DICTIONARY
OF
SCIENTIFIC BIOGRAPHY

PUBLISHED UNDER THE AUSPICES OF
THE AMERICAN COUNCIL OF LEARNED SOCIETIES

The American Council of Learned Societies, organized in 1919 for the purpose of advancing the study of the humanities and of the humanistic aspects of the social sciences, is a nonprofit federation comprising forty-five national scholarly groups. The Council represents the humanities in the United States in the International Union of Academies, provides fellowships and grants-in-aid, supports research-and-planning conferences and symposia, and sponsors special projects and scholarly publications.

MEMBER ORGANIZATIONS

AMERICAN PHILOSOPHICAL SOCIETY, 1743
AMERICAN ACADEMY OF ARTS AND SCIENCES, 1780
AMERICAN ANTIQUARIAN SOCIETY, 1812
AMERICAN ORIENTAL SOCIETY, 1842
AMERICAN NUMISMATIC SOCIETY, 1858
AMERICAN PHILOLOGICAL ASSOCIATION, 1869
ARCHAEOLOGICAL INSTITUTE OF AMERICA, 1879
SOCIETY OF BIBLICAL LITERATURE, 1880
MODERN LANGUAGE ASSOCIATION OF AMERICA, 1883
AMERICAN HISTORICAL ASSOCIATION, 1884
AMERICAN ECONOMIC ASSOCIATION, 1885
AMERICAN FOLKLORE SOCIETY, 1888
AMERICAN DIALECT SOCIETY, 1889
AMERICAN PSYCHOLOGICAL ASSOCIATION, 1892
ASSOCIATION OF AMERICAN LAW SCHOOLS, 1900
AMERICAN PHILOSOPHICAL ASSOCIATION, 1901
AMERICAN ANTHROPOLOGICAL ASSOCIATION, 1902
AMERICAN POLITICAL SCIENCE ASSOCIATION, 1903
BIBLIOGRAPHICAL SOCIETY OF AMERICA, 1904
ASSOCIATION OF AMERICAN GEOGRAPHERS, 1904
HISPANIC SOCIETY OF AMERICA, 1904
AMERICAN SOCIOLOGICAL ASSOCIATION, 1905
AMERICAN SOCIETY OF INTERNATIONAL LAW, 1906
ORGANIZATION OF AMERICAN HISTORIANS, 1907
AMERICAN ACADEMY OF RELIGION, 1909
COLLEGE ART ASSOCIATION OF AMERICA, 1912
HISTORY OF SCIENCE SOCIETY, 1924
LINGUISTIC SOCIETY OF AMERICA, 1924
MEDIAEVAL ACADEMY OF AMERICA, 1925
AMERICAN MUSICOLOGICAL SOCIETY, 1934
SOCIETY OF ARCHITECTURAL HISTORIANS, 1940
ECONOMIC HISTORY ASSOCIATION, 1940
ASSOCIATION FOR ASIAN STUDIES, 1941
AMERICAN SOCIETY FOR AESTHETICS, 1942
AMERICAN ASSOCIATION FOR THE ADVANCEMENT OF SLAVIC STUDIES, 1948
METAPHYSICAL SOCIETY OF AMERICA, 1950
AMERICAN STUDIES ASSOCIATION, 1950
RENAISSANCE SOCIETY OF AMERICA, 1954
SOCIETY FOR ETHNOMUSICOLOGY, 1955
AMERICAN SOCIETY FOR LEGAL HISTORY, 1956
AMERICAN SOCIETY FOR THEATRE RESEARCH, 1956
SOCIETY FOR THE HISTORY OF TECHNOLOGY, 1958
AMERICAN COMPARATIVE LITERATURE ASSOCIATION, 1960
AMERICAN SOCIETY FOR EIGHTEENTH-CENTURY STUDIES, 1969
ASSOCIATION FOR JEWISH STUDIES, 1969

DICTIONARY OF SCIENTIFIC BIOGRAPHY

CHARLES COULSTON GILLISPIE
Princeton University

EDITOR IN CHIEF

Volume 7

IAMBLICHUS – KARL LANDSTEINER

CHARLES SCRIBNER'S SONS · NEW YORK

First publication in an eight-volume edition 1981.

Library of Congress Cataloging in Publication Data

Main entry under title:

Dictionary of scientific biography.

"Published under the auspices of the American Council of Learned Societies."
Includes bibliographies and index.
1. Scientists—Biography. I. Gillispie, Charles Coulston. II. American Council of Learned Societies Devoted to Humanistic Studies.
Q141.D5 1981 509'.2'2 [B] 80-27830
ISBN 0-684-16962-2 (set)

ISBN 0-684-16963-0 Vols. 1 & 2
ISBN 0-684-16964-9 Vols. 3 & 4
ISBN 0-684-16965-7 Vols. 5 & 6
ISBN 0-684-16966-5 Vols. 7 & 8
ISBN 0-684-16967-3 Vols. 9 & 10
ISBN 0-684-16968-1 Vols. 11 & 12
ISBN 0-684-16969-X Vols. 13 & 14
ISBN 0-684-16970-3 Vols. 15 & 16

Published simultaneously in Canada
by Collier Macmillan Canada, Inc.

7 9 11 13 15 17 19 B/C 20 18 16 14 12 10 8

Printed in the United States of America

Editorial Staff

Panel of Consultants

Contributors to Volume 7

The following are the contributors to Volume 7. Each author's name is followed by the institutional affiliation at the time of publication and the names of articles written for this volume. The symbol † indicates that an author is deceased.

MARK B. ADAMS
University of Pennsylvania
A. O. KOVALEVSKY

S. MAQBUL AHMAD
Aligarh Muslim University
AL-IDRĪSĪ; IBN KHURRADĀDHBIH

MICHELE L. ALDRICH
Smith College
KING

TOBY A. APPEL
Kirkland College
LACAZE-DUTHIERS; LACÉPÈDE

A. ALBERT BAKER, JR.
University of Oklahoma
A. LADENBURG

MARGARET E. BARON
IVORY; W. JONES

IRINA V. BATYUSHKOVA
Academy of Sciences of the U.S.S.R.
KARPINSKY

JOSEPH BEAUDE
Centre National de la Recherche Scientifique
B. LAMY

SILVIO A. BEDINI
National Museum of History and Technology
JEFFERSON

LUIGI BELLONI
University of Milan
LANDRIANI

MICHAEL BERNKOPF
Pace College
H. VON KOCH; LAGUERRE

KURT-R. BIERMANN
German Academy of Sciences
JOACHIMSTHAL; L. KRONECKER; KUMMER

P. W. BISHOP
JARS

A. BLAAUW
European Southern Observatory
KAPTEYN

L. J. BLACHER
Academy of Sciences of the U.S.S.R.
P. P. IVANOV; V. O. KOVALEVSKY

HERMANN BOERNER
University of Giessen
KNESER

MARTIN BOPP
Botanical Institute, University of Heidelberg
KLEBS

WALLACE A. BOTHNER
University of New Hampshire
JAGGAR; A. JOHANNSEN

GERT H. BRIEGER
Duke University
KELSER

T. A. A. BROADBENT †
LANDEN

BARUCH BRODY
Massachusetts Institute of Technology
W. E. JOHNSON

K. E. BULLEN
University of Sydney
IMAMURA; KNOTT; LAMB

VERN L. BULLOUGH
California State University, Northridge
KNOX

IVOR BULMER-THOMAS
ISIDORUS OF MILETUS

WERNER BURAU
University of Hamburg
C. F. KLEIN; J. KOENIG; KÖNIGSBERGER

DEAN BURK
National Cancer Institutes, Bethesda
R. KUHN

LESLIE J. BURLINGAME
Mount Holyoke College
LAMARCK

J. C. BURNHAM
Ohio State University
JENNINGS

H. L. L. BUSARD
State University of Leiden
LA FAILLE

JOHN T. CAMPBELL
KAUFMANN

H. B. G. CASIMIR
KRAMERS

JAMES H. CASSEDY
National Library of Medicine
E. O. JORDAN

CARLO CASTELLANI
LANCISI

SEYMOUR L. CHAPIN
California State University, Los Angeles
JEAURAT

ROBERT A. CHIPMAN
University of Toledo
JENKIN

R. J. CHORLEY
University of Cambridge
D. W. JOHNSON; W. D. JOHNSON

MARSHALL CLAGETT
Institute for Advanced Study, Princeton
JOHN OF PALERMO

EDWIN CLARKE
University College, London
J. H. JACKSON

WILLIAM COLEMAN
Northwestern University
KIELMEYER

D. E. COOMBE
University of Cambridge
T. JOHNSON

PIERRE COSTABEL
École Pratique des Hautes Études
LAGNY

J. K. CRELLIN
Wellcome Institute of the History of Medicine
JONSTON

EDWARD E. DAUB
University of Wisconsin
KRÖNIG

SALLY H. DIEKE
The Johns Hopkins University
M. J. JOHNSON; KEELER; H. J. KLEIN

J. DIEUDONNÉ
C. JORDAN

YVONNE DOLD-SAMPLONIUS
AL-JAYYĀNĪ; AL-KHĀZIN

CLAUDE E. DOLMAN
University of British Columbia
H. H. R. KOCH

J. G. DORFMAN
Institute for the History of Science and Technology, Moscow
KURCHATOV; G. S. LANDSBERG

HAROLD DORN
Stevens Institute of Technology
KATER

OLLIN J. DRENNAN
Western Michigan University
F. W. G. KOHLRAUSCH; R. H. A. KOHLRAUSCH

L. C. DUNN
Columbia University
W. L. JOHANNSEN

J. M. EDMONDS
University of Oxford
KIDD

CONTRIBUTORS TO VOLUME 7

JAMES W. ELLINGTON
University of Connecticut
KANT

VASILY A. ESAKOV
Institute for the History of Science and Technology, Moscow
KRUBER

FOCKO EULEN
Ruhr University, Bochum
KELLNER

JOSEPH EWAN
Tulane University
JEPSON; A. KELLOGG

JOAN M. EYLES
JAMESON; JUKES

V. A. EYLES
JOLY

A. S. FEDOROV
Academy of Sciences of the U.S.S.R.
KRASHENINNIKOV

I. A. FEDOSEEV
Institute for the History of Science and Technology, Moscow
KRASNOV

J. FELDMANN
University of Paris
LAMOUROUX

E. A. FELLMANN
Institut Platonaeum, Basel
J. S. KOENIG

EUGENE S. FERGUSON
University of Delaware
KENNEDY

SARAH FERRELL
JAMES

MARTIN FICHMAN
York University
JUNCKER; E. KÖNIG

BERNARD S. FINN
National Museum of History and Technology
IVES; M. H. VON JACOBI; J. KERR

W. S. FINSEN
INNES

JAROSLAV FOLTA
Czechoslovak Academy of Sciences
JONQUIÈRES; KLÜGEL

MICHAEL FORDHAM
JUNG

H. C. FREIESLEBEN
KAYSER; KONKOLY THEGE

HANS FREUDENTHAL
State University of Utrecht
KERÉKJÁRTÓ; KNOPP

JOSEPH S. FRUTON
Yale University
KEILIN

TSUNESABURO FUJINO
Osaka University
KITASATO

PATSY A. GERSTNER
Dittrick Museum of Historical Medicine
J. T. KLEIN

GEORGE E. GIFFORD, JR.
C. T. JACKSON

JEAN GILLIS
University of Ghent
KEKULE VON STRADONITZ

CHARLES C. GILLISPIE
Princeton University
KOYRÉ

OWEN GINGERICH
Smithsonian Astrophysical Observatory
KEPLER; LACAILLE

GEORGE GOE
New School for Social Research
KAESTNER

RAGNAR GRANIT
Karolinska Institutet, Stockholm
KALM

EDWARD GRANT
Indiana University, Bloomington
JORDANUS DE NEMORE

FRANK GREENAWAY
Science Museum, London
KIRKALDY

JOSEPH T. GREGORY
University of California, Berkeley
G. F. JAEGER

SAMUEL L. GREITZER
Rutgers University
LAMÉ

NORMAN T. GRIDGEMAN
National Research Council of Canada
JEVONS; KIRKMAN

A. T. GRIGORIAN
Institute for the History of Science and Technology, Moscow
KOCHIN; KOLOSOV; KOTELNIKOV; A. N. KRYLOV; N. M. KRYLOV; L. D. LANDAU

HENRY GUERLAC
Cornell University
LA BROSSE

V. GUTINA
Institute for the History of Science and Technology, Moscow
IVANOVSKY

MARIE BOAS HALL
Imperial College of Science and Technology
KUNCKEL

ROBERT E. HALL
The Queen's University of Belfast
AL-KHĀZINĪ

THOMAS L. HANKINS
University of Washington
LALANDE

ROY FORBES HARROD
Fellow of Royal Economic Society, London
KEYNES

FREDERICK HEAF
Welsh National School of Medicine
LAENNEC

JOHN L. HEILBRON
University of California, Berkeley
KINNERSLEY; KLEIST

DAVID HEPPELL
Royal Scottish Museum
JEFFREYS

ARMIN HERMANN
University of Stuttgart
W. KUHN; KURLBAUM

HEINRICH HERMELINK
AL-JAYYĀNĪ

DIETER B. HERRMANN
Archenhold Observatory, Berlin
LAMONT

EDWARD HINDLE
The Royal Society
J. G. KERR

ERICH HINTZSCHE
Medizinhistorische Bibliothek, University of Bern
KOELLIKER

TETU HIROSIGE
Nihon University
ISHIWARA

MARJORIE HOOKER
U.S. Geological Survey
A. LACROIX

WLODZIMIERZ HUBICKI
Marie Curie-Skłodowska University
KOSTANECKI

KARL HUFBAUER
University of California, Irvine
KLAPROTH

AARON J. IHDE
University of Wisconsin, Madison
KAHLENBERG

JEAN ITARD
Lycée Henri IV
KRAMP; S. F. LACROIX; LAGRANGE

BØRGE JESSEN
University of Copenhagen
J. L. W. V. JENSEN

HANS KANGRO
University of Hamburg
JUNGIUS; C. KHUNRATH; H. KHUNRATH; KIRCHER

GEORGE B. KAUFFMAN
California State University, Fresno
JØRGENSEN; W. KOSSEL

CONTRIBUTORS TO VOLUME 7

BRIAN B. KELHAM
KANE

HENRY S. VAN KLOOSTER
Rensselaer Polytechnic Institute
F. M. JAEGER

BRONISLAW KNASTER
JANISZEWSKI

HIDEO KOBAYASHI
Hokkaido University
KOTŌ

MANFRED KOCH
Bergbau Bucherei, Essen
JUSTI; KARSTEN

SHELDON J. KOPPERL
Grand Valley State College
KHARASCH; KIPPING; KRAUS

HANS-GÜNTHER KÖRBER
Zentralbibliothek des Meteorologischen Dienstes, Potsdam
JOLLY; A. KÖNIG; KUNDT

EDNA E. KRAMER
Polytechnic Institute of Brooklyn
S. KOVALEVSKY

DAVID KUBRIN
Liberation School, San Francisco
JOHN KEILL

P. G. KULIKOVSKY
Academy of Sciences of the U.S.S.R.
IDELSON; KAVRAYSKY; KOSTINSKY; KOVALSKY

V. I. KUZNETSOV
Institute for the History of Science and Technology, Moscow
IPATIEV; KONDAKOV

YVES LAISSUS
Bibliothèque Centrale du Muséum National d'Histoire Naturelle
J. DE JUSSIEU

WILLIAM LEFANU
Royal College of Surgeons of England
KEITH

HENRY M. LEICESTER
University of the Pacific
H. C. JONES; KOLBE; KOPP

MARTIN LEVEY †
IBRĀHĪM IBN YAʿQŪB

JACQUES R. LÉVY
Paris Observatory
JANSSEN

A. C. LLOYD
University of Liverpool
IAMBLICHUS

J. M. LÓPEZ DE AZCONA
Comisión Nacionale de Geologia, Madrid
IBÁÑEZ; JUAN Y SANTACILLA

RICHARD P. LORCH
University of Manchester
JĀBIR IBN AFLAḤ

MARVIN W. MCFARLAND
Library of Congress
LANCHESTER

PATRICIA P. MACLACHLAN
W. JOHNSON

BRIAN G. MARSDEN
Smithsonian Astrophysical Observatory
D. KIRKWOOD

LORENZO MINIO-PALUELLO
University of Oxford
JAMES OF VENICE

M. G. J. MINNAERT †
JULIUS; KAISER

A. G. MOLLAND
University of Aberdeen
JOHN OF DUMBLETON

PIERCE C. MULLEN
Montana State University
KOFOID

LETTIE S. MULTHAUF
KIRCH FAMILY

SHIGERU NAKAYAMA
University of Tokyo
INŌ; KIMURA

G. V. NAUMOV
Academy of Sciences of the U.S.S.R.
KROPOTKIN

W. NIEUWENKAMP
State University of Utrecht
KRAYENHOFF

LOWELL E. NOLAND †
JUDAY

J. D. NORTH
Museum of the History of Science, Oxford
JERRARD

L. NOVÝ
Czechoslovak Academy of Sciences
JONQUIÈRES

WILFRIED OBERHUMMER
Austrian Academy of Sciences
JACQUIN

HERBERT OETTEL
JUEL; LALOUVÈRE

ROBERT OLBY
University of Leeds
KOELREUTER; KÖHLER; K. M. L. A. KOSSEL

C. D. O'MALLEY †
INGRASSIA

GEORGE F. PAPENFUSS
University of California, Berkeley
KÜTZING; KYLIN

KURT MØLLER PEDERSEN
University of Aarhus
KRAFT

OLAF PEDERSEN
University of Aarhus
JOHANNES LAURATIUS DE FUNDIS; JOHN SIMONIS OF SELANDIA

FRANCIS PERRIN
Collège de France
JOLIOT; JOLIOT-CURIE

MOGENS PIHL
University of Copenhagen
KLINGENSTIERNA; KNUDSEN

DAVID PINGREE
Brown University
JAGANNĀTHA; JAYASIṂHA; KAMALĀKARA; KANAKA; KEŚAVA; KṚṢṆA; LALLA

JACQUES PIQUEMAL
Université Paul Valery, Montpellier
A. JORDAN

A. F. PLAKHOTNIK
Academy of Sciences of the U.S.S.R.
KNIPOVICH

L. PLANTEFOL
Laboratoire de Botanique, University of Paris
G. LAMY

M. PLESSNER
Hebrew University
JĀBIR IBN ḤAYYĀN; AL-JĀḤIẒ

J. B. POGREBYSSKY †
KOROLEV

EMMANUEL POULLE
École Nationale des Chartes
JOHN OF LIGNÈRES; JOHN OF MURS; JOHN OF SAXONY; JOHN OF SICILY

J. A. PRINS
KEESOM

HANS QUERNER
Institut für Geschichte der Medizin, University of Heidelberg
KLEINENBERG; KÜHN

SAMUEL X. RADBILL
College of Physicians of Philadelphia
ISAACS

JOHN B. RAE
Harvey Mudd College
KETTERING

VARADARAJA V. RAMAN
Rochester Institute of Technology
KALUZA

ROSHDI RASHED
Centre National de la Recherche Scientifique
IBRĀHĪM IBN SINĀN; KAMĀL AL-DĪN; AL-KARAJĪ

RUTH GIENAPP RINARD
Kirkland College
LANDOLT

GLORIA ROBINSON
Yale University
JENKINSON

FRANCESCO RODOLICO
University of Florence
ISSEL

PAUL LAWRENCE ROSE
New York University
KECKERMAN

CONTRIBUTORS TO VOLUME 7

EDWARD ROSEN
City College, City University of New York
JANSEN

B. A. ROSENFELD
Institute for the History of Science and Technology, Moscow
AL-KĀSHĪ; AL-KHAYYĀMĪ

L. ROSENFELD
Nordic Institute for Theoretical Atomic Physics, Copenhagen
JOULE; G. R. KIRCHHOFF

FRANZ ROSENTHAL
Yale University
IBN KHALDŪN

JOHN ROSS
Massachusetts Institute of Technology
J. G. KIRKWOOD

K. E. ROTHSCHUH
University of Münster/Westphalia
H. KRONECKER; KÜHNE

A. I. SABRA
Harvard University
AL-JAWHARĪ

A. S. SAIDAN
University of Jordan
KUSHYĀR

S. SAMBURSKY
Hebrew University
JOHN PHILOPONUS

CECIL J. SCHNEER
University of New Hampshire
KNORR

BRUNO SCHOENEBERG
University of Hamburg
C. F. KLEIN; E. LANDAU; G. LANDSBERG

E. L. SCOTT
Stamford High School, Lincolnshire
KEIR; KIRWAN

CHRISTOPH J. SCRIBA
Technical University, Berlin
C. G. J. JACOBI; LAMBERT

A. N. SHAMIN
Institute for the History of Science and Technology, Moscow
K. S. KIRCHHOF

ROBERT S. SHANKLAND
Case Western Reserve University
K. R. KOENIG

WILLIAM D. SHARPE
ISIDORE OF SEVILLE

NABIL SHEHABY
Warburg Institute
ISḤĀQ IBN ḤUNAYN

A. G. SHENSTONE
Princeton University
R. W. LADENBURG

OSCAR B. SHEYNIN
Academy of Sciences of the U.S.S.R.
KRASOVSKY

ELIZABETH NOBLE SHOR
Scripps Institution of Oceanography
D. S. JORDAN; V. L. KELLOGG

DIANA M. SIMPKINS
Polytechnic of North London
KNIGHT

P. N. SKATKIN
I. I. IVANOV

PIETER SMIT
Catholic University Nijmegen
KLUYVER

CYRIL STANLEY SMITH
Massachusetts Institute of Technology
JEFFRIES

E. SNORRASON
Rigshospitalet, Copenhagen
C. O. JENSEN; KROGH

Y. I. SOLOVIEV
Institute of the History of Natural Science and Engineering, Moscow
KABLUKOV; KONOVALOV; KURNAKOV

PAUL SPEISER
University of Vienna
LANDSTEINER

FRANZ A. STAFLEU
State University of Utrecht
A. H. L. DE JUSSIEU; A. DE JUSSIEU; A.-L. DE JUSSIEU; B. DE JUSSIEU

S. M. STERN †
ISAAC ISRAELI

D. J. STRUIK
Massachusetts Institute of Technology
KORTEWEG

CHARLES SÜSSKIND
University of California, Berkeley
JEWETT; KÁRMÁN; KENNELLY

FERENC SZABADVÁRY
Technical University, Budapest
IRINYI; JAHN; KITAIBEL; KJELDAHL

RENÉ TATON
École Pratique des Hautes Études
KOENIGS; G.-P. LA HIRE; P. DE LA HIRE; LANCRET

KENNETH L. TAYLOR
University of Oklahoma
LAMÉTHERIE

SEVIM TEKELI
Ankara University
AL-KHUJANDĪ

V. V. TIKHOMIROV
Academy of Sciences of the U.S.S.R.
KEYSERLING

HEINZ TOBIEN
University of Mainz
JACCARD; JAEKEL

G. J. TOOMER
Brown University
AL-KHWĀRIZMĪ

ANDRZEJ TRAUTMAN
University of Warsaw
INFELD

SHERWOOD D. TUTTLE
University of Iowa
KAY

G. UBAGHS
University of Liège
L.-G. DE KONINCK

GEORG USCHMANN
University of Jena
KRAUSE

F. M. VALADEZ
JAMES KEILL

J. VAN DEN HANDEL
KAMERLINGH ONNES

PETER W. VAN DER PAS
South Pasadena, California
INGEN-HOUSZ; JOBLOT; KAEMPFER; KNUTH; KUENEN

ARAM VARTANIAN
New York University
LA METTRIE

L. VEKERDI
Library, Hungarian Academy of Sciences
KÜRSCHÁK

JUAN VERNET
University of Barcelona
IBN JULJUL; AL-KHUWĀRIZMĪ

KURT VOGEL
University of Munich
JOHN OF GMUNDEN; KÖBEL

A. I. VOLODARSKY
Institute for the History of Science and Technology, Moscow
K. S. KIRCHHOF

A. R. WEILL
JACQUET

RALPH H. WETMORE
JEFFREY

LEONARD G. WILSON
University of Minnesota
JENNER

A. E. WOODRUFF
JEANS

HATTEN S. YODER, JR.
Carnegie Institution of Washington, Geophysical Laboratory
IDDINGS

A. A. YOUSCHKEVITCH
Institute for the History of Science and Technology, Moscow
KHINCHIN

A. P. YOUSCHKEVITCH
Institute for the History of Science and Technology, Moscow
KAGAN; AL-KĀSHĪ; AL-KHAYYĀMĪ

S. Y. ZALKIND
Institute for the History of Science and Technology, Moscow
KOLTZOFF

DICTIONARY OF SCIENTIFIC BIOGRAPHY

DICTIONARY OF SCIENTIFIC BIOGRAPHY

IAMBLICHUS—LANDSTEINER

IAMBLICHUS (*b.* Chalcis, Syria, *ca.* A.D. 250; *d. ca.* A.D. 330), *philosophy.*

Iamblichus' parents were Syrian and he taught philosophy in Syria. Otherwise almost nothing is known of his life. It is clear from later writers that he was of major importance in the elaborate systematization of Neoplatonism that occurred after Plotinus. He wrote an encyclopedic work on Neopythagorean philosophy which included arithmetic, geometry, physics, and astronomy. But what has survived of this work has virtually no philosophical or scientific interest. We have only the traditional "Pythagorean" claims that all sciences are based on Limit and the Unlimited, that numbers are generated from the One and a principle of plurality, and that geometric solids are generated from unit points, lines, and surfaces. Because this "procession" of the One generates also beauty and then goodness, the study of mathematics and of the sciences based on mathematics is said to be the path to true virtue; individual numbers moreover are symbols of individual gods of the Greek pantheon. But Iamblichus makes all these claims in only a compressed and dogmatic manner. His *Life of Pythagoras* has no historical value.

Iamblichus' commentaries on Aristotle are lost. They contained some acute, if unoriginal, defenses of Aristotelian doctrine as well as some less well-judged attempts to incorporate Neoplatonic metaphysics. For example, he correctly defended Aristotle's definition of motion as incomplete or potential against the claims of both Plotinus and the Stoics that motion was activity and actuality (*energeia*). He argued that Aristotle was concerned with the concept of "being potentially (possibly) so and so" as opposed to "being actually so and so," while his opponents confused this notion with the concept of possibility itself, as opposed to actuality or activity (see Simplicius, "In Aristotelis Categorias," pp. 303 ff.). Historians have perhaps failed to notice the significance of Iamblichus' influential commentary on Aristotle's psychology. By sharpening the distinctions that his Neoplatonic forerunners had blurred between soul and intellect and between a human soul and a pure or disembodied soul, he showed how Platonic prejudices need not stand in the way of separating psychology from metaphysics.

Like most Platonists of his age, Iamblichus was attracted by the prevalent Gnostic and sometimes magical practices that were supposed to lead to salvation. His work *On the Egyptian Mysteries* is a characteristic attempt to reconcile these practices with Platonic philosophy. Although it rationalizes them (more than has been recognized), the work is without scientific significance. In the Renaissance, however, it was partly responsible for the fascination with hieroglyphics and other supposed symbols of the East.

BIBLIOGRAPHY

I. ORIGINAL WORKS. E. Des Places, ed., *Les mystères d'Égypte* (Paris, 1966), contains Greek text with French trans. and intro. *De vita pythagorica*, *Protrepticus*, *Theologumena arithmeticae*, and *De communi mathematica scientia* are in various eds. of Teubner Classics (Leipzig–Hildesheim).

II. SECONDARY LITERATURE. Simplicius' commentaries on Aristotle's *On the Soul* and *Categories* are in *Commentaria in Aristotelem graeca*, II (Berlin, 1882), and VIII (Berlin, 1907), respectively—see indexes under Iamblichus.

See also A. H. Armstrong, ed., *Cambridge History of Later Greek and Early Medieval Philosophy* (Cambridge, 1967), pp. 295–301.

A. C. LLOYD

IBÁÑEZ E IBÁÑEZ DE IBERO, CARLOS (*b.* Barcelona, Spain, 14 April 1825; *d.* Nice, France, 28 January 1891), *geodesy.*

Ibáñez' father, Martín Ibáñez de Prado, was a soldier and mathematician: a national hero for his participation in the sieges of Zaragoza and one of the first postulators of non-Euclidean geometry. In 1832, when Ibáñez was barely seven, his father was assassinated for political reasons. The boy entered the

Academy of Army Engineers in 1839, receiving training in both military and scientific subjects.

Ibáñez' interest in geodesy was awakened by the practical courses he taught at the Academy of Engineers. In 1853 he joined the recently created commission for drawing up a national map. As a member he studied and planned a Geodimeter, known as the "Spanish rule," to measure the base at Madridejos. In 1859 he devised a method for carrying out the census of rural and urban real property and for its conservation. He was cofounder (1866) and later president (until his death) of the International Geodesic Association. In 1875, as plenipotentiary envoy of the king, he attended the inauguration of the International Office of Weights and Measures, which he actively promoted in order to achieve a worldwide system of units of measure and decimal currency. He was also its first president.

At Ibáñez' initiative the Census and Geographic Institute was created in 1870—he was its first director, as he was of the Corps of Geodesists (today known as the Geographical Engineers)—and in 1877 he was responsible for creation of the Statistics Corps. An early advocate of international scientific collaboration, he was a member of the Royal Academy of Sciences of Madrid, as well as the corresponding organizations of Barcelona, Paris, Berlin, Rome, Belgium, the United States, Buenos Aires, and Egypt.

Ibáñez was concerned mainly with precision in measurement and with scientific organization. He obtained a probable error of $\pm 1/5{,}800{,}000$ in geodesic bases, compared with the $\pm 1/1{,}200{,}000$ achieved until then. The fifteen-kilometer base measured in La Mancha was of particular note. At the request of the Swiss Confederation he carried out in 1880 the measurement of the central base of Aarberg at 2.4 kilometers.

As a result of Ibáñez' initiative and eagerness to measure the globe, it was agreed in 1860 to remeasure the arc of meridian from Dunkirk to Formentera. He projected the geodesic union of Europe with Africa, with an interruption of 270 kilometers, from Shetland to the Sahara; this had never been achieved by observations of a geodesic landmark. He carried out these observations in 1878 between the inhospitable peaks of Mulhacén and the Teticas in Spain and the Filhaoussen and the M'Sabiha in Algeria. The error in closing the triangles was on the order $\pm$ one second of arc, and the precision brought him the Poncelet Prize of the Academy of Sciences of Paris in 1889. Measurements of horizontal angles with $\pm 1/10{,}000$ of degree in the centesimal system were achieved through his techniques.

BIBLIOGRAPHY

I. Original Works. The following are outstanding among Ibáñez' twenty-eight publications: *Manual del pontonero* (Madrid, 1853); *Aparato de medir bases* (Madrid, 1859); *Historia de los instrumentos de observación en astronomía y geodesía* (Madrid, 1863); *Nivelación geodésica* (Madrid, 1864); *Base central de la triangulación geodésica de España* (Madrid, 1865); *Nuevo aparto para medir bases geodésicas* (Madrid, 1869); *Determinación del metro y kilógramo internacionales* (Madrid, 1875); *Enlace geodésico y astronómico de Europa y Africa* (Madrid, 1880); and *Jonction geodésique et astronomique de l'Algérie avec l'Espagne* (Paris, 1886).

II. Secondary Literature. See the following, listed chronologically: A. Hirsch, *Le Général Ibáñez;* C. I. de P. and M., necrological note (Neuchâtel, 1891); *Commemoration du centenaire de la naissance du Général Ibáñez de Ibero* (Paris, 1925); *Inauguración del monumento en memoria del General de Ingenieros Carlos Ibáñez de Ibero, Marqués de Mulhacén* (Madrid, 1957); and Carlos Ibáñez de Ibero, *Biografía del General Ibáñez de Ibero, Marqués de Mulhacén* (Madrid, 1957); and *Episodios de la guerra de la independencia* (Madrid, 1963).

J. M. López de Azcona

IBN. See next element of name.

IBRĀHĪM IBN SINĀN IBN THĀBIT IBN QURRA (*b.* Baghdad [?], 908; *d.* Baghdad, 946), *mathematics, astronomy.*

Born into a family of celebrated scholars, Ibn Sinān was the son of Sinān ibn Thābit, a physician, astronomer, and mathematician, and the grandson of Thābit ibn Qurra. Although his scientific career was brief—he died at the age of thirty-eight—he left a notable body of work, the force and perspicuity of which have often been underlined by biographers and historians. This work covers several areas, such as tangents of circles, and geometry in general; the apparent motions of the sun, including an important optical study on shadows; the solar hours; and the astrolabe and other astronomical instruments.

Since it would hardly be feasible to give even a summary sketch of Ibn Sinān's entire work in a brief article, the best course will be to concentrate attention on two important contributions: his discussions of the quadrature of the parabola and of the relations between analysis and synthesis.

His study of the parabola followed directly out of the treatment given the problem in the work of his grandfather. Thābit ibn Qurra had already resolved this problem in a different way from that of Archimedes. Although his method may have been equivalent to that of summing integrals, the approach was

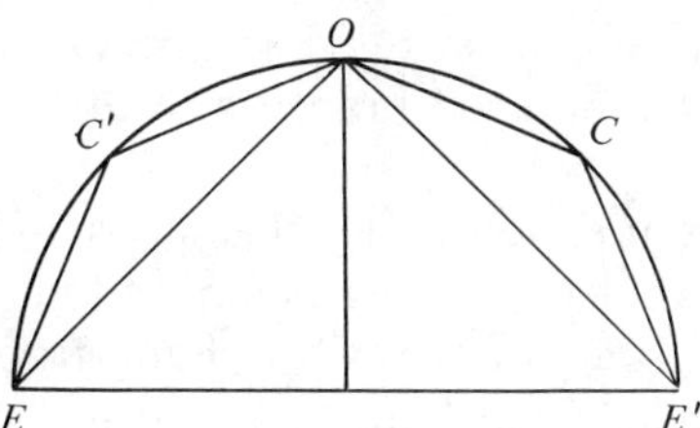

FIGURE 1

more general than that of Archimedes in that the intervals of integration were no longer divided into equal subintervals. Thābit's demonstration was lengthy, however, containing twenty propositions. Another mathematician, one al-Mahānni, had given a briefer one but Ibn Sinān felt it to be unacceptable that (as he wrote) "al Mahānni's study should remain more advanced than my grandfather's unless someone of our family (the Ibn Qurra) can excel him" (*Rāsa'ilu Ibn-i-Sinān*, p. 69). He therefore sought to give an even more economical demonstration, one that did not depend upon reduction to the absurd. The proposition on which Ibn Sinān founded his demonstration, and which he took care to prove beforehand, is that the proportionality of the areas is invariant under affine transformation.

His method considers the polygon a_n to be composed of $2^n - 1$ triangles and inscribed in the area a of the parabola. The polygon a_1 is the triangle EOE', a_2 is the polygon $ECOC'E'$, etc. Ibn Sinān demonstrated that if a_n and a'_n are two polygons inscribed, respectively, in the two areas a and a' of the parabola, then

$$\frac{a_n}{a'_n} = \frac{a_1}{a'_1}.$$

Actually, he derived an expression equivalent to

$$\frac{a}{a'} = \lim_{n\to\infty} \frac{a_n}{a'_n} = \frac{a_1}{a'_1},$$

from which he obtained

$$\frac{\frac{1}{2}(a - a_1)}{a} = \frac{\frac{1}{2}(a_2 - a_1)}{a_1} = \frac{1}{8},$$

and finally derived

$$a = \frac{4}{3} a_1.$$

Ibn Sinān's originality in this investigation is manifest. It was with that same independence of mind that he intended to revive classical geometric analysis in order to develop it in a separate treatise. By virtue of that study, the author may be considered one of the foremost Arab mathematicians to treat problems of mathematical philosophy. His attempt has the form of a critique of the practical geometry in his own times. "I have found," he wrote, "that contemporary geometers have neglected the method of Apollonius in analysis and synthesis, as they have in most of the things I have brought forward, and that they have limited themselves to analysis alone in so restrictive a manner that they have led people to believe that this analysis did not correspond to the synthesis effected" (*ibid.*, p. 66).

In this work, Ibn Sinān proposed two tasks simultaneously, the one technical and the other epistemological. On the one hand, the purpose was to provide those learning geometry with a method (*tarīq*) which could furnish what they needed in order to solve geometrical problems. On the other hand, it was equally important to think about the procedures of geometrical analysis itself and to develop a classification of geometrical problems according to the number of the hypotheses to be verified, explaining the bearing, respectively, of analysis and synthesis on each class of problems.

Considering both the problem of infinitesimal determinations and the history of mathematical philosophy, it is obvious that the work of Ibn Sinān is important in showing how the Arab mathematicians pursued the mathematics they had inherited from the Hellenistic period and developed it with independent minds. That is the dominant impression left by his work.

BIBLIOGRAPHY

I. ORIGINAL WORKS. The *Rāsa'ilu Ibn-i-Sinān* (Hyderabad, 1948) comprises *Fī'l astrolāb* ("On the Astrolabe"), *Al-Tahlīl wa'l-Tarkīb* ("Analysis and Synthesis"), *Fī Ḥarakati'š-Šams* ("On Solar Movements"), *Rasm al-quṭū' aṯṯalāṯa* ("Outline of Three Sections"), *Fī misāḥat qaṭ'al-Maḥrūṭ al-mukāfī* ("Measurement of the Parabola"), and *Al-Handasa wa'n-Nujūm* ("Geometry and Astronomy").

II. SECONDARY LITERATURE. See Ibn al-Qifti, *Ta'riḥ al-Ḥukamā'*, J. Lippert, ed. (Leipzig, 1903); Ibn al-Nadīm, *Kitāb al-Fihrist*, Flugel, ed. (Leipzig, 1871–1872), p. 272; C. Brockelmann, *Geschichte der arabischen Literatur*, I (Leiden, 1943), 245; H. Suter, *Die Mathematiker und Astronomer der Araber und ihre Werke* (Leipzig, 1900), pp. 53–54; "Abhandlung über die Ausmessung der Parabel von Ibrahim ben Sinan ben Thabit ben Kurra," in *Vierteljahrsschrift der Naturforschenden Gesellschaft in Zürich*, **63** (1918), 214 ff.; A. P. Youschkevitch, "Note sur les déterminations infinitésimales chez Thabit ibn Qurra," in *Archives internationales d'histoire des sciences*, no. 66 (January–March 1964), pp. 37–45.

ROSHDI RASHED

IBRĀHĪM IBN YA'QŪB

IBRĀHĪM IBN YA'QŪB AL-ISRĀ'ĪLĪ AL-TURṬUSHI (*b.* Tortosa, Spain; *fl.* second half of the tenth century), *geography.*

Ibrāhīm ibn Ya'qūb, a Spanish Jewish merchant, was known for his travels through eastern, central, and western Europe on business and, probably, diplomatic missions. At that time, about 965, there were colonies of Jews throughout Europe who would be apt to receive a fellow Jew. Only fragments remain of his one or more geographical writings.

Because of the difficulty in identifying the names of towns and places from those who quoted Ibrāhīm's description of his journey, the itinerary may be given only tentatively. He crossed the Adriatic, traveled through Bordeaux, Noirmoutier, St.-Malo, Rouen, Utrecht, Aix-la-Chapelle (Aachen), Mainz, Fulda, Soest, Paderborn, Schleswig, Magdeburg (where he met Bulgarian ambassadors at the court of Emperor Otto I), then along the right bank of the Elbe and through Prague, Cracow, Augsburg, Cortona, and Trapani. This route led Ibrāhīm through the Slavic and Frankish regions of Europe.

Such later geographers as al-Bakrī (*d.* 1094), *Kitāb al-masālik wa'l-mamālik* ("Routes and Kingdoms"; Paris, B.N. MSS 2218, 5905), and al-Qazwīnī (*b.* 1203), author of *'Ajā'ib al-makhlūqāt wa'āthār al-bilād* ("The Marvels of Creation and Peculiarities of Existing Things"; Paris, B.N. MS 2775)—possibly through the intermediary of al-'Udhrī (*fl.* 1213), author of *Siyar al-Nāṣir lil-ḥaqq*, a history of dynasties which contains notes on towns of southern and western Europe—quoted Ibrāhīm for his particularly rich account of the Slav areas. Since Ibrāhīm was able to learn much at first hand from the Jewish natives, he gave a reliable description of the articles of commerce, their prices, local manufactures, the military situation, customs of the people, Jewish life and merchants, Old Russian history and tribes, agriculture, health conditions, food and drinks, the salt pans of Soest, Sāmānid dirhams struck at Samarkand (914–915) but found in the German region, and much else.

Most of Ibrāhīm's material was obtained directly by observation, some of it from the natives' oral literature, and some from written works. It is recognized as being one of the best accounts of the period. His narrative is of great interest in the history of medieval cultures. He writes that the Slavs have no bathhouses but that they erect a wooden house and caulk it with a green marsh moss they call *mokh*, a substance used on their boats instead of tar. Then a stone is placed in a corner under a high window which lets out the smoke from the fire built to heat the stone. Basins of water are poured over this red-hot stone to create steam. Each person stirs the air with a bundle of dried branches to draw the heat to himself.

Ibrāhīm's attention to detail is shown in his description of the vehicle of kings as having four wheels and posts. The frame of the cab is hung from the posts by strong chains which are wrapped with silk so that the shaking inside the cab is reduced. Such vehicles were made also for the sick and wounded. The fragments on the Slavs have been much studied in eastern Europe, and the portions on western Europe have been translated into French by A. Miquel. A complete edition is planned by M. Canard.

BIBLIOGRAPHY

The following works may profitably be consulted: S. M. Ali, *Arab Geography* (Aligarh, 1959), p. 97; C. Brockelmann, "Zur arabischen Handschriftenkunde," in *Zeitschrift für Semitisik*, **10** (1935), 230–233; and *Geschichte der arabischen Literatur*, I (Leiden, 1937), 410; M. Conrad, "Ibrāhīm ibn Ya'qūb et sa relation de voyage en Europe," in *Études d'orientalisme Lévi-Provençal*, II (Paris, 1962), 503–508; R. Frye, "Remarks on Some New Islamic Sources of the Rūs," in *Byzantion*, **18** (1948), 119–125; I. Hrbek, "Arabico-Slavica," in *Archiv für Orientforschung*, **23** (1955), 109–135; G. Jacob, *Ein arabischer Berichterstatter aus dem 10. Jahrhundert über Fulda, Schleswig, Soest, Paderborn und andere Städte* (Berlin, 1891); *Studien in arabischen Geographen* (Berlin, 1892–1896); and *Arabische Berichte von Gesandten an germanische Fürstenhöfe aus dem 9. und 10. Jahrhundert* (Berlin, 1927); T. Kowalski, *Relacja Ibrāhīma ibn Ja'ḳūba z podrózy do Krajów slowiańskich w przekazie al-Bekrīego* (Cracow, 1946); I. Y. Krachkovsky, *Arabskaya geograficheskaya literatura* (Moscow, 1957), pp. 190–193; A. Kunik and V. Rosen, *Izvestiya al-Bekri i drugikh avtorov o Rusi i Slavyanakh* (St. Petersburg, 1878–1903); A. Miquel, "L'Europe occidentale dans la relation arabe d'Ibrāhīm b. Ya'qūb," in *Annales de l'École supérieure des sciences* (1966), 1048–1064; *La géographie humaine du monde musulman jusqu'au milieu de XIe siècle* (Paris, 1967), pp. 146–148; and "Ibrāhīm b. Ya'kūb," in *Encyclopaedia of Islam* (Leiden, 1969), III, 991; V. Minorsky, *Ḥudūd al-'Ālam "The Regions of the World"—A Persian Geography* (London, 1937), pp. 191, 427–428, 442; Ramhumar Chaube, "India as Described by an Unknown Early Arab Geographer of the Tenth Century," in *Proceedings of the Third Indian Hist. Congress* (1939), pp. 661–671; Semen Rapoport, "On the Early Slavs. The Narrative of Ibrāhīm-ibn-Yakub," in *Slavonic and East European Review*, **8** (1929–1930), 331–341; and F. Westberg, "Ibrāhīm's Reisebericht," in *Mémoires de l'Académie des sciences de St. Pétersbourg*, **3** (1898).

MARTIN LEVEY

IDDINGS, JOSEPH PAXSON (*b.* Baltimore, Maryland, 21 January 1857; *d.* Brinklow, Maryland, 8 September 1920), *petrology*.

Iddings was the son of William Penn Iddings and the former Almira Gillet. His father encouraged him to become a mining engineer, and he graduated from the Sheffield Scientific School of Yale University in 1877. Following a year as an instructor at Yale, he spent a year at the School of Mines, Columbia University, eventually turning to geology as a result of the influence of Clarence King. During 1879–1880 Iddings studied at Heidelberg under K. H. F. Rosenbusch, an experience which led to his career as a petrographer. On his return to America in 1880 he joined the U.S. Geological Survey, at about the same time as C. Whitman Cross, and served through 1892. He then joined the new department of geology at the University of Chicago, leaving abruptly in 1908 when he learned of the death of an aunt; an inheritance was presumed to be involved. The remainder of his bachelor life was spent collecting, writing, lecturing, traveling, conversing with his scientific friends in Washington, or residing at his ancestral home in Maryland.

One of the early few to study thin rock sections by means of the microscope, Iddings became one of the foremost petrographers of his time through his detailed, worldwide collecting and study of igneous rocks. His early broad surveys of rocks were done mainly in conjunction with fieldworkers such as Arnold Hague, C. D. Walcott, and G. F. Becker. Participation in the exploration and mapping of the geology of Yellowstone National Park was the most rewarding of Iddings' field studies. He concluded from these studies that the textural and chemical variation of igneous rocks depends on the variety of physical conditions imposed by the geological environment; that the consanguinity of some igneous rocks can be attributed to descent from a common parental magma; that mineralogical and structural variations were due largely to the rate of cooling of the magma; and that volatile constituents played a special role in rock magmas. His physicochemical approach to rocks, still valid today, greatly influenced the Carnegie Institution of Washington in the establishment of the Geophysical Laboratory as well as the course of petrology.

Iddings' teaching duties led to his need for a satisfactory classification of rocks. He enlisted the help of his friends C. W. Cross, L. V. Pirsson, G. H. Williams, and later, on Williams' death, H. S. Washington. They collaborated in revising rock nomenclature and expressing the compositional variations among rocks on a quantitative basis. The widely used normative method of calculating simple theoretical minerals from a chemical analysis of a rock is referred to, after its authors (alphabetically arranged), as the C.I.P.W. system. This classification served as the basis for a unique and original two-volume work, the first volume dealing with the physical chemistry of magmas and the second, a compilation of the geographic distribution of igneous rocks and the related problem of petrographic provinces.

Iddings' course of Silliman lectures at Yale in 1914 was published as *The Problem of Volcanism.* This honor was preceded by election to the National Academy of Sciences in 1907, honorary membership in the Société Française de Minéralogie in 1914, and an honorary D.Sc. from Yale (1907). Iddings was described as a very reserved and shy man for a world traveler, yet possessing personal charm, broad culture, and a poetic view of his surroundings—whether they were the rock sections of his profession, the butterflies of his hobby, or the landscapes of his travels.

BIBLIOGRAPHY

I. Original Works. Iddings' U.S. Geological Survey publications are *On the Development of Crystallization in the Igneous Rocks of Washoe, Nevada, With Notes on the Geology of the District*, Bulletin no. 17 (Washington, D.C., 1885), written with A. Hague; a description of rocks in J. S. Diller, *The Educational Series of Rock Specimens*, Bulletin no. 150 (Washington, D.C., 1898), *passim;* "Microscopical Petrography of the Eruptive Rocks of the Eureka District, Nevada," app. B of Arnold Hague, *Geology of the Eureka District, Nevada*, Monograph no. 20 (Washington, D.C., 1892), pp. 337–404; and the following articles in monograph no. 32, *Geology of Yellowstone National Park, Part II* (Washington, D.C., 1899): "Descriptive Geology of the Gallatin Mountains," pp. 1–59, written with W. A. Weed; "The Intrusive Rocks of the Gallatin Mountains," pp. 60–88; "The Igneous Rocks of Electric Peak and Sepulchre Mountain, Yellowstone National Park," pp. 89–148; "Descriptive Geology of the Northern End of the Teton Range, Yellowstone National Park," pp. 149–164, written with W. H. Weed; "The Dissected Volcano of Crandall Basin, Wyoming," pp. 215–268; and "The Igneous Rocks of the Absaroka Range, Yellowstone National Park," pp. 269–439.

Journal articles include "Notes on the Change of Electric Conductivity Observed in Rock Magmas of Different Composition on Passing From Liquid to Solid," in *American Journal of Science*, 3rd ser., **44** (1892), 242–249, written with Carl Barus; "The Origin of Igneous Rocks," in *Bulletin of the Philosophical Society of Washington*, **12** (1897), 89–216; "Chemical and Mineral Relationships in Igneous Rocks," in *Journal of Geology*, **6** (1898), 219–237; "A Quantitative Chemico-Mineralogical Classification and Nomenclature of Igneous Rocks," *ibid.*, **10** (1902), 555–690, written with C. W. Cross, L. V. Pirsson, and H. S. Washington; "The Isomorphism and Thermal Properties of the Feldspars. Part II. Optical Study," in *Publications. Carnegie Institution of Washington*, **31** (1905), 77–95, written with A. L. Day and E. T. Allen; "The Texture of Igneous Rocks," in *Journal of Geology*, **14** (1906), 692–707, written with C. W. Cross, L. V. Pirsson, and H. S. Washington; and "Some Examples of Magmatic Differentiation and Their Bearing on the Problem of Petrographical Provinces," in *Comptes rendus. XIIe session du Congrès géologique international* (Toronto, 1914), pp. 209–228.

Iddings' books are *Quantitative Classification of Igneous Rocks, Based on Chemical and Mineral Characters, With Systematic Nomenclature* (Chicago, 1903); *Rock Minerals, Their Chemical and Physical Characters and Their Determination in Thin Sections* (New York, 1906); *Igneous Rocks:* I, *Composition, Texture, and Classification* (New York, 1909), and II, *Description and Occurrence* (New York, 1913); and *The Problem of Volcanism* (New Haven, 1914).

II. SECONDARY LITERATURE. On Iddings and his work, see "Memorial of Joseph Paxson Iddings," in *Bulletin of the Geological Society of America*, **44** (1933), 352–374.

H. S. YODER, JR.

IDELSON, NAUM ILICH (*b*. St. Petersburg, Russia, 13 March 1885; *d*. near Riga, Latvian S.S.R., 14 July 1951), *astronomy, history of astronomy*.

Idelson's father, a mathematician, wished his son to become a lawyer. After graduating from the Gymnasium he entered the law faculty of St. Petersburg University, studying mathematics at the same time in the physical-mathematical faculty. His brilliance enabled him to graduate in 1909 from both faculties. For a while he was assistant to a barrister, but his interest in mathematics led to his teaching that subject in a secondary school. Obviously a born teacher, he devoted all his free time to serious scientific studies—theoretical astronomy and celestial mechanics.

In 1914 Idelson was elected a member of the Russian Astronomical Society and of the Society of Amateur Naturalists. The latter society brought together many professional and amateur astronomers. In 1918 Idelson was invited to join the computational bureau of the astronomical section of the P. F. Lesgaft Scientific Institute, headed by the extraordinary scientist and revolutionary N. A. Morozov (named Shlisselburgsky). This bureau computed the astronomical tables known as "The Canon of Solar Eclipses" (similar to Oppolzer's well-known "Canon of Eclipses"), which are indispensable for the study of the chronology of Russian history.

Although a project for compiling a Russian astronomical yearbook had been proposed in April 1917 at the First All-Russian Astronomical Congress in Petrograd, not until 1919 was a special institution created—the State Computing Institute—to satisfy the new nation's need for precise astronomical data for both scientific and practical use. Idelson became the head of the group computing the basic tables of the yearbook, for which project he studied in depth the theory and technique of compiling astronomical ephemerides.

In 1923 the Astronomical-Geodesic Institute was merged with the Computing Institute to form the Leningrad Astronomical Institute (from 1943 the Institute of Theoretical Astronomy of the Academy of Sciences of the U.S.S.R.). Idelson became the head of its astrometrical section and, in 1924, assistant to the director, B. V. Numerov. In 1924 Idelson visited the Berlin Computing Institute, the Computing Institute at Frankfurt-am-Main, the Paris Bureau of Longitudes, and a number of French observatories. His familiarity with the activity of foreign astronomical computing institutions aided the progress of corresponding work at the new institute, which besides the basic *Astronomical Yearbook* began publication in 1929 of the *Marine Astronomical Yearbook* and, later, the *Aviation Yearbook* and *Ephemerides of 500 Zinger Pairs*.

At the end of 1920 the Pulkovo Observatory invited Idelson to direct the Petrograd section of its computing bureau, where he organized the compilation of tables of Besselian values *A*, *B*, *C*, *D*, and *E* for 1920–1960, necessary for processing meridional observations. Idelson also conducted a huge project for deriving corrections of the equinox from the series of Pulkovo observations of the sun's position in 1904–1915.

In 1926 Idelson, who had taught mathematics, mechanics, and geophysics at various higher educational institutions, was invited to Leningrad University. In 1933 he became professor of astronomy there, giving courses in theoretical astrometry, theory of tides, potential theory, theory of the shape of the earth, theory of mathematical analysis of observations, general mechanics, history of astronomy, and history of mechanics. From 1930 to 1937 he also occupied the chair of theoretical mechanics at the Leningrad Institute of Precision Mechanics and Optics.

In December 1941, Idelson was evacuated from blockaded Leningrad to Kazan, where he lectured on celestial mechanics at Kazan University, occupied the chair of geophysics, and headed the gravimetry laboratory. After his return to Leningrad he renewed his work at the university, at the Institute of Precision Mechanics and Optics, and at the Pulkovo Observatory, where he directed the astrometrical section.

Idelson's areas of basic scientific interest were fundamental astrometry, celestial mechanics, and the history of astronomy. His works related to the theory of ephemerides, published in the appendixes to the *Astronomichesky ezhegodnik* ("Astronomical Yearbook") in 1941 and 1942, gained wide recognition, as did those in potential theory and the theory of the shape of the earth, the subject of his basic monograph *Teoria potentsiala s prilozheniem k teorii figury Zemli i geofizike* ("Potential Theory With an Application to the Theory of the Shape of the Earth and to Geophysics").

Idelson left a deep mark on the history of astronomy. Particularly notable are his excellent book *Istoria kalendarya* ("History of the Calendar," 1925) and his articles on this subject in the *Great Soviet Encyclopedia;* his discerning commentaries and articles on Clairaut, Appell, Lobachevsky, Copernicus, and Galileo; and his sketches of Newton, Laplace, and Le Verrier. It is curious that one of Idelson's first publications in the history of astronomy, *Istoria i astronomia* ("History and Astronomy," 1925), was devoted to a criticism of the "horoscopical method" of investigating the facts of world history developed by Morozov. In it he showed a broad knowledge of the scientific literature of antiquity and great feeling for the spirit of the epoch. He established the authenticity of ancient scientists' astronomical observations and the special importance of Ptolemy's *Almagest* in the history of science.

Idelson's most important and original work in the history of astronomy was *Etyudy po istorii planetnykh teory*, in which he masterfully investigated the development from Hipparchus to Kepler of mathematical methods of representing the movement of the planets. The precision of his succinct description of the scientists of the past and the depth of his scientific analysis make Idelson an unsurpassed authority on problems in the history of astronomy. Unfortunately, all his work in this field has been published only in Russian and has not attracted the foreign attention that it deserves.

BIBLIOGRAPHY

I. Original Works. The complete list of Idelson's works in Yakhontova's biography (see below) includes more than 70 titles. The following is a representative list: "Tables auxiliaires pour le calcul des quantités Besseliennes A, B, C, D, E, pour 0^h temps sidéral Poulkovo et 12^h temps sidéral Greenwich pour les époques 1920–1960," in *Izvestiya Glavnoi astronomicheskoi observatorii v Pulkove*, no. 91 (1921), pp. 235–274; *Istoria kalendarya* ("History of the Calendar"; Leningrad, 1925); *Uravnitelnye vychislenia po sposobu naimenshikh kvadratov* ("Equalizing Calculations for the Method of Least Squares"; Leningrad, 1926; 2nd ed., rev. and enl., 1932); "Die Stokesche Formel in der Geodäsie als Lösung einer Randwertaufgabe," in *Beiträge zur Geophysik*, **29**, no. 2 (1931), 156–160, written with N. R. Malkin; *Teoria potentsiala i ee prilozhenia k geofizike* ("The Theory of Potential and Its Application to Geophysics"), 2 pts. (Leningrad–Moscow, 1931–1932), 2nd ed., rev. and enl. (Moscow–Leningrad, 1936); "Über die Bestimmung der Figur der Erde aus Schwerkraftmessungen," in *Comptes rendus de la 7e séance de la Commission géodésique baltique* (Helsinki, 1935), pp. 9–23; "Reduktsionnye vychislenia v astronomii" ("Calculations of Reductions in Astronomy"), in *Astronomichesky ezhegodnik SSSR na 1941 g.* ("Astronomical Yearbook of the U.S.S.R. for 1941"; Moscow, 1940), pp. 379–432; "Zamechania po povodu teorii Lomonosova o kometnykh khvostakh" ("Remarks on Lomonosov's Theory of Comet Tails"), in *Lomonosov. Sbornik statey i materialov* ("Lomonosov. Collected Papers and Materials"; Moscow–Leningrad, 1940), pp. 66–116; "Galiley v istorii astronomii" ("Galileo in the History of Astronomy"), in *Galileo Galilei, 1564–1642* (Moscow, 1943), pp. 68–141; "Zakon vsemirnogo tyagotenia i teoria dvizhenia Luny" ("The Universal Law of Gravity and the Theory of the Moon's Movement"), in *Isaac Newton, 1643–1727* (Moscow, 1943), pp. 161–210; *Sposob naimenshikh kvadratov i teoria matematicheskoy obrabotki nablyudeny* ("The Method of Least Squares and the Theory of the Mathematical Treatment of Observations"; Moscow, 1947); "Zhizn i tvorchestvo Kopernika" ("The Life and Work of Copernicus"), in *Nikolay Kopernik* (Moscow, 1947), pp. 5–42; "Etyudy po istorii planetnykh teory" ("Studies in the History of the Theory of Planetary Movement"), *ibid.*, pp. 84–179; "Lobachevsky—Astronom," in *Istoriko-matematicheskie issledovaniya*, no. 2 (1949), pp. 137–167; and "Raboty A. N. Krylova po astronomii" ("The Works of A. N. Krylov in Astronomy"), in *Trudy Instituta istorii estestvoznaniya i tekhniki. Akademiya nauk SSSR*, **15** (1956), 24–31.

II. Secondary Literature. An obituary is in *Astronomichesky tsirkular SSSR*, no. 117 (1951), p. 14. Biographies include N. S. Yakhontova in *Istorikoastronomicheskie issledovaniya*, no. 4 (1958), pp. 387–405, with complete bibliography; and S. N. Korytnikov, *ibid.*, pp. 407–431, dealing with Idelson's work in the history of astronomy. See also *Bolshaya sovetskaya entsiklopedia*, 2nd ed., XVII, 327.

P. G. Kulikovsky

AL-IDRĪSĪ, ABŪ ʿABD ALLĀH MUḤAMMAD IBN MUḤAMMAD IBN ʿABD ALLĀH IBN IDRĪS, AL-SHARĪF AL-IDRĪSĪ (*b.* Ceuta, Morocco, 1100; *d.* Ceuta, 1166), *geography, cartography.*

Al-Idrīsī belonged to the house of the ʿAlavī Idrīsīs, claimants to the caliphate who ruled in the region around Ceuta from 789 to 985; hence his title "al-

Sharīf" (the noble) al-Idrīsī. His ancestors were the nobles of Málaga but, unable to maintain their authority, they migrated to Ceuta in the eleventh century. Al-Idrīsī was educated in Córdoba, then an important European center of learning. He started his travels when he was hardly sixteen years old with a visit to Asia Minor. He later traveled along the southern coast of France, visited England, and journeyed widely in Spain and Morocco. In 1138 he received an invitation from Roger II, the Norman king of Sicily, to visit Palermo where he remained at the request of the king, who told him: "You belong to the house of the Caliphs. If you live among the Muslims, their kings will contrive to kill you, but if you stay with me you will be safe" (al-Ṣafadī, cited from *Arabskaya geograficheskaya literatura*, Arabic trans., I, 282–283). Al-Idrīsī lived in Palermo until the last years of his life, returning to Ceuta after Roger's death in 1154.

An important meeting ground of Arab and European culture, Sicily was at this time the political seat of the Normans, who were keenly interested in the promotion of the arts and sciences. It was in this atmosphere that al-Idrīsī, under the patronage of Roger II, collaborated with Christian scholars and made important contributions to geography and cartography. Roger himself displayed great interest in these subjects and wished to have a world map constructed and a comprehensive world geography produced that would present detailed information on various regions of the world. It is possible that his objective was political, but in any case he was unsatisfied with existing Greek and Arabic works. After sending envoys to collect firsthand information in various regions, he ordered the construction of a large circular map of the world in silver. Al-Idrīsī, with the assistance of technicians and other scholars, constructed the relief map, depicting on it the seven climes, rivers and gulfs, seas and islands, mountains, towns and ports, and other physical features. The data utilized were drawn from Greek and Arabic sources as well as from the accounts of Roger's envoys and other travelers.

Only the geographical compendium with sectional maps, entitled *Kitāb nuzhat al-mushtāq fī ikhtirāq al-āfāq*, is extant today. The sectional maps are, in all probability, reproductions of the silver map; their basic framework is Ptolemaic. The inhabited world (*oikoumenē*), mainly of the northern hemisphere, is divided into seven latitudinal climes (*iqlīm*), parallel to the equator. Each clime is subdivided longitudinally into ten sections; for each of the seventy sections there is a separate map. By joining the sections a total picture of the world known to the Arabs and the Normans may be obtained. But the placing of a vast amount of information, both ancient and contemporary, on a 1,000-year-old map by Ptolemy produced a somewhat distorted picture of the relative geographical positions of certain places. Again, the maps and their descriptions do not always concur in their details, probably because the two were compiled separately. It is also evident that the author's knowledge of Europe, the Mediterranean region, and the Middle East was more accurate and reliable than that of other parts of the world. In addition, the maps were not drawn mathematically, and latitudes and longitudes known to Arab and Greek astronomers and geographers were not used in determining the geographical positions of place names.

The extremely rich and varied information presented in the text pertains to various countries of Europe, Asia, and North Africa and not only includes topographical details, demographic information, and reports of descriptive and physical geography but also describes socioeconomic and political conditions. It is thus a rich encyclopedia of the medieval period. The material is cataloged and indexed by section and then inserted into its proper place along with the sectional map. Al-Idrīsī's geographical conceptions are based mainly on the theoretical works of Greek and early Arab geographers and astronomers: he displays no originality of thought in mathematical and astronomical geography. Among his important sources are Ptolemy's *Geography*, Abu'l-Qāsim ibn Ḥawqal's *Kitāb ṣūrat al-arḍ*, Abu'l-Qāsim ʿUbayd Allāh ibn ʿAbd Allāh ibn Khurradādhbih's *Kitāb al-masālik wa'l-mamālik*, and Abū ʿAbd Allāh Muḥammad ibn Aḥmad al-Jayyānī's *Kitāb al-masālik wa'l-mamālik*.

Al-Idrīsī's work represents the best example of Arab-Norman scientific collaboration in geography and cartography of the Middle Ages. For several centuries the work was popular in Europe as a textbook; a number of abridgments were also produced, the first being published at Rome in 1592. A Latin translation by Joannes Hesronite and Gabriel Sionita was published at Paris in 1619 under the rather misleading title *Geographia Nubiensis*. Not until the middle of the nineteenth century was a complete two-volume French translation produced, by P. Amédée Jaubert, *Géographie d'Edrisi* (Paris, 1836–1840). Many scholars have edited and translated sections of the work pertaining to various countries. A complete edition of the text with translation and commentary is being prepared in Italy under the auspices of Istituto Universitario Orientale di Napoli and Istituto

Italiano per il Medio e l'Estremo Oriente. Two fascicules of the text have appeared under the title *Opus Geographicum.*

BIBLIOGRAPHY

See I. Y. Krachkovsky, *Arabskaya geograficheskaya literatura* (Moscow–Leningrad, 1957), translated into Arabic by Ṣalāḥ al-Dīn ʿUthmān Hāshim as *Taʾrikh al-adab al-jughrāfī al-ʿArabī*, I (Cairo, 1963); Konrad Miller, *Mappae Arabicae*, 5 vols. (Stuttgart, 1926–1930); and Giovanni Oman, "Notizie bibliografiche sul geografo arabo al-Idrīsī (XII secolo) e sulle sue opere," in *Annali dell'Istituto universitario orientale di Napoli*, n.s. **11** (1961), 25–61.

S. MAQBUL AHMAD

IKHWĀN AL-ṢAFĀʾ (also known as the **Brethren of Purity**). Secret association founded at Basra *ca.* 983.

For a complete study see Supplement.

IMAMURA, AKITUNE (*b.* Kagoshima, Japan, 14 June 1870; *d.* Tokyo, Japan, 1 January 1948), *seismology.*

Imamura was the second son of Akikiyo Imamura, a member of the Shimazu feudal clan. Among his ancestors was Eisei Imamura, a Dutch scholar and pioneer of veterinary science in Japan. The Meiji restoration of 1868 resulted in reducing most members of the feudal clans to near poverty, and Imamura was brought up in straitened circumstances. He early showed scholarly brilliance, and his family made great sacrifices to send him to Tokyo Imperial University. In 1894 he graduated in physics and became a university research assistant.

Two major earthquakes in 1891 and 1894 kindled in Imamura an intense desire to study seismology. He made important contributions to the scientific side of the subject but is noted above all for his concern with the human aspects. Dedicated to the problem of predicting earthquakes and mitigating their effects, he devoted his research principally, although not exclusively, to that end and was a member of numerous civil committees concerned with earthquakes.

In 1905 Imamura received the D.Sc. for work on the travel times of seismic waves. In 1923 he became a professor at Tokyo Imperial University and president of the Earthquake Investigation Committee of Japan. The following year he established at the Faculty of Science a department of seismology which has become world-famous. He also held posts at the imperial universities of Kyoto, Kyushu, and Tohoku. He was a founding member of the Japanese Academy of Sciences and of the Earthquake Disaster Prevention Society of Japan, and was president of the Seismological Society of Japan.

In his more purely scientific work Imamura contributed to the development of the seismograph and of other instruments. He carried out special studies of tiltmeter records with a view to obtaining clues on impending earthquakes and was among the first to show systematic connections between ground tilting and earthquake occurrence. He made important related studies of earthquake foreshocks and aftershocks. Imamura carried out intensive field studies and analyses of macroseismic effects of earthquakes—effects observed in damaged areas—and drew up maps of expected earthquake intensities in specific regions of Japan.

Passionately convinced that his mission in life was to mitigate disastrous earthquake effects, Imamura held to this vision; his last paper on the subject was dictated from his deathbed. His efforts were on the widest scale, including scientific efforts to predict earthquakes, steps to have seismic areas of Japan finely surveyed, efforts to have buildings made earthquake-proof, measures to improve Japan's fire-fighting facilities, and campaigns to educate the public on earthquake precautions. Examples of the detail to which Imamura went are his advocacy that kerosene lamps be abolished in Tokyo, his success in including a section on earthquakes in primary school syllabuses, and his public advocacy—to the extent of addressing a public meeting at a street corner at the age of seventy-seven. He was responsible for the reinforcement of the Imperial Diet Building in Tokyo, following his own experiments, and the erection of protective barriers against tsunami (seismic sea waves) along vulnerable coastlines. Knowledge of the Ainu language enabled him to determine the present locations of many obsolete Japanese place names and thereby to interpret more fully accounts of ancient earthquakes.

BIBLIOGRAPHY

I. ORIGINAL WORKS. Imamura's principal publication is *Theoretical and Applied Seismology*, published in Japanese (Tokyo, 1936), trans. into English by D. Kennedy (Tokyo, 1937). This book incorporates his most important contributions to seismology, including summarized accounts of the content of many earlier papers published (in Japanese and English) in Japanese seismological journals.

II. SECONDARY LITERATURE. References to Imamura's more important findings and published papers are included in C. F. Richter, *Elementary Seismology* (San Francisco, 1958); and Takeo Matuzawa, *Study of Earthquakes* (Tokyo, 1964). An obituary article is H. Kawasumi, in *Jishin*, **18** (May–Dec. 1948), 1–11, in Japanese.

K. E. BULLEN

INFELD, LEOPOLD (*b.* Cracow, Poland, 20 August 1898; *d.* Warsaw, Poland, 15 January 1968), *theoretical physics.*

Infeld was the son of Salomon and Ernestyna Infeld, his father being a merchant in the leather business. Against his will Leopold was sent to a commercial school. He became interested in the theory of relativity as a student of physics at the Jagiellonian University in Cracow. In 1920 he went to Berlin for eight months, where he met Einstein and worked on his doctoral dissertation, which was devoted to the problem of light waves in general relativity. In 1929 Infeld was given an appointment at the University of Lvov. After a short visit to Leipzig in 1932, he undertook research on spinor analysis in Riemannian spaces, a spinor being a mathematical concept used to describe particles possessing intrinsic angular momentum (spin). In a paper published in 1933, Infeld and Bartel L. van der Waerden showed how spinor calculus can be extended to take into account the influence of gravitation on spinning particles.

As a fellow of the Rockefeller Foundation, from 1933 to 1934, Infeld spent over a year in Cambridge, England. He gave there a new interpretation of a theory of nonlinear electrodynamics and, together with Born, worked on how to describe particles and quanta within that theory, which is now known as the Born-Infeld theory. In 1936, at Einstein's suggestion, Infeld was offered a fellowship at the Institute for Advanced Study in Princeton. He accepted it and, at Princeton from 1936 to 1938, joined Einstein in research on the problem of motion of heavy bodies according to the theory of general relativity. Together with Banesh Hoffmann, they laid the foundations of a new approximation method, today known as the EIH (Einstein-Infeld-Hoffmann) method. The method is well suited for the solution, within the framework of general relativity, of all problems related to the motion of slowly moving, gravitating bodies. One of the principal results is that the motion of bodies, described by singularities of the field, is determined by the equations of the gravitational field. In this respect, general relativity differs considerably from other physical theories, where the equations of motion usually have to be postulated separately.

Questions connected with the motion of bodies, such as the problem of gravitational radiation and the structure of sources, dominated the subsequent scientific activity of Infeld and his students. Infeld showed that gravitational radiation is strongly inhibited by the nature of Einstein's equations, and he simplified the derivation of post-Newtonian corrections to the motion of celestial bodies. These results are collected in *Motion and Relativity* (1960), a unique book on this subject, written by Infeld in collaboration with Jerzy Plebański.

At the end of Infeld's stay in Princeton, he wrote with Einstein *The Evolution of Physics* (1938), today a widely read book on modern physics for the layman. Later, at the University of Toronto (1938–1950), Infeld had many students and did research in several fields: he developed the factorization method of solving differential equations and, in collaboration with Alfred Schild, he formulated a new approach to relativistic cosmology, based on the similarity of propagation of light in cosmological models and in flat space. During the war, he worked on waveguides and antennas for a Canadian defense project. While in Canada, Infeld wrote *Whom the Gods Love* (1948), a biographical novel about Evariste Galois.

After a short visit to Poland in 1949, Infeld left Canada and, in the fall of 1950, returned to Warsaw to become professor at the university. He subsequently played a leading role in the development of theoretical physics in his country, becoming director of the new Institute of Theoretical Physics at the university and heading the theoretical division of the Institute of Physics of the Polish Academy of Sciences. Within a short time Infeld gathered around himself a large group of young physicists, whom he inspired with enthusiasm for scientific research and encouraged in new domains of physics and new approaches to science teaching.

Infeld's own research during the Warsaw period was devoted mostly to the classical theory of fields. But he also followed with fascination the rapid development of other areas of physics and encouraged Polish students to do theoretical work in nuclear, high-energy, and solid-state physics.

In recognition of his achievements, Infeld was awarded the highest distinctions by the Polish government. He was a member not only of the Polish Academy of Sciences but of several other academies. His activities were never restricted to the domain of science alone. Infeld did a great deal for international scientific cooperation, for physics in Poland, and especially for theoretical physics in Warsaw. His feeling of responsibility for the world prompted him to sign the Einstein–Russell appeal that gave birth to the Pugwash movement, in which he became very active.

BIBLIOGRAPHY

I. ORIGINAL WORKS. Infeld published more than 100 scientific papers and several books; a comprehensive list of his scientific writings may be found in *General Relativity and Gravitation*, I (1970), 191–208. In addition to the books mentioned in the text, Infeld wrote *Albert Einstein* (New York, 1950), a book describing Einstein's main ideas in simple terms.

II. SECONDARY LITERATURE. Infeld's *Quest* (New York, 1941) is an autobiography written in Canada and covering the period until 1939. *Szkice z przeszłości* ("Sketches From the Past") and *Kordian i ja* ("Kordian and I"), both published in Warsaw in 1965 and 1968, respectively, are two autobiographical books written in Polish by Infeld during the last years of his life. The former has also been published in German as *Leben mit Einstein* (Vienna, 1969). A third book, *W służbie cesarza i fizyki*, is in preparation.

ANDRZEJ TRAUTMAN

INGEN-HOUSZ, JAN (*b*. Breda, Netherlands, 8 December 1730; *d*. Bowood Park, near Calne, Wiltshire, England, 7 September 1799), *medicine, plant physiology, physics*.

Ingen-Housz[1] was the second son of Arnoldus Ingen-Housz and Maria Beckers. His father[2] was a leather merchant and is also mentioned as having been a pharmacist after 1755. The family was Roman Catholic.

Ingen-Housz was educated at the Breda Latin school where he excelled in classical languages. During the War of the Austrian Succession, British troops were encamped in Terheijden, a nearby village. The physician-general of the British forces, John Pringle, often visited in Breda and became a friend of the Ingen-Housz family. He took a great interest in the bright young Jan and later persuaded him to come to England, where he guided his career.

In the eighteenth century, Catholics in the Netherlands, rather than attend the national universities, preferred the Catholic University of Louvain, from which Ingen-Housz received the M.D. degree *summa cum laude* on 24 July 1753.[3] Probably at the advice of Pringle, he then matriculated at Leiden on 21 December 1754, where he most likely remained no longer than a year. There he continued his medical studies under H. D. Gaubius, a former pupil of Boerhaave and his successor in the chair of medicine and chemistry. He also studied anatomy under B. S. Albinus and physics under van Musschenbroek. He is said to have subsequently studied at Paris and Edinburgh.[4]

Ingen-Housz settled in Breda, where perhaps he had earlier started the family pharmacy.[5] While living at his father's house, he established a successful medical practice and began experimenting in physics and chemistry; he probably built his electrostatic machine during this period. Upon his father's death in 1764 the elder brother Ludovicus inherited the leather business and the pharmacy. Jan used his share of the estate to go to England.

In Edinburgh he became friends with the chemist W. Cullen, the physician A. Dick, and the anatomist A. Monro. He soon moved to London and under Pringle's guidance became acquainted with the anatomists W. Hunter, J. Hunter, the elder A. Monro, and the pediatrician G. Armstrong. He also met Priestley, who was then interested in electricity and optics, and Franklin, who has just then arrived in London. Franklin, especially, became a lifelong friend.

There was at this time a controversy raging in England over Daniel Sutton's revival of the use of inoculation against smallpox,[6] despite the charge that such inoculation had in the past proved dangerous and often lethal. It was known that Sutton had a secret regimen and special prescriptions and because his casualties were few, there were physicians who adopted his medical practices. Among them was William Watson, with whom Ingen-Housz worked from 1766 at the Foundling Hospital, where inoculation was mandatory. Watson soon entrusted Ingen-Housz with all inoculations, and in addition to working at other hospitals, Ingen-Housz developed a private inoculation practice. In early 1768 there was a serious smallpox epidemic of the dreaded confluent type in the villages of Berkamstead and Bayford. Ingen-Housz accompanied Thomas Dimsdale, another inoculation proponent, and together they inoculated 700 persons.

In the same year, through a dispute over method, Ingen-Housz' name became known in his native land. A. Sutherland planned to bring Sutton's method of inoculation to Holland. When his qualifications were challenged, C. Chais, pastor of the Walloon community in The Hague and author of a book defending inoculation against theological objections,[7] published a brochure that posed questions, together with answers he had received from Sutherland.[8] Ingen-Housz thereupon addressed an open letter to Chais in which he defended Dimsdale's method of inoculation. These brochures spurred an avalanche of pamphlets which led Fagel, secretary of the States General, to ask Ingen-Housz to settle as an inoculator in The Hague.

Ingen-Housz retained his interest in smallpox inoculation until his death and lived to see the introduction of Jenner's vaccination. He disagreed with Jenner on several points and correponded with him.[9]

In 1768 Ingen-Housz was sent by George III on

the advice of a commission[10] to the Austrian court to inoculate the royal family. After successfully inoculating the archdukes Ferdinand and Maximilian and the Archduchess Therese, Ingen-Housz was showered with gifts and honors. Empress Maria Theresa appointed him court physician with a lifelong annual income of 5,000 gulden. While the empress was disappointed in her hopes that the shy, kind man would develop into an interesting courtier, Ingen-Housz' use of his financial independence—for research—proved of inestimable value. He traveled throughout the empire, inoculating relatives of the imperial family and practicing and teaching inoculation. In January 1771 he went to Paris and then to London, where he was admitted to the Royal Society on 21 March 1771.[11] He returned to Vienna in May 1772. In 1775 Ingen-Housz married Agatha Maria Jacquin, the daughter of the famous botanist N. Jacquin; they had no children.

In 1777 and 1778 he was again in Holland and in England, where he delivered the Bakerian lecture before the Royal Society in June 1778. A year later he made his famous experiments on photosynthesis at a country house near London and wrote his book on this subject; it was hurriedly printed in London and Ingen-Housz took copies with him when he returned to Vienna. In July 1780 he was in Paris, where he visited Franklin. Ingen-Housz stayed in Vienna until 1789. He then returned to Paris, arriving on the day the Bastille was stormed. The violence so alarmed him that he immediately left for the Netherlands, and from there returned to England. There is some indication that he actually planned to immigrate to the United States (he had already bought land near Philadelphia), but after Franklin died (17 April 1790) he decided against this move. He remained in London, often staying for long periods at Bowood Park, near Calne, the manor of the marquess of Lansdowne, the former patron of Priestley. It was at Bowood Park that Ingen-Housz died.

Ingen-Housz is most widely known for his discovery of photosynthesis.[12] In the summer of 1771 Priestley had discovered that plants can restore air that has been made unfit for respiration by combustion or putrefaction; that is, by having a candle burn out in it or an animal die in it. (In these early experiments Priestley found only a few cases in which plants did not "improve" the quality of spoiled air; in his experiments of 1779, however, he found many such cases.) Some investigators outside of England, notably the Swedish chemist C. W. Scheele, were unable to confirm Priestley's observations. Others, and particularly the Dutch chemists J. R. Deiman and Paets van Troostwijk, were more fortunate. They were the first to use the endiometric method in research on photosynthesis, and Ingen-Housz must have been familiar with their work.

Since there is no evidence that Ingen-Housz had worked on photosynthesis previously, it must have been Priestley's publications on the subject that motivated his own investigations.[13] The book in which Ingen-Housz reported his results, *Experiments Upon Vegetables, Discovering Their Great Power of Purifying the Common Air in the Sunshine and of Injuring it in the Shade and at Night*, advanced the understanding of the phenomenon considerably. He established that only the green parts of a plant can "restore" the air, that they do this only when illuminated by sunlight, and that the active part of the sun's radiation is in the visible light and not in the heat radiation. In addition he found that plants, like animals, exhibit respiration, that respiration continues day and night, and that all parts of the plant—green as well as nongreen, flowers and fruit as well as roots—take part in the process.

Felice Fontana had found that during respiration animals produce fixed air (CO_2), and Ingen-Housz determined that plant respiration produces the same gas. He proved these and several less important points through a series of well-designed experiments. His results also explained why Scheele's and some of Priestley's experiments had failed. (It is interesting to note that Priestley's failures in 1771 occurred during the dark month of December.)

In the years that followed, partly inspired by controversies with Priestley and Senebier, Ingen-Housz continued to study plant assimilation and respiration. He knew the amount of dephlogisticated air (oxygen) consumed by plants through respiration to be far smaller than the amount of the gas produced through photosynthesis. This finding led him to believe that plants and animals mutually support each other, the animals consuming dephlogisticated air and producing fixed air, the plants doing exactly the reverse. He also felt that the phenomenon of photosynthesis would enable scientists to establish the demarcation between plants and animals of the lower forms of life.

Because of their attacks on Ingen-Housz, Priestley and Senebier are often given credit for the discovery of photosynthesis. But it is beyond doubt that neither of the two had even a vague understanding of this process before the publication of Ingen-Housz' book in 1779; even in their later writings the process is not entirely clear. Therefore the discovery of both photosynthesis and plant respiration belongs to Ingen-Housz alone.

During this period Ingen-Housz met Sir John Sinclair, president of the Board of Agriculture, who encouraged his studies of plant nutrition. The chemistry of plant growth was not yet understood, especially the origin of carbon in plants. The French chemist Hassenfratz had proposed the theory that the carbon was taken up from the soil by the roots of the plants, the so-called humus theory; the humus, decayed remains of animals and plants in the soil, being the storehouse of the carbon. In a contribution intended for publication by the Board of Agriculture, "on the Food of Plants and the Renovation of Soils," Ingen-Housz declared that carbon dioxide in air is the source of carbon in plants, thus explaining the disappearance of the gas and the production of oxygen in photosynthesis.

The work on photosynthesis had given Ingen-Housz an understanding of the influence of air quality on the performance of organisms. As a physician he conjectured that respiratory ailments could be relieved if purer air could be administered. He consequently devised an apparatus to produce and administer to the patient pure, dephlogisticated air. Although there is no evidence that he employed this respiratory treatment himself, others soon followed his suggestion. Ingen-Housz can thus be credited with the initiation of oxygen therapy.

Ingen-Housz' interest in photosynthesis may well have derived from his interest in the chemistry of gases. He was especially interested in "inflammable air," at that time the name for combustible gases. He experimented with explosive gas mixtures and used a mixture of air and ethyl ether vapor as the propellant for a pistol, which was fired electrically. He also studied the production of dephlogisticated air from metal oxides, and was the first to try the demonstration-experiment of burning a steel wire in pure oxygen.[14] To replace the cumbersome tinderbox, he designed a hydrogen-fueled, electrically ignited lighter, and he substantially improved the phosphor matches invented by Peibla (1781).

Ingen-Housz' earliest scientific interest was in the field of physics, especially electricity. He was the first to use disks instead of revolving cylinders or globes in electrostatic generators.[15] In the Bakerian Lecture of 1778, he explained the phenomena of Volta's electrophore by means of Franklin's theory of positive and negative electricity, a crucial demonstration of the correctness of Franklin's view. He also entered the controversy on blunt lightning rods (advocated by Wilson) versus pointed ones (advocated by Franklin). While living in Austria, he often advised the government on the placement of lightning rods on powder houses. He experimentally refuted the statement that plant growth is promoted by electricity.

In the field of magnetism, Ingen-Housz experimented with artificial magnets, made according to the ideas of Gowin Knight, and devised methods of dampening the vibrations of magnetic needles. He also discovered the paramagnetism of platinum.

Ingen-Housz' research on algae led to three other significant contributions. Although he could not identify the "green matter" that Priestley spoke of (1779), he discovered its swarm spores, and his investigation prompted him to suggest the use of the very thin glass resulting from glassblowing operations as cover plates for liquid microscopic preparations, which greatly facilitated observation. But perhaps the most important consequence of his algae research was his discovery and correct description of Brownian motion.

Brown, in his discovery paper of 1827, had himself suggested the names of some possible precursors; but an examination of their work has shown that none of these scientists was aware that lifeless particles might show the motion.[16] Ingen-Housz demonstrated that finely powdered charcoal suspended in alcohol shows the irregular motion, just as minute organisms such as Infusoria will.[17] Although Ingen-Housz has not been properly credited with this discovery, his name did become associated with an experiment that Franklin inspired, the demonstration of the difference of heat conductivity of different metals. Ingen-Housz carried out Franklin's ideas and obtained correct, although only qualitative, results.[18]

NOTES

1. Ingen-Housz himself used this form of the name (although he sometimes signed his letters with Housz only) and was so referred to by his contemporaries. Descendants of Ingen-Housz' older brother Ludovicus usually write the name without the hyphen.
2. On Arnoldus and Ludovicus Ingen-Housz see G. J. Rehm, *De Bredasche apothekers van de 15e tot het begin van de 19e eeuw* (Breda, 1961), pp. 84–90.
3. Although biographers have generally stated that Ingen-Housz started his studies at Louvain in 1746, this is unlikely since the town was under enemy occupation after May 1746. It is probable that he did not go to Louvain before the autumn of 1748. The earliest matriculation record of Ingen-Housz falls between August 1750 and February 1751, allowing for only three years to obtain his medical degree. A printed dissertation is not known; Ingen-Housz was probably promoted after the defense of propositions. A broadside poem in Latin, issued at his promotion, establishes the date. A facsimile of this poem is reproduced by H. S. Reed, facing p. 31.
4. V. Flint, p. 9, states that she has a letter from the University of Edinburgh which confirms that Ingen-Housz was there between 1755 and 1757 but did not matriculate. He probably attended a *privatissimum* of W. Cullen.
5. Because, in Breda, an M.D. had the right to dispense prescriptions, it is logical to assume that the pharmacy was operated under the authority of Jan Ingen-Housz, rather than his father, who would have had to go through a four-year apprenticeship.
6. Inoculation consisted of introducing the live virus of the

INGEN-HOUSZ

disease under the skin of the patient and must be distinguished from vaccination, introduced later by Jenner (1798).

7. C. Chais, *Essai apologétique sur la méthode de communiquer la petite vérole par inoculation* (The Hague, 1754).
8. C. Chais, *Lettre . . . à Mr. Sutherland . . . sur la nouvelle méthode d'inoculer la petite vérole* (The Hague, 1768).
9. See P. W. van der Pas, "The Ingen-Housz-Jenner Correspondence."
10. *Oordeel van de genees—en heelmeesters van den koning van Engeland aangaande de manier van inentinge der kinderpokjes der heeren Suttonianen* (The Hague, 1767).
11. Ingen-Housz had been elected a fellow of the Royal Society on 25 May 1769.
12. In this brief discussion of Ingen-Housz' work, the old chemical nomenclature is used as it appeared in his papers and books on photosynthesis. It is not certain whether Ingen-Housz agreed with Lavoisier's views on phlogiston, but on 5 January 1788 he wrote a letter to J. H. Hassenfratz to thank him for "l'exemplaire de la nouvelle nomenclature chimique, que je n'approuve pas."
13. It would seem from a letter to Franklin dated 25 May 1779 (see I. M. Hays, *Calendar of the Franklin Papers*, II, 38) that Ingen-Housz had originally intended to retire to the country to finish his book on smallpox, which was never completed.
14. E. Cohen, "Wie heeft de verbranding van een horlogeveer in zuurstof het eerst uitgevoerd?," in *Chemisch weekblad*, **8** (1911), 87–92.
15. In the 1st ed. (1767) of *The History and Present State of Electricity*, Priestley attributed this invention to J. Ramsden; in the 2nd ed. (1769) he credits Ingen-Housz, who must have convinced him of his priority. Ingen-Housz' machine was still a friction and not an induction machine, which was the forerunner of the modern high-voltage electrostatic generators.
16. P. W. van der Pas, "The Early History of the Brownian Motion," in *Proceedings. Twelfth International Congress of the History of Science and Technology* (August 1968), *Actes XIIe Congrès International d'Histoire des Sciences* (Paris, 1968), VIII, 143–158.
17. See P. W. van der Pas, "The Discovery of the Brownian Motion."
18. Max Jacob, *Heat Transfer*, 2 vols (New York, 1949), I, 207–209, gives the mathematical theory of the experiment.

BIBLIOGRAPHY

I. Manuscripts. Wiesner (see below) mentions many letters and other personal writings preserved at various institutions. In addition he presents abstracts of documents concerning Ingen-Housz, preserved in a number of Austrian archives.

II. Original Works. As Wiesner's bibliography, while having few omissions, is wanting in accuracy, a complete bibliography is given here in the hope of providing more correct information; only the primary references are included.

(1) "Extract of a Letter From Dr. J. Ingenhousz to Sir John Pringle, Bart., PRS, Containing Some Experiments on the Torpedo, Made at Leghorn, January 1, 1773 (After Having Been Informed of Those by Mr. Walsh). Dated Salzburg, March 27, 1773," in *Philosophical Transactions of the Royal Society*, **65** (1775), 1–4.

(2) "Easy Methods of Measuring the Diminution of Bulk, Taking Place on the Mixture of Common Air and Nitrous Air, With Experiments on Platina," *ibid.*, **66** (1776), 257–267.

(3) "A Ready Way of Lighting a Candle, by a Very Moderate Electric Spark," *ibid.*, **68** (1778), 1022–1026.

(4) "Electrical Experiments, to Explain How Far the Phenomena of the Electrophorus May Be Accounted for by Dr. Franklin's Theory of Positive and Negative Electricity, Being the Annual Lecture, Instituted by the Will of Henri Baker, Esq., FRS.," *ibid.*, 1027–1048.

(5) "Account of a New Kind of Inflammable Air or Gas, Which Can Be Made in a Moment Without Apparatus, and Is as Fit for Explosion as Other Inflammable Gases in Use for That Purpose; With a New Theory of Gun Powder," *ibid.*, **69** (1779), 376–418.

(6) "On Some New Methods of Suspending Magnetic Needles," *ibid.*, 537–546.

(7) "Improvements in Electricity," *ibid.*, 659–673.

(8) "On the Degree of Salubrity of the Common Air at Sea, Compared With That of the Sea-shore and That of Places, Far Removed From the Sea," *ibid.*, **70** (1780), 354–377.

(9) "Exposition de plusieurs lois qui paroissent s'observer constamment dans les divers mouvements du fluide électrique et auxquelles les physiciens n'avoient pas fait une suffisante attention," in *Journal de physique théorique et appliquée*, **16** (1780), 117–126.

(10) "Uitslag der proefnemingen op de planten, strekkende ter ontdekking van derzelver zonderlinge eigenschap om de gemeene lucht te zuiveren op plaatsen waar de zon schijnt, en dezelve te bederven in de schaduwe en gedurende den nacht," in *Algemeene Vaderlandsche Letteroefeningen*, **2** (1780), 247–249.

(11) "Verhandeling over de gedephlogisteerde lucht en de manier hoe men dezelve kan bekomen en tot de ademhaling kan doen dienen," in *Verhandelingen Bataafsch genootschap der proefondervindelijke wijsbegeerte te Rotterdam*, **6** (1781), 107–160.

(12) "Some Farther Considerations on the Influence of the Vegetable Kingdom on the Animal Creation," in *Philosophical Transactions of the Royal Society*, **72** (1782), 426–439.

(13) "Observations sur la vertu de l'eau impregnée d'air fixe, de differens acides et de plusieurs autres substances, pour en obtenir par le moyen des plantes et de la lumière du soleil de l'air déphlogistiqué," in *Journal de physique théorique et appliquée*, **24** (1784), 337–348.

(14) "Réflexions sur l'économie des végétaux," *ibid.*, 443–455.

(15) "Remarques sur l'origine et la nature de la matière verte de M. Priestley, sur la production de l'air déphlogistiqué par le moyen de cette manière et sur le changement de l'eau en air déphlogistiqué," *ibid.*, **25** (1784), 3–12.

(16) "Remarques de M. Ingen Housz sur la lettre précédente avec quelques observations ultérieurs sur la vertu de l'eau impregnée d'air fixe," *ibid.*, 78–91, reply to a letter of Senebier (see below).

(17) "Lettre de M. Ingen Housz à M. Jan van Breda au sujet de la quantité d'air déphlogistiqué que les végétaux répandent dans l'atmosphère pendant le jour; au sujet des raisons de l'inexactitude de la quantité d'air déphlogistiqué qu'on obtient par les végétaux, exposés au soleil dans

l'eau imbibée d'air fixe, ainsi que sur la véritable cause de l'influence méphitique nocturne des végétaux dans l'air," *ibid.*, 437–450.

(18) "Observations sur la construction et l'usage de l'eudiomètre de M. Fontana et sur quelques propriétés particulières de l'air nitreux, adressé à M. Dominique Beck . . . ," *ibid.*, **26** (1785), 339–359.

(19) "Lettre de M. Ingen Housz à M. N. C. Molitor, . . . au sujet de l'effet particulier qui ont sur la germination des sémences et sur l'accroissement des plantes formés, les differentes espèces d'airs, les differens degrés de lumière et de la chaleur et de l'électricité," *ibid.*, **28** (1786), 81–92.

(20) "Lettre de M. Ingen Housz . . . à M. Molitor . . . au sujet de l'influence de l'électricité atmosphérique sur les végétaux," *ibid.*, **32** (1788), 321–337.

(21) "Lettre de M. Ingen Housz à M. de la Métherie sur les métaux comme conducteurs de la chaleur," *ibid.*, **34** (1789), 68–69.

(22) "Expériences qui prouvent: 1e, que les plantes évaporent une quantité plus grande d'air vital pendant le jour à l'air libre que nous en voyons répandre étant couvert d'eau pure; 2e, que leur évaporation nocturne d'un air méphitique, qui est très petite lorsqu'elles sont couvertes d'eau et très considerable dans l'état naturel, qu'il y a un mouvement et déplacement continuel du fluide aerien dans les végétaux," *ibid.*, 436–446.

(23) "Aqua mephitica alcalina of loogzoutig luchtzuur water, een nieuw ontdekt en uitmuntend geneesmiddel in het graveel en den steen," in *Scheikundige Bibliotheek*, **1**, no. 1 (1792), 41; no. 2, 95; no. 3, 175.

(24) "Brief aan Jan van Breda, behelzende eenige proeven met wormdoodende vogten genomen," *ibid.*, **2** (1794), 153.

Of the following books by Ingen-Housz, some are collections of his papers, but in most cases these reprints are considerably modified:

(25) *Lettre de Monsieur Ingenhousz, Docteur en médecine à Monsieur Chais, Pasteur de l'église Wallone de la Haye, au sujet d'un brochure contenant sa lettre à M. Sutherland, et une réponse de M. Sutherland à M. Chais, sur la nouvelle méthode d'inoculer la petite vérole* (Amsterdam, 1768).

(26) *Nova, tuta, facilisque methodus curandi calculum, scorbutum, podagram, destruendique vermes in humano corpore nidulantes. Cui addita est methodus extemporanea impregnandi aquam aliosque liquores aëre fixo etc. Latino sermone ab J. Ingenhousz* (Leiden, 1778; 2nd ed., Louvain, 1797), Latin trans. of a book by N. Hulme (see below), with appended notice by Ingen-Housz on the use of carbonated water; Dutch trans., *Nieuwe, veilige en gemakkelijke manier om den steen . . .* (Rotterdam, 1778); German trans., *Neue, sichere Methode der Heilung des Steins . . .* (Vienna, 1781).

(27) *The Baker Lecture for Year 1778, Read at the Royal Society, June 4, 1778. Experiments on the Electrophorus* (London, 1779); Dutch trans., *Proeve over den electrophorus . . .* (Delft, 1780); German trans., *Anfangsgründe der Elektrizität* (1781).

(28) *Experiments Upon Vegetables Discovering Their Great Power of Purifying the Common Air in the Sunshine and of Injuring it in the Shade and at Night, to Which is Joined, a new Method of Examining the Accurate Degree of Salubrity of the Atmosphere* (London, 1779); Dutch trans., *Proeven op plantgewassen . . .* (Delft, 1780), a promised second vol. never having been published; French trans., *Expériences sur les végétaux . . .* (Paris, 1780); 2nd ed. (1785), see G. A. Pritzel, *Thesaurus literaturae botanicae* (Milan, 1871; repr. 1950), no. 4435; 3rd ed., 2 vols.: I (Paris, 1787); II (Paris, 1789); no evidence of a third vol., which is sometimes mentioned, has been found; German trans., *Versuche mit Pflanzen, wodurch entdeckt worden dasz sie die Kraft besitzen, die atmosphärische Luft beim Sonnenschein zu reinigen, und im Schatten und des Nachtsüber zu verderben* (Leipzig, 1780); 2nd ed., 3 vols. (Vienna, 1786–1790), the best and most complete ed.

(29) *Nouvelles expériences et observations sur divers objets de physique*, 2 vols.: I (Paris, 1785); II (Paris, 1789); German trans., *Vermischte Schriften physisch-medizinischen Inhalts . . . nebst einigen Bemerkungen über den Einfluss der Pflanzen auf das Tierreich* (Vienna, 1782); 2nd ed., 2 vols.: I (Vienna, 1785); II (Vienna, 1784 [sic]); Dutch trans., *Verzameling van verhandelingen over verschillende natuurkundige onderwerpen*, 2 vols.: I (The Hague, 1784); II (The Hague, 1785).

(30) *Epistola ad J. A. Scherer* (Vienna, 1794).

(31) *Miscellana physico-medica* (Vienna, 1795).

(32) *Additional Appendix to the Outlines of the Fifteenth Chapter of the Proposed General Report From the Board of Agriculture; On the Subject of Manures* (London, 1796).

(33) *An Essay on the Food of Plants and the Renovation of Soils* (n.p., 1796); Dutch trans., *Proeve over het voedsel der planten en de vrugtbaarmaking van landerijen* (Delft, 1796); German trans., *Ueber die Ernährung der Pflanzen und Fruchtbarkeit des Bodens* (Leipzig, 1798).

III. Secondary Literature. The following items are from the contemporary literature which pertains to Ingen-Housz:

(34) N. Hulme, *A Safe and Easy Remedy Proposed for the Relief of the Stone, the Scurvy, Gout, etc., and for the Destruction of Worms in the Human Body . . . Together With an Extemporaneous Method of Impregnating Water . . . With Fixed Air* (London, 1778).

(35) W. Henley, "Observations and Experiences, Tending to Confirm Dr. Ingen Housz' Theory of the Electrophorus and to Show the Impermeability of Glass to the Electric Fluid," in *Philosophical Transactions of the Royal Society*, **68** (1778), 1049–1058.

(36) J. Senebier, "Mémoire sur la matière verte, ou plutôt sur l'espèce de Conserve, qui croît dans les vaisseaux pleins d'eau exposés à l'air et sur l'influence singulière de la lumière pour la déveloper," in *Journal de physique théorique et appliquée*, **17** (1781), 209–216.

(37) "Lettre de M. Senebier à M. Ingen Housz, (sur ses observations sur l'eau impregnée d'air fixe, et de differens acides)," *ibid.*, **25** (1784), 76–77.

(38) B. Franklin, "A Letter From Dr. B. Franklin to Dr. Ingenhausz . . . (Throughts Upon the Construction and Use of Chimneys)," in *Transactions of the American Philosophical Society*, **2** (1786), 1–27.

(39) "Lettre de M. Fontana au célèbre M. Ingen Housz . . . sur la décomposition de l'eau," in *Journal de physique théorique et appliquée*, **28** (1786), 310–315.

(40) "Lettre à M. Ingen Housz sur la décomposition de l'eau, par M. Adet," *ibid.*, 436–441.

(41) M. de la Métherie, "Réflexions sur la lettre précédente de M. Adet, relativement à la décomposition de l'eau," *ibid.*, **28** (1786), 442–446.

(42) "Effets de l'électricité sur les plantes, Réflexions ultérieurs sur le contenu du mémoire de M. Ingen Housz," *ibid.*, **35** (1789), 81–83.

(43) "Lettre de M. van Marum à Jean Ingen Housz, . . . contenant la description d'une machine électrique, construite d'une manière nouvelle et simple et qui réunit plusieurs avantages sur la construction ordinaire," *ibid.*, **38** (1791), 447–459.

(44) "Seconde lettre de M. van Marum à M. Jean Ingen Housz, . . . contenant quelques expériences et des considérations sur l'action des vaisseaux des plantes qui produit l'ascension et le mouvement de leur sève," *ibid.*, **41** (1792), 214–220.

(45) "Lettre de A. F. Humboldt à Ingen Housz sur l'absorption de l'oxygène par les terres," *ibid.*, **47** (1795), 377–378.

The following writings are of interest for data on the life of Ingen-Housz:

(46) M. J. Godefroi, "Het leven van Dr. Jan Ingen-Housz, geheimraad en lijfarts van Z. M. Keizer Josef II van Oostenrijk," in *Handelingen van het Provinciaal Genootschap van Kunsten en Wetenschappen in Noord Brabant* ('s Hertogenbosch, 1875).

(47) M. Treub, "Jan Ingen-Housz," in *De Gids*, **18** (1880), 478–500.

(48) H. W. Heinsius, "Jan Ingen-Housz," in *Album der Natuur*, **46** (1897), 1–15.

(49) J. Wiesner, *Jan Ingen-Housz, sein Leben und sein Wirken als Naturforscher und arzt* (Vienna, 1905), the best biography of Ingen-Housz.

(50) E. Mortreux, "Johannes Ingen-Housz," in *Nieuw Nederlandsch Biografisch woordenboek*, **6** (1912), 832–837.

(51) H. S. Reed, "Jan Ingen-Housz, Plant Physiologist. With a History of the Discovery of Photosynthesis," in *Chronica botanica*, **11** (1950), 285–396.

(52) *Dictionary of National Biography*, **10** (1903), 433.

(53) M. Speter, "Jan Ingenhousz' 'verbessertes' Sauerstoff Inhalierungs Apparat (1783–84) und dessen Ausgestaltung durch Paskal Joseph Ferro," in *Wissenschaftliche Mitteilungen des Drägerwerks*, no. 5 (1936).

(54) Vera Flint, "The Botanical Studies of Jan Ingen-Housz and the Influence of his Work on his Contemporaries and Successors," diss. (Univ. of London, 1950).

(55) P. W. van der Pas, "The Ingenhousz-Jenner Correspondence," in *Proceedings. Tenth International Congress of History of Science and Technology* (1964), 957–960; also in *Janus*, **51** (1964), 202–220, more complete.

(56) P. W. van der Pas, "The Discovery of the Brownian Motion," in *Scientiarium historia*, **13** (1971), 27–35.

P. W. VAN DER PAS

INGRASSIA, GIOVANNI FILIPPO (*b.* Regalbuto, Sicily, *ca.* 1510; *d.* Palermo, Sicily, 6 November 1580), *medicine.*

Nothing appears to be known with certainty about Ingrassia's family, and of his early education we know only that he first studied medicine in Palermo with Giovanni Battista di Pietra. Attracted by the fame of the medical faculty of the University of Padua, he went there to continue his studies and received the M.D. degree in 1537. Thereafter his activities are obscure until 1544, when he was invited to teach anatomy and the practice of medicine at the University of Naples. In 1556, on the recommendation of the Spanish viceroy of Sicily and by decree of Philip II of Spain, he was called to Palermo as *protomedicus.*

Little is known likewise of Ingrassia's medical practice in Palermo except for his celebrated case involving Giovanni d'Arragona, marquis of Terranova, who had received a penetrating wound of the left chest in a tournament. When the marquis failed to respond to his treatment, Ingrassia circularized the leading physicians of Europe for suggestions and ultimately elicited, in 1562, Vesalius' remarkable description of his surgical procedure for treatment of empyema. Ingrassia acknowledged the advice in the following year but declared that he found it unnecessary to employ it since the marquis had finally recovered. Nevertheless he published Vesalius' description of his procedure in *Quaestio de purgatione per medicamentum* (1568, pp. 92–98), as he declared, for the sake of posterity.

As *protomedicus*, Ingrassia was concerned for the most part with problems of hygiene, epidemiology, and the general administration of Sicilian medicine. His activities included efforts to suppress quackery, to control the pharmaceutical trade, and to improve the conditions in hospitals. He was able with some success to control the endemic malaria of Palermo through drainage of marshes, and his greater use of isolation hospitals (*lazzaretti*) was instrumental in decreasing the severity of the plague of 1575. It was Ingrassia's belief that there ought to be three kinds of hospitals: for those suspected to be infected, for the infected, and for the convalescent. The whole subject of plague and infection was discussed, with other matters of public health, in his *Informazione del pestifero morbo* (1576). Ingrassia was responsible for the establishment of one of the first sanitary codes and a council of public health. He was also a founder of the study of legal medicine, for which he composed his *Methodus dandi relationes* in 1578; owing to his death two years later, the book was not published until 1914.

Ingrassia is best known for his anatomical studies, admittedly based upon the methods and procedures of Vesalius, for whom he expressed the greatest admiration. These studies were for the most part the result of his period of teaching anatomy at Naples, but were only published posthumously under the title of *In Galenum librum de ossibus doctissima commentaria* (Palermo, 1603; Venice, 1604). This is a Latin version of Galen's work on osteology accompanied by Ingrassia's commentary, and demonstrated both Galen's dependence upon the study of nonhuman material and Ingrassia's own discoveries from his investigation of human osteology. Because of the long delay in publication of the book, whatever claims he may have had to certain discoveries were preempted by other scientists whose findings were printed during the second half of the sixteenth century. Ingrassia must nevertheless be recognized as having investigated and described the sutures of the skull in minute detail; as having provided a precise description of the sphenoid bone and its sinuses, as well as of the ethmoid; and as having displayed an excellent knowledge of the bony structure of the auditory apparatus. According to Falloppio in *Observationes anatomicae* (1561), Ingrassia in 1546 described orally to his students in Naples the third auditory ossicle or stapes, actually calling it *stapha* because of its resemblance in shape to the stirrup commonly used in Sicily; since his account of the ossicle did not appear in print until 1603, priority of published description must be awarded to the Spanish anatomists Pedro Jimeno (1549) and Luis Collado (1555).

Building upon the work of Vesalius, Ingrassia also provided an excellent description of the atlas and the atlanto-occipital articulation, and drew attention to the distinguishing differences between the male and female pubic bones. In addition to his work on osteology, he also made note of the existence of some of the blood vessels in the cerebral substance and in the walls of the ventricles that had not been described in the *Fabrica* of Vesalius.

Upon his death Ingrassia was entombed in the chapel of Santa Barbara in Palermo.

BIBLIOGRAPHY

The earliest study of Ingrassia of any value appears to have been that of A. Spedalieri, *Elogio storico di Giovanni Filippo Ingrassia* (Milan, 1817). Recognition of the four-hundredth anniversary (1910) of Ingrassia's birth led to two publications by G. G. Perrando: "Festiggiamenti commemorativi," in *Rivista di storia critica delle scienze mediche e naturali*, **1** (1910–1912), 75–79; and "La storia e le vicende di un prezioso codice ms. di Gianfilippo Ingrassia," *ibid.*, 29–41. Other articles include G. Petrè, "Pel IV centenario della nascita di G. F. Ingrassia," in *Atti della R. Accademia delle scienze mediche* (1913–1915), pp. 150–167; and G. Bilancioni, "L'opera medico-legale di Ingrassia," in *Cesalpino*, **11** (1915), 249–271.

A fairly sound but brief account of Ingrassia is to be found in P. Capparoni, "Giovan Filippo Ingrassia," in *Profili bio-bibliografici di medici e naturalisti celebri italiani dal sec. XV al sec. XVIII*, I (Rome, 1926), 42–44, which also contains an extensive short-title bibliography of Ingrassia's writings; and A. Piraino, "G. F. Ingrassia, l'Ippocrate siciliano del '500 e la sua opera," in *Cultura medica moderna*, **15** (1936), 270–278.

There is reference to the case of the marquis of Terranova in Harvey Cushing, *Bio-bibliography of Andreas Vesalius*, 2nd ed. (Hamden, Conn., 1962); and in C. D. O'Malley, *Andreas Vesalius of Brussels 1514–1564* (Berkeley–Los Angeles, 1964), in which the correspondence between Ingrassia and Vesalius is given in English trans.

C. D. O'MALLEY

INNES, ROBERT THORBURN AYTON (*b.* Edinburgh, Scotland, 10 November 1861; *d.* Surbiton, England, 13 March 1933), *astronomy.*

Innes, the eldest of twelve children of John Innes and Elizabeth Ayton, left school at the age of twelve; although thereafter he was entirely self-taught, this was apparent only in his unprejudiced and often unconventional outlook. His proficiency as a mathematician, even in his earlier years, is shown by his published contributions to celestial mechanics. Always preferring the direct approach, he tended to favor numerical methods such as Cowell's (he would have been in his element in the computer age) and his arithmetical adroitness was legendary. Yet it is by his outstanding ability as a practical astronomer and observer that Innes is chiefly remembered. He was elected a fellow of the Royal Astronomical Society at the age of seventeen, of the Royal Society of Edinburgh in 1904, and of several other learned societies; his doctorate of science from the University of Leiden was conferred in 1923 *honoris causa.* Throughout his life he took a leading part in civic cultural activities, where his wide range of interests, his persuasive diplomacy, and his unfailing urbanity found a natural outlet.

Shortly after his marriage in 1884 to Anne Elizabeth Fennell, by whom he had three sons, Innes emigrated to Sydney, Australia, where he prospered as a wine merchant; his leisure was devoted, as before, to astronomy. His success in a search for new double stars led Sir David Gill to offer him the post of secretary at the Cape Observatory, South Africa, at a very modest salary. Here Innes somehow found time to continue his double-star observations, to compile a catalog of southern double stars, and to revise the *Cape Photographic Durchmusterung*. In 1903 he was appointed director of the newly established Transvaal Observatory in Johannesburg. Although his official duties were meteorological, by 1907 he had acquired a nine-inch telescope; and in 1909, three years before the renamed Union Observatory became a purely astronomical institute, he persuaded the government to order a 26.5-inch refractor. Unfortunately, he had the use of it for only two years before his retirement in 1927.

Innes was the first to place double-star research in the southern hemisphere on a sound modern footing. He had unusually acute eyesight, discovering with small telescopes doubles that are difficult to observe with much larger instruments. Altogether Innes is credited with 1,628 new doubles; in addition he made many thousands of measurements that drew attention to the excellence of the astronomical "seeing" on the high veld. His second general catalog of southern double stars appeared in 1927, and in 1926 his interest in practical computation led to his proposal of the orbital parameters now known as the Thiele-Innes constants.

But double-star astronomy was not enough for a man of Innes' versatility; he also found time to do important work in such diverse fields as proper motions, variable stars, lunar occultations, and the Galilean satellites of Jupiter. He always insisted that the proper function of an observatory was first and foremost to observe, especially in the southern hemisphere; theoretical work could be left to the more numerous observatories in the north, which are often situated in less favorable climates. But his self-confessed personal preference for theoretical work was not always to be denied, and he was probably the first to offer a definite proof of the variability of the earth's rotation. Innes was a pioneer in the use of the blink microscope in astronomy—in the face, surprisingly, of some criticism. It was with this instrument that he made his celebrated discovery, as the result of a deliberate search, of Proxima Centauri, still the nearest known star to the solar system.

BIBLIOGRAPHY

I. Original Works. Innes' writings include "Reference Catalogue of Southern Double Stars," in *Annals of the Royal Observatory, Cape Town*, **2**, pt. 2 (1899); "Revision of the Cape Photographic Durchmusterung," *ibid.*, **9** (1903); "Discovery of Variable Stars, etc., With Pulfrich's Blink-Mikroskop, and Remarks Upon Its Use in Astronomy," in *Union Observatory Circular*, no. 20 (1914)—see also nos. 28 (1915), and 35 (1916); "A Faint Star of Large Proper Motion," *ibid.*, no. 30 (1915), which concerns Proxima Centauri—see also no. 40 (1917); "Transits of Mercury, 1677–1924," *ibid.*, no. 65 (1925), on the variability of the earth's rotation; "Orbital Elements of Binary Stars," *ibid.*, no. 68 (1926), written with W. H. van den Bos; and *Southern Double Star Catalogue* $-19°$ *to* $-90°$ (Johannesburg, 1927), written with B. H. Dawson and W. H. van den Bos. Many other papers are in *Transvaal* (later *Union*) *Observatory Circular*, *Monthly Notices of the Royal Astronomical Society*, and *Astronomische Nachrichten*.

II. Secondary Literature. An obituary notice with *curriculum vitae* by W. de Sitter is in *Monthly Notices of the Royal Astronomical Society*, **94** (1934), 277. See also D. Brouwer, "Discussion of Observations of Jupiter's Satellites Made at Johannesburg in the Years 1908–1926"' in *Annalen van de Sterrewacht te Leiden*, **16**, pt. 1 (1928); and J. Hers, "R. T. A. Innes and the Variable Rotation of the Earth," in *Monthly Notes of the Astronomical Society of Southern Africa*, **30** (1971), 129.

W. S. Finsen

INŌ, TADATAKA (*b.* Ozekimura, Yamabegun, Kazusanokuni, Japan, 11 February 1745; *d.* Kameshimachō, Hacchōbori, Tokyo, Japan, 17 March 1818), *astronomy, surveying, cartography.*

The son of Jinpo Rizeamon Sadatsune, Tadataka had an unhappy childhood. His mother died in 1751, and because his father and his stepmother could not support him he stayed with various relatives. It is believed that as a boy he studied mathematics and medicine.

In 1762 he married a girl four years his senior, the daughter of a wealthy landowner and brewer named Inō. (Since the Inōs had no son, he was adopted by them and took their surname.) The adopted Inō proved himself an able businessman, managing a brewery, buying and selling grain, and setting up a firewood warehouse in Tokyo. His wife died in 1784, and he remarried in 1790.

In 1794 Inō officially retired and the next year left for Edo (Tokyo), where he studied astronomy under Takahashi Yoshitoki, an official astronomer. His formal scientific studies began only at the age of fifty, and from then until the age of seventy-three, two years before his death, he worked energetically in astronomical surveying. (After his retirement Inō called himself Kageyu.) At the time that Inō became Takahashi's pupil, the Asada school was the most prominent in Japanese astronomy. Although Asada Gōryū himself was past his prime, his students Takahashi and Hazama Shigetomi were revising the calendar based upon such Sino-Jesuit works as *Li-hsiang K'ao-ch'eng*. In 1897 the Kansei revision of the calendar was completed.

A major astronomical and geodetic problem of the

time in Japan was the finding of the length of a meridian by Japanese measure. Since *Li-hsiang K'ao-ch'eng* had set zero longitude at Peking, that of Japan had to be accurately measured so that, in predicting a solar eclipse, the Sino-Jesuit method could be employed for the Japanese longitude.

In order to find the length of a meridian Inō volunteered to undertake a geodetic survey. Takahashi negotiated for him with the government, and in 1800 official permission for the survey was received. The Asada school was interested in the project from the point of view of astronomical geodesy, but the government permitted the private survey in hopes that it would contribute to the defense of northern Japan against possible Russian encroachments. The Russians had been active in the north since the end of the eighteenth century and the Japanese now wanted a coastal survey of Hokkaido, there being no satisfactory marine chart of that coast.

With several followers, Inō set out for Hokkaido via the northern part of the main island of Honshu. During the day they measured distances by number of steps (sometimes using a pedometer) and the bearings of distant mountains. At night, using a quadrant, they observed the altitude of a fixed star as it crossed the meridian. After compiling the results of their survey, Inō produced a map and presented it to the government. He subsequently conducted many successful surveys in northeastern Japan. His success aroused enthusiasm for surveying among many of his followers, especially Honda Rimei.

In 1804 Inō undertook a government project to survey the western seacoast of Japan. In comparison with the privately done, somewhat inexact survey of northeastern Japan, which had been carried out with insufficient personnel and funds, this better supported, government-sponsored survey of western Japan was very accurate and detailed; there was a larger budget, and personnel were also allowed various privileges on the site. After making over 2,000 measurements of latitude, Inō calculated the length of a meridian which agreed (within several tenths of a second of a degree) with the figure given in the Dutch translation of Lalande's *Astronomie* (Amsterdam, 1775), which source Takahashi had obtained in 1803.

On Inō's maps zero longitude is through Kyoto. Inō tried to utilize celestial observation to measure longitude, as by noting, for instance, the solar and lunar eclipses from two different points and by observing an eclipse of a satellite of Jupiter. He had to revert, however, to fixing longitude by measuring distances along the earth's surface. This procedure affected the accuracy of his maps, especially that of Hokkaido, in which there was a systematic error of several tenths of a minute.

Inō was an energetic field observer but did not excel in devising new methods or new theories in either astronomy or geodesy. While he was active, knowledge of Western astronomy was available through Dutch translations and Sino-Jesuit works and, later, through the works of Lalande; but Inō had no knowledge of Dutch or dynamics and little understanding of astronomical theories. When calculating the length of the meridian, he considered the earth as a perfect sphere rather than a spheroid. Moreover, when observing the positions of fixed stars, he did not take into account the effects of refraction, parallax, or nutation. In his surveying, Inō did not use modern triangulation but relied upon the old traverse method. His mapmaking approach resembled the Sanson-Flamsteed method (it is presumed that his method was developed independently), which is appropriate only for small areas; Inō nonetheless used the method for an area as large as all of Japan.

Despite Inō's scientific failings, his map of Japan, based upon surveys covering the length and breadth of the land, has an important place in geographical history. George Sarton compares his contribution with that of Ferdinand Hassler, founder of the U.S. Coast and Geodetic Survey.

Historically, the only scientific technique used in Japanese mapmaking and surveying had been that of the plane survey, adopted from China in ancient times and used mainly for measuring fields. Astronomical observation had been restricted to city planning, and used for establishing the north-south axis of the checkerboard grid plans copied from the cities of the ancient Chinese dynasties. In the Middle Ages, when the influence of the Chinese civilization weakened and Japan was constantly engaged in internal wars, techniques of mapmaking and surveying improved somewhat, since they were necessary for military purposes in measuring terrain and laying out fortresses; but, judging from extant maps, these techniques were only good enough to make crude sketch maps.

In the sixteenth and seventeenth centuries Westerners came to Japan, bringing with them European surveying techniques and instruments such as the astrolabe; but after Japan virtually closed its doors to the outside world in the seventeenth century, it lost any direct contact with Western countries. The only surveying school, the Shimizu in Nagasaki, was secretive about its methods, which were never published in book form or developed much further. On the other hand, in the seventeenth century the world map of Matteo Ricci (in Sino-Jesuit works) was introduced, and in 1720 the ban on publication of non-Christian works in Western languages was lifted. Sino-Jesuit books on astronomy and surveying were increasingly studied. The first Japanese map showing latitude and longitude was published by Nagakubo Sekisui in 1779. This map, although it went through many revisions and was widely published, was deficient in interpreting the basically Western concept of longitude and latitude. (The first government-appointed astronomer, Shibukawa Harumi, and his follower Tani Jinzan had tried in the seventeenth century to determine the latitude of various places, but their observation error was well over ten minutes of a degree.)

Hence Inō's map of Japan was far superior to maps then in use, and to an amateur, his results look almost like modern maps. It was a revolutionary step forward. But since his map was produced by government order, it was not published or made available to the public; thus its influence was very limited.

In 1826 the German natural historian Philipp Franz von Siebold came to Edo. Takahashi Kageyasu, the official astronomer of the time and a son of Takahashi, gave Inō's map of Japan to Siebold in exchange for his maps and books. Knowledge of this reached the government in 1828, when Siebold was about to leave Japan. Takahashi Kageyasu was arrested and died in prison, and Siebold was subsequently deported. This incident amply illustrates the government's treatment of Inō's map as a top-secret document.

Because Inō's brilliant work was not known to the rest of the world, the Europeans depended on a map produced in 1827 by a Russian admiral, Adam Johann von Krusenstern, which was clearly inferior to Inō's. Although the original of Inō's map was confiscated, Siebold succeeded in smuggling out a copy. After revising it on the basis of the Mercator projection, he published it in 1840 as *Karte vom Japanischen Reiche*. Since Inō's map was not published in Japan, this revision was reimported to Japan, where it was copied.

Under the Edo Treaty of 1858, H.M.S. *Acteon* came to Japan in 1861 and asked the shogunate for permission to survey the coastline. In Japan xenophobia was at its peak, and the government thought it unsafe to grant the permission. Instead it gave the British a copy of Inō's map. The British found the coastline described with sufficient accuracy for them to be satisfied with measuring only the depth of the surrounding seas.

After the Meiji Restoration, with the new government anxious to build a modern nation, an accurate map of Japan became a necessity for reasons both of prestige and of foreign trade. All the Japanese maps produced during the 1870's and 1880's by various government departments and the military were based on Inō's pioneering map.

BIBLIOGRAPHY

I. Original Works. Inō's works consist mainly of maps, observations, records of his surveys, field notes, and diaries. Most of these are in the Inō Memorial Hall in Sahara. Among them there is a copybook entitled *Bukkoku rekishōhen sekimō* (1816 or 1817), a thesis strongly criticizing Entsū's *Bukkoku rekishōhen*, 5 vols. (1810), in which Entsū disputed the Western astronomical cosmology, basing his rebuttal upon the Buddhist theory of Shumisen.

II. Secondary Literature. Ōtani Ryokichi, *Inō Tadataka* (Tokyo, 1917), which was published in commemoration of the centenary of Inō's death, is the standard biography at present. It is an exhaustive critical study. There are many biographies of Inō, including some aiming for popularity, but all of them are either excerpts from Ōtani's book or partial additions to it.

In commemoration of the 150th anniversary of Inō's death, the Tokyo Geographical Society published many articles (in Japanese) on Inō in its *Chigaku zasshi.* Many of them are partial amendments or additions to Ōtani's book. Significant among them are "The Life of Tadataka Inō, the First Land Surveyor in the Yedo Period and his Contribution to the Modernization of Japan Since the Meiji Restoration," **76**, no. 1 (1967), 1–21; "Significance and Essential Features of Inō's Map in the History of Japanese Science and Cartography," **77**, no. 4 (1968), 193–222; Hoyanagi Mutsumi, "British Preliminary Chart of Japan and Part of Korea Compiled From Inō's Map," **79**, no. 4 (1970), 224–236; Masumura Hiroshi, "Some Criticism on the Surveying Trips in 'Tadataka Inō,' Written by Professor Ryokichi Ōtani," **77**, no. 1 (1968), 24–36; Nakamura Hiroshi, "Appreciation of Maps of Japan Made by Land Survey in the Edo Period Seen From the Standpoint of Cartographers in Europe and America," **78**, no. 1 (1969), 1–18; Akioka Takejiro, "Notes on Some of Inō's Maps Preserved in Japan," **76**, no. 6 (1967), 313–321; and Hirose Hideo, "On the Value of Longitude of Kyoto Appearing on Inō's Map Introduced to Europe by P. Siebold," **76**, no. 3 (1967), 150–153.

The publications in English are Ōtani Ryokichi, *Tadataka Inō* (Tokyo, 1932), rev. for foreign readers and trans. into English; George Sarton, in *Isis*, **26**, no. 1 (1936), 196–200, a comment on Ōtani's book; and Norman Pye and W. G. Beasley, "An Undescribed Manuscript Copy of Inō Chukei's Map in Japan," in *Geographical Journal*, **117** (1951), 178–187.

On the history of cartography, see Fujita Motoharu, *Japanese Geographical History*, rev. ed. (Katanae, 1942); and Akioka Takejiro, *History of Japanese Mapmaking* (Kawaide, 1955), which evaluates Inō's contribution to geographical history.

Shigeru Nakayama

IPATIEV, VLADIMIR NIKOLAEVICH (*b.* Moscow, Russia, 9 November 1867; *d.* Chicago, Illinois, 29 November 1952), *chemistry.*

Ipatiev received a military secondary and higher education, graduating from the Artillery School and then, in 1892, from the Artillery Academy in St. Petersburg. Yet his calling was not military but scientific. In 1899 he became professor at the Artillery Academy. In 1914 he was elected associate member, and in 1916 member, of the Russian Academy of Sciences. After the October Revolution, Ipatiev held many high administrative posts and was a member of the Presidium of the Supreme Soviet of the National Economy, exerting leadership over the chemical industry and scientific research. From 1926 he was simultaneously a consultant to many chemical enterprises in Germany, particularly to the Bavarian central laboratory for nitrogen-producing factories. From 1930 Ipatiev was director of the Catalytic High Pressure Laboratory (now bearing his name) at Northwestern University, Evanston, Illinois.

At the beginning of his scientific career (1892–1896), while studying action of bromine upon tertiary alcohols and of hydrogen bromine upon acetylene and allene hydrocarbons in acetic acid solution, at the suggestion of his teacher, A. Y. Favorsky, Ipatiev established new means for the synthesis of unsaturated hydrocarbons and obtained isoprene. This was the first synthesis of the substance which is the basic monomeric component of natural rubber. Before Ipatiev's work this was separated only from the products of the pyrogenic decomposition of rubber or terpenes. The synthesis of isoprene immediately made Ipatiev's name well-known.

After 1900 Ipatiev began to depart from the classic methods of organic synthesis in his development of heterogeneous catalysis. Studying the various directions of the thermocatalytic decomposition of alcohols, he was able to prepare aldehydes, esters, and olefin and diene hydrocarbons by the catalytic dehydrogenation and dehydration of alcohol with various catalysts and under various physical reaction conditions. At this time Ipatiev and his colleagues began a systematic investigation of the catalytic properties of alumina, one of the most widely used catalysts in contemporary chemistry.

In 1904 Ipatiev introduced high pressures—400–500 atmospheres and higher—into heterogeneous catalysis. A device he constructed, the "Ipatiev bomb," introduced into chemical practice the use of a new type of reactor: the autoclave. Using such a device, Ipatiev was the first to synthesize methane from carbon and hydrogen; to obtain changes of reaction equilibriums in the dehydrogenation and dehydration of alcohol, the process being interrupted at intermediate stages; and to demonstrate the possibility of hydrogenating compounds which do not take up hydrogen in the presence of the same catalysts at normal atmospheric pressure. The introduction of high pressures into organic synthesis, which at first met with skepticism (for example, from Sabatier), allowed the kinetics of chemical reactions to be radically changed—to accelerate them a thousandfold or increase the equilibrium relationships of the product concentrations of the reaction a millionfold. In addition, Ipatiev promulgated the application of high pressure to inorganic reactions; in particular he proposed methods of separating metals from water solutions of their salts using hydrogen at high pressure (1909). These methods permit pure metals and minerals, as well as new modifications of element metalloids, to be obtained.

In 1912 Ipatiev was the first to use multicomponent catalysts. He demonstrated the possibility of combining oxidation-reduction reactions with dehydration reactions in one process which proceeds with the aid of the two-component catalyst Ni_2O_3/Al_2O_3. Multifunctional catalysts have come to occupy a leading position in cracking and re-forming processes and in other branches of petrochemical synthesis.

Ipatiev was the author of one of the most effective theories of catalysis, according to which the basic role in the heterogeneous catalytic reaction belongs to metallic oxides. In this connection he examined the catalytic activity of many oxides—FeO, Fe_2O_3, Cr_2O_3, TiO, Mo_2O_5, WO_3, and WO—and selected from among them catalysts for processes that found wide application in the petrochemical industry and often bore his name: (1) the synthesis of benzene polymers from the gas by-products of cracking by means of "solid phosphoric acid"; (2) the dehydrogenation of C_4- and C_5-alkanes, obtaining olefins and diene monomers of synthetic rubber; (3) alkylation of aromatic and paraffin hydrocarbons by means of olefins; (4) the synthesis of isooctane; (5) the isomerization of paraffins with the aim of increasing the octane number of gasoline; (6) many cracking and re-forming processes; (7) the dehydrocyclization of paraffins, obtaining alicyclic hydrocarbons and aromatics.

Following unsuccessful attempts by many, including Butlerov, Ipatiev was the first to achieve the polymerization of ethylene in reduction (1913), indicating the possibility of obtaining polyethylene of various molecular weights. He discovered a series of reactions which exemplify selective catalysis, such as the reaction of hydrodemethylation.

Combining the qualities of researcher, engineer,

and administrator, Ipatiev found industrial applications for all the results of his scientific research—even for the most unexpected. Many factories throughout the world use technology he developed or produce goods according to his methods.

Ipatiev trained many chemists, including, in the Soviet Union, G. A. Razuvaev, B. L. Moldavsky, E. I. Shpitalsky, A. D. Petrov, A. V. Frost, B. N. Dolgov, and V. V. Ipatiev, and, in the United States, H. Pines, R. Barvell, and L. Schmerling.

Ipatiev wrote 350 papers and took out 200 patents.

BIBLIOGRAPHY

I. ORIGINAL WORKS. Ipatiev's writings include *Kurs organicheskoy khimii* ("A Course in Organic Chemistry"; St. Petersburg, 1903); *Kurs neorganicheskoy khimii* ("A Course in Inorganic Chemistry"; St. Petersburg, 1909), written with A. V. Sapozhnikov; *Neft i ee proiskhozhdenie* ("Petroleum and Its Origin"; Moscow, 1922); *Kataliticheskie reaktsii pri vysokikh temperaturakh i davleniakh* ("Catalytic Reactions at High Temperatures and Pressures"; Moscow–Leningrad, 1936); and *The Life of A Chemist: Memoirs of Vladimir N. Ipatieff*, X. J. Eudin *et al.*, eds. (Stanford, Calif., 1946).

II. SECONDARY LITERATURE. See *K 35-letiyu nauchnoy deyatelnosti akademika V. N. Ipatieva* ("On the Thirty-Fifth Year of Academician V. N. Ipatiev's Scientific Career"; Moscow, 1929), an anthology; V. I. Komarewsky, ed., *Advances in Catalysis and Related Subjects*, V (1948), 9; and V. I. Kuznetsov, *Razvitie kataliticheskogo organicheskogo sinteza* ("The Development of Catalytic Organic Synthesis"; Moscow, 1964).

V. I. KUZNETSOV

IBN ʿIRĀQ. See **Manṣūr ibn ʿIrāq.**

IRINYI, JÁNOS (*b.* Nagyléta, Hungary, 17 May 1817; *d.* Vértes, Hungary, 17 December 1895), *chemistry.*

Irinyi was the son of an agronomist and estate agent, also named János, who set up Hungary's first alcohol factory equipped with steam engines. He studied chemistry at the Vienna Polytechnikum and agriculture at the Agricultural Academy in Hohenheim. Irinyi is often called the inventor of the safety match, but this is only partially true, as many researchers contributed to its development. In 1805 Jean Chancel, a Frenchman, invented the "dip lighter," and the Englishmen John Walker (1827) and Samuel Jones (1832) also have individual claims as pioneers of the friction match. Since the ignition materials in these primitive matches were potassium chlorate and antimony trisulfide, they ignited violently and explosively. The suggestion of adding white phosphorus was contributed by István Rómer, a Hungarian manufacturer, who in 1832 applied in Vienna for a patent for this process.

In 1835, Irinyi, while still a student, had the idea of substituting lead oxide for the potassium chlorate. He thereby obtained an explosionless, noiseless, and smoothly igniting match whose head consisted of white phosphorus, lead oxide, and sulfur. Irinyi sold his invention to Rómer, who thereafter manufactured the new type of match in Austria. Irinyi himself established a match factory in Buda (today Budapest), Hungary, but the volume of business did not meet his expectations. He soon fell into financial difficulties and had to give up the factory. This failure was probably caused in part by his many scientific and public activities.

Irinyi wrote several books and worked to create an artificial Hungarian technical language in which all chemical terms would be "Magyarized"; this language prevailed in scientific usage only for a very short time. Irinyi also participated in the revolutionary events of the year 1848, and during the Hungarian war of independence he was charged with the organization and supervision of the Hungarian manufacture of arms. Upon the defeat of the uprising he was imprisoned. Following his release he worked in various steam-powered corn mills and sugar factories. He spent the last years of his life in retirement in Vértes, cultivating a small plot of land he had inherited.

BIBLIOGRAPHY

I. ORIGINAL WORKS. Irinyi's most important works are *Über die Theorie der Chemie im allgemeinen und die der Schwefelsäure insbesondere* (Berlin, 1838); and *A vegytan elemei* ("Principles of Chemistry"; Nagyvarad, 1847).

II. SECONDARY LITERATURE. On Irinyi's contribution to the development of the match, see *Ullmans Encyklopädie der technischen Chemie*, XIX (Munich, 1969), 263; J. R. Partington, *A History of Chemistry*, IV (London, 1964), 197. Two biographical treatments, both in Hungarian, are by J. Nyilasi in *Természettudományi közlöny* (1960), pp. 516–518; and by Z. Szökefalvi Nagy and E. Táplányi in *Magyar Vegyészeti Muzeum Közleményei* (1971), no. 1, pp. 3–31.

FERENC SZABADVÁRY

ISAAC ISRAELI (*b.* Egypt, *fl.* ninth-tenth century), *medicine, philosophy.*

Nothing is known of Isaac Israeli's early life or his education. A Jewish physician and philosopher, he immigrated to Ifriqiya (now Tunisia) sometime after 900 and became the court physician to the last Aghlabid emir and, after he was ousted, to the

Fatimid caliph who succeeded him. Although the date of Israeli's death is uncertain, there is some ground for placing it about 955.

Of his medical works, the *Book on Fevers* and the *Book on Urine* were highly regarded textbooks. An edition of the Arabic original of the *Book on Fevers* is in preparation; a comparison of Constantine the African's Latin version with the original has shown it to be more a condensed paraphrase than a literal translation. Constantine also prepared Latin versions of the *Book on Urine* and the *Book on Foodstuffs and Drugs*. There are also Hebrew translations of the medical works.

Israeli's philosophy, purely Neoplatonic in character, is mainly based on a treatise in Arabic that, like other similar texts, was attributed to Aristotle, and on the writings of the Muslim philosopher al-Kindī. His themes were the process of emanation, the elements, and the soul and its return to the upper world. He wrote a number of short treatises on philosophy, of which the *Book of Definitions and Descriptions*, largely based on al-Kindī, was widely used by the Schoolmen in a Latin version by the twelfth-century translator Gerard of Cremona. Whereas the *Book of Substances*, of which only part is extant, is a kind of commentary on the pseudo-Aristotelian text, the *Book on Spirit and Soul* supports its doctrines with biblical quotations. In addition, there is a treatise called *Chapter on the Elements*, extant only in the Hebrew version, and a lengthier *Book on the Elements*, which exists in Latin and Hebrew editions, the former by Gerard of Cremona.

BIBLIOGRAPHY

There is a biographical note in A. Altmann and S. M. Stern, *Isaac Israeli. A Neoplatonic Philosopher of the Early Tenth Century* (Oxford, 1958), which also contains references to the editions of the philosophical treatises, the English translation, with commentary, of the philosophical treatises, and a systematic exposition of Israeli's philosophy that supersedes J. Guttmann, *Die philosophischen Lehren des Isaak b. Solomon Israeli* (Münster, 1911). Constantine the African's Latin versions were printed in *Opera omnia Ysaac* (Lyons, 1515). The relation between the Arabic original of the *Book on Fevers* and the Latin (as well as the Castilian) version is studied by J. D. Latham in *Journal of Semitic Studies* (1969).

Among the bibliographic references in Altmann and Stern, the most important are M. Steinschneider, *Die hebräischen Übersetzungen des Mittelalters* (Berlin, 1893), sec. 479; and *Die arabische Literatur der Juden* (Frankfurt, 1902), sec. 28; and G. Sarton, *Introduction to the History of Science* (Baltimore, 1927), pp. 639 ff.

S. M. STERN

ISAAC JUDAEUS. See **Isaac Israeli.**

ISAACS, CHARLES EDWARD (*b.* Bedford, New York, 24 June 1811; *d.* Brooklyn, New York, 16 June 1860), *medicine.*

Isaacs was the first American to do work in experimental kidney physiology. By use of painstaking techniques in these researches Isaacs settled the controversy concerning the connection of the Malpighian bodies with the uriniferous tubules of the kidneys (until then strongly maintained by Bowman and as strenuously denied by Müller, Hyrtl, and others), and ingeniously demonstrated the presence of nucleated cells on the surface of the Malpighian tuft, as well as the selective ability of the Malpighian tuft to separate many products of the urine from the blood. He introduced into the intestinal tract dyes which were absorbed into the blood; he was thus able to demonstrate conclusively that the Malpighian bodies separated these coloring matters from the blood, excreting them into the urine.

He was the youngest of four children of William and Mary Isaacs; his father, a merchant and farmer, died when Charles was only seven. Educated at a parish school run by Samuel Holmes, he could read Latin and Greek by the age of twelve and became fluent as well in French and German. With a Dr. Belcher, a relative living in New York City, as his preceptor, Isaacs attended his first course of medical lectures at the College of Physicians and Surgeons of New York. He then entered the office of John J. Graves, at that time one of the editors of the *New York Medical Journal;* with Graves he moved to Baltimore about 1831, where he graduated M.D. from the University of Maryland in 1833. Moving to North Carolina soon afterward, Isaacs was appointed surgeon to the Cherokee Indians when they were removed to the West, and traveled with them through the southern states to their place of relocation west of the Mississippi. He entered the army in 1841 but resigned in 1845 to join William H. Van Buren's private medical school on Greene Street in New York City. Two years later he joined his friend T. G. Catlin in private practice for six months in Youngstown, New York; he then accepted the position of deputy health officer on Staten Island, New York, but remained for only a month before rejoining Catlin.

In September 1848, Isaacs was appointed demonstrator of anatomy at the College of Physicians and Surgeons of New York City, where he remained for several years. He next moved to the University Medical College of New York City as demonstrator and adjunct professor of anatomy. By serving between school terms as surgeon on transatlantic steamers he was able to visit anatomy departments of schools in Paris and other European cities. His last move, in 1857, was to Brooklyn, where he finally achieved financial success in practice and was invited to lecture on surgical anatomy at the Brooklyn City Hospital.

One of the founders of the New York Pathological Society, Isaacs served it as both vice-president and president. In 1850 he was elected a member of the New York Academy of Medicine; his papers on the structure and physiology of the kidney, presented to the Academy in 1856 and 1857, attracted worldwide attention for the first time to this important medical group. A monument to Isaacs' patient industry and scientific zeal, they were the only papers considered worthy of publication by the Academy in that year; and following their publication they were acclaimed, translated, and republished in France by Brown-Séquard in the *Journal de l'anatomie et de la physiologie normales*, and in Germany by Karl Christian Schmidt.

In 1858 Isaacs served as one of the New York Academy of Medicine's vice-presidents; after his death the Academy paid a striking tribute to his memory at a joint special meeting with the New York Pathological Society, at which Van Buren read a fine memoir. Isaacs was also a member of the Kings County Medical Society; one of the surgeons to the Brooklyn City Hospital; and consulting surgeon to the Kings County Hospital, to the Sailor's Snug Harbor seaman's retreat on Staten Island, and to the Municipal Hospital on Blackwell's (Welfare) Island. Having suffered from "malarious and camp exposure of military life" since his army days, he died unexpectedly from pleuropneumonia with renal complications.

BIBLIOGRAPHY

I. ORIGINAL WORKS. Isaacs' "Researches Into the Minute Anatomy of the Kidney" were reported as an abstract of the proceedings of the New York Academy of Medicine in the *New York Journal of Medicine*, 3rd ser., **1** (1856), 60–64. His "Researches Into the Structure and Physiology of the Kidney," in *Transactions of the New York Academy of Medicine*, **1** (1857), 377–435; and "On the Function of the Malpighian Bodies of the Kidney," *ibid.*, pp. 437–457, were critically reviewed in Schmidt's *Jahrbücher der in- und ausländischen gesamten Medizin*, **96** (1857), 155–156, and **104** (1859), 3. Other publications by Isaacs are "Extent of the Pleura Above the Clavicle," in *Transactions of the New York Academy of Medicine*, **2** (1863), 3–19; and "Remarks on Chylous or Milky Urine," *ibid.*, pp. 77–96.

The Anatomical Remembrancer or Complete Pocket Anatomist, a pocket compendium, was originally published in London and republished with corrections and additions by Isaacs (New York, 1850 and many subsequent eds.). He also edited (with W. H. Van Buren) Claude Bernard and C. Huette's *Illustrated Manual of Operative Surgery and Surgical Anatomy*, "adapted to the use of the American medical student" (New York, 1864).

II. SECONDARY LITERATURE. See "The Late Charles E. Isaacs, M.D." [editorial], in *American Medical Times*, **1** (1860), 26; Raymond N. Bieter, "Charles Edward Isaacs: A Forgotten American Kidney Physiologist," in *Annals of Medical History*, n.s. **1** (1929), 363–377, which reviews in detail Isaacs' papers on renal function and structure; and Joseph C. Hutchison, "An Address on the Life and Character of the Late Charles Edward Isaacs, M.D., Delivered to the Graduates of the Long Island College Hospital July 14, 1862," in *American Medical Monthly*, **18** (1862), 81–97, the chief source for biographical information.

SAMUEL X. RADBILL

ISḤĀQ IBN ḤUNAYN, ABŪ YAʿQŪB (*d.* Baghdad, 910 or 911), *medicine, scientific translation.*

The son of Ḥunayn ibn Isḥāq, and like him a physician, Isḥāq was trained under his father's supervision in the Greek sciences and the discipline of translation. A Nestorian Christian from al-Ḥīra (Iraq) and probably of Arab descent, his first language was Syriac, but he knew Greek and al-Qifṭī considered his Arabic to be superior to that of his father,[1] who, although bilingual, preferred to write in Syriac. Isḥāq's brother, Dāwūd ibn Ḥunayn, was a physician. Of his two sons, Dāwūd ibn Isḥāq became a translator and Ḥunayn ibn Isḥāq ibn Ḥunayn a physician.

Isḥāq is associated with the translation movement in Baghdad, which continued to flourish after the decline of the academy founded by the Caliph al-Ma'mūn for the purposes of scientific translation. Both Isḥāq and his father were court physicians; Isḥāq found special favor with the caliphs al-Muʿtamid (who reigned from 870 to 892) and al-Muʿtaḍid (892–902) and with the latter's vizier, Qāsim ibn ʿUbaydallāh. He is sometimes connected with the group of scholars who met with the Shīʿite theologian al-Ḥasan ibn al-Nawbakht, and al-Bayhaqī is among those who claim he converted to Islam.[2]

Isḥāq's original works are few. His books *On Simple Medicines* and *Outline of Medicine* are not extant. His *History of Physicians*, which does survive, is based, as Isḥāq indicates, on the work of the same name

ISḤĀQ IBN ḤUNAYN

by John Philoponus. Isḥāq supplements the original author's list with the names of the philosophers who lived during the lifetime of each physician, adding very little chronological matter. The account of medical practitioners is not continued beyond Philoponus' time. The epitome of Aristotle's *De Anima*, although attributed to Isḥāq, is unlikely to be his.[3]

Isḥāq's most notable contributions are his translations from Greek and Syriac. Here his work is associated with his first cousin Ḥubaysh ibn al-Ḥasan al-Aʿṣam and with ʿĪsā ibn Yaḥyā (neither of whom knew Greek), but especially with his father, with whom he translated medical works, and with Thābit ibn Qurra, who independently revised several of Isḥāq's translations, particularly those of mathematical treatises. Ḥunayn credits Isḥāq with the translation of several of Galen's books, mostly into Arabic but also into Syriac; he translated epitomes of Galenic works as well.[4]

Among Isḥāq's translations of philosophical works are Galen's *The Number of the Syllogisms* and *On Demonstration*, books XII–XV; three books of the epitome of Plato's *Timaeus*, and the *Sophist* (with the commentary by Olympiodorus). He translated into Arabic Aristotle's *Categories*, *On Interpretation*, *Physics*, *On Generation and Corruption*, *On the Soul*, book α and other parts of the *Metaphysics* (with Themistius' commentary on book Λ), *Nicomachean Ethics* and perhaps *On Sophistical Refutations*, *Rhetoric*, and *Poetics*. His Syriac translations include part of the *Prior Analytics*, all of the *Posterior Analytics*, and the *Topics* (with Ammonius' commentary on books I–IV and the commentary of Alexander of Aphrodisias on the remainder, with the exception of the last two chapters of book VIII). Other translations are Alexander of Aphrodisias' *On the Intellect;* Nicholas of Damascus' *On Plants* (revised by Thābit ibn Qurra); and Nemesius of Emesa's *On the Nature of Man* (*Kitāb al-Abwāb ʿalā Raʾy al-Ḥukamāʾ waʾl-falāsifa*), which is not a work by Gregory of Nyssa as is sometimes stated.

Of special consequence are Isḥāq's mathematical translations: Euclid's *Elements*, *Optics*, and *Data;* Ptolemy's *Almagest;* Archimedes' *On the Sphere and the Cylinder;* Menelaus' *Spherics;* and works by Autolycus and Hypsicles. The *Elements*, *Optics*, and *Almagest* were revised and presumably improved mathematically by Thābit ibn Qurra. The influence of the several versions and recensions of the Arabic *Elements* and *Almagest* is a basic and virtually unstudied problem in the history of Islamic mathematics and astronomy. Because so few texts have been established, the sorting out of separate traditions is not yet possible.

NOTES

1. Ibn al-Qifṭī, p. 80.
2. ʿAlī ibn Zayd al-Bayhaqī, p. 5.
3. See M. S. Hasan, in *Journal of the Royal Asiatic Society* (1956), p. 57; R. Walzer, in *Oriens*, **6** (1953), 126; and R. M. Frank, in *Cahiers de Byrsa*, **8** (1958–1959), 231 ff. The text is in A. F. al-Ahwānī, pp. 125–175.
4. On the question of attribution for the medical translations, see the articles on Ḥunayn ibn Isḥāq listed in the bibliography.

BIBLIOGRAPHY

I. Original Works. For information on Isḥāq's MSS, see the works by C. Brockelmann and F. Sezgin (listed below); see also H. Suter, "Die Mathematiker und Astronomen der Araber und ihre Werke," in *Abhandlungen zur Geschichte der Mathematik*, **10** (1900); and "Nachträge und Berichtigungen," *ibid.*, **14** (1902); *cf.* H. J. P. Renaud, "Additions et corrections à Suter, 'Die Mathematiker . . .,' " in *Isis*, **17** (1932), 166–183; and M. Krause, "Stambuler Handschriften islamischen Mathematiker," in *Quellen und Studien zur Geschichte der Mathematik, Astronomie und Physik*, Sec. B. Studien, **3** (1936), 437–532.

Works by Isḥāq are in F. Rosenthal, ed. and translator, "Isḥāq b. Ḥunayn's 'Taʾrīkh al-Aṭibbāʾ'," in *Oriens*, **7** (1954), 55–80; and A. F. al-Ahwānī, *Talkhīṣ Kitāb al-Nafs l'Ibn Rushd* (Cairo, 1950).

II. Translations. Isḥāq's translations of Galen's works are bound up with those of Ḥunayn ibn Isḥāq. For the Arabic versions of Galen's medical books, see the bibliography in G. Strohmaier, "Ḥunayn b. Isḥak," in B. Lewis *et al.*, eds., *Encyclopaedia of Islam*, new ed. (Leiden–London, in press); *cf.* "Djālīnūs," *ibid.* The translations of Galen's mathematical works and the work of Plato are in "Galeni compendium Timaei Platonis," in P. Kraus and R. Walzer, eds., *Plato Arabus*, vol. I (London, 1951). The Arabic translations of the Greek physicians are to be included in the *Corpus Medicorum Graecorum: Supplementum Orientale* (in press). For Isḥāq's translations of Aristotle, see F. E. Peters, *Aristoteles Arabus: The Oriental Translations and Commentaries of the Aristotelian Corpus* (Leiden, 1968).

For the trans. of Nicholas of Damascus' *On Plants*, see A. J. Arberry, "An Early Arabic Translation From the Greek," in *Bulletin of the Faculty of Arts* (Cairo University), **1** (1933), 48 ff.; **2** (1934), 72 ff.; and R. P. Bouyges, "Sur le 'de Plantis' d'Aristote-Nicolas à propos d'un manuscrit arabe de Constantinople," in *Mélanges de la Faculté orientale, Université St.-Joseph*, **9** (1924), 71 ff. Isḥāq's trans. of Alexander of Aphrodisias' work is in J. Finnegan, "Texte arabe de 'peri nou' d'Alexandre d'Aphrodise," *ibid.*, **33** (1956), 157 ff.

III. Secondary Literature. Medieval biobibliographies are included in ʿAlīb. Zayd al-Bayhaqī, *Tatimmat Ṣiwān al-Ḥikma*, M. Shafīʿ, ed. (Lahore, 1935); Ibn Juljul, *Ṭabaqāt al-Aṭibbāʾ waʾl-Ḥukamāʾ*, F. Sayyid, ed. (Cairo, 1955); Ibn Khallikān, *Wafayāt al-Aʿyān*, F. Wüstenfeld, ed., 2 vols. (Göttingen, 1835–1843), English trans. by MacGuckin de Slane as *Ibn Khallikan's Biographical Dic-*

tionary, 4 vols. (Paris, 1842–1871); Ibn al-Nadīm, *Al-Fihrist*, G. Flügel, ed., 2 vols. (Leipzig, 1871–1872), English trans. by B. Dodge as *The Fihrist of al-Nadim*, 2 vols. (New York, 1970); Ibn al-Qifṭī, *Ta'rīkh al Ḥukamā'*, J. Lippert, ed. (Leipzig, 1903); Sā'id al-Andalusī, *Ṭabaqāt al-Umam*, L. Cheikho, ed. (Beirut, 1912), French trans. by R. Blachère as *Livre des Catégories des Nations* (Paris, 1935); and Ibn Abī Uṣaybi'a, *'Uyn al-Anba' fi Ṭabaqāt al-Aṭibbā'*, A. Müller, ed., 2 vols. (Cairo-Königsberg, 1882–1884). See also three works by Barhebraeus: *Chronicon Ecclesiasticum*, J. B. Abbeloos and T. J. Lamy, eds. (Louvain, 1872–1877); *Chronicon Syriacum*, P. Bedjan, ed. (Paris, 1890), Latin trans. by P. J. Bruns and G. Kirsch (Leipzig, 1789); and *Ta'rīkh Mukhtaṣar al-Duwal*, A. Sālhānī, ed. (Beirut, 1890).

Modern biobibliographies are in A. Baumstark, *Geschichte der syrischen Literatur* (Bonn, 1922); C. Brockelmann, *Geschichte der arabischen Literatur*, 2 vols. and 3 suppl. vols. (Leiden, 1937–1949); G. Graf, *Geschichte der christlichen-arabischen Literatur*, 5 vols. (Rome, 1944–1953); G. Sarton, *Introduction to the History of Science*, 3 vols. (Baltimore, 1927–1928); F. Sezgin, *Geschichte des arabischen Schrifttums*, vol. I (Leiden, 1967); M. Ullmann, "Die medizin im Islam," in B. Spuler, ed., *Handbuch der Orientalistik* (Leiden, 1970), sec. 1, suppl. vol. VI, 119, 128; and the article on Isḥāq in T. Houtsma *et al.*, eds., *Encyclopaedia of Islam*, 4 vols. (Leiden–London, 1913–1938), and in new ed. (in press).

Literature on the translations is in M. Steinschneider, *Die arabischen Übersetzungen aus den Griechischen* (Graz, 1960), repr.: G. Bergsträsser, *Ḥunain b. Isḥâq u. seine Schule* (Leiden, 1913); and "Ḥunain über die syrischen und arabischen Galenübersetzungen," in *Abhandlungen für die Kunde des Morgenlandes*, **17** (1925); M. Meyerhof, "New Light on Ḥunain b. Isḥāq and his Period," in *Isis*, **8** (1926), 685–724; J. Kollesch, "Das 'Corpus medicorum graecorum'—Konzeption und Durchführung," in *Medizinhistorisches Journal*, **3** (1968), 68–73; M. Plessner, "Diskussion über das 'Corpus Medicorum Graecorum,' speziell das 'Supplementum Orientale.' Einleitendes referat," in *Proceedings. International Congress of the History of Medicine*, **19** (1966), 238–248; F. Rosenthal, "On the Knowledge of Plato's Philosophy in the Islamic World," in *Islamic Culture*, **14** (1940), 387 ff. (*cf.* "Aflāṭūn," in *Encyclopaedia of Islam*, new ed. [in press]); H. Gätje, "Studien zur Überlieferung der aristotelischen Psychologie im Islam," in *Annales Universitatis saraviensis*, **11** (1971); J. Murdoch, "Euclid: Transmission of the Elements," in C. C. Gillispie, ed., *Dictionary of Scientific Biography*, IV (New York, 1971), 437–459; M. Clagett, *Archimedes in the Middle Ages* (Madison, 1964), I, "The Arabo-Latin Tradition"; and the Ptolemy article in the *Encyclopaedia of Islam*, new ed. (in press).

For additional information, see A. Badawi, *La transmission de la philosophie grecque au monde arabe* (Paris, 1968); F. E. Peters, *Aristotle and the Arabs* (New York, 1968); and F. Rosenthal, *Das Fortleben der Antike im mittelalterlichen Islam* (Zurich–Stuttgart, 1965).

NABIL SHEHABY

ISHIWARA, JUN (*b.* Tokyo, Japan, 15 January 1881; *d.* Chiba prefecture, Japan, 19 January 1947), *physics.*

Ishiwara was the son of Ryo Ishiwara, a minister of a Japanese Christian church, and of Chise Ishiwara. He was graduated from the department of theoretical physics of the College of Science of the Imperial University of Tokyo in July 1906 and continued his studies at the graduate school of the college. He became a teacher at the Army School of Artillery and Engineers in April 1908 and in April 1911 was appointed assistant professor at the College of Science, Tohoku Imperial University. From April 1912 to May 1914 he studied in Munich, Berlin, and Zurich and was greatly influenced by Sommerfeld and Einstein. In May 1914 he became full professor at the Tohoku Imperial University, and in May 1919 he was awarded an Imperial Academy prize for his study on the theory of relativity and the quantum theory.

In August 1921 a love affair forced Ishiwara to resign his post at the university, and, ending his scientific career, he subsequently devoted himself to writing. He edited four volumes of a complete edition of Einstein's works in Japanese translation (1922–1924) and wrote many popular books and articles introducing and explaining the latest developments in physics. Shortly before the outbreak of World War II he wrote many essays criticizing the government's control over the study of science.

Ishiwara's fields of study included the electron theory of metal, the special and general theories of relativity, and the quantum theory. Between 1909 and 1911 he wrote numerous papers dealing with the theory of relativity: on propagation of light within moving objects, cavity radiation, dynamics of electrons, and the energy-momentum tensor of the electromagnetic field. He concentrated particularly on the principle of least action; and in 1913, using this principle, he drew the energy-momentum tensor, which was also done by Minkowski. Ishiwara tried to revise the concept of a constant velocity of light within the theory of relativity, arguing that a variable time scale, such that the product *cdt* remained constant, would produce equivalent results. From this point of view, between 1913 and 1915 he investigated the interrelationship among the theories of gravity as set up by Gunnar Nordström, Abraham, and Einstein and proved that each of their theories can be derived from Ishiwara's theory.

Ishiwara later tried to develop the five-dimensional theory unifying the gravitational and electromagnetic fields. As suggested by Sommerfeld's paper (1911) proposing the quantization of the aperiodic process in terms of the action integral, Ishiwara presented, in 1915, an interpretation of the quantum by relating it to the elementary cell in the phase space. That is, he assumed that the motion of a material system is such that we may divide its phase space into elementary cells of equal probability, whose extension is

$$h = \frac{1}{j} \sum_{i=1}^{j} \int q_i \, dp_i.$$

He utilized this assumption in discussing the spectra of hydrogen and helium and also the spectra of characteristic X rays.

BIBLIOGRAPHY

Ishiwara's papers were published in *Proceedings of the Tokyo Mathematico-Physical Society* and *Science Reports of Tohoku Imperial University*. His major works are "Über das Prinzip der kleinsten Wirkung in der Elektrodynamik bewegter ponderabler Körper," in *Annalen der Physik*, 6th ser., **42** (1913), 986–1000; "Zur relativistischen Theorie der Gravitation," in *Science Reports of Tohoku Imperial University*, **4** (1915), 111–160; "Universelle Bedeutung des Wirkungsquantums," in *Proceedings of the Mathematico-Physical Society*, **8** (1915), 106–116; and *Sōtaisei Genri* ("Principle of Relativity"; Tokyo, 1921).

Obituaries are in *Kagaku*, **22** (1947), 93–99.

TETU HIROSIGE

ISIDORE OF SEVILLE (*b.* Spain [?], *ca.* 560; *d.* Seville, Spain, 4 April 636), *dissemination of knowledge.*

An encyclopedist, confessor-bishop, and Doctor of the Church, Isidore was educated by his elder brother Leander (a friend of Gregory the Great) and in monastery schools. He succeeded Leander as bishop of Seville and Catholic primate of Spain in 599. Much concerned with the reformation of church discipline and with the establishment of schools, he exerted an influence on science entirely through writings intended as textbooks.

Isidore wrote extensively on Scripture, canon law, systematic theology, liturgy, general and Spanish history, and ascetics. His scientific writings are chiefly to be found as parts of the glossary *Libri duo differentiarum* (*De differentiis verborum*, and *De differentiis rerum*), two short works on cosmology (*De natura rerum* and *De ordine creaturarum*) and his great encyclopedic dictionary, the *Etymologiae* or *Origines*. This last work briefly defines or discusses terms drawn from all aspects of human knowledge and is based ultimately on late Latin compendia and gloss collections. The books of greatest scientific interest deal with mathematics, astronomy, medicine, human anatomy, zoology, geography, meteorology, geology, mineralogy, botany, and agriculture. Isidore's work is entirely derivative—he wrote nothing original, performed no experiments, made no new observations or reinterpretations, and discovered nothing—but his influence in the Middle Ages and Renaissance was great, and he remains an interesting and often authoritative source for Latin lexicography, particularly in technical, scientific, and nonliterary fields.

His sources seem to have included, apart from Scripture, the Servian Vergil commentaries, gloss collections, grammars, cookbooks, and technical manuals, Ambrose, Augustine, Boethius, an abridgment of Caelius Aurelianus, Cassiodorus, Cassius Felix, Cicero, some form of Dioscorides, Donatus, a Latin digest of Galen, Gargilius Martialis, Gregory the Great, Hegesippus, Horace, Hyginus, Jerome, Lactantius, Lucan, Lucretius, Macrobius, Orosius, Ovid, Palladius, Placidus, Pliny the Younger, Pseudo-Clement, Sallust, Seneca, Solinus, Suetonius, Tertullian, Varro, Vergil, Verrius Flaccus, Victorinus, and doubtless other writers at first or second hand.

Isidore's universe was composed of a primordial substance which, by itself, possessed neither quality nor form but was given shape by four elemental qualities: coldness, dryness, wetness, and hotness. Isidore followed Lucretius and many Greek cosmographers in regarding these elements as in constant flux between the earth and the solar fire at the center of the universe. Although all elemental qualities are present in all created things, the elemental name assigned in any specific case depends upon those qualities which are most prominent. Isidore shared the microcosmic theory which views each individual human being as a microcosm paralleling the macrocosm, on a smaller scale, and regards man as the central link in this chain of being. The elements shade into each other and are arranged in the solar system by weight, each stratum of the concentric spheres having its proper inhabitants: angels in the fiery heavens, birds in the air, fish in the water, and man and animals on solid earth.

Isidore summarizes this view in the *Etymologiae*, (13.3.1–3; see also his *De natura rerum*, 11.1):

> *Hylê* is the Greek word for a certain primary material of things, directly formed in no shape but capable of all bodily forms, from which these visible elements are shaped, and it is from this derivation that they get their name. This *hylê* the Latins call "matter," because being altogether formless from which anything is to be made, it is always termed "matter." . . . The Greeks, however, have named the elements *stoicheia*, because they come together by a certain commingling and concordance of association. They are thus said to be joined among

> themselves by a certain natural ratio, so that something originating in the form of fire returns again to earth, and from earth to fire just as, for example, fire ends in air, air is condensed into water, water thickens into earth, and earth again is dissolved into water, water evaporates into air, air is reduced into fire. . . [Sharpe, *Isidore of Seville: The Medical Writings*, p. 23].

The same distribution of elements occurs in the human body: blood, like air, is hot and moist; yellow bile, like fire, hot and dry; black bile, like earth, cold and dry; and phlegm, like water, is cold and wet. Individual temperaments are determined by the dominant humoral qualities, and health depends upon their balance. Disease arises from excess or defect among them: acute diseases from the hot, and chronic diseases from the cold elemental humors. Therapy attempts to restore their normal balance. The living organism is governed by the soul but animated by the *pneuma*, which is assigned various names as it assumes various functions within the organism. Isidore rejects the pantheistic notion that the individual soul is either part of or indistinguishable from the world *pneuma*. His psychology follows late classical views of cerebral localization of function (sensation anteriorly, memory centrally, and thought posteriorly) and of the traditional faculties of the soul: intellect, will, memory, reason, judgment, sensation, and the like. The soul is distinct both from the mind and from the vital spirit; sensation and thought are distinguished, as are illusion and error.

Western Europe in Isidore's time had little direct contact with the Greek scientific tradition and derived both science and philosophy at second hand. The bulk of early Latin scientific writing was severely practical or anecdotal and descriptive. Most of Isidore's scientific passages merely define words or phrases. A man of his time, Isidore was more concerned with analogy than with analysis, with the unusual than with the typical. An encyclopedic dictionary is too disconnected to present a scientific world view; but Isidore carefully and quite accurately preserved much of the scientific lore current late in the Roman period, when original work had long since ceased and facility in Greek had perished. If he was no Aristotle, he was a great improvement on Pliny, and—considerations of style apart—his scientific content compares very favorably with that of Lucretius.

BIBLIOGRAPHY

I. Original Works. Editions of Isidore are Faustinus Arevalo, *Isidori Hispalensis opera omnia*, 7 vols. (Rome, 1797–1803), in J. P. Migne, *Patrologia latina*, LXXXI–LXXXIV; W. M. Lindsay, *Isidori Hispalensis Etymologiarum sive Originum libri*, 2 vols. (Oxford, 1911); and Jacques Fontaine, *Isidore de Seville: Traité de la nature* (Bordeaux, 1960).

II. Secondary Literature. See Ernest Brehaut, *An Encyclopedist of the Dark Ages* (New York, 1912), which is unreliable; R. B. Brown, *Printed Works of Isidore of Seville* (Lexington, Ky., 1949), useful but incomplete and confuses Isidore of Seville with other Isidores; Luis Cortés y Góngora, *Etimologías: Versión castellana* (Madrid, 1951); Jacques Fontaine, *Isidore de Seville et la culture classique dans l'Espagne visigothique*, 2 vols. (Paris, 1959), the best general study; F. S. Lear, "St. Isidore and Mediaeval Science," in *Rice Institute Pamphlets*, **23** (1936), 75–105; and W. D. Sharpe, *Isidore of Seville: The Medical Writings* (Philadelphia, 1964), which translates *Etymologiae* 4 and 11 with an intro. and bibliography.

William D. Sharpe

ISIDORUS OF MILETUS (*b.* Miletus; *fl.* Constantinople, sixth century), *architecture*, *mathematics.*

Isidorus of Miletus was associated with Anthemius of Tralles (a neighboring town of Asia Minor) in the construction of the church of Hagia Sophia at Constantinople. The church begun by Constantine was destroyed in the Nika sedition on 15 January 532.[1] Justinian immediately ordered a new church to be built on the same site, and it was begun the next month.[2] Procopius names Anthemius as the man who organized the tasks of the workmen and made models of the future construction, adding: "With him was associated another architect, Isidorus by name, a Milesian by birth, an intelligent man and in other ways also worthy to execute Justinian's designs."[3] Paul the Silentiary concurs in his labored hexameters: "Anthemius, a man of great ingenuity and with him Isidorus of the all-wise mind—for these two, serving the wills of lords intent on beauty, built the mighty church."[4] It is commonly held that Anthemius died in or about 534,[5] when Isidorus was left in sole charge, but this must be regarded as unproved. The church was dedicated on 27 December 537.[6]

In the astonishing space of five years Anthemius and Isidorus erected one of the largest, most ingenious, and most beautiful buildings of all time. The ground plan is a rectangle measuring seventy-seven by seventy-one meters, but the interior presents the appearance of a basilica terminating in an apse, flanked by aisles and galleries, and surmounted by a dome greater than any ecclesiastical dome ever built. The dome rests upon four great arches springing from four huge piers; the pendentives between the arches were at that time a novel device. As in the church of SS. Sergius and Bacchus in the same city, the stresses of the central dome are shared by half domes to the west and east, and the general similarity of plan has led to conjectures that the same architects built the earlier church. The dome nevertheless exerted a greater outward thrust on the piers supporting it than was safe, and when it had to be reconstructed after an earthquake twenty years later it was made six meters higher; but in general the applied mathematics of the architects (no doubt applied instinctively rather than consciously) have proved equal to the exacting demands of fourteen centuries. The decoration of the building was worthy of its artifice; the empire was ransacked to adorn it with gold, silver, mosaics, fine marbles, and rich hangings. Its ambo excited particular admiration.

Anthemius and Isidorus were consulted by Justinian when the fortifications at Daras in Mesopotamia were damaged by floods; but on this occasion the advice of Chryses, the engineer in charge, was preferred.[7]

Isidorus probably died before 558, for when a section of the dome and other parts of Hagia Sophia were destroyed by an earthquake at the end of the previous year, it was his nephew, called Isidorus the Younger, who carried out the restoration.[8] No doubt he had learned his art in his uncle's office. Essentially what is extant is the church of Anthemius and Isidorus, as repaired by the latter's nephew and patched after no fewer than thirty subsequent earthquakes, in addition to the ordinary ravages of time.

Isidorus was a mathematician of some repute as well as an architect. Notes at the end of Eutocius' commentaries on Books I and II of Archimedes' *On the Sphere and the Cylinder* and *Measurement of the Circle* indicate that Isidorus edited these commentaries.[9] The first such note reads, "The commentary of Eutocius of Ascalon on the first of the books of Archimedes *On the Sphere and the Cylinder*, the edition revised by Isidore of Miletus, the engineer (μηχανικός), our teacher"; and, *mutatis mutandis*, the other two are identical. It was formerly supposed on the strength of these notes that Eutocius was a pupil of Isidorus; but other considerations make this impossible, and it is now agreed that the three notes must be interpolations by a pupil of Isidorus.[10] A similar note added to Eutocius' second solution to the problem of finding two mean proportionals—"The parabola is traced by the *diabetes* invented by Isidorus of Miletus, the engineer, our teacher, having been described by him in his commentary on Hero's book *On Vaultings*"—must also be regarded as an interpolation by a pupil of Isidorus.[11] The nature of the instrument invented by Isidorus can only be guessed—the Greek word normally means "compass"—and nothing is otherwise known about Hero's book or Isidorus' commentary on it.

The third section of the so-called Book XV of Euclid's *Elements* shows how to determine the angle of inclination (dihedral angle) between the faces meeting in any edge of any one of the five regular solids. The procedure begins with construction of an isosceles triangle with vertical angle equal to the angle of inclination. Rules are given for drawing these isosceles triangles, and the rules are attributed to "Isidorus our great teacher."[12] It may therefore be presumed that at least the third section of the book was written by one of his pupils.

The above passages are evidence that Isidorus had a school, and it would appear to have been in this school that Archimedes' *On the Sphere and the Cylinder* and *Measurement of the Circle*—in which Eutocius had revived interest through his commentaries—were translated from their original Doric into the vernacular, with a number of changes designed to make them more easily understood by beginners. It is evident from a comparison of Eutocius' quotations with the text of extant manuscripts that the text of these treatises which Eutocius had before him differed in many respects from that which we have today, and the changes in the manuscripts must therefore have been made later than Eutocius.[13]

NOTES

1. "Chronicon Paschale," in *Corpus scriptorum historiae Byzantinae*, X (Bonn, 1832), 621.20–622.2.
2. Zonaras, *Epitome historiarum*, XIV.6, in the edition by Dindorf, III (Leipzig, 1870), 273.23–29.
3. Procopius, *De aedificiis*, I.1.24, in his *Opera omnia*, Haury, ed., IV (Leipzig, 1954), 9.9–16. In another passage Procopius says that "Justinian and the architect Anthemius along with Isidorus employed many devices to build so lofty a church with security" (*ibid.*, I.1.50; *Opera omnia*, IV, 13.12–15) and in yet another reference he relates how Anthemius and Isidorus, alarmed at possible collapse, referred to the emperor, who in one instance ordered an arch to be completed and in another ordered the upper parts of certain arches to be taken down until the moisture had dried out—in both cases with happy results (*ibid.*, I.1.66–77; *Opera omnia*, IV, 15.17–17.7). The word translated "architect" in these passages (μηχανοποιός) might equally be rendered "engineer." There was no sharp distinction in those days. Perhaps "master builder" would be the best translation.
4. Paul the Silentiary, *Description of the Church of the Holy Wisdom*, ll. 552–555, Bekker, ed., *Corpus scriptorum historiae Byzantinae*, XL (Bonn, 1837), 28. Agathias, *Historiae*, V.9, R. Keydell, ed. (Berlin, 1967), 174.17–18, mentions Anthemius alone, but this is not significant; in his account of the church, Evagrius Scholasticus—*Ecclesiastical History*, Bidez and Parmentier, eds. (London, 1898), 180.6–181.14—mentions neither.
5. F. Hultsch, "Anthemius 4," in Pauly-Wissowa, I (Stuttgart, 1894), col. 2368, "um 534"; followed more precisely by G. L. Huxley, *Anthemius of Tralles* (Cambridge, Mass., 1959), "in A.D. 534." But Agathias, V.9, on which Hultsch

relies, cannot be made to furnish this date; and the latest editor, R. Keydell, in his *Index nominum*, merely deduces from the passage *pridem ante annum 558 mortuus*.
6. Marcellinus Comes, "Chronicon," in J. P. Migne, ed., *Patrologia latina*, LI (Paris, 1846), col. 943D.
7. Procopius, *op. cit.*, II.3.1–15; *Opera omnia*, IV, 53.20–55.17.
8. Agathias, *op. cit.*, 296. Procopius records that the younger Isidorus had previously been employed by Justinian, along with John of Byzantium, in rebuilding the city of Zenobia in Mesopotamia (*op. cit.*, II.8.25; *Opera omnia*, IV, 72.12–18).
9. *Archimedis opera omnia*, J. L. Heiberg ed., 2nd ed., III (Leipzig, 1915), 48.28–31, 224.7–10, 260.10–12. The Greek will bear the interpretation that it was the treatises of Archimedes, rather than the commentaries by Eutocius, which Isidorus revised. This was the first opinion of Heiberg—*Jahrbuch für classische Philologie*, supp. **11** (1880), 359—but he was converted by Tannery to the view given in the text: *Archimedis opera omnia*, III, xciii.
10. Paul Tannery, "Eutocius et ses contemporains," in *Bulletin des sciences mathématiques*, 2nd ser., **8** (1884), 315–329, repr. in *Mémoires scientifiques*, II (Toulouse–Paris, 1912), 118–136.
11. *Archimedis opera omnia*, III, 84.8–11.
12. *Euclidis opera omnia*, J. L. Heiberg and Menge, eds., V (Leipzig, 1888), 50.21–22. See also T. L. Heath, *The Thirteen Books of Euclid's Elements*, 2nd ed., III (Cambridge, 1926), 519–520.
13. J. L. Heiberg, "Philologische Studien zu griechischen Mathematikern II. Ueber die Restitution der zwei Bücher des Archimedis *περὶ σφαίρας καὶ κυλίνδρου*," in *Neues Jahrbuch für Philologie und Pädagogik*, supp. **11** (1880), 384–385; *Quaestiones Archimedeae* (Copenhagen, 1879), pp. 69–77; *Archimedis opera omnia*, III, xciii. The delight with which Eutocius found an old book which preserved in part Archimedes' beloved Doric dialect—*ἐν μέρει δὲ τὴν Ἀρχιμήδει φίλην Δωρίδα γλῶσσαν απέσωζον*—shows that there had been a partial loss of Doric forms even before his time.

BIBLIOGRAPHY

I. Original Works. Isidorus edited the commentaries of Eutocius on Archimedes' *On the Sphere and the Cylinder* and *Measurement of the Circle*. These survive—with subsequent editorial changes—and are in *Archimedis opera omnia*, J. L. Heiberg, ed., 2nd ed., III (Leipzig, 1915). A commentary which Isidorus wrote on an otherwise unknown book by Hero, *On Vaultings*, has not survived.

II. Secondary Literature. The chief ancient authorities for the architectural work of Isidorus are Procopius, *De aedificiis*, in *Opera omnia*, Haury, ed., IV (Leipzig, 1954); Paul the Silentiary, *Description of the Church of the Holy Wisdom*, Bekker, ed., *Corpus scriptorum historiae Byzantinae*, XL (Bonn, 1837); and Agathias Scholasticus, *Historiae*, R. Keydell, ed. (Berlin, 1967). One of the best modern books is W. R. Lethaby and Harold Swainson, *The Church of Sancta Sophia Constantinople* (London, 1894). A more recent monograph is E. H. Swift, *Hagia Sophia* (New York, 1940). There are good shorter accounts in Cecil Stewart, *Simpson's History of Architectural Development*, II (London, 1954), 66–72; and Michael Maclagan, *The City of Constantinople* (London, 1968), pp. 52–62.

For Isidorus' contribution to the study of the five regular solids, see T. L. Heath, *The Thirteen Books of Euclid's Elements*, 2nd ed. (Cambridge, 1926; repr. New York, 1956), III, 519–520.

Ivor Bulmer-Thomas

ISSEL, ARTURO (*b.* Genoa, Italy, 11 April 1842; *d.* Genoa, 27 November 1922), *geology.*

Issel was the son of Raffaele and Elisa Sonsino Issel. He studied under Giuseppe Meneghini and graduated with a degree in natural science from the University of Pisa in 1863. From 1866 to 1917, he taught geology, mineralogy, paleontology, and geography at Genoa. Issel was a skillful geologist, but the most characteristic aspect of his talent was the versatility that enabled him to work in an astonishing variety of fields. Based on a solid foundation of learning and far from being any sort of dilettantism, this versatility was particularly oriented toward the study of living mollusks, a field in which Issel soon became a master, and ethnology, especially the then emerging paleethnology.

Issel belonged to that elite group of traveling Italian naturalists who, in the second half of the nineteenth century, contributed much to the scientific knowledge of distant regions, especially of Africa. In 1865 he traversed a considerable portion of the coast and the islands of the Red Sea, collecting living and fossil mollusks, both on the present shores and in the sediments deposited since Miocene times. Thus, he simultaneously carried out zoological and paleontological researches, which are presented in the excellent *Malacologia del Mar Rosso* (1869), in which 804 species, eighty-five of them previously unknown, are described and discussed in relation to the fauna of the neighboring seas. Also worthy of mention are Issel's minor contributions to the malacology of Italy and of regions outside Europe (including Tunisia, Persia, and Borneo). He returned to the coast of the Red Sea in 1870, also making an expedition to Cheren in the Ethiopian highlands. Shortly thereafter he published a lively diary entitled *Viaggio nel Mar Rosso e tra i Bogos* (1872). He reflected at length on his geological observations of the two voyages, and not until almost thirty years later did he issue his valuable paleogeographic study "Essai sur l'origine et la formation de la Mer Rouge" (1899).

Much of Issel's research concerns the recent geological events of the Mediterranean basin, which he traversed from the Greek archipelago to Malta, from the Ligurian coast to the Tunisian coast. He contributed to the geological study of Malta, Zante, and Galite (Jezīret Jālita); he established a geological stage of the Pleistocene marine series, the Tyrrhenian, which followed the Calabrian and the Sicilian; he devoted particular attention to the valleys which, in Liguria, continue below sea level. This last subject is related to Issel's research on the slow oscillations of the ground, for which he proposed the name "bradyseisms." Collecting and synthesizing not only his own observations made along the shores of the

Red Sea and the Mediterranean but also those made previously by other investigators, he wrote the now classic "Le oscillazioni lente del suolo o bradisismi" (1883). The criteria set forth there were further used in the detection and exact determination of bradyseisms on the shore. Issel also concerned himself with the present-day conditions in the Mediterranean through the study of samples from its bottom.

Born and raised in Liguria, Issel devoted much time to the study of this region, which is small but very interesting from a naturalist's point of view. In addition to geologic surveys made in collaboration with L. Mazzuoli and D. Zaccagna, he made dozens of contributions in geography, seismology, geology, petrography, paleethnology, and paleontology, which were collected in two large volumes: *Liguria geologica e preistorica* (1892). The title of this work invites closer consideration of the science of paleethnology, in which Issel was already interested at the age of twenty; in fact, there are two articles, dated 1864 and 1866, devoted to the ossiferous caverns which he visited near Finale (Liguria) and on Malta. Throughout his life he conducted fruitful research on the remains of prehistoric man in Liguria, maintaining close contact with the famous paleethnologist L. Pigorini, on whose *Bullettino di paletnologia italiana* he collaborated. The 1892 analytical work was followed by the synthetic "Liguria preistorica" (1908). He also conducted some more purely ethnological investigations, studying African (Bogos, Niam Niam) and Burmese populations.

BIBLIOGRAPHY

I. Original Works. Issel's writings include *Malacologia del Mar Rosso: Ricerche zoologiche e paleontologiche* (Pisa, 1869); *Viaggio nel Mar Rosso e tra i Bogos* (Milan, 1872); "Le oscillazioni lente del suolo o bradisismi: saggio di geologia storica," in *Atti della R. Università di Genova*, **5** (1883), 1–422; *Liguria geologica e preistorica*, 2 vols. (Genoa, 1892); "Essai sur l'origine et la formation de la Mer Rouge," in *Bulletin de la Société belge de géologie, de paléontologie et d'hydrologie*, **13** (1899), 65–84; and "Liguria preistorica," in *Atti della Società ligure di storia patria*, **30** (1908), 1–775.

II. Secondary Literature. Biographical and bibliographical notes are M. Canavari, "Commemorazione di Arturo Issel," in *Memorie dell'Accademia dei Lincei*, classe di scienze fisiche, matematiche e naturali, 5th ser., **14** (1923), 679–697, with complete bibliography; P. Principi, "Arturo Issel," in *Bollettino della Società geologica italiana*, **42** (1923), xx–xxiv; and F. Sacco, "Arturo Issel," in *Bollettino del R. Ufficio geologico*, **49** (1922–1923), 1–25. On Issel's voyages, see F. Rodolico, *Naturalisti esploratori dell'Ottocento italiano* (Florence, 1967), pp. 151–171, with selections from Issel's work.

Francesco Rodolico

IVANOV, ILYA IVANOVICH (*b.* Shigry, Kursk guberniya, Russia, 1 August 1870; *d.* Alma-Ata, Kazakh S.S.R., 20 March 1932), *biology*.

Ivanov's father, a clerk in the district treasury, came from the lower middle class; his mother, from a minor landowning family. After graduating from the Sumskaya Gymnasium (Ukraine) in 1890, Ivanov studied at the biology faculties of the University of Moscow and, later, the University of Kharkov. After graduation he worked in the biochemistry and microbiology laboratories of the universities of St. Petersburg and Geneva, and in 1897–1898 he completed a course of theoretical and practical study at the Pasteur Institute, Paris.

Ivanov was distinguished by good health, abundant energy, exceptional single-mindedness, and persistence in overcoming difficulties. He was active in the work of the Petersburg Society of Natural Scientists and Physicians and, after the October Revolution, in various scientific societies.

From his student years Ivanov manifested an interest in problems of reproductive biology, interspecies hybridization, and the artificial insemination of domestic animals, which had been little studied at that time. In 1899 he published a detailed historical essay, "Iskusstvennoe oplodotvorenie u mlekopitayushchikh" ("Artificial Impregnation of Mammals"), which was incorporated into his monograph of the same title (1906). Using the data of Spallanzani, Jakobi, Remy, Coste, and Vrassky and the results of experiments by dog breeders, horse breeders, veterinarians, and medical doctors, he believed that "the artificial impregnation of domestic mammals is not only possible but also must become one of the powerful forces of progress in the practice of livestock breeding" ("Iskusstvennoe oplodotvorenie u mlekopitayushchikh" [1903], p. 456).

Ivanov stressed that the method was widespread in fish breeding at the end of the nineteenth century, owing to the application of the "Russian methods" for the artificial insemination of fish roe described by Vrassky, while in livestock breeding it was not used at all. In fact, there was a negative attitude toward it, many people believing that, in mammals, exclusion of the sex act and human interference in the complex physiological process of reproduction would destroy the full biological value of the offspring and the health of the animals used for artificial insemination. "As long as the question of the viability and strength of the offspring obtained from artificial impregnation remained unresolved," wrote Ivanov, "this method had no right to wide application" ("Iskusstvennoe oplodotvorenie domashnikh zhivotnykh" [1910], p. 8). It was also necessary to develop a method, suitable in practice and safe for the animal,

IVANOV

which would permit insemination of a significant number of females with the semen of one sire, for "only with such a technical setup does artificial impregnation acquire its significance and can it count on widespread practical application" ("Iskusstvennoe oplodotvorenie u mlekopitayushchikh" [1906], p. 411). For this reason Ivanov, in 1898, formulated a program of extensive research on the biology of mammalian reproduction and on the formulation of the theoretical and technical problems involved in artificial insemination of domestic animals, subsequently publishing it in an article (1903) and monographs (1906, 1907, 1910).

On his return to Russia in 1898, Ivanov set about the realization of this program in the special zoological laboratory of the Academy of Sciences, directed by A. O. Kovalevsky; in the physiology laboratory, directed by Pavlov; and in the biochemistry laboratory, directed by M. V. Nentsky, of the Institute of Experimental Medicine. In 1901 he founded the world's first center for the artificial insemination of horses (Dolgoe village, Orlovskaya guberniya); in 1908, the physiology section of the veterinary laboratory of the ministry of internal affairs (St. Petersburg); and in 1910, a zootechnical station (at Askania-Nova, the estate of F. E. Falzfein in Taurida guberniya). In these establishments he investigated the peculiarities of the sexual physiology of male and female domestic mammals, the biology of their sexual cells, and especially the role of secretions of accessory sexual glands during impregnations.

The results of these studies led Ivanov to conclude that the sole necessary condition for the impregnation of domestic mammals and poultry is the possibility for the meeting and union of spermatozoon and egg; the sex act, with its complex processes of the engorgement and hardening of the sexual apparatus, and even the natural liquid medium of the semen, are not absolutely necessary. They can be replaced by the artificial introduction of semen—or even spermatozoa in an artificial medium—into the female's sexual organs. His second fundamental conclusion was that spermatozoa could retain not only their motility but also their capability for causing conception for a certain period of time outside the organism if the conditions in which they were kept were favorable.

Starting from these prerequisites, Ivanov developed a method for the artificial insemination of domestic mammals and poultry by spermatozoa in their natural medium, intended for the use of pure-strain breeders on farms, and a method of insemination by spermatozoa in an artificial medium, to use the testicles of castrated or killed pure-strain stock or of wild animals. The results of testing these methods under laboratory and farm conditions demonstrated the practical suitability of the techniques, their great effectiveness, and their safety for the breed animals used. The full biological value of the offspring was established by prolonged observations of their growth, development, and quality. Ivanov therefore proposed that the method of artificial insemination be employed in livestock raising, with the goals of more effective use of pure-strain stock and interspecies hybridization of domestic mammals and poultry with wild varieties. He organized the production of special equipment for centers involved in the artificial insemination of mares; wrote a practical textbook (1910) and technical instructions; and, in the courses he created, prepared veterinarians for the practical realization of artificial insemination, which permitted the artificial insemination of around 8,000 mares from 1908 to 1917 on Russian farms.

Ivanov's results became more widely used after the October Revolution, when he became director of the section of animal reproductive biology of the State Institute of Experimental Veterinary Medicine and the Artificial Insemination Bureau of the All-Union State Organization of Beef Sovkhozes (Skotovod) and Sheep-Raising Sovkhozes (Ovtsevod), as well as consultant to the National Commissariat of Agriculture. At the same time he taught a course on the reproductive biology of farm animals at the Moscow and Alma-Ata zootechnical institutes. Through research carried out at the Skotovod and Ovtsevod sovkhozes and at other farms, and with the first mass experiments of their kind, Ivanov devised the basic directions for treating problems of the reproductive biology of farm animals (sexual periodicity and ovulation in females, impregnation, sperm formation, and the biology and biochemistry of sex cells) as well as for dealing with theoretical and practical problems of artificial insemination (methods of obtaining, evaluating, diluting, preserving, and disinfecting semen). These methods were later successfully developed by the biological-zootechnical school that he created.

By 1932, over 180,000 mares, 385,000 cows, and 1,615,000 ewes had been artificially inseminated on the Skotovod and Ovtsevod sovkhozes. Artificial insemination has since become the fundamental method of reproduction of farm animals in the Soviet Union.

Ivanov began the practice in livestock raising of interspecies hybridization with wild animals by artificial insemination in order to obtain economically usable hybrids as well as to develop new breeds of animals that can endure more severe conditions and are more resistant to illness. He obtained hybrids of a domestic horse by crossbreeding a zebra and

Przhevalski's horse, and produced hybrids of cattle with aurochs, bison, yak, and other hybrids. He organized experiments on the mass interspecies hybridization of cattle on the Skotovod sovkhozy. With A. Filipchenko, Ivanov gave a zoological description of interspecies hybrids and determined their economically useful characteristics and the degree of fertility in various generations. Using the program outlined by Ivanov, including interspecies hybridization and artificial insemination with spermatozoa in an artificial medium, his students and followers produced a new fine-haired breed of arkharo-merino sheep which is now widely distributed in the Kazakh and Kirgiz republics.

Ivanov also began work on preserving species of wild animals that are becoming extinct (the aurochs, bison, Przhevalski's horse). He was one of the organizers of the Sukhumsky Monkey Nursery, which in 1926 conducted the African expedition of the Soviet Academy of Sciences for the interspecies hybridization of monkeys and the delivery of them to the nursery.

BIBLIOGRAPHY

Among Ivanov's writings are "Iskusstvennoe oplodotvorenie u mlekopitayushchikh i primenenie ego v skotovodstve i v chastnosti v konevodstve" ("Artificial Impregnation of Mammals and Its Use in Cattle Raising, Especially in Horse Breeding"), in *Trudy Sankt-Petersburgskogo obshchestva estestvoispytatelei*, **30**, pt. 1 (1899), 341–343; "Iskusstvennoe oplodotvorenie u mlekopitayushchikh (predvaritelnoe soobshchenie)" ("Artificial Impregnation of Mammals [Preliminary Report]"), in *Russkii vrach*, **2**, no. 12 (1903), 455–457; "Iskusstvennoe oplodotvorenie u mlekopitayushchikh" ("Artificial Impregnation of Mammals"), in *Arkhiv biologicheskikh nauk*, **12**, pts. 4–5 (1906), 376–509, also in *Archives des sciences biologiques* (St. Petersburg), **12**, nos. 4–5 (1907), 377–511; *Iskusstvennoe oplodotvorenie u mlekopitayushchikh. Eksperimentalnoe issledovanie* ("Artificial Impregnation of Mammals. Experimental Investigation"; St. Petersburg, 1907); *Iskusstvennoe oplodotvorenie domashnikh zhivotnykh* ("Artificial Impregnation of Domestic Animals," St. Petersburg, 1910); *Die künstliche Befruchtung der Haustiere* (Hannover, 1912); *Kratky otchet o deyatelnosti Fiziologicheskogo otdelenia Veterinarnoy laboratorii pri Veterinarnom Upravlenii Ministerstva vnutrennikh del za 1909–1913 gg.* ("Brief Account of the Activities of the Physiological Section of the Veterinary Laboratory Attached to the Veterinary Department of the Ministry of Internal Affairs During the Period 1909–1913"; St. Petersburg, 1913); "The Application of Artificial Insemination in the Breeding of Silver and Black Foxes," in *Veterinary Journal*, **79**, no. 5 (1923), 164–173; and "Iskusstvennoe osemenenie mlekopitayushchikh, kak zootekhnichesky metod" ("Artificial Insemination of Mammals as a Zootechnical Method"), in *Trudy Pyatogo Sezda zootekhnikov Moskovskogo zootekhnicheskogo instituta* (Moscow, 1929), "Conference plenum," pp. 57–67. See also *Isbrannye Trudy* ("Selected Works," Moscow, 1970).

P. N. Skatkin, *Ilya Ivanovich Ivanov—vydayushchysya biolog* ("Ilya Ivanovich Ivanov—A Distinguished Biologist"; Moscow, 1964), has a complete bibliography of works by and concerning Ivanov, as well as information on archival sources.

P. N. SKATKIN

IVANOV, PIOTR PAVLOVICH (*b.* St. Petersburg [now Leningrad], Russia, 24 April 1878; *d.* Kostroma, U.S.S.R., 15 February 1942), *embryology.*

Ivanov graduated from St. Petersburg University in 1901 and, from 1903, was an assistant to the professor of invertebrate zoology. In 1906–1907 he traveled through the islands of the Malay Archipelago, where he collected materials on embryonic development of *Xiphosura* and *Scolopendra.* In 1909 and 1911 he worked at the zoological station in Naples, studying the embryology and regeneration of annelids. In 1912 at St. Petersburg University he defended his master's thesis, devoted to regeneration of annelids, and as associate professor gave a course on theoretical embryology. In 1922 Ivanov was appointed head of the embryological laboratory at the university. At the same time he occupied the chair of zoology at the Psychoneurological Institute (later the Second Leningrad Medical Institute), where from 1924 to 1942 he headed first the department of zoology and later the department of general biology, supervising at the same time the work of the embryological laboratory at the All-Union Institute of Experimental Medicine.

In addition to a number of special works Ivanov published two manuals on general and comparative embryology (1937, 1945). The basic morphological generalization that he formulated, comparable in its importance with the theory of germinal layers, was called the theory of the larval body or theory of the primary heteronomy of the bodies of segmented animals. Through his experiments on regeneration of *Oligochaeta* and *Polychaeta*, Ivanov established that after the fore end is amputated, only the segments whose structure is characteristic for the larval stage regenerate; during ontogeny the postlarval segments are formed by budding on the hind end of the larval body. The theory of the larval body allows the establishment of the relationship between various types of animals; generalization of the theory is given in Ivanov's article published posthumously in 1945.

Ivanov's theory was used in E. Korschelt's *Ver-*

gleichende Entwicklungsgeschichte der Tiere (2nd ed., Jena, 1936) and in V.N. Beklemishev's *Osnovy sravnitelnoy anatomii bezpozvonochnykh* ("Principles of Comparative Anatomy of Invertebrates," 1944, 1952); it has been further developed in the research conducted by P. G. Svetlov.

BIBLIOGRAPHY

I. Original Works. Ivanov's writings include "Die Regeneration von Rumpf- und Kopfsegmenten bei Lumbriculus variegatus Gr.," in *Zeitschrift für wissenschaftliche Zoologie*, **75**, no. 3 (1903), 327–390; "Die Regeneration der Segmente bei Polichaeten," *ibid.*, **85**, no. 1 (1907), 1–47; "Die Regeneration des vorderen und des hinteren Körperendes bei Spirographis spallanzanii Viv.," *ibid.*, **91**, no. 4 (1908), 511–558; *Regenerativnye protsessy u mnogoshchetinkovykh chervey i otnoshenie ikh k ontogenezu i morfologii annelid* ("Regenerative Processes in Oligochaete Annelids and Their Relation to Ontogeny and Morphology of Annelids"; St. Petersburg, 1912), his master's thesis; "Die Entwicklung der Larvalsegmente bei den Anneliden," in *Zeitschrift für Morphologie und Ökologie der Tiere*, **10**, no. 1 (1928), 62–161; "Die Embryonale Entwicklung von Limulus mollucanus," in *Zoologische Jahrbücher*, Morph. Abt., **56**, no. 2 (1933), 163–348; *Obshchaya i sravnitelnaya embriologia* ("General and Comparative Embryology"; Moscow, 1937); *Embrionalnoe razvitie skolopendry v svyazi s embriologiey i morfologiey Tracheata* ("Embryonic Development of Scolopendra in Connection With Embryology and Morphology of Tracheata"), in *Izvestiya Akademii nauk SSSR, otdel biologicheskikh nauk*, no. 2 (1940), pp. 831–860; *Rukovodstvo po obshchey i sravnitelnoy embriologii* ("Manual of General and Comparative Embryology"; Moscow, 1945); and "Pervichnaya i vtorichnaya metameria tela" ("Primary and Secondary Metamerism of the Body"), in *Zhurnal obshchei biologii*, **5**, no. 2 (1945), 61–94.

II. Secondary Literature. See P. G. Svetlov, "Zhizn i tvorchestvo P. P. Ivanova" ("The Life and Work of P. P. Ivanov"), in *Trudy Instituta istorii estestvoznaniya i tekhniki. Akademiya nauk SSSR*, **24** (1958), 151–176; and L. N. Zhinkin, "Piotr Pavlovich Ivanov," in *Uchenye zapiski Leningradskogo ordena Lenina gosudarstvennogo universiteta im A. A. Zhdanova. Seria biologicheskikh nauk*, no. 20 (1949), pp. 5–17.

L. J. Blacher

IVANOVSKY, DMITRI IOSIFOVICH (*b.* Gdov, Russia, 9 November 1864; *d.* U.S.S.R., 20 June 1920), *botany, microbiology.*

Ivanovsky was the son of Iosif Antonovich Ivanovsky, a landowner in Kherson guberniya. He was educated at the Gymnasium of Gdov, then that of St. Petersburg, from which he graduated as gold medalist in the spring of 1883. In August of that year he enrolled at St. Petersburg University in the natural science department of the physics and mathematics faculty. Among his teachers were I. M. Sechenov, N. E. Vvedensky, D. I. Mendeleev, V. V. Dokuchaev, A. N. Beketov, and A. S. Famintsyn—the leading representatives of contemporary Russian science.

In 1887 Ivanovsky and V. V. Polovtsev, a fellow student in the department of plant physiology, were commissioned to investigate the causes of a disease which had struck the tobacco plantations of the Ukraine and Bessarabia. During 1888 and 1889 they studied this disease, called "wildfire," and concluded that it was not infectious and arose from an abrupt change by the plants from weak to more intensive transpiration, producing light blemishes on the leaves. This work determined Ivanovsky's future scientific interests.

On 1 February 1888, having defended his graduation thesis "O dvukh boleznyakh tabachnykh rasteny" ("On Two Diseases of Tobacco Plants"), Ivanovsky graduated from St. Petersburg University, receiving the degree of candidate of science. On the recommendation of two professors at the university—A. N. Beketov and K. Y. Gobi—he was retained at the university in order to prepare for a teaching career. In 1891 he joined the staff of the botanical laboratory of the Academy of Sciences.

In 1890 another disease appeared in the tobacco plantations of the Crimea, and the directors of the Department of Agriculture suggested to Ivanovsky that he study it. He left for the Crimea that summer. The first results of his investigations of mosaic disease in tobacco—*O dvukh beloznyakh tabaka* ("On Two Diseases of Tobacco")—were published in 1892. This was the first study containing factual proof of the existence of new infectious pathogenic organisms—viruses.

To continue his scientific career Ivanovsky needed the secure position in scientific circles which could be attained only after defending a dissertation. He was for this reason compelled to turn to the study of a more specific problem. On 22 January 1895 he defended his master's dissertation, *Issledovania nad spirtovym brozheniem* ("An Investigation Into the Fermentation of Alcohol"), a study of the vital activity of yeast under aerobic and anaerobic conditions. He thereby earned the degree of master of botany and was subsequently assigned to give a course of lectures on the physiology of lower plants. He was further confirmed as assistant professor.

By this time Ivanovsky had married E. I. Rodionova and had a son, Nikolai. Straitened financial conditions compelled him to seek a better-paying position. In October 1896 he joined the Technological Institute as an instructor in plant anatomy and physiology, remaining there until 1901. During this period Ivanovsky returned to his early interest and became deeply involved in the study of the etiology of tobacco mosaic disease.

In August 1908 Ivanovsky moved to Warsaw: in October 1901 he had been named extraordinary professor at Warsaw University. His *Mozaichnaya bolezn tabaka* ("Mosaic Disease in Tobacco"), in which his investigations of the etiology of mosaic disease were summed up, was published in 1902. In 1903 he presented this book as his doctoral dissertation, defending it at Kiev. He received a D.Sc. and the title of full professor.

After defending his doctoral dissertation, Ivanovsky abandoned the study of viruses. Apparently he took this step because of both the unusual complexity of the problem itself and also the indifference and lack of understanding that most scholars showed toward his work. Neither his contemporaries nor Ivanovsky himself properly evaluated the consequences of his discovery. Either his work went unnoticed or it was simply ignored. A possible reason for this was Ivanovsky's uncommon modesty; he never publicized his discoveries.

In Warsaw Ivanovsky studied plant photosynthesis in relation to the pigments of green leaves. The choice of this topic was the result of his interest in the chlorophyll-bearing structures (chloroplasts) in plants, a problem which had arisen during his work on mosaic disease. During these investigations Ivanovsky made a study of the adsorption spectra of chlorophyll in a living leaf and in solution and demonstrated that chlorophyll in solution is quickly destroyed by light. He also propounded the theory that the yellow pigments of a leaf—xanthophyll and carotene—act as a screen to protect the green pigment from the destructive action of blue-violet rays.

Ivanovsky's chief fame, however, is as the discoverer of viruses. He discovered a new type of pathogenic source, which M. W. Beijerinck rediscovered in 1893 and named "virus." He established that the sap of a diseased plant remains infectious after filtration through a Chamberland candle, even though the bacteria visible under a microscope have been filtered out. Ivanovsky believed that this pathogenic source had the form of discrete particles—exceedingly small bacteria or bacteria spores. His point of view here differed from that of Beijerinck, who considered a virus to be *contagium vivum fluidum*. Ivanovsky repeated the experiments which had led Beijerinck to believe that a virus is liquid and became convinced of the rightness of his own conclusions. After following Ivanovsky's methods, Beijerinck agreed.

As the result of exhaustive histoanatomical investigations of tissue preparations from healthy and diseased plants, Ivanovsky discovered crystalline particles. He associated their presence with the onset of tobacco mosaic disease and simultaneously posed the question of a connection between the crystals that he had discovered and the minuscule living bacteria which he considered to be the pathogenic organisms of tobacco mosaic disease. Ivanovsky maintained that this pathogenic agent could exist only in the body of a living organism, that is, that it was a parasite.

Almost all the fundamental tenets of Ivanovsky's discovery have been confirmed and developed in modern virology. The sole exception is his proposition that the source of infection for tobacco mosaic disease was a minuscule bacterium, but Ivanovsky himself had not been fully convinced of its validity. Even during his lifetime progress was being made by filtering a contagious source through a Chamberland candle, the method he had used: dozens of viral diseases of plant and animals were discovered. Ivanovsky's hypothesis of the existence of a direct connection between the crystals he had found and the pathogenic source was confirmed in 1935 in the work of Wendell Stanley, who obtained crystals in a test tube of the virus that causes mosaic disease in tobacco and confirmed the infectious nature of the crystals that were separated.

The parasitic nature and corpuscularity of viruses, noted by Ivanovsky, have been confirmed during the seventy-year development of virology. Ivanovsky's view that viruses are living parasitic microorganisms is shared by many scientists, who are influenced by the consideration that viruses possess the properties of pathogenic microorganisms: specialized parasitism, a cyclical infectional process, and immunization formation.

BIBLIOGRAPHY

I. Original Works. Ivanovsky's writings include "Iz deyatelnosti mikroorganismov v pochve" ("On the Activity of Microorganisms in the Soil"), in *Trudy Volnogo Ekonomicheskogo Obshchestva*, **2**, no. 6 (1891), 222; *O dvukh boleznyakh tabaka* ("On Two Diseases of Tobacco"; St. Petersburg, 1892); *Issledovania nad spirtovym brozheniem* ("Investigations Into the Fermentation of Alcohol"; St. Petersburg, 1894), his master's diss.; *Mozaichnaya bolezn tabaka* ("Mosaic Disease in Tobacco"; Warsaw, 1902), his doctoral diss.; and *Fiziologia rasteny* ("The Physiology of Plants"; Moscow, 1924). His writings were brought together in *Izbrannye proizvedenia* ("Selected Works"; Moscow, 1953).

II. Secondary Literature. See M. A. Novikova,

"D. I. Ivanovsky," in *Lyudi russkoy nauki* ("Men of Russian Science"; Moscow, 1963), p. 319; K. E. Ovcharov, *Dmitry Iosifovich Ivanovsky* (Moscow, 1952); *Pamyati Dmitria Iosifovicha Ivanovskogo* ("In Memory of ... Ivanovsky"; Moscow, 1952); Wendell M. Stanley, "Soviet Studies on Viruses," in *Science*, **99**, no. 2564 (1944), 136–138; *O prirode virusov* ("On the Nature of Viruses"; Moscow, 1966); and G. M. Vayndrakh and O. M. Knyazhansky, *D. I. Ivanovsky i otkrytie virusov* ("D. I. Ivanovsky and the Discovery of Viruses"; Moscow, 1952).

V. GUTINA

IVES, HERBERT EUGENE (*b.* Philadelphia, Pennsylvania, 31 July 1882; *d.* New York, N.Y., 13 November 1953), *physics.*

The course of Ives's career was strongly influenced by his father, Frederic Eugene Ives, who developed several processes connected with color photography and halftone printing. Much of the elder Ives's experimentation was done at home and must inevitably have influenced his son. In the period 1898–1901 Herbert worked for his father in the Ives Kromskop Company, designing and constructing apparatus for color photography. He then attended the University of Pennsylvania, receiving the B.S. in 1905. He obtained a Ph.D. from Johns Hopkins in 1908, working under R.W. Wood and writing a dissertation on a study of standing light waves in the Lippmann photographic process.

Ives was employed by the National Bureau of Standards (1908–1909), the National Electric Lamp Association in Cleveland, Ohio (1909–1912), and the United Gas Improvement Company in Philadelphia (1912–1917). He volunteered for the army, working on aerial photography for the Signal Corps (1918–1919). One result of his service was the book *Airplane Photography*, published in 1920. After the war Ives went to work for the Bell Telephone Laboratories, where he remained until his retirement in 1947.

Ives received numerous awards during his lifetime, including medals from the Franklin Institute and the Optical Society of America, and the Rumford Medal from the American Academy of Arts and Sciences. He was president of the Optical Society in 1924–1925 and was elected to the National Academy of Sciences in 1933.

On 14 November 1908 Ives married Mabel Agnes Lorenz; they had three children. His avocations included coin collecting, and he was president of the American Numismatic Society in 1942–1946. A portrait painter of some talent, he developed a three-color palette.

Ives's early involvement with photography led him into a long association with problems in colorimetry and photometry, and papers on these subjects dominate the period of his life prior to World War I. He was especially concerned with the design of photometric instruments. His papers are credited with having been largely responsible for introducing tristimulus colorimetry into the United States.

After moving to the Bell Laboratories, Ives became more interested in photoelectric effects and in television. He measured in great detail the photoelectric effect of alkali metal films as a function of polarization, angle of incidence, and alloy composition. Changes in these variables produced some striking anomalies, which Ives eventually concluded were due to standing wave patterns formed in the film. Another series of experiments led him to conclude that the photoelectric and thermionic work functions were identical.

Ives spent a considerable amount of time from 1924 to 1930 on the development of television. Using Nipkov disks with photoelectric cells at the transmitter and neon lamps at the receiver, he performed a series of successful demonstrations, beginning in 1927 with a transmission between Washington and New York.

In 1938 and again in 1941 Ives, together with G. R. Stilwell, described a series of experiments on the transverse Doppler effect. It had been suggested by Einstein in 1907 that this effect—which could confirm the Lorentz transformations as applied in the special theory of relativity—might be discovered by observing hydrogen canal rays. A special tube developed by A. J. Dempster in 1932 produced spectral lines narrow enough so that the small displacement could be observed. Other specialized experimental techniques enabled Ives and Stilwell to find the effect, which was consistent with prediction. Nevertheless, Ives was an opponent of Einstein's theory and attacked it in several of his publications.

Ives received more than 100 patents for a variety of inventions, most of them related to his interests in photography, photoelectricity, and television. All but half a dozen of them were issued during the period of his employment at the Bell Laboratories.

BIBLIOGRAPHY

An essentially complete bibliography of more than 250 papers plus 100 patents is given in the memoir by Buckley and Darrow (see below). Ives's work on the photoelectric effect, covered in a series of papers published from 1922 to 1938, is summarized in his Rumford Medal lecture, "Adventures in Standing Waves," in *Proceedings of the American Academy of Arts and Sciences*, **81** (1951), 1–32. The canal ray experiments are reported in "Experimental Study of the Rate of a Moving Atomic Clock," in *Journal of the*

Optical Society of America, **28** (1938), 215–226, and **31** (1941), 369–374. Criticisms of Einstein include "Revisions of the Lorentz Transformations," in *Proceedings of the American Philosophical Society*, **95** (1951), 125–131; and "Derivation of the Mass-Energy Relation," in *Journal of the Optical Society of America*, **42** (1952), 540–543. Ives's papers are preserved at the Library of Congress and the Smithsonian Institution. Experimental apparatus is at the Smithsonian and at the Bell Telephone Laboratories.

A biography of Ives by Oliver E. Buckley and Karl K. Darrow appears in *Biographical Memoirs. National Academy of Sciences*, **29** (1953), 145–189.

BERNARD S. FINN

IVORY, JAMES (*b.* Dundee, Scotland, 17 February 1765; *d.* London, England, 21 September 1842), *mathematics.*

The son of James Ivory, a watchmaker, Ivory was educated at the universities of St. Andrews (1779–1785) and Edinburgh (1785–1786). After taking the M.A. degree (1783) he studied theology, with a view to entering the Church of Scotland. His studies in divinity were not pursued further, for immediately on leaving the university he was appointed teacher of mathematics and natural philosophy in Dundee. After three years he became the manager of a flax-spinning company in Forfarshire (now Angus). In 1804 the company was dissolved, and Ivory took up a mathematical professorship at the Royal Military College at Great Marlow (subsequently at Sandhurst). He held this office until 1819, when ill health compelled an early retirement. During the remainder of his life Ivory lived in London, devoting himself entirely to mathematical investigations, the results of which he made available in a long series of articles published in scientific journals. Sixteen of his papers were printed in the *Philosophical Transactions of the Royal Society* (he was elected a fellow of the Society in 1815). He was awarded the Copley Medal in 1814 and received the Royal Medal in 1826 and 1839.

Ivory's interests lay mainly in the application of mathematics to physical problems, and his principal contributions may be summarized under six categories.

1. The attraction of homogeneous ellipsoids upon points situated within or outside them. His paper "On the Attractions of Homogeneous Ellipsoids," containing the well-known theorem which bears his name, in which the attraction of an ellipsoid upon a point exterior to it is made to depend upon the attraction of another ellipsoid upon a point interior to it, was printed in the *Philosophical Transactions* for 1809 (pp. 345–372). Although Laplace had already reduced this problem to a similar form, Ivory's solution was regarded as simpler and more elegant.

2. Critical commentaries on the methods used by Laplace in the third book of the *Mécanique céleste* for computing the attraction of spheroids differing little from spheres and the substitution of analytical methods for some of Laplace's geometrical considerations (1812, 1822). Although some of Ivory's criticisms seem to have been unjustified, Laplace himself paid tribute to Ivory's work.

3. The investigation of the orbits of comets (1814).

4. Atmospheric refraction (1823, 1838).

5. The equilibrium of fluid bodies (1824, 1831, 1834, 1839).

6. The equilibrium of a homogeneous ellipsoid with three unequal axes rotating about one of its axes, based on a theorem of Jacobi and Liouville (1838).

Ivory's scientific reputation, for which he was accorded many honors during his lifetime, including knighthood of the Order of the Guelphs, Civil Division (1831), was founded on the ability to understand and comment on the work of the French analysts rather than on any great originality of his own. At a time when few in England were capable of understanding the work of Laplace, Ivory not only grasped its significance but also showed himself capable, in many cases, of substituting a clearer and more direct process for the original. Ivory's work, conducted with great industry over a long period, helped to foster in England a new interest in the application of analysis to physical problems.

BIBLIOGRAPHY

A list of ninety papers published by Ivory is in the Royal Society, *Catalogue of Scientific Papers*, III, 502–505. These include brief notes, comments and corrections, correspondence from the *Philosophical Magazine* (1821–1828), and his most important papers in the *Philosophical Transactions of the Royal Society*.

Biographical notices include R. E. Anderson in *Dictionary of National Biography*, XXIX, 82–83; and W. Norrie in *Dundee Celebrities* (Dundee, 1878), pp. 70–73. An informed critique of Ivory's work is in *Proceedings of the Royal Society*, n.s. **55** (1842), 406–513. Isaac Todhunter discusses Ivory's contribution to the theory of attraction in *A History of the Mathematical Theories of Attraction and the Figure of the Earth*, 2 vols. (London, 1873), II, 221–224, and *passim*.

MARGARET E. BARON

JĀBIR IBN AFLAḤ AL-ISHBĪLĪ, ABŪ MUḤAMMAD (*fl.* Seville, first half of the twelfth century), *astronomy, mathematics.*

Usually known in the West by the Latinized name

Geber, Jābir has often been confused with the alchemist Jābir ibn Ḥayyān and occasionally with the astronomer Muḥammad ibn Jābir al-Battānī. He should also be distinguished from Abū Aflaḥ ha-Saraqosṭī, the author of the mystical *Book of the Palm*, and from the Baghdad poet Abu'l Qāsim ʿAlī ibn Aflaḥ. Almost nothing is known of Jābir ibn Aflaḥ's life. He can be roughly dated by Maimonides' citation in his *Guide of the Perplexed:* ". . . Ibn Aflaḥ of Seville, whose son I have met . . . ."[1] That he came from Seville is deduced from the name "al-Ishbīlī" in manuscripts of his works and in the above quotation from Maimonides.

Jābir's most important work was a reworking of Ptolemy's *Almagest* in nine books. Its title in one Arabic manuscript (Berlin 5653) is *Iṣlāḥ al-Majisṭī* ("Correction of the *Almagest*"), but it had no fixed title in the West—Albertus Magnus calls it *Flores*, presumably short for *Flores Almagesti*, in his *Speculum astronomiae.*[2] According to the contemporary historian Ibn al-Qifṭī,[3] the text was revised by Maimonides and his pupil Joseph ibn ʿAqnīn. This revision seems to have been done about 1185, and so it was almost certainly from the unrevised text that Gerard of Cremona made his Latin translation. The *Iṣlāḥ* was translated from Arabic into Hebrew by Moses ibn Tibbon in 1274 and again by his nephew Jacob ben Māḥir; the latter translation was revised by Samuel ben Judah of Marseilles in 1335.

Jābir describes the principal differences between the *Iṣlāḥ* and the *Almagest* in the prologue: Menelaus' theorem is everywhere replaced by theorems on right spherical triangles, so that a proportion of four quantities is substituted for one of six; further, Jābir does not present his theorems in the form of numerical examples, as Ptolemy did. So far the changes seem to be the same as those made by Abu'l Wafā', but Jābir's spherical trigonometry is less elaborate. It occupies theorems 12–15 of book I and follows a theorem giving criteria for the sides of a spherical triangle to be greater or less than a quadrant (so that the sides may be known from their sines). In modern notation it may be summarized as follows:

Theorem 12. If all the lines in the figure are arcs of great circles, then

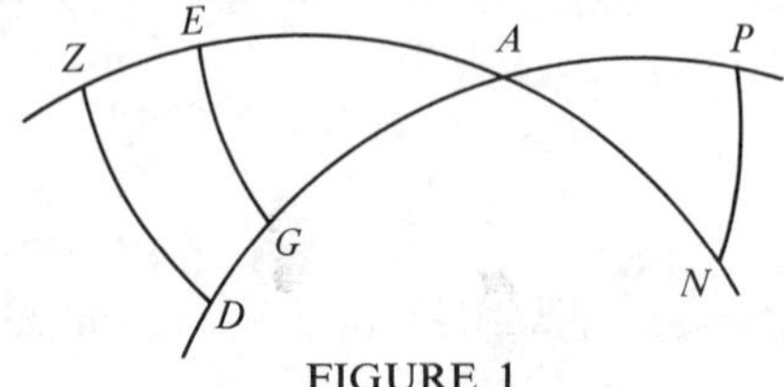

FIGURE 1

$$\text{Sin } AG{:}\text{Sin } GE = \text{Sin } AD{:}\text{Sin } DZ = \text{Sin } AN{:}\text{Sin } NP.$$

Theorem 13. In any spherical triangle ABG, $\text{Sin } BG{:}\text{Sin } \hat{A} = \text{Sin } GA{:}\text{Sin } \hat{B} = \text{Sin } AB{:}\text{Sin } \hat{G}$.

Theorem 14. If in spherical triangle ABG, $\hat{B}$ is right, then $\text{Sin } \hat{A}{:}\text{Sin } \hat{B} = \text{Cos } \hat{G}{:}\text{Cos } AB$.

Theorem 15. If in spherical triangle ABG, $\hat{B}$ is right, then $\text{Cos } AG{:}\text{Cos } BG = \text{Cos } AB{:}\text{Sin (quadrant)}$.

Theorems 13 and 15 are the most frequently used. Because of the differences in treatment it is unreasonable to suppose that Jābir copied directly from Abu'l Wafā', whose writings have survived. They may both have derived their fundamental theorems from Thābit ibn Qurra's tract on Menelaus' theorem, or all three may depend upon some source that in turn depends upon the third book of Menelaus' *Spherics.* As a trigonometer Jābir is important only because he was translated into Latin, whereas works such as Abu'l Wafā''s—which carried an equivalent, or a better, trigonometry—were not.

Jābir criticized Ptolemy—sometimes very violently—on a number of astronomical matters. Ptolemy's "errors" are listed in the prologue of the *Iṣlāḥ.* The most substantial, and most famous, deviation from the *Almagest* concerns Venus and Mercury. Ptolemy placed them beneath the sun, claiming that they were never actually on the line joining the eye of the observer and the sun. Jābir contradicted this justification, putting Venus and Mercury above the sun. The *Iṣlāḥ* is the work of a theorist. The demonstrations are free of all numbers and there are no tables. Jābir does, however, describe a torquetum-like instrument, which he says replaces all the instruments of the *Almagest.*

Although Jābir was quoted in the twelfth century by al-Biṭrūjī and by the author of the compendium of the *Almagest* ascribed to Ibn Rushd, and although the *Iṣlāḥ* was epitomized by Quṭb al-Dīn al-Shīrāzī in the thirteenth century, Jābir was better known in the West through Gerard of Cremona's translation. His name was used as that of an authority who criticized Ptolemy. But more serious was his influence on Western trigonometry. For instance, Richard of Wallingford cited him several times in the *Albion* and in the *De sectore* (a variant of the *Quadripartitum*); Simon Bredon took a great deal from Jābir in his commentary on the *Almagest;* and part of a commentary on the *Iṣlāḥ* in which Jābir's theorems are made more general is extant. But his most important influence was upon Regiomontanus' *De triangulis*, written in the early 1460's and printed in 1533, which systematized trigonometry for the Latin West. The core of the fourth book of this treatise is taken from Jābir without acknowledgment; the plagiarism was the subject of several pungent remarks by Cardano. Jābir was still quoted in the sixteenth and seventeenth centuries—for instance, by Sir Henry Savile and Pedro Nuñez. Copernicus' spherical trigonometry is of the same general type, but we have no reason to believe it was taken straight from the *Iṣlāḥ*. He called Jābir an "egregious calumniator of Ptolemy."

NOTES

1. Pt. II, ch. 9. See *The Guide of the Perplexed*, Schlomo Pines, trans. (Chicago, 1963), p. 268.
2. See Erfurt, Wissenschaftliche Bibliothek, MS Q223, fols. 106r–106v, and other MSS. The 1891 ed. is somewhat corrupt at this point.
3. *Ta'rīkh al-ḥukamā'*, J. Lippert, ed. (Leipzig, 1903), pp. 319, 392–393. The text is the abridgment by Muḥammad ibn ʿAlī al-Zawzānī (1249); the original is lost.

BIBLIOGRAPHY

I. Original Works. *Iṣlāḥ al-Majisṭī* is in the following Arabic MSS: Berlin 5653; Escorial 910 and 930; Paris, B.N. héb. 1102 (fragment of bk. V in Arabic but in Hebrew script). Hebrew MSS are Moses ibn Tibbon, trans., Bodleian Opp. Add. fol. 17 (Neubauer 2011); and Jacob ben Māḥir, trans., rev. by Samuel ben Judah, Paris, B.N. héb. 1014, 1024, 1025, 1036. At the end of the text Samuel describes the circumstances of the translation. This passage is transcribed by Renan with a French paraphrase in "Les écrivains juifs français," in *Histoire littéraire*, **31** (Paris, 1893), 560–563.

There are some 20 Latin MSS plus five fragments; the text was published by Peter Apian (Nuremberg, 1534) together with his *Instrumentum primi mobilis*. There is a description of a different but similar instrument in the Latin version, but the original diagrams remain. Jacob ben Māḥir describes both instruments.

The commentary on Thābit ibn Qurra's tract on Menelaus' theorem and the commentary on Menelaus' *Spherics* (fragment) occur together and are extant only in Hebrew. MSS are Bodleian Hunt. 96 (Neubauer 2008), fols. 40v–42v, and Bodleian Heb. d.4 (Neubauer 2773), fols. 165r–177v. Berlin Q 747 (Steinschneider catalog no. 204) contains part of this text.

The anonymous Latin commentary on the *Iṣlāḥ* is in Paris, B.N. Lat. 7406, fols. 114ra–135rb.

The six-book *Parvum Almagestum*, which exists only in Latin, is almost certainly not by Jābir, as has sometimes been held—see Lorch (below), ch. 3, pt. 1.

II. Secondary Literature. See H. Bürger and K. Kohl, "Zur Geschichte des Transversalensatzes des Ersatztheorems, der Regel der vier Grössen und des Tangentensatzes," in *Abhandlungen zur Geschichte der Naturwissenschaften und der Medizin*, **7** (1924), a substantial article following Axel Björnbo's ed. of Thābit ibn Qurra's tract on Menelaus' theorem; J. B. J. Delambre, *Histoire de l'astronomie du moyen âge* (Paris, 1819; repr. 1965), esp. pp. 179–185—Delambre is very hostile to Jābir; and R. P. Lorch, "Jābir ibn Aflaḥ and His Influence in the West" (Manchester, 1970), a Ph.D. thesis concerned mainly with spherical trigonometry.

For further references, see G. Sarton, *Introduction to the History of Science*, II (Washington, D.C., 1931), 206.

R. P. Lorch

JĀBIR IBN ḤAYYĀN (*fl.* late eighth and early ninth centuries?), *alchemy.*

Jābir ibn Ḥayyān is the supposed author of a very extensive corpus of alchemical and other scientific writings in Arabic. The earliest mention of him in a historical work is that in the *Notes* (*taʿālīq*) of Abū Sulāimān al-Manṭiqī al-Sijistānī (*d. ca.* 981), the head of a scholarly circle in Baghdad, who disputed Jābir's authorship of the corpus and asserted that the true author, a certain al-Ḥasan ibn al-Nakad from Mosul, was personally known to him. Shortly after Abū Sulāimān's death the bibliographer Ibn al-Nadīm presented a biography and bibliography of Jābir in his *Fihrist* of 987; in the same work he opposed the already present doubt concerning Jābir's existence. Al-Nadīm, a Shiite, supported the identity of a man named Jaʿfar, whom Jābir often called his teacher, with the sixth Shiite imam, Jaʿfar ibn Muḥammad al-Ṣādiq (*ca.* 700–765) and opposed his identification with the Barmakid vizier Jaʿfar ibn Yaḥyā (executed in 803 under Hārūn al-Rashīd).

In any case Jābir was, as a student or favorite of one of the two Jaʿfars, a personality of the eighth and of a part of the ninth century. E. J. Holmyard believed that Jābir's father was an apothecary named Ḥayyān who lived in Kūfa and was sent as a Shiite agent to Khurasan at the beginning of the eighth century. At the time this thesis was presented (1925), only the few writings that M. Berthelot had published in the third volume of his *Chimie au moyen âge* (1893) and in *Archéologie et histoire des sciences* (1906) were known. Holmyard published in 1928 a not very extensive volume of Jābir's writings based on an Indian lithograph. The immense list of Jābir's works in the *Fihrist*, which previously had been considered fantastic, was at least partially confirmed by these publications. Holmyard's detailed reconstruction of Jābir's biography on the basis of the supposition that he was the son of the above-mentioned Ḥayyān need not be repeated here; for even granting his historicity, Jābir is by no means the author of all the writings which bear his name.

JĀBIR IBN ḤAYYĀN

The study of all the printed texts of Jābir and of the manuscripts partly discovered by M. Meyerhof in Cairo libraries led Paul Kraus to the following conclusions: that the corpus of the writings attributed to Jābir is not the work of a single man but, rather, that of a school; that the degree of scientific knowledge shown and the terminology employed presuppose the translations from the Greek produced by Ḥunayn ibn Isḥāq (*d.* 874) and his school; that the references in these works to the theology of Muʿtazila suggest that the writings are of the same period, at the earliest; that the earliest mention of the books appears in Ibn Waḥshīya (first half of the tenth century) and in Ibn Umail ("Senior Zadith," *ca.* 960); and, above all, that the writings reveal the same more or less veiled Ismaʿili propaganda as that of the epistles of the Brethren of Purity, who were closely connected with the establishment of the Fatimid caliphate in Egypt—in short, that the works were not written until the tenth century and that Jaʿfar was really the Shiite imam, the name of whose eldest son, Ismail, was the eponym of Ismaʿiliya.

Kraus assembled a huge collection of Jābir manuscripts which enabled him to produce a relative chronology of the works enumerated in the *Fihrist*, with the help of citations contained in the texts themselves. The list of the *Fihrist* and the manuscripts form the basis of his numerical critical bibliography of Jābir's writings (works will be cited below according to this numeration). We also are indebted to Kraus for a comprehensive historical presentation of the scientific teaching of the corpus; he died before he could complete his account of Jābir's place in the religious history of Islam.

The publication of the writings belonging to the Arabic *Corpus Jābirianum* already had enabled Berthelot definitively to separate the author or authors of these works from the Latin alchemical writings appearing in the thirteenth and fourteenth centuries under the name of Geber, who had been considered identical with Jābir. These Latin writings—*Summa perfectionis magisterii*, *Liber de investigatione perfectionis*, *Liber de inventione veritatis*, *Liber fornacum*, and a *Testamentum Geberi*—had long confused researchers. Hermann Kopp gathered the material indicating that from the standpoint of literary history the author of these Latin writings was not the same as the author of the Arabic. Berthelot compared the contents of the Arabic texts with that of the Latin Geber texts and concluded that the latter displayed a level of chemical knowledge well beyond that shown by the Arabs. The keystone of the proof was provided by Berthelot's recognition that the *Liber de septuaginta* by Geber, which he edited, was a translation of Geber's *Book of Seventy* (Kraus nos. 123–192) listed in the *Fihrist;* hence, their omission from the Arabic list of the five Latin Geber texts mentioned serve as additional proof of their being spurious.

Recently a further demonstration of the correctness of Berthelot's thesis has been offered: in the *Occulta philosophia* of Agrippa von Nettesheim the comparison of the alchemists with God is spoken of in connection with the name Geber—which is in striking opposition to the Geber of the Latin writings mentioned above, who feels himself to be a poor creature whose success depends on scrupulous adherence to the instructions in his formulas and on God's grace. We will therefore completely disregard the Latin Geber texts and consider the Arabic texts exclusively. Moreover, for brevity we will speak of Jābir as if the name applied to a single author.

The corpus of Arabic Jābir texts comprises both individual books and groups of books. The latter are in part designated according to the number of individual writings they contain; the largest are entitled *Seventy*, *One Hundred Twelve*, and *Five Hundred.* The *Seventy* includes seven groups of ten, well distinguished from each other; *One Hundred Twelve* (Kraus nos. 6–122) is the product of four—the number of the basic qualities, the elements, the humors—and twenty-eight, the number of the mansions of the moon and of the letters of the Arabic alphabet, which is itself equal to four times seven. The *Five Hundred* (Kraus nos. 447–946) are not, in contrast with the *One Hundred Twelve* and the *Seventy*, individually enumerated in the *Fihrist;* since only some of them are known and shown to belong to the group, there exists no certainty concerning their character or the meaning of the number in the title.

Other writings gathered into groups are the ten supplementary books (*muḍāfa*) to the *Seventy* (Kraus nos. 193–202); ten books of corrections (*muṣaḥḥahāt*) of the teachings of mostly ancient philosophers (including Homer) and physicians (Kraus nos. 203–212); the *Twenty* (Kraus nos. 213–232); the *Seventeen* (the sacred number of the Ismaʿiliya) (Kraus nos. 233–249); the 144 *Books of the Balances* (Kraus nos. 303–446); and the books of the seven metals (*al-ajsād al-sabʿa;* Kraus nos. 947–956).

Alchemy takes a commanding position both in theory and in actuality in the corpus, but all the writings belonging to it are by no means concerned with this subject. Rather, all the sciences—philosophy, linguistics, astrology, the science of talismans, the sciences of the quadrivium, metaphysics, cosmology, and theology—are represented, as are fields which do not belong to the sciences, such as medicine, agriculture, and technology. The philosophical writings include notes on various pre-Socratics, a work attacking Plato's *Laws*, another against Aristotle's *De anima*, and a commentary on his *Rhetoric* and *Poetics*. A series of writings is based on Balīnās or Balīnūs, that is, Apollonius of Tyana. A book attributed to him, *The Secret of Creation*, is preserved in many manuscript copies; it was written about 820, in the time of al-Ma'mūn, and is a cosmological-alchemical commentary on the *Tabula Smaragdina* which concludes the work and which appears there for the first time. The writings of the corpus are full of quotations from ancient authors whose works are partly preserved elsewhere in Arabic translations.

Besides the writings published by Berthelot (1893) and Holmyard (1928), Kraus edited, during his investigation of the corpus, a volume of texts which contains several complete writings and others in extract (1935). A facsimile of the Arabic manuscript of the *Poison Book* (Kraus no. 2145) was published in 1958 with a German translation by A. Siggel. A French translation by H. Corbin of the *Livre du glorieux* (Kraus no. 706) appeared in 1950.

Besides the Latin translation of the *Liber de septuaginta*, only one other book is available in Latin, the *Liber misericordiae* (Kraus no. 5), edited by E. Darmstaedter in 1925.

The writings of the corpus are copiously cited in the Arabic literature, and long sections from them have been copied by other authors. The long list of quotations from Jābir in Arabic works given in Kraus (I, 189–196) has long been out of date. Since most books of the corpus are still unedited, the presentation of Jābir's doctrines must be based on the previously published material and on Kraus, who gives many quotations from the manuscripts.

For full appreciation of Jābir's achievement, a philologically based elucidation of his relationship to the Greek alchemists is the foremost requirement. In contrast with the later works, the *One Hundred Twelve* contains many quotations from the Greek alchemists and references to them. Since a portion of the Greek alchemical corpus must have been translated into Arabic, as is apparent from the numerous word-for-word quotations from such writings, especially in the *Turba philosophorum*, one of the chief problems is how Jābir freed himself from the confusion in the occult writings of the Greek authors and succeeded in constructing the system represented in the *Seventy*. It is also not inconceivable that the quotations from the Greek could contribute to the textual criticism of the original passages.

According to Jābir's developed theory, the ingredients of the elixir are not exclusively mineral; rather, some are vegetable and animal. The substances of all three kingdoms of nature can be combined so as to arrive at a mixture in which the basic qualities contained in all natural objects are represented in the proposition sought. The theoretical interest of this procedure is at least as strong as the practical interest in the transmutation of metals. The ideal goal is a catalog of all natural objects in which the basic qualities and peculiar properties of each substance, which are to be determined experimentally, are numerically specified. The scientific principle of such research Jābir called *mīzān* (balance); its all-encompassing importance results from the wealth of its applications. The word represents the Greek *zygon* (balance), also in the sense of specific weight; but it also stands for the *stathmos* (weight) of the Greek alchemists, in the sense of the measurement in a mixture of substances.

Beyond this purely scientific meaning, the term constitutes a basic principle of Jābir's world view: *mīzān al-ḥurūf*, the balance of the letters, concerns the relationship of the twenty-eight letters of the Arabic alphabet (four times seven) to the four qualities (warm, cold, moist, and dry), a relationship which also embraces the metaphysical hypostases of Neoplatonism—intellect, world-soul, matter, space, and time. The concept thus becomes a principle of Jābir's scientific monism, in opposition to the dualistic world view of Manichaeism—the struggle against this religion was one of the chief concerns of Islam at that time. This religious side of Jābir's world view is based on the appearance of the word *mīzān* in the Koran, where it is used both in the sense of the balance in which deeds are weighed at the Last Judgment (for example, 21, 47), and an eternal, essential part of heaven itself, along with the stars (55, 7–9). The allegorical interpretation of the Koranic balance, which also appears in Islamic gnosis, unites Jābir's scientific system with his religious doctrine.

Jābir finds an expression of this world view in the theory, already developed by the Greeks, of the specific properties (*proprietates*) of things, of their sympathies and antipathies, and of their specific suitabilities in practical applications, especially in medicine. Finally, this theory leads him to conceive of the possibility of the artificial production of natural objects, and therefore also of the homunculus; this conception expressly places the activity of his ideal scholar in parallel with that of the Demiurge.

Jābir's rationalism is not obscured by this theory; rather, it is here that he finds the working of natural law, as he sees it. The same is true of his treatment of arithmetic; the significance of number in nature, a notion developed by the Pythagoreans and Plato, is for him at once an empirical fact and a principle.

JĀBIR IBN ḤAYYĀN

The number 28 is not only the product of 4 and 7 but also the seventh number in the arithmetical series 1-3-6-10-15-21-28. It is a "perfect number" in that it is equal to the sum of its divisors (1, 2, 4, 7, 14). Besides this series Jābir readily used the series 1-3-5-8, which defines the relationships between the degree of the basic qualities and their intensity. It should be observed that the sum of these numbers is seventeen, a religiously significant number to the Isma'ilis. He considers this number to be the basis of the theory of the balance; it indicates the equilibrium that governs the constitution of every object in the world.

For authors of Jābir's time it is obvious that the astrological world view played a prominent role in the entire theory. The stars are not only a constituent of the world of which they are a part; their unique position in the cosmos also makes them of decisive importance in terrestrial events. This view is expressed in Jābir's very detailed talisman theory. The talisman bears the powers of the stars and, according to him, is for this reason called *ṭilasm* (*ṭlsm* in vowelless Arabic script), because it is given domination (*musallaṭ*, without vowels, *msl*ṭ) over events in the world. But Jābir did not stop with the importance of the stars for the creation of talismans; rather, he believed that they can be made directly subservient through sacrifice and prayer. The character of such sacrifices and prayers can be gathered from the extensive chapters dealing with similar matters in the *Picatrix* (Arabic, *Ghāyat al-ḥakīm*), which is traditionally attributed, incorrectly, to the Spanish mathematician and astronomer Maslama al-Majrīṭī). The author of the *Picatrix* expressly names Jābir as one of his intellectual leaders. This portion of his teachings is one of the most prominent pieces of evidence for the survival of the belief in the stars as divine beings, as they were originally viewed (hence the names of the planets), even though the monotheistic religions had officially removed them from this status. The Hebrew and Latin translations of the *Picatrix* show that the lingering on of this "idolatry" was not confined to the Islamic world.

Two of the writings contained in the corpus, which Kraus edited and placed at the beginning of his volume of texts, permitted him to reconstruct Jābir's system of the sciences: *Book of the Transformation of the Potential Into the Actual* (Kraus no. 331, no. 29 of the *Books of the Balances Texts*, pp. 1–95) and *Book of Definitions* (Kraus no. 780, belonging to the *Five Hundred Texts*, pp. 97–114). This system is divided into sections on the religious and the secular sciences. Within the secular sciences, alchemy and its dependent sciences occupy one side of the genealogical tree; all the others constitute the second side. Among the former, medicine plays a major role; Jābir's remarkable knowledge in this area is displayed in his book on poisons. As for the religious sciences, it is remarkable that here the intellectual disciplines (the science of letters, the science of the senses, philosophy, metaphysics, and others) are on the same footing. A valuable project would be the comparison of this ranking with that developed by al-Fārābī, who in his *De scientiis* likewise sought to place the religious sciences into a total system.

Jābir is among the pioneers of the completion of the "spirits," that is, of the volatile substances sulfur, mercury, and arsenic, through a fourth one, sal ammoniac, which was unknown to the Greeks. He knew mineral ammonia and other kinds that can be prepared chemically. Hair, blood, and urine served him as its material bases. The Arabic word for ammonia, *nushādir*, is of Persian origin and hence it is reasonable to suppose that it was discovered in the Sassanid kingdom.

Holmyard, using Kraus's analytical presentation as a basis, attempted a complete synthetic description of Jābir's alchemical system. Such an attempt will not be repeated here, because there is hope that a translation of the *Seventy*, a work central to Jābir's theory, prepared over forty years ago by the present writer, may yet be published.

The acquaintance of the author of the corpus with the best works of Greek science is astounding. Kraus gathered the evidence for his knowledge of the works of Aristotle, Alexander of Aphrodisias, Galen, Archimedes, and of the *Placita philosophorum* of pseudo-Plutarch. This evidence is not complete, nor is the number of quotations from ancient authors exhausted by it. That this schooling in Greek science is completely compatible with the gnostic speculations of the Sābians of Ḥarrān has been demonstrated; for from their midst have emerged some of the most outstanding mathematicians, astronomers, and physicians of Islam.

Jābir's importance for the history not only of alchemy but also of science in general, and for the intellectual history of Islam, has by no means been sufficiently examined. Future study of the writings contained in the corpus will no doubt provide many surprises concerning the position of their authors in the intellectual culture of the Middle Ages.

BIBLIOGRAPHY

The following works may be consulted: M. Berthelot, *La chimie au moyen âge*, I (Paris, 1893), 320–350; and

Archéologie et histoire des sciences (Paris, 1906), pp. 308–363; H. Corbin, "Le livre du glorieux de Jābir ibn Ḥayyān," in *Eranos-Jahrbuch*, **18** (1950), 47–114; E. Darmstaedter, *Die Alchemie des Geber* (Berlin, 1922); and "Liber misericordiae Geber, eine lateinische Übersetzung des grösseren *kitāb ąl-raḥma*," in *Archiv für Geschichte der Medizin*, **18** (1925), 181–197; J. W. Fück, "The Arabic Literature on Alchemy According to al-Nadim," in *Ambix*, **4** (1951), 81–144; E. J. Holmyard, *The Works of Geber, Englished by Richard Russell, 1678, a New Edition* (London–New York, 1928); and *Alchemy* (London, 1957), pp. 66–80; B. S. Jørgensen, "Testamentum Geberi," in *Centaurus*, **13** (1968), 113–116; H. Kopp, *Beiträge zur Geschichte der Chemie*, III (Brunswick, 1875), 13–54; P. Kraus, "Studien zu Jābir ibn Ḥayyān," in *Isis*, **15** (1931), 7–30; *Jābir ibn Ḥayyān, essai sur l'histoire des idées scientifiques dans l'Islam*, I, *Textes choisis* (Paris, 1935); and *Jābir ibn Ḥayyān, contribution à l'histoire des idées scientifiques dans l'Islam. I. Le corpus des écrits jabiriens*, and *II. Jābir et la science grecque*, in *Mémoires. Institut d'Egypte*, **44** (1943) and **45** (1942); P. Kraus and M. Plessner, "Djabir b. Hayyan," in *Encyclopaedia of Islam*, new ed. (1965); H. M. Leicester, *The Historical Background of Chemistry* (New York–London, 1956), pp. 63–70; *Picatrix, Das Ziel des Weisen von Pseudo-Mağrīṭī*, 2 vols.: I, Arabic text, H. Ritter, ed. (Leipzig, 1933), II, German trans. by H. Ritter and M. Plessner (London, 1962); M. Plessner, "The Place of the Turba philosophorum in the Development of Alchemy," in *Isis*, **45** (1954), 331–338; "Ğābir ibn Ḥayyān und die Zeit der Entstehung der arabischen Ğābir-Schriften," in *Zeitschrift der Deutschen morgenländischen Gesellschaft*, **115** (1965), 23–35; "Balīnūs," in *Encyclopaedia of Islam*, new ed. (1960); "Geber and Jābir ibn Ḥayyān: An Authentic Sixteenth-Century Quotation From Jābir," in *Ambix*, **16** (1969), 113–118; and "Medicine and Science," in *The Legacy of Islam* (1972); J. Ruska, "Sal ammoniacus, Nušādir und Salmiak," in *Sitzungsberichte der Heidelberger Akademie der Wissenschaften*, Phil.-hist. Kl. (1923), 5; *Tabula Smaragdina* (Heidelberg, 1926); and *Turba philosophorum* (Berlin, 1931); J. Ruska and P. Kraus, "Der Zusammenbruch der Dschabir-Legende: I. Die bisherigen Versuche, das Dschabir-Problem zu lösen, II. Dschabir Ibn Hajjan und die Isma'ilijja," in *Jahresbericht der Forschungsinstituts für Geschichte der Naturwissenschaften in Berlin*, **3** (1930); F. Sezgin, "Das Problem des Ğābir ibn Ḥayyān im Lichte neu gefundener Handschriften," in *Zeitschrift der Deutschen morgenländischen Gesellschaft*, **114** (1964), 255–268; and A. Siggel, *Das Buch der Gifte des Ğābir ibn Ḥayyān, arabischer Text, übersetzt und erläutert* (Wiesbaden, 1958).

M. PLESSNER

JACCARD, AUGUSTE (*b.* Culliairy, Neuchâtel, Switzerland, 6 July 1833; *d.* Le Locle, Neuchâtel, Switzerland, 5 January 1895), *geology, paleontology*.

Auguste Jaccard came from a strongly Protestant family; his parents belonged to the Mährische Brüder community. His father made music boxes and was a small farmer. Hoping to better themselves financially the family resettled in 1840 in Ste. Croix, where Auguste attended the village school. In 1845 the family moved to Le Locle. The father was by now pursuing the watchmaker's trade, and Auguste helped him in this while continuing with his schooling.

The revolutionary ideas and political events of 1845–1848, in which the students of Le Locle actively participated, induced the intensely religious, conservative father to take Auguste and his three brothers out of school. Until 1875 the father and his sons ran a family watchmaking business. Auguste then became independent, operating his own watchmaking business until 1885.

In 1849 there appeared the first symptoms of pulmonary tuberculosis, of which he was to die. He married Marie Joly in 1857; they had four children.

Even as a boy Jaccard collected fossils. At first he did not recognize their importance, since he had no books and no guidance in identifying the specimens. But in the winter of 1848–1849 a Ste. Croix physician named Campiche took an interest in him and therewith began his passion for geology and paleontology. In 1853 he came under the influence of Desor, professor of geology at the Academy of Neuchâtel. Through Desor, Jaccard came to know the important geologists of the Swiss Jura, including Gressly, Tribolet, Pictet de la Rive, Loriol, and Renevier.

In 1868 Desor named Jaccard as his assistant, and in 1873 Jaccard succeeded him as professor of geology; he held this chair until his death. In 1883 the University of Zurich conferred a doctorate *honoris causa* on him. Over the years Jaccard held various administrative and public posts. His colleagues and friends saw in him a tireless worker, a passionate collector, and an upstanding, ever-helpful human being.

Jaccard's geological and paleontological works are concerned almost exclusively with the Swiss Jura and adjacent parts of France. His first publication, in 1856, dealt with 140 subtropical types of Tertiary flora from the freshwater limestone of Le Locle. He was aided and encouraged by Oswald Heer in Zurich. His later publications treated the tortoises of this locality, as well as reptiles, fish, and invertebrates of the Jurassic and Cretaceous systems in the Jura.

Geological mapping constituted a great segment of Jaccard's lifework. In 1861 he became a collaborator on the geological map of Switzerland, and he published the following year three sheets in the scale 1 : 100,000, with commentary, dealing with parts of the Jura mountains. In 1877 he published a geological map of the canton of Neuchâtel, and in 1892 there appeared a geological mapping and description of the French Chablais Alps south of the Lake of Geneva—the result of six year's work.

Along with his mapping and the consequent paleontological, stratigraphic, and tectonic projects, Jaccard busied himself over many years with questions and problems of practical geology. There were numerous publications on hydrogeology, hydrology, and the springs and groundwaters of the Jura. He provided expert advice on water supplies, often as the member of a commission, to various cities of his home region. In 1882 he reported on a hydrological map of Neuchâtel canton. His other publications treated the possibility of establishing factories utilizing Jura limestone for cement production.

As a geologist and member of the board of directors for asphalt mining in Val de Travers (southwest of Neuchâtel), he became involved in investigating the occurrence and origin of bitumen, petroleum, and asphalt in Switzerland. He was among the first to declare petroleum to be of organic origin—as opposed to the then widely held view proposing an inorganic, hydrothermal origin from the depths of the earth. Surface oil traces in the Tertiary Molasse sediments of the Neuchâtel environs, as well as the investigation of geological circumstances there, prompted Jaccard to propose deep drilling in the hope of finding petroleum. Drilling and exploration in the Swiss Tertiary Molasse sediments continued up to the middle of the twentieth century.

BIBLIOGRAPHY

I. Original Works. "Notes sur la flore fossile du terrain d'eau douce supérieur du Locle," in *Bulletin de la Société des sciences naturelles de Neuchâtel*, **2** (1856); "Description de quelques débris de reptiles et de poissons fossiles trouvés dans l'étage jurassique supérieur (virgulien) du Jura neuchâtelois," in *Matériaux pour la paleontologie suisse*, 3rd ser., no. 1 (1860), 20 pls., written with F. J. Pictet de la Rive; "Description géologique du Jura vaudois et de quelques districts adjacents du Jura français et de la plaine suisse," in *Beiträge zur geologischen Karte der Schweiz*, **6** (1869), 8 pls., 2 maps; "Supplément à la description du Jura vaudois et neuchâtelois (sixième livraison)," *ibid.*, **7**, pt. 1 (1870), 4 pls., map; "Observations sur les roches utilisées par la fabrique de ciment de Saint-Sulpice et sur les terres à briques du Jura," in *Bulletin de la Société des sciences naturelles de Neuchâtel*, **9** (1879); "Étude sur les massifs du Chablais compris entre l'Arve et la Drance (feuilles de Thonon et d'Annecy)," in *Bulletin du Service de la carte géologique de la France et Topographie souterraine*, **3**, no. 26 (1892); "Le pétrole, l'asphalte et le bitume au point de vue géologique," in *Bibliothèque scientifique international*, **81** (1895).

II. Secondary Literature. M. de Tribolet, "Notice sur la vie et les travaux d'Auguste Jaccard," in *Bulletin de la Société des sciences naturelles de Neuchâtel*, **23** (1895), 210–242 (with bibl. and portrait); and "Auguste Jaccard 1833–1895," in *Actes de la Société helvétique des sciences naturelles* (Zermatt, 1895), pp. 205–211; H. Rivier, "La Société neuchâteloise des Sciences naturelles 1832–1932. Notice historique publiée à l'occasion de son centenaire," in *Bulletin de la Société des sciences naturelles de Neuchâtel*, **56** (1932), 5–83, 7 pls.

Heinz Tobien

JACKSON, CHARLES THOMAS (*b.* Plymouth, Massachusetts, 21 June 1805; *d.* Somerville, Massachusetts, 28 August 1880), *medicine, chemistry, mineralogy, geology.*

Jackson was descended from the original settlers of Plymouth. His sister Lydia (later renamed Lydian) was Ralph Waldo Emerson's second wife. He had an irritable personality and it is difficult to avoid putting the label of "paranoid" on his behavior. He died insane.

Jackson received his early education in the town school and in the private school of Dr. Allyne of Duxbury. His health failed and he made a walking expedition through New York and New Jersey with Baron Lederer, William McClure, Lesueur, and Troost, who were to foster his interest in natural history and geology. He returned to Boston and prepared privately for Harvard Medical School. During medical school, he received the Boylston Prize for a dissertation on paruria mellita, and in 1829 he was "authorized to give instruction in chemistry during the absence of Dr. [John White] Webster, and at his expense," at Harvard College. He was graduated in 1829.

Jackson's interest in geology began when he found chiastolite crystals in a glacial drift schist in Lancaster, Massachusetts. In 1827 he visited Nova Scotia with his friend, Francis Alger, and he returned again in in the summer of 1829 to geologize. His observations were published in the 1828 *American Journal of Science.*

In 1829 Jackson went to Europe where he studied medicine at the University of Paris, attended lectures of the École de Médecine, the Collège de France, and the scientific lectures at the Sorbonne, as well as those on geology given by L. Élie de Beaumont at the École Royale des Mines. He made a walking tour of Europe, and, in Vienna, did 200 autopsies with Dr. John Fergus of Scotland and Dr. Johannes Glaisner of Poland on cholera victims. In Paris he became acquainted with the statistical methods then being introduced into medicine by Pierre Louis and his empirical school. He returned to Boston in 1832.

He abandoned the practice of medicine completely in 1836 and established a laboratory in Boston for instruction in analytical chemistry, the first laboratory of its kind in America to receive students.

Jackson was the first state geologist of Maine, Rhode Island, and New Hampshire. The movement for geological surveys was on, but when Great Britain claimed 10,000 square miles of Maine, it served to stimulate the legislature of Massachusetts to cooperate with that of Maine in the survey of the public lands owned by the former in the latter's territory. Three reports were issued (1837, 1838, 1839). In 1839 Jackson was employed to make a geological examination of Rhode Island; the report was published in 1840. In 1839 he was appointed state geologist for New Hampshire and his results were published in a quarto volume in 1844. In 1847 he was appointed U.S. geologist to report on the public lands in the Lake Superior region. After two seasons' work, conflicts with his fellow geologists J. D. Whitney and J. W. Foster led to his discharge. After Jackson was dismissed as U.S. geologist, there appeared the pamphlet *Full Exposure of the Conduct of Charles T. Jackson, Leading to His Discharge From Government Service, and Justice to Messrs. Foster and Whitney* (Washington, 1851). Thereafter, he made frequent reports for mining companies.

Jackson was a descriptive geologist who was interested in the economic advantages that might come from his work. He was more interested in mineralogical geology than stratigraphical geology, and he minimized the importance of fossils in determining the ages of rocks. His reliance on mineralogy alone obscured possible correlations of strata. Everywhere, Jackson saw igneous causes. He also considered "the causes formerly in action vastly more energetic than they are now." He was not interested in the question of chronology and held on to the Wernerian mineralogical school's terminology, such as "transition" rather than using the terms "Cambrian" and "Silurian" as introduced by Sedgwick and Murchison. In 1843 he stated that "this country presents no proofs of the glacial theory as taught by Agassiz, but on the contrary the general bearing of the facts is against the theory." Jackson appears to have been the first to observe the occurrence of tellurium and silenium in America.

In 1836 Jackson claimed discovery of guncotton after it had been announced by C. F. Schönbein. Jackson had returned to America in 1832 on the same ship with Samuel F. B. Morse and some years later claimed to have pointed out to Morse the principles of the electric telegraph which Morse patented in 1840. (For a defense of Morse see A. Kendall, *Morse's Patent. Full Exposure of Dr. Charles T. Jackson's Pretensions to the Invention of the American Electro-Magnetic Telegraph.*)

William Thomas Green Morton, who had been a student of Jackson's, demonstrated to a group of students and physicians the use of ether as a general anesthetic on 16 October 1846 at the Massachusetts General Hospital. Morton applied for a patent on 28 October 1846 and, upon the advice of others, Jackson's name was included as a patentee. The patent was granted (no. 4848) on 12 November 1846, but predictably Jackson claimed the discovery, although he had assumed no responsibility. He addressed two letters to Élie de Beaumont (1 December and 20 December 1846), to be read at the French Academy, and in which, without mentioning Morton, he announced himself as the discoverer of surgical anesthesia. On 2 March 1847, Jackson made a similar announcement at the American Academy of Arts and Sciences.

He spent much time and effort trying to prove his primary role in the discovery of the use of ether as a general anesthetic. In 1873 he was committed to McLean Hospital as insane. He died in 1880. His valid scientific reputation rests upon his geological and mineralogical work.

BIBLIOGRAPHY

I. Original Works. An incomplete bibliography of Jackson's writings is to be found in J. B. Woodworth, "Charles Thomas Jackson," in *American Geologist*, **20** (Aug. 1897), 87–110.

II. Secondary Literature. There has been no adequate biography of Jackson; the available biographical accounts are William Barber, "Dr. Jackson's Discovery of Ether," in *National Magazine* (Oct. 1896), 46–58; Edward Waldo Emerson, "A History of the Gift of Painless Surgery," in *Atlantic Monthly*, **78**, 679–686; Anonymous, "Charles Thomas Jackson," in *Proceedings of the American Academy of Arts and Sciences*, **16** (1881), 430–432; T. T. Bouve, "Biographical Notice of C. T. Jackson," in *Proceedings of the Boston Society of Natural History*, **21** (1881), 40–47; Thomas Edward Keys, *The History of Surgical Anesthesia* (New York, 1945); Bruce Rogers, *The Semi-Centennial of Anaesthesia Oct. 16, 1846–Oct. 16, 1896* (Boston, 1897); L. J. Ludovici, *The Discovery of Anaesthesia* (New York, 1961); John F. Fulton and Madeline Stanton, *An Annotated Catalogue of Books and Pamphlets Bearing on the Early History of Surgical Anaesthesia* (New York, 1946); William Henry Welch, *A Consideration of the Introduction of Surgical Anesthesia* (Boston, 1909); Richard Manning Hodges, *A Narrative of Events Connected With the Introduction of Sulphuric Ether Into Surgical Use* (Boston, 1891); Thomas Frances Harrington, *The Harvard*

Medical School 1782–1905, II (New York, 1905), 604–635; R. H. Dana, Jr., ed., "Ether Discovery," in *Littell's Living Age* (1848), pp. 529–571; Victor Robinson, *Victory Over Pain* (New York, 1946).

On the ether controversy, the Boston Athenaeum holds "clippings and correspondence relating to etherization, Dr. W. T. G. Morton and Dr. C. T. Jackson (1848–1861)." There are about twenty letters relating to the ether discovery, including an undated, six-page letter from C. T. Jackson to the U.S. Congress. The Countway Medical Library at Harvard Medical School has the correspondence of Dr. Stanley Cobb (1936–1937) regarding a portrait of Jackson and a memorial tablet to be placed in the Ether Dome at the Massachusetts General Hospital, in order to give Jackson recognition for the discovery of ether. This includes copies of letters (1849–1873) verifying Dr. Jackson's discovery. The Countway also holds three volumes, "letters, affidavits and other papers relating to the discovery of the use of ether," deposited in the Boston Medical Library in March 1917 by Mrs. Bridges of Medford, Jackson's daughter. There are approximately fifteen letters in the Countway. The Houghton Library at Harvard holds about twenty letters, including nine letters to Emerson (1845–1856). The Boston Museum of Science (with holdings of the old Boston Society of Natural History) has three lectures and two letters. The Massachusetts Historical Society has about thirty letters relating to Jackson and the ether controversy. The Concord Public Library and the Boston Public Library each has one. There are also a few letters in the William Sharswood correspondence in the American Philosophical Society and one letter in the Association of American Geologists and Naturalists papers of the Academy of Natural Science in Philadelphia. There are 65 items in the Manuscript Division of the Library of Congress.

GEORGE EDMUND GIFFORD, JR.

JACKSON, JOHN HUGHLINGS (*b.* Providence Green, Green Hammerton, Yorkshire, England, 4 April 1835; *d.* London, England, 7 October 1911), *clinical neurology, neurophysiology.*

Jackson's father, Samuel, was a yeoman who owned and farmed his land; his mother, the former Sarah Hughlings, was of Welsh extraction. There were three sons and a daughter besides John; his brothers emigrated to New Zealand and his sister married a Dr. Langdale. Jackson attended small country schools, and little is known of this period of his life except that his general education was limited. At the age of fifteen he was apprenticed to a Dr. Anderson, a lecturer at the now defunct York Hospital Medical School, and completed his medical education at St. Bartholomew's Hospital Medical School in London from 1855 to 1856, when he gained the form of English medical qualification usual at that time: Member of the Royal College of Surgeons and Licentiate of the Society of Apothecaries. He then went back to York Hospital as house surgeon to the dispensary, a post he held until 1859, when he returned to London, intending to study philosophy rather than medicine.

Jonathan Hutchinson, also from York, dissuaded Jackson, and he was appointed to the staff of the Metropolitan Free Hospital in 1859 and at the same time was made lecturer on pathology at the London Hospital. In 1860 he took the M.D. degree of the University of St. Andrews and in 1861 was admitted Member of the Royal College of Physicians of London. By 1863 he was assistant physician at the London Hospital and, in 1874, physician; he remained on the active staff until 1894, thereafter until 1911 serving as consulting physician. Jackson's other main attachment, to the National Hospital for the Paralysed and Epileptic, Queen Square, began in 1862 as assistant physician; elevated to full physician in 1867, he retired in 1906 as consulting physician. Jackson also had other appointments to London hospitals, the most consequential being that to Moorfields Eye Hospital from 1861; in addition he had a private practice.

In 1868 Jackson was elected a Fellow of the Royal College of Physicians of London, and over the ensuing years he was invited to give several of the important named lectures of the College and to assist in its administration. He was appointed Fellow of the Royal Society in 1878 and in 1885 was the first president of the Neurological Society, later absorbed by the Royal Society of Medicine.

He married his cousin Elizabeth Dade Jackson in 1865; they had no children, and she died of a cerebrovascular disorder in 1876. Both Jackson and his wife died at 5 Manchester Square, and his residence there is today recorded on a blue plaque affixed to the house. He suffered from a chronic form of vertigo and died of pneumonia.

Jackson was an intensely shy, modest, and obsessional man who disliked most social activities. He was gentle, ever-courteous, and had a subtle, if somewhat hidden, sense of humor. He never exhibited excesses of passion, and his intellectual honesty allowed him always readily to acknowledge the efforts of others. He had no religious convictions and denied the existence of life after death.

Much has been written on the fertility of Jackson's genius, which basically depended upon his ability to diverge and converge. Like his contemporary Charles Darwin, he had the keenest powers of observation and at the same time an ability to manipulate philosophical concepts, together with a comprehensiveness of view. He could on the one hand give laborious attention to the minutest local clinical neurological detail, which he analyzed with infinite care, and on the other he could synthesize and propound wide generalizations and fundamental doctrines involving the whole nervous system. His writings are usually considered difficult to read because of an involved and tedious style, dictated by his obsessive need to qualify and document statements and to repeat himself.

Jackson's first paper was published in 1861, his last in 1909. Unfortunately he never recorded a systematic account of his views, but his scattered contributions to the neurological sciences can be considered under six specific areas. His writings are voluminous and teeming with ideas, only some of which can be presented here. For most of Jackson's ideas it is impossible to indicate in a brief summary either their breadth and subtlety or the rich accumulation of diverse clinical phenomena supporting them.

When Jackson first entered the field of clinical neurology, it was a chaotic mass of clinical and pathological data concerning diseases that were usually of unknown etiology. Neurophysiology was also in its infancy. During his lifetime and as a result of the combined activity of neurological clinics mainly in Paris, Heidelberg, Vienna, New York, and Philadelphia and of his own institution in Queen Square, order was established, methods of clinical examination were introduced, etiologies (usually based on advances in physiology or pathology) were discovered, advances in physiology were made, and the roots of modern neurology were created. It was in this milieu that Jackson excelled because of his versatility in the collection of data and the enunciation of basic principles. Although acclaimed as the greatest British scientific clinician of the nineteenth century, he carried out no experiments and rarely employed the microscope; but he was acutely aware of their importance and was widely read in the literature. He was devoted to clinical observation, to the analysis of facts, and to philosophic reasoning.

Thus armed, Jackson profoundly influenced all of the neurological sciences. He applied the data of abnormal functioning to the elucidation of the normal action of the nervous system. Jackson's creed was as follows: "I should be misunderstood if I were supposed to underrate the physiological study of disease of the nervous system. Indeed, I think that to neglect it shows want of education, but to neglect the clinical shows want of experience and sagacity. Never forget that we may run the risk of being over-educated and yet under-cultivated." This approach remains acceptable today. He was less concerned with the etiology of disease and its cataloging than with the interpretation of disorders in terms of physiology. Jackson always wished to know how a symptom could reveal normal function and how knowledge of a disease could illuminate the normal dynamic properties of the nervous system. In sum, he profoundly influenced the development of clinical neurology, neurophysiology, and psychology in the nineteenth century and continues to do so in the twentieth.

In early life Jackson was influenced by four men: Thomas Laycock, who introduced him to the Paris school of clinicopathologic correlation and to clinical neurology; Jonathan Hutchinson, who imbued him with the Hunterian tradition of scrupulous observation and biological and pathophysiologic principles; Charles Brown-Séquard, who also influenced him to devote his attention to the nervous system; and Herbert Spencer, whose positivistic evolutionary theory led Jackson to certain basic neurological doctrines.

Jackson's earlier papers deal primarily with ophthalmological problems. The ophthalmoscope had been invented in 1851 by Helmholtz; and when Jackson began his neurological career ten years later he emphasized its importance as a diagnostic instrument and the need for neurologists to study eye diseases. In London he began a tradition, which still exists, of neurologists being affiliated with ophthalmological hospitals. He was one of the first to insist upon the relationships between ocular and cerebral disease and wrote extensively on papilledema and "optic neuritis"; he published a summary of this work in 1871 and also wrote many papers on ocular palsies. Jackson's contribution to the growth of ophthalmology and neuro-ophthalmology was therefore a significant one, and it continued throughout his career.

Before Jackson's time it was thought that epileptic phenomena originated in the medulla oblongata, and it was his work more than anyone else's that initiated progress toward the present concept. He considered each attack to be an experimental situation from which, by close observation and deduction, he could add to the knowledge of the functioning of the normal brain. At a physiological level he considered a convulsion to be a symptom, not a disease per se: "an occasional, an excessive and a disorderly discharge of nerve tissue on muscle." The word "discharge" was an inspired forecast of electrical phenomena. Jackson was the first to place the site of this disturbance in the cerebral cortex; and from his careful analysis of the localized type of epilepsy, with its "march" of events, as he put it, he deduced that in this case it was limited to a certain area of one cerebral hemisphere. This focal fit, or convulsion beginning unilaterally, had been described earlier by L. F. Bravais (*Recherches sur les symptômes et le traitement de l'épilepsie hémiplégique*, Thèse de Paris, no. 118 [1827]) and by others, but its significance had remained unknown; it is therefore appropriately termed "Jacksonian epilepsy."

Each local epileptic manifestation received Jackson's attention; and he is also well known for the first adequate account of the uncinate fit, which involves a "dreamy state," a hallucination of smell or taste, and involuntary chewing or tasting. The loss of consciousness associated with the major (grand mal) and the minor (petit mal) varieties of epilepsy also intrigued him; he was able to demonstrate that it was due to a wide spread of the discharge. Concerning the etiology of epilepsy he often found no pathological change and suggested a nutritional disturbance of brain cells, which may well be the explanation in certain cases. Jackson's interest in epilepsy led him to several important discoveries in neurophysiology, the most significant being cortical localization and speech function.

One of Jackson's most outstanding contributions to the neurosciences was his contention that function is localized to areas of the cerebral cortex. This he deduced from his studies on focal epilepsy and from philosophical considerations, here clearly being influenced by Herbert Spencer and supplementing the work of Broca. By 1870 he was certain of his contention, and it was in the same year that Fritsch and Hitzig showed experimentally that electrical stimulation of the cortex produced contralateral limb movement. Others quickly verified this; but the most important confirmatory study was that of Ferrier, whose aim was to test Jackson's suggestion fully. Jackson believed in neither precise cortical mosaics nor diffuse representation; but in the case of the representation of movements, he postulated a motor area and worked out the cortical pattern accepted today, noting its gradations and overlapping. His views were not widely accepted at the time but are nearer to modern beliefs than are those of his contemporaries.

Jackson's study of epileptics whose attacks involved a disturbance of speech allowed him to investigate the physiology of speech, particularly its central mechanisms. Unlike his contemporaries, who were seeking precise cerebral centers for the various manifestations of speech function, Jackson preferred to elucidate its physiological basis. He concluded that there were two components of speech. First was intellectual or "propositioning" speech, the unit of which is the proposition with precise word order, and not the word, although a single word may have a propositional value if, when it is uttered, other words are understood. Second was emotional speech, which expresses feelings but has no propositional value. As Walshe has pointed out, this simple analysis may be nearer the truth than many modern approaches. Jackson's suggestion concerning the respective roles of the two cerebral hemispheres in the production and control of speech is no longer acceptable.

The various motor defects manifested by neurological disease provided Jackson with further data to expand his basic concepts of nervous system activity. The focal motor epileptiform attack, the multifarious phenomenon of hemiplegia, and the complex, involuntary movements of chorea were each employed to contribute in particular to his ideas on the functioning of the cerebral cortex and the cerebellum. He considered that the latter controlled continuous (tonic) movements, whereas the cerebrum controlled changing (clonic) movements, and thereby formulated the modern doctrine of the tonic nature of cerebellar activity. In addition, Jackson envisaged functional localization in the cerebellum, which has been demonstrated only very recently. His so-called "cerebellar attacks," occurring with tumors of the cerebellum, were of brainstem origin.

Jackson's most basic neurological concept, the evolution and dissolution of the nervous system, again reveals the influences of Spencer and Laycock; the best account of it is in his Croonian lectures (1884). The nervous system evolves from the lowest, simplest, and most automatic center, present at birth, to the highest, most complex, and voluntary. There are three morphological and functional levels: the spinal cord and brainstem, the long afferent and efferent tracts (including their cortical connections), and the prefrontal cerebral cortex. Disease results in dissolution, the reverse of evolution; and there are always two elements present in the clinical picture: the negative, due to local destruction, and the positive, produced by the surviving tissue, which may have been released from higher control.

BIBLIOGRAPHY

I. Original Works. Jackson's writings comprise approximately 320 articles which relate exclusively to clinical neurology and neurophysiology. He attempted no synthesis or *opera omnia*, despite the entreaties of Osler, Weir Mitchell, and J. J. Putnam (see *Neurological Fragments* [*below*], p. 25). There are, however, two collections of his works, both edited by J. Taylor—*Selected Writings of John Hughlings Jackson*, 2 vols. (London, 1931); I, *On Epilepsy and Epileptiform Convulsions*, and II, *Evolution and Dissolution of the Nervous System. Speech. Various Papers, Addresses and Lectures*, with bibliography in II, 485–498; and *Neurological Fragments of J. Hughlings Jackson* (London, 1925), a series of brief articles published originally in *Lancet* over a period of sixteen years, all dealing with clinical and physiological neurology and each written in Jackson's characteristic style. In *Medical Classics*, **3** (1939), 889–971, there is a bibliography on pp. 890–913; and four of his papers are reprinted on pp. 918–971.

The important contributions are as follows, arranged by topic:

General features of his work are presented in *Neurological Fragments* (see above); "Notes on the Physiology and Pathology of the Nervous System," in *Selected Writings*, II, 215–237; "Certain Points in the Study and Classification of Diseases of the Nervous System," *ibid.*, pp. 246–250, the Goulstonian lectures, 1869 (abstract only); and "On the Study of Diseases of the Nervous System," in *Brain*, **26** (1903), 367–382.

Secondary literature in this area is E. F. Buzzard, "Jackson and His Influence on Neurology," in *Lancet* (1934), **2**, 909–913; and S. H. Greenblatt, "The Major Influences on the Early Life and Work of John Hughlings Jackson," in *Bulletin of the History of Medicine*, **39** (1965), 345–376, with a bibliography of his writings from 1861 to 1864.

Ophthalmology is the subject of "Lectures on Optic Neuritis From Intracranial Disease," in *Selected Writings*, II, 251–264; "Ophthalmology in Its Relation to General Medicine," *ibid.*, pp. 300–319; and "Ophthalmology and Diseases of the Nervous System," *ibid.*, pp. 346–358.

See B. Chance, "Short Studies on the History of Ophthalmology. III. Hughlings Jackson, the Neurologic Ophthalmologist, With a Summary of His Works," in *Archives of Ophthalmology*, **17** (1937), 241–289, with a bibliography of his ophthalmological papers; and J. Taylor, "The Ophthalmological Observations of Hughlings Jackson and Their Bearing on Nervous and Other Diseases," in *Proceedings of the Royal Society of Medicine*, **9** (1915), 1–28.

"A Study of Convulsions," in *Selected Writings*, I, 8–36, is the best collection of Jackson's views on the underlying mechanisms of epilepsy. See also "On the Anatomical, Physiological, and Pathological Investigations of Epilepsies," *ibid.*, pp. 90–111; "On the Scientific and Empirical Investigation of Epilepsies," *ibid.*, pp. 162–273; and "Epileptic Attacks With a Warning of a Crude Sensation of Smell and With the Intellectual Aura (Dreamy State) . . .," *ibid.*, pp. 464–473.

Commentaries are H. L. Parker, "Jacksonian Convulsions: An Historical Note," in *Journal-Lancet*, **49** (1929), 107–111; O. R. Langworthy, "Hughlings Jackson: Opinions Concerning Epilepsy," in *Journal of Nervous and Mental Diseases*, **76** (1932), 574–585; G. Jefferson, "Jacksonian Epilepsy: Background and Postscript," in *Post-Graduate Medical Journal*, **11** (1935), 150–162; and O. Temkin, *The Falling Sickness: A History of Epilepsy From the Greeks to the Beginnings of Modern Neurology* (Baltimore, 1945), pp. 288–323.

Cortical localization is discussed in "Loss of Speech," in *Clinical Lectures and Reports of the Medical and Surgical Staff, London Hospital*, **1** (1864), 388–471; "On the Anatomical and Physiological Localisation of Movement in the Brain," in *Selected Writings*, I, 37–76; "Observations on the Localisation of Movements in the Cerebral Hemispheres, as Revealed by Cases of Convulsion, Chorea, and 'Aphasia,' " *ibid.*, pp. 77–89; and "On Some Implications of Dissolution of the Nervous System," *ibid.*, II, 29–44.

See E. Hitzig, "Hughlings Jackson and the Cortical Motor Centres in the Light of Physiological Research," in *Brain*, **23** (1900), 544–581; and E. Clarke and C. D. O'Malley, *The Human Brain and Spinal Cord* (Berkeley–Los Angeles, 1968), pp. 499–505.

On speech disorders, see "On the Nature of the Duality of the Brain," in *Selected Writings*, II, 129–145; and "On Affections of Speech From Disease of the Brain," *ibid.*, pp. 155–204.

Secondary writings are Sir Henry Head, "Hughlings Jackson on Aphasia and Kindred Affections of Speech," in *Brain*, **38** (1915), 1–190; and "Hughlings Jackson," in *Aphasia and Kindred Disorders of Speech*, I (Cambridge, 1926), 30–53; and McD. Critchley, "Jacksonian Ideas and the Future With Special Reference to Aphasia," in *British Medical Journal* (1960), **2**, 6–12.

Motor functions are discussed in "Observations on the Physiology and Pathology of Hemi-Chorea," in *Selected Writings*, II, 238–245; "On Certain Relations of the Cerebrum and Cerebellum . . .," *ibid.*, pp. 452–458; on the cerebellar attitude and attacks, see *Brain*, **29** (1906), 425–440, 441–445.

See Sir Victor Horsley, "On Dr. Hughlings Jackson's Views of the Functions of the Cerebellum as Illustrated by Recent Research," in *Brain*, **29** (1906), 446–466, the Hughlings Jackson lecture (1906); and E. Clarke and C. D. O'Malley, *The Human Brain and Spinal Cord* (Berkeley–Los Angeles, 1968), pp. 689–690.

On evolution and dissolution of the nervous system, see "The Croonian Lectures on Evolution and Dissolution of the Nervous System," in *Selected Writings*, II, 45–75; "Remarks on . . .," *ibid.*, pp. 76–118; and "Relations of Different Divisions of the Central Nervous System to One Another and to Parts of the Body," *ibid.*, pp. 422–443, the first Hughlings Jackson lecture (1897).

Secondary writings are O. Sittig, "Dr. Hughlings Jackson's Principles of Cerebral Pathology," in *Post-Graduate Medical Journal*, **11** (1935), 135–138; W. Riese, "The Sources of Jacksonian Neurology," in *Journal of Nervous and Mental Diseases*, **124** (1956), 125–134; and Walshe (see below), pp. 127–130.

II. Secondary Literature. In addition to the works listed above, the following can be especially recommended: "Obituary. John Hughlings Jackson, M.D. St. And., F.R.C.P. Lond., L.L.D., D.Sc., F.R.S.," in *Lancet* (1911), **2**, 1103–1107, with portrait; "Obituary. John Hughlings Jackson, M.D., F.R.C.P., F.R.S.," in *British Medical Journal* (1911), **2**, 950–954, with portrait; J. Hutchinson, "The Late Dr. Hughlings Jackson: Recollection of a Lifelong Friendship," *ibid.*, pp. 1551–1554, repr. in *Neurological Fragments*, pp. 28–39; C. A. Mercier, "The Late Dr. Hughlings Jackson: Recollections," in *Neurological Fragments*, pp. 40–46; J. Taylor, "Jackson, John Hughlings (1835–1911)," in *Dictionary of National Biography, Supp., 1901–1911*, II, 356–358; and "Biographical Memoir," in *Neurological Fragments*, pp. 1–26.

See also (listed chronologically) W. Broadbent, "Hughlings Jackson as Pioneer in Nervous Physiology and Pathology," in *Brain*, **26** (1903), 305–366 (some of discussion is out of date); A. W. Campbell, "Dr. John Hughlings Jackson," in *Medical Journal of Australia* (1935), **2**, 344–347; W. G. Lennox, in W. Haymaker, ed., *The Founders of Neurology* (Springfield, Ill., 1953), pp. 308–311; W. Riese and W. Gooddy, "An Original Clinical Record of Hughlings Jackson With an Interpretation," in *Bulletin of the History of Medicine*, **29** (1955), 230–238, a case of *grande hystérie* seen in 1881; Gordon Holmes, in K. Kolle, *Grosse Nervenärzte; Lebensbilder*, I (Stuttgart, 1956), 135–144; McD. Critchley, "Hughlings Jackson, the Man: and the Early Days of the National Hospital," in *Proceedings of the Royal Society of Medicine*, **53** (1960), 613–618; F. M. R. Walshe, "Contributions of John Hughlings Jackson to Neurology: A Brief Introduction to His Teaching," in *Archives of Neurology and Psychiatry* (Chicago), **5** (1961), 119–131; and E. Stengel, "Hughlings Jackson's Influence in Psychiatry," in *British Journal of Psychiatry*, **109** (1963), 348–355.

In *Medical Classics*, **3** (1939), 913–914, there is a list of thirty-seven biographical articles, fourteen of which are cited above.

EDWIN CLARKE

JACOB BEN MĀḤIR IBN TIBBON. See **Ibn Tibbon.**

JACOBI, CARL GUSTAV JACOB (*b.* Potsdam, Germany, 10 December 1804; *d.* Berlin, Germany, 18 February 1851), *mathematics.*

The second son of Simon Jacobi, a Jewish banker, the precocious boy (originally called Jacques Simon) grew up in a wealthy and cultured family. His brother Moritz, three years older, later gained fame as a physicist in St. Petersburg. His younger brother, Eduard, carried on the banking business after his father's death. He also had a sister, Therese.

After being educated by his mother's brother, Jacobi entered the Gymnasium at Potsdam in November 1816. Promoted to the first (highest) class after a few months in spite of his youth, he had to remain there for four years because he could not enter the university until he was sixteen. When he graduated from the Gymnasium in the spring of 1821, he excelled in Greek, Latin, and history and had acquired a knowledge of mathematics far beyond that provided by the school curriculum. He had studied Euler's *Introductio in analysin infinitorum* and had attempted to solve the general fifth-degree algebraic equation.

During his first two years at the University of Berlin, Jacobi divided his interests among philosophical, classical, and mathematical studies. Seeing that time would not permit him to follow all his interests, he decided to concentrate on mathematics. University lectures in mathematics at that time were at a very elementary level in Germany, and Jacobi therefore in private study mastered the works of Euler, Lagrange, and other leading mathematicians. (Dirichlet, at the same time, had gone to Paris, where Biot, Fourier, Laplace, Legendre, and Poisson were active. Apart from the isolated Gauss at Göttingen, there was no equal center of mathematical activity in Germany.)

In the fall of 1824 Jacobi passed his preliminary examination for *Oberlehrer*, thereby acquiring permission to teach not only mathematics but also Greek and Latin to all high school grades, and ancient and modern history to junior high school students. When—in spite of being of Jewish descent—he was offered a position at the prestigious Joachimsthalsche Gymnasium in Berlin in the following summer, he had already submitted a Ph.D. thesis to the university. The board of examiners included the mathematician E. H. Dirksen and the philosopher Friedrich Hegel. Upon application he was given permission to begin work on the *Habilitation* immediately. Having become a Christian, he was thus able to begin a university career as *Privatdozent* at the University of Berlin at the age of twenty.

Jacobi's first lecture, given during the winter term 1825–1826, was devoted to the analytic theory of curves and surfaces in three-dimensional space. He greatly impressed his audience by the liveliness and clarity of his delivery, and his success became known to the Prussian ministry of education. There being no prospect for a promotion at Berlin in the near future, it was suggested that Jacobi transfer to the University of Königsberg, where a salaried position might be available sooner. When he arrived there in May 1826, the physicists Franz Neumann and Heinrich Dove were just starting their academic careers, and Friedrich Bessel, then in his early forties, occupied the chair of astronomy. Joining these colleagues, Jacobi soon became interested in applied problems. His first publications attracted wide attention among mathematicians. On 28 December 1827 he was appointed associate professor, a promotion in which Legendre's praise of his early work on elliptic functions had had a share. Appointment as full professor followed on 7 July 1832, after a four-hour disputation in Latin. Several months earlier, on 11 September 1831, Jacobi had married Marie Schwinck, the daughter of a formerly wealthy *Kommerzienrat* who had lost his fortune in speculative transactions. They had five sons and three daughters.

For eighteen years Jacobi was at the University of

Königsberg, where his tireless activity produced amazing results in both research and academic instruction. Jacobi created a sensation among the mathematical world with his penetrating investigations into the theory of elliptic functions, carried out in competition with Abel. Most of Jacobi's fundamental research articles in the theory of elliptic functions, mathematical analysis, number theory, geometry, and mechanics were published in Crelle's *Journal für die reine und angewandte Mathematik*. With an average of three articles per volume, Jacobi was one of its most active contributors and quickly helped to establish its international fame. Yet his tireless occupation with research did not impair his teaching. On the contrary—never satisfied to lecture along trodden paths, Jacobi presented the substance of his own investigations to his students. He would lecture up to eight or ten hours a week on his favorite subject, the theory of elliptic functions, thus demanding the utmost from his listeners. He also inaugurated what was then a complete novelty in mathematics—research seminars—assembling the more advanced students and attracting his nearest colleagues.

Such were Jacobi's forceful personality and sweeping enthusiasm that none of his gifted students could escape his spell: they were drawn into his sphere of thought, worked along the manifold lines he suggested, and soon represented a "school." C. W. Borchardt, E. Heine, L. O. Hesse, F. J. Richelot, J. Rosenhain, and P. L. von Seidel belonged to this circle; they contributed much to the dissemination not only of Jacobi's mathematical creations but also of the new research-oriented attitude in university instruction. The triad of Bessel, Jacobi, and Neumann thus became the nucleus of a revival of mathematics at German universities.

In the summer of 1829 Jacobi journeyed to Paris, visiting Gauss in Göttingen on his way and becoming acquainted with Legendre (with whom he had already been in correspondence), Fourier, Poisson, and other eminent French mathematicians. In July 1842 Bessel and Jacobi, accompanied by Marie Jacobi, were sent by the king of Prussia to the annual meeting of the British Association for the Advancement of Science in Manchester, where they represented their country splendidly. They returned via Paris, where Jacobi gave a lecture before the Academy of Sciences.

Early in 1843 Jacobi became seriously ill with diabetes. Dirichlet, after he had visited Jacobi for a fortnight in April, procured a donation (through the assistance of Alexander von Humboldt) from Friedrich Wilhelm IV, which enabled Jacobi to spend some months in Italy, as his doctor had advised. Together with Borchardt and Dirichlet and the latter's wife, he traveled in a leisurely manner to Italy, lectured at the science meeting in Lucca (but noticed that none of the Italian mathematicians had really studied his papers), and arrived in Rome on 16 November 1843. In the stimulating company of these friends and of the mathematicians L. Schläfli and J. Steiner, who also lived in Rome at that time, and further blessed by the favorable climate, Jacobi's health improved considerably. He started to compare manuscripts of Diophantus' *Arithmetica* in the Vatican Library and began to resume publishing mathematical articles. By the end of June 1844 he had returned to Berlin. He was granted royal permission to move there with his family because the severe climate of Königsberg would endanger his health. Jacobi received a bonus on his salary to help offset the higher costs in the capital and to help with his medical expenses. As a member of the Prussian Academy of Sciences, he was entitled, but not obliged, to lecture at the University of Berlin. Because of his poor health, however, he lectured on only a very limited scale.

In the revolutionary year of 1848 Jacobi became involved in a political discussion in the Constitutional Club. During an impromptu speech he made some imprudent remarks which brought him under fire from monarchists and republicans alike. Hardly two years before, in the dedication of volume I of his *Opuscula mathematica* to Friedrich Wilhelm IV, he had expressed his royalist attitude; now he had become an object of suspicion to the government. A petition of Jacobi's to become officially associated with the University of Berlin, and thus to obtain a secure status, was denied by the ministry of education. Moreover, in June 1849 the bonus on his salary was retracted. Jacobi, who had lost his inherited fortune in a bankruptcy years before, had to give up his Berlin home. He moved into an inn and his wife and children took up residence in the small town of Gotha, where life was considerably less expensive.

Toward the end of 1849 Jacobi was offered a professorship in Vienna. Only after he had accepted it did the Prussian government realize the severe blow to its reputation which would result from his departure. Special concessions from the ministry and his desire to stay in his native country finally led Jacobi to reverse his decision. His family, however, was to remain at Gotha for another year, until the eldest son graduated from the Gymnasium. Jacobi, who lectured on number theory in the summer term of 1850, joined his family during vacations and worked on an astronomical paper with his friend P. A. Hansen.

JACOBI

Early in 1851, after another visit to his family, Jacobi contracted influenza. Hardly recovered, he fell ill with smallpox and died within a week. His close friend Dirichlet delivered the memorial lecture at the Berlin Academy on 1 July 1852, calling Jacobi the greatest mathematician among the members of the Academy since Lagrange and summarizing his eminent mathematical contributions.

The outburst of Jacobi's creativity at the very beginning of his career, combined with his self-conscious attitude, early caused him to seek contacts with some of the foremost mathematicians of his time. A few months after his arrival at Königsberg he informed Gauss about some of his discoveries in number theory, particularly on cubic residues, on which he published a first paper in 1827. Jacobi had been inspired by Gauss's *Disquisitiones arithmeticae* and by a note on the results which Gauss had recently presented to the Göttingen Academy, concerning biquadratic residues. Obviously impressed, Gauss asked Bessel for information on the young mathematician and enclosed a letter for Jacobi, now lost—as are all subsequent letters from Gauss to Jacobi. No regular correspondence developed from this beginning.

Another contact, established by a letter from Jacobi on 5 August 1827, initiated an important regular mathematical correspondence with Legendre that did not cease until Legendre's death. Its topic was the theory of elliptic functions, of which Legendre had been the great master until Abel and Jacobi came on the scene. Their first publications in this subject appeared in September 1827—Abel's fundamental memoir "Recherches sur les fonctions elliptiques" in Crelle's *Journal* (**2**, no. 2) and Jacobi's "Extraits de deux lettres . . ." in *Astronomische Nachrichten* (**6**, no. 123). From these articles it is clear that both authors were in possession of essential elements of the new theory. They had developed these independently: Abel's starting point was the multiplication, Jacobi's the transformation, of elliptic functions; both of them were familiar with Legendre's work.

The older theory centered on the investigation of elliptic integrals, that is, integrals of the type $\int R(x, \sqrt{f(x)})\,dx$, where R is a rational function and $f(x)$ is an integral function of the third or fourth degree. Examples of such integrals had been studied by John Wallis, Jakob I and Johann I Bernoulli, and in particular G. C. Fagnano. Euler continued this work by investigating the arc length of a lemniscate, $\int \frac{dx}{\sqrt{1-x^4}}$; by integrating the differential equation

$$\frac{dx}{\sqrt{1-x^4}} + \frac{dy}{\sqrt{1-y^4}} = 0$$

he was led to the addition formula for this integral (elliptic integral of the first kind). When he extended these investigations—for example, to the arc length of an ellipse (elliptic integral of the second kind)—he concluded that the sum of any number of elliptic integrals of the same kind (except for algebraic or logarithmic terms, which may have to be added) may be expressed by a single integral of this same kind, of which the upper limit depends algebraically on the upper limits of the elements of the sum. This discovery shows Euler to be a forerunner of Abel.

The systematic study of elliptic integrals and their classification into the first, second, and third kinds was the work of Legendre, who had cultivated this field since 1786. The leading French mathematicians of his day were interested in the application of mathematics to astronomy and physics. Therefore, although Legendre had always emphasized the applicability of his theories (for instance, by computing tables of elliptic integrals), they did not appreciate his work. Gauss, on the other hand, was well aware of the importance of the subject, for he had previously obtained the fundamental results of Abel and Jacobi but had never published his theory. Neither had he given so much as a hint when Legendre failed to exploit the decisive idea of the inverse function.

It was this idea, occurring independently to both Abel and Jacobi, which enabled them to take a big step forward in the difficult field of transcendental functions. Here Abel's investigations were directed toward the most general question; Jacobi possessed an extraordinary talent for handling the most complicated mathematical apparatus. By producing an almost endless stream of formulas concerning elliptic functions, he obtained his insights and drew his conclusions about the character and properties of these functions. He also recognized the relation of this theory to other fields, such as number theory.

When Legendre first learned of the new discoveries of Abel and Jacobi, he showed no sign of envy. On the contrary, he had nothing but praise for them and expressed enthusiasm for their creations. He even reported on Jacobi's first publications (in the *Astronomische Nachrichten*) to the French Academy and wrote to Jacobi on 9 February 1828:

> It gives me a great satisfaction to see two young mathematicians such as you and him [Abel] cultivate with success a branch of analysis which for so long a time has been the object of my favorite studies and which has not been received in my own country as well as it would

> deserve. By these works you place yourselves in the ranks of the best analysts of our era.

Exactly a year later Legendre wrote in a letter to Jacobi:

> You proceed so rapidly, gentlemen, in all these wonderful speculations that it is nearly impossible to follow you—above all for an old man who has already passed the age at which Euler died, an age in which one has to combat a number of infirmities and in which the spirit is no longer capable of that exertion which can surmount difficulties and adapt itself to new ideas. Nevertheless I congratulate myself that I have lived long enough to witness these magnanimous contests between two young athletes equally strong, who turn their efforts to the profit of the science whose limits they push back further and further.

Jacobi, too, was ready to acknowledge fully the merits of Abel. When Legendre had published the third supplement to his *Traité des fonctions elliptiques et des intégrales eulériennes*, in which he presented the latest developments, it was Jacobi who reviewed it for Crelle's *Journal* (**8**[1832], 413–417):

> Legendre to the transcendental functions $\int \frac{f(x)\,dx}{\sqrt{X}}$, where X exceeds the fourth degree, gives the name "hyperelliptical" [*ultra-elliptiques*]. We wish to call them *Abelsche Transcendenten* (Abelian transcendental functions), for it was Abel who first introduced them into analysis and who first made evident their great importance by his far-reaching theorem. For this theorem, as the most fitting monument to this extraordinary genius, the name "Abelian theorem" would be very appropriate. For we happily agree with the author that it carries the full imprint of the depth of his ideas. Since it enunciates in a simple manner, without the vast setup of mathematical formalism, the deepest and most comprehensive mathematical thought, we consider it to be the greatest mathematical discovery of our time although only future—perhaps distant—hard work will be able to reveal its whole importance.

Jacobi summarized his first two years' research, a good deal of which had been obtained in competition with Abel, in his masterpiece *Fundamenta nova theoriae functionum ellipticarum*, which appeared in April 1829. His previous publications in *Astronomische Nachrichten* and in Crelle's *Journal* were here systematically collected, greatly augmented, and supplemented by proofs—he had previously omitted these, thereby arousing the criticism of Legendre, Gauss, and others.

The *Fundamenta nova* deals in the first part with the transformation, and in the second with the representation, of elliptic functions. Jacobi took as his starting point the general elliptic differential of the first kind and reduced it by a second-degree transformation to the normal form of Legendre. He studied the properties of the functions U (even) and V(odd) in the rational transformation $Y = U/V$ and gave as examples the transformations of the third and fifth degrees and the pertinent modular equations. By combining two transformations he obtained the multiplication of the elliptic integral of the first kind, a remarkable result. He then introduced the inverse function $\varphi = am\, u$ into the elliptic integral

$$u(\varphi, k) = \int_0^\varphi \frac{d\varphi}{\sqrt{1 - k^2 \sin^2 \varphi}};$$

hence

$$x = \sin \varphi = \sin am\, u.$$

Further introducing $\cos am\, u = am\,(K - u)$ $\left(\text{with } K = u\left[\frac{\pi}{2}, k\right]\right)$,

$$\Delta\, am\, u = \sqrt{1 - k^2 \sin^2 am\, u},$$

he collected a large number of formulas. Using the substitution $\sin \varphi = i \tan \psi$, he established the relation

$$\sin am\,(iu,k) = i \tan am\,(u,k');$$

the moduli k and k' are connected by the equation $k^2 + k'^2 = 1$. He thus obtained the double periodicity, the zero values, the infinity values, and the change of value in half a period for the elliptic functions. This introduction of the imaginary into the theory of elliptic functions was another very important step which Jacobi shared with Abel. Among his further results is the demonstration of the invariance of the modular equations when the same transformation is applied to the primary and secondary moduli. Toward the end of the first part of his work Jacobi developed the third-order differential equation which is satisfied by all transformed moduli.

The second part of the *Fundamenta nova* is devoted to the evolution of elliptic functions into infinite products and series of various kinds. The first representation of the elliptic functions $\sin am\, u$, $\cos am\, u$, $\Delta\, am\, u$, which he gave is in the form of quotients of infinite products. Introducing $q = e^{-\frac{\pi K'}{K}}$, Jacobi expressed the modulus and periods in terms of q, as for instance

$$k = 4\sqrt{q}\left\{\frac{(1+q^2)(1+q^4)(1+q^6)\cdots}{(1+q)(1+q^3)(1+q^5)\cdots}\right\}^4.$$

Another representation of the elliptic functions and their nth powers as Fourier series leads to the sums (in terms of the moduli) of various infinite series in q. Integrals of the second kind are treated after the function

$$Z(u) = \frac{F^1 E(\varphi) - E^1 F(\varphi)}{F^1} \quad (\varphi = am\ u)$$

has been introduced. Jacobi reduced integrals of the third kind to integrals of the first and second kinds and a third transcendental function which also depends on two variables only. In what follows, Jacobi's function

$$\Theta(u) = \Theta(0) \cdot \exp\left(\int_0^u Z(u)\, du\right)$$

played a central role. It is then supplemented by the function $H(u)$ such that $\sin am\, u = \frac{1}{\sqrt{k}} \cdot \frac{H(u)}{\Theta(u)}$. $\Theta(u)$ and $H(u)$ are represented as infinite products and as Fourier series. The latter yield such remarkable formulas as

$$\sqrt{\frac{2kK}{\pi}} = 2\sqrt[4]{q} + 2\sqrt[4]{q^9} + 2\sqrt[4]{q^{25}} + 2\sqrt[4]{q^{49}} + \cdots.$$

After a number of further summations and identities Jacobi closed this work with an application to the theory of numbers. From the identity

$$\begin{aligned}\left(\frac{2K}{\pi}\right)^2 &= (1 + 2q + 2q^4 + 2q^9 + 2q^{16} + \cdots)^4 \\ &= 1 + 8\sum \varphi(p)(q^p + 3q^{2p} + 3q^{4p} \\ &\qquad + 3q^{8p} + \cdots),\end{aligned}$$

where $\varphi(p)$ is the sum of the divisors of the odd number p, he drew the conclusion that any integer can be represented as the sum of at most four squares, as Fermat had suggested.

Jacobi lectured on the theory of elliptic functions for the first time during the winter term 1829–1830, emphasizing that double periodicity is the essential property of these functions. The theta function should be taken as foundation of the theory; the representation in series with the general term $e^{-(an+b)2}$ ensures convergence and makes it possible to develop the whole theory. In his ten hours a week of lecturing in the winter of 1835–1836 Jacobi for the first time founded the theory on the theta function, proving the famous theorem about the sum of products of four theta functions and defining the kinds of elliptic functions as quotients of theta functions. He continued this work in his lectures of 1839–1840, the second part of which is published in volume I of his *Gesammelte Werke*. Volume II contains a historical summary, "Zur Geschichte der elliptischen und Abel'schen Transcendenten," composed by Jacobi probably in 1847, which documents his view of his favorite subject toward the end of his life.

Some of Jacobi's discoveries in number theory have already been mentioned. Although he intended to publish his results in book form, he was never able to do so. The theory of residues, the division of the circle into n equal parts, the theory of quadratic forms, the representation of integers as sums of squares or cubes, and related problems were studied by Jacobi. During the winter of 1836–1837 he lectured on number theory, and some of his methods became known through Rosenhain's lecture notes. In 1839 Jacobi's *Canon arithmeticus* on primitive roots was published; for each prime and power of a prime less than 1,000 it gives two companion tables showing the numbers with given indexes and the index of each given number.

Most of Jacobi's work is characterized by linkage of different mathematical disciplines. He introduced elliptic functions not only into number theory but also into the theory of integration, which in turn is connected with the theory of differential equations where, among other things, the principle of the last multiplier is due to Jacobi. Most of his investigations on first-order partial differential equations and analytical mechanics were published posthumously (in 1866, by Clebsch) as *Vorlesungen über Dynamik*. Taking W. R. Hamilton's research on the differential equations of motion (canonical equations) as a starting point, Jacobi also carried on the work of the French school (Lagrange, Poisson, and others). He sought the most general substitutions that would transform canonical differential equations into such equations. The transformations are to be such that a canonical differential equation (of motion) is transformed into another differential equation which is again canonical. He also developed a new theory for the integration of these equations, utilizing their relation to a special Hamiltonian differential equation. This method enabled him to solve several very important problems in mechanics and astronomy. In some special cases Clebsch later improved Jacobi's results, and decades later Helmholtz carried Jacobi's mechanical principles over into physics in general.

Among Jacobi's work in mathematical physics is research on the attraction of ellipsoids and a surprising discovery in the theory of configurations of rotating liquid masses. Maclaurin had shown that a homogeneous liquid mass may be rotated uniformly about a fixed axis without change of shape if this shape is an ellipsoid of revolution. D'Alembert, Laplace, and Lagrange had studied the same problem; but it was left for Jacobi to discover that even an ellipsoid of three different axes may satisfy the conditions of equilibrium.

The theory of determinants, which begins with Leibniz, was presented systematically by Jacobi early in 1841. He introduced the "Jacobian" or functional determinant; a second paper—also published in Crelle's *Journal*—is devoted entirely to its theory, including relations to inverse functions and the transformation of multiple integrals.

Jacobi was also interested in the history of mathematics. In January 1846 he gave a public lecture on Descartes which attracted much attention. In the same year A. von Humboldt asked him for notes on the mathematics of the ancient Greeks as material for his *Kosmos* and Jacobi readily complied—but Humboldt later confessed that some of the material went beyond his limited mathematical knowledge. In the 1840's Jacobi became involved in the planning of an edition of Euler's works. He corresponded with P. H. von Fuss, secretary of the St. Petersburg Academy and great-grandson of the famous mathematician, who had discovered a number of Euler's unpublished papers. Jacobi drew up a very detailed plan of distributing the immense number of publications among the volumes of the projected edition. Unfortunately, the project could be realized only on a much reduced scale. It was not until 1911 that the first volume of *Leonhardi Euleri opera omnia*—still in progress—appeared.

Jacobi's efforts to promote an edition of Euler were prompted by more than the ordinary interest a mathematician might be expected to take in the work of a great predecessor. Jacobi and Euler were kindred spirits in the way they created their mathematics. Both were prolific writers and even more prolific calculators; both drew a good deal of insight from immense algorithmical work; both labored in many fields of mathematics (Euler, in this respect, greatly surpassed Jacobi); and both at any moment could draw from the vast armory of mathematical methods just those weapons which would promise the best results in the attack on a given problem. Yet while Euler divided his energies about equally between pure and applied mathematics, Jacobi was more inclined to investigate mathematical problems for their intrinsic interest. Mathematics, as he understood it, had a strong Platonic ring. For the disputation at his inauguration to a full professorship in 1832 Jacobi had chosen as his first thesis "Mathesis est scientia eorum, quae per se clara sunt."

BIBLIOGRAPHY

I. Original Works. Jacobi's works have been collected twice. *Opuscula mathematica* is in 3 vols. (Berlin, 1846–1871). Vol. I was edited by Jacobi himself; vol. II, also prepared by him, was published posthumously by Dirichlet; vol. III was published by his pupil C. W. Borchardt.

The standard ed., 7 vols. and supp., was issued by the Prussian Academy of Sciences as *C. G. J. Jacobi's Gesammelte Werke*, C. W. Borchardt, A. Clebsch, and K. Weierstrass, eds. (Berlin, 1881–1891). Vol. I contains, among other works, the *Fundamenta nova theoriae functionum ellipticarum* (Königsberg, 1829). The supp. vol. is *Vorlesungen über Dynamik*, first published by A. Clebsch (Leipzig, 1866). *Gesammelte Werke* has been repr. (New York, 1969).

Jacobi's only other publication in book form, the *Canon arithmeticus* (Berlin, 1839), is not in the *Gesammelte Werke* but appeared in a 2nd ed. recomputed by W. Patz and edited by H. Brandt (Berlin, 1956).

Kurt-R. Biermann has published "Eine unveröffentlichte Jugendarbeit C. G. J. Jacobis über wiederholte Funktionen," in *Journal für die reine und angewandte Mathematik*, **207** (1961), 96–112.

A list of Jacobi's publications and of his lectures is in *Gesammelte Werke*, VII, 421–440. See also Poggendorff, I, 1178–1181, 1576; III, 681; IV, 688; VIIa, Supp. 302–303.

Brief information on 16 vols. of manuscript material, in the archives of the Deutsche Akademie der Wissenschaften in Berlin, is in *Gelehrten- und Schriftstellernachlässe in den Bibliotheken der DDR*, I (Berlin, 1959), 50, no. 315, "Jakobi" [*sic*].

II. Secondary Literature. The main secondary sources are J. P. G. Lejeune Dirichlet, "Gedächtnisrede" (1852), repr. in *Gesammelte Werke*, I; and Leo Koenigsberger, *Carl Gustav Jacob Jacobi. Festschrift zur Feier der hundertsten Wiederkehr seines Geburtstages* (Leipzig, 1904); and *Carl Gustav Jacob Jacobi. Rede zu der von dem Internationalen Mathematiker-Kongress in Heidelberg veranstalteten Feier der hundertsten Wiederkehr seines Geburtstages, gehalten am 9. August 1904* (Leipzig, 1904), also in *Jahresbericht der Deutschen Mathematikervereinigung*, **13** (1904), 405–433. For further secondary literature see Poggendorff, esp. VIIa Supp.

Christoph J. Scriba

JACOBI, MORITZ HERMANN VON (*b.* Potsdam, Germany, 21 September 1801; *d.* St. Petersburg, Russia, 27 February 1874), *physics.*

At the urging of his parents Jacobi studied architecture at Göttingen and in 1833 set up practice in Königsberg, where his younger brother Carl was a professor of mathematics. He also began to turn his attention to physics and chemistry. In 1835 he went to the University of Dorpat as a professor of civil engineering, and in 1837 he moved to St. Petersburg. There he became a member of the Imperial Academy of Sciences (adjunct in 1839, extraordinary in 1842, and ordinary in 1847) and devoted his energies to research on electricity and its various practical applications, his interest in this subject having developed since his days in Göttingen.

Jacobi engaged in a number of studies of great interest in the fast-developing subject of electricity, dealing especially with its possible technical applications. Although most of the results of his work were published and were generally available, their impact was minimal. One reason for this certainly lies in his physical isolation from the centers of development in electricity in France and England. Another can probably be found in that most of his practical applications proved to be premature; that is, the technology had not developed enough to sustain them.

Jacobi's most interesting work, reported to the St. Petersburg Academy in 1838 and to the British Association two years later, was his investigation of the power of an electromagnet as a function of various parameters; electric current, thickness of wire, number of turns on the helix, diameter of the helix, and thickness of the iron core. Of great practical value in the design of motors and generators, this work was pursued in greater detail by Henry Rowland and John Hopkinson almost half a century later.

In May 1834 Jacobi built one of the first practical electric motors. He performed a variety of tests on it, for instance measuring its output by determining the amount of zinc consumed by the battery. In 1838 his motor drove a twenty-eight-foot boat carrying a dozen Russian officials on the Neva River at a speed of one and one-half miles per hour. His hopes of covering the Neva with a fleet of magnetic boats were doomed from the beginning, however, by the cost of battery-powered operation and by the fumes that such batteries emitted.

In a separate enterprise Jacobi was asked to continue the work of Baron Pavel Schilling, who had demonstrated the needle (electromagnetic) telegraph to the Russian government in 1837 but who had died that year before an experimental line could be set up. Jacobi improved on Schilling's design and by 1839 had constructed an instrument quite similar to Morse's first, and earlier, receiver. Various experimental lines were run in succeeding years, but practical telegraphy did not come to Russia until the 1850's, with the introduction of the Siemens and Halske system.

In 1838 Jacobi announced his discovery of the process he called "galvanoplasty" (now called electrotyping), the reproduction of forms by electrodeposition. In subsequent publications he described his techniques in great detail.

BIBLIOGRAPHY

I. Original Works. Jacobi's articles are listed in the Royal Society's *Catalogue of Scientific Papers*, III, 517–518; VIII, 8. Among the more important are "Expériences électromagnétiques," in *Bulletin de l'Académie impériale des sciences de St. Pétersbourg*, **2** (1837), 17–31, 37–44, trans. in R. Taylor *et al.*, eds., *Scientific Memoirs*, **2** (1841), 1–19; and "Galvanische und electromagnetische Versuche. Ueber electro-telegraphische Leitungen," in *Bulletin de l'Académie impériale des sciences de St. Pétersbourg*, **4** (1845), 113–135, **5** (1847), 86–91, 97–113, 209–224, **6** (1848), 17–44, **7** (1849), 1–21, 161–170, and **8** (1850), 1–17.

Published separately as pamphlets were *Mémoire sur l'application de l'électro-magnétisme au mouvement des machines* (Potsdam, 1835), trans. in *Scientific Memoirs*, **1** (1837), 503–531; and *Die Galvanoplastik* (St. Petersburg, 1840), trans. in *Annals of Electricity, Magnetism and Chemistry*, **7** (1841), 323–328, 337–344, 401–448, and **8** (1842), 66–74, 168–173.

Jacobi's papers are preserved in the archives of the Soviet Academy of Sciences in Leningrad. His correspondence with his brother was published by Wilhelm Ahrens as *Briefwechsel zwischen C. G. J. Jacobi und M. H. Jacobi* (Leipzig, 1907).

II. Secondary Literature. A biographical account, with portrait, appears in German in the *Bulletin de l'Académie impériale des sciences de St. Pétersbourg*, **21** (1876), 262–279. A much shorter sketch is Ernest H. Huntress, in *Proceedings of the American Academy of Arts and Sciences*, **79** (1951), 22–23.

BERNARD S. FINN

JACQUET, PIERRE ARMAND (*b.* St. Mande, France, 7 April 1906; *d.* at sea, off the coast of Spain, 6 September 1967), *chemical engineering*.

Jacquet received his diploma from the École Nationale Supérieure de Chimie in 1926 and quickly turned his attention to electrochemical research. In 1929, at the research laboratory of Le Matériel Téléphonique Society, he was given the task of finding a method for preparing a perfectly smooth nickel surface. He solved the problem while making a test on electrolytic deposition, in which he reversed the polarity, and in this experiment he discovered electrolytic polishing.

He continued his research under Charles Marie at the École Nationale Supérieure de Chimie and then at F. Joliot-Curie's nuclear chemistry laboratory at the Collège de France, before returning to industry (1939–1945). In 1945 he was engaged as an engineer by the French navy; he worked independently in this capacity until his retirement in 1966, serving as consultant to the Office National des Études et Recherches Aéronautiques (ONERA), the Commissariat à l'Energie Atomique, and various industries.

Most of Jacquet's approximately 200 publications refer to electrolytic polishing. By 1940 he had established the conditions for polishing most of the ordinary metals, and the precautions to be taken in handling them. In 1956 he published a summary of his results, along with the practical operating conditions for most metals.

Certain of Jacquet's micrographs were epochmaking: the precipitates in Duralumin (1939), the dislocations in alpha brass (1954), the dislocation mills in aluminum-copper alloys (1956), and polygonization in industrial metals.

He contributed to the clarification of various metallurgical problems, such as the chemical nature of surfaces, their corrosion and passivity; temper brittleness, fatigue, and fracture of steels; and stress corrosion and intergranular corrosion of light alloys.

During the last ten years of Jacquet's life, he developed, with E. Mencarelli, an idea of A. Van Effenterre: to render metallography nondestructive. They discovered that after local polishing with a buffer and application of nitrocellulose varnish, a replica of any region of a large piece reveals its structure, thus saving the removal of any portion of the object. A piece may then be placed (or replaced) in service after direct testing, whatever its dimensions. The process was later extended to the study of fractured surfaces and the preparation of thin films for transmission electron microscopy.

Jacquet retired to Banyuls, in southern France, in 1966. He died a year later in a boating accident.

BIBLIOGRAPHY

I. Original Works. Jacquet's works in electrolytical polishing technique and applications include *Brevet français* (Paris, 1930), written with H. Figour; *Contribution à l'étude expérimentale de la structure cristalline des dépôts électrolytiques* (1938), thesis; "The Principles and Scientific Applications of the Electrolytic Polishing of Metals," in *Proceedings. Third International Conference on Electrodeposition* (1947); and *Le polissage électrolytique des surfaces métalliques et ses applications*, vol. I, *Aluminium, magnésium, alliages légers* (Paris, 1947); "Contribution du polissage électrolytique à la physique et à la chimie des métaux," in *Revue de métallurgie*, **48** (1951), 1–16; "Electrolytic and Chemical Polishing," in *Metallurgical Reviews*, **1**, pt. 2 (1956), 157–238; and "Le polissage électrolytique dans les techniques, réalisations et perspectives," in *Revue de métallurgie*, **49** (1962), 1056–1069.

Other of his papers are "The Age-Hardening of a Copper-Aluminum Alloy of Very High Purity," in *Journal of the Institute of Metals*, **65**, no. 2 (1939), 121–137, written with J. Calvet and A. Guinier; "Recherches expérimentales sur la microstructure de la solution solide cuivre-zinc 65–35 polycristalline faiblement déformée par traction et sur son évolution au recuit entre 200 et 600°C.," in *Acta Metallurgica*, **2** (1954), 752–790; "Sur un type de sous-structure en spirales dans un laiton bêta à l'aluminium," in *Comptes-rendus hebdomadaires des séances de l'Académie des sciences*, **239** (1954), 1799–1801; "Sur quelques cas de polygonisation non provoquée d'alliages industriels," *ibid.*, 1384–1386, written with A. R. Weill; and "Influence des traitements thermiques sur le seuil de précipitation de la phase thêta prime dans alliage aluminium-cuivre à 4 % de cuivre," in *Mémoires scientifiques de la Revue de métallurgie*, **63** (1961), 97–106, written with A. R. Weill.

For Jacquet's research in nondestructive metallography, see *Technique non destructive pour les observations, en particulier de nature métallographique, sur les surfaces métalliques*, Note technique de l'Office National des Études et Recherches Aéronautiques, no. 54 (Oct. 1959); and "An Innovation in Inspection and Control. Non-Destructive Metallography and Microfractography," in *Metal Progress* (Feb. 1964), p. 114.

II. Secondary Literature. For information on Jacquet, see "P. A. Jacquet. Ingénieur contractuel des constructions et armes navales. Chevalier de la Légion d'honneur (1906–1967)," in *Mémorial de l'artillerie navale*, **42** (1968); and Paul Lacombe, "P. A. Jacquet (1906–1967)," in *Metallography*, **1**, no. 1 (1968), 1–3.

A. R. Weill

JACQUIN, NIKOLAUS JOSEF (*b.* Leiden, Netherlands, 16 February 1727; *d.* Vienna, Austria, 26 October 1817), *botany, chemistry.*

Jacquin was the grandson of a Frenchman who had immigrated to the Netherlands in the second half of the seventeenth century. His mother came from a noble Dutch family; his father, a distinguished cloth and velvet manufacturer, was an admirer of the great writers of antiquity and was responsible for his son's attaining a thorough classical education, even though the latter was expected to become a merchant. Jacquin therefore completed the course at the highly reputed Jesuit Gymnasium in Antwerp. His father's business failure and death obliged Jacquin to direct his studies toward a specific profession. He began to study theology but soon changed to medicine. In Paris, where he continued his studies, Jacquin realized that his lack of funds would never permit him to obtain the M.D. In need of help, he turned to his father's friend, Gerard van Swieten, who as *Protomedicus* and Director of the Medical Faculty of the University of Vienna, invited him to Vienna.

At Vienna, Jacquin continued his medical studies, supported by Van Swieten, on whose recommendation he was sent by Francis I on a trip to the West Indies and South America for the purpose of enlarging

JACQUIN

the imperial natural history collections. The trip lasted from the end of 1754 to the middle of 1759, with stops at Martinique, Curaçao, Santo Domingo, Jamaica, Cuba, Venezuela, and Colombia. In 1763 Jacquin was appointed professor of "practical mining and chemical knowledge" at the Mining School in Schemnitz, Hungary, and in 1768 he was appointed to the chair of chemistry and botany in the Medical Faculty of the University of Vienna. He occupied this chair until 1796 and in 1809 was appointed rector of the university. His home, at which Mozart was a frequent visitor, played a not insignificant role in Viennese scientific and social life. In 1774 he was elevated to the nobility and in 1806 he was made a baron. The Royal Society, the Academy of Sciences in Paris, and the Netherlands Academy of Sciences elected him to foreign membership.

Jacquin's significance for chemistry lay in his acceptance of Joseph Black's revolutionary concepts concerning the chemical events occurring in the burning of lime. At this time the balance was not yet universally used in the interpretation of a chemical reaction, and chemists were far from unanimous in accepting the idea that "air" can enter into combination with a solid body, thereby markedly changing it. The predominant interpretation was that something was added to lime upon combustion and that this something gave the "fiery" property of slaked lime. Black, however, came to the conclusion, based on careful weighings of the initial and final products and on collection of the escaping air, that the change of lime on burning should be explained by the escape of some special type of air. This concept provoked opposition because of the then almost universal acceptance of the phlogiston theory.

Among the writings opposed to Black, a work by I. C. Meyer gained special significance. Meyer sought to prove, by examining Black's experiments, that the changes in lime during combustion are caused by the addition of a substance consisting of fire material with some other substance. Meyer's work was the occasion for a careful experimental investigation by Jacquin, in which he proved conclusively that Meyer was wrong and that Black's interpretation was correct.

A further contribution by Jacquin to chemistry is a chemistry textbook which he designed specifically for the instruction of pharmacists and physicians; enlarged and modified by his son and successor at Vienna, Josef Franz, Baron von Jacquin, the work became a widely known textbook of general chemistry. It appeared in several editions and determined the direction of chemical instruction in Austria for two generations; it was also translated into English and Dutch.

As a botanist Jacquin was the most important of the younger contemporaries of Linnaeus. He was the first writer in German to utilize to any large extent Linnaeus' system of binary nomenclature, and was foremost in his time with respect to the number of new species described precisely and in a consistent way. His descriptions are still valid today. Jacquin's interest in botany had been stimulated while he was a student at Leiden by Theodor Gronovius, of the scholarly Gronovius family who were acquainted with Jacquin's family; and also his seeing a blooming of Zingiber, a pharmaceutical plant then known as *Costus speziosus* or *Costus arabicus*. His monumental floral works, containing colored illustrations by him and by other artists using his models, are among the most beautiful of their kind. At this time Antoine and Bernard de Jussieu were developing the natural system of botanical classification. Jacquin had known the Jussieus during his Paris sojourn, although he did not contribute to the development of the natural system.

BIBLIOGRAPHY

I. Original Works. A list of Jacquin's botanical works is in G. A. Pritzel, *Thesaurus literaturae botanicae* (Leipzig, 1872). Jacquin's botanical writings include *Enumeratio systematica plantarum quas in insulis Caribaeis, vicinaque Americes continente detexit novas, aut iam cognitas, emendavit* (Leiden, 1760); *Selectarum stirpium americanarum historia* (Vienna, 1763; 2nd ed., *ca.* 1780); *Flora Austriacae sive plantarum selectarum in Austriae Archiducatu sponte crescentium icones*, 5 vols. (Vienna, 1773–1778); *Hortus botanicus Vindobonensis*, 3 vols. (Vienna, 1773–1776); *Icones plantarum rariorum*, 3 vols. (Vienna, 1781–1793); *Oxalis, monographia iconibus illustrata* (Vienna, 1784); *Collectanae ad botanicam, chemiam et historiam naturalem spectantia*, 5 vols. (Vienna, 1786–1796); *Plantarum rariorum horti caesarei Schoenbrunnensis descriptiones et icones*, 4 vols. (Vienna, 1797–1804); *Stapeliarum in hortis Vindobonensibus cultarum descriptiones* (Vienna, 1806); and *Fragmenta botanica figuris coloratis illustrata* (Vienna, 1809).

There is a Jacquin MS, "Genera ex Cryptogamia, Linnaei figuris ad vivum expressis illustrata," in the Botanical Institute of the University of Vienna.

A chemical work is *Examen chemicum doctrinae Meyerianae de acido pingui, et Blackianae de aero fixo, respectu calcis* (Vienna, 1769).

Two textbooks are *Anfangsgründe der medizinisch-praktischen Chemie zum Gebrauch seiner Vorlesung* (Vienna, 1783; 2nd ed., 1785); and *Anleitung zur Pflanzenkenntnis nach Linne's Methode* (Vienna, 1785; 3rd ed., 1840).

II. Secondary Literature. For a list of unpublished material and for older biographical literature, see Wilfried Oberhummer, "Die Chemie an der Universität Wien in der

Zeit von 1749 bis 1848 und die Inhaber des Lehrstuhls für Chemie und Botanik," in *Studien zur Geschichte der Universität Wien*, III (Graz–Cologne, 1965), 126–202. On Jacquin's botanical activities, see J. H. Barnhart, *Biographical Notes Upon Botanists* (Boston, 1965); August Neilreich, "Geschichte der Botanik in Niederoesterreich," in *Verhandlungen der Zoologisch-botanischen Vereins in Wien*, **5** (1855), 23; and Ignatius Urban, *Symbolae Antillanae seu fundamenta florae Indiae occidentalis*, I (Berlin, 1898).

WILFRIED OBERHUMMER

JAEGER, FRANS MAURITS (*b.* The Hague, Netherlands, 11 May 1877; *d.* Haren, near Groningen, Netherlands, 2 March 1945), *crystallography, physical chemistry.*

Jaeger was the oldest of three sons of an officer who left the army at twenty-eight and then taught mathematics at the Gymnasium in The Hague. After completing his primary and secondary education in his native town, Jaeger entered the nearby University of Leiden in 1895, working as a part-time assistant at the geological museum. After his final examination in 1900 he obtained a stipend for two years of study at the University of Berlin, where he obtained practical experience in the laboratories of E. Fischer, E. Warburg, and J. F. C. Klein. While in Berlin he married the sister of his classmate B. R. de Bruijn. He began to teach chemistry in the fall of 1902 at a secondary school in Zaandam, near Amsterdam. On 9 October 1903 he obtained his Ph.D. from Leiden University with a thesis suggested by his promoter, A. P. Franchimont, entitled *Kristallografische en moleculaire symmetrie van plaatsings-isomere benzolderivaten.*

Under the supervision of H. W. Bakhuis Roozeboom, the successor to J. H. van't Hoff, Jaeger continued his studies at the municipal university in Amsterdam. On the recommendation of Roozeboom, he was allowed to teach as *privaat-docent* at the university while retaining his salaried position at Zaandam. In 1908 he was appointed lecturer, and the following year professor, of physical chemistry at the University of Groningen; he held this position until his dismissal by the Nazis in November 1944. One of his first tasks was to plan a new chemical laboratory to replace the one destroyed by fire in 1906.

Through his friend H. R. Kruyt, professor of colloid science at the University of Utrecht, Jaeger became acquainted with A. L. Day, director of the Geophysical Division of the Carnegie Institution of Washington. He spent one semester (September 1910–March 1911) with Day and thereby learned how to equip his new laboratory (opened in 1912) for the study of silicates and other materials with high melting points. In 1929 Jaeger returned to the United States as nonresident George Fisher Baker lecturer at Cornell University, where he gave a summary of his laboratory work from 1912 to 1929.

Throughout his active life as a scientist Jaeger was fascinated by the study of crystals. His first book on this topic, *Lectures on the Principle of Symmetry and Its Applications in All Natural Sciences*, was published in Amsterdam in 1917. Jaeger's most important work was his study of molten salts and silicates at extremely high temperatures. It involved the determination of viscosity, surface tension, conductivity, and specific heat at temperatures ranging from $-50°$ to 1,600°C. Jaeger's bent for writing on the history of chemistry was probably inherited from his father, who wrote in his spare time under the pseudonym of Maurits Smit. This hobby became a welcome necessity during the two world wars, when laboratory work was limited or entirely halted, particularly after the German invasion of Holland in 1940. Most of Jaeger's contributions to science are recorded in English in *Proceedings. K. Nederlandse akademie van wetenschappen*, of which academy he became a member in 1915.

BIBLIOGRAPHY

I. ORIGINAL WORKS. Jaeger's writings include *Anleitung zur Ausführung exakter physiko-chemischer Messungen bei höheren Temperaturen* (Groningen, 1913); and *The George Fisher Baker Nonresident Lecturership in Chemistry at Cornell University*, VII (New York, 1930): I, "Spatial Arrangements of Atomic Systems and Optical Activity"; II, "Methods, Results and Problems of Precise Measurements at High Temperatures"; III, "The Construction and Structure of Ultramarines."

II. SECONDARY LITERATURE. A résumé, by his collaborators, of 25 years of Jaeger's activity as professor appeared in *Chemisch weekblad*, **31** (1934), 182–212; an obituary by his colleague J. M. Bijvoet is in *Jaarboek der K. Nederlandsche akademie van wetenschappen* (1945).

HENRY S. VAN KLOOSTER

JAEGER, GEORG FRIEDRICH (*b.* 25 December 1785, Stuttgart, Germany; *d.* 10 September 1866, Stuttgart), *paleontology, medicine.*

The youngest son of Christian Friedrich Jaeger, a court physician, Jaeger attended the Stuttgart Gymnasium, studied medicine at Tübingen (M.D., 1808), and then spent a year traveling, during which he studied osteology and fossil skeletons under Georges Cuvier at Paris. He returned to Stuttgart and established a successful medical practice, becoming a member and eventually senior councillor of the Medicinal Collegium.[1] From 1817 until 1856 he held the post of inspector of the royal natural history cabinet,[2] and from 1822 until 1842 was also professor of chemistry and natural history in the Stuttgart Obergymnasium. A large man of robust health, he married twice[3] and had thirteen children.

Jaeger wrote on abnormal growth and anatomy of man and animals, physiological effects of poisons on plants, parasitism, mammalian systematics and distribution, geology, and anthropology. His principal contributions were to paleontology.

In 1822 he discovered in the collections of the Stuttgart Gymnasium a slab of limestone containing a large reptile skeleton, and within this skeleton a much smaller one.[4] Although it was unlike any reptile described by Cuvier,[5] Jaeger recognized its similarity to the fossil remains recently described in England and named Ichthyosaurus. His monograph contains careful observations (considering the unexposed condition of the fossil) and judicious inferences. He pointed out that the structure of the limbs is more like the paddles of a porpoise than the feet of either salamanders or crocodiles, the animals between which it had been placed in classifications. In 1842 and 1852 he suggested that the small skeleton might be that of a fetus, and that ichthyosaurs may have given birth to living young.

Jaeger's monograph on fossil plants of the Triassic sandstones near Stuttgart (1827) was followed in 1828 by a more extensive study of fossil reptiles, which contains the earliest descriptions of labyrinthodont amphibians and of the crocodile-like phytosaurs of the Triassic. He then turned his attention to the fossil mammals from fissures in the Schwabian Alb, in Germany, and in 1835 and 1839 provided the first detailed account of this material. This account was important for Jaeger's recognition of considerable differences between the faunas of different localities. In the absence of any super-positional relationships between fissures, he failed to grasp the implication of these differences for the relative ages of the faunas; instead he sought to explain them by varying circumstances of deposition and accumulation of the bones. Although he attempted to fit them into the Cuvierian concept of an ancient fauna (that of the Paris gypsum) and a more recent assemblage of still living animals mixed with not long extinct species such as were found in caves and river alluvium, he repeatedly expressed doubts about this interpretation, and arranged the various faunas in their proper time sequence. He also described the tusks of mammoths and associated fossils found near Stuttgart in 1700 and 1816.[6] In addition to these monographs Jaeger wrote many shorter articles on various vertebrate fossils.

Jaeger actively promoted science, medicine, and natural history. He was highly regarded by his contemporaries, a member of thirty-five academies and learned societies, and a recipient of state and national honors.

NOTES

1. The highest health authority in Württemberg.
2. His brother, Carl Christian Friedrich Jaeger, had held this post from 1798–1817.
3. Jaeger married Charlotte Hoffmann, who died in 1818, leaving two sons and two daughters. He later married Charlotte Schwab, who bore four sons and five daughters.
4. These specimens, from Boll, in Württemberg, had been collected in 1749 by Christian Albert Mohr and described as fishes in an unpublished dissertation. Boll had been known as an important locality for fossils since the 1598 memoir of J. Bauhin.
5. *Recherches sur les ossemens fossiles de quadrupèdes, ou l'on rétablit les caractères de plusieurs espèces d'animaux que les révolutions du globe paroissent avoir détruites*, 4 vols. (Paris, 1812); later eds. include a section on ichthyosaurs.
6. The caches of mammoth ivory at Cannstatt, near Stuttgart, attracted great attention. Those found in 1700 were largely sold for medicine; a piece of "unicorn horn" from this find was given by Duke Eberhard Ludwig of Württemberg to the citizens of Zurich to help them fight a plague epidemic. King Frederick I of Württemberg personally supervised the excavations in the winter of 1816–1817; and he died of pneumonia contracted during this work.

BIBLIOGRAPHY

I. Original Works. Jaeger's works include *De effectibus Arsenici albi in varios organismos* (Tübingen, 1808), dissertation; *Anleitung zur Gebirgskunde* (Stuttgart, 1811); *Über die Missbildungen der Gewächse* (Stuttgart, 1814); *Das wissenwürdigste aus der Gebirgskunde* (Stuttgart, 1815); *De Ichthyosauri sive proteosauri fossilis speciminibus in agro Bollensis in Würtembergia repertis commentatur Georgius Friedericus Jaeger* (Stuttgart, 1824); *Ueber die Pflanzenversteinerungen des Bausandsteins in Stuttgart* (Stuttgart, 1827); *Über die fossilen Reptilien, welche in Württemberg aufgefunden worden sind* (Stuttgart, 1828); *Über die fossilen Säugethiere, welche in Württemberg aufgefunden worden sind* (Stuttgart, 1835), continued as *Über die fossilen Säugethiere, welche in Württemberg in verschiedenen Formationen aufgefunden worden sind, nebst geognostischen Bemerkungen über diese Formationen* (Stuttgart, 1839); *Beobachtungen und Untersuchungen über die regelmässigen Formen der Gebirgsarten* (Stuttgart, 1846); and *Ueber die Wirkung des Arseniks auf Pflanzen im Zusammenhang mit Physiologie, Landwirtschaft und Medicinalpolizei* (Stuttgart, 1864).

II. Secondary Literature. Citations of Jaeger's publications dealing with fossil vertebrates are given in A. S. Romer *et al.*, "Bibliography of Fossil Vertebrates

Exclusive of North America 1509–1927," in *Memoirs of the Geological Society of America*, **87** (1962), II, 685–687 (contains 44 titles). J. G. von Kurr lists 28 articles dealing with natural history, including the major paleontological monographs, in *Württembergische naturwissenschaftliche Jahreshefte*, **23** (1867), 34–36; 27 short papers are noted in the index, **20** (1864), 315–316. A. Hirsch lists a few medical papers in *Biographisches Lexikon der hervorragenden Ärzte aller Zeiten und Völker*, III (Vienna–Leipzig, 1886), 372–373.

Biographical notices have been published by C. G. Carus in *Leopoldina*, **5** (1866), 138; and J. G. von Kurr (see above), who says that Jaeger published 143 articles, presumably about half of which were on medical subjects. For an appreciation of Jaeger's paleontological work see K. Staesche, "Ein Jahrhundert Paläontologie in Württemberg," in *Jahresheft des Vereins für vaterländische Naturkunde in Württemberg*, **113** (1958), 24.

JOSEPH T. GREGORY

JAEKEL, OTTO (*b.* Neusalz an der Oder, Germany [now Nowa Sól, Poland], 21 February 1863; *d.* Peking, China, 6 March 1929), *paleontology*.

Jaekel, whose parents ran a butcher shop in Neusalz, attended the Gymnasium in Liegnitz. He then studied geology under Ferdinand Roemer at Breslau (1883) and paleontology under K. A. R. von Zittel at Munich (1885–1886), where he received his doctorate. He became an assistant at the Geological-Paleontological Institute of the University of Strasbourg, and following a short stay in London was made a *Privatdozent* at the University of Berlin in 1890; four years later he was appointed an assistant professor and curator of the Geological-Paleontological Institute and museum. Although a desired nomination to the chair of paleontology at Vienna did not materialize, in 1906 he obtained this post at the University of Greifswald (Pomerania). Upon his retirement in 1928, he accepted an invitation from Sun Yat-sen University, Canton, China, to assume a professorship in paleontology. He died of pneumonia six months later in the German Hospital in Peking; he was survived by his son and daughter.

Outside of a few geological works, Jaekel's field of research was primarily paleontology. He concentrated on the echinoderms and the vertebrates and wrote several major monographs on the Paleozoic stalked echinoderms (pelmatozoans) in which he established, among other things, the new class Carpoidea. Among the vertebrates, his chief interests were fish and reptiles, and general questions concerning the origin and descent of vertebrates. With regard to fish, he investigated primarily the Paleozoic groups, especially the placoderms and the Elasmobranchii. Through field collections made at the Upper Devonian locality of Wildungen, near Kassel, he provided, in large part, the specimens used in these investigations, which he published only in provisional form. His study of reptiles included the placodonts, turtles, and dinosaurs of the Triassic, as well as other Mesozoic reptilian groups. Here, too, he carried out his own excavations in the Upper Triassic at Halberstadt, near Magdeburg.

In the last years of his life Jaekel was occupied with general problems concerning the origin of vertebrates, the morphogenesis of the teeth and the skeleton, and descent and phylogenetic interrelationships of the great vertebrate groups. Like many of his other studies, his works on these subjects contain numerous new ideas and insights. But they were often insufficiently substantiated by the evidence, and therefore provoked criticism from his contemporaries. Much of this criticism must be attributed to his artistic inclinations and to his personal temperament. A passionate, gifted painter and connoisseur of Far Eastern art, his approach to paleontology was sometimes that of the artist rather than that of the scientist, and thus his ideas, often brilliant and stimulating, were not always verified by the critical scientific method. Advocating a greater independence of paleontology from geology, he founded, in 1912, the Paleontological Society and its publication *Paläontologische Zeitschrift*.

BIBLIOGRAPHY

I. ORIGINAL WORKS. Jaekel's works include "Die Selachier aus dem oberen Muschelkalk Lothringens," in *Abhandlungen zur geologischen Spezialkarte von Elsass-Lothringen*, **3** (1889), 275–332; *Die eozänen Selachier vom Monte Bolca, ein Beitrag zur Morphologie der Wirbeltiere* (Berlin, 1894); *Stammesgeschichte der Pelmatozoen* (Berlin, 1899); "Neue Wirbeltierfunde aus dem Devon von Wildungen," in *Sitzungsberichte der Gesellschaft naturforschender Freunde zu Berlin*, no. 3 (1906), pp. 73–85; *Die Wirbeltiere, eine Übersicht über die fossilen und lebenden Formen* (Berlin, 1911); "Die Wirbeltierfunde aus dem Keuper von Halberstadt," in *Paläontologische Zeitschrift*, **2** (1918), 88–214; "Phylogenie und System der Pelmatozoen," *ibid.*, **3** (1921), 1–128; "Das Problem der chinesischen Kunstentwicklung," in *Zeitschrift für Ethnologie*, no. 6 (1920–1921), pp. 493–518; "Das Mundskelett der Wirbeltiere," in *Gegenbaurs morphologisches Jahrbuch*, **55** (1925), 402–484; "Der Kopf der Wirbeltiere," in *Zeitschrift für die gesamte Anatomie*, **27**, sec. 3 (1927), 815–974; and "Die Morphogenie der ältesten Wirbeltiere," in *Monographien zur Geologie und Paläontologie*, 1st ser., no. 3 (1929).

II. SECONDARY LITERATURE. See O. Abel, "Otto Jaekel," in *Palaeobiologica*, **2** (1929), 143–186, with complete bibliography and portrait; E. Hennig, "Otto Jaekel," in *Zentralblatt für Mineralogie, Geologie und Paläontologie*, sec. B (1929), pp. 268–271; S. von Bubnoff, "Otto Jaekel als Forscher," in *Mitteilungen des naturwissenschaftlichen Vereins für Neu-Vorpommern und Rügen in Greifswald*, **57–58** (1929–1930), 1–9; F. Krüger, "Otto Jaekel als Persönlichkeit," *ibid.*, 10–17; W. Gross, "Herkunft und Entstehung der Wirbeltiere in der Sicht Otto Jaekel's," in *Paläontologische Zeitschrift*, **37** (1963), 32–48; and H. Wehrli, "Otto Jaekel (Greifswald, 1906–1928)," in *Festschrift zur 500-Jahrfeier der Universität Greifswald* (Greifswald, 1956), pp. 498–503.

HEINZ TOBIEN

JAʿFAR AL-BALKHĪ. See **Abū Maʿshar.**

JAGANNĀTHA (*fl.* India, *ca.* 1720–1740), *astronomy, mathematics.*

According to legend, Jagannātha Samrāṭ was discovered by Jayasiṃha of Amber during a campaign against the Marāṭha chief Śivājī in 1664–1665; Jagannātha was then supposed to be twenty years old. Unfortunately for the story, it was Jayasiṃha I, known as Mirjā, who was involved with Śivājī; the patron of Jagannātha was Jayasiṃha II, known as Savāī, who ruled Amber from 1699 to 1743. For Jayasiṃha II, Jagannātha translated Euclid's *Elements* and Ptolemy's *Syntaxis Mathēmatikē* (both in the recensions of Naṣīr al-Dīn al-Ṭūsī) from Arabic into Sanskrit as a part of Jayasiṃha's program to revitalize Indian astronomy and Indian culture in general.

Jagannātha translated Euclid's *Elements* under the title *Rekhāgaṇita* shortly before 1727, the date of the earliest manuscript copied at his command by Lokamaṇi. He translated Ptolemy's *Syntaxis Mathēmatikē* in 1732 under the title *Siddhāntasamrāṭ*. This contains not only a translation of al-Ṭūsī's Arabic recension but also notes of his own referring to Ulugh Beg and al-Kāshī of Samarkand as well as to Muḥammad Shāh, the Mogul emperor to whom Jayasiṃha dedicated his *Zīj-i-jadīd-i Muḥammad-Shāhī* in 1728; these additions closely link Jagannātha's translation with the work of the other astronomers assembled by Jayasiṃha. (See essays on Indian science in Supplement.)

BIBLIOGRAPHY

The *Rekhāgaṇita* was edited by H. H. Dhruva and K. P. Trivedi as Bombay Sanskrit series no. 61–62, 2 vols. (Bombay, 1901–1902); the *Siddhāntasamrāṭ* was edited by Rāmasvarūpa Śarman, 3 vols. (New Delhi, 1967–1969).

Secondary literature includes Sudhākara Dvivedin, *Gaṇakataraṅginī* (Benares, 1933), repr. from *Pandit*, n.s. **14** (1892), 102–110; and L. J. Rocher, "Euclid's Stoicheia and Jagannātha's Rekhāgaṇita," in *Journal of the Oriental Institute, Baroda*, **3** (1953–1954), 236–256.

DAVID PINGREE

JAGGAR, THOMAS AUGUSTUS, JR. (*b.* Philadelphia, Pennsylvania, 24 January 1871; *d.* Honolulu, Hawaii, 17 January 1953), *geology, volcanology.*

Jaggar was the son of the Reverend Thomas Augustus and Anna Louisa (Lawrence) Jaggar. As he later wrote, a love of the outdoors was instilled in him by his father, and at an early age he was tramping the backwoods of Maine and the Maritime Provinces. At fourteen, while in Europe with his family, he was schooled in French and Italian, climbed Vesuvius, and became committed to natural science. Jaggar earned three degrees (B.A., 1893; M.A., 1894; Ph.D., 1897) at Harvard University, and spent two postgraduate years in Europe at the University of Munich, with K. A. von Zittel (1894), and at the University of Heidelberg, with K. H. Rosenbusch, V. Goldschmidt, and E. Osann (1895). In 1903 he married Helen Kline, later marrying, in 1917, Isabel P. Maydwell, a valued assistant and companion throughout his career.

The need for careful field observation was impressed upon Jaggar at Harvard by Nathaniel Shaler, and also by Arnold Hague, with whom Jaggar worked as an assistant in the Rocky Mountain volcanic province (1893). The latter experience introduced Jaggar to volcanology, the field of geology that was to dominate his scientific career. While a graduate student, he studied intrusive rocks with R. A. Daly, and for his doctoral dissertation he invented a "mineral hardness instrument" (microsclerometer) and studied the xenoliths in the dikes in Boston.

After completing his formal education Jaggar worked for the U.S. Geological Survey (1898–1901). He participated with S. F. Emmons, J. D. Irving, Bailey Willis, and N. H. Darton in studies on the laccoliths and economic resources of the Black Hills in South Dakota, and with Charles Palache on the geology of the Precambrian granites of the Bradshaw Mountains in Arizona. In 1899 Jaggar provided for Charles D. Walcott, director of the survey, the first estimate for a geological survey of Hawaii, a project which eventually led Jaggar to that area of the Pacific. During the next several years Jaggar and his associates in Boston and at Harvard experimented with various devices, including stream tables and model geysers. The experimental geyser was in effect a miniature

"Old Faithful," yielding (although scale was not a serious consideration) some mechanistic understanding and, by application, more knowledge of phreatic volcanic eruptions. In addition, squeeze-box experiments by Jaggar and his associates extended the earlier work of Bailey Willis, and experiments by the team on the crystallization of basalt melts and artificial mixtures of their constituent minerals yielded observations on textural variations and on the rate of cooling.

In 1901 Jaggar left the U.S. Geological Survey to become an assistant professor at Harvard and in 1906 head of the geology department at the Massachusetts Institute of Technology, where he remained in close association with the survey until 1910. It was during this time (1901–1912) that Jaggar began to make his volcanic expeditions to the island of Martinique after the Mount Pelée eruption (1902); to Vesuvius, where he became acquainted with F. A. Perret (1906), the noted volcanic photographer and volcanologist; to the Aleutian Islands (1907); and ultimately to sixty of 450 still-active volcanoes.

In 1911, with financial help from the Massachusetts Institute of Technology and the Volcano Research Association of Honolulu, Jaggar established what was to become the Hawaii Volcano Observatory. He served as director until 1940. Under his charge all measurable parameters of Hawaiian volcanic activity were recorded—shape, character of flow, height of lava in craters (particularly Halemaumau, since this crater remains filled for extended times during eruption), temperature (his early attempts used pyrometer, iron pipe, and immersed thermocouples), and eruptive periods. Seismic data (some recorded on Jaggar's "shock recorder") and accurate surveying yielded important evidence of swelling of volcanic edifices and the first hints of eruptive predictability. The cyclic nature of volcanism was substantiated by the continuous, careful records kept by Jaggar and his associates. The work at the Hawaii Volcano Observatory and Jaggar's experience at other volcanoes culminated in the publication of "Origin and Development of Craters" (1947).

Jaggar was responsible for a classification of volcanoes by viscosity (1910) and for early direct temperature measurements of basaltic lava. He also formulated nomenclature, although it is not used frequently today, for specific lava conditions. "Pyromagma," for example, referred to hot, gas-charged lava in lakes; "epimagma" to partially solidified and crystallized lava; and "hypomagma" to the subsurface source magma. Much of Jaggar's work, based on firsthand observation and experience, was qualitative. He made significant contributions on the development of volcanoes and on the role of groundwater in explosive eruptions. He always demonstrated concern for the people who lived in volcanic areas. In 1936 he recommended a plan of bombing from aircraft that succeeded in diverting the flows endangering Hilo.

After Jaggar retired from the Hawaii Volcano Observatory in 1940, he became a research associate in geophysics at the University of Hawaii. Up to his death he continued to impart his knowledge of volcanology through travels with his wife and through his writing.

BIBLIOGRAPHY

I. Original Works. A large number of Jaggar's works, including his early publications and important journal articles through 1945, are listed in the reference section of his major work "The Origin and Development of Craters," *Memoirs. Geological Society of America*, **21** (1947). Significant among these are "Japanese Volcanoes and Volcano Classification," in *M.I.T. Bulletin of the Society of the Arts* (Feb. 1910); "Seismometric Investigation of the Hawaiian Lava Column," in *Bulletin of the Seismological Society of America*, **10** (1920), 155–275; and "Protection of Harbors From Lava Flows," in *American Journal of Science*, **243A** (1945), 333–351, with reference to a number of reports written by Jaggar for the *Volcano Letter of the Hawaiian Volcano Research Association*, published by the Hawaii Volcano Observatory.

Additional works include "The Mechanism of Volcanoes," in *Volcanology*, National Research Council Bulletin no. 77 (1931), 49–71; *Volcanoes Declare War* (Honolulu, 1945); *Steam Blast Volcanic Eruptions*, Hawaiian Volcano Observatory, 4th spec. report, a significant work; and *My Experiments With Volcanoes* (Honolulu, 1956), an autobiography that provides an interesting look into Jaggar's experiences through 1952 and contains information on his early inclination toward natural science and his developing interest in Hawaii.

II. Secondary Literature. A brief but important sketch of Jaggar's life is found in F. M. Bullard, *Volcanoes —in History, in Theory, in Eruption* (Austin, 1962), pp. 27–30; and in *World Who's Who in Science* (Hannibal, Mo., 1968), p. 870. Reference to much of Jaggar's professional work is in A. Rittmann, *Volcanoes* (New York, 1962), pp. xiii, 18, 55, 63, 167, 188; F. A. Perrett, "Volcanological Observations," in *Publications, Carnegie Institution of Washington*, **549** (1950), 50, 133; and most recently in G. A. MacDonald, *Volcanoes* (Englewood Cliffs, N.J., 1972), pp. xii, 37–39.

Wallace A. Bothner

AL-JĀḤIẒ, ABŪ ʿUTHMĀN ʿAMR IBN BAḤR (*b.* Basra, Iraq, *ca.* 776; *d.* Basra, 868/869), *natural history.*

Al-Jāḥiẓ is a nickname that means "the goggle-eyed." His ugliness is further attested to by sources that mention it as the reason he lost his post as tutor to the children of Caliph al-Mutawakkil. Although ardently devoted to Basra, al-Jāḥiẓ spent extended periods in Baghdad and Sāmarrā. His teachers were the philologists and men of letters al-Aṣmaʿī, Abū ʿUbayda, and Abū Zayd. Among other things he studied translations from the Greek that had recently become available.

A tireless reader, al-Jāḥiẓ also obtained a great deal of oral information from the sailors, bedouins, and men of all classes who could be found in Basra. In politics and religion he adhered to the rational theology of the Muʿtazila school. This allegiance can be seen in, for example, a number of writings he devoted especially to defending the legitimacy of the Abbasid dynasty. He also wrote polemical works against the Jews and Christians. He earned so much money from his books that he was able to support himself even when he was not holding an office.

Of the long list of writings attributed to al-Jāḥiẓ by literary historians, approximately 200 are genuine, of which less than thirty are extant. Many contain noteworthy remarks pertaining to the various sciences; but a group of them is devoted specifically to scientific themes. Among the shorter writings, mostly lost, are *Of the Lion and the Wolf; On the Mule and Its Uses; Dogs; Grain, Dates, Olives, and Grapes; Minerals; Man; On the Difference Between Jinn and Men; Refutation of He Who Considers Man to Be an Indivisible Entity* (*Atom*)*; On Cripples, Lepers, and the Poor; On the Difference Between Men and Women; Contest Between Female Slaves and Young Men; The Limbs; The Bedouin Diet; On the Drinker and Drink* (*on the Types of Date Wine*)*; Critique of Medicine; The Grocer's Shop; Against Alchemy; Countries* (*Geography*)*;* and *Contest Between Winter and Summer.*

By far the most important of these works, and one of al-Jāḥiẓ's most extensive, is his book on animals (*Kitāb al-Ḥayawān*) in seven parts. As yet no satisfactory edition of it exists, but much of it has been translated into European languages, particularly English and Spanish; the most recent edition contains a very detailed name and subject index. The book is not a systematic account of zoology but, rather, a literary work meant to entertain, the arrangement of which is based on certain groups of animals. For this reason it treats far fewer animals than the total number known to al-Jāḥiẓ, who considers only the larger mammals, some important birds, and, with special enthusiasm, the insects, such as flies, gnats, scorpions, and lice. Al-Jāḥiẓ describes the animals and relates, with many literary digressions, what the Arabs knew about them. The work is therefore a kind of national zoology in which he includes the results of his own scientific studies. He is acquainted with and eagerly draws on Aristotle's *Historia animalium* but he is not dependent on it. Other Greek writers are cited as well.

Al-Jāḥiẓ distinguishes running, flying, swimming, and crawling animals and opposes the carnivores to the herbivores. He likewise differentiates doglike animals, catlike animals, and ruminants. He divides the birds into birds of prey, defenseless birds, and small birds. For lack of reliable material he does not discuss fishes. He rejects the division into useful and harmful animals, since even the animals harmful to man have their uses in the divine plan of the universe and the opposition between good and bad in general is one of the foundations of the organization of the universe. Al-Jāḥiẓ displays an interest in the adaptation of certain animals, accepts the possibility of the spontaneous generation of some animals (for example, of frogs from ice), and considers such special problems as the language of animals. He also discusses the effects of intoxication and castration on animals, as well as their sexual anomalies, including sodomy. For al-Jāḥiẓ man is a microcosm that unites within itself the attributes of numerous animals.

Al-Jāḥiẓ did not slavishly accept the material he found in the writings of his predecessors. He formed his own judgments and even conducted his own investigations, some of which are remarkable for their methodology. He was critical of tradition, even of the Koran. A book on zoology of this scope never appeared again in the Islamic world.

In 1946 Oscar Löfgren published the illustrations preserved in a manuscript of this work in the Biblioteca Ambrosiana in Milan. Some of them represent coitus between animals, a subject that was very seldom depicted. In one picture an act of sodomy is illustrated. The illustrations are monochromatic, but R. Ettinghausen has reproduced in color the image of an ostrich sitting on its eggs.

Al-Jāḥiẓ held that alchemy was not impossible in principle but spoke out against its practice, since in the course of thousands of years so many great scholars had achieved no practical results.

BIBLIOGRAPHY

I. ORIGINAL WORKS. Two of al-Jāḥiẓ's books are *Kitāb al-Ḥayawān*, ʿAbd al-Salām Hārūn, ed., 2nd ed., 7 vols. (Cairo, 1938–1945); and *Livre des mulets*, ed. and with notes by Charles Pellat (Cairo, 1955). Translations from the "quasi-scientific works" of al-Jāḥiẓ are in Charles Pellat, *The Life and Works of Jāḥiẓ* (London, 1969), pp. 126–199. Other translations are in Oskar Rescher, *Excerpte und Übersetzungen aus den Schriften des Philologen und Dogmatikers Ǧâḥiẓ aus Baçra (150–250 H.)*, I (Stuttgart, 1931). See also Oscar Löfgren, *Ambrosian Fragments of an Illuminated Manuscript Containing the Zoology of al-Jāḥiẓ* (Uppsala, 1946); and R. Ettinghausen, *Arab Painting* (Paris, 1962), pl. p. 157.

II. SECONDARY LITERATURE. See the following, listed chronologically: G. van Vloten, *Ein arabischer Naturphilosoph im 9. Jahrhundert (el-Dschâḥiẓ)* (Stuttgart, 1918); M. Asin Palacios, "El 'Libro de los animales' de Jâḥiẓ," in *Isis*, **14** (1930), 20–54; and L. Kopf, "The 'Book of Animals' (Kitāb al-Ḥayawān) of al-Jāhiz (ca. 767–868)," in *Actes du VIIe Congrès international d'histoire des sciences* (Jerusalem, 1953), pp. 395–401. See also Charles Pellat, *Le milieu baṣrien et la formation de Ǧāḥiẓ* (Paris, 1953); and his article "Djāḥiẓ," in *Encyclopaedia of Islam*, II, 2nd ed. (London–Leiden, 1965), 384–387; and George Sarton, *Introduction to the History of Science*, I (Baltimore, 1927), 597.

M. PLESSNER

JAHN, HANS MAX (*b.* Küstrin, Germany [now Kostrzyn, on the Oder, Poland], 4 July 1853; *d.* Berlin, Germany, 7 August 1906), *physical chemistry.*

Jahn studied chemistry and mathematics at the universities of Berlin and Heidelberg. During his student years he was a private assistant to A. W. von Hofmann at Berlin, where in 1875 he earned his doctorate with a work on the derivatives of secondary octyl alcohol. He then went to Athens, where he was an instructor and then a professor at the university. He moved to Vienna in 1877 and worked as *Privatdozent* in association with Ernst Ludwig. In 1883 he married Sophie von Sichrovsky.

In 1884 Jahn went to Graz University in Austria and, in 1889, to the Landwirtschaftliche Hochschule in Berlin to work with Hans Landolt. When the latter was appointed to the University of Berlin, Jahn went with him and taught at his institute, first as an instructor and later as an assistant professor, lecturing on electrochemistry. But his hearing steadily deteriorated, hindering his career. He died unexpectedly of complications stemming from an appendectomy.

Jahn's earliest scientific works dealt with organic chemistry, as did most German writings of that period on chemistry. His most important publications concerned the decomposition of simple organic compounds by means of zinc dust. While at the university in Graz, he turned his attention to the field of electrochemistry. In numerous works he considered various thermodynamic problems raised by electrochemical phenomena. In his studies on the relationship of the chemical energy and electrical energy of galvanic cells and on the equivalence of these two kinds of energy, he presented experimental evidence (1886) suggesting that the total chemical energy is modified only when the electromotive force does not change with the temperature. He thereby provided the quantitative demonstration of the validity for electrochemical phenomena of the so-called Gibbs-Helmholtz equation (1878–1882).

Jahn also demonstrated the existence of reversible Peltier heat effects in voltaic cells and, with Otto Schönrock, investigated electrochemical polarization. In his last years Jahn was concerned with electrolytic dissociation, attempting to calculate the relationship between degree of dissociation and electrical conductance in weak and strong electrolytes; he was unsuccessful in this endeavor. His last works are devoted to improvements in cryoscopic methods.

BIBLIOGRAPHY

Jahn's major works are: *Die Grundsätze der Thermochemie und ihre Bedeutung für die theoretische Chemie* (Vienna, 1882); *Die Elektrolyse und ihre Bedeutung für die theoretische und angewandte Chemie* (Vienna, 1883); and *Grundriss der Elektrochemie* (Vienna, 1895). In addition to those Jahn produced about fifty other publications.

For a biographical treatment of Jahn, see H. Landolt, "Hans Jahn," in *Berichte der Deutschen chemischen Gesellschaft*, **39** (1906), 4463, with portrait.

FERENC SZABADVÁRY

JAMES OF VENICE, also known as **IACOBUS VENETICUS GRECUS** (*d.* after 1147), *philosophy, law, Aristotelian translations.*

The available evidence suggests that James was born in Venetia—not necessarily in Venice—and the qualification "Grecus" could mean either that he spent much of his life in some Greek-speaking part of the Byzantine Empire or that he was of Greek descent. There are only three known dates relevant to his life. On 3 April 1136, he attended, in the Pisan quarter of Constantinople, a theological debate between Anselm, Catholic bishop of Havelberg, and Nicetas, Orthodox archbishop of Nicomedia. When Moses of Vercelli, archbishop of Ravenna, claimed in Cremona, 7 July 1148, the privilege of sitting at the right hand of the Pope, he was supported by a legal advice written for him by James. In 1159 John of Salisbury quoted James's translation of Aristotle's *Posterior Analytics* as being older than another version of the same treatise.

James was probably the most important of the scholars on whose work the knowledge of Aristotle's writings in the Latin Middle Ages depended. He was the first to translate the *Physics*, *De anima*, *Metaphysics* (at least books I–IV.4, possibly all fourteen books), and most of the shorter treatises which go under the title of *Parva naturalia*, that is, *De memoria, De longitudine et brevitate vitae, De iuventute, De respiratione,* and *De vita et morte*. He was perhaps the first to translate the epistemological treatise *Posterior Analytics;* Boethius' translation, if it was ever made, does not seem to have been known by anybody. He translated anew the *Sophistici elenchi* and probably the *Prior Analytics* and *Topics.* Fragments of Alexander of Aphrodisias' commentary on *Posterior Analytics* and *Sophistici elenchi,* translated by James, still survive, as does his version of an anonymous introduction to the *Physics* (published under the title *De intelligentia Aristotelis*) and of scholia to *Metaphysics I*. He himself wrote a commentary on the *Sophistici elenchi* and perhaps on other Aristotelian works.

James provided the link between the Greek philosophical schools in Constantinople and those of the Latin West. At this time the study of Aristotle was prospering in Constantinople after the revivals of the ninth and eleventh centuries, which in turn were based on the work done in the schools of the second to sixth centuries. The philosophy masters in Constantinople frequented the same circles as James, whose commentary on the *Sophistici elenchi* contains clear evidence of its connection with the Greek teaching on this subject; there is no other place in the Greek world where, at that time, it would have been possible to have access to so many works of Aristotle. James's work on sophisms, extensively quoted and discussed in logical treatises, was most probably written in northern Italy in the second half of the twelfth century. At the same time, his translations reached Normandy; copies of some of them, written in Mont-Saint-Michel before the end of the century, still survive. John of Salisbury knew at least one of them and asked for others.

James's translations (particularly of the works on philosophy of science) and Boethius' translations of most of the logical works formed the main body of work, to which were added, during the next four generations, all the other latinized texts of Aristotle. Some of these translations, either in an unaltered form like the *Posterior Analytics*, or in a form slightly revised by William of Moerbeke (?)—like the *Physics, De anima,* and *De memoria*—were the recognized "authentic" texts for over three centuries. They contributed in considerable measure to the formation or establishment of the technical language of philosophy and, indirectly, of the scientific and common language of the Western world. The several hundred manuscripts, the dozens of printed editions of the original or revised translations, and the vast number of commentaries, elaborations, and *quaestiones* by Roger Bacon, Grosseteste, Albertus Magnus, Aquinas, Ockham, and many other philosophers testify to the importance of James's work.

BIBLIOGRAPHY

I. Original Works. The translation of *Posterior Analytics*, anonymous until 1500, was printed either by itself (Louvain, 1476), or with commentaries (the first being with Grosseteste's commentary [Naples, before 1479]), or as part of Aristotle's so-called *Organon*, 1st ed. (Augsburg, 1479). From 1503 to 1891 it was either wrongly ascribed to Boethius or still anonymous in a text revised by Jacques Lefèvre d'Étaples, repr. in Migne, *Patrologia Latina*, LXIV, cols. 711–762. The translation is ascribed to James in the critical ed. by L. Minio-Paluello and B. G. Dod in *Aristoteles Latinus* IV.1–4 (Bruges–Paris, 1968), 5–107.

James's translations of *De anima* and *De longitudine et brevitate vitae*, edited anonymously by M. Alonso, are in *Pedro Hispano, obras filosóficas*, III (Madrid, 1952), 89–395, 405–411.

Translation of the *Metaphysics* I–IV.4 appears anonymously in R. Steele, *Opera adhuc inedita Rogeri Baconi*, XI (Oxford, 1932), 255–312, and ascribed to James in the critical ed. by G. Vuillemin-Diem in *Aristoteles Latinus* XXV.1–1*a* (Brussels–Paris, 1970), 5–73.

The translation of the introduction to the *Physics* was printed twice, anonymously, in Venice (1482, 1496).

The translation of the scholia to the *Metaphysics I* has been edited twice: by L. Minio-Paluello, "Note," VII (below), 491–495, and G. Vuillemin-Diem, *op. cit.*, 74–82.

For the extant fragments of James's commentary on the *Sophistici elenchi* and of his translations from Alexander of Aphrodisias, see L. Minio-Paluello, "Note," XI and XIV, and De Rijk, *Logica modernorum* (see below). See also R. W. Hunt, "Studies on Priscian in the Twelfth Century," in *Mediaeval and Renaissance Studies*, **2** (1950), 43.

No edition exists of the original texts of James's translations of the *Physics*, *De memoria*, and other minor treatises, but there are several editions of these texts as revised by William of Moerbeke (?).

James's "Legal Advice to Archbishop Moses" was edited by A. Gaudenzi in *Bullettino dell'Istituto Storico Italiano*, **39** (1919), 54–55; E. Franceschini, "Il contributo . . .," in *Atti della XXVI Riunione della Società Italiana per il Progresso delle Scienze*, 1937 (Rome, 1938), pp. 307–308; and L. Minio-Paluello, "Iacobus" and "Il chronicon" (see below).

II. Secondary Literature. All the scanty information and speculation on James previous to 1952 is critically reviewed in L. Minio-Paluello, "Iacobus Veneticus Grecus: Canonist and Translator of Aristotle," in *Traditio*, **8** (1952), 265–304.

See also L. Minio-Paluello, "Il chronicon altinate e

Giacomo Veneto," in *Miscellanea in onore di Roberto Cessi*, I (Rome, 1958), 153–169; "Giacomo Veneto e l'Aristotelismo Latino," in A. Pertusi, ed., *Venezia e l'Oriente fra tardo medioevo e rinascimento* (Florence, 1966), pp. 53–74; and "Note sull'Aristotele Latino medievale," I, in *Rivista di Filosofia Neo-Scolastica*, **42** (1950), 222–226; VI–VII, **44** (1952), 398–411, 485–495; IX, **46** (1954), 223–231; XIV, **54** (1962), 131–137. All of these articles have been reprinted in L. Minio-Paluello, *Opuscula: The Latin Aristotle* (Amsterdam, 1972).

For additional information, see the introductions to vols. IV.1–4 and XXV.1–1*a* of *Aristoteles Latinus;* and L. M. De Rijk, *Logica modernorum*, I (Assen, 1962), esp. 83–100, and passages listed in the *index nominum*, p. 643.

LORENZO MINIO-PALUELLO

JAMES, WILLIAM (*b.* New York, N. Y., 11 January 1842; *d.* Chocorua, New Hampshire, 26 August 1910), *psychology, philosophy.*

James was the first of five children of Mary Robertson Walsh and Henry James, Sr.; their second was the novelist Henry James. Although he studied with tutors and in schools in the United States and throughout Europe, James may most properly be said to have received his early education at the family dinner table. The elder Henry James was a man of private means who had turned to travel and Swedenborgianism as perhaps the ultimate result of a childhood accident by which he had lost a leg. Having found the consolations of intellect and philosophy, he encouraged his children in critical investigation and discussion; it is probably significant that William James's first published book (1885) was his edition of *The Literary Remains of Henry James*, a work which rises above mere filial piety in containing, in the introduction, an early statement of some of his own religious views. The *Remains* themselves show their author to have been something rather more than the usual nineteenth-century American religious crank, and certainly his sons seem to have benefited from his tutelage.

James's first ambition was to become an artist, and in 1860 the entire family relocated from Paris to the United States, so that he could study painting with William Morris Hunt in Newport, Rhode Island. John La Farge, a fellow student, noticed his talent, but James soon changed his mind about his vocation and took up the study of chemistry, enrolling in the Lawrence Scientific School of Harvard in 1861. At Lawrence, James attended Agassiz's lectures, which led him from chemistry into the biological sciences. In 1864 he entered the Harvard Medical School, which he left in April 1865 to join Agassiz on an expedition up the Amazon. It was not a happy journey. James found that he had no skill as a field naturalist—indeed, he recorded that he hated collecting—and he became ill. He resumed his medical studies in 1866, but discontinued them again shortly thereafter because of lingering poor health. The following year he went to Germany to take a course of water cures and to study the physiology of the nervous system. He returned after two years, still sick, but able to take the M.D. from Harvard in 1869.

James never practiced medicine. The three years immediately following the award of his degree he remained at home, too unwell for regular employment, reading, writing occasional literary reviews, and apparently undergoing the shattering spiritual experience that he later described in "The Sick Soul" in *The Varieties of Religious Experience*. His recovery came in part through his reading of the *Essais de critique générale* of Charles Renouvier, from which he formulated the belief in volitional free will that shook him from his moral lethargy. By 1873 he was well enough to accept enthusiastically an appointment as instructor in anatomy and physiology at Harvard, where he was subsequently assistant professor of physiology (1876), assistant professor of philosophy (1880), and professor of philosophy (1885).

In 1878 James married Alice Howe Gibbens, of Cambridge; the four of their five children who lived past infancy were brought up in the Jamesian tradition of travel, familial affection, and abstract discussion. In the same year he contracted to write a textbook of psychology, to be brought out in two years' time. The book was published only in 1890, but it was definitive—*The Principles of Psychology*.

The intent of the *Principles* was descriptive and antimetaphysical; it marks one of the earliest attempts to treat psychology as a natural science. James conceived of the mind as being subject to both Darwinian evolutionary principles and to acts of the will. Consciousness exists for practical results, and its characteristics are conditioned by such results; it flows—"the stream of consciousness" is one of James's many felicitous phrases—and the perception of a fact is represented as a brief halt in the flow. An innovation is James's recognition of the significance of transitive as well as substantive processes; he includes the fringe areas of thought, dimly if at all perceived, as "the free water of consciousness." He further treated of the will, defining it as the relation of the mind to concepts, or attention, and described pathological states of mind, drawing on the work of the European psychologists Charcot, Janet, and Binet. (That James was working along the same lines as European scientists is further shown by the James-Lange theory of the physiological bases of the emotions, formulated at about this time, independently and almost simultaneously, by James and the Danish physiologist C. G. Lange.) The *Principles* was an immediate success, and an abridgment of the original two-volume work, the *Briefer Course*, was published in 1892.

James's next book, *The Will to Believe and Other Essays on Popular Philosophy* (1897), contains his dedication to C. S. Peirce, "To whose philosophic comradeship in old times and to whose writings in more recent years I owe more incitement and help than I can express or repay." The first four essays are concerned with what James called "the legitimacy of religious faith," while others take up determinism, the moral life, great men (including a discussion of their place in Darwinian theory), individuality, Hegel, and psychic research (James was a member of an association for that purpose). To these religious arguments he added, in 1898, the Ingersoll lecture, given at Harvard, *Human Immortality: Two Supposed Objections to the Doctrine*, in which he held the compatibility of immortality with "our present mundane consciousness." None of these essays gives any sort of metaphysical formula; all suggest cheerfully that belief is probably not a bad thing.

In the summer of 1898 James sustained an irreparable heart lesion while on a strenuous hike in the New Hampshire mountains. He continued to philosophize and write, however: *Talks to Teachers on Psychology: and to Students on Some of Life's Ideals* was published in 1899, while 1902 saw the publication of his major work of descriptive psychology, *The Varieties of Religious Experience*, being the Gifford lectures on natural religion delivered at the University of Edinburgh. In the *Varieties*, James approached the religious impulse in man largely through individual documents, presenting a full panoply of its forms. In a postscript he posited the necessity of such pluralism, and set out a brief statement of the pragmatic value of religion. Although the book contains no notable synthesis, its wit and style give it a special place in American letters.

In 1906 James lectured at Stanford University for a half term (a tenure that was cut short by the San Francisco earthquake, which largely destroyed the campus). In 1907 he gave the Lowell Institute lectures, choosing as his subject "Pragmatism," the theory with which his name is most closely linked. These lectures gave a system to ideas apparent in all of his previously published work and were themselves published in 1907 as *Pragmatism: A New Name for Some Old Ways of Thinking*. James gave credit for the invention of pragmatism as an entity to Peirce, although it may more accurately be said to have grown out of their association in the Metaphysical Club that they had founded in Cambridge in the 1870's. James extended Peirce's notion of pragmatism and, indeed, refashioned it. Peirce was concerned with practical results as an empirical tool; James moved them into the moral realm of the good and the true. Thus, he was able to define good as the plurality of practical results beneficial to conduct and could state that a theory is true insofar as it "works" (thereby leaving his own theory open to the ready criticism that it is self-justifying). He insisted that the same flexibility must be granted to metaphysics. Such extensions would seem to have appalled Peirce, but James's book became startlingly popular and influential in the United States, perhaps because of its essential Americanness.

James resigned from all teaching duties at Harvard in 1907. In 1908 he gave the Hibbert lectures at Manchester College, Oxford, which were collected as *A Pluralist Universe* (1909). These, in effect, develop the idea of a multiplicity of standards of truth and rationality that is suggested in the postscript to *The Varieties of Religious Experience*. He died at his summer house in New Hampshire, leaving incomplete *Some Problems of Philosophy: A Beginning of an Introduction to Philosophy*, which work nevertheless contains some important formulations of his ideas, in particular those regarding perception. Another especially significant work, the essay "Does Consciousness Exist?," was also published posthumously. In it, James speculates on a single primal material, which he calls "pure experience." The essay was published in *Essays in Radical Empiricism*, a term James had invented and used in the preface of *The Will to Believe*.

James is buried in Cambridge Cemetery, next to his novelist brother Henry—with whom his lifelong relationship had been complex, mutually and advantageously critical, affectionate, and epistolary—and near his novelist friend William Dean Howells.

BIBLIOGRAPHY

I. Original Works. Ralph Barton Perry, *Annotated Bibliography of the Writings of William James* (New York, 1920), lists more than 300 items and may be considered definitive. See also his son Henry James, ed., *The Letters of William James*, 2 vols. (Boston, 1920); and F. O. Matthiessen, *The James Family, Including Selections from the Writings of Henry James, Senior, William, Henry, and Alice James* (New York, 1947), *passim*.

II. Secondary Literature. Charming personal recollections may be found in the autobiographical sketches of Henry James, *A Small Boy and Others* (New York, 1913);

and *Notes of a Son and Brother* (New York, 1914). Although a good short treatment, especially of James as a teacher, is Lloyd Morris, *William James. The Message of a Modern Mind* (New York–London, 1950), the best formal biography remains Ralph Barton Perry, *The Thought and Character of William James*, 2 vols. (Boston, 1935).

For a brief general discussion of pragmatism, its beginnings, its influence, and James's part in it, see Philip P. Wiener, "Pragmatism," in *Dictionary of the History of Ideas* (New York, 1973), which includes a useful bibliography.

SARAH FERRELL

JAMESON, ROBERT (*b.* Leith, Scotland, 11 July 1774; *d.* Edinburgh, Scotland, 19 April 1854), *geology, natural history.*

Jameson was the third son of Thomas Jameson, a prosperous soap manufacturer, and the former Catherine Paton, daughter of a brewer. At school he showed a preference for natural history, frequently playing truant to follow his hobby and collect insects and shells. He wished to follow a maritime career but was persuaded by his father to accept an apprenticeship to a Leith surgeon, John Cheyne. He also attended classes in medicine, botany, chemistry, and natural history at Edinburgh University.The professor of natural history was John Walker, who lectured on geology and mineralogy as well as on botany, zoology, and meteorology. Jameson's enthusiasm for these subjects led to his becoming a favorite pupil, and he was soon given charge of the university museum. Edinburgh was at this time a center of geological thought, and geology and mineralogy soon became his principal interest; in consequence he gave up his post as assistant to Cheyne.

In 1793 Jameson went to London for two months, meeting prominent naturalists, visiting museums, and making lengthy notes on all he saw. The following year he spent three months in the Shetland Islands and was "zealously occupied in exploring their geology, mineralogy, zoology and botany." In 1795 he was elected to the Royal Medical Society, a student organization in Edinburgh, and a year later he read to the Society two papers on current geological topics: "Is the Volcanic Opinion of the Formation of Basaltes Founded on Truth?" and "Is the Huttonian Theory of the Earth Consistent With Fact?" In these papers he enumerated his reasons for replying in the negative to both queries, quoting Kirwan and Werner as authorities as well as describing his own observations in the Edinburgh district and the Shetlands. Jameson had probably already received direct information about Werner's theories from two students, E. F. da Camera de Bethencourt, a Portuguese, and A. Deriabin, a Russian. They had been at the Bergakademie at Freiberg, Saxony, in 1792–1793, and subsequently visited Edinburgh and became Jameson's friends.

In 1797 Jameson went to Ireland and met Kirwan; before he returned to Edinburgh he spent some time on the island of Arran. In 1798 his first book, published at Edinburgh, was *An Outline of the Mineralogy of the Shetland Islands, and of the Island of Arran.* He spent the summer of 1798 exploring the Hebrides and the Western Isles, and in 1799 he investigated the Orkneys and revisited Arran. As a result of these journeys he issued a much larger two-volume work, *Mineralogy of the Scottish Isles* (Edinburgh, 1800).

For at least five years Jameson had been advocating the theories of Werner, and in September 1800 he went to the Bergakademie to study under the master himself. He stayed over a year, returning to Scotland early in 1802. Later that year, after reading Playfair's *Illustrations of the Huttonian Theory*, he wrote several articles, published in Nicholson's *Journal of Natural Philosophy, Chemistry, and the Arts*, which expounded the Wernerian view of granite and basalt.

John Walker had been nearly blind and very ill for some years. In 1801 both Kirwan and the mineralogist Charles Hatchett had written to Sir Joseph Banks, president of the Royal Society, indicating Jameson's qualifications for the Regius chair of natural history at Edinburgh should it become vacant. With this support, it is not surprising that soon after Walker's death on 31 December 1803, Jameson was elected to the chair, which he occupied with great distinction for fifty years.

In 1804 Jameson published at Edinburgh the first volume of his *System of Mineralogy;* in 1805 the second volume appeared there, as did *A Treatise on the External Characters of Minerals* and *A Mineralogical Description of the County of Dumfries.* The third volume of the *System*, issued in 1808 and subtitled *Elements of Geognosy*, contains the first detailed account in English of Werner's geognostic theories and his classification of the rock strata.

By this time Jameson was the acknowledged leader of the Scottish Wernerians, or Neptunists; and in 1808 he and eight other scientists and laymen interested in natural history founded the Wernerian Natural History Society, which attracted many members and remained in existence for nearly fifty years, with Jameson president until his death. During this period eight volumes of memoirs were published, to which Jameson contributed over a dozen papers on geological and mineralogical topics, as well as a few on botany and zoology.

In 1819 Jameson and David Brewster founded the

Edinburgh Philosophical Journal, and from 1824 to 1854 Jameson was sole editor. The *Journal* was highly esteemed, and many leading men of science contributed to it. Jameson also edited and provided notes for translations of Cuvier's *Essay on the Theory of the Earth* and Buch's *Travels Through Norway and Lapland* (London, 1813), and for an Edinburgh edition of Wilson and Bonaparte's *American Ornithology*, as well as other works.

Jameson took every opportunity to increase the university museum collections, which were very small when he became professor. During his fifty years' tenure, he instigated many direct purchases of collections by the university, and also successfully urged his former students journeying abroad to bring back specimens. By 1852 there were over 74,000 zoological and geological specimens, and in Great Britain the natural history collection was second only to that of the British Museum. Shortly after Jameson's death it was transferred by the university to the crown and became part of what is now the Royal Scottish Museum.

Although Jameson made no considerable direct contributions to geology, either in theory or in fieldwork, many of his field observations are still of interest. But his interpretations of various rock junctions in terms of Wernerian concepts present a strangely unreal picture to a modern geologist, just as the modern chemist finds it difficult to understand the chemical ideas of the phlogistic period. Nevertheless, Jameson earned a place in the history of geology by his influence on the progress of geology and of natural history in general, both through his teaching and through the manner in which he undoubtedly inspired a large number of naturalists and naturalist travelers. His lectures may have been dull, but Robert Christison, a student in 1816, wrote:

> The lectures were numerously attended in spite of a dry manner, and although attendance on Natural History was not enforced for any University honour or for any profession, the popularity of his subject, his earnestness as a lecturer, his enthusiasm as an investigator, and the great museum he had collected for illustrating his teaching, were together the cause of his success (*The Life of Sir Robert Christison* [London–Edinburgh, 1885–1886]).

Edward Forbes, his successor in the chair of natural history and one of Jameson's most distinguished pupils, stated:

> A large share of the best naturalists of the day received their first instruction . . . from Professor Jameson. Not even his own famous master, the eloquent and illustrious Werner, could equal him in this genesis of investigators (G. Wilson and A. Geikie, *Memoir of Edward Forbes, F.R.S.* [Cambridge–London, 1861], p. 554).

In this connection it should be noted that whereas the small mining school at Freiberg admitted only some twenty new students a year, during 1800–1820 there were over 1,500 students annually at Edinburgh; and between fifty and one hundred attended Jameson's classes each year. Hence in the first two decades of the century far more students instructed in Wernerian doctrines must have emerged from the portals of Edinburgh University than from the Bergakademie.

It seems fairly certain that Jameson gradually gave up the more controversial parts of Werner's teaching in the decade following the latter's death in 1817. Jameson's own former students, in particular Ami Boué, writing for the *Edinburgh Philosophical Journal*, must have done much to convince him that Werner's belief that basalts were of aqueous origin was no longer tenable. In 1826, for example, in an article on countries discovered by J. C. Ross and W. E. Parry (*Edinburgh Philosophical Journal,* **16**, 105), Jameson mentions "secondary trap-rocks, such as basalts" as "intimations of older volcanic action." A note "On Primitive Rocks," which appeared in the first four editions of Cuvier's *Essay* and described granite "as far as we know at present" as the oldest and first-formed of all primitive rocks, was replaced in the fifth edition (1827) by a long extract from a memoir by Mitscherlich discussing the igneous origin of mountains. There are other indications of Jameson's progressive acceptance of new ideas. His early interest in glacial phenomena has been described by G. L. Davies (*The Earth in Decay* [London, 1969]), quoting papers printed by Jameson in the *Edinburgh Philosophical Journal* in 1827 and 1836–1839.

BIBLIOGRAPHY

I. Original Works. Jameson's early essay "Is the Huttonian Theory of the Earth Consistent With Fact?" was printed in *Dissertations by Eminent Members of the Royal Medical Society* (Edinburgh, 1892), pp. 32–39. Most of Jameson's books have been mentioned above. Unless otherwise indicated, the following works were all published at Edinburgh. A 2nd ed. of *System of Mineralogy*, 3 vols., but without the *Elements of Geognosy* (the original vol. III), appeared in 1816. A 3rd ed. in 1820 was completely revised and based on a different system of classification which owed much to F. Mohs. *A Treatise on the External Characters of Minerals* (1805) appeared in a 2nd ed., considerably enlarged, in 1816, with the title *A Treatise on the External, Chemical, and Physical Characters of Minerals;*

there was a 3rd ed. in 1817. In 1821 Jameson published another work, *Manual of Mineralogy*, containing a long section entitled "Description and Arrangement of Mountain Rocks"; this included some of the material found in *Elements of Geognosy*.

Cuvier's *Discours sur les révolutions de la surface du globe*, translated by R. Kerr into English as *Essay on the Theory of the Earth*, includes Jameson's "Appendix Containing Mineralogical Notes and an Account of Cuvier's Discoveries." This appeared first in 1813; and subsequent eds., in which the notes were enlarged, came out in 1815, 1817, and 1822. The 3rd ed. also appeared with a New York imprint in 1818, and this had a long additional section by Samuel L. Mitchill, "Observations on the Geology of North America." The 5th ed. of the *Essay* (Edinburgh–London, 1827) was a much larger work, translated from a new and revised ed. by Cuvier; the notes by Jameson are considerably revised. Since Kerr died in 1813, the new translation may have been prepared by Jameson himself.

A list of Jameson's scientific papers is given in Royal Society, *Catalogue of Scientific Papers (1800–1863)*, **3** (1869), 531–532.

The library of Edinburgh University has a collection of Jameson's MSS and lecture notes.

II. Secondary Literature. A biographical memoir by Jameson's nephew Laurence Jameson, in *Edinburgh New Philosophical Journal*, **57** (1854), 1–49, provides the fullest contemporary account. Further information can be found in V. A. Eyles, "Robert Jameson and the Royal Scottish Museum," in *Discovery* (Apr. 1954), pp. 155–162; and J. Ritchie, "A Double Centenary. Robert Jameson and Edward Forbes," in *Proceedings of the Royal Society of Edinburgh*, **66B** (1956), 29–58. An account of his ancestors and relations is given by Jessie M. Sweet in "Robert Jameson and Shetland: A Family History," in *Scottish Genealogist*, **16** (1969), 1–18; the same author has published excerpts from Jameson's MS journals: "Robert Jameson in London, 1793," in *Annals of Science*, **19** (1963), 81–116; and "Robert Jameson's Irish Journal, 1797," *ibid.*, **23** (1967), 97–126. His earliest papers are discussed in J. M. Sweet and C. D. Waterston, "Robert Jameson's Approach to the Wernerian Theory of the Earth, 1796," *ibid.*, 81–95; and the Wernerian Natural History Society in Edinburgh has been described by J. M. Sweet in "Abraham Gottlob Werner Gedenkschrift," *Freiberger Forschungshefte*, **223C** (1967), 205–218. Interesting sidelights on Jameson's teaching can be found in *Scottish Universities Commission of 1826 and 1830, Evidence*, I (London, 1837), 613–617, 632–637, and *passim*.

There are various portraits of Jameson, some reproduced by Ritchie (1956) and Sweet (1963, 1967); a fine bust, executed when Jameson was seventy-one, is in the library of Edinburgh University.

Joan M. Eyles

JAMSHĪD IBN MAḤMŪD AL-KĀSHĪ. See **Al-Kāshī.**

JANISZEWSKI, ZYGMUNT (*b.* Warsaw, Poland, 12 June 1888; *d.* Lvov, Poland [now U.S.S.R.], 3 January 1920), *mathematics.*

Janiszewski founded, with Stefan Mazurkiewicz and Wacław Sierpiński, the contemporary Polish school of mathematics and its well-known organ *Fundamenta mathematicae*, devoted to set theory and allied fields (topology, foundations of mathematics, and other areas).

Janiszewski's father Czesław, a licentiate of the University of Warsaw and a financier by profession, was director of the Société du Crédit Municipal in Warsaw; his mother was Julia Szulc-Chojnicka. He completed his secondary education in his native city in 1907 and immediately began studying mathematics at Zurich. There, along with several of his colleagues, including Stefan Straszewicz, he organized a group of Polish students. He continued his studies in Munich, Göttingen, and Paris. Among his professors were the mathematicians Burkhardt, H. K. Brunn, Hilbert, H. Minkowski, Zermelo, Goursat, Hadamard, Lebesgue, Picard, and Poincaré, and the philosophers Foerster, Bergson, and Durkheim.

Janiszewski received a doctorate from the Sorbonne in 1911 for his thesis on a topic proposed by Lebesgue. The bold notions that he introduced in it and the results it contained became an important part of set theory (see, for example, F. Hausdorff, *Mengenlehre*, 2nd ed. [Berlin–Leipzig, 1927]). Beginning in the same year he taught mathematics at the Société des Cours des Sciences, which had replaced the Polish university in Warsaw, banned by the czarist regime. In 1913 he obtained the *agrégation* in mathematics from the University of Lvov, where until World War I he lectured on the theory of analytic functions and functional calculus.

At the outbreak of the war Janiszewski enlisted in the legion fighting for Polish independence. A soldier in the artillery, he participated in the costly winter campaign (1914–1915) in the Carpathians. A year later, refusing with a substantial part of the legion to swear allegiance to the Central Powers, he took refuge under the pseudonym of Zygmunt Wicherkiewicz at Boiska, near Zwoleń, and at Ewin, near Włoszczowa. At Ewin he directed a refuge for homeless children, which he founded and supported. In 1918, when the University of Warsaw, which had again become Polish, offered him a chair in mathematics, he began to engage in notable scientific, teaching, and editorial activities. But these were suddenly cut short by his death two years later at Lvov, following a brief illness.

For Janiszewski teaching was a mission and the student a comrade, and his attitude was shared by the other mathematicians of the Polish school. In order to better prepare his courses, he took up residence in a small isolated house in Klarysew, near Warsaw. By applying mathematical logic, he wished methodically to unmask the defects and confusions in the structure of fundamental mathematical concepts. His first research works (1910–1912) dealt with the concepts of arc, curve, and surface, which had not yet been defined precisely. In 1912, in a communication to the International Congress of Mathematicians in Cambridge, England, he sketched the first construction of a curve without arcs (that is, without homeomorphic images of the segment of a straight line).

Three topological theorems are especially associated with his name:

1. If a continuum C has points in common with a set E and with the complement of this set, then each component of the set $C \cdot \bar{E}$ (where $\bar{}$ designates closure) has points in common with the boundary $Fr(E)$.

2. If a continuum is irreducible between two points and does not contain subcontinua which are nondense on it, then it is an arc. This intrinsic topological characterization of the notion of arc is due to Janiszewski.

3. In order that the sum of two continua, neither of which is a cut of the plane which contains them, be a cut of this plane, it is necessary and sufficient that their common part is not connected (that is, that it has more than one component). This theorem abridged and simplified considerably the demonstration of the Jordan curve theorem. Moreover, it constitutes the most essential part of the topological characterization of the plane—a success all the more remarkable because the problem of a topological characterization of Euclidean spaces of more than two dimensions still remains unsolved.

When Poland became independent in 1918, the Committee of the Mianowski Foundation in Warsaw, an important social institution patronizing scientific research, invited Polish scientists to give their views on the needs of the various disciplines in Poland. In his article in *Nauka polska*, the organization's yearbook, Janiszewski advocated the concentration of mathematical research in a special institution (now the Institute of Mathematics of the Polish Academy of Sciences) and the foundation of a periodical devoted solely to a single branch of mathematics having in Poland sufficiently numerous and capable practitioners; the latter criterion would assure its value and worldwide importance and, at the same time, create a favorable mathematical climate for youth. Such was the origin of *Fundamenta mathematicae*.

Through a series of articles on philosophy and the various branches of mathematics in volume I of *Poradnik dla samouków* ("Adviser for Autodidacts"), of which he was the principal author, Janiszewski exerted an enormous influence on the development of mathematics in Poland. He was aware of social problems. As a student in Paris he had been strongly influenced by Marc Sangnier, founder of the "Sillon" group, a Christian-democrat movement, and author of *Vie profonde*. Thus when chevrons were initiated in the Polish Legion, Janiszewski refused to accept this distinction for himself, contending that it introduced inequality. He donated for public education all the money he received for scientific prizes and an inheritance from his father. Before he died he willed his possessions for social works, his body for medical research, and his cranium for craniological study, desiring even to be "useful after his death."

BIBLIOGRAPHY

I. Original Works. Janiszewski's works include "Contribution à la géométrie des courbes planes générales," in *Comptes rendus hebdomadaires des séances de l'Académie des sciences*, **150** (1910), 606–609; "Sur la géométrie des lignes cantoriennes," *ibid.*, **151** (1910), 198–201; "Nowy kierunek w geometryi," in *Wiadomości matematyczne*, **14** (1910), 57–64; *Sur les continus irréductibles entre deux points* (Paris, 1911), also in *Journal de l'École polytechnique*, **16** (1912), 79–170, and *Comptes rendus hebdomadaires des séances de l'Académie des sciences*, **152** (1911), 752–755, his thesis; "Über die Begriffe 'Linie' und 'Fläche,'" in *Proceedings. International Congress of Mathematicians* (Cambridge, 1912), pp. 1–3; "Démonstration d'une propriété des continus irréductibles entre deux points," in *Bulletin de l'Académie des sciences de Cracovie* (Cracow, 1912), pp. 906–914; and "Sur les coupures du plan faites par des continus," in *Prace matematyczno-fizyczne*, **26** (1913), in Polish with French summary.

Among his articles in *Poradnik dla samouków*, I (Warsaw, 1915), are "Wstep ogólny" ("General Introduction"), 3–27; "Topologia," 387–401; and "Zakończenie" ("Conclusion"), 538–543. These have also been published, with French trans., in Janiszewski's *Oeuvres choisies* (Warsaw, 1962). See also "O realizmie i idealizmie w matematyce," in *Przegląd filozoficzny*, **19** (1916), 161–170; and "Sur les continus indécomposables," in *Fundamenta mathematicae*, **1** (1920), 210–222, written with C. Kuratowski.

II. Secondary Literature. For information on Janiszewski, see S. Dickstein, "Przemówienie ku uczczeniu Zygmunta Janiszewskiego," in *Wiadomości matematyczne*, **25** (1921), 91–98, with portrait; B. Knaster, "Zygmunt Janiszewski," *ibid.*, **74** (1960), 1–9, with portrait and bibliography; K. Kuratowski, "10 lat Instytutu Matematycznego," in *Nauka polska*, **7**, no. 3 (1959), 29–48, English trans. in *Review of the Polish Academy of Sciences*, **4**, no. 3 (1959), 16–32; H. Lebesgue, "À propos d'une nouvelle revue mathématique: *Fundamenta Mathematicae*," in *Bulletin des sciences mathématiques*, 2nd ser., **46** (1921), 1–3; and E. Marczewski, *Rozwój matematyki w Polsce*, Historia Nauki Polskiej, I (Cracow, 1948), 18–21, 33–34, 40; "Uwagi o środowisku naukowyn" ("Remarks on the Scientific Milieu"), in *Życie nauki*, no. 4 (1951), 352–370, Czech trans. in *Časopis pro pěstování matematiky a fysiky*, **78** (1953), 31–45; and "Zygmunt Janiszewski," in *Polski Słownik Biograficzny*, X (Warsaw, 1962–1964), 527–530.

See also obituaries by J. Ryglówna, "Dr. Zygmunt Janiszewski," in *Dziennik Ludowy*, no. 7 (1920); W. Sierpiński, "Śp. profesor Zygmunt Janiszewski," in *Kurier warszawski* (7 Jan. 1920); H. Steinhaus, "Wspomnienie pośmiertne o Zygmuncie Janiszewskim," in *Przegląd filozoficzny*, **22** (1920), 113–117, and obituary in *Fundamenta mathematicae*, **1** (1920), p. v, with list of nine works and portrait, repr. (1937), with preface, pp. iv–vi; and J. D. Tamarkin, "Twenty-five Volumes of *Fundamenta Mathematicae*," in *Bulletin of the American Mathematical Society*, **42** (1936), 300.

B. KNASTER

JANSEN, ZACHARIAS (*b*. The Hague, Netherlands, 1588; *d*. Amsterdam, Netherlands, 1628–1631), *optics*.

As besieged Antwerp was about to fall to the Spaniards, Jansen's parents fled to Middelburg. From there they often traveled to nearby fairs, where the husband plied his trade as an optician. Thus it happened that Zacharias Jansen was born in The Hague in 1588, four years before his father died. His mother taught him how to manage his father's shop in Middelburg, and on 6 November 1610 approved his marriage. His son, Johannes Sachariassen, was baptized on 25 September 1611.

Through his neighbor Willem Boreel (1591–1668), the mintmaster's son, Jansen learned how to counterfeit Spanish copper quarters. Although the nominal penalty was death and confiscation of property, Jansen was merely fined on 22 April 1613 for performing this patriotic act harmful to the former oppressors of the Dutch people. He moved to nearby Arnemuiden and escalated his counterfeiting to gold and silver coins. Condemned to death in 1618, he evaded the penalty by returning to Middelburg. There his first wife died in 1624, and in the following year he remarried. Sued for nonpayment of the interest on his mortgage, he leased a house in Amsterdam. His failure to meet the installment due on 1 May 1628 was followed by bankruptcy and auction of his property. On 17 April 1632 his son's marriage banns described the bridegroom as fatherless.

Although twice convicted for counterfeiting, Jansen never pretended that he invented the telescope. Long after his death that false claim was made by his son on 30 January 1655, when the Middelburg authorities were taking testimony about the disputed invention to comply with the request of Willem Boreel, then Dutch ambassador to France. His request had been prompted by Pierre Borel, who was writing a book about the true inventor of the telescope.

In order to assert his father's priority over all claimants, Johannes Sachariassen lyingly testified in 1655 that Jansen invented the telescope in 1590 (at the age of two!). Why 1590? Descartes's friend Isaac Beeckman had visited Sachariassen's shop in Middelburg to improve his lens-polishing technique. In 1634, before 30 April, Sachariassen privately told Beeckman that Jansen had made the first telescope in Holland in 1604, after an Italian model marked 1590. This had presumably been brought to Middelburg by one of the many immigrant Italian craftsmen.

BIBLIOGRAPHY

The invention of the telescope was wrongly attributed to Jansen by Borel, *De vero telescopii inventore* (The Hague, 1655–1656). Jansen's actual contribution to the recognition and utilization of the telescope was discovered by Cornelis de Waard, *De uitvinding der verrekijkers* (The Hague, 1906), summarized in French by de Waard in *Ciel et terre*, **28** (1907–1908), 81–88, 117–124, English trans. by Albert van Helden. See also Antonio Favaro, "La invenzione del telescopio," in *Atti del Istituto veneto di scienze, lettere ed arti*, **66**, pt. 2 (1906–1907), 19–46; Hendrik Fredrik Wijnman, "Sacharias Jansen te Amsterdam," in *Amstelodamum*, **20** (1933), 125–126; *ibid*., **21** (1934), 82–83; André Danjon and André Couder, *Lunettes et télescopes* (Paris, 1935), pp. 592–604; *Journal tenu par Isaac Beeckman de 1604 à 1634*, de Waard, ed., 4 vols. (The Hague, 1939–1953), I, 209; II, 210, 295; III, 249, 308, 376; Gerard Doorman, *Patents for Inventions in the Netherlands During the 16th, 17th and 18th Centuries*, abridged English version (The Hague, 1942), pp. 71–72; J. H. Kruizinga, "De strijd om een Veerekijker," in *Historia*, **13** (1948), 140–144; Henry Charles King, *The History of the Telescope* (Cambridge, Mass., 1955), pp. vii, 30–33.

The portrait of Jansen in Borel was engraved by Jacob van Meurs after a drawing by Hendrick Berckman, who was in Middelburg in 1655 when the testimony concerning the invention of the telescope was being taken. Berckman drew Jansen's portrait from imagination long after the subject's death.

EDWARD ROSEN

JANSSEN, PIERRE JULES CÉSAR (*b*. Paris, France, 22 February 1824; *d*. Meudon, France, 23 December 1907), *physical astronomy, spectroscopy, photography*.

Janssen was born into a cultivated family. His

father was a musician of Belgian descent, and his maternal grandfather was the architect Paul-Guillaume Le Moyne. An accident in his early childhood left him permanently lame. He was thus kept at home and never attended school. Financial difficulties obliged him to go to work at an early age. While working for a bank from 1840 to 1848, he devoted himself to completing his education and earned the baccalaureate at the age of twenty-five.

Janssen attended the University of Paris, receiving his *licence ès sciences* in 1852. He then obtained a post as substitute teacher in a lycée. In 1857 he was sent on an official mission to study the position of the magnetic equator in Peru. He contracted a severe case of dysentery and had to return to France, where he agreed to become a tutor for the Schneider family, who owned iron and steel mills in Le Creusot.

Janssen's first scientific work was a study of the absorption of radiant heat in the mediums of the eye ("Sur l'absorption de la chaleur rayonnante obscure dans les milieux de l'œil," in *Annales de physique et de chimie*, 3rd series, **60**, 71–93). He showed that the mediums are transparent only for visual rays and that the focalization of the thermal radiation has no harmful effect on the retina because nine-tenths of the radiation is absorbed. This carefully executed work earned him a doctorate of science in 1860. The work actually had no real scientific significance; the conclusion could be expected because the absorbing mediums, being aqueous, have precisely the properties of water. Yet it is of interest because of the way in which Janssen was led to undertake it. In it he wrote:

> . . . having often had the opportunity to be present during the tapping of blast furnaces, I noticed that the radiation from the bath of molten metal . . . in no way affects the eyes; thus one can follow without fatigue the various phases of the operation if one takes the precaution of protecting the face with a mask that exposes only the eyes. This absorption by the mediums of the eye having appeared to me to be an important physiological fact, I proposed to verify and measure it by precise experiments.

Throughout his life, when a phenomenon aroused Janssen's curiosity, he immediately studied it. In the case of the blast furnace he perfected a delicate experimental device that permitted, in particular, the measurement of weak radiations by suitable adaptation of Melloni's thermopile; and in less than six months he carried out the essential portion of his study.

In October 1859, in a celebrated report, Kirchhoff demonstrated the presence of terrestrial elements in the constitution of the sun. Janssen immediately realized that the study of the radiation from the Le Creusot furnaces would never allow him to make discoveries so splendid as that of solar radiation, and he decided to direct his career toward physical astronomy.

In Paris, where in 1862 he had come to work with E. Follin of the Faculty of Medicine on the construction of an ophthalmoscope, Janssen mounted a small observatory on the flat roof of the house that his wife owned north of Montmartre. Here he began work on a problem posed by Brewster in 1833, that of the nature of certain dark bands in the solar spectrum, bands irregular in presence and most noticeable at sunrise and sunset. For this purpose Janssen constructed a spectroscope possessing a high dispersive power and furnished with a device for regulating the luminous intensity. He was able to establish in 1862 that these spectral bands resolve into rays and that their presence is permanent. On a mission to Italy in the following year he demonstrated with precision that the intensity of the rays varies in the course of the day as a function of the density of the terrestrial atmosphere traversed. The terrestrial origin of the phenomenon was demonstrated—thus Janssen proposed for them the name "telluric rays."

In 1864 Janssen moved to the Bernese Alps, at an altitude of 2,700 meters, to verify that the intensity of the telluric rays is lessened in the mountains. The weakening exceeded that expected on the basis of the decrease in density of atmosphere traversed, and Janssen attributed it partly to the dryness of the air. In order to verify his supposition he studied, at Geneva, the telluric rays of the spectrum of an artificial source (in this case a wood fire) situated twenty-one kilometers away on the shores of Lake Leman. His results confirmed the effect of humidity on the intensity of the rays. Janssen then turned to the direct study of the absorption rays of water vapor, carrying out his experiment in a gasworks near his home and using an iron tube thirty-seven meters long to hold the water vapor under a pressure of seven atmospheres. The source, illuminating gas, which ordinarily yields a continuous spectrum, furnished a spectroscopic image displaying most of the telluric rays.

Janssen had stated at the beginning of his research that the existence of telluric rays entailed the possibility of making a chemical analysis of the atmospheres of the planets. His determination of the water vapor spectrum was a major step in this direction. As early as 1867 he was able to announce the presence of water vapor in the atmosphere of Mars.

In the following year Janssen made another im-

JANSSEN

portant contribution to knowledge of solar structure. In order to observe the total eclipse of 18 August 1868, which was visible in India, Janssen went to the city of Guntur, near the Bay of Bengal. His aim was to study the solar prominences. Keeping the slit of the spectroscope on the lunar limb, he was able to observe highly luminous spectra while the sun was in eclipse. Visual observation in a finder showed that these spectra came from two great prominences. Janssen measured the position of the brightest rays: they corresponded to rays C and F of the solar spectrum, which are produced by hydrogen.

The brightness of the rays led Janssen to suspect the possibility of observing the prominences even when there was no eclipse. The next day he resumed his observation of the solar limb, admitting only the red portion of the spectrum. He ascertained first that a bright line appeared in the exact extension of a dark line of the solar spectrum, the C ray. Exploring the contour of the sun, Janssen observed the variations in the intensity of the line and the modifications in its structure. He also made other bright lines appear, all of them corresponding exactly to the dark lines of the absorption spectrum. "Thus was demonstrated the possibility of observing the lines of the prominences outside of eclipses, and of finding therein a method for studying these bodies" (*Annuaire du Bureau des longitudes* for 1869, p. 596).

From 18 August to 4 September Janssen worked on establishing maps of prominences. He continued his observations at Simla, in the Himalayas. On 25 December he wrote that the solar photosphere is surrounded

> . . . by an incandescent atmosphere, the general, if not exclusive, base of which is formed by hydrogen. . . . The atmosphere in question is low, [its] level very uneven and broken; often it does not rise above the projections of the photosphere, but the remarkable phenomenon is that it forms a continuous whole with the prominences, the composition of which is identical and which appear to be simply raised portions of it, projected and often detached in isolated clouds [*Comptes rendus hebdomadaires des séances de l'Académie des sciences*, **68** (1869), 181].

Several other astronomers worked in these areas during this period, notably Lockyer, who arrived at the same method of analyzing the prominences as Janssen did; but it was the latter who, in the course of his stay at Simla, created the first spectrohelioscope. He described the essential device of this apparatus as follows:

> . . . a metallic diaphragm, placed at the focus of the spectroscope, and pierced by a slit at the precise point where one of the bright lines of the light from the prominences appears, permits complete separation of this light from that of the photosphere, which lacks the bundle of rays of precisely this degree of refrangibility. . . . This focal slit . . . , when combined with a rotary movement imparted to the spectroscope, makes it possible to obtain the series of monochromatic images that a luminous body is capable of furnishing [*ibid.*, pp. 713–714].

This discovery facilitated daily examination of the sun.

For carrying out his investigations Janssen received official subsidies at the times of his missions. In addition, in 1865 he was appointed professor of physics at the École Spéciale d'Architecture. Yet, in France at least, he had little more than his home to use for his technical and experimental work. In 1869 the minister of education, Victor Duruy, tried to find him an observatory. Janssen had had the opportunity to make several measurements at the Paris observatory but was not able to install himself there permanently, since the director, Le Verrier, considered the establishment to be his personal property. Janssen was offered the pavilion of Breteuil in Sèvres, but the Franco-Prussian War prevented his using it.

In 1874 the French government decided to establish an observatory for physical astronomy. Janssen had the choice of two sites: Malmaison (the former residence of Empress Josephine) or Meudon, better located in terms of climate. Janssen chose Meudon and it was granted to him in 1876.

The estate at Meudon was in ruins when Janssen moved there in October 1876. He commenced repairs on the buildings and began to prepare the astronomical equipment. The offices and laboratories were lodged in the principal part of the estate, which formerly consisted of a modest chateau, stables, and other outbuildings. A separate building, the Chateau Neuf, built by Mansart in 1706, was restored and topped by an astronomical dome 18.5 meters in diameter.

Janssen had hoped he would rapidly acquire the means necessary to extend his investigations, which until then he had conducted with small instruments and improvised devices. Soon, however, there were financial difficulties. It became necessary to use funds budgeted for research for the completion of the buildings, a task that required twenty years. The staff was insufficient moreover—until 1906 there were only two astronomers. Nevertheless Janssen was able to endow the observatory with two large instruments: a double refractor of sixteen meters with a visual objective of eighty-three centimeters (the largest in Europe) and a photographic objective of sixty-two centimeters; and a telescope with an aperture of one meter and a focal length of three meters. For the spectral investigation of gases, he also set up a large laboratory with a steel tube sixty meters long, closed by thin transparent plates and capable of supporting a pressure of 200 atmospheres.

The most famous of Janssen's projects at Meudon during this period was the atlas of solar photographs. Composed of a selection of exposures made between 1876 and 1903, it summarized the history of the surface of the sun during these years. Janssen employed a photoheliograph of his own design. Its telescope was achromatized for violet radiations; and its shutter, which had a movable and variable slit, permitted exposures on the order of 1/3,000 of a second.

It was not possible to make all the solar observations at Meudon. Janssen was well aware of the advantages of observing at high altitudes. Wishing to know whether the dark rays of oxygen are entirely telluric or whether certain of them are present before the radiation reaches the earth's atmosphere, he went in October 1888 to the Mont Blanc massif, to the refuge of the Grands-Mulets at an altitude of 3,000 meters. His age and his lameness did not allow him to make the climb on foot, especially at that season. Thus he invented a conveyance to be borne by porters. It consisted of a seat fixed under a horizontal ladder: the upper part of his body emerged between two rungs in such a way that his arms were supported by the uprights. The ascent, which lasted thirteen hours, was as exhausting for Janssen as for his porters. But the instruments were installed immediately; and the observations, which he was able to make during the whole of the third day, were sufficient to provide a solution to the problem under study. The dark rays were either nonexistent or so weak that it could be deduced that they would not exist for an observer at the limit of the terrestrial atmosphere.

Encouraged by this experiment, Janssen repeated it in 1890, this time at the summit of Mont Blanc (4,800 meters). The measurements confirmed the earlier results. Despite the difficulties encountered (the caravan left Chamonix on 17 August but did not arrive at the summit until 22 August), Janssen decided to erect an observatory there for conducting studies in physical astronomy, terrestrial physics, and meteorology. By July 1891 he had gathered the necessary funds and equipment, and two years later the observatory was completed. The initial stages were completed at Meudon, where a fifteen-ton building had been set up, which was then transported to Mont Blanc in pieces. Each piece had to weigh less than thirty kilograms so that it could be carried by porter to the summit.

Although his observatory did not withstand the rigors of the weather, Janssen had set a splendid example by his energy and unfailing courage. In 1897 the annual Mont Blanc expedition set out to determine the solar constant. Janssen had broken a leg on the staircase of the large dome at Meudon and was unable to manage the climb. He nevertheless arranged to be carried on a stretcher to Chamonix in order to organize the work of his collaborators. He was then seventy-three.

The most famous instance of Janssen's adventurous spirit occurred during the Franco-Prussian War. He had planned to observe the eclipse of 22 December 1870 in Algeria. On 24 October, while Paris was under siege, he wrote to the Academy of Sciences:

> Despite the very critical circumstances . . . that prevail in our country at this moment, it does not seem that France should abdicate and renounce taking part in the observation of this important phenomenon. Despite the siege . . . an observer would be able, at an opportune moment, to head toward Algeria by the aerial route; he would carry with him only the most indispensable parts of his instruments.

A balloon, the *Volta*, was placed at his disposal, and Janssen left Paris with an assistant on 2 December. He headed west at an altitude of 2,000 meters and descended in sight of the Atlantic coast. In spite of a violent wind he succeeded in making a good landing, and his instrument cases remained intact. The weather proved to be unfavorable for the observation of the eclipse.

Janssen later profited from his experience as a balloonist to think out an aeronautic compass designed to furnish instantly direction and speed of flight by observation of the apparent movement of the ground. He never ceased to be interested in aeronautical problems, the importance of which he foresaw. Opening the International Aeronautical Congress held at Paris in 1889, he declared,

> . . . the twentieth century . . . will see the realization of great applications of aerial navigation and the terrestrial atmosphere navigated by apparatuses that will take possession of it to make a daily and systematic study of it or to establish among nations communications and relations that will take continents, seas, and oceans in their stride.

A prophet of aviation, Janssen was perhaps also the precursor of observations from outside the atmosphere. He was interested, for example, in the meteor shower of the Leonids, which appears about mid-November, a period generally unfavorable for observations in Europe. The determination of its intensity, which attains a maximum every thirty-three years, was of great interest to celestial physics. Janssen had the idea of undertaking the observations in a balloon above the cloud layer. He obtained a balloon for this exploit in 1898 and for several others in the following years (in the course of which balloon observations were also carried out abroad at his request). In 1900 he wrote, "The application of balloons to astronomical observations is destined to render to this science services whose extent is difficult to measure today" (*Comptes rendus hebdomadaires des séances de l'Académie des sciences*, **131** [1900], p. 128).

Janssen, who thought that "the photographic plate is the retina of the scientist" (*L'astronomie*, **2** [1883], 128), was one of the first to understand that a photograph can do more than record what the eye perceives: "I realized that photography ought to have distinct advantages over optical observation in bringing out effects and relationships of light that are imperceptible to sight" (*Association française pour l'avancement des sciences* [Le Havre–Paris, 1877], p. 328). He subsequently made photographs and, through the brevity of the exposures and a special method of developing, he obtained images which disclosed a new and scientifically true aspect of the solar surface. The technique of short exposures led Janssen in 1873 to conceive of a device of historical interest, the photographic revolver.

In planning for the observation of the transit of Venus, which he was to observe in Japan on 9 December 1874, Janssen decided to substitute for visual observation at the time of transit a series of photographs taken in rapid succession, which would permit him to measure the successive positions of the planet in relation to the solar limb. He ordered the construction of an apparatus consisting of three circular disks with the same axis: the first, pierced by twelve slits, served as the shutter; the second contained a window; the photographic plate, which was circular, was fixed to the third. The first two disks turned with a synchronized movement, the shutter disk continuously and the other irregularly in the intervals of time in which the window was not swept by a slit. A series of separate images laid out on a circle was thus obtained on the plate. In a general manner the apparatus provided an analysis of a motion on the basis of the sequence of its elemental aspects. Here Janssen realized one of the operations necessary for cinematography, which was invented twenty years later, and which required, besides analysis, the synthesis of images.

Even in fields in which he was not a specialist, Janssen displayed astonishing insight. In 1865, in a course designed for architects, he set forth the principles of effective illumination: "In retail stores light is squandered in the least intelligent manner; . . . instead of seeking power, . . . would it not be preferable to adapt the luminous intensity to the objects, . . . and to avoid above all those radiant points that are so fatiguing for our sight?" (*Oeuvres scientifiques*, I (1929), 111–112). At the same time he recommended the soundproofing of apartment buildings:

> Since the increasing value of space in our big cities has imposed the necessity of joining, under the same roof, a large number of families . . . ; since the rooms have diminished in area and in height in order to increase in number, and all the common walls . . . have become thinner, our dwellings present an intolerable resonance to which the promptest remedy urgently needs to be applied. On this point, almost everything remains to be done [*ibid.*, p. 110].

Janssen had acquired an international reputation from the start of his scientific work. He was elected to the Academy of Sciences in 1873 and to the Bureau of Longitudes in 1875 and was also a member of the academies of Rome, Brussels, St. Petersburg, Edinburgh, and the United States. He carried out his duties as director of the Meudon observatory until his death, which was the result of a pulmonary congestion.

It is not surprising that Janssen was able to accomplish so much that he undertook. As he himself wrote, "There are very few difficulties that cannot be surmounted by a firm will and a sufficiently thorough preparation."

BIBLIOGRAPHY

I. Original Works. Between 1859 and 1907 Janssen published approximately 350 items, almost all of which were reprinted in *Oeuvres scientifiques de Jules Janssen*, H. Dehérain, ed., 2 vols. (Paris, 1929–1930). Janssen's scientific works originally appeared mainly in the *Comptes rendus hebdomadaires des séances de l'Académie des sciences* and in the *Annuaire du Bureau des longitudes*. Janssen collected about thirty of his speeches in *Lectures académiques. Discours* (Paris, 1903). In addition, in 1896 he founded the *Annales de l'Observatoire d'astronomie physique de Paris, sis au Parc de Meudon*. The first volume, which he himself wrote, is devoted to the establishment of the Meudon observatory.

II. Secondary Literature. A list of biographies and obituaries is included in the *Oeuvres scientifiques*, II, 632–634. The most important accounts are G. Bigourdan, "J. Janssen," in *Bulletin astronomique*, **25** (1908), 49–58; A. de la Baume Pluvinel, "Jules César Janssen," in *Astrophysical Journal*, **28** (1908), 88–99; and R. Radau and H. Deslandres, "Discours prononcés aux funérailles de

Janssen," in *Annuaire du Bureau des longitudes* (1909), pp. C1–C11.

See also G. Bigourdan, H. Deslandres, Prince R. Bonaparte, C. Flammarion, P. Renard, and M. Dubuisson, *Inauguration de la statue de Jules Janssen* (Paris, 1920).

JACQUES R. LÉVY

JARS, ANTOINE GABRIEL (*b.* Lyons, France, 26 January 1732; *d.* Clermont-Ferrand, Auvergne, France, 20 August 1769), *mining engineering, metallurgy*.

The second son of Gabriel Jars and Jeanne-Marie Valioud, Jars began his studies in chemistry at the College of Lyons. After working for some years in his father's copper mines at St.-Bel and Chessy, Lyonnais, he attracted the attention of Joseph Florent, the marquis de Vallière, who arranged for him to enter the École des Ponts-et-Chaussées at Paris about 1754. There Jars designed and built a furnace to refine the Chessy ores.

While still students, Jars and Duhamel visited the lead mines of Britanny and the mines of Pontpéan and Ste.-Marie-aux-Mines in Alsace. In 1757 the French government sent them to inspect Central European mines, particularly those of Saxony and of several provinces of Austria, including Bohemia, Hungary, Tirol, Carinthia, and Styria. After two years Jars returned to Chessy, where, with the exception of a year at the coal mines of Franche-Comté, he remained until 1764. He was then sent to study the English coal mines and the manufacture and use of coke in metallurgical work.

In addition to a thorough examination of the more advanced English and Scottish technology, Jars visited lead mines, observed the preparation of white and red lead, the making of steel by cementation, and the manufacture of oil of vitriol. He was accorded unusually generous treatment by the proprietors of the establishments he visited and was honored by election to the Royal Society of Arts as a corresponding member (1765). After fifteen months Jars returned to France in September 1765. Although reports gave a most valuable account of contemporary British industrial practice, they were not published by the French government (perhaps, in the opinion of Charles Ballot, to avoid making Jars's information available to other countries).

The following year he visited the Low Countries, Germany, and Scandinavia. A correspondent of the French Royal Academy of Sciences since 1761, Jars became a member on 18 May 1768, when he shared a tie vote with Lavoisier. Soon after he toured east-central France from Champagne to Franche-Comté, with government orders to examine factory operations and advise the proprietors on methods that would bring their manufacturing "to the degree of perfection of which they are capable" (Ballot, p. 439). His success led to a similar survey of central France from Orléans to Auvergne. Unhappily, his mission was not completed; he suffered a sunstroke and died after a short illness.

In spite of his long and arduous journeys, Jars found time to experiment at St.-Bel. By applying coke to the melting of copper he demonstrated, for the first time in France, the melting of iron with coke (January 1769). A few months later he conducted the experiment again at the plant of the Wendel family at Hayange, where, although this process was not adopted immediately, ". . . the English procedures were successfully naturalized in France."

Jars, probably the first professional French metallurgist, was an important element in the French government's endeavors to bring about the modernization of industrial practices to meet the challenge offered by the drastic developments occurring in England. His early death may, indeed, have retarded the changes if only because he seems to have been the only person, until 1773, to have direct knowledge and experience of English methods, especially of using coke in the smelting of iron. The reports published by his brother between 1774 and 1781, coupled with the importation of English specialists, accelerated the change.

BIBLIOGRAPHY

I. ORIGINAL WORKS. For a compilation of Jars's works from 1757 to 1769, see *Voyages métallurgiques ou recherches et observations sur les mines et forges de fer . . .*, Gabriel Jars, ed., 3 vols. (Lyons, 1774–1781). The following were published in the *Mémoires* of the Paris Academy of Sciences: "Observations sur la circulation de l'air dans les mines" (1768), pp. 218–235; "D'un grand fourneau à raffiner le cuivre" (1769), pp. 589–606; "Procédé des Anglois pour convertir le plomb en minium" (1770), pp. 68–72; "Observations métallurgiques sur la séparation des métaux," *ibid.*, pp. 423–436, 514–525; and "Observations sur les mines en général," *ibid.*, pp. 540–557.

II. SECONDARY LITERATURE. On Jars and his work, see (listed chronologically) Grandjean de Fouchy, "Éloge de M. Jars," in *Histoires de l'Académie Royale des Sciences* (1769), p. 173; *ibid.* (1770), p. 59; Charles Ballot, *L'introduction du machinisme dans l'industrie française* (Paris–Lille, 1923), pp. 437 ff.; and Jean Chevalier, "La mission de Gabriel Jars dans les mines et les usines britanniques en 1764," in *Transactions. The Newcomen Society for the Study of the History of Engineering and Technology*, **26** (1947–1949), 57.

P. W. BISHOP

AL-JAWHARĪ, AL-ʿABBĀS IBN SAʿĪD (*fl.* Baghdad, *ca.* 830), *mathematics, astronomy.*

Al-Jawharī was one of the astronomers in the service of the ʿAbbāsid Caliph al-Ma'mūn (813–833). He participated in the astronomical observations which took place in Baghdad in 829–830 and in those which took place in Damascus in 832–833. Ibn al-Qifṭī (*d.* 1248) describes him as an expert in the art of *tasyīr* (ἄφεσις, "prorogation"), the complex astrological theory concerned with determining the length of life of individuals (Ptolemy, *Tetrabiblos* III, 10), and adds that he was in charge of (*qayyim ʿlā*) the construction of astronomical instruments. According to Ibn al-Nadīm (*fl.* 987), he worked mostly (*al-ghālib ʿalayh*) in geometry.

Ibn al-Nadīm lists two works by al-Jawharī: *Kitāb Tafsīr Kitāb Uqlīdis* ("A Commentary on Euclid's Elements") and *Kitāb al-Ashkāl allatī zādahā fi 'l-maqāla 'l-ūlā min Uqlīdis* ("Propositions Added to Book I of Euclid's Elements"). To this list Ibn al-Qifṭī adds *Kitāb al-Zīj* ("A Book of Astronomical Tables"), which, he says, was well known among astronomers, being based on the observations made in Baghdad. None of these works has survived.

Naṣīr al-Dīn al-Ṭūsī (*d.* 1274), in his work devoted to Euclid's theory of parallels, *al-Risāla 'l-shāfiya ʿan al-shakk fi 'l-khuṭūṭ al-mutawāziya*, ascribes to al-Jawharī an "Emendation of the Elements" (*Iṣlāḥ li-Kitāb al-Uṣūl*), which may be identical with the "Commentary" (*Tafsīr*) mentioned by Ibn al-Nadīm and Ibn al-Qifṭī. According to al-Ṭūsī, this work included additions by al-Jawharī to the premises and the theorems of the *Elements*, the added theorems totaling "nearly fifty propositions." From among these al-Ṭūsī quotes six propositions constituting al-Jawharī's attempt to prove Euclid's parallels postulate.

Al-Jawharī's is the earliest extant proof of the Euclidean postulate written in Arabic. As a premise (which his book included among the common notions) al-Jawharī lays down a rather curious version of the so-called Eudoxus-Archimedes axiom: If from the longer of two unequal lines a half is cut off, and from the [remaining] half another half is cut off, and so on many times; and if to the shorter line an equal line is added, and to the sum a line equal to it is added, and so on many times, there will remain of the halves of the longer line a line shorter than the multiples (*aḍʿāf*) of the shorter line. The axiom, which in different forms became a common feature of many Arabic proofs of the postulate, had already been applied in the same context in a demonstration attributed by Simplicius to an associate (*ṣāḥib*) of his named Aghānīs or Aghānyūs (Agapius [?]). This demonstration was known to mathematicians in Islam through the Arabic translation of a commentary by Simplicius on the premises of Euclid's *Elements*. The exact date of this translation is unknown, but it was available to al-Nayrīzī (*fl.* 895) and could have been made early in the ninth century.

The six propositions making up al-Jawharī's proof are the following:

(1) If a straight line falling on two straight lines makes the alternate angles equal to one another, then the two lines are parallel to one another; and if parallel to one another, then the distance from every point on one to the corresponding (*naẓīra*) point on the other is always the same, that is, the distance from the first point in the first line to the first point in the second line is the same as that from the second point in the first line to the second point in the second line, and so on.

(2) If each of two sides of any triangle is bisected and a line is drawn joining the dividing points, then the remaining side will be twice the joining line.

(3) For every angle it is possible to draw any number of bases (sing. *qāʿida*).

(4) If a line divides an angle into two parts (*bi-qismayn*) and a base to this angle is drawn at random, thereby generating a triangle, and from each of the remainders of the sides containing the angle a line is cut off equal to either side of the generated triangle, and a line is drawn joining the dividing points, then this line will cut off from the line dividing the given angle a line equal to that which is drawn from the [vertex of the] angle to the base of the generated triangle.

(5) If any angle is divided by a line into two parts and a point is marked on that line at random, then a line may be drawn from that point on both sides [of the dividing line] so as to form a base to that given angle.

(6) If from one line and on one side of it two lines are drawn at angles together less than two right angles, the two lines meet on that side.

Proposition (6) is, of course, Euclid's parallels postulate. Proposition (5) is, essentially, an attempt to prove a statement originally proposed by Simplicius, as we learn from a thirteenth-century document, a letter from ʿAlam al-Dīn Qayṣar to Naṣīr al-Dīn al-Ṭūsī, which is included in manuscripts of the latter's *al-Risāla 'l-shāfiya*. The attempted proof, which makes use of the Eudoxus-Archimedes axiom, rests on proposition (4) and ultimately depends on (1) and (2). Proposition (3), used in the deduction of (4), also formed part of Simplicius' attempted demonstration. The first part of proposition (1) is the same as Euclid I, 27, and does not depend on the parallels postulate. To prove the second part al-Jawharī takes

$HO = TQ$ on the two parallel lines cut by the transversal HT (Figure 1). The alternate angles AHT,

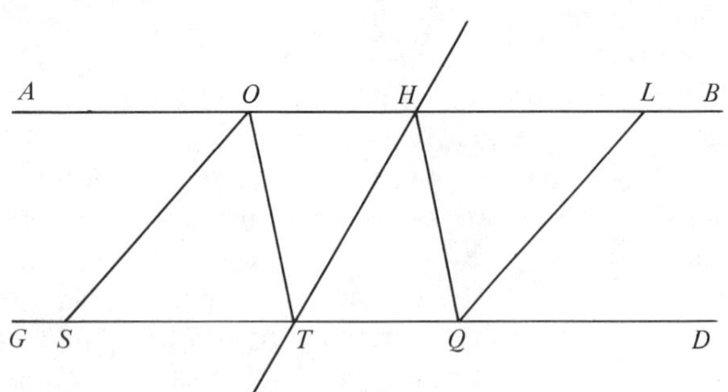

FIGURE 1

HTD being equal, it follows that the corresponding angles and sides in the triangles OHT, HTQ are equal. He then takes $HL = TS$ and similarly proves the congruence of the triangles OST, QLH, and hence the equality of the corresponding sides OS, QL. As the lines OS, QL join the extremities of the equal segments OL, SQ, they may be said to join "corresponding points" of the latter, parallel lines, and they have been shown to be equal.

As al-Ṭūsī remarked, the proof fails to establish the intended general case: it does not establish the equality of lines joining "corresponding points" on the same side of the transversal or at unequal distances on either side of the transversal, nor does it show the equality of either OS or LQ to the transversal HT itself, even if one takes $HL = TQ = OH = ST$. It is this failure which is overlooked in proving proposition (2), which, in turn, forms the basis of (4).

It seems clear that al-Jawharī took his starting point from Simplicius, although he himself appears to have been responsible for propositions (1) and (2). His attempt should therefore be grouped with those Arabic proofs clustered round Simplicius' propositions. Another proof belonging to this group was one proposed in the thirteenth century by Muḥyi 'l-Dīn al-Maghribī, and still another is the anonymous treatise on parallel lines in Istanbul MS Carullah 1502, fols. 26v-27r, dated A.H. 894 (A.D. 1488–1489).

Also extant by al-Jawharī are some "additions" (*ziyādāt*) to book V of the *Elements*. Istanbul MS Feyzullah 1359, fols. 239v-240v, dated A.H. 868 (A.D. 1464–1465), contains only a fragment consisting of three propositions taken either from a longer work on that part of Euclid's book or, probably, from al-Jawharī's comprehensive commentary on or emendation of the *Elements*. The first of these propositions "proves" Euclid's definition of proportionals (book V, def. 5), the second is the counterpart of the first, and the third is the same as Euclid's definition of "to have a greater ratio" (book V, def. 7). Further, al-Ṭūsī quotes from the "Emendation" one proposition which al-Jawharī added after Euclid I, 13: If three straight lines are drawn from any point in different directions, the three angles thus contained by the three lines are together equal to four right angles.

BIBLIOGRAPHY

Naṣīr al-Dīn al-Ṭūsī's *al-Risāla 'l-shāfiya 'an al-shakk fi 'l-khuṭūṭ al-mutawāziya*, which contains al-Jawharī's proof of Euclid's parallels postulate and the letter of 'Alam al-Dīn Qayṣar to al-Ṭūsī, is published in *Majmū' Rasā'il al-Ṭūsī*, II (Hyderabad, A.H. 1359), *Risāla* no. 8, see esp. pp. 17–26. A Russian trans. of al-Ṭūsī's *Risāla* by B. A. Rosenfeld with introduction and notes by B. A. Rosenfeld and A. P. Youschkevitch is in *Istoriko-matematicheskie issledovaniya*, **13** (1960), 475–532. MS Feyzullah 1359, containing what is left of al-Jawharī's additions to bk. V of the *Elements*, is listed in M. Krause, "Stambuler Handschriften islamischer Mathematiker," in *Quellen und Studien zur Geschichte der Mathematik, Astronomie und Physik*, Abt. B, Studien, **3** (1936), 446. See also C. Brockelmann, *Geschichte der arabischen Literatur*, Supp. I (Leiden, 1937), 382.

Brief biobibliographical notices on al-Jawharī are in Ibn al-Nadīm, *al-Fihrist*, G. Flügel, ed., I (Leipzig, 1871), 266, 272; and in Ibn al-Qifṭī, *Ta'rīkh*, J. Lippert, ed. (Leipzig, 1903), pp. 64, 219. On al-Jawharī's participation in the astronomical observations under al-Ma'mūn, see Ibn Yūnus, *al-Zīj al-Ḥākimī*, in *Notices et extraits des manuscrits de la Bibliothèque Nationale...*, VII (Paris, 1803), pp. 57, 167.

For questions of the identity of "Aghānīs," and for references to his demonstration of Euclid's postulate, see A. I. Sabra, "Thābit ibn Qurra on Euclid's Parallels Postulate," in *Journal of the Warburg and Courtauld Institutes*, **31** (1968), 12–32, esp. 13. On the proof attributed to Simplicius and the related proof by al-Maghribī, see A. I. Sabra, "Simplicius's Proof of Euclid's Parallels Postulate," in *Journal of the Warburg and Courtauld Institutes*, **32** (1969), 1–24.

A. I. SABRA

JAYASIṂHA (*b.* Amber, Rajasthan, India, 1686; *d.* Jaipur, Rajasthan, 2 October 1743), *astronomy.*

Savāī Jayasiṃha II, a Kachwāha Rajput, succeeded to the throne of Amber in 1699—not, of course, as a sovereign monarch but as a subordinate of the Mogul emperor. As maharaja he patronized a revival of Brahman culture, his most notable act in this regard being the performance of an *aśvamedha*, or "horse sacrifice," in 1742. But his most fascinating effort was the attempt to restore Indian astronomy by introducing Islamic and European scientific works and instruments into the traditional astronomy.

In pursuit of his goal Jayasiṃha had many parts of the traditional Islamic course of study of mathematical astronomy translated into Sanskrit. Naṣīr al-Dīn al-Ṭūsī in 1255 had divided this course into three parts: the first contained Euclid's *Elements*; the second, the "Middle Mathematics" or "Little Astronomy," consisting of various works by Euclid, Theodosius, Autolycus, Aristarchus, Hypsicles, Archimedes, and Menelaus; and the third, Ptolemy's *Syntaxis Mathēmatikē*. The first and third parts of this course were translated by Jagannātha: the Euclid as the *Rekhāgaṇita* shortly before 1727 and the Ptolemy as the *Siddhāntasamrāṭ* in 1732. In 1725 or 1730 Nayanasukhopādhyāya translated from Arabic into Sanskrit Theodosius' *Spherics* (*Ukara*) (the second treatise in al-Ṭūsī's "Middle Mathematics"), and it seems likely that Jayasiṃha was his patron. Jayasiṃha also patronized the preparation of a set of astronomical tables with instructions for their use in Persian, the *Zīj-i-jadīd-i Muḥammad-Shāhī*, which was probably written largely by Abū al-Khayr Khayr Allāh Khān and which Jayasiṃha dedicated to the Mogul emperor Muḥammad Shāh in 1728. (The star catalog is dated 1725–1726; the preface was written after 1734.) This *zīj* contains tables for calendars, for oblique ascensions in the seven climes, and for planetary positions; it was intended to be an improvement on the *zīj* of Ulugh Beg and the *Zīj-i Khāqānī* of al-Kāshī.

Also imitative of Ulugh Beg was Jayasiṃha's construction of five astronomical observatories—at Delhi, Jaipur (the capital city he founded in 1728), Ujjain, Benares, and Mathurā. The story related in the preface to his *zīj* that he had observations made at these observatories for seven years before publishing the *zīj* must be false—at least in the case of Jaipur, which was founded in the same year that the *zīj* was finished —and the observatories at Benares and Mathurā seem to have been built after 1734. Jayasiṃha did determine the obliquity of the ecliptic to be 23;28° (in 1729) and the latitude of Ujjain to be 23;10°, both of which values are very close to the truth; but what relationship there might be between his observations and the parameters employed in his *zīj* is not yet apparent. Jayasiṃha also used the *Tabulae Astronomicae* of Philippe de la Hire (published 1687–1702) and the *Historia Coelestis Britannica* of John Flamsteed (published 1712–1725), although his emissaries were not sent to Europe (Portugal) until 1728 or 1729, after publication of the *zīj*.

The instruments constructed for Jayasiṃha were of metal and stone. The metal instruments include astrolabes (*yantrarāja*), a graduated brass circle 17.5 feet in diameter, equatorial circles, and a declination circle (*krāntivṛttiyantra*). The masonry instruments, which are the most spectacular remains of Jayasiṃha's observatories, include huge equinoctial dials (*samrāṭyantra*), hemispherical dials (*jayaprakāśa*), azimuth instruments (*digaṃśayantra*), meridian circles (*dakṣiṇavṛttiyantra*), cylindrical dials (*rāmayantra*), fixed sextants (*ṣaṣṭāṃśayantra*), "mixed instruments" (*miśrayantra*), and zodiacal dials (*rāśiyantra*).

Jayasiṃha wrote a work in Sanskrit, the *Yantrarājaracanā*, describing the astrolabe. His other work in Sanskrit is the *Jayavinoda*, a set of tables for computing *tithis*, *nakṣatras*, and *yogas*, written in 1735.

James Tod thus concludes his account of Jayasiṃha: "Three of his wives and several concubines ascended his funeral pyre, on which science expired with him." In fact, his grand design to revitalize Indian astronomy failed. Although his *zīj* and other works and the translations of Jagannātha and Nayanasukhopādhyāya were copied, the observatories were abandoned. European advances in astronomy were ignored by Indian scientists until 1835, when the high school at Sihora in Mālwā, under the direction of Lancelot Wilkinson (the editor of many classical Sanskrit texts on astronomy and mathematics), began to teach and expound Western astronomy in Sanskrit.

(See essays on Indian science in Supplement.)

BIBLIOGRAPHY

I. Original Works. The *Yantrarājaracanā* was published in *Pandit* (Jaipur), **1**, no. 1 (1924), together with a note by A. ff. Garrett; and again with Kedāranātha's own commentary, *Yantrarājaprabhā* and the *Yantrarājaprabhā* of Śrīnātha by Kedāranātha, Rajasthan Oriental series no. 5 (Jaipur, 1953). An incomplete version of Jayasiṃha's *Jayavinoda* is described in D. Pingree, "Sanskrit Astronomical Tables in the United States," in *Transactions of the American Philosophical Society*, n.s. **58**, no. 3 (1968), 66*b*–67*a*.

II. Secondary Literature. The best treatment of Jayasiṃha's life remains James Tod, *Annals and Antiquities of Rajasthan*, 3 vols. (Oxford, 1920), vol. 3, pp. 1341–1356. Among numerous articles dealing with various aspects of his life are D. C. Sircar, "Sewai Jaysingh of Amber, A.D. 1699–1743," in *Indian Culture*, **3** (1936–1937), 376–379; P. K. Gode, "The Aśvamedha Performed by Sevai Jayasing of Amber (1699–1744 A.D.)," in *Poona Orientalist*, **2** (1937), 166–180; V. S. Bhatnagar, "The Date of Aśvamedha Performed by Sawai Jai Singh of Jaipur," in *Journal of the Bihar Research Society*, **46** (1960), 151–154; and P. D. Pathak, "A Further Evidence on Sawai Jai Singh and the New City of Jaipur Founded by Him With Reference to Buddhi-vilāsa—a Contemporary Jain Work," in *Journal of the Oriental Institute, Baroda*, **13** (1963–1964), 281–284.

An article on Jayasiṃha as an astronomer is in Ś. B. Dīkṣita, *Bhāratīya Jyotiḥśāstra* (Poona, 1896; repr. Poona, 1931), pp. 292–295. A bibliography of the unpublished *Zīj-i-jadīd-i Muḥammad-Shāhī* is in C. A. Storey, *Persian Literature*, II, pt. 1 (London, 1958), 93–94; W. Hunter's article (see below) is most informative regarding the contents of the *zīj*. But the observatories have received the most attention; on them see R. Barker, "An Account of the Brahmin's Observatory at Benares," in *Philosophical Transactions of the Royal Society*, **67** (1777), pt. 2, 598–607; J. L. Williams, "Further Particulars Respecting the Observatory at Benares," in *Philosophical Transactions of the Royal Society*, **88** (1793), pt. 1, 45–49; W. Hunter, "Some Account of the Astronomical Labours of Jayasinha, Rajah of Ambhere or Jaynagar," in *Asiatic Researches*, **5** (1799), 177–211, 424; A. ff. Garrett and C. Guleri, *The Jaipur Observatory and Its Builder* (Allahabad, 1902); S. Noti, *Land und Volk des königlichen Astronomer Dschaisingh II Maharadscha von Dschaipur* (Berlin, 1911); and A. P. Stone, "Astronomical Instruments at Calcutta, Delhi and Jaipur," in *Archives internationales d'histoire des sciences*, **42** (1958), 159–162. The standard work is G. R. Kaye, *The Astronomical Observatories of Jai Singh*, Memoirs of the Archaeological Survey of India, Imperial Series 40 (Calcutta, 1918); a short version of this is Kaye, *A Guide to the Old Observatories at Delhi; Jaipur; Ujjain; Benares* (Calcutta, 1920).

DAVID PINGREE

AL-JAYYĀNĪ, ABŪ ʿABD ALLĀH MUḤAMMAD IBN MUʿĀDH (*b.* Córdoba[?], Spain, *ca.* 989/990; *d.* after 1079), *mathematics, astronomy.*

"Jayyānī" means from Jaén, the capital of the Andalusian province of the same name. The Latin form of his name is variously rendered in the manuscripts as Abenmoat, Abumadh, Abhomadh, or Abumaad, corresponding to either Ibn Muʿādh or Abū . . . Muʿādh.

Very little is known about al-Jayyānī. Ibn Bashkuwāl (*d.* 1183) mentions a Koranic scholar of the same name who had some knowledge of Arabic philology, inheritance laws (*farḍ*), and arithmetic. Since in his treatise *Maqāla fī sharḥ al nisba* (*On Ratio*) al-Jayyānī is called *qāḍī* (judge) as well as *faqīḥ* (jurist), he is thought to be identical with this scholar, who was born in Córdoba in 989/990 and lived in Cairo from the beginning of 1012 until the end of 1017. The date of al-Jayyānī's death must be later than 1079, for he wrote a treatise ("On the Total Solar Eclipse") on an eclipse which occurred in Jaén on 1 July 1079. This means that he took the real astronomical and not the average date according to the ordinary Islamic lunar calendar (3 July 1079). In *Tabulae Jahen* he explains that the difference between these two dates may amount to as much as two days.

"On the Total Solar Eclipse" was translated into Hebrew by Samuel ben Jehuda (*fl. ca.* 1335), as was a treatise entitled "On the Dawn." A Latin translation of the latter work, *Liber de crepusculis*, was made by Gerard of Cremona. The Arabic texts of these two works are not known to be extant.

The *Liber de crepusculis*, a work dealing with the phenomena of morning and evening twilights, was for a long time attributed to Ibn al-Haytham, probably because in some manuscripts it comes immediately after his *Perspectiva* or *De aspectibus*, sometimes without any mention of the name of the author of the second work. In it al-Jayyānī gives an estimation of the angle of depression of the sun at the beginning of the morning twilight and at the end of the evening twilight, obtaining the reasonably accurate value of 18°. On the basis of this and other data he attempts to calculate the height of the atmospheric moisture responsible for the phenomena of twilights. The work found a wide interest in the Latin Middle Ages and in the Renaissance.

Liber tabularum Iahen cum regulis suis, the Latin version of the *Tabulae Jahen*, was also translated from the Arabic by Gerard of Cremona. A printed edition of the *Regulae*, lacking the tables, appeared in 1549 at Nuremberg as *Saraceni cuiusdam de Eris*. These tables were based on the tables of al-Khwārizmī, which were converted to the longitude of Jaén for the epoch of midnight, 16 July 622 (the date of the *hijra*), completed and simplified. For the daily needs of the *qāḍī* a practical handbook without much theory was sufficient. The *Tabulae Jahen* contains clear instructions for determining such things as the direction of the meridian, the time of day, especially the time and direction of prayer, the calendar, the visibility of the new moon, the prediction of eclipses, and the setting up of horoscopes. Finally al-Jayyānī deals critically with previous astrological theories. He rejects the theories of al-Khwārizmī and Ptolemy on the division of the houses and the theory of Abū Maʿshar on ray emission (ἀκτινοβολία; *emissio radiorum*); his astrological chronology refers to Hindu sources.

In the *Libros del saber* (II, 59, 309), al-Jayyānī is quoted as considering the twelve astrological houses to be of equal length. Other astronomical works by al-Jayyānī are the *Tabula residuum ascensionum ad revolutiones annorum solarium secundum Muhad Arcadius*, preserved in Latin translation (possibly a fragment of the *Tabulae Jahen*), and *Maṭraḥ shuʿāʿāt al-kawākib* ("Projection of the Rays of the Stars").

Several mathematical works by al-Jayyānī are extant in Arabic. His treatise *Kitāb majhūlāt qisiyy al-kura* ("Determination of the Magnitudes of the Arcs on the Surface of a Sphere"), which is also cited in *Saraceni cuiusdam de Eris*, is a work on spherical trigonometry.

Ibn Rushd mentions the Andalusian mathematician Ibn Muʿādh as one of those who consider the angle to be a fourth magnitude along with body, surface, and line (*Tafsīr* II, 665). Although he finds the argument not very convincing, he regards Ibn Muʿādh as a progressive and high-ranking mathematician. This Ibn Muʿādh is presumably al-Jayyānī, especially since in *On Ratio* an even more elaborate point of view is found. Here al-Jayyānī defines five magnitudes to be used in geometry: number, line, surface, angle, and solid. The un-Greek view of considering number an element of geometry is needed here because al-Jayyānī bases his definition of ratio on magnitudes.

The treatise *On Ratio* is a defense of Euclid. Al-Jayyānī, a fervent admirer of Euclid, says in his preface that it is intended "to explain what may not be clear in the fifth book of Euclid's writing to such as are not satisfied with it." The criticism of Euclid, to which al-Jayyānī objected, was a general dissatisfaction among Arabic mathematicians with Euclid V, definition 5. The cool, abstract form in which the Euclidean doctrine of proportions was presented did not appeal to the Arabic mind, since little or nothing could be deduced regarding the way in which it had come into being. So from the ninth century on, the Arabs tried either to obtain equivalent results more in accord with their own views, or to find a relation between their views and the unsatisfying theory. Those who chose the second way, such as Ibn al-Haytham, al-Khayyāmī, and al-Ṭūsī, tried to explain the Greek technique of equimultiples in terms of more basic, better-known concepts and methods.

The most successful among them was al-Jayyānī. To establish a common base he assumes that a right-thinking person has a primitive conception of ratio and proportionality. From this he derives a number of truths characteristic of proportional magnitudes, without proofs, since "There is no method to make clear what is already clear in itself." He then makes the connection by converting Euclid's multiples into parts, so that magnitudes truly proportional according to his own view also satisfy Euclid's criterion. The converse is proved by an indirect proof much resembling the one of Ibn al-Haytham, being based on the existence of a fourth proportional and the unlimited divisibility of magnitudes. In the third part al-Jayyānī deals with unequal ratios.

Al-Jayyānī shows here an understanding comparable with that of Isaac Barrow, who is customarily regarded as the first to have really understood Euclid's Book V.

BIBLIOGRAPHY

I. Original Works. Lists of MSS, which may not be complete, are in C. Brockelmann, *Geschichte der arabischen Literatur*, supp. I (Leiden, 1937), 860; and F. J. Carmody, *Arabic Astronomical and Astrological Sciences in Latin Translation* (Berkeley, 1956), p. 140. Extant MSS are "On the Total Solar Eclipse"—see Hermelink (below); "Tabula residuum ascensionum," Madrid 10023, fol. 66r—see J. M. Millás Vallicrosa, *Las traducciones orientales de la Biblioteca Catedral de Toledo* (Madrid, 1942), p. 246; and "Kitāb majhūlāt qisiyy al-kura" ("Determination of the Magnitudes of the Arcs on the Surface of a Sphere"), Escorial 955 and in a codex in the Biblioteca Medicea-Laurenziana, Or. 152—the latter also contains "Maṭraḥ shuʿāʿ [marg., shuʿāʿāt] al-kawākib" ("Projection of the Ray[s] of the Stars").

Published works are *De crepusculis* (Lisbon, 1541); *On Ratio*, with English trans., in E. B. Plooij, *Euclid's Conception of Ratio* (Rotterdam, 1950); and *Tabulae Jahen* (Nuremberg, 1549)—see H. Hermelink, "Tabulae Jahen," in *Archives for History of Exact Sciences*, **2**, no. 2 (1964), 108–112.

II. Secondary Literature. A survey of al-Jayyānī's work is in A. I. Sabra, "The Authorship of the *Liber de Crepusculis*, an Eleventh-Century Work on Atmospheric Refraction," in *Isis*, **58** (1967), 77–85. Biographical references are in M. Steinschneider, *Die hebräischen Übersetzungen des Mittelalters* (Berlin, 1893), pp. 574–575; H. Suter, *Die Mathematiker und Astronomen der Araber und ihre Werke* (Leipzig, 1900), pp. 96 and 214, *n*. 44; and "Nachträge und Berichtigungen zur *Die Mathematiker* . . .," in *Abhandlungen zur Geschichte der Mathematik*, **14** (1902), 170; and Sabra and Hermelink (see above). On the Arabic title of the *Liber de crepusculis*, see P. Kunitzsch, "Zum Liber Alfadhol eine Nachlese," in *Zeitschrift der Deutschen morgenländischen Gesellschaft*, **118** (1968), 308, *n*. 28. The conception of ratio in the Middle Ages is discussed in J. E. Murdoch, "The Medieval Language of Proportions," in A. C. Crombie, ed., *Scientific Change* (Oxford, 1961), pp. 237–272. See also Ibn Rushd, *Tafsīr Mā baʿd aṭ-ṭabiʿat*, M. Bouyges, ed., II (Beirut, 1942), 665.

YVONNE DOLD-SAMPLONIUS
HEINRICH HERMELINK

AL-JAZARĪʿ, BADĪʿAL-ZAMĀN, ABU-L-ʿIZZ ISMĀʿĪL IBN AL-RAZZĀZ (*fl. ca.* 1181–*ca.* 1206), *mechanics*.

For a complete study of his life and works, see Supplement.

JEAN. See **John.**

JEANS, JAMES HOPWOOD (*b.* Ormskirk, Lancashire, England, 11 September 1877; *d.* Dorking, Surrey, England, 16 September 1946), *physics, astronomy.*

Jeans was the son of William Tullock Jeans, a parliamentary journalist who wrote two books of lives of scientists. The family moved to London when Jeans was three. His childhood was not very happy and played an important role in forming his rather shy, apparently aloof personality. Jeans was a precocious child. His early passion was clocks, which he would dismantle, boil, and reassemble; he wrote a little booklet on clocks at the age of nine. He attended the Merchant Taylors' School from 1890 to 1896 and entered Trinity College, Cambridge, in 1896; he was second wrangler on the mathematical tripos in 1898. While recovering from a tubercular infection of the joints, Jeans took a first class on part two of the tripos in 1900 and was awarded a Smith's Prize. In the following year he was elected a fellow of Trinity College, and he obtained his M.A. in 1903. His first treatise, *Dynamical Theory of Gases,* was published in 1904. It became a standard textbook, both because of its clarity and elegance and because Jeans incorporated into it the results of his own research. "It is all a joyous adventure," wrote E. A. Milne, the astrophysicist and Jeans's biographer.

From 1905 to 1909 Jeans was professor of applied mathematics at Princeton University, where he wrote two textbooks, *Theoretical Mechanics* (1906) and *Mathematical Theory of Electricity and Magnetism* (1908). The latter work, written in Jeans's fluent style, was widely used and went through many editions. In 1907 he was elected a fellow of the Royal Society and married an American from a wealthy family, Charlotte Tiffany Mitchell. They had one daughter.

Jeans was Stokes lecturer in applied mathematics at Cambridge from 1910 to 1912, when he retired from university duties, devoting himself to research and writing. His *Report on Radiation and the Quantum Theory* appeared in 1914 and helped to spread acceptance of the early quantum theory. From this time his interest turned more exclusively to astronomy, culminating in his Adams Prize essay, *Problems of Cosmogony and Stellar Dynamics* (1919), and his book *Astronomy and Cosmogony* (1928). He was elected a secretary of the Royal Society for 1919–1929, during which time he was instrumental in improving the quality of the physical section of the *Proceedings.* He was vice-president of the Royal Society for 1938–1940, president of the Royal Astronomical Society for 1925–1927, and president of the British Association for the Advancement of Science meeting at Aberdeen in 1934. In 1923 he was made a research associate of the Mt. Wilson Observatory; from its establishment in 1935 until the year of his death he held the chair of astronomy of the Royal Institution. He was knighted in 1928. Other honors included the Royal Medal of the Royal Society (1919), the Hopkins Prize of the Cambridge Philosophical Society (for 1921–1924), the gold medal of the Royal Astronomical Society (1922), and the Franklin Medal of the Franklin Institute (1931).

In 1928 Jeans ended his career in scientific research and devoted himself to the popular exposition of science for which he became so famous. His first wife died in 1934, and in 1935 he married Suzanne Hock, a concert organist. An accomplished organist himself, Jeans had a deep interest in music, as shown by his book on musical acoustics, *Science and Music* (1938), published in the year in which he became a director of the Royal Academy of Music. The couple had three children. Jeans died of coronary thrombosis in 1946.

Jeans's biographer E. A. Milne divides his scientific life into four parts. During the first of these, from the taking of his degree to 1914, Jeans devoted his major attention to problems of molecular physics. After an initial student work, with the assistance of J. J. Thomson, on electrical discharges in gases, he turned to the foundations of kinetic molecular theory. In his attempt to provide a new derivation of the theorems of kinetic theory avoiding the assumption of "molecular disorder," he was challenged by S. H. Burbury, with whom he carried on a controversy. His first book, *The Dynamical Theory of Gases,* includes his treatment of the persistence of molecular velocities after collisions. When a molecule undergoes a collision in gas there is, statistically, a tendency for it to maintain some motion in the direction that it took before collision. If account is taken of this favoring of forward motion over rebounding, correction factors must be included in the derivation of the coefficients of viscosity, heat conduction, and diffusion of gases. His major efforts, though, were devoted to the problems posed by the classical theorem of equipartition of energy in its application to specific heats and particularly to blackbody radiation.

In 1905, in connection with this work, Jeans corrected a numerical error in Rayleigh's derivation of the classical distribution of blackbody radiation, so that the law has become known as the Rayleigh-Jeans law. This law states that if a hot body is placed inside a reflecting cavity, nearly all the heat energy will be associated with high-frequency radiation in the cavity when equilibrium is reached, the body cooling off and approaching absolute zero in temperature. But the facts, which were well presented by Planck's law, showed a reasonable distribution of energy between matter and radiation, with the highest concentrations of radiant energy associated with a finite frequency and relatively little of such energy concentrated in the high frequencies. Jeans hoped to preserve classical physical ideas, according to which the Rayleigh-Jeans law represented the true ultimate equilibrium distribution, by arguing that such a distribution would not be reached for an extremely long time under most conditions, because the usual processes generating radiation (such as collisions involving charged bodies) would produce very little high-frequency radiation at low or moderate temperatures. Jeans held that a steady-state distribution of radiant energy would quickly set in, which would not be the true equilibrium distribution because energy would be dissipated into the high frequencies at a very slow rate. He believed that Planck's law represented such a steady-state distribution.

In 1914, when Jeans presented his *Report on Radiation and the Quantum Theory* to the Physical Society, he had abandoned these ideas. This report was strongly influenced by Poincaré's important memoir of 1912, "Sur la théorie des quanta," which demonstrated the near-impossibility of circumventing the quantum hypothesis by classical arguments. Jeans constructed arguments convincing himself that Planck's law could not result as a steady-state distribution in classical physics; and in this report he stressed as sharply as possible the break which the early quantum theory represented with classical principles, and the inadequate state of the quantum theory of that time. Yet he did not contribute to the development of this theory, turning at this time almost exclusively to astrophysics.

As early as 1902–1903 Jeans had occupied himself with the forms and stability of rotating liquid masses, inspired in this by the work of George Darwin. Poincaré had traced the evolution of a rotating incompressible fluid mass slowly contracting gravitationally through ellipsoidal figures to a pear-shaped figure but was unable to decide the stability of the latter. By an incomplete argument Darwin concluded that the pear-shaped figure was stable, but in 1905 Lyapunov demonstrated the opposite. Jeans's earliest work in this field had been to compute the equilibrium figures of rotating liquid cylinders; this simplified problem allowed him to refine the calculations to a much higher degree of accuracy while still showing characteristics of the more complex three-dimensional case. He returned to the general problem in papers published in 1914–1916, demonstrating that Darwin had not gone to a sufficiently high approximation in his calculations to be able to decide the stability of the pear-shaped figure and that, if this were done, the figure was indeed shown to be unstable.

Jeans went beyond his predecessors by treating compressible as well as incompressible fluids. His results, summarized in his Adams Prize essay, *Problems of Cosmogony and Stellar Dynamics* (1919), led him to distinguish two cases, represented in their extremes, respectively, by an incompressible mass of fluid and by a gas of negligible mass surrounding a mass concentrated at its center (Roche's model). Upon contraction or, equivalently, upon attaining an increasing angular momentum, the incompressible mass underwent the evolution described above; but when the unstable pear-shaped configuration was reached, the furrow deepened and the mass split in two. In such a fashion double stars could be formed. On Roche's model, on the other hand, the gas evolved through similar ellipsoidal figures to a lenticular shape, which ejected matter from its sharp, equatorial edge, a process which Jeans associated with the formation of spiral nebulae.

Fluid masses with properties intermediate between the incompressible fluid and Roche's model behaved in either of these two ways, with the conditions for fissional or equatorial breakup well specified. As a result, Jeans concluded that rotation of a contracting mass evidently could not give rise to the formation of a planetary system. For this reason he rejected the theory of origin of the solar system of Kant and Laplace and favored a tidal theory somewhat like that of T. C. Chamberlin and F. R. Moulton, in which planetary systems were created during the close passage of two stars. Jeans had developed a formula giving the approximate distance between gravitational condensations in a gaseous medium, which he had applied to the matter in the arms of spiral nebulae. The same formula indicated that in the tidal case several small masses could be formed from the material drawn out in the cataclysm. Since such near collisions were extremely unlikely, this would mean that planetary systems were very rare. Jeans also treated a third situation (besides the rotation of a single mass and the near collision of two passing masses), the evolution of a double-star system, here drawing mainly on the classical work of Roche and George Darwin.

His masterful work on the equilibrium of rotating masses, culminating in the Adams Prize essay, constitutes the second phase of Jeans's career. After this he continued to work on astrophysical problems for another decade, until 1928. In connection with his work on rotating stars, he introduced in 1926 the concept of radiative viscosity (viscosity mediated by the action of radiation on matter) and computed its coefficient. From 1913 he applied kinetic theory arguments to the stars making up a star cluster or a galaxy. An association of stars should approach a Maxwellian distribution of velocities over a very long period of time, as a result of their mutual gravitational interactions when they pass each other at moderate distances. Jeans developed this idea mathematically and used it to attempt estimates of the ages of stellar systems. Beginning in 1917, A. S. Eddington developed a theory of the internal constitution of stars. Jeans contributed the observation that because the internal matter of the stars should be highly ionized, the mean molecular weight, which enters into Eddington's equations, should be much smaller than it would be if the atoms were not ionized. Eddington's theory, treating the radiative equilibrium of a gaseous star, gave as a result a unique relation between mass and luminosity. Jeans believed that such a unique relation was spurious because it ignored the source of stellar energy, which he concluded to be a type of radioactive process involving massive atoms and independent of the temperature of the star's interior. According to Jeans, the interior matter of these stars would progressively become more ionized and denser, causing the star to evolve from a red giant through a main sequence stage (in which most stars, including the sun, are found) to a white dwarf. With the aid of stability arguments he concluded that the material in stars could not obey the ideal gas law in their interiors, and he investigated the structure and stability of "liquid" stars, the substance of which does not behave like a gas. Much of this work on stellar interiors, based on assumptions which could not then be tested, has not held up with the passage of time. Jeans's astrophysical investigations were gathered together in *Astronomy and Cosmogony* (1928), which was practically his last research work.

From 1928, Jeans occupied himself with the popularization of science. In that year he gave a series of radio lectures which served as a source for *The Universe Around Us* (1929). By impressive analogies Jeans conveyed to his readers some idea of the immense differences in scale from the atomic nucleus to the galaxies, then proceeded to sketch his ideas concerning the evolution of stars and the universe. The Rede lecture in 1930 led to *The Mysterious Universe*, in which, after a discussion of modern physics and astronomy, he propounded his rather uncritical idealistic speculations, picturing the universe as "the thought of . . . a mathematical thinker." The book was immensely popular and appeared in at least fourteen languages. Further works followed: *The Stars in Their Courses* and *Through Space and Time*, popularizing astronomy, and *The New Background of Science*, treating modern physics, all written in Jeans's fluent and exciting style. His final books, *Physics and Philosophy* (1942) and *The Growth of Physical Science* (1947), were written in a more historical and restrained manner.

BIBLIOGRAPHY

I. Original Works. A list of Jeans's books and papers is included in Milne's biography and in his obituary notice (see below). Among the most important technical books are *The Dynamical Theory of Gases* (Cambridge, 1904); *The Mathematical Theory of Electricity and Magnetism* (Cambridge, 1908); *Problems of Cosmogony and Stellar Dynamics* (Cambridge, 1919); and *Astronomy and Cosmogony* (Cambridge, 1928). The popular books include *The Universe Around Us* (Cambridge, 1929); *The Mysterious Universe* (Cambridge, 1930); and *The Growth of Physical Science* (Cambridge, 1947).

II. Secondary Literature. See E. A. Milne, *Sir James Jeans, a Biography* (Cambridge, 1952); an obituary notice in *Obituary Notices of Fellows of the Royal Society of London*, **5** (1945–1948), 573–589; and Sydney Chapman, in *Dictionary of National Biography (1941–1950)*, pp. 430–432.

A. E. Woodruff

JEAURAT, EDME-SÉBASTIEN (*b.* Paris, France, 14 September 1724; *d.* Paris, 8 March 1803), *astronomy.*

Jeaurat's maternal grandfather was the noted engraver Sébastien Leclerc. Because his father also pursued that art form and his uncle was a painter, it was appropriate that Jeaurat's early years were spent in painting and engraving (he won a medal for drawing from the Academy of Painting), although he also studied mathematics under Lieutaud. His subsequent efforts in mapmaking, including work under Cassini de Thury, and in preparing a treatise aimed at rendering artistic perspective more rigorously geometric soon led him into scientific pursuits, a professorship in mathematics at the École Militaire in Paris, and astronomical studies.

The most important of his first works was concerned with perfecting planetary tables, especially those of Jupiter and Saturn, by comparing recorded oppositions with tabular predictions. In 1760 he pierced the roof of an attic at the École Militaire and began his own observations of those phenomena. He became a member of the Paris Academy of Sciences in 1763; and the results of these efforts were presented to that body, first as memoirs and then, in 1766, as completed tables of the motions of Jupiter.

Although a more permanent and better-equipped observatory was established at the École Militaire in 1769 under his direction, Jeaurat went to live at the royal observatory in 1770. Some of his subsequent work, such as his invention of a "double-image" telescope, was devoted to the improvement of instruments. But far more important was his assumption of the editing of the *Connaissance des temps*. The twelve volumes that he prepared, for the years 1776–1787, were enriched by a number of tables and catalogs: reimpressions, such as J. T. Mayer's catalog of zodical stars in the volume for 1778; original efforts by others, such as Messier's list of nebulae (1783); and his own contributions, such as his calculations of the principal towns, cities, and landmarks of France according to the operations forming the basis of Cassini's map (1787).

Attainment of pensioner rank in 1784 led Jeaurat to abandon his work on the *Connaissance des temps*. Indeed, although he later served the Academy as vice-director (1791) and director (1792), his subsequent contributions consisted largely of protests that his work was being slighted or unjustly criticized.

Partly because of this record of jealousy but also because he had risen within the Academy as a geometer while pursuing almost solely astronomical work, Jeaurat was not among the original members of the First Class of the Institut de France and was not elected to its astronomical section until late in 1796. Neither this recognition nor his regaining of lost lodgings at the observatory, pitiably and unsuccessfully requested many times between 1793 and 1796, restored him to the status of contributing scientist.

BIBLIOGRAPHY

I. ORIGINAL WORKS. Since there exist several discrepancies in the list of Jeaurat's works as recorded in such standard references as Poggendorff, Quérard, and even the catalog of the Bibliothèque Nationale, only those works mentioned by Jeaurat himself in the *Indication succincte des travaux scientifiques publiés à Paris . . . par le citoyen Edme-Sébastien Jeaurat . . .* are indicated here. Nevertheless, minor problems remain, because late in his career he brought out at least three such *Indications* or *Notices* in support of various appeals—such as for a "national recompense" and for membership in both the Institut National and the Bureau des Longitudes; these lesser questions, involving only his contributions to the Academy's *Mémoires* and not his separately published works, have been satisfactorily resolved.

Jeaurat's first work was his *Traité de perspective à l'usage des artistes . . .* (Paris, 1750). Prior to his election to the Academy, he presented several memoirs there, four of which were printed in *Mémoires de mathématique et de physique, présentés à l'Académie royale des sciences . . .*, **4** (1763): "Observations de la comète de 1682, 1607 et 1531, faites en mai 1759," pp. 182–187; "Projection géométrique des éclipses de soleil, assujétie aux règles de la perspective ordinaire," pp. 318–335; "Mémoire sur le mouvement des planètes, et moyen de calculer leur équation du centre pour un temps donné," pp. 524–540; and "Détermination directe de la distance d'une planète au soleil de sa parallaxe et de son diamètre horizontal pour un temps donné," pp. 601–611. His 1761 observation of the transit of Venus was published in the Academy's regular *Mémoires* (1762), pp. 570–577, perhaps because the diagrams accompanying it were placed on the same plate as those bearing upon Lalande's simultaneous observation.

His presentations of the next several years were concerned primarily with oppositions of Saturn and, particularly, Jupiter, although those from 1769, 1772, and 1779 also included his observations, respectively, of the Venus transit and solar eclipse of 1769, of Venus in its greatest elongation in 1772, and a determination of the position of 64 stars of the Pleiades. See the *Mémoires*, as follows: (1763), 85–120, 241–251, 252–259; (1765), 376–388, 435–438; (1766), 100–119; (1767), 252–255, 266–267, 340–342, 484–486; (1768), 91–92; (1769), 147–152; (1772), pt. 2, 35–43; (1779), 505–525. The results of the earlier parts of these investigations were brought together in *Essai sur la théorie des satellites de Jupiter, avec les tables de Jupiter* (Paris, 1766).

Although Jeaurat had occasionally dealt with eclipses and occultations prior to assuming the editing of *Connaissance des temps*—for instance, *Mémoires* (1766), 407–415, 417–422, it became a major subject of his contributions to the *Mémoires* thereafter: (1776), 268–272, 438–440; (1777), 487–490; (1778), 39–43; (1781), 9–20; (1785), 229–232; (1787), 5–6; (1788), 742–746. More important were his "instrumental" memoirs: one on experimentation with various types of glass and combinations of lenses for making achromatic objectives (1770), 461–486; two on his "double-image" telescope (1779), 23–50 and (1786), 562–571; and one describing his *astéréomètre* (1779), 502–504. An observation of Uranus in opposition (1787), 1–4; and a study of the nonapplication of the aberrational correction in the calculation of transits (1786), 572–573, complete his contributions to the *Mémoires*.

Virtually all of Jeaurat's mathematical works and astronomical observations of the 1790's remain in MS. One exception is a 1793 memoir, "Méthode graphique de la trisection de l'angle, suivie de la relation des sinus, tangentes et secantes de 10°, de 20°, de 40°, de 50°, de 70°, et de 80°," submitted to the Committee of Public Instruction; Jeaurat published an extract of this work, *Relations Géométriques*, at his own expense.

II. SECONDARY LITERATURE. Although brief, the account of Jeaurat by J. M. Quérard in his *La France littéraire ou Dictionnaire bibliographique des savants . . .*, IV (Paris, 1830), 222–223, includes a complete listing of his articles in the *Mémoires* but, apparently erroneously, credits him with at least one nonexistent "separately published" work. Later short accounts, depending on Quérard, repeat that error while adding little of importance: J. F. Michaud, ed., *Biographie universelle*, XXI (Paris, 1858), 27–28; and

Niels Nielsen, *Géomètres français du dix-huitième siècle* (Paris, 1935), pp. 212–215. The best account of his astronomical work remains J. B. J. Delambre, *Histoire de l'astronomie au dix-huitième siècle* (Paris, 1826), pp. 748–755. On his first observational locales, see G. Bigourdan, "Histoire des observatoires de l'École Militaire," in *Bulletin astronomique*, **4** (1887), 497–504, and **5** (1888), 30–40.

SEYMOUR L. CHAPIN

JEFFERSON, THOMAS (*b.* Goochland [now Albemarle] County, Virginia, 13 April 1743; *d.* Albemarle County, 4 July 1826), *agriculture, botany, cartography, diplomacy, ethnology, meteorology, paleontology, surveying, technology.*

Jefferson was the elder son of Peter Jefferson, a land developer and surveyor, and of Jane Randolph, a member of a distinguished Virginia family. He was born at Shadwell, Peter Jefferson's home on the edge of western settlement, and spent part of his early life there. Seven years of his boyhood were spent at Tuckahoe, William Randolph's estate, which Peter Jefferson administered after Randolph's death.

Jefferson received his early schooling, including instruction in Latin, Greek, French, and mathematics, from the Reverend William Douglas and later from the Reverend James Maury, grandfather of Matthew Fontaine Maury. In March 1760 he entered the school of philosophy of the College of William and Mary, where he continued his interest in mathematics as well as other sciences in his course of studies with the Reverend William Small, professor of natural philosophy. Small exercised considerable influence over Jefferson's academic interests as well as the direction of his future. The warm friendship that developed between them brought Jefferson into close association with George Wythe, a lawyer, and with the lieutenant-governor of the province, Francis Fauquier. He shared many common interests, including science, with each of them.

In 1762 Jefferson left the College of William and Mary to read for the law in Wythe's law office at Williamsburg for the next five years. The close contact that he maintained with his three friends during this period was marred only by Small's return to England in 1764; and it was this association which led Jefferson into his pursuit of the sciences. He was admitted to the bar in 1767 and successfully practiced law at Williamsburg until 1769, when he was elected to the House of Burgesses. In 1770 he was appointed county lieutenant of Albemarle, and on 1 January 1772 he was married to Martha Wayles Skelton. During the next decade they had six daughters, only three of whom survived their mother, who died in 1782. Jefferson never remarried.

In 1773 Jefferson was appointed county surveyor of Albemarle County, with the right to name a deputy. This appears to have been a political appointment, for although he made a number of surveys of his own properties and those of friends during his life, there is no evidence that he practiced as a professional surveyor.

Jefferson served in the House of Burgesses until it ceased to function in 1775, and he was gradually drawn into the historic events that finally led to a call to arms. He was among those who drew up the resolves forming the provincial Committee of Correspondence, of which he was a member. Although he was unable to attend the Virginia Convention in 1774, he prepared a paper entitled "A Summary View of the Rights of British America," which was later published and widely distributed although not adopted by the Convention. He was appointed to the Continental Congress in 1775 as Peyton Randolph's alternate. Following the introduction of the resolution in the Congress by Richard Henry Lee on 7 June 1776, Jefferson was delegated with four others to draft a declaration of independence. Although it was subsequently modified by Franklin, Adams, and the Congress itself, the language of the document remained primarily Jefferson's. In September 1776 he left the Congress to enter the House of Delegates, in which he served until 1 June 1779, when he was elected governor of Virginia. He served as governor for two terms, until 1 June 1781, then resigned and retired to his estate at Monticello. The following period of frustration and sadness was marked by the invasion of Monticello by British troops under Colonel Tarleton, an investigation by the Assembly into his administration as governor, the death of his wife in 1782, and finally his fall from a horse.

During this period Jefferson compiled his *Notes on the State of Virginia* from memoranda that he had been assembling for some time in reply to a series of questions about Virginia prepared by Barbé de Marbois, French representative to the United States. In his work he included statistics and descriptions of the geography, climate, flora and fauna, topography, ethnology, population, commercial production, and other aspects of the region. He took this opportunity also to combat assumptions made by the French naturalist Buffon concerning American animals and aborigines. Jefferson presented facts and arguments to refute Buffon's conclusions and had the *Notes* printed in France in 1784–1785. An unauthorized French version was published in 1786; and the work was published in London and Philadelphia in 1788. The *Notes* received wide acclaim as the first comprehensive study of any part of the United States and as one of the most important works derived from America to that time.

In June 1783 Jefferson was elected to the Continental Congress, in which he was active for the next two years. Among his most significant contributions of this period were a proposal entitled "Notes on the Establishment of a Money Unit" and his reports on the western territory, which were subsequently used as a basis for the Northwest Ordinance of 1787.

In 1784 Jefferson was appointed with John Adams to assist Franklin in the negotiation of commerce treaties at Paris; and upon Franklin's retirement in 1785 he succeeded him as minister to France, a post he retained until late 1789. During this period he was able to fulfill a dream of his youth and tour Europe. He observed the state of the sciences and new advances in technology, noting agricultural and mechanical innovations and labor-saving devices, all of which he reported to correspondents in America and a number of which he adapted for his own use at Monticello. He reported to James Madison the new "phosphoretic matches," the invention of the Argand lamp, and various applications of steam power that had come to his attention. He envisaged steam not as the means to achieve an industrial revolution but rather as a supplementary source of power. He considered its primary application to be in navigation and for powering gristmills and small manufactures which would liberate manpower for increased agricultural pursuits.

The type of plough used by French peasants led Jefferson to design an improved moldboard, which he subsequently had constructed and tested successfully at Monticello. His moldboard achieved distinction in France and England and was widely used in America. He introduced dry rice to North Carolina and brought the olive tree and Merino sheep to America—he considered these among his major achievements. He was intrigued with the first French balloon ascensions and manufacture with interchangeable parts, and wrote of them to correspondents in America.

In 1790 Jefferson was appointed secretary of state by President Washington. He became the leader of the Republicans, a group opposed to Alexander Hamilton's policies affecting foreign policy, national banking, and other major issues. During this period he was closely involved with President Washington in the survey of the Federal Territory for a national capital at Washington. He initiated measures for the establishment of a decimal system for a standard coinage and a system of weights and measures. He was instrumental in developing a system for granting patents; and when the law authorizing the issuance of patents was passed in 1790, he served as a member of the tribunal reviewing applications and was the keeper of records of patents granted.

Elected a member of the American Philosophical Society in 1780, Jefferson served on its committee to study the Hessian fly and in 1781 was elected a councillor. He became the Society's vice-president in 1791 and in 1797 was elected its third president, succeeding David Rittenhouse. He was reelected each year until 1815, and remained a member of the Society until his death. His interest in paleontological studies developed with his acquisition and study of the fossilized bones of a ground sloth, which he named the *Megalonyx*; he presented a paper on them to the Society in 1797. The following year he read a paper on his moldboard. Both papers were subsequently published in the Society's *Transactions*.

Jefferson became involved in the development of a plan for distribution of public lands in the west that was subsequently embodied in the Land Act of 1796.

In 1797 Jefferson became vice-president; and in 1801 he became the third president of the United States, the first to be inaugurated in Washington. During his two terms in office he repeatedly sponsored governmental support of science for the common good. Following the Louisiana Purchase, which he negotiated in 1803 and which nearly doubled the national area, he supported several expeditions to explore and report on unsettled lands. In 1803 he was responsible for a survey of Mississippi by Isaac Briggs, and in the same year he launched the Lewis and Clark expedition. Jefferson was personally involved in many aspects of the preparations for the expedition, specifying scientific training for Meriwether Lewis and William Clark and providing detailed instructions for the selection of scientific equipment and its use in the field. The success of the expedition provided Jefferson and his administration with the support required to sponsor a second expedition, under Zebulon M. Pike, to explore the sources of the Mississippi River and western Louisiana. These projects, followed by other expeditions, led to the formation of the U.S. Geological Survey in 1879.

Shortly after Lewis and Clark returned, Jefferson directed his attention to a survey of American coasts, for which he submitted a recommendation to Congress in 1806. The Congress authorized the survey, which resulted in the formation of the United States Coast Survey (later the U.S. Coast and Geodetic Survey).

At the end of his second presidential term, Jefferson retired to Monticello. He dedicated himself to the improvement of education in Virginia, advocating a statewide system based on a proposal that he had initiated many years earlier. He worked to create the

University of Virginia, which was finally chartered in 1819 and which opened in 1825. Jefferson played an important role in defining the university, and his efforts were a decisive factor in its establishment. He served on the first board of visitors and was elected rector, a position which he retained until his death. He was responsible in large part for the planning of the buildings and grounds, the organization of the schools within the university, and its curriculum of studies, in which the practical sciences were emphasized.

Jefferson died at the age of eighty-three, at 12:50 P.M. on 4 July 1826, several hours before the death of his friend John Adams. He was survived by only one of his daughters, Martha Randolph, and by eleven grandchildren and their progeny. He was interred in the family burial ground at Monticello, in a grave marked by a stone obelisk inscribed with words of his own choosing: "Here was buried Thomas Jefferson Author of the Declaration of American Independence, of the Statute of Virginia for religious freedom & Father of the University of Virginia."

BIBLIOGRAPHY

I. ORIGINAL WORKS. Jefferson was the author of one book and of two papers, which were read before the American Philosophical Society and published in its *Transactions: Notes on the State of Virginia* . . . (Paris, 1782; not published until 1784–1785), French trans. by M. J. as *Observations sur la Virginie* (Paris, 1786); later eds. of English version were: (London, 1787, 1788), (Philadelphia, 1788, 1792, 1794, 1801, 1812, 1815, 1825), (Baltimore, 1800), (New York, 1801, 1804), (Newark, 1801), (Boston, 1801, 1802, 1829, 1832), (Trenton, 1803, 1812), (Richmond, 1853); "A Memoir of the Discovery of Certain Bones of an Unknown Quadruped, of the Clawed Kind, in the Western Part of Virginia," in *Transactions of the American Philosophical Society*, **4** (1799), 246–260; and "The Description of a Mould-Board of the Least Resistance and of the Easiest and Most Certain Construction," *ibid.*, 313–322.

Two volumes based on Jefferson's notes on specific subjects have also been published: Edwin M. Betts, ed., *Thomas Jefferson's Garden Book*, Memoirs of the American Philosophical Society, XXII (Philadelphia, 1944); and *Thomas Jefferson's Farm Book*, Memoirs of the American Philosophical Society, XXXV (Philadelphia, 1953).

Collected writings include Julian P. Boyd and Lyman C. Butterfield, eds., *The Papers of Thomas Jefferson*, 18 vols. (Princeton, 1950–); Paul Leicester Ford, ed., *The Works of Thomas Jefferson*, 12 vols. (New York, 1904); Andrew A. Lipscomb and Albert Ellery Bergh, eds., *Writings of Thomas Jefferson*, 20 vols. (Washington, D.C., 1903–1905); and H. A. Washington, ed., *Writings of Thomas Jefferson*, 9 vols. (Washington, D.C., 1853–1854).

II. SECONDARY LITERATURE. Works on Jefferson's scientific pursuits include Daniel J. Boorstin, *The Lost World of Thomas Jefferson* (New York, 1948); William Elerey Curtis, *The True Thomas Jefferson* (Philadelphia, 1901), ch. 12; Edwin T. Martin, *Thomas Jefferson, Scientist* (New York, 1952); and Saul K. Padover, ed., *Thomas Jefferson and the Nation's Capital* (Washington, D.C., 1946).

SILVIO A. BEDINI

JEFFREY, EDWARD CHARLES (*b.* St. Catherines, Ontario, 21 May 1866; *d.* Cambridge, Massachusetts, 19 April 1952), *botany*.

Jeffrey was the son of Andrew and Cecilia Mary Walkingshaw Jeffrey, both of Calvinistic border-Scotch ancestry. He earned his B.A. degree at the University of Toronto in 1888, and a gold medal with honors in modern languages and English. Having audited courses in biology and finding high school teaching of languages not to his liking, he returned to the university for graduate study in biology, where such study was essentially zoological. Almost immediately he obtained a three-year appointment as fellow in biology (1889–1892). Because of his interest in plants, stemming from the floristic environs of Toronto and of eastern Quebec where his family spent summers, he directed his fellowship time to wide reading in botanical literature. He later stated to his classes that no single work had had so profound an influence on him as did Darwin's *Origin of Species*.

In 1892 Jeffrey received a permanent appointment as lecturer at Toronto, with the suggestion that he build a program in botany comparable to that in zoology. During his ten years in this lectureship he decided on his future program of study—the evolutionary history and sequence of vascular plants in geological time and their interrelationships. His Darwin-motivated intent included not only the assembling of available knowledge but also the exploitation of comparative morphology and anatomy for new evolutionary evidence. He devised technical methods which enabled him to make thin microscopic sections of refractory plant materials such as wood and fossilized remains.

In addition to beginning work, on his own initiative, on a Ph.D. thesis problem, he developed some of his most important work while at Toronto. The series of original papers published between 1899 and 1905 established Jeffrey's reputation. An example of his quickly acquired maturity in comparative or evolutionary morphology is his reclassification of vascular plants as a whole in 1899 into the Lycopsida and the Pteropsida; this change from the classical system won worldwide acceptance and, with little alteration, has withstood the test of increased knowledge of fossil as well as of living plants.

Taking a leave of absence for a year, Jeffrey completed his Ph.D. in botany at Harvard in 1899. In 1901 he married Jennette Atwater Street of Toronto and a year later accepted an assistant professorship in vegetable histology at Harvard. From 1907 until his retirement in 1933 he was professor of plant morphology, and as emeritus professor he continued daily use of his laboratory until shortly before his death in 1952.

In the early years of the twentieth century, indomitable will and firm convictions were needed to face the controversies brought on by the rediscovery of Mendelism and the interpretation of genetic change in the causal interpretation of evolution. Jeffrey's Scottish background, his absolute faith in the doctrine of evolution, his love of battle, his skill in writing, and his vigorous health enabled him to maintain the same direction and intensity of effort throughout his life.

Two facets of Jeffrey's program deserve special comment: his cytological studies, unfinished at his death, and his studies on coal, published in final form in 1925. The former have been questioned for their putative evolutionary mechanisms and their lack of genetic checks. But in regard to the latter studies, although geologists have added to his concept of a single origin of coal, they have supported his demonstration of its vegetable origin.

BIBLIOGRAPHY

A complete list of Jeffrey's publications, with a biographical sketch and a photographic portrait, appeared in *Phytomorphology*, **3** (1953), 127–132; other bibliographic information is in Ralph H. Wetmore, "Edward Charles Jeffrey," in *Microscope*, **8** (1953), 145–146. His 115 titles include two books: *The Anatomy of Woody Plants* (Chicago, 1917), an anatomical evolutionary overview of representatives from the different groups of vascular plants—a recognizedly important contribution; and *Coal and Civilization* (New York, 1925), a semipopular treatise on the origin and nature of coal and the rationale behind the industrial uses of different grades of coal. It was accompanied by a monograph, "The Origin and Organization of Coal," in *Memoirs of the American Academy of Arts and Sciences*, n.s. **15** (1925), 1–52, which was acclaimed by botanists and geologists alike.

Jeffrey's reputation was established early in his career by the series of publications (1899–1910) on comparative anatomy and phylogeny of the different groups of vascular plants. Of special note are papers 5, 9, 10, 15, 25, and 34 in the *Phytomorphology* bibliography mentioned above.

Jeffrey's intended third book was never completed; the MS as he left it, entitled "Chromosomes," is in the archives of the Harvard University Library. He expressed uncertainty about some of its chapters on controversial material; therefore referees decided a posthumous edition could not, in fairness to Jeffrey, be completed for publication.

RALPH H. WETMORE

JEFFREYS, JOHN GWYN (*b*. Swansea, Wales, 18 January 1809; *d*. London, England, 24 January 1885), *marine zoology*.

Jeffreys was the eldest son of John Jeffreys, a solicitor. He was articled to a solicitor at seventeen but, as his tastes were scientific rather than legal, he spent his holidays dredging from a rowboat in Swansea Bay. When only nineteen he submitted "A Synopsis of the Testaceous Pneumonobranchous Mollusca of Great Britain" to the Linnean Society and was elected a fellow the following year. In 1840 Jeffreys married Anne Nevill, and in the same year he was elected fellow of the Royal Society and received an honorary LL.D. from St. Andrews University.

Jeffreys practiced as a solicitor until 1856, when he was called to the bar at Lincoln's Inn. Although he could spare only short holidays, each summer from 1861 to 1868 was spent dredging, mostly to the north and west of Scotland. In 1866 he retired from the legal profession to devote all his time to the study of the European Mollusca. His discoveries early led him to suspect that the present-day malacofauna is directly descended from that of the late Tertiary deposits, as he found many crag mollusks, formerly supposed to be extinct, still living in the seas around Shetland and the Hebrides. After publishing numerous short papers on the results of his explorations, Jeffreys brought out a five-volume systematic treatise, *British Conchology*, which remains a standard work of reference on the subject to this day.

In 1843 Edward Forbes had postulated that no life would be found in the sea below a limit of about 300 fathoms. This was still generally believed in the 1860's, although by then enough evidence had already come to light to have made the hypothesis no longer tenable. In 1868 a successful haul had been made from 650 fathoms during the experimental cruise of H.M.S. *Lightning*, demonstrating the possibility of exploring depths rather greater than the 200 fathoms to which most previous dredging had been confined. That cruise also cast doubt on the then current belief that the temperature of seawater is a constant 4°C. below a certain depth. In 1869 and 1870 the Admiralty survey ship *Porcupine* was made available for further oceanographic investigations, and Jeffreys was given charge of the scientific work on two of her cruises.

A great number of new species, especially of mollusks, were collected, with others previously known only as Tertiary fossils. The dredge was successfully worked to a maximum depth of 2,435 fathoms and life was found to be present at all levels. The existence of cold-water and warm-water areas in close proximity at similar depths was also confirmed. Thus two prevalent ideas—the azoic zone and the universal minimum temperature—were shown conclusively to be false. From a conchological point of view, the cruises of the *Lightning* and *Porcupine* yielded considerably more material than the subsequent and much more extensive voyage of the *Challenger*; and the mollusks obtained occupied Jeffreys for the rest of his life. In 1875 he superintended the deep-sea explorations of H.M.S. *Valorous*, which accompanied the Arctic expedition of Captain Sir George S. Nares as far as Baffin Bay; and in 1880, by invitation of the French government, he took part in dredging the deep water of the Bay of Biscay on board the *Travailleur*. This was Jeffreys' last active participation in marine research. In 1884 he was one of the founders of the Marine Biological Association of the United Kingdom.

Jeffreys appreciated more than any conchologist before him the necessity for careful comparison with good series and actual specimens of types. For this purpose he visited all the principal European collections and added extensively to his personal collection by exchange and purchase. This collection was unrivaled for British mollusks and also contained a very extensive series of Mediterranean, Scandinavian, and Arctic species. His exact knowledge of recent European mollusks made his opinions on those of the late Tertiary deposits of particular value, and the latter too were well represented in his collection. Jeffreys' collection was intended for the British Museum but, as a result of a disagreement with those in authority there, it was sold to the Smithsonian Institution for a thousand guineas a few years before Jeffreys' death.

BIBLIOGRAPHY

I. ORIGINAL WORKS. Jeffreys' major work is *British Conchology, or an Account of the Mollusca Which Now Inhabit the British Isles and the Surrounding Seas*, 5 vols. (London, 1862–1869), vol. I (only) repr. 1904. More than 100 papers on European mollusks are listed in the Royal Society's *Catalogue of Scientific Papers*, **3** (1869), 541–542; **8** (1879), 20; **10** (1894), 332–333; **12** (1902), 366; **16** (1918), 93, of which the most important are "Preliminary Report of the Scientific Exploration of the Deep Sea in H. M. Surveying-Vessel '*Porcupine*' During the Summer of 1869," in *Proceedings of the Royal Society*, **18** (1870), 397–492, written with W. B. Carpenter and C. Wyville Thomson; "New and Peculiar Mollusca . . . Procured in the '*Valorous*' Expedition," in *Annals and Magazine of Natural History*, 4th ser., **18** (1876), 424–436, 490–499, and **19** (1877), 153–158, 231–243, 317–339; "On the Mollusca Procured During the '*Lightning*' and '*Porcupine*' Expeditions, 1868–70," in *Proceedings of the Zoological Society of London* (1878), 393–416; (1879), 553–588; (1881), 693–724, 922–952; (1882), 656–687; (1883), 88–115; (1884), 111–149, 341–372; (1885), 27–63; 5 further parts were published by E. R. Sykes in *Proceedings of the Malacological Society of London*, **6** (1904), 23–40; **6** (1905), 322–332; **7** (1906), 173–190; **9** (1911), 331–348; **16** (1925), 181–193.

II. SECONDARY LITERATURE. The most detailed obituary of Jeffreys is probably W. B. Carpenter, in "Obituary Notices of Fellows Deceased," in *Proceedings of the Royal Society*, **38** (1885), xiv–xviii. Numerous others are listed in the Royal Society's *Catalogue of Scientific Papers*, XVI, p. 93. A most readable account of the general results of the dredging cruises of the *Lightning* and *Porcupine* is C. Wyville Thomson, *The Depths of the Sea* (London, 1873). The contents of Jeffreys' collection are detailed in a letter to W. H. Dall published in *Smithsonian Miscellaneous Collections*, **104**, no. 15 (1946), 9.

DAVID HEPPELL

JEFFRIES, ZAY (*b*. Willow Lake, South Dakota, 22 April 1888; *d*. Pittsfield, Massachusetts, 21 May 1965), *industrial metallurgy*.

Jeffries graduated in mechanical engineering from the South Dakota School of Mines and Technology in 1910. His first employment was in 1911, as instructor at the Case School of Applied Science, Cleveland, Ohio. In the same year he became a consultant for the Aluminum Company of America and the General Electric Company, beginning a lifelong association with these companies. His first research was the development of new methods for the measurement of the grain size of metals and its relation to their properties. Industrial work on tungsten for electric lamp filaments led him to basic studies of secondary recrystallization and the role of inclusions, which he used as a thesis topic at Harvard (Ph.D., 1918) and which stimulated a decade of research on grain growth. In 1924–1926, with Robert S. Archer, Jeffries developed strong aluminum alloys for casting and forging, exploiting the recently discovered phenomenon of precipitation hardening. This work led them to the slip-interference theory of hardening, the first theory of metal hardening to be based realistically upon crystal structure and was the immediate precursor of dislocation theory. Highly regarded by his professional colleagues for his combination of scientific theory and industrial realities, Jeffries served on many committees; and in his last years he took more pride in his managerial recommendations that enabled others to do research than in his own scientific accomplishments. The report of the committee under his chairmanship, "Prospectus on Nucleonics" (1944), was the first comprehensive study of the probable impact of nuclear energy on industry and society and served to trigger scientists' wider concern in public affairs before the atomic bombing of Hiroshima.

BIBLIOGRAPHY

I. ORIGINAL WORKS. Jeffries' writings include "Metallography of Tungsten," in *Transactions of the American Institute of Mining and Metallurgical Engineers*, **60** (1919), 588–656; "The Slip Interference Theory of the Hardening of Metals," in *Chemical and Metallurgical Engineering*, **24** (1921), 1057–1067; *The Science of Metals* (New York, 1924), written with Robert S. Archer; *The Aluminum Industry*, 2 vols. (New York, 1930), written with J. D. Edwards and F. C. Frary; "Autobiographical Notes of a Metallurgist," in C. S. Smith, ed., *Sorby Centennial Symposium on the History of Metallurgy* (New York, 1965), pp. 109–119; and "Prospectus on Nucleonics," report of the Jeffries Committee to the Metallurgical Laboratory, Manhattan District, University of Chicago (Nov. 1944), classified material published in part in A. K. Smith, *A Peril and a Hope* (Chicago, 1965), app. A, pp. 539–559.

Many of Jeffries' professional papers have been deposited with the American Philosophical Society, Philadelphia.

II. SECONDARY LITERATURE. A biographical article by C. G. Suits and a list of publications will appear in *Biographical Memoirs. National Academy of Sciences* (in press). A full biography of Jeffries written by William Mogerman is to be published by the American Society for Metals (Cleveland, in press).

CYRIL STANLEY SMITH

JENKIN, HENRY CHARLES FLEEMING (*b.* near Dungeness, Kent, England, 25 March 1833; *d.* Edinburgh, Scotland, 12 June 1885), *engineering.*

Fleeming (pronounced "Fleming") Jenkin was the only child of Charles Jenkin, a naval officer, and the former Henrietta Camilla Jackson, a political liberal and popular novelist. He received his early schooling in Edinburgh. His family, in reduced financial circumstances, lived in Frankfurt, Paris, and Genoa during the period 1846–1851; he received the M.A. degree from the University of Genoa in the latter year.

After ten years of employment in various British engineering firms, mainly in the design and manufacture of the earliest long submarine cables (such as that under the Red Sea) and the associated cable-laying equipment, Jenkin in 1861 formed a consulting engineering partnership in London. In the same year his close friend William Thomson initiated the Committee on Electrical Standards of the British Association for the Advancement of Science, of which Jenkin was appointed reporter. Jenkin's lasting reputation in electrical science rests largely on his contributions to the work of this committee, through policy direction, participation in experiments, and the writing or editing of six reports between 1862 and 1869. Of major importance was the establishment of the ohm as an absolute unit of resistance, including the preparation of materials for construction of reliable resistance units, and the development of precision methods (0.1 percent) for resistance measurement. In 1867 he made the first absolute measurement of capacitance. Collaborators in the committee's activities included Thomson, James Clerk Maxwell, Carey Foster, Latimer Clark, and Charles Wheatstone.

Taking part in numerous cable-laying expeditions after 1861, Jenkin often shared the consultant duties with Thomson. Of his thirty-five British patents, many on cable-laying inventions were held jointly with Thomson. The patents and consulting work eventually made him financially independent.

Jenkin was a man of extremely broad interests. In a long essay written in 1867 he advanced detailed arguments—based on animal breeding experiments, genetic probabilities, and contemporary estimates of the geological time scale—for rejecting the two principal evolution mechanisms (indefinite variation and natural selection) proposed by Darwin in the first four editions of *The Origin of Species.* Jenkin asserted that a large body of available evidence dictated two conclusions opposing Darwin's views. First, the possible variations of an existing species must be considered as quite limited and "contained within a sphere of variation" centered on a norm. Second, the probability of favorable variations in a single individual becoming incorporated in a population must be slight, since such variations are infrequent and their effect is diluted by the breeding of the rest of the population. The remainder of his essay questioned Darwin's implicit assumption of the indefinite age of the earth. In the fifth (1869) edition, and in correspondence with others, Darwin acknowledged that he had modified some of his opinions substantially after reading Jenkin.

After 1876 Jenkin waged a vigorous campaign against unsanitary plumbing practices in Edinburgh and elsewhere, and he actively promoted the automated electric transport of industrial raw materials by monorail and cable car (telpherage.)

The Royal Society (London) elected Jenkin a fellow in 1865; the Royal Society of Edinburgh followed suit in 1869, and he was its vice-president in 1879. He was also a member of the Institution of Civil Engineers and held an honorary LL.D. from the University of Glasgow (1883).

BIBLIOGRAPHY

Papers Literary, Scientific, etc. by the Late Fleeming Jenkin, F.R.S., Sidney Colvin and J. A. Ewing, eds., 2 vols. (London, 1887), contains, principally, the Stevenson memoir (see below) and Jenkin's nontechnical writings. Of interest are a short note by Kelvin on Jenkin's contributions to electricity, a longer note by A. Ferguson on Jenkin's contributions to sanitary reform, a concise list of Jenkin's patents, and brief abstracts of all of his scientific and engineering papers. *Electricity and Magnetism* (London, 1873) is an elementary textbook that went through many English and foreign-language eds. Jenkin edited *Reports of the Committee on Electrical Standards* (London, 1873), which contains the six reports of the committee, a summary report by Jenkin, and Jenkin's five lectures on submarine telegraphy to the Royal Society of Arts in 1866. His arguments for rejecting Darwin's two main evolution mechanisms are in "The Origin of Species," in *North British Review*, **46**, no. 92 (June 1867), 277–318.

Robert Louis Stevenson, *Memoir of Fleeming Jenkin*, was written, out of friendship for Jenkin and his wife, as a preface for the collected papers edited by Colvin and Ewing. Its first separate appearance was an American ed. (New York, 1887). Stevenson had studied under Jenkin at Edinburgh and became a lifelong friend. This book-length biography is an unusual account of an unusual man. Its principal technical information is in Jenkin's letters written during cable-laying expeditions. It is included in most eds. of Stevenson's collected works.

A modern view of Jenkin's influence on Darwin's thought is Peter Vorzimmer, "Charles Darwin and Blending Inheritance," in *Isis*, **54** (Sept. 1963), 371–390. See also Loren Eisley, *Darwin's Century: Evolution and the Men Who Discovered It* (New York, 1958), ch. 8, which contains a summary of Jenkin's opposition to Darwin's theory of evolution and states that Mendel's work eventually proved Darwin correct.

ROBERT A. CHIPMAN

JENKINSON, JOHN WILFRED (*b.* London, England, 31 December 1871; *d.* Gallipoli, Turkey, 4 June 1915), *comparative embryology, experimental embryology.*

A pioneering experimental embryologist in England in the first part of this century, Jenkinson, through his researches and teaching, stimulated interest in what was then a relatively new field of developmental biology. He was the second son of William Wilberforce Jenkinson, a surveyor, and the former Alice Leigh Bedale. As a schoolboy at Bradfield College he was an avid botanist, and his records of a number of the plants to be seen near Bradfield were cited in George Claridge Druce's *Flora of Berkshire* (1897), which mentions a catalog that Jenkinson had made of the plants found in that vicinity.

When Jenkinson matriculated at Exeter College, Oxford, in 1890, it was with a classical scholarship; and during the next several years his studies centered on the classics, in which he attained honors, receiving his degree in *Litterae humaniores.* But he had managed to hear some lectures in biology; and in 1894 he turned to zoological studies with characteristic enthusiasm, entering University College, London, and gaining the necessary scientific background under the guidance of W. F. R. Weldon.

In zoology Jenkinson was drawn to embryology, and his study of normal comparative developmental biology led to an interest in experimental embryology. He was soon engaged in original investigation as he faced the difficulties involved in understanding growth and the causes of differentiation. Several stays in the Utrecht laboratory of the embryologist A. A. W. Hubrecht during vacation periods afforded Jenkinson further direction and facilities for researches which were published in 1900 in his first paper, on the early embryology of the mouse.

He returned to Oxford to assist in the teaching of comparative anatomy and embryology. In 1905 he added the D.Sc. (Oxon.) to the M.A. and was married to Constance Stephenson. The next year he was named university lecturer in comparative and experimental embryology, and Exeter College elected him a research fellow in 1909. But the course of his researches was not to be completed. With the outbreak of World War I he volunteered for service; sent to the Dardanelles, he was killed within days at Gallipoli.

His classical studies had given Jenkinson a broad perspective in his approach to the problems of the biologist. He particularly examined the concepts of vitalism from the Aristotelian to those of Driesch, whose neovitalism he strongly contested. Repeatedly, in his writings and lectures, he returned to the issue of vitalism, while in his own researches he experimented to clarify some of the conditions, both internal and external, determining embryological development. Both as a scientist and as a philosopher Jenkinson examined the premises on which his experimentation was based, convinced that the processes of growth and change so remarkably evident during embryogeny bore investigation and that it was possible to isolate physical and chemical factors that interacted and influenced the mechanics of development.

Among the subjects of Jenkinson's studies was the development of the mammalian placenta. Over several years, too, he sought to determine the relationship between the symmetry of the egg and that of the embryo in the frog. In his continuing investigations of the factors affecting the plane of symmetry, he realized their complexity; he demonstrated that in certain forms light and gravity might play some role. Jenkinson experimented to define the effects of various chemical agents upon the development of the embryo; using different isotonic solutions, he found that in some cases segmentation, gastrulation, or the course of formation of the medullary folds was affected, and chemical environmental factors were shown to change the rate or affect the normalcy of development.

To his contemporaries Jenkinson's work was a reference point. His texts on experimental embryology (1909) and vertebrate embryology (1913) were compendiums and are of interest for their discussions of vitalism. They present some of the views of the day on embryology and on the germ cells. Jenkinson himself, for example, considered that the nucleus played only a limited role in inheritance; he thought the chromosomes were concerned with the transmission of generic, varietal, or individual characters; he assigned the larger part in heredity to cytoplasmic factors in the ovum. His papers in *Archiv für Entwicklungsmechanik der Organismen*, *Biometrika*, and other journals focused attention on the problems of the embryologist.

BIBLIOGRAPHY

I. Original Works. Among Jenkinson's articles, defining some of the questions confronting embryology and describing his own work, are "Remarks on the Germinal Layers of Vertebrates and on the Significance of Germinal Layers in General," in *Memoirs and Proceedings of the Manchester Literary and Philosophical Society*, **50** (1906), 1–89; "On the Effect of Certain Solutions Upon the Development of the Frog's Egg," in *Archiv für Entwicklungsmechanik der Organismen*, **21** (1906), 367–460; "On the Relation Between the Symmetry of the Egg, the Symmetry of Segmentation, and the Symmetry of the Embryo in the Frog," in *Biometrika*, **7** (1909), 148–209, one of several communications on the subject over several years' work; and "On the Effect of Certain Isotonic Solutions on the Development of the Frog," in *Archiv für Entwicklungsmechanik der Organismen*, **32** (1911), 688–699. In vertebrate morphology his papers include "The Development of the Ear-Bones in the Mouse," in *Journal of Anatomy and Physiology*, **45** (1911), 305–318. Many of Jenkinson's observations and experimental results appear in his *Experimental Embryology* (Oxford, 1909) and *Vertebrate Embryology* (Oxford, 1913). *Three Lectures on Experimental Embryology* (Oxford, 1917) appeared posthumously. His views on vitalism are published as a separate essay, "Vitalism," in Charles Singer, ed., *Studies in the History and Method of Science*, I (Oxford, 1917), 59–78, and in *Hibbert Journal*, **9** (1911), 545–559; see also his essay, "Science and Metaphysics," in *Studies . . .*, II (1921), 447–471.

II. Secondary Literature. Singer provides a biography and a bibliography in *Studies . . .*, I, 57–58. A biographical note by R. R. Marett prefaces Jenkinson's *Three Lectures*, pp. xi–xvi. Obituary notices are "Dr. J. W. Jenkinson," in *Nature*, **95** (1915), 456; and "Captain J. W. Jenkinson, M.A., D.Sc.," in *Proceedings of the Royal Society*, **89B** (1917), xlii–xliii.

Gloria Robinson

JENNER, EDWARD (*b*. Berkeley, Gloucestershire, England, 17 May 1749; *d*. Berkeley, 26 January 1823), *natural history, immunology, medicine*.

Edward Jenner was the sixth and youngest child of the Reverend Stephen Jenner, rector of Rockhampton and vicar of Berkeley, a small market town in the Severn Valley. His mother was a daughter of the Reverend Henry Head, a former vicar of Berkeley. In addition to his church offices, Jenner's father owned a considerable amount of land in the vicinity of Berkeley.

In 1754, when Edward was five years old, both parents died within a few weeks of each other and he came under the guardianship of his elder brother, the Reverend Stephen Jenner, who had succeeded their father as rector of Rockhampton. Jenner's first schooling was received from the Reverend Mr. Clissold at the nearby village of Wotton-under-Edge. Later he was sent to a grammar school at Cirencester. One of his favorite boyhood activities was searching for fossils among the oolite rocks of the countryside.

In 1761 Jenner was apprenticed to Daniel Ludlow, a surgeon of Sodbury, with whom he worked until 1770, when he went to London to study anatomy and surgery under John Hunter. Hunter had just taken over the large house of his brother William in Jermyn Street, and Jenner was one of Hunter's first boarding pupils. Jenner also served as Hunter's anatomical assistant and, while with Hunter, arranged the zoological specimens brought back by Joseph Banks from the first voyage of H. M. S. *Endeavour*. In 1773 Jenner returned to Berkeley, where he lived with his elder brother and began to practice medicine.

Jenner's medical practice at Berkeley left him enough leisure time for activity in local medical societies, making observations in natural history, playing the flute, and now and then writing verse. His poetry has sometimes a simple beauty, his best poems being "Address to a Robin" and "The Signs of Rain."

Hunter continually encouraged Jenner's studies in natural history, for instance, by asking him to obtain particular specimens and to investigate temperatures of hibernating animals. Hunter incorporated many of Jenner's observations in his own papers, published in the *Philosophical Transactions of the Royal Society* and in his book *Observations on . . . the Animal Oeconomy* (London, 1786).

On 6 March 1788 Jenner married Katherine Kingscote and moved to Chantry Cottage, a comfortable Georgian country house at Berkeley, where he resided, except for intervals at London and Cheltenham, for the rest of his life. His wife, who bore him four children, died on 13 September 1815.

In 1786 Jenner wrote a paper on the breeding habits of the cuckoo and Hunter submitted it to the Royal Society. Jenner had shown that when a cuckoo's egg, laid in the nest of another bird such as the hedge sparrow, was hatched, the eggs or nestlings of the foster parent were thrown out of the nest, apparently by their own parents. Jenner had no explanation for this seemingly unnatural behavior. The paper was read before the society on 29 March 1787 and was accepted for publication in the *Philosophical Transactions*. Then, on 18 June 1787, Jenner discovered that it was the newly hatched cuckoo which ejected from the nest of its "foster parent" the hedge sparrow's own unhatched eggs and nestlings. Accordingly, Jenner withdrew his original paper before publication and revised it. On 27 December 1787 he sent the revised report to Hunter, and it was read before the Royal Society on 13 March 1788.

When Jenner began medical practice at Berkeley, he was frequently asked to inoculate persons against smallpox. Smallpox inoculation had been introduced into England early in the eighteenth century. A person in good health was inoculated with matter from smallpox pustules and was thus given what was usually a mild case of the disease in order to confer immunity against further smallpox infection. The practice was dangerous, however, since smallpox thus induced could be severe or fatal, and it tended to spread smallpox among the population.

Such inoculation was evidently not a common practice in the English countryside until about 1768, when it was improved by Robert Sutton of Debenham, Suffolk. Sutton required the patient to rest and maintain a strict diet for two weeks before inoculation. He inoculated by taking, on the point of a lancet, a very small quantity of fluid from an unripe smallpox pustule and introducing it between the outer and inner layers of the skin of the upper arm without drawing blood. He used no bandage to cover the incision.

Jenner began to inoculate against smallpox using Sutton's method, but he soon found some patients to be completely resistant to the disease. On inquiry he found that these patients had previously had cowpox, the disease which produced a characteristic eruption on the teats of milk cows and was frequently transmitted to people who milked the cows. Jenner also found that among milkmen and milkmaids it was generally believed that contraction of cowpox prevented subsequent susceptibility to smallpox, although there had apparently been instances where this had not been the case. His fellow medical practitioners in the countryside did not agree that cowpox prevented smallpox with certainty.

As early as 1780 Jenner learned that the eruptions on the teats of infected cows differed. All were called cowpox and all could be communicated to the hands of the milkers, but only one kind created a resistance to smallpox. He called this type "true cowpox." Jenner subsequently found that even true cowpox conferred immunity against smallpox only when matter was taken from the cowpox pustules before they were too old (as had been the case with Sutton's smallpox fluid). Jenner thought (mistakenly) that true cowpox was identical with a disease of the feet of horses known as "grease," and that the pox was carried from horses to cattle on the hands of milkmen who also cared for horses. He also believed at that time that the cowpox could be transmitted from person to person, serving to protect them from smallpox. But he was not able to confirm his opinions for another sixteen years.

On 14 May 1796 Jenner inoculated James Phipps, an eight-year-old boy, with matter taken from a pustule on the arm of Sarah Nelmes, a milkmaid suffering from cowpox. The boy contracted cowpox and recovered within a few days. On 1 July 1796 Jenner inoculated him with smallpox, but the inoculation produced no effect. In June 1798 Jenner published at his own expense a slender volume of seventy-five pages, *An Inquiry Into the Causes and Effects of the Variolae Vaccinae*. In this work he described twenty-three cases in which cowpox had conferred a lasting immunity to smallpox. Jenner described "grease," the disease on the heels of horses, and suggested that it could cause cowpox in cows. He showed that cowpox was transmitted to the milkmaids, giving rise to pustules on their hands and arms but not to systemic disease. They recovered after a few days of mild illness and were thereafter immune to smallpox.

To describe the matter producing cowpox Jenner introduced the term "virus," contending that the cowpox virus had to be acquired from the cow and that it gave permanent protection from smallpox. In Case IV of the *Inquiry* Jenner also describes a kind of reaction now known as anaphylaxis. In 1791 one Mary Barge, who had had cowpox many years before, was inoculated with smallpox. A pale red inflammation appeared around the inoculation site and spread extensively, but it disappeared within a few days. Jenner noted how remarkable it was that the smallpox virus should produce such inflammation more rapidly than it could produce smallpox itself. He also observed that although cowpox gave immunity to smallpox, it did not confer similar immunity to the cowpox itself.

Following the publication of Jenner's book the practice of vaccination was adopted and spread with astonishing speed. It was taken up not only by medical practitioners but also by country gentlemen, clergymen, and schoolmasters. Jenner found that lymph taken from smallpox pustules might be dried in a glass tube or quill and kept for as long as three months without losing its effectiveness. The dried vaccine could thus be sent long distances. Jean de Carro, a Swiss physician living in Vienna, introduced vaccination on the continent of Europe and was instrumental in sending vaccine virus into Italy, Germany, Poland, and Turkey. In 1801 Lord Elgin, British ambassador at Constantinople, sent vaccine virus received from de Carro overland to Bussora (Basra) on the Persian Gulf, and thence to Bombay. The marquis of Wellesley, governor general of India, actively promoted the distribution of the vaccine and many thousands of people were vaccinated in India during the next few years. In Massachusetts, Benjamin Waterhouse introduced vaccination to America with vaccine received from Jenner. Jenner also sent vaccine to President Jefferson, who vaccinated his family and his neighbors at Monticello.

After 1798 Jenner's life was taken up almost entirely by the question of vaccination. He had to provide vaccine to those who requested it, explain the details of the procedure, and defend the practice against ill-informed criticism. He had to answer first the somewhat casual criticisms of Ingen-Housz and in 1799 the more serious attack of William Woodville, head of the London Smallpox Hospital. Woodville had inadvertently inoculated his patients with smallpox when he attempted to vaccinate them, a misfortune which produced serious cases of smallpox and at least one death. Jenner wrote a number of pamphlets in defense of vaccination. He was obliged to spend extended periods of time at London and to carry on a vast correspondence. In 1802 the British Parliament voted him a grant of £10,000 in recognition of his discovery and in 1806 an additional grant of £20,000. In 1803 the Royal Jennerian Society was founded at London to promote vaccination and Jenner took a large part in its affairs; it was superseded in 1808 by a national vaccination program.

Although Great Britain and France were at war, in 1804 Napoleon had a medal struck in honor of Jenner's discovery and in 1805 he made vaccination compulsory in the French army. At Jenner's request he also released certain Englishmen who had been interned in France. In 1813 the University of Oxford awarded Jenner an honorary M.D. degree.

After his wife's death in 1815, Jenner rarely left Berkeley and never visited London. He resumed his studies in natural history and completed a paper on the migration of birds, published after his death, in which he showed that birds appeared to migrate into England in summer for the purpose of reproduction, and that the ovaries of the female and testes of the male were enlarged at that time. Jenner also served as a justice of the peace at Berkeley.

The day before his death he walked to a neighboring village, where he ordered that fuel be provided for certain poor families. In 1820 he had suffered a mild stroke, and on 25 January 1823, a severe one. He died early the next morning.

Jenner's discovery of vaccination made possible the immediate control of smallpox and the saving of untold lives. It also made possible, as Jenner realized, the ultimate eradication of smallpox as a disease, an end which is only now (1972) within sight for the whole world. Jenner must be considered the founder of immunology; in vaccination he made the first use of an attenuated virus for immunization. For his coining of the term "virus," his effort to describe the natural history of the cowpox virus, and his description of anaphylaxis, he must be considered the first pioneer of the modern science of virology.

BIBLIOGRAPHY

A complete description of Jenner's published work and a survey of sources concerning his life is provided in W. R. Lefanu, *A Bio-Bibliography of Edward Jenner* (London, 1951). All later biographical accounts of Jenner have been based on John Baron, *The Life of Edward Jenner* (London, 1827).

Jenner has also been the subject of criticism by anti-vaccinationists. E. M. Crookshank, *History and Pathology of Vaccination*, 2 vols. (London, 1889); and Charles Creighton, *Jenner and Vaccination* (London, 1889), are writings of this order. There is no modern detailed biography of Jenner and his life awaits critical reexamination. The rapid development of immunobiology gives an ever-increasing historical interest to his work.

LEONARD G. WILSON

JENNINGS, HERBERT SPENCER (*b.* Tonica, Illinois, 8 April 1868; *d.* Santa Monica, California, 14 April 1947), *zoology*.

Jennings was the son of George Nelson Jennings, a physician, and the former Olive Taft Jenks. After attending local schools he studied at Illinois Normal (now Illinois State University) and taught school near Tonica. In 1889–1890 (at the age of twenty-one and without a degree), through the influence of a former teacher he was appointed assistant professor of botany and horticulture at the Agricultural and Mechanical College of Texas. From 1890 to 1893 he attended the University of Michigan, where he met Jacob Reighard, a young zoologist and ichthyologist. After obtaining the B.S. from Michigan and spending a year in graduate study there, Jennings went on to work with Reighard's teacher, E. L. Mark, at Harvard. There he took the M.A. in 1895 and the Ph.D. in 1896, submitting a thesis on the morphogenesis of a rotifer. After finishing his doctorate Jennings held the Parker traveling fellowship, which took him to Jena in the winter of 1896–1897, where he worked with Max Verworn, a pioneer student of the behavior of protozoans. In the spring he went to the zoological station at Naples (to which he returned in 1903–1904 as one of the first scientists subsidized by the Carnegie Institution of Washington).

Upon returning to the United States, Jennings was without a position. In August 1897, he was called to the Agricultural College of Montana in Bozeman (now Montana State University) as professor of botany and horticulture. The next summer he married Mary Louise Burridge, who did many of the illustrations in his publications. He spent the academic year 1898–1899 at Dartmouth as instructor in zoology; and in 1899 he rejoined Reighard at Michigan as instructor in zoology. By 1901 he had advanced to assistant professor, but in 1903 he departed for an identical appointment at the University of Pennsylvania. Finally, in 1906, Jennings accepted an associate professorship of zoology at Johns Hopkins University. In 1907 the title was changed to professor of experimental zoology. He was named Henry Walters professor and director of the zoological laboratory in 1910. After his retirement in 1938, he became research associate at the University of California, Los Angeles. His wife died in 1937; and in 1939 he married Lulu Plant Jennings, the widow of his brother. At Johns Hopkins, Jennings was noted for his dedication to research and graduate instruction. Although he did not have a great number of students, among them were T. M. Sonneborn, William Taliaferro, and Karl Lashley.

Jennings concentrated his attention upon only two types of microorganisms, the Rotifera and the Protozoa; but his research nevertheless mirrored the changes that were taking place in the shifting mainstream of biology. His first works were descriptive and systematic; he then turned his attention to physiology and adaptation; and finally took up the question of variation and reproduction and made major contributions to genetics.

The earliest phase of Jennings' work is reflected in his publications for the Michigan State Board of Fish Commissioners and the U. S. Fish Commission, working under the direction, in each case, of Reighard, who saw to his employment during several summers. Jennings' descriptions and classifications of Rotifera, including new species, hold a respectable place in the systematic and morphological literature. At Harvard, C. B. Davenport helped kindle his enthusiasm for experimental methods; and Jennings' work with the paramecium, undertaken with Verworn at Jena in 1897, began a decade of study of the behavior of the very simple organisms. This research resulted in a number of papers and, ultimately, in a book, *Behavior of the Lower Organisms* (1906), a classic of zoology and comparative psychology.

Behavior of the Lower Organisms had an impact in three areas particularly. First, Jennings, unlike previous workers on the protozoa, studied the reactions of individual organisms rather than generalized or group behavior. He thus was able to raise questions about the specific processes and patterns involved when stimuli were followed by responses. Second, the book (and an earlier report of 1904) presented the first clear challenge, with experimental evidence, to the theory of physicochemical tropisms, of which Jacques Loeb was the chief exponent. Since most of the workers supporting the existence of tropistic behavior had utilized metazoic organisms, Jennings' demonstration in even more primitive one-celled animals of phenomena that the concept of tropistic responses could not encompass was devastating to the theory, which eventually languished.

Finally, through his book Jennings did much to bring very simple and one-celled organisms into the realm of psychology. Jennings now adapted to paramecia the use of experiment in comparative psychology, largely pioneered by E. L. Thorndike in 1898; and he asserted the identity of the basic nature of activity and reactivity in all animals, from protozoans to man. The idea was not new but Jennings' experimental evidence was, and it appeared at the time to be conclusive.

With the completion of the book, Jennings turned away from the subject of animal adaptation and functioning. Even the course on animal behavior at Johns Hopkins was turned over to his new colleague and ally in the field, S. O. Mast. Jennings now embarked on four decades of research in the recently opened field of genetics, still utilizing the very smallest animals. Although again basing his work upon characteristics of individual organisms (aggregated statistically rather than studied in groups or swarms), he devoted himself "to what happens in the passage of generations in these creatures; to a study of the biology of races rather than of individuals, to life, death, mating, generation, heredity, variation and evolution in the Protista" (*Life and Death*, p. 19).

In the course of his inquiry Jennings made numerous important contributions. He contended, chiefly against Gary Calkins, that conjugation was not necessarily essential in maintaining the vitality of strains of one-celled animals, an opinion that he reversed at the end of his life. As he systematically applied Mendelian theory, he helped to found mathematical genetics through his calculations of expectable ratios of traits in various types of inheritance. Of all of Jennings' work in genetics, however, the most momentous was that in which he investigated the questions of variation and evolution.

Between 1908 and 1916 Jennings and his students published a number of papers on the constancy and variability of protozoan lines of inheritance. He was able to show that within a given species there exist a number of distinctive strains whose traits persist over many generations—essentially the kinds of variations that Darwin had discussed originally as providing the basis for the effects of natural selection. But Jennings also observed the spontaneous development of this type of variation, usually only a very slight—but a persisting—alteration. A university publication characterized Jennings—not entirely hyperbolically—as the first scientist "actually to see and control the process of evolution among living things." This work did much to modify the theory of mutations, because the "saltations" that Jennings reported were very slight indeed and suggested that evolution must proceed gradually by very small changes.

After 1916 Jennings did much less laboratory work, instead producing a notable series of works popularizing genetics and discussing philosophical questions raised by the newer methods and discoveries in experimental biology. He discussed particularly the fundamental finding of his own lifework, that life processes are identical throughout all of the animal kingdom, for he had extended his contention from areas of reactivity to include inheritance. In the 1940's he even wrote about social phenomena among unicellular beings, a conception that grew out of the writings of W. C. Allee as well as his own renewed burst of laboratory research, an inquiry into the nature of sexuality in the paramecium.

As a youngster Jennings had at first been interested in the humanities, and this interest appeared later in his published philosophical discussions. As early as the 1910's he had written in opposition to Driesch's vitalism because, he asserted, he was afraid that biological experimentation that was purely scientific would be inhibited by vitalistic beliefs. He later wrote in a more positive vein about broader questions. For example, in the Terry lectures (published in 1933) on the relation between biology and religion, he affirmed the finality of death but also stressed the purposiveness of life and the compatibility of ethics with a strictly biological viewpoint.

Jennings' popularizations of experimental biology were very widely read and cited, but the precise influence of *The Biological Basis of Human Nature* (1930) and other writings on Western culture is very hard to determine. The impact of Jennings upon science, however, can be measured, at least in part, by his place in the literature. Experimental methods that he introduced were still in use a generation or two later. Lines of investigation in genetics that he started or fostered were exciting and productive for at least half a century. For years he was the most conspicuous figure in a new genetics developed in protozoology, complementary to but different from the classic *Drosophila* work. His results in the *Behavior of the Lower Organisms* were still fundamental in that field in the 1960's. His thorough experimental procedures and clear thinking gained him respect among his colleagues that was reflected in prizes, honorary degrees, lectureships, and membership in the American Philosophical Society, the National Academy of Sciences, and similar bodies. Jennings assisted in editing a number of major journals and, through his influence within the powerful inner circles of American science, helped to develop the best traditions of experimental work in the United States.

BIBLIOGRAPHY

I. Original Works. Two books sum up most of Jennings' contributions: *Behavior of the Lower Organisms* (New York, 1906) and *Life and Death, Heredity and Evolution in Unicellular Organisms* (Boston, 1920). He reviewed his own work in *Genetics of the Protozoa* (The Hague, 1929). The most important popularizations are *The Biological Basis of Human Nature* (New York, 1930) and *The Universe and Life* (New Haven, 1933). Other major books include *Genetic Variations in Relation to Evolution, A Critical Inquiry Into the Observed Types of Inherited Variation, in Relation to Evolutionary Change* (Princeton, 1935) and *Genetics* (New York, 1935). No complete bibliography is known to exist, but a good list can be compiled from standard bibliographical guides and citations in Jennings' works. The Jennings papers in the American Philosophical Society Library contain a number of unpublished materials, the most important of which are a diary covering 1906–1943 and a lecture, "History of Zoology in America, Partly as Reminiscences" (1929). Few letters exist in that collection, but a number may be found in the Robert M. Yerkes papers in the Yale University Medical Library. An autobiographical fragment is "Stirring Days at A. and M.," in *Southwest Review*, **31** (1946), 341–344.

II. SECONDARY LITERATURE. The best biographical sketch is T. M. Sonneborn, "Herbert Spencer Jennings, April 8, 1868–April 14, 1947," in *Genetics*, **33** (1948), 1–4. See also Samuel Wood Geiser, "Herbert Spencer Jennings, Apostle of the Scientific Spirit," in *Bios* (Iowa), **16** (1946), 3–18; *National Cyclopaedia of American Biography*, XLVII, 92–93; and Donald D. Jensen, "Foreword to the 1962 Edition," in H. S. Jennings, *Behavior of the Lower Organisms* (Bloomington, Ind., 1962), pp. ix–xvii. Newspaper obituaries are not helpful. For Jennings' work at Johns Hopkins, in addition to materials there and official university reports, see Carl Pontius Swanson, "A History of Biology at the Johns Hopkins University," in *Bios* (Iowa), **22** (1951), esp. 245–248.

J. C. BURNHAM

JENSEN, CARL OLUF (*b.* Frederiksberg, Copenhagen, Denmark, 18 March 1864; *d.* Middelfart, Denmark, 3 September 1934), *veterinary medicine.*

The son of Peter Jensen, a pipe fitter, and Dorthea Rasmusdatter, Jensen in his boyhood was avidly interested in mathematics and astronomy. In 1882, when only eighteen years old, chance led him to become a veterinary surgeon. He practiced veterinary medicine at Nimtofte in 1883, but he could not earn enough and this fact and illness compelled him to move to Copenhagen, where he began studies under the bacteriologist Bernhard Bang at the Copenhagen State Agricultural College and the physician C. J. Salomonsen. In 1886, with G. Sand, Jensen published a study concerning gas gangrene and the edema bacillus, an anaerobic form which Pasteur in 1877 had erroneously stated to be a septic vibrio.

During the autumn of 1887, Jensen accepted an opportunity to study bacteriology at Koch's institute in Berlin, and after his return he became an assistant to Bang. In 1889 he was appointed lecturer at the Royal Veterinary and Agricultural College, where he remained for the next forty-five years; in 1903 he became full professor of pathology and pathological anatomy. In 1890 he married Maria Magdalene Schmit. From 1892 to 1898 Jensen also supervised the small-animal clinic and from 1898 he lectured on serology and serotherapy. Besides developing these new disciplines, he mastered abdominal surgery in the clinic and became the first veterinarian to utilize X-ray investigations in surgery.

In 1909 he founded a serum laboratory, which in 1932 was placed officially under the Ministry of Agriculture, and he continued as its superintendent until his death. From 1898 he was a member of the Veterinarian Board of Health and from 1928 its chairman. He was appointed *Veterinaerfysicus*, the highest veterinary office in Denmark, in 1922, succeeding Bang; in 1931 his title became veterinary director, a post he held until 1933. In these official capacities he was able to make effective contributions to the defeat of foot-and-mouth disease in Denmark and toward the export of safer Danish bacon to England.

When he became assistant at the college, Jensen continued the investigations begun with Sand. Together they demonstrated in 1888 that strangles, an infectious horse disease affecting the windpipe, was the result of a special type of pyogenic streptococcus, *Streptococcus equi.* During the following years he took up the problem of defects in buttermaking which led to an unpleasant taste and poor consistency in cream and butter. Inspired by the investigations of Pasteur and Emil K. Hansen of yeast, Jensen demonstrated the necessity of milk pasteurization and of pure bacterial cultures in butter production; he showed that the noxious bacteria originated in the vessels and tools used in the purification process. In 1891 he published his book on milk and butter defects which enabled farmers to save large sums of money.

In the following year Jensen demonstrated that the bacteria causing swine erysipelas can be found in both the throat and the digestive tract of healthy animals but that with reduced somatic resistance swine contract the disease; disinfection and isolation are then ineffective. In 1892 he found that vaccination with weakened cultures was the only effective prevention. During these years Jensen also investigated infectious diarrhea in cattle and braxy (bradsot), a severe infectious disease in Icelandic sheep. He developed effective serum preparations for these diseases, as well as methods for identification and differentiation of the various bacterial types and races (1897).

With the Nobelist J. Fibiger, Jensen undertook several significant investigations concerning the relation between human and animal tuberculosis—demonstrating, in opposition to the ideas of Koch, that many children are mortally stricken by animal tuberculosis bacteria (1902–1908).

Beginning in 1901 Jensen did experimental work

with mice and rats on cancer transplantation; the studies showed that cancer cells survive through generations by transplantation. In 1910 he demonstrated a type of cancerous tumor in turnips, produced by *Bacterium tumefaciens* (described by Erwin F. Smith). The tumors could be transplanted and survive through generations of turnips not infected with the bacteria.

From 1916 to 1921 Jensen took up an endocrinological investigation, showing that the axolotl, a salamander which normally lives in the larval state and even breeds in this condition, during feeding with a thyroid substance develops to a degree never found in nature. Jensen worked out a standardized method for the effective preparation of thyroid hormone for medical use (1920).

In 1903 Jensen became a member of the Royal Danish Society of Sciences. He later received the Walker Prize (1906) for his cancer investigations and was created honorary doctor of medicine at the Copenhagen University (1910) and honorary doctor of veterinary medicine at the Berlin Veterinary College (1912). He became a corresponding or honorary member of veterinary societies all over the world. In 1928 he was made president of the Danish Cancer Committee. He founded *Maanedsskrift for Dyrlaeger* ("Monthly Review for Veterinary Surgeons") in 1885 and was its coeditor until his death. He died suddenly from apoplexy while on a vacation.

BIBLIOGRAPHY

I. ORIGINAL WORKS. "Ueber malignes Oedem beim Pferde," in *Deutsche Zeitschrift für Thiermedizin und vergleichenden Pathologie*, **12** (1886), 31–45, written with G. Sand; "Die Aetiologie der Druse," *ibid.*, **14** (1888), 437–467, written with Sand; "Die Aetiologie des Nesselfiebers und diffusen Hautnekrose des Schweines," *ibid.*, **18** (1892), 278–305; "Ueber Bradsot und derer Aetiologie," *ibid.*, **21** (1897), 249–274; "Uebertragung der Tuberculose des Menschen auf das Rind," in *Berliner klinische Wochenschrift*, **39** (1902), 881–886; **41** (1904), 129–133, 171–174; **45** (1908), 1876–1883, 1926–1936, 1977–1980, 2026–2031, written with J. Fibiger.

See also "Von echten Geschwülsten bei Pflanzen," *12 conférence internationale pour l'étude du cancer. Rapport* (Paris, 1910), pp. 254 ff.; "Om standardisering af thyreoidea-praeparater ved anvendelse af Axolotlr," in *Hospitalstidende*, **63** (1920), 505–515; *Selected Papers*, I (Copenhagen, 1948), covering his work from 1886 to 1908; and II (Copenhagen, 1964), covering 1908 to 1934.

Full catalogues of Jensen's works are printed in *Kongelige Veterinaer- og Landbohøiskole 1858–1908* (Copenhagen, 1908), pp. 555–556, listing his works published between 1883 and 1907; *Kongelige Veterinaer- og Landbohøiskoles Aarsskrift* (1919), pp. 295–296, listing works between 1908 and 1918; and *Den danske Dyrlaegestand* (Copenhagen, 1934), pp. 150–152, listing works between 1919 and 1933.

See also *Selected Papers*, II, 207–212; *Forelaesninger over maelk og maelkekontrol* (Copenhagen, 1903), pp. 214 ff.; *Grundriss der Milchkunde und Milchhygiene* (Stuttgart, 1903), pp. 235 ff., trans. and amplification by Leonard Pearson as *Essentials of Milk Hygiene* (Philadelphia–London, 1907), pp. 275 ff.

II. SECONDARY LITERATURE. See M. Christiansen, *Selected Papers*, I (Copenhagen, 1948), ix–xvi; Oluf Thomsen, "Carl Oluf Jensen," in *Acta pathologica et microbiologica scandinavica*, **18**, supp. (1934), 9–27; H. M. Høyberg, *Den danske Dyrlaegestand* (Copenhagen, 1934), pp. 149–152; L. Bahr, "Carl Oluf Jensen," in *Deutsche tierärztliche Wochenschrift*, **42** (1934), 161–163; and E. Gotfredsen, *Medicinens historie* (Copenhagen, 1964), pp. 452, 464.

E. SNORRASON

JENSEN, JOHAN LUDVIG WILLIAM VALDEMAR (*b.* Nakskov, Denmark, 8 May 1859; *d.* Copenhagen, Denmark, 5 March 1925), *mathematics.*

Jensen's career did not follow the usual pattern for a mathematician. He was essentially self-taught and never held an academic position. His father was an educated man of wide cultural interests but unsuccessful in a series of ventures, and pecuniary problems were frequent. At one time the family moved to northern Sweden, where for some years the father managed an estate, and Jensen considered these years the most wonderful of his life. After they returned to Denmark he finished school in Copenhagen and, at the age of seventeen, entered the College of Technology, where he studied mathematics, physics, and chemistry. He soon became absorbed in mathematics and decided to make it his sole study; his first papers date from this time.

In 1881 Jensen's life took an unexpected turn. In order to support himself, he became an assistant at the Copenhagen division of the International Bell Telephone Company, which in 1882 became the Copenhagen Telephone Company. His exceptional gifts and untiring energy soon made Jensen an expert in telephone technique, and in 1890 he was appointed head of the technical department of the company. He held this position until the year before his death. He was extremely exacting, and it was largely through his influence that the Copenhagen Telephone Company reached a high technical level at an early time. He continued his mathematical studies in his spare time and acquired extensive knowledge, in particular, of analysis. Weierstrass was his ideal, and his papers are patterns of exact and concise exposition.

Jensen's most important mathematical contribution is the theorem, named for him, expressing the mean value of the logarithm of the absolute value of a holomorphic function on a circle by means of the distances of the zeros from the center and the value at the center. This was communicated in a letter to Mittag-Leffler, published in *Acta mathematica* in 1899. Jensen thought that by means of this theorem he could prove the Riemann hypothesis on the zeros of the zeta function. This turned out to be an illusion, but in occupying himself with the Riemann hypothesis he was led to interesting results on algebraic equations and, from such results, to generalizations on entire functions.

Another important contribution by Jensen is his study of convex functions and inequalities between mean values, published in *Acta mathematica* in 1906; he showed there that a great many of the classical inequalities can be derived from a general inequality for convex functions. Among other subjects studied by Jensen is the theory of infinite series. In 1891 he published an excellent exposition of the theory of the gamma function, an English translation of which appeared in *Annals of Mathematics* in 1916.

BIBLIOGRAPHY

A list of Jensen's papers is contained in the obituaries mentioned below.

Obituaries by N. E. Nørlund appear in *Oversigt over det K. Danske Videnskabernes Selskabs Forhandlinger Juni 1925–Maj 1926* (1926), pp. 43–51, and in *Matematisk Tidsskrift B* (1926), pp. 1–7 (in Danish). See also G. Pólya, "Über die algebraisch-funktionentheoretischen Untersuchungen von J. L. W. V. Jensen," in *Mathematisk-fysiske Meddelelser*, VII, **17** (1927), 1–33.

BØRGE JESSEN

JEPSON, WILLIS LINN (*b.* Little Oak Ranch, Vacaville, California, 19 August 1867; *d.* Berkeley, California, 7 November 1946), *botany*.

Jepson became the "high priest of the California Flora." Of Scottish-English ancestry, he was the son of William and Martha Potts Jepson, originally from New England. Early in his schooling he was fascinated by the identification of plants through keys in Volney Rattan's local floras. On visits to the California Academy of Sciences he was cordially received by Albert Kellogg. His student years at the University of California at Berkeley coincided with the professorship of Edward Lee Greene, whose influence on Jepson was deep and lasting. Although Jepson never shrank from the defense of Greene's frailties, he disavowed Greene's views on the fixity of species. Jepson's decision to follow the great nineteenth-century British systematists in making taxonomic judgments as he organized the scattered writings on the California flora proved to be fortunate for the future of botany in California (Keck, 1948).

Jepson took his bachelor's (1889) and doctor's (1899) degrees at Berkeley, interspersed with terms at Cornell (1895) and Harvard (1896–1897). His association at Harvard with the conservative taxonomist Benjamin L. Robinson was critical. Jepson's teaching was enriched by field excursions to redwood wilderness, mountains, and deserts. He was at ease with lumberjacks and prospectors and was admitted into a tribe of the Hupa Indians.

Jepson's collections and voluminous notes, totaling nearly sixty closely written field books, provided the data for his *Trees of California* (1909), *Silva of California* (1910), and *Manual of the Flowering Plants of California* (1925). The latter, the leading book in its field, described 4,019 species and was extensively illustrated with line drawings by closely supervised artists. *A Flora of California*, monographic documentation to the *Manual*, with references, notes on types, citations of supporting specimens, and ecological notes, shows Jepson the taxonomist at his best. He followed his manuscripts through the press with fanatical care, yet in his zeal for priority of publication at times maneuvered unscrupulously.

Jepson studied the significance of revegetation after fire, rainpool ecology, endemism, and floristic provinces. He founded the California Botanical Society in 1913, was a founder and spokesman for the Save-the-Redwoods League, and implemented the Point Lobos Reserve, near Monterey. His scientific prose was sensitive and facile, as attested by his articles in *Erythea*, *Madroño*, and the *Dictionary of American Biography*.

He never married. His personality ranged from that of charming host to implacable hermit. He delighted in esoteric allusion—designating a plant collected by Katherine Brandegee, whom he intensely disliked, as "Viper Parsnip"—in the dramatic, and in rigorous tests of loyalty.

BIBLIOGRAPHY

I. ORIGINAL WORKS. Jepson's *A Flora of Western Middle California* (1901) was based on his doctoral thesis; first-edition stock was destroyed in the San Francisco earthquake-fire of 1906. The introduction (pp. 1–32) and concluding parts (vol. 1, pt. 8 and vol. 3, pt. 3) of *A Flora of California* are as yet unpublished.

The first two volumes of *Madroño*, an undetermined number of issues of *Erythea*, and all issues of *Nemophila*, "meeting and field guide" of the California Botanical Society, were edited and wholly or partly written by Jepson, although some articles are unsigned.

His correspondence (1887–1946), in 62 vols., is in the Jepson Library, department of botany, University of California at Berkeley, and was arranged and indexed annually under his supervision. A bibliography of his scientific writings (1891–1962), compiled by L. R. Heckard, J. T. Howell, and R. Bacigalupi, in *Madroño*, **19** (1967), 97–108, is provided with topical indexes of plant genera and biographical sketches.

II. SECONDARY LITERATURE. Obituaries and historical appraisals are listed by Heckard, Howell, and Bacigalupi, *op. cit.*, pp. 97–98. Unlisted there are animadversions by Marcus E. Jones in his intermittent publication *Contributions to Western Botany* (privately printed, Claremont, California), **15** (1929), 2–8; **18** (1933–1935), 9–10. Jepson's estimate of Jones appeared in *Madroño*, **2** (1934), 152–154. See also D. D. Keck, "Place of Willis Linn Jepson in California Botany," in *Madroño*, **9** (1948), 223–228. For recent commentary on Jepson see J. Ewan, ed., *A Short History of Botany in the United States* (New York, 1969), passim.

JOSEPH EWAN

JERRARD, GEORGE BIRCH (*b.* Cornwall, England, 1804; *d.* Long Stratton, Norfolk, England, 23 November 1863), *mathematics.*

George Jerrard was the son of Major General Joseph Jerrard. Although he entered Trinity College, Dublin, where he was a pupil of T. P. Huddart, on 4 December 1821, he did not take his B.A. until the spring of 1827. Jerrard's name is remembered for an important theorem in the theory of equations, relating to the reduction of algebraic equations to normal forms. In 1824 Abel had shown that the roots of the general quintic equation cannot be expressed in terms of its coefficients by means of radicals. E. W. Tschirnhausen had previously generalized the technique of Viète, Cardano, and others of removing terms from a given equation by a rational substitution. Then, in 1786, E. S. Bring reduced the quintic to a trinomial form

$$x^5 + px + q = 0$$

by a Tschirnhausen-type transformation with coefficients expressible by one cube root and three square roots (that is, the coefficients defined by equations of degree three or less). Jerrard also obtained this result, independently, and in a more general form, reducing any equation of degree n to an equation in which the coefficients of x^{n-1}, x^{n-2}, and x^{n-3} are all zero.

When Hermite found a solution for quintic equations in terms of elliptic modular functions, he cited only Jerrard. Unaware that Bring had found the result for the case $n = 5$, Hermite stated that Jerrard's theorem was the most important step taken in the algebraic theory of equations of the fifth degree since Abel. Bring's partial priority, later brought to light by C. J. D. Hill in 1861, did not entirely detract from the importance of Jerrard's research.

BIBLIOGRAPHY

Apart from his *Mathematical Researches*, 3 vols. (Bristol–London, 1832–1835), in which his theorem was first given, Jerrard wrote *An Essay on the Resolution of Equations*, 2 vols. (London, 1858). On his earlier work, see W. R. Hamilton, "Inquiring Into the Validity of a Method Recently Proposed by George B. Jerrard . . .," in *British Association Report*, **5** (1837), 295–348. On the question of Bring's priority, see Felix Klein, *Vorlesungen über die Ikosaeder* (Leipzig, 1884), pt. 2, ch. 1, sec. 2, Eng. trans. (London, 1913), pp. 156–159, English repr. (New York, 1956).

Jerrard wrote extensively in the *Philosophical Magazine*; references to these articles are in Klein, *op. cit.*; and the Royal Society, *Catalogue of Scientific Papers, 1800–1863*, **3** (1869), 547–548. For a useful bibliography on the theory of equations and its history, see G. Loria, *Bibliotheca mathematica*, **5** (1891), 107–112. A short obituary is in *Gentleman's Magazine*, **1** (1864), 130; sparse personal details are in the registers of Trinity College, Dublin.

J. D. NORTH

JEVONS, WILLIAM STANLEY (*b.* Liverpool, England, 1 September 1835; *d.* Hastings, England, 13 August 1882), *logic, economics, philosophy of science.*

Jevons' name is purportedly of Welsh origin, akin to Evans. His father, Thomas, was a notable iron merchant with an inventive trait, and his mother, Mary Anne Roscoe, belonged to an old Liverpool family of bankers and lawyers with a literary bent. The ninth of eleven children, Jevons was brought up, and always in spirit remained, a Unitarian. He was a timid, clever boy and not narrowly studious, showing unusual mechanical aptitude. At University College, London, he took science courses, and his prowess in chemistry was such that he was recommended, while still an undergraduate and only eighteen years of age, for the job of assayer at the newly established Australian mint. In part to help ease a shrunken family budget, he decided to interrupt his education and to accept the post, which carried the handsome remuneration of over £600 a year. In

Sydney he cultivated his interests in meteorology, botany, and geology and published papers in these fields. After five years he renounced a prosperous future in Australia and went back to England to further his education with a view to becoming an academic; he was already saying that "my forte will be found to lie . . . in the moral and logical sciences."

The first subject Jevons concentrated on was political economy, which he felt he could transform. His early and sustained interest in economics must have owed something to two disjoint features of his youth: a family bankruptcy caused by a trade slump, and his having been involved in the physical creation of money. The fluctuations of the national economy always fascinated him.

Having earned his master's degree at University College in 1863, Jevons was appointed junior lecturer at Owens College, Manchester, thus beginning a long association with that city. Two years later he became a part-time professor of logic and political economy at Queen's College, Liverpool, and in 1866 the Manchester institution raised him to a full-time double professorship of "logic and mental and moral philosophy," and of political economy. The following year Jevons married Harriet Ann Taylor, daughter of the founder and first editor of the *Manchester Guardian*, and the couple was soon enjoying the lively intellectual atmosphere of Victorian Manchester. They had three children, of whom one, Herbert Stanley, became a well-known economist. Jevons was made a fellow of the Royal Society in 1872.

In 1876, tired of his teaching chores (he was a poor lecturer and hated to speak in public), Jevons moved to a less onerous but more prestigious professorship at University College. Four years later, he resigned, anxious to spend all his time writing, but his health had begun to deteriorate, despite many long recuperative vacations in England and Norway. A few weeks before his forty-seventh birthday he drowned (possibly as the result of a seizure) while swimming off the Sussex coast.

Jevons had a strong and almost visionary sense of his own destiny as a thinker, and he worried about its fulfillment. A prodigious worker, he sustained many side interests and was passionately fond of music. Among his tributary writings are articles on the Brownian movement, the spectrum, communications, muscular exertion, pollution, skating, and popular entertainment. He tried hard and successfully to improve his literary style, and his later writings are more concise and readable than his earlier ones.

Whether economics or the rationale of scientific methodology benefited more from Jevons' attention is still arguable, but it was certainly as an economist that he became a public figure. Ironically, his popular fame rested on two achievements that now seem slight or even misguided. The first was his book *The Coal Question* (1865), a homily about dwindling English fuel supplies in relation to rocketing future demands. The work was Malthusian in the sense that he discussed industry and coal in much the terms that the earlier author had discussed population and food. It was a tract and obviously, as Keynes remarked, *épatant*. Gladstone, then chancellor of the Exchequer, was deeply impressed by the book, whose "grave conclusions" influenced his fiscal policy. A royal commission was subsequently appointed to look into the matter. Jevons' second achievement was the thesis, developed in the late 1870's from tentative suggestions made by earlier writers, that trade cycles could be correlated with sunspot activity through agrometeorology and, at a further remove, through the price of wheat. It was an ingenious and inherently plausible argument, but the data could not be manipulated to yield convincing evidence and the theory has no standing today.

Jevons brought to economic theory a fruitful insistence on a mathematical framework with an abundance of statistical material to fill out the structure. He was a diligent collector and sifter of statistics, and his methods of presenting quantitative data showed insight and skill. He strongly advocated the use of good charts and diagrams (colored, if possible), which, he said, were to the economist what fine maps are to the geographer. (At one time he toyed with a project to sell illustrated statistical information bulletins to businessmen.) He clarified certain concepts, particularly that of value, which he regarded as a function of utility, a property he wrote about with luminance. Indeed, this was his major contribution, and one that led to the explicit use of the calculus and other mathematical tools by economists.

Utility, said Jevons, "is a circumstance of things arising out of their relation to man's requirements and that normally diminishes." One of his favorite examples concerns bread, a daily pound of which, for a given person, has the "highest conceivable utility," and he went on to show that extra pounds have progressively smaller utilities, illustrating that the "final degree of utility" declines as consumption rises. In effect he was arguing that the second derivative of the function $U_a = U(Q_a)$, where U_a is the total utility or satisfaction derived from the consumption of a in the amount Q_a, is negative. This simple proposition was quite new, for the classical theorists, including Marx, had analyzed value from the supply side only, whereas Jevons' analysis was from demand. Later writers were to recognize the necessity of both approaches. Incidentally, Jevons' approach was the forerunner of the idea of "marginal utility," the first of the "marginal" concepts on which modern economics may be said to rest.

Jevons' work on price index-numbers deserves mention, as his setting them on a sound statistical footing enormously advanced understanding of changes in price and in the all-important value of gold.

Although some of Jevons' ideas can be found inchoate in the works of his predecessors Augustus Cournot and H. H. Gossen, and of his contemporaries Karl Menger and Léon Walras, he arrived at his theories independently and can be seen as a pathbreaker in modern economics. His stress on the subject as essentially a mathematical science was judicious, and he held no exaggerated notions about the role of mathematics—he had observed, he said slyly, that mathematical students were no better than any others when faced with real-life problems.

Economics and logic have traditionally been associated in England, and Jevons belongs to the chain of distinguished thinkers, from John Stuart Mill to Frank Ramsey, who are linked to both disciplines. He was opposed to Mill in divers particulars, and at times he looked upon the older man's deep influence on the teaching of logic as hardly short of disastrous. Mill's mind, he once averred, was "essentially illogical," and Jevons eagerly seized on Boole's remarkable new symbolic logic to show up what he deemed the vastly inferior warmed-over classical logic of Mill. Jevons actually improved on Boole in some important details, as, for instance, in showing that the Boolean operations of subtraction and division were superfluous. Whereas Boole had stuck to the mutually exclusive "either-or," Jevons redefined the symbol + to mean "either one, or the other, or both." This change, which was at once accepted and became permanent, made for greater consistency and flexibility. The expression $a + a$, which was an uninterpretable nuisance in the Boolean scheme, now fell into place, the sum being a.

At the same time Jevons deprecated certain aspects of Boole's work. He thought it too starkly symbolic and declared that "the mathematical dress into which he [Boole] threw his discoveries is not proper to them, and his quasi-mathematical processes are vastly more complicated than they need have been"—an extravagant statement, even in the light of Jevons' own logic. Indeed, some of Jevons' writings about Boole's system, and especially his worries about the discrepancies between orthodox and Boolean algebras, suggest that he did not fully grasp Boole's originality, potential, and abstractness. Nevertheless, he can certainly be reckoned a leading propagandist for Boole, particularly among those who could not understand, or who would not brook, Boole's logic. Moreover, Jevons was led through Boole's ideas to some original work on a logical calculus.

Jevons' logic of inference was dominated by what he called the substitution of similars, which expressed "the capacity of mutual replacement existing in any two objects which are like or equivalent to a sufficient degree." This became for him "the great and universal principle of reasoning" from which "all logical processes seem to arrange themselves in simple and luminous order." It also allowed him to develop a special equational logic, with which he constructed various truth-tablelike devices for handling logical problems. He did not foresee that a truth-table calculus could be developed as a self-contained entity, but he was able to devise a logic machine—a sort of motional form of the later diagrammatic scheme of John Venn. Jevons' "logical piano" (as he eventually called it in preference to his earlier terms "abacus," "abecedarium," and "alphabet") was built for him by a Salford clockmaker. It resembled a small upright piano, with twenty-one keys for classes and operations in an equational logic. Four terms, *A*, *B*, *C*, and *D*, with their negations, in binary combinations, were displayed in slots in front and in back of the piano; and the mechanism allowed for classification, retention, or rejection, depending upon what the player fed in via the keyboard. The keyboard was arranged in an equational form, with all eight terms on both left and right and a "copula" key between them. The remaining four keys were, on the extreme left, "finis" (clearance) and the inclusive "or," and, on the extreme right, "full stop" (output) and the inclusive "or again." In all 2^{16} (65,536) logical selections were possible.

The machine earned much acclaim, especially after its exhibition at the Royal Society in 1870. At present it is on display in the Oxford Museum of the History of Science. Although its principal value was as an aid to the teaching of the new logic of classes and propositions, it actually solved problems with superhuman speed and accuracy, and some of its features can be traced in modern computer designs.

Jevons' various textbooks on logic sold widely for many decades, and his *Elementary Lessons in Logic* (London, 1870) was still in print in 1972. Moreover, he considered his ambitious work on the rationale of science to be an extension of his logic into a special field of human endeavor. His biggest and most celebrated book, *The Principles of Science*, is firmly rooted in Jevonian logic and contains practically all his ideas on, and contributions to, the subject.

Inescapably, the matter of induction, the basis of scientific method and the bugbear of scientific philosophy, is lengthily explored and analyzed. Jevons confidently declared that "induction is, in fact, the inverse operation of deduction." Such a statement—in one sense a truism and in another a travesty—might be thought a feeble beginning for a study of the how and why of science, but Jevons acquits himself admirably. He does not confuse the formal logic of induction with the problems of inductive inference in the laboratory, and he is obviously under no illusions about the provisional nature of all scientific "truth."

Nineteenth-century English scholars, inspired by the phenomenal explicatory and material success of science, had been taking an increasing interest in its philosophy. John Stuart Mill and William Whewell are particularly associated with these early studies; and in Jevons' view their work contained serious flaws. He thought that Mill, first, expected too much of science as a key to knowledge of all kinds and, second, that he was overly respectful of Bacon's view of science as primarily the collection and sortation of data. His criticism of Whewell centered on that writer's apparent assumption that exact knowledge is a reality attainable by scientific patience.

Jevons was perhaps the first writer to insist that absolute precision, whether of observation or of correspondence between theory and practice, is necessarily beyond human reach. Taking a thoroughly modern position, Jevons held that approximation was of the essence, adding that "in the measure of continuous quantity, perfect correspondence must be accidental, and should give rise to suspicion rather than to satisfaction." He also felt that causation was an overrated if not dangerous concept in science, and that what we seek are logically significant interrelations. All the while, he said, the scientist is framing hypotheses, checking them against existing information, and then designing experiments for further support. There can be no cut-and-dried conclusion to most investigations and no guarantee that correct answers can be issued. The scientist must act in accordance with the probabilities associable with rival hypotheses, which probabilities, or, as many would prefer to say today, likelihoods, constitute the decision data. Thus "the theory of probability is an essential part of logical method, so that the logical value of every inductive result must be determined consciously or unconsciously, according to the principles of the inverse method of probability." An entire chapter of *The Principles* is devoted to direct probability and another to inverse probability.

As a probabilist Jevons was fundamentally a disciple of Laplace, or at least of Laplace as reshaped by Jevons' own college teacher and mentor Augustus De Morgan—that is to say, he was a subjectivist. Probability, he maintained, "belongs wholly to the mind" and "deals with quantity of knowledge." Yet he was careful to emphasize that probability is to be taken as a measure, not of an individual's belief, but of rational belief—of what the perfectly logical man would believe in the light of the available evidence. In espousing this view, Jevons sidestepped Boole's disturbing reservations about subjectivism—mainly because he had difficulty grasping them. Writing to Herschel, he stated, frankly: "I got involved in Boole's probabilities, which I did not thoroughly understand. . . . The most difficult points ran in my mind, day and night, till I got alarmed. The result was considerable distress of head a few days later, and some signs of indigestion." In general Jevons was silent about the movement toward a frequential theory of probability that was growing out of the work of Leslie, Ellis, and Poisson, as well as that of his contemporary, Venn.

To the modern reader, however, Jevons may seem altogether too self-assured in this notoriously treacherous field. For example, he wholeheartedly accepted Laplace's controversial rule of succession and offered a naive illustration of its applicability: Observing that of the sixty-four chemical elements known to date (Jevons was writing in 1873) fifty are metallic, we say that the quantity $(50 + 1)/(64 + 2) = 17/22$ is the probability that the next element discovered will be a metal. To the frequentist, insistent on a clearly delineated sample space, this statement is almost wholly devoid of meaning. Another, more bizarre example of his naive Laplaceanism is his contention that the proposition "a platythliptic coefficient is positive" has, because of our complete ignorance, a probability of correctness of 1/2. Boole had rightly objected to this sort of thing and would agree with Charles Terrot that such a probability has the numeric but wholly indeterminate value of 0/0. To understand Jevons' position, however, we must bear in mind that the prestige of Laplace, especially in this area, was then enormous.

Curiously, in discoursing on what he calls "the grand object of seeking to estimate the probability of future events from past experience," Jevons made only one casual and unenlightening reference to Thomas Bayes, who, a century earlier, had been the first to attempt a coherent theory of inverse probability. Today the implications of Bayes's work form the subject of lively discussion among probabilists.

Some of the most illuminating sections of *The*

Principles of Science are those dealing with technical matters, such as the methodology of measurement, the theory of errors and means, and the principle of least squares. Yet the book offers only a shallow treatment of the logic of numbers and arithmetic, and it has been criticized for the absence of any serious discussion of the social and biological sciences. By and large, however, *The Principles* is something of a landmark in the bleak country of nineteenth-century philosophy of science.

BIBLIOGRAPHY

I. ORIGINAL WORKS. According to Harriet Jevons' bibliography (see below), Jevons' first appearance in print was a weather report (24 Aug. 1856) in the *Empire*, a Sydney, Australia, newspaper, for which he wrote weekly reports until 1858. His first publication in a scholarly journal was "On the Cirrus Form of Cloud, With Remarks on Other Forms of Cloud," in *London, Edinburgh and Dublin Philosophical Magazine*, **14** (July 1857), 22–35. His account of the logical piano is "On the Mechanical Performance of Logical Inference," in *Philosophical Transactions of the Royal Society*, **160** (1870), 497–518.

His books, all published in London, include *Pure Logic* (1863); *The Coal Question* (1865); *The Theory of Political Economy* (1871); *The Principles of Science* (1874); *Studies in Deductive Logic* (1880); *The State in Relation to Labour* (1882); *Methods of Social Re-Form* (1883); *Investigations in Currency and Finance* (1884); and *Principles of Economics* (1905). The last three were published posthumously, and the very last is a fragment of a large work that he was writing at the time of his death. *The Principles of Science, A Treatise on Logic and Scientific Method*, the frontispiece of which is an engraving of the logical piano, is available as a paperback reprint (New York, 1958).

For information on the 1952 exhibition of Jevons' works at the University of Manchester, see *Nature*, **170** (1952), 696. The library of that university has Jevons' economic and general MSS, and Wolfe Mays of the Department of Philosophy owns the philosophic and scientific MSS; much of this material is still unpublished.

II. SECONDARY LITERATURE. No full biography exists. The primary source is Harriet A. Jevons, *Letters and Journal of W. Stanley Jevons* (London, 1886), with portrait and bibliography. See also the obituary by the Reverend Robert Harley in *Proceedings of The Royal Society*, **35** (1883), i–xii. Jevons' granddaughter, Rosamund Könekamp, contributed "Some Biographical Notes" to *Manchester School of Economics and Social Studies*, **30** (1962), 250–273. For a modern view of his logic, see W. Mays and D. P. Henry, "Jevons and Logic," in *Mind*, **62** (1953), 484–505. The logic contrivances are described and placed in historic perspective in Martin Gardner, *Logic Machines and Diagrams* (New York, 1958). Jevons' economics is reviewed in J. M. Keynes, *Essays in Biography*, 2nd ed. (London, 1951); and, more formally, in E. W. Eckard, *Economics of W. S. Jevons* (Washington, D.C., 1940); both of the foregoing also contain biographical material. His contemporary influence is discussed in R. D. C. Black, "W. S. Jevons and the Economists of His Time," in *Manchester School of Economics and Social Studies*, **30** (1962), 203–221. Ernest Nagel has written a preface on Jevons' philosophy of science for the paperback ed. of *Principles of Science*. See also W. Mays, "Jevons's Conception of Scientific Method," in *Manchester School of Economics and Social Studies*, **30** (1962), 223–249.

NORMAN T. GRIDGEMAN

JEWETT, FRANK BALDWIN (*b.* Pasadena, California, 5 September 1879; *d.* Summit, New Jersey, 18 November 1949), *telecommunications*.

The descendant of a long line of New England ancestors going back to Pilgrim days, Jewett was the son of Stanley Jewett and Phebe Mead. Originally from Cincinnati, his parents moved to California, where his father became a pioneering railroad man. Jewett attended Throop Polytechnic Institute in Pasadena (now the California Institute of Technology); after graduation in 1898 he went to the University of Chicago to work with Albert Michelson, who was to become the first American scientist to receive the Nobel Prize (1907). Before obtaining a Ph.D. in physics in 1902, Jewett helped develop a laboratory machine for the manufacture of diffraction gratings used in spectrographic analysis and also formed a lifelong friendship with another future Nobel laureate, Robert A. Millikan, then a young instructor in the same department.

Jewett next spent two years teaching at the Massachusetts Institute of Technology. In 1904 he joined the engineering department of the American Telephone and Telegraph Company, then headquartered in Boston. There his excellent technical and managerial judgment quickly attracted attention. When the department moved to New York in 1907, its laboratory was consolidated with the engineering department of Western Electric, the company's manufacturing branch. Jewett continued his work as transmission engineer in New York until 1912, when he became assistant chief engineer and was put in charge of the newly formed research department. His main responsibility was to develop the transcontinental telephone line that would link New York with San Francisco in time for the Panama-Pacific International Exposition of 1915.

Jewett believed that the crucial component was a more reliable amplifier for the repeater circuits inserted at regular intervals in all long-distance transmission lines. On Millikan's suggestion, he hired H. D. Arnold to do research in this new area of electron physics, and Arnold soon solved the problem by devising a production model of de Forest's three-electrode vacuum tube (triode) amplifier. Arnold's improvements of the device (notably high vacuum) permitted production of electronic amplifiers with predictable performance and long life, and the New York–San Francisco service was initiated on schedule.

When the Bell Telephone Laboratories were organized in 1925, Jewett became the first president and Arnold the director of research. Together they established the laboratories as one of the foremost centers of industrial research and supervised Bell's contributions to radio communication systems, carrier telephony, the talking motion picture, the electric phonograph, the transmission of photographs by telephone, the transatlantic telephone, and the high-speed cable. Jewett was an early proponent of the idea of a nearly autonomous research department within an industrial organization, and under his leadership as president and chairman of the board until his retirement in 1944, his vision was most successfully realized.

In addition to his role within the Bell System, Jewett was commissioned in the U.S. Army Signal Corps in World War I and served in several important posts. From 1933 to 1935 he was a member of President Roosevelt's Science Advisory Board and in World War II he was one of the eight members of the National Defense Research Committee of the Office of Scientific Research and Development. After the atomic explosions, he spoke up in favor of civilian control of scientific research.

He received many professional and scientific honors, including the presidency of the American Institute of Electrical Engineers (1922–1923) and of the National Academy of Sciences (1939–1947)—the first time an industrial scientist had been elected to that post.

Jewett married Fannie C. Frisbie of Rockford, Illinois, a fellow graduate student in physics at the University of Chicago, in 1905; she died in 1948. They had two sons.

BIBLIOGRAPHY

"The Career of Frank Baldwin Jewett," a biography written by John Mills, his friend and collaborator, on the occasion of Jewett's retirement, appears in *Bell Laboratories Record*, **22** (1944), 541–549, followed by an announcement of five annual Jewett Fellowships in the Physical Sciences begun in his honor. For a biography, followed by a bibliography, see O. E. Buckley, in *Biographical Memoirs. National Academy of Sciences*, **27** (1952), 238–264. Obituaries are in the *Bell Laboratories Record*, **27** (1949), 442–445; *New York Times* (19 November 1949); and *Proceedings of the Institute of Radio Engineers*, **38** (1950), 189.

CHARLES SÜSSKIND

JOACHIM, GEORG. See **Rheticus.**

JOACHIMSTHAL, FERDINAND (*b.* Goldberg, Germany [now Złotoryja, Poland], 9 March 1818; *d.* Breslau, Germany [now Wrocław, Poland], 5 April 1861), *mathematics.*

Son of the Jewish merchant David Joachimsthal and Friederike Zaller, Joachimsthal attended school in Liegnitz (now Legnica, Poland), where he had the good fortune to have Kummer as his teacher. In 1836 after completing his studies there, he went to Berlin for three semesters, where his mathematics teachers were Dirichlet and Jacob Steiner.

Starting with the summer semester of 1838, he spent another four semesters as a student at the University of Königsberg (now Kaliningrad, U.S.S.R.). Most notable among his teachers there were Jacobi, Germany's leading mathematician after Gauss, and Bessel, foremost German astronomer of his day. Joachimsthal thus received the best available mathematical training.

After completion of these studies, he went to Halle to work for his doctorate under Otto Rosenberger, obtaining his Ph.D. on 21 July 1840. His dissertation, "De lineis brevissimis in superficiebus rotatione ortis," was graded *docta*, and he passed his oral examinations *cum laude*.

As was then still generally customary, Joachimsthal took the *Examen pro facultate docendi* and in 1844 joined the teaching staff of the Königliche Realschule in Berlin. Starting in 1847, he taught in Berlin at the Collège Royal Français, after 1852 with the rank of full professor. In 1845 he applied to the school of philosophy at the University of Berlin for accreditation to teach there. Prior to this time, after taking his doctor's degree, Joachimsthal had become a convert from Judaism to the Protestant faith.

After being accepted as an applicant for university teaching credentials, on 7 August 1845 he delivered before the assembled department a trial lecture entitled *Über die Untersuchungen der neueren Geometrie, welche sich der Lehre von den Brennpunkten anschliessen* and on 13 August a public trial lecture entitled "De curvis algebraicis." The *Venia legendi* was conferred upon him that day.

From the winter semester of 1845–1846 to that of 1852–1853, in addition to his teaching, Joachimsthal lectured as *Privatdozent* at the University of Berlin to beginners in analytic geometry and differential and integral calculus; to advanced students of the theory of surfaces and calculus of variations; and to special students on statics, analytic mechanics, and the theory of the most important curves encountered in architecture. Profiting from his experience in Jacobi's seminar, he also held mathematical drill sessions, a relative novelty at Berlin. His effective teaching won the unanimous approval of the department. His lectures attracted more students than did those of Eisenstein, his brilliant colleague in the same field.

Meanwhile, Joachimsthal's prospects in Berlin were not at all promising. On 7 May 1853 he was finally promoted to full professor by the Prussian Ministry of Culture, after repeated urgings and commendations from his department at the university, and received an appointment in Halle as successor to Sohnke. By 1855 he had received a new offer and went as Kummer's successor to Breslau, where his lectures were very popular. He taught, among other things, analytic geometry, differential geometry, and the theory of surfaces, in which—exceptional for the time—he operated with determinants and parameters. He gave special lectures on geometry and mechanics for students of mining engineering and metallurgy.

The average number of his listeners exceeded that of Kummer, who with Weierstrass was later to become one of the most sought-after teachers of mathematics. In 1860, for example, Joachimsthal had an audience of sixty-six attending his mechanics lectures. By the time of his death, at the age of forty-three, he had acquired a wide reputation as an excellent teacher and kind person.

His *Cours de géométrie élémentaire à l'usage des élèves du Collège Royal Français* (1852) had demonstrated his talent as a textbook writer through its clear logical structure and insight (rarely found in accomplished mathematicians) into the difficulties facing beginners, and he was naturally expected to turn out equally valuable university texts.

Jacobi persuaded him to write an *Analytische Geometrie der Ebene* as a supplement to the *Geometrie des Raumes* that he himself was planning. It was published posthumously in 1863. A printed version of his lecture during the winter semester of 1856–1857 on the application of infinitesimal calculus to surfaces and lines of double curvature was also published posthumously in 1872. Reprinted several times, both books were in use for some thirty years, due largely to their clear, simple exposition and to the general applicability of their conclusions.

But the reputation of a teacher tends to be transitory, and Joachimsthal's contributions would have receded into oblivion had it not been for his outstanding original research. Those qualities of clarity, rigor, and elegance that made him one of the most eminent teachers of his day were also characteristic of his own work. One of his favorite fields of study was the theory of surfaces. He dealt repeatedly with the problem of normals to conic sections and second-degree surfaces.

His published writings (the first of which appeared in 1843) show him to have been influenced primarily by Jacobi, Dirichlet, and Steiner. Although never prolific, he always went deeply into a subject, seeking to discover connections between isolated, ostensibly unrelated phenomena. In treating a problem of attraction, for example, he gave a solution constituting an application of the Abel method for defining a tautochrone.

His striving for general validity and his critical acumen emerged also in treatises in which he took up problems dealt with by other mathematicians, such as Bonnet, La Hire, Carl Johann Malmsten, Heinrich Schröter, Steiner, Jacques Charles Sturm, whose solutions he strengthened. As in his lectures, he was primarily concerned in his writings with analytic applications in geometry. Not only did he use determinants himself but explicitly indicated their possible applications in geometry. His use of oblique coordinates also deserves special mention.

Today his name is associated with Joachimsthal surfaces, which possess a family of plane lines of curvature within the planes of a pencil; the Joachimsthal theorem concerning the intersection of two surfaces in three-dimensional real Euclidean space along a common line of curvature; and a theorem on the four normals to an ellipse from a point inside it.

Joachimsthal's contributions were substantial and lucid. His marked predilection for mature, polished exposition was expressed in constant recasting, revising, and rewriting; so that many planned works never reached completion. In addition, during the few years of his greatest potential, when he was teaching in Berlin as *Privatdozent*, he lived surrounded by an unprecedented galaxy of luminaries within his field (Dirichlet, Jacobi, Steiner, Eisenstein, and Borchardt).

BIBLIOGRAPHY

I. Original Works. Most of Joachimsthal's writings not published independently appeared in the *Journal für die reine und angewandte Mathematik* between 1843 and 1871. A bibliography of most of his published works is given in Poggendorff, I, 1196, and III, 692. Works not mentioned there are found in *Nouvelles annales de mathématiques*, **6** (1847); **9** (1850); and **12** (1853); and in *Abhandlungen zur Geschichte der Mathematik*, **20** (1905), 76–79. See also Royal Society, *Catalogue of Scientific Papers*, III, 548–549, and VIII, 27. Also worthy of mention is his "Mémoire sur les surfaces courbes," in *Collège Royal Français, Programme* (Berlin, 1848), pp. 3–20; and his foreword in Friedrich Engel, *Axonometrische Projectionen der wichtigsten geometrischen Flächen* (Berlin, 1854).

Some of the works listed in Poggendorff appeared as preprints in school and university publications before reaching a wider public in the *Journal für die reine und angewandte Mathematik*.

A collection of Joachimsthal's papers is to be found partly in the archives of Humboldt University, Berlin, German Democratic Republic, and in those of the Martin Luther University of Halle-Wittenberg in Halle, German Democratic Republic.

II. SECONDARY LITERATURE. Borchardt published a brief obituary in the *Journal für die reine und angewandte Mathematik*, **59** (1861), 124. The only detailed biography, by Moritz Cantor, in *Allgemeine deutsche Biographie*, XIV (Leipzig, 1881), 96–97, is the source of most biographic references to Joachimsthal, such as those in Karl Gustav Heinrich Berner, *Schlesische Landsleute* (Leipzig, 1901), p. 221; and S. Wininger, *Grosse jüdische National-Biographie*, III (Chernovtsy, 1928), 317.

Additional information is given in Rudolf Sturm, "Geschichte der mathematischen Professuren im ersten Jahrhundert der Universität Breslau 1811–1911," in *Jahresberichte der Deutschen Matematikervereinigung*, **20** (1911), 314–321; this appeared in virtually the same form in *Festschrift zur Feier des hundertjährigen Bestehens der Universität Breslau*, II, *Geschichte der Fächer* (Breslau, 1911), 434–440.

Joachimsthal is mentioned also in Wilhelm Lorey, *Das Studium der Mathematik an den deutschen Universitäten seit Anfang des 19. Jahrhunderts* (Leipzig–Berlin, 1916), pp. 87, 88, 120; Max Lenz, *Geschichte der Königlichen Friedrich-Wilhelms-Universität zu Berlin*, II, pt. 2 (Halle, 1918), pp. 155 and 156; Heinrich Brandt, "Mathematiker in Wittenberg und Halle," in *450 Jahre Martin-Luther-Universität Halle-Wittenberg*, II (Halle, 1952), 449–455.

KURT-R. BIERMANN

JOBLOT, LOUIS (*b.* Bar-le-Duc, Meuse, France, 9 August 1645; *d.* Paris, France, 27 April 1723), *microscopy, physics.*

Joblot was the fourth of six children of Nicolas Joblot, probably a moderately well-to-do merchant, and Anne Guilly. The family was Roman Catholic.[1] Nothing certain is known of Joblot's life prior to his thirty-fifth year, and subsequent data are meager.[2] He was probably educated at the Collège Gilles de Trèves, founded at Bar-le-Duc in the second half of the sixteenth century. In 1680 he was appointed assistant professor of mathematics (geometry and perspective) at the École Nationale des Beaux-Arts, which was part of the Royal Academy of Painting and Sculpture. Neither this position nor the rank of academician associated with it was salaried, although Joblot was obliged to attend the meetings of the Academy. In 1697 Joblot obtained an eighteen-month leave of absence to travel to Italy. Soon after his return, Sébastien le Clerc resigned his position as professor of mathematics and in May 1699 Joblot succeeded him at a salary of 300 livres. During the summer of 1702 Joblot read to the Academy a series of lectures on optics and on the anatomy of the eye. In December 1716 he presented the outline of his book on microscopy to the Academy, who decided to allow the book to be printed. Although the illustrations were ready at that time, this work was not completed until 1718. Joblot resigned his professorship on 26 April 1721 and died two years later at the age of seventy-seven.

It is probable that Joblot was led to his research on microscopy by the arrival in Paris of Huygens and Hartsoeker in the summer of 1678. In July of that year, Huygens showed microscopes that he had brought from Holland and demonstrated Infusoria before the Academy of Sciences. An account of this demonstration was published in the *Journal des sçavans* (15 August 1678), which, two weeks later, published a description of Hartsoeker's microscope.[3] Probably inspired by these events, Joblot began his own research at about this time, and his *Descriptions et usages de plusieurs nouveaux microscopes* makes particular reference to Hartsoeker's visit and to his work on Infusoria.

The publication of *Descriptions* established Joblot as the first French microscopist. The first part of the book described several microscopes and their construction and introduced some improvements, including the use of stops (diaphragms) in compound microscopes to correct for chromatic aberration. Joblot designed the first *porte loupe*, a simple preparation microscope in which the lens is supported by a string of "Musschenbroek nuts," forming a ball-and-socket jointed arm.

The second part of the book discusses Joblot's microscopical observations. With the exception of his work on the vinegar-eels (nematodes), which dates from 1680,[4] Joblot's observations mainly concern Protozoa and were made between 1710 and 1716. Leeuwenhoek had observed the Protozoa previously,[5] but Joblot's is the earliest treatise on them. He described and illustrated a large number of new types and, according to Cole, was the first to observe the contractile vacuole, while Oudemans states that he was the first to picture the larva of a hydrachnid[6] and the nymph of *Unionicola ypsilophorus*, a parasite of the pond mussel.

Joblot believed that Infusoria originated from germs already present in the material from which the cold infusion was made or germs that were floating in the air. He proved his supposition by preparing two identical cold infusions of hay, one of which was shielded from the air; both developed Infusoria. In a similar experiment, but with infusions that had been boiling for fifteen minutes, the unshielded preparation developed Infusoria, while, even after a considerable time, the shielded one did not. When the latter infusion was exposed to the air, Infusoria developed. Because Joblot's use of sterilized infusions anticipated the later work of Spallanzani and Pasteur, he should be included, after Redi and Leeuwenhoek, in the list of investigators who disproved the doctrine of spontaneous generation.

In addition to mathematics (to which science he did not contribute) and microscopy, Joblot was interested in physics and especially magnetism. He associated with Amontons and G. Homberg,[7] the physicians J. Méry and Bourdelot and regularly attended meetings at the home of M. F. Geoffroy devoted to discussions of physics. In 1701 he constructed the first artificial magnet, using thin strips of magnetized iron, an arrangement which is usually attributed to Pierre Lemaire (1740).

Challenging the Cartesian theory of magnetism, Joblot formulated his own theory in which he denied, among other hypotheses, the existence of a dual flow of magnetic matter moving in opposite directions; he maintained instead that only a single flow exists, a theory which sparked a controversy with L. de Puget. Further, Joblot refuted experimentally Descartes's idea that with magnetic bars of equal weight and equal degree of magnetization, the longest will support the greatest weight. Although he is remembered for these contributions, Joblot's role in the development of magnetism has been insufficiently explored.

NOTES

1. After an extensive search in the archives of Bar-le-Duc and surrounding communities, Konarski, Joblot's biographer, found only the name of Nicolas Joblot, with no mention of his profession or circumstances. The godparents of his children (all of whom were baptized in the church of the parish of Notre Dame) were either merchants or belonged to the upper classes of the city.
2. Most of the data are to be found in the minutes of the meetings of the Royal Academy of Painting and Sculpture.
3. This description was written by Huygens.
4. Leeuwenhoek described vinegar-eels in two letters to H. Oldenburg. The first (11 Feb. 1675) was not published at the time; the second (9 Oct. 1676), in which more particulars were given, was published in part in the *Transactions of the Royal Society*, **12** (1677), 821–831; and in the *Journal des sçavans*, **6** (1678), 106–110, 132–135. The observations on the vinegar-eel were omitted in these early publications; hence Joblot's observations were probably the first published. Leeuwenhoek's complete letter was published in *The Collected Letters of Antoni van Leeuwenhoek* (Amsterdam, 1941), II, 60–161.
5. Described in a letter to H. Oldenburg (7 Sept. 1674) that was published in the *Philosophical Transactions of the Royal Society*, **9** (1674), 178–182.
6. A curious picture (pl. 6, no. 12) that resembles a human head.
7. Joblot (*Descriptions* . . ., II, 7) has Hombert, but this must be a printing error; no person by that name was a member of the French Academy.

BIBLIOGRAPHY

I. Original Works. Joblot's works include "Lettre de M. Joblot, Professeur en mathématiques dans l'Académie Royale de Peinture et Sculpture, à Paris, à M. de Puget, à Lyon," in *Mémoires pour servir à l'histoire des sciences et des beaux-arts* (July 1703); "Extrait d'une nouvelle hypothèse sur l'Aiman," *ibid.* (Sept. 1703); *Descriptions et usages de plusieurs nouveaux microscopes, tant simples que composez; avec de nouvelles observations faites sur une multitude innombrable d'insectes, & d'autres animaux de diverses espèces, qui naissent dans des liqueurs préparés, & dans celles qui ne le sont point* (Paris, 1718); 2nd ed., enl. and with entomological notes, *Observations d'histoire naturelle, faites avec le microscope sur un grand nombre d'insectes et sur les animalcules qui se trouvent dans les liqueurs . . . . Avec la description et les usages des différens microscopes*, 2 vols. (Paris, 1754–1755).

II. Secondary Literature. An anonymous work concerning Joblot is "Description d'un aimant artificiel qui est dans le cabinet de M. Chamard," in *Mémoires pour servir à l'histoire des sciences et des beaux-arts* (Nov.–Dec. 1702). Two very rare booklets by L. de Puget, announced in *Journal des sçavans* (31 July 1702), are *Lettres écrites à un philosophe sur le choix d'une hypothèse propre à expliquer les effets d'un aimant* (Lyons, 1702), and *Lettre de M. Puget, de Lyon, à M. Joblot sur l'aimant* (Lyons, 1702).

See also the following works (listed chronologically): B. de Fontenelle, *Histoire de l'Académie royale des sciences, année 1703* (Amsterdam ed., 1739), sec. 7, pp. 26–27; L. Lémery, "Diverses expériences et observations chimiques et physiques sur le fer et l'aimant," in *Mémoires de l'Académie royale des sciences, année 1706* (Amsterdam ed., 1708), pp. 148–169; J. M. Fleck, "Quels sont les premiers observateurs des infusoires?," in *Mémoires de l'Académie de Metz*, **56** (3rd ser. **4**) (1876), 651–652; P. Cazeneuve, "La génération spontanée d'après les livres d'Henri Baker et de Joblot," in *Revue scientifique*, **31** (4th ser. **1**) (1894), 161–166; J. Boyer, "Joblot et Baker," *ibid.*, 283–284; W. Konarski, "Un savant barrisien, précurseur de M. Pasteur, Louis Joblot (1645–1723)," in *Mémoires de la Société des lettres, sciences et arts, Bar-le-Duc*, 3rd ser., **4** (1895), 205–333; H. Brocard, *Louis de Puget, François*

Lamy, Louis Joblot, leur action scientifique d'après de nouveaux documents (Bar-le-Duc, 1905); C. Dobell, "A Protozoological Bicentenary, Antony van Leeuwenhoek (1632–1723) and Louis Joblot (1645–1723)," in *Parasitology*, **15** (1923), 308–319; F. J. Cole, *The History of Protozoology* (London, 1926); A. C. Oudemans, "Kritisch historisch overzicht der acarologie," in *Tijdschrift voor entomologie*, **69** (1926), supp.; **72** (1929), supp.; and L. L. Woodruff, "Louis Joblot and the Protozoa," in *Scientific Monthly*, **44** (1937), 41–47.

P. W. VAN DER PAS

JOHANNES. See **John.**

JOHANNES LAURATIUS DE FUNDIS (*fl.* Bologna, 1428–1473), *astronomy, astrology.*

Most of what we know about Johannes Lauratius (or Paulus) de Fundis is found in Thorndike, according to whom he is mentioned during most of the period 1428–1473 as lecturer in astrology at the University of Bologna. In April 1433 he observed Mars on several consecutive nights; and in the same year he published the first of his many writings, *Questio de duracione . . . huius aetatis* (MS Brit. Mus. Reg. 8.E.VII). Other works are *Tacuinus astronomico-medicus*, a prediction dated 7 February 1435 (MS Bologna Univ. Libr. 2); *Rescriptus super tractatum de spera*, dated 1437 (MS Paris, B.N. lat. 7273), a commentary on Sacrobosco; *Questio de fine seu durabilitate mundi* (a revised version of *Questio de duracione . . .*), dating from 1445 (MS Paris, B.N. lat. 10271, 204r–227v); and *Tractatus reprobacionis eorum que scripsit Nicolaus Orrem . . . contra astrologos*, dated 30 Oct. 1451 (*ibid.*, 63r–153v), a defense of astrology against the anti-astrological writings of Nicole Oresme.

Nova spera materialis, dated 10 Aug. 1456 (MSS Utrecht 724; Venice, San Marco VIII, 33), is a complete but brief exposition of astronomy, dealing with the system of the world, spherical astronomy, and planetary theory. Presumably it was written as an alternative to Sacrobosco's standard exposition. Finally, there is *Nova theorica planetarum* (incipit, "*Theorica speculativa dicitur scientia motuum planetarum*"), known from the same codices as *Nova spera materialis* but undated. It quotes a treatise entitled *De sphera rotunda*, which seems to be identical to *Rescriptus super tractatum de spera* and accordingly must be later than 1437 but earlier than 1456, when the Utrecht MS was copied at Bologna. It is meant as a substitute for the thirteenth-century *Theorica planetarum* wrongly ascribed to Gerard of Cremona, Gerard of Sabioneta, and others.

The first six works in this list were examined by Thorndike, who determined that the author was an astrologer and astronomer of no exceptional qualities and without any outstanding ideas. *Nova theorica planetarum* is interesting only insofar as it reveals how planetary theory was taught around the middle of the fifteenth century at one of the major chairs of astronomy. It is almost contemporary with Peurbach's famous work of the same title and exhibits many of the same features. Thus the geometric models of the old *Theorica* (and the *Almagest*) are embedded in the system of "physical" spheres known from Ptolemy's *Planetary Hypotheses* but are moved by separate "intelligences." But where Peurbach maintains the Alfonsine theory of precession, Johannes is skeptical toward trepidation and prefers the Ptolemaic theory of a constant rate of precession. Like the Arabs, he is much interested in conjunctions and in the initial conditions of the planetary system.

BIBLIOGRAPHY

See L. Thorndike, *History of Magic and Experimental Science*, IV (New York, 1934), 232–242; O. Pedersen, "The Theorica Planetarum Literature of the Middle Ages," in *Classica et mediaevalia*, **23** (1962), 225–232; and a roneotyped ed. of *Nova theorica planetarum* by O. Pedersen and B. Dalsgaard Larsen, "A 15th Century Planetary Theory" (Aarhus, 1961).

OLAF PEDERSEN

JOHANNES LEO. See **Leo the African.**

JOHANNSEN, ALBERT (*b.* Belle Plaine, Iowa, 3 December 1871; *d.* Winter Park, Florida, 11 January 1962), *petrology, petrography.*

Johannsen was educated at the universities of Illinois (B.S., 1894), Utah (B.A., 1898), and Johns Hopkins (Ph.D., 1903). He worked for the Maryland Geological Survey (1901–1903) and for the U.S. Geological Survey (1903–1925), during which time he served as acting chief of the petrology section (1907–1910).

In 1910 Johannsen went to the University of Chicago where he rose in eight years to the rank of full professor. Principally a petrographer, his early contributions were mainly improvements of the polarizing microscope and methods of optical analysis of minerals (1918). About this time Johannsen began work on a quantitative mineralogical classification of the igneous rocks primarily because of "errors introduced by loose usage of [petrologic] terms" ("Suggestions for a . . . Classification of Igneous Rocks" [1917]; *Descriptive Petrography* [1931], vol. I, p. 129). His classification is "strictly mineralogical, quantitative and modal" and used as its base the "double tetrahedron" with quartz, potassium feldspar, sodium feldspar, calcium feldspar, and the feldspathoids as end members (*Descriptive Petrography,* p. 141). It is this classification that is used in his major scientific work, the four-volume *Descriptive Petrography of the Igneous Rocks.* It includes complete petrographic descriptions, chemical and modal analyses, and the historical background of virtually all known igneous rocks, as well as biographical sketches of the early petrologists responsible for identifying and studying many of those rocks. Although Johannsen's classification is not in use today, the information contained in this work serves as a standard reference in the field of petrography and as a monument to the thoroughness and meticulous attention to detail so characteristic of his scientific work.

Johannsen retired in 1937. The remaining twenty-five years of his life were spent in nonscientific pursuits, yet the two works produced during this period are indicative of his scientific approach. From his keen interest in the "dime and nickel novels" of the late nineteenth century came the history and biographies of the *House of Beadle and Adams.* His collection of the first editions of the works of Charles Dickens provided the source for a detailed examination of the plates in Dickens' novels and the publication of *Hablot Knight Browne (1815–1822): Phiz—Illustrations From the Novels of Charles Dickens.* These works attest to Johannsen's unwavering eye for detail, a trait revealed throughout his entire career.

BIBLIOGRAPHY

I. Original Works. Johannsen wrote twenty-two scientific papers dealing with improvements of the petrographic microscope, optical analysis of minerals, rock classification, and reviews of early petrologists' published works. A complete bibliography is in D. J. Fisher, "Memorial to Albert Johannsen," in *Bulletin of the Geological Society of America,* **73** (1962), 109–114. His paper on rock classification is "Suggestions for a Quantitative Mineralogical Classification of Igneous Rocks," in *Journal of Geology,* **25** (1917), 63–97. Several books of importance are *Determination of Rock-Forming Minerals* (New York, 1908); *Manual of Petrographic Methods* (New York, 1918); and *Essentials for the Microscopic Determination of Rock-Forming Minerals and Rocks in Thin Section* (Chicago, 1922). His most important scientific work, still in print, is *Descriptive Petrography of the Igneous Rocks,* 4 vols. (Chicago, 1931–1938).

Johannsen's nonscientific works are *The House of Beadle and Adams,* 2 vols. (Norman, 1950); and *Hablot Knight Browne (1815–1882): Phiz—Illustrations From the Novels of Charles Dickens* (Chicago, 1956).

II. Secondary Literature. The most complete discussion of Johannsen's life is in D. J. Fisher's article mentioned above. There is also a brief biographical sketch in J. Cattell, ed., *American Men of Science,* 10th ed. (Tempe, 1960), p. 2007.

A review of vols. III and IV of Johannsen's major work, *Descriptive Petrography,* including additional insights into the man and scientist, is in T. T. Quirke, "Reviews," in *Journal of Geology,* **47** (1939), 774–776.

WALLACE A. BOTHNER

JOHANNSEN, WILHELM LUDVIG (*b.* Copenhagen, Denmark, 3 February 1857; *d.* Copenhagen, 11 November 1927), *biology.*

Johannsen, one of the founders of the science of genetics, was the son of a Danish army officer, Otto Julius Georg Johannsen, and the former Anna Margrethe Dorothea Ebbesen. His father's family included many civil servants; the interest of his mother and maternal grandmother in German culture influenced Johannsen's childhood experience. He attended a good elementary school in Copenhagen, but his father's means did not permit him to enter the university, so in 1872 he was apprenticed to a pharmacist. He worked in pharmacies in Denmark, where he taught himself chemistry, and in Germany, where he became interested in botany.

Johannsen returned to Denmark in 1879, passed the pharmacist's examination, and continued his studies in botany and chemistry. In 1881 he became assistant in the department of chemistry of the newly established Carlsberg laboratory. His chief was the chemist Johan Kjeldal, who developed the method for determining nitrogen in organic substances. Here he was given freedom to work independently. His research centered on the metabolic processes connected with ripening, dormancy, and germination in plants, especially in seeds, tubers, and buds. These were the formative years of Johannsen's scientific life, both with respect to the problems which he chose and stated with great clarity and especially to the exact quantitative methods which he sought and used. Essentially self-taught, Johannsen was widely read in several languages, philosophy, aesthetics, and belles lettres as well as in science.

In 1887 Johannsen resigned his post at the Carlsberg laboratory but with the aid of stipends continued some of his own research there and discovered a method of breaking the dormancy of winter buds.

(First demonstrated in 1893, this work is described in *Das Aether-Verfahren.*) He traveled to Zurich, Darmstadt, and Tübingen for further work in plant physiology. In 1892 he became lecturer, and in 1903 professor, of botany and plant physiology at the Copenhagen Agricultural College.

In 1905 Johannsen was appointed professor of plant physiology at the University of Copenhagen and in 1917 became rector of the university, although he had no university education. His scientific eminence was recognized by the award of several honorary degrees and membership in the Royal Danish Academy of Sciences and in academies and learned societies outside Denmark.

Johannsen was an interesting figure in many ways: as theorist and analytical logician, as discoverer of a major concept in genetics ("pure line theory"), and as critic, clarifier, popularizer, and historian of scientific ideas. His claim on the attention of historians of science has steadily increased since his death, even though, with a few exceptions, his publications were written in Danish. The reason may be that genetics, which in its formative stages (1900–1915) was strongly influenced by Johannsen, has grown in the direction he emphasized. His main concern was with the analysis of the heredity of normal characters which vary quantitatively, such as size, fertility, and degree of response to environmental factors. These provide much of the variety upon which natural selection acts and with which breeders of useful plants and animals are mainly concerned. Through the development of population genetics, the genetic structure of populations, as Johannsen was one of the first to recognize, occupies an increasingly important place in evolutionary biology.

Moreover, Johannsen's view of the unit of heredity, to which he first gave the name "gene" (1909), has survived the changes brought about by the discovery of the physical basis of heredity, first in the chromosomes and then more precisely in the structure of the nucleic acids. He conceived of genes as symbols: *Rechnungseinheiten*, units of calculation or accounting.

Proof of the existence of such elements was of course due to Mendel (1866), but it was Johannsen who first stated clearly the fundamental distinction between the symbolic view of the hereditary constitution of the organism—its genotype, consisting of the totality of its genes—and its phenotype, how it appears and acts. The latter, as Johannsen pointed out, is the observed reality, representing the responses of the organism, as determined by its genotype, to the conditions encountered by the individual during its life history. Introduced by Johannsen in 1909, these terms embody concepts that remain essential in the interpretation of processes of heredity and of evolution.

Johannsen began his study of variability in relation to heredity in the early 1890's. He was strongly influenced by Darwin's work and especially by that of Francis Galton, whose "Theory of Heredity" (*Journal of the Anthropological Institute of Great Britain and Ireland* [1876]) suggested both ideas and quantitative, statistical methods which Johannsen improved. Galton had derived his first so-called law of regression to the mean from the self-fertilizing sweet pea plant. Johannsen chose the common princess bean and discovered, as Galton had, that the seeds of the offspring have the same average weight as the seeds of the parent plant, in spite of selection for higher or for lower weight.

This seemed a remarkable discovery to Johannsen, and he dedicated his paper of 1903, "Über Erblichkeit in Populationen und in reinen Linien," to "the creator of the exact science of heredity, Francis Galton F.R.S. in respect and gratitude." Yet Johannsen had shown that Galton's "law of regression" was wrong when applied as Galton had done—to impure or mixed populations. He proved that selection was ineffective—that is, regression to parental averages was complete—only in the offspring derived by self-fertilization from a single parent, as in peas, beans, and other "selfers." The offspring and descendants from such a plant Johannsen referred to as a "pure line." His theory was that the offspring of a pure line were genetically identical and that fluctuating variability among such offspring was due to effects of chance and of environmental factors. These effects were not heritable and hence were not subject to the action of selection, either natural or artificial. The differences between the different pure lines composing a "population" were inherited and had arisen by mutation, a process then recently invoked (1900) by the Dutch botanist Hugo de Vries as the source of inherited variations which led to the origin of new "elementary species." Johannsen was the first to attribute to mutation the origin of the small differences in the continuous kind of variability characteristic of normal heredity. His proof of the existence of two kinds of variability—heritable and nonheritable—eliminated the need felt by Darwin and other nineteenth-century naturalists for invoking the inheritance of acquired characters as an evolutionary process.

Johannsen had first set forth his views on the nature of the evolutionary process, including the part played by discontinuous variation (mutation) in the little book *Om arvelighed og variabilitet* ("On Heredity and Variation"), issued by the student organization of the Copenhagen Agricultural College in 1896. It proved to contain a preview, not only of the direction of Johannsen's own future work but also of the new attitudes toward Darwin's theory of the mechanism of evolution which arose after the elaboration of de Vries's mutation theory and the rediscovery of Mendel's theory of heredity beginning in 1900. Johannsen's main ideas about heredity, including the assumption of particulate elements, antedated the rediscovery of Mendel's work, of which Johannsen first took public notice on 27 November 1903. Johannsen was clearly prepared to appreciate the significance of Mendel's theory and immediately set about incorporating it into his theory of the evolutionary process. This book was enlarged and reissued in 1905 as *Arvelighedslaerens elementer* ("The Elements of Heredity"). Greatly expanded and rewritten by Johannsen in German, it became in 1909 the first and most influential textbook of genetics on the European continent. About half of the book was devoted to the mathematical and statistical methods needed in the analysis of the quantitative data arising from experiments in genetics. This was what "exakten" in the title meant, not pretensions on the part of a new biology but, rather, delimitation of that part of a heterogeneous field which could be dealt with by quantitative methods applied to verifiable facts. In this book Johannsen defined the basic concepts of a new science—"gene," "genotype," "phenotype"—and forecast the effects to be expected from it upon the central problem of biology, that of the mechanism of organic evolution. The *Elemente* appeared in revised editions, the third (and last) in 1926. Many European biologists owed to it their introduction to genetics.

After his "pure line" work, Johannsen gave up experimentation. His work as critic and historian of science showed the same lively mind of a free-lancer in science as his scientific contributions. *Falske Analogier* (1914) revealed that the kind of logic which Johannsen introduced into genetics was in his hands a tool in a wider crusade to banish obscurantism, teleology, and mysticism from biology. In one chapter he analyzed Henri Bergson's *Évolution créatrice* as "a whole system of false analogies" based on "unverified speculation" and concluded: "It is a pure waste of time to lose oneself in such an author's 'positive' views; they are just not worth a bean" (pp. 102–103). (Beans were Johannsen's chief research material.)

His 1923 book *Arvelighed i historisk og experimentel Belysning* ("Heredity in the Light of History and Experimental Study"), although it appeared only in Danish, went through four editions. It is a lively account of ideas on heredity from the Greeks to T. H. Morgan, author of *The Theory of the Gene*, of whose views Johannsen had been sharply critical.

Johannsen's place in the history of biology may come to be seen as a bridge over which nineteenth-century ideas of heredity and evolution passed to be incorporated, after critical purging, into modern genetics and evolutionary biology.

BIBLIOGRAPHY

I. Original Works. Johannsen's chief writings are *Laerebog i plantefisiologi med henblik paa plantedyrkningen* (Copenhagen, 1892); *Om arvelighed og variabilitet* (Copenhagen, 1896), enl. as *Arvelighedslaerens elementer* (Copenhagen, 1905), rewritten and enl. as *Elemente der exakten Erblichkeitslehre* (Jena, 1909; 3rd ed., 1926); *Das Aether-Verfahren beim Frühtreiben mit besonderer Berücksichtigung der Fliedertreiberei* (Jena, 1900); *Über Erblichkeit in Populationen und in reinen Linien* (Jena, 1903), first published in *K. Danske Videnskabernes Selskabs Forhandlinger*, no. 3 (1903), abridged English trans. in James A. Peters, ed., *Classic Papers in Genetics* (Englewood Cliffs, N.J., 1959); *Falske analogier, med henblik paa lighed, slaegtskab, arv, tradition og udvikling* (Copenhagen, 1914); and *Arvelighed i historisk og experimentel belysning*, 4th ed. (Copenhagen, 1923).

II. Secondary Literature. The following treat Johannsen's life and work: Jean Anker, "Wilhelm Johannsen," in *Danmark*, **5** (1946–1947), 295–300, in Danish; P. Boysen Jensen, "Wilhelm Ludvig Johannsen," in *Københavns Universitets Festskrift* (Nov. 1928), 105–118, in Danish, with a bibliography of over 100 of Johannsen's publications (1883–1927); L. Kolderup Rosenvinge, "Wilhelm Ludvig Johannsen. I. Liv og personlighed," in *Oversigt over det K. Danske Videnskabernes Selskabs Forhandlinger* (1927–1928), 43–68; and Øjvind Winge, "Wilhelm Ludvig Johannsen. II. Videnskabelig Virksomhed," *ibid.*, 64–69; and "Wilhelm Johannsen," in *Journal of Heredity*, **49** (1958), 82–88.

Unpublished materials on Johannsen are in the archives of the Carlsberg laboratory, Copenhagen, and in the library of the American Philosophical Society, Philadelphia.

L. C. Dunn

JOHN BURIDAN. See **Buridan, Jean.**

JOHN DANKO OF SAXONY. See **John of Saxony.**

JOHN DE' DONDI. See **Dondi, Giovanni.**

JOHN OF DUMBLETON (*b.* England; *d. ca.* 1349), *natural philosophy.*

Virtually nothing is known of John of Dumbleton's life. His name suggests that he may have come from the Gloucestershire village of Dumbleton. He is mentioned as a fellow of Merton College, Oxford, at various dates between 1338 and 1348, and in 1340 he was named as one of the original fellows of Queen's College, but he could not have been there long.

His work should be considered in relation to that of other fellows of Merton College, notably Bradwardine, Heytesbury, and Richard Swineshead; but whereas the extant writings of these three deal only with particular problems of natural philosophy, Dumbleton's huge *Summa logicae et philosophiae naturalis* attempts a fairly complete coverage of the topic, providing an invaluable source for opinions current at Oxford in his time. Of the nine extant parts of the *Summa*, the first deals with logic and the remaining eight with natural philosophy. These take their starting point in the Aristotelian writings and consider such subjects as matter and form; intension and remission of qualities; the definition and measure of motion; time; elements and mixtures; light; maxima and minima in physical actions; natural motions; the first mover; whether motion is eternal; the generation of animals; the soul; and the senses. The promised tenth part, which was to deal more fully with the rational soul and to consider the Platonic forms, is not extant, and quite probably Dumbleton died without having written it. In fact, there are indications throughout the treatise of a lack of careful editing.

Most of the subject matter of the *Summa* was, of course, commonplace in medieval discussions; but the techniques that Dumbleton employed were strongly influenced by the more mathematical scientific language for which Merton has become famous, and Dumbleton was always aware of the quantitative aspect of the problems he faced. In the first part of the *Summa*, for example, he considered at length the intension and remission of knowledge and doubt with respect to the evidence available and seemed to be exploring the possibility of a quantitative grammar of assent.

Although Dumbleton favored the Ockhamist definition of motion, this preference did not prevent him from applying to it a thoroughly mathematical treatment; by keeping a very close analogy between the speeds of motions and geometric straight lines, he was able to couch his discussion firmly within the language of the latitudes of forms. He accepted Bradwardine's "law of motion" (which he regarded as being the view of Aristotle and Ibn Rushd) and devoted some space to expounding its consequences. His proof of the "Merton mean speed theorem" is interesting and in some ways reminiscent of the geometric method of exhaustion. He also considered how mathematical techniques could be applied to motions other than local motion. On the less mathematical side he tied his discussion of the continuance of projectile motion to his view that every body has a twofold natural motion: one upward or downward depending on its elementary composition, and another, more primitive motion arising from the desire of "every body. . . to be with another and follow it naturally lest a vacuum be left." He suggested that after a projectile has left the hand it follows the air in front of it by virtue of the second type of natural motion.

One complete part of the *Summa* is devoted to "spiritual action" and more particularly to light, "through which spiritual action is made most apparent to us." This part is based solidly on Aristotle and Ibn Rushd, with the addition of a long mathematical discussion of the intensity of spiritual action. Geometrical optics does not appear here, but in discussing vision, Dumbleton considered how the "lines, triangles, and visual rays" used by writers on perspective are to be reconciled with the Aristotelian interpretation of vision. He concluded that they are mere fictions useful for calculating the position of the image when an object is viewed by reflection.

A full appreciation of Dumbleton's work and its relation to that of his predecessors, contemporaries, and successors awaits much further research. As so often happens, we cannot easily ascertain how much of Dumbleton's discussion is strictly original. He has, nevertheless, left us with much precious evidence relating to a period of intense intellectual activity, and the number of extant manuscript copies (at least twenty-one) testifies to the influence of his *Summa.*

BIBLIOGRAPHY

I. Original Works. For a list of manuscript copies of the *Summa logicae et philosophiae naturalis* see J. A. Weisheipl, "Repertorium Mertonense," in *Mediaeval Studies*, **31** (1969), 174–224. Dumbleton's rather banal *Compendium sex conclusionum* and a small portion of the *Summa* have been edited in J. A. Weisheipl, *Early Fourteenth Century Physics of the Merton "School,"* D.Phil. thesis (Oxford, 1956), pp. 392–436.

II. Secondary Literature. On Dumbleton and his work, see M. Clagett, *The Science of Mechanics in the Middle Ages* (Madison, Wis.–London, 1959); A. C. Crombie, *Robert Grosseteste and the Origins of Experimental Science* (Oxford, 1953); P. Duhem, *Le système du monde*, VII and VIII (Paris, 1956–1958); A. Maier, *Zwei Grundprobleme der scholastischen Naturphilosophie* (Vienna–

Leipzig, 1939–1940; 3rd ed., Rome, 1968); *An der Grenze von Scholastik und Naturwissenschaft* (Essen, 1943; 2nd ed., Rome, 1952); *Zwischen Philosophie und Mechanik* (Rome, 1958); and J. A. Weisheipl, "The Place of John Dumbleton in the Merton School," in *Isis*, **50** (1959), 439–454; "Ockham and Some Mertonians," in *Mediaeval Studies*, **30** (1968), 163–213.

A. G. MOLLAND

JOHN DUNS SCOTUS. See **Duns Scotus, Johannes.**

JOHN OF GMUNDEN (*b.* Gmunden am Traunsee, Austria, *ca.* 1380–1384; *d.* Vienna, Austria, 23 February 1442), *astronomy, mathematics, theology.*

John of Gmunden's origins were long the subject of disagreement. Gmunden am Traunsee, Gmünd in Lower Austria, and Schwäbisch Gmünd were all thought to be possible birthplaces; and Nyder (Nider), Schindel, Wissbier, and Krafft possible family names. Recent research in the records of the Faculty of Arts of the University of Vienna, however, appears to have settled the question. The Vienna matriculation register records the entrance, on 13 October 1400, of an Austrian named "Johannes Sartoris de Gmundin," that is, of the son of a tailor from Gmunden.[1] He was surely the "Johannes de Gmunden" who was admitted to the baccalaureate examination on 13 October 1402.[2] If he was the astronomer John of Gmunden, who was accepted as master into the Faculty of Arts on 21 March 1406, along with eight other candidates, then he spent all his student years at Vienna.[3] His birthplace could only have been Gmunden am Traunsee, since Gmünd and Schwäbisch Gmünd were then known only as "Gamundia"; the locality on the Traunsee, which even in Latin sources, is called "Gmunden" (Gemunden until 1350).[4] Schwäbisch Gmünd must be eliminated from consideration because our John of Gmunden was an examiner of Austrian students, which means he had to be of Austrian birth. The family names Nyder and Schindel can be excluded, but that of Krafft is better established. In his writings and the records of his deanship of the Arts faculty, carefully written in his own hand, he calls himself Johannes de Gmunden exclusively, thus clearly he never used a family name.[5]

John of Gmunden's career can be divided into four periods. In the first (1406–1416) his early lectures—besides one given in 1406 on "Theorice"—were devoted to nonmathematical subjects: "Physica" (1408), "Metheora" (1409, 1411), "Tractatus Petri Hyspani" (1410), and "Vetus ars" (1413).[6] On 25 August 1409 he became *magister stipendiatus* and received an appointment at the Collegium Ducale.[7] He gave his first mathematics lecture in 1412. He was also interested in theology, the study of which he completed in 1415 as "Baccalaureus biblicus formatus in theologia." Two lectures in this field concerned the Exodus (1415) and the theology of Peter Lombard (1416).

In 1416–1425 John of Gmunden lectured exclusively on mathematics and astronomy, which led to the first professorship in these fields at the University of Vienna; the position became permanent under Maximilian I. When John became ill in 1418, he lost his salary, since only someone actively teaching (*magister stipendiatus legens*) could be paid; but, at the request of the faculty, the duke removed this hardship. John obtained permission to hold lectures in his own house—a rarely granted privilege.[8] During his years at the university he held many honorary offices. He was dean twice (1413 and 1423) and examiner of Saxon (1407), Hungarian (1411), and Austrian (1413) students.[9] In 1410 he was named *publicus notarius.*[10] In 1414 he was receiver (bursar) of the faculty treasury and member of the dormitory committee[11] for the bylaws for the *burse.* In 1416 he was "Conciliarius of the Austrian nation," and from 1423 to 1425 he was entrusted with supervising the university's new building program.[12]

The third period (1425–1431) began when John of Gmunden retired from the Collegium Ducale and, on 14 May 1425, became canon of the chapter of St. Stephen.[13] Previously he had been ordained priest (1417) and delivered sermons.[14] He was also vice-chancellor of the university, which had long been closely associated with St. Stephen's Gymnasium.[15] Henceforth John devoted himself to writings on astronomy, astronomical tables, and works on astronomical instruments. He also lectured on the astrolabe.[16]

In the last period (1431–1442) John became *plebanus* in Laa an der Thaya, an ecclesiastical post that yielded an income of 140 guldens.[17] In 1435 he wrote his will, in which he bequeathed his books (particularly those he himself had written) and instruments to the library of the Faculty of Arts. He also gave precise instructions for their use.[18] We note the absence from his list of those books on which his own works were based, but these undoubtedly were in the Faculty of Arts library. John died on 23 February 1442 and was probably buried in St. Stephen's cathedral; no monument indicates where he was laid to rest. Moreover, we possess no likeness of him except for an imaginative representation that shows him wearing a full-bottomed wig.[19]

John of Gmunden's work reflects the goal of the instruction given in the Scholastic universities: to teach science from existing books, not to advance it. He was above all a teacher and an author. His mathematics lectures were entitled "Algorismus de minutiis" (1412, 1416, 1417), "Perspectiva" (1414), "Algorismus de integris" (1419), and "*Elementa* Euclidis" (1421).[20] In this series one does not find a lecture on *latitudines formarum*, which was already part of the curriculum and a topic on which John's teacher, Nicolaus of Dinkelsbühl, had lectured in 1391. John's main concerns, however, lay in astronomy, and thus even in his mathematical writings he treated only questions of use to astronomers. Tannstetter cites three mathematical treatises by him: an arithmetic book with sexagesimal fractions, a collection of tables of proportions, and a treatise on the sine.[21] Only the first of these was printed, appearing in a compendium containing a series of writings that provided the basis for mathematics lectures.[22] In this area John introduced no innovations.[23] For example, in extracting the square root of a sexagesimal number, he first transformed the latter into seconds, quarters, and so forth (therefore into minutes with even index), added an even number of zeros (*cifras*) in order to achieve greater accuracy, then extracted the root, divided through the *medietates cifrarum*, and expressed the result sexagesimally (thus $\sqrt{a} = \frac{1}{10^n}\sqrt{a \cdot 10^{2n}}$). Even the summation of two zodiac signs of thirty degrees into a *signum physicum* of sixty degrees had appeared earlier.[24] The treatise on the sine was recently published.

John's other mathematical works are contained in manuscript volumes that he himself dated (Codex Vindobonensis 5151, 5268).[25] From his writings on angles, arcs, and chords it is clear that Arabic sine geometry was known in Vienna. Yet it is doubtful whether—as has been asserted[26]—John was also acquainted with the formula (corresponding to the cosine law) $\frac{\sin h}{\sin H} = \frac{\sin \operatorname{vers} b - \sin \operatorname{vers} t}{\sin \operatorname{vers} b}$.[27] The formula, which was employed in calculating the sun's altitude for every day of the year, was discovered by Peurbach with "God's help."[28]

John of Gmunden's work in astronomy was of greater importance than his efforts in mathematics. It was through his teaching and writings that Vienna subsequently became the center of astronomical research in Europe. His astronomy lectures were entitled "Theoricae planetarum" (1406, 1420, 1422, 1423), "Sphaera materialis" (1424), and "De astrolabio" (1434).[29] He probably did not himself make any systematic observations of the heavens, but his students are known to have done so. The instruments that he had constructed according to his own designs were used only in teaching and to determine time.[30] He was no astrologer, as can be seen from a letter of September 1432 to the prior Jacob de Clusa, who had made predictions on the basis of planetary conjunctions.[31] Although his library contained numerous astrological writings, the stringent directions in his will pertaining to the lending of these dangerous works show what he thought of this pseudoscience. If occasionally he spoke of the properties of the zodiac signs and of bloodletting, it was because these subjects were of particular interest to purchasers of almanacs.[32]

The great number of extant manuscripts of John of Gmunden's works attests his extensive literary activity, which began in 1415 and steadily intensified until his death. Many of the manuscripts are in his own hand, such as those in Codex Vindobonensis 2440, 5151, 5144, and 5268.[33] Most of those done by students and other scribes date from the fifteenth century. Only a small portion (about twenty of the total of 238 manuscripts that Zinner located in the libraries of Europe) are from a later period. This indicates that his works were superseded by those of Peurbach and Regiomontanus. John's writings can be divided into tables, calendars, and works on astronomical instruments.

John of Gmunden produced five versions of his tabular works, which contained tables of the motions of the sun, the moon, and the planets, as well as of eclipses and new and full moons. They also included explanatory comments (*tabulae cum canonibus*). They are all enumerated in his deed of gift. Regiomontanus studied the first of the tabular works (in Codex Vindobonensis 5268) and found an error in it that he noted in the margin.[34] John was also the author of many individual writings on astronomy that were not made part of the tabular works. Shorter than the latter (with the exception of the tables of eclipses), they contained tables of planetary and lunar motions, of the true latitudes of the planets (with explanations), and of the true new and full moons, as well as tables of eclipses.[35] Further astronomical writings can be found in his works on astronomical instruments.

Along with elaborating and improving the values of his tables, John of Gmunden was especially concerned with the preparation of calendars, which provided in a more usable form the information contained in the tabular works. In addition to such astronomical data, they included the calendar of the first year of a cycle with saints' days and feast days, dominical letter, and the golden numbers, so that the calendar could be used during all nineteen years of a cycle.[36] He brought out four editions of the calendar: the first covered 1415–1434; the second, 1421–1439; the third, 1425–1443; and the fourth, 1439–1514.[37] The fourth edition was printed on Gutenberg's press in 1448.[38] Two other calendars bearing John's name were published later. From one of them, a xylographic work, there remains only a woodblock; of the other, a peasants' calendar, only a single copy is extant.[39]

John of Gmunden's third area of interest was astronomical instruments; he explained how they operated and gave directions for making them. In his deed of gift he mentioned two works in this field: a volume bound in red parchment containing *Astrolabium Alphonsi* and a little book written by himself, entitled *Astrolabii quadrantes.*[40] In his will he lists the following instruments: a celestial globe (*sphaera solida*), an "equatorium" of Campanus with models taken from the *Albion* (devised and written about by Wallingford), an astrolabe, two quadrants, a *sphaera materialis*, a large cylindrical sundial, and four "*theorice lignee*."[41] He stipulated that these instruments should be kept well and seldom loaned out—the equatorium only very seldom (*rarissime*). Of all this apparatus nothing remains in Vienna.[42] On the other hand, about 100 manuscripts of his treatises on astronomical instruments have been preserved. To date no one has made a study of these manuscripts (which contain other works on astronomy) thorough enough to establish, in detail, what John took from his predecessors and what he himself contributed. The instruments he discussed, with regard to both their theoretical basis and their production and use, were the following:

1. The astrolabe. The text is composed of fourteen manuscripts.[43] The star catalog joined to one of them indicates that the first version dates from 1424.[44]

2. The quadrant. John's treatise on the quadrant exists in three versions.

Quadrant I: Fifteen manuscripts are extant of this version of the work, which dates from 1424–1425.[45] Here he drew on a revision from 1359 of the *Quadrans novus* of Jacob Ibn Tibbon (also known as Jacob ben Maḥir or Profatius Judaeus) from 1291–1292 and on a revision from 1359.[46] To these John appended an introduction and remarks on measuring altitudes.[47] Several of the manuscripts also contain additional data that he presented in 1425: a table of the true positions of the sun at the beginnings of the months, star catalogs, and a table of the sun's entrance into the zodiac signs.[48]

Quadrant II: A second, more elaborate version of the work on the quadrant exists in only one manuscript; it no doubt stems from a student, who speaks of a "tabula facta a Johanne de Gmunden, 1425."[49]

Quadrant III: This version, which is independent of John's other writings on the topic, is known in thirteen manuscripts.[50] One of them is dated 1439.[51] Many of the manuscripts contain tables of solar altitudes for every half month and for various localities (calculated or taken from the celestial globe) as well as tables for the shadow curves of cylindrical sundials.[52]

3. The albion. This universal device ("all-by-one"), which combined the properties of the instruments used for reckoning time and location, was devised and built by Richard of Wallingford.[53] His treatise on it (1327) was revised by John of Gmunden, to whom it is often incorrectly attributed. Several manuscripts contain further additions by John, such as a star catalog for 1430 and also (most probably by him) instructions (1433) on the use of the albion in the determination of eclipses.[54] He had earlier used the instrument for this purpose for 1415–1432.[55]

4. The equatorium. This instrument, made of either metal or paper, could represent the motions of the planets.[56] It is found in the thirteenth century in the writings of Campanus and in those of John of Lignères and Ibn Tibbon.[57] John called the device "instrumentum solemne."[58] Following Campanus, he set forth its theoretical basis and described how to make and use it in a work published at the University of Vienna that was highly regarded by Peurbach and Regiomontanus.[59] The manuscripts occasionally also present tables of the mean motion of the sun and moon for 1428.[60]

5. The torquetum. This instrument, whose origin is uncertain, was the subject of a treatise by a Master Franco de Polonia (Paris, 1284); it exists in manuscripts of the fourteenth and fifteenth centuries.[61] John completed the treatise with an introduction and a conclusion. In the latter he stated that with the "turketum" one can determine the difference in longitude between two localities.

6. The cylindrical sundial. The origin of this instrument is likewise unknown; it is described as early as the thirteenth century in an Oxford manuscript.[62] John introduced it to Vienna. His work on it, *Tractatus de compositione et usi cylindri*, exists in nineteen manuscripts.[63] From these it can be inferred that he composed his treatise between 1430 and 1438. He calculated the shadow curves at Vienna, taking the latitude as $\varphi = 47°46'$; in the Oxford manuscript φ is taken as $51°50'$. Shadow curves for other localities also appear in the manuscripts.[64]

A further work on sundials and nocturnals was written by John or by one of his students.[65]

The study of mathematics and astronomy beyond what was offered in the quadrivium first became possible at Vienna through the efforts of Henry of Hesse, who brought back from Paris knowledge of the recent advances in mathematics (as is reported by Petrus Ramus[66]). The first evidence of this is found in the work of Nicolaus of Dinkelsbühl, who taught, besides Sacrobosco's astronomy and Euclid's *Elements*, Oresme's "latitudines formarum." His last lecture (1405), on theories of the planets,[67] may have stimulated the young John of Gmunden to study astronomy. In any case, John studied the relevant available writings, transcribed them, and frequently added to them. Although he seldom mentioned his predecessors, his sources can be inferred to some extent. He was acquainted with the Alphonsine and Oxford tables and knew Euclid from the edition prepared by Campanus, to whose ideas on planetary theories he subscribed.[68] In his first tabular work John cited Robert the Englishman.[69] Moreover, his *Algorismus de minuciis phisicis* undoubtedly follows the account of John of Lignères. In 1433 he transcribed and completed John of Murs's treatise on the tables of proportions; hence many manuscripts name him as the author.[70] It is not clear to what extent his treatise on the sine, arc, and chord depended on the work of his predecessors (Levi ben Gerson, John of Murs, John of Lignères, and Dominic de Clavasio). His dependence on earlier authors is most evident in his writings on astronomical instruments (Campanus' equatorium, Ibn Tibbon's new quadrants, Richard of Wallingford's albion, and Franco de Polonia's torquetum). In his will he also mentions a work entitled *Astrolabium Alphonsi.*

John of Gmunden was influential both through his teaching and, long after his death, through his writings. Among his students Tannstetter mentions especially Georg Pruner of Ruspach. There are transcriptions (in London) by the latter of John's works with remarks by Regiomontanus.[71] John's co-workers included Johann Schindel, Ioannes Feldner, and Georg Müstinger, prior of the Augustinian monastery in Klosterneuburg.[72] Fridericus Gerhard (*d.* 1464–1465), of the Benedictine monastery of St. Emmeran in Regensburg, also had connections with the Vienna school; however, they were indirect, being based on his contacts with Master Reinhard of Kloster-Reichenbach, who worked at Klosterneuburg. Gerhard was a compiler of manuscript volumes that reflected the mathematical knowledge of the age; in them he included works by John of Gmunden, drawn partially from a lecture notebook of 1439.[73] Gerhard was particularly interested in geography; and geographical coordinates play a role in astronomy. Thus it is quite possible that his knowledge in this area also came from Vienna.[74]

John of Gmunden greatly influenced Peurbach. To be sure, the latter cannot be considered a direct student of John's since Peurbach was nineteen when John died. Nevertheless, he undoubtedly knew John personally, and he studied his writings thoroughly. The same is known of Regiomontanus, who in his student years at Vienna made critical observations in copies of John's works.[75] He also bought a copy of the treatise on the albion.[76]

The outstanding achievements of Peurbach and Regiomontanus resulted in John of Gmunden's being overshadowed, as can be seen from the small number of manuscripts of his works that date from after the fifteenth century. Yet it was he who initiated the tradition that was established with Peurbach, whose scientific reputation caused the young Regiomontanus to come to Vienna (1450) instead of staying longer at Leipzig.

NOTES

1. *Matrikel der Universität Wien*, I, p. 57. Another John Sartoris de Gmunden is listed in the register for 14 April 1403 (I, p. 65). He cannot yet have been a master in 1406. He may have been the son of another tailor, or it may be a second matriculation.
2. *Acta Facultatis artium Universitatis Viennensis 1385–1416*, p. 212. (Cited below as *AFA*.)
3. *AFA*, p. 261. John of Gmunden is the only one of all the candidates for whom no family name is given; but if it is true (letter from P. Uiblein) that the lectures were distributed to the *magistri legentes* in the same sequence in which they received their *magisterium*, then John is identical with Krafft. Johannes Wissbier of Schwäbisch Gmünd studied at Ulm in 1404. See M. Curtze, "Über Johann von Gmunden," in *Bibliotheca mathematica*, **10**, no. 2 (1896), 4.
4. R. Klug, *Johann von Gmunden*, p. 14. Several variants exist in the MSS, including Gmund, Gmunde, Gmundt, and Gmundia. Concerning the two dots over the "u" in "Gmünden" see *ibid.*, p. 16. It is not an *Umlaut* but a vowel mark; on this point see H. Rosenfeld, in *Studia neophilologica*, **37** (1965), 126 ff., 132 f.
5. The name Nyder does not appear, as has been asserted, in the obituaries of the canons of St. Stephen. See J. Mundy, "John of Gmunden," p. 198, n. 29. Regarding Schindel, it is a matter of a change of name in the Vienna MSS. See Klug, *op. cit.*, p. 17, Curtze, *op. cit.*, p. 4. In 1407 this scholar left Prague for Vienna, where he taught privately for several years. He is mentioned with praise along with Peurbach and Regiomontanus by Kepler in the preface to the *Rudolphine Tables.* A Master Johannes Krafft lectured on the books of Euclid in 1407 (*AFA*, p. 281). Others also lectured on Euclid around this time (*AFA*, pp. 253, 292, 453).
6. *AFA*, pp. 292, 325, 365, 338, and 401.
7. *AFA*, p. 324: "magistro Iohanni de Gmunden data fuit regencia"; Klug, pp. 18 f.
8. J. Aschbach, *Geschichte der Wiener Universität im ersten Jahrhundert ihres Bestehens*, p. 457.
9. On his deanship see *AFA*, p. 405; Aschbach, *op. cit.*, p. 458. On his posts as examiner see *AFA*, pp. 284, 370, 402. In 1413 he was also examiner of candidates for the licentiate (*AFA*, p. 391).
10. *AFA*, p. 345.
11. *AFA*, p. 421.
12. *AFA*, p. 472; Aschbach, *loc. cit.*
13. *AFA*, p. 530.
14. Mundy, *op. cit.*, p. 199; Klug, *op. cit.*, pp. 20 f.
15. Aschbach, *op. cit.*, p. 459; *AFA*, p. 530.
16. Klug, *op. cit.*, p. 18. Even if he retired in 1434, he was already a clergyman in Laa.
17. *AFA*, p. 530; Klug, *op. cit.*, p. 21.

18. The text of the will is given in Mundy, *op. cit.*, p. 198; Klug, *op. cit.*, pp. 90 ff.
19. E. Zinner, *Leben und Wirken des Johann Müller von Königsberg, genannt Regiomontanus*, p. 196, n. 16. (Cited below as *ZR*.)
20. *AFA*, pp. 381, 430; Klug, *op. cit.*, p. 18. See also Note 5.
21. G. Tannstetter, in *Tabulae eclypsium Magistri Georgii Peurbachii*, wrote: "Libellum de arte calculandi in minuciis phisicis, Tabulas varias de parte proporcionali, Tractatum sinuum" (fol. aa 3[v]).
22. The title of the compendium is *Contenta in hoc libello*; on this point see D. E. Smith, *Rara arithmetica*, p. 118.
23. See Simon Stevin, *De thiende*, H. Gericke, K. Vogel, ed., in Ostwald's Klassiker der Exacten Wissenschaften, n.s. 1 (Braunschweig, 1965), pp. 47 ff.
24. Mundy, *op. cit.*, p. 199.
25. The treatise on tables of proportions, along with explanatory comments, is cited in E. Zinner, *Verzeichnis der astronomischen Handschriften des deutschen Kulturgebietes*, nos. 3585, 3586, 3695, and 3696 (dated for 1433 and 21 May 1440); two treatises on the sine, chord, and arc are noted in this work by Zinner (cited below as *ZA*) as nos. 3591 and 3592; the letter (from Codex Vindobonensis 5268) was published by Busard (Vienna, 1971); then follows a work on Euclid, also by John of Gmunden.
26. Klug, *op. cit.*, p. 50.
27. Here H = meridian altitude, b = semidiurnal arc, and t = horary angle.
28. Codex Vindobonensis 5203, fol. 54r; *ZR*, p. 25—here $\cos b$ should be altered to $(\cos b - 1)$.
29. Klug, *op. cit.*, p. 18.
30. *ZR*, p. 15. The *Tabula de universo* undoubtedly was also conceived for use at the university; on this point see Mundy, *op. cit.*, p. 206; Klug, *op. cit.*, pp. 63 ff.
31. *ZA*, no. 3584.
32. *ZA*, no. 3732; Mundy, *op. cit.*, p. 204.
33. *ZR*, p. 16.
34. The MSS of the tabular works are as follows:
    Version I: Codex Vindobonensis 5268, fols. 1v–34r (*ZA*, no. 3587; *ZA*, no. 3588 is an extract), with explanation in *ZA*, no. 3589 (from 10 August 1437); the tables were valid for Vienna during 1433, 1436, and 1440, among other years. For a marginal notation by Regiomontanus ("non valet. Nam in alio circulo sumitur declinatio et in alio latitudo") see *ZR*, p. 43.
    Version II: *ZA*, no. 3709 (for 1400).
    Version III: *ZA*, nos. 3710–3716, also *ZA*, no. 3719. MS 3711 (in Codex Vindobonensis 5151) was written by John of Gmunden himself, as was *ZA*, no. 3694 (of 20 May 1440).
    Version IV: *ZA*, nos. 3717, 3718. A student's transcription in a MS at the British Museum is dated 1437; on this point see Mundy, *op. cit.*, p. 202, n. 73.
    Version V: *ZA*, nos. 3691–3693, with explanatory material at nos. 3688–3690 (for 1440, 1444, 1446).
35. On planetary and lunar motions see *ZA*, nos. 3496, 3720, 3721, 3723; moreover, the tables in *ZA*, nos. 11203, 11204, and 11207 ("quamvis de motibus mediis . . .") stem from John of Gmunden (*ZA*, p. 523). On the true planetary latitudes see *ZA*, nos. 3697, 3699 (from 21 and 25 May 1440), 3698, 3700. On the true new and full moons see *ZA*, nos. 3702, 3707, 3708, 3733, 3734. Tables of eclipses are at *ZA*, nos. 3498, 3701, 3703–3706, 3735 (for 1433 and 1440). Further works by him are undoubtedly *ZA*, nos. 3590 (astronomy), 3725 (position of the heavenly spheres), and 3729 (intervals between heavenly bodies). Codex latinus monaiensis 10662, fols. 99v–102r, contains a "Tabula stellarum per venerabilem Joh. de Gmunden" for 1430, and Codex latinus monaiensis 8950, fols. 81r–92v, a treatise on the "radices" of the sun, moon, and "Caput draconis." See Mundy, *op. cit.*, p. 201, n. 72.
36. Tannstetter (fol. aa 3[v]) records that he left to the library a "Kalendarium quod multis sequentibus annis utile erat et jucundissimum" (perhaps *ZA*, no. 3606).
37. 1st calendar: *ZA*, nos. 3499–3502 (four MSS).
    2nd calendar (nine MSS): *ZA*, nos. 3503–3511, 5378. This was the calendar for whose publication John of Gmunden obtained permission. See Mundy, *op. cit.*, p. 201, n. 71; Klug, *op. cit.*, p. 91.
    3rd calendar (fifteen MSS): *ZA*, nos. 3512–3526; no. 3513 dates from 1431.
    4th calendar: Exists in eighty MSS (Zinner, in 1938, knew of ninety-nine copies [*ZR*, p. 15]): *ZA*, nos. 3606–3687, of which three date from 1439; this calendar was announced at the University of Vienna (*ZA*, p. 425). An extract with explanation by John of Gmunden exists in MS 12118 (*ZA*, p. 536).
38. J. Bauschinger and E. Schröder, "Ein neu entdeckter astronomischer Kalender für das Jahr 1448."
39. Mundy, *op. cit.*, p. 203; Klug, *op. cit.*, pp. 79 ff. (with illustrations on p. 81 and plates VIII and IX); Aschbach, *op. cit.*, pp. 465 ff.
40. Each astronomical instrument served to "grasp the stars" (λαμβάνειν τὰ ἄστρα).
41. Tannstetter (fol. aa 3[v]) simply groups Campanus' instrument ("equatorium motuum planetarum ex Campano transsumptum") and almost all the others as "Compositio Astrolabii & utilitates eiusdem & quorundam aliorum instrumentorum."
42. An ivory quadrant in the Kunsthistorisches Museum in Vienna was undoubtedly designed by John of Gmunden. See *ZR*, p. 16; Klug, *op. cit.*, p. 26.
43. *ZA*, nos. 3593, 3593a–3605.
44. *ZA*, no. 3593; *ZA*, no. 3602 contains still another star catalog as well as tables for the rising of the signs and for the entrance of the sun into the signs for the year 1425.
45. *ZA*, nos. 3555–3569.
46. *ZA*, p. 424. On Ibn Tibbon see R. T. Gunther, *Early Science in Oxford*, II, 164.
47. *ZA*, p. 468.
48. *ZA*, nos. 3556, 3557, 3569 for the table of true positions; *ZA*, nos. 3557, 3559, 3561, 3562 for the star catalogs; and for the sun's entrance into the zodiac signs *ZA*, nos. 3557 and 3564 for 1424; no. 3568 for 1425; and no. 3569.
49. *ZA*, no. 3570; additional material in *ZA*, no. 3724 contains tables of equatorial altitudes. In the same volume of MSS there is also an essay on solar quadrants that is by either John of Gmunden or a student of his (*ZA*, no. 3731).
50. *ZA*, nos. 3571–3583.
51. *ZA*, no. 3578.
52. For solar attitudes see *ZA*, nos. 3572, 3578, 3580 for Vienna, Nuremberg, Klosterneuburg, Prague, Venice, Rome, and the town of "Cöppt"; for tables of shadow curves see *ZA*, no. 3578 for the places named and for Regensburg. In addition, nos. 3577, 3579, 3583 have tables for Vienna; and no. 3576 has them for Vienna and Nuremberg.
53. Gunther, *op. cit.*, pp. 49 f., 349 ff.; *ZA*, nos. 11584–11586; p. 52 g.
54. *ZA*, nos. 11590–11593, 11596, and p. 529.
55. *ZA*, no. 3498.
56. Perhaps the wooden instruments (*theorice lignee*) named in the will are such equatoria.
57. Gunther, *op. cit.*, p. 234.
58. *ZA*, no. 3527–3535; s. 2A p. 423.
59. John of Gmunden's explanation of Campanus' work is *ZA*, no. 1912; construction and use of the instrument is at *ZA*, p. 423.
60. *ZA*, nos. 3527, 3531.
61. *ZA*, nos. 2787–2800 and p. 416. See also G. Sarton, *Introduction to the History of Science*, II, 1005 and III, 1846; and Gunther, *op. cit.*, pp. 35, 370 ff.
62. Gunther, *op. cit.*, p. 123. There had been a MS in Germany since the fourteenth century.
63. *ZA*, nos. 3536–3554.

64. Venice, Rome, Nuremberg, Prague, and Klosterneuburg; they stem in part from John of Gmunden and in part from Prior Georg. Moreover, tables of solar altitudes for Vienna, Nuremberg, and Prague are appended to some of the MSS (*ZA*, p. 424).
65. *ZA*, nos. 3722, 3726, 3727, 3730.
66. Petrus Ramus, *Mathematicarum scholarum, libri duo* (Basel, 1569), p. 64: "Henricus Hassianus . . . primo mathematicas artes Lutetia Viennam transtulit" (Aschbach, *op. cit.*, p. 386, n. 3).
67. *AFA*, p. 253.
68. A star catalog is completed in *ZA*, no 452. See also *ZA*, p. 390. The tables of the planetary latitudes in *ZA*, no. 3700 were taken from the Oxford tables (*ZA*, p. 426). On Campanus' edition of Euclid, see *ZA*, no. 1912 and p. 405.
69. *ZR*, p. 14; Mundy, *op. cit.*, p. 200.
70. *ZA*, no. 7423; on his being considered the author, see *ZA*, p. 475.
71. *ZR*, pp. 15, 43; Mundy, *op. cit.*, pp. 197, 202, n. 73.
72. *ZA*, p. 529: "Selder"; Mundy, *op. cit.*, p. 197. Tannstetter, fol. aa 3[v]: Schinttel.
73. *ZR*, pp. 50 f.; see also *ZA*, nos. 3565, 3578, 10979, 11198, 11205, 11206.
74. D. B. Durand, "The Earliest Modern Maps of Germany and Central Europe," p. 498; Mundy, *Eine Schrift über Orts Koordinaten*, in *ZA*, 3728.
75. *ZR*, p. 43.
76. *ZR*, pp. 53, 218; here one can find references to other works by John of Gmunden that Regiomontanus studied (and in part copied).

BIBLIOGRAPHY

I. Original Works. The only works by John of Gmunden to be printed, except for the posthumous calendars mentioned above, were the treatise on the sine (see below on Busard) and the "Algorithmus Magistri Joannis de Gmunden de minuciis phisicis," which appeared in a compendium prepared by Joannes Sigrenius, entitled *Contenta in hoc libello* (Vienna, 1515). A facsimile of the title page is given in Smith, *Rara arithmetica*, p. 117. An extract on the finding of roots is in C. J. Gerhardt, *Geschichte der Mathematik in Deutschland* (Munich, 1877), pp. 7 f. John of Gmunden's will is published in Mundy, "John of Gmunden," p. 198; and in Klug, "Johann von Gmunden," p. 90 ff., with a facsimile of the first page. All his other writings are preserved only in MS; they were compiled by Zinner in *ZA*.

Three MSS are obtainable in microfilm or photostatic reproduction (Document 1645 of the American Documentation Institute, 1719 N Street, N.W., Washington, D.C.; in this regard see Mundy, *op. cit.*, p. 196):

1. Codex latinus Monaiensis 7650, fols. 1r–8r (the calendar cited at *ZA*, no. 3524).

2. Cod. lat. Mon. 8950, fol. 81v ("Proprietates signorum," *ZA*, no. 3732).

3. Cod. St. Flor. XI, 102, 1rv (letter to Jacob de Clusa, *ZA*, no. 3584; this letter was published in Klug, *op. cit.*, pp. 61 ff.).

II. Secondary Literature. See *Acta Facultatis artium Universitatis Viennensis 1385–1416*, P. Uiblein, ed. (Vienna, 1968), a publication of the Institut für Österreichische Geschichtsforschung, 6th ser., Abt. 2; J. Aschbach, *Geschichte der Wiener Universität im ersten Jahrhundert ihres Bestehens* (Vienna, 1865), pp. 455–467; J. Bauschinger and E. Schröder, "Ein neu entdeckter astronomischer Kalender für das Jahr 1448," in *Veröffentlichungen der Gutenberg-Gesellschaft*, **1** (1902), 4–14; D. B. Durand, "The Earliest Modern Maps of Germany and Central Europe," in *Isis*, **19** (1933), 486–502; R. T. Gunther, *Early Science in Oxford*, II, *Astronomy* (Oxford, 1923); R. Klug, "Johann von Gmunden, der Begründer der Himmelskunde auf deutschem Boden. Nach seinen Schriften und den Archivalien der Wiener Universität," Akademie der Wissenschaften Wien, Phil.-hist. Kl., Sitzungsberichte, **222**, no. 4 (1943), 1–93; *Die Matrikel der Universität Wien*, I, *1377–1435*, ed. by the Institut für Österreichische Geschichtsforschung (Vienna, 1956); J. Mundy, "John of Gmunden," in *Isis*, **34** (1942–1943), 196–205; G. Sarton, *Introduction to the History of Science*, III (Baltimore, 1948), 1112 f.; D. E. Smith, *Rara arithmetica* (Boston–London, 1908, 117, 449); G. Tannstetter, *Tabulae eclypsium Magistri Georgii Peurbachii. Tabula primi mobilis Joannis de Monteregio* (Vienna, 1514), which contains *Viri mathematici quos inclytum Viennense gymnasium ordine celebres habuit*, fol. aa 3[v]; L. Thorndike and P. Kibre, *A Catalogue of Incipits of Mediaeval Scientific Writings in Latin* (London, 1963), index, p. 1838; and E. Zinner, *Verzeichnis der astronomischen Handschriften des deutschen Kulturgebietes* (Munich, 1925), 119–126; and *Leben und Wirken des Johann Müller von Königsberg, genannt Regiomontanus* (Munich, 1938), pp. 14 ff. and index, p. 284.

See also H. H. Busard, "Der Traktat De sinibus, chordis et arcubus von Johannes von Gmunden," in *Denkschriften der Akademie der Wissenschaften*, **116** (Vienna, 1971), 73–113.

Kurt Vogel

JOHN OF HALIFAX. See **Sacrobosco, Johannes de.**

JOHN OF HOLYWOOD. See **Sacrobosco, Johannes de.**

JOHN LICHTENBERGER. See **Lichtenberger, Johann.**

JOHN OF LIGNÈRES, or **Johannes de Lineriis** (*fl.* France, first half of fourteenth century), *astronomy, mathematics.*

Originally from the diocese of Amiens, where any of several communes could account for his name, John of Lignères lived in Paris from about 1320 to 1335. There he published astronomical and mathematical works on the basis of which he is, with justice, credited with diffusion of the Alfonsine tables in the Latin West.[1]

JOHN OF LIGNÈRES

In astronomy the work of John of Lignères includes tables and canons of tables, a theory of the planets, and treatises on instruments. The tables and the canons of tables have often been confused among themselves or with the works of other contemporary Paris astronomers named John: John of Murs, John of Saxony, John of Sicily, and John of Montfort. There are three canons by John of Lignères.

1. The canons beginning *Multiplicis philosophie variis radiis* . . . are sometimes designated as the *Canones super tabulas magnas;* they provide the daily and annual variations of the mean motions and mean arguments of the planets in a form which, although not the most common, is not exceptional. The tables of equations, on the other hand, are completely original: one enters them with both mean argument and mean center at the same time and reads off directly a single compound equation, the sum of the equation of center and the corrected equation of argument; it is sufficient to add this compound equation to the mean motion in order to obtain the true position. The tables also permit the calculation of the mean and true conjunctions and oppositions of the sun and the moon; but there is no provision for the determination of the eclipses and for planetary latitudes. The radix of the mean motions is of the time of Christ, but the longitude is not specified (in all probability that of Paris); nevertheless, since the list of apogees is established by reckoning from 1320, *Tabule magne* may be dated approximately to that year.

The canons are dedicated, as is the treatise on the saphea (see below), to Robert of Florence, dean of Glasgow in 1325.[2] It is not certain whether these tables were calculated on the basis of the Alfonsine tables, which John of Lignères would therefore already have known.[3] Although certain characteristics (the use of physical signs of thirty degrees and not of natural signs of sixty degrees) are not decisive, the tables of the equations of Jupiter and Venus appear to be calculated following neither those of the Alfonsine tables nor the Ptolemaic eccentricities:[4] insofar as one can judge on the basis of tables from which it is difficult to derive the equation of center and the equation of argument, the eccentricities used are not the customary ones.

The text of the canons of John of Lignères is very concise. It was the object of an explanatory effort by John of Speyer, *Circa canonem de inventione augium*. . ., which may have been written in 1348, since it contains an example calculated for that year.[5]

2. In 1322[6] John of Lignères composed a set of tables completely different from the preceding. (One cannot tell, however, if they are earlier or later than the *Tabule magne*.) These tables, and especially their canons, are in three parts often found separately, particularly the first. The canons of the *primum mobile* (*Cujuslibet arcus propositi sinum rectum* . . .), in forty-four chapters, correspond to the trigonometric part of the tables and consider the problems linked to the daily movement of the sun: trigonometric operations, the determination of the ascendant and of the celestial houses, of equal and unequal hours, and so on. Three of the canons describe the instruments used in astronomical observation: Ptolemaic parallactic rulers and a quadrant firmly fixed in the plane of the meridian. The corresponding part of the tables is thus made up of a table of sines, a table of declinations (the maximum declination is 23°33′30″), and tables of right and oblique ascensions for the latitude of Cremona and of Paris (seventh clima). This portion of the canons is not dated, and the contents do not provide any chronological information whatever; but it is reasonable to suppose that the canons were published at the same time as the tables and the canons of the movements of the planets, with which they form a harmonious ensemble.

The canons of the *primum mobile* were the object of a commentary, accompanied by many worked-out examples, by John of Saxony (*ca.* 1335): *Quia plures astrologorum diversos libros fecerunt* . . . . One of the canons of the *primum mobile*, no. 37, concerning the equation of the celestial houses, was printed at the end of the canons by John of Saxony that were appended to the edition of the Alfonsine tables published by Erhard Ratdolt in 1483.

The canons of the movements of the planets form the second part of the treatise whose first part comprises the canons of the *primum mobile: Priores astrologi motus corporum celestium* . . . . They treat the conversion of eras (very briefly), the determination of the true positions of the planets and of their latitudes, the mean and true conjunctions and oppositions of the sun and the moon and their eclipses, the coordinates of the stars, and the revolution of the years. The corresponding tables give (*a*) chronological schemes which permit the conversion from one era to another with a sexagesimal computation of the years; (*b*) the mean motions and mean arguments of the planets for groups of twenty years (*anni collecti*), single years (*anni expansi*), months, and days—both for the epoch of the Christian era and 31 December 1320 at the meridian of Paris; and (*c*) the tables of equations according to the usual presentation.

These canons, which are usually dated by their explicit references to 1322, allude to the Alfonsine tables, to which the tables of John of Lignères are certainly related, if only by the adoption of the

JOHN OF LIGNÈRES

motion compounded from precession and accession and recession for the planetary auges. Yet the longitudes of the stars, established by adding a constant to Ptolemy's longitudes, are not the same as those of Alfonso X. Nor are the tables of equations of Jupiter and Venus those ordinarily found in the Alfonsine tables or those of the *Tabule magne* but, rather, the equations of the Toledan tables. On the other hand, John of Lignères used the physical signs of thirty degrees and not the natural signs of sixty degrees. J. L. E. Dreyer observed that the Castilian canons of the Alfonsine tables in their original version, as they were published by Rico y Sinobas, do not correspond at all to the tables commonly designated as Alfonsine; he believed he had found them in a state closer to the original in the Oxford tables constructed by William Reed with the year 1340 as radix.[7] It is reasonable to suppose that the tables of John of Lignères represent, twenty years before Reed, an analogous effort to reduce the Castilian tables to the meridian of Paris; he preserved their general structure, and notably three characteristics: the compound motion of the auges (precession and accession and recession), physical signs of thirty degrees, and tables of periodic movements presented on the basis of twenty-year *anni collecti.*

The part of the canons of 1322 dealing with the determination of eclipses frequently appears separately (*Diversitatem aspectus lune in longitudine et latitudine*), despite the many references made to the preceding canons—references that thereby lose all significance.

In 1483 some of the canons on the planetary motions were included in the edition of the Alfonsine tables and with the canons of John of Saxony: canons 21–23, concerning the determination of the latitudes of the three superior planets and of Venus and Mercury, and canons 38 and 40, on eclipses.

3. Finally, John of Lignères wrote canons beginning *Quia ad inveniendum loca planetarum . . .* in order to treat the tabular material ordinarily designated as Alfonsine tables.[8] The signs are the natural signs of sixty degrees, and the tables of the mean motions and mean arguments of the planets consist of the sexagesimal multiples of the motions during the day. The chronological portion has, obviously, had to be increased since with this system, in order to enter the tables of mean motions, it is necessary to transform a date expressed in any given calendar into a number of days in sexagesimal numeration. As it is certain that the sexagesimal form of the Alfonsine tables does not represent their original state, there is firm evidence for believing that the transformation which they underwent was carried out at Paris in the 1320's, either by John of Lignères himself, under his direction, or under the direction of John of Murs. The date of these very succinct new canons cannot be determined from the text; but it is certainly later than that of the canons *Priores astrologi . . .* (1322) and may perhaps be earlier than 1327, when John of Saxony produced a new version of the canons of the Alfonsine tables.

For those who are aware of the almost universal diffusion of the Alfonsine tables at the end of the Middle Ages, almost to the exclusion of any other tables, there is no need to emphasize John of Lignères's exceptional role in the history of astronomy. The magnitude of the work that he and his collaborators accomplished in so few years is admirable. Although there is no formal proof of the existence of a team of workers, the terms in which John of Saxony expressed his admiration for his "maître" bear witness to the enthusiasm that John of Lignères evoked.

In order to complete the account of John of Lignères's work on astronomical tables, we must notice the execution of an almanac conceived, like that of Ibn Tibbon (Profacius), on the principle of the "revolutions" of each planet and therefore theoretically usable in perpetuity, provided a correction is applied based on the number of revolutions intervening since the starting date (1321) of the almanac. The work appears to be preserved in only one manuscript, unfortunately incomplete, with a short canon: *Subtrahe ab annis Christi 1320 annos Christi . . . .*

John of Lignères's theory of the planets, *Spera concentrica vel circulus concentricus dicitur . . .*, represents the theoretical exposition of the principles of the astronomy of planetary motions, the application of which is furnished by the Alfonsine tables. In particular this theory provides a detailed justification of the compound motion of the eighth sphere; in it the author strives to demonstrate at length the inanity of the solution recommended by Thābit ibn Qurra (a motion of simple accession and recession). Furthermore, he promises to return, in a work which it is not known whether he wrote, to certain difficulties remaining under the Alfonsine theory. John of Lignères provided no indication of the values of the planetary eccentricities, of the lengths of the radii of the epicycles, or of the values of the various motions at any particular date. The only precise information, the reference to the position of the star Alchimech in 1335, allows the text to be dated about that year.

John of Lignères's astronomical work also included treatises on three instruments: the saphea, the equatorium, and the directorium. The saphea is an astrolabe with a peculiar system of stereographic projection: the pole of projection is one of the points of intersection of the equator and the ecliptic, and the plane of projection is that of the colure of the solstices.

 JOHN OF LIGNÈRES

Following a rather clumsy effort by William the Englishman in 1231 to reconstruct the principle of an instrument attributed to al-Zarqāl that he no doubt had never seen, the saphea was introduced in the West by the translation, done by Ibn Tibbon in 1263, of al-Zarqāl's treatise. The saphea described by John of Lignères (*Descriptiones que sunt in facie instrumenti notificate . . .*) presents technically several improvements over al-Zarqāl's instruments.[9] The most notable is the use of a kind of rete, the *circulus mobilis*, consisting of a graduated circle of the same diameter as the face and an arc of circle bearing the stereographic projection of the northern half of the zodiac, as in the classic astrolabe. The diameter that subtends this projection of the zodiac carries a graduation similar to almucantars on the meridian line of the astrolabe's tablet; a rule graduated in the same manner can be mounted on the *circulus mobilis*, forming a given angle with the diameter of the latter. On such an instrument one may consider either (1) one of the diameters of the face as a horizon, in the projection which characterizes the saphea: the diameter of the *circulus mobilis* then serves to refer to this horizon every position located in the unique system of the almucantars and of the azimuths traced on the instrument for the diameter of the horizon (this is the principle used in al-Zarqāl's canons); or (2) the almucantars of the face as the horizons of a tablet of the horizons in a classic stereographic projection bounded by the equator: the half of the ecliptic traced on the *circulus mobilis* then plays the same role as the ordinary rete. The judicious alternate use of both systems allowed John of Lignères to offer simpler and more rapid solutions to the problems dealt with in al-Zarqāl's canons without losing any of the saphea's advantages.

John of Lignères wrote two treatises on the equatorium. The first, *Quia nobilissima scientia astronomie non potest . . .*, is an adaptation of Campanus' instrument.[10] In order to find the true positions of the planets, Campanus recommended a series of three disks, that is, six "instruments" (one on each face of one disk), which reproduced fairly closely the schema of the geometric analysis of the planetary motions. John of Lignères maintained this principle but simplified the construction by adopting a common disk to bear the equants of all the planets (but not the moon). To avoid difficulties in reading, the equants are represented by a circle without graduations; and these ones, which begin at a different point for each planet, are replaced by a graduated ring which is superposed on the equant in the position suitable to the planet for which one is operating. The radii of the deferents are represented by a small rule bearing, on one side, a nail to be fixed in the center of the deferent of the planet, and on the other, an epicycle at the center of which is turning another small rule bearing, at appropriate distances, the "bodies" of the planets. Two threads represent the radius of the equant which measures the mean center and the radius of the zodiac which passes through the planet.

The other treatise on the equatorium (uses: *Primo linea recta que est in medio regule . . .*; construction: *Fiat primo regula de auricalco seu cupro . . .*[11]) is fundamentally different from the first. The problem is no longer to reproduce the geometric construction of a planet's true position but, rather, to calculate graphically, so to speak, the angular corrections (equation of center and equation of argument) that, added to the mean motion of a planet, determine its true position. The sole function of the instrument's five parts (the so-called ruler of the center of the epicycle, the disk of the centers, the epicycle, the square carrying the "bodies" of the planets, and the rule for reading off) therefore is to furnish and to position in relation to each other the parameters of the planets (eccentricities, epicycle radii). Successively determined—exactly as in a calculation carried out with the Alfonsine tables—are the equation of center on the basis of the mean center, then the equation of argument on the basis of the true center and the true argument; the true position is obtained by adding the two equations to the mean motion.

The astrologer Simon de Phares, whose account of John of Lignères is otherwise fairly correct, attributes to him a directorium the incipit of which ("*Accipe tabulam planam rotundam cujus . . .*") corresponds very closely to that of a text on this instrument preserved in at least four anonymous manuscripts: "Accipe tabulam planam mundam super cujus extremitatem . . . ."[12] The directorium was used only for astrology: it served to "direct" a planet or a point in the zodiac having a particular astrological value, that is, to lead it to another point in the zodiac by counting the degrees of the equator corresponding to this course. In fact, it is very similar to the astrolabe, except that the fixed celestial reference sphere, represented only by the horizon of the place and by the meridian line, is made to turn above the sphere of the stars and of the zodiac. Since, in a good calculation of "direction," the latitude of the planets must be taken into account, the zodiac is represented by a wide band on which are traced its almucantars and its azimuths, as far as six degrees on either side of the ecliptic. John of Lignères's directorium presents no special features.

Finally, a Vatican manuscript attributes to John of Lignères an "armillary instrument" that is difficult to define (*Rescriptiones* [read *Descriptiones*] *que sunt in facie instrumenti notificare. Trianguli equilateri ex tribus quartis arcus circuli magni . . .*): in the absence of a section on its construction, the uses and the brief description that precede them give a very imperfect idea of the instrument, which appears to derive from the new quadrant. John of Lignères's idea seems to have been to replace the rotation of the margarites (which, in the new quadrant, compensated for the immobility in which this instrument held the rete of the astrolabe because of the reduction of the latter to one of its quarters)[13] by the rotation of another quarter-disk bearing the oblique horizon. This conception amounted to a return to what had constituted the justification of the stereographic projection characterizing the astrolabe, that is, to the rotation of the sphere of the stars and of the zodiac on the celestial sphere used for reference but with a reversal, as in the directorium, of the respective roles of the spheres.[14]

None of John of Lignères's treatises on astronomical instruments is dated or contains information from which a date can be established. Nevertheless, the preface to the canons of the *Tabule magne*, addressed to Robert of Florence, notes the simultaneous sending, along with the tables, of an equatorium and a "universal astrolabe." The latter should be identified with the saphea; as for the equatorium, defined as suitable to furnish "easily and rapidly the equations of the planets," it is more likely to be the second of the instruments described above.[15]

We have seen the development that John of Lignères gave to sexagesimal numeration in the astronomical tables, since the tables of the regular movements of the planets in the Alfonsine tables have been modified so as to permit the systematic use of this type of numeration. He was so aware of the astronomer's need for its use that he introduced, at the beginning of the canons of the *Tabule magne*, a long section on the technique of working on the "physical minutes." He took up the question again and expounded it in the *Algorismus minutiarum*, in which he simultaneously treated physical fractions and vulgar fractions. Its great success is attested to by the number of manuscripts in which the *Algorismus* is preserved.

NOTES

1. P. Duhem, *Le système du monde*, IV (Paris, 1916), 578–581, following G. Bigourdan, maintains that John of Lignères was alive after 1350; he bases this on a letter from Wendelin to Gassendi that mentions the positions of the stars determined by John of Lignères and reproduced by John of Speyer in his *Rescriptum super canones J. de Lineriis*. As long as John of Speyer's work had not been found, one could—just barely—give credence to this tale. But the *Rescriptum* of John of Speyer, identified through MS Paris lat. 10263—see E. Poulle, *La bibliothèque scientifique d'un imprimeur humaniste au XV*[e] *siècle* (Geneva, 1963), p. 49—and dating from about 1348, makes no reference to any table of stars.
2. G. Sarton, *Introduction to the History of Science*, III (Baltimore, 1947), 649*n*.
3. Despite the title of the MS Paris lat. 7281, fol. 201v: *Canones super tabulas magnas per J. de Lineriis computatas ex tabulis Alfonsii* (in another hand: *ad meridianum Parisiensem*).
4. The equations of Jupiter and Venus given in the Alfonsine tables use simultaneously two values for the eccentricities of these planets. See E. Poulle and O. Gingerich, "Les positions des planètes au moyen âge: application du calcul électronique aux tables alphonsines," in *Comptes rendus des séances de l'Académie des inscriptions et belles-lettres* (1967), pp. 531–548, esp. 541.
5. See note 1.
6. Some MSS, notably MS Paris lat. 7281, fol. 201v, which Duhem used in constructing his account, give the date as 1320: it corresponds to the epoch of the tables (31 Dec. 1320), that is, to the beginning of the first year following the closest leap year to the date of composition of the tables and canons.
7. J. L. E. Dreyer, "On the Original Form of the Alfonsine Tables," in *Monthly Notices of the Royal Astronomical Society*, **80** (1919–1920), 243–262. M. Rico y Sinobas, *Libros del saber del rey d. Alfonso X de Castilla*, IV (Madrid, 1866), 111–183; the tables actually published by Rico y Sinobas, in facs. (*ibid.*, pp. 185 ff.), are spurious, as J.-M. Millás Vallicrosa has shown in *Estudios sobre Azarquiel* (Madrid–Granada, 1943–1950), pp. 407–408.
8. The Alfonsine tables reorganized at the time of John of Lignères are shorter than those published in 1483, which were completed by tables of ascensions, by tables of proportion and by tables for the calculation of eclipses. A portion of this supplement, but not the whole of it, is borrowed from John of Lignères's tables of the *primum mobile* and from the part of the tables of 1322 dealing with the calculation of eclipses; but the canons *Quia ad inveniendum* . . . are silent on the use of this part of the tables and give no special attention to eclipses. In the medieval MSS the list of the tables forming the Alfonsine tables varies considerably from one MS to another, and it is very difficult to reconstruct the original core of the text; one can rely on little more than the uses specified by the canons.
9. John of Lignères's treatise on the saphea contains only uses, preceded by a chapter of description. But the MS Paris lat. 7295, which preserves the text of the treatise (fols. 2–14), also included (fols. 18v–19) two incomplete and unidentified drawings; these must be compared with John of Lignères's text, which they illustrate most pertinently.

   On the saphea, see G. García Franco, *Catalogo crítico de astrolabios existentes en España* (Madrid, 1945), pp. 64–65; M. Michel, *Traité de l'astrolabe* (Paris, 1947), pp. 95–97; and E. Poulle, "Un instrument astronomique dans l'occident latin, la saphea," in *A Giuseppe Ermini* (Spoleto, 1970), pp. 491–510, esp. pp. 499–502.
10. In MS Oxford, Digby 57, fols. 130–132v, the same incipit introduces another treatise on the equatorium, composed at Oxford with 31 December 1350 as the radix. MS Paris fr. 2013, fols. 2–8v, preserves a text in French ("Pour composer l'equatoire des sept planètes . . .") presented as the translation, in 1415, of a treatise on the equatorium by John of Lignères written in 1360; besides the fact that the date cannot be accepted, the instrument, although similar to John of Lignères's first equatorium, is not identical.
11. The part dealing with its construction is found in only one of the two MSS of the texts, and there it is placed after the uses. The incipit "Descriptiones (eorum) que sunt in equatorio . . ." noted by L. Thorndike and P. Kibre in *Catalogue of Incipits*, 2nd ed. (Cambridge, Mass., 1963),

JOHN OF LIGNÈRES

col. 402, is the title of the descriptive chapter that broaches the section on the uses.

12. E. Wickersheimer, ed., *Recueil des plus célèbres astrologues et quelques hommes doctes faict par Symon de Phares* (Paris, 1929), p. 214.
13. On the new quadrant, see E. Poulle, "Le quadrant nouveau médiéval," in *Journal des savants* (1964), pp. 148–167, 182–214.
14. MS Berlin F. 246, fol. 155, preserves extracts from a *Tractatus de mensurationibus* by John of Lignères: they are actually several of the chapters from the section on geometric uses in the treatise on the armillary instrument, a section extremely similar, in terms of its contents, to the treatise on the ancient quadrant by Robert the Englishman.
15. Paris lat. 7281, fol. 202: "Post multas excogitatas vias, feci instrumentum modici sumptus, levis ponderis, quantitate parvum et continentia magnum quod planetarium equatorium nuncupavi, eo quod in eo faciliter et prompte eorum equationes habetur; . . . unum composui instrumentum omnium predictorum instrumentorum [astrolabe, saphea, solid sphere] vires et excellentias continens quod merito universale astrolabium nuncupatur, eo quod unica superficie tota celi machina continetur et illa eadem cunctis regionibus applicatur . . .. Suscipiatis, o domine decane, instrumenta et tabulas que vobis . . . offero."

BIBLIOGRAPHY

I. ORIGINAL WORKS. Of the canons and tables written by John of Lignères, only the canons of the *primum mobile* (the first part of the canons of 1322) have been published in part: M. Curtze, "Urkunden zur Geschichte der Trigonometrie im christlichen Mittelalter," in *Bibliotheca mathematica*, ser. 3, **1** (1900), pt. 7, 321–416, pp. 390–413: "Die canones tabularum primi mobilis des Johannes de Lineriis"; there are the first nineteen canons (pp. 391–403), followed by the titles of the succeeding canons, as well as by the tables of sines and chords and of shadows and the *tabula proportionis* (pp. 411–413); canon 9 of the *canones super tabulas latitudinum planetarum et etiam eclipsium* (the second part of the canons of 1322) is also included (pp. 403–404). See also J.-M. Millás Vallicrosa (note 7), p. 414. To study John of Lignères's work on astronomical tables recourse to the MSS is therefore necessary.

The almanac of 1321 is in MS Philadelphia Free Library 3, fols. 3–10 (the beginning is incomplete).

The *Tabule magne* are very rare. The canons are in Erfurt 4° 366, fols. 28–32v; Paris lat. 7281, fols. 201v–205v; and Paris lat. 10263, fols. 70–78. The tables are in Erfurt F.388, fols. 1–42; and (tables of equations only) Lisbon Ajuda 52-VI-25, fols. 67–92v.

The tables and the canons of 1322, on the other hand, are fairly common; but the tables themselves are seldom complete, probably because those among them that duplicated the Alfonsine tables were not so well accepted as the latter and hence only the tables for the *primum mobile*, those for the latitudes, and those for the eclipses were preserved: Basel F.II.7, fols. 38–57v, 62–77v (incomplete canons and tables in part); Catania 85, fols. 144–173 (canons), 192–201v (partial tables); Erfurt 4° 366, fols. 1–25v (canons only); Paris lat. 7281, fols. 178v–201v (canons only); Paris lat. 7282, fols. 46v–52v (canons), 113–128v (partial tables); Paris lat. 7286 C, fols. 9–58v (tables and canons), etc. The canons of the *primum mobile* often appear alone: Paris lat. 7286, fols. 35–42v (unfinished canons and partial tables); Paris lat. 7290 A, fols. 66–75v; Paris lat. 7292, fols. 1–12v; Paris lat. 7378 A, fols. 46–52. The canons of the planetary movements likewise are frequently found by themselves: Cusa 212, fols. 74–108 (with tables); Paris lat. 7295 A, fols. 155–181v (with tables); Paris lat. 7407, fols. 40–63, etc. Those of John of Lignères's tables and canons that, for the latitudes and the eclipses, complete the Alfonsine tables are sometimes integrated into the latter, as in Paris lat. 7432, fols. 224–358v. The portion of the canons that treats eclipses can be found separately: Paris lat. 7329, fols. 127–131v.

See also Cracow 557, fols. 58–96 v (canons and partial tables).

John of Lignères's canons on the Alfonsine tables, while much less common than those of John of Saxony, are nevertheless not rare: Cusa 212, fols. 65–66v; Oxford, Digby 168, fols. 145–146; Oxford, Hertford College 4, fols. 148v–155; Paris lat. 7281, fols. 175–178; Paris lat. 7286, fols. 1–3v; Paris lat. 7405, fols. 1–4v. Moreover, they often duplicate those of John of Saxony.

John of Lignères's other astronomical works do not seem to have had as great a diffusion. The theory of the planets is preserved in Cambridge Mm.3.11, fols. 76–80v; Paris lat. 7281, fols. 165–172. Another Cambridge MS, Gg.6.3, fols. 237v–260, also preserves this text in a version that appears to be quite different, but this MS is very mutilated and practically unusable.

The saphea can be found (the incipit of which is very similar to the one in Ibn Tibbon's translation of al-Zarqāl's treatise) in Erfurt 4° 355, fols. 73–81v; Erfurt 4° 366, fols. 40–49; Paris lat. 7295, fols. 2–14. The first chapter (description) was published in L. A. Sédillot, "Mémoire sur les instruments astronomiques des Arabes," in *Mémoires présentés par divers savants à l'Académie des inscriptions et belles-lettres*, ser. 1, **1** (1844), 1–220, see 188–189n.

The first treatise on the equatorium was published by D. J. Price as an appendix to the treatise attributed to Chaucer: *The Equatorie of the Planetis* (Cambridge, 1955), pp. 188–196, but the text is very defective and it is still necessary to refer to the MSS: Cambridge Gg.6.3, fols. 217v–220v; Cracow 555, fols. 11–12v; Cracow 557, fols. 11–12v; Oxford, Digby 168, fols. 65v–66; and Vatican Palat. 1375, fols. 8v–10v. The treatise on the equatorium preserved in Oxford, Digby 57, fols. 130–132v, under the same incipit, is not the one by John of Lignères. The second equatorium is unpublished: Vatican Urbin. lat. 1399, fols. 16–21 (uses and construction); Oxford, Digby 228, fols. 53v–54v (uses only).

The treatise on the directorium, *Accipe tabulam planam* . . ., is found only anonymously in Florence, Magl. XX.53, fols. 35–37; Oxford, Digby 48, fols. 91v–94; Salamanca 2621, fols. 21v–23; Wolfenbuttel 2816, fols. 125–126v. The armillary instrument is attributed to John of Lignères in Vatican Urbin. lat. 1399, fols. 2–15.

The *Algorismus minutiarum* (*Modum representationis minutiarum vulgarium* . . .) was published very early: Padua, 1483 (Klebs 167.1) and Venice, 1540. See A. Favaro,

"Intorno alla vita ed alle opere di Prosdocimo de' Beldomandi," in *Bullettino di bibliografia e di storia delle scienze matematiche e fisiche*, **12** (1879), 115–125; D. E. Smith, *Rara arithmetica* (Boston, 1908), pp. 13–15; and H. L. L. Busard, "Het rekenen met breuken in de middeleeuwen, in het bijzonder bij Johannes de Lineriis," in *Mededelingen van de K. academie voor wetenschappen, letteren en schoone kunsten van België* (1968). There are a great many MSS of this work.

II. SECONDARY LITERATURE. Pierre Duhem, *Le système du monde*, IV (Paris, 1916), 60–69, 578–581; and L. Thorndike, *A History of Magic and Experimental Science*, III (New York, 1934), 253–262, although they supersede most of the earlier works—see G. Sarton, *Introduction to the History of Science*, III (Baltimore, 1947), 649–652— do not really bring John of Lignères's work into clear focus; the canons of the tables, especially, have been confused with each other and with the treatises on the instruments. Moreover, Duhem's hypothesis that the *Algorismus minutiarum* ought to be attributed to John of Sicily rather than to John of Lignères is not based on any serious evidence: the medieval attribution is unanimously to John of Lignères.

EMMANUEL POULLE

JOHN MARLIANI. See **Marliani, Giovanni.**

JOHN OF MURS (*fl.* France, first half of the fourteenth century), *mathematics, astronomy, music.*

Originally from the diocese of Lisieux in Normandy, John of Murs was active in science from 1317 until at least 1345. He wrote most of his works in Paris, at the Sorbonne, where he was already a master of arts in 1321. Between 1338 and 1342 he was among the clerks of Philippe III d'Évreux, king of Navarre, and in 1344 he was canon of Mézières-en-Brenne, in the diocese of Bourges.[1] The date of his death is not known. His letter to Clement VI on the conjunctions of 1357 and 1365 must have been sent before the pope's death in 1352; on the other hand, the chronicler Jean de Venette prefaced his account of the year 1340 with two prophecies, one for the year 1315 and the other, no date given, attributed to John of Murs, of whom he speaks in the past tense.[2] But this prophecy is probably not by John of Murs.[3] Moreover, Jean de Venette, whose information is not necessarily firsthand, wrote his chronicle at different times and probably made corrections and additions which do not permit the assignment of a definite year to the composition of the account of the year 1340.[4] There has been an attempt to argue that John of Murs's life extended beyond the accession of Philippe de Vitry to the see of Meaux in 1351, but there is no ground for accepting this assertion.

John of Murs wrote a great deal, but certain of his works appear not to have been preserved. Among the missing are one on squaring the circle and a "genealogia astronomie," both cited at the end of the *Canones tabule tabularum* as composed in 1321. The other writings are devoted to music, mathematics, and astronomy.

John of Murs's musical works include *Ars nove musice*, composed in 1319, according to the explicit of one of the manuscripts; *Musica speculativa secundum Boetium*, dating from 1323 and written at the Sorbonne; *Libellus cantus mensurabilis;* and *Questiones super partes musice*, which takes up again, in the form of questions and answers, the material of the *Libellus.* We do not know whether to this list should be added the *Artis musice noticia* cited by the *Canones tabule tabularum* among the works composed in 1321, or whether this text is the same as the *Ars nove musice* mentioned above. The *Musica speculativa* is a commentary on Boethius. The other treatises bear witness to a scientific conception of music, new at the beginning of the fourteenth century: it is as a mathematician that John of Murs views musical problems. In addition to its fundamental originality, his work reveals the pedagogic qualities that assured his musical writings a wide diffusion until the end of the Middle Ages.

It was in mathematics that John of Murs's learned work received its greatest development. The *Canones tabule tabularum* mentions a squaring of the circle which does not seem to have been preserved; it is therefore not known whether he was acquainted at that time with Archimedes' *De mensura circuli* in the translation of William of Moerbeke, which he mentions knowing twenty years later. This quadrature aside, the earliest mathematical work of John of Murs is the *Tabula tabularum* with its canons "Si quis per hanc tabulam tabularum proportionis . . ." It is a table giving, for the numbers one to sixty inscribed as both abscissas and ordinates, the product of their multiplication expressed directly in sexagesimal notation. The year of this table, 1321, and its title clearly reveal the preoccupations which led John of Murs to construct it, since he was associated at that time with the project of recasting the astronomical tables of Alfonso X of Castile in a strictly sexagesimal presentation. This systematic conversion of the chronological elements into the number of days expressed in sexagesimal numeration presupposed great suppleness in the mental gymnastics involved in such a conversion.

In addition to calculations in sexagesimal numeration, knowledge of trigonometry was necessary in astronomy. Hence it is not surprising to find, under the name of John of Murs, a short treatise on trigonometry entitled *Figura inveniendi sinus kardagarum*

("Omnes sinus recti incipiunt a dyametro orthogonaliter . . ."), which concerns the construction of a table of sines.

Yet it would be wholly incorrect to consider John of Murs's mathematical work as only a sort of handmaiden to astronomy. About 1344[5] he completed *De arte mensurandi* ("Quamvis plures de arte mensurandi inveniantur tractatus . . ."), in twelve chapters —of which the first four chapters and the beginning of the fifth had already been written by another author and deal precisely with the mathematical knowledge necessary for astronomy (operations on sexagesimal fractions and trigonometry). Going beyond these elementary notions, John of Murs utilized Archimedes' treatises on spirals, on the measurement of the circle, on the sphere and the cylinder, and on the conoids and spheroids, which he knew in the translations of William of Moerbeke. Moreover, he inserted in this work, as the eighth chapter, a squaring of the circle which is sometimes found separately ("Circulo dato possibile est accipere . . .") and which is dated 1340. The propositions of the *De arte mensurandi* appeared, without the demonstrations, under the title *Commensurator* or as *Problemata geometrica omnimoda*, long attributed to Regiomontanus.[6]

John of Murs's most famous mathematical work is his *Quadripartitum numerorum* ("Sapiens ubique sua intelligit . . ."), which takes its name from its division into four books. They are preceded by a section in verse ("Ante boves aratrum res intendens. . .") and completed by a *semiliber* interpolated between books III and IV. The arithmetical portions of this treatise derive from al-Khwārizmī, with no evidence of any great advance over the original. Yet the appearance, in book III, of the use of decimal fractions in a particular case, that of the extraction of square roots, is noteworthy; but reference to their use is almost accidental and is not developed. The sections on algebra, both in the versified portion and in book III, draw on the *Flos super solutionibus* of Leonardo Fibonacci. Since book IV is devoted to practical applications of arithmetic, John of Murs uses this occasion to introduce a discussion on music (*De sonis musicis*) and two treatises on mechanics (*De movimentis et motis* and *De ponderibus*), the second of which reproduces long extracts from the *Liber Archimedis de incidentibus in humidum*.[7]

The *Quadripartitum* is dated 13 November 1343, and the versified part is addressed to Philippe de Vitry. Since the Paris manuscripts of this text note that this celebrated poet and musician was also the bishop of Meaux,[8] it has been claimed that the versified part cannot be prior to 1351, the year in which Philippe de Vitry assumed his episcopal functions; in fact, the part in verse was indeed written after the prose part, but the date of the former is certainly not much later than that of the latter (see note 5). The reference to the bishopric of Meaux is made by the copyist of the Paris manuscript, not by John of Murs.

In astronomy John of Murs's name is associated, as is that of John of Lignères, with the introduction of the Alfonsine tables into medieval science. Yet his first astronomical writing, a critique of the ecclesiastical computation of the calendar ("Autores calendarii nostri duo principaliter tractaverunt . . ."), in 1317, is that of a convinced partisan of the Toulouse tables, which he declares to be the best. The attribution of this text to John is proposed only by a fifteenth-century manuscript, but there is no reason to contest it; moreover, the author's style, very critical and impassioned, is definitely that of John of Murs when, later on, he attacked the defects of the calendar. The reference to the Toulouse tables would then demonstrate that, whatever P. Duhem may have believed, the introduction of the Alfonsine tables among the Paris astronomers was not yet complete in 1317.

Nor is that introduction established for 1318. In fact, we possess the report of the observation of the equinox and of the calculation of the hour of the entry of the sun into Aries, both made in that year at Évreux by John of Murs. Since the report invokes the authority of Alfonso X and his tables, Duhem saw in it proof that those tables were then in current use; but his account rests on an erroneous subdivision of a poorly identified text, the *Expositio intentionis regis Alfonsii circa tabulas ejus*, preserved in the manuscript Paris lat. 7281 ("Alfonsius Castelle rex illustris florens . . ."). Duhem made two different texts from it, dating the first 1301 and proposing to attribute it to William of Saint-Cloud, and assigning to John of Murs only the second, reduced to the account of the observation of 1318. In truth, the references to 1300 (*anno perfecto*, that is to say 1301) are found in both texts, and therefore cannot signify the year in which the texts were composed, for they accompany the results of the observation of 1318; the latter, moreover, is not described as a very recent event but as evidence invoked a posteriori to confirm the excellence of the Alfonsine tables. This *Expositio*, including the account of 1318, must correspond to the *Expositio tabularum Alfonsi regis Castelle* mentioned in John's *Canones tabule tabularum* as being among the works that he composed in 1321. It must, consequently, have been between 1317 and 1321 that John learned of the Alfonsine tables. These dates may be compared with those of the first two tables of John of Lignères: those from around 1320, which appear to be independent of the Alfonsine tables, and those from 1322, which present the Alfonsine tables in a first draft. This *Expositio* is presented as a technical study of the values given by the Alfonsine tables for the composite movement of the apogees of the planets and for the mean movement of the sun; as the copyist of manuscript Paris lat. 7281 remarks in a final note, nothing appears about the eccentricities of the planets.[10] It was not until 1339 that John of Murs composed, after John of Lignères and John of Saxony, canons of the Alfonsine tables in their definitive version: "Prima tabula docet differentiam unius ere . . ."[11]

We have seen that John of Murs had observed the sun at Évreux in 1318, on the occasion of the vernal equinox. This was not his only observation: a manuscript in the Escorial preserves abundant autograph notes by him dealing with his observations at Bernay, Fontevrault, Évreux, Paris, and Mézières-en-Brenne between 1321 and 1344, notably at the time of the solar eclipse of 3 March 1337.[12] They attest to the scientific character of an outstanding mind, for the records of medieval astronomical observations are quite exceptional.

An informed practitioner very closely associated with the diffusion of the Alfonsine tables, John of Murs was not unaware of the extent to which astronomical tables based on the calculation of the mean movements and mean arguments of the planets, and on the corresponding equations, however satisfying they might be theoretically, contained snares and difficulties when put to practical use. An important part of his work was therefore devoted to perfecting the tables and the calculating procedures in order to lighten the task of determining planetary positions on a given date.

Thus the tables of 1321, bearing the canons "Si vera loca planetarum per presentes tabulas invenire...," represent one of the most original productions of medieval astronomy. They are based on the generalization to all the planets of the principle ordinarily applied in calculating solar and lunar conjunctions and oppositions. This calculation rests on the determination of a mean conjunction or opposition, a unique moment in which the two bodies have the same mean movement and, consequently, the equation of the center of the moon is null. Likewise, John of Murs provided, for the sixty years beginning on 1 January 1321, the list of dates on which the sun and each of the planets have the same mean movement; the argument of the planet and the equation of the argument are then null. Next, a *contratabula* gives directly the equation to be added to the mean movement in order to obtain the true position, partly as a function of the difference between the date for which the true position of the planet and that of its "mean conjunction" with the sun are sought and partly as a function of the mean center of the planet at the moment of the "mean conjunction."

For the particular case of the sun and the moon, John of Murs proposed to simplify further the calculation of their conjunctions and oppositions by means of new tables, termed *tabule permanentes*, and of their canon "Omnis utriusque sexus armoniam celestem . . .": knowing the date of a mean conjunction or opposition of the two bodies (it is determined very easily with the aid of the table of mean elongation of the sun and the moon, which is included among the tables of mean movements and mean arguments of the planets), John of Murs presented directly the difference in time which separates the mean conjunction or opposition from the true conjunction or opposition in a double-entry table, where the sun's argument is given as the abscissa and that of the moon as the ordinate.

Maintaining the goal of a rapid determination of the conjunctions and oppositions of the sun and the moon, the *Patefit* (so designated after the first word of its canon: "Patefit ex Ptolomei disciplinis in libro suo . . .") offers a complete solution that is limited to the period 1321–1396.[13] A series of tables gives, without the necessity of calculation, the dates of the mean conjunctions and oppositions, the true positions of the two bodies at the times of the mean conjunctions, and the data needed to calculate rapidly, from this information, their actual positions at the times of the true conjunctions. Other tables deal with the determination of those conjunctions and oppositions which eclipse one of the two bodies and also with the calculation of the duration of the eclipse. All these tables form an annex to a calendar of which the originality consists in providing, in addition to the true daily position of the sun during the years of a bissextile cycle, the correction to be employed after the years 1321–1324 of the first cycle. Here John of Murs's concern to replace the ecclesiastical calendar, frozen in a nonscientific conservatism (the faults of which already were revealed in 1317), by a chronological instrument conforming to astronomical reality becomes fully apparent.

John of Murs expressed that concern again on two occasions in texts on the calendar and on the reforms that should be made in it. One of these ("De regulis computistarum quia cognite sunt a multis . . ."), by the violence of its style, almost seems to be a pamphlet against the traditional *computus* and the computists;[14] it nevertheless offers some constructive solutions, such as suppressing, for forty years, the intercalation of the bissextile or shortening eleven months of any

JOHN OF MURS

given year by one day each, so that at the end of the period thus treated the calendar will have lost the eleven-day advance that it then would have recorded over the astronomical phenomena whose rhythm it should have reproduced. Another of its suggestions was to adopt a lunar cycle of four times nineteen years, a better one than the ordinary cycle of nineteen years. One of the manuscripts of the *De regulis computistarum* preserved at Erfurt assigns to the text the date of 1337.[15]

The other text on the calendar has a more official character; in fact, in 1344, John of Murs and Firmin de Belleval were called to Avignon by Pope Clement VI to give their opinion on calendar reform.[16] The result of this consultation was, in 1345, a memoir ("Sanctissimo in Christo patri ac domino . . .") in which the experts proposed two arrangements: the suppression of a bissextile year every 134 years to correct the solar calendar (after applying a suitable correction to compensate for the gap of eleven days between the date of the equinox of the computists and the true date), and the adoption of a new table of golden numbers to correct the lunar calendar.[17] It was suggested that the reform begin in 1349, which offered the advantage of being the first year after a bissextile and of having "1" for its golden number according to the ancient *computus*. This advice was not followed, and the Julian calendar retained its errors for more than two centuries.[18]

It was perhaps to follow up on these matters that John of Murs again sent to Clement VI, at an unknown date but necessarily before the pope's death in 1352, an opinion concerning the anticipated conjunction of Saturn and Jupiter on 30 October 1365 and of Saturn and Mars on 8 June 1357 ("Sanctissimo et reverendissimo patri et domino . . ."). In it he informed the pope of the particularly favorable conditions which were to conjoin in 1365 for the success of a crusade against the Muslims, but he beseeched him at the same time to use the weight of his authority to prevent the wars between the Christian states inscribed in the very unfavorable conjunction of 1357. Analogous astrological concern had elicited, at the time of the triple conjunction of 1345, parallel commentaries by Leo of Balneolis (his commentary was translated into Latin by Peter of Alexandria), by Firmin de Belleval, and by John of Murs ("Ex doctrina mirabili sapientium qui circa noticiam . . ."); the conjunctions were predicted for 1 March between Jupiter and Mars, for 4 March between Saturn and Mars, and for 20 March between Saturn and Jupiter, all in the sign of Aquarius. An autograph note by John of Murs on the same conjunction is found in one of the manuscripts of *De arte mensurandi*.[19]

NOTES

1. L. Gushee, "New Sources for the Biography of Johannes de Muris," in *Journal of the American Musicological Society*, **22** (1969), 3–26, esp. 19, 26.
2. "Quam, ut fertur, fecit magister Johannes de Muris qui temporibus suis fuit magnus astronomus," in H. Géraud, *Chronique latine de Guillaume de Nangis de 1113 à 1300 avec les continuations de cette chronique de 1300 à 1368*, II (Paris, 1843; Société de l'histoire de France), 181. This prophecy is completely independent of the texts on the conjunction of 1345 and the conjunctions of 1357 and 1365.
3. This prophecy appears elsewhere than in Jean de Venette's chronicle: see H. L. D. Ward, *Catalogue of Romances in the Department of Manuscripts in the British Museum*, I (London, 1883), 302, 314, 316–319, 321. It is taken up again by the fifteenth-century historian Thedericus Pauly, in *Speculum historiale*, edited by W. Focke in his inaugural dissertation, *Theodericus Pauli ein Geschichtsschreiber des XV. Jahrhunderts* (Halle, 1892), pp. 47–48, but only Jean de Venette attributes it to John of Murs; it is generally given under the name of Hemerus, the equivalent of Merlin.
4. A. Coville, "La chronique de 1340 à 1368 dite de Jean de Venette," in *Histoire littéraire de la France*, **38** (1949), 333–354, esp. 344–346.
5. The *De arte mensurandi* was completed after the prose part of the *Quadripartitum numerorum*, to which it alludes in several places, but before the epistle in verse which accompanies the *Quadripartitum* and in which there is an allusion to the *De arte mensurandi.*
6. M. Clagett, "A Note on the Commensurator Falsely Attributed to Regiomontanus," in *Isis*, **60** (1969), 383–384.
7. E. A. Moody and M. Clagett, *The Medieval Science of Weights* (Madison, Wis., 1960), pp. 35–53. It was published by Clagett in *The Science of Mechanics in the Middle Ages*, pp. 126–135.
8. The allusion to the bishopric of Meaux is not found in either of the two Vienna MSS.
9. The announcement of the observation is made in a quite solemn and perhaps parodic manner, according to a formulation borrowed from the charters: "Noverint preterea presentes et futuri . . ."; similarly at the end there is a prohibitive clause against the ignorant and the jealous.
10. Paris lat. 7281, fol. 160: after the explicit of the *Expositio* the copyist has added: "Per Joh. de Muris credo; mirum videtur quod iste non determinavit de quantitate eccentricitatum deferentis solis et aliorum planetarum et de quantitate epiciclorum, consequenter de quantitate equationum argumenti solis, centri et argumenti etc. ceteris planetis convenientium secundum intentionem regis Alfonsii quia alias et differentes posuit ab antiquis, prospecto quod de istis fuit semper diversitas inter consideratores."
11. These canons are not very frequently found in the MSS and often appear only in a fragmentary state, which explains why John of Murs is constantly credited with canons on the eclipses that Duhem assigned to the year 1339, distinguishing them from the canons of the Alfonsine tables that he thought dated from 1321, having confused them with the *Canones tabule tabularum* that he had not read; in fact, the canons on the eclipses form the last part of the canons of the Alfonsine tables. MS Oxford Hertford Coll. 4, fols. 140–147, appears to preserve the totality of these canons, but its text is constantly interrupted by explicits, anonymous or referring to John of Murs.
12. G. Beaujouan (who is preparing an ed. of these notes), in *École pratique des hautes études, IV*[e] *section, Sciences Historiques et Philologiques, Annuaire*, 1964–1965, pp. 259–260; these notes were partially used by L. Gushee (see note 1).
13. In the London MS, the *Patefit* is designated as *Calendarium Beccense* and includes a long explicit in which the author, who does not identify himself, dedicates his work to

Geoffroy, abbot of Bec-Hellouin. A problem results from the fact that the abbot of Bec in 1321 was Gilbert de St.-Étienne; Geoffroy Fare did not become abbot until 1327. It is perhaps for this reason that an annotator of the Metz MS, in which the tables are attributed to John of Murs, has corrected them thus: "Falsum, et quidam dicunt quia fuit cujusdam monachi Beccensis."

14. The computists were reproached in particular for never stating whether their dates were "completo" or "incompleto anno" and for calculating the life of Christ in solar years rather than in lunar years.
15. Erfurt 4° 371, fol. 45. It is this MS, which is undoubtedly the source of the information on John of Murs's calendrical work before 1345, on which Duhem relied—*Le système du monde*, IV (Paris, 1916), 51—following a work by Schubring (1883) cited by M. Cantor, *Vorlesungen über die Geschichte der Mathematik*, II, 2nd ed. (Leipzig, 1900), 125, which no one has been able to locate.
16. The papal letters addressed to John of Murs and Firmin de Belleval were published in E. Deprez, "Une tentative de réforme du calendrier sous Clément VI: Jean de Murs et la chronique de Jean de Venette," in *École française de Rome, Mélanges d'archéologie et d'histoire*, **19** (1889), 131–143, republished in Clement VI, *Lettres closes, patentes et curiales se rapportant à la France*, E. Deprez, ed., I (Paris, 1901–1925), nos. 1134, 1139, 1140.
17. The summary found at the end of the text is an integral part of it and is in all the MSS.
18. A London MS—Sloane 3124, fols. 2–8v—preserves a calendar whose brief canon ("Canon autem tabule ita scripte ut supra apparet est de renovatione lune . . .") attributes it to John of Murs and to the other experts who composed it at the request of Clement VI; but this calendar was established for a classical cycle of nineteen years beginning in 1356.
19. Paris lat. 7380, fol. 38v.

BIBLIOGRAPHY

I. ORIGINAL WORKS. Almost all of John of Murs's musical work has been published: The *Ars nove musice* was included by M. Gerbert in his *Scriptores ecclesiastici de musica sacra*, III (St.-Blaise, 1784), but it was fragmented under various titles (pp. 256–258, 312–315, 292–301), as were the *Musica speculativa* (*ibid.*, pp. 249–255, 258–283; also printed in Cologne, *ca.* 1500, in a collection entitled *Epitoma quadrivii practica* [Klebs 554.1]) and the *Questiones super partes musice* (*ibid.*, pp. 301–308). The *Questiones* was reproduced, under the title of *Accidentia musice*, by E. De Coussemaker in *Scriptorum de musica medii aevi nova series*, III (Paris, 1869), 102–106; Coussemaker also published the *Libellus cantus mensurabilis* (*ibid.*, pp. 46–58). U. Michels, "Die Musiktraktate des Johannes de Muris," in *Beihefte zum Archiv für Musikwissenschaft*, **8** (1970). The *Summa musice*, published under the name of John of Murs by Gerbert (*op. cit.*, pp. 190–248), and the *Speculum musice*, published in part by Coussemaker (*op. cit.*, II [Paris, 1867], 193–433), although long attributed to John, are not by him.

Of John of Murs's mathematical works, the only ones which have been published are an abridgment of Boethius' *Arithmetica* (Vienna, 1515; Mainz, 1538), dealt with in A. Favaro, "Intorno alla vita ed alle opere di Prosdocimo de' Beldomandi," in *Bullettino di bibliografia e di storia delle scienze matematiche e fisiche*, **12** (1879), 231, D. E. Smith, *Rara arithmetica* (Boston, 1908), pp. 117–119, and H. L. L. Busard, "Die 'Arithmetica speculativa' des Johannes de Muris," in *Scientiarum historia*, **13** (1971), 103–132; and the short treatise on trigonometry, M. Curtze, ed., "Urkunden zur Geschichte der Trigonometrie im christlichen Mittelalter," in *Bibliotheca mathematica*, 3rd ser., **1** (1900), 321–416, no. 8, pp. 413–416: "Die Sinusrechnung des Johannes de Muris." A partial ed. of *De arte mensurandi* is in preparation: M. Clagett, *Archimedes in the Middle Ages*, III; it will be based on MS Paris lat. 7380, of which the parts composed by John of Murs are autograph—see S. Victor, "Johannes de Muris' Autograph of the *De Arte Mensurandi*," in *Isis*, **61** (1970), 389–394. Other MSS are Florence, Magliab. XI-2, fols. 1–89, and XI-44, fols. 2–26v.

The *Canones tabule tabularum* are in the following MSS: Berlin F.246, fols. 79v–81; Brussels 1022–47, fols. 41–43v, 154v–158v; Erfurt F.377, fols. 37–38; Paris lat. 7401, pp. 115–124; Vienna 5268, fols. 35–39. Of the MSS cited, only those of Paris and Vienna contain the table itself.

Extracts of bk. II of the *Quadripartitum* were published in A. Nagl, "Das Quadripartitum des Johannes de Muris," in *Abhandlungen zur Geschichte der Mathematik*, **5** (1890), 135–146; and extracts of the versified portion and of bk. III were published in L. C. Karpinski, "The Quadripartitum numerorum of John of Meurs," in *Bibliotheca mathematica*, 3rd ser., **13** (1912–1913), 99–114. The second tract of bk. IV was published in M. Clagett, *The Science of Mechanics in the Middle Ages* (Madison, Wis., 1959; 1961), pp. 126–135. The *Quadripartitum* is preserved in four MSS: Paris lat. 7190, fols. 21–100v; Paris lat. 14736, fols. 23–108; Vienna 4770, fols. 174–324v; Vienna 10954, fols. 4–167. MS Paris lat. 14736, which begins with bk. II and has a lacuna in bk. IV, was completed by its copyist with the *De elementis mathematicis* of Wigandus Durnheimer, which replaces bk. I, and with the text of the versified portion, inserted in the middle of bk. IV. This MS served as the model for MS Paris lat. 7190; but since Durnheimer's text was incomplete in it, it was completed, in the sixteenth century, by the MS now cited as Paris lat. 7191, where it was wrongly baptized "Residuum primi libri Quadripartiti numerorum Johannis de Muris." In MS Vienna 10954, the epistle in verse appears after bk. IV.

The only text of John of Murs's astronomical *oeuvre* which has been published is that on the triple conjunction of 1345: H. Pruckner, *Studien zu den astrologischen Schriften des Heinrich von Langenstein* (Leipzig, 1933), pp. 222–226. The letter to the pope on the conjunctions of 1357 and 1365 is translated in P. Duhem, *Le système du monde*, IV (Paris, 1916), 35–37; the original text can be found in MS Paris lat. 7443, fols. 33–34v.

The criticism of the *computus* of 1317 is in MSS Vienna 5273, fols. 91–102; and Vienna 5292, fols. 199–209v. The treatise on the calendar, *De regulis computistarum*, is in MSS Brussels 1022–47, fols. 40–40v, 203–204v; Erfurt 4° 360, fols. 51v–52; Erfurt 4° 371, fols. 44v–45. The letter to Clement VI on calendar reform is in Paris lat. 15104, fols. 114v–121v (formerly fols. 50v–58v, or fols. 208v–215v, the MS having three simultaneous foliations); Vienna

5226, fols. 73–77v; Vienna 5273, fols. 111–122; and Vienna 5292, fols. 221–230.

The *Expositio tabularum Alfonsii* is preserved in only one MS, Paris lat. 7281, fols. 156v–160.

The tables of 1321 and their canons are in MSS Lisbon, Ajuda 52-VI-25, fols. 24–66; Oxford, Canon. misc. 501, fols. 54–106v. The *Canones tabularum permanentium* are in MSS Munich lat. 14783, fols. 198v–200v; London, Add. 24070, fols. 55, 57v; Vatican, Palat. lat. 1354, fols. 60–60v; Vienna 5268, fols. 45v–48v. None of the MSS cited in L. Thorndike and P. Kibre, *A Catalogue of Incipits*, 2nd ed. (London, 1963), col. 1004, appears to contain the tables, which are found only in the Vienna MS. The Alfonsine canons of 1339 are in MSS Erfurt 4° 366, fols. 52–52v; Oxford, Hertford Coll. 4, fols. 140–147; Paris lat. 18504, fols. 209–209v.

The *Patefit* is in MSS Erfurt 4° 360, fols. 35–51, 52–55; Erfurt 4° 371, fols. 2–42v; London, Royal 12.C.XVII, fols. 145v–190, 203–210; Metz 285. In MS Lisbon, Ajuda 52-VI-25, fols. 1–14v, is an extract of the *Patefit:* the list of mean and true conjunctions and oppositions for 1321–1396, with the canon "In canone hujus operis continentur medie et vere conjonctiones . . . Deus dat bona hominibus qui sit benedictus . . ." This extract seems to have been printed in 1484; see O. Mazal, "Ein unbekannter astronomischer Wiegendruck," in *Gutenberg Jahrbuch*, 1969, 89–90.

Duhem, *op. cit.*, p. 33, has called attention to a MS of *Fractiones* or *Arbor Boetii*, written in 1324; and L. Thorndike, *A History of Magic and Experimental Science*, III (New York, 1934), 301, mentions a *Figura maris aenei Salomonis*, also of 1324, the nature of which is uncertain.

A Cambridge MS attributes to John of Murs a short memoir refuting the Alfonsine tables in 1347–1348, "Bonum mihi quidem videtur omnibus nobis . . .," in Cambridge, Trinity Coll. 1418, fols. 55–57v; this attribution, which contradicts John of Murs's actions during the same period, cannot be upheld. This text is sometimes also attributed to Henri Bate, despite the chronological improbability. Duhem (*op. cit.*, pp. 22–24) resolved this difficulty by very subtle but unconvincing artifices. Also very suspect are the attributions to John of Murs of a geomancy according to a Venetian MS—see Thorndike, *op. cit.*, III, 323–324—and of a poem in French on the philosophers' stone, the "Pratique de maistre Jean de Murs parisiensis"—Florence, Laurenz. Acq. e Doni 380, fols. 83–86v.

II. Secondary Literature. John of Murs's work has interested historians of music. The article in *Grove's Dictionary of Music and Musicians*, 5th ed., V (London, 1954), 1005–1008, is now completely outdated; that by H. Besseler, *Die Musik in Geschichte und Gegenwart*, VII (Kassel, 1958), cols. 105–115, is excellent and contains an abundant bibliography. For John of Murs's astronomical work, however, it is dependent on Duhem, *op. cit.*, pp. 30–38, 51–60; and Thorndike, *op. cit.*, pp. 268–270, 294–324, which should be used—especially the former—with caution. L. Gushee, "New Sources for the Biography of Johannes de Muris," in *Journal of the American Musicological Society*, **22** (1969), 3–26, is presented as a restatement, with new documentation, of Besseler's article but likewise remains tied to Duhem's information.

Emmanuel Poulle

JOHN OF PALERMO (*fl.* Palermo, Sicily, 1221–1240), *translation of scientific works.*

John of Palermo, translator from Arabic to Latin, worked at the court of Emperor Frederick II. Little is known of his life. He was designated as Frederick's "philosopher" by the well-known mathematician Leonardo Fibonacci in the introduction to the latter's *Flos.* John is also mentioned in the introduction to Fibonacci's *Liber quadratorum* (dated 1225). He appears to be identical with the Johannes de Panormo mentioned in diplomatic documents of Frederick II ranging in date from 1221 to 1240.

The only known work by John of Palermo is a Latin translation of an Arabic tract on the hyperbola entitled, in Latin, *De duabus lineis semper approximantibus sibi invicem et nunquam concurrentibus.* The original Arabic may be related to a work by Ibn al-Haytham of similar title. The tract consists of five propositions. Its overall objective is to show that the hyperbola and one of its asymptotes have the desired relationship between a straight line and a curve that always, on extension, come closer together but never meet. That is, its purpose is to demonstrate the asymptotic property of the hyperbola. The author makes free use of Apollonius' *Conics* but does not use Apollonius' special parameter, the *latus rectum*; rather, he employs the fundamental axial property to which Archimedes was accustomed to refer. The *De duabus lineis* was one of the few works available in Latin in the Middle Ages that treated conic sections in a nonoptical context. A somewhat later Latin treatise entitled *De sectione conica orthogona, quae parabola dicitur* shares three propositions with the *De duabus lineis.* A version of the *De sectione conica* was published in 1548 and both tracts appear to have influenced a variety of authors, including Johann Werner, *De elementis conicis* (Nuremberg, 1522); Oronce Fine, *De speculo ustorio* (Paris, 1551); Jacques Peletier, *Commentarii tres* (Basel, 1563); and Francesco Barozzi, *Geometricum problema tredecim modis demonstratum* (Venice, 1586).

BIBLIOGRAPHY

The text and an English translation of the *De duabus lineis*, and a collection of references to John of Palermo, are in M. Clagett, "A Medieval Latin Translation of a Short Arabic Tract on the Hyperbola," in *Osiris*, **11** (1954),

359–385. Since the appearance of this text, which was based on the earliest and best MS, Oxford, Bodl., D'Orville 70, 61v–62v, three further MSS have been discovered: Paris, B.N. lat. 7434, 79v–81r (colophon missing); and Vienna, Nationalbibliothek 5176, 143v–146r (colophon missing), and 5277, 276v–277r (proem. proofs of propositions I–IV, and colophon missing). The text will be republished and related to the sixteenth-century authors in volume IV of M. Clagett, *Archimedes in the Middle Ages*.

The *De sectione conica orthogona, quae parabola dicitur* was published in an altered version by Antonius Gongava Gaviensis in an ed. of Ptolemy's *Quadripartitum* (Louvain, 1548). For a comparison of this printed text with a sixteenth-century MS, Verona, Bibl. Capitolare, cod. 206, 1r–8v, see J. L. Heiberg and E. Wiedemann, "Eine arabische Schrift über die Parabel und parabolische Hohlspiegel," in *Bibliotheca mathematica*, 3rd ser., **11** (1910–1911), 193–208. There is a further copy of this work in Regiomontanus' hand: Vienna, Nationalbibliothek 5258, 27r–38v. Other copies are in Oxford, Bodl., Canon. Misc. 480, 47r–54r, 15c; and Florence, Bibl. Laur. Medic. Ashb. 957, 95r–110v, 15–16c. In both of these the tract is attributed to Roger Bacon.

On the work of Ibn al-Haytham that may be related to *De duabus lineis*, see F. Woepcke, *L'algèbre d'Omar al-Khâyyamî* (Paris, 1851), pp. 73 ff.; and L. Leclerc, *Histoire de la médecine arabe*, I (Paris, 1876), 515. Woepcke translates the title given by Ibn al-Haytham (through Ibn abī Uṣaibi'a) as "18. Mémoire sur la réfutation de la démonstration que l'hyperbole et ses deux asymptotes s'approchent indéfiniment l'une des autres, sans cependant jamais se rencontrer."

M. CLAGETT

JOHN PECKHAM. See **Peckham, John.**

JOHN PHILOPONUS (*b.* Caesarea [?], late fifth century; *d.* Alexandria, second half of sixth century), *philosophy*, *theology*.

Most of what is known about Philoponus is found in a few remarks made by him and by some of his contemporaries. He gives the dates of two of his books: his commentary on Aristotle's *Physica* was written in 517 and his book against Proclus in 529. One of his last works, *De opificio mundi*, was dedicated to Sergius, who was patriarch of Antioch from 546 to 549. Philoponus was one of the last holders of the chair of philosophy in Alexandria, succeeding Ammonius the son of Hermias. His philosophical background was Neoplatonic; but he was—probably from birth—a member of the Monophysite sect, which was declared heretical in the seventh century.

Philoponus' main significance for the history of science lies in his being, at the close of antiquity, the first thinker to undertake a comprehensive and massive attack on the principal tenets of Aristotle's physics and cosmology, an attack unequaled in thoroughness until Galileo. The essential part of his criticism is in his commentary on Aristotle's *Meteorologica*, in his book *De aeternitate mundi contra Proclum*, and in excerpts from his book against Aristotle's doctrine of the eternity of the world. This last work has been lost, but Philoponus' pagan adversary Simplicius quoted from it extensively in his commentaries on Aristotle's *Physica* and *De caelo*.

Philoponus' philosophy of nature was the first to combine scientific cosmology and monotheism. The monotheistic belief in the universe as a creation of God and the subsequent assumption that there is no essential difference between things in heaven and on earth, as well as the rejection of the belief in the divine nature of the stars, had already been expressed in the Old Testament and was taken over by Christianity and later by Islam. The unity of heaven and earth had been accepted as a fact, but Philoponus was the first to interpret it in the framework of a scientific conception and to explain it in terms of a world view differing from myth or pagan beliefs. His point of departure was a criticism, supported by physical arguments, of Aristotle's doctrine of the eternity of the universe and the invariable structure of the celestial region. The physical basis of Aristotle's dichotomy of heaven and earth was his assumption that the celestial bodies are made of the indestructible fifth element, the ether. As early as the first century B.C. an attack on the concept of ether was made by the Peripatetic philosopher Xenarchus. His book *Against the Fifth Element* is lost, but fragments are extant in quotations by Simplicius in his commentary on Aristotle's *De caelo*. No doubt Xenarchus' book was also known to Philoponus; but from a remark by Simplicius it appears that Philoponus' arguments against the ether went much further than those of Xenarchus, particularly those concerned with his physical proofs in favor of the fiery nature of the sun and stars. Aristotle had claimed that "the stars are neither made of fire nor move in fire" (*De caelo*, 289a34) and that the celestial stuff "is eternal, suffers neither growth nor diminution, but is ageless, unalterable, and impassive" (*ibid.*, 270b1). Heat and light emitted from the celestial bodies are produced, according to him, by friction resulting from their movements, a case similar to that of flying projectiles. This is what makes us think that the sun itself possesses the quality of fire, but even the color of the sun does not suggest a fiery constitution: "The sun, which appears to be the hottest body, is white rather than fiery in appearance" (*Meteorologica*, 341a36).

JOHN PHILOPONUS

Philoponus denied Aristotle's statement regarding the color of the sun and emphasized that the color of a fire depends on the nature of the fuel: "The sun is not white, of the kind of color which many stars possess; it obviously appears yellow, like the color of a flame produced by dry and finely chopped wood. However, even if the sun were white, this would not prove that it is not of fire, for the color of fire changes with the nature of the fuel" (*In Meteorologica*, 47, 18).

Philoponus expressed this idea elsewhere, explicitly comparing celestial and terrestrial sources of light: "There is much difference among the stars in magnitude, color, and brightness; and I think the reason for this is to be found in nothing else than the composition of the matter of which the stars are constructed. . . . Terrestrial fires lit for human purposes also differ according to the fuel, be it oil or pitch, reed, papyrus, or different kinds of wood, either humid or in a dry state" (*De opificio mundi*, IV, 12). If the different colors of the stars indicate their different constitutions, it follows that stars are composite bodies; and since composite things imply decomposition and things implying decomposition imply decay, one must conclude that celestial bodies are subject to decay. But, Philoponus argued, even those who believe the stars to be made of ether must assume them to be composed of both the matter of the fifth element and their individual form, different for each star. "However, if one abstracts the forms of all things, there obviously remains the three-dimensional extension only, in which respect there is no difference between any of the celestial and the terrestrial bodies" (Philoponus, *apud*: Simplicius, *In Physica*, 1331, 10). Thus, anticipating Descartes, Philoponus arrived at the conclusion that all bodies in heaven, as well as on earth, are substances whose common attribute is extension. Against the objection raised by Simplicius that no change can be observed in the celestial bodies, Philoponus adduced arguments from physics, stressing that the greater the mass of a body, the slower its rate of decay. Furthermore, the slowness of change is a function not only of the mass but also of certain physical properties, such as hardness; moreover, it is well known, for instance, that different animals have different life spans and that some parts of them are more resistant than others to change.

The monotheistic dogma of the creation of the universe *ex nihilo* by the single act of a God who transcends nature implied, for Philoponus, the creation of matter imbued with all the physical faculties for its independent development according to the laws of nature, a development that he conceived of as extending from the primary chaotic state to the present organized structure of the universe. This deistic conception of a world that, once created, continues to exist automatically by natural law, was completely foreign to the classical Greek view, which never considered the gods to be "above nature" but associated them with nature, reigning not above it but within it. The shock created by this conception of Philoponus' is reflected in the words of Simplicius, who is bewildered by the idea of a god who acts only at the single moment of creation and then hands over his creation to nature.

Philoponus' anti-Aristotelian views were not restricted to problems of cosmology and to the removal of the barriers between heaven and earth. He also took strong exception to some of the main tenets of Aristotle's dynamics. According to Aristotle, movement is not possible without a definite medium in which it can take place; thus, statements on the movement of bodies must always be related to a certain medium. Aristotle asserted, for instance, that in a given medium the velocities of falling bodies are proportional to their weights, and that the velocities of a given body in different media are inversely proportional to the densities of these media. Furthermore, one of the many reasons given by Aristotle for denying the existence of a void was that it would be like a medium of zero density; and thus the velocity of a falling body *in vacuo* would reach infinity, regardless of its weight. Philoponus, in opposition to Aristotle, did not exclude the feasibility of movement in a void. However, against the view held by the Epicureans (proved to be correct), he assumed that in the void Aristotle's law of the proportionality of the velocities and weights of falling bodies would be exact. Against Aristotle he stressed that the impeding influence of a medium on a falling body consists in an additional increase of the body's time in motion over and above that of the natural motion *in vacuo*, depending on the density of the medium. This additional time will be directly proportional to the density of the interfering medium. In a lengthy argument Philoponus refuted Aristotle's statements and emphasized that experience shows that "if one lets fall simultaneously from the same height two bodies differing greatly in weight, one will find that the ratio of their times of motion does not correspond to the ratio of their weights, but that the difference in time is a very small one" (*In Physica*, 683, 17).

Philoponus had his doubts about the essence and the causes of the natural motion of light and heavy bodies. For instance, he wrote that one cannot agree with Aristotle that air tends to move only upward. Air may move downward for some physical reason, such as the removal of earth or water beneath it; in this case it will rush down, filling the void thus created. On the other hand, it may well be that the so-called natural motion upward has a similar cause, if there happens to be an empty space in the upper region.

Of special importance is Philoponus' criticism of Aristotle's theory of forced motion. He rejected the main contention of the Peripatetics that in every forced motion there must always be an immediate contact between the mover and the body forced to move in a direction other than that of its natural motion. In particular Philoponus denied Aristotle's hypothesis that besides the push given to a missile by the thrower, the air behind the missile is set in motion and continues to push it. He argued convincingly that if string and arrow, or hand and stone, are in direct contact, there is no air behind the missile to be moved, and that the air which is moved along the sides of the missile can contribute nothing, or very little, to its motion. Philoponus concluded that "some incorporeal kinetic power is imparted by the thrower to the object thrown" and that "if an arrow or a stone is projected by force in a void, the same thing will happen much more easily, nothing being necessary except the thrower" (*ibid.*, 641, 29). This is the famous theory of the impetus, the precursor of the modern vectorial term "momentum" or scalar term "kinetic energy." The impetus was rediscovered by Philoponus 700 years after it had been conceived of by Hipparchus (see Simplicius, *In De caelo*, 264, 25). In the physics of medieval Islamic philosophers and Western Schoolmen the concept of impetus was developed further, mainly as a consequence of a tradition following Philoponus.

Philoponus returned to his idea of the impetus in his anti-Aristotelian theory of light, which he developed in the guise of an interpretation of Aristotle's doctrine that centers on the basic categories of potentiality (*dynamis*) and actuality (*energeia*). According to Aristotle, light is the state of actual transparency in a potentially transparent medium; by such an actualization any potentially colored body found in this medium becomes actually colored and thus visible. Light is therefore a static phenomenon whose emergence and disappearance are instantaneous and have nothing to do with locomotion. Philoponus raised the fundamental question of how Aristotle's view can be compatible with both the laws of geometrical optics, developed in the Hellenistic period, and the thermic effects of light, which are so strongly enhanced by its concentration through burning glasses. He emphasized that light must be a directional phenomenon and that visual rays move in straight lines and are reflected according to the law of equal angles. However, at the same time he pointed out that these rays are not projected from our eyes to the luminous object, as was formerly assumed, but that they move in the opposite direction, from the luminous object to the eye. He clearly stated the principle of reversibility of the path of light for the case of reflection: "It makes no difference whether straight lines proceed from the eye toward the mirror or whether they are reflected from the mirror toward the eye" (*In De anima*, 331, 27). Making this assumption, Philoponus interpreted Aristotle's term *energeia* (actuality) as a kinetic phenomenon proceeding from the luminous object to the eye. He attempted to reconcile Aristotle's conception of light as actualization of a state with geometrical optics by identifying the visual rays with the *energeia* light, interpreting *energeia* as "force" and conceiving the emission of light in terms of the doctrine of impetus. Light is "an incorporeal kinetic force [*energeia kinetikē*]" emitted from the luminous object, similar to the force imparted by the thrower to the body thrown (*In Physica*, 642, 11).

Even when Philoponus accepted Aristotle's tenets, he was most remarkable in the originality and ingenuity of his exposition or amplification of Aristotle's physical doctrines. Sometimes he posed questions never raised before, anticipating much later developments; and some of the solutions he offered are evidence of the great acuity of his mind. Conspicuous examples are his discussion of the functional dependence of one set of variable quantities on another and his clear recognition of the course of a function—in modern language its first derivative. Assuming with Aristotle that the physical properties of a substance ultimately depend on the mixture of the four elementary qualities—hot, cold, dry, and moist—he asked how a reasonable explanation can be given of the fact that one of the physical properties of a given substance may remain practically unaltered while the other is undergoing a visible change. Two examples are the sweetness of honey remaining constant while its color changes from yellow to white and the color of wine remaining the same while its taste changes to sour. If all the properties derive from the primary qualities, one should expect them to change together with the qualities. Philoponus' answer is given in what can be defined as a verbal description of a graphic representation (unknown before the late Middle Ages). He explained that every physical property is a variable depending on the four primary qualities, so that if the qualities are diminishing, the physical properties are also being reduced. However, the rate of change is different for each of the properties; and thus, "if the mixture of the independent variables is slightly varied, the sweetness of the honey, e.g., will not alter appreciably, but its color may change completely" (*In De generatione et corruptione*, 170, 32).

Another very acute remark of Philoponus' is his comment on Aristotle's statement in the *Physica* that "all things that exist by nature seem to have within themselves a principle of movement and of rest" (Aristotle, *Physica*, 192b13). Many Aristotelian commentators have pondered the question of how to include the heavens in this definition of nature, since they are never at rest but move eternally in a circle. Philoponus answered this question by interpreting the uniform and circular motion of the celestial bodies as inertial motion: "Rest is found in all things. For the perpetually moving heavens partake in rest, because the very persistence of perpetual motion is rest" (*In De anima*, 75, 11). Elsewhere he repeated his definition of inertial motion, adding that "the celestial bodies are, if I may say so, motionless in their motion" (*In Meteorologica*, 11, 31).

The concepts of potentiality and actuality, which Aristotle used extensively in his physical treatises, were occasionally supplemented, from the second century on, by a term expressing the capacity of a body to actualize a certain property or state that exists only potentially. The Greek word for this was *epitedeiotes*, meaning "fitness," "appropriateness," or "suitability"; it was sometimes used as a synonym for potentiality but later came to signify the sufficient condition for actuality, thus restricting potentiality to a necessary condition for actuality. In several of his writings Philoponus makes frequent use of this meaning of "fitness," occasionally in order to amplify Aristotle's doctrine of the basic requirements for physical action, whereby it is supposed that both the thing acting and the thing acted upon must be alike in kind but contrary in species. One of the examples given by Aristotle is the change of color, which he regarded as a process in which the object acted upon changes into the acting object by assimilation. Philoponus, commenting on this, remarked that such processes require the fitness of the active object to accomplish the assimilation. The black ink of a cuttlefish, he said, will overpower the whiteness of milk; but the black of a piece of ebony, when put into the milk, will not affect its color because of its lack of fitness. In the same way, brass or silver or similar metals will resound for some time after having been struck—i.e., they are capable of turning potential sound into actual sound—because they have a fitness for producing sound, in contrast with wood or other nonmetallic substances.

On another occasion Philoponus made use of the concept of fitness in order to defend against Aristotle's criticism Plato's doctrine of the soul as the mover of the body. Aristotle in his *De anima* argued that if Plato were right, it would be possible for the soul that had left the body to enter it again, and thus resurrection of the dead could be feasible, although it had never been observed. Philoponus emphasized that the soul keeps the body moving only so long as the body has the mechanical fitness to be worked on, and it loses that fitness when death occurs. Characteristically, he adduced mechanical similes for his view: "A stick pushed against a door cannot move it when it has not the fitness necessary for being moved. . . . It will not do so when fastened by nails or when the hinges are loose. Everything set in motion by something else generally needs a certain specific fitness" (*In De anima*, 108, 24). One interesting aspect of Philoponus' treatment of this problem is the way in which, anticipating Descartes, he looked at the human body as a mechanism capable of functioning only if its parts have the necessary mechanical fitness.

Philoponus' Neoplatonic background, depending largely on Stoic conceptions, is also evident in his manner of discussing a problem that in modern terms can be defined as resonance; it also shows his keen powers of observation. He described the ripples produced in the water in a metal cup when the cup is brought into a state of vibration. He assumed that these vibrations are not transferred directly to the water but that the air enclosed in the metal acts as an intermediate agent. This assumption shows influences of the Aristotelian theory of metals (*Meteorologica*, III, 6) as well as of the Stoic doctrine of *pneuma*.

> If we pass a wet finger round the rim [of the cup], a sound is created by the air squeezed out by the finger, which air is ejected into the cavity of the cup, producing the sound by striking against the walls. Experimental evidence for it can be brought in the following way: If one fills a cup with water, one can see how ripples are produced in the water when the finger moves round the rim [*In De anima*, 355, 34].

If the cup itself is held by the hand, no sound is produced, because, as Philoponus explained, "the body struck must vibrate softly, so that the air . . . is emitted continuously into the upper part, striking the walls of the cup and being reflected toward all of its parts" (*ibid.*).

A very ingenious physical illustration was given by Philoponus to explain the perturbation of a system by external forces. He discussed the Aristotelian concepts "according to nature" and "contrary to nature" in the context of explanations given of an illness or a congenital deformity. Such phenomena, according to his view, have to be regarded in a wider framework, as parts of a whole, in order to be considered natural. This is basically the Stoic idea that if something goes wrong, the event or object in question must be seen as a partial phenomenon. In the framework of a wider system, taken as a totality, the wrong is compensated in some way and the harmony of the whole is restored. Philoponus introduced a more physical notion into this trend of thought. When something "contrary to nature" happens to a physical object, one has to regard it as a perturbation caused by outside factors. The intervention of these factors, taken together with the resulting perturbation, restores the phenomenon as a "natural" one, as something in accordance with nature. Part of Philoponus' example is worth quoting:

> I will give you an illustration that will explain what happens with things contrary to nature: Suppose that a lyre player tunes his instrument according to one of the musical scales and is then ready to begin his music. . . . Let us assume for the sake of this illustration that the strings are affected by the state of humidity of the environment and thus get out of tune. . . . When the player strikes the lyre, the substance of the strings does not perform the melody that he had in mind; but instead an unmusical, distorted, and indefinite sound is produced [*In Physica*, 201, 28].

Philoponus then went on to say that the harmony of the whole is restored by taking into account the climactic changes and the perturbation of the strings caused by them.

On several occasions Philoponus discussed the problem of the infinite. He rejected the use of the infinite in the sense of the unlimited in extension; and in his rejection he went even further than Aristotle, not only denying, as Aristotle did, the existence of the infinite as an actual entity but also excluding the potentially infinite. Aristotle had admitted the possibility of entities that can be increased *in infinitum* without ever reaching actual infinity, but he did this mainly in order to reconcile his doctrine of the eternity of the universe and the infinite duration of the human race with the concept of infinity.

From his opposite position, believing in the beginning of the world at a finite point in the past, Philoponus argued that acquiescence in the existence of the potentially infinite will perforce lead to the admission of the actually infinite. Once one admits the infinite as a never-ending process, he said, the existence of an infinite magnitude existing by itself, or of a number that cannot be passed through to the end, cannot be excluded. From this, in his view, obvious absurdities would follow. A few sentences from his argument may be quoted here:

> If the universe were eternal, it is obvious that the number of men up to now would be infinite, i.e., actually infinite—since obviously they all have actually come into existence—and thus an infinite number would be possible. For if all human individuals have become actual up to now—and we, for instance, will be the limit of the actually infinite number of men who have been before us—then the infinite will actually have been passed through to the end [*In Physica*, 428, 25].

Philoponus went on to say that if we extend this definite limit to a future generation, the infinite will be further increased:

> This increase will tend toward infinity, if the universe is incorruptible, and thus the infinite will be infinitely increased. . . . For each generation, e.g., my own, will have an infinite number of men who were born before it. . . . Since it is impossible for the actually infinite to have been passed through to its end, and for something to be greater and more infinite than the infinite itself, it is impossible for time or for the universe to be eternal [*ibid.*].

Another argument of Philoponus' against the eternity of the universe is worth noting because it was later used by Islamic philosophers, e.g., al-Ghazālī. It is quoted by Simplicius (who wrote a polemic against it) from Philoponus' lost work against Aristotle. Philoponus, by a *reductio ad absurdum*, set out to prove that a universe without a beginning would necessarily involve the existence of different actual infinities, representing the relative numbers of the revolutions of the planets:

> Since the spheres do not move with equal periods of revolution, but one in thirty years, the other in twelve years, and others in shorter periods . . ., and if the celestial motion were without beginning, then necessarily Saturn must have revolved an infinity of times, but Jupiter nearly three times more, the sun thirty times more, the moon 360 times more, and the sphere of the fixed stars are more than 10,000 times as often. Is it not beyond any absurdity to suppose a ten-thousandfold infinity or even an infinite time of infinity, while the infinite cannot be comprised even once. Thus necessarily the revolution of the celestial bodies must have had a beginning [Philoponus, *apud*: Simplicius, *In Physica*, 1179, 15].

This passage is of interest to the historians of mathematics, since Philoponus, although he rejected altogether the notion of the infinite, here, for the first time in a specific case, made use of infinite cardinal numbers, anticipating modern concepts by more than 1,300 years.

BIBLIOGRAPHY

I. ORIGINAL WORKS. Editions of Philoponus' writings include *In Physica*, H. Vitelli, ed. (Berlin, 1887); *In De anima*, M. Hayduck, ed. (Berlin, 1897); *In De generatione et corruptione*, H. Vitelli, ed. (Berlin, 1897); *De opificio mundi*, G. Reichardt, ed. (Leipzig, 1897); *In Categoria*, A. Busse, ed. (Berlin, 1898); *De aeternitate mundi contra Proclum*, H. Rabe, ed. (Leipzig, 1899); *In Meteorologica*, M. Hayduck, ed. (Berlin, 1901); *In Analytica priora*, M. Wallies, ed. (Berlin, 1905); and *In Analytica posteriora*, M. Wallies, ed. (Berlin, 1909).

II. Secondary Literature. Editions of Simplicius' works are *In Physica*, H. Diels, ed. (Berlin, 1882); and *In De caelo*, J. L. Heiberg, ed. (Berlin, 1894). See also A. H. Armstrong, ed., *The Cambridge History of Later Greek and Early Medieval Philosophy* (Cambridge, 1967), pp. 477–483; E. Evrard, "Les convictions religieuses de Jean Philopon et la date de son Commentaire aux Météorologiques," in *Bulletin de l'Académie royale de Belgique*, classe de lettres, **6** (1955), 299 ff.; "Ioannes Philoponus," in Pauly-Wissowa, IX, cols. 1764–1793; H. D. Saffrey, "Le Chrétien J. Philopon et la survivance de l'école d'Alexandrie," in *Revue des études grecques*, **67** (1954), 396–410; Walter Böhm, *Johannes Philoponus, Ausgewählte Schriften* (Munich, 1967); Michael Wolff, *Fallgesetz und Massebegriff: zwei wissenschaftshistorische Untersuchungen zur Kosmologie des Johannes Philoponus* (Berlin, 1971); and S. Sambursky, *The Physical World of Late Antiquity* (London–New York, 1962); and "Note on John Philoponus' Rejection of the Infinite," in *Festschrift for Richard Walzer* (Oxford, 1972).

S. Sambursky

JOHN OF SACROBOSCO. See **Sacrobosco, Johannes de.**

JOHN OF SAXONY (*fl.* France, first half of the fourteenth century), *astronomy*.

Probably from Germany, John Dank, Danco, Danekow, or Danekow of Saxonia was active in science at Paris between 1327 and 1335;[1] but his scientific career may possibly have begun as early as 1297. John of Saxony, who considered himself a student of John of Lignères, composed various works on the Alfonsine tables or works that employed them and a commentary on the astrological treatise of al-Qabisi (Alcabitius).

In 1327 John of Saxony published canons on the Alfonsine tables: "Tempus est mensura motus ut vult Aristoteles. . . ."[2] An exact appreciation of the place of these canons in the history of astronomy is dependent on knowledge of the introduction of the Alfonsine tables among the Paris astronomers. P. Duhem thought that the tables were already known to William of Saint-Cloud in 1300;[3] but his conclusions are based on an unsound subdivision of poorly identified texts (see the article on John of Murs), and it seems unlikely that the tables were known in Paris before about 1320. Their first appearance in medieval science may have been in the *Expositio tabularum Alphonsi regis Castelle*, written by John of Murs in 1321, and in the canons of the tables (1322) by John of Lignères. These canons, however, do not apply to the Alfonsine tables in the form known in the Latin West at the end of the Middle Ages. A short time later, in fact, the Alfonsine tables underwent a considerable transformation affecting both form and substance—the form through substitution of a sexagesimal representation of the mean movements of the planets for the traditional mode employing *anni collecti* and *anni expansi*, the substance through adoption of a double eccentricity for the equation of Venus and Jupiter. It is to this new drafting of the Alfonsine tables that the following canons apply: the undated "Quia ad inveniendum loca planetarum . . ." of John of Lignères; the canons of John of Saxony of 1327; and the canons "Prima tabula docet differentiam . . ." of John of Murs (1339).

It may be wondered why these three astronomers, who very likely worked together, produced texts on the same subject that duplicate one another. Basically, these texts deal with the same tabular material and defend the same principles, particularly in regard to the movement of planetary apogees. John of Lignères's very succinct account deals only with changes of calendar and with determining the mean solar and lunar conjunctions and oppositions and computing the true places of the planets. John of Saxony developed this account; his canons are clearer, and he added chapters on finding a "revolution" (the moment when the sun returns to a previously occupied position); calculating the date and hour of a true conjunction of the sun and moon and of their positions "in quarter aspect"; determining the time of the entrance of the sun into one of the signs of the zodiac; establishing the date of the conjunction of two planets. John of Saxony's canons enjoyed considerable success, attested to by the number of extant manuscripts and by their inclusion in the first printed edition of the Alfonsine tables (1483); the canons of John of Lignères, like those of John of Murs, were never printed.

Produced through the efforts of Erhard Ratdolt, this first printed edition bears, following John of Saxony's canons, the words "Expliciunt canones et quod sequitur est additio." This supplement comprises a general remark on interpolation in the tables of equations, canons of the eclipses, and several chapters —preceded by a separate title page—on the latitudes of the planets. The canons of the eclipses ("Eclypsis solis quantitatem et durationem . . .") are also credited by the manuscripts to John of Saxony. Consequently, they complete the chapter on determining the true conjunctions of the sun and moon and are designed to accompany particular tables which appear at the end of the Alfonsine tables and were not part of the first group. John of Saxony's canons of eclipses duplicate those of John of Lignères for the tables of 1322, not the canons "Quia ad inveniendum . . .," which do not treat eclipses. A manuscript of the canons of John of Saxony attributes the date 1330 to them.[4]

Another way in which John of Saxony participated in efforts to spread knowledge of the astronomical tables was in his working out of examples in the *Exempla super tabulas et canones primi mobilis* of John of Lignères. The work is not, properly speaking, a commentary but a collection of numerical applications, developed in a pedagogical fashion, of the canons of John of Lignères on the canons of the *primum mobile*, "Cujuscumque arcus propositi . . .," that is, of the portion of the canons of 1322 dealing with astronomical trigonometry. According to a note in MS Paris lat. 7281,[5] it was believed that these *exempla* appeared simultaneously under two incipits, "Non fuit mortuus qui scientiam vivificavit . . ." and "Quia plures astrologorum diversos libros . . .," with the date 1355; but this note has been misinterpreted. The author of the manuscript, which in the middle of the fifteenth century constituted a collection of thirteenth- and fourteenth-century astronomical texts, merely wished to indicate that there existed, under the incipit "Non fuit mortuus . . .," a collection of examples at Paris dated 1356 (and not 1355), dealing with the canons of astronomical tables; the information is correct, the text thus explained being the canons of the Alfonsine tables published by John of Saxony in 1327. These examples are found in particular in the manuscripts Paris lat. 7407, fols. 1–26v, and Paris lat. 15104, fols. 122v–137 and 122. Nothing, however, indicates that the examples of 1356 are also by John of Saxony.[6] As for the applications of the canons of the *primum mobile* ("Quia plures astrologorum . . .")—the only ones formally attributable to John of Saxony—they do not include dated examples, since, for the purpose of these canons, they were not needed. The only example from which chronological information might be drawn is one which concerns the star Aldebaran, whose movement since the determination of the Alfonsine coordinates is estimated at 51′: the movement of the apogees and of the sphere of the fixed stars is too slow for this information to be interpreted with precision; it corresponds approximately to the year 1335.

The astronomical tables provide a general means of determining the positions of the planets at all times and in all locations, but they do not give these positions themselves. The calculations for establishing the latter are, moreover, long and tedious. In order to prevent their character from turning young people away from astronomy, John of Saxony did the calculations in advance, compiling an ephemeris for the period 1336–1380 and for the meridian of Paris with a short canon: "Cum animadverterem quamplurimos magistros et scholares in studio Parisiensi . . . ."[7] The basis for the calculations is obviously the Alfonsine tables.

In 1331 John of Saxony wrote a commentary on the great astrological treatise of al-Qabisi known under the title *Liber isagogicus*. Printed in 1485, at the same time as al-Qabisi's work, which had already reached its third edition, this commentary is found very frequently in the manuscripts. Although he himself is confident of the possibilities of astrology, especially in meteorology, John of Saxony distinguishes between his own specialty and the domain of faith, taking care not to encroach upon the latter.

Medieval manuscripts and modern scholars have credited John of Saxony with all kinds of astronomical and astrological texts; the majority are only extracts of canons on the Alfonsine tables or on tables of the eclipses. For the remainder, the attribution is suspect, to say the least: there is no evidence for affirming, for example, that he wrote the astrological commentary concerning a person born on 10 March 1333 at a place situated at 52° latitude and 3° east of Paris.[8] The *exempla* of 1356 on the canons of the Alfonsine tables had no relation to him. The treatise on the astrolabe attributed to John of Saxony by MS Erfurt 4°366, fols. 82–85v ought to be assigned to John of Seville. Finally, another John of Saxony was responsible for texts of a medical nature—which, moreover, date from a later period.[9]

Two manuscripts attribute to John of Saxony a computus dated 1297 ("Omissis preternecessariis cum intentionis sit . . .") and a commentary on it. Since the date appeared incompatible with a chronology which would have extended his period of activity to 1355, Duhem and Thorndike concluded that a homonymous author was involved, unless the date is corrected to 1397 (which settles nothing) or 1357. But the date of 1297 is confirmed by technical data furnished by the text (year of indiction 10 and golden number 6); and since his career is not known to extend beyond 1335, there is no major objection to supposing that John of Saxony wrote on the *computus* thirty years before publishing canons on the Alfonsine tables, the introduction of which had occurred during the intervening period. The author of the *computus*, who identifies himself as Johannes Alemanus, is concerned with the longitudes of Paris and Magdeburg; the latter is given as the native city of John of Saxony by one of the manuscripts of his canons on the Alfonsine tables.[10]

If the attribution of the *computus* of 1297 to John of Saxony is accepted, then it follows that he was considerably older than John of Murs, whose scientific activity extended from 1317 (his first work also concerns the calendar) to after 1345. Moreover, since John of Saxony acknowledged that he was the student of John of Lignères, who wrote between 1320 and 1335, the three Johns who made such a profound mark on fourteenth-century astronomy may be arranged in the following sequence of birth: John of Lignères, John of Saxony, and, about twenty years later, John of Murs.

NOTES

1. MS Berlin F. 246, fol. 121, which dates from 1458, calls John of Saxony "magister J. de Saxonia alio nomine magister Johannes Danekow de Magdeborth." MS Paris lat. 7281 calls him both John of Saxony and John of Counnout; P. Duhem (*Le système du monde*, IV, 78) has erected daring hypotheses on the basis of this. It is sufficient to observe that, while the MS Paris lat. 7281 is an exceptionally valuable document for the history of astronomy because of the variety and quality of the texts that it unites, and while it demonstrates its author's fine curiosity with regard to texts, many of which were then out of date, it is still a late testimony (mid-fifteenth century) and contains certain misinformation.
2. The date, furnished by many explicits in the MSS, is confirmed by the example of 3 July 1327, given in connection with the expression of dates in sexagesimal numeration.
3. Duhem, *op. cit.*, pp. 20–24.
4. Brussels 1022–47, fols. 37–39v; the incipit is somewhat different—"Ad eclipsim solis inveniendam quere primo conjonctionem . . ."—but the text is the usual one.
5. Paris lat. 7281, fol. 222: "Canones cum exemplis particularibus ad longum 'Non fuit mortuus qui scientiam vivificavit etc.'; ponuntur exempla in omnibus canonibus super radicem anni Christi 1355 completi et super Parisius" (note misread by Duhem, *op. cit.*, p. 78, and by Thorndike, *A History of Magic* . . ., III, 255).
6. Despite the account, erroneous on this point, of Simon de Phares (E. Wickersheimer, ed., p. 216): "et fist une declaracion bien ample sur les mouvemens des planetes qui se commence 'Non fuit mortuus.' "
7. In a phrase which appears to make of it a doublet from "Tempus est mensura motus . . .," Simon de Phares (*ibid.*) points out this incipit, thus deformed: "Quamplures astrorum diversos . . .."
8. Thorndike, *op. cit.*, p. 267. The text is in Oxford, Hertford Coll. 4, fols. 126–130v: "Investigationis gradus ascendentis nativitatis . . .," followed, in fols. 131–133v, by another commentary: "In hac nativitate sic processi . . .," not dated, but in which the planetary positions allow us to refer to 17 March 1308 for 55° latitude; they concern two particular applications of a general rule for determining the ascendant at the moment of birth. The other MS cited by Thorndike, Vienna 5296, fols. 23–25, contains another application, to a person born on 28 September 1444, as does MS Catania 85, fols. 251–253.
9. E. Wickersheimer, *Dictionnaire biographique des médecins en France au moyen âge* (Paris, 1936), p. 475.
10. See note 1.

BIBLIOGRAPHY

I. Original Works. The 1327 canons on the Alfonsine tables and the canons of the eclipses composed by John of Saxony were printed at Venice in 1483, with the Alfonsine tables (Klebs 50.1); the MSS containing these texts number in the hundreds. The *Exempla super . . . canones primi mobilis* may be found in the following MSS: Erfurt F. 386, fols. 26–32; Paris lat. 7281, fols. 222–232; Paris lat. 7285, fols. 30–36; Paris lat. 7407, fols. 27–40. The ephemeris of 1336–1380 is preserved in MSS Erfurt F. 386, fols. 62–108; Erfurt F. 387; Erfurt 4° 360, fols. 77v–78v (only the canons).

The commentary on the *Liber isagogicus* of al-Qabisi was printed with the latter in 1485 (Klebs 41.3) and several times afterward. See B. Boncompagni, "Intorno alle vite inedite di tre matematici . . .," in *Bullettino di bibliografia e di storia delle scienze matematiche e fisiche*, **12** (1879), 373–374. There are numerous MSS.

The *computus* of 1297 is preserved in MS Florence Plut. 30.24, fols. 78-86; the commentary on this *computus* ("Sicut dicit Ptolomeus in Almagesti disciplina . . .") is found following the above text in the same MS, fols. 87–96v; and in Erfurt 4° 365, fols. 132–139.

II. Secondary Literature. P. Duhem, *Le système du monde*, IV (Paris, 1916), 76–90; and L. Thorndike, *A History of Magic and Experimental Science*, III (New York, 1934), 253–267, must be corrected on many points.

Emmanuel Poulle

JOHN SCOTTUS ERIUGENA. See **Eriugena, Johannes Scotus.**

JOHN OF SICILY (*fl.* France, second half of the 13th century), *astronomy.*

Nothing is known of John of Sicily's life except that he was part of the Paris scientific community at the end of the thirteenth century. His only extant work is a commentary on the "Quoniam cujusque actionis quantitatem . . .," Gerard of Cremona's translation of the canons of the tables of al-Zarqāl. This commentary is generally dated 1290 but is more probably from September 1291 (*anno Domini 1290 completo*), to judge from the numerous examples calculated for that date.

After citing the opening words of each chapter of the canons, John of Sicily very methodically summarizes its purpose, states its plan, the different sections of which are carefully indicated by appropriate citations, and then comments at length following the plan of the original. For the most part, these developments give John of Sicily's work the character of a treatise on trigonometry and planetary astronomy. In accordance with al-Zarqāl's canons, the commentary is divided into three parts. First it discusses what Gerard of

Cremona called the "measure of time," that is, everything concerning the divisions of the year in the lunar and solar calendars and converting each of the four standard calendars, that is, the Arab, Christian, Persian, and Greek, to the others. It should be noted that the version of the Toledan tables of which the treatise *Quoniam cujusque actionis*... forms the canons has replaced the tables giving the equivalence between Christian years and Arab years with tables for reducing the years of all calendars to days expressed in sexagesimal numeration. The Alfonsine tables later adopted this principle and systematized its applications. The astronomy of the *primum mobile*, which constitutes the second part, includes the application of trigonometry to astronomy. Everything relating to the movements of the planets forms the third part.

The canons of al-Zarqāl discuss the consequences of the motion of "accession and recession" only at the very end of the section on planetary motions: instead of considering the effect of the motion of the eighth sphere on the positions of the planetary auges with respect to the ninth sphere, the true places of the planets are determined with respect to the ecliptic of the eighth sphere, as if the auges were fixed, and the correction resulting from the motion of the eighth sphere is applied to get the final result. This procedure permits the calculator not to commit himself as to the motion he assigns to the eighth sphere until he reaches the very last step in the calculation.

In commenting on al-Zarqāl, a partisan of "accession and recession," John of Sicily recalled the various hypotheses that had been proposed: a simple movement of precession, estimated by Ptolemy at one degree in 100 years and by al-Battānī at one degree in sixty-six years; a back-and-forth movement (one degree in eighty years with an amplitude of eight degrees) disproved by al-Battānī; and the movement of "accession and recession" suggested by Thābit ibn Qurra. John of Sicily rejected the last of these movements, invoking arguments the origin of which P. Duhem traced to Roger Bacon. He adhered to precession as Ptolemy presented it, conceding, however, that its exact quantity was uncertain and would become determinable only by extensive observations.

Two types of star table are included in the Latin version of the Toledan tables (P. Kunitzsch, *Typen von Sternverzeichnissen in astronomischen Handschriften des 10. bis 14. Jahrhunderts* [Wiesbaden, 1966], pp. 73–94). One contains forty stars with their ecliptic coordinates; the other, thirty-four stars with both their ecliptic and equatorial coordinates. This second table is declared verified for the Arab year 577 (A.D. 1181–1182). The canons do not specify to which type of star table they refer. Without indicating the number of stars, John of Sicily alluded to the double coordinates: he therefore had before him the table verified for A.H. 577.

BIBLIOGRAPHY

I. ORIGINAL WORKS. John of Sicily's commentary is found under two incipits: "Inter cetera veritatis philosophice documenta ..." (Paris lat. 7266, fols. 136–220v; the text preserved in Erfurt 4° 366, fols. 74–79v, contains, under the same incipit, only extracts; this is, in all probability, the case with Oxford Laud. misc. 594, fols. 22–40v) and "Cum inter cetera philosophice documenta ..." (Paris lat. 7281, fols. 46–138; Paris lat. 7406, fols. 1–9v, mentioned under this incipit by L. Thorndike and P. Kibre, *A Catalogue of Incipits*, 2nd ed. [London, 1963], col. 311, contains only the canons of al-Zarqāl: their title nevertheless indicates that John of Sicily composed a commentary on them). The version of the canons that are the subject of the commentary appears to correspond to no. 31.1.a of F. Carmody and not to that in Paris lat. 7281 (fols. 30–45; none of the three other MSS cited contains the canons). See F. J. Carmody, *Arabic Astronomical and Astrological Sciences in Latin Translation* (Berkeley, Calif., 1956), p. 159.

II. SECONDARY LITERATURE. The only account of John of Sicily is found in P. Duhem, *Le système du monde*, IV (Paris, 1916), 6–10. G. Sarton, *Introduction to the History of Science*, II (Baltimore, 1931), 987–988, rejected, on good grounds, the identification of John of Sicily with John of Messina, one of the translators employed by Alfonso X in the preparation of his tables.

EMMANUEL POULLE

JOHN SIMONIS OF SELANDIA (*fl.* France, fifteenth century), *astronomy*.

It is still uncertain whether John Simonis of Selandia was Danish, as supposed by P. Lehmann, or Dutch, as maintained by G. Beaujouan. In fact, nothing is known of his life except that he was a *doctor artium* who in 1417 wrote his only known work at Vienne. It is entitled *Speculum planetarum* and has the incipit "Ad utilitatem communem studentium in astrologia et specialiter medicorum." The treatise was quite well known, and both Regiomontanus and Arnald of Brussels copied it. At least twenty manuscripts are extant; besides those listed by E. Zinner there are Darmstadt 780; Paris, B.N. lat. 10266; Vatican Palat. 1340; Vatican lat. 5006; and Yale De Ricci Supp. 25. The text (an edition of which is in progress) describes the construction and use of an equatorium, or analogue computer for planetary longitudes, with several features that are new and interesting compared to the previous instruments of Campanus of Novara, Peter Nightingale, John of Lignères, and Chaucer.

The new features are on that part of the instrument related to the motion of the sun. Here all earlier equatoria had been based on the theory of Hipparchus (no epicycle and an eccentric deferent); John of Selandia based his speculum on the equivalent theory of Apollonius (one epicycle and a concentric deferent). He represented all epicycles by imaginary circles produced by points on a ruler turning about the epicycle center and gave all the planets a deferent of the same size, produced by a knot on a thread revolving about a pin that was placed in holes on the *mater* (ground plate) of the instrument.

Engraved upon the *mater* was a system of graduated circles, the outermost of which represented the ecliptic of the sun and moon, as well as the equant circles of the five other planets. With regard to the latter the center of the *mater* accordingly represents all the equant centers, but not the center of the earth, as in previous instruments. This conception entails some curious consequences. First, each planet must have its individual "center of the earth" represented by a hole properly placed on the *mater*. Second, the instrument must be provided with five particular circles, each representing the ecliptic of a particular planet. Third, since all these ecliptics are concentric with the *mater*, they must be divided into 360 unequal degrees in such a way as to appear equal when seen from the corresponding "center of the earth." This construction, highly ingenious and sophisticated, presumably was unique in the history of instrument making until this time; but the cumbersome division of scales into unequal degrees may well be the reason why no specimen of the actual instrument is known and why most of the manuscripts are without illustrations.

BIBLIOGRAPHY

The following works may be consulted: G. Beaujouan, *Manuscrits scientifiques médiévaux de l'Université de Salamanque* (Bordeaux, 1962), p. 166; Paul Lehmann, "Skandinaviens Anteil an der lateinischen Literatur des Mittelalters," in *Sitzungsberichte der Bayerischen Akademie der Wissenschaften*, Philosophisch-Historische Abteilung, 1936, **2** (Munich, 1936), 55; O. Pedersen, "Two Mediaeval Equatoria," in *Actes du XIe Congrès international d'histoire des sciences*, III (Warsaw, 1968), 68–72; E. Poulle, *La bibliothèque scientifique d'un imprimeur humaniste au XVe siècle* (Geneva, 1963), p. 64; and E. Zinner, *Verzeichnis der astronomischen Handschriften* . . . (Munich, 1925), nos. 9629–9642; and *Leben und Wirken des . . . Regiomontanus* (Munich, 1938), p. 52.

OLAF PEDERSEN

JOHNSON, DOUGLAS WILSON (*b.* Parkersburg, West Virginia, 30 November 1878; *d.* Sebring, Florida, 24 February 1944), *geomorphology*.

Johnson's father, a farmer-turned-lawyer, died when the boy was twelve, leaving his upbringing to his mother, an intellectual who was a leader of the Woman's Christian Temperance Union and an advocate of women's suffrage. It was from this background that Johnson developed the sharp legalistic mind, love of order, self-discipline, and emotional austerity which characterized both his dealings with his colleagues and his scholarship.

Johnson entered Denison University at the age of eighteen. He had never been robust and, fearing tuberculosis, transferred to the University of New Mexico, where he assisted its president, Clarence Luther Herrick, himself late of Denison University, in his summer geological fieldwork. Herrick's humane scientific influence led Johnson to take up geology; he subsequently pursued graduate work at Columbia University, where he received his doctorate in 1903, the year in which he married Alice Adkins, daughter of a Baptist preacher. The deep love between these two sustained them through thirty-five years of marriage. During most of this time Alice was totally blind, and all five of their children died within a few hours of birth as did a foster child within a few days. Triumphing over her affliction, Alice was his companion on worldwide travels and Johnson never recovered from her death in 1938.

Taking up an instructorship in geology at the Massachusetts Institute of Technology, Johnson continued his studies in physical geography at Harvard, where he came under the influence of W. M. Davis, whose upbringing and intellect were much like his own. Inspired by Davis' sharpness of reasoning and by the clarity and effectiveness of his written and graphic exposition, Johnson later wrote to him (April 1921): "I have always felt that no one of the teachers with whom it was my fortune to be associated did so much for me in the way of development of correct methods of investigation and exposition as did you." When, toward the end of his life, Johnson wrote his penetrating but unfinished "Studies in Scientific Method" (1938–1941), he gave pride of place to the "analytical method of presentation" (also called the "method of multiple working hypotheses") as exemplified by Davis' "Rivers and Valleys of Pennsylvania" (1889). Johnson's last major work, *The Origin of the Carolina Bays* (1942), employed this analytical method to arrive at the suggestion that the Carolina bays were caused by a combination of submarine artesian spring action, lacustrine solution, and beach processes.

In 1907 Johnson moved to Harvard as assistant professor of geology; two years later he edited an important collection of Davis' works, *Geographical Essays* (1909), thereby perpetuating Davis' earlier writings, on which later generations of geomorphologists were to draw, to the virtual exclusion of Davis' important later work. Johnson was an unswerving disciple of Davis, disagreeing with him only on the spelling of the word "peneplain" (1916). It is rather ironic that one of his last students was Arthur N. Strahler, who later did much to propagate modern "quantitative geomorphology" to the detriment of the "classical" work of Davis and Johnson. Speaking of his teacher, Strahler wrote: "Even as recently as 1943, Douglas Johnson presented his graduate classes in geomorphology with subject matter faithfully reproducing the principles and details as written by Davis 45 years earlier" (*Annals of the Association of American Geographers*, **40** [1950], 209–213). Johnson was, nevertheless, an energetic and meticulous teacher much influenced at Harvard by the mercurial Nathaniel Southgate Shaler. He conducted field trips with almost regimental efficiency and produced a constant stream of Ph.D.'s beginning with Armin K. Lobeck (1917).

In 1911 Johnson received a grant from the Shaler Memorial Fund to study the whole eastern shoreline of the United States and that of parts of western Europe. He had already published work on beach processes and sea-level changes, topics he pursued until World War II. After moving to Columbia University, first as associate professor in 1912 and then as professor in 1919, he devoted much time to these topics and in 1919 published his important book *Shore Processes and Shoreline Development*, notable for the completeness of its review of the literature and, particularly, for its detailed elaboration of the idea of F. P. Gulliver, Davis' only Ph.D. student, that the cyclic concept should be applied to shoreline evolution and classification. This work was followed by a regional application of these principles in *The New England–Acadian Shoreline* (1925). These marine interests were largely instrumental in the setting up of the National Research Council's Committee on Shoreline Investigations in 1923 with Johnson as chairman; it concerned itself with studies of mean sea level (on which Johnson published some fourteen papers between 1910 and 1930) and with coastal protection. He developed an interest in the formation and correlation of marine terraces: his "Principles of Marine Level Correlation" (1932) was followed by some seven other papers on the study of Pleistocene and Pliocene terraces. He also served as president of the International Geographical Union's commission on the subject from 1934 to 1938. In 1939 Johnson published *Origin of Submarine Canyons*, reviewing the large number of hypotheses which had been proposed and tentatively suggesting a working hypothesis that involved the sapping action of submarine artesian springs.

Johnson's orderly, authoritarian outlook led him to take a profound interest in the course of World War I, particularly in the way in which military operations were affected by terrain; in this he followed Davis' interest in the influence of the Appalachian topography on the course of the American Civil War. His anti-German views were reflected in his election as chairman in 1916 of the executive committee of the American Rights League, which sought American entry into the war; and there is little doubt that his influence on the elderly W. M. Davis did much to widen the academic breach which had developed between the latter and Albrecht and Walther Penck.

Johnson's political and scholarly views were reflected in his many contributions to what was then termed "military geography." In 1917 he published *Topography and Strategy in the War* and received a commission as major in the intelligence division of the U. S. Army before proceeding to Europe to study firsthand the influence of landforms on modern warfare. This interest culminated in his *Battlefields of the World War* (1921). He returned to Columbia University in 1920, and in 1923–1924 he lectured on American geomorphology at several French universities; he published the substance of his lectures in *Paysages et problèmes géographiques de la terre américaine* (1927), and in his extensive review of European geography (1929), in which he argued forcefully that geomorphology be viewed as part of geology rather than geography.

It is for his work in geomorphology, fluvial as well as coastal, that Johnson is remembered most. Early in his career he wrote "The Tertiary History of the Tennessee River" and he continued to contribute articles on a wide variety of geomorphic topics virtually until his death. Particularly notable are "Baselevel" (1929), "Geomorphologic Aspects of Rift Valleys" (1931), "Streams and Their Significance" (1932), and several ascribing the origin of pediments primarily to lateral stream corrasion (1931, 1932). In his more advanced years, particularly after Davis' death in 1934, Johnson occupied the position of America's most influential geomorphologist, passing critical judgments on his contemporaries.

Johnson chose for his major and most lasting contribution a return to the denudation chronology of the central and northern Appalachians made classic forty years before by Davis' two brilliant papers, "The Rivers and Valleys of Pennsylvania" and "The Rivers of Northern New Jersey." Two circumstances permitted him to produce a more streamlined and satisfying theoretical history of Appalachian geomorphology than had Davis. First he was free from the necessity of assuming the existence of Appalachia with an initial east-west drainage, the subsequent reversal of which had to be accounted for. Second, he saw that the situation would be greatly simplified if it could be assumed that the sub-Cretaceous unconformity of the coastal plain (the Fall Zone peneplain) was of different age from the summit peneplain of the Appalachians further west (the Schooley peneplain). Johnson's *Stream Sculpture on the Atlantic Slope* (1931) ranks with the work of H. Baulig on the Massif Central and of S. W. Wooldridge and D. L. Linton on Southeast England as one of the masterpieces of denudation chronology. In it, with highly effective writing and use of block diagrams, he pictures the development of the complex relief and drainage of the region through a series of rational steps. His masterstroke was to postulate a widespread Cretaceous marine cover over the Fall Zone peneplain which was subsequently upwarped and from which eastward-flowing rivers were superimposed on the underlying structures; there followed a series of discontinuous diastrophic uplifts which were responsible for the successive Schooley, Harrisburg, and Somerville surfaces.

Johnson was president of both the Geological Society of America and the Association of American Geographers; he held six honorary degrees, three of them from French universities; and he received many medals and two decorations, one of which was that of chevalier of the Legion of Honor.

BIBLIOGRAPHY

I. ORIGINAL WORKS. A comprehensive bibliography of Johnson's writings is in the obituary by W. H. Bucher in *Biographical Memoirs. National Academy of Sciences*, **24** (1947), 197–230. The works mentioned in the text include "The Tertiary History of the Tennessee River," in *Journal of Geology*, **13** (1905), 194–231; W. M. Davis, *Geographical Essays* (Boston, 1909), of which Johnson was editor; "Beach Cusps," in *Bulletin of the Geological Society of America*, **21** (1910), 599–624; "The Supposed Recent Subsidence of the Massachusetts and New Jersey Coasts," in *Science*, **32** (1910), 721–723; "Plains, Planes and Peneplanes," in *Geographical Review*, **1** (1916), 443–447; *Topography and Strategy in the War* (New York, 1917); *Shore Processes and Shoreline Development* (New York, 1919); *Battlefields of the World War: A Study in Military Geography*, American Geographical Society Research series, no. 3 (New York, 1921); *The New England–Acadian Shoreline* (New York, 1925); *Paysages et problèmes géographiques de la terre américaine* (Paris, 1927); "The Central Plateau of France," in *Geographical Review*, **19** (1929), 662–667; "The Geographic Prospect," in *Annals of the Association of American Geographers*, **19** (1929), 167–231; "Baselevel," in *Journal of Geology*, **37** (1929), 775–782; "Geomorphologic Aspects of Rift Valleys," in *Comptes rendus. 15th International Geological Congress*, II (1931), 354–373; "Planes of Lateral Corrasion," in *Science*, **73** (1931), 174–177; *Stream Sculpture on the Atlantic Slope* (New York, 1931); "Streams and Their Significance," in *Journal of Geology*, **40** (1932), 480–497; "Rock Fans of Arid Regions," in *American Journal of Science*, 5th ser., **137** (1932), 389–416; "Rock Planes of Arid Regions," in *Geographical Review*, **22** (1932), 656–665; "The Role of Analysis in Scientific Investigation," in *Science*, **77** (1933), 569–576, also in *Bulletin of the Geological Society of America*, **44** (1933), 461–494; "Development of Drainage Systems and the Dynamic Cycle," in *Geographical Review*, **23** (1933), 114–121; "Available Relief and Texture of Topography: A Discussion," in *Journal of Geology*, **41** (1933), 293–305; obituary of W. M. Davis, in *Science*, **79** (1934), 445–449; "Studies in Scientific Method," in *Journal of Geomorphology*, **1** (1938), 64–66, 147–152; **2** (1939), 366–372; **3** (1940), 59–66, 256–262, 353–355; **4** (1941), 145–149, 328–332; **5** (1942), 73–77, 171–173; *Origin of Submarine Canyons* (New York, 1939); "Memorandum . . . on the Mimeographed Outline of the Proposed Symposium on the Geomorphic Ideas of Davis and Walther Penck," in *Annals of the Association of American Geographers*, **30** (1940), 228–232; "The Function of Meltwater in Cirque Formation," in *Journal of Geomorphology*, **4** (1941), 253–262; and *The Origin of the Carolina Bays* (New York, 1942).

II. SECONDARY LITERATURE. On Johnson and his work, see R. J. Chorley, "Diastrophic Background to Twentieth-Century Geomorphological Thought," in *Bulletin of the Geological Society of America*, **74** (1963), 953–970; R. J. Chorley, R. P. Beckinsale, and A. J. Dunn, *The History of the Study of Landforms:* vol. II, *The Life and Work of William Morris Davis* (1973); A. K. Lobeck, "Douglas Johnson," in *Annals of the Association of American Geographers*, **34** (1944), 216–222; A. N. Strahler, "Davis' Concepts of Slope Development Viewed in the Light of Recent Quantitative Investigations," in *Annals of the Association of American Geographers*, **40** (1950), 209–213; and F. J. Wright and A. Z. Wright, "Memorial to Douglas Johnson," in *Proceedings of the Geological Society of America, Annual Report for 1944* (New York, 1945), pp. 223–239.

R. J. CHORLEY

JOHNSON, MANUEL JOHN (*b.* Macao, China, 23 May 1805; *d.* Oxford, England, 28 February 1859), *astronomy.*

While stationed on the island of St. Helena, Johnson measured the positions of southern hemisphere stars and compiled a catalogue of them. Later, while serving as Radcliffe observer at Oxford, he made a catalogue of northern hemisphere stars as well.

His father was an Englishman named John William Johnson. As a boy, Manuel attended Addiscombe College, a military school located near Croydon, just south of London, where cadets were prepared for service in the British East India Company. At age sixteen, with the rank of lieutenant, he was assigned to an artillery division stationed on St. Helena.

With time on his hands at this isolated place, Johnson began to study the heavens. He was encouraged in this pursuit by the governor of St. Helena, General Alexander Walker, to whom he had been assigned as aide-de-camp. Funds to build and equip an observatory were provided by the East India Company; and Johnson was sent twice to Cape Town, in 1825 and again in 1829, to get advice from Fearon Fallows, then royal astronomer at the recently established Cape of Good Hope observatory.

The observatory on St. Helena was completed in 1829. During the next four years Johnson made the observations that formed the basis of his *Catalogue of 606 Principal Fixed Stars of the Southern Hemisphere*, published in 1835 at the expense of the East India Company.

St. Helena was returned to the British crown in 1834. The artillery unit was disbanded, and Johnson returned to England on a pension. In February 1835 he was awarded the gold medal of the Royal Astronomical Society for his star catalogue, and in December of that same year he enrolled as an undergraduate in Magdalen College, Oxford. He graduated B.A. in 1839 and M.A. in 1842.

When the Radcliffe observer, Stephen Peter Rigaud, died early in 1839, Johnson applied for the position and obtained it; in October 1839 he took up residence in the Radcliffe Observatory (then in the northwest suburbs of Oxford but transferred about a century later to Pretoria, South Africa). He remained there until his death twenty years later.

As Radcliffe observer, Johnson, with the initial help of Sir Robert Peel (one of the Radcliffe trustees), reequipped the observatory, buying telescopes and a heliometer, and—starting out only to revise Groombridge's catalogue—assembled material for his *Radcliffe Catalogue of 6317 Stars, Chiefly Circumpolar*. He also continued the meteorological observations begun by Rigaud and made many differential measurements with the heliometer.

In 1850 Johnson married Caroline Ogle. He was elected to fellowship in the Royal Society in 1856 and served as president of the Royal Astronomical Society in 1856–1857. His death followed a period of declining health because of heart disease.

BIBLIOGRAPHY

I. Original Works. Johnson's writings include *A Catalogue of 606 Principal Fixed Stars in the Southern Hemisphere; Deduced From Observations Made at the Observatory, St. Helena, From November 1829 to April 1833* (London, 1935) and *Astronomical Observations Made at the Radcliffe Observatory, Oxford*, I–XIX (Oxford, 1842–1861) —with vol. XIV the title was changed to *Astronomical and Meteorological Observations . . .*, and in that volume is a summary of the meteorological records kept at the Radcliffe Observatory for twenty-five years, extending back into Rigaud's tenure. Johnson's northern hemisphere star observations appeared as *The Radcliffe Catalogue of 6317 Stars, Chiefly Circumpolar; Reduced to the Epoch 1845.0; Formed From the Observations Made at the Radcliffe Observatory, Under the Superintendence of Manuel John Johnson, M.A., Late Radcliffe Observer* (Oxford, 1860); this was complete and in the hands of the printer before Johnson died.

In addition, Johnson published thirteen papers, listed in the Royal Society's *Catalogue of Scientific Papers*, III (London, 1869), 556–557; one of these was written with Norman Pogson as junior author (Pogson, who devised the scale of stellar magnitudes still in use today, served as Johnson's assistant for some years, beginning in 1851).

II. Secondary Literature. Johnson's certificates of marriage (16 July 1850) and death are in the General Register Office, London. Francis Baily's citation when he presented Johnson the gold medal of the Royal Astronomical Society appeared in *Memoirs of the Royal Astronomical Society*, **8** (1835), 298–301. Other contemporary accounts of Johnson's life and accomplishments can be found in the (London) *Times* (2 Mar. 1859, p. 5, col. 6), and (4 Mar. 1859, p. 5, col. 1); *Monthly Notices of the Royal Astronomical Society*, **20** (1860), 123–130; and *Proceedings of the Royal Society*, **10** (1860), xxi–xxiv.

Later sources include C. André and G. Rayet, *L'astronomie pratique*, I (Paris, 1874), 57–60; Agnes Mary Clerke, in *Dictionary of National Biography*, XXX (London, 1892), 22–23; and Joseph Foster, *Alumni Oxoniensis, 1715–1886* (Oxford, 1888), II, 757, col. 2.

SALLY H. DIEKE

JOHNSON, THOMAS (*b.* Selby [?], Yorkshire, England, *ca.* 1600, *d.* Basing, Hampshire, England, September 1644), *botany*.

The year of Thomas Johnson's birth and his parentage are unknown. He was certainly born in Yorkshire, probably at Selby; and, if there, possibly in either 1600 or 1605. He lived in Lincolnshire before 1620; and his travels in 1626 in Lincolnshire, Yorkshire, and County Durham suggest connections with the bourgeoisie and lesser landed gentry there.

On 28 November 1620 Johnson was apprenticed to the London apothecary William Bell and on 28 November 1628 was made a free brother of the Society of

Apothecaries. His laudatory Latin contribution to John Parkinson's *Paradisi in sole, Paradisus terrestris,* published in 1629, shows that he rapidly gained eminence in his profession. In 1629 George Johnson, son of Marmaduke Johnson of Rotsea in Holderness, Yorkshire, was bound apprentice to Thomas Johnson, who had certainly been to Rotsea in 1626, presumably to visit his relatives there. By 1633 Johnson was established at his apothecary's shop on Snow Hill, London, where it is likely that his kinsman George continued the practice after Thomas' death. By 1630 he had become acquainted with Dr. George Bowles and by 1631 with John Goodyear, who with John Parkinson and Johnson himself were the ablest British botanists of the first half of the seventeenth century. On 12 December 1633 Johnson was admitted liveryman of the Society of Apothecaries, and on 3 August 1640 he was sworn to the Society's Court of Assistants.

By 1642, with the outbreak of civil war, the position of Royalist citizens in London was precarious; and although Johnson attended the Apothecaries' Court on 8 December 1642, at about that time he apparently joined other prominent London Royalists with Charles I at Oxford. Johnson figures among the notorious "Caroline creations" at Oxford University. He was made bachelor of physic on 31 January 1643 and doctor of physic on 9 May 1643, no doubt in recognition of his loyalty, although he was academically more deserving than most of the recipients. As Lieutenant Colonel Johnson he was in the king's army at Basing House on 7 November 1643, when it was besieged by Sir William Waller. On 14 September 1644, still under siege, Johnson was "shot in the shoulder, whereby contracting a Feaver he dyed a fortnight after." He probably left a widow and a son, Thomas, who was bound apprentice to the Society of Apothecaries on 10 May 1649.

Johnson's *Iter plantarum* (1629) is a lively description of a botanical journey into Kent and of a visit to Hampstead Heath. The lists of plants observed establish him as the foremost British field botanist of his day, and the *Iter* is the first local flora of Britain. *Descriptio itineris* (1632) brought the total number of species recorded for the first time in Kent to over 300; it includes an additional list of species for Hampstead Heath.

In 1632 Johnson was commissioned to produce within a year a revised edition of John Gerard's *Herball* (1597). The changes which Johnson made in his edition of 1633 are quite remarkable. A new set of 2,766 wood-block illustrations was incorporated into the text of 1,634 folio pages (with about forty leaves of additional matter), and many of Gerard's mistakes were corrected. Johnson wrote on p. 1114: "Our author here (as in many other places) knit knots somewhat intricate to loose." Passages which Johnson substantially emended were marked with a dagger, and completely new ones with a double cross. Contributions by his friends John Parkinson, George Bowles, John Goodyear, and others are acknowledged by name. Many of the additions are based on Johnson's own journeys. In all, about 120 plants were recorded for the first time in Britain in the *Iter*, the *Descriptio*, and his edition of Gerard. Another excellent addition by Johnson to Gerard is a survey of the history of botany, the first such in English. The 1636 reprint of the *Herball* contains only minor corrections but adds Johnson's intention, with the help of some of his friends, to travel over the greater part of England to discover the native plants.

An account of a botanical journey to Bath and Bristol, including the already famous Avon gorge, appeared in *Mercurius botanicus* in 1634. *Mercurii botanici pars altera* (1641) describes Johnson's last and longest journey into North Wales, made in 1639. In this he clearly stated his intentions: having published a catalogue of all the plants found on his previous excursions, he proposed to add the discoveries of his friends and records from the literature to his own observations; he hoped to add accurately drawn figures and eventually, with his friend Goodyear, to publish their histories; and he hoped that others would notify him of their records. Johnson intended the two parts of the *Mercurius* to be a complete catalogue of all known British plants, a total of about 900, of which nearly fifty were new to Britain. After Johnson's death William How brought together the two parts of the *Mercurius* in *Phytologia Britannica* (1650), a hasty and defective compilation which earned How undeserved credit.

Thomas Johnson, apothecary, soldier, and botanist, was almost certainly the same Thomas Johnson whose translation of the massive works of the French surgeon Ambroise Paré first appeared in 1634; this English edition had a profound influence on British surgery until at least the end of the seventeenth century. The editorial method, prose style, and botanical emendations strongly suggest the author.

Johnson was an amiable companion, a brave soldier, a successful apothecary, and an industrious and scholarly editor. Although he contributed nothing to the principles or ideas of scientific botany, he stands high among the pioneers of the study of the British flora. His botanical work was respected by John Ray, inspired Sir Joseph Banks to an interest in plants, and continues to give pleasure to many botanists. Furthermore, his edition of the works of

Paré had a profound and beneficial influence on British surgery.

He is commemorated in the name given by Robert Brown to the liliaceous genus *Johnsonia.*

BIBLIOGRAPHY

I. ORIGINAL WORKS. A complete bibliography and references to related works are given in the excellent biography by H. Wallis Kew and H. E. Powell, *Thomas Johnson, Botanist and Royalist* (London, 1932). Johnson's principal works are *Iter plantarum investigationis ergo susceptum. A decem sociis, in agrum Cantianum. Anno Dom. 1629. Julii 13. Ericetum Hampstedianum . . . 1 Augusti* (London [1629]); *Descriptio itineris plantarum investigationis ergo suscepti, in agrum Cantianum Anno Dom. 1632. Et enumeratio plantarum in ericeto Hampstediano . . .* ([London], 1632); *The Herball or Generall Historie of Plantes. Gathered by John Gerarde . . . . Very Much Enlarged and Amended . . .* (London, 1633; repr. with minor alterations, 1636); *Mercurius botanicus. Sive plantarum gratia suscepti itineris, anno 1634 descriptio . . .* (London, 1634); *The Workes of That Famous Chirurgion Ambrose Parey. Translated out of Latine and Compared With the French* (London, 1634; reiss., 1649, 1665, 1678); and *Mercurii botanici pars altera sive Plantarum gratia suscepti itineris in Cambriam sive Walliam descriptio . . .* (London, 1641). The *Iter*, *Descriptio*, *Mercurius*, and *Mercurii . . . pars altera* were reprinted by T. S. Ralph in *Opuscula omnia botanica Thomae Johnsoni* (London, 1847). Facs. repr. of the exceedingly rare *Iter* and *Descriptio*, with English trans. by C. E. Raven, appear in *Thomas Johnson. Botanical Journeys in Kent and Hampstead*, J. S. L. Gilmour, ed. (Pittsburgh, Pa., 1972).

II. SECONDARY LITERATURE. No significant biographical details have come to light since the account of Kew and Powell. The best appreciations of Johnson's botanical work are in Agnes Arber, *Herbals*, 2nd ed. (Cambridge, 1938), pp. 134–135, more extensively in C. E. Raven, *Early English Naturalists From Neckham to Ray* (Cambridge, 1947), ch. 16, "Thomas Johnson and His Friends"; and, for the early journeys, by several contributors to Gilmour's ed. of the *Iter* and *Descriptio*. A vivid account of the importance of Johnson's trans. of the works of Ambroise Paré is given by Sir D'Arcy Power, "Epoch-making Books in British Surgery. VI. Johnson's Ambrose Parey," in *British Journal of Surgery*, **16** (1928), 181–187.

D. E. COOMBE

JOHNSON, WILLARD DRAKE (*b.* Brooklyn, New York, 1859; *d.* Washington, D.C., 13 February 1917), *geomorphology.*

When Johnson was two years old, his family moved to Washington, D.C. After graduating from the Sheffield Scientific School of Yale, Johnson joined the topographical division of the U.S. Geological Survey in 1879, receiving a permanent appointment as topographer in 1883. During his probationary period he was fortunate to work with G. K. Gilbert, surveying the abandoned shorelines of Lake Bonneville and the ancient delta of the Logan River, and then with Israel C. Russell on a geological reconnaissance of southern Oregon, where he carried out soundings of Silver Lake and recognized that Guano Valley had contained a shallow Quaternary lake. The detailed maps which he prepared for both the publications reporting on these surveys marked Johnson as a master of the then artistic science of topographical surveying. In 1883 he mapped the Mono Basin for Russell; in the course of this mapping he occupied a survey point on the summit of Mt. Dana in the Sierras, transporting his instruments there on muleback. The results of his survey, notably maps of the Mono Basin and Mono Lake on scales of 1:250,000 and 1:125,000, respectively, appeared in Russell's "Quaternary History of Mono Valley, California." In view of his later contribution to glacial geomorphology, it is interesting that at this time Johnson explored the existing snowfields and glaciers on Mt. Couness, Mt. McClure, and Mt. Ritter, providing information which later appeared in Russell's *Glaciers of North America* (1897). In 1884 Johnson completed the survey of the shorelines of ancient Lake Lahontan in western Nevada. He was then assigned to Massachusetts and remained in the East for the next three years. During this time he was in charge of a number of survey parties mapping chiefly in western Massachusetts (notably the Becket, Sandesfield, Chesterfield, and Glanville quadrangles) and in one season surveyed fully 460 square miles with the aid of two assistants.

After a short period of surveying in the vicinity of the Delaware Water Gap, Johnson was sent to Colorado in 1888 in charge of a large party to survey some 1,500 square miles in the valley of the Arkansas River near Pueblo, with a view to assessing the possibilities for irrigation. At this time J. W. Powell personally obtained a significant promotion for him from the secretary of the interior by pointing out that Johnson was in charge of five or six independent survey parties. This survey was completed at the end of 1889, and Johnson was then transferred to the Irrigation Survey, which was responsible for mapping in Colorado, Wyoming, and the Dakotas. In 1891 Johnson was ordered to California to set up an office of the U.S. Geological Survey in the Gold Belt section of the central part of the state. Shortly after his arrival he became a charter member and a director of the Sierra Club. Although the mountain named after him by the Geological Survey in 1917 (12,250 feet on the crest of the Sierras, 1.5 miles northeast of Parker Pass) was subsequently renamed Mt. Lewis, after a former Yosemite Park superintendent, the peak recommended by the Sierra Club in 1926 to commemorate him still bears his name (12,850 feet on the crest of the Sierras, between Mt. Gilbert and Mt. Goode).

Johnson was granted a year's leave of absence in 1895 to accompany W J McGee, whom he had met in 1883 when the latter was a geological assistant to Russell, on a hazardous expedition for the Bureau of Ethnology to Tiburon Island in the Gulf of California, in order to study the Seri Indians. He produced a fine map on a scale of 1:380,160 for McGee's report and joined McGee in producing a general article on Seriland (1896). On his return Johnson was appointed hydrographer and was assigned to the division of hydrography, continuing in this capacity until 1913. Between 1897 and 1900 he worked in the high plains on water supply problems, becoming increasingly concerned with geological investigation. His most extensive publication, "The High Plains and Their Utilization" (1901–1902), contains the definitive description of the origin and structure of the depositional surface of the plains, emphasizing the preponderance of silt content and the occurrence of depressions, as well as a survey of groundwater occurrence, with special reference to underflow in the river valleys, and a description of water utilization in the region.

Between 1900 and 1904 Johnson worked as a topographer and geologist under Gilbert, mapping the Wasatch, Fish Spring, and Swasey ranges and studying the Pleistocene glacial features west of the Wasatch Range. In 1904 he returned to the Sierras and published "The Profile of Maturity in Alpine Glacial Erosion," which was to become his most enduring scholarly legacy. Drawing on observations he had made in 1883 north of Mt. Lyell, Johnson formally presented his theory of cirque backcutting by nivation along the bergschrund; it had been foreshadowed by three previously published abstracts (1896–1899). Johnson's outstanding contribution to glacial geomorphology rested on his association of cirque-floor truncation with processes going on at the base of the curving ice crevasse occurring at the head of a cirque. He wrote: ". . . my instant surmise, therefore, was that this curving great schrund penetrated to the foot of the [cirque head] wall, or precipitous rock-slope, and that a causal relation determined the coincidence in position of the line of deep crevassing and the line of assumed basal undercutting" ("The Profile of Maturity in Alpine Glacial Erosion," p. 573). Johnson had had himself lowered 150 feet into a bergschrund and had found that the lower twenty to thirty feet coincided with the base of the cirque headwall which was riven by freeze and thaw. It seemed clear that erosion is concentrated in this zone and that the glacier removes the resulting debris, so that the cirque floor is constantly extended backward and downward into the range, giving a "down-at-heel" effect. "The ultimate effect, upon a range of high-altitude glaciation, would be rude truncation. The crest would be channeled away, down to what might be termed the base-level of glacial generation" (*ibid.*, p. 577). It is difficult to overemphasize the importance of this ten-page paper.

Johnson later investigated the effects of the earthquake of 1872 in the Owens Valley with W. H. Hobbs (1908, 1910), but he began to be dogged by ill health, which eventually forced him to become a part-time employee of the U.S. Geological Survey on a *per diem* basis. His remaining professional life was a series of tragedies. His transfer to Portland, Oregon, in 1913 to take charge of the geography section of the U.S. Forest Service was not a success; the following year he was confined to a private sanatorium, suffering from "colitis and paranoia with suicidal tendencies." His friend Gilbert assumed the expenses. Johnson's slow recovery was hampered by an appendectomy, although he did carry out triangulation work in the Cascade Range. In 1915 he suddenly returned to Washington, dispirited and mentally and physically ill, and again entered a sanatorium. After discharge he worked part-time classifying the photograph collection of the Survey. He took his own life, according to Gilbert, after having ". . . learned of the collapse of an enterprise by which he hoped to reestablish himself."

Johnson possessed above all "an eye for country." His survey method was to use a plane table for trial sketching and correction by intersection, and his maps show that he was an artistic genius of landscape. He was an untiring surveyor and teacher of surveying, doing much to improve mapmaking procedures. In 1887 he patented the Johnson tripod head, waiving all royalties, and established the design still used today. An impoverished, lonely, and sensitive man, "he seemed to have no thought for anything but the advancement of scientific work of the Geological Survey, sacrificing his health, pleasures, and means to the end" (*Geographical Review* [1917], p. 330). Shortly after his death, Gilbert wrote to the U.S. Forest Service in Portland, asking if any notebooks or other records of Johnson's geologic work remained there. The assistant district forester replied: "He had accumulated throughout his life a vast store of information which was carried in his mind and never reduced to writing. . . . the great store of his learning is forever lost."

BIBLIOGRAPHY

I. ORIGINAL WORKS. Johnson's publications include "An Early Date for Glaciation in the Sierra Nevada," in *American Geologist*, **18** (1896), 61–62, also in *Science*, n.s. **3** (1896), 823, an abstract; "Seriland," in *National Geographic Magazine*, **7** (1896), 125–133, written with W J McGee; "An Unrecognized Process in Glacial Erosion," in *American Geologist*, **23** (1899), 99–100, also in *Science*, n.s. **9** (1899), 106, an abstract; "The Work of Glaciers in High Mountains," in *Science*, n.s. **9** (1899), 112–113, an abstract; "Subsidence Basins of the High Plains," *ibid.*, pp. 152–153, an abstract; "The High Plains and Their Utilization," in *Report of the United States Geological Survey*, **21**, pt. 4 (1901), 601–741, and **22**, pt. 4 (1902), 631–669; "The Profile of Maturity in Alpine Glacial Erosion," in *Journal of Geology*, **12** (1904), 569–578; "The Grade Profile in Alpine Glacial Erosion (Sierra Nevada, California)," in *Sierra Club Bulletin*, **5** (1905), 271–278; "The Earthquake of 1872 in the Owens Valley, California," in *Science*, n.s. **27** (1908), 723, an abstract written with W. H. Hobbs; and "Recent Faulting in Owens Valley, California," in *Science*, n.s. **32** (1910), 31, also in *Bulletin of the Geological Society of America*, **21** (1910), 792, an abstract.

II. SECONDARY LITERATURE. Biographical sources are his service record at the U.S. Geological Survey; the unsigned "Willard D. Johnson," in *Geographical Review*, **3** (1917), 329–330; and "W. D. Johnson Dies by His Own Hand," in Washington *Evening Star* (13 Feb. 1917).

R. J. CHORLEY

JOHNSON, WILLIAM (*b. ca.* 1610; *d.* London, England, September 1665), *chemistry.*

Born into the gentry and probably educated for the clergy, Johnson began his career as a chemist about 1648. In June 1648 the College of Physicians of London had decided to erect a laboratory for the preparation of chemical medicines by the doctors. Soon afterward, however, Johnson was allowed to fit up and man a laboratory at the west end of the College garden at his own expense. His effort to make this a commercial venture in which both the public and the College were served with his chemical preparations was unanimously condemned by the College; nevertheless Johnson soon became known as the operator to the Royal College of Physicians, and his career as such is in many respects typical of a seventeenth-century chemical operator—a competent technician versed in the manipulations of chemistry.

As operator to the College, Johnson prepared chemical medicines and ingredients as samples and for sale. (His occupation as a dispenser of chemical medicines was recognized when he was granted the freedom of the Society of Apothecaries in 1654.) He instructed Collegians and possibly the public in chemical operations and analyzed suspicious medicines for the College, using, in part, a rudimentary comparison by weights.

Johnson served his profession with the publication of *Lexicon chemicum* (1652), which he freely admitted was simply gleaned and rearranged from such German authors as Basilius Valentinus, J. B. van Helmont, and especially Ruland. At least five printings attest to the usefulness of such a dictionary, in which the dark phrases of chemists were ordered and classified.

Because of the *Lexicon* and a less orderly publication, issued in the same year, Johnson was placed among the early followers of Helmont but later was considered a traitor by the dogmatic iatrochemists, who soon were challenging the established legal medical practice of London. One of the more important of these chemists was George Thomson, in reply to whose *Galeno-pale* (London, 1665) Johnson wrote *Some Brief Animadversions* on behalf of the College of Physicians. The College expressed its pleasure with a gift of £100. While the medical theories at stake are of great interest, for Johnson they were secondary to more immediate questions of the technical competence and professional status of the writers. His defense of chemical Galenism rarely attempted to tackle the philosophical issues raised.

The urgency of this debate was increased by the outbreak of the plague, which took the lives of some of the chief participants, including Johnson, who had taken part in the dissection of the body of a plague victim.

BIBLIOGRAPHY

I. ORIGINAL WORKS. Johnson's writings include *Lexicon chemicum* (London, 1652, 1657, 1660); and *'Αγυρτο-Μαστιξ or Some Brief Animadversions Upon Two Late Treatises . . .* (London, 1665). Johnson was the editor of some parts of *The Excellence of Physik and Chirurgery . . . Short Animadversions Upon a Work of Noah Biggs, Three Exact Pieces of Leonard Phiororant, etc.* (London, 1652).

II. SECONDARY LITERATURE. G. N. Clark, *History of the Royal College of Physicians* (London, 1964); Gerard Eis, "Vor und nach Paracelsus," in *Medizin in Geschichte und Kultur*, **8** (1965), 141; Patricia P. MacLachlan, "Scientific Professionals in the Seventeenth Century," (Ph.D. thesis, Yale University, 1968), pp. 61–83; and C. Wall, H. C. Cameron, and E. A. Underwood, *A History of the Worshipful Society of Apothecaries of London*, I (London, 1963), 335. Appropriate extracts from the unpublished *Annals* of the Royal College of Physicians and Farre's MS *History of the Royal College of Physicians* were furnished by Mr. L. M. Payne, librarian of the R.C.P., London.

PATRICIA PETRUSCHKE MACLACHLAN

JOHNSON, WILLIAM ERNEST (*b.* Cambridge, England, 23 June 1858; *d.* Northampton, England, 14 January 1931), *logic.*

Johnson's father, William Henry Johnson, was headmaster of a school in Cambridge, and Johnson first studied there. In 1879 he entered King's College, Cambridge, where he was eleventh wrangler in the mathematics tripos of 1882 and placed in the first class in the moral sciences tripos of 1883. For the next nineteen years he held a variety of temporary positions around Cambridge. During that period he published three technical papers on Boolean logic and one on the rule of succession in probability theory. In 1902 Johnson was appointed to the Sidgwick lectureship in moral science and was awarded a fellowship at King's College. He held these positions until shortly before his death. Although shy and sickly, he was a popular, respected teacher. Indeed, it was his students, especially Naomi Bentwich, who persuaded him to publish his three-volume *Logic* (1921–1924). A fourth volume, on probability, was never finished, but the first few chapters were published posthumously in *Mind.* This book won Johnson fame, honorary degrees from Manchester (1922) and Aberdeen (1926), and election as a fellow of the British Academy (1923).

Johnson made some technical contributions to logic. In "On the Logical Calculus" he developed an elegant version of Boolean propositional and functional logic, using conjunction and negation as his primitive symbols. He even attempted to define the quantifiers in terms of these connectives. In "Sur la théorie des équations logiques" and in his later writings on probability, he developed various rules of succession for the theory of probability. His primary contributions were, however, in the foundations of logic and of probability theory.

Johnson made many worthwhile, although not major, contributions to the philosophy of logic. Perhaps the most important were his distinction between determinables and determinates, his theory of ostensive definition, and his distinction between primary and secondary propositions. On all of these topics his ideas influenced, directly or indirectly, many contemporary logicians.

Johnson was one of the first to expound the view that probability claims should be interpreted as expressing logical relationships between evidence propositions and hypothesis propositions, relationships determined in each case by the content of these propositions. This view, also adopted by J. M. Keynes, Harold Jeffreys, and Rudolf Carnap, is one of the main contemporary alternatives to the frequency interpretation of probability claims. Although Keynes and Jeffreys published their books before the appearance of Johnson's "Probability," Keynes freely admitted his indebtedness to Johnson's ideas.

BIBLIOGRAPHY

Johnson's main writings are "The Logical Calculus," in *Mind,* **1** (1892), 3–30, 235–250, 340–347; "Sur la théorie des équations logiques," in *Bibliothèque du Congrès international de philosophie* (1901); *Logic,* 3 vols. (Cambridge, 1921–1924); and "Probability," in *Mind* (1932).

BARUCH BRODY

JOLIOT, FRÉDÉRIC (*b.* Paris, France, 19 March 1900; *d.* Paris, 14 August 1958), *nuclear physics.*

Joliot's father, Henri Joliot, took part in the Commune of Paris at the end of the Franco-Prussian War and was obliged to spend several years in Belgium to escape the subsequent repression. Upon returning to France he settled in Paris, where he became a well-to-do tradesman. He was an ardent fisherman and hunter, and a virtuoso performer on the hunting horn, an instrument for which he composed numerous calls. Joliot's mother, Émilie Roederer, came from a petit bourgeois Alsatian Protestant family. She married Henri Joliot in 1879; Frédéric was the last of their six children. Raised in a completely nonreligious family, Joliot never attended any church and was a thoroughgoing atheist all his life. At the age of ten he became a boarder at the Lycée Lakanal, located in a southern suburb of Paris but, following the death of his father and family financial difficulties, he transferred to the École Primaire Supérieure Lavoisier in Paris. There he prepared for the entrance examination for the École Supérieure de Physique et de Chimie Industrielle of the City of Paris, to which he was admitted in 1920.

In this prestigious school, which ordinarily would have prepared him for a career in engineering, Joliot was introduced to basic science and scientific research pursued solely to satisfy the passion for knowledge, with no concern for practical application. The director of studies at the time was the physicist Paul Langevin, who had a decisive influence on Joliot: he oriented the young man not only toward scientific research but also toward a pacifist and socially conscious humanism that eventually led him to socialism. During these years at the École Supérieure de Physique et de Chimie Industrielle, where a large portion of the curriculum was devoted to laboratory work, Joliot developed his talents as an experimenter. He graduated first in his class, but after fifteen months of military service he was still undecided on a career.

A summer job in a large steel mill in Luxembourg (1922) left a strong impression on him, and the value of his engineering degree from one of the most highly regarded of the *grandes écoles* assured him a brilliant position in industry; but he felt himself drawn to scientific research. He discussed his situation with Paul Langevin, who, recognizing his exceptional gifts and sensing his true aspirations, advised him to accept a stipend that Mme. Curie had at her disposal to pay for a personal assistant.

Joliot took this advice and began work at the Institut du Radium of the University of Paris in the spring of 1925, under the guidance of Mme. Curie. At first he undertook further studies in modern physical chemistry, particularly radioactivity, at the laboratory itself; he also took courses at the university that enabled him to earn his *licence*. At the same time he successfully concluded his first personal research, on the electrochemical properties of polonium. He presented the results of this work, in the course of which he displayed great skill in handling difficult techniques, in his doctoral thesis (defended in 1930).

When Joliot entered the Radium Institute, Mme. Curie's elder daughter, Irène, was already an assistant there. Brought in contact with her through his work, he was rapidly attracted by her remarkable personality, which was wholly different from his own. They were married in 1926 but for most of the time continued to work separately. Only in 1931 did they begin the four years' close collaboration that so successfully united their complementary qualities.

Frédéric and Irène Joliot-Curie had a daughter and a son, both of whom became distinguished scientists. Hélène Joliot (b. 1927), who, like her father, graduated first in her class from the École Supérieure de Physique et de Chimie Industrielle of Paris, became a researcher in nuclear physics. In 1949 she married a grandson of Paul Langevin, who likewise was engaged in research. Thus in the third generation there was to be a married couple each of whom was a research physicist working in the field opened by Pierre and Marie Curie through the discovery of radium and extended by Frédéric and Irène Joliot-Curie through that of artificial radioactivity. Pierre Joliot (b. 1932) chose to specialize in biophysics. He too maintained the family tradition, for he and his scientist wife Anne both chose the same area of research.

After Joliot defended his thesis, no academic post was available at the Radium Institute and he had to consider leaving scientific research and taking a job as an engineer in industry. Fortunately for the advancement of science, Jean Perrin, who directed the Laboratory of Physical Chemistry, located near the Radium Institute, and who had already appreciated Joliot's great abilities as a researcher, arranged for him to receive a scholarship from the Caisse Nationale des Sciences, whose creation by the government Perrin had only recently obtained.

This scholarship permitted Joliot to continue research and to select freely his area of study. First he assembled the equipment that would enable him to observe under the best possible conditions the ionizing radiations emitted directly or secondarily by radioactive substances. His training as an engineer enabled him to draw up the plans and supervise in detail the construction of a greatly improved Wilson chamber. This device, also called a cloud chamber, makes it possible to see and to photograph the trajectories of electrically charged particles passing through a gas saturated with water vapor: a sudden expansion produces a supersaturation and causes tiny droplets of water to condense around each of the ions formed along its trajectory by every charged particle (electron, proton, α particle, and so on). Joliot called the direct, detailed view of individual corpuscular phenomena provided by this apparatus "the most beautiful experience in the world," and the cloud chamber was always his favorite tool of research. The one that he had constructed in 1931 could function at various pressures, from the low pressure of pure saturant water vapor at room temperature up to a pressure of several atmospheres. It had a diameter of about fifteen centimeters and could operate in a magnetic field of 1,500 gauss produced by large coils surrounding the cylindrical chamber in which the expansion took place. This arrangement permitted Joliot to determine the energy of the electronic rays (β rays) by measuring the curvature of their trajectories on the photographs. In addition Joliot set up devices to count the ionized particles detected by a thin-walled Geiger counter and an ionization chamber connected to an electrometer of high sensitivity. Finally, taking advantage of a large stock of radium D patiently accumulated at the Radium Institute by Mme. Curie, Joliot, in collaboration with his wife, prepared very strong sources of α rays emitted by polonium that had been deposited as thin layers possessing a high surface density of activity. The preparation of these sources was both difficult and dangerous because of the very high toxicity of polonium.

Joliot used all this equipment with a fertile imagination and a keen sense of those experiments which might lead to the observation of unexpected phenomena. In little more than two years of intense activity, alone or in collaboration with Irène Curie, he made a series of remarkable discoveries that culminated in the discovery of artificial radioactivity at the beginning of 1934. An enthusiastic innovator, Joliot constantly devised new experiments the significance of which was so immediately obvious that they always appeared extremely simple. The elegance of their conception and execution belied the laborious work with complex apparatus that had gone into their preparation.

The first experiment Joliot carried out with this equipment well demonstrates the manner in which he worked in order to increase his chances of observing unforeseen phenomena. He decided to study, in collaboration with Irène Curie, the strangely penetrating radiation emitted—as the German physicists W. Bothe and J. Becker had discovered in 1930—by certain light elements, notably beryllium and boron, when they are bombarded with α rays. The Joliot-Curies set up an intense source of this mysterious radiation by placing one of their very strong polonium preparations against a beryllium plate; and they used their highly sensitive ionization chamber to detect, after filtration through fifteen millimeters of lead, the penetrating radiation issuing from this source. Thinking that the ionization measured might be the result of easily absorbable secondary radiations produced in the wall of the ionization chamber by the very penetrating radiation (as was the case for γ rays) and wishing to be able to vary the nature of the last solid plates traversed by the radiation, they made the ultrapenetrating radiation enter the ionization chamber through a window covered by a sheet of aluminum only five microns thick; in front of this sheet they were able to interpose plates of various substances.

The analogy of the γ rays with ionizing secondary radiations may have suggested the use of a much thicker sheet of aluminum to cover the entrance window, but the more difficult option of an extremely thin sheet made possible the observation of secondary rays of very low penetrating power and thus permitted the Joliot-Curies to make an important discovery. With this arrangement they ascertained that interposing plates of most of the substances under examination (carbon, aluminum, copper, silver) between the source and the ionization chamber left the measured current virtually unaffected but that placing a screen made of a hydrogen-containing substance (paraffin, cellophane, water) in front of the window of the ionization chamber caused a large increase in the current. The secondary radiation responsible for this unexpected increase was completely absorbed by a sheet of aluminum 0.20 millimeter thick. It appeared that this radiation, produced only in hydrogen-containing substances, consisted of protons ejected by the penetrating Bothe-Becker radiation. This hypothesis was confirmed by various experiments that the Joliot-Curies reported when they announced the discovery of this surprising phenomenon in a note presented to the Academy of Sciences on 18 January 1932. A short time later they were in fact able to observe, with the aid of their Wilson chamber, the easily identifiable trajectories of the protons thus ejected and to prove, by filling their ionization chamber with helium, that the Bothe-Becker radiation also ejected helium nuclei (*Comptes rendus ... de l'Académie des sciences* **97**, 708 [22 Feb. 1932]).

Joliot would not have been able to make these unforeseen discoveries if he had closed his ionization chamber with a sheet of aluminum several tenths of a millimeter thick instead of making the effort required by the use of a foil a hundred times thinner. Analyzing the conditions of his success, he wrote in 1954: "I have always attached great importance to the manner in which an experiment is set up and conducted. It is, of course, necessary to start from a preconceived idea; but whenever it is possible, the experiment should be set up to open as many windows as possible on the unforeseen."

Immediately following the Joliot-Curies' first publication on this subject, the English physicist James Chadwick began to study the ejection of atomic nuclei by Bothe-Becker radiation. He employed a proportional pulse amplifier that allowed him to compare the energies of the ejected helium or nitrogen nuclei with that of the protons. He concluded that these ejections were the result of collisions between the nuclei and fast-moving uncharged particles that possessed a mass of the same order of magnitude as that of the protons and undoubtedly were torn from the nuclei of beryllium or boron by the α particles. Chadwick had discovered the neutrons; he published his results in *Nature* on 27 February 1932.

Joliot devoted the years 1932 and 1933 to studying, generally in collaboration with his wife, Bothe-Becker radiation and the phenomena accompanying its production. They proved that this radiation is complex, consisting not only of the neutronic rays that cause the ejection of light nuclei but also of γ rays of several million electron volts; when the latter pass through matter, they tear away electrons and eject them at high speeds.

The Joliot-Curies found that these high-energy γ rays also eject positive electrons—their existence had been predicted by Dirac and they had just been discovered in cosmic radiation by the American physicist C. D. Anderson. Furthermore, by operating their Wilson chamber in a magnetic field, they were able to make the first photographs of the creation of an electron pair (one positive and one negative) by materialization of a γ photon.

Resuming the study of radiation emitted by light elements bombarded by α particles, the Joliot-Curies discovered that when certain of these elements, notably boron and aluminum, are submitted to such bombardment there occurs an emission not only of protons or neutrons but also of positive electrons, the origin of which they attributed to some induced transmutations. They showed that the energies of the positive electrons created in this manner form a continuous spectrum analogous to that formed by the energies of the negative electrons emitted in β radioactivity, suggesting that the emission of a positive electron during transmutation is accompanied by that of a neutrino bearing a variable fraction of the available energy. In retrospect it seems that the observation of artificial radioactivity could have immediately followed this last discovery, which was presented and discussed in October 1933 at the Solvay Physics Conference, a gathering of the world's greatest nuclear physicists. As a matter of fact, Joliot did not continue his research on the emission of positive electrons by aluminum bombarded by α particles until he had successfully completed, at the end of December 1933, a study of the annihilation of positive electrons stopped by matter, in which he proved—as Dirac had foreseen—that this annihilation is accompanied by the emission of two γ photons of approximately 500 KEV.

Resuming the earlier investigations of emission phenomena, Joliot covered the window of his cloud chamber with a thin sheet of aluminum foil, against which he placed a strong source of polonium. He was surprised to observe that the emission of positive electrons, induced by the polonium, continued for several minutes after it had been removed and, therefore, after all irradiation of the aluminum had ceased. Realizing the significance of this observation and the importance of rapidly deducing from it every possible consequence, Joliot called in his wife; he wanted her to take part in the experiments that had to be carried out immediately in order to furnish decisive proof of the creation of new radioelements. Their observations, repeated with a thin-walled Geiger counter, confirmed that radioactive atoms with a half-life of a little more than three minutes are formed in aluminum irradiated by α rays. The radioactivity was analogous to the β radioactivity of the natural radioelements but was of a new type, since the electrons emitted were positive. The formation of atoms emitting delayed positive electrons appeared to be associated with the emission of neutrons, which had been previously observed, according to the nuclear reaction

$$^{27}_{13}\mathrm{Al} + {}^{4}_{2}\mathrm{He}(\alpha) \rightarrow {}^{30}_{15}\mathrm{P} + {}^{1}_{0}n.$$

These atoms therefore had to be atoms of a radioactive isotope of phosphorus that would be transformed by the emission of positive electrons and neutrinos into atoms of one of the known stable isotopes of silicon:

$$^{30}_{15}\mathrm{P} \rightarrow {}^{30}_{14}\mathrm{Si} + e^{+} + \nu.$$

The similar production, by the irradiation of boron with α rays, of a radioelement emitting positive electrons possessing a period of more than ten minutes was also established; this radioelement had to be an isotope of nitrogen. Frédéric Joliot and Irène Curie announced their discovery of a new type of radioactivity and of the artificial formation of light radioelements in a note to the Academy of Sciences on 15 January 1934. Within less than two weeks after their announcement they were able to conceive and skillfully execute radiochemical experiments proving that the radioelement formed in aluminum bombarded with α rays had exactly the same chemical properties as phosphorus and that the one formed in boron had those of nitrogen.

These elegant experiments, which provided the first chemical proof of induced transmutations and showed the possibility of artificially creating radioisotopes of known stable elements, were repeated and extended in the major nuclear physics laboratories of various countries. In Italy, Enrico Fermi demonstrated that neutron bombardment of most elements, even those of high atomic mass, gave rise to radioelements emitting negative electrons, often isotopes of the initial element. The next year, in November 1935, Frédéric Joliot and Irène Curie were awarded the Nobel Prize in chemistry for "their synthesis of new radioactive elements."

Thirty-five years old and at the height of his scientific career, Joliot had fully developed his personality. Slightly taller than average, with black hair and black eyes, a lively expression, and an athletic appearance, he possessed considerable charm. He was a brilliant conversationalist who loved to please and to be admired. An avid and exceptionally good skier, sailor, and tennis player, he was also an enthusiastic and skillful hunter and fisherman. Joliot had a taste for certain luxuries that wealth brought but was deeply attracted to the common people; he enjoyed spending time with workers and sailors, with whom he was able to communicate easily. Politically a socialist, he was active in antifascist organizations.

The fame that came with the Nobel Prize brought Joliot numerous responsibilities that interrupted his research for several years. Named professor at the Collège de France in 1937, he sought to equip its new laboratories with the instruments needed for the study of nuclear reactions; he directed the construction of a 7 MeV cyclotron and a 2,000,000-volt electric pulse accelerator.

At the beginning of 1939 the great German radiochemist Otto Hahn published chemical data proving that the nucleus of a uranium atom can be split into two nuclei of similar mass by the impact of a neutron. Within a few days Joliot furnished a direct physical proof of the explosive character of this bipartition of the uranium atom, subsequently called fission. In an elegantly simple experiment he demonstrated that the radioactive atoms produced in a thin layer of uranium by a flux of neutrons are ejected with a velocity sufficient to permit them to pass through a thin sheet of cellophane. The great kinetic energy of the fission fragments was established independently by this experiment and by the one done in Copenhagen by O. R. Frisch, who employed a proportional pulse amplifier. Pursuing the study of this phenomenon, Joliot was incontestably the first, in collaboration with Hans von Halban and Lew Kowarski, to prove that the fission of uranium atoms is accompanied or followed by an emission of neutrons (uranium submitted to a flux of slow neutrons emits rapid neutrons) and subsequently that the fission of a uranium atom induced by one neutron produces, on the average, an emission of several neutrons (March–April 1939). It was now possible to envision, as Joliot immediately did, a process in which uranium atoms would undergo successive fissions linked in divergent chains by neutrons and consequently developing like an avalanche. Hence an immense number of atoms, constituting a ponderable mass of uranium, might be disintegrated within a relatively short time by the minute excitation due to a single neutron.

The principle of the liberation of the internal energy of uranium atoms had thus been discovered, and the conditions in which the nuclear chain reactions could develop were rapidly determined. In particular it became apparent that it would be necessary to slow down the neutrons emitted during fission by joining to the uranium a "moderator" containing light atoms absorbent of neutrons and that the best moderator would be heavy water. For this reason Joliot, who had obtained about six tons of uranium oxide from the Belgian Congo, ordered from Norway the only sizable stock of heavy water then existing. The heavy water arrived safely in Paris even though World War II had begun, but there was too little time before the invasion of France for it to be used there. Joliot decided to remain in France but had Halban and Kowarski carry the precious substance with them to England to continue the group's investigations. In Paris, Joliot discontinued all his work on atomic energy that might benefit Germany. While continuing research in pure physics, he became increasingly involved in dangerous resistance activities, working closely with militant Communists; in 1942 he joined the then clandestine Communist party.

Following the liberation of France and the explosion of the first atomic bombs, Joliot, foreseeing the potential industrial importance of atomic energy and convinced of the impossibility of obtaining sufficient money for any fundamental research in nuclear physics not linked with practical applications, persuaded General de Gaulle, president of the provisional government, to create an atomic energy commission. Established in October 1945, this commission was endowed with broad powers and substantial funds. The new organization was headed by Joliot, who as high commissioner was responsible for scientific and technical activities, and by a chief administrator responsible for administrative and financial matters. Joliot assembled a dynamic group, and under his vigorous direction France's first atomic pile began operation in December 1948; in the same year a valuable uranium deposit had been discovered near Limoges. The first laboratories, which were installed in a former fort, became inadequate. Joliot persuaded the government to build a major nuclear research center on the plateau of Saclay, twelve miles south of Paris, and he supervised the construction of its first equipment.

Under pressure from the Communist party Joliot publicly took positions irritating to the government, although they were not of the sort to cast doubt on the loyalty with which he performed his duties. Using as a pretext a declaration of Joliot's in which he stated that he would never participate, in his capacity as a scientist, in a war against the Soviet Union, the president of the Council, Georges Bidault, removed him from his functions as high commissioner in April 1950. Although he had partially provoked it, Joliot suffered from this dismissal. Since the war, in fact, he had often seemed a tormented spirit, plagued by deep self-doubt despite his brilliant successes and seeking in the adulation of crowds compensation for the reserve that he sometimes perceived among his peers.

After 1950 Joliot once again gave most of his time to his laboratory and to his teaching at the Collège de France, but he felt he should dedicate his best efforts to what seemed to him the most effective struggle against the threat of war; he lent his great prestige to the World Organization of the Partisans of Peace, whose president he had become. He was greatly shaken by the death of his wife in March 1956, at a time when he had just survived a very serious attack of viral hepatitis. He succeeded Irène Joliot-Curie as head of the Radium Institute, where, with her, he had done his finest work. He carried out these new duties with devotion and enthusiastically supervised the relocation of the Institute in its new laboratories, then under construction in Orsay. His health remained delicate; he died on 14 August 1958, at the age of fifty-eight, following an operation made necessary by an internal hemorrhage. General de Gaulle, who had again become head of the government, decided that Joliot, whom thirteen years earlier he had appointed High Commissioner for Atomic Energy, should receive a state funeral.

BIBLIOGRAPHY

I. Original Works. The works of Frédéric and Irène Joliot-Curie are collected in *Oeuvres scientifiques complètes* (Paris, 1961). A selection of his work is in *Textes choisis* (Paris, 1959). His principal scientific publications include "Sur une nouvelle méthode d'étude du dépôt électrolytique des radio-éléments," in *Comptes rendus hebdomadaires des séances de l'Académie des sciences*, **184** (1927), 1325; (with Irène Curie) "Sur le nombre d'ions produits par les rayons alpha du RaC′ dans l'air," *ibid.*, **186** (1928), 1722; **187** (1928), 43; (with Irène Curie) "Sur la nature du rayonnement absorbable qui accompagne les rayons alpha du polonium," *ibid.*, **189** (1929), 1270; "Étude électrochimique des radioéléments," in *Journal de chimie physique*, **27** (1930), 119; (with Irène Curie) "Rayonnements associés à l'émission des rayons alpha du polonium," in *Comptes rendus*, **190** (1930), 627; "Sur la détermination de la période du Radium C′ par la méthode de Jacobsen. Expérience avec le thorium C′," *ibid.*, **191** (1930), 1292; (with Irène Curie) "Étude du rayonnement absorbable accompagnant les rayons alpha du polonium," in *Journal de physique et le radium*, **2** (1931), 20.

For further reference, see "Sur la projection cathodique des éléments et quelques applications" and "Propriétés électriques des métaux en couches minces préparées par projection thermique et cathodique," in *Annales de physique*, **15** (1931), 418; "Le phénomène de recul et la conservation de la quantité de mouvement," in *Comptes rendus hebdomadaires des séances de l'Académie des sciences*, **192** (1931), 1105; (with Irène Curie) "Préparation des sources de polonium de grande densité d'activité," in *Journal de chimie physique*, **28** (1931), 201; "Sur l'excitation des rayons gamma nucléaires du bore par les particules alpha. Energie quantique du rayonnement gamma du polonium," in *Comptes rendus . . . des sciences*, **193** (1931), 1415; (with Irène Curie) "Émission de protons de grande vitesse par les substances hydrogénées sous l'influence des rayons gamma très pénétrants," *ibid.*, **194** (1932), 273; (with Irène Curie) "Effet d'absorption de rayons gamma de très haute fréquence par projection de noyaux légers," *ibid.*, **194** (1932), 708; (with Irène Curie) "Projection d'atomes par les rayons très pénétrants excités dans les noyaux légers," *ibid.*, **194** (1932), 876; (with Irène Curie) "Sur la nature du rayonnement pénétrant excité dans les noyaux légers par les particules alpha," *ibid.*, **194** (1932), 1229; (with Irène Curie and P. Savel) "Quelques expériences sur les rayonnements excités par les rayons alpha dans les corps légers," *ibid.*, **194** (1932), 2208; (with Irène Curie) "New Evidence for the Neutron," in *Nature*, **130** (1932), 57; (with Irène Curie) "L'existence du neutron," in *Actualités scientifiques et industrielles* (Paris, 1932); (with Irène Curie) "Sur les conditions d'émission des neutrons par actions des particules α sur les éléments légers," in *Comptes rendus . . . des sciences*, **196** (1933), 1105; (with Irène Curie) "Contribution à l'étude des électrons positifs," *ibid.*, **196** (1933), 1105; and (with Irène Curie) "Sur l'origine des électrons positifs," *ibid.*, **196** (1933), 1581.

Other of Joliot's papers are: (with Irène Curie) "Preuves expérimentales de l'existence du neutron," in *Journal de physique et le radium*, **4** (1933), 21; (with Irène Curie) "Électrons positifs de transmutation," in *Comptes rendus . . . des sciences*, **196** (1933), 1885; (with Irène Curie) "Nouvelles recherches sur l'émission des neutrons," *ibid.*, **197** (1933), 278; (with Irène Curie) "La complexité du proton et la masse du neutron," *ibid.*, **197** (1933), 237; (with Irène Curie) "Mass of the Neutron," in *Nature*, **133** (1934), 721; (with Irène Curie) "Électrons de matérialisation et de transmutation," in *Journal de physique*, **4** (1933), 494; (with Irène Curie) "Rayonnement pénétrant des atomes," *7ème Conseil de physique Solvay, 22 octobre 1933* (Paris, 1934), p. 121; "Preuve expérimentale de l'annihilation des électrons positifs," in *Comptes rendus . . . des sciences*, **197** (1933), 1622; "Sur la dématérialisation de paires d'électrons," *ibid.*, **198** (1934), 81; "Preuve expérimentale de l'annihilation des électrons positifs," in *Journal de physique*, **5** (1934), 299; "Le neutron et le positron," in *Helvetica acta*, **81** (1934), 211; (with Irène Curie) "Un nouveau type de radioactivité," in *Comptes rendus . . . des sciences*, **198** (1934), 254; (with Irène Curie) "Artificially Produced Radioelements," *Joint Conference of the International Union of Pure and Applied Physics, and the Physical Society*, **1** (Cambridge, 1934); (with Irène Curie) "Production artificielle d'éléments radioactifs" and "Preuve chimique de la transmutation des éléments," in *Journal de physique*, **5** (1934), 153; "Réalisation d'un appareil Wilson pour pressions variables (1 cm de Hg à plusieurs atmosphères)," in *Journal de physique*, **5** (1934), 216; "Étude des rayons de recul radioactifs par la méthode des détentes de Wilson," *ibid.*, **5** (1934), 219; (with Irène Curie and P. Preiswerk) "Radioéléments créés par bombardement de neutrons. Nouveau type de radioactivité," in *Comptes rendus . . . des sciences*, **198** (1934), 2089.

Other works include "Les nouveaux radioéléments. Preuves chimiques des transmutations," in *Journal de chimie physique*, **31** (1934), 611; (with Irène Curie) "L'électron positif," in *Actualités scientifiques et industrielles* (Paris, 1934); (with L. Kowarski) "Sur la production d'un rayonnement d'énergie comparable à celle des rayons cosmiques mous," in *Comptes rendus . . . des sciences*, **200** (1935), 824; (with A. Lazard and P. Savel) "Synthèse de radioéléments par des deutérons accélérés au moyen d'un générateur d'impulsions," *ibid.*, **201** (1935), 826; (with Irène Curie) "Radioactivité artificielle," in *Actualités scientifiques et industrielles* (Paris, 1935); (with I. Zlotowski) "Sur l'énergie des groupes de protons émis lors de la transmutation du bore par les rayons α," in *Comptes rendus*, **206** (1938), 750; (with I. Zlotowski) "Formation d'un isotope stable de masse 5 de l'hélium lors des collisions entre hélions et deutérons," in *Journal de physique*, **9** (1938), 403; (with I. Zlotowski) "Sur la détermination par la méthode Wilson de la nature et de l'énergie des particules émises lors des transmutations. Application à la réaction $^{10}_{5}B\ (\alpha, p)\ ^{13}_{6}C$," *ibid.*, **9** (1938), 393; "Preuve expérimentale de la rupture explosive des noyaux d'uranium et de thorium sous l'action des neutrons," in *Comptes rendus . . . des sciences*, **208** (1939), 341; "Observations par la méthode Wilson des trajectoires de brouillard des produits de l'explosion des noyaux d'uranium," *ibid.*, **208** (1939), 647; "Sur la rupture explosive des noyaux U and Th sous l'action des neutrons," in *Journal de physique*, **10** (1939), 159; (with L. Dodé, H. von Halban, L. Kowarski) "Sur l'énergie des neutrons libérés lors de la partition nucléaire de l'uranium," in *Comptes rendus . . . des sciences*, **208** (1939), 995; and (with H. von Halban and L. Kowarski) "Liberation of Neutrons in the Nuclear Explosion of Uranium," in *Nature*, **143** (1939), 470.

See also: (with H. von Halban and L. Kowarski) "Number of Neutrons Liberated in the Nuclear Fission of Uranium," in *Nature*, **143** (1939), 680; (with H. von Halban and L. Kowarski) "Energy of Neutrons Liberated in the Nuclear Fission of Uranium Induced by Thermal Neutrons," *ibid.*, **143** (1939), 939; (with H. von Halban, L. Kowarski, and F. Perrin) "Mise en évidence d'une réaction nucléaire en chaîne au sein d'une masse uranifère," in *Journal de physique*, **10** (1939), 428; (with B. Lacassagne) "Cancer du foie apparu chez un lapin irradié par les neutrons," in *Comptes rendus . . . des sciences*, **138** (1944), 50; "Sur une méthode de mesure de parcours des radioéléments de nature chimique déterminée, projetés lors de la bipartition de l'uranium," *ibid.*, **218** (1944), 488; (with Irène Curie) "Sur la bipartition de l'ionium sous l'action des neutrons," in *Annales de physique*, **19** (1944), 107; "Sur une méthode physique d'extraction des radioéléments de bipartition des atomes lourds et mise en évidence d'un radiopraséodyme de période 13 j," in *Comptes rendus . . . des sciences*, **218** (1944), 733; (with R. Courrier, A. Horeau, and P. Süe) "Sur l'obtention de la thyroxine marquée par le radioiode et son comportement dans l'organisme," *ibid.*, **218** (1944), 769; (with H. von Halban and L. Kowarski) "Sur la possibilité de produire dans un milieu uranifère des réactions nucléaires en chaîne illimitée. 30 octobre 1939," *ibid.*, **299** (1949), 19; and (with Irène Curie) "Sur l'étalonnage des sources de radioéléments," *Commission Mixte des Unions Internationales de Physique et de Chimie, Juillet 1953*.

II. SECONDARY LITERATURE. Louis de Broglie, *La vie et l'oeuvre de Frédéric Joliot* (Paris, 1959); P. M. S. Blackett, "Jean-Frédéric Joliot," in *Biographical Memoirs of Fellows of the Royal Society*, **6** (Nov. 1960), 87; and Pierre Biquard, *Frédéric Joliot-Curie* (Paris, 1961).

FRANCIS PERRIN

JOLIOT-CURIE, IRÈNE (*b.* Paris, France, 12 September 1897; *d.* Paris, 17 March 1956), *radioactivity, nuclear physics.*

Irène Joliot-Curie's fame stems principally from the discoveries she made with her husband, Frédéric Joliot, particularly that of artificial radioactivity, for which they shared the Nobel Prize in chemistry in 1935. Yet her own investigations on the radioelements produced by the irradiation of uranium with neutrons were sufficiently important to secure her a position among the great modern scientists.

Her father, Pierre Curie, married the brilliant Polish student Marie Skłodowska in July 1895. Their marriage marked the beginning of a close collaboration between two dedicated scientific researchers which culminated in the discovery of radium hardly more than a year after the birth of Irène, their first child. Marie Curie's devotion to her laboratory work left her little time to spend with her daughter. Young Irène would have had scarcely any company other than her governesses had not her grandfather, Eugène Curie, come to live with Pierre and Marie Curie in 1898. Eugène Curie, a doctor, had distinguished himself by treating the wounded during the uprising in Paris of June 1848 and of the Commune of 1871. Until his death in 1910 he exerted a great influence on Irène's personality, especially after her father's death in 1906. It was to her grandfather, a convinced freethinker, that Irène owed her atheism, later politically expressed as anticlericalism. He was also the source of her attachment to the liberal socialism to which she remained faithful all her life.

Marie Curie did, however, very early take charge of Irène's scientific education. Irène did not attend school until the age of twelve, but for the two preceding years she studied at the teaching cooperative established by some of Marie's colleagues and friends for their own children: Marie Curie taught physics; Paul Langevin, mathematics; and Jean Perrin, chemistry. Irène next went to the Collège Sévigné; she received her *baccalauréat* just before the outbreak of World War I. From then until 1920 she studied at the Sorbonne and took the examinations for a *licence* in physics and mathematics. During the war she served for many months as an army nurse, assisting her mother in setting up apparatus for the radiography of the wounded; at the age of eighteen she had sole responsibility for installing radiographic equipment in an Anglo-Canadian hospital a few miles from the front in Flanders.

In 1918 Irène Curie became an assistant at the Radium Institute, of which her mother was the director, and in 1921 she began scientific research. Her first important investigation concerned the fluctuations in the range of α rays. She determined these variations by photographing the tracks that the rays formed in a Wilson cloud chamber. Presented in her doctoral thesis in 1925, this work was followed by a series of studies on classical radioactivity, some of which were in collaboration with Frédéric Joliot, whom she had married on 26 October 1926. Not until 1931, however, did they begin the constant collaboration, lasting several years, that brought them the Nobel Prize. It is worth noting that for their Nobel addresses Frédéric, considered to be the physicist, chose to deal with the chemical identification of the artificially created radioisotopes, while Irène, the chemist, recounted the discovery of a new type of radioactivity, the positive β decay. Marie Curie had died of acute leukemia in July 1934 and thus could not witness the triumph of her daughter and son-in-law, which duplicated her own accomplishment with Pierre Curie thirty-two years earlier.

Honors did not change Irène Joliot-Curie, who retained throughout her life a great simplicity and a thorough uprightness. Her pensive attitude made her appear somewhat slow and aloof, but she could be quite lively with her few close friends. She loved to be close to nature and enjoyed rowing, sailing, and especially swimming during vacations in Brittany. She was also fond of taking long walks in the mountains, where she was often obliged to go because of a tubercular condition. Although her interest in science was preeminent, she deeply loved the writings of certain French and English authors, especially Victor Hugo and Rudyard Kipling; she even translated some of Kipling's poems. She found great joy in motherhood and, despite the hours spent in the laboratory, devoted much time to her children until their adolescence. Both Hélène and Pierre became brilliant researchers: the former, like her mother and grandmother, in nuclear physics; the latter, in biophysics.

After serving for four months in 1936 as secretary of state in Léon Blum's Popular Front government, Irène Joliot-Curie was elected professor at the Sorbonne in 1937. She continued to work at the Radium Institute, while Frédéric Joliot transferred his research activities to the Collège de France, where he had received a professorship.

It was during these years preceding World War II that Irène Joliot-Curie did her most remarkable individual work. Aided by her great experience in radiochemistry, she sought to analyze the complex phenomena that result from bombarding uranium with neutrons. First brought to light by Enrico Fermi, these phenomena were subsequently studied by Otto Hahn and Lise Meitner, who demonstrated that in uranium submitted to a neutron flux there appear a rather large number of β radioactivities, displaying different periods associated with diverse chemical properties. This discovery led them to suppose the formation not only of several transuranic radioelements but also of new radioisotopes of elements preceding uranium (down to radium itself). Irène Joliot-Curie, in collaboration with the Yugoslav physicist P. P. Savic, showed that, among the radioisotopes formed, a radioelement with a period of 3.5 hours could be carried away by adding actinium to the solution of irradiated uranium and then separating it out again through precipitation. But this radioelement was not an isotope of actinium, since by adding lanthanum to the actinium extract and then separating it out again through fractional precipitations, the new radioelement was shown to follow lanthanum, its chemical properties therefore being closer to those of lanthanum than to those of actinium.

Reproducing these experiments, the result of which he found surprising, Otto Hahn proved that the bombardment of uranium with neutrons produces not only radioactive atoms possessing chemical properties very similar to those of the lanthanides but also, undoubtedly, atoms of a radioactive isotope of barium. This was the proof that a neutron can induce the bipartition of a uranium atom into two atoms of a comparable mass—a phenomenon soon afterward termed "fission." Irène Joliot-Curie had instigated this important discovery—which she herself would probably have made had a fortuitous complication not concealed the formation of a true radioisotope of lanthanum in the uranium irradiated by neutrons. The former existed in association with a radioisotope of promethium with a similar period, which explains why the fractional precipitation of the lanthanum separated from the actinium results in the appearance in the top fractions of an increase in the 3.5-hour activity period.

At the time of the German invasion in 1940 Irène Joliot-Curie decided to remain in France with the researchers in her laboratory. In 1944, a few months before the liberation of Paris, the Communist resistance organization, fearing that she might suffer reprisals for the resistance activities of her husband, who had gone underground, had her smuggled into Switzerland with her children. In 1946 she was named director of the Radium Institute, created for her mother some thirty years before, in which she conducted all her own research. From 1946 to 1950 she was also one of the directors of the French Atomic Energy Commission, of which Frédéric Joliot was the high commissioner.

Irène Joliot-Curie divided her efforts in the following years between the creation of the Radium Institute's large, new laboratories at Orsay, a southern suburb of Paris, and working for women's pacifist movements. She died at the age of fifty-eight, a victim, like her mother, of acute leukemia. The disease was undoubtedly a consequence of the X and γ radiations to which she had been exposed, first as an inadequately protected nurse-radiologist during World War I and then in the laboratory, when the dangers of radioactivity were still not fully realized.

BIBLIOGRAPHY

See Frédéric and Irène Joliot-Curie, *Oeuvres scientifiques complètes* (Paris, 1961). Irène Joliot-Curie's publications in collaboration with her husband are listed in the preceding article "Frédéric Joliot-Curie." Her principal scientific publications include "Sur le poids atomique du chlore dans quelques minéraux," in *Comptes rendus hebdomadaires des séances de l'Académie des sciences*, **172** (1921), 1025; "Sur la vitesse d'émission des rayons α du polonium," *ibid.*, **175** (1922), 220; "Sur la distribution de longueur des rayons α," in *Journal de physique et le radium*, **4** (1923), 170; "Sur le rayonnement γ du radium *D* et du radium *E*," in *Comptes rendus*, **176** (1923), 1301; "Sur la constante radioactive du radon," in *Journal de physique et le radium*, **5** (1924), 238, written with C. Chamié; "Sur la distribution de longueur des rayons α du polonium dans l'oxygène et dans l'azote," in *Comptes rendus*, **179** (1924), 761, written with N. Yamada; "Sur l'homogénéité des vitesses initiales des rayons α du polonium," *ibid.*, **180** (1925), 831; "Recherches sur les rayons α du polonium. Oscillation de parcours, vitesse d'émission, pouvoir ionisant," in *Annales de physique*, **2** (1925), 403, diss.; "Sur les particules de long parcours émises par le polonium," in *Journal de physique et le radium*, **6** (1925), 376, written with N. Yamada; "Sur le spectre magnétique des rayons α du radium *E*," in *Comptes rendus*, **181** (1925), 31; "Extraction et purification du dépôt actif à évolution lente du radium," in *Journal de physique et le radium*, **22** (1925), 471; "Étude de la courbe de Bragg relative aux rayons du radium *C′*," *ibid.*, **7** (1926), 125, written with F. Béhounck; "Sur la distribution de longueur des rayons α du radium *C* et du radium *A*," *ibid.*, 289, written with P. Mercier; "Sur l'oscillation de parcours des rayons α dans l'air," *ibid.*, **8** (1927), 25; "Sur la mesure du dépôt actif du radium par le rayonnement γ pénétrant," in *Comptes rendus*, **188** (1929), 64; "Sur la quantité de polonium accumulée dans d'anciennes ampoules de radon et sur la période du radium *D*," in *Journal de physique et le radium*, **10** (1929), 388; "Sur la décroissance du radium *D*," *ibid.*, 385, written with Marie Curie; "Sur la complexité du rayonnement α du radioactinium," in *Comptes rendus*, **192** (1931), 1102; "Sur un nouveau composé gazeux du polonium," *ibid.*, 1453, written with M. Lecoin; and "Sur le rayonnement γ nucléaire excité dans le glucinium et dans le lithium par les rayons α du polonium," *ibid.*, **193** (1931), 1412.

See also "Sur le rayonnement α du radioactinium, du radiothorium et de leurs dérivés. Complexité du rayonnement α du radioactinium," in *Journal de physique et le radium*, **3** (1932), 52; "Sur la création artificielle d'éléments appartenant à une famille radioactive inconnue, lors de l'irradiation du thorium par les neutrons," *ibid.*, **6** (1935), 361, written with H. von Halban and P. Preiswerk; "Remarques sur la stabilité nucléaire dans le domaine des radioéléments naturels," *ibid.*, 417; "Sur les radioéléments formés par l'uranium irradié par les neutrons," *ibid.*, **8** (1937), 385, written with P. Savic; "Sur le radioélément de périod 3,5 h. formé dans l'uranium irradié par les neutrons," in *Comptes rendus*, **206** (1938), 1643, written with P. Savic; "Sur les radioéléments formés dans l'uranium irradié par les neutrons, II," in *Journal de physique et le radium*, **9** (1938), 355, written with P. Savic; "Sur le rayonnement du corps de période 3,5 h. formé par irradiation de l'uranium par les neutrons," *ibid.*, 440, written with P. Savic and A. Marquès da Silva; "Sur les radioéléments formés dans l'uranium et le thorium irradiés par les neutrons," in *Comptes rendus*, **208** (1939), 343, written with P. Savic; "Comparaison des isotopes radioactifs des terres rares formés dans l'uranium et le thorium," in *Journal de physique et le radium*, **10** (1939), 495, written with Tsien San-tsiang; "Détermination de la période de l'actinium," in *Cahiers de physique*, nos. **25–26** (1944), 25–67, written with G. Bouissières; "Parcours des rayons α de l'ionium," in *Journal de physique et le radium*, **6** (1945), 162, written with Tsien San-tsiang; "Détermination empirique du nombre atomique *Z*, correspondant au maximum de stabilité des atomes de nombre de masse *A*," *ibid.*, 209; "Sur la possibilité d'étudier l'activité des roches par l'observation des trajectoires des rayons alpha dans l'émulsion photographique," *ibid.*, **7** (1946), 313; *Les radioéléments naturels. Propriétés chimiques. Préparation. Dosage* (Paris, 1946); "Sur le rayonnement gamma de l'ionium," in *Journal de physique et le radium*, **10** (1949), 381; "Autoradiographie par neutrons. Dosage séparé de l'uranium et du thorium," in *Comptes rendus*, **232** (1951), 959, written with H. Faraggi; "Sélection et dosage du carbone dans l'acier par l'emploi de la radioactivité artificielle," in *Journal de physique et le radium*, **13** (1952), 33, also in *Bulletin. Société chimique de France*, **20** (1952), 94; "Détermination de la proportion de mésothorium, radium, radiothorium dans une ampoule de mésothorium commercial," in *Journal de physique et le radium*, **15** (1954), 1; and "Sur une nouvelle méthode pour la comparaison précise du rayonnement des ampoules de radium," *ibid.*, 790.

Details on the biography of Irène Joliot-Curie can be found in Eugénie Cotton, *Les Curie* (Paris, 1963).

FRANCIS PERRIN

JOLLY, PHILIPP JOHANN GUSTAV VON (*b.* Mannheim, Germany, 26 September 1809; *d.* Munich, Germany, 24 December 1884), *physics.*

Jolly became well-known as an experimental physicist primarily through his instruments and methods for making measurements. He was of Huguenot descent and his father, an army captain who later became a merchant, was for many years the mayor of Mannheim. Jolly attended the Gymnasium in Mannheim and, from 1829 to 1834, the universities of Heidelberg and Vienna, where he also studied technology and mechanics, and then concentrated on mathematics and physics in Berlin. He received his doctorate in 1834 from Heidelberg and qualified there as privatdocent in mathematics, physics, and technology. In 1839 he became professor of mathematics, and in 1846 he obtained the chair of physics. At Heidelberg, Jolly was often consulted by J. R. von Mayer, the discoverer of the law of the conservation of energy. At that time Jolly was concerned with questions of osmosis. Although his idea that the same amounts by weight of salt and water are exchanged through a membrane did not prove to be correct, he nevertheless contributed substantially to the elucidation of this process.

An outstanding experimenter, Jolly was also able to present the fundamentals of the physics of the period in a readily understandable manner in his *Prinzipien der Mechanik* (1852). In 1854 he was called to the University of Munich, where he was a popular teacher until his retirement shortly before his death. In addition, Jolly was an expert consultant for the reorganization of Bavaria's technical schools, a member of the Bavarian commission on standards, and German representative at the international conference on the meter held at Paris in 1872. He was also a member of the Bavarian Academy of Sciences and longtime chairman of the Munich Geographical Society.

With his mechanic A. Berberich, Jolly constructed and improved various measuring devices, including the spring balance, the air thermometer for determining the coefficients of expansion of gases (1874), the eudiometer, and the mercury air pump. In addition, he greatly increased the accuracy of balances. With the aid of his precision balances Jolly determined the earth's gravitation (that is, the acceleration due to gravity) and density by means of a remarkable experimental procedure (1878–1881). Through measurements based on various methods he also succeeded in demonstrating the variability of the oxygen content of the air, which is important in meteorology. He found (1879) that the oxygen content of the "North Wind"—masses of polar air—is greater than that of the "South Wind"—masses of tropical air.

BIBLIOGRAPHY

I. ORIGINAL WORKS. Bibliographies of Jolly's works are in Böhm's biography (see below); Poggendorff, I, 1199, and III, 695–696; and G. Hellmann, *Repertorium der deutschen Meteorologie* (Leipzig, 1883), pp. 221–222, meteorological papers only.

Among his writings are *De Euleri meritis de functionibus circularibus* (Heidelberg, 1834), the prize question of the University of Heidelberg for 1830; "Experimental-Untersuchungen über Endosmose," in *Annalen der Physik und Chemie*, **78** (1849), 261–271; *Die Principien der Mechanik gemeinfasslich dargestellt* (Stuttgart, 1852); *Ueber die Physik der Molecularkräfte* (Munich, 1857); *Das Leben Fraunhofers* (Munich, 1866); "Ausdehnungs-Koefficienten einiger Gase und über Luftthermometer," in *Annalen der Physik und Chemie*, jubilee vol. (1874), 82–101; "Die Anwendung der Waage auf Probleme der Gravitation," *ibid.*, n.s. **5** (1878), 112–134, and **14** (1881), 331–355, also in *Abhandlungen der Bayerischen Akademie der Wissenschaften*, math.-phys. Kl., **13** (1880), I, 3, 155–176, and **14** (1883), II, 1, 1–26; and "Die Veränderlichkeit in der Zusammensetzung der atmosphärischen Luft," in *Annalen der Physik und Chemie*, n.s. **6** (1879), 520–544, also in *Abhandlungen der Bayerischen Akademie der Wissenschaften*, math.-phys. Kl., **13** (1880), II, 2, 49–79.

II. SECONDARY LITERATURE. See G. Böhm, *Philipp von Jolly, ein Lebens- und Charakterbild* (Munich, 1886), with a bibliography of his works; and C. von Voit, "Philipp Johann Gustav von Jolly," in *Sitzungsberichte der Bayerischen Akademie der Wissenschaften zu München*, math.-phys. Kl., **15** (1885), 118–136. Shorter biographies are "Jolly, Philipp von," in *Allgemeine deutsche Biographie*, LV (1910), 807–810; and "Philipp Jolly," in *Badische Biographien*, pt. 4 (Karlsruhe, 1891), 199–204.

HANS-GÜNTHER KÖRBER

JOLY, JOHN (*b.* Holywood, King's County [now Offaly], Ireland, 1 November 1857; *d.* Dublin, Ireland, 8 December 1933), *geology, experimental physics, chemistry, mineralogy.*

Joly was the third son of Rev. J. P. Joly, who was of French extraction, and Anna Comtesse de Lusi, who came from a mixed German-Italian family. He was educated at Rathmines School and Trinity College, Dublin. He graduated in 1882 with a degree in engineering, physics, chemistry, geology, and mineralogy; he then became assistant to the professor of engineering. In 1897 Joly was appointed professor of geology at Trinity College, a post he held until his death. He had both a fertile mind and the ability to apply the fundamental principles of physics and chemistry to the explanation of new facts; and he often devised new forms of apparatus for his researches.

One of Joly's earliest inventions was the steam calorimeter, which he used to determine the specific heats of minerals. Using this apparatus, he also determined, for the first time, the specific heats of gases at constant volume. In 1895 he devised a new method for the production of photographs in natural colors. Using an ordinary camera with an isochromatic plate, Joly placed a glass screen ruled with closely spaced alternating lines of red, blue, and violet between the lens and plate. The resulting negative, viewed through a similar screen with red, blue, and violet lines, gave a picture reproducing natural colors.

In 1899 Joly estimated the age of the earth by a method originally suggested by Halley, based on the rate of increase in the sodium content of the oceans. His estimate of 80–90 million years represented the time elapsed since moisture first condensed upon the earth. Although now of only historic interest, at the time his estimate was of some importance because it supported the views of geologists and evolutionists who were unwilling to accept the much lower estimates of contemporary physicists, notably Lord Kelvin, which were based on the supposed rate at which the earth had cooled by radiation, assuming only the sources of heat then known.

The discovery of radioactivity, and particularly the possibility of its application to the solution of geological problems, aroused Joly's interest. In 1903 he drew attention to the probable importance of radioactivity as a source of terrestrial heat and the effect it would have on calculations of the age of the earth made by Kelvin's method. In 1907 Joly, using his knowledge of mineralogy, made a discovery that proved of great importance in connection with the new method of calculating the earth's age by radioactive methods. Mineralogists had long known that certain rock-forming minerals, especially biotite, when viewed under the microscope, were often characterized by the presence of small circular dark spots or concentric rings, known as pleochroic halos, which were centered on minute inclusions of other minerals, for example, zircon. Joly demonstrated that these halos were spherical in form and proved that they had been formed by radioactive emanations from the mineral at their center. Subsequently he and others carried out exact measurements of halos present in rocks of differing geological ages, establishing by this means that the rate of decomposition of radioactive minerals must have been constant throughout geological time, an assumption necessary in all subsequent calculations of the age of geological formations by radioactive methods. Joly's studies in radioactivity were incorporated in his books *Radioactivity and Geology* (1909) and *The Surface History of the Earth* (1925).

Joly's interest in radioactivity also extended to its use for therapeutic purposes. It was on his suggestion that the Royal Dublin Society founded a radium institute; and, in collaboration with Walter Stevenson, he invented a hollow needle for use in deep-seated radiotherapy (the "Dublin method"), which came into worldwide use.

Joly carried out much experimental research into the physical and chemical properties of minerals, publishing many papers on the subject. He was a pioneer in the microscopical study of rock-forming minerals in relation to their suitability as road metal.

During his lifetime Joly did much to improve the facilities for teaching science and for carrying out scientific research at Trinity College. He was elected fellow of the Royal Society in 1892 and was awarded its Royal Medal in 1910. He received the Boyle Medal from the Royal Dublin Society in 1911 and the Murchison Medal from the Geological Society of London in 1923.

BIBLIOGRAPHY

A complete list of Joly's numerous publications, mainly contributions to learned societies and journals, is in *Obituary Notices of Fellows of the Royal Society of London*, **1**, no. 3 (1932–1935), 259. His two books are *Radioactivity and Geology* (London, 1909); and *The Surface History of the Earth* (Oxford, 1925).

No biography has been published. The most complete account of Joly's life and scientific work is contained in the obituary notice cited above, which also reproduces his portrait. The following obituary notices are also worth consulting: *Nature*, **133** (1934), 90; and *Quarterly Journal of the Geological Society of London*, **90** (1934), "Proceedings," lv–lvii.

V. A. EYLES

JONES, HAROLD SPENCER. See **Spencer Jones, Harold.**

JONES, HARRY CLARY (*b.* New London, Maryland, 11 November 1865; *d.* Baltimore, Maryland, 9 April 1916), *physical chemistry.*

Jones was professor of physical chemistry at Johns Hopkins University and one of the pioneer promoters of this subject in the United States. The son and grandson of farmers, he always considered farming his avocation and, whenever he had the time, he spent it in managing and improving his three farms. His scientific career was determined during his elementary school years, by the reading of one of Tyndall's books on science. Jones's tremendous driving energy first showed itself in this decision: although poorly prepared for a scientific education, his enthusiasm enabled him to enter Johns Hopkins as a special student in 1887 and to secure his bachelor's degree two years later. He received his doctorate in June 1892.

During his graduate study Jones became fascinated by the newly developing field of physical chemistry. He spent the next two years studying with the masters in this field: Ostwald at Leipzig, Arrhenius at Stockholm, and van't Hoff at Amsterdam. Ostwald and Arrhenius remained close personal friends of Jones's for the rest of his life.

In 1894 Jones returned to Johns Hopkins as an honorary fellow and in the following year he became instructor in physical chemistry. He rose to full professor in 1903 and remained in this position until his death. In 1902 he married Harriet Brooks, of an old Baltimore family.

While studying with Arrhenius, Jones had investigated hydrates of sulfuric acid; his interest in solutions developed from this work. All of his later researches related in some way to an attempt to develop a general theory, a modification of Mendeleev's concept that solution came from the formation of a series of solvates. In support of his theory Jones developed at least sixteen lines of evidence, the chief of which came from his studies of solubility, absorption spectra of solutions, electrolytic conductivity, and the influence of solvent and solute on each other. While van't Hoff had concerned himself with the theory of ideal solutions, Jones studied the behavior of actual solutions. Much of his work was supported by the Carnegie Institution of Washington. In 1913 he received the Langstreth Medal of the Franklin Institute for his work.

In addition to his scientific papers Jones wrote twelve textbooks and semipopular scientific works. His most successful book was *Elements of Physical Chemistry* (1902), which was translated into Russian and Italian. He served on the editorial boards of *Zeitschrift für physikalische Chemie*, *Journal de chimie physique*, and *Journal of the Franklin Institute*. He was a man of strong opinions, with enormous energy and an insatiable desire for work, both in the laboratory and in preparing books and papers. These activities eventually led to a breakdown and his death at the age of fifty.

BIBLIOGRAPHY

There is a bibliography of 158 papers and twelve books by Jones and his co-workers in his posthumously published *The Nature of Solution* (New York, 1917), pp. 359–370.

There are short, appreciative obituaries in *Nature*, **97** (1916), 283; and *Journal de chimie physique*, **14** (1916), 488. A longer biographical sketch by E. Emmet Reid is in *The Nature of Solution*, pp. vii–xiii.

HENRY M. LEICESTER

JONES, WILLIAM (*b.* Llanfihangel Tw'r Beird, Anglesey, Wales, 1675; *d.* London, England, 3 July 1749), *mathematics.*

According to Welsh custom Jones, the son of a small farmer, John George, took the Christian name of his father (John) as his own surname (Jones). His mother was Elizabeth Rowland. Although Jones has little claim to eminence as a mathematician in his own right, his name is well-known to historians of mathematics through his association with the correspondence and works of many seventeenth-century mathematicians, particularly Newton.

In his early schooling Jones showed enough promise to secure the patronage of a local landowner (Bulksley of Baron Hill) who helped him to enter the countinghouse of a London merchant. Subsequently he traveled to the West Indies and taught mathematics on a man-of-war. Upon his return to London, Jones established himself as a teacher of mathematics; tutorships in great families followed. One of his pupils, Philip Yorke (afterward first earl of Hardwicke), later became lord chancellor; Jones traveled with him on circuit and was appointed "secretary for peace." He also taught Thomas Parker, afterward first earl of Macclesfield, and his son George, who became president of the Royal Society. For many years Jones lived at Shirburn Castle, Tetsworth, Oxfordshire, with the Parker family. There he met and married Maria Nix, daughter of a London cabinetmaker; they had two sons and a daughter.

In 1702 Jones published *A New Compendium of the Whole Art of Navigation*, a practical treatise concerned with the application of mathematics to astronomy and seamanship. His second book, *Synopsis palmariorum matheseos* (1706), attracted the attention of Newton and Halley. Although the book was designed essentially for beginners in mathematics, it contained a fairly comprehensive survey of contemporary developments, including the *method of fluxions* and the *doctrine of series*. Of the binomial theorem he wrote: ". . . and in a word, there is scarce any *Inquiry* so Sublime and Intricate, or any *Improvement* so

Eminent and Considerable, in *Pure Mathematics*, but by a *Prudent application* of this *Theorem*, may easily be exhibited and deduced." Although all the symbols used by Jones are sensible and concise, in only one respect does he appear to have been an innovator: he introduced π for the ratio of the circumference of a circle to the diameter.

From 1706 on, Jones remained in close touch with Newton and was one of the privileged few who obtained access to his manuscripts. About 1708 he acquired the papers and correspondence of John Collins, a collection that included a transcript of Newton's *De analysi* (1669). In 1711 Newton permitted Jones to print the tracts *De analysi per aequationes numero terminorum infinitas* and *Methodus differentialis* (along with reproductions of his tracts on quadratures and cubics) as *Analysis per quantitatum series, fluxiones ac differentias; cum enumeratione linearum tertii ordinis*. In the same year Jones was appointed a member of the committee set up by the Royal Society to investigate the invention of the calculus. With John Machin and Halley, he was responsible for the preparation of the printed report. On 30 November 1712 he was elected a fellow of the Royal Society and subsequently became vice-president. He contributed sundry papers to the *Philosophical Transactions*, mostly of a practical character.

At his death Jones left a voluminous collection of manuscripts and correspondence which he had assembled mainly through his connections with Newton and the Royal Society. It seems that he intended to publish an extensive work on mathematics and, to this end, made copious notes and transcripts from manuscripts lent by Newton. This material became inextricably mixed with the original manuscripts and the transcripts of others, including those of John Collins and James Wilson. John Coulson (1736) used a transcript made by Jones as the basis for an English version of Newton's 1671 tract, *The Method of Fluxions and Infinite Series*. Subsequently Samuel Horsley (Newton's *Opera omnia*, I [1779]) retained Jones's title for the tract on fluxions (1671) and copied the "dot" notation inserted by Jones. D. T. Whiteside (*Newton Papers*, I, xxxiii) remarks that the sections of the Portsmouth collection relating to fluxions are "choked with irrelevant, fragmentary transcripts by Jones and Wilson." After Jones's death most of the manuscript collection passed into the hands of the second earl of Macclesfield. Two volumes of correspondence from this collection were published by Rigaud in 1841. The task of separating the mass of material compiled by Jones from Newton's original manuscripts has only recently been completed by Whiteside.

BIBLIOGRAPHY

I. Original Works. Jones's books are *A New Compendium of the Whole Art of Navigation* (London, 1702), with tables by J. Flamsteed; and *Synopsis palmariorum matheseos, or a New Introduction to the Mathematics* (London, 1706). Charles Hutton, *The Mathematical and Philosophical Dictionary*, 2 vols. (London, 1795), I, 672, lists the papers (mostly slight) published by Jones in the *Philosophical Transactions of the Royal Society* and gives some account of the disposal of his library of MSS after his death. F. Maseres, *Scriptores logarithmici* (London, 1791), contains a paper by Jones on compound interest. D. T. Whiteside, *The Mathematical Papers of Isaac Newton*, I–II (Cambridge, 1967–1968), makes numerous references to Jones and his connection with the Newton MSS. A number of letters written by and received by Jones were printed in S. J. Rigaud, *Correspondence of Scientific Men of the Seventeenth Century*, 2 vols. (Oxford, 1841).

II. Secondary Literature. Biographical material is available in Hutton's *Mathematical Dictionary* (see above) and in John Nichols, *Biographical and Literary Anecdotes of William Bowyer, Printer, F.S.A.* (London, 1782), pp. 73–74. See also Lord Teignmouth, *Memoirs of the Life, Writings and Correspondence of Sir William Jones* (London, 1804); and David Brewster, *Memoirs of the Life, Writings and Discoveries of Sir Isaac Newton*, 2 vols. (Edinburgh, 1855), I, 226, II, 421.

M. E. Baron

JONQUIÈRES, ERNEST JEAN PHILIPPE FAUQUE DE (*b.* Carpentras, France, 3 July 1820; *d.* Mousans-Sartoux, near Grasse, France, 12 August 1901), *mathematics*.

Jonquières entered the École Navale at Brest in 1835 and subsequently joined the French navy, in which he spent thirty-six years. He achieved the rank of vice-admiral in 1879, and retired in 1885. He traveled all over the world, particularly to Indochina. In 1884 he was named member of the Institut de France.

In the 1850's Jonquières became acquainted with the geometric work of Poncelet and Chasles, which stimulated his own work in the field of synthetic geometry. In 1862 he was awarded two-thirds of the Grand Prix of the Paris Academy for his work in the theory of fourth-order plane curves. Geometry remained his main scientific interest. He was outstanding in solving elementary problems, for which, besides traditional methods, he used projective geometry. In addition to elementary problems Jonquières studied then-current questions of the general theory of plane curves, curve beams, and the theory of algebraic curves and surfaces, linking his own work with that of Salmon, Cayley, and Cremona.

In his studies he generalized the projective creation of curves and tried to obtain higher-order curves with projective beams of curves of lower order. In 1859–1860 (before Cremona), he discovered the birational transformations (called by him "isographic"), which can be considered as a special case of Cremona's transformations; in nonhomogeneous coordinates they have the form:

$$x' = x \qquad y' = \frac{\alpha y + \beta}{\gamma y + \delta}$$

where α, β, γ, δ are functions of x and $\alpha\delta - \beta\gamma$ does not equal zero.

A number of Jonquières's results were in the field of geometry which Schubert called "abzählende Geometrie."

Besides geometry, Jonquières studied algebra and the theory of numbers, in which he continued the tradition of French mathematics. Here again his results form a series of detailed supplements to the work of others and reflect Jonquières's inventiveness in calculating rather than a more profound contribution to the advancement of the field.

BIBLIOGRAPHY

An autobiographical work is *Notice sur la carrière maritime, administrative et scientifique du Vice-Amiral de Jonquières, Grand officier de la Légion d'honneur, Directeur général du Dépôt des cartes et plans de la marine, Vice-Président de la Commission des phares, Membre de la Commission de l'Observatoire* (Paris, 1883).

On Jonquières's work, see Gino Loria, "L'oeuvre mathématique d'Ernest de Jonquières," in *Bibliotheca mathematica*, 3rd ser., **3** (1902), 276–322, and "Elenco delle pubblicazioni matematiche di Ernesto de Jonquières," in *Bullettino di bibliografia e storia delle scienze matematiche e fisiche*, **5** (1902), 72–82. See also H. G. Zeuthen, "Abzählende Methoden"; L. Berzolari, "Allgemeine Theorie der höheren ebenen algebraischen Kurven"; and L. Berzolari, "Algebraische Transformationen und Korrespondenzen"; all in *Encyklopädie der mathematischen Wissenschaften*, III, *Geometrie*.

L. Nový
J. Folta

JONSTON, JOHN (*b.* Sambter, Poland, 3 September 1603; *d.* Liegnitz, Poland, 8 June 1675), *natural history, medicine.*

Of Scottish extraction, Jonston gained an extensive education while traveling (sometimes as a private tutor) in Germany, Scotland, England, and Holland. He attended St. Andrews, Cambridge, Leiden, and Frankfurt universities, obtaining M.D. degrees in 1632 at Cambridge (ad eundem) and Leiden, where he later practiced medicine. He refused the chair in medicine at Leiden in 1640, but in 1642 he did become, for a short while, professor of medicine at Frankfurt.

Jonston's widespread education is reflected in his prolific and wide-ranging writings, which comprise natural history, medicine, and miscellaneous works. Commentators on his books have tended to dismiss them as mere compilations, exhibiting more learning than judgment. There is some justice in this view, especially with regard to his extensive publications on natural history, in which he often relied heavily on the writings of others, for example, Aldrovandi. But that Jonston's works failed to reach the standard of critical organization set by some of his contemporaries should not overshadow the significant contribution his works made to the growing interest in natural history during the first half of the seventeenth century. For example, four of his dictionary-style works on fish, birds, quadrupeds, and insects—published between 1650 and 1653 with excellent illustrations—were widely read and translated.

Of Jonston's many medical writings, his best known is *Idea Universae Medicinae Practicae* (Amsterdam, 1644), which was published in five editions. The book also appeared with commentaries by J. Michaelis and T. Bonnet, and was translated into English by Nicholas Culpeper (1652). Jonston's book emphasized the teaching of clinical medicine to students, and therefore represented an interesting choice for Culpeper, about half of whose work was undertaken in response to the needs of the English apothecaries, then increasingly numerous and influential.

Since Jonston made it clear that his text owed much to Daniel Sennert, it is no surprise that the work is both systematic and Galenic in outlook. Despite this debt to Sennert, Jonston's own conciseness and care in preparing the text made it eminently suitable for students. A wide-ranging work, it dealt not only with clinical conditions, but also provided summaries on, for instance, materia medica and on the importance of non-naturals (which he listed as air, meat, drink, motion and rest, sleep and watching) in the preservation of health. Jonston's emphasis on signs and symptoms undoubtedly contributed to a growing empirical outlook in clinical medicine, an influence enhanced by Michaelis' commentary on Jonston, which has already been mentioned.

Jonston's miscellaneous works include items of general scientific interest, such as his *De Naturae Constantiae* (Amsterdam, 1652; English translation, 1657). Through many examples Jonston indicated that both natural phenomena and human nature had not changed since classical times. His theme perhaps reflected conservatism in his own views; but concerning the theory of matter, he considered that there were three elements (rather than four); he believed that fire was the supreme part of pure air and asked, "since the Scripture doth no where speak of fire . . . why should we maintain it?" He accepted Paracelsian views that salt, sulfur, and mercury were the fundamental constituents of matter, and favored this *tria prima* theory because of the prominence of the number three —as in, for instance, the Trinity, the three spirits of man (animal, vital, and natural), and the three types of vessels (nerves, arteries, and veins). He also spoke of three humors in the blood, although it is not clear whether he dismissed the traditional four-humor theory, which he held in his *Idea Universae Medicinae Practicae*. Jonston's writings were a useful contribution to seventeenth-century thought, although he was not in the forefront of changing concepts of the time.

BIBLIOGRAPHY

Lists of Jonston's writings may be found in J. P. Niceron, *Mémoires pour servir a l'histoire des hommes illustres*, **41** (1740), 269–276; and the *Dictionary of National Biography*.

Other valuable sources are T. Bilikiewicz, "Johann Jonston (1603–1675) und seine Tätigkeit als Artz," in *Sudhoffs Archiv*, **23** (1930), 357–381; and T. Bilikiewicz, *Jan Jonston* (Warsaw, 1931).

J. K. CRELLIN

JORDAN, (CLAUDE THOMAS) ALEXIS (*b.* Lyons, France, 29 October 1814; *d.* Lyons, 7 February 1897), *botany*.

Jordan belonged to one of the most distinguished families of the Lyons bourgeoisie. His father, César, was a rich merchant; his mother, Jeanne-Marie (called Adèle) Caquet d'Avaize, was the daughter of a lawyer. Camille Jordan, the writer and politician, was his uncle and the mathematician Camille Jordan was a cousin. A Catholic and a royalist, Jordan was a member of the Société Botanique de France, the Imperial Society of Naturalists of Moscow, and the Royal Botanical Society of Belgium, among others.

Only botany could induce Jordan to leave Lyons. He received his secondary education there and, renouncing a career in commerce, turned to the natural sciences. He frequented a group of cultivated amateurs who enlivened the Linnaean Society of Lyons and he soon specialized in botany under the direction of Nicolas Seringe, a military surgeon who became director of the Lyons Botanical Garden and whose assistant he was for a time.

Apparently Jordan's task originally was purely descriptive and modest: to complete or correct the existing French floras on certain points. Between 1836 and 1846 he was essentially an observer and a researcher in the field. Each year he made several long botanical journeys, usually beyond the southeast but rarely outside of France (except to Corsica and Italy). These trips and the samples sent him by the many botanists with whom he corresponded allowed Jordan to assemble one of the most important private herbaria in Europe. His personal fortune enabled him to amass a library and experimental equipment comparable with those of the great professional botanists of his time.

Between the ages of twenty-five and thirty, Jordan concentrated on certain plants difficult to classify that had been brought to his attention by his friend Marc-Antoine Timeroy. These plants did not coincide exactly with the descriptions of the floras, and yet they could not be treated as simple varieties. Jordan soon became convinced, on these grounds, that the classic method of description itself was too schematic: a single species name was almost always applied to a multiplicity of forms that were quite similar, yet distinct and stable. This was particularly the explanation of the so-called polymorphous species, to which the very special faculty of "varying" was attributed. As early as 1846 Jordan concluded that at least five "easy to distinguish although very closely related" species are designated by the name *Viola tricolor* L. (the pansy); each of them, moreover, proves to be invariable after several years of cultivation.

Jordan announced these general conclusions in 1846 and 1847 to the Linnaean Society of Lyons in a series of seven monographs entitled *Observations sur plusieurs plantes nouvelles rares ou critiques*. From then on, the concept of species itself was in question. Beginning in the 1850's, Jordan became a theoretician and virtually *chef d'école*. The Linnaean notion of species corresponded, according to him, not to the real boundaries of the plant forms but to a crude division suggested by simple, practical convenience: it retained only those characteristics that are easily distinguished with the naked eye by a botanist of average experience and that are preserved when the plant is dried in a herbarium. Experiment shows that a great many essential traits are not necessarily of this sort. Furthermore, the rigorous invariability of the species, largely underestimated by classical botanists, is confirmed by the facts and is already contained in the pure concept of species in general, which is logically prior to experiment and is the basis on which experiment can be methodically conducted and interpreted. "The observer who wishes to proceed on sure ground . . . should take philosophy for his guide and theology for his compass" (*Remarques. . .*, p. 23). Every authentic being was "conceived by thought as absolutely one and indivisible . . . as immutable and unalterable" in that which is proper to it (*De l'origine . . .*, p. 5). The living species was a being of this kind: its "substantial" character was confirmed by its permanence during the course of generations, and every type that showed itself to be variable in its lineage was hybrid. Varieties, properly speaking, originate among the plants from superficial, environmentally determined modifications, which are not transmitted to the descendants; they do not affect the "substance" of the species.

This break with traditional concepts necessitated a change in method. "Closely related" species (*espèces affines*) exist everywhere; a given species, Jordan noted, was almost always surrounded, in a single location, by several analogous forms. All the true species could be counted only by controlling the descendants, that is, by cultural experimentation. Giving up his botanical expeditions, Jordan now limited his activity as an investigator to his experimental garden; the first one measured only about 400 square meters. About 1850 an inheritance from an uncle enabled him to buy a plot at Villeurbanne that he gradually increased to 12,000 square meters. At its best it had about 400 flower beds, grouped in equal squares and containing approximately 50,000–60,000 plants. A series of related forms belonging to the same Linnaean species was placed in the same flower bed, and their complete stability from year to year was verified. Thus, twenty-five kinds of *Scabiosa succisa*, thirty-five of *Sempervivum tectorum*, and so on, coexisted without ever intermingling. The record was established by *Draba verna* (a crucifer with rosette leaves and small flowers that is frequently found in spring on walls and embankments), from which as many as 200 distinct forms were obtained. These "genuine species" were distinguished not by a major difference limited to one characteristic (Jordan considered the sudden modifications that sometimes disturb a characteristic in a line to be an accident, a *lusus naturae* that may be disregarded) but rather by a series of small but very stable details: for example, by bifurcated or trifurcated hairs, petals that are more or less narrow, fruits varying in size in relation to the length of the stem, and so on.

Yet if true species are rigorously invariable, how can one interpret the effects of cultivation, which seems to create and determine the quantity of new forms? In reply to this question Jordan published *De l'origine des diverses variétés ou espèces* (1853). In fields, gardens, and orchards, he explains, many plants may be seen that were unknown, say, in the seventeenth century. But either they are purely individual variations due to the environment—and it is the permanence of this environment, not heredity, that creates the uniformity of the successive generations—or else it is a question of true species; but then they are new only for us, since they already existed in earlier times.

Jordan's career—he was both a bachelor with a difficult nature and an increasingly intransigent theorist—ended in growing isolation. In 1864 he entrusted the management of his garden to Jules Fourreau. Some deplored the influence that Fourreau immediately began to exert on him: he was more Jordanian than his master. He incited Jordan, it was said, to multiply species without limit by arbitrarily purging his flower beds in order to make the lines more homogeneous, sometimes dispensing with the cultural criterion in the process. Jordan maintained this orientation after Fourreau's death. The last years of this obstinate monarchist saw his involvement in a naïve political enterprise that estranged a great many of his former friends.

When Jordan died, his reputation was at its lowest ebb. The conservative botanists reproached him for "pulverizing" the species and ruining systematics; on the other side, a triumphant Darwinism drew the younger botanists away from this "ultra" of fixity. Nevertheless, French and foreign researchers repeated certain of his experiments and were able to confirm them. Beginning in 1900 it was, paradoxically, the neo-Darwinians who rediscovered the radical separation of hereditary characteristics and variations due simply to the environment and who emphasized the idea that the Linnaean species is only a convenient category, a subgroup in a discontinuous series of elementary types. Hugo de Vries reappraised Jordan's work, reinterpreting it in terms of mutation; and in 1916 J. P. Lotsy introduced the term "Jordanon," in opposition to "Linneon," in order to designate the species in Jordan's sense.

Thus, Jordan's fixity, now completely outdated, survives only in the conclusions that this unusual theorist drew and confirmed from it: the intransmissibility of acquired characteristics and the traditional concept of species considered as the blending of separate but closely related homogeneous types, since they appear genetically stable at first consideration. And the *lusus naturae* or mutations that he eliminated from botany have become for contemporary biology a way of conceiving of the origin of "Jordanon species."

BIBLIOGRAPHY

I. ORIGINAL WORKS. Jordan's works in systematic botany include *Observations sur plusieurs plantes nouvelles rares ou critiques*, 7 pts. (Paris, 1846–1847); *Pugillus plantarum novarum praesertim gallicarum* (Paris, 1852); *Diagnoses d'espèces nouvelles ou méconnues pour servir de matériaux à une flore réformée de la France et des contrées voisines* (Paris, 1864); and his masterwork, *Icones ad floram Europae novo fundamento instaurandam spectantes*, 3 vols. (Paris, 1866–1903), written with J. Fourreau.

His theoretical writings include *De l'origine des diverses variétés ou espèces d'arbres fruitiers et autres végétaux généralement cultivés pour les besoins de l'homme* (Paris, 1853); and *Remarques sur le fait de l'existence en société à l'état sauvage des espèces végétales affines et sur d'autres faits relatifs à la question de l'espèce* (Lyons, 1873). There is also the critical study "Rapport sur l'*Essai de phytostatique appliquée à la chaîne du Jura et aux contrées voisines* par M. Thurmann," in *Annales des sciences physiques et naturelles, d'agriculture et d'industrie publiées par la Société nationale d'agriculture, d'histoire naturelle et des arts utiles de Lyon* (1850), pp. 7–30.

II. SECONDARY LITERATURE. An excellent source is C. Roux and A. Colomb, "Alexis Jordan et son oeuvre botanique," in *Annales de la Société linnéenne de Lyon*, n.s. **54** (1908), 181–258. See also A. Magnin, *Prodrome d'une histoire des botanistes lyonnais* (Lyons, 1906), pp. 97–107; and Viviand-Morel, "Histoire abrégée des cultures expérimentales du jardin d'A. Jordan," in *Lyon horticole et horticulture nouvelle* (1907), nos. 3, pp. 57–59; 4, pp. 77–78; 7, pp. 137–140; 21, pp. 415–418. Recent studies on Jordan's thought are M. Breistoffer, "Sur la nomenclature botanique de quelques botanistes lyonnais," in *Comptes rendus du 89° Congrès des sociétés savantes* (Paris, 1965); and J. Piquemal, "Alexis Jordan et la notion d'espèce," *Conférences du Palais de la découverte* (Paris, 1964).

JACQUES PIQUEMAL

JORDAN, CAMILLE (*b.* Lyons, France, 5 January 1838; *d.* Paris, France, 22 January 1921), *mathematics.*

Jordan was born into a well-to-do family. One of his granduncles (also named Camille Jordan) was a fairly well-known politician who took part in many events from the French Revolution in 1789 to the beginning of the Bourbon restoration; a cousin, Alexis Jordan, is known in botany as the discoverer of "smaller species" which still bear his name ("jordanons"). Jordan's father, an engineer, was a graduate of the École Polytechnique; his mother was a sister of the painter Pierre Puvis de Chavannes. A brilliant student, Jordan followed the usual career of French mathematicians from Cauchy to Poincaré: at seventeen he entered the École Polytechnique and was an engineer (at least nominally) until 1885. That profession left him ample time for mathematical research, and most of his 120 papers were written before he retired as an engineer. From 1873 until his retirement in 1912 he taught simultaneously at the École Polytechnique and the Collège de France. He was elected a member of the Academy of Sciences in 1881.

Jordan's place in the tradition of French mathematics is exactly halfway between Hermite and Poincaré. Like them he was a "universal" mathematician who published papers in practically all branches of the mathematics of his time. In one of his first papers, devoted to questions of "analysis situs" (as combinatorial topology was then called), he investigated symmetries in polyhedrons from an exclusively combinatorial point of view, which was then an entirely new approach. In analysis his conception of what a rigorous proof should be was far more exacting than that of most of his contemporaries; and his *Cours d'analyse*, which was first published in the early 1880's and had a very widespread influence, set standards which remained unsurpassed for many years. Jordan took an active part in the movement which started modern analysis in the last decades of the nineteenth century: independently of Peano, he introduced a notion of exterior measure for arbitrary sets in a plane or in n-dimensional space. The concept of a function of bounded variation originated with him; and he proved that such a function is the difference of two increasing functions, a result which enabled him to extend the definition of the length of a curve and to generalize the known criteria of convergence of Fourier series. His most famous contribution to topology was to realize that the decomposition of a plane into two regions by a simple closed curve was susceptible of mathematical proof and to imagine such a proof for the first time.

Although these contributions would have been enough to rank Jordan very high among his mathematical contemporaries, it is chiefly as an algebraist that he reached celebrity when he was barely thirty; and during the next forty years he was universally regarded as the undisputed master of group theory.

When Jordan started his mathematical career, Galois's profound ideas and results (which had remained unknown to most mathematicians until 1845) were still very poorly understood, despite the efforts of A. Serret and Liouville to popularize them; and before 1860 Kronecker was probably the only first-rate mathematician who realized the power of these ideas and who succeeded in using them in his own algebraic research. Jordan was the first to embark on a systematic development of the theory of finite groups and of its applications in the directions opened by Galois. Chief among his first results were the concept of composition series and the first part of the famous Jordan-Hölder theorem, proving the invariance of the system of indexes of consecutive groups in any composition series (up to their ordering). He also was the first to investigate the structure of the general linear group and of the "classical" groups over a prime finite field, and he very ingeniously applied his results to a great range of problems; in particular, he was able to determine the structure of the Galois group of equations having as roots the parameters of some well-known geometric configurations (the twenty-seven lines on a cubic surface, the twenty-eight double tangents to a quartic, the sixteen double points of a Kummer surface, and so on).

Another problem for which Jordan's knowledge of these classical groups was the key to the solution, and to which he devoted a considerable amount of effort from the beginning of his career, was the general study of solvable finite groups. From all we know today (in particular about p-groups, a field which was started, in the generation following Jordan, with the Sylow theorems) it seems hopeless to expect a complete classification of all solvable groups which would characterize each of them, for instance, by a system of numerical invariants. Perhaps Jordan realized this; at any rate he contented himself with setting up the machinery that would automatically yield all solvable groups of a given order n. This in itself was no mean undertaking; and the solution imagined by Jordan was a gigantic recursive scheme, giving the solvable groups of order n when one supposes that the solvable groups of which the orders are the exponents of the prime factors of n are all known. This may have no more than a theoretical value; but in the process of developing his method, Jordan was led to many important new concepts, such as the minimal normal subgroups of a group and the orthogonal groups over a field of characteristic 2 (which he called "hypoabelian" groups).

In 1870 Jordan gathered all his results on permutation groups for the previous ten years in a huge volume, *Traité des substitutions,* which for thirty years was to remain the bible of all specialists in group theory. His fame had spread beyond France, and foreign students were eager to attend his lectures; in particular Felix Klein and Sophus Lie came to Paris in 1870 to study with Jordan, who at that time was developing his researches in an entirely new direction: the determination of all groups of movements in three-dimensional space. This may well have been the source from which Lie conceived his theory of "continuous groups" and Klein the idea of "discontinuous groups" (both types had been encountered by Jordan in his classification).

The most profound results obtained by Jordan in algebra are his "finiteness theorems," which he proved during the twelve years following the publication of the *Traité*. The first concerns subgroups G of the symmetric groups $\mathfrak{S}_n$ (group of all permutations of n objects); for such a group G, Jordan calls "class of G" the smallest number $c > 1$ such that there exists a permutation of G which moves only c objects. His finiteness theorem on these groups is that there is an absolute constant A such that if G is primitive and does not contain the alternating group $\mathfrak{A}_n$, then $n \leqslant Ac^2 \log c$ (in other words, there are only finitely many primitive groups of given class c other than the symmetric and alternating groups).

The second, and best-known, finiteness theorem arose from a question which had its origin in the theory of linear differential equations: Fuchs had determined all linear equations of order 2 of which the solutions are all algebraic functions of the variable. Jordan reduced the similar problem for equations of order n to a problem in group theory: to determine all finite subgroups of the general linear group $GL(n, \mathbf{C})$ over the complex field. It is clear that for $n \geqslant 1$ there are infinitely many such groups, but Jordan discovered that for general n the infinite families of finite subgroups of $GL(n, \mathbf{C})$ are of a very special type. More precisely, there exists a function $\varphi(n)$ such that any finite group G of matrices of order n contains a normal subgroup H which is conjugate in $GL(n, \mathbf{C})$ to a subgroup of diagonal matrices, and such that the index $(G : H)$ is at most $\varphi(n)$ (equivalently, the quotient group G/H can only be one of a finite system of groups, up to isomorphism).

Jordan's last finiteness theorem is a powerful generalization of the results obtained earlier by Hermite in the theory of quadratic forms with integral coefficients. Jordan considered, more generally, the vector space of all homogeneous polynomials of degree m in n variables, with complex coefficients; the unimodular group $SL(n, \mathbf{C})$ operates in this space, and Jordan considered an orbit for this action (that is, the set of all forms equivalent to a given one F by unimodular substitutions). Within that orbit he considered the forms having (complex) integral coefficients (that is, coefficients which are Gaussian integers), and he placed in the same equivalence class all such forms which are equivalent under unimodular substitutions having (complex) integral coefficients. His fundamental result was then that the number of these classes is finite, provided $m > 2$ and the discriminant of F is not zero.

BIBLIOGRAPHY

Jordan's papers were collected in *Oeuvres de Camille Jordan*, R. Garnier and J. Dieudonné, eds., 4 vols (Paris, 1961–1964). His books are *Traité des substitutions et des équations algébriques* (Paris, 1870; repr. 1957) and *Cours d'analyse de l'École Polytechnique*, 3 vols. (3rd ed., 1909–1915).

A detailed obituary notice is H. Lebesgue, in *Mémoires de l'Académie des sciences de l'Institut de France*, 2nd ser., **58** (1923), 29–66, repr. in Jordan's *Oeuvres*, IV, x–xxxiii.

J. DIEUDONNÉ

JORDAN, DAVID STARR (*b.* Gainesville, New York, 19 January 1851; *d.* Stanford, California, 19 September 1931), *ichthyology, education.*

A childhood in rural New York State provided young Jordan ample opportunity to indulge his early interests in plants, stars, maps, and reading. His parents, Hiram Jordan and the former Huldah Lake Hawley, had both been teachers as well as owners of a prosperous farm, where Jordan, the fourth of five children, took charge of a flock of sheep and later the making of maple sugar. His pre-college schooling was, by special exemption, at the nearby Gainesville Female Seminary. Intending to specialize in botany or animal husbandry, he entered Cornell University, to which he had received a scholarship, in March 1869. Of the staff he was most impressed by C. Frederick Hartt in geology, Burt G. Wilder in zoology, and Albert N. Prentiss in botany. Because of undergraduate work as an instructor in botany, he was awarded the M.S. instead of the B.S. in 1872.

Jordan entered the field of education by teaching natural science for one year at Lombard College in Galesburg, Illinois, and the next year he was principal and teacher at Appleton Collegiate Institute in Wisconsin. He moved on to teach science at Indianapolis High School (1874–1875) and then became professor of biology at Butler University, Indianapolis (1875–1879). That position led to his becoming professor of natural history at Indiana University (1879) and later president of the university (1885–1891).

Always ahead of his time, Jordan instituted electives and a major field at Indiana, on the premise that "the duty of real teachers is to adapt the work to the student, not the student to the work" (*Days of a Man*, I, 237). His successful theories of education attracted the attention of Leland Stanford, and in 1891 Jordan became the first president of Leland Stanford Junior University. In 1913, in order to devote more time to outside interests, Jordan became chancellor.

Jordan was inspired to enter ichthyology by Louis Agassiz in the summer of 1873, at the Anderson School of Natural History on Penikese Island, Massachusetts. At Butler University he turned to local fish fauna as the most rewarding undeveloped field in which a young scientist could distinguish himself. He chose well, for from his first paper on fishes in 1874 he dominated ichthyology and drew the best science students to it.

Descriptive ichthyology was then in its infancy in the United States. The eccentric Constantine Samuel Rafinesque essentially founded it with his descriptions of fishes of the Ohio River frontier country (1820), which were modified by Jared Potter Kirtland twenty years later. In 1850 David Humphries Storer published a *Synopsis of the Fishes of North America*, and government explorations of the western territories provided a wealth of new material, the fishes of which were mostly described by Charles Frederic Girard and his coauthor Spencer Fullerton Baird. Individual regions were under study by various workers, one of the most significant investigations being Louis Agassiz's 1850 report on Lake Superior.

Jordan began in Indiana but soon went farther afield. From 1876 he customarily spent each summer collecting, the earliest trips being largely along the rivers of the Allegheny Mountains and in much of the South. He spent three summers on extensive walking and collecting tours in Europe. In 1876 he studied the fishes of Ohio for that state's fish commission. Later, for the U.S. Fish Commission he collected and presented taxonomic monographs on fishes of the Pacific coast, the Gulf coast, Florida, and Cuba, and the fish faunas of the major American rivers. While at Stanford, besides making many trips within California, Jordan visited Mazatlán, Mexico; the Bering Sea, while investigating the fur seal dispute between the United States and Great Britain (1896); the interior of Mexico; Japan; Hawaii; Samoa; Alaska; and Europe. From 1908 to 1910 he served as the U.S. International Commissioner of Fisheries for the conservation of fisheries along the Canadian border.

The result of Jordan's work was the naming of 1,085 genera and more than 2,500 species of fishes, as well as synopses of the classification. An uncanny ability to distinguish similar species, an unfailing intuition of diagnostic characters, and a phenomenal memory made Jordan an outstanding taxonomist.

Unlike his mentor Agassiz, Jordan was an early adherent of and contributor to the theory of Darwinian evolution. From his early trips in the southern United States he derived Jordan's law: The species most closely related to another is found just beyond a barrier to distribution. From his worldwide studies of fishes he concluded that extreme specialization along a given line of development is followed by progressive degeneration. Enlarging on observations by Albert Günther and Theodore Gill, he also found that, almost universally, equatorial fishes have considerably fewer and larger vertebrae than do their polar relatives.

A prolific writer, in addition to his many papers on fish collections and areal faunas, Jordan published thirteen editions of *Manual of Vertebrates* (1876–1929); several valuable manuals on fish classification; with C. H. Gilbert the useful "Synopsis of the Fishes of North America" (1883), which gave the first great impetus to American ichthyology; and with B. W. Evermann the indispensable "Fishes of North and Middle America" (1896–1900), which for many years almost ended the study, since he and many others considered the subject largely completed.

Jordan's honors were legion. He received half a dozen honorary degrees; was president of the American Association for the Advancement of Science in 1909–1910; president of the California Academy of Sciences three times; and a member of the International Commission of Zoological Nomenclature from 1904 until his death. Among other societies, he was a member of the American Philosophical Society and the Zoological Society of London. The Smithsonian Institution made him an honorary associate in zoology in 1921.

BIBLIOGRAPHY

I. ORIGINAL WORKS. In 1883 a fire at Indiana University destroyed some of Jordan's collections and incomplete MSS. From then on, he published promptly. A list of his works compiled by Alice N. Hays, "David Starr Jordan. A Bibliography of His Writings," *Stanford University Publications*, University series, **1** (1952), contains 1,372 general writings and 645 on ichthyology. Mentioned in the text are his most valuable references on ichthyology: *Manual of the Vertebrates of the Northern United States* (Chicago, 1876), 13th ed. entitled *Manual of the Vertebrate Animals of the Northeastern United States Inclusive of Marine Species* (Yonkers, N.Y., 1929); "A Synopsis of the Fishes of North America," *Bulletin. United States National Museum*, **16** (1883), written with C. H. Gilbert; and "The Fishes of North and Middle America," pts. 1–4, *ibid.*, **47** (1896–1900), written with B. W. Evermann. In addition, "The Genera of Fishes," *Stanford University Publications*, monograph series, **27**, **36**, **39**, **43** (1917–1920), and "A Classification of Fishes," *ibid.*, Biological Sciences, **3** (1923), reissued in book form (Stanford, 1963), are standard tools of ichthyologists.

Monographs on the fishes of specific regions are catalogued in Bashford Dean, *A Bibliography of Fishes* (New York, 1916), pp. 643–661. Jordan's law is expounded in "The Law of Geminate Species," in *American Naturalist*, **42** (1908), 73–80. His conclusions on degeneration after specialization appear in *Evolution and Animal Life* (New York, 1907), written with V. L. Kellogg. His observations on numbers of vertebrae are in "Temperature and Vertebrae: A Study in Evolution . . .," in *Wilder Quarter-Century Book* (Ithaca, N.Y.), pp. 13–36.

Jordan's general works, ranging from international relations, philosophy, evolution, and education to poetry and children's books, can be found in Hays (see above) and in *Days of a Man*.

The life of an unbelievably busy man is presented in Jordan's *The Days of a Man, Being Memories of a Naturalist, Teacher and Minor Prophet of Democracy*, 2 vols. (Yonkers, N.Y., 1922).

II. SECONDARY LITERATURE. Insights on Jordan as a leader and teacher are given in B. W. Evermann, "David Starr Jordan, the Man," in *Copeia* (Dec. 1930), pp. 93–105. An excellent analysis of his influence on ichthyology is Carl. L. Hubbs, "History of Ichthyology in the United States After 1850," *ibid.* (Mar. 1964), pp. 42–60.

ELIZABETH NOBLE SHOR

JORDAN, EDWIN OAKES (*b.* Thomaston, Maine, 28 July 1866; *d.* Lewiston, Maine, 2 September 1936), *bacteriology*.

Jordan spent much of his first three years at sea with his parents. His father, Joshua Lane Jordan, was a captain of merchant vessels; his mother, Eliza Bugbee Jordan, had taught school. After his secondary schooling in Maine and Massachusetts, Jordan attended the Massachusetts Institute of Technology, where he became one of the early protégés of William Thompson Sedgwick. Following his graduation in 1888 Jordan worked with Sedgwick and Allen Hazen for two years at the new Lawrence Experiment Station of the Massachusetts Board of Health, investigating the bacteria of water and sewage. In 1890 he began graduate studies in zoology at Clark University under Charles Otis Whitman and received his Ph.D. in 1892, in time to accompany Whitman to the University of Chicago. Jordan remained at Chicago for the next forty-one years, moving up from instructor in "sanitary biology" to professor of bacteriology and, from 1914 to 1933, serving as chairman of the department of hygiene and bacteriology. He married Elsie Fay Pratt in 1893; they had three children.

Modest and soft-spoken, Jordan nevertheless was one of the energizers of the second generation of American bacteriologists. As his many graduate students moved to laboratories and teaching positions around the country, his influence within the fields of bacteriology and public health expanded similarly. He helped organize the Society of American Bacteriologists in 1899 and during the 1920's helped found the American Epidemiological Society. As a trustee and staff member Jordan played an active role in the work of the John McCormick Institute for Infectious Diseases. In particular, he did much to raise the quality of American scientific writing in his capacity as joint editor, beginning in 1904, of the McCormick Institute's *Journal of Infectious Diseases* and as editor of its *Journal of Preventive Medicine* from 1926 to 1933.

An authority on waterborne diseases as well as other aspects of sanitation, Jordan frequently served as consultant to local, national, and international health agencies. Notable were his studies of the self-purification of streams, which he made between 1899 and 1903 for the Sanitary District of Chicago in connection with the controversial Chicago drainage canal and its pollution of the Illinois River. From 1920 to 1927 Jordan was a member of the International Health Board of the Rockefeller Foundation and from 1930 to 1933 served on the Board of Scientific Directors of the Rockefeller Foundation's International Health Division.

In 1899 Jordan translated Ferdinand Hueppe's *Die Methoden der Bakterien-Forschung* into English. Later he wrote *Textbook of General Bacteriology* (1908). This standard work went through eleven American editions before Jordan's death and was translated into several foreign languages. Another major publication was his authoritative *Food Poisoning* (1917), much expanded in 1931. Of comparable significance was *Epidemic Influenza* (1927). This exhaustive study, which grew out of the frustration experienced by scientists and health officials during the 1918–1919 pandemic, failed to establish the etiology of influenza; but its organization of the voluminous literature proved invaluable for subsequent research efforts on the disease.

Jordan's authority on public health matters derived at least partly from the continuing basic research in bacteriology which he and his associates conducted. His broad biological view of bacteriology produced a variety of studies on host and parasite populations and on the mechanism of transmitting infective agents. Among his many other studies, he was one of the earliest investigators concerned with the problem of the variation of bacteria. For Jordan, however, the pursuit of fundamental scientific knowledge was never a wholly abstract matter but, rather, an activity which often related intimately to the demands and unsolved problems of practical sanitation.

BIBLIOGRAPHY

I. Original Works. A collection of Jordan's professional correspondence and other papers is deposited in the library of the University of Chicago. Much of his personal library, including books, reprints, photographs, and other materials, is in the archives of the American Society for Microbiology, Washington, D.C.

A complete bibliography of Jordan's writings, arranged by year of publication, was prepared by William Burrows in 1939 (see below). The list includes several hundred scientific papers. It also includes two book-length works not mentioned in the text: *Textbook of General Bacteriology* (Philadelphia–London, 1908); and *Food Poisoning* (Chicago, 1917; enl., 1931). See also *A Pioneer in Public Health, William Thompson Sedgwick* (New Haven, 1924), written with G. C. Whipple and C. E. A. Winslow; and *The Newer Knowledge of Bacteriology and Immunology* (Chicago, 1928), edited with I. S. Falk.

II. Secondary Literature. The fullest account to date of Jordan's life and work is William Burrows, in *Biographical Memoirs. National Academy of Sciences*, **20** (1939), 197–228. There are informative sketches by Stanhope Bayne-Jones, in *Dictionary of American Biography*, supp. II, 352–354; and by Paul F. Clark, in his *Pioneer Microbiologists of America* (Madison, Wis., 1961), pp. 255–261. Among the most useful obituaries are N. Paul Hudson, in *Journal of Bacteriology*, **33** (1937), 242–248; and Ludvig Hektoen, in *Science*, **84** (1936), 411–413. Hektoen's account also appeared in *Proceedings of the Institute of Medicine of Chicago*, **11** (1936), 182–185. See also brief accounts in *Journal of the American Medical Association*, **107** (1936), 2051; and *Who Was Who in America*, I, 652.

James H. Cassedy

JORDANUS DE NEMORE (*fl. ca.* 1220), *mechanics, mathematics.*

Although Jordanus has been justly proclaimed the most important mechanician of the Middle Ages and one of the most significant mathematicians of that period, virtually nothing is known of his life. That he lived and wrote during the first half of the thirteenth century, and perhaps as early as the late twelfth century, is suggested by the inclusion of his works in the *Biblionomia*, a catalogue of Richard de Fournival's library compiled sometime between 1246 and 1260.[1] In all, twelve treatises are ascribed to Jordanus de Nemore, whose name is cited four times in this form.[2] Since most of his genuine treatises are included, it seems reasonable to infer that Jordanus' productive career antedated the *Biblionomia*.

The appellation "Jordanus de Nemore" is also found in a number of thirteenth-century manuscripts of works attributed to Jordanus. The meaning and origin of "de Nemore" are unknown. It could signify "from" or "of Nemus," a place as yet unidentified (the oft-used alternative "Nemorarius," frequently associated with Jordanus, is apparently a later derivation from "Nemore"), or it may have derived

JORDANUS DE NEMORE

from a corruption of "de numeris" or "de numero" from Jordanus' arithmetic manuscripts.[3]

Identification of Jordanus de Nemore with Jordanus de Saxonia (or Jordanus of Saxony), the master general of the Dominican order from 1222 to 1237, has been made on the basis of a statement by Nicholas Trivet (in a chronicle called *Annales sex regum Angliae*) that Jordanus of Saxony was an outstanding scientist who is said to have written a book on weights and a treatise entitled *De lineis datis*.[4] Although a late manuscript of a work definitely written by Jordanus de Nemore is actually ascribed to "Jordanus de Alemannia" (Jordanus of Germany, and therefore possibly Jordanus of Saxony), no mathematical or scientific works can be assigned to Jordanus of Saxony, whose literary output was seemingly confined to religion and grammar. At no time, moreover, was Jordanus of Saxony called Jordanus de Nemore or Nemorarius. Finally, if Jordanus de Nemore lectured at the University of Toulouse, as one manuscript indicates,[5] this could have occurred no earlier than 1229, the year of its foundation. As master general of the Dominican order during the years 1229–1237, the year of his death, Jordanus of Saxony could hardly have found time to lecture at a university. For all these reasons it seems implausible to suppose that Jordanus of Saxony is identical with Jordanus de Nemore.

It was in mechanics that Jordanus left his greatest legacy to science. The medieval Latin "science of weights" (*scientia de ponderibus*), or statics, is virtually synonymous with his name, a state of affairs that has posed difficult problems of authorship. So strongly was the name of Master Jordanus identified with the science of weights that manuscripts of commentaries on his work, or works, were frequently attributed to the master himself. Since the commentaries were in the style of Jordanus, original works by him are not easily distinguished. At present only one treatise, the *Elementa Jordani super demonstrationem ponderum,* may be definitely assigned to Jordanus. Whether he inherited the skeletal frame of the *Elementa* in the form of its seven postulates and the enunciations of its nine theorems, for which he then supplied proofs, is in dispute.[6] Indisputable, however, is the great significance of the treatise. Here, under the concept of "positional gravity" (*gravitas secundum situm*), we find the introduction of component forces into statics. The concept is expressed in the fourth and fifth postulates, where it is assumed that "weight is heavier positionally, when, at a given position, its path of descent is less oblique" and that "a more oblique descent is one in which, for a given distance, there is a smaller component of the vertical."[7] In a constrained system the effective weight of a suspended body is proportional to the directness of its descent, directness or obliquity of descent being measured by the projection of any segment of the body's arcal path onto the vertical drawn through the fulcrum of the lever or balance. It is implied that the displacement which measures the positional gravity of a weight can be infinitely small. Thus, by means of a principle of virtual displacement (since actual movement cannot occur in a system in equilibrium, positional gravity can be measured only by "virtual" displacements) Jordanus introduced infinitesimal considerations into statics.

These concepts are illustrated in Proposition 2, where Jordanus demonstrates that "when the beam of a balance of equal arms is in horizontal position, then if equal weights are suspended from its extremities, it will not leave the horizontal position; and if it should be moved from the horizontal position, it will revert to it."[8] If the balance is depressed on the side of *B* (see Figure 1), Jordanus argues that it will return to a

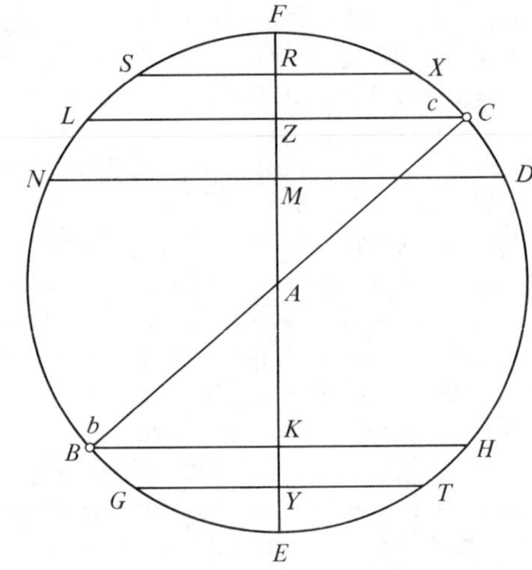

FIGURE 1

horizontal position because weight *c* at *C* will be positionally heavier than weight *b* at *B*, a state of affairs which follows from the fact that if any two equal arcs are measured downward from *C* and *B*, they will project unequal intercepts onto diameter *FRZMAKYE*. If the equal arcs are *CD* and *BG*, Jordanus can demonstrate (by appeal to his *Philotegni*, or *De triangulis*, as it was also called) that the intercept of arc *CD*—*ZM*—is greater than the intercept of arc *BG*—*KY*—and the "positionally" heavier *c* will cause *C* to descend to a horizontal position. The concept of positional heaviness, although erroneous when applied to arcal paths, may have derived ultimately from application of an idea in the pseudo-Aristotelian

JORDANUS DE NEMORE

Mechanica, where it was argued that the further a weight is from the fulcrum of a balance, the more easily it will move a weight on the other side of the fulcrum, since "a longer radius describes a larger circle. So with the exertion of the same force the motive weight will change its position more than the weight which it moves, because it is further from the fulcrum."[9] It was by treating the descent of *b* independently from the ascent of *c* that Jordanus fell into error. A comparison of the ratio of paths formed by a small descent of *b* and an equal ascent of *c* with the ratio of paths formed by a small descent of *c* and an equal ascent of *b* would have revealed the equality of these ratios and demonstrated the absence of positional advantage. As we shall see below, however, when the concept of positional gravity was applied to rectilinear, rather than arcal, constrained paths, perhaps by Jordanus himself, it led to brilliant results.

More important than positional gravity is Jordanus' proof of the law of the lever by means of the principle of work. In Figure 2, *ACB* is a balance beam and

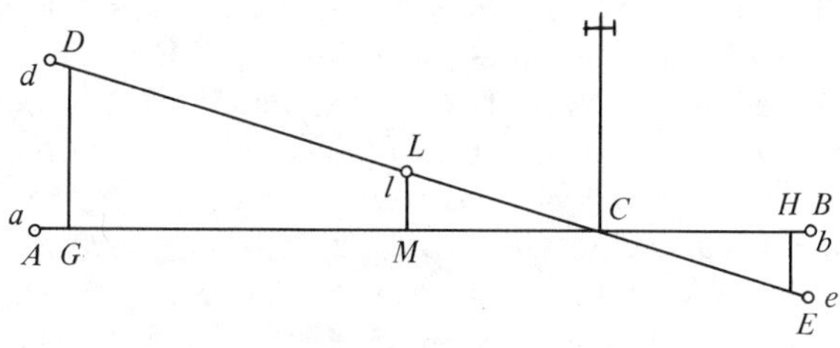

FIGURE 2

a and *b* are suspended weights. If we assume that $b/a = AC/BC$, no movement of the balance will occur. The demonstration takes the form of an indirect proof. It is assumed that the balance descends on *B*'s side so that as *b* descends through vertical distance *HE*, it lifts *a* through vertical distance *DG*. If a weight *l*, equal to *b*, is now suspended at *L*, Jordanus shows, on the basis of similar triangles *DCG* and *ECH*, that $DG/EH = b/a$. On the assumption that $CL = CB$ and drawing perpendicular *LM*, he concludes that $LM = EH$. Therefore $DG/LM = b/a = l/a$. At this point the principle of work is applied, for "what suffices to lift *a* to *D*, would suffice to lift *l* through the distance *LM*. Since therefore *l* and *b* are equal, and *LC* is equal to *CB*, *l* is not lifted by *b*; and, as was asserted, *a* will not be lifted by *b*."[10] If a weight is thus incapable of lifting an equal weight the same distance that it descends, it cannot raise a proportionally smaller weight a proportionally greater distance.

The principles of positional gravity and work were superbly employed in the *De ratione ponderis*, which contains forty-five propositions and is probably the most significant of all medieval statical treatises. If, as the manuscripts indicate, it was by Jordanus himself[11] (although there is some doubt about this),[12] not only did Jordanus extend his own concept of positional gravity to rectilinear paths (the incorrect application to arcal paths was, however, retained in a few propositions) but he also applied that concept, in conjunction with the principle of work, to a formulation of the first known proof—long before Galileo—of the conditions of equilibrium of unequal weights on planes inclined at different angles. Paradoxically, in Book I, Proposition 2, the *De ratione ponderis* included reasoning which, if rigorously applied, would have destroyed the notion that an elevated weight has greater positional gravity with which to restore the equilibrium of a balance.[13]

In Book I, Proposition 9, Jordanus (for convenience we shall assume his authorship) shows that positional gravity—the heaviness or force of a weight—along an inclined plane (see Figure 3) is the same at any

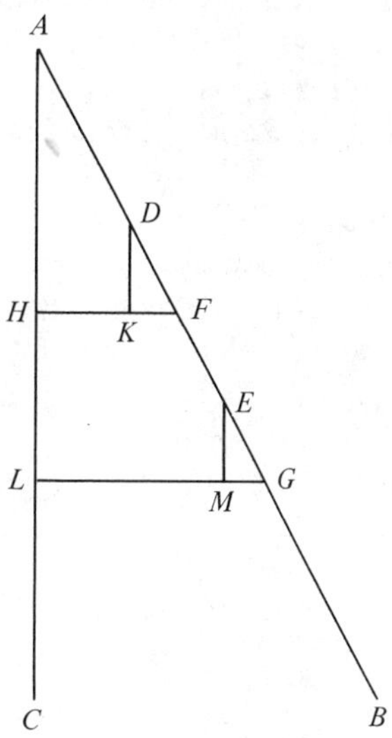

FIGURE 3

point. Thus a given weight at *D* or *E* will possess equal force, since for equal segments of the inclined path *AB*—*DF* and *EG*—equal segments of the vertical *AC* will be intercepted—*DK* and *EM*.

On the basis of Postulates 4 and 5 of the *Elementa*, which are also Postulates 4 and 5 of the *De ratione ponderis*, and Book I, Proposition 9, the inclined plane proof is enunciated in Book I, Proposition 10, as follows: "If two weights descend along diversely inclined planes, then, if the inclinations are directly proportional to the weights, they will be of equal force in descending."[14] Jordanus demonstrates that weights *e* and *h*, on differently inclined planes, are of

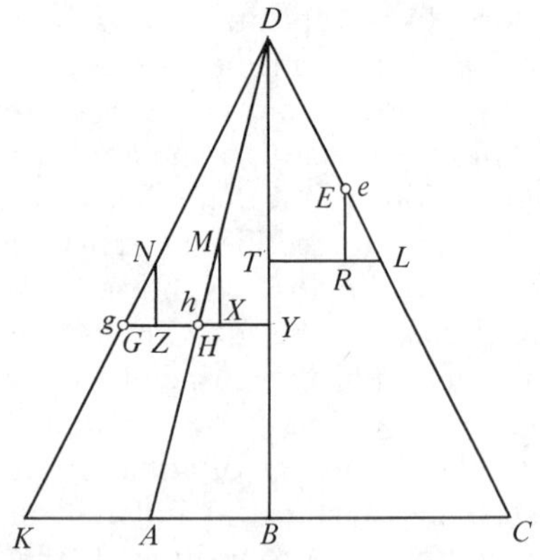

FIGURE 4

equal force. He first assumes that a weight g, equal to e, is on another plane, DK, whose obliquity is equal to that of DC and then assumes that if e moves to L through vertical distance ER, it will also draw h up to M. Should this occur, however, it would follow by the principle of work that what is capable of moving h to M can also move g to N, since it can be shown that $MX/NZ = g/h$. But g is equal to e and at the same inclination; hence, by Book I, Proposition 9, they are of equal force because they will intercept equal segments of vertical DB. Therefore e is incapable of raising g to N and, consequently, unable to raise h to M. By substituting a straight line for an arcal path and utilizing Postulate 5 of the *Elementa* (see above), Jordanus was, in modern terms, measuring the obliquity of descent, or ascent, by the sine of the angle of inclination. The force along the rectilinear oblique path, or incline, is thus equivalent to

$$F = W \sin a,$$

where W is the free weight and a is the angle of inclination of the oblique path.

The principle of work, which was but a vague concept prior to Jordanus, was not only used effectively in the proof of the inclined plane and, as indicated above, in the proof of the law of the lever in the *Elementa*, a proof repeated in Book I, Proposition 6, of the *De ratione ponderis*, but was also applied successfully to the first proof of the bent lever in Book I, Proposition 8, of the *De ratione ponderis*, which reads: "If the arms of a balance are unequal, and form an angle at the axis of support, then, if their ends are equidistant from the vertical line passing through the axis of support, equal weights suspended from them will, as so placed, be of equal heaviness."[15] In this proof there is also an anticipation of the concept of static moment, that the effective force of a weight is dependent on the weight and its horizontal distance from a vertical line passing through the fulcrum.[16]

Over and above his specific contributions to the advance of statics, Jordanus marks a significant departure in the development of that science. He joined the dynamical and philosophical approach characteristic of the dominant Aristotelian physics of his day with the abstract and rigorous mathematical physics of Archimedes. Thus the postulates of the *Elementa* and *De ratione ponderis* were derived from, and consistent with, Aristotelian dynamical concepts of motion but were arranged in a manner that permitted the derivation of rigorous proofs within a mathematical format modeled on Archimedean statics and Euclidean geometry.

The extensive commentary literature on the statical treatises ascribed to Jordanus began in the middle of the thirteenth century and continued into the sixteenth. Through printed editions of the sixteenth century, the content of this medieval science of weights, identified largely with the name of Jordanus, became readily available to mechanicians of the sixteenth and seventeenth centuries. Dissemination was facilitated by works such as Peter Apian's *Liber Iordani Nemorarii ... de ponderibus propositiones XIII et earundem demonstrationes* (Nuremberg, 1533); Nicolo Tartaglia's *Questii ed invenzioni diverse* (Venice, 1546, 1554, 1562, 1606; also translated into English, German, and French), which contained a variety of propositions from Book I of the *De ratione ponderis*; and *Jordani Opusculum de ponderositate* (Venice, 1565), a version of the *De ratione ponderis* published by Curtius Trojanus from a copy owned by Tartaglia, who had died in 1557.

Concepts such as positional gravity, static moment, and the principle of work, or virtual displacement, were now available and actually influenced leading mechanicians, including Galileo, although some preferred to follow the pure Greek statical tradition of Archimedes and Pappus of Alexandria. In commenting on Guido Ubaldo del Monte's *Le mechaniche* (1577), which he himself had translated into Italian, Filippo Pigafetta remarked that Guido Ubaldo's more immediate predecessors

> ... are to be understood as being the modern writers on this subject cited in various places by the author [Guido Ubaldo], among them Jordanus, who wrote on weights and was highly regarded and to this day has been much followed in his teachings. Now our author [Guido Ubaldo] has tried in every way to travel the road of the good ancient Greeks, ... in particular that of Archimedes of Syracuse ... and Pappus of Alexandria ...[17]

JORDANUS DE NEMORE

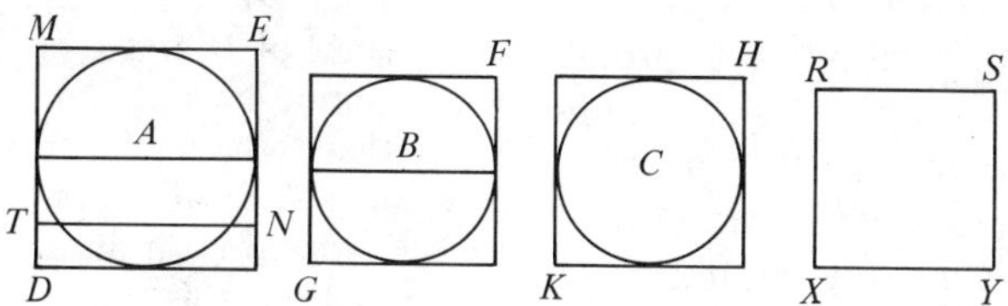

FIGURE 5

Guido Ubaldo's attitude was costly, for it led him to reject Jordanus' correct inclined-plane theorem in favor of an erroneous explanation by Pappus.

Since few editions of his mathematical treatises have been published, and critical studies and evaluations are largely lacking, Jordanus' place in the history of medieval and early modern mathematics has yet to be determined. Treatises on geometry, algebra, proportions, and theoretical and practical arithmetic have been attributed to him.

The *Liber Philotegni de triangulis*, a geometrical work extant in two versions, represents medieval Latin geometry at its highest level. In the four books of the treatise we find propositions concerned with the ratios of sides and angles; with the division of straight lines, triangles, and quadrangles under a variety of given conditions; and with ratios of arcs and plane segments in the same circle and in different circles. The fourth book contains the most significant and sophisticated propositions. In IV.20 Jordanus presents three solutions for the problem of trisecting an angle, and IV.22 offers two solutions for finding two mean proportionals between two given lines. A proof of Hero's theorem on the area of a triangle—$A = \sqrt{s(s-a)(s-b)(s-c)}$, where s is the semiperimeter and a, b, and c are the sides of the triangle—may also have been associated with the *De triangulis*. Jordanus drew his solutions largely from Latin translations of Arabic works, which were themselves based on Greek mathematical texts. He did not always approve of these proofs and occasionally displayed a critical spirit, as when he deemed two proofs of the trisection of an angle based on mechanical means inadequate and uncertain (although no source is mentioned, they were derived from the *Verba filiorum* of the Banū Mūsā) and offered what is apparently his own demonstration,[18] in which a proposition from Ibn al-Haytham's *Optics* is utilized. In IV.16 a non-Archimedean proof of the quadrature of the circle may have been original. It involves finding a third continuous proportional. Here is the proof.[19]

To Form a Square Equal to a Given Circle.

For example, let the circle be A [see Figure 5].

Disposition: Let another circle B with its diameter be added; let a square be described about each of those circles. And the circumscribed square [in each case] will be as a square of the diameter of the circle. Hence, by [Proposition] XII.2 [of the *Elements*], circle A/circle B = square DE/square FG. Therefore, by permutation, $DE/A = FG/B$. Let there be formed a third surface C, which is a [third] proportional [term] following DE and A. Now C will either be a circle or a surface of another kind, like a rectilinear surface. In the first place, let it be a circle which is circumscribed by square HK. And so, $DE/A = A/C$ but also, by [Proposition] XII.2 [of the *Elements*], $DE/A = HK/C$. Therefore, HK as well as A is a mean proportional between DE and C. Therefore, circle A and square HK are equal, which we proposed.

Next, let C be some [rectilinear] figure other than a circle. Then let it be converted into a square by the last [proposition] of [Book] II [of the *Elements*], with its angles designated as R, S, Y, and X. And so, since DE is the larger extreme [among the three proportional terms], DE is greater than square RY, and a side [of DE] is greater than a side [of RY]. Therefore, let MT, equal to RX, be cut from MD. Then a parallelogram MN—contained by ME and MT—is described. Therefore, MN is the mean proportional between DE and RY, which are the squares of its sides, since a rectangle is the mean proportional between the squares of its sides. But circle A was the mean proportional between them [i.e., between square DE and C (or square RY)]. Therefore, circle A and parallelogram MN are equal. Therefore, let MN be converted to a square by the last [proposition] of [Book] II [of the *Elements*], and this square will be equal to the given circle A, which we proposed.

The *De numeris datis*, Jordanus' algebraic treatise in four books, which was praised by Regiomontanus, was more formal and Euclidean than the algebraic treatises derived from Arabic sources. Indeed it has recently been claimed[20] that in the *De numeris datis* Jordanus anticipated Viète in the application of analysis to algebraic problems. This may be seen in Jordanus' procedure, where he regularly formulated problems in terms of what is known and what is to be found (this is tantamount to the construction of an equation), and subsequently transforms the initial equation into a final form from which a specific computation is made with determinate numbers that meet the general conditions of the problem.

JORDANUS DE NEMORE

The general pattern of every proposition is thus (1) formal enunciation of the proposition; (2) proof; and (3) a numerical example, which is certainly non-Euclidean and was perhaps patterned after Arabic algebraic treatises. In this wholly rhetorical treatise, Jordanus used letters of the alphabet to represent numbers. An unknown number might be represented as *ab* or *abc*, which signify $a + b$ and $a + b + c$ respectively. Occasionally, when two unknown numbers are involved, one would be represented as *ab*, the other as *c*.

Typical of the propositions in the *De numeris datis* are Book I, Proposition 4, and Book IV, Proposition 7. In the first of these, a given number, say s,[21] is divided into numbers x and y, whose values are to be determined. It is assumed that g, the sum of x^2 and y^2, is also known. Now $s^2 - g = 2xy = e$ and $g - e = h = (x - y)^2$. Therefore $(x - y) = \sqrt{h} = d$. Since d is the difference between the unknown numbers x and y, their values can be determined by Book I, Proposition 1, where Jordanus demonstrated that "if a given number is divided in two and their difference is given, each of them will be given." The numbers are found from their sum and difference. Since $x + y = s$ and $x - y = d$, it follows that $y + d = x$ and, therefore, $2y + d = x + y = s$; hence $2y = s - d$, $y = (s - d)/2$, and $x = s - y$. Should we be given the ratio between x and y, say r, and the product of their sum and difference, say p, the values of x and y are determinable by Book IV, Proposition 7, as follows: since $x/y = r$, $x^2/y^2 = r^2$; moreover, since $(x + y)\ (x - y) = p$, therefore $x^2 - y^2 = p$. Now $x^2 = r^2y^2$, so that $(r^2 - 1)y^2 = p$ and $y = \sqrt{p/(r^2 - 1)}$. In the numerical example $r = 3$ and $p = 32$, which yields $y = 2$ and $x = 6$.

In the *Arithmetica*, the third and probably most widely known of his three major mathematical works, Jordanus included more than 400 propositions in ten books which became the standard source of theoretical arithmetic in the Middle Ages. Proceeding by definitions, postulates, and axioms, the *Arithmetica* was modeled after the arithmetic books of Euclid's *Elements*, a treatise which Jordanus undoubtedly used, although the proofs frequently differ. Jordanus' *Arithmetica* contrasts sharply with the popular, nonformal, and often philosophical *Arithmetica* of Boethius. A typical proposition, which has no counterpart in Euclid's *Elements*, is Book I, Proposition 9:

> *The* [*total sum or*] *result of the multiplication of any number by however many numbers you please is equal to* [*est quantum*] *the result of the multiplication of the same number by the number composed of all the others.*
>
> Let A be the number multiplied by B and C to produce D and E [respectively]. I say that the composite [or sum] of D and E is produced by multiplying A by the composite of B and C. For it is obvious by Definition [7] that B measures [*numerat*] D A times and that C measures E by the same number, namely, A times. By the sixth proposition of this book, you will easily be able to argue this.[22]

Thus Jordanus proves that if $A \cdot B = D$ and $A \cdot C = E$, then $D + E = A(B + C)$. By Definition 7, $D/B = A$ and $E/C = A$. And since D and E are equimultiples of B and C, respectively, then, by Proposition 6, it follows that $B + C = \frac{1}{n}(D + E)$ and, assuming $n = A$, we obtain $A(B + C) = D + E$.

The Arabic number system also attracted Jordanus' attention—if the *Demonstratio Jordani de algorismo* and a possible earlier and shorter version of it, the *Opus numerorum*, are actually by Jordanus. Once again Jordanus proceeded by definitions and propositions in a manner that differed radically from Johannes Sacrobosco's *Algorismus vulgaris*, or *Common Algorism*. Unlike Sacrobosco, Jordanus described the arithmetic operations and extraction of roots succinctly and formally and without examples. Among the twenty-one definitions of the *Demonstratio Jordani* are those for addition, doubling, halving, multiplication, division, extraction of a root (these definitions are illustrated as propositions), simple number, composite number, digit, and article (which is ten or consists of tens). Propositions equivalent to the following are included:

3. If $a : b = c : d$, then $a \cdot 10^n : b = c \cdot 10^n : d$
12. $1 \cdot 10^n + 9 \cdot 10^n = 1 \cdot 10^{n+1}$
19. $a \cdot 10^n + b \cdot 10^n = (a + b)10^n$
32. If $a_1 = a \cdot 10$, $a_2 = a \cdot 100$, $a_3 = a \cdot 1{,}000$, then $(a \cdot a_1)/a^2 = (a_1 \cdot a_2)/a_1^2 = (a_2 \cdot a_3)/a_2^2 = \cdots$.

An algorithm of fractions, called *Liber* or *Demonstratio de minutiis* in some manuscripts, may also have been written by Jordanus. It describes in general terms arithmetic operations with fractions alone and with fractions and integers. He also composed a *Liber de proportionibus*, a brief treatise containing propositions akin to those in Book V of Euclid's *Elements*.

NOTES

1. Marshall Clagett, *The Science of Mechanics in the Middle Ages*, pp. 72–73.
2. Leopold Delisle, *Le cabinet des manuscrits de la Bibliothèque nationale*, II (Paris, 1874), 526, 527.
3. This has been suggested by O. Klein, "Who Was Jordanus Nemorarius?," in *Nuclear Physics*, **57** (1964), 347.
4. Maximilian Curtze, "Jordani Nemorarii Geometria vel De triangulis libri IV," in *Mitteilungen des Coppernicus-Vereins*, **6** (1887), iv, n. 2.
5. *Ibid.*, p. vi.

JORDANUS DE NEMORE

6. Clagett, *op. cit.*, p. 73.
7. Translated by E. A. Moody and M. Clagett, *The Medieval Science of Weights*, p. 129.
8. *Ibid.*, p. 131.
9. 850b.4–6 in the translation of E. S. Forster (Oxford, 1913).
10. Moody and Clagett, trans., *op. cit.*, pp. 139, 141.
11. A position adopted by Moody, *ibid.*, pp. 171–172.
12. Joseph E. Brown, *The Scientia de ponderibus in the Later Middle Ages*, pp. 64–66.
13. Clagett, *op. cit.*, pp. 76–77.
14. Moody and Clagett, trans., *op. cit.*, p. 191.
15. *Ibid.*, pp. 185, 187.
16. Clagett, *op. cit.*, p. 82.
17. Translated by Stillman Drake in *Mechanics in Sixteenth-Century Italy*, trans. and annotated by Stillman Drake and I. E. Drabkin (Madison, Wis., 1969), p. 295.
18. Marshall Clagett, *Archimedes in the Middle Ages*, I, 675.
19. Trans. by Clagett, *ibid.*, pp. 573–575.
20. Barnabas B. Hughes (ed. and trans.), *The De numeris datis of Jordanus de Nemore*, pp. 50–52.
21. In his ed. of the *De numeris datis*, Curtze altered the letters in presenting the analytic summaries of the propositions; in a few instances I have altered Curtze's letters.
22. My translation from Edward Grant, ed., *A Source Book in Medieval Science* (in press).

BIBLIOGRAPHY

I. ORIGINAL WORKS. In Richard Fournival's *Biblionomia* twelve works are attributed to Jordanus. In codex 43 we find (1) *Philotegni*, or *De triangulis*; (2) *De ratione ponderum*; (3) *De ponderum proportione*; and (4) *De quadratura circuli*. In codex 45 three works are listed: (5) *Practica*, or *Algorismus;* (6) *Practica de minutiis;* and (7) *Experimenta super algebra*. Codex 47 contains the lengthy (8) *Arithmetica*. Codex 48 includes (9) *De numeris datis;* (10) *Quedam experimenta super progressione numerorum;* and (11) *Liber de proportionibus*. Codex 59 includes a treatise called (12) *Suppletiones plane spere*.

Numbers (2) and (3) are obviously statical treatises. The *De ratione ponderum* is probably the *De ratione ponderis* edited and translated by E. A. Moody in E. A. Moody and M. Clagett, *The Medieval Science of Weights* (*Scientia de ponderibus*) (Madison, Wis., 1952); its attribution to Jordanus has been questioned by Joseph E. Brown, *The Scientia de ponderibus in the Later Middle Ages* (Ph.D. diss., University of Wis., 1967), pp. 64–66. In the same volume Moody has also edited and translated the *Elementa Jordani super demonstrationem ponderum*, a genuine work of Jordanus, which perhaps corresponds to the *De ponderum proportione* of the *Biblionomia*.

The *Philotegni*, or *De triangulis*, exists in two versions. The longer, and apparently later, version was published by Maximilian Curtze, "Jordani Nemorarii Geometria vel De triangulis libri IV," in *Mitteilungen des Coppernicus-Vereins für Wissenschaft und Kunst zu Thorn*, **6** (1887), from MS Dresden, Sächsische Landesbibliothek, Db 86, fols. 50r–61v. Utilizing additional MSS, Marshall Clagett reedited and translated Props. IV.16 (quadrature of the circle), IV.20 (trisection of an angle), and IV.22 (finding of two mean proportionals) in his *Archimedes in the Middle Ages*, I, *The Arabo-Latin Tradition* (Madison, Wis., 1964), 572–575, 672–677, and 662–663, respectively. A shorter version, which lacks Props. II.9–12, 14–16, and IV.10 and terminates at IV.9 or IV.11, has been identified by Clagett. Both versions will be reedited by Clagett in vol. IV of his *Archimedes in the Middle Ages*. Of the 17 MSS of the two versions which Clagett has found thus far, we may note, in addition to the Dresden MS used by Curtze, the following: Paris, BN lat. 7378A, 29r–36r; London, Brit. Mus., Sloane 285, 80r–92v; Florence, Bibl. Naz. Centr., Conv. Soppr. J. V. 18, 17r–29v; and London, Brit. Mus., Harley 625, 123r–130r.

The *De quadratura circuli* attributed to Jordanus as a separate treatise in Fournival's *Biblionomia* may be identical with Bk. IV, Prop. 16 of the *De triangulis*, which bears the title "To Form a Square Equal to a Given Circle" (quoted above; for the Latin text, see Clagett, *Archimedes in the Middle Ages*, I, 572, 574). In at least one thirteenth-century MS (Oxford, Corpus Christi College 251, 84v) the proposition stands by itself, completely separated from the rest of the *De triangulis*, an indication that it may have circulated independently (for other MSS, see Clagett, *Archimedes*, I, 569).

The *De numeris datis* has been edited three times. It was first published on the basis of a single fourteenth-century MS, Basel F.II.33, 138v–145v, by P. Treutlein, "Der Traktat des Jordanus Nemorarius 'De numeris datis,' " in *Abhandlungen zur Geschichte der Mathematik*, no. 2 (Leipzig, 1879), pp. 125–166. Relying on MS Dresden Db86, supplemented by MS Dresden C80, Maximilian Curtze reedited the *De numeris datis* and subdivided it into four books in "Commentar zu dem 'Tractatus de numeris datis' des Jordanus Nemorarius," in *Zeitschrift für Mathematik und Physik*, hist.-lit. Abt., **36** (1891), 1–23, 41–63, 81–95, 121–138. In MS Dresden C80 Curtze found additional propositions (IV.16–IV.35) beyond the concluding proposition in Treutlein's ed. The additional propositions included no proofs but only the enunciations of the propositions followed immediately by a single numerical example for each. That these extra propositions formed a genuine part of the *De numeris datis* was verified by MSS Vienna 4770 and 5303, which included not only the additional propositions but also their proofs. Using MS Vienna 4770, from which 5303 was copied, R. Daublebsky von Sterneck published complete versions of Props. IV.15–IV.35 and also supplied corrections and additions to a few propositions in Bk. I in "Zur Vervollständigung der Ausgaben der Schrift des Jordanus Nemorarius: 'Tractatus de numeris datis,' " in *Monatshefte für Mathematik und Physik*, **7** (1896), 165–179. A third ed., with the first English trans., has been completed by Barnabas Hughes: *The De numeris datis of Jordanus de Nemore, a Critical Edition, Analysis, Evaluation and Translation* (Ph.D. diss., Stanford University, 1970). Hughes's thorough study also includes (pp. 104–126) a history of previous editions, as well as a description of twelve MSS, whose relationships are discussed in detail.

A Russian translation from Curtze's edition was made by S. N. Šreĭder, "The Beginnings of Algebra in Medieval Europe in the Treatise De numeris datis of Jordanus de Nemore," in *Istoriko-Matematicheskie issledovaniya*, **12** (1959), 679–688.

As yet there is no ed. of the ten-book *Arithmetica*, although the enunciations of the propositions were published by Jacques Lefèvre d'Étaples (Jacobus Faber Stapulensis), who supplied his own demonstrations and comments in *Arithmetica (Iordani Nemorarii) decem libris demonstrata* (Paris, 1496, 1503, 1507, 1510, 1514). At least sixteen complete or partial MSS of it are presently known, among which are two excellent and complete thirteenth-century MSS: Paris, BN 16644, 2r–93v; and Vat. lat., Ottoboni MS 2069, 1r–51v.

The Latin text of the definitions and enunciations of the 34 propositions of the *Demonstratio Jordani de algorismo* were published by G. Eneström, "Über die 'Demonstratio Jordani de algorismo,'" in *Bibliotheca mathematica*, 3rd ser., **7** (1906–1907), 24–37, from MSS Berlin, lat. 4° 510, 72v–77r (Königliche Bibliothek, renamed Preussische Staatsbibliothek in 1918; the fate of this codex after World War II, when the basic collection was divided between East and West Germany, is unknown to me) and Dresden Db 86, 169r–175r. The *Demonstratio* appears to be an altered version of a similar and earlier work beginning with the words "Communis et consuetus . . .," which Eneström called *Opus numerorum*. The Latin text of its introduction and a comparison of its propositions with those of the *Demonstratio Jordani* were published by Eneström as "Über eine dem Jordanus Nemorarius zugeschriebene kurze Algorismusschrift," in *Bibliotheca mathematica*, 3rd ser., **8** (1907–1908), 135–153. He relied primarily on MS Vat. lat. Ottob. 309, 114r–117r, supplemented by MSS Vat. lat. Reg. Suev. 1268, 69r–71r; Florence, Bibl. Naz. Centr., Conv. Soppr. J.V. 18 (cited by Eneström as San Marco 216, its previous designation), 37r–39r; and Paris, Mazarin 3642, 96r and 105r. Since the *Demonstratio Jordani* was definitely ascribed to Jordanus, and the *Opus numerorum* seemed an earlier version of it, Eneström conjectured that the *Opus* was a more likely candidate for Jordanus' original work, while the *Demonstratio Jordani*, which omits most of the introduction but expands the text itself, may have been revised by Jordanus or someone else.

Each of these two treatises has associated with it a brief work, attributed in some MSS to Jordanus, on arithmetic operations with fractions. The treatise associated with the *Opus numerorum*, which Eneström calls *Tractatus minutiarum*, contains an introduction and 26 highly abbreviated propositions; the work on fractions associated with the *Demonstratio Jordani de algorismo*, which Eneström calls *Demonstratio de minutiis*, consists of an introduction and 35 propositions. Although the introductions differ, all 26 propositions of the *Tractatus minutiarum* have, according to Eneström, identical counterparts in the longer *Demonstratio de minutiis*. In "Das Bruchrechnen des Jordanus Nemorarius," in *Bibliotheca mathematica*, 3rd ser., **14** (1913–1914), 41–54, Eneström includes a list of MSS for both treatises (pp. 41–42), the Latin texts of the introductions, the texts of the enunciations of the propositions, and analytic representations of the propositions. By analogy with his reasoning about the relations obtaining between the *Opus numerorum* and *Demonstratio Jordani de algorismo*, Eneström conjectures that Jordanus is the author of the *Tractatus minutiarum*, the briefer treatise associated with the *Opus numerorum*. One of the MSS is Bibl. Naz. Centr., Conv. Soppr. J.V. 18, 39r–42v, which follows immediately after the *Opus numerorum* in the same codex cited above; correspondingly, MS Berlin, lat. 4° 510, 72v–77r, of the *Demonstratio Jordani de algorismo* is followed immediately by a version of the *Demonstratio de minutiis* on fols. 77r–81v, a relation which also seems to obtain in Bibl. Naz. Centr., Conv. Soppr. J.I. 32, 113r–118v, 118v–124r. Whether the two algorithm treatises and the two associated treatises on fractions bear any relation to works (5), (6), (7), or (10), cited above from the *Biblionomia*, has yet to be determined and may, indeed, be impossible to determine. The *Algorismus demonstratus* published in 1534 by J. Schöner and formerly ascribed to Jordanus, was composed by a Master Gernardus, who is perhaps identical with Gerard of Brussels.

The *Liber de proportionibus*, mentioned in the *Biblionomia*, is probably a brief work by Jordanus beginning with the words "Proportio est rei ad rem determinata secundum quantitatem habitudo . . . ." A seemingly complete MS of it is Florence, Bibl. Naz. Centr., Conv. Soppr. J.V. 30, 8r–9v. Other MSS are listed in L. Thorndike and P. Kibre, *A Catalogue of Incipits of Mediaeval Scientific Writings in Latin*, rev. ed. (Cambridge, Mass., 1963), col. 1139. The *Suppletiones plane spere* of the *Biblionomia* is probably a commentary on Ptolemy's *Planisphaerium*. According to G. Sarton, *Introduction to the History of Science*, 3 vols. in 5 pts., II, pt. 2 (Baltimore, 1931), 614, it is "a treatise on mathematical astronomy, which contains the first general demonstration of the fundamental property of stereographic projection—i.e., that circles are projected as circles (Ptolemy had proved it only in special cases)." In Thorndike and Kibre, *op. cit.*, Jordanus' *Planisphaerium* is listed under three separate and different incipits (see cols. 1119, 1524, and 1525, where MSS are listed for each). An edition appeared at Venice in 1558, under the title *Ptolemaei Planisphaerium: Iordani Planisphaerium; Federici Commandi Urbinatis in Ptolemaei Planisphaerium commentarius.* A work on isoperimetric figures, *De figuris ysoperimetris*, is also attributed to Jordanus: MSS Florence, Bibl. Naz. Centr., Conv. Soppr. J.V. 30, 12v (a fragment) and Vienna 5203, 142r–146r, the latter actually copied by Regiomontanus, who was also acquainted with Jordanus' *De triangulis*, *Planisphaerium*, *Arithmetica*, *De numeris datis*, and *De proportionibus;* the enunciations of the eight propositions in the Vienna MS were published by Maximilian Curtze, "Eine Studienreise," in *Zentralblatt für Bibliothekswesen*, **16** (1899), 264–265.

II. Secondary Literature. The most significant studies on Jordanus are monographic in character and have been cited above, since they are associated with editions and translations of his works. No general appraisal and evaluation of his scientific works has yet been published. To what has already been cited, the following may be added: O. Klein, "Who Was Jordanus Nemorarius?," in *Nuclear Physics*, **57** (1964), 345–350; Benjamin Ginzberg, "Duhem and Jordanus Nemorarius," in *Isis*, **25** (1936), 341–362, which seeks to refute Duhem's claims for medieval science and for Jordanus' statics in particular (Ginzberg seriously misread Duhem and was also ignorant of Jordanus' subsequent impact on later statics, believing mistakenly that all of it was rediscovered); M. Clagett, *The Science of Mechanics in the Middle Ages* (Madison, Wis., 1959), ch. 2, which is a summary of medieval contributions in statics, including source selections from the works of Jordanus; Joseph E. Brown, *The Scientia de ponderibus in the Later Middle Ages* (Ph.D. diss., University of Wis., 1967), which includes summaries and evaluations of the major principles in Jordanus' statical treatises and their subsequent influence in the commentary literature; and G. Wertheim, "Über die Lösung einiger Aufgaben in *De numeris datis*," in *Bibliotheca mathematica*, **1** (1900), 417–420. Additional bibliography is given in Sarton, *op. cit.*, II, pt. 2, 614–616.

EDWARD GRANT

JØRGENSEN, SOPHUS MADS (*b.* Slagelse, Denmark, 4 July 1837; *d.* Copenhagen, Denmark, 1 April 1914), *chemistry.*

The son of Jens Jørgensen, a tailor, and Caroline Grønning, Jørgensen attended school in Slagelse and later studied at the Sorø *Velvillie*. In 1857 he entered the University of Copenhagen, from which he received his master's degree in chemistry (1863) and his doctorate (1869) with the dissertation *Overjodider af Alkaloiderne* ("Polyiodides of Alkaloids"). At the university he became assistant (1864) to Edward Augustus Scharling and director of the chemical laboratories (1867). In 1867 he was also appointed *Laerer* at the Technical University. In 1871, after an engagement of five years, he married Louise Wellmann and also became *Lektor* at the university. In 1887 he was appointed professor of chemistry, a position which he held until his retirement in 1908.

Except for some early isolated organic and inorganic research, Jørgensen devoted himself exclusively to investigating the coordination compounds of cobalt, chromium, rhodium, and platinum; this work, upon which his fame rests, forms an interconnected and continuous series from 1878 to 1906. Jørgensen created no new structural theory of his own. His interpretations of the luteo (hexaammines), purpureo (halopentaammines), roseo (aquopentaammines), praseo (*trans*-dihalotetraammines), violeo (*cis*-dihalotetraammines), croceo (*trans*-dinitrotetraammines), flavo (*cis*-dinitrotetraammines), and other series of coordination compounds were made in light of his logical extensions and modifications of the famous chain theory proposed by the Swedish chemist Christian Wilhelm Blomstrand.

For fifteen years Jørgensen's views remained the most acceptable of the numerous theories advanced to explain the properties and reactions of the so-called molecular compounds, which were not explicable under the contemporary valence theory. In 1893 Alfred Werner, an unknown twenty-six-year-old privatdocent at the Eidgenössische Polytechnikum in Zurich, challenged the old system with his radically new coordination theory [*Zeitschrift für anorganische Chemie*, **3** (1893), 267–330]. The ensuing controversy between Jørgensen and Werner constitutes an excellent example of the synergism so often encountered in the history of science. Jørgensen regarded Werner's theory as an ad hoc explanation insufficiently supported by experimental evidence and an unwarranted break in the development of theories of chemical structure. In their scholarly rivalry each chemist did the utmost to prove his views, and in the process a tremendous amount of fine experimental work was performed by both. Although not all of Jørgensen's criticisms [*ibid.*, **19** (1899), 109–157] were valid, Werner, in a number of cases, was forced to modify various aspects of his theory.

Werner's ideas eventually triumphed, yet Jørgensen's experimental observations were in no way invalidated. On the contrary, his experiments, performed with extreme care, have proved completely reliable in most cases and form the foundation not only of the now obsolete Blomstrand–Jørgensen theory but also of Werner's coordination theory (a debt acknowledged by Werner). It is perhaps not an exaggeration to state that Werner's theory might never have been propounded had not Jørgensen's experimental work provided the observations requiring such explanation. Ironically enough, Jørgensen's work bore the seeds that undid the Blomstrand–Jørgensen theory, for many of the compounds first prepared by him later proved instrumental in demonstrating the validity of Werner's views. When Werner finally succeeded in 1907 in preparing the long-sought *cis*-dichlorotetraamminecobalt(III) salts [*Berichte der Deutschen Chemischen Gesellschaft*, **40** (1907), 4817–4825], whose existence was a necessary consequence of his theory but not of Blomstrand's, Jørgensen graciously capitulated.

A solitary research worker, Jørgensen was methodical and painstaking. Although he could have delegated much routine work to assistants, he insisted on personally performing all his analyses, a task for which he reserved one day a week. In terms of Wilhelm Ostwald's twofold classification of scientific genius, classic vis-à-vis romantic [E. Farber, *Journal of Chemical Education*, **30** (1953), 600–604], Jørgensen would seem to be the classic type—the slow and deep-digging scientist who proceeds with careful deliberation and completes a traditional theory or develops it to new consequences. Considering his passion for perfection, his research output was tremendous; and we are indebted to him for many of the basic experimental facts of coordination chemistry.

BIBLIOGRAPHY

The majority of Jørgensen's papers appeared in the *Journal für praktische Chemie* (Leipzig), until the founding of the *Zeitschrift für anorganische Chemie* (1892), after which they began to appear in the latter journal. For a complete list of Jørgensen's publications (seventy-six papers and nineteen volumes), see Stig Veibel, *Kemien i Danmark*, II (Copenhagen, 1953).

The details of Jørgensen's life and a critical discussion of his work and controversy with Alfred Werner are given in G. B. Kauffman, "Sophus Mads Jørgensen (1837–1914): A Chapter in Coordination Chemistry History," in *Journal of Chemical Education*, **36** (1959), 521–527, repr. in A. J. Ihde and W. F. Kieffer, eds., *Selected Readings in the History of Chemistry* (Easton, Pa., 1965), pp. 185–191. For a fuller account see G. B. Kauffman, "Sophus Mads Jørgensen and the Werner–Jørgensen Controversy," in *Chymia*, **6** (1960), 180–204.

A little-known obituary and evaluation of Jørgensen's work by his chief scientific adversary is given in A. Werner, *Chemiker-Zeitung*, **38** (1914), 557–564. Biographical data in Danish can be found in S. P. L. Sørensen, *Fysisk Tidsskrift*, **12** (1913–1914), 217; and *Oversigt over det K. Danske Videnskabernes Selskabs Forhandlinger*, **46–49** (1914); and S. Veibel, *Dansk biografisk leksikon*, XII (Copenhagen, 1937), 253–256.

GEORGE B. KAUFFMAN

JOULE, JAMES PRESCOTT (*b.* Salford, near Manchester, England, 24 December 1818; *d.* Sale, England, 11 October 1889), *physics.*

Joule's ancestors were Derbyshire yeomen; his grandfather had become wealthy as the founder of a brewery at Salford. James was the second of five children of Benjamin and Alice Prescott Joule. Together with his elder brother, James received his first education at home. From 1834 to 1837 the two brothers were privately taught elementary mathematics, natural philosophy, and some chemistry by John Dalton, then about seventy years old.

James never took part in the management of the brewery or engaged in any profession. He shared his father's Conservative allegiance and entertained conventional Christian beliefs. He married Amelia Grimes, of Liverpool, in 1847, but she died in 1854. He spent the rest of his life with his two children in various residences in the neighborhood of Manchester. He had a shy and sensitive disposition, and his health was delicate.

Joule's pioneering experiments were carried out in laboratories he installed at his own expense in his successive houses (or in the brewery). Later, owing to financial losses, he could no longer afford to work on his own and received some subsidies from scientific bodies for his last important investigations. His friends eventually procured him a pension from the government, in 1878, but by then his mental powers had begun to decline. He died after a long illness.

Joule's scientific career presents two successive periods of very different character. During the decade 1837–1847, he displayed the powerful creative activity that led him to the recognition of the general law of energy conservation and the establishment of the dynamical nature of heat. After the acceptance by the scientific world of his new ideas and his election to the Royal Society (1850), he enjoyed a position of great authority in the growing community of scientists.

Joule carried on for almost thirty years a variety of skillful experimental investigations; none of them, however, was comparable to the achievements of his youth. His insufficient mathematical education did not allow him to keep abreast of the rapid development of the new science of thermodynamics, to the foundation of which he had made a fundamental contribution. Here Joule's fate was similar to that of his German rival Robert Mayer. By the middle of the century, the era of the pioneers was closed, and the leadership passed to a new generation of physicists who possessed the solid mathematical training necessary to bring the new ideas to fruition.

Joule began independent research at the age of nineteen under the influence of William Sturgeon, a typical representative of those amateur scientists whose didactic and inventive activities were supported by the alert tradesmen of the expanding industrial cities of England. Taking up Sturgeon's interest in the development of electromagnets and electromagnetic engines, the young Joule at once transformed a rather dilettantish effort into a serious scientific investigation by introducing a quantitative analysis of the "duty," or efficiency, of the designs he tried. This was a far from trivial step, since it implied defining, for the various magnitudes involved, the standards and units that were still almost entirely lacking in voltaic electricity and magnetism. Joule's preoccupation with this fundamental aspect of physical science is apparent throughout his work and culminated with the precise determination of the mechanical equivalent of heat.

At first Joule was so far removed from any idea of equivalence between the agencies of nature that for a while he hoped that electromagnets could become a source of indefinite mechanical power. He found their mutual attraction to be proportional to the square of the intensity of the electric current, whereas the chemical power necessary to produce the current in the batteries was simply proportional to the intensity. But he soon learned of the counter-induction effect discovered by M. H. Jacobi, which set a limit to the efficiency of electromagnetic engines. Subjecting the question to quantitative measurement, he realized, much to his dismay, that the mechanical effect of the current would always be proportional to the expense of producing it, and that the efficiency of the electromagnetic engines that he could build would be much lower than that of the existing steam engines. He presented this pessimistic conclusion in a public lecture (1841) at the Victoria Gallery in Manchester (one of Sturgeon's short-lived educational ventures).

Joule's early work, although rather immature, exhibited features that persisted in all his subsequent investigations and that unmistakably revealed Dalton's influence. Adopting Dalton's outlook, Joule believed that natural phenomena are governed by "simple" laws. He designed his experiments so as to discriminate among the simplest relations which could be expected to connect the physical quantities describing the effect under investigation; in fact, the only alternative that he ever contemplated was between a linear or a quadratic relation. This explains the apparent casualness of his experimental arrangements, as well as the assurance with which he drew sweeping conclusions from very limited series of measurements. In the search for simple physical laws, Joule necessarily relied on theoretical representations. We find the first explicit mention of these in the Victoria Gallery lecture, where Joule operated with a crude, but quite effective, atomistic picture of matter. His views embodied then-current ideas about the electric nature of the chemical forces and the electrodynamic origin of magnetization, as well as the concept of heat as a manifestation of vibratory motions on the atomic scale.

Abandoning hope of exploiting electric current as a source of power, Joule decided to study the thermal effects of voltaic electricity. Indirect evidence strongly suggests that this choice was motivated by the wish to enter a field of investigation made "respectable" by Faraday's example. Yet whatever expectations he had in this respect were quickly dashed by the Royal Society's frigid reception of his first paper and he turned again to the more sympathetic audience he found in the Manchester Literary and Philosophical Society.

Joule derived the quantitative law of heat production by a voltaic current—its proportionality with the square of the intensity of the current and with the resistance—from a brief series of measurements of the simplest description: he dipped a coiled portion of the circuit into a test tube filled with water and ascertained the slight changes of temperature of the water for varying current intensity and resistance (December 1840). The critical step in these, as well as in all his further experiments, was the measurement of small temperature variations; Joule's success crucially depended on the use of the best available thermometers, sensitive to about a hundredth of a degree. To outsiders, who could not be aware of his extraordinary skill and accuracy, and failed to appreciate the logic underlying the design of his experiments, Joule's derivation of statements of utmost generality from a few readings of minute temperature differences was bound to appear too rash to be readily trusted. Joule's self-confidence may be understood only by realizing that his experimental work was deliberately directed toward testing the theoretical conceptions gradually taking shape in his mind.

During the next two years Joule made a systematic study of all the thermal effects accompanying the production and passage of the current in a voltaic circuit. From this study, completed by January 1843, he obtained a clear conception of an equivalence between each type of heat production and a corresponding chemical transformation or resistance to the passage of the current. Regarding the nature of heat, no conclusion could be derived from the phenomena of the voltaic circuit: voltaic electricity was "a grand agent for carrying, arranging and converting chemical heat"; but this heat could either be some substance simply displaced and redistributed by the current, or arise from modifications of atomic motions inseparable from the flow of the current.

Joule saw the possibility of settling this last question—and at the same time of subjecting the equivalence idea to a crucial test—by extending the investigation to currents not produced by chemical change but induced by direct mechanical effect. This brilliant inference led him to the next set of experiments, among the most extraordinary ever conceived in physics. He enclosed the revolving armature of an electromagnetic engine in a cylindrical container filled with a known amount of water and rotated the whole apparatus during a given time between the poles of the fixed electromagnet, ascertaining the small change of temperature of the water: the heat produced in this way could only be dynamical in origin. Moreover, by studying the heating effects of the induced current, to which a voltaic one was added or subtracted, he established, by a remarkably rigorous argument, the strict equivalence of the heat produced on revolving the coil and the mechanical work spent in the operation. He thus obtained a first determination of the coefficient of equivalence (1843).

After this accomplishment, his last series of experiments concerned with the mechanical equivalent of heat—those described in every elementary textbook—appear rather pedestrian by comparison, although they offer further examples of Joule's virtuosity as an experimenter. They consist in direct measurements of the heat produced or absorbed by mechanical processes: the expansion and compression of air (1845) and the friction of rotating paddle wheels in water and other liquids (1847). The experiments with air are of special interest because they were based on the same argument used by Mayer in his own derivation of the equivalent (letter to Baur, September 1841). But while Joule performed all the necessary experiments himself, Mayer made an extremely skillful use of available experimental results—most notably the difference of the specific heats at constant pressure and constant volume, and Gay-Lussac's little-known demonstration (1806) that if a gas expands without doing work, its temperature remains constant. This law (which, strictly speaking, applies only to ideal gases) is usually ascribed to Joule—not without justification, since his experiment was much more accurate than Gay-Lussac's.

Joule did not announce his momentous conclusions to a wider audience before he had completed single-handed all his painstaking measurements. Significantly, he did not venture outside his familiar Manchester environment. He simply gave a public lecture in the reading room of St. Ann's Church (May 1847) and was content to have the text of his address published in the *Manchester Courier* (a newspaper for which his brother wrote musical critiques). This synthetic essay, entitled "On Matter, Living Force, and Heat," gave the full measure of his creative imagination. In a few pages of limpid, straightforward description, he managed to draw a vivid picture of the transformation of "living force" into work and heat and to pass on to the kinetic view of the nature of heat and the atomic constitution of matter.

At the same time, he did not neglect to present a more technical account of his work before the scientific public. In particular, he reported his final determinations of the equivalent to the French Academy of Sciences, and presented this learned body with the iron paddle-wheel calorimeter he had used in the case of mercury. In contrast to previous occasions, Joule's report to the British Association meeting at Oxford (June 1847) met with a lively response from the twenty-two-year-old William Thomson, an academically trained physicist who was better prepared than his elders to receive fresh ideas. How this dramatic encounter stimulated Thomson to formulate his own theory of thermodynamics is a story that no longer belongs to Joule's biography. Indeed, the very moment of Joule's belated recognition marked the end of his influence on scientific progress. Although Thomson had the highest regard for Joule's experimental virtuosity, and repeatedly enlisted him in undertakings that required measurements of high accuracy, the scope of Thomson's research was no longer within Joule's full grasp.

The only substantial contribution to thermodynamics to which the joint names of Joule and Thomson are attached belongs to an idea conceived by Thomson, who saw the possibility of analyzing the deviations of gas properties from the ideal behavior. In particular, a non-ideal gas, made to expand slowly through a porous plug (so as to approximate a specified mathematical condition—constant enthalpy), would in general undergo a cooling (essentially a transformation of atomic motion into work spent against the interatomic attractions). For the delicate test of this effect Thomson required Joule's unsurpassed skill (1852). But the application of the Joule-Thomson effect to the technology of refrigeration belongs to a later stage in the development of thermodynamics.

In 1867 Joule was induced to carry out two high-precision determinations of the equivalent on behalf of the British Association Committee on Standards of Electrical Resistance. The first experiment, based on the thermal effect of currents, was designed by Thomson to test the proposed resistance standard. Because his result showed a 2 percent discrepancy from the original paddle-wheel calorimeter determination, Joule was asked to repeat the latter. He did so in painstaking experiments from 1875 to 1878 and fully confirmed his previous value. Joule's results thus displayed the necessity of improving the resistance standard. This was Joule's last contribution to the science his pioneering work had initiated.

BIBLIOGRAPHY

I. Original Works. See *The Scientific Papers of James Prescott Joule*, 2 vols. (London, 1884–1887).

II. Secondary Literature. Information on Joule may be found in Osborne Reynolds, "Memoir of James Prescott Joule," in *Memoirs and Proceedings of the Manchester Literary and Philosophical Society*, 4th ser., **6** (1892); and J. C. Crowther, *British Scientists of the Nineteenth Century* (London, 1935), ch. 3.

L. Rosenfeld

JUAN Y SANTACILLA

JUAN Y SANTACILLA, JORGE (*b.* Novelda, Alicante, Spain, 5 January 1713; *d.* Madrid, Spain, 21 July 1773), *geodesy.*

Juan's parents, Bernardo Juan y Cancia and Violante Santacilla y Soler, were hidalgos, the lower aristocracy. Orphaned at the age of three, he first attended school in Zaragoza and at the age of twelve went to Malta, where he joined the Knights of Malta with the rank of commander of Gracia de Aliaga. At seventeen he enlisted as a midshipman at the Compañía of Cádiz, where he completed his higher studies. His comrades nicknamed him "Euclid" because of his aptitude for the exact sciences. He took part in various privateering campaigns against the Moors and in the expedition against Oran. Juan never married.

After Cassini's measurements of the meridians seemed to show that the earth was a spheroid elongated at the poles, in clear opposition to Newton's theory, the French Academy of Sciences proposed that two series of measurements of one degree of an arc of meridian should be made, one near the North Pole, the other near the equator. Louis XV designated a Hispano-French commission for the measurement at the equator, in which, by appointment of Philip V, Juan and Antonio de Ulloa would participate, on behalf of Spain, with La Condamine, Godin, Bouguer, and Joseph de Jussieu.

In 1736 the commission's expedition began its work, principally in the regions of Quito and Guayaquil. Complementary scientific observations were made of the speed of sound and of various aspects of astronomy, physics, geography, biology, and geology. Great effort went into achieving precision and accuracy for the measurements of the Peruvian meridian. The measurement made by the French members of the expedition gave a longitude of 56,750 toises, while that of the Spaniards gave 56,758; that is, a minor difference on the order of 14:100,000. These measurements confirmed the Newtonian theory of the shape of the earth and were extraordinary for their precision.

In 1745, nine years after the inception of the expedition, Juan and his colleague Ulloa returned to Spain, each taking a different route as a precaution for safe arrival of the data.

Juan subsequently designed and directed the shipyards at El Ferrol and La Carraca, took part in the improvement of the working and development of the mines of Almadén, founded the astronomical observatory of Cádiz, and carried out several diplomatic and special missions. He was squadron commander of the Royal Armada and director, at the age of fifty-seven, of the Royal Seminary of Nobles. During his captaincy of the company of midshipmen of Cádiz he established, in his house there, the Friendly Literary Society, which is considered the forerunner of the Royal Society of Sciences of Madrid. For several years this group met each Thursday to consider questions of mathematics, physics, geography, hygiene, history, and archaeology.

His combination of theoretical learning and practical experience enabled Juan, in his *Examen marítimo*, to provide a considerable base for the improvement of naval science, to refute several empirical theories of navigation, and to establish the fundamental principles of naval architecture. The book is a valuable application of mechanics to naval science.

Among the societies in which Juan held membership were the Royal Society, the French Academy of Sciences, the Royal Academy of Sciences of Berlin, and the Spanish Academy of San Fernando.

BIBLIOGRAPHY

I. ORIGINAL WORKS. Juan y Santacilla's writings are *Relación histórica del viaje a América meridional* . . ., 4 vols. (Madrid, 1747), written with Antonio de Ulloa—vol. II, para. 1,026 tells of the discovery of platinum by the Spaniards in Peru; *Disertación histórica y geográfica sobre el meridiano de demarcación entre los dominios de España y Portugal y los parajes por donde pasa en la América meridional* . . . (Madrid, 1749), written with Ulloa; and *Examen marítimo teórico práctico* . . . (Madrid, 1771, 1793), translated into English (London, 1784) and French (Nantes, 1783; Paris, 1793), also republished in Spanish (Madrid, 1968).

See also "Reglamento para la construcción de lonas" (1751), MS in the collection of Vargas Ponce; *Compendio de navegación para el uso de los Caballeros Guardiasmarinas* (Cadiz, 1757); "Informe a S. M. sobre los perjuicios de la construcción francesa de los bajeles" (1773), MS; *Observaciones astronómicas y fichas hechas por O. de S. M. en los reinos del Peru* (Madrid, 1773, 1778), written with Ulloa.

Other works are *Estado de la astronomía en Europa* . . . (Madrid, 1773); "Reflexiones sobre la fábrica y uso del cuarto-de-circulo" (Dirección de Hidrografia, 1809); "Las observaciones del paso de Venus por el disco del sol (Memoir Deposito Hidrografico, 1809); "Metodo de levantar o dirigir el mapa o plano general de España" (Memoir Deposito Hidrografico, 1809); *Noticias secretas de América sobre el estado moral y militar y político de los reinos del Perú y provincias de Quito, cosas de Nueva Granada y Chile* . . . (London, 1826), written with Ulloa; and "Relox o crónometro inventado por Juan Harrison."

A fair number of Juan's notes and papers on cosmography and navigation exist, the majority of them unpublished. Among these are two volumes with ten blueprints for naval construction.

II. SECONDARY LITERATURE. See *Diccionario enciclopédico hispano-americano*, XI (Barcelona, 1892), 217; Francisco Cervera y Jíménez Alfaro, *Jorge Juan y la colonización española en America . . .* (Madrid, 1927).

J. M. LÓPEZ DE AZCONA

JUDAY, CHANCEY (*b.* Millersburg, Indiana, 5 May 1871; *d.* Madison, Wisconsin, 29 March 1944), *limnology*.

Juday, the son of Elizabeth Heltzel and Baltzer Juday, received the B.A. from the University of Indiana in 1896. He took the M.A. (1897) under Carl Eigenmann, who aroused his interest in aquatic biology. Having taught high school for two years, Juday returned to science to accept a position as biologist with the Wisconsin Geological and Natural History Survey. His work was interrupted by an attack of tuberculosis, but he returned to the survey in 1905 and remained there until 1930. At the same time, he assumed academic positions at the University of California (1904), the University of Colorado (1905), and finally the University of Wisconsin, where he collaborated with E. A. Birge in researches on Wisconsin lakes. (Their major monograph, "The Inland Lakes of Wisconsin. The Dissolved Gases," was published in 1911.)

From October 1907 to June 1908, Juday traveled in England and on the Continent, visiting universities, biological stations, and lakes and meeting leading European aquatic biologists. On his return to the United States he began giving the courses on limnology and plankton organisms that he was to continue until his retirement from teaching in 1941. In February of 1910 he traveled in Central America, studying four semitropical lakes in Guatemala and El Salvador. In the same year Juday made studies on the Finger Lakes of New York.

From 1925 until 1941 Juday was director of the limnological laboratory at Trout Lake, Wisconsin, spending two months of each summer there. The station attracted biologists from the United States and Europe, and work done there was the subject of many important monographs. In his duties at the University of Wisconsin, Juday similarly encouraged research relations among departments whose work touched on lake studies. He was made professor of limnology, within the department of zoology, in 1931.

Juday published more than one hundred papers, including works on plankton, hydrography and morphometry of the inland lakes of Wisconsin, crustaceans, anaerobiosis, productivity of lakes, mineral content of waters and muds, hydrogen ion concentration, the effects of fertilizing lakes, and the growth of game fish and photosynthesis as indexes of the productivity of lakes. In addition to his teaching and research he worked with the Wisconsin Conservation Department, for whom he directed studies of the game fish of that state, and served as a consultant to the U.S. Bureau of Fisheries.

Juday was elected president of the Limnological Society of America upon its foundation in 1935; secretary and president of the Wisconsin Academy of Sciences, Arts and Letters; and president of both the American Microscopical Society and the Ecological Society of America. He received an honorary doctorate from the University of Indiana (1933) and the Leidy Medal of the Academy of Natural Sciences of Phildelphia (1943). He married Magdalen Evans on 6 September 1910; they had three children.

Upon his retirement from teaching, Juday was retained by the University of Wisconsin as a research associate. He died three years later, without completing the comprehensive treatment of Wisconsin limnology that he had begun.

BIBLIOGRAPHY

I. ORIGINAL WORKS. A complete list of Juday's writings is Arthur D. Hasler, "Publications of Chancey Juday," in *Special Publications. Limnological Society of America*, no. 16 (1945), pp. 4–9. Among the most important of these are three book-length reports, "The Inland Lakes of Wisconsin. I. The Dissolved Gases and Their Biological Significance," in *Bulletin of the Wisconsin Geological and Natural History Survey*, sci. ser. 7, no. 22 (1911), written with E. A. Birge; "The Inland Lakes of Wisconsin. II. The Hydrography and Morphometry of the Lakes," *ibid.*, no. 27 (1914); and "The Inland Lakes of Wisconsin. The Plankton. I. Its Quantity and Chemical Composition," *ibid.*, sci. ser. 13, no. 64 (1922), written with E. A. Birge.

II. SECONDARY LITERATURE. On Juday and his work, see Lowell E. Noland, "Chancey Juday," in *Special Publications. Limnological Society of America*, no. 16 (1945), pp. 1–3.

LOWELL E. NOLAND

JUEL, SOPHUS CHRISTIAN (*b.* Randers, Denmark, 25 January 1855; *d.* Copenhagen, Denmark, 24 January 1935), *mathematics*.

Juel's father, a judge, died the year after his son was born. The boy spent his youth in the country and attended the Realschule in Svendborg. At the age of fifteen he went to Copenhagen, where in 1871 he entered the Technical University. In January 1876, being more interested in pure science, he took the examinations for admission to the University of Copenhagen. Completing his university studies in 1879 with the state examination, he received his doctor's degree in 1885. From 1894 he was lecturer at the Polytechnic Institute, where in 1907 he became full professor. He occasionally lectured at the University of Copenhagen.

From 1889 to 1915 he was editor of the *Matematisk Tidsskrift*. In 1925 he became an honorary member of the Mathematical Association and in 1929 received an honorary doctorate from the University of Oslo. He married a daughter of T. N. Thiele, professor of mathematics and astronomy. Failing eyesight plagued him in later years.

Juel's writings include schoolbooks, textbooks, and essays. He made substantial contributions to projective geometry for the cases of one and two complex dimensions, and to the theory of curves and surfaces. His book on projective geometry is very similar in approach to that of Staudt but is easier to understand; his treatment of autocollineations goes beyond Staudt's. Segré arrived at similar results independently.

In 1914 Juel devised the concept of an elementary curve, which is in the projective plane without straight-line segments and has the topological image of a circle and a tangent at every point. Outside these points a convex arc can be described on each side. Thus an elementary curve consists of an infinite number of convex arcs passing smoothly one into another.

Juel, whose treatment of his subject was loose and incomplete, dealt mainly with fourth-order curves, developing the concept of the order of an elementary curve and setting up a correspondence principle and theory of inflection points. His third-order elementary curve is very close to a third-order algebraic curve but no longer has three points of inflection on one straight line.

Juel worked also on the theory of finite equal polyhedra, on cyclic curves, and on oval surfaces.

BIBLIOGRAPHY

I. ORIGINAL WORKS. Juel's textbooks include *Vorlesungen über Mathematik für Chemiker* (Copenhagen, 1890); *Elementar stereometri* (Copenhagen, 1896); *Analytisk stereometri* (Copenhagen, 1897); *Ren og anvendt aritmetik* (Copenhagen, 1902); *Forlaesinger over rational mekanik* (Copenhagen, 1913; enl. ed., 1920); and *Vorlesungen über projektive Geometrie mit besonderer berücksichtigung der von staudtschen Imaginärtheorie* (Berlin, 1934).

His articles and essays include *Inledning i de imaginaer linies og den imaginaer plans geometrie* (Copenhagen, 1885), his dissertation; "Grundgebilde der projektiven geometrie," in *Acta mathematica*, **14** (1891); and "Parameterbestimmung von Punkten auf Kurven 2. und 3. Ordnung," in *Mathematische Annalen*, **47** (1896), written with R. Clebsch; and three that appeared in *Kongelige Danske Videnskabernes Selskabs Skrifter:* "Inledning i laeren om de grafiske kurver" (1899); "Caustiques planes" (1902); "Égalité par addition de quelques polyèdres" (1902).

The following articles appeared in *Matematisk Tidsskrift:* "Kegelsnitskorder des fra et fast punkt ses under ret vinkel" (1886); "Korder i en kugel, der fra et fast punkt ses under ret vinkel" (1887); "Vivianis theorem" (1891); "Transformationer af Laguerre" (1892); "Polyeder, der ere kongruente med deres speilbilleder uden at vaere selvsymmetriske" (1895); "Konstrukter af dobbelpunktstangenterne ved en rumkurve af 4. order" (1897); and "Arealer ot voluminere" (1897).

II. SECONDARY LITERATURE. Details concerning Juel's work can be found in David Fog, "The Mathematician C. Juel—Commemorative Address Delivered Before the Mathematical Association on March 18, 1935," in *Matematisk Tidsskrift*, **B** (1935), 3–15.

HERBERT OETTEL

JUKES, JOSEPH BEETE (*b.* Summerhill, near Birmingham, England, 10 October 1811; *d.* Dublin, Ireland, 29 July 1869), *geology, geomorphology*.

The only son of John Jukes, a Birmingham manufacturer, Beete Jukes studied geology at Cambridge under Sedgwick, graduating in 1836. For several years he traveled about central and northern England, studying geology and giving lectures on the subject. In 1839 he accepted the post of geological surveyor in the colony of Newfoundland. He spent a year and a half there, completed his report, and returned to England at the end of 1840. In 1842 he set off again as naturalist on H.M.S. *Fly*, sent to survey part of the Great Barrier Reef and the Torres Strait. Jukes returned to England in June 1846; subsequently he published an account of the voyage, as well as several papers on the geology of Australia.

Soon after his return Jukes was appointed to the Geological Survey of Great Britain and began field work in North Wales, later working in the South Staffordshire coalfield. In 1850 he was promoted director of the Irish branch of the Geological Survey, a post he held until his death in 1869. His final illness followed a head injury received in 1864 but was undoubtedly exacerbated by overwork.

It was in Ireland that Jukes carried out the work for which he is best known, a study of river action. The Huttonian concept of intense denudation by the agency of rain and rivers, although maintained by Scrope in 1827, had been eclipsed by Lyell's advocacy of marine action and earthquakes as the major factors in the shaping of rising land. Darwin, too, had followed Lyell in emphasizing marine erosion. In North America, J. D. Dana was almost alone in stressing the greater powers of subaerial denudation and stream erosion.

Jukes himself had supported the popular marine erosion theory until 1862, when he seems quite suddenly to have changed his views and pronounced in favor of fluvial action as the principal agent in producing land relief.

In 1862 he read, first in Dublin and then in London, a paper which was a careful study of the pattern of river valleys in southern Ireland and their relation to the underlying geological structure. This paper has become a classic in geomorphology. Jukes's views on the importance of river action were quickly accepted by several leading geologists, particularly Ramsay, Geikie, and Croll. As a result, within a few years Darwin had changed his views and Lyell had modified his.

BIBLIOGRAPHY

I. ORIGINAL WORKS. A complete list of Jukes's works is given in *Letters and Extracts From the Addresses and Occasional Writings of J. Beete Jukes*, Mrs. C. A. Browne (his sister), ed. (London, 1871), pp. 591–596. This is also the main source for biographical information. There is a list of his scientific papers in the Royal Society's *Catalogue of Scientific Papers*, III (1869), 588.

His classic paper is "On the Mode of Formation of the River-Valleys in the South of Ireland," in *Quarterly Journal of the Geological Society of London*, **18** (1862), 378–403. A valuable and little-known historical pamphlet by Jukes is his address *Her Majesty's Geological Survey of the United Kingdom, and Its Connection With the Museum of Irish Industry in Dublin and That of Practical Geology in London* (Dublin, 1867).

II. SECONDARY LITERATURE. Jukes's work in geomorphology is fully discussed in R. J. Chorley, A. Dunn, and R. P. Beckinsale, *The History of the Study of Landforms* (London–New York, 1964), pp. 391–401; and in G. L. Davies, *The Earth in Decay* (London, 1969), pp. 319–333.

Obituary notices of Jukes are in *Geological Magazine*, **6** (1869), 430–432; and in *Quarterly Journal of the Geological Society of London*, **26** (1870), xxxii. See also H. B. Woodward, *The History of the Geological Society of London* (London, 1907), pp. 187, 228–232. A photograph of Jukes is reproduced in the latter work.

There are letters from Jukes to W. B. Clarke in the Mitchell Library, Sydney, Australia; some of these are quoted in James Jervis, "Rev. W. B. Clarke . . . the Father of Australian Geology," in *Royal Australian Historical Society, Journal and Proceedings*, **30** (1944), 345–358.

JOAN M. EYLES

JULIUS, WILLEM HENRI (*b.* Zutphen, Netherlands, 4 August 1860; *d.* Utrecht, Netherlands, 15 April 1925), *solar physics.*

The son of Willem Julius and Maria Margareta Dumont, Julius studied at the University of Utrecht. He became professor of physics at the University of Amsterdam in 1890 and at the University of Utrecht in 1896. A modest man, he lived well and in the traditional manner, showing full devotion to both science and the arts. Among his friends were Einstein, Ehrenfest, Zeeman, Eykman, and Einthoven.

Julius studied the infrared radiation of flames with a radiometer he had constructed. In order to avoid tremors he mounted this instrument in such a way that its center of gravity was supported, a technique known as the Julius suspension. His observation of the solar eclipse of 1901 was the turning point in Julius' activity; from then on he devoted all of his work to solar physics. August Schmidt had stressed the effects of refraction in the solar gaseous sphere; Julius modified this conception and gave more consideration to the refraction in irregular inhomogeneities and the anomalous refraction of rays having wavelengths quite near to the wavelength of an absorption line. He explained solar prominences as regions with strong inhomogeneities, where the light of the sun is refracted toward us. The darkness of sunspots was explained by regular refraction. In these conceptions the importance of refraction was vastly exaggerated. Later, however, Julius extended his argument, stating that the rays of a Fraunhofer line would also show anomalous scattering, which would explain the darkness inside the line. This view was later developed independently by Unsöld, and even today anomalous scattering is assumed to be the mechanism by which strong solar resonance lines are formed.

During the eclipses of 1905 and 1912 Julius applied a new method for determining the distribution of brightness over the sun's disk by recording the variation of the total intensity during the partial phases.

BIBLIOGRAPHY

A survey of Julius' theories and a complete bibliography of his works are in his *De Natuurkunde van de Zon* (Groningen, 1927).

Individual works include "Solar Phenomena, Considered in Connection With Anomalous Dispersion," in *Astrophysical Journal*, **12** (1900), 185; "Hypothese over den oorsprung der zonneprotuberanties," in *Proceedings of the Academy of Amsterdam*, **5** (1902–1903), 162, and *Physikalische Zeitschrift*, **4** (1902–1903), 85; "A New Method for Determining the Rate of Decrease of the Radiating Power From the Center Toward the Limb of the Solar Disk," in *Astrophysical Journal*, **23** (1906), 312; "Selective Absorption and Anomalous Scattering of Light in Extensive Masses of Gas," in *Proceedings of the Academy of Amsterdam*, **13** (1910–1911), 881, and *Physikalische*

Zeitschrift, **12** (1911), 329; "The Total Solar Radiation During the Annular Eclipse of April 17, 1912," in *Astrophysical Journal*, **37** (1913), 225; "Anomalous Dispersion and Fraunhofer Lines. Reply to Objections," *ibid.*, **43** (1916), 43; and "How to Utilize Actinometric Results Obtainable During Solar Eclipses," in *Bulletin of the Astronomical Institute of the Netherlands*, no. 33 (1923), p. 189.

M. G. J. MINNAERT

IBN JULJUL, SULAYMĀN IBN ḤASAN (*b.* Córdoba, Spain, 944; *d. ca.* 994), *medicine, pharmacology.*

Ibn Juljul's course of studies is known through his autobiography, preserved by Ibn al-Abbār. He studied medicine from the age of fourteen to twenty-four with a group of Hellenists that had formed in Córdoba around the monk Nicolas and was presided over by the Jewish physician and vizier of ʿAbd al-Raḥmān III, Ḥasdāy ibn Shaprūṭ. Later he was the personal physician of Caliph Hishām II (976–1009). The famous pharmacologist Ibn al-Baghūnish was his disciple.

Among Ibn Juljul's works is *Ṭabaqāt al aṭibbāʾ waʾl-ḥukamāʾ* ("Generations of Physicians and Wise Men"). It is the oldest and most complete extant summary in Arabic—except for the work on the same subject written by Isḥāq ibn Ḥunayn, which is inferior to that of Ibn Juljul—on the history of medicine. It is of particular interest because Ibn Juljul uses both Eastern sources (Hippocrates, Galen, Dioscorides, Abū Maʿshar) and Western ones. The latter had been translated into Arabic from Latin at Córdoba in the eighth and ninth centuries and include Orosius, St. Isidore, Christian physicians, and anonymous authors who served the first Andalusian emirs. The work has frequent chronological mistakes, especially when it deals with the earliest periods, but it never lacks interest.

The *Ṭabaqāt* contains fifty-seven biographies grouped in nine generations. Thirty-one are of oriental authors: Hermes I, Hermes II, and Hermes III, Asclepiades, Apollon, Hippocrates, Dioscorides, Plato, Aristotle, Socrates, Democritus, Ptolemy, Cato, Euclid, Galen, Al-Ḥārith al-Thaqafī, Ibn Abi Rumtha, Ibn Abḥar, Masarjawayhi, Bakhtīshūʿ, Jabril, Yuḥannā ibn Māsawayhi, Yuḥannā ibn al-Biṭrīq, Ḥunayn ibn Isḥāq, al-Kindī, Thābit ibn Qurra, Qusṭa ibn Lūqā, al-Rāzī, Thābit ibn Sinān, Ibn Waṣīf, and Nasṭās ibn Jurayḥ. The rest of the biographies are of African and Spanish scholars, who generally are less well-known than the Eastern ones. Since he knew many of the latter and possibly attended some of them, there is no reason to question the details given concerning their behavior or illnesses. The remarks on these topics are not real clinical histories but transmit details (allergic asthma, dysentery, and so on) that give a clear idea of life in Córdoba in the tenth century.

Ibn Juljul also provides interesting information about the oldest Eastern translations into Arabic, in the time of Caliph ʿUmar II (717–719), when he states that the latter ordered the translation from Syriac of the work of the Alexandrian physician Ahran ibn Aʿyan (*fl.* seventh century). One should not disdain his reflections on the causes hindering the development of science when, referring to the East, he justifies not mentioning more scholars from this region after al-Rāḍī's caliphate (*d.* 940), saying:

> In later reigns there was no notable man known for his mastery or famous for his scientific contributions. The Abbasid empire was weakened by the power of the Daylamites and Turks, who were not concerned with science: scholars appear only in states whose kings seek knowledge [*Ṭabaqāt*, p. 116].

Tafsīr asmāʾ al-adwiya al-mufrada min kitāb Diyusqūrīdūs, written in 982, may concern a copy of Dioscorides' *Materia medica.* In it is a text, quite often copied, on the vicissitudes of the Arabic translation of the famous Greek work. *Maqāla fī dhikr al-adwiya al-mufrada lam yadhkurhā Diyusqūrīdūs* is a complement to Dioscorides' *Materia medica. Maqāla fī adwiyat al-tiryāq* concerns theriaca. *Risālat al-tabyīn fī ma ghalaṭa fīhi baʿḍ al-mutaṭabbibīn* probably dealt with errors committed by quacks.

Ibn Juljul may be the author of the *De secretis* quoted by Albertus Magnus in his *De sententiis antiquorum et de materia metallorum* (*De mineralibus* III, 1, 4), which is attributed to a certain Gilgil.

The work of Ibn Juljul must have remained popular in Muslim Spain for a long time; otherwise we could not account for the frequent references given by a botanist such as the unnamed Spanish Muslim studied by Asín Palacios.

BIBLIOGRAPHY

I. ORIGINAL WORKS. A list of MSS is in C. Brockelmann, *Geschichte der arabischen Litteratur*, I (Weimar, 1898), 237, and *Supplementband*, I (Leiden, 1944), 422. The text of *Ṭabaqāt* . . . is available in a good ed. by Fu'ād Sayyid (Cairo, 1955); the last chapter of this work has appeared in Spanish trans. by J. Vernet in *Anuario de estudios medievales* (Barcelona), **5** (1968), 445–462.

II. SECONDARY LITERATURE. See G. Sarton, *Introduction to the History of Science*, I (Baltimore, 1927), 682; Ibn al-Abbār, *Takmila*, A. González Palencia and M. Alarcón, eds. (Madrid, 1915), p. 297; Ibn Abī Uṣaybiʿa, *ʿUyūn al-anbāʾ fī ṭabaqāt al-aṭibbāʾ*, edited and translated into French by H. Jahier and A. Noureddine (Algiers, 1958), pp. 36–41; and Miguel Asín Palacios, *Glosario de voces romances registradas por un botánico anónimo hispano-musulmán* (*siglos XI–XII*) (Madrid–Granada, 1943), index.

J. VERNET

JUNCKER, JOHANN (*b.* Londorf, Germany, 23 December 1679; *d.* Halle, Germany, 25 October 1759), *chemistry, medicine.*

Born in modest circumstances, Juncker received his primary education in Giessen. He was a student of philosophy at the University of Marburg in 1696 and then went to Halle, where he studied theology and followed a program in literature under the classical scholar Christopher Cellarius. Juncker taught in Halle from 1701 to 1707 and then left to study medicine in Erfurt, where he received his M.D. degree in 1717. In 1716 he had returned to Halle as physician to the Royal Pedagogical Institute and Orphanage, beginning a distinguished career in that city which culminated in his appointment to the chair of medicine in 1729 and ultimately his selection as Prussian privy councillor. Juncker married three times.

The University of Halle, a Pietistic stronghold, possessed two of the outstanding chemical and medical theorists of the early eighteenth century, Georg Ernst Stahl and Friedrich Hoffmann. Juncker benefited from his close association with these brilliant colleagues and became one of Stahl's most gifted and prominent disciples. When Stahl went to Berlin in 1716 Juncker corresponded with him and published numerous dissertations and books that expounded and developed Stahlian ideas in chemistry and medicine. He reiterated Stahl's admonition to keep these two disciplines distinct, arguing that chemical theory had little to offer medical practice at that time. His medical treatises censured both the iatrochemical and iatromathematical traditions and elaborated Stahl's vitalist theories. Juncker and another colleague in Halle, Michael Alberti, disseminated Stahlian vitalism throughout Europe and assisted in establishing an alternative in medical thought to the mechanical theories of Boerhaave.

Juncker's most important chemical text was the *Conspectus chemiae theoretico-practicae* (1730), a systematic exposition of the ideas and experiments of Becher and Stahl. By providing a critical and coherent treatment of Stahl's studies on chemical composition and reaction, the *Conspectus* offered a more intelligible version of Stahl's work that gave it a greater audience. Juncker stressed the necessity for grounding chemical theory in accurate and extensive experimental data and, after establishing the definition, aims, and utility of chemistry, applied the Becher-Stahl hierarchy of matter as the fundamental schema for chemical explanation. He adopted Stahl's ideas on the nature of the elements, including phlogiston (which he emphasized was a material principle and not merely the property of burning) and, like Stahl, denied air a chemically active role, maintaining that it acted only as a physical instrument during reactions. Thus in combustion and calcination, air served to expedite the release of phlogiston from compounds, without itself entering into any chemical combination.

By effecting a clarification in Stahlian theory and methodology, Juncker played a significant part in the development of his mentor's ideas as a major force for reform in eighteenth-century chemistry. His concern with the broader implications of Stahl's work, transcending other, more narrow approaches that focused on phlogiston, prefigured the orientation of important groups of chemists in Germany and France at mid-century.

BIBLIOGRAPHY

I. ORIGINAL WORKS. Many of Juncker's writings include Stahl's name and method in their titles. Representative medical texts are *Conspectus chirurgiae tam medicae methodo Stahliana conscriptae* (Halle, 1721; 2nd ed., 1731); and *Conspectus formularum medicarum . . . ex praxi Stahliana potissimum desumta* (Halle, 1723; 2nd ed., 1730; 4th ed., 1753). Juncker's major chemical work is *Conspectus chemiae theoretico-practicae in forma tabularum repraesentatus . . . e dogmatibus Becheri et Stahlii potissimum explicantur* (Halle, 1730; 2nd ed., 2 vols., 1742–1744), French trans. by J. F. Demachy, *Élémens de chymie, suivant les principes de Becker et de Stahl*, 6 vols. (Paris, 1757). A complete list of Juncker's writings, including numerous dissertations on chemical and medical subjects, is given in Johann Georg Meusel, *Lexikon der vom Jahr 1750 bis 1800 Verstorbenen Teutschen Schriftsteller*, VI (1806; repr. Hildesheim, 1967), 340–347.

II. SECONDARY LITERATURE. Details concerning Juncker's life appear in F. Hoefer, ed., *Nouvelle biographie générale*, XXVII (Paris, 1858), 238–240; and Johann C. Adelung, *Fortsetzung und Ergänzungen zu Christian G. Jöchers Allgemeinem Gelehrten-Lexicon*, II (Leipzig, 1787; repr. Hildesheim, 1960), 2347–2348. Brief assessments of Juncker's scientific work are J. F. Gmelin, *Geschichte der Chemie seit dem Wiederaufleben der Wissenschaften bis an das Ende der achtzehnten Jahrhunderts*, II (Göttingen, 1797–1798), 681; and James R. Partington, *A History of Chemistry*, II (London, 1961), 688–689.

MARTIN FICHMAN

JUNG, CARL GUSTAV (*b.* Kesswil, Switzerland, 26 July 1875; *d.* Küsnacht, Switzerland, 6 June 1961), *analytical psychology.*

Jung's father was a pastor of the Basel Reformed Church; eight of his uncles, as well as his maternal grandfather, were also pastors; and the atmosphere of religious tradition and practice in which he grew up had an all-pervading influence on his life. It is also significant, in view of his choice of career, that his paternal grandfather, an imposing figure, had identical Christian names and was a physician.

Jung's mother, who suffered from ill health during his childhood, was warm and down-to-earth. Although not an intellectual, she was well enough read to introduce her son to Goethe. She was superstitious and communicated with her son, in whom she confided extensively, on two levels: one was conventional; the other, primitive, superstitious and very direct. These two levels of communication became important to Jung—especially a "voice" that "told the truth." Jung's father had had a successful university career, studying philology and linguistics. A kind, generous man who evoked his son's affection, he developed intellectual doubts about religion but overinsisted on the need for belief—an attitude that Jung could not accept—and a rift was created between father and son. He gradually became hypochondriacal, irritable, and ill-tempered, and family quarrels occurred. Before he died, Jung's father started reading a book on psychology, perhaps in an attempt to solve his doubts.

Jung was an only child until he was nine years old. Secretive and highly imaginative, he spent considerable time with the peasants, among whom his family lived, and assimilated much of their folklore and easy acceptance of nature.

In his third or fourth year, Jung suffered intense anxieties focusing on the Jesuits, of whom he heard his parents talking, and came to distrust Jesus although he was presented as gentle and mild. His rich imagination began to focus on religious themes and led him to experience God. Together with a strong feeling for dreams this imagination was to develop into a lifelong sense of purpose. The need to understand unconventional and even shocking religious experiences during childhood—especially a vision of God defecating on a cathedral—was a powerful element in Jung's drive to study relevant and often abstruse topics. This inner life was revealed only in his autobiography, written in his last years.

Jung's education was typical for a child of his circumstances. At the age of six he entered the village school, which made a considerable impression on him. Forced to adapt in a way that was out of keeping with his home and inner life, he became aware of the need for living as if he were two persons, "personalities number one and two," as he called them. An intellectually precocious boy, Jung soon read fluently, and his father started teaching him Latin when he was six. He also prepared his son for confirmation, but in an unsatisfying way, because he did not answer his son's questioning mind. His father's emphasis on belief became suspect and was a contributing factor in Jung's turning away from formal religion.

At the age of thirteen he entered the Gymnasium in Basel, where he met the sons of well-to-do families; his acute awareness of his poverty eventually had significant bearing on his choice of a career.

While still at school, Jung read extensively in religion and philosophy, including Goethe's *Faust*, the Scholastic theologians, and to his great fascination the mystical writings of Meister Eckhart. He also studied the Greek philosophers, Hegel, and Schopenhauer, whose relation to Kant drew his attention; later, at the University of Basel, he became fascinated by Nietzsche. In all this he was working and exploring on his own; indeed, there was nobody with whom he could discuss his ideas freely. Jung used to go weekly to an uncle whose family would discourse on theology; he enjoyed the intellectual ingenuity but did not find satisfaction in it.

Jung's entry into university life was thus enormously liberating. To his excitement and delight he found others with similar interests and comparable intellectual gifts, with whom he could enjoy a free and broad exchange of ideas. He had difficulty in deciding which subject to take when he won a place and a bursary. Although attracted to the humanities, he had developed a strong interest in science, especially zoology, paleontology, and geology. Since none of these subjects would enable him to earn the good living that he desired, Jung chose medicine, for which subject he showed considerable aptitude. After graduation, he was offered the post of assistant to his chief, Frederick Müller. A successful career as physician lay before him, but this was not to be.

While still a student Jung had become interested in spiritualism. One day a wood table in his home suddenly split with a bang. After the shock of surprise and incredulity had subsided, his mother remarked, "That means something." Soon afterward a steel knife broke into several pieces in a way that he could not rationally explain; his mother looked meaningfully at him. It then appeared that some relatives had been engaged in table turning and the group had been thinking of asking Jung to join them. The possibility that the séances and broken objects were related began

Jung's serious interest in parapsychology and seems to have been the prototype for his theory of synchronicity. But apart from these considerations, noting irrational events and following them up was characteristic of Jung, who always struggled to use his powerful intellectual drive to try to understand them instead of explaining them away.

The data Jung collected from the séances was the subject of his doctoral thesis, delivered at the faculty of medicine of the University of Zurich, in which the influence of Pierre Janet, from whom he took a number of ideas, is first recorded. His thesis "Zur Psychologie und Pathologie sogenannter occulter Phänomene" was published in 1902. The data from the séances also marked a turning point in his scientific and professional life, for he discovered that psychiatry held the best hope of understanding what he had observed. In Krafft-Ebing's *Lehrbuch der Psychiatrie* he read that the psychoses could be considered as diseases of the personality rather than of the central nervous system and that a subjective factor was a significant part of psychiatry. These two notions led Jung to a concept of science based upon the interaction of two psychical systems. He was so powerfully affected by this idea that it "wiped out philosophy" as a method of explaining his religious and parapsychological experiences; they could now be replaced by the psychological point of view. His subsequent decision to become a psychiatrist was the first of two decisive steps (the second was his break with Freud). Each threw into relief his wholehearted way of pursuing his interests regardless of the consequences. Psychiatry was then very much a backwater in the medical profession, and Jung seemed to be sacrificing his career altogether. His contemporaries were astounded that he should be willing to do so.

Fortunately, Eugen Bleuler was conducting research into schizophrenia at the Burghölzli Asylum. Jung found in him support for applying association tests to the psychology of normal persons and the mentally diseased. The technique had been initiated by Francis Galton and developed in Emil Kraepelin's laboratory by Gustav Aschaffenburg. By ingeniously studying the irregularity in responses to stimulus words, Jung developed a theory of complexes and grasped that they could be explained by Freud's theory of repression. He started corresponding with Freud, to whom he sent a small volume on schizophrenia in which he unraveled the meaning of a patient's delusions, hallucinations, and stereotypes. Freud was much interested, and in 1907 a close but complex relationship began which lasted for seven years.

From the outset Jung was greatly impressed by Freud, although he increasingly came to have doubts about the sexual theory to which Freud attached such importance. Jung began to think of it as a concealed religion but kept this view private, as he had kept his religious convictions secret from his father. There were also differences over parapsychology, and the positive importance that Jung gave to religion was unacceptable to Freud. Jung's doubts increased especially in 1909, when he and Freud lectured at Clark University in Worcester, Massachusetts.

In 1909 Jung resigned his lectureship at the University of Zurich, which he had held since 1905, thus sacrificing his academic prospects. He claimed that this act was due to the pressure of work, but a contributing factor may have been the attacks and threats made because of his promotion of psychoanalysis. At this time Jung was active in the psychoanalytic movement: he became editor of the *Jahrbuch für psychoanalytische und psychopathologische Forschungen*, the main psychoanalytic journal, and later the first president of the International Psychoanalytical Association.

With the publication of his large and erudite *Wandlungen und Symbole der Libido* (1912), in which he applied psychoanalytic theory to the study of myths, Jung's relations with Freud and psychoanalysis had become very strained and the work was heavily attacked by Sandor Ferenczi. In 1913 Karl Abraham issued a parallel and devastating criticism of Jung's lectures entitled "Theory of Psychoanalysis" at Fordham University, New York. He was also criticized for his handling of the International Congress for Psychoanalysis in 1913 when he delivered a short paper on psychological types. At that time the serious conflicts among schools of psychoanalysis evoked intense and even personal animosities; in particular a group differing with Freud on scientific issues began to center on Alfred Adler, who eventually formed his own school.

Jung's attempt to resolve the conflict by introducing his theory of types was not appreciated, and in 1914 he formally severed all connections with psychoanalysis to form his own school of analytical psychology. The break was the second essential, seemingly catastrophic step Jung took in his personal development and in his scientific and professional career. He was left virtually isolated, although a few colleagues remained interested in the development of his concepts and practices.

This period was characterized by profound disorientation. From his own dreams Jung had already derived the idea of a substratum of historical structures in the psyche, which he thought existed in the unconscious beneath the personal level that Freud had investigated. It was as if the personal psyche of man was founded in archaic and historical roots which, expressed in myths, both determined the course of, and gave meaning to, his life. Having developed this concept in *The Psychology of the Unconscious*, in 1913 Jung began to explore its personal implications; he wanted to discover his own myth. The decision was not entirely voluntary; he experienced a horrifying vision of Europe covered with a sea of blood which lost its spell only at the outbreak of World War I. He came to think of it as an intimation of what was to come, but it also seemed that he had perceived the unconscious processes latent in European man. At this time his dreams became especially significant.

Jung's intensive exploration of his own inner life began, however, from childhood games played with stones. As his games developed his imagination grew more intense and a stream of imagery, sometimes of visionary quality, began to emerge; with the persons of his fantasy he held an inner dialectic that he later called "active imagination." This period of "creative illness" sometimes threatened to become a mahifest psychosis, but once the process was under way, he succeeded in controlling and confining it so that it did not seriously disrupt his family life and analytic practice with patients.

In 1903 Jung had married Emma Rauschenbach, a comparatively rich, intelligent, and devoted woman who kept his life running smoothly and assured his material security. Related to this turbulent period was his building of a small house at Bollingen, on a remote part of the Lake of Zurich; it was to be a "representation in stone of my inner thoughts and of the knowledge I had acquired." Started without detailed plan but based on the huts of primitive people, it was enlarged over the years and Jung often retired to it. His simple life there combined cooking and looking after himself with painting and stone carving. Thus Jung made concrete the two personalities that he had discovered when he went to the village school.

The period of Jung's intense inner life ended in 1917, when the stream of imagery faded. His careful records of his imagery, dreams, and visions were to form the basis of his conceptual framework. In the theory of conscious systems that he developed, the ego was at the center. Its constituents were arranged by types: there were two attitude types, introversion and extroversion. Of the four function types, thinking and feeling were rational; sensation and intuition were irrational. Within the psychic organization there were combinations of types, but the rational and irrational functions were arranged in opposites and so could not combine. Thinking was thus incompatible with feeling, and sensation with intuition. The unconscious was also relatively organized by inherited archetypes, which Jung inferred from the tendency of fantasy images to show regularities around particular personifications, such as the parent images, the hero, and the child. The unconscious forms compensated the ego and interacted with it to produce, under favorable circumstances, increased consciousness of the self or personality as a whole. The process that brought this about Jung termed "individuation." As this took place in the person so did it occur in society, and he developed a theory of history and social change.

Important in Jung's formulation was a special theory of symbols in which inner imaginative life, having a validity of its own, was expressed. The irrational nature of symbols made possible the combination of opposites; consequently they had an integrative function, forming, in Jung's view, the basis of religion. This structural theory required a concept of energy to account for the manifestly dynamic relation of the elements he had defined. As an abstract concept, psychic energy is inevitably neutral; but in relation to structures it operates in terms of gradients. Higher and lower energy potentials exist like the positive and negative poles of an electrical system.

Jung found that his techniques of dream analysis and active imagination applied to those—like himself—in the "second half of life," to near-psychotics, or in selected cases to the clinically insane, especially to those schizophrenics in whose psychotherapy he had pioneered at the Burghölzli Asylum. He also found that a number of normal persons for whom life had lost its meaning could benefit from his findings. In these cases Jung concluded that the solution was "religious" in nature, although his meaning of the term was essentially refined and psychological.

In developing his theories and practices Jung proceeded empirically, using comparative methods. He compared clinical material, carefully obtained first from himself and then gathered from others, with relevant ethnological material. Thus he followed Freud's method, in that self-analysis was concurrent with the scientific investigation of patients.

After Jung had tested his theories and begun to publish his conclusions, his practice became international; this expansion was important for testing his concepts of archetypes and the collective unconscious. At one time more English and Americans than Swiss and Germans came to study with him—Jung's proficiency in languages facilitated this interchange. His method of teaching was to combine personal analytic treatment with seminars on dreams, visions, or a long series on Nietzsche's *Thus Spake Zarathustra.* His remarkable capacity for exposition carried his audience with him. Here his extensive knowledge of philosophy, comparative religion, myths, and other ethnological material was impressively displayed. Pupils who gathered round him returned to their own countries to practice what they had learned.

One path that Jung had been following attracted much attention: the idea that a serious disturbance existed in the European unconscious because of the one-sided development of consciousness. He was particularly struck by the threatening archetypal themes in his German patients. In 1918 Jung published a paper in which he stated that World War I would not be the end of the matter and he thought that Germany would again be a danger to western culture. The assumption of power by the Nazis was therefore no surprise to him—indeed, he was fascinated by this confirmation of his prediction. There is reason to think that Jung hoped his ideas would be of use in understanding the events taking place and that they might even influence their course. He published a number of articles and became president of the International General Medical Society for Psychotherapy in 1934, of which the German National Society was a member; it was a stormy period in which he was attacked either for being a Nazi sympathizer or inimical to the Nazi regime: he was put on their blacklist. Although his political influence was insignificant, it gained Jung the reputation of being a commentator on national and international affairs. It was his second and last excursion into the politics of psychotherapy.

Until World War II, Jung traveled widely, mainly in response to the invitations of scientific and other societies. He went several times to the United States; often to Germany, France, and Great Britain; and once to India—in each country he delivered lectures and seminars. Other travels were made to study primitive cultures: one was to meet the Pueblo Indians; the other longer excursion was to Kenya and Uganda, where he especially studied the life of the Elgonyi tribe. Jung's account of these travels, which he considered more a personal test than a scientific expedition, appeared only in his autobiography.

Jung's work attracted a number of specialists, including the sinologist Richard Wilhelm, the student of mythology Carl Kerényi, and the physicist Wolfgang Pauli, with each of whom he published the results of combined study. These intellectual interchanges were enhanced by the annual Eranos conferences at Ascona on Lake Maggiore in Switzerland, which he attended between 1932 and 1951. It attracted scholars in a variety of disciplines who discussed a common topic of psychological relevance. Jung was the focal point of these gatherings, and he used this platform to introduce a series of researches on alchemy and on the psychology of religious themes.

After World War II, Jung ceased traveling. He concentrated intensively on organizing and developing his research on alchemy, gnosticism, and early Christianity and its development (in which many heretical movements seemed a logical outcome). In addition Jung developed his theory of the self as a motive force in history and linked the Judaic and Christian traditions in a highly original way. Acutely concerned with the state of humanity, he developed the theme that man himself must mature in self-realization if he is not to become the victim of his scientific achievements. Finally he developed a theory of parapsychological data that challenged scientific thinking and method by stressing not the causes but the meaningfulness of random occurrences—the essence of his controversial theory of synchronicity, which also postulated the relativization of time and space.

During his productive last years Jung became mythologized as the "Sage of Zurich"; many traveled to consult him and gain illumination. Interviews often became treasured, recorded, and sometimes published. His profound knowledge of people enabled him to help them, but the role cast for him was not one that he liked or fostered.

Jung was averse to becoming a leader of a school or of anything resembling a sect and took active although not entirely successful steps to prevent it. He never founded an organized school of analytical psychology or trained therapists in a formal sense. He was anxious that his ideas and practices be considered part of the general development of psychological science and that they not be taken dogmatically.

Jung's recognition came first in a long series of honorary foreign degrees; in Switzerland recognition came relatively late. In 1935 he was named professor at the Eidgenössische Technische Hochschule; not until 1943 did he become professor at his old University of Basel. By then he was too old to take up his duties.

Jung's life was essentially identified with his work, which had a dual aspect. On the one hand he studied others first as a psychiatrist and later as an analytical therapist; on the other, he worked on himself and his own development. For the rest, his outer life was stable and calm. His marriage gave him a basic security, and he lived a rather typical Swiss family life with his wife and five children. His wife, who died in 1955, was also his collaborator and contributed useful research of her own.

Jung was widely known as a pioneer, with Freud and

Adler, in the early stages of dynamic psychology. Outside psychological circles his influence has been significant not only in religion and art, through his rehabilitation of symbolic expression, but also in history, economics, and the philosophy of science. The influence of Jung's researches has not yet been fully felt, and before his death groups of analysts formed to develop and modify his theory and practice. His theory of types has been subjected to further experimental investigation, with more support for the basic attitude types than for the function types. Jung's concept of archetypes has been refined, developed, and applied to childhood; his concepts of the self and individuation have been independently developed by psychoanalysts. In the process of assimilation the mixture of original theorizing and discovery will no doubt influence and change psychological and psychiatric thinking and will itself be changed by it.

BIBLIOGRAPHY

I. Original Works. Jung's collected works, Herbert Read, Michael Fordham, and Gerhard Adler, eds., are being published as Bollingen Series no. 20 (Princeton, 1953–); 18 vols. have appeared as of 1973. Other recent publications of his works are *Memories, Dreams, Reflections* (London–New York, 1963); and *Letters. Volume I, 1906–1950*, Gerhard Adler and Aniela Jaffé, eds., Bollingen Series no. 95 (Princeton, 1972).

II. Secondary Literature. On Jung and his work see Joseph Campbell, ed., *Papers From the Eranos Year Books*, Bollingen Series no. 30, 3 vols. (Princeton, 1954–1957); Henri F. Ellenberger, "Carl Gustav Jung and Analytical Psychology," in his *The Discovery of the Unconscious* (New York, 1970), pp. 657–748; F. Fordham, *Introduction to Jung's Psychology* (Harmondsworth, 1953); M. Fordham, *Children as Individuals* (New York–London, 1969); and W. Pauli, "The Influence of Archetypal Ideas on the Scientific Theories of Kepler," in C. G. Jung and W. Pauli, *The Interpretation and Nature of the Psyche* (London–New York, 1955), pp. 151–240.

Michael Fordham

JUNGIUS, JOACHIM (*b.* Lübeck, Germany, 22 October 1587; *d.* Hamburg, Germany, 23 September 1657), *natural science*, *mathematics*, *logic.*

Jungius was the son of Nicolaus Junge, a professor at the Gymnasium St. Katharinen in Lübeck who died in 1589, and Brigitte Holdmann, who later married Martin Nortmann, another professor at St. Katharinen. Jungius attended that Gymnasium, where he commented on the *Dialectic* of Petrus Ramus, as well as writing on logic and composing poetry, then entered the Faculty of Arts of the University of Rostock in May 1606.

At Rostock Jungius studied with Johann Sleker, from whom he learned metaphysics in the tradition of Francisco Suarez and his school. In general, however, he preferred to concentrate on mathematics and logic. In May 1608 Jungius went to the new University of Giessen to continue his studies. He took the M.A. at Giessen on 22 December 1608, and remained there until 1614 as professor of those disciplines then generally designated as mathematics. His inaugural dissertation was the famous oration on the didactic significance, advantage, and usefulness of mathematics for all disciplines, which he later repeated at Rostock and Hamburg and which revealed the idea that guided his lifework. He ardently pursued mathematical studies. He copied a book by F. Viète, although which one is not known, and in 1612 and 1613, while on a journey to Frankfurt, observed sunspots, the existence of which had been confirmed by Johann Fabricius and Christoph Scheiner.

At this time Jungius was attracted to pedagogy. In 1612 he traveled to Frankfurt with Christoph Helvich of the University of Giessen to attend the coronation of the emperor Matthias; there he met Wolfgang Ratke, who was trying to revive the "Lehrkunst." Jungius resigned his post at Giessen in 1614 and devoted himself to educational reform in Augsburg and Erfurt, but by the time of his return to Lübeck, on 27 July 1615, he had changed his mind in favor of the natural sciences. He began to study medicine at the University of Rostock in August 1616 and received the M.D. at Padua on 1 January 1619.

The years between 1619 and 1629 were a peak in Jungius' scientific life. He deepened his knowledge in the natural sciences while practicing medicine at Lübeck (1619–1623) and at Brunswick and Wolfenbüttel (1625) and during his tenure as a professor of medicine at the University of Helmstedt. He improved his abilities in mathematics as a professor of mathematics at Rostock in 1624–1625 and again from 1626 until 1628. He utilized this practical experience in the intensive private research that he conducted at the same time. This is particularly apparent in his "Protonoeticae philosophiae sciagraphia" and in his "Heuretica." In addition, he founded in about 1623 the Societas Ereunetica, a short-lived group dedicated to scientific research and perhaps modeled on the Accademia dei Lincei, with which Jungius had become acquainted in Italy. Finally, he was appointed professor of natural science and rector of the Akademisches Gymnasium at Hamburg, a post he held until his death.

Two tragic features characterized this last period of Jungius' life. His wife, Catharina, the daughter of Valentin Havemann of Rostock, whom he had married on 10 February 1624, died on 16 June 1638. During the 1630's, too, he became subject to the envy of his colleagues and even to attacks by the clergy, despite his devout Protestantism. He was thereafter reluctant to publish his writings and left some 75,000 pages in manuscript at the time of his death, of which two-thirds were destroyed in a fire in 1691, while the remainder have been little studied. Indeed, the primary source of Jungius' influence on his disciples and contemporaries must be sought in his correspondence and in his composition of some forty disputations.

Jungius tried to apply his mathematical training in two ways. First, he used it to solve problems, as, for example, in proving that the catenary is not, as Galileo had assumed, a parabola. Many of his problems in arithmetic and geometry, including those set out in his *Geometria numerosa* and *Mathesis specialis*, have not been found. He was one of the first to use exponents to represent powers. His experiments and views on the laws of motion are also mathematical in nature, as was explicit in the *Phoranomica*, which in part set out the instruction given by Jungius to Charles Cavendysshe, Jr., of Newcastle-upon-Tyne, when the latter mathematician was staying at Hamburg, from 8 July 1644 to February 1645, as a refugee from Cromwell's regime. In this complete, but lost, *Phoranomica* Jungius also wrote on such topics as "De impetu," "De intensione motus" (on velocity), "De tempore," and "De tendentia motuum." A specimen of this work, containing the titles of single chapters, was sent to the Royal Society of London in December 1669. Astronomy was at that time comprised in mathematics, and an account of Jungius' observations of the variable star Mira (Omicron) Ceti, made in 1647, was also sent to the Royal Society by Heinrich Sivers in a letter of 23 June 1673. While Jungius made other astronomical observations and calculations, they remained unpublished, as did his optical researches.

Second, Jungius used mathematics as a model on which to base a theory of science in general. He outlined this principle in the "Protonoeticae philosophiae sciagraphia," of which a copy was sent by Samuel Hartlib to Robert Boyle in 1654. In this paper and in his orations in praise of mathematics and his "Analysis heuretica," Jungius worked out a scientific method analogous to the mathematical mode of proof that he called "ecthesis." These works were composed more than eight years before Descartes's *Discours* appeared. In other writings Jungius rejected such Scholastic devices as single syllogisms and consequences and advocated the "clear and distinct" methodological principle of Galen. He further elaborated a theory of mathematical operations ("zetetica") that continued in more detail the "general mathematics" of the school of Proclus, Conrad Dasypodius, and Johann Heinrich Alsted. Jungius thought that this methodology was closely connected with the logical doctrine of proof that he presented in 1638 in the fourth book of his *Logica Hamburgensis*, in which he for the first time also treated such mathematical principles as "problems," "regulas," and "theorems"; abandoned distinctions in favor of exact nominal definitions; recommended a "geometric style" ("stylus protonoeticus"); and defined a systematic science ("scientia totalis"). His method of scientific inference as here set forth was based upon "demonstrations" from principles (including definitions) and upon both complete and incomplete induction.

Jungius' taste for systematizing led him to morphological studies in botany and to a corpuscular theory of chemistry, among other things. All his arguments were based on observations that he put in writing as "protonoetical papers." In botany he built his system on what Andrea Cesalpino had defined as plant morphology; some of his work was incorporated by John Ray in *Catalogus plantarum circa Cantabrigiam nascentium* (1660) and was communicated to the Royal Society of London by John Beale on 6 May 1663.

Jungius' chemical system was elaborated before 1630 and was published in two *Disputationes* (1642) and in the *Doxoscopiae physicae minores* (1662). It was based upon planned experiment and closely related to the medical tradition of the corpuscular hypothesis, as opposed to atomism. Jungius explained the apparent homogeneity of a natural body, the mechanism of chemical reaction, and the conservation of matter and weight through the assumption of invisible particles of no fixed size or shape. This enabled him to elucidate the precipitation of copper by iron in solution as an exchange of individual particles at the metal and in the solution, as opposed to the "transubstantiation" suggested by Andreas Libavius, the mere extraction from solution proposed by Nicolas Guibert and Angelo Sala, and the simple disappearance of iron particles in the solution postulated by J. B. van Helmont.

Jungius stressed that the parts of a body should be reducible to their original states with the same weights that they had originally had. In keeping with his analytical point of view he defined an element a posteriori, that is as experimentally separable. He found that gold, silver, sulfur, mercury, saltpeter, common salt, soda, and some other substances had existed as discrete elements before separation. He distinguished the bodies arrived at after separation, that is, those "exactly simple bodies," from the substantial parts, that is, "elements," in the natural body. He chose to emphasize the former, and stated that each consisted of like particles—although he did not specify how the particles of one such exactly simple body might be told from those of another. He further recognized spontaneous reactions, but referred them to attraction, and so he did not believe that any motion is inherent to the corpuscles. Like Galileo, he tried to objectify the properties of bodies and studied the transitions between their solid, liquid, and vapor phases. He was opposed to the Peripatetic notions of substantial forms and inseparable matter and fought strongly against the ideas of inherent qualities and a single principle of combustion.

Jungius' systems for botany and chemistry—cited here as an example—were products of his methodological program for all sciences, with its emphasis on observation and mathematical demonstration.

BIBLIOGRAPHY

I. Original Works. A complete bibliography of Jungius' printed works is in Hans Kangro, *Joachim Jungius' Experimente und Gedanken zur Begründung der Chemie als Wissenschaft, ein Beitrag zur Geistesgeschichte des 17. Jahrhunderts* (Wiesbaden, 1968), pp. 350–394, with photographic reproductions of nearly all title pages.

Important works published during Jungius' lifetime are *Kurtzer Bericht von der Didactica, oder Lehrkunst Wolfgangi Ratichii . . . durch Christophorum Helvicum . . . und Joachimum Jungium* (Frankfurt am Main, 1613); *Geometrica empirica* (Rostock, 1627); *Logica Hamburgensis*, bks. I–III (Hamburg, 1635), bks. I–VI (Hamburg, 1638); *Verantwortung wegen desjenigen was neulich vor und in den Pfingsten wegen des griechischen Neuen Testaments und anderer Schulsachen von öffentlicher Kanzel fürgebracht*, in Johannes Geffcken, "Joachim Jungius, Über die Originalsprache des Neuen Testaments vom Jahre 1637," in *Zeitschrift des Vereines für hamburgische Geschichte*, **5** (n.s. **2**) (1866), 164–183; *Candido lectori salutem* (Hamburg, 1639), with the incipit "Pervenit tandem hestierno die . . .," Jungius' answer to an attack by Johannes Scharff of Wittenberg; *De stilo sacrarum literarum, et praesertim Novi Testamenti Graeci* (n.p., 1639); *Compendium Logicae Hamburgensis* (Hamburg, 1641); a pamphlet (Hamburg, 1642), with the incipit "L. S. P. Philosophiae studium . . .," Jungius' invitation to the oration of his disciple Caspar Westermann; some forty *Disputationes* printed between 1607 and 1652, in which Jungius was "respondens," afterward "praesidens," the exact dates of which may be found in Kangro's bibliography (cited above); *Dokt. Joach. Jungius Reisskunst* (n.p., n.d.), only fols. A1–D4 plus one page, a free German translation from the Latin *Geometria empirica*.

Important works published after Jungius' death are *Doxoscopiae physicae minores*, Martin Fogel, ed. (Hamburg, 1662); 2nd ed., entitled *Praecipuae opiniones physicae*, M. Fogel and Johann Vaget, eds. (Hamburg, 1679), also contains *Harmonica* (n.p., n.d.) and *Isagoge phytoscopica* (preface dated 1678); *Germania Superior*, J. Vaget, ed. (Hamburg, 1685); *Mineralia*, Christian Buncke and J. Vaget, eds. (Hamburg, 1689); *Historia vermium*, J. Vaget, ed. (Hamburg, 1691); and *Phoranomica, id est De motu locali* (n.p., n.d., but perhaps not earlier than 1699, since it first appears in Johann Adolph Tassius, *Opuscula mathematica* [Hamburg, 1699]). Selections from Jungius' voluminous correspondence were published—although the collection is not perfect, some letters being presented only in extract or translation—by Robert C. B. Avé-Lallement, *Des Dr. Joachim Jungius aus Lübeck Briefwechsel mit seinen Schülern und Freunden* (Lübeck, 1863); the incipit "Quod iis evenire solet . . ." to Jungius' oration on the propaedeutic use of mathematics in studying liberal arts, presented 19 March 1629 at his inauguration in Hamburg, was edited by J. Lemcke and A. Meyer [-Abich] in *Beiträge zur Jungius-Forschung* (*Festschrift der Hamburgischen Universität anlässlich ihres zehnjährigen Bestehens*), A. Meyer [-Abich], ed. (Hamburg, 1929), pp. 94–120, with German trans. There is also "Protonoeticae philosophiae sciagraphia," the first four sheets of a copied or dictated MS, edited by H. Kangro in *Joachim Jungius' Experimente . . .*, pp. 256–271, with German trans.

A reprint, arranged by Jungius, of *Auctarium epitomes physicae . . . Dn. Danielis Sennerti* (author unknown; Wittenberg, 1635) appeared at Hamburg in 1635.

The main collection of MSS, including orations and correspondence, is in the Staats- und Universitätsbibliothek Hamburg; these include nearly all the MSS on botany, as well as part of Jungius' correspondence with John Pell. The rest of the Jungius–Pell letters are in London, BM Sloane 4279 and 4280. Other MSS are "Phoranomica" ("praelecta . . . 1644"), perhaps addressed to Charles Cavendysshe, in the Niedersächsische Landesbibliothek Hannover, MS IV, 346; "Definitiones geometricae inservientes Phoranomicae," written down by Cavendysshe, London BM Harl. 6083, fols. 246–265; and "Isagoge phytoscopica," copied or dictated before 1660, in MSS of Samuel Hartlib in the possession of Lord Delamere. Also in the Niedersächsische Landesbibliothek Hannover are Jungius' "Heuretica," partly copied by Leibniz under the title "Logica did. [actica]," LH Phil. VII C, fols. 139r–145r; "Texturae contemplatio," LH XXXVIII, fols. 26–29; "De dianoea composita lectiones," LH Phil. VII C, fols. 149r–150r, which is fragmentary; and sheets on various topics interspersed in MSS XLII 1923 of Jungius' disciple Martin Fogel.

II. Secondary Literature. The best biography, although an old one, is Martin Fogel, *Memoriae Joachimi Jungii mathematici summi . . .* (Hamburg, 1657), 2nd ed., entitled *Historia vitae et mortis Joachimi Jungii . . .* (Strasbourg, 1658). There are relevant additions by J. Moller in his *Cimbria literata* (Copenhagen, 1744), III, 342–348.

On Jungius' philosophy see G. E. Guhrauer, *Joachim*

Jungius und sein Zeitalter (Stuttgart–Tübingen, 1850), original but in need of updating. His corpuscular hypothesis and chemistry are discussed in E. Wohlwill, "Joachim Jungius und die Erneuerung atomistischer Lehren im 17. Jahrhundert," in *Festschrift zur Feier des fünfzigjährigen Bestehens des Naturwissenschaftlichen Vereins in Hamburg* (Hamburg, 1887), paper II, which presents a positivistic point of view; a new view of Wohlwill's theses is given in R. Hooykaas, "Elementenlehre und Atomistik im 17. Jahrhundert," in *Die Entfaltung der Wissenschaft* (Hamburg, n. d. [1958]), pp. 47–65; and H. Kangro, *Joachim Jungius' Experimente und Gedanken zur Begründung der Chemie als Wissenschaft, ein Beitrag zur Geistesgeschichte des 17. Jahrhunderts* (Wiesbaden, 1968). An original sketch of Jungius' botany is W. Mevius, "Der Botaniker Joachim Jungius und das Urteil der Nachwelt," in *Die Entfaltung der Wissenschaft* (Hamburg, n. d. [1958]), pp. 67–77. Texts of original MSS concerning Jungius' conflict with the clergy are Erich von Lehe, "Jungius-Archivalien aus dem Staatsarchiv," in *Beiträge zur Jungius-Forschung*, A. Meyer[-Abich], ed. (Hamburg, 1929), pp. 62–87.

On other topics see the following by H. Kangro: "Heuretica (Erfindungskunst) und Begriffskalkül—ist der Inhalt der Leibnizhandschrift Phil. VII C 139r–145r Joachim Jungius zuzuschreiben?," in *Sudhoffs Archiv, Vierteljahrsschrift für Geschichte der Medizin und der Naturwissenschaften, der Pharmazie und der Mathematik*, **52** (1968), 48–66; "Joachim Jungius und Gottfried Wilhelm Leibniz, ein Beitrag zum geistigen Verhältnis beider Gelehrten," in *Studia Leibnitiana*, **1** (1969), 175–207; "Die Unabhängigkeit eines Beweises: John Pells Beziehungen zu Joachim Jungius und Johann Adolph Tassius (aus unveröffentlichten MSS)," in *Janus*, **56** (1969), 203–209; "Martin Fogel aus Hamburg als Gelehrter des 17. Jahrhunderts," in *Ural-Altaische Jahrbücher*, **41** (1969), 14–32, containing many relations between Fogel and Jungius; and "Organon Joachimi Jungii ad demonstrationem Copernici hypotheseos Keppleri conclusionibus suppositae," in *Organon* (in press).

HANS KANGRO

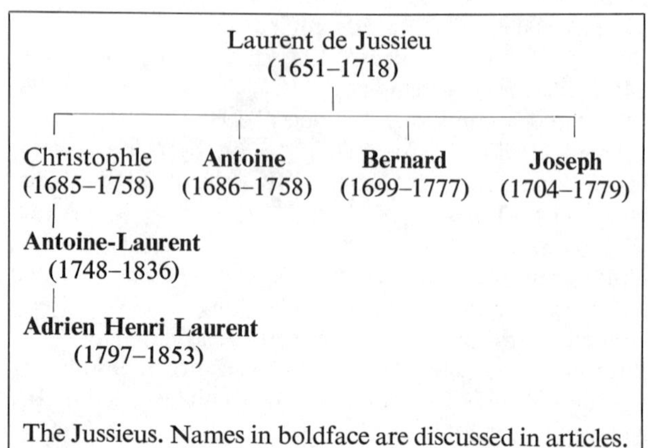

The Jussieus. Names in boldface are discussed in articles.

JUSSIEU, ADRIEN HENRI LAURENT DE (*b.* Paris, France, 23 December 1797; *d.* Paris, 29 June 1853), *botany*.

Adrien de Jussieu, the last in a long familial line of botanists, was the son of Antoine-Laurent de Jussieu. As a third-generation botanist he was able to follow his vocation with considerably less initial difficulty than his father and granduncles. He received a thorough classical training and developed a predilection for belles lettres, befriending Prosper Mérimée, Stendhal, and Victor Jacquemont. He turned to both medicine and botany, specializing from the beginning in the latter. His thesis, written in Latin (quite uncharacteristic at the time in France), consisted of a monograph on the Euphorbiaceae (1824), in which he pursued work his father had begun in tracing natural affinities on the basis of morphological relationship with attention to pharmaceutical and chemical detail.

In 1826 Jussieu succeeded his father as professor of botany at the Muséum National d'Histoire Naturelle and, judging from the number of persons who took his courses, his teaching was brilliant; his *Botanique. Cours élémentaire d'histoire naturelle* went through no fewer than twelve editions between 1842 and 1884. Jussieu's botanical work, mainly monographic, showed an increasing emphasis on the provision of new characteristics from related fields, especially anatomy and developmental morphology. His main contributions to the general theory of taxonomy were his articles "Taxonomie (végétale)" and "Géographie botanique" in d'Orbigny's *Dictionnaire universel des sciences naturelles*.

At the Muséum Jussieu, in collaboration with his friend Adolphe Brongniart, built up a large herbarium, which was supplemented, at Jussieu's death, by the family herbarium. His inestimable library was dispersed at public auction.

BIBLIOGRAPHY

I. ORIGINAL WORKS. Jussieu's publications include *De Euphorbiacearum generibus medicisque earundem viribus tentamen* (Paris, 1824); *Mémoires sur les Rutacées, ou considération de ce groupe de plantes, suivies de l'exposition des genres qui les composent* (Paris, 1825), repr. with independent pagination, from *Mémoires du Muséum d'histoire naturelle*, **12** (1825), 384–542; *Mémoire sur le groupe des Méliacées* (Paris [1832]), repr. from *Mémoires du Muséum d'histoire naturelle*, **19** (1830), 153–304; *Botanique. Cours élémentaire d'histoire naturelle à l'usage des collèges* (Paris, 1842; 12th ed., 1884); *Monographie des Malpighiacées, ou exposition des caractères de cette famille de plantes, des genres et espèces qui la composent* (Paris, 1843), repr. from *Archives du Muséum d'histoire naturelle*, **3** (1843), 5–151,

255–616, pt. 2 with independent pagination; and *Taxonomie. Coup d'oeil sur l'histoire et les principes des classifications botaniques* (Paris, 1848), repr. from C. d'Orbigny, *Dictionnaire universel de l'histoire naturelle* (Paris, 1841–1849).

II. SECONDARY LITERATURE. For information on Jussieu, see J. Decaisne, "Notice historique sur M. Adrien de Jussieu," in *Bulletin Société botanique de France*, **1** (1854), 384–400; A. Lacroix, "Notice historique sur les cinq Jussieu," in *Mémoires de l'Académie des sciences de l'Institut de France*, 2nd ser., **63** (1941), 59–62; and F. A. Stafleu, *Taxonomic Literature* (Utrecht, 1967), pp. 236–237.

FRANS A. STAFLEU

JUSSIEU, ANTOINE DE (*b.* Lyons, France, 6 July 1686; *d.* Paris, France, 22 April 1758), *botany, paleontology.*

Antoine de Jussieu was the son of Laurent de Jussieu, a Lyonnais apothecary. He was the first in the botanical dynasty that included his younger brothers Bernard and Joseph and his nephew Antoine-Laurent. Jussieu studied medicine and botany at Montpellier under Pierre Magnol, the first French botanist to attempt a natural classification of plants and the originator of the family concept in botany. Having obtained the M.D. degree on 15 December 1707, Jussieu went to Paris to study under Tournefort, who died shortly after Jussieu's arrival. Tournefort's successor as professor of botany at the Jardin du Roi, Danty d'Isnard, resigned in 1710 and Jussieu, then twenty-four, was appointed to this post, which he occupied until his death. During his first years as professor he traveled in France, Spain, and Portugal; his later years were spent in Paris.

Jussieu's published contributions to the natural sciences were numerous but relatively modest with respect to content. His main activities were the development of the Jardin du Roi and the training of pupils. As a teacher he was a faithful follower of Tournefort, and his brothers were among his students. Jussieu was also a successful physician, laying the foundation for the fortune that enabled the other family members to pursue their scientific careers.

Jussieu was the first (1715) to give a scientific description of the coffee plant, which he grew from seed obtained from the Amsterdam Botanic Garden. Although his description was detailed and precise, he did not recognize the plant, as did Linnaeus in 1737, as a genus of its own. In later years Jussieu tried to stimulate the cultivation of coffee in several of the French colonies, especially on the Île de Bourbon (now the island of Réunion).

Jussieu was responsible for a posthumous edition of Jacques Barrelier's important *Plantae per Galliam, Hispaniam et Italiam observatae* (1714) and the third edition of Tournefort's *Institutiones.* He also wrote numerous memoirs, the most noteworthy of which is his treatise (1728) on the need to establish the fungi as a separate class of plants, the *Plantae fungosae.* Discovering the fungal nature of the nongreen component of lichens, Jussieu proposed that lichens and fungi be classified together. He regarded the spores of the higher basidiomycetes as seeds.

Possibly influenced by Scheuchzer's *Herbarium diluvianum* (1709), Jussieu in 1718 was one of the first to give a correct interpretation of fossil remains of ferns found in the coal mines of the Lyons region.

He also recognized the animal nature of ammonites, and his interest in archaeology led him to publish on the various uses of flint by prehistoric tribes, insisting on the extreme patience and care with which some of these early instruments and tools had been made.

Jussieu's universality, together with an open-minded inductive approach to nature, made him a forerunner of the *philosophes*, a pioneer in colonial agriculture, and the originator of botanical hypotheses that would not be accepted until much later in the century.

BIBLIOGRAPHY

I. ORIGINAL WORKS. Jussieu's works include "Description du Coryspermum hyssipifolium," in *Mémoires de l'Académie royale des sciences* (1712), pp. 187–189; "Histoire du café," *ibid.*, 1713 (1716), pp. 291–299; "Histoire du Kali d'Alicante," *ibid.* (1717), pp. 73–78; *Discours sur le progrès de la botanique au jardin royal de Paris, suivi d'une introduction à la connaissance des plantes* (Paris, 1718); "Examen des causes des impressions de plantes marquées sur certaines pierres des environs de Saint-Chaumont dans le Lyonnais," in *Mémoires de l'Académie royale des sciences* (1718), pp. 287–297; "Réflexions sur plusieurs observations concernant la nature du Gyps," *ibid.* (1719), pp. 82–93; "Appendix," in J. P. de Tournefort, *Institutiones rei herbariae*, 3rd ed. (Paris, 1719); "The Analogy Between Plants and Animals, Drawn From the Difference of Their Sexes," in Richard Bradley, *A Philosophical Account of the Works of Nature* (London, 1721), pp. 25–32; "Recherches physiques sur les pétrifications qui se trouvent en France de divers parties de plantes et d'animaux étrangers," in *Mémoires de l'Académie royale des sciences* (1721), pp. 69–75, 322–324; "De l'origine et de la formation d'une sorte de pierre figurée que l'on nomme corne d'ammon," *ibid.* (1722), pp. 235–243; "De l'origine et des usages de la pierre de foudre," *ibid.* (1723), pp. 6–9; "Observations sur quelques ossements d'une tête d'hippopotame," *ibid.* (1724), pp. 209–215; "Description d'un champignon qui peut-être nommé champignon-lichen," *ibid.* (1728), pp. 268–272; "De la nécessité d'établir dans la méthode nouvelle des plantes, une classe particulière pour les fungus, à laquelle doivent se rapporter non seulement les champignons, les agarics, mais encore les lichen," *ibid.*, pp. 377–383; and *Traité des vertus des plantes* (Nancy, 1771, 2nd ed., Paris, 1772), a posthumous work edited and enlarged by Gandoger de Foigny.

II. SECONDARY LITERATURE. For an index to Jussieu's publications, see J. Dryander, *Catalogus bibliothecae historico-naturalis Josephi Banks*, V (London, 1800), 299. Additional references include A. Lacroix, "Notice historique sur les cinq Jussieu," in *Mémoires de l'Académie des sciences de l'Institut de France*, 2nd ser., **63** (1941), 8–21; W. J. Lütjeharms, *Zur Geschichte der Mykologie. Das XVIII. Jahrhundert* (Gouda, 1936), pp. 131–133; A. Magnin, "Prodrome d'une histoire des botanistes Lyonnais," in *Bulletin de la Société botanique de Lyon*, **31** (1906), 28; and F. A. Stafleu, *Introduction to Jussieu's Genera plantarum* (Weinheim, 1964), pp. iv–viii.

FRANS A. STAFLEU

JUSSIEU, ANTOINE-LAURENT DE (*b.* Lyons, France, 12 April 1748; *d.* Paris, France, 17 September 1836), *botany.*

Antoine-Laurent de Jussieu's father, Christophle, was the elder brother of Antoine, Bernard, and Joseph de Jussieu, and himself a dedicated amateur botanist. In 1765 Antoine-Laurent went to Paris to finish his studies at the Medical Faculty, from which he obtained a doctorate in 1770 with a thesis comparing animal and vegetable physiology. Soon afterward Jussieu became deputy to L. G. Le Monnier, professor of botany at the Jardin du Roi. In his first botanical publication (1773), a reexamination of the taxonomy of the Ranunculaceae, Jussieu developed his ideas on plant classification in general. In 1774 he published a paper on the new arrangement of plants adopted at the Jardin du Roi, an arrangement which was essentially that used by his uncle Bernard at the Trianon garden, Versailles. The paper dealt mainly with the units of classification above the family level and stressed that for the purposes of taxonomy certain characteristics had been given unequal importance.

Jussieu's thorough study of the genera and families of flowering plants (1774–1789) resulted in the publication of his epoch-making *Genera plantarum* (1789). For this work Jussieu had at his disposal not only the rich collections of living plants at the royal garden, but also his uncle's and his own rich herbarium, as well as the collections made by Philibert Commerson on his world voyage with Bougainville; the Commerson collections proved to be of critical importance for the inclusion of many tropical angiosperm families. Through an exchange of specimens, Jussieu also had access to part of Sir Joseph Banks's collections from Cook's first voyage, and another valuable London contact was with James Edward Smith, owner of Linnaeus's herbarium.

The *Genera plantarum* soon found its way to centers of botanical research. With excellent generic descriptions, the book presented a thorough summary of current knowledge of plant taxonomy. The genera were arranged in a natural system based upon the correlation of a great number of characteristics, a system which proved to be so well designed that within a few decades it was accepted by all leading European botanists, the most active proponents being Robert Brown and A. P. de Candolle. Jussieu's arrangement of families is among those elements of the *Genera plantarum* that remain a part of the contemporary system of classification.

During the French revolution Jussieu occupied a civil post in the municipal government, but in 1793, with the reorganization of the Jardin du Roi as the Muséum National d'Histoire Naturelle, he was appointed professor of botany and was charged with field courses. One of his first tasks was to set up an institutional herbarium at the museum, making use of herbaria captured by the French revolutionary armies in Belgium, Italy, and the Netherlands, and of collections and libraries confiscated from monasteries and private homes. Until 1802 much of the scientists' time at the museum was directed toward organizational duties characteristic of the revolutionary change. Jussieu signed the declaration of "hate to royalty and anarchy" presented by the Welfare Committee of the Division of Plants, and in 1800 succeeded Daubenton as director of the museum. After the resumption of scientific activity in 1802, Jussieu published six memoirs on the history of the Paris botanical garden in the new *Annales* of the museum, to which journal he continued to contribute regularly. Most of his later articles, which numbered fifty-nine, and those on plants that he wrote for the sixty-volume *Dictionnaire des sciences naturelles* (1816–1830) elaborated the principles he had set down in *Genera plantarum*, and although he made many notes, a second edition of this classic work never appeared.

Jussieu resigned from his post at the museum in 1826 and went to live with his family, where he remained until his death.

BIBLIOGRAPHY

I. ORIGINAL WORKS. Jussieu's works are "Examen de la famille des Renoncules," in *Mémoires de l'Académie royale des sciences* (1773), pp. 214–240; *Principes de la méthode naturelle des végétaux* (Paris, 1824), repr. from G. Cuvier, ed., *Dictionnaire des sciences naturelles*, III (1805); "Exposition d'un nouvel ordre de plantes adopté dans les démonstrations du jardin royal," in *Mémoires de l'Académie royale des sciences* (1774), pp. 175–197; *Genera plantarum* (Paris, 1789); other eds. are P. Usteri, ed. (Zurich, 1791); and facs. of 1789 ed. (Weinheim, 1964); "Notice historique sur le Muséum d'histoire naturelle," in *Annales du Muséum d'histoire naturelle*, **1** (1802), 1–14; **2** (1803), 1–16; **3** (1804), 1–17; **4** (1804), 1–19; **6** (1805), 1–20; **11** (1808), 1–41; and *Introductio in historiam plantarum* (Paris [1838]), posthumously published by his son Adrien de Jussieu in the *Annales des sciences naturelles*.

II. Secondary Literature. Information on Jussieu is in A. Brongniart, "Notice historique sur Antoine-Laurent de Jussieu," in *Annales des sciences naturelles, botanique*, 2nd ser., **7** (1837), 5–24; P. Flourens, *Éloge historique d'Antoine Laurent de Jussieu* (n.p., n.d.); A. Lacroix, "Notice historique sur les cinq Jussieu," in *Mémoires de l'Académie des sciences de l'Institut de France*, 2nd ser., **63** (1941), 34–47; G. A. Pritzel, *Thesaurus literaturae botanicae* (Leipzig, 1871–1877), with a list of nearly all of Jussieu's memoirs (p. 160); and F. A. Stafleu, *Introduction to Jussieu's Genera plantarum* (Weinheim, 1964).

Frans A. Stafleu

JUSSIEU, BERNARD DE (*b.* Lyons, France, 17 August 1699; *d.* Paris, France, 6 November 1777), *botany*.

At the invitation of his elder brother Antoine de Jussieu, Bernard came to Paris in 1714 to finish his botanical and medical studies. He accompanied Antoine, who was then professor of botany at the Jardin du Roi, on his travels through France, Portugal, and Spain, and then took a degree in medicine at Montpellier and another at Paris in 1726. He was appointed *sous-démonstrateur de l'extérieur des plantes* at the Jardin du Roi on 30 September 1722, filling the vacancy created by the death of Sébastien Vaillant. Jussieu was charged with field courses and supervision of the gardens and greenhouses, and it was at the royal gardens that he developed his greatest gift, that of teaching, in which he exhibited profound botanical knowledge and great personal charm. Jussieu's field trips were famous, and the list of those botanists inspired by his course includes Adanson, Guettard, Poivre, Duhamel, L. G. Le Monnier, Thouin, Claude and Antoine Richard, his brother Joseph, and his nephew Antoine-Laurent. Among those attending his classes were Buffon, Malesherbes, and foreign visitors such as Carl Linnaeus in 1738.

Although he published little, Jussieu's influence on eighteenth-century French botany was unequaled. He wrote on *Pilularia*, *Lemna*, and *Litorella* but never prepared his botanical lectures or ideas on classification for publication. Rather, it was through the arrangement of the botanical garden of the Trianon, near Versailles, that Jussieu became known as one of the great protagonists of a natural classification of plants. Louis XV, interested in horticulture and forestry, wanted a living collection of as many species of cultivated plants as possible for his Trianon garden. Jussieu was charged with the arrangement, on the recommendation of one of his amateur pupils, Louis de Noailles, Duc d'Ayen.

With the help of the gardener Claude Richard, Jussieu designed part of the garden as a "botany school" to illustrate his natural system, the layout of the grounds reflecting his ideas on the natural relationships of plants and the circumscription of his plant families. It was not until 1789 that this system appeared in print. At that time Antoine-Laurent published in his *Genera plantarum* a simple list of genera, which, according to his uncle, constituted certain natural families. Although in many ways comparable to similar enumerations published by Linnaeus in his *Genera plantarum* (1737) and *Philosophia botanica* (1751), Jussieu's work played an important role in the development of the concept of natural relationship as the main basis for classification, opposing the prevalent artificial sexual system of Linnaeus.

Jussieu had a decisive influence on the expansion of the botanical garden of the Jardin du Roi, a task which Antoine left almost entirely to him. Under Jussieu's care the collections of living plants grew rapidly, and the garden evolved from a simple apothecary's establishment to one of the best botanical gardens of the time. Jussieu corresponded with many of his colleagues abroad, and his letters to Linnaeus, published in 1855, provide perhaps the greatest source of information on his life, ideas, and character.

BIBLIOGRAPHY

I. Original Works. Jussieu reviewed and enlarged J. P. Tournefort's *Histoire des plantes qui naissent aux environs de Paris avec leur usage dans la médecine*, 2nd ed., 2 vols. (Paris, 1725). Among his other works are "Histoire d'une plante, connue par les botanistes sous le nom de Pilularia," in *Mémoires de l'Académie royale des sciences* (1739), pp. 240–256; "Histoire du Lemma," *ibid.* (1740), pp. 263–275; "Observations sur les fleurs de Plantago palustris gramineo folio monanthos parisiensis," *ibid.* (1742), pp. 121–138; "Examen de quelques productions marines, qui ont été mises au nombre des plantes, et qui sont l'ouvrage d'une sorte d'insectes de mer," *ibid.*, pp. 290–302; "Dissertation sur le Papyrus," in *Mémoires de l'Académie des inscriptions et belles-lettres*, **26** (1759), 267–320, written with A. P. de Caylus; and "Bernardi

Jussieu ordines naturales in Ludovici XV horto Triano-nensi dispositi, anno 1759," in A. L. Jussieu, *Genera plantarum* (1789), pp. lxiii–lxx.

II. SECONDARY LITERATURE. Information on Jussieu and his work is in Adrien de Jussieu, "Caroli a Linné ad Bernardum de Jussieu ineditae, et mutuae Bernardi ad Linnaeum epistolae," in *Memoirs of the American Academy of Arts and Sciences*, 2nd ser., **5** (1855), 179–234; A. Lacroix, "Notice historique sur les cinq Jussieu," in *Mémoires de l'Académie des sciences de l'Institut de France*, 2nd ser., **63** (1941), 21–34; A. Magnin, "Prodrome d'une histoire des botanistes Lyonnais," in *Bulletin de la Société botanique de Lyon*, **31** (1906), 28–29; and F. A. Stafleu, *Introduction to Jussieu's Genera plantarum* (Weinheim, 1964), vi–viii.

FRANS A. STAFLEU

JUSSIEU, JOSEPH DE (*b.* Lyons, France, 3 September 1704; *d.* Paris, France, 11 April 1779), *natural history.*

After hesitating between the medical and engineering professions, Jussieu opted for the former, thus following his elder brothers Antoine and Bernard. Like them he took an interest in botany, and his life would undoubtedly have paralleled theirs but for an event that occurred in 1735. Godin, Bouguer, and La Condamine were charged with measuring an arc of meridian in Peru near the equator; and the minister of the Navy and of the Colonies, Maurepas, an enlightened patron of scientific explorations, was seeking a physician to accompany them and to double as naturalist, collecting and describing the natural products of the countries visited. Jussieu accepted the post.

The mission, which left from La Rochelle on 16 May 1735, in 1743 completed its geodesic work between Quito, in the north, and Cuenca, in the south, in what is now Ecuador. During this period cinchona, known until then in Europe only in the form of quinine, was observed for the first time by La Condamine (1737) near Loja, while he was traveling from Quito to Lima. Jussieu returned to Loja in 1739, repeating and completing La Condamine's observations and gathering valuable data on cinchona. Their mission concluded, the party separated. Godin went to Lima, where he had accepted the chair of mathematics at the university and the post of first cosmographer of his Catholic majesty; La Condamine and Bouguer went back to France. Jussieu, left ill and penniless, was forced to earn a living and to save for his return passage by practicing medicine.

By 1745 Jussieu had saved nearly enough to pay for his return, but in the meantime his abilities and devotion had become so well known that when an epidemic of smallpox broke out in Quito, a formal order of the royal court forbade him to leave the city and threatened anyone who aided in his departure with serious penalties. Resigned to remaining, Jussieu was torn between the desire to see his friends and family and the passion for discovery. He soon found it impossible to give up a visit to a new region, even if it meant missing a chance to return home. Moreover, for want of money, he was even under pressure to abandon natural history in order to practice medicine for a living.

When he was finally able to leave the province of Quito, Jussieu began, at the request of Maurepas, a long journey to Lima, where the astronomer Godin was living. At the outset he made a detour to examine the canella tree in its natural habitat; its bark is the source of cinnamon. He arrived in Lima in 1748 and left on 27 August with Godin. The following summer they arrived at La Paz, having inspected the mercury mines of Huancavelica, crossed the Great Cordillera of the Andes, and followed the Rio Urubamba as far as Lake Titicaca, where Jussieu assembled a collection of aquatic birds.

At La Paz, once again seized by the passion to explore, Jussieu let Godin continue to Europe and traveled to the Las Yungas Mountains in the eastern Cordillera, to study the cultivation of coca. From there, continuing northeast, he entered the swampy Majos region. He then recrossed the Andes and visited Santa Cruz de la Sierra, Oruro, and Chuquisaca (Sucre), reaching Potosí in July 1749. He stayed in that city four years, studying the famous silver mines while practicing medicine and even serving as engineer in 1754 when, at the command of the governor of the province, he supervised the construction of a bridge.

Exhausted, Jussieu returned to Lima in 1755. Despondent after the death of his mother and two of his brothers and in a deteriorating physical state, he remained in Lima, caring for the poor and the rich, without the means or the will to tear himself from this draining existence. His family pleaded with him to return, and several French friends, alarmed by his state, finally convinced him to leave. He embarked in October 1770 and sailed for Spain by way of Panama before continuing to France. On 10 July 1771 he reached Paris after an absence of thirty-six years. Affectionately welcomed by his brother Bernard and his nephew Antoine-Laurent, he lived with them in the family residence. Their care partially restored his health but not his taste for life. He published nothing and no longer went out, not even to the Academy of Sciences, to which he had been elected in 1743 but which he never visited in his thirty-six years as member. Venerated by those close to him as a martyr to science, he lived in a state of despondency for eight more years in the house on the rue des Bernadins from which he had set out, young and enthusiastic, in 1735.

When he set out on his voyage home, Jussieu had left the majority of his scientific papers in Lima; they were destroyed following the death of the man to whom they were entrusted.

BIBLIOGRAPHY

I. Original Works. Jussieu's surviving scientific papers are preserved mainly at the library of the Muséum National d'Histoire Naturelle, Paris, as MSS 111, 179, 779, 1152, and 1625–1627.

See also Amédée Boinet, "Manuscrits de la Bibliothèque du Muséum d'histoire naturelle," in *Catalogue général des manuscrits des bibliothèques publiques de France, Paris*, II (Paris, 1914), 19, 26–27, 131, 191, 242; and Yves Laissus, "Note sur les manuscrits de Joseph de Jussieu, 1704–1779, conservés à la Bibliothèque centrale du Muséum national d'histoire naturelle," in *Comptes-rendus du 89e Congrès national des sociétés savantes, Lyon 1964, Section des sciences*, III, *Histoire des sciences* (Paris, 1965), 9–16.

II. Secondary Literature. On Jussieu and his work, see Condorcet, "Éloge de M. de Jussieu," in *Histoire de l'Académie royale des sciences* for 1779 (Paris, 1782), pp. 44–53; Charles-Marie de La Condamine, *Journal du voyage fait par ordre du roi à l'équateur, servant d'introduction historique à la mésure des trois premiers degrés du méridien . . .* (Paris, 1751); and Alfred Lacroix, *Notice historique sur les cinq de Jussieu membres de l'Académie des sciences* (Paris, 1936), pp. 48–59, repr. in Lacroix, *Figures de savants*, IV (Paris, 1938), 159–173, with portrait.

Yves Laissus

JUSTI, JOHANN HEINRICH GOTTLOB VON (*b.* Brücken, Thuringia, Germany, 25 December 1720; *d.* Küstrin, Germany, 21 July 1771), *political economy, mining.*

The son of a tax assessor, Justi spent his earliest years, concerning which we have only dubious information, in modest circumstances. His career can be followed with some certainty from the commencement of his legal studies at the University of Wittenberg. He interrupted these studies after a short time to enter the Prussian military service, and he participated in the first Silesian War in 1741–1742. At the end of the war he continued his education in Jena and Leipzig. In 1747 he received a prize from the Prussian Academy of Sciences for his work on monads, and following the conclusion of his studies in 1747 he took the post of estates manager for the duchess of Sachsen-Eisenach at Sangerhausen.

Justi married twice and had several sons and daughters from each marriage.

In 1751 Justi accepted a professorship of cameralistics at the newly established Theresian Academy (Theresianum) in Vienna, where he gave lectures on financial and fiscal science and where he also occupied the chair of German eloquence and rhetoric. Along with this teaching activity, he was entrusted in Vienna with administrative tasks because of his extensive knowledge of government and finance, and he was appointed imperial counselor of finances and mines (K. K. Finanz- und Bergrat). Justi's good fortune did not last, however; as a result of failures in mining ventures, he lost the respect and trust of his superiors and in 1754 resigned from the Austrian civil service. From that time on he wrote his name as von Justi, asserting that Emperor Francis I had ennobled him.

Thereafter Justi led an unsettled life which took him to Leipzig, Erfurt, and Göttingen. He did not remain long in Göttingen, even though in 1755 he was given the position of mining counselor and chief of police (*Polizeidirektor*) there, with the prerogative of delivering lectures at the university. In 1757 he again left the electorate of Brunswick, brought his family to Altona, and went himself to Copenhagen, where he obtained a commission from Count Bernstorff to inspect the Jutland heath region and to submit proposals for its cultivation.

After completing this assignment and rejecting a highly paid position as Norwegian superintendent of mines, Justi left Denmark in 1758 for Berlin on the advice of the Prussian state official Hecht—in order to seek employment in the state administration. The government held out hope of an appointment as soon as the Seven Years' War ended. In the subsequent period of involuntary leisure (1758–1766), Justi displayed an uncommonly varied literary activity. He worked on two prize questions posed by the Bavarian Academy of Sciences in Munich and won both prizes in 1761. In recognition of these extraordinary and important achievements, he was offered the position of president of that scientific society. Justi refused, on personal grounds, whereupon in 1762 the Academy elected him an honorary member.

By 1766 Frederick the Great, honoring his promise of an appointment, made Justi superintendent of mines and inspector general of the state mines as well as of glass and steel works in Prussia. Unfortunately, excessive writing had so weakened Justi's eyes that he could only do his work with the help of an assistant appointed by the king. A short time later Justi became completely blind. A growing sense of grief and distrust at his fate led him to be headstrong and injudicious in certain actions and brought him into discredit for supposedly squandering state funds. In 1768 he was dismissed from his position on this charge, although it had not been proved. In order to exonerate himself and win his reinstatement, Justi called for an investigatory commission and voluntarily entered state custody at Küstrin. He died there of a stroke on 21 July 1771 before the end of the trial.

Justi was respected for his extraordinary abilities and diligence, which were evident in both his scientific and his literary work. He wrote as effortlessly as he grasped things mentally. Although his style lacks polish, it has something original and naïve, and Justi's friends compared him in character and style to Buffon.

Justi's scientific importance undoubtedly lies—as his numerous publications show—in the field of political science. Blessed with a rich knowledge of the real conditions of public life and public administration, he independently brought a new direction to this field. In other areas such as mineralogy and mining, however, he was reproached for his ignorance. The first German systematist of the political sciences, including police science, financial science, and industrial organization, Justi originally based his political economics on mercantilism. Later the influence on him of Montesquieu and the Encyclopedists became noticeable, and Justi came to represent more the newly founded physiocratic doctrines than the previously reigning mercantilist viewpoint.

BIBLIOGRAPHY

I. Original Works. Justi's works include *Von den römischen Feldzügen in Deutschland* (Copenhagen, 1748); *Das entdeckte Geheimnis der neuen sächsichen Farben . . .* (Vienna, 1750); *Von der Abtretung des Reichslehns im Frieden mit auswärtigen Mächten* (Vienna, 1751); *Gutachten von dem vernünftigen Zusammenhange und praktischen Vortrage aller ökonomischen und Kameralwissenschaften . . .* (Leipzig, 1754); *Neue Wahrheiten zum Vorteil der Naturkunde und des gesellschaftlichen Lebens der Menschen*, 2 vols. (Leipzig, 1754–1758); *Abhandlung von den Mitteln, die Erkenntnis in den ökonomischen und Kameralwissenschaften dem gemeinen Wesen recht nützlich zu machen* (Göttingen, 1755); *Entdeckte Ursachen des verderbten Münzwesens in Deutschland* (Leipzig, 1755); *Staatswirtschaft, oder systematische Abhandlung aller ökonomischen und Kameralwissenschaften . . .* (Leipzig, 1755; 2nd ed., 1758); *Grundsätze der Polizeiwissenschaft in einem vernünftigen . . .* (Göttingen, 1756; 2nd ed., 1759; 3rd ed., 1782); *Der handelnde Adel, welchem der kriegerische Adel entgegengesetzt wird* (Göttingen, 1756); *Göttingische Polizeiamtsnachrichten . . .* (Göttingen, 1757); *Grundriss des gesamten Mineralreiches worinnen alle Fossilien in einem, ihren wesentlichen Beschaffenheiten gemässen, Zusammenhange vorgestellt und beschrieben werden* (Göttingen, 1757); *Rechtliche Abhandlung von den Ehen, die ungültig und nichtig sind* (Leipzig, 1757); *Anweisung zu einer guten deutschen Schreibart* (Leipzig, 1758); *Die Chimäre des Gleichgewichts von Europa . . .* (Altona, 1758); *Die Chimäre des Gleichgewichts der Handlung und Schiffahrt* (Altona, 1759); *Die Folgen der wahren und falschen Staatskunst in der Geschichte des Psammitichus, Königs von Egypten und der damaligen Zeiten*, 2 pts. (Frankfurt am Main, 1759–1760); and *Der Grundriss einer guten Regierung* (Frankfurt am Main, 1759).

In the 1760's Justi published *Abhandlung von der Macht, Glückseligkeit und Kredit eines Staates* (Ulm, 1760); *Historisch-juristische Schriften*, 2 vols. (Frankfurt am Main, 1760); *Die Natur und das Wesen der Staaten als die Grundwissenschaft der Staatskunst, der Polizei und aller Regierungswissenschaften . . .* (Berlin, 1760; 2nd ed., Mietau–Leipzig, 1771); *Von der Vollkommenheit der Landwirtschaft* (Ulm, 1760); *Abhandlung von der Vollkommenheit der Landwirtschaft und der höchsten Kultur der Länder* (Ulm, 1761); *Fortgesetzte Bemühungen zum Vorteil der Naturkunde und des gesellschaftlichen Lebens des Menschen* (Berlin–Stettin, 1759–1761); *Die Grundveste zu der Macht und Glückseligkeit der Staaten . . .*, 2 vols. (Königsberg–Leipzig, 1760–1761); *Moralische und philosophische Schriften*, 3 vols. (Berlin, 1760–1761); *Oekonomische Schriften über die wichtigsten Gegenstände der Stadt- und Landwirtschaft*, 2 vols. (Berlin, 1760–1761; 2nd ed., 1766–1767); *Vollständige Abhandlung von den Manufakturen und Fabriken*, 2 vols. (Copenhagen, 1758–1761), 2nd ed., J. Beckmann, ed. (Berlin, 1780), 3rd ed., Beckmann, ed. (Berlin, 1788); *Abhandlung von den Steuern und Abgaben* (Königsberg, 1762); *Von dem Manufaktur- und Fabrikreglement* (Berlin, 1762); *Vergleich der europäischen Regierung mit der asiatischen* (Berlin, 1762); *La chimère de l'équilibre du commerce et de la navigation* (Copenhagen, 1763); *Gesammelte politische und Finanzschriften über wichtige Gegenstände der Staatskunst, Kriegswissenschaft und des Kameral- und Finanzwesens*, 3 pts. (Copenhagen, 1761–1764); *Die Kunst, Silber zu raffinieren* (Königsberg, 1765); *Scherzhafte und satyrische Schriften*, 3 vols. (Berlin, 1765); *System des Finanzwesens nach vernünftigen . . .* (Halle, 1766); *Gesammelte chymische Schriften . . .*, 3 vols. (Berlin–Leipzig, 1760–1771).

Works of the 1770's are *Geschichte des Erdkörpers . . .* (Berlin, 1771); and *Gekrönte Abhandlung über die Frage . .*, (Leipzig, 1776). Justi was editor of *Deutsche Memoires, oder Sammlung verschiedener Anmerkungen, die Staatsklugheit und das Kriegswesen betreffend*, 3 vols. (Vienna, 1750). He also translated the first four vols. of the encyclopedia *Description des arts et métiers*, Diderot and d'Alembert, eds., under the title *Schauplatz der Künste und Handwerke . . .* (Berlin, 1762–1765).

II. Secondary Literature. See the following articles and works: D. M., "Précis historique sur la vie de Mr. Justi," in *Journal des Sçavans combiné avec les meilleurs journaux anglois* (Amsterdam, Sept. 1777), p. 460; Johann Beckmann, in *Physikalisch ökonomische Bibliothek*, **10** (1779), 458; J. S. Putter, *Akademische Gelehrtenge-*

schichte von der Universität Göttingen (Göttingen, 1765–1788), I, 113; II, 68; J. D. A. Hock, in *Magazin der Staatswirtschaft und Statistik*, **1** (1797), 29 ff.; C. G. Salzmann, *Denkwürdigkeiten aus dem Leben ausgezeichneter Deutschen des 18. Jahrhunderts* (Schnepfenthal, 1802), pp. 681 ff.; J. Beckmann, *Vorrat kleiner Anmerkungen*, III (Göttingen, 1806); J. G. Meusel, *Lexikon der von 1750–1800 verstorbenen deutschen Schriftsteller*, VI (Leipzig, 1806); J. S. Ersch and J. G. Gruber, *Allgemeine Enzyklopädie der Wissenschaften und Künste*, 2nd ed., pt. 30 (Leipzig, 1853), pp. 15–16; J. Kautz, *Theorie und Geschichte der Nationalökonomik*, II (Vienna, 1860), 292–293; W. Roscher, "J. H. G. Justi," in *Archiv für sächsische Geschichte*, **6** (1867), 76 ff.; K. Walcker, *Schutzzölle, laissez-faire und Freihandel* (Leipzig, 1880), pp. 568–569; "Johann Heinrich Gottlob von Justi," in *Allgemeine deutsche Biographie*, XIV (Leipzig, 1881), 747–753; G. Marchet, *Studien über die Entwicklung der Verwaltungslehre in Deutschland* (Munich–Leipzig, 1885); G. Deutsch, "Justi und Sonnefels," in *Zeitschrift für die gesamte Staatswissenschaft*, **44** (1888), 135 ff.; "J. H. G. von Justi," *ibid.*, **45** (1889), 554 ff.; and "Johann Gottlob von Justi, der erste Lehrer der Kameralwissenschaft in Oesterreich," in *Oesterreichisch-ungarische Revue* (Jan. 1890); J. K. Ingram, "Justi," in Palgrave, *Dictionary of Political Economy*, II (London, 1896), 499; *Festschrift zur Feier des 150 jährigen Bestehens der K. Gesellschaft der Wissenschaften zu Göttingen. Beiträge zur Gelehrtengeschichte Göttingens* (Berlin, 1901), pp. 495 ff.; F. Frensdorff, "J. H. G. von Justi," in *Nachrichten der kgl. Gesellschaft der Wissenschaften zu Göttingen*, phil.-hist. Kl., no. 4 (1903); W. Stieda, "Die Nationalökonomie als Universitätswissenschaft," in *Abhandlungen der K. Sächsischen Gesellschaft der Wissenschaften*, phil.-hist. Kl., **25** (1906), no. 2; A. Jaeger, *Vergleichende Darstellung der Ansichten von R. Price und J. H. G. von Justi über die Staatsschuldentilgung* (Diessen, 1910), phil. diss. (Erlangen, 1910); C. Meitzel, "Johann Heinrich Gottlob von Justi," in *Handwörterbuch der Staatwissenschaften*, 4th ed., V (Jena, 1923), 535–536; A. Tautscher, "Johann Heinrich Gottlob von Justi," in *Handwörterbuch der Sozialwissenschaften*, V (Stuttgart–Tübingen–Göttingen, 1956), 452–454; J. Remer, *Johann Heinrich Gottlob von Justi. Ein deutscher Volkswirt des 18. Jahrhunderts* (Stuttgart, 1938); M. Koch, *Geschichte und Entwicklung des bergmännischen Schrifttums* (Clausthal–Zellerfeld, 1960), diss.; and Walter Serlo, *Männer des Bergbaus* (Berlin, 1937), pp. 81–82.

M. KOCH

KABLUKOV, IVAN ALEXSEVICH (*b.* Selo Prussi, Moskovskaya Guberniya, Russia, 2 September 1857; *d.* Tashkent, U.S.S.R., 5 May 1942), *chemistry.*

Kablukov was the son of A. F. Kablukov, a doctor who came from a family of serfs, and E. S. Storozhevaya, who also came from a peasant family. He completed courses at the School of Physics, Mathematics, and Sciences of the University of Moscow, where he studied under V. V. Markovnikov. In 1885 he became a privatdocent, and in 1903 he became a professor at the University of Moscow. For many years, beginning in 1899, he conducted a course in inorganic chemistry at the Moscow Agricultural Institute and at the Academy of Trade and Industry (1933–1941). In 1928 he became a corresponding member of the Academy of Sciences of the U.S.S.R. and in 1932 he was elected an honorary member.

The development of physical chemistry in Russia owes much to Kablukov's work as a scientist and teacher. One of the first to investigate the electrical conductivity of nonaqueous solutions (methyl, ethyl, and isobutyl alcohols) Kablukov discovered the effect of anomalous conductivity, namely, that molecular electroconductivity of ethereal solutions of HCl diminishes with dilution. He is one of the founders of the theory of ionic hydration. In 1891 he reached the conclusion that "water, in disintegrating the molecules of a dissolved substance, enters with the ions into unstable compounds which are in a state of dissociation." These concepts served as a basis for amalgamation of Mendeleev's chemical theory of solutions and Arrhenius' theory of electrolytic dissociation. For many years Kablukov was a close friend of Arrhenius, whose theory he defended and promoted in Russia.

In the field of thermochemistry, Kablukov demonstrated (1887) that the heats of formation of isomeric organic molecules are dissimilar. Using the results of his thermochemical research as a basis, Kablukov formulated a number of laws on the reaction capacity of organic compounds:

(1) When organic oxides combine with halides, acid halogen joins the most hydrogenated carbon atom and hydroxyl the least hydrogenated.

(2) The heat of combination of bromine with unsaturated hydrocarbons of the ethylene series increases on transition from the lower to the higher homologues.

(3) The substitution in an unsaturated hydrocarbon of one atom of hydrogen for one of bromine retards the combining reaction of bromine.

Kablukov was the author of many study manuals in organic and physical chemistry.

BIBLIOGRAPHY

Kablukov's writings include *Glitseriny, ili trekhatomnye spirty i ikh proizvodnye* ("Glycerines, or the Triatomic Alcohols and Their Derivatives"; Moscow, 1887); "Über die elektrische Leitfähigkeit von Chlorwasserstoff in verschiedenen Lösungsmitteln," in *Zeitschrift für physikalische*

Chemie, **4** (1889), 429–434; *Sovremennye teorii rastvorov (van't Hoff, Arrhenius) v svazi c ycheniem o khimicheskom ravnovesii* ("Modern Theories of Solutions"; Moscow, 1891); "Sur la chaleur dégagée dans la combinaison du brome avec quelques substances non saturées de la série grasse," in *Journal de chimie physique et de physico-chemie biologique*, **4** (1906), 489–506, and **5**, nos. 4–5 (1907), 186–202, written with V. F. Luginin; *Osnovnye nachala neorganicheskoi khimii* ("Fundamentals of Inorganic Chemistry"; Moscow, 1900; 13th ed., 1936); *Osnovnye nachala fisicheskoi khimii* ("Fundamentals of Physical Chemistry"; Moscow, 1900; 2nd ed., 1902, 3rd ed., 1910); and *Pravilo faz v primenii k nakyshchennym rastvoram solei* ("The Phase Rule in Its Application to Saturated Salt Solutions"; Leningrad, 1933).

A secondary source is Y. I. Soloviev, M. I. Kablukov, and E. V. Kolesinikov, *Ivan Aleksevich Kablukov* (Moscow, 1957).

Y. I. SOLOVIEV

KAEMPFER, ENGELBERT (*b.* Lemgo, Germany, 16 September 1651; *d.* Lemgo, 2 November 1716), *geography, botany*.

Kaempfer's father, Johannes Kemper (Engelbert later changed the spelling of the family name) was a Lutheran minister, first pastor of the Nicolai church in Lemgo. His mother, Christine Drepper, the daughter of Kemper's predecessor as pastor, died young; his father's second wife, Adelheid Pöppelmann, bore him six more children. Kaempfer's oldest brother, Joachim, who studied law in Leiden, later became city mayor of Lemgo.

Kaempfer felt an urge to travel from an early age, and this is reflected to some extent in his schooling. He attended the Latin schools of Lemgo (1665) and Hameln (1667), the Gymnasia of Lüneburg (1668–1670) and Lübeck (1670–1672), and the Athenaeum of Danzig (1672–1674), where his first book was published, *Exercitatio politica de majestatis divisione* (1673).

For his university studies, Kaempfer went to Thorn (1674–1676); to Cracow (1676–1680), where he studied languages, history, and medicine, and obtained a master's degree; and finally to Königsberg (1680–1681), where he studied physics and medicine. After completing his studies, he traveled by way of Lemgo to Sweden, where he lived in Uppsala and Stockholm until 1683.

His wish to undertake a great journey was fulfilled when he was invited to join the embassy sent by King Charles XI of Sweden to the shah of Persia; Ludwig Fabritius was the ambassador, Kaempfer his secretary and also physician to the embassy. The group left Stockholm in March 1683 and traveled via Helsingfors, Narva, Novgorod, Moscow, and Saratov to Astrakhan, before crossing the Caspian Sea and arriving at Isfahan, capital of Persia, in March 1684.

During the journey through Persia, Kaempfer made several side trips. He climbed Mount Barmach (not identified, but possibly Mount Babadag, northwest of Baku), and visited the "burning earth" (from oil or gas seepage) near Baku and the Apsheron peninsula. Obliged to wait with the embassy for a year and a half before being received at court, Kaempfer used this time to study the Persian language, the geography of Isfahan and its surroundings, and the flora of the country. Wishing to continue his voyage instead of returning with the embassy, he joined the Dutch East India Company and was stationed as a physician at Bandar Abbas, from which he explored the surrounding area. In 1688 and 1689, he served as ship physician traveling between Indian ports.

Kaempfer arrived in Java in October 1689. The following year, he was appointed to accompany the annual voyage to Japan of the East India Company as a physician. He remained in Nagasaki from September 1690 to October 1692 and twice accompanied the chief of the factory at Deshima on his embassy to Edo (now Tokyo). In Nagasaki he made a profound study of Japanese history, geography, customs, and flora. Soon after his return to Java in March 1693 he left for Holland, arriving there in October 1693.

Arrived in Holland, Kaempfer visited many prominent scientists and earned his doctorate in medicine at the University of Leiden. In 1694 he returned home to Lemgo, where he settled on the estate "Steinhof" in the neighboring village of Lieme. He intended to spend his remaining years writing about his ten-year travels; unfortunately, these plans were only partly realized. He was soon appointed court physician to Friedrich Adolf, count of Lippe, and the post left him little free time. He held this position until his death in 1716.

In December 1700, Kaempfer married Maria Sophia Wilstach, who was much younger than he. The marriage, which was far from successful, may have hampered his literary production even more than did his occupation as court physician.

Apart from Kaempfer's doctoral dissertation, which contained observations made during his travels, only one book resulting from his journeys was published during his lifetime *Amoenitatum exoticarum* (1712). In this work Kaempfer presents his observations on Persia and adjacent countries; information on Japanese paper-making and a brief discussion of Japan; a number of discussions on various topics of natural history; a long chapter on the date palm; and, finally, a catalog of Japanese plants that must have been intended as a prodromus for a more complete flora of Japan. The description of the nearly 500 plants is often brief and cryptic. In most cases, Kaempfer gives the Japanese names, both *kun* and *on* readings, and in many cases the Chinese characters of these names. (In Japanese, a Chinese character has at least two pronunciations, analogous to Latin and vernacular names in Western usage. The *on* reading is associated with the old Chinese pronunciation, while the *kun* is the true Japanese pronunciation.) But, because of his orthography of Japanese names and his imperfect rendering of Chinese characters, it is difficult to determine the identity of many plants. This is probably the reason why the work did not attract much attention at the time: Linnaeus in his *Species plantarum* (1753) mentions only a few of them. Attempts at identifying Kaempfer's plants have been made by Karl Peter Thunberg, J. G. Zuccarini, and the Japanese scientists Ishida Chō and Katagiri Kazuo.

After Kaempfer's death, his manuscripts passed into the hands of Sir Hans Sloane, who had the German manuscript on Japan translated and published. The resulting *History of Japan* (1727) was for more than a century the chief source of Western knowledge of the country. It contains the first biography of Kaempfer, an account of his journey, a history and description of Japan and its fauna, a description of Nagasaki and Deshima; a report on two embassies to Edo with a description of the cities which were visited on the way; and six appendixes, on tea, Japanese paper, acupuncture, moxa, ambergris, and Japan's seclusion policy.

It is regrettable that Kaempfer did not document more of his experiences and observations and that so few of those that were documented appeared in print. But even the small portion of his work that was published is sufficient to insure Kaempfer the gratitude of the Orientalist, and the student of Tokugawa Japan.

BIBLIOGRAPHY

I. Original Works. After Kaempfer's death, his MSS were bought by Sir Hans Sloane; they are now in the British Museum. For an index see E. J. L. Scott, *Index to the Sloane Manuscripts in the British Museum* (London, 1904), p. 286. Karl Meier Lemgo, a lifelong student of Kaempfer's career, mentions on p. 42 of his 1960 book a holograph flora of Persia, which is not mentioned by Scott. A holograph copy of the *Geschichte* was later found in Germany in the estate of a niece of Kaempfer; the MS served as the basis for the German edition of this book, which therefore contains Kaempfer's own text.

Among Kaempfer's published works see *Exercitatio politica de Majestatis divisione* (Danzig, 1673); *Decas miscellanearum observationem* (Leiden, 1694).

Amoenitatum exoticarum politico-physico-medicarum, fasciculi V, etc. (Lemgo, 1712). The five parts are: Relationes de aulae Persiae statu hodiernis; Relationes et observationes historico-physicas de rebus variis; Observationes physico-medicas curiosas; Relationes botanico-historicas de palme dactylifera in Perside cressante; and Plantarum Japonicarum, quas regnum peragranti solum natale conspiciendas objectit, nomina et characteres sinices, intermixtis, pro specimine, quarandam plenis descriptionibus, unà cum iconibus.

Geschichte und Beschreibung von Japan, Aus den Originalhandschriften des Verfassers herausgegeben von Christian Wilhelm Dohm, 2 vols. (Lemgo, 1777–1779); facsimile repr. (Stuttgart, 1964); the *Geschichte* is translated into English by J. G. Scheuchzer as *The History of Japan, etc.*, 2 vols. (London, 1727); 2nd ed., T. Woodward and C. Davis, eds., 2 vols. (London, 1728), with additional material trans. from the *Amoenitatum;* 3rd ed., 3 vols. (Glasgow, 1906). There are abstracts of his trans. in several later works. The French trans. is *Histoire naturelle, civile et ecclesiastique de l'empire du Japon*, 2 vols. (The Hague, 1729); 2nd ed., 3 vols. (The Hague, 1732); and the Dutch trans., *De beschrijving van Japan etc.* (The Hague, 1729). The French and Dutch trans. are based on the English, since the original German text came into print only fifty years later. For complete titles, additional bibliographical data, and early authors discussing Kaempfer, see Cordier's *Bibliotheca Japonica*.

There is a partial Japanese trans. by the Nagasaki interpreter Shitsuku Tadao 志筑忠雄 (dates unknown), *Sakoku Ron* 鎖國論 ("Essay on National Isolation") Kyōwa 1 (1801). This book served as the basis for a discussion of the advantages and disadvantages of the national isolation policy by Kurosawa Okinamaro 黒澤翁滿 (1795–1859), *Ijin Kyōfu Fu* 異人恐怖傳("Thoughts on Fear of Foreigners"), Kaei 3 (1850). Another partial trans. was made by Kure Shūzō 呉秀三(1865–1932): *Kemperu Edo Bakufu Kikō* ケンペル江戸幕府紀行("Journal of a Trip to the Court in Edo").

According to Meier Lemgo, a Japanese trans. was published in 1937 and a copy given to the museum in the city of Lemgo by Shigetomo Koda. I have not succeeded in identifying this book.

A number of Kaempfer's drawings of Japanese plants, which are in the Sloane collection, were published by Joseph Banks as *Icones selectae plantarum, quas in Japonica collegit et delineavit Engelbertus Kaempfer, ex architypis in Musea Brittannica asservatis* (London, 1791).

II. Secondary Literature. Authors who have attempted to identify Kaempfer's plants include J. P. Thunberg, "Kaempferus illustratus I," in *Nova acta Regiae Societatis scientiarum upsaliensis*, **3** (1780), 196–209; "Kaempferus illustratus II," *ibid.*, **4** (1783), 31–40; *Flora Japonica etc.* (Leipzig, 1784), containing a repr. of *Kaempferus illustratus*, pp. 371–391; J. G. Zuccarini, "Weitere Notitzen über die Flora von Japan etc.," in *Gelehrte Anzeigen*, **18** (1844), 430–472; Ishida Chō 石田肇 and Katagiri Kazuo

片桐一男 *Kemperu no shokubutsu kenkyũ* ケンペル日本植物研究 ("Kaempfer's Research on Japanese Botany"), in *Rangaku Shiryõ Kenkyũ Kai* 蘭学資料研究会 (Society for Research on Dutch Studies), Report no. 98, 18 November 1961.

With the publications during the last thirty-five years of Meier Lemgo, previous literature on Kaempfer has become obsolete. Meier Lemgo's works on Kaempfer include *Engelbert Kämpfer: Seltsames Asien* (Detmold, 1933), containing trans. of selected chs. from the *Amoenitatum; Engelbert Kämpfer, der erste Deutsche Forschungsreisende 1651–1716* . . . (Stuttgart, 1937), with a 2nd, corrected and augmented ed. entitled *Engelbert Kaempfer erforscht das seltsame Asien* (Hamburg, ca. 1960); "Ueber die echte Mumie," in *Archiv für Geschichte der Medizin*, **30** (1937), 62–77; "Das *Stammbuch* Engelbert Kaempfers," in *Mitteilungen aus der Lippischen Geschichte und Landeskunde*, **21** (1952), 192–200. The *Stammbuch*, a *liber amicorum* which Kaempfer carried on his travels to collect mottoes and signatures of interesting people he met, is now in the Lippische Landesbibliothek in Detmold.

Also by Lemgo, see "Aus E. Kaempfers Leben und Forschung," *ibid.*, **26** (1957), 264–276; "Die Wirkung und Geltung Engelbert Kaempfers bei der Nachwelt," *ibid.*, **34** (1965), 192–228; "Die Briefe Engelbert Kaempfers," in *Abhandlungen. Mathematisch-naturwissenschaftliche Klasse. Akademie der Wissenschaften und der Literatur, Mainz*, **9** (1965), 265–314; "Engelbert Kaempfer, 1651–1716," *Mitteilungen aus dem Engelbert-Kaempfer-Gymnasium, Lemgo*, no. 15 (1967), *Die Reisetagebücher Engelbert Kaempfers* (Wiesbaden, 1968). Excerpts from Kaempfer's letters and diaries, as well as excerpts from the *Amoenitatum*, are preserved in the Sloane collection.

A novel based on Kaempfer's life is H. S. Thielen, *Der Medicus Engelbert Kaempfer entdeckt das unterhimmliche Reich* (Leipzig, *ca.* 1935). This book contains both *Dichtung* and *Wahrheit*.

PETER W. VAN DER PAS

KAESTNER, ABRAHAM GOTTHELF (*b.* Leipzig, Germany, 27 September 1719; *d.* Göttingen, Germany, 20 June 1800), *mathematics.*

Kaestner's father, a professor of jurisprudence, began early preparing him to enter that field but the young man's interests turned to philosophy, mathematics, and physics. After his *Habilitation* at the University of Leipzig in 1739, Kaestner lectured there on mathematics, logic, and natural law, as privatdocent until 1746, and then as extraordinary professor. In 1756 he was appointed professor of mathematics and physics at the University of Göttingen, where he remained for the rest of his life, becoming an influential figure through his teaching and writing; Göttingen's reputation as a center of mathematical studies dates from that time. Kaestner is also known in German literature, notably for his epigrams. He was a devout Lutheran. Kaestner married twice and had a daughter by his second wife.

Kaestner owes his place in the history of mathematics not to any important discoveries of his own but to his great success as an expositor and to the seminal character of his thought. His output as a writer in mathematics and its applications (optics, dynamics, astronomy), in the form of long works and hundreds of essays and memoirs, was prodigious. Most popular was his *Mathematische Anfangsgründe*, which appeared in four separately titled parts, each going through several editions (Göttingen, 1757–1800). Of lesser significance was his other four-volume work, *Geschichte der Mathematik* (Göttingen, 1796–1800).

From today's point of view Kaestner's historical significance lies mostly in the interest he promoted in the foundations of parallel theory. His own search for a proof of Euclid's parallel postulate culminated in his sponsorship of, and contribution of a postscript to, a dissertation by G. S. Klügel (1763) in which thirty purported proofs of that postulate are examined and found defective. This influential work prompted J. H. Lambert's important researches on parallel theory. The three men who independently founded non-Euclidean (hyperbolic) geometry in the early nineteenth century were all directly or indirectly influenced by Kaestner: Gauss had studied at Göttingen during Kaestner's tenure there; Johann Bolyai's father, Wolfgang, who personally taught his son geometry, had studied under Kaestner and had tried his own hand at proving Euclid's postulate; Lobachevsky studied mathematics at the University of Kazan under J. M. C. Bartels, a former student of Kaestner's.

As a student, Gauss is said to have shunned Kaestner's lectures as too elementary. Yet the *princeps mathematicorum* shows the influence of Kaestner, not only in the matter of parallelism but in other areas as well. Kaestner opposed, as did Gauss, the concept of actual infinity in mathematics (see, for example, Kaestner and G. S. Klügel, *Philosophische-mathematische Abhandlungen* [Halle, 1807]); and he felt the need, later clearly expressed by Gauss (*Werke* [Göttingen, 1870–1927], VIII, 222), for postulates of order in geometry. Indeed, Kaestner anticipated M. Pasch in explicitly postulating the division of the plane, by a line, into two parts, and in enunciating the needed assumptions concerning the intersections of a circle with a line or another circle (*Anfangsgründe*, I).

BIBLIOGRAPHY

I. ORIGINAL WORKS. Most of Kaestner's scientific publications are listed in the article on him in Poggendorff, I, cols. 1217–1219. Also valuable is the bibliography in the article on Kaestner in the *Biographie universelle* (Paris, 1852–1868), XXI, which includes literary works. Neither of these two bibliographies cites Kaestner's sponsorship of and contribution to the dissertation by G. S. Klügel, *Conatuum praecipuorum theoriam demonstrandi recensio, quam publico examini submittent Abrah. Gotthelf Kaestner et auctor respondens Georgius Simon Klügel* (Göttingen, 1763). For details of Kaestner's life, see his autobiography, *Vita Kestneri* (Leipzig, 1787).

II. SECONDARY LITERATURE. References to Kaestner's preparatory role in the development of non-Euclidean geometry are found in Friedrich Engel and Paul Stäckel, *Theorie der Parallelinien von Euclid bis auf Gauss* (Leipzig, 1895), pp. 138–140; and in Roberto Bonola, *Non-Euclidean Geometry: A Critical and Historical Study of Its Developments*, trans. by H. S. Carslaw (New York, 1955), pp. 50, 60, 64, 66. For Kaestner's anticipations of Pasch, see George Goe, "Kaestner, Forerunner of Gauss, Pasch, Hilbert," in *Proceedings of the 10th International Congress of the History of Science*, II (Paris, 1964), 659–661.

GEORGE GOE

KAGAN, BENJAMIN FEDOROVICH (*b.* Shavli, Kovno [Kaunas] district [now Siauliai, Lithuanian S.S.R.], 10 March 1869; *d.* Moscow, U.S.S.R., 8 May 1953), *mathematics.*

The son of a clerk, Kagan entered Novorossysky University, Odessa, in 1887, but was expelled in 1889 for participating in the democratic students' movement and was sent to Ekaterinoslav (now Dnepropetrovsk). In 1892 he passed the examinations in the department of physics and mathematics of Kiev University. He passed the examinations for the master's degree at St. Petersburg (1895), becoming lecturer at Novorossysky in 1897 and professor in 1917. Besides teaching at Novorossysky, Kagan gave higher education classes for women and presented courses at a Jewish high school. He edited *Vestnik opytnoi fiziki i elementarnoi matematiki* ("Journal of Experimental Physics and Elementary Mathematics") in 1902–1917 and was a director of a large scientific publishing house, Mathesis.

Kagan's first important work was devoted to a very original and ingenious exposition of Lobachevsky's geometry. Next he considered problems of the foundations of geometry, proposing in 1902 a system of axioms and definitions considerably different from all previously suggested, and particularly different from that of Hilbert. This system was based on the notion of space as a set of points in which to every two points there corresponds a nonnegative number—distance—invariant in respect to a system of point transformations (movements) in this space; the point, the principal element from which other figures are generated, is not defined. A very complete construction of Euclid's geometry on such a basis is in the first volume of Kagan's master's thesis, defended in 1907; the second volume contains a detailed history of the doctrines of the foundations of geometry. In 1903 Kagan presented a new demonstration, remarkable in its simplicity, of Dehn's well-known theorem on equal polyhedrons (1900). Since he was interested in Einstein's theory of relativity, Kagan also began studies in tensor differential geometry which he pursued intensively in Moscow, to which he moved in 1922.

For almost ten years Kagan was in charge of the science department of the state publishing house, and for many years he supervised the department of mathematical and natural sciences of the *Great Soviet Encyclopedia.* But his principal efforts were directed to Moscow University, where he was elected professor in 1922; in 1927 he organized a seminar on vector and tensor analysis, and from 1934 he held the chair of differential geometry. At Moscow, Kagan created a large scientific school with considerable influence on the development of contemporary geometrical thought; his disciples include Y. S. Dubnov, P. K. Rashevsky, A. P. Norden, and V. V. Wagner. Kagan himself was concerned mainly with the theory of subprojective spaces, a generalization of Riemannian spaces of constant curvature.

Kagan also wrote studies on the history of non-Euclidean geometry and published a detailed biography of Lobachevsky. He was the general editor of the five-volume edition of Lobachevsky's complete works (1946–1951).

In 1926 Kagan was raised to the rank of honored scientist of the Russian Federation; in 1943 he was awarded the U.S.S.R. State Prize.

BIBLIOGRAPHY

I. ORIGINAL WORKS. A bibliography of Kagan's writings is in Lopshitz and Rashevsky (see below). They include "Ocherk geometricheskoy systemy Lobachevskogo" ("Outline of Lobachevsky's Geometrical System"), in *Vestnik opytnoi fiziki i elementarnoi matematiki* (1893–1898), also published separately (Odessa, 1900); "Ein System von Postulaten, welche die euklidische Geometrie definieren," in *Jahresbericht der Deutschen Mathematikervereinigung*, **11** (1902), 403–424; "Über die Transformation der Polyeder," in *Mathematische Annalen*, **57** (1903), 421–424; *Osnovania geometrii* ("Foundations of Geometry"), 2 vols. (Odessa, 1905–1907); "Über eine Erweiterung des Begriffes vom projectiven Raume und dem zugehörigen Absolut," in *Trudy seminara po vektornomu i tensornomu analysu* ("Transactions of the Seminar on Vector and

Tensor Analysis"), I (Moscow–Leningrad, 1933), 12–101, repr. in Kagan's *Subproektivnye prostranstva* ("Subprojective Spaces"; Moscow, 1960); *Lobachevsky* (Moscow–Leningrad, 1944; 2nd ed., 1948); *Osnovy teorii poverkhnostey v tensornom izlozhenii* ("Foundations of the Theory of Surfaces Exposed by Means of Tensor Calculus"), 2 vols. (Moscow–Leningrad, 1947–1948); *Osnovania geometrii* ("Foundations of Geometry"), 2 vols. (Moscow–Leningrad, 1949–1956); and *Ocherki po geometrii* ("Essays on Geometry"; Moscow, 1963), a volume of collected papers and discourses.

II. SECONDARY LITERATURE. See A. M. Lopshitz and P. K. Rashevsky, *Benjamin Fedorovich Kagan* (Moscow, 1969); I. Z. Shtokalo, ed., *Istoria otechestvennoy matematiki* ("History of Native Mathematics"), II-III (Kiev, 1967–1968), see index; and A. P. Youschkevitch, *Istoria matematiki v Rossii do 1917 goda* ("History of Mathematics in Russia Until 1917"; Moscow, 1968), see index.

A. P. YOUSCHKEVITCH

KAHLENBERG, LOUIS ALBRECHT (*b.* Two Rivers, Wisconsin, 27 January 1870; *d.* Sarasota, Florida, 18 March 1941), *chemistry.*

Kahlenberg was the son of Albert Kahlenberg, a butcher who had been a sailor in his youth, and Bertha Albrecht, both immigrants from Germany. He received his early education at the local German Lutheran school and at Two Rivers High School. A short course at Oshkosh Normal School prepared him to teach in a country school near Two Rivers. After two years Kahlenberg attended Milwaukee Normal School for a year, then transferred to the University of Wisconsin in 1890 and received the B.S. with a chemistry major in 1892. A fellowship enabled him to complete his M.S. in 1893.

His interest in the newly developing field of physical chemistry led Kahlenberg to Leipzig, where he studied in Ostwald's laboratory. His dissertation dealt with the solubility of copper and lead salts in organic acids such as tartrates, a subject he had first studied at Wisconsin. The Ph.D. was granted *summa cum laude* in 1895. On returning to Wisconsin, Kahlenberg became instructor in pharmaceutical technique and physical chemistry. A year later he moved from the pharmacy school to the chemistry department, where he became instructor in physical chemistry. He rapidly climbed the academic ladder, becoming a full professor in 1901 and department chairman in 1908.

Kahlenberg began an active research program on his return to Wisconsin. Over the years he studied solutions, dialysis, gas electrodes, and the activation of gases by metals, potentiometric titration, boric acid in the treatment of blood poisoning, the use of colloidal gold in treatment of malignancies, and the use of dichloroacetic acid in medicine. He pioneered in the establishment of graduate studies in chemistry at the University of Wisconsin, the first Ph.D. being granted to Azariah T. Lincoln in 1899; the second was awarded to Kahlenberg's boyhood friend Herman Schlundt in 1901. By the time of his retirement in 1940 Kahlenberg had directed the studies of some twenty doctoral candidates.

Kahlenberg's research on nonaqueous solutions soon led him to doubt the worth of Arrhenius' theory of ionization. He became a leading opponent of the theory and for many years took issue with its supporters, who constituted a sizable majority of American chemists. His opposition was doubtless a factor in the ultimate reexamination of solution theory, which led to such modifications of Arrhenius' theory as those of Debye and Hückel. Kahlenberg never accepted such variants and as a consequence of his rigid opposition to ions lost influence in chemical circles.

Although a loyal American, Kahlenberg had a deep love for Germany and was an outspoken opponent of America's entry into World War I. This position was unpopular at the University of Wisconsin during the war years, and in 1919 Kahlenberg was demoted from his chairmanship of the chemistry department. He continued his professorship, teaching introductory chemistry to engineers, a course in solution chemistry, and courses in the history of chemistry.

Kahlenberg married Lillian Belle Heald, a fellow student at the university, in 1896. They had a daughter and two sons. During the 1920's Kahlenberg and his son Herman, one of his Ph.D. candidates, opened the Kahlenberg Laboratories at Two Rivers, Wisconsin, to produce Equisetene, a skin suture material, and certain other pharmaceuticals developed in the course of his research. The company was later moved to Sarasota, Florida.

BIBLIOGRAPHY

I. ORIGINAL WORKS. A full bibliography of Kahlenberg's publications is in N. F. Hall's biography (see below). The State Historical Society of Wisconsin holds twelve file boxes of Kahlenberg papers relevant to his activities between 1900 and 1939. One box contains articles and addresses; the rest contain correspondence. The University of Wisconsin archives also contain Kahlenberg papers, mostly dealing with his chairmanship of the chemistry department. One file box contains materials relevant to his demotion. There are also three bound volumes of his reprints.

His books are *Laboratory Exercises in General Chemistry* (Madison, Wis., 1907; 9th ed., 1938); *Outlines of Chemistry* (New York, 1909; rev. ed., 1915); *Chemistry and Its Relation to Everyday Life* (New York, 1911), written with E. B. Hart; and *Qualitative Chemical Analysis* (Madison, Wis., 1911; 3rd ed., 1932), written with J. H. Walton.

II. SECONDARY LITERATURE. The best biography of Kahlenberg is that by his colleague Norris F. Hall, "A Wisconsin Chemical Pioneer—The Scientific Work of Louis Kahlenberg," in *Transactions of the Wisconsin Academy of Sciences, Arts and Letters*, **39** (1949), 83–96, and **40** (1950), 173–183. See also A. J. Ihde and H. A. Schuette, "Early Days of Chemistry at the University of Wisconsin," in *Journal of Chemical Education*, **29** (1952), 67–72; and A. T. Lincoln, "Louis Kahlenberg," in *Industrial and Engineering Chemistry. News Edition*, **16** (1938), 336–337. There is also *Encyclopedia of American Biography*, new ed., XV (New York, 1942), 166–167. An obituary appeared in *The Capital Times* (Madison, Wis., 19 Mar. 1941).

AARON J. IHDE

KAISER, FREDERIK (*b.* Amsterdam, Netherlands, 10 June 1808; *d.* Leiden, Netherlands, 28 July 1872), *astronomy.*

Known chiefly for his reorganization of the Leiden observatory and his work on the fundamental coordinates of stars, Kaiser was the son of Johann Wilhelm Kaiser, a teacher of German, and Anna Sibella Liernur. His father died when he was eight years old and his uncle, who educated him, died when he was fourteen. By then Kaiser had already published a computation of the occultation of the Pleiades by the moon. In 1831 he married Aletta Rebecca Maria Barkey, who bore him one daughter and three sons. Although Kaiser was given the name Friedrich at birth, he preferred the Dutch form, Frederik.

In 1826 Kaiser became observer at the Leiden observatory, but the instruments were inferior and his relationship with the director Uylenbroek was tense. Kaiser left the observatory in 1831 and in the same year graduated from the university. In 1835 he gained some prominence by calculating the orbit of Halley's comet and predicting its return more accurately than any of his contemporaries. In the same year he was awarded a doctoral degree *honoris causa* by the University of Leiden. He became a lecturer in astronomy and director of the observatory in 1837, and three years later a professor. After years of strenuous observational work and a year-long campaign for a new observatory building, for which appreciable funds had been raised through a national subscription, he succeeded in inaugurating the new Leiden observatory (1861–1862), where the meridian circle was the main instrument. In planning this building he had been considerably inspired by the Pulkovo observatory; although he himself had never visited Russia, he acquired a detailed description of the Pulkovo observatory in 1854. (A history of the Leiden observatory and of the new building is found in *Annalen der Sternwarte in Leiden*, **1** [1868], intro.) Kaiser's staff was extremely small and he was overburdened by his administrative and teaching duties. Nervous and sensitive, he struggled throughout his life with bad health; he nevertheless continued to be productive and thorough in his work.

Kaiser is noted primarily for his observations and measurements of fundamental stellar positions; certainly the most precise made at that time, they became the basis for the international reputation of the Leiden observatory. Applying Bessel's classical precepts, Kaiser carefully determined any errors in his instruments or observations. In volume 1 of the *Annalen* (1868) he fully explained his methods and recorded about 16,000 meridian observations of 190 stars, which were not fully reduced. The reduced declinations for those stars, used in European triangulation, and the results for the polar height at Leiden appeared in volume 2 (1869).

Kaiser also devoted special attention to the theory of the equatorially mounted telescope, to time determination, and to a critical investigation of Airy's double-image micrometer. He advised the government on nautical instruments, becoming inspector of instruments for the navy, and on methods for position determination in the Dutch East Indies. For such purposes he invented the fluid compass and improved Steinheil's prismatic circle, which was more precise than the sextant. Kaiser represented the Netherlands on the Commission for the Triangulation of Europe and played an important role in this enterprise (1864–1871). He made numerous drawings of Mars (1862, 1864) and of the comets 1861 (II) and 1864 (II) which were posthumously published in the *Annalen*, volume 3 (1872).

Kaiser contributed in an important way to the diffusion of astronomical knowledge in the Netherlands by his popular book *De Sterrenhemel*, which had several editions; by his popular account of planet discoveries (1851); and by his *Populair Sterrekundig Jaarboek*.

BIBLIOGRAPHY

I. ORIGINAL WORKS. See *Annalen der Sternwarte in Leiden*, **1** (1868), **2** (1870), and **3** (1872). See also *De inrigting der Sterrewachten, beschreven naar de Sterrewacht op den heuvel Pulkowa en het ontwerp eener Sterrewacht* *voor de Hoogeschool te Leiden* (Leiden, 1854); *De Sterrenhemel* (Amsterdam, 1843–1844), which had several eds.; *De geschiednis der ontdekkingen van planeten* (Amsterdam, 1859); and *Populair Sterrekundig Jaarboek* (Amsterdam, 1845–1863).

II. SECONDARY LITERATURE. A biography and bibliography covering Kaiser's career up to 1868 is found in *Annalen der Sternwarte in Leiden*, **1** (1868), intro. For a general biography and complete bibliography see J. A. C. Oudemans, *Jaarboek van de K. akademie van wetenschappen gevestigd te Amsterdam* (1875), pp. 39–104. Shorter biographies appear in *Vierteljahrsschrift der astronomischen Gesellschaft*, **7** (1872), 266–273, and *Monthly Notices of the Royal Astronomical Society*, **33** (1873), 209–211.

M. G. J. MINNAERT

KALBE, ULRICH RÜLEIN VON. See **Rülein von Calw, Ulrich.**

KALM, PEHR (*b.* Ångermanland, Sweden, 6 March 1716; *d.* Turku, Finland, 16 November 1779), *natural history*.

The defeat of Charles XII of Sweden and Finland left the latter country open to a reign of Russian terror during which many people fled. Among them were Gabriel Kalm, curate of Korsnäs Chapel in Närpes parish, county of Ostrobothnia, and his wife Catharina Ross, who escaped to Sweden. Their son Pehr was born somewhere in the county of Ångermanland; the father died there and the widow returned to Finland after the Treaty of Nystad (1721). Pehr was educated at the Gymnasium in Vaasa and matriculated (1735) at the University of Åbo (founded by Queen Christina in 1640 as Åbo Academy and shifted to Helsinki when Åbo [Finnish Turku] was destroyed by fire in 1827).

The poor but gifted and well-connected boy found influential supporters among the university professors, including the professor of physics Johan Browallius and Carl Fredrik Mennander, both later to become bishops. The vice-president of the Åbo Law Court, Baron Bielke, then took him to his estate, Löfstad, near Uppsala, where for seven years Kalm served as superintendent of his experimental plantation. Bielke introduced him to his rich library of natural history and to his famous friend Linnaeus, under whose guidance Kalm completed his studies at the University of Uppsala. Bielke sent him on botanical expeditions to the south of Sweden and to Finland, and took him as a companion on a journey to St. Petersburg and Moscow. Kalm became a learned and well-trained naturalist in the pattern of his great teacher and in 1747 was named *professor oeconomiae* ("economy" here meant the utilitarian aspects of natural science) at the University of Åbo.

The great event in his life was his journey, sponsored by the Royal Swedish Academy of Sciences, to North America and Canada to discover useful plants capable of withstanding the Scandinavian climate. Kalm landed in Philadelphia in September 1748. Benjamin Franklin and two correspondents of Linnaeus, John Bartram and Cadwallader Colden, the latter lieutenant-governor of the New York colony, became helpful friends. Both of them were keen botanists admired by Kalm and Linnaeus. When this part of the country had been explored, Kalm departed in May 1749 for New York, Albany, Lake Champlain, and Canada, where French officials received him in princely fashion and paid his traveling expenses within the colony. He returned to Philadelphia in October. A second journey to Canada was undertaken in 1750 (the diary from which has been lost). In February 1751 Kalm left Philadelphia for Stockholm going thence to Åbo, where he remained for the rest of his life.

Kalm's biographer, the eminent Swedish botanist Carl Skottsberg, calls him a descriptive naturalist of the first rank, cautious, penetrating, and precise as an observer. In the *Species plantarum* of Linnaeus, Kalm was cited for ninety species, sixty of them new. The mountain laurel genus *Kalmia* was named for him. Extreme utilitarian that he was, at Åbo Kalm spent his time trying to grow economically useful plants.

Kalm's description of his American journey does not constitute a complete picture, but what remains is important enough to make it an informative source on eighteenth-century American colonial life, customs, agriculture, politics (Kalm predicted American independence), and Indian tribes. Kalm's diary (5 October 1747–31 December 1749), from which he selected material for the three volumes published in his lifetime, was discovered by Georg Schauman, chief librarian in the university library at Helsinki, and the Society for Swedish Literature in Finland has included part of it in its republication of Kalm's book on North America. To posterity, the diary itself is the most interesting part of his writings because of its wealth of cultural and ethnographic detail, and also because of the reliability of the observer. Kalm's contemporaries had little interest in these aspects of the journey.

BIBLIOGRAPHY

Peter Kalm's Travels in North America, Adolph Benson, ed., 2 vols. (New York, 1937). The Swedish original,

Fredrik Elfving and Georg Schauman, eds., 4 vols., appears in the series *Skrifter Utgivna av Svenska Litteratursällskapet i Finland:* I as vol. LXVI (Helsinki, 1904); II as vol. XCIII (Helsinki, 1910); III as vol. CXX (Helsinki, 1915); and IV (from the diary), as vol. CCX (Helsinki, 1929). The first complete publication of the diary has now begun. Vol. I (M. Kerkkonen, ed.) has appeared as vol. CDXIX (Helsinki, 1966) in the *Skrifter* series. This covers his stay in England en route to America.

See also Carl Skottsberg, "Pehr Kalm" in *Kungliga Svenska vetenskapsakademiens levnadsteckningar*, no. 139 (Stockholm and Uppsala, 1951), pp. 221–503.

RAGNAR GRANIT

KALUZA, THEODOR FRANZ EDUARD (*b.* Ratibor, Germany [now Raciborz, Poland], 9 November 1885; *d.* Göttingen, Germany, 19 January 1954), *mathematical physics.*

Theodor Kaluza was the only child of the German Anglicist Max Kaluza, whose works on phonetics and Chaucer were widely read in his day. The Kaluza family tree may be traced back to 1603, the family having been in Ratibor for over three centuries.

Kaluza was a bright student at school. Beginning his mathematical studies at the age of eighteen at the University of Königsberg, he prepared a doctoral dissertation on Tschirnhaus transformation[1] under Professor F. W. F. Meyer and qualified to lecture there in 1909. He married in the same year and remained a meagerly remunerated privatdocent in Königsberg for two decades.[2]

By the time Kaluza was past forty, Einstein, recognizing his worth and finding him in a position far below his merits, recommended him warmly for something better.[3] At last, in 1929, Kaluza obtained a professorship at the University of Kiel. In 1935 he moved to the University of Göttingen, where he became a full professor. Two months before he was to be named professor emeritus, Kaluza died after a very brief illness.

By the close of the nineteenth century, the concept of ether had become an integral part of physics. It was generally expected that the ether, and perhaps even the electromagnetic equations themselves, would explain all of physics, including gravitation. But when Einstein developed his general relativity theory (1910–1920), in which gravitational effects are traced to changes in the structure of a four-dimensional Riemannian manifold, the question arose as to whether the electromagnetic field could be incorporated into such a manifold. The aim was to give a unified picture of the gravitational and electromagnetic phenomena. This was referred to as the unitary problem.

Kaluza's essentially mathematical mind was attracted to the problem. He initiated a line of attack by introducing into the structure of the universe a fifth dimension which would account for the electromagnetic effects. When he communicated his ideas to Einstein, the latter encouraged him to pursue such an approach, submitting that this was an entirely original point of view.[4] Kaluza's major paper on this question appeared in 1921.[5] Here he combined the ten gravitational potentials, which arise in Einstein's general relativity theory as the components of the metric tensor of a four-dimensional space-time continuum, with the four components of the electromagnetic potential. He did this by means of his fifth dimension, which had the characteristic restriction that in it the trajectory of a particle is always a closed curve. This makes the universe essentially filiform with respect to the fifth dimension.

Mathematically, the five-dimensional manifold may be defined in terms of the metric

$$d\sigma^2 = \gamma^{mn} dx^m dx^n \qquad (m, n = 1, 2, 3, 4, 5),$$

in which the coefficients γ^{mn} are assumed to be independent of the fifth coordinate x^5. With the additional restriction that γ^{55} is a constant, Kaluza could deduce that the charge-mass ratio is a constant for the electron. The motion of electrically charged particles in an electromagnetic field is described by the equations of the geodesics in such a space.

If one were to assume that the periodicity of the fifth dimension is a "quantum effect"—indeed, that it is the physical source of Planck's constant—then the radius of the curves in the fifth dimension which would give the empirical value of the electron's charge would be on the order of 10^{-30} cm, and would thus be beyond the reach of experiment. (This result is due to O. Klein.)

Kaluza's theory was criticized on the ground that the fifth dimension is a purely mathematical artifice, with only a formalistic significance and no physical meaning whatever. Nevertheless, the five-dimensional idea was explored by several mathematical physicists.[6]

Kaluza also worked on models of the atomic nucleus, applying the general principles of energetics. He wrote on the epistemological aspects of relativity and was sole author of or collaborator on several mathematical papers. In 1938 a text on higher applied mathematics written by Kaluza and G. Joos was published; in this work he showed himself as a mathematician rather than as a mathematical physicist.[7]

Kaluza was a man of wide-ranging interests. Although mathematical abstraction delighted him tremendously, he was also deeply interested in languages, literature, and philosophy. He studied more than fifteen languages, including Hebrew, Hungarian, Arabic, and Lithuanian. He had a keen sense of humor. A nonswimmer, he once demonstrated the power of theoretical knowledge by reading a book on swimming, then swimming successfully on his first attempt (he was over thirty when he performed this feat). Kaluza loved nature as much as science and was also fond of children.

He was liked and respected by his students and had extremely good relations with his colleagues. He never used notes while lecturing, except on one occasion, when he had to copy down a fifty-digit number which showed up in number theory.

NOTES

1. The dissertation was published in *Archiv der Mathematik und Physik*, **16** (1910), 197–206.
2. Privatdocents did not have a definite salary; they were merely allowed the privilege of giving lectures. If a privatdocent gave x hours of lectures a week and had y students, he would earn about $5xy$ gold marks per semester, an inconsiderable sum.
3. In a note written to a colleague in November 1926 Einstein praised Kaluza's "schöpferische Begabung." He considered it unfortunate that "Kaluza unter schwierigen äusseren Bedingungen arbeitet" and added, "Es würde mich sehr freuen, wenn er einen passenden Wirkungskreis bekäme."
4. In his first reaction to Kaluza's private communication of the five-dimensional idea, Einstein wrote, ". . . der Gedanke, dies (elektrischen Feldgrössen) durch eine fünfdimensionale Zylinderwelt zu erzielen, ist mir nie gekommen und dürfte überhaupt neu sein. Ihr Gedanke gefällt mir zunächst ausserordentlich" (letter dated 21 Apr. 1919).
5. "Zum Unitärsproblem der Physik," in *Sitzungsberichte der Preussischen Akademie der Wissenschaften*, **54** (1921), 966–972. The communication was delivered by Einstein on 8 December 1921.
6. The most important of these were O. Klein, L. de Broglie, Einstein, E. P. Jordan, and Y. R. Thiry. For a detailed bibliography on these extensions the reader may consult the treatise by Tonnelat cited in the bibliography.
7. *Höhere Mathematik für die Praktiker* (Leipzig, 1938).

BIBLIOGRAPHY

I. Original Works. A bibliography of Kaluza's works is found in Poggendorff, VIIA, pt. 2 (1958), 684. I am indebted to Theodor Kaluza, Jr., for letting me see the scientific correspondence of his father, especially that with Einstein, and for relating personal details.

II. Secondary Literature. Good discussions of Kaluza's five-dimensional theory may be found in P. G. Bergmann, *An Introduction to the Theory of Relativity* (New York, 1942); and M. A. Tonnelat, *Les théories unitaires de l'électro-magnétisme et de la gravitation* (Paris, 1965).

Varadaraja V. Raman

KAMĀL AL-DĪN ABU'L ḤASAN MUḤAMMAD IBN AL-ḤASAN AL-FĀRISĪ (*d.* Tabrīz [?], Iran, 1320), *optics, mathematics.*

Kamāl al-Dīn was the disciple of the famous Qutb al-Dīn al-Shīrāzī, mathematician, astronomer, and commentator on Ibn Sīnā.[1] Scholars since Wiedemann and Sarton have linked the names of the two, and some questions of priority have arisen, as will be seen below. Although Kamāl al-Dīn produced a number of writings in different branches of mathematics—particularly arithmetic and geometry—his essential contribution was in optics. It was in response to a question addressed to him on the principles of refraction that al-Shīrāzī recommended to Kamāl al-Dīn that he study the *Kitāb al-manāẓir* ("Book of Optics") of Ibn al-Haytham. Once Kamāl al-Dīn had undertaken this study, al-Shīrāzī, who was at this time occupied in commenting on the *Canon* of Ibn Sīnā, suggested further that Kamāl al-Dīn write his own commentary on Ibn al-Haytham's book.

Kamāl al-Dīn chose to extend the task set him to other works of Ibn al-Haytham as well, so that his *Tanqīḥ al-manāẓir li-dhawi 'l-abṣār wa'l-basā'ir* contains, in addition to the originally planned study of the *Kitāb al-manāẓir*, essays on Ibn al-Haytham's *The Burning Sphere*, *The Halo and the Rainbow*, *Shadows*, *The Shape of Eclipse*, and the *Discourse on Light*. He was also led, in the course of this work, to study Ibn al-Haytham's *The Solar Rays*, although he did not comment upon it. Kamāl al-Dīn was thus dealing with the essential optical works of Ibn al-Haytham, and with this group we must also consider his own work on optics, *Al-baṣā'ir fī 'ilm al-manāẓir* ("Insights Into the Science of Optics"). This is basically a textbook for students of optics, presenting the conclusions of the *Tanqīḥ* without the proofs or experiments.

In order to grasp the meaning and scope of Kamāl al-Dīn's contribution, it must first be understood that his work was more properly a revision (*tanqīḥ*) than a commentary (*sharḥ*), as the title itself indicates. To Kamāl al-Dīn "to comment" meant a reconsideration and reinterpretation, rather than the medieval notion of a return to the original sources for a more faithful reading. In the course of his revision, Kamāl al-Dīn did not hesitate to refute certain of Ibn al-Haytham's theories, such as the analogy between impact and the propagation of light, an essential element of the explanation of reflection and refraction. He further had no reluctance in developing other of Ibn al-Haytham's ideas, notably the example of the camera obscura, refraction in two transparent spheres, and the numerical tabulation of refraction (air to glass); indeed, from time to time he simply set aside Ibn al-Haytham's doctrine to substitute one of his own. An important instance is the theory of the rainbow.

This profound change in the notion of a commentary is directly attributable to the new stage reached by Ibn al-Haytham in his optics, which may be briefly characterized as the systematic introduction of new norms—mathematical and experimental—to treat traditional problems in which light and vision are united. Until then light had been considered to be the instrumentality of the eye and to see an object was to illuminate it. In order to construct a theory of light, it was necessary to begin with a theory of vision; but to establish a theory of vision required taking a position on the propagation of light. Each task immediately involved the other and each theory borrowed the language of the other. The optics of Aristotle, like that of Euclid and even that of Ptolemy, comprised both factors. In order to introduce the new norms systematically, a better differentiation forced itself on Ibn al-Haytham. But the distinction between seeing and illuminating had to allow the transfer of the notions of a physical doctrine to an experimental situation and thus to bring about a realization of the initial project.

The essential and most representative part of Kamāl al-Dīn's work, however, is his study of the rainbow. The question of Kamāl al-Dīn's originality here has been raised; recalling that Kamāl al-Dīn had borrowed the idea of studying the rainbow from his teacher, Carl Boyer writes, "Hence the discovery of the theory presumably should be ascribed to the latter [al-Shīrāzī], its elaboration to the former [Kamāl al-Dīn]."[2] Although the same notion is supported by Crombie and many subsequent authors, it remains unconvincing, despite a manuscript text on the rainbow attributed to al-Shīrāzī (at the end of his commentary on Ibn Sīnā's *Canon*, in a manuscript kept at Paris). The manuscript, written before 1518, is incomplete, and the text dealing with the rainbow occurs after several pages on alchemy that are irrelevant to the rest of the book, and in a different hand. The text on the rainbow itself is in yet another hand; after examining this manuscript and comparing it with one of the same book in the National Library at Cairo, Naẓīf suggested that the passage is an interpolation.[3] The Cairo manuscript has in turn been compared with a complete version of the same book, copied in an elegant handwriting and dating from 1785.[4] In confirmation of Naẓīf's theory, this last altogether lacks the passage on the rainbow.

Even were this text on the rainbow to be accepted as being by al-Shīrāzī, no doubt would be cast on Kamāl al-Dīn's originality, since we have seen that Kamāl al-Dīn drew upon a new interpretation of Ibn al-Haytham's optics. The theory of the rainbow elucidated in the text in question deals with the reflection of light on droplets of water dispersed in the atmosphere, a traditional conception that does not agree with Kamāl al-Dīn's (although it is not too unlike al-Shīrāzī's, since the latter, following in the path of such geometers as al-Ṭūsī, was still concerned with visual rays). The disputed manuscript reveals a further fundamental difference from the work of Kamāl al-Dīn in its optical terminology.

Ibn al-Haytham, on the other hand, had in his discussion of the rainbow dealt specifically with the problem of reflection; that is, in order to explain the form of the arc, he had proposed that the light from the sun is reflected on the cloud before reaching the eye. He sought the condition under which a ray emanating from a source of light—the sun—and reflected on a concave spherical surface, outside the axis, passes through the eye after its reflection. Admitting, as did the Aristotelian tradition before him, the possibility of a direct study of the arc, Ibn al-Haytham did not attempt to construct an experimental situation in order to verify the geometrical hypotheses. But the direct study of the rainbow did not lend itself to this sort of proof, even though Ibn al-Haytham called for it.

Kamāl al-Dīn took up Ibn al-Haytham's project at this point. Despite Ibn al-Haytham's authority, Kamāl al-Dīn began by submitting his predecessor's attempt to a severe criticism that, essentially, showed the need of a better physics which, when joined with geometry, would allow him to reach the goal formulated but unattained by Ibn al-Haytham.

Thus Kamāl al-Dīn returned to the doctrine of the rainbow proposed by Ibn Sīnā, who conceived of the arc as being produced by reflection from a totality of the water droplets dispersed in the atmosphere at the moment when the clouds dissolve into rain. Ibn Sīnā's improvement justified an analogy—important for the explanation of the rainbow—between a drop of water and a transparent sphere filled with water.

Having stated the analogy, Kamāl al-Dīn wished to introduce two refractions between which one or several reflections occur. He benefited here from the results obtained by Ibn al-Haytham in *The Burning Sphere*, in which the latter showed that the paths followed by the light propagated between the two refractions are a function of the relationships of the increase in the angles of incidence and those of the increase in the angles of deviation.

Ibn al-Haytham established that for two rays to intersect inside the circle—that is, for the points of the second refraction to approach O' instead of moving away from each other—it is necessary that

$D' - D > 1/2 \ (i' - i)$ (compare Kamāl al-Dīn's diagram, Figure 1). While it is true that this relationship is valid for the passage from air to glass, it can be easily demonstrated that it is independent of n. Drawing upon this relationship, however, Ibn al-Haytham was able to show by a simple geometric demonstration that the angle beginning with which this intersection occurs is 50° for the case in which $n = 3/2$ (from air to glass). This can be verified by the relation $\frac{dD}{di} = 1 - \frac{\cos i}{n \cos r}$. It should be noted that Ibn al-Haytham thought that with the incident ray at 90°, the second point of refraction was on the same side of the axis as the point of the first refraction; this was not verified in the air-to-glass case that he was considering. In the water-to-air case that Kamāl al-Dīn studied, on the other hand, this was easily verifiable, so that in taking up Ibn al-Haytham's results, Kamāl al-Dīn did not encounter the same difficulty.

Kamāl al-Dīn thus considered the incident rays to be parallel to the axis OO'. These rays intersect the sphere at points increasingly removed from O and are refracted in it at points distant from O' on the opposite portion of the sphere up to the angle of incidence of 50°. For an angle of incidence greater than 50°, the points of the second refraction successively approach O'. Concerning the propagation of rays at their exit from the sphere, Ibn al-Haytham had already demonstrated spherical aberration.

With these results Kamāl al-Dīn attempted to show how, following double refraction in the sphere and depending on whether rays near to or distant from the axis are considered, one or several images of a luminous object as well as different forms can be obtained—an arc or a ring in the case of a circular object. Before treating in detail double refraction in the sphere, however, Kamāl al-Dīn eliminated a difficulty resulting from the fact that, unlike the sphere, the drop does not have a glass envelope and that there are therefore four refractions, not two, in the sphere. In order to guarantee the correspondence between the manufactured object—the sphere—and the natural object—the drop of water—Kamāl

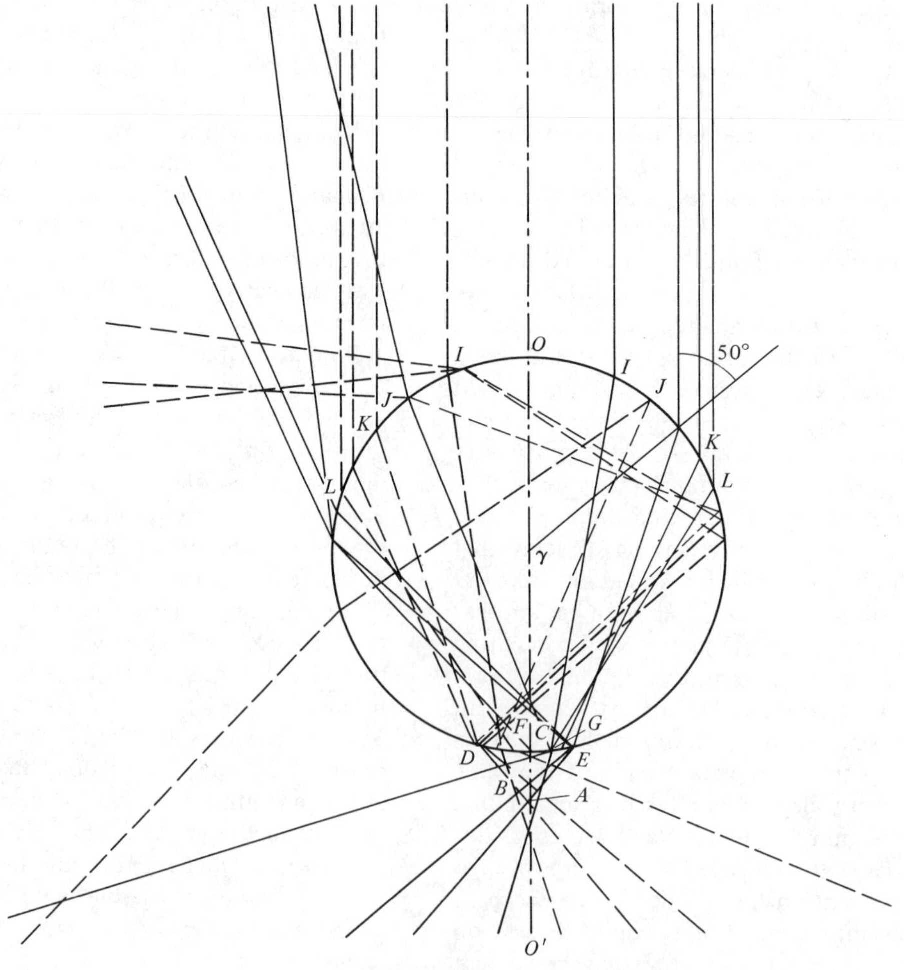

FIGURE 1

KAMĀL AL-DĪN

al-Dīn employed an approximation furnished by the study of refraction and justified by the consideration that the indexes of the two mediums are quite close, which allowed him, finally, to disregard the glass envelope.

Kamāl al-Dīn considered the circle of center γ and the rays that form angles of incidence of 10°, 20°,..., 90° with it. He divided the rays into two groups. The first five form angles of incidence of less than 50°; the four others, of more than 50°. (See Figure 2.) He divided the arc *DE* into two equal parts at O' and took *F* and *G* equidistant from O'. Let *SJ* be the ray with the angle of incidence 50° and SJ' its symmetric counterpart in relation to the axis OO'. These two rays are refracted along the lines *JE* and $J'D$ and meet after the second refraction at point *A*, exterior to the sphere on the axis. Following the first refraction, all the rays of incidence of less than 50° are contained in the interior of the trunk of the cone generated by *JE* and $J'D$, called the "central cone" by Kamāl al-Dīn. Following the second refraction, these same rays are contained within the cone generated by *EA* and *DA*, the "burning cone." The rays that constitute the second group—with angle of incidence greater than 50°—are refracted, some between *JE* and *LG* and others symmetrically between $J'D$ and $L'F$, which generate the two exterior cones, or "hollow cones." These rays are refracted a second time, some between *GB* and *EA* and some between *FC* and *DA*; they generate the exterior refracted cones or "hollow opposites." These rays intersect on the axis at points *H* and *A*.

At this stage, Kamāl al-Dīn's problem was to produce, under certain conditions, several possible images of the same object placed before the sphere. He could then vary their respective positions, causing them to become more distant from each other or superimposing them. Kamāl al-Dīn sought, in fact, to place himself outside what are today called Gauss's approximation conditions in order to produce this multiplicity of images.

He then returned to his model and complicated it with new, precise details. He examined the propagation of rays inside the sphere between two refractions and also treated the different types of reflection. Kamāl al-Dīn believed that a bundle of parallel rays falling on the drop of water is transformed, following a certain number of reflections in the sphere, into a divergent bundle. He knew, moreover, that the rays refracted in the drop of water after one or several reflections in its interior are not sent equally in all directions but produce a mass of rays in certain regions of space. This mass—and Kamāl al-Dīn's text allows no doubt on this point—is in the vicinity of the point of emergence of the ray which corresponds to the maximum (actually maximum or minimum) of deviation.[6] He stated, in addition, that the intensities of the lights join together, producing a greater illumination. He expressed these ideas in the complicated language of "cones" of rays that have been refracted after having undergone one or two reflections in the interior of the sphere and also in the concept of a greater illumination at the edges of the "cones." In the case of one reflection between two refractions, he distinguished two bundles of rays coming from the exterior cones and the central cone (see Figure 2); in the case of two reflections, he obtained two groups of rays that were more divergent than in the case of one reflection and that also gave one or two images. If the eye receives the rays coming from the central cone, Kamāl al-Dīn stated, a single image will be seen in a single position; and if the eye is placed in the region where the rays issuing from the central cone and the exterior cone intersect, two images will be seen in two positions.

In order to test the completed model, Kamāl al-Dīn employed an experimental procedure that was independently rediscovered by Descartes. He constructed a dark chamber with one opening, and placed inside it a transparent sphere illuminated by the rays of the sun. He masked half of the sphere with a dense white body and observed the face on the side toward the sphere: on it he saw an arc whose center was on the axis leading from the center of the sphere to the sun. This arc was formed from light rays that had undergone a refraction, a reflection, and another refraction. The inside of the arc was brighter than the outside because it contained rays emitted by both the central cone and the exterior cone. Kamāl al-Dīn next placed another white body, less dense than the first, before the sphere and again observed the face turned toward the sphere. This time he saw a complete ring that always displayed the colors of the rainbow. This ring was formed from the rays refracted a second time after having been reflected in the sphere. He noted the variation in the intensity of the colors according to the position of the screen, then employed the same dark chamber to consider the case of two reflections between two refractions.

This introduced into the study an important possibility that had not been considered then: the transfer through geometry of a physical doctrine of this phenomenon—essentially that of Ibn Sīnā—into the realm of experiment. It was in fact a question of restoration, contrary to Ibn al-Haytham, of the latter's own style of optics. The new optics promised to respect the norms of the combination of geometry and physics. But to follow the new norms with some

KAMĀL AL-DĪN

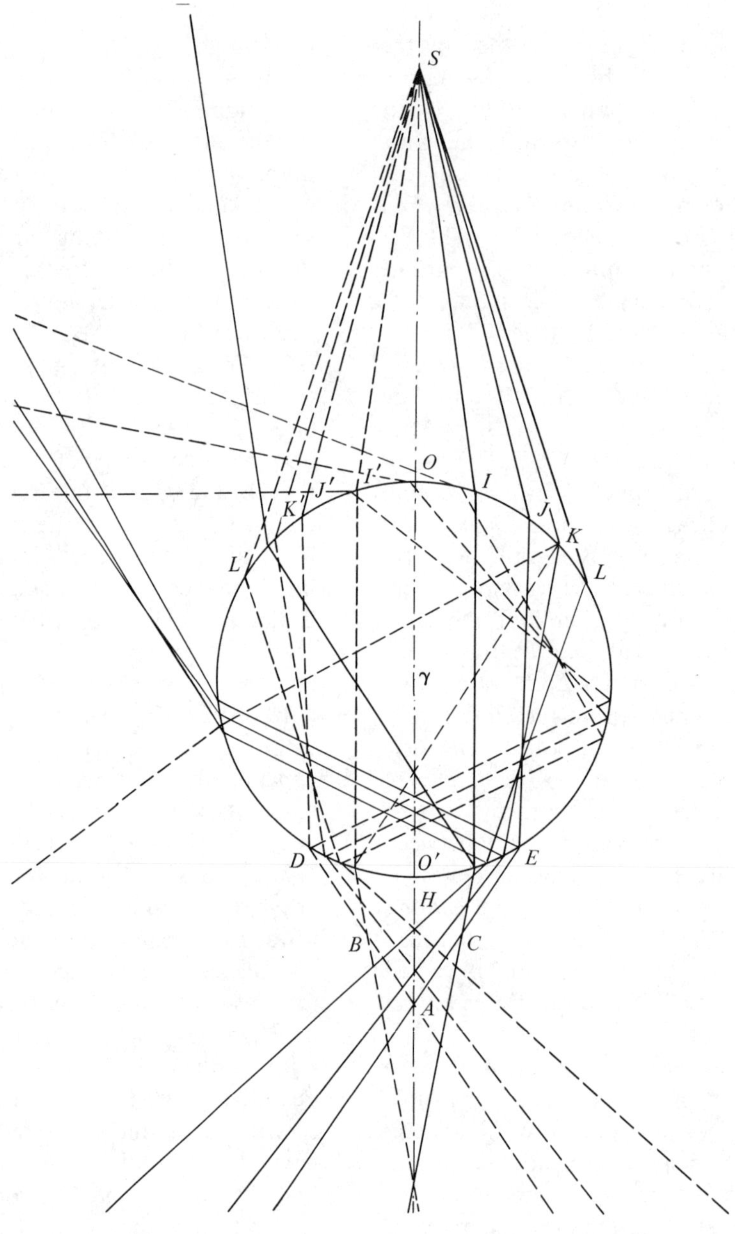

FIGURE 2

prospect of success necessarily led, in the case of a phenomenon as complicated as the rainbow, to the abandonment of direct study. This abandonment led to research on phenomena better mastered by the contemporary optical knowledge and more accessible to experimental verification—to the use of practical analogy. The analogue could be subjected to objective observation, and the resulting data applied to the study of the proposed natural object. Thus, Kamāl al-Dīn's spherical glass vial filled with water served to demonstrate the natural phenomenon of refraction.

On the problem of color, Kamāl al-Dīn turned to a commentary by al-Shīrāzī on the text of Ibn Sīnā's *Canon.*[5] His work soon began to diverge from its older model, however. In particular, Kamāl al-Dīn chose to treat four colors instead of three and to treat the problem of color by a reformulation of al-Shīrāzī's method. Kamāl al-Dīn set forth the doctrine of color, then limited its scope so as to consider only the colors formed on the screen in front of the sphere after the combination of reflections and refractions. He wrote:

> The colors of the arc are different but related, between the blue, the green, the yellow, and the dark red, and come from a strong luminous source reaching the eye by a reflection or a refraction or a combination of the two [*Tanqīḥ* . . ., p. 337].

KAMĀL AL-DĪN

Thus varying the respective positions of the images in the different cones formed by the refracted rays, Kamāl al-Dīn declared that he perceived the different colors gradually as the two images were superposed. The bright blue was produced by the approach, without superposition, of two images; the bright yellow resulted from the superposition of two images; and the darkish red appeared at the edge of the bundle of rays. It was no longer, therefore—as in a traditional doctrine of color—the mixture of light and darkness that produced color, but the bringing together or the superposition of two or more images—or, still better, "forms"—of light on a background of darkness that explained the formation and diversity of colors.

Kamāl al-Dīn thought that he finally could explain how the rainbow should be observed. He showed that when the sphere was moved up and down along the perpendicular to the axis from the eye to the center of the sun (see Figure 3), then according to the position

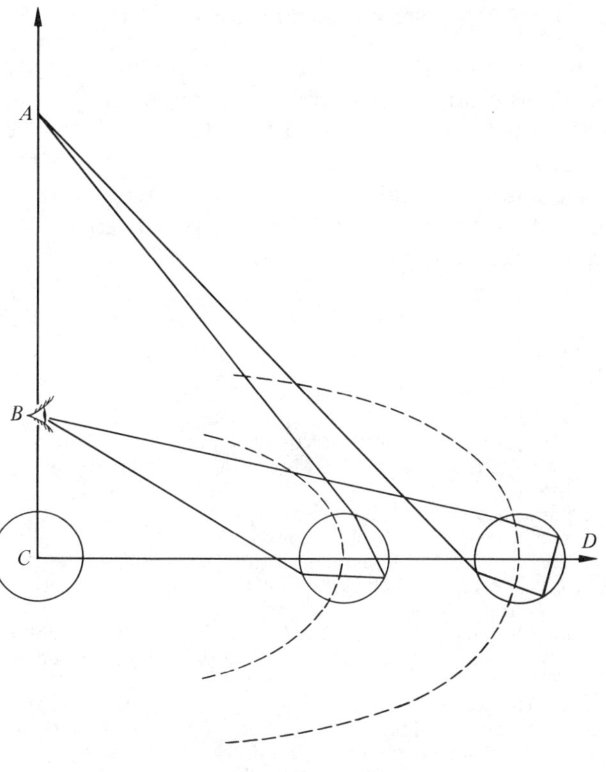

FIGURE 3

of the sphere the image of the sun could be produced by simple reflection between two refractions. In other words, depending on the angle formed by the rays of sun meeting the sphere, the well-placed observer will perceive either the rays refracted after one reflection or the rays refracted after two reflections. Then the colors of the first arc and those of the second are obtained successively. It must be noted that Kamāl al-Dīn employed here—as elsewhere—the principle of reversibility. Thus he imagined the cones of the rays refracted after one or two reflections, by putting, in the first step, the light source where the eye had been. In the second step he reversed the situation in order to consider the displacement of the sun in relation to these cones of rays, the eye being returned to its initial position. He wrote:

> Let us suppose that *B*, the center of the eye, is between *A*, the center of the sun, and *C*, the center of a polished transparent sphere. *ABC* is a straight line. Draw a perpendicular, *CD*, from *C* and suppose that the sphere is moved away from the line *ABC* in such a manner that its center remains on the perpendicular. If its center is moved away from *ABC*, the cone of rays refracted after one reflection will incline toward the sun while the latter, proportionally to the displacement of the sphere from *ABC*, will continue to approach the edge of the cone in the direction of the movement of the sphere and will appear in two images, at two positions on the sphere. . . . To the extent that the sphere is displaced, the two images draw closer until they become tangent. It is then that the light becomes stronger and produces an *isfanjūnī* blue if it blends with the darkness or with the green. If the images then interpenetrate, the light is again intensified and produces a bright yellow. Next, the blended image diminishes and becomes a darker and darker red until it disappears when the sun is outside the cone of rays refracted after one reflection.
>
> If the sphere continues to become more distant from the line *ABC*, the cone of rays refracted after two reflections approaches closer and closer to the sun until the sun is contained within this cone, and then what had disappeared in the beginning reappears in inverse order, beginning with the purple red, then the bright yellow, then the pure blue, and finally a light that is not really perceived because of the disappearance of one of the images or because of their mutual separation. If there are a great many drops of water massed in the air, these, arranged in a circle—each drop giving one of the images mentioned according to its size—produce the image of two arcs, as one may see: the small one is red on its exterior circumference, then yellow, then blue. The same colors appear in inverse order on the superior arc, hiding what is behind it by the colors and lights that appear in it. The air between the two arcs is darker than the air above and below them, because the portions between the two arcs are screened from the light of the sun [*Tanqīḥ* . . ., pp. 340–342].

In order to bring the combination of geometry and physics as in Ibn al-Haytham's optics to the study of the rainbow—that is, to arrive at a valid proof through geometrical deduction and experimental verification—Kamāl al-Dīn was led to reject as a starting point the notion of direct study, used by Ibn al-Haytham and by a whole tradition. He therefore elaborated a mode of explanation by reduction by establishing a group of correspondences between a natural object and a synthetic object, which he then systematically reduced by the geometry of the propagation of light in the first object to its propagation in the second.

Appearing in the wake of Ibn al-Haytham's reform, this achievement was a means of extending that reform to an area where it was not yet operative. It is in this way that the importance of Kamāl al-Dīn's contribution is to be understood.

It remains for us to consider Kamāl al-Dīn's work on the rainbow in conjunction with that of Dietrich von Freiberg. Dietrich's *De iride et radialibus impressionibus* was written between 1304 and 1311;[6] Krebs found the direct influence of Ibn al-Haytham in this work: "However, it seems very likely," he wrote, "that Dietrich used fully the great work of the Arabic father of modern optics. . . ."[7] Würschmidt, too, stated, "that Dietrich, by his own testimony, used in the treatment of this problem of the rainbow ... the optics of Ibn al-Haytham."[8]

Wiedemann concluded that Kamāl al-Dīn completed the definitive version of his work between 1302 and 1311,[9] during Dietrich's lifetime. In support of this thesis he offered the arguments that the book was written during al-Shīrāzī's lifetime (that is, before 1311), and that in it Kamāl al-Dīn refers to a lunar eclipse that, according to Wiedemann, occurred in 1302. This evidence has been accepted by other historians; Naẓīf, however, took exception to it.

> In his research on the rainbow included in the appendix to the *Tanqīḥ*, al-Fārisī [Kamāl al-Dīn] borrowed from al-Shīrāzī's commentary to the *Canon* the latter's conception of the manner in which colors originate; the passage containing al-Shīrāzī's remarks clearly indicates that the commentary had not been completed. This is tantamount to saying that al-Fārisī had completed the *Tanqīḥ* before al-Shīrāzī finished the commentary to the *Canon*. As for the lunar eclipse that Wiedemann emphasizes, if the year 1304 is accepted for its occurrence (Wiedemann gives 1302), the fact remains that the eclipse is not mentioned either in the main portion of the *Tanqīḥ*, in its conclusion, or in the appendix. The eclipse is referred to only in al-Fārisī's commentary on one of Ibn al-Haytham's treatises that al-Fārisī appended to his own book. This is *Shadows*, and it is conceivable that these treatises were added to the book after publication or that the reference to the eclipse was added at a later date.
>
> At least one can speculate; I do not believe that it is mistaken to say that al-Fārisī had completed the research on which he would base his two theories of the rainbow before al-Shirazi had finished his commentary. This is not to generalize and include the entire *Tanqīḥ*—corpus, conclusion, appendixes, and excursus—in this chronology. Thus, I am not suggesting what is probable but, rather, what is certain, in alleging that al-Fārisī had completed the research on the rainbow that is included in the appendix to the *Tanqīḥ* at least ten years before Theodoricus [Dietrich] wrote his treatise between the years 1304 and 1311 [M. Naẓīf, "Kamāl al-Dīn al-Fārisī . . .," p. 94].

Naẓīf went on to posit the possibility of Kamāl al-Dīn's influence upon Dietrich. Such influence would seem tenuous at best, however; no trace has been found of Kamāl al-Dīn's work in Latin, and Dietrich himself did not cite him. The influence of Ibn al-Haytham on Dietrich is another matter. As Würschmidt wrote:

> . . . a comparison of these works [those of Kamāl al-Dīn] with those of Master Dietrich indicates that the latter definitely did not know Kamāl al-Dīn's commentary; Kamāl al-Dīn avoided a succession of errors which occur with Dietrich as well as with earlier Arab scholars, and saw clearly especially the returned rays so important later in Descartes's rainbow theory.[10]

It may thus be seen that Kamāl al-Dīn's priority in no way implies his influence upon Dietrich, but, rather, that both Kamāl al-Dīn and Dietrich were disciples of Ibn al-Haytham and, relying upon the same source for their essential ideas, independently arrived at the model of the transparent sphere to explain the rainbow.

NOTES

1. See R. Rashed, "Le modèle de la sphère . . .," p. 114.
2. *The Rainbow: From Myth to Mathematics*, p. 125.
3. The Paris MS is Bib. Nat., Fonds arabe, MS 2517; that at Cairo, written in 1340 at Mossoul, is National Library MS 7797.
4. Paris, Bib. Nat., Fonds arabe, MS 2518.
5. *Tanqīḥ*, p. 331 *et seq.*
6. See E. Krebs, "Meister Dietrich (Theodoricus Teutonicus de Vriberg), sein Leben, seine Werke, seine Wissenschaft," in *Beiträge zur Geschichte der Philosophie und Theologie des Mittelalters*, V, pts. 5–6 (Münster in Westfalen, 1906), 105 ff.; and P. Duhem, *Le système du monde*, III (Paris, 1915), 383 ff.
7. See Krebs, *op. cit.*, p. 40.
8. "Dietrich von Freiberg; Über den Regenbogen und die durch Strahlen erzeugten Eindrücke," in *Beiträge zur Geschichte der Philosophie und Theologie des Mittelalters*, XII, pts. 5–6 (Münster in Westfalen, 1906), p. 1.
9. "Zu Ibn al-Haitams Optik," in *Archiv für die Geschichte der Naturwissenschaften und der Technik*, no. 3 (1912), pp. 3-4.
10. *Op. cit.*, p. 2, note 8.

BIBLIOGRAPHY

I. ORIGINAL WORKS. The *Tanqīḥ al-manāẓir*, 2 vols. (Hyderabad, 1928–1929), contains, at the end of vol. II, commentaries on the following works of Ibn al-Haytham: *The Halo and the Rainbow*, trans. by E. Wiedemann as "Theorie des Regenbogens von Ibn al-Haitam," in *Sitzungsberichte der Physikalisch-medizinische Sozietät in Erlangen*, **46** (1919), 39–56; *The Burning Sphere*, trans. by Wiedemann as "Über die Brechung des Lichtes in Kugeln nach Ibn al-Haitam und Kamal al-Din al-Farisi," *ibid.*, **42** (1910), 15–58; *Shadows*, trans. by Wiedemann as "Über eine Schrift von Ibn al-Haitam, über die Beschaffenheit der Schatten," *ibid.*, **39** (1907), 226–248; *The Shape of Eclipses*; and *Discourse on Light*, trans. by R. Rashed in *Revue d'histoire des sciences*, **21**, no. 3 (1968), 197–224.

Works still in MS are *Al-Basā'ir fī 'ilm al-manāẓir fi'l ḥikma*; *Asās al-gawā'id fī uṣūl al'fawa'id*; *Taḏhirat al-aḥhāb fī bayān al-tahābb*; and "Treatise on a Geometrical Proposition of Nasīr al-Dīn al-Ṭūsi." See C. Brockelmann, *Geschichte der arabischen Literatur*, supp. II (Leiden, 1938), p. 295; and H. Suter, *Die Mathematiker und Astronomen der Araber und ihre Werke* (Leipzig, 1900), p. 159.

II. SECONDARY LITERATURE. On Kamāl al-Dīn or his work, see Carl Boyer, *The Rainbow: From Myth to Mathematics* (New York, 1959), pp. 127–129; M. Schramm, "Steps Towards the Idea of Function: A Comparison Between Eastern and Western Science of the Middle-Ages," in *History of Science*, **4** (1956), 70–103, esp. 81–85; M. Naẓīf, "Kamāl al-Dīn al-Fārisī wa ba'ḍ buḥūṭuhu fī 'ilm al-ḍaw'," in *La société égyptienne et histoire des sciences*, no. 2 (Dec. 1958), 63–100 (in Arabic); R. Rashed, "Le modèle de la sphère transparente et l'explication de l'arc-en-ciel: Ibn al-Haytham—al-Fārisī," in *Revue d'histoire des sciences*, **22** (1970), 109–140; and J. Würschmidt, "Über die Brennkugel," in *Monatshefte für den naturwissenschaftlichen Unterricht*, **4** (1911), 98–113.

ROSHDI RASHED

KAMALĀKARA (*b.* Benares, India, *ca.* 1610), *astronomy.*

Kamalākara was a scion of a family of astronomers whose origin is traced back to a Mahārāṣṭra Brāhmaṇa, Rāma of the Bhāradvājagotra, who lived in Gologrāma on the west bank of the Godāvarī River (near Pathri, Mahārāṣṭra). Rāma's son was Bhaṭṭācārya, and Bhaṭṭācārya's son was Divākara of Golagrāma, a pupil of Gaṇeśa of Nandod (*b.* 1507). Divākara moved to Benares, where his five sons were born between about 1560 and 1570. Viṣṇu, Mallāri, and Viśvanātha were the main commentators on Gaṇeśa's principal astronomical works. The eldest, Kṛṣṇa, had two sons, of whom Nṛsiṃha (*b.* 1586) continued the family tradition of commenting on Gaṇeśa; the other was Śiva. Nṛsiṃha had four sons: of these Divākara (*b.* 1606), Kamalākara, and Raṅganātha were all noted astronomers in Benares between 1625 and 1675.

Kamalākara, who studied under his elder brother Divākara (himself a pupil of his uncle Śiva), was the leading rival of Munīśvara Viśvarūpa among Benares astronomers. He combined traditional Indian astronomy with Aristotelian physics and Ptolemaic astronomy as presented by Islamic scientists (especially Ulugh Bēg). Following his family's tradition he wrote a commentary, *Manoramā*, on Gaṇeśa's *Grahalāghava* and, like his father, Nṛsiṃha, another commentary on the *Sūryasiddhānta*, called the *Vāsanābhāṣya*, both of which are still unpublished. His chief claims to fame are his *Siddhāntatattvaviveka* (see essay in supplement), written in Benares in 1658, and the later supplement, *Śeṣavāsanā*.

The *Siddhāntatattvaviveka* contains fifteen chapters:

1. On the units of time measurement.
2. On the mean motions of the planets.
3. On the true longitudes of the planets.
4. On the three problems related to diurnal motion.
5. On the diameters and distances of the planets.
6. On the earth's shadow.
7. On the lunar crescent.
8. On heliacal risings and settings.
9. On the syzygies.
10. On lunar eclipses.
11. On solar eclipses.
12. On planetary transits.
13. On the *pātas* of the sun and moon.
14. On the "great problems."
15. Conclusion.

The *Siddhāntatattvaviveka* has been edited with his own notes by Sudhākara Dube (Dvivedin) (5 vols., Benares Sanskrit Series 1, 2, 3, 6, and 14 [Benares, 1880–1885]; 2nd ed. revised by Muralīdhara Jhā [Benares, 1924–1935]). It was also published with his own commentary by Gaṅgādhara Miśra (Lucknow, 1929). The *Śeṣavāsanā* is a collection of additional discussions of various topics in the *Siddhāntatattvaviveka*, and is published as an appendix to volume 5 of the Benares Sanskrit Series edition of that work.

BIBLIOGRAPHY

Articles on Kamalākara are Sudhākara Dube (Dvivedin), *Gaṇakataraṅginī* (Benares, 1933), pp. 98–99, repr. from *The Pandit*, n.s. **14** (1892); Ś. B. Dīkṣita, *Bhāratīya Jyotiḥśāstra* (Poona, 1896; repr. Poona, 1931), pp. 287–288; Padmākara Dvivedin, "Kamalākarabhaṭṭa," in *Proceedings of the Benares Mathematical Society*, **2** (1920), 67–80; and D. Pingree, *Census of the Exact Sciences in Sanskrit*, ser. A, II (Philadelphia, 1971), pp. 21–23.

DAVID PINGREE

KAMERLINGH ONNES, HEIKE (*b.* Groningen, Netherlands, 21 September 1853; *d.* Leiden, Netherlands, 21 February 1926), *physics.*

Kamerlingh Onnes was the son of a well-known manufacturer in Groningen. After attending secondary school, he was admitted in 1870 to the University of Groningen, where he studied physics and mathematics. In November 1871 he passed the intermediate examination for the bachelor's degree, whereupon he spent some time at Heidelberg. There he studied for three semesters with Bunsen and Kirchhoff, a tenure that was made possible by the Seminarpreis. Earlier he had won two other competition prizes, the gold medal of the University of Utrecht and the silver medal of the University of Groningen, both for research on the chemical bond. In April 1873 he returned to Groningen to complete his studies under R. A. Mees. In June 1876 he passed his doctoral examination, and on 10 July 1879 he defended his dissertation, entitled "Nieuwe bewijzen voor de asweteling van de aarde" ("New Proofs for the Axial Rotation of the Earth"), a subject which, stimulated by Kirchhoff, he had started to study at Heidelberg. On the basis of this dissertation, in which he showed that he was also an excellent mathematician, he was awarded the doctorate *magna cum laude*.

In 1878 Kamerlingh Onnes was appointed assistant to Johannes Bosscha, who was then the director of the Polytechnic School (later the Technical University) at Delft. In 1880–1881 and 1881–1882 he lectured there for Snijders and Bosscha. During this time he was in close contact with van der Waals, who was then professor of physics in Amsterdam, and thus he became acquainted with problems related to the molecular theory of matter. An indication of this is found in his article "Théorie générale de l'état fluide" (1884).

In 1882 P. L. Rijke, professor of physics at Leiden, retired and Kamerlingh Onnes was appointed his successor at the age of twenty-nine. He held this professorship, which included the directorship of the physics laboratory, for forty-two years.

The period in which Kamerlingh Onnes worked can be characterized as transitional for physics. The increasing importance of experimental physics is demonstrated by his appointment to the first chair of experimental physics in the Netherlands. Before then experimental and theoretical physics were not separated. On the other hand, the mechanistic image of physics was gradually being abandoned under the influence of Maxwell's theory of electromagnetism; physicists were also gradually coming to believe that matter is not a continuum but has a corpuscular nature. When Kamerlingh Onnes came to Leiden most physicists still adhered to the idea of continuity, but Boltzmann and van der Waals in particular were promoting the corpuscular theory.

In his inaugural address at Leiden (11 November 1882), "The Significance of Quantitative Research in Physics," Kamerlingh Onnes stated: "In my opinion it is necessary that in the experimental study of physics the striving for quantitative research, which means for the tracing of measure relations in the phenomena, must be in the foreground. I should like to write 'Door meten tot weten' ['Through measuring to knowing'] as a motto above each physics laboratory." This motto was a declaration of principle to which he always remained loyal.

In conducting his research and developing the necessary facilities Kamerlingh Onnes showed an enormous capacity for work, the more admirable because he was in very delicate health. His strong will and the great devotion and care of his wife, Elisabeth, enabled him to achieve what he did.

When Kamerlingh Onnes received his appointment at Leiden, he made it his purpose to give experimental support to van der Waals's theory of the behavior of gases and especially to the "law of corresponding states." This theory is based on the hypothesis that a gas consists of molecules circulating and exerting forces on each other. The law of corresponding states, which van der Waals had derived from his equation of state but which had a wider validity than for this form of the equation alone, says that all gases behave in exactly the same way and obey the same equation of state, when the units in which pressure, volume, and temperature are measured are adapted to the gas under specific consideration.

Kamerlingh Onnes was greatly interested in this theory, for he had concluded that the conformity in the behavior of gases could be found in "the stationary mechanical similarity of the substances," as he stated in his Nobel address. He was "mightily attracted" by the idea of carrying out precise measurements in order to verify the results of this theory. For this purpose he would have to consider the behavior of gases with simple molecules having low condensation temperatures. Moreover, since it would be important to have a large range of temperatures at his disposal, it was desirable to use the lowest temperatures possible. Just five years earlier (December 1877) Cailletet and Pictet, using different methods, had liquefied air for the first time and so opened this new temperature region. It was necessary for Kamerlingh Onnes first to build an apparatus for the liquefaction of air in large quantities. Here the advantage of his method became evident. He did his work with great accuracy and perseverance, systematically, and with attention to all details, thus obtaining important results and advancing far ahead of all other researchers in this field. In 1892 his apparatus for the "cascade method" (using liquid methyl chloride and ethylene) for the liquefaction of oxygen and air was ready. (Boiling points of oxygen and air of normal composition are −183° C. and −193° C., respectively.) In the meantime much information was obtained about the behavior of pure gases and binary gas mixtures, a study that could be extended, after 1892, to lower temperatures.

The research of Kamerlingh Onnes and his collaborators followed two lines, one related to van der Waals's theories (equation of state, viscosity, capillarity), and the other to the theoretical work of Lorentz (magnetorotation of the plane of polarization, Kerr effect, Hall effect). In a volume commemorating Kamerlingh Onnes' forty years as a professor Lorentz referred to an earlier book (1904), similarly in his honor:

> Many a physicist would be glad if, at the end of his career, he could look back at researches of the quantity and importance of those which are described in that book. But in the following years all this proved to be only an overture to a higher flight in which results and points of view were reached of which originally even the most daring imagination had not been able to dream.

This "higher flight" became possible by the liquefaction of hydrogen and helium, which have boiling points of −252.7° C. and −268.9° C. (20.4 K. and 4.2 K.). In February 1906 the hydrogen liquefier was ready, and on 10 July 1908 helium was first liquefied. Construction of the helium liquefier was facilitated by knowledge of the law of corresponding states. With this liquefaction a vast new temperature region was opened for research—a field in which, until his retirement in 1923, Kamerlingh Onnes remained absolute monarch.

It was evident that the simple van der Waals law, $p = \frac{RT}{V-b} - \frac{a}{V^2}$, where p, T, and V are pressure, temperature, and volume and R, a, and b are constants, could not represent the results of the measurements quantitatively. Therefore Kamerlingh Onnes set up the experimental law

$$pV = A\left(1 + \frac{B}{V} + \frac{C}{V^2} + \frac{D}{V^4} + \frac{E}{V^6} + \frac{F}{V^8}\right),$$

where $A(=RT$ for one mole), B, C, . . ., which he called the virial coefficients, depend on the temperature. For this dependence he wrote

$$B = b_1 + \frac{b_2}{T} + \frac{b_3}{T^2} + \frac{b_4}{T^4} + \frac{b_5}{T^6},$$

with similar expressions for C, D, E, and F. In this way he had twenty-five coefficients to describe the measured values. This did not lead to a better formulation of the law of corresponding states, but the second virial coefficient B and sometimes C as well are commonly used to represent deviations from the ideal gas law.

The study of the resistance of metals was Kamerlingh Onnes' second major field. Originally accepting the idea expressed in 1902 by Kelvin, he expected that with decreasing temperature the resistance, after reaching a minimum value, would become infinite as electrons condensed on the metal atoms. Later, when this proved to contradict experimental results, he supposed that the resistance, caused by Planck vibrators which lose their energy at low temperatures, would become zero. This proved to be true although, for certain metals, in a way different from that he expected. In order to diminish the influence of impurities, very pure mercury resistors were prepared. To Kamerlingh Onnes' great surprise the resistance showed a discontinuous decrease to zero. Discovered in 1911, this phenomenon, which he called supraconductivity (later superconductivity), was found for various metals having different transition points, all at very low temperatures. J. Bardeen, J. N. Cooper, and J. R. Schrieffer gave a theoretical explanation of this phenomenon in 1957.

Kamerlingh Onnes had originally hoped that this property would allow him to establish strong magnetic fields without cooling difficulties; but he soon found that the superconductive state disappears in a magnetic field of a temperature-dependent value, never very high in the cases he studied. Also, a current sent through a superconducting wire destroys the superconductive state by its own magnetic field. Only today, after the discovery of alloys that can support strong fields, is it possible to take advantage of superconductivity for cheap production of very intense magnetic fields.

In 1913 Kamerlingh Onnes received the Nobel Prize in physics "for researches on the properties of matter at low temperatures, which researches have among others also led to the liquefaction of helium." He received honors from the Dutch and foreign governments and was a member of many academies and societies. An especial honor was his election to membership in the Royal Academy of Sciences in Amsterdam before he was thirty.

Kamerlingh Onnes was also concerned with the application of low temperatures in everyday matters such as food preservation, refrigerated transport, and the production of ice. In 1908, at the opening ceremonies of the first International Congress of Refrigeration in Paris, he "formally proposed the creation of an international organization of refrigeration which would further the work of the congress." He insisted that one of the commissions be devoted to scientific problems. In the Netherlands he stimulated the foundation of the Nederlandsche Vereeniging voor Koeltechniek, of which he was president until his death. Another organization for which Kamerlingh Onnes was responsible was the Vereeniging tot Bevordering van de Opleiding tot Instrumentmaker (1901). The workshops of his laboratory were organized as a school, the Leidsche Instrumentmakersschool. This establishment has been of great importance in training instrument makers, glassblowers, and glass polishers in the Netherlands.

BIBLIOGRAPHY

I. Original Works. Most of Kamerlingh Onnes' writings can be found in the *Proceedings of the Section of Sciences* of the Royal Netherlands Academy of Sciences in Amsterdam and the *Verhandelingen* (later *Verslagen*) of the meetings of the Academy. Reprints of these works are in *Communications from the Physical Laboratory at the University of Leiden.* Review articles on most of his research can also be found in the *Reports* and *Communications* presented by the president of the First Commission of the International Institute of Refrigeration to the third and fourth International Congresses of Refrigeration.

Among his works are "Algemeene theorie der vloeistoffen," in *Verhandelingen der K. akademie van wetenschappen*, **21** (1881), and the following writings that appeared in *Communications from the Physical Laboratory at the University of Leiden:* "On the Cryogenic Laboratory at Leiden and on the Production of Very Low Temperatures," in *Communication*, **14** (1894), which gives a review of the work of the first twelve years, including the liquefaction of oxygen; "The Importance of Accurate Measurements at Very Low Temperatures," in *Communication*, supp. 9 (1904), his address as *rector magnificus* of the University of Leiden on its 329th anniversary; "Die Zustandsgleichung," in *Communication*, supp. 23 (1912), repr. from *Encyklopädie der mathematischen Wissenschaften*, V, pt. 10 (1912), written with W. H. Keesom; "Untersuchungen über die Eigenschaften der Körper bei niedrigen Temperaturen, welche Untersuchungen unter anderen auch zur Herstellung von flüssigem Helium geführt haben," in *Communication*, supp. 35 (1913), his Nobel address; and "On the Lowest Temperature Yet Obtained," in *Communication*, **159** (1922), a paper read at the joint meeting of the Faraday Society and the British Cold Storage and Ice Association, repr. from *Transactions of the Faraday Society*, **18** (1922).

The liquefaction of hydrogen is described in *Communication*, **94** (1906), and that of helium in *Communication*, **108** (1908). The empirical equation of state is introduced and discussed in *Communication*, **74** (1901). The superconductivity of mercury is treated in *Communication*, **122B** and **124C** (1911).

II. Secondary Literature. *In Memoriam Heike Kamerlingh Onnes* (Leiden, 1926) contains many of the newspaper obituaries and addresses at scientific societies, a review of his work, and a sketch of his personality; most of these are in Dutch. A similar work is Ernst Cohen, "Kamerlingh Onnes Memorial Lecture," in *Journal of the Chemical Society* (1927), p. 1193. See also *Gedenkboek aangeboden aan H. Kamerlingh Onnes 10 Juli 1904* (Leiden, 1904), and *Het Natuurkundig laboratorium der Rijksuniversiteit te Leiden in de jaren 1904–1922* (Leiden, 1922).

J. van den Handel

KANAKA (*b.* India; *fl.* Baghdad [?], *ca.* 775–820), *astronomy*.

Kanaka appears in the Arabic bibliographical tradition as Kankah al-Hindī. In the astrological compendium *Kitāb al-Mughnī*, written by Ibn Hibintā about 950, there is a passage (Munich Arab. 852, fols. 69v–70) which is alleged to be a quotation from Kankah. Al-Bīrūnī in his *Chronology* (ed., p. 132; trans., p. 129) states that Kankah was an astrologer at the court of Hārūn al-Rashīd (786–809) but attributes to him two specific predictions concerning the fall of the ʿAbbāsids and the rise of the Buwayhids, the first of which was in fact made by Māshāllāh about 810 and the second by Māshāllāh's epitomizer, Ibn Hibintā (see E. S. Kennedy and D. Pingree, *The Astrological History of Māshāʾallāh* [Cambridge, Mass., 1971], pp. 56–59, 67–68). It is possible, then, that al-Bīrūnī had only Ibn Hibintā's work before him and somehow confused the references in it to Māshāllāh and Kankah, and that we have no right to connect Kankah with al-Rashīd. But al-Jahānī (*fl.* 1079) attributes to Kankaraf the same beginning of various cycles used in astrological history that was employed by Māshāllāh (f. Zl); perhaps, then, they were associates. It is true that Abū Maʿshar, in his *Kitāb al-ulūf* (see D. Pingree, *The Thousands of Abū Maʿshar* [London, 1968], p. 16), which was written between 840 and 860, states that Kankah was an authority on astronomy among Indian scientists "in ancient times"—that is, long before al-Rashīd's caliphate—but Abū Maʿshar is a notorious liar, so the question cannot be answered on the basis of his statement. One may tentatively conclude, then, that Kanaka was in Baghdad during the reign of al-Rashīd and was an associate of Māshāllāh. There has recently been located in Ankara a manuscript of an astrological history of the caliphs entitled *Kitāb Kankah al-Hindī* (*Book of Kankah the Indian*). This history stops in the reign of al-Maʾmūn (813–833).

Among the biographers Ibn al-Nadīm (*Fihrist*, p. 270) contents himself with listing four astrological treatises which he claims were written by Kankah: *Kitāb al-nāmūdār fī al-āʿmār* (Book of the Nāmūdār [Used for Determining the Lengths of] Lives"), *Kitāb asrār al-mawālīd* ("Book of the Secrets of Nativities"), *Kitāb al-qirānāt al-kabīr* ("Great Book of Conjunctions"), and *Kitāb al-qirānāt al-ṣaghīr* ("Small Book of Conjunctions"). These works all dealt with topics of great interest to early ʿAbbāsid astrologers.

In addition the Indian astrologer Kalyāṇavarman, who wrote his *Sārāvalī* in Bengal about 800, refers (*Sārāvalī* 53, 1) to Kanakācārya as an authority on the nativities of plants and animals. If this Kanakācārya is identical with Kankah, as is suggested by Ramana-Sastrin, he must have written something on astrology in Sanskrit about 750–775 and may subsequently have traveled to Baghdad. There is, however, no real evidence to connect the two.

If, then, one is willing to accept the traditions of the ninth and tenth centuries as referring to a historical personality, Kankah emerges as an Indian astrologer who practiced his art at Baghdad toward the end of the eighth and in the early ninth centuries but whose works in Arabic fall within the ʿAbbāsid tradition of astrology (derived from Greek and Iranian sources); and the existing fragments appear to display no specifically Indian traits.

Later Arab scholars, especially in Spain, constructed elaborate theories regarding the role of Kankah in the history of science; because their fables have been accepted by modern Western historians an article on Kanaka is included here. There were two sources for the development of the Kankah legend: the story of an Indian embassy to the court of al-Manṣūr as related by Ibn al-Adamī (*ca.* 920) in his *Niẓām al-ʿiqd* (see fragment Z1 of al-Fazārī), and that of Mankah al-Hindī, a physician who is alleged to have traveled from India to Iraq and to have translated Shānāq (Cāṇakya) from an Indian language into Persian (or Arabic) during the time of Hārūn al-Rashīd for Isḥaq ibn Sulaymān ibn ʿAlī al-Hāshimī (the most complete account seems to be that of Ibn abī Uṣaybiʿa, III, 51–52).

Ibn al-Adamī associates the translation of the *Zīj al-Sindhind* that serves as the basis of the works of al-Fazārī, Yaʿqūb ibn Ṭāriq, and others with an unnamed member of an embassy sent from Sind to Baghdad in 773. This passage from Ibn al-Adamī is quoted by Ṣāʿid al-Andalusī of Toledo (*Kitāb ṭabaqāt al-umam*, ed., pp. 49–50; trans., p. 102) in 1067–1068; his next biographer, Ibn al-Qifṭī (pp. 265–267), who died at Aleppo in 1248–1249, quotes some of Ibn al-Adamī's story in his article on Kankah but without actually connecting Kankah with the *Zīj al-Sindhind*. Apparently Abraham ibn Ezra (*ca.* 1090–1167) was the first to identify Kankah with Ibn al-Adamī's unnamed scholar (in the preface to his translation of Ibn al-Muthanna's *Fī ʿilal zīj al-Khwārizmī* [see fr. Z2 of Yaʿqūb ibn Ṭāriq] and in *Liber de rationibus tabularum*, p. 92); there is no real basis for this invention, although it is dutifully repeated by Steinschneider, Suter, and Sarton.

The confusion of Kankah with Mankah sometimes leads to the attribution of medical knowledge and writings to the former—for instance, by Ibn abī Uṣaybiʿa (III, 49). This tradition also is without basis. Finally, pure fancy has produced a fabulous Kankah al-Hindī in alchemical literature. His fantastic exploits are recounted in pseudo-al-Majrīṭī's *Ghāyat al-ḥakīm* (ed., pp. 278 ff.; trans., pp. 285 ff.). These stories have no place in serious history.

Kanaka's significance, then, is as a name to which either serious accounts of the transmission of Indian science to the Arabs or alchemists' dreams of ancient philosopher-kings can be conveniently attached. He was so easily subjected to this treatment because, in fact, nothing reliable was known about him.

BIBLIOGRAPHY

Standard reference works on the history of science contain notices of Kanaka, but they follow the fictions of Ibn Ezra. The authorities to which I have referred are the following: Abraham ibn Ezra, *De rationibus tabularum*, edited by J. M. Millás Vallicrosa as *El libro de los fundamentos de las Tablas astronómicas* (Madrid–Barcelona, 1947); and his trans. of Ibn al-Muthanna, edited by B. Goldstein as *Ibn al-Muthanna's Commentary on the Astronomical Tables of al-Khwārizmī* (New Haven, 1967); Abū Maʿshar, "Kitāb al-ulūf," in D. Pingree, *The Thousands of Abū Maʿshar* (London, 1968); al-Bīrūnī, *Chronology*, edited by C. E. Sachau as *Chronologie orientalischer Voelker von Albērūnī* (Leipzig, 1878), translated into English by Sachau as *The Chronology of Ancient Nations* (London, 1879); al-Fazārī, in D. Pingree, "The Fragments of the Works of al-Fazārī," in *Journal of Near Eastern Studies*, **29** (1970), 103–123; Ibn abī Uṣaybiʿa, *ʿUyūn al-anbāʾ*, 3 vols. (Beirut, 1956–1957); Ibn Hibintā, "Kitāb al-Mughnī," MS Munich Arab. 852; Ibn al-Nadīm, *Fihrist*, G. Flügel, ed., 2 vols. (Leipzig, 1871–1872); Ibn al-Qifṭī, *Taʾrīkh al-ḥukamāʾ*, J. Lippert, ed. (Leipzig, 1903); al-Jahānī, Latin trans. by Gerard of Cremona, J. Heller, ed., in Māshāʾallāh's *De elementis et orbibus coelestibus* (Nuremberg, 1549), fols. Ni-Zii. Kalyāṇavarman, *Sārāvalī*, V. Subrahmanya Sastri, ed., 3rd ed. (Bombay, 1928); pseudo-al-Majrīṭī, *Ghāyat al-ḥakīm*, H. Ritter, ed. (Hamburg, 1933), translated into German as "*Picatrix.*" *Das Ziel des Weisen von Pseudo-Mağrīṭī*, by H. Ritter and

M. Plessner (London, 1962); Māshāllāh, in E. S. Kennedy and D. Pingree, *The Astrological History of Māshā'allāh* (Cambridge, Mass., 1971); V. V. Ramana-Sastrin, "Kanaka," in *Isis*, **14** (1930), 470; Ṣā'id al-Andalusī, *Kitāb ṭabaqāt al-umam*, L. Cheikho, ed. (Beirut, 1912), translated into French as *Ṭabaqāt alumam (Succession des Communautés religieuses)* by R. Blachère (Paris, 1935); and Ya'qūb ibn Ṭāriq, in D. Pingree, "The Fragments of the Works of Ya'qūb ibn Ṭāriq," in *Journal of Near Eastern Studies*, **27** (1968), 97–125.

DAVID PINGREE

KANE, ROBERT JOHN (*b.* Dublin, Ireland, 24 September 1809; *d.* Dublin, 16 February 1890), *chemistry*.

Kane's father was a Dublin chemical manufacturer, and the son's exposure to this business nurtured in him a precocious interest in chemistry. While still a schoolboy, Kane attended the lectures of the Royal Dublin Society and, when only twenty years old, described the natural arsenide of manganese, which was named Kaneite in his honor.

Kane also had an early interest in medicine; in 1829 he became a licentiate of the Apothecaries' Hall, where in 1831 he became professor of chemistry. In 1829 he also enrolled at Trinity College, Dublin, from which he received his B.A. in 1835. He published the *Elements of Practical Pharmacy* and founded the *Dublin Journal of Medical and Chemical Science*.

In 1834 Kane became lecturer (and subsequently professor) of natural philosophy at the Royal Dublin Society. He retained the post until 1847 and during this period carried out much research. In 1836 he spent three months at Liebig's laboratory at Giessen, where he isolated acetone from wood spirit; in the following year he transformed this compound into a ring compound, which he called mesitylene. The significance of this reaction was not realized at the time. For his work on the metallic compounds of ammonia and on the chemical history of archil and litmus, he received medals from the Royal Irish Academy and the Royal Society of London. His writing also prospered. In 1840 he was appointed an editor of the *Philosophical Magazine*, and in 1840–1841 he published in three parts his *Elements of Chemistry*, which was successful in both Great Britain and America.

In 1844 Kane published *Industrial Resources of Ireland* and in the following year was appointed director of the newly formed Museum of Economic Geology in Dublin. Under his guidance the museum evolved into the Royal College of Science for Ireland. Other appointments and honors followed. In 1845 Kane became president of Queen's College, Cork; in 1846 he was knighted; and in 1849 he became a fellow of the Royal Society of London. Because of the burden of administrative work his research lapsed, but he retained the editorship of the *Philosophical Magazine* until his death.

BIBLIOGRAPHY

I. ORIGINAL WORKS. A comprehensive list of Kane's work is in the Royal Society *Catalogue of Scientific Papers*. These include *Elements of Practical Pharmacy* (Dublin, 1831); "Research on the Combinations Derived From Pyroacetic Spirit," in *Proceedings of the Royal Irish Academy*, **1** (1837), 42–44; "On the Chemical History of Archil and Litmus," in *Philosophical Transactions of the Royal Society*, **130** (1840), 273–324; *Elements of Chemistry* (Dublin, 1841); and *Industrial Resources of Ireland* (Dublin, 1844).

II. SECONDARY LITERATURE. For biographical treatments of Kane see D. Reilly, *Sir Robert Kane* (Cork, 1942); and "Robert John Kane (1809–1890), Irish Chemist and Educator," in *Journal of Chemical Education*, **32** (1955), 404–406; T. S. Wheeler, "Sir Robert Kane," in *Endeavour*, **4** (1945), 91–93; and B. B. Kelham, "Royal College of Science for Ireland," in *Studies*, **56** (1967), 297–309.

BRIAN B. KELHAM

KANT, IMMANUEL (*b.* Königsberg, Germany [now Kaliningrad, R.S.F.S.R.], 22 April 1724; *d.* Königsberg, 12 February 1804), *philosophy of science*.

Kant was the fourth child of Johann Georg Cant and Anna Regina (Reuter) Cant. His paternal grandfather was an immigrant from Scotland, where the name Cant is still not uncommon in the northern parts. Immanuel changed the spelling to Kant in order that the name might conform more comfortably with the usual practices of German pronunciation. His father was a saddle maker of modest means. His mother was much given to Pietism, a Protestant sect (not unlike the Quakers and early Methodists) which flourished in northern Germany at the time.

When Kant was ten, he entered the Collegium Fridericianum, intending to study theology. But he actually spent more time with classics, and he became quite adept in Latin. In 1740 he entered the University of Königsberg and studied mainly mathematics and physics with Martin Knutzen and Johann Teske. These years doubtless influenced him much in his interest in the philosophy of science. In 1746 Kant's father died, and he was forced to interrupt his studies to help care for his brother and three sisters by being a private tutor in three different families successively for a period of some nine years. Finally, in 1755 he was able to resume his studies at the university and received his doctorate in philosophy in the autumn of that year. For the next fifteen years he earned a meager living as a *Privatdozent* lecturing on physics and nearly all aspects of philosophy. In 1770 he was given the chair of logic and metaphysics at the University of Königsberg, a position which he held until he retired in 1797.

Although Kant was brought up in Pietistic surroundings and in his youth even considered becoming a minister, in his maturity he became the one who, above all others, liberated philosophy and science from theology. His single-minded devotion to both philosophy and science also accounts largely for the fact that he never married. He was slightly built and gave the appearance of having a delicate constitution, but his careful attention to the laws of health and the regularity of his habits enabled him to live to be almost eighty.

Kant was very modest in his style of living. In 1783 he purchased a house in the center of town and quite regularly thereafter entertained friends at dinner. The number of his table companions was never large because his dinner service could accommodate but six persons. These companions were for the most part men of great culture and learning, and his dinners were widely known for the liveliness and diversity of the conversation.

The German writer Johann Gottfried von Herder said that Kant in his prime had the happy sprightliness of a boy and that he continued to have much of it even as an old man. He had a broad forehead, Herder continued, that was built for thinking and that was the seat of an imperturbable cheerfulness and joy. Speech rich in thought issued from his lips. He also possessed playfulness, wit, and humor. He enlivened his lectures and conversations by drawing on the history of men and peoples and on natural history, science, mathematics, and his own observations. He was indifferent to nothing that was worth knowing, concluded Herder.

Even though Kant is widely considered to be one of the two or three greatest philosophers that Western civilization has produced, he was also much interested in science and especially in the philosophy of science. He was not an experimental scientist and did not contribute to the body of scientific knowledge, but he was much concerned with the foundations of science and made significant contributions to that field. He has sometimes been accused (as by Erich Adickes in his *Kant als Naturforscher*) of being an armchair scientist. He might more accurately be called an armchair philosopher speculating on the fundamental bases of science. He was not interested in gleaning facts and data; rather, he speculated concerning the grand scheme in which the facts gleaned by others are arrayed.

The two main influences on Kant in his philosophical reflections on science were Leibniz and Newton. During his first period of study at the University of Königsberg, from 1740 to 1746, Knutzen taught that version of Leibniz's metaphysics which the German philosopher Christian von Wolff had made popular. He also taught the mathematical physics which Newton had developed. He revealed to the young Kant the various oppositions, puzzles, and contradictions of these two great natural philosophers.

The nature of space and time was what interested the young Kant most in these disputes between Leibniz and Newton. He studied the famous exchange of letters between Leibniz and Samuel Clarke, a defender of Newton's philosophy. Leibniz claimed that the universe is made up of an infinitude of monads, which are simple, immaterial (spiritual) substances. Every monad is endowed with some degree of consciousness. He conceived of space as a set of relations which the monads have to one another; it is the order of coexistent things. He thought of time as the relations of the successive states of consciousness of a single monad. Physical bodies, on this theory, are groups of monads. Mathematically considered, every monad is a dimensionless point. Length, breadth, and position can be represented as relations of monads. Space, then, is a continuous, three-dimensional system of mathematical points corresponding to the order of a plenum of distinct monads. Time has but one dimension; succession and coexistence are the only temporal relations, corresponding, as they do, to the order of perceptions in the consciousness of a monad. For Leibniz, then, space and time were relations among things (monads) which would have no existence whatever if there were no monads.

By contrast, Newton held that space and time are infinite and independent of the physical bodies that exist in space and time. For him space and time were things, and they would exist even if there were no bodies. He held that there are absolute positions in space and time that are independent of the material entities occupying them and, furthermore, that empty space (void) and empty time are possible. Leibniz denied both tenets. Neither Leibniz nor Clarke was able fully to undermine the position of the other, and the result was an impasse.

In his early years Kant pondered the nature of space and time first from the point of view of Leibniz and then of Newton, but eventually he found both positions unsatisfactory. In his *Thoughts on the True Estimation of Living Forces* (1747) he took Leibniz's view and tried to explain the nature of space by means of the forces of unextended substances (monads) that cause such substances to interact. He attempted to account for the threefold dimensionality of space by appealing to the laws that govern such interactions; but he was not very successful, as he himself admitted.

In his *On the First Ground of the Distinction of Regions in Space* (1768) Kant took Newton's view that space is absolute and argued against Leibniz's relational theory of space. He used the example of a pair of human hands. They are perfect counterparts of one another, yet they are incongruent (like left- and right-hand spirals). The two hands are identical as far as their spatial relations are concerned, but they are, nevertheless, spatially different. Therefore space is not just the relationship of the parts of the world to one another.

When Kant was inaugurated as professor of logic and metaphysics in 1770, he submitted a dissertation, in accordance with the custom of the time. It was entitled *On the Form and Principles of the Sensible and Intelligible World.* Here his views on space and time had developed to a point that was very close to the views enunciated in the *Critique of Pure Reason* (1781). Space and time are the schemata and conditions of all human knowledge based on sensible intuition. Our concepts of space and time are acquired from the action of the mind in coordinating its sensa according to unchanging laws. The sensa are produced in the mind by the presence of some physical object or objects. Space and time are now based epistemologically on the nature of the mind rather than ontologically on the nature of things, either as a relation among monads (Leibniz) or as a thing (Newton's absolute space). Kant had turned from modes of being to ways of knowing. This new epistemological view of space and time provided him with a way of reconciling the opposed views of Leibniz and Newton. Space and time are indeed the relational orders of contemporaneous objects and successive states, inasmuch as space and time are the conditions of intuitive representations of objects, rather than being mere relations of independent substances (monads). Space and time are indeed absolute wholes in which physical objects are located, inasmuch as they are forms of sensible intuition lying ready in the mind, rather than being independently existing containers for physical objects.

Kant's views in the *Critique* differ from those of the *Dissertation* in that space and time are held in the former to be passive forms of intuition by means of which a manifold of sensa are presented to the understanding, which has the active function of synthesizing this manifold. Space is the form of all appearances of the external senses, just as time is the form of all appearances of the internal sense. As such, space and time are nothing but properties of the human mind. Everything in our knowledge that belongs to spatial intuition contains nothing but relations: locations in an intuition (extension), change of location (motion), and the laws of moving forces according to which change of location is determined. The representations of the external senses are set in time, which contains nothing but relations of succession, coexistence, and duration.

Geometry is based on the pure intuition of space. To say that a straight line is the shortest distance between two points involves an appeal to spatial intuition. The concept of straight is merely qualitative. The concept of shortest is not already contained in the concept of straight but is an addition to straight through recourse to the pure intuition of space. Accordingly, the propositions of geometry are not analytic but a priori synthetic. So are the propositions of arithmetic. The concept of number is achieved by the successive addition of units in the pure intuition of time. Leibniz had claimed that the propositions of mathematics are analytic. For Kant even some of the propositions of mechanics are a priori synthetic because pure mechanics cannot attain its concepts of motion without employing the representation of time.

As we have seen, Kant rejected Newton's absolute space conceived as an independently existing whole containing all physical objects. In the *Metaphysical Foundations of Natural Science* (1786) he pointed out a meaning for "absolute space" which makes it a legitimate idea. At the beginning of "Phoronomy" in that treatise, he distinguishes relative space from absolute space. Relative (or material or empirical) space is the sum total of all objects of experience (bodies). Such space is movable because it is defined by material entities (bodies). If the motion of a movable space is to be perceived, that space must be contained in another, larger space in which it is to move. This larger space must be contained in another, still larger one, and so on to infinity. Absolute space is merely that largest space which includes all relative ones and in which the relative ones move. As such, absolute (empty) space cannot be perceived because it is not defined by material entities, as relative (empirical) spaces are, and so exists merely in idea, with no actual ontological status. Kant claimed that Newton mistakenly endowed such absolute space with ontological significance.

The terms "physical entities," "material objects," "bodies," and similar ones have been used from time to time in the foregoing discussion without any exact definitions being given for them. They are all roughly analogous terms and involve what Kant usually calls "matter" or "body." Toward the end of the "Dynamics" in the *Metaphysical Foundations of Natural Science* he does say that a body is matter between determinate boundaries and thus has definite shape. "Matter" is therefore a more general term than "body," but he often uses the two interchangeably. What, indeed, is matter for Kant? In the development of his thought at the stage of the *Dissertation*, he distinguished a sensible world from an intelligible one. The former is the world which sense reveals, and the latter is what the intellect reveals. He called the former world phenomenal and the latter noumenal. The former is the world of things as they appear, while the latter consists in things as they are. Sensibility with its two pure forms, space and time, provides the foundation for the validity of physics and geometry; however, the scope of the application of these two sciences is restricted to phenomena. Intellect with its pure concepts of substance, cause, possibility, existence, and necessity provides the foundation for the validity of the metaphysics of monads; this science yields an intellectual knowledge of such substances (monads) as they are in themselves, but there is no sensible knowledge of monads. The concepts of matter and body are empirical, sensitive ones belonging to physics but not to metaphysics.

By the time Kant's thought attained full maturity in the *Critique of Pure Reason*, the pure concepts of substance, cause, possibility, existence, and necessity had become coterminous with the two pure forms of intuition, space and time, in having a valid application to nothing but phenomena. The intelligible world of monads, conceived in the *Dissertation* as a known realm of things-in-themselves, becomes in the *Critique* an unknowable realm of noumena underlying the knowable realm of phenomena. One has a detailed and actual knowledge of matter and body but only a problematic knowledge of monads and noumena. The noumena did not wither away completely in the *Critique*. If they had, Kant would have been an idealist like Berkeley or a phenomenalist like Hume. Rather, Kant was a type of realist not unlike Descartes or Locke in his claim that appearances are not all that there is but are all that one has an actual and detailed knowledge of. There is a reality behind the appearance, but one has only a problematic concept of this reality. He often characterized this position of the *Critique* as transcendental idealism in order to distinguish this brand of idealism from the extreme form typified by Berkeley.

And so matter can be defined, in most general terms, as an appearance given in space. When one turns to a more particular characterization of matter, one finds that Kant's mature theory of matter developed as an opposition to the atomist view of matter held by Newton and the monadist view of matter held by Leibniz. For Newton matter is composed of physical atoms, which are things-in-themselves. He seems to espouse some form of simple realism and doubtless would have held that these atoms would move about in empty space even if there were no sentient beings anywhere to perceive them. The atoms are absolutely impenetrable, and this means that the matter constituting an atom coheres with a force that cannot be overcome by any existing force in nature. Atoms are absolutely homogeneous as to density. They differ from one another only in size and shape. Bodies are aggregates of such atoms and differ in density according to how much empty space, or void, is interspersed among the atoms.

In the "Dynamics" of the *Metaphysical Foundations* Kant objected to such absolute impenetrability as being an occult quality that no experiment or experience whatsoever could substantiate. We have seen earlier what Kant thought about absolute, empty space. Newton thus regarded matter as an interruptum. So also did Kant in his early work entitled *Physical Monadology* (1756). But in the Critical thought of the *Metaphysical Foundations*, he rejected all forms of atomism and monadology. He maintained that matter is a continuum, as we shall see.

Motive forces were for him the fundamental attributes of matter, a position which he held even in the days of the *Physical Monadology*. By contrast, Newton had taken a different view on the relation between forces and matter. For him atoms are inert but mobile. Since inertia is an entirely passive property of the atoms, they must be moved by an active principle external to them. God is the ultimate cause of gravitational motion by virtue of His acting through the immaterial medium of absolute space, as one can infer from various scholia in the *Principia* and queries in the *Opticks*. Accordingly, Newton did not regard attraction (as Kant did) as a basic property of matter itself. For Kant only two kinds of moving forces are possible: repulsive and attractive. If two bodies (regarded as mathematical points) are being considered, then any motion which the one body can impress on the other must be imparted in the straight line joining the two points. They either recede from one another or approach one another; there are no other possibilities. Since forces are what cause bodies to move, the only kinds of forces are therefore repulsive and attractive.

When one body tries to enter the space occupied by another body, the latter resists the intrusion and the former is moved in the opposite direction. The repulsive (or expansive) force exerted here is also called elastic. For Kant all matter is originally elastic, infinitely compressible but impenetrable—one body cannot compress another to the extent that the first occupies all the space of the second. He called such elasticity "relative impenetrability" and contrasted it with the absolute impenetrability posited by atomism. The relative kind has a degree that can be ascertained by experience—for instance, gold is more penetrable than iron—whereas the absolute kind is open to no experience whatsoever. On the atomic theory, bodies are compressed when the empty space among the atoms constituting bodies is eliminated and the atoms stand tightly packed. But once so packed, they admit of no further compression.

Unless there were another force acting in an opposite direction to repulsion, that is, acting for approach, matter would disperse itself to infinity. By means of universal attraction all matter acts directly on all other matter and so acts at all distances. This force is usually called gravitation, and the endeavor of a body to move in the direction of the greater gravitation is called its weight. If matter possessed only gravitational force, it would all coalesce in a point. The very possibility of matter as an entity filling space in a determinate degree depends on a balance between repulsion and attraction. Sensation makes us aware of repulsion when we feel or see some physical object and ascertain its size, shape, and location. Repulsion is directly attributed to matter. Attraction is attributed to matter by inference, since gravitation alone makes us aware of no object of determinate size and shape but reveals only the endeavor of our body to approach the center of the attracting body.

True attraction is action at a distance. The earth attracts the moon through space that may be regarded as wholly empty. And so gravity acts directly in a place where it is not. Descartes and others thought this to be a contradictory notion and tried to reduce all attraction to repulsive force in contact. Attraction is therefore nothing but apparent attraction at a distance. Descartes propounded the theory of a plenum with fourteen vortices to account for the celestial motions of the planets about the sun and the moons about the planets. Newton objected to such plenum and vortices because he thought that the friction between the celestial bodies and this hypothesized swirling fluid medium would slow down the celestial motions and eventually terminate them. He, like Kant, espoused a true attraction rather than an apparent one.

If Kant had ever critically examined Newton's suggestions as to the ultimate cause of gravitation, he doubtless would have had emphatic objections. He showed in the *Critique of Pure Reason* that God's existence cannot be established by theoretic reason. For him attraction is a property of matter itself. He argued against Descartes's apparent attraction by pointing out that such attraction operates by means of the repulsive forces of pressure and impact so as to produce the endeavor to approach, just as in the case when one billiard ball approaches another after the first has been hit by a cue. But there would not even be any impact or pressure unless matter cohered in such a way as to make such impact and pressure possible. Matter would disperse itself to infinity if it possessed nothing but repulsive force. Hence there must be a true attraction acting contrary to repulsion in order for impact and pressure to bring about even apparent attraction.

Thus matter in general was reduced by Kant to the moving forces of repulsion and attraction. He appealed to these forces to account for the specific varieties of matter. Attraction depends on the mass of the matter in a given space and is constant. Repulsion depends on the degree to which the given space is filled; this degree can vary widely. For example, the attraction of a given quantity of air in a given volume depends on its mass and is constant, while its elasticity is directly proportional to its temperature and varies accordingly. This means that repulsion can, with regard to one and the same attractive force, be originally different in degree in different matters. Consequently, a spectrum of different kinds of matter each having the same mass (and therefore having the same attraction) can vary widely in repulsion—running, for instance, from the density of osmium to the rarity of the ether. And so every space can be thought of as full and yet as filled in varying measure.

Kant claimed that matter is continuous quantity involving a proportion between the two fundamental forces of attraction and repulsion. For an atomist like Newton matter is discrete quantity, and the force of attraction in his theory is superadded through the agency of God. The varying densities of elements and compounds of matter are for Newton a function of the amount of empty space interspersed among the atoms. Empty space, according to Kant, is a fiction that can be discerned by no sense experience whatever. The senses reveal to us only full spaces. Kant's theory of matter committed him to accept the existence of an ether.

The ether was mentioned in many of his writings. In his doctoral dissertation, entitled *A Brief Account of Some Reflections on Fire* (1755), Kant said in proposition VIII that the matter of heat (or the caloric) is nothing but the ether (or the matter of light) which is compressed within the interstices of bodies by means of their strong attraction. In the *Metaphysical Foundations of Natural Science* he accepted the existence of the ether cautiously, as a hypothesis that he found more plausible than the atomists' hypothesis of the reality of absolutely impenetrable atoms and absolutely empty space. Toward the end of chapter 2 the ether is characterized as a matter that entirely fills space, leaving no void. It is so rarefied that it fills its space with far less quantity of matter than any of the bodies known to us fill their spaces. In relation to its attractive force the repulsive force of the ether must be incomparably greater than in any other kind of matter known to us.

Between 1790 and 1803 Kant worked on what is now called the *Opus postumum*. At his death this unfinished work survived as a stack of handwritten pages, which were eventually gathered by editors into thirteen fascicles (*Convoluten*). Sections of it constitute coherent wholes, others provide illustrations, and still others are repetitions of earlier works. The *Opus* appears in Volumes XXI and XXII of the Royal Prussian Academy edition of Kant's works. Part of the *Opus* contains the *Transition From the Metaphysical Foundations of Natural Science to Physics*. The theory of the ether figures in almost all parts of the *Transition* but especially in *Convolute* X, XI, and XII of the *Opus*. There the ether is characterized as a matter that occupies absolutely every part of space, that penetrates the entire material domain, that is identical in all its parts, and that is endowed with a spontaneous and perpetual motion.

Kant based his proof of the ether's existence upon the unity of experience. Space, which is unitary, is the form of all experience; hence experience is unitary. Experience is a system made up of a manifold of sense perceptions synthesized in space by the intellect. These perceptions are caused by the actions of the material forces which fill space. Accordingly, the motive forces of matter must collectively be capable of constituting a system in order to conform to the unity of possible experience. Such a system is possible only if one admits, as the basis of these forces, the existence of an ether that has the properties listed above. Therefore, the existence of the ether is the a priori condition of the system of experience. Many critics have found this proof unconvincing, as well they might. The *Transition*, as it has come down to us, is merely a series of sketches for a work that was never finished. Accordingly, it suggests about as many unanswered questions as it provides solutions.

Kant in his mature period opposed not only atomism but also all forms of monadology. He was like Leibniz and unlike Newton in that he put the emphasis, both in his youth and in his maturity, on force rather than on atomic particles of impenetrable mass. In the *Physical Monadology* (1756) he even claimed (following Leibniz) that bodies are composed of monads, which are indivisible, simple substances. The space which bodies fill is infinitely divisible and is not composed of original, simple parts because space does not have any substantiality; it is only the appearance of the external relations bound up with the unity of the monad. So conceived, the infinite divisibility of space is not opposed to the simplicity of monads. Matter is not infinitely divisible, while space is. But by the time his thought had arrived at the critical phase represented by the *Critique of Pure Reason* and the *Metaphysical Foundations*, Kant had repudiated both the view (derived from Leibniz) that bodies are composed of monads and the view (espoused by Leibniz) that space is a relation among monads. Perceptible matter was now continuous quantity, and space was a form of sensible intuition. What did he think now about the infinite divisibility of matter and space? One must turn to the Second Antinomy in the "Dialectic" of the *Critique* and the "Dynamics" of the *Metaphysical Foundations* to learn the answer.

The outcome of these discussions is this: Matter as appearance is infinitely divisible and therefore consists of infinitely many parts, but matter as appearance does not consist of the simple (either atoms or monads); matter as thing-in-itself does not consist of infinitely many parts (either atoms or monads), but matter as thing-in-itself does consist of the simple. It was pointed out earlier that according to Kant's position of transcendental idealism, we have actual cognition of things as appearances and a problematic concept of the reality behind the appearance. Accordingly, matter can be regarded as appearance or as the reality (thing-in-itself) behind the appearance.

Intuitive space is divisible to infinity. Any matter filling such space is also divisible to infinity. But this means that matter as appearance is infinitely divisible; it does not mean that matter as thing-in-itself consists of infinitely many parts (as an atomist or a monadist might claim). Only the division of the appearance can be infinitely continued, not the division of the thing-in-itself. Any whole as thing-in-itself must already contain all the parts into which it can be divided. But the division process can never be finished. And so the thought that matter as thing-in-itself contains infinitely many parts is self-contradictory.

Furthermore, it cannot be maintained that matter as appearance is made up of the simple. The composite of things-in-themselves must consist of the simple,

KANT

since the parts must be given before all composition. But the composite in the appearance does not consist of the simple, because an appearance can never be given in any way other than as composite (extended in space); its parts can be given only through the process of division, and therefore not before the composite but only in it (and thus the atomist and monadist are foiled again).

Even though matter as appearance is mathematically divisible into infinitely many parts, no real distance of parts is to be assumed. Physicists usually represent the repulsive forces of the parts of elastic matters (for instance, a gas) when these matters are in a state of greater or lesser compression as increasing or decreasing in a certain proportion to the distance of their parts from one another. This is necessary for the mathematical construction of the concept corresponding to such a state of elastic matters; in this construction all contact of parts is represented as an infinitely small distance. The posited spatial distance of the parts should be understood, however, as nothing but a mathematical convenience, necessary convenience though it is. In reality, matter is continuous quantity, and there is no spatial distance between its parts; they are always in contact.

Time, space, matter, force—Kant's views on these fundamental concepts of natural science have now been examined; but motion has not yet been considered in any detail. In contrast with Newton, Kant claimed in the "Phenomenology" of the *Metaphysical Foundations*, as well as in the earlier *New Conception of Motion and Rest* (1758), that all motion is relative. The motion of a body is the change of its external relations to a given space. If a ball rolls on a table top, it changes its position relative to various points on the top. But we have the same change of positions if the ball remains at rest and the table moves under it in the opposite direction with the same velocity. Hence the rectilinear motion of a body with regard to an empirical space can be viewed as either the body moving by reference to the space at rest or as the body at rest and the space moving relative to it. It is impossible to think of a body in rectilinear motion relative to no material space outside of it. Matter can be thought of as moved or at rest only in relation to matter and never by reference to mere space without matter. Furthermore, there is no fixed empirical point by reference to which absolute motion and absolute rest can be determined. The center of the sun might be fixed as the center of our solar system, but our solar system moves relative to other solar systems in the Milky Way, and the Milky Way moves relative to other galaxies, and so on.

Accordingly, there is no empirical space defined by matter that can provide a reference system for all possible rectilinear motions of bodies in the universe. Therefore, all motion or rest is merely relative, and neither is absolute. But the empirical space (for instance, the table top) in relation to which a body (for instance, the ball) moves or remains at rest must itself be referred to another (absolute) space at rest within which this given empirical space is movable. If one did not invoke such immovable, absolute space, one would be claiming that the given empirical space is immovable and hence absolute; but by experience all material spaces are movable. This ultimate absolute space by reference to which all empirical spaces are movable (and hence relative) exists merely in thought and not in fact, since only empirical (material) spaces actually exist. But such absolute (immovable, immaterial) space is nevertheless a necessary idea that serves as a rule for considering all motion therein as relative. Everything empirical is movable in such ideal absolute space; and all such motions in it are valid as merely relative to one another, while none is valid as absolute. And so the rectilinear motion of a body in relative space is reduced to absolute space when one thinks of the body as in itself at rest but thinks of the relative space as moved in the opposite direction in absolute space.

The circular motion of a body might seem, at first glance, to be an absolute motion. In contrast with the foregoing case of rectilinear motion, it is not all the same whether the earth is regarded as rotating on its axis while the heavens remain still (Copernicus) or the earth is regarded as staying still while the heavens rotate about it (Ptolemy). Both give the same appearance of motion. But the former case is the true one, while the second one is false. To prove that the earth rotates, Kant says that if one puts a stone at some distance from the surface of the earth and drops it, then the stone will not remain over the same point of the earth's surface in its fall but will wander from west to east. Accordingly, the rotation of the earth on its axis (or the rotation of any other body) is not to be represented as externally relative. But does this mean that the motion is absolute? Even though circular motion exhibits no real change of place with regard to the space outside of the rotating body, such motion does exhibit a continuous dynamic change of the relation of matter within its space. If the earth were to stop spinning, it would contract in size. The present size of the earth involves a balance between centrifugal forces and attracting ones. Hence the actuality of the earth's rotation rests upon the tendency of the parts of the earth on opposite sides of the axis of rotation to recede from one another. The rotation is actual in absolute space, since this

KANT

rotation is referred to the space within, and not to that outside of, the rotating body. And so rotation is not absolute motion but is a continuous change of the relations of matters (or parts of the rotating body) to one another; this change is represented in absolute space (the space within the rotating body), and for this reason such change is actually only relative motion.

The case of the translation of a body relative to a material reference system and the case of a rotating body have now been considered. What about the third and last case, in which one body hits another? Is this motion absolute? In this case both the matter and the (relative) space must necessarily be represented as moved at the same time: in every motion of a body whereby it is moving with regard to another body, an opposite and equal motion of the other body is necessary. One body cannot by its motion impart motion to another body that is absolutely at rest; this second body must be moved (together with its relative space) in the opposite direction with just that quantity of motion which is equal to that quantity of motion which it is to receive through the agency of the first body and in the direction of this first one. Both bodies, subsequently, put themselves relative to one another—that is, in the absolute space lying between their two centers—in a state of rest. But with reference to the relative space outside of the impacting bodies, the bodies move after impact with equal velocity in the direction in which the first body is moving. This same law holds if the impact involves a second body that is not at rest but moving. There is no absolute motion in this third case, even though a body in absolute space is thought of as moved with regard to another body. The motion in this case is relative not to the space surrounding the bodies but only to the space between them. When this latter space is regarded as absolute, it alone determines their external relation to one another. And so this motion is merely relative.

In the case of rectilinear motion, the change of place may be attributed either to the matter (that is, space at rest and matter moving with respect to it) or to the space (that is, matter at rest and space moving with respect to it). In the case of rotatory motion, the change of place must be attributed to the matter. In the case of colliding bodies, both the matter and the (relative) space must necessarily be represented as moved at the same time. Motion is relative in all three cases by reference to absolute space: in the first case by reference to absolute space outside of the body, in the second to absolute space inside of the body, and in the third to absolute space between two bodies.

In a pre-Critical work entitled *Universal Natural History and Theory of the Heavens* (1755) Kant was much more favorably disposed toward Newton than he was in his mature period. In fact, the full title of the work has this addition: *An Essay on the Constitution and Mechanical Origin of the Whole Universe Treated According to Newtonian Principles*. In this book Kant went far beyond anything that is to be found in Newton's own writings. Newton aimed at nothing more than describing and explaining the regularities that exist in the world in its present state of evolution. He paid special attention to the regularities of the planets and their motions around the sun, but nowhere did he try to explain, on the basis of his own mechanical principles, how the solar system originated and reached its present state of uniformity. In his *Theory of the Heavens*, Kant, by a series of bold strokes, anticipated astronomical facts that were later confirmed by very powerful observational techniques and with the help of relativistic cosmological theory. He conjectured that our solar system is a part of a vast system of stars making up a single galaxy, that the so-called nebulous stars are galactic systems external to but similar to our own galaxy (a fact that was not confirmed until the twentieth century), and that there are many such galaxies making up the universe as a whole. Much of this thought was stimulated by the work of an Englishman, Thomas Wright of Durham, entitled *Original Theory or New Hypothesis of the Universe* (1750). Kant read an abstract of this book in a Hamburg newspaper of January 1751. Wright gave the first essentially correct interpretation of the Milky Way and suggested that the nebulae are systems of stars much like our own galaxy. Newton provided Kant with the fundamental physical principles to help him in the development of his cosmogony, while Wright gave suggestions for working out the particulars of the spatial organization of the main components of the universe as a whole.

At this pre-Critical stage Kant claimed that the world had a beginning in time and is infinite in spatial extent. The universe came into existence through an act of creation on the part of a transcendent deity. In the *Critique of Pure Reason* these claims were presented as unresolvable antinomies. In the First Antinomy the thesis claims that the world has a beginning in time and is also limited spatially, while the antithesis claims that the world has no beginning and is not limited spatially. In the Fourth Antinomy the thesis claims that an absolutely necessary being belongs to the world as its part or as its cause, while the antithesis claims that no absolutely necessary being exists anywhere in the world, nor does such a being exist outside the world as its cause. However, in his *Universal Natural History and Theory of the*

Heavens, Kant adopted the theistic view (espoused also by Newton and Leibniz) that the cosmos as a whole owes its genesis to a Mosaic deity.

But once God had created time, space, and matter and had endowed them with the very laws that Newton eventually discovered, how did the universe evolve? The term "natural history" in the title of the work indicates that Kant was interested in an evolutionary account of the universe. On this subject he made suggestions that resemble those later set forth by Laplace. Kant claimed that the planets of our solar system arose from the condensation of primordial diffused matter. This position contrasts with one claiming that some celestial body passed near our sun and set up cataclysmic actions that caused the planets to be born of our sun through the agency of tidal forces. For Kant the sun, the planets, and their moons all originated by a process of condensation of a diffused mass of widely distributed, thin matter (or, in other words, a nebulous mass of matter). He appealed to the Newtonian attraction and repulsion of the various material particles of the original nebulous mass as being the causes of the solar system's flattening out into a disk. Other astronomical systems in the universe developed in a similar way.

On Kant's view the universe is not a static mechanism; it undergoes a fundamental change. Various regions of the universe undergo cyclical changes, being born as just described and dying when the train of planets associated with each star tends to run down and eventually falls back into its respective sun. The sun then heats up with this new matter and eventually explodes into a nebular cloud of matter. Our own developed world is midway between the ruins of the nature that has been destroyed and the chaos of the nature that is still unformed. He claimed that this Phoenix of Nature burns itself only in order to revive from its ashes in restored youth through all the infinity of times and places.

Cosmological theory has made great advances since Kant's day, but his theory has nonetheless inspired proponents of recent theories. C. F. von Weizsäcker, in his *History of Nature* (1949) and again in his Gifford lectures for 1959–1960 (entitled *The Relevance of Science*), has espoused the view that the planets of our solar system were formed from a nebula that surrounded the sun. G. Kuiper of Yerkes Observatory thinks that some denser parts of the nebula condensed further under the influence of their own gravity—the very view that Kant advanced. Kant has also been the object of much criticism. The claim has been made by many that he put forth bold conjectures before the experimental evidence was in and that he even twisted some evidence that was in so that it would conform with his theories. Perhaps he did make too many bold conjectures; but if such conjectures were never made, would science ever make much progress? In his *Theory of the Heavens*, Kant borrowed the principles of Newton's system of the world and by a sort of thought experiment used them to extend and deepen man's picture of the universe. It is a wonder that his thought experiment turned out to be so close to much of subsequent cosmological theory.

In the middle of the preface to the *Theory of the Heavens*, Kant said that it would be possible to grasp the origin of the whole present constitution of the universe by means of mechanical (efficient) causes before it would be possible to grasp the production of even a single herb or a caterpillar by means of mechanical causes alone. He did not categorically deny the possibility that organisms might some day be completely explained mechanically. But in the *Critique of Teleological Judgment* (1790), Kant did deny the possibility of such an explanation. All the phenomena of inanimate nature can be explained in terms of the motion of matter in space and enduring through time, while for living things such efficient causes are not enough—they must be explained in terms of an end and thus require final causes in addition to efficient ones. In more modern terms, biology, for Kant, cannot in the final analysis be explained solely in terms of physics and chemistry.

Kant was opposed to the Cartesian conception of animal machines; no one can or ever will be able to produce a caterpillar from a given bit of matter. An organism exists as a physical end and is both cause and effect of itself. The parts of an organism, both in their existence and in their form, are possible only by their relation to the whole; furthermore, the parts combine spontaneously to constitute the unity of a whole by being reciprocally cause and effect of their form. A machine has only motive power, while an organized being possesses inherent formative power, which it imparts to raw materials devoid of form.

An oak tree, for example, prepares the matter that it assimilates; and it bestows upon this matter a specifically distinctive quality which mere mechanical nature cannot supply. The tree develops itself by means of a material which the tree itself produces through an original capacity to select and construct the raw material that it derives from nature outside of it. It is as though the tree were itself a supremely ingenious artisan in the building of itself, and the subjunctive mood must be emphasized here. One cannot claim that there really is an end in the thing that is operating in its production; to do so would be to foist an unprovable anthropomorphism on organisms. A person coughs when some water goes down his windpipe while he is taking a drink. Substances heavier than air trigger the coughing mechanism. Such a mechanism is said to have adaptive significance for the organism, in that it enables the organism to live. But it cannot be said that the cough mechanism is intended to keep the organism alive, except on an analogy with artistic production. Final causes are merely regulative concepts which human beings use to comprehend biological organisms, which differ essentially from inorganic entities. One deals with them by means of an analogy with artifacts; it is as though a tree organized itself in a way not unlike the way an artisan forms his product.

Finality is read into the facts, and teleological principles are nothing but heuristic maxims whose justification resides in their fruitfulness in providing systematic comprehension of living organisms. When finality is so viewed, the question of whether, in the unknown inner basis of nature itself, efficient causes and final ones may cohere in a single principle is left open. Accordingly, teleology and mechanism in no way contradict one another, as Kant explains with the greatest epistemological subtlety in §77 of the *Critique of Teleological Judgment*. Efficient causes are concepts that determine (and do not merely regulate) our knowledge of phenomena. Therefore, the investigations of biology must be pushed as far as possible in the direction of efficient causes; the simple mechanism of nature must be the basis of research in all investigation of biological phenomena. But this does not mean that such phenomena are possible as entities solely on the basis of efficient causation. The principle of teleology directs one to continue research as far as possible on the basis of efficient causation. The processes of digestion are not understood by any appeal to the principle that such processes enable the organism to live and thrive. Such processes are understood by accounting for the passage of chemical substances through membranes, but why such membranes permit the passage only of certain chemical substances and no others may not be able to be accounted for on the basis of mechanical causes alone.

Efficient causes are progressive, while final causes are reciprocal. In the former the connection constitutes a series (the so-called causal chain) which is always such that the things which, as effects, presuppose other things as their causes cannot themselves also be causes of these other things. Final causes are such that the series involves regressive as well as progressive dependency (a house is the cause of one's receiving rent money, but the house was built in the first place so that one might receive such rent). Kant calls efficient causes a nexus of real ones and final causes a nexus of ideal ones. The former are also said to determine our knowledge of phenomena. He treats of efficient causation in the Second Analogy of Experience in the "Transcendental Analytic" of the *Critique of Pure Reason*, where he distinguishes between a subjective connection of cognitions and an objective connection of them. One walks into a warm room and sees a glowing stove. As far as the subjective order of cognitions is concerned, one first feels warm and only later spies the stove concealed behind a screen in the corner. But yet one says that it is the stove which causes the room to be warm and not that it is the warm room which causes the stove to glow. In order to have knowledge through perceptions, one must connect them in their objective time relations. There is no necessity in the subjective order of cognitions, but there certainly is in one's synthetic reorganization of that order. If event A (glowing stove) precedes event B (warm room) objectively, then one must think of A as preceding B or else be wrong. It makes no difference whether one perceives A first and then B, or B first and then A in his subjective consciousness.

Kant worked out his theory of efficient causation largely in opposition to Hume's position on the subject. Hume claimed that there are three conditions which two events must fulfill in order for one event to be considered the cause of another: the cause precedes the effect in time, cause and effect are contiguous in space, and cause and effect are found constantly conjoined in experience. Kant leveled his attack mainly against the third condition, but other people have found objections to the first two conditions as well. In the realm of colliding billiard balls, the cause does temporally precede the effect; but in the case of boiling water, the boiling occurs just when the water reaches 100° C.; and so cause and effect are simultaneous. As for spatial contiguity, the moon through empty space attracts the waters of the earth's seas and oceans to produce tides. In the case of such action at a distance, there is no contiguity in space. As for constant conjunction, some have pointed out that night and day are always conjoined, but night is not the cause of day. Kant objected to the third condition by claiming that on Hume's view there is no way to distinguish the subjective order of cognitions from the objective order of them.

Hume held that the idea of necessary connection between cause and effect arises when we develop a habit of association from a repeated subjective succession of perceptions (fire always burns). He thus based causation entirely on sensible experience. In contrast, Kant claimed that the objective reordering of the subjective succession of cognitions (which is based on sense perception and imagination) is actually

KANT

a synthetic reorganization of the a posteriori order of perception. This synthetic reorganization is an a priori act of the human understanding. In other words, the causal ordering of cognitions is an act of the intellect that is brought to experience (or, even better, that makes experience) and is not an ordering derived from experience (as Hume claimed). For Kant in his Critical period the pure concepts of substance, cause, possibility, existence, and necessity were a priori concepts that are coterminous with the pure forms of intuition, time and space. Experience is the result of the synthetic activity of the intellect by means of such pure concepts in organizing empirically given sense perceptions that are arrayed in time and space. The history of theories of efficient causation did not end with Kant. The controversy between the a posteriori and a priori views broke out again in the middle of the nineteenth century with the debates between John Stuart Mill and William Whewell. Mill disagreed with Kant, while Whewell agreed. In the twentieth century Bertrand Russell and A. C. Ewing have continued the debate.

Time, space, matter, force, motion, cause—the major portions of Kant's philosophy of science have now been examined in terms of these key concepts. This philosophy is rationalistic in comparison with the views of such later thinkers as Mill and Russell but seems almost empiricistic in comparison with such of Kant's immediate German idealist successors as Fichte, Schelling, and Hegel. According to Kant, the human mind supplies the form of experience (time, space, and the categories of the understanding); but the content of experience is empirically given in sensation from a source outside the human self (the real material world). The German idealists claimed that the self is the source not only of the form of experience but also of its content. On this view nature becomes a sort of external symbol or image of the self. Nature is the self taken as object. Accordingly, Schelling thought that the whole of physics could be spun out of the mind itself. If so, what need is there for experiment?

The accusation of armchair scientist which Erich Adickes leveled at Kant might more appropriately be directed against these Romantic idealists. Apart from the most general, formal aspects of nature (matter is a continuum and not an interruptum, there is no absolute motion, the changes in nature are causally connected, and so on), all the rest of nature in its particular aspects (temperature of Venus, strength of the gravitational pull of the moon, the cause of diabetes, and so on) must for Kant be learned by experiment. To be sure, he was more interested in the formal aspects than in the particular, but he never claimed that the particular aspects could be dealt with in any way other than by observation and experiment.

The Kantian emphasis on causality in conformity with law and on mathematical rigor in conformity with experience contributed important elements to the philosophical depth and seriousness that animated the German scientific movement from the middle of the nineteenth century on and that distinguished it from the scientific traditions of other cultures. Two quotations from classic texts will illustrate the way in which creative German scientists formed their expectations of what a scientific explanation does from their knowledge of Kant. The first, concerning physics, is from Helmholtz's *Ueber die Erhaltung der Kraft* (1847), the famous memoir on the conservation of energy which in its philosophical aspect Helmholtz attributed expressly to Kant's inspiration:

> The final goal of the theoretical natural sciences is to discover the ultimate invariable causes of natural phenomena. Whether all processes may actually be traced back to such causes, in which case nature is completely comprehensible, or whether on the contrary there are changes which lie outside the law of necessary causality and thus fall within the region of spontaneity or freedom, will not be considered here. In any case it is clear that science, the goal of which is the comprehension of nature, must begin with the presupposition of its comprehensibility and proceed in accordance with this assumption until, perhaps, it is forced by irrefutable facts to recognize limits beyond which it may not go [from *Selected Writings of H. L. F. von Helmholtz*, Russell Kahl, ed. (Middletown, Conn., 1971), p. 4].

The second quotation, concerning biology, is from Schwann's *Mikroskopische Untersuchungen über die Uebereinstimmung in der Struktur und dem Wachstume der Tiere und Pflanzen* (1839), the equally famous treatise concerning the cellular structure of the living organism. It concludes with a regulatory discussion of methodology deriving from the Kantian distinction between mechanistic and teleological explanation in the *Critique of Teleological Judgment*:

> Teleological views cannot be discarded for the time being since not all phenomena are to be clearly explained by the physical view. Discarding them is not necessary, however, for a teleological explanation is admissible if and only if a physical explanation can be shown to be unattainable. Certainly it brings science closer to the goal to try at least to formulate a physical explanation. I should like to repeat that when I speak of a physical explanation of organic phenomena, I do not necessarily mean that an explanation in terms of known physical forces like that universal resort, electricity, is to be understood, but rather an explanation in terms of

> forces which operate like physical forces in service to strict laws of blind necessity, whether or not such forces be found in inorganic nature [Ostwalds Klassiker der Exacten Wissenschaften (Leipzig, 1910), p. 187].

In general, it was in service to the rigorous idealism of Kant that leaders of the first great generation of German science—including Müller, Schleiden, Mayer, du Bois-Reymond, and Virchow—repudiated the literary and speculative *Naturphilosophie* of the Romantic idealists, which they dismissed as an episode of cultural wild oats in the adolescence of the German spirit.

In the twentieth century Kant's thought has not had the direct influence on experimental scientists that it had earlier on Helmholtz and Schwann. But in the philosophy of science Ernst Cassirer gave a Kantian interpretation of the metaphysical and epistemological foundations of relativity in his *Einstein's Theory of Relativity Considered From the Epistemological Standpoint* and of quantum theory in his *Determinism and Indeterminism in Modern Physics.* C. F. von Weizsäcker gave a Kantian interpretation of quantum theory in his *The World View of Physics.*

BIBLIOGRAPHY

I. ORIGINAL WORKS. The best ed. of Kant's works is *Kants gesammelte Schriften*, published by the Royal Prussian Academy of Sciences (Berlin, 1902–). This ed. already runs to some 27 vols. and, when complete, will contain not only all the published works, letters, and fragments but also all the extant transcripts of lectures.

English translations of some of Kant's works mentioned in this essay follow, in the order of the publication of the original German works: *Universal Natural History and Theory of the Heavens*, trans. by W. Hastie (Ann Arbor, Mich., 1969), a repr. of Hastie's *Kant's Cosmogony* (Edinburgh, 1900); *Kant's Inaugural Dissertation and Early Writings on Space*, trans. by J. Handyside (Chicago, 1929); *Critique of Pure Reason*, trans. by Norman Kemp Smith, 2nd ed., rev. (London, 1933); *Metaphysical Foundations of Natural Science*, trans. by James W. Ellington (Indianapolis, 1970); *Critique of Teleological Judgement*, trans. by J. C. Meredith (Oxford, 1928).

II. SECONDARY LITERATURE. The best biography of Kant is Karl Vorländer, *Immanuel Kant, der Mann und das Werk*, 2 vols. (Leipzig, 1924). The best in English is J. W. H. Stuckenberg, *The Life of Immanuel Kant* (London, 1882).

The literature about Kant is enormous. Erich Adickes, "Bibliography of Kant," in *The Philosophical Review* (1893), lists 2,832 titles and goes only to 1802. The *Literaturverzeichnisse* preceding Kant's various writings on science and the philosophy of science that are in vol. VII of Karl Vorländer's ed. of Kant's works in the Philosophische Bibliothek series, entitled *Kants sämtliche Werke*, 10 vols. (Leipzig, 1913–1922), are recommended for books specifically on Kant's works on science and the philosophy of science.

A brief list of the works especially relevant to this essay is Erich Adickes, *Kants Opus postumum*, which is *Kant-Studien*, supp. vol. no. 50 (Berlin, 1920); and *Kant als Naturforscher*, 2 vols. (Berlin, 1924–1925); Ernst Cassirer, *Substance and Function and Einstein's Theory of Relativity* (Chicago, 1923), pp. 347–456; and *Determinism and Indeterminism in Modern Physics* (New Haven, 1956); James W. Ellington, introduction and supplementary essay entitled "The Unity of Kant's Thought in His Philosophy of Corporeal Nature," in the trans. of the *Metaphysical Foundations of Natural Science* (see above); Irving Polonoff, *Force, Cosmos, Monads, and Other Themes of Kant's Early Thought*, which is *Kant-Studien*, supp. vol. no. 106 (Bonn, 1972); Lothar Schäfer, *Kants Metaphysik der Natur* (Berlin, 1966); and C. F. von Weizsäcker, *The World View of Physics* (Chicago, 1952); and *The Relevance of Science* (London, 1964).

JAMES W. ELLINGTON

KAPTEYN, JACOBUS CORNELIUS (*b.* Barneveld, Netherlands, 19 January 1851; *d.* Amsterdam, Netherlands, 18 June 1922), *astronomy.*

Kapteyn was the ninth of fifteen children of G. J. Kapteyn and E. C. Koomans, who conducted a boarding school for boys in Barneveld. Many of these children were extraordinarily gifted in science. As a boy Kapteyn showed outstanding intellectual ability and curiosity. At the age of sixteen he passed the entrance examination for the University of Utrecht, which, however, his parents judged him too young to enter until the following year. He studied mathematics and physics and obtained the doctor's degree with a thesis on the vibration of a membrane. Kapteyn was interested in many branches of science, and it was mostly through his accepting (1875) a position as observer at the Leiden observatory that he began his career in astronomy. In 1878, at the age of twenty-seven, he was elected to the newly instituted chair of astronomy and theoretical mechanics at the University of Groningen, which he held until his retirement in 1921 at the age of seventy.

Kapteyn's major contributions are in the field of stellar astronomy, particularly in research on the space distribution and motions of the stars. At the time of these studies, the problem of the space distribution of the stars was still tantamount to the problem of the structure of the universe. Kapteyn's work presents the first major step in this field after the great works of William and John Herschel. It culminated in the views presented in the article "First Attempt at a Theory of the Arrangement and Motion of the Sidereal System," published in the May 1922 issue of the *Astrophysical Journal*, shortly before Kapteyn's death. Seeking to resolve the structure and kinematic properties of the stellar system, Kapteyn devoted his efforts both to the problem of the methods to be developed and their mathematical aspects, and to that of obtaining proper observational data. For the latter purpose he established and participated in extensive international observational projects of many kinds. These have served astronomical research in fields remote from Kapteyn's own. Thus Kapteyn deeply influenced astronomy not only by his analyses of (sometimes rather meager) observational material available at his time but also by providing the framework for future observational programs and more detailed analyses.

Kapteyn possessed the ability to conduct several large-scale undertakings at once, and to handle with great care and ability the more detailed matters essential to their successful completion. He thereby made significant contributions to several special fields of astronomy. This article will concentrate first on his major contributions.

Kapteyn's first major achievement was the compilation, together with David Gill, of the *Cape Photographic Durchmusterung*, a catalogue of stars in the southern hemisphere. Their approach to this project was quite untraditional according to astronomical practice of that time. Since the University of Groningen, in spite of Kapteyn's requests, could not provide him with a telescope, he looked for other ways to contribute to the observational work. In 1885 he contacted Gill, then director of the Royal Observatory in Cape Town, South Africa, to offer to measure the photographic plates, covering the whole southern sky, which Gill had taken at the Cape with a Dallmeyer objective. These would be measurements of the position and apparent brightness of the stars, down to magnitudes comparable with those of the *Bonner Durchmusterung* (Bonn, 1859–1862), and thus would supplement, by photographic means, what had been accomplished by Argelander, by visual means, for the northern sky.

The extremely laborious work was finished after thirteen years of excellent collaboration. For the measurement of the plates, done from 1886 to 1896, Kapteyn devised an unconventional method. Instead of measuring x and y coordinates, he used a theodolite, observing the plate from a distance equal to the focal length of the telescope that had produced the plates. He thus obtained equatorial coordinates directly; the catalogue gives the right ascensions to $0^{s}.1$ and the declinations to $0'.1$. Approximate apparent (photographic) magnitudes were also given. The catalogue was published as volumes III, IV, and V of the *Annals of the Royal Observatory*, *Cape Town* (1896–1900) and lists the positions of 454,875 stars, between the South Pole and declination $-18°$, down to the tenth magnitude. An invaluable reference work on the southern sky, it is remarkably free of errors because of the painstaking care with which Kapteyn himself participated in much of the routine work. Assistance in the routine included labor by certain convicts of the state prison at Groningen, who were put at Kapteyn's disposal by the prison authorities. The measurements were all carried out in two small rooms of the physiological laboratory of the University of Groningen. Thus started Kapteyn's "astronomical laboratory," a kind of institute unique at that time, which soon would become world famous and recognized a much-needed complement to institutes equipped with telescopes. After being housed in various other "guest" institutes, the "laboratory" in 1913 acquired the whole of the building of the original physiological laboratory.

Structure of the Stellar System. The principal unknowns which Kapteyn tried to solve were the function $D(r)$, that is, the space density of stars as a function of the distance r from the sun; and the function $\varphi(M)$, or the distribution of the stars according to brightness per unit volume. In a series of investigations by Kapteyn and his collaborators, extending over several decades, these functions became defined in more and more detail. Thus, the function $D(r)$, initially determined only for stars generally without regard to their spectral type or galactic latitude or longitude, could later be determined separately for different types of stars, and $\varphi(M)$ could be distinguished more and more according to both distance from the galactic plane and spectral type.

Kapteyn's approach to these problems was basically different from that of contemporaries such as Hugo von Seeliger and Karl Schwarzschild. The latter proposed certain analytical expressions for the aforementioned functions, as well as for the distribution of observed quantities, and then tried to solve for the parameters involved by means of integral equations. Kapteyn, on the other hand, preferred the purely numerical approach, allowing full freedom for the form of the solution.

In principle the procedure applied was the following. Statistics could be obtained on the numbers of stars of given apparent brightness $N(m)$ and, to some extent, on these numbers subdivided according to the size of the proper motion, μ, of the stars, $N(m, \mu)$. The proper motion was introduced as an auxiliary quantity because, like m, it is a measure of distance; hence, if the velocity distribution of the stars is independent of r, then knowledge of μ should help in unraveling the distance distribution. Moreover, largely through his own efforts, Kapteyn obtained for stars of a given apparent magnitude the mean value of the trigonometric parallax $\langle\pi(m)\rangle$ and, to some extent, this mean subdivided according to the stars' proper motions, $\langle\pi(m, \mu)\rangle$. If one assumes that all stars have the same intrinsic brightness—a very special case of $\varphi(M)$—and that the space density is uniform (that is, $D(r)$ is constant), then a certain, predictable form of the statistics $N(m)$ results. The actually observed shape of $N(m)$ is different, and the problem is to find by proper adjustments the true $D(r)$ and $\varphi(M)$. For these adjustments the values $\langle\pi(m)\rangle$ and (more refined) $\langle\pi(m, \mu)\rangle$ appeared the most adequate quantities.

For a detailed account of Kapteyn's procedure, we refer to the literature cited below, especially to the relevant chapters in the books by von der Pahlen and by de Sitter. In a long series of papers, mostly in *Publications. Astronomical Laboratory at Groningen* (referred to below as *Gron. Publ.*), Kapteyn and his co-workers gave ever more complete tables for the quantities and the resulting solutions. Data on $\langle\pi(m)\rangle$ published in *Astronomische Nachrichten*, no. 3487 (1898); on $\langle\pi(m, \mu)\rangle$ in *Gron. Publ.*, no. 8 (1901); and on $N(m)$ and $N(m, \mu)$ in *Gron. Publ.*, no. 11 (1902), were analyzed for a provisional solution of $D(r)$ and $\varphi(M)$ in *Gron. Publ.*, no. 11 (1902). The final, more precise and detailed solutions were published by Kapteyn and van Rhijn in 1920 in *Contributions from the Mount Wilson Solar Observatory*, no. 188 (also in *Astrophysical Journal*, **52**) and, after Kapteyn's death, by van Rhijn in *Gron. Publ.*, no. 38 (1925). These were based on the improved data for

$N(m)$, in *Gron. Publ.*, no. 18 (1908) and no. 27 (1917);
$N(m, \mu)$, *ibid.*, no. 30 (1920) and no. 36 (1925);
$\langle\pi(m)\rangle$, *ibid.*, no. 29 (1918); see also no. 45 (1932); and for
$\langle\pi(m, \mu)\rangle$, *ibid.*, no. 8 (1901), *Contributions from the Mount Wilson Solar Observatory*, no. 188, and *Gron. Publ.*, no. 34 (1923).

Some of the results of these investigations have been of lasting value and some have been superseded. As to those for $D(r)$, we should distinguish between the change of densities with distance in the direction of the galactic plane (or at adjacent, low, galactic latitudes) and in the directions away from the plane. Whereas the latter results have proved to be essentially correct, the former are now known to be spurious. Kapteyn and van Rhijn found that at low galactic latitudes the star density in all directions diminishes with increasing distance from the sun. Thus, at 600 parsecs (2,000 light-years) it was found to be about 60 percent of that near the sun, at 1,600 parsecs about 20 percent, and at 4,000 parsecs only 5 percent. This apparent decrease, which gave rise to interpretation in terms of a more or less isolated, flattened, and spheroidal local stellar system (the "Kapteyn system"), is due to the fact that Kapteyn assumed starlight to pass through space without being dimmed on its way to the earth.

Actually, as is now known, interstellar absorption by small grains does cause a dimming effect, hence in the numerical solutions the stars appear too distant. This results in an apparent decrease of the derived star density with distance. Kapteyn was fully aware of the interstellar absorption as a possible cause of inaccuracies in his results, and therefore he made several attempts to detect its existence. He correctly assumed that interstellar absorption should be accompanied by reddening of the starlight, the expected absorption in the blue being stronger than in the yellow and the red. But investigations of this reddening did not lead to positive results, and accordingly absorption could not be taken into account. (It was only in 1930 that Trümpler could prove its existence.)

Kapteyn's results for high galactic latitudes were hardly affected by the dimming because the absorbing matter is concentrated close to the galactic plane. At 100 parsecs the star density appeared to be about 55 percent of that near the sun, at 250 parsecs 40 percent, at 600 parsecs 12 percent, and at 1,600 parsecs less than 2 percent. The sun was found to be close to the plane of symmetry. These results apply to the combined population of all spectral types. Results for the luminosity function $\varphi(M)$ were essentially correct because they had been derived mostly from the nearest, unobscured stars. It was shown that the frequency of stars per unit volume increases from the most luminous objects (with intrinsic brightnesses about 1,000 times that of the sun) to those of about solar brightness, which are about 1,000 times more prevalent, and that the frequency subsequently tends to level off. The results included stars with a luminosity down to about half solar brightness.

Discovery of the Star Streams. Reference has been made to the use of the proper motion, μ, as well as the apparent magnitude, m, as an indicator of distance. In the early stages of his work, after having explored the use of the trigonometric parallaxes, Kapteyn emphasized the use of μ rather than that of m because of the large spread known to exist among the absolute magnitudes. As a prerequisite to the application of the method, an attempt was made to determine the distribution of the stars according to their peculiar velocities; that is, of the velocities with respect to the mean motion of the stars, the latter, in turn, being the reflex of the motion of the sun. A basic assumption was that the stellar motions have a random character, like those of the molecules of a gas, without preferred direction.

When tests of the method using μ as a distance indicator gave unsatisfactory results, Kapteyn found that the assumption of random motion was incorrect: preferred directions did exist. It appeared that the stars belong to two different, but intermingled, groups having different mean motions with respect to the sun.

This phenomenon, termed "the two star streams," was announced by Kapteyn at the International Congress of Science at St. Louis in 1904 and before the British Association in Cape Town in 1905 (*Report of the British Association for the Advancement of Science*, Sec. A) and deeply impressed the astronomical world. It demonstrated that a certain order, rather than the hitherto assumed random motion, dominated stellar motions. During the subsequent decades, numerous investigations were devoted to the subject and alternative interpretations presented. Of these latter by far the most significant is that of K. Schwarzschild, who, in 1907, instead of assuming the existence of two intermingling populations, postulated an undivided population; however, he conceived this population to have an ellipsoidal distribution of peculiar velocities, with the largest peculiar velocity components in the direction of the largest axis of the velocity ellipsoid. This interpretation appeared to accord with the observational data equally well. The discovery of the two star streams—and especially the hypothesis of ellipsoidal distribution—was of fundamental importance for the theory of the dynamics of the stellar system.

The Plan of Selected Areas. During the early stages of Kapteyn's investigations, the approximate position and apparent brightness were known for somewhat less than a million stars; proper motions were known with varying degrees of accuracy for several thousand, and trigonometric parallaxes for fewer than 100 stars. Kapteyn encouraged efforts of many observatories to procure more data on trigonometric parallaxes, radial velocities, spectra, and proper motions; wherever possible, he assisted in the measurements of plates by means of the facilities of his growing laboratory. A carefully planned undertaking appeared in order, however, particularly because from the fainter stars (there are about ten million down to the fourteenth magnitude) a selection had to be made.

In order to make sure that this selection would involve as much as possible the same stars for each observational program, Kapteyn devised a plan which was proposed to the international astronomical world in the booklet *Plan of Selected Areas* (Groningen, 1906). His plan resulted from many letters and discussions with colleagues abroad (and was a study topic for an Astronomical Society of the Atlantic, created ad hoc aboard the ship on which Kapteyn and other astronomers made a voyage to South Africa for their meeting in 1905).

The *Plan* proposed to concentrate work on 206 stellar areas, uniformly demarcated over the sky and at declinations $+90°$, $+75°$, $+60°$ . . . to $-90°$. Photographic and visual magnitudes were to be measured for all stars in these areas ($\pm$200,000); and, for more limited numbers, the quantities more difficult to measure, such as proper motion, parallax, spectral type, and radial velocity. This observational material would provide a proper sampling of the stellar system for the purpose of revealing its main structural features. At the instigation of the astronomer E. C. Pickering, a supplementary program of forty-six areas was proposed, chosen where the Milky Way shows particularly striking features, such as excessive star density and dark or bright nebulae. Pickering's program became known as the "Special Plan." Methods for observing and for evaluation of the material and the prospects for analyses were extensively discussed in the booklet.

Astronomical institutes throughout the world responded most favorably to the proposal—not least because of the cooperative spirit Kapteyn and his laboratory had shown on many occasions when their help was solicited by others. Work on Kapteyn's plan, and to a lesser degree on Pickering's special plan, progressed during the first half of the twentieth century and continues to be an outstanding example of international scientific collaboration. To date forty-three observatories have in one way or other collaborated. Shortly after international agreement on the plan had been reached, its supervision was placed in the hands of an international committee of prominent astronomers; W. S. Adams, F. W. Dyson, Gill, Hale, Küstner, E. C. Pickering, K. Schwarzschild, and Kapteyn himself. The committee was later incorporated into the International Astronomical Union as one of its commissions, and progress reports on the plan are to be found in the *Transactions* of the union.

Dynamical Theory of the Stellar System. With the newly obtained results on stellar density distribution (the "Kapteyn system") and the new knowledge of stellar kinematics (the peculiar motions, solar motion, and star streams), Kapteyn toward the end of his career developed a dynamical theory of the system. Such a theory aimed at explaining both observed density distribution and motions in terms of gravitational forces, and it would do this on the assumption that the system is in a state of equilibrium.

Kapteyn's theoretical results were communicated in the 1922 paper already quoted. In considering his results we may again distinguish between two basic directions: the one toward the "pole" of the galaxy, that is, along the minor axis of the spheroidal system, and the one perpendicular to this axis, in the galactic plane.

In the first direction the galactic situation may be compared with that in the earth's atmosphere: its scale height is such as to balance those gravitational forces that tend to flatten the atmosphere with the force of thermal motions perpendicular to the earth's surface, which tend to increase the thickness of the atmosphere. For a given gravitational field, increased thermal velocities would lead to increased scale height. Similarly, considerations of equilibrium allowed Kapteyn to derive, from the known distribution of the components of the velocities in the direction perpendicular to the galactic plane and, from the observed "scale height" in the same direction, the strength of the gravitational field at various distances from the plane. This calculation led in turn to an estimate of the total mass density per volume, a very fundamental quantity. Kapteyn expressed the results in mean masses per star—knowing the number of stars per unit volume—and found values between 2.2 and 1.6 solar masses, well in agreement with later determinations.

The larger extension of the stellar system in the direction of its equatorial plane was explained by the occurrence of a general rotation around the polar axis. This hypothesis was related to the phenomenon of the star streams, the assumption being that the system is composed of two subsystems with opposing directions of rotation; in that case, centrifugal force plus random motions must be balanced by the gravitational field. Here, too, Kapteyn succeeded in arriving at a coherent picture. But the concept of the spheroidal system could not be upheld, and the phenomenon of star streams has since been given a different explanation by B. Lindblad.

Apart from these main achievements, Kapteyn made essential contributions in many other fields. Among these are his attempts, in his early years at the Leiden observatory, to improve upon the measurement of trigonometric parallaxes with the meridian telescope and his later efforts to apply photographic methods for this purpose as well as for the measurement of stellar magnitudes. In his early years he also devised a method to find the altitude of the equatorial pole which would be free of errors in the declinations of the stars and would be independent of errors in the atmospheric refraction. Kapteyn emphasized on many occasions the great need for improvement of the fundamental system of declinations and proposed observational methods to eliminate systematic errors. He demonstrated certain relations between the various spectral types of the stars and their kinematic properties and pursued especially the properties of the earliest types (the "helium stars"), for which the small ratio between peculiar velocity and solar motion allowed the determination of accurate individual parallaxes. The accounts of this latter work, in which Kapteyn's approach to the handling of such delicate quantities as small proper motions is quite remarkable, are given in two extensive papers (*Astrophysical Journal*, **40** [1914] and **47** [1918]; repr. in *Contributions from the Mount Wilson Solar Observatory*, nos. 82 and 147).

Kapteyn's interest in statistical properties of natural phenomena outside astronomy is shown by his thorough studies of tree growth and other phenomena in the booklet *Skew Frequency Curves in Biology and Statistics* (Groningen, 1903) and in the article "Tree-Growth and Meteorological Factors (1889–1908)," in *Recueil des travaux botaniques néerlandais* (1914). In the course of his researches he introduced many concepts that have come into common acceptance in astronomy, including those of absolute magnitude and color index.

Kapteyn had an almost inexhaustible capacity for scientific activity. In his attitude toward research he was extremely critical, with respect both to his own work and to that of others. He never sacrificed clarity of treatment or exposure of essential details for elegance of presentation; and, although a mathematician himself by his early training, he strongly disliked treatises in which emphasis lay more on the form of the mathematical expression than on proper evaluation of the basic observations. It was only through his thorough knowledge of their strengths and weaknesses that he was able to draw proper conclusions from what were sometimes limited data.

In his relation to friends and colleagues, Kapteyn was very sensitive to friendship and cordiality. Having suffered in early youth from a lack of warmth and protection in his family life—his parents being fully occupied with their boarding school and perhaps having aimed at equal treatment of all their "children" —he later responded all the more readily to human relations. From his collaboration with many colleagues grew close ties of friendship, such as that with Gill (with whom a regular correspondence developed over three decades).

Kapteyn had a keen sense of justice and suffered

deeply when World War I disrupted the international communication between scientists. He firmly believed in the duty of scientists to bridge the gaps caused by political developments, and therefore he urged that upon termination of the war—at least in the scientific world—reconciliation between Germany and the Allies be reestablished. He was thus deeply shocked, and protested violently, when in 1919 the Interallied Association of Academies was founded with Germany excluded. When, in spite of his and a few others' protests, the Royal Netherlands Academy of Sciences and Letters decided to join the International Research Council (from which Germany was again excluded), he resigned his long-standing membership in the academy.

Kapteyn had a keen sense of humor and was a celebrated lecturer to audiences of all kinds. In the town of Groningen, where he lived for more than forty years with his wife and family (he married Elise Kalshoven in 1879 and had two daughters and one son), he was well remembered more than thirty years after his death.

BIBLIOGRAPHY

I. ORIGINAL WORKS. Numerous papers by Kapteyn, some of them collaborations, appeared in the main astronomical journals, especially the *Astrophysical Journal*, *Astronomische Nachrichten*, and *Astronomical Journal*, and in the reports of the *Koninklijke Akademie van Wetenschappen* of Amsterdam. A list of the principal papers is given in an appendix to the obituary by W. de Sitter (see below).

The series Publications of the Astronomical Laboratory at Groningen, created by Kapteyn, contains both treatises on the analyses of observational data and catalogues of measurements. Other important catalogues besides the *Cape Photographic Durchmusterung* (see text) are "Durchmusterung of the Selected Areas," in *Annals of Harvard College Observatory*, nos. 101, 102, and 103 (1918–1924), compiled with E. C. Pickering and P. J. van Rhijn; and the *Mount Wilson Catalogue of Photographic Magnitudes in Selected Areas 1–139* (Washington, D.C., 1930), compiled with F. H. Seares and P. J. van Rhijn.

II. SECONDARY LITERATURE. Many obituaries appeared in scientific journals after Kapteyn's death. Of special note are A. Pannekoek, "J. C. Kapteyn und sein astronomisches Werk," in *Naturwissenschaften*, **10**, no. 45 (1922), 967–980; J. J. (J. Jackson?), in *Monthly Notices of the Royal Astronomical Society*, **83** (1923), 250–255; W. de Sitter, "Jacobus Cornelius Kapteyn †," in *Hemel en dampkring*, **20** (1922), 98–110, in Dutch; C. Easton, "Persoonlijke herinneringen aan Kapteyn," *ibid.*, 112–117, and **21** (1922), 151–164, in Dutch; and A. S. Eddington, "Jacobus Cornelius Kapteyn," in *Observatory*, **45** (1922), 261–265.

An excellent chapter describing Kapteyn's work in the context of developing historical insight into the structure of the universe appears in W. de Sitter, *Kosmos* (Cambridge, Mass., 1932), ch. 4, a lecture series at the Lowell Institute in Boston. A good detailed account of Kapteyn's statistical treatments is given by E. von der Pahlen in *Lehrbuch der Stellarstatistik* (Leipzig, 1937), ch. 8, sec. ID, pp. 434–479. A biography in Dutch, *J. C. Kapteyn, zijn leven en werken* (Groningen, 1928), by Kapteyn's daughter, H. Hertzsprung-Kapteyn (wife of the famous astronomer E. Hertzsprung), gives a fine account of Kapteyn's personal life, his relations with colleagues, and his scientific achievements as experienced by his family.

A. BLAAUW

AL-KARAJĪ (or **AL-KARKHĪ), ABŪ BAKR IBN MUḤAMMAD IBN AL ḤUSAYN** (or **AL-ḤASAN)** (*fl.* Baghdad, end of tenth century/beginning of eleventh), *mathematics*.

Virtually nothing is known of al-Karajī's life; even his name is not certain. Since the translations by Woepcke and Hochheim he has been called al-Karkhī, a name adopted by historians of mathematics.[1] In 1933, however, Giorgio Levi della Vida rejected this name for that of al-Karajī.[2] This debate would have been pointless if certain authors had not attempted to use the name of this mathematician to deduce his origins: Karkh, a suburb of Baghdad, or Karaj, an Iranian city. In the present state of our knowledge della Vida's argument is plausible but not decisive. On the basis of the manuscripts consulted it is far from easy to decide in favor of either name.[3] Turning to the "commentators" does not take us any further.[4] For example, the *al-Bāhir fī'l jabr* of al-Samaw'al cites the name al-Karajī, as indicated in MS Aya Sofya 2718. On this basis some authors have sought to derive a definitive argument in favor of this name.[5] On the other hand, another hitherto unknown manuscript of the same text (Esat Efendi 3155) gives the name al-Karkhī.[6] Because the use of the name al-Karajī is beginning to predominate—for no clear reasons—and because we do not wish to add to the already great confusion in the designation of Arab authors, we shall use the name al-Karajī—refraining from any speculation designed to infer our subject's origins from this name. It is sufficient to know that he lived and produced the bulk of his work in Baghdad at the end of the tenth century and the beginning of the eleventh and that he probably left that city for the "mountain countries,"[7] where he appears to have ceased writing mathematical works in order to devote himself to composing works on engineering, as indicated by his book on the drilling of wells.

Al-Karajī's work holds an especially important place in the history of mathematics. Woepcke remarked that it "offers first the most complete or rather the only theory of algebraic calculus among the Arabs known to us up to the present time."[8] It is true that al-Karajī employed an approach entirely new in the tradition of the Arab algebraists—al-Khwārizmī, Ibn al-Fatḥ, Abū Kāmil—commencing with an exposition of the theory of algebraic calculus.[9] The more or less explicit aim of this exposition was to find means of realizing the autonomy and specificity of algebra, so as to be in a position to reject, in particular, the geometric representation of algebraic operations. What was actually at stake was a new beginning of algebra by means of the systematic application of the operations of arithmetic to the interval $[0, \infty]$. This arithmetization of algebra was based both on algebra, as conceived by al-Khwārizmī and developed by Abū Kāmil and many others, and on the translation of the *Arithmetica* of Diophantus, commented on and developed by such Arab mathematicians as Abu'l Wafā' al-Būzjānī.[10] In brief, the discovery and reading of the arithmetical work of Diophantus, in the light of the algebraic conceptions and methods of al-Khwārizmī and other Arab algebraists, made possible a new departure in algebra by al-Karajī, the author of the first account of the algebra of polynomials.

In his treatise on algebra, *al-Fakhrī*, al-Karajī first presented a systematic study of algebraic exponents, then turned to the application of arithmetical operations to algebraic terms and expressions, and concluded with a first account of the algebra of polynomials. He studied[11] the two sequences $x, x^2, \cdots, x^9, \cdots; 1/x, 1/x^2, \cdots, 1/x^9, \cdots$ and, successively, formulated the following rules:

$$(1)\quad \frac{1}{x} : \frac{1}{x^2} = \frac{1}{x^2} : \frac{1}{x^3} = \cdots$$

$$(2)\quad \frac{1}{x} : \frac{1}{x^2} = \frac{x^2}{x} \cdots = \frac{1}{x^{n-1}} : \frac{1}{x^n} = \frac{x^n}{x^{n-1}}$$

$$\left.\begin{array}{ll} (3) & \frac{1}{x}\cdot\frac{1}{x} = \frac{1}{x^2}, \\ & \frac{1}{x^2}\cdot\frac{1}{x} = \frac{1}{x^3}, \cdots, \\ & \frac{1}{x^n}\cdot\frac{1}{x^m} = \frac{1}{x^{n+m}} \\ (4) & \frac{1}{x}\cdot x^2 = \frac{x^2}{x}, \\ & \frac{1}{x}\cdot x^3 = \frac{x^3}{x}, \cdots, \\ & \frac{1}{x^n}\cdot x^m = \frac{x^m}{x^n} \end{array}\right\} \begin{array}{l} m = 1, 2, 3, \cdots \\ n = 1, 2, 3, \cdots \end{array}$$

In order to appreciate the importance of this study, it is necessary to see how al-Karajī's more or less immediate successors exploited it. For example, al-Samaw'al[12] was able, on the basis of al-Karajī's work, to utilize the isomorphism of what would now be called the groups $(Z, +)$ and $([x^n; n \in Z], \times)$ in order to give for the first time, in all its generality, the rule equivalent to $x^m x^n = x^{m+n}$, where $m, n \in Z$.

In applying arithmetical operations to algebraic terms and expressions, al-Karajī first considered the application of these rules to monomials before taking up "composed quantities," or polynomials. For multiplication he thus demonstrated the following rules: (1) $(a/b) \cdot c = ac/b$ and (2) $a/b \cdot c/d = ac/bd$, where a, b, c, and d are monomials. He then treated the multiplication of polynomials, for which he gave the general rule. He proceeded in the same manner and with the same concern for the symmetry of the operations of addition and subtraction. Yet this algebra of polynomials was uneven. In division and the extraction of roots al-Karajī did not achieve the generality already attained for the other operations. Hence he considered only the division of one monomial by another and of a polynomial by a monomial. Nevertheless, these results permitted his successors—notably al-Samaw'al—to study, for the first time to our knowledge, divisibility in the ring $[Q(x) + Q(1/x)]$ and the approximation of whole fractions by elements of the same ring.[13] As for the extraction of the square root of a polynomial, al-Karajī succeeded in giving a general method—the first in the history of mathematics—but it is valid only for positive coefficients. This method allowed al-Samaw'al to solve the problem for a polynomial with rational coefficients or, more precisely, to determine the root of a square element of the ring $[Q(x) + Q(1/x)]$.[14] Al-Karajī's method consisted in giving first the development of $(x_1 + x_2 + x_3)^2$—where x_1, x_2, and x_3 are monomials—for which he proposed the canonical form

$$x_1^2 + 2x_1x_2 + (x_2^2 + 2x_1x_3) + 2x_2x_3 + x_3^2.$$

This last expression is itself, in this case, a polynomial ordered according to decreasing powers. Al-Karajī then posed the inverse problem: finding the root of a five-term polynomial. He therefore considered this polynomial to be of the canonical form and proposed two methods. The first consisted in taking the sum of the roots of two extreme terms—if these exist—and the quotient of either the second term divided by twice the root of the first or of the fourth term divided by twice the root of the last.[15] The second method consisted in subtracting from the third term twice the product of the root of the first term times the root of the last term, then the root of the remainder from the subtraction is added to the roots of the extreme terms. Great care must be exercised here. This form is not restricted to the particular example, and al-Karajī's method, as can be seen in *al-Badī*ʿ, is general.[16]

Again with a view to extending algebraic computation al-Karajī pursued the examination of the application of arithmetical operations to irrational terms and expressions.

"How multiplication, division, addition, subtraction, and the extraction of roots may be used [on irrational algebraic quantities]."[17] This was the problem posed by al-Karajī and used by al-Samaw'al as the title of the penultimate chapter of his work on the use of arithmetical tools on irrational quantities. The problem marked an important stage in al-Karajī's whole project and therefore also in the extension of the algebraic calculus. Just as he had explicitly and systematically applied the operations of elementary arithmetic to rational quantities, al-Karajī, in order to achieve his objectives, wished to extend this application to irrational quantities in order to show that they still retained their properties. This project, while conceived as purely theoretical, led to a greatly increased knowledge of the algebraic structure of real numbers. Clear progress indeed, but to make it possible it was necessary to risk a setback—a risk at which some today would be scandalized—in that it did not base the operation on the firm ground of the theory of real numbers. The arithmetician-algebraists were only interested in what we might call the algebra of R and did not attempt to construct the field of real numbers. Here progress was made in another algebraic field, that of geometrical algebra, later revived by al-Hayyām and Šaraf al-Dīn al-Tusī.[18] In the tradition of this algebra al-Karajī and al-Samaw'al could extend their algebraical operations to irrational quantities without questioning the reasons for their success or justifying the extension. Because an unfortunate lack of any such justification gave the sense of a setback al-Karajī simultaneously adopted the definitions of books VII and X of the *Elements*. While he borrowed from book VII the definition of number as "a whole composed of unities" and of unity—not yet a number—as that which "qualifies by an existing whole," it is in conformity with book X that he defined the concepts of incommensurability and irrationality. For Euclid, however, as for his commentators, these concepts apply only to geometrical objects or, in the expression of Pappus, they "are a property which is essentially geometrical."[19] "Neither incommensurability nor irrationality," he continued, "can exist for numbers. Numbers are rational and commensurable."[20]

Since al-Karajī explicitly used the Euclidean definitions as a point of departure, it would have been useful if he could have justified his use of them on incommensurable and irrational quantities. His works may be searched in vain for such an explanation. The only justification to be found is extrinsic and indirect and is based on his conception of algebra. Since algebra is concerned with both segments and numbers, the operations of algebra can be applied to any object, be it geometrical or arithmetical. Irrationals as well as rationals may be the solution of the unknown in algebraic operations precisely because they are concerned with both numbers and geometrical magnitudes. The absence of any intrinsic explanation seems to indicate that the extension of algebraic calculation—and therefore of algebra—needed for its development to forget the problems relative to the construction of R and to surmount any potential obstacle, in order to concentrate on the algebraic structure. An unjustified leap, indeed, but a fortunate one for the development of algebra. This is the exact meaning of al-Karajī when he writes, without transition immediately after referring to the definitions of Euclid, "I show you how these quantities [incommensurables, irrationals] are transposed into numbers."[21]

One of the consequences of this project, and not the least important, is the reinterpretation of book X of the *Elements*.[22] This had until then been considered by most mathematicians, even by one so important as Ibn al-Haytham, as merely a geometry book. For al-Karajī its concepts concerned magnitudes in general, both numerical and geometric, and by algebra he classified the theory in this book in what was later to be known as the theory of numbers. To extend the concepts of book X of the *Elements* to all algebraic quantities al-Karajī began by increasing their number. "I say that the monomials are infinite: the first is absolutely rational, five for example, the second is potentially rational, as the root of ten, the third is defined by reference to its cube as the *côté* of twenty, the fourth is the *médiale* defined by reference to the square of its square, the fifth is the square of the quadrato-cube, then the *côté* of the cubo-cube and so on to infinity."[23] In the same way binomials can also be split infinitely. In this field, as in so many others, al-Samaw'al is continuing the work of al-Karajī. At the same time one contribution belongs to him alone and that is his generalization of the division of a polynomial with irrational terms.[24] He thus developed the calculus of radicals introduced by his predecessors. At the beginning of *al-Badī*ʿ[25] is a statement—for the monomials x_1, x_2 and the strictly positive natural integers m, n—of the rules that make it possible to calculate the following:

AL-KARAJĪ

$$x_1 \sqrt[n]{x_2};\ \sqrt[n]{x_1}/\sqrt[m]{x_2};\ \sqrt[n]{x_1} \cdot \sqrt[m]{x_2}$$
$$\sqrt[n]{x_1}/\sqrt[n]{x_2};\ \sqrt[n]{x_1}/\sqrt[m]{x_2}$$
$$\sqrt[n]{x_1} \pm \sqrt[n]{x_2}.$$

Al-Karajī next discussed the same operations carried out on polynomials and gave, among others, rules that allow calculation of expressions such as

$$\frac{\sqrt{x_1}}{\sqrt{x_2} - \sqrt{x_3}};\ \frac{x_1}{4\sqrt{x_2} + 4\sqrt{x_3}};$$
$$\sqrt{x_1 + \sqrt{x}};\ \sqrt{\sqrt{x_1} + \sqrt{x_2}}.$$

In addition he attempted, unsuccessfully, to calculate

$$\frac{x_1}{\sqrt{x_2} + \sqrt{x_3} + \sqrt{x_4}}.$$

In the same spirit al-Karajī took up binomial developments. In *al-Fakhrī*[26] he gives the development of $(a + b)^3$, and in *al-Badī'*[27] he presents those of $(a - b)^3$ and $(a + b)^4$. In a long text of al-Karajī reported by al-Samaw'al are the table of binomial coefficients, its formation law $C_n^m = C_{n-1}^{m-1} + C_{n-1}^m$, and the expansion $(a + b)^n = \sum_{m=0}^{n} C_n^m\, a^{n-m}\, b^n$ for integer n.[28]

To demonstrate the preceding proposition as well as the proposition $(ab)^n = a^n b^n$, where a and b are commutative and for all $n \in N$, al-Samaw'al uses a slightly old-fashioned form of mathematical induction. Before proceeding to demonstrate the two propositions he shows that multiplication is commutative and associative—$(ab)(cd) = (ac)(bd)$—and recalls the distributivity of multiplication with respect to addition —$(a + b)\lambda = a\lambda + b\lambda$. He then uses the expansion of $(a + b)^{n-1}$ to prove the identity for $(a + b)^n$ and that of $(ab)^{n-1}$ to prove the identity for $(ab)^n$. It is the first time, as far as we know, that we find a proof that can be considered the beginning of mathematical induction.

Turning to the theory of numbers, al-Karajī pursued further the task of extending algebraic computation. He demonstrated the following theorems:[29]

$$\sum_{i=1}^{n} i = (n^2 + n)/2 = n(\tfrac{1}{2} + n/2) \qquad (1)$$

$$\sum_{i=1}^{n} i^2 = \sum_{i=1}^{n} i(2n/3 + \tfrac{1}{3}). \qquad (2)$$

Actually al-Karajī did not demonstrate this theorem; he only gave the equivalent form

$$\sum_{i=1}^{n} i^2 \Big/ \sum_{i=1}^{n} i = (2n/3 + \tfrac{1}{3}).$$

The algebraic demonstration appeared for the first time in al-Samaw'al:[30]

$$\sum_{i=1}^{n-1} i(i+1) = \left(\sum_{i=1}^{n} i\right)(2n/3 - \tfrac{2}{3}) \qquad (3)$$

$$\sum_{i=1}^{n} i^3 = \left(\sum_{i=1}^{n} i\right)^2 \qquad (4)$$

$$\sum_{i=0}^{n-1} (2i+1)(2i+3) + \sum_{i=1}^{n} 2i(2i+2) \qquad (5)$$
$$= \left(\sum_{i=1}^{2n+2} i\right)(\tfrac{2}{3}[2n+2] - \tfrac{5}{3}) + 1$$

$$\sum_{i=1}^{n-2} i(i+1)(i+2) = \sum_{i=1}^{n-1} i^3 - \sum_{i=1}^{n-1} i \qquad (6)$$
$$= \left(\sum_{i=1}^{n-1} i\right)^2 - \sum_{i=1}^{n-1} i.$$

For al-Karajī, the "determination of unknowns starting from known premises" is the proper task of algebra.[31] The aim of algebra is to show how unknown quantities are determined by known quantities through the transformation of the given equations. This is obviously an analytic task, and algebra was already identified with the science of algebraic equations. One can thus understand the extension of algebraic computation and why al-Karajī's followers[32] did not hesitate to join algebra to analysis and, to a certain extent, to oppose it to geometry, thus affirming its autonomy and its independence. Since al-Khwārizmī the unity of the algebraic object was no longer founded in the unity of mathematical entities but in that of operations. It was a question, on the one hand, of the operations necessary to reduce an arbitrary problem to one form of equation—or, more precisely, to one of the canonical types stated by al-Khwārizmī—and, on the other hand, of the operations necessary to give particular solutions, that is, the "canons." In the same fashion al-Karajī took up the six canonical equations[33]—$ax = b$, $ax^2 = bx$, $ax^2 = b$, $ax^2 + bx = c$, $ax^2 + c = bx$, $bx + c = ax^2$ —in order to solve equations of higher degree: $ax^{2n} + bx^n = c$, $ax^{2n} + c = bx^n$, $bx^n + c = ax^{2n}$, $ax^{2n+m} = bx^{n+m} + cx^m$.

AL-KARAJĪ

Next, following Abū Kāmil in particular, al-Karajī studied systems of linear equations[34] and solved, for example, the system $x/2 + w = s/2$, $2y/3 + w = s/3$, $5z/6 + w = s/6$, where $s = x + y + z$ and $w = 1/3(x/2 + y/3 + z/6)$.

The translation of the first five books of Diophantus' *Arithmetica* revealed to al-Karajī the importance of at least two fields. Yet, unlike Diophantus, he wished to elaborate the theoretical aspect of the fields under consideration. Therefore al-Karajī benefited from both a conception of algebra renewed by al-Khwārizmī and a more developed theory of algebraic computation, and he was able, through his reading of Diophantus, to state in a general form propositions still implicit in Diophantus and to add to them others not initially foreseen. In *al-Fakhrī*, as in *al-Badī*ʿ, by indeterminate analysis (*istiqrā*ʾ)[35] al-Karajī meant "to put forward a composite quantity [that is, a polynomial or algebraic expression] formed from one, two, or three successive terms, understood as a square but the formulation of which is nonsquare and the root of which one wishes to extract."[36] By the solution in q of a polynomial with rational coefficients al-Karajī proposed to find the values of x in q such that $P(x)$ will be the square of a rational number. In order to solve in this sense, for example, $A(x) = ax^{2n} + bx^{2n-1}$, where $n = 1, 2, 3, \cdots$, divide by x^{2n-2} to arrive at the form $ax^2 + b$, which should be set equal to a square polynomial of which the monomial of maximum degree is ax^2, such that the equation has a rational root.

Al-Karajī noted that problems of this type have an infinite number of solutions and proposed to solve many of them, some of which were borrowed from Diophantus while others were of his own devising. An exhaustive enumeration of these problems cannot be given here. We shall present only the principal types of algebraic expressions or polynomials that can be set equal to a square.[37]

1. Equations in one unknown:

$$ax^n = u^2$$

$ax^2 + bx = u^2$ and in general $ax^{2n} + bx^{2n+1} = u^2$

$ax^2 + b = u^2$ and in general $ax^{2n} + bx^{2n-2} = u^2$

$ax^2 + bx + c = u^2$ and in general $ax^{2n} + bx^{2n-1} + cx^{2n-2} = u^2$

$ax^3 + bx^2 = u^2$ and in general $ax^{2n+1} + bx^{2n} = u^2$ for $n = 1, 2, 3 \cdots$.

2. Equations in two unknowns:

$$x^2 + y^2 = u,\ x^3 \pm y^3 = u^2,\ (x^2)^{2m} \pm (y^3)^{2m+1} = u^2$$

$$(x^{2m+1})^{2m+1} - (y^{2m})^{2m} = u^2.$$

3. Equation in three unknowns:

$$x^2 + y^2 + z^2 \pm (x + y + z) = u^2.$$

4. Two equations in one unknown:

$$\begin{cases} a_1x + b_1 = u_1^2 \\ a_2x + b_2 = u_2^2 \end{cases} \text{and in general} \begin{cases} a_1x^{2n+1} + b_1x^{2n} = u_1^2 \\ a_2x^{2n+1} + b_2x^{2n} = u_2^2 \end{cases}$$

$$\begin{cases} a_1x^2 + b_1x + c_1 = u_1^2 \\ a_2x^2 + b_2x + c_2 = u_2^2. \end{cases}$$

5. Two equations in two unknowns:

$$\begin{cases} x^2 + y = u^2 \\ x + y^2 = v^2 \end{cases} \quad \begin{cases} x^2 - y = u^2 \\ x^2 - x = v^2 \end{cases} \quad \begin{cases} x^3 + y^2 = u^2 \\ x^3 - y^2 = v^2 \end{cases}$$

$$\begin{cases} x^2 - y^3 = u^2 \\ x^2 + y^3 = v^2 \end{cases} \quad \begin{cases} x^2 + y^2 = u^2 \\ x^2 + y^2 \pm (x + y) = v^2 \end{cases}$$

$$\begin{cases} x + y + x^2 = u^2 \\ x + y + y^2 = v^2. \end{cases}$$

6. Two equations in three unknowns:

$$\begin{cases} x^2 + z = u^2 \\ y^2 + z = v^2. \end{cases}$$

7. Three equations in two unknowns:

$$\begin{cases} x^2 + y^2 = u^2 \\ x^2 + y = v^2 \\ x + y^2 = w^2. \end{cases}$$

8. Three equations in three unknowns:

$$\begin{cases} x^2 + y = u^2 \\ x + z = v^2 \\ z^2 + x = w^2 \end{cases} \quad \begin{cases} x^2 - y = u^2 \\ y^2 - z = v^2 \\ z^2 - x = w^2 \end{cases}$$

$$\begin{cases} (x + y + z) - x^2 = u^2 \\ (x + y + z) - y^2 = v^2 \\ (x + y + z) - z^2 = w^2. \end{cases}$$

In al-Karajī's work there are other variations on the number of equations and of unknowns, as well as a study of algebraic expressions and of polynomials that may be set equal to a cube. From a comparison of the problems solved by al-Karajī and those of Diophantus it was found that "more than a third of the problems of the first book of Diophantus, the problems of the second book starting with the eighth, and virtually all the problems of the third book were included by al-Karajī in his collection."[38] It should be noted that al-Karajī added new problems.

Two sorts of preoccupations become evident in al-Karajī's solutions: to find methods of ever greater generality and to increase the number of cases in which the conditions of the solution should be examined. Hence, for the equation $ax^2 + bx + c =$

AL-KARAJĪ

u^2—although he supposed that its solution requires that a and c be positive squares—he considered the various possibilities: a is a square, b is a square, neither a nor b is a square in $ax^2 + b = u^2$ but $-b/a$ is a square. In addition he showed that $\pm(bx - c) - x^2 = u^2$ has no rational solution unless $b^2/4 \pm c$ is the sum of two squares.[39] Another example is that of the solution of the system $ax + b = u^2$ and $ax + c = v^2$ where he set up $b - c = a \cdot (b - c)/\alpha$ and took $ax + b = (a + [b - c]/a)^2/4$.

The same preoccupation appears in his solution of the system $x^2 + y = u^2$ and $y^2 + x = v^2$, where he sought first to transform $x = at$ and $y = bt$, $a > b$, in order to posit $(a - b)t = \lambda$; $a^2 + t^2 + bt = u$; $b^2t^2 + at = v$, and to solve the problem by means of the demonstrated identity

$$\frac{1}{4}\left[\left(\frac{u-v}{\lambda} + \lambda\right)^2 - \left(\frac{u-v}{\lambda} - \lambda\right)^2\right] = u - v.$$

This concern with generality is also evident in the following two examples: (1) $x^3 + y^3 = u^2$, where he set $y = mx$ and $u = nx$, with $n, m \in q$ and derived $x = n^2/1 + m^2$—a method applicable to more general rational problems of the form $ax^n + by^n = cu^{n-1}$—and $x^3 + ax^2 = u^2$; $x^3 - bx^2 = v^2$, where a and b are integers; he set $u = mx$, $v = nx \Rightarrow x = m^2 - a = n^2 + b$, from which he showed that the condition that m and n should fulfill is $m^2 - n^2 = a + b$. He set $m = n + t$ and obtained $2nt + t^2 = a + b \Rightarrow n = a + b - t^2/2t$.

A great many other examples could be cited to illustrate al-Karajī's incontestable concern with generality and with the study of solutions, as well as a considerable number of other mathematical investigations and results. His most important work, however, remains this new start he gave to algebra, an arithmetization elicited by the discovery of Diophantus by a mathematician already familiar with the algebra of al-Khwārizmī. This new impetus was understood perfectly and extended by al-Karajī's direct successors, notably al-Samaw'al. It is this tradition, as all the evidence indicates, of which Leonardo Fibonacci had some knowledge, as perhaps did Levi ben Gerson.[40]

NOTES

1. F. Woepcke, *Extrait du Fahri, traité d'algèbre* (Paris, 1853); A. Hochheim, *Al-Kāfī fīl Ḥisāb*, 3 pts. (Halle, 1877–1880).
2. G. Levi della Vida, "Appunti e quesiti di storia letteraria araba, IV," in *Rivista degli studi orientali*, **14** (1933), 264 ff.
3. No claim for completeness is made for this table, because of the dispersion of the Arabic MSS and their insufficient classification.

Title	al-Karkhī	al-Karajī
al-Fakhrī	BN Paris 2495 Esat Efendi Istanbul 3157 Cairo Nat. Lib., 21	Köprülü Istanbul 950
al-Kāfī	Gotha 1474 Alexandria 1030	Topkapi Sarayī, Istanbul A. 3135 Damat, Istanbul no. 855 Sbath Cairo 111
al-Badī'		Barberini Rome 36, 1
'ilal-ḥisāb al-jabr	Hūsner pasha, Istanbul 257	Bodleian Library I, 968, 3
Inbat al-miyāh al-khafiyyat	Publ. Hyderabad, 1945, on the basis of the MSS. of the library of Aya Sofya and of the library of Bankipore.	

4. One encounters the same difficulties when one considers the MSS of the later Arab commentators and scholars. Thus in the commentaries of al-Shahrazūrī (Damat 855) and of Ibn al-Shaqqāq (Topkapi Sarayī A. 3135), both of which refer to *al-Kāfī*, one finds the name al-Karajī, whereas in MS Alexandria 1030 one finds al-Karkhī.
5. See A. Anbouba, *L'algèbre al-Badī' d'al-Karajī* (Beirut, 1964), p. 11; this work has an introduction in French.
6. This MS was classified as anonymous until the present author identified it as being the *al-Bāhir* of al-Samaw'al. See R. Rashed, "L'arithmétisation de l'algèbre au 11ème siècle," in *Actes du Congrès de l'histoire des sciences* (Moscow, in press); and R. Rashed and S. Ahmad, *L'algèbre al-Bāhir d'al-Samaw'al* (Damascus, 1972).
7. In Arab dictionaries the "mountain countries" include the cities located between "Ādharbayjān, Arab Iraq, Khourestan, Persia, and the land of Deīlem (a land bordering the Caspian Sea)."
8. Woepcke, *op. cit.*, p. 4.
9. See R. Rashed, "Algèbre et linguistique: L'analyse combinatoire dans la science arabe," in R. Cohen, ed., *Boston Studies in the Philosophy of Science*, X (Dordrecht).
10. See M. I. Medovoi, "Mā yaḥtāj ilayh al-Kuttāb wa'l-'ummāl min sinā'at al-ḥisab," in *Istoriko-mathematicheskie issledovaniya,* **13** (1960), pp. 253–324.
11. *Al-Fakhrī*; see Woepcke, *op. cit.*, p. 48.
12. See al-Samaw'al, *op. cit.*, pp. 20 ff. of the Arabic text.
13. *Ibid.*
14. *Ibid.*, p. 60 of the Arabic text.
15. For example, for the first method, to find the root of $x^6 + 4x^5 + (4x^4 + 6x^3) + 12x^2 + 9$; one takes the roots of x^3 and of 9; one then divides $4x^5$ by x^3 or $12x^2$ by 3; in both cases one obtains $4x^2$. The root sought is thus $(x^3 + 2x^2 + 3)$. For the second method, take
$$x^8 + 2x^6 + 11x^4 + 10x^2 + 25.$$
One finds the roots of x^8 and of 25; x^4 and 5, then subtracts as indicated to obtain x^4, the root of which is x^2. The root sought is thus $(x^4 + x^2 + 5)$. See *al-Fakhrī*, p. 55; and *al-Badī'*, p. 50 of the Arabic text.
16. Al-Samaw'al, *op. cit.*
17. *Al-Badī'*, p. 31 of the Arabic text.
18. See Šaraf al-Dīn al Tusī, MSS India office 80th 767 (I.O. 461) and the important work on decimal numbers.
19. See *The Commentary of Pappus on Book X of Euclid's Elements*, W. Thomson, ed. (Cambridge, Mass., 1930), p. 193.
20. *Ibid.*
21. Al-Karajī, *op. cit.*, p. 29 of the Arabic text.

22. For Euclid, book X, see Van der Waerden, *Erwachende Wissenschaft* (Basel-Stuttgart, 1956), J. Vuillemin, *La philosophie de l'algèbre* (Paris, 1962), and P. Dedron and J. Itard, *Mathématiques et mathématisation* (Paris, 1959).
23. *Al-Badī'*, p. 29 of the Arabic text.
24. See the introduction to the present author's edition of al-Bāhir, cited above (note 7).
25. See Anbouba, *op. cit.*, pp. 32 ff. of the Arabic text and pp. 36 ff. of the French intro.
26. See *al-Fakhrī*, in Woepcke, *op. cit.*, p. 58.
27. See *al-Badī'*, in Anbouba, *op. cit.*, p. 33 of the Arabic text.
28. See the chapter on numerical principles in al-Samaw'al, *op. cit.*
29. See *al-Fakhrī*, in Woepcke, *op. cit.*, pp. 59 ff.
30. See al-Samaw'al, *op. cit.*, pp. 64 ff.
31. See *al-Fakhrī*, in Woepcke, *op. cit.*, p. 63, with the trans. improved by comparison with MSS of the Bibliothèque Nationale, Paris.
32. See al-Samaw'al, *op. cit.*, pp. 71 ff. of the Arabic text.
33. See *al-Fakhri*, in Woepcke, *op. cit.*, pp. 64 ff.
34. *Ibid.*, pp. 90–100.
35. *Ibid.*, p. 72; *Al-Badī'*, in Anbouba, *op. cit.*, p. 62 of the Arabic text.
36. *Al-Fakhrī*, with trans. improved by comparison with the MSS of the Bibliothèque Nationale.
37. See *al-Fakhrī* and *al-Badī'*.
38. See *al-Fakhrī*, *op. cit.*, p. 21.
39. *Ibid.*, p. 8.
40. See the comparison made by Woepcke, *op. cit.*; and G. Sarton, *Introduction to the History of Science (1300–1500)*, p. 596.

BIBLIOGRAPHY

I. Original Works. In addition to the works cited in note 3, all of which have been published except *'ilal ḥisāb al-jabr*, the Arabic bibliographies and al-Karajī himself mention other texts that seem to have been lost. Those mentioned in the bibliographies are *Kitāb al 'uqūd wa'l abniyah* ("Of Vaults and Buildings") and *Al-madkhal fi 'ilm al-nujūm* ("Introduction to Astronomy"). Cited by Karajī in *al-Fakhrī* are *Kitāb nawādir al-ashkāl* ("On Unusual Problems") and *Kitāb al dūr wa'l wiṣāyā* ("On Houses and Wills"); and in *al-Badi'*, "On Indeterminate Analysis" and *Kitāb fi'l-hisāb al-hindi* ("On Indian Computation"). Finally, al-Samaw'al mentions a book by al-Karajī from which he has extracted his text on binomial coefficients and expansion.

II. Secondary Literature. Besides the works cited in the notes, see Amir Moez, "Comparison of the Methods of Ibn Ezra and Karhī," in *Scripta mathematica*, **23** (1957); and L. E. Dickson, *History of the Theory of Numbers* (New York, 1952).

See also R. Rashed, "L'induction mathématique-al-Karajī et As-Samaw'al," in *Archive for History of Exact Sciences*, **1** (1972), 1–21.

Roshdi Rashed

KÁRMÁN, THEODORE VON (*b.* Budapest, Hungary, 11 May 1881; *d.* Aachen, Germany, 7 May 1963), *aerodynamics*.

Theodore (in Hungarian, Todor) von Kármán was the son of Maurice (Mór) Kármán, a university professor of education who was knighted by Francis Joseph of Austria-Hungary in 1907 for reorganizing Hungarian secondary education, and Helen Konn, descendant of a distinguished Bohemian-Jewish family. The third of four sons, he also had a younger sister, Josephine (Pipö), to whom he remained devoted until her death in 1951. Having attended the Minta, a model Budapest Gymnasium organized according to his father's ideas, Kármán won the Eötvös Prize for Hungarian secondary students in science and mathematics before entering the Palatine Joseph Polytechnic (now the Technical University of Budapest), where he first became interested in developing a theoretical basis for the solution of problems in mechanics. After a year of compulsory military service as an artillery cadet, he spent three years as an instructor at the Palatine Joseph.

In 1906 Kármán received a two-year fellowship for postgraduate work from the Hungarian Academy of Sciences and decided to go to Göttingen, where he worked with Ludwig Prandtl, the "father of aerodynamics" and only six years Kármán's senior; he also came under the influence of the great mathematicians David Hilbert and Felix Klein. Apart from short terms at Berlin and Paris, Kármán remained at Göttingen for six years, serving as privatdocent during the last three. Besides working in aerodynamics, he collaborated with Max Born in an attempt to explain the temperature dependence of specific heat. Their theory, based on the assumption that atoms were arranged in a regular lattice, proved to be more general than the theory of atomic heats of Peter Debye, which was published first. (Both Born and Debye later received Nobel prizes.) Göttingen also pioneered some of the ideas that later became crucial to the development of technical education, notably that engineering and other applied sciences must rest firmly on a scientific foundation if they are to advance other than by trial and error—a viewpoint that Kármán defended throughout his long career.

Before leaving Göttingen, Kármán attracted considerable attention with his first important work, the elucidation of a phenomenon that had been observed by others (notably Henri Bernard) but had not been previously analyzed. He found that when a fluid flows at a velocity V past a cylindrical obstacle of diameter d at right angle to the axis of the cylinder, a separation of the wake into two rows of periodic vortices occurs, alternating in position between the two sides like street lights. The phenomenon is known as the Kármán vortex street (or "Karmansche Wirbelstrasse") or simply Kármán vortices. It leads to self-induced vibrations at a frequency $n = 0.207V/d$ cycles per second that can build up to destructively large

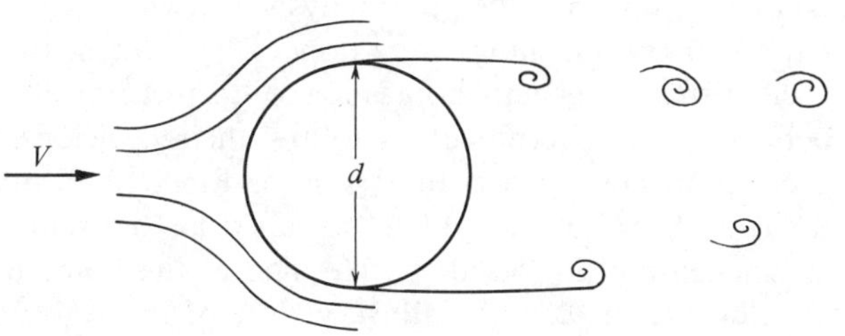

FIGURE 1

magnitudes when a structure designed on the basis of static considerations is subjected to dynamic conditions, as in aircraft wing flutter or the behavior of a bridge in a high wind. (The failure of the first suspension bridge across the Tacoma Narrows at Puget Sound, Washington, in a fresh gale in 1940 was later shown by Kármán to have been caused by Kármán vortices created when the wind reached a velocity of $V = 42$ miles per hour.)

After a term of teaching at a mining college in Schemnitz (now Baňská Štiavnica) in Slovakia, under conditions that were not conducive to a research career, Kármán returned to Göttingen and presently secured a professorship at the Technische Hochschule in Aachen, where he remained (except for military service during World War I) until 1929. There he became heavily involved in the development of aviation.

Kármán's interest in flying dated back to a demonstration he observed in Paris in 1908. In World War I he was assigned to the nascent Austro-Hungarian air force and worked on problems relating to propeller design, synchronized guns, and fire protection of fuel tanks. He also experimented with helicopters and demonstrated the superiority of two counterrotating propellers from the viewpoints of vibration and control. When he returned to Aachen after the war, he became interested in helping the students design glider planes, little realizing that he was laying the foundations of a new German air force, the *Luftwaffe* of World War II. He counted such pioneers as Hugo Junkers, Ernst Heinkel, and A. H. G. Fokker among his friends and associates.

Kármán's most important contribution stemming from this period was a new law of turbulence, a field in which he continued a friendly rivalry with his former mentor Prandtl, still at Göttingen. Once again, the underlying mathematical theory proved to be of considerable practical importance, not only in aeronautical engineering (in predicting drag on the surface of aircraft and—later—rockets) but also in describing flow through pipes, an aspect of his research of great benefit to the oil industry and to other hydraulics applications.

During 1926–1927 Kármán made an extensive trip to the United States and Japan. In the United States he spent some time at the California Institute of Technology in Pasadena, where a new aeronautics laboratory had been endowed by Daniel Guggenheim; while in Japan, he helped to modernize the Kawanishi Company in Kobe, for which he designed a new wind tunnel. Soon after his return to Aachen, Kármán received an offer to become the director of the Guggenheim Aeronautical Laboratory from Caltech's chief, R. A. Millikan, and accepted. He left Aachen at the end of 1929.

Kármán's long tenure at Caltech saw its emergence as one of the top aeronautical research centers in the world. But perhaps even more important were his contributions to the teaching of aeronautical engineering, which he put on a scientific (especially mathematical) basis and which he greatly extended in scope to postgraduate and postdoctoral studies. A substantial number of professorial chairs in aeronautics in the United States and in other countries came to be occupied by Kármán's students. One such student was Hsue-shen Tsien, an extraordinarily talented engineer from Shanghai, who later taught at the Massachusetts Institute of Technology and at Caltech, before falling under false suspicion during the anti-Communist crusades led by Senator Joseph McCarthy. Tsien ultimately returned to China, where he achieved a high position and contributed substantially to that nation's technological development. Another of Kármán's students was Francis H. Clauser, later a well-known aircraft designer (DC-6) and engineering educator.

Kármán's presence at Caltech also played a part in the development of the aircraft (and later space) industry in southern California. He helped to found the Aerojet Engineering Corporation, which later (after Kármán had liquidated his interest in it) grew into one of the industrial giants of the jet age, as Aerojet-General Corporation, a subsidiary of the General Tire and Rubber Company.

Another development deriving from Kármán's activities was the organization of the institute's Jet Propulsion Laboratory (JPL), a government-funded center of rocket research and space communications techniques that is acknowledged as a principal contributor to America's preeminence in space technology. Originally concerned primarily with rocket research arising from military requirements during and after World War II, the laboratory was greatly expanded after the United States entered the field of space exploration and shifted its emphasis to propellants and to remote control; but the aerodynamics research started by Kármán remained a concern of the laboratory throughout.

At Caltech, Kármán and his students laid the foundations for aerodynamic design leading to supersonic flight, an area that at one point appeared to be stymied by vehicle-design (rather than propellant or engine) considerations. At the same time Kármán continued to be a valued consultant of the U.S. Air Force and played a part in its emergence from a subsidiary position as a branch of the U.S. Army to autonomous status. He was chairman of a committee that produced the report *Toward New Horizons*, which became the blueprint for the new air force; he subsequently served on the force's scientific advisory board. In addition, he had a hand in the organization of the Rand Corporation, the first of the "think tanks" or quasi-independent civilian research organizations that work under contract to a government department.

When the North Atlantic Treaty Organization (NATO) was organized in 1949, Kármán proposed the organization of the Advisory Group for Aeronautical Research and Development (AGARD) to review aeronautical advances, exchange information among the treaty members, and generally help solve defense problems of mutual interest. Under his leadership the terms of reference of AGARD were very broadly interpreted and led to the establishment of the international aerodynamics school known as the Training Center for Experimental Aerodynamics (later named the Von Kármán Center). He also played an important part in the formation of the Advanced Research Projects Agency of the U.S. Department of Defense.

Kármán's multiple roles as aeronautical engineer, university professor, and industrial and government consultant brought him frequently to public notice, a role that he did not seek. After the Nazi government took over the Junkers firm in Germany, it was discovered that some American aircraft manufacturers were infringing inventions belonging to Junkers, which resulted in a patent suit in a United States court. Although Jewish, Kármán testified in support of the Junkers contention, which many thought was carrying scientific detachment too far. He protested vigorously when Hsue-shen Tsien was persecuted for alleged Communist ties, placed under a deportation order, and then detained for five years, presumably to allow his knowledge of secret projects to become obsolete.

Kármán remained quite unabashed about his lifelong association with military authorities, first in Austria-Hungary, then in Germany, and finally in the United States and NATO. His viewpoint was that of an engineer of an earlier era who may be considered to have discharged his debt to society once he has contrived to "provide an analysis of what would happen if certain things were done"; he thought that "scientists as a group should not try to force or even persuade the government to follow their decisions."

Kármán never married. His sister Pipö, first with their mother and then alone, managed his household in Aachen and in Pasadena. He died at the home of his close friend Bärbel Talbot, widow of the Aachen manufacturer Georg Talbot.

Kármán received many honorary degrees and medals, including the U.S. Medal for Merit (1946), the Franklin Gold Medal (1948), and the National Medal of Science (from President Kennedy, 1963), as well as most awards given in aeronautics and fluid mechanics. He was a commander of the French Legion of Honor, a member of the Pontifical Academy of Science, and the recipient of similar decorations from Germany, Greece, Spain, and the Netherlands.

Despite his many public activities, he never became a great public figure in the way of many inventors, perhaps because theoretical aerodynamics is not a very accessible field to the layman; nevertheless, his work in that field and in rocket research has helped shape both scientific and political history.

BIBLIOGRAPHY

I. Original Works. Kármán was the author or coauthor of 171 papers and articles. His five books are *General Aerodynamic Theory*, 2 vols. (Berlin, 1924), written with J. M. Burgers; *Mathematical Methods in Engineering* (New York, 1940), written with M. A. Biot, translated into French, Spanish, Portuguese, Italian, Turkish, Japanese, Polish, and Russian; *Aerodynamics: Selected Topics in Light of Their Historical Development* (New York, 1954), translated into Spanish, Italian, German, French, and Japanese; and *From Low-speed Aerodynamics to Astronautics* (London, 1961). For his other publications, see the four-volume *Collected Works of Dr. Theodore von Kármán* (London, 1956), which contains the papers published until then; a complete bibliography, including twenty-five later papers, appears in his biography (see below).

II. Secondary Literature. In his late seventies, Kármán contracted with a journalist, Lee Edson, to publish an autobiography contrived from dictation and taped interviews. The work was about three-quarters finished when Kármán died, but the U.S. Air Force underwrote its completion and it was ultimately published as *The Wind and Beyond: Theodore von Kármán, Pioneer in Aviation and Pathfinder in Space* (New York, 1967). It is largely a personal biography, which mentions Kármán's scientific work only in passing, but it does contain a complete bibliography of his writings and a list of his degrees, decorations, and awards. Successive bibliographies also appear in Poggendorff, V, 612; VI, 1282; and VIIa, 692–693.

Charles Süsskind

KARPINSKY, ALEXANDR PETROVICH (*b.* Bogoslovsk [now Karpinsk], Russia, 7 January 1847; *d.* Moscow, U.S.S.R., 15 July 1936), *geology.*

Karpinsky's grandfather, Mikhail Mikhaylovich, and his father, Petr Mikhaylovich Karpinsky, were mining engineers; his mother, Maria Ferdinandovna Grasgof, was the daughter of a mining engineer. His childhood, spent in the Urals, awakened in him a permanent love for the region and determined his future profession. In 1858, after the death of his father, Karpinsky was sent to study at the Mining Corps in St. Petersburg (later the Mining Institute), from which he graduated in 1866 with a gold medal and a diploma as a mining engineer. In 1868 he began his teaching career, which continued for twenty-eight years. In 1869, after defending his dissertation, Karpinsky was made adjunct; and from 1877 to 1896 he was professor at the Mining Institute. Every year he did fieldwork, the greatest part of it in the Urals.

Karpinsky's general scientific activity was extremely broad in scope. From 1885 to 1903 he was director of the central geological institution of the country, the Geological Committee; in 1886 he was elected adjunct of the St. Petersburg Academy of Sciences, and in 1896 he became an academician; in 1916 he was elected vice-president, and from May 1917 to 1936 he was president of the Soviet Academy of Sciences. In addition, from 1899 to 1936 Karpinsky was president of the Mineralogical Society. From 1881 he was present at all the sessions of the International Geological Congress and was president of the Seventh, which was held in St. Petersburg. Karpinsky remained active even in his declining years. In 1933 he was a member of an Academy of Sciences expedition to the northern region; and in 1936, shortly before his death, he took part in a series of meetings and conferences.

Karpinsky was a charming and warm family man. Even when extremely busy, he never refused scientific help to anyone. He loved music and was an excellent singer, and held musical evenings in his home that were attended by eminent musicians.

Karpinsky's first works were in petrography. In 1869 he defended his dissertation on the augitite rocks of the Urals (from the village of Muldakaeva), which he called "muldakaite." In the same year this work was published both as an article and as a book. In preparing this work Karpinsky was one of the first to use the microscope for research on metamorphic rock. Subsequently he studied the principal metamorphic rocks of the Urals. The study of beresovite (a quartz-rich aplite) with microscope and chemical analysis showed its similarity to greisen (1875, 1877). His research on the rock listwanite of the southern Urals revealed that it could be regarded as the result of the transformation of limestone. Karpinsky also investigated the pegmatites of the Urals with carbonatite inclusions. The alkali rocks of the Ilmen Mountains next drew his attention. He described these rocks and presented a taxonomy of pegmatite lodes in a guidebook for the Seventh International Geological Congress (1897).

In 1902 Karpinsky described in detail the nepheline syenites of the Ilmen Mountains. He considered that for these rocks—consisting of orthoclase, nepheline, and semiprecious minerals—it was necessary to keep "miaskite" as a generic term, and thus these rocks have entered petrographic literature under that name. In brief communications to the St. Petersburg Society of Natural Scientists (1874, 1909) Karpinsky described the rare Urals rock associated with syenite—kyschtymite, which consists of plagioclase and corundum.

In other articles devoted to geological research on the Urals, Karpinsky often returned to questions of petrography: he described uralitic and actinolite rocks, effusive rocks, tuffs and serpentine of the southern Urals, and others. But not only the Urals attracted his attention—he wrote many notes on rocks from various regions of Russia, including breccia of diabase composition from the Olonets region (1882), crystal shale of the Kaninsky Range (1892), diorite from the Yenisey (1888), and basalt and porphyry from the Far East (1897).

The study of various rocks led Karpinsky to a number of important conclusions and generalizations. Investigating the metamorphic rock epidosite, he suggested its formation from limestone by means of contact metamorphism and discussed the formation of rocks by metamorphism (1871). In his early works Karpinsky had already dealt with petrographic laws (1870) and laws of association of feldspars (1876). For feldspars, the principal rock-forming minerals, Karpinsky established the regularity of the association of plagioclase and orthoclase.

Karpinsky was also interested in methods of petrographic research, especially in separations by heavy liquid and in determining the free quartz in rock by means of chemical processing. In addition, the classification of rocks that he compiled in his lecture course for students at the Mining Institute was important for its time. At the end of the nineteenth century the chemical composition of rock received special attention. Karpinsky opposed the classification of rock by chemical composition alone, considering mineralogical composition more important. He presented a report on the principles of classification in 1900 to the Eighth International Geological Congress at Paris.

Karpinsky lectured on petrography at the Mining Institute for almost thirty years. At meetings of the

Russian Mineralogical Society he delivered a number of reports and maintained an interest in petrography until his death. At the end of the century, however, the geological-paleontological orientation began to predominate in his work.

At the jubilee meeting of the St. Petersburg Academy of Sciences, 29 December 1886, Karpinsky gave a speech in connection with his election as an active member; it was published in 1887 as "Ocherk fizikogeograficheskikh uslovy Evropeyskoy Rossii v minuvshie geologicheskie periody" ("Sketch of the Physical-Geographical Conditions of European Russia in Past Geological Periods"). In 1893 Karpinsky, with S. N. Nikitin and F. N. Chernyshev, compiled a new geological map of European Russia on the scale of 1:2,500,000. In 1894 he published "Obshchy kharakter kolebany zemnoy kory v predelakh Evropeyskoy Rossii" ("The General Character of the Movements of the Earth's Crust Within the Boundaries of European Russia"). These three works represented a generalization of the tremendous amount of material on the geology of Russia which had been accumulated by the end of the nineteenth century. Karpinsky attempted, on the basis of the available factual material, to present a sketch of the ancient oceans and dry land and their changes in the course of geological history, and to explain the character of the movement of the earth's crust.

Karpinsky's work in paleogeography was based on the principle that all geological phenomena are stages in the historical process of the development of the earth and can be understood only in relation to associated phenomena. Karpinsky proposed to distinguish two main periods in the history of the earth: the prehistoric-prepaleozoic, which, he believed, could not be deciphered; and the historical, from the Cambrian to today, for which paleogeographical reconstruction was possible. In 1880 he attempted to determine the location of dry land and sea for the Russian platform in the Devonian, Carboniferous, Permian, and Triassic periods. In 1887 and 1894 he did the same reconstruction for the whole "historical" period.

The history of the development of the section under consideration can be clearly seen on the paleogeographical maps attached to Karpinsky's works. In the Cambrian period the western part of the Baltic massif slowly broke away and the sea entered Scandinavia. In the Devonian the dry land at first rose again, with the continental red sandstones on its surface; then the advance of the sea flooded almost the whole Russian platform. In the bays of the Devonian seas organic sediment accumulated, leading to formation of great oil deposits. In the Carboniferous period the sea contracted, and the tropical vegetation on its shores turned into coal deposits. At the end of this period the Urals rose, and the outlines of the ocean basins were drawn into a north-south alignment by this movement. A period of drought began in the Permian period. The map shows a closed basin in the salty lagoons of which rock and potassium salts were deposited.

The continental conditions continued in the Triassic period. The marine transgression began again only in the Jurassic period, as a consequence of the sinking of the Baltic massif in the upper Jurassic and Cretaceous periods. A broad basin came into existence, extending along an east-west axis. Seas of the Tertiary period are shown on other maps.

Finally the southern seas took their present outline. The northern Russian platform was covered with ice, under the weight of which this region slowly sank and a northern transgression occurred. After the melting of the ice, the Baltic massif again rose, and the northern seas took their present form.

Karpinsky's construction was later confirmed by the research of Soviet geologists. Only minor corrections were made in his maps; the map of the lower Silurian alone has been substantially modified. Developing Karpinsky's ideas, Soviet geologists produced analogous paleogeographic constructions for earlier geological periods not investigated by Karpinsky and introduced greater detail into his sketches. In addition the areas of the continents and continental deposits were studied, since Karpinsky's descriptions dealt for the most part with ocean basins.

Karpinsky's paleogeographic work was closely connected with his tectonic conclusions. In works on the geological structure of European Russia (1880, 1883) he showed that in the structure of the platform there were two clearly distinguished elements: the folded base of crystalline rock and a cover of sedimentary deposits. Until Karpinsky's research Murchison's idea of the existence of an anticlinal fold, "a Devonian axis" between the basin near Moscow and the Donets basin, was accepted. Karpinsky showed that there was no such axis but that there were lower Devonian strata from north to south. This conclusion had great significance for the determination of the depth of the Kursk beds of magnetic ore. Especially important was Karpinsky's discovery of the belts of sedimentary rock in the south of Russia (1883) displaced parallel to the Caucasus. Karpinsky called this folded region, not expressed orographically, which also involves the Donets basin, a "vestigial range." His work on this newly defined tectonic structure became widely known. Suess accepted Karpinsky's view, and in his *Das Antlitz der Erde* he called the lines bounding this

KARPINSKY

structure "Karpinsky lines." The study of the vestigial range occupied many Soviet geologists, including D. N. Sobolev and A. D. Arkhangelsky. N. S. Shatsky proposed that this structure originated with the intrusion of neighboring folded zones into the body of the platform massif—a process similar to that proposed for the formation of the Wichita system in the United States.

The articles "Ocherk fiziko-geograficheskikh uslovy Evropeyskoy Rossii v minuvshie geologicheskie periody" (1887) and "Obshchy kharakter kolebany zemnoy kory v predelakh Evropeyskoy Rossii" (1894) were important contributions to tectonics. Karpinsky developed a method of using paleogeographic analysis to obtain tectonic information, using the structures of the outlines of the basins, not at the point of maximum transgression but only the outline of the part that was most submerged and thus preserved from later denudations. This method of studying the position of "mean basins" combined with the analysis of paleogeographic maps made possible a clearer representation of the movement of the earth's crust and of the history of the tectonic development of the Russian platform.

The 1887 work was descriptive. At the end Karpinsky concluded that the distribution of oceans is closely connected with displacement processes. The tectonic pattern of the platform, its fold trends, and the sequence of their formation were also described. In the 1894 work, which was a continuation of the first, conclusions were drawn from the factual material presented in the first. Karpinsky saw the contraction of the crust because of the cooling of the earth as the reason for all the tectonic movements—that is, he subscribed to the contraction hypothesis and considered it to be among the "most successful achievements of science." He distinguished two types of structure in the earth's surface: the "plicate," in which folds are formed, and the "disjunctive," where displacements and settling occur. The second type of structure is the platform, the region of plains with undisturbed accumulations of strata. From his research on this type of structure on the Russian platform, Karpinsky concluded that high and low portions of the platform arose from oscillations of the earth's crust. Analysis of paleogeographic maps showed that basins in different geological periods were elongated sometimes in an east-west, sometimes in a north-south, direction. Consequently the oscillations occurred in these directions. Only in the early Paleozoic era did basins reach the Baltic shield. Near it other parts of the platform rose and fell. Thus since earliest geological times the northeast of the Russian platform has remained dry land almost continuously, while since the upper Devonian the southeast has been sea. In the southern and central parts of the Russian platform east-west depressions have predominated; in the eastern part, north-south depressions.

This was connected with the orogenic process in neighboring geosynclinal regions, since in the times of mountain-forming movement in the Urals during the Devonian, Carboniferous, and Permian periods, depressions extended by these movements in a north-south direction predominated on the platform. In the Jurassic, upper Cretaceous, Paleocene, and Eocene, the elongation was east-west, corresponding to intense activity in the Caucasus. The formation of all other dislocations of the platform—gentle folds, displacements, and so on—depended directly on these oscillations and appeared especially when there was a shift of the north-south and east-west axes at the intersection of gently sloping synclinal and anticlinal curves. With the formation of mountain ranges belts of depressions were formed in the foothill areas. The association of orogenic motion in geosynclines with oscillations of the platform is a basic regularity discovered by Karpinsky.

In the second edition of his works, published in an anthology in 1919, and in the article "K tektonike Evropeyskoy Rossii" ("Toward a Tectonics of European Russia"; 1919) Karpinsky expanded his basic tectonic and paleogeographical conclusions with data obtained through further study of the geological structure of Russia. In particular he placed great importance on the depressions at the edges of the Russian platform, which served as a stop to tangential pressure.

Karpinsky's works in tectonics were of major importance for Russian geology. These short articles represented not only a synthesis but also a methodology for research on the platform. The paleogeographic method he proposed was extremely important for the clarification of the geological structure and the history of the development of a large part of the earth's crust. After the translation and publication of "Obshchy kharakter kolebany zemnoy kory v predelakh Evropeyskoy Rossii" in *Annales de géographie* (1896) Karpinsky's ideas became widely known and greatly influenced the development of geology. His views were developed by A. D. Arkhangelsky, N. S. Shatsky, and V. V. Belousov.

To learn the geological history of any part of the earth's crust it is necessary to know not only the direction of the orogenic movement but also the character of the fauna that inhabited the seas in various geological periods. But not all organic forms are obvious, and the geological chronicle often remains incomplete. Precisely because of this Karpinsky took a lively interest in paleontology, especially the identification of mysterious forms. In this area he wrote substantial general works of major importance: on the ammonoids, on the characteristics of the fossils of the family Edestidae, and on the study of the Devonian algae, the charophytes.

Exceptionally important was Karpinsky's "Ob ammoneyakh artinskogo yarusa i nekotorykh skhodnykh s nimi kamennougolnykh formakh" ("On the Ammonoids of the Artinsk Stage [Permo-Carboniferous] and Certain Carboniferous Formations Similar to Them"; 1891) and his further works on this subject (1896, 1922, 1928). Karpinsky related ontogenesis and phylogenesis to the historical development of organisms. The morphology of the shells of the ammonoids is extraordinarily complex, so that the sequence of various stages is very noticeable. Careful research enabled Karpinsky to construct the genealogical tree of the ammonoids and thus to determine their phylogenetic relationships. The detailed method of research had been used before Karpinsky; the novelty of his work consisted in the fact that the method of studying the ontogenesis of the ammonoids was applied to the study of the fauna of a whole geological horizon (the Artinsk Stage).

Karpinsky was also interested in the origin of the geometrical regularity of the ammonoids' spiral shells. In his opinion this mathematical regularity was necessitated by economy of matter and energy, because such regular forms also occur in other organisms, such as the foraminifers and cephalopods.

Karpinsky's research on the ammonoids and the application of the ontogenetic method are considered classic. Many of his contemporaries noted the great value of his work, including A. A. Chernov, A. A. Borisyak, J. Perrin Smith, E. Haug, and K. von Zittel. For his research Karpinsky received the Cuvier Prize of the French Academy.

Also important was Karpinsky's research on the upper Paleozoic fossil sharks of the family Edestidae, especially of the genus *Helicoprion*. In 1899 "Ob ostatkakh edestid i o novom ikh rode Helicoprion" ("On the Remains of the Edestidae and the New Genus *Helicoprion*") was published, and he often returned to this question (in the period 1903–1930 he published ten articles on *Helicoprion*). The fossil remains of *Helicoprion* were preserved in strange forms: flat spirals with separated turns.

These remains were studied before Karpinsky by many scientists, including E. Hitchcock, J. S. Newberry, L. Agassiz, R. Owen, H. Woodward, and K. von Zittel. There were, however, contradictory opinions on their origin. Before Karpinsky's research the idea was widespread that these remains were ichthyodorulites—spines from the backs of sharklike fishes. Through careful analysis Karpinsky showed the unsoundness of this hypothesis and suggested that the fossil forms were the dental apparatus of *Helicoprion*. In his opinion, "the teeth of the middle row of the edestid, forced out of the mouth cavity, did not fall away but, closely touching the teeth moving in behind them, were gradually moved to the ends of the jaw" ("Ob ostatkakh edestid i o novom ikh rode Helicoprion," p. 64). In continuing his investigations, Karpinsky concluded that the inner teeth of the spiral were smaller because they belonged to a younger animal of smaller size; then, in proportion to the animal's growth, the teeth became bigger. Knit into an arc, the teeth went beyond the limits of the mouth and formed an organ of defense or attack, similar to what can be observed in the sawfish (*Pristis*).

Karpinsky's research was carried out with great thoroughness. He subjected the fossil remains to detailed morphological and comparative anatomical study, compared them with the remains of other fossil and contemporary animals, studied the rock in which the fossil was found and the process of fossilization, and took into account paleogeographical data on the specific part of the earth, the stratigraphic position of the horizon, and the geological history of the time at which the animals lived. This thoroughness made Karpinsky's conclusions indisputable and ensured the success of his hypothesis. All other suggestions concerning the origin of the remains of *Helicoprion* proposed before and even after the appearance of Karpinsky's work were gradually discarded, and it has retained its significance.

Of great interest is Karpinsky's research on Devonian algae, the so-called charophytes, which were long considered mysterious. The small spherical or ellipsoid fossils of the little bodies were described by various researchers as seeds, spores, fish eggs, echinoderms, foraminifers, and so on. Becoming interested in these forms, Karpinsky conducted a thorough study in terms of comparative morphology, paleontological history, evolutionary development, and the processes of fossilization. As a result of these investigations in 1906 he published "O trokhiliskakh" ("On Trochiliscids"). He showed that the mysterious fossils were lime shells of *sporophydia oogons*, belonging to Devonian algae. He made a thorough study of the anatomy and taxonomy of contemporary charophytes and showed their closeness to the extinct Devonian forms; both had developed from common ancestors. These ancient forms were distinguished from the contemporary by their living in brackish and ocean waters. The development of lime shells was connected with their adaptation to the environment and thus could be explained in terms of natural selection. Karpinsky showed that the charophytes had a long evolutionary history and that their ancient forms were far more varied than the modern forms.

Karpinsky's work on trochiliscids is a classic in paleobotany. Karpinsky was interested in this subject for many years, as his articles published in 1909, 1927, and 1932 show. Further investigations of these interesting fossils confirmed all of his conclusions.

A comparison of Karpinsky's paleontological works demonstrates that they are united by the desire to discover the unknown pages of organic development in past geological periods. He approached this project as a Darwinist.

Karpinsky's works in stratigraphy are related to his paleontological, tectonic, and paleogeographic research. Deposits of all geological ages have developed on the Russian platform, and they were carefully studied by Karpinsky. In "Zamechania ob osadochnykh obrazovaniakh Evropeyskoy Rossii" ("Notes on Sedimentary Formations of European Russia"; 1880), Karpinsky solved tectonic and paleogeographic problems and also made important stratigraphic generalizations: the deposits of the Carboniferous period were described more precisely and the Triassic age of the rock on the east of the Russian platform was established. On the basis of his research on the ammonoids of the Artinsk Stage (1891) the possibility of determining the stratigraphic position of these strata was demonstrated: they are transitional between the Carboniferous and the Permian systems. Karpinsky's proposal for the classification of sedimentary formations of the earth's crust was accepted at the Second International Geological Congress at Bologna in 1881. Karpinsky did stratigraphic research on the eastern slope of the Urals for the compilation of the geological map published in 1884. In 1909 he gave a general characterization of the Mesozoic deposits of the Urals and studied the stratigraphy and geological structure of many other regions of the country.

Karpinsky's geological research always had a practical cast, even though some problems he studied appeared to be of only theoretical importance. The paleogeographic maps he compiled, and his tectonic and stratigraphic research, served as a basis for finding useful fossils. Many of Karpinsky's works were devoted to the study of deposits and theoretical questions of ore formation.

In 1870 Karpinsky published an article emphasizing the possibility of finding rock salt in the Donets coal basin. This prediction was confirmed by drilling. In 1881 he published a long summary work on the deposits of useful fossils in the Urals. His research on the deposits of coal in the eastern slope of the Urals were of great importance, and the results were published in 1908 and 1909, and in 1913 in the *Proceedings* of the Twelfth International Geological Congress in Toronto. Later works were devoted to the origin of deposits of platinum. All these works influenced the development of Russian industry.

BIBLIOGRAPHY

I. Original Works. Karpinsky's works are listed in full in *Aleksandr Petrovich Karpinsky. Bibliografichesky ukasatel trudov* (". . . Bibliographical List of Works"; Moscow–Leningrad, 1947); they were collected in *Sobranie sochineny* ("Collected Works"), 4 vols. (Moscow–Leningrad, 1939–1949). His most important works are "Ob Avgitovykh porodakh derevni Muldakaevoy i gory Kachkanar na Urale" ("On the Augitite Rocks of the Village of Muldakaeva and Kachkanar Mountain in the Urals"), St. Petersburg, 1869; "O petrograficheskikh zakonakh" ("On Petrographic Laws"), *ibid.*, **2**, no. 4 (1870), 63–79; "O vozmozhnosti otkrytia zalezhey kamennoy soli v Kharkovkoy gub" ("On the Possibility of Discovering Deposits of Rock Salt in Kharkov Province"), *ibid.*, **3**, no. 9 (1870), 449–466; "Zakony sovmestnogo nakhozhdenia polevykh shpatov" ("Laws of Association of Feldspars"), *ibid.*, **3**, no. 7 (1874), 46–60; "Zamechania ob osadochnykh obrazovaniakh Evropeyskoy Rossii" ("Notes on Sedimentary Formations of European Russia"), *ibid.*, **4**, nos. 11–12 (1880), 242–260; *Zamechania o kharaktere dislokatsii porod v Yuzhnoy polovine Evropeyskoy Rossii* ("Notes on the Character of Rock Dislocations in the Southern Half of European Russia"; St. Petersburg, 1883); *Geologicheskaya karta vostochnogo sklona Urala* ("Geological Map of the Eastern Slope of the Urals"; St. Petersburg, 1884); "Ocherk fiziko-geograficheskikh uslovy Evropeyskoy Rossii v minuvshie geologicheskie periody" ("Sketch of the Physical-Geographical Conditions of European Russia in Past Geological Periods"), in *Zapiski Imperatorskoi akademii nauk* ("Notes of the Academy of Sciences"), **55**, app. 8 (1887), 1–36; "Ob ammoneyakh artinskogo yarusa i nekotorykh skhodnykh s nimi kamennougolnykh formakh" ("On the Ammonoids of the Artinsk Stage and Certain Carboniferous Formations Similar to Them"), in *Zapiski S.-Peterburgskogo mineralogicheskago obshchestva* ("Notes of the St. Petersburg Mineralogical Society"), 2nd ser., **27** (1891), pp. 15–208; "Obshchy kharakter kolebany zemnoy kory v predelakh Evropeyskoy Rossii" ("The General Character of the Movements of the Earth's Crust Within the Boundaries of European Russia"), in *Izvestiya Imperatorskoi akademii nauk*, 5th ser., **1**, no. 1 (1894), pp. 1–19; "Ob ostatkakh edestid i o novom ikh rode Helicoprion" ("On the Remains of Edestidae and the New Genus Helicoprion"), in *Zapiski Imperatorskoi Akademii nauk*, **8**, no. 7 (1899), 1–67; "Mezozoyskie uglenoskii otlozhenia vostochnogo sklona Urala" ("Mesozoic Coal Deposits of the Eastern Slope of the Urals"), in *Gornyi zhurnal*, **3**, no. 7 (1909), 53–86; "O trokhiliskakh" ("On Trochiliscids"), in *Trudy*

Geologicheskago komiteta, n.s. no. 27 (1906), 1–166; "K tektonike Evropeyskoy Rossii" ("Toward a Tectonics of European Russia"), in *Izvestiya Rossiiskoi akademii nauk*, 6th ser., **13**, nos. 12–15, pp. 573–590 (1919); and *Ocherk geologicheskogo proshlogo Evropeyskoy Rossii* ("Sketch of the Geological Past of European Russia"; Petrograd, 1919).

II. SECONDARY LITERATURE. On Karpinsky or his work, see A. A. Borisyak, "A. P. Karpinsky," in I. V. Kuznetsov, ed., *Lyudi russkoy nauki. Geologia i geografia* ("Men of Russian Science. Geology and Geography"; Moscow, 1962), pp. 46–53; and V. A. Obruchov, "Akademik Aleksandr Petrovich Karpinsky," in *Izvestiya Akademii nauk SSSR*, Ser. geolog., no. 3, pp. 3–7 (1951); *ibid.*, no. 1 (1947) was dedicated to Karpinsky.

IRINA V. BATYUSHKOVA

KARSTEN, KARL JOHANN BERNHARD (*b.* Bützow, Germany, 26 November 1782; *d.* Berlin, Germany, 22 August 1853), *metallurgy, mining.*

Karsten received his early education in Bützow and later in Rostock, where his father, Franz Christian Lorenz Karsten, was professor of political economy at the University of Rostock. At the age of seventeen Karsten matriculated at that university to study law and medicine. His friendly relations there with the later renowned botanist Heinrich Link, who was then lecturing on the natural sciences, awakened in Karsten an interest in physics and chemistry.

After attending the university for only one year, Karsten published *Vollständiges Register über Green's neues Journal der Physik*. He was called to Berlin in 1801 to collaborate in editing *Scherer's Journal*, while continuing his medical and scientific studies. He devoted himself with special zeal to mineralogy and metallurgy and from 1805 to 1810 published with S. Weiss a German edition of René Haüy's *Traité de minéralogie*. At about the same time, he independently produced a translation of Beaume's chemical system.

After Karsten earned his doctoral degree with the dissertation *De affinitate chemica* and parted with Scherer, he gained experience at ironworks in Brandenburg and Upper Silesia. On the basis of several excellent field reports, Karsten received a ministerial commission in 1804 to erect a plant for extracting coal tar at the metalworks in Gleiwitz (now Gliwice, Poland); the plant was the first of its kind in Germany. At the end of 1804 he was accepted in the government service as a *Referendar* (assistant mining inspector). The following year he was promoted to *Assessor* (associate inspector) and was entrusted with the technical supervision of all Upper Silesian metallurgical works. He was named *Bergrat* (mining inspector) in 1810 and in 1811 *Oberhüttenrat* (senior foundry inspector) and *Oberhüttenverwalter* (senior foundry manager) for Upper and Lower Silesia. Karsten won special recognition for his part in the growth of the Silesian zinc industry. He constructed the Lydognia metalworks where, for the first time, zinc was prepared directly from calamine.

In 1815 Karsten was asked to provide his expert opinion on the Siegerland ore mines situated in territory conquered during the Napoleonic wars, in the interests of establishing the boundary between Prussia and Nassau. He subsequently returned to Breslau for a short time, leaving in 1819 to accept an offer as *Geheimer Bergrat*, a prestigious post in the ministry of the interior in Berlin. In 1821 he became *Geheimer Oberbergrat* (privy councilor). In this position he successfully administered the entire metallurgical and salt-mining industry in Prussia for thirty years.

A prolific writer, Karsten published a German edition (1814–1815) of Rinman's history of iron, *Geschichte des Eisens*, a preparatory work for Karsten's own *Handbuch der Eisenhüttenkunde*. Two years later he published *Grundriss der Metallurgie und der metallurgischen Hüttenkunde*; its brilliant success led Karsten to expand it into a large handbook which appeared in 1831 as *System der Metallurgie*. With this work Karsten achieved fame as a founder of scientific metallurgy. His literary activity was not confined to metallurgy, however. In 1828 he produced an important source book of mining law in his *Grundriss der deutschen Bergrechtslehre*. In 1843 his *Philosophie der Chemie* appeared and in 1846–1847 his excellent two-volume *Lehrbuch der Salinenkunde*.

Karsten also established a reputation as an editor of the mining and metallurgical journal *Archiv für Bergbau und Hüttenwesen*, the title of which was changed in 1829 to *Archiv für Mineralogie, Geognosie, Bergbau und Hüttenkunde*. Many of his shorter papers were published in this journal.

Karsten resigned in December 1850 after forty-six years in government service. The two years preceding his death were occupied with scientific research and political activities, for Karsten was at this time a deputy in the Prussian Upper Chamber.

BIBLIOGRAPHY

I. ORIGINAL WORKS. Karsten's major writings include *Revision der chemischen Affinitätslehre mit Rücksicht auf Berthollets neue Theorie* (Leipzig, 1803); *Handbuch der Eisenhüttenkunde*, 2 vols. (Halle, 1816; trans. into French; 2nd ed., 4 vols., Berlin, 1827–1828; 3rd ed., 5 vols., Berlin, 1841, with atlas); *Grundriss der Metallurgie und der metallurgischen Hüttenkunde* (Breslau, 1818).

Metallurgische Reise durch einen Theil von Baiern und durch die süddeutschen Provinzen Oesterreichs (Halle, 1821); *Grundriss der deutschen Bergrechtslehre mit Rücksicht auf die französische Bergwerksgesetzgebung* (Berlin, 1828); *System der Metallurgie; geschichtlich, statistisch, theoretisch und technisch*, 5 vols. (Berlin, 1831), with atlas; *Philosophie der Chemie* (Berlin, 1843); *Über den Ursprung des Berg-Regals in Deutschland* (Berlin, 1844); *Lehrbuch der Salinenkunde*, 2 vols. (Berlin, 1846–1847).

Karsten edited the journals *Archiv für Bergbau und Hüttenwesen*, 20 vols. (1818–1828) and *Archiv für Mineralogie, Geognosie, Bergbau und Hüttenkunde*, 26 vols. (1829–1855), for both of which he wrote articles. A great many short articles by Karsten also appear in Scherer's *Allgemeines Journal der Chemie*, and in *Abhandlungen der K. Preussischen Akademie der Wissenschaften*, and *Monatsberichte der K. Preussischen Akademie der Wissenschaften*.

II. Secondary Literature. See "Umrisse zu Karl Johann Bernhard Karsten's Leben und Wirken," in *Archiv für Mineralogie, Geognosie, Bergbau und Hüttenkunde*, **26**, no. 2 (1855), 195–372; "Karl Johann Bernhard Karsten," in *Allgemeine Deutsche Biographie*, XV (Leipzig, 1882), 427–430; and "Karl Johann Bernhard Karsten," in Walter Serlo, *Männer des Bergbaus*, pp. 82–84.

M. Koch

AL-KĀSHĪ (or **AL-KĀSHĀNĪ**), **GHIYĀTH AL-DĪN JAMSHĪD MAS'ŪD** (*b.* Kāshān, Iran; *d.* Samarkand [now in Uzbek, U.S.S.R], 22 June 1429), *astronomy, mathematics.*

The biographical data on al-Kāshī are scattered and sometimes contradictory. His birthplace was a part of the vast empire of the conqueror Tamerlane and then of his son Shāh Rukh. The first known date concerning al-Kāshī is 2 June 1406 (12 Dhū'l-Hijja, A.H. 808), when, as we know from his *Khaqānī zīj*, he observed a lunar eclipse in his native town.[1] According to Suter, al-Kāshī died about 1436; but Kennedy, on the basis of a note made on the title page of the India Office copy of the *Khaqānī zīj*, gives 19 Ramaḍān A.H. 832, or 22 June 1429.[2] The chronological order of al-Kāshī's works written in Persian or in Arabic is not known completely, but sometimes he gives the exact date and place of their completion. For instance, the *Sullam al-samā'* ("The Stairway of Heaven"), a treatise on the distances and sizes of heavenly bodies, dedicated to a vizier designated only as Kamāl al-Dīn Maḥmūd, was completed in Kāshān on 1 March 1407.[3] In 1410–1411 al-Kāshī wrote the *Mukhtaṣar dar ʿilm-i hay'at* ("Compendium of the Science of Astronomy") for Sultan Iskandar, as is indicated in the British Museum copy of this work. D. G. Voronovski identifies Iskandar with a member of the Tīmūrid dynasty and cousin of Ulugh Bēg, who ruled Fars and Iṣfahān and was executed in 1414.[4] In 1413–1414 al-Kāshī finished the *Khaqānī zīj*. Bartold assumes that the prince to whom this *zīj* is dedicated was Shāh Rukh, who patronized the sciences in his capital, Herat;[5] but Kennedy established that it was Shāh Rukh's son and ruler of Samarkand, Ulugh Bēg. According to Kennedy, in the introduction to this work al-Kāshī complains that he had been working on astronomical problems for a long time, living in poverty in the towns of Iraq (doubtless Persian Iraq) and mostly in Kāshān. Having undertaken the composition of a *zīj*, he would not be able to finish it without the support of Ulugh Bēg, to whom he dedicated the completed work.[6] In January 1416 al-Kāshī composed the short *Risāla dar sharḥ-i ālāt-i raṣd* ("Treatise on . . . Observational Instruments"), dedicated to Sultan Iskandar, whom Bartold and Kennedy identify with a member of the Kārā Koyunlū, or Turkoman dynasty of the Black Sheep.[7] Shishkin mistakenly identifies him with the above-mentioned cousin of Ulugh Bēg.[8] At almost the same time, on 10 February 1416, al-Kāshī completed in Kāshān *Nuzha al-ḥadāiq* ("The Garden Excursion"), in which he described the "Plate of Heavens," an astronomical instrument he invented. In June 1426, at Samarkand, he made some additions to this work.

Dedicating his scientific treatises to sovereigns or magnates, al-Kāshī, like many scientists of the Middle Ages, tried to provide himself with financial protection. Although al-Kāshī had a second profession—that of a physician—he longed to work in astronomy and mathematics. After a long period of penury and wandering, al-Kāshī finally obtained a secure and honorable position at Samarkand, the residence of the learned and generous protector of science and art, Sultan Ulugh Bēg, himself a great scientist.

In 1417–1420 Ulugh Bēg founded in Samarkand a *madrasa*—a school for advanced study in theology and science—which is still one of the most beautiful buildings in Central Asia. According to a nineteenth-century author, Abū Ṭāhir Khwāja, "four years after the foundation of the *madrasa*," Ulugh Bēg commenced construction of an observatory; its remains were excavated from 1908 to 1948.[9] For work in the *madrasa* and observatory Ulugh Bēg took many scientists, including al-Kāshī, into his service. During the quarter century until the assassination of Ulugh Bēg in 1449 and the beginning of the political and ideological reaction, Samarkand was the most important scientific center in the East. The exact time of al-Kāshī's move to Samarkand is unknown. Abū Ṭāhir Khwāja states that in 1424 Ulugh Bēg discussed with al-Kāshī, Qāḍī Zāde al-Rūmī, and another scientist from Kāshān, Muʿin al-Dīn, the project of the observatory.[10]

In Samarkand, al-Kāshī actively continued his mathematical and astronomical studies and took a great part in the organization of the observatory, its provision with the best equipment, and in the preparation of Ulugh Bēg's *Zīj*, which was completed after his (al-Kāshī's) death. Al-Kāshī occupied the most prominent place on the scientific staff of Ulugh Bēg. In his account of the erection of the Samarkand observatory the fifteenth-century historian Mirkhwānd mentions, besides Ulugh Bēg, only al-Kāshī, calling him "the support of astronomical science" and "the second Ptolemy."[11] The eighteenth-century historian Sayyīd Raqīm, enumerating the main founders of the observatory and calling each of them *maulanā* ("our master," a usual title of scientists in Arabic), calls al-Kāshī *maulanā-i ālam* (*maulanā* of the world).[12]

Al-Kāshī himself gives a vivid record of Samarkand scientific life in an undated letter to his father, which was written while the observatory was being built. Al-Kāshī highly prized the erudition and mathematical capacity of Ulugh Bēg, particularly his ability to perform very difficult mental computations; he described the prince's scientific activity and once called him a director of the observatory.[13] Therefore Suter's opinion that the first director of the Samarkand observatory was al-Kāshī, who was succeeded by Qāḍī Zāde, must be considered very dubious.[14] On the other hand, al-Kāshī spoke with disdain of Ulugh Bēg's nearly sixty scientific collaborators, although he qualified Qāḍī Zāde as "the most learned of all."[15] Telling of frequent scientific meetings directed by the sultan, al-Kāshī gave several examples of astronomical problems propounded there. These problems, too difficult for others, were solved easily by al-Kāshī. In two cases he surpassed Qāḍī Zāde, who misinterpreted one proof in al-Bīrūnī's *al-Qānūn al-Masʿūdī* and who was unable to solve one difficulty connected with the problem of determining whether a given surface is truly plane or not. Nevertheless his relations with Qāḍī Zāde were amicable. With great satisfaction al-Kāshī told his father of Ulugh Bēg's praise, related to him by some of his friends. He emphasized the atmosphere of free scientific discussion in the presence of the sovereign. The letter included interesting information on the construction of the observatory building and the instruments. This letter and other sources characterize al-Kāshī as the closest collaborator and consultant of Ulugh Bēg, who tolerated al-Kāshī's ignorance of court etiquette and lack of good manners.[16] In the introduction to his own *Zīj* Ulugh Bēg mentions the death of al-Kāshī and calls him "a remarkable scientist, one of the most famous in the world, who had a perfect command of the science of the ancients, who contributed to its development, and who could solve the most difficult problems."[17]

Al-Kāshī wrote his most important works in Samarkand. In July 1424 he completed *Risāla al-muḥīṭīyya* ("The Treatise on the Circumference"), a masterpiece of computational technique resulting in the determination of 2π to sixteen decimal places. On 2 March 1427 he finished the textbook *Miftāḥ al-ḥisāb* ("The Key of Arithmetic"), dedicated to Ulugh Bēg. It is not known when he completed his third chef d'oeuvre, *Risāla al-watar wa'l-jaib* ("The Treatise on the Chord and Sine"), in which he calculated the sine of 1° with the same precision as he had calculated π. Apparently he worked on this shortly before his death; some sources indicate that the manuscript was incomplete when he died and that it was finished by Qāḍī Zāde.[18] Apparently al-Kāshī had developed his method of calculation of the sine of 1° before he completed *Miftāḥ al-ḥisāb*, for in the introduction to this book, listing his previous works, he mentions *Risāla al-watar wa'l-jaib*.

As was mentioned above, al-Kāshī took part in the composition of Ulugh Bēg's *Zīj*. We cannot say exactly what he did, but doubtless his participation was considerable. The introductory theoretical part of the *Zīj* was completed during al-Kāshī's lifetime, and he translated it from Persian into Arabic.[19]

Mathematics. Al-Kāshī's best-known work is *Miftāḥ al-ḥisāb* (1427), a veritable encyclopedia of elementary mathematics intended for an extensive range of students; it also considers the requirements of calculators—astronomers, land surveyors, architects, clerks, and merchants. In the richness of its contents and in the application of arithmetical and algebraic methods to the solution of various problems, including several geometric ones, and in the clarity and elegance of exposition, this voluminous textbook is one of the best in the whole of medieval literature; it attests to both the author's erudition and his pedagogic ability.[20] Because of its high quality the *Miftāḥ al-ḥisāb* was often recopied and served as a manual for hundreds of years; a compendium of it was also used. The book's title indicates that arithmetic was viewed as the key to the solution of every kind of problem which can be reduced to calculation, and al-Kāshī defined arithmetic as the "science of rules of finding numerical unknowns with the aid of corresponding known quantities."[21] The *Miftāḥ al-ḥisāb* is divided into five books preceded by an introduction: "On the Arithmetic of Integers," "On the Arithmetic of Fractions," "On the 'Computation of the Astronomers'" (on sexagesimal arithmetic), "On the Measurement of Plane Figures and Bodies," and "On the Solution of Problems by Means of Algebra [linear and quadratic equations] and of the Rule of Two False Assumptions, etc." The work comprises many interesting problems and carefully analyzed numerical examples.

In the first book of the *Miftāḥ*, al-Kāshī describes in detail a general method of extracting roots of integers. The integer part of the root is obtained by means of what is now called the Ruffini–Horner method. If the root is irrational, $a < \sqrt[n]{a^n + r} < a + 1$ (a and r are integers), the fractional part of the root is calculated according to the approximate formula $\frac{r}{(a+1)^n - a^n}$.[22] Al-Kāshī himself expressed all rules of computation in words, and his algebra is always purely "rhetorical." In this connection he gives the general rule for raising a binomial to any natural power and the additive rule for the successive determination of binomial coefficients; and he constructs the so-called Pascal's triangle (for $n = 9$). The same methods were presented earlier in the *Jāmiʿ al-ḥisāb biʾl takht waʾl-tuzāb* ("Arithmetic by Means of Board and Dust") of Naṣīr al-Dīn al-Ṭūsī (1265). The origin of these methods is unknown. It is possible that they were at least partly developed by al-Khayyāmī; the influence of Chinese algebra is also quite plausible.[23]

Noteworthy in the second and the third book is the doctrine of decimal fractions, used previously by al-Kāshī in his *Risāla al-muḥīṭīyya*. It was not the first time that decimal fractions appeared in an Arabic mathematical work; they are in the *Kitāb al-fuṣūl fīʾl-ḥisāb al-Hindī* ("Treatise of Arithmetic") of al-Uqlīdisī (mid-tenth century) and were used occasionally also by Chinese scientists.[24] But only al-Kāshī introduced the decimal fractions methodically, with a view to establishing a system of fractions in which (as in the sexagesimal system) all operations would be carried out in the same manner as with integers. It was based on the commonly used decimal numeration, however, and therefore accessible to those who were not familiar with the sexagesimal arithmetic of the astronomers. Operations with finite decimal fractions are explained in detail, but al-Kāshī does not mention the phenomenon of periodicity. To denote decimal fractions, written on the same line with the integer, he sometimes separated the integer by a vertical line or wrote in the orders above the figures; but generally he named only the lowest power that determined all the others. In the second half of the fifteenth century and in the sixteenth century al-Kāshī's decimal fractions found a certain circulation in Turkey, possibly through ʿAlī Qūshjī, who had worked with him at Samarkand and who sometime after the assassination of Ulugh Bēg and the fall of the Byzantine empire settled in Constantinople. They also appear occasionally in an anonymous Byzantine collection of problems from the fifteenth century which was brought to Vienna in 1562.[25] It is also possible that al-Kāshī's ideas had some influence on the propagation of decimal fractions in Europe.

In the fifth book al-Kāshī mentions in passing that for the fourth-degree equations he had discovered "the method for the determination of unknowns in . . . seventy problems which had not been touched upon by either ancients or contemporaries."[26] He also expressed his intention to devote a separate work to this subject, but it seems that he did not complete this research. Al-Kāshī's theory should be analogous to the geometrical theory of cubic equations developed much earlier by Abu'l-Jūd Muḥammad ibn Laith, al-Khayyāmī (eleventh century), and their followers: the positive roots of fourth-degree equations were constructed and investigated as coordinates of points of intersection of the suitable pairs of conics. It must be added that actually there are only sixty-five (not seventy) types of fourth-degree equations reducible to the forms considered by Muslim mathematicians, that is, the forms having terms with positive coefficients on both sides of the equation. Only a few cases of fourth-degree equations were studied before al-Kāshī.

Al-Kāshī's greatest mathematical achievements are *Risāla al-muḥīṭīyya* and *Risāla al-watar waʾl-jaib*, both written in direct connection with astronomical researches and especially in connection with the increased demands for more precise trigonometrical tables.

At the beginning of the *Risāla al-muḥīṭīyya* al-Kāshī points out that all approximate values of the ratio of the circumference of a circle to its diameter, that is, of π, calculated by his predecessors gave a very great (absolute) error in the circumference and even greater errors in the computation of the areas of large circles. Al-Kāshī tackled the problem of a more accurate computation of this ratio, which he considered to be irrational, with an accuracy surpassing the practical needs of astronomy, in terms of the then-usual standard of the size of the visible universe or of the "sphere of fixed stars."[27] For that purpose he assumed, as had the Iranian astronomer Quṭb al-Dīn al-Shīrāzī (thirteenth-fourteenth centuries), that the radius of this sphere is 70,073.5 times the diameter of the earth. Concretely, al-Kāshī posed the problem of calculating the said ratio with such precision that the error in the circumference whose diameter is equal to 600,000 diameters of the earth will be smaller than the thickness of a horse's hair. Al-Kāshī used the following old Iranian units of measurement: 1 parasang (about 6 kilometers) = 12,000 cubits, 1 cubit = 24 inches (or fingers), 1 inch = 6 widths of a medium-size grain

AL-KĀSHĪ

of barley, and 1 width of a barley grain = 6 thicknesses of a horse's hair. The great-circle circumference of the earth is considered to be about 8,000 parasangs, so al-Kāshī's requirement is equivalent to the computation of π with an error no greater than $0.5 \cdot 10^{-17}$. This computation was accomplished by means of elementary operations, including the extraction of square roots, and the technique of reckoning is elaborated with the greatest care.

Al-Kāshī's measurement of the circumference is based on a computation of the perimeters of regular inscribed and circumscribed polygons, as had been done by Archimedes, but it follows a somewhat different procedure. All calculations are performed in sexagesimal numeration for a circle with a radius of 60. Al-Kāshī's fundamental theorem—in modern notation—is as follows: In a circle with radius r,

$$r(2r + crd\ \alpha^\circ) = crd^2\left(\alpha^\circ + \frac{180^\circ - \alpha^\circ}{2}\right),$$

where $crd\ \alpha^\circ$ is the chord of the arc α° and $\alpha^\circ < 180^\circ$. Thus al-Kāshī applied here the "trigonometry of chords" and not the trigonometric lines themselves. If $\alpha = 2\varphi$ and $d = 2$, then al-Kāshī's theorem may be written trigonometrically as

$$\sin\left(45^\circ + \frac{\varphi^\circ}{2}\right) = \sqrt{\frac{1 + \sin\varphi^\circ}{2}},$$

which is found in the work of J. H. Lambert (1770). The chord of 60° is equal to r, and so it is possible by means of this theorem to calculate successively the chords $c_1, c_2, c_3, \cdots$ of the arcs 120°, 150°, 165°, $\cdots$; in general the value of the chord c_n of the arc $\alpha_n^\circ = 180^\circ - \frac{360^\circ}{3 \cdot 2^n}$ will be $c_n = \sqrt{r(2r + c_{n-1})}$. The chord c_n being known, we may, according to Pythagorean theorem, find the side $a_n = \sqrt{d^2 - c_n^2}$ of the regular inscribed $3 \cdot 2^n$-sided polygon, for this side a_n is also the chord of the supplement of the arc α_n° up to 180°. The side b_n of a similar circumscribed polygon is determined by the proportion $b_n : a_n = r : h$, where h is the apothem of the inscribed polygon. In the third section of his treatise al-Kāshī ascertains that the required accuracy will be attained in the case of the regular polygon with $3 \cdot 2^{28} = 805{,}306{,}368$ sides.

He resumes the computation of the chords in twenty-eight extensive tables; he verifies the extraction of the roots by squaring and also by checking by 59 (analogous to the checking by 9 in decimal numeration); and he establishes the number of sexagesimal places to which the values used must be taken. We can concisely express the chords c_n and the sides a_n by formulas

$$c_n = r\sqrt{2 + \sqrt{2 + \cdots + \sqrt{2 + \sqrt{3}}}}$$

and

$$a_n = r\sqrt{2 - \sqrt{2 + \cdots + \sqrt{2 + \sqrt{3}}}},$$

where the number of radicals is equal to the index n. In the sixth section, by multiplying a_{28} by $3 \cdot 2^{28}$, one obtains the perimeter p_{28} of the inscribed $3 \cdot 2^{28}$-sided polygon and then calculates the perimeter P_{28} of the corresponding similar circumscribed polygon. Finally the best approximation for $2\pi r$ is accepted as the arithmetic mean $\frac{p_{28} + P_{28}}{2}$, whose sexagesimal value for $r = 1$ is $6\ 16^{\mathrm{I}}\ 59^{\mathrm{II}}\ 28^{\mathrm{III}}\ 1^{\mathrm{IV}}\ 34^{\mathrm{V}}\ 51^{\mathrm{VI}}\ 46^{\mathrm{VII}}\ 14^{\mathrm{VIII}}\ 50^{\mathrm{IX}}$, where all places are correct. In the eighth section al-Kāshī translates this value into the decimal fraction $2\pi = 6.2831853071795865$, correct to sixteen decimal places. This superb result far surpassed all previous determinations of π. The decimal approximation $\pi \approx 3.14$ corresponds to the famous boundary values found by Archimedes, $3\frac{10}{71} < \pi < 3\frac{1}{7}$; Ptolemy used the sexagesimal value $3\ 8^{\mathrm{I}}\ 30^{\mathrm{II}}$ (≈ 3.14166), and the results of al-Kāshī's predecessors in the Islamic countries were not much better. The most accurate value of π obtained before al-Kāshī by the Chinese scholar Tsu Ch'ung-chih (fifth century) was correct to six decimal places. In Europe in 1597 A. van Roomen approached al-Kāshī's result by calculating π to fifteen decimal places; later Ludolf van Ceulen calculated π to twenty and then to thirty-two places (published 1615).

In his *Risāla al-watar wa'l-jaib* al-Kāshī again calculates the value of sin 1° to ten correct sexagesimal places; the best previous approximations, correct to four places, were obtained in the tenth century by Abu'l-Wafā' and Ibn Yūnus. Al-Kāshī derived the equation for the trisection of an angle, which is a cubic equation of the type $px = q + x^3$—or, as the Arabic mathematicians would say, "Things are equal to the cube and the number." The trisection equation had been known in the Islamic countries since the eleventh century; one equation of this type was solved approximately by al-Bīrūnī to determine the side of a regular nonagon, but this method remains unknown to us. Al-Kāshī proposed an original iterative method of approximate solution, which can be summed up as follows: Assume that the equation

$$x = \frac{q + x^3}{p}$$

possesses a very small positive root x; for the first approximation, take $x_1 = \frac{q}{p}$; for the second approximation, $x_2 = \frac{q + x_1^3}{p}$; for the third, $x_3 = \frac{q + x_2^3}{p}$, and generally

$$x_n = \frac{q + x_{n-1}^3}{p}, \qquad x_0 = 0.$$

It may be proved that this process is convergent in the neighborhood of values of x, $\frac{3x^2}{p} < r < 1$. Al-Kāshī used a somewhat different procedure: he obtained x_1 by dividing q by p as the first sexagesimal place of the desired root, then calculated not the approximations $x_2, x_3, \cdots$ themselves but the corresponding corrections, that is, the successive sexagesimal places of x. The starting point of al-Kāshī's computation was the value of sin 3°, which can be calculated by elementary operations from the chord of 72° (the side of a regular inscribed pentagon) and the chord of 60°. The sin 1° for a radius of 60 is obtained as a root of the equation

$$x = \frac{900 \sin 3° + x^3}{45 \cdot 60}.$$

The sexagesimal value of sin 1° for a radius of 60 is $1\ 2^{I}\ 49^{II}\ 43^{III}\ 11^{IV}\ 14^{V}\ 44^{VI}\ 16^{VII}\ 26^{VIII}\ 17^{IX}$; and the corresponding decimal fraction for a radius of 1 is 0.017452406437283571. All figures in both cases are correct.

Al-Kāshī's method of numerical solution of the trisection equation, whose variants were also presented by Ulugh Bēg, Qāḍī Zāde, and his grandson Maḥmūd ibn Muḥammad Mīrīm Chelebī (who worked in Turkey),[28] requires a relatively small number of operations and shows the exactness of the approximation at each stage of the computation. Doubtless it was one of the best achievements in medieval algebra. H. Hankel has written that this method "concedes nothing in subtlety or elegance to any of the methods of approximation discovered in the West after Viète."[29] But all these discoveries of al-Kāshī's were long unknown in Europe and were studied only in the nineteenth and twentieth centuries by such historians of science as Sédillot, Hankel, Luckey, Kary-Niyazov, and Kennedy.

Astronomy. Until now only three astronomical works by al-Kāshī have been studied. His *Khāqānī Zīj,* as its title shows, was the revision of the *Īlkhānī Zīj* of Naṣīr al-Dīn al-Ṭūsī. In the introduction to al-Kāshī's *zīj* there is a detailed description of the method of determining the mean and anomalistic motion of the moon based on al-Kāshī's three observations of lunar eclipses made in Kāshān and on Ptolemy's three observations of lunar eclipses described in the *Almagest.* In the chronological section of these tables there are detailed descriptions of the lunar Muslim (Hijra) calendar, of the Persian solar (Yazdegerd) and Greek-Syrian (Seleucid) calendars, of al-Khayyāmī's calendar reform (Malikī), of the Chinese-Uigur calendar, and of the calendar used in the Il-Khan empire, where Naṣīr al-Dīn al-Ṭūsī had been working. In the mathematical section there are tables of sines and tangents to four sexagesimal places for each minute of arc. In the spherical astronomy section there are tables of transformations of ecliptic coordinates of points of the celestial sphere to equatorial coordinates and tables of other spherical astronomical functions.

There are also detailed tables of the longitudinal motion of the sun, the moon, and the planets, and of the latitudinal motion of the moon and the planets. Al-Kāshī also gives the tables of the longitudinal and latitudinal parallaxes for certain geographic latitudes, tables of eclipses, and tables of the visibility of the moon. In the geographical section there are tables of geographical latitudes and longitudes of 516 points. There are also tables of the fixed stars, the ecliptic latitudes and longitudes, the magnitudes and "temperaments" of the 84 brightest fixed stars, the relative distances of the planets from the center of the earth, and certain astrological tables. In comparing the tables with Ulugh Bēg's *Zīj,* it will be noted that the last tables in the geographical section contain coordinates of 240 points, but the star catalog contains coordinates of 1,018 fixed stars.

In his *Miftāḥ al-ḥisāb* al-Kāshī mentions his *Zīj al-tashīlāt* ("Zīj of Simplifications") and says that he also composed some other tables.[30] His *Sullam al-samā',* scarcely studied as yet, deals with the determination of the distances and sizes of the planets.

In his *Risāla dar sharḥ-i ālāt-i raṣd* ("Treatise on the Explanation of Observational Instruments") al-Kāshī briefly describes the construction of eight astronomical instruments: triquetrum, armillary sphere, equinoctial ring, double ring, Fakhrī sextant, an instrument "having azimuth and altitude," an instrument "having the sine and arrow," and a small armillary sphere. Triquetra and armillary spheres were used by Ptolemy; the latter is a model of the celestial sphere, the fixed and mobile great circles of which are represented, respectively, by fixed and mobile rings. Therefore the armillary sphere can represent positions of these circles for any moment; one ring has diopters for measurement of the altitude of a star, and the direction of the plane of this ring determines the azimuth. The third and seventh instruments consist of several rings of armillary spheres. The equinoctial ring (the circle in the plane of the celestial equator), used for observation of the transit of the sun through the equinoctial points, was invented by astronomers who worked in the tenth century in Shīrāz, at the court of the Buyid sultan ʿAḍūd al-Dawla. The Fakhrī sextant, one-sixth of a circle in the plane of the celestial meridian, used for measuring the altitudes of stars in this plane, was invented about 1000 by al-Khujandī in Rayy, at the court of the Buyid sultan Fakhr al-Dawla. The fifth instrument was used in the Marāgha observatory directed by Naṣīr al-Dīn al-Ṭūsī. The sixth instrument, al-Kāshī says, did not exist in earlier observatories; it is used for determination of sines and "arrows" (versed sines) of arcs.

In *Nuzha al-ḥadāiq* al-Kāshī describes two instruments he had invented: the "plate of heavens" and the "plate of conjunctions." The first is a planetary equatorium and is used for the determination of the ecliptic latitudes and longitudes of planets, their distances from the earth, and their stations and retrogradations; like the astrolabe, which it resembles in shape, it was used for measurements and for graphical solutions of problems of planetary motion by means of a kind of nomograms. The second instrument is a simple device for performing a linear interpolation.

NOTES

1. See E. S. Kennedy, *The Planetary Equatorium* . . ., p. 1.
2. H. Suter, *Die Mathematiker und Astronomen* . . ., pp. 173–174; Kennedy, *op. cit.*, p. 7.
3. See M. Krause, "Stambuler Handschriften . . .," p. 50; M. Ṭabāṭabā'i, "Jamshīd Ghiyāth al-Dīn Kāshānī," p. 23.
4. D. G. Voronovski, "Astronomy Sredney Azii ot Muhammeda al-Havarazmi do Ulugbeka i ego shkoly (IX–XVI vv.)," pp. 127, 164.
5. See V. V. Bartold, *Ulugbek i ego vremya*, p. 108.
6. Kennedy, *op. cit.*, pp. 1–2.
7. Bartold, *op. cit.*, p. 108; Kennedy, *op. cit.*, p. 2.
8. V. A. Shishkin, "Observatoriya Ulugbeka i ee issledovanie," p. 10.
9. See T. N. Kary-Niyazov, *Astronomicheskaya shkola Ulugbeka*, 2nd ed., p. 107; see also Shishkin, *op. cit.*
10. See Kary-Niyazov, *loc. cit.*
11. See Bartold, *op. cit.*, p. 88.
12. *Ibid.*, pp. 88–89.
13. E. S. Kennedy, "A Letter of Jamshīd al-Kāshī to His Father," p. 200.
14. Suter, *op. cit.*, pp. 173, 175; E. S. Kennedy, "A Survey of Islamic Astronomical Tables," p. 127.
15. Kennedy, "A Letter . . .," p. 194.
16. See Bartold, *op. cit.*, p. 108.
17. See *Zīj-i Ulughbeg*, French trans., p. 5.
18. See Kennedy, *The Planetary Equatorium* . . ., p. 6.
19. See *Taʿrīb al-zīj*; Kary-Niyazov, *op. cit.*, 2nd ed., pp. 141–142.
20. See P. Luckey, *Die Rechenkunst* . . .; A. P. Youschkevitch; *Geschichte der Mathematik im Mittelalter*, p. 237 ff.
21. al-Kāshī, *Klyuch arifmetiki* . . ., p. 13.
22. See P. Luckey, "Die Ausziehung des *n*-ten Wurzel . . . ."
23. P. Luckey, "Die Ausziehung des *n*-ten Wurzel . . ."; Juschkewitsch, *op. cit.*, pp. 240–248.
24. See A. Saidan, "The Earliest Extant Arabic Arithmetic . . ."; Juschkewitsch, *op. cit.*, pp. 21–23.
25. H. Hunger and K. Vogel, *Ein byzantinisches Rechenbuch des 15. Jahrhunderts*, p. 104.
26. al-Kāshi, *Klyuch arifmetiki* . . ., p. 192.
27. *Ibid.*, p. 126.
28. Kary-Niyazov, *op. cit.*, 2nd ed., p. 199; Qāḍī Zāde, *Risāla fī istikhrāj jaib daraja wāhida*; Mīrīm Chelebī, *Dastūr al-ʿamal wa taṣḥīḥ al-jadwal.*
29. H. Hankel, *Zur Geschichte der Mathematik* . . ., p. 292.
30. al-Kāshī, *Klyuch arifmetiki* . . ., p. 9.

BIBLIOGRAPHY

I. Original Works. Al-Kāshī's writings were collected as *Majmūʾ* ("Collection"; Teheran, 1888), an ed. of the original texts; "Matematicheskie traktaty," in *Istoriko-matematicheskie issledovaniya*, **7** (1954), 9–439, Russian trans. by B. A. Rosenfeld and commentaries by Rosenfeld and A. P. Youschkevitch; and *Klyuch arifmetiki. Traktat of okruzhnosti* ("The Key of Arithmetic. A Treatise on Circumference"), trans. by B. A. Rosenfeld, ed. by V. S. Segal and A. P. Youschkevitch, commentaries by Rosenfeld and Youschkevitch, with photorepros. of Arabic MSS.

His individual works are the following:

1. *Sullam al-samāʾ fī ḥall ishkāl waqaʿa liʾl-muqaddimīn fiʾl-abʿād waʾl-ajrām* ("The Stairway of Heaven, on Resolution of Difficulties Met by Predecessors in the Determination of Distances and Sizes"; 1407). Arabic MSS in London, Oxford, and Istanbul, the most important being London, India Office 755; and Oxford, Bodley 888/4.

2. *Mukhtaṣar dar ʿilm-i hayʾat* ("Compendium on the Science of Astronomy") or *Risāla dar hayʾat* ("Treatise on Astronomy"; 1410–1411). Persian MSS in London and Yezd.

3. *Zīj-i Khaqānī fī takmīl-i Zīj-i Īlkhānī* ("Khaqānī Zīj—Perfection of Īlkhānī Zīj"; 1413–1414). Persian MSS in London, Istanbul, Teheran, Yezd, Meshed, and Hyderabad-Deccan, the most important being London, India Office 2232, which is described in E. S. Kennedy, "A Survey of Islamic Astronomical Tables," pp. 164–166.

4. *Risāla dar sharḥ-i ālāt-i raṣd* ("Treatise on the Explanation of Observational Instruments"; 1416). Persian MSS in Leiden and Teheran, the more important being Leiden, Univ. 327/12, which has been pub. as a supp. to V. V. Bartold, *Ulugbek i ego vremya;* and E. S. Kennedy, "Al-Kāshī's Treatise on Astronomical Observation Instruments," pp. 99, 101, 103. There is an English trans. in Kennedy, "Al-Kāshī's Treatise . . .," pp. 98–104; and a Russian trans. in V. A. Shishkin, "Observatoriya Ulugbeka i ee issledovanie," pp. 91–94.

5. *Nuzha al-ḥadāiq fī kayfiyya ṣanʿa al-āla al-musammā bi ṭabaq al-manāṭiq* ("The Garden Excursion, on the Method of Construction of the Instrument Called Plate of Heavens"; 1416). Arabic MSS are in London, Dublin, and Bombay, the most important being London, India Office Ross 210. There is a litho. ed. of another MS as a supp. to the Teheran ed. of *Miftāḥ al-ḥisāb;* see also *Risāla fiʾl-ʿamal bi ashal āla min qabl al-nujūm;* G. D. Jalalov, "Otlichie 'Zij Guragani' ot drugikh podobnykh zijey" and "K voprosu o sostavlenii planetnykh tablits samar-

kandskoy observatorii"; T. N. Kary-Niyazov, *Astronomicheskaya shkola Ulugbeka;* and E. S. Kennedy, "Al-Kāshī's 'Plate of Conjunctions.' "

6. *Risāla al-muḥīṭīyya* ("Treatise on the Circumference"; 1424). Arabic MSS are in Istanbul, Teheran, and Meshed, the most important being Istanbul, Ask. müze. 756. There is an ed. of another MS in *Majmū'* and one of the Istanbul MS with German trans. in P. Luckey, *Der Lehrbrief über den Kreisumfang von Gamšīd b. Mas'ūd al-Kāšī*. Russian trans. are in "Matematicheskie traktaty," pp. 327–379; and in *Klyuch arifmetiki*, pp. 263–308, with photorepro. of Istanbul MS on pp. 338–426.

7. *Ilkaḥāt an-Nuzha* ("Supplement to the Excursion"; 1427). There is an ed. of a MS in *Majmū'*.

8. *Miftāḥ al-ḥisāb* ("The Key of Arithmetic") or *Miftāḥ al-ḥussāb fi 'ilm al-ḥisāb* ("The Key of Reckoners in the Science of Arithmetic"). Arabic MSS in Leningrad, Berlin, Paris, Leiden, London, Istanbul, Teheran, Meshed, Patna, Peshawar, and Rampur, the most important being Leningrad, Publ. Bibl. 131; Leiden, Univ. 185; Berlin, Preuss. Bibl. 5992 and 2992a, and Inst. Gesch. Med. Natur. I.2; Paris, BN 5020; and London, BM 419 and India Office 756. There is a litho. ed. of another MS (Teheran, 1889). Russian trans. are in "Matematicheskie traktaty," pp. 13–326; and *Klyuch arifmetiki*, pp. 7–262, with photorepro. of Leiden MS on pp. 428–568. There is an ed. of the Leiden MS with commentaries (Cairo, 1968). See also P. Luckey, "Die Ausziehung des *n*-ten Wurzel . . ." and "Die Rechenkunst bei Ǧamšīd b. Mas'ūd al-Kāšī . . . ."

9. *Talkhīṣ al-Miftāḥ* ("Compendium of the Key"). Arabic MSS in London, Tashkent, Istanbul, Baghdad, Mosul, Teheran, Tabriz, and Patna, the most important being London, India Office 75; and Tashkent, Inst. vost. 2245.

10. *Risāla al-watar wa'l-jaib* ("Treatise on the Chord and Sine"). There is an ed. of a MS in *Majmū'*.

11. *Ta'rīb al-zīj* ("The Arabization of the Zīj"), an Arabic trans. of the intro. to Ulugh Bēg's *Zīj*. MSS are in Leiden and Tashkent.

12. *Wujūh al-'amal al-ḍarb fī'l-takht wa'l-turāb* ("Ways of Multiplying by Means of Board and Dust"). There is an ed. of an Arabic MS in *Majmū'*.

13. *Natā'ij al-ḥaqā'iq* ("Results of Verities"). There is an ed. of an Arabic MS in *Majmū'*.

14. *Miftāḥ al-asbāb fī 'ilm al-zīj* ("The Key of Causes in the Science of Astronomical Tables"). There is an Arabic MS in Mosul.

15. *Risāla dar sakht-i asṭurlāb* ("Treatise on the Construction of the Astrolabe"). There is a Persian MS in Meshed.

16. *Risāla fī ma'rifa samt al-qibla min dāira hindiyya ma'rūfa* ("Treatise on the Determination of Azimuth of the Qibla by Means of a Circle Known as Indian"). There is an Arabic MS at Meshed.

17. Al-Kāshī's letter to his father exists in 2 Persian MSS in Teheran. There is an ed. of them in M. Ṭabāṭabā'ī, "Nāma-yi pisar bi pidar," in *Amūzish wa parwarish*, **10**, no. 3 (1940), 9–16, 59–62. An English trans. is E S. Kennedy, "A Letter of Jamshīd al-Kāshī to His Father"; English and Turkish trans. are in A. Sayili, "Ghiyāth al-Dīn al-Kāshī's Letter on Ulugh Bēg and the Scientific Activity in Samarkand," in *Türk tarih kurumu yayinlarinden*, 7th ser., no. 39 (1960).

II. SECONDARY LITERATURE. See the following: V. V. Bartold, *Ulugbek i ego vremya* ("Ulugh Bēg and His Time"; Petrograd, 1918), 2nd ed. in his *Sochinenia* ("Works"), II, pt. 2 (Moscow, 1964), 23–196, trans. into German as "Ulug Beg und seine Zeit," in *Abhandlungen für die Kunde des Morgenlandes*, **21**, no. 1 (1935); L. S. Bretanitzki and B. A. Rosenfeld, "Arkhitekturnaya glava traktata 'Klyuch arifmetiki' Giyas ad-Dina Kashi" ("An Architectural Chapter of the Treatise 'The Key of Arithmetic' by Ghiyāth al-Dīn Kāshī"), in *Iskusstvo Azerbayjana*, **5** (1956), 87–130; C. Brockelmann, *Geschichte der arabischen literatur*, 2nd ed., II (Leiden, 1944), 273 and supp. II (Leiden, 1942), 295; Mīrīm Chelebī, *Dastūr al-'amal wa taṣḥīḥ al-jadwal* ("Rules of the Operation and Correction of the Tables"; 1498), Arabic commentaries to Ulugh Bēg's *Zīj*, contains an exposition of al-Kāshī's *Risāla al-watar wa'l-jaib*—Arabic MSS are in Paris, Berlin, Istanbul, and Cairo, the most important being Paris, BN 163 (a French trans. of the exposition is in L. A. Sédillot, "De l'algèbre chez les Arabes," in *Journal asiatique*, 5th ser., **2** [1853], 323–350; a Russian trans. is in *Klyuch arifmetiki*, pp. 311–319); A. Dakhel, *The Extraction of the n-th Root in the Sexagesimal Notation. A Study of Chapter 5, Treatise 3 of Miftāḥ al Ḥisāb*, W. A. Hijab and E. S. Kennedy, eds. (Beirut, 1960); H. Hankel, *Zur Geschichte der Mathematik im Altertum und Mittelalter* (Leipzig, 1874); and H. Hunger and K. Vogel, *Ein byzantinisches Rechenbuch des 15. Jahrhunderts* (Vienna, 1963), text, trans., and commentary.

See also G. D. Jalalov, "Otlichie 'Zij Guragani' ot drugikh podobnykh zijey" ("The Difference of 'Gurgani Zīj' from Other Zījes"), in *Istoriko-astronomicheskie issledovaniya*, **1** (1955), 85–100; "K voprosu o sostavlenii planetnykh tablits samarkandskoy observatorii" ("On the Question of the Composition of the Planetary Tables of the Samarkand Observatory"), *ibid.*, 101–118; and "Giyas ad-Din Chusti (Kashi)—krupneyshy astronom i matematik XV veka" ("Ghiyāth al-Dīn Chūstī [Kāshī]—the Greatest Astronomer and Mathematician of the XV Century"), in *Uchenye zapiski Tashkentskogo gosudarstvennogo pedagogicheskogo instituta*, **7** (1957), 141–157; T. N. Kary-Niyazov, *Astronomicheskaya shkola Ulugbeka* (Moscow–Leningrad, 1950), 2nd ed. in his *Izbrannye trudy* ("Selected Works"), VI (Tashkent, 1967); and "Ulugbek i Savoy Jay Singh," in *Fiziko-matematicheskie nauki v stranah Vostoka*, **1** (1966), 247–256; E. S. Kennedy, "Al-Kāshī's 'Plate of Conjunctions,' " in *Isis*, **38**, no. 2 (1947), 56–59; "A Fifteenth-Century Lunar Eclipse Computer," in *Scripta mathematica*, **17**, no. 1–2 (1951), 91–97; "An Islamic Computer for Planetary Latitudes," in *Journal of the American Oriental Society*, **71** (1951), 13–21; "A Survey of Islamic Astronomical Tables," in *Transactions of the American Philosophical Society*, n.s. **46**, no. 2 (1956), 123–177; "Parallax Theory in Islamic Astronomy," in *Isis*, **47**, no. 1

(1956), 33–53; *The Planetary Equatorium of Jamshid Ghiyāth al-Dīn al-Kāshī* (Princeton, 1960); "A Letter of Jamshīd al-Kāshī to His Father. Scientific Research and Personalities of a Fifteenth Century Court," in *Commentarii periodici pontifici Instituti biblici, Orientalia*, n.s. **29**, fasc. 29 (1960), 191–213; "Al-Kāshī's Treatise on Astronomical Observation Instruments," in *Journal of Near Eastern Studies*, **20**, no. 2 (1961), 98–108; "A Medieval Interpolation Scheme Using Second-Order Differences," in *A Locust's Leg. Studies in Honour of S. H. Tegi-zadeh* (London, 1962), pp. 117–120; and "The Chinese-Uighur Calendar as Described in the Islamic Sources," in *Isis*, **55**, no. 4 (1964), 435–443; M. Krause, "Stambuler Handschriften islamischer Mathematiker," in *Quellen und Studien zur Geschichte der Mathematik, Astronomie und Physik*, Abt. B, **3** (1936), 437–532; P. Luckey, "Die Ausziehung des *n*-ten Wurzel und der binomische Lehrsatz in der islamischen Mathematik," in *Mathematische Annalen*, **120** (1948), 244–254; "Die Rechenkunst bei Ǧamšīd b. Mas'ūd al-Kāšī mit Rückblicken auf die ältere Geschichte des Rechnens," in *Abhandlungen für die Kunde des Morgenlandes*, 31 (Wiesbaden, 1951); and *Der Lehrbrief uber den Kreisumfang von Ǧamšīd b. Mas'ūd al-Kāšī*, A. Siggel, ed. (Berlin, 1953); *Risāla fi'l-'amal bi ashal āla min qabl al-nujūm* ("Treatise on the Operation With the Easiest Instrument for the Planets"), a Persian exposition of al-Kāshī's *Nuzha*—available in MS as Princeton, Univ. 75; and in English trans. with photorepro. in E. S. Kennedy, *The Planetary Equatorium;* B. A. Rosenfeld and A. P. Youschkevitch, "O traktate Qāḍī-Zāde ar-Rūmī ob opredelenii sinusa odnogo gradusa" ("On Qāḍī-Zāde al-Rūmī's Treatise on the Determination of the Sine of One Degree"), in *Istoriko-matematicheskie issledovaniya*, **13** (1960), 533–556; and Mūsā Qāḍī Zāde al-Rūmī, *Risāla fī istikhrāj jaib daraja wāhida* ("Treatise on Determination of the Sine of One Degree"), an Arabic revision of al-Kāshī's *Risāla al-watar wa'l-jaib*—MSS are Cairo, Nat. Bibl. 210 (ascribed by Suter, p. 174, to al-Kāshī himself) and Berlin, Inst. Gesch. Med. Naturw. I.1; Russian trans. in B. A. Rosenfeld and A. P. Youschkevitch, "O traktate Qāḍī-Zāde . . ." and descriptions in G. D. Jalalov, "Giyas ad-Din Chusti (Kashi) . . ." and in Ṣālih Zakī Effendī, *Athār bāqiyya*, I.

Also of value are A. Saidan, "The Earliest Extant Arabic Arithmetic. *Kitāb al-fuṣūl fi al-ḥisāb al-Hindī* of . . . al-Uqlīdisī," in *Isis*, **57**, no. 4 (1966), 475–490; Ṣālih Zakī Effendī, *Athār bāqiyya*, I (Istanbul, 1911); V. A. Shishkin, "Observatoriya Ulugbeka i ee issledovanie" ("Ulugh Bēg's Observatory and Its Investigations"), in *Trudy Instituta istorii i arkheologii Akademii Nauk Uzbekskoy SSR*, V, *Observatoriya Ulugbeka* (Tashkent, 1953), 3–100; S. H. Sirazhdinov and G. P. Matviyevskaya, "O matematicheskikh rabotakh shkoly Ulugbeka" ("On the Mathematical Works of Ulugh Bēg's School"), in *Iz istorii epokhi Ulugbeka* ("From the History of Ulugh Bēg's Age"; Tashkent, 1965), pp. 173–199; H. Suter, *Die Mathematiker und Astronomen der Araber und ihre Werke* (Leipzig, 1900); M. Ṭabāṭabā'ī, "Jamshīd Ghiyāth al-Dīn Kāshānī," in *Amuzish wa Parwarish*, **10**, no. 3 (1940), 1–8 and no. 4 (1940), 17–24; M. J. Tichenor, "Late Medieval Two-Argument Tables for Planetary Longitudes," in *Journal of Near Eastern Studies*, **26**, no. 2 (1967), 126–128; D. G. Voronovski, "Astronomy Sredney Azii ot Muhammeda al-Havarazmi do Ulugbeka i ego shkoly (IX–XVI vv.)" ("Astronomers of Central Asia from Muḥammad al-Khwārizmī to Ulugh Bēg and His School, IX–XVI Centuries"), in *Iz istorii epokhi Ulugbeka* (Tashkent, 1965), pp. 100–172; A. P. Youschkevitch, *Istoria matematiki v srednie veka* ("History of Mathematics in the Middle Ages"; Moscow, 1961), trans. into German as A. P. Juschkewitsch, *Geschichte der Mathematik in Mittelalter* (Leipzig, 1964); and *Zīj-i Ulughbēg* ("Ulugh Bēg's Zīj") or *Zīj-i Sulṭānī* or *Zīj-i jadīd-i Guragānī* ("New Guragānī Zīj"), in Persian, the most important MSS being Paris, BN 758/8 and Tashkent, Inst. Vost. 2214 (a total of 82 MSS are known)—an ed. of the intro. according to the Paris MS and a French trans. are in L. A. Sédillot, *Prolegomènes des tables astronomiques d'Oloug-Beg* (Paris, 1847; 2nd ed., 1853), and a description of the Tashkent MS is in T. N. Kary-Niyazov, *Astronomicheskaya shkola Ulugbeka* (2nd ed., Tashkent, 1967), pp. 148–325.

A. P. YOUSCHKEVITCH
B. A. ROSENFELD

KATER, HENRY (*b*. Bristol, England, 16 April 1777; *d*. London, England, 26 April 1835), *geodesy*.

Although his formal scientific background was limited, Kater, the son of a bakery proprietor, rose to eminence in the Royal Engineers and generally in the world of British science. He was encouraged by his father to become an attorney, but after his father's death in 1794 he abandoned his legal training and served with the British army in India, where he assisted William Lambton in surveying a region of Madras.

His higher education was confined to a brief period, when he was thirty-one, at the Royal Military College at Sandhurst. Kater's most significant scientific contributions consisted of improvements in geodetic instruments, refinements of geodetic measurements, and in the standardization of weights and measures. After 1815, when he was elected a fellow of the Royal Society, he became active in the society's affairs; he served on its council and as vice-president, was the first scientist to become treasurer, and won its Copley Medal in 1817 for his pendulum experiments.

On the basis of the principle enunciated by Huygens that the centers of suspension and oscillation are interchangeable, Kater devised a reversible pendulum (which became known as "Kater's pendulum") with knife edges accurately adjusted to lie at the conjugate points. By using the distance between these points as the "length" in the formula for a simple pendulum, he was able to determine with great accuracy the length of a pendulum beating seconds under specified conditions. He thereby obtained accurate values for *g*, the acceleration due to gravity, at several stations of the Trigonometrical Survey of Great Britain and estimated the ellipticity of the earth. Kater performed these experiments as a member of a committee appointed by the Royal Society in response to a request by the government for assistance in standardizing weights and measures.

He also contributed to the improvement of telescopes by devising floating collimators, and in 1821 he reported the appearance of volcanic action on the moon. His experiments on the relative illuminating powers of Cassegrainian and Gregorian telescopes led him to conjecture on the nature of light (1813). It was one of his rare ventures into the realm of theory. The German astronomer Wilhelm Olbers, attempting to account for the darkness of the night sky, cited Kater's reports to support his belief that space is not perfectly transparent.

BIBLIOGRAPHY

I. Original Works. The papers in which Kater described his pendulum experiments are "An Account of Experiments for Determining the Length of the Pendulum Vibrating Seconds in the Latitude of London," in *Philosophical Transactions of the Royal Society*, **108** (1818), 33–102; and "An Account of Experiments for Determining the Variation in the Length of the Pendulum Vibrating Seconds, at the Principal Stations of the Trigonometrical Survey of Great Britain," *ibid.*, **109** (1819), 337–508. The papers cited by Olbers are "On the Light of the Cassegrainian Telescope, compared with that of the Gregorian," *ibid.*, **103** (1813), 106–212; and "Further Experiments on the Light of the Cassegrainian Telescope compared with that of the Gregorian," *ibid.*, **104** (1814), 231–247. A list of many of Kater's publications is given in the Royal Society's *Catalogue of Scientific Papers, 1800–1863*, III (London, 1869). He also wrote, in collaboration with Dionysius Lardner, *A Treatise on Mechanics* (London, 1830), contributing the chapter "On Balances and Pendulums."

II. Secondary Literature. Many of Kater's papers are summarized in *Abstracts of the Papers. Royal Society of London*, I (London, 1832), II (London, 1833), and III (London, 1837). The fullest biographical articles on Kater are in *Dictionary of National Biography* and Charles Knight's *The English Cyclopaedia*, Biography, III (London, 1856).

HAROLD DORN

IBN KAṬĪR AL-FARGHĀNĪ. See **Al-Farghānī.**

KAUFMANN, NICOLAUS. See **Mercator, Nicolaus.**

KAUFMANN, WALTER (or **Walther**) (*b.* Elberfeld [Wuppertal], Germany, 5 June 1871; *d.* Freiburg, Germany, 1 January 1947), *physics.*

Kaufmann studied at Berlin and Munich, receiving the doctorate at Munich in 1894. In 1896 he was assistant in the Physics Institute at Berlin; three years later he accepted a similar position at Göttingen, later being promoted to *Privatdozent*. Kaufmann became associate professor at Bonn in 1903 and full professor and director of the Physics Institute of Königsberg in 1908; he retired in 1935 as professor emeritus. He then moved to Freiburg, where until his death he served occasionally as visiting professor.

While at Berlin in 1896–1898 Kaufmann began research on the magnetic deflection of cathode rays, attempting a first approximation of the ratio of electron charge to mass (e/m). His most accurate determination of this ratio was 1.865×10^7 cgs/gm.

During this period a controversy arose over whether electrons, believed to be the ultimate constituents of matter, could have "apparent" mass in addition to "real" (material) mass. Apparent mass would be the "electromagnetic mass" gained from the interaction of the moving charge with its own field. Kaufmann's major works were concerned primarily with attempts to measure and characterize this electromagnetic mass of electrons.

During the Göttingen years, 1899–1902, Kaufmann conducted research on the magnetic and electric deflection of radium emanations—then known as Becquerel rays. From the Curies he obtained several radioactive particles of radium chloride and set about measuring the e/m ratio. Since these newly discovered rays had velocities approaching the speed of light, it was assumed that the maximum possible electromagnetic charge was imparted to them. On the basis of his initial e/m measurements in 1901, Kaufmann asserted that the apparent mass was appreciably larger than the real mass—by an estimated magnitude of at least three to one. His successful measurements apparently were made possible by his experimental apparatus, which attained a more complete vacuum than other experimenters could produce in their vacuum tubes.

About the same time a fellow professor at Göttingen, Max Abraham, had formulated a theory of electrons assuming the electromagnetic mass as the total mass of rigid, spherical electrons. Kaufmann adopted this hypothesis. By 1902 Kaufmann produced experimental evidence that the mass of electrons was entirely electromagnetic, that is, that electromagnetic mass constituted the total mass of electrons. More importantly, in these same investigations he presented evidence that the mass of electrons was dependent on their velocity, noting that this dependence was accurately calculated by Abraham's theoretical formula. Thus, a sacrosanct Newtonian principle—that mass was invariant with velocity—was contradicted by Kaufmann's experimental data! By March 1903 Kaufmann confidently declared that not only the Becquerel rays but also the cathode rays consisted of electrons having a mass entirely electromagnetic.

By May 1904 H. A. Lorentz had developed a theory of electrons as being contractable with velocity and in the direction of motion. This view of electrons later became associated with Einstein's theory of relativity. In the same year Alfred Bucherer advanced a view intermediate between Abraham's theory and that of Lorentz. He believed electrons were elastic and could be deformed or contracted in the direction of motion but would maintain constant volume.

During his years at Bonn, Kaufmann undertook a new series of measurements in an attempt to corroborate one of the three rival theories. Upon completion of this work, and after requesting a thorough review by Sommerfeld, he published his results in 1906. He found that both Abraham's and Bucherer's theories were within the limits of experimental error for his measurements, but that the Lorentz-Einstein theory was not. He concluded that Lorentz's theory was thus refuted and that Einstein's theory of relativity was faulty in this respect.

Near the end of 1906 the significance of Kaufmann's measurements was challenged by Max Planck. Developing his own mathematical calculations, Planck reached the tentative conclusion that neither Lorentz's nor Abraham's theory conformed closely to Kaufmann's data. He contended that a different interpretation of Kaufmann's measurements might conceivably place Lorentz's theory in a more favorable position. In 1907 Einstein reviewed Kaufmann's data, noting that these data could conform to relativity theory. He objected to the theoretically limited scope of Abraham's theory—it could not explain as many phenomena as could the theory of relativity. By 1908 Bucherer published experimental data, of greater accuracy than Kaufmann's measurements; these new data supported the Lorentz-Einstein viewpoint.

After 1906 Kaufmann apparently abandoned further investigations in this area. He progressed academically, performing other types of research until his retirement.

As early as 1901 Kaufmann reviewed the history of electron theory in his address "Die Entwicklung des Elektronenbegriffs," delivered at the seventy-third Naturforscher Versammlung at Hamburg. He noted the fruitless efforts in the past to reduce electrical phenomena to mechanical phenomena and advocated reversing the process by attempting to reduce mechanics to electrical principles. Acknowledging the contributions of Lorentz, J. J. Thomson, and W. Wein in this direction, Kaufmann reasoned that if atoms consisted of conglomerates of electrons, then their inertia resulted as a matter of course. In this sense, at so early a date, Kaufmann may be considered a pioneer of twentieth-century physics. The significance of Kaufmann's experimental evidence that electron mass varied with velocity, coupled with his belief that mass could be expressed as essentially electromagnetic phenomena, has rarely been recognized. He outlined a major pathway along which research in twentieth-century physics would be directed.

BIBLIOGRAPHY

I. Original Works. Kaufmann's major works are "Über die Deflexion der Kathodenstrahlen," in *Annalen der Physik und Chemie*, **62** (1897), 588–595, written with Emil Aschkinass; "Die magnetische Ablenkbarkeit electrostatisch beeinflusster Kathodenstrahlen," *ibid.*, **65** (1898), 431–439; "Grundzüge einer elektrodynamischen Theorie der Gasenladungen" (pts. 1 and 2), in *Nachrichten von der Gesellschaft der Wissenschaften zu Göttingen*, Math.-phys. Kl., **1** (1899), 243–259; "Die magnetische und electrische Ablenbarkeit der Bequerelstrahlen und die scheinbare Masse der Elektronen," *ibid.*, **2** (1901), 143–155, translated into English as "Magnetic and Electric Deflectability of the Becquerel Rays and the Apparent Mass of the Electron" (editors' translation), in *The World of the Atom* (edited by Henry A. Boorse and Lloyd Motz), I (New York, 1966), 502–512; "Die Entwicklung des Elektronenbegriffs," in *Physikalische Zeitschrift*, **3** (1901), 9–15, translated into English as "The Development of the Electron Idea," in *Electrician* (8 November 1901), 95–97; "Die elektromagnetische Masse des Elektrons," in *Physikalische Zeitschrift*, **4** (1902), 54–57; Kaufmann's letter (microfilm) to Arnold Sommerfeld (4 Nov. 1905), in the Archive for the History of Quantum Physics of the American Philosophical Society Library, Philadelphia; and "Über die Konstitution des Elektrons," in *Annalen der Physik*, **19** (1906), 487–553.

II. Secondary Literature. See the following, listed chronologically: Max Abraham, "Prinzipien der Dynamik des Elecktrons," in *Annalen der Physik*, **10** (1903), 105–179; Max Planck, "Die Kaufmannschen Messungen der Ablenbarkeit der β-Strahlen in ihrer Bedeutung für die Dynamik der Elecktronen," in *Physikalische Zeitschrift*, **7** (1906), 753–761; Max Planck, "Nachtrag zu der Besprechung der Kaufmannschen Ablenkungsmessungen," in *Verhandlungen der Deutschen Physikalischen Gesellschaft*, **9** (1907), 301–305; Albert Einstein, "Über das Relativitätsprinzip und die aus demselben gezogenen Folgerungen," in *Jahrbuch der Radioaktivität und Elektronik*, **3** (1907), 411–439; and John T. Campbell, "Walter Kaufmann and the Electromagnetic Mass of Electrons," unpub. research paper (The Johns Hopkins University, 1967). Obituaries are W. Kossel, "Walter Kaufmann," in *Naturwissenschaften*, **34** (1947), 33–34; and (author unknown) "Walther Kaufmann," in *Physikalische Blätter*, **3** (1947), 17.

JOHN T. CAMPBELL

KAVRAYSKY, VLADIMIR VLADIMIROVICH (*b.* Zherebyatnikovo, Simbirsk province [now Ulyanovsk oblast], Russia, 22 April 1884; *d.* Leningrad, U.S.S.R., 26 February 1954), *astronomy, geodesy, cartography.*

Kavraysky was born into a family of landed gentry and government officials. In 1903 he graduated from the Simbirsk Gymnasium with a gold medal and entered the mathematical section of Moscow University, where one of his teachers was the mathematician Boleslaw Mlodzeewski, who greatly influenced Kavraysky's development as a scientist. His involvement in the revolution of 1905 forced Kavraysky to leave the University of Moscow; not until 1916 did he graduate with distinction from Kharkov University, where one of his chief professors was Ludwig Struve, director of the astronomical observatory. In the interim Kavraysky had earned his living by teaching mathematics and physics in various educational institutions in Saratov and later in Kharkov. In 1915–1916 he worked as a calculator at the Kharkov University observatory. Even before 1916 he showed a deep interest in astronomy and geodesy and (starting in 1910) published ten articles in various scientific publications. Among them was "Graficheskoe reshenie astronomicheskikh zadach" ("A Graphic Solution of Astronomical Problems"; 1913). In this work he proposed the so-called "Kavraysky grid," a transverse grid of equally spaced azimuthal projection, which was widely distributed and was awarded a prize by the Russian Astronomical Society. In 1915 Kavraysky received the V. Pavlovsky Prize from the Faculty of Physics and Mathematics of Kharkov University for his research on the polarization of light from clear daylight sky.

After graduating from the university Kavraysky entered naval service in Petrograd as assistant chief, and later chief, on the workshop producing nautical instruments for the Main Hydrographical Administration of the Navy. From 1918 to 1926 he served as astronomer at the Administration's observatory. In 1921 he began teaching at the Faculty of Hydrographics of the Naval Academy and in 1922 at the Mining Institute—first mining surveying and later the theory of instruments, astronomy, and mathematical cartography. From 1926 through 1930 Kavraysky was extraordinary astronomer at the Pulkovo observatory, conducting studies in astronomy and geodesy. In 1930–1938 Kavraysky was a member of the newly created Leningrad Institute of Geodesy and Cartography (now the Central Institute of Geodesy, Aerial Photography, and Cartography in Moscow). His scientific authority was so great that he was asked to participate in all important cartographic and geodesic projects. Kavraysky retired in 1949 with the title of engineer–rear admiral. He was awarded the degree of doctor of physico-mathematical sciences (1934) and the title of professor (1935), as well as four orders and several medals.

A full list of Kavraysky's published works includes eighty-six titles; manuscripts of a number of his major unpublished works are preserved in various archives. He was responsible for a number of inventions and treatments of original cartographic projections and nomograms. For the invention of two new nautical instruments, a tiltmeter and a direction finder, he received the State Prize in 1952.

Kavraysky's scientific activity covered many fields, each of which must be considered separately. All of his publications are distinguished by unusual clarity and mastery of exposition. After a statement of the history of a problem and survey of the literature, Kavraysky expressed the mathematical essence of the problem and gave not only an exhaustive solution but also all the information necessary for practical use of the solution, including tables, a list of possible variants, and an estimate of error. In this manner he worked out in detail the mathematical aspects of the introduction into the Soviet Union of a unified system of two-dimensional rectangular coordinates on a Gauss projection for all geodesic and cartographic work. Kavraysky's work for many years was connected with mathematical cartography, the establishment of strict criteria for evaluating cartographic projects, and the development of the most useful projects for various problems. Volume II of the *Izbrannye trudy* ("Selected Works") includes many —but not all—of his works in this area. These works gave him the opportunity to propose original projections for maps of the world and of individual sections of the earth's surface. Many maps and atlases have been published with these projections. Kavraysky also worked on the computations associated with the making of globes. He gave an extraordinarily clear and strict statement of the complex problem of cartographic projection in his articles in *Bolshaya sovetskaya entsiklopedia* (1st ed., XLVII [1940]; 2nd ed., XX [1953]).

A special group of Kavraysky's works is connected with the solution of practical problems of navigation, including the major investigation, *Graficheskie sposoby opredelenia mesta korablya po radiopelengam* ("Graphic Methods of Determining Positions of Ships by Radio Bearings"), in which all problems are solved by the construction of "position lines." Many works in this group are in volume II of *Izbrannye trudy*. This group also includes Kavraysky's numerous inventions and his improvement of nautical instruments.

Part of Kavraysky's theory of astronomical observations is the improvement and simplification of known methods of solving problems in practical astronomy and geodesy, as is a series of original methods of simultaneous determination of time and latitude, which proved highly effective in the high northern latitudes (from 60° to 80°). Kavraysky developed methods of determining locations near the pole for the first Soviet expedition to the North Pole and developed the necessary tables and nomograms for this project. He contributed to the Pulkovo observatory's *Vvedenie v prakticheskuyu astronomiyu* ("Introduction to Practical Astronomy"; 1936). Related to this work is a group of works on the theory and practice of the use of astronomical and geodesic instruments.

In essence, all of Kavraysky's scientific works were devoted to the solution of problems of navigation.

Kavraysky devoted the last years of his life to the preparation of *Rukovodstva po matematicheskoy kartografii* ("Guide to Mathematical Cartography"), which remained unfinished. But all that he had written was included in volume II of the *Izbrannye trudy*.

BIBLIOGRAPHY

I. ORIGINAL WORKS. Most of Kavraysky's writings are in *Izbrannye trudy* ("Selected Works"), 2 vols. (Moscow, 1956). Vol. I, *Astronomia i geodezia* has a complete bibliography (pp. 355–358) and contains the following works: "Graficheskoe reshenie astronomicheskikh zadach" ("A Graphic Solution of Astronomical Problems"; 1913), pp. 13–138; "Zapiski po sferoidicheskoy geodezii" ("Notes on Spheroidal Geodesy"; 1944), pp. 139–248; "Obobshchenny sposob liny polozhenia" ("A Generalized Method of Lines of Position"; 1943), pp. 249–282; "Linii polozhenia i ikh primenenie" ("Lines of Position and Their Use"; 1939), pp. 283–306; and "Teoria opredelenia polozhenia tochki na poverkhnosti" ("Theory of Determination of Positions of Points on a Surface"; 1956), pp. 307–350.

Vol. II, *Matematicheskaya kartografia*, is in 3 pts.: "Obshchaya teoria kartograficheskikh proektsy" ("General Theory of Cartographical Projections"; 1958); "Konicheskie i tsilindricheskie proektsii, ikh primenenie" ("Conic and Cylindrical Projections, Their Use"; 1959); and "Perspektivnye, krugovye i drugie vazhneyshie proektsii. Navigatsionnye zadachi" ("Perspective, Circular and Other Major Projections. Navigational Problems"; 1960).

He also contributed to *Sovmestnoe opredelenie vremeni i shiroty po sootvetstvuyushchim vysotam zvyezd s efemeridami yarkikh zvyezd dlya shiroty ot +60° do +80°, vychislennymi Astronomicheskim Institutom* ("Simultaneous Determination of Time and Latitude From Altitudes of Stars Corresponding to Ephemerides of Bright Stars in Latitudes From +60° to +80°, Computed by the Astronomical Institute"; Moscow–Leningrad, 1936).

II. SECONDARY LITERATURE. See M. K. Venttsel, "Kratky ocherk istorii prakticheskoy astronomii v Rossii i v SSSR: Razvitie metodov opredelenia vremeni i shiroty" ("Brief Sketch of the History of Practical Astronomy in Russia and the U.S.S.R.: Development of Methods of Determining Time and Latitude"), in *Istoriko-astronomicheskie issledovaniya*, **2** (1956), 7–137, see 113–119; A. P. Yushchenko, "Vladimir Vladimirovich Kavraysky," in Kavraysky's *Izbrannye trudy*, 5–12; and K. A. Zvonarev, "Vladimir Vladimirovich Kavraysky," in P. G. Kulikovsky, ed., *Istoriko-astronomicheskie issledovaniya*, **9** (1966), 261–285.

P. G. KULIKOVSKY

KAY, GEORGE FREDERICK (*b.* Virginia, Ontario, 14 September 1873; *d.* Iowa City, Iowa, 19 July 1943), *geology*.

Kay's forebears were English and Scotch-Irish pioneers in Ontario. Born on the family farm, he was the fifth of seven children of Joseph Sidney and Elizabeth Marshall Rae Kay. He married Bethea Hopper of Paisley, Ontario; they had two sons and a daughter.

At the University of Toronto (B.A., 1900; M.S., 1901) and the University of Chicago (Ph.D., 1914, under Joseph Iddings), Kay was trained in mineralogy, petrology, and economic geology. He did exploration work in Ontario and—for the U.S. Geological Survey—in Colorado, Oregon, California, and Alaska. Although his first faculty post was at the University of Kansas (1904–1907), he was associated with the University of Iowa for thirty-six years as professor of geology (1907–1943), head of the department of geology and state geologist (1911–1934), and dean of the College of Liberal Arts (1917–1941). He was also an officer of the Geological Society of America and the American Association for the Advancement of Science.

Kay's interests were scientific, practical, and philosophical in nature, reflecting his Calvinist upbringing and his lifelong commitment to education (even while a dean he continued teaching both graduate and undergraduate courses). His scientific contributions in both his early and later years were significant.

Before 1910, with J. S. Diller, Kay mapped nickel, copper, and gold deposits in Oregon. Then of little interest, the host rocks (Franciscan) are now of primary importance in new concepts of global tectonics. In Iowa, Kay became interested in Pleistocene deposits. He proposed the term "gumbotil" for dark clays he believed were formed by chemical weathering of till (some he described are now thought to be accretion gleys). Using gumbotils and time estimates based on weathering rates, Kay correlated drift sheets and developed a series of absolute dates for glacial and interglacial stages. His work is still evident in the stratigraphic nomenclature of the Pleistocene, especially that of the upper Mississippi Valley. His final scientific effort was the preparation of a three-part monograph, based on his own work and that of others, on the Pleistocene geology of Iowa.

BIBLIOGRAPHY

I. Original Works. Kay's major writings are "Nickel Deposits of Nickel Mountain, Oregon," in *Bulletin of the United States Geological Survey*, no. 315 (1907), 120–127; "Gold-quartz Mines of the Riddle Quadrangle, Oregon," *ibid.*, no. 340 (1908), 134–147; "Notes on Copper Prospects of the Riddle Quadrangle, Oregon," *ibid.*, no. 340 (1908), 152; "Mineral Resources of the Grants Pass Quadrangle and Bordering Districts, Oregon," *ibid.*, no. 380 (1909), 48–79, written with J. S. Diller; "Gumbotil, a New Term in Pleistocene Geology," in *Science*, n.s. **44** (1916), 637–638; "The Origin of Gumbotil," in *Journal of Geology*, **28** (1920), 89–125, written with J. N. Pearce; "Description of the Riddle Quadrangle, Oregon," in *Geologic Atlas of the United States*, no. 218 (1924), written with J. S. Diller; "The Relative Ages of the Iowan and Illinoian Drift Sheets," in *American Journal of Science*, 5th ser., **16** (1928), 497–518; "Classification and Duration of the Pleistocene Period," in *Bulletin of the Geological Society of America*, **42** (1931), 425–466; "Eldoran Epoch of the Pleistocene Period," *ibid.*, **44** (1933), 669–674, written with M. M. Leighton; *The Pleistocene Geology of Iowa:* pt. 1 (repr.), "The Pre-Illinoian Pleistocene Geology of Iowa," written with E. T. Apfel, in *Report of the Iowa Geological Survey*, **34** (1929), 1–304; pt. 2 (repr.), "The Illinoian and Post-Illinoian Pleistocene Geology of Iowa," written with J. B. Graham, *ibid.*, **38** (1943), 11–262; pt. 3, "The Bibliography of the Pleistocene of Iowa," in *Report of the Iowa Geological Survey* (1943), 1–55.

II. Secondary Literature. See M. M. Leighton, "The Naming of the Subdivisions of the Wisconsin Glacial Stage," in *Science*, **77** (1933), 168; R. V. Ruhe, *Quaternary Landscapes in Iowa* (Ames, Iowa, 1969); A. C. Trowbridge, "Memorial to George Frederick Kay," in *Proceedings. Geological Society of America* (1944), 169–176; "George Frederick Kay, 1873–1943," in *Proceedings of the Iowa Academy of Science*, **51** (1944), 109–111; and "Discussion, Accretion-Gley and the Gumbotil Dilemma," in *American Journal of Science*, **259** (1961), 154–157; H. E. Wright, Jr., and David G. Frey, eds., *The Quaternary of the United States* (Princeton, 1965), pp. 8, 29–41, 527, 759–762.

Sherwood D. Tuttle

KAYSER, HEINRICH JOHANNES GUSTAV (*b.* Bingen, Germany, 16 March 1853; *d.* Bonn, Germany, 14 October 1940), *physics.*

Kayser was the son of Heinrich Kayser, a former lord of the manor, and the former Amelie von Metz. He attended the *Pädagogium* in Halle and the Sophie Gymnasium in Berlin, where he received a diploma on 4 March 1872. After a year of traveling Kayser studied from April 1873 until March 1879 in Strasbourg, Munich, and Berlin, mainly with Kundt, Helmholtz, and Kirchhoff. On 13 March 1879 he graduated as Ph.D. with the thesis "Der Einfluss der Intensität des Schalles auf seine Fortpflanzungsgeschwindigheit."

In the previous year Kayser, initially together with Heinrich Hertz, had become assistant to Helmholtz at the Berlin Physical Institute. He remained there until the fall of 1885. On 26 September 1881 he gained qualification as academic lecturer with the dissertation "Über die Verdichtung von Gasen an Oberflächen in ihrer Abhängigkeit von Druck und Temperatur." In 1885 he was appointed professor of physics at the Hannover Technical University. Here, working with Carl Runge, he began his investigations in the field of spectroscopy. In his *Handbuch der Spektroskopie* (1900) he described the purpose of his investigations:

> It is certain that the light is produced by the motions of the molecules or of the particles or of their electrical charges. It was expected that chemical elements would be similar to a certain extent in the structure of their spectra according to their periodic classification. Balmer was the first who derived a real result of regularity in the distribution of the wave numbers of the spectral lines of hydrogen. It was hoped that similar laws would be detected for other elements.

Kayser and Runge began these investigations at about the same time that Rydberg began working along the same lines. Kayser and Runge determined the spectra anew, using a Rowland concave grating, and found the results to be much more reliable from this method; Rydberg evaluated the existing older measurements anew. The investigations showed that for many elements a regular structure could indeed be demonstrated. For the alkali metals (lithium, sodium, potassium, rubidium, cesium), all known spectra lines could be settled at three series described very accurately by equations of the same structure; these formulas were also interrelated. Furthermore, an important relation between the atomic weight and the structure of the spectra was discovered. Kayser and his co-worker learned not only that the spectra of these five related elements were ordered by the same plan, but also that they changed with perfect regularity, according to the increasing atomic weight.

Kayser and Runge next investigated, with similar results, the alkali earths and also some metals of the groups IB and IIB of the periodic table of elements. The regularity was not so perfect in this case, however. Kayser said that the number of irregular lines grows as one proceeds in the natural system of elements. He emphasized that the formulas he and his associate had found in Hannover, as well as those of Rydberg, were only empirical and far from the discovery, through the structure of the spectra, of the behavior of the atoms. In his criticism of Rydberg, Kayser was always willing to acknowledge the merits of Rydberg's work. From the vantage of today, the work of Rydberg and of Kayser and Runge was indispensable to the atomic theory brought forth twenty-five years later by Rutherford and Bohr. Although Kayser provided the solid experimental foundation for this theory with his experiments—he was the experimenter, Runge the theorist—Rydberg, full of ideas and speculations, was more successful in formulating the spectra equations; hence the name Rydberg constant. Nevertheless, Kayser and Runge's lists of the exact frequencies of many spectral lines guarantee their place in the history of science.

Kayser and Rydberg also collaborated in the discussion with Pickering concerning the spectrum of ζ Puppis. The latter discovered there some spectral lines very near to the Balmer series. Bohr was later able to explain that these lines are produced by ionized helium. At the end of the nineteenth century, the lines occurring in O stars were attributed to hydrogen and called protohydrogen.

In 1894 Kayser was appointed successor to Heinrich Hertz as professor of physics at the University of Bonn, at this time an outstanding professorial chair for this discipline. During Kayser's tenure, the Institute of Physics in Bonn became a center of spectroscopic investigations, and Kayser obtained a new and modern building after more than a decade. Previous to his retirement, Kayser wrote his *Handbuch der Spektroskopie*, comprising eight volumes. Although he was greatly interested in astrophysics, the field is not treated in this work. Even today, the *Handbuch* is a remarkably comprehensive achievement for a single author, compiling countless facts and aspects of spectroscopy beginning with Newton. The first astrophysical investigations by Kayser were of spectra of comets and variable stars. He later published his ideas concerning the temperature of stars, and was one of the first to attempt an explication of novae in terms of radiation processes (1912). In addition to teaching, he wrote a textbook for students which had several editions.

Kayser was a member of the International Union for Solar Research and an honorary member of the Royal Institution of Great Britain, and he belonged to several foreign academies. In 1912 he received an honorary doctorate of jurisprudence from the University of St. Andrews. He had widespread interests in different fields, notably in Greek and Roman art, and he made excellent photographs during journeys to Italy and Greece. In 1887 Kayser married Auguste Hofmann, surviving her by nearly twenty-five years.

BIBLIOGRAPHY

Kayser's original works include *Lehrbuch der Spektralanalyse* (Berlin, 1883); "Spektren der Elemente," in *Abhandlungen der Konigl. Preussischen Akademie der Wissenschaften in Berlin* (1888–1893), written with Carl Runge; "Die Dispersion der Luft," *ibid.* (1893), written with Runge; "Bogenspektren der Elemente der Pb-Gruppe," *ibid.* (1897); *Handbuch der Spektroskopie*, 8 vols. (Leipzig, 1900–1932), vols. VII and VIII written with Heinrich Konen; and *Lehrbuch der Physik für Studierende* (Stuttgart, 1890; 6th ed., 1921). See also various articles by Kayser in *Astronomische Nachrichten*, **134** (1894), **135** (1894), **162** (1903), **191** (1912); and *Astrophysical Journal*, **1**, **4**, **5**, **7**, **13**, **14**, **19**, **20**, **26**, **32**, **39** (1895–1914).

For obituaries see R. Frerichs, in *Naturwissenschaften*, **29** (1941), 153–155; F. Paschen, in *Physikalische Zeitschrift*, **41** (1941), 429–433; and H. Crow, in *Astrophysical Journal*, **94** (1941), 5–11.

H. C. Freiesleben

KECKERMANN, BARTHOLOMEW (*b.* Danzig [now Gdansk], Poland, 1571/73; *d.* Danzig, 25 July 1609), *astronomy, mathematics, methodology.*

Keckermann, the son of George and Gertrude Keckermann, was educated by Jacob Fabricius, rector of the Danzig Gymnasium, who imbued him with strict Calvinist doctrine and a detestation of Anabaptists and Catholics. In 1590 he was sent to Wittenberg University, then to Leipzig for a semester (1592), and finally to Heidelberg (1592). In the latter city he obtained his M.A. in 1595, afterward being appointed tutor and then lecturer in philosophy. The chair of Hebrew was conferred on him in 1600. Keckermann's growing reputation had resulted in an invitation in 1597 from the Danzig senate to return to that city's Gymnasium. Although he declined this offer, preferring to work toward his doctorate of divinity at Heidelberg (obtained 1602), Keckermann accepted a later invitation and became professor of philosophy at Danzig in 1602. There he remained until he died, "worn out with mere scholastic drudgery," in 1609.

At the Danzig Gymnasium, Keckermann tried to implement a Ramist reform of the curriculum with a scheme intended to give youths an encyclopedic education within three years. In this new *cursus philosophicus* the first year was devoted to logic and physics, the second year to mathematics and metaphysics, and the third to ethics, economics, and politics. The key to this syllabus was Keckermann's systematic method, which was influenced by the view of Petrus Ramus that the correct approach to a discipline is topical and analytical, rather than merely historical or narrative.

Keckermann was not a pure Ramist, however, and was most sympathetic to the progressive Aristotelian views outlined in Jacopo Zabarella's *De methodis.* Like Zabarella, Keckermann believed that much of the effort being devoted to the textual analysis of Aristotle (effort that led to the prolonging of the *cursus philosophicus*) could be better diverted to developing new Aristotelian methods and analytical systems. He thus drew heavily on both Aristotelian and Ramist ideas for his philosophical and logical *Praecognita*, in which he gave the first theoretical discussion of systems (the set of precepts characterizing each science).

In his lectures at Danzig, he made abundant use of his systematic method. In its published form the typical lecture course is entitled *Systema* . . .. Among the published *systeme* are treatments of logic, politics, physics, metaphysics, ethics, theology, Hebrew, geography, geometry, astronomy, and optics. These works are philosophical and pedagogical in character and contain little material of any scientific value; certainly there is no scientific originality in them. Their main interest lies perhaps in their illustration of the content of university courses in mathematics and natural philosophy in the early years of the seventeenth century.

Keckermann's *Systema physicum*, a set of lectures delivered in 1607 and published in 1610, discussed physics, astronomy, and natural philosophy, all in largely Aristotelian terms. The author differed from most Peripatetics by describing the four elements as less complete and perfect in form than the mixed bodies. Since elements are not completely and individually sui generis, Keckermann found it plausible that they should be capable of rapid transmutation into one another.

The long discussion of comets has a theological flavor, which is not surprising in view of Keckermann's religious training and devotion. Comets are conventionally defined as terrestrial exhalations produced by action of the planets in the supreme aerial region. God then encourages angels, or permits demons, to join with the comet in producing extraordinary terrestrial effects. God's unpredictable choice of angels or demons for the task explains the good and bad effects of comets, although some allowance must be made for the comet's relation to the stars and planets. Predominantly, however, the effects of comets are malign and indicate divine wrath.

There are serious gaps and errors in the *Systema physicum*. The vacuum is not adequately discussed in terms of Aristotelian motion and place. Keckermann also maintained that water contracts when frozen. This mistake was criticized in 1618 by Isaac Beeckman, who remarked that either a simple experiment or common sense would have exposed the fallacy. Keckermann's use of experiment—or, rather, of experience—is in fact very crude and imprecise.

The *Systema compendiosum totius mathematices* (1617) consists of lectures, read in 1605 and other unspecified years, on geometry, optics, astronomy, and geography; it was intended to form the second year of the *cursus philosophicus.* The geometry section is elementary, although it describes the duplication of the cube and other problems. The main influence of this section—and of the whole *Systema*—seems to have been Ramus, *Scholarum mathematicarum* (1569). Among the geometrical authors cited are Regiomontanus, Albrecht Dürer, and Wilhelm Xylander. For the section on optics Keckermann drew on Arab writers, Witelo, and Peter Apian. The astronomical section follows Regiomontanus and Georg Peurbach but also cites Copernicus, Erasmus Reinhold's *Prutenic Tables*, and Tycho Brahe. Keckermann remarks that while the Ptolemaic theory of the *primum mobile* is certain, the theory of the planets has defects which compelled Copernicus and Brahe to try to reduce the planetary motions to "greater certainty and superior method" (1621 ed., p. 349). Keckermann disappointingly failed to follow up this interesting statement, although later (p. 357) he cites with approval Copernicus' criticism (*De revolutionibus*, III, cap. 13) of the Ptolemaic treatment of the solar year. There is, however, no real examination of the Copernican system.

BIBLIOGRAPHY

I. Original Works. The collected ed. of Keckermann's works is *Operum omnium quae extant*, 2 vols. (Geneva,

1614), which includes his religious works as well as the *Systeme* and *Praecognita*. The *Systema physicum septem libris adornatum* . . . first appeared at Danzig in 1610. A 3rd ed. was published at Hanau in 1612. The various parts of the *Systema mathematices* were published separately soon after Keckermann's death. They were collected into the *Systema compendiosum totius mathematices* . . . (Hanau, 1617, 1621; Oxford, 1661).

II. SECONDARY LITERATURE. The main source for Keckermann's life is the nearly contemporary biography (1615) by Melchior Adam, *Vitae Germanorum philosophorum* (3rd ed., Frankfurt, 1706), pp. 232–234. The work of Keckermann is surveyed in Bronisław Nadolski, *Zycie i działalność naukowa uczenego gdańskiego Bartlomieja Keckermanna; studium z dziejów Odrodzenia na Pomorzu* (Torun, 1961). W. H. van Zuylen, *Bartholomäus Keckermann: Sein Leben und Wirken*, Tübingen dissertation (Leipzig, 1934), concentrates on the theological works.

For notes on Keckermann's physics, see Lynn Thorndike, *A History of Magic and Experimental Science*, 8 vols. (New York, 1923–1958), VII, 375–379. For Keckermann as a systematist, see Otto Ritschl, *System und systematische Methode in der Geschichte des wissenschaftlichen Sprachgebrauchs und der philosophischen Methodologie* (Bonn, 1906), pp. 26–31; and Neal W. Gilbert, *Renaissance Concepts of Method* (New York, 1960), pp. 214–220. Beeckman's criticism appears in Cornelis de Waard, ed., *Journal tenu par Isaac Beeckman de 1604 à 1634*, I (The Hague, 1939), 215; see also II, 253.

PAUL LAWRENCE ROSE

KEELER, JAMES EDWARD (*b.* La Salle, Illinois, 10 September 1857; *d.* San Francisco, California, 12 August 1900), *astronomy*.

When he died at the age of forty-two, Keeler was the leading astrophysicist in the United States. He is best remembered today for his spectroscopic proof that the rings of Saturn are composed of small particles moving independently, and for his discovery of the abundance of spirals among the nebulae.

His father was William F. Keeler, who served as a paymaster in the U.S. Navy during the Civil War; his mother, Anna, was the daughter of Henry Dutton, onetime governor of Connecticut. Keeler attended public schools in La Salle until 1869, when the family moved to Mayport, Florida. In this small settlement a few miles east of Jacksonville, Keeler helped his father and older brother to build the house they lived in. He had no formal secondary education. At age eighteen, Keeler sent away for two lenses and made a telescope. This was the beginning of his "Mayport Astronomical Observatory"; other equipment included a quadrant, chronometer, and meridian circle—all homemade.

Providentially, Charles H. Rockwell, of Tarrytown, New York, learned of Keeler's interest in astronomy and made it possible for him—by then twenty years old—to enroll as a freshman at the newly opened Johns Hopkins University in Baltimore, Maryland. Keeler earned part of his expenses there as assistant to Charles S. Hastings, professor of physics, and with him took part in the U.S. Naval Observatory expedition to Central City, Colorado, to observe the solar eclipse of 29 July 1878.

Upon receiving his B.A. degree in June 1881, Keeler went to Pittsburgh, Pennsylvania, as assistant to Samuel P. Langley, who was then director of the Allegheny Observatory. Keeler arrived just in time to take part in the expedition to Mount Whitney, California (July, 1881), when Langley's new bolometer was used to measure the infrared radiation of the sun.

In 1883 Keeler went abroad for a year, to study at Heidelberg under G. H. Quincke and at the University of Berlin under H. L. F. von Helmholtz. He then returned to Allegheny, to remain until 1886, when he became the first professional astronomer to reside on Mount Hamilton, where the Lick Observatory was under construction; his main job was to set up a time service for distribution from there to various commercial interests.

When the University of California took formal possession of Lick in 1888, Keeler remained, with the title of astronomer. Here it was that he used the thirty-six-inch refracting telescope and a spectroscope incorporating one of Henry A. Rowland's concave gratings to measure (1890) the wavelengths of the bright lines in nebular spectra. His accuracy was sufficient to show that—like stars—gaseous nebulae have measurable motions toward or away from the earth. The precision of these measurements also helped to show that some of the wavelengths did not correspond to any atomic transitions known to occur on earth; this led to Keeler's involvement in the early stages of the "nebulium" controversy, which was finally resolved by Ira S. Bowen in 1927.

In June of 1891 Keeler married Cora Slocomb Matthews, niece of the president of the Lick board of trustees. That same year he left Lick for seven years, having been appointed successor to Langley as director of the Allegheny observatory. Langley had become secretary of the Smithsonian Institution. During this period of his life Keeler designed a spectrograph—differing from a spectroscope in that spectral lines are recorded photographically rather than being located by eye—and with it obtained (1895) the classic proof of James Clerk Maxwell's theoretical prediction that the rings of Saturn are meteoritic in nature.

Returning to Lick in 1898 to succeed Edward S. Holden as director, Keeler was able to put into use the thirty-six-inch Crossley reflecting telescope, which had defied earlier astronomers (it was difficult to operate because of an unusual mounting, designed, furthermore, for its original location in England). With the Crossley, Keeler took a series of photographs that revealed how greatly spiral nebulae—later identified as exterior galaxies—outnumbered all the other hazy objects detectable in the sky. He was awaiting the completion of a slitless spectrograph he had designed for use with this telescope when he had a heart attack and died.

Keeler was granted an honorary Sc.D. by the University of California in 1893. He was elected a fellow and foreign associate of the Royal Astronomical Society (London) in 1898, and awarded the Rumford Medal of the American Academy of Arts and Sciences that same year, for his applications of spectroscopy to astronomy. On the same basis he received the Henry Draper medal from the National Academy of Sciences in 1899 and was elected to membership in 1900. He was coeditor with George Ellery Hale of the *Astrophysical Journal* from its inception.

BIBLIOGRAPHY

I. Original Works. Abstracts from Keeler's "Records of Observations Made at the Mayport Observatory" are included in the second obituary notice by Campbell (see below); the original work seems to have disappeared.

Keeler's first published work, a description of the solar corona during the eclipse of 29 July 1878, appeared as "Addendum E of Appendix III," to *Astronomical and Meteorological Observations Made During the Year 1876 at the U.S. Naval Observatory* (Washington, 1880), pp. 170–173; his second work, describing what he saw during the transit of Venus on 5 December 1882, was "The Ring of Light Surrounding Venus," in *Sidereal Messenger*, **1** (1882–1883), 292–294.

Products of Keeler's early days at Lick were "The Time Service of the Lick Observatory," *ibid.*, **6** (1887), 233–248; and "First Observations of Saturn With the 36-Inch Equatorial of Lick Observatory," *ibid.*, **7** (1888), 79–83; the latter records the very first use of that great refracting telescope.

Keeler's second solar eclipse expedition was one he led from Lick to Bartlett Springs, California; it is described in "Total Eclipse of the Sun of January 1, 1889," in *Contributions From the Lick Observatory*, **1**, pt. 2 (1889), 31–55. His work on the radial velocities of nebulae with bright line spectra appeared as "On the Motions of the Planetary Nebulae in the Line of Sight," in *Publications of the Astronomical Society of the Pacific*, **2** (1890), 265–280, with an expanded version, "Spectroscopic Observations of Nebulae," in *Publications of the Lick Observatory*, **3** (1894), 161–229. Criticism of these results can be found in the verbatim account of the Royal Astronomical Society's meeting of 8 May 1891, in *Observatory*, **14** (1891), 209–213; for Keeler's reply, see "Elementary Principles Governing the Efficiency of Spectroscopes for Astronomical Purposes," in *Sidereal Messenger*, **10** (1891), 433–453.

While he served as director of the Allegheny observatory, Keeler published forty-eight papers, including "Physical Observations of Mars, Made at the Allegheny Observatory in 1892," in *Memoirs of the Royal Astronomical Society*, **51** (1892–1895), 45–52, with 12 sketches tipped in: "A Spectroscopic Proof of the Meteoritic Constitution of Saturn's Rings," in *Astrophysical Journal*, **1** (1895), 416–427; and "The Importance of Astrophysical Research and the Relation of Astrophysics to the Other Physical Sciences [Address Delivered at the Dedication of Yerkes Observatory, 21 Oct. 1897]," *ibid.*, **6** (1897), 271–288, reprinted in *Science*, n.s. **6** (19 Nov. 1897), 745–755; this address provides a good summary of the current state of astrophysics and also displays the clarity of Keeler's thinking.

Among Keeler's publications while director of Lick are "The Crossley Reflector of the Lick Observatory," in *Astrophysical Journal*, **11** (1900), 325–349, reprinted in *Publications of the Astronomical Society of the Pacific*, **12** (1900), 146–167, and also in *Publications of the Lick Observatory*, **8** (1908), see below; and "Photograph of the Trifid Nebula, in *Sagittarius*," in *Publications of the Astronomical Society of the Pacific*, **12** (1900), 89–90, with photogravure repro. facing p. 89. Keeler's program for work with the Crossley was completed after his death by Charles Dillon Perrine and appeared as "Photographs of Nebulae and Clusters, Made with the Crossley Reflector, by James Edward Keeler, Director of the Lick Observatory, 1898–1900," in *Publications of the Lick Observatory*, **8** (1908), 1–46, followed by 70 plates.

A list of 126 publications by Keeler is included in Campbell's first obituary notice and reprinted in Hastings' biographical memoir (see below for both).

II. Secondary Literature. Charles Sheldon Hastings wrote the entry on Keeler in *Biographical Memoirs. National Academy of Sciences*, **5** (1905), 231–246, which includes a portrait facing p. 231 and a list of publications, pp. 241–246. This memoir is based on obituaries by John Alfred Brashear (covering Keeler's days at Allegheny) in *Popular Astronomy*, **8** (1900), 476–481; William Wallace Campbell (Lick) in *Astrophysical Journal*, **12** (1900), 239–253, including a list of publications; and George Ellery Hale in *Science*, n.s. **12** (1900), 353–357, reminiscences of their long association.

Other obituaries are those by William Wallace Campbell, in *Publications of the Astronomical Society of the Pacific*, **12** (1900), 139–146, with excerpts from the Mayport Observatory records; and by Charles Dillon Perrine in *Popular Astronomy*, **8** (1900), 409–417.

SALLY H. DIEKE

KEESOM, WILLEM HENDRIK (*b.* Texel, Netherlands, 21 June 1876; *d.* Oegstgeest, Netherlands, 24 March 1956), *physics.*

The son of a farmer, Keesom studied physics at Amsterdam University, where J. D. van der Waals was one of his teachers. An excellent student, he received the doctorate with a thesis on the isotherms of oxygen and carbon dioxide mixtures in 1904. He became a close collaborator of Kamerlingh Onnes at the University of Leiden, assisting him, for instance, in the liquefaction of helium (1908) and in the writing of a comprehensive treatise on the equation of state for the *Enzyklopädie der mathematischen Wissenschaften* (1912).

In 1917 Keesom became a teacher, and the next year a professor, of physics at the veterinary school in Utrecht (later incorporated into the university). In 1923 he returned to Leiden to occupy one of the two chairs of experimental physics, the other being occupied by W. J. de Haas in 1924.

At Utrecht, Keesom succeeded in finding a connection between the X-ray diffraction pattern and the intermolecular distance in liquids. This is a good example of his tendency to combine theoretical and experimental methods in order to clarify the picture of liquids and compressed gases without mathematical sophistications. (The introduction of the radial distribution function for the intermolecular distance is due to others.)

As a director of the Kamerlingh Onnes laboratory at Leiden, Keesom continued the tradition of low-temperature research, especially (although by no means exclusively) on helium. He was the first to solidify it, by applying external pressure to overcome "repulsion" between the atoms, which in all other crystals is overcome by mutual attraction. Since the saturated vapor pressure is much lower than the pressure needed for crystallization, there is no triple point and the liquid state extends down to absolute zero. (See Figure 1.)

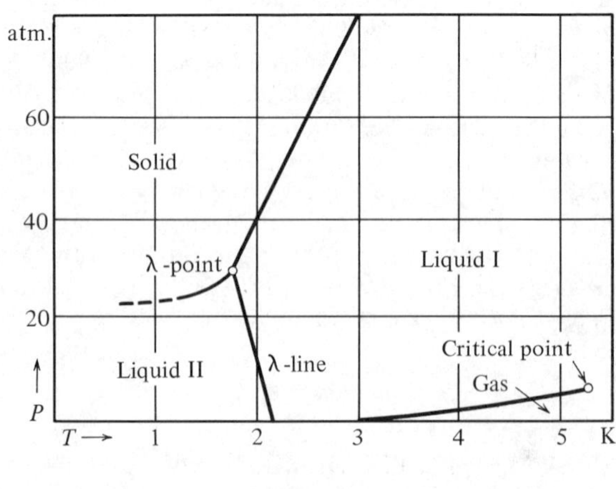

FIGURE 1

In the course of further work in 1927–1933 it gradually became clear that at very low temperatures the liquid (II) differed fundamentally in its behavior from ordinary liquids and from high-temperature liquid helium (I). The two liquid states are separated by the λ line indicated in Figure 1. At this line, for instance, the heat capacity changes abruptly ("second-order transition") and the internal friction disappears, leading to "superfluidity" (more or less like the superconductivity in a number of metals). Theoretical explanations, founded on quantum statistics and the third law of thermodynamics, were presented later. The essence is that below a certain temperature both the liquid and the crystal become so fixed in the ground state that entropy is practically zero, and no further changes occur in it. Such a "thermally degenerate" system no longer behaves as an (irreversible) thermal system but as a (reversible) mechanical system.

In 1942 Keesom wrote a standard work on helium, and in 1945 he retired. Most of his work is to be found in *Physica* (The Hague), *Proceedings of the Section of Sciences* of the Royal Netherlands Academy of Sciences, and *Communications from the Kamerlingh Onnes Laboratory of the University of Leiden*. He received foreign as well as national honors and was elected a member of the Royal Netherlands Academy in 1924.

BIBLIOGRAPHY

Keesom's writings include *Isothermen van mengsels van zuurstof en koolzuur* (Amsterdam, 1904), his thesis; "Die Zustandsgleichung," in *Enzyklopädie der mathematischen Wissenschaften*, V, 1 (1912), 615–945, written with Kamerlingh Onnes; and *Helium* (Amsterdam, 1942).

A biography is in *Jaarboek van de K. Akademie van wetenschappen gevestigd te Amsterdam* (1956–1957), p. 225.

J. A. PRINS

KEILIN, DAVID (*b.* Moscow, Russia, 21 March 1887; *d.* Cambridge, England, 27 February 1963), *biochemistry, parasitology.*

Born of Polish parents temporarily residing in Moscow, Keilin received his early schooling at home and then attended the Gorski Gymnasium in Warsaw from 1897 to 1904. After graduation he embarked on premedical studies at the University of Liège, but in 1905 he decided to continue his studies at Paris, where he came under the influence of Maurice Caullery, the distinguished parasitologist at the Laboratoire d'Évolution des Êtres Organisés. In this laboratory Keilin began research dealing with the life cycle of the fly *Pollenia rudis* and published his first paper in 1909, which formed a major part of his doctoral dissertation, presented at the Sorbonne in 1917.

By 1914 Keilin's observations had become well-known, and after the outbreak of World War I he was invited to become research assistant in the laboratory of G. H. F. Nuttall, Quick professor of biology at the University of Cambridge. Keilin occupied this post from 1915 to 1920, when he was elected to a Beit fellowship. In 1921 the Quick laboratory was incorporated into the newly established Molteno Institute for parasitology and Keilin accompanied Nuttall in this move. In 1925 Keilin became university lecturer in parasitology and in 1931 he succeeded Nuttall as Quick professor and as director of the Molteno Institute. Keilin retired from these posts in 1952, but continued active work until his death. Among his many honors was the Copley Medal of the Royal Society (1952).

As a parasitologist Keilin became known through his work on the life cycle of parasitic and free-living Diptera. In his entomological papers (he published about seventy up to 1931), he described eleven protists that were placed in new species or new genera. His interest in the adaptation of the dipterous larvae to the parasitic mode of life led him to a closer study of their respiration, which brought him to biochemical studies on respiratory mechanisms. After 1930 his entomological efforts gave way almost entirely to biochemical work. But as editor (1934–1963) of the journal *Parasitology*, he continued to influence that field.

Keilin's entry into biochemistry came during the course of studies (1922–1924) on the life cycle of the horse botfly (*Gasterophilus intestinalis*). In 1924 he observed with the aid of a microspectroscope the presence of a four-banded absorption spectrum in the muscles of this insect and others and also found a similar spectrum in aerobic microorganisms. The observation that the four-banded spectrum disappeared on shaking the cell suspension with air, and reappeared shortly afterward, led Keilin to conclude that he was dealing with an intracellular respiratory pigment that is widely distributed in nature. He named this pigment cytochrome and in his first paper (1925) on this subject he suggested that the three components of cytochrome (*a*, *b*, and *c*) are iron-porphyrin compounds that serve as catalysts of oxidation-reduction processes in respiring cells, being alternately oxidized by oxygen and reduced by the action of enzymes that dehydrogenate metabolites.

This concept of the role of cytochrome provided a synthesis of the conflicting views of T. Thunberg and H. Wieland, working with dehydrogenases, and those of O. H. Warburg, who emphasized the activation of oxygen by an iron-containing oxidase he termed "respiratory enzyme." During the period 1925–1935 Keilin's work provided clear experimental evidence in support of his view that cytochrome is the link between the dehydrogenases and an oxidase. In subsequent investigations (1937–1939) he isolated one of the cytochrome components (cytochrome *c*) and characterized more fully the oxidase system (cytochromes a and a_3).

During the early years of these studies Keilin came upon the work of C. A. MacMunn, who had reported in 1884 the spectroscopic observation of a muscle pigment which had a four-banded spectrum, and to which MacMunn assigned a role in intracellular respiration. Except for a strong criticism in 1890 by Hoppe-Seyler, MacMunn's work attracted relatively little attention at the time. Keilin's finding represented more than a rediscovery of MacMunn's pigment, however, since Keilin firmly established the chemical position of cytochrome and clearly delineated its role as an electron-carrier system in the metabolic process whereby metabolites are oxidized by oxygen.

The recognition that cytochrome is a protein containing an iron-porphyrin unit led Keilin to study the enzymes catalase and peroxidase; between 1934 and 1958 he described a series of important studies on the mechanism of action of these two iron-porphyrin enzymes. He also examined several copper-containing enzymes and in 1938 discovered hemocuprein, a copper protein of red blood corpuscles. Keilin's investigation of the enzyme carbonic anhydrase (1939–1944) showed that it is a zinc protein and that it is strongly inhibited by sulfanilamide; the latter result served as a basis for the subsequent development of a valuable drug for the treatment of glaucoma. Other enzymes fruitfully studied by Keilin included the flavoproteins glucose oxidase, D-amino acid oxidase, and xanthine oxidase.

The breadth of Keilin's biological interests manifested itself in many ways. He conducted studies on the comparative biochemistry of hemoglobin, especially in relation to its presence and function in microorganisms and in the root nodules of leguminous plants. He was interested in the problem of anabiosis, that is, the suspended animation of living things after desiccation and freezing (he preferred to call it cryptobiosis), and his Leeuwenhoek lecture to the Royal Society (published in 1959) on the history of the problem gave clear evidence of his thoroughness and perception as a historian of biology. In the latter years of his life, Keilin began to write a history of intracellular respiration, but did not complete the book. His daughter, Joan, prepared the manuscript for the press and added much new material; the book was published in 1966.

Keilin resided in Cambridge from 1915 to the end

of his life. During that time, and especially after 1931, he exerted a profound influence on the development of science at the university, notably in the encouragement of younger men to embark on new lines of research. A few days after his death, M. F. Perutz wrote in *The Times* of London: "J. C. Kendrew and I owe Keilin a tremendous debt, for he was one of the first to see the potentialities of our physical approach to biochemistry." In 1962 Perutz and Kendrew were awarded the Nobel Prize in chemistry for their work on the X-ray crystallography of proteins.

BIBLIOGRAPHY

I. Original Works. Keilin's only book was *The History of Cell Respiration and Cytochrome* (Cambridge, 1966). He published about 200 scientific articles; a list of his publications follows the article by T. Mann in *Biographical Memoirs* (see below).

II. Secondary Literature. See the obituary articles by M. Dixon and P. Tate, in *Journal of General Microbiology*, **45** (1966), 159–185; E. F. Hartree, in *Biochemical Journal*, **89** (1963), 1–5; T. Mann, in *Biographical Memoirs of Fellows of the Royal Society*, **10** (1964), 183–197; and in *Nature*, **198** (1963), 736–737; and P. Tate, in *Parasitology*, **55** (1965), 1–28.

Joseph S. Fruton

KEILL, JAMES (*b.* Edinburgh, Scotland, 27 March 1673; *d.* Northampton, England, 16 July 1719), *physiology, anatomy*.

James Keill was the younger brother of John Keill, the distinguished Newtonian mathematician. They both entered the University of Edinburgh, although for James it can be specified only that he registered in 1688 for the philosophy course of Andrew Massey. He later went to Paris, where he attended the chemistry lectures of Nicolas Lemery and perhaps the anatomical demonstrations of Joseph Duverney; finally he matriculated at the University of Leiden on 16 October 1696 but did not receive a degree. Upon his return to England, Keill found a ready use for his Continental education as an unofficial anatomy lecturer at Oxford and Cambridge, whose students were wholly dependent on private teachers for any instruction in the basic medical sciences.

In May 1699 Keill obtained an M.D. degree from Aberdeen which probably reflected little more than the payment of a fee. He also received an honorary M.D. degree from Cambridge on 16 April 1705, yet apparently on the strength of his earlier qualifications Keill's medical practice had already begun to prosper in Northampton. Henceforth he combined research with a successful career as a country physician, in which capacity he numbered several members of the nobility among his patients. That he was a conscientious if not a very innovative practitioner can be drawn from an extensive medical correspondence with Sir Hans Sloane, whose friendship was an important factor in Keill's election to the Royal Society in 1712. Aside from an attack of bladder stones, Keill generally enjoyed good health until 1716, when he developed a tumor in his mouth. His death in 1719 resulted possibly from cancer or septic lymphadenopathy.

The first edition of Keill's popular *Anatomy of the Humane Body Abridged* (1698) was largely copied from the contemporary French compendium of Amé Bourdon. Subsequent, more original editions successively reflected not only Keill's increasing anatomical knowledge but also his own physiological research, in which he showed a definite iatromechanical bias as early as the second edition (1703)—presenting, for example, the rudiments of his theory of glandular secretion.

The fourth edition of his *Anatomy* (1710) represented an extensive and final revision with summaries of the physiological theories first described extensively in *An Account of Animal Secretion, the Quantity of Blood in the Humane Body, and Muscular Motion* (1708). In this work Keill examined the problems suggested in the title by using measurement and mathematics in general, and more particularly by positing an attractive force between particles of matter. This concept, admittedly derived from the Newtonian-inspired theories of attraction developed by his brother, led James to propose, among other things, that glandular secretions consisted of cohesions of particles in the blood; and that these particles had united through forces of attraction and were mechanically filtered by various glands according to size. Muscle contraction involved the presence in muscle fibers of blood globules, which had compressed air molecules that could expand when the blood globules were pulled apart by the attraction of animal spirits.

In contrast to Stephen Hales, Keill was not an ingenious experimentalist. Often he simply took a few anatomical measurements and then retreated into mathematical abstractions—which in one case led to extravagant results regarding the rate of blood flow. Yet even here Keill may be credited with discerning a new problem, since he claimed the first calculations of the absolute velocity at which blood travels through the aorta and smaller vessels; he also recognized that the blood's velocity must decrease as the number of arterial branches increases. Keill would also appear to have been one of the first to study the ratio of the fluid to the solid portions of the body, partly through experiments involving tissue desiccation. Finally, he deserves praise for stressing the value of physiological studies in response to his more empirically minded contemporaries. The second edition of Keill's physiological treatise, *Essays on Several Parts of the Animal Oeconomy* (1717), contained a study of the force of the heart which provoked a debate with the physician James Jurin, who believed that Keill had not sufficiently understood the Newtonian principles he had used to obtain his result.

In summary, Keill's anatomical texts provided sound basic knowledge to generations of students, and his physiology may at least be considered a rational attempt at quantification. In his own century, however, his reputation declined as vitalistic trends overshadowed the quantitative approach in English physiology.

BIBLIOGRAPHY

I. Original Works. For the most complete listing of Keill's anatomical and physiological works, see K. F. Russell, *British Anatomy 1525–1800 a Bibliography* (Melbourne, 1963), pp. 146–150. He lists 18 English eds. of Keill's *Anatomy*, a French ed., and possibly Dutch and Latin translations, making the work the most popular anatomical epitome of its time. In the *Philosophical Transactions of the Royal Society* Keill published "An Account of the Death and Dissection of John Bayles, of Northampton, Reputed to Have Been 130 Years Old," **25** (1706), 2247–2252; and "De viribus cordis epistola," **30** (1719), 995–1000. Keill was responsible for Nicolas Lemery's *A Course of Chymistry . . . the Third Edition, Translated from the Eighth Edition in the French* (London, 1698).

II. Secondary Literature. The only recent review of Keill's life is F. M. Valadez and C. D. O'Malley, "James Keill of Northampton, Physician, Anatomist and Physiologist," in *Medical History*, **15** (1971), 317–335. A more detailed consideration of Keill's physiology and its general relation to post-Harveian English physiology is offered by T. M. Brown in his dissertation, *The Mechanical Philosophy and the "Animal Oeconomy"—a Study in the Development of English Physiology in the Seventeenth and Early Eighteenth Century* (Princeton, 1968).

F. M. Valadez

KEILL, JOHN (*b*. Edinburgh, Scotland, 1 December 1671; *d*. Oxford, England, 31 August 1721), *physics, mathematics*.

Keill's early education was at Edinburgh, where he also attended the university, studying under David Gregory, the first to teach pupils on the basis of the newly published Newtonian philosophy. He graduated M.A. before going to Oxford with Gregory, who had been made Savilian professor of astronomy there. Keill was incorporated M.A. at Balliol in 1694 and in 1699 became deputy to Thomas Millington, Sedleian professor of natural philosophy. After a short absence from Oxford he became Savilian professor of astronomy there in 1712, and a year later a public act made him doctor of physic. He remained as Savilian professor until his death.

Keill was one of the very important disciples gathered around Newton who transmitted his principles of philosophy to the scientific and intellectual community, thereby influencing the directions and emphases of Newtonianism. As one of the few around Newton with High Church patronage, Keill apparently tried to counter the Low Church influences of such spokesmen as Richard Bentley and William Whiston. While agreeing with them that the discoveries and doctrine of universal attraction of Newtonianism should play a crucial role in fighting "atheistic" Cartesianism and mechanical thinking, he rejected the notion that this should be accomplished exclusively or primarily by means of natural theology. Rather, natural theology should be subordinated to the Scripture, while natural philosophy should acknowledge the important role played not only by Providence but also by outright miracles. These arguments are made in Keill's first work, *An Examination of Dr Burnet's Theory of the Earth. Together With Some Remarks on Mr. Whiston's New Theory . . .* (1698). This was probably written before he had met Newton, and was an attack on the cosmogonical treatises about the world's creation then being widely debated by many members of the Royal Society. Although supposedly written specifically against the unscientific methods of the theories of Thomas Burnet and William Whiston, in substance it amounted to a very hostile attack—in the name of orthodoxy—on the delusions of "world-making" which were caused, Keill claimed, by Cartesian natural philosophy. As an antidote Keill prescribed the more modest and exact Newtonian philosophy, based solidly on mathematical reasoning, even though Newton himself was known at the time to have sympathies with the cosmogonical theories. Besides those of Burnet and Whiston, Keill attacked the ideas of Richard Bentley, who had tried to use Newtonian principles as the foundation for his physicotheology in his famous Boyle lectures in 1692.

In effect, Keill's work offered itself to Newton as an alternative Newtonian theology, different from that of the Low Church disciples. Newton's public acceptance of Keill's basic criticism against "world-making" was incorporated in 1706 in what was to be the famous 31st Query of the *Opticks*.

Keill's role as propagator of Newtonian philosophy

was carried out primarily through his major work, *Introductio ad veram physicam* . . . (1701), based on the series of experimental lectures on Newtonian natural philosophy he had been giving at Oxford since 1694. The first such lectures ever given, their attempt to derive Newton's laws experimentally did much to influence later publications. Although Keill makes the decidedly anti-Newtonian principle of the infinite divisibility of matter in nature a fundamental axiom, the *Introductio* again unfavorably contrasts Cartesian mechanism, with its dangers of atheism, and Newtonianism. Descartes's insufficient use of geometry, his attempt to define the essences of things rather than being content merely to describe their major properties, and his desire to explain the complex before he can adequately deal with the simple distinguish his fictions from the true principles of Newton. An appendix to the *Introductio* gives a proof for the law of centrifugal "force," whose magnitude had been announced in 1673 by Christiaan Huygens. Several years after the *Introductio*, Keill published an article on the laws of attraction, dealing mainly with short-range forces between small particles, in which he elaborated on Newtonian hypotheses that Newton himself had been unable to pursue.

Some of Keill's writings also brought hostile attacks against Newtonianism from the Continent. For example, his charge that Leibniz had plagiarized from Newton's invention of the calculus gave rise to a major dispute between English and Continental natural philosophers, in which Keill served as Newton's "avowed Champion." Keill's article on the laws of attraction also brought criticisms from the Continent against the employment in Newtonianism of such dubious philosophical concepts as attraction.

In 1700 Keill was elected fellow of the Royal Society. Support from Henry Aldrich, dean of Christ Church College, Oxford, helped Keill's preferment, particularly in becoming deputy to Millington in 1699, just after the attack on Burnet, Whiston, and Bentley. In 1709 Robert Harley helped Keill become treasurer for the refugees from the Palatinate, in which connection he traveled to New England. From 1712 to 1716, with Harley's help, he was a decipherer to Queen Anne.

Keill's uncle was John Cockburn, a controversial Scottish clergyman with Jacobin sympathies. His brother, James, with help from John, tried to apply Newtonian principles to medicine; at his death James left a large sum of money to John. John's marriage in 1717 to Mary Clements, many years his junior and of lesser social standing, was the cause of some scandal. Besides her, Keill was survived by a son, who became a linen draper in London.

BIBLIOGRAPHY

I. ORIGINAL WORKS. *Introductio ad veram physicam, accedunt Christiani Hugenii theoremata de vi centrifuga et motu circulari demonstrata* . . . (Oxford, 1701) was translated as *An Introduction to Natural Philosophy, or Philosophical Lectures Read in the University of Oxford* . . . (London, 1720); when Newtonianism began to make inroads in France, it was translated into French. *An Examination of Dr Burnet's Theory of the Earth. Together With Some Remarks on Mr Whiston's New Theory of the Earth* (Oxford, 1698) includes, in the 1734 ed., Maupertuis's *Dissertation on the Celestial Bodies.* Keill answered Burnet's and Whiston's defenses in *An Examination of the Reflections on the Theory of the Earth. Together With a Defence of the Remarks on Mr Whiston's New Theory* (Oxford, 1699). *Introductio ad veram astronomiam, seu lectiones astronomicae* . . . (Oxford, 1718) was translated as *An Introduction to the True Astronomy; or, Astronomical Lectures* . . . (London, 1721) and also appeared in French. "On the Laws of Attraction and Other Principles of Physics" is in *Philosophical Transactions of the Royal Society*, no. 315 (1708), p. 97. "Response aux auteurs des remarques, sur le différence entre M. de Leibnitz et M. Newton," in *Journal litéraire de la Haye*, **2** (1714), 445–453, is one of several articles by Keill on the calculus controversy. He edited the *Commercium epistolicum D. Johannis Collins, et aliorum, de analysi promota* . . . (London, 1712), which contains the original documents bearing on the Newton–Leibniz controversy. Samuel Halkett and John Laing, *Dictionary of Anonymous and Pseudonymous English Literature*, II (Edinburgh, 1926), 202, cite a contemporary MS note in attributing authorship of Martin Strong [pseud.], *An Essay on the Usefulness of Mathematical Learning. In a Letter From a Gentleman in the City to His Friend at Oxford* (London, 1701), to John Arbuthnot and Keill. "Theoremata quaedam infinitam materiae divisibilitatem spectantia, quae ejusdem raritatem et tenuem compositionem demonstrans, quorum ope plurimae in physica tolluntur difficultates" is in *Philosophical Transactions of the Royal Society*, no. 339 (1714), p. 82. There are letters from Keill in *Correspondence of Sir Isaac Newton and Professor Cotes*, J. Edleston, ed. (London, 1850). Two boxes of Keill MSS, including some letters, drafts of lectures, notebooks, and an inventory of his library are in the Lucasian Papers at Cambridge University Library.

II. SECONDARY LITERATURE. There has been very little attention given to Keill by historians of science, and mention of him generally is found only in connection with the controversy over the calculus. Among Newton's biographers, Sir David Brewster, *Memoirs of the Life, Writings, and Discoveries of Sir Isaac Newton*, I (Edinburgh, 1855), pp. 335, 341–342, II, pp. 43–44, 53, 69; and Frank Manuel, *Portrait of Isaac Newton* (Cambridge, Mass., 1968), pp. 271–278, 321–323, 329, 335–338, 351, 399, 456, discuss Keill. There is a section on Keill's approach to natural philosophy in E. W. Strong, "Newtonian Explications of Natural Philosophy," in *Journal of the History of Ideas*, **18** (1957), 49–83. Ernst Cassirer, *Das Erkenntnisproblem* *in der Philosophie und Wissenschaft der neueren Zeit*, II (Berlin, 1907), pp. 404–406, has a brief discussion of Keill. Pierre Brunet, *L'introduction des théories de Newton en France au XVIIIe siècle. Avant 1738* (Paris, 1931), p. 79 f., briefly deals with Keill. See Arnold Thackray, " 'Matter in a Nut-Shell': Newton's Opticks and Eighteenth Century Chemistry," in *Ambix*, **15** (1968), 29–53, for Keill's ideas on the infinite divisibility of matter and *Atoms and Powers. An Essay on Newtonian Matter-Theory and the Development of Chemistry* (Cambridge, Mass., 1970). A chapter on Keill in David Kubrin, "Providence and the Mechanical Philosophy: The Creation and Dissolution of the World in Newtonian Thought," unpub. diss. (Cornell University, 1968), discusses Keill's attack on Burnet and Whiston. See also Robert Schofield, *Mechanism and Materialism. British Natural Philosophy in an Age of Reason* (Princeton, 1969), pp. 15n, 25–30, 42, 42n, 43–44, 55, 80.

DAVID KUBRIN

KEIR, JAMES (*b.* Edinburgh, Scotland, 29 September 1735; *d.* West Bromwich, England, 11 October 1820), *chemistry.*

A pioneer industrial chemist, Keir developed the first commercially successful process for making synthetic alkali and did much to disseminate chemical knowledge. The youngest of the eighteen children of John Keir and the former Magdalene Lind, both from prominent Scottish families, he was educated in Edinburgh, at the Royal High School and Edinburgh University's medical school; at the latter, during the session 1754–1755, he met Erasmus Darwin and became his lifelong friend. Wishing to travel, he left without taking a degree, purchased a commission in the army, and served during and after the Seven Years' War; he resigned with the rank of captain in 1768.

Keir had retained an interest in chemistry acquired at Edinburgh and had corresponded with Darwin on scientific matters; through the latter, who had settled in practice at Lichfield, near Birmingham, he was soon drawn into the group (which included Matthew Boulton, Josiah Wedgwood, James Watt, and later Joseph Priestley) which constituted the Lunar Society and exercised such a profound influence on the course of the industrial revolution.[1] In 1770 he married Susanna Harvey and settled in West Bromwich.

Shortly before, Keir had begun translating P. J. Macquer's *Dictionnaire de chymie* (Paris, 1766),[2] adding notes and new articles, particularly on the recent work of Black and Cavendish. He also translated the second edition, adding an appendix (later published separately) that summarized recent work on gases. To keep up with the rapidly accelerating development of the science, he prepared a new dictionary of his own, of which only the first part was published.[3] At this time (1789) Keir, like his friend Priestley, was a phlogistonist; unlike Priestley, however, he later abandoned the theory.

From 1771 to 1778 Keir managed a glass factory at Stourbridge. The first and most important of his three papers read to the Royal Society (he became a fellow in 1785) was based on his observations of the crystallization of glass during slow cooling; it included an early and reasoned suggestion that basalt was of volcanic origin.

At least as early as 1771 Keir had, in common with many others, begun to experiment on the production of soda from common salt.[4] The course of his experiments is not known because of the destruction of most of his papers in a fire in 1845; but in 1780, in partnership with a former fellow officer, Alexander Blair, he founded the Tipton Chemical Works, where (more than forty years before the establishment of the Leblanc process in Britain) alkalies were manufactured from sodium and potassium sulfates, waste products from the manufacture of hydrochloric acid. The process was never published but has recently been elucidated by a descendant of Keir.[5] The sinking of a coal mine in 1794 to supply the Tipton works led to a paper on the geology of Staffordshire.

Liberal in politics like many of his associates, Keir supported the French Revolution—until dismayed by its excesses. A well-informed man of great common sense, his advice was frequently sought; his tact and diplomacy were valuable attributes in keeping the Lunar Society together.

NOTES

1. See R. E. Schofield, *The Lunar Society of Birmingham* (Oxford, 1963), pp. 75–82 and *passim*; for opposed views on the membership and duration of the Lunar Society see E. Robinson, "The Lunar Society: Its Membership and Organization," in *Transactions. Newcomen Society for the Study of the History of Engineering and Technology*, **35** (1962–1963 [1964]), 153–177.
2. See D. McKie, "Macquer, the First Lexicographer of Chemistry," in *Endeavour*, **16** (1957), 133–136; R. G. Neville, "Macquer and the First Chemical Dictionary," in *Journal of Chemical Education*, **43** (1966), 486–490.
3. For a discussion of these works (details in bibliography), see W. A. Smeaton, "The Lunar Society and Chemistry: A Conspectus," in *University of Birmingham Historical Journal*, **11** (1967), 51–64.
4. For the background of these early experiments see R. Padley, "The Beginnings of the British Alkali Industry," *ibid.*, **3** (1951–1952), 64–78.
5. See J. L. Moilliet, "Keir's Caustic Soda Process."

BIBLIOGRAPHY

I. ORIGINAL WORKS. Keir's books include his trans. (although his name is not given) of P. J. Macquer, *A Dic-*

tionary of Chemistry (London, 1771; 2nd ed., London, 1777), prepared from sheets supplied by Macquer and published the year before the French ed.; *A Treatise on the Various Kinds of Permanently Elastic Fluids or Gases* (London, 1777; 2nd ed., 1779), in which the use of the word "gases" instead of "airs" was a break with convention; *The First Part of a Dictionary of Chemistry* (Birmingham, 1789); and *An Account of the Life and Writings of Thomas Day, Esq.* (London, 1791)—Day (1748–1789) was a social and political reformer and member of the Lunar Society.

His articles include "On the Crystallizations Observed in Glass," in *Philosophical Transactions of the Royal Society*, **66** (1776), 530–542; "Experiments on the Congelation of the Vitriolic Acid," *ibid.*, **77** (1787), 267–281 (he discovered the crystalline hydrate $H_2SO_4 \cdot H_2O$); "Experiments and Observations on the Dissolution of Metals in Acids; and Their Precipitations; With an Account of a New Compound Acid Menstruum, Useful in Some Technical Operations of Parting Metals," *ibid.*, **80** (1790), 359–384; and "Mineralogy of the South-West Part of Staffordshire," in S. Shaw, *The History and Antiquities of Staffordshire*, I (London, 1798), 116–125.

An unpublished MS, a chemistry "primer" for Keir's only child, Amelia, is in the possession of his descendants (see Moilliet, 1964).

II. SECONDARY LITERATURE. The main biographical source is Amelia Moilliet, *Sketch of the Life of James Keir, Esq., With a Selection From His Correspondence* (n.d.; preface dated 1868), compiled by his daughter and edited after her death in 1857 by her grandson, J. K. Moilliet. See also J. L. Moilliet, "Keir's 'Dialogues on Chemistry'—an Unpublished Masterpiece," in *Chemistry and Industry* (1964), 2081–2083; and "Keir's Caustic Soda Process—an Attempted Reconstruction," *ibid.* (1966), 405–408; B. M. D. Smith and J. L. Moilliet, "James Keir of the Lunar Society," in *Notes and Records. Royal Society of London*, **22** (1967), 144–154; and S. Timmins, "James Keir, F.R.S., 1735–1820," in *Transactions of the Birmingham and Midland Institute*, Archaeological Section, **24** (for 1898; pub. 1899), 1–5.

E. L. SCOTT

KEITH, ARTHUR (*b.* Persley, Aberdeen, Scotland, 5 February 1866; *d.* Downe, Kent, England, 7 January 1955), *anatomy, anthropology.*

Arthur Keith was the fourth son of John Keith, a small farmer, and the former Jessie Macpherson. He received a bachelor of medicine degree from the University of Aberdeen in 1888 and the following year went as physician to a goldmining project in Siam. The mine failed and many laborers died of malaria, but Keith collected plants for Kew Gardens, dissected monkeys, and became interested in racial types. Returning to Britain in 1892, he studied anatomy under G. D. Thane at University College, London, and under R. W. Reid at Aberdeen, where he won the first Struthers prize (1893) with a demonstration of the ligaments of man and ape. In 1894 he became a fellow of the Royal College of Surgeons of England and received his M.D. degree from Aberdeen, but his thesis on primate muscles was not published. During 1895 Keith worked under Wilhelm His at Leipzig. In 1899 he married Celia Gray, who died in 1934.

Keith was appointed senior demonstrator of anatomy at the London Hospital in 1895, and became head of the department in 1899. He was an excellent teacher and inspired many students to research. He edited two anatomy textbooks and also wrote his successful *Human Embryology and Morphology* (1902). He made extensive research on malformations, particularly those of the heart, on which he was helped by James Mackenzie; and with his pupil Martin Flack he first described in 1906 the sinoatrial node, or pacemaker, of the heart. This observation was of much value to cardiology, especially when heart surgery was developed forty years later. Keith resigned from the hospital in 1908 to become conservator of the Royal College of Surgeons Museum, where a vast and somewhat heterogeneous medical collection had grown round the nucleus of John Hunter's museum of comparative anatomy and pathology. He revived the scientific side of the college's work, gave stimulating demonstration-lectures, and encouraged surgeons and anatomists to use the museum; but there were then no students at the college and little facility for research.

Keith's interest now reverted to anthropology. He had studied primate skulls in 1895 and had published *An Introduction to the Study of Anthropoid Apes* (London, 1897); he had also written a monograph, *Man and Ape*, which his publisher refused in 1900. In 1911 he published in London a short book, *Ancient Types of Man*, on the theme that the modern type was as old as the extinct primitive types. He followed this with *The Antiquity of Man* (1915), an anatomical survey of all important human fossil remains, which urged the same theme; he enlarged it in 1925 but "with diminishing conviction." In *New Discoveries* (1931), Keith admitted that evidence really suggested that modern races arose from types already separate in the early Pleistocene. Between 1919 and 1939, when he completed his study of the Palestinian Stone Age remains, he published many reports on human fossils and became the principal arbiter in discussing them.

Keith believed that it was a curator's first duty to make the resources of his museum available to research workers, and that a scientist ought to awaken the public to the message of his work and ideas. He thus became a successful popularizer in the tradition of Huxley, and published two semipopular books in 1919. *Engines of the Human Body* offered "fresh interpretations" of structure and function; Keith recorded that it met "fair success, but was soon out of date." His *Menders of the Maimed* was a historical critique of orthopedic surgery, combined with an exposition of the natural powers of living bone.

Keith was active in several societies, becoming president of the Royal Anthropological Institute (1914–1917), president of the Anatomical Society (1918) and editor of the *Journal of Anatomy* (1916–1936), honorary secretary of the Royal Institution (1922–1926), and president of the British Association for the Advancement of Science (1927). He was elected a fellow of the Royal Society in 1913 and was knighted in 1921.

At Keith's instigation and with the financial support of Buckston Browne, a retired surgeon, the Royal College of Surgeons founded in 1932 a research institute at Downe, the country village south of London where Darwin had once lived; Keith was appointed the master of the new institute when he retired from the college in 1933, and held the post until his death in 1955 at the age of eighty-eight. During his twenty-one years at Downe, besides advising and inspiring successive young researchers there, Keith continued his work in anthropology and wrote his autobiography, a life of Darwin, and several books which sought to correlate the physical and moral evolution of man.

BIBLIOGRAPHY

I. Original Works. Keith bequeathed his MSS, diaries, and other papers to the Royal College of Surgeons Library, London, which has a complete bibliography of his writings; the scientific writings are listed in the memoirs named below. His chief books and papers comprise *Human Embryology and Morphology* (London, 1902; 6th ed., 1948); "The Auriculo-Ventricular Bundle of the Human Heart," in *Lancet* (1906), **2**, 359, written with Martin Flack; "The Form and Nature of the Muscular Connections Between the Primary Divisions of the Vertebrate Heart," in *Journal of Anatomy*, **41** (1907), 172, written with Flack; *The Antiquity of Man* (London, 1915; 2nd ed., 2 vols., 1925); *The Engines of the Human Body* (London, 1919; rev. ed., 1925); *Menders of the Maimed* (London, 1919; repr. 1952); *New Discoveries Relating to the Antiquity of Man* (London, 1931); "A New Theory of the Origin of Modern Races of Mankind," in *Nature*, **138** (1936), 194; *The Stone Age of Mount Carmel—the Fossil Human Remains* (Oxford, 1939), written with T. D. McCown; *An Autobiography* (London, 1950); *Darwin Revalued* (London, 1955).

II. Secondary Literature. For two informative memoirs, see W. Le Gros Clark, in *Biographical Memoirs of Fellows of the Royal Society*, **1** (1955), 145–162; and J. C. Brash and A. J. E. Cave, in *Journal of Anatomy*, **89** (1955), 403–418.

William LeFanu

KEKULE VON STRADONITZ, (FRIEDRICH) AUGUST (*b*. Darmstadt, Germany, 7 September 1829; *d*. Bonn, Germany, 13 July 1896), *chemistry*.

Kekulé was descended from the Czech line of an old Bohemian noble family, Kekule ze Stradonič, Stradonice being a village northeast of Prague. The family can be traced to the end of the fourteenth century; a branch emigrated to Germany during the Thirty Years' War and in the eighteenth century became established in Darmstadt. Kekulé's father, *Oberkriegsrat* Ludwig Carl Emil Kekule, added the accent to the family name following Napoleon's inclusion of Hesse-Darmstadt in the Confederation of the Rhine. When Kekulé himself was ennobled by William II of Prussia, in March 1895, the terminal accent was dropped in the full style.

Kekulé attended the Gymnasium in Darmstadt, where he distinguished himself by his studiousness, aptitude for languages, and talent for drawing. His family intended him to be an architect, and he began the appropriate studies at the University of Giessen in the winter semester of 1847–1848. During the second semester, however, he so enjoyed Liebig's chemistry course that he decided to become a chemist. Kekulé's father had died, and the family council did not give its immediate consent to his new plan, although it was agreed that he might attend the Höhere Gewerbeschule in Darmstadt to study science and mathematics. He accordingly spent the winter semester there and, remaining resolute, was allowed to study chemistry at Giessen, beginning in the summer semester of 1849.

At Giessen, Kekulé first worked under the direction of Heinrich Will, undertaking a study on the ester of amylsulfuric acid and its salts. In the winter of 1850–1851 he began to work in Liebig's laboratory. Liebig was at that time devoting his energies to enlarging his *Chemische Briefe;* he entrusted Kekulé with research on the composition of gluten and wheat bran, and cited Kekulé's results in his twenty-seventh letter. He offered Kekulé an assistantship, but Kekulé found practical laboratory work unsympathetic to his speculative mind and decided to continue his studies abroad.

In 1851 Kekulé, upon Liebig's advice, went to Paris, where he took courses in physics and chemistry and, in particular, became the student and friend of

Charles Gerhardt. He thus came to know Gerhardt's unitary theory of chemistry, his theory of radicals, and his systematization of organic compounds into four types: water (H_2O), hydrogen (H_2), hydrogen chloride (HCl), and ammonia (NH_3); Gerhardt further made the manuscript of his *Traité de chimie organique* available to Kekulé. It was at this time, too, that Kekulé became interested in the problems of the philosophy of chemistry that were to concern him for some time.

Kekulé returned to Germany when his mother died. At Giessen he defended a thesis on the ester of amylsulfuric acid and was awarded the doctorate on 25 June 1852. He then became assistant to Adolf von Planta at Reichenau, Switzerland, where he remained for a year and a half before taking up a similar position, on Liebig's recommendation, with John Stenhouse at St. Bartholomew's Hospital in London. Kekulé stayed in London from the end of 1853 until the autumn of 1855. During this time he met several other of Liebig's former students, including A. W. Williamson, who had shortly before synthesized simple and mixed ethers that corresponded exactly to Gerhardt's water type. (Gerhardt himself had just discovered the anhydrides of organic acids, thereby confirming the significance of this same type.)

Williamson and Kekulé became friends, and Williamson was influential in the development of Kekulé's theoretical views. It was at Williamson's instigation, moreover, that Kekulé began his work on the reaction of phosphorus pentasulfide on acetic acid. From this reaction Kekulé was able to isolate thioacetic acid, which he classified as a new type, hydrogen sulfide, corresponding to Gerhardt's water and hydrogen chloride types. This work, published in 1854, marks the beginning of Kekulé's scientific maturity. At the same time Kekulé had begun to consider, in the Gerhardt types of organic molecules, not only the radicals, but more and more the atoms themselves; he himself gave an account of a vision that he had on top of a London omnibus, in which he saw the atoms "gambolling" before his eyes. This fantasy, which was perhaps influenced by his early training in architecture, was soon to result in his theory of valence and in his structure theory.

Kekulé was, however, eager to begin a university career, and at the suggestion of Liebig and Bunsen he enrolled in the University of Heidelberg in order to be admitted there as a privatdocent. Having passed the requisite examinations, in the summer semester of 1856 Kekulé began teaching organic chemistry. He further installed, at his own expense, a lecture room and a laboratory in the first two floors of a house on the main street of Heidelberg, and it was in this private laboratory that he carried out his experiments on the chemical constitution of fulminate of mercury. Here, too, Adolf von Baeyer studied compounds of arsenic trimethyl.

During these years, too, Kekulé arrived at the concept of polyvalent radicals and introduced multiple and mixed types in a single formula of a particular compound. He introduced also the marsh gas type and worked out the theory of the tetravalence of carbon, as may be seen from an article that he published in *Justus Liebigs Annalen der Chemie* in 1857; in a more extensive publication of the following year he was able to state not only that the carbon atom is tetravalent in such simple compounds as CH_4, CH_3Cl, CCl_4, $CHCl_3$, and CO_2, but also that in compounds containing more than one carbon atom, the carbon atoms can link together in chains which can, in turn, form various polyvalent radicals. An ordered classification of organic compounds thus becomes possible. Indeed, by creating the new type CH_4 and by stating the ability of carbon atoms to join up with each other, Kekulé laid the foundation of structural chemistry. He based his courses at Heidelberg on these principles, illustrating his lectures with models of individual atoms and of molecular groupings. By projecting the shadows of these models on a blackboard or on paper, Kekulé obtained the "graphic formulas" that were one of his favorite teaching aids. His innovative course was a great success, and Kekulé began to consider publishing a treatise on organic chemistry.

Before he could do so, however, a chair of chemistry became vacant at the University of Ghent. The Belgian chemist Jean Servais Stas, wishing to revivify the teaching of chemistry in Belgium, strongly urged Kekulé's nomination as full professor. Kekulé accepted the position and, at the age of twenty-nine, moved to Ghent. Stas had obtained a promise that practical chemistry would be introduced into the curriculum at Ghent and Kekulé was promised a new laboratory for both teaching and research. He was also given permission to accept private students, of whom Baeyer, one of the first, became his personal research assistant.

Despite the difficulties of adjusting to a foreign environment and of teaching in French, Kekulé soon established himself in a scholarly mode of life. He spent the entire day in the laboratory, dedicated the evening to composing the first sections of his *Lehrbuch der organischen Chemie* (of which the first fascicle was printed in June 1859), and, in the hours after midnight, prepared his courses for the next day. He also found time to take the initiative in organizing the first International Congress of Chemists, which met at Karlsruhe in September 1860. The purpose of the

Congress was to reduce confusion in chemical nomenclature—Kekulé was as aware as anyone of discrepancies in defining such basic concepts as the atom, the molecule, and equivalence—and to promote greater uniformity of terminology in the world chemical literature. It served a further important end as well, since it was here that Cannizzaro reestablished the importance of the Avogadro-Ampère molecular hypothesis, which had lain neglected for nearly fifty years.

Kekulé achieved some significant experimental work even before his new laboratory was ready. In particular he was concerned with the chemical structure of the organic acids and carried out, in sealed tubes, the bromination of succinic acid; from the silver salt of dibromosuccinic acid he prepared optically inactive tartaric acid and from the silver salt of the monobromosuccinic acid he obtained maleic acid. He further demonstrated that the same family relationship exists between salicylic acid and benzoic acid as between glycolic acid and acetic acid. These researches led Kekulé to recognize the isomerism of the phenolic aromatic acids, but he was unable to account for it.

The new laboratory, constructed according to plans drawn up by Stas and Kekulé, was inaugurated in 1861, and Kekulé began to study the unsaturated dibasic acids. He was aided in this undertaking by Théodore Swarts and Eduard Linnemann, his assistants, and by one of his students, Hermann Wichelhaus. His attention had been drawn to the subject by his discovery of fumaric acid and maleic acid, two unsaturated dibasic isomers, related to succinic acid, each of which contains four carbon atoms. These acids further readily fix bromine to form two different dibromide derivatives. Having identified these entities, Kekulé was unable to interpret their structure, and the problem became more complex when he discovered three other unsaturated isomeric dibasic acids with five carbon atoms each. Since Kekulé had long held the tetravalence of carbon to be as invariable as its atomic weight, it was necessary for him to create a new theory to acknowledge the presence, in unsaturated isomers, of lacunae or double bonds between two neighboring carbon atoms. This theory of unsaturates was published in 1862; by means of it Kekulé was able to account for both the two isomers with four carbon atoms and the three acids with five carbon atoms.

The problem of unsaturated substances almost immediately came again to Kekulé's attention in the following year, since he was writing the second part of his *Lehrbuch*, in which he planned to deal with the chemical structure of the aromatic compounds. The solution in this instance came to Kekulé in a vision—half awake, he saw before his eyes the animated image of a chain of carbon atoms, closing upon itself like a snake biting its own tail. He was instantly aware of the significance of such a closure, and spent the rest of the night determining the consequences of his inspired hypothesis. He arrived at a closed chain of six carbon atoms, linked alternately by three single and three double bonds and constituting the common nucleus of all the aromatic substances. He then set himself the task of experimental confirmation, but his work toward this end was delayed by various events.

At the time of his arrival in Ghent, Kekulé had met George William Drory, inspector general of the Continental Gas Association. Like Kekulé, Drory was a Protestant; they soon became close friends and Kekulé became a frequent visitor to Drory's house. There he met and fell in love with Drory's youngest daughter, Stéphanie, whom he married on 24 June 1862. Kekulé was thirty-two, Stéphanie nineteen. Their son, Stephan, was born the following May, and two days later Stéphanie Kekulé died. Kekulé was unable to take up his creative work for several months following her death.

He returned to his research in 1864, again taking up the search for confirmation of his benzene theory, which he had already set down in manuscript form. He first tried to do the necessary work by himself, but soon recognized the actual extent of his project and hired two assistants, Karl Glaser and Wilhelm Körner, both trained at Giessen. All the activity of the laboratory was for some time thereafter concentrated upon the derivatives of benzene and their isomers, but Kekulé still did not publish his theory. It was only after Tollens and Fittig brought out their excellent work on the synthesis of the hydrocarbons of the benzene series that he decided to make his own work known. Thus Wurtz presented Kekulé's benzene theory to the Société chimique de Paris on 27 January 1865, in a session presided over by Pasteur. It was subsequently published in the *Bulletin de la Société chimique de Paris* under the title "Sur la constitution des substances aromatiques," and concluded with a table of formulas for benzene and similar compounds.

On 11 May 1865, Kekulé presented to the Académie Royale de Belgique, of which he had been elected an associate member, a "Note sur quelques produits de substitution de la benzine," in which he considered the geometry of the benzene nucleus and used it to determine the number of its possible monosubstituted, disubstituted, and trisubstituted isomeric derivatives. He and his associates then set out to prove these figures experimentally, and succeeded after several years' work. They found the most diverse substituents to be those fixed on the ring or onto the lateral

KEKULE VON STRADONITZ

chains—namely the halogens and the NO_2, NH_2, diazo, CO_2H, SO_3H, OH, and SH groups—and attempted to localize these substituents in each of the benzene isomers. On 3 August 1867, Kekulé presented to the Academy a remarkable work on this subject by Körner, "Faits pour servir à la détermination du lieu chimique dans la série aromatique." (Körner himself stated his "absolute" method, which provided an elegant means for establishing unambiguously the ortho, meta, and para positions of the disubstituted derivatives of benzene, some seven years later.)

In addition to his work on the structure of aromatic substances, beginning in 1865, Kekulé took up the study of their azo and diazo derivatives. He began this research with a view toward incorporating the results of it in the second volume of his *Lehrbuch;* in addition, the subject had assumed considerable industrial importance once the potential of the intermolecular transformation of diazobenzene into aminobenzene became known. In 1866 Kekulé provided a masterful interpretation of this transformation and of the catalytic role of the aniline salts, drawing upon his new theories of the constitution of the diazo group and its mode of fixation on the benzene ring. In his wonderful researches on diazocompounds Griess prepared a new compound, called phenylendisulfuric acid, formed by interaction of concentrated sulfuric acid and diazobenzensulfate. Kekulé's interest was aroused and he proved theoretically as well as experimentally that the product was in reality a disulfonic derivative of phenol. He turned then to the study of the sulfonic derivatives of phenol and was able to clarify the double mode of action of sulfuric acid on organic matter, showing that it produces both readily decomposable sulfuric esters and highly stable sulfonic derivatives; he further emphasized the striking analogy in this respect between the sulfonyl and carbonyl groups fixed on the benzene ring.

Kekulé also discovered that sulfonic derivatives of benzene fuse with potash to create their corresponding phenols. This discovery was to become important in the industrial production of phenols. In a variation of an earlier experiment, made in London, in which he used phosphorus pentasulfide, Kekulé succeeded in transforming phenol into thiophenol by substituting sulfur for the oxygen of the former. He demonstrated thereby that the oxygen of phenol is more strongly bonded to the carbon of the benzene ring than to the OH group of the fatty alcohols.

Throughout this strenuous period of research Kekulé did not neglect his teaching duties. (In 1867, for example, he published and recommended as a teaching aid a new model of the carbon atom.) But he wished to be able to teach in German again, and when he was offered the chair of chemistry at the University of Bonn—vacant since A. W. Hofmann had gone to Berlin—he accepted it gladly. He was additionally assured the directorship of a new chemical institute, the construction of which was virtually complete. In September 1867, the Belgian government accepted Kekulé's resignation and he left Ghent for Bonn.

The new chemical institute was officially opened in 1868; the inaugural ceremonies coincided with those in celebration of the fiftieth anniversary of the university itself. Kekulé was awarded an honorary M.D. on this occasion in recognition of his contributions to theoretical chemistry. Many students were drawn to Bonn to hear his lectures and observe his class experiments and laboratory work; one of them, in 1873, was J. H. van't Hoff, to whom Kekulé's model of the carbon atom suggested the concept of the asymmetric carbon atom of his *La chimie dans l'espace* of 1875. Other students became Kekulé's direct collaborators, among them Theodor Zincke (in work on condensation of aldehydes), Hermann Wichelhaus —who had followed him from Ghent—and Thomas Edward Thorpe (on aromatic compounds), Nicolas Franchimont (on triphenylmethane and anthraquinone), Otto Strecker (on the constitution of benzene), and Richard Anschütz (on oxyderivatives of fumaric and maleic acids).

At Bonn, Kekulé found it necessary to delegate some of his teaching responsibilities to others in order to concentrate on his own research, in which he had the aid of several private assistants. His first projects were continuations of work he had begun in Ghent; he resumed his study of the sulfonic derivatives of phenol and nitrophenol, and extended his earlier investigations of camphor and oil of turpentine to include cymol, thymol, and carvacrol (the latter work was completed in 1874). More important, however, was the resumption of his attempt to provide experimental evidence for his benzene theory, particularly for the presence in the ring of three alternating double bonds. Having observed that trimethylbenzene is formed through the condensation of three acetone molecules, Kekulé hoped to synthesize the benzene ring through the condensation of aldehyde. He was unable to obtain such a synthesis; his attempts to do so, however, resulted in an elegant series of works (published between 1869 and 1872) on the condensation of acetaldehyde. These studies treat the formation of crotonaldehyde and some of its derivative products, as well as dealing with polymerization products of aldehyde.

Kekulé also wished to demonstrate the superiority of his own formula for benzene over those put forth by A. Claus, H. Wichelhaus, and A. Ladenburg. By 1872 he had created the complementary "oscillation theory," which took into account the existence of only one bisubstituted derivative in the ortho position, rather than two. He thus permitted the delocalization of single and double bonds, which he had considered to be fixed in his earlier theory.

In the same year Kekulé and Franchimont succeeded in synthesizing triphenylmethane, the fundamental hydrocarbon in rosaniline dyes, and also obtained anthraquinone in the course of preparing benzophenone. The elucidation of the structure of these compounds proved crucial to the development of synthetic dyes; the subsequent rapid growth of the German aniline dye industry, based on the triphenylmethane group and anthraquinone, was its direct result.

The growing number of Kekulé's students and co-workers soon necessitated an expansion of the chemical institute, to which a number of new workrooms were added in 1874 and 1875. During this period Kekulé was offered the chair of chemistry at the University of Munich, which had become vacant with the death of Liebig; he declined the post, however, and recommended Adolph von Baeyer in his stead. At the same time, his health had begun to fail. Twenty years of overexertion had begun to take their toll, as had an unfortunate second marriage to his former housekeeper, a woman much younger than he, who was incapable of relieving him of his cares. A month after this marriage, too, Kekulé contracted measles from his son and suffered prolonged aftereffects. He nonetheless continued to serve the university, being elected rector in 1877, on which occasion he gave an address on the scientific goals and accomplishments of chemistry. Upon completion of his office the following year, he spoke upon the principles of higher education and educational reform.

At about the same time Kekulé resumed work on the *Lehrbuch*, in collaboration with Gustav Schultz, Richard Anschütz, and, slightly later, Wilhelm La Coste; but the rapid growth of chemistry at that time did not allow them to maintain the original plan of composition, and the work was never completed. Although volume III appeared in 1882, volume IV, published in 1887, consisted of only one of the planned sections. From 1879 to 1885 Kekulé also engaged in research, primarily experiments designed to support his own benzene theory against the prismatic formula advocated by August Ladenburg. Ludwig Barth had become a partisan of the latter theory, arguing from the formation of carboxytartronic acid from pyrocatechol. In 1883 Kekulé was able to show that this acid is simply tetraoxysuccinic acid, the formation of which from pyrocatechol was better explained by his own hexagonal theory. In a series of investigations on trichlorophenomalic acid made with Otto Strecker in the following year, Kekulé again corroborated his own thesis and confirmed the superiority of his own formula, which provided an atom-by-atom explanation of the formation of β-trichloracetylacrylic acid through the oxydochlorination of quinone.

A high point in Kekulé's career occurred in 1890, when he read his paper "Ueber die Konstitutionen des Pyridins" to the general assembly of the Deutsche Chemische Gesellschaft in Berlin on 10 March. The communication summed up the investigations on pyridine, of which the formula is comparable to that of benzene, that he had carried out since 1886. The day after this presentation, a great celebration was held to honor Kekulé on the occasion of the twenty-fifth anniversary of his benzene theory. Kekulé, in thanks, gave a remarkable speech in which he reviewed his life's work and made public for the first time the details of his visionary solution of the benzene ring.

Although he had grown deaf by this time, Kekulé continued to teach and carry out administrative duties. In 1892 he also prepared formic aldehyde in the pure state, thus extending his earlier work on the condensation of the aldehydes. His health was again seriously impaired following an attack of influenza, and he died shortly thereafter. He was buried in the family vault in the cemetery of Poppelsdorf; a bronze statue of him, paid for largely by subscription from the German dyestuffs industry, was erected, facing his chemical institute, in 1903.

BIBLIOGRAPHY

Kekulé's works were collected, with a biography, by Richard Anschütz to honor him at the centenary of his birth: *August Kekulé*, I, *Leben und Wirken*, II, *Abhandlungen, Berichte, Kritiken, Artikel, Reden* (Berlin, 1929).

Secondary literature includes G. V. Bykov, *August Kekulé* (Moscow, 1964), in Russian; J. Gillis, "Auguste Kekulé et son oeuvre, réalisée à Gand de 1858 à 1867," in *Mémoires de l'Académie royale de Belgique. Classe des sciences*, **37** (1966), 1–40; Francis R. Japp, "Kekulé Memorial Lecture," in *Journal of the Chemical Society*, **73** (1898), 97–138; and R. Wizinger-Aust *et al.*, *Kekulé und seine Benzolformel . . .* (Weinheim, 1966).

JEAN GILLIS

KELLNER, DAVID (*b.* Gotha, Germany, mid-seventeenth century; *d.* [?]), *medicine, chemistry.*

Little is known about the life of the physician and metallurgist David Kellner; indeed, even eighteenth-century reference works were unable to furnish biographical information concerning him. Although Kellner wrote a comedy that in form and content quite met the standards of German baroque theater, he is nowhere mentioned by historians of German literature. Furthermore, there is no secondary literature of any value that deals with his importance as a physician or—and here the omission is more surprising—with his contribution to the development of specialized literature in the field of metallurgy.

Kellner studied medicine in Helmstedt when the renowned physician and polymath Hermann Conring taught there. He was undoubtedly influenced by Conring, who waged violent battles against alchemy and esoteric medicine.

Kellner received his doctor's degree in Helmstedt in 1670 (accordingly, his date of birth must be set in the mid-seventeenth century). He wrote two surgical dissertations, *De ossium constitutione naturali et praeternaturali* and *De empyemate.* He dedicated the second of these, a work on festering wounds, to Johann Langguth, a physician in the service of Duke Ernst of Saxony. It is possible that Langguth advanced him in his scientific work.

Kellner later worked in Nordhausen. In the majority of his own writings, as well as in those he edited, he signed himself as "Practitioner in the Imperial Free City of Nordhausen, and Body and Court Physician of Royal Prussia, Princely Saxony, and the County of Stolberg." Beyond this he left no references to himself, except for a remark in the dedication to the reader in his *Schenkeldiener* (1690). He relates there that he wrote the book in 1683 when he was with Duke Heinrich, his prince and overlord, in Römhild (Franconia). It was dedicated to the surgeon and barber of Gotha, the city of the prince's residence, Johann Scheib, whom Kellner calls his friend and patron. The *Schenkeldiener* is a reference work on bone injuries and includes prescriptions as well as advice on diagnosis and therapy.

The names of the scholars with whom Kellner associated are not known. In his works he cites, in the traditional manner, only ancient or older German authors, with the exception of famous *Kameralisten* such as Johann Joachim Becher and Wilhelm von Schroder. The latter's *Fürstliche Schatz- und Rentkammer* ("Princely Treasury and Revenue Office") is the opening chapter of a work on mining and saltworks that Kellner edited.

Kellner's interest in scientific writing manifested itself mainly in the field of metallurgical chemistry. He wished above all to free this literature, and indeed all scientific publication, from the fantasies of alchemists. Toward this end he wrote for a lay audience and for future scientists, rather than for an exclusive circle of initiates. In all, the number of writings by other authors that he collected and edited exceeded that of his own published works.

Kellner's comedy about the "harmful Society of Alchemists" (1700) displays a fertile inventiveness that is typical of the baroque. In the play he excoriates alchemy. The climax is a scene in which seven alchemists mix, cook, and toil—to no apparent purpose—in the kitchen of a baron who has been taken in by their promises.

Kellner himself once fell under suspicion: he was accused of being one of the "chemical heretics." He was obliged to defend himself in an apologetic, but none the less polemical, "Epistle to the Unnamed Authors of the German Purgatory of Refining." Nonetheless, among those who wrote on science in his time, Kellner was one of the more serious authors and was certainly so considered by his contemporaries.

This judgment is justified by the tenor of most of Kellner's writings. They were meant to be, as their titles indicate, contributions to the science of assaying. Kellner sought to state, as clearly as possible, prescriptions and methods for experimentation. He asserted, however, that "it is highly necessary for all who are devoted to chemistry and medicine, and not just for those whose own profession is metal assaying, to know what is contained in the mineral kingdom, and how it might be purified, smelted, and even improved."

BIBLIOGRAPHY

I. Original Works. Among them are *De ossium constitutione naturali et praeternaturali. De empyemate* (Helmstedt, 1670), medical diss.; *Curieuser Schenkeldiener* (Frankfurt–Leipzig, 1690); *Die durch seltsame Einbildung und Betriegerei schaden bringende Alchymisten-Gesellschaft in einem nützlichen Lustspiele vorgestellet* (Frankfurt–Leipzig, 1700); *Hochnutzbar und bewahrte edle Bierbraukunst, mit einem Anhang über Wein und Essig* (Leipzig–Gotha, 1690; 2nd ed., Leipzig–Eisenach, 1710); *Ars separatoria oder Scheidekunst* (Leipzig, 1693; 2nd ed., enlarged by several new experiments, entitled *Erneuerte Scheidekunst*, Chemnitz, 1710; 3rd ed., Chemnitz, 1727).

II. Secondary Literature. See also Johann Bernhard Horn, *Synopsis metallurgica oder Anleitung zur Probierkunst* (Gotha, 1690); *Praxis metallica curiosa oder Schmelzproben* (Nordhausen, 1701); Ulysses Aldrovandus, *Synopsis musaei metallici* (Leipzig, 1701); *Kurz abgefasstes Berg- und Salzwerks-Buch* (Frankfurt–Leipzig, 1702); and L. Martin Schmuck *et al.*, *Chymische Schatzkammer* (Leipzig, 1702).

Focko Eulen

KELLOGG, ALBERT (*b.* New Hartford, Connecticut, 6 December 1813, *d.* Alameda, California, 31 March 1887), *botany.*

The first resident botanist of California, Kellogg came from a line of pioneer English farmers. He was the son of Isaac and Aurill Barney Kellogg. His boyhood was spent on the farm and he showed an early interest in plants. He began studying medicine as an apprentice to a Middletown, Connecticut, physician but his health failed. He resumed his medical studies at Charleston, South Carolina, but tuberculosis necessitated his removal to the interior. After obtaining his M.D. degree at Transylvania College, Lexington, Kentucky, he practiced in Kentucky, Georgia, and Alabama; those who knew him say he never requested payment.

Kellogg was in San Antonio, Texas, in 1845 but shortly returned to Connecticut. In search of new botanical fields and intending to practice, he joined a party heading for the gold fields by way of the Horn. Kellogg arrived in Sacramento 8 August 1849 but moved on to San Francisco where with six others he founded what is now known as the California Academy of Sciences in 1853. By his encouragement of the Academy's beginnings, its collections, library, and publications, by his own reports and exceptional artistic talents, Kellogg influenced natural sciences in California. He described 215 species of plants, of which about fifty are today recognized in the manuals.

In 1867, the year of the Alaska purchase, Kellogg accompanied the Coast Survey cutter *Lincoln* as far as Unalaska in the Aleutians. He made about 500 plant collections in three sets, destined for the Smithsonian Institution, the Academy of Natural Sciences of Philadelphia, and the California Academy (the latter almost wholly destroyed in the 1906 fire). His *Forest Trees of California* was the state's first dendrological report. He finished 400 drawings, principally of woody plants; those of the oaks were published posthumously. He never married. His biographer E. L. Greene said Kellogg would not have claimed to be "a scientific botanist" and that his writings were "a commingling of matters, poetical, theological and botanical." Others described him as a "dreamy imaginative man," with "childlike enthusiasm and unworldliness," who lived "a happy life and died respected."

BIBLIOGRAPHY

I. ORIGINAL WORKS. Kellogg's most important writings, some in collaboration with H. H. Behr, were published in the *Bulletin* and the *Proceedings of the California Academy of Natural Sciences*, including first reports on the singular plant forms of Baja California. His *Forest Trees of California*, appendix to *Second Report of the California State Mineralogist, 1880–1882*, 1–116, was reprinted separately (Sacramento, 1882). *Illustrations of West American Oaks . . . the text by Edward Lee Greene* (San Francisco, 1889) includes 24 of Kellogg's line drawings.

II. SECONDARY LITERATURE. Edward Lee Greene wrote the principal sketch of Kellogg for *Pittonia*, **1** (1887), 145–151; this has been used by subsequent authors in their biographies, including W. L. Jepson in the *Dictionary of American Biography*, V, pt. 1 (New York, 1933), 300–301. Greene's contention that Kellogg met Audubon has been proved erroneous by S. W. Geiser, *Naturalists of the Frontier*, 2nd ed. (Dallas, 1948), p. 276.

Kellogg's Alaskan itinerary is summarized by Eric Hulteń, *Botaniska Notiser*, **50** (1940), 302. Kellogg's friend of twenty years, George Davidson, contributed a eulogistic preface to *West American Oaks.*

JOSEPH EWAN

KELLOGG, VERNON LYMAN (*b.* Emporia, Kansas, 1 December 1867; *d.* Hartford, Connecticut, 8 August 1937), *entomology, zoology.*

Kellogg was the son of a college professor, Lyman Beecher Kellogg, and Abigail Homer Kellogg. Although he had shown a considerable interest in the animals of his native Kansas, Kellogg intended to become a journalist when he entered the University of Kansas. At the university he worked on the local newspaper with his close friend and fellow student William Allen White. But the persuasive influence of entomologist Francis Huntington Snow, chancellor of the university, impelled him to follow a scientific career. Kellogg received the B.A. at Kansas in 1889 and the M.A. from the same university in 1892, by which time he was already assistant professor of entomology (1890) and secretary to Snow. He became associate professor in 1893.

In 1894 Kellogg went to Stanford University at the urging of the prominent entomologist John Henry Comstock, who spent three months there each year. Kellogg became professor of entomology and head of the department at Stanford in 1895. He took leaves of absence to study at Cornell University, and at Leipzig and Paris. During World War I he served with the Commission for Relief in Belgium, headed by his former student Herbert Hoover, and in other relief and peace activities. Kellogg resigned his professorship in 1920 to become permanent secretary of the National Research Council.

Under Snow's leadership, the University of Kansas developed a fine entomological center, and Kellogg was one of its significant contributors. He and Samuel Wendell Williston added numerous specimens to the insect collection. Stanford, also, was an active entomological site, where David Starr Jordan's forceful personality influenced many branches of science. Kellogg accumulated the largest collection of *Mallophaga* in the United States and published extensively on them. He observed that closely related species of these bird lice on different hosts indicated a close relationship between the hosts. His work on silkworms was a pioneer study in genetics in America. He classified the Dipteran family Blepharoceridae, and he also investigated Lepidoptera scales and the morphology and development of mouth parts in insects. He wrote, alone or with Jordan, about a dozen books on evolution and general biology.

BIBLIOGRAPHY

I. Original Works. A list of Kellogg's most significant publications is presented in McClung's memorial, cited below. Outstanding entomological contributions are "List of North American Species of Mallophaga," in *Proceedings of the United States National Museum*, **22** (1900), 39–100; "Mallophaga," in *Genera insectorum*, fasc. 66 (1908), pp. 1–87; "Diptera family Blepharoceridae," *ibid.*, fasc. 56 (1907), pp. 1–15; and *Inheritance in Silkworms*, Stanford University Publications, ser. 1 (Stanford, Calif., 1908). His three highly esteemed eds. of *American Insects* (New York, 1904; 3rd ed., 1914) were especially valuable in popularizing the field.

Among his other significant books are *Darwinism Today* (New York, 1907); *Evolution and Animal Life* (New York, 1907), written with David Starr Jordan; *Mind and Heredity* (Princeton, N.J., 1923); and *Evolution* (New York, 1924).

II. Secondary Literature. C. E. McClung presented Kellogg's personality and accomplishments in "Biographical Memoir of Vernon Lyman Kellogg," in *Biographical Memoirs. National Academy of Sciences*, **20** (1939), 243–257. A volume entitled *Vernon Kellogg, 1867–1937* (Washington, D.C., 1939), C. C. Fisher, ed., which includes excerpts from his writings, tributes from associates, and some biographical material, was published by the Belgian American Education Foundation. Kellogg is included in E. O. Essig, *A History of Entomology* (New York, 1965), a facs. of the 1931 ed.; in Herbert Osborn, *Fragments of Entomological History* (Columbus, Ohio, 1937; pt. 2, 1946); and in Arnold Mallis, *American Entomologists* (New Brunswick, N.J., 1971).

Elizabeth Noble Shor

KELSER, RAYMOND ALEXANDER (*b.* Washington, D.C., 2 December 1892; *d.*, Philadelphia, Pennsylvania, 16 April 1952), *veterinary medicine, microbiology*.

Raymond Kelser was the first of eight children born to Charles Kelser, a skilled mechanic, and Josie Potter Kelser. Educated in the public schools of Washington, D.C., he took a business course in high school and became a messenger for the Bureau of Animal Industry of the Department of Agriculture. Here he came under the friendly and helpful influence of John R. Mohler, chief of the pathological division of the bureau. The subsequent rise and achievements of the poor but gifted young Kelser were typical of the American success story.

While working at the bureau, Kelser enrolled in night classes at the School of Veterinary Medicine of George Washington University and received the D.V.M. degree in 1914. After his marriage in that year to Eveline Harriet Davison and some brief experience as a commercial bacteriologist, Kelser took the civil service examination and returned to the Bureau of Animal Industry. From 1915 to 1918 he and his colleagues studied anthrax immune serum and improved the vaccine for the disease.

In 1918 Kelser joined the Veterinary Corps of the U.S. Army. He served with distinction in the army for twenty-eight years, achieving the rank of brigadier general. In 1942 he became the first general officer in the corps; and while he was chief veterinary officer of the army, the number of officers in the corps grew from 126 in 1938 to over 2,200 during World War II.

Kelser's abilities in research, evident even prior to his military service, led the army to assign him to various laboratories for periods of time sufficiently long to enable him to carry out some highly successful researches. From 1921 to 1925 he was chief of the Veterinary Laboratory Division of the Army Medical School in Washington, D.C. During this period he perfected a better test for detecting botulinus toxin in canned foods. He also pursued graduate studies at American University, earning an M.A. degree in 1922 and a Ph.D. in 1923.

From 1925 to 1928 Kelser served in the Philippine Islands, where he made a major scientific contribution. Cattle plague, or rinderpest, was at the time a serious problem in many parts of the world, and Philippine agriculture was suffering severely from its effects. Kelser developed an effective means of inactivating the virus with chloroform without destroying its immunizing properties. The resulting vaccine led to the eventual control of the disease.

From 1928 to 1933 Kelser again headed the Veterinary Laboratory Division in Washington, where he completed several important studies. The most notable of these was his elucidation of the mechanism of transmission of the virus of equine encephalomyelitis. Kelser, who was among the early virologists in the 1930's, making important studies in virus characteristics and transmission, showed that the agent of equine encephalomyelitis could be passed from guinea pig to guinea pig and from horse to horse by mosquitoes. Kelser also showed that the mosquito acted not merely as a mechanical agent of viral transfer from animal to animal, but served as a necessary incubating host for the virus. The virus multiplied while in the mosquito, thus increasing its infective powers. This finding was of great interest to those trying to understand the nature of virus diseases. It also had practical application—in mosquito control—in dealing with widespread encephalomyelitis among horses.

Kelser retired from the army in 1946 but continued to advise the Department of Defense on matters of biological warfare. Upon his retirement he became dean and professor of bacteriology of the School of Veterinary Medicine at the University of Pennsylvania. In the six years before his sudden death from a stroke, he helped to expand and improve the school's facilities and research activities. He received numerous awards and was a member of many scientific organizations, including the National Academy of Sciences, to which he was elected in 1948.

BIBLIOGRAPHY

I. ORIGINAL WORKS. Kelser's numerous scientific articles have never been collected in a single bibliography. Among them is "Carriers of Organisms Pathogenic for Both Man and the Lower Animals," in Henry J. Nichols, *Carriers in Infectious Diseases* (Baltimore, 1922), pp. 121–180, an early work. Of much greater importance was Kelser's *Manual of Veterinary Bacteriology* (Baltimore, 1927). This was a standard work in its field and appeared in successive rev. and enl. eds. in 1933, 1938, and 1943.

II. SECONDARY LITERATURE. The most informative article on Kelser is Richard E. Shope, "Raymond Alexander Kelser 1892–1952," in *Biographical Memoirs. National Academy of Sciences*, **28** (1952–1954), 199–217. For obituaries see *Veterinary Medicine*, **47** (1952), 250; *Military Surgeon*, **110** (1952), 460–461; and *Journal of the American Veterinary Medical Association*, **120** (1952), 398–400.

GERT H. BRIEGER

KELVIN. See **Thomson, William.**

KENNEDY, ALEXANDER BLACKIE WILLIAM (*b.* Stepney, London, England, 17 March 1847; *d.* London, 1 November 1928), *kinematics of mechanisms, testing of materials and machines.*

As a young man Kennedy contributed significantly to the kinematics of mechanisms, which is a theoretical treatment in machine design of relative displacements of machine members, and to laboratory testing of machines as a part of engineering training. The latter half of his life, from 1889, was devoted to engineering aspects of electric power systems.

The eldest son of John Kennedy, a Congregational minister, and the former Helen Blackie, Kennedy attended the City of London School and the School of Mines until age sixteen. Successively as apprentice, draftsman, and consultant, he was for ten years concerned with marine steam engine design and construction in their embryonic stages. From 1874 to 1889 he was professor of engineering at University College, London, where in 1878 he organized the first mechanical testing laboratory intended primarily for the instruction of undergraduate students. He was anxious to give students experience in the laboratory so that they might recognize the problems and importance of precise measurements and thus use critically the data tabulated in reports and handbooks. Also during his teaching career he published his English translation of Franz Reuleaux's *Theoretische Kinematik* (1876). Ten years later he published his own textbook of kinematics, in which appeared Kennedy's law of three centers, which is fundamental to kinematic analysis employing instantaneous centers.

In 1889 Kennedy turned from mechanical to electrical engineering. As consulting engineer, he quickly became a leading authority in the design and construction of electric generating and distribution systems for both domestic and railway service. He was responsible for the design of numerous systems installed in the principal British cities.

After 1900 Kennedy was a member of several government technical boards and commissions. In 1894 he was president of the Institution of Mechanical Engineers and of Section G (Engineering) of the British Association for the Advancement of Science and in 1906 president of the Institution of Civil Engineers; he was also a member of the Institution of Electrical Engineers. Kennedy was elected fellow of the Royal Society in 1887 and was knighted in 1905. He received honorary degrees from the universities of Glasgow (1894), Birmingham (1909), and Liverpool (1913). He was married in 1874 to Elizabeth Verralls, eldest daughter of William Smith of Edinburgh.

Throughout Kennedy's life there was a consistent thread of controlled quantitative testing of materials and machines, apparently stemming from his conviction, stated obliquely when he was president of Section G of the British Association, that "the essence of science may be rightly summed up in [the] one word 'measurement.' "

BIBLIOGRAPHY

I. ORIGINAL WORKS. Kennedy translated Franz Reuleaux's *Theoretische Kinematik* as *The Kinematics of Machinery: Outlines of a Theory of Machines* (London, 1876; repr. New York, 1963). Among his writings are "The Use and Equipment of Engineering Laboratories," in *Minutes of Proceedings of the Institution of Civil Engineers*, **88**, pt. 2 (1886–1887), 1–153, including over 70 pp. devoted to discussion of the paper; "Experiments Upon the Transmission of Power by Compressed Air in Paris," in F. E. Idell, ed., *Compressed Air* (New York, 1892), pp. 7–52; and *The Mechanics of Machinery* (London, 1886; 4th ed., 1902).

II. SECONDARY LITERATURE. See *Dictionary of National Biography, 1922–1930*, pp. 464–466. The best biographical sketch is *Minutes of Proceedings of the Institution of Civil Engineers*, **227** (1929), 269–275, which cites a number of minor papers and addresses. A portrait appears as the frontispiece in *Minutes of Proceedings of the Institution of Civil Engineers*, **167** (1907).

EUGENE S. FERGUSON

KENNELLY, ARTHUR EDWIN (*b*. Colaba, near Bombay, India, 17 December 1861; *d*. Boston, Massachusetts, 18 June 1939), *electrical engineering*.

Kennelly was the son of David Joseph Kennelly, an Irish-born employee of the East India Company who later became a barrister and practiced in England and Canada, and Kathrine Heycock Kennelly, English-born daughter of a Bombay cotton-mill owner. She died when Kennelly and his older sister were small children; the father later remarried twice and had ten more children. The boy was educated in Britain and on the Continent but did not attend a university.

His interest in engineering having been aroused by a lecture by Latimer Clark on submarine telegraphy, Kennelly left school at fourteen to become an office boy at the Society of Telegraph Engineers (predecessor of the Institution of Electrical Engineers). At fifteen he became a telegraph operator for the Eastern Telegraph Company, whose employee he remained for ten years, acquiring an engineering education through practice and independent study.

In 1887 Kennelly immigrated to the United States, where he became an assistant to Thomas A. Edison and a consulting engineer; in 1894 he founded his own consulting firm with Edwin J. Houston but continued to be active in his own specialty, submarine cables. In 1902 he was appointed professor of engineering at Harvard University, a post he held until he retired in 1930. Between 1913 and 1924 Kennelly had a second appointment at the Massachusetts Institute of Technology. During the remainder of his career he made important contributions in three areas: the theory and practice of electrical engineering, the study of the ionosphere, and the evolution of electrical units and standards.

Kennelly's principal contribution to electrical engineering arose from an early interest (contemporaneously with C. P. Steinmetz) in the representation of alternating-current quantities by complex variables; his first publication on that subject appeared in 1893. A little later another great contemporary with whom his name was to be linked on several occasions, Oliver Heaviside, proposed the representation of the distribution of current and voltage in a cable by hyperbolic functions; Kennelly extended that notion by the use of complex hyperbolic functions and also introduced polar notation for the complex quantities—that is, using $re^{i\theta}$ instead of $x + iy$, where $r = +\sqrt{x^2 + y^2}$ and $\theta = \arctan(y/x)$—an innovation of considerable pedagogical and practical value.

Kennelly's best-known contribution is his suggestion, following Marconi's success in bridging the Atlantic by a radiotelegraphic signal in 1901, that radio waves must be reflected from a discontinuity in the ionized upper atmosphere. Soon thereafter the same explanation occurred independently to Heaviside, and the name Kennelly-Heaviside layer was given to the region; it is now known as the ionosphere.

In his third major activity Kennelly's interests again overlapped with Heaviside's: both were interested in the evolution of electrical notation, units, and standards. Kennelly served as president of the American Metric Society, officer of the Metric Association, secretary of the standards committee of the American Institute of Electrical Engineers (AIEE), and secretary and president of the U.S. National Committee of the International Electrotechnical Commission, which he had helped found in 1904. He was instrumental in the adoption of a uniform nomenclature and of the meter-kilogram-second (mks) system as an international standard. Kennelly also was president of the AIEE (1898–1900), the Illuminating Engineering Society (1911), the Institute of Radio Engineers (1916), and the International Radio Scientific Union (honorary, 1935). He received several honorary degrees and many medals, including the AIEE's Edison Medal, and was elected to membership of the U.S. National Academy of Sciences and the Swedish Academy.

In 1903 Kennelly married Julia Grice, a physician. They had a daughter, who died in infancy, and a son.

BIBLIOGRAPHY

Kennelly published more than 350 papers and 28 books (18 as coauthor). A bibliography follows the biography by

Vannevar Bush in *Biographical Memoirs. National Academy of Sciences*, **22** (1943), 83–119. Another biography, by C. L. Dawes, is in *Dictionary of American Biography*, XXII (1958), 357–359. An appreciation by the same author appears in *Science*, **90** (1939), 319–321; and by others in *American Philosophical Society Yearbook* for 1939, pp. 453–457, and in *Transactions of the Illuminating Engineering Society*, **34** (1939), 661. Some of Kennelly's MSS and correspondence are in the Harvard University archives.

CHARLES SÜSSKIND

KEPLER, JOHANNES (*b.* Weil der Stadt, Germany, 27 December 1571; *d.* Regensburg, Germany, 15 November 1630), *astronomy, physics.*

Although Kepler is remembered today chiefly for his three laws of planetary motion, these were but three elements in his much broader search for cosmic harmonies and a celestial physics. With the exception of Rheticus, Kepler became the first enthusiastic Copernican after Copernicus himself; he found an astronomy whose clumsy geocentric or heliostatic planetary mechanisms typically erred by several degrees and he left it with a unified and physically motivated heliocentric system nearly 100 times more accurate.

When Kepler was twenty-five and much occupied with astrology, he compared the members of his family with their horoscopes.[1] His grandfather Sebald, mayor of Weil in 1571, when Kepler was born, was "quick-tempered and obstinate." His grandmother was "clever, deceitful, blazing with hatred, the queen of busybodies." His father, Heinrich, was described as "criminally inclined, quarrelsome, liable to a bad end" and destined for a "marriage fraught with strife." When Kepler was three years old, his father joined a group of mercenary soldiers to fight the Protestant uprising in Holland, thereby disgracing his family. Soon after his return in 1576, he again joined the Belgian military service for a few years; and in 1588 he abandoned his family forever.

Although Kepler describes his mother, the former Katharina Guldenmann, as "thin, garrulous, and bad-tempered," he adds that "treated shabbily, she could not overcome the inhumanity of her husband." Katharina showed her impressionable son the great comet of 1577. Later, Kepler spent many months between 1617 and 1620 preparing a legal defense when his aged but meddlesome mother was accused of and tried for witchcraft.

Kepler first attended the German Schreibschule in Leonberg, where his family had moved in 1576; shortly after, he transferred to the Latin school, there laying the foundation for the complex Latin style displayed in his later writings. In 1584 he entered the Adelberg monastery school; and two years later enrolled at Maulbronn, one of the preparatory schools for the University of Tübingen. In October 1587 Kepler formally matriculated at Tübingen; but because no room was available at the Stift, the seminary where, as a scholarship student supported by the duke of Württemberg, he was expected to lodge, he continued at Maulbronn for another two years. On 25 September 1588 he passed the baccalaureate examination at Tübingen, although he did not actually take up residence there until the following year.

At Tübingen, Kepler's thought was profoundly influenced by Michael Maestlin, the astronomy professor. Maestlin knew Copernican astronomy well; the 1543 *De revolutionibus* he owned is probably the most thoroughly annotated copy extant; he edited the 1571 edition of the *Prutenicae tabulae*, and he used them to compute his own *Ephemerides.* Although Maestlin was at best a very cautious Copernican, he planted the seed that with Kepler later blossomed into a full Copernicanism. The ground was fertile. Kepler's quarterly grades at the university, still preserved, show him as a "straight A" student; and when he applied for a scholarship renewal at Tübingen, the senate noted that he had "such a superior and magnificent mind that something special may be expected of him." Nevertheless, Kepler himself wrote concerning the science and mathematics of his university curriculum that "these were the prescribed studies, and nothing indicated to me a particular bent for astronomy."[2]

On 11 August 1591 Kepler received his master's degree from Tübingen and thereupon entered the theological course. Halfway through his third and last year, however, an event occurred that completely altered the direction of his life. Georgius Stadius, teacher of mathematics at the Lutheran school in Graz, died; and the local authorities asked Tübingen for a replacement. Kepler was chosen; and although he protested abandoning his intention to become a clergyman, he set out on the career destined to immortalize his name.

Graz and the Mysterium Cosmographicum. On 11 April 1594, the twenty-two-year-old Kepler arrived in southern Austria to take up his duties as teacher and as provincial mathematician. In the first year he had few pupils in mathematical astronomy and in the second year none, so he was asked to teach Vergil and rhetoric as well as arithmetic. But the young Kepler made his mark in another way; soon after coming to Graz, he issued a calendar and prognostication for 1595, which contained predictions of bitter cold, peasant uprisings, and invasions by the

KEPLER

Turks. All were fulfilled, to the great enhancement of his local reputation. Five more calendars followed in annual succession, and later in Prague he issued prognostications for 1602 to 1606. These ephemeral items are now extremely rare, some surviving in unique copies; and all the copies of nearly half the editions are totally lost.

Kepler's personal reaction to astrology was mixed. He rejected most of the commonly accepted rules, and he repeatedly referred to astrology as the foolish little daughter of respectable astronomy. In *De fundamentis astrologiae certioribus* (1601) he wrote: "If astrologers sometimes do tell the truth, it ought to be attributed to luck."[3] Nevertheless, his profound feeling for the harmony of the universe included a belief in a powerful concord between the cosmos and the individual. These views found their fullest development in the *Harmonice mundi.* Furthermore, his astrological opinions continually provided welcome supplementary income and later became a significant justification for his office as imperial mathematician. At least 800 horoscopes are still preserved in his manuscript legacy. Included are many for himself; if we are to believe the deduced time of his conception (16 May 1571, at 4:37 A.M. on his parents' wedding night), then he was a seven-month baby.

Concerning the calendars, Kepler later wrote: "Because astrology has no language other than that used by common man, so the common man will not understand otherwise, knowing nothing of the generalities of abstractions and seeing only the concrete, will often praise a calendar in an accidental case that the author never intended or blame it when the weather doesn't come as he expects: so much trouble have I brought upon myself, that I finally have given up writing calendars."[4] Nevertheless, Kepler later produced a series from 1618 to 1624, excusing himself with the remark that when his salary was in arrears, writing calendars was better than begging.

Meanwhile, just over a year after his arrival in Graz, Kepler's fertile imagination hit upon what he believed to be the secret key to the universe. His own account, here greatly abridged, appears in the introduction to the resulting work, the *Mysterium cosmographicum* of 1596.

> When I was studying under the distinguished Michael Maestlin at Tübingen six years ago, seeing the many inconveniences of the commonly accepted theory of the universe, I became so delighted with Copernicus, whom Maestlin often mentioned in his lectures, that I often defended his opinions in the students' debates about physics. I even wrote a painstaking disputation about the first motion, maintaining that it happens because of the rotation of the earth. I have by degrees—partly out of hearing Maestlin, partly by myself—collected all the advantages that Copernicus has over Ptolemy. At last in the year 1595 in Graz when I had an intermission in my lectures, I pondered on this subject with the whole energy of my mind. And there were three things above all for which I sought the causes as to why it was this way and not another—the number, the dimensions, and the motions of the orbs.[5]

After describing several false attempts, Kepler continues:

> Almost the whole summer was lost with this agonizing labor. At last on a quite trifling occasion I came nearer the truth. I believe Divine Providence intervened so that by chance I found what I could never obtain by my own efforts. I believe this all the more because I have constantly prayed to God that I might succeed if what Copernicus had said was true. Thus it happened 19 July 1595, as I was showing in my class how the great conjunctions [of Saturn and Jupiter] occur successively eight zodiacal signs later, and how they gradually pass from one trine to another, that I inscribed within a circle many triangles, or quasi-triangles such that the end of one was the beginning of the next. In this manner a smaller circle was outlined by the points where the lines of the triangles crossed each other [see Fig. 1].

The proportion between the circles struck Kepler's eye as almost identical with that between Saturn and Jupiter, and he immediately initiated a vain search for similar geometrical relations.

> And then again it struck me: why have plane figures among three-dimensional orbits? Behold, reader, the invention and whole substance of this little book! In memory of the event, I am writing down for you the sentence in the words from that moment of conception: The earth's orbit is the measure of all things; circumscribe around it a dodecahedron, and the circle containing this will be Mars; circumscribe around Mars a tetrahedron, and the circle containing this will be Jupiter; circumscribe around Jupiter a cube, and the circle containing this will be Saturn. Now inscribe within the earth an icosahedron, and the circle contained in it will be Venus; inscribe within Venus an octahedron, and the circle contained in it will be Mercury. You now have the reason for the number of planets.

Kepler of course based his argument on the fact that there are five and only five regular polyhedrons.

> This was the occasion and success of my labors. And how intense was my pleasure from this discovery can never be expressed in words. I no longer regretted the time wasted. Day and night I was consumed by the computing, to see whether this idea would agree with the Copernican orbits, or if my joy would be carried away by the wind. Within a few days everything worked, and I watched as one body after another fit precisely into its place among the planets.

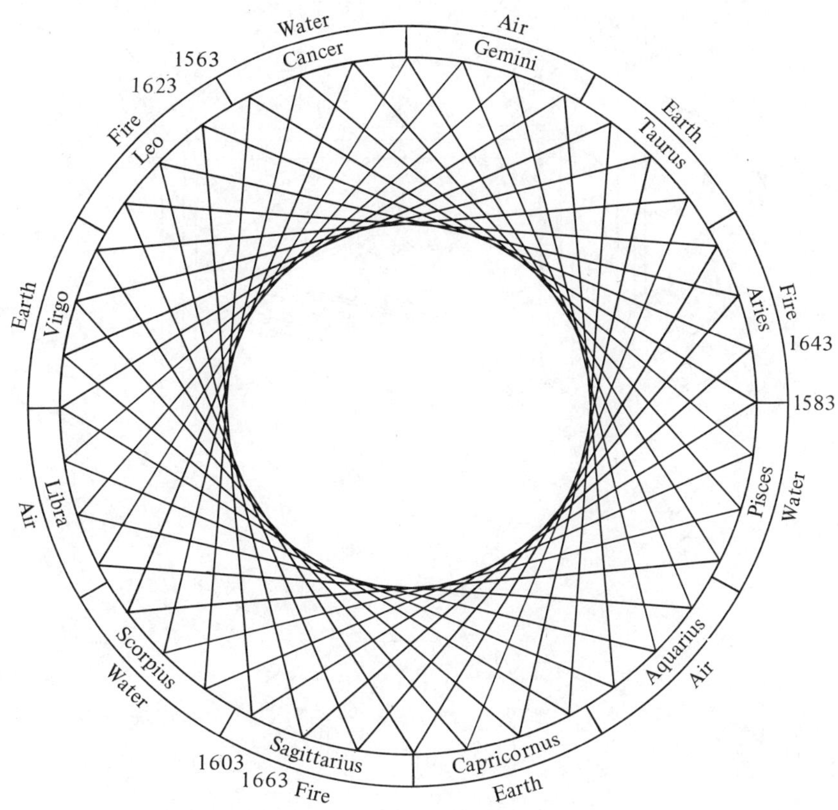

FIGURE 1. The pattern of Jupiter–Saturn conjunctions with the astrological elements added (*Mysterium cosmographicum*).

Astonishingly, Kepler's scheme works with fair accuracy when space is allowed for the eccentricities of the planetary paths. The numbers are given in Table I. Kepler was obliged to compromise the elegance of his system by adopting the second value for Mercury, which is the radius of a sphere inscribed in the square formed by the edges of the octahedron, rather than in the octahedron itself. With this concession, everything fits within 5 percent—except Jupiter, at which "no one will wonder, considering such a great distance."

TABLE I. Ratios of adjacent planetary orbits (assuming the innermost part of the outer orbit to be 1000).

Planet	Intervening Regular Solid	Computed by Kepler	From Copernicus
Saturn			
	Cube	577	635
Jupiter			
	Tetrahedron	333	333
Mars			
	Dodecahedron	795	757
Earth			
	Icosahedron	795	794
Venus			
	Octahedron	577 or 707	723
Mercury			

Quixotic or chimerical as Kepler's polyhedrons may appear today, we must remember the revolutionary context in which they were proposed. The *Mysterium cosmographicum* was essentially the first unabashedly Copernican treatise since *De revolutionibus* itself; without a sun-centered universe, the entire rationale of his book would have collapsed. Moreover, even the inquiry about the basic causes for the number and motions was itself a novel break with the medieval tradition, which considered the "naturalness" of the universe sufficient reason. For Kepler, the theologian-cosmologist, nothing was more reasonable than to search for the architectonic principles of creation. "I wanted to become a theologian," he wrote to Maestlin in 1595; "for a long time I was restless. Now, however, behold how through my effort God is being celebrated in astronomy."[6]

Furthermore, Kepler demanded to know how God the architect had set the universe in motion. He recognized that although in Copernicus' system the sun was near the center, it played no physical role. Kepler argued that the sun's centrality was essential, for the sun itself must provide the driving force to keep the

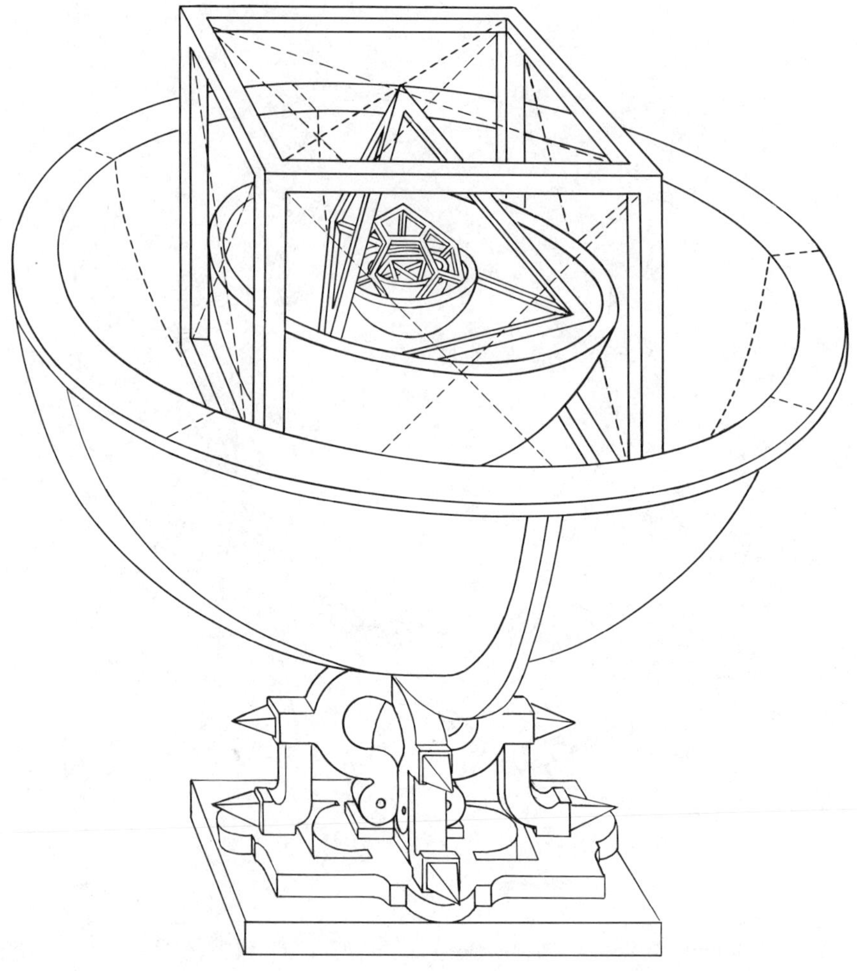

FIGURE 2. Kepler's nested polyhedrons and planetary spheres (*Mysterium cosmographicum*, 1596).

planets in motion. This physical reasoning, which characterizes Kepler's astronomy, makes its appearance in the latter part of the *Mysterium cosmographicum*. After announcing his celebrated nest of spheres and regular solids, which to him explained the spacing of the planets, he turned to the search for the basic cause of the regularities in the periods.

Kepler knew that the more distant a planet was from the sun, the longer its period—indeed, this was one of the most important regularities of the heliocentric system, already noted by Copernicus, that had appealed so strongly to Kepler's aesthetic sense. Kepler believed that the longer periods directly reflected the diminution with distance of the sun's driving force. Thus, he sought to relate the planetary periods ($P_1, P_2, \cdots$) to the intervals between the planets; with this step he had gone from the heliostatic scheme of Copernicus to a physically heliocentric system. After several trials he formulated a relation for the ratios of the distances equivalent to $(P_1/P_2)^{1/2}$ rather than the correct $(P_1/P_2)^{2/3}$, but this gave a sufficiently satisfactory first result, as seen in Table II.

TABLE II. Mean ratios of the planetary orbits.

Planets	Computed by Kepler	From Copernicus
Jupiter/Saturn	574	572
Mars/Jupiter	274	290
Earth/Mars	694	658
Venus/Earth	762	719
Mercury/Venus	563	500

Although the principal idea of the *Mysterium cosmographicum* was erroneous, Kepler established himself as the first, and until Descartes the only, scientist to demand physical explanations for celestial phenomena. Seldom in history has so wrong a book been so seminal in directing the future course of science.

As an impecunious young instructor, Kepler submitted his manuscript to the scrutiny of Tübingen University because his publisher would go ahead only with the approval of the university authorities. Without dissent the entire senate endorsed the publication of Kepler's militantly pro-Copernican treatise, but they requested that he explain his discovery and also Copernicus' hypotheses in a clearer and more popular style. In the actual publication the reasons for abandoning the Ptolemaic in favor of the Copernican system are set forth in the first chapter with remarkable lucidity. J. L. E. Dreyer has noted that "it is difficult to see how anyone could read this chapter and still remain an adherent of the Ptolemaic system."[7]

The Tübingen senate also recommended that Kepler delete his "discussion of the Holy Writ in several theses." This Kepler did, but he later incorporated his arguments into the introduction of his *Astronomia nova:*

> But now the Sacred Scriptures, speaking to men of vulgar matters (in which they were not intended to instruct men) after the manner of men, that so they might be understood by men, do use such expressions as are granted by all. . . . What wonder is it then, if the scripture speaks according to man's apprehension, at such time when the truth of things doth dissent from the conception [of] all men?

This version, from Thomas Salusbury in 1661, is a part of the first and only seventeenth-century translation of any of Kepler's works. The passage was also repeatedly reprinted as an appendix to the Latin translation of Galileo's *Dialogo.* In the words of Edward Rosen, "Kepler's clarion call, trumpeted to receptive ears, echoed and reechoed down the corridors of the seventeenth century and thereafter. It demonstrated how unswerving allegiance to the scientific quest for truth could be combined in one and the same person with unwavering loyalty to religious tradition: accept the authority of the Bible in questions of morality, but do not regard it as the final work in science."[8]

As soon as the *Mysterium cosmographicum* arrived from the printer early in 1597, Kepler sent copies to various scholars. By return courier Galileo sent a few civil sentences saying that he had as yet read only the preface. Kepler, unsatisfied, sent a spirited reply urging Galileo to "believe and step forth." Tycho Brahe offered a detailed critique, calling the nest of inscribed spheres and polyhedrons a clever and polished speculation. Kepler's book, notwithstanding its faults, had thrust him into the front rank of astronomers. Looking back as a man of fifty, Kepler remarked that the direction of his entire life and work took its departure from this little book.

Kepler had entitled his book *Prodromus dissertationum cosmographicarum continens mysterium cosmographicum . . .* ("A Precursor to Cosmographical Treatises . . ."), thus implying a continuation. Following the publication of his first book, Kepler plunged into studies for not one but four cosmological treatises. His interests ranged from the observation of lunar and solar eclipses—he first found the so-called annual equation of the moon's motion—to chronology and harmony. By 1599 he had outlined the plan for one of his principal works, the *Harmonice mundi.* Yet fate, in the form of the gathering storm of the Counterreformation, once more diverted the course of Kepler's life; and the *Harmonice* was not completed until 1619.

Meanwhile, another discovery molded Kepler's life: the eldest daughter of a wealthy mill owner, Barbara Müller, had "set his heart on fire." Two years younger than Kepler, she had been widowed twice. Early in 1596 Kepler sought her hand, but his seven-month absence on a trip to Tübingen almost scuttled the courtship. The wedding took place 27 April 1597, under ominous constellations, as Kepler noted in his diary. The initial happiness of his marriage gradually dissolved as he realized that his wife understood nothing of his work—"fat, confused, and simple-minded" was Kepler's later description of her. The early death of his first two children grieved him deeply. His wife's fortune was tied into estates, so it was difficult to transfer their assets when the Lutheran Kepler was forced to abandon Catholic Graz and move to Prague. There Kepler was eventually to find an exhilarating freedom, but his wife, out of her depth in court circles, found only homesickness and monetary worries.

Prague and the Astronomia Nova. The numerous Protestants in Graz remained unmolested by their Catholic rulers until mid-September 1598. On 28 September, all the teachers, including Kepler, were abruptly ordered to leave town before sunset. Although, unlike his colleagues, he was allowed to return, conditions remained tense; and Kepler tried vainly to secure a position at Tübingen. In August 1599 he learned that Tycho Brahe had gone to the court of Rudolph II in Bohemia, so he set out in January 1600 for an exploratory visit to the great Danish astronomer, arriving at Tycho's Benatky Castle observatory outside Prague early in February.

Although Tycho welcomed him "not so much as a guest but as a highly desirable participant in our observations of the heavens," he promptly treated the sensitive Kepler as a beginner. Kepler at first had little opportunity to participate except at meals, "where one day Tycho mentioned the apogee of one planet, the next day the nodes of another."[9] Conscious of his own genius, Kepler expected to be regarded as an independent investigator; plagued by the financial worries as well as the uncertainties of his position, either in Graz or in Prague, he brought the matter to a heated crisis early in April. Happily, a reconciliation followed, and Kepler worked another month at Benatky before going to Prague and thence back to Graz.

Kepler had quickly perceived the quality of Brahe's treasure of observations, but he realized that Tycho lacked an architect for the erection of a new astronomical structure. By Divine Providence, as he was later to view it, Kepler was assigned to the theory of Mars; and in his three months at Benatky he established two fundamental points: first, the orbital place of Mars must be referred to the true sun, and not to the center of the earth's orbit, as previous astronomers had assumed; and second, the traditional mechanism for the earth-sun relation had to be modified to include an equant. The equant, a seat of uniform angular motion within a circular orbit, satisfied Kepler's physical intuition that a planet must move proportionally more quickly when it is closer to the sun (see Fig. 3). Although the other planetary mechanisms

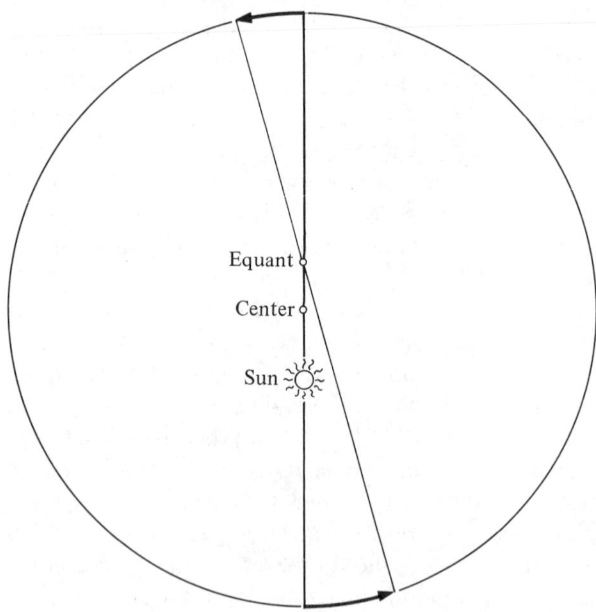

FIGURE 3. The unequally spaced equant of the vicarious hypothesis.

had traditionally employed the equant, the earth-sun system did not. Hence, it was of paramount importance to Kepler's physics to prove that the earth's motion resembled those of the other planets, and this he accomplished by an ingenious triangulation from the earth's orbit to Mars.

Kepler continued his astronomical studies after his return to Graz, working on geometrical theorems relating to the Mars problem and building a projection device for observing the solar eclipse of 10 July 1600. Shortly after, his work was interrupted by a commission of the Counter-reformation; Kepler was examined on 2 August and was among the sixty-one men banished from Graz for refusing to change their faith. Although he was uncertain which way to turn, the deadline for leaving allowed little time for negotiations; consequently, Kepler left Graz with his family on 30 September. At Linz a hoped-for letter from Maestlin had not arrived. Depressed and in poor health, Kepler arrived in Prague on 19 October. Tycho gladly took Kepler back, especially because his chief assistant, Longomontanus, had just resigned.

Kepler resumed his work on Mars, notably his attempt to fit the observations with a circular orbit and equant. He departed from the traditional procedure by allowing the equant to fall at an arbitrary point along the line joining the sun and the center of the orbit (see Fig. 3); in principle the minimum error in heliocentric longitude from this model is only about one minute of arc. By the spring of 1601 he had achieved a far more accurate solution for the longitudes than had any of his predecessors, but the latitudes were not satisfactory.

Meanwhile, Tycho had assigned to Kepler the unhappy task of composing a defense against Nicolaus Raymarus Ursus, whom Tycho accused of plagiarism. Kepler's contacts with Ursus dated back to 1595, when, as a still-unknown youth, he had written Ursus a letter praising him as the leading mathematician of the age. In 1597 Ursus incorporated the letter into a venomous attack on Brahe. The embarrassed Kepler blamed the extravagance of his letter on his own immaturity, but the incident continued to rankle Tycho, who must have found grim satisfaction in requiring Kepler to write the rebuttal. Kepler, however, took the opportunity to analyze the nature of scientific hypotheses and to sharpen his own arguments on the truth of the Copernican premises. "If in their geometrical conclusions two hypotheses coincide, nevertheless in physics each will have its own peculiar additional consequence."[10] Thus the stage was set for the critical distinction between the "vicarious" and the "physical" hypotheses in his Mars researches the following spring.

Kepler returned to Graz in April 1601, on a futile trip to look after his wife's inheritance. The visit dragged on, his wife wrote from Prague of her financial worries, and Kepler responded indignantly to Brahe. Kepler returned to Prague at the end of August, and the differences with Brahe were patched up. Nevertheless, Kepler continued to chafe under the secretive jealousy with which Tycho guarded his observations. Then, suddenly, the Danish astronomer fell ill; and on 24 October 1601 he died. On his deathbed Tycho urged Kepler to complete the proposed *Rudolphine Tables* of planetary motion, adding his hope that they would be framed according to the Tychonic hypothesis. Within two days Kepler received the appointment to Tycho's post of imperial mathematician, although five months passed before he received his first salary.

Kepler's encounter with Tycho had been a fateful one—"God let me be bound with Tycho through an unalterable fate and did not let me be separated from him by the most oppressive hardships,"[11] he wrote—yet he had worked with the Danish master altogether less than ten months. Kepler always spoke of Tycho with high esteem; but clearly Tycho's unexpected death freed Kepler to work out the planetary theory without the continual strain that had characterized their relationship.

As the first step toward the construction of the *Rudolphine Tables*, Kepler continued to perfect the quasi-traditional circular orbit with its equant, which yielded heliocentric longitudes accurate to 2′; "If you are wearied by this tedious procedure," he later implored his readers, "take pity on me who carried out at least seventy trials."[12] From the predicted latitudes, however, he realized that his model gave erroneous distances; unlike previous astronomers, who were satisfied with separate mathematical mechanisms for the longitudes and latitudes, Kepler sought a unified, physically acceptable model. Thus, by the spring of 1602 he began to distinguish between the "vicarious hypothesis" that he had achieved and the desired "physical hypothesis." To obtain the correct distances that a physical model demanded, he was obliged to reposition his circular orbit with its center midway between the sun and the equant (unlike Fig. 3). With this bisected eccentricity, the error in heliocentric longitude rose to 6′ or 8′ in the octants. "Divine Providence granted us such a diligent observer in Tycho Brahe," wrote Kepler, "that his observations convicted this Ptolemaic calculation of an error of 8′; it is only right that we should accept God's gift with a grateful mind. . . . Because these 8′ could not be ignored, they alone have led to a total reformation of astronomy."[13]

Kepler now revised his earlier speculations on the planetary driving force emanating from the sun. Jean Taisner's book on the magnet (1562) and, later, William Gilbert's convinced him that the force might be magnetic. Kepler envisioned a rotating sun with a rotating field of magnetic emanations that continuously drove the planets in their orbits. He supposed that such a force would act only in the planes of the orbits, and consequently (unlike light) would diminish inversely with distance. Kepler's new model with bisected eccentricity, especially of the earth's orbit, enabled him to formulate what we can call his distance law: that the orbital velocity of a planet is inversely proportional to its distance from the sun. Although this holds strictly only at aphelion and perihelion, Kepler promptly generalized the relation to the entire orbit. Controlling the angular motion by his distance law immediately raised a difficult quadrature problem that could be solved only by tedious numerical summations. Here he had the fortunate inspiration to replace the sums of the radius vectors (that is, the lines from the sun to the planet) required by the distance law with the area within the orbit. Thus the radius vector swept out equal areas in equal times. Kepler recognized that this was mathematically objectionable, but like a miracle the predicted longitudes matched the observations. Today it is called his second law, but nowhere in his great book on Mars is the area rule clearly stated. Kepler properly understood its fundamental nature only later, when he based the calculations of the *Rudolphine Tables* on it; and both the area law and a revised distance law are correctly stated in book V of his *Epitome astronomiae Copernicanae* (1621).

Whereas the area law worked well for the earth's orbit, when it was applied to Mars the eight-minute discrepancy again appeared. Kepler recognized at once that a noncircular orbit could provide a solution, although the area law itself was still suspect. A triangulation to three points on the Martian orbit confirmed that Mars's path bowed in from a circle, but the exact amount was difficult to establish. Kepler now resumed an exploration of the effects of a small epicycle, which he had started in 1600. He knew from the traditional model for Mercury that its epicycle produced an oval or, more properly, an ovoid curve. His attempts to find a quadrature for the ovoid and to confirm the area law led, in Kepler's own words, to a veritable labyrinth of calculation. In fact, the difference between the longitudes generated from the distance law and the area law reaches 4′, precariously close to the eight-minute discrepancy that had driven him to a renewed assault on the problem. Writing to David Fabricius in July 1603, Kepler noted, "I lack only a knowledge of the geometric generation of the oval or face-shaped curve. . . . If the figure were a perfect ellipse, then Archimedes and Apollonius would be enough."[14] As shown with exaggerated eccentricity in Figure 4, the ovoid is quite similar to an ellipse; but since the approximating ellipse has an eccentricity of $\sqrt{2}\,e$ (where e is the eccentricity of the true ellipse), it has no physical connection with the sun.

KEPLER

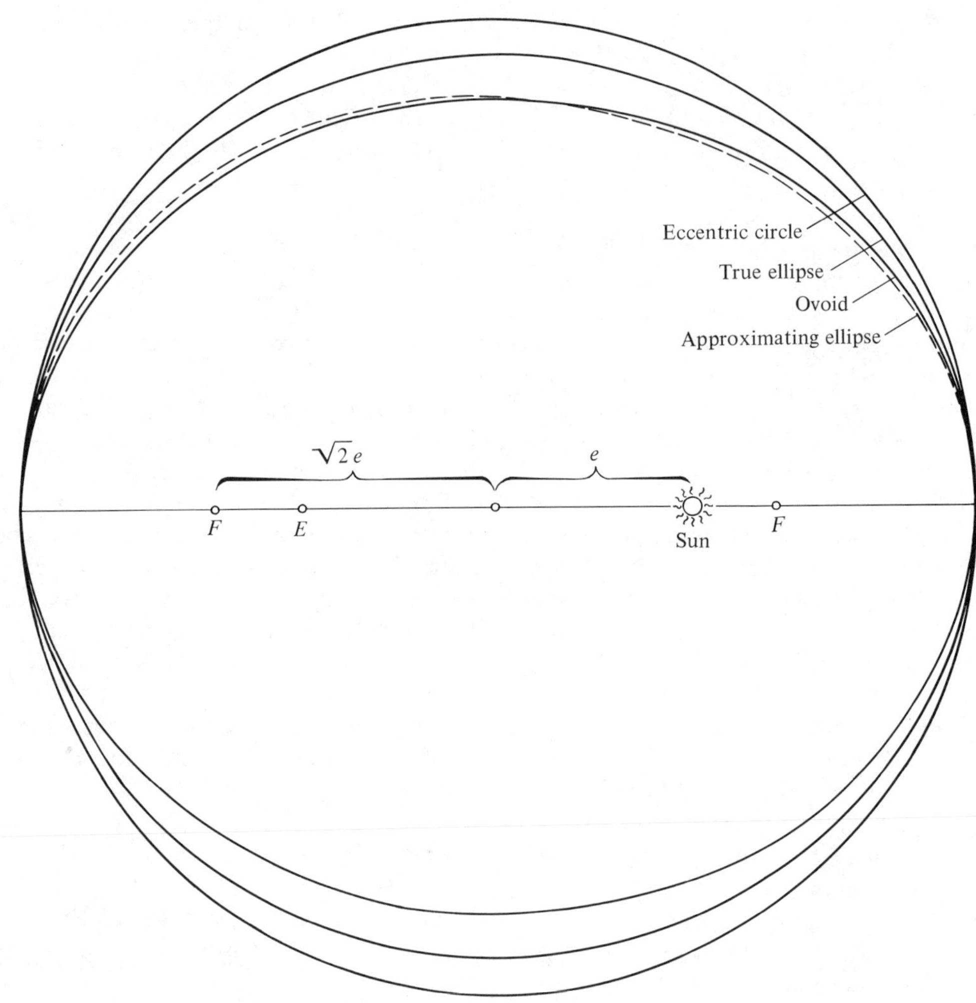

FIGURE 4. The ovoid, its approximating ellipse (broken line), and the true ellipse, with greatly exaggerated eccentricity. *F* designates the foci of the approximating ellipse, *E* the "empty focus" of the true ellipse.

Kepler spent most of his effort from September 1602 to the end of 1603 on his *Astronomiae pars optica* (see below), but in 1604 he began in earnest to prepare his *Astronomia nova* or *Commentarius de stella martis.* In December he wrote to Fabricius: "I am now completely immersed in the *Commentarius*, so that I can hardly write fast enough." By early 1605 he had completed fifty-one chapters without yet discovering the ellipse. By now, however, Kepler must have realized that his problems lay not with an inadequate quadrature but in his defective knowledge of the size of the crescent-shaped lunula by which the path of Mars departed from a circle.

His renewed assault included a revised triangulation to the Martian orbit, carried out by coupling observed geocentric position angles with the heliocentric longitudes predicted by his accurate but physically unacceptable vicarious hypothesis. The results showed that the true orbit lay midway between the oval and circle (see Fig. 4), and thus his elaborately reasoned physical causes for the ovoid itself "went up in smoke."

These physical causes were an extension of the magnetic emanation that drove the planets around the sun: The sun's presumed unipolar magnetism could act on a planet's magnetic axis with a fixed direction in space, and could alternately attract and repel the planet from the sun. Thus the same mechanism might, he hoped, account for both the varying speed of a planet in its orbit and its varying distance from the sun. His goal, as he wrote J. G. Herwart von Hohenburg in February 1605, was "to show that the celestial machine is not so much a divine organism but rather a clockwork . . . inasmuch as all the variety of motions are carried out by means of a single very simple magnetic force of the body, just as in a clock all the motions arise from a very simple weight."[15]

Kepler had invested so much speculative energy justifying the "librations" produced by the epicycle (even though he was simultaneously repelled by the epicycle's lack of physical properties) that he refused to abandon his idea. Thus, he now proceeded to another epicyclic construction, the *via buccosa* or "puffy-cheeked" (as distinguished from the oval, "face-shaped") curve. Although this curve matches within the accuracy of Tycho's observations, Kepler made a conceptual error in his calculation and therefore found a disagreement; thus he felt justified in rejecting it. By this time he realized that an ellipse would satisfy the observations but he could not at first connect an ellipse with the magnetic hypothesis. In chapter 58 he writes:

> I was almost driven to madness in considering and calculating this matter. I could not find out why the planet would rather go on an elliptical orbit. Oh, ridiculous me! As if the libration on the diameter could not also be the way to the ellipse. So this notion brought me up short, that the ellipse exists because of the libration. With reasoning derived from physical principles agreeing with experience, there is no figure left for the orbit of the planet except a perfect ellipse.

Kepler had thus arrived at what we now call his first law, that the planetary orbits are ellipses with the sun at one focus. With justifiable pride he could call his book "The New Astronomy." Its subtitle emphasizes its repeated theme: "Based on Causes, or Celestial Physics."

Unfortunately, publication of the work did not proceed promptly, partly from the lack of imperial financial support but mostly because of interference from the "Tychonians." Emperor Rudolph II had offered the heirs 20,000 talers for Tycho's observations but had actually paid only a few thousand. Consequently the heirs, particularly the nobleman Franz Gansneb Tengnagel (who had married Tycho's daughter), held a vested interest in the data. Tengnagel, although unqualified, promised his own publication and, after failing to produce, threatened to suppress Kepler's commentary. Then a compromise was reached: Tengnagel was allowed a preface to warn the readers not to "become confused by the liberties that Kepler takes in deviating from Brahe in some of his expositions, particularly those of a physical nature." The printing finally began at Heidelberg in 1608, and the *Astronomia nova* was published in the summer of 1609. Although the distribution of the large and magnificent folio was a privilege held by the emperor, Kepler eventually sold the edition to the printer in an attempt to recover part of his back salary.

The Nova of 1604. Of the various astrological matters on which Rudolph II sought Kepler's opinion, the great conjunction of Jupiter and Saturn in 1603 was particularly remarkable. Such conjunctions occur every twenty years; only comets were considered more ominous. As shown in Figure 1, the conjunctions fall in a regular pattern, so that a series of ten occurs within a particular zodiacal "trigon," one of the four sets designated by astrologers according to the Aristotelian elements; in an 800-year cycle the conjunctions pass through all four trigons. The conjunction at the end of 1603 marked the beginning of the 200-year series within the fiery trigon. Excitement reached a peak in October 1604, when a brilliant new star unexpectedly appeared within a few degrees of Jupiter, Saturn, and Mars.

Kepler at once published in German an eight-page tract on the new star, describing its appearance and comparing it to Tycho's nova of 1572. Then he allowed himself a frivolous prediction: the nova portended good business for booksellers, because every theologian, philosopher, physician, mathematician, and scholar would have his own ideas and would want to publish them.

Kepler's extensive collection of observations and opinions appeared in a longer work, *De stella nova*, in 1606. A subtitle announced it as "a book full of astronomical, physical, metaphysical, meteorological and astrological discussions, glorious and unusual." That it was. Early chapters described the nova's appearance, astrological significance, and possible origin. He rejected the possibility of a chance configuration of atoms, and in a charming passage presented

> . . . not my own opinion, but my wife's. Yesterday, when weary with writing, I was called to supper, and a salad I had asked for was set before me. "It seems then," I said, "if pewter dishes, leaves of lettuce, grains of salt, drops of water, vinegar, oil and slices of eggs had been flying about in the air from all eternity, it might at last happen by chance that there would come a salad." "Yes," responded my lovely, "but not so nice as this one of mine."[16]

In chapter 21 Kepler argued that stars are not suns, but his well-reasoned case rested on an erroneously large angular diameter of the stars. Finally, he meditated on the astrological interpretations of the nova: conversion of the Indians in America, a universal migration to the new world, the downfall of Islam, or even the return of Christ. His speculations broke off abruptly; as a "good and peaceful German," as he called himself, he had avoided controversy; and he urged his readers, in the presence of a celestial sign, to examine their sins and repent.

De stella nova is a monument of its time but the least significant of Kepler's major works. It broke no new astronomical ground, although twentieth-century astronomers have preferred its faithful descriptions over numerous other accounts when searching the literature to help distinguish supernovae from ordinary novae.

In terms of Kepler's own scholarly effort, the appendix was the most important part. In its dedicatory epistle he spoke of entering a "chronological forest," and the printer seized upon *Sylva chronologica* as the running title. In 1605 Kepler had come upon a tract by Lawrence Suslyga that argued for 5 B.C. as the date of Christ's birth. Noting that an initial conjunction in the fiery trigon, comparable with the one in 1603, presumably had occurred in 5 B.C., Kepler drew an analogy between the nova of 1604 and the star of the Magi; following Suslyga's arguments in part, he settled upon 5 B.C. as the year of Christ's birth. (A similar adjustment is commonly accepted today for chronological reasons.) Afterward, Kepler elaborated his arguments in several works, including his definitive account, *De vero anno . . .* (1614).

Optical Researches. Kepler's interest in optics arose as a direct result of his observations of the partial solar eclipse of 10 July 1600. Following instructions from Tycho Brahe, he constructed a pinhole camera; his measurements, made in the Graz marketplace, closely duplicated Brahe's and seemed to show that the moon's apparent diameter was considerably less than the sun's. Kepler soon realized that the phenomenon resulted from the finite aperture of the instrument (see Fig. 5); his analysis, assisted by actual threads, led to a clearly defined concept of the light ray, the foundation of modern geometrical optics.

Kepler's subsequent work applied the idea of the light ray to the optics of the eye, showing for the first time that the image is formed on the retina. He introduced the expression "pencil of light," with the connotation that the light rays draw the image upon the retina; he was unperturbed by the fact that the image is upside down.

Kepler constructed an expression for the traditional "angle of refraction," that is, the difference between the angles of the incident and refracted rays, as

$$i - r = n \cdot i \cdot \sec r,$$

where n is the index of refraction. He arrived at this result at least partly by theoretical considerations of the resistance offered by the denser medium. His formulation matched the somewhat erroneous data given by Witelo just as well as the correct sine law of refraction did. Descartes, the discoverer of the sine law in its modern form, acknowledged to Mersenne that "Kepler was my principal teacher in optics, and I think that he knew more about this subject than all those who preceded him."[17]

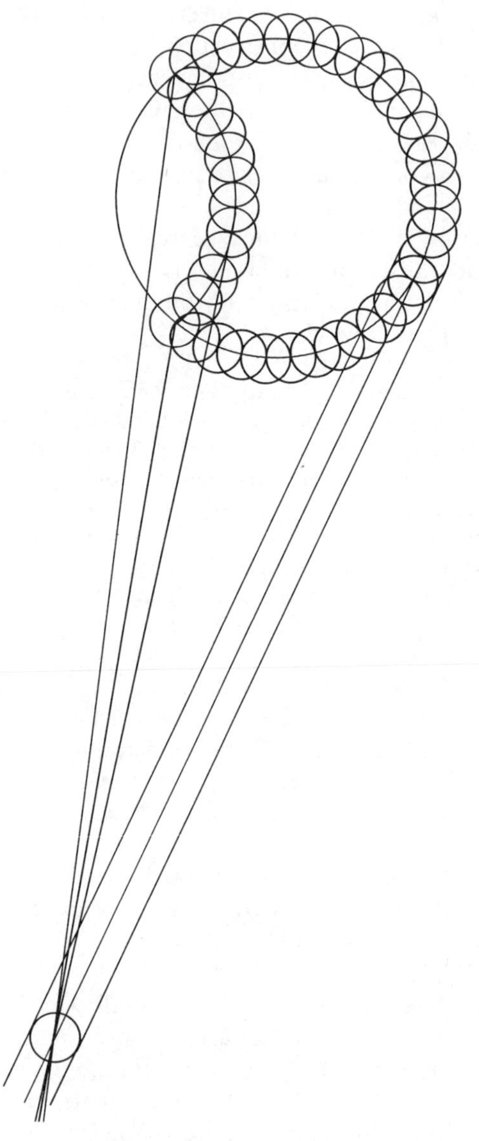

FIGURE 5. The formation of an eclipse image by a round aperture (Leningrad Kepler MS, vol. XV, fol. 250r).

Kepler intended at first to publish his optical analyses merely as *Ad Vitellionem paralipomena*, but by 1602 this "Appendix to Witelo" had taken second place to the broader program of *Astronomiae pars optica*. The book was published in 1604 with both titles, but Kepler regularly referred to his work by the latter. The six astronomical chapters include not only a discussion of parallax, astronomical refraction, and his eclipse instruments but also the annual variation in the apparent size of the sun. Since the changing size of the solar image is inversely proportional to the sun's distance, this key problem was closely related to his planetary theory; unfortunately, his observational results were not decisive.

The immediate impact of Kepler's optical work was not great; but ultimately it changed the course of optics, especially after his *Dioptrice* (1611), which applied these principles to the telescope. "Optical tubes" had been discussed in Giambattista della Porta's *Magia naturalis* (1589); but Kepler confessed that "I disparaged them most vigorously, and no wonder, for he obviously mixes up the incredible with the probable."[18] Thus Kepler, who himself used spectacles, discussed lenses only in passing in his *Astronomiae pars optica*. Nevertheless, he had set forth the essential background by which the formation of images with lenses could be explained, and so he was able to complete his *Dioptrice* within six months after he had received Galileo's *Sidereus nuncius* (1610). With great thoroughness Kepler described the optics of lenses, including a new kind of astronomical telescope with two convex lenses. The preface declares, "I offer you, friendly reader, a mathematical book, that is, a book that is not so easy to understand," but his severely mathematical approach only serves to place the *Dioptrice* all the more firmly in the mainstream of seventeenth-century science.

Minor Works. On 8 April 1610 Kepler received a copy of Galileo's *Sidereus nuncius*, and a few days later the Tuscan ambassador in Prague transmitted Galileo's request for an opinion about the startling new telescopic discoveries. What a contrast with 1597, when Kepler, an unknown high-school teacher, had sought in vain Galileo's reaction to his own book! Kepler was now the distinguished imperial mathematician, whose opinion mattered; he responded generously and quickly with a long letter of approval.

He promptly published his letter as *Dissertatio cum Nuncio sidereo;* in accepting the new observations with enthusiasm, he also reminded his readers of the earlier history of the telescope, his own work on optics, his ideas on the regular solids and on possible inhabitants of the moon, and his arguments against an infinite universe. A few months later, in the second of the only three known letters that Galileo wrote directly to Kepler, the Italian astronomer stated, "I thank you because you were the first one, and practically the only one, to have complete faith in my assertions."[19]

Not until August 1610 was Kepler's great desire to use a telescope satisfied, when Elector Ernest of Cologne lent him one. From 30 August to 9 September, Kepler observed Jupiter; he published the results in *Narratio de Jovis satellitibus* (1611), a booklet that was quickly reprinted in Florence. It provided a strong witness to the authenticity of the new discoveries.

Three other short works from Kepler's Prague period deserve mention. In *Phaenomenon singulare* (1609) he reported on a presumed transit of Mercury that he had observed on 29 May 1607. Unfortunately, Kepler had caught between clouds only two glimpses of a "little daub" that appeared on the solar image projected through a crack in the roof. After Galileo's discovery of sunspots, Maestlin pointed out the error and Kepler ultimately printed a retraction in his *Ephemerides* for 1617, noting that unwittingly he had been the first of his century to observe a sunspot.

In *Tertius interveniens* (1610) Kepler played the role of the "third man in the middle" both against those who uncritically accepted grotesque astrological predictions and against those critics who would "throw out the baby with the bath." Writing in the vernacular German interspersed with numerous scraps of Latin, Kepler argued: "No one should consider unbelievable that there should come out of astrological foolishness and godlessness also cleverness and holiness . . . out of evil-smelling dung a golden corn scraped for by an industrious hen."[20] As part of the dung he counted most astrological rules, including the distinctions of the zodiacal signs and the meanings of the twelve houses. Kepler insisted, however, on the harmonic significance of the configurations of the planets among themselves and with ecliptic points such as the ascendant. The stars do not compel, he said, but they impress upon the soul a special character.

In his *Strena* (1611), "A New Year's Gift, or On the Six-Cornered Snowflake," Kepler ponders the problem of why snowflakes are hexagonal. Composed as a present for his friend at court, Counselor J. M. Wackher, it is not only a charming letter, light-hearted and full of puns, but also a perceptive, pioneering study of the regular arrangements and the close packing that are fundamental in crystallography.

Linz. In the spring of 1609 Kepler journeyed to the Frankfurt Book Fair; to Heidelberg, where the *Astronomia nova* was being printed; and also, for the first time in thirteen years, to Tübingen. While in Swabia he attempted to pave the way for a permanent return to his homeland by sending a petition to the duke of Württemberg, reminding him how quickly troubles might arise, leaving Kepler unemployed. Nevertheless, under Rudolph's patronage his position in Prague seemed secure despite the increasing religious turmoil. But in 1611 Kepler's world suddenly collapsed. His wife became seriously ill; and his three children were stricken with smallpox. His favorite son died. Prague became a scene of bloodshed; and in May, Rudolph was forced to abdicate.

Kepler turned once more to the duke of Württemberg, but his cherished hopes of returning to Tübingen as a professor were finally dashed when the Württemberg theologians objected to his friendliness with Calvinists and to his reservations about the Formula of Concord. Earlier, Kepler had been invited to Linz as provincial mathematician, a post created specially for him; his decision to go there was motivated in part by a desire to find a town more congenial to his wife. Before the move took place, however, Barbara Kepler was infected by the typhus carried into Prague by the troops; and on 3 July 1611 she died. Meanwhile, the deposed Rudolph demanded Kepler's presence, and not until the monarch's death in January 1612 was the lonely astronomer free to leave Prague. Simultaneously his appointment as imperial mathematician was renewed.

Although Kepler's most creative period lay behind him, his fourteen-year sojourn in Linz eventually saw the production of his *Harmonice mundi* and *Epitome astronomiae Copernicanae* and the preparation of the *Tabulae Rudolphinae*. His stay in Linz started badly, however, for the local Lutheran pastor, who knew the opinion of the Württemberg theologians, excluded him from communion when he refused to sign the Formula of Concord. Kepler did not accept the exclusion willingly and produced repeated appeals to the Württemberg consistory and to his Tübingen teacher, Matthias Hafenreffer, but always in vain. While his coreligionists considered him a renegade, the Catholics tried to win him to their side. When the Counter-reformation swept into Linz in 1625, an exception was made so that he was not banished; but his library was temporarily sealed and his children were forced to attend the Catholic services. Thus Kepler, a peaceful and deeply religious man, suffered greatly for his conscience' sake throughout his life and especially in Linz.

One bright spot in his Linz career was his second marriage, to Susanna Reuttinger, a twenty-four-year-old orphan, on 30 October 1613. In an extraordinary letter to an unidentified nobleman, Kepler details his slate of eleven candidates for marriage and explains how God had led him back to number five, who had evidently been considered beneath him by his family and friends. The marriage was successful, far happier than the first; but of their seven children, five died in infancy or childhood. Likewise, only two of the five children of his first marriage survived to adulthood.

That Kepler, engulfed in a sea of personal troubles, published no astronomical works from 1612 through 1616 is not surprising. Yet he did produce the *Stereometria doliorum vinariorum* (1615), which is generally regarded as one of the significant works in the prehistory of the calculus. Desiring to outfit his new household with the produce of a particularly good wine harvest, Kepler installed some casks in his house. When he discovered that the wine merchant measured only the diagonal length of the barrels, ignoring their shape, Kepler set about computing their actual volumes. Abandoning the classical Archimedean procedures, he adopted a less rigorous but productive scheme in which he considered that the figures were composed of an infinite number of thin circular laminae or other cross sections. Captivated by the task, he extended it to other shapes, including the torus.

In 1615 Kepler brought the first printer to Linz, and thus the *Stereometria* was the first work printed there. He had hoped to profit from his book; and when sales lagged, he edited a considerably rearranged popular German version, the *Messekunst Archimedis* (1616). Incidentally, his work proved that the simple gauging rod was valid for Austrian wine casks.

Harmonice Mundi. The Linz authorities were not entirely pleased that Kepler had taken three-quarters of a year to produce the *Stereometria*, for his appointment had charged him first of all to complete the astronomical tables long since proposed by Tycho. They urged him to get on with the important *Tabulae Rudolphinae*. But to a correspondent, Kepler responded, "Don't sentence me completely to the treadmill of mathematical calculations—leave me time for philosophical speculations, my sole delight."[21]

In 1599 Kepler had drafted a plan for a work on the harmony of the universe; but since he had left Graz, his cosmological studies had lain comparatively dormant. Meanwhile, in 1607 he had finally obtained a Greek manuscript of Ptolemy's *Harmony*, and he was further stimulated by the Neoplatonic views of Proclus' commentary on Euclid. Alongside his studies of mathematical and musical harmony, he was formulating an astrological world view consonant with the laws of harmony. Thus, after he had completed his *Ephemerides* for 1617 in November 1616, he began to work intermittently on his *Harmonice mundi* but devoted most of his time to his tables and to the first part of his *Epitome*.

In the fall of 1617 Kepler was forced to journey to Württemberg to arrange the defense for his mother's witchcraft trial. In September the first daughter of his second marriage had died; and in February 1618, several weeks after his return to Linz, his second infant daughter died. Distraught and oppressed, he wrote, "I set the tables aside since they required peace, and turned my mind to refining the *Harmonice*."[22] A major work of 255 pages, the five parts of the *Harmonice mundi* were swiftly completed by 27 May

1618, although the final section was further revised while the type was being set.

Max Caspar, in his biography *Kepler*, gives an extended and perceptive summary of the *Harmonice*, concluding:

> Certainly for Kepler this book was his mind's favorite child. Those were the thoughts to which he clung during the trials of his life and which brought light to the darkness that surrounded him. . . . With the accuracy of the researcher, who arranges and calculates observations, is united the power of shaping of an artist, who knows about the image, and the ardor of the seeker for God, who struggles with the angel. So his *Harmonice* appears as a great cosmic vision, woven out of science, poetry, philosophy, theology, mysticism. . . .[23]

Kepler developed his theory of harmony in four areas: geometry, music, astrology, and astronomy. In the short book I he examined the geometry of polygons with an eye to their constructability. In the second book he investigated both polygons and polyhedrons, especially for their properties of filling a plane or space. He was here the first to introduce the great and small stellated dodecahedrons. (In 1810 Louis Poinsot rediscovered these, along with the two other examples of this class; the four polyhedrons are now known as the Kepler-Poinsot solids.)

Kepler's notion that the archetypal principles of the universe were based on geometry rather than on number found confirmation in book III, on musical harmony. He could not see any sufficient reason why God should have chosen the numbers 1, 2, 3, 4, 5, 6 for generating musical consonances and have excluded the numbers 7, 11, 13, and so on. He knew, of course, that the regular polygons produced only five regular polyhedrons and he sought some other procedure with polygons to yield the seven ratios that had only within his lifetime become commonly accepted as the basis for the "just" scale: 1/2, 2/3, 3/4, 4/5, 5/6, 3/5, and 5/8. By appropriately dividing those polygons that could be easily constructed with a rule and compass, he convinced himself of a rationale for these ratios and no others. Had he stuck to the ancient Greek system of intonation, with only the consonances 1/2, 2/3, and 3/4, his geometrical rationalization would have been much simpler; but apparently he had adopted the just intonation by actually listening to the harmonies. (One authority on modern music that Kepler cites is Galileo's father, Vincenzo, whose *Dialogo della musica antica e moderna* [1581] he read during his 1617 journey to Württemberg.)

Kepler's astrological views, already expounded in the *De stella nova* and *Tertius interveniens*, found their fully organized expression in book IV of the *Harmonice mundi*. In his theory of aspects, the geometrically formed human soul created the zodiac as a projection of itself. Thus, the heavenly bodies traveling on the zodiac produced an excitement within the soul whenever they formed angles corresponding to those of the regular polygons. The form of the planetary configurations at one's birth remained impressed on his soul for the whole of his life. As the spur was to the horse, or the trumpet to the soldier, so was the so-called influence of the heavens to the soul. For the benefit of other astrologers, Kepler gave his own nativity, adding:

> My stars were not Mercury rising in the seventh angle in quadrature with Mars, but Copernicus and Tycho Brahe, without whose observation books everything that I have brought into the clearest light would have remained in darkness; my rulers were not Saturn predominating over Mercury, but the emperors Rudolph and Matthias; not a planetary house, Capricorn with Saturn, but Upper Austria, the house of the emperor.[24]

The earth, too, participated in the cosmic harmony; its soul held a bond of sympathy with the entire firmament. Although in the introduction to the *Astronomia nova* Kepler had already explained the tides by the moon's attraction, as he noted here, he nevertheless now proceeded to interpret them as the breathing of the earth like an enormous living animal. Swept on by his fantasy, Kepler found animistic analogies everywhere. Yet in his *Epitome astronomiae Copernicanae*, written at almost the same time, he stated: "No soul oversees this revolution (of the planet . . .), but there is the one and only solar body, situated in the middle of the entire universe, to which this motion of the primary planets about the sun can be ascribed."[25]

In the *Mysterium cosmographicum* the young Kepler had been satisfied with the rather approximate planetary spacings predicted by his nested polyhedrons and spheres; now, imbued with a new respect for data, he could no longer dismiss its 5 percent error. In the astronomical book V of the *Harmonice mundi*, he came to grips with this central problem: By what secondary principles did God adjust the original archetypal model based on the regular solids? Indeed, he now found a supposed harmonic reason not only for the detailed planetary distances but also for their orbital eccentricities. The ratios of the extremes of the velocities of the planets corresponded to the harmonies of the just intonation. Of course, one planet would not necessarily be at its perihelion when another was at aphelion. Hence, the silent harmonies did not sound simultaneously, but only from time to time as the planets wheeled in their generally dissonant courses around the sun. Swept on by the grandeur of his vision, he exclaimed:

> It should no longer seem strange that man, the ape of his Creator, has finally discovered how to sing polyphonically, an art unknown to the ancients. With this symphony of voices man can play through the eternity of time in less than an hour and can taste in small measure the delight of God the Supreme Artist by calling forth that very sweet pleasure of the music that imitates God.[26]

In the course of this investigation, Kepler hit upon the relation now called his third or harmonic law: The ratio that exists between the periodic times of any two planets is precisely the ratio of the 3/2 power of the mean distances. Neither here nor in the few later references to it does he bother to show how accurate the relation really is. Yet the law gave him great pleasure, for it so neatly linked the planetary distances with their velocities or periods, thus fortifying the a priori premises of the *Mysterium* and the *Harmonice*. So ecstatic was Kepler that he immediately added these rhapsodic lines to the introduction to book V:

> Now, since the dawn eight months ago, since the broad daylight three months ago, and since a few days ago, when the full sun illuminated my wonderful speculations, nothing holds me back. I yield freely to the sacred frenzy; I dare frankly to confess that I have stolen the golden vessels of the Egyptians to build a tabernacle for my God far from the bounds of Egypt. If you pardon me, I shall rejoice; if you reproach me, I shall endure. The die is cast, and I am writing the book —to be read either now or by posterity, it matters not. It can wait a century for a reader, as God himself has waited six thousand years for a witness.[27]

At the instigation of a third party, Kepler appended a comparison of the "colossal difference" between his theory and that of Robert Fludd, the Oxford physician and Rosicrucian. The ensuing controversy at least illuminates the intellectual climate of the early 1600's, when the new, quantitative mathematical approach to nature collided with the qualitative, symbolical, alchemical tradition. Fludd counterattacked in an arrogant, polemical pamphlet, to which Kepler replied in his *Pro suo opere Harmonice mundi apologia* (1622). The *Apologia* was appended to a reissue of his *Mysterium cosmographicum.* Although republished in a larger format, the 1596 text of the *Mysterium* was unchanged. Numerous new footnotes called attention to the subsequent work, especially in the *Harmonice mundi.*

On Comets. In *De Cometis libelli tres* (1619) Kepler discussed in detail the bright comets of 1607 and 1618. Reflecting on the ephemeral nature of comets, he proposed a strictly rectilinear trajectory, which of course appeared more complex because of the earth's motion. Some decades later Edmond Halley made extensive use of the observations recorded in this book when he showed the seventy-six-year periodicity of the comet of 1607. The brief second section of Kepler's trilogy concerned the "physiology of comets": they fill the ether as fish fill the sea but are dissipated by the sun's light, forming the tail that points away from the sun. The final section treated the significations of the comets. Although he asserted that the common astrological beliefs rested on superstition, Kepler was still convinced that comets announced evil and misfortune. In those politically uncertain times, however, he wisely refrained from any specific prognostications.

A few years later Kepler published the *Hyperaspistes* (1625), a polemical defense of Tycho's comet theories against the Aristotelian views expressed by Scipione Chiaramonti in his *Antitycho.* As Delambre remarked, one regrets that Kepler took such pains with a point-by-point refutation, because the book is difficult to read in its entirety. More interesting is the appendix, which takes Galileo to task for some of the same erroneous views on comets. Kepler brings to Galileo's attention the fact that the observed phases of Venus can be as easily explained by the Tychonic as by the Copernican system.

The Epitome. Despite its title, Kepler's *Epitome astronomiae Copernicanae* was more an introduction to Keplerian than to Copernican astronomy. Cast in a catechetical form of questions and answers typical of sixteenth-century astronomy textbooks, it treated all of heliocentric astronomy in a systematic way, including the three relations now called Kepler's laws. Its seven books were issued in three installments. Taken together, they constitute a squat, unprepossessing octavo volume whose physical appearance scarcely marks it as Kepler's longest and most influential work. J. L. Russell has maintained that from 1630 to 1650 the *Epitome* was the most widely read treatise on theoretical astronomy in Europe.[28]

The composition of the *Epitome* was closely intertwined with the personal vicissitudes of its author's life. Although he had been pressed for a more popular book on Copernican astronomy when his very technical *Astronomia nova* appeared, not until the spring of 1615 were the first three books ready for the printer. This part finally appeared in 1617, having been delayed a year because, even though he had previously signed a contract with an Augsburg publisher, Kepler wanted the work done by his new Linz printer. By that time his seventy-year-old mother had been charged with witchcraft, and the astronomer felt obliged to go to Württemberg to aid in her legal defense. Afterward, the writing of the *Harmonice mundi* interrupted progress on the *Epitome*, so that the second installment, book IV, did not appear until 1620. The printing was barely completed when Kepler again journeyed to Württemberg, this time for the actual witchcraft trial. During pauses in the proceedings, he consulted with Maestlin at Tübingen about the lunar theory and arranged the printing of the last three books in Frankfurt. The publisher completed his work in the autumn of 1621, just as Kepler's mother won acquittal after enduring the threat of torture.

The first three books of this compendium deal mainly with spherical astronomy. Occasionally Kepler went beyond the conventional subject matter, considering, for example, the spatial distribution of stars and atmospheric refraction. Of special interest are the arguments for the motions of the earth; in describing the relativity of motion, he went considerably further than Copernicus and correctly formulated the principles later given more detailed treatment in Galileo's *Dialogo* (1632). Because of these arguments, and as a result of the anti-Copernican furor stirred up by Galileo's polemical writings, the *Epitome* was placed on the *Index Librorum Prohibitorum* in 1619. In spite of assurances that his works would be read all the more attentively in Italy, Kepler was alarmed; fearing that the circulation of his *Harmonice mundi* might also be restricted, he urged Italian book dealers to sell his works only to the highest clergy and the most important philosophers.

The most remarkable section of the *Epitome* was book IV, on theoretical astronomy, subtitled, "Celestial Physics, That Is, Every Size, Motion and Proportion in the Heavens Is Explained by a Cause Either Natural or Archetypal." In conception this installment came after books V–VI, and to a great extent it epitomized both the *Harmonice mundi* and the new lunar theory that Kepler completed in April 1620.

Book IV opened with one of his favorite analogies, one that had already appeared in the *Mysterium cosmographicum* and that stressed the theological basis of his Copernicanism: The three regions of the universe were archetypal symbols of the Trinity—the center, a symbol of the Father; the outermost sphere, of the Son; and the intervening space, of the Holy Spirit. Immediately thereafter Kepler plunged into a consideration of final causes, seeking reasons for the apparent size of the sun, the length of the day, and the relative sizes and the densities of the planets. From first principles he attempted to deduce the distance of the sun by assuming that the earth's volume is to the sun's as the radius of the earth is to its distance from the sun. Nevertheless, his assumption was tempered by a perceptive examination of the observations. In their turn the nested polyhedrons, the harmonies, the magnetic forces, the elliptical orbits, and the law of areas also found their place within Kepler's astonishing organization.

The harmonic law, which Kepler had discovered in 1619 and announced virtually without comment in the *Harmonice mundi*, received an extensive theoretical justification in the *Epitome*, book IV, part 2, section 2. His explanation of the $P = a^{3/2}$ law, in modern form, was based on the relation

$$\text{Period} = \frac{L \times M}{S \times V},$$

where the longer the path length L, the longer the period; the greater the strength S of the magnetic emanation, the shorter the period (this magnetic "species," emitted from the sun, provided the push to the planet); the more matter M in the planet, the more inertia and the longer the period; the greater the volume V of the planet, the more magnetic emanation could be absorbed and the shorter the period. According to Kepler's distance rule, the driving force S was inversely proportional to the distance a, and hence L/S was proportional to a^2; thus the density M/V had to be proportional to $1/a^{1/2}$ in order to achieve the 3/2 power law. Consequently, he assumed that the density (as well as both M and V) of each planet depended monotonically on its distance from the sun, a requirement quite appropriate to his ideas of harmony. To a limited extent he could defend his choice of V from telescopic observations of planetary diameters, but generally he was obliged to fall back on vague archetypal principles.

The lunar theory, which closed book IV of the *Epitome*, had long been a preoccupation of its author. In Tycho's original division of labor, Kepler had been assigned the orbit of Mars and Longomontanus that of the moon; but not long after Tycho's death Kepler applied his own ideas of physical causes to the lunar motion. To Longomontanus' angry remonstrance Kepler replied that it was not the same with astronomers as with smiths, where one made swords and another wagons.[29] He believed that the moon would undergo magnetic propulsion from the sun as well as from the earth, but the complicated interrelations gave much difficulty. In 1616 Maestlin wrote to him:

> Concerning the motion of the moon, you write that you have traced all the inequalities to physical causes; I do not quite understand this. I think rather that one should leave physical causes out of account, and should explain astronomical matters only according to astronomical method with the aid of astronomical, not

> physical, causes and hypotheses. That is, the calculation demands astronomical bases in the field of geometry and arithmetic . . . .[30]

In other words, the circles, epicycles, and equants that Kepler had ultimately abandoned in his *Astronomia nova.*

Kepler persisted in seeking the physical causes for the moon's motion and by 1620 had achieved the basis for his lunar tables. The fundamental form of his lunar orbit was elliptical, but the positions were further modified by the evection and by Tycho's so-called variation. Kepler's lunar theory, as given in book IV of the *Epitome*, failed to offer much foundation for further advances; nevertheless, his very early insight into the physical relation of the sun to this problem had enabled him to discover the annual equation in the lunar motion, which he handled by modifying the equation of time.

Books V–VII of the *Epitome* dealt with practical geometrical problems arising from the elliptical orbits, the law of areas, and his lunar theory; and together with book IV they served as the theoretical explanation to the *Tabulae Rudolphinae*. Book V introduced what is now called Kepler's equation,

$$E = M - e \sin E,$$

where e is the orbital eccentricity, M is the mean angular motion about the sun, and E is an auxiliary angle related to M through the law of areas; Kepler named M and E the mean and the eccentric anomalies, respectively. Given E, Kepler's equation is readily solved for M; the more useful inverse problem has no closed solution in terms of elementary trigonometric functions, and he could only recommend an approximating procedure. In the *Tabulae Rudolphinae*, Kepler solved the equation for a uniform grid of E values and provided an interpolation scheme for the desired values of M (see below).

Book VI of the *Epitome* treated problems of the apparent motions of the sun, the individual planets, and the moon. The short book VII discussed precession and the length of the year. To account for the changing obliquity, Kepler placed the pole of the ecliptic on a small circle, which in turn introduced a minor variation in the rate of precession (one last remnant of trepidation); because he was not satisfied with the ancient observations, he tabulated alternative rates in the *Tabulae Rudolphinae*. Such problems, he proposed, could be left to posterity "if it has pleased God to allot to the human race enough time on this earth for learning these left-over things."[31]

Tabulae Rudolphinae. In his own eyes Kepler was a speculative physicist and cosmologist; to his imperial employers he was a mathematician charged with completing Tycho's planetary tables. He spent most of his working years with this task hanging as a burden as well as a challenge; ultimately it provided the chief vehicle for the recognition of his astronomical accomplishments. In excusing the long delay in publication, which finally took place in 1627, he mentioned in the preface not only the difficulties of obtaining his salary and of the wartime conditions but also "the novelty of my discoveries and the unexpected transfer of the whole of astronomy from fictitious circles to natural causes, which were most profound to investigate, difficult to explain, and difficult to calculate, since mine was the first attempt."[32]

Although the rudiments of the tables must have been finished by 1616, when he calculated the first of the annual *Ephemerides* for 1617–1620, Kepler was still wrestling with the form of the lunar theory; in fact, the double-entry table for lunar evection ultimately determined the page size of the printed edition. But before he cast these tables in a final form, his project was overtaken by what he called a "happy calamity"—his initiation into logarithms.

Kepler had seen John Napier's *Mirifici logarithmorum canonis descriptio* (1614) as early as 1617; but he did not study the new procedure carefully until by chance, the following year, he saw Napier's tables reproduced in a small book by Benjamin Ursinus. Kepler then grasped the potentialities offered by the logarithms; but lacking any description of their construction, he re-created his own tables by a new geometrical procedure. He tried to base his theory of logarithms on a Eudoxian theory of general proportion; but he could not resolve the problem of limit increments, which he concealed in the guise of numerical approximation. The form of his logarithms differed both from Napier's and from Briggs's; in modern notation Kepler's log x was

$$\lim_{n\to\infty} 10^5 \cdot 2^n \left[1 - \left(\frac{x}{10^5}\right)^{\frac{1}{2^n}}\right],$$

so that his $\log x = 10^5 \ln(10^5/x)$. His tables and theory were published in the *Chilias logarithmorum* (1624), and numerous examples of their use appeared in the *Supplementum* . . . (1625).

Unlike the *Ephemerides*, the *Tabulae Rudolphinae* did not contain sequential positions of planets for specified days; rather, it provided perpetual tables for calculating such positions for any date in the past or future. To compute the longitude for a particular planet on a specified date, the user must first find the mean longitude and the position of the aphelion by adding the appropriate angles from the mean motion tables for that planet to the starting positions tabulated for the beginning of the preceding century. The difference between the mean longitude and the aphelion angle is the mean anomaly, which is entered in the "Tabula aequationum": for the planet in question the user extracts from the table the true anomaly (called by Kepler the "anomalia coequata"), that is, the angle at the sun measured from the aphelion. The table in effect provides a tabulated solution of Kepler's equation coupled with the conversion from the eccentric anomaly to the true anomaly. (Kepler calls the term $e \sin E$ the "physical equation," and the remainder of the conversion the "optical equation.") It is here that he first exploited the logarithms, as an aid to interpolation. (As stated above in the discussion of the *Epitome*, he solved his equation for a uniform grid of eccentric anomaly angles, which led to a set of nonuniformly spaced mean anomalies.) The true anomaly obtained from the interpolation, added to the aphelion angle, gives the heliocentric longitude.

Previous planetary tables yielded geocentric planetary positions directly from a single procedure. In Kepler's more exact version, the heliocentric positions of the earth and the planets are calculated separately and then combined to produce the geocentric position—essentially a problem in vector addition. Thus, the second important use of logarithms arose from this thoroughly heliocentric basis of the *Tabulae Rudolphinae*. Kepler tabulated the logarithms of the heliocentric distances and provided a convenient double-entry "Tabula Anguli" for combining the heliocentric longitudes into geocentric ones. He explained all these procedures, including the manipulation of logarithms, in a series of precepts that preceded the tables. There the ellipse was introduced, but for the full astronomical theory the reader was urged to consult the *Epitome*.

The *Tabulae Rudolphinae* gave planetary positions far more accurate than those of earlier methods; for example, the predictions for Mars previously erred up to 5°, but Kepler's tables kept within $\pm 10'$ of the actual position. In calculating his *Ephemeris* for 1631, Kepler realized that the improved accuracy of his tables enabled him to predict a pair of remarkable transits of Mercury and of Venus across the disk of the sun. These he announced in a small pamphlet, *De raris mirisque anni 1631 phenomenis* (1629). Although he did not live to see his predictions fulfilled, the Mercury transit was observed by Pierre Gassendi in Paris on 7 November 1631; this observation, the first of its kind in history, was a tour de force for Kepler's astronomy, for his prediction erred by only 10′ compared to 5° for tables based on Ptolemy, Copernicus, and others. (The transit of Venus in 1631 was not visible in Europe because it took place at night.)

The printed volume of the *Tabulae Rudolphinae* contains 120 folio pages of text in the form of precepts and 119 pages of tables. Besides the planetary, solar, and lunar tables and the associated tables of logarithms it includes Tycho Brahe's catalog of 1,000 fixed stars, a chronological synopsis, and a list of geographical positions. In some of the copies there is also a foldout map of the world, measuring 40 × 65 centimeters; the map was engraved in 1630 but apparently was not distributed until many years later. This work stands alone among Kepler's books in having an engraved frontispiece—filled with intricate baroque symbolism, it represents the temple of Urania, with the Tychonic system inscribed on the ceiling. Hipparchus, Ptolemy, Copernicus, and Tycho are at work within the temple, and Kepler himself is depicted in a panel below. The dome of the allegorical edifice is adorned with goddesses whose paraphernalia subtly remind the readers of Kepler's scientific contributions.

As with Kepler's other great books, the printing history of the tables was intricately linked with his personal odyssey. The material was ready for printing in 1624, but he believed that the Linz press was inadequate for this great work. Furthermore, his printer, Johannes Plank, intended to leave Linz because of the religious turmoil.

Since the tables were to be named after Rudolph, Kepler hoped to use their potential publication as leverage to collect 6,300 guldens in back pay. Consequently he spent the autumn of 1624 promoting his affairs at the imperial court in Vienna. Emperor Ferdinand II approved a scheme to impose the back payment on the cities of Nuremberg, Memmingen, and Kempten but insisted that the tables be printed in Austria. The following spring Kepler visited these cities; although he concluded satisfactory arrangements with the latter two towns, which supplied the paper for the printing, he never collected the 4,000 guldens imposed on Nuremberg. In order not to delay the work further, Kepler finally financed the printing from his own funds, even importing his own type for the numerical tables, a fact referred to on the final elaborate title page.

Barely had the typesetting begun when the Counterreformation struck Linz. Although Kepler received a concession for himself and his Lutheran printers to remain, the printing progressed very slowly. By the summer of 1626 Linz was blockaded, and Plank's house and press went up in flames. In a letter Kepler described his own circumstances:

> You ask me what I have been doing during the long siege? First ask me what I have been *able* to do in the middle of the soldiers. It appeared to be a favor from the commissioner when a year ago I moved into a gov-

ernment house. This house lies along the city wall. All the towers had to be kept open for the soldiers, who by their going in and out disturbed my sleep by night and my studies by day. An entire reserve detachment settled in our house. The ear was incessantly exhausted by the noise of the cannons, the nose by the stench, the eye by the glare of fire.

> Nevertheless, in these evil circumstances I myself undertook against Scaliger the same thing that our garrison undertook against the peasants. I have composed a splendid treatise on chronology. . . . This pugnacious sort of writing wiped out for me much boredom from the inconveniences of the siege and impediments to work. If I had not happened upon this there would have been something else for me to do in making the tables more useful.[33]

As soon as the siege was lifted, Kepler petitioned the emperor for permission to move to Ulm. Although he had worked in Linz longer than he had in any other place, the astronomer was glad to leave. He packed up his household, books, manuscripts, and type and traveled by boat up the Danube to Regensburg. After finding accommodations there for his wife and children he continued on to Ulm, where the printing was soon under way. Kepler spent many hours supervising the typesetting in order to guarantee a neat, aesthetic result. By September 1627, the large edition of 1,000 copies was at last completed.

Last Years in Sagan and Regensburg. Even before the *Tabulae Rudolphinae* was printed, Kepler began to search for a new residence. Some years earlier he had dedicated his *Harmonice mundi* to James I; in 1619 the English poet John Donne visited him, and in 1620 the English ambassador Sir Henry Wotton had called on him in Linz and had invited him to England. To his Strasbourg friend Matthias Bernegger he confided, "Therefore shall I cross the sea, where Wotton calls me? I, a German? A lover of firm land, who dreads the confinement of an island?"[34] But in 1627 he wrote to Bernegger, "As soon as the *Rudolphines* are published, I desire to find a place where I can lecture about them to a large audience, if possible in Germany, otherwise in Italy, France, Holland or England, provided the salary is adequate for a traveler."[35]

Torn between his desire to find religious toleration and his reluctance to lose his salaries as provincial and imperial mathematician, Kepler journeyed to Prague at the end of 1627 to arrange further employment. Ferdinand III has just been crowned king of Bohemia; and the imperial commander-in-chief, Albrecht von Wallenstein, was at the height of his power. Ferdinand received Kepler graciously and awarded him 4,000 guldens (to be paid by Nuremberg and Ulm) for the dedication of the tables, but it was made clear that the astronomer should become a Catholic to remain in imperial service.

On the other hand, Wallenstein, a superstitious man who had earlier applied anonymously to Kepler for a horoscope, favored peaceful coexistence of the various creeds. The general had just received the duchy of Sagan as a fief and, anxious to raise the status of his new possession as well as to have close access to an astrologer, agreed to support both Kepler and a printing press there. Because Protestants were not yet restricted in Sagan, Kepler accepted. But he was unwilling to "let himself be used like an entertainer" and was reluctant to compromise his own scientific convictions to satisfy the "quite visibly erroneous delusion" of his astrologically minded patron. The general arranged everything to his satisfaction, however, by employing Kepler to calculate the precise positions and then obtaining the predictions from less inhibited astrologers.

Kepler collected his family in Regensburg, settled his affairs in Linz, and finally reached Sagan in July 1628. Apart from his ceaseless work, he found little of interest there. "I am a guest and stranger," he complained to Bernegger, "almost completely unknown, and I barely understand the dialect so that I myself am considered a barbarian."[36] Shortly after his arrival religious strife broke out, and for political reasons Wallenstein pressed Catholicism onto his subjects. Although Kepler was not personally affected, the persecutions made it difficult to attract printers there for the intended publication of Brahe's observations. It took months to find a press and workmen, and Kepler himself acted as printer.

Eager to reap the fruits of his own astronomical tables, he set to work calculating ephemerides. In this he was assisted by Jacob Bartsch, a young scholar who had studied astronomy and medicine at Strasbourg and who calculated the positions for 1629–1636. *Ephemerides pars III* (for 1629–1639) was printed first, then the second part (for 1621–1629), which also contained Kepler's daily weather observations. (The first part, for 1617–1620, had been printed year by year at Linz.)

At Sagan, Kepler finally began to print a short book of a far different sort, whose beginnings went back to his school days at Tübingen: his *Somnium seu astronomia lunari*. The "Dream" is a curiously interesting tract for two reasons. First, its fantasy framework of a voyage to the moon made it a pioneering and remarkably prescient piece of science fiction. Second, its perceptive description of celestial motions as seen from the moon produced an ingenious polemic on behalf of the Copernican system.

Kepler had written out the *Somnium* in 1609, and

copies circulated in manuscript. Because the work took the form of an interview with a knowledgeable "daemon" who explained how a man could be transported to the moon, there were overtones of witchcraft that later played an embarrassing role in his mother's trial. Kepler himself remarked about this in one of the 223 notes that he added when he returned to the *Somnium* in 1621: "Would you believe that in the barber shops there was chatter about this story of mine? When this gossip was taken up by senseless minds, it flared up into defamation, fanned by ignorance and superstition. If I am not mistaken, you will judge that my family could have gotten along without that trouble for six years"[37]—a clear reference to his mother's legal entanglement between 1615 and 1621.

In Sagan, Kepler waited in vain for the payment of his salary claims, which had been transferred to Wallenstein. In June 1630, Ferdinand III summoned an electoral congress at Regensburg; and in August, Wallenstein lost his position as commander-in-chief. On 8 October the fifty-eight-year-old Kepler set out for Regensburg, taking with him all his books and manuscripts. Although his ultimate goal was Linz, where he hoped to collect interest on two Austrian district bonds, he must have intended to consult with the emperor and his court friends in Regensburg about a new residence. A few days after reaching Regensburg, Kepler became sick with an acute fever; the illness became worse, and on 15 November 1630 he died. He was buried in the Protestant cemetery; the churchyard was completely demolished during the Thirty Years War.

Jacob Bartsch, who had married Kepler's daughter Susanna in March 1630, became a faithful protector of the bereaved and penniless family. He pressed on with the printing of the *Somnium*, and he tried in vain to collect the 12,694 guldens still owed by the state treasury. He recorded the epitaph that Kepler himself has composed:

Mensus eram coelos, nunc terrae metior umbras:
Mens coelestis erat, corporis umbra jacet.

I used to measure the heavens,
now I shall measure the shadows of the earth.
Although my soul was from heaven,
the shadow of my body lies here.

Evaluation. Kepler was a small, frail man, nearsighted, plagued by fevers and stomach ailments, yet nonetheless resilient. In his youth he had compared himself to a snappish little house dog who tried to win the favor of his masters but who drove others away, but his later years in part belied this self-image. He never rid himself of a feeling of dependence, nor could he exhibit the imperious self-assurance of Tycho or of Galileo. Nevertheless, his ready wit, modest manner, and scrupulous honesty, as well as his wealth of knowledge, won him many friends. In the dedication to the *Epitome* he wrote, "I like to be on the side of the majority"; but in his Copernicanism and in his deep-felt religious convictions he learned the role of a staunch, lonesome minority.

Delambre has aptly summarized Kepler's persistent approach to scientific achievement:

> Ardent, restless, burning to distinguish himself by his discoveries, he attempted everything, and when he had glimpsed something, nothing was too hard for him in following or verifying it. All his attempts did not have the same success, and indeed that would have been impossible. . . . When in search of something that really existed, he sometimes found it; when he devoted himself to the pursuit of a chimera, he could only fail; but even there he revealed the same qualities and that obstinate perseverance that triumphed over any difficulties that were not insurmountable.[38]

Kepler's scientific thought was characterized by his profound sense of order and harmony, which was intimately linked with his theological view of God the Creator. He saw in the visible universe the symbolic image of the Trinity. Repeatedly, he stated that geometry and quantity are coeternal with God and that mankind shares in them because man is created in the image of God. (In this framework Kepler can be called a mystic.) From these principles flowed his ideas on the cosmic link between man's soul and the geometrical configurations of the planets; they also motivated his indefatigable search for the mathematical harmonies of the universe.

Contrasting with Kepler's mathematical mysticism, and yet growing out of it through the remarkable quality of his genius, was his insistence on physical causes. Many examples illustrate his physical insight: his embryonic ideas of universal gravitation as articulated in the introduction to the *Astronomia nova*; his trailblazing (but not fully correct) use of "inertia"; the statement "If the word *soul* is replaced by *force*, we have the very principle on which the celestial physics of the Mars Commentaries is based."[39] In Kepler's view the physical universe was not only a world of discoverable mathematical harmonies but also a world of phenomena explainable by mechanical principles.

Kepler wrote prolifically, but his intensely personal cosmology was not very appealing to the rationalists of the generations that followed. A much greater audience awaited a more gifted polemicist, Galileo, who became the persuasive purveyor of the new cosmology. Kepler was an astronomer's astronomer. It was the astronomers who recognized the immense superiority of the *Tabulae Rudolphinae*. For the professionals the improvement in planetary predictions was a forceful testimony to the efficacy of the Copernican system.

Tables copied after Kepler's were published by N. Durret (Paris, 1639), V. Renieri (Florence, 1639), J. B. Morin (Paris, 1650, 1657), M. Cunitz (Oels, 1650), H. Coley (London, 1675), N. Mercator (London, 1676), and T. Streete (1705); many others based ephemerides on Kepler's work. Use of the tables sometimes also generated an interest in the physical bases; Durret, for example, described both the elliptical orbits and an equivalent form of the law of areas. In England an early and influential disciple was Jeremiah Horrocks; at the time of his early death in 1641, he was working on a book, published posthumously, that strongly supported Kepler's theories. Descartes apparently was ignorant of his work on planetary motions, but in 1642 Pierre Gassendi mentioned the ellipses and physical theories with apparent approval. In his *Astronomica Philolaica* (1645) Ismael Boulliau accepted the elliptical orbit although he rejected Kepler's celestial physics.

Newton's early student notebooks show that he learned Kepler's first and third laws from Streete's *Astronomia Carolina* (1661). The first author after Kepler to state all three of Kepler's laws was G. B. Riccioli in his *Almagestum novum* (1651); somewhat later Mercator did so in his *Institutionum astronomicarum* (1676). Isaac Newton's well-thumbed copy of this latter work was undoubtedly the source of his information about Kepler's second law, which played a crucial role in the development of his physics. Although nowhere in book I of the *Principia* is Kepler's name mentioned (Newton attributes the harmonic law to him in book III), the work was introduced to the Royal Society as "a mathematical demonstration of the Copernican hypothesis as proposed by Kepler." Perhaps the most just evaluation of Kepler has come from Edmond Halley in his review of the *Principia*; Newton's first eleven propositions, he wrote, were "found to agree with the *Phenomena* of the Celestial Motions, as discovered by the great Sagacity and Diligence of *Kepler*."[40]

NOTES

KGW stands for *Johannes Kepler Gesammelte Werke*; full references are found in the bibliography.

1. C. Frisch, ed., *Kepleri opera omnia*, VIII, 671–672.
2. *KGW*, III, 108.
3. *Ibid.*, IV, 12.
4. *Ibid.*, 253.
5. *Ibid.*, I, 9 ff.
6. *Ibid.*, XIII, 40.
7. J. L. E. Dreyer, *History of the Planetary Systems*, p. 373.
8. E. Rosen, in *Johannes Kepler Werk und Leistung*, p. 148.
9. *KGW*, XIV, 130.
10. Frisch, ed., *op. cit.*, I, 240.
11. *KGW*, XIV, 203.
12. *Ibid.*, III, 156.
13. *Ibid.*, 178.
14. *Ibid.*, XIV, 410.
15. *Ibid.*, XV, 146.
16. *Ibid.*, I, 285.
17. *Ibid.*, IV, 519.
18. *Ibid.*, 293.
19. *Ibid.*, XVI, 327.
20. *Ibid.*, IV, 161.
21. *Ibid.*, XVII, 327.
22. *Ibid.*, 254.
23. M. Caspar, *Kepler* (1959), p. 290.
24. *KGW*, VI, 280.
25. *Ibid.*, VII, 297.
26. *Ibid.*, VI, 328.
27. *Ibid.*, 290.
28. J. L. Russell, "Kepler's Laws of Planetary Motion: 1609–1666."
29. *KGW*, XV, 136.
30. *Ibid.*, XVII, 187.
31. *Ibid.*, III, 408.
32. *Ibid.*, X, 42–43.
33. *Ibid.*, XVIII, 272–273.
34. *Ibid.*, 63.
35. *Ibid.*, 277–278.
36. *Ibid.*, 402.
37. E. Rosen, *Kepler's Somnium*, p. 40.
38. J. B. J. Delambre, *Histoire de l'astronomie moderne*, I, 358.
39. *KGW*, VIII, 113.
40. E. Halley, in *Philosophical Transactions of the Royal Society*, no. 186 (1687), 291.

BIBLIOGRAPHY

Max Caspar's *Bibliographia Kepleriana* (Munich, 1936; rev. by Martha List, Munich, 1968) gives the definitive annotated listing of Kepler's printed works, later eds., and trans. It includes 574 secondary writings about Kepler to 1967. Since at least 100 articles and books were produced in connection with the 1971 anniversary, the following list is necessarily highly selective.

I. ORIGINAL WORKS. A. **Printed Books.** *Joannis Kepleri astronomi opera omnia*, Christian Frisch, ed., 8 vols. (Frankfurt–Erlangen, 1858–1871; reprinted, Hildesheim, 1971–) includes all of the major printed works and also extensive excerpts from Kepler's correspondence, copious editorial notes, a 361-page Latin *vita*, and an index as yet unsurpassed. It contains the initial publication of, and is still the only printed source for, the "Apologia Tychonis contra Nicolaum Ursum," "Hipparchus," and other MSS; these are listed by Caspar, *op. cit.*, pp. 111–113.

In the twentieth century a monumental Kepler *opera* was planned by Walther von Dyck and Max Caspar and carried out under the auspices of the Deutsche Forschungsgemeinschaft and the Bayerische Akademie der Wissenschaften; Caspar and Franz Hammer have served successively as eds. The extensive notes and commentaries make this ed. the single most valuable source for any Kepler scholar. As now envisioned, the *Johannes Kepler Gesammelte Werke* (Munich, 1937–) will encompass 22 vols. when complete. Concerning this edition see Franz Hammer, "Problems and Difficulties in Editing Kepler's Collected Works," in *Vistas in Astronomy*, **9** (1967), 261–264. An abridged table of contents, with short titles and the dates of original publication, is given here for reference: I, *Mysterium cosmographicum* (1596), *De stella nova* (1606); II, *Ad Vitellionem paralipomena* or *Astronomiae pars optica* (1604); III, *Astronomia nova* (1609); IV, *Phaenomenon singulare* (1609), *Tertius interveniens* (1610), *Strena* (1611), *Dissertatio cum Nuncio sidereo* (1610), *Dioptrice* (1611); V, *De vero anno* (1614), *Bericht vom geburtsjahr Christi* (1613), *Eclogae chronicae* (1615); VI, *Harmonice mundi* (1619); VII, *Epitome astronomiae Copernicanae* (1618–1621); VIII, *Mysterium cosmographicum* (1621), *De cometis libelli tres* (1619), *Hyperaspistes* (1625); IX, *Nova stereometria doliorum vinariorum* (1615), *Messekunst Archimedis* (1616), *Chilias logarithmorum* (1624); X, *Tabulae Rudolphinae* (1627); XI, *Ephemerides novae* (1617–1619), *Ephemerides* (1630), and the small calendars; XII, *Somnium* (1634), theological writings, and trans.; XIII–XVIII, letters, in chronological order; XIX, documents; XX–XXII, synopsis of the MS material, and index.

Except for the two collected eds., Kepler's works have been reprinted rather rarely; most of the other reprinting occurred in connection with the greater interest in Galileo: *Dioptrice* (London, 1653, 1683); *Dissertatio cum Nuncio sidereo* (Modena, 1818; Florence, 1846, 1892); *Perioche ex introductione in Martem* (Strasbourg, 1635; London, 1641, 1661 [in English], 1663; Modena, 1818; Florence, 1846). Recent facsimiles include *Dioptrice* (Cambridge, 1962); *Dissertatio cum Nuncio sidereo* (Munich, 1964); *Astronomia nova*, *Harmonice mundi*, and *Astronomiae pars optica* (Brussels, 1968); and *Somnium* (Osnabrück, 1969).

Nor have trans. been frequent. Outstanding exceptions are the three published in German by Max Caspar: *Das Weltgeheimnis* (*Mysterium cosmographicum*) (Augsburg, 1923; Munich–Berlin, 1936); *Neue Astronomie* (Munich–Berlin, 1929); *Weltharmonik* (Munich–Berlin, 1939; repr. 1967). In addition, Caspar and Walther von Dyck published *Johannes Kepler in seinem Briefen*, 2 vols. (Munich–Berlin, 1930); this served as the basis for Carola Baumgardt, *Johannes Kepler: Life and Letters* (New York, 1951). See also *Johannes Kepler—Selbstzeugnisse*, sel. by Franz Hammer and trans. by Esther Hammer (Stuttgart–Bad Cannstatt, 1971).

The chief English trans. are the secs. of the *Epitome* (bks. IV and V) and the *Harmonice* (bk. V), prep. by Charles Glenn Wallis for Great Books of the Western World, XVI (Chicago, 1952). Several smaller works have been trans. in full, including John Lear, ed., *Kepler's Dream*, trans. by Patricia Frueh Kirkwood (Berkeley, 1965); Edward Rosen, *Kepler's Conversation With Galileo's Sidereal Messenger* (New York, 1965); and *Kepler's Somnium* (Madison, Wis., 1967); and Colin Hardie, *The Six-Cornered Snowflake* (Oxford, 1966). An English trans. of the *Astronomia nova* by Owen Gingerich and Ann Wegner Brinkley is nearing completion. Extended quotations from the *Astronomia nova* are available in French in Alexandre Koyré, *La révolution astronomique* (Paris, 1961) and in its English trans. by E. W. Maddison (Paris, 1972).

B. **Manuscripts.** The thousands of MS sheets left at Kepler's death went to his son Ludwig, who promised publication but lacked both the time and the scientific knowledge for the undertaking. After Ludwig's death the Danzig astronomer Johannes Hevelius acquired the collection and published a brief inventory in *Philosophical Transactions of the Royal Society*, **9** (1674), 29–31. In 1707 Michael Gottlieb Hansch obtained the material with the intention of publishing it, and in 1718 he produced *Joannis Kepleri aliorumque epistolae mutuae*, a large folio vol. containing 77 letters by Kepler and 407 to him. Hansch had the MSS bound in vellum in 22 vols., which he cataloged briefly in *Acta eruditorum*, no. 57 (1714), 242–246; in 1721 financial difficulties forced him to pawn 18 of the vols. The other four—VII, VIII, and XII, which had formed the basis for his *Epistolae*, and VI, which was used for *De calendario Gregoriano* (Frankfurt, 1726)—eventually found their way to the Österreichische Nationalbibliothek in Vienna, where VI is codex 10704 and the three original vols. of letters have apparently been rebound into codices 10702 and 10703. Not until about 1765 were the 18 vols. rediscovered, and in 1773 Catherine II purchased them for the Academy of Sciences in St. Petersburg. They are still preserved in Leningrad. Details of this odyssey are chronicled by Martha List, *Der handschriftliche Nachlasz der Astronomen Johannes Kepler und Tycho Brahe* (Munich, 1961).

The 18 Leningrad MS vols. contain roughly the following: I, the unfinished "Hipparchus" dealing with planetary and especially lunar theory, partially published by Frisch (*op. cit.*, III); II, lunar theory and tables; III, nova of 1604, including letters; IV, musical theory, including notes to Vincenzo Galilei's *Dialogo della musica antica e della moderna* and Kepler's trans. of bk. III of Ptolemy's *Libri harmonicorum;* V, geometry, studies of Euclid, "Apologia Tychonis" [VI–VIII, XII, in Vienna]; IX–XI, letters; XIII, motion of Mercury, Venus, Jupiter, Saturn; XIV, workbook on Mars, dating from 1600–1601, plus early drafts of some chs. of *Astronomia nova*; XV, observations and theory of solar and lunar eclipses; XVI, "Chronologia ab origine rerum usque ad annum ante Christum"; XVII, chronological notes on Scaliger and Petavius, German trans. and annotations of ch. 13 from Aristotle's *De caelo*, bk. II, on the position and shape of the earth; XVIII, horoscopes, examination of observations of Regiomontanus and Bernhard Walther; XIX, length of year, Greek astronomy, horoscopes; XX, MS printer's copy of the *Tabulae Rudolphinae;* XXI, trigonometry, planetary tables, about 300 horoscopes; XXII, Biblical chronology, including three tracts published by Frisch (*op. cit.*, VII).

From 1839 to 1937 the MSS were housed at the Pulkovo observatory, and during that time they were made available to Frisch and to Dyck and Caspar for the preparation of the two eds. Unfortunately, some of the material copied by Frisch is no longer to be found, for example, part of the remarkable "self-analysis" formerly in XXI. At the Royal Observatory in Edinburgh is a page of a Kepler letter given to Lord Lindsay when he visited the Pulkovo observatory, and undoubtedly other leaves were dispersed in a similar manner. A complete set of photocopies of the currently available pages (together with photographs of virtually all the other known Kepler MSS) is found at the Kepler Commission of the Bavarian Academy of Sciences at the Deutsches Museum in Munich. Secondary sets are located at the Württembergische Landesbibliothek in Stuttgart and at the Bayerische Staatsbibliothek in Munich.

Besides the four vols. already mentioned, many additional MSS are found in Vienna, especially with the Tycho Brahe material in codices 10686–10689. These are listed, rather unreliably, in *Tabulae codicum manu scriptorum . . . in Bibliotheca palatina Vindobonensi asservatorum*, VI (Vienna, 1871). The third most important repository of Kepler MSS is the Württembergische Landesbibliothek in Stuttgart, with about 60 letters and some notes; in addition, the Hauptstaatsarchiv in Stuttgart contains rich documentation on the witchcraft trial of Kepler's mother. Some of the latter material was published by Frisch (*op. cit.*, VIII). Other significant collections, especially of letters, are in Oxford, Munich, Graz, Paris, Florence, Wolfenbüttel, and Tübingen.

Some of the letters were originally printed in the Nova Kepleriana series in the *Abhandlungen der Bayerischen Akademie der Wissenschaften* (1910–1936); a new series is continuing the publication of other MS material (Munich, 1969–), beginning with Jürgen Hübner's ed. of *Unterricht vom h. Sacrament.*

Kepler's rather sparsely annotated copy of Copernicus' *De revolutionibus*, now in the Leipzig University Library, has been reproduced in facsimile (Leipzig, 1965).

II. SECONDARY LITERATURE. Still useful among many early biographical accounts are John Drinkwater Bethune, *Life of Kepler* (London, 1830), a pamphlet in the Library of Useful Knowledge and later in the *Penny Cyclopaedia* (London, 1839); Edmund Reitlinger's uncompleted *Johannes Kepler* (Stuttgart, 1868); and Christian Frisch, "Kepleri vita," in his *Kepleri opera omnia*, VIII (Frankfurt, 1871), 668–1028. The standard biography by Max Caspar, *Johannes Kepler* (Stuttgart, 1948), English trans. by C. Doris Hellman (New York, 1959), is based on an extensive familiarity with the sources, including the letters; from a scholarly viewpoint, its lack of citations and its inadequate index are disappointing. Arthur Koestler, *The Sleepwalkers* (New York, 1959), is a compellingly written account with an emphasis on the psychology of discovery; the part specifically on Kepler has been published separately in an expurgated version, *The Watershed* (New York, 1960).

Two particularly well-illustrated books are Walther Gerlach and Martha List, *Johannes Kepler, Dokumente zu Lebenszeit und Lebenswerk* (Munich, 1971); and Justus Schmidt, *Johann Kepler, sein Leben in Bildern und eigenen Berichten* (Linz, 1970).

The quadricentennial of Kepler's birth in 1971 provided the occasion for several collections of papers; the contents of these are listed and selected papers reviewed by Robert S. Westman, "Continuities in Kepler Scholarship: The European Kepler Symposia, 1971, in Historiographical Perspective." This paper, together with the others presented at a symposium in Philadelphia, is found, edited by Arthur Beer, in *Vistas in Astronomy*, **18** (1975). Thirteen papers appear in the catalog for the Kepler exhibition at Linz organized by Wilhelm Freh, *Johannes Kepler, Werk und Leistung*, Katalog des Oberösterreichen Landesmuseums no. 74, Katalog des Stadtmuseums Linz no. 9 (Linz, 1971). Papers on "little-known aspects of Kepler" appear in Fritz Krafft, Karl Meyer, and Bernhard Sticker, eds., *Internationales Kepler-Symposium Weil der Stadt 1971*, Beiträge zur Wissenschaftsgeschichte, ser. A, I (Hildesheim, 1972). Twelve papers appear in Ekkehard Preuss, ed., *Kepler Festschrift 1971* (Regensburg, 1971), sponsored by the Naturwissenschaftlichen Verein Regensburg as a successor to Karl Stöckl, ed., *Kepler Festschrift* (Regensburg, 1930). Papers presented at the International Union for the History and Philosophy of Science's International Kepler Symposium in Leningrad will be published both in a Russian ed. and in Western languages, edited by Arthur Beer, in *Vistas in Astronomy,* **18** (1975).

Three outstanding review papers given at these symposia are Walther Gerlach, "Johannes Kepler, Leben, Mensch und Werk" (Leningrad); Edward Rosen, "Kepler's Place in the History of Science" (Philadelphia); and I. Bernard Cohen, "Kepler's Century" (Philadelphia).

The printing history of Kepler's work is presented in Friedrich Seck, "Johannes Kepler und der Buchdruck," in *Archiv für Geschichte des Buchwesens*, **11** (1971), 610–726.

An early post-Newtonian appreciation and critique of Kepler is David Gregory, *Astronomiae physicae et geometricae elementa* (Oxford, 1702); the rev. English ed. of 1726 has been reprinted (New York, 1972). A succinct discussion of Kepler's astronomical achievements is found in J. L. E. Dreyer, *History of the Planetary Systems From Thales to Kepler* (Cambridge, 1906; repr. New York, 1953). An earlier account is E. F. Apelt, *Johann Keppler's astronomische Weltansicht* (Leipzig, 1849).

For fuller details, one must turn to more technical analyses such as Robert Small, *An Account of the Astronomical Discoveries of Kepler* (London, 1804; repr. Madison, Wis., 1963); and J. B. J. Delambre, *Histoire de l'astronomie moderne* (Paris, 1821), I, 314–615. These writers were interested in Kepler's techniques, not his philosophical foundations; and they were frequently embarrassed by his metaphysical reasoning—in fact, Small's chapter-by-chapter account of the *Astronomia nova* completely omits ch. 57, on the magnetic forces underlying Kepler's formulation. For a more sympathetic treatment of the technical aspects of Kepler's laws within the framework of the metaphysical background, see Alexandre Koyré, *La révolution astronomique* (Paris, 1961), pp. 119–458; Curtis Wilson, "Kepler's Derivation of the Elliptical Path," in *Isis*, **59** (1968), 5–25; E. J. Aiton, "Kepler's Second Law of Planetary Motion," *ibid.*, **60** (1969), 75–90; and Owen Gingerich, "Johannes Kepler and the New Astronomy," in *Quarterly Journal of* *the Royal Astronomical Society*, **13** (1972), 346–373; and "The Origins of Kepler's Third Law," Leningrad Kepler Symposium, 1971. See also E. J. Aiton, *The Vortex Theory of Planetary Motions* (London, 1972), ch. 2, pp. 12–19; and Curtis Wilson, "How Did Kepler Discover His First Two Laws?," in *Scientific American*, **226** (Mar. 1972), 92–106.

Analyses of Kepler's planetary tables are Owen Gingerich, "A Study of Kepler's Rudolphine Tables," in *Actes du XI Congrès international d'histoire des sciences*, III (Wrocław, 1968), 31–36; Volker Bialas, "Die Rudolphinischen Tafeln von Johannes Kepler," in *Nova Kepleriana*, n.s. **2** (1969); and "Jovialia, Die Berechnung der Jupiterbahn nach Kepler," *ibid.*, n.s. **4** (1971); see also Owen Gingerich, "Johannes Kepler and the Rudolphine Tables," in *Sky and Telescope*, **42** (1971), 328–333. The impact of Kepler's tables and laws on subsequent astronomers has been described by J. L. Russell, "Kepler's Laws of Planetary Motion: 1609–1666," in *British Journal for the History of Science*, **2**, no. 5 (1964), 1–24; and by Curtis Wilson, "From Kepler's Laws, So-called, to Universal Gravitation: Empirical Factors," in *Archive for History of Exact Sciences*, **6** (1970), 89–170.

Other aspects of Kepler's astronomy are found in J. A. Ruffner, "The Curved and the Straight: Cometary Theory From Kepler to Hevelius," in *Journal for the History of Astronomy*, **2** (1971), 178–194; A. Koyré, "Johannes Kepler's Rejection of Infinity," in his *From the Closed World to the Infinite Universe* (Baltimore, 1957), ch. 3, pp. 58–87; and Victor E. Thoren, "Tycho and Kepler on the Lunar Theory," in *Publications of the Astronomical Society of the Pacific*, **79** (1967), 482–489.

An awareness of Kepler as a philosopher of science, and of the unitary nature of his thought, is found in Ernst Cassirer, *Das Erkenntnisproblem in der Philosophie und Wissenschaft der neuren Zeit*, I (Berlin, 1906), 328–377. See also Shmuel Sambursky, "Kepler in Hegel's Eyes," in *Proceedings of the Israel Academy of Sciences and Humanities*, **5** (1971), 92–104.

A seminal critique of the metaphysical foundations of Kepler's scientific concepts is Gerald Holton, "Johannes Kepler's Universe: Its Physics and Metaphysics," in *American Journal of Physics*, **24** (1956), 340–351. See also E. J. Dijksterhuis, *De Mechanisering van het Wereldbeeld* (Amsterdam, 1950); German trans. by H. Habicht (Berlin, 1956); English trans. by C. Dikshoorn (Oxford, 1971), pp. 337–359; Max Jammer, *Concepts of Force* (Cambridge, Mass., 1957), pp. 81–93; and Fritz Krafft, "Kepler's Beitrag zur Himmelsphysik," in *Internationales Kepler-Symposium, Weil der Stadt 1971* (Hildesheim, 1972).

Aspects of the intellectual background of Kepler are discussed by Harry A. Wolfson, "The Problem of the Souls of the Spheres From the Byzantine Commentaries on Aristotle . . . to Kepler," in *Dumbarton Oaks Papers*, **16** (1962), 67–93; J. O. Fleckenstein, "Kepler and Neoplatonism," in *Vistas in Astronomy,* **18** (1975); Robert Westman, "The Comet and the Cosmos: Kepler, Mästlin and the Copernican Hypothesis," in *Studia Copernicana* (Warsaw, in press); and K. Hujer, "Kepler in Prague—1600–1612," Leningrad Kepler Symposium, 1971.

Kepler's methodology is discussed by Robert Westman, "Johannes Kepler's Adoption of the Copernican Hypothesis" (doctoral diss., Univ. of Michigan, 1971); and "Kepler's Theory of Hypothesis and the 'Realist Dilemma,'" in *Internationales Kepler-Symposium, Weil der Stadt 1971* (Hildesheim, 1972). In the Weil der Stadt symposium, see also J. Mittelstrass, "Wissenschaftstheoretische Elemente der Keplerschen Astronomie"; and G. Buchdahl, "Methodological Aspects of Kepler's Theory of Refraction." These last three papers are also repr. in *Studies in History and Philosophy of Science*, **3**, no. 3 (1972).

Kepler's theory of light is discussed extensively in Stephen Straker, "Kepler's Optics, a Study in the Development of Seventeenth-Century Natural Philosophy" (doctoral diss., Indiana Univ., 1971); for other aspects, see Vasco Ronchi, *The Nature of Light* (Cambridge, Mass., 1970), pp. 87–98; Huldrych M. Koelbing, "Kepler und die physiologische Optik," in *Internationales Kepler-Symposium, Weil der Stadt 1971* (Hildesheim, 1972); and Johannes Lohne, "Zur Geschichte des Brechungsgesetzes," in *Sudhoffs Archiv*, **47** (1963), 152–172.

A succinct outline of Kepler's contribution to mathematics is D. J. Struik, "Kepler as a Mathematician," in *Johann Kepler 1571–1630*, Special Publication no. 2 of the History of Science Society (Baltimore, 1931). Aspects of the background of his mathematics are found in E. J. Aiton, "Infinitesimals and the Area Law," in *Internationales Kepler-Symposium, Weil der Stadt 1971;* and in Joseph Ehrenfried Hofmann, "Über einige fachliche Beiträge Keplers zur Mathematik," *ibid.;* and in his "Johannes Kepler als Mathematiker," in *Praxis der Mathematik*, **13** (1971), 287–293, 318–324. See also H. S. M. Coxeter, "Kepler and Mathematics," in *Vistas in Astronomy*, **18** (1975), on contributions to the geometry of polyhedrons; Charles Naux, *Histoire des logarithmes*, I (Paris, 1966), ch. 6, pp. 128–158; and Kuno Fladt, "Das Keplerische Ei," in *Elemente der Mathematik*, **17** (1962), 73–78. J. A. Belyj and D. Trifunovic, "Zur Geschichte der Logarithmentafeln Keplers," in *NTM-Schriftenreihe für Geschichte der Naturwissenschaften, Technik und Medizin*, **9** (1972), 5–20, includes an extensive trans. of a previously unpublished Kepler MS.

For Kepler's musical harmonies, see D. P. Walker, "Kepler's Celestial Music," in *Journal of the Warburg and Courtauld Institutes*, **30** (1967), 228–250; Eric Werner, "The Last Pythagorean Musician: Johannes Kepler," in *Aspects of Mediaeval and Renaissance Music*, Jan La Rue, ed. (New York, 1966), pp. 867–882; and Michael Dickreiter, "Dur und Moll in Keplers Musiktheorie," in *Johannes Kepler, Werk und Leistung* (Linz, 1971), pp. 41–50. See also Ulrich Klein, "Johannes Kepler's Bemühungen um die Harmonieschriften des Ptolemaios und Porphyrios," *ibid.*, pp. 51–60; and Francis Warrain, *Essai sur l'Harmonice mundi*, Actualités Scientifiques et Industrielles no. 912 (Paris, 1942). Paul Hindemith wrote a five-act opera based on Kepler's life, *Die Harmonie des Welt* (text; Mainz, 1957).

For Kepler's mysticism and astrology, see W. Pauli, "Der Einflusz archetypischer Vorstellungen auf die Bildung naturwissenschaftlicher Theorien bei Kepler," in *Naturerklärung und Psyche* (Zurich, 1952), 109–194, trans. by Priscilla Silz as "The Influence of Archetypal Ideas on the Scientific Theories of Kepler," in *The Interpretation of Nature and the Psyche*, Bollingen Series, **51** (New York, 1955), 147–240; Franz Hammer, "Die Astrologie des Johannes Kepler," in *Sudhoffs Archiv*, **55** (1971), 113–135; and Arthur Beer, "Kepler's Astrology and Mysticism," in *Vistas in Astronomy*, **16** (1973). See also Martha List, "Das Wallenstein-Horoskop von Johannes Kepler," in *Johannes Kepler, Werk und Leistung* (Linz, 1971), pp. 127–136.

Aspects of Kepler's religious influences are treated in Edward Rosen, "Kepler and the Lutheran Attitude Toward Copernicanism," *ibid.*, pp. 137–158 and also Leningrad Kepler Symposium; Jürgen Hübner, "Naturwissenschaft als Lobpreis des Schöpfers," in *Internationales Kepler-Symposium, Weil der Stadt 1971;* M. W. Burke-Gaffney, *Kepler and the Jesuits* (Milwaukee, 1944); J. Hübner, "Johannes Kepler als theologischer Denker," in *Kepler-Festschrift 1971* (Regensburg, 1971), pp. 21–44; and Martha List, "Kepler und die Gegenreformation," *ibid.*, pp. 45–63.

OWEN GINGERICH

KERÉKJÁRTÓ, BÉLA (*b.* Budapest, Hungary, 1 October 1898; *d.* Gyöngyos, Hungary, 26 June 1946), *mathematics.*

In 1920 Kerékjártó took the Ph.D. degree at Budapest University. He became a privatdocent at Szeged University in 1922, extraordinary professor in 1925, and professor ordinarius in 1929. In 1938 he became professor ordinarius at Budapest University. From 1922 to 1926 he had also traveled abroad: in 1922–1923 he stayed at Göttingen University where he gave lectures on topology and mathematical cosmology; in 1923 he taught geometry and function theory at the University of Barcelona; and from 1923 to 1925 he was at Princeton University, where he lectured on topology and continuous groups. When he returned to Europe he lectured in Paris. Kerékjártó was a corresponding member of the Hungarian Academy of Sciences from 1934 and a full member from 1945. He was a coeditor of *Acta litterarum ac scientiarum Regiae Universitatis hungarica . . .*, Sectio scientiarum mathematicarum, beginning with volume 6 (1932–1934).

After Max Dehn and P. Heegaard's article "Analysis situs" (1907), in *Encyklopädie der mathematischen Wissenschaften* and Schönflies' *Die Entwicklung der Lehre von den Punktmannigfaltigkeiten* (1908), the first three monographs on topology to appear were Veblen's "Analysis situs," in American Mathematical Society Colloquium Publications (1922), Kerékjártó's (1923), and S. Lefschetz's "L'analysis situs et la géométrie algébrique" (1924). Kerékjártó's probably sold best and was the most widely known, but it has exerted much less (if any) influence than the two others. One reason is the restriction of subject and method to two dimensions, at a time when all efforts were directed to understanding higher dimensions. The other, decisive reason is that everyone knew Kerékjártó's *Vorlesungen über Topologie* was not a good book and, therefore, nobody read it. The present author has held this opinion for many years and now feels obliged to make a closer examination of Kerékjártó's works.

The book opens with a proof which is unintelligible and probably wrong. This, indeed, is the worst possible beginning, but it continues the same way. The greater part of Kerékjártó's own contributions are hardly intelligible and most are apparently wrong. The work of others is often taken over almost literally or in a way which proves that Kerékjártó had not really assimilated the material. The level of the book is far below that of topology at that time, and the organization is chaotic. When referring to a concept, a notation, or an argument, he often quotes a proof, a page, or an entire chapter—but often the material quoted is not found where he cites it; sometimes footnotes serve to fill gaps in arguments. For many years this also was the style of Kerékjártó's papers. They are full of mistakes or gaps which should have been filled by other papers which never appeared.

Kerékjártó's papers written around 1940 make a more favorable impression. In general they are correct. They deal with his earlier problems on topological groups but use methods which in the meantime had become obsolete. It is quite probable that he did not know the developments in topology after 1923. The strangest feature is that he never used set-theory symbols, such as the signs for belonging to a set, inclusion, union, and intersection. Apparently he did not know of their existence.

Kerékjártó mainly continued the work of Brouwer and Hilbert on mappings of surfaces and topological groups acting upon surfaces. The undeniable merits of his work are obscured by the manner of presentation. The classification of open surfaces is usually ascribed to Kerékjártó, but the exposition of this subject in his book hardly justifies this claim. It was probably his greatest accomplishment that he became interested in groups of locally equicontinuous mappings (of a surface), although his definition of this notion did not match the way in which it was applied; strangely enough, he did not notice that this notion had already been fundamental in Hilbert's work. The best result in studying such groups with Kerékjártó's methods has recently been achieved by I. Fary, who proved that equicontinuous, orientation-preserving groups of the plane are essentially subgroups of the Euclidean or of the hyperbolic group.

In addition to the work on topology in German and a work on foundations of geometry in Hungarian, which has been translated into French, Kerékjártó wrote some sixty papers, most of them comprising only a few pages. The bibliography is restricted to the more mature ones.

BIBLIOGRAPHY

Kerékjártó's works include *Vorlesungen über Topologie* (Berlin, 1923); "Sur le caractère topologique du groupe homographique de la sphère," in *Acta mathematica*, **74** (1941), 311–341; "Sur le groupe des homographies et des antihomographies d'une variable complexe," in *Commentarii mathematici helvetici*, **13** (1941), 68–82; "Sur les groupes compacts de transformations topologiques des surfaces," in *Acta mathematica*, **74** (1941), 129–173; "Sur le caractère topologique du groupe homographique de la sphere," in *Journal de mathématiques pures et appliquées*, 9th ser., **2** (1942), 67–100; and *A geometria alapjai*, 2 vols. (I, Szeged, 1937; II, Budapest, 1944), translated into French as *Les fondements de la géométrie euclidienne*, 2 vols. (Budapest, 1955–1956).

See also I. Fary, "On a Topological Characterization of Plane Geometries," in *Cahiers de topologie et géométrie différentielle*, **7** (1964), 1–33. There is also an obituary in *Acta Universitatis szegediansis, Acta scientiarum mathematicarum*, **11** (1946–1948), v–vii.

HANS FREUDENTHAL

KERR, JOHN (*b*. Ardrossan, Ayrshire, Scotland, 17 December 1824; *d*. Glasgow, Scotland, 18 August 1907), *physics*.

The second son of Thomas Kerr, a fish dealer, Kerr received part of his early education at a village school on Skye. He attended the University of Glasgow, beginning in 1841, and received his M.A. with "highest distinction in Physical Science" in 1849. He studied natural philosophy under David Thomson in 1845–1846 and then under William Thomson. He was a member of the first group to work with the latter in the laboratory that was converted from a wine cellar and was known among the students as the "coal hole."

A divinity student, Kerr completed the courses in theology at the Free Church College in Glasgow but did not take up clerical duties. In 1857 he was appointed lecturer in mathematics at the Free Church Normal Training College for Teachers in Glasgow, remaining in this post for forty-four years. The facilities for research at this institution were limited, as was the time that Kerr could devote to it. Therefore the paucity of his publications is not surprising; their quality, however, is high.

Kerr is remembered primarily for two discoveries. The first, which he announced in 1875, was the birefringence developed in glass in an intense electric field. He bored holes into the ends of a piece of glass two inches thick until they were about a quarter of an inch apart. An intense electric field was applied to electrodes placed in these holes. The effect on a beam of polarized light shining perpendicular to the electric field was to give it elliptical polarization. The effect was strongest when the plane of polarization was at an angle of 45° to the field and zero when it was parallel or perpendicular to the field. In subsequent papers Kerr extended his findings to other materials, including a large number of organic liquids. He also found that the size of the effect was proportional to the square of the electric force.

Kerr's second discovery, which bears his name, was announced at the meeting of the British Association in Glasgow in 1876, and an account was published the following year. The Kerr effect is detected when a beam of plane polarized light is reflected from the pole of an electromagnet. When the magnet is activated, the beam becomes elliptically polarized, with the major axis rotated from the direction of the original plane. Extended by Kerr and others, these experiments were first treated theoretically by George F. Fitzgerald in "On the Electromagnetic Theory of the Reflection and Refraction of Light" (*Philosophical Transactions of the Royal Society*, **171** [1880], 691–711) and in more general terms by Joseph Larmor in "The Action of Magnetism on Light" (*Report of the British Association* . . . [1893], 335–372).

Kerr received an honorary LL.D. degree from the University of Glasgow in 1868, in recognition of his achievements in teaching, and was elected to the Royal Society in 1890. He was married to Marion Balfour and had three sons and four daughters.

BIBLIOGRAPHY

I. ORIGINAL WORKS. Kerr's books include *The Metric System; Its Prospects in This Country* (London, 1863); *An Elementary Treatise on Rational Mechanics* (London, 1867); *Memories Grave and Gay; Forty Years of School Inspection* (Edinburgh–London, 1902); and *Scottish Education, School and University, From Early Times to 1908* (Cambridge, 1910). His articles are listed in the *Royal Society Catalogue of Scientific Papers;* most of them appeared in *Philosophical Magazine*. The two discoveries cited above were announced in "On a New Relation Between Electricity and Light: Dielectrified Media Birefringent," in *Philosophical Magazine*, **50** (1875), 337–348,

446–458; and "On the Rotation of the Plane of Polarization by Reflection From the Pole of a Magnet," *ibid.*, n.s. **3** (1877), 321–343. Some of Kerr's original apparatus is preserved at the University of Glasgow.

II. SECONDARY LITERATURE. Short biographical accounts include those by C. G. Knott, in *Nature*, **76** (1907), 575–576; by Andrew Gray, in *Proceedings of the Royal Society*, **82A** (1909), i–v; and, by Robert Steele, in *Dictionary of National Biography*, supp. II, 394.

BERNARD S. FINN

KERR, JOHN GRAHAM (*b.* Arkley, Hertfordshire, England, 18 September 1869; *d.* Barley, Hertfordshire, England, 21 April 1957), *zoology.*

Graham Kerr, as he was known during most of his life, was the son of James Kerr, a former principal of Hoogly and Hindu College, Calcutta, and Sybella Graham. He attended the Collegiate School, Edinburgh, and then the Royal High School, before enrolling in Edinburgh University to study medicine. A new career opened to him when in 1889 he was selected to join the staff of an expedition of the Argentine navy for the survey of the Pilcomayo River from the Paraná to the Bolivian frontier. An account of this famous expedition is contained in his book *A Naturalist in the Gran Chaco* (Cambridge, 1950).

Kerr returned to England in 1891 and entered Christ's College, Cambridge. After graduation he led a second expedition to Paraguay (1896-1897) with the object of obtaining material for the study of the Dipnoi group of fishes, to which he had decided to dedicate his research career. He succeeded in collecting many specimens of the eel-shaped dipnoan or lungfish, *Lepidosiren paradoxa.* On his return he was appointed demonstrator in animal morphology at Cambridge and elected a fellow of Christ's College.

In 1902 Kerr became regius professor of natural history (changed to zoology in 1903) at the University of Glasgow, where he remained until 1935. Throughout his professorship he was especially interested in the teaching of medical students and his lectures were famous. His approach was largely morphological and embryological and is summarized in his textbook *Zoology for Medical Students* (London, 1921). Apart from his heavy teaching and administrative duties, he continued research work on various aspects of dipnoan anatomy and embryology, and he and his colleagues published a whole series of papers based on the material he had collected. In addition he was president of the Scottish Marine Biological Association and helped in the organization of the marine biological station at Millport, on Great Cumbrae Island in southwestern Scotland.

He was twice married, first to a cousin, Elizabeth Mary Kerr, daughter of Thomas Kerr, who died in 1934 and by whom he had two sons and one daughter; he later married a widow, Isobel Clapperton.

He was elected a fellow of the Royal Society of London in 1909 and was a member of numerous other learned societies, serving as president of many of them. Later in life he took an increasing interest in politics and in 1935 was elected Conservative member of Parliament for the Scottish universities. He was knighted in 1939. Other public recognition included honorary LL.D. degrees at Edinburgh (1935) and St. Andrews (1950); an honorary fellowship at Christ's College, Cambridge (1935); associate membership of the Royal Academy of Belgium (1946); and the Linnean Gold Medal (1955).

Graham Kerr was one of the last of the famous zoologists of the nineteenth century, men who for the most part were widely traveled, good naturalists, and who possessed an almost encyclopedic knowledge of their subject. Kerr's output of zoological papers was considerable, especially in his early days, and included studies on the anatomy not only of the Dipnoi but of the pearly nautilus (genus *Nautilus*). His outlook was essentially morphological and phylogenetic, and he tended to mistrust the experimental approach.

Kerr also took a great interest in devising camouflage according to natural and biological, or so-called dazzle, principles and upon the outbreak of World War I advised the British admiralty to use obliterative shading and disruptive patterns to make warships less conspicuous. His pioneering suggestion was eventually adopted and was used during World War II, even for land installations.

BIBLIOGRAPHY

Kerr's books include *Primer of Zoology* (London, 1912); *Lectures on Sex and Heredity Delivered in Glasgow 1917–1918* (London, 1919), written with F. O. Bower and W. E. Agar; *Textbook of Embryology* (London, 1919); *Zoology for Medical Students* (London, 1921); *Evolution* (London, 1926); *An Introduction to Zoology* (London, 1929); and *A Naturalist in the Gran Chaco* (Cambridge, 1950).

An obituary appears in *Biographical Memoirs of Fellows of the Royal Society*, **4** (London, 1958), 155–166.

EDWARD HINDLE

KEŚAVA (*fl.* Nandod, Gujarat, India, 1496), *astronomy.*

Keśava, the son of Kamalākara of the Kauśikagotra and a pupil of Vaidyanātha, was the first of a line of astronomers at Nandigrāma (Nandod) that includes his sons Ananta, Rāma, and Gaṇeśa (*b.* 1507) and his grandson Nṛsiṃha (*b.* 1548). Gaṇeśa lists his father's works in his *Muhūrtadīpikā*, a commentary on Keśava's *Muhūrtatattva:*

1. *Grahakautuka*
2. *Tithisiddhi*
3. *Jātakapaddhati*
4. *Jātakapaddhativivṛti*
5. *Tājikapaddhati*
6. *Siddhāntopapattipāṭhanicaya*
7. *Muhūrtatattva*
8. *Kāyasthādidharmapaddhati.*

Like Gaṇeśa, then, Keśava wrote on Hindu law as well as on astronomy and astrology.

The *Grahakautuka* is a treatise on astronomy, apparently following the Brāhmapakṣa (see essay in Supplement), written in 1496. It is accompanied by astronomical tables. Although several manuscripts of this work survive, it has not been studied or published. There is a commentary on it by Viśvanātha (*fl.* 1612–1634), the son of Divākara of Golagrāma, a pupil of Keśava's son Gaṇeśa.

The *Tithisiddhi* presumably contained tables for computing *tithis*, *nakṣatras*, and *yogas*. No manusscripts are known.

The *Jātakapaddhati* or *Keśavīpaddhati* is a short treatise on horoscopy which has been immensely popular in India. It is usually accompanied by a commentary which includes extensive astronomical tables. The commentaries are the following (for editions see the list of editions of the *Jātakapaddhati*):

1. *Vivṛti* of Keśava himself.
2. *Udāharaṇa* of Viśvanātha (*fl.* 1612–1634), the son of Divākara of Golagrāma.
3. *Prauḍhamanoramā* of Divākara (1626), the great-grandson of Divākara of Golagrāma. Published.
4. *Vāsanābhāṣya* of Dharmeśvara (*fl. ca.* 1600–1650).
5. *Udāharaṇa* of Nārāyaṇa (1678).
6. *Subodhinī* of Umāśaṅkara Miśra (1857). Published.
7. *Udāharaṇa* of Apūcha Śarman (Jhā) (1858). Published.
8. *Sarvamanoramā* of Sītārāma Jhā (1924). Published.
9. *Udāharaṇadarśinī* of Gopīkānta Śarman. Published.
10. *Udāharaṇa* of Gurudāsa.
11. *Udāharaṇadīpikā* of Rāmadhīna Śarman. Published.

The *Jātakapaddhati* has frequently been published:

1. Edited with a Marāṭhī translation by A.D.S. Vadikara and V. L. J. Kannaḍakara (Bombay, 1872).
2. Edited with a Hindī commentary by B. Prabhuṇe (Benares, 1877).
3. Edited with the *Prauḍhamanoramā* of Divākara by Vāmanācārya (Benares, 1882).
4. Edited with the *Subodhinī* of Umāśaṅkara Miśra (n.p., 1890).
5. Edited with a Hindī commentary by Jagadīśaprasāda Tripāṭhin (Bombay, 1899; 2nd ed., Bombay, 1924).
6. Edited with a Gujarātī commentary by D. K. Mayāśaṅkara (Bombay, 1909).
7. Edited with the *Udāharaṇadarśinī* of Gopīkānta Śarman (Ayodhyā, 1924).
8. Edited with the *Udāharaṇa* of Apūcha Jhā, the *Udāharaṇadīpikā* of Rāmadhīna Śarman, and his own *Sarvamanoramā* by Sītārāma Jhā (Benares, 1925; 2nd ed., Benares, 1948).

The *Jātakapaddhativivṛti* has been mentioned above. Several manuscripts exist, but it has not yet been published.

The *Tājikapaddhati* or *Varṣaphalapaddhati* is a work on annual predictions based on the Islamic doctrine of the revolution of years of the world. It has been commented on by two of the sons of Divākara of Golagrāma, Mallāri (*fl. ca.* 1600) and Viśvanātha (*fl.* 1612–1634). There are two editions: one with Viśvanātha's *Udāharaṇa* (Benares, 1869) and the other with a Telegu translation (Madras, 1916).

The *Siddhāntopapattipāṭhanicaya* seems to be "a collection of readings on the origin (of statements in) the (astronomical) Siddhāntas"; nothing more is known of it.

The *Muhūrtatattva* is a well-known work on catarchic astrology. There are commentaries by Keśava's son Gaṇeśa (the *Muhūrtadīpikā*), and perhaps by Kṛpārāma. The *Muhūrtatattva* has been edited twice: at Benares in 1856 and with a Marāṭhī translation by V. V. Śāstrī Jośī (3rd ed., Poona, 1927).

The *Kāyasthādidharmapaddhati* is a work on the religious duties of Kāyasthas (members of the scribal caste) and others. No manuscripts are known.

In addition to these works a *Gotrapravaramaṅgalāṣṭaka* has been published by D. A. Sāvanta in *Maṅgalāṣṭakasaṅgraha* [Belgaum, 1924], as by Keśava the astrologer, but the attribution of this text to Keśava of Nandigrāma is doubtful. The subject of the work is the system of exogamous lineages prevalent in India.

BIBLIOGRAPHY

There are articles on Keśava by Sudhākara Dvivedin in *Gaṇakataraṅginī* (Benares, 1933), repr. from *The Pandit*, n.s. **14** (1892), 53–55; Ś. B. Dīkṣita in *Bhāratīya Jyotiḥśāstra* (Poona, 1896; repr. Poona, 1931), pp. 258–259; and D. Pingree, *Census of the Exact Sciences in Sanskrit*, Series A, **2**, *Memoirs of the American Philosophical Society*, **86** (Philadelphia, 1971), 65–74.

DAVID PINGREE

KETTERING, CHARLES FRANKLIN (*b.* Loudonville, Ohio, 25 November 1876; *d.* Loudonville, 25 November 1958), *engineering, invention.*

Kettering's genius lay in his ability to adopt new methods and concepts in solving technical problems. He was a questioner with a fondness for challenging the apparently obvious: Why is grass green? Have you ever been a piston in a diesel engine?

Kettering was one of five children of Jacob Kettering, a farmer and carpenter, and the former Mary Hunter. As a boy his schooling was interrupted by poor eyesight, but he eventually graduated from the Ohio State University in 1904 with a degree in electrical engineering. He joined the inventions staff of the National Cash Register Company, working there for five years; then he left to found the Dayton Engineering Laboratories Company (Delco) with Edward A. Deeds. The company subsequently became part of General Motors.

Kettering's first great achievement was the electric starter for automobiles, installed on the Cadillac car in 1912. His contribution was a motor powerful enough to turn the engine over but small enough to fit in a motor vehicle. The concept originated when he was working on an electric cash register and realized that the motor he required need not carry a constant load but merely had to deliver an occasional surge of power. Kettering was invited to work on the starter because his company had designed a successful wet-battery ignition system for Cadillac. This system was alleged to be a cause of engine knock; and to refute this criticism Kettering undertook to find the real cause of knock—which, with remarkable insight, he decided was imperfect combustion of the fuel. After long research in cooperation with Thomas Midgley, Jr., and T. A. Boyd, he found a remedy in the addition of tetraethyl lead to gasoline. The resulting product, ethyl gasoline, was put on the market in 1922. At the same time Kettering was working with Du Pont chemists to develop a quick-drying lacquer finish for motor vehicle bodies, thereby eliminating a troublesome source of delay in production.

In 1919 Kettering became head of the General Motors Research Corporation, an assignment designed to give greater scope and resources for his talents. It was in this capacity that he completed his work on antiknock gasoline and quick-drying finishes. He also tried to design an engine with an air-cooling system using copper fins on the cylinders over which air was blown by a fan. Some cars using this system were built in 1922 but were subsequently withdrawn, and the experiment was discontinued as impractical in the existing state of the art. A search for a nontoxic refrigerant had a happier outcome, the result being the fluorine compounds known as Freon.

During the 1930's Kettering was active in the refinement and improvement of the diesel engine. The specific steps involved using a two-cycle system, better fuel injection, and prevention of piston overheating. The result was the replacement of steam by diesel power on railroads and the use of diesel engines in trucks and buses. Subsequently, through the Charles F. Kettering Foundation and the Sloan-Kettering Institute for Cancer Research, Kettering encouraged and took an active interest in various fields of medical research, as well as basic research in photosynthesis and magnetism. He was not, however, a pure research scientist; his interest in photosynthesis and magnetism lay in a hope of finding new sources of energy for human use.

BIBLIOGRAPHY

Kettering delivered many lectures and addresses, but they have not been collected. During the 1930's he wrote frequently for the *Saturday Evening Post.*

Secondary sources are T. A. Boyd, *Professional Amateur. The Biography of Charles F. Kettering* (New York, 1957); Mrs. Wilfred C. Leland, *Master of Precision. Henry M. Leland* (Detroit, 1960), written with Minnie D. Millbrook, which challenges Kettering's claim to be the inventor of the electric starter; Arthur Pound, *The Turning Wheel* (New York, 1934); John B. Rae, *American Automobile Manufacturers: The First Forty Years* (Philadelphia, 1959), pp. 109–114, 155–157, 198–199; and Alfred P. Sloan, Jr., *My Years with General Motors* (Garden City, N.Y., 1964), pp. 108–111, 222–225, 249–261, 341–359.

JOHN B. RAE

KEULEN, LUDOLPH VAN. See **Ceulen, Ludolf van.**

KEYNES, JOHN MAYNARD (*b.* Cambridge, England, 5 June 1883; *d.* Firle, Sussex, England, 21 April 1946), *economics, mathematics.*

His father, John Nevile Keynes, was the author of *Formal Logic* (1884) and *Scope and Method of Political*

Economy (1890). Both were thoroughly up-to-date in their day and remained standard texts for a number of years. John Nevile was a lifelong fellow of Pembroke College, Cambridge, and registrary (chief administrative officer) of Cambridge University from 1910 to 1925. Maynard's mother was Florence Ada, daughter of John Brown, who wrote what was for long regarded as the standard life of John Bunyan, the author of *The Pilgrim's Progress.* She was an authoress and an ardent worker for social causes and in local government, eventually becoming the mayor of Cambridge.

Keynes went to the Perse School Kindergarten, Cambridge (1890), and to St. Faith's Preparatory School, Cambridge (1892). He won a scholarship at Eton College (1897) and at Kings College, Cambridge (1902). For his degree he studied mathematics only and in 1905 was twelfth wrangler (i.e., twelfth on the list of those offering mathematics in that year). This was sufficiently distinguished but not eminently so. The fact is that he did little work at academic studies when an undergraduate, devoting his time to wide reading, some political activity (he was president of the Cambridge Union) and, more particularly, to the cultivation of literary friends (Lytton Strachey and others) who were destined to play a notable part in the intellectual life of England.

He spent the next year (1905–1906) as a graduate at Cambridge, not working for a degree but enlarging his reading, including that in economics, in which he had instruction from Alfred Marshall and A. C. Pigou, and which he had also imbibed in early boyhood from his father.

In 1906 he took the British Civil Service examination and was second on the list for all England. There happened to be only one vacancy in the Treasury in that year, and he opted for the India Office as his second preference. He remained there for two years. During those years and in the three years that followed, he devoted the greater part of his time to work on the theory of probability.

His *Treatise on Probability* was not published until 1921, owing to the interruption of the war, but had been almost completed by 1911. This was at once a work of great learning and also an exposition of important original ideas. Its bibliography of the literature is one of the most comprehensive that has ever been made.

In regard to the original ideas, his ambition was to provide a firm mathematical basis for the probability theory on lines comparable to those of the *Principia Mathematica*, in which Russell and Whitehead laid the foundations of symbolic deductive logic. Of Keynes's book Bertrand Russell afterward wrote, "The mathematical calculus is astonishingly powerful, considering the very restricted premises which form its foundation . . . the book as a whole is one which it is impossible to praise too highly" (*Mathematical Gazette* [July 1922]).

While Keynes was an innovator in expressing probability theory in terms of modern-type symbolism, and in this respect his book constituted a landmark, two of its central doctrines have not been widely accepted since.

(1) Keynes thought it proper to postulate that probability is a concept that is capable of being apprehended by direct intuition and requires no definition. This approach was due to the influence of the Cambridge philosopher G. E. Moore but no longer finds favor.

(2) Keynes translated into his own symbolism the central proposition of Bayes. The Bayes-type approach demonstrates how favorable instances can increase the probability of a given premise. For this reasoning to work, the premise must have some prior probability of its own. The trouble is that in the very beginning of the inductive process there are no empirical propositions with any intrinsic probability of their own. So how to make a start? This is, of course, the crux of the problem of induction. Keynes thought it proper to overcome this difficulty by postulating the principle of limited independent variety, meaning that there is a finite number of "ultimate generator properties" in the universe. This would enable one to assign a positive probability to the proposition that one (or another) of the ultimate properties was operating in a given case. The objection was then made that to get any significant probability for a conclusion it would not be enough to postulate a finite number of ultimate generator properties, but a specific number. The impossibility of doing this is clearly a stumbling block for the Keynes-type approach.

In 1908 he resigned from the India Office and went to Cambridge, without official appointment, on the invitation of Alfred Marshall to assist the new Department of Economics there. Shortly afterwards he was awarded a fellowship at Kings College, open to competition, on the strength of his thesis on probability.

Meanwhile he was also at work on his book *Indian Currency and Finance* (1913), which included a description of what is called a "gold exchange standard." He had also been invited to serve on a royal commission on Indian finance and currency. He contributed much to the report and also appended an annex of his own, recommending a central bank. At that time this was a revolutionary idea for a less-developed country.

He was in the British Treasury from 1915 to 1919,

KEYNES

in the later years as head of the department looking after foreign exchange controls. In 1919 he went to the Paris Peace Conference as principal representative of the British Treasury and deputy for the chancellor of the exchequer. In June 1919 he resigned, on the ground that the proposals put forward for German reparations payments were impractical and unjust. In December 1919 he published *The Economic Consequences of the Peace*, which won him a worldwide reputation for its brilliant writing and character sketches, its humane and liberal outlook, and the cogency of its arguments about the German reparations problem.

He returned to Kings College, Cambridge. For a period his primary interest was in German reparations. *A Revision of the Treaty* was published in 1922. Other economic matters began to engage his attention, namely the evils of deflation, which became severe both in the United States and the United Kingdom in 1920, and the unemployment question.

On the monetary side he published *A Tract on Monetary Reform* (1923), which was a lucid exposition of monetary theory partly on traditional lines. He departed from those lines, however, in advocating that there should not be a return to a fixed parity between the pound and the dollar but that a floating exchange rate should be regarded as normal. When, despite his advocacy, the United Kingdom returned to the gold standard in 1925, he wrote a devastating pamphlet entitled *The Economic Consequences of Mr. Churchill.*

On the side of unemployment he began at an early date to advocate public works, mainly in articles in *The Nation*. Orthodoxy claimed that public works would not decrease unemployment, on the ground that money spent on them would entail that private enterprise had that much less money to spend, so that there would be no net gain of employment at all. This was sometimes known as the "Treasury view." It was the intellectual challenge presented by this view which drove him to the conclusion that quite a considerable part of economics would have to be rethought, and to that he devoted his main powers for the next dozen years. The fruits of his thinking were published in *A Treatise on Money* (December 1930), when his intellectual journey was half complete, and in *The General Theory of Employment, Interest and Money* (January 1936). Note should also be made of his membership (1929–1931) of the official Macmillan committee of enquiry into finance and industry. He gave that committee the benefit of a statement of his views on money which lasted for five days. A rescript of this is due eventually to be published in his collected works.

Of the two books, the *Treatise* is the more comprehensive volume and contains much vital material not to be found elsewhere. For knowledge of Keynes the *General Theory* is compulsory reading, because it contains his final synthesis; but this has had the unfortunate effect that the *Treatise* has not been read as much as it should be by those who wish to understand Keynes in depth over a wide range of subjects.

This is not the place to summarize Keynes's theory. His position in the history of economic thought may be described by saying that he was the first economist to provide a systematic "macrostatics." Traditional economics had had a considerable measure of success in what is known as "microstatics"; this refers to the analysis of the supply and demand for particular commodities, the allocation of productive resources among different uses, the distribution of income, decisions by firms, etc.; but there was lacking a systematic account of what determines the level of activity in the economy as a whole, and the balance between saving and investment requirements. Whatever criticisms have been or may in due course be made in detail, Keynes's work will remain a landmark in the history of theories relating to these topics.

Mention should be made of his influence during World War II in getting the British government first to compile, and later to publish, national income statistics, which give the factual material required for the practical application of his theories. Most countries have come to think it needful to compile such statistics.

Keynes had a serious illness in 1937 and was never thereafter restored to full health. In 1940 he was invited into the British Treasury in an honorary capacity. Although he did not have responsibilities such as he had in World War I, his advice was constantly sought on all matters relating to the economics of the war. Then he began, as early as 1941, to acquire a position of leadership in matters relating to Anglo-American cooperation for postwar world reconstruction. At this point mention should be made of his booklet *The Means to Prosperity* (1933), which he published shortly before the World Economic Conference in London and which contained some of the ideas to which he began to give a more elaborate form in his Clearing Union plan in 1941, which was the British contribution to the Anglo-American-Canadian effort leading to the Bretton Woods Conference (1944) and the foundation of the International Monetary Fund.

Prior to his illness he devoted much time to practical finance, on his own behalf, on that of Kings College, Cambridge, of which he was bursar for many years, and of certain insurance and investment companies with which he was associated. He was joint editor of the *Economic Journal* from 1912 to 1945. He also made important collections of old books and of modern paintings. He was the founder of the Arts Theatre in Cambridge. During the war he became chairman of the British Arts Council. In 1925 he married Lydia Lopokova, the famous Russian ballerina.

In his book collecting, he specialized in the philosophers and thinkers of the seventeenth and eighteenth centuries and, later, in the general English literature, including drama and poetry, of the sixteenth century. He had an exceptionally important collection of Newton manuscripts. It was this, doubtless, that caused him to prepare an essay on "Newton, the Man" for the tricentenary celebrations (1942).

He pays tribute to Newton's world preeminence as a scientist. "His peculiar gift was his power of holding continuously in his mind a mental problem until he had seen straight through it. I fancy his pre-eminence is due to his muscles of intuition being the strongest and most enduring with which a man has ever been gifted." But he gives more space to Newton's other interests—alchemy and apocalyptic writings, to which, so the manuscripts suggest, he devoted as much time as he did to physics.

Had Keynes completed his work, he would doubtless have inserted the fact (of relevance to Keynes's own work!) that Newton became Master of the Mint and established a new bimetallic parity for Britain (1717). Alexander Hamilton was responsible for the original parity of the U. S. gold and silver dollars, and expressed indebtedness to Newton's writings on the topic.

In 1942 Keynes was made a member of the House of Lords, where he sat on the Liberal benches.

The Royal Society paid him the honor, rare for a nonscientist, of making him a fellow, doubtless in recognition of his basically scientific approach to all things.

He died in his country home, Tilton, on 21 April 1946, shortly after his return from a meeting in Savannah, which was concerned with details relating to the setting up of the International Monetary Fund and the International Bank for Reconstruction and Development.

BIBLIOGRAPHY

For a list of Keynes's writings see British Museum *General Catalogue of Printed Books*, CXXII, cols. 706–710.

Recent works in English on Keynes are Dudley Dillard, *The Economics of John Maynard Keynes* (New York, 1948), with bibliography, pp. 336–351; Seymour E. Harris, *The New Economics* (London, 1960), with bibliography, pp. 665–686; Roy Forbes Harrod, *The Life of John Maynard Keynes* (New York–London, 1951; repr. 1969); and the obituary by A. C. Pigou in *Proceedings of the British Academy*, **32** (1946), 395–414, with portrait.

ROY FORBES HARROD

KEYS, JOHN. See **Caius, John.**

KEYSERLING, ALEXANDR ANDREEVICH (*b.* Kabillen farm, Courland, Latvia, 15 August 1815; *d.* Raikül estate, Estonia, 8 May 1891), *geology, paleontology, botany.*

Keyserling was the fifth son of seven children of Count Heinrich Dietrich Wilhelm Keyserling and the former Anne Nolde. He received a good education and in 1834 began to study law at Berlin University. Under the influence of Buch and Humboldt, whom he met there, he became interested in natural sciences and chose geology as his specialty. In 1840 Keyserling returned to Russia and a year later became an official handling special missions in the Mining Department. In 1842 Berlin University awarded him a doctorate.

In 1844 Keyserling married Zinaida Kankrina, the daughter of Russia's minister of finance. Being financially secure, he did not have to consider permanent employment; and in 1850 he left government service. He settled down on his estate in Estonia and continued his scientific research. In 1858 the St. Petersburg Academy of Sciences elected him corresponding member and in 1887 granted him the title of honorary academician.

Keyserling began his scientific research while still a student. With the zoologist J. H. Blasius, later a professor at Brunswick, he made a number of excursions in the Carpathians and the Alps, collecting material for his first scientific paper (1837). Later they studied vertebrates of Europe.

Keyserling returned to Russia with Blasius, and they participated in A. K. Meyendorff's expedition that studied the natural resources and industry of European Russia. In 1841 Keyserling joined the special expedition conducted by Murchison to study the geological structure of European Russia and the Urals; and he studied a vast area—the Kirghiz steppes—along the left bank of the Volga southwest of Orenburg.

To process the material collected there Keyserling visited Paris and London in 1842. A year later he took part in another expedition, to study the geological structure of the Pechora basin, the northern Urals, and Timan. Previously this area was virtually unknown geologically, and Keyserling's research provided extensive new material on the geological structure and the paleontology of Paleozoic and Jurassic deposits developed there. The paper he published on the basis of the data collected was awarded the Demidoff Prize by the St. Petersburg Academy of Sciences. The paper was included in the second volume of a large summary of the geology of Russia published by Murchison, P. E. de Verneuil, and Keyserling in 1845, in which the greater part of the paleontological section is Keyserling's. In general he paid great attention to the study of extinct organisms, and many of his papers are devoted to descriptions of fossils collected by other researchers during their Siberian investigations. After he retired from his official posts in 1850 Keyserling practically gave up traveling until 1860, when he made several crossings of the Pyrenees with Verneuil.

Along with the usual descriptions Keyserling's geological papers contain elements of facies analysis, which was quite new at that time. Thus in 1842, on the basis of the lithology and color of the rocks, he reconstructed the changing paleogeographical conditions in the Carboniferous sea of the Moscow basin.

Keyserling had an active interest in botany and worked out the systematics of the fern genus *Adiantum*. This research provided him with abundant material for theoretical deductions in biology. He came to the conclusion that the entire complex of plants and animals inhabiting the earth originated through evolution of primitive cellular elements, or protoplasts. In 1853 he suggested that under the chemical effects of various elements the embryos of living beings undergo a transformation that leads to the creation of new species. In this process only the most adaptable survive, the others becoming extinct. At that time such ideas were very daring and new.

Charles Darwin praised Keyserling's views, referring to him in *The Origin of Species* (1859) as one of his predecessors. Darwin's theory of evolution had a marked effect upon Keyserling; under its influence he changed his views substantially but never became a consistent evolutionist, believing that the changes of a species take place abruptly rather than by gradual modification.

Keyserling contributed substantially to the progress of culture and education in the Baltic provinces. From 1862 to 1869 he was a trustee of the Dorpat (now Tartu) educational region. With J. F. Schmidt he founded a museum of natural history in Reval (Tallinn), which has the world's richest collections of Ordovician and Silurian fauna of the Baltic provinces.

Keyserling was an honorary or corresponding member of numerous Russian and foreign scientific societies, including the mineralogical society of St. Petersburg and the geological societies of London and Paris; and for many years he was president of the Agricultural Society of Estonia. Keyserling's titles at the court were gentleman in attendance and court tutor, as well as land counselor of Estland.

BIBLIOGRAPHY

I. Original Works. Keyserling's major writings are "Bemerkungen während des Überganges von Latsch nach Bormio durch das Marterthal," in *Neues Jahrbuch der Mineralogie* . . . (1837), pp. 389–502, written with J. H. Blasius; *The Geology of Russia and the Ural Mountains*, 2 vols. (London–Paris, 1845), written with R. I. Murchison and P. E. de Verneuil; *Wissenschaftliche Beobachtungen auf einer Reise in das Petschoraland im Jahre 1843* (St. Petersburg, 1846); and "Genus Adiantum recensuit L.," in *Mémoires de l'Académie des sciences de St. Pétersbourg*, 7th ser., **22**, no. 2 (1875), 1–44.

II. Secondary Literature. See J. F. Schmidt and S. N. Nikitin, "Aleksandr Keyserling," in *Izvestiya Geologicheskogo Komiteta*, **10**, no. 15 (1891), 1–11, which includes a bibliography of 21 titles; and Helene von Taube von der Issen (Keyserling's daughter), ed., *Graf Alexander Keyserling. Ein Lebensbild aus seinen Briefen und Tagebüchern* . . ., 2 vols. (Berlin, 1902), with portrait.

V. V. Tikhomirov

IBN KHALDŪN (*b.* Tunis, 27 May 1332; *d.* Cairo, 17 March 1406), *history, sociology.*

ʿAbd-al-Raḥmān, the son of Muḥammad, derived his family name, Ibn Khaldūn, from a remote ancestor named Khaldūn, who is said to have settled in Spain in the eighth century, not long after the Muslim conquest. His family gave up its patrician home in Seville before the Christian conquest of the city in 1248, crossing over to northwest Africa and eventually establishing itself in Tunis. Close ties with the ruling circles in northwest Africa and excellence in legal and religious scholarship were a family tradition. After the ravages of the Black Death, which killed his parents, Ibn Khaldūn entered the service of the Ḥafṣid ruler of Tunis. He soon was dissatisfied and, in 1352, left to look for a more flourishing intellectual atmosphere and more promising career opportunities. Accepting an invitation of Abū ʿInān, the Merinid ruler of Fez, Ibn Khaldūn arrived at Fez in 1354. There he completed his education, profiting from contact with the able scholars assembled at Abū ʿInān's court. Early in 1357 his Tunisian connections made him suspect to Abū ʿInān; he was imprisoned and was not released until Abū ʿInān's death, twenty-one months later.

IBN KHALDŪN

Realizing that Fez was heading toward increasing political instability, Ibn Khaldūn succeeded, with difficulty, in obtaining permission to leave the city. In December 1362 he reached Granada, where he was cordially welcomed by its ruler, Muḥammad V. A diplomatic mission undertaken for Muḥammad V in 1364 offered Ibn Khaldūn the opportunity for a brief visit to his ancestral city, Seville. In the spring of 1365 an invitation extended to him by the Ḥafṣid Abū ʿAbdallāh of Bougie made it possible for him to return to northwest Africa. During the following decade his activities had to be adapted to the turbulent course of northwest African power politics.

An attempt to escape to Spain did not succeed, and ultimately Ibn Khaldūn took refuge in Qalʿat Ibn Salāma, a small village in the Algerian hinterland northwest of Biskra. Here, during a stay of over three years (1375–1378), he started work on his world history. In November 1377 he completed the first draft of its "Introduction" (*Muqaddima*), which was to bring him lasting fame. By then, however, he was becoming restless in the isolation of Qalʿat Ibn Salāma. He asked for and obtained permission from the ruler of Tunis, the Ḥafṣid Abū 'l-ʿAbbās, to return to the city of his birth. According to Ibn Khaldūn himself he soon gained the confidence of Abū 'l-ʿAbbās but thereby aroused the envy of other courtiers and officials, who undermined the ruler's trust in him. Feeling uncertain of his situation, he thought it advisable to escape from Tunis. To this end he used the time-honored subterfuge of declaring his intention to make a pilgrimage to Mecca. He arrived by boat at Alexandria on 8 December 1382 and proceeded to Cairo, where he arrived on 6 January 1383. He remained in Egypt for the rest of his life, except for periods of travel in Syria and Palestine and an eight-month trip to Arabia in connection with the pilgrimage.

Ibn Khaldūn's talents as a scholar and diplomat were soon utilized by the new ruler of Egypt, Barqūq (1382–1399), and his career finally reached fulfillment. Ibn Khaldūn was given various academic positions; and the coveted appointment to a judgeship came to him for the first time on 8 August 1384, when he was made Egyptian chief judge of the Mālikites, many of whom had—like himself—strong ties to northwest Africa. Judgeships were highly sensitive positions dependent on political circumstances and the ruler's favor. Thus, it is not astonishing that Ibn Khaldūn was deposed from the judgeship and reappointed to it five more times, the last reappointment coming a few days before his death. During his twenty-three years in Egypt, Ibn Khaldūn took an active part in internal Egyptian politics and in international affairs, capitalizing, in particular, upon his northwest African connections.

In 1401, during a military expedition to Syria initiated by Barqūq's young successor, Faraj, a memorable meeting took place between Ibn Khaldūn and the great Tamerlane, who was then laying siege to Damascus. Despite all his judicial, political, and academic activities, Ibn Khaldūn found the time to continue his scholarly research, mainly with a view to improving and expanding his great historical work. Since he wrote an autobiography, possibly the longest work of its kind until then, and was also much noticed by contemporary biographers, the story of his life is known in considerable detail; but the larger questions of the sources of his genius and the psychological motivation for his many extraordinary activities can only, if at all, be answered by uncertain and unsatisfactory speculation.

Ibn Khaldūn entitled his world history "Book of the Lessons and Archive of Early and Subsequent History, Dealing With the Political Events Concerning the Arabs, Non-Arabs, and Berbers, and the Supreme Rulers Who Were Contemporary With Them." Its introduction and first book became known as an independent work, entitled "The Introduction" (*Muqaddima*), even during the lifetime of Ibn Khaldūn, who was convinced of his work's originality and significance. The *Muqaddima* was indeed the first large-scale attempt to analyze the group relationships that govern human political and social organization on the basis of environmental and psychological factors. Against a background of Muslim legal thought and Islamized Greek philosophy, human society is described as following a constantly repeated rhythm of growth and decay, which includes a continuous slight forward movement provided by the retention of certain cultural achievements of earlier generations.

Human beings require cooperation for the preservation of the species, and they are by nature equipped for it. Their labor is the only means at their disposal for creating the material basis for their individual and group existence. Where human beings exist in large numbers, a division of activities becomes possible and permits greater specialization and refinement in all spheres of life. The result is *ʿumrān* ("civilization" or "culture"), with its great material and intellectual achievements, but also with a tendency toward luxury and leisure which carries within itself the seeds of destruction.

Large concentrations of people are possible only in urban environments, which therefore present the opportunity for the highest flowering of civilization. The force that makes people cohere and cooperate and

IBN KHALDŪN

then aspire toward the achievement of political control is called *ʿaṣabiyya*, which may be translated roughly as "group feeling." Originally a negative term with the connotation of "unfair bias" and as such generally condemned, it was applied in a positive sense by Ibn Khaldūn. *ʿAṣabiyya* is man's psychological attraction to those of the same blood and racial origins. Outsiders may come to share in the *ʿaṣabiyya* of a group through long and close contact.

Political leaders and dynasties attain their eminence by virtue of the ability to concentrate the group feeling upon themselves and thereby profit from its natural bent for the acquisition of power. The achievement of political predominance sets in motion a process of territorial overexpansion that dilutes the group support of the dynasty. More important, it also marks the beginning of an inevitable three-generation cycle of weakening the dynasty's moral fiber. The dynasty becomes alienated from its supporters, and its realm falls prey to others who are fired by a strong and unspoiled group feeling.

All the factors of environment and human psychology operate without the direct intervention of the superhuman, divine establishment. Ibn Khaldūn accepts as a fact that God created these factors and saw to it that they would operate as they do. He also acknowledges the existence of occasions on which God directly intervenes in history—for instance, by sending a prophet with a divine message to mankind. But unusual events of this sort bring about only an intensification of the normal situation or an abnormal interruption that soon comes to an end, permitting resumption of the normal development of human affairs.

In the course of reviewing the totality of the institutions of Muslim society in order to illustrate his sociological views, Ibn Khaldūn shows himself an able and effective historian of science and scholarship. In the sixth chapter of the *Muqaddima*, he presents a number of usually brief sketches of the various religious and legal disciplines, of the natural sciences, and of the functions of language and literature, together with instructive essays on the methodology of scholarship. Although he neglects certain recondite and remarkable achievements of Muslim science, he gives an acceptable picture of the more obvious elements in the development of each science and successfully captures the general flavor of medieval scholarship and the broad outlines of its history. The reality of some of the occult sciences was not denied by Ibn Khaldūn, who believed in the legitimacy of white magic and the existence of black magic. Yet, taking sides in an old controversy, he rather lengthily refuted the claims of astrology and alchemy. He was especially concerned with showing the harm that belief in the reality of astrology and alchemy is able to do to human society. A treatise on elementary arithmetic written in his earlier years is not preserved. It was probably of no scientific importance.

As a descriptive historian Ibn Khaldūn ranks among the greatest of the Muslim world. He saw no need to spell out constantly and in detail how the ideas expounded in the *Muqaddima* applied to individual historical situations; but clearly he was convinced that his historical exposition bore out these ideas, as in fact it does to some degree. In large portions of his history he was naturally obliged to follow one or another of the older standard works; but he also utilized unusual sources, some of them of Christian or Jewish origin. He searched for the best information available concerning events of his own time and attempted to make his history of the world as balanced and complete as possible in his time and place. He helped to increase the intense interest in history current in the Egypt of the fourteenth and fifteenth centuries and among later Ottoman historians and statesmen.

In nearly every individual instance the sources of Ibn Khaldūn's information and ideas can be traced, with the noteworthy exception of the origin of the concept of *ʿaṣabiyya* in the sense that he uses it. But the synthesis, according to all we know, is entirely his own. It stands out boldly in the Muslim context, no matter how greatly it is indebted to Muslim scholarly tradition. It is a summing up of Muslim medieval civilization, but it also points beyond its own time to fundamental problems of the modern science of sociology.

BIBLIOGRAPHY

I. ORIGINAL WORKS. W. McG. de Slane translated the *Muqaddima* as *Prolégomènes historiques d'Ibn Khaldoun*, 3 vols. (Paris, 1862–1868); there is also an English trans. by F. Rosenthal, *The Muqaddimah* (New York, 1958), 2nd ed., 3 vols. (Princeton, 1967), published as vol. XLIII of the Bollingen Series; and a French trans. by V. Monteil, *Ibn Khaldûn, Discours sur l'histoire universelle*, 3 vols. (Beirut, 1967–1969).

The section from the *History* on northwest Africa was translated by de Slane as *Histoire des Berbères et des dynasties musulmanes de l'Afrique septentrionale*, 4 vols. (Algiers, 1852–1856; new ed., Paris, 1925–1956).

The Arabic text of the complete *Autobiography* was edited by M. Ibn Tāwīt al-Ṭanjī, as *al-Taʿrīf bi-Ibn Khaldūn* (Cairo, 1951); an English trans. of the report on the meeting with Tamerlane is by W. J. Fischel, *Ibn Khaldūn and Tamerlane* (Berkeley–Los Angeles, 1952).

Further works are *Lubāb al-Muḥaṣṣal*, L. Rubio, ed.

(Tetuán, 1952); and *Shifā' al-sā'il li-tahdhīb al-masā'il*, M. Ibn Tāwīt al-Ṭanjī, ed. (Istanbul, 1958).

II. SECONDARY LITERATURE. A bibliography by W. J. Fischel is appended to Rosenthal's trans. of the *Muqaddima* and is further updated by Fischel in his *Ibn Khaldūn in Egypt* (Berkeley–Los Angeles, 1967). See also 'Abd al-Raḥmān Badawī, *Mu'allafāt Ibn Khaldūn* (Cairo, 1962); and M. Talbi, "Ibn Khaldūn," in *Encyclopaedia of Islam*, 2nd ed. (Leiden–London, 1968 [1969]), III, 825–831. Some of the more recent works on Ibn Khaldūn are (listed chronologically) Muhsin Mahdi, *Ibn Khaldūn's Philosophy of History* (London, 1957); H. Simon, *Ibn Khaldūns Wissenschaft von der menschlichen Kultur* (Leipzig, 1959); W. J. Fischel, *Ibn Khaldūn in Egypt* (Berkeley–Los Angeles, 1967); M. Nassar, *La pensée réaliste d'Ibn Khaldūn* (Paris, 1967); and M. M. Rabī', *The Political Theory of Ibn Khaldūn* (Leiden, 1967).

FRANZ ROSENTHAL

KHARASCH, MORRIS SELIG (*b.* Kremenets, Ukraine, Russia [formerly Krzemieniec, Poland], 24 August 1895; *d.* Chicago, Illinois, 7 October 1957), *organic chemistry.*

Although his parents, Selig and Louise Kneller Kharasch, were in relatively comfortable circumstances, Kharasch and his brothers immigrated to the United States when he was thirteen years old. An older brother in Chicago helped care for the children upon their arrival. Kharasch received his B.S. degree from the University of Chicago in 1917 and his Ph.D. from the same institution two years later, despite his service in 1918 with the Gas Flame Division of the Army. There he worked on toxic gases at Johns Hopkins and the Edgewood Arsenal. He was later (1926) a consultant with the Chemical Warfare Service.

After graduation Kharasch became a research fellow in organic chemistry at Chicago. From 1922 until 1924 he was at the University of Maryland, first as an associate professor and then as a full professor. He returned to the University of Chicago as associate professor of chemistry in 1928 and remained with that institution until his death; in 1935 he was promoted to professor. Just prior to his death in 1957 he became director of the Institute of Organic Chemistry, which had been created in his honor by the university. He was one of the founders of the *Journal of Organic Chemistry* and late in life became American editor of *Tetrahedron*, an international journal of organic chemistry. He married Ethel May Nelson in 1923 and had two children.

Kharasch is best known for his studies, begun in 1930, on the addition of hydrogen bromide to unsaturated organic compounds. He and his student Frank Mayo carefully determined that in the absence of peroxides, "normal" Markovnikov addition occurs —that is, the hydrogen adds to that carbon of a double bond which already has the largest number of hydrogen atoms. When peroxides are present, however (even in small amounts such as in the use of old reagents), reverse addition occurs. An understanding of the free radical mechanism of this peroxide effect soon followed, and Kharasch applied these ideas to other chemical systems.

During World War II Kharasch collaborated with Frank Westheimer, his colleague at the University of Chicago, in the investigation of reaction mechanisms of polymer formation, greatly aiding the U.S. government's synthetic rubber program. For his efforts he received the Presidential Merit Award in 1947. For his synthesis of alkyl mercury compounds (developed in 1929) he received the John Scott Award in 1949. His varied chemical interests are shown by his patents for a treatment of fungus disease in small grains and his isolation of the active principle (ergotocine) in ergot. He also assisted in preparation of the thermochemical section of the International Critical Tables.

BIBLIOGRAPHY

I. ORIGINAL WORKS. The first of Kharasch's important papers on the peroxide effect is M. S. Kharasch and Frank R. Mayo, "Peroxide Effect in the Addition of Reagents to Unsaturated Compounds. I. The Addition of Hydrogen Bromide to Allyl Bromide," in *Journal of the American Chemical Society*, **55** (1933), 2468–2490. Many of his other important research results appear in the patent literature.

II. SECONDARY LITERATURE. Details of Kharasch's personal life can be found in early editions of *American Men of Science*. His scientific contributions are discussed in detail in Cheves Walling, "The Contributions of Morris S. Kharasch to Polymer Chemistry," in *Tetrahedron*, supp. 1 (1959), pp. 143–150; this article also cites his pertinent papers. An elementary discussion of his discovery of the peroxide effect is R. D. Billinger and K. Thomas Finley, "Morris Selig Kharasch, a Great American Chemist," in *Chemistry*, **38** (June 1965), 19–20.

SHELDON J. KOPPERL

AL-KHAYYĀMĪ (or KHAYYĀM), GHIYĀTH AL-DĪN ABU'L-FATḤ 'UMAR IBN IBRĀHĪM AL-NĪSĀBŪRĪ (or AL-NAYSĀBŪRĪ), also known as **Omar Khayyam** (*b.* Nīshāpūr, Khurasan [now Iran], 15 May 1048 [?]; *d.* Nīshāpūr, 4 December 1131 [?]), *mathematics, astronomy, philosophy.*

As his name states, he was the son of Ibrāhīm; the epithet "al-Khayyāmī" would indicate that his father or other forebears followed the trade of making tents. Of his other names, "ʿUmar" is his proper designation, while "Ghiyāth al-Dīn" ("the help of the faith") is an honorific he received later in life and "al-Nīsābūrī" refers to his birthplace. Arabic sources of the twelfth to the fifteenth centuries[1] contain only infrequent and sometimes contradictory references to al-Khayyāmī, differing even on the dates of his birth and death. The earliest birthdate given is approximately 1017, but the most probable date (given above) derives from the historian Abu'l-Ḥasan al-Bayhaqī (1106–1174), who knew al-Khayyāmī personally and left a record of his horoscope. The most probable deathdate is founded in part upon the account of Niẓāmī ʿArūḍī Samarqandī (1110–1155) of a visit he paid to al-Khayyāmī's tomb in A.H. 530 (A.D. 1135/1136), four years after the latter's death.[2] This date is confirmed by the fifteenth-century writer Yār-Aḥmed Tabrīzī.[3]

At any rate, al-Khayyāmī was born soon after Khurasan was overrun by the Seljuks, who also conquered Khorezm, Iran, and Azerbaijan, over which they established a great but unstable military empire. Most sources, including al-Bayhaqī, agree that he came from Nīshāpūr, where, according to the thirteenth/fourteenth-century historian Faḍlallāh Rashīd al-Dīn, he received his education. Tabrīzī, on the other hand, stated that al-Khayyāmī spent his boyhood and youth in Balkh (now in Afghanistan), and added that by the time he was seventeen he was well versed in all areas of philosophy.

Wherever he was educated, it is possible that al-Khayyāmī became a tutor. Teaching, however, would not have afforded him enough leisure to pursue science. The lot of the scholar at that time was, at best, precarious, unless he were a wealthy man. He could undertake regular studies only if he were attached to the court of some sovereign or magnate, and his work was thus dependent on the attitude of his master, court politics, and the fortunes of war. Al-Khayyāmī gave a lively description of the hazards of such an existence at the beginning of his *Risāla fi'l-barāhīn ʿalā masā'il al-jabr wa'l-muqābala* ("Treatise on Demonstration of Problems of Algebra and Almuqabala"):

> I was unable to devote myself to the learning of this *al-jabr* and the continued concentration upon it, because of obstacles in the vagaries of Time which hindered me; for we have been deprived of all the people of knowledge save for a group, small in number, with many troubles, whose concern in life is to snatch the opportunity, when Time is asleep, to devote themselves meanwhile to the investigation and perfection of a science; for the majority of people who imitate philosophers confuse the true with the false, and they do nothing but deceive and pretend knowledge, and they do not use what they know of the sciences except for base and material purposes; and if they see a certain person seeking for the right and preferring the truth, doing his best to refute the false and untrue and leaving aside hypocrisy and deceit, they make a fool of him and mock him.[4]

Al-Khayyāmī was nevertheless able, even under the unfavorable circumstances that he described, to write at this time his still unrecovered treatise *Mushkilāt al-ḥisāb* ("Problems of Arithmetic") and his first, untitled, algebraical treatise, as well as his short work on the theory of music, *al-Qāwl ʿalā ajnās allatī bi'l-arbaʿa* ("Discussion on Genera Contained in a Fourth").

About 1070 al-Khayyāmī reached Samarkand, where he obtained the support of the chief justice, Abū Ṭāhir, under whose patronage he wrote his great algebraical treatise on cubic equations, the *Risāla* quoted above, which he had planned long before. A supplement to this work was written either at the court of Shams al-Mulūk, *khaqan* of Bukhara, or at Isfahan, where al-Khayyāmī had been invited by the Seljuk sultan, Jalāl al-Dīn Malik-shāh, and his vizier Niẓām al-Mulk, to supervise the astronomical observatory there.

Al-Khayyāmī stayed at Isfahan for almost eighteen years, which were probably the most peaceful of his life. The best astronomers of the time were gathered at the observatory and there, under al-Khayyāmī's guidance, they compiled the *Zīj Malik-shāhī* ("Malik-shāh Astronomical Tables"). Of this work only a small portion—tables of ecliptic coordinates and of the magnitudes of the 100 brightest fixed stars—survives. A further important task of the observatory was the reform of the solar calendar then in use in Iran.

Al-Khayyāmī presented a plan for calendar reform about 1079. He later wrote up a history of previous reforms, the *Naurūz-nāma*, but his own design is known only through brief accounts in the astronomical tables of Naṣīr al-Dīn al-Ṭūsī and Ulugh Beg. The new calendar was to be based on a cycle of thirty-three years, named "Malikī era" or "Jalālī era" in honor of the sultan. The years 4, 8, 12, 16, 20, 24, 28, and 33 of each period were designated as leap years of 366 days, while the average length of the year was to be 365.2424 days (a deviation of 0.0002 day from the true solar calendar), a difference of one day thus accumulating over a span of 5,000 years. (In the Gregorian calendar, the average year is 365.2425 days long, and the one-day difference is accumulated over 3,333 years.)

Al-Khayyāmī also served as court astrologer,

although he himself, according to Niẓāmī Samarqandī, did not believe in judicial astrology. Among his other, less official activities during this time, in 1077 he finished writing his commentaries on Euclid's theory of parallel lines and theory of ratios; this book, together with his earlier algebraical *Risāla*, is his most important scientific contribution. He also wrote on philosophical subjects during these years, composing in 1080 a *Risāla al-kawn wa'l-taklīf* ("Treatise on Being and Duty"), to which is appended *Al-Jawab ʿan thalāth masāʾil: ḍarūrat al-taḍadd fi'l-ʿālam wa'l-jabr wa'l baqāʾ* ("An Answer to the Three Questions: On the Necessity of Contradiction in the World, on the Necessity of Determinism, and on Longevity"). At about the same time he wrote, for a son of Mu'ayyid al-Mulk (vizier in 1095–1118), *Risāla fi'l kulliyat al-wujūd* ("Treatise on the Universality of Being"). (His two other philosophical works, *Risāla al-ḍiyāʾ al-ʿaqlī fi mawḍūʿ al-'ilm al-kullī* ["The Light of Reason on the Subject of Universal Science"] and *Risāla fi'l wujūd* ["Treatise on Existence"] cannot be dated with any certainty.)

In 1092 al-Khayyāmī fell into disfavor, Malik-shāh having died and his vizier Niẓām al-Mulk having been murdered by an Assassin. Following the death of Malik-shāh his second wife, Turkān-Khātūn, for two years ruled as regent, and al-Khayyāmī fell heir to some of the hostility she had demonstrated toward his patron, Niẓām al-Mulk, with whom she had quarreled over the question of royal succession. Financial support was withdrawn from the observatory and its activities came to a halt; the calendar reform was not completed; and orthodox Muslims, who disliked al-Khayyāmī because of the religious freethinking evident in his quatrains, became highly influential at court. (His apparent lack of religion was to be a source of difficulty for al-Khayyāmī throughout his life, and al-Qifṭī [1172–1239] reported that in his later years he even undertook a pilgrimage to Mecca to clear himself of the accusation of atheism.)

Despite his fall from grace al-Khayyāmī remained at the Seljuk court. In an effort to induce Malik-shāh's successors to renew their support of the observatory and of science in general, he embarked on a work of propaganda. This was the *Naurūz-nāma*, mentioned above, an account of the ancient Iranian solar new year's festival. In it al-Khayyāmī presented a history of the solar calendar and described the ceremonies connected with the Naurūz festival; in particular, he discussed the ancient Iranian sovereigns, whom he pictured as magnanimous, impartial rulers dedicated to education, building edifices, and supporting scholars.

Al-Khayyāmī left Isfahan in the reign of Malik-shāh's third son, Sanjar, who had ascended the throne in 1118. He lived for some time in Merv (now Mary, Turkmen S.S.R.), the new Seljuk capital, where he probably wrote *Mizān al-ḥikam* ("Balance of Wisdoms") and *Fi'l-qusṭas al-mustaqīm* ("On Right *Qusṭas*"), which were incorporated by his disciple al-Khāzinī (who also worked in Merv), together with works of al-Khayyāmī's other disciple, al-Muẓaffar al-Isifīzarī, into his own *Mīzān al-ḥikam*. Among other things, al-Khayyāmī's *Mizān* gives a purely algebraic solution to the problem (which may be traced back to Archimedes) of determining the quantities of gold and silver in a given alloy by means of a preliminary determination of the specific weight of each metal. His *Fī'l-qusṭas* deals with a balance with a mobile weight and variable scales.[5]

Arithmetic and the Theory of Music. A collection of manuscripts in the library of the University of Leiden, Cod. or. 199, lists al-Khayyāmī's "Problems of Arithmetic" on its title page, but the treatise itself is not included in the collection—it may be surmised that it was part of the original collection from which the Leiden manuscript was copied. The work is otherwise unknown, although in his algebraic work *Risāla fi'l-barāhīn ʿalā masāʾil al-jabr wa'l-muqābala*, al-Khayyāmī wrote of it that:

> The Hindus have their own methods for extracting the sides of squares and cubes based on the investigation of a small number of cases, which is [through] the knowledge of the squares of nine integers, that is, the squares of 1, 2, 3, and so on, and of their products into each other, that is, the product of 2 with 3, and so on. I have written a book to prove the validity of those methods and to show that they lead to the required solutions, and I have supplemented it in kind, that is, finding the sides of the square of the square, and the quadrato-cube, and the cubo-cube, however great they may be; and no one has done this before; and these proofs are only algebraical proofs based on the algebraical parts of the book of Elements.[6]

Al-Khayyāmī may have been familiar with the "Hindu methods" that he cites through two earlier works, *Fī uṣul ḥisāb al-hind* ("Principles of Hindu Reckoning"), by Kushyār ibn Labbān al-Jīlī (971–1029), and *Al-muqniʿ fi'l-ḥisāb al-hindī* ("Things Sufficient to Understand Hindu Reckoning"), by ʿAlī ibn Aḥmad al-Nasawī (*fl.* 1025). Both of these authors gave methods for extracting square and cube roots from natural numbers, but their method of extracting cube roots differs from the method given in the Hindu literature and actually coincides more closely with the ancient Chinese method. The latter was set out as early as the second/first centuries B.C., in the "Mathematics in Nine Books," and was used by medieval

Chinese mathematicians to extract roots with arbitrary integer exponents and even to solve numerical algebraic equations (it was rediscovered in Europe by Ruffini and Horner at the beginning of the nineteenth century). Muslim mathematics—at least the case of the extraction of the cube root—would thus seem to have been influenced by Chinese, either directly or indirectly. Al-Jīlī's and al-Nasawī's term "Hindu reckoning" must then be understood in the less restrictive sense of reckoning in the decimal positional system by means of ten numbers.

The earliest Arabic account extant of the general method for the extraction of roots with positive integer exponents from natural numbers may be found in the *Jāmiʿ al-ḥisāb bi'l-takht wa'l-turāb* ("Collection on Arithmetic by Means of Board and Dust"), compiled by al-Ṭūsī. Since al-Ṭūsī made no claims of priority of discovery, and since he was well acquainted with the work of al-Khayyāmī, it seems likely that the method he presented is al-Khayyāmī's own. The method that al-Ṭūsī gave, then, is applied only to the definition of the whole part a of the root $\sqrt[n]{N}$, where

$$N = a^n + r, \qquad r < (a+1)^n - a^n.$$

To compute the correction necessary if the root is not extracted wholly, al-Ṭūsī formulated—in words rather than symbols—the rule for binomial expansion

$$(a+b)^n = a^n + na^{n-1} + \cdots + b^n,$$

and gave the approximate value of $\sqrt{a^n + r}$ as $a + \dfrac{r}{(a+1)^n - a^n}$, the denominator of the root being reckoned according to the binomial formula. For this purpose al-Ṭūsī provided a table of binomial coefficients up to $n = 12$ and noted the property of binomials now expressed as

$$C_n^m = C_{n-1}^{m-1} + C_{n-1}^m.$$

Al-Khayyāmī applied the arithmetic, particularly the theory of commensurable ratios, in his *al-Qawl ʿalā ajnās allatī bi'l-arbaʿa* ("Discussion on Genera Contained in a Fourth"). In the "Discussion" al-Khayyāmī took up the problem—already set by the Greeks, and particularly by Euclid in the *Sectio canonis*—of dividing a fourth into three intervals corresponding to the diatonic, chromatic, and enharmonic tonalities. Assuming that the fourth is an interval with the ratio 4:3, the three intervals into which the fourth may be divided are defined by ratios of which the product is equal to 4:3. Al-Khayyāmī listed twenty-two examples of the section of the fourth, of which three were original to him. Of the others, some of which occur in more than one source, eight were drawn from Ptolemy's "Theory of Harmony"; thirteen from al-Fārābī's *Kitāb al-mūsikā al-kabīr* ("Great Book of Music"); and fourteen from Ibn Sīnā, either *Kitāb al-Shifāʾ* ("The Book of Healing") or *Dānish-nāmah* ("The Book of Knowledge"). Each example was further evaluated in terms of aesthetics.

Theory of Ratios and the Doctrine of Number. Books II and III of al-Khayyāmī's commentaries on Euclid, the *Sharḥ ma ashkala min muṣādarāt kitāb Uqlīdis*, are concerned with the theoretical foundations of arithmetic as manifested in the study of the theory of ratios. The general theory of ratios and proportions as expounded in book V of the *Elements* was one of three aspects of Euclid's work with which Muslim mathematicians were particularly concerned. (The others were the theory of parallels contained in book I and the doctrine of quadratic irrationals in book X.) The Muslim mathematicians often attempted to improve on Euclid, and many scholars were not satisfied with the theory of ratios in particular. While they did not dispute the truth of the theory, they questioned its basis on Euclid's definition of identity of two ratios, $a/b = c/d$, which definition could be traced back to Eudoxus and derived from the quantitative comparison of the equimultiples of all the terms of a given proportion (*Elements*, book V, definition 5).

The Muslim critics of the Euclid-Eudoxus theory of ratios found its weakness to lie in its failure to express directly the process of measuring a given magnitude (a or c) by another magnitude (b or d). This process was based upon the definition of a proportion for a particular case of the commensurable quantities a, b, and c, d through the use of the so-called Euclidean algorithm for the determination of the greatest common measure of two numbers (*Elements*, book VII). Beginning with al-Māhānī, in the ninth century, a number of mathematicians suggested replacing definition 5, book V, with some other definition that would, in their opinion, better express the essence of the proportion. The definition may be rendered in modern terms by the continued fraction theory: if $a/b = (q_1, q_2, \cdots, q_n, \cdots)$ and $c/d = (q_1', q_2', \cdots, q_n', \cdots)$, then $a/b = c/d$ under the condition that $q_k' = q_k$ for all k up to infinity (for commensurable ratios, k is finite). Definitions of inequality of ratios $a/b > c/d$ and $a/b < c/d$, embracing cases of both commensurable and incommensurable ratios and providing criteria for the quantitative comparison of rational and irrational values, are introduced analogously. In the Middle Ages it was known that this "anti-phairetical" theory of ratios existed in Greek mathematics before Eudoxus; that it did was discovered only by Zeuthen

AL-KHAYYĀMĪ

and Becker. The proof that his theory was equivalent to that set out in the *Elements* was al-Khayyāmī's greatest contribution to the theory of ratios in general. Al-Khayyāmī's proof lay in establishing the equivalence of the definitions of equality and inequalities in both theories, thereby obviating the need to deduce all the propositions of book V of the *Elements* all over again. He based his demonstration on an important theorem of the existence of the fourth proportional d with the three given magnitudes a, b, and c; he tried to prove it by means of the principle of the infinite divisibility of magnitudes, which was, however, insufficient for his purpose. His work marked the first attempt at a general demonstration of the theorem, since the Greeks had not treated it in a general manner. These investigations are described in book II of the *Sharḥ*.

In book III, al-Khayyāmī took up compound ratios (at that time most widely used in arithmetic, as in the rule of three and its generalizations), geometry (the doctrine of the similitude of figures), the theory of music, and trigonometry (applying proportions rather than equalities). In the terms in which al-Khayyāmī, and other ancient and medieval scholars, worked, the ratio a/b was compounded from the ratios a/c and c/b—what would in modern terms be stated as the first ratio being the product of the two latter. In his analysis of the operation of compounding the ratios, al-Khayyāmī first set out to deduce from the definition of a compound ratio given in book VI of the *Elements* (which was, however, introduced into the text by later editors) the theorem that the ratio a/c is compounded from the ratios a/b and b/c and an analogous theorem for ratios $a/c, b/c, c/d$, and so on. Here, cautiously, al-Khayyāmī had begun to develop a new and broader concept of number, including all positive irrational numbers, departing from Aristotle, whose authority he nonetheless respectfully invoked. Following the Greeks, al-Khayyāmī properly understood number as an aggregate of indivisible units. But the development of his own theory—and the development of the whole of calculation mathematics in its numerous applications—led him to introduce new, "ideal" mathematical objects, including the divisible unit and a generalized concept of number which he distinguished from the "absolute and true" numbers (although he unhesitatingly called it a number).

In proving this theorem for compound ratios al-Khayyāmī first selected a unit and an auxiliary quantity g whereby the ratio $1/g$ is the same as a/b. He here took a and b to be arbitrary homogeneous magnitudes which are generally incommensurable; $1/g$ is consequently also incommensurable. He then described the magnitude g:

> Let us not regard the magnitude g as a line, a surface, a body, or time; but let us regard it as a magnitude abstracted by reason from all this and belonging in the realm of numbers, but not to numbers absolute and true, for the ratio of a to b can frequently be non-numerical, that is, it can frequently be impossible to find two numbers whose ratio would be equal to this ratio.[7]

Unlike the Greeks, al-Khayyāmī extended arithmetical language to ratios, writing of the equality of ratios as he had previously discussed their multiplication. Having stated that the magnitude g, incommensurable with a unit, belongs in the realm of numbers, he cited the usual practice of calculators and land surveyors, who frequently employed such expressions as half a unit, a third of a unit, and so on, or who dealt in roots of five, ten, or other divisible units.

Al-Khayyāmī thus was able to express any ratio as a number by using either the old sense of the term or the new, fractional or irrational sense. The compounding of ratios is therefore no different from the multiplication of numbers, and the identity of ratios is similar to their equality. In principle, then, ratios are suitable for measuring numerically any quantities. The Greek mathematicians had studied mathematical ratios, but they had not carried out this function to such an extent. Al-Khayyāmī, by placing irrational quantities and numbers on the same operational scale, began a true revolution in the doctrine of number. His work was taken up in Muslim countries by al-Ṭūsī and his followers, and European mathematicians of the fifteenth to seventeen centuries took up similar studies on the reform of the general ratios theory of the *Elements*. The concept of number grew to embrace all real and even (at least formally) imaginary numbers; it is, however, difficult to assess the influence of the ideas of al-Khayyāmī and his successors in the East upon the later mathematics of the West.

Algebra. Eastern Muslim algebraists were able to draw upon a mastery of Hellenistic and ancient Eastern mathematics, to which they added adaptations of knowledge that had come to them from India and, to a lesser extent, from China. The first Arabic treatise on algebra was written in about 830 by al-Khwārizmī, who was concerned with linear and quadratic equations and dealt with positive roots only, a practice that his successors followed to the degree that equations that could not possess positive roots were ignored. At a slightly later date, the study of cubic equations began, first with Archimedes' problem of the section by a plane of a given sphere into two segments of which the volumes are in a given ratio. In the second half of the ninth century, al-Māhānī expressed the problem as an equation of the type $x^3 + r = px^2$

(which he, of course, stated in words rather than symbols). About a century later, Muslim mathematicians discovered the geometrical solution of this equation whereby the roots were constructed as coordinates of points of intersection of two correspondingly selected conic sections—a method dating back to the Greeks. It was then possible for them to reduce a number of problems, including the trisection of an angle, important to astronomers, to the solution of cubic equations. At the same time devices for numerical approximated solutions were created, and a systematic theory became necessary.

Al-Khayyāmī's construction of such a geometrical theory of cubic equations may be accounted the most successful accomplished by a Muslim scholar. In his first short, untitled algebraic treatise he had already reduced a particular geometrical problem to an equation, $x^3 + 200x = 20x^2 + 2{,}000$, and had solved it by an intersection of circumference $y^2 = (x - 10) \cdot (20 - x)$ and equilateral hyperbola $xy = 10\sqrt{2}\,(x - 10)$. He also noted that he had found an approximated numerical solution with an error of less than 1 percent, and he remarked that it is impossible to solve this equation by elementary means, since it requires the use of conic sections. This is perhaps the first statement in surviving mathematical literature that equations of the third degree cannot be generally solved with compass and ruler—that is, in quadratic radicals—and al-Khayyāmī repeated this assertion in his later *Risāla*. (In 1637 Descartes presented the same supposition, which was proved by P. Wantzel in 1837.)

In his earlier algebraic treatise al-Khayyāmī also took up the classification of normal forms of equations (that is, only equations with positive coefficients), listing all twenty-five equations of the first, second, and third degree that might possess positive roots. He included among these fourteen cubic equations that cannot be reduced to linear or quadratic equations by division by x^2 or x, which he subdivided into three groups consisting of one binomial equation ($x^3 = r$), six trinomial equations ($x^3 + px^2 = r$; $x^3 + r = qx$; $x^3 + r = px^2$; $x^3 + qx = r$; $x^3 = px^2 + r$; and $x^3 = qx + r$), and seven quadrinomial equations ($x^3 = px^2 + qx + r$; $x^3 + qx + r = px^2$; $x^3 + px^2 + r = qx$; $x^3 + px^2 + qx = r$; $x^3 + px^2 = qx + r$; $x^3 + qx = px^2 + r$; and $x^3 + r = px^2 + qx$). He added that of these four types had been solved (that is, their roots had been constructed geometrically) at some earlier date, but that "No rumor has reached us of any of the remaining ten types, neither of this classification,"[8] and expressed the hope that he would later be able to give a detailed account of his solution of all fourteen types.

Al-Khayyāmī succeeded in this stated intention in his *Risāla*. In the introduction to this work he gave one of the first definitions of algebra, saying of it that, "The art of *al-jabr* and *al-muqābala* is a scientific art whose subject is pure number and measurable quantities insofar as they are unknown, added to a known thing with the help of which they may be found; and that [known] thing is either a quantity or a ratio . . ."[9] The "pure number" to which al-Khayyāmī refers is natural number, while by "measurable quantities" he meant lines, surfaces, bodies, and time; the subject matter of algebra is thus discrete, consisting of continuous quantities and their abstract ratios. Al-Khayyāmī then went on to write, "Now the extractions of *al-jabr* are effected by equating . . . these powers to each other as is well known."[10] He then took up the consideration of the degree of the unknown quantity, pointing out that degrees higher than third must be understood only metaphorically, since they cannot belong to real quantities.

At this point in the *Risāla* al-Khayyāmī repeated his earlier supposition that cubic equations that cannot be reduced to quadratic equations must be solved by the application of conic sections and that their arithmetical solution is still unknown (such solutions in radicals were, indeed, not discovered until the sixteenth century). He did not, however, despair of such an arithmetical solution, adding, "Perhaps someone else who comes after us may find it out in the case, when there are not only the first three classes of known powers, namely the number, the thing, and the square."[11] He then also repeated his classification of twenty-five equations, adding to it a presentation of the construction of quadratic equations based on Greek geometrical algebra. Other new material here appended includes the corresponding numerical solution of quadratic equations and constructions of all the fourteen types of third-degree equations that he had previously listed.

In giving the constructions of each of the fourteen types of third-degree equation, al-Khayyāmī also provided an analysis of its "cases." By considering the conditions of intersection or of contact of corresponding conic sections, he was able to develop what is essentially a geometrical theory of the distribution of (positive) roots of cubic equations. He necessarily dealt only with those parts of conic sections that are located in the first quadrant, employing them to determine under what conditions a problem may exist and whether the given type manifests only one case—or one root (including the case of double roots, but not multiple roots, which were unknown)—or more than one case (that is, one or two roots). Al-Khayyāmī went on to demonstrate that some types of equations are characterized by a diversity of cases, so that they may possess no roots at all, or one root, or two roots. He also investigated the limits of roots.

As far as it is known, al-Khayyāmī was thus the first to demonstrate that a cubic equation might have two roots. He was unable to realize, however, that an equation of the type $x^3 + qx = px^2 + r$ may, under certain conditions, possess three (positive) roots; this constitutes a disappointing deficiency in his work. As F. Woepcke, the first editor of the *Risāla*, has shown, al-Khayyāmī followed a definite system in selecting the curves upon which he based the construction of the roots of all fourteen types of third-degree equations; the conic sections that he preferred were circumferences, equilateral hyperbolas of which the axes, or asymptotes, run parallel to coordinate axes; and parabolas of which the axes parallel one of the coordinate axes. His general geometrical theory of distribution of the roots was also applied to the analysis of equations with numerical coefficients, as is evident in the supplement to the *Risāla*, in which al-Khayyāmī analyzed an error of Abū'l-Jūd Muḥammad ibn Layth, an algebraist who had lived some time earlier and whose work al-Khayyāmī had read a few years after writing the main text of his treatise.

His studies on the geometrical theory of third-degree equations mark al-Khayyāmī's most successful work. Although they were continued in oriental Muslim countries, and known by hearsay in Moorish countries, Europeans began to learn of them only after Descartes and his successors independently arrived at a method of the geometrical construction of roots and a doctrine of their distribution. Al-Khayyāmī did further research on equations containing degrees of a quantity inverse to the unknown ("part of the thing," "part of the square," and so on) including, for example, such equations as $1/x^3 + 3\ 1/x^2 + 5\ 1/x = 3\ 3/8$, which he reduced by substituting $x = 1/z$ in the equations that he had already studied. He also considered such cases as $x^2 + 2x = 2 + 2\ 1/x^2$, which led to equations of the fourth degree, and here he realized the upper limit of his accomplishment, writing, "If it [the series of consecutive powers] extends to five classes, or six classes, or seven, it cannot be extracted by any method."[12]

The Theory of Parallels. Muslim commentators on the *Elements* as early as the ninth century began to elaborate on the theory of parallels and to attempt to establish it on a basis different from that set out by Euclid in his fifth postulate. Thābit ibn Qurra and Ibn al-Haytham had both been attracted to the problem, while al-Khayyāmī devoted the first book of his commentaries to the *Sharḥ* to it. Al-Khayyāmī took as the point of departure for his theory of parallels a principle derived, according to him, from "the philosopher," that is, Aristotle, namely that "two convergent straight lines intersect and it is impossible that two convergent straight lines should diverge in the direction of convergence."[13] Such a principle consists of two statements, each equivalent to Euclid's fifth postulate. (It must be noted that nothing similar to al-Khayyāmī's principle is to be found in any of the known writings of Aristotle.)

Al-Khayyāmī first proved that two perpendiculars to one straight line cannot intersect because they must intersect symmetrically at two points on both sides of the straight line; therefore they cannot converge. From the second statement the principle follows that two perpendiculars drawn to one straight line cannot diverge because, if they did, they would have to diverge on both sides of the straight line. Therefore, two perpendiculars to the same straight line neither converge nor diverge, being in fact equidistant from each other.

Al-Khayyāmī then went on to prove eight propositions, which, in his opinion, should be added to book I of the *Elements* in place of the proposition 29 with which Euclid began the theory of parallel lines based on the fifth postulate of book I (the preceding twenty-eight propositions are not based on the fifth postulate). He constructed a quadrilateral by drawing two perpendicular lines of equal length at the ends of a given line segment AB. Calling the perpendiculars AC and BD, the figure was thus bounded by the segments AB, AC, CD, and BD, a birectangle often called "Saccheri's quadrilateral," in honor of the eighteenth-century geometrician who used it in his own theory of parallels.

In his first three propositions, al-Khayyāmī proved that the upper angles C and D of this quadrilateral are right angles. To establish this theorem, he (as Saccheri did after him) considered three hypotheses whereby these angles might be right, acute, or obtuse; were they acute, the upper line CD of the figure must be longer than the base AB, and were they obtuse, CD must be shorter than AB—that is, extensions of sides AC and BD would diverge or converge on both ends of AB. The hypothetical acute or obtuse angles are therefore proved to be contradictory to the given equidistance of the two perpendiculars to one straight line, and the figure is proved to be a rectangle.

In the fourth proposition al-Khayyāmī demonstrated that the opposite sides of the rectangle are of equal length, and in the fifth, that it is the property of any two perpendiculars to the same straight line that any perpendicular to one of them is also the perpendicular to the other. The sixth proposition states that if two straight lines are parallel in Euclid's sense—that is, if they do not intersect—they are both perpendicular to one straight line. The seventh proposition adds that if two parallel straight lines are intersected by a third straight line, alternate and corresponding angles are equal, and the interior angles of one side are two right angles, a proposition coinciding with Euclid's book I, proposition 29, but one that al-Khayyāmī reached by his own, noncoincident methods.

Al-Khayyāmī's eighth proposition proves Euclid's fifth postulate of book I: two straight lines intersect if a third intersects them at angles which are together less than two right angles. The two lines are extended and a straight line, parallel to one of them, is passed through one of the points of intersection. According to the sixth proposition, these two straight lines—being one of the original lines and the line drawn parallel to it—are equidistant, and consequently the two original lines must approach each other. According to al-Khayyāmī's general principle, such straight lines are bound to intersect.

Al-Khayyāmī's demonstration of Euclid's fifth postulate differs from those of his Muslim predecessors because he avoids the logical mistake of *petitio principi*, and deduces the fifth postulate from his own explicitly formulated principle. Some conclusions drawn from hypotheses of acute or obtuse angles are essentially the same as the first theorems of the non-Euclidean geometries of Lobachevski and Riemann. Like his theory of ratios, al-Khayyāmī's theory of parallels influenced the work of later Muslim scholars to a considerable degree. A work sometimes attributed to his follower al-Ṭūsī influenced the development of the theory of parallels in Europe in the seventeenth and eighteenth centuries, as was particularly reflected in the work of Wallis and Saccheri.

Philosophical and Poetical Writings. Although al-Khayyāmī wrote five specifically philosophical treatises, and although much of his poetry is of a philosophical nature, it remains difficult to ascertain what his world view might have been. Many investigators have dealt with this problem, and have reached many different conclusions, depending in large part on their own views. The problem is complicated by the consideration that the religious and philosophical tracts differ from the quatrains, while analysis of the quatrains themselves is complicated by questions of their individual authenticity. Nor is it possible to be sure of what in the philosophical treatises actually reflects al-Khayyāmī's own mind, since they were written under official patronage.

His first treatise, *Risālat al-kawn wa'l-taklīf* ("Treatise on Being and Duty"), was written in 1080, in response to a letter from a high official who wished al-Khayyāmī to give his views on "the Divine Wisdom in the Creation of the World and especially of Man and on man's duty to pray."[14] The second treatise, *Al-Jawab ʿan thalāth masā'il* ("An Answer to the Three Questions"), closely adheres to the formula set out in the first. *Risāla fi'l kulliyat al-wujūd* ("Treatise on the Universality of Being") was written at the request of Mu'ayyid al-Mulk, and, while it is not possible to date or know the circumstances under which the remaining two works, *Risālat al-ḍiyāʾ al-ʿaqlī fi mawḍūʿ al-ʿilm al-kullī* ("The Light of Reason on the Subject of Universal Science") and *Risāla fi'l wujūd* ("Treatise on Existence"), were written, it would seem not unlikely that they had been similarly commissioned. Politics may therefore have dictated the contents of the religious tracts, and it must be noted that the texts occasionally strike a cautious and impersonal note, presenting the opinions of a number of other authors, without criticism or evaluation.

It might also be speculated that al-Khayyāmī wrote his formal religious and philosophical works to clear his name of the accusation of freethinking. Certainly strife between religious sects and their common aversion to agnosticism were part of the climate of the time, and it is within the realm of possibility that al-Khayyāmī's quatrains had become known to the religious orthodoxy and had cast suspicion upon him. (The quatrains now associated with his name contain an extremely wide range of ideas, ranging from religious mysticism to materialism and almost atheism; certainly writers of the thirteenth century thought al-Khayyāmī a freethinker, al-Qifṭī calling the poetry "a stinging serpent to the Sharīʿa" and the theologian Abū Bakr Najm al-Dīn al-Rāzī characterizing the poet as "an unhappy philosopher, materialist, and naturalist.")[15]

Insofar as may be generalized, in his philosophical works al-Khayyāmī wrote as an adherent of the sort of eastern Aristotelianism propagated by Ibn Sīnā—that is, of an Aristotelianism containing considerable amounts of Platonism, and adjusted to fit Muslim religious doctrine. Al-Bayhaqī called al-Khayyāmī "a successor of Abū ʿAli [Ibn Sīnā] in different domains of philosophical sciences,"[16] but from the orthodox point of view such a rationalistic approach to the dogmas of faith was heresy. At any rate, al-Khayyāmī's philosophy is scarcely original, his most interesting works being those concerned with the analysis of the problem of existence of general concepts. Here al-Khayyāmī—unlike Ibn Sīnā, who held views close to Plato's realism—developed a position similar to that which was stated simultaneously in Europe by Abailard, and was later called conceptualism.

AL-KHAYYĀMĪ

As for al-Khayyāmī's poetical works, more than 1,000 quatrains, written in Persian, are now published under his name. (Govinda counted 1,069.) The poems were preserved orally for a long time, so that many of them are now known in several variants. V. A. Zhukovsky, a Russian investigator of the poems, wrote of al-Khayyāmī in 1897:

> He has been regarded variously as a freethinker, a subverter of Faith, an atheist and materialist; a pantheist and a scoffer at mysticism; an orthodox Musulman; a true philosopher, a keen observer, a man of learning; a bon vivant, a profligate, a dissembler, and a hypocrite; a blasphemer—nay, more, an incarnate negation of positive religion and of all moral beliefs; a gentle nature, more given to the contemplation of things divine than the wordly enjoyments; an epicurean skeptic; the Persian Abū'l-ʿAlā, Voltaire, and Heine. One asks oneself whether it is possible to conceive, not a philosopher, but merely an intelligent man (provided he be not a moral deformity) in whom were commingled and embodied such a diversity of convictions, paradoxical inclinations and tendencies, of high moral courage and ignoble passions, of torturing doubts and vacillations?[17]

The inconsistencies noted by Zhukovsky are certainly present in the corpus of the poems now attributed to al-Khayyāmī, and here again questions of authenticity arise. A. Christensen, for example, thought that only about a dozen of the quatrains might with any certainty be considered genuine, although later he increased this number to 121. At any rate, the poems generally known as al-Khayyāmī's are one of the summits of philosophical poetry, displaying an unatheistic freethought and love of freedom, humanism and aspirations for justice, irony and skepticism, and above all an epicurean spirit that verges upon hedonism.

Al-Khayyāmī's poetic genius was always celebrated in the Arabic East, but his fame in European countries is of rather recent origin. In 1859, a few years after Woepcke's edition had made al-Khayyāmī's algebra—previously almost unknown—available to Western scholars, the English poet Edward FitzGerald published translations of seventy-five of the quatrains, an edition that remains popular. Since then, many more of the poems have been published in a number of European languages.

The poems—and the poet—have not lost their power to attract. In 1934 a monument to al-Khayyāmī was erected at his tomb in Nīshāpūr, paid for by contributions from a number of countries.

NOTES

1. V. A. Zhukovsky, *Omar Khayyam i "stranstvuyushchie" chetverostishia*; Swami Govinda Tirtha, *The Nectar of Grace*; and Niẓāmī ʿArūḍī Samarqandī, *Sobranie redkostei ili chetyre besedy*.
2. Samarqandī, *op. cit.*, p. 97; in the Browne trans., p. 806, based on the later MSS, "four years" is "some years."
3. Govinda, *op. cit.*, pp. 70–71.
4. *Risāla fi'l-barāhīn ʿalā masāʾil al-jabr waʾl-muqābala*, Winter-ʿArafat trans., pp. 29–30.
5. I. S. Levinova, "Teoria vesov v traktatakh Omara Khayyama i ego uchenika Abu Hatima al-Muzaffara ibn Ismaila al-Asfizari."
6. *Risāla*, Winter-ʿArafat trans., pp. 34 (with correction), 71.
7. *Omar Khayyam, Traktaty*, pp. 71, 145.
8. First algebraic treatise, Krasnova and Rosenfeld trans., p. 455; omitted from Amir-Moéz trans.
9. *Risāla*, Winter-ʿArafat trans., p. 30 (with correction).
10. *Ibid.*, p. 31.
11. *Ibid.*, p. 32 (with correction).
12. *Ibid.*, p. 70.
13. *Omar Khayyam, Traktaty*, pp. 120–121; omitted from *Sharḥ mā ashkala min muṣādarāt kitāb Uqlīdis*, Amir-Moéz trans.
14. *Omar Khayyam, Traktaty*, p. 152.
15. Zhukovsky, *op. cit.*, pp. 334, 342.
16. Govinda, *op. cit.*, pp. 32–33.
17. Zhukovsky, *op. cit.*, p. 325.

BIBLIOGRAPHY

I. ORIGINAL WORKS. The following are al-Khayyāmī's main writings:

1. The principal ed. is *Omar Khayyam, Traktaty* ("... Treatises"), B. A. Rosenfeld, trans.; V. S. Segal and A. P. Youschkevitch, eds.; intro. and notes by B. A. Rosenfeld and A. P. Youschkevitch (Moscow, 1961), with plates of the MSS. It contains Russian trans. of all the scientific and philosophical writings except the first algebraic treatise, *al-Qawl ʿalā ajnās allātī bi'l-arbaʿa*, and *Fī'l-qusṭas al-mustaqīm*.

2. The first algebraic treatise. MS: Teheran, Central University library, VII, 1751/2. Eds.: Arabic text and Persian trans. by G. H. Mossaheb (see below), pp. 59–74, 251–291; English trans. by A. R. Amir-Moéz in *Scripta mathematica*, **26**, no. 4 (1961), 323–337; Russian trans. with notes by S. A. Krasnova and B. A. Rosenfeld in *Istoriko-matematicheskie issledovaniya*, **15** (1963), 445–472.

3. *Risāla fi'l-barāhīn ʿalā masāʾil al-jabr wa'l-muqābala* ("Treatise on Demonstration of Problems of Algebra and Almuqabala"). MSS: Paris, Bibliothèque Nationale, Ar. 2461, 2358/7; Leiden University library, Or. 14/2; London, India Office library, 734/10; Rome, Vatican Library, Barb. 96/2; New York, collection of D. E. Smith.

Eds.: F. Woepcke, *L'algèbre d'Omar Alkhayyâmî* (Paris, 1851), text of both Paris MSS and of the Leiden MS, French trans. and ed.'s notes—reedited by Mossaheb (see below), pp. 7–52, with Persian trans. (pp. 159–250) ed. by the same author earlier in *Jabr-u muqābala-i Khayyām* (Teheran, 1938); English trans. by D. S. Kasir, *The Algebra of Omar Khayyam* (New York, 1931), trans. from the Smith MS, which is very similar to Paris MS Ar. 2461, and by H. J. J. Winter and W. ʿArafat, "The Algebra of ʿUmar Khayyam," in *Journal of the Royal Asiatic Society of Bengal Science*, **16** (1950), 27–70, trans. from the London MS; and Russian trans. and photographic repro. of Paris

MS 2461 in *Omar Khayyam, Traktaty*, pp. 69–112; 1st Russian ed. in *Istoriko-matematicheskie issledovaniya*, **6** (1953), 15–66.

4. *Sharḥ mā ashkala min muṣādarāt kitāb Uqlīdis* ("Commentaries to Difficulties in the Introductions to Euclid's Book"). MSS: Paris, Bibliothèque Nationale, Ar. 4946/4; Leiden University library, Or. 199/8.

Eds.: T. Erani, *Discussion of Difficulties of Euclid by Omar Khayyam* (Teheran, 1936), the Leiden MS, reed. by J. Humai (see below), pp. 177–222, with a Persian trans. (pp. 225–280); *Omar Khayyam, Explanation of the Difficulties in Euclid's Postulates*, A. I. Sabra, ed. (Alexandria, 1961), the Leiden MS and text variants of Paris MS; an incomplete English trans. by A. R. Amir-Moéz, in *Scripta mathematica*, **24**, no. 4 (1959), 275–303; and Russian trans. and photographic repro. of Leiden MS in *Omar Khayyam, Traktaty*, pp. 113–146; 1st Russian ed. in *Istoriko-matematicheskie issledovaniya*, **6** (1953), 67–107.

5. *Al-Qawl ʿalā ajnās allatī bi'l-arbaʿa* ("Discussion on Genera Contained in a Fourth"). MS: Teheran, Central University library, 509, fols. 97–99.

Ed.: J. Humai (see below), pp. 341–344.

6. *Mīzān al-ḥikam* ("The Balance of Wisdoms") or *Fī ikhtiyāl ma'rafa miqdāray adh-dhahab wa-l-fiḍḍa fī jism murakkab minhumā* ("On the Art of Determination of Gold and Silver in a Body Consisting of Them"). Complete in Abdalraḥmān al-Khāzinī, *Kitāb mīzān al-ḥikma* ("Book of the Balance of Wisdom"). MSS: Leningrad, State Public Library, Khanykov collection, 117, 57b–60b; also in Bombay and Hyderabad. Incomplete MS: Gotha, State Library, 1158, 39b–40a.

Eds. of the Bombay and Hyderabad MSS: Abdalraḥmān al-Khāzinī, *Kitāb mīzān al-ḥikma* (Hyderabad, 1940), pp. 87–92; S. S. Nadwi (see below), pp. 427–432. German trans. by E. Wiedemann in *Sitzungsberichte der Physikalisch-medizinischen Sozietät in Erlangen*, **49** (1908), 105–132; Russian trans. and repro. of the Leningrad MS in *Omar Khayyam, Traktaty*, pp. 147–151; 1st Russian ed. in *Istoriko-matematicheskie issledovaniya*, **6** (1953), 108–112.

Eds. of the Gotha MS: Arabic text in Rosen's ed. of the *Rubā'ī* (see below), pp. 202–204), in Erani's ed. of the *Sharḥ* (see above), and in M. ʿAbbasī (see below), pp. 419–428; German trans. by F. Rosen in *Zeitschrift der Deutschen morgenländischen Gesellschaft*, **4(79)** (1925), 133–135; and by E. Wiedemann in *Sitzungsberichte der Physikalisch-medizinischen Sozietät in Erlangen*, **38** (1906), 170–173.

7. *Fī'l-qusṭas al-mutaqīm* ("On Right *Qusṭas*"), in al-Khāzinī's *Mīzān* (see above), pp. 151–153.

8. *Zīj Malik-shāhī* ("Malik-shāh Astronomical Tables"). Only a catalogue of 100 fixed stars for one year of the Malikī era is extant in the anonymous MS Bibliothèque Nationale, Ar. 5968.

Eds.: Russian trans. and photographic repro. of the MS in *Omar Khayyam, Traktaty*, pp. 225–235; same trans. with more complete commentaries in *Istoriko-astronomicheskie issledovaniya*, **8** (1963), 159–190.

9–11. *Risāla al-kawn wa'l-taklīf* ("Treatise on Being and Duty"), *Al-Jawab ʿan thalāth masā'il: ḍarūrat al-taḍadd fī'l-ʿālam wa'l-jabr wa'l-baqā'* ("Answer to Three Questions: On the Necessity of Contradiction in the World, on Determinism and on Longevity"), *Risāla al-ḍiyā' al-ʿaqlī fī mawḍūʿ al-'ilm al-kullī* ("The Light of Reason on the Subject of Universal Science"). MSS belonging to Nūr al-Dīn Muṣṭafā (Cairo) are lost.

Arabic text in *Jāmiʿ al-badā'iʿ* ("Collection of Uniques"; Cairo, 1917), pp. 165–193; text of the first two treatises published by S. S. Nadwī (see below), pp. 373–398; and S. Govinda (see below), pp. 45–46, 83–110, with English trans.; Persian trans., H. Shajara, ed. (see below), pp. 299–337; Russian trans. of all three treatises in *Omar Khayyam, Traktaty*, pp. 152–171; 1st Russian ed. in S. B. Morochnik and B. A. Rosenfeld (see below), pp. 163–188.

12. *Risāla fī'l-wujūd* ("Treatise on Existence"), or *al-Awṣāf wa'l-mawṣūfāt* ("Description and the Described"). MS: Berlin, former Prussian State Library, Or. Petermann, B. 466; Teheran, Majlis-i Shurā-i Millī, 9014; and Poona, collection of Shaykh ʿAbd al-Qādir Sarfaraz.

The Teheran MS is published by Saʿīd Nafīsī in *Sharq* ("East"; Shaʿbān, 1931); and by Govinda (see below), pp. 110–116; Russian trans. in *Omar Khayyam, Traktaty*, pp. 172–179; 1st Russian ed. in S. B. Morochnik and B. A. Rosenfeld (see below), pp. 189–199.

13. *Risāla fī kulliyat al-wujūd* ("Treatise on the Universality of Existence"), or *Risāla-i silsila al-tartīb* ("Treatise on the Chain of Order"), or *Darkhwāstnāma* ("The Book on Demand"). MSS: London, British Museum, Or. 6572; Paris, Bibliothèque Nationale, Suppl. persan, 139/7; Teheran, Majlis-i Shurā-i Millī, 9072; and al-Khayyāmī's library. London MS reproduced in B. A. Rosenfeld and A. P. Youschkevitch (see below), pp. 140–141; the Paris MS is reproduced in *Omar Khayyam, Traktaty*; the texts of these MSS are published in S. S. Nadwi (see below), pp. 412–423; the Majlis-i Shurā-i Millī MS is in Nafīsī's *Sharq* (see above) and in M. ʿAbbasī (see below), pp. 393–405; the al-Khayyāmī library MS is in *ʿUmar Khayyām, Darkhwāstnāma*, Muḥammad 'Alī Taraqī, ed. (Teheran, 1936). Texts of the London MS and the first Teheran MS are published by Govinda with the English trans. (see below), pp. 47–48, 117–129; French trans. of the Paris MS in A. Christensen, *Le monde orientale*, I (1908), 1–16; Russian trans. from the London and Paris MSS, with repro. of the Paris MS in *Omar Khayyam, Traktaty*, pp. 180–186—1st Russian ed. in S. B. Morochnik and B. A. Rosenfeld (see below), pp. 200–208.

14. *Naurūz-nāma*. MS: Berlin, former Prussian State Library, Or. 2450; London, British Museum, Add. 23568.

Eds. of the Berlin MS: *Nowruz-namah*, Mojtaba Minovi, ed. (Teheran, 1933); by M. ʿAbbasī (see below), pp. 303–391; Russian trans. with repro. of the Berlin MS in *Omar Khayyam, Traktaty*, pp. 187–224.

15. *Rubāiyāt* ("Quatrains"). Eds. of MS: *Rubāiyāt-i hakīm Khayyām*, Sanjar Mirzā, ed. (Teheran, 1861), Persian text of 464 *ruba'i*; Muhammad Sadīq 'Alī Luknawī, ed. (Lucknow, 1878, 1894, 1909), 762 (1st ed.) and 770 (2nd and 3rd eds.) *ruba'i*; Muḥammad Raḥīm Ardebili, ed. (Bombay, 1922); Husein Danish, ed. (Istanbul, 1922, 1927), 396 quatrains with Turkish trans.; Jalāl al-Dīn Aḥmed Jafrī, ed. (Damascus, 1931; Beirut, 1950), 352

quatrains with Arabic trans.; Sa'īd Nafīsī, ed. (Teheran, 1933), 443 quatrains; B. Scillik, ed., *Les manuscrits mineurs des Rubaiyat d'Omar-i-Khayyam dans la Bibliothèque National* (Paris-Szeged, 1933–1934)—1933 MSS containing 95, 87, 75, 60, 56, 34, 28, 8, and 6 *ruba'i* and 1934 MSS containing 268, 213, and 349 *ruba'i*; Maḥfūz al-Ḥaqq, ed. (Calcutta, 1939) repro. MS containing 206 *ruba'i* with minatures; Muḥammad 'Ali Forughī, ed. (Teheran, 1942, 1956, 1960), 178 selected *ruba'i* with illustrations; R. M. Aliev, M. N. Osmanov, and E. E. Bertels, eds. (Moscow, 1959), photographic repro. of MS containing 252 *ruba'i* and Russian prose trans. of 293 selected *ruba'i*.

English trans.: Edward FitzGerald (London, 1859, 1868, 1872, 1879) a poetical trans. of 75 (1st ed.) to 101 (4th ed.) quatrains, often repr. (best ed., 1900); E. H. Whinfield (London, 1882, 1883, 1893), a poetical trans. of 253 (1st ed.), 500 (2nd ed.), and 267 (3rd ed.) *ruba'i* from the MS published by Luknawi, in the 2nd ed. with the Persian text; E. Heron-Allen (London, 1898), a prose trans. and repro. of MS containing 158 *ruba'i;* S. Govinda (see below), pp. 1–30, a poetical trans. and the text of 1,069 *ruba'i;* A. J. Arberry (London, 1949), a prose trans. and the Persian text of MS containing 172 *ruba'i* with FitzGerald's and Whinfield's poetical trans., 1952 ed., a poetical trans. of 252 *ruba'i* from the MS published in Moscow in 1959. French trans.: J. B. Nicolas (Paris, 1867), prose trans. and the Persian text of 464 *ruba'i* from the Teheran ed. of 1861; German trans.: C. H. Rempis (Tübingen, 1936), poetical trans. of 255 *ruba'i*; Russian trans.: O. Rumer (Moscow, 1938), poetical trans. of 300 *ruba'i;* V. Derzhavin (Dushanbe, 1955), verse trans. of 488 *ruba'i;* and G. Plisetsky (Moscow, 1972), verse trans. of 450 *ruba'i*, with commentaries by M. N. Osmanov.

II. Secondary Literature. The works listed below provide information on al-Khayyāmī's life and work.

1. Muḥammad 'Abbasī, *Kulliyāt-i athār-i parsī-yi hakīm 'Umar-i Khayyām* (Teheran, 1939), a study of al-Khayyāmī's life and works. It contains texts and translations of *Mizān al-ḥikam, Risālat al-kawn wa'l-taklīf, Al-Jawab 'an thalāth masā'il, Risālat al-ḍiyā' . . ., Risāla fi'l-wujūd*, and *Risāla fi kulliyat al-wujūd* and the quatrains.

2. C. Brockelmann, *Geschichte der arabischen Literatur*, I (Weimar, 1898), 471; supp. (Leiden, 1936), 855–856; III (Leiden, 1943), 620–621. A complete list of all Arabic MSS and their eds. known to European scientists; supp. vols. mention MSS and eds. that appeared after the main body of the work was published.

3. A. Christensen, *Recherches sur les Rubâiyât de 'Omar Hayyâm* (Heidelberg, 1904), an early work in which the author concludes that since there are no criteria for authenticity, only twelve quatrains may reasonably be regarded as authentic.

4. A. Christensen, *Critical Studies in the Rubaiyát of 'Umar-i-Khayyám* (Copenhagen, 1927). A product of prolonged study in which a method of establishing the authenticity of al-Khayyāmī's quatrains is suggested; 121 selected quatrains are presented.

5. J. L. Coolidge, *The Mathematics of Great Amateurs* (Oxford, 1949; New York, 1963), pp. 19–29.

6. Hâmit Dilgan, *Büyük matematikci Omer Hayyâm* (Istanbul, 1959).

7. F. K. Ginzel, *Handbuch der mathematischen und technischen Chronologie*, I (Leipzig, 1906), 300–305, information on al-Khayyāmī's calendar reform.

8. Swami Govinda Tirtha, *The Nectar of Grace, 'Omar Khayyām's Life and Works* (Allahabad, 1941), contains texts and trans. of philosophical treatises and quatrains and repros. of MSS by al-Bayhaqī and Tabrīzī giving biographical data on al-Khayyāmī.

9. Jamāl al-Dīn Humāī, *Khayyām-nāmah*, I (Teheran, 1967). A study of al-Khayyāmī's commentary to Euclid; text and Persian trans. of *Sharḥ mā ashkala min muṣādarāt kitāb Uqlīdis* and text of *al-Qawl 'alā ajnās allatī bi'l-arba'a* are in the appendix.

10. U. Jacob and E. Wiedemann, "Zu Omer-i-Chajjam," in *Der Islam*, **3** (1912), 42–62, critical review of biographical data on al-Khayyāmī and a German trans. of al-Khayyāmī's intro. to *Sharḥ mā ashkala min muṣādarāt kitāb Uqlīdis*.

11. I. S. Levinova, "Teoria veso v traktatakh Omara Khayyama i ego uchenika Abu Hatima al-Muzaffara ibn Ismaila al-Asfizari," in *Trudy XV Nauchnoy Konferencii . . . Instituta istorii estestvoznaniya i tekhniki, sekoiya istorii matematiki i mekhaniki* (Moscow, 1972), pp. 90–93.

12. V. Minorsky, "'Omar Khayyām," in *Enzyklopädie des Islams*, III (Leiden–Leipzig, 1935), 985–989.

13. S. B. Morochnik, *Filosofskie vzglyady Omara Khayyama* ("Philosophical Views of Omar Khayyam"; Dushanbe, 1952).

14. S. B. Morochnik and B. A. Rosenfeld, *Omar Khayyam—poet, myslitel, uchenyi* (". . . Thinker, Scientist"; Dushanbe, 1957).

15. C. H. Mossaheb, *Hakim Omare Khayyam as an Algebraist* (Teheran, 1960). A study of al-Khayyāmī's algebra; text and trans. of the first algebraic treatise and *Risāla fi'l-barāhīn 'alā masā'il al-jabr wa'l muqābala* are in appendix.

16. Seyyīd Suleimān Nadwī, *Umar Khayyam* (Azamgarh, 1932), a study of al-Khayyāmī's life and works, with texts of *Mizān al-ḥikam Risālat al-kawn wa'l taklīf, Al-Jawab 'an thalāth masā'il, Risālat al-ḍiyā' . . ., Risāla fi'l-wujūd*, and *Risāla fi kulliyat al-wujūd* in appendix.

17. B. A. Rosenfeld and A. P. Youschkevitch, *Omar Khayyam* (Moscow, 1965), consisting of a biographical essay, analysis of scientific (especially mathematical) works, and detailed bibliography.

18. Niẓāmī 'Arūḍī Samarqandī, *Sobranie redkostei ili chetyre besedy* ("Collection of Rarities or Four Discourses"), S. I. Bayevsky and Z. N. Vorosheikina, trans., A. N. Boldyrev, ed. (Moscow, 1963), pp. 97–98; and "The Chahár Maqála" ("Four Discourses"), E. G. Browne, English trans., in *Journal of the Royal Asiatic Society*, n. s. **31** (1899), 613–663, 757–845, see 806–808. Recollections of a contemporary of al-Khayyāmī's regarding two episodes in the latter's life.

19. G. Sarton, *Introduction to the History of Science*, I (Baltimore, 1927), 759–761.

20. Husein Shajara, *Tahqīq-i dar rubā'iyāt-i zindagānī-i*

Khayyām (Teheran, 1941). A study of al-Khayyāmī's life and work; Persian trans. of *Risālat al-kawn wa'l-taklīf* and *Al-Jawab ʿan thalāth masā'il* are in appendix.

21. D. E. Smith, "Euclid, Omar Khayyam and Saccheri," in *Scripta mathematica*, **3**, no. 1 (1935), 5–10, the first critical investigation of al-Khayyāmī's theory of parallels in comparison with Saccheri's.

22. D. J. Struik, "Omar Khayyam, Mathematician," in *Mathematical Teacher*, no. 4 (1958), 280–285.

23. H. Suter, *Die Mathematiker und Astronomen der Araber und ihre Werke* (Leipzig, 1900), pp. 112–113.

24. A. P. Youschkevitch, "Omar Khayyam i ego Algebra," in *Trudy Instituta istorii estestvoznaniya*, **2** (1948), 499–534.

25. A. P. Youschkevitch, *Geschichte der Mathematik im Mittelalter* (Leipzig, 1964), pp. 251–254, 259–269, 283–287.

26. A. P. Youschkevitch and B. A. Rosenfeld, "Die Mathematik der Länder des Osten im Mittelalter," in G. Harig, ed., *Sowjetische Beiträge zur Geschichte der Naturwissenschaften* (Berlin, 1960), pp. 119–121.

27. V. A. Zhukovsky, "Omar Khayyam i 'stranstvuyuschie' chetverostishiya" ("Omar Khayyam and the 'Wandering' Quatrains"), in *al-Muẓaffariyya* (St. Petersburg, 1897), pp. 325–363. Translated into English by E. D. Ross in *Journal of the Royal Asiatic Society*, n. s. **30** (1898), 349–366. This paper gives all principal sources of information on al-Khayyāmī's life and presents the problem of "wandering" quatrains, that is, *ruba'i* ascribed to both al-Khayyāmī and other authors.

A. P. YOUSCHKEVITCH
B. A. ROSENFELD

AL-KHĀZIN, ABŪ JAʿFAR MUḤAMMAD IBN AL-ḤASAN AL-KHURĀSĀNĪ (*d.* 961/971), *astronomy, mathematics.*

Al-Khāzin, usually known as Abū Jaʿfar al-Khāzin, was a Sabaean of Persian origin. The *Fihrist* calls him al-Khurāsānī, meaning from Khurāsān, a province in eastern Iran. He should not be confused with ʿAbd al-Raḥmān al-Khāzinī (*ca.* 1100), the probable author of *Kitāb al-ālāt alʿajība al-raṣdiyya*, on observation instruments, often attributed to al-Khāzin. (E. Wiedemann attributed this work, inconsistently, to al-Khāzin in the *Enzyklopaedie des Islam*, II [Leiden–Leipzig, 1913], pp. 1005–1006, and to al-Khāzinī in *Beiträge*, **9** [1906], 190. De Slane confounded these two astronomers in his translation of Ibn Khaldūn's *Prolegomena*, I, 111.)

Abū Jaʿfar al-Khāzin, said to have been attached to the court of the Buwayhid ruler Rukn al-Dawla (932–976) of Rayy, was well known among his contemporaries. In particular his *Zīj al-ṣafā'iḥ* ("Tables of the Disks [of the astrolabe]"), which Ibn al-Qifṭī calls the best work in this field, is often cited; it may be related to manuscript "Liber de sphaera in plano describenda," in the Laurentian library in Florence (Pal.-Med. 271).

Al-Bīrūnī's *Risāla fī fihrist kutub Muḥammad b. Zakariyyā' al-Rāzī* ("Bibliography") of 1036 lists several texts (written in cooperation with Abū Naṣr Manṣūr ibn ʿIrāq), one of which is *Fī taṣḥīḥ mā waqaʿa li Abī Jaʿfar al-Khāzin min al-sahw fī zīj al-ṣafā'iḥ* ("On the Improvement of What Abū Jaʿfar Neglected in His Tables of the Disks"). In *Tamhīd al-mustaqarr li-taḥqīq maʿnā al-mamarr* ("On Transits"), al-Bīrūnī criticizes Abū Jaʿfar al-Khāzin for not having correctly handled two equations defining the location of a planet but remarks that the *Zīj al-ṣafā'iḥ* is correct on this matter. Abū Jaʿfar al-Khāzin criticized the claim of Abū Maʿshar that, unlike many others, he had fully determined the truth about the planets, which he had included in his *Zīj*. Abū Jaʿfar al-Khāzin regarded this work as a mere compilation. Al-Bīrūnī compared Abū Jaʿfar al-Khāzin very favorably with Abū Maʿshar, and in his *al-Āthār al-bāqiya min al-qurūn al-khāliya* ("Chronology of Ancient Nations") he refers to *Zīj al-ṣafā'iḥ* for a good explanation of the progressive and retrograde motion of the sphere.

An anonymous manuscript in Berlin (*Staatsbibliothek, Ahlwardt Cat. No.* 5857) contains two short chapters on astronomical instruments from a work by Abū Jaʿfar al-Khāzin, probably the *Zīj al-ṣafā'iḥ*. The MS Or. 168(4) in Leiden by Abū'l-Jūd quotes Abū Jaʿfar al-Khāzin's remark in *Zīj al-ṣafā'iḥ* that he would be able to compute the chord of an angle of one degree if angle trisection were possible.

In *Kitāb fī istīʿāb*, dealing with constructions of astrolabes, al-Bīrūnī cites Abū Jaʿfar al-Khāzin's work "Design of the Horizon of the Ascensions for the Signs of the Zodiac." And in his *Chronology* he describes two methods for finding the *signum Muḥarrami* (the day of the week on which al-Muḥarram, the first month of the Muslim year, begins) described by Abū Jaʿfar al-Khāzin in *al-Madkhal al-kabīr fīʿilm al-nujūm* ("Great Introduction to Astronomy"). Neither work is extant.

Also treated in al-Bīrūnī's *Chronology* is Abū Jaʿfar al-Khāzin's figure, different from the eccentric sphere and epicycle, in which the sun's distance from the earth is always the same, independent of the rotation. This treatment gives two isothermal regions, one northern and one southern. Ibn Khaldūn gives a precise exposition of Abū Jaʿfar al-Khāzin's division of the earth into eight climatic girdles.

Al-Kharaqī (*d.* 1138/1139), in *al-Muntahā*, mentions Abū Jaʿfar al-Khāzin and Ibn al-Haytham as having the right understanding of the movement of the spheres. This theory was perhaps described in Abū Ja'far al-Khāzin's *Sirr al-ʿālamīn* (not extant).

In *Taḥdīd nihāyāt al-amākin* . . ., al-Bīrūnī criticizes the verbosity of Abū Ja'far al-Khāzin's commentary on the *Almagest* and objects to Ibrāhīm ibn Sīnān and Abū Ja'far al-Khāzin's theory of the variation of the obliquity of the ecliptic; al-Bīrūnī himself considered it to be constant. The obliquity was measured by al-Harawī and Abū Ja'far al-Khāzin at Rayy (near modern Teheran) in 959/960, on the order of Abu'l Faḍl ibn al-ʿAmīd, the vizier of Rukn al-Dawla. The determination of this quantity by "al-Khāzin and his collaborators using a ring of about 4 meters" is recorded by al-Nasawī.

Abū Ja'far al-Khāzin was, according to Ibn al Qifṭī, an expert in arithmetic, geometry, and *tasyīr* (astrological computations based on planetary trajectories). According to al-Khayyāmī, he used conic sections to give the first solution of the cubic equation by which al-Māhānī represented Archimedes' problem of dividing a sphere by a plane into two parts whose volumes are in a given ratio (*Sphere and Cylinder* II, 4) and also gave a defective proof of Euclid's fifth postulate.

Abū Ja'far al-Khāzin wrote a commentary on Book X of the *Elements*, a work on numerical problems (not extant), and another (also not extant) on spherical trigonometry, *Maṭālib juz'iyya mail al-muyūl al-juz' iyya wa 'l-maṭāliʿ fi'l-kura al-mustaqīma*. From the latter, al-Ṭūsī, in *Kitāb šakl al-qaṭṭāʿ* ("On the Transversal Figure"), quotes a proof of the sine theorem for right spherical triangles. Al-Ṭūsī also added another proof of Hero's formula to the *Verba filiorum* of the Banū Mūsā (in *Majmūʿ al-rasā'il*, II [Hyderabad, 1940]), attributing it to one al-Khāzin. This proof, closer to that of Hero than the proof by the Banū Mūsā, and in which the same figure and letters are used as in Hero's *Dioptra*, is not found in the Latin editions of the *Verba filiorum*.

BIBLIOGRAPHY

I. ORIGINAL WORKS. Not many of al-Khāzin's writings are extant. The available MSS are listed in C. Brockelmann, *Geschichte der arabischen Literatur, Supplementband*, I (Leiden, 1943), 387. The commentary on Book X of the *Elements* is discussed by G. P. Matvievskaya in *Uchenie o chisle na srednevekovom Blizhnem i Srednem Vostoke* ("Studies About Number in the Medieval Near and Middle East"; Tashkent, 1967), ch. 6.

II. SECONDARY LITERATURE. Biographical and bibliographical references can be found in Ya'qub al-Nadim, *al-Fihrist*, G. Flügel, ed. (Leipzig, 1871–1872), pp. 266, 282; Ibn al-Qifṭī, *Ta'rīkh-al-ḥukamā'*, J. Lippert, ed. (Leipzig, 1903), 396; Hājjī Khalifa, *Lexicon bibliographicum* (repr. New York, 1964), I, 382, II, 584, 585, III, 595, VI, 170; H. Suter, *Die Mathematiker und Astronomen der Araber ubd ihre Werke* (Leipzig, 1900), p. 58, and *Nachträge*, p. 165; and A. Sayili, *The Observatory in Islam* (Ankara, 1960), pp. 103–104, 123, 126, which emphasizes the observations at Rayy. For Abū Ja'far al-Khāzin's astronomical theories and activities, see Ibn Khaldūn, *Prolegomena*, I, M. de Slane, trans. (repr. Paris, 1938), p. 111; and al-Bīrūnī, *Chronology of Ancient Nations*, C. E. Sachau, ed. (London, 1879), pp. 183, 249; *On Transits*, M. Saffouri and A. Ifram, trans. with a commentary by E. S. Kennedy (Beirut, 1959), pp. 85–87, and *Taḥdīd nihāyāt al-amākin* (Cairo, 1962), pp. 57, 95, 98, 101, 119.

M. Clagett, *Archimedes in the Middle Ages*, I, *The Arabo-Latin Tradition* (Madison, Wis., 1964), p. 353; and H. Suter, "Über die Geometrie der Söhne des Mūsā ben Schākir," in *Bibliotheca mathematica*, 3rd ser., **3**, no. 1 (1902), p. 271, mention the proof of Hero's formula. For the cubic equation of al-Māhānī, see F. Woepcke, *L'algèbre d'Omar Alkhayyāmī* (Paris, 1851), pp. 2–3; for the sine theorem, see Naṣir al-Dīn al-Ṭūsī, *Traité du quadrilatère*, A. Carathéodory, ed. (Constantinople, 1891), pp. 148–151; for the fifth postulate, see G. Jacob and E. Wiedemann, "Zu ʿOmer-i-Chajjâm," in *Der Islam*, **3** (1912), p. 56. Other articles by E. Wiedemann containing information on Abū Ja'far al-Khāzin are in *Beiträge* **60** (1920–1921) and **70** (1926–1927), of *Sitzungsberichte der Physikalisch-Medizinischen Sozietät zu Erlangen*. Now available in E. Wiedemann, *Aufsätze zur arabischen Wissenschaftsgeschichte*, II (Hildesheim, 1970), pp. 498, 503, 633.

YVONNE DOLD-SAMPLONIUS

AL-KHĀZINĪ, ABU'L-FATḤ ʿABD AL-RAḤMĀN [sometimes **Abū Manṣūr ʿAbd al-Raḥmān** or **ʿAbd al-Raḥmān Manṣūr**] (*fl.* Merv, an Iranian city in Khurāsān [now Mary, Turkmen S.S.R.], *ca.* 1115–*ca.* 1130), *astronomy, mechanics, scientific instruments.*

A slave-boy of Byzantine origin (a *castrato*, according to the edition of al-Bayhaqī by Shafīʿ, who reads *majbūb* for *maḥbūb*), al-Khāzinī was owned by Abu'l-Ḥusayn (Abu'l-Ḥasan, according to Shafīʿ) ʿAlī ibn Muḥammad al-Khāzin al-Marwazī, whose name indicates that he was treasurer of the court at Merv and who seems to have been sometime chancellor there (or, according to Meyerhof's translation of al-Bayhaqī, a religious judge, *qāḍī* being read for *māḍī*). Because of the owner's rank the form "al-Khāzinī," which denotes a relationship to the *khāzin*, should probably be preferred to "al-Khāzin," a form which, however, is encountered very often. His master gave the young man the best possible education in mathematical and philosophical (*ʿaqliyya*) disciplines. Al-Khāzinī "became perfect" in the geometrical sciences and pursued a career as a mathematical practitioner under the patronage of the Seljuk court. His work seems to have been done at Merv.[1] That city was then a capital of Khurāsān and from 1097 to 1157 was a seat of the Seljuk ruler Sanjar ibn Malikshāh, who held power first as emir of Khurāsān, then as sultan of the Seljuk empire. It became a brilliant center of literary and scientific activity and by the end of this period was renowned for its libraries. Al-Khāzinī's book of astronomical tables was composed for Sanjar, and his balance was constructed for Sanjar's treasury.

Noted for his asceticism, al-Khāzinī dressed as a Ṣūfī mystic and ate "the food of pious men "—meat but three times a week and otherwise two cakes of bread a day. Rewards he refused: he handed back 1,000 dinars sent him by the wife of the emir Lājī Ākhur Beg al-Kabīr; the same amount, presented to him by Sanjar through the emir Shāfiʿ ibn ʿAbd al-Rashīd (a pupil of al-Ghazālī, *d.* 1146/1147), presumably on the occasion of his completing the astronomical tables, was also returned. He had, he said, ten dinars already and lived on three a year, for in his household there was only a cat. Al-Khāzinī had students, but only one name has survived, an otherwise unknown al-Ḥasan al-Samarqandī.

Scarcely anything else is known of al-Khāzinī's life (although his own works have not been fully searched). The basic biographical account is that by al-Bayhaqī (*d.* 1169), who seems to have been personally acquainted with al-Khāzinī. (Meyerhof's translation of the notice must be preferred to that by Wiedemann, who wrote before the publication of Shafīʿ's critical edition.) Al-Shahrazūrī adds nothing significant and subtracts a good deal; Ḥājjī Khalīfa has only a few lines with nothing new. Ṭāshköprüzāde merely mentions an "al-Khāzinī" in connection with astronomical instruments. He does not appear among the 266 "ʿAbd al-Raḥmāns" in al-Ṣafadī.[2]

At various times al-Khāzinī has been mistakenly identified with Alhazen (*i.e.*, Ibn al-Haytham), Abū Jaʿfar al-Khāzin (especially in connection with the treatise on astronomical instruments; see below), and Abu'l-Fatḥ al-Khāzimī [or al-Hāzimī] (a twelfth-century astronomer of Baghdad).[3] There is no evidence that al-Khāzinī ever worked in Baghdad; assertions that he did must be based on the false assumption that the Seljuk court would be there.

One doubtful passage (Quṭb al-Dīn al-Shīrāzī [*d.* 1311], *Nihāyat al-idrāk* . . .) indicates that he made astronomical observations at Iṣfahān; "at Iṣfahān," however, seems to be an addition of unknown origin or authority.[4] Chronology makes it extremely unlikely that al-Khāzinī was a member of the staff of the observatory which was established by the Seljuk Sultan Malikshāh in Iṣfahān and which lasted but a short while after the founder's death in 1092; ʿUmar al-Khayyāmī (Omar Khayyam; *d.* 1131[?]) and al-Muẓaffar ibn Ismāʿīl al-Asfizārī (mentioned below in connection with al-Khāzinī's balance), both a generation older than al-Khāzinī, had in fact been there.[5] Indeed, no evidence shows al-Khāzinī to have been associated with any observatory, that is, as a member of a group of researchers attached to an actual astronomical institution.[6] In calculating his *zīj* (book of astronomical tables) al-Khāzinī was said to have worked with Ḥusām al-Dīn Sālār (otherwise dated only as writing between the times of al-Bīrūnī [*d.* 1051 or after] and Naṣīr al-Dīn al-Ṭūsī [*d.* 1274]); but the source is the sixteenth-century Persian historian Ḥasan-i Rūmlū, who also associates al-Khāzinī with the poet Anwarī. But Anwarī, astronomically learned though he was, and patronized by Sanjar, almost certainly lived at least a generation later.

Al-Khāzinī, al-Khāzimī, and Anwarī are also among those variously reported to have been involved in the unfortunate astrological prediction of devastating windstorms in 1186 (the entire year was so calm in Khurāsān that the grain crop could not be properly winnowed); but al-Khāzinī's involvement, again on chronological grounds, is hardly likely.[7]

Al-Khāzinī's Scientific Accomplishments. The known works of al-Khāzinī, seemingly all extant, are the following: *al-Zīj al-Sanjarī* ("The Astronomical Tables for Sanjar"), also in a summary (*wajīz*) by the author; *Risāla fi'l-ālāt* ("Treatise on [Astronomical] Instruments"), which actually may not be the work mentioned by the biobibliographers (see below; al-Bayhaqī does not refer to it); and *Kitāb mīzān al-ḥikma* ("Book of the Balance of Wisdom"), a wide-ranging work that deals primarily with the science of weights and the art of constructing balances. To the manuscripts listed by Brockelmann should be added 1) Sipahsālār Mosque [madrasa] Library (Teheran) 681–682 (cataloged as "Zīj-i Sanjarī" but containing a collection of al-Khāzinī's works including *Risāla fi'l-ālāt* but not the complete *zīj*)[8] and 2) the manuscript used for the Cairo edition of *Kitāb mīzān al-ḥikma* (see below). The contents of the works are discussed later.

It is hard to assess the importance of al-Khāzinī. His hydrostatic balance can leave no doubt that as a maker of scientific instruments he is among the greatest of any time. As a student of statics and hydrostatics, even in their most practical aspects, he is heavily dependent upon earlier workers and borrows especially from al-Bīrūnī and al-Asfizārī; but his competence is not to be denied, and *Kitāb mīzān al-ḥikma* is of outstanding importance to the historian of mechanics, whatever its claims to originality or comprehensiveness may prove to be. In astronomy, as in mechanics, al-Khāzinī's direct predecessors are ʿUmar al-Khayyāmī and al-Asfizārī. His *zīj* takes its place in the Eastern Islamic astronomical tradition after those of al-Bīrūnī and ʿUmar al-Khayyāmī and is succeeded by those produced by the labors of the Marāgha Observatory (Naṣīr al-Dīn al-Ṭūsī and Quṭb al-Dīn al-Shīrāzī) and the Samarkand observatory (al-Kāshī [*d. ca.* 1430] and Ulugh Beg [the sultan; *d.* 1449]). Al-Khāzinī is one of twenty-odd Islamic astronomers known to have performed original observations.[9] Kennedy rates his *zīj* very highly and, in suggesting eclipse and visibility theory as subjects that would particularly reward monographic treatment, names topics—particularly visibility theory—for which al-Khāzinī's tables are an especially rich source.[10]

In mechanics no works are known that follow in the tradition of *Kitāb mīzān al-ḥikma*; treatments of balances or the science of weights become mere manuals for craftsmen who make simple scales or steelyards, or for merchants or inspectors who use them or check them. That branch of learning ceases to be a part of the scientific tradition.

Although al-Khāzinī's publications were well-known in the Islamic world, and particularly in the Iranian part of it, they do not seem to have been used elsewhere save in Byzantium. The *Sanjarī zīj* (*ζῆζι Σαντζαρῆς*) was utilized, at least for its tables of stars, by George Chrysococces (*fl.* Trebizond, *ca.* 1335–*ca.* 1346), an astronomer and geographer, and through him by Theodore Meliteniotes, an astronomer in Constantinople (*fl. ca.* 1360–*ca.* 1388).[11]

Works: the Astronomical Tables. The *Sanjarī Zīj*, whose full title is *al-Zīj al-muʿtabar al-Sanjarī al-Sulṭānī* ("The Compared [or "Tested"] Astronomical Tables Relating to Sultan Sanjar") is also called by shorter forms of the same title (*al-Zīj al-sulṭānī* refers to other works, however); and by the name *Jāmiʿ al-tawārīkh li'l-Sinjarī* ("Collection of Chronologies for Sanjar," if Sanjar can be called al-Sinjarī, after his native town)—the last title resulting from the large amount of calendrical material and the tables of holidays and fasts and rulers and prophets.[12] The known manuscripts are Vatican Library cod. Ar. 761 and British Museum cod. Or. 6669; the work runs to 192 folios (32 × 20.5 cm.) in the Vatican manuscript, which is sometimes considered an autograph. Ḥamdallāh al-Qazwīnī, in *Nuzhat al-qulūb*, presents a table to use in conjunction with the Indian dial for determining the *qibla* (direction of Mecca) for most places in Iran. He indicates that it was produced by al-Khāzinī on the order of Sultan Sanjar. One would expect to find such a table in the *zīj*, but it is missing—as are geographical coordinates of cities—from both the Vatican and British Museum manuscripts (the latter being nearly complete, despite LeStrange's remark—the table of contents at the beginning of the codex, however, omits many sections).[13]

In 1130/1131 (A.H. 525) al-Khāzinī wrote an abridgement of his tables called *Wajīz al-Zīj al-muʿtabar al-sulṭānī*;[14] that year presumably marks a *terminus ante quem* for the tables themselves. The British Museum and Vatican manuscripts of the *zīj* have no date in the obvious places. The year A.H. 530 is assigned by Suter and taken over by Sayili without basis.[15] Kennedy, Destombes, and Nallino have produced no precise dating. Nallino, using the Vatican manuscript (folios 191v–192r), describes the star tables as having longitudes and latitudes of forty-three fixed stars for A.H. 509 (1115/1116);[16] Kennedy, using the same manuscript, describes the same table as providing latitudes and longitudes, temperaments, and magnitudes of forty-six stars for A.H. 500 (1106/1107), presumably on the basis of the parameters.[17] Destombes says—as Nallino does, but using the British Museum manuscript—that the star table is for forty-three stars for A.D. 1115.[18] Thence Destombes presumes that the *zīj* was written in 1115 and "corrects" the date of *ca.* 1120 attached to the tables by Kennedy without discussion.[19] The tables are, however, dedicated to Sanjar, who was sultan of the empire only from 1118; but he had been emir of Khurāsān since 1097, and the use of the title "sulṭān" in the *zīj* is in any case problematical. Sayili does cite a report that the *zīj* had been finished before Sanjar's coronation.[20] Nallino, however, had long since pointed out a reference to the caliph Mustarshid bi'llāh, who occupied the office in 1118–1135.[21] One is left, then, with the interval 1118–1131 for the completion of the *zīj*, all subsidiary evidence pointing to the beginning of this period.

That al-Khāzinī made a certain number of actual observations is not questioned; probably they were done at Merv independently of any observatory. Quṭb al-Dīn al-Shīrāzī discusses the measurements of the obliquity of the ecliptic by al-Khāzinī and indicates that they were very careful—praise which suggests high technical competence and good intruments.[22] In the *Wajīz* (fol. 1v) al-Khāzinī states that he compared observed and calculated positions for all planets (including sun and moon) at conjunctions and eclipses and found disagreement for all of them.[23] In fact the word *muʿtabar* in the title suggests just such a comparison, indeed a testing or "experimental verification." But al-Bayhaqī in his biographical notice says that the mean motions (*awsāṭ*) and equations (*ta'dīlāt*) determined by al-Khāzinī need further study—except in the case of Mercury, especially in its retrograde motion, for which the positions had been observed and tried.

The Indian theory of cycles (i.e., those which culminate in the "world day," the period which the cosmos takes to return to any given state) as reported in the *Sindhind* and in Abu Ma'shar's *al-Hazārāt* ("The Thousands") greatly interested al-Khāzinī despite al-Bīrūnī's unequivocal strictures against that sort of astronomy.[24] It is possible to deduce those cycles from the motions one observes, al-Khāzinī claims, but difficult because of the amount of calculation.[25] The *Sanjarī zīj* has a fair amount of such material, but al-Khāzinī keeps all his computations strictly within the Islamic Ptolemaic tradition (as far as can be said).

Among his predecessors in astronomy, apart from al-Bīrūnī it is Thābit ibn Qurra and al-Battānī whose *zīj*'s seem to have concerned him most.[26] He reproduces Thābit's work on lunar visibility before presenting his own exceptionally detailed treatment, and frequently he reports the methods or conclusions of Thābit or al-Battānī in other connections. For his value of the obliquity of the ecliptic al-Khāzinī, like al-Battānī, chooses 23°35′—but only after discussing the discrepancies among the results obtained by others, mentioning difficulties due to refraction, and then rejecting both decreasing and alternately increasing and decreasing values of the obliquity.[27] Unlike any other Islamic astronomer except Ḥabash al-Ḥāsib al-Marwazī, al-Khāzinī uses the canonical religious date for the Hijrī epoch.[28]

Al-Khāzinī's *zīj* in general is very rich. The chronologies and the section on visibility have already been mentioned.[29] The latter, besides tabulating the arcs of visibility for the five planets (perhaps calculated in an original fashion) as well as those for the moon, also presents differences according to clime and incorporates historical material. Tables of trigonometric functions, of astronomical parameters generally, and especially of planetary mean motions (including those of sun, moon, and lunar nodes) are thorough and highly precise—the planetary mean motions, for example, are given in degrees or revolutions per day to eight or more significant sexagesimal (fourteen or more decimal) figures; and the tables relating to eclipse theory are also greatly elaborated. The absence of material on terrestrial geography has been noted, and the star tables have been described in connection with the dating of the *zīj*. There are, finally, a number of tables of astrological quantities. Positions are recorded here for "al-Kayd," perhaps a comet.[30]

Treatise on Instruments. The *Treatise on Instruments* (*Risāla fi'l-ālāt*), found by Sayili in codices 682 and 681 of the library of the Sipahsālār Mosque in Teheran, is a short work, occupying seventeen folios in the manuscript.[31] It is probably the same as *al-Ālāt al-'ajība* (*al-raṣadiyya*) ("The Remarkable [Observational] Instruments"), which was noted by Ibn al-Akfānī, Ṭāshköprüzāde, and Ḥājjī Khalīfa.[32] Sayili ascribes the work to 'Abd al-Raḥmān al-Khāzinī. So does Brockelmann, following Wiedemann, "Beiträge . . . IX"; Wiedemann repeats this ascription in "Beiträge . . . LVII," but in his articles for the *Encyclopaedia of Islam* on "al-Khāzin, Abu Dja'far . . ." and "al-Khāzinī . . ." he allots the work without comment (although, indeed, with citation of the passages in the "Beiträge . . .") to al-Khāzin, an astronomer, mathematician, and instrument-maker of the mid-tenth century.[33] Ibn al-Akfānī, Tāshköprüzādeh, and Ḥajjī Khalīfa (at both places) ascribe a treatise of that title to "al-Khāzinī" without further identification; but this carries no weight, for Ḥājjī Khalīfa refers to "Abu Ja'far al-Khāzinī" four times, to "Abu'l-Fatḥ 'Abd al-Raḥmān al-Khāzin" once, and otherwise to "al-Khāzinī"—so that Flügel's index assigns *al-Ālāt al-'ajība* to Abū Ja' far. De Slane's note, following upon Ibn Khaldūn's mention of "Abu Ja' far al-Khāzinī," should no longer be misleading, for not only is *al-Ālāt al-'ajība* attributed by de Slane to Abu Ja' far, but so also are both *Kitāb mīzān al-ḥikma* and *Zīj al-safā' iḥ* (the former now known to be by 'Abd al-Raḥmān, the latter, by Abu Ja'far).[34] Since the treatise on instruments is a minor one al-Bayhaqī's failure to note it as a work of 'Abd al-Raḥmān al-Khāzinī means nothing; similarly, the absence of the title from the frequent references of al-Bīrūnī to works by Abu Ja'far al-Khāzin can produce no certainty in the other direction. The incidental mentions by the biobibliographers seem to suggest the later man, but the only concrete evidence for assigning the work to 'Abd al-Raḥmān al-Khāzinī is that of the Teheran manuscript, mentioned above, which was copied in A.H. 683 (1284/1285).

The *Risāla* has seven parts, each devoted to a different instrument: a triquetrum, a dioptra, a "triangular instrument," a quadrant, devices involving reflection, an astrolabe, and simple helps for the naked eye. The quadrant is in fact called a *suds*, or "sextant," and performs the functions of the sextant, although its arc is 90°. Apart from describing the devices and their use, the treatise also demonstrates their geometrical basis.

Kitāb Mīzān al-Ḥikma. The most interesting and

important of al-Khāzinī's writings, both in itself and as a source of information on earlier work—if only because it is a much rarer sort of book than a *zīj*—is *Kitāb mīzān al-ḥikma*, the *Book of the Balance of Wisdom*. A long treatise (the Hyderabad ed. has 165 large octavo pages of Arabic text, exclusive of figures and tables), it studies the hydrostatic balance, its construction and uses, and the theories of statics and hydrostatics that lie behind it, as well as other topics both related and unrelated. Written in A.H. 515 (1121/1122) for Sultan Sanjar's treasury,[35] *Kitāb mīzān al-ḥikma* has survived in four manuscripts, of which three are independent. The treatise has been published, partially edited and largely translated.

Study of the *Kitāb mīzān al-Ḥikma* may begin from either the edition of selected parts, accompanied by sometimes inaccurate English translations, that was produced in 1859 by Khanikoff and the editors of the *Journal of the American Oriental Society*—or from the uncritical but serviceable text of the Hyderabad edition, which was made on the basis of the two related Indian manuscripts and a photocopy of the one used by Khanikoff.[36] (It is Khanikoff's manuscript that seems to be the oldest.) Variants can be sought from the Cairo edition, which is a rather unprofessional transcription of an additional manuscript, from East Jerusalem;[37] the text, of which up to half is missing, seems closer to Khanikoff's copy of the work than to the Indian ones.[38]

Those parts of his not fully complete manuscript left untranslated by Khanikoff were almost entirely rendered into German by Wiedemann,[39] who, however, provided no Arabic text and occasionally abridged and paraphrased without sufficient indication. Of the long studies, that by Ibel is helpful; Bauerreiss' thesis demands caution save when he is describing the apparatus. The commentary in Khanikoff's article is by now badly dated.

An elaborate literary conceit (three pages in the Hyderabad edition) on the name *mīzān al-ḥikma*—thus far translated as "the balance of wisdom"—opens the book; and the phrase does indeed repay consideration. The hydrostatic balance built by al-Asfizārī (who was a generation older than al-Khāzinī) had been called *mīzān al-ḥikma*;[40] an improvement upon earlier instruments of the type first constructed by Archimedes, it was likewise intended to detect alloys passing for gold, and other frauds. Created for Sultan Sanjar, the scales was destroyed, out of fear by his treasurer (not the one who was al-Khāzinī's master); and al-Asfizārī "died of grief."[41] Al-Khāzinī subsequently built a similar balance, further refined, for Sanjar's treasury; this he called *al-mīzān al-jāmiʿ* (the "comprehensive" or "combined balance") and *mīzān al-ḥikma*, in honor, presumably, of al-Asfizārī.[42] The primary meaning, then, of *mīzān al-ḥikma* is "balance of true judgment," of accurate discrimination between pure and adulterated metals, between real gems and fakes. The name in fact consciously echoes the Koranic balance with the long beam that is to be erected on the Day of Judgment.[43]

The first words of *Kitāb mīzān al-ḥikma* are praises to God the Wise (*al-Ḥakīm*), the Just (*al-ʿAdl*)—or, in variants, the Judgment (*al-Ḥukm*), the Truth (*al-Ḥaqq*), the Justice (*al-ʿAdl*; lexically distinct from the form above).[44] Words derived from the root Ḥ-K-M are then cleverly woven into the text, together with forms from the root ʿ-D-L (which denotes justice in the sense of equitability and evenhandedness, and one of whose derivatives, *iʿtadala*, means "to balance" and is specifically applied to weights on a scales). "Justice," says al-Khāzinī, "is the support of all virtues and the foundation of all excellencies. For perfect virtue is wisdom and has two parts, knowledge (*ʿilm*) and action (*ʿamal*), and two halves, religion and the world, perfect knowledge and proficient (*muḥkam*) activity (*fiʿl*); and justice is the combination of [those] two and the union of the two perfections of it [wisdom], by which is conferred the limit of every greatness and by means of which is attained precedence in every excellence." God in his Mercy, continues al-Khāzinī, has set up among men three arbiters [*ḥukkām*] of justice: the glorious Koran, to which the Traditions of the Prophet are the sequel; the rightly guided and well-versed scholars (*ʿulamāʾ*), among whom is the just governor, alluded to in the words of the Blessed, "the *sulṭān*, the shadow of the Most-High God upon Earth, the refuge of the injured, and the judge (*ḥākim*)";[45] and the balance, which is the tongue of justice, the just judgment whose decision satisfies all, the order and justice in human conduct and transactions—the balance which God Himself has associated with his very Koran (as al-Khāzinī shows with a surprising number of strongly worded and explicit textual proofs from the Koran).[46]

But these religious (and political) themes must not obscure the fact that for contemporary students of the sciences *ḥikma* meant not only wisdom but particularly philosophy (that is to say, Islamic Peripateticism), with its two divisions, theoretical and practical, answering to the two virtues σοφία and φρόνησις, which may reasonably if not perfectly be associated with the divisions of knowledge and action, religion and the world, stated by al-Khāzinī. Certainly he proceeds to describe what a later age called a "philosophical balance"—the other possible translation of *mīzān al-ḥikma*. Al-Khāzinī writes:

> This just balance is founded upon geometrical demonstrations and deduced from physical causes, in two aspects: 1) as regards centers of gravity, the most elevated and noble division of the mathematical sciences, which is knowledge that the weights of heavy things differ according to the distances they are placed [from a fulcrum]—the foundation of the steelyard; and 2) [as regards] knowledge that the weights of heavy things differ according to the rarity or density of the fluids in which the thing weighed is immersed—the foundation of the *mīzān al-ḥikma*.[47]

When al-Khāzinī lists the advantages of his balance, "which is something worked out by the human intellect and perfected by trying out and testing" and which "performs the functions of skilled craftsmen," he names benefits variously theoretical and practical—precision, ability to distinguish pure metal from alloy and to determine the content of binary alloys, usefulness in calculations relating to a treasury, gains due to ease and versatility in use (for instance, the possibility of recourse to any reference liquid—from the broad scope of its applications comes its other name, "the comprehensive balance"), and the seventh and last advantage, "the gain above all others," that it enables judging true gems from false.[48] His is a philosophical balance desirable both for the superior theory in its construction and the range and excellence of uses to which it can be put; and of its practical virtues the greatest is the ability to judge genuine from fraudulent. Among al-Khāzinī's great Islamic precursors in this art al-Rāzī (Lat., Rhazes; the famous physician) had called his water balance *al-mīzān al-ṭabīʿī* ("the physical balance," that is, as pertaining to physical principles), whereas ʿUmar al-Khayyāmī had designated his highly developed steelyard *al-qusṭās al-mustaqīm* ("the upright [or "honest"] balance"); al-Asfizārī and al-Khāzinī had found a name which included both aspects: *mīzān al-ḥikma*—the balance of wisdom—meaning the "balance of right judgment" and "the philosophical balance."

Al-Khāzinī is perfectly explicit in stating what sort of book he is composing. As a preliminary he divides the fundamental principles of any art into three classes: those which are acquired in early childhood and youth, after one sensation or several sensations, spontaneously, and which are called first things and common knowledge; those which are demonstrated in other sciences; and those which are obtained by trying out and by assiduous investigation (in the area of the art itself). So it is, then, with the art of the balance, which has principles both geometrical and physical (considering as it does the categories both of quantity and of quality); but the author will not mention the obvious principles belonging to common knowledge and will refer only in passing, as necessary, to principles taken over from other disciplines or obtained by investigation.[49]

Even though he presents propositions and general theorems of statics and hydrostatics in books I and II, al-Khāzinī supplies no proofs and frequently no explanations; he employs demonstration in his treatise only when it is required in connection with designing or using the balance of wisdom or another instrument. It is not a deductive work of mathematical science but rather a technical presentation of the art of the philosophical (or scientific) balance.

Contrary to what is assumed about most medieval authors, al-Khāzinī was well aware of the historical progress made in his art—the introduction to *Kitāb mīzān al-ḥikma* contains two sections[50] which report the invention of the hydrostatic balance by Archimedes (following Menelaus' account) and the modifications and perfections introduced by later workers up to al-Khāzinī himself. He states, indeed, that "the knowledge of the relations [in specific weight] of one metal to another depends upon that perfecting of the balance through delicate and detailed devising by all who have studied it, or developed it by fixing the marks for specific gravities of metals relative to a particular sort of water."[51] Hence he had seen fit "to assemble on this subject whatever we have gained from the works of the ancients and of later philosophers who have followed them, in addition to what [our own] thought, with God's aid and giving of success had granted."[52] In fact much of *Kitāb mīzān al-ḥikma* is composed of extracts; most of what is original relates to the "balance of wisdom" itself or to its applications.

Contents of Kitāb Mīzān al-Ḥikma. The *Book of the Balance of Wisdom* comprises eight books (*maqālāt*) divided into fifty chapters (*abwāb*); larger, intermediate, and smaller divisions of the text are also indicated, but inconsistently. In particular, the initial summary and table of contents by al-Khāzinī, as given by the manuscripts, differ from each other and from the headings in the actual text.

(Because cross-references or headings are often missing from the translations of *Kitāb mīzān al-ḥikma*, making them hard to use, page references are supplied here. All the translations follow the Khanikoff manuscript. Wied. = Wiedemann; B = Wiedemann, "Beiträge . . ." [numbers in parentheses refer to reprint]; Khan. = Khanikoff edition. I: 1.1 = book I, chapter 1, section 1. All numbers are inclusive.)

Al-Khāzinī's long introduction and his own summary and table of contents precede the body of the treatise.

Introduction: Khan., 3–16; extracts in Clagett, *The Science of Mechanics in the Middle Ages,* 56-58. *Summary of contents*: Khan., 16–18. *Table of contents*: Khan., 18–24; Ibel, 80–83; compare table of contents drawn up according to headings in the text, at end of Hyderabad edition.

Book I sets forth geometrical and physical principles underlying the hydrostatic balance: theorems on centers of gravity from works by Ibn al-Haytham and Abu Sahl al-Qūhī (ch. 1); theorems from Arabic translations of works entitled "On the Heavy and the Light"—by Archimedes (a fragment of "On Floating Bodies") (ch. 2), by Euclid (ch. 3), and by Menelaus (ch. 4); repetition or summary of important theorems (ch. 5); and propositions on sinking and floating (ch. 6), following Archimedes (?).[53] Thus far no proofs or discussions.

Chapter 7 is a detailed description of the construction and use of Pappus' araeometer,[54] an instrument for determining specific gravities of liquids; here geometrical demonstrations are indicated, in accordance with al-Khāzinī's aims.

I:1, Khan., 25–33, retranslated in Ibel, 85–88, reprinted in Clagett, 58–61. *I:2,* B VII; compare Clagett, 52–55. *I:3,* Ibel, 37–39; see also Ernest A. Moody and Marshall Clagett, *The Medieval Science of Weights* (Madison, Wis., 1960), pp. 23–31. *I:4,* not translated, but see Ibel, 77–78, 181–185. *I:5,* Khan., 34–38; largely reprinted in Clagett, 61–63; compare, for *I:5. 3,* Wied., "Inhalt . . .". *I:6,* B XVI, 133–135 (I, 492-494). *I:7,* Khan., 40–52 (and the notes); compare Bauerreiss, 95–108.

Book II begins with a discussion of the balancing of weights and its various causes, taken from a work of Thābit ibn Qurra.[55] The rest of this book derives from al-Asfizārī and treats, without demonstration, the following topics: constrained motion of the centers of gravity of bodies; the equilibrium of a balance beam, geometrical or physical, with application of the results to a spear held in the hand; the construction of a steelyard, the graduation of its beam, and the methods of weighing with it; and the conversion of steelyards from one system of weights to another.

II, entire, B XVI, 136–158 (I, 495–517).

Book III has three parts. Part 1 (chapters 1–3) comes from al-Bīrūnī's [*Maqāla*] *fī'l-nisab* [*allātī*] *bayna'l-filizzāt wa'l-jawāhir fī'l-ḥajm* ("On the Relations [in Weight] Among Metals and Precious Stones With Respect to [a Given] Volume"): specific gravities —or water-equivalents (weights of water equal in volume to reference weights of the given materials)— of metals, precious stones, and other substances of interest. Al-Bīrūnī here describes his "cone-shaped instrument" (*al-āla al-makhrūṭīya*)—a pycnometric metal vessel shaped like an Erlenmeyer flask, with a handle and a spout that is a narrow tube projecting out and down from the neck—and explains its use in measuring the weight of a volume of water equal to the sample, which has been introduced into the flask and displaces water through the spout into one of the pans of a balance. The neck of the flask is about the diameter of a man's little finger; the spout is perforated all along its sides to minimize the effects of surface tension.

This part of book III and the remaining two parts, also by al-Bīrūnī, have elaborate tables of values and detailed indications of procedure.

Part 2 records how the weight was obtained of a cubic cubit of water by making an exact hollow cube of brass, determining its internal volume through a precise measurement of its dimensions, weighing the water necessary to fill it, and multiplying the result by the appropriate ratio. The weights of a cubic cubit of several metals are then found, using their water equivalents. This part ends with the calculation of the weight of gold required to fill the volume of the earth (chapter 4, section 3). Part 3 (which is chapter 5) continues in similar vein with problems about dirhams doubled successively on each square of a chessboard, starting with a single dirham on the first square—their total number, the number of chests to hold them, the length of time to spend them.

III:1.1–.3 and III*:1.5–.6,* missing in Khanikoff's manuscript, now in the Hyderabad edition; no translation, but compare Khan., 53–56; Wied., ". . . Bêrûnîsche Gefäss . . ."; and Wied., "Mīzān", p. 534. *III:1.4,* Khan., 56–58. *III:2–III:4.2,* Khan., 58–78; on *III:4.1–.2* see B XXXIV. *III:4.3–III:5,* B XIV; see also Julius Ruska, "Ḳazwīnīstudien," 254–257.

Book IV is historical. First come descriptions of the hydrostatic balances of Archimedes and of Menelaus, an explanation of the latter's methods of analyzing alloys, and a summary of the values he found for specific gravities (thus far according to Menelaus). Then follow presentations of the "physical balance" of Muḥammad ibn Zakariyyâ al-Râzī and of the water balance of ʿUmar al-Khayyāmī, with detailed diagrams, based on works by those authors.

IV:1, Archimedes, according to Menelaus, Ibel, 185–186. *IV:2–IV:3,* Menelaus; B XV, 107–112 (I, 466–471). *IV:4,* al-Râzī, Ibel, 153–156. *IV:3* and *IV:4* are reversed in the Hyderabad edition. In the Khanikoff manuscript *IV:4.2–.3* was displaced to the end of the book, section 3 being abridged; the full text of section 3 is in the Hyderabad edition, pages 85–86, without abridgement or disordering. *III:5,*

B XV, 113–117 (I, 472–476); *III:5.1* also in Ibel, 158–159.

In Books V and **VI**, *Kitāb mīzān al-ḥikma* becomes a manual of the "balance of wisdom." Starting with the instruction al-Asfizārī had left, the discussion becomes al-Khāzinī's own after the first chapter of book V (or perhaps after the first section of that chapter). Diagrams, illustrations, and tables are outstandingly rich. Book V explains the fabrication of the parts of the balance, their arrangement and assembly, and the adjustment and checking of the balance, pointing out defects that may be found or mistakes that may be made. Book VI, the longest of the work (although only a fifth of the whole), sets forth the operation of the balance: selecting the counterpoises, leveling the beam, and weighing, then graduating the balance in order to use it for measuring specific gravities. After that has been done, a number of special procedures can be exploited: testing the genuineness of metals and precious stones by use of two movable scale pans (a method restricted to the "balance of wisdom") and discovering the ratio of the constituents of a two-element alloy or other mixture (perhaps to determine a correct monetary value); assaying and appraising by another technique, that of *tajrīd* ("isolation"), which involves a single movable bowl and algebraic calculation; and finding specific gravities of substances by computation from their weights in air and in water. Other, related procedures are also mentioned, and some special theorems are introduced. The end of book VI (chapter 10) is an appendix on the prices of gems in times past, taken from al-Bīrūnī's *Kitāb al-jamāhir fi maʿrifat al-jawāhir* ("Book of Gatherings on Knowledge of Precious Stones").

Many of the methods of using the balance are provided with geometrical proofs. The suspension of the instrument, its indicator tongue, and the placing of the marks on the beam are treated with special care.

V: Ibel, 112–136. *V:1.4*, also Khan., 88–94. *VI:1–VI:4*, Ibel, 136–151 (also: *VI:2.5*, Khan., 98–99, and *VI:4.1–.2*, Khan., 100–104). *VI:5–V:9*, B XV, 117–132 (I, 476–491) (beginning with the method of *tajrīd*). *VI:10*, Wied., ". . . Wert von Edelsteinen . . . ."

Books VII and **VIII** treat special modifications of the "balance of wisdom" and other specialized balances. There are abundant diagrams and tables. Differently graduated and without the extra scale pans that make it a hydrostatic balance, al-Khāzinī's instrument can be applied to exchanging among different coinages. To his discussion of the adjustment and use of the exchanging balance (*VII:2–VII:4*) the author prefixes an unattributed analysis of proportion (*VII:1.1–.5*) and other mathematical material (*VII:1.6–.7*). The text on proportions is nearly identical to that by al-Bīrūnī in *Kitāb al-tafhīm*.[56] In *VII:5* al-Khāzinī adds propositions and theorems relating to the mint and to exchange.

He then (*VII:6* to the end of book VIII) considers the other, special-purpose balances: (*VII:6*) a scales for weighing dirhams and dinars without counterpoises; (*VII:7*) balances for use in leveling, measuring differences in level, and smoothing vertical surfaces;[57] (*VII:8*) the "righteous steelyard" of ʿUmar al-Khayyāmī, which can weigh from a grain to a thousand dirhams or a thousand dinars by using an indicator tongue and three counterpoises associated with three different graduations of the beam; and the clock-balance, to which al-Khāzinī devotes a rather long notice (thirteen pages plus figures, all of book VIII in the Hyderabad edition) that pays special attention to the water or sand reservoir and to the measurement of short intervals (for instance, for astronomical purposes), down to seconds of time.

VII and *VIII*: the Hyderabad text and al-Khāzinī's own list of chapters assign eight chapters to book VII; that enumeration is followed here. Al-Khāzinī's brief summary of the work and Khanikoff's altered table of contents give five chapters to book VII, moving the next three to the beginning of book VIII.

VII:1–VII:6 (except *VII:1.1–.5*), B LXVIII, 6–15 (II, 220–229); *VII:1.1–.5*, *ibid.*, 3–6 (II, 217–220). *VII:7*, Ibel, 159–160. *VII:8*, Ibel, 107–110. *VIII:1–VIII:4* and first paragraph of *VIII:5* (according to al-Khāzinī's list of chapters; VIII, part 1, chapters 1–4 and first paragraph of part 2, chapter 1, in Hyderabad edition and in Khanikoff manuscript), B XXXVII. The Khanikoff MS ends there; the Hyderabad edition, pp. 164–165, gives the other two paragraphs of VIII, part 2, chapter 1, and the short chapter 2, which seems to end the book properly. VIII, part 2, chapters 1 and 2, seems to contain what is suggested by the title of *VIII:5* in al-Khāzinī's list of chapters.

Kitāb mīzān al-ḥikma is also well-stocked with miscellaneous incidental statements of interest—on the rising and sinking of mountains, for example, and the natural production of gold out of lead.

Al-Khāzinī and the Science of Weights in Islam. Much of what is most interesting in the work comes from other authors, Greek or Islamic. For his theorems in geometry and physics al-Khāzinī draws upon Euclid, Archimedes, Menelaus, and, without citation and perhaps indirectly, Pseudo Aristotle's *Mechanica problemata*, chapters 1 and 2,[58] among the Greeks, and from Thābit ibn Qurra, Abu Sahl al-Qūhī, Ibn al-Haytham, al-Bīrūnī, and al-Asfizārī. Exactly what al-Khāzinī does to the extracts he incorporates is a matter for detailed study, but there seems to be nothing in the way of basic physical theory that is his own.

He is especially indebted for significant material to al-Bīrūnī. The very careful explanations (in book III) of refined instruments and methods for determining specific gravities come from al-Bīrūnī's "On the Relations Among Metals . . .," and, when discussing the determination of specific gravities by "isolation" (in *VI:5*), al-Khāzinī uses data from a lost work by al-Bīrūnī.[59] The treatment of (mathematical) proportion (beginning of book VII) almost certainly comes, as was noted, from the *Kitāb al-tafhīm*; and the numerical problems involving large numbers (end of book III) are taken from al-Bīrūnī's writings. How much (if any) of al-Khāzinī's material on the exchanging balance or the chronometric balance derives from al-Bīrūnī's lost treatises on those subjects cannot, of course, be known.[60] The historically valuable notice on the prices of gems at various times and places (*VI:10*) is also from al-Bīrūnī.

When al-Khāzinī traces his scientific lineage in the art of the hydrostatic balance, he first names Archimedes (giving Menelaus' account of the assay of the crown presented to Hiero II, tyrant of Syracuse) and Menelaus, who is said to have been attempting to solve the problem of a three-component alloy. As his first precursors in the Muslim world he lists Sanad ibn ʿAlī, Yūḥannā ibn Yūsuf, and Aḥmad [ibn] al-Faḍl al-Massāḥ al-Bukhārī, who were contemporaries in the mid-ninth century. (Those three and only they are mentioned as his Islamic predecessors by al-Bīrūnī, in "On the Relations Among Metals . . .," in the study of specific gravities. Only Sanad ibn ʿAlī is otherwise known, although there was a Yūḥannā ibn Yūsuf al-Qass, a scientist who died in 980/981.) Then he mentions al-Rāzī (who included in one of his works—not extant save insofar as it is presented in al-Khāzinī —a chapter on his water balance; this is cited by al-Bīrūnī in the study just mentioned); and, surprisingly, Ibn al-ʿAmīd (*d.* 969/970) and Ibn Sīnā (Avicenna; *d.* 1037), neither of whom is known to have worked in this art; next, al-Bīrūnī, then Abu Ḥafṣ [*sic*; usually Abu'l-Fatḥ] ʿUmar al-Khayyāmī; and last, al-Asfizārī, who had died before al-Khāzinī composed *Kitāb mīzān al-ḥikma* and before "reducing all his views on the subject to writing."

It was al-Rāzī who added the indicator tongue to the hydrostatic balance[61] (see Figure 1), and a third scale pan was attached not later than the time of al-Bīrūnī. Al-Asfizārī put on the two movable scale pans and indicated the possibility of cutting specific-gravity marks into the beam. Al-Khāzinī made further refinements, mainly, it seems, in marking the beam for specific gravities of various substances for more than one reference liquid.[62]

The grounds for excluding, as practitioners in this art, Euclid, Pappus, the Banū Mūsā, al-Kindī, Thābit ibn Qurra, Abu Sahl al-Qūhī, and Ibn al-Haytham may be that al-Khāzinī regards them as not actually having worked with a hydrostatic balance. (The only balances considered in books IV and V are the ones of Archimedes, Menelaus, al-Rāzī, ʿUmar al-Khayyāmī, and al-Asfizārī, besides his own.) That can scarcely be said, however, about Abū Manṣūr al-Nayrīzī, whose work on the determination of specific gravities was used by al-Bīrūnī and thus by al-Khāzinī.[63]

No real successors to al-Khāzinī in the art of the balance seem to have arisen in the Islamic world. The *Book of the Balance of Wisdom* is used as a source, however, in several encyclopedias and mineralogical compilations. Fakhr al-Dīn al-Rāzī (*d.* 1209) has long extracts in one of his Persian encyclopedias of the sciences, *Jāmiʿ al-ʿulūm*.[64] In the lapidary (*Azhār al-afkār fi jawāhir al-aḥjār*) by Aḥmad ibn Yūsuf al-Tīfāshī (*d.* 1253) and in the mineralogical section of the *Cosmography* (*ʿAjā'ib al-makhlūqāt*) by Zakariyyā ibn Muḥammad al-Qazwīnī (*d.* 1283) are passages parallel to ones found in both al-Bīrūnī and al-Khāzinī.[65] Many later mineralogical works are heavily indebted to material by al-Bīrūnī; how many of the authors were familiar with al-Khāzinī's supplementary endeavors on the specific gravities of gems is unclear. The later medieval Islamic literature concerning specific gravities, in either the mathematical or the lapidary tradition, is treated by Bauerreiss, J. J. Clément-Mullet, and Wiedemann.[66]

Al-Khāzinī's Archimedean World Picture. *Kitāb mīzān al-ḥikma* has no integrated exposition of the theories of mechanics, but the theorems and excerpts on physical fundamentals that compose books I and II have a very definite cast. Most important for determining al-Khāzinī's theoretical framework is chapter 5 of book I, for he himself seems to have selected the theorems that are repeated there for emphasis. Certain conceptual foundations, however, must be found in chapter 1 of book I. Heaviness (*al-thiqal*), one is told, is the force (*al-quwwa*), an inherent force by which any heavy body is moved toward the center of the world, and in no other direction, without cease, until (and only until—compare section 6) it reaches the center (*I:1.1*). The force of a heavy body varies according to its density (*al-kathāfa*) or rarity (*al-sakhāfa*)(*I:1.2*). Also missing from the recapitulation on chapter 5 are the theorems in *I:1.4–.9* on centers of gravity and the law of the lever, notably the axiomatization in section 5 of the balancing of two heavy bodies (relative to a given point or plane) according to the pattern of book I of Euclid's *Elements*. The whole of *I:1* is discussed in Clagett's commentary;[67] although Clagett does not reedit the text or revise Khanikoff's deficient translation (sections 7 and 8 are particularly troubling), his analysis has not greatly suffered.

Two subjects of great interest emerge from the presentation of book I, chapter 5. The first is al-Khāzinī's idea of "gravity." His conception of heaviness, which is obviously Aristotelian, is here fitted to a picture of the subcelestial world that is purely "hydrostatic" and, in this sense, Archimedean. A heavy body becomes heavier in a rarer medium, lighter in a denser medium; two bodies of different substances but of the same weight in some given medium differ in weight elsewhere, the body of smaller volume being the heavier in a denser, and the lighter in a rarer, medium (*I:5.1*). In section 2, theorems 6 and 7 make explicit that heavy bodies are essentially heavier than they are found to be in air and are heavier in a rarer air, lighter in a denser one.[68]

A general relationship is then stated (*I:5.3*): the weight of any heavy body varies according to its distance from the center of the world, and the relation of weight (*al-thiqal*; meaning precisely the weight as measured in the medium at that distance) to weight is as the relation of distance from the center to distance from the center. A body thus has its maximum and essential weight where there is no interfering medium, and has zero weight at the center. (No concept of "essential lightness" is found; nor is there any question of "mass," although "density" [*al-kathāfa*] is in one way closely related.)

Clagett blames al-Khāzinī for saying that (quoting Clagett) "gravity [i.e., *al-thiqal*] varies directly as distance from the center of the world" after he has said that "gravity depends on the density of the medium in accordance with Archimedes' principle."[69] The assumption behind this criticism is that al-Khāzinī must have the density of the medium vary directly as the distance from the center; yet al-Khāzinī is more likely to hold the opposite opinion—that the density of the medium varies roughly as the distance from the periphery of the world—for the density of the medium is the cause of the reduction in weight of a body weighed in it as that body is brought nearer the center. The weight of the body at a given distance from the center of the world less its weight at another distance must be equal to the weight of an equal volume of the medium at the second distance less the weight of the same volume of the (different) medium at the first distance.

The relationships stated by al-Khāzinī in *I:5.3* can hardly have been intended as continuous and exact ones. But if he be granted any reasonable assumptions—for example, finite densities at the surface of the earth, zero density at the periphery of the cosmos, finite weights at the periphery, zero weight at the center—then in the idealized continuous case (earth shading off into water, water into air, etc.) weight will be directly proportional to distance from the center of the world, although in the form $W = a + br$, where a and b are constants; and the difference in weight of a body at two distances from the center will be a constant multiple of the difference in distance: $\Delta W = b\Delta r$. Thus al-Khāzinī's statement that weight to weight is as distance to distance, although strictly wrong, is not an unreasonable brief presentation of the results of his physical picture of the sublunar cosmos.

In connection with that view of weight, one discovers an important corollary, seen most directly from theorems 3 and 4 of *I:1.9* but implicit throughout chapters 1 and 5. The weight of an object, to summarize the text, has (at least) two possible manifestations: through its inherent force that tends toward the center of the world and acts against the interference of the ambient medium; and through the force with which it acts against the interference of another body when they are turning about a fulcrum. The rule for instances of the second sort is that, when any two bodies balance each other with reference to a determined point, the weight of one to the weight of the other is inversely as the ratio between the two segments of a horizontal line cut off by vertical lines that pass through the centers of gravity of the two bodies and through the pivot point. Or, weight is to weight as distance from the center is to distance from the center (to wit, from the fulcrum, which is also the combined center of gravity of the balancing bodies). For both kinds of "heaviness," then, weight is to weight as distance from the center to distance from the center; the symmetry is absolute. There is no doubt that it is intentional, and no doubt that al-Khāzinī intends his statement of proportionality about gravity (or heaviness) with regard to distance from the center of the world to be taken as strictly as possible.

Hydrodynamical Ideas. The second topic to be considered is the movement of heavy bodies through a liquid. In *I:1.3* their motion is said to be proportional to the fluidity (*al-ruṭūba*) of the medium; further, if two bodies unequal in density (*al-kathāfa*) or rarity (*al-sakhāfa*) but having the same shape and the same volume move (that is, fall) in the same medium, the denser is faster. Then comes an addition to the

AL-KHĀZINĪ

Archimedean analysis: in the case of equal volumes and densities, the body of smaller surface moves faster in a given medium. But the treatment cannot be successfully completed in a symmetric manner: if two bodies of the same density but different volumes move in a given medium, one is now told, the larger moves more quickly (all manuscripts). Glossators, Khanikoff, and Clagett prefer "more slowly"[70]—and indeed a greater volume tends to have a greater surface, and certainly does so if the shape is the same; but the next section, *I:1.4,* states the expected law, that it is the heavier body that moves faster. So the effects of total weight versus those of specific weight need to have been analyzed further. And it is clear that theory has not fully assimilated the effects of shape; nor could it have done so.

Al-Khāzinī takes over into *I:5.2* the non-Archimedean notion that a cause of the differing forces of the motions of bodies, in liquids and in air, is their difference in shape. A liquid medium interferes (*ʿāwaqa*) with the motion of a heavy body through it; it also reduces the body's force and heaviness in proportion (*bi qadr*) to its volume, that is, in proportion to the weight of an equal volume of the liquid medium (Archimedes' Principle). Whenever the moving body is increased in size, the interference (*al-muʿāwaqa*) becomes greater. "Interference" here refers to both the kinetic and the static effects. But the interference as regards weight is known to be due to the density (*al-kathāfa*) of the medium, and the interference in motion to its liquidity (*al-ruṭūba*), inversely—compare the theorems of *I:1.3,* reviewed in the preceding paragraph. So a good start has been made on separating static and kinetic effects and distinguishing viscosity from density, even if it has not been carried through completely and consistently. From this vantage al-Khāzinī's observations farther on, in section 2 of chapter 3, become revealing. He is reporting what happens to the beam of the hydrostatic balance during its actual operation: "Yet," he says, "when a body lies at rest in the water-bowl [the bowl filled with the reference liquid], the beam of the balance rises according to the measure of the volume of the body, not according to its shape"; whereas "the rapidity of the motion of the beam is in proportion to the force of the body [and hence its shape], not to its volume."

Al-Asfizārī on Mechanics. Al-Asfizārī's discussions of mechanical topics, included in book II of *Kitāb mīzān al-ḥikma,* have notable examples and make interesting points, even though they supply no proofs. First to be treated is the problem of several heavy bodies simultaneously seeking the center of the world. Al-Asfizārī mentions in passing that that center must in fact be a natural place and not a geometrical point, and asserts that it is the common center of gravity of those bodies that must come into coincidence with the middle of the cosmos. As heuristics for this idea, al-Asfizārī considers the cases of one and two spheres free to roll in a concave spherical bowl and of one and two spherical bobs freely suspended (for the case of two, from a single point by cords of equal length). In the most difficult instances that he treats, two spheres equal in size but unequal in weight are employed (this must be the intention of the text, which, however, is inexplicit); in both such situations a particular vertical line (the one passing through the bowl perpendicularly to the plane tangent at its lowest point, in the example with the bowl, through the point of suspension in the other) cuts the line joining the centers of gravity of the two spheres at a point such that the two line segments thus formed are in inverse ratio to the weights of the spheres.[71]

The second chapter by al-Asfizārī investigates the conditions for equilibrium of a balance. He distinguishes between the constrained motion occurring around the point of suspension of the beam and the natural motion of falling. To achieve equilibrium, *two independent causes of motion* of the balance must be made proportionate—the distance of the weights from the center of suspension, which al-Asfizārī prefers to treat in terms of the arcs described, and the natural heaviness of the weights, their tendency toward the center of the world. However, the case of a physical balance-beam (one that has weight) suspended from a point away from its center and kept even by unequal weights hung from its ends is then handled according to the method of Euclid in "On the Balance," a procedure related closely to the one used in Thābit ibn Qurra's *Risāla fī'l-qarasṭūn.*[72] In that proof Peripatetic concepts are ignored, and the analysis diverges widely from Archimedes' as well.

The results thus derived for physical levers are applied to the problem of the forces acting on the hand of someone holding up a spear, well back along the shaft, in a horizontal position. Al-Asfizārī recognizes two components, the (natural) weight of the spear and the unbalanced weight (of the other sort, the kind that produces forced motion about a center) of its front portion. His explanation (*bayān,* not *burhān* ["demonstrative proof"], as throughout these chapters by al-Asfizārī) is well advanced, although wrong—he does not distinguish moments from forces; it was too advanced for the scribes, who have muddled the text.[73]

Specific Gravities. Much of the scholarly attention that has been paid to *Kitāb mīzān al-ḥikma* has been stimulated by its tables of specific gravities. The literature has concerned itself especially with their accuracy and with the relationship between al-Bīrūnī's investigations and al-Khāzinī's.[74]

Precautions are taken by al-Bīrūnī and al-Khāzinī to assure the purity of the substances whose water equivalents are being measured, and the difficulties of entrapped air, especially in cavities in gems, are dealt with. But knowledge of chemical identities and physical states (for instance, of alloys) is quite rudimentary.[75] Temperature effects, at least on the reference liquid, are reasonably well known.[76] (The change of density with temperature is explicitly recognized, but apparently not a change of volume; the discussions consequently deserve study from the standpoint of theories of matter.)[77] The need for standardization is accepted, although nothing effective could be done—the lack of a sufficient institutional basis for science prevented it. The specific gravities that are recorded are rarely correct to within 1 percent in cases where precision is possible; whatever their source, these uncertainties are two orders of magnitude larger than is necessitated by the balance itself.

As regards metals, the values for mercury, lead, and tin are excellent (0.06 to 0.3 percent off); for gold and iron, reasonably correct (about 1 percent); brass and bronze are surprisingly well done, and the value for copper is right, although there, as for other elemental metals as well as alloys, the description of the physicochemical state of the substance is insufficient. The same is true for glass, and more acutely so for the special earths that are tried. Nor are measurements on precious stones successful, except for emeralds, where the result falls in the middle of the actual range (specific gravity 2.68–2.78). The figure for salt is 1.5 percent high. The specific gravity of blood of a healthy man is about 2 percent low. Water at the boil is exact, as far as can be told; but ice is 5 percent high and saturated salt water, 6 percent low. The gravimetric precision attained is more than sufficient, however, to discriminate reliably among hot and cold water, hot and cold human urine, and fresh water, sea water, and saturated salt water.

The Balance of Wisdom. In the Islamic world al-Khāzinī's treatise was particularly valued for its descriptions of the instruments themselves—the araeometer of Pappus, al-Bīrūnī's pycnometer flask, the earlier hydrostatic balances of book IV, the specialized balances and steelyards of books VII and VIII, and al-Khāzinī's own balance. His instrument must now be explained. (In the account that follows the capital letters refer to Figure 1.)[78]

The "balance of wisdom" is a hydrostatic balance of standard form with five scale pans, a rather complicated polyfilar suspension, and a sensitive indicator tongue. The overall length of the beam is four cubits (about two meters or six feet), the length of the tongue, about fifty centimeters. The extraordinary limit of precision of this balance arises from its long beam, very accurate construction, and nearly frictionless suspension, of which the center of gravity and axis of oscillation turn out to be very close together. The double suspension increases stability but is such also as to magnify the motion of the tongue (*D*), which is much more carefully designed than the illustration suggests. Al-Khāzinī claims, and Wiedemann and Bauerreiss leave unchallenged, a sensitivity of one part in 60,000 (or up to 1 in 100,000, depending on the value of the *ḥabba*[79]) for a weight of about 4.5 kilograms. This would mean a noticeable deflection for a change in weight of about forty-five to seventy-five milligrams.[80] Khanikoff and Wiedemann point out that specific gravities measured with the "balance of wisdom" are as precise as those obtained up to the eighteenth or beginning of the nineteenth century;[81] in fact the "balance of wisdom" had the precision of an analytical balance, although it required quite large samples. It was too far in advance of chemical knowledge to be of service to chemistry, however, and belonged to researchers in a largely separate tradition.

The beam, *A*, is six centimeters thick, strengthened at the middle by the brace *C*. Crossbar *B* is set in through *C*; corresponding to it are the two lower crosspieces, *F*, of the fork. The fork is hung by its upper crosspiece on rings encircling a rod, the rod itself being anchored in any convenient manner. The beam is suspended from many parallel threads running between exactly opposite points on the crosspieces *B* and *F*. The knob below the center of the beam fastens and adjusts the tongue, a peg in the base of the tongue extending through crosspiece *B* and the beam. On both halves the beam is graduated with marks cut into its top; into the marks fit the points of the precisely made rings of steel from which the scale pans hang. Up to five pans, all of equal weight, are employed for measuring specific gravities, analyzing binary alloys, or detecting fraudulent gems. Pan *H*, whose bottom is drawn out into a point to facilitate its sinking into a liquid, is called the "cone-shaped" or "the judge" (*al-ḥākim*), for it is there that a suspected object is placed. Pan *J*, called "the winged" (*al-mujannaḥ*), has deeply indented sides so that it can be brought close to other pans. *K* is a running weight (*rummāna sayyāra*) used, when necessary, to adjust the leading of that end of the beam. In many kinds of measurements, after the balance has been brought into equilibrium the desired magnitudes can be read

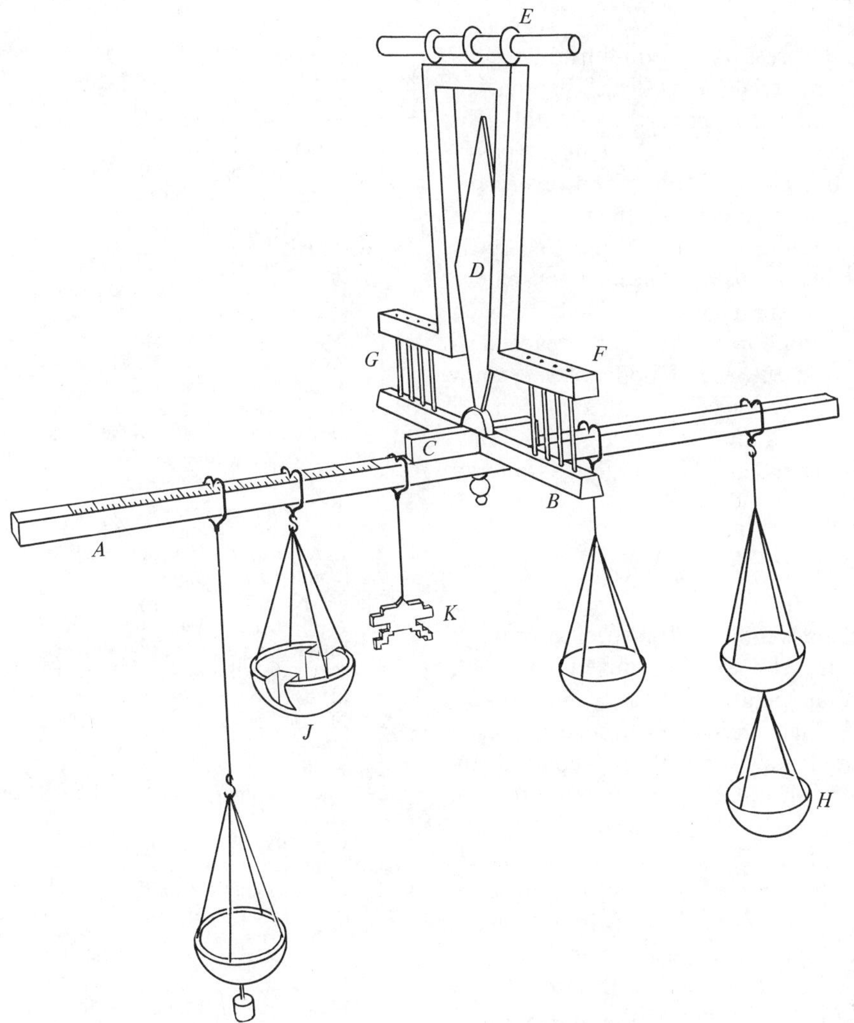

FIGURE 1. "Balance of Wisdom" (or "Comprehensive Balance") of ʿAbd al-Raḥmān al-Khāzinī. Marks for waterweights of several substances with respect to one or more reference liquids are cut into the top of the beam. The diagram follows a photograph of a 1 : 4 scale model constructed in Erlangen about 1913 under the supervision of H. Bauerreiss.

directly from the divisions on the beam (hence al-Khāzinī's talk of locating the marks for given substances and for given liquid media).

The problem of suspending a balance beam is treated in detail (*V:1.4*). Al-Khāzinī's table showing the stability or instability of the instrument for different positions of the axis of suspension is reproduced in translation in Khanikoff and in Ibel.[82]

In determining the composition of alloys or testing the genuineness of precious stones, the great advantage of the "balance of wisdom," apart from its sensitivity, is the directness of the measurement and the consequent avoidance of calculations with dubious parameters. Thus the most immediate way of testing a suspected gem or metal object or measuring the constituents of a binary alloy—a method possible only with an instrument as advanced as the "balance of wisdom"—is the procedure set forth by al-Khāzinī in *VI:4.1–.2*.[83]

The substance is first weighed, the sample being placed in the air pan at the left end of the beam and the *mithqāl*'s (the weights) in the air pan at the right end. The movable pans are both hung from the right half of the beam, at distances corresponding to the two assumed components of the sample. (The distance, d', for any such constituent, measured from the center of suspension, is given by the formula $d'/l = W'_w/W'$, where l is the distance from the center of suspension to the point where the air pan at the left end of the beam is hung, W'_w is the weight in water of an arbitrary volume of the given substance, and W' is the weight in air of the same volume of that substance.)

The sample is then placed in the filled water pan, sufficient care being taken to assure that water (or, more generally, the reference liquid) reaches all its parts. The *mithqāl*'s from the air pan at the right-hand end are transferred to one of the movable air pans, then to the other, if necessary; if equilibrium results in either case, the material being tested is that to which the given movable pan corresponds—for instance, the pure gem or its likeness in colored paste. For an alloy or other mixed body, however, the *mithqāl*'s must be distributed between the two movable pans to create equilibrium (with sand or sifted seeds substituted for the last *mithqāl*, if needed to achieve an exact balance). The weight contained in each movable pan when equilibrium has been reached is equal to the weight in the mixed body of the component corresponding to the position of that pan.

Khanikoff provides a formula by which that procedure may be expressed (the notation is altered here for clarity):[84]

$$W'' = W \frac{1/s' - 1/s}{1/s' - 1/s''},$$

where W denotes air weight and s specific gravity; unprimed quantities refer to the mixed material, primes to one component, seconds to the other. Specific gravities, in fact, are neither needed nor used in the method, and they should be replaced by water weights and air weights according to the identity $1 - 1/s = W_w/W$ (for unprimed, primed, and seconded quantities). The right-hand ratio is, of course, the one involved in the placing of the movable pans. This equation for the weights of the two components is correct upon the assumption that the volume of the mixed body is equal to the sum of the volumes of its constituents, a relation that holds for all mechanical mixtures and is closely true for many alloys.

NOTES

1. For most of the scant evidence see E. S. Kennedy, "A Survey of Islamic Astronomical Tables," pp. 159–160, secs. 12C and 12J; and C. A. Nallino, *Al-Battānī*, pt. 1, lxvii.
2. Listed by G. Gabrieli, in *Rendiconti dell'Accademia nazionale dei Lincei*, Classe di Scienze Morale, Storiche e filologiche, ser. 5, **22** (1913), 596–620.
3. For al-Khāzinī in this connection, see Aydin Sayili, *The Observatory in Islam*, p. 178; M. Mīnawī, "Conjonction des planètes . . ." pp. 28–31, 50–53; compare Thomas Ibel, *Die Wage im Altertum und Mittelalter*, pp. 75–76.
4. Sayili, *Observatory* . . ., p. 177.
5. *Ibid.*, pp. 160–166.
6. *Ibid.*, pp. 165, 177; see also Kennedy, "Parallax Theory in Islamic Astronomy," in *Isis*, **47** (1956), 46, and *idem*, "Survey . . .," p. 159.
7. On Anwarī: Sayili, *Observatory* . . ., p. 178; Mīnawī, "Conjonction . . .," 24–26, 28–31, 38–41, 50–53; and E. G. Browne, *A Literary History of Persia*, II (London, 1906, and many reprs.), pp. 365–371 in repr. of 1964.
8. The manuscript is discussed in Sayili, "Al-Khâzini's Treatise."
9. Kennedy, "Survey . . .," p. 169.
10. *Ibid.*, pp. 172–173.
11. See Hermann Usener, *Ad historiam astronomiae symbola*, p. 15.
12. Kennedy, "Survey . . .," p. 129.
13. Al-Qazwīnī, LeStrange trans., II, 27; *cf.* Kennedy, "Survey . . .," p. 160.
14. Contained in Istanbul, MS Hamidiye 859; see Max Krause, "Stambuler Handschriften islamischer Mathematiker," p. 487.
15. Sayili, *Observatory* . . ., p. 177—misinterpreted as the date for the positions recorded in the star tables; compare Heinrich Suter, *Die Mathematiker und Astronomen* . . ., p. 122.
16. Nallino, *Al-Battānī*, pt. 1, lxvii.
17. Kennedy, "Survey . . .," pp. 160–161.
18. Marcel Destombes, "L'Orient et les catalogues d'étoiles au Moyen Âge," p. 343.
19. Destombes, in his review of Kennedy's "Survey . . .", p. 272; Kennedy, "Survey . . .," p. 129.
20. Sayili, *Observatory* . . ., p. 178; he follows Mīnawī, "Conjonction . . .," p. 29.
21. The mention appears on fol. 121v. of the Vat. MS.
22. In *Nihāyat al-idrāk*; see Sayili, *Observatory* . . ., pp. 177–178.
23. Sayili, *Observatory* . . ., p. 177.
24. See Kennedy, "Survey . . .," pp. 133–134. 160.
25. Nallino, *Raccolta di scritti editi e inediti*, V, pt. 2, 227–228, following Vat. MS, fol. 49r.
26. Kennedy, "Survey . . .," pp. 159–161; Nallino, *Al-Battānī*, pt. 1, 269–271, 279–282.
27. Nallino, *Al-Battānī*, pt. 1, 159–161, following Vat. MS, fol. 10v.
28. Nallino, *Al-Battānī*, pt. 2, 199.
29. The summary now follows Kennedy, "Survey . . .," pp. 159–161.
30. Kennedy, "Survey . . .," pp. 132 (no. 54), 145.
31. Reported in Sayili, "Al-Khâzinî's Treatise . . ."
32. Ibn al-Akfānī, *Irshād al-qāṣid* (Beirut, A.H. 1322; author's name given in the form ". . . Muḥammad ibn Ibrāhīm . . . al-Anṣārī al-Sanjārī"), p. 119; Ṭāshköprüzāde, *Mevzūʿāt ül-Ulūm*, Turkish trans. by the author's son, Kemālüddīn Mehmed, 2 vols. (Istanbul [?], A.H. 1313), I, 413; and Ḥājjī Khalīfa, Flügel ed., in notices no. 1122 (I, 394–398), p. 394, and no. 9887 (V, 48–49).
33. Carl Brockelmann, *Geschichte der arabischen Litteratur, Supplementband* I, 902; Eilhard Wiedemann, "Beiträge zur Geschichte der Naturwissenschaften. IX," in *Sitzungsberichte der Physikalisch-medizinischen Sozietät zu Erlangen*, **38** (1906), 190, n. 3 (I, 267); "Beiträge . . . LVII," *ibid.*, **50–51** (1918–1919), 26 (II, 456); "al-Khāzin, Abu Djaʿfar . . .," in *Encyclopaedia of Islam* II, 937; "al-Khāzinī . . .," *ibid.*, pp. 937–938.
34. Ibn Khaldūn, *Prolégomènes historiques d'Ibn Khaldoun*, trans. W. M. de Slane, 3 vols. [*Notices et extraits des manuscrits de la Bibliothèque impériale, XIX–XXI*], (Paris, 1862–1868), I, 111, n. 1.
35. Khanikoff ed., p. 16.
36. See the publisher's postface, p. 169.
37. The MS is discussed in the foreword, p. 4.
38. MSS of *Kitāb mīzān al-ḥikma* are the following: Khanikoff 117, Leningrad, Gosudarstvennaya Publikhnaya Biblioteka, listed in the catalog by B. Dorn (St. Petersburg, 1865); Bombay, Jāmeʿ Masjid (no number [?]); Hyderabad, Deccan, Āṣafīya Mosque, cat. I, 125 (Hyderabad, A.H. 1333 [1914–1915])—copied from the preceding (?). On the Indian MSS see al-Nadwī, *Tadhkirat al-nawādir*, nos. 267, 282. The East Jerusalem MS seems to remain uncataloged.
39. In Wiedemann's "Beiträge . . ." and elsewhere; the trans. in Ibel are also by Wiedemann. See under "Original Works" in the bibliography, below.
40. Khanikoff ed., p. 14.
41. al-Bayhaqī, notice no. 68.
42. Khanikoff ed., pp. 14–15.
43. See Khanikoff ed., pp. 5, 8.

44. Hyderabad ed., p. 2, for the variant readings.
45. True "Oriental hyperbole" in honor of Sanjar is found further along, in Khanikoff ed., pp. 15–16.
46. The preceding parts of the introduction are retranslated from Khanikoff ed., pp. 3–8.
47. *Ed. cit.*, p. 10.
48. *Ed. cit.*, pp. 8–10, 15.
49. *Ed. cit.*, pp. 11–12.
50. *Ed. cit.*, pp. 12–16; compare bk. IV and bk. V, ch. 1.
51. *Ed. cit.*, p. 15.
52. *Ed. cit.*, p. 10.
53. Discussed by Wiedemann, in "Beiträge . . . VII."
54. Analysis in Heinrich Bauerreiss, *Zur Geschichte des spezifischen Gewichtes im Altertum und Mittelalter*, pp. 99–102.
55. Presumably as excerpts from the *Risāla fi'l-qarasṭūn*. See Wiedemann, "Beiträge . . . XVI," 136 (I, 495).
56. *The Book of Instruction in the Elements of the Art of Astrology, by . . . al-Bīrūnī*, trans. R. Ramsay Wright (London, 1934), pp. 11–16. See Wiedemann, "Beiträge . . . XLVIII," 1–6 (II, 215–220).
57. Wiedemann, "al-Mīzān," pp. 537–539, deals with leveling, etc.
58. In the discussion of the "balance of wisdom," *V:1.4*; see Ibel, *Die Wage . . .*, p. 123.
59. Wiedemann, "al-Mīzān," p. 534; and "Beiträge . . . XV," p. 119 (I, 478).
60. These works are noted by Wiedemann, "Beiträge . . . XLVIII," p. 1 (II, 215), n. 1.
61. Khanikoff ed., p. 86.
62. For al-Khāzinī's list of his forerunners and their modifications of the water balance see Khanikoff ed., pp. 12–15; compare Ibel, *Die Wage . . .*, pp. 77–80; and Wiedemann, "al-Mīzān," pp. 531, 534. Wiedemann, "Beiträge . . . VI, pt. 1, Ueber arabische Literatur über Mechanik," in *Sitzungsberichte . . . Erlangen*, **38** (1906), 2–16 [I, 174–188], outlines the history of Islamic mechanics of the Archimedean and Heronian types.
63. On al-Nayrīzī, otherwise unknown and not the Euclid commentator, but whose writings on the specific gravities of mixed bodies have survived, see in Wiedemann, "Beiträge . . . VIII, Ueber Bestimmung der spezifischen Gewichte," in *Sitzungsberichte . . . Erlangen*, **38** (1906), 163–180 (I, 240–257), pp. 166–170 (I, 243–247); the treatises on specific gravities by ʿUmar al-Khayyāmī and pseudo-Plato are translated in the remainder of the article.
64. Discussed in Wiedemann, "Ueber die Kenntnisse der Muslime . . .."
65. Passages and commentary in Julius Ruska, "Ḳazwīnīstudien."
66. Bauerreiss, *Zur Geschichte . . .*, pp. 44–46; J. J. Clément-Mullet, "Pesanteur spécifique de diverses substances minérales, procédé pour l'obtenir d'après Abou'l-Raihan Albirouny. Extrait de l'*Ayin Akbery*," in *Journal asiatique*, 5th ser., **11**(1858), 379–406; and Wiedemann: "Beiträge . . . VIII" (see note 63, above); "Beiträge . . . XXX, Zur Mineralogie im Islam," in *Sitzungsberichte . . . Erlangen*, **44** (1912), 205–256 (I, 829–880); "Beiträge . . . XXXI, Ueber die Verbreitung der Bestimmungen des spezifischen Gewichtes nach Bīrūnī," in *Sitzungsberichte . . . Erlangen*, **45** (1913), 31–34 (II, 1–4); "Beiträge . . . XXXIV"; and "al-Mīzān," p. 534.
67. *The Science of Mechanics in the Middle Ages*, pp. 58–61.
68. Compare *VI:2.5*, Khanikoff ed., pp. 98–99.
69. Clagett, *The Science of Mechanics*, p. 68, n. 42.
70. Khanikoff ed., p. 28; Clagett, *The Science of Mechanics*, p. 58; Hyderabad ed., p. 17, for readings of the MSS.
71. II: part 2, ch. 1, also numbered as *II:2*; trans. in Wiedemann, "Beiträge . . . XVI," pp. 141–144 (I, 500–503).
72. For Euclid, see Clagett, *The Science of Mechanics*, pp. 24–30, and Ibel, *Die Wage . . .*, pp. 32–36; compare E. A. Moody and M. Clagett eds. and trans., *The Medieval Science of Weights* (Madison, Wis., 1952; repr. 1960), pp. 57–75, on Thābit.
73. II, part 2, chapter 2, also numbered as *II:3;* Wiedemann, "Beiträge . . . XVI," pp. 144–150 (I, 503–509).
74. See Clément-Mullet, "Pesanteur spécifique . . ."; Khanikoff ed., pp. 55–58, 65–78, 83–85; H. C. Bolton, " 'The Book of the Balance of Wisdom.' An Essay on Determination of Specific Gravity"; Wiedemann, "Arabische spezifische Gewichtsbestimmungen"; and Bauerreiss, *Zur Geschichte . . .*, pp. 28–33, 41–44.
75. On purification techniques see, for example, Khanikoff ed., pp. 55–56.
76. *VI:2.5*; Khanikoff ed., pp. 98–99; and Ibel, *Die Wage . . .*, pp. 140–142. The remarks on thermometry by Khanikoff, *loc. cit.* and p. 106, cannot be accepted; determination of temperature was purely by sense—compare Ibel, *loc. cit.*
77. Compare Khanikoff ed., pp. 83, 98–99.
78. Other presentations are in Wiedemann, "al-Mīzān," pp. 532–533; Bauerreiss, *Zur Geschichte . . .*, pp. 50–58; and Khanikoff ed., pp. 87–98. Al-Khāzinī's own account occurs in *V:1–V:2*.
79. On the possible metric equivalents, see Walther Hinz, *Islamische Masse und Gewichte . . .*, pp. 12–13.
80. For the sensitivity, Khanikoff ed., p. 8; Wiedemann, "al-Mīzān," p. 533; and Bauerreiss, *Geschichte . . .*, p. 54. Bauerreiss had tried out two working models of the scales.
81. Khanikoff ed., p. 85; Wiedemann, "al-Mīzān," p. 535.
82. Khanikoff ed., p. 94; Ibel, *Die Wage . . .*, p. 122.
83. Khanikoff ed., pp. 100–104; Ibel, *Die Wage . . .*, pp. 145–148.
84. Khanikoff ed., p. 104; quoted in Clagett, *The Science of Mechanics*, p. 65.

BIBLIOGRAPHY

I. ORIGINAL WORKS. Of al-Khāzinī's works only the *Kitāb mīzān al-ḥikma* has been printed: *Kitāb mīzān al-ḥikma* (Hyderabad, Deccan, A.H. 1359 [A.D. 1940–1941]); and *Mīzān al-ḥikma*, Fu'ād Jamīʿān, ed. (Cairo, [1947]) which is incomplete and gives the author's name in the form "al-Khāzin." N. Khanikoff, "Analysis and Extracts of *Kitāb mīzān al-ḥikma*, 'Book of the Balance of Wisdom,' an Arabic work on the Water-Balance, written by al-Khāzinī in the Twelfth Century," in *Journal of the American Oriental Society*, **6** (1859), 1–128, contains selected texts with English trans. and notes. Eilhard Wiedemann provides German trans., with notes, of other portions of *Kitāb mīzān al-ḥikma* in "Beiträge . . ." nos. VII, XIV, XV, XVI, XXXVII, and XLVIII, and in "Über den Wert von Edelsteinen . . ." (see below). Thomas Ibel, *Die Wage im Altertum . . .* (see below), contains German trans., with notes, of additional parts of *Kitāb mīzān al-ḥikma*, on pp. 37–39, 80–83, 84–88, 107–110, 112–151, 153–156, 158–160, and 185–186. The trans. by Khanikoff and by Ibel, and those by Wiedemann, except the ones in "Beiträge . . ." XXXVII and XLVIII and in "Über den Wert . . .," are tabulated by Wiedemann in "Beiträge . . . XVI," pp. 158–159, according to the folio numbers of the Khanikoff MS; see also the complete listing by book and chapter in the text of this article.

II. SECONDARY LITERATURE. *Biobibliographical sources* (modern): Carl Brockelmann, *Geschichte der arabischen Litteratur*, 2 vols. and 3 supp. vols. (Leiden, 1937–1949), *Supplementband I*, 902; Max Krause, "Stambuler Handschriften islamischer Mathematiker," in *Quellen und Stu-*

dien zur Geschichte der Mathematik, Astronomie und Physik, sec. B, *Studien*, **3** (1936), 437–532—see p. 487 [no. 293]; George Sarton, *Introduction to the History of Science*, 3 vols. in 5 pts. (Baltimore, 1927–1948), II, pt. 1, 216–217; and Heinrich Suter, *Die Mathematiker und Astronomen der Araber und ihre Werke* [*Abhandlungen zur Geschichte der mathematischen Wissenschaften*, X] (Leipzig, 1900), 122 (no. 293) and 226. See also Wiedemann's "al-Khazini"; the notice in Ibel, *Die Wage im Altertum . . .*, at pp. 73–80; and Sayili, *The Observatory in Islam* (*loc. cit.*), all listed below.

Biobibliographical sources (medieval; listed chronologically): al-Bayhaqī, *Tatimma ṣiwān al-ḥikma of 'Alī b. Zaid al-Baihaḳi*, Moḥammad Shafī', ed., Fasciculus I, Arabic Text (Lahore, 1935), 161–162 (no. 103)—English trans. of Bayhaqī's notice on al-Khāzinī in Max Meyerhof, "'Alī al-Bayhaqī's 'Tatimmat ṣiwān al-ḥikma': A Biographical Work on Learned Men of the Islam," in *Osiris*, **8** (1948), 122–127, at pp. 196–197 (written in 1939; the trans. follows the Shafī' ed.); also a German translation by Wiedemann in "Beiträge . . . XX" (see below). Shams al-Dīn Muḥammad b. Maḥmūd al-Shahrazūrī, *Kitāb Kanz al-Ḥikma*, Persian trans. by Diyā' al-Dīn Durrī of *Nuzhat al-arwāḥ wa rawḍat al-afrāḥ fi tawārīkh al-ḥukamā'*, 2 vols. (Teheran, A.H. (solar) 1316 [A.D. 1937–1938]), II, 66; Ḥamdullāh Mustawfī Aḥmad b. abi Bakr al-Qazwīnī, *The Geographical Part of the 'Nuzhat al-Qulūb,' composed by Ḥamdallah Mustawfi of Qazwin in 740 (1340)*, Persian text and English trans. by G. LeStrange, 2 vols. (Leiden, 1915, 1919)—see vol. I (Persian text), pp. 22–26 and table opposite p. 26, and II (English trans.), 24–31; Ḥājjī Khalīfa, [*Kashf al-ẓunūn.*] *Lexicon bibliographicum et encyclopaedicum a . . . Haji Khalfa . . . compositum*, ed. with Latin trans. by Gustav Flügel, 7 vols. (Leipzig, London, 1835–1858), III, 564 (no. 6945).

Studies: Heinrich Bauerreiss, *Zur Geschichte des spezifischen Gewichtes im Altertum and Mittelalter*, Inaug.-Diss. Univ. Erlangen (Erlangen, 1914)—much of the work, which is topically arranged and lacks an index, is devoted to al-Khāzinī's *Kitāb mīzān al-ḥikma;* see esp. pp. 50–58, 99–102. H. Carrington Bolton, " 'The Book of the Balance of Wisdom.' An Essay on Determination of Specific Gravity," in *The American Chemist* (May, 1876), 20 pp. in the offprint; Marshall Clagett, *The Science of Mechanics in the Middle Ages* (Madison, Wis., 1961), 56–68 and *passim* (see index)—short passages of *Kitāb mīzān al-ḥikma*, taken from Khanikoff, are provided with a commentary; Thomas Ibel, *Die Wage im Altertum und Mittelalter*, Inaug. Diss. Univ. Erlangen (Erlangen, 1908) (rev. from first pub. version, *Die Wage bei den Alten. Programm des königlichen Luitpoldprogymnasiums Forchheim, 1906*), pp. 73–80 and *passim*—see also the list of trans. from *Kitāb mīzān al-ḥikma*, above. E. S. Kennedy, "A Survey of Islamic Astronomical Tables," in *Transactions of the American Philosophical Society*, n.s. **46**, pt. 2 (1956), 121–177—on al-Khāzinī's *zīj* (no. 27), see pp. 129 and 159-161, and see also pp. 169, 172–173; Kennedy's study is reviewed by Marcel Destombes in *Isis*, **50** (1959), 272–273—also consult Destombes's "L'Orient et les catalogues d'étoiles au Moyen Âge," in *Archives internationales d'histoire des sciences*, **35** (1956), 339–344. M. Mīnawī, "Conjonction des planètes en 582 de l'Hégire," in *Revue de la Faculté des lettres de l'Université de Téhéran*, **2**, no. 4 (1955), 16–53 [in Persian]. Carlo Alfonso Nallino, ed. and trans., *Al-Battānī sive Albattenii opus astronomicum*, 3 vols. [*Pubblicazioni del Reale Osservatorio di Brera in Milano*, no. 40, pts. 1–3] (Milan, 1899–1907)—pt. 1, pp. lxvii, 269–271, 279–282, and see *s.v.* "Khāzinī," in index, pt. 2, p. 392; *idem*, *Raccolta di scritti editi e inediti*, V: *Astrologia, astronomia, geografia* (Rome, 1944)—see pt. 2 (pp. 88–329), "Storia dell'astronomia presso gli Arabi nel Medio Evo," trans. by Maria Nallino, pp. 227–228 [original ed., *'Ilm al-falak: ta'rīkhuhu 'inda'l-'arab fī'l-qurūn al-wusṭā* (Rome, 1911–1912), p. 179]. Julius Ruska, "Ḳazwīnīstudien," in *Der Islam*, **4** (1913), 14–66, 236–262—see pp. 247–252, on precious stones (parallel passages, etc., in Bīrūnī, Khāzinī, Qazwīnī, and Tīfāshī), and 254–257 on mathematical problems involving large numbers (parallels, etc., in Bīrūnī, Khāzinī, and Qazwīnī); Aydin Sayili, "Al-Khâzinî's Treatise on Astronomical Instruments," in *Ankara üniversitesi Dil ve tarin-coğrafya fakültesi dergisi*, **14**, nos. 1–2 (1956), 18–19—compare the longer Turkish version, "Khâzinî'nin rasat aletleri üzerindeki risalesi," *ibid.*, 15–17; *idem*, *The Observatory in Islam*, [*Publications of the Turkish Historical Society*, 7th ser., no. 38] (Ankara, 1960), pp. 177–178, also 160–166; and Hermann Usener, *Ad historiam astronomiae symbola. Programm der Universität Bonn, 1876*—see p. 15.

Much of the literature relating to al-Khāzinī is by Eilhard Wiedemann: "Arabische spezifische Gewichtsbestimmungen," in *Annalen der Physik*, n.s. **20** (1883), 539–541; "Inhalt eines Gefässes in verschiedenen Abständen vom Erdmittelpunkte nach al Khâzinî und Roger Baco," *ibid.*, n.s. **39** (1890), 319; cf. Khanikoff, ed., p. 38; "Über das al Bêrûnîsche Gefäss zur spezifischen Gewichtsbestimmung," in *Verhandlungen der Deutschen Physikalischen Gesellschaft*, **10** (1908), 339–343—a comparison of al-Bīrūnī's and al-Khāzinī's descriptions; "Über die Kenntnisse der Muslime auf dem Gebiete der Mechanik und Hydrostatik," in *Archiv für die Geschichte der Naturwissenschaften und der Technik*, **2** (1909–1910), 394–398, on the use of al-Khāzinī's work by Fakhr al-Dīn al-Rāzī; "Über den Wert von Edelsteinen bei den Muslimen," in *Der Islam*, **2** (1911), 345–358—pp. 347–353 have passages from al-Khāzinī, *Kitāb mīzān al-ḥikma*, that derive from al-Bīrūnī; and the following articles from *Encyclopaedia of Islam*, M. T. Houtsma *et al.*, eds., 4 vols. and Supp. (Leiden–London, 1913–1938): "al-Ḳarasṭūn," II, 757–760; "al-Khāzinī," II, 937–938; and "al-Mīzān," III, 530–539.

Also by Eilhard Wiedemann are the "Beiträge zur Geschichte der Naturwissenschaften. I–LXXIX," in *Sitzungsberichte der Physikalisch-medizinischen Sozietät zu Erlangen*, **34** (1902)–**60** (1928), repr. (with 3 additional articles by Wiedemann, the bibliography of Wiedemann's works by H. J. Seemann, and indices of terms and personal names by W. Fischer) as *Aufsätze zur arabischen Wissenschaftsgeschichte*, 2 vols. (Hildesheim, 1970). The following parts of the "Beiträge" are relevant for al-Khāzinī (in the reprint, see also the index of personal names, *s.v.* "Khāzinī"; vol. and page refs. given in parentheses refer to the reprint): *VII*, "Über arabische Auszüge aus der Schrift des Archimedes über die schwimmenden Körper," **38** (1906), 152–162 (I, 229–239), with additions in XIV (see below), pp. 60–62 (I, 459–461); *XIV*, **40** (1908), 1–64 (I, 400–463), sec. 4.2: "Über das Schachspiel und dabei vorkommende Zahlenprobleme: Stellen aus dem Werk 'Die Wage der Weisheit,' " pp. 45–54 (I, 444–453); *XV*, "Über die Bestimmung der Zusammensetzung von Legierungen," *ibid.*, 105–132 (I, 464–491); *XVI*, "Über die Lehre vom Schwimmen, die Hebelgesetze und die Konstruktion des *Qarasṭûn*," *ibid.*, 133–159 (I, 492–518)—XV and XVI are trans., with discussion, from the *Kitāb mīzān al-ḥikma; XX*, "Einige Biographien nach al Baihaqî," **42** (1910), 59–77 (I, 641–659)—the biography of al-Khāzinī (no. 103) is on pp. 73–74 (I, 655–656); *XXXIV*, "Über die Gewichte der Kubikelle u.s.w. verschiedener Substanzen nach arabischen Schriftstellern," **45** (1913), 168–173 (II, 39–44)—comparison of the values of al-Bīrūnī and al-Khāzinī with those from a later, anonymous compilation; *XXXVII*, "Über die Stundenwage," **46** (1914), 27–38 (II, 57–68)—trans. from the *Kitāb mīzān al-ḥikma*, with which compare "Beiträge . . . X," **38** (1906), 349 (I, 314); and *XLVIII*, "Über die Wage des Wechselns von al Châzinî und über die Lehre von den Proportionen nach al Bîrûnî," **48/49** (1916–1917), 1–15 (II, 215–229).

On Islamic metrology, see Walther Hinz, *Islamische Masse und Gewichte umgerechnet ins metrische System*, [*Handbuch der Orientalistik, Abt. I: Der Nahe und der mittlere Osten, Ergänzungsband I, Heft 1*] (repr. with additions and corrections, Leiden, 1970).

ROBERT E. HALL

KHINCHIN, ALEKSANDR YAKOVLEVICH (*b.* Kondrovo, Kaluzhskaya guberniya, Russia, 19 July 1894; *d.* Moscow, U.S.S.R., 18 November 1959), *mathematics.*

The son of an engineer, Khinchin graduated from a technical high school in Moscow in 1911 and, from 1911 until 1916, studied at the Faculty of Physics and Mathematics of Moscow University. In 1916 he was retained by the university to prepare for professorship. From 1918 Khinchin taught at various colleges in Moscow and Ivanovo; in 1927 he became a professor at Moscow University. He was elected an associate member of the Soviet Academy of Sciences in 1939 and a member of the Academy of Pedagogical Sciences of the R.S.F.S.R. in 1944. He received the State Prize in 1940 for his scientific achievements. With A. N. Kolmogorov, Khinchin was one of the founders of the Moscow school of probability theory, one of the most influential in the twentieth century.

Khinchin's interest in mathematics was awakened in high school. Other strong interests of his youth were poetry and the theater. At the university Khinchin became an active member of the group of gifted young mathematicians guided by N. N. Luzin, the passionate propagandist of the modern theory of functions. In this group Khinchin began to work on the metric theory of functions. His first paper (1916), on a generalization of the Denjoy integral, began a series of works dealing with the properties of functions which remain after the removal of a set of density 0 at a given point (asymptotic derivative, asymptotic monotonicity).

After 1922 Khinchin turned to the theory of numbers and to probability theory. First he studied metric problems of the theory of Diophantine approximations and of the theory of continuous fractions. These problems, which deal with properties true for almost all real numbers, are naturally connected with the asymptotic properties of functions mentioned above. Later Khinchin studied classical Diophantine approximations, which hold true for all numbers; in particular he established the so-called principle of transposition. Another topic of the theory of numbers was studied in his works on the density of sequences.

In 1923 Khinchin established the so-called law of the iterated logarithm, strengthening the results obtained by G. H. Hardy and John Littlewood on the frequency of zeros in the binary expansion of real numbers. In the probabilistic interpretation this law improves the strengthened law of large numbers established by Borel. Probability theory proved to be an auspicious field for the application of the methods of the metric theory of functions, and Khinchin was drawn more and more into the problems of the summation of independent random variables. During the 1920's and 1930's this classical branch of probability theory assumed its present form in the closely related works of Kolmogorov, P. Lévy, Khinchin, and others. Khinchin's contribution included results on the applicability of the law of large numbers to equally distributed random variables with finite mathematical expectations, on the coincidence of the class of all limit distributions with the class of all infinitely divisible laws, on the convergence of series of random variables (jointly with Kolmogorov), and on the structure of stable laws (jointly with Lévy).

In a series of papers written between 1932 and 1934, Khinchin laid the foundation of the general theory of stationary random processes, revealed the spectral representation of their correlation functions, and generalized G. D. Birkhoff's ergodic theorem, which is a strengthened law of large numbers for such processes.

In other works Khinchin dealt with the convergence of discrete Markov chains to continuous diffusion, with large deviations, with the arithmetic of distribution laws, and with the method of arbitrary functions. In the 1940's Khinchin's interest shifted to statistical mechanics. With the aid of local limit theorems, he substantiated the possibility of replacing means in time by means in the phase space both for classical and quantum statistics. In the last years of his life Khinchin studied information theory and queuing theory.

Khinchin also wrote several popular books on the theory of numbers and published articles devoted to pedagogic and philosophic questions of mathematics.

BIBLIOGRAPHY

I. Original Works. Khinchin's writings include "Über dyadische Brüche," in *Mathematische Zeitschrift*, **18** (1923), 109–116, on the law of the iterated logarithm; "Recherches sur la structure des fonctions mesurables," in *Fundamenta mathematica*, **9** (1927), 212–279, a summary work on the theory of functions; *Osnovnye zakony teorii veroyatnostey* ("Basic Laws of Probability Theory"; Moscow, 1927, 2nd ed., rev., 1932), on the summation of independent random variables; *Asymptotische Gesetze der Wahrscheinlichkeitsrechnung* (Berlin, 1933), a monograph on the convergence of Markov chains to diffusion processes; "Korrelationstheorie der stationären stochastischen Prozesse," in *Mathematische Annalen*, **109** (1934), 604–615, the principal work on stationary processes; *Predelnye raspredelenia dlya summ nezavisimykh sluchaynykh velichin* ("Limit Distributions for Sums of Independent Random Variables"; Moscow, 1938); *Matematicheskie osnovania statisticheskoy mekhaniki* ("Mathematical Foundations of Statistical Mechanics"; Moscow, 1943), also in English (New York, 1949); *Matematicheskie osnovania kvantovoy statistiki* ("Mathematical Foundations of Quantum Statistics"; Moscow, 1951); *Pedagogicheskie stati* ("Pedagogical Articles"; Moscow, 1963), English trans., *The Teaching of Mathematics* (London, 1968); and *Raboty po matematicheskoy teorii massovogo obsluzhivania* ("Works on the Mathematical Theory of Queuing"; Moscow, 1963).

II. Secondary Literature. A biography of Khinchin by B. V. Gnedenko is in *Pedagogicheskie stati* (see above, pp. 180–196); there is also an article by A. I. Markushevich in the same volume (pp. 173–179; both are in the English ed.). See also Gnedenko's article in *Uspekhi matematicheskikh nauk*, **10**, no. 3 (1955), 197–212; and the obituary by Gnedenko and Kolmogorov, *ibid.*, **15**, no. 4 (1960), 97–110. Each of these articles has a full bibliography of Khinchin's works up to the time of publication.

See also *Nauka v SSR za pyatnadtsat let. Matematika* ("Fifteen Years of Science in the U.S.S.R. Mathematics"; Moscow–Leningrad, 1932), 150–151, 166–169; *Matematika v SSSR za tridtsat let* ("Thirty Years of Mathematics in the U.S.S.R."; Moscow–Leningrad, 1948), 57, 60–61, 259–260, 509, 706–713, 724–727; and *Matematika v SSR za sorok let* ("Forty Years of Mathematics in the U.S.S.R."), I (Moscow–Leningrad, 1959), 129–130, 789, 795.

A. A. Youschkevitch

AL-KHUJANDĪ, ABŪ MAḤMŪD ḤĀMID IBN AL-KHIḌR (*d.* 1000), *mathematics, astronomy.*

Little is known of al-Khujandī's life. Nāṣir al-Dīn al-Ṭūsī states that he had the title of khan, which would lead one to believe that he was one of the khans of Khujanda on the Syr Darya, or Jaxartes, in Transoxania. For a time he lived under the patronage of the Buwayhid ruler Fakhr al-Dawla (976–997). He died in 1000.

Ḥājjī Khalīfa, Suter, and Brockelmann ascribe the following scientific works to al-Khujandī: *Risāla fi'l mayl wa'arḍ balad* ("On the Obliquity of the Ecliptic and the Latitude of the Lands"), a text on geometry, and *Fi'amal al-āla al-'amma* or *al-āla al-shāmila* ("The Comprehensive Instrument").

According to Nāṣir al-Dīn al-Ṭūsī, al-Khujandī discovered *qānūn al-haiya*, the sine theorem relative to spherical triangles; it displaced the so-called theorem of Menelaus. Abu'l-Wafā' and Abū Naṣr ibn 'Alī ibn 'Irāq (tenth century) also claimed to have discovered the sine theorem.

Al-Ṭūsī, in his *Shakl al-qaṭṭāʿ*, gives al-Khujandī's solution related to the sine theorem.

Given the spherical triangle *ABC* whose sides *AC* and *AB* are completed into quadrants. *RA*, *RD*, *RE*, and *RB* are joined and form radii of the sphere.

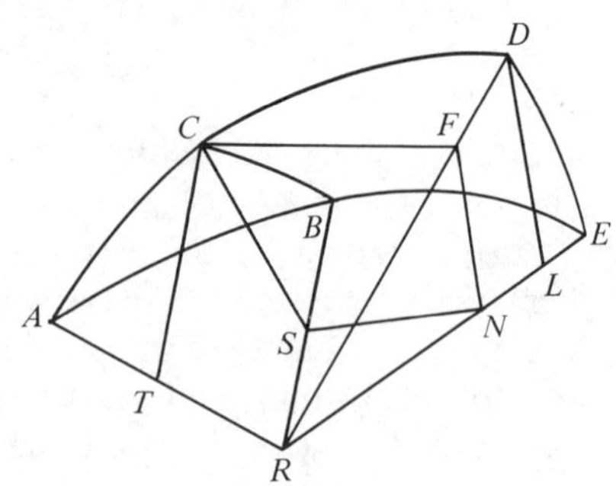

FIGURE 1

$RA \perp$ plane of the circle DE

At the same time

$RA \perp$ radii RE and RD

Erect the perpendicular CF on the plane of the circle DE. The perpendiculars FN and CS are erected on the plane ABE. $CFNS$ is a rectangle, and $DE \parallel FN$.

$$\overset{\triangle}{DER} \sim \overset{\triangle}{FNR}.$$

The perpendicular CT is erected on the plane of the circle AR and is parallel to RS,

$$\text{angle } R = \text{angle } T = 90^\circ$$

$$CF \perp RH$$

$$\text{angle } CFR = 90^\circ$$

Therefore

$$CFRT \text{ forms a rectangle}$$

$$\frac{RF = CT = \sin AC}{FN = CS = \sin CB} = \frac{RH = \sin 90^\circ}{FE = \sin A}$$

In geometry al-Khujandī proved (imperfectly) that the sum of two cubic numbers cannot be a cubic number.

Under the patronage of Fakhr al-Dawla, al-Khujandī constructed, on a hill called *jabal Tabrūk*, in the vicinity of Rayy, an instrument called *al-suds al-Fakhrī* ("sixth of a circle") for the measurement of the obliquity of the ecliptic. The instrument can be described as follows.

Two walls, parallel to the meridian and 40 *zira*ʿ in height, are constructed. Near the southern wall, there is an arched ceiling with an aperture about three inches in diameter.

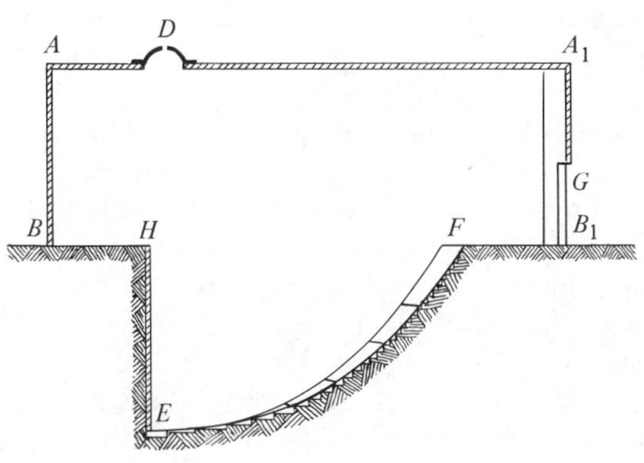

FIGURE 2

The floor directly underneath this aperture is excavated to a depth of forty *zira*ʿ. A wooden arc of 60°, forty *zira*ʿ in diameter and covered with sheets of copper, is placed between the two walls. Each degree of the arc is divided into sixty minutes and each minute into ten parts.

Since the sun's rays projected through the aperture form a cone, an instrument is needed to find the center of its base. This instrument, a circle with two diameters intersecting at right angles, coincides with the base of the cone. It is moved as the cone moves until its center is at the meridian. The arc between the plumb-line and the altitude of the sun is equal to the cosine of the altitude of the sun.

Al-Khujandī says that this instrument is his own invention and adds, "We have attained to the degrees, minutes, and seconds with this instrument." According to Al-Bīrūnī, on this instrument each degree was subdivided into 360 equal parts and each ten-second portion was distinguished on the scale. It should be noted that before al-Khujandī the instruments did not indicate the seconds.

Before al-Khujandī a domed building was used to make solar measurements. According to Al-Bīrūnī, Abū Sahl al-Kūhī (tenth century), at the Sharaf al-Dawla observatory (built in 988) constructed a domed building with an aperture on the top. This structure was a section of a sphere with a radius of 12.5 meters. Solar rays entered through the aperture and traced the daily trajectory of the sun.

After al-Khujandī, an instrument like *suds al-Fakhrī* was constructed at the Marāgha observatory (built in 1261). The huge meridian arc of the Samarkand observatory (built in 1420) apparently was similar to al-Khujandī's *suds al-Fakhrī.*

The astronomers of Islam tried to increase the precision of their instruments and to make it possible to read the smaller fractions of a degree. For this purpose they increased the size of the instruments. Al-Khujandī and Ulugh Beg represent the extreme examples of this tendency. Increased size, however, causes slight displacement. Al-Bīrūnī says that the aperture of *suds al-Fakhrī* sank by about one span because of the weight of the instrument. Experience with several large instruments proved disappointing and may have led to some doubts about the advisability of continuing to build them.

For the observations of the planets al-Khujandī constructed an armillary sphere and other instruments. He also built a universal instrument called *al-āla al-shāmila* (comprehensive instrument), which was used instead of the astrolabe or the quadrant. It could, however, be used for only one latitude. Al-Badī al-Aṣṭurlābī al-Baghdadī al-Isfahanī (first half of the twelfth century) constructed an astrolabe used for all latitudes.

Al-Khujandī observed the sun and the planets and determined the obliquity of the ecliptic and the latitude of Rayy. He says that these observations were made in the presence of a group of distinguished astronomers and that they gave their written testimony concerning the observations. Using these observations, he compiled his *Zīj al-Fakhrī*. There is in the Library of the Iranian Parliament (Teheran MS 181) an incomplete copy of a *zīj* written in Persian about two centuries after the death of al-Khujandī, which may have been based on his observations.

Al-Khujandī observed the meridian altitude of the sun on two consecutive days, 16 and 17 June 994, and found it to be 77°57′40″. According to this result the entrance of the sun into the summer solstice must have taken place at midnight.

He then observed the sun on 14 December 994 and found the meridian altitude to be 30°53′35″. On the following two days the weather was cloudy, and on the third day he found the meridian altitude of the sun to be 30°53′32″. The entrance of the sun into the winter solstice must have taken place between these two observations. But the second observation is 3″ less than the first. Al-Khujandī calculated from them that its least meridian altitude must have been 30°53′2.30″ (the least altitude of the sun).

Half of the difference between the greatest and the least altitudes of the sun is equal to the obliquity of the ecliptic:

$$1/2(77°57'40'' - 30°53'2'') = 23°32'19''.$$

Al-Khujandī says that the Indians found the greatest obliquity of the ecliptic, 24°; Ptolemy, 23°51′; and he himself, 23°32′19″. These divergent values cannot be due to defective instruments. Actually the obliquity of the ecliptic is not constant; it is a decreasing quantity.

Al-Khujandī calculated the latitude of Rayy by adding the obliquity of the ecliptic (23°32′18.45″) to the least altitude of the sun (30°53′2.30″) and subtracting the result from 90°(90° − 54°25′21.15″ = 35°34′38.45″).

BIBLIOGRAPHY

I. Original Works. Editions of al-Khujandī's writings are L. Cheiko, *Risāla al-Khujandī fi'l mayl wa'ard balad, al-machriq*, II (Beirut, 1908), 60–68; E. Wiedemann, "Über den Sextant des al-Chogendī," in *Archiv für die Geschichte der Naturwissenschaften*, **2** (1919), 148–151; and "Avicennas Schrift über ein von ihm ersonnenes Beobachtungsinstrument," in *Acta orientalia*, **5** (1926), 81–167; and O. Schirmer, "Studien zur Astronomie der Araber," in *Sitzungsberichte der Physikalisch-medizinische Sozietät in Erlangen*, **58–59** (1926–1927), 43–79.

II. Secondary Literature. See C. Brockelmann, *Geschichte der arabischen Literatur*, supp. I (Leiden, 1937), 390; M. Cantor, *Vorlesungen über Geschichte der Mathematik*, 2 vols. (Leipzig, 1880–1892); A. P. Carathéodory, *Traité du quadrilatère attribué à Nassiruddin-el-Toussy* (Constantinople, 1891), pp. 108–120; J. Frank, "Über zwei astronomische arabische Instrumente," in *Zeitschrift für Instrumentenkunde*, **41** (1921), 193–200; Ḥājjī Khalīfa, *Kashf al-zunuh*, S. Yaltkaya, ed., 2 vols. (Istanbul, 1941–1943); E. S. Kennedy, "A Survey of Islamic Astronomical Tables," in *Transactions of the American Philosophical Society*, n.s. **46** (1956), 123–177; J. A. Repsold, *Zur Geschichte der astronomischen Messwerkzeuge von Purbach bis Reinach, 1450 bis 1830*, I (Leipzig, 1908), 8–10; G. Sarton, *Introduction to the History of Science*, I (Baltimore, 1927), 667; A. Sayili, *The Observatory in Islam and Its Place in the General History of the Observatory* (Ankara, 1960), pp. 118–120; A. L. Sédillot, "Mémoire sur les instruments astronomiques des Arabes," in *Mémoires de l'Académie des inscriptions et belles-lettres*, **1** (1884), 202–206; H. Suter, *Die Mathematiker und Astronomen der Araber und ihre Werke* (Leipzig, 1900), p. 74; S. Tekeli, "Nasirüddin, Takiyüddin ve Tycho Brahe'nin Rasataletlerinin Mukayesesi," in *Ankara üniversitesi Dil ve tarih-cografya fakültesi dergisi*, **16**, no. 3–4 (1958), 301–393; E. Wiedemann, "Al-Khujandī," in *Encyclopédie de l'Islam*, II; and Sālih Zakī, *Athār-ī bāqiyya*, I (Istanbul, 1911), 165.

Sevim Tekeli

KHUNRATH (or **KUNRATH** or **KUHNRAT[H]** or **CUNRADIUS** or **CONRATHUS), CONRAD** (*b.* Leipzig, Germany; *d.* not later than 1614), *medicine, preparation of medicines, chemistry.*

Little is known of Khunrath's life. It has not yet been proved that he is identical with a person of the same name and place of birth who enrolled at Leipzig University in the winter of 1562. Khunrath lived in the duchy of Holstein, then Danish, for several years and in 1594 at Schleswig.

Khunrath's main work is *Medulla Destillatoria et Medica*, which seems to have been successful, for it passed through at least eight editions, the last in 1703. The two volumes, each of 650 pages, with detailed subject indexes, contain descriptions of many kinds of illnesses, prescriptions for curing them, and manifold applications of the process of distillation, including instructions for constructing the appropriate apparatus. Distilling is defined as "a dissolution or separation of a composed body into its simpler parts by help of heat." This operation, subsequently improved by others, was used in the *Chymische Kunst*. All is based on practical experience, the result of which—the medicament—is therefore called "experiment." Numerous medicines are treated: fruits, plants, salts, natural waters, stones, and animals—in short, as much as God has provided in the three natural kingdoms. Since human nourishment is considered to be essential for health, descriptions of grain, olives, and other foods are included.

Khunrath was a follower of Paracelsus, whom he knew personally, but his reference to Georg Agricola and Conrad Gesner confirms the influence also of metallurgical knowledge on the development of chemistry. Although Khunrath, as his contemporaries were doing, occasionally attributed cures to God's blessing or the effects of the signs of the zodiac and the planets, his book represents a masterpiece of clear, practical prescriptions and in the 17th century appears to have been considered a storehouse of information on curing.

It is impossible to give here an impression of his fine work, which is cited by several contemporaries, such as Libau and Joachim Jungius. It is certain that to historians of science he and his work are virtually unknown, being overshadowed by the writings of the theosopher and alchemist Heinrich Khunrath.

BIBLIOGRAPHY

I. ORIGINAL WORKS. Personal examination of the books and a check of catalogs in modern libraries reveals the following: *Medulla Destillatoria et Medica. Das ist, Warhafftiger eigentlicher gründtlicher bericht* ... (Schleswig [1594]), in German; *Chirurgia Vulnerum: Das ist, von Heylung der Wunden: Philippi Theophrasti, Paracelsi ... Durch einen fleissigen Zuhörer aus seinem Munde auffgezeichnet* ... (Schleswig [1595]); *Vier Schöne Medicische Tractat, vor nie in Truck kommen. De Elleboro. De Rore solis. De saccaro. Von der Schlangen.* ... (n.p., 1597)—J. C. Adelung (in *Fortsetzung und Ergänzungen zu ... Jöchers allgemeinem Gelehrten-Lexico*, 7 vols. (Leipzig, later Delmenhorst, Bremen, 1784–1897), gives an ed. of five papers, adding *De Absinthio*. See J. Ferguson, *Bibliotheca Chemica*, I (London, 1954); C. G. Jöcher, *Allgemeines Gelehrten Lexicon* (Leipzig, 1750–1751); W. Heinsius, *Allgemeines Bücherlexikon oder vollständiges alphabetisches Verzeichnis aller von 1700 bis 1892 erschienenen Bücher*, 19 vols. (Leipzig, 1812–1894); and F. Ferchl, *Chemisch-Pharmazeutisches Bio-Bibliographikon*, 2 vols. (Mittenwald, 1937–1938); *Eigentliche Beschreibung Derer fürnembsten Virtutes ... des ... Olei Succini* ... (n.p., 1598); *Relatio oder Erzehlung, wie der Grossmechtigste Herr Christianus Quartus, zu Dennemarck ... in ... Engellandt angelanget* ... (Hamburg, 1607), a trans. of two pamphlets by Henry Roberto: "The Most Royall and Honourable Entertainement of the Famous and Renowned King Christian the Fourth" and "England's Farewell to Christian the Fourth"; and *Medulla Destillatoria et Medica, quartum aucta et renovata. Das ist: Gründliches und Vielbewehrtes Destillier und ArtzneyBuch*, vol. I (Hamburg, 1614), vol. II (Hamburg, 1619); vol. II also Hamburg, 1615—for other eds. see catalogs of the British Museum, the Bibliothèque Nationale, W. Heinsius (1812), T. Georgi, *Allgemeines europäisches Bücherlexicon*, 5 vols. and supps. 1–3 (Leipzig, 1742–1758); and Ferguson; the 1703 ed. was pub. at "Franckfurt and Leipzig" (Ferguson and Georgi are mistaken here); the work also appeared under the title *Edelstes Kleinod Menschlicher Gesundheit; das ist: ... Destillier- und Artzeney-Kunst* (Frankfurt am Main–Leipzig, 1680).

A further work deduced from bibliographies is *Verteutschte Oration Königs Jacobi I. Anno 1605 von der Pulver-Verschwörung zu London* (Hamburg, 1606)—see T. Georgi, supp. 3.

II. SECONDARY LITERATURE. No full presentation of Khunrath's life or work is known. Only remarks are included in the following works (listed chronologically): E. O. von Lippmann, *Abhandlungen und Vorträge* ..., II (Leipzig, 1913), 383; J. R. Partington, *A History of Chemistry*, II (London–New York, 1961), 88; H. Kangro, *Joachim Jungius' Experimente und Gedanken zur Begründung der Chemie als Wissenschaft, ein Beitrag zur Geistesgeschichte des 17. Jahrhunderts* (Wiesbaden, 1968), pp. 198, 220, 235, 307; and R. J. Forbes, *A Short History of the Art of Distillation* (Leiden, 1970), pp. 153, 158, 380—bibliographically incorrect. Older sources of remarks can be found in Ferguson's catalog.

HANS KANGRO

KHUNRATH (or **KUNRATH**), **HEINRICH** (*b.* Leipzig, Germany, ca. 1560 [?]; *d.* Dresden [Leipzig?], Germany, 9 September 1605), *theosophy, alchemy, medicine.*

It has not yet been proved that Heinrich was a brother of Conrad Khunrath; nor is it known whether he is identical with a certain Henricus Conrad Lips, who enrolled at Leipzig University in the winter of 1570. He matriculated at the University of Basel in May 1588 and received the M.D. on 3 September 1588. On several title pages of his books he is called *medicinae utriusque Doct.* (doctor of theoretical and practical medicine). Later he practiced medicine at least at Hamburg (in 1598) and Dresden.

It is difficult to give even an outline of Heinrich Khunrath's work, which is largely unknown, in brief compass. He believed himself to be an adept of various spiritual traditions of alchemy dominated by the Paracelsian belief in the divine science of medicine as a privilege of the initiated (*scientia arcanorum*). Beginning with his twenty-eight doctoral theses (defended at Basel on 24 August 1588) he used the then modern Neoplatonic combination of heavenly and earthly processes to develop his ideas of Christianized natural magic (*Divino-Magicum*). This irrational component in his works attracted readers in the 17th century as well as in later times: an edition of his *Amphitheatrum sapientiae* appeared as late as 1900. On the other hand, he held experience alone to be decisive against the opinions of the great philosophers. With this union of theosophical experience and experience from natural observations he tried to shape what was then called *physicochemica*, that is, a chemical art grounded on general principles and practiced by the *physicochemici*. They used the artificial method, that is, the application of chemical means to obtain reactions from natural substances, and took as a basis the three principles sulfur, salt, and mercury, together with essentia. The last he made equivalent to the Mosaic "Ruach Elohim," to the Hermetic *anima mundi*, to the philosophers' *forma*, and to the *essentia quinta* of the doctrine of elements. He explicitly distinguished the mercury as a philosophical principle mentioned from quicksilver, as did Michael Potier in his time. Needless to say, Khunrath based his ideas on a profound knowledge of the *Chymia* of his time and of medicine.

Khunrath ardently sought to find and demonstrate the secret, divine primary matter endowed with the universal virtue of ruling all natural processes (*hylealischer, pri-materialischer catholischer Chaos*; *Chaos physico-chymicorum catholicus*; *Allgemeiner, naturlicher, dreieiniger ... allergeheimbster Chaos*; *materia prima mundi*). These efforts would, he claimed, afford eternal wisdom. Medical and chemical operations have to be undertaken, he stated, first under the guidance of God, then of Jesus Christ, and finally through the three methods called *Christiano-Kabalice*, *Divino-Magice*, and *Physico-Chymice*. Through prayer and work (*orando et laborando*) the theosopher has to implement them. The relation to God was an essential feature of the chemistry of Khunrath's time (see, for instance, Clemens Timpler).

On 1 February 1625 Khunrath's *Amphitheatrum* was condemned by the Sorbonne for its mixture of Christianity and magic. It must be admitted that to appreciate the symbolic character of his works requires enthusiasm and admiration for the mystical; a modern reader looking for real relations to medical and chemical reactions would scarcely understand it. Nevertheless Khunrath deserves more than admiration from enthusiasts of the occult. He still awaits the kind of understanding that would derive from an appreciation of the reciprocal role of thought in mediating between the two ranges of spiritual striving and experimental knowledge. The theosopher himself invited his readers to heed the motto of the owl: "Was helffn Fackeln, Liecht oder Brilln, Wann die Leute nicht sehen wölln." (What good are torches, light, or spectacles, to folk who won't see.)

BIBLIOGRAPHY

I. Original Works. Mentioned in the text is *Amphiatheatrum sapientiae aeternae* ... (Frankfurt am Main, 1653; other eds. are listed in bibliographical sources quoted below); an ed. not mentioned in the literature appeared at Magdeburg in 1608. The many other works and eds. that are not fictitious can be found in the catalogs of the Bibliothèque Nationale, Library of Congress, and British Museum; also consult those of T. Georgi, *Allgemeines europäisches Bücherlexicon*, 8 vols. (Leipzig, 1742–1758); W. Heinsius, *Allgemeines Bücherlexikon oder vollständiges alphabetisches Verzeichnis aller von 1700 bis 1892 erschienenen Bücher*, 19 vols. (Leipzig, 1812–1894); and J. Ferguson, *Bibliotheca chemica*, 2 vols. (London, 1954). Other works extant but not listed in these bibliographies are *Naturgemes-alchymisch Symbolon, oder, gahr kurtze Bekentnus, Henrici Khunrath Lips.: Beyder Artzney Doct.* ... (Hamburg, 1598); *Warhafftiger Bericht von philosophischen Athanor* (Magdeburg, 1599); and *Confessio de chao physico-chemicorum Catholico* (Strassburg, 1699). F. Ferchl (in *Chemisch-pharmazeutisches Bio-Bibliographikon*, 2 vols., (Mittenwald, 1937–1938)) mentions *Tractat von gründlicher Curation Tartari, Grieses, Sandes, Steins, Zipperleins an Händen und Füssen* (Hof, 1611)—see W. Heinsius (1812). Heinsius and J. F. Gmelin, in *Geschichte der Chemie*, I (Göttingen, 1797), 570, report on a MS at the library of the University of Jena, entitled "Die Kunst, den Lapidem Philosophorum nach dem hohen Liede Salomon's zu verfertigen."

II. Secondary Literature. See Claude K. Deischer and Joseph L. Rabinowitz, "The Owl of Heinrich Khunrath: Its Origin and Significance," in *Chymia*, **3** (1950), 243–250. No detailed historical biography or interpretation of Khunrath's work is known. A few notes can be found in the following books (listed chronologically): Lynn Thorndike, *A History of Magic and Experimental Science*, VII (New York, 1958), 273–275—from the few remarks a distorted picture of Khunrath is inferred; J. R. Partington, *A History of Chemistry*, II (London–New York, 1961), 645; Walter Pagel, *Das medizinische Weltbild des Paracelsus, seine Zusammenhänge mit Neuplatonismus und Gnosis*, which is *Kosmosophie*, I (Wiesbaden, 1962), pp. 71–72, 93, pls 2 and 3; and John Read, *Prelude to Chemistry* (Cambridge, Mass., 1966), pp. 81–82, 251–252, pl. 52. For older sources of remarks, see Ferguson.

Hans Kangro

IBN KHURRADĀDHBIH (or **IBN KHURDĀDHBIH), ABU'L-QĀSIM ʿUBAYD ALLĀH ʿABD ALLĀH** (*b. ca.* 820; *d. ca.* 912), *geography, history, music.*

Ibn Khurradādhbih was of Persian descent; his grandfather was a Zoroastrian who later accepted Islam. His father was the governor of Ṭabaristān. Ibn Khurradādhbih occupied the high office of chief of posts and information in al-Jibāl (Media). In his later years he became a close companion of Caliph al-Muʿtamid at Sāmarrā. He wrote on such subjects as history, genealogy, geography, music, and wines and cookery, thus showing the scope of his knowledge and erudition and his keen interest in the social and cultural life of his time. At least nine of his works have been mentioned in Arabic biobibliographical literature. Al-Nadīm gives the following list: *Kitāb adab al-samā'* ("On the Art of Music"); *Kitāb jumhurat ansāb al-Furs wa'l-nawāfil* ("On the Genealogies of the Ancient Persians"); *Kitāb al-masālik wa'l-mamālik* ("On Geography"); *Kitāb al-ṭabīkh* ("On Cookery"); *Kitāb al-lahw wa'l-malāhī* ("On Entertainments and Musical Instruments); *Kitāb al-sharāb* ("On Wines"); *Kitāb al-anwā'* ("On the Appearance of the Stationary Stars"); and *Kitāb al-nudamā' wa'l-julasā'* ("On Royal Companionship"). Al-Mas'ūdī attributes to Ibn Khurradādhbih a large work on history dealing with the non-Arabs of pre-Islamic times and admires it for its methodology and vast information (*Murūj*, I, 12–13). It is not unlikely that this book was the one on ancient Persian genealogies. For the caliph, Ibn Khurradādhbih had Ptolemy's *Geography* translated from a "foreign language" (probably Greek or Syriac) into the "correct language," but this may have been a simple adaptation of the work into Arabic. Some excerpts from his work on music are preserved by al-Mas'ūdī (*Murūj*, VIII, 88–100). Ibn Khurradādhbih's main contribution, however, was in geography. His geographical compendium, *Kitāb al-masālik wa'l-mamālik*, has come down to us in an abridged form. The original work, according to de Goeje, was prepared about 846–847 and the second draft was made not later than 885–886. The extant abridged version deals with regional, descriptive, economic, and political geography and covers not only the "Islamic kingdom" under the 'Abbāsid rule but also the non-Islamic world. The portions on mathematical and physical geography are insufficient, although the material seems to have been drawn from Ptolemy's work and from contemporary Arabic writings on the subject. The work also deals with stories of exploration and adventure and the wonders of the world. The major portion of it is devoted to detailed and accurate descriptions of the itineraries and road systems of the *oekumene* (*al-rub' al-ma'mūra*, the inhabited portion of the earth). It is here that Ibn Khurradādhbih displays his ability to handle scientifically the vast material at his disposal. Besides ancient Persian source books on the subject he seems to have used government records and firsthand accounts of Arab merchants, travelers, and sailors. The arrangement and presentation of the subject matter, the use of Persian terminology for the subdivisions, districts, and regions, and the use of Persian names shows a distinct Persian influence.

The very fact that Ibn Khurradādhbih assigns Iraq a central position vis-à-vis other provinces and takes Baghdad as the starting point to describe the itineraries shows that he equated it with the Irānshahr (Iraq) of the ancient Persians. He begins his description with al-Sawād, for, he says, the ancient Persian kings considered it *dil-i Irānshahr*, "the heart of Iraq." The land and sea routes pass out of Baghdad and lead in the four directions. To the east they reach as far as Central Asia and the sea route to India and China; to the west they go as far as North Africa and Spain; to the north to Azerbaijan and the Caucasus; and to the south they extend to southern Arabia. This material formed an indispensable source of knowledge for later geographers and travelers.

BIBLIOGRAPHY

Ibn Khurradādhbih's *Kitāb al-masālik wa'l-mamālik* is in J. de Goeje, ed., *Bibliotheca geographorum arabicorum*, VI (Leiden, 1889).

See also *Ḥudud al-'ālam, The Regions of the World*, trans. and explained by V. Minorsky (London, 1937), with preface by V. V. Barthold; C. Brockelmann, *Geschichte der arabischen Literatur*, I (Leiden, 1943), p. 225; I. I. Krachkovsky, *Istoria arabskoi geograficheskoi literatury* (Moscow–Leningrad, 1957), pp. 155–158, trans. into Arabic by Ṣalāḥ al-Dīn Uthmān Hāshim as *Ta'rīkh al-adab al-jughrāfī al-'Arabī* (Cairo, 1963); al-Mas'ūdī, *Murūj al-dhahab wa ma'ādin al-jauhar, Les prairies d'or*, 9 vols., Arabic text and French trans. by C. Barbier de Maynard and P. de Courteille (Paris, 1859–1877), I (1859), pp. 12–13, VIII (1874), pp. 88–100; and al-Nadīm, *Fihrist* (Cairo, n.d.), pp. 218–219.

S. MAQBUL AHMAD

AL-KHUWĀRIZMĪ, ABŪ 'ABD ALLĀH MUḤAMMAD IBN AḤMAD IBN YŪSUF (*fl.* in Khuwarizm *ca.* 975), *transmission of science.*

We know little of al-Khuwārizmī other than the biographical data he provides incidentally. He should not be confused with the mathematician Muḥammad ibn Mūsā al-Khwārizmī (see below) and the secretary Abu Bakr al-Khuwārizmī (*d.* 993). The subject of this article wrote a book entitled *Mafātīḥ al-'ulūm* ("Keys of the Sciences"), which he dedicated to Abu'l-Ḥasan al-'Utbī, vizier of the Samanid sovereign Nūḥ II (976–997). Analysis of its contents indicates that it was composed shortly after 977.

Intended as a manual for the perfect secretary, the *Mafātīḥ al-'ulūm* contains all the knowledge possessed by a cultured person living at that time in eastern Persia. Its purpose was to provide exact definitions of the technical terms in common use. The *Mafātīḥ* consists of two parts: the first analyzes the sciences traditionally considered of Arab or Muslim origin (theology, grammar, state administration, poetry, and others); and the second analyzes those imported from the Hellenic world. In the latter part much information on the history of the sciences can be found, for it contains chapters on philosophy, logic, medicine, arithmetic, geometry, astronomy, music, mechanics, and alchemy. Of special note is the author's interest in giving the correct etymologies of the terms he defines, their equivalents in Persian or Greek, and, in some cases, numerical examples (for example, when speaking of *jabr* and *muqābala*) that avoid misunderstandings.

Al-Khuwārizmī rarely states his sources and, when they do appear, they are not, in the case of scientific subjects, the best ones. Yet the latter were undoubtedly known to him; otherwise the good information that he does transmit could not be explained. On the other hand, he seems to have some points of contact with the *Rasā'il* ("Epistles") of the Ikhwān al-Ṣafā'.

BIBLIOGRAPHY

An inventory of al-Khuwārizmī's MSS is in C. Brockelmann, *Geschichte der arabischen Literatur*, I (Weimar, 1898), 282, and supp. I (Leiden, 1944), 434. The text has been published by B. Carra de Vaux under the title *Liber Mafātīḥ al-olum explicans vocabula technica scientiarum* (Leiden, 1895; repr. 1968). Many chs. of the second part have been trans. by E. Wiedemann and others. Details of these trans. can be found in C. E. Bosworth, "A Pioneer Arabic Encyclopedia of the Sciences: Al-Khuwārizmī, Keys of the Sciences," in *Isis*, **54** (1963), 97–111. See also G. Sarton, *Introduction to the History of Science*, I (Baltimore, 1927), 659–660; and E. Wiedemann, "Al-Khuwārizmī," in *Encyclopédie de l'Islam*, II (Leiden–Paris, 1927), 965.

J. VERNET

AL-KHWĀRIZMĪ, ABŪ JA'FAR MUḤAMMAD IBN MŪSĀ (*b.* before 800; *d.* after 847), *mathematics, astronomy, geography*.

Only a few details of al-Khwārizmī's life can be gleaned from the brief notices in Islamic bibliographical works and occasional remarks by Islamic historians and geographers. The epithet "al-Khwārizmī" would normally indicate that he came from Khwārizm (Khorezm, corresponding to the modern Khiva and the district surrounding it, south of the Aral Sea in central Asia). But the historian al-Ṭabarī gives him the additional epithet "al-Quṭrubbullī," indicating that he came from Quṭrubbull, a district between the Tigris and Euphrates not far from Baghdad,[1] so perhaps his ancestors, rather than he himself, came from Khwārizm; this interpretation is confirmed by some sources which state that his "stock" (*aṣl*) was from Khwārizm.[2] Another epithet given to him by al-Ṭabarī, "al-Majūsī," would seem to indicate that he was an adherent of the old Zoroastrian religion. This would still have been possible at that time for a man of Iranian origin, but the pious preface to al-Khwārizmī's *Algebra* shows that he was an orthodox Muslim, so al-Ṭabarī's epithet could mean no more than that his forebears, and perhaps he in his youth, had been Zoroastrians.

Under the Caliph al-Ma'mūn (reigned 813–833) al-Khwārizmī became a member of the "House of Wisdom" (Dār al-Ḥikma), a kind of academy of scientists set up at Baghdad, probably by Caliph Harūn al-Rashīd, but owing its preeminence to the interest of al-Ma'mūn, a great patron of learning and scientific investigation. It was for al-Ma'mūn that al-Khwārizmī composed his astronomical treatise, and his *Algebra* also is dedicated to that ruler. We are told that in the first year of his reign (842) Caliph al-Wāthiq sent al-Khwārizmī on a mission to the chief of the Khazars, who lived in the northern Caucasus.[3] But there may be some confusion in the source here with another "Muḥammad ibn Mūsā the astronomer," namely, one of the three Banū Mūsā ibn Shākir. It is almost certain that it was the latter, and not al-Khwārizmī, who was sent, also by al-Wāthiq, to the Byzantine empire to investigate the tomb of the Seven Sleepers at Ephesus.[4] But al-Khwārizmī survived al-Wāthiq (*d.* 847), if we can believe the story of al-Ṭabarī that he was one of a group of astronomers, summoned to al-Wāthiq's sickbed, who predicted on the basis of the caliph's horoscope that he would live another fifty years and were confounded by his dying in ten days.

All that can be said concerning the date and order of composition of al-Khwārizmī's works is the following. The *Algebra* and the astronomical work, as we have seen, were composed under al-Ma'mūn, in the earlier part of al-Khwārizmī's career. The treatise on Hindu numerals was composed after the *Algebra*, to which it refers. The treatise on the Jewish calendar is dated by an internal calculation to 823–824. The *Geography* has been tentatively dated by Nallino ("al-Khuwārizmī," p. 487) to soon after 816–817, since one of the localities it mentions is Qiman, an Egyptian village of no importance whatever except that a battle was fought there in that year; but the inference is far from secure. The *Chronicle* was composed after 826, since al-Ṭabarī quotes it as an authority for an event in that year.[5]

The *Algebra* is a work of elementary practical mathematics, whose purpose is explained by the author

(Rosen trans., p. 3) as providing "what is easiest and most useful in arithmetic, such as men constantly require in cases of inheritance, legacies, partition, lawsuits, and trade, and in all their dealings with one another, or where the measuring of lands, the digging of canals, geometrical computations, and other objects of various sorts and kinds are concerned." Indeed, only the first part of the work treats of algebra in the modern sense. The second part deals with practical mensuration, and the third and longest with problems arising out of legacies. The first part (the algebra proper) discusses only equations of the first and second degrees. According to al-Khwārizmī, all problems of the type he proposes can be reduced to one of six standard forms. These are (here, as throughout, we use modern notation, although al-Khwārizmī's exposition is always rhetorical) the following:

(1) $ax^2 = bx$

(2) $ax^2 = b$

(3) $ax = b$

(4) $ax^2 + bx = c$

(5) $ax^2 + c = bx$

(6) $ax^2 = bx + c,$

where a, b, and c are positive integers. Such an elaboration of cases is necessary because he does not recognize the existence of negative numbers or zero as a coefficient. He gives rules for the solution of each of the six forms—for instance, form (6) is solved by

$$x^2 = (b/a)\,x + c/a,$$

$$x = \sqrt{\left[\frac{1}{2}\left(\frac{b}{a}\right)\right]^2 + \frac{c}{a}} + \frac{1}{2}\left(\frac{b}{a}\right).$$

He also explains how to reduce any given problem to one of these standard forms. This is done by means of the two operations *al-jabr* and *al-muqābala*. *Al-jabr*, which we may translate as "restoration" or "completion," refers to the process of eliminating negative quantities. For instance, in the problem illustrating standard form (1) (Rosen trans., p. 36), we have

$$x^2 = 40x - 4x^2.$$

By "completion" this is transformed to

$$5x^2 = 40x.$$

Al-muqābala, which we may translate as "balancing," refers to the process of reducing positive quantities of the same power on both sides of the equation. Thus, in the problem illustrating standard form (5) (Rosen trans., p. 40), we have

$$50 + x^2 = 29 + 10x.$$

By *al-muqābala* this is reduced to

$$21 + x^2 = 10x.$$

These two operations, combined with the arithmetical operations of addition, subtraction, multiplication, and division (which al-Khwārizmī also explains in their application to the various powers), are sufficient to solve all types of problems propounded in the *Algebra*. Hence they are used to characterize the work, whose full title is *al-Kitāb al-mukhtaṣar fī ḥisāb al-jabr wa'l-muqābala* ("The Compendious Book on Calculation by Completion and Balancing"). The appellation *al-jabr wa'l-muqābala*, or *al-jabr* alone, was commonly applied to later works in Arabic on the same topic; and thence (via medieval Latin translations from the Arabic) is derived the English "algebra."

In his *Algebra* al-Khwārizmī employs no symbols (even for numerals) but expresses everything in words. For the unknown quantity he employs the word *shay'* ("thing" or "something"). For the second power of a quantity he employs *māl* ("wealth," "property"), which is also used to mean only "quantity." For the first power, when contrasted with the second power, he uses *jidhr* ("root"). For the unit he uses *dirham* (a unit of coinage). Thus the problem

$$(x/3 + 1)(x/4 + 1) = 20$$

and the first stage in its resolution,

$$x^2/12 + x/3 + x/4 + 1 = 20,$$

appear, in literal translation, as follows:

> A quantity: I multiplied a third of it and a *dirham* by a fourth of it and a *dirham:* it becomes twenty. Its computation is that you multiply a third of something by a fourth of something: it comes to a half of a sixth of a square (*māl*). And you multiply a *dirham* by a third of something: it comes to a third of something; and [you multiply] a *dirham* by a fourth of something to get a fourth of something; and [you multiply] a *dirham* by a *dirham* to get a *dirham*. Thus its total, [namely] a half of a sixth of a square and a third of something and a quarter of something and a *dirham*, is equal to twenty *dirhams*.[6]

After illustrating the rules he has expounded for solving problems by a number of worked examples, al-Khwārizmī, in a short section headed "On Business Transactions," expounds the "rule of three," or how to determine the fourth member in a proportion sum where two quantities and one price, or two prices and one quantity, are given. The next part concerns practical mensuration. He gives rules for finding the area of various plane figures, including the circle, and for finding the volume of a number of solids, including cone, pyramid, and truncated pyramid. The third part, on legacies, consists entirely of solved problems. These involve only arithmetic or simple linear equations but require considerable knowledge of the complicated Islamic law of inheritance.

We are told that al-Khwārizmī's work on algebra was the first written in Arabic.[7] In modern times considerable dispute has arisen over the question of whether the author derived his knowledge of algebraic techniques from Greek or Hindu sources. Both Greek and Hindu algebra had advanced well beyond the elementary stage of al-Khwārizmī's work, and none of the known works in either culture shows much resemblance in presentation to al-Khwārizmī's. But, in favor of the "Hindu hypothesis," we may note first that in his astronomical work al-Khwārizmī was far more heavily indebted to a Hindu work than to Greek sources; second, that his exposition is completely rhetorical, like Sanskrit algebraic works and unlike the one surviving Greek algebraic treatise, that of Diophantus, which has already developed quite far toward a symbolic representation; third that the "rule of three" is commonly enunciated in Hindu works but not explicitly in any ancient Greek work; and fourth that in the part on mensuration two of the methods he gives for finding the circumference of the circle from its diameter are specifically Hindu.[8]

On the other hand, in his introductory section al-Khwārizmī uses geometrical figures to explain equations, which surely argues for a familiarity with Book II of Euclid's *Elements.* We must recognize that he was a competent enough mathematician to select and adapt material from quite disparate sources in order to achieve his purpose of producing a popular handbook. The question of his sources is further complicated by the existence of a Hebrew treatise, the *Mishnat ha-Middot*, which is closely related in content and arrangement to the part of al-Khwārizmī's work dealing with mensuration. If we adopt the conclusion of Gandz, the last editor of the *Mishnat ha-Middot*, that it was composed about A.D. 150,[9] then al-Khwārizmī must be the borrower, either through an intermediary work or even directly—his treatise on the Jewish calendar (see below) shows that he must have been in contact with learned Jews. But the Hebrew treatise may be a later adaptation of al-Khwārizmī's work. Gad Sarfatti (*Mathematical Terminology in Hebrew Scientific Literature of the Middle Ages*, Jerusalem, 1968, 58–60) argues on linguistic grounds that the *Mishnat ha-Middat* belongs to an earlier Islamic period.

Al-Khwārizmī wrote a work on the use of the Hindu numerals, which has not survived in Arabic but has reached us in the form of a Latin translation (probably much altered from the original). The Arabic title is uncertain; it may have been something like *Kitāb ḥisāb al-ʿadad al-hindī* ("Treatise on Calculation With the Hindu Numerals"),[10] or possibly *Kitāb al-jamʿ waʾl-tafrīq bi ḥisāb al-hind* ("Book of Addition and Subtraction by the Method of Calculation of the Hindus").[11] The treatise, as we have it, expounds the use of the Hindu (or, as they are misnamed, "Arabic") numerals 1 to 9 and 0 and the place-value system, then explains various applications. Besides the four basic operations of addition, subtraction, multiplication, and division, it deals with both common and sexagesimal fractions and the extraction of the square root (the latter is missing in the unique manuscript but is treated in other medieval works derived from it). In other words, it is an elementary arithmetical treatise using the Hindu numerals. Documentary evidence (eighth-century Arabic papyri from Egypt) shows that the Arabs were already using an alphabetic numeral system similar to the Greek (in which 1, 2, 3, . . . 9, 10, 20, 30, . . . 90, 100, 200, . . . 900 are each represented by a different letter). The sexagesimal modified place-value system used in Greek astronomy must also have been familiar, at least to learned men, from the works such as Ptolemy's *Almagest* which were available in Arabic before 800. But it is likely enough that the decimal place-value system was a fairly recent arrival from India and that al-Khwārizmī's work was the first to expound it systematically. Thus, although elementary, it was of seminal importance.

The title of al-Khwārizmī's astronomical work was *Zīj al-sindhind.*[12] This was appropriate, since it is based ultimately on a Sanskrit astronomical work brought to the court of Caliph al-Manṣūr at Baghdad soon after 770[13] by a member of an Indian political mission. That work was related to, although not identical with, the *Brāhmasphuṭasiddhānta* of Brahmagupta. It was translated into Arabic under al-Manṣūr (probably by al-Fazārī), and the translation was given the name *Zīj al-sindhind. Zīj* means "set of astronomical tables"; and *sindhind* is a corruption of the Sanskrit *siddhānta*, which presumably was part of the title of the Hindu source work. This translation formed the basis of astronomical works (also called *Zīj al-sindhind*) by al-Fazārī and Yaʿqūb ibn Ṭāriq in the late eighth century. Yet these astronomers also used other sources for their work, notably the *Zīj al-shāh*, a translation of a Pahlavi work composed for the Sassanid ruler Khosrau I (Anūshirwān) about 550, which was also based on Hindu sources.

Al-Khwārizmī's work is another "revision" of the *Zīj al-sindhind.* Its chief importance today is that it is the first Arabic astronomical work to survive in anything like entirety. We are told that there were two editions of it; but we know nothing of the differences between them, for it is available only in a Latin translation made by Adelard of Bath in the early twelfth century. This translation was made not from the original but from a revision executed by the Spanish Islamic astronomer al-Majrīṭī (*d.* 1007–1008) and perhaps further revised by al-Majrīṭī's pupil Ibn al-Ṣaffār (*d.* 1035).[14] We can, however, get some notions of the original form of the work from extracts and commentaries made by earlier writers.[15] Thus from the tenth-century commentary of Ibn al-Muthannā we learn that al-Khwārizmī constructed his table of sines to base 150 (a common Hindu parameter), whereas in the extant tables base 60 (more usual in Islamic sine tables) is employed. From the same source we learn that the epoch of the original tables was era Yazdegerd (16 June 632) and not the era Hijra (14 July 622) of al-Majrīṭī's revision.[16]

The work as we have it consists of instructions for computation and use of the tables, followed by a set of tables whose form closely resembles that made standard by Ptolemy. The sun, the moon, and each of the five planets known in antiquity have a table of mean motion(s) and a table of equations. In addition there are tables for computing eclipses, solar declination and right ascension, and various trigonometrical tables. It is certain that Ptolemy's tables, in their revision by Theon of Alexandria, were already known to some Islamic astronomers; and it is highly likely that they influenced, directly or through intermediaries, the form in which al-Khwārizmī's tables were cast.

But most of the basic parameters in al-Khwārizmī's tables are derived from Hindu astronomy. For all seven bodies the mean motions, the mean positions at epoch, and the positions of the apogee and the node all agree well with what can be derived from the *Brāhmasphuṭasiddhānta*. The maximum equations are taken from the *Zīj al-shāh*. Furthermore, the method of computing the true longitude of a planet by "halving the equation" prescribed in the instructions is purely Hindu and quite alien to Ptolemaic astronomy.[17] This is only the most notable of several Hindu procedures found in the instructions. The only tables (among those that can plausibly be assigned to the original *Zīj*) whose content seems to derive from Ptolemy are the tables of solar declination, of planetary stations, of right ascension, and of equation of time. Nowhere in the work is there any trace of original observation or of more than trivial computation by the author. This appears strange when we learn that in the original introduction (the present one must be much altered) al-Khwārizmī discussed observations made at Baghdad under al-Ma'mūn to determine the obliquity of the ecliptic.[18] The value found, 23° 33′, was fairly accurate. Yet in the tables al-Khwārizmī adopts the much worse value of 23° 51′ from Theon. Even more inexplicable is why, if he had the Ptolemaic tables available, he preferred to adopt the less accurate parameters and obscure methods of Hindu astronomy.

The *Geography*, *Kitāb ṣūrat al-arḍ* ("Book of the Form of the Earth"), consists almost entirely of lists of longitudes and latitudes of cities and localities. In each section the places are arranged according to the "seven climata" (in many ancient Greek geographical works the known world was divided latitudinally into seven strips known as "climata," each clima being supposed to enjoy the same length of daylight on its longest day), and within each clima the arrangement is by increasing longitude. Longitudes are counted from an extreme west meridian, the "shore of the western ocean." The first section lists cities; the second, mountains (giving the coordinates of their extreme points and their orientation); the third, seas (giving the coordinates of salient points on their coastlines and a rough description of their outlines); the fourth, islands (giving the coordinates of their centers, and their length and breadth); the fifth, the central points of various geographical regions; and the sixth, rivers (giving their salient points and the towns on them).

It is clear that there is some relationship between this work and Ptolemy's *Geography*, which is a description of a world map and a list of the coordinates of the principal places on it, arranged by regions. Many of the places listed in Ptolemy's work also occur in al-Khwārizmī's, with coordinates that are nearly the same or systematically different. Yet it is very far from being a mere translation or adaptation of Ptolemy's treatise. The arrangement is radically different, and the outline of the map which emerges from it diverges greatly from Ptolemy's in several regions. Nallino is surely right in his conjecture that it was derived by reading off the coordinates of a map or set of maps based on Ptolemy's but was carefully revised in many respects. Nallino's principal argument is that al-Khwārizmī describes the colors of the mountains in a way which could not possibly represent their physical appearance but might well represent their depiction on a map. To this we may add that in those areas where al-Khwārizmī agrees, in general, with Ptolemy, the coordinates of the two frequently differ by 10, 15, 20, or more minutes, up to one degree of arc; such discrepancies cannot be explained by scribal errors but are plausible by supposing a map as intermediary. The few maps accompanying the sole manuscript of al-Khwārizmī's *Geography* are crude things; but we know that al-Maʾmūn had had constructed a world map, on which many savants worked. According to al-Masʿūdī, the source of this information, al-Maʾmūn's map was superior to Ptolemy's.[19] Nallino makes the plausible suggestion that al-Khwārizmī's *Geography* is based on al-Maʾmūn's world map (on which al-Khwārizmī himself had probably worked), which in turn was based on Ptolemy's *Geography*, which had been considerably revised.

The map which emerges from al-Khwārizmī's text is in several respects more accurate than Ptolemy's, particularly in the areas ruled by Islam. Its most notable improvement is to shorten the grossly exaggerated length of the Mediterranean imagined by Ptolemy. It also corrects some of the distortions applied by Ptolemy to Africa and the Far East (no doubt reflecting the knowledge of these areas brought back by Arab merchants). But for Europe it could do little more than reproduce Ptolemy; and it introduces errors of its own, notably the notion that the Atlantic is an inland sea enclosed by a western continent joined to Europe in the north.

The only other surviving work of al-Khwārizmī is a short treatise on the Jewish calendar, *Istikhrāj taʾrīkh al-yahūd* ("Extraction of the Jewish Era"). His interest in the subject is natural in a practicing astronomer. The treatise describes the Jewish calendar, the 19-year intercalation cycle, and the rules for determining on what day of the week the first day of the month Tishrī shall fall; calculates the interval between the Jewish era (creation of Adam) and the Seleucid era; and gives rules for determining the mean longitude of the sun and moon using the Jewish calendar. Although a slight work, it is accurate, well informed, and of importance as evidence for the antiquity of the present Jewish calendar.

Al-Khwārizmī wrote two works on the astrolabe, *Kitāb ʿamal al-asṭurlāb* ("Book on the Construction of the Astrolabe") and *Kitāb al-ʿamal biʾl-asṭurlāb* ("Book on the Operation of the Astrolabe"). Probably from the latter is drawn the extract found in a Berlin manuscript of a work of the ninth-century astronomer al-Farghānī. This extract deals with the solution of various astronomical problems by means of the astrolabe—for instance, determination of the sun's altitude, of the ascendant, and of one's terrestrial latitude. There is nothing surprising in the content, and it is probable that al-Khwārizmī derived it all from earlier works on the subject. The astrolabe was a Greek invention, and we know that there were once ancient Greek treatises on it. Astrolabe treatises predating al-Khwārizmī survive in Syriac (by Severus Sebokht, seventh century) and in Arabic (now only in Latin translation, by Māshāʾllāh, late eighth century).

The *Kitāb al-taʾrīkh* ("Chronicle") of al-Khwārizmī does not survive, but several historians quote it as an authority for events in the Islamic period. It is possible that in it al-Khwārizmī (like his contemporary Abū Maʿshar) exhibited an interest in interpreting history as fulfilling the principles of astrology.[20] In that case it may be the ultimate source of the report of Ḥamza al-Iṣfahānī about how al-Khwārizmī cast the horoscope of the Prophet and showed at what hour Muḥammad must have been born by astrological deduction from the events of his life.[21] Of a book entitled *Kitāb al-rukhāma* ("On the Sundial") we know only the title, but the subject is consonant with his other interests.

Al-Khwārizmī's scientific achievements were at best mediocre, but they were uncommonly influential. He lived at a time and in a place highly favorable to the success of his works: encouraged by the patronage of the caliphs, Islamic civilization was beginning to assimilate Greek and Hindu science. The great achievements of Islamic science lay in the future, but these early works which transmitted the new knowledge ensured their author's lasting fame. Between the ninth and twelfth centuries algebra was developed to a far more sophisticated level in Islamic lands, aided by the spread of knowledge of Diophantus' work. But even such advanced algebraists as al-Karajī (*d.* 1029) and ʿUmar al-Khayyāmī (*d.* 1123–1124) still used the rhetorical exposition popularized by al-Khwārizmī.

Al-Khwārizmī's *Algebra* continued to be used as a textbook and praised highly (see, for instance, the quotations in Hajī Khalfa, V, 67–69). The algebraic part proper was twice translated into Latin in the twelfth century (by Robert of Chester and by Gerard of Cremona) and was the chief influence on medieval European algebra, determining its rhetorical form and some of its vocabulary (the medieval *cossa* is a literal translation of Arabic *shayʾ*, and *census* of *māl*). The treatise on Hindu numerals, although undoubtedly important in introducing those useful symbols into more general use in Islamic lands, achieved its greatest success only when introduced to the West through Latin translation in the early twelfth century (occasional examples of the numerals appeared in the West more than a century earlier, but only as isolated curiosities). The work quickly spawned a number of adaptations and offshoots, such as the *Liber alghoarismi* of John of Seville (*ca.* 1135), the *Algorismus* of John of Sacrobosco (*ca.* 1250), and the *Liber ysagogarum Alchorizmi* (twelfth century). In fact, al-Khwārizmī's name became so closely associated with the "new arithmetic" using the Hindu numerals that the Latin form of his name, *algorismus*, was given to any treatise on that topic. Hence, by a devious path, is derived the Middle English "augrim" and the modern "algorism" (corrupted by false etymology to "algorithm").

The other works did not achieve success of such magnitude; but the *Zīj* continued to be used, studied, and commented on long after it deserved to be superseded. About 900 al-Battānī published his great astronomical work, based on the *Almagest* and tables of Ptolemy and on his own observations. This is greatly superior to al-Khwārizmī's astronomical work in nearly every respect, yet neither al-Battānī's opus nor the other results of the prodigious astronomical activity in Islamic lands during the ninth and tenth centuries drove al-Khwārizmī's *Zīj* from the classroom. In fact it was the first such work to reach the West, in Latin translation by Adelard of Bath in the early twelfth century. Knowledge of this translation was probably confined to England (all surviving manuscripts appear to be English), but many of al-Khwārizmī's tables reached a wide audience in the West via another work, the *Toledan Tables*, a miscellaneous assembly of astronomical tables from the works of al-Khwārizmī, al-Battānī, and al-Zarqāl which was translated into Latin, probably by Gerard of Cremona, in the late twelfth century and which, for all its deficiencies, enjoyed immense popularity throughout Europe for at least 100 years.

The *Geography* too was much used and imitated in Islamic lands, even after the appearance of good Arabic translations of Ptolemy's *Geography* in the later ninth century caused something of a reaction in favor of that work. For reasons rather obscure the medieval translators of Arabic scientific works into Latin appear to have avoided purely geographical treatises, so al-Khwārizmī's *Geography* was unknown in Europe until the late nineteenth century. But some of the data in it reached medieval Europe via the lists of longitudes and latitudes of principal cities, which were commonly incorporated into ancient and medieval astronomical tables.[22]

NOTES

1. Al-Ṭabarī, de Goeje ed., III, 2, 1364.
2. E.g., *Fihrist*, Flügel ed., I, 274; followed by Ibn al-Qifṭī, Lippert, ed., p. 286.
3. Al-Muqaddasī, de Goeje ed., p. 362.
4. The story is found in several sources, all of which call the envoy "Muḥammad ibn Mūsā the astronomer." Only one, al-Mas'ūdī, *Kitāb al-tanbīh,* de Goeje ed., p. 134, adds "ibn Shākir." For the full story see Nallino, "Al-Khuwārizmī," pp. 465–466.
5. Al-Ṭabarī, de Goeje ed., III, 2, 1085.
6. Rosen, text, p. 28, somewhat emended. The translation is mine.
7. E.g., Hajī Khalfa, Flügel ed., V, 67, no. 10012.
8. These are $c = \sqrt{10d^2}$ and $c = 62832d/20000$. See Rosen's note on pp. 198–199 of his ed. The second value, which is very accurate, is attested for the later *Pauliśasiddhānta* and also, significantly, for Ya'qūb ibn Ṭāriq, al-Khwārizmī's immediate predecessor, by al-Bīrūnī, *India*, Sachau trans., I, 168–169. Pauliśa presumably derived it from the *Āryabhaṭīya* (see the *Āryabhaṭīya*, Clark ed., p. 28).
9. See Gandz's ed. of the *Mishnat ha-Middot*, pp. 6–12.
10. Some such title seems to be implied by Ibn al-Qifṭī, Lippert ed., pp. 266–267.
11. As conjectured by Ruska, "Zur ältesten arabischen Algebra," pp. 18–19.
12. *Fihrist*, Flügel ed., I, 274.
13. See, e.g., al-Bīrūnī, *India*, Sachau trans., II, 15.
14. For the latter revision see Ibn Ezra, *Libro de los fundamentos*, p. 109.
15. For a list of these see Pingree and Kennedy, commentary on al-Hāshimī's *Book of the Reasons Behind Astronomical Tables*, sec. 11; see also in biblio. (below).
16. For the value 150 see, e.g., Goldstein, *Ibn al-Muthannâ*, p. 178. For the epoch, *ibid.*, p. 18.
17. On "halving the equation" see Neugebauer, *al-Khwārizmī*, pp. 23–29.
18. Ibn Yūnus, quoted by Nallino, "Al-Khuwārizmī," p. 469.
19. Al-Mas'ūdī, *Kitāb al-tanbīh*, de Goeje ed., p. 33.
20. On Abū Ma'shar see especially Pingree, *The Thousands of Abū Ma'shar*.
21. Ḥamza, *Ta'rikh*, Beirut ed., p. 126. However, Ḥamza quotes this not directly from the *Chronicle* (which he uses elsewhere, *ibid.*, p. 144) but from Shādhān's book of Abū Ma'shar's table talk, so the ultimate source might be a conversation between al-Khwārizmī and Abū Ma'shar.
22. The list in the *Toledan Tables*, which is certainly in part related to al-Khwārizmī's *Geography*, is printed with commentary in Toomer, "Toledan Tables," pp. 134–139.

BIBLIOGRAPHY

The principal medieval Arabic sources for al-Khwārizmī's life and works are Ibn al-Nadīm, *Kitāb al-fihrist*, Gustav Flügel, ed., 2 vols. (Leipzig, 1872; repr. Beirut, 1964), I, 274—trans. by Heinrich Suter, "Das Mathematiker-Verzeichniss im Fihrist des Ibn Abî Ja'ḳûb an-Nadîm," which is Abhandlungen zur Geschichte der Mathematik, VI, 29, supp. to *Zeitschrift für Mathematik und Physik*, 37 (1892); Ibn al-Qifṭī, *Ta'rīkh al-ḥukamā'*, Julius Lippert, ed. (Leipzig, 1903; repr. Baghdad, n.d.), p. 286, a mere repetition of the *Fihrist* but with more information under the entry "Kanka," p. 266; Ṣâ'id al-Andalusî, *Kitâb ṭabaḳât al-umam* (*Livre des catégories des nations*), which is Publications de l'Institut des Hautes Études Marocaines, XXVIII, Régis Blachère, trans. (Paris, 1935), pp. 47–48, 130; Hajī Khalfa, *Lexicon bibliographicum*, G. Flügel, ed., V (London, 1850; repr. London–New York, 1964), 67–69, no. 10012; *Annales quos scripsit Abu Djafar Mohammed ibn Djarir at-Tabari*, M. J. de Goeje, ed., III, 2 (Leiden, 1881; repr. Leiden, 1964), 1364; *Descriptio imperii moslemici auctore al-Mokaddasi*, M. J. de Goeje, ed. (Leipzig, 1876–1877), p. 362; al-Mas'ūdī, *Kitāb al-tanbīh wa'l-ishrāf*, which is Bibliotheca Geographorum Arabicorum, VIII, M. J. de Goeje, ed. (Leiden,

1894; repr. 1967), pp. 33, 134. The best modern account of his life is C. A. Nallino, "Al-Khuwārizmī e il suo rifacimento della Geografia di Tolomeo," in his *Raccolta di scritti editi e inediti*, V (Rome, 1944), 458–532 (an amended repr. of his article in *Atti dell'Accademia nazionale dei Lincei. Memorie*, Classe di scienze morali, storiche e filologiche, 5th ser., II, pt. 1), and sec. 2, 463–475, where references to further source material may be found.

The Arabic text of the *Algebra* was edited with English trans. by Frederic Rosen as *The Algebra of Mohammed ben Musa* (London, 1831; repr. New York, 1969). Editing and trans. are careless. A somewhat better Arabic text is provided by the ed. of ʿAlī Muṣṭafā Masharrafa and Muḥammad Mursī Aḥmad (Cairo, 1939), which is Publications of the Faculty of Science, no. 2. Both eds. are based only on the MS Oxford Bodleian Library, I 918, 1, but other MSS are known to exist. I owe the refs. to the following to Adel Anbouba: Berlin 5955 no. 6, ff. 60r–95v; also a MS at Shibin el-Kom (Egypt) mentioned in *Majalla Maʿhad al-Makhṭūṭāt al-ʿArabiyya* (Cairo, 1950), no. 19. The section of the *Algebra* concerning mensuration is published with an English trans. by Solomon Gandz, together with his ed. of the *Mishnat ha-Middot*, which is *Quellen und Studien zur Geschichte der Mathematik, Astronomie und Physik*, Abt. A, **2** (1932). A useful discussion of the *Algebra* is given by Julius Ruska, "Zur ältesten arabischen Algebra und Rechenkunst," in *Sitzungsberichte der Heidelberger Akademie der Wissenschaften*, Phil.-hist. Kl. (1917), sec. 2, where further bibliography will be found. On the section dealing with legacies, see S. Gandz, "The Algebra of Inheritance," in *Osiris*, **5** (1938), 319–391. The Latin trans. by Robert of Chester was edited with English trans. by Louis Charles Karpinski, *Robert of Chester's Latin Translation of the Algebra of al-Khowarizmi* (Ann Arbor, 1915); repr. as pt. I of Louis Charles Karpinski and John Garrett Winter, *Contributions to the History of Science* (Ann Arbor, 1930). The editor perversely chose to print a sixteenth-century reworking rather than Robert's original translation, but his introduction and commentary are occasionally useful. The anonymous Latin version printed by G. Libri in his *Histoire des sciences mathématiques en Italie*, I (Paris, 1858), 253–297, is probably that of Gerard of Cremona, but the problem is complicated by the existence of another Latin text which is a free adaptation of al-Khwārizmī's *Algebra*, whose translation is expressly ascribed to Gerard of Cremona. This is printed by Baldassarre Boncompagni in *Atti dell'Accademia pontificia dei Nuovi Lincei*, **4** (1851), 412–435. A. A. Björnbo, "Gerhard von Cremonas Übersetzung von Alkwarizmis Algebra und von Euklids Elementen," in *Bibliotheca mathematica*, 3rd ser., **6** (1905), 239–241, argues that the version printed by Libri is the real Gerard translation. On al-Karaji's *Algebra* see Adel Anbouba, *L'algèbre al-Bādiʿ d'al-Karagī* (Beirut, 1964), which is Publications de l'Université Libanaise, Section des Études Mathématiques, II. On ʿUmar al-Khayyāmī's *Algebra* see F. Woepcke, *L'algèbre d'Omar al-Khayyâmî* (Paris, 1851); and, for discussion and further bibliography, Hâmit Dilgan, *Büyük matematikci Ömer Hayyâm* (Istanbul, 1959), in the series Istanbul Technical University Publications. On Hindu values of π see *Alberuni's India*, Edward C. Sachau, trans., I (London, 1910), 168–169; and *The Āryabhaṭīya of Āryabhaṭa*, Walter Eugene Clark, trans. (Chicago, 1930), p. 28.

The Latin text of the treatise on Hindu numerals was first published, carelessly, *Algoritmi de numero indorum* (Rome, 1857), which is Trattati d'aritmetica, B. Boncompagni, ed., I. A facs. text of the unique MS was published by Kurt Vogel, *Mohammed ibn Musa Alchwarizmi's Algorismus* (Aalen, 1963), which is Milliaria, III. Vogel provides a transcription as inaccurate as his predecessor's and some useful historical information. Of the numerous medieval Latin works named *Algorismus* the following have been published: John of Seville's *Alghoarismi de practica arismetrice*, B. Boncompagni, ed. (Rome, 1857), which is Trattati d'aritmetica, II; John of Sacrobosco's *Algorismus*, edited by J. O. Halliwell as "Joannis de Sacro-Bosco tractatus de arte numerandi," in his *Rara mathematica*, 2nd ed. (London, 1841), pp. 1–31; and Alexander of Villa Dei (*ca.* 1225), "Carmen de algorismo," *ibid.*, pp. 73–83. See also M. Curtze, "Über eine Algorismus-Schrift des XII Jahrhunderts," in *Abhandlungen zur Geschichte der Mathematik*, **8** (1898), 1–27.

The Latin version of al-Khwārizmī's *Zīj* was edited by H. Suter, *Die astronomischen Tafeln des Muḥammed ibn Mūsā al-Khwārizmī* (Copenhagen, 1914), which is Kongelige Danske Videnskabernes Selskabs Skrifter, 7. Raekke, Historisk og filosofisk Afd., III, 1. Suter has a useful commentary, but an indispensable supplement is O. Neugebauer, *The Astronomical Tables of al-Khwārizmī* (Copenhagen, 1962), Kongelige Danske Videnskabernes Selskabs, Historisk-filosofiske Skrifter, IV, 2, which provides a trans. of the introductory chapters and an explanation of the basis and use of the tables. Important information on al-Khwārizmī's *Zīj* will be found in the forthcoming ed. of al-Hāshimī's *Book of the Reasons Behind Astronomical Tables* (*Kitāb fī ʿIlal al-Zījāt*), ed. and trans. by Fuad I. Haddad and E. S. Kennedy, with a commentary by David Pingree and E. S. Kennedy. The Arabic text of Ibn al-Muthannā's commentary is lost, but one Latin and two Hebrew versions are preserved. The Latin version has been miserably edited by E. Millás Vendrell, *El comentario de Ibn al-Mutannā a las Tablas astronómicas de al-Jwārizmī* (Madrid–Barcelona, 1963). It is preferable to consult Bernard R. Goldstein's excellent ed., with English trans. and commentary, of the Hebrew versions, *Ibn al-Muthannâ's Commentary on the Astronomical Tables of al-Khwârizmî* (New Haven–London, 1967). On the origin of the *Sindhind* and early versions of it, see David Pingree, "The Fragments of the Works of al-Fazārī," in *Journal of Near Eastern Studies*, **29** (1970), 103–123; "The Fragments of the Works of Yaʿqūb ibn Ṭāriq," *ibid.*, **26** (1968), 97–125; and *The Thousands of Abū Maʿshar* (London, 1968). On Maslama and Ibn al-Ṣaffār's revision of al-Khwārizmī's *Zīj* see Ibn Ezra, *El libro de los fundamentos de las tablas astronómicas*, J. M. Millás Vallicrosa, ed. (Madrid-Barcelona, 1947), pp. 75, 109–110. The relationship of the mean motions in al-Khwārizmī's *Zīj* to the *Brāhmasphuṭasid-*

dhānta was demonstrated by J. J. Burckhardt, "Die mittleren Bewegungen der Planeten im Tafelwerk des Khwârizmî," in *Vierteljahrsschrift der Naturforschenden Gesellschaft in Zürich*, **106** (1961), 213–231; and by G. J. Toomer, review of O. Neugebauer's *The Astronomical Tables of al-Khwārizmī*, in *Centaurus*, **10** (1964), 203–212. Al-Battānī's *Zīj* was edited magisterially by C. A. Nallino, *Al-Battānī sive Albatenii opus astronomicum*, 3 vols. (Milan, 1899–1907), which is Pubblicazioni del Reale Osservatorio di Brera in Milano, XL (vols. I and II repr. Frankfurt, 1969; vol. III repr. Baghdad [?], 1970 [?] [n.p., n.d.]). The *Toledan Tables* have never been printed in their entirety, but they are extensively analyzed by G. J. Toomer, "A Survey of the Toledan Tables," in *Osiris*, **15** (1968), 5–174.

The text of the *Geography* was published from the unique MS by Hans von Mžik, *Das Kitāb Ṣūrat al-Arḍ des Abū Ǧaʿfar Muḥammad ibn Mūsā al-Ḫuwārizmī* (Leipzig, 1926). The classic study of the work is that by C. A. Nallino mentioned above. See also Hans von Mžik, "Afrika nach der arabischen Bearbeitung der *Γεωγραφικὴ ὑφήγησις* des Claudius Ptolemaeus von Muḥammad ibn Mūsā al-Ḫwārizmī," which is *Denkschriften der K. Akademie der Wissenschaften* (Vienna), Phil.-hist. Kl., **59**, no. 4 (1916); and "Osteuropa nach der arabischen Bearbeitung der *Γεωγραφικὴ ὑφήγησις* des Klaudios Ptolemaios von Muḥammad ibn Mūsā al-Ḫuwārizmī," in *Wiener Zeitschrift für die Kunde des Morgenlandes*, **43** (1936), 161–193; and Hubert Daunicht, *Der Osten nach der Erdkarte al-Ḫuwārizmīs* (Bonn, 1968), with further bibliography.

The treatise on the Jewish calendar is printed as the first item in *al-Rasāʾil al-mutafarriqa fiʾl-hayʾa* (Hyderabad [Deccan], 1948). See E. S. Kennedy, "Al-Khwārizmī on the Jewish Calendar," in *Scripta mathematica*, **27** (1964), 55–59. The extract from the treatise on the astrolabe survives in MSS Berlin, Arab. 5790 and 5793. A German trans. and commentary was given by Josef Frank, *Die Verwendung des Astrolabs nach al Chwârizmî* (Erlangen, 1922), which is Abhandlungen zur Geschichte der Naturwissenschaften und der Medizin, no. 3. Severus Sabokht's treatise was edited by F. Nau, "Le traité sur l'astrolabe plan de Sévère Sabokt," in *Journal asiatique*, 9th ser., **13** (1899), 56–101, 238–303, also printed separately (Paris, 1899). The Latin trans. of Māshāʾllāh's treatise was printed several times in the sixteenth century; a modern ed. is in R. T. Gunther, *Chaucer and Messehalla on the Astrolabe*, which is *Early Science in Oxford*, V (Oxford, 1929), 133–232. For the *Chronicle* the principal excerptor is Elias of Nisibis, in his *Chronography*, written in Syrian and Arabic. See the ed. of the latter, with trans. and commentary, by Friedrich Baethgen, *Fragmente syrischer und arabischer Historiker*, which is Abhandlungen für die Kunde des Morgenlandes, VIII, 3 (Leipzig, 1884; repr. Nendeln, Lichtenstein, 1966), esp. pp. 4–5. A fuller ed., with Latin trans., is given in E. W. Brooks and J.-B. Chabot, *Eliae metropolitae Nisibeni opus chronologicum*, 2 vols. (Louvain, 1910), which is Corpus Scriptorum Christianorum Orientalium, Scriptores Syri, vols. XXIII and XXIV. See also the French trans. by L.-J. Delaporte, *La chronographie d'Élie bar-Šinaya* (Paris, 1910). See also Ḥamza al-Ḥasan al-Iṣfahānī, *Taʾrīkh sinī mulūk al-arḍ wa l-anbiyāʾ* (Beirut, 1961), pp. 126, 144. Other excerptors are listed by Nallino, "Al-Khuwārizmī," pp. 471–472.

G. J. TOOMER

KIDD, JOHN (*b.* London, England, 10 September 1775; *d.* Oxford, England, 17 September 1851), *chemistry, anatomy.*

John Kidd was the son of John Kidd, a captain of a merchant ship; and his mother was the daughter of Samuel Burslem, vicar of Etwall, near Derby. He married Fanny Savery, daughter of the chaplain of St. Thomas's Hospital, London, and they had four daughters.

In 1789 Kidd entered Westminster School, London, with a king's scholarship. He was elected to a studentship at Christ Church, Oxford, in 1793, and graduated B.A. in 1797, M.A. in 1800. He then studied at Guy's Hospital, London, from 1797 to 1801, and took his medical degrees at Oxford, M.B. in 1801 and M.D. in 1804. In 1801 he returned to Oxford as reader in chemistry, and in 1803 he became the first Aldrichian professor of chemistry.

Kidd began his teaching career just when the reformed system of examinations, introduced at Oxford in 1800, was requiring greater concentration by students on the classical and mathematical syllabus. Undergraduates, especially those aiming at an honors degree, were thus discouraged by their college tutors from attending lectures on scientific subjects. This system, which did not allow a formal study of the sciences, was attacked from many quarters, but the science teachers at Oxford tacitly acquiesced during the first three decades of the nineteenth century.

In his pamphlet *An Answer to a Charge Against the English Universities* (1818), Kidd discussed to what extent chemistry and other sciences ought to be taught in Oxford, where the training of students was almost exclusively for the church, the law, and the diplomatic service. He concluded that it would be unreasonable to introduce any general requirement regarding the study of science. Nevertheless, Kidd noted that his chemistry course, consisting of thirty lectures a year, was equal in length to that at Guy's Hospital, which was regarded as "the best school in London for Physical Sciences." Kidd also lectured on mineralogy and geology and published the textbooks *Outlines of Mineralogy* (1809) and *A Geological Essay* (1815). Among his pupils was William Buckland, who took over his teaching in these subjects when the readership in mineralogy was instituted in 1813.

After his election to the readership in anatomy in

1816, Kidd became increasingly occupied with the teaching of anatomy and in improving the osteological material in the college museum of Christ Church. He resigned from the chair of chemistry on his appointment as Regius professor of medicine in 1822.

Kidd was always conscious that his audiences were mainly nonscientists and senior members of the university who were genuinely interested in the rapid and varied advances then proceeding in science. They were also most anxious to reconcile their own religious dogmas and beliefs with findings of science which apparently contradicted them. In 1824 Kidd published *An Introductory Lecture to a Course in Comparative Anatomy, Illustrative of Paley's "Natural Theology."* In addition, Kidd was one of eight prominent scientists chosen to contribute to the *Bridgewater Treatises on the Power, Wisdom, and Goodness of God as Manifested in the Creation.* These volumes, financed from a bequest by the eighth Earl of Bridgewater, were intended to maintain the Paley standpoint. Kidd's *On the Adaptation of External Nature to the Physical Condition of Man* (1833) examined the physical nature of man in relation to his whole environment. He went beyond Paley to embrace the idea of a universe designed and adapted to the physical and intellectual requirements of man, chosen by the Creator to be the supreme being. Although these studies were to become somewhat discredited later by the discoveries of Darwin and his followers, they were fully representative of the best thought of the time, both in the level of their learning and in the variety of their religious ideas.

In practical administration, Kidd gave much attention to the content and scope of medical education and to the introduction of a single licensing authority for practitioners, and thus contributed to the modernization of the medical profession in Britain.

BIBLIOGRAPHY

I. Original Works. In addition to the references in the text, Kidd also published a number of papers in the *Philosophical Transactions of the Royal Society*. His most substantial work was the Bridgewater Treatise on *The Physical Condition of Man* (London, 1833). It seems that no manuscript material of Kidd's has survived.

II. Secondary Literature. Apart from obituary notices appearing shortly after his death, little has been written about Kidd's life and work. R. T. Gunther, *Early Science in Oxford*, III (Oxford, 1925), p. 118, confines his reference to a few anecdotes about Kidd in later life.

J. M. Edmonds

KIELMEYER, CARL FRIEDRICH (*b.* Bebenhausen, Württemberg, Germany, 22 October 1765; *d.* Stuttgart, Germany, 24 September 1844), *comparative physiology, anatomy, chemistry.*

Born in a small Swabian town near Tübingen, Kielmeyer was the son of Georg Friedrich and Anna Maria Oberreuter Kielmeyer. His father was an important official in the ducal forest and hunting service. In 1774 he entered the Karlsschule near Stuttgart, recently organized to prepare the most promising youths of Württemberg for state service. Here Kielmeyer and other students—among his contemporaries were Friedrich von Schiller and Georges Cuvier—received comprehensive instruction in the classics, modern languages, public administration and law, mathematics, and all aspects of the natural sciences. Upon completion of the philosophical course Kielmeyer turned to that in medicine; he received medical certification in 1786, upon completing his formal studies, but never practiced. With ducal support he then undertook his *Wanderjahre*. Study at Göttingen with Johann Friedrich Blumenbach, Johann Friedrich Gmelin, and, most important, the physicist Georg Christoph Lichtenberg, was followed by a tour of museums and chemical laboratories in northern Germany.

In 1790 Kielmeyer was made teacher (*Lehrer*) of zoology and associate curator of the natural history collections at the Karlsschule. He assumed direction of chemical instruction and the small chemical laboratory in 1792. Upon suppression of the Karlsschule (1794) he again undertook scientific travels: on the Baltic and North Sea coasts he pursued exacting anatomical studies of marine invertebrates. Kielmeyer's instructional role resumed in 1796 with appointment as professor of chemistry at the University of Tübingen; five years later he was charged with the additional responsibilities of the chair of botany, materia medica, and pharmacy. After leaving Tübingen in 1816, he assumed direction of the Württemberg state art and scientific collections and of the state library, all located in Stuttgart. He retained this position until his retirement in 1839.

Kielmeyer's professorial activity is of crucial importance in understanding the influence of the man and his views. Ever devoted to the broadest viewpoint and a master of the imaginative yet controlled development of an argument, he published very little. Apart from one celebrated essay, his great contemporary reputation and influence rested upon his decisive effect upon his friends and students. This effect resulted as much from exceptional personal character as from the boldness and cogent presentation of doctrine. Kielmeyer's biographer, unfortunately, cannot experience this personality and must deal with Kielmeyer's scientific views only through quite inadequate published material.

Almost no direct record remains of Kielmeyer's anatomical investigations, which began at the Karlsschule and continued at least through the 1790's. He was, in the opinion of qualified observers, an extraordinarily able and diligent practitioner of dissection and inspired many others to follow his interest. Foremost among these was Cuvier, who warmly acknowledged his indebtedness to Kielmeyer and whose accomplishments the latter proudly but not uncritically recorded.[1] Kielmeyer's plan for his celebrated course on comparative anatomy at the Karlsschule (1790–1793) survives in manuscript and deals with important generalities; it is possible that discovery of copies of students' notes from this course—known to have been circulated and read in Germany—may cast light on the obviously great factual foundations upon which he built.

This lack of evidence is extremely harmful to a just assessment of Kielmeyer's overall scientific endeavor. He was deeply concerned with problems of method. Although certain traits appear to link him closely to *Naturphilosophie*—in the 1790's a body of doctrine just taking shape—it is clear that his great regard for Immanuel Kant and tenacious adherence to concrete evidence precluded his ever tumbling into the abyss of radical idealism. Friedrich Schelling may have admired Kielmeyer[2]; the latter certainly could not applaud the extravagances of Schelling and his followers. Kantian criticism was the first step in Kielmeyer's scientific inquiry. "Prior to any research," he wrote, "the human mind must first find agreement on the extent and limitations within which, with the undivided and reciprocal support of all of its powers, it may advance the inquiry."[3] Kielmeyer then found that space and time are the fundamental categories of all understanding. But what we may deduce from them about the external world will find credit only insofar as concrete evidence supports our inferences. Kielmeyer always felt that his conclusions rested on sound empirical foundations; he stressed, moreover, that one's first principles should necessarily be tentative in character. These prescriptions were difficult to observe faithfully.

Kielmeyer proposed a dynamic view of nature. Force and its modalities underlie all phenomena. Force induces effects not only in the present but in the past and future; time is the decisive measure of all things. Kielmeyer's debt to Johann Gottfried von Herder was great. Herder had posited the total historicity of existence and of understanding; he explored the forces by which, ostensibly, changes occurred; he focused closely upon plants and animals and explicitly called for a "philosophical dissector" who would prosecute "comparative physiology" and establish, "through the determination of distinct and identifiable forces," the relations of animals to man and set these relations in close connection to the "whole organization of creation."[4] Here was Kielmeyer's program. He sought to articulate an all-inclusive system of nature. Organisms, which he viewed from the dynamic stance of the comparative physiologist and not the static view of traditional anatomy, were the product of a developmental force—and that force was strictly analogous to (and perhaps identical with) the predominant forces of chemical transformation and, more fundamentally, the forces of physical change in general. Physics and chemistry were presumably the primary bases for interpreting biological change, and Kielmeyer devoted considerable attention to the possible relationship between attraction, chemical affinity, and the developmental force of organisms. The imponderables—light, electricity, and, above all, heat and magnetism—were emphasized and a coherent doctrine of the interaction of opposites (*Polarität*) advocated.

Nevertheless, while deriving the foundation of his general system from physics and devoting a major portion of his professional activity to chemistry, Kielmeyer exerted his greatest influence in biology. All anatomical and classificatory evidence suggested, he believed, the existence of a graduated scale of organisms (*Stufenfolge*) in the present world. The history of the earth as a whole, according to Kielmeyer's "concept of natural history," must deal "not only with its present condition, but also with that which has gone before and perhaps with that which will follow after, that is, [with the earth] as it is, as it was and as it will be."[5]

These incessant transformations are guided by a developmental force, which constituted the heart of Kielmeyer's biological doctrines (it stands as the primary force in the great *Rede* of 1793; irritability and sensibility are later acquisitions of the developing higher organism). Unity of phenomena is dictated by unity of cause, that is, a common developmental force: "I hold that *the force* which, in previous times, brought forth on our earth the series of organisms, is, in its essence and laws [of action], *one and the same* as the force by which today are produced in each organized *individual* the series of its developmental stages."[6] The sequence in both stages was comparable; and thus it was on the dual grounds of observed similarities and the commonality of the developmental force, the latter being decisive, that Kielmeyer gave early expression to what subsequently became known as the doctrine of ontogenetic recapitulation. He also recognized the expression of this force in the characteristic stages of a man's lifetime and hoped to extend its power, through analogy with terrestrial magnetism, to the evolution of the earth itself.

The concept of a common developmental force satisfied Kielmeyer's keen ambition to introduce "unity into all human knowledge," to create "a genealogy of our knowledge" and to do so without the self-deception and arrogance characteristic of contemporary *Naturphilosophen.*[7] As for the nature of the developmental force, Kielmeyer shrewdly offered no inflexible opinions. He accepted it as testimony and concomitant to the essential fact of organic existence, the self-sufficient directedness of vital processes. Following Kant, he declared that the "organs stand in a purposeful relationship to one another. . . . Each is the effect and cause of the other—and for us, therefore, the relationship is purposeful" and not mechanical.[8] And here the analysis must terminate: the author of force in nature is surely also the source of nature's purposefulness. On these matters Kielmeyer maintained a spiritualistic, anti-mechanistic, and probably Christian outlook.

Kielmeyer's instruction and methodological cautions bore rich fruit. Few if any of his students participated in the more excessive forms of *Naturphilosophie*, and some became biologists of great distinction. Cuvier, the celebrated experimental plant hybridizer Carl Friedrich von Gärtner, the anatomist and paleontologist Georg Jäger, Christian Heinrich Pfaff, George Louis Duvernoy, Johann von Autenrieth, and numerous others found their inspiration in Kielmeyer's teaching and informal instruction. To a man they celebrated his exceptional powers and compassionate personality; many looked to him also as a leader of the German people and an outspoken advocate of human dignity and freedom in a period of war, reaction, and national self-definition. He was perhaps the preeminent teacher of physiology in Germany in the generation before Johannes Müller and fully deserved both Alexander von Humboldt's designation as the "foremost physiologist of Germany" and Müller's own tribute: "Germans may proudly claim that it was Kielmeyer who first viewed comparative anatomy from this, its inner side; he who first called it to life and endowed it with [its] intellectual orientation."[9]

NOTES

1. G. Cuvier, "Mémoires pour servir à celui qui fera mon éloge," in P. Flourens, *Recueil des éloges historiques. Première série* (Paris, 1856), pp. 173–174; Kielmeyer, "Einige Notizen über . . . G. Cuviers," pp. 163–186, esp. 177–178.
2. F. Schelling, *Von der Weltseele* (1798); cited in Balss, "Kielmeyer als Biologe," p. 269.
3. Kielmeyer, *Gesammelte Schriften*, pp. 112–113.
4. J. G. von Herder, "Ideen zur Geschichte der Menschheit. Erste Theil (1784)," in *Sämmtliche Werke. Zur Philosophie und Geschichte*, pt. 4 (Stuttgart–Tübingen, 1827), pp. 101–102.
5. Kielmeyer, *Gesammelte Schriften*, p. 228.
6. *Ibid.*, p. 205.
7. *Ibid.*, pp. 239–240, 236.
8. *Ibid.*, p. 180. See I. Kant, *Critique of Teleological Judgement*, translated by J. C. Meredith (Oxford, 1928), § 66 (pp. 24–25).
9. Humboldt, cited in Balss, *op. cit.*, p. 270; Müller, cited in Kielmeyer, *Gesammelte Schriften*, p. 6.

BIBLIOGRAPHY

I. Original Works. Kielmeyer's publications are few in number, diverse in nature, and exceedingly difficult to obtain. Two items are indispensable for understanding his thought. At the birthday celebration (11 Feb. 1793) for Duke Karl Eugen of Württemberg, Kielmeyer pronounced his most important single statement of doctrine. This was soon published as *Ueber die Verhältnisse der organischen Kräfte unter einander in der Reihe der verschiedenen Organisationen, die Gesetze und Folgen dieser Verhältnisse* (Stuttgart, 1793; repr. Tübingen, 1814), extracts apparently trans. into French by Oelsner (Paris, 1815), modern repr. by Heinrich Balss in *Sudhoffs Archiv für Geschichte der Medizin und der Naturwissenschaften*, **23** (1930), 247–267. Kielmeyer's widow left her husband's unpublished scientific MSS to the Württemberg state library (today the Württembergische Landesbibliothek). A selection of exceptional importance and interest from these MSS was published as Kielmeyer's *Gesammelte Schriften*, F. H. Holler, ed., with the collaboration of Julius Schuster (Berlin, 1938). This work is of extraordinary rarity (the publisher's stock was destroyed during World War II), and no copy could be located in the United States; the principal contents are therefore listed here (the titles are the editor's): I. "Selbstbiographie"; II. "Das älteste Programm der deutschen vergleichenden Zoologie"; III. "Naturforschung. Infusionstierchen"; IV. "Die Bewegungslehre. Dynamik"; V. "Organische Kräfte" [the 1793 *Rede* published from the original MS, and including material not found in Balss's ed.]; V. [*sic*] "Die Natur. Gesprochene Urfassung"; VI. [*sic*] "Geschichte und Theorie der Entwicklung"; VII. "Über den Organismus"; VIII. "Über Erde und Leben"; IX. "Über Naturgeschichte"; X. "Über Kant und die deutsche Naturphilosophie. Ein Schreiben an Cuvier"; X*a*. "Lethe. Ein Gedicht"; XI. "Die Württembergischen Botaniker"; XII. "Zu Dissertationen."

Gesammelte Schriften, pp. 12, 254, lists Kielmeyer's other publications; these include his medical dissertation (an analysis of mineral waters) and lectures and reports on chemistry, plant development, and animal magnetism. There is no census or any certain ed. of lecture notes kept by Kielmeyer's students. Some authors (most notably Balss, "Kielmeyer als Biologe") have accepted Gustav Wilhelm Münther's *Allgemeine Zoologie oder Physik der organischen Körper* (Halle, 1840) as an authentic representation of his teacher's viewpoint. There seems to be little evidence either to support or to deny this claim.

II. SECONDARY LITERATURE. While no comprehensive biography of Kielmeyer exists, there are several valuable notices concerning both his life and his doctrines. Foremost among these are Carl Friedrich von Martius, "Carl Friedrich Kielmeyer" [1845], in *Akademische Denkreden* (Leipzig, 1866), pp. 181–209; and Georg Friedrich von Jäger, "Ehrengedächtniss des königl. württembergischen Staatsraths von Kielmeyer," in *Nova acta physico-medica Academiae Caesareae Leopoldino Carolinae germanicae naturae curiosorum*, **21**, pt. 2 (1845), xvii–xcii; Jäger presents often quite full discussion of the content of Kielmeyer's various lecture courses. See also "Selbstbiographie," in *Gesammelte Schriften*, pp. 7–12. Heinrich Balss, "Kielmeyer als Biologe," in *Sudhoffs Archiv für Geschichte der Medizin und der Naturwissenschaften*, **23** (1930), 268–288, describes Kielmeyer's scientific work but bases his analysis exclusively on the contestable ground that Münther fairly restated the master's views. Balss also (p. 288) provides a good bibliography of publications dealing with Kielmeyer. Among these are Max Rauther, "Ungenütze Quellen zur Kenntnis K. F. Kielmeyers," in *Besondere Beilage des Staatsanzeiger für Württemberg* (Stuttgart), no. 6 (1921), 113–122, which examines MSS relating to Kielmeyer's students; and J. H. F. Kohlbrugge, "G. Cuvier and K. F. Kielmeyer," in *Biologische Centralblatt*, **32** (1912), 291–295, a brief analysis of Cuvier's unbroken communication with his Karlsschule friends and based on the Fonds Cuvier of the Institut de France. Cuvier's letters to his German friends are printed in *George Cuvier's Briefe an C. H. Pfaff*, W. F. G. Behn, ed. (Kiel, 1845), trans. into French by L. Marchant (Paris, 1858); see index for references to Kielmeyer. See also Max Rauther, "Carl Friedrich Kielmeyer zu Ehren," in *Sudhoffs Archiv für Geschichte der Medizin und der Wissenschaften*, **31** (1938), 345–350; Felix Buttersack, "Karl Friedrich Kielmeyer (1765–1844). Ein vergessenes Genie," *ibid.*, **23** (1930), 236–246; and the derivative Klüpfel, "Karl Friedrich Kielmeyer," in *Allgemeine deutsche Biographie*, XV (Leipzig, 1882), 721–723.

The basic work for comprehending the context of Kielmeyer's scientific views, above all those on the developmental force and historical understanding, is Owsei Temkin, "German Concepts of Ontogeny and History Around 1800," in *Bulletin of the History of Medicine*, **24** (1950), 227–246. On the Karlsschule and intellectual life in Kielmeyer's Württemberg, see E. Stübler, *Johann Heinrich Ferdinand Autenrieth. 1775–1835. Professor der Medizin und Kanzler der Universität Tübingen* (Stuttgart, 1948). The rise of *Naturphilosophie* and the emergence of "romantic medicine" is brilliantly chronicled by Ernst Hirschfeld, "Romantische Medizin. Zu einer künftigen Geschichte der naturphilosophischen Ära," in *Kyklos*, **3** (1930), 1–89, see pp. 10–11; Hirschfield's bibliography of primary and secondary materials dealing with these issues is outstanding. On the background of Schelling's scientific work, see Rudolf Haym, *Die romantische Schule. Ein Beitrag zur Geschichte des deutschen Geistes*, 3rd ed. (Berlin, 1914), pp. 636–645. On Kielmeyer's doctrinal relations with Schelling see Ludwig Noack, *Schelling und die Philosophie der Romantik. Ein Beitrag zur Culturgeschichte des deutschen Geistes* (Berlin, 1859), I, 216 ff., and the very full and careful discussion, including an accurate restatement of the argument of the 1793 *Rede*, by Kuno Fischer, *Schellings Leben, Werke und Lehre*, which is his *Geschichte der neuern Philosophie*, 3rd ed., VII (Heidelberg, 1902), pp. 342–347.

WILLIAM COLEMAN

KIMURA, HISASHI (*b.* Kanazawa, Japan, 9 September 1870; *d.* Setagaya-ku, Tokyo, Japan, 26 September 1943), *astronomy*.

Hisashi Kimura graduated from the department of astronomy, College of Science, Tokyo University, in 1892. During the following six years, while attending graduate school at the same university, he began to study latitude variation through observation. In 1899 he became director of the Mizusawa Latitude Observatory and devoted the rest of his life to the observation of latitude variation.

The question of latitude variation was of considerable interest to astronomers at the end of the nineteenth century because of Euler's work. In 1765 Euler's theory had predicted that a difference of motion between the principal axis and the rotational axis of the earth's moment of inertia would cause the latter to rotate around the former in a cycle of 305 days. This fact was substantiated 120 years later by Seth Chandler and Friedrich Küstner.

Kimura discovered that in latitude variation there is an annual term independent of the motion of the earth's axis which can be observed regardless of the observer's position ("A New Annual Term in the Variation of Latitude, Independent of the Components of the Pole's Motion," in *Astronomical Journal*, 22 [1902], 197). The observation of the value became an important problem in international observation. Even today the cause for this annual term is not clear but it is presumed that it is a geophysical phenomenon rather than an astronomical one.

Kimura became the chairman of the Latitude Variation Committee of the International Astronomical Union in 1918, and when the Mizusawa Observatory was established as the Central Bureau of International Latitude Observation in 1922, Kimura was elected director; he held both positions until his resignation in 1936.

BIBLIOGRAPHY

His works are published in *Results of International Latitude Service*, **7** (Mizusawa, 1935) and **8** (Mizusawa,

1940). His research achievements are published in *Astronomical Journal,* **36** (1943), 117–119, and in "An Interim Report on an International Research Project: the Wandering of the North Pole," in Harlow Shapley, ed., *Source Book in Astronomy, 1900–1950* (Cambridge, Mass., 1960), pp. 64–66.

S. NAKAYAMA

AL-KINDĪ, ABŪ YŪSUF YA'QŪB IBN ISḤĀQ AL-ṢABBĀḤ (*b.* Basra, beginning ninth century; *d. ca.* 873), *philosophy.*

For a detailed study of his life and work, see Supplement.

KING, CLARENCE RIVERS (*b.* Newport, Rhode Island, 6 January 1842; *d.* Phoenix, Arizona, 24 December 1901), *geology.*

Clarence King's ancestors included Rhode Islanders distinguished in politics, business, and the arts. He was the son of Caroline Florence Little and James Rivers King, a Canton trader. Mrs. King raised her son to be a Congregationalist and helped him with classical languages as a youth. His father died in 1848, but the family business prospered until the Panic of 1857. In 1859 Mrs. King married a merchant who paid for her son's college education. Clarence King took the intensive chemistry course, which included James Dwight Dana's geology lectures, at Yale's Sheffield Scientific School from September 1860 to July 1862, when he graduated with a bachelor of philosophy degree. Between graduation and April 1863, King read further in geology and audited Louis Agassiz's lectures on glaciers.

King joined the California Geological Survey as an assistant from the fall of 1863 to the fall of 1866. He briefly studied the geology of Arizona as a scientific escort to a military road survey in the winter of 1865. From 1867 to 1878 he directed the U.S. Geological Exploration of the Fortieth Parallel, a study of topography, petrology, and geological history along the Union Pacific and Central Pacific railroad lines. The twenty-five-year-old King had obtained this responsibility over the heads of four major generals. During May 1879 to March 1881 he was the first director of the U.S. Geological Survey, winning the appointment with the support of John Wesley Powell, who became his successor in 1881. King also led the mining investigations for the tenth census from May 1879 to May 1882. After resigning he worked as a mining geologist. Despite rheumatism, malaria, and a spinal affliction, King was robust enough for strenuous fieldwork until 1893, when financial and personal worries culminated in a nervous breakdown.

King and Ada Todd, a Negro, were married in New York in September 1888 by a Negro Methodist minister, but fear of scandal kept them from filing the certificate that would have legalized the ceremony. They had five children. Apart from his secret life with his family, King lived in genteel, affluent, literary society when in San Francisco or the East. An intimate friend of John Hay and Henry Adams, King influenced the latter with his enthusiastic adoption of Lord Kelvin's concept of a short age for the earth. He was elected to the National Academy of Sciences (1876) and was a founding member of the Geological Society of America, in addition to joining the American Philosophical Society and other scientific groups.

King's scientific work may be arbitrarily divided into practical, descriptive, and theoretical units. In 1870 he outlined the Green River coal deposits and correctly predicted greater silver strikes in the Comstock lode, and in 1872 he exposed a diamond fraud. His descriptive work included mapping parts of the Sierra Nevada while on the California survey. In 1863 and 1864 he found fossils that dated the Mariposa gold-bearing slates as Jurassic, and in 1870 he discovered glaciers on Mount Shasta. King hired the microscopic petrographer Ferdinand Zirkle for the fortieth-parallel survey to prepare the first extensive monograph (1876) on American rocks studied in thin section. He also instructed his topographers to use the new triangulation methods developed on the California survey and to record their data on contour maps. In 1880 King established Carl Barus' laboratory as part of the U.S. Geological Survey to measure physical constants of rocks, and in 1893 he used Barus' data for diabase to calculate the age of the earth's crust. King's calculation of twenty-four million years, based on Kelvin's theory of the cooling of the earth, was a much shorter time than the uniformitarians had assumed. This figure was widely accepted until the concept of radioactive energy upset the basis of King and Kelvin's work.

In 1877 King, faced with having to explain the geological past of the American West, promulgated a new catastrophist theory that employed more rapid rates of geological change than those operating on the present landscape. In 1878 he extended his theory to account for the source of volcanic lava: when very rapid erosion took place, decreased pressure allowed subcrustal local melting. He refined Ferdinand von Richthofen's law of the succession of volcanic rocks by adding acid, neutral, and basic phases resulting from gravity separation in the magma chamber. His neocatastrophism led him to propose a modification of Darwin's theory of biological evolution: natural selection explained biological change in geologically quiet times, but in revolutions only flexible organisms adapted and survived the rapid change in environment, while the others died out. Given King's prestige, his ideas eventually helped theories such as diastrophism and neo-Lamarckianism to gain a hearing.

BIBLIOGRAPHY

Thurman Wilkins' scholarly, well-written *Clarence King, A Biography* (New York, 1958), includes a bibliography of King's popular and scientific publications, a list of MS collections which have material on King, and citations to printed secondary works.

Treatments of scientific institutions with which King was associated have appeared since Wilkins' book. See, for example, Gerald Nash, "The Conflict Between Pure and Applied Science in Nineteenth Century Public Policy: The California Geological Survey, 1860–1874," in *Isis*, **54** (1963), 217–228; Thomas Manning, *Government in Science: The U.S. Geological Survey, 1867–1894* (Lexington, Ky., 1967); Gerald White, *Scientists in Conflict: The Beginnings of the Oil Industry in California* (San Marino, Calif., 1968), ch. 2, on the California survey; and the section on the fortieth-parallel survey in Richard Bartlett, *Great Surveys of the American West* (Norman, Okla., 1962).

William Goetzmann reviews King's entire scientific career in *Exploration and Empire: The Explorer and the Scientist in the Winning of the American West* (New York, 1968), chs. 10, 12, 16; Loren Eiseley, in *Darwin's Century: Evolution and the Men Who Discovered It* (Garden City, N.Y., 1958), ch. 9, discusses the importance of the problem of the age of the earth; and Edward Pfeifer, in "The Genesis of American Neolamarckianism," in *Isis*, **56** (1965), 156–167, comments on King's relevance for other American scientists.

MICHELE L. ALDRICH

KINNERSLEY, EBENEZER (*b.* Gloucester, England, 30 November 1711; *d.* Pennepack, Pennsylvania, March 1778), *electricity*.

Kinnersley intended to follow the career of his father, an English Baptist minister who settled near Philadelphia in 1714. But his period of probation coincided with that evangelical rush known as the Great Awakening, which offended his rationalistic sensibilities; and in 1740, after preaching in his minister's absence against "whining, roaring Harangues, big with affected Nonsense," he found himself temporarily outside his communion. Although the squabble was composed and he ordained (1743), he never received a pulpit; his religious career effectively ended in 1747 with the last of his published polemics against the elders of his church.

At just this time Franklin, who had printed some of Kinnersley's tracts, began to study electricity. The unoccupied minister, "being honoured with Mr. Franklin's intimacy," became his principal collaborator. A flair for striking demonstrations and a ministerial facility of speech fitted Kinnersley for public lecturing; and, with Franklin's help and encouragement, he successfully toured the colonies from 1749 to 1753, spreading the Philadelphia system and the truth about lightning. In 1753, again with Franklin's help, he became professor of English and rhetoric at the Philadelphia Academy (forerunner of the University of Pennsylvania), a position he retained until 1772.

Kinnersley did his best work in electricity away from Franklin. For example, in 1752 while on a lecture tour in Boston, he rediscovered the two electricities of Dufay and forced upon Franklin the problem of deciding which was truly positive. Kinnersley raised another fundamental point in 1761. Orthodox Franklinian theory ascribed the mutual repulsion between positively charged bodies to their "atmospheres" of springy electrical matter. The reciprocal recession of negatively charged bodies therefore became a problem of principle, since they lacked by definition the necessary repulsive mechanism. Kinnersley suggested that no electrical repulsion existed; the air, he said, draws apart objects similarly charged via the usual attraction between neutral and electrified bodies. Although Franklin rejected the idea, it had important adherents, such as Beccaria (who had proposed it in his *Lettere al Beccari* [Turin, 1758]) and Volta. It was one of several eighteenth-century theories that avoided macroscopic actions at a distance by overtaxing the air with chores later entrusted to the electromagnetic ether. Kinnersley's best-known contribution is the so-called electrical air thermometer, a device for estimating the increase in pressure caused by the passage of a spark through a confined volume of air. It is representative of the plight of colonial American philosophers that here again he was anticipated by Beccaria.

BIBLIOGRAPHY

Kinnersley's important paper of 1761, which includes a description of his thermometer and the theory of repulsion, appeared as "New Experiments in Electricity," in *Philosophical Transactions*, **53** (1763), 84–97. Franklin printed it, as well as the letter of 1752 regarding the two electricities, in later editions of his *Experiments and Observations on Electricity*; see I. B. Cohen, *Benjamin Franklin's Experiments* (Cambridge, Mass., 1941), 250–252, 348–358; the text of a hitherto unpublished lecture on electricity that Kinnersley wrote out in 1752 is on pp. 409–421. A full bibliography appears in J. A. L. Lemay, *Ebenezer Kinnersley, Franklin's Friend* (Philadelphia, 1964), 123–124. On Kinnersley's life see Lemay; *The Papers of Benjamin Franklin*, L. W. Labaree *et al.*, eds. (New Haven, Conn., 1959); and Cohen, *op. cit.*, 401–408. On his work see Lemay, "Franklin and Kinnersley," in *Isis*, **52** (1961), 575–581; M. Gliozzi, "Giambatista Beccaria nella storia dell'elettricità," in *Archeion*, **17** (1935), 15–47; and I. B. Cohen, *Franklin and Newton* (Philadelphia, 1956), 492–494, 531–534.

JOHN L. HEILBRON

KIPPING, FREDERICK STANLEY (*b.* Manchester, England, 16 August 1863; *d.* Criccieth, England, 1 May 1949), *chemistry*.

Kipping became interested in chemistry through the influence of a neighbor, the district's public analyst, who convinced Kipping's banker father that a career in chemistry was honorable. The boy's chemistry teacher in grammar school further whetted his appetite for the subject. He enrolled at the University of London in 1879 but actually attended Owens College in Manchester and graduated with a degree in chemistry three years later. Realizing that his position as chemist at the Manchester Gas Department held little promise of advancement, Kipping entered the University of Munich in 1886 to begin graduate work in Adolf von Baeyer's laboratory. His work there was supervised by W. H. Perkin, Jr., who became a close friend. Kipping received his doctorate with highest honors in 1887 and was awarded a doctor of science degree in the same year from the University of London, the first person to be awarded this degree from that institution solely on the basis of research.

Kipping's first position after graduation was that of demonstrator under Perkin at Heriot-Watt College in Edinburgh. Two years later he was promoted to assistant professor of chemistry and lecturer in agricultural chemistry. In 1890 he was appointed chief demonstrator of the chemistry department at what is now Imperial College of Science and Technology in London. In 1897, the year of his election as a fellow of the Royal Society, Kipping was appointed to the chair of chemistry at University College in Nottingham. His resignation accepted in 1936, Kipping became emeritus professor of chemistry and continued to work regularly in his laboratory until the outbreak of World War II caused him to move with his wife and daughter to the seaside town of Criccieth, where he died at the age of eighty-five. His younger son, Frederick Barry, became a well-known chemist, and his elder son a noted composer of chess problems.

Kipping's early studies involved the preparation and properties of optically active camphor derivatives and nitrogen compounds. He then turned his attention to asymmetric organosilicon compounds and began his most important research. In his preparative reactions he discovered the value of Grignard reagent to substitute organic groups onto silicon atoms. Although his stereoisomer studies were only partially successful, Kipping did report the preparation of the forerunners of the organosilicon polymers. Turning to a detailed investigation of these condensation products, he attempted to prepare the silicon analogues of simple carbon compounds, particularly those with a double bond. The polymeric materials obtained in his endeavor to prepare the analogue of ketones were named "silicones"—the common name now given to the entire class of oxygen-containing organosilicon polymers. Although Kipping was unable to prepare any double-bonded silicon compounds, his extensive studies led to the synthesis of many silicone polymers and to a clear exposition in his fifty-odd published papers of the laboratory techniques necessary to obtain others.

Ironically, Kipping saw absolutely no practical value for the polymeric materials he had laboriously prepared, and in 1937 he lamented that "the prospect of any immediate and important advance in this section of organic chemistry does not seem to be very hopeful." Within four years the first patents for silicone polymers had been issued, and a rapidly growing industry had been born from a marriage of Kipping's experimental procedures and the war's pressure on industry to develop new products.

BIBLIOGRAPHY

I. ORIGINAL WORKS. Kipping's name was familiar to two generations of organic chemistry students as coauthor of the popular textbook *Organic Chemistry*, written with Perkin. The first edition appeared in London in 1894; and after Perkin's death in 1929 Kipping himself made periodic revisions, the last of which is dated 1949. His studies on silicones account for fifty-one numbered papers (and two outside the main series) published in the *Journal of the Chemical Society;* the last (1944) concludes by thanking the society for publishing so much of his work. His 1937 Bakerian Lecture, from which the quote in the text was taken, was published as "Organic Derivatives of Silicon," in *Proceedings of the Royal Society*, **159A** (1937), 139–147.

II. SECONDARY LITERATURE. A detailed biographical

sketch is Frederick Challenger, "Frederick Stanley Kipping," in Eduard Farber, ed., *Great Chemists* (New York, 1961), pp. 1157–1179. His silicone studies are discussed in relation to later developments in Eugene Rochow, *An Introduction to the Chemistry of the Silicones*, 2nd ed. (New York, 1951), pp. 15–87, and in Howard W. Post, *Silicones and Other Silicon Compounds* (New York, 1949).

SHELDON J. KOPPERL

KIRCH. A family of scientists who flourished in Germany in the seventeenth and eighteenth centuries. Four of its members were astronomers:
Gottfried Kirch; his wife, **Maria Margarethe Winkelmann;** his son **Christfried Kirch;** and daughter **Christine Kirch.**
Kirch, Gottfried (*b.* Guben, Germany, 18 December 1639; *d.* Berlin, Germany, 25 July 1710), *astronomy.*
Kirch, Maria Margarethe Winkelmann (*b.* Panitsch near Leipzig, Germany, 25 February 1670; *d.* Berlin, 29 December 1720), *astronomy.*
Kirch, Christfried (*b.* Guben, 24 December 1694; *d.* Berlin, 9 March 1740), *astronomy.*
Kirch, Christine (*b. ca.* 1696; *d.* Berlin, 6 May 1782), *astronomy.*

Gottfried Kirch was the son of a tailor. Because of the unrest of the times, his parents were forced to flee to Poland; on the way, the enemy took all their belongings, and apparently Gottried had to provide for himself while continuing his education. He studied at Jena under the then famous polyhistorian Erhard Weigel, who recommended him to Hevelius in Danzig, one of the most careful observers of his time. After his apprenticeship Kirch returned to Guben, but he also lived for periods of time in Leipzig and Coburg.

In 1692 he married Maria Margarethe Winkelmann (apparently his second wife), the daughter of a Protestant minister. It is thought that she had become interested in astronomy through Christoph Arnold of Sommerfeld, the so-called "astronomical peasant." Arnold was a self-taught astronomer from near Leipzig who observed, among other astronomical phenomena, the great comet of 1683 and the transit of Mercury in 1690. The council of Leipzig was so impressed with these accomplishments that they gave him a sum of money and lifelong freedom from taxes. After his death in 1697 his picture was placed in the library of the city council. Arnold willed part of his manuscripts to Kirch and the rest to the library of the Leipzig city council.

Kirch had fourteen children, two of whom became astronomers. He made his living by computing and publishing calendars and ephemerides. His first calendar appeared in 1667 in Jena and Helmstedt. Calendars were published yearly from 1685 until 1728 in Nuremberg; after Gottfried's death in 1710 his son Christfried continued to publish them under his father's name. The calendars were made with care and became very popular. His ephemerides were calculated for the years 1681–1702. They were based essentially on Kepler's Rudolphine tables and were well-known throughout Europe. Kirch was the first to introduce Halley's *Catalogus stellarum Australium* (1679) to Germany by publishing it as a supplement to his ephemerides for 1681. He was one of the earliest astronomers to search the skies systematically with a telescope and thus discovered several comets, among them the large one of 1680. This comet became an important link in cometary theory, and on the basis of observations of it, Newton indicated the method which Halley used to calculate the parabolic orbits of twenty-four comets.

In 1638 Holwarda of Franeker had found that the magnitude of a star in Cetus fluctuated with a periodicity of 11 months; it was named Mira Ceti ("the miraculous one in the Whale"). In 1667 Montanari at Bologna noticed variations in Algol, and in 1672, a variable star in the constellation Hydra. Kirch had also described Mira Ceti in 1678, and in 1685 he discovered a variable star in the neck of the constellation Cygnus ("in collo Cygni"). He calculated that it had a periodicity of 404½ days. Kirch also observed sunspots, eclipses, and the transit of Mercury in 1707, and designed a new, circular micrometer.

The acceptance of the Gregorian calendar by the Protestant estates of Germany at the end of the seventeenth century gave Frederick III, elector of Brandenburg (later Frederick I of Prussia), the impetus to found an astronomical observatory as well as an academy of science. Kirch was called to Berlin in May 1700 as the first astronomer. The official date of the founding was 11 July 1700, the king's birthday. Because of the meager allotment of funds by the spendthrift king, the building, which was designed to house the observatory as well as the offices of the academicians, proceeded slowly, and the official ceremonies dedicating the building were not held until 19 January 1711, half a year after Kirch's death. While waiting for the building to be finished, Kirch made his observations in his own house and in the private observatory of Baron von Krosigk. A wealthy nobleman and amateur astronomer, Krosigk sent his secretary Kolbe to the Cape of Good Hope in 1705 to make observations corresponding to those being made by Kirch in his observatory in Berlin.

Apparently Kirch had a rather phlegmatic, not too cheerful character and a sickly constitution. While living in Leipzig under difficult financial circumstances, some of his friends, without his knowledge, obtained a stipend from the elector; when Kirch heard about it, he refused to accept it for fear that some poor students, for whom the funds available for scholarship had originally been intended, would be deprived.

Kirch's wife, Maria Margarethe, worked regularly with her husband making observations and especially doing calculations for calendars. After his death she continued to publish on her own. She discovered the comet of 1702, and in 1709 she published a pamphlet about the 1712 conjunction of the sun with Saturn and Venus. In 1712 she wrote about the coming conjunction of Saturn and Jupiter and also communicated rather extensive astrological prognostications, although she admitted that no great value should be given to her interpretation. She went to work in Krosigk's well-equipped observatory in 1712, and upon his death in 1714 moved to Danzig. Peter the Great wanted her to come to Russia, but when her son Christfried became the astronomer of the Berlin observatory she joined him there. She continued to calculate calendars for Breslau, Nuremberg, Dresden, and Hungary until her death in 1720.

Christfried Kirch started his astronomical studies in Leipzig and then journeyed to Königsberg and Danzig to work in Hevelius' old observatory. He put in order the manuscripts Hevelius had left behind and repaired his instruments so that they could be used for his own observations. He remained in Danzig for eighteen months; he returned to Berlin in 1716 to succeed J. H. Hoffman as astronomer of the Berlin observatory; this was the same position his father had held, and Christfried occupied it until his death in 1740. He was a careful observer who noted eclipses of the moon, occultations, and the transit of Mercury in 1720. By observing the eclipses of the satellites of Jupiter, he tried to calculate the differences between the meridians of Berlin, St. Petersburg, and Paris. Together with his assistant Grischon and the famous astronomer Celsius, he observed the solar eclipse of May 1733. He was also interested in ancient astronomical observations, especially those made in China, and Eastern calendars and chronologies. The French Academy of Sciences elected him as a regular correspondent in 1723, and he maintained an extensive correspondence with most of the European astronomers. Christfried was a hardworking, serious man who never married; he lived with his three sisters in complete harmony for almost twenty years. He died of a heart attack.

Christine Kirch assisted her brother Christfried with his observations and calculations. For many years she calculated the calendar for Silesia.

BIBLIOGRAPHY

I. Original Works. According to Lalande, *Bibliographie astronomique*, J. de l'Isle acquired manuscripts and correspondence from the Kirch family for his collection of astronomical material. This collection was bought by the French government and placed in the Dépôt de la Marine. The Berlin Observatory also has manuscript observations. Aside from the ephemerides and calendars, most of their articles were published in *Miscellanea berolinensia*, **1–5** (Berlin, 1710–1737). See Poggendorff I for complete listing.

II. Secondary Literature. See J. E. Bode, *Astronomisches Jahrbuch für das Jahr 1816* (Berlin, 1813); and *Allgemeine deutsche Biographie*, XV, 787–788.

Lettie S. Multhauf

KIRCHER, ATHANASIUS (*b.* Geisa at the Ulster, Germany, 2 May 1602 [or 1601]; *d.* Rome, Italy, 28 November 1680), *polymathy, dissemination of knowledge.*

Kircher was the youngest of six sons (there were also three daughters) of Johannes Kircher of Mainz, D.D. and bailiff of the abbey of Fulda, and Anna Gansek of Fulda. He was educated at the Jesuit Gymnasium in Fulda (1614–1618), where he learned Greek and Hebrew, and entered the Society of Jesus in 1616.

He soon afterwards was at Paderborn, where, until 1622, he studied humanities, natural science, and the various disciplines of seventeenth-century mathematics. After fleeing to Münster and Neuss from the volunteer corps of Christian of Brunswick, he continued his studies at Cologne, where he completed his education in philosophy. At Koblenz (1623), he took up humanities and languages and taught Greek. The following year he taught grammar at Heiligenstadt in Saxony, also studying languages and, especially, physical curiosities.

From 1625 to 1628 Kircher studied theology at Mainz; he was ordained a priest in 1628. His surveying work for the elector during this time contributed to his later interest in geography. At Mainz he first used the telescope for his observations, chiefly of sunspots. He then spent the year of probation at Speyer, where he became interested in hieroglyphics after reading a book on obelisks of Rome. In 1628 Kircher was appointed professor of philosophy and mathematics, as well as of Hebrew and Syriac, at the University of Würzburg. Here he had his first exposure to professional medicine. He also wrote and edited his first book, *Ars magnesia*, which was based on his own experiments.

In 1631, because of the Thirty Years' War, Kircher fled Würzburg and took his disciple, Caspar Schott, with him. Kircher reached Lyons and was appointed to lecture at Avignon, on papal territory. At Avignon he was engaged in different fields, including astronomy, deciphering hieroglyphics, and surveying. His efforts to design a planetarium, using mirrors to direct the light of the sun and moon into the De La Motte tower of the Jesuit college, resulted in a book on astronomical observations by reflected light and another one on catoptrics.

In Avignon, Kircher met the young J. Höwelcke (Hevelius) and corresponded with Christoph Scheiner. In 1633 he was introduced by N. C. Fabri to Gassendi in Aix. It was also Fabri who advised Kircher to attempt an interpretation of Egyptian hieroglyphics. Kircher later edited an improved Coptic grammar (Rome, 1643), recognizing the importance of Coptic in deciphering hieroglyphics.

In 1633 Kircher was appointed by Ferdinand II to a professorship of mathematics in Vienna. But after several shipwrecks, he arrived by chance at Rome, only to learn that he had in the meanwhile been called there—with the intercession of Fabri—by Pope Urban VIII and Cardinal Barberini. In his first years in Rome he worked independently. Later, perhaps in 1638, he was appointed professor of mathematics at the College of Rome. He resigned this post after some eight years. On the whole, he devoted himself to independent studies in this cultural center for some forty-six years until his death.

Some forty-four books and more than 2,000 extant letters and manuscripts attest to the extraordinary variety of his interests and to his intellectual endowments. His studies covered practically all fields both in the humanities and the sciences. This is in harmony with the style of the period, in which polymathy was highly praised. A tendency to deal with curious questions led him to study orientology, including the culture of the Far East. Kircher enjoyed the privilege of living in Rome, the center of a worldwide network of Jesuit missionaries and others who reported on their journeys. Kircher sought to disseminate the knowledge that was at his disposal. His printed works, comprehensive and illustrative, became very popular. Guericke, for example, was greatly indebted to Kircher's *Magnes, sive de arte magnetica* (1643), *Ars magna lucis et umbrae* (1646), *Itinerarius exstaticum* (1656), and *Mundus subterraneus* (1665). Schott exchanged letters with Kircher about Guericke's discoveries, and Jungius and Leibniz quoted from Kircher's works.

Like Manfredo Settala at Milan, Kircher collected various objects and rarities of nature, art, and superstition. In 1663 Martin Fogel of Hamburg visited "Kircher's Museum," as he called it (although it had been founded by Alfonso Donmines in 1650). He recorded his amazement at, for example, a rare piece of wood dug out of the earth; an *Automatum musicum organum*; a "Daimunculus in a liquid, ascending, descending, or remaining in the middle of it depending on man's direction"; and a supposed rib and tail from one of the legendary Sirens. In 1913, relics of Kircher's museum were divided between the Museo Nazionale Romano, Museo Nazionale di Castel Sant'Angelo, and Museo Paleoetnografico del Collegio Romano.

Kircher's diverse studies—including magnetism, optics, astronomy, philology, music theory, acoustics, physics, geology, chemistry, geography, archeology, arithmetic, geometry, theology, philosophy, and medicine—have been only partially explored by specialists as they exist only today, in terms of their own respective sciences.

Certain specific studies by Kircher, however, are worthy of mention. In *Ars magnesia* he described a device for measuring magnetic power by means of a balance. Later, he carefully compiled measurements of magnetic declination from several places around the world, as reported by Jesuit scholars, and particularly by his disciple Martin Martini (1638), who in a letter suggested the possibility of determining longitudes by the declination of a magnetic needle. Recognizing the importance of this method, Kircher brought it to the attention of the scientific world. Kircher also drew lines from the pole of a terella to the magnetic needle moved around it. In short, magnetism, on which Kircher published five books, was for him an omnibus of scientific and also fantastic theories.

Equally interested in optical phenomena, Kircher was one of the first to report on the fluorescence produced by a tincture of wooden pieces called by the Mexicans "Tlapazalli"; that is, *lignum nephriticum*. The tincture was used for curing nephritis. Related to the fluorescence was Kircher's artificial production of the phosphorescent substance first described as *Lapis Bononiensis* (stone of Bologna) by J. C. La Galla in 1612; its basic material was heavy spar. It was Kircher, too, who first reported on sea phosphorescense of organic origin and on afterimages.

In *Ars magna lucis et umbrae*, Kircher applied "magna" to "magnes" of his first work. He argued that light—the "attracting magnes of all things" and connected with the heavens by an unknown chain—behaves exactly like the magnes. He discussed the projecting of sunlight or candlelight on plane mirrors, which were painted with colored pictures, or through an illustrated glass sphere. Only Thomas Rasmussen Walgensten (to whom Kircher referred in the second edition in 1671) succeeded in uniting this principle of projecting translucent pictures by rays having a pointlike source with G. Porta's projection through a hole, to the true magic lantern. Exploring the myth of Archimedes' burning mirrors, Kircher stated that the more times light is reflected between several plane mirrors, the more burning power the rays will obtain. He thus supported the story of Archimedes' purported device.

Optics and horology, both allied to astronomy, also interested Kircher. As a young man he had erected sundials at the Jesuit colleges in Koblenz, Heiligenstadt, and Würzburg. Later he collected various kinds of clocks, including those powered by "heliotropic revolutions of the seed of the solanum plant."

In astronomy Kircher made some progress with telescopes. For the first time he depicted Jupiter and Saturn. His chief interest lay in observations of solar and lunar eclipses and of comets. He acted as a clearinghouse, particularly for reports on eclipses, and supplied astronomers including G. B. Riccioli, G. D. Cassini, and Hevelius with valuable information. Kircher was aware of the improbability of any advance on or deviation from the theory of Copernicus but was enthusiastic about Tycho's system, and he believed—his censors notwithstanding—in the existence of similar worlds created by an omnipotent God.

Kircher's third scientific work worthy of mention was *Mundus subterraneus*. This book on the "subterranean world" is a mixture of odd, partly true speculation. Kircher pointed out a hydrologic circle of water by evaporation, geysers, creeks, cold-water springs, and oozing through the seabed back to the abyss. He assumed the existence of vast underground reservoirs. From subterranean naphtha springs, he suggested, there might be an ever-burning lamp fed on the way through channels; he traced this idea back to the Egyptians of antiquity. He saw hot springs and volcanic eruptions as the consequence of subterranean regions of fire (Kircher witnessed the eruptions in 1638 of Stromboli, Etna, and Vesuvius).

The *Mundus subterraneus* comprised many branches of science, including physics, geography, and chemistry. Kircher described in it a graduated aerometer, unaware that Hypatia had already used this principle around the year 400. Like Magiotti, Kircher explained in his *Magnes* (3rd ed., 1654) the measuring of temperatures in terms of buoyancy by immersing small glass balls in a liquid. Although his geography remained on the general level of sixteenth-century knowledge, unrelated to recent views such as those of Bernhard Varen, his description of the influence of weathering, which he ascribed to a kind of chemical process and to cold, was sound, as was that of the geological action of water and wind. He opposed fraudulent alchemy but supported the transmutation of metals, particularly of iron into copper.

As a "mathematician," Kircher dealt not only with "the hidden mysteries of numbers" and geometry—inventing a "pantometrum" for solving problems of practical geometry—but also, as indicated, with optics, statics, hydrostatics, astronomy, and acoustics. He suggested in 1638 in his *Specula Melitensis* ("a watch tower of Malta") a machine for reading scientific data and, in 1668, a device resembling an organ for teaching methods of solving mathematical problems. Designed to teach all disciplines systematically Kircher's *Ars magna sciendi* (1669) was in the mainstream of the didactic and encyclopedic movement of the century. His *Polygraphia nova et universalis* (1663) contains a system to reduce by the art of combination all languages into one universal tongue; Kircher built on the tradition of Lull but provoked the criticism of Leibniz, Martin Fogel, and the linguist Andreas Müller of Berlin.

It is not surprising that Kircher recorded the first —though imperfect—description of a speaking trumpet (*Musurgia universalis*, 1650). He also reported on remarkable echoes and on the sound of a bell in a *vacuum campana* ringed by a magnet but did not reach a genuine interpretation of the phenomenon. On Kircher's alleged discovery of classical Greek music notes from Pindaric odes, see P. Friedlaender, in *Hermes*, **70** (1935), 463–471. In biology he stated only a relative constancy of the species appealing to God and the history of creation. He frequently used the microscope in his medical investigations (*Scrutinium physico-medicum*, 1658).

Despite particular contributions in specific scientific fields, it should be kept in mind that by far the most of what Kircher described in his works was already known and was due rather to amusement and dissemination of news than to reasonable demonstration of knowledge.

BIBLIOGRAPHY

I. Original Works. A fairly complete bibliography of Kircher's printed works is in Carlos Sommervogel, *Bibliothèque de la Compagnie de Jésus*, part 1, vol. 9 (Brussels–Paris, 1893), cols. 1046–1077. See also the catalogues of the Bibliothèque Nationale, Paris; Library of Congress; and the British Museum. For catalogues of printed writings listed in Kircher's works, see Sommervogel, col. 1072.

Works by Kircher treated above are *Ars magnesia, hoc est disquisitio bipartita empeirica seu experimentalis, physico-mathematica de natura, viribus et prodigiosis effectibus magnetis* (Würzburg, 1631); *Horologium Aven-astronomico-catoptricum* (Avignon, 1634); *Primitiae gnomonicae catoptricae* (Avignon, 1635); *Prodromus coptus sive aegyptiacus* (Rome, 1636); *Specula Melitensis encyclica, hoc est syntagma novum instrumentorum physico-mathematicorum* (Naples, 1638); *Magnes, sive de arte magnetica* (Rome, 1641; 2nd ed., Cologne, 1643; 3rd ed., Rome, 1654); *Ars magna lucis et umbrae in mundo* (Rome, 1646; 2nd ed., Amsterdam, 1671); *Musurgia universalis* (Rome, 1650; facs. repr. Hildesheim, 1970); *Itinerarium exstaticum* (Rome, 1656); *Scrutinium physico-medicum contagiosae luis, quae pestis dicitur* (Rome, 1658); *Pantometrum Kircherianum, hoc est, instrumentum geometricum novum, à . . . Kirchero antehac inventum, nunc . . . explicatum . . . à Gaspare Schotto* (Würzburg, 1660); and *Polygraphia nova et universalis, ex combinatoria arte detecta* (Rome, 1663).

Subsequent works include *Mundus subterraneus* (Amsterdam, 1665); *Arithmologia, sive de abditis numerorum mysterijs* (Rome, 1665); *Magneticum naturae regnum* (Rome, 1667); *Organum mathematicum, libris IX. explicatum a P. Gaspare Schotto* (Würzburg, 1668); *Ars magna sciendi* (Amsterdam, 1669); *Phonurgia* (Kempten, 1673; facs. repr., New York, 1966); and *Tariffa Kircheriana, sive mensa Pythagorica expansa* (Rome, 1679).

The main collection of Kircher's letters and MSS is preserved in the archives of the Pontificia Università Gregoriana, Rome. On other letters, see the papers of John E. Fletcher quoted and Sommervogel, cols. 1070–1077. The Staats- und Universitätsbibliothek, Hamburg, has an original letter of Kircher to C. Schott (30 May 1672); a copy of a letter from Kircher to Schott (12 Feb. 1675); and an extract of a letter from Kircher to Joannes Monrath (9 Apr. 1660).

A letter from Leibniz to Kircher (16 May 1670) in the Pontificia Università Gregoriana has been published by Paul Friedländer in *Atti della Pontificia Accademia Romana di Archeologia*, 3rd ser., *Rendiconti* **13** (1937), 229–231; Kircher's answer (23 June 1670), in the Niedersächsische Landesbibliothek, Hannover, is in G. W. Leibniz, *Sämtliche Schriften und Briefe*, ser. 2, **1** (Darmstadt, 1926), 48–49, and in Friedländer (see above), pp. 232–233. The Hannover library also has other MSS concerning Kircher.

An early collection of Kircher's correspondence has been edited by H. A. Langenmantel (see below), comprising 100 separately numbered pages. A letter by Kircher to an unknown person (7 Dec. 1664) has been reprinted in facsimile in Seng (see below).

II. SECONDARY LITERATURE. A nearly complete bibliography of papers edited from 1913 to 1965, is in M. Whitrow, ed., *Isis Cumulative Bibliography*, **2** (London, 1971), 21; for a list of earlier literature on Kircher see J. Ferguson, *Bibliotheca chemica*, **1** (London, 1954), 466–468.

The best, although older, biographies are A. Behlau, "Athanasius Kircher, eine Lebensskizze," in *Programm des Königlichen katholischen Gymnasiums zu Heiligenstadt* (Heiligenstadt, 1874), 1–18; and Karl Brischar, "P. Athanasius Kircher, ein Lebensbild," Katholische Studien, III, no. 5 (1877). A detailed, although not faultless biography is G. J. Rosenkranz, "Aus dem Leben des Jesuiten Athanasius Kircher 1602–1680," in *Zeitschrift für vaterländische Geschichte und Alterthumskunde* (Verein für Geschichte und Alterthumskunde Westfalens ed.), **13**, n.s. **9** (1852), 11–58.

Kircher's autobiography, *Vita admodum Reverendi P. A. Kircheri* (to be used with some reserve) is in H. A. Langenmantel, ed., *Fasciculus epistolarum* (Augsburg, 1684), 1–78. There is a German trans. with a few relevant additions by Nikolaus Seng, *Selbstbiographie des P. Athanasius Kircher aus der Gesellschaft Jesu* (Fulda, 1901); an account of Kircher's life for young students is J. Leonhardt Pfaff, in *Examina autumnalia in Lyceo et Gymnasio Fuldensi, DD. 20–30 M. Sept. 1831 celebranda* (Fulda, n.d.), 4–39.

A study on Kircher as polymath is Conor Reilly, "Father A. Kircher, S.J., Master of an Hundred Arts," in *Studies*, **44** (Dublin, 1955), 357–468: it is not always accurate and is based on a distorted picture of seventeenth-century Germany.

On Leibniz' relations with Kircher see Paul Friedländer, "Athanasius Kircher und Leibniz, ein Beitrag zur Geschichte der Polyhistorie im XVII. Jahrhundert," in *Atti della Pontificia Accademia Romana di Archeologia*, 3rd ser., *Rendiconti*, **13** (1937), 229–247. On letters of Joannes Marcus Marci von Kronland to Kircher, see J. Marek, "Neznámé dopisy Jana Marka Marci z Kronlandu," in *Dějiny věd a techniky*, **3** (1970), 43–45; and Josef Smolka, "Nové pohledy na J. Marka Marci a jeho dobu," *ibid.*, 45–49. For works describing Kircher's museum, see Sommervogel, col. 1076.

Single topics, apart from those listed in *Isis* (see above) are the following: Three worthwhile analyses by John Fletcher based on new knowledge from Kircher's MSS are "Athanasius Kircher and the Distribution of His Books," in *The Library*, 5th ser. **23** (1969), 108–117; "Medical Men and Medicine in the Correspondence of Athanasius Kircher," in *Janus*, **56** (1969), 259–277; and "Astronomy in the Life and Correspondence of Athanasius Kircher," in *Isis*, **61** (1970), 52–67. Kircher's thoughts on the hydrologic cycle are touched on in a short note by Asit K. Biswas, in *Civil Engineering*, **35** (Apr. 1965), 72.

On Kircher's biology, see Joseph Gutmann, *Athanasius Kircher (1602–1680) und das Schöpfungs- und Entwicklungsproblem* (Fulda, 1938). For aspects of Kircher's geography see Karl Sapper, "Athanasius Kircher als Geograph," in M. Buchner, ed., *Aus der Vergangenheit der Universität Würzburg* (Berlin, 1932), 355–362; a rough sketch on Kircher as music scholar is Oskar Kaul, "Athanasius Kircher als Musikgelehrter," *ibid.*, 363–370.

On medicine, apart from the article by Fletcher (see above), see Georg Sticker, "Die medica facultas Wirtzeburgensis im siebzehnten Jahrhundert," in *Festschrift zum 46. deutschen Ärztetag in Würzburg* (Würzburg, 1927), 75–87. A. Erman gives a critical treatment of Kircher's deciphering of hieroglyphics in *Allgemeine Deutsche Biographie*, XVI (Leipzig, 1882), 1–4. On Martin Fogel's opinion of Kircher's comparison of languages and on Fogel's travels, see Hans Kangro, "Martin Fogel aus Hamburg als Gelehrter des 17. Jahrhunderts," in *Ural-Altaische Jahrbücher*, **41** (1969), 14–32.

On Kircher's chemistry see J. R. Partington, *A History*

of Chemistry, II (London, 1961), 328–333 (the author's emphasis on Kircher's criticism of alchemy is more than historically justifiable). On his physics see Edmund Hoppe, *Geschichte der Physik* (Brunswick, 1926; repr. 1965), *passim;* Ernst Gerland, *Geschichte der Physik* (Munich–Berlin, 1913), *passim;* and Ferdinand Rosenberger, *Die Geschichte der Physik*, I (Brunswick, 1882; repr. Hildesheim, 1965), *passim.*

HANS KANGRO

KIRCHHOF, KONSTANTIN SIGIZMUNDOVICH (GOTTLIEB SIGISMUND CONSTANTIN) (*b.* Teterow, Mecklenburg-Schwerin [now D.D.R.], 19 February 1764; *d.* St. Petersburg, Russia, 14 February 1833), *chemistry.*

Kirchhof's father, Johann Christof Kirchhof, owned a pharmacy until 1783 and at the same time was a postmaster. His mother, the former Magdalena Windelbandt, was the daughter of a tin smelter.

In his youth Kirchhof helped his father run the pharmacy; after the latter's death in 1785 he worked in various pharmacies in the duchy of Mecklenburg-Schwerin, qualifying as a journeyman apothecary. In 1792 he moved to Russia and worked in the same capacity at the St. Petersburg Chief Prescriptional Pharmacy. From 1805 he was a pharmacist and became a member of the Fizikat Medical Council, a scientific and administrative group that supervised the checking of the quality of medicaments and certain imported goods. Kirchhof began his chemical studies under Tobias Lowitz, the manager of the pharmacy, and A. A. Musin-Pushkin. A few of his works were undertaken jointly with A. N. Scherer, and all of his scientific activity was carried out in Russia. In 1805 he was elected a corresponding member, in 1809 an adjunct, and in 1812 an academician adjunct of the St. Petersburg Academy of Sciences. In 1801 Kirchhof was elected a member of the Mecklenburg Natural Science Society, in 1806 a member of the Russian Independent Economical Society, in 1812 a member of the Boston Academy of Sciences, in 1815 a member of the Vienna Economical Society, and in 1816 a member of the Padua Academy of Sciences.

Kirchhof's first major discovery was the decomposition of barite with water, which Lowitz reported in "Vermischte chemische Bemerkungen" (*Chemische Annalen* [1797], 179–181), explicitly mentioning the discoverer. Klaproth had discovered this reaction much earlier. In 1797 Kirchhof reported two important results: the bleaching of shellac, which had an appreciable significance for the production of sealing wax, and a wet process that made it possible to begin industrial production of cinnabar. Cinnabar was produced of such high quality that it supplanted imported cinnabar, and some was exported. In 1805 Kirchhof developed a method for refining "heavy earth" (barite) by allowing caustic potash to react with barium salts. In 1807 he entered a competition organized by the Independent Economical Society to develop a method for refining vegetable oil. In collaboration with Alexander Crichton he worked out the sulfuric acid method of refining oil and received a prize of 1,000 rubles. The two men founded an oil purifying plant in St. Petersburg on Aptekarskiy Island, the largest factory at that time, with an output of about 4,400 pounds of oil per day. In many respects (for example, in the method of adding acid and the clarification of oil by glue) Kirchhof's method is closer to modern methods than that of Thénard (1801).

In 1809 Kirchhof resigned from the Chief Prescriptional Pharmacy but continued to carry out the assignments of the Fizikat Medical Council in his laboratory there; he also conducted investigations in his home laboratory. During this period he began prolonged research to find a method for producing gum from starch in order to supplant the imported products; he then began investigating the optimal conditions for obtaining sugar from starch.

Kirchhof studied the action of mineral and organic acids (sulfuric, hydrochloric, nitric, oxalic and so on) on starch and found that these acids inhibit the jelling of starch and promote the formation of sugar from starch. He also studied the effect of acids on the starches of potatoes, wheat, rye, and corn as well as the effect of acid concentration and temperature on the rate of hydrolysis. At the same time he was searching for new raw materials for producing sugar by the hydrolysis of starch. In 1811 Kirchhof presented to the St. Petersburg Academy of Sciences the samples of sugar and sugar syrup obtained by hydrolysis of starch in dilute acid solutions. He advanced a technological method for producing sugar that was based on his investigations published in 1812. Best results were obtained by adding 1.5 pounds of sulfuric acid in 400 parts of water to 100 pounds of starch. The duration of reaction was between twenty-four and twenty-five hours at 90–100° C. The bulk of the acid did not enter into the reaction with starch, because after completion of the reaction, Kirchhof neutralized it with a specific amount of chalk. This was the first controlled catalytic reaction.

In 1814 Kirchhof submitted to the Academy of Sciences his report "Über die Zucker bildung beim Malzen des Gestreides und beim Bebrühen seines Mehl mit kochendem Wasser," which was published the following year in Schweigger's *Journal für Chemie und Physik*. This report describes the biocatalytic (amylase) action, discovered by Kirchhof, of gluten and of malt in saccharifying starch in the presence of these agents. He showed that gluten induces saccharification of starch even at 40–60° C. in eight to ten hours. During the first hour or two the starch paste was converted into liquid, which after filtration became as transparent as water. Mashed dry barley malt saccharified the starch at 30° R. in one hour. Similarly, Kirchhof studied the starch contained in the malt, separating starch from gluten by digesting it with a 3 percent aqueous solution of caustic potash. The starch treated in this manner could not be converted into sugar. Thus he proved that malt gluten is the starting point for the formation of sugar, while starch is the source of sugar.

The catalytic enzyme hydrolysis of starch discovered by Kirchhof laid the foundation for the scientific study of brewing and distilling and resulted in the creation of the theory of the formation of alcohol.

In his last years of scientific activity Kirchhof developed a method of producing unglazed pottery by treating it with drying oils; a method to refine *chervets* (a substitute for cochineal) from oily substances; and a method for rendering wood, linen, paper, and other substances nonflammable. For refining *chervets* he suggested the regeneration of turpentine by mixing it with water and then distilling the mixture.

Kirchhof also conducted research assigned by the Academy of Sciences, including analysis of gunpowders, William Congreve's rocket fuel, mineral samples, and mineral and organic substances.

BIBLIOGRAPHY

I. Original Works. Kirchhof's writings include "O privedenii chistoy tyazheloy zemli v kristaly posredstvom prostogo sredstva" ("On Crystallizing Pure Barite by Simple Means"), in *Tekhnologichesky zhurnal*, **4**, no. 2 (1807), 116–119; "Über Zucker und Syrop aus Kartoffelmehle Getreide und Andere," in *Bulletin des Neusten und Wissenwürdigsten aus der Naturwissenschaft . . .*, **10** (1811), 88; "O prigotovlenii sakhara iz krakhmala" ("On the Preparation of Sugar From Starch"), in *Tekhnologichesky zhurnal*, **9**, no. 1 (1812), 3–26; and "Über die Zuckerbildung beim Malzen des Getreides und beim Bebrühen seines Mehl mit kochendem Wasser," in Schweigger's *Journal für Chemie und Physik*, **14** (1815), 389–398.

II. Secondary Literature. On Kirchhof's life or work, see S. N. Danilov, "Khimia sakharov v Rossii" ("Chemistry of Sugar in Russia"), in *Materialy po istorii otechestvennoy khimii* ("Papers on the History of Russian Chemistry"; Moscow, 1951), pp. 282–283; A. Mittasch and E. Theiss, *Von Davy und Dobereiner bis Deacon, ein halbes Jahrhundert Grenzflächenkatalyse* (Berlin, 1932); pp. 14, 97, 243; A. A. Osinkin, "Zhizn i deyatelynost akademika K. Kirkhgofa" ("The Life and Work of Academician K. Kirchhof"), in *Trudy Instituta istorii Estestvoznaniya i tekhniki. Akademiya nauk SSSR*, **30** (1960), 253–287; and A. N. Shamin, *Biokataliz i biokatalizatory* (*istorichesky ocherk*) ("Biocatalysis and Biocatalysts [A Historical Note]"; Moscow, 1971), 51–54, 102–117, 171–183.

A. N. Shamin
A. I. Volodarsky

KIRCHHOFF, GUSTAV ROBERT (*b.* Königsberg, Germany, 12 March 1824; *d.* Berlin, Germany, 17 October 1887), *physics.*

Kirchhoff's major contribution to physics was his experimental discovery and theoretical analysis in 1859 of a fundamental law of electromagnetic radiation: for all material bodies, the ratio of absorptive and emissive power for such radiation is a universal function of wavelength and temperature. Kirchhoff made this discovery in the course of investigating the optical spectra of chemical elements, by which, in collaboration with Bunsen, he laid the foundation of the method of spectral analysis (1860). Outstanding among his other contributions were his early work on electrical currents (1845–1849) and on the propagation of electricity in conductors (1857). A master in the mathematical analysis of the phenomena, he insisted on the clear-cut logical formulation of physical concepts and relations, directly based on observation and leading to coherent systems free of hypothetical elements. His teaching had a considerable influence on the development in Germany of a flourishing school of theoretical physics during the first three decades of the twentieth century.

Kirchhoff's uneventful life and career afford a typical example of the somewhat parochial but substantial comfort which the academic profession enjoyed in Germany during a period of unprecedented economic expansion. Latecomers in the industrial revolution, the Germans, more than their wealthier English and French competitors, had to rely on scientific methods for the improvement of technology. The paradoxical result was that physics and chemistry found more favorable conditions of development under the multitude of petty feudal governments in Germany than in the progressive environments of prosperous manufacturing centers in England and France.

In Königsberg, Kirchhoff's birthplace, a nucleus of enterprising tradesmen and able officials had fostered a thriving intellectual circle. Kirchhoff's father, a law councillor (*Justizrat*), belonged to the strongly disciplined body of state functionaries which also included university professors. He regarded it as a matter of course that his sons keep up, according to their diverse talents, the family's allegiance to the service of the Prussian state. Gustav, the most gifted, upheld this tradition which still determined the careers of his own children. Yet, like other prominent figures of the German intelligentsia, he does not seem to have had difficulty in reconciling submission to authority in political matters with liberal opinions in other respects. The Manchester chemist Henry Roscoe, who knew both Kirchhoff and Bunsen well, relates an incident from a visit they paid him, which shows them taking quite a Voltairean view of the church.

Boltzmann described Kirchhoff, at the height of his powers, as being not easily drawn out but of a cheerful and obliging disposition. A disability from an accident, which compelled him to use crutches or a wheelchair, did not alter his cheerfulness, and he bore with patience the long illness of his last years.

At the university in his native city, Kirchhoff came under the influence of Franz Neumann; in the new science of electromagnetism, Neumann introduced and further developed in Germany the ideas and methods of the leading French school of mathematical physics. In 1847 Kirchhoff graduated from the university and married Clara Richelot, the daughter of another of his teachers—thus fulfilling the two prerequisites for a successful academic career. In 1848 he obtained in Berlin the *venia legendi* (the right to lecture privately in a university) and two years later became extraordinary professor in Breslau. In 1851 Bunsen, Kirchhoff's senior by thirteen years, came to Breslau, only to leave again the next year for Heidelberg. This brief period was sufficient to create a lasting friendship between the two men.

In 1854, on Bunsen's proposal, Kirchhoff was called to Heidelberg. He found there a congenial environment for his talents as teacher and investigator; and it was there that, partly in collaboration with Bunsen, he made his greatest contributions to science. This was the heyday of the university of Heidelberg, where the academic circle gathered around Helmholtz and, dominated by him, led a showy social life.

Kirchhoff's wife died in 1869, leaving him with two sons and two daughters; in 1872 he married Luise Brömmel, the superintendent in the ophthalmological clinic. On two occasions he turned down calls to other universities; only when his failing health hindered his experimental work did he accept a chair of theoretical physics offered him in Berlin (1875). He took up this new task with great devotion, until illness forced him to give up his teaching activity in 1886. A year later, physically weak but intellectually alert he died peacefully, presumably of a cerebral congestion.

Kirchhoff's first scientific work dates from the time when he was studying under Neumann. One of the results he then arrived at has become, on account of its practical importance, a classical part of the theory of stationary electric currents: it is the formulation of the laws governing the distribution of tension and current intensity in networks of linear conductors (1845–1846). The derivation of these laws was essentially a simple application of Ohm's law, but generalizing it fully, as the twenty-one-year-old student did, demanded uncommon mathematical skill.

Ohm's theory of electric current (1828) was based on the hypothetical analogy between the flow and distribution of current in a conductor of any shape to which a "tension" is applied, and the flow of heat in a body at whose boundary some inequality of temperature is established. But apparently neither Ohm himself nor others had realized that the failure to follow up the analogy consistently had led to erroneous results. Thus Ohm thought that a uniform distribution of electricity could subsist at rest inside a conductor.

Kirchhoff's turn of mind was such that he was not long in discovering such a logical flaw and finding the way to mend it. In 1849 he was induced to look into the matter when confronted with some experiments by Kohlrausch on a closed circuit including a condenser, which involved both a static distribution and a flow of electricity. Kirchhoff pointed out that a consistent formulation of Ohm's theory required (at least for stationary currents) the identification of the tension with the electrostatic potential. Thus a correct mathematical unification of electrostatics and the theory of voltaic currents was achieved after more than twenty years of neglect.

The theory of variable currents raised more difficult problems. The law of dynamical interactions between currents had been formulated by Ampère (1826) in the spirit of the concept of action at a distance. The followers of the French school in Germany, Franz Neumann and Wilhelm Weber, concentrated their efforts on the search for an extension of the law of electrostatic interaction between charges, which would embody the new forces at play when the charges are in motion. Although their first attempts in this direction date from 1845–1846, progress was slow, owing above all to the technical difficulty of the experiments required for checking the validity of the necessarily speculative hypotheses on the nature of the electric current, from which the theoretical developments had to start.

The field was still open when Kirchhoff entered it in 1857 with his own general theory of the motion of electricity in conductors. His first paper, in which he treated linear conductors from the same premises as Weber, turned out to coincide in all essentials with an investigation carried out by Weber shortly before but delayed in publication. Both physicists noticed a remarkable implication of their theory: in a perfectly conducting circuit, oscillating currents could be propagated with a constant velocity, independent of the nature of the conductors, and numerically equal to the velocity of light. Both Kirchhoff and Weber, however, pointing to the extreme character of the condition of infinite conductivity, dismissed this result as a mere accidental coincidence.

In a second paper Kirchhoff presented a generalization of the theory to conductors of arbitrary shape. Although his equations purporting to give the local distribution of current and electromotive force were fundamentally wrong, they did yield for the total current the approximate equation already derived by William Thomson, and known as the "telegraphists' equation" on account of its application to the propagation of current in the transatlantic cable then being laid.

The element lacking in Kirchhoff's analysis was obviously the displacement current, or in equivalent terms, the introduction of retarded potentials. It is highly instructive to observe that this decisive step was indeed taken in 1866 by Ludvig Lorenz. Starting from Kirchhoff's equations (modified in order to express the finite velocity of propagation of electromagnetic forces), Lorenz demonstrated the existence of purely transversal current waves which, in a perfectly transparent medium (a medium of vanishing conductivity), are propagated at the velocity of light. Lorenz was quite clear about the far-reaching consequences of his analysis: it spelled the end of the conception of action at a distance and opened the way to an identification of optical and electromagnetic waves.

Kirchhoff, aiming at a neat mathematical theory complete in itself, was operating with limited sets of concepts and relations directly suggested by experience. That he thus narrowly missed a great discovery illustrates the weakness inherent in his phenomenological method: emphasizing logical consistency entails the risk of closing the logical construction too soon and of overlooking possible connections between qualitatively different phenomena. In the case of voltaic currents, the closure of the theory demanded an extension of the scope of the potential concept, and the method led—by good luck—to a unification of two hitherto separated domains. But in electrodynamics the opposite happened. The ideal program of a physics in which the various forces of nature would be ascribed to specific, sharply separated types of action at a distance blinded its adherents to the strong hint of a possible similarity between the dynamics underlying optical and electromagnetic phenomena. Lorenz' success, by contrast, resulted from his firm belief in the essential unity of all physical phenomena.

The events leading to the foundation and elaboration of the method of spectral analysis have been described by Bunsen (whose testimony is related by W. Ostwald in his edition of the classical paper by Bunsen and Kirchhoff). Bunsen was exploring the possibility of analyzing salts on the basis of the distinctive colors they gave to flames containing them; he had tried with some success to use colored pieces of glass or solutions to distinguish similarly colored flames. Kirchhoff pointed out that a much finer and surer distinction could be obtained from the characteristic spectra of such colored flames; unknown to him, the approach had been tried before, if only in a dilettantish way.

By rigorous experimentation, however, Bunsen and Kirchhoff soon put the method on a firm basis. The burner invented by Bunsen gave a flame of very high temperature and low luminosity, which emitted line spectra of great sharpness. The salts they investigated were prepared in a state of highest purity, and a spectroscope was specially designed to allow the positions of the lines to be accurately determined. By testing an extensive variety of chemical compounds, the ascription to each metal of its characteristic line spectrum was uniquely established (1860). The power and importance of spectral analysis became immediately apparent: its very first systematic application to alkali compounds led Bunsen to the discovery of two new alkaline elements, cesium and rubidium (1860).

In the course of his preparatory work in the autumn of 1859, Kirchhoff made an unexpected observation. It had long been known that the dark D lines, noticed in the solar spectrum by Fraunhofer (1814), coincided with the yellow lines emitted by flames containing sodium. (This effect could be accurately checked by allowing sunlight to reach the spectroscope after traversing a sodium flame; if the sunlight was sufficiently dimmed, the dark Fraunhofer lines were replaced by the bright lines from the flame.) Kirchhoff's unexpected discovery was that if the intensity of the solar spectrum increased above a certain limit, the dark D lines were made much darker by the interposition of the sodium flame. He instantly felt that he had got hold of "something fundamental," even though he was at a loss to suggest an explanation.

On the day following the surprising observation, Kirchhoff found the correct interpretation, which was soon confirmed by new experiments: a substance capable of emitting a certain spectral line has a strong absorptive power for the same line. In particular, the interposition of a sodium flame of low temperature is sufficient to produce artificially the dark D lines in the spectrum of an intense light source which did not show them originally. The dark D lines in the solar spectrum could accordingly be ascribed to absorption by a solar atmosphere containing sodium. Immense prospects thus opened up of ascertaining the chemical composition of the sun and other stars from the study of their optical spectra.

A few more weeks sufficed for Kirchhoff to elaborate a quantitative theory of the relationship between emissive and absorptive power. He attacked the problem directly by a wonderfully simple and penetrating argument. He considered the balance of radiative exchanges between bodies with appropriately chosen properties of absorption and emission. From the sole condition of radiative equilibrium at a given temperature, he was able to conclude that the ratio of absorptive and emissive powers, for each wavelength, must be independent of the nature of the bodies, and hence that it was a universal function of wavelength and temperature. In a later elaboration of the argument (1862), he introduced the conception of a "black body," which absorbs completely every radiation incident on it. Since by definition the absorptive power of such a body has its maximum value, unity, for all wavelengths, its emissive power directly represents the universal function whose existence is asserted by Kirchhoff's law. Hence, this function expresses the spectral distribution of the energy of radiation in equilibrium with a black body of given temperature; moreover, the empirical determination of this universal distribution is reduced to the practical problem of devising a material system with properties approximating those of a black body, and of measuring its emissive power.

Thus, Kirchhoff's law was the key to the whole thermodynamics of radiation. In the hands of Planck, Kirchhoff's successor to the Berlin chair, it proved to be the key to the new world of the quanta, well beyond Kirchhoff's conceptual horizon.

Kirchhoff's derivation of the fundamental law of radiative equilibrium is the triumph of his phenomenological method. He was fully aware of this methodological aspect and attached great importance to it. About ten years before the events just related, Stokes had commented to William Thomson on the coincidence of the Fraunhofer D lines and the bright lines of the sodium flame. Stokes suggested resonance as a mechanical explanation of this phenomenon: the sodium atom would have a proper frequency of vibration corresponding to that of the yellow light it emits and would accordingly absorb most intensively light of the same frequency. Now, Stokes's suggestion, which appears to us a striking anticipation of the atomic basis of Kirchhoff's law, did not appeal to Kirchhoff. When called upon to express an opinion on it (1862), Kirchhoff firmly asserted that the truth of the law had been established only by his own theoretical considerations and the supporting experiments; he thus implicitly denied Stokes's argument any demonstrative value.

Yet, Kirchhoff was not averse to atomistic ideas. Whenever he judged the atomic substratum of phenomena to be sufficiently accessible to analysis, as in the kinetic theory of gases, he readily adopted the proposed atomistic picture. Fully sharing the common ideal of a purely mechanical description of the universe, he realized that such a description could be achieved only on the atomic scale; but he thought—with some reason—that the time was not ripe for it. For him, arguments depending on detailed and unwarranted assumptions about the structure and properties of atoms were without cogency in spite of their suggestiveness. Kirchhoff's fidelity to the phenomenological point of view was thus dictated solely by methodological reasons; if this viewpoint sometimes proved too narrow, it nevertheless inspired not only his discoveries but his no less original attempt at a systematic exposition of the whole of physics. The historical importance of this attempt should not be underestimated.

In a period of expanding scientific horizons, the need soon arises for ordering and logical analysis of new knowledge. Among the leading physicists of the nineteenth century, it was Kirchhoff whose temperament was best suited to this task. In all his work he strove for clarity and rigor in the quantitative statement of experience, using a direct and straightforward approach and simple ideas. His mode of thinking is as conspicuous in his contributions of immediate practical value (the laws of electrical networks) as in those with wide implications (the method of spectral analysis). The excellence of Kirchhoff as a teacher can be inferred from the printed text of his lectures (he managed to publish only those on mechanics, the others being edited posthumously). They set a standard for the teaching of classical theoretical physics in German universities, at a time when they were taking a leading position in the development of science.

BIBLIOGRAPHY

I. Original Works. Kirchhoff edited his collected works as *Gesammelte Abhandlungen* (Leipzig, 1882); they were completed with *Nachtrag*, L. Boltzmann, ed. (Leipzig, 1891). *Vorlesungen über mathematische Physik* was published at Leipzig in 4 vols.: I. *Mechanik* (1876; 4th ed., 1897); II. *Mathematische Optik*, K. Hensel, ed. (1891); III. *Electricität und Magnetismus*, M. Planck, ed. (1891); and IV. *Theorie der Wärme*, M. Planck, ed. (1894).

See also *Chemische Analyse durch Spectralbeobachtungen*, Ostwalds Klassiker der Exakten Wissenschaften, no. 72 (Leipzig, 1895), written with R. Bunsen and edited with notes by W. Ostwald.

II. SECONDARY LITERATURE. On Kirchhoff and his work, see L. Boltzmann, "Gustav Robert Kirchhoff," in *Populäre Schriften* (Leipzig, 1905), 51–75; H. Roscoe, *The Life and Experiences of Sir Henry Enfield Roscoe Written by Himself* (London, 1906), 68–74; L. Rosenfeld, "The Velocity of Light and the Evolution of Electrodynamics," in *Nuovo cimento*, **4**, supp. 5 (1956), 1630–1669; and W. Voigt, *Zum Gedächtniss von G. Kirchhoff* (Göttingen, 1888), also in *Abhandlungen der K. Gesellschaft der Wissenschaften zu Göttingen*, **35** (1888).

L. ROSENFELD

KIRKALDY, DAVID (*b.* Dundee, Scotland, 4 April 1820; *d.* London, England, 25 January 1897), *metallurgy, mechanical engineering.*

Kirkaldy, who was poor in health as a boy, was educated by a Dr. Low at Dundee and at Merchiston Castle, Edinburgh. He worked first in his father's shipping office and then as an apprentice in the shipbuilding works of Robert Napier in Govan (Glasgow) in 1843. He was transferred to the drawing office where he developed skills which, in 1855, brought him the signal honor of having his engineering drawings exhibited at the Royal Academy of Arts in London. He received many other awards for the excellence of his drawings, among them a gold medal at the 1862 International Exhibition in London. He took an early interest, entirely novel in his day for its quantitative approach, in the performance of ships on trial. He showed how to rectify defects in a vessel by the comparative study of performance figures.

Kirkaldy's main contribution to science and engineering, however, proved to be in the field of testing. He became engaged in this in 1858 when Napier asked him to compare the merits of iron and steel for some high-pressure boilers. He studied testing methods for some years and published his findings in papers and in *Results of an Experimental Enquiry* (1862), a book of great importance. Some of his tests gave rise to new methods of improving steel, for example, an oil-hardening process which he patented.

He eventually became independently established in his profession and spent the rest of his life as a testing consultant and manufacturer of testing machinery. Kirkaldy's first machine was publicly demonstrated on 1 January 1866, and one of his first public commissions was for testing the materials for the Blackfriars Bridge. He was able to attract many commissions, for testing and for supplying testing machinery, from foreign governments, construction firms, and even a laboratory at University College, London.

Forthright in manner, Kirkaldy commanded respect but fell foul of more conservative professional colleagues. Nevertheless he succeeded in establishing the testing of materials as an essential factor in civil engineering construction.

BIBLIOGRAPHY

Kirkaldy's originality is readily seen in *Results of an Experimental Enquiry Into the Comparative Strength and Other Properties of Various Kinds of Wrought Iron and Steel* (Glasgow, 1862), and in *Result of an Experimental Enquiry Into the Relative Properties of Wrought Iron Plates Manufactured at Essen and in Yorkshire* (London, 1876). For a summary of his professional experience, written by his son and supervised by himself in the latter part of his life, see W. G. Kirkaldy, *Illustration of D. Kirkaldy's System of Mechanical Testing as Originated and Carried On by Him* (London, 1891). Kirkaldy also contributed supplementary matter to Peter Barlow, *Treatise on the Strength of Materials*, 3rd ed. (London, 1867).

On Kirkaldy's contribution to knowledge of the crystal structure of metals, see C. S. Smith, *History of Metallography* (Chicago, 1960), pp. 161–163. Memoirs of Kirkaldy are found in *Proceedings of the Institution of Civil Engineers*, **128** (1897), 351–356, and in *Engineer*, **83** (1897), 147, with portrait.

FRANK GREENAWAY

KIRKMAN, THOMAS PENYNGTON (*b.* Bolton, England, 31 March 1806; *d.* Croft, near Warrington, England, 3 February 1895), *mathematics.*

Raised in an unscholastic mercantile family, Kirkman had to struggle for a decent education, and even so he received no instruction in mathematics at any level. He earned an arts degree at Dublin University in 1833 (M.A., 1850), and was ordained into the Church of England, becoming rector at Croft, Lancashire. Nominally, this was his life's work, for there the Reverend Mr. Kirkman tended his parish and defended his creed by sermon and pamphlet for more than fifty years. But he also taught himself mathematics with a thoroughness and insight that propelled him swiftly to the frontiers of current research and earned him the admiration and friendship of Cayley, De Morgan, and William Rowan Hamilton. He was elected to the Royal Society in 1857. Kirkman was a good linguist and an individualistic writer, if perhaps overly fond of neologisms and stylistic gimmickry. He delighted in versifying problems and in devising mnemonics for troublesome formulas—in fact he wrote a whole book on this topic.

Kirkman's interests extended to the controversies of the times, and he was fierce in his opposition to the new materialistic trends. Herbert Spencer's philosophy aroused his especial contumely, and his satiric paraphrase of Spencer's definition of evolution is a notable example of a Kirkmannerism. Spencer, he wrote, was really defining the concept as "a change from a nohowish untalkaboutable all-likeness, to a somehowish and in-general-talkaboutable not-all-likeness, by continuous somethingelseifications and sticktogetherations."

His mathematical work contributed to five topics then in infancy: topology, group theory, hypercomplex numbers, combinatorics, and knots. He also wrote lengthily on a very old topic: polyhedra (or, as he insisted, on calling them, *polyedra*). Hamilton's discovery of quaternions stimulated Kirkman to one of the earliest attempts to extend the notion further, and he named his new numbers *pluquaternions*. It is, however, in combinatorics that Kirkman's name is now best known, and his Fifteen Schoolgirls Problem and its variations became and remained famous. (Essentially, it concerns ways of rearranging a sevenfold 5×3 array of distinct objects, with the restriction that the triples are individually unique and collectively comprehensive.) Many other problems of this nature were first enunciated and solved by Kirkman.

BIBLIOGRAPHY

Fifty-nine of Kirkman's chief papers are listed in the Royal Society *Catalogue of Scientific Papers*. Three of note are "On Pluquaternions and Homoid Products of *n* Squares," in *Philosophical Magazine*, **33** (1848), 447–459, 494–509; "Application of the Theory of Polyedra to the Enumeration and Registration of Results," in *Proceedings of the Royal Society*, **12** (1862–1863), 341–380; and "The Complete Theory of Groups, Being the Solution of the Mathematical Prize Question of the French Academy for 1860," in *Memoirs and Proceedings of the Manchester Literary and Philosophical Society*, **4** (1865), 171–172.

His Fifteen Schoolgirls Problem was first posed in the *Lady's and Gentleman's Diary* (1850), p. 48, and it is thoroughly discussed in W. W. R. Ball, *Mathematical Recreations and Essays*, revised by H. S. M. Coxeter, 11th ed. (London, 1939).

Only recently has the generalized problem (for $6n + 3$ girls) been solved: "Solution of Kirkman's Schoolgirl Problem," in D. K. Ray-Chaudhuri and R. M. Wilson, *Combinatorics*, Vol. XIX of *Proceedings of Symposia in Pure Mathematics* (Providence, R. I., 1971).

Some miscellaneous publications of Kirkman's are cited in *Memoirs and Proceedings of the Literary and Philosophical Society*, **9** (1894–1895), 241–243, preceded by a short memoir on the author. A fuller account of the man and his work is Alexander Macfarlane, *Lectures on Ten British Mathematicians of the Nineteenth Century* (New York, 1916), pp. 122–133.

NORMAN T. GRIDGEMAN

KIRKWOOD, DANIEL (*b.* Harford County, Maryland, 27 September 1814; *d.* Riverside, California, 11 June 1895), *astronomy*.

Of Scots-Irish descent, Kirkwood was the son of John Kirkwood, a farmer, and Agnes Hope. He spent his formative years on his father's farm, and his early education was limited to a nearby country school. In 1833, having little interest or aptitude in farming, he took a teaching post at a small school in Hopewell, Pennsylvania. There he encountered a student who wished to study algebra, and together they mastered Bonnycastle's *Algebra*. In the spring of 1834, Kirkwood entered York County Academy, where he continued his mathematical studies; in 1838 he became first assistant and instructor in mathematics. In 1843 he was appointed principal of Lancaster High School, and two years later he married Sarah A. McNair, of Newtown, Pennsylvania. He became principal of Pottsville Academy in 1849. From 1851 to 1856 he was professor of mathematics at Delaware College, after 1854 also serving as college president.

Never one to relish public notice or the assumption of authority, Kirkwood preferred to devote his energies to teaching and research and eagerly accepted the chair of mathematics at Indiana University. Except for the interval 1865–1867 (when he was professor of mathematics and astronomy at Washington and Jefferson College in Pennsylvania), he remained at Indiana University for thirty years, retiring in 1886. He then moved to California and at the age of seventy-seven became nonresident lecturer on astronomy at Stanford University. Kirkwood was an immensely popular, enthusiastic, and inspiring teacher. He was a deeply religious and serene person who never found any conflict between his cosmogonical studies and his Presbyterian faith.

Kirkwood's research and many writings were generally devoted to an understanding of the nature, origin, and evolution of the solar system; he studied, in particular, the role of the lesser members of the system—asteroids, comets, and meteoric and meteoritic bodies. His first publication (1849) consisted of a demonstration that the square of the number of rotations per orbital revolution of a planet is proportional to the cube of the radius of the sphere of attraction given by the Laplace nebular hypothesis. His subsequent research revealed that both this "Kepler-type" law and the nebular hypothesis required severe modification, although he never abandoned the hypothesis completely.

Kirkwood's most important astronomical discovery was of the "gaps" or "chasms" in the distribution of the mean distances of the asteroids from the sun. He made this discovery as early as 1857, when only about fifty asteroids were known. He published two short lists of asteroidal resonances in the *Astronomical Journal* (1860); the lists revealed an obvious lack of asteroids in simple resonance with Jupiter, while asteroids in resonance with Mars were commonplace. His first formal publication of the discovery was not made until 1866, by which time the number of asteroids known had risen to eighty-seven. He remarked on discontinuities in the distribution at distances corresponding to periods of revolution of one-third, two-fifths, and two-sevenths that of Jupiter. He later mentioned gaps at one-half, three-fifths, four-sevenths, five-eighths, three-sevenths, five-ninths, seven-elevenths, and four-ninths. Several more gaps have since been recognized. Kirkwood also pointed out in 1866 that the Cassini division between rings A and B of Saturn exhibits the same phenomenon, for any particles on the division would have periods of revolution one-third that of the satellite Enceladus. Soon afterward he noted that the periods would also be close to one-half that of Mimas and that there were also resonances with Tethys and Dione; furthermore, the Encke division in ring A also corresponded to resonances with Saturn's satellites.

A simple qualitative explanation of the Kirkwood gaps is that the repetitive gravitational action of Jupiter (or Mimas) would quickly remove any asteroid (or ring-particle) away from resonance. The effect would be most pronounced for the resonances of lowest order, where the difference between the numerators and denominators of the fractions is small. Some of Kirkwood's contemporaries doubted that this was really an explanation. In addition Kirkwood himself consistently maintained that the regularly increasing orbital eccentricities of particles initially in resonance would lead to the possibility of collisions with nearby non-resonant particles, and that this was really the cause of the elimination. In the asteroid problem he felt that mutual collisions might be too infrequent, and elimination could be achieved instead by collisions with the sun: according to the nebular hypothesis, the radius of the sun would be only slightly smaller than the mean distance of a newly formed asteroid.

A completely satisfactory explanation of the gaps is still lacking. It is not clear whether they represent merely a statistical underpopulation, with the instantaneous mean distances of objects near resonance tending at any time to be near the extremes of long-term oscillations about the critical values, or whether the gaps still exist if one considers the distribution of mean distances averaged over a long period of time. Current thinking is that the latter is the case, and therefore that nongravitational effects such as collisions, even among the asteroids, play an important role. The problem was complicated for Kirkwood toward the end of his life by his realization that the only asteroids existing at large mean distances *were* in resonance, having periods two-thirds and three-fourths that of Jupiter; he therefore conjectured that these orbits were unstable. Asteroids are especially numerous at the two-thirds resonance—the Hilda group—and it is now known that they are stable, avoiding encounters with Jupiter precisely because they oscillate about exact resonance.

Kirkwood also anticipated the asteroid "families," listing thirty-two groups with similar orbits (1892), and he surmised that the members of each group had separated from one another soon after their formation. Hirayama put the concept on a firmer basis with the use of "proper orbital elements."

From about 1869 onward Kirkwood began to question Laplace's nebular hypothesis. Although Kirkwood recognized that the existence of the asteroids and Saturn's rings suggested that the contracting sun had thrown off a continuous succession of rings, he thought it curious that the orbits of the major planets should be so widely separated. He also objected to the nebular hypothesis because the planets would have to be nearly cold by the time they had formed; furthermore, the satellites could not be explained, and the time required for production of the planets would be much longer than what was then believed to be the age of the solar system. He concluded (1880–1885) that planets and satellites were formed not by accumulation from complete rings but from limited arcs expelled in the equatorial plane of the shrinking sun. In order to explain the revolution of the inner satellite of Mars in a period shorter than that of the rotation of Mars, he supposed that the satellite's motion had been accelerated by passage through the resisting medium of the solar atmosphere. The possibility that the satellite's motion was being accelerated was still under active discussion in the mid-twentieth century.

When discussing Saturn's rings in 1884, Kirkwood remarked that "planets and comets have not formed from rings, but rings from planets and comets." He had considered this possibility for comets alone as early as 1861, when he gave the first convincing demonstration of an association between meteors and comets; confirmation was provided during the following few years, with the realization that the orbits of several of the best-known meteor streams were virtually identical with those of particular comets. He discussed the consequences of collisions of comets with meteoric rings, and his idea that this could explain the nongravitational acceleration of Encke's comet was subsequently adopted by Backlund. Kirkwood also seems to have been the first to suggest (1880) that there exists a genetically connected group of sun-grazing comets.

Kirkwood was the first to consider (1866–1867) the possible relationship of comets and asteroids and of shower meteors and stony meteorites. Although he was forced to withdraw his supposition that meteorites have a tendency to fall and bright fireballs to appear during meteor showers, the asteroidal versus cometary nature of bright fireballs is still an unresolved issue. As for the possible asteroid-comet relationship, more evidence in favor of this was forthcoming as Kirkwood's life advanced; with the discovery in the twentieth century of the Apollo group of asteroids, Hidalgo, and two short-period comets of remarkably asteroidal appearance, connection in some cases seems to be assured.

Kirkwood received an honorary master of arts degree from Washington (now Washington and Jefferson) College, Pennsylvania, in 1848 and that of doctor of laws from the University of Pennsylvania in 1852. His name has been appended to asteroid number 1578, which, appropriately, is a member of the Hilda group.

BIBLIOGRAPHY

I. Original Works. Kirkwood wrote three books: *Meteoric Astronomy* (Philadelphia, 1867), *Comets and Meteors* (Philadelphia, 1873), and *The Asteroids* (Philadelphia, 1888). His numerous papers include "On a New Analogy in the Periods of Rotation of the Primary Planets," in *Proceedings of the American Association for the Advancement of Science for 1849* (1850), 207; see also the letter from S. C. Walker communicating Kirkwood's information to the editor of the *Astronomische Nachrichten*, **30** (1850), 11–14; "Instances of Nearly Commensurable Periods in the Solar System," in *Mathematical Monthly*, **2** (1860), 126–132; "On the Nebular Hypothesis," in *American Journal of Science and Arts*, 2nd ser.; **30** (1860), 161–181; the obscure article that first discussed the association of comets and meteors, in *The Danville Quarterly Review* (Dec. 1861); "On Certain Harmonics in the Solar System," in *American Journal of Science and Arts*, 2nd ser., **38** (1864), 1–17.

See also "On the Theory of Meteors," in *Proceedings of the American Association for the Advancement of Science for 1866* (1867), pp. 8–14, for the first discussion of the gaps in the distribution of the asteroids and in Saturn's rings; "On the Nebular Hypothesis, and the Approximate Commensurability of the Planetary Periods," in *Monthly Notices of the Royal Astronomical Society*, **29** (1869), 96–102; "On the Periodicity of the Solar Spots," in *Proceedings of the American Philosophical Society*, **11** (1871), 94–101; "On the Formation and Primi-Structure of the Solar System," *ibid.*, **12** (1872), 163–166; "The Asteroids Between Mars and Jupiter," in *Annual Report of the Smithsonian Institution for 1876*, pp. 358–371; "On Some Remarkable Relations Between the Mean Motions of the Primary Planets," in *Astronomische Nachrichten*, **88** (1876), 77–78; "The Satellites of Mars and the Nebular Hypothesis," in *The Observatory*, **1** (1878), 280–282; "On Croll's Hypothesis of the Origin of Solar and Sidereal Heat," *ibid.*, **2** (1879), 116–118.

For further reference, see "On the Aerolitic Epoch of November 12th–13th," *ibid.*, **2** (1879), 118–121; "The Cosmogony of Laplace," *ibid.*, **3** (1880), 409–412; "On the Origin of the Planets," *ibid.*, **3** (1880), 446–447; "On the Great Southern Comet of 1880," *ibid.*, **3** (1880), 590–592; "The Divisions in Saturn's Rings," *ibid.*, **6** (1883), 335–336; "The Limits of Stability of Nebulous Planets, and the Consequences Resulting From Their Mutual Relations," in *The Sidereal Messenger*, **4** (1885), 65–77; "The Relation of Short-Period Comets to the Zone of Asteroids," *ibid.*, **7** (1888), 177–181; "On the Age of Periodic Comets," in *Publications of the Astronomical Society of the Pacific*, **2** (1890), 214–217; "Groups of Asteroids," in *The Sidereal Messenger*, **11** (1892), 785–789; "On the Relations Which Obtain Between the Mean Motions of Jupiter, Saturn and Certain Minor Planets," *ibid.*, **12** (1893), 302–303.

II. Secondary Literature. Biographical information is contained in W. W. Payne, "Daniel Kirkwood," in *Popular Astronomy*, **1** (1893), 167–169; and J. Swain, "Daniel Kirkwood," in *Publications of the Astronomical Society of the Pacific*, **13** (1901), 140–147.

For modern works on the Kirkwood gaps and related problems see D. Brouwer, "The Problem of the Kirkwood Gaps in the Asteroid Belt," in *Astronomical Journal*, **68** (1963), 152–159; P. J. Message, "On Nearly-Commensurable Periods in the Restricted Problem of Three Bodies, With Calculations of the Long-Period Variations in the Interior 2:1 Case," in G. Contopoulos, ed., *International Astronomical Union Symposium No. 25: The Theory of Orbits in the Solar System and in Stellar Systems* (London–New York, 1966), pp. 197–222.

W. H. Jeffreys, "Nongravitational Forces and Resonances in the Solar System," in *Astronomical Journal*, **72** (1967), 872–875; J. Schubart, "Long-Period Effects in the Motion of Hilda-Type Planets," *ibid.*, **73** (1968), 99–103; G. Colombo, F. A. Franklin, and C. M. Munford, "On a Family of Periodic Orbits of the Restricted Three-Body Problem and the Question of the Gaps in the Asteroid Belt and in Saturn's Rings," *ibid.*, **73** (1968), 111–123; F. Schweizer, "Resonant Asteroids in the Kirkwood Gaps and Statistical Explanations of the Gaps " *ibid.* **74** (1969), 779–788; A. T. Sinclair, "The Motions of Minor Planets Close to Commensurabilities with Jupiter," in *Monthly Notices of the Royal Astronomical Society,* **142** (1969), 289–294; B. G. Marsden, "On the Relationship Between Comets and Minor Planets," in *Astronomical Journal,* **75** (1970), 206–217; and F. A. Franklin and G. Colombo, "A Dynamical Model for the Radial Structure of Saturn's Rings," in *Icarus,* **12** (1970), 338–347.

BRIAN G. MARSDEN

KIRKWOOD, JOHN GAMBLE (*b.* Gotebo, Oklahoma, 30 May 1907; *d.* New Haven, Connecticut, 9 August 1959), *theoretical chemistry.*

Kirkwood received his early education in Wichita, Kansas. He studied for two years at the California Institute of Technology before enrolling at the University of Chicago in 1925, where he graduated B.S. in 1926. From February 1927 to June 1929 he did graduate work at the Massachusetts Institute of Technology. After several postdoctoral years in the United States and abroad, he assumed academic positions at Cornell University (1934), at the University of Chicago (1937), at the California Institute of Technology (1947), and at Yale University (1951), where he was Sterling professor of chemistry and head of the department.

Kirkwood married Gladys Danielson in 1930; their son John Millard Kirkwood, was born in 1935. They were divorced in 1951. In 1958 Kirkwood married Platonia Kaldes.

Kirkwood received many distinctions and honors, among them the American Chemical Society Award in Pure Chemistry, the Theodore William Richards Medal, a Presidential Certificate of Appreciation, and honorary degrees from the University of Chicago and the Free University of Brussels. He was a member of the National Academy of Sciences and its foreign secretary from 1955 to 1958.

Kirkwood's scientific contributions were in the theory of chemical physics and range over a wide variety of topics: polarizability and long-range interactions of molecules; dielectric properties of fluids; molecular distributions in fluids; systematic treatment of solutions, including order-disorder problems; the separation of proteins; the theory of shock and detonation waves; quantum statistics; and irreversible processes. In each of these fields his research was incisive and dominant. His topics were varied but his approach remained much the same. As an illustrative example, in his theories of irreversible processes he began his analysis with fundamentals (Liouville's equation), carried his derivation as far as possible without further assumptions or approximations, then introduced constraints based on physical reasoning to obtain working equations (Boltzmann and Fokker–Planck equations). His penetrating derivations put the field of his interests on a firm foundation which pointed out the limitations of models and guided the way to future development. Kirkwood had many students and co-workers whom he inspired by his insight and high standards.

BIBLIOGRAPHY

See *John Gamble Kirkwood Collected Works,* 8 vols. (New York, 1965–1968): *Dielectrics–Intermolecular Forces –Optical Rotation* (1965), R. H. Cole, ed.; *Quantum Statistics and Cooperative Phenomena* (1965), F. H. Stillinger, Jr., ed.; *Molecules* (1967), P. L. Auer, ed.; *Proteins* (1967), G. Scatchard, ed.; *Shock and Detonation Waves* (1967), W. W. Wood, ed.; *Selected Topics in Statistical Mechanics* (1967), R. W. Zwanzig, ed.; *Theory of Liquids* (1968), B. J. Alder, ed.; *Theory of Solutions* (1968), Z. W. Salsburg, ed.

JOHN ROSS

KIRWAN, RICHARD (*b.* Cloughballymore, County Galway, Ireland, 1733[?]; *d.* Dublin, Ireland, 1 June 1812), *chemistry, mineralogy, geology, meteorology.*

The second son of Martin Kirwan, of Cregg Castle near Corrandulla, and Mary French, of Cloughballymore near Kilcolgan, Richard Kirwan was descended from two prominent landed families. Little is known for certain about his early years; he was apparently a precocious, bookish child. After his father's death in 1741 he was brought up at his mother's home, where he had been born.

In about 1750 Kirwan enrolled at the University of Poitiers, to which his elder brother, Patrick, had gone earlier, their religion virtually excluding them from British universities. A letter from his mother written at about this time indicates that he had already developed an interest in chemistry, for which he was neglecting his specified curriculum; he nevertheless seems to have become proficient in Latin. His mother died soon afterwards, and in 1754 Kirwan left the university but remained in Europe and became a Jesuit novice.

Upon the death of Patrick and his consequent inheritance of the family estates Kirwan abandoned his novitiate and returned to Ireland. In 1757 he married Anne Blake, daughter of Sir Thomas Blake of Menlough Castle, just north of Galway; they had two daughters. Kirwan seems to have spent most of his married life at the home of the Blakes, where he fitted up a laboratory and amassed a library. His wife died in 1765 while he was in London studying law.

The motives for Kirwan's turning to law are not clear. He renounced his Catholic beliefs in 1764 as a prerequisite of his being called to the Irish bar in 1766, but practiced for only about two years, finding the profession uncongenial and the rewards inadequate. During the next eight or nine years, about three of which were spent in London (*ca.* 1769–1772), he increased his knowledge of science and languages; and in 1777 he returned to London, where he stayed for ten years. His house became a well-known meeting place for those distinguished by birth, position, or achievement—particularly in science.

In 1787 Kirwan returned to Ireland. He was one of the original members of the Royal Irish Academy, to which he presented a large number of papers on a remarkable variety of subjects, and of which he was president from 1799 until his death. Much of Kirwan's work in chemistry was done during the decade he spent in London. He was admitted to the Royal Society in 1780 and awarded the Copley Medal for his work on chemical affinity.

Although by some the term "affinity" was used in preference to "attraction" to avoid the implications of the latter, Kirwan had no such reservations: "Chymical affinity or attraction is that power by which the invisible particles of different bodies intermix and unite with each other so intimately as to be inseparable by mere mechanical means." It differed from nonchemical forms of attraction in that it caused "a body already united to another to quit that other and unite with a third, and hence it is called *elective* attraction." He coined the terms "quiescent" and "divellent" (afterwards generally adopted) to denote, respectively, those affinities which resisted decomposition and those which tended to effect decomposition and bring about a new union.

Kirwan began his three-part paper (1781–1783) by referring to the "recent great improvement in the subject by the excellent Mr. Bergman."[1] But up until then, he felt, only the *order* of affinities had been attended to (the first table showing this had been published by E. F. Geoffroy in 1718); and none except "Mr. Morveau of Dijon" had thought of ascertaining the actual *degrees* of attraction between one substance and others, or between the same two substances in different circumstances. Kirwan, however, had already "bestowed much pains" on this problem.[2]

He set out to determine the weights of various bases and metals that neutralized or dissolved in a given weight of each of the three mineral acids, believing these weights to be proportional to the affinities of the particular acid with the given bases or metals. His preliminary problem, however, was to find the weight of "real acid" in an aqueous solution. He believed that Priestley's discovery of "marine acid air" (hydrogen chloride)—that is, the acid freed from all water—had shown him how to do this. Knowing the specific gravity of the air and that of any solution, the weight of real acid in the solution could be found. For other acids, he made use of the assumption (presumably derived from Homberg) that the same weights of different acids neutralized a given weight of a particular alkali.

The measurements of affinity which Kirwan obtained were, of course, equivalents. His findings, translated and published in France and Germany, must have contributed to the formulation of the law of reciprocal proportions explicated by Richter. In two later papers on the composition of salts, read in 1790 and 1797 and published in Ireland, he revised some of his results and criticized some of those obtained by Richter and others. The first of these papers was translated by Mme. Lavoisier, who earlier had translated Kirwan's best-known work, the *Essay on Phlogiston.*[3]

First published in 1787, the *Essay* defended the phlogiston theory against the views then being promulgated in France by Lavoisier and his followers. Kirwan identified phlogiston with "inflammable air" (hydrogen), comparing it with "fixed air" (carbon dioxide); the latter, Black had shown, could exist "fixed, concrete and unelastic" in solids and in a "fluid, elastic and aëriform" state. He did not deny the observations on which Lavoisier had based his rejection of phlogiston, but believed them to be explicable in terms of the older theory, which, on the whole, accorded best with known chemical facts.

Kirwan's arguments hinged on his belief that inflammable air and "dephlogisticated air" (oxygen), although they had been shown to form water at red heat, formed fixed air at lower temperatures, and that the fixed air contained a greater proportion of phlogiston than did water. Thus, for example, he was able to explain the gain in weight of metals when calcined: this was due to the fixed air formed from dephlogisticated air and the phlogiston in the metal, and then absorbed. In the French translation of the *Essay* (1788), lengthy comments on each section were added by leading "antiphlogistians" (including Lavoisier); these, translated by William Nicholson, were included, with brief replies by Kirwan, in the second English edition (1789). A detailed refutation of the *Essay* was published in the same year by Kirwan's countryman William Higgins, whom he afterwards befriended.[4]

Kirwan abandoned the phlogiston theory in 1791 because he failed to show conclusively the formation of fixed air from phlogiston and oxygen. In spite of his conversion, Kirwan was not overly enthusiastic about the new nomenclature. He forbore using it in the second edition of his *Elements of Mineralogy* and employed some terms of his own, earning a reproof from Guyton de Morveau.[5] Following Cronstedt, Kirwan based his classification on qualitative chemical tests. Commenting on the vast increase of mineralogical information in Europe during the decade following his first edition, Kirwan said he would have despaired of assimilating it had it not been for the acquisition of the Leskean collection of minerals. He negotiated the purchase of this collection for the Royal Dublin Society and used his influence to obtain a grant from the Irish Parliament of £1,200 toward its cost.

Much has been written regarding Kirwan's intemperate attack on James Hutton's geomorphological theory. Although he was by no means alone in opposing Hutton, it was apparently his paper (1793), "Examination of the Supposed Igneous Origin of Stony Substances," that led Hutton to expand his ideas in his *Theory of the Earth* (1795). It has often been overlooked, however, that Kirwan's attack, if forthright, was less acrimonious than either Hutton's reply in the second chapter of his book or John Playfair's later defense of Hutton.[6] Kirwan also challenged the experimental support of Hutton's theories given by Sir James Hall.[7] In his *Geological Essays* and papers published in the *Transactions of the Royal Irish Academy*, he attempted to reconcile his observations with the history of the earth as related in Genesis.[8]

Kirwan's defense of outmoded ideas has tended to overshadow his more positive contributions to science. In particular, his pioneering work in meteorology has only recently claimed much attention. From records of Irish weather covering forty-one years, compiled by John Rutty, he worked out a system of probabilities with a view to forecasting the weather for seasons ahead. His predictions were more often right than wrong and were much valued by farmers; though the sequences he observed have not persisted in Ireland, methods of autocorrelation similar to Kirwan's are today proving successful elsewhere. Noteworthy, too, is his concept of air masses (redeveloped independently during the present century), his terms for which—"polar" and "equatorial," "supra-marine" and "supra-terrene"—anticipated modern classification.[9]

Kirwan was interested in the application of science to industry and wrote informatively on coal mining, manures, and bleaching (he probably introduced chlorine bleaching to Ireland).[10] In the last decade of his life, however, he virtually abandoned science for other interests. He published volumes on logic and metaphysics, and in one paper tried to prove that man's first language was a primitive form of Greek. He developed a number of eccentricities, particularly in later life, which were the subject of many anecdotes in the literature of the period.

NOTES

1. Bergman had a high opinion of Kirwan as a chemist, and on his proposal Kirwan was elected a member of both the Royal Society of Sciences in Uppsala and the Royal Academy of Sciences in Stockholm (see the introduction [pp. xlvii–li] to G. Carlid and J. Nordström, eds., *Torbern Bergman's Foreign Correspondence* (Stockholm, 1965), which contains eleven letters from Kirwan to Bergman (1782–1784)).
2. He later mentioned the attempts by Guyton de Morveau and C. F. Wenzel to quantify affinities, but pointed out their limited application. For an account of their methods see W. A. Smeaton, "Guyton de Morveau and Chemical Affinity," in *Ambix*, **11** (1963), 55–64.
3. See D. I. Duveen, "Madame Lavoisier, 1758–1836," in *Chymia*, **4** (1953), 13–29.
4. T. S. Wheeler and J. R. Partington, *The Life and Work of William Higgins* (Oxford, 1960), which contains a facs. repr. of *A Comparative View of the Phlogistic and Antiphlogistic Theories*, 2nd ed. (London, 1791), analyzes the arguments and gives much information about Kirwan.
5. In a review of the first volume (*Annales de chimie*, **23** [1797], 102–106). Kirwan thought highly of Guyton, and they were regular correspondents (see Smeaton, "L. B. Guyton de Morveau and His Relations with British Scientists," in *Notes and Records. Royal Society of London*, **22** (1967), 113–130.
6. In *Illustrations of the Huttonian Theory of the Earth* (Edinburgh, 1802), *passim.* Playfair disparaged Kirwan's reputation as a mineralogist (p. 481).
7. See C. S. Smith, "Porcelain and Plutonism," in C. J. Schneer, ed., *Towards a History of Geology* (Cambridge, Mass.–London, 1969), pp. 317–338.
8. A good account of his theory is in G. L. Davies, *The Earth in Decay* (London, 1969), pp. 142–145.
9. For an appraisal by meteorologists see F. E. Dixon, in *Dublin Historical Record*, **24** (1971), 58–59, and in *Weather*, **5** (1950), 63–65; also W. E. Knowles Middleton, *A History of the Theories of Rain* (London, 1965), p. 86.
10. See A. E. Musson & E. Robinson, *Science and Technology in the Industrial Revolution* (Manchester, 1969), pp. 186, 259–260, 289, 319–320.

BIBLIOGRAPHY

I. Original Works. Kirwan's books are *Elements of Mineralogy* (London, 1784; 2nd ed., rev. and much enl.: I, London, 1794; II, Dublin, 1796); a 3rd ed., virtually a repr. of the 2nd ed., was published in Dublin in 1810, apparently "against his approbation": Kirwan had "declined for some time previously the further cultivation of the science" (see R. Bakewell, *An Introduction to Mineralogy* [London, 1819], p. iv); *An Essay on Phlogiston and the Constitution of the Acids* (London, 1787; 2nd ed. [with notes from the French ed.], London, 1789; a facs. repr. of 2nd ed., London, 1968).

See also *An Estimate of the Temperature of Different Latitudes* (London, 1787), an early work in comparative climatology; *An Essay on The Analysis of Mineral Waters* (London, 1799), a good account of the qualitative and quantitative methods available; *Geological Essays* (London, 1799); *Logick*, 2 vols. (London, 1807); *Metaphysical Essays* (London, 1809) was styled "Vol. I" in its 1st ed., but this was omitted from the title page of a repr. in 1811.

The list of Kirwan's papers in the Royal Society's *Catalogue of Scientific Papers*, **3** (1969), 665–667, omits those in *Philosophical Transactions of the Royal Society;* almost complete lists are given by M. Donovan in *Proceedings of the Royal Irish Academy*, **4** (1850), xcv–xcvii, and by J. Reilly and N. O'Flynn in *Isis*, **13** (1930), 316–317, although a few articles and reprs. in periodicals are omitted.

In order of their treatment in the text, Kirwan's most important papers on chemistry are "Experiments and Observations on the Specific Gravities and Attractive Powers of Various Saline Substances," in *Philosophical Transactions of the Royal Society*, **71** (1781), 7–41; **72** (1782), 179–236; and **73** (1783), 15–84; "Experiments on Hepatic Air," *ibid.*, **76** (1786), 118–154 (this gives a good account of hydrogen sulfide, and describes Kirwan's discovery—independently of P. Gengembre—of hydrogen phosphide, which he called "phosphoric hepatic air").

"Of the Strength of Acids, and the Proportions of Ingredients in Neutral Salts," in *Transactions of the Royal Irish Academy*, **4** (n.d.), 3–84, read in 1790; "Additional Observations on the Proportion of Real Acid in the Three Ancient Known Mineral Acids, and on the Ingredients in Various Neutral Salts and Other Compounds," *ibid.*, **7** (1800), 163–297.

Kirwan's geological publications include "Examination of the Supposed Igneous Origin of Stony Substances," *ibid.*, **5** (n.d.), 51–81, read in 1793, his first attack on Hutton's theory; "On the Primitive State of the Globe and its Subsequent Catastrophe," *ibid.*, **6** (1797), 233–308; "Observations on the Proofs of the Huttonian Theory of the Earth, Adduced by Sir James Hall, Bart.," *ibid.*, **8** (1802), 3–27; "An Illustration and Confirmation of Some Facts Mentioned in an Essay on the Primitive State of the Globe," *ibid.*, 29–34; "An Essay on the Declivities of Mountains," *ibid.*, 35–52.

On meteorology, see "A Comparative View of Meteorological Observations Made in Ireland Since the Year 1788, With Some Hints Towards Forming Prognostics of the Weather," *ibid.*, **5** (n.d.), 3–29, read in 1793; "Essay on the Variations of the Barometer," *ibid.*, **2** (1788), 43–72; "Of the Variations of the Atmosphere," *ibid.*, **8** (1802), 269–507 (his concept of "air masses" is in these last two).

Kirwan's papers on applied science include "Observations on Coal Mines," *ibid.*, **2** (1788), 157–170; "On the Composition and Proportion of Carbon in Bitumens and Mineral Coal," *ibid.*, **6** (1797), 141–167; "What are the Manures Most Advantageously Applicable to the Various Sorts of Soils, and What Are the Causes of Their Beneficial Effect in Each Particular Instance," *ibid.*, **5** (n.d.), 129–198, read in 1794 (this was repr. separately and rev. several times); "Experiments On the Alkaline Substances Used in Bleaching, and On the Colouring Matter of Linen-Yarn," *ibid.*, **3** (1789), 3–47.

II. Secondary Literature. The main biography of Kirwan is by M. Donovan, who said he had access to family records, in *Proceedings of the Royal Irish Academy*, **4** (1850), lxxxi–cxviii; biographies since 1850 have relied heavily on this. Using it in conjunction with other sources, and providing a historical background, P. J. McLaughlin gives a very readable account in *Studies*, **28** (1939), 461–474, 593–605, and **29** (1940), 71–83, 281–300.

Other biographies are J. O'Reardon, "The Life and Works of Richard Kirwan," in *National Magazine* [Dublin], **1** (1830), 330–342, 469–475; J. R. O'Flanagan, in *Dublin Saturday Magazine*, **2** (1865), 242–244, 254–256, 266–269; C. J. Brockman, "Richard Kirwan, Chemist, 1733–1812," in *Journal of Chemical Education*, **4** (1927), 1275–1282; J. Reilly and N. O'Flynn, "Richard Kirwan, An Irish Chemist of the Eighteenth Century," in *Isis*, **13** (1930), 298–319; F. E. Dixon, "Richard Kirwan the Dublin Philosopher," in *Dublin Historical Record*, **24** (1971), 53–64.

On Kirwan's chemistry, see J. R. Partington, *History of Chemistry*, III (1962), 660–671 and *passim*. On Kirwan and Hutton, see C. C. Gillispie, *Genesis and Geology* (Cambridge, Mass., 1951), pp. 49–56 and *passim;* and E. B. Bailey, *James Hutton, the Founder of Modern Geology* (London–New York, 1967), 69–73.

E. L. Scott

KITAIBEL, PÁL (or **PAUL)** (*b.* Nagymarton, Hungary [now Mattersdorf, Austria], 3 February 1757; *d.* Pest, Hungary, 13 December 1817), *botany, chemistry, mineralogy.*

Kitaibel studied theology, jurisprudence, and finally medicine at the University of Pest. Although he qualified as a physician (1785), he never practiced medicine. He remained at the university as assistant to Jacob Joseph Winterl, at the Institute for Chemistry and Botany.

Kitaibel spent almost his entire life traveling through Hungary, studying the plant and animal life, collecting minerals, and analyzing mineral waters. His travel journals reveal a wide spectrum of interests, extending as far as folklore. Following Winterl's death, the institute was divided into two sections and Kitaibel was appointed to the professorship of botany. Because of his constant travels he published little and almost never lectured.

Kitaibel classified more than fifty unknown plants, many of which bear his name (for example, *Kitaibelia vitifolia*), and published a three-volume flora of

Hungary (1799–1812). He also recorded and named several species of animals. His botanical and mineralogical collection became the basis for the natural history collection of the Hungarian National Museum. In addition he wrote a description of Hungarian mineral springs, published after his death by a colleague. His travel accounts were also published posthumously.

Kitaibel was the discoverer of chloride of lime, and he even observed its bleaching effect (1795). But he did not consider its potential industrial use for textile bleaching, probably because at the time there was no important textile factory in Hungary. In 1798, unaware of Franz Müller's earlier discovery (1784), Kitaibel independently discovered the element later named tellurium in a mineral from the Borzsony Mountains. As a result he became involved in a heated and unjustified priority dispute with H. M. Klaproth.

BIBLIOGRAPHY

I. ORIGINAL WORKS. See *Descriptiones et icones plantarum rariorum Hungariae*, F. Waldstein, ed., 3 vols. (Vienna, 1812); *Topographische Beschreibung von Ungarn* (Pest, 1803); *Hydrographia Hungariae*, J. Schuster, ed., 2 vols. (Pest, 1829); "P. Kitaibelii Addimenta ad floram Hungariae," A. Kanitz, ed., in *Linnaea*, **32** (1863), 305–642; and *Diaria itinerarum Pauli Kitaibelii*, E. Gombocz, ed., 2 vols. (Budapest, 1945). Kitaibel's papers are partially enumerated in Poggendorff I.

II. SECONDARY LITERATURE. See S. Javorka, *Kitaibel Pál* (Budapest, 1937), which has a complete biblio. On the Kitaibel–Klaproth controversy, see M. E. Weeks, *Discovery of the Elements* (Easton, Pa., 1956), pp. 326–377; on the discovery of chlorinated lime, see L. Szathmáry, "Paul Kitaibel entdeckt den Chlorkalk," in *Chemiker Zeitung*, **55** (1931), 645; see also L. Szathmáry, "Einige chemisch-physikalische Apparate des ungarischen Chemikers Paul Kitaibel," in *Chemische Apparatur*, **19** (1932), 49, for a discussion of Kitaibel's apparatus.

FERENC SZABADVÁRY

KITASATO, SHIBASABURO (*b.* Oguni, Kumamoto, Japan, 20 December 1852; *d.* Nakanojo, Gumma, Japan, 13 June 1931), *bacteriology.*

Kitasato, one of the foremost Japanese bacteriologists, was born sixteen years before the Meiji Restoration in a country village in the mountains of Kumamoto prefecture. He was the eldest son of Korenobu Kitasato, the mayor of the village. In 1872 he enrolled at a newly founded medical college in the city of Kumamoto, where he met a Dutch physician, C. G. van Mansvelt, who had been invited to the school as the principal advisor for medical education. Kitasato became greatly devoted to the Dutch scholar, who in turn recognized his pupil's ability. Mansvelt invited Kitasato to his own home almost every evening, and gave private tutoring to Kitasato not only in the medical sciences but in Western history and culture.

When Mansvelt left Kumamoto, he suggested to Kitasato that he pursue a medical education at the University of Tokyo and then go to Europe for further study. Kitasato received his medical degree from the University of Tokyo in 1883. He then worked as a government officer at the new Public Health Bureau. There he became a research assistant to Masanori Ogata, associate professor at the University of Tokyo, who had just returned from Germany and had opened a new laboratory of bacteriology at the bureau; the first such laboratory in Japan, it was equipped with German microscopes and other apparatus.

Kitasato married Torako Matsuo in 1884. In the same year during an outbreak of cholera at Nagasaki, Kitasato is said to have demonstrated the presence of comma bacillus, the causative bacteria of the disease, under a microscope. Following the recommendation of the chief of the Public Health Bureau, Kitasato went to Germany for further study at Robert Koch's laboratory (1886–1891).

Many of Kitasato's papers are milestones in the history of bacteriology. In 1889 he published a paper on his method of culturing the anaerobic bacterium *Clostridium chauvoei*, the causative agent of blackleg in cattle; Kitasato found that the bacterium could grow in solid media surrounded by a hydrogen atmosphere. In the same year he published a paper on the bacillus that causes tetanus. It had been thought impossible to get a pure culture of *Clostridium tetani*, which had hitherto been grown in symbiosis with other bacteria. But Kitasato thought otherwise and discussed his belief with Koch and other colleagues. He found that the spores of the bacillus, strongly heat-resistant, could be heated to 80° C. without perishing. He utilized this property to obtain a pure culture: he heated a mixed culture of *Clostridium tetani* and other bacteria at 80° C. for forty-five to sixty minutes and then cultivated them in a hydrogen atmosphere. He thereby derived the first pure culture of *Clostridium tetani.*

In 1890, with Behring, Kitasato published a paper on immunity to diphtheria and tetanus, the section on diphtheria being written by Behring and the greater part of the paper, on tetanus, by Kitasato. This report opened a new field of science—that of serology—and provided the first evidence that immune serum can serve in the curing of an infectious disease.

The existence of tetanus toxin in the culture filtrate of *clostridium tetani* was unknown until Kitasato found it. By diluting the toxin and injecting it into rabbits, he established the minimum lethal dose. He then injected nonlethal doses and found that the animals contracted no symptoms of tetanus, and that repeated injections with an increasing amount of toxin made them immune. Moreover, Kitasato found that subsequent injection of a large amount of toxin—much more than the normal minimum lethal dose—did not kill the immune animals. He also demonstrated that serum containing antitoxin taken from immune animals could neutralize (inactivate) the toxin, and that injection of such serum in nonimmune animals had a prophylactic and therapeutic effect against tetanus infection.

In 1894 there occurred an outbreak of bubonic plague at Hong Kong, and Kitasato was dispatched to the city by the Japanese government. He identified there the causative bacterium of the plague, *Pasteurella pestis*. In one paper, in collaboration with James A. Lowson, a British naval surgeon, he presented several photographs of the isolated bacterium. In a later paper he described its nature in detail. Kitasato sent his Hong Kong strain to Koch's laboratory and in 1897 published a third paper, in Japanese, on plague bacteria. Unknown to him, however, the bacterium described in this paper was quite different from that which he had isolated at Hong Kong. Two years later, Kitasato realized his error and published a correction. Based upon his papers published in *Lancet* and on the strain he sent to Koch's laboratory, the strain he isolated (at Hong Kong), *Pasteurella pestis*, has been generally accepted as being the causative bacterium of bubonic plague. During the same Hong Kong outbreak, Yersin discovered the same bacterium independently.

During a final period of his stay in Germany, Kitasato worked on tuberculin, which had been discovered by Koch in 1890. (During this period Kitasato's stay was supported by the Imperial household of Japan, not by the Japanese government.) When he returned home from Germany in 1892, he found no laboratory where he could work satisfactorily. But two people proved helpful to Kitasato at this juncture: Yukichi Fukuzawa, founder of Keio University and president of a large newspaper company, and Ichizaemon Morimura, a businessman. Together they founded the Institute for Infectious Diseases, and Kitasato became its director. In 1899, this institute became part of the Public Health Bureau of the government. In 1914, when it was suddenly transferred from the Bureau to the Ministry of Education, Kitasato and the entire research staff left the institute protesting the sudden change in its bureaucratic affiliation. Kitasato founded the Kitasato Institute the same year and most of his researchers rejoined him there.

In 1917 Kitasato became the first dean of the school of medicine of Keio University in Tokyo. In the same year he was appointed a member of the House of Peers by the Japanese government. In 1923, when the Japanese Medical Association was founded, he was elected its first president. The following year, he was created baron by the emperor, then a supreme honor for a Japanese scientist. In 1925 he was awarded the Harben Gold Medal by the Royal Institute of Public Health.

Among Kitasato's notable disciples were Kiyoshi Shiga, the discoverer of *Shigella dysenteriae*, the cause of bacillary dysentery; and Sahachiro Hata, who found with Ehrlich the antisyphilitic effect of Salvarsan. The Prussian government made Kitasato a professor; he was decorated by the governments of Prussia, Norway, and France, and was elected an honorary member of national academies and scientific societies of various countries.

In 1908 Koch visited Japan at the invitation of Kitasato and was officially welcomed by the Japanese government. After Koch's death on 27 May 1910, Kitasato built a small shrine in front of his laboratory in honor of the German bacteriologist and deposited there a strand of Koch's hair and a fingernail, which he had secretly obtained during Koch's stay in Japan. In 1931 Kitasato died of a stroke and was laid to rest in the shrine of his respected teacher. Each year, on the anniversaries of Koch's and Kitasato's deaths, many people pay their respects at the shrine. The notable friendship between Koch and Kitasato is well remembered in Japan as an example of the close bond possible between teacher and pupil.

Kitasato left behind him the memory of a pioneer, of gratitude to his teachers and colleagues, of wisdom carried into practice, and of indomitability. Thus inspired, the Kitasato Institute has remained active in the fields of bacteriology, serology, and virology. In 1967, on the fiftieth anniversary of the Institute, Kitasato University was founded.

BIBLIOGRAPHY

Among Kitasato's important papers are "Über den Rauschbrandbacillus und sein Culturfahren," in *Zeitschrift für Hygiene und Infektionskrankheiten*, **6** (1889), 105–116; "Über dem Tetanusbacillus," *ibid.*, **7** (1889), 225–234; "Über das Zustandekommen der Diptherie-Immunität und der Tetanus-Immunität bei Thieren," in *Deutsche Medizinische Wochenschrift*, **16** (1890), 1113–

1114, written with Emil von Behring; "The Plague at Hong Kong," in *Lancet* (11 August 1894), p. 325; and "The Bacillus of Bubonic Plague," *ibid.* (25 August 1894), p. 428-430.

TSUNESABURO FUJINO

KJELDAHL, JOHANN GUSTAV CHRISTOFFER (*b.* Jagerpris, Denmark, 16 August 1849; *d.* Tisvildeleje, Denmark, 18 July 1900), *analytical chemistry*.

Kjeldahl's father was district physician in his native village on the island of Sjaelland; his mother was Johanne Lohmann. He received his schooling at the Gymnasium in Roskilde and then studied chemistry at the Technological Institute in Copenhagen. He passed his state examination "in applied science" with distinction in 1873 and became an instructor at the Agricultural College. There he became acquainted with J. C. Jacobsen, the owner of the Carlsberg brewery, who hired him in 1875 to set up a laboratory for making various technical analyses. Jacobsen soon established the Carlsberg Foundation which helped to found the Carlsberg Laboratory, a scientific research institution. In 1876 Kjeldahl was appointed the director of the laboratory, and he held this position until his death.

Kjeldahl's name is known above all for the "Kjeldahl method" for the estimation of nitrogen in organic substances. This discovery, of such great value in analytical chemistry, was first made as an auxiliary step in his efforts to develop a method for use in experiments in agricultural chemistry. Yet his method of nitrogen determination is of far greater importance than all of his results in agricultural chemistry. Kjeldahl realized the problem involved in nitrogen determination while investigating protein transformation in beer fermentation.

Lavoisier thought that organic substances consisted solely of carbon, hydrogen, and oxygen, while Berthollet found in 1786 that certain substances of animal origin contained nitrogen. After the experiments of Gay-Lussac and Thénard, Dumas finally succeeded in 1831 in creating a practical method of nitrogen determination, one which required burning and gasometric measurement. As is evident from the contemporary literature, Dumas's method was "a torment for everyone" because it was so complicated. A suitable wet method had long been sought to replace Dumas's combustion method. In 1841 Franz Varrentrapp and Heinrich Will created a method in which the substance was heated with barium hydroxide, nitrogen was converted into ammonia, and the latter was conducted into hydrochloric acid and precipitated with platinic chloride. Although this procedure was indeed an advance, it was also rather inexact, time-consuming, and costly.

Kjeldahl attacked the problem systematically, starting from James Wanklyn's observation that under certain conditions potassium permanganate converts the nitrogen in organic bodies into ammonia. This phenomenon, however, occurs only very irregularly. Kjeldahl found that in concentrated sulfuric acid the oxidation and conversion through permanganate take place regularly and quantitatively, and he determined the amount of ammonia by titration. He reported his method to the Chemical Society of Copenhagen on 7 March 1883 (*Zeitschrift für analytische Chemie*, **22** [1883], 366).

He later added phosphoric acid in order to convert those nitrogenous compounds that had resisted the original method, but this measure was only partially successful. The "Kjeldahl flask" which he constructed in 1888 to simplify the method is still in use today.

Kjeldahl's method is extremely important for agriculture, medicine, and drug manufacturing, and it is still universally employed in its essentially unmodified original form.

Mention should also be made of Kjeldahl's work on sugar-forming enzymes. He reported on a previously unrecognized polysaccharide (amylan) in barley, and conducted experiments to distinguish among various kinds of sugar. Although the latter attempt was unsuccessful, he improved the values that were assigned to the reducing capacities of various sugars in the tables employed in analytic calculations.

In 1890 Kjeldahl was elected to the Danish Society of Sciences. Although subject to morbid depressions, he was nevertheless capable of undertaking steady, although increasingly less extended, research, and in his last years he occasionally had to stop working completely. He traveled a great deal for reasons of health, and died of a heart attack while bathing in the sea.

BIBLIOGRAPHY

I. ORIGINAL WORKS. Kjeldahl's publications have not been compiled. His most important paper is "Neue Methode zur Bestimmung des Stickstoffs in organischen Körpern" in *Zeitschrift für analytische Chemie*, **22** (1883), 366; for his work on sugar-forming enzymes see *Meddelelser fra Carlsberg Laboratoriet*, **4** (1895), 1.

II. SECONDARY LITERATURE. See W. Johannsen, "Johann Kjeldahl," in *Berichte der Deutschen Chemischen Gesellschaft*, **33** (1900), 3881; R. E. Oesper, "Johan Kjeldahl and the Determination of Nitrogen," in *Journal of Chemical Education*, **11** (1934), 457; St. Veibel, "Johan Kjeldahl," in *Journal of Chemical Education*, **26** (1949),

459; H. Lund, "Scandinavian Contributions to Chemistry," in *Selecta chimica*, **12** (1953), 3; A. J. Ihde, *The Development of Modern Chemistry* (New York, 1964), p. 296; and F. Szabadváry, *History of Analytical Chemistry* (Oxford, 1966), p. 298.

FERENC SZABADVÁRY

KLAPROTH, MARTIN HEINRICH (*b.* Wernigerode, Germany, 1 December 1743; *d.* Berlin, Germany, 1 January 1817), *chemistry.*

"Suffer and hope"—with these words Klaproth in 1765 captured the essence of his youth. The third son of Johann Julius Klaproth, a poor but respected tailor with pietistic leanings, he had been intended for the clergy. Shortly after his fifteenth birthday, however, an unpleasant incident apparently forced him to drop out of Wernigerode's Latin school. Deciding to take up pharmacy, probably because of its connection with the natural sciences, Klaproth became an apprentice in a Quedlinburg apothecary shop in 1759. His master worked him hard, giving him little, if any, theoretical training and less spare time. In 1766, two years after becoming a journeyman, he moved, in the same capacity, to Hannover. There, at last, he had the opportunity to begin transcending pharmacy. Choosing chemistry, he read the texts of J. F. Cartheuser and J. R. Spielmann and conducted many minor investigations. After two years in Hannover, followed by two and a half years in Berlin and a few months in Danzig, Klaproth settled at Berlin in 1771. During his first decade there he supported himself by managing the apothecary shop of a deceased friend, the minor chemist Valentin Rose the elder. In 1780 he finally gained self-sufficiency—a fortunate marriage to A. S. Marggraf's wealthy niece enabled him to purchase his own shop.

In the meantime Klaproth had continued his pursuit of chemistry, studying not only by himself but also, it seems, with Marggraf. He ventured into print for the first time in 1776 when a friend persuaded him to contribute a chapter on the chemical properties of copal to a book on the natural history of that resinous substance. By 1780 he felt sufficiently knowledgeable to request permission to give private lectures on chemistry under the auspices of Berlin's Medical-Surgical College. The college's professors, who were eager to avoid such competition for student fees, blocked his request. In 1782, after publishing several articles on chemical topics and securing the backing of influential Masonic brothers, Klaproth was in a stronger position. That year he was named to the second seat for pharmacy on Prussia's highest medical board and soon afterward was granted permission to lecture on chemistry. Thus, at the relatively advanced age of thirty-nine, he embarked on his administrative and teaching career.

Over the years Klaproth moved up in the Prussian medical bureacracy from assessor (1782–1797) to councillor (1797–1799), to high councillor (1799–1817). Meanwhile he secured teaching posts, serving as private lecturer at the Medical-Surgical College (1782–1810); teacher of chemistry at the Mining School (1784–1817); professor of chemistry at G. F. von Tempelhoff's Artillery School and its successors, the Royal Artillery Academy and the General War School (1787–1812?); and full professor of chemistry in the University of Berlin's Philosophical Faculty (1810–1817). In 1800 Klaproth was appointed to succeed F. K. Achard as the Berlin Academy's representative for chemistry. No longer needing his apothecary shop, he sold it at a handsome profit and moved into the academy's new laboratory-residence complex in 1803. Here Klaproth, the tailor's son who once could only "suffer and hope," worked until his death from a stroke on New Year's Day 1817.

Although his wealthy wife and influential friends had helped Klaproth launch his career, it was his accomplishments as a chemist that propelled his subsequent rise. His most important work was in analytical chemistry. Indeed, he was the leading analytical chemist in Europe from the late 1780's, when he established himself as Bergman's intellectual successor, until the early 1800's, when Berzelius gradually took his place. Working with minerals from all parts of the globe, Klaproth discovered or co-discovered zirconium (1789), uranium (1789), titanium (1792), strontium (1793), chromium (1797), mellitic acid (1799), and cerium (1803) and confirmed prior discoveries of tellurium (1798) and beryllium (1798). More consequential than these specific results were Klaproth's new techniques. For instance, he found that many particularly insoluble minerals could be dissolved if they were first ground to a fine powder and then fused with a carbonate. With his student Valentin Rose the younger he introduced the use of barium nitrate in the decomposition of silicates. He constantly drew attention to the necessity of either avoiding or making allowances for contamination from apparatus and reagents. Most significant, he broke with the tradition of ignoring "small" losses and gains in weight in analytical work. Instead, he used discrepancies over a few percentage points as a means of detecting faulty and incomplete analyses. Once satisfied with his procedure for analyzing a mineral, he reported his final results—including the remaining discrepancy. This practice became a convention with the next generation of analysts.

Besides his influence as an analyst, Klaproth played a role of some consequence in the German acceptance of Lavoisier's theory. In the spring of 1792, after studying his friend S. F. Hermbstädt's manuscript translation of Lavoisier's *Traité* and repeating some of its main experiments, he announced his tentative support for the antiphlogistic system. During the ensuing year he often joined with Hermbstädt in repeating the reduction of mercuric oxide before skeptical and important witnesses. By the summer of 1793 they had discredited F. A. C. Gren and other phlogistonists who denied the accuracy of Lavoisier's account of the experiment, thereby preparing the way for the success of the antiphlogistic revolution in Germany. In the remaining decades of his life, however, Klaproth avoided taking an active part in the theoretical development of chemistry.

Klaproth's aversion to theory in no way dampened international enthusiasm for his work. Among the numerous honors that he received were membership in the Royal Society of London (1795) and, far more important, membership as one of six foreign associates in the Institut de France (1804).

BIBLIOGRAPHY

A complete list of Klaproth's many publications appears in Georg Edmund Dann, *Martin Heinrich Klaproth (1743–1817): Ein deutscher Apotheker und Chemiker: Sein Weg und seine Leistung* (Berlin, 1958). That Klaproth was Marggraf's student is revealed by Lorenz von Crell, "Lebensgeschichte Andreas Sigismund Marggraf's . . .," in *Chemische Annalen*, no. 1 (1786), 181–192. For an appreciative assessment of Klaproth's work by a member of the next generation of analytical chemists, see Thomas Thomson, *The History of Chemistry*, II (London, 1831), 191–210. For Klaproth's role in the German antiphlogistic revolution, see Karl Hufbauer, "The Formation of the German Chemical Community, 1700–1795" (University of California, Berkeley, 1970), diss., chs. 6–7.

KARL HUFBAUER

KLEBS, GEORG ALBRECHT (*b.* Neidenburg, Germany, 23 October 1857; *d.* Heidelberg, Germany, 15 October 1918), *botany.*

Klebs was the third son of Emil Klebs, a Prussian Consistory Councillor. He received his education in the small East Prussian city of Wehlau, entering elementary school in 1864 and graduating from the Realgymnasium in 1874. He then studied the natural sciences, philosophy, and art history at the University of Königsberg. During his first semester he composed a prize essay in philosophy; but in preparing a study of the Desmidiaceae, an algae family, his interest was drawn to botany. This work (1879) brought him a position as an assistant to Anton de Bary at the University of Strasbourg (1878–1880). Subsequently he became an assistant to the leading plant physiologists of the time, Julius Sachs at Würzburg and Wilhelm Pfeffer at Tübingen. In 1883 he qualified as lecturer at Tübingen. Klebs was appointed a full professor at Basel in 1887 and obtained the same position in 1898 at Halle and in 1907 at Heidelberg, where he was simultaneously named privy councillor. In 1913 he received the Knight's Cross of the Zähringer Löwen. Klebs was rector of the University of Basel for a year and twice dean of the Science and Mathematics Faculty at Heidelberg. Shortly before his death he was elected rector of the University of Heidelberg. He was a member of many scientific academies. On 20 March 1888 he married Luise Charlotte von Sigwart of Tübingen; they had three children.

In his scientific work, which was stimulated primarily by de Bary, Klebs at first concentrated on systematics among the algae and fungi, but he soon turned his attention to the cellular and reproductive physiology of these plants. His important discoveries in these areas include the fact that the presence of the nucleus is necessary for the formation of a cell wall.

Klebs's principal work began with his extensive investigations of developmental variation among both lower and higher plants, particularly of the way in which variation is brought about through alteration of environmental factors. This research was set forth in two books that appeared in 1896 and 1903. Starting with his first works in this field, Klebs not only presented interesting facts but also furnished, through suitable definitions, the correct points of departure for mastering developmental physiology through experimentation. This was his most lasting achievement. Specifically, he was the first to make a logically consistent division of the influences affecting development into external conditions, internal conditions, and specific structure. He defined their combined activity as follows: "All variations of [a species] are generated by the *external environment* in that it materializes, through its effect on the *inner conditions*, the powers lying dormant in the *specific structure*" ("Uber das Verhältnis der Aussenwelt zur Entwicklung der Pflanzen," p. 12). The crucial innovation of this conception is the distinction between the unchangeable specific structures (the genetic endowment) of the cell and the internal factors that are altered in response to external conditions. Klebs demonstrated the validity of this point of view in many experiments on fungi, ferns, and flowering plants. Moreover, with this definition he gave developmental physiology its own methodology and was thus its real founder.

Although Klebs's view was accepted only hesitantly —undoubtedly because of certain errors it contained, including the belief that it is not necessary to distinguish between mutations and variations—many contemporaries praised him unreservedly.

The full importance of Klebs's conception could not be appreciated, however, until the "arbitrary," and therefore often indiscriminate, developmental variations which he had posited were replaced by regular variations.

BIBLIOGRAPHY

I. Original Works. A complete bibliography of Klebs's writings can be found in Küster (see below). They include *Über die Fortpflanzungsphysiologie der niederen Organismen der Protobionten* . . . (Jena, 1896); *Willkürliche Entwicklungsänderungen bei Pflanzen* . . . (Jena, 1903); and "Über das Verhältnis der Aussenwelt zur Entwicklung der Pflanzen. Eine theoretische Betrachtung," in *Sitzungsberichte der Heidelberger Akademie der Wissenschaften*, Sec. B (1913), 5th essay, 1–47.

II. Secondary Literature. On Klebs or his work, see M. Bopp, "Georg Klebs und die heutige Entwicklungsphysiologie," in *Naturwissenschaftliche Rundschau*, **22** (1969), 97–101; E. Küster, "Georg Klebs 1857–1918," in *Berichte der Deutschen botanischen Gesellschaft*, **36** (1918), 90–116, with a complete bibliography; G. Lakon, "Über den rhythmischen Wechsel von Wachstum und Ruhe bei den Pflanzen," in *Biologisches Zentralblatt*, **35** (1915), 401–471; G. Melchers, "Einführung" in W. Ruhland, ed., *Handbuch der Pflanzenphysiologie*, XVI (Berlin–Göttingen–Heidelberg, 1961), xix–xxvi; and E. Ungerer, "Die Beherrschung der pflanzlichen Form. Eine Einführung in die Forschungen von Georg Klebs," in *Naturwissenschaften*, **6** (1918), 683–691.

M. Bopp

KLEIN, CHRISTIAN FELIX (*b.* Düsseldorf, Germany, 25 April 1849; *d.* Göttingen, Germany, 22 June 1925), *mathematics.*

Klein graduated from the Gymnasium in Düsseldorf. Beginning in the winter semester of 1865–1866 he studied mathematics and physics at the University of Bonn, where he received his doctorate in December 1868. In order to further his education he went at the start of 1869 to Göttingen, Berlin, and Paris, spending several months in each city. The Franco-Prussian War forced him to leave Paris in 1870. After a short period of military service as a medical orderly, Klein qualified as a lecturer at Göttingen at the beginning of 1871. In the following year he was appointed a full professor of mathematics at Erlangen, where he taught until 1875. From 1875 to 1880 he was professor at the Technische Hochschule in Munich, and from 1880 to 1886 at the University of Leipzig. From 1886 until his death he was a professor at the University of Göttingen. He retired in 1913 because of poor health. During World War I and for a time thereafter he gave lectures in his home. In August 1875 Klein married Anne Hegel, a granddaughter of the philosopher; they had one son and three daughters.

One of the leading mathematicians of his age, Klein made many stimulating and fruitful contributions to almost all branches of mathematics, including applied mathematics and mathematical physics. Moreover, his extensive activity contributed greatly to making Göttingen the chief center of the exact sciences in Germany. An opponent of one-sided approaches, he possessed an extraordinary ability to discover quickly relationships between different areas of research and to exploit them fruitfully.

On the other hand, he was less interested in work requiring subtle and detailed calculations, which he gladly left to his students. In his later years Klein's great organizational skill came to the fore, enabling him to initiate and supervise large-scale encyclopedic works devoted to many areas of mathematics, to their applications, and to their teaching. In addition Klein became widely known through his many books based on his lectures dealing with almost all areas of mathematics and with their historical development in the nineteenth century.

Klein's extraordinarily rapid development as a mathematician was characteristic. At first he wanted to be a physicist, and while still a student he assisted J. Plücker in his physics lectures at Bonn. At that time Plücker, who had returned to mathematics after a long period devoted to physics, was working on a book entitled *Neue Geometrie des Raumes, gegründet auf der geraden Linie als Raumelement*. His sudden death in 1868 prevented him from completing it, and the young Klein took over this task. Klein's dissertation and his first subsequent works also dealt with topics in line geometry. The new aspects of his efforts were that he worked with homogeneous coordinates, which Plücker did only occasionally; that he understood how to apply the theory of elementary divisors, developed by Weierstrass a short time before, to the classification of quadratic straight line complexes (in his dissertation); and that he early viewed the line geometry of P_3 as point geometry on a quadric of P_5, which was a completely new conception.

In 1870 Klein and S. Lie (see *Werke*, I, 90–98) discovered the fundamental properties of the asymptotic lines of the famous Kummer surface, which, as the surface of singularity of a general quadratic straight-line complex, occupied a place in algebraic line geometry. Here and in his simultaneous investigations of cubic surfaces (*Werke*, II, 11–63) there is evidence of Klein's special concern for geometric intuition, whether regarding the forms of plane curves or the models of spatial constructions. A further result of his collaboration with Lie was the investigation, in a joint work, of the so-called W-curves (*Werke*, I, 424–460). These are curves that admit a group of projective transformations into themselves.

Klein's most important achievements in geometry, however, were the projective foundation of the non-Euclidean geometries and the creation of the "Erlanger Programm." Both of these were accomplished during his enormously productive youth.

Hyperbolic geometry, it is true, had already been discovered by Lobachevsky (1829) and J. Bolyai (1832); and in 1868, shortly before Klein, E. Beltrami had recognized that it was valid on surfaces of constant negative curvature. Nevertheless, the non-Euclidean geometries had not yet become common knowledge among mathematicians when, in 1871 and 1873, Klein published two works entitled *Über die sogenannte nicht-euklidische Geometrie* (*Werke*, I, 254–351). His essential contribution here was to furnish so-called projective models for three types of geometry: hyperbolic, stemming from Bolyai and Lobachevsky; elliptic, valid on a sphere on which antipodal points have been taken as identical; and Euclidean. Klein based his work on the projective geometry that C. Staudt had earlier established without the use of the metric concepts of distance and angle, merely adding a continuity postulate to Staudt's construction. Then he explained, for example, plane hyperbolic geometry as a geometry valid in the interior of a real conic section and reduced the lines and angles to cross ratios. This had already been done for the Euclidean angle by Laguerre in 1853 and, more generally, by A. Cayley in 1860; but Klein was the first to recognize clearly that in this way the geometries in question can be constructed purely projectively. Thus one speaks of Klein models with Cayley-Klein metric.

The conceptions grouped together under the name "Erlanger Programm" were presented in 1872 in "Vergleichende Betrachtungen über neuere geometrische Forschungen" (*Werke*, I, 460–498). This work reveals the early familiarity with the concept of group that Klein acquired chiefly through his contact with Lie and from C. Jordan. The essence of the "Erlanger Programm" is that every geometry known so far is based on a certain group, and the task of the geometry in question consists in setting up the invariants of this group. The geometry with the most general group, which was already known, was topology; it is the geometry of the invariants of the group of all continuous transformations—for example, of the plane. Klein then successively distinguished the projective, the affine, and the equiaffine or principal group of the particular dimension; in certain cases the succeeding group is a subgroup of the previous one. To these groups belong the projective, affine, and equiaffine geometries with their invariants, whereby the equiaffine geometry is the same as the Euclidean elementary geometry.

The non-Euclidean geometries accounted for with the aid of the Cayley-Klein models, as well as the various types of circular and spherical geometries devised by Moebius, Laguerre, and Lie, could likewise be viewed as the invariant theories of certain subgroups of the projective groups. In his later years Klein returned to the "Erlanger Programm" and, in a series of works (*Werke*, I, 503–612), showed how theoretical physics, and especially the theory of relativity, which had emerged in the meantime, can be understood on the basis of the ideas presented there. The "Programm" was translated into six languages and guided much work undertaken in the following years: for example, the analytic geometry of Lothar Heffter, school instruction, and the lifelong efforts of W. Blaschke in differential geometry. Only later in the twentieth century was it superseded.

Klein considered his work in function theory to be the summit of his work in mathematics. He owed some of his greatest successes to his development of Riemann's ideas and to the intimate alliance he forged between the latter and the conceptions of invariant theory, of number theory and algebra, of group theory, and of multidimensional geometry and the theory of differential equations, especially in his own fields, elliptic modular functions and automorphic functions.

For Klein the Riemann surface is no longer necessarily a multisheeted covering surface with isolated branch points on a plane, which is how Riemann presented it in his own publications. Rather, according to Klein, it loses its relationships to the complex plane and then, generally, to three-dimensional space. It is through Klein that the Riemann surface is regarded as an indispensable component of function theory and not only as a valuable means of representing multivalued functions.

Klein provided a comprehensive account of his conception of the Riemann surface in 1882 in *Riemanns Theorie der algebraischen Funktionen und ihre Integrale.* In this book he treated function theory as geometric function theory in connection with potential theory and conformal mapping—as Riemann had done. Moreover, in his efforts to grasp the actual relationships and to generate new results, Klein deliberately worked with spatial intuition and with concepts that were borrowed from physics, especially from fluid dynamics. He repeatedly stressed that he was much concerned about the deficiencies of this method of demonstration and that he expected them to be eliminated in the future. A portion of the existence theorems employed by Klein had already been proved, before the appearance of the book by Klein, by H. A. Schwarz and C. Neumann. Klein did not incorporate their results in his own work: He opposed the spirit of the reigning school of Berlin mathematicians led by Weierstrass, with its abstract-critical, arithmetizing tendency; Riemann's approach, which inclined more toward geometry and spatial representation, he considered more fruitful. The rigorous foundation of his own theorems and the fusion of Riemann's and Weierstrass' concepts that Klein hoped for and expected found its expression—still valid today—in 1913 in H. Weyl's *Die Idee der Riemannschen Fläche.*

A problem that greatly interested Klein was the solution of fifth-degree equations, for its treatment involved the simultaneous consideration of algebraic equations, group theory, geometry, differential equations, and function theory. Hermite, Kronecker, and Brioschi had already employed transcendental methods in the solution of the general algebraic equation of the fifth degree. Klein succeeded in deriving the complete theory of this equation from a consideration of the icosahedron, one of the regular polyhedra known since antiquity. These bodies sometimes can be transformed into themselves through a finite group of rotations. The icosahedron in particular allows sixty such rotations into itself. If one circumscribes a sphere about a regular polyhedron and maps it onto a plane by stereographic projection, then to the group of rotations of the polyhedron into itself there corresponds a group of linear transformations of the plane into itself. Klein demonstrated that in this way all finite groups of linear transformations are obtained, if the so-called dihedral group is added. By a dihedron Klein meant a regular polygon with n sides, considered as a rigid body of null volume.

Through the relationships of the fifth-degree equations to linear transformations and through the joining of his investigations with H. A. Schwarz's theory of triangular functions, Klein was led to the elliptic modular functions, which owe their name to their occurrence in elliptic functions. He dedicated a long series of basic works to them and, with R. Fricke, presented the complete theory of these functions in two extensive volumes that are still indispensable for research. Individual aspects of the theory were known earlier. It was a question here of holomorphic functions in the upper half-plane $\mathscr{H}$ with a pole at infinity, which remain invariant under the transformations of the modular group Γ:

$$z \to \frac{az+b}{cz+d};\ a, b, c, d \text{ integers};\ ad - bc = 1.$$

If one sets $z = x + iy$, then the set F of points z, with

$$-\tfrac{1}{2} \leqslant x < \tfrac{1}{2},\ x^2 + y^2 \geqslant 1,\ y > 0;$$

and additionally, $x \leqslant 0$ if $x^2 + y^2 = 1$, has at every point in $\mathscr{H}$ exactly one point equivalent to that point under Γ. F is a fundamental domain for Γ relative to $\mathscr{H}$. It had already been recognized as such by Gauss. In 1877, somewhat later than Dedekind and independently of him, Klein discovered the fundamental invariant $J(\tau)$, which assumes each value in F exactly once and by means of which all modular functions are representable as rational functions.

Klein next investigated the subgroups Γ_1 of Γ with finite index, their fundamental domains, and the related functions. He thus arrived at algebraic function fields, which he investigated with the concepts and methods of Riemann's function theory. The Abelian integrals and differentials, and thereby the modular forms, as a generalization of the modular functions, lead to the modular functions on Γ_1. We also owe to Klein the congruence groups. These are subgroups Γ_1 of Γ that contain the group of all transformations

$$z \to \frac{az-b}{cz+d};\ a \equiv d \equiv \pm 1,\ b \equiv c \equiv 0 \bmod m$$

for fixed natural number m. The least possible m for a group Γ_1 Klein designated as the level of the group. The congruence groups are intimately related to basic theorems of number theory. The theory of modular functions was further developed by direct students of Klein, such as A. Hurwitz and R. Fricke, and most notably by Erich Hecke; its application to several variables was due especially to David Hilbert and Carl Ludwig Siegel.

From the modular functions Klein arrived at the automorphic functions, which, along with the former, include the singly and doubly periodic functions. Automorphic functions are based upon arbitrary groups Γ of linear transformations that operate on the Riemann sphere or on a subset thereof; they have interior points in their domain of definition that have neighborhoods in which no two points are equivalent under Γ. They also possess a fundamental domain F. Klein studied the various types of networks produced from F by the action of Γ. A primary role is played by the *Grenzkreisgruppen*, by means of which the net fills the interior of a circle that goes into itself under Γ; and under them there are again finitely many generators. The groups lead to algebraic function fields, and thus Klein could apply the ideas of Riemann that he had further developed. At the same time as Klein and in competition with him, Poincaré developed a theory of automorphic functions. In opposition to Klein, however, he established his theory in terms of analytic expressions—called, accordingly, Poincaré series. The correspondence between the two mathematicians during 1881 and 1882, which was beneficial to both of them, can be found in volume III of Klein's *Gesammelte mathematische Abhandlungen.*

The path from automorphic functions to algebraic functions may be traveled in both directions—that is the essence of the statements that Klein termed the "fundamental theorems," which were set forth by both himself and Poincaré in reciprocally influential works. Among the fundamental theorems, for example, is the following portion of the *Grenzkreistheorem*: Let $f(w, z)$ be an irreducible polynomial in w and z in the field of the complex numbers. Then one obtains all solution pairs of the equation $f(w, z) = 0$ in the form $w = g_1(t)$, $z = g_2(t)$, where $g_1(t)$ and $g_2(t)$ are rational functions in t, or doubly periodic functions, or automorphic functions under a *Grenzkreis* group, according to whether the Riemann surface corresponding to $f(w, z) = 0$ is of genus 0, 1, or higher than 1. The variable t is said to be *Grenzkreis* uniformizing; it is well defined up to a linear transformation. Klein, like Poincaré, worked with the fundamental theorems without being able to prove them fully. This was first accomplished at the beginning of the twentieth century by Paul Koebe. The progress made in the theory of automorphic functions since the 1930's is due primarily to W. H. H. Petersson (*b.* 1902).

In the 1890's Klein was especially interested in mathematical physics and engineering. One of the first results of this shift in interest was the textbook he composed with A. Sommerfeld on the theory of the gyroscope. It is still the standard work in this field of mechanics.

Klein was not pleased with the increasingly abstract nature of contemporary mathematics. His long-standing concern with applications was further strengthened by the impressions he received during two visits to the United States. He sought, on the one hand, to awaken a greater feeling for applications among pure mathematicians and, on the other, to lead engineers to a greater appreciation of mathematics as a fundamental science. The first goal was advanced by the founding, largely through Klein's initiative, of the Göttingen Institute for Aeronautical and Hydrodynamical Research; at that time such institutions were still uncommon in university towns. Moreover, at the turn of the century he took an active part in the major publishing project *Encyklopädie der mathematischen Wissenschaften mit Einschluss ihrer Anwendungen.* He himself was editor, along with Konrad Müller, of the four-volume section on mechanics.

What a fruitful and stimulating teacher Klein was can be seen from the number—forty-eight—of dissertations prepared under his supervision. Starting in 1900 he began to take a lively interest in mathematical instruction below the university level while continuing to pursue his academic functions. An advocate of modernizing mathematics instruction in Germany, in 1905 he played a decisive role in formulating the "Meraner Lehrplanentwürfe." The essential change recommended was the introduction in secondary schools of the rudiments of differential and integral calculus and of the function concept. In 1908 at the International Congress of Mathematicians in Rome, Klein was elected chairman of the International Commission on Mathematical Instruction. Before World War I, the German branch of the commission published a multivolume work containing a detailed report on the teaching of mathematics in all types of educational institutions in the German empire.

BIBLIOGRAPHY

I. Original Works. Klein's papers were brought together in *Gesammelte mathematische Abhandlungen*, 3 vols. (Berlin, 1921–1923). His books include *Über Riemanns Theorie der algebraischen Funktionen und ihrer Integrale* (Leipzig, 1882); *Vorlesungen über das Ikosaeder und die Auflösung der Gleichungen vom 5. Grade* (Leipzig, 1884); *Vorlesungen über die Theorie der elliptischen Modulfunktionen*, 2 vols. (Leipzig, 1890–1892), written with R. Fricke; *Über die Theorie des Kreisels*, 4 vols. (Leipzig, 1897–1910), written with A. Sommerfeld; *Vorlesungen über die Theorie der automorphen Funktionen*, 2 vols. (Leipzig, 1897–1912), written with R. Fricke; *Elementarmathematik vom höheren Standpunkt aus*, 3 vols. (Berlin, 1924–1928); *Die Entwicklung der Mathematik im 19. Jahrhundert*, 2 vols. (Berlin, 1926); *Vorlesungen über höhere Geometrie* (Berlin, 1926); *Vorlesungen über nicht-euklidische Geometrie* (Berlin, 1928); and *Vorlesungen über die hypergeometrische Funktion* (Berlin, 1933).

II. SECONDARY LITERATURE. See Richard Courant, "Felix Klein," in *Jahresberichte der Deutschen Mathematikervereinigung*, **34** (1925), 197–213; the collection of articles by R. Fricke, A. Voss, A. Schönflies, C. Carathéodory, A. Sommerfeld, and L. Prandtl, "Felix Klein zur Feier seines 70. Geburtstages," which is *Naturwissenschaften*, **7**, no. 17 (1919); G. Hamel, "F. Klein als Mathematiker," in *Sitzungsberichte der Berliner mathematischen Gesellschaft*, **25** (1926), 69–80; W. Lorey, "Kleins Persönlichkeit und seine Verdienste um die höhere Schule," *ibid.*, 54–68; and L. Prandtl, "Kleins Verdienste im die angewandte Mathematik," *ibid.*, 81–87.

WERNER BURAU
BRUNO SCHOENEBERG

KLEIN, HERMANN JOSEPH (*b.* Cologne, Germany, 14 September 1844; *d.* Cologne, 1 July 1914), *astronomy, meteorology.*

Klein's scientific curiosity was first aroused by the changing patterns of cloud cover over the Rhine Valley. From studies of the weather he branched out into observational astronomy, spending considerable time looking for changes in surface features of the moon.

In the course of an early career as a bookdealer in Cologne, Klein met Eduard Heis, professor of mathematics and astronomy in Münster and editor of the *Wochenschrift für Astronomie, Meteorologie und Geographie* from 1857 to 1875. Under Heis's tutelage, Klein obtained the necessary background in mathematics and astronomy to become a doctoral candidate at the University of Giessen. Here he was granted a Ph.D. in 1874, with a dissertation on the size and shape of the earth. He had already published a number of brief papers, based on astronomical work done in his own private observatory in Cologne; the first such, in 1867, dealt with the lunar crater Linné.

In 1879 Klein reported the discovery of a newly formed crater near the Hyginus rille, which became known as Hyginus N (N for *Nova*). And in 1882 he described a bright flash he had seen, close to another rille inside the crater Alphonsus. These and other observations convinced Klein that at least some of the circular lunar structures referred to as craters had resulted from vulcanism, which process he believed was still occurring on the moon. Volcanic activity inside Alphonsus was confirmed in 1958 by the Russian astronomer Kozyrev. More recently Klein's viewpoint seems to have been verified by manned exploration of the moon.

In 1880 Klein became director of a combined meteorological and astronomical observatory located in Lindenthal, a western suburb of Cologne, and sponsored by the newspaper *Kölnischen Zeitung*. Here he continued writing, observing the moon, and studying cirrus clouds for the rest of his life. In 1882 he began editing *Sirius*, a semipopular astronomical journal. He also wrote a number of books that were widely read throughout Europe and the United States, making him one of the foremost popularizers of meteorology and astronomy of his day.

A lunar crater, formerly known as Albategnius A, was renamed Klein in his honor.

BIBLIOGRAPHY

I. ORIGINAL WORKS. Klein's article "Ueber den Mondcrater Linne" appeared in *Astronomische Nachrichten*, **69** (1867), cols. 35–36; his discovery of Hyginus N was reported in "Ueber die Neubildungen beim Hyginus auf dem Monde," *ibid.*, **95** (1879), cols. 297–300. His account of a bright flash inside Alphonsus is included, with a halftone illustration, in "Über eine vulcanische Formationen auf dem Monde," in *Petermanns geographische Mitteilungen*, **28** (1882), 207–210.

Klein wrote his first book on meteorology when he was twenty-one years old: it was *Wetterpropheten und Wetterprophezeiung* (Neuwied, 1865). His first study of cirrus clouds was "Ueber die Periodicität der Cirruswolken," in *Zeitschrift für Meteorologie* (Vienna), **7** (1872), 209–212; an overall account of his observations and conclusions appeared in two articles entitled "Cirrus-Studien," in *Meteorologische Zeitschrift*, **18** (1901), 157–172, and **23** (1906), 67–82.

Poggendorff, III (Leipzig, 1898), 723–724; IV (Leipzig, 1904), 756; and V (Leipzig, 1926), 636, lists a total of twenty-four articles by Klein and twenty-four books. Notable among the latter are *Anleitung zur Durchmusterung des Himmels*, 2nd ed. (Brunswick, 1882); *Astronomischen Abende*, 6th ed. (Leipzig, 1905); *Allgemeine Witterungskunde*, 2nd ed. (Leipzig, 1905); *Star Atlas*, new ed. with 18 charts (New York, 1910), a translation of *Sternatlas* (Leipzig, 1886); and *Allgemeinverstandlische Astronomie* 10th ed. (Leipzig, 1911).

II. SECONDARY WORKS. An obituary notice on Klein, written by Hans Hermann Kritzinger, appeared in *Astronomische Nachrichten*, **199** (1914), cols. 15–16. Other details about his life can be found in Poggendorff (see above), and in *Meyers grosses Konversations-Lexikon*, 6th ed., XI (Leipzig, 1905), 114.

The lunar crater Klein is described in H. Percy Wilkins and Patrick Moore, *The Moon* (London, 1955), p. 143. The spectroscopic detection of transient gases over Alphonsus is described by Nikolai A. Kozyrev in "Observations of a Volcanic Process on the Moon," in *Sky and Telescope*, **18** (1958–1959), 184–186, translated by Luigi G. Jacchia.

SALLY H. DIEKE

KLEIN, JACOB THEODOR (*b.* Königsberg, East Prussia, 15 August 1685; *d.* Danzig [now Gdansk, Poland], 27 February 1759), *zoology.*

Klein's *Naturalis dispositio Echinodermatum* (1734) was one of the earliest monographic treatments of the sea urchins. It includes descriptions, illustrations, and a classification of both recent and fossil sea urchins. Klein called these the Echinodermata and divided them into three classes according to the position of the vent. The classes were then divided into nine sections, corresponding to the genera of later authors, and twenty-two species. Although altered and enlarged, this work was a major source of information on the Echinoidea for zoologists and paleontologists throughout the eighteenth century and remained a point of departure in discussions by such early nineteenth-century authors as James Parkinson.

Klein, who studied law at the University of Königsberg and served as court secretary in Danzig from 1714, had many and diverse interests in natural history besides sea urchins. He developed a botanical garden in Danzig, founded and directed a naturalist's society there, made extensive collections, and published about two dozen monographs, including studies of birds, fishes, reptiles, and invertebrates other than the sea urchins, particularly the mollusks. Fossils are dealt with in various publications, and Klein edited the *Sciagraphia lithologica curiosa, seu lapidum nomenclator* (1740) of J. J. Scheuchzer, which was published after Scheuchzer's death.

A principal concern in his monographs is classification. Klein's taxonomic method was based entirely on external characteristics, such as the number and position of limbs and the mouth; and he vigorously opposed any method, including the Linnaean system, based on characters not visible externally.

Klein was a member of the St. Petersburg Academy of Sciences and the Royal Society of London. He was a frequent contributor to the latter's *Philosophical Transactions* between 1730 and 1748.

BIBLIOGRAPHY

I. Original Works. Klein's writings include *Naturalis dispositio Echinodermatum. Accessit lucubratiuncula de aculeis echinorum marinorum, cum spicilegio de belemnitis* (Danzig, 1734), trans. into French as *Ordre natural des oursins de mer et fossiles; avec des observations sur les figures des oursins de mer, et quelques remarques sur les bélemnites* (Paris, 1754), Latin ed. rev. by N. G. Leske (Leipzig, 1778); and *Summa dubiorum circa classes quadrupedum et amphibiorum in celebris domini Caroli Linnaei systemate naturae, sive naturalis quadrupe dum historiae promovendae prodromus, cum praeludio de crustatis* (Leipzig, 1743), which summarizes Klein's feelings about a taxonomic method. A list of Klein's works is in Johann Georg Meusel, *Lexikon der vom Jahr 1750 bis 1800 verstorbenen teutschen Schriftsteller*, VII (Leipzig, 1808), pp. 53–60.

II. Secondary Literature. The brief biographical sketch in *Allgemeine deutsche Biographie*, XVI (repr. Berlin, 1969), pp. 92–94, is based on Christof Sendel, *Lobrede auf Herrn Jacob Klein* (Danzig, 1759).

Patsy A. Gerstner

KLEINENBERG, NICOLAUS (NICOLAI) (*b.* Liepaja, Russia, 11 March 1842; *d.* Naples, Italy, 5 November 1897), *biology.*

Kleinenberg was the son of Friedrich Kleinenberg, a municipal official. He studied medicine from 1860 to 1867 at the University of Dorpat, but botany became his primary interest. In 1863–1864 he attended the lectures of Matthias Jacob Schleiden. To further his botanical education he went to Jena in 1868, to study with Ernst Hallier. He soon came under the influence of Ernst Haeckel and became closely involved with Anton Dohrn, Ernst Abbe, and Karl Snell. In 1869 and 1870 Kleinenberg was an assistant to Haeckel at the Zoological Institute. During this period he prepared his dissertation, on the development of the freshwater polyp. He received his doctorate in 1871, having passed the oral examination on 5 January 1869.

In 1873 Anton Dohrn persuaded Kleinenberg to go to Naples to help him establish the first marine biological station. At first, he was very close to Dohrn, even accompanying him on a vacation in St. Moritz. The Stazione Zoologica di Napoli opened in October 1873. Kleinenberg, along with the English writer Charles Grant and the artists Hans von Marées and Adolf von Hildebrand, belonged to the first intimate circle around Dohrn (which is portrayed in the famous frescoes by von Marées in the library of the station at Naples). Yet Kleinenberg possessed certain character traits that led him to give up his assistant's post as early as 1875. For the next few years he lived on the island of Ischia, and he remained in Italy until his death. In 1879 he was appointed, principally through the intercession of Dohrn, to the chair of zoology and comparative anatomy at the University of Messina. In 1895 he became professor of the same subjects at Palermo.

Although considered an original and versatile scientist, Kleinenberg was a difficult person and unsuited to teamwork because of pride and a strong need for independence. Descriptions of his personality can be found in Theodor Heuss's work on Anton Dohrn and in the biography of Ernst Abbe by Felix

Auerbach. Like other students of Haeckel—such as Dohrn, Oscar Hertwig, and Hans Driesch—Kleinenberg soon found himself sharply opposing his teacher, especially the latter's ideologically interpreted Darwinism. Kleinenberg became friendly with one of the first visiting scientists at the Naples station, the young English physiologist Francis Maitland Balfour, and translated a textbook on embryology to which Balfour had made a major contribution, Michael Foster's *The Elements of Embryology* (London, 1874), a description of the development of the chick embryo. In the translator's preface he makes a criticism of scientific journalism that is interesting in the context of the period. He also characteristically includes a philosophical and epistemological excursus.

Kleinenberg's actual scientific work in zoology is not extensive. His dissertation of 1871 contains a chapter ("Die Furchung des Eies von Hydra. Ein Beitrag zur Kenntnis der Plasmabewegungen") from *Hydra. Eine anatomisch-entwicklungsgeschichtliche Untersuchung*, one of his two publications that are still of interest. Kleinenberg dedicated this work, the first detailed investigation of the development of the freshwater polyp, to Haeckel. In it he also took up the then very important theoretical question of the comparability, from the ontogenetic point of view, of the coelenterate ectoderm and endoderm with those of the vertebrate. He began by establishing the correspondence—postulated by Huxley in 1849—of the physiological functions of both germinal layers in these phyla.

In 1881 Kleinenberg published a first report on the embryology of the polychaete *Lopadorhynchus*. His comprehensive "Die Entstehung des Annelids aus der Larve von Lopadorhynchus. Nebst Bemerkungen über die Entwicklung anderer Polychaeten" appeared in 1886. The author wished "to evaluate the development of an animal according to new principles" and presented a "justification" of this endeavor in the first chapter, "Etwas von den Keimblättern." In a vigorous controversy in which Kleinenberg disputed Haeckel's gastraea theory—of which a kernel of truth stemmed from Huxley—and the views concerning the early development of animals held by R. Hertwig, A. von Kölliker, W. His, and B. Hatscheck, he sought to show that there was no mesoderm. This meant, in effect, demonstrating that "a reasonable [*verständiges*] system of tissues is possible only on a physiological basis." Kleinenberg thus belonged, like Alexander Goette, to the group of zoologists who wished to give up the purely phylogenetic evaluation of embryological processes in favor of a physiological approach. The latter ultimately led to the evolutionary physiology of W. Roux.

Kleinenberg again discussed the theoretical standpoint of ontogenetic research in the final chapter of "Die Entstehung . . .." This time the problem under consideration was the origin of organs performing similar functions—for example, among animals that exhibit metamorphosis. In this context Kleinenberg set forth his "substitution theory," with which he entered the dispute over the concept of homology. He drew attention to organ systems which develop by the substitution of organs which are unrelated from the point of view of embryological development —interpreted either ontogenetically or phylogenetically. The example he used was that of the larval and imaginal nervous systems of the polychaetes. The significance of Kleinenberg's distinction between homology and substitution has been emphasized by Adolf Remane in his discussion of the concept and criteria of homology in *Die Grundlagen des natürlichen Systems, der vergleichenden Anatomie und der Phylogenetik* (Leipzig, 1952).

Kleinenberg paid tribute to the work of Charles Darwin in a short piece published in the year of the latter's death. Besides the translation of Balfour and Foster's embryological text, Kleinenberg produced a German translation of Foster's textbook of physiology. He also published a work on the difference between art and science.

BIBLIOGRAPHY

I. ORIGINAL WORKS. Kleinenberg's writings include *Die Furchung des Eies von Hydra viridis. Ein Beitrag zur Kenntnis der Plasmabewegungen* (Jena, 1871), his dissertation; *Hydra. Eine anatomisch-entwicklungsgeschichtliche Untersuchung* (Leipzig, 1872); *Sullo sviluppo des Lumbricus trapezoides* (Naples, 1878), also in English in *Quarterly Journal of Microscopical Science*, **19** (1879), 206–244; "Über die Entstehung der Eier bei Eudendrium," in *Zeitschrift für wissenschaftliche Zoologie*, **35** (1881), 326–332; *Carlo Darwin e l'opera sua* (Messina, 1882); "Die Entstehung des Annelids aus der Larve von Lopadorhynchus. Nebst Bemerkungen über die Entwicklung anderer Polychaeten," in *Zeitschrift für wissenschaftliche Zoologie*, **44** (1886), 1–227; *Intorno alla differenza essenziale fra arte e scienza* (Messina, 1892); "Sullo sviluppo del sistema nervoso periferico nei Molluschi," in *Monitore zoologico italiano*, **5** (1894), 75; and *Cenno biografico e catalogo delle opere di Pietro Doderlein* (Palermo, 1896).

His translations are of Michael Foster and F. M. Balfour, *Grundzüge der Entwicklungsgeschichte der Thiere* (Leipzig, 1876); and Michael Foster, *Lehrbuch der Physiologie* (Heidelberg, 1881), with a foreword by W. Kühne.

II. SECONDARY LITERATURE. An obituary is Paul Mayer, in *Anatomischer Anzeiger*, **14** (1898), 267–271. See also Felix Auerbach, *Ernst Abbe. Sein Leben, sein Wirken, seine Persönlichkeit* (Leipzig, 1918); Theodor Heuss,

Anton Dohrn in Neapel (Berlin–Zürich, 1940; 2nd ed., Stuttgart–Tübingen, 1948); and Georg Uschmann, *Geschichte der Zoologie und der zoologischen Anstalten in Jena 1779–1919* (Jena, 1959).

HANS QUERNER

KLEIST, EWALD GEORG VON (*b.* probably Prussian Pomerania, *ca.* 1700; *d.* Köslin [?], Pomerania [now Koszalin, Poland], 11 December 1748), *physics.*

Kleist's father, a district magistrate (*Landrat*), sent him to the University of Leiden to prepare for his place in the Prussian administrative squirearchy. He returned, with an interest in science, to become dean of the cathedral chapter at Cammin and a member of the high court of justice (*Hofgericht*) at Köslin. Kleist's only recorded researches concern electricity, which he began to study in the mid-1740's, inspired by the electrical flare (the ignition of spirits by sparks) and the spectacular displays introduced by G. M. Bose.

Kleist began the experiments which culminated in the invention of the condenser with attempts to increase the strength and reliability of the flare. It appears that he tried to build a portable model, and to this end he placed a nail in a "narrow-necked medicine glass" containing alcohol as fuel. He was quite unprepared for the shock he received when he grasped the nail after touching it to his electrical machine. "What really surprises me," he wrote to J. G. Krüger, a professor at Halle, in December 1745, "is that the powerful effect occurs only [when the bottle is held] in the hand. . . . No matter how strongly I electrify the phial, if I set it on the table and approach my finger to it, there is no spark, only a fiery hissing. If I grasp it again, without electrifying it anew, it displays its former strength." Apparently Kleist had held (that is, grounded) the bottle while charging it, thereby transforming a simple conductor (the nail) into the positive coating of a condenser.

In the winter of 1745–1746 Kleist reported his discovery to several German savants, but without specifying clearly the necessity of grounding the bottle's exterior during charging. None of his correspondents succeeded in reproducing his results. The general theory of electricity accepted at the time, which assumed that electrical matter could traverse glass of the thickness of bottle bottoms, counterindicated Kleist's arrangement for concentrating electrical force; it was not until Pieter van Musschenbroek described more exactly a similar chance experiment done at Leiden toward the beginning of 1746 that others could confirm Kleist's claims. Their subsequent discovery that the shock from the Leyden jar (or Kleist vial) was the greater the thinner the bottle was (that is, the larger the theoretical leak), dealt a deathblow to the traditional approach, cleared the way for the Franklinian system, and won Kleist a foreign membership in the Berlin Academy of Sciences.

BIBLIOGRAPHY

Kleist's letters describing the invention of the condenser appear in J. C. Krüger, *Geschichte der Erde* (Halle, 1746), pp. 177–181; and D. Gralath, "Geschichte der Elektricität [II]," in *Versuche und Abhandlungen der naturforschenden Gesellschaft zu Danzig*, **2** (1754), 402–411.

Biographical details will be found in *Allgemeine deutsche Biographie* (repr. Berlin, 1969), XVI, 112–113; and A. von Harnack, *Geschichte der königlichen preussischen Akademie der Wissenschaften*, I, pt. 1 (Berlin, 1900), 474. On the condenser see J. L. Heilbron, "A propos de l'invention de la bouteille de Leyde," in *Revue d'histoire des sciences*, **19** (1966), 133–142; and "G. M. Bose: The Prime Mover in the Invention of the Leyden Jar?," in *Isis*, **57** (1966), 264–267, and literature cited there.

JOHN L. HEILBRON

KLINGENSTIERNA, SAMUEL (*b.* near Linköping, Sweden, 18 August 1698; *d.* Stockholm, Sweden, 26 October 1765), *physics.*

Klingenstierna was the son of a Swedish army officer and the grandson of two bishops. After having studied at the University of Uppsala and pursuing various activities, including service as a secretary to the Swedish treasury, in 1727 he began a study tour of great importance for his scientific development that led him to Marburg, Basel, Paris, and London. After his return in 1731 Klingenstierna was professor of geometry at the University of Uppsala, a position to which he had been appointed in 1728. He contributed decisively to the development of teaching and research in mathematics and physics at Uppsala, and in 1750 he took over the newly created chair of physics. In 1754 Klingenstierna was appointed teacher to the Swedish crown prince (later Gustavus III) and became a highly respected member of the Swedish court. He held this difficult position, one rather inappropriate for a man of his qualifications, until 1764, the year before his death.

Klingenstierna was an able mathematician and physicist. In the history of physics he is remembered mainly for his important contributions to geometrical optics, having been the first to give a comprehensive theory for achromatic and aplanatic optical systems (systems without color dispersion and spherical aberration).

According to Newton, achromatic refraction is impossible: the dispersion is proportional to the refraction. This was denied by Euler, who cited the human eye as a seemingly achromatic lens system and asked the famous optical instrument maker John Dollond to conduct new experiments in this field; but Euler's proposals did not lead to successful results.

At this stage Klingenstierna, in a paper published in *Kungliga Svenska vetenskapsakademiens Handlingar* in 1754, gave a complete theoretical proof that Newton's assertion disagrees with the fundamental law of refraction. Through Mallet he informed Dollond of this proof, and Dollond immediately began to investigate the problem experimentally. He first found that an experiment mentioned by Newton in his *Opticks* could not be correct: A glass prism was placed in a prism filled with water so that the light was refracted in opposite directions by the two prisms; for a suitable angle of the water prism it was then found that there was no refraction but that the color dispersion did not vanish, as postulated by Newton. Dollond further showed that by increasing the angle of the water prism it was possible to obtain refraction without dispersion, because the dispersion of glass is greater than that of water. Although these findings were in complete agreement with the results presented in Klingenstierna's paper, he was not mentioned in Dollond's account of his own investigations, published in 1758 in *Philosophical Transactions of the Royal Society*.

In this paper Dollond also described his successful construction of achromatic lenses, first by using water and glass, and later crown glass and flint glass, as refractive media. He also showed how he succeeded in the approximate elimination of the spherical aberration.

After reading Dollond's paper Klingenstierna took up the problem of achromatic and aplanatic lens systems for theoretical investigation. In 1760 he published in *Kungliga Svenska vetenskapsakademiens Handlingar* a comprehensive theory for such lens systems, referring to Dollond's experimental results and mentioning that he had made Dollond aware of his proof that Newton's assertion was wrong. Klingenstierna's important 1760 paper is now considered a classic contribution to geometrical optics, paving the way for the later works of Abbe and Gullstrand.

Dollond continued to deny the importance of Klingenstierna's contributions for his own research, and consequently Klingenstierna in a letter to the secretary of the Royal Society protested against this attitude. He also challenged Dollond to enter a competition, arranged by the St. Petersburg Academy of Sciences, concerning the removal of the imperfections in optical instruments caused by color dispersion and spherical aberration. Dollond did not accept the challenge, and Klingenstierna was awarded the prize. His paper, mainly of the same content as that of 1760, was published in 1762 in the *Proceedings* of the St. Petersburg Academy.

BIBLIOGRAPHY

I. ORIGINAL WORKS. Klingenstierna's proof that achromatism is not impossible is given in "Anmärkning vid Brytnings-Lagen af särskilta slags Ljus-strålar, da de gå ur ett genomskinande medel in i åtskilliga andra," in *Kungliga vetenskapsakademiens Handlingar* (1754), pp. 297–300. His general theory of achromatic and aplanatic lens systems is "Om ljusstrålars aberration efter deras brytning genom Spheriske Superficier och Lentes," *ibid.*, **21** (1760), 79–125. His treatise for the St. Petersburg Academy competition is "Tentamen de definiendis et corrigendis aberrationibus radiorum lumines in lentibus sphaericis refracta et de perficiendo telescopio Dioptrico" (1762).

II. SECONDARY LITERATURE. Works on Klingenstierna include G. Gezelli, *Biographiska lexicon*, II (Stockholm, 1779), 38–41; Harald J. Heymann, "Samuel Klingenstierna," in Sten Lindroth, ed., *Swedish Men of Science* (Stockholm, 1952), pp. 59–65; H. Hildebrand Hildebrandsson, *Samuel Klingenstiernas levnad och verk*, I, *Levnadsteckning* (Stockholm, 1919), which includes an account of the controversy between Klingenstierna and Dollond; C. G. Nordin, *Minnen öfver namnkundiga svenska män*, I (Stockholm, 1818), 232–254; and Mårten Strömer, "Åminnelsetal över S. Klingenstierna," in *Kungliga Vetenskapsakademiens Åminnelsetal*, III (1763–1768).

MOGENS PIHL

KLÜGEL, GEORG SIMON (*b.* Hamburg, Germany, 19 August 1739; *d.* Halle, Germany, 4 August 1812), *mathematics, physics.*

Klügel was the first son of a businessman and received a solid mathematical education at the Hamburg Gymnasium Academicum, which he attended after completing the local Johanneum. In 1760 he entered the University of Göttingen to study theology; but he soon came under the influence of A. G. Kaestner, who interested him in mathematics and induced him to devote himself exclusively to that science. At Kaestner's suggestion Klügel took as the subject of his thesis, which he defended on 20 August 1763 (*Conatuum praecipuorum theoriam parallelarum demonstrandi recensio*), a critical analysis of the experiments made thus far to prove the parallel postulate. His criticism of the errors provided a new incentive to investigate this problem. (For instance, Lambert, one of the outstanding forerunners of non-Euclidean geometry, expressly referred to Klügel, who is also cited by most later critics of the problem of parallel lines.)

After five years at Göttingen, Klügel went in 1765 to Hannover, where he edited the scientific contributions to the *Intelligenzblatt*; two years later he was appointed professor of mathematics at Helmstedt. His extensive work in mathematics and physics began then and increased after his transfer to the chair of mathematics and physics at the University of Halle. His work included papers, textbooks, and handbooks on various branches of mathematics.

Despite the generally encyclopedic character of his works, in some respects Klügel put forward new ideas. His most important contribution was in trigonometry. His *Analytische Trigonometrie* analytically unified the hitherto separate trigonometric formulas and introduced the concept of trigonometric function, which in a coherent manner defines the relations of the sides in a right triangle. He showed that the theorems on the sum of the sines and cosines already "contain all the theorems on the composition of angles" and extended the validity of six basic formulas for a right spherical triangle. Not even Euler, who returned to the problem of extending Euclid's trigonometry nine years after Klügel, was able to achieve Klügel's results in certain respects. Klügel's trigonometry was very modern for its time and was exceptional among the contemporary textbooks. Other work in advance of its time concerned stereographic projection, where the properties of this transformation of a spherical surface onto a plane were geometrically derived and the ideas were also applied to spherical trigonometry and gnomonics.

In 1795 in the small publication *Über die Lehre von den entgegengesetzten Grössen*, Klügel dealt with questions of formal algebra and tried to define formal algebraic laws. His most popular and useful work was his mathematical dictionary, *Mathematisches Wörterbuch oder Erklärung der Begriffe, Lehrsätze, Aufgaben und Methoden der Mathematik*, to which three volumes by Mollweide and Grunert were added in 1823–1836; it was used throughout much of the nineteenth century.

While at Halle, Klügel was elected member of the Berlin Academy on 27 January 1803. In 1808 he became seriously ill, and he died four years later.

BIBLIOGRAPHY

I. Original Works. A complete list of Klügel's works can be found in *Hamburger Schriftstellerlexikon* (see below). They include *Conatuum praecipuorum theoriam parallelarum demonstrandi recensio* . . . (Göttingen, 1763); *Analytische Trigonometrie* (Brunswick, 1770); *Von den besten Einrichtung der Feuerspritzen* (Berlin, 1774); *Geschichte und gegenwärtiger Zustand der Optik nach der Englischen Priestleys bearbeitet* (Leipzig, 1776); *Analytische Dioptrik*, 2 vols. (Leipzig, 1778); *Enzyklopädie oder zusammenhängender Vortrag der gemeinnützigsten Kenntnisse*, 3 pts. (Berlin–Strettin, 1782–1784, 2nd ed. in 7 pts., 1792–1817); "Über die Lehre von den entgegengesetzten Grössen," in Hindenburg's *Archiv der reinen und angewandten Mathematik*, **3** (1795); and *Mathematisches Wörterbuch oder Erklärung der Begriffe, Lehrsätze, Aufgaben und Methoden der Mathematik*, 3 vols. (Leipzig, 1803–1808).

II. Secondary Literature. On Klügel or his work, see M. Cantor, *Vorlesungen über Geschichte der Mathematik*, IV (Leipzig, 1908), especially pp. 27, 88, 389, 406, 412 ff., 424 ff., 616, which evaluates Klügel's mathematical work; *Lexikon der hamburgischen Schriftsteller* . . ., IV (Hamburg, 1858), 65–73, which includes a complete list of Klügel's writings; and A. H. Niemeyer, obituary, in *Hallisches patriotisches Wochenblatt* (5 September 1812), 561–569.

Jaroslav Folta

KLUYVER, ALBERT JAN (*b.* Breda, Netherlands, 3 June 1888; *d.* Delft, Netherlands, 14 May 1956), *microbiology, biochemistry.*

Kluyver was the second child and only son of Marie Honingh and Jan Cornelis Kluyver, an engineer who was later professor of mathematics at Leiden. He entered the Technical University of Delft in 1905 and received a degree in chemical engineering in 1910. In the latter year he became assistant to professor G. van Iterson in the Technical Botany Laboratory of the University and began work on a thesis on biochemical sugar determinations, published in 1914 as *Biochemische Suikerbepalingen*. In his thesis Kluyver reported on his measurement of the amount of carbon dioxide produced by yeast during aerobic incubation with sugar solutions. He further suggested that it might be possible, through the use of selected types of yeast, to determine quantitatively any one of a number of sugars in such mixtures. (He returned to this subject in a paper of 1935, "Die bakteriellen Zuckervergärungen.")

In 1916 Kluyver went to the Dutch East Indies as an adviser to the Netherlands Indies Government on the promotion of native industries. He returned to the Netherlands to succeed, in 1922, M. W. Beyerinck in the chair of general and applied microbiology at the Technical University of Delft. He remained there for the rest of his life, his tenure being interrupted only in 1923, when he made a long journey to study the coconut-fiber and copra yarn industries of Ceylon and the Malabar coast of India to determine the practicality of establishing these industries in Java.

In 1922, when Kluyver took up his duties at Delft, the study of the chemical activities of microorganisms was still in its infancy. The microbiologist C. B. van Niel, Kluyver's first student at Delft and later his collaborator, gave this assessment of the state of the science at the time:

> . . . the knowledge of the chemical activities of microorganisms was virtually restricted to an awareness of a large number of more or less specific transformations that can be brought about by the diverse and numerous representative types. Except in a few institutions, the study of biochemistry itself appeared to consist in little more than the development and application of methods for the analysis of urine and blood. Within ten years Kluyver succeeded in welding together a vast amount of detailed information into a coordinated picture, whose strong and simple outlines encompassed the totality of the chemical manifestations of all living organisms, and whose structure brought into strong relief the dynamic aspects of these processes. In another ten years the direction of biochemical research throughout the world had been guided into the paths mapped out by Kluyver. And the spectacular successes scored in enlarging and intensifying biochemical understanding through investigations with appropriate micro-organisms, an approach repeatedly practiced and advocated by Kluyver, had convinced an increasing number of biochemists of the potentialities of such studies, with the logical result that the post-1940 biochemical literature has become predominantly occupied by publications on various aspects of microbiological chemistry [A. F. Kamp *et al.*, eds., *Albert Jan Kluyver* (Amsterdam, 1959), p. 69].

Nor was Kluyver's work only theoretical; in his inaugural address he had already proposed that some of the chemical activities of microorganisms might be utilized on a commercial scale. In 1924 he made the first investigation of *Acetobacter suboxydans* and recognized its importance in the production of sorbose, an intermediate stage in the commercial manufacture of ascorbic acid. Kluyver continued to develop the ties between theoretical and applied microbiology; by the end of the 1920's, he had begun a significant collaboration with the Netherlands Yeast and Alcohol Manufacturing Company.

Kluyver's study of *Acetobacter suboxydans* led him to recognize that the vast diversity of metabolic processes can be reduced to a relatively simple and unified principle of gradual oxidation. He applied this discovery, over the next two decades, to his studies of alcoholic fermentation, phosphorylation, the assimilatory processes, the nature and mechanism of biocatalysis, and cellulose decomposition in the rumen of cattle, and further utilized it in his classification of microorganisms (he was a member of the International Commission for Nomenclature and Classification of the International Congress for Microbiology). In 1933, in collaboration with L. H. C. Perquin, he reported on their development of submerged cultivation of molds, by which it became possible to obtain physiologically uniform cell material for use in studying the oxidative metabolism of these organisms. In 1942, in collaboration with A. Manten, he published a paper entitled "Some Observations on the Metabolism of Bacteria Oxidizing Molecular Hydrogen," in which he showed that biochemical properties might be used advantageously to subdivide a genus already partly characterized by physiological data. Of his various contributions to science, however, his statement of the principle of hydrogen transfer as the fundamental feature of all metabolic processes has had the most far-reaching effect.

Although the chief orientation of Kluyver's work was biochemical, he was also profoundly interested in problems concerning the morphology and development of microorganisms. During World War II, the department of technological physics at Delft was able to provide him with an electron microscope. He thus took up new techniques in studying microbial morphology.

After the war, Kluyver's scientific concerns became more general; at the same time he took up new commercial duties. The Netherlands Yeast Factory greatly expanded its manufacture of pharmaceuticals, necessitating the appointment of a group of medical advisers, and Kluyver took his place at the head of this group. He further became an adviser to the Royal Dutch Shell Laboratory and to the Governmental Fiber Institute and was a trustee of the Central Organization for Applied Natural Scientific Research (T.N.O.) and a member of many of its committees. Following his example, an increasing number of Kluyver's students went to work in industry.

Kluyver received a wide spectrum of honors and was a member of many scientific societies. Of particular interest was his membership in the Royal Netherlands Academy of Sciences, to which he was elected in 1926. He served as president of the natural sciences section of that organization from 1947 until 1954; during this period he was also appointed chairman of the commission bearing his name, the purpose of which was to prepare for the government recommendations concerning investigations of atomic energy. He became a member of the executive council of the Netherlands Reactor Center. He further remained active in the Biophysical Research Group Delft-Utrecht, an organization dedicated to investigating biophysical problems, which Kluyver had founded with L. S. Ornstein, professor of physics at the University of Utrecht. This group published some eighty-seven papers between 1936 and 1956.

Kluyver married Helena Johanna van Lutsenburg Maas in 1916; they had two sons and three daughters. His character has been described as sensitive, restrained, and always courteous.

BIBLIOGRAPHY

I. ORIGINAL WORKS. Kluyver's writings include *Biochemische Suikerbepalingen*, his thesis (Leiden, 1914); *Microbiologie en Industrie*, his inaugural address (Delft, 1922), English trans. as "Microbiology and Industry," in A. F. Kamp *et al.* (see below), pp. 165–185; "Klappervezel- en Klappergarennijverheid," *Mededelingen van het Koloniaal instituut te Amsterdam*, vol. **20** (1923), written with J. Reksohadiprodjo; "*Acetobacter suboxydans*, eine merkwürdige Eisenbakterie," in *Deutsche Essigindustrie*, **29** (1925), 175 ff., written with F. J. G. de Leeuw; "The Catalytic Transference of Hydrogen as the Basis of the Chemistry of Dissimilation Processes," in *Proceedings. K. Nederlandse akademie van wetenschappen*, **28** (1925), 605–618; "Die Einheit in der Biochemie," in *Chemisch Zelle Gewebe*, **13** (1926), 134–190, written with H. J. R. Donker, repr. in A. F. Kamp *et al.*, below, pp. 211–267; "Über das sogenannte Coenzym der alkoholischen Gärung. Ein Versuch zu einer synthetischen Betrachtung des Coenzymproblems auf experimenteller Grundlage," in *Biochemische Zeitschrift*, **201** (1928), 212–258, written with A. P. Struyk; "Atmung, Gärung und Synthese in ihrer gegenseitigen Abhängigkeit," in *Archiv für Mikrobiologie*, **1** (1930), 181–196; *The Chemical Activities of Micro-organisms* (London, 1931); "Zur Methodik der Schimmelstoffwechseluntersuchung," in *Biochemisches Zeitschrift*, **266** (1933), 68–95, written with L. H. C. Perquin; "Die bakteriellen Zuckervergärungen," in *Ergebnisse der Enzymforschung*, **4** (1935), 230–273; "Die gegenseitigen Beziehungen zwischen dem Oxydationszustande in den Zellen und den Stoffwechselvorgängen," in *Proceedings of the VIth International Botanical Congress*, II (Amsterdam, 1935), 273–274; "Beziehungen zwischen den Stoffwechselvorgängen von Hefen und Milchsäurebakterien und dem Redox-Potential im Medium," pt. 1, in *Enzymologia*, **1** (1936), 1–21; pt. 2, in *Biochemisches Zeitschrift*, **272** (1936), 197–214, both parts written with J. G. Hoogerheide; "Prospects for a Natural System of Classification of Bacteria," in *Zentralblatt für Bakteriologie, Parasitenkunde, Infektionskrankheiten und Hygiene*, Abt. II, **94** (1936), 369–403, written with C. B. van Niel, repr. in A. F. Kamp *et al.*, below, pp. 289-328; "Intermediate Carbohydrates Metabolism of Micro-organisms," in *Proceedings of the IInd International Congress for Microbiology* (London, 1937), pp. 459–461; "Beyerinck, the Microbiologist," in *Martinus Willem Beyerinck: His Life and His Work* (The Hague, 1940), pp. 97–154; "Microbial Metabolism and Its Significance to the Microbiologist," in *Proceedings of the IIIrd International Congress for Microbiology* (New York, 1940), pp. 73–86; "Some Observations on the Metabolism of Bacteria Oxidizing Molecular Hydrogen," in *Antonie van Leeuwenhoek*, **8** (1942), 71–85, written with A. Manten; "Microbial Metabolism and Its Industrial Implications," read to the inaugural meeting of the Microbiology Group of the Royal Society, 7 March 1951, in *Chemistry and Industry* (1952), pp. 136–145, repr. in A. F. Kamp *et al.*, below, pp. 424–472; "The Changing Appraisal of the Microbe," Leeuwenhoek lecture, in *Proceedings of the Royal Society*, **141B** (1953), 147–161; "From Dutch Settlements to the Rutgers Institute of Microbiology," in S. A. Wakman, ed., *Perspectives and Horizons in Microbiology, a Symposium* (New Brunswick, N.J., 1955), pp. 213–220; and *The Microbe's Contribution to Biology* (Cambridge, Mass., 1956), written with C. B. van Niel.

II. SECONDARY LITERATURE. On Kluyver's life and work, see the Jubilee Volume of the Netherlands Society for Microbiology, issued on the occasion of his twenty-fifth anniversary at the University of Delft as *Antonie van Leeuwenhoek*, vol. XII; and A. F. Kamp, J. W. M. la Rivière, and W. Verhoeven, eds., *Albert Jan Kluyver: His Life and Work* (Amsterdam–New York, 1959), which contains, among other things, a list of papers from Kluyver's laboratory published between 1922 and 1956, a list of doctor's theses published under his direction, and a complete bibliography of his own publications.

PIETER SMIT

KNESER, ADOLF (*b.* Grüssow, Germany, 19 March 1862; *d.* Breslau, Germany [now Wrocław, Poland], 24 January 1930), *mathematics.*

One of the most distinguished German mathematicians of the years around 1900, Kneser was the son of a Protestant clergyman who died when the boy was one year old. His mother moved to Rostock in order to educate her four sons. There Kneser completed his secondary schooling and studied for a year at the university. As early as this (1880) he published his first paper, on the refraction of sound waves. He then went to Berlin. Of the great Berlin mathematicians Kronecker was above all his teacher, but certainly Kneser was also influenced by Weierstrass. In 1884 he received his doctorate and began his teaching activity. In 1889 he became associate professor, and in 1890 full professor, at Dorpat. In 1900 he went to the Bergakademie at Berlin; and in 1905 he received a professorship at the University of Breslau which he held for the rest of his life. Kneser was "Dr. e. h." (honorary doctor in engineering) of the Technische Hochschule at Breslau and a corresponding member of the Prussian and Russian Academies of Sciences. In 1894 he married Laura Booth; they had four children. Their son Helmuth was professor at the University of Tübingen, and Helmuth's son Martin became professor at the University of Göttingen. So Kneser may be considered as the founder of a mathematical dynasty.

Although Kneser appears in the history of mathematics primarily as a master of analysis, he was, at first, more concerned with algebra. His dissertation and some subsequent papers are dedicated to algebraic functions and algebraic equations. He next turned to geometry, with a series of interesting works on space curves (1888–1894). Much later Kneser made another important discovery in the theory of curves: the so-called four-vertex theorem (1912). In 1888 he had begun his analytical investigations, the first of which involved elliptic functions, a subject still of interest to him in later years. Soon, however, he turned his attention to one of the two main subjects of his lifework: linear differential equations, and especially the group of ideas associated with the so-called Sturm-Liouville problem (from 1896). Since 1906, integral equations were added, after Fredholm's fundamental works had appeared, the two subjects being closely connected. Kneser's decisive achievement was to bring the theory of developing arbitrary functions into series with respect to the eigenfunctions of a Sturm-Liouville differential equation to the same level of generality that Dirichlet had achieved in the special case of Fourier series. Kneser's treatment of all this, which found final expression in his book on integral equations (1911), is characterized by a very intensive consideration of the theory's applications to mathematical physics, the theory of heat conduction, for instance.

The calculus of variations, the other main subject of Kneser's research (from 1897)—in fact the most important—is also of value to physics. His engagement in this classical topic was not the direct result of his studies with the field's great master, Weierstrass, but, rather, of his teaching experience at Dorpat. Kneser brought the theory of the so-called second variation to a certain conclusion. Especially, he favored one of its geometric aspects, the theory of families of resolution curves and their envelopes, closely connected with the Jacobian theory of conjugate points. But above all, the decisive advances toward the solution of the so-called Mayer Problem, recently introduced to the calculus of variations, are due to Kneser. His textbook on the calculus of variations (1900) had an enduring influence on later research. Many of the technical terms nowadays usual in calculus of variations were created by Kneser, e. g. "extremal" (for a resolution curve), "field" (for a family of extremals), "transversal," and "strong" and "weak" extremum.

An example of Kneser's interest in the history of mathematics is his booklet *Das Prinzip der kleinsten Wirkung von Leibniz bis zur Gegenwart* (1928).

BIBLIOGRAPHY

I. Original Works. A bibliography of Kneser's works can be found in Koschmieder (see below). They include *Lehrbuch der Variationsrechnung* (Brunswick, 1900; 2nd ed., 1925); "Variationsrechnung," in *Encyklopädie der mathematischen Wissenschaften*, II, pt. 1 (1904), 571–625; *Die Integralgleichungen und ihre Anwendungen in der mathematischen Physik* (Brunswick, 1911; 2nd ed., 1922); and *Das Prinzip der kleinsten Wirkung von Leibniz bis zur Gegenwart* (Leipzig, 1928).

II. Secondary Literature. See *Zur Erinnerung an Adolf Kneser* (Brunswick, 1930), reprint of the commemorative addresses delivered at Breslau in Feb. 1930; and L. Koschmieder, "Adolf Kneser," in *Sitzungsberichte der Berliner mathematischen Gesellschaft*, **29** (1930), 78–102, which includes a bibliography of 81 items; and "El profesor Adolfo Kneser," in *Revista matemática hispano-americana*, 2nd ser., **5** (1930), 281–288.

Hermann Boerner

KNIGHT, THOMAS ANDREW (*b.* near Ludlow, Herefordshire, England, 12 August 1759; *d.* London, England, 11 May 1838), *botany, horticulture.*

T. A. Knight was born into an old Shropshire family with independent means and a substantial estate. His father, Thomas Knight, died when he was young; his elder brother, Richard Payne Knight, the numismatist, helped him by establishing him at Elton with a farm and hothouses, and later by handing over to him the management of 10,000 acres at Downton Castle, his home in Herefordshire.

As a young man, Knight had some education at Ludlow grammar school, but he learned more by observing and asking questions in the gardens at his home. He graduated at Balliol College, Oxford, and in 1791 married Frances Felton. They had a son who died in youth and three daughters. As a result of his early experiments on breeding fruit, vegetables, and cattle, he was recommended by his brother to Sir Joseph Banks as a correspondent to the Board of Agriculture. He met Banks in London in 1795 and began a correspondence in which he recorded his most important scientific work, read by Banks to the Royal Society. He was elected to the Royal Society in 1805. When the Horticultural Society was founded in 1804 Knight was a member, and from 1811 he was its president.

The main impetus to Knight's research was practical: He wanted to improve the culture and yield of produce from farm and garden. His most important contribution to horticulture and agriculture was in the application of scientific principles and techniques to practical situations of the grower or breeder. He worked on the design of hothouses, control of pests, and cider making as carefully as he did on theoretical research. His first published letter to Banks (1795) was on the gradual decline of stock propagated by grafting and the need to develop new varieties from seed, particularly by cross fertilization. The latter would produce both greater vigor and a wider range of offspring from which the most useful could be chosen. He wrote in 1799: "In promoting this sexual intercourse between neighbouring plants of the same species, nature appears to me to have an important purpose in view."

For practical purposes he bred apples, pears, and Herefordshire cattle, but he also worked from 1787 on peas selected as suitable because they were annuals with clearly differentiated paired characteristics. In a paper of 1799 he described how in the crossing of a gray pea with a white pea all the first generation are gray, but the white reappear in the second generation. He had observed dominance, but had done none of the careful statistical work that made Mendel's experiments of half a century later so significant.

Study of the developing fruit led him to design careful experiments on the translocation of sap. He used colored infusions to trace the ascent of sap through the outer layers of alburnum (xylem), branching to petioles and passing to the developing fruit; he also used ringing to trace the descent of sap, disproving Hales's theory that bark was formed from the alburnum. He believed that the sap circulated, but showed that sap in the bark differs from aqueous sap in the alburnum, as a result of nutrients received from the leaves. Sap which is carried to leaves and air "seems to acquire (by what means I shall not attempt to decide) the power to generate the various inflammable substances that are found in the plant. It appears to be brought back again . . . to add new matter." If leaves were shaded the quantity of alburnum deposited was small. He discussed at length the forces causing this circulation, certain only that the explanation must be mechanical. He knew of capillarity and demonstrated transpiration and observed the spiral vessels, but he took no account of root pressure or cohesion of liquid columns, so he suggested a process involving the contraction and expansion of "silver grains." Sap descended, he thought, by gravitation.

His most famous work was on what are now called geotropisms. In a letter read by Banks to the Royal Society in 1806, Knight described how he eliminated the influence of gravitation on germinating seeds: He attached them at various angles to the rim of a vertical wheel which was driven by a stream in his garden to revolve continuously at a rate of 150 r.p.m. As the germinating plants grew, each shoot was directed to the center of the wheel; when a shoot passed the center of the wheel its tip turned back so that growth was still centripetal; the roots grew away from the center. Next he set up a similar structure with the wheel horizontal and rotating at 250 r.p.m. so that the seedlings were influenced by both gravitation of the earth and the centrifugal force. In this case, growth was at an angle of 80° to the vertical, the shoot upward and inward, and the root downward and out. Reducing the rotation to 80 r.p.m. decreased the centrifugal force to such an extent that the plants grew at an angle of 45° to the vertical.

Here he ran into the philosophical problem of how the plants "perceived" the force acting on them, as he was himself "wholly unable to trace the existence of anything like sensation or intellect in the plant." Yet these seedlings clearly reacted to gravitational force, for he had shown in 1801 that vine leaves always move so as to present the upper surface to the sun; and in 1811 he was to demonstrate that in certain conditions roots may be deflected from the vertical by moisture. The plants were making adaptive responses, but his explanation was typically mechanical: the roots bend down by their own weight, while in the shoot, nutrient sap moves to the lower side and there stimulates differential growth and curvature to the vertical. He was aware that this was not entirely satisfactory as he had not explained the "weeping" tree.

Adaptive response to the physical environment by differential growth was his only major discovery, although he came surprisingly near several others. His papers, readily accessible in *Philosophical Transactions*, were known to Darwin, who worked on light sensitivity in climbers as well as on variation and selection. Knight was friendly with many scientists of his day apart from Banks. Dutrochet visited him in 1827 and repeated some of his experiments. Sir Humphry Davy quoted his experiments with wheels in his lectures on agricultural chemistry of 1802 to 1812, and included an elegant plate of the apparatus in the published version. Knight added some notes to the third edition, and in the fourth edition Davy altered the dedication from the Board of Agriculture to Knight.

BIBLIOGRAPHY

I. ORIGINAL WORKS. Knight's scientific work was published in *Philosophical Transactions of the Royal Society of London*, and is listed in the Society's *Catalogue of Scientific Papers*. The most significant of the papers referred to in the text are "Observations of the Grafting of Fruit Trees," **85** (1795), 290–295; "An Account of Some Experiments on the Fecundation of Vegetables," **89** (1799), 195–204; a series of papers on sap, **91** (1801), 333–353; **93** (1803), 277–289; **94** (1804), 183–190; and **95** (1805), 88–103; and "On the Direction of the Radicle and Germen During the Vegetation of Seeds," **96** (1806), 99–108.

The above and some eighty other papers were reprinted in *A Selection From the Physiological and Horticultural Papers, Published in the Transactions of the Royal and Horticultural Societies by the Late Thomas Andrew Knight, to Which is Prefixed a Sketch of His Life* (London, 1841).

His two most important books are *Pomona Herefordiensis* (London, 1811) and *A Treatise on the Culture of the Apple and Pear, and on the Manufacture of Cider and Perry* (Ludlow, 1797; 2nd ed., 1801; 3rd ed., 1808).

II. Secondary Sources. The most useful personal source is the *Sketch of His Life*, written by his daughter Mrs. Frances Acton and published in the *Selection* already cited. The best scientific evaluation is C. A. Shull and J. F. Stanfield, "Thomas Andrew Knight: in Memoriam," in *Plant Physiology*, **14** (1939), 1–8. A brief obituary notice is in *Abstracts of the Papers Printed in the Philosophical Transactions of the Royal Society of London From 1837 to 1843* (later the *Proceedings*), **4** (1843), 92–93.

There is an account of his practical work in "Thomas Andrew Knight and His Work in the Orchard," in Woolhope Naturalists' Field Club, *Herefordshire Pomona*, **1** (1876), 29–46.

Darwin does not appear to have referred to Knight's work on geotropisms, but he mentioned his work on hybrid vigor several times in *The Effects of Cross and Self Fertilisation in the Vegetable Kingdom* (London, 1876). He also quoted Knight's definition of domestication in his essay first exploring the ideas of the *Origin* in 1844.

See also Sir Humphry Davy, *Elements of Agricultural Chemistry* (London, 1813, 3rd ed., 1821).

There are numerous references to Knight's place in the history of scientific thought in the standard textbooks and in biographies of his many friends. Many letters remain from his voluminous correspondence, the largest collection being in the British Museum (Natural History). Those written to Banks may be traced through W. R. Dawson, ed., *The Banks Letters, a Calendar of the Manuscript Correspondence of Sir Joseph Banks* (London, 1958).

A good bibliography of source material is included in the entry for Knight in *A Biographical Index of Deceased British and Irish Botanists*, compiled by J. Britten and G. S. Boulger, 2nd ed. (London, 1931), pp. 176–177.

Diana M. Simpkins

KNIPOVICH, NIKOLAI MIKHAILOVICH (*b.* Suomenlinna, Finland, 25 March 1862; *d.* Leningrad, U.S.S.R., 23 February 1939), *marine biology, marine hydrology.*

In 1886 Knipovich graduated from the Physics and Mathematics Faculty of St. Petersburg University. A year earlier he was a member of O. A. Grimm's expedition to the Lower Volga to evaluate the long-range prospects for herring fishery in the Volga–Caspian basin. This experience aroused his interest in hydrology and the study of marine life, thereby defining the subjects of his subsequent scientific activity.

Beginning in 1887 Knipovich studied the biology of the White Sea. Soon his work expanded to the Barents Sea, where he organized and headed the Murmansk expedition for research on natural resources (1898–1901). This expedition, which was carried out on the *Andrey Pervozvanny*, the first vessel especially constructed for such work, brought back the first substantial information on the nature and rich natural resources of the northern seas of European Russia. Knipovich organized three Caspian expeditions for research on natural resources (1904, 1912–1913, and 1914–1915). The expeditions made it possible to present in great detail the peculiar hydrological features of the Caspian Sea and the distribution and annual cycle of marine life. In 1921 Knipovich summarized and published the results of the Caspian expeditions. They provided the basis for the subsequent regulation of commerce and the protection of the resources of the Caspian Sea. From 1905 to 1911 Knipovich organized and led research expeditions on the natural resources of the Baltic Sea. After processing the results of the third Caspian expedition, he led similar expeditions to the Sea of Azov and the eastern part of the Black Sea (1922–1927). In 1931–1932 he headed a research expedition to study the natural resources of the entire Caspian Sea.

Before Knipovich, such research expeditions to the Russian seas merely described the objects of possible commercial exploitation, without considering the water's influence on them. Knipovich was a pioneer in the study of marine fishes and their close connection with hydrological conditions. "Productivity of reservoirs," he wrote in *Gidrologia morey i solonovatykh vod* (1938), "is always restricted by certain limits which are determined by the aggregate of the physical and chemical conditions. . . ." Knipovich's half-century of work is summed up in the same work: "As there is no scientific hydrobiology without calculation of hydrological conditions, there is no scientific hydrology without calculation of biological factors."

BIBLIOGRAPHY

I. Original Works. Knipovich wrote 164 works. The principal publications are *Osnovy gidrologii Evropeyskogo Ledovitogo okeana* ("Principles of Hydrology in the

European Arctic Ocean"; St. Petersburg, 1906); *Gidrologicheskie issledovania v Kaspyskom more* ("Hydrological Explorations in the Caspian Sea"; Petrograd, 1914–1915); *Trudy Kaspyskoy ekspeditsii 1914–1915* ("Transactions of the Caspian Expedition . . ."; Petrograd, 1921); "Opredelitel ryb Chernogo: Azovskogo morey" ("Checklist of Fishes of the Black and Azov Seas"; Moscow, 1923); "Opredelitel ryb morey Barentseva, Belogo i Karskogo" ("Checklist of Fishes of the Barents, White, and Kara Seas"), in *Trudy Instituta po izucheniyu Severa*, no. 27 (1926); "Gidrologicheskie issledovania v Azovskom more" ("Hydrological Explorations in the Azov Sea"), in *Trudy Azovo-chernomorskoi nauchno-promyslovoi ekspeditsii*, no. 5 (1932); "Gidrologicheskie issledovania v Chernom more" ("Hydrological Explorations in the Black Sea"), *ibid.*, no. 10 (1933); and *Gidrologia morey i solonovatykh vod (v primenenii k promyslovomu delu)* ("Hydrology of the Seas and of Saline Waters [Applied to Production]"; Moscow–Leningrad, 1938).

II. SECONDARY LITERATURE. On Knipovich or his work, see "Polveka nauchnoy i obshchestvennoy deyatelnosti N. M. Knipovicha" ("Half a Century of the Scientific and Public Activity of N. M. Knipovich"), in *Sbornik posvyashchenny nauchnoy deyatelnosti N. M. Knipovicha* ("Collection Dedicated to the Scientific Activity of N. M. Knipovich"; Moscow–Leningrad, 1939), an anthology; V. K. Soldatov, "Nikolai Mikhailovich Knipovich," in *Sbornik v chest professora N. M. Knipovicha* ("Collection of Articles in Honor of Professor N. M. Knipovich"; Moscow, 1927), pp. 1–14; and P. Sushkin and A. Kapinsky, "Zapiska ob uchenykh trudakh N. M. Knipovicha" ("A Note on the Scholarly Works of N. M. Knipovich"), in *Izvestiya Akademii nauk SSSR*, ser. 5, no. 18 (1927), 1485–1488.

A. F. PLAKHOTNIK

KNOPP, KONRAD (*b.* Berlin, Germany, 22 July 1882; *d.* Annecy, France, 20 April 1957), *mathematics.*

After one semester at the University of Lausanne in 1901, Knopp returned to Berlin to study at its university. He passed the teacher's examination in 1906 and received the Ph.D. in 1907. In the spring of 1908 he went to Nagasaki to teach at the Commercial Academy; he also traveled in China and India. In the spring of 1910 he returned to Germany, where he married the painter Gertrud Kressner. They moved to Tsingtao, where Knopp taught at the German-Chinese Academy. Back in Germany in 1911, he received his *habilitation* at Berlin University while teaching at the Military Technical Academy and at the Military Academy. An officer in the German army, he was injured early in World War I. In the fall of 1914 Knopp resumed teaching at Berlin University, in 1915 he became extraordinary professor at Königsberg University and in 1919 ordinary professor. In 1926 he was appointed to Tübingen University, where he remained until retiring in 1950.

Knopp was a specialist in generalized limits. He not only contributed many details but also clarified the general concept and the aims of the theory of generalized limits. He is well known for his extremely popular books on complex functions, which have often been republished and translated. He was responsible for the sixth through tenth editions of H. von Mangoldt's popular *Einführung in die höhere Mathematik*. He was a cofounder of *Mathematische Zeitschrift* in 1918 and from 1934 to 1952 was its editor.

BIBLIOGRAPHY

Knopp's writings include *Grenzwerte von Reihen bei der Annäherung an die Konvergenzgrenze* (Berlin, 1907), his diss.; "Neuere Untersuchungen in der Theorie der divergenten Reihen," in *Jahresbericht der Deutschen Mathematiker-Vereinigung*, **32** (1923), 43–67; "Zur Theorie der Limitierungsverfahren I. II," in *Mathematische Zeitschrift*, **31** (1929), 97–127, 276–305; *Theorie und Anwendung der unendlichen Reihen*, which is Grundlehren der mathematischen Wissenschaften, II, 4th ed. (Berlin, 1947); *Funktionentheorie*, 2 vols., which is Sammlung Göschen, nos. 668 and 703, 9th ed. (Berlin, 1957); *Aufgabensammlung zur Funktionentheorie*, 2 vols., which is Sammlung Göschen, nos. 877 and 878, 5th ed. (Berlin 1957–1959); and *Elemente der Funktionentheorie*, which is Sammlung Göschen, no. 1109, 7th ed. (Berlin, 1966).

An obituary is E. Kamke and K. Zeller, "Konrad Knopp †," in *Jahresbericht der Deutschen Mathematiker-Vereinigung*, **60** (1958), 44–49.

HANS FREUDENTHAL

KNORR, GEORG WOLFGANG (*b.* Nuremberg, Germany, 30 December 1705; *d.* Nuremberg, 17 September 1761), *engraving, paleontology.*

Knorr was one of the protogeologists of the eighteenth century who is intermediate between the collectors of cabinets of natural history and those who first made use of fossils for the identification and mapping of stratigraphic succession. This was the generation that finally established the organic origin of fossils and accumulated sufficient descriptive material to classify their finds within the biological kingdom, thus providing the paleontologic basis for the law of faunal succession.

Apprenticed to his father's craft of turner, Knorr at the age of eighteen became an engraver of copperplates for Leonhard Blanc, working with Martin Tyroff on the illustrations for Jacob Scheuchzer's *Physica sacra* (1731). This work and his acquaintance with J. A. Beurer, a mineralogist-correspondent of the Royal Society, sharpened Knorr's interest in natural history. He was greatly influenced by Dürer's work, some of which he later reproduced, publishing "Albert Dürer, Opera Omnia" as an appendix to his *General History of Artists or The Lives, Works, and Accomplishments of Famous Artists* (1759).

Knorr's earliest independent works were views of Nuremberg and its environs after sketches by J. C. Dietzsch (1737), *Historische Künstler* (1738). His first scientific work was the preparation of 301 colored copperplate engravings for *Thesaurus rei herbariae hortensisque universalis* (1750), with Latin and German text by P. F. Gmelin and G. R. Boehmer. The remainder of Knorr's life was devoted to the preparation and publication of these costly folio volumes of mixed text and engravings.

Knorr's geological concerns culminated in 1755 with the publication of his *Sammlung* . . . ("Collection of Natural Wonders and Antiquities of the Earth's Crust"), comprising about 125 handsome plates in folio with a descriptive text. The copperplates are in several colors of ink (more than one impression), as well as being hand-washed with watercolor and possibly aquatinted. They depict *lusus naturae* (dendrites, Florentine *paesina*, moss agate), a *Landkarten-Stein* (a variety of dendrite), leaves, crustaceans, ammonites, crinoids, fish, medusae, corals, echinoderms, brachiopods (terebratulids), mollusks of various species, ferns, bark, seeds—in short, the contents (excluding mineral and strictly rock specimens) typical of the fossil cabinets of the eighteenth century. Credit for the scientific content, which, Zittel wrote, places Knorr's *Sammlung* far ahead of all other eighteenth-century paleontological works, is usually given to J. E. I. Walch, who undertook to continue and complete the work for Knorr's heirs, expanding it ultimately to four volumes.

Nevertheless, the volume published by Knorr alone, although modestly entitled ". . . nach der Mëynung der berühmsten Männer . . .," makes it clear that he was thoroughly familiar with the literature of the period and that he was an experienced enough observer to exercise good judgment in weighing the views of others. Knorr classified as diluvialists John Woodward, John Ray, John Morton, Tenzel, Buttner, Johann Bayer, Scheuchzer, Liebknecht, G. A. Volkmann, Christlob Mylius, Nicolaus Steno, Linnaeus, Wohlfarth, Johann Sulzer, Lesser, and Leibniz. He subscribed to the primitive diluvialism which made no distinction between fossils in the rock and the drift: ". . . a man cannot contradict his eyes" (*Sammlung*, 1755, p. 2).

Knorr also briefly referred in his introduction to the views of Voszius, Stillingfleet, Clericus, Plot, Lang, Camerarius, Conring, and Moro. There are additional references to Bernard de Jussieu, who classified the plant fossils depicted by Knorr, as well as to Peter Collinson, F. Bayer's "Sublim. Oryctograph." (1730), Pontopiddan, and Wallerius. An essay by Mylius in the form of a letter to A. v. Haller on the zoophytes of Greenland is included. An engraved title page and a double folio frontispiece in watercolors, showing the quarry of Solnhofen (Bavaria), completed the first volume of the work (the three later volumes were published by Walch).

The extraordinary quality of the plates, representing the eighteenth-century continuation of the tradition of Dürer, led to expansion of the work by Walch, as well as to French and Dutch editions. It is scarely an exaggeration to say that the beauty of some of Knorr's illustrations exceeds that of their models and that in all cases the artist's eye has transformed neutral, natural objects into permanent, formal aspects of humanism.

The detail and accuracy of Knorr's engravings not only made possible zoological classification but firmly established the distinction between fossils of organic origin and sports of nature.

BIBLIOGRAPHY

I. Original Works. The value of Knorr's copperplates ensured successive impressions, which have yet to be classified.

See *Historische Künstler—Belustigung, oder Gespräche in dem Reiche derer Todten, zwischen* . . . (Nuremberg, 1738); *Monumentorum et aliarum quae ad sepulcra veterum pertinent rerum, imagines* (Nuremberg, 1753); *Deliciae naturae selectae, oder auserlesenes Natüralien-Cabinet, welches aus den drey Reichen der Natur zeiget was von curiösen Liebhabern aufbehalten und gesammlet zu werden verdienet . . . herausgegeben von Georg Wolfgang Knorr* (Nuremberg, 1754), with color plates; French trans., 6 pts. in 4 vols. (Nuremberg, 1757–1773), with plates and color frontispiece; *ibid.*, 3 vols. (1760–1775); 2nd ed. (1766–1777), with 88 plates; 3rd ed., 2 vols. (1769–1776); *Sammlung von Merckwürdigkeiten der Natur und Alterthümern des Erdbodens, welch petrificirte Cörper enthält . . . beschrieben von Georg Wolffgang Knorr* (Nuremberg, 1755), with 2nd engraved title page in Latin and 126 color plates. *Les délices des yeux et de l'esprit, à la représentation d'une collection universelle des coquilles et des autres corps qui sont à trouver dans la mer, produite par Georg Guelphe Knorr*, 6 pts., in 4 vols. (Nuremberg, 1757–1773), with 2 plates and color frontispiece; 2nd ed., 6 pts. in 3 vols. (1760–1773); *Allgemeine Künstler-Historie, oder berühmter Künstler Leben, Werke und Verrichtungen, mit vielen Nachrichten von raren, alten und neuen Kupferstichen beschrieben von Georg Wolfgang Knorr* (Nuremberg, 1759), with portrait, engravings; and *Regnum florae, das Reich der Blumen mit allen seinen Schönheiten nach der Natur und ihren Farben vorgestellt* (Nuremberg, n.d.), with 101 color plates.

See also *Nürnbergischer Prospecten, ...* (*ca.* 1745), written with J. A. Delsenback and others; *Thesaurus rei herbariae hortensisque universalis . . . Allgemeines Blumen-, Kräuter-, Frucht- und Garten-Buch* (Nuremberg, 1750), with color plates, written with P. F. Gmelin and G. R. Boehmer; and *Die Naturgeschichte der Versteinerungen zur Erläuterung der Knorrischen Sammlung von Merkwürdigkeiten der Natur, herausgegeben von J. E. I. Walch* . . . (Nuremberg, 1755–1773); Dutch trans., 3 pts. in 4 vols. (Amsterdam, 1773); French trans., 4 pts. in 6 vols. (Nuremberg, 1768–1778).

Subsequent works are *Délices physiques choisies, ou choix de tout ce que les trois règnes de la nature renferment de plus digne des recherches d'un amateur curieux pour en former un cabinet . . . par Georg Wolfgang Knorr . . . Continué par ses héritiers, avec les descriptions de Philippe Louis Stace Müller, et traduit en français par Mathieu Verdier de La Blaquière* . . ., 2 vols. (Nuremberg, 1766–1767), in German and French with color plates; 2nd ed. (Nuremberg, 1779), in French alone; and *Recueil choisi de desseins de fleurs, à l'usage des dames, à l'aide duquel on peut apprendre très facilement et sans beaucoup d'instruction l'art de dessiner les fleurs*, 2 pts. (Nuremberg, n.d.), with color plates.

II. SECONDARY LITERATURE. Biographical notices of Knorr appear in Poggendorff, I, 1284; G. K. Nagler, *Künstler-Lexikon;* Ulrich Thieme and Felix Becker, *Künstler-Lexikon*, XXI, 30–31; and *Nouvelle biographie générale*, XXVII–XXVIII, 915. There are brief notices in *La grande encyclopédie*, XXI, 575, and Larousse's *Dictionnaire du XIX Siècle*, IX, 1232. The best notice of Knorr is by Gümbel in *Deutsche Biographie*, XVI, 326–327. Details of Knorr's life in these notices are from J. G. Meusel's *Lexikon vom jahr 1750 bis 1800 verstorbeuen teutschen schriftsteller*, VII, 142, and G. A. Will's *Nürnbergisches gelehrten-lexicon.*

CECIL J. SCHNEER

KNOTT, CARGILL GILSTON (*b.* Penicuik, Scotland, 30 June 1856; *d.* Edinburgh, Scotland, 26 October 1922), *natural philosophy, seismology.*

Cargill Gilston Knott was the son of Pelham Knott, author of a volume of poetry, who died young. Cargill was brought up by an uncle and aunt. He entered Edinburgh University in 1872, and after gaining his first degree became assistant to P. G. Tait in the Department of Natural Philosophy. In 1883 he left Edinburgh to become a professor of physics at the Imperial University of Japan. In Tokyo he married a Scottish lady, Mary Dixon, sister of the professor of English at the Imperial University. He returned to Edinburgh in 1891 as lecturer, later reader, in applied mathematics, and died in his office from a heart attack in 1922.

On arriving in Japan, Knott became a member of the famous group (J. Milne, M. Ewing, and T. Gray from Britain, and several Japanese) who inaugurated the modern era in earthquake study. Knott's contributions were chiefly on the mathematical side. He utilized the observational results of his colleagues to pioneer the application of seismology to determine the internal mechanical properties of the earth. In the course of this work, he traced the connection between the times taken by earthquake waves in traveling from the center of the disturbance through the earth's interior to seismic recording stations, and worked out many of the detailed physical characteristics of the waves. These and other pioneering results formed the basis of many later researches on the interior of the earth.

While in Japan, Knott also organized and supervised the first comprehensive magnetic survey of Japan. Many of his pupils and their pupils became Japan's leading investigators of the earth's magnetic field. In addition he investigated Japanese volcanic eruptions.

His work on seismology and geophysics generally earned him many honors, including the award by the Japanese emperor of the Order of the Rising Sun, election to the Royal Society of London, and the award of the Keith Prize by the University of Edinburgh. He was a principal founder of the Edinburgh Mathematical Society, an energetic secretary of the Royal Society of Edinburgh, and contributed many articles to the *Encyclopaedia Britannica.* He was also a zealous officeholder in the United Free Church of Scotland.

BIBLIOGRAPHY

Knott wrote five books, was editor of several others, and was the author of 100 published papers. The papers cover not only his seismological and other geophysical researches, but also researches in physics and pure mathematics, and include a few general and biographical articles. A full list of his publications is given in the *Proceedings of the Royal Society of Edinburgh*, **48** (1923), 242–248.

Among his most important publications are *Physics of Earthquake Phenomena* (Oxford, 1908); *Electricity and Magnetism* (Edinburgh, 1893); a rev. 3rd ed. of *Kelland and Tait's Quaternions* (Edinburgh, 1904); and "The Propagation of Earthquake Waves Through the Earth, and Connected Problems," in *Proceedings of the Royal Society of Edinburgh*, **39**, 157–208.

K. E. BULLEN

KNOX, ROBERT (*b.* Edinburgh, Scotland, 4 September 1793; *d.* Hackney, London, England, 20 December 1862), *anatomy*.

Knox was the eighth child and fifth son of Robert Knox and Mary Sherer or Scherer, of German extraction. His father, who claimed kinship with John Knox, was a schoolmaster at George Heriot School in Edinburgh. For a time, following the outbreak of the French Revolution, the senior Knox had been connected with liberal prorevolutionary groups, but he broke with them before the government instituted repressive measures.

Young Robert was tutored at home until he entered Edinburgh High School at age twelve, where he was an honor student. After his graduation in 1810, he enrolled as a medical student at the University of Edinburgh. He stood for examination three years later but failed in anatomy, in part because Alexander Monro *tertius*, the official professor in the subject, was so bored with it that he was content to read his grandfather's century-old lecture notes. To remedy his deficiency, Knox turned to studying with John Barclay, who ran an extramural school of anatomy. He not only passed the examination in 1814 but decided to make anatomy his area of special interest.

Knox's doctoral dissertation on the effects of alcohol and other stimulants on the human body led him to conclude that alcohol was detrimental to long life. In later years he stated that he had only three rules of health: temperance, early rising, and frequent changes of linen. A case of smallpox in his youth left his face scarred and destroyed one of his eyes, leaving an ugly, raised cicatrix in the cornea. As a result he always wore glasses. He also had a reputation as an ornate and fastidious dresser, perhaps, as one biographer has suggested, to distract attention from his face.

After completing his studies at Edinburgh he went to London for further study, and was soon assigned as a hospital assistant with the British forces in Europe. He arrived in Brussels in time to administer to casualties from the Battle of Waterloo. Knox later publicly expressed shock and distress at the treatment of the wounded there, and went so far as to argue that it would have been preferable to tend the wounded in the open field since the mortality in the hospitals was overwhelming. He was soon sent back to England in charge of a party of wounded. In 1817 he set sail with the 72nd Regiment for the Cape of Good Hope.

During the voyage Knox measured the temperature of the ocean and of the "Superincumbent Atmosphere" three times daily, dissected sharks and dolphins, and studied the "action of the heart in fishes." Results of his studies and experiments were published in the *Edinburgh Philosophical Journal* as well as the *Edinburgh Medical and Surgical Journal;* the latter journal had also published an article based upon his doctoral thesis.

In South Africa Knox participated in the fifth Kaffir War (with the Bantu), which ended in 1819, and spent much of his spare time studying the fauna and shooting and dissecting numerous animals. It was in South Africa that he demonstrated his tendency to irritate so many of his compatriots. Knox apparently had a very high opinion of his own talents and little tolerance for the views of others who he felt lacked his ability and dedication. He was also outspoken, acquiring a reputation as a radical and an atheist. In South Africa he was regarded as pro-Bantu.

For reasons which are not clear (his own accounts are lost), Knox was censured by a court of inquiry in 1820 for alleged actions against a fellow officer. He was also horsewhipped by a citizen who felt the censure was insufficient. The ambiguities of the case are compounded by the army's having continued to keep him on half-pay after his return to Edinburgh until his retirement in 1832, even though he was never again on active duty.

After his return to Edinburgh, he published a series of articles in Scottish medical journals on his experiences in South Africa, dealing with various topics, for example, the Bantu, tapeworms, and necrosis and regeneration of the bone. In 1821 he went to Paris for further study with Georges Cuvier and Geoffroy Saint-Hilaire. His stay confirmed him in his admiration for Napoleon and things French, deepening his contempt for most of his Scottish predecessors and contemporaries.

Back in Edinburgh, by the end of 1822, Knox concentrated on the anatomy of the eye, and in 1823 published a study entitled "Observations on the Comparative Anatomy of the Eye," in which he reasoned that it was a muscle and not a ligament that received the nerves governing vision. He also took steps to establish a museum of comparative anatomy, a project for which he received the backing of the Royal College of Surgeons. In 1825 Knox also made an agreement with John Barclay to take over his extramural school of anatomy.

In 1824 Knox married a woman named Mary, some four years younger than himself, by whom he had six children; her maiden name is not known. Henry Lonsdale, who eventually became his partner as well as his biographer, regarded Knox's wife as of "inferior social rank," and held that this marriage imperiled his career. Few of Knox's friends and companions ever met her, and many thought him unmarried. His "acknowledged residence" (Lonsdale's term) was with his mother and sisters, and his elder sister Mary acted as hostess for him. On the testimony of his son-in-law, however, Knox was devoted to his wife and family. She died in 1841 of puerperal fever following the birth of her sixth child. Most of Knox's children died before him.

Knox was tremendously successful as an anatomy teacher. During the academic year 1828–1829 he had 504 students, and so popular were his lectures that he gave special Saturday lectures to the public. He demonstrated considerable showmanship in his lectures (and some of his peers were shocked when they discovered that he rehearsed them).

It was Knox's very success that caused him trouble since, basing his opinion on his experience in France, he came to believe it essential for students to have their own cadavers to dissect. Cadavers were in short supply in Edinburgh, however, and had been for a century. Gangs of "Resurrectionists" robbed graves to sell bodies, and Edinburgh anatomists imported bodies from Ireland and even from London. While Knox attempted to gain Parliamentary sanction to acquire the unclaimed bodies of paupers, in the meantime he offered premium prices for cadavers regardless of source—once unknowingly paying for a corpse stolen from another dissecting room. He kept students on duty at night to receive the bodies, instructing them to ask no questions, and to pay the agreed fees in cash.

In November 1827 William Burke and William Hare, tenant and owner, respectively, of an Edinburgh rooming house, found the body of a tenant in one of their rooms and sold to it Knox. They soon decided to make this their business, but instead of robbing graves, which they considered too dangerous, they turned to murder. During the next year they brought some sixteen bodies of victims whom they had murdered (through suffocation) to Knox's anatomical school. In October 1828 a suspicious tenant reported their activities, and authorities learned that a recently murdered corpse had been delivered to Knox's school.

Burke was convicted, hanged, and turned over to Monro for dissection, but Hare, who had testified against him, was granted immunity by the prosecution. When Hare left the city, angry Edinburgh citizens, unable to get at him, hanged Knox in effigy; many of them regarded Knox as the patron and instigator of the crimes. Some professional colleagues, already antagonistic toward him, took the opportunity to dissociate themselves. A committee of his friends sponsored a private investigation which cleared him of any duplicity, and Knox announced the results in a letter to the press, his only public statement. His students supported him, giving him a gold vase as evidence of their support. The affair served to mobilize Parliament, which in 1832 finally passed an anatomy act allowing the use of unclaimed bodies of paupers for study.

But Knox remained under a cloud of suspicion, and although he was soon busy dissecting a whale, he apparently ceased to do any basic research on human anatomy. He continued to lecture, but his reputation was now such that in 1831 the Royal College of Surgeons encouraged his resignation as conservator of their museum; his attempts to gain a university appointment were in vain. The university began to apply restrictions against all extramural schools and his in particular. With his school declining he turned it over to Lonsdale and departed for London in 1842.

His notoriety having preceded him, Knox received a cold shoulder from the English surgical profession. In order to support himself he turned to medical journalism and to public lectures. He published extensively in the *Lancet* and lectured widely on such topics as the human races, publishing a book, *Races of Men*, in 1850.

In 1844 Knox went to Glasgow to lecture at the Portland Street School but left by the middle of the school year, apparently because of public hostility. His troubles deepened when in 1847 he was accused of signing a statement of attendance in 1839–1840 for a student who could not possibly have been in his anatomy classes. Although he may simply have made an error, and although others involved in the case received no punishment, the Edinburgh College of Surgeons withdrew any accreditation from lectures given by Knox after 1847 until such time as they should be satisfied that he had answered the charges. His attempt to become an anatomy lecturer in London at a school set up by one of his supporters was thwarted by the actions of the Edinburgh College of Surgeons, as were his attempts to gain various government positions.

Knox finally managed, in 1856, to gain an appointment as pathological anatomist to the London Cancer Hospital, and settled down with his only surviving son, Edward, and his devoted sister, Mary, in the East End section of London. Here he built up a small practice and partially overcame his bitterness, and here he died.

BIBLIOGRAPHY

I. ORIGINAL WORKS. Knox wrote numerous books and hundreds of articles. Isobel Rae, his most recent biographer, calculated that his medical journalistic writings would fill several vols.

Several of his books went through more than one ed.

His most successful were probably *Races of Men* (London, 1850, 2nd enlarged ed., 1862), *Manual of Human Anatomy, Descriptive, General, and Practical* (London, 1853), and *Fish and Fishing in the Lone Glens of Scotland* (London, 1854). He also translated several French anatomical and medical works, including those of P. A. Béclard, Hippolyte Cloquet, L. A. J. Quetelet, H. Milne Edwards, and J. Fau.

Some of Knox's early papers were published together under the title *Memoirs Chiefly Anatomical and Physiological* (Edinburgh, 1837). Other works include *Great Artists and Great Anatomists* (London, 1852), *A Manual of Artistic Anatomy for the Use of Sculptors, Painters and Amateurs* (London, 1852), and *Man: His Structure and Physiology* (London, 1857–1858). His greatest influence was probably through his students, who included John Goodsir and William Fergusson.

II. Secondary Literature. As one of the central figures in a *cause célèbre*, Knox was the focus of a play by James Bridie entitled *The Anatomist*, first produced in 1930 (London, 1931), and of a screenplay (published but never filmed) by Dylan Thomas entitled *The Doctor and the Devils*, Much about the Burke and Hare affair appears in *Burke and Hare* (London–Edinburgh, 1921, enlarged 1948), a volume in the *Notable British Trial Series*, William Roughead, ed. Of the many other accounts of the "Resurrectionists," see James Moores Ball, *The Sack'em-up Men* (Edinburgh–London, 1928).

There are two biographies of Knox, one by his former partner, Lonsdale, *A Sketch of the Life and Writings of Robert Knox, the Anatomist* (London, 1870); and, more recently, Isobel Rae, *Knox: The Anatomist* (Edinburgh, 1964). See also James A. Ross and Hugh W. Y. Taylor, "Robert Knox's Catalogue," in *Journal of the History of Medicine and Allied Sciences*, **10** (1955), 269–276.

For a brief personal sketch by one of his students, see Lloyd G. Stevenson, "E. D. Worthington on Student Life in Edinburgh, with A Character Sketch of Robert Knox," in *Journal of the History of Medicine and Allied Sciences*, **19** (1964), 71–73. See also Douglas Guthrie, *A History of Medicine* (London, 1945).

Vern L. Bullough

KNUDSEN, MARTIN HANS CHRISTIAN (*b.* Hansmark, Denmark, 15 February 1871; *d.* Copenhagen, Denmark, 27 May 1949), *physics, hydrography.*

Knudsen's parents owned a small estate, and he led the healthy and simple existence of a country boy. His outstanding abilities soon became evident; and upon entering the University of Copenhagen in 1890, he began to study physics, mathematics, astronomy, and chemistry. In 1896 he was granted the M.S., with physics as his main subject. In 1895 Knudsen had answered one of the prize questions posed by the University of Copenhagen on electrical sparks and had received the university's gold medal. In 1901 the position of docent was established so that he could teach physics to medical students. When C. Christiansen retired in 1912, Knudsen succeeded him as professor of physics, a position he held until 1941, when he retired. He also taught at the Technical University.

Knudsen's scientific work was centered mainly on the properties of gases at low pressure; and using simple methods he obtained important results for further development as well as for the technology of the vacuum. In one of his earliest projects he examined the escape of gases through a small hole and obtained a confirmation of the correctness of the predictions of the kinetic molecular theory; in particular he was responsible for the first indirect experimental confirmation of Maxwell's law of the distribution of velocity, the experimental determination of the flow of gases through the small hole being in accordance with the formula calculated on the basis of Maxwell's law. At the same time Knudsen conducted a study of the flow of gases through narrow tubes and thus arrived at the laws of molecular diffusion. He used an elegant application of diffusion to describe the vapor pressure of mercury at low temperature. Continuing this research, he next examined the behavior of a gas at low pressure in a container in which there is a temperature gradient. This led to a quantitative theory of the "radiometer forces" at low pressure and the discovery of the absolute manometer, now commonly called the Knudsen manometer. Knudsen was also responsible for an extensive series of basic investigations into the behavior of gases at very low pressures and found that it is not the mean free path of the molecules following collisions, but the dimension of the container, which is decisive.

Knudsen also occupies an important place in hydrography. He developed methods to define the various properties of seawater and was very active as an administrator. It was mainly through his initiative that the Central Committee for Oceanic Research of the International Council for Exploration of the Sea was based in Copenhagen. From 1902 to 1947 he was the Danish delegate to the council, the last fourteen years serving as vice-president, and he edited the *Bulletin hydrografique* from 1908 to 1948.

Besides his work as a researcher and administrator Knudsen carried a heavy load as teacher of physics at the University of Copenhagen and the Polytechnical Institute. Well liked by all his colleagues, he received many honors, both Danish and foreign, for his research and administrative achievements.

BIBLIOGRAPHY

I. Original Works. A complete bibliography is given in *Fysisk Tidsskrift*, **47** (1949), 159–164. The main results of Knudsen's experimental research in the field of kinetic molecular theory of gases are summarized in his small book *The Kinetic Theory of Gases* (London, 1934). He was editor of *Hydrological Tables* (*Hydrographische Tabellen*) (Copenhagen–London, 1901). A short autobiography is in *Innbjudning til filosofie doktorpromotion vid Lunds universitets 250 årsfest* (1918), p. xliii.

II. SECONDARY LITERATURE. A biography of Knudsen by H. M. Hansen is in *Dansk Biografisk Leksikon*, XII (Copenhagen, 1937), 615–618. Obituaries by Niels Bohr and E. R. H. Rasmussen are in *Fysisk Tidsskrift*, **47** (1949), 145–159. See also Mogens Pihl, *Betydningsfulde danske bidrag til den klassiske fysik* (Copenhagen, 1972).

MOGENS PIHL

KNUTH, PAUL ERICH OTTO WILHELM (*b.* Greifswald, Germany, 20 November 1854; *d.* Kiel, Germany, 30 October 1900), *botany*.

Knuth was the son of a municipal official (*Privat Sekretär*), and the former Sophie Bremer.[1] The family was Lutheran. Until August 1873 he attended the Realschule and then the University of Greifswald, where he was awarded the doctorate on 30 December 1876. On 1 October of the same year, he accepted the position of provisional teacher at the Realschule in Iserlohn; he became a full member of the staff one year later, after obtaining his teaching certificate, with high honors, on 28 July 1877. After five years at Iserlohn, Knuth moved to Kiel as teacher at the Oberrealschule, beginning his duties on 1 October 1881. He remained associated with this school until his death in 1900. In 1891 he was promoted to senior teacher (*Oberlehrer*); on 15 December 1895 he was given the title of professor; and on 8 December 1898 he became a councillor of the fourth degree.

From 1891 Knuth was in ill health and often had to interrupt his teaching for long periods. In August 1898 he asked for permission to study at the botanical garden in Buitenzorg, Java; he stayed there from November 1898 until March 1899. On 20 March 1899 he wrote to the authorities in Kiel, seeking an extension of his leave of absence; when their refusal reached Buitenzorg in May, Knuth had already left, probably because of ill health. He returned to Kiel via Japan, California, and New York, arriving in July 1899. He died in Kiel about a year later.

Although Knuth apparently was educated as a chemist, his scientific work was in botany. This interest was probably awakened by the flora in the vicinity of Kiel and the North Sea islands. Soon he began writing books and papers on the flora of Schleswig-Holstein and the North Sea islands. He also wrote on the history of botany in Schleswig-Holstein. All this would have earned Knuth the reputation of a good regional botanist; it is his work on the fertilization of flowers that brought him fame. This field had been opened up by C. Sprengel, whose *Das entdeckte Geheimnis der Natur* (1783) Knuth prepared for a new edition in Ostwald's Klassiker der Exakten Wissenschaften. The topic became of special interest after the publication of Darwin's *Origin of Species* (1859) and Darwin's studies on forms and fertilization of flowers.[2]

While doing work in systematics Knuth started collecting data on the identity of the insects that visit various flowers, and later he set up a network of observers to enlarge the area covered. This type of research had been started in Germany by the brothers Fritz and Hermann Müller; the latter had written a book on the subject in 1873.[3] The numerous observations collected by Knuth made Müller's book obsolete, so Knuth decided to publish his *Handbuch der Blütenbiologie*, which was planned on such an elaborate scale that it still remains the definitive handbook on the subject. Death prevented Knuth from completing more than two volumes of this work; it was continued by Otto Appel and Ernst Loew, who wrote the three subsequent volumes.

The journey to Buitenzorg had been undertaken to collect material on the fertilization of exotic plants that was to be used in later volumes. At Buitenzorg, Knuth studied the pollination history of more than 200 plants and worked on a great variety of related problems.[4] His untimely death prevented him from working out and publishing the results of these studies.

NOTES

1. I am indebted to Miss H. Sievert of the city archives of Kiel for information on Knuth.
2. *On the Various Contrivances by Which British and Foreign Orchids Are Fertilized by Insects, and on the Good Effects of Intercrossing* (London, 1862); *The Effects of Cross and Self Fertilisation in the Vegetable Kingdom* (London, 1876); *The Different Forms of Flowers on Plants of the Same Species* (London, 1877).
3. H. Müller, *Die Befruchtung der Blumen durch Insekten* (Leipzig, 1873); English trans. by d'Arcy W. Thompson, *The Fertilisation of Flowers* (London, 1883).
4. P. Honig and F. Verdoorn, *Science and Scientists in the Netherlands Indies* (New York, 1945), see p. 66.

BIBLIOGRAPHY

I. ORIGINAL WORKS. A list of Knuth's writings is found in Otto Appel's obituary and in the Royal Society *Catalogue of Scientific Papers*, XVI, 349–350. Both should be consulted. His writings include *Ueber eine neue Tribrombenzolsulfosaüre und einige ihrer Derivate* (Greifswald, 1876), his dissertation, also in *Justus Liebig's Annalen der*

Chemie, **186** (1877), 290–306; *Lehrbuch der Chemie für Maschinisten und Torpeder* (Kiel, 1884); *Flora der Provinz Schleswig-Holstein, des Fürstenthums Lübeck, sowie des Gebietes der freien Städte Hamburg und Lübeck* (Leipzig, 1887); *Einige Bemerkungen, meine Flora von Schleswig-Holstein betreffend* (Leipzig, 1888); *Schulflora der Provinz Schleswig-Holstein* (Leipzig, 1888); *Botanische Wanderungen auf der Insel Sylt-Tondern und Westerland* (n.p., 1890); *Geschichte der Botanik in Schleswig-Holstein*, 2 vols. (Kiel, 1890–1892); *Ueber blütenbiologische Beobachtungen* (Kiel–Leipzig, 1893); *Blumen und Insekten auf den nordfriesischen Inseln* (Kiel–Leipzig, 1894); *Christian Konrad Sprengel, Das entdeckte Geheimnis der Natur . . .*, edited by Knuth for Ostwald's Klassiker der Exakten Wissenschaften, nos. 48–51 (Leipzig, 1894); *Grundriss der Blütenbiologie* (Kiel–Leipzig, 1894); *Flora der nordfriesischen Inseln* (Kiel–Leipzig, 1895); *Flora der Insel Helgoland* (Kiel, 1896); and *Handbuch der Blütenbiologie unter Zugrundelegung von Hermann Müller's Werk "Die Befruchtung der Blumen durch Insekten*, 2 vols. in 3 pts. (Leipzig, 1898–1899); English trans. by J. R. Ainsworth Davis, *Handbook of Flower Pollination, Based on Hermann Müller's Work "The Fertilisation of Flowers by Insects*" (Oxford, 1906–1909).

Knuth wrote many papers, most of which are found in *Botanisches Centralblatt; Botanisch Jaarboek, uitgegeven door het kruidkundig genootschap Dodonaea te Gent; Die Heimath, Monatschrift des Vereines zur Pflege der Natur- und Landeskunde in Schleswig-Holstein, Hamburg und Lübeck; Humboldt; Kieler Zeitung;* and *Deutsche botanische Monatsschrift*.

II. Secondary Literature. On Knuth or his work, see the anonymous "Biographische Mittheilungen (Paul Knuth)," in *Leopoldina*, **35** (1899), 180; Otto Appel, "Paul Knuth," in *Berichte der Deutschen botanischen Gesellschaft*, **18** (1900), 162–170; I. H. B., "Paul Knuth," in *Nature*, **61** (1899–1900), 205; L. István, "Nekrológ Paul Knuth," in *Természettudományi közlöny*, **32** (1900), 693–694; F. Ludwig, "Nekrolog. Das Leben und Wirken Prof. Dr. Paul Knuths," in *Illustrierte Zeitschrift für Entomologie*, **4** (1899), 365–367; and E. Wunschmann, "Paul Knuth," in *Allgemeine deutsche Biographie*, LI (1906), 274–275.

Peter W. van der Pas

KÖBEL (or **KOBEL** or **KOBILIN** or **KIBLIN** or **CABALLINUS), JACOB** (*b.* Heidelberg, Germany, 1460/1465; *d.* Oppenheim, Germany, 31 January 1533), *mathematics, law, publishing, municipal administration.*

Köbel was the son of Klaus Köbel, a goldsmith. He began his studies at the University of Heidelberg on 20 February 1480[1] and earned his bachelor's degree from the Faculty of Arts in July 1481. Concerning the following years it is known only that "Jacobus Kiblin" was active in the book trade in 1487. Simultaneously he studied law, receiving the bachelor's degree on 16 May 1491. He appears to have gone then to Cracow, where he studied mathematics, a subject then flourishing at the Jagiellonian University. He is also reported to have been a fellow student of Copernicus, who had enrolled there in 1491 under the rectorship of Mathias de Cobilyno (perhaps a relative of Köbel).[2] In 1494 Köbel was in Oppenheim, where on 8 May 1494 he married Elisabeth von Gelthus, the daughter of an alderman. They are known to have had one son and two daughters. Köbel worked as town clerk and official surveyor, as well as manager of the municipal wine tavern. A scholar of manifold interests, he wrote extensively and was also a printer and publisher. As a member of the Sodalitas Litteraria Rhenana he was friendly with many humanists.[3] In the religious conflicts he stood with the Catholic reformers. He died after suffering greatly in his last years from gout and was buried in Oppenheim, in the Church of St. Katherine. A portrait of him can be found in his 1532 essay on the sundial.[4]

Between 1499 and 1532 Köbel published ninety-six works, at first those of others and then his own. Among the authors whose writings he published were Albertus Magnus, Virdung, and especially his friend Johann Stöffler.[5] Köbel's publishing activity decreased markedly after 1525, no doubt as a result of his poor health.

Köbel wrote three arithmetic books of varied content, all of which were well received. They appeared during the period in which the algorithm, with new numerals and methods—propagated especially through the writings of Sacrobosco—was gradually supplanting the traditional computation with the abacus and with Roman numerals (which Köbel called "German" numerals). Köbel's first book was *Rechenbüchlein vf den Linien mit Rechenpfenigen* because such a book was the easiest sort for beginners, who had to know only the corresponding Roman letters. The book was widely read and went through many editions, most of them under altered titles, and was continually revised and enlarged. In it Köbel treated the manipulation of the abacus, computational operations (with duplication, mediation, and progression but without the roots, since they are "unsuitable for domestic use"), the rule of three, fractions (also with Roman numerals), and a few problems of recreational mathematics.

The next to appear was *Eyn new geordent Vysirbuch* (1515), which dealt with the calculation of the capacity of barrels. Köbel presented the new methods of calculation with Arabic numerals in *Mit der Kryden oder Schreibfedern durch die Zeiferzal zu rechnen* (1520).

Köbel's writings were most widely disseminated in a collection that he himself had prepared, *Zwey*

Rechenbüchlin: uff der Linien und Zipher. Mit eym angehenckten Visirbuch (1537). It contained almost verbatim the line arithmetic book of 1525 (now without Roman numerals), as well as the *Vysirbuch* and *Mit der Kryden oder Schreibfedern.* In the editions after 1544 a chapter was added on the commercially important measures and coins of many foreign lands.

A *Geometrei* appeared posthumously (1535) and was in print until 1616. This work consisted of three papers by Köbel: "Von vrsprung der Teilung . . ." (1522), which contained formulas for the surveyor;[6] an essay on the Jacob's staff, written in February and May 1531,[7] and "Feldmessung durch Spiegel," which was first published in the *Geometrei.*

As an astronomer Köbel was concerned with the astrolabe and with the publication of numerous popular calendars. His *Astrolabii declaratio* (1532) went through several editions. He also published a treatise entitled *Eyn künstliche sonn-Uhr inn eynes yeden menschen Lincken handt gleicht wie in eynem Compass zu erlernen . . .* (1532), which later appeared under the titles *Bauren Compas* and *LeyenCompas.*[8]

Besides informative handbooks and many poems, Köbel wrote works on law—for example, on inheritance cases and rules of the court. He also wrote on imperial history and continued the chronicle of Steinhöwel from the time of Frederick III to his own day.

The high esteem accorded to Köbel's writings is reflected in the numerous editions that appeared until the beginning of the seventeenth century. Today his importance lies principally in his dissemination of mathematical knowledge, especially of the new Hindu numerals and methods, among broad segments of the population. He accomplished this through the use of German in his work, a practice he was the first to adopt since the publication of the arithmetic books of Bamberg (1482, 1483) and Widmann (1489)—which, moreover, were basically collections of problems. Adam Ries, who replaced Köbel as teacher of the nation, used his books, having become acquainted with them while in Erfurt (1518–1522) through the humanist Georg Sturtz.[9]

NOTES

1. The printed register of Heidelberg University, I, 367, gives the name Johannes Köbel; the original has the correct form.
2. Starowolsky, *Scriptorum Polonicorum . . .*, p. 88; in it Köbel is cited as Cobilinius in *Catalogus illustrium Poloniae scriptorum* (p. 133). Benzing, in *Jakob Köbel zu Oppenheim, 1494–1533*, p. 8, remarks that we have no proof that Köbel studied at Cracow.
3. Vigilius, who was a guest of Köbel (Caballinus), reports in a letter to Conradus Celtis (Heidelberg, 19 Apr. 1496) that Köbel was estranged from Johann von Dalberg because Köbel had, without permission, given Celtis a book he had borrowed from Dalberg. See Rupprich, *Der Briefwechsel des Konrad Celtis*, pp. 178–227 ff.; and Morneweg, *Johann von Dalberg*, pp. 196 ff.
4. The portrait can be found in Benzing, *op. cit.*, p. 6.
5. For example, Stöffler's *Calendarium Romanorum magnum* (1518); *Der newe grosz Römisch Calender* (1522); and *Elucidatio fabricae ususque astrolabii* (1513).
6. With regard to the errors criticized by Kaestner (*Geschichte der Mathematik*, I [1796], 655), it should be said that Köbel intended to provide the surveyor only with formulas, such as the ancient Egyptian approximation formula for quadrangles.
7. Köbel explained this work by stating that he himself was now obliged to use a staff.
8. See Benzing, *op. cit.*, pp. 79 ff.
9. Köbel never realized his intention to write on algebra. See Unger, *Die Methodik der praktischen Arithmetik*, p. 45.

BIBLIOGRAPHY

I. Original Works. A list of all of Köbel's writings, with full titles, subsequent eds., and present locations, is in Joseph Benzing, *Jakob Köbel zu Oppenheim, 1494–1533. Bibliographie seiner Drucke und Schriften* (Wiesbaden, 1962). Most of his works were published "under such varied titles and in such different combinations with his other books, that it is difficult to say whether a given edition is a new work or merely a revision" (Smith, *Rara Arithmetica*, p. 102). Among them are *Rechenbüchlein vf den Linien mit Rechenpfenigen* (Oppenheim–Augsburg, 1514; 2nd ed., 1517; 3rd ed., 1518); *Eyn new geordent Vysirbuch* (1515; 1527), later issued with line arithmetic book (also with square and cube roots) (1531; 1532); *Über die Pestilenz* (1519); *Was Tugend und Geschicklichkeit ein Oberster regirer an ynn haben soll* (1519), an exhortation to Charles V; *Mit der Kryden oder Schreibfedern durch die Zeiferzal zu rechnen* (1520); the line arithmetic book, *Eyn new Rechenbüchlin Jacob Köbels Stadtschreiber zu Oppenheym auff den Linien vnd Spacien gantz leichtlich zu lernen mit Vyelen zusetzen* (Oppenheim, 1525); *Astrolabii declaratio* (1532), also published with Stöffler's *Elucidatio* and later trans. into German as *Von gerechter Zubereitung des Astrolabiums . . .* (1536); *Eyn künstliche sonnUhr inn eynes yeden menschen Lincken handt gleich wie in eynem Compass zu erlernen . . .* (1532); *Geometrei* (1535; 1550; 1570; 1584; 1598; 1616), with a treatise on the quadrant by Johann Dryander, who also published a work on the *Nachtuhr* begun by Köbel (1535); and *Zwey Rechenbüchlin: Uff der Linien und Zipher. Mit eym angehenckten Visirbuch* (1537; 1543; 1564; 1584).

II. Secondary Literature. On Köbel or his work, see *Allgemeine deutsche Biographie*, XVI, 345–349, and XIX, 1827; M. Cantor, *Vorlesungen über Geschichte der Mathematik*, 2nd ed., II (Leipzig, 1913), 419 f.; S. Günther, *Geschichte des mathematischen Unterrichts im deutschen Mittelalter bis zum Jahre 1525* (Berlin, 1887), p. 386; K. Haas, "Der Rechenmeister Jakob Köbel," in *Festschrift zum 125-jährigen Jubiläum des Helmholtz-Gymnasiums in Heidelberg* (Heidelberg, 1960), pp. 151–155; K. Morneweg, *Johann von Dalberg* (Heidelberg, 1887); F. W. E. Roth, "Jakob Köbel, Verleger zu Heidelberg, Buchdrucker und

Stadtschreiber zu Oppenheim am Rhein 1489–1533," in *Neues Archiv zur Geschichte der Stadt Heidelberg*, **4** (1901), 147–179; H. Rupprich, *Der Briefwechsel des Konrad Celtis* (Munich, 1934); D. E. Smith, *Rara arithmetica* (Boston–London, 1908), pp. 100–114; Szymon Starowolsky, *Scriptorum Polonicorum* 'ΕΚΑΤΟΝΤΆΣ (Frankfurt, 1625); and F. Unger, *Die Methodik der praktischen Arithmetik* (Leipzig, 1888), pp. 44–46.

KURT VOGEL

KOCH, HEINRICH HERMANN ROBERT (*b.* Clausthal, Oberharz, Germany, 11 December 1843; *d.* Baden-Baden, Germany, 27 May 1910), *bacteriology, hygiene, tropical medicine.*

Many of the basic principles and techniques of modern bacteriology were adapted or devised by Koch, who therefore is often regarded as the chief founder of that science. His isolation of the causal agents of anthrax, tuberculosis, and cholera brought him worldwide acclaim as well as leadership of the German school of bacteriology. Directly or indirectly he influenced authorities in many countries to introduce public health legislation based on knowledge of the microbic origin of various infections, and he stimulated more enlightened popular attitudes toward hygienic and immunologic measures for controlling such diseases.

Robert was the third son of Hermann Koch, a third-generation mining official, and his wife, Mathilde Julie Henriette Biewend, daughter of an iron-mine inspector. The parents were blood relatives, Mathilde being her husband's grandniece. They were Lutherans and natives of an old mountain town of some 10,000 inhabitants, situated about fifty miles south of Hannover. The family tree can be traced to Hermann's grandfather, Johann Wilhelm Koch (1730–1808), a mine foreman who married the daughter of the Clausthal town clerk. Hermann Koch began his career as a miner but broadened his horizons as a young man by visiting several European countries. His wife bore him thirteen children. Two died in infancy, leaving Robert with six brothers and two sisters younger than himself and two elder brothers. When the boy was about ten years old, his father became overseer of all local mines and acquired the title of *Bergrat*. Industrious, methodical, and dutiful, he encouraged in Robert a desire for travel and respect for nature's beauties and wonders.

Management of the large household devolved upon Mathilde Koch, an industrious, thrifty, and selfless woman who promoted a harmonious, untrammeled atmosphere in which the children became self-reliant and polished each other's manners. Particularly fond of animals and flowers, she fostered similar sentiments in Robert, her favorite child. Knowledge of plant and animal life was also stimulated by the maternal grandfather, a humorous, sensitive nature lover, and by his highly educated son, Eduard Biewend, who befriended Robert, conducted him frequently on natural history excursions, and imparted his wide interests in scientific subjects and the new art of photography. The boy avidly collected mosses and lichens, insects, and mineralized stones, identifying them with a lens. Later he dissected and mounted larger animals and prepared their skeletons.

Robert Koch taught himself to read and write before entering the local primary school in 1848. A rapid learner, he was transferred to the Clausthal Gymnasium in 1851 and headed his class four years later. Thereafter his school progress slowed, perhaps because of marked adolescent emotional turbulence, as indicated in letters collected by K. Kolle. After repeating the final grade, he graduated in 1862, with good standing in mathematics, physics, history, geography, German, and English. Despite only "satisfactory" rating in Latin, Greek, Hebrew, and French, he declared an intention to study philology; but the school principal suggested his aptitudes were for either medicine or mathematics and natural sciences. Other alternatives were apprenticeship to a shoe merchant and emigration to America. His father's promotion to *Geheimer Bergrat* and improved family finances facilitated the youth's decision to study natural sciences at Göttingen University, some fifteen miles distant, where he enrolled at Easter 1862.

After studying botany, physics, and mathematics for two semesters, Koch transferred to medicine because of a disinclination for professional teaching and the realization that natural science interests were compatible with a medical career. Several distinguished professors appreciated his unobtrusive industry and notable curiosity about vital phenomena. Many years later Koch acknowledged that his sense for scientific investigation had been awakened by the anatomist Jacob Henle, the physiologist Georg Meissner (a master of animal experimentation), and Karl Hasse, a clinician who taught him pathology and therapeutics. Especially important was his close association with Henle, whose classic essay, "Von den Miasmen und Contagien," had appeared over twenty years before. Henle lectured only on anatomy. As Koch later emphasized, no bacteriology was taught in his student days at Göttingen; but the final section of Henle's *Handbuch der rationellen Pathologie* (Brunswick, 1846–1853) reaffirms his previous convictions regarding the living nature of contagious agents. Moreover, in the early 1860's university circles fiercely debated Louis

Pasteur's assertions about the specific fermentative properties of the lower fungi and the myth of spontaneous generation. With so promising and inquisitive a student Henle could scarcely have failed to discuss how the *contagium animatum* of an infective disease might be identified.

In his fifth semester Koch began a prize task set by Henle, concerning the existence and distribution of uterine nerve ganglia. He completed it during his sixth semester, while assisting Wilhelm Krause in the Pathological-Anatomical Institute. Entitled "Ueber das Vorkommen von Ganglienzellen an den Nerven des Uterus" and dedicated to his father, the report—bearing the motto "Nunquam otiosus"—won the first prize. Koch then investigated, at Meissner's Physiological Institute, the mechanism of succinic acid development in the body and its urinary excretion. The project entailed ingestion of discomforting amounts of certain foodstuffs, such as a half-pound of butter daily for several days. Koch's findings, entitled "Ueber das Entstehen der Bernsteinsäure im menschlichen Organismus," appeared in 1865 in the *Zeitschrift für rationelle Medizin*, a journal founded by Henle. This report was accepted as his doctoral dissertation. In the final examinations at Göttingen, in January 1866, he obtained highest distinction; two months later he passed the state examination at Hannover. Upon graduating, Koch visited Berlin to attend the Charité clinics and Rudolf Virchow's course in pathology. Finding the hospital and lectures overcrowded, he returned home after a few months and became engaged to the youngest daughter of the general superintendent of Clausthal, Emmy Adolfine Josefine Fraatz.

The next six years were very unsettled. Koch's aspirations to become a military physician, or to see the world as a ship's doctor, were stifled by lack of opportunities and by his fiancée's refusal to travel abroad. In 1866 an assistantship at Hamburg General Hospital, during a cholera outbreak, familiarized him with this scourge; but since the position was unsuited to his prospective marriage, he resigned after three months. He then became assistant at an institution for retarded children in Langenhagen, a large village near Hannover where private practice was allowed, prospering sufficiently to acquire a riding horse and a large apartment and to marry Emmy Fraatz in July 1867.

Within two years Koch resigned this post because an economy drive threatened to reduce his salary. He made attempts to establish a small-town practice, mostly in the province of Posen (now Poznán, Poland). Following a brief, disappointing stay alone at Braetz, he moved with his wife to Niemegk, near Potsdam, where his only daughter, Gertrud, was born in September 1868; but the family suffered economic hardship and after ten months migrated eastward again, settling in Rakwitz. Here Koch's quiet efficiency was recognized, his practice flourished and he became a popular figure. This idyllic interlude was interrupted by the Franco-Prussian War.

Despite severe myopia, Koch volunteered for service as a field hospital physician. He gained invaluable experience, especially while attached later to a typhoid hospital at Neufchâteau and a hospital for wounded near Orléans. Early in 1871, responding to a petition from Rakwitz citizens, he left the army and resumed practice. Shortly afterward he passed the qualifying examinations for district medical officer (*Kreisphysicus*) and was advised by the influential Baron von Unruhe-Bomst, an appreciative patient, to apply for a vacant position at Wollstein (now Wolsztyn, Poland). He was appointed in August 1872. For eight years the family lived happily in this lakeside town set in forested countryside. Koch became highly respected locally and started on his path to international fame.

Despite increasing professional activities, Koch found time for hobbies as well as scientific pursuits. He excavated ancient Teutonic graves in the neighborhood, developing an interest in anthropology; inquired into occupational diseases, such as lead poisoning; and intensified the studies of algae and infusoria begun at Langenhagen and Rakwitz. Because he strongly favored the parasitic over the miasmatic theory in the recurrent controversies about the etiology of infection, he extended the scale of microscopic investigations to include bacteria. The consulting area of his four-room house was divided by a curtain, and the rear part served as laboratory. Besides a good Hartnack microscope, an incubator, glass jars for mice, and sundry smaller apparatus, it contained microphotographic devices and a converted wardrobe for darkroom. At forty years of age Koch turned his attention to anthrax, then enzootic in the district.

After verifying C.-J. Davaine's contention of ten years before, that anthrax was caused by rodlike microorganisms seen in the blood of infected sheep, Koch invented techniques for culturing them in drops of cattle blood or aqueous humor on the warm stage of his microscope, under varied conditions of moisture, temperature, and air access. He traced accurately their mode of growth and life cycle, including the phenomena of spore formation and germination, which Davaine neither observed nor suspected. Koch's cultures developed no transitional forms or other evidence of acquired pleomorphism. Sporogenesis required adequate moisture and oxygen and temperatures above 15°C.; and whereas the bacilli were relatively short-lived, the spores withstood prolonged drying and remained infective for years. Koch showed that anthrax developed in mice only when the inoculum contained viable rods or spores of *Bacillus anthracis*, grown either *in vitro* or in infected animals. Characteristically, he sought to correlate these laboratory findings with the peculiar recurrences and seasonal incidence of anthrax, and with sound preventive measures.

Before publishing these observations, Koch sought an interview with Ferdinand Cohn, the famous botanist at Breslau (now Wrocław, Poland), who in his pioneering *Untersuchung über Bacterien* (1872–1876) had stressed the fixity of bacterial species and anticipated the spore-forming properties of *Bacillus anthracis*. In the spring of 1876 Koch demonstrated his methods and preparations to Cohn and to the pathologist Julius Cohnheim and his assistants. After personally confirming the results, Cohn included Koch's classic report on the etiology of anthrax in the next issue of his journal, *Beiträge zur Biologie der Pflanzen*. In 1877 the *Beiträge* contained another paper by Koch, "Verfahren zur Untersuchung, zum Conservieren und Photographieren der Bakterien." This described techniques for dry-fixing thin films of bacterial culture on glass slides, for staining them with aniline dyes (according to information received from Carl Weigert in Breslau), and for recording their structure by microphotography. Koch used apparatus built to his own specifications, following expert advice from the physiologist Gustav Fritsch; and through extraordinary patience and ingenuity he transmuted the early inspiration from his uncle into pictures of astonishing fidelity. These clearly revealed the flagella of motile bacteria, and the morphological distinctions between harmless spirochetes in marsh water or tooth slime and those just reported by Otto Obermeier in relapsing fever.

Koch revisited Breslau in 1877 and in Cohnheim's laboratory showed his latest findings to a group that included Paul Ehrlich and John Burdon Sanderson. He was welcomed there again in 1878; but in Berlin shortly afterward Virchow, the outstanding pathologist and quasi-miasmatist, received him ungraciously. Nevertheless, by now Koch's self-confidence was such that in 1878 he published an aggressively critical review of Carl von Naegeli's *Die niederen Pilze in ihren Beziehungen zu den Infektionskrankheiten und der Gesundheitspflege* (1877), attacking its pleomorphist doctrines. In that same year appeared Koch's first monograph, *Untersuchungen über die Aetiologie der Wundinfectionskrankheiten*, an English translation of which followed in 1880.

This work reported his findings on the bacteriology of infected wounds—a problem still unsettled more than a decade after Joseph Lister introduced antisepsis. To avoid confusion from imprecise clinical terms such as "septicemia" and "pyemia" applied to human patients, Koch induced artificial infections in mice and rabbits by injecting putrid fluids. He declared that a "thoroughly satisfactory proof" of the parasitic nature of traumatic infective diseases would be forthcoming only "when the parasitic micro-organisms are successfully found in all cases of the disease in question; further, when their presence is demonstrable in such numbers and distribution that all the symptoms of the disease thereby find their explanation; and finally, when for every individual traumatic infective disease, a micro-organism with well-marked morphological characters is established." Thus he first explicitly stated the criteria implicit in Henle's essay on contagion, which after modification became known as "Koch's postulates." He equipped his microscope with Ernst Abbe's new condenser and oil-immersion system (manufactured by Carl Zeiss) so that he could detect organisms appreciably smaller than *B. anthracis*. Koch identified six transmissible infections, two in mice and four in rabbits, that were pathologically and bacteriologically distinctive. He deduced that human traumatic infections would prove similarly due to specific parasites and concluded that his experiments illustrated the diversity and immutability of pathogenic bacteria. Among favorably impressed surgeons were Lister and Theodor Billroth, who thereafter strongly supported him.

Early in 1879 the Breslau medical faculty unsuccessfully petitioned the minister of public instruction (*Kultus Minister*) to create a professorship in hygiene for Koch in a proposed new institute. That summer, on Cohn's urging, he was appointed city physician at Breslau; but the small salary and negligible practice made the position untenable and within three months the family was welcomed back in Wollstein. Koch's reputation and ambitions, however, had outgrown his environment. Obligations to patients conflicted with keeping abreast of specialized literature and with conducting laboratory researches that demanded more apparatus and experimental animals than the household could accommodate. The anthrax studies required innumerable mice, in addition to guinea pigs, rabbits, frogs, dogs, a partridge, and a sparrow; and he had recently transmitted relapsing fever spirochetes to two monkeys. His wife, who had previously collected algae specimens and helped with photographic procedures, now shared with their daughter such duties as feeding animals and cleaning microscope slides: some disenchantment was understandable. In 1880, on Cohnheim's recommendation, Koch was appointed government adviser (*Regierungsrat*) with the Imperial Department of Health (*Kaiserlichen Reichsgesundheitsamt*) in Berlin. His home henceforth was in the capital city.

The Health Department, established in 1876, occupied a former apartment house near the Charité hospital. At first Koch shared one small laboratory with his assistants, Friedrich Loeffler and Georg Gaffky, both army staff doctors, whose competence, industry, and loyalty eased his transition from solitary worker to team leader. Their main assignments, under the Health Department's director, Heinrich Struck, were to develop reliable methods for isolating and cultivating pathogenic bacteria and to gather bacteriological data and establish scientific principles bearing on hygiene and public health. Koch's disciples worked tirelessly beside him while, in Loeffler's words, "almost daily new miracles of bacteriology displayed themselves before our astonished eyes." As the program expanded, Ferdinand Hueppe and Bernhard Fischer were seconded from the army medical staff and chemists Georg Knorre and Bernhard Proskauer recruited. All were destined for distinguished careers, but Koch was the undisputed leader.

As publishing medium for the Health Department's scientific findings, Struck instituted in 1881 the *Mittheilungen aus dem Kaiserlichen Gesundheitsamt*. In the first article, "Zur Untersuchung von pathogenen Organismen," Koch extended his earlier account of bacteriological methods. He stressed the importance of avoiding contamination through use of strictly sterile techniques and advocated nutrient gelatin as a solid medium that allowed individual colonies to be selected, thus ensuring pure cultures. He also specified that newly isolated pathogens should be investigated for transferability to animals, portals of entry and localizations in the host, natural habitats, and susceptibility to harmful agents. This work, illustrated with numerous microphotographs, long remained the basic instructional manual for bacteriological laboratories. Koch then studied disinfectant substances and processes, comparing their inhibitory or destructive action on certain bacterial species, mainly anthrax bacilli and spores. His declaration in "Ueber Desinfection" that carbolic acid was inferior to mercuric chloride hastened dethronement of Lister's "carbolic spray," and the reports, written with associates, that live steam surpassed hot air in sterilizing power revolutionized hospital operating room practices.

In 1881 Koch's preoccupation with methodology culminated and began to yield a rich harvest. In August, while attending the International Medical Congress at London, he demonstrated his pure-culture techniques in Lister's laboratory and there met Louis Pasteur, who magnanimously termed the methods "un grand progrès." Upon returning to Berlin, Koch launched experiments on tuberculosis, convinced of its chronic infectious nature. In six months, working alone and without hint to colleagues, he fully verified the still-disputed claims of J.-A. Villemin, J. Cohnheim, C. J. Salomonsen, H. E. von Tappeiner, and P. C. Baumgarten that the disease was transmissible. Further, a bacillus of exacting cultural and staining properties was demonstrated and isolated from various tuberculous specimens of human and animal origin; and tuberculosis was induced by inoculating several species of animals with pure cultures of this bacterium.

Identification of the tubercle bacillus was rendered exceptionally difficult by its small size and often scanty distribution, restricted stainability (due to a waxy coat), fastidious nutritional requirements, and very slow growth *in vitro*. Eventually Koch found it would retain alkaline methylene blue in tissues counterstained with Bismarck brown. Inconspicuous colonial growth appeared on test-tube "slopes" of heat-coagulated cattle or sheep serum during the second week of incubation at 37°C. His resolute, single-handed ingenuity in surmounting difficulties was matched by the thoroughness and completeness of his lecture, entitled simply "Ueber Tuberculose," delivered before the Physiological Society in Berlin on 24 March 1882—a red-letter day in bacteriological history. Although no orator, Koch presented his evidence with such logic and conviction that the audience was too spellbound to applaud or engage in official discussion. Paul Ehrlich, who later recalled that evening as his "greatest scientific event," developed overnight an improved method of staining tubercle bacilli, which Koch adopted. These two very different characters became firm friends. Virchow, who on pathological grounds upheld belief in the nontuberculous nature of phthisis, was absent. Inter alia, Koch's demonstrations of tubercle bacilli in the caseous material from phthisical lungs, as well as in specimens from miliary and other forms of tuberculosis, refuted this doctrine of duality.

Within three weeks Koch's paper appeared in the *Berliner klinische Wochenschrift* as "Die Aetiologie der Tuberculose." Although unsurpassed in lucidity of style and directness of statement, in thoroughness and precision of experimentation, and in stringency of requirements for proof, the report contains some errors. For example, the granular staining displayed by many cultures was misinterpreted as sporulation. Differences between cultures of human and cattle origin were overlooked, leading Koch to assert that bovine tuberculosis is identical with human tuberculosis and thus transmissible to man—a contention he later denied. Because of the susceptibility of experimental animals to tuberculosis he modified his criteria for establishing the causal relationship of these bacteria to the disease. He stipulated that the bacilli had to be isolated from the body and "cultivated in pure culture until freed of all adherent products of disease originating in the animal organism" and that tuberculosis must be reproduced in animals injected with the isolated bacilli. This last clause in his "postulates" presented awkward obstacles when the disease in question was not transmissible to animals. The general luster of the report nevertheless remains unblemished.

Koch's chief findings were confirmed wherever his techniques were carefully followed—in the United States, for example, by Theobald Smith and E. L. Trudeau. The demonstration of tubercle bacilli in the sputum was soon accepted as of crucial diagnostic significance, and his co-workers began investigating such problems as the disinfection of tuberculous sputum. Koch himself continued to amass evidence for converting those who clung to the belief that tuberculosis was dyscrasic rather than infective. By 1883 he had induced the disease in over 500 animals of ten species, of which more than 200 succumbed to pure cultures of the bacillus administered by various routes, and had obtained new data on the cultural properties and modes of spread of the causal bacillus. Publication of an expanded version of "Die Aetiologie der Tuberculose" in the *Mittheilungen* was delayed until 1884. Meanwhile, in the *Deutsche medizinische Wochenschrift*, Koch deplored the "incorrect and clumsy technique" and the "altogether empty literature" of those who disputed the importance of the tubercle bacillus. In 1883 he received the title *Geheimer Regierungsrat*.

Koch's propensity for aggressive criticism became conspicuous in 1881, when he attacked P. G. Grawitz, a pupil of Virchow, for espousing Naegeli's theory of the transformation of fungi and took issue, in "Zur Aetiologie des Milzbrandes," with Hans Buchner's and Pasteur's researches on anthrax. Buchner allegedly produced the disease in animals with cultures derived from the hay bacillus; but Koch exposed several sources of error, including use of unsterilized blood as nutrient medium. The dispute with Pasteur was more complex, profound, and sustained, for personal jealousy and national pride aggravated their disagreements. Koch's indictment appeared in the *Mittheilungen*, supplemented by separate contributions from Gaffky and Loeffler. They disparaged much of Pasteur's four years' work on anthrax as plagiaristic or inaccurate and impugned the purity of his cultures. Koch disputed Pasteur's contentions that farm animals acquired the disease through mouth abrasions caused by thistle prickles and that anthrax spores were brought to the surface by earthworms.

These attacks from Berlin, and Pasteur's sensational demonstration at Pouilly-le-Fort that sheep could be protected against virulent anthrax cultures by vaccination with attenuated strains, were reported almost simultaneously. Pasteur was unaware of the former when he met Koch two months later in London and, amid great acclaim, addressed the Medical Congress on vaccination against chicken cholera and anthrax. In September 1882, however, before the International Congress of Hygiene and Demography at Geneva, he concluded an invited address on bacterial attenuation by repudiating the "disagreeable diatribes" of Koch and his pupils and offered to enlighten anyone who shared the opinions of his "stubborn contradictors." Responding briefly, Koch expressed disappointment at having heard nothing new, termed the occasion inappropriate for dealing with Pasteur's attacks, and reserved his reply for medical journals. Three months later he published an acerbic critique, whose most cogent complaint was that Pasteur offered no explicit details of his method of attenuating anthrax bacilli. Koch now conceded the feasibility of attenuation but still doubted Pasteur's immunization claims. The latter's eloquently scornful retort took the form of a lengthy open letter to Koch dated Christmas Day 1882. The controversy flared intermittently for another five years.

These polemics and Koch's tuberculosis researches were interrupted by the 1883 Hygiene Exhibition in Berlin, which he helped to organize. In the Health Department's pavilion he enjoyed demonstrating bacterial preparations to many distinguished visitors, including the crown prince. A more challenging diversion was an outbreak of cholera in the Nile delta that summer. The French government, warned by Pasteur that the epidemic could invade Europe and that the cause of cholera was probably microbial, dispatched a four-man scientific mission which reached Alexandria in mid-August. Nine days later Koch arrived, heading a German government commission that included Gaffky and Fischer. Within three weeks he had observed large numbers of tiny rods in sectioned walls of the small intestines from ten autopsied cholera cases and had isolated a gelatin-liquefying organism from the intestinal contents of about twenty cholera cadavers and patients. This organism, although unassociated with other diseases, failed to induce choleraic effects when fed to or injected into monkeys, dogs, chickens, and mice. The French team, meanwhile, suffered disappointment and tragedy. Their bouillon cultures yielded a confusing assortment of intestinal bacteria from cholera victims, whose blood, however, contained suspicious bodies. These bodies proved to be merely blood platelets, as Koch later pointed out. One month after the French mission's arrival in Egypt the epidemic had waned; but Louis Thuillier, their youngest member, contracted fatal cholera. The German commission paid appropriate homage: Koch visited the dying man and served as pallbearer.

While awaiting governmental permission to proceed to India (where cholera persisted) to continue his investigations, Koch drew attention to the regional prevalence of amoebic dysentery and various helminthic manifestations. He also reported that "Egyptian ophthalmia" included two different disease processes, one probably gonococcal and the other due to a minute organism later known as the Koch-Weeks bacillus. His interests in sanitation and in world travel were exercised by visits to quarantine stations, a pilgrim camp, and ancient monuments.

The Egyptian findings were confirmed in Bengal. Two months after arriving at Calcutta with Gaffky and Fischer, Koch had observed the same nonsporulating, comma-shaped bacillus in seventy cholera victims. Despite inability to provoke the disease therewith in experimental animals, he asserted that it was the specific cause of cholera. His final communication from India (4 March 1884) designated village ponds, used for drinking water and all domestic purposes, as sources of localized outbreaks. He had isolated cholera bacilli from one such pond. The commission returned triumphant in May. The Kaiser awarded Koch the Order of the Crown, the Reichstag voted him 100,000 marks, and the Berlin Medical Society tendered a festive banquet in his honor at which Ernst von Bergmann lavishly praised him.

Koch's six letter-reports to the minister during the commission's nine months abroad were supplemented in 1887 by Gaffky's complete account of their activities, constituting volume 3 of the Health Department's *Arbeiten*. At Koch's instigation two conferences of experts considered cholera problems at the Health Department in July 1884 and May 1885, under Virchow's chairmanship. On the first occasion Koch detailed the properties of his comma bacillus, including its susceptibility to various disinfectants and to desiccation. Max von Pettenkofer, Germany's senior hygienist, who believed other factors besides a microbial agent governed cholera epidemics, was invited only to the five-day second conference. Koch repudiated Pettenkofer's hazy arguments and fallacious claims with facts and straightforward logic; and he adumbrated control measures that were adopted successfully for the German empire, although not by the International Sanitary Conference at Rome in July 1885, to which he was an official delegate.

Koch's advisory duties became very extensive. As Health Department representative on a Reichstag commission on smallpox vaccination, he vigorously opposed antivaccinationists and initiated regulations for improving calf lymph. He reported to state and municipal authorities on problems that ranged from water supply, sewage disposal, and canal purification to testing disinfection apparatus, denaturing alcohol, and reuse of cotton wool. In addition the Health Department sponsored a training course in cholera diagnosis, and visiting doctors clamored to learn the methods that yielded so many discoveries. (Between 1882 and 1884 the bacilli of swine erysipelas, glanders, and diphtheria had been isolated by Loeffler, and the typhoid bacillus by Gaffky.) At this juncture the Minister of Public Instruction, Gustav Gossler and his adviser Friedrich Althoff decided that additional institutes of hygiene should be established in Prussia. In 1885 Koch accepted the new chair of hygiene at the University of Berlin and directorship of a prospective institute, while retaining honorary membership in the Health Department. He also received the title *Geheimer Medizinalrat*.

After consulting Carl Flügge, director of the Hygiene Institute at Göttingen (with whom he founded and for twenty-five years coedited the *Zeitschrift für Hygiene*), Koch conscientiously prepared lecture courses and organized field excursions and discussion groups for students, practitioners, and public health officials. Notable assistants and trainees of this period included Carl Fraenkel, Wilhelm Dönitz, Richard Pfeiffer, and Emil von Behring from Germany; Shibasaburo Kitasato from Japan; and William Welch and Mitchell Prudden from the United States. Late in 1886, despite oppressive teaching and administrative duties in improvised quarters, Koch informed Flügge that he had resumed experimental work "with the greatest zeal" on long-term problems. Working alone and secretively, he sought a specific remedy for tuberculosis.

On 4 August 1890, at the tenth International Medical Congress in Berlin, Koch ended a pedestrian address on bacteriological research by announcing that after testing many chemicals, he had "at last hit upon a substance which has the power of preventing the growth of tubercle bacilli," both *in vitro* and *in vivo*. Injections of the substance into guinea pigs rendered normal animals resistant to tuberculosis and arrested the generalized disease. Hopes aroused by these incomplete experiments were enhanced in mid-

November, when Koch reported excellent results in clinical trials of the agent, prepared and administered by two physicians—E. Pfuhl, his son-in-law, and A. Libbertz, of the Höchst pharmaceutical firm. Emphasizing its destructive effects upon human tuberculous tissues, Koch urged caution in treating advanced pulmonary tuberculosis but asserted that early phthisis "can be cured with certainty by this remedy." Several distinguished clinicians, including Bergmann, also issued optimistic reports.

Koch's name was now on all lips: doctors and patients made pilgrimages to Berlin, filling hospitals, clinics, and hotels, clamoring for his "lymph." He received the honorary freedom of Clausthal, Wollstein, and Berlin; awards from foreign rulers and societies; Pasteur's congratulations; and the Grand Cross of the Red Eagle from the Kaiser. Minister of Public Instruction Gossler informed the Prussian legislature that Koch disclosed to him in October his discovery of a specific against tuberculosis, which he wished to investigate outside the state service. Although he had relinquished direction of the Hygiene Institute, the government intended to build hospital facilities near the Charité and to provide an adjacent bacteriological research institute which Koch would direct. Meanwhile, in the Moabit municipal hospital, 150 beds were reserved for specific treatment of tuberculous paupers under Ehrlich's supervision. Gossler discounted rumors of exorbitant charges for injections and undertook to safeguard the remedy's manufacture: to discourage imitations, he had persuaded Koch to postpone revealing its nature.

The government's plan to monopolize production of Koch's fluid was viewed unfavorably at home and abroad. Moreover, despite Ehrlich's good results from small dosages in early phthisis, and Lister's endorsement after visiting Berlin, mounting evidence of the drug's toxicity—particularly Virchow's postmortem demonstrations of intense local inflammatory reactions in treated cases—intensified demands for revelation of its nature. In January 1891, Koch published the long-awaited formula in "Fortsetzung der Mittheilungen über ein Heilmittel gegen Tuberkulose"; but the report proved anticlimactic. His definition, "a glycerine extract of pure cultures of tubercle bacilli," lacked essential details and was somewhat misleading. Besides, although his paper correctly described the contrasting responses of normal and previously exposed guinea pigs to injections of tubercle bacilli (subsequently termed "Koch's phenomenon"), which he claimed instigated his discovery, he attempted to explain the agent's curative action in terms of its necrotizing rather than its allergenic properties.

Widespread doubts now arose about the remedy. Its merits were debated at length in the Berlin Medical Society, and some centers banned its use. Koch's projected hospital and research institute seemed in jeopardy. Disturbed by these developments and also by increasing marital infelicity, he journeyed to Egypt to recover equanimity. (He had become infatuated with Hedwig Freiburg, a comely minor actress some thirty years his junior. A consequent divorce and remarriage in 1893 provoked more censure than sympathy, but the childless union lasted and was happy.) Koch stayed away from Berlin until his buildings were assured. Gossler resigned as Minister in March, having misguidedly pressed Koch to announce his discovery and having recently boasted that it was unique for a secret remedy to be "accepted by the entire world on the strength of one man's name." Despite the domestic scandal and professional skepticism temporarily clouding Koch's reputation, the government felt honor-bound to support him. Althoff secured funds for converting a three-story edifice (the "Triangle") into the Institute for Infectious Diseases, and he persuaded the city of Berlin to complete a multipavilion barrack hospital (later known as "Koch's sheds") accommodating over 100 patients. In the legislature Virchow protested without avail the hasty approval of Koch's 20,000-mark salary as overall director and of a budget equaling the total research funds for all scientific departments at Berlin University.

In October 1891, shortly after occupying the new quarters, Koch reported certain chemical characteristics of "tuberculin." Previous allusions to complicated methods of preparation had been misleading, for now crude tuberculin was revealed as a filtrate of tubercle bacilli grown for six to eight weeks in a glycerol-containing medium (described by E. Nocard and E. Roux in 1887) evaporated to one-tenth volume. Koch chided fellow bacteriologists for neither following nor developing his method but did not mention the unsuccessful attempts of several French workers, and of Trudeau and others in the United States, to immunize animals with derivatives of tubercle bacilli. Hueppe bitterly criticized his former chief for graver errors than those he had condemned unjustly in Pasteur. When others disputed Koch's findings, Bergmann sought his autopsy records on the experimentally protected animals. These were unavailable: autopsies had not been performed.

Still undaunted, Koch and such followers as L. Brieger, F. Neufeld, and J. Petruschky endeavored to improve tuberculin's safety and efficacy and to determine its optimal dosage. In "Ueber neue Tuberkulinpräparate" (1897), describing three new forms of tuberculin, Koch asserted that "nothing better of this kind can be produced." Eventually the specific diagnostic value of the hypersensitive response of tuberculous patients and cattle to tuberculin injections helped to restore the prestige lost through excessive confidence in the agent's curative powers. Koch realized that these local reactions were of different significance from the antitoxic immunity of tetanus and diphtheria, and from the specific bacteriolytic phenomenon involving *V. cholerae*, observed in his institute by Behring and Kitasato (1890) and by Pfeiffer (1894), respectively; but the "allergy" concept, which illuminated the connection between immune mechanisms and tuberculin hypersensitivity, was first proposed in 1906 by C. von Pirquet.

Others now working with Koch included P. Ehrlich, H. Kossel, B. Proskauer, and A. von Wasserman. Ehrlich, allowed to choose his field of work, conducted brilliant studies on active and passive immunity and was an indispensable help to Behring in producing potent antitoxins. Visitors flocked to the Institute, some attracted by the sudden fame of Behring, who, on leaving the Institute in 1894, began exploring aggressively the antitoxin treatment of tuberculosis. Unable to produce a serum effective in cattle, he developed a bewildering succession of vaccines for "Jennerization" against bovine tuberculosis. Koch, whose prime lifelong quest was tuberculosis control, considered Behring an interloper and resented this challenge to his superstar status. Their worsening relationship culminated in 1898–1899 when the Höchst Farbwerke and Behring obtained patents for two different extracts of tubercle bacilli, despite Koch's formal opposition.

Cholera reached Hamburg in August 1892. Within ten weeks 18,000 cases occurred, including about 8,200 fatalities. Koch responded to the city's pleas and, in collaboration with Gaffky and W. P. Dunbar, stressed early bacteriological detection and isolation of ambulant cases, disinfection of patients' excreta, and scrupulous sanitation of water supplies. Pettenkofer disparaged these measures; and to dramatize his conviction that comma bacilli alone could not induce cholera, swallowed some freshly isolated culture. Rudolf Emmerich and a few other disciples followed his example. Two developed a choleraic syndrome; but since no experimenter died and he himself suffered only mild diarrhea, Pettenkofer claimed his theory was verified. In three subsequent papers (1893) Koch reported on the bacteriological diagnosis of cholera, the control of sand filtration of water, and the waterborne origins of the Hamburg and related Prussian epidemics. His views won widespread support, while Pettenkofer's adherents dwindled. Berlin's water supplies had been tested regularly at the Hygiene Institute since 1885. During and after these outbreaks, Koch's new Institute examined countless fecal, sewage, and water samples. The undertaking verified the existence of cholera carriers and of nonpathogenic vibrios resembling *V. cholerae* which could be differentiated through Pfeiffer's bacteriolytic reaction.

Koch's tireless leadership in the cholera emergency brought him increased public responsibilities. He established stations in the Institute for Pasteurian treatment of rabies and for diphtheria antitoxin assay. The communicable diseases control law promulgated in 1900 incorporated his recommendations of the early 1890's. Following his leprosy survey in Memel in 1896, the disease became notifiable in Prussia and a leprosarium was established. Then the Cape Colony government engaged him to investigate rinderpest, ravaging cattle north of the Orange River. His thirst for foreign travel revived, and his microbiological interests were redirected. Arriving at Kimberley in December 1896, accompanied by his wife and staff surgeon Paul Kohlstock, Koch assembled a menagerie of experimental animals and within four months found that the infective agent was nonbacterial, transmissible by infected blood, and unattenuated by passage through animals. He achieved active immunization by inoculating susceptible cattle with a mixture of blood serum from recovered animals and virulent rinderpest-infected blood. Inoculation with bile from cattle freshly dead of the disease was even more protective. These procedures, outlined in succinct reports, were implemented by Kohlstock and veterinary officer G. Turner and further developed by W. Kolle (from the Institute), who replaced Koch after his departure for India to head a German government plague commission made up of his disciples Gaffky, Pfeiffer, G. Sticker, and A. Dieudonné.

By May 1897, when Koch reached Bombay, bubonic plague was epidemic in upper India; and other European governments had sent scientific missions. Under Gaffky's direction the German commission had confirmed the etiologic role of the plague bacillus (discovered in 1894 by A. Yersin and by Kitasato) and had launched epidemiological inquiries. Koch organized laboratory tests of Yersin's serum and W. Haffkine's vaccine against plague. He designated rats as plague source and urged reoriented control measures, but (overlooking the flea as vector) he presumed the reservoir to be maintained by cannibalism. Visiting the North-West Frontier Province, he and Gaffky recognized a local disease as endemic plague.

Koch left India for Dar-es-Salaam when invited to

German East Africa to curb rinderpest. Instead he found two protozoan diseases—surra, a trypanosomiasis affecting horses, and Texas cattle fever, identified as a piroplasmosis by Theobald Smith. He began to study malaria and blackwater fever—soon attributing the latter to quinine intoxication—and detected an endemic plague focus at Kisiba on Lake Victoria. Returning to Berlin in May 1898, after eighteen months' absence, Koch delivered to the German Colonial Society an address entitled "Aertzliche Beobachtungen in den Tropen." He described four types of malaria, favored the mosquito-borne theory, compared immunity in malaria and Texas fever, and asserted that he had "pioneered new routes and set new goals in malaria research." The various accomplishments were recounted in *Reise-Bericht über Rinderpest, Bubonenpest in Indien und Afrika . . .* (1898).

Koch recommended to the government that further malaria studies would foster colonial development and improve military hygiene. He proposed another visit to Italy, followed by an extensive tropical expedition. That autumn, working with Pfeiffer and Kossel in the Lombardy plains, the Campagna di Roma, and other Italian malarial districts, he confirmed Ronald Ross's discovery of the avian malaria parasite's life cycle. The main expedition started in April 1899, halted in Tuscany (where Koch and P. Frosch correlated mosquito activities and the incidence of estivo-autumnal fever), and proceeded to Java. Although quinine had abated the disease there, Koch noted the high susceptibility of young children, particularly Europeans, and the apparent immunity of native adults in endemic areas. Orangutans and gibbons resisted experimental human malaria. He found no mosquito-free localities harboring malaria and averred "no mosquitoes, no endemic malaria."

In German New Guinea, where his wife became ill and was sent home, the disease was prevalent. Because mosquito eradication seemed hopeless, Koch evolved a control policy based on destruction of the parasite within its exclusive host. This entailed microscopic blood examinations of the population concerned, with systematic quininization of all parasite carriers until they were symptomless and relapse-free, and had negative blood films. The regimen, subsequently adopted throughout the German empire, was intrinsically handicapped by the imperfect specificity and potential toxicity of quinine; but it was successful when drug supplies and trained physicians were freely available and the population disciplined. The owner and 300 inhabitants of the Istrian islet of Brioni erected a monument to Koch for liberating them from malaria. In 1901 the Kaiser Wilhelm Academy acknowledged the value of his discoveries to military hygiene by electing him to its senate, with the rank of major general.

Koch returned to Berlin in October 1900, having spent only nine months there in four years. Pfeiffer's loyalty as acting director could not compensate for these prolonged absences. The Institute's transfer to planned larger quarters in north Berlin, adjoining the Rudolf Virchow Hospital, needed supervision. Indigenous public health problems demanded attention. In July 1901, Koch presented to the first British Tuberculosis Congress in London an address entitled "The Fight Against Tuberculosis." He specified sputum as main source of infection in man but cited cattle experiments (conducted with W. Schütz) that indicated human bacilli could not infect cattle. The converse possibility was so negligible that he deemed countermeasures inadvisable. Lord Lister, who chaired the meeting, disputed this "doctrine of the immunity of man to bovine tubercle." Koch's assertion caused consternation in Britain. Two royal commissions had declared ingestion of tuberculous matter in food to be dangerous, and milk from cows with tuberculous udders had recently been banned for human consumption. Yet in October 1902, at the International Congress on Tuberculosis in Berlin, Koch reaffirmed this position—in stark contrast with the etiological opinion he had expressed twenty years before.

Typhoid fever was prevalent in Prussia. From endemic foci serious waterborne epidemics arose, such as that involving 2,000 cases at Beuthen, Silesia, in 1887. A similar outbreak occurred in 1901 at Gelsenkirchen, in the Ruhr. Koch was requested to report on this and subsequent outbreaks in the Trier vicinity. Besides emphasizing sanitary water supplies and sewage disposal, he stressed the importance of contact infections. Key control measures were early detection and isolation of cases and "bacillary carriers," disinfection of their excreta, and bacteriological investigation of their surroundings. This program required several new regional laboratories, bacteriologically trained health officers, and special efforts at the Institute by W. von Drigalski and H. Conradi to improve culture media and techniques. Koch's broad experience, thoroughness, and zealous leadership halted the epidemics and the typhoid morbidity fell.

Early in 1903 Koch and his wife departed for Bulawayo, Southern Rhodesia, accompanied by F. Neufeld and F. W. Kleine. He had been invited to investigate "Rhodesian redwater," another cattle epizootic, a tick-borne piroplasmosis resembling Texas fever. Since he had noted the same blood parasite in East African coastal cattle, Koch termed the disease African coast fever. After painstaking field and laboratory research he recommended control measures that included immunization by repeated injections of parasite-infected blood. Similar studies on "horse-sickness" (*Pferdesterbe*) failed to disclose a causal bacterium or protozoon, for—like rinderpest—this is a viral disease; but he could induce protection by alternately injecting serum from recovered horses and infected blood.

As his sixtieth birthday approached, Koch proposed to retire from state service. The authorities offered generous working privileges at the Institute, with appointment as consulting hygienist. On his return to Berlin in June 1904, disciples and admirers presented a *Festschrift*, with contributions from over forty pupils, and a marble bust. In voicing appreciation, tinged with bitterness at increasing competition and "passionate opposition," Koch undertook to serve science as long as strength permitted. With Gaffky, his favorite disciple and successor, he maintained an unblemished relationship to the end. Retirement was sweetened with an annual honorarium of 10,000 marks, besides the statutory pension. The Kaiser awarded him the Order of Wilhelm.

Although the battle against tuberculosis remained his prime concern, Koch spent most of the next three years in equatorial Africa. Early in 1905 he arrived at Dar-es-Salaam. On this expedition he investigated the life cycles of the piroplasmas of Coast and Texas fevers, and of sleeping sickness trypanosomes in *Glossina* (tsetse flies). He also showed African relapsing fever to be a spirochetosis transmitted by *Ornithodoros moubata*, a tick infesting caravan routes and native huts. Monkeys exposed to infected tick progeny hatched in isolation acquired the disease, indicating transovarian passage of spirochetes in the arachnid. Similar findings made independently in the Belgian Congo by J. E. Dutton and J. L. Todd had prior publication. Koch returned to Berlin the following October. Three months later, in Stockholm, he received the 1905 Nobel Prize for physiology or medicine for his work on tuberculosis. His lecture on current control measures against this disease scarcely mentioned tuberculin but reasserted that bovine bacilli were harmless to man.

In April 1906, Koch led the German Sleeping Sickness Commission, comprising M. Beck, F. W. Kleine, O. Panse, and R. Kudicke, to East Africa. After visiting regional stations, Koch sent his wife home from Entebbe, to spare her the hardships and dangers of the Sesse Islands, Uganda—an area of rampant trypanosomiasis in northwestern Lake Victoria, where the expedition established straw hut headquarters and patients' camp. They studied indefatigably all aspects of the disease, from symptomatology to laboratory diagnosis and prevention. Therapeutic trials of atoxyl indicated large doses were effectively trypanocidal; but of 1,633 patients treated, 23 became permanently blind from optic atrophy. Koch proposed some drastic alternatives, such as eliminating tsetse fly harborages through clearance of undergrowth and tree-cutting on the littoral, and exterminating crocodiles, on whose blood *Glossina palpalis* fed. In November 1907 he and Beck returned to Berlin: the others continued investigations.

Many honors were now bestowed on Koch. To previous awards, such as the 1901 Harben Medal and foreign membership in the Paris Academy of Sciences (he succeeded Virchow in 1902), were added in 1906 the Prussian order Pour le Mérite and—following his latest tropical exploits—the title *Wirklicher Geheimer Rat* with the predicate *Excellenz*. Early in 1908, Berlin physicians attended a festive evening to witness his receiving the first Robert Koch Medal—starting a series intended to commemorate the greatest living physicians. Proposals for a Robert Koch Foundation to combat tuberculosis won official approval, and over a million marks quickly accumulated. The Kaiser contributed 100,000 and Andrew Carnegie 500,000 marks.

Koch and his wife now embarked upon a journey planned as a restful world tour. In April the New York German Medical Society feted him at a sumptuous banquet, where he was eulogized by W. H. Welch, his best-known American disciple, and by Carnegie. After visiting his brothers and other relatives in Chicago and St. Louis, Koch traveled via Honolulu to Japan, to be welcomed with solemn honors by Kitasato, presented to the mikado, and escorted like a demigod around the country. This idyll was disrupted by instructions to lead Germany's delegates to the Sixth International Congress on Tuberculosis at Washington, D.C., at the end of September. Koch gave an address entitled "The Relationship Between Human and Cattle Tuberculosis." After belatedly acknowledging that Theobald Smith (who was present) had first drawn attention to differences between human and bovine tubercle bacilli, he again defended his entrenched position, somewhat equivocally. He disputed some key findings on the potential dangers of the bovine bacillus to man, documented in the very thorough report of the royal commission (1904) appointed after the 1901 London Congress. An informal conference to resolve these issues was held *in camera* a few days later under the chairmanship of Hermann Biggs, whose antituberculosis program for New York City had been praised by Koch. However, the latter's intransigence frustrated this attempt by leading international experts to reach common ground with him.

The remainder of Koch's life was devoted to tuberculosis control. He worked daily at the Institute, supervising production and clinical trials of new tuberculins. In 1910 earlier intimations of cardiac trouble became insistent. On April 9, three nights after lecturing on the epidemiology of tuberculosis before the Berlin Academy of Sciences, he suffered a severe anginal attack. He failed to recuperate and died peacefully in his chair at a sanatorium. His ashes were deposited in a mausoleum at the Institute, which the Kaiser ordered named after Robert Koch; a shrine was dedicated to him in Japan; and Metchnikoff brought from the Pasteur Institute a plaque of gilded laurel and palm.

The former *Kreisphysicus* of Wollstein was ranked by Ehrlich among "the few princes of medical science"; and Theobald Smith—not given to loose praise—called him "the master of us all in bacteriology." Such tributes were evoked by a remarkable combination of qualities—extraordinarily methodical technique, dogged perseverance in verifying theories and fearless logic in applying findings, and tireless industry. "Nicht locker lassen" (don't let up) was a favorite exhortation to himself and associates. These characteristics emerged at a crucial time, as Koch modestly admitted in New York in 1908: "I have worked as hard as I could and have fulfilled my duty and obligations. If the success really was greater than is usually the case, the reason for it is . . . that in my wanderings through the medical field I came upon regions where gold was still lying by the wayside." Considering how much territory he prospected, the true gold was seldom confused with base metals.

Koch was thoroughly German, a senior civil servant and government consultant, and accustomed to assistants with military background. His consequent hierarchical concepts and attitudes partly account for such faults attributed to him as pugnacity, arrogance, failure to acknowledge borrowed ideas or to give credit where due, and reluctance to admit mistakes. He was also accused of self-interest, particularly in connection with his sojourns abroad. These doubtless satisfied his yearnings for travel, furnished escape from social disapproval in the capital, and offered fieldwork opportunities in beguiling environments where his wide knowledge of botany and zoology could be fully exercised. Nevertheless, he wore himself out in the imperial service, helping to elucidate intricate medical and veterinary mysteries in a period of national rivalries so intense that his verdict was necessary to stimulate government action on public health issues.

Koch was liable to be suspicious and aloof with strangers; but to friends and colleagues he was kind and considerate, and with his daughter he remained on affectionate terms. In congenial company he reminisced entertainingly, revealing the wide scope of his secondary interests. These ranged from the arts to astronomy and mathematics; from anthropology, ethnology, and geography to the dilemmas of missionaries on furlough. He was a great admirer of Goethe and addicted to chess. Although unattracted by didactic lecturing, on special occasions he enjoyed the role of *praeceptor mundi*. An unfortunate tendency to use the pen as sword and cudgel sometimes marred the lucidity and persuasiveness of his earlier writings.

Koch expressed his career's basic motivation in his first paper on tuberculosis: "I have undertaken my investigations in the interests of public health and I hope the greatest benefits will accrue therefrom." Less impersonal compassion was displayed in his determination to maintain contacts with patients and in his continuing quests for specific remedies—diphtheria antitoxin, tuberculin, quinine, and atoxyl. In disclosing the causes of disease and expounding the means of prevention, Robert Koch at his best was unexcelled. The Faustian weaknesses and perplexities he carried do not diminish the lasting benefits that his aspirations bestowed upon mankind.

BIBLIOGRAPHY

I. Original Works. The only ed. of Koch's collected works is *Gesammelte Werke von Robert Koch*, 2 vols. in 3 pts., J. Schwalbe, ed., in association with G. Gaffky and E. Pfuhl (Leipzig, 1912), containing repros. of 99 of his published monographs and scientific papers, as well as 92 previously unpublished reports to national, state, and municipal authorities. Vol. I includes his early classic reports on anthrax, wound infections, disinfection, and methods of isolating bacteria, and 18 papers on tuberculosis that appeared between 1882 and 1910. Vol. II, pt. 1, contains 9 publications on cholera and 30 on tropical diseases, chiefly malaria and sleeping sickness. Among the reports to governments in vol. II, pt. 2, 53 deal with zoonoses, acute infectious diseases (including cholera and typhoid fever), and tropical diseases; 8 with tuberculosis; 5 with vaccination regulations and procedures; 14 with sewage disposal and water supplies; and 12 with miscellaneous topics, ranging from the denaturing of alcohol to smoke nuisances. The entire text is German, except for two short articles in English. A few additional addresses and reports first appeared in English, and several others were translated and republished in English journals and texts. Some of his best-known articles were also translated into French, Italian, and other languages. A detailed bibliography of 80 items in *Medical Classics*, **2** (1937–1938), 720–731, has many minor inaccuracies and lacks several publications reproduced in *Gesammelte Werke* but includes some unimportant works omitted from the latter. Appended to the obituary by W. W. Ford (see below) is a list of 61 of Koch's chief publications.

Monographs by Koch include *Untersuchungen über die Aetiologie der Wundinfectionskrankheiten* (Leipzig, 1878), trans. by W. W. Cheyne as *Investigations Into the Etiology of Traumatic Infective Diseases* (London, 1880); *Die Cholera auf ihren neuesten Standpunkte* (Berlin, 1886); *Bericht über die im hygienischen Laboratorium der Universität Berlin ausgeführten Untersuchungen des Berliner Leitungswassers in der Zeit vom 1. Juni 1885 bis 1. April 1886* (Berlin, 1887); *Die Bekämpfung der Infektionskrankheiten, insbesondere der Kriegsseuchen* (Berlin, 1888); *Reise-Berichte über Rinderpest, Bubonenpest in Indien und Afrika, Tsetse- oder Surrakrankheit, Texasfieber, tropische Malaria, Schwarzwasserfieber* (Berlin, 1898); *Interim Report on Rhodesian Redwater or African Coast Fever* (Salisbury, 1903); and *Bericht über die Tätigkeit der deutschen Expedition zur Erforschung der Schlafkrankheit im Jahre 1906/07 nach Ostafrika entsandten Kommission* (Berlin, 1909), written with M. Beck and F. Kleine, which also appeared in *Arbeiten aus dem Kaiserlichen Gesundheitsamt*, **31** (1911), 1–320.

Among his more important and characteristic early works are "Ueber das Vorkommen von Ganglienzellen an den Nerven des Uterus," prize dissertation, Medical Faculty (Göttingen, 1865); "Ueber das Entstehen der Bernsteinsäure im menschlichen Organismus," in *Zeitschrift für rationelle Medizin*, 3rd ser., **24** (1865), 264–274, published while he was still a medical student; "Die Aetiologie der Milzbrand-Krankheit, begründet auf die Entwicklungsgeschichte des Bacillus Anthracis," in *Beiträge zur Biologie der Pflanzen*, **2** (1876), 277–311, repro. in Karl Sudhoff's *Klassiker der Medizin*, no. 9 (1910), trans. as "The Etiology of Anthrax, Based on the Ontogeny of the Anthrax Bacillus," in *Medical Classics*, **2** (1937–1938), 787–820, and abstr. as "The Etiology of Anthrax, Based on the Life History of *Bacillus Anthracis*," in *Milestones in Microbiology*, T. Brock, ed. (Englewood Cliffs, N.J., 1961), pp. 89–95; and "Entgegnung auf den von Dr. Grawitz in der Berliner medizinischen Gesellschaft gehaltenen Vortrag über die Anpassungstheorie der Schimmelpilze," in *Berliner klinische Wochenschrift*, **18** (1881), 769–774.

Koch's fundamental contributions to bacteriological techniques are in "Verfahren zur Untersuchung, zum Conservieren und Photographieren der Bakterien," in *Beiträge zur Biologie der Pflanzen*, **2** (1877), 399–434, trans. and abstr. as "Methods for Studying, Preserving, and Photographing Bacteria," in *Microbiology: Historical Contributions from 1776 to 1908*, R. N. Doetsch, Jr., ed. (New Brunswick, N.J., 1960), pp. 67–73; "Zur Untersuchung von pathogenen Organismen," in *Mittheilungen aus dem Kaiserlichen Gesundheitsamt*, **1** (1881), 1–48, trans. by V. Horsley as "On the Investigation of Pathogenic Organisms," in *Microparasites in Disease. Selected Essays*, W. W. Cheyne, ed. (London, 1886), pp. 3–64, and abstr. by T. Brock as "Methods for the Study of Pathogenic Organisms," in *Milestones in Microbiology*, pp. 101–108; "Ueber die neuen Untersuchungsmethoden zum Nachweis der Mikroorganismen in Boden, Luft and Wasser," in *Aerztliches Vereinblatt für Deutschland*, no. 237 (1883), 244–250, trans. by R. N. Doetsch, Jr., in *Microbiology: Historical Contributions from 1776 to 1908*, pp. 122–131.

With associates Koch published a series of papers on disinfection, including "Ueber Desinfection," in *Mittheilungen aus dem Kaiserlichen Gesundheitsamt*, **1** (1881), 234–282; "Untersuchungen über die Desinfection mit heisser Luft," *ibid.*, 301–321, written with G. Wolffhügel; and "Versuche über die Vermerthbarkeit heisser Wasserdämpfe zu Desinfectionszwecken," *ibid.*, 322–340, written with G. Gaffky and F. Loeffler. These were trans. and abstr. by B. A. Whitelegge in *Microparasites in Disease. Selected Essays*, as "On Disinfection," pp. 493–518, "Disinfection by Hot Air," pp. 519–525, and "Disinfection by Steam," pp. 526–533.

Koch's dispute with Pasteur over anthrax is covered in "Zur Aetiologie des Milzbrandes," in *Mittheilungen aus dem Kaiserlichen Gesundheitsamt*, **1** (1881), 49–79; *Ueber die Milzbrandimpfung. Eine Entgegnung auf den von Pasteur in Genf gehaltenen Vortrag* (Leipzig, 1882); "Experimentelle Studien über die künstliche Abschwächung der Milzbrandbazillen und Milzbrandinfection durch Fütterung," in *Mittheilungen aus dem Kaiserlichen Gesundheitsamt*, **2** (1884), 147–181, written with G. Gaffky and F. Loeffler; and "Ueber die Pasteurschen Milzbrandimpfungen," in *Deutsche medizinische Wochenschrift*, **13** (1887), 722.

His pioneer report, "Die Aetiologie der Tuberculose," in *Berliner klinische Wochenschrift*, **19** (1882), 221–230, trans. by B. Pinner and M. Pinner, appears as "The Aetiology of Tuberculosis," with foreword by A. K. Krause, in *American Review of Tuberculosis*, **25** (1932), 285–323, and as a pamphlet published by the National Tuberculosis Association (New York, 1932). Another version of this report appears in *Medical Classics*, **2** (1937–1938), 853–880, trans. by W. de Rouville. There followed "Kritische Besprechung der gegen die Bedeutung der Tuberkelbazillen gerichteten Publicationen," in *Deutsche medizinische Wochenschrift*, **9** (1883), 137–141; and "Die Aetiologie der Tuberkulose," in *Mittheilungen aus dem Kaiserlichen Gesundheitsamt*, **2** (1884), 1–88, trans. by S. Boyd as "The Etiology of Tuberculosis," in *Microparasites in Disease*, pp. 67–201, and abstr. in H. A. Lechevalier and M. Solotorovsky, *Three Centuries of Microbiology* (New York, 1965), pp. 69–79.

His chief earlier works on cholera are "Cholera-Berichte aus Egypten und Indien," in *Deutsche Vierteljahrsschrift für öffentliche Gesundheitspflege*, **16** (1884), 493–515; "Conferenz zur Erörterung der Cholerafrage am 26. Juli 1884," in *Berliner klinische Wochenschrift*, **21** (1884), 478–483, 493–503 (trans. by G. L. Laycock in *Microparasites in Disease. Selected Essays*, pp. 327–369), followed by discussion, pp. 509–521; "Ueber die Cholerabakterien," in *Deutsche medizinische Wochenschrift*, **10** (1884), 725–728; "Conferenz zur Erörterung der Cholerafrage. (Zweites Jahr)," *ibid.*, **11** (1885), 1–60, of which Koch's opening address, pp. 1–8, at the Second Conference on Cholera, appears in part trans. by G. L. Laycock in *Microparasites in Disease*, pp. 370–384, and also trans. in full as "Further

Researches on Cholera," in *British Medical Journal* (1886), **1**, 6–8, 62–66.

Koch's first paper on tuberculin was "Ueber bakteriologische Forschung," in *Verhandlungen des X. internationalen medizinische Kongresses*, I (Berlin, 1890), 35–47, trans. as "An Address on Bacteriological Research," in *British Medical Journal* (1890), **2**, 380–383. Three other papers—"Weitere Mittheilungen über ein Heilmittel gegen Tuberkulose," in *Deutsche medizinische Wochenschrift*, **16** (1890) 1029–1032; "Fortsetzung der Mittheilungen über ein Heilmittel gegen Tuberkulose," *ibid.*, **17** (1891), 101; and "Weitere Mittheilungen über das Tuberkulin," *ibid.*, 1189–1192—are repro. in Sudhoff's *Klassiker der Medizin*, no. 19 (Leipzig, 1912), and are also abstr. and trans. as "A Further Communication on a Remedy for Tuberculosis," in *British Medical Journal* (1890), **2**, 1193–1195; (1891), **1**, 125–127; and (1891), **2**, 966–968. For comments on some of the issues involved see "Correspondence From Berlin," in *British Medical Journal* (1890), **2**, 1197–1198, 1327–1328; (1891), **1**, 1096–1097; and editorial, *ibid.* (1891), **2**, 954–955. His last report in this field was "Ueber neue Tuberkulinpräparate," in *Deutsche medizinische Wochenschrift*, **23** (1897), 209–213.

The later reports on cholera, following the Hamburg epidemic of 1892, include "Ueber den augenblicklichen Stand der bakteriologischen Choleradiagnose," in *Zeitschrift für Hygiene und Infektionskrankheiten*, **14** (1893), 319–338; "Wasserfiltration und Cholera," *ibid.*, 393–426; and "Die Cholera in Deutschland während des Winters 1892 bis 1893," *ibid.*, **15** (1893), 89–165. These appear trans., respectively, in *Practitioner*, **51** (1893), 466–476; *ibid.*, 146–160, 218–240; and *Lancet* (1893), **2**, 828–830, 891. The three reports were repub. in book form as *Professor Koch on Cholera*, trans. by G. Duncan (Edinburgh, 1894).

Koch's views on the innocuousness of bovine tuberculosis to man began with "The Fight Against Tuberculosis in the Light of Experience Gained in the Successful Combat of Other Infectious Diseases," in *British Medical Journal* (1901), **2**, 189–193, and with slightly modified title in *Journal of State Medicine*, **9** (1901), 441–457. The German version appeared later as "Die Bekämpfung der Tuberkulose unter Berücksichtigung der Erfahrungen, welche bei der erfolgreichen Bekämpfung anderer Infektionskrankheiten gemacht sind," in *Deutsche medizinische Wochenschrift*, **27** (1901), 549–554. His opinions were reiterated in "Uebertragbarkeit der Rindertuberkulose auf den Menschen," *ibid.*, **28** (1902), 857–862, trans. as "The Transference of Bovine Tuberculosis to Man," in *British Medical Journal* (1902), **2**, 1885–1889; and again in "The Relations of Human and Bovine Tuberculosis," in *Journal of the American Medical Association*, **51** (1908), 1256–1258, with discussion pp. 1258–1260, followed by "Conference in Camera on Human and Bovine Tuberculosis," pp. 1262–1268; and in "Das Verhältnis zwischen Menschen- und Rindertuberkulose," in *Berliner klinische Wochenschrift*, **45** (1908), 2001–2003, with discussion pp. 2003–2006.

Koch's continuing interest in tuberculosis is further exemplified by "Ueber die Agglutination der Tuberkelbazillen und über die Verwerthung dieser Agglutination," in *Deutsche medizinische Wochenschrift*, **27** (1901), 829–834; "Ueber die Immunisierung von Rindern gegen Tuberkulose," in *Zeitschrift für Hygiene und Infektionskrankheiten*, **51** (1905), 300–327, written with W. Schütz, F. Neufeld, and H. Miessner; "Ueber den derzeitigen Stand der Tuberkulosebekämpfung," Nobel Prize address, 12 Dec. 1905, in *Deutsche medizinische Wochenschrift*, **32** (1906), 89–92, trans. as "How the Fight Against Tuberculosis Now Stands," in *Lancet* (1906), **1**, 1449–1451; "Ueber therapeutische Verwendung von Tuberkulin," in *Medizinische Woche*, **7** (1906), 493–496; "Zur medikamentösen Behandlung der Lungentuberkulose," in *Therapeutische Rundschau*, **3** (1909), 101–103; and "Epidemiologie der Tuberkulose," in *Zeitschrift für Hygiene und Infektionskrankheiten*, **67** (1910), 1–18, trans. in *Smithsonian Institution Annual Report for 1910*, no. 2049 (1911), 659–674.

The investigations of rinderpest in South Africa were reported in English to the Secretary for Agriculture, in *Cape of Good Hope Agricultural Journal*, **10** (1897), 94–96, 96–101, 216–219, 220–221, 413–418, 418–419; as "Researches Into the Cause of Cattle Plague," in *British Medical Journal* (1897), **1**, 1245–1246, trans. of two letters to the editor; and as "Berichte des Prof. Dr. Koch über seine in Kimberley gemachten Versuche bezüglich Bekämpfung der Rinderpest," in *Centralblatt für Bakteriologie*, Abt. 1, **21** (1897), 526–537.

Reports on malaria included "Berichte des Geheimen Medicinalrathes Professor Dr. R. Koch über die Ergebnisse seiner Forschungen in Deutsch-Ostafrika," in *Arbeiten aus dem Kaiserlichen Gesundheitsamt*, **14** (1898)—"Die Malaria," 292–304, "Das Schwarzwasserfieber," 304–308; "Ergebnisse der wissenschaftlichen Expedition des Geheimen Medicinalrathes Professor Dr. Koch nach Italien zur Erforschung der Malaria," in *Deutsche medizinische Wochenschrift*, **25** (1899), 69–70. The findings of the 1899–1900 expedition to the Dutch East Indies and New Guinea appeared in six articles, followed by a summary, in *Deutsche medizinische Wochenschrift*, the first being "Erste Bericht über die Thätigkeit der Malariaexpedition. Aufenthalt in Grosseto von 25. April bis 1. August 1899," *ibid.*, **25** (1899), 601–604. The final report appears as "Zusammenfassende Darstellung der Ergebnisse der Malariaexpedition," *ibid.*, 781–783, 801–805. Other papers on malaria problems are "Ueber die Entwicklung der Malariaparasiten," in *Zeitschrift für Hygiene und Infektionskrankheiten*, **32** (1899), 1–24; "Die Bekämpfung der Malaria," *ibid.*, **43** (1903), 1–4; and "Address on Malaria to the Congress at Eastbourne," in *Journal of State Medicine*, **9** (1901), 613–625.

Koch's extensive activities in tropical medicine are further illustrated by "Ein Versuch zur Immunisierung von Rindern gegen Tsetsekrankheit (Surra)," in *Deutsches Kolonialblatt*, **12** (1901), 1–4; "Framboesia tropica und Tinea imbricata," in *Archiv für Dermatologie und Syphilis*, **59** (1902), 3–8; "On Rhodesian Redwater or African Coast Fever," a series of four reports, of which the first two were published separately in Salisbury (1903) and the third in

Bulawayo (1903), and also in *Journal of Comparative Pathology and Therapeutics*, **16** (1903), 273–280, 280–284, 390–398, and **17** (1904), 175–181. The first three reports, trans. by R. Hollandt, appear as "Ueber das Rhodesische Rotwasser oder 'Afrikanische Küstenfieber,'" in *Archiv für wissenschaftliche und praktische Tierheilkunde*, **30** (1904), 281–319; and the fourth is included in English as "Fourth Report on African Coast Fever," in *Gesammelte Werke*, II, pt. 2, 787–798. Also noteworthy are "Ueber die Trypanosomenkrankheiten," in *Deutsche medizinische Wochenschrift*, **30** (1904), 1705–1711, trans. as "Remarks on Trypanosome Diseases," in *British Medical Journal* (1904), **2**, 1445–1449; "Untersuchungen über Schutzimpfungen gegen Horse-Sickness (Pferdesterbe)," in *Deutsches Kolonialblatt*, **15** (1904), 420–424, 459–463; "Vorläufige Mittheilungen über die Ergebnisse einer Forschungsreise nach Ostafrika," *ibid.*, **31** (1905), 1865–1869, trans. by P. Falcke as "Preliminary Statement on the Results of a Voyage of Investigation to East Africa," in *Journal of Tropical Medicine*, **9** (1906), 43–45, 75–76, 104–105, 137–138; "Beiträge zur Entwicklungsgeschichte der Piroplasmen," in *Zeitschrift für Hygiene und Infektionskrankheiten*, **54** (1906), 1–9; "Ueber afrikanischen Rekurrens," in *Berliner klinische Wochenschrift*, **43** (1906), 185–194, trans. by H. T. Brooks as "African Recurrent Fever," in *Post-Graduate* (New York), **21** (1906), 770–789; and "Ueber den bisherigen Verlauf der deutschen Expedition zur Erforschung der Schlafkrankheit in Ostafrika," in *Deutsche medizinische Wochenschrift*, **32** (1906), 1–8.

Miscellaneous publications that further illustrate the scope of Koch's versatility include "Versuche über die Desinfection des Kiel- oder Bilgeraumes von Schiffen," in *Arbeiten aus dem Kaiserlichen Gesundheitsamt*, **1** (1886), 199–221; "Beobachtungen über Erysipel-Impfungen am Menschen," in *Zeitschrift für Hygiene und Infektionskrankheiten*, **23** (1896), 477–489, written with J. Petruschky; "Die Lepra-Erkrankungen im Kreise Memel," in *Klinisches Jahrbuch*, **6** (1897), 239–253; "Ueber die Verbreitung der Bubonenpest," in *Deutsche medizinische Wochenschrift*, **24** (1898), 437–439; "Typhusepidemie in Gelsenkirchen. Berlin, 21. Oktober 1901," in *Gesammelte Werke*, II, pt. 2, 910–915; and "Berichte über die Wertbestimmung des Pariser Pestserums," in *Klinisches Jahrbuch*, **9** (1902), 643–704, written with E. von Behring, R. Pfeiffer, W. Kolle, and E. Martini.

Koch's personal scientific library of just over 300 vols. was bequeathed to the Robert Koch Institute in Berlin, where it has been kept intact. His ashes, a bust, and a memorial tablet citing his accomplishments are in a mausoleum at the Institute. In an adjacent small museum are personal memorabilia, including a few handwritten letters and photographs, a diary kept during his journey to South Africa, and a map depicting his various travel routes; there are also sample vials of tuberculin labeled by him, some early laboratory apparatus, and other relics. A collection made by H. B. Jacobs of about 60 handwritten letters from Koch to Carl Flügge, dating from 1879 to 1907, is in the Welch Library, Institute of the History of Medicine, Johns Hopkins University, Baltimore.

II. SECONDARY LITERATURE. Obituaries in German include P. Ehrlich, "Robert Koch, 1843–1910," in *Frankfurter Zeitung* (Erstes Morgenblatt), **54**, no. 150 (2 June 1910), 1–3, trans. by S. Klein in *Chicago Medical Recorder*, **32** (1910), 443–450; and "Robert Koch †," in *Zeitschrift für Immunitätsforschung*, **6** (1910), preface; C. Fraenkel, "Robert Koch," in *Münchener medizinische Wochenschrift*, **57** (1910), 1345–1349; G. Gaffky, "Gedächtnisrede auf Robert Koch," in *Deutsche medizinische Wochenschrift*, **36** (1910), 2321–2324; M. Kirchner, "Robert Koch," in *Zeitschrift für Tuberkulose*, **16** (1910), 105–114; and R. Pfeiffer, "Robert Koch †," in *Berliner klinische Wochenschrift*, **47** (1910), 1045–1048.

Obituaries in English include W. W. Ford, "The Life and Work of Robert Koch," in *Bulletin of the Johns Hopkins Hospital*, **22** (1911), 415–425; S. A. Knopf, "Robert Koch (December 11, 1843–May 27, 1910). The Father of the Modern Science of Tuberculosis," *ibid.*, 425–428; C. J. Martin, "Robert Koch, M.D.," in *British Medical Journal* (1910), **1**, 1386–1388; and "Robert Koch, 1843–1910," in *Proceedings of the Royal Society*, **83** (1910), xviii–xxiv; G. S. Woodhead, "Robert Koch," in *Journal of Pathology and Bacteriology*, **15** (1911), 108–114; J. A. Wyeth, "Memorial Address on Doctor Robert Koch," in *Medical Record*, **79** (1911), 95–97; and the following unsigned tributes: "Robert Koch, M.D.," in *British Medical Journal* (1910), **1**, 1384–1386; "Professor Robert Koch," in *Lancet* (1910), **1**, 1583–1588; and "Robert Koch and His Achievements," in *Journal of the American Medical Association*, **54** (1910), 1872–1876.

Other references in German to Koch's life and work are R. Bassenge, M. Beck, L. Brieger, *et al.*, *Festschrift zum sechzigsten Geburtstage von Robert Koch* (Jena, 1903), containing contributions from former colleagues and pupils; P. Boerner, "R. Koch's Polemik gegen Buchner und Pasteur," in *Deutsche medizinische Wochenschrift*, **8** (1882), 40–41; L. Brieger and F. Kraus, "Krankheitsgeschichte Robert Kochs," *ibid.*, **36** (1910), 1045–1046; W. Bulloch, "Robert Koch und England," *ibid.*, **58** (1932), 508–509; W. von Drigalski, "Planmässige Seuchenbekämpfung nach Robert Koch, mit besonderer Berücksichtigung der Typhus Bekämpfung," *ibid.*, 503–505; and "Robert Koch und die Entwicklung der kommunalen Gesundheitspflege," in *Medizinische Welt*, **6** (1932), 348–350; P. Ehrlich, "Erinnerung aus der Zeit der ätiologische Tuberculoseforschung Robert Kochs," in *Deutsche medizinische Wochenschrift*, **39** (1913), 2444–2446; J. Fibiger and C. O. Jensen, "Untersuchungen über die Beziehungen zwischen der Tuberkulose und den Tuberkelbacillen des Menschen und der Tuberkulose und den Tuberkelbacillen des Rindes," in *Berliner klinische Wochenschrift*, **45** (1908), 1977–1980, 2026–2031; I. Fischer, "Robert Koch," in *Biographisches Lexikon der hervorragenden Ärzte der letzten fünfzig Jahre*, I (Berlin–Vienna, 1932), 784–786; G. Gaffky, "Dem Andenken Robert Kochs," in *Deutsche medizinische Wochenschrift*, **42** (1916), 653–655; S. Guttmann, ed., *Robert Koch's Heilmittel gegen die Tuberculose* (Berlin–Leipzig, 1890); L. Haendel, "Robert Koch und das Reichsgesundheitsamt," in *Medizinische Welt*, **6** (1932),

351–353; R. Harms, *Robert Koch, Arzt und Forscher. Ein biographische Roman* (Hamburg, 1966); B. Heymann, "Zur Fünfzig-Jahr-Feier der Entdeckung des Tuberkelbacillus," in *Klinische Wochenschrift*, **12** (1932), 489–490; and *Robert Koch* (Leipzig, 1932), covering the period to 1882; F. Hueppe, "R. Koch's Mittheilungen über Tuberkulin," in *Berliner klinische Wochenschrift*, **28** (1891), 1121–1122; G. Jaeckel, *Die Charité*, 2nd ed. (Bayreuth, 1965), pp. 276–296; J. Kathe, *Robert Koch und sein Werk* (Berlin, 1961); M. Kirchner, "Robert Koch," in *Meister der Heilkunde*, M. Neuberger, ed., V (Vienna–Berlin, 1924), 7–84; K. Kisskalt, "Robert Kochs Gedächtnis. Die Entdeckung des Tuberkelbazillus," in *Münchener medizinische Wochenschrift*, **79** (1932), 497–501; and "Die ersten Beurteilungen Robert Kochs durch die Schule Pettenkofers," in *Archiv für Hygiene und Bakteriologie*, **112** (1934), 167–180; F. K. Kleine, "Ein Tagebuch von Robert Koch während seiner deutsch-ostafrikanischen Schlafkrankheitsexpedition i. J. 1906/07," in *Deutsche medizinische Wochenschrift*, **50** (1924), 21–24, 55–56, 88–89, 121–122, 152–153, 184–185, 216–217, 248–249; and "Der Anteil R. Kochs an der Erforschung tropischer Seuchen," *ibid.*, **58** (1932), 505–508; K. Kolle, ed., *Robert Koch Briefe an Wilhelm Kolle* (Stuttgart, 1959); W. Kolle, "Zur Erinnerung an Robert Koch. Gedenkrede, gehalten zur 70. Wiederkehr seines Geburtstags am 11. Dezember 1913," in *Medizinische Klinik*, **9** (1913), 2137–2138, 2159–2161; "Robert Koch und das Spezifizitätsproblem," in *Deutsche medizinische Wochenschrift*, **39** (1913), 2446–2448; and "Robert Koch," in *Zentralblatt für Bakteriologie*, **127** (1932), 3–10; H. Kossel, "Zeitliche und örtliche Disposition bei Infektionskrankheiten im Lichte experimenteller Forschung," in *Deutsche medizinische Wochenschrift*, **39** (1913), 2448–2450; W. Leibbrand, "Robert Koch, 1843–1910," in *Die grossen Deutschen*, IV (Berlin, 1957), 93–102; F. Loeffler, "Zur Immunitätsfrage," in *Mittheilungen aus dem Kaiserliche Gesundheitsamt*, **1** (1881), 137–187; and "Zum 25jährigen Gedenktage der Entdeckung des Tuberkelbacillus," in *Deutsche medizinische Wochenschrift*, **33** (1907), 449–451, 489–495; R. Maresch, N. Jagié and F. Hamburger, "Zum 24. März 1882," in *Wiener klinische Wochenschrift*, **45** (1932), 417–422; P. Martell, "Robert Koch," in *Zeitschrift für ärztliche Fortbildung*, **32** (1935), 332–335; M. Miyajima, "Robert Koch in Japan," in *Deutsche medizinische Wochenschrift*, **58** (1932), 509–511; B. Möllers, *Dr. med. Robert Koch: Persönlichkeit und Lebenswerk. 1843–1910* (Hannover, 1950); R. Paltauf, "Robert Koch," in *Wiener klinische Wochenschrift*, **16** (1903), 1377–1381; E. Pfuhl, "Privatbriefe von Robert Koch," in *Deutsche medizinische Wochenschrift*, **37** (1911), 1399–1400, 1443–1444, 1483–1485, 1524–1526; and "Robert Kochs Entwicklung zum bahnbrechenden Forscher," *ibid.*, **38** (1912), 1101–1102, 1148–1150, 1195–1197; L. Roudolf, "Bemerkungen zu den Forschungsreisen Robert Kochs mit besonderer Berücksichtigung Afrikas," *ibid.*, **87** (1962), 1680–1686; and "Die wissenschaftliche Bibliothek Robert Kochs," in *Zentralblatt für Bakteriologie, I. Referate*, **175** (1960), 447–472; F. Sauerbruch, "Robert Koch," in *Zeitschrift für Tuberkulose*, **64** (1932), 7–9; J. Schwalbe, "Robert Koch zum Gedächtnis," in *Deutsche medizinische Wochenschrift*, **39** (1913), 2441; and "Die Enthüllung des Robert Koch Denkmals," *ibid.*, **42** (1916), 704–705; G. Seiffert, "Die Tuberkulose als übertragbare Krankheit und ihre Bekämpfung vor Robert Koch," in *Münchener medizinische Wochenschrift*, **79** (1932), 501–506; H. Unger, *Robert Koch. Roman eines grossen Lebens* (Berlin–Vienna, 1936); W. von Waldeyer-Harz, *Lebenserinnerungen* (Bonn, 1921), pp. 283–285; F. A. Weber, "Robert Koch und die Bekämpfung der Tuberkulose. Zur Erinnerung an die Entdeckung des Tuberkelbazillus vor 50 Jahren," in *Zeitschrift für Tuberkulose*, **64** (1932), 399–415; K. Wezel, *Robert Koch* (Berlin, 1912); and H. Zeiss and R. Bieling, *Behring. Gestalt und Werk* (Berlin, 1940).

English and French references to Koch's life and work include E. R. Baldwin, "A Call Upon Robert Koch in His Laboratory in 1902," in *Journal of the Outdoor Life*, **16** (1919), 302–303; E. von Bergmann, "Demonstration of Cases Treated by Koch's Anti-Tubercular Liquid," in *Lancet* (1890), **2**, 1120–1122; (1891), **1**, 50–51; L. Brown, "Robert Koch," in *Bulletin of the New York Academy of Medicine*, **8** (1932), 549–584; and "Robert Koch (1843–1910). An American Tribute," in *Annals of Medical History*, n.s. **7** (1935), 99–112, 292–304, 385–401; editorial, "Koch's Work Upon Tuberculosis, and the Present Condition of the Question," in *Science*, **4** (1884), 59–61; editorial, "The Debate on Koch's Remedy at the Berlin Medical Society," in *Lancet* (1891), **1**, 215–217, 271–272, 328–330, 389, 450–452, 506–507, 567–568, 630–631; H. C. Ernst, "Robert Koch (1843–1910)," in *Proceedings of the American Academy of Arts and Sciences*, **53** (1918), 825–827; S. Flexner and J. T. Flexner, *William Henry Welch and The Heroic Age of American Medicine* (New York, 1941), pp. 146–149; A. P. Hitchens and M. C. Leikund, "The Introduction of Agar-Agar Into Bacteriology," in *Journal of Bacteriology*, **37** (1939), 485–493; T. James, "Professor Robert Koch in South Africa," in *South African Medical Journal*, **44** (1970), 621–624; L. S. King, "Dr. Koch's Postulates," in *Journal of the History of Medicine and Allied Sciences*, **7** (1952), 350–361; A. K. Krause, "Essays on Tuberculosis. IV. The Tubercle Bacillus," in *Journal of the Outdoor Life*, **15** (1918), 129–137, and "XVI. The First Experiments in Resistance. The Discovery of Tuberculin: Trudeau and Koch," *ibid.*, **16** (1919), 129–132, 150–152; E. Lagrange, *Robert Koch: Sa vie et son oeuvre* (Tours, 1938); H. R. M. Landis, "The Reception of Koch's Discovery in the United States," in *Annals of Medical History*, n.s. **4** (1932), 531–537; Sir Joseph Lister, "Koch's Treatment of Tuberculosis," in *Lancet* (1890), **2**, 1257–1260; E. Metchnikoff, *The Founders of Modern Medicine: Pasteur. Koch. Lister*, D. Berger, trans. (New York, 1939), pp. 60–75, 112–124; G. H. F. Nuttall, "Biographical Notes Bearing on Koch, Ehrlich, Behring and Loeffler, With Their Portraits and Letters From Three of Them," in *Parasitology*, **16** (1924), 214–238—"Robert Koch, 1843–1910," pp. 214–223; L. Pasteur, "De l'atténuation des virus," in *Revue scientifique*, 2nd ser., **4** (1882), 353–361, written with C. Chamberland, E. Roux, and L. Thuillier; and "La vaccination charbonneuse.

Réponse à un mémoire de M. Koch," *ibid.*, **5** (1883), 74–84; Sir Robert Philip, "Koch's Discovery of the Tubercle Bacillus, Some of Its Implications and Results," in *British Medical Journal* (1932), **2**, 1–5; J. Plesch, *Janos, the Story of a Doctor*, E. Fitzgerald, trans. (London, 1947), pp. 50–51; V. Robinson, "Robert Koch," in *Pathfinders in Medicine* (New York, 1929), pp. 714–746; T. Smith, "Koch's Views on the Stability of Species Among Bacteria," in *Annals of Medical History*, n.s. **4** (1932), 524–530; D. A. Stewart, "The Robert Koch Anniversary—the Man and His Work," in *Canadian Medical Association Journal*, **26** (1932), 475–478; E. L. Trudeau, "An Experimental Study of Preventive Inoculation in Tuberculosis," in *Medical Record*, **38** (1890), 565–568; "Some Personal Reminiscences of Robert Koch's Two Greatest Achievements in Tuberculosis," in *Journal of the Outdoor Life*, **7** (1910), 189–192; and *An Autobiography* (Philadelphia–New York, 1916), pp. 212–216; R. Virchow, "On the Action of Koch's Remedy Upon Internal Organs in Tuberculosis," in *Lancet* (1891), **1**, 130–132; M. E. M. Walker, "Robert Koch, M.D., F.R.S., 1843–1910," in *Pioneers of Public Health* (Edinburgh–London, 1930), pp. 178–192; G. B. Webb, "Robert Koch (1843–1910)," in *Annals of Medical History*, n.s. **4** (1932), 509–523; W. Welch, "Tribute to Robert Koch," in *Journal of the Outdoor Life*, **5** (1908), 165–167; and C.-E. A. Winslow, *The Life of Hermann Biggs* (Philadelphia, 1929), pp. 54–56, 176–180, 216–220.

Besides the 60th birthday *Festschrift*, various medical journals honored Koch's work through commemorative numbers or special contributions. To celebrate the 70th anniversary of his birth, several papers on tuberculosis, by W. Kolle, H. Kossel, F. Loeffler, and others, appeared in the *Deutsche medizinische Wochenschrift*, **39** (1913), 2442–2466. To mark the 50th anniversary of the discovery of the tubercle bacillus many journals issued memorial numbers, including the *Deutsche medizinische Wochenschrift*, **58** (1932), 475–511, with contributions by W. Bulloch, W. von Drigalski, F. K. Kleine, and others; *Medizinische Klinik*, **28** (1932), 387–424, with contributions by P. Uhlenhuth, L. Aschoff, T. Burgsch, and others; *Medizinische Welt*, **6** (1932), 325–364, with contributions by O. Lenz, R. Otto, R. Pfeiffer, and others; and *Zeitschrift für Tuberkulose*, **64** (1932), 1–126, 476–499, with 17 contributions. Some articles from these anniversary publications are cited above.

CLAUDE E. DOLMAN

KOCH, HELGE VON (*b.* Stockholm, Sweden, 25 January 1870; *d.* Stockholm[?], 11 March 1924), *mathematics.*

Von Koch is known principally for his work in the theory of infinitely many linear equations and the study of the matrices derived from such infinite systems. He also did work in differential equations and in the theory of numbers.

The history of infinitely many equations in infinitely many unknowns is long; special cases of infinite systems were studied by Fourier, who used them naïvely in his celebrated *Théorie analytique de la chaleur;* and there are even earlier examples. Yet despite the many applications in differential equation theory and in geometry, the rigorous study of infinite systems began only in 1884–1885 with the publication by Henri Poincaré of a few special results.

Von Koch's interest in infinite matrices came from his investigations in 1891 into Fuchs's equation:

$$D^n + P_2(x)\,D^{n-1} + \cdots + P_n(x)\,y = 0,$$

where

$$D^r = \frac{d^r y}{dx^r} \text{ and } P_r(x) = \sum_{k=-\infty}^{\infty} a_{rk}x^k,$$

all of which converge in some annulus A with center at the origin. It was known that there existed a solution

$$y = \sum_{k=-\infty}^{\infty} b_k x^{k+\rho}$$

which also converged in A; but in order explicitly to calculate the coefficient b_k and the exponent ρ, von Koch was led to an infinite system of linear equations. Here he used Poincaré's theory, which forced him to assume some unnaturally restrictive conditions on the original equation.

To remove the restrictions, von Koch published another paper in 1892 which was concerned primarily with infinite matrix theory. He considered the infinite array or matrix

$$A = \{A_{ik} : i, k = \cdots, -2, -1, 0, 1, 2, \cdots\}$$

and set

$$D_m = \det\{A_{ik} : i, k = -m, \cdots, m\}.$$

The determinant D of A was defined to be $\lim_{m\to\infty} D_m$ if this limit existed. He then noted that the same array could give rise to denumerably many different matrices—by the use of different systems of enumeration—each with a different main diagonal. He was, however, able to prove that if $\prod_{i=-\infty}^{\infty} A_{ii}$ converged absolutely and $\sum_{i,k=-\infty;i\neq k}^{\infty} A_{ik}$ also converged absolutely, then D existed and was independent of the enumeration of A. A matrix which satisfied the above hypotheses was said to be in normal form.

Various methods to evaluate D were then given by von Koch, all of them analogous to the evaluation of finite determinants. Minors of finite and infinite order were defined, and it was proved that D could be evaluated by the method of expansion by minors in a direct generalization to infinite matrices of the Laplace expansion. Finally, he showed that

$$D = 1 + \sum_{p=-\infty}^{\infty} a_{pp} + \sum_{p<q} \det \begin{pmatrix} a_{pp} & a_{pq} \\ a_{qp} & a_{qq} \end{pmatrix}$$

$$+ \sum_{p<q<r} \det \begin{pmatrix} a_{pp} & a_{pq} & a_{pr} \\ a_{qp} & a_{qq} & a_{qr} \\ a_{rp} & a_{rq} & a_{rr} \end{pmatrix} + \cdots .$$

Here, $a_{pq} = A_{pq} - \delta_{pq}(\delta_{jk} = 1$ if $j = k$, $\delta_{jk} = 0$ if $j \neq k)$; the largest summation index in each term is to range over all integers; and the others are to range over all integers as indicated. This is particularly interesting because it was the form used by Fredholm in 1903 to solve the integral equation

$$\phi(x) + \int_0^1 f(x, y)\, \phi(y)\, dy = \psi(x)$$

for the unknown function ϕ, the other functions being supposed known.

Von Koch then went on to prove that if A and B are in normal form, then the usual product matrix $C = AB$ can be formed. The matrix C will also be in normal form and det C = (det A)(det B). He also was able to show that the property of being in normal form is not a necessary condition for D to exist and indicated how his theory could be extended to matrices whose entries are functions all analytic in the same disk.

Finally, von Koch applied his results to systems of infinitely many linear equations in infinitely many unknowns. Although he claimed a certain amount of generality, he actually considered only the homogeneous case

$$\sum_{k=-\infty}^{\infty} A_{ik}x_k = 0 \qquad (i = -\infty, \cdots, \infty).$$

Here the matrix $\{A_{ik}\}$ was supposed to be in normal form, and the only solutions sought were those for which $|x_k| \leq M$ for $k = -\infty, \cdots, \infty$. He then established that if det$\{A_{ik}\}$ is different from zero, then the only such solution for the above equation is $x_k = 0$ for $k = -\infty, \cdots, \infty$. He then showed that if $D = 0$ but $A_{ik} \not\equiv 0$, there will always exist a minor of smallest order m which is not zero. Then if the nonvanishing minor is obtained from $\{A_{ik}\}$ by deleting columns $k_1, k_2, \cdots, k_m$, a solution $\{x_k\}$ can be obtained by assigning arbitrary values to $x_{k_1}, x_{k_2} \cdots, x_{k_m}$ and expressing each of the remaining x_k's as a linear combination of $x_{k_1}, x_{k_2} \cdots, x_{k_m}$. This is similar to the finite case. Von Koch then asserted that analogous results could be obtained for unhomogeneous systems, which is now known to be false unless further restrictions are placed on $\{A_{ik}\}$.

Von Koch's work cannot be called pioneering. His results were all fairly readily accessible, although many of the calculations are lengthy. He was aware, through a knowledge of Poincaré's work, of the possibility of obtaining pathological results but did little to explore them. Yet this work can be said to be the first step on the long road which eventually led to functional analysis, since it provided Fredholm with the key for the solution of his integral equation.

BIBLIOGRAPHY

A complete bibliography of von Koch's papers is in *Acta mathematica*, **45** (1925), 345–348. Of particular interest is "Sur les déterminants infinis et les équations différentielles linéaires," in *Acta mathematica*, **16** (1892–1893), 217–295.

A secondary source is Ernst Hellinger and Otto Toepletz, "Integralgleichungen und Gleichungen mit unendlichenvielen Unbekannten," in *Encyklopädie der mathematischen Wissenschaften*, II, pt. C (Leipzig, 1923–1927), 1335–1602, also published separately.

MICHAEL BERNKOPF

KOCHIN, NIKOLAI YEVGRAFOVICH (*b.* St. Petersburg, Russia [now Leningrad, U.S.S.R.], 1901; *d.* Moscow, U.S.S.R., 31 December 1944), *physics, mathematics.*

Kochin's father was a clerk in a dry goods store. After graduating from Petrograd University in 1923, Kochin gave courses in mechanics and mathematics there from 1924 to 1934 and then at Moscow University until 1944. From 1932 to 1939 he worked in the Mathematics Institute of the Soviet Academy of Sciences, and from 1939 to 1944 he was head of the mechanics section of the Mechanics Institute of the Academy.

Kochin's work covered a wide range of scientific problems. At the beginning of his career he published a number of very important works in meteorology. He made significant contributions in the development of gas dynamics. His research on shock waves in compressed liquids was of great importance in the development of this area of science. In hydrodynamics he was responsible for a number of classical investigations. His "K teorii voln Koshi-Puassona" ("Towards a Theory of Cauchy-Poisson Waves," 1935) gives the solution of the problem of small-amplitude free waves on the surface of an uncompressed liquid. In 1937 Kochin published "O volnovom soprotivlenii i podyomnoy sile pogruzhennykh v zhidkosty tel" ("On the Wave Resistance and Lifting Strength of Bodies Submerged in Liquid"), in which he proposed a general method of solving the two-dimensional problem of an underwater fin, the formulas for the resistance of a body (a ship), forms of a wave surface, and lifting force. Using this method, Kochin in 1938 solved the two-dimensional problem of the hydroplaning of a slightly curved contour on the surface of a heavy uncompressed liquid. "Teoria voln, vynuzhdaemykh kolebaniami tela pod svobodnoy poverkhnostyu tyazheloy neszhimaemoy zhidkosti" ("Theory of Waves Created by the Vibration of a Body Under a Free Surface of Heavy Uncompressed Liquid," 1940) provided a basis for a new theory of the pitch and roll of a ship, taking into account the mutual influence of the hull of the ship and the water.

In aerodynamics Kochin was the first (1941–1944) to give strict solutions for the wing of finite span; he introduced formulas for aerodynamic force and for the distribution of pressure.

Kochin also produced important works on mathematics and theoretical mechanics. He wrote textbooks on hydromechanics and vector analysis, was coauthor and editor of a two-volume monograph on dynamic meteorology, and was the editor of the posthumous edition of the works of A. M. Lyapunov.

BIBLIOGRAPHY

Kochin's works were brought together as *Sobranie sochineny* ("Collected Works"), 2 vols. (Moscow–Leningrad, 1949). There is also a bibliography: *Nikolai Yevgrafovich Kochin. Bibliografia sost. N. I. Akinfievoy* ("Bibliography Compiled by N. I. Akinfieva"; Moscow–Leningrad, 1948).

See also P. I. Polubarinova-Kochina, *Zhizn i deyatelnost N. Ye. Kochina* ("Life and Work of N. Y. Kochin"; Leningrad, 1950).

A. T. GRIGORIAN

KOELLIKER, RUDOLF ALBERT VON (*b.* Zurich, Switzerland, 6 July 1817; *d.* Würzburg, Germany, 2 November 1905), *comparative anatomy, histology, embryology, physiology.*

Koelliker's father, Johannes Koelliker, was a bank officer; his early death left his widow, Anna Maria Katharina Füssli, responsible for the education of their two sons. The family was of the upper middle class, with strong ties to letters and the arts. In his memoirs, published in 1899, Koelliker was to recall his carefree youth and his affection for his mother.

Koelliker attended the Gymnasium in Zurich, receiving supplementary private tuition in foreign languages. His interest in natural sciences, particularly botany, was early manifest and led him to study medicine when he entered the University of Zurich in the spring of 1836. His teachers included the botanist Oswald Heer; Lorenz Oken, whose lectures on zoology and *Naturphilosophie* he attended; and Friedrich Arnold, the anatomist, who instructed him in the basic tenets of the subject that he was to make his lifework. In 1839 Koelliker studied for a semester in Bonn; then, the following autumn, he went to Berlin, where he remained for three semesters. In Berlin he was strongly influenced by Johannes Müller's lectures on comparative anatomy and physiology and was instructed in microscopy by F. G. J. Henle. Robert Remak introduced him to the study of embryology. Koelliker was singularly fortunate in his teachers; and the course of his career was then determined.

In the fall of 1840, Koelliker, with his close friend Naegeli and two other Swiss students, undertook a journey to the islands of Föhr and Helgoland to collect and study seabirds and marine animals. He continued to do independent research; the following winter, having bought a microscope, he began to investigate the spermatozoa of invertebrates. In refutation of the parasitic theory, Koelliker recognized the origin of the spermatozoa to be in the sperm-producing cells and thereby deduced their cellular nature. For these investigations he was awarded the Ph.D. at Zurich in spring of 1841. He took the state medical examination there in the summer of the same year, and then went on to study the development of two types of fly larvae. His results on this subject formed the basis for his M.D. dissertation. He received the degree from Heidelberg in 1842.

At about the same time Koelliker became assistant to Henle, who had assumed the professorship of anatomy at Zurich in 1840; their association was to develop into a long and fruitful friendship. In summer of 1842, Koelliker made another expedition with Naegeli to investigate the flora and fauna of the Mediterranean at Naples and Messina. Returning to Zurich, he became Henle's prosector; in 1843 a discourse on the development of invertebrates qualified him for the post of lecturer in the university. In 1844 Henle accepted an appointment at Heidelberg. His professorship in Zurich was divided between two successors, Koelliker becoming associate professor of physiology and comparative anatomy and continuing to lecture on embryology and general anatomy.

In 1844, too, Koelliker published his work on the development of the cephalopods, a continuation of his earlier investigations of Mediterranean fauna. He had also by this time begun his researches on nerve tissue and demonstrated, in a paper on relative independence of the sympathetic nerve system, that among the vertebrates certain ramifications of the nerve cells exist as medullary nerve fibers. With Henle, he studied the lamellar corpuscles, for which they introduced the name "Pacinian corpuscles." The following year he published an important paper on single-celled animals, particularly the gregarines, and in 1846 he brought out his studies on the formation of mammalian red blood corpuscles, with special emphasis on their nucleus-bearing first stages. The latter work is notable in that there Koelliker discussed the formation of blood in the embryo, localizing the site of hematopoiesis in the liver. In 1846 he also studied the structure of the smooth muscles, isolating smooth muscle fibers for the first time; he further recognized the cellular nature of such muscle fibers and ascertained their wide distribution throughout the body.

In 1847 Koelliker was called to the University of Würzburg as full professor of physiology and comparative anatomy. Before accepting this new post, he stipulated that he was also to be made professor of anatomy as soon as that chair became vacant; he duly received that appointment in 1849. He was assigned to teach courses on human tissues and organs, in preparation for the textbook, *Mikroskopische Anatomie*, that he planned to base on his own investigations. In 1848 he married Maria Schwarz, of Mellingen, Aargau; his mother came to join their household shortly thereafter.

It was typical of Koelliker's method of working that he sought to verify, if not personally study, each subject treated in his textbooks. Often before publishing a general work, he brought out individual treatises on his observations of specific details. Thus, while working on the *Mikroskopische Anatomie*, Koelliker in 1849 published his important findings on the formation of the skull, in which he made a distinction between the preformed, cartilaginous structure and the bony plates that develop from the connective tissue. He further continued his investigations of the central nervous system, and published an article on the course of the nerve fibers in the human spinal cord the following year. This procedure brought him a reputation for detailed research; he maintained it by being wary of hasty generalizations.

The publication of Koelliker's *Mikroskopische Anatomie oder Gewebelehre des Menschen* provides a good illustration of this wariness. The second volume, bearing the separate title *Spezielle Gewebelehre*, appeared in three parts, between 1850 and 1854; the first volume, projected as a general treatment, was never published. That Koelliker was at this time working toward a generalization may, however, be seen in the chapter "Allgemeine Gewebelehre" of the textbook *Handbuch der Gewebelehre des Menschen* of 1852. This chapter might almost be considered a draft for the planned larger work, since it contains, in addition to a section on cytology, both detailed and general descriptions of ten different tissues. It is interesting to note that in the second edition, published three years later, Koelliker enumerated only five kinds of tissue, while in the third edition of 1859, he dealt with four. All the tissues he wrote of bore the names still in use, with the important exception of the epithelium, which he called *Zellengewebe*—cellular tissue.

Koelliker was thus one of the first to utilize the cellular elements of tissue structure descriptively. Indeed, his breakthrough lay in presenting the study of tissue in terms of the cell theory. His *Handbuch* was translated and had many editions; by this means his classification of tissues became known and accepted throughout central Europe.

While his microscopical work was receiving wide currency, Koelliker continued to teach and do research in physiology. Among other projects he confirmed the existence of the musculus dilator pupillae (1855). The following year he published his findings on the effect of various poisons on nerves and muscles; he also demonstrated that electric current is produced by muscle contraction. In 1857, Koelliker brought out his study on the light organs of the lamprey, in which he cited the dependence of these organs upon the nervous system. He extended his studies of the spinal cord to the lower invertebrates in 1858.

At the same time, Koelliker continued to lecture on embryology, a subject that had been of interest to him since his student days in Berlin. There he employed the cell theory in interpreting the development of the embryo, as he also did in his histological studies of tissues. His early studies of spermatozoa—in which he emphasized their cellular nature—were important in helping him to achieve this viewpoint. By 1856, in an investigation of their motility in various organic and inorganic substances, he was able to predicate that the motion of flagella and that of the sperm is essentially the same, noting in particular that both become more mobile in alkaline solutions. In a further series of studies he demonstrated that the same types of developmental processes occur in both invertebrates and vertebrates, although he based his work on an inaccurate notion of fertilization. He viewed the egg as a single cell, and correctly regarded its segmentation as a continuous production of daughter cells, which he interpreted as material for the developing tissues and organs. Koelliker thus opposed Schwann's doctrine of free-cell formation in the cytoblast, although he did not exclude it in every instance, and fully rejected it only in 1859.

For his lectures on embryology Koelliker made

himself thoroughly acquainted with the existing literature and also had illustratory drawings made from specimens that he himself had prepared. He further drew upon the findings that he had made in regard to specific organs—as, for example, his results on the eye, ear, spinal cord, brain, and the olfactory and sex organs. In this way he came to compile a vast amount of information on the subject, which he decided to gather into as comprehensive a work as possible. His first publication, *Entwicklungsgeschichte des Menschen und der höheren Thiere* (1861), represented a collection of his classroom lectures. A second edition, published in 1879, was so fully revised as to constitute an entirely new book, while an abridgment for students, *Grundriss der Entwicklungsgeschichte des Menschen und der höheren Thiere*, appearing in 1880, required a second edition by 1884. Here again, as in his work on tissues, Koelliker sought to incorporate all new data, subject, in so far as was possible, to his own investigation and verification.

In 1864 Koelliker gave up the chair of physiology at Würzburg. Two separate departments were organized for the teaching of anatomy and embryology, one comprising systematic and topographical anatomy, and the other, comparative anatomy, microscopy, and embryology. Koelliker and his co-workers taught alternate courses in macroscopic and microscopic anatomy, including related lectures on ontogeny.

As part of his work on the development of the embryo Koelliker was led to consider the origin of species and the laws of heredity. Although he carried out no special researches in this field, he had, as early as 1841, suggested that a particularly important function of the cell nucleus—in addition to its participation in the metabolism of the cell—was its agency in the transmission of inherited characteristics. In 1864, Koelliker made known his objections to Darwinian natural selection, which he thought too teleological. He pointed out that variations in certain characteristics are more apt to appear suddenly than gradually, and he emphasized the significance of such abrupt changes, thereby closely foreshadowing De Vries's theory of mutations.

Koelliker returned to his study of nerve tissue with a treatise on nerve endings in the cornea in 1866. His work on the central nervous system took a new direction after 1884, when he heard of Golgi's discoveries and adopted his methods of research, extending them to a study of parts of the brain. Koelliker was thus able to make important contributions toward substantiating the doctrine of the neuron as the basic unit of the nervous system. (Some of his results were to be incorporated into the sixth edition of the *Handbuch der Gewebelehre*.)

In 1873 Koelliker took up the study of the processes involved in the absorption of bone. He identified the large multinucleated cells that are active in osseous absorption and removal and named them "osteoclasts." His findings were published, with illustrations, in his memoirs (pp. 315–323). In 1884, the same year in which he espoused Golgi's work on nerve tissue, Koelliker also rejected His's embryological theories. Having stated his objections to His's parablast theory, Koelliker lived to see himself proved correct. He took further exception—again correctly—to His's notion that the processes involved in the formation of the embryo might be understood through the mechanical model of an unevenly stretched elastic plate.

In 1897 Koelliker retired from teaching, but not from research. In 1899 he demonstrated the presence of uncrossed fibers in the optic chiasm, while in 1902 he made further exact studies of the nuclei (which he named for his anatomy demonstrator Hofmann) of the avian spinal cord. In 1903, when he was eighty-six years old, Koelliker conducted investigations into the origin of the vitreous body of the eye. He had retained his post as director of the microscopical institute until fall of 1902; three years later he died of a lung infarct.

In addition to his textbooks, Koelliker published about 300 separate items during his lifetime. A list of these appears in Ehlers' memoir of him ("Albert von Koelliker. Zum Gedächtnis," pp. x-xxvi); it is arranged chronologically, but lacks fully adequate documentation. Koelliker himself mentioned almost all his published works in his *Erinnerungen aus meinen Leben* of 1899. He was also the founder and, with Theodor von Siebold, the editor of the *Zeitschrift für wissenschaftliche Zoologie*, which has been issued continuously from 1848 to the present.

Koelliker was instrumental in the founding of the Physikalisch-Medizinische Gesellschaft of Würzburg; he was active in the Anatomische Gesellschaft, of which he was the first chairman and later honorary president; and he worked ceaselessly to promote the international cooperation of scientists. He was, as Waldeyer wrote (in "Albert von Koelliker zum Gedächtnis," p. 543), a "member of all the learned societies for which his knowledge qualified him," and received international recognition in the form of honorary degrees from the universities of Utrecht, Bologna, Glasgow, and Edinburgh, as well as numerous medals and special awards. He was also a knight of the Maximiliansorden für Wissenschaft und Kunst and thereby personally ennobled.

The effect of Koelliker's work was widespread and long lasting. His books set high standards for subsequent texts in histology and embryology, and his students included Haeckel and Gegenbaur. During his tenure, Würzburg became an important center for medical education. He was the first to recognize the cellular nature of tissue and extended the cell theory into new areas; his extensive investigations of histogenesis and comparative tissue theory helped to establish histology as an independent branch of science. Cytology too became a subject for separate study after Koelliker pointed out the significance of the nucleus in the physiology of the cell and began to study cell structure in detail.

BIBLIOGRAPHY

I. ORIGINAL WORKS. Koelliker's most important scientific publications are *Beiträge zur Kenntnis der Geschlechtsverhältnisse und der Samenflüssigkeit wirbelloser Thiere . . .* (Berlin, 1841), his doctoral diss.; *Observationes de prima insectorum genesi . . .* (Zurich, 1842), his M.D. diss.; "Beiträge zur Entwicklungsgeschichte wirbelloser Thiere I. Ueber die ersten Vorgänge im befruchteten Ei," in Müller's *Archiv für Anatomie, Physiologie und wissenschaftliche Medicin* (1843), 66–141; *Entwicklungsgeschichte der Cephalopoden* (Zurich, 1844); *Die Selbständigkeit und Abhängigkeit des sympathischen Nervensystems . . .* (Zurich, 1844); "Die Lehre von der tierischen Zelle," in *Zeitschrift für wissenschaftliche Botanik* (1845), 46–102; "Ueber die Struktur und die Verbreitung der glatten oder unwillkürlichen Muskeln," in *Mitteilungen der Naturforschenden Gesellschaft in Zürich*, **1** (1847), 18–28; "Ueber den Faserverlauf im menschlichen Rückenmarke," in *Sitzungsberichte der Physikalisch-medizinischen Gesellschaft zu Würzburg*, **1** (1850), 189–207; *Mikroskopische Anatomie oder Gewebelehre des Menschen*, II, *Spezielle Gewebelehre*, 3 pts. (Leipzig, 1850–1854), vol. I never published; *Handbuch der Gewebelehre des Menschen* (Leipzig, 1852; 2nd ed., 1855; 3rd ed., 1859; 4th ed., 1863; 5th ed., 1867; 6th ed., 1889–1902), translated into French (Paris, 1856; 1872), English (London, 1853–1854; Philadelphia, 1854), and Italian (Milan, 1856); "Experimenteller Nachweis von der Existenz eines Dilatator pupillae," in *Zeitschrift für wissenschaftliche Zoologie*, **6** (1855), 143; "Physiologische Studien über die Samenflüssigkeit," *ibid.*, **7** (1856), 201–273; "Physiologische Untersuchungen über die Wirkung einiger Gifte," in *Virchows Archiv für pathologische Anatomie*, **11** (1856), 3–77; "Ueber die Leuchtorgane von Lampyris," in *Verhandlungen der Physikalisch-medizinischen Gesellschaft zu Würzburg*, **8** (1857), 217–224; "Vorläufiger Bericht über den Bau des Rückenmarkes der niederen Wirbelthiere," in *Zeitschrift für wissenschaftliche Zoologie*, **9** (1858), 1–12; *Entwicklungsgeschichte des Menschen und der höheren Thiere* (Leipzig, 1861; 2nd ed., 1879), also translated into French (Paris, 1882); *Grundriss der Entwicklungsgeschichte des Menschen und der höheren Thiere* (Leipzig, 1880; 2nd ed., 1884); "Ueber das Chiasma," in *Anatomischer Anzeiger*, **16**, supp. (1899), 30–31; "Weitere Beobachtungen über die Hofmannschen Kerne am Mark der Vögel," *ibid.*, **21** (1902), 81–84; and "Ueber die Entwicklung und Bedeutung des Glaskörpers," *ibid.*, **23**, supp. (1903), 49–51.

His autobiography, *Erinnerungen aus meinem Leben* (Leipzig, 1899), contains a bibliography and analysis of his works on pp. 188–396.

II. SECONDARY LITERATURE. See E. Ehlers, "Albert von Koelliker. Zum Gedächtnis," in *Zeitschrift für wissenschaftliche Zoologie*, **84** (1906), i–xxvi, with bibliography on pp. x–xxvi; and Wilhelm Waldeyer, "Albert von Koelliker zum Gedächtnis," in *Anatomischer Anzeiger*, **28** (1906), 539–552.

ERICH HINTZSCHE

KOELREUTER, JOSEPH GOTTLIEB (*b.* Sulz, Germany, 27 April 1733; *d.* Karlsruhe, Germany, 12 November 1806), *botany*.

Koelreuter was the son of an apothecary. At fifteen he went to the nearby University of Tübingen to read medicine and graduated in 1755. He spent the next six years as keeper of the natural history collections belonging to the Imperial Academy of Sciences in St. Petersburg. Although his chief duties concerned the classification of fish, he began his study of flower and pollen structure, pollination, and fertilization. When the Academy offered a prize for an essay on the experimental demonstration of the sexuality of plants, Koelreuter set out to produce plant hybrids.

In 1761 Koelreuter returned to Germany. In Leipzig and in Calw (Swabia), as the guest of Achatius Gaertner, he continued his hybridization experiments until his appointment as professor of natural history and director of the gardens in Karlsruhe, which belonged to the margrave of Baden, Karl Friedrich. Caroline, wife of the margrave, was an enthusiast for botany and protected Koelreuter from the jealousy of the gardeners, who resented the intrusion of so much experimental botany in the margrave's fine gardens. On her death in 1786 Koelreuter was dismissed. In 1775 he had married the daughter of a local judge; she bore him six daughters and one son. The son was given a good education and sent to St. Petersburg to read medicine, but the rest of the family lived in Karlsruhe in straitened circumstances. In his latter years Koelreuter complained of lack of recognition and financial support. He died embittered.

Koelreuter's strength lay in his brilliant experimentation, which was combined with great curiosity. His deep commitment to the concept of the harmony of nature and to the purposive character of all organic structures led him to inquire where others had merely described. His enthusiasm for the current interest in alchemical notions colored his interpretation of the facts of fertilization and led him to undertake his famous experiments in the "transmutation" of plant species.

Two theories of plant fertilization were current in the eighteenth century among those who accepted the concept of plant sexuality. Those who adhered to the doctrine of preformation and were spermists identified the germ of the new organism with the granules in the fluid which was expelled from pollen grains immersed in water. Ovists, on the other hand, denied a genetic role to the pollen. Those who denied preformation tended to think of the agents of fertilization as fluids. These male and female "seeds" had to mix in order to generate the offspring. Koelreuter thought this granular fluid was too crude to be the male "seed." Instead, he imagined that it was perfected to yield an oil which passed through the system of excretion canals in the wall of the pollen grain in order to reach the stigma. These canals were not really canals, and the oil he observed did not come from the interior of the grains.

The then current conception of fertilization as a mixing of liquids was further supported by Koelreuter's apparent demonstration that more than one pollen grain was required to fertilize one ovule. A certain minimum number of pollen grains had to be supplied to the stigma before any seeds were formed. Only in the case of *Mirabilis* did Koelreuter find that one grain sufficed to fertilize its uniovulate flowers; but because of the large size of these grains he rejected this first evidence that fertilization is a unitary and discrete process.

His study of the curious architecture of pollen grains led Koelreuter to inquire into their function. He soon perceived that pollen, stamens, and stigmas are designed to insure efficient pollination. He drew attention to the agency of wind and insects in this process and described the special sensitive stamens of *Berberis* and the sensitive stigmas of *Martynia.* He perceived the significant fact that many hermaphrodite flowers fail to be self-pollinated because the stamens and stigmas ripen at different times. These observations were extended by C. K. Sprengel and led to the overthrow of the old view that the role of insects in visiting flowers was to remove the harmful waxy and sugary secretions which would otherwise prevent seed formation. They also raised a new question: why are many hermaphrodite species adapted for cross-pollination?

R. Camerarius had described experiments in support of the sexuality of plants in 1691, but doubt on this subject continued long after Koelreuter's student days in Tübingen, when his teacher J. G. Gmelin reprinted Camerarius' account. Koelreuter perceived that if he could produce plant hybrids and show analogies between them and animal hybrids, he would achieve powerful support for the theory of plant sexuality. His first success was with the cross *Nicotiana rustica* × *N. paniculata.* Because his approach to the study of this tobacco hybrid was thoroughly scientific, the account he published in 1761 constitutes the first of its kind in the literature. Over the next five years he published further reports on such experiments, in which he discovered the uniformity and almost complete sterility of these plant hybrids, the identity of reciprocal crosses, and the contrast between these hybrids and their progeny. The latter were not uniform but tended to return to one of the parental species.

Because he believed in the preestablished harmony of nature, Koelreuter was delighted to find how infertile these plant hybrids were, for he saw this as a mechanism for preventing the confusion which unbridled crossing would yield. Unfortunate for this view was his subsequent discovery of fertile hybrids, especially in the genus *Dianthus.* To overcome this counterevidence he made a distinction between the natural world as it came from the hand of God and the artificial world of man's making. The latter was to be seen in zoological and botanical gardens, where species had been brought together which in nature had been deliberately separated to prevent their cross-breeding. He saw a corresponding distinction between "perfect" and "imperfect" hybrids. The former were all exactly intermediate between the two originating species and were almost completely sterile—mule plants. Imperfect hybrids, if fertile, were the product of crossing garden varieties; and if nonintermediate, they must have arisen from a fertilization between two species in which a tincture of pollen from the mother plant had also been active. Such products he termed "half-hybrids."

Koelreuter drew an analogy between the pollen and the sulfur of the alchemist, on the one hand, and between the female seed material and the mercury of the alchemist, on the other. By successive pollinations of a hybrid and its progeny he saw a biological means of effecting a transmutation. His first success in 1763, when he "transmuted" *Nicotiana rustica* into *N. paniculata,* spurred him on to further efforts. Only a lack of facilities prevented him from attempting the transmutation of the canary into the goldfinch. These experiments furnished the champions of bisexual heredity with strong evidence in their favor, evidence which told against the preformation theory. At the same time Koelreuter saw these experiments as a demonstration of the impossibility of producing new species by hybridization. Hybrids tended to revert to one of the stem species. By hybridization one species could be changed to another, but no new combination of existing species could reproduce its kind indefinitely. Where such new and permanent forms did arise, they were not to be given the status of new species.

Koelreuter may be said to have initiated the scientific study of plant hybridization. Mendel, Sprengel, and Darwin owed him a debt. If the world view which underlay his modern approach to nature belonged to that of the seventeenth century, his experimentation was not surpassed until the time of Mendel.

BIBLIOGRAPHY

I. ORIGINAL WORKS. Koelreuter's doctoral thesis was *Dissertatio inauguralis medica de Insectis Coleopteris, necnon de plantis quibusdam rarioribus* (Tübingen, 1755). His famous experiments on plant sexuality and hybridization are described in his *Vorläufige Nachricht von einigen das Geschlecht der Pflanzen betreffenden und Beobachtungen, nebst Fortsetzungen 1, 2 und 3* (Leipzig, 1761–1766), reprinted by W. Pfeffer in *Ostwald's Klassiker der exakten Wissenschaften*, no. 41. His work on the cryptogams is *Das entdeckte Geheimniss der Cryptogamie* (Carlsruhe, 1777).

II. SECONDARY LITERATURE. The best account of Koelreuter's life and work is by J. Behrens, "Joseph Gottlieb Koelreuter. Ein Karlsruhe Botaniker des achtzehnten Jahrhunderts," in the *Verhandlungen des Naturwissenschaftlichen Vereins in Karlsruhe*, **11** (1895), 268–320. For more recent accounts see R. C. Olby, "Joseph Koelreuter, 1733–1806," in Olby, ed., *Late Eighteenth Century European Scientists* (Oxford, 1966), pp. 33–65, and *Origins of Mendelism* (London, 1966), pp. 20–36.

See also H. F. Roberts, *Plant Hybridization Before Mendel* (1929; repr. New York, 1965).

ROBERT OLBY

KOENIG (KÖNIG), JOHANN SAMUEL (*b.* Büdingen, Germany, July 1712; *d.* Zuilenstein, near Amerongen, Netherlands, 21 August 1757), *mathematics, physics.*

Koenig was the son of the theologian, philologist, and mathematician Samuel Koenig (1671–1750), who after a very active existence spent his last twenty years as a professor of Oriental studies in his native city of Bern. Koenig received his first instruction in science from his father, whose enthusiasm he shared. After studying for a short time in Bern, in 1729 he attended the lectures of Frédéric de Treytorrens in Lausanne. In 1730 he left for Basel to study under Johann I Bernoulli and, beginning in 1733, under the latter's son Daniel as well—thus receiving the best mathematical training possible. During his stay of more than four years in Basel, Koenig, along with Clairaut and Maupertuis, studied the whole of mathematics, particularly Newton's *Principia mathematica.* Koenig was introduced to Leibniz' philosophical system by Jakob Hermann, who returned from St. Petersburg in 1731. He was so impressed by it that in 1735 he went to Marburg to further his knowledge of philosophy and law under the guidance of Leibniz' disciple Christian von Wolff.

Koenig's first mathematical publications appeared in 1735. In 1737 he returned to Bern to compete for the chair at Lausanne left vacant by the death of Treytorrens (the position went to Crousaz). Koenig then began to practice law in Bern and was so successful that he seriously intended to give up mathematics, which he had found something less than lucrative. First, however, he wanted to write on dynamics; two articles appeared in 1738. Before the start of the new year Koenig was in Paris, where in March 1739 Maupertuis introduced him to the marquise du Châtelet, Voltaire's learned friend. During the following months Koenig instructed the marquise du Châtelet in mathematics and Leibnizian philosophy. He also went to Charenton with Voltaire and the marquise to visit Réaumur, who inspired Koenig to write his paper on the structure of honeycombs. On the basis of this work Koenig was named a corresponding member of the Paris Academy of Sciences. Following the break with the marquise—the result, according to René Taton, of a disagreement about money—Koenig remained in Paris for a year and a half and then settled in Bern. By this time, after repeated unsuccessful attempts, he had given up hope of obtaining a chair in Lausanne. Besides conducting his legal practice, he studied the works of Clairaut and Maupertuis, whose influence is evident in his book on the shape of the earth (1747, 1761).

In 1744 Koenig was exiled from Bern for ten years for having signed a political petition that was considered too liberal, although it was in fact very courteously written. Through the intervention of Albrecht von Haller, Koenig finally obtained a suitable position as professor of philosophy and mathematics at the University of Franeker, in the Netherlands, and had considerable success there. Under the patronage of Prince William IV of Orange he moved to The Hague in 1749 as privy councillor and librarian. He became a member of the Prussian Academy on Maupertuis's nomination.

While still in Franeker, Koenig wrote the draft of his important essay on the principle of least action, which was directed against Maupertuis. The controversy touched off by this work, which was published in March 1751, resulted in perhaps the ugliest of all the famous scientific disputes.[1] Its principal figures were Koenig, Maupertuis, Euler, Frederick II, and Voltaire; and, as is well known, it left an unseemly stain on Euler's otherwise untarnished escutcheon. The quarrel occupied Koenig's last years almost completely; moreover, he had been ill for several years before it started. Koenig emerged the moral victor from this affair, in which all the great scientists of Europe—except Maupertuis and Euler—were on his side. The later finding of Kabitz[2] testifies to Koenig's irreproachable character.

Koenig never married. A candid and amiable man, he was distinguished by erudition of unusual breadth even for his time. He was a member of the Paris Academy of Sciences, the Royal Prussian Academy, the Royal Society, and the Royal British Society of Sciences in Göttingen. The opinion is occasionally voiced that were it not for the controversy over the principle of least action, Koenig would be completely forgotten in the history of science. His formulation of the law (named for him) of the kinetic energy of the motion of a mass point system relative to its center of gravity[3] is sufficient in itself to refute this view. According to Charles Hutton, Koenig "had the character of being one of the best mathematicians of the age." It is most regrettable that Koenig never accomplished his favorite project, publication of the correspondence between Leibniz and Johann Bernoulli.

NOTES

1. See *Dictionary of Scientific Biography*, IV, 471.
2. Willy Kabitz, "Ueber eine in Gotha aufgefundene Abschrift des von S. Koenig in seinem Streite mit Maupertuis und der Akademie veröffentlichten, seiner Zeit für unecht erklärten Leibnizbriefes," in *Sitzungsberichte der K. Preussischen Akademie der Wissenschaften zu Berlin*, **2** (1913), 632–638.
3. The law states that the kinetic energy of a system of mass points is equal to the sum of the kinetic energy of the motion of the system relative to the center of gravity and of the kinetic energy of the total mass of the system considered as a whole, which moves as the center of gravity of the system; therefore

$$\Sigma m_i v_i^2 = MV^2 + \Sigma m_i v_i'^2.$$

See A. Masotti, "Sul teorema di Koenig," in *Atti dell' Accademia pontificia dei Nuovi Lincei*, **85** (1932), 37–42. Koenig's original formulation of the law can be found in "De universali principio aequilibrii et motus . . . ."

BIBLIOGRAPHY

I. Original Works. Koenig's writings include *Animadversionem rhetoricarum specimen subitum quod cessante professoris rhetorices honore Academico d. 17. Nov. 1733 propon. Ant. Birrius respondente lectissimo juvene J. Sam. Koenigio, J. Sam. Bernate, philos. imprimisque mathesi sublimiori studioso* (Basel, 1733); "Epistola ad geometras," in *Nova acta eruditorum* (Aug. 1735), 369–373; "De nova quadam facili delineatu trajectoria, et de methodis, huc spectantibus, dissertatiuncula," *ibid.* (Sept. 1735), 400–411; "De centro inertiae atque gravitatis meditatiuncula prima," *ibid.* (Jan. 1738), 34–48; "Demonstratio brevis theorematis Cartesiani," *ibid.*, p. 33; "Lettre de Monsieur Koenig à Monsieur A. B., écrite de Paris à Berne le 29 novembre 1739 sur la construction des alvéoles des abeilles, avec quelques particularités littéraires," in *Journal helvétique* (Apr. 1740), 353–363; and *Figur der Erden bestimmt durch die Beobachtungen des Herrn von Maupertuis . . .* (Zurich, 1741; 2nd ed., 1761).

Subsequent works are *De optimis Wolfianae et Newtonianae philosophiae methodis earumque consensu* (Franeker, 1749; Zurich, 1752); the MS of the 2nd pt. of this history of philosophy must have been in existence at Koenig's death but appears to have been lost; "Mémoire sur la véritable raison du défaut de la règle de Cardan dans le cas irréducible des équations du troisième degré et de sa bonté dans les autres," in *Histoire de l'Académie Royale de Berlin* (1749), pp. 180–192, on which see M. Cantor, *Geschichte der Mathematik*, 2nd ed. (Leipzig, 1901), III, 599 ff.; "De universali principio aequilibrii et motus, in vi viva reperto, deque nexu inter vim vivam et actionem, utriusque minimo dissertatio," in *Nova acta eruditorum* (Mar. 1751), 125–135, 162–176; *Appel au publique du jugement de l'Académie royale de Berlin sur un fragment de lettre de Monsieur de Leibnitz cité par Monsieur Koenig* (Leiden, 1752); *Défense de l'Appel au publique* (Leiden, 1752); *Recueil d'écrits sur la question de la moindre action* (Leiden, 1752); *Maupertuisiana* (Hamburg, 1753), published anonymously (see *Mitteilungen der Naturforschenden Gesellschaft in Bern* [1850], 138); and *Élémens de géométrie contenant les six premiers livres d'Euclide mis dans un nouvel ordre et à la portée de la jeunesse sous les directions de M. le prof. Koenig et revus par M. A. Kuypers* (The Hague, 1758).

Miscellaneous mathematical works can be found in *Feriis Groningianis*. Correspondence with Haller was published by R. Wolf in *Mitteilungen der Naturforschenden Gesellschaft in Bern*, nos. 14, 20, 21, 23, 29, 34, and 44 (1843–1853). A portrait of Koenig by Robert Gardelle (1742) is in the possession of Dr. Emil Koenig, Reinach, Switzerland; it is reproduced in the works by E. Koenig and I. Szabó (see below). MSS and unpublished letters are scattered in the libraries of Basel, Bern, Franeker, The Hague, Leiden, Paris, and Zurich. Two unpublished MSS, "Demonstrationes novae nonnullarum propositionum principiorum philosophiae naturalis Isaaci Newtoni" and "De moribus gysatoriis," appear to have been lost.

II. Secondary Literature. See *Frieslands Hoogeschool und das Rijksathenaem zu Franeker*, II, 487–491; J. H. Graf, *Geschichte der Mathematik und der Naturwissenschaften in Bernischen Landen*, no. 3, pt. 1 (Bern–Basel, 1889), pp. 23–62; E. Koenig, *400 Jahre Bernburgerfamilie Koenig* (Bern, 1968), pp. 31–35, and *Gestalten und Geschichten der Bernburger Koenig* (Bern, 1972), pp. 6–8; O. Spiess, *Leonhard Euler* (Frauenfeld–Leipzig, 1929), pp. 126 ff.;

and R. Wolf, *Biographien zur Kulturgeschichte der Schweiz*, II (Zurich, 1858–1862), pp. 147–182.

On the principle of least action, see P. Brunet, *Étude historique sur le principe de la moindre action*, Actualités Scientifiques, no. 693 (Paris, 1938), with bibliography; *Leonhardi Euleri opera omnia*, J. O. Fleckenstein, ed., 2nd ser., V, intro. (Zurich, 1957), pp. vii–xlvi, including a bibliography by P. Brunet; and I. Szabó, "Prioritätsstreit um das Prinzip der kleinsten Aktion an der Berliner Akademie im XVIII. Jahrhundert," in *Humanismus und Technik*, **12**, no. 3 (Oct. 1968), 115–134.

E. A. FELLMANN

KOENIG, JULIUS (*b.* Györ, Hungary, 16 December 1849; *d.* Budapest, Hungary, 8 April 1914), *mathematics.*

Koenig studied at Vienna and Heidelberg, where he earned his Ph.D. in 1870. He qualified as a lecturer at Budapest in 1872 and became a full professor only two years later at the city's technical university. He remained in Hungary; and during his last years he was involved, as a senior civil servant in the Ministry of Education, with the improvement of training in mathematics and physics. He was also a secretary of the Royal Hungarian Academy of Sciences in Budapest.

Koenig's two years at Heidelberg (1868–1870) were of decisive importance for his scientific development. Helmholtz was still active there, and under his influence Koenig began working on the theory of the electrical stimulation of the nerves. But the mathematician Leo Königsberger, who was very well known at that time, soon persuaded Koenig to devote himself to mathematics; Koenig therefore wrote his dissertation on the theory of elliptic functions.

In Hungary, Koenig progressed very rapidly in his academic career. He was also productive in various fields of mathematics, chiefly analysis and algebra. Some of his works appeared simultaneously in German and Hungarian; others published only in Hungarian were naturally less influential. Among Koenig's writings is the prize essay for the Royal Hungarian Academy of Sciences, which was published in German in *Matematische Annalen* under the title "Theorie der partiellen Differentialgleichungen Ordnung mit 2. unabhängigen Veränderlichen." In it Koenig specified when the integration of a second-order differential equation can be reduced to the integration of a system of total differential equations, for which there already existed the integration methods devised by Jacobi and Clebsch.

Koenig's most important work is the voluminous *Einleitung in die allgemeine Theorie der algebraischen Grössen*, published in German and Hungarian in 1903. This book draws heavily on a fundamental study by Kronecker, *Grundzüge einer arithmetischen Theorie der algebraischen Grössen* (1892), although Koenig had had very little personal contact with Kronecker. In his work Kronecker had set forth the principles of the part of algebra later called the theory of polynomial ideals. Koenig developed Kronecker's results and presented many of his own results concerning discriminants of forms, elimination theory, and Diophantine problems. He also employed Kronecker's notation and added some of his own terms, but these did not gain general acceptance. The theory of polynomial ideals later proved to be a highly important topic in modern algebra and algebraic geometry. To be sure, many of Kronecker's and Koenig's contributions were simplified by later writers, notably Hilbert, Lasker, Macaulay, E. Noether, B. L. van der Waerden, and Gröbner; and their terminology was modified extensively. Hence, despite its great value, Koenig's book is now of only historical importance.

In the last eight years of his life Koenig took great interest in Cantor's set theory and the discussion that it provoked concerning the foundations of mathematics. The result of his investigations was the posthumous *Neue Grundlagen der Logik, Arithmetik und Mengenlehre* (1914) published by his son Dénes. The title originally planned was *Synthetische Logik;* and in it Koenig intended to reduce mathematics to a solidly established logic, hoping in this way to avoid the many difficulties generated by the antinomies of set theory. Dénes Koenig (*b.* 1884) also has become known in the literature of mathematics through his *Theorie der endlichen und unendlichen Graphen* (Leipzig, 1936).

BIBLIOGRAPHY

Koenig's writings include *Zur Theorie der Modulargleichungen der elliptischen Funktionen* (Heidelberg, 1870); "Theorie der partiellen Differentialgleichungen 2. Ordnung mit 2 unabhängigen Veränderlichen," in *Matematische Annalen*, **24** (1883), 465–536; *Einleitung in die allgemeine Theorie der algebraischen Grössen* (Leipzig, 1903); and *Neue Grundlagen der Logik*, *Arithmetik und Mengenlehre* (Leipzig, 1914), with a portrait of Koenig.

WERNER BURAU

KOENIG, KARL RUDOLPH (*b.* Königsberg, East Prussia [now Kaliningrad, R.S.F.S.R.], 1832; *d.* Paris, France, 1901), *acoustics.*

Koenig's father was on the faculty of the University of Königsberg. He took his Ph.D. in physics there and studied with Helmholtz, although at that time the latter was not primarily interested in acoustics. Upon completing his studies, Koenig moved to Paris in 1851 to become an apprentice to Vuillaume, one of the most famous violin makers of the time. Upon completing his apprenticeship in 1858 Koenig started his own business as a designer and maker of original acoustical apparatus of the highest quality. For the remainder of his life he produced equipment used for acoustical research throughout the world and renowned for the precision and skill of its workmanship. Every piece of equipment was tested by Koenig himself and usually employed in his own basic researches before it was sold. Much of the fundamental research in acoustics before the advent of modern electronic methods was done with Koenig's equipment; and even today Koenig organ pipes, tuning forks, and other apparatus are still used. Koenig never developed a large and lucrative business, preferring to produce instruments to be sold to scientists who he knew would appreciate their precision.

At the London International Exposition in 1862 Koenig displayed his equipment, including his new manometric-flame apparatus. For this he was awarded a gold medal, a recognition which first attracted wide public attention to his apparatus. Koenig went to Philadelphia in 1876 to exhibit a large collection of his acoustical apparatus at the Centennial Exposition. His exhibit was given the highest rating by the awards committee, and he received another gold medal. His hopes of developing business relations in the United States were not realized, however; and despite the efforts of Joseph Henry and other influential American scientists, Koenig's equipment was not sold. Finally, a part of the extensive collection was purchased by subscription and presented to the U.S. Military Academy at West Point. Another part of the Philadelphia exhibit was purchased by the University of Toronto for research work, and the remainder was ultimately returned to Paris—to Koenig's great disappointment and financial loss.

One of Koenig's most famous instruments was the clock tuning fork used to determine the absolute frequency of sound sources by direct reference to a standard clock. It employed a variable-frequency tuning fork of sixty-two to sixty-eight vibrations per second, which served the clock escapement much as the pendulum does in an ordinary clock; and by comparing the rate of the Koenig clock tuning fork with that of an unknown sound, the latter's frequency was determined. This instrument was of great accuracy and was employed in many important researches in acoustics. Koenig himself used his clock tuning fork to establish in 1859 the standard of pitch for music known as the "diapason normal." This was adopted in 1891 as the international pitch of $A = 435$ cycles per second. A pioneer in the graphic recording of sound, Koenig greatly improved the phonautograph invented by Leon Scott in 1857. The method was to focus sound through a horn onto a diaphragm attached to a stylus, thus producing a visual record of the sound wave on a revolving drum. This instrument and method were well known to Edison in his work on the phonograph, invented in 1877, although there is no specific reference to Koenig's pioneer invention.

Koenig developed special acoustical apparatus for the study of vowels, for the analysis of tone quality, for the synthesis of speech sounds, and for many other acoustical studies demanding high-precision measurement. His largest tuning forks were eight feet long with resonators twenty inches in diameter. One of Koenig's precision tuning forks was used by Albert A. Michelson in a stroboscopic comparison to determine precisely the speed of the revolving mirror used in his measurements of the velocity of light at Case School of Applied Science in 1882–1884.

In addition to producing precision instruments for other workers in acoustics, Koenig conducted important fundamental research. His achievements include studies of the physical characteristics of vowels, the nature of tone quality in sound, the effect of the phases of the several components of a complex sound on tone quality, the nature and characteristics of combination tones, and the frequency limits of audibility of sound. In 1882 he published a number of his researches in a book entitled *Quelques expériences d'acoustique*, which summarizes scientific work that had appeared previously in *Annalen der Physik* and in *Comptes rendus hebdomadaires des séances de l'Académie des sciences.*

Koenig was a lifelong bachelor and lived in the same apartment in which he built his equipment and carried on his researches, located on the Quai d'Anjou, facing the Seine on the Île St. Louis. This was one of the quietest places in Paris, where he could carry on his acoustical work under ideal conditions. The walls of his rooms were lined with tuning forks, resonators, and other apparatus. Koenig contributed a great deal to the development of the science of sound during the nineteenth century. Primarily an experimentalist and instrument maker, he was a man of great intellectual power with a deep physical understanding of the nature of sound and music. His attention to detail was phenomenal, and the quality of his finished apparatus was superb. Some of Koenig's finest equipment is now maintained in the Conservatoire des Arts et Métiers in Paris, a fitting memorial to his great contributions to the science of acoustics.

BIBLIOGRAPHY

Among Koenig's most important works are *Quelques expériences d'acoustique* (Paris, 1882); and an article in *Comptes rendus hebdomadaires des séances de l'Académie des Sciences*, **70** (1870), 931. There are also articles in the following issues of *Annalen der Physik:* **146** (1872), 161; **9** (1880), 394–417; **57** (1896), 339–388, 555–566; and **69** (1899), 626–660, 721–738.

ROBERT S. SHANKLAND

KOENIGS, GABRIEL (*b*. Toulouse, France, 17 January 1858; *d*. Paris, France, 29 October 1931), *differential geometry, kinematics, applied mechanics.*

After achieving a brilliant scholarly record, first at Toulouse and then in Paris at the École Normale Supérieure, which he entered in 1879, Koenigs passed the examination for the *agrégation* in 1882 and in the same year defended his doctoral thesis, "Les propriétés infinitésimales de l'espace réglé." After a year as *agrégé répétiteur* at the École Normale he was appointed a deputy lecturer in mechanics at the Faculty of Sciences of Besançon (1883–1885) and then of mathematical analysis at the University of Toulouse. In 1886 he was named lecturer in mathematics at the École Normale and deputy lecturer at the Sorbonne, which post he held until 1895. In addition he taught analytical mechanics on a substitute basis at the Collège de France.

Appointed assistant professor (1895) and professor (1897) of physical and experimental mechanics at the Sorbonne, Koenigs henceforth devoted himself to the elaboration of a method of teaching mechanics based on integrating theoretical studies and experimental research with industrial applications. He created a laboratory of theoretical physical and experimental mechanics designed especially for the experimental study of various types of heat engines and for perfecting different testing procedures. This laboratory, which began operations in new quarters in 1914, played a very important role during World War I. Koenigs won several prizes from the Académie des Sciences and was elected to that organization, in the mechanics section, on 18 March 1918.

A disciple of Darboux, Koenigs directed his first investigations toward questions in infinitesimal geometry, especially, following Plücker and F. Klein, toward the study of the different configurations formed by straight lines: rules surfaces and straight-line congruences and complexes. In analysis he was one of the first to take an interest in iteration theory, conceived locally; and in analytic mechanics he applied Poincaré's theory of integral invariants to various problems and advanced the study of tautochrones.

His *Leçons de cinématique* (1895–1897) enjoyed considerable success. They were characterized by numerous original features, including a definite effort to apply recent progress in various branches of geometry to kinematics. This work also contains a thorough investigation of articulated systems, an area in which Koenigs made several distinctive contributions. He demonstrated, in particular, that every algebraic surface can be described by an articulated system, and he produced various devices for use in investigating gyrations. His interest in the study of mechanisms is also reflected in his important memoir on certain types of associated curves, called conjugates.

Starting about 1910, however, Koenigs, working in his laboratory of physical and experimental physics, increasingly concentrated on research in applied thermodynamics and on the development of more precise test methods. Despite his successes in these areas it is perhaps regrettable that this disciple of Darboux thus abandoned his initial approach, the originality of which appeared potentially more fruitful.

BIBLIOGRAPHY

I. ORIGINAL WORKS. Koenigs' books are *Sur les propriétés infinitésimales de l'espace réglé* (Paris, 1882), his dissertation; *Leçons de l'agrégation classique de mathématiques* (Paris, 1892); *La géométrie réglée et ses applications. Coordonnées, systèmes linéaires, propriétés infinitésimales du premier ordre* (Paris, 1895); *Leçons de cinématique* . . . (Paris, 1895); *Leçons de cinématique . . . Cinématique théorique* (Paris, 1897), with notes by G. Darboux and E. Cosserat; *Introduction à une théorie nouvelle des mécanismes* (Paris, 1905); and *Mémoire sur les courbes conjuguées dans le mouvement relatif le plus général de deux corps solides* (Paris, 1910).

Koenigs published some 60 papers, most of which are listed in Poggendorff, IV, 778–779; V, 652–653; and VI, 1354; and in the Royal Society *Catalogue of Scientific Papers*, X, 429; and XVI, 376–377.

Koenigs analyzed the main points of his work in his *Notice sur les travaux scientifiques de Gabriel Koenigs* (Tours, 1897; new ed., Paris, 1910).

II. SECONDARY LITERATURE. Besides the bibliographies and his *Notice* (see above), Koenigs' life and work have been treated in only a few brief articles: A. Buhl, in *Enseignement mathématique*, **30** (1931), 286–287; L. de Launay, in *Comptes rendus hebdomadaires des séances de l'Académie des sciences*, **193** (1931), 755–756; M. d'Ocagne, in *Histoire abrégée des sciences mathématiques* (Paris, 1955), pp. 338–339; and P. Sergescu, in *Tableau du XXe siècle (1900–1933)*, II, *Les sciences* (Paris, 1933), pp. 67–68, 98, 117, 177.

RENÉ TATON

KOFOID, CHARLES ATWOOD (*b.* Granville, Illinois, 11 October 1865; *d.* Berkeley, California, 30 May 1947), *zoology.*

Kofoid's career reflects the changing nature of institutional support for science, and he clearly exemplifies the increasingly professional nature of American scientific endeavor. Born a Midwesterner, the son of Nelson Kofoid and the former Janet Blake, he completed his baccalaureate at Oberlin in 1890. He immediately took up graduate work at Harvard, completing his doctorate in 1894. He taught for one year at the University of Michigan and in 1895 was appointed director of the Biological Station at the University of Illinois. He investigated plankton and suspended life systems in the Illinois River and developed new techniques of biological survey to study these systems.

In 1901 he became a member of the zoology department of the University of California. He became chairman of the department in 1910 and retained that post, with a leave from 1919 to 1923, until his retirement in 1936. He trained about sixty doctoral students, edited the *University of California Publications in Zoology* for twenty-six years and bequeathed a valuable library to the university.

Anyone who has had the privilege of using the great resources of the Kofoid Collection in the Biology Library of the University of California at Berkeley is aware of the debt of the life sciences to Kofoid. His personal bookplate, which depicts the world of marine biology, protozoology, and parasitology, testifies to his wide interests and enthusiasms. At present the Kofoid Collection consists of about 31,000 volumes and 46,000 pamphlets; a third of these volumes can be classified as rare books. During his career he supported from his own purse a number of deserving graduate students.

Kofoid's research centered on the plankton and pelagic life of the Pacific ocean. He published many articles and monographs on dinoflagellates and tintinnids. New collection techniques and a more systematic attack upon the problems of marine biology resulted from his work. He accompanied Alexander Agassiz on one voyage (1904–1905) and retained a lifelong interest in the implications of a Darwinian approach to biology. After a visit to the Far East in 1915–1916, he became interested in rumen ciliates and general parasitology. Although his system of classifying ciliates is not universally accepted, distinguished workers in the field still follow it. In World War I he served as a major in the Sanitary Corps and worked on hookworm and general parasitology. Later he served as director of the California State Parasitological Division, and from research in this area he published a long series of papers on the parasitic protozoa in man.

Applied biology interested Kofoid, and he studied shipworms and termites in the San Francisco Bay area. In the interests of a more systematic approach to biological research, he played an instrumental role in founding what is now called the Scripps Institution of Oceanography at La Jolla, California. He helped establish *Biological Abstracts* and served as editor of its general biology section for many years. He served as an associate editor of *Isis* and other journals. He is credited with the development of the plankton net, a deep sea water sampler and a self-closing plankton net for horizontal towing.

Kofoid was a vigorous and dedicated scientist. The very range of his research interests may have blunted his ability to achieve outstanding prominence in any one field. He exemplifies the inherent difficulties of American biology at a time when it was maturing; and he gave it dignity and helped shape its direction.

BIBLIOGRAPHY

The bulk of Kofoid's published work is in article form, and the long titles preclude full citations here. He wrote extensively on marine protozoology up to about 1930. Thereafter parasitic protozoology tended to occupy his interests, with specific focus on amebiasis, trypanosomiasis, and rumen ciliates. He published in a wide variety of journals, but the majority of his important works after 1911 are in *University of California Publications in Zoology;* representative samples can be found in nearly every issue, many of his articles being co-authored with others in his department.

Kofoid edited *Termites and Termite Control* (Berkeley, 1934). See also *Marine Borers and Their Relation to Marine Construction on the Pacific Coast* (San Francisco, 1927), ch. 12 in G. N. Calkins and F. M. Summers, eds., *Protozoa in Scientific Research* (New York, 1941).

For an obituary by Harold Kirby, see *Science*, **106** (1947), 462–463.

PIERCE C. MULLEN

KÖHLER, AUGUST KARL JOHANN VALENTIN (*b.* Darmstadt, Germany, 4 March 1866; *d.* Jena, Germany, 12 March 1948), *microscopy.*

Köhler was the son of Julius Köhler, accountant to the grand duke of Hesse. After attending the Gymnasium and the Technische Hochschule in Darmstadt, he studied at the universities of Heidelberg and Giessen; when he took the state examination for teachers in 1888, he had studied zoology, botany, mineralogy, physics, and chemistry. He taught in Gymnasiums in Darmstadt and Bingen until 1891, when he was appointed assistant at the Institute of Comparative Anatomy of the University of Giessen.

Köhler's new design in 1893 for microscope illumination, which was to replace the existing condenser system, attracted the attention of the firm of Carl Zeiss in Jena; and six years later, Siegfried Czapski came to Bingen and invited Köhler to work in Jena for six months. In 1900 Köhler joined the Zeiss firm. He spent the rest of his life in Jena. In 1922 the university made him professor of microphotometry and projection, a post he held until 1945. He received honorary degrees from Edinburgh and Jena in 1934.

Köhler's boyhood passion had been for geology, but his contact with the life sciences drew him to zoology; his earliest papers were on freshwater and land mollusks and his doctoral thesis (1893) on *Siphonaria*. The paper which brought him fame was "Ein neues Beleuchtungsverfahren für mikrophotographische Zwecke" (1893). From the start, Köhler's aim had been to raise the standard and ease of microphotography. He began in 1893 by introducing the "collector" lens, which focused a magnified image of the light source in the plane of the condenser iris. By movement of the iris diaphragm, the size of the cone of illumination could be altered at will. This became known as the Köhler principle of illumination. Much later, at the Zeiss works, Köhler overcame the imperfections in microphotographs caused by the curvature of the image. The resulting fuzzy margins were avoided in Köhler's negative "Homal" system (1922).

As a Gymnasium teacher in Bingen, Köhler had wanted to improve the resolving power (R) of the light microscope. Abbe and Zeiss had raised R by increasing the numerical aperture (NA) and in 1886 had introduced their apochromatic lenses, which pushed the optical microscope to the apparent limit of its resolving power. But Köhler wanted to go yet further by reducing the wavelength (λ) of the light source. From Abbe's theory, the relation $R = 0.61\lambda/NA$ follows, and it can be seen that conversion from visual light of $\lambda = 5500Å$ to ultraviolet of $\lambda = 2750Å$ should yield a twofold increase in resolving power.

In the summer of 1900 Köhler, Moritz von Rohr, and Hans Boegehold began to work on the ultraviolet microscope. They began by attempting to make an objective suitable for light of short wavelengths. By 1902 they succeeded, but only for the green mercury line. A further two years passed before Köhler succeeded in designing a lens suited to the ultraviolet spectrum of cadmium. This objective, known as the "monochromator," was used in the ultraviolet microscope which he described in 1904. Although Köhler, von Rohr, and their colleagues had overcome numerous difficulties, from the design of the cadmium arc light to that of the fused quartz lenses, the instrument was still difficult to use. Direct focusing and viewing were, of course, impossible.

Although their microscope was shown to the medical profession in Vienna in 1905 and offered for trial, there was little enthusiasm. Köhler's fine pictures of the chromosomes in the epithelial cells of salamander gill buds, in which he noted the strongly ultraviolet absorbing character of the chromatin, were forgotten; and it was not until Tobjörn Caspersson made a thorough study of the absorption spectra of cell constituents some thirty years later that the ultraviolet microscope became popular. Since that time, the establishment of phase and fluorescent microscopy and closed circuit television conversion of the ultraviolet image have transformed the instrument which Köhler designed into a useful tool for the student of the living, unfixed cell.

BIBLIOGRAPHY

I. ORIGINAL WORKS. Seventy-two papers by Köhler are listed in Reinert's obituary notice (see below). The majority of his papers appeared in the *Zeitschrift für wissenschaftliche Mikroskopie und für mikroscopische Technik*, and of these the most important are "Ein neues Beleuchtungsverfahren für mikrophotographische Zwecke," **10** (1893), 433–440; "Beleuchtungsapparat für gleichmässige Beleuchtung mikroskopischer Objekte mit beliebigem einfarbigem Licht," **16** (1899), 1–28; "Mikrophotographische Untersuchungen mit ultraviolettem Licht," **21** (1904), 129–165, 273–304.

Reports on his later studies, found in *Naturwissenschaften*, are "Einige Neuerungen auf dem Gebiet der Mikrophotographie mit ultraviolettem Licht," **21** (1933), 165–172; and "Das Phasenkontrastverfahren und seine Anwendung in der Mikroskopie," **29** (1941), 49–61, written with W. Loos.

Köhler contributed three essays to *Handbuch der biologischen Arbeitsmethoden*, pts. 1–13 (Berlin–Vienna, 1921–1927): "Das Mikroskop und seine Anwendung," pt. 1, sec. 2 (1925), 171–352; "Die Verwendung des Polarisationsmikroskops für biologische Untersuchungen," pt. 2, sec. 2 (1928), 907–1108; "Mikrophotographie," pt. 2, sec. 2 (1931), 1691–1978. His suggestion for making ultraviolet microscopy quantitative will be found in "Mikroskopische Untersuchungen einiger Augenmedien mit ultraviolettem und mit polarisiertem Licht," in *Archiv für Augenheilkunde*, **99** (1928), 263–280, written with A. F. Togby.

II. SECONDARY LITERATURE. For personal details, a full bibliography, and photograph, see G. G. Reinert's obituary notice in *Mikroskopie*, **4** (1949), 65–70. On Köhler's scientific contributions the best accounts are by K. Michel, "August Köhler siebzig Jahre alt," in *Naturwissenschaften*, **24** (1936), 145–150; and by M. von Rohr, "Persönliche Erinnerungen an A. Köhler," in *Zeitschrift für Instrumentenkunde*, **56** (1936), 93–97. Köhler's work is discussed in F. Schomerus, *Geschichte der Jenaer Zeisswerkes 1846–1946* (Stuttgart, 1952), pp. 76–79. Further obituary notices are mentioned in Poggendorff, VIIa, pt. 2, 830–831.

On the way in which Köhler's ultraviolet microscope was developed, see T. Caspersson's essay, "Ueber den chemischen Aufbau der Strukturen des Zellkernes," in *Skandinavisches Archiv für Physiologie*, **73**, supp. 8 (1936), 1–151.

ROBERT OLBY

KOHLRAUSCH, FRIEDRICH WILHELM GEORG (*b.* Rinteln, Germany, 14 October 1840; *d.* Marburg, Germany, 17 January 1910), *chemistry, physics.*

Kohlrausch is best known for his experiments on the electrical conductivity of solutions. The son of Rudolph Kohlrausch, he was educated at the Polytechnikum at Kassel and at the universities of Marburg, Erlangen, and Göttingen, receiving his doctor's degree at Göttingen in 1863 under Wilhelm Weber. He then acted as assistant in the astronomical observatory at Göttingen and in the laboratory of the Physical Society at Frankfurt before being appointed extraordinary professor at the University of Göttingen (1866–1870).

Kohlrausch held the professorship of physics in the Polytechnikum at Zurich (1870–1871), at Darmstadt (1871–1875), and at the University of Würzburg (1875–1888), and then succeeded Kundt as director of the physical laboratory at Strasbourg. On the death of Helmholtz in 1894, he left Strasbourg to accept the appointment of director of the Physikalisch Technische Reichsanstalt at Charlottenburg. He was elected a member of the Academy of Sciences in Berlin in 1895 and was a member of scientific societies in many countries.

Kohlrausch's contributions to physical science were characterized by a high degree of precision. They included research on the electrical conductivity of electrolytes, on elasticity (begun in 1866), on magnetic measurements (begun in 1869), and on the determination of the electrochemical equivalent of silver in 1886 with his brother Wilhelm.

When Kohlrausch began his research in conductivity of solutions, the structure of a solution was controversial. The determination of whether or not Ohm's law applied to electrolytic solutions was confused by the question of polarization of the electrodes. When a direct current was forced through the electrolyte, ions gathered around the electrodes and partially neutralized the electric potential, decreasing the current; the effect produced inconsistent values for the conductivity of the solution being measured.

In 1868 Kohlrausch began to study the problem, developing the technique of using an alternating current rather than a direct current. In this way the decomposition which took place at the electrodes was reversed many times each second. The alteration of the solution was thus kept at a minimum while conductivity measurements were being made. At the same time the products of decomposition were not allowed to collect at the electrodes, and polarization was thus also reduced to a minimum. In a paper published in 1870 with W. A. Nippoldt, Kohlrausch showed that there was a maximum in the conductivity curve of sulfuric acid diluted with increasing amounts of water. He concluded that there was something fundamentally associated with the act of mixing itself that imparted conductivity to solutions. In a later paper, written with Otto Grotrian (1874), Kohlrausch showed that the conductivity of solutions increased with increasing temperature.

In 1876 Kohlrausch pointed out that, following the work of Hittorf on the migration of ions, the ions in very dilute solutions did not encounter appreciable resistance to their movement from other similar ions, and that the water in which the ions were dissolved provided the only friction serving to retard their motion. He concluded that "in a dilute solution every electrochemical element has a perfectly definite resistance pertaining to it, independent of the compound from which it is electrolyzed" (*The Fundamental Laws of Electrolytic Conduction*, p. 86). Thus the conductivity of electrochemically equivalent solutions of two electrolytes which have a component in common would vary inversely with the transference numbers of the common component. Kohlrausch was able to substantiate his conclusion by a comparison of the transference numbers measured by Hittorf with his own values for the conductivity of the same solutions.

The work produced by Kohlrausch on the conductivity of electrolytic solutions was important in leading to the eventual statement by Arrhenius postulating the electrolytic dissociation theory of solution structure.

Kohlrausch was also one of the first teachers to prepare an instructive work on physical laboratory methods, *Leitfaden der praktischen Physik* (1870). It was widely used and republished, being translated into four languages, including English.

BIBLIOGRAPHY

I. ORIGINAL WORKS. Kohlrausch's works have been collected and published under the title *Gesammelte*

Abhandlungen, 2 vols. (Leipzig, 1910–1911). He summarized his contributions and their place in the field in other books: *Das Leitvermögen der Elektrolyte, Methode, Resultate, und Anwendungen* (Leipzig, 1898) and *Die Energie oder Arbeit und die Anwendungen des elektrische Stromes* (Leipzig, 1900).

The first eight eds. of his laboratory manual were entitled *Leitfaden der praktischen Physik* (1st ed., Leipzig, 1870), while the 9th through the 16th eds. were published under the title *Lehrbuch der praktischen Physik* (9th ed., Leipzig, 1901). The memoir in which Kohlrausch stated his final conclusions with respect to conductivity and ions was "Ueber das Leitungsvermogen der in Wasser gelosten Electrolyte in Zusammenhang mit der Wanderung ihrer Bestandtheile," in *Göttingen Nachrichten* (1876), p. 213; it was republished in Harry Manly Goodwin, *The Fundamental Laws of Electrolytic Conduction* (New York–London, 1899), with memoirs of Faraday and Hittorf.

II. Secondary Literature. For discussions of Kohlrausch's work on conductivity of solutions, see Wilhelm Ostwald, *Elektrochemie, ihre Geschichte und Lehre* (Leipzig, 1896), or Harry C. Jones, *The Theory of Electrolytic Dissociation and Some of its Applications* (New York, 1900).

Ollin J. Drennan

KOHLRAUSCH, RUDOLPH HERRMANN ARNDT (*b.* Göttingen, Germany, 6 November 1809; *d.* Erlangen, Germany, 9 March 1858), *physics.*

Kohlrausch is best remembered for showing, with Wilhelm Weber, that the ratio of the absolute electrostatic unit of charge to the absolute electromagnetic unit of charge equals the speed of light.

Kohlrausch taught mathematics and physics, successively, at the Ritterakademie at Lüneburg, the Gymnasium at Rinteln, the Polytechnikum in Kassel, and the Gymnasium at Marburg. He became professor at the University of Marburg in 1853 and at the University of Erlangen in 1857, a year before his death. Kohlrausch was the father of the physicists Friedrich Wilhelm Kohlrausch and Wilhelm Friedrich Kohlrausch and the grandfather of the physiologist Arnt Ludwig Friedrich Kohlrausch.

Kohlrausch improved the operation of the Dellmann electrometer (1847–1848) and measured the electromotive force of various cells (1849–1853). He verified Ohm's law in electric circuits in 1848 when he showed that the electromotive force produced by a cell was proportional to the electroscopic tension of the same cell.

In 1856 Kohlrausch and Weber used the tangent galvanometer, developed by Weber, to determine experimentally the electromagnetic value of the discharge current when a Leyden jar is discharged through the galvanometer. They compared this value with the value, determined experimentally, of the electrostatic charge contained in the Leyden jar before discharge. Kohlrausch and Weber found that the ratio of the two measurements—electrostatic to electromagnetic—equalled 3.107×10^{10} cm. sec., a figure close to the accepted value for the velocity of light. This result was the continuation of Weber's measurements, with Gauss, of the absolute units of terrestrial magnetism.

The coincidence of the ratio and the speed of light led Kirchhoff to state in 1857 that an electric disturbance was propagated along a perfectly conducting wire at the velocity of light.

BIBLIOGRAPHY

Kohlrausch's works have not yet appeared in collected form. Original articles can be found in various journals, including Poggendorff's *Annalen der Physik und Chemie*, in which the major paper with Weber, "Ueber die Electricitätsmenge, welche bei galvanischen Strömen durch den Querschnitt der Kette fliest," was published (**99** [1856], 10–25). It was reprinted with Friedrich Kohlrausch's paper on conductivity of solutions, in *Ostwald's Klassiker der exakten Wissenschaften*, **142** (1904).

Ollin J. Drennan

KOLBE, ADOLF WILHELM HERMANN (*b.* Eliehausen near Göttingen, Germany, 27 September 1818; *d.* Leipzig, Germany, 25 November 1884), *chemistry.*

Hermann Kolbe was the oldest of fifteen children of a Lutheran pastor and was raised in the towns of Eliehausen and Stockheim, in the vicinity of Göttingen, where his father held pastorates. His mother, Auguste, was the daughter of A. F. Hempel, professor of anatomy at the University of Göttingen. Kolbe showed an early interest in science. When he entered the Göttingen Gymnasium, at the age of fourteen, he was introduced to chemistry by a fellow student who had studied this subject with Robert Bunsen, then a privatdocent at the university. Kolbe later said that this encounter led him to choose chemistry as his career. In 1838 he entered the University of Göttingen, where Wöhler had recently begun to teach chemistry. While he was a student he met Berzelius, who was visiting Wöhler, and was deeply impressed by him; Berzelius later took a great interest in Kolbe's first major research. It is not surprising that the young chemist accepted Berzelius' theories wholeheartedly and founded his later theoretical ideas upon them.

In 1842 Kolbe published his first short paper, on fusel oil, and began work on his doctoral dissertation. While this dissertation was in progress he was offered an assistantship with Bunsen, who had been called to Marburg. He accepted and completed his dissertation at Marburg. While there he perfected his knowledge of Bunsen's methods of gas analysis.

In 1845 Lyon Playfair, at the School of Mines in London, was studying firedamp in coal mines and needed a chemist qualified to perform gas analyses. He asked Bunsen to recommend someone. Bunsen proposed Kolbe, who went to London in the autumn of that year and remained until 1847. He met most of the English chemists, and became a close friend of Edward Frankland, who was beginning the studies that led him to the theory of valence. Together Kolbe and Frankland began a study of the conversion of nitriles to fatty acids. Kolbe himself investigated the action of the galvanic current on organic compounds; the results of these studies led him directly to the development of his chemical system. He returned to Marburg in the spring of 1847, accompanied by Frankland, and they continued their joint studies for a time.

In the autumn of the same year, Kolbe undertook a new activity. The publishing firm of Vieweg and Son had been bringing out a *Handwörterbuch der Chemie*, edited by Liebig, Wöhler, and Poggendorff. Kolbe was asked to continue this work and moved to Brunswick for the purpose. Temporarily abandoning most of his experimental work, he began the literary activity which he continued for the rest of his life. During this period he developed a number of theoretical ideas which he became anxious to test in the laboratory; when offered a professorship at Marburg, he gladly accepted.

Kolbe returned to the university in 1851. Since he had never served as a privatdocent and was only thirty-two years old, he was received with some jealousy by a few of the older professors, but his ability was soon recognized. With the aid of a number of talented students he established a solid reputation. In 1853 he married the youngest daughter of Major General von Bardeleben. During the next fourteen years he developed his theoretical ideas and wrote a comprehensive textbook of chemistry.

In 1865 he was called to Leipzig. Here he constructed the largest and best equipped chemical laboratory of its day, completed in 1868. It attracted so many students that in spite of its size, which had been criticized by Liebig, it was soon filled. Kolbe carried out most of his instruction in the laboratory rather than in lectures. In 1870 he took over the editorship of the *Journal für praktische Chemie*, which he used to express his very personal opinions of the state of chemistry. In his violent criticism of many of his contemporaries, Kolbe used terms that were outspoken even for his time, when polemical arguments were frequent and vigorous. The death of his wife in 1876 was a severe blow to him, and his health began to fail soon afterwards. He continued his writings until 1884, when he died at the age of sixty-six.

Kolbe was a brilliant experimenter, and his laboratory work in organic chemistry resulted in the discovery of many important compounds and reactions. He was also interested in the nature of chemical composition and developed his own system for representing the structure of the compounds with which he worked. Although this system involved a number of incorrect ideas, he stubbornly refused to abandon them until the evidence against them became overwhelming. Nevertheless, his chemical intuition was so keen that in spite of his conservatism he was able to make important predictions about the chemical behavior of many compounds. His own method of representing structure eventually gave way to the much simpler structural theory based on the work of Kekulé, but his unorthodox formulas actually embodied many of the ideas which Kekulé developed.

One of the major difficulties in Kolbe's formulas was that he refused to abandon equivalent weights for atomic weights until 1869, long after other chemists had adopted them. He still followed Berzelius in using the values $C = 6$ and $O = 8$, so that he had to double the number of atoms of these elements in his formulas —thus, his notation for the methyl group was C_2H_3, and hydrated carbonic acid became $2HO \cdot C_2O_4$.

Kolbe's early work was strongly influenced by the copula theory of Berzelius. The latter had been forced to abandon his original radical theory, expressed in terms of the dualistic electrochemical theory, when studies of substitution showed that positive hydrogen could be replaced by negative chlorine in organic compounds. Berzelius then assumed that in acetic acid the methyl radical was copulated with oxalic acid and water, so that his formula was written

$$C_2H_3 + C_2O_3 + HO.$$

The methyl group was a passive partner in which substitution by chlorine could produce the radical C_2Cl_3 without altering the properties of the compound greatly, since these depended chiefly on the active C_2O_3 radical. This was the theory that Kolbe adopted in his dissertation, and from which most of his later speculations were derived. His profound admiration for Berzelius made it impossible for him to abandon the concept of radicals, although he eventually modified his view. Precisely because he thought in terms of radicals and of their relative positions in the molecule, he was able to avoid the difficulties encountered by the adherents of the type theory, who were unable to conceive of a general reaction affecting specific parts within the molecule.

In his doctoral investigation Kolbe studied the action of moist chlorine on carbon disulfide. Among other products he obtained trichloromethylsulfonic acid (CCl_3SO_2OH), which he formulated as $HO + C_2Cl_3S_2O_5$. He at once saw the similarity to trichloroacetic acid, which he called trichlorocarbon oxalic acid, $HO + C_2Cl_3 \cdot C_2O_3$. Each of these compounds contained a group C_2Cl_3, which could be reduced to methyl, C_2H_3. Kolbe's attention was thus focused on organic acids and these became the basis for his later studies. In the course of this work he described the synthesis of acetic acid from its elements, the second time an organic compound had been so synthesized. (The first instance had been Wöhler's synthesis of urea.)

Kolbe was now convinced that methyl groups actually existed in his compounds and could be isolated. By the electrolysis of potassium acetate he obtained a gas which met the analytical criteria for methyl (although it was really ethane). Frankland had obtained "free ethyl" (butane) through the action of zinc on ethyl iodide; and Kolbe and Frankland were now sure that they had proved the existence of radicals in organic compounds. Their study of the conversion of nitriles to fatty acids seemed to prove that these acids must consist of the acidic group joined to the proper radical, since methyl cyanide gave methyl oxalic acid, ethyl cyanide gave ethyl oxalic acid, and so on, thus confirming the copula formulas. Kolbe had actually recognized what is known today as carboxyl, a single group joined to a hydrocarbon radical in all the fatty acids, and his copula formulas thus contained an essential truth that made many of his further speculations fruitful. The adherents of the type theory, with their formal attempts to squeeze all compounds into a few rigid types, missed this point completely, for which Kolbe criticized them, pointing out that any number of different types could be assumed to fit any number of special cases.

By 1857 Kolbe had worked out all the essentials of his system. Since the controlling group in his acids was "oxalic acid," all acids could be derived from hydrated carbonic acid, $2HO \cdot C_2O_4$, by replacing an OH (and an O to keep the equivalent balance) with another radical such as methyl. Thus he wrote acetic acid $HO(C_2H_3)\ C_2O_3$. Other organic compounds, however, notably aldehydes and alcohols, contain less oxygen, and the other oxygens in his formula could therefore also be replaced by methyl radicals. This consideration led him to accept as the fundamental radical "acetyl," which he wrote as the oxygen-free group $(C_2H_3)\ C_2$. In this new radical the point of attack by other elements was the double carbon atom attached to the methyl group. When an oxygen of acetic acid (that is, an HOO group) was replaced by hydrogen, the product was $\left.\begin{matrix}C_2H_3\\H\end{matrix}\right\} C_2O_2$, which must therefore represent the first substance produced in the reduction of an acid, an aldehyde. Here a new phenomenon could be observed. A replaceable hydrogen appeared, and if a methyl group replaced it, the product $\left.\begin{matrix}C_2H_3\\C_2H_3\end{matrix}\right\} C_2O_2$ was acetone. The relationship of aldehydes and ketones was thus explained, and a new group, carbonyl, was identified.

To go on a step, reduction of the aldehyde to an alcohol could lead only to the formula $HO \left\{\begin{matrix}C_2H_3\\H_2\end{matrix}\right\} C_2O$, and another important fact thus emerged. Either one or both of the hydrogens could be replaced, leading to a "singly or doubly methylated alcohol," and Kolbe could predict their behavior on oxidation. The discovery of secondary alcohols by Friedel in 1862 and of tertiary alcohols by Butlerov in 1864 fully confirmed Kolbe's predictions.

Although Kolbe's system involved many of the same ideas that were expressed by Kekulé in his famous paper of 1858, Kolbe was never able to see the similarity. He bitterly opposed the whole idea of structural formulas and kept some of his most scathing and sarcastic invective for the theories of Kekulé and their development by others. He ridiculed the theory of stereochemical isomerism of van t'Hoff and Le Bel. The pages of the *Journal für praktische Chemie* were filled with his diatribes.

In spite of his literary ferocity, however, he was a delightful companion, and his students thought highly of him and remembered the personal interest he took in them. He claimed that in attacking what he felt to be false theory he was defending the science he loved against "inexact scientific principles," rather than making personal attacks on any chemists. His criticisms, however, do not sound as if this had been the case.

In his later years Kolbe worked with the nitroparaffins and developed a method for large-scale synthesis of salicylic acid. He was impressed by the antiseptic and food-preserving power of this acid and founded an industry on its manufacture.

During the first part of his life, Kolbe's outstanding experimental work won him the respect and admiration of his colleagues. He was always highly regarded as a chemist, although his later refusal to accept new chemical theories and his bitter attacks on other chemists somewhat isolated him from the rest of his profession. Nevertheless, his criticisms of the type theory helped to weaken it and prepare the way for Kekulé. With the passage of time, however, it has become possible to see that Kolbe's system was basically sound; and the greater simplicity of Kekulé's structural system should not blind us to Kolbe's acute chemical insight.

BIBLIOGRAPHY

I. Original Works. The complete exposition of Kolbe's system was given by him in "Ueber den natürlichen Zusammenhang der organischen mit den unorganischen Verbindungen; die wissenschaftliche Grundlage zu einer natürgemässen Classification der organischen chemischen Körper," in *Annalen der Chemie*, **113** (1860), 293–332.

Kolbe's account of how he developed his theories, with an extensive bibliography and a full-scale attack on the structural theory of Kekulé and his successors, is given in a series of articles entitled "Meine Betheiligung an der Entwicklung der theoretischen Chemie," in *Journal für praktische Chemie*, n.s. **23** (1881), 305–323, 353–379, 497–517, and **24** (1881), 375–425. These were collected and published as *Zur Entwicklungsgeschichte der theoretischen Chemie* (Leipzig, 1881).

II. Secondary Literature. There is a rather sympathetic obituary of Kolbe in *Journal of the Chemical Society*, **47** (1885), 323–327, and a very cool and restrained obituary in *Berichte der deutschen chemischen Gesellschaft*, **17** (1884), 2809–2810, which probably reflects the resentment that Kolbe aroused. The most complete account of his life and work is given by his son-in-law, E. von Meyer, in "Zur Erinnerung an Hermann Kolbe," in *Journal für praktische Chemie*, n.s. **30** (1884), 417–466. There is also a useful biography by G. Lockeman, in G. Bugge, *Das Buch der grossen Chemiker*, **2** (Berlin, 1930), 124–135.

Henry M. Leicester

KOLOSOV, GURY VASILIEVICH (*b.* Ust, Novgorod guberniya, Russia, 25 August 1867; *d.* Leningrad, U.S.S.R., 7 November 1936), *theoretical physics, mechanics, mathematics.*

Kolosov graduated from the Gymnasium in St. Petersburg with a gold medal in 1885 and in that year joined the faculty of physics and mathematics of St. Petersburg (now Leningrad) University. He graduated from the university in 1889 and remained there to prepare for a teaching career.

In 1893 Kolosov passed his master's examination and was named director of the mechanics laboratory of the university and teacher of theoretical mechanics at the St. Petersburg Institute of Communications Engineers. From 1902 to 1913 he worked at Yurev (now Tartu) University, as privatdocent and then as professor. In 1913 he returned to St. Petersburg, where he became head of the department of theoretical mechanics at the Electrotechnical Institute; in 1916 he also became head of the department of theoretical mechanics at the university. Kolosov worked in these two institutions until the end of his life. In 1931 he was elected a corresponding member of the Academy of Sciences of the U.S.S.R.

Kolosov's scientific work was devoted largely to two important areas of theoretical mechanics: the mechanics of solid bodies, with which he began his career; and the theory of elasticity, on which he worked almost exclusively from 1908.

Kolosov's first important achievement in the mechanics of solid bodies was his discovery of a new "integrated" case of motion for a top on a smooth surface, related to the turning of a solid body about a fixed point. This result was published by Kolosov in 1898 in "Ob odnom sluchae dvizhenia tyazhelogo tverdogo tela, . . ." ("On One Case of the Motion of a Heavy Solid Body Supported by a Point on a Smooth Surface"). His basic results in the mechanics of solid bodies are discussed in his master's dissertation, "O nekotorykh vidoizmeneniakh nachala Gamiltona . . ." ("On Certain Modifications of Hamilton's Principle in its Application to the Solution of Problems of Mechanics of Solid Bodies" [1903]).

Kolosov's main results in the theory of elasticity are contained in his classic work *Ob odnom prilozhenii teorii funktsy kompleksnogo peremennogo* . . . ("On One Application of the Theory of Functions of Complex Variables to the Plane Problem of the Mathematical Theory of Elasticity," 1909). Kolosov's most important achievement was his establishment of formulas expressing the components of the tensor of stress and of the vector of displacement through two functions of a complex variable, analytical in the area occupied by the elastic medium. In 1916 Kolosov's method was applied to heat stress in the plane problem of the theory of elasticity by his student N. I. Muskhelishvili. Specialists in the theory of elasticity still use Kolosov's formulas.

Many of Kolosov's more than sixty works in mechanics and mathematics were published in major German, English, French, and Italian scientific journals.

BIBLIOGRAPHY

I. Original Works. Kolosov's most important works are "Ob odnom sluchae dvizhenia tyazhelogo tverdogo tela, opirayushchegosya ostriem na gladkuyu ploskost" ("On One Case of the Motion of a Heavy Solid Body Supported by a Point on a Smooth Surface"), in *Trudy*

Obshchestva lyubiteley estestvoznania, Otd. fiz. nauk, **9** (1898), 11–12; *O nekotorykh vidoizmeneniakh nachala Gamiltona v primenenii k resheniyu voprosov mekhaniki tverdogo tela* ("On Certain Modifications of Hamilton's Principle in Its Application to the Solution of Problems of Mechanics of Solid Bodies"; St. Petersburg, 1903); *Ob odnom prilozhenii teorii funktsy kompleksnogo peremennogo k ploskoy zadache matematicheskoy teorii uprugosti* ("On One Application of the Theory of Functions of Complex Variables to the Plane Problem of the Mathematical Theory of Elasticity"; Yurev [Tartu], 1909); and *Primenenie kompleksnoy peremennoy k teorii uprugosti* ("Application of the Complex Variable to the Theory of Elasticity"; Moscow–Leningrad, 1935).

II. SECONDARY LITERATURE. See N. I. Muskhelishvili, "Gury Vasilievich Kolosov," in *Uspekhi matematicheskikh nauk*, no. 4 (1938), 279–281; and G. Ryago, "Gury Vasilievich Kolosov," in *Uchenye zapiski Tartuskogo gosudarstvennogo universiteta*, no. 37 (1955), 96–103.

A. T. GRIGORIAN

KOLTZOFF, NIKOLAI KONSTANTINOVICH (*b.* Moscow, Russia, 15 July 1872; *d.* Leningrad, U.S.S.R., 2 December 1940), *zoology, cytology, genetics.*

Koltzoff's father, Konstantin Stepanovich Koltzoff, was an accountant for a large furrier; his mother, Varvara Ivanovna Bykhovskaya, came from an educated family of merchants. Koltzoff married his pupil and co-worker Maria Polievktovna Sadovnikova-Shorygina.

After graduating from the Gymnasium with a gold medal in 1890, Koltzoff entered the natural sciences section of the faculty of physics and mathematics of Moscow University, from which he graduated in 1894 with a first-class diploma and a gold medal. While a student Koltzoff worked in the department of comparative anatomy under the direction of M. A. Menzbir; his second teacher was the gifted embryologist and histologist V. N. Lvov. His close friends were A. N. Severtsov and P. P. Sushkin. Koltzoff began his scientific work in comparative anatomy, studying the origin and development of the paired limbs of vertebrates. His first published work was devoted to the development of the pelvis in frogs; and for his thesis, "Taz i zadnie konechnosti pozvonochnykh" ("Pelvis and Posterior Extremities of Vertebrates"), he received a gold medal.

Following his graduation from the university, Koltzoff remained to prepare for a teaching career. After three years of work and passing six master's examinations, he went abroad for two years, working in the laboratories of Flemming at Kiel and of Otto Bütschli in Heidelberg, and at the biological stations in Naples, Villefranche, and Roscoff. From this trip Koltzoff brought back material for his master's thesis, "Razvitie golovy minogi. K ucheniyu o metamerii golovy pozvonochnykh" (published as "Metamerie des Kopfes von Petromyzon planeri"), which he defended in the fall of 1901. In 1902–1904 he again worked abroad in the laboratories of Bütschli and O. Hertwig and in the biological stations at Naples and Villefranche. Koltzoff returned to Moscow with his doctoral dissertation, "Issledovania o spermiakh desyatinogikh rakov" ("Research on the Spermatozoa of the Decapoda"), the defense of which was set at Moscow University for January 1906. In connection, however, with the severe repression of the first Russian revolution by the czarist government, Koltzoff, who belonged to the radical and revolutionary-minded wing of the younger faculty of the university, refused to defend his dissertation. (He did not receive the doctorate until 1935.)

In 1903, Koltzoff began teaching at Moscow University and at the Women's University. In 1911 Koltzoff left the university with a large group of progressive professors and teachers, in protest against the reactionary politics of the czarist minister Kasso. The center of his scientific and teaching activity shifted to the Shanyavsky People's University, which was free from government control. Here Koltzoff organized an important biological laboratory. Students who later became outstanding scientists worked there, among them M. M. Zavadovsky, A. S. Serebrovsky, S. N. Skadovsky, G. V. Epstein, G. I. Roskin, and P. I. Zhivago. Koltzoff returned to Moscow University in 1918 and remained there until 1930, heading the department of experimental biology. Koltzoff's outstanding ability as a scientific administrator developed after the Revolution; in 1917 he was named head of the Institute of Experimental Biology, which he then directed for twenty-two years. The Institute was the first Russian biological research institute (not including the small zoological laboratories at the Academy of Sciences). It played a leading role in the development of new experimental areas in biological sciences—genetics, cytology, protozoology, hydrobiology, physicochemical biology, endocrinology, experimental embryology, and animal psychology. About 1,000 investigations were carried out under Koltzoff's direction, and the majority of the leading workers in new areas of experimental biology studied under him.

Koltzoff's career, which lasted more than forty-five years, may be divided into several periods that reflect his evolving scientific interests. Koltzoff quickly lost interest in comparative anatomy, which had become somewhat narrow and static, and even during his student years he had transferred his attention to experimental cytology. He was especially interested in the biology of the cell, particularly its formative structures. Koltzoff advanced the idea of the existence of the fibrillary elastic skeleton, which determines not only the anatomy of the cell but also, in a more general way, its entire organization. Starting from this theoretical principle (which has come to be called the Koltzoff principle), he moved to the study of nonmotile spermatozoa, such as that of the Decapoda, which had been very little studied. His research on this subject has become classic in the study of the structure, development, physicochemical properties, and physiology of these peculiar, highly specialized cells and their homology with motile spermatozoa. In physiology Koltzoff's observations on the method of penetration of the Decapoda spermatozoon into the egg cell were especially important. Experimentally verifying his proposed principle of skeletal structure in the cell, Koltzoff carried out a series of basic studies on the form of the cell. The first part was published in Russian in 1905 and contained the data on the decapod spermatozoa. The second part, which appeared in German in 1908, is devoted to the comparative study of the skeleton of the spermatazoon head in a number of animals.

On the basis of this research Koltzoff concluded that in the cell each contractile fiber must consist of a firm skeleton and a surrounding liquid protoplasm. To confirm this hypothesis Koltzoff devoted the third part of his major work, which appeared in German in 1911, to the statics and dynamics of the contractile stem of the sea infusorion *Zoothamnium*. On the basis of the data obtained Koltzoff expressed certain hypotheses about the mechanism of the contractile processes in such highly specialized elements as the muscle fibers. The fourth part of his research on the form of cells consisted of the study of the physicochemical properties, morphology, and functions of cells of the effector organs, particularly the pigment cells of the skin. This research was carried out during his last years and remained unfinished; his works on the morphology and nerve and hormone regulation of melanophores were not published until 1940, the year of his death.

The necessity for careful physicochemical analysis of the structures and processes that determine the form of the cell led Koltzoff to the second area of his research—physicochemical and colloidal biology. Having mastered the principles and methods of this new area of research, Koltzoff carried out several important works in it, resulting in such publications as "Über die physiologische Kationenreihe" (1912), "Über die Wirkung von H-Ionen auf die Phagocytose von Carchesium lachmani" (1914), and "Les principes physico-chimiques de l'irritabilité des cellules pigmentaires, musculaires et glandulaires" (1929). To some degree "Über die künstliche Partenogenese des Seidenspinners" (1932) also belongs to this area because of its use of chemical methods of stimulating egg cells. In another sense this research, because of its great general importance, must be related to the experimental and theoretical analysis of general biological problems.

In this third area Koltzoff was a pioneer in and apologist for genetics. His institute became the center for important work in general and applied genetics, such as the pioneer research of the group headed by S. S. Chetverikov on the genetic structure of the *Drosophila* population, research on artificial mutations, analysis of research on coloration of guinea pigs and the chemical properties of blood groups in man, research on genetics of farm animals and fish and on the genetic method in silk culture. Koltzoff's theoretical ideas—which proved to be prophetic—on the submicroscopic structure and template process of reproduction of the chromosome's macromolecular structure were especially important. This hypothesis, which undoubtedly had a strong influence on the development of theoretical genetics, was expressed as early as 1927. Koltzoff postulated the existence of "hereditary molecules," gigantic polymerous protein macromolecules which constitute an axial, genetically active structure of the chromosomes; the genes are amino acid radicals connected with these molecules. The replication of these gigantic molecules occurs according to the principles of self-reproduction—"omnis molecula ex molecula."

The sole essential difference between the views of Koltzoff and those of contemporary genetics is in the idea that genetic information is coded not by sequence of nucleotides of DNA but by the sequence of amino acids in the highly polymerous protein molecule; it must be noted, however, that at the end of the 1920's almost nothing was known of the significance of nucleic acids. Koltzoff gave his attention to one other cardinal question of genetics, the mechanism for realization of the influence of the genes on the characteristics depending on it. This question was examined in particular detail through consideration of oocytes of certain vertebrates in "Struktura khromosom i obmen veshchestv v nikh" ("Structure of Chromosomes and Exchange of Substances in Them" [1938]). In this work he introduced ideas of the exchange of chromosomal substances and of the chemical influence on the cytoplasm of the egg and of the formed organism. One of Koltzoff's basic theoretical ideas was that of the necessity of synthesis and mutual exchange in the new areas of experimental biology:

genetics, cytology, experimental embryology, and biochemistry. He spoke of this in many papers the titles of which emphasize the importance of the relations between these disciplines: "Fiziko-khimicheskie osnovy morfologii" ("Physicochemical Bases of Morphology" [1928]); *Ob eksperimentalnom poluchenii mutatsy* ("On the Experimental Obtaining of Mutations" [1930]), *Physiologie du développement et génétique* (1935), *Rol gena v fiziologii razvitia* ("The Role of the Gene in the Physiology of Development" [1935]), and *Les molecules héréditaires* (1939). Koltzoff tried to use all the achievements of experimental biology in medicine and in agriculture: his institute carried out a wide range of research in endocrinology, applied genetics, silk culture, and other fields.

An excellent teacher, Koltzoff introduced a course in experimental biology at Moscow University and taught it for thirty years, continually improving it. At Shanyavsky University he introduced a two-year major practicum during which the students carried out various independent research projects. Koltzoff was one of the founders and, for many years, an editor of the journals *Priroda* ("Nature"), *Zhurnal eksperimentalnoy biologii* ("Journal of Experimental Biology"), *Uspekhi sovremennoy biologii* ("Progress in Contemporary Biology"), and *Biologicheskii zhurnal* ("Biological Journal"). He organized several biological stations for his institute and aided the development of theoretical and applied biological research in various areas of the Soviet Union.

Koltzoff was a corresponding member of the Russian (Soviet) Academy of Sciences from 1916, president of the biological section of the Association of Natural Scientists and Physicians, an active member of the V. I. Lenin All-Union Academy of Agricultural Sciences (1935), an honorary member of the Leningrad Society of Amateurs of Natural Science, Anthropology and Ethnography (1928), the Moscow Society of Experimenters With Nature (1936), and the Royal Society of Edinburgh (1933), and an Honored Worker in Science of the R.S.F.S.R. (1934).

BIBLIOGRAPHY

I. ORIGINAL WORKS. Koltzoff's writings include "Metamerie des Kopfes von Petromyzon planeri," in *Anatomischer Anzeiger*, **16**, no. 20 (1899), 510–523; *Entwicklungsgeschichte des Kopfes von Petromyzon planeri. Ein Beitrag zur Lehre über Metamerie des Wirbelthierkopfes* (Moscow, 1902); *Issledovania o spermiakh desyatinogikh rakov v svyazi s obshchimi soobrazheniami otnositelno organizatsii kletki* ("Research on the Spermatozoa of Decapoda in Relation to Considerations of the Organization of the Cell"; Moscow, 1905); "Studien über die Gestalt der Zelle. I. Untersuchungen über die Spermien der Decapoden als Einleitung in das Problem der Zellengestalt," in *Archiv für mikroskopische Anatomie und Entwicklungsmechanik*, **67** (1906), 365–572; II. "Untersuchungen über das Kopfskelett des tierischen Spermiums," in *Archiv für Zellforschung*, **2** (1908), 1–65; III. "Untersuchungen über Kontraktilität des Stammes von Zoothamnium alternans," *ibid.*, **7** (1911), 244–423; "Über die physiologische Kationenreihe," in *Pflügers Archiv für die gesamte Physiologie*, **149** (1912), 327–363; "Über die Wirkung von H-Ionen auf die Phagocytose von Carchesium lachmani," in *Internationale Zeitschrift für physikalisch-chemische Biologie*, **1**, nos. 1–2 (1914), 82–107; "O nasledstvennikh khimicheskikh svoistvakh krovi" ("On Inherited Chemical Characteristics of the Blood"), in *Uspekhi eksperimentalnoy biologii*, **1**, nos. 3–4 (1922), 333–361; "Über erbliche chemische Bestandteile des Blutes," in *Zeitschrift für induktive Abstammungs- und Vererbungslehre*, supp. (1928), 931–935; "Fiziko-khimicheskie osnovy morfologii" ("Physicochemical Bases of Morphology"), in *Uspekhi eksperimentalnoy biologii*, ser. B, **7**, no. 1 (1928), 3–31; "Les principes physico-chimiques de l'irritabilité des cellules pigmentaires, musculaires et glandulaires," in *Revue générale des sciences pures et appliquées*, **40**, no. 6 (1929), 165–171; "Ob eksperimentalnom poluchenii mutatsy" ("On the Experimental Obtaining of Mutations"), in *Zhurnal eksperimentalnoy biologii*, **6**, no. 4 (1930), 237–268; "Über die künstliche Partenogenese des Seidenspinners," in *Biologisches Zentralblatt*, **52**, nos. 11–12 (1932), 626–642; "Rol gena v fiziologii razvitia" ("The Role of the Gene in the Physiology of Development"), in *Biologicheskii zhurnal*, **4**, no. 5 (1935), 753–774; *Physiologie du développement et génétique*, Actualités Scientifiques et Industrielles no. 254 (Paris, 1935); *Organizatsia kletki . . .* ("Organization of the Cell"; Moscow–Leningrad, 1936); "Issledovania po razdrozhimosti effektornykh khromatoforov" ("Research on the Divisibility of the Effector Chromatophores," in *Biologicheskii zhurnal*, **7**, nos. 5–6 (1938), 895–936; "Struktura khromosom i obmen veshchestv v nikh" ("Structure of Chromosomes and Exchange of Substances in Them"), *ibid.*, no. 1 (1938), 3–46; "O vozmozhnosti planomernogo sozdania novykh genotipov putem karioklasticheskikh vozdeystvy" ("On the Possibility of the Planned Creation of New Genotypes by Means of Karyoclastic Action"), *ibid.*, no. 3 (1938), 679–697; *Les molecules héréditaires*, Actualités Scientifiques et Industrielles no. 776 (Paris, 1939); "Amikroskopicheskaya morfologia melanofora" ("Amicroscopic Morphology of the Melanophore"), in *Doklady Akademii nauk SSSR*, **28**, no. 6 (1940), 554–558; "Gormonalnaya regulyatsia melanoforov" ("Hormonal Regulation in Melanophores"), *ibid.*, 548–553; "Nervnaya regulyatsia melanoforov" ("Nerve Regulation of Melanophores"), *ibid.*, no. 5 (1940), 463–469; and "Mikroskopicheskaya morfologia melanoforov" ("Microscopic Morphology of Melanophores"), *ibid.*, 458–462.

II. SECONDARY LITERATURE. See B. L. Astaurov, "Pamyati N. K. Koltsova" ("Recollections of N. K. Koltsov"), in *Priroda* (1941), no. 5, 108–117; and "Dve vekhi v razvitii geneticheskikh predstavleny" ("Two Landmarks in the Development of Genetic Ideas"), in *Byulleten Moskovskago obshchestva ispytatelei prirody*, biological sec., **70**, no. 4 (1965), 23–32; *N. K. Koltsov, Materialy k bio-bibliografii uchyenykh SSSR* ("Material for a Biobibliography of Scientists of the U.S.S.R."), biological science ser. (Moscow, in press), with intro. by B. L. Astaurov; V. Polynin, *Prorok v svoem otechestve* ("Prophet in His Country"; Moscow, 1969), a popular work; and S. Y. Zalkind, "Tsitologia" ("Cytology"), in *Sovetskaya nauka i tekhnika za 50 let. Razvitie biologii v SSSR* ("Soviet Science and Technology After 50 Years. Development of Biology in the U.S.S.R."; Moscow, 1967), pp. 408–426.

S. Y. ZALKIND

KOMENSKY, JAN AMOS. See **Comenius, Johannes.**

KONDAKOV, IVAN LAVRENTIEVICH (*b.* Vilyuisk, Yakutia, Russia [now Yakut A.S.S.R.], 8 October 1857; *d.* Elva, Estonia, 14 October 1931), *chemistry.*

Kondakov's scientific activity began at St. Petersburg University, from which he graduated in 1884. Continuing the traditional research of Butlerov, Kondakov thoroughly studied the transformation of trimethylethylene, establishing the possibility of a transition to a diene hydrocarbon, difficult to obtain at that time:

$$\underset{\displaystyle\overset{|}{C}}{C - C} = C - C \rightarrow C - \underset{\displaystyle\overset{|}{C}}{C} = C = C.$$

This was the first step toward the synthesis of isoprene, and it determined all of Kondakov's further creative work.

From 1886 to 1895 Kondakov worked at Warsaw University, systematically studying the syntheses of C_5 olefins and their transformations. From 1895 to 1918 he was professor at the University of Yurev (Tartu), where he continued his research, chiefly on the polymerization of unsaturated hydrocarbons. Kondakov's main contribution was his discovery of a series of very important regularities followed by the processes of polymerization; it aided the development of modern methods for the industrial synthesis of rubber.

In 1900 and 1901 Kondakov concluded that it was possible to synthesize rubber on the basis not only of isoprene but also of other diene hydrocarbons, including butadiene and diisopropenyl. He showed that the latter could be polymerized in three ways: by the catalytic action of alcoholic alkali, by raising the temperature, and by the action of light. In 1901, through the photopolymerization of diisopropenyl, Kondakov obtained rubber that was stable under the influence of hydrocarbon solvents. "We have here a product that undoubtedly must be accepted as the first known homologue of rubber . . . impervious to solvents and oils," K. O. Weber wrote (*Gummi Zeitung*, B. **17** [1902], p. 207). These methods of polymerization of diisopropenyl provided the basis of the industrial production of synthetic "methyl rubber," accomplished in 1915 in Germany.

Kondakov was one of the first to discover that metallic sodium can serve as a catalyst for the polymerization of dienes. He is to be credited with the development of methods for synthesizing spirits, ethers, acyl chlorides, and other difficult-to-obtain compounds that are bases of olefins by means of zinc chloride (1890–1895).

BIBLIOGRAPHY

Kondakov's writings include "K voprosu o polimerizatsii etilenovykh uglevodorodov" ("On the Question of the Polymerizations of Ethylene Hydrocarbons"), in *Zhurnal Russkago fiziko-khimicheskago obshchestva . . .*, **28** (1896), 784; "Ein bemerkenswerter Fall der Polimerisation des Dimethyl-2, 3-Butadien-1, 3," in *Journal für praktische Chemie*, **64** (1901), 109; and *Sintetichesky kauchuk, ego gomologi i analogi* ("Synthetic Rubber, Its Homologues and Analogues"; Yurev, 1912).

A secondary source is N. Y. Ryago, "Iz istorii khimicheskogo otdelenia Tartusskogo gosudarstvennogo universiteta" ("On the History of the Chemical Department of the Imperial University of Tartu"), *Trudy Instituta istorii estestvoznania i tekhniki . . .*, **12** (1956), 105–134.

V. I. KUZNETSOV

KÖNIG, ARTHUR (*b.* Krefeld, Germany, 13 September 1856; *d.* Berlin, Germany, 26 October 1901), *physics.*

König, one of Helmholtz's most prominent students was a leading representative of physiological optics. The son of a teacher, he attended the Realgymnasium in Krefeld; after graduating in 1874, he became a merchant. In 1878 he began scientific studies at the universities of Bonn, Heidelberg, and Berlin. He became an assistant to Helmholtz in 1882 at the physics institute of the University of Berlin, where he earned his doctorate in 1882 and qualified for lecturing in physics in 1884. In 1889 he became a full professor and head of the physics division of the physiological institute of the University of Berlin. König devoted himself entirely to physiological optics, especially to psychophysics and the physiology of the sense organs. In 1891, with H. Ebbinghaus, he founded his own journal covering these fields.

An excellent experimenter, König improved the Helmholtz leukoscope and constructed a spectrophotometer. He worked on the theory of colors and was a zealous defender of the Young-Helmholtz theory of color perception. He investigated the blending of colors, the brightness distribution of colors in the spectrum, and the significance of visual purple in sight. König also developed new data in his studies on visual acuity and color blindness. For example, he demonstrated that those who are totally color-blind have no visual perceptions in the center of the retina and hence are blind there. By using the Young-Helmholtz color theory, with its basic perceptions of red, green, and blue, König showed that in cases where one of these basic perceptions is lacking, the color confusions of red-blind and green-blind persons can be explained in terms of the normal trichromatic color system. With Conrad Dieterici, König investigated the structure of this abnormal, dichromatic color system (blue-yellow or red-green blindness).

Along with his works on physiological optics, König conducted psychophysical studies, particularly on Weber's law. He also considered other experimental and theoretical questions in physics. In the first years of his scientific activity he worked on galvanic polarization, developed a new method of determining the modulus of elasticity, and, with Franz Richarz, made a new determination of the gravitational constants. After the death of Helmholtz, König became coeditor of his manuscripts and supervised the second edition of his *Handbuch der physiologischen Optik*.

König was also very active in the editing of periodicals. Beginning in 1889 he was the sole editor of the *Verhandlungen der Deutschen physikalischen Gesellschaft* of Berlin, and from 1891 to 1901 he edited the *Älteren Beiträge*, later called *Beiträge zur Physiologie der Sinnesorgane*. With H. Ebbinghaus he edited *Zeitschrift für Psychologie und Physiologie der Sinnesorgane* (from 1830).

BIBLIOGRAPHY

I. ORIGINAL WORKS. Bibliographies can be found in Poggendorff, III, 735, and IV, 777; and A. Harnack, *Geschichte der Königlichen Preussischen Akademie der Wissenschaften zu Berlin*, III (Berlin, 1900), 154 (König's academic papers). König wrote about 40 scientific papers, including "Ueber die Beziehungen zwischen der galvanischen Polarisation und der Oberflächenspannung des Quecksilbers," in *Annalen der Physik und Chemie*, n.s. **16** (1882), 1–38, his doctoral dissertation; "Das Leukoskop und einige mit demselben gemachten Beobachtungen," *ibid.*, **17** (1882), 990–1008; "Zur Kenntniss dichromatischer Farbsysteme," *ibid.*, **22** (1884), 567–578; "Ueber die Empfindlichkeit des normalen Auges für Wellenlängenunterschiede des Lichts," *ibid.*, 579–589, written with C. Dieterici; "Eine neue Methode zur Bestimmung der Gravitationsconstante," *ibid.*, **24** (1885), 664–668, written with F. Richarz; "Modern Development of Thomas Young's Theory of Colour-Vision," in *Report of the British Association for the Advancement of Science* (1886); "Experimentelle Untersuchungen über die psychophysische Fundamentalformel in Bezug auf den Gesichtssinn," in *Sitzungsberichte der Preussischen Akademie der Wissenschaften zu Berlin* (1888), **2**, 917–931, and (1889), **2**, 641–644, written with E. Brodhun; "Die Grundempfindungen in normalen und anormalen Farbsystemen und ihre Intensitäts-Vertheilung im Spectrum," in *Zeitschrift für Psychologie und Physiologie der Sinnesorgane*, **4** (1893), 241–347, written with C. Dieterici (first results published in *Sitzungsberichte der Preussischen Akademie der Wissenschaften zu Berlin* [1886], **2**, 805–829); "Über den menschlichen Sehpurpur und seine Bedeutung für das Sehen," *ibid.* (1894), **2**, 577–598; and "Über 'Blaublindheit,'" *ibid.* (1897), **2**, 718–731.

II. SECONDARY LITERATURE. See H. Ebbinghaus, "Arthur König," in *Zeitschrift für Psychologie und Physiologie der Sinnesorgane*, **27** (1901), 145–147; W. Uhthoff, "Arthur König," in *Klinische Monatsblätter für Augenheilkunde*, **39** (1901), 950–953; and the unsigned "Arthur König," in *Leopoldina*, no. 37 (1901), 109–110.

HANS-GÜNTHER KÖRBER

KÖNIG, EMANUEL (*b.* Basel, Switzerland, 1 November 1658; *d.* Basel, 30 July 1731), *natural history, medicine.*

The son of a bookseller, König was educated in his native city. He pursued a comprehensive program of studies at the University of Basel, receiving his M.D. degree on 31 October 1682. In that year, through the efforts of his friend Georg Wolfgang Wedel, professor of medicine at Jena, König joined the German Academia Naturae Curiosorum (after 1687 the Academia Caesarea Leopoldina) and subsequently contributed a number of papers to its *Miscellanea.* After traveling for several years in Italy and France, he returned to Basel in 1695 to become professor of Greek. He remained in that city until his death, becoming professor of physics in 1703 and professor of medicine in 1711. It is as popularizer rather than innovator that König is important in the history of science. His writings are marked by clarity and draw upon a broad range of contemporary as well as classical scientific and medical literature, including the proceedings of major scientific societies.

König's most important published works are three excellent texts: *Regnum animale* (1682), *Regnum minerale* (1686), and *Regnum vegetabile* (1688). In the first treatise, which was praised in the *Acta eruditorum* as the best of its kind that had yet appeared, he presents a detailed analysis of animal internal structure and physiology. König emphasizes the causal role of animal spirits in physiological activity, frequently citing John Mayow's theory and experiments concerning nitroaerial particles (or spirits) to explain respiration, muscle action, and disease. Adopting Descartes's concept of animals as automatons, he divides them into five classes—quadrupeds, flying animals, swimming animals, serpents, and insects—with the proviso that this classification be understood as approximate rather than precise. Although critical of astrology and the "cures" which infested the medical literature of the seventeenth century, König was nonetheless intrigued with the use of animal parts and products as medicines; and he occasionally credited somewhat extravagant claims for the efficacy of skulls, elephants' tusks, and the like.

König's writings on the vegetable and mineral realms parallel his treatment of animals: they are well-reasoned books which critically utilize the results of considerable reading and research. The *Regnum minerale* contains much chemical information and employs rational and convenient symbols for which König supplies a clear explanatory plate. His analysis of metals, gems, salts, sulfurs, and earths is accurate, although he shows credulity with respect to the magic virtue of gems and relies upon the common, albeit erroneous, analogy made between the generation, nutrition, and augmentation of metals and that of animals. His work on plants is reliable and repudiates Van Helmont and Boyle's opinion that plants are nourished by water alone, by demonstrating that saline and nitrous ingredients play a role in vegetative growth. König was attracted to the new corpuscular philosophy and asserted that the reputed occult virtues of plants could be explained mechanically; his treatment of the doctrine of signatures is an example of the attempt to apply corpuscular theory.

An astute observer of contemporary developments, König was able to modify traditional ideas. His writings, while not completely uncontaminated by superstition, were successful in disseminating major ideas in natural history and medicine during the last decades of the seventeenth and the early years of the eighteenth centuries.

BIBLIOGRAPHY

I. Original Works. A complete list of König's writings is in Heinrich Rotermund, *Fortsetzung und Ergänzungen zu Christian G. Jöchers Allgemeinem Gelehrten-Lexicon* (Delmenhorst, 1810; repr. Hildesheim, 1961), III, 641–643. König's major scientific writings are *Regnum animale . . . physice, medice, anatomice, mechanice, theoretice, practice . . . enumeratum et emedullatum, hominis scilicet et brutorum, machinam hydraulico-pneumaticam comparate* (Basel, 1682; 3rd ed., 1703); *Regnum minerale . . . metallorum, lapidum, salium, sulphurum, terrarum . . . praeparationes selectissimas ususque multiplices candide sistens* (Basel, 1686); *Regnum minerale generale et speciale, quorum illud naturalem et artificialem mineralium productionem cum parallelismo alchymico verorum philosophorum* (Basel, 1703); and several different works with the general title *Regnum vegetabile*, the most important being *Regnum vegetabile . . . vegetabilium nimirum naturam, ortum, propagandi modum,. . . colorem, figuram, signaturam* (Basel, 1688).

II. Secondary Literature. Details concerning König's life can be found in Christian G. Jöcher, *Allgemeinem Gelehrten-Lexicon* (Leipzig, 1750; repr. Hildesheim, 1961), II, 2136; and F. Hoefer, ed., *Nouvelle biographie générale*, XXVIII (1859), 7. Recent assessments of König's scientific work are J. R. Partington, *A History of Chemistry*, II (London, 1961), 318, 616, 713–714; and Lynn Thorndike, *A History of Magic and Experimental Science* (New York, 1958), VII, 266–267, 690, 693; VIII, 43–47, 79, 426.

Martin Fichman

KÖNIGSBERGER, LEO (*b.* Posen, Germany [now Poznań, Poland], 15 October 1837; *d.* Heidelberg, Germany, 15 December 1921), *mathematics.*

The son of a wealthy merchant, Königsberger began to study mathematics and physics at the University of Berlin in 1857. After graduating in 1860, he taught mathematics and physics to the Berlin cadet corps from 1861 to 1864. In the latter year his academic career commenced at the University of Greifswald, as an associate professor; in 1869 he became a full professor at Heidelberg. After teaching at the Technische Hochschule in Dresden (1875–1877) and at the University of Vienna (1877–1884), he returned in 1884 to Heidelberg, where he remained until his death. He retired in 1914.

Königsberger was one of the most famous mathematicians of his time, member of many academies, and universally respected. He contributed to several fields of mathematics, most notably to analysis and analytical mechanics.

Königsberger's mathematical work was early influenced by his teacher Weierstrass. In 1917 he published a historically important account of Weierstrass' first lecture on elliptic functions, which he had heard in 1857, during his first semester at Berlin. Königsberger also was extremely skillful in treating material from the Riemannian point of view, as can be seen from his textbooks on elliptic functions (1874) and hyperelliptic integrals (1878). In addition he worked intensively on the theory of differential equations. This subject, which grew out of function theory, is associated especially with Lazarus Fuchs, with whom Königsberger was friendly during his youth. Königsberger was the first to treat not merely one differential equation, but an entire system of such equations in complex variables.

In Heidelberg, Königsberger maintained close friendships with the chemist Bunsen and the physicists Kirchhoff and Helmholtz. These contacts undoubtedly provided the stimulation both for his series of works on the differential equations of analytical mechanics and his biography of Helmholtz (1902). The latter and the biographical *Festschrift* for C. G. J. Jacobi (1904) have proved to be his best-known works, despite his many other publications.

BIBLIOGRAPHY

Königsberger's writings include *Vorlesungen über elliptische Funktionen* (Leipzig, 1874); *Vorlesungen über die Theorie der hyperelliptischen Integrale* (Leipzig, 1878); *Lehrbuch der Theorie der Differentialgleichungen mit einer unabhängigen Veränderlichen* (Leipzig, 1889); *H. v. Helmholtz*, 2 vols. (Brunswick, 1902); *C. G. J. Jacobi, Festschrift zur 100. Wiederkehr seines Geburtstages* (Leipzig, 1904); "Weierstrass' erste Vorlesung aus der Theorie der elliptischen Funktionen," in *Jahresberichte der Deutschen Mathematikervereinigung*, **25** (1917), 393–424; and *Mein Leben* (Heidelberg, 1919).

WERNER BURAU

KONINCK, LAURENT-GUILLAUME DE (*b.* Louvain, Belgium, 3 May 1809; *d.* Liège, Belgium, 15 July 1887), *chemistry, paleontology*.

De Koninck studied at the University of Louvain, from which he graduated at the age of twenty-two with a doctorate in medicine, pharmacy, and natural sciences. He practiced medicine for only a short time; he was named a *préparateur* at the University of Louvain in 1831. In 1834 and 1835 he frequented the laboratories of several great chemists of this period and visited Germany's most famous professors. Upon his return to Belgium, he was placed in charge of a course in industrial chemistry at the University of Ghent (1835), and was then transferred at his own request to the University of Liège (1836), where he taught various branches of chemistry until his retirement in 1876. Although his principal scientific activity was in paleontology, he was never authorized to give more than an optional course in this field.

De Koninck's reputation was considerable during his lifetime. He was named a member of the Belgian Royal Academy and of many foreign academies and scientific societies as well. In 1875 he received the Wollaston Medal.

Although he tackled very diverse subjects in paleontology, his chief work was concerned with the fauna of the Carboniferous limestone. The limestone of Visé, the type section of the Visean stage, is located a few kilometers from Liège. The rich fossil content of this formation was already known when De Koninck came to settle in that city. There he made important collections, which he completed by means of fossils from the limestone of Tournai, the type section of the Tournaisian stage. These collections (preserved in large part at the Museum of Comparative Zoology of Harvard University) revealed to him the importance of a subject that F. McCoy at the same period, and John Phillips before him, had approached in Great Britain. In his first major work on this question (1842–1844), he described and illustrated 434 species, of which he considered 208 to be new. In this work he demonstrated the complexity, not then appreciated, of the Carboniferous fauna, attempted to establish the relative age of the sedimentary deposits by means of the fossils, and compared the Belgian Carboniferous fauna with that of other regions.

De Koninck concentrated his efforts around the systematic inventory of the fossil fauna, their chronological signification, and their geographical extension. Besides writing monographs on particular genera and groups, De Koninck revealed the existence of the Devonian system in China (1846), made substantial contributions to the knowledge of the Paleozoic fossils of Spitsbergen (1846, 1849), India (1863), and New South Wales (1877), as well as of various countries in Europe. From 1878 until his death he worked on a monumental study of the *Faune du calcaire carbonifère de la Belgique*, treating successively the fishes and the genus *Nautilus* (1878), the Cephalopoda (1880), the Gastropoda (1881, 1883), the Lamellibranchia (1885, in collaboration with J. Fraipont), and finally the Brachiopoda (1887)—a total of 1,302 species described and illustrated, of which he judged 891 to be new.

He changed his views on the relative chronology of the Carboniferous limestone. In his first monograph he thought he was able to explain the differences of the fauna of the Tournai and Visé limestones by supposing that they had belonged to different basins. Later, he considered the Visean deposits to be slightly older than the Tournaisian ones. Then, having recognized his error and having placed these formations back in their natural order, he accepted the existence of a third division of the Carboniferous limestone, the Waulsortian, intermediate between the Tournaisian and the Visean. (The Waulsortian is actually not a stage but a facies of the Belgian Carboniferous limestone.)

Convinced of the fixity of species, De Koninck remained faithful to the school of Cuvier and d'Orbigny until the end of his life. Refusing to admit that a species might cross the boundary between stages, he was led to exaggerate the number of species. His work was essentially analytical but nevertheless is valuable for the precision of the descriptions and for the number of fossil forms that it helped to make known.

BIBLIOGRAPHY

I. ORIGINAL WORKS. De Koninck published twenty-one works on chemistry, and seventy articles, memoirs, and reports on paleontology and geology. The two main works are *Description des animaux fossiles qui se trouvent dans le terrain carbonifère de Belgique*, 2 vols. (Liège, 1842–1844); and *Faune du calcaire carbonifère de la Belgique*, in *Annales du Musée royale d'histoire naturelle de Belgique*, 6 pts. (1878–1887).

II. SECONDARY LITERATURE. Several obituary notices were published on De Koninck with a complete listing of his publications. See, in particular, J. Fraipont, "Laurent-Guillaume De Koninck, sa vie et ses oeuvres," in *Annales de la Société géologique de Belgique* (*Bulletin*), **14** (1889), 189–255; and E. Dupont, "Notice sur Laurent-Guillaume De Koninck," in *Annuaire de l'Académie royale de Belgique*, **57** (1891), 437–483.

G. UBAGHS

KONKOLY THEGE, MIKLÓS VON (*b*. Budapest, Hungary, 20 January 1842; *d*. Budapest, 17 February 1916), *astronomy, geophysics.*

Konkoly Thege was the son of Elek Konkoly Thege and Klára Földváry. He first studied law but at the same time eagerly attended lectures in science at Budapest. After a short time in the civil service he moved to Berlin, where he received the Ph.D. in astronomy in 1862. He then returned to Hungary and earned a captain's certificate on the Danube steamship line.

In 1869 Konkoly Thege established a small astronomical observatory at his country estate at Ógyalla. In 1874 he expanded it to include a mechanical workshop. His largest instrument was a ten-inch refractor.

Konkoly Thege was chiefly interested in the new methods of celestial photography and astrophysics—especially spectroscopy. At this stage in these disciplines, each observer had to construct his own instruments and prepare the photographic materials. Konkoly Thege's works contain valuable notes and comments on the conditions and procedures of the time.

Konkoly Thege not only corresponded with the leading scientists, but his personal fortune enabled him to travel widely and visit most of the European observatories and instrument workshops. He in turn entertained many famous astronomers at Ógyalla. Among the young astronomers who worked at Ógyalla was Kobold.

In 1898, Konkoly Thege presented his observatory to the Hungarian government, together with the funds necessary to ensure its continuation. The observatory remained at Ógyalla until 1919, when the instruments became the basis for the new Budapest observatory.

From 1890 Konkoly Thege directed the Hungarian Meteorological Service, a task he undertook with his characteristic energy. After 1891, forecasts were sent out by telegraph. In September 1900 a meteorological-magnetic observatory was in operation, and by 1910 a central station was established in Budapest. He retired in 1911.

Konkoly Thege married Erzsébet Madarassy, who died in 1919. Their two children died in early childhood.

BIBLIOGRAPHY

See *Praktische Anleitung zur Anstellung astronomischer Beobachtungen mit besonderer Berücksichtigung der Astrophysik nebst moderner Instrumentenkunde* (Brunswick, 1883); *Beobachtungen am Astrophysikalischen Observatorium in Ógyalla in Ungarn*, 16 vols. (Halle, 1879–1893); *Praktische Anleitung zur Himmelsphotographie* (Halle, 1887); *Handbuch für Spectroscopiker im Cabinet und am Fernrohr* (Halle, 1890).

Many short notes appear in *Astronomische Nachrichten*, **80–190** (1873–1913), concerning spectroscopic observations of meteorites and comets; there are also many Hungarian papers, especially *160 állócsillag szinképs* (Budapest, 1877).

H. C. FREIESLEBEN

KONOVALOV, DMITRY PETROVICH (*b*. Ivanovka, Ekaterinoslav guberniya [now Dnepropetrovsk oblast], Russia, 22 March 1856; *d*. Leningrad, U.S.S.R., 6 January 1929), *chemistry.*

Konovalov graduated from the Institute of Mines at St. Petersburg in 1878; from 1886 he was a professor at St. Petersburg University. From 1890 he studied under Mendeleev and his successor in the department of inorganic chemistry at St. Petersburg. Konovalov was the director of the St. Petersburg Institute of Mines from 1904 and from 1907 director of the government's Department of Mines. He was deputy minister of trade and industry from 1908 until 1915. On 13 January 1923 he was elected a member of the Soviet Academy of Sciences. From 1922 until 1929 Konovalov was president of the Bureau of Weights and Measures in Leningrad and a member of the International Bureau of Weights and Measures.

Konovalov's basic works are in the theory of solutions, kinetics, and catalysis. Developing Mendeleev's idea of the interaction between the solute and the solvent, he studied the vapor pressure of solutions of liquids in liquids. Prior to Konovalov's work, science had only fragmentary information concerning the vapor pressure of liquid systems, provided by Regnault and Roscoe. Konovalov defined the distillation conditions of a mixture of liquids in relationship to the shape of general vapor pressure curves for mixtures. In 1884 he established that, compared with the solution, the vapor contains an abundance of that component which, when added to the solution, increases the general vapor pressure of the latter. At the points corresponding to the maximum and minimum of the curve expressing vapor pressure as a function of the percentage composition of a liquid, the vapor has precisely the same composition as the liquid. These laws, which entered the chemical literature as Konovalov's laws, were confirmed in the later work of Duhem, Margules, Planck, and van der Waals.

Konovalov's book *Ob uprugosti para rastvorov* ("On the Vapor Pressure of Solutions" [1884]) stated the scientific bases for the theory of the distillation of solutions, which made possible the industrial processes associated with the distillation of solutions.

In 1890 Konovalov gave a general thermodynamic definition of osmotic pressure, according to which osmotic equilibrium is "equality of the vapor pressure on both sides of a membrane." This definition provides the basis for calculating the value of osmotic pressure in modern thermodynamics. Konovalov introduced the method of electroconductivity in the study of the interaction of the components of two-liquid systems (1890–1898). He discovered a special class of electrolytes, the solvoelectrolytes, which include aniline and acetic acid.

Konovalov initiated work on the physicochemical theory of catalysis. He introduced (1885) the concept of active surface area, which played an important role in the development of the theory of heterogeneous catalysis. A study of the formation and decomposition of complex esters in their liquid phase led Konovalov to a conclusion expressed in the formula for the rate of autocatalytic reactions,

$$dx/dt = K(1 - x)(x + x_0),$$

where dx/dt represents the relative quantity of x ester decomposed in time t and x_0 is the initial concentration of acetic acid. An analogous formula was deduced by Ostwald (1888) for the saponification of methylacetate. The Ostwald-Konovalov formula, which expressses the fundamental law of autocatalysis, has become firmly fixed in the literature of chemical kinetics.

In 1923 Konovalov deduced a formula for calculating the heat of combustion of organic substances.

From 1890 to 1904 Konovalov was chairman of the chemistry division of the Russian Technical Society. He participated in the organization of the chemical section of the Russian pavilion at the Columbian Exposition at Chicago (1893). Konovalov's *Promyshlennost Soedinenykh Shtatov Severnoy Ameriki i sovremennye priemy khimicheskoy tekhnologii* ("Industry in the United States of North America and Modern Methods of Chemical Technology"; St. Petersburg, 1894) resulted from his trip to the United States.

BIBLIOGRAPHY

I. Original Works. Konovalov's writings include *Ueber die Dampfspannungen der Flüssigkeitsgemische* (Leipzig, 1881), his inaugural diss.; *Ob uprugosti para rastvorov* ("On the Vapor Pressure of Solutions"; St. Petersburg, 1884; 3rd ed., Leningrad, 1928); *Rol kontaktnykh deystvy v yavleniakh dissotsiatsii* ("The Role of Contact Action in Dissociation Phenomenae"; St. Petersburg, 1885); "Nekotorye soobrazhenia, kasayushchiesya teorii zhidkostey" ("Some Considerations Concerning the Theory of Liquids"), in *Zhurnal Russkogo fiziko-khimicheskogo obshchestva*, **18** (1886), 395–404; "O razlozhenii uksusnogo efira tretichnogo amilovogo spirta v zhidkom sostoyanii" ("On the Decomposition of Acetic Ester of Tertiary Amyl Alcohol in the Liquid State"), *ibid.*, 346–350; "O deystvii kislot na uksusny efir tretichnogo amilovogo spirta" ("On the Action of Acids on Acetic Ester of Tertiary Amyl Alcohol"), *ibid.*, **20** (1888), 586–594; "O prirode osmoticheskogo davlenia" ("On the Nature of Osmotic Pressure"), *ibid.*, **22** (1890), 71–72; "Ob elektroprovodnosti rastvorov" ("On the Electroconductivity of Solutions"), *ibid.*, **24** (1892), 336–338, 440–450, and **25** (1893), 192–201; "O teplotvornoy sposobnosti uglerodistykh veshchestv" ("On the Calorific Value of Carbon Substances"), *ibid.*, **50** (1918), 81–105; "On the Calorific Value of Carbon Compounds," in *Journal of the Chemical Society*, **123** (1923), 2184–2202; and *Materialy i protsessy khimicheskoy tekhnologii* ("Materials and Processes of Chemical Technology"), 2 vols. (Petrograd, 1924; Leningrad, 1925).

II. Secondary Literature. See A. A. Baykov, *Dmitry Petrovich Konovalov* (Leningrad, 1928); and Y. I. Soloviev and A. Y. Kipnis, *Dmitry Petrovich Konovalov (1856–1929)* (Moscow, 1964).

Y. I. Soloviev

KOPP, HERMANN (*b.* Hanau, Electoral Hesse, 30 October 1817; *d.* Heidelberg, Germany, 20 February 1892), *chemistry.*

Kopp's father, Johann Heinrich Kopp, was a practicing physician who had a strong interest in science. He taught chemistry, physics, and natural history in the local lyceum and possessed an outstanding mineralogical collection. He occasionally published papers on mineralogy and physiological chemistry. His son was thus exposed early in life to chemistry and crystallography, which later were his major scientific concerns. During his studies at the Hanau Gymnasium, however, Kopp's subjects were chiefly Latin and Greek. When he entered the University of Heidelberg in 1836, he intended to study philology. His interest in chemistry was aroused by the lectures of Leopold Gmelin, and he therefore decided to devote himself to this subject.

Because there was little opportunity then for individual experimental work at Heidelberg, in 1837 Kopp went to Marburg. He received his doctorate on 31 October 1838 with the dissertation *De oxydorum densitatis calculo reperiendae modo*, which revealed his early interest in the physical properties of substances. After a short period at Hanau, Kopp moved to Giessen in 1839 to work with Liebig, remaining there for twenty-four years. He became a privatdocent in 1841, lecturing on theoretical chemistry, crystallography, meteorology, and physical geography. In 1843 he was appointed extraordinary professor; and in 1852, when Liebig left Giessen for Munich, Kopp and Heinrich Will were jointly appointed to succeed him. Kopp did not like administrative work, and after a year he turned control of the laboratory over to Will. His relations with Liebig remained close, and he corresponded with him for the rest of his life. Most of Kopp's experimental work was carried out at Giessen, and he also began to collect materials on the history of chemistry while there. He taught this subject at intervals early in his career, and in later life he made this one of his main teaching activities.

Kopp's wife, Johanna, whom he married in 1852, came from Bremen. They had two sons, who died in infancy, and a daughter, Therese. Both Kopp and his wife suffered from poor health much of their lives, and his letters to Liebig are full of complaints concerning illnesses.

In 1863 Kopp left Giessen for Heidelberg, where he spent the rest of his life. He was called three times to Berlin but always preferred to remain at Heidelberg. During his years there he gave courses only in crystallography and the history of chemistry. He retired in 1890 and died two years later.

As his dissertation had shown, Kopp's chief interest lay in the study of the physical properties of substances. Under the influence of Liebig when he first came to Giessen, Kopp studied the action of nitric acid on mercaptans; but this was his only excursion into the customary organic chemistry of his day. His real concern was an attempt to establish a connection between the physical properties and the chemical nature of substances. Since he seldom worked with students, he had to carry out most of his experiments by himself. Kopp's first published paper (1837) had concerned the construction of a differential barometer; he delighted in designing and building the apparatus needed for his many accurate determinations of physical constants. His work involved many tedious purifications and much laborious calculation, but he enjoyed this type of study. In the course of his work he accurately measured the boiling points of many organic substances for the first time.

Beginning in 1839, Kopp studied the specific gravity of a number of compounds. In developing formulas for calculating such values he used the concept of specific volume, which he defined as the molecular weight divided by the specific gravity. He showed the similarity of this value in similar elements and isomorphous compounds and related it to their crystal structures, although he was unable to generalize the work to the extent he desired. In 1841 he observed the relations between chain length and boiling point in various classes of organic compounds. He pointed out the generally constant increase in this value as the chain length in a homologous series is increased by addition of a methylene group, but he stressed that the exact value for the increase varies in different types of compounds. He concluded that the boiling point of a liquid was a function partly of molecular weight and partly of chemical constitution.

In 1864 Kopp undertook the study of specific heats of a large number of elements and compounds, in an attempt to verify Neumann's law that the product of molecular weight and specific heat is a constant, regardless of the nature of the substance. He found that in fact the relation was much more complicated and involved a large number of factors. He was, however, able to show that each element has the same specific heat in its free solid state as in its solid compounds. The specific heats of compounds could be calculated from those of their elements.

Kopp's general conclusions often had to be modified later, but in many cases the evidence he presented of a relation between physical properties and chemical structure opened the way for advances in both organic and physical chemistry.

Many of Kopp's researches were paralleled to

some extent by the work of H. G. F. Schröder. Heated disputes between the two chemists frequently took place, and a considerable polemical literature resulted as Kopp asserted his priority in certain discoveries or the correctness of his interpretations. Although essentially mild-mannered, Kopp never hesitated to express his views when he felt aggrieved.

While carrying on his laboratory studies, Kopp was also engaged in literary activities. With Liebig he continued publication of the *Jahresbericht über die Fortschritte der Chemie* after the death of Berzelius, its founder. As editors they changed the plan of the publication, making it a general review of chemistry and related subjects, and enlisted the aid of their university colleagues, so that almost all the members of the philosophical faculty took part in the work. Kopp handled the details of general editorial management. He was also an editor for many years of *Justus Liebigs Annalen der Chemie*. As a result of teaching courses in crystallography he was able to prepare a well-known text on the subject, although he did little laboratory work in this field. It was through his historical books that he established his greatest reputation.

Kopp's linguistic training was undoubtedly of great value to him in this work. He gave his first course of lectures on the history of chemistry when he was only twenty-four; two years later he published the first volume of *Geschichte der Chemie*, which appeared in four volumes between 1843 and 1847. It is clear that he must have been collecting materials for a longer period than that required for giving his course. The first complete, accurate, and readable history of chemistry, the book was notable for its success in relating the development of chemistry to contemporary cultural events.

The first volume contained a general history of the science; the second consisted of individual histories of special branches of chemistry; the last two gave histories of individual substances, elements, and organic compounds. The style was simple and direct, in contrast with the very involved sentences characteristic of Kopp's later historical writings. Ruska has suggested that this complex style came from continued reading of Latin authors. Since Kopp was writing in relative isolation from large libraries, he was not able in his first historical work to utilize source material from early Greek and Arabic authors. The other major contemporary historians of chemistry, Ferdinand Hoefer and Marcellin Berthelot, working in Paris, were able to produce more complete surveys of these periods; but Ruska believes Kopp's treatment of later eras was superior to theirs. His later works rectified the lack of Greek and Arabic material.

Shortly after his call to Heidelberg, Kopp was asked to prepare a history of recent chemical developments in Germany. The resulting *Die Entwicklung der Chemie in der neueren Zeit* (1873) went far beyond the original plan. It discussed the development of chemistry to about 1858, not only in Germany but in all the major countries. An internationalist in outlook, Kopp differed from the French historians, who tended to regard chemistry as a French science. As a result of this approach, Kopp was led in the later years of his life into polemical disputes with Berthelot. The latter very grudgingly mentioned the publications of Kopp and Hoefer on alchemy but remarked that he had reconstituted the whole science which others had neglected. Once more Kopp's priority was challenged, and he responded with a strong defense of his work.

Kopp always intended to revise his *Geschichte der Chemie*, and during most of his life he collected materials for this purpose; but he was never able to put these in final form, and only his scattered publications on alchemy reveal the richness of his historical thought. After his death his material for revision of his *Geschichte* was lost, and so the four volumes of his history remain his chief monument.

BIBLIOGRAPHY

I. Original Works. Kopp's summary of his lifetime of experimental work is "Ueber die Molecularvolume von Flüssigkeiten," in *Justus Liebigs Annalen der Chemie*, **250** (1889), 1–117. His chief historical works are *Geschichte der Chemie*, 4 vols. (Brunswick, 1843–1847); *Beiträge zur Geschichte der Chemie*, 3 pts. (Brunswick, 1869–1875); *Die Entwicklung der Chemie in der neueren Zeit* (Munich, 1873); and *Die Alchemie in alterer und neuerer Zeit* (Heidelberg, 1886). There is a bibliography of his scientific papers by T. E. Thorpe, in *Journal of the Chemical Society*, **63** (1893), 782–785.

II. Secondary Literature. An appreciative obituary is by A. W. von Hofmann, in *Berichte der Deutschen chemischen Gesellschaft*, **25** (1892), 505–521. T. E. Thorpe, "Kopp Memorial Lecture," in *Journal of the Chemical Society*, **63** (1893), 775–815, devotes particular attention to Kopp's experimental work. Julius Ruska, "Hermann Kopp, Historian of Chemistry," in *Journal of Chemical Education*, **14** (1937), 3–12, critically evaluates the historical writings. Max Speter, " 'Vater Kopp' Bio-, Biblio- und Psychographisches von und über Hermann Kopp (1817–1892)," in *Osiris*, **5** (1938), 392–460, gives many personal details, largely drawn from Kopp's letters to Liebig.

Henry M. Leicester

KOROLEV, SERGEY PAVLOVICH (*b.* Zhitomir, Russia, 12 January 1907; *d.* Moscow, U.S.S.R., 14 January 1966), *mechanics, rocket and space technology.*

Korolev's parents were teachers who were divorced soon after his birth. The boy was then brought up in the family of his maternal grandfather, in Nezhin and Kiev. He was ten years old when his mother remarried, moved with her husband to Odessa, and sent for her son. His stepfather, an engineer named Balanin, became both father and friend to Korolev and supported his early interest in technology.

In 1922, after completing his general education, Korolev entered the first Odessa Professional School, which offered specialized training in construction work. He graduated in 1924. It was during this period that he was first attracted to aviation. He began to construct gliders and worked as an instructor for glider clubs in Odessa.

In 1924 Korolev entered the Faculty of Mechanics at Kiev Polytechnic Institute for professional training in aviation. He transferred in 1926 to the Faculty of Aeromechanics of the Bauman Higher Technical School in Moscow. He graduated in 1929, having defended his thesis on the SK-4 airplane, which he had built himself under the direction of A. N. Tupolev (the letters S.K. from his initials and the 4 because it was the fourth plane which he had built).

In 1930 Korolev graduated from the Moscow Summer School and received a pilot's diploma. In 1931 he married a childhood friend, Oksana Maksimilianovna Vintsentina, a physician. They had one daughter. Both in Kiev and in Moscow Korolev combined study with the construction of gliders and airplanes, and made test flights.

In 1930–1931 Korolev became acquainted with the work of Tsiolkovsky and decided to devote himself to rocket and space technology. He participated in the organization at Moscow in 1931 of a group formed to study jet propulsion; he also organized and directed experimental workshops affiliated with the central Moscow group. After the creation in 1933 of the Institute for Jet Research, Korolev moved there and directed both construction and the scientific research. During this period he designed new gliders for carrying large loads. One, the SK-9, with a capacity of about twenty-six kilograms per square meter, was equipped with a liquid-fuel rocket engine. In 1940 V. P. Fedorov made what apparently was the first flight on such a plane.

In 1941–1945 Korolev concerned himself mainly with military uses of rocket planes, and in the years immediately after the war he turned totally to the construction of long-distance rockets.

Korolev combined theoretical research with construction work and teaching. His building a very powerful rocket led Korolev to become a major builder of space rocket vehicles that resulted in outstanding achievements of Soviet space technology, beginning with the launching of the first artificial earth satellite on 4 October 1957. The first Soviet launches of interplanetary probes to Venus (1961–1965) and Mars (1962) were carried out under Korolev's direction.

For his scientific achievements Korolev was elected a corresponding member of the Academy of Sciences of the U.S.S.R. in 1953; in 1958 he became an academician and received the Lenin Prize.

BIBLIOGRAPHY

I. Original Works. Korolev's writings include *Raketny polet v stratosfere* ("Rocket Flight to the Stratosphere"; Moscow, 1934); "Polet reaktivnykh apparatov v stratosfere" ("The Flight of Reactive Apparatuses to the Stratosphere"), in *Trudy Vsesoyuznogo konferentsii po izucheniyu stratosfery* (Moscow, 1935), 849–855; "O prakticheskom znachenii nauchnykh i tekhnicheskikh predlozheny K. E. Tsiolkovskogo v oblasti raketnoy tekhniki" ("On the Practical Importance of the Scientific and Technical Proposals of K. E. Tsiolkovsky in the Field of Rocket Technology"), in *Iz istorii aviatsii i kosmonavtiki*, no. 4 (1966), 7–21; and "O nekotorykh problemakh osvoenia kosmicheskogo prostranstva" ("On Certain Problems in the Conquest of Cosmic Space"), *ibid.*, no. 5 (1967), 3–5. There are also articles and notes in *Samolyot* (1931), nos. 1 and 12; (1932), no. 4; (1935), no. 11; *Vestnik vozdushnogo flota* (1931), no. 2; and *Tekhnika vozdushnogo flota* (1935), no. 7.

II. Secondary Literature. See O. Apenchenko, *Sergey Korolev* (Moscow, 1969); P. T. Astashenkov, *Akademik S. P. Korolev* (Moscow, 1969); and A. Romanov, *Konstruktor kosmicheskikh korabley* ("Constructor of Spaceships"; Moscow, 1969).

J. B. Pogrebyssky

KORTEWEG, DIEDERIK JOHANNES (*b.* 's Hertogenbosch, Netherlands, 31 March 1848; *d.* Amsterdam, Netherlands, 10 May 1941), *mathematics.*

Korteweg studied at the Polytechnical School of Delft, but before graduation as an engineer he turned to mathematics. After teaching in secondary schools at Tilburg and Breda, he entered the University of Amsterdam, where he received his doctorate in 1878. From 1881 until his retirement in 1918 he was professor of mathematics at the same university, where, with P. H. Schoute at Groningen and J. C. Kluyver at Leiden, he did much to raise mathematics in the Netherlands to the modern level.

The subject of Korteweg's dissertation was the velocity of wave propagation in elastic tubes. His sponsor was the physicist J. D. van der Waals, with whom Korteweg subsequently worked on several papers dealing with electricity, statistical mechanics, and thermodynamics. His main scientific work was thus in applied mathematics, including rational mechanics and hydrodynamics; but through his work on Huygens he also contributed greatly to the history of seventeenth-century mathematics. Korteweg established a criterion for stability of orbits of particles moving under a central force (1886), investigated so-called folding points on van der Waals's thermodynamic ψ-surface (1889), and discovered a type of stationary wave advancing in a rectangular canal given by $y = h \text{ cn}^2 (ax)$, the "cnodoil wave" (1895).

From 1911 to 1927 Korteweg edited the *Oeuvres* of Christiaan Huygens, especially volumes XI–XV. He was an editor of the *Revue semestrielle des publications mathématiques* (1892–1938) and of the *Nieuw archief voor wiskunde* (1897–1941).

BIBLIOGRAPHY

I. Original Works. Most of Korteweg's papers appeared in *Verhandelingen* and *Mededelingen der K. nederlandsche akademie van wetenschappen* (Amsterdam) and in *Archives néerlandaises des sciences exactes et naturelles* between 1876 and 1907, the latter often publishing French translations of the Dutch papers appearing in the former. Other papers include "Über Stabilität periodischer ebener Bahnen," in *Sitzungsberichte der K. Akademie der Wissenschaften in Wien*, Math.-naturwiss. Kl., **93**, sec. 2 (1886), 995–1040; "Ueber Faltenpunkte," *ibid.*, **98** (1889), 1154–1191; *Het bloeitydperk der wiskundige wetenschappen in Nederland* (Amsterdam, 1894); and "On the Change of Form of Long Waves Advancing in a Rectangular Canal, and on a New Type of Stationary Waves," in *Philosophical Magazine*, 5th ser., **39** (1895), 422–443, written with G. de Vries.

II. Secondary Literature. For biographies see H. J. E. Beth and W. van der Woude, "Levensbericht van D. J. Korteweg," in *Jaarboek der koninklyke nederlandsche akademie van wetenschappen 1945–1946* (Amsterdam, 1946), pp. 194–208; and L. E. J. Brouwer in *Euclides* (Groningen), **17** (1941), 266–267. Appreciations of his work are by H. A. Lorentz in *Algemeen Handelsblad* (Amsterdam), *Avondblad* (12 July 1918); and an unsigned article, *ibid.* (30 March 1928), p. 13, on the occasion of Korteweg's eightieth birthday.

D. J. Struik

KOSSEL, KARL MARTIN LEONHARD ALBRECHT (*b.* Rostock, Germany, 16 September 1853; *d.* Heidelberg, Germany, 5 July 1927), *nucleoprotein chemistry.*

Kossel, the only son of Albrecht Kossel, a merchant and consul, and of the former Clara Jeppe, was a keen botanist as a schoolboy in Rostock. Only his father's influence persuaded him to read medicine, but Kossel chose to go to Strasbourg in order to attend the lectures of the mycologist Anton de Bary. There he came under the influence of Germany's foremost physiological chemist, Felix Hoppe-Seyler, to whom he returned as assistant in 1877 after passing the state medical examination at Rostock. His *Habilitationsschrift* was accepted in 1881, and in 1883 du Bois-Reymond appointed him director of the chemical division of the Berlin physiological institute, where he became an assistant professor in 1887. After ten years in Berlin, Kossel received the chair of physiology and directorship of the physiological institute in Marburg. In 1901 he succeeded Willy Kuhne in the chair of physiology at Heidelberg. On his retirement in 1924 he directed the new institute for the study of proteins at Heidelberg.

Kossel married Luise Holtzmann in 1886. He received the Nobel Prize in physiology or medicine in 1910, as well as many honorary degrees. He was survived by his son, Walther, professor of theoretical physics at Kiel, and a daughter. Among his students were W. J. Dakin, A. P. Mathews, and P. A. Levene.

After studying salt diffusion and the pepsin digestion of fibrin, Kossel turned in 1879 to the nucleins (nucleoproteins) discovered by J. F. Miescher in 1869, taking the chemical characterization of these compounds to a much deeper level than had Miescher. Between 1885 and 1901 Kossel and his students discovered adenine, thymine, cytosine, and uracil. He demonstrated that these, together with xanthine, hypoxanthine, and guanine (sarcine), are breakdown products of nucleic acids, which can be used to distinguish between the true nucleins of the cell nucleus and the spurious nucleins found in milk and egg yolk, which he termed "paranucleins." His suggestion that hypoxanthine is a secondary product of adenine, and therefore not a primary constituent of nucleic acid, was correct; and his belief in the presence of a hexose sugar in nucleic acid from the thymus was not far from the truth (2-deoxyribose). In 1893 he also suggested, correctly, that the carbohydrate in yeast nucleic acid is a pentose.

Kossel's invaluable distinction between true nucleins and paranucleins gained acceptance only slowly; its impact was rendered less decisive by his advocacy of the mistaken view that in the synthesis of nucleins the xanthine bases are simply added to preexisting paranuclein molecules, such as are abundant in the egg yolk. Kossel's recognition of chromatin as nucleic acid with varying proportions of histone (1893) did not lead him into the discussions current at that time on the identity of the hereditary substance.

From physiological studies Kossel correctly concluded that the function of nuclein is neither to act as a storage substance nor to furnish energy for muscular contraction; rather, it must be associated with the formation of fresh tissue. He found embryonic tissue to be especially rich in nuclein. Also from physiological studies he showed that uric acid is more closely associated with the breakdown of nucleins than with that of proteins. Although he believed that the nuclein molecule consists of some twelve subunits, or a multiple of twelve, he left the task of formulating its structure to others.

In 1884 Kossel turned to the basic component of nuclein; and from the nuclei of goose erythrocytes he isolated a substance like Miescher's protamine, which he named histone. He regarded it as a peptone and demonstrated that the amino acids leucine and tyrosine are among its decomposition products. When he examined the basic component of fish spermatozoa —Miescher's protamine—he found that, like histone, it is protein in character and on decomposition yields arginine, lysine, and a new amino acid, which he named histidine. Using his own quantitative methods, Kossel made comparative studies of the protamines of the sperm of various fish species which showed varying proportions of monoamino and diamino acids. He tried to formulate sequences of amino acids with the aid of his identification of decomposition products as small as arginylarginine.

Always anxious to unite chemical description with physiological function, and seeking to move from the static world of protein chemistry to the dynamic world of physiology, Kossel formulated a scheme of protein synthesis based on the idea that all proteins possess a nucleus of the diamino acids to which monoamino acids are added progressively during embryogeny. In the reverse process of gametogenesis, diamino acids from protein breakdown are selectively utilized to form the nuclear protamine of the gametes. He followed this process quantitatively in the male salmon.

As one of Hoppe-Seyler's most successful students, Kossel continued to develop the tradition of physiological chemistry in Germany. He was reserved, modest, unexcitable, very conscientious, an unimpressive speaker, and dominated by the vision of a biological meaning for his chemical discoveries. Thus he believed that the reactivity of proteins depends upon that of the residues in exposed positions on the molecule. In a given reaction of a protein certain characteristic groups will be involved. In his Herter Foundation lecture (1912) Kossel clearly recognized the potential diversity of polypeptides and saw in the structure of proteins the chemical basis of biological specificity. The inadequacy of the techniques then available to him prevented him from carrying these essentially modern ideas further.

BIBLIOGRAPHY

I. Original Works. Kossel's papers up to 1900 are listed in the Royal Society *Catalogue of Scientific Papers*, XII, pp. 404–405, and XVI, p. 427. His later papers are in his book *The Protamines and Histones* (London, 1928), also in German as *Protamine und Histone* (Leipzig–Vienna, 1929).

He also wrote *Leitfaden für medicinisch-chemische Kurse* (Berlin, 1888; 8th ed., 1921). His Heidelberg vice-rectoral address appeared as a booklet, *Die Probleme der Biochemie* (Heidelberg, 1908). With W. Behrens and P. Schiefferdecker he wrote *Das Mikroskop und die Methoden der mikroskopischen Untersuchungen*, which is vol. I of their *Die Gewebe des menschlichen Körpers und ihre mikroskopische Untersuchung* (Brunswick, 1889). He also contributed an essay, "Beziehungen der Chemie zur Physiologie," to E. von Meyer, ed., *Die Kultur der Gegenwart ihre Entwicklung und ihre Ziele: Chemie* (Leipzig–Berlin, 1913), pp. 376–412.

Kossel's early interest in proteins is found in "Ein Beitrag zur Kenntniss der Peptone," in *Pflüger's Archiv für die gesamte Physiologie*, **13** (1876), 309–320; "Ueber die Peptone und ihre Verhältniss zu den Eiweisskörpern," *ibid.*, **21** (1888), 179–184. His interest in the clinical aspects of his chemical researches is found in "Dosage de l'hypoxanthine et de la xanthine," in *Journal de pharmacie*, **7** (1883), 325–326.

Many of his papers on nucleins appeared in the journal he edited for over thirty years, *Hoppe-Seyler's Zeitschrift für physiologische Chemie*, including "Ueber das Nuclein der Hefe," **3** (1879), 284–291, and **4** (1880), 290–295; "Ueber die Herkunft des Hypoxanthins in der Organismen," **5** (1881), 152–157; "Ueber Guanin," **8** (1883–1884), 404–410; "Ueber einen peptonartigen Bestandtheil des Zellkerns," *ibid.*, 511–515; "Weitere Beiträge zur Chemie des Zellkerns," **10** (1886), 248–264; "Ueber das Adenin," **12** (1888), 241–253; with A. P. Mathews he published "Zur Kenntniss der Trypsinwirkung," **25** (1898), 190–194; with F. Kutscher, "Beiträge zur Kenntniss der Eiweisskörper," **31** (1900), 165–214; with H. Steudel, "Ueber einen basischen Bestandtheil thierischen Zellen," **37** (1902–1903), 177–189.

Most of his publications on nuclein in the *Berichte der Deutschen chemischen Gesellschaft* concern adenine and thymine: "Ueber eine neue Base aus dem Thier-Körper," **18** (1885), 79–81; "Ueber das Adenin," *ibid.*, 1928–1930, and **20** (1887), 3356–3358; "Ueber eine neue Base aus dem

Pflanzenreich," **21** (1888), 2164–2167; and two papers with A. Neumann—"Ueber das Thymin, ein Spaltungsproduct der Nucleinsäure," **26** (1893), 2753–2756; and "Darstellung und Spaltungsproducte der Nucleinsäure (Adenylsäure)," **27** (1894), 2215–2222. His important comparisons of true nucleins and paranucleins are found in "Ueber das Nuclein im Dotter des Hühnereies," in *Archiv für Anatomie und Physiologie* (1885), 346–347; "Ueber die chemische Zusammensetzung der Zelle," *ibid.* (1891), 181–186; and "Ueber die Nucleinsäure," *ibid.* (1893), 157–164.

Kossel's Nobel Prize lecture, "Ueber die Beschaffenheit des Zellkerns" (1910), is translated in *Nobel Lectures Including Presentation Speeches and Laureates' Biographies: Physiology or Medicine 1901–1921* (Amsterdam, 1967), pp. 394–405. His best general lectures are "The Chemical Composition of the Cell," in *Harvey Lectures* (1911–1912), 33–51; and "Lectures on the Herter Foundation" in *Johns Hopkins Hospital Bulletin* **23** (1912), 65–76.

II. SECONDARY LITERATURE. The best source of biographical information is S. Edlbacher, "Albrecht Kossel zum Gedächtnis," in *Hoppe-Seyler's Zeitschrift für physiologische Chemie*, **177** (1928), 1–14. For a more up-to-date assessment of his work, Kurt Felix's centenary essay should be consulted: "Albrecht Kossel: Leben und Werk," in *Naturwissenschaften*, **42** (1955), 473–477. The greater part of this essay has been translated in Eduard Farber, ed., *Great Chemists* (New York–London, 1961), pp. 1033–1037. Obituary notices are in *Science*, **66** (1927), 293; *Deutsche medizinische Wochenschrift*, **53** (1927), 1441; *Journal of the American Medical Association*, **89** (1927), 524–525; *Nature*, **120** (1927), 233; and *Berichte der Deutschen chemischen Gesellschaft*, **60** (1927), A159–A160.

Kossel's portrait appears in the 60th anniversary vol. of *Hoppe-Seyler's Zeitschrift für physiologische Chemie*, **130** (1923), and also in **169** (1927). For a personal account of research in Kossel's institute which is far from complimentary to Kossel as a director of research, see Sir Ernest Kennaway, "Some Recollections of Albrecht Kossel, Professor of Physiology in Heidelberg, 1901–1924," in *Annals of Science*, **8** (1952), 393–397. The best reviews of Kossel's work on the nucleic acids are P. A. Levene and L. W. Bass, in *Nucleic Acids* (New York, 1931), ch. 8; and R. Markham and J. D. Smith, "Nucleoproteins and Viruses," in H. Neurath and K. Bailey, eds., *The Proteins: Chemistry, Biological Activity, and Methods* (New York, 1954 [1st ed. only]), IIa, ch. 12. For brief critical comments on Kossel's contribution to protein chemistry, see J. M. Luck, "Histone Chemistry: The Pioneers," in J. Bonner and P. Ts'o, eds., *The Nucleohistones* (San Francisco, 1964), ch. 1.

ROBERT OLBY

KOSSEL, WALTHER (LUDWIG JULIUS PASCHEN HEINRICH) (*b.* Berlin, Germany, 4 January 1888; *d.* Kassel, Germany, 22 May 1956), *physics.*

Kossel was descended from an old family of distinguished scholars. His father, Albrecht Kossel, longtime professor of physiology at the University of Heidelberg, received the Nobel Prize in physiology or medicine in 1910 for "contributions to the chemistry of the cell through his work on proteins, including the nucleic substances." His mother's maiden name was Holtzmann. The atmosphere in the parental home fostered in young Kossel two characteristics that always appeared in his work—his delight in a careful, well-ordered style and his thorough clarity of presentation. He considered these traits crucial to a scientist, and he took for his life's work an interest in "that which most intrinsically holds physics together."

After attending the Gymnasiums at Marburg and Heidelberg, Kossel studied physics under Philipp Lenard at the University of Heidelberg. There he was an assistant in physics from 1910. He received his doctorate in 1911 with a dissertation on an experimental investigation of the character and quantity of secondary cathode rays produced in different gases by primaries of diverse velocities (1). In the same year, he married Hedwig Kellner.

In order to advance his knowledge of physics he then moved to Munich, where Roentgen and Sommerfeld at the University, and Zenneck at the Technische Hochschule, presided over thriving institutes where P. S. Epstein, P. P. Ewald, W. Friedrich, M. von Laue, P. Knipping, P. P. Koch, and E. Wagner were working. In the spring of 1912, Laue, Friedrich and Knipping had shown that X rays could be diffracted by crystals, thereby opening new avenues in the physics of X rays and providing a new method for exploring the structure of crystals. Soon after, Sommerfeld, under whom Kossel had worked, began his explorations of atomic structure and spectral lines within the framework of Bohr's quantum theory of the atom. Besides remaining in touch with these developments, Kossel, who became Zenneck's assistant in 1913, mastered the then emerging field of electronics, which enabled him to develop radio amplifying tubes for use in World War I.

Kossel's first important contribution to physics was his extension of Bohr's theory to the mechanism of X-ray emission (2). According to Kossel, who here succeeded where both Bohr and Moseley had failed, characteristic high-frequency (X) radiation accompanies the binding of an electron into a prior vacancy within the atom. The deeper the hole and the greater the distance through which the electron falls the higher the frequency of the emitted quantum. This picture, as developed especially by Sommerfeld, brought a general understanding of the X-ray spectrum, an estimate of the number of electrons n_j supposed arranged in concentric rings about the nucleus, and the recognition (in contrast to Bohr's original system) that the normal atom contains electrons characterized by more than one quantum of angular momentum $h/2\pi$. But Kossel's theory also had its difficulties, especially conflicts between (*a*) the calculated n_j and their values as inferred from chemical evidence and (*b*) computed and observed frequencies of X-ray lines. Kossel later recognized (3) that (*a*) arose from the working of a selection principle which Coster and Wentzel then fully specified; as for (*b*), Kossel left it to others, and contented himself with establishing n_j on chemical grounds.

Kossel discussed the n_j in the context of the theories of valence and bonding, to which he made central contributions (4). Physicists had already tried to relate the electronic structure of atoms to two fundamental chemical phenomena—the chemical bond, that is, the attraction between atoms in a molecule, and valence, the quality that determines the number of atoms or groups with which any single atom or group will unite chemically, and also expresses this combining capacity relative to the hydrogen atom. In particular, J. J. Thomson, in his well-known *Corpuscular Theory of Matter* (1907), had given a theory of heteropolar bonding based upon the transfer of electrons from one molecule partner to another. The subsequent work of Rutherford, Bohr, and Moseley, which established the doctrine of atomic number, made possible more precise correspondences than Thomson had been able to suggest, and inspired Kossel to update the theory.

According to Kossel, who here followed Bohr and not Thomson, the number of electrons in the outer ring, the so-called valence electrons, determines the chemical properties of an atom. Kossel postulated that the extraordinary stability of the noble gases (then called inert gases) was due to the "closed" or complete nature of their outer ring or shell; there are eight electrons in the outer shell of all these gases except for helium, which has two. He believed, with Thomson, that metals achieve the stable electronic configuration of the nearest noble gas by losing electrons, and nonmetals achieve it by gaining electrons. The electrons lost by the metal are transferred to the nonmetal, and the resulting ions—cation (or positive ion) in metals and anion (or negative ion) in nonmetals—are held together by electrostatic (Coulombic) attraction. The theory has since been completely confirmed.

A month after the publication of Kossel's paper (1916), the American chemist G. N. Lewis, working independently of him, published a paper (5) dealing with electrovalent compounds and especially covalent compounds, that is, those formed by the sharing of electrons. The ideas of Kossel and Lewis did not achieve the immediate success among chemists that they deserved, possibly because of the interference of World War I with scientific activity. Their ideas achieved general acceptance in 1919 largely because of Langmuir's systematizing efforts. Since then the theory has been developed and extended by many scientists, in particular, Sidgwick (6) and Linus Pauling (7). Kossel also helped extend it in an important paper written in collaboration with Sommerfeld (8). They showed that the spark spectrum of a given element (that is, the spectrum of its positive ion) has the same structure as the arc spectrum of the element one below it; apparently the electronic superstructure of the ion of the first element is identical in form to that of the atom of the second.

In 1920 Kossel became privatdocent at the Technische Hochschule, Munich. The following year he was called to the University of Kiel, where an excellent tradition in physics already existed, both Lenard and Dieterici having held chairs there. At Kiel he became professor ordinarius of theoretical physics and director of the Institute for Theoretical Physics. In addition to his scientific activities, Kossel offered his services to the academic administration of the university; in 1926 he served as dean of the Faculty of Mathematics, and in 1929–1930 he was appointed rector of the university. He continued his interest in administrative affairs throughout his life and later represented the University of Tübingen in the union of universities (*Hochschulverband*).

At Kiel, Kossel continued his work on valency and published a second edition of his book *Valenzkräfte und Röntgenspektren* (9). He further developed the idea that the chemical bond was due for the most part to electrostatic forces, and he applied this idea to investigating the growth of ionic crystals. He turned increasingly to X-ray exploration of crystals and their growth, a subject which appealed to him esthetically. In 1927 he completed his first work on crystal growth (10), in which he dealt only with simple structures but in a way that furnished the basis for much recent work.

In 1932 Kossel went to Danzig (now Gdańsk, Poland), where he became professor ordinarius of experimental physics and director of the Institute for Experimental Physics at the Technische Hochschule. He made the Institute a place of lively, productive activity. In Danzig he found better opportunities to explore new fields, and he soon acquired a large school of students and co-workers, with whom he maintained friendly contacts throughout his life. Among his most important work at this time was his discovery of the interference effects (the so-called Kossel effect, first announced in 1935) produced by characteristic Roentgen rays excited in a single crystal. He also demonstrated the interference of electrons in converging beams, which gave great impetus to the theory of electron diffraction. For these and previous work Kossel was awarded in 1944 the Deutsche Physikalische Gesellschaft's highest honor—the Max Planck Medal. In 1955 he was made an honorary member of the Society. Additional honors followed including offers, which he declined, from the Universities of Berlin (1939) and Strasbourg (1942).

Kossel's activity at Danzig was halted by the Russian occupation in 1945. He left with his family and most of the Physics Institute. Kossel moved the Institute's costly equipment to the West. After a troubled time he found a new base of operations for his research in 1947, when he was appointed professor ordinarius of experimental physics and director of the Experimental Physics Institute at the University of Tübingen.

At Tübingen, Kossel and his colleagues worked in electron and optical diffraction, electrical discharge in gases, solid-state physics, acoustics, and crystal structure. He used large spherical single crystals of metal, of the kind first produced by his old Heidelberg friend Wilhelm Hausser. In the gas discharge he discovered continuous Lichtenberg figures, a phenomenon of gas discharge physics unique in color and symmetry. In acoustics he continued the classical work of Helmholtz with the help of electronic methods. In collaboration with medical colleagues, Kossel developed a technique for measuring Roentgen dosage within the body. With his students he developed electrostatic band generators with field voltages of up to 1.5 million volts and small disk generators with potentials up to 100,000 volts. In addition, Kossel continued to expend much time and effort on his experimental lectures.

Kossel was already suffering from the prolonged liver ailment which eventually claimed his life when he was grieved by his wife's death in 1953. But he was mentally clear and active until the very end and as late as 1955 he lectured by special invitation in Paris. His last work, *Individuation in der unbelebten Welt* (11), published during his final illness, gives some idea of the range of his scientific activities.

Kossel was a corresponding member of the Göttingen and Halle Academies, an honorary member of the Deutsche Chemische Gesellschaft, the Deutsche Mineralogische Gesellschaft, the Verband Deutscher Physikalischer Gesellschaften, and the Bremen Naturwissenschaftliche Gesellschaft. He was also an honorary citizen of the Christian-Albrechts University of Kiel.

BIBLIOGRAPHY

The works referred to in the text are

(1) "Über die sekundäre Kathodenstrahlung in der Nähe des Optimums der Primärgeschwindigkeit," in *Annalen der Physik*, **37** (1912), 393–424.

(2) "Bemerkung zur Absorption homogener Röntgenstrahlen," in *Verhandlungen der deutschen physikalischen Gesellschaft*, **16** (1914), 898–909.

(3) "Zum Bau der Röntgenspektren," in *Zeitschrift für Physik*, **1** (1920), 119–134.

(4) "Über Molekülbindung als Frage des Atombaus," in *Annalen der Physik*, **49** (1916), 229–362 (Received Dec., 1915). Partial English translations appear in H. M. Leicester, *Source Book in Chemistry, 1900–1950* (Cambridge, Mass., 1968), pp. 94–100, and W. G. Palmer, *A History of the Concept of Valency to 1930* (Cambridge, 1965), pp. 129–132.

(5) G. N. Lewis, "The Atom and the Molecule," in *Journal of the American Chemical Society*, **38** (1916), 762–785 (Received Jan., 1916). Further details and developments can be found in Lewis' book *Valence and the Structure of Atoms and Molecules* (New York, 1923; paperbound rep. ed., 1966).

(6) N. V. Sidgwick, *The Electronic Theory of Valency* (London, 1929).

(7) L. Pauling, *The Nature of the Chemical Bond*, 3rd ed. (Ithaca, N.Y., 1960). Additional information can be found in G. V. Bykov, "Historical Sketch of the Electron Theories of Organic Chemistry," in *Chymia*, **10** (1965), 199–253.

(8) With A. Sommerfeld, "Auswahlprinzip und Verschiebungssatz bei den Serienspektren," in *Verhandlungen der deutschen physikalischen Gesellschaft*, **21** (1919), 240.

(9) *Valenzkräfte und Röntgenspektren* (Berlin, 1924).

(10) "Zur Theorie des Kristallwachstoms," in *Nachrichten der Akademie der Wissenschaften zu Göttingen*, Mathematisch-physikalische Klasse (1927), 135–143.

(11) *Individuation in der unbelebten Welt* (Berlin, 1956).

Most of the biographical information in this article is taken from two obituaries: "Zum Tode von Walther Kossel," in *Tübinger Chronik* (26 May 1956), and E. N. da C. Andrade, "Prof. Walther Kossel," in *Nature*, **178** (1956), 568–569. Details about Kossel's fundamental work on X-ray spectra may be found in papers by J. L. Heilbron, from which the present account has been taken: "The Kossel–Sommerfeld Theory and the Ring Atom," in *Isis*, **58** (Winter, 1967), 450–485; "The Work of H. G. J. Moseley," in *Isis*, **57** (Fall, 1966), 336–364.

GEORGE B. KAUFFMAN

KOSTANECKI, STANISŁAW (*b.* Myszakow, Poznań province, Poland, 16 April 1860; *d.* Würzburg, Germany, 15 November 1910), *chemistry.*

Kostanecki was the oldest son of Nepomucen Kostanecki, a small landowner, and Michalina Dobrowolska. From 1871 to 1883 he attended the nonclassical secondary school in Poznań, where he studied chemistry under Teodor Krug. In 1883 Kostanecki graduated with distinction and entered the Faculty of Philosophy of the University of Berlin. He also attended the lectures of R. H. Finkener and Liebermann at the Gewerbeakademie.

In 1884 Kostanecki became Liebermann's assistant, with whom he jointly published two papers on azo compounds. These papers, together with those written on the compounds of the hydroxyanthraquinone group, formed the basis of the so-called Liebermann-Kostanecki rule; it stated that the only technically satisfactory dyestuffs are those hydroxyanthraquinone ones with two hydroxyl groups attached, as in alizarin. Over a two-year period in Liebermann's laboratory, Kostanecki published thirteen scientific papers. Several were published jointly with Stefan Niementowski on cochineal dyestuffs, and others with Augustyn Bistrzycki on euxanthone.

From 1886 to 1889 Kostanecki was *chef de travaux* at the École de Chimie in Mulhouse. Emil Noelting, a leading dye specialist, was director of the school, and Kostanecki developed a lasting friendship with him. While he was in Mulhouse, Kostanecki experimented with derivatives of resorcinol, especially nitroso compounds, and wrote thirteen papers.

In March 1888 Kostanecki was invited to become professor of organic chemistry at the Jagiellonian University in Cracow, but the Ministry of Education in Vienna did not consent to his appointment. On 7 May 1890, after the death of Valentin von Schüpfen Schwarzenbach, Kostanecki accepted the chair of organic and theoretical chemistry at the University of Bern. He received his doctorate from the University of Basel in 1890 after his nomination as professor at Bern.

At Bern, Kostanecki built up the chemical laboratory with the help of Marceli Nencki, professor of physiological chemistry at the university and later director of the department of chemistry of the Institute of Experimental Medicine in St. Petersburg. During his twenty years at Bern, Kostanecki published 182 papers and supervised 161 doctoral dissertations.

Kostanecki himself did all analyses of newly discovered compounds. His research was concentrated on the structural problems of vegetable dyestuffs, especially of the flavone group. He carried out the synthesis of chrysin, trihydroxyflavone dyes derived from *Reseda luteola*, fisetin, quercetin, kaempferol, galangin, and morin. He also investigated the structure of brazilin and hematoxylon. In his last years he conducted research on the structure of curcumin dyestuffs obtained from *Curcuma tinctoria*.

Kostanecki's many-sided investigations on dyestuffs provided the basis for his classification of dyes and for his formulation of the relationship between the structure of compounds, their color, and dyeing ability. Among his pupils was Casimir Funk. Kostanecki died of chronic appendicitis in Würzburg Hospital and was buried in Kazimierz, near Łódź.

BIBLIOGRAPHY

I. Original Works. Most of Kostanecki's papers were published in *Berichte der Deutschen chemischen Gesellschaft*, **17–43** (1884–1910). Complete listings, with full titles, are in the obituary notice by E. Noelting in *Verhandlungen der Schweizerischen naturforschenden Gesellschaft* (1911), 74–128; and in W. Lampe, "Prace ś.p. St. Kostaneckiego," in *Chemik polski*, **11**, no. 2 (1911), 1–25; "Kostanecki, Stanisław," in *Wielka encycklopedia powszechna*, VI (Warsaw, 1965), 87; and *Stanisław Kostanecki życie i działalność naukowa* (Warsaw, 1958).

II. Secondary Literature. On Kostanecki and his work see *Gedächtnisreden für Herrn Professor Dr. St. v. Kostanecki gehalten an der Trauerfeier* . . . (Bern, 1911); A. Bistrzycki, "Stanislaus von Kostanecki," in *Chemikerzeitung*, **142** (1910), 1261; T. Estreicher, *Stanisław Kostanecki, wspomnienie pośmiertne* (Cracow, 1910), 1–25; M. Sarnecka Keller, "Kostanecki Stanisław," in *Polski Slownik Biograficzny*, XIV (Cracow, 1969), 334–335; S. Niementowski, "Życie i naukowe prace Prof. Dr. St. Kostaneckiego," in *Kosmos*, **37** (1912), 1–63; A. Szlagowski, "Mowa na nabożeństwie żałobnym za duszę ś.p. Prof. St. Kostaneckiego 1.XII.1910 r.," in *Chemik polski*, **10** (1910), 550–552; J. Tambor, "St. v. Kostanecki, Nachruf," in *Berichte der Deutschen chemischen Gesellschaft*, **45** (1912), 1683; and J. S. Turski and B. Więcławek, *Barwniki roślinne i zwierzęce* (Warsaw, 1952).

Włodzimierz Hubicki

KOSTINSKY, SERGEY KONSTANTINOVICH (*b.* Moscow, Russia, 12 August 1867; *d.* Pulkovo, near Leningrad, U.S.S.R., 21 August 1936), *astronomy.*

Kostinsky graduated from the first Moscow Gymnasium and in 1890 from the Faculty of Physics and Mathematics of Moscow University, where his teachers in astronomy were Bredikhin and Ceraski. When Bredikhin was elected director of the Pulkovo observatory, Kostinsky was one of the first Russian astronomers invited to work there. In 1890 Kostinsky was supernumerary astronomer of the observatory, from 1894 adjunct astronomer, and from 1902 senior astronomer. In 1895 the new director, O. A. Baklund, commissioned Kostinsky to organize a section of astrophotography and to set up the normal astrograph at Pulkovo for the application of photography to precise measurements in astronomy.

In 1896 he obtained an excellent photograph of the

solar corona during observation of a total solar eclipse on the island of Novaya Zemlya in the northern Arctic Ocean. In 1899–1901 Kostinsky made astronomical determinations and trigonometrical measurements on Spitsbergen in connection with the works of the Russian–Swedish expedition for the measurement of an arc of meridian.

At Pulkovo, from the end of 1899, Kostinsky lectured and taught scientific photography to young astronomers, geodesists, and hydrographers who were sent to the observatory for practical work. In 1915 he was elected corresponding member of the Academy of Sciences and honorary doctor of Moscow University. In 1919 Kostinsky began to teach at Petrograd-Leningrad University as professor.

Kostinsky's scientific activity at Pulkovo began with the study of the fluctuation of astronomical latitudes based on observations with a transit instrument on the first vertical. The method he proposed for computing the curved motion of the pole has been widely used. This method was based on the analysis of the variations in latitude of several observatories located at various geographical longitudes.

Working in a new branch of science, astrophotography, in 1898, together with F. F. Renz, Kostinsky studied in detail an instrument designed to make very precise measurement of astronegatives. In doing this he discovered the effect of mutual repulsion of very close stellar images on the negative. This phenomenon is now known as the Kostinsky effect. Kostinsky then attempted to determine the proper motion of the stars by intensive photography with the normal astrograph. The first series of photographs ("first epoch") was compared with a second series taken a decade later ("second epoch") in order to obtain the precise measures of the yearly changes of positions of the stars ("proper motions").

Having developed a method of measuring negatives and deducing reduction formulas from them, Kostinsky turned to the determination by photography of stellar parallaxes. For the comparison of negatives, Kostinsky began to use, instead of the blink microscope, the stereocomparator, which, by superimposing the images of two negatives, stereoscopically marks out an object with a noticeable proper movement from the background stars. Kostinsky developed a method of using the stereocomparator and became a strong advocate of it. His method and results were praised by A. Y. Orlov, director of the Odessa observatory, and by Van Rhijn, who was amazed at the precision of proper motions that Kostinsky had achieved.

Over a ten-year period Kostinsky obtained remarkable photographs of star clusters, nebulae, the satellites of Mars, and the planets Uranus and Saturn; made precise measurements of the positions of the planets; and became a recognized authority on photographical astrometry. He obtained about 3,000 astronegatives in all. His numerous photographs of selected areas of Kapteyn permitted Kostinsky's pupils in the 1930's to compile a catalog of the proper motions of 18,000 stars, a valuable contribution to the study of stellar kinematics. Kostinsky's photographs of Eros during the "great opposition" in 1900 were used in England by A. R. Hincks for a new determination of the solar parallax. In 1914 Kostinsky published in Russian his later well-known work on the parallax of Mira Ceti ($0.02'' \pm 0.02''$); the currently accepted value is $0.013'' \pm 0.011''$.

At Pulkovo, Kostinsky established a school of specialists in photographic astrometry, now headed by A. N. Deutsch. Kostinsky contributed a valuable chapter to the two-volume monograph compiled mainly by the Pulkovo astronomers, *Kurs astrofiziki i zvezdnoy astronomii* ("A Course in Astrophysics and Stellar Astronomy," 1934). He also contributed greatly to popularizing astronomical science in Russia. In 1916–1917 Kostinsky was one of the most active organizers of the All-Russian Astronomical Society and its first congress.

BIBLIOGRAPHY

I. ORIGINAL WORKS. Among Kostinsky's papers are "Ob izmenenii astronomicheskikh shirot" ("On Variations of Astronomical Latitudes") in *Zapiski Akademii nauk*, **73**, app. 10 (1893), 1–101; "Po povodu odnoy lichnoy oshibki pri izmerenii fotograficheskikh snimkov" ("Concerning One Personal Error in the Measurement of Photographs"), in *Izvestiya Akademii nauk*, **3** (1895), 491–498; and "O Bredikhinskoy teorii kometnykh form" ("On Bredikhin's Theory of Cometary Forms"), in S. A. Vengerov, ed., *Kritiko-biografichesky slovar russkikh pisateley i uchenykh* ("Critical-Biographical Dictionary of Russian Writers and Scientists"), V (St. Petersburg, 1897), 279–290.

Subsequent works include "Zur Frage über die Parallaxe von β Cassiopejae," in *Astronomische Nachrichten*, **163** (1903), 350; "Untersuchungen auf dem Gebiete der Sternparallaxen mit Hilfe der Photographie," in *Publications de l'Observatoire Central à Poulkovo*, **17**, no. 2 (1905); "Über die Einwicklung zweier Bilder auf einander bei astrophotographischen Aufnahmen," in *Mitteilungen der Sternwarte zu Pulkowo*, **2**, no. 14 (1907), 15–28; "Durchmusterung der Eigenbewegungen in der Umgebund des Sternhaufens NGC 7209," in *Astronomische Nachrichten*, **238** (1935), 245–248; "The Star-Streamings in the Region of Spiral Nebula Messier 51 [NGC 5194]," in *Tsirkulyar Glavnoi astronomicheskoi observatorii v Pulkove*, no. 15 (1935), 18–21; and "Stereoscopic Durchmusterung of Proper Motions in Four Regions of the Sky," *ibid.*, no. 20 (1936), 22–31.

II. SECONDARY LITERATURE. Kostinsky's autobiography

is in S. A. Vengerov, ed., *Kritiko-biografichesky slovar russkikh pisateley i uchenykh*, VI (1904), 49–50; for further information on Kostinsky's life, see the obituaries by A. N. Deutsch in *Astronomicheskii zhurnal*, **13**, no. 6 (1936), 505–507; and by M. S. Eigenson in *Priroda*, no. 9 (1936), 128–132; and Y. G. Perel, *Vydayushchiesya russkie astronomy* ("Outstanding Russian Astronomers"; Moscow–Leningrad, 1951), pp. 178–193.

See also "K biografii S. K. Kostinskogo" ("Toward a Biography of S. K. Kostinsky"), in *Istoriko-astronomicheskie issledovaniya*, no. 3 (1957), 531–540.

P. G. KULIKOVSKY

KOTELNIKOV, ALEKSANDR PETROVICH (*b.* Kazan, Russia, 20 October 1865; *d.* Moscow, U.S.S.R., 6 March 1944), *mechanics, mathematics.*

Kotelnikov was the son of P. I. Kotelnikov, a colleague of Lobachevsky, and the only one to publicly praise Lobachevsky's discoveries in geometry during the latter's lifetime. In 1884, upon graduation from Kazan University, Kotelnikov taught mathematics at a Gymnasium in Kazan. Later he was accepted by the department of mechanics of Kazan University in order to prepare for the teaching profession. He began his teaching career at the university in 1893, and in 1896 he defended his master's dissertation, "Vintovoe ischislenie i nekotorie primenenia ego k geometrii i mekhanike" ("The Cross-Product Calculus and Certain of Its Applications in Geometry and Mechanics"). Kotelnikov's calculus is a generalization of the vector calculus, describing force moments in statics and torques in kinematics. In his many years of teaching theoretical mechanics, Kotelnikov was an advocate of vector methods.

In 1899 Kotelnikov defended his doctoral dissertation, "Proektivnaya teoria vektorov" ("The Projective Theory of Vectors"), for which he simultaneously received the doctorate in pure mathematics and the doctorate in applied mathematics. Kotelnikov's projective theory of vectors is a further generalization of the vector calculus to the non-Euclidean spaces of Lobachevsky and Riemann and the application of this calculus to mechanics in non-Euclidean spaces.

Kotelnikov served as professor and head of the department of pure mathematics at both Kiev (1899–1904) and Kazan (1904–1914). He headed the department of theoretical mechanics at Kiev Polytechnical Institute (1914–1924) and at the Bauman Technical College in Moscow (1924–1944).

Among his many works, special mention must be made of his paper "Printsip otnositelnosti i geometria Lobachevskogo" ("The Principle of Relativity and Lobachevsky's Geometry"), on the relationship between physics and geometry, and "Teoria vektorov i kompleksnie chisla" ("The Theory of Vectors and Complex Numbers"), in which generalizations of the vector calculus and questions of non-Euclidean mechanics are again examined.

His papers on the theory of quaternions and complex numbers in application to geometry and mechanics are of considerable significance.

Kotelnikov edited and annotated the complete works of both Zhukovsky and Lobachevsky.

In 1934 Kotelnikov was named an Honored Scientist and Technologist of the R.S.F.S.R. In 1943 he was awarded the State Prize of the U.S.S.R.

BIBLIOGRAPHY

I. ORIGINAL WORKS. Among Kotelnikov's papers are *Vintovoe ischislenie i nekotorie primenenia ego k geometrii i mekhanike* ("The Cross-Product Calculus and Certain of its Applications in Geometry and Mechanics"; Kazan, 1885); *Proektivnaya teoria vektorov* ("The Projective Theory of Vectors"; Kazan, 1899); *Vvedenie v teoreticheskuyu mekhaniku* ("Introduction to Theoretical Mechanics"; Moscow–Leningrad, 1925); "Printsip otnositelnosti i geometria Lobachevskogo" ("The Principle of Relativity and Lobachevsky's Geometry"), in *In Memoriam N. I. Lobatschevskii*, II (Kazan, 1927); and "Teoria vektorov i kompleksnie chisla" ("The Theory of Vectors and Complex Numbers"), in *Nekotorie primenenia geometrii Lobachevskogo k mekhanike i fizike* ("Certain Applications of Lobachevsky's Geometry to Mechanics and Physics"; Moscow–Leningrad, 1950).

II. SECONDARY LITERATURE. See A. T. Grigorian, *Ocherki istorii mekhaniki v Rossi* ("Essays on the History of Mechanics in Russia"; Moscow, 1961); and B. A. Rosenfeld, "Aleksandr Petrovich Kotelnikov," in *Istoriko-matematicheskie issledovania*, IX (Moscow, 1956).

A. T. GRIGORIAN

KOTŌ, BUNJIRO (*b.* Tsuwano, Iwami [now Shimane prefecture], Japan, 4 March 1856; *d.* Tokyo, Japan, 8 March 1935), *geology, seismology.*

Born into a feudal samurai clan from Tsuwano, Kotō was the eldest son of Jisei Kotō. He went as a recommended student to Tokyo to study Western science. In 1879 he graduated from Tokyo Imperial University, where he was the first student to specialize in geology.

In 1881 Kotō entered the University of Leipzig, where he studied geology under Credner and petrography under Zirkel, a pioneer of microscopic petrology. Kotō entered the University of Munich in 1882 but received his Ph.D. from Leipzig in 1884. He became professor at Tokyo University in 1885 and directed teaching and research in the department of geology together with Tsunashiro Wada and Toyokichi Harada. He held this professorship until 1921.

In the early stage of his research Kotō studied mainly metamorphic rocks of Japan, and his papers introduced uniquely Japanese varieties of rock to geologists throughout the world. In these works Kotō proposed the terms Sambagawa system, Mikabu system, Takanuki system, and Gozaisho system.

In 1892, following the great Nōbi earthquake (28 October 1891) in which 7,000 persons were killed, an investigatory committee was formed by the ministry of education. Kotō was an active member and directed research on volcanoes. About this time his interest turned from the petrology of metamorphic rocks to volcanoes, earthquakes, and geotectonics. The photograph of the Neo-dani Fault, taken and published by Kotō in his paper on the Nōbi earthquake, was reprinted in many geology textbooks. Between 1893 and 1931 Kotō published many articles on earthquakes, volcanoes, morphology, and geotectonics. In "The Scope of the Volcanological Survey of Japan" (p. 93), he wrote:

> So-called tectonic fracture lines play a most important part. It is the key with which the structure and the origin of continents and oceans, mountains and lands, tablelands and basins, etc., are disclosed and explained. I must say plainly, that the chains of volcanoes, the system of mountains, and the nonvolcanic earthquake appear to me to have very intimate and fundamental relation with the so-called tectonic lines.

During the first years of the twentieth century Kotō published many papers on the geology and topography of the Korean peninsula. His works on the geotectonics of the Japanese islands of the Pacific Ocean date from about 1930.

Kotō had considerable administrative authority at Tokyo; and although at times apt to be dictatorial, he was professionally respected. Not until 1921 did he allow a lectureship of applied geology (in ore deposit) to be established.

The petrology of Kotō was descriptive. His geotectonic theory was essentially static and was based on the concept of the fracture-line. Late in his life geological research underwent a kind of revolution: experimental petrology was developed, and comparative tectonics, treating the movement of the earth's crust, appeared.

BIBLIOGRAPHY

I. Original Works. Among Kotō's more important papers are "Studies on Some Japanese Rocks," in *Quarterly Journal of the Geological Society of London*, no. 159 (1884), 431; "A Note on Glaucophane," in *Journal of the College of Science, Imperial University of Tokyo*, **1** (1887), 85; "Some Occurrence of Piedmontite in Japan," *ibid.*, 303; "On the So-Called Crystalline Schist of Chichibu," *ibid.*, **2** (1888), 77; "On the Cause of the Great Earthquake in Central Japan 1891," *ibid.*, **5** (1893), 295; "The Archaean Formation of the Abukuma Plateau," *ibid.*, 197; "The Scope of the Volcanological Survey of Japan," Earthquake Investigation Committee, no. 3 (1900), p. 89; "Topography of the Southern Part of Korea," in *Geological Magazine of Japan*, **13**, no. 150 (1901), 342, and no. 151 (1901), 413; "Topography of the Northern Part of Korea," *ibid.*, **14**, no. 162 (1902), 399, and no. 163 (1902), 467; "An Orographic Sketch of Korea," in *Journal of the College of Science, Imperial University of Tokyo*, **19** (1903), 1; "The Great Eruption of Sakurajima in 1914," *ibid.*, **38**, art. 3 (1916), 3; "On the Volcanoes of Japan," in *Journal of the Geological Society of Japan*, **23** (1916), 1, 17, 29, 77, 95; "The Rocky Mountain Arcs in Eastern Asia," in *Journal of the Faculty of Science, Tokyo University*, **3**, pt. 3 (1931), 131; and "The Seven Islands of Izu Province: A Volcanic Chain," *ibid.*, pt. 5 (1931).

II. Secondary Literature. There is a biography of Kotō in Japanese by Matajiro Yokoyama in *Journal of the Geological Society of Japan*, **42** (Apr. 1935), 39. A brief English biography with complete bibliography and portrait is T. A. Jaggar, "Memorial of Bunjiro Koto," in *Proceedings. Geological Society of America* (1936), pp. 263–272.

Hideo Kobayashi

KOVALEVSKY, ALEKSANDR ONUFRIEVICH (*b.* Daugavpils district, Vitebsk region, Russia [now Latvian S.S.R.], 19 November 1840; *d.* St. Petersburg, Russia, 22 November 1901), *embryology*.

Kovalevsky, the leading Russian embryologist of the late nineteenth century, was an adherent of Darwin's evolutionary theory. His numerous embryological studies of vertebrates and invertebrates established the occurrence of gastrulation by blastular invagination in a wide range of organisms and made a major contribution to the theory of germinal layers. His father, Onufry Osipovich Kovalevsky, was a Russianized Polish landowner of modest means; his mother, Polina Petrovna, was Russian. In 1856 Kovalevsky entered an engineering school in St. Petersburg; but in 1859, against the wishes of his father, he left it and enrolled in the natural sciences division of the Physico-Mathematics Faculty of St. Petersburg University, where he studied histology and microscopy under L. A. Tsenkovsky and zoology under S. S. Kutorga.

In the fall of 1860 Kovalevsky went to Heidelberg, where he worked in the laboratory of Ludwig Carius, publishing two works in organic chemistry, and attended lectures in zoology by G. K. Bronn. He spent three semesters at Tübingen before returning to St. Petersburg in 1862 to take his examinations and to prepare a thesis. He returned in August 1863 to Tübingen, where he studied microscopy and histology with F. Leidig.

In the summer of 1864 Kovalevsky traveled to Naples to begin the embryological investigations on amphioxus, tunicates (simple and complex ascidians), holothurians, Chaetognatha, *Phoronis,* and Ctenophora that launched the studies in comparative embryology which were to be almost his sole scientific concern for the next thirty-five years and which formed the basis for both his master's thesis (on amphioxus, 1865) and his doctoral dissertation (on *Phoronis*, 1866). These and later studies proved that a wide variety of organisms—coelenterates, echinoderms, worms, ascidians, and amphioxi—develop from a bilaminar sac (gastrula) produced by invagination. His work also showed that later developmental stages of the larvae of ascidians and amphioxi are similar (which finding contributed to their revised classification as chordates rather than mollusks), as are the mode of origin of equivalent organs in the embryos of worms, insects, and vertebrates, and that the nerve layers of insects and vertebrates are homologous. Theoretically, his work was seen as providing embryological evidence for the descent theory and as refuting the widely accepted view, implicit in Cuvier's work, that the organs of organisms from different *embranchements* cannot be homologous.

Kovalevsky apparently reached Naples in 1864 with a detailed plan of research which he subsequently followed. How this plan was formulated and how his intellectual outlook was formed are unclear: the relative importance of Tsenkovsky, N. D. Nozhin, Bronn, Leidig, Pagenstecher, Karl Ernst von Baer, Darwin's *Origin of Species*, and Fritz Müller's *Für Darwin* is disputed in the literature. But the importance of Kovalevsky's studies was quickly recognized by Baer, who nonetheless criticized their evolutionary tone; by Haeckel, who was greatly excited and generalized them well beyond Kovalevsky's conclusions into his own theory of the gastrula; and by Darwin, who saw them as providing embryological proofs for his theory of descent.

In the fall of 1866 K. F. Kessler, zoologist and rector of St. Petersburg University, appointed Kovalevsky curator of the zoological cabinet and privatdocent. He subsequently served on the faculties of Kazan University (1868–1869); the University of St. Vladimir in Kiev (1869–1873); Novorossisk University in Odessa (1873–1890), where for a time he served as prorector; and St. Petersburg University (1891–1894).

Kovalevsky is described by contemporaries as a shy man who had almost no social life, a man totally dedicated to science, a demanding and thorough teacher who much preferred research. His only nonscientific interest seemed to be his family; in 1867, the year his father died, he married Tatiana Kirillovna Semenova; they had three daughters. He also maintained close contacts with his younger brother, Vladimir, a paleontologist, and his sister-in-law Sonya, the mathematician.

Kovalevsky was active as a scientific organizer. He used his research trips to Naples, Trieste, Messina, Villefranche, Marseilles, the Red Sea, Algeria, and Sevastopol—which were almost annual—to make collections for Russian universities. At every university where he taught, he helped to found or was active in a natural history society; and he was instrumental in promoting Russian biological stations at Villefranche and Sevastopol and in furthering Russian participation in the Naples Station and at Messina.

During his lifetime Kovalevsky published nothing about politics; but privately he was not totally apolitical, especially in his youth, when a number of his closest friends were politically active. Both at Kiev (1873) and at Odessa (1881) he was distressed by the government's increasing interference in faculty appointments and university affairs; and in the 1880's Kovalevsky seriously considered leaving Russia to join A. F. Marion at Marseilles or A. Dohrn at Naples, where he hoped to find less interference and greater appreciation of his talents. By 1886 he was an honorary member of the Cambridge Philosophical Society and the Society of Naturalists of Modena; a corresponding member of the academies of sciences of Brussels and Turin; and a foreign member of the Royal Society; and he had won two prizes (1882, 1886) awarded by the French Academy of Sciences. He became a member of the Russian Academy of Sciences in 1890 and had to teach at St. Petersburg University as a professor of histology (1891–1894) in order to receive a pension.

BIBLIOGRAPHY

I. Original Works. During his lifetime Kovalevsky published over 100 monographs on vertebrate and invertebrate embryology. His writings include *Istoria razvitia Amphioxus lanceolatus ili Branchiostoma lumbricum* ("The Developmental History of *Amphioxus lanceolatus* or *Branchiostoma lumbricum*"), his master's thesis (St. Petersburg, 1865); "Beiträge zur Anatomie und Entwickelungsgeschichte des *Loxosoma neapolitanum,*" in *Mémoires de l'Académie impériale des sciences de St.-Pétersbourg,*

7th ser., **10**, no. 2 (1866); "Anatomie des Balanoglossus delle Chiaje," *ibid.*, no. 3 (1866), 1–18; "Entwickelungsgeschichte der Rippenquallen," *ibid.*, no. 4 (1866); "Entwickelungsgeschichte der einfachen Ascidien," *ibid.*, no. 15 (1866); "Entwickelungsgeschichte des *Amphioxus lanceolatus*," *ibid.*, **11**, no. 4 (1867); "Beiträge zur Entwickelungsgeschichte der Holothurien," *ibid.*, no. 6 (1867); "Anatomia i istoria razvitia *Phoronis*" ("Anatomy and Developmental History of *Phoronis*"), his doctoral dissertation, preface to *Zapiski Akademii nauk* (St. Petersburg), **11**, no. 1 (1867); "Untersuchungen über die Entwickelung der Coelenteraten," in *Nachrichten von der Gesellschaft der Wissenschaften zu Göttingen* (1868), no. 7, 154–159; "Beiträg zur Entwickelungsgeschichte der Tunikaten," *ibid.*, no. 19, 401–415; *Kratkii uchebnik zoologii* ("Short Textbook of Zoology"), 2nd rev. ed. (St. Petersburg, 1869); "Embryologische Studien an Würmern und Arthropoden," in *Mémoires de l'Académie impériale des sciences de St.-Pétersbourg*, 7th ser., **16**, no. 12 (1871), pp. 1–70; "Weitere Studien über die Entwickelung der einfachen Ascidien," in *Archiv für mikroskopische Anatomie*, **7** (1871), 101–130; "Sitzungsberichte der zoologischen Abtheilung der III. Versammlung russischer Naturforscher in Kiew," in *Zeitschrift für wissenschaftliche Zoologie*, **22**, no. 3 (1872), 283–304; "Nabliudenia nad razvitiem *Coelenterata*" ("Observations on the Development of *Coelenterata*"), in *Izvestia Obshchestva liubitelei estestvoznanii, antropologii i etnografii*, **10**, no. 2 (1873), 1–36; "Nabliudenia nad razvitiem *Brachiopoda*" ("Observations on the Development of *Brachiopoda*"), *ibid.*, **14** (1874), 1–40; "Ueber die Knospung der Ascidien," in *Archiv für mikroskopische Anatomie*, **10** (1874), 441–470; "Ueber die Entwickelungsgeschichte der *Pyrosoma*," *ibid.*, **11** (1875), 598–635; "Weitere Studien über die Entwickelungsgeschichte des *Amphioxus lanceolatus*, nebst einem Beitrag zur Homologie des Nervensystems der Würmer und Wirbelthiere," *ibid.*, **13** (1876), 181–208; "Documents pour l'histoire embryogénique des Alcyonaires," in *Annales du Musée d'histoire naturelle de Marseilles*, **1**, no. 4 (1883), 7–43, written with A. F. Marion; "Embryogénie du *Chiton Polii* (*Philipii*) avec quelques remarques sur le développement des autres Chitons," *ibid.*, Zoologie, **1**, no. 5 (1883), 5–37; "Étude sur l'embryologie du Dentale," *ibid.*, no. 7 (1883), 7–46; "Matériaux pour servir à l'histoire de l'Anchinie," in *Journal de l'anatomie et de la physiologie*, **19** (1883), 1–22, written with J. Barrois; "Beiträge zur Kenntnis der nachembryonalen Entwicklung der Musciden. 1 Theil," in *Zeitschrift für wissenschaftliche Zoologie*, **45** (1887), 542–588; "Ein Beitrag zur Kenntnis der Excretionsorgane," in *Biologisches Zentralblatt*, **9**, no. 2 (1889), 33–47; no. 3 (1889), 65–76; no. 4 (1889), 127–128; "Études expérimentales sur les glandes lymphatiques des invertébrés (communication préliminaire)," in *Mélanges biologiques, Bulletin de l'Académie impériale des sciences de St.-Pétersbourg*, **13** (1894), 437–459; "Étude des glandes lymphatiques de quelques Myriapodes," in *Archives de zoologie expérimentale et générale*, **3** (1896), 591–614; "Étude biologique de l'Haementeria costata Müller," in *Mémoires de l'Académie impériale des sciences de St.-Pétersbourg*, 8th ser., **11**, no. 10 (1900), 1–19; "Études anatomiques sur le genre Pseudovermis," *ibid.*, **12**, no. 4 (1901), 1–32; "Les Hedilidées, étude anatomique," *ibid.*, no. 6 (1901).

II. SECONDARY LITERATURE. See *Biograficheskii slovar' professorov i prepodavatelei imp. S.-Peterburgskago universiteta za istekshuiu tret'iu chetvert' veka ego sushchestvovaniia, 1869–1894* ("Biographical Dictionary of Professors and Teachers of the Imperial St. Petersburg University During the Third Quarter Century of its Existence, 1869–1894"), I (St. Petersburg, 1896), 320–324, with a bibliography of works by Kovalevsky; L. I. Bliakher, *Istoria embriologii v Rossii* (*s serediny XIX do serediny XX veka*). *Bespozvonochnye* ("The History of Embryology in Russia [From the Mid-nineteenth to the Mid-twentieth Century]. Invertebrates"; Moscow, 1959), with bibliographies of works by and on Kovalevsky; P. Buchinskii, "A. O. Kovalevskii. Ego nauchnye trudy i ego zaslugi v nauke" ("A. O. Kovalevskii. His Scientific Works and His Services to Science"), in *Zapiski Novorossiiskago obshchestva estestvoispytatelei*, **24**, no. 2 (1901–1902), 1–23; K. N. Davydov, "A. O. Kovalevskii i ego rol' v sozdanii sravnitel'noi embriologii" ("A. O. Kovalevskii and His Role in the Creation of Comparative Embryology"), in *Priroda* (1916), no. 4, 463–467; nos. 5/6, 579–598; and "A. O. Kovalevskii kak chelovek i kak uchenyi (Vospominaniia uchenika)" ("A. O. Kovalevskii as a Person and as a Scientist [Memoirs of a Student]"), in *Trudy Instituta istorii estestvoznaniia i tekhniki. Akademii nauk SSSR*, **31**, no. 6, 326–363; V. A. Dogel, "Embriologicheskie raboty A. O. Kovalevskogo v 60-80kh godakh XIX v" ("Embryological Works by A. O. Kovalevskii From the 1860's Through the 1880's"), in *Nauchnoe nasledstvo*, nat. sci. ser., **1** (1948), 206–218; and *A. O. Kovalevskii* (*1840–1901*) (Moscow–Leningrad, 1945), with bibliographies of works by and on Kovalevsky; A. E. Gaisinovich, "A. O. Kovalevskii i ego rol' v vozniknovenii evoliutsionnoi embriologii v Rossii" ("A. O. Kovalevskii and His Role in the Origin of Evolutionary Embryology in Russia"), in *Uspekhi sovremennoi biologii*, **36** (1953), 252–272; L. L. Gelfenbein, *Russkaia embriologia vtoroi poloviny XIX veka* ("Russian Embryology in the Second Half of the Nineteenth Century"; Kharkov, 1956); V. S. Ikonnikov, ed., *Biograficheskii slovar' professorov i prepodavatelei imp. Universiteta Sv. Vladimira* (*1834–1884*) ("Biographical Dictionary of Professors and Teachers of the University of St. Vladimir [1834–1884]"; Kiev, 1884), 264–268, with a bibliography of Kovalevsky's writings; P. P. Ivanov, "A. O. Kovalevskii i znachenie ego embriologicheskikh rabot" ("A. O. Kovalevskii and the Significance of His Embryological Works"), in *Izvestia Akademii nauk SSSR*, biological ser. (1940), no. 6, 819–830; A. G. Knorre, "A. O. Kovalevskii—osnovopolozhnik sravnitel'noi embriologii (k 100-letiiu so dnia rozhdeniia)" ("A. O. Kovalevskii—Founder of Comparative Embryology [for the 100th Anniversary of His Birth]"), in *Uspekhi sovremennoi biologii*, **13**, no. 2 (1940), 195–206; V. A. Kovalevskaia-Chistovich, "Aleksandr Onufrievich Kovalevskii. Vospominaniia docheri" ("Aleksandr Onufrievich Kovalevskii. Memoirs of a Daughter"), in *Priroda* (1926), nos. 7–8,

5–20; T. V. Makarova, "Aleksandr Onufrievich Kovalevskii v Peterburgskom universitete" ("Aleksandr Onufrievich Kovalevskii at Petersburg University"), in *Trudy Instituta istorii estestvoznaniia i tekhniki, Akademii nauk SSSR*, **24**, 222–254; A. I. Markevich, *Dvadtsatipiatiletie imp. Novorossiiskogo universiteta. Istoricheskaia zapiska* ("Twenty-fifth Anniversary of the Imperial Novorossisk University. Historical Note"; Odessa, 1890), pp. 457–661; V. F. Mirek, "Aleksandr Onufrievich Kovalevskii (1840–1901)," in *Liudi Russkoi nauki* ("People of Russian Science"), II (Moscow-Leningrad, 1948), 705–715, with a bibliography of works on Kovalevsky; E. Ray-Lankaster, "Alexander Kowalevsky," in *Nature*, **66**, no. 1712 (1902), 394–395; V. Shimkevich, "A. O. Kovalevskii (nekrolog)," in *Obrazovanie* (1901), no. 11, 107–114; A. D. Nekrasov and N. M. Artemov, eds., *A. O. Kovalevskii. Izbrannye raboty* ("A. O. Kovalevskii. Selected Works"; Moscow–Leningrad, 1951), with biographical essay, pp. 536–621, commentary by the eds., and bibliographies of works by and on Kovalevsky; V. L. Omelianskii, "Razvitie estestvoznaniia v Rossii v posledniuiu chetvert' veka" ("The Development of Science in Russia in the Last Quarter-Century"), in *Istoria Rossii v XIX veke*, IX, 116–142; Iu. I. Polianskii, I. I. Sokolov, and L. K. Kuvanova, eds., *Pis'ma A. O. Kovalevskogo k I. I. Mechnikovu (1866–1900)* ("A. O. Kovalevskii's Letters to I. I. Mechnikov [1866–1900]"; Moscow–Leningrad, 1955); I. I. Puzanov, "Aleksandr Onufrievich Kovalevskii, ego zhizn' i znachenie v mirovoi nauke" ("Aleksandr Onufrievich Kovalevskii, His Life and Significance for World Science"), in *Trudy Odesskogo derzhavnogo universiteta*, **145** (1955), 5–19; S. Ia. Shtraikh, *Sem'ia Kovalevskikh* ("The Kovalevskii Family"; Moscow, 1948); and "Iz perepiski V. O. Kovalevskogo" ("From the Correspondence of V. O. Kovalevskii"), in *Nauchnoe nasledstvo*, nat. sci. ser., **1** (1948), 219–423; and V. V. Zalenskii, "A. O. Kovalevskii," in *Izvestia Akademii nauk*, **15** (1901), xci–xciv; and "Spisok sochinenii Akademika A. O. Kovalevskago" ("List of the Works of A. O. Kovalevskii"), in *Izvestia imperatorskoi Akademii nauk*, **22**, no. 1 (1905), 1–4.

MARK B. ADAMS

KOVALEVSKY, SONYA (or **Kovalevskaya, Sofya Vasilyevna**) (*b.* Moscow, Russia, 15 January 1850; *d.* Stockholm, Sweden, 10 February 1891), *mathematics*.

Sonya Kovalevsky was the greatest woman mathematician prior to the twentieth century. She was the daughter of Vasily Korvin-Krukovsky, an artillery general, and Yelizaveta Shubert, both well-educated members of the Russian nobility. The general was said to have been a direct descendant of Mathias Korvin, king of Hungary; Soviet writers believe that Krukovsky's immediate background was Ukrainian and that his family coat of arms resembled the emblem of the Polish Korwin-Krukowskis.

In *Recollections of Childhood* (and the fictionalized version, *The Sisters Rajevsky*), Sonya Kovalevsky vividly described her early life: her education by a governess of English extraction; the life at Palabino (the Krukovsky country estate); the subsequent move to St. Petersburg; the family social circle, which included Dostoevsky; and the general's dissatisfaction with the "new" ideas of his daughters. The story ends with her fourteenth year. At that time the temporary wallpaper in one of the children's rooms at Palabino consisted of the pages of a text from her father's schooldays, namely, Ostrogradsky's lithographed lecture notes on differential and integral calculus. Study of that novel wall-covering provided Sonya with her introduction to the calculus. In 1867 she took a more rigorous course under the tutelage of Aleksandr N. Strannolyubsky, mathematics professor at the naval academy in St. Petersburg, who immediately recognized her great potential as a mathematician.

Sonya and her sister Anyuta were part of a young people's movement to promote the emancipation of women in Russia. A favorite method of escaping from bondage was to arrange a marriage of convenience which would make it possible to study at a foreign university. Thus, at age eighteen, Sonya contracted such a nominal marriage with Vladimir Kovalevsky, a young paleontologist, whose brother Aleksandr was already a renowned zoologist at the University of Odessa. In 1869 the couple went to Heidelberg, where Vladimir studied geology and Sonya took courses with Kirchhoff, Helmholtz, Koenigsberger, and du Bois-Reymond. In 1871 she left for Berlin, where she studied with Weierstrass, and Vladimir went to Jena to obtain his doctorate. As a woman, she could not be admitted to university lectures; consequently Weierstrass tutored her privately during the next four years. By 1874 she had completed three research papers on partial differential equations, Abelian integrals, and Saturn's rings. The first of these was a remarkable contribution, and all three qualified her for the doctorate *in absentia* from the University of Göttingen.

In spite of Kovalevsky's doctorate and strong letters of recommendation from Weierstrass, she was unable to obtain an academic position anywhere in Europe. Hence she returned to Russia where she was reunited with her husband. The couple's only child, a daughter, "Foufie," was born in 1878. When Vladimir's lectureship at Moscow University failed to materialize, he and Sonya worked at odd jobs, then engaged in business and real estate ventures. An unscrupulous company involved Vladimir in shady speculations that led to his disgrace and suicide in 1883. His widow turned to Weierstrass for assistance and, through the efforts of the Swedish analyst

Gösta Mittag-Leffler, one of Weierstrass' most distinguished disciples, Sonya Kovalevsky was appointed to a lectureship in mathematics at the University of Stockholm. In 1889 Mittag-Leffler secured a life professorship for her.

During Kovalevsky's years at Stockholm she carried on her most important research and taught courses (in the spirit of Weierstrass) on the newest and most advanced topics in analysis. She completed research already begun on the subject of the propagation of light in a crystalline medium. Her memoir, *On the Rotation of a Solid Body About a Fixed Point* (1888), won the Prix Bordin of the French Academy of Sciences. The judges considered the paper so exceptional that they raised the prize from 3,000 to 5,000 francs. Her subsequent research on the same subject won the prize from the Swedish Academy of Sciences in 1889. At the end of that year she was elected to membership in the Russian Academy of Sciences. Less than two years later, at the height of her career, she died of influenza complicated by pneumonia.

In mathematics her name is mentioned most frequently in connection with the Cauchy-Kovalevsky theorem, which is basic in the theory of partial differential equations. Cauchy had examined a fundamental issue in connection with the existence of solutions, but Sonya Kovalevsky pointed to cases that neither he nor anyone else had considered. Thus she was able to give his results a more polished and general form. In short, Cauchy, and later Kovalevsky, sought necessary and sufficient conditions for the solution of a partial differential equation to exist and to be unique. In the case of an ordinary differential equation the general solution contains arbitrary constants and therefore yields an infinity of formulas (curves); in the general solution of a partial differential equation, arbitrary functions occur and the plethora of formulas (surfaces or hypersurfaces) is even greater than in the ordinary case. Hence additional data in the form of "initial" or "boundary" conditions are needed if a unique particular solution is required.

The simplest form of the Cauchy-Kovalevsky theorem states that any equation of the form

$$p = f(x, y, z, q)$$

where $p = \partial z/\partial x$, $q = \partial z/\partial y$, and the function f is analytic (has convergent power series development) in its arguments for values near (x_0, y_0, z_0, q_0), possesses one and only one solution $z(x, y)$ which is analytic near (x_0, y_0) and for which

$$z(x_0, y) = g(y)$$

where $g(y)$ is analytic at y_0 with

$$g(y_0) = z_0 \quad \text{and} \quad g'(y_0) = q_0$$

In the general theorem, the simple case illustrated is generalized to functions of more than two independent variables, to derivatives of order higher than the first, and to systems of equations.

To place Sonya Kovalevsky's second doctoral paper and some of her later research in a proper setting, one must examine analytic concepts developed gradually in the work of Legendre, Abel, Jacobi, and Weierstrass. It is a familiar fact of elementary calculus that the integral,

$$\int f(x, y)\, dx,$$

can be expressed in terms of elementary functions (algebraic, trigonometric, inverse trigonometric, exponential, logarithmic) if y^2 is a polynomial of degree 1 or 2 in x, and $f(x, y)$ is a rational function of x and y. If the degree of the polynomial for y^2 is greater than 2, elementary expression is not generally possible. If the degree is 3 or 4, the integral is described as *elliptic* because a special case of such an integral occurs in the problem of finding the length of an arc of an ellipse. If the degree is greater than 4, the integral is called *hyperelliptic*. Finally, one comes to the general type that includes the others as special cases. If y is an algebraic function of x, that is, if y is a root of $P(x, y) = 0$, where P is a polynomial in x and y, the above integral is described as *Abelian*, after Abel, who carried out the first important research with such integrals. Abel's brilliant inspiration also clarified and simplified the theory of elliptic integrals (just after Legendre had given some forty years to investigating their properties).

If the integral

$$u = \int_0^x \frac{dt}{\sqrt{1 - t^2}} = \sin^{-1} x$$

is "inverted," one obtains $x = \sin u$, which elementary trigonometry indicates to be easier to manipulate than its inverse, $u = \sin^{-1} x$. Therefore it occurred to Abel (and subsequently to Jacobi) that the inverses of elliptic integrals might have a simpler theory than that of the integrals themselves. The conjecture proved to be correct, for the inverses, namely the *elliptic functions*, lend themselves to a sort of higher trigonometry of doubly periodic functions. For example, while the period of $\sin x$ is 2π, the corresponding elliptic function, sn z, has two periods whose ratio is a complex number, a fact indicating that the theory of elliptic functions belongs to complex (rather than real) analysis. Inversion of Abelian integrals leads to *Abelian functions* which, in the first generalization beyond the elliptic functions, have two independent complex variables and four periods.

Abel died within a year of the research he started in that area, and there was left to Weierstrass and his pupils the stupendous task of developing the theory of general Abelian functions having k complex variables and $2k$ periods and of considering the implications for the inverses, the corresponding Abelian integrals. Kovalevsky's doctoral research contributed to that theory by showing how to express a certain species of Abelian integral in terms of the relatively simpler elliptic integrals.

Complex analysis and nonelementary integrals were also a feature of the Kovalevsky paper which won the Bordin Prize. In her paper she generalized work of Euler, Poisson, and Lagrange, who had considered two elementary cases of the rotation of a rigid body about a fixed point. Her predecessors had treated two symmetric forms of the top or the gyroscope, whereas she solved the problem for an asymmetric body. This case is an exceedingly difficult one and she was able to solve the differential equations of motion by the use of hyperelliptic integrals. Her solution was so general that no new case of rotatory motion about a fixed point has been researched to date.

In her study of the form of Saturn's rings, as in her other research, she had great predecessors—Laplace, in particular, whose work she generalized. Whereas, for example, he thought certain cross sections to be elliptical, she proved that they were merely egg-shaped ovals symmetric with respect to a single axis. Although Maxwell had proved that Saturn's rings could not possibly be continuous bodies—either solid or molten—and hence must be composed of a myriad of discrete particles, Kovalevsky considered the general problem of the stability of motion of liquid ring-shaped bodies; that is, the question of whether such bodies tend to revert to their primary motion after disturbance by external forces or whether deviation from that motion increases with time. Other researchers completed her task by establishing the instability of such motion.

Her concern for Saturn's rings caused the British algebraist Sylvester to write a sonnet (1886) in which he named her the "Muse of the Heavens." Later, Fritz Leffler, the mathematician's brother, stated in a poetic obituary,

> While Saturn's rings still shine,
> While mortals breathe,
> The world will ever remember your name.

She was remembered by the eminent Russian historian Maxim Kovalevsky (who was unrelated to her husband) who dedicated several works to her. She had met him when he came to lecture at Stockholm University in 1888 after he had been discharged from Moscow University for criticizing Russian constitutional law. It was believed that they were engaged to be married but that she hesitated because his new permanent position was in Paris, and joining him there would have meant sacrificing the life professorship for which she had worked so long and hard.

She was remembered, too, by her daughter who, at the age of seventy-two, was guest speaker when the centenary of her mother's birth was celebrated in the Soviet Union. After her mother's death, Foufie had returned to Russia to live at the estate of her godmother Julia Lermontov, a research chemist and agronomist, and a good friend from Sonya's Heidelberg days. Foufie studied medicine and translated major foreign literary works into Russian.

An unusual aspect of Sonya Kovalevsky's life was that, along with her scientific work, she attempted a simultaneous career in literature. The titles of some of her novels are indicative of their subject matter: *The University Lecturer*, *The Nihilist* (unfinished), *The Woman Nihilist*, and, finally, *A Story of the Riviera.* In 1887 she collaborated with her good friend and biographer, Mittag-Leffler's sister, Anne Charlotte Leffler-Edgren (later Duchess of Cajanello), in writing a drama, *The Struggle for Happiness*, which was favorably received when it was produced at the Korsh Theater in Moscow. She also wrote a critical commentary on George Eliot, whom she and her husband had visited on a holiday trip to England in 1869.

BIBLIOGRAPHY

I. Original Works. Among Kovalevsky's papers are "Zur Theorie der partiellen Differential-gleichungen," in *Journal für die reine und angewandte Mathematik*, **80** (1875), 1–32; "Zusätze und Bemerkungen zu Laplaces Untersuchungen über die Gestalt der Saturnsringe," in *Astronomische Nachrichten*, **3** (1883), 37–48; "Über die Reduction einer bestimmten Klasse Abelscher Integrale dritten Ranges auf elliptische Integrale," in *Acta Mathematica*, **4** (1884), 393–414; and "Sur le problème de la rotation d'un corps solide autour d'un point fixe," in *Acta Mathematica*, **12** (1889), 177–232.

II. Secondary Literature. See E. T. Bell, *Men of Mathematics* (New York, 1937), 423–429; J. L. Geronimus, *Sofja Wasilyevna Kowalewskaja—Mathematische Berechnung der Kreiselbewegung* (Berlin, 1954); E. E. Kramer, *The Main Stream of Mathematics* (New York, 1951), 189–196, and *The Nature and Growth of Modern Mathematics* (New York, 1970), 547–549; A. C. Leffler-Edgren, duchessa di Cajanello, *Sonia Kovalevsky, Biography and Autobiography*, English trans., L. von Cossel (New York, 1895); O. Manville, "Sophie Kovalevsky," in *Mélanges scienti-*

fiques offerts à M. Luc Picart (Bordeaux, 1938); G. Mittag-Leffler, "Sophie Kovalevsky, notice biographique," in *Acta Mathematica*, **16** (1893), 385–390; and P. Polubarinova-Kochina, *Sophia Vasilyevna Kovalevskaya, Her Life and Work*, English trans., P. Ludwick (Moscow, 1957).

EDNA E. KRAMER

KOVALEVSKY, VLADIMIR ONUFRIEVICH (*b.* Dünaberg, Vitebsk region, Russia [now Daugavpils, Latvian S.S.R.], 14 August 1842; *d.* Moscow, Russia, 28 April 1883), *paleontology*.

One of the founders of evolutionary paleontology, Kovalevsky graduated in 1861 from the School of Jurisprudence. Thereafter he was engaged in publishing, doing translations and editing books of Alfred Braem, Darwin, Lyell, L. Agassiz, and many others. In 1869 he married Sonya (Sofya) Korvin-Krukovsky (see article above). From 1869 to 1874 he attended lectures on various aspects of natural science in Heidelberg, Munich, Würzburg, and Berlin; made geological observations and collected fossils in northern Italy and southern France; and studied paleontological collections in the museums of Germany, France, Holland, and Great Britain.

In 1872 Kovalevsky passed his doctoral examinations in Jena and submitted a thesis on the paleontological history of horses; this was later the subject of his master's degree (1875). He was associate professor at Moscow University from 1880 to 1883.

The paleontological researches of Kovalevsky deal with the evolution of morphological characteristics of the teeth apparatus and skull of mammals as related to change of plant food composition; and with the phylogeny of ungulates, particularly of horses and pigs. Basing his evolutionary argumentation on Darwin's theory, Kovalevsky established the conception of inadaptive and adaptive evolution in the special case of the extremities of the ungulates. He suggested that adaptive reduction ensured survival, but that nonadaptive reduction—of the fingers of the *Entelodon* giant pig, for example—could not save a species from extinction. Kovalevsky was the first to attempt to construct the genealogy of hoofed animals, in particular the horses. Developing Darwin's views on divergency, he advanced the idea of adaptive radiation as a means of evolutionary transformation.

The opinion put forward (by E. Koken, R. Hoernes, C. Diener, and O. Abel) that Kovalevsky is a forerunner of E. Cope and H. Osborn, the founders of Neo-Lamarckism in paleontology, is groundless. Kovalevsky was a consistent Darwinist and attributed evolutionary changes in fossil forms not to autogenesis, nor to use or disuse of parts, but to natural selection.

BIBLIOGRAPHY

I. ORIGINAL WORKS. Kovalevsky's works include "Sur l'Anchitherium aurelianense Cuv. et sur l'histoire paléontologique des chevaux," in *Zapiski Imperatorskoi akademii nauk*, 7th ser., **20**, no. 5 (1873), 1–73; "On the Osteology of the Hyopotamidae," in *Philosophical Transactions of the Royal Society*, **163** (1873), 19–94; "Monographie der Gattung Antracotherium Cuv. und Versuch einer natürlichen Classification der fossilen Hufttier," in *Paleontographica*, no. 3 (1873), 131–210, no. 4 (1874), 211–290; and "Ostéologie des Genus Gelocus Aym.," in *Paleontographica* (1876), 415–450, (1877), 145–162.

II. SECONDARY LITERATURE. See A. A. Borisiak, *V. O. Kovalevsky, His Life and Scientific Works* (Moscow, 1928); L. Sh. Davitashvili, *V. O. Kovalevsky* (Moscow, 1946).

L. J. BLACHER

KOVALSKY, MARIAN ALBERTOVICH (VOYTEKHOVICH) (*b.* Dobrzhin, Russia [now Dobrzyn nad Wisła, Poland], 15 August 1821; *d.* Kazan, Russia, 28 May 1884), *astronomy*.

The son of a minor official, Kovalsky graduated from the Gymnasium in the city of Płock in 1840. From the fall of 1841 he studied mathematics at St. Petersburg University, supporting himself and his younger brother by giving private lessons. In 1845 Kovalsky graduated from the university with the degree of candidate and a gold medal for his work "O printsipakh mekhaniki" ("On the Principles of Mechanics"). In 1847 he defended his dissertation for the master's degree, "O vozmushcheniakh v dvizhenii komet" ("On Perturbations in the Motion of Comets"). Working in 1846 at Pulkovo Observatory, Kovalsky made astronomical observations and calculations, in addition to studying the basic works on celestial mechanics of Laplace, Lagrange, Poisson, and P. A. Hansen.

In 1847 Kovalsky was invited by the Russian Geographical Society to join an expedition to the Urals to determine astronomical coordinates from Cherdyn to the Arctic Ocean. Over a two-year period Kovalsky determined the coordinates of 186 geographical points and the altitudes of seventy-two points. He determined for the first time the elements of earth magnetism for five points in the Northern Urals. The result was Kovalsky's work *Severny Ural i beregovoy khrebet Pay-Khoy* ("The Northern Urals and the Pay-Khoy Coastal Range," 1853). On the recommendation of W. Struve, director of the

Pulkovo Observatory, Kovalsky was invited to Kazan as assistant in the department of astronomy, and in September 1850 he began lecturing on astronomy and geodesy.

For nearly thirty-five years Kovalsky almost singlehandedly did all teaching of astronomy at the university. In 1852, having defended his doctoral dissertation, "Teoria dvizhenia Neptuna" ("Theory of the Motion of Neptune"), he became extraordinary professor and in 1854, ordinary professor. From 1855 he was also director of the Kazan Observatory, and from 1862 to 1868 and 1871 to 1882 he was dean of the faculty of physics and mathematics of the university. In 1863 Kovalsky was elected corresponding member of the Academy of Sciences and foreign member of the Royal Astronomical Society in London.

In 1867 the first congress of the Astronomische Gesellschaft was held in Bonn; Kovalsky had participated in its organization since 1864. At this congress it was decided to coordinate observations on the meridian circles of all stars of the well-known Bonner Durchmusterung catalog. The zone from $-75°$ to $+80°$ was assigned to Kazan, and Kovalsky set up an extensive program for these observations. A catalog of 4,218 stars to magnitude 9.5 was published in 1887 by D. I. Dubyago. In 1869 St. Petersburg University elected Kovalsky an honorary member, and Kazan (1875) and Kiev (1884) universities followed suit. In 1875 Kovalsky received the title of distinguished professor.

In 1856 Kovalsky married the daughter of a Nizhny Novgorod physician, Henriette Serafimovna Gatsisskaya. Their son Aleksandr became an astronomer at Pulkovo.

Kovalsky's contributions were especially important in the areas of celestial mechanics, astronomy, and stellar astronomy. His first important work (1852) on celestial mechanics was his doctoral dissertation on the theory of motion of Neptune, the existence of which had been predicted in 1846 by Le Verrier and J. C. Adams. In 1852 Kovalsky conducted a detailed study of perturbations from the large planets, and in 1853 he obtained on the meridian circle a series of observations for a more accurate definition of the orbit of Neptune. Kovalsky's complete theory of Neptune's motions (1855), including positional predictions for the planet (ephemerides) served as a source for Newcomb in his reexamination (1864) of the theory of planetary motion for the entire solar system. Two other works of Kovalsky also deal with celestial mechanics: "O vozmushcheniakh v dvizhenii komet" ("On Perturbations in the Motion of Comets," 1847) and "Développement de la fonction perturbatrice en série" (1859).

Theoretical astronomy was represented by Kovalsky's work on the improvement of the elliptical orbit based on many observations by means of the method of differential corrections (1860), and by the memoir "Ob opredelenii ellipticheskoy orbity . . ." ("On the Determination of the Elliptical Orbit of the Planets . . .," 1873). In this work Kovalsky uses, instead of the classical method of Gauss, the theorem of Euler-Lambert. This theorem makes it possible to obtain the major axis from a simple expression, which includes a rapidly converging series permitting any desired degree of precision in determining the unknown quantity.

Of great interest is Kovalsky's report on the well-known Bertrand's problem (published from the manuscript copy only in 1951 in the Martynov edition of Kovalsky's works). In "O zatmenniakh" ("On Eclipses," 1856), Kovalsky substantially simplified and improved the computation of all the circumstances of solar eclipses and occultation of stars by the moon, by means of a theory that was much simpler and more precise than Bessel's. Published only in Russian, it did not receive wide recognition. Since the advent of electronic computers, however, Kovalsky's method for computing occultations has proved to be the most satisfactory.

Kovalsky's analytical method of determining the elements of the orbits of double stars, presented in his official opinion on V. N. Vinogradsky's dissertation, is also widely known. Kovalsky's fundamental work on the theory of refraction (1878), partially based on his own observations of stars at very low altitudes over the horizon, included new tables of refraction.

Kovalsky's important theoretical work (1860) on the analysis of the proper motion of 3,136 stars of Bradley's catalog presented the first practical method of discovering the rotation of the Galaxy from the proper motion of the stars. (Final confirmation of the rotation was not obtained until 1927 by J. H. Oort.) Kovalsky showed the impossibility of the existence of a massive central body in the Galaxy, that is, one which would play a role analogous to that of the sun in our planetary system; J. Mädler had spoken persistently of such a central body since 1846. At the same time, Kovalsky developed a method of determining the elements of the motion of the sun in space among the stars; although this method is named after Airy, it could fairly be called the Kovalsky-Airy method. One of Kovalsky's methods of analyzing stellar motions was the compilation of so-called polar diagrams, later used successfully by J. Karteyn in his treatment (1904) of his well-known theory of two star streams.

Although much of Kovalsky's work did not receive wide recognition in his time, his influence on the development of astronomy in the nineteenth century is indisputable.

BIBLIOGRAPHY

I. ORIGINAL WORKS. Kovalsky's selected works in astronomy were published as *Izbrannye raboty po astronomii*, D. Y. Martynov, ed. (Moscow, 1951). Separate works include *Teoria dvizhenia Neptuna* ("Theory of the Motion of Neptune"; Kazan, 1852), his doctoral diss.; *Severny Ural i beregovoy khrebet Pay-Khoy: Geograficheskie opredelenia mest i magnitnye nablyudenia M. Kovalskogo ekstraordinarnogo professora astronomii v Kazanskom universitete* ("The Northern Urals and the Pay-Khoy Coastal Range: Geographical Determinations of Locations and Magnetic Observations by M. Kovalsky, Extraordinary Professor of Astronomy at Kazan University"; St. Petersburg, 1853); "O zatmeniakh" ("On Eclipses"), in *Sbornik uchenykh statey, napisannykh professorami Imp. Kazanskogo universiteta v pamyat pyatidesyatiletia ego sushchestvovania*, I (Kazan, 1856), 341–478, also separately published (Kazan, 1856), 1–138.

Subsequent works are "Développement de la fonction perturbatrice en série," in *Uchenye zapiski izdavaemye Imperatorskim Kazanskim universitetom* (1860), 94–155, repr. in *Recherches astronomiques de l'observatoire de Kasan*, no. 1 (1859), 107–168; his short paper on this work is "Développement de la fonction perturbatrice en série (Abstract)," in *Monthly Notices of the Royal Astronomical Society*, **21** (1861), 37–38; "Sur les lois du mouvement propre des étoiles de Bradley," in *Uchenye zapiski izdavaemye Imperatorskim Kazanskim universitetom*, no. 1 (1860), 47–136, repr. in *Recherches astronomiques de l'observatoire de Kasan*, no. 1 (1859), 1–90; "Sur le calcul de l'orbite elliptique ou parabolique d'après un grand nombre d'observations," in *Uchenye zapiski izdavaemye Imperatorskim Kazanskim universitetom*, no. 1 (1860), 166–181, repr. in *Recherches astronomiques de l'observatoire de Kasan*, no. 1 (1859), 91–106.

Kovalsky's review of V. N. Vinogradsky's diss., "Ob opredelenii elementov dvoynykh zvezd" ("On the Determination of the Elements of Binary Stars"), in *Izvestiya i uchenye zapiski Kazanskago universiteta*, **10**, no. 2 (1873), 329–339, contains a statement of Kovalsky's method, which was widely used—see, for example, S. P. Glazenapp, "On a Graphical Method for Determining the Orbit of a Binary Star," in *Monthly Notices of the Royal Astronomical Society*, **49** (1889), 276–280; B. P. Modestoff, "Sur la méthode de Kowalski pour le calcul des orbites des étoiles doubles," in *Annales de l'observatoire astronomique de Moscou*, **3**, no. 2 (1896), 82–87; and W. M. Smart, "On the Derivation of the Elements of a Visual Binary Orbit by Kowalsky's Methods," in *Monthly Notices of the Royal Astronomical Society*, **90** (1930), 534–538.

See also *Recherches sur la réfraction astronomique* (Kazan, 1878); "Ob opredelenii ellipticheskoy orbity planet pomoshchiyu dvukh dannykh radiusov-vektorov, ugla, mezhdu nimi zaklyuchayushchegosya, i vremeni, upotreblennogo na opisanie etogo ugla" ("On the Determination of the Elliptical Orbit of the Planets With the Aid of Two Given Radius Vectors, the Angle Between Them, and the Time Required to Describe This Angle"), in *Izvestiya i uchenye zapiski Kazanskago universiteta*, **12**, no. 2 (1875), 289–312, and in French in *Bulletin de l'Académie impériale des sciences de St. Petersbourg*, **20** (1875), 559–571.

A series of eight lithographed courses in various problems of astronomy and geodesy were published from 1859 to 1882.

II. SECONDARY LITERATURE. On Kovalsky and his work see the obituaries in *Vierteljahrsschrift der Astronomischen Gesellschaft*, **19** (1884), 172–179; and *Monthly Notices of the Royal Astronomical Society*, **45** (1885), 208–211.

Other works are D. Y. Martynov, "Ob odnoy zabytoy rabote M. A. Kovalskogo" ("On One Forgotten Work of M. A. Kovalsky"), in *Astronomichesky zhurnal SSSR*, **27**, no. 3 (1950), 169–176; and "Marian Albertovich Kovalsky. Biografichesky ocherk" ("Biographical Sketch"), in *Izbrannye raboty po astronomii* (Moscow, 1951), 7–40, with complete annotated bibliography of 34 titles on pp. 40–48; A. A. Mikhaylovsky, "Marian Albertovich Kovalsky," in *Biografichesky slovar professorov i prepodavateley Kazanskogo Universiteta 1804–1904*, pt. 1 (Kazan, 1904), 358–365; Yu. G. Perel, "Marian Albertovich Kovalsky," in *Vydayushchiesya russkie astronomy* (Moscow, 1951), 108–122; P. Rybka, "M. Kowalsky," in *Problemy* (Warsaw), **14**, no. 2 (1958), 837–838, in Polish; and O. Struve, "M. A. Kovalsky and His Work on Stellar Motions," in *Sky and Telescope*, **23**, no. 5 (1962), 250–252.

P. G. KULIKOVSKY

KOWALEWSKY. See **Kovalevsky.**

KOYRÉ, ALEXANDRE (*b.* Taganrog, Russia, 29 August 1892; *d.* Paris, France, 28 April 1964), *history of science, of philosophy, and of ideas.*

Koyré's work was threefold. First, he exercised a formative influence upon an entire generation of historians of science, and especially in the United States. In France, secondly, where his circle was mainly philosophical, he also initiated the revival of Hegelian studies in the 1930's and published important studies of other pure philosophers, most notably Spinoza [6]. Thirdly, his essays on Russian thought and philosophical sensibility were important contributions to the intellectual history of his native country [4, 11]. A strong vein of philosophical idealism inspired all his writings, which proceeded from the assumption that the object of philosophical reasoning is reality, even when the subject is religious.

A remark in the preface to his study of Jacob Boehme might equally well be applied to any of his books: "We believe . . . that the system of a great philosopher is inexhaustible, like the very reality of which it is an expression, like the master intuition that dominates it."[1]

For Koyré was ever a Platonist. Indeed, the best introduction to the unity of view and value inspiring the whole body of his work is his beautiful essay *Discovering Plato* [9], published in 1945 in French and English editions in New York, and originally composed in the form of lectures given in Beirut after the fall of France in 1940. Koyré never despaired of European civilization, however Hellenic its apparent disintegration. It was always his inner belief that mind might yet prevail. The contemplative tone disarms resistance to the hortatory discourse, which, mingling jest with seriousness in true Platonic style, opens to the reader the implications of philosophy for personality and of personality for politics, those being the themes that invest the dialogues with dramatic tension.

Koyré said little here of Platonism in the development of science, but the relation of intellect to character and of personal excellence to civic responsibility that this essay brings out explains his sympathy for the Platonic inspiration that he detected (and in other writings perhaps exaggerated) in the motivations of the founders of modern science, particularly Galileo.

Koyré began his secondary education at Tiflis and completed it at the age of sixteen at Rostov-on-Don. His father, Vladimir, was a prosperous importer of colonial products and successful investor in the Baku oil fields. Husserl was the idol of Koyré's schooldays, and in 1908 he went to Göttingen, where, besides the master of phenomenology he had come to follow, he also met Hilbert and attended his lectures in higher mathematics. In 1911 he moved on to Paris and the Sorbonne, where he listened to Bergson, Victor Delbos, André Lalande, and Léon Brunschvicg. Although he did not become as familiar with any of his teachers in Paris as he had with Husserl and his family (Frau Husserl had mothered him a bit), he felt at ease in the cooler climate of French civilization.

Before the war he had already begun work on a thesis on Saint Anselm under the direction of François Picavet, then teaching at the École Pratique des Hautes Études. In 1914 Koyré, though not yet a citizen, enlisted in the French army and fought in France for two years. Then he transferred his service to a Russian regiment when a call came for volunteers and went back to Russia, where he continued to fight on the southwestern front until the collapse in October 1917. During the civil war that followed, Koyré found himself among opposition groups which can be best compared to resistance forces, fighting against both Reds and Whites. After a time, he decided to disengage himself from the melee, and, the war being over, he made his way back to Paris. There he was married with great happiness to Dora Rèybermann, daughter of an Odessa family. Her sister also married his elder brother. In Paris he resumed a life of scholarship and philosophy, finding to his astonishment that the proprietor of the hotel where he had lodged in his student days had faithfully preserved the manuscript of his thesis on Anselm throughout the war.

Always a philosopher in his own sense of professional identity, Koyré began his career in the study of religious thought, though it was in the history of science that he later did his deepest work. His first books were theological: *Essai sur l'idée de Dieu et les preuves de son existence chez Descartes* (1922), *L'idée de Dieu dans la philosophie de St. Anselme* (1923), and *La philosophie de Jacob Boehme* (1929). Completion of the first qualified him for the diploma of the École Pratique and won him election as *chargé de conférences*, or lecturer, in that institution, with which he remained associated throughout his life. The work on Anselm, completed earlier, was published later and satisfied the Sorbonne's requirements for the university doctorate, a degree elevated into the *doctorat d'État* by virtue of the Boehme thesis.

Students of Koyré's later writings on the history of science will recognize characteristic motifs and methods in the analysis he gave these early subjects. The theological tradition that appealed to him was that most highly intellectualized of apologetic strategies, the ontological argument for the existence of God. In the versions given both by its originator, Anselm, and by Descartes, mind rather than religious experience made the connection between personal existence apprehended subjectively and external reality, of which the important aspect in this context was God—though it could as easily be nature when Koyré turned his interest to the natural philosophers. His central proposition in regard to Descartes was that the philosopher of modernity owed much to medieval predecessors. It is one that would no longer need to be argued. Neither would his more interesting, supporting assertion of the philosophic value of scholastic reasoning, "subtleties" being a word that Koyré never thought pejorative.

To historians of science the most interesting feature of the discussion is the use that Koyré found Descartes making of the concepts of perfection and infinity. In handling the latter, he showed how the mathematician in Descartes had fortified the philosopher and invested the ontological argument with a sophistication unattainable by the reasoning of Anselm. Occasional asides presaged the direction in which Koyré's own interests would afterwards develop: for example, "we consider that the most notable achievement of Descartes the mathematician was to recognize the continuity of number. In assimilating discrete number to lines and extended magnitudes, he introduced continuity and the infinite into the domain of finite number."[2] In this book, however, Koyré had his attention on the *Meditations* and on Descartes the theologian and metaphysicist. Only later, in the beautiful and lucid *Entretiens sur Descartes* [8], did Koyré handle instead the *Discourse on Method*, emphasizing that it was the preface to Descartes's treatises of geometry, optics, and meteorology. Koyré would then no longer have agreed with his own youthful statement to the effect that, although Descartes altered the whole course of philosophy, the history of science would have been little different if he had never lived.[3]

Indeed, Koyré's own natural predilections emerge from the contrast in tone between his two major writings on Descartes. The *Entretiens* is an enthusiastic book, sympathetic and almost affectionate in its treatment of Descartes. Not so the thesis, a little stilted in its quality, wherein the author does not seem quite at ease with his subject. The constraint comes out overtly in passages concerning Descartes's want of candor, but the reader is left with a more general feeling of artificiality about the very enterprise of treating Descartes theologically. Koyré's having been a candidate in the division of the École Pratique concerned with "sciences réligieuses," the Ve Section, may quite naturally have affected his choice of a subject. What is surprising, however, is that Koyré remained associated with that section throughout a life devoted largely to the history of science, there being no appropriate provision for the latter subject in the academic structures of Paris. It was a circumstance bespeaking both the rigidity of institutions in the French capital and the flexibility of their administrators, despite whose generosity Koyré felt some difficulty over his commitments in his later years.

No ambiguities beclouded the simplicity and serenity of Anselm's commitments, and though Koyré's monograph on the founder of the ontological proof of God's being is less suggestive of his later interests in its thematics than the thesis on Descartes, it is more so in its treatment, specifically on the score of sympathy and penetration of the man through the texts.

Perhaps Koyré's most characteristic gift as a scholar (it was the manifestation in scholarship of his personal quality) was his ability to enter into the world of his subject and evoke for the reader the way in which things were then seen: in this case, the spiritual and intellectual reality in which Anselm perceived both beatifically and logically the necessity of God's being; in other instances, Aristotle's world of physical objects apprehended by common sense and ranged into an orderly philosophy; Jacob Boehme's tissue of signatures and correspondences between man and nature; the Copernican globes spinning and revolving for the simple and sufficient reason that they are round; Kepler's vision of numerical form and Pythagorean solidity; Galileo's abstract reality of quantifiable bodies kinematically related in geometrical space; and finally Newton's open universe, with consciousness situated in infinite space instead of in the cosmos of ancient Greek philosophy.

It was through meticulous analysis of essential texts, however, and not through general summary or paraphrase that Koyré thus opened spacious implications out of the intellectual constructions of his subjects. He liked to print extensive passages from the text to accompany his analysis in order that the reader might see what he was about. Indeed, his writing adapted the French instructional technique of *explication de texte* to the highest purposes of scholarship. Most of his later works derived from courses, often from individual lectures, given in the many institutions in France, Egypt, and the United States, where he taught regularly or was a guest. In later years his knowledge sometimes made him seem severe to younger scholars unsure of their own. The effect was altogether unintended. Fundamentally his was a deeply humane intellectual temperament, critical in the analytical and never in the denigrating or destructive sense. He wished to bring out the value in the subjects that he studied, not to expose what might be found of hollowness or falsity in them. Easy targets never tempted him. His own self-assurance was thus compatible with the most serious humility, for he subordinated his gifts to enhancing the merits of those who by mind, daring, imagination, and taste had contributed to civilizing our culture, and who had thereby aroused his admiration.

Such qualities of empathy animated the important studies he made of Hegelian philosophy and of the intellectual culture of nineteenth-century Russia. Neither of those concerns bore directly on the history of science, but perhaps a word may be said. His knowledge of Hegel derived from youthful immersion in Husserl's phenomenology. In the early 1930's he thought to convey the interest it held to his circle

KOYRÉ

of philosophical friends in Paris—formed for the most part in the École Normale Supérieure—to whom it was largely alien, not to say terra incognita. These papers were well received,[4] and readers whose case is similar may find particularly illuminating his "Note sur la langue et la terminologie hégéliennes."[5] Similarly, the papers in his two volumes on Russian intellectual history developed for a French learned public a subject for which Koyré had special competence: the dilemma of Russian writers torn between the necessity for assimilating European culture if their country were to become civilized, and resisting it if Russia were to establish its own national identity [4, 11]. Admirers of Koyré's writings in the history of science would do well to read the most considerable of those studies, a monograph on Tchaadaev.[6] Although it has nothing to do with their subject, it is one of the finest, most sympathetic and revealing pieces that he wrote.

By contrast, Koyré's work on the German mystics did have an important if somewhat enigmatic bearing on his historiography of science, for although he was never more earnest than in mediating between this inaccessible tradition and his modern reader's sensibility, his own reaction to it was to turn from theological subjects back to the scientific interests of his student days at Göttingen. His major doctoral thesis remains the most considerable and reliable study of Boehme, a lucid book on an obscure writer. Koyré also gathered into a little book four short pieces on Schwenkfeld, Sebastian Franck, Paracelsus, and Valentin Weigel, Boehme's most important sources [12]. Its reissue in 1971 coincided with a revival of the occult that the author would have deplored. True, it might be held that Boehme took an interest in the natural world even as did Galileo, Descartes, and Kepler, his contemporaries. Any resemblance is only apparent, however, for Boehme's sense of nature was altogether symbolic, the reality of phenomena residing for him in the signatures they bear of the divine. It is true that Koyré's awareness of how the world had impinged on consciousness before modern science destroyed these symbolic meanings sensitized his later writings on the scientific revolution. But he came to feel a certain futility in the enterprise of exploring the experiences of mystics, which by definition could be known only by him to whom they happened. Boehme was consistent in always seeking to read the correspondence between man and the world out of what he often called the book of himself, whereas the Koyré of *Études galiléennes* observed in the opening lines that only the history of science invests the idea of progress with meaning since it records the conquests won by the human mind at grips with reality.[7]

However that may be, the leitmotif of Koyré's work in history of science was the problem of motion; and he first identified it in a philosophical essay, "Bemerkungen zu den Zenonischen Paradoxen," published in 1922, prior to these theological writings.[8] In this, his first substantial publication,[9] Koyré argued that understanding Zeno's puzzles required analysis not merely of motion but of the manner in which its conceptualization in parameters of time and space involved ideas of infinity and continuity. After reviewing the Zenonian contributions of Brochard, Noël, Evelyn, and Bergson, Koyré (no doubt thinking back to his studies with Hilbert) invoked the findings of Bolzano and Cantor on the infinite and the nature of limits, and distinguished between motion as a process involving bodies in their nature and motion as a relation to which they are indifferent in themselves. A footnote anticipated Koyré's lifework in a single sentence: "All the disagreement between ancient and modern physics may be reduced to this: whereas for Aristotle, motion is necessarily an action, or more precisely an actualization (*actus entis in potentia in quantum est in potentia*), it became for Galileo as for Descartes a state."[10] Towards the end of his life, Koyré was sometimes asked how he happened to turn from theology to science, and once said, "I returned to my first love."[11]

His own career was full of movement. He had prepared his materials on Russian intellectual history in the first instance for a course at the Institut d'Études Slaves of the University of Paris. In 1929, the year *Jacob Boehme* appeared, he was appointed to a post in the Faculty of Letters at Montpellier and taught there from September 1930 until December 1931, enjoying the climate and quality of life in the Midi while regretting the inaccessibility of libraries. In January 1932 he was elected a *directeur d'études* at the École Pratique and returned to Paris, where his course treated of science and faith in the sixteenth century. Having read Copernicus for that purpose, and found how little was really known of his epochal accomplishment, Koyré prepared a translation of book I of *De revolutionibus*, its theoretical and cosmological part, together with a historical and interpretative introduction [5]. It was his initial contribution to history of science proper. In it Copernicus stands forth a thinker about the universe and no mere manipulator of epicycles, a thinker at once archaic and revolutionary. He was archaic in his addiction to the Platonic aesthetic of circularity, making it into a cosmic kinematics. He was revolutionary in his conviction that geometric form must comport with physical reality, and that no hypothesis joining the two was too daring to adopt, let the consequences be what they might for tradition and common sense. By implication, form itself became geometric, instead of substantial, and down that road lay modern science.

When Koyré published his Copernicus edition in 1934, he was teaching on a visiting basis at the University of Cairo. Finding his colleagues and students most congenial, he returned there in 1936–1937 and again in 1937–1938. For that audience he prepared lectures later developed into the *Entretiens sur Descartes*. Having turned from Copernicus to Galileo, it was also in Cairo that he settled down with the great Favaro edition of the latter's works, a set of which he had brought to Egypt, and there composed his masterpiece, *Études galiléennes*.[12] The title page gives 1939 for the date of publication. Actually it appeared in Paris in April 1940, just prior to the German invasion. Koyré and his wife were once again in Cairo. He wished to serve amid the disasters, and they hurried back to France, reaching Paris just as the city was surrendered. Thereupon they turned about, making their way first to Montpellier, and then by way of Beirut back to Cairo. Koyré had already determined to rally to the Free French and offered his services to De Gaulle when the General came to Cairo. Since Koyré held an American visa, De Gaulle felt that the Free French cause might benefit from the presence in the United States of a man of intellectual prominence able to express the Gaullist point of view in a country where government policy was favorable to Pétain. Somehow, the Koyrés found transportation by way of India, the Pacific crossing, and San Francisco to New York. There he joined a group of French and Belgian scientists and scholars in creating the École Libre des Hautes Études, and he taught there as well as in the New School for Social Research throughout the war, making one trip to London in 1942 to report to De Gaulle. In New York he developed the familiarity with American life that made it natural for him to spend in his later years something like half of his professional life in the United States.

It was in the United States in the immediate postwar years that *Études galiléennes*, not much noticed amid the distraction of scholarship by war, found its widest and most enthusiastic public, a case of the right book becoming known at the right time. A new generation of historians of science, the first to conceive of the subject in a fully professional way, was just then finding an opportunity in the expanding American university system, which more than made up in flexibility and enthusiasm for science whatever it may have lacked in scholarly sophistication and philosophical depth. Casting about through the literature in search of materials, they came upon *Études galiléennes* as upon a revelation of what exciting intellectual interest their newly found subject might hold, a book which was no arid tally of discoveries and obsolete technicalities, nor a sentimental glorification of the wonders of the scientific spirit, nor yet (despite the author's Platonism) a stalking horse for some philosophical system, whether referring to science like the positivist outlook or to history like the Marxist.

Instead, they found a patient, analytical, and still a tremendously exciting history of the battle of ideas waged by the great protagonists, Galileo and Descartes, in their struggle to win through to the most fundamental concepts of classical physics, formulations that later seemed so simple that schoolchildren could learn them with ease and without thought. It was a struggle waged not against religion, nor superstition, nor ignorance, as the received folklore of science would have it, but against habit, against common sense, against the capacity of the greatest of minds to commit error amid the press of their own commitments. Koyré sometimes observed, indeed, that the history of error is as instructive as that of correct theory, and in some ways more so, for although nothing to be celebrated—he was no irrationalist—it does exhibit the force and nature of the constraints amid which intellect needs must strive in order to create knowledge. (The more strictly philosophical problem of the false was one that he developed intensively in its classical context in a charmingly ironic essay, *Epiménide le menteur* [10].)

Koyré's technique was to study problems both intensively and broadly, intensively for themselves and broadly in the awareness of their widest significance. *Études galiléennes* consists of three essays published in separate fascicles. The first is entitled *À l'aube de la science classique*, the latter phrase meaning classical physics. The theme that unites all three is the emergence of that science (without which the rest of modern science is unthinkable) from the effort to formulate the law of falling bodies and the law of inertia, the subjects respectively of the second and third fascicles. The subtitle of the first fascicle, "La jeunesse de Galilée," implies that Galileo's early education and first researches recapitulated the main stages in the history of physics from its origins in antiquity. Koyré's sympathetic summary of Aristotelian physics emphasizes the anomaly of the cause attributed to motion in projectiles and explicates the reasoning of Benedetti and Bonamico, from whom Galileo learned physics and who developed the fourteenth-century impetus theory into a scheme for explaining the flight of missiles and fall of heavy bodies. Only when Galileo abandoned the idea of causal impetus, however, did he begin to lead the way from a physics of quality to a physics of quantity. He first attempted that step in the analysis in his youthful *De motu*, left in manuscript. There he substituted Archimedean for Aristotelian methods and formulated the relation between a body and its surrounding medium in terms of relative density.

In Koyré's view, geometrization of physical quantity in the Archimedean sense was the crux of the scientific revolution. The intellectual drama, becoming at times a comedy of errors in *Études galiléennes*, is made to consist of a counterpoint between Galileo and Descartes striving to disengage the law of falling bodies and the law of inertia, respectively the earliest and the most general laws of modern dynamics, from concealment by the gross behavior of ordinary bodies throughout the everyday world. In the end Galileo achieved the law of fall and Descartes the concept of inertia. Galileo began in 1604 in private correspondence with a correct statement of the former law—that the distance traversed in free fall from rest is proportional to the square of the elapsed time—and simultaneously attributed it to an erroneous principle—that the velocity acquired at any point is proportional to the distance fallen.

In fact, velocity is proportional to time in constant acceleration, and the irony that reveals the depth of the mistake is that Descartes independently repeated these same confusions fifteen years later in his correspondence with Beeckman. The specific trouble lay in the mutual unfamiliarity of mathematics and dynamics. However clearly Galileo saw the need for formulating the latter in terms of the former, his only tools for mathematicizing motion were arithmetic and geometry. Analytical though his mind was, proportion had to do the work of functional interdependence, and it was not intuitively clear to him at the outset that lapse of time could naturally be expressed in geometric magnitudes. His instinct having been eminently that of a physicist, Galileo eventually worked through to a resolution of his error. The *Discorsi* incorporates a fully mathematical derivation of the law from the principle of uniform acceleration, followed by the famous experimental verification on the inclined plane (which Koyré in his own excessive skepticism about the experimental component of early physics dismissed as a thought experiment).

Less fortunate with this problem was Descartes. Committed to identifying physics with geometry, he never did perceive that his formulation of fall was inconsistent with the physical description of the phenomenon. But if this tendency to "géometrisation à outrance" concealed the elements of the physical problem from Descartes, it was on the other hand just such mathematical radicalism that led him to the law of inertia, unconcerned to say where motion would stop and what could hold the world together if bodies tended to move in straight lines to infinity. Before this physical problem, Galileo finally drew back into the traditional conception that on the cosmic scale motion endures in circles, and left it to Descartes to enunciate the more general, the universal law of motion. Attributing the law of inertia to Descartes was certainly one of the most original and surprising of Koyré's findings in *Études galiléennes*, and it is central to the argument. In consequence of that principle, the ancient notion of a finite cosmos centered around man and ordered conformably to his purposes disappeared into the comfortless expanse of infinite space. In Koyré's view, the scientific revolution entailed a more decisive mutation in man's sense of himself in the world than any intellectual event since the beginnings of civilization in ancient Greece, and it came about because of the change that solving the basic problems of motion required in conceiving their widest boundaries and parameters.

In the postwar years Koyré resumed his post in Paris while lecturing from time to time at Harvard, Yale, Johns Hopkins, Chicago, and the University of Wisconsin. Western Reserve University awarded him its honorary doctorate of L.H.D. in 1964. In 1955 he came to the Institute for Advanced Study in Princeton, where he was appointed to permanent membership in the following year. From then until his health began failing in 1962, he spent six months of the year in Princeton, returning to Paris each spring to give his annual course at the École Pratique. The tranquillity of the Institute, and specifically its Rosenwald collection of first editions in the history of science, were essential to the completion of his further works. He was greatly stimulated and encouraged in their composition by his association with Harold Cherniss and Erwin Panofsky, and also by the acumen and criticism of Robert Oppenheimer, then the director of the Institute, in whose bracing company Koyré was one of the very few people with the intellectual self-possession to feel at ease.

Those works carry further the main themes that Koyré discerned in the scientific revolution, its history and philosophical aspects. *La révolution astronomique* was the last book he left in finished form, and consists of a very substantial treatise on Kepler's transformation of astronomy, preceded by a resume of Koyré's earlier discussion of Copernicus and followed by an essay on the celestial mechanics of Borelli. This last was one of his most original contributions to the literature, for although Borelli has been well known to scholars for his mechanistic physiology, the intricate rationalities of his world machine had been very little studied in modern times. As for the main part of the book, Kepler was always one of Koyré's favorite figures, appreciated for his boldness, for his imagination, for his Platonism, finally for his accuracy. Koyré distinguished his touch from that of Copernicus by making him out an astrophysicist needing a physical explanation of the planetary motions, in search of which he came upon his mathematical laws. In no way did Koyré underplay the fantastic and Pythagorean aspects of Kepler's thought. Indeed, it might be said that Koyré's earlier interest in German mysticism met his later commitment to science in his study of Kepler. Ultimately, however, we have Kepler making his mark through the fertility of an imagination controlled by fidelity to physical fact.

The themes that interested Koyré reached their dénouement in the Newtonian synthesis, and his essay on its significance is one of the most lucid, serene, and comprehensive of his writings. It opens the volume of *Newtonian Studies* published after his death. Perhaps it is a pity that he did not see fit to include "A Documentary History of the Problem of Fall From Kepler to Newton" [13], for that meticulous monograph exhibits at his scholarly best his gift for treating the ramifications of a single problem in detail and in generality as they appeared to the succession of analytical minds that handled it. For the rest, Koyré was not given the time to establish the same degree of coherence among his several studies of Newton that he did in *Études galiléennes*. In the last years of his life, he was collaborating with I. Bernard Cohen on the preparation of a variorum edition of the *Principia*, currently in press. The essay in *Newtonian Studies* on "Hypothesis and Experiment in Newton" translates the famous "hypotheses non fingo" to mean "feign" not "frame," and takes issue with the attribution of a positivistic philosophy to Newton himself. The most substantial essay in the volume contrasts Newtonian with Cartesian doctrines of space, and carefully explores the theological implications of the difference, a theme worked out more fully in Koyré's *From the Closed World to the Infinite Universe*.

Completed earlier than *Newtonian Studies*, this important work follows the metaphysical course of the transition epitomized in the title, beginning with the cosmology of Nicolas of Cusa and culminating in the Newtonian assertion of the absoluteness of infinite space and the omnipotence of a personal God distinct from nature. Theologically, the critical issue throughout was the relation of God to the world, for it appeared, and most subtly so to Henry More, that Cartesian science escaped atheism only by falling into pantheism. In regard to these issues Koyré's discussion may seem a little bodiless to readers whose sensibilities are less finely attuned to the metaphysical and theological implications of the old ontologies. The problems will come alive, however, if they are transposed from a metaphysical into a psychological key. It is the sort of reading that would be consonant with his own admiration for the writings of Émile Meyerson, to whose memory he dedicated *Études galiléennes*,[3] and that would place *From the Closed World* alongside that work as its more philosophical complement or companion, concerned with what Koyré now calls "world-feelings"[14] in contrast to world views.

The central theme is that of alienation, the alienation of consciousness from nature by its own creation of science. Put in those terms the metaphysical anxieties about God and the world will take on reality in modern eyes, and that is precisely what the destruction of the Greek cosmos entailed:

> The substitution for the conception of the world as a finite and well-ordered whole, in which the spatial structure embodied a hierarchy of perfection and value, that of an indefinite and even infinite universe no longer united by natural subordination, but unified only by the identity of its ultimate and basic components and laws; and the replacement of the Aristotelian conception of space—a differentiated set of innerworldly places—by that of Euclidean geometry—an essentially infinite and homogeneous extension—from now on considered as identical with the real space of the world.[15]

Yet if this emphasis in Koyré might give aid to the current fashion for deploring science as something set against humanity, his treatment gives protagonists of antiscientism no comfort. It is significant that of all the great minds of the seventeenth century, the only one apart from Bacon with whom Koyré felt little sympathy was Pascal [18*q*]. For he always held the creations of intelligence to be triumphs in the long battle between mind and disorder, not burdens to be lamented.

NOTES

1. *La philosophie de Jacob Boehme*, p. viii.
2. *L'idée de Dieu et les preuves de son existence chez Descartes*, p. 128.
3. *La philosophie de Jacob Boehme*, p. vi.
4. Jean Wahl, "Le rôle de A. Koyré dans le développement des études hégéliennes en France," in *Archives de philosophie*, **28** (July-Sept. 1965), 323–336.
5. *Études d'histoire de la pensée philosophique*; originally published in *Revue philosophique*, **112** (1931), 409–439.
6. *Études sur l'histoire des idées philosophiques*, pp. 19–102.
7. *Études galiléennes*, p. 6.

KOYRÉ

8. *Jahrbuch für Philosophie und phänomenologische Forschung*, **5** (1922), 603–628; published in French in [17*a*].
9. He had published one small note prior to World War I, "Remarques sur les nombres de M. B. Russell," in *Revue de metaphysique et de morale*, **20** (1912), 722–724.
10. *Études d'histoire de la pensée philosophique*, p. 30, n. 1.
11. Koyré left among his papers a curriculum vitae of 1951 which sets out his own sense of the interconnectedness of the work that he had accomplished and that he then projected; see *Études d'histoire de la pensée scientifique*, pp. 1–5.
12. Two articles containing parts of the work had already appeared: "Galilée et l'expérience de Pise," in *Annales de l'Université de Paris*, **12** (1937), 441–453; "Galilée et Descartes," in *Travaux du IXe Congrès international de Philosophie*, **2** (1937), 41–47.
13. Cf. Koyré's "Die Philosophie Émile Meyersons," in *Deutsch-Französische Rundschau*, **4** (1931), 197–217, and his "Les essais d'Émile Meyerson," in *Journal de psychologie normale et pathologique* (1946), 124–128.
14. *From the Closed World to the Infinite Universe*, p. 43.
15. *Ibid.*, p. viii.

BIBLIOGRAPHY

A Festschrift entitled *Mélanges Alexandre Koyré*, 2 vols. (Paris, 1964), was organized on the occasion of Koyré's seventieth birthday. The second volume opens with the list of his principal publications, comprising some seventy-five titles. We limit the present article to identifying his books, together with the more important of his articles, those mentioned in the footnotes above and under items [17], [18], and [19] below. It is a testimonial to the continuing interest in Koyré's specialized studies that in his later years and after his death, associates and publishers thought it important to collect and reissue these writings in book form. Readers may find it helpful to know the contents of those collections.

[1] *L'idée de Dieu et les preuves de son existence chez Descartes* (Paris, 1922; German trans., Bonn, 1923).

[2] *L'idée de Dieu dans la philosophie de S. Anselme* (Paris, 1923).

[3] *La philosophie de Jacob Boehme; Étude sur les origines de la métaphysique allemande* (Paris, 1929).

[4] *La philosophie et le mouvement national en Russie au début du XIXe siècle* (Paris, 1929).

[5] *N. Copernic: Des Révolutions des orbes célestes, liv. 1, introduction, traduction et notes* (Paris, 1934; repub. 1970).

[6] *Spinoza: De Intellectus Emendatione, introduction, texte, traduction, notes* (Paris, 1936).

[7] *Études galiléennes* (Paris, 1939): I, *À l'aube de la science classique;* II, *La loi de la chute des corps, Descartes et Galilée;* III, *Galilée et la loi d'inertie.*

[8] *Entretiens sur Descartes* (New York, 1944); repub. with [9] (Paris, 1962).

[9] *Introduction à la lecture de Platon* (New York, 1945); English trans., *Discovering Plato* (New York, 1945); Spanish trans. (Mexico City, 1946); Italian trans. (Florence, 1956); repub. in combination with [8] (Paris, 1962).

[10] *Epiménide le menteur* (Paris, 1947).

[11] *Études sur l'histoire des idées philosophiques en Russie* (Paris, 1950).

[12] *Mystiques, spirituels, alchimistes du XVIe siècle allemand: Schwenkfeld, Šeb. Franck, Weigel, Paracelse* (Paris, 1955; repub. 1971).

[13] "A Documentary History of the Problem of Fall From Kepler to Newton: De motu gravium naturaliter cadentium in hypothesi terrae motae," in *Transactions of the American Philosophical Society*, **45**, pt. 4 (1955), 329–395. A French translation is in press (Vrin) under the title *Chute des corps et mouvement de la terre de Kepler à Newton: Histoire et documents du problème.*

[14] *From the Closed World to the Infinite Universe* (Baltimore, 1957; repub. New York, 1958); French trans. (Paris, 1961).

[15] *La révolution astronomique: Copernic, Kepler, Borelli* (Paris, 1961).

[16] *Newtonian Studies* (Cambridge, Mass., 1965); French trans. (Paris, 1966).

[17] *Études d'histoire de la pensée philosophique* (Paris, 1961).

(*a*) "Remarques sur les paradoxes de Zénon" (1922).
(*b*) "Le vide et l'espace infini au XIVe siècle" (1949).
(*c*) "Le chien, constellation céleste, et le chien, animal aboyant" (1950).
(*d*) "Condorcet" (1948).
(*e*) "Louis de Bonald" (1946).
(*f*) "Hegel à Iena" (1934).
(*g*) "Note sur la langue et la terminologie hégéliennes" (1934).
(*h*) "Rapport sur l'état des études hégéliennes en France" (1930).
(*i*) "De l'influence des conceptions scientifiques sur l'évolution des théories scientifiques" (1955).
(*j*) "L'évolution philosophique de Martin Heidegger" (1946).
(*k*) "Les philosophes et la machine" (1948).
(*l*) "Du monde de l''à-peu-près' à l'univers de précision" (1948).

[18] *Études d'histoire de la pensée scientifique* (Paris, 1966).

(*a*) "La pensée moderne" (1930).
(*b*) "Aristotélisme et platonisme dans la philosophie du Moyen Age" (1944).
(*c*) "L'apport scientifique de la Renaissance" (1951).
(*d*) "Les origines de la science moderne" (1956).
(*e*) "Les étapes de la cosmologie scientifique" (1952).
(*f*) "Léonard de Vinci 500 ans après" (1953).
(*g*) "La dynamique de Nicolo Tartaglia" (1960).
(*h*) "Jean-Baptiste Benedetti, critique d'Aristote" (1959).
(*i*) "Galilée et Platon" (1943).*
(*j*) "Galilée et la révolution scientifique du XVIIe siècle" (1955).*
(*k*) "Galilée et l'expérience de Pise: à propos d'une légende" (1937).
(*l*) "Le 'De motu gravium' de Galilée: de l'expérience imaginaire et de son abus" (1960).**
(*m*) " 'Traduttore-traditore,' à propos de Copernic et de Galilée" (1943).
(*n*) "Une expérience de mesure" (1953).*

(o) "Gassendi et la science de son temps" (1957).**
(p) "Bonaventura Cavalieri et la géométrie des continus" (1954).
(q) "Pascal savant" (1956).**
(r) "Perspectives sur l'histoire des sciences" (1963).
* English original republished in [19].
** English translation published in [19].
[19] *Metaphysics and Measurement* (London, 1968). English versions of [18] *i*, *j*, *l*, *n*, *o*, and *q*.

II. Secondary Literature. Accounts of Koyré and his work have appeared as follows: Yvon Belaval, *Critique*, nos. 207–208 (1964), 675–704; Pierre Costabel and Charles C. Gillispie, *Archives internationales d'histoire des sciences*, no. 67 (1964), 149–156; Suzanne Delorme, Paul Vignaux, René Taton, and Pierre Costabel in *Revue d'histoire des sciences*, **18** (1965), 129–159; T. S. Kuhn, "Alexander Koyré and the History of Science," in *Encounter*, **34** (1970), 67–69; René Taton, *Revue de synthèse*, **88** (1967), 7–20.

Charles C. Gillispie

KRAFT, JENS (*b.* Fredrikstad, Norway, 2 October 1720; *d.* Sorø, Denmark, 18 March 1765), *mathematics, physics, anthropology, philosophy.*

Kraft's mother, Severine Ehrensfryd Scolt, died when he was only two, and his father, Anders Kraft, a senior lieutenant in the Norwegian army, died when he was five years old. He was privately educated in Denmark at the manor of his uncle, Major Jens Kraft, and took the master's degree in Copenhagen in 1742. Kraft was married twice, to Bodil Cathrine Evertsen, who died in 1758, and to Sophie Magdalene Langhorn, who survived him.

A traveling grant enabled him to study philosophy with Christian Wolff in Germany, and mathematics and physics in France. Later he often expressed his admiration for Wolff, Daniel Bernouilli, Clairaut, and d'Alembert, whose works changed his general scientific outlook. On his return in 1746, he was admitted as a fellow of the Royal Danish Academy of Science and Letters. The following year he became the first professor of mathematics and philosophy in the reestablished academy for the nobility at Sorø, where he remained until his death. An eminent teacher, Kraft's lectures and private colloquia helped to diminish the prevailing influence of Cartesianism, and to bring Danish science back into the mainstream of the eighteenth century.

Kraft's best-known work is a textbook on theoretical and technical mechanics (1763–1764). The book, written in an easy and fluent style, contains a series of lectures based on Newtonian principles. Each lecture is provided with a supplement giving a more advanced mathematical exposition of the subject matter. In Denmark this work gave theoretical physics a firm basis as an academic subject, while its large section on machines stimulated the expansion of industry. The book was favorably received abroad and was translated into Latin and German.

Kraft's broad cultural interests were also reflected in a book on the life and manners of primitive peoples which is regarded as a pioneer work in social anthropology. It was written in the belief that a study of savage cultures would reveal the general origin of human institutions and beliefs.

His first paper, presented to the Royal Danish Academy of Science and Letters in 1746, was a clear exposition of the systems of Descartes and Newton. In opposition to his admired teacher, Christian Wolff, Kraft sided with Newton by showing that the Cartesian vortex theory was incompatible with accepted mechanical principles. Kraft did write several textbooks, nevertheless, on logic, ontology, cosmology, and psychology, inspired primarily by Wolffian philosophy.

Mathematics was one of Kraft's major areas of interest. Two early theses (written in 1741 and 1742) present no really new contributions to mathematics, but they show Kraft to have been a skilled and well-read mathematician. For example, the theses contain discussions of equations which are solved by means of Descartes's method of cuts between parabolas and circles. In 1748–1750 Kraft published two mathematical treatises. In the first he proved that if

$$y = \sum_{i=0}^{n} \beta_i x^i$$

has two equal roots α, then α is also a root in dy/dx. In the second paper Kraft discusses the following problem: Given an equation

$$A = \alpha x^r y^f + \beta x^m y^t + \gamma x^p y^l + \delta x^q y^h + \cdots$$

with rational exponents, y can be found as a series

$$y = Bx^n + Cx^{n+k} + Dx^{n+2k} + \cdots$$

with rational exponents. In his introduction, Kraft mentioned Newton, Leibniz, Maclaurin, Sterling, and 's Gravesande as examples of mathematicians who had treated this problem before, and Kraft's own method is a refinement of that of 's Gravesande.

Furthermore, in two small treatises from 1751–1754 Kraft argued that the concepts of infinitely large and infinitely small do not exist in an absolute sense in mathematics and physics, and that they must be conceived as relative quantities.

BIBLIOGRAPHY

I. Original Works. Kraft's textbook on mechanics was published in two volumes. The first, *Forelæsninger over mekanik med hosføiede tillæg* (Sorø, 1763) was translated into Latin, *Mechanica Latine* (Wismar, 1773), and into German, *Mechanik, aus Lateinischen mit Zusätzen vermehrten Uebersetzung Tetens ins Deutsche übersetzt und hin und wieder verbessert von Joh. Chr. Aug. Steingrüber* (Dresden, 1787); the other appeared as *Forelæsninger over statik og hydrodynamik med Maskin-Væsenets theorier* (Sorø, 1764).

His book on ethnology, *Kort fortælning af de vilde folks fornemmeste indretninger, skikke og meninger, til oplysning af det menneskeliges oprindelse og fremgang i almindelighed* (Sorø, 1760), was translated into German, *Die Sitten der Wilden zur Aufklärung des Ursprungs und Aufnahme der Menschheit* (Copenhagen, 1766), and Dutch, *Verhandeling over de zeden en gewoontens der oude en hedendaagsche wilde volker* (Utrecht, 1779).

Kraft's paper on the systems of Descartes and Newton, "Betænkning over Neutons og Cartesii systemer med nye Anmærkninger over Lyset," was published in *Det Kiøbenhavnske Selskabs Skrifter*, **3** (1747), 213–296. Kraft's mathematical papers include *Explicationum in Is. Neutoni Arithmeticam universalem particulam primum* (Copenhagen, 1741); *Theoria generalis succincta construendi aeqvationes analyticas* (Copenhagen, 1742); "Anmerkning over de Liigheder, i hvilke af flere Værdier af den ubekiendte Størrelse er lige store" in *Det Kiøbenhavnske Selskabs Skrifter*, **5** (1750), 303–309; and "Metode at bevise, hvorledes man i alle Tilfælde kand bestemme den ene Ubekiendte ved en u-endelig Følge af Terminis, som gives ved den anden, i de algebraiske Liigheder, som indeholde to Ubekiendte," *ibid.*, 324–354.

His most important philosophical papers are *Systema mundi deductum ex principiis monadis, Dissertation, qui a remporté le prix proposé par l'Académie des sciences et belles lettres sur le système des monades avec les pièces, qui ont concouru* (Berlin, 1748); and "Afhandling om en Deel Contradictioner, som findes i det sædvanlige Systema over Materien og de sammensatte Ting," in *Det Kiøbenhavnske Selskabs Skrifter*, **6** (1754), 189–216.

II. Secondary Literature. See *Dansk Biografisk Leksikon.* An account of Kraft's contribution to ethnology is given by Kaj Birket-Smith, "Jens Kraft, A Pioneer of Ethnology in Denmark" in *Folk, Dansk etnografisk tidsskrift*, **2** (Copenhagen, 1960), 5–12.

Kurt Møller Pedersen

KRAMERS, HENDRIK ANTHONY (*b.* Rotterdam, Netherlands, 17 December 1894; *d.* Oegstgeest, Netherlands, 24 April 1952), *theoretical physics.*

Kramers, the third of five sons of a physician, received his early schooling at Rotterdam. In 1912 he went to Leiden and studied theoretical physics, mainly with P. Ehrenfest, who in 1912 had succeeded H. A. Lorentz. In 1916, after passing his *doctoraal* (roughly equivalent to obtaining a master's degree), he taught for a few months in a secondary school and in September set out for Copenhagen, where he became a close collaborator of Niels Bohr. In 1920 Bohr's Institute of Theoretical Physics was opened; Kramers was first an assistant and in 1924 became a lecturer. In 1926 he accepted the chair of theoretical physics at Utrecht and in 1934 returned to Leiden as a successor to Ehrenfest, who had died in September 1933. From 1934 until his death Kramers taught at Leiden and paid numerous visits to other countries, including the United States.

During his years at Copenhagen, Kramers worked mainly on the further development of the quantum theory of the atom. It was a surprising feature of Bohr's theory that the frequency of a spectral line determined by the equation

$$h\nu = E_n - E_m$$

did not coincide with a kinetic frequency of electrons. The situation was mitigated by Bohr's correspondence principle: The frequency ν is an average of kinetic frequencies of electrons in the initial and in the final states, and in the limit of high quantum numbers these two frequencies and the frequency of the emitted radiation approach each other. Bohr further concluded that polarizations and intensities should, in the limit of high quantum numbers, be given by the Fourier components of the quantized motion and that even at low quantum numbers the Fourier components in the initial and final states should give an indication of the intensities to be expected. In his doctoral thesis at Leiden in 1919 (published by the Royal Danish Academy of Sciences) Kramers developed the mathematical formalism required to apply these ideas; he also carried out detailed calculations for the case of a hydrogen atom in an external electric field. This led to a satisfactory interpretation of the intensities of Stark components.

Other papers from this period deal with the relativistic theory of the Stark effect in hydrogen (1920), the continuous X-ray spectrum (1923), and the quantization of the rotation of molecules when there is a "built-in flywheel" (an electronic angular momentum around an axis fixed with respect to the molecule [1923]). His paper on the helium atom (1923) was of special importance for the development of quantum theory. In this paper Kramers showed that application of the theory of quantization of classical orbits to the fundamental state of helium does not lead to a stable state and gives far too low a value for the binding energy. He pointed out that this revealed the fundamental inadequacy of the provisional quantum theory. From then on, the helium atom became a test case for a new theory. Eventually this challenge was successfully met by the new quantum mechanics, as was shown by W. Heisenberg and, with greater numerical precision, by Hylleraas.

Kramers was coauthor of the famous paper by Bohr, Kramers, and J. C. Slater (1924) which suggested that conservation of energy might not hold in elementary processes. Although this idea was not substantiated by subsequent experimental and theoretical work, the paper had a profound influence. It emphasized the notion of virtual oscillators associated with quantum transitions. This concept formed the basis for Kramers' theory of dispersion. In classical theory an isotropic harmonic oscillator with charge e, mass m, and frequency ν_1, in an alternating electric field with amplitude E and frequency ν, would acquire an induced polarization.

$$P = E(e^2/m)/4\pi^2(\nu_1{}^2 - \nu_2).$$

Kramers showed that a similar formula should hold in quantum theory. To each possible transition there corresponds a virtual oscillator with an effective value $(e^2/m)^*$ that can be calculated from the transition probabilities. For a transition to a higher level this value is positive but for a transition to a lower level it is negative, an entirely new feature closely related to Einstein's stimulated emission.

In Kramers' subsequent paper with Heisenberg (1924) the theory was developed in more detail; it was shown that one must expect the scattered radiation also to contain frequencies $\nu \pm \nu_1$. This paper thus described quantitatively the effect that was later found experimentally by Raman and that had already been predicted by Smekal on the basis of considerations on light quanta. The notion of virtual oscillators was the starting point of Heisenberg's quantum mechanics—the virtual oscillators became the matrix elements of the coordinates. In connection with the theory of dispersion, Kramers also wrote two later papers (1927, 1929) in which he established the now well-known relations between the real and the imaginary part of the polarizability (Kramers-Kronig relations).

Kramers' later work, produced after his departure from Copenhagen, may be divided into four groups. There were a number of papers dealing with the mathematical formalism of quantum mechanics. One of his earliest and best-known papers in this field dealt with what became later known as the W(entzel)-K(ramers)-B(rillouin) method (1926). It is a method to obtain approximate solutions of a one-dimensional Schrödinger equation of the form

$$U'' + (\lambda - W(x))\,U = 0.$$

One approximate solution is

$$(\lambda - W)^{-1/4} \exp \int (\lambda - W)^{1/2}\,dx,$$

but this solution breaks down near the zeros of $\lambda - W$. In this region Kramers replaces $\lambda - W$ with αx. Then the solution becomes a Bessel function of order 1/3, the behavior of which can easily be discussed. A solution can then be "patched" together from the solution shown above and the Bessel function. Kramers showed that this leads to the quantization rule of the older quantum theory but with quantum numbers:

$$n + 1/2 \qquad (n = 0, 1, 2, \cdots).$$

Although this method yielded quite satisfactory approximate wave functions—compare, for example, the paper that Kramers wrote with E. M. van Engers on the ion of molecular hydrogen (1933)—its practical value has diminished since the arrival of modern computers; it nevertheless remains valuable in elucidating the relations between quantum mechanics, classical mechanics, and the older methods of quantization.

Kramers developed a special formalism for dealing with the theory of the multiplet structure of spectra (1930). It was based on Weyl's treatment of the rotation group combined with notations current in the theory of invariants. By this powerful method he derived a general formula for the quantum mechanical analogon of the classical expression $P_l(\cos AB)$, where P_l is a Legendre polynomial (1931). For $l = 1$ one obtains the well-known Landé cosine; for $l = 2$ and $l = 3$ the expressions for quadrupole and octopole coupling. Later (1943) Kramers also gave a treatment of multipole radiation.

With G.P. Ittmann, Kramers studied the Schrödinger equation of the asymmetric top and made several additions to the theory of Lamé functions (1933, 1938). In a very elegant paper (1935) he dealt with the solutions and eigenvalues of the Schrödinger equation for a particle in a one-dimensional periodic force field. Kramers showed in a very general way that there exists an infinite number of zones of allowed energy values separated by forbidden regions. In many of these papers Kramers is as much a mathematician as a physicist. Also his textbook on quantum mechanics (1933, 1938) contains a wealth of mathematical detail not found elsewhere. It is even more valuable, however, because it analyzes very carefully the basic principles and assumptions of quantum mechanics.

A second group of papers dealt with paramagnetism, magneto-optical rotation, and ferromagnetism. Several of these papers were the result of Kramers' collaboration with Jean Becquerel, who regularly came to Leiden to perform low-temperature measurements on magneto-optic rotation in crystals of the rare earths. Kramers' calculations of the behavior of magnetic ions were essentially straightforward and concur in many cases with experimental results. Mention should be made of "Kramers' theorem": If an ion containing an odd number of electrons is placed in an arbitrary static electric field, then every state remains $2p$-fold degenerate ($p = 1, 2, 3, \cdots$). In particular the lowest state is always at least doubly degenerate. Kramers was coauthor of the first papers on cooling by adiabatic demagnetization published by the Leiden school (1933, 1934).

Two of Kramers' papers (1934, 1936) dealt with ferromagnetism and the theory of spin waves; they formed the transition to a third group of papers—those dealing with statistical and kinetic theory. With Wannier, Kramers studied the two-dimensional Ising model. He was unable to find a complete analytical solution—that was done later by L. Onsager, who was much influenced by Kramers' work—but he was able to show that the Curie temperature T_c, if it exists, is related to the coupling constant J by the equation $J/kT_c = 0.8814$; and he worked out approximate solutions for high and low temperatures.

Kramers and J. Kistemaker made an important contribution to the kinetic theory of gases (1943, 1949). Maxwell had already shown that the aerodynamic boundary condition, according to which the velocity of a gas at the surface of a wall is equal to the velocity of the wall, is not strictly valid when there is a velocity gradient perpendicular to the wall or a temperature gradient along the wall; Maxwell had calculated in 1879 both this viscosity slip and the thermal slip. Kramers noticed that there should also be a diffusion slip which occurs when there is a concentration gradient along the wall. This would lead to a pressure gradient's arising in a stationary state of diffusion through a capillary, and experiments confirmed this prediction.

Mention should also be made of an early contribution to the theory of strong electrolytes (1927), of a paper on the behavior of macromolecules in inhomogeneous flow (1946), and a very instructive paper on the use of Gibbs's "grand ensemble" (1938).

In his treatment of Brownian motion in a field of force, Kramers dealt specifically with the escape of a particle over the edge of a potential-hole. Although the most important factor in the probability of escape is

$$\exp(-Q/kT),$$

where Q is the height of the potential barrier, Kramers found quite different factors in front of the exponential, depending on whether the viscosity was large or small. The model is used to discuss chemical reactions—in 1923 Kramers had already written on chemical reactions with J. A. Christiansen—but it can also be used in connection with the Bohr-Wheeler theory of fission and has many other applications.

Finally there were also a number of papers on relativistic formalisms in particle theory and on the theory of radiation. Kramers' report to the 1948 Solvay Congress, entitled "Nonrelativistic Quantum Electrodynamics and Correspondence Principle" summarized ideas that had already been presented in his textbook on quantum mechanics (1933, 1938). His aim was to arrive at structure-independent results, and his method involved a separation between the proper field of the electron and the external field. To a certain extent these considerations have been superseded by later developments of quantum electrodynamics.

Kramers' work, which covers almost the entire field of theoretical physics, is characterized both by outstanding mathematical skill and by careful analysis of physical principles. It also leaves us with the impression that he tackled problems because he found them challenging, not primarily because they afforded chances of easy success. As a consequence his work is somewhat lacking in spectacular results that can easily be explained to a layman; but among fellow theoreticians he was universally recognized as one of the great masters. He played an important part in the scientific life of his country and in the world of physics.

In 1946 Kramers was elected chairman of the Scientific and Technological Committee of the United Nations Atomic Energy Commission, and he presented a unanimous report on the technological feasibility of control of atomic energy. From 1946 to 1950 he was president of the International Union of Pure and Applied Physics.

Kramers received honorary degrees from the universities of Oslo, Lund, Stockholm, and the Sorbonne and was a member of many learned societies. He is also remembered as a gifted musician and an excellent linguist.

BIBLIOGRAPHY

I. ORIGINAL WORKS. Many of Kramers' writings were brought together in *Collected Scientific Papers* (Amsterdam, 1956). His books include *Die Grundlagen der Quantentheorie, Hand- und Jahrbuch der chemischen Physik*, I (Leipzig, 1933); and *Quantentheorie des Elektrons und der Strahlung* (Leipzig, 1938); an English trans. of these two volumes by D. ter Haar was published as *Quantum Mechanics* (Amsterdam, 1956). A complete bibliography, also mentioning popular or general articles not reprinted in the *Collected Scientific Papers* was published in *Nederlands tijdschrift voor natuurkunde*, **18** (1952), 173.

II. SECONDARY LITERATURE. A number of obituary notices have appeared, among them J. Becquerel, in *Comptes-rendus hebdomadaires des séances de l'Académie des sciences*, **234** (1952), 2122–2126; F. J. Belinfante and D. ter Haar, *Science*, **116** (1952), 555; N. Bohr, "Hendrik Anthony Kramers," in *Nederlands Tijdschrift voor natuurkunde*, **18** (1952), 161; and H. B. G. Casimir, "The Scientific Work of H. A. Kramers," *ibid.*, 167; and in *Jaarboek der Koninklijke Nederlandsche akademie van wetenschappen* (1952–1953), pp. 302–305.

The best account of Kramers' life and personality was given by his friend J. Romein, in *Jaarboek van de maatschappij der Nederlandse letterkunde* (1951–1953), 82–91; see also J. A. Wheeler, *Year Book. American Philosophical Society* (1953).

H. B. G. CASIMIR

KRAMP, CHRÉTIEN *or* **CHRISTIAN** (*b.* Strasbourg, France, 8 July 1760, *d.* Strasbourg, 13 May 1826), *physics, astronomy, mathematics.*

Kramp's father, Jean-Michel, was a teacher (*professeur régent*) at the Gymnasium in Strasbourg. Brought up speaking French and German, Kramp studied medicine and practiced in several Rhineland cities that were contained in the region annexed to France in 1795. Turning to education, Kramp taught mathematics, chemistry, and experimental physics at the École Centrale of the department of the Ruhr in Cologne. Following Napoleon's reorganization of the educational system, whereby the Écoles Centrales were replaced by lycées and faculties of law, letters, medicine, and science were created, Kramp, around 1809, became professor of mathematics and dean of the Faculty of Science of Strasbourg. A corresponding member of the Berlin Academy since 1812, he was elected a corresponding member of the geometry section of the Academy of Sciences of Paris at the end of 1817.

In 1783, the year the Montgolfier brothers made the first balloon ascension, Kramp published in Strasbourg an account of aerostatics in which he treated the subject historically, physically, and mathematically. He wrote a supplement to this work in 1786. In 1793 he published a study on crystallography (in collaboration with Bekkerhin) and, in Strasbourg, a memoir on double refraction.

Kramp published a medical work in Latin in 1786 and another, a treatise on fevers, in German in 1794. His critique of practical medicine appeared in Leipzig in 1795. Moreover, in 1812 he published a rather mediocre study on the application of algebraic analysis to the phenomenon of the circulation of the blood. He corresponded with Bessel on astronomy and made several calculations of eclipses and occultations in the years before 1820; his most important astronomical work, however, is the *Analyse des réfractions astronomiques et terrestres* (1798), which was very favorably received by the Institut de France. He wrote several elementary treatises in pure mathematics, as well as numerous memoirs, and the *Éléments d'arithmétique universelle* (1808). A disciple of the German philosopher and mathematician K. F. Hindenburg, Kramp also contributed to the various journals that Hindenburg edited. He may thus be considered to be one of the representatives of the combinatorial school, which played an important role in German mathematics.

In the *Analyse des réfractions astronomiques* Kramp attempted to solve the problem of refraction by the simplifying assumption that the elasticity of air is proportional to its density. He also presented a rather extensive numerical table of the transcendental function

$$\varphi(x) = \int_0^x e^{-t^2}\,dt,$$

which is so important in the calculus of probabilities, and which sometimes is called Kramp's transcendental. In this same work he considered products of which the factors are in arithmetic progression. He indicated the products by $a^{n|d}$; hence

$$a(a+d)(a+2d)\cdots[a+(n-1)\,d] = a^{n|d}.$$

He called these products "facultés analytiques," but he ultimately adopted the designation "factorials," proposed by his fellow countryman Arbogast.

Although Kramp was not aware of it, his ideas were in agreement with those of Stirling (1730) and especially those of Vandermond. The notation $n!$ for the product of the first n numbers, however, was his own. Like Bessel, Legendre, and Gauss, Kramp extended the notion of factorial to non-whole number arguments, and in 1812 he published a numerical table that he sent to Bessel. In his *Arithmétique universelle* Kramp developed a method that synthesizes the fundamental principles of the calculus of variations as stated by Arbogast with the basic procedures of combinatorial analysis. He thus strove to create an intimate union of differential calculus and ordinary algebra, as had Lagrange in his last works.

BIBLIOGRAPHY

I. ORIGINAL WORKS. Kramp's writings include *Geschichte der Aërostatik, historisch, physisch und mathematisch ausgefuehrt*, 2 vols. (Strasbourg, 1783); *Anhang zu der Geschichte der Aërostatik* (Strasbourg, 1786); *De vi vitali Arteriarum diatribe. Addita nova de Febrium indole generali Conjectura* (Strasbourg, 1786); *Krystallographie des Mine-*

ralreichs (Vienna, 1794), written with Bekkerhin; *Fieberlehre, nach mecanischen Grundsaetzen* (Heidelberg, 1794); *Kritik der praktischen Arzneykunde, mit Ruecksicht auf die Geschichte derselben und ihre neuern Lehrgebaeude* (Leipzig, 1795); *Analyse des réfractions astronomiques et terrestres* (Strasbourg–Leipzig, 1798); *Éléments d'arithmétique* (Cologne–Paris, 1801); *Éléments de géométrie* (Cologne, 1806); and *Éléments d'arithmétique universelle* (Cologne, 1808). He also translated into German Lancombe's *Art des Accouchements* (Mannheim, 1796) and contributed to Hindenburg's *Sammlung combinatorisch-analytischer Abhandlungen* and *Archiv der reinen und angewandte Mathematik* (1796); the *Nova Acta* of the Bayerische Akademie der Wissenschaften (1799); and Gergonne's *Annales des mathématiques pures et appliquées* (from 1810 to 1821).

II. SECONDARY LITERATURE. Poggendorff, I, col. 1313 contains a partial list of Kramp's work. See also Gunther, in *Allgemeine deutsche Biographie*, XVII (Leipzig, 1883), 31–32; L. Louvet, in Hoefer, *Nouvelle Biographie générale*, XXVIII (Paris, 1861), 191–192; Niels-Nielsen, *Géomètres français sous la Révolution* (Copenhagen, 1929), pp. 128–134; and Royal Society of London, *Catalogue of Scientific Papers*, III (1869), 743–744, which lists 32 memoirs published after 1799.

JEAN ITARD

KRASHENINNIKOV, STEPAN PETROVICH (*b.* Moscow, Russia, 11 November 1711; *d.* St. Petersburg, Russia [now Leningrad, U.S.S.R.], 8 March 1755), *geography, ethnography, botany, history.*

Krasheninnikov, the son of a soldier, studied at the Moscow Slavonic-Greek-Latin Academy (1724–1732) and then at the University of the St. Petersburg Academy of Sciences (1732–1733). From 1733–1743 he took part in the second Kamchatka expedition of the Academy. During the first three years Krasheninnikov studied the history, geography, and ethnography of Siberia under the direction of J. G. Gmelin and G. F. Muller. The extensive material that he and other members of the expedition gathered provided a basis for the general geographical description of Siberia.

From 1737–1741 Krasheninnikov traveled through Kamchatka. At first he studied the warm springs and the flora and fauna of the western shore of the peninsula; then he studied the geographical peculiarities of the eastern part of Kamchatka in the area of the Avachinskaya volcano. Finally, he carefully investigated many central areas of the peninsula. He described the rocks and minerals, observed the Avacha and Kliuchevskoi volcanoes, and reported on earthquakes. The articles based on these observations were the first scientific works devoted to Kamchatka.

In the following years Krasheninnikov prepared the "Description of the Land of Kamchatka" (1756). This major work contains a detailed geographical description of Kamchatka and information on its natural resources and animal and plant life. Krasheninnikov also described, for the first time, the life, customs, and language of the local populations—Kamchadals and Kurils—and supplied dictionaries of their languages. The last part of the book describes the history of the peoples who settled the peninsula and the discovery of Kamchatka by the Russian traveler V. V. Atlasov in 1697–1699. Krasheninnikov's work was soon translated into European languages, and there have been several subsequent editions.

In February of 1743 Krasheninnikov returned to St. Petersburg, where he worked in the Academy of Sciences on the systematization of the extensive material gathered on his ten-year journey. In 1745, after defending his dissertation in ichthyology, he received the title of adjunct of the St. Petersburg Academy. In the same year he began to work in the Academy botanical garden. In April 1750 Krasheninnikov was appointed professor of natural history and botany and also a full member of the Academy. At the same time he was named rector of the University and inspector of the Gymnasium, both associated with the Academy.

In the last years of his life, despite his teaching and administrative activities, Krasheninnikov continued his scientific work on the material from his Siberian-Kamchatka expedition and did botanical research in St. Petersburg province. Named after Krasheninnikov are an island off Kamchatka, a volcano in Kamchatka, and a point, cape, and bay in the Kuril islands.

BIBLIOGRAPHY

I. ORIGINAL WORKS. Major works by Krasheninnikov are *Opisanie zemli Kamchatki* ("Description of the Land of Kamchatka"; St. Petersburg, 1756); "Rech o polze nauk i khudozhestv" ("Speech on the Usefulness of the Sciences and Arts") in *Torzhestvo Akademii nauk 6 sentyabrya 1750 g* ("Festival of the Academy of Sciences, 6 September 1750"; St. Petersburg, 1750).

II. SECONDARY LITERATURE. For reference see N. G. Fradkin, *S. P. Krasheninnikov* (Moscow, 1954); L. S. Berg, "Stepan Petrovich Krasheninnikov," in *Otechestvennye fiziko-geografy i puteshestvenniki* ("Russian Physical-Geographers and Travellers"; Moscow, 1959); and A. I. Andreev, "Stepan Petrovich Krasheninnikov," in *Lyudi russkoy nauki. Geologia, geografia* ("People of Russian Science. Geology, Geography"; Moscow, 1962).

A. S. FEDOROV

KRASNOV, ANDREY NIKOLAEVICH (*b.* St. Petersburg, Russia, 8 November 1862; *d.* Tbilisi, Russia, 1 January 1915), *geography, geobotany.*

Krasnov's father was a Cossack general and his mother came from a family of the St. Petersburg intelligentsia. In 1880 he graduated from the Gymnasium and in 1885 from the natural sciences section of the Faculty of Physics and Mathematics at St. Petersburg University. His scientific education was especially influenced by the lectures of Beketov and Dokuchaev. In 1889 he defended his master's dissertation on development of flora of the southern part of the Eastern Tien Shan. In 1894 Krasnov defended a dissertation at Moscow University and received the degree of doctor of geography. The subject was grassland steppes of the northern hemisphere.

Krasnov's scientific and teaching career was connected with Kharkov University, where he was professor of geography from 1889 to 1912. He spent the last two years of his life, when he was already seriously ill, creating the botanical garden in Batum, on the Black Sea.

Krasnov's love of nature began in his childhood, and from his student years he was an ardent world traveler. His travels to foreign countries were substantially aided by his knowledge of modern and ancient languages.

Krasnov considered the main purpose of geography to be the clarification of the mutual relationships of natural phenomena that constitute particular geographical complexes (landscapes) and the explanation of the evolution of the latter through the use of the comparative geographical method. Krasnov found similarities between the flora of the Tien Shan and of Central Russia, and believed contemporary flora of the mountains, steppes, and the Arctic to be the product of the regeneration of a single Palaearctic flora. He presented the flora of each country by the formula

$$F = f_1 + f_2 + f_3,$$

with F as the totality of the forms of plants now living; f_1 the Palaearctic species that have survived without change to our times; f_2 the Palaearctic species that have changed under the influence of the changes in the conditions of life in a given country; and f_3 the species that have migrated to the given country in later times. Krasnov distinguished three types of flora, relating them to definite geographic areas; ancient ($F = f_1$), migrating ($F = f_3$), and transformed ($F = f_2$).

Krasnov's research on the nature of steppe and tropical and subtropical plants occupied an important place in his scientific work. He advanced an original geomorphological hypothesis to explain the absence of forest in the steppes. The flatness of the land causes poor drainage, which in turn is responsible for a surplus accumulation of harmful salts.

Krasnov drew on his wide-ranging knowledge of the countries of the world and his many years of teaching to create the first original Russian university textbooks in geography.

BIBLIOGRAPHY

I. Original Works. See "Opyt istorii razvitia flory yuzhnoy chasti vostochnogo Tyan-Shanya" ("An Attempt at a History of the Development of the Flora of the Southern Part of the Eastern Tian-Shan"), in *Zapiski Russkogo Geograficheskogo Obshchestva*, **19** (1888); "Geografia kak novaya universitetskaya nauka" ("Geography as a New University Science"), in *Zhurnal Ministerstva Narodnogo Prosveshchenia*, **261**, sec. 2 (1890); *Relef, rastitelnost i pochvy Kharkovskoy gubernii* ("Topography, Vegetation and Soil of Kharkov Province"; Kharkov, 1893); and "Iz poezdki na Dalny Vostok Azii. Zametki o rastitelnosti Yavy, Yaponii i Sakhalina" ("From a Trip to the Far East of Asia. Notes on the Vegetation of Java, Japan, and Sakhalin"), in *Zemlevedenie*, bks. 2-3 (1894).

For further reference, see *Travyanye stepi severnogo polusharia* ("The Grassland Steppes of the Northern Hemisphere"); in *Izvestia Obshchestva Liubiteley Yestestvoznania Antropologii i Ethnogorafii*, **81** (1894); *Osnovy zemlevedenia* ("Bases of Soil Science"), nos. 1–4 (Kharkov, 1895–1899); *Chaynye okrugi subtropicheskikh oblastev Azii* ("Tea Regions of the Subtropical Areas of Asia"), nos. 1-2 (Kharkov, 1897–1898); *Iz koybeli tsivilizatsii* ("From the Cradle of Civilization"; St. Petersburg, 1898); *Pisma iz krugosvetnogo plavania* ("Letters From a Voyage Around the World"; St. Petersburg, 1898); and *Pod tropikami Azii* ("In the Tropics of Asia"; Moscow, 1956).

II. Secondary Literature. See *Professor A. N. Krasnov* (Kharkov, 1916); D. N. Anuchin, *O Lyudyakh russkoy nauki i kultury, izdani* ("On People of Russian Science and Culture"; 2nd ed., Moscow, 1952); and Milkov, F. N., *A. N. Krasnov—geograf i puteshestvennik* ("A. N. Krasnov —Geographer and Traveler"; Moscow, 1955).

I. A. Fedoseev

KRASOVSKY, THEODOSY NICOLAEVICH (*b.* Galich, Kostroma guberniya, Russia, 26 September 1878; *d.* Moscow, U.S.S.R., 1 October 1948), *earth sciences, mathematics.*

Krasovsky graduated from the Moscow Geodetic Institute in 1900. Until 1903 he studied physics and mathematics at Moscow University and astronomy at the Pulkovo observatory. An instructor at the Geodetic Institute from 1902, he became professor in 1916 and chairman of higher geodesy in 1921. He was also a corresponding member of the Academy of Sciences of the U.S.S.R., an Honored Scientist and Technologist of the R.S.F.S.R., and in the mid-1930's, vice-president of the Baltic Geodetic Commission.

Krasovsky contributed considerably to the study of the geometry of the figure of the earth—the "spheroid." He devised an efficient method of adjusting primary triangulation, deduced the parameters of the earth's spheroid, and drew up scientific specifications for triangulation and subsequent geodetic work for the U.S.S.R. subcontinental territory. With M. S. Molodenski, Krasovsky was a pioneer in his emphasis on geodetic gravimetry rather than isostatic theory.

He reorganized the institutions of higher geodetic study in the U.S.S.R., and his pupils were future Soviet specialists. In his last years Krasovsky studied the earth's interior by combining geodetic and other geophysical as well as geological data. The figure of the earth now generally accepted differs only slightly from the "Krasovsky spheroid."

BIBLIOGRAPHY

I. ORIGINAL WORKS. The most important of Krasovsky's more than 120 published works are in *Izbrannye sochinenia* ("Selected Works"), 4 vols. (Moscow, 1953–1956). Vol. I is devoted to the figure of the earth and to the adjustment of primary triangulation, and contains an essay on Krasovsky's life and works; vol. II to various branches of geodesy, field astronomy, and map projections; vol. III to geodetic control; and vol. IV to the geometry of the spheroid. A few of Krasovsky's papers and reports were published in German in *Comptes rendus de la Commission géodésique baltique*, sessions 5, 6, 7, 8, and 9 (Helsinki, 1931–1937).

II. SECONDARY LITERATURE. The most comprehensive source is G. V. Bagratuni, *T. N. Krasovsky* (Moscow, 1959), which includes a bibliography. There is a short obituary by V. V. Danilov and M. S. Molodenski in *Izvestiya Akademii nauk SSSR*, Geograf.-geofiz. ser., **13**, no. 1 (1949), 3–4; and brief information on Krasovsky is in *Bolshaya sovetskaya entsiklopedia* ("Greater Soviet Encyclopedia"), XXIII (1953), 281, with portrait.

O. B. SHEYNIN

KRAUS, CHARLES AUGUST (*b.* Knightsville, Indiana, 15 August 1875; *d.* East Providence, Rhode Island, 27 June 1967), *chemistry*.

Although he spent most of his childhood on a Kansas farm, Kraus developed a strong interest in physics and electrical engineering. He entered the University of Kansas in 1894. In his junior year he coauthored a paper on spectroscopy, and he received his bachelor's degree in engineering in 1898. After a year's further study at Kansas and a year as research fellow at Johns Hopkins University, he served as an instructor of physics at the University of California from 1901 until 1904. He then became a research assistant at the Massachusetts Institute of Technology, where he received his doctorate in chemistry in 1908. After graduation he remained at MIT as research associate (1908–1912) and assistant professor of physical chemical research (1912–1914).

In 1914 he was named professor of chemistry and head of the chemical laboratory at Clark University, where he remained for nine years. He became professor of chemistry and director of chemical research at Brown University in 1924. Although he retired from the position in 1946, he continued his research and was working on a book when he died in 1967 at the age of ninety-one. During World War I he was active in government service. Kraus received five honorary degrees, the Willard Gibbs Medal (1935), and the Priestley Medal (1950) of the American Chemical Society.

Kraus's studies in chemical research began at the University of Kansas, where Hamilton P. Cady and Edward C. Franklin became interested in liquid ammonia solutions. Kraus was attracted by the opportunities that appeared in the field, especially by the study of solutions of the alkali metals. Using the glassblowing experience he had gained in Germany, Franklin, in conjunction with Kraus, published eight papers between 1898 and 1905, dealing with such problems as the solubility of substances in liquid ammonia, the molecular rise in the boiling point of a solution of a solute in liquid ammonia as the molal concentration of the solution is increased; and various physical properties of the solutions.

Continuing his research on liquid ammonia at MIT, Kraus published a series of articles (beginning in 1907). He reported that alkali and alkaline earth metals could form solutions with ammonia. Highly concentrated solutions would behave like metals while dilute solutions would resemble ionic solutions of salts. The dilute solutions ionize to give the normal positive metal ion and negative electrons, the latter associated with large amounts of ammonia. Gilbert N. Lewis remarked on Kraus's "extraordinary experimental skill" in designing and constructing glass apparatus, which was largely responsible for his successful studies.

At Clark University, Kraus studied other physical properties of metal ammonia solutions, notably density, conductivity, and vapor pressure. His results indicated beyond any doubt that alkali metals do not form compounds with ammonia, but alkaline earths do form compounds by combining with six molecules of ammonia to produce metal-like ammoniates.

His interest in radicals that exhibit metallic properties led to studies of alkyl mercury and ammonium and substituted ammonium groups. From these groups he then turned to work on organic radicals bonded to group four metals. He published over twenty-five papers on these metal-organic systems; many of the studies were carried out in liquid ammonia solutions because of the water or air sensitivity of most of the compounds.

Around 1920 Thomas Midgley, Jr., and T. A. Boyd of General Motors discovered that tetraethyl lead would reduce engine knocking when added to gasoline. As a consultant to Standard Oil Company of New Jersey, Kraus was asked to design an economical quantity synthesis of this substance. After a three-month intensive study with Conrall C. Callis late in 1922, Kraus succeeded in producing a reaction between a lead-sodium alloy and ethyl chloride under high pressures and relatively low temperatures followed by steam distillation. This discovery permitted the automobile industry to develop high compression engines.

Further studies at Clark and Brown concerned the electrical properties of substances dissolved in solvents of very low dielectric constant such as benzene, dioxane, and ethylene dichloride. Of particular interest are his studies of large anions and cations, which have led to a clearer picture of the nature of electrolyte solutions.

During World War II he worked with the navy and served as a consultant to the Manhattan Project. Specifically he was instrumental in working out a process for purifying uranium salts.

BIBLIOGRAPHY

I. Original Works. Among Kraus's more significant publications are "Solutions of the Metals in Non-metallic Solvents. I. General Properties of Solutions of Metals in Liquid Ammonia," in *Journal of the American Chemical Society*, **29** (1907), 1556–1571; "Solutions . . . II. On the Formation of Compounds between Metals and Ammonia," *ibid.*, **30** (1908), 653–668; and "General Relations Between the Concentration and the Conductance of Ionized Substances in Various Solvents," *ibid.*, **35** (1913), 1315–1434, written with W. C. Bray. These papers give sufficient references to Kraus's earlier work. His tetraethyl lead studies received U.S. Patent Numbers 1,612,131 (28 Dec. 1926); 1,694,268 (4 Dec. 1928); and 1,697,245 (1 Jan. 1929).

II. Secondary Literature. R. M. Fuoss published a memoir of Kraus in the *Biographical Memoirs of the National Academy of Sciences*, **42** (1971), 119–159. A detailed, but incomplete, discussion of Kraus's early achievements is Warren C. Johnson, "The Scientific Work of Charles A. Kraus," in *The Chemical Bulletin*, **22** (1935), 123–127. A helpful obituary notice appeared in *Chemical and Engineering News*, **45** (July 17, 1967), p. 59. In this latter article, his birth year is incorrectly given as 1865. The G. N. Lewis quotation in the text comes from Johnson's article.

Sheldon J. Kopperl

KRAUSE, ERNST LUDWIG, also known as **Carus Sterne** (*b.* Zielenzig, Germany [now Sulęcin, Poland], 22 November 1839; *d.* Eberswalde, Germany, 24 August 1903), *scientific popularization.*

After attending the Realschule in Meseritz (now Międzyrzecz, Poland, Krause trained to be an apothecary. In 1857 he began to study science at the University of Berlin, where he attended the lectures of Alexander Braun on botany, of Gustav Rose on mineralogy, and of Johannes Müller on comparative anatomy. Krause never worked as an apothecary but educated himself in a variety of fields and wrote popular scientific works. After receiving the doctorate from the University of Rostock in 1874, he lived in Berlin. In 1899, near the end of his life, he moved to Eberswalde.

Krause's first publications (1862–1863) were directed against spiritualism. He early became an enthusiastic adherent of Darwin's theory, which he made the basis of his own "natural system" for plants (1866). In this connection he criticized the hypothesis of the inheritance of acquired adaptations (*Die botanische Systematik*, p. 154). From 1866 until his death Krause was friendly with Haeckel and defended the latter's monistic world view in numerous popular essays. The great success of Krause's *Werden und Vergehen* (1876) is understandable in the context of the vehement ideological disputes provoked by Darwin's theory. In this period Krause introduced many readers to the basic ideas of the theory of evolution. Simultaneously, he showed that the new views had thoroughly shaken the traditional anthropomorphic conception of God and of God's actions.

Belief in progress, a doctrine founded on the achievements of science, led to the creation of the journal *Kosmos*, of which Krause was an editor from 1877 to 1883. Among his co-workers on *Kosmos*, which advocated a unified world view, were Darwin, Haeckel, Arnold Lang, Strasburger, and a group of philosophers. Krause's essay on Erasmus Darwin was translated into English (1879) at the urging of Charles Darwin, who wrote a biographical introduction for it. This work is still indispensable for studying the history of the theory of evolution.

On the other hand, many of Krause's other papers, dealing with such subjects as the scientific basis of myths and with topics in prehistory and ethnography, were of little significance.

BIBLIOGRAPHY

I. ORIGINAL WORKS. The works preceded by (C. S.) were published under Krause's pseudonym, Carus Sterne: *Die Naturgeschichte der Gespenster. Physikalisch-physiologisch-psychologische Studien* (Weimar, 1863); *Die botanische Systematik in ihrem Verhältnis zur Morphologie. Kritische Vergleichung der wichtigsten älteren Pflanzensysteme nebst Vorschlägen zu einem natürlichem Pflanzensystem nach morphologischen Grundsätzen* (Weimar, 1866); (C. S.) *Werden und Vergehen. Eine Entwicklungsgeschichte des Naturganzen in gemeinverständlicher Fassung* (Berlin, 1876), 6th ed., rev., 2 vols. (1905–1906); *Erasmus Darwin und seine Stellung in der Geschichte der Descendenz-Theorie. Mit seinem Lebens- und Charakterbilde von Charles Darwin* (Leipzig, 1880). English trans. by W. S. Dallas as *Erasmus Darwin. With a preliminary notice by Charles Darwin* (London, 1879); (C. S.) *Die Krone der Schöpfung. Vierzehn Essays über die Stellung des Menschen in der Natur* (Vienna, 1884); "Charles Darwin und sein Verhältnis zu Deutschland," in E. Krause, ed., *Gesammelte kleinere Schriften von Charles Darwin*, I (Leipzig, 1885), 1–236; (C. S.) *Die alte und die neue Weltanschauung. Studien über die Rätsel der Welt und des Lebens* (Stuttgart, 1887); (C. S.) *Die allgemeine Weltanschauung in ihrer historischen Entwickelung. Charakterbilder aus der Geschichte der Naturwissenschaften* (Stuttgart, 1889); and *Die Trojaburgen Nordeuropas. Ihr Zusammenhang mit der indogermanischen Trojasage* (Glogau, 1893).

For Krause's editorial contributions, see *Kosmos* **1–13** (1877–1883).

II. SECONDARY LITERATURE. See Wilhelm Bölsche, "Zur Erinnerung an Carus Sterne," in the 6th ed. of Krause's *Werden und Vergehen*, with portrait; and Victor Hantzsch, "Krause, Ernst Ludwig," in *Biographisches Jahrbuch und deutscher Nekrolog*, VIII (Berlin, 1905), p. 305–307.

GEORG USCHMANN

KRAYENHOFF, CORNELIS RUDOLPHUS THEODORUS (*b.* Nijmegen, Netherlands, 2 June 1758; *d.* Nijmegen, 24 November 1840), *geodesy, engineering.*

Krayenhoff was the son of C. J. Krayenhoff, a lieutenant colonel of engineers, and Clara Jacoba de Man. Sensitive to disappointments in his own military career, his father intended him for the bar. Thus, in 1777, after completing grammar school, Krayenhoff matriculated as a law student at Harderwijk, but his interest in physics drew him to other faculties, where he took the not unusual two degrees, in science (1780) and in medicine (1784). He mounted the first lightning rod on a public building in the Netherlands, on the bell tower at Doesburg, near Arnhem, in 1782. The following year he published a Dutch elaboration of a French treatise on electricity.

During ten years of medical practice in Amsterdam, Krayenhoff continued his scientific pursuits, giving well-attended lectures on physical subjects. His proficiency impressed scientists, including Van Swinden, Paets van Troostwijk, and Van Marum. When the French army occupied, or liberated, Holland in 1795 the patriot Krayenhoff was persuaded to give up medicine for a military career. After commanding the forces in Amsterdam, he was appointed general of engineers. He became minister of war under King Louis Bonaparte, but he was relieved within the year and gladly returned to the more palpable tasks of engineering.

Krayenhoff zealously studied the arts of war, especially that of fortification. He strengthened to the utmost the defenses of Amsterdam until Napoleon ordered the king to cease further extensions. At the impending incorporation of Holland into the French empire, Krayenhoff urged using his forts for armed resistance, but the king had to clear the way for his imperial brother. In 1811 on a visit to Amsterdam, the third city of his empire, Napoleon severely rebuked Krayenhoff, who shouted at the emperor that he was responsible for his former conduct to nobody but King Louis. During a tour of inspection a week later, Napoleon admired Krayenhoff's defense works and invited him to lunch in the fortress of Naarden. The general was later summoned to Paris to join the commission on fortifications, which, happily, also gave him time to visit each session of the Société des Sciences for "solid instruction" and to complete his *Précis historique*.

Under King William I, Krayenhoff was fully occupied with fortifications (for which he visited Curaçao in 1825) and with hydraulic engineering, regulating the great rivers and draining lakes. Pensioned in 1830, he devoted his time fully to physics and astronomy. He was buried at Fort Krayenhoff near Nijmegen.

In 1798 it was decided to follow the French example and divide the "Republique Batave, une et indivisible" into *départements* and *arrondissements*, for which purpose a committee with broad powers was installed. The first need was a suitable map of the whole territory. All kinds of maps were collected, but their poor quality made it impossible to fit them together. Krayenhoff then wrung from the committee an order for effecting a triangulation, and during most of 1799 he measured, with a sextant, angles from and between church towers. In February 1800, with a specially forged chain, he measured a base on the ice of the Zuider Zee about 5.6 kilometers long. This survey enabled him to complete two of the nine sheets planned for the great map. On showing them to Van Swinden, then a member of the Directory, he was amply praised for having achieved so much by such simple means. All the same, it was to be regretted that this occasion had not been seized upon to obtain an extension of the triangulation done by Delambre and Méchain.

As a member of the committee for weights and measures (about 1796 to 1799), Van Swinden had had to recalculate the 115 triangles involved and thus was completely familiar with this famous endeavor. He explained to Krayenhoff the method and necessary instruments, and what precautions were needed. The general's enthusiasm was boundless. His committee objected that with this new project all observations with the sextant would then be thrown to the winds, but Krayenhoff pleaded that these would be highly useful as a preliminary for the great undertaking. If the sextant work was not followed up, what a poor opinion the world would form of the state of science in the land of Huygens and Snellius!

Krayenhoff began by remeasuring fifteen of the twenty-two triangles observed in Flanders and Zeeland by the French astronomer J. Perny de Villeneuve. Some discrepancies, however, made him distrustful and he started again, taking as his base a side of the northernmost triangle determined by Delambre with its vertex at Dunkerque. Van Swinden put at his disposal an excellent repeating circle of Borda. Krayenhoff's military duties caused frequent interruptions, once of two years in succession, but by 1811 the whole territory was covered by 162 triangles.

Krayenhoff in his *Précis historique* gives a full account of the measured angles and of the reductions applied. The measurements are of the highest quality, and still more credit is due Krayenhoff as the first to adjust completely an extensive network of triangles. Here he made an original contribution by introducing "the rule of sines in a polygon" (Fig. 1). This elementary proposition was in itself nothing new—Krayenhoff states that he came upon it in Lazare Carnot's *Géométrie de position.* But his demonstration of its easy and useful application in geodesy was a memorable feat.

The then standard triangulation method generally comprised simple chains of triangles, limited only by the obvious condition that the angles of a triangle must add up to $180° + \epsilon$ (ϵ being the spherical excess, 1″ for every 198 square kilometers of the triangle's surface). In the network the angular tour of the horizon about a central point has to equal 360°. The

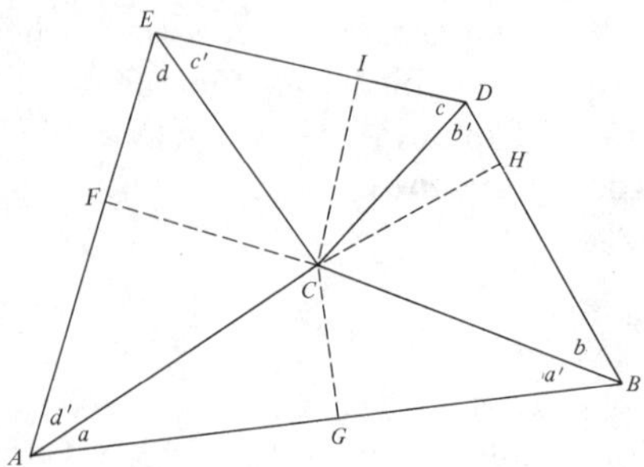

FIGURE 1. The theorem of sines in a polygon as applied by Krayenhoff in reducing the results of his triangulation (*Précis historique*, p. 38n.).

Demonstration of the theorem that the product of the sines of the angles a, b, c, d is equal to the product of the sines of the angles a', b', c', d'.

In the triangle ACB, $CG = AC \times \sin a = BC \times \sin a'$.
" " DCB, $CH = BC \times \sin b = DC \times \sin b'$.
" " DCE, $CI = DC \times \sin c = EC \times \sin c'$.
" " ACE, $CF = EC \times \sin d = AC \times \sin d'$.

Therefore $AC \times BC \times DC \times EC \times \sin a \times \sin b \times \sin c \times \sin d = BC \times DC \times EC \times AC \times \sin a' \times \sin b' \times \sin c' \times \sin d'$.

Dividing this equation by AC, BC, DC, EC yields $\sin a \times \sin b \times \sin c \times \sin d = \sin a' \times \sin b' \times \sin c' \times \sin d'$ and $\log \sin a + \log \sin b + \log \sin c + \log \sin d = \log \sin a' + \log \sin b' + \log \sin c' + \log \sin d'$.

This theorem is equally true for triangles forming the planes of any pyramid; it is also true for all the scalene triangles that can be formed on the faces of the pyramid if only the sides opposite to the angles all begin at the top of the pyramid and are coincident for two consecutive triangles; and finally for all the scalene triangles that can be formed according to the same conditions on a sphere about an arbitrary point.

high accuracy obtained is most clearly shown by the minor difference in length Krayenhoff was confronted with when a common side of two triangles was calculated along two different paths through the network. Of the fifty-two sides involved, twenty-nine showed a difference of less than 2.10^{-5} and only five one of more than 5.10^{-5}. For the determination of the azimuth, Krayenhoff introduced a simple method consisting in a determination of the time when the sun is passing the vertical plane through the line of vision.

Krayenhoff's work had been abundantly extolled in the first half of the nineteenth century, especially in Holland but also by Delambre. Then grudging, adverse criticism came forth from Dutch scientists. Their disapproval followed upon an unfair treatment by Gauss, who had taken as representative the two triangles with the largest errors out of the list of 162 triangles in Krayenhoff's *Précis historique*, intimating in this way that the adjacent Dutch triangulation was of a yet poorer quality than his own work in Hannover (see *Dictionary of Scientific Biography*, V, 303). He must have considered the rule of sines in a polygon so elementary that he forgot to mention Krayenhoff's priority. Since then Gauss's overwhelming influence has thrust Krayenhoff into oblivion. In Holland, however, an active study of Krayenhoff's achievements has been carried on by geodesists and astronomers.

Another example of Krayenhoff's inventiveness is a contrivance for gauging average stream velocity. A hollow cylinder, partly emergent and held in a vertical position by disks of lead, will float along at mean velocity in a certain tract, when the adjusted load keeps its lower end just clear of the bottom.

BIBLIOGRAPHY

I. Original Works. Krayenhoff's works on geodesy include "Batavische Vermessung, Schreiben des Oberst-Leutenant Krayenhoff an Freiherrn von Zach," in *Monatliche Correspondenz*, **9** (Feb. 1804), 168, 264, with map; original Dutch text in *Algemene Konst en Letterbode* (Apr. 1804), pp. 225–240, with map; *Instructie voor de Geographische Ingenieurs door Generaal-Majoor Krayenhoff*, Ministry of War (The Hague, 1808), with maps and drawings; *Hydrographische en Topographische Waarnemingen in Holland door den Oud-minister van Oorlog C. R. T. Krayenhoff* (Amsterdam, 1813).

Précis historique des opérations géodésiques et astronomiques, faites en Hollande, pour servir de base à la topographie de cet état, exécutées par le Lieutenant-Général Krayenhoff (The Hague, 1815), with map scale 84:100,000,000; 2nd ed. (1827), with a report by Delambre; *Chorotopographische Kaart der Noordelijke Provinciën*.

An original written protocol of Krayenhoff's observations, and of his calculations in 18 vols., is preserved at the University of Leiden, codex 241.

Krayenhoff's autobiography is *Levensbijzonderheden van den Leutenant-Generaal Baron Krayenhoff, door hem zelven in schrift gesteld en op zijn verlangen in het licht gegeven door Mr. H. W. Tydeman prof. jur.* (Nijmegen, 1844).

II. Secondary Literature. J. D. van der Plaats, "Overzicht van de Graadmetingen in Nederland," in *Tijdschrift voor Kadaster en Landmeetkunde*, **5** (1889), 217–243, 257–306, with map, and **7** (1891), 65–101, 109–133; W. Koopmans, "De laatste arbeid van Krayenhoff," in *Geodesia* (1962), pp. 144–147; N. van der Schraaf, "Historisch overzicht van het Nederlandse driehoeksnet," in *Nederlands geodetisch tijdschrift*, no. 4 (1972); N. D. Haasbroek, *Investigation of the Accuracy of Krayenhoff's Triangulation 1802–1811 in Belgium, the Netherlands and a Part of North Western Germany*, publication of the Netherlands Geodetic Commission (1973).

W. Nieuwenkamp

KROGH, SCHACK AUGUST STEENBERG (*b.* Grenå, Jutland, Denmark, 15 November 1874; *d.* Copenhagen, Denmark, 13 September 1949), *zoology, physiology.*

His family had lived for 300 years in southern Jutland, and Krogh never lost his affection for this area. His father, a brewer who had been educated as a naval architect, inspired his son's love of ships and the sea.

In 1893 Krogh graduated from the Århus cathedral school—in 1889 he had tried a voluntary apprenticeship in the Royal Navy to become an officer—and entered the University of Copenhagen as a medical student. But a friend of the family, the zoologist William Sørensen, inspired Krogh to study zoology and recommended that he attend the physiology lectures of Christian Bohr. After obtaining his M.Sc. degree in 1899, Krogh became an assistant to Bohr in his laboratory. In 1905 he married Marie Jørgensen, a physician. They had three daughters and one son.

In 1896 Krogh had begun studying the hydrostatic mechanism of the air bladders in *Corethra* larvae. He demonstrated that these organs "function like the diving tanks of a submarine." For his studies Krogh used a microscopic method of analysis on air in minimal bubbles. During these years he also published critical works on physical properties in biology, topics rather uncommon at that time. All his life Krogh was more interested in physical than in chemical problems in biology, and he explained his critical attitude thus:

> When experimental results are found to be in conflict with those of an earlier investigator, the matter is often taken too easily and disposed of for an instance by pointing out a possible source of error in the experiments of the predecessor, but without inquiring whether the error, if present, would be quantitatively sufficient to explain the discrepancy. I think that disagreement with former results should never be taken easily, but every effort should be made to find a true explanation. This can be done in many more cases than it actually is; and as a result, it can be done more easily than anybody else by the man "on the spot" who is already familiar with essential details. But it may require a great deal of imagination, and very often it will require supplementary experiments ["August Krogh," in *Festkrift Københavns Universitet 1950* (Copenhagen, 1950), p. 185].

In 1903, at the University of Copenhagen, Krogh defended a thesis on the cutaneous and pulmonary respiration of the frog, which was a synthesis of Bohr's work in respiratory physiology and his own in zoology. Krogh demonstrated that the exchange of oxygen takes place essentially through the thin pulmonary walls, with their short diffusion path, while the more diffusible carbon dioxide is expired through the skin. The study showed his abilities for quantitative and very accurate work and for "visual thinking," which enabled him to foresee and construct his often simple experimental tools without drawing them beforehand. This showed up especially in the first of his works to gain international fame, his paper on the pulmonary exchange of nitrogen, for which he was awarded the Seegen Prize of the Vienna Academy of Sciences. Through a new, very accurate temperature-control apparatus, he was able to demonstrate that free nitrogen takes no part in respiratory exchanges.

During his first years with Bohr, Krogh believed that pulmonary air exchanges took place mainly through secretory processes regulated by the nervous system. Since the great problem concerning passive or active moments for pulmonary exchange could not be solved with his instruments, Krogh invented a tonometer and a device for microanalysis of gases. Only 10 cu. mm. was necessary for the microtonometer to obtain equilibrium between the tension of air in blood and in the bubble, and dissolved gases in the pulmonary alveoli and in the bubble could be contrasted with sufficient accuracy.

Through his investigations Krogh concluded that the pulmonary gas exchange depended on diffusion; in 1904 he published with Bohr and K. A. Hasselbalch a study on the relation between the carbon dioxide tension and the oxygen association of blood. In 1910 he published his seven famous articles on the mechanism of gas exchange, prepared in collaboration with his wife. Krogh wrote:

> I shall be obliged in the following pages to combat the views of my teacher Prof. Bohr on certain essential points and also to criticize a few of his experimental results. I wish here not only to acknowledge the debt of gratitude which I personally owe to him, but also to emphasize the fact, obvious to everybody familiar with the problems here discussed, that the real progress made during the last twenty years in the knowledge of the processes in the lungs is mainly due to his labors and to that refinement of methods which he has introduced. The theory of the lung as a gland has justified its existence and done excellent service in bringing forward facts, which shall survive any theoretical construction that has been or may hereafter be put upon them ["The Mechanism of Gas Exchange" (1910), p. 257].

Krogh summed up his results as follows: "The absorption of oxygen and the elimination of carbon dioxide in the lungs take place by diffusion and by diffusion alone. There is no trustworthy evidence of any regulation of this process on the part of the organism." Against criticism from followers of the secretion theory, Krogh's further investigations confirmed that the partial tension of oxygen was always greater in the alveolar air than in the blood, so that conditions for diffusion had to be equal.

This new point of view created several other problems, and during the following years Krogh published works concerning the blood flow through the lungs and the influence of the venous supply upon the output of the heart. With Johannes Lindhard, later professor of gymnastic theory, he took up the topics of dead space, respiration and muscle work, and variations in the composition of alveolar air. His wife, a physician, inspired him. In 1913 they published a study of the diet and metabolism of the Eskimo, a result of travels in 1902 and 1908 to Greenland. Krogh's work on the tension of carbonic acid in natural waters was a result of the Greenland journey.

In 1908 a special associate professorship of zoophysiology was created for Krogh at the University of Copenhagen, and he left Bohr's laboratory even without the prospect of a laboratory for himself. In 1910, however, he acquired a laboratory in Ny Vestergade, which originally had been used by the bacteriologist Salomonsen. It was simply equipped, so Krogh set up a laboratory in his official residence and lived in the small rooms of the top story. Here he developed his many instruments for evaluating the function of blood flow and respiration: the rocker spirometer, the electromagnetic bicycle ergometer, and a gas analysis apparatus accurate to 0.001 percent. He demonstrated the influence of fat and carbohydrate as sources of muscle energy and found the oxygen deficit created through muscle work. With Lindhard he worked out the nitrous oxide method for determination of the heart's volume per minute.

From these studies Krogh early concluded that the capillaries of the muscles were partially closed during rest and, for the most part, open during work; using intensive microscopical and histological methods, he was able to demonstrate the truth of his ideas and in 1920 was awarded the Nobel Prize in physiology or medicine. Several others had conducted capillary investigations before Krogh, but none had addressed the problem specifically nor understood its significance for the entire circulation.

During these years many foreign scientists flocked to Krogh's laboratory: Goeran Liljestrand (Sweden), Joseph Barcroft and Thomas Lewis (England), and Edward Churchill, Cecil Drinker, Eugene Landis, Alfred N. Richards, and A. H. Turner (United States). Among the problems studied was the evaluation of hormonal, chemical, and nervous regulation of the capillaries. The Danish physician Bjovolf Vimtrup, who worked with Krogh, demonstrated in a doctoral thesis, "Über contractile Elemente in der Gefässwand der Blutcapillaren" (Copenhagen, 1922), that Rouget cells constricted frog capillaries. Later investigations have confirmed the appearance of an epithelial mechanism in mammals.

With his co-workers Krogh now demonstrated that capillary movements were influenced by both nerves and hormones, and in 1922 he published his monograph on the anatomy and physiology of capillaries. In 1916 he had written on respiratory exchange in animals and man, but this was mainly a reference book. It opened the way for tracheal insufflation in medicine and paved the way for the use of hypothermia in heart surgery through its demonstration of the slowing down of gaseous exchange at low temperatures. His later monograph, on the other hand, reads like an exciting novel; it reached five editions and was translated into German in two editions (Krogh wrote perfect English).

After 1920 Krogh worked with problems concerning edema and stasis, and during a lecture trip to the United States in 1922 he was able to study the newly discovered insulin; his wife had diabetes mellitus. On his return to Denmark, he and the internist H. C. Hagedorn organized the fabrication of Danish insulin and sold it privately without profit. Two famous institutions, Nordisk Insulinlaboratorium and Nordisk Insulinfond, were established largely through the efforts of these two foresighted men. With A. M. Hemmingsen, Krogh worked on the standardization of insulin.

Two years later the Rockefeller Foundation offered to provide better working facilities for Krogh, and a new institution with room for five other university laboratories was erected in Juliane Mariesvej; Krogh worked out plans for every room (even considering the shadows cast by the trees surrounding the building). In 1928 the institute, named the Rockefeller Institute, was officially opened. Krogh had tried to obtain special dustless rooms, but the building materials at that time made it impossible. During the first years of the institute, he continued his studies on heavy muscle work. He created new methods for determination of total osmotic tension of blood and studied the balance of insensible perspiration. Krogh also became interested in the physiological problems of heating houses. As chairman of a committee of the Academy for the Technical Sciences, he eagerly took part in the planning of technical and physiological investigations; and his microclimatograph proved to have many applications. He worked in this area until his death.

Besides his physiological studies of vertebrates, Krogh never lost his interest in zoophysiology. He started with the investigations of the *Corethra* larvae, and with his microanalysis apparatus for measuring air tensions in blood he demonstrated the essential importance of diffusion for insect respiration. Later, with his microrespirometer apparatus, he showed the influence of temperature on the metabolism of insects; in this way he also established that metabolism, even in other animals, follows van't Hoff's theorem. Between these studies he carried out investigations with J. Leitch on the oxygenation of fishes and with H. O. Schmit-Jensen on the fermentation of cellulose.

Another question which interested Krogh was the metabolism of sea animals: whether these creatures could live on dissolved material. With the zoologist R. Spärck he studied plankton and dissolved substances as food for aquatic organisms, as well as the aforementioned investigations on the relationship of carbon dioxide in air and seawater. Refuting the theories of A. Pütter, Krogh turned his attention to the exchange of water and inorganic ions through the surfaces of living cells and membranes, utilizing isotopes as indicators. He continued his osmoregulatory investigations and in 1939 set forth his ideas in a monograph—like the book on capillaries, a classic. One year later he published his fascinating and lucid monograph on respiratory mechanisms, with up-to-date knowledge of comparative physiology.

During his last years—after his retirement in 1945 as ordinary professor, a chair which had been created for him in 1916—Krogh took up a problem from his youth: the flight of insects and birds. In his private laboratory in Gentofte, with T. Weis-Fogh, he worked with a merry-go-round (a circular revolving platform) of thirty-two grasshoppers. Krogh even took up the development of buds in trees in collaboration with H. Burstrøm.

A superb lecturer, Krogh eventually tired of preparing his talks and in 1934 withdrew from academic duties. But since 1908 he has been known to Danish schoolchildren from his *Laerebog i Menneskets Fysiologi*, which went through eleven editions. Krogh was a prolific popular author as well as a scholar in the history of science. He enjoyed Kipling and contributed citations to the *Oxford English Dictionary*.

As a scholar Krogh was gifted, kindly, sympathetic, and inspiring. Physically he was a small man and wore a mustache and goatee. Although universally respected, during World War II he was apparently slated for "liquidation" by the Germans. Forced to go underground, he escaped to Sweden, where he stayed incognito until the end of the war.

Krogh could at times be argumentative, but he was always loyal in his friendships. In 1929, in Boston, he said: "We may fondly imagine that we are impartial seekers after truth, but with a few exceptions, to which I know that I do not belong, we are influenced—and sometimes strongly—by our personal bias; and we give our best thoughts to those ideas which we have to defend."

Although Krogh declined decorations, he appreciated other tokens of esteem. He was Silliman lecturer at Yale University in 1922, Charles Mickle fellow of the Toronto Medical Faculty in 1925, Cooper lecturer at Swarthmore College in 1939, and Croonian lecturer at the Royal Society, London, in 1945, the same year in which he was awarded the Baly Medal. He held numerous honorary degrees. In 1916 he became a member of the Royal Danish Society of Sciences, but he resigned when he found its scientific attitude too unprogressive. On his fiftieth birthday Krogh received a Festschrift, *Physiological Papers Dedicated to August Krogh* (1926), from twenty-two foreign and Danish pupils. In 1939 he was declared an honorary citizen of Grenå.

BIBLIOGRAPHY

I. ORIGINAL WORKS. A complete catalog of Krogh's writings is in *Meddelelser fra Akademiet for de Tekniske Videnskaber*, **1** (1949), 39–50.

II. SECONDARY LITERATURE. On Krogh and his work see H. C. Hagedorn, "August Krogh," in *Meddelelser fra Akademiet for de Tekniske Videnskaber*, **1** (1949), 33–38; G. Liljestrand, "August Krogh," in *Acta physiologica scandinavica*, **20** (1950), 109–120; P. Brandt Rehberg, "August Krogh," in *Festskrift, Københavns Universitet 1950* (Copenhagen, 1950), pp. 182–215; and L. G. Rowntree, *Amid Masters of Twentieth Century Medicine* (Chicago, 1958), pp. 171–174.

E. SNORRASON

KRONECKER, HUGO (*b.* Liegnitz, Germany [now Legnica, Poland], 27 January 1839; *d.* Bad Nauheim, Germany, 6 June 1914), *physiology*.

Hugo Kronecker was the brother of the mathematician Leopold Kronecker. After attending the grammar school at Liegnitz, he went to Berlin, where he began studying medicine around 1859–1860. He soon took a great interest in physiology, taught at that time by du Bois-Reymond. He continued his studies at Heidelberg, where at the institute he took part in investigations of the physiology of the muscles carried out by Hermann von Helmholtz and W. Wundt. After further studies at Pisa, he returned to Berlin. There he took the doctorate in 1863 under du Bois-Reymond with a thesis on the problem of the fatigue of the muscles. In 1865 he was registered as a medical practitioner. He received his clinical training at Berlin as the personal assistant of his friend, Ludwig Traube, who was a follower of the so-called physiological school in medicine. He also worked under Wilhelm Kühne. Kühne was at that time in charge, under Virchow, of the chemico-physiological laboratory in the Charité Hospital in Berlin.

In 1868 Kronecker moved to Leipzig, where he worked in the Physiologische Anstalt directed by Carl Ludwig. After a temporary absence, during which he participated as a medical officer in the Franco-Prussian war of 1870–1871, Kronecker became Ludwig's assistant; in the same year, he qualified as a lecturer with a significant dissertation in the field of the physiology of muscles. In 1875 he was appointed senior lecturer at the Physiologische Anstalt, where he stayed until 1877. In that year young physiologists from many countries were working at Ludwig's laboratory; among them were S. von Basch, Baxt, Bowditch, Buchner, Flechsig, M. von Frey, W. Gaskell, Gaule, Merunowicz, A. Mosso, G. Schwalbe, and Scäfer. Kronecker was of great aid to them, in part because of his outstanding knowledge of foreign languages and his congeniality. From 1878 to 1884 Kronecker was in charge of the "special physiological department" of the Institute of Physiology at Berlin, directed by du Bois-Reymond. Once again he helped the foreigners studying at the Berlin institute, among them the Russians N. E. Wedenskij, Th. Openchowski, D. von Ott, N. W. Jastrebov, and D. V. Kireev. In addition S. von Basch from Vienna and Francis Gotch from England worked under him.

When he was appointed successor to Paul Grützner as full professor for physiology at Bern in 1884, many foreigners worked at his institute, especially after the "Hallerianum," a new building attached to the institute had been constructed. He took a great part in founding the International Physiological Congress. The first one took place at Basel in 1889. Kronecker held his chair at Bern until his sudden death at Bad Nauheim of a perforated aortic aneurysm. He was seventy-five. After his death obituaries appeared throughout the world. His extreme amiability, his noble character, his generosity as a host, his sociability, and his merits as a scientist were praised in all of them.

Kronecker's scientific investigations were mainly centered around questions concerning the muscles, the heart and circulation, the mechanism of deglutition, saline infusion, and mountain sickness. Because most of the works appeared under his name and those of his collaborators, and some only under the name of the latter, it is often difficult to assess the share contributed by Kronecker himself. His methodological talents were spoken of very highly. He improved the proofs for the all-or-none law applied to the heart (1873) as well as the method of the isolated heart (1874). Almost simultaneously with E. J. Marey, he described the refractory period of the heart (1874). A device for irrigating and measuring the pressure of the isolated heart is named after him. He also discovered a blood substitute in saline solution. He reported on the life-saving NaCl injection in 1884.

Kronecker made an essential contribution to the development of the first clinically applicable indirect method of measuring the blood pressure, carried out at Berlin by S. von Basch (1880). Together with S. J. Meltzer he worked on the deglutition reflex, its speed, and its dependence on the nervous system. Throughout his life Kronecker defended the theory of the neurogenic origin of the automatism of the heart beat, which later was shown to be mistaken. He constructed a much-used calibrated induction coil for the excitation of living organs and on several occasions dealt with questions concerning the excitability of the skeletal muscle and of the heart muscle.

When he was at Bern, most of his research dealt with the problem of mountain sickness. Kronecker argued against the oxygen-deficiency theory put forward by Paul Bert, favoring a mechanical theory—that is, he regarded a congestion of the lung vessels by pulmonary edema as its cause. This theory, too, was mistaken and Bert's theory of oxygen-deficiency was proven right. In 1894 Kronecker led an expedition to Zermatt for the investigation of these pulmonary disturbances. The purpose of the expedition was to draw up an expert's report on the possible dangers of the quick ascent up Jungfrau mountain on the cable-railway line, which was then in the planning stage (1892). The monograph "Mountain Sickness" (1903) presents the results of his work.

BIBLIOGRAPHY

I. Original Works. Among Kronecker's more important publications are "De ratione qua musculorum defatigatio ex labore eorum pendeat," Ph.D. dissertation, University of Berlin, 1863; "Über die Ermüdung und Erholung der quergestreiften Muskeln" (Habilitation thesis), in *Arbeiten aus der Physiologischen Anstalt zu Leipzig 1871* (Leipzig, 1872), pp. 177–265; "Das charakteristische Merkmal der Herzmuskelbewegung," in *Beiträge zur Anatomie und Physiologie als Festgabe für Carl Ludwig zum 15. Oktober 1874 gewidmet von seinen Schuelern* (1874), pp. CLXXIII–CCIV; "Über die Speisung des Froschherzens," in *Archives für Anatomie und Physiologie*, Physiologische Abteilung (1878), 321–322; "Die Genesis des Tetanus," *ibid.* (1878), 1–40, written with William Stirling; "Über die Form des minimalen Tetanus," *ibid.* (1877), pp. 571–573; "Über den Mechanismus der Schluckbewegung," *ibid.* (1880), 296–299; *ibid.* (1880), 446–447; *ibid.* (1881), 465–466; and *ibid.* (1883) (Suppl. Band), 328–362, written with S. J. Meltzner.

For further reference, see Kronecker and Schmey, "Das Coordinationszentrum der Herzkammerbewegungen," in *Monatsberichte der Preussischen Akademie der Wissenschaften zu Berlin* (1884), 87–89; "Ueber graphische Methoden in der Physiologie. Z. Instrumentenkunde," **1** (1881), 26–28; and "Ueber Störungen der Coordination des Herzkammerschlages," in *Zeitschrift für Biologie*, n.s., **34** (1896), 529–603.

II. Secondary Literature. For detailed obituaries, see S. J. Meltzner, "Professor Hugo Kronecker," in *Science*, **40** (1914), 441–444; and Paul Heger, "Hugo Kronecker," in *Münchener medizinische Wochenschrift*, (1914), pp. 1629–1631. Short obituaries include A. Loewi, "Hugo Kronecker," in *Deutsche medizinische Wochenschrift*, (1914), pp. 1437–1438; and anonymous works in *Lancet* (1914), **92**, 270–271; *British Medical Journal* (1914), **2**, 491; Corr. Blatt für Schweizer Ärzte, Jg. XLIV (1914), 848–850; E. A. S. (E. A. Schäfer?), "Hugo Kronecker 1839–1914," in *Proceedings of the Royal Society*, Ser. B, **89** (1917), 14–50. A short biography in *Biographisches Lexikon der hervorragenden Ärzte aller Zeiten und Völker*, II, repr. by Haberling, III (Berlin–Vienna, 1931), 616–617.

For a bibliography of Kronecker's works consult Royal Society, *Catalogue of Scientific Papers*, **8** (1879), 128, **10** (1894), 467–468. See also H.-J. Marseille, "Das physiologische Lebenswerk von Emil du Bois-Reymond mit besonderer Berücksichtigung seiner Schüler," diss. (Münster, 1968).

K. E. Rothschuh

KRONECKER, LEOPOLD (*b.* Liegnitz, Germany [now Legnica, Poland], 7 December 1823; *d.* Berlin, Germany, 29 December 1891), *mathematics.*

Kronecker's parents were Isidor Kronecker, a businessman, and his wife, Johanna Prausnitzer. They were wealthy and provided private tutoring at home for their son until he entered the Liegnitz Gymnasium. At the Gymnasium, Kronecker's mathematics teacher was E. E. Kummer, who early recognized the boy's ability and encouraged him to do independent research. He also received Evangelical religious instruction, although he was Jewish; he formally converted to Christianity in the last year of his life.

Kronecker matriculated at the University of Berlin in 1841. He attended lectures in mathematics given by Dirichlet and Steiner; in astronomy, by Encke; in meteorology by Dove; and in chemistry, by Mitscherlich. Like Gauss and Jacobi, he was interested in classical philology, and heard lectures on this subject. He also attended Schelling's philosophy lectures; he was later to make a thorough study of the works of Descartes, Spinoza, Leibniz, Kant, and Hegel, as well as those of Schopenhauer, whose ideas he rejected.

Kronecker spent the summer semester of 1843 at the University of Bonn, having been attracted there by Argelander's astronomy lectures. He also became acquainted with such democrats as Eduard Kinkel, and was active in founding a *Burschenschaft*, a student association. Kronecker's career might thus have been endangered by his political associations. The following autumn he went to Breslau (now Wrocław, Poland) because Kummer had been appointed professor there. He remained for two semesters, returning to Berlin in the winter semester of 1844-1845 to take the doctorate.

In his dissertation, "On Complex Units," submitted

to the Faculty of Philosophy on 30 July 1845, Kronecker dealt with the particular complex units that appear in cyclotomy. He thereby arrived at results and methods closely related to the theory of "ideal numbers" that Kummer was to propound a short time later. (In 1893 Frobenius, in a memorial address on Kronecker, compared this dissertation to a work of "chemistry without the atomic hypothesis.") In evaluating the dissertation, Dirichlet said that in it Kronecker demonstrated "unusual penetration, great assiduity, and an exact knowledge of the present state of higher mathematics."

Kronecker took his oral examination on 14 August 1845. Encke questioned him on the application of the calculus of probabilities to observations and to the method of least squares; Dirichlet, on definite integrals, series, and differential equations; August Boeckh, on Greek; and Adolf Trendelenburg, on the history of legal philosophy. He was awarded the doctorate on 10 September.

Dirichlet, his professor and examiner, was to remain one of Kronecker's closest friends, as was Kummer, his first mathematics teacher. (On the occasion of the fiftieth anniversary of the latter's doctorate, in 1881, Kronecker said that Kummer had provided him with the "most essential portion" of his "intellectual life.") In the meantime, in Berlin, Kronecker was also becoming better acquainted with Eisenstein and with Jacobi, who had recently returned from Königsberg (now Kaliningrad, U.S.S.R.) for reasons of health. During the same period Dirichlet introduced him to Alexander von Humboldt and to the composer Felix Mendelssohn, who was both Dirichlet's brother-in-law and the cousin of Kummer's wife.

Family business then called Kronecker from Berlin. In its interest he was required to spend a few years managing an estate near Liegnitz, as well as to dissolve the banking business of an uncle. In 1848 he married the latter's daughter, his cousin Fanny Prausnitzer; they had six children. Having temporarily renounced an academic career, Kronecker continued to do mathematics as a recreation. He both carried on independent research and engaged in a lively mathematical correspondence with Kummer; he was not ambitious for fame, and was able to enjoy mathematics as a true amateur. By 1855, however, Kronecker's circumstances had changed enough to allow him to return to the academic life in Berlin as a financially independent private scholar.

This was a momentous time for mathematics in Germany. In 1855 Dirichlet left Berlin to go to Göttingen as successor to Gauss; Kummer succeeded Dirichlet in Berlin; and Carl Wilhelm Borchardt became editor of the *Journal für die reine und angewandte Mathematik*, following the death of its founder Crelle. In 1856 Weierstrass was called to Berlin and Kronecker and Kummer soon became friends with Borchardt and Weierstrass.

Although Kronecker had published some scientific articles before he returned to Berlin, he soon brought out a large number of mathematical tracts in rapid succession. Among other subjects he wrote on number theory (one of his earliest interests, instilled in him by Kummer), the theory of elliptical functions, algebra, and, particularly, on the interdependence of these mathematical disciplines. In 1860 Kummer, seconded by Borchardt and Weierstrass, nominated Kronecker to the Berlin Academy, of which he became full member on 23 January 1861.

In the winter semester of the following year Kronecker, at Kummer's suggestion, made use of a statutory right held by all members of the Academy to deliver a series of lectures at the University of Berlin. His principal topics were the theory of algebraic equations, the theory of numbers, the theory of determinants, and the theory of simple and multiple integrals. He attempted to simplify and refine existing theories and to present them from new perspectives. His teaching and his research were closely linked and, like Weierstrass, he was most concerned with ideas that were still in the process of development. Unlike Weierstrass—and for that matter, Kummer—Kronecker did not attract great numbers of students. Only a few of his auditors were able to follow the flights of his thought, and only a few persevered until the end of the semester. To those students who could understand him, however, Kronecker communicated something of his joy in mathematical discusssion. The new ideas that he offered his colleagues and students often received their final formulation in the course of such scholarly exchanges. He was allowed a considerable degree of autonomy in his teaching at Berlin, so much so that when in 1868 he was offered the chair at Göttingen that had been held successively by Gauss, Dirichlet, and Riemann, he refused it.

Kronecker was increasingly active and influential in the affairs of the Academy, particularly in recruiting the most important German and foreign mathematicians for it. Between 1863 and 1886 he personally helped fifteen mathematicians in becoming full, corresponding, or honorary members, or in obtaining a higher degree of membership. The names of these men constitute a formidable catalog; they were, in the order in which Kronecker assisted them, Heine, Riemann, Sylvester, Clebsch, E. Schering, H. J. Stephen Smith, Dedekind, Betti, Brioschi, Beltrami, C. J. Malmsten, Hermite, Fuchs, F. Carorati, and L. Cremona. The formal nominations that Kronecker made during this period are of great interest, not least because of their subjectivity. Thus, to give one example, in his otherwise comprehensive evaluation of Dedekind's work (1880), Kronecker, who was then seeking to reduce all mathematical operations to those dealing in positive whole numbers, ignored Dedekind's *Stetigkeit und irrationale Zahlen* of 1872.

Kronecker's influence outside Germany also increased. He was a member of many learned societies, among them the Paris Academy, of which he was elected a corresponding member in 1868, and the Royal Society of London, of which he became a foreign associate in 1884. He established other contacts with foreign scientists in his numerous travels abroad and in extending to them the hospitality of his Berlin home. For this reason his advice was often solicited in regard to filling mathematical professorships both in Germany and elsewhere; his recommendations were probably as significant as those of his erstwhile friend Weierstrass.

Kronecker's relations with Weierstrass had been disintegrating since the middle of the 1870's. They continued to work together, however; in 1880, following Borchardt's death, Kronecker took over the editorship of the *Journal für die reine und angewandte Mathematik*, in which Weierstrass for a time assisted him. In 1883 Kummer retired from the chair of mathematics, and Kronecker was chosen to succeed him, thereby becoming the first person to hold the post at Berlin who had also earned the doctorate there. He was simultaneously named codirector of the mathematics seminar that Kummer and Weierstrass had founded in 1861. Kronecker continued to lecture, as he had done for twenty years, but now, as a member of the faculty, was able to assume all the rights thereof, including participation in the granting of degrees, the nomination of professors, and the qualifying examinations for university lecturers. He was enabled, too, to sponsor his own students for the doctorate; among his candidates were Adolf Kneser, Paul Stäckel, and Kurt Hensel, who was to edit his works and some of his lectures.

The cause of the growing estrangement between Kronecker and Weierstrass was the following. The very different temperaments of the two men must have played a large part in it, and their professional and scientific differences could only have reinforced their personal difficulties. Since they had long maintained the same circle of friends, their friends, too, became involved on both levels. A characteristic incident occurred at the new year of 1884-1885, when H. A. Schwarz, who was both Weierstrass' student and Kummer's son-in-law, sent Kronecker a greeting that included the phrase: "He who does not honor the Smaller [Kronecker], is not worthy of the Greater [Weierstrass]." Kronecker read this allusion to physical size—he was a small man, and increasingly self-conscious with age—as a slur on his intellectual powers and broke with Schwarz completely. (Other scholars, among them Hofmann and Helmholtz, maintained lasting good relations with Kronecker by displaying more tact toward his special sensitivities.)

At any rate, personal quarrel became scholarly polemic. Weierstrass, for example, believed (perhaps rightly) that Kronecker's opposition to Cantor's views on "transfinite numbers" reflected opposition to his own work.

The basis of Kronecker's objection to Weierstrass' methods of analysis is revealed in his well-known dictum that "God Himself made the whole numbers—everything else is the work of men." Kronecker believed that all arithmetic could be based upon whole numbers, and whole numbers only; he further classified all mathematical disciplines except geometry and mechanics as arithmetical, a category that specifically included algebra and analysis. He never actually stated his intention of recasting analysis without irrational numbers, however, and it is possible that he did not take his radical notions altogether seriously himself. Weierstrass could not afford to regard Kronecker's demands as merely whimsical; in 1885 he claimed indignantly that for Kronecker it was an axiom that equations could exist only between whole numbers, while he, Weierstrass, granted irrational numbers the same validity as any other concepts.

Kronecker's remarks that arithmetic could put analysis on a more rigorous basis, and that those who came after him would recognize this and thereby demonstrate the falseness of so-called analysis, angered and embittered Weierstrass. He saw in these words an attempt by Kronecker not only to invalidate his whole life's work, but also to seduce the younger generation of mathematicians to an entirely new theory. The two men were further at odds over a Swedish mathematics prize contest and over the editing of Borchardt's works. By 1888, Weierstrass had confided to a few close friends that his break with Kronecker was complete; Kronecker, for his part, apparently did not realize how gravely his opinions and activities had wounded Weierstrass, since on several later occasions he still referred to himself as being his friend.

Weierstrass at this time even considered leaving Germany for Switzerland to avoid the constant conflict with Kronecker, but one consideration kept him in Berlin. Kronecker had remained on good terms with Kummer and with Kummer's successor,

Fuchs; it was therefore likely that Kronecker would have considerable influence in the choice of Weierstrass' own successor. Weierstrass believed that all his work would be undone by a successor acceptable to Kronecker; for this reason he stayed where he was. In the meantime, new sources of antagonism arose, among them Weierstrass' scruple about the qualifications as a lecturer of Kronecker's protégé Hensel and Kronecker's stated objection to granting an assistant professorship to Weierstrass' pupil Johannes Knoblach. These new difficulties never reached a crucial point, however, since Kronecker's wife died on 23 August 1891, and he survived her by only a few months.

Kronecker's greatest mathematical achievements lie in his efforts to unify arithmetic, algebra, and analysis, and most particularly in his work on elliptical functions. His boundary formulas are particularly noteworthy in this regard, since they laid bare the deepest relationships between arithmetic and elliptical functions and provided the basis for Erich Hecke's later analytic-arithmetical investigations. Kronecker also introduced a number of formal refinements in algebra and in the theory of numbers, and many new theorems and concepts. Among the latter, special mention should be made of his theorem in regard to the cyclotomic theory, according to which all algebraic numbers with Abelian and Galois groups (over the rational number field) are rational combinations of roots of unity. His theorem on the convergence of infinite series is also significant.

The most important aspects of Kronecker's work were manifest as early as his dissertation of 1845. In his treatment of complex units, Kronecker sought to present a theory of units in an algebraic number field, and, indeed, to present a whole system of units as a group. Twenty-five years later he succeeded in constructing an implicit system of axioms to rule finite Abelian groups, although he did not at that time apply it explicitly to such groups. His work thus lay clearly in the line of development of modern algebra.

For this reason it might be useful to assess Kronecker's position with respect to other mathematicians. One criterion that suggests itself is the application of the algorithm, and while few mathematicians have held unalloyed opinions on this matter, two sharply differentiated positions may be distinguished. One group of mathematicians then—of whom Gauss, Dirichlet, and Dedekind are representative—found the algorithm to be most useful as a concept, rather than a symbol; their work centered on ideas, not calculations. The other group, which includes Leibniz, Euler, Jacobi, and—as Kneser demonstrated—Kronecker, stressed the technical use of the algorithm, employing it as a means to an end. Kronecker's goal was the perfection of the technique of calculation and he employed symbols to avoid the repetition of syllogisms and for clarity. He termed Gauss's contrasting method of presenting mathematics "dogmatic," although he retained a great respect for Gauss and for his work.

Kronecker's mathematics lacked a systematic theoretical basis, however, and for this reason Frobenius asserted that he was not the equal of the greatest mathematicians in the individual fields that he pursued. Thus, Frobenius considered Kronecker to be inferior to Cauchy and Jacobi in analysis; to Riemann and Weierstrass in function theory; to Kummer and Dirichlet in arithmetic; and to Abel and Galois in algebra.

Kronecker was nevertheless preeminent in uniting the separate mathematical disciplines. Moreover, in certain ways—his refusal to recognize an actual infinity, his insistence that a mathematical concept must be defined in a finite number of steps, and his opposition to the work of Cantor and Dedekind—his approach may be compared to that of intuitionists in the twentieth century. Kronecker's mathematics thus remains influential.

BIBLIOGRAPHY

I. Original Works. Kronecker's writings, including collected editions, are listed in Poggendorff, I, 1321, 1579; III, 752–753; IV, 807–808; and VI, 1412. See also Ernst Schering, "Briefwechsel zwischen G. Lejeune Dirichlet und Leopold Kronecker," in various issues of *Nachrichten der Königlichen Gesellschaft der Wissenschaften zu Göttingen* beginning with that of 4 July 1885; Emil Lampe, "Schriften von L. Kronecker," in *Jahresbericht der Deutschen Mathematiker-vereinigung*, **2** (1893), 23–31; and "Brief Leopold Kroneckers an Ernst Eduard Kummer vom 9. September 1881," in *Abhandlungen der Geschichte der mathematischen Wissenschaften*, **29** (1910), 102–103. Additional material may be found in the archives of the Deutschen Akademie der Wissenschaften of Berlin and Humboldt University, Berlin.

Kronecker also edited the first vol. of Dirichlet, *Werke* (Berlin, 1889).

II. Secondary Literature. On Kronecker and his work see Kurt-R. Biermann, "Vorschläge zur Wahl von Mathematikern in die Berliner Akademie," in *Abhandlungen der Deutschen Akademie der Wissenschaften zu Berlin*, Klasse für Mathematik, Physik und Technik, no. 3 (1960), 29–34; "Karl Weierstrass," in *Journal für die reine und angewandte Mathematik*, **223** (1966), 191–220; "Die Mathematik und ihre Dozenten an der Berliner Universität 1910–1920" (MS, Berlin, 1966); "Richard Dedekind im Urteil der Berliner Akademie," in *Forschungen und Fort-*

schritte, **40** (1966), 301–302, which contains a reprint of Kronecker's memorial address for Dedekind; Georg Frobenius, "Gedächtnisrede auf Leopold Kronecker," in *Abhandlungen der Königlich Preussischen Akademie der Wissenschaften zu Berlin* (1893); Lotte Kellner, "The Role of Amateurs in the Development of Mathematics," in *Scientia*, **60** (1966), 1–5 (see especially p. 4); Adolf Kneser, "Leopold Kronecker," in *Jahresbericht der Deutschen Mathematiker-vereinigung*, **33** (1925), 210–228; Emil Lampe, "Leopold Kronecker," in *Annalen der Physik und Chemie*, **45** (1892), 595–601; Heinrich Weber, "Leopold Kronecker," in *Jahresbericht der Deutschen Mathematiker-vereinigung*, **2** (1893), 5–23; and Hans Wussing, "Zur Entstehungsgeschichte der abstrakten Gruppentheorie," in '*NTM*,' *Schriftenreihe für Geschichte der Naturwissenschaften, Technik und Medizin*, **2**, no. 5 (1965), 1–16, esp. pp. 7–9.

KURT-R. BIERMANN

KRÖNIG, AUGUST KARL (*b.* Schildesche, Westphalia, Germany, 20 September 1822; *d.* Berlin, Germany, 5 June 1879), *physics.*

Remarkably little is known about Krönig, who is commonly recognized as the originator of the kinetic theory of gases. The sixth of the seven children of a country pastor, he entered the University of Bonn in 1839. He chose to study the physical sciences only after transferring to Berlin where he completed his doctoral dissertation (*De acidi chromici salibus crystallinus*) in 1845. One of the fifty-three original members gathered by Gustav Magnus to form the Berlin Physikalische Gesellschaft in 1845, Krönig served as secretary in 1848. He edited the three-volume *Journal für Physik und physikalische Chemie des Auslandes*, which appeared in 1851 and introduced significant foreign scientific work in German translation. He also was editor of the annual literature survey, *Die Fortschritte der Physik*, between 1855 and 1859. Declining health caused him to relinquish the editorship and led to his early retirement in 1861 from his position as professor at the Berlin Königliche Realschule. Krönig soon became a forgotten man, and only a brief obituary notice by his wife and son in *Die National-Zeitung* marked his passing.

Krönig was not a significant figure in the Berlin scientific community. His books, *Neue Methode zur Vermeidung and Auffindung von Rechenfehlern* (1855) and *Die Chemie bearbeitet als Bildungsmittel*, were elementary treatises addressed to students and the general reader. The former presented techniques for testing the accuracy of various calculations, using the numbers 9, 11, 37, and 101; the latter was an overly simplified and somewhat unorthodox version of chemistry which became the target for polemical reviews. Apart from the well-known "Grundzüge einer Theorie der Gase" (1856), his first paper, none of his papers was of great import. Most of them appeared in 1864—some being only a few pages in length—and ranged over a variety of topics from how to locate the position of real images using a pinpoint to the explanation of the Davy safety lamp in terms of radiation rather than conduction of heat by the metal screen.

In the absence of any other significant theoretical work, the paper on the kinetic theory of gases seems strangely anomalous. His approach was an exceedingly simple one; he assumed that atoms move unhindered between the walls of a container, with exactly one-third of them moving in each of the three Cartesian coordinate directions. From this minimal model, however, he reached a number of significant conclusions. After first demonstrating the usual relations for ideal gas behavior, Krönig went on to establish the proposition that different gases should contain equal numbers of atoms in equal volumes at the same temperature and pressure. He also suggested that lighter gases should diffuse more rapidly, if temperature is a measure of molecular *vis viva*, and explained why gases are warmed by compression—the moving piston imparts additional velocity to the rebounding atoms.

It is likely that Krönig's theory was not altogether original with him. During his years as editor of *Die Fortschritte der Physik*, an abstract of Waterston's ideas came under review in which every one of these conclusions was succinctly stated, and Krönig may well have inadvertently received the general guidelines of his theory from reading that brief résumé.

Crippled by blindness and paralysis during his last years, Krönig published in 1874 a collection of theological and philosophical fragments written over the years, *Das Dasein Gottes und das Glück der Menschen.* While recognizing that the teleological behavior of organic nature suggests a personal God of incomparable intelligence, Krönig could not accept the Christian image of God. He posited a God who structured the universe intelligently but was little involved in what ensued. The world, he said, would have to be somewhat different were God omnipotent, omniscient, and perfectly good.

BIBLIOGRAPHY

The only complete biographical source is Grete Ronge, "Biographische Notizen zu August Karl Krönig," in *Gesnerus*, **18** (1961), 67–70; see also Poggendorff, I, 1320–1321, and II, 752.

On the roles of Waterston, Krönig, and Clausius in the

development of the kinetic theory of gases, see Stephen Brush, "The Development of the Kinetic Theory of Gases, II. Waterston," in *Annals of Science*, **13** (1957), 273–282; "III. Clausius," *ibid.*, **14** (1958), 185–196; E. E. Daub, "Waterston's Influence on Krönig's Kinetic Theory of Gases," in *Isis*, **62** (1971), 512–515; Grete Ronge, "Zur Geschichte der kinetischen Wärmetheorie," in *Gesnerus*, **18** (1961), 45–67.

EDWARD E. DAUB

KROPOTKIN, PETR ALEKSEEVICH (*b.* Moscow, Russia, 9 December 1842; *d.* Dmitrov, Moscow oblast, U.S.S.R., 8 February 1921), *geography*.

Kropotkin's father was Prince Aleksei Petrovich Kropotkin, a general and large landowner. His mother, Ekaterina Nikolaevna Kropotkina, was the daughter of General N. Sulim.

Kropotkin was first educated at home, and from 1857 until 1862 he studied in St. Petersburg at the School of Pages. Having completed his schooling, Kropotkin enlisted in the Amur Cossack army in order to study Siberia and further the development of this territory. In Irkutsk, Kropotkin was appointed adjutant to General Kukel, chief of staff of the governor-general of eastern Siberia.

Kropotkin traveled widely and wrote a number of outstanding scientific works on geography and geomorphology. He was one of the founders of the paleography of the Quaternary period and an author of the doctrine of ancient glaciation. Kropotkin believed that geography was the only subject that could unite all the natural sciences.

In the 1870's Kropotkin became one of the ideologists of anarchism. On 23 March 1874 he was arrested and imprisoned in the Peter and Paul Fortress in St. Petersburg. On 30 June 1876 he escaped and went to Edinburgh. He then lived in Switzerland for some time before moving to France, where he was arrested in 1883 for disseminating anarchist propaganda and was imprisoned until 1886, first in Lyons and then in Clairvaux. After his release Kropotkin lived in London. In 1917, after the February Revolution, he returned to Russia.

Kropotkin officially adhered to Russian Orthodoxy. He was married and had one daughter, Aleksandra.

Kropotkin first took part in military missions for geographical investigations to the Transbaikal region. In 1863 he accompanied a barge carrying provisions to Cossack outposts located along the shores of the Amur River. During the spring of 1864 he carried out the reconnaissance of a previously unknown route from southeastern Transbaikalia, through northern Manchuria, to the Amur. He crossed the Greater Khingan Range and discovered a region of inactive volcanoes. In *Zapiski Sibirskogo otdelo Russkogo geograficheskogo obshchestva*, **8** (1865), he described this journey and his steamship voyage from the mouth of the Sungari River to the town of Kirin.

During the summer of 1865, Kropotkin traveled at his own expense to the Eastern Sayan Mountain Range. During this trip he scaled the Tunkinskiye Goltsy Mountains, a range in the Eastern Sayan system. On the dome-shaped summits, which are above the timber line, he discovered clear traces of glaciers. In the basin of the Dzhanbulak River (a tributary of the Oka River) Kropotkin found volcanic craters; he surveyed their environs and collected samples of volcanic rock. His account of this trip shed new light on the geography of eastern Siberia.

In 1866 Kropotkin headed an expedition of the Siberian section of the Russian Geographic Society to the basins of the Vitim and Olekma rivers. The expedition traveled in boats downstream along the Lena River to the village of Krestovaya. On packhorses, it followed a complex route—from the basin of the Chara River to the Vitim River, crossing it below the Parama rapids; once past the confluence of the Vitim with the Parama and Muya rivers, the expedition moved on to the Tsipa River and its tributary, the Tsipikan. Next it crossed the Big and Little Amalat, Dzhelinde, and Mangoi rivers and arrived at Chita. On this route the expedition surmounted the mountain ranges of the Olekmo-Charskoye Nagorye uplands (the northern and southern Muya range) and a number of hills of the Vitim plateau. As a result of the expedition, a route for cattle drives was found. The scientific results, published in "Otchet ob Olekminsko-Vitimskoy ekspeditsii" ("Account of the Olekma-Vitim Expedition," 1873), are of special interest.

From his observations, Kropotkin found sufficient proof to "accept as an indisputably proven fact that the limestone appearing in the Lena valley between Kirensk and Olekminsk is older than the red sandstone that lies in horizontal layers between Kachug and Kirensk" (*ibid.*, p. 181).

In the Olekma River basin Kropotkin found further evidence to support his conjecture concerning the past glaciation of eastern Siberia—the smoothed surfaces of gneiss mountains heights; the presence of typical marine deposits, which form the gold-bearing sands located high above the present level of rivers; and striated boulders and other signs of the effects of glaciers. Kropotkin wrote: "At first, having faith in the authority of geologists who had visited Siberia and having been convinced that Siberia did not present traces of the glacial period, I slowly but surely had to retreat before the obviousness of the facts and had to arrive at the opposite conviction—that glacial phenomena did extend into eastern Siberia, at least into its northeast sector" (*ibid.*, p. 223).

Kropotkin noted that the climate of eastern Siberia in the post-Pliocene period had been moist enough that glaciers could be formed. This observation was new in the paleogeography of eastern Siberia. Since then, traces of ancient glaciation in the mountains north of Lake Baikal have been found, and modern glaciers—including more than thirty in the Kodar Range—have been discovered.

From the Lena River to the town of Chita, Kropotkin distinguished four geomorphological regions: the Lena River valley, the flat Lena heights, the Olekma-Vitim mountainous country, and the Vitim plateau. He gave a detailed characterization of each region. After his account of the Olekma-Vitim expedition, Kropotkin began studying the orography of eastern Siberia. He used earlier accounts of a number of expeditions which described their routes and noted elevations. Kropotkin published in 1875 "Obshchy ocherk orografii Vostochnoy Sibiri" ("A General Essay on the Orography of Eastern Siberia"). In 1904 he published an extract from this work in Brussels and in the *Geographical Journal.*

From his observations Kropotkin concluded that the erosion of plateaus (high and low) into dome-shaped summits and rounded crests played a major role in forming the relief of eastern Siberia. Kropotkin refuted the old concept of the Stanovoy Khrebet Mountain range as the watershed between the waters of the Arctic Ocean and the Pacific Ocean. He pointed out that in the area of the Zeya River's upper reaches, the tributaries of the Lena and the Amur rivers begin not in the watershed range but on the high plateau, and then rush into the Lena and the Amur. Kropotkin's orographic map of eastern Siberia was not superseded until 1950.

After the completion of the Olekma-Vitim expedition, Kropotkin left the military service and in the autumn of 1867 enrolled in the mathematics department of St. Petersburg University. In November 1868 he was elected secretary of the physical geography section of the Russian Geographical Society. While working in this post, Kropotkin compiled "Doklad komissii po snaryazheniyu ekspeditsii v severnye morya" ("Report of the Commission for the Outfitting of an Expedition to the Northern Seas"), in which he substantiated Shilling's hypothesis concerning the existence of the then unknown Zemlya Frantsa Iosifa, which was discovered in 1873 by the Austro-Hungarian expedition of Julius von Payer and Karl Weiprecht.

Kropotkin's major scientific work was the corroboration of the theory of ancient continental glaciation. In 1871 he visited Finland and Sweden on behalf of the Russian Geographic Society in order to study glacial phenomena. On 21 March 1874 he reported his conclusions to the Society and refuted the prevailing view that huge boulders had been transported to European fields by floating ice floes. Kropotkin proved that the plains of northern Europe, Asia, and America had once been covered by powerful continental ice which had (in the case of Europe) spread from the heights of Fennoscandia. His conclusions were printed in "Issledovanie o lednikovom periode" ("An Investigation of the Glacial Period"), which proved, for the first time in scientific literature, the presence of an ancient period of glaciation in the plains areas. The work also contained interesting comments on glacial relief forms and accurate notions on the formation of loess from ground glacial debris carried by water from under the glacial cap. Kropotkin wrote: "We can conclude that under certain climatic conditions extensive ice caps 1,000, 2,000, and 3,000 meters in thickness must have been formed on the continents, and that these ice caps (and even those which were ten and twenty times thinner) must have crawled along the land, regardless of the existing relief" (*ibid.*, p. 539).

In exile Kropotkin continued to deal with questions of the geography of Russia. In 1880 he participated in the compilation of the sixth volume, *Asiatic Russia,* of Reclus' *General Geography.*

While visiting the United States in 1897, Kropotkin agreed to write his memoirs for the *Atlantic Monthly.* The work appeared in English (London, 1899) and was later translated into Russian (London, 1902) as a separate book entitled *Memoirs of a Revolutionist.* This work received wide circulation in western countries and was frequently reprinted in the Soviet Union (7th ed., Moscow, 1966). From 1892 to 1901 Kropotkin also wrote scientific reviews for the English magazine *Nineteenth Century.*

In 1912 the world scientific community celebrated Kropotkin's seventieth birthday. The Royal Geographic Society wrote in its congratulatory address: "Your service in the field of the natural sciences, your contribution to geography and geology, your amendments to Darwin's theory have gained you worldwide fame and have broadened our understanding of nature." In the Soviet Union a range on the southern perimeter of the Patom uplands, a town in the Krasnodar region, and a settlement in the Irkutsk oblast have been named for Kropotkin. His scientific works still serve as references for young investigators of the geography of the Soviet Union.

BIBLIOGRAPHY

I. Original Works. Kropotkin's writings include "Doklad komissii po snaryazheniyu ekspeditsii v severnye morya" ("Report of the Commission for the Outfitting of an Expedition to the Northern Seas"), in *Izvestiya Russkago geograficheskago obshchestva*, no. 3 (1871), 29–117; "Otchet ob Olekminsko-Vitimskoy ekspeditsii" ("Account of the Olekma-Vitim Expedition"), in *Zapiski Russkogo geograficheskogo obshchestva po obshchey geografii*, **3** (1873), 1–482; "Obshchy ocherk orografii Vostochnoy Sibiri" ("A General Essay on the Orography of Eastern Siberia"), *ibid.*, **5** (1875), 1–91; "Issledovanie o lednikovom periode" ("An Investigation of the Glacial Period"), 2 pts., *ibid.*, **7**, no. 1 (1876); *Memoirs of a Revolutionist*, 2 vols. (London, 1899), also in Russian, *Zapiski revolyutsionera* (London, 1902; St. Petersburg, 1906); *Mutual Aid, a Factor of Evolution* (London, 1902); "The Orography of Asia," in *Geographical Journal*, **23** (1904), 176–207, 331–361; and *Orographie de la Sibirie, précédée d'une introduction et d'un aperçu de l'orographie de l'Asie* (Brussels, 1904).

II. Secondary Literature. On Kropotkin and his work see *Petr Kropotkin, sbornik statey, posvyashchennykh pamyati P. A. Kropotkina* (Moscow, 1922), a collection of articles dedicated to the memory of Kropotkin; L. S. Berg, "Petr Alekseevich Kropotkin kak geograf" ("Petr Alekseevich Kropotkin as Geographer"), in *Otechestvennye fiziko-geografy* ("National Physical Geographers"; Moscow, 1959), 352–359; N. N. Sokolov, "Petr Alekseevich Kropotkin kak geograf," in *Trudy Instituta istorii estestvoznaniya i tekhniki. Akademiya nauk SSSR*, **4** (1952), 408–442; and A. A. Velichko, "P. A. Kropotkin kak sozdatel uchenia o lednikovom periode" ("P. A. Kropotkin as the Creator of the Theory of the Glacial Period"), in *Izvestiya akademii nauk SSSR, Seria geografia*, no. 1 (1957).

G. V. Naumov

KṚṢṆA (*fl.* 1653 at Taṭāka, Mahārāṣṭra, India), *astronomy*.

Kṛṣṇa, the son of Mahādeva of the Kāśyapagotra and his wife Barvāī, lived at an unidentified town called Taṭāka in the Koṅkaṇa, that is, the coastal region north and south of the city of Bombay. Kṛṣṇa wrote the *Karaṇakaustubha* at the command of the bhūpati Śiva (presumably this is Śivājī the Marāṭha monarch whose career began in 1646 though his coronation took place only in 1674; he died in 1680). The epoch of the *Karaṇakaustubha* is 1653. Though Śivājī occupied Kalyāna in the Koṅkaṇa in 1652, he was forced by his father's imprisonment to be inactive for five years thereafter; he acquired the rest of the Koṅkana only between 1657 and 1662. If Kṛṣṇa wrote in 1653, then Taṭāka is most probably near Kalyāṇa, but it is also possible that he wrote elsewhere between 1657 and 1680 but used an earlier epoch.

The *Karaṇakaustubha* (see essay in Supplement) is based on the *Grahalāghava* of Gaṇeśa and contains fourteen chapters, accompanied by numerous tables:

1. On the mean motions of the planets.
2. On the true longitudes of the Sun and Moon.
3. On the true longitudes of the five "star planets."
4. On the three problems relating to diurnal motion.
5. On lunar eclipses.
6. On solar eclipses.
7. On the two eclipses from a table of *tithis*.
8. On heliacal risings and settings.
9. On planetary latitudes.
10. On the lunar crescent.
11. On planetary conjunctions.
12. On the fixed stars.
13. On the *pātas* of the Sun and Moon.
14. Conclusion.

The text was edited by V. G. Āpṭe, Ānandāśrama Sanskrit Series 96 (Poona, 1927).

BIBLIOGRAPHY

The only articles on Kṛṣṇa are by Ś. B. Dīkṣita, *Bhāratīya Jyotihśāstra* (Poona, 1931; reprint of the first edition, Poona, 1896), pp. 290–291, and by D. Pingree, *Census of the Exact Sciences in Sanskrit*, series A, **2** (Philadelphia 1971), 55–56.

David Pingree

KRUBER, ALEKSANDR ALEKSANDROVICH (*b.* Voskresensk [now Istra], Moscow guberniya, Russia, 10 August 1871; *d.* Moscow, U.S.S.R., 15 December 1941), *geography*.

Kruber's father was a schoolteacher. From 1891 to 1896 Kruber studied in the natural science section of the physics and mathematics faculty of Moscow University. Upon graduation he was retained in the department of geography and ethnography. In 1897 he began to teach geography and to lead field investigations, studying mainly lakes, swamplands, and karst phenomena in the territory of European Russia and the Caucasus. In 1915 he was awarded the degree of master of geography for his work "Karstovaya oblast gornogo Kryma" ("The Karstic Region of Mountainous Crimea"). In 1919 he occupied the newly created separate chair of geography and from 1923 to 1927 was director of the Moscow University Scientific Research Institute of Geography. Kruber was coeditor with Anuchin of the journal *Zemlevedenie* ("Geography") from 1917 to 1923 and sole editor from 1923 to 1927. The result of a severe illness forced him to retire in 1927.

Kruber belonged to the Russian school of geography founded by Anuchin and distinguished by a "complexion" approach to the study of nature. This approach regarded the geographic field as a complex of interrelated phenomena, the result of diverse but interacting forces. Kruber's *Fiziko-geograficheskie oblasti Evropeyskoy Rossii* ("Physiographic Regions of European Russia"; 1907) and *Ocherki geografii Rossii* ("Essays on the Geography of Russia"; 1910) are clear examples of this approach applied to the study of geographic regions. Kruber ascribed great significance to analysis of the interaction of distinct phenomena and processes on the earth's surface and created the university text *Obshchee zemlevedenie* ("General Geography"; 1912–1922) and many other works that played an important role in training Soviet geographers.

"The task of geography," Kruber wrote, "has always been and remains the description of the earth's surface and the interpretation of its peculiarities" (*Obshchee zemlevedenie* [1923], p. 31). In this regard he emphasized the study of land masses as natural, historical complexes and focused on all the forces that had participated in their creation.

Kruber was one of the founders of the scientific study of karstic phenomena in Russia, which gained many followers among Soviet geographers. He gave the first summary of karstic phenomena in European Russia (1900) and presented a model study of the karstic phenomena of the mountainous Crimea. He also elucidated the question of water circulation in a karst (1913). Kruber approached the study of karstic relief forms from the evolutionary viewpoint, taking into account the diversity of physiogeographic conditions and processes. He considered corrosion (the dissolution of rock) to be the leading process in the creation of most karstic forms, with erosion also playing a significant role. Kruber's papers on physiographic classification, anthropogeography, and demography are also of interest. His services in creating geography textbooks for secondary schools are also considerable.

BIBLIOGRAPHY

I. Original Works. Kruber's works include "K voprosu ob izuchenii bolot Evropeyskoy Rossii" ("On the Question of the Study of the Swamplands of European Russia"), in *Zemlevedenie*, **4**, bks. 3–4 (1897), 99–115; "Opyty razdelenia Evropeyskoy Rossii na rayony" ("Experiments in Regional Subdivision of European Russia"), *ibid.*, **5**, bks. 3–4 (1898), 175–184; "O karstovykh yavleniakh v Rossii" ("On Karstic Phenomena in Russia"), *ibid.*, **7**, bk. 4 (1900), 1–34; and "I. Fiziko-geograficheskie oblasti Evropeyskoy Rossii; II. Ocherki reliefa i prirody Evropeyskoy Rossii, Kavkaza i Sibiri" ("I. Physiographic Regions of European Russia; II. Essays on the Relief and Nature of European Russia, the Caucasus and Siberia"), in S. G. Grigoriev, A. S. Barkov, and S. V. Chefranov, *Ocherki po geografii Rossii* ("Essays on the Geography of Russia"; Moscow, 1910).

See also *Obshchee zemlevedenie* ("General Geography"), 3 pts. (Moscow, 1912–1922); "Gidrografia karsta" ("The Hydrography of Karst"), in *Sbornik v chest 70-letia D. N. Anuchina* ("Collection Honoring the Seventieth Birthday of D. N. Anuchin"; Moscow, 1913), pp. 215–299; *Karstovaya oblast gornogo Kryma* ("The Karstic Region of the Mountainous Crimea"; Moscow, 1915), issued as app. to *Zemlevedenie* (1915); *Kurs geografii Rossii* ("A Course in the Geography of Russia"; Moscow, 1917); and *Chelovecheskie rasy i ikh rasprostranenie* ("Human Races and Their Distribution"; Moscow–Petrograd, 1923).

II. Secondary Literature. On Kruber and his work see A. S. Barkov, M. S. Bodnarsky, and S. V. Chefranov, "Pamyati A. A. Krubera" ("In Memory of A. A. Kruber"), in *Zemlevedenie*, n.s. **2** (1948), 11–15; and N. A. Gvozdetsky, "Vydayushchysya deyatel russkogo karstovedenia" ("The Outstanding Figure in the Russian Study of Karstic Phenomena"), *ibid.*, n.s. **4** (1957), 276–278; and "Aleksandr Aleksandrovich Kruber," in *Otechestvennye fiziko-geografy i puteshestvenniki* ("National Physical Geographers and Travellers"; Moscow, 1959), pp. 619–625, which includes a bibliography.

Vasily A. Esakov

KRYLOV, ALEKSEI NIKOLAEVICH (*b.* Visyaga, Simbirskoy province [now Ulyanovskaya oblast], Russia, 15 August 1863; *d.* Leningrad, U.S.S.R., 26 October 1945), *mathematics, mechanics, engineering.*

Krylov was born on the estate of his father, Nikolai Aleksandrovich Krylov, a former artillery officer. In 1878 he entered the Maritime High School in St. Petersburg. When he left in 1884 he was appointed to the compass unit of the Main Hydrographic Administration, where he began research on a theory of compass deviation, a problem to which he often returned. In 1888 Krylov joined the department of ship construction of the Petersburg Maritime Academy where he received a thorough mathematical grounding under the guidance of A. N. Korkin, a distinguished disciple of Chebyshev. In 1890 Krylov graduated first in his class from the Maritime Academy and at Korkin's suggestion remained there to teach mathematics. He taught various theoretical and engineering sciences for almost fifty years at this military-maritime institute, creating from among his students a large school of shipbuilders who were both engineers and scientists. From 1900 to 1908, he directed the experimental basin, where he engaged in extensive research and tested models of various vessels. Krylov's work covered an unusually wide spectrum of the problems of what Euler referred to as naval science: theories of buoyancy, stability, rolling and pitching, vibration, and performance, and compass theories. His investigations always led to a numerical answer. He proposed new and easier methods of calculating the structural elements of a ship, and his tables of seaworthiness quickly received worldwide acceptance. From 1908 to 1910 Krylov, who had attained the rank of general, served as chief inspector for shipbuilding and was a president of the Maritime Engineering Committee. His courage and integrity led to conflicts with officials of the Maritime Ministry and to his refusal to do further work for them.

In 1914, Moscow University awarded Krylov the degree of doctor of applied mathematics, *honoris causa*, and the Russian Academy of Sciences elected him a corresponding member. He was elected to full membership in 1916.

After the October Revolution, Krylov sided with the Soviet government. During this period he continued to be both active and productive. From 1927 to 1932 he was director of the Physics and Mathematics Institute of the Soviet Academy of Sciences. He also played an important role in the organization, in 1929, of the division of engineering sciences of the Soviet Academy. The title of honored scientist and engineer of the Russian Soviet Federated Socialist Republic was conferred upon Krylov in 1939, and in 1943 he was awarded the state prize (for his work in compass theory) and the title of hero of socialist labor.

While using mathematics and mechanics to work out his theory of ships, Krylov simultaneously improved the methods of both disciplines, especially that in the theory of vibrations and that of approximate calculations. In a paper on forced vibrations of fixed-section pivots (1905), he presented an original development of Fourier's method for solving boundary value problems, pointing out its applicability to a series of important questions: for example, the theory of steam-driven machine indicators, the measurement of gas pressure in the conduit of an instrument, and the twisting vibrations of a roller with a flywheel on its end. Closely related to this group of problems was his ingenious and practical method for increasing the speed of convergence in Fourier and related series (1912). He also derived a new method for solving the secular equation that serves to determine the frequency of small vibrations in mechanical systems (1931). This method is simpler than those of Lagrange, Laplace, Jacobi, and Leverrier. In addition, Krylov perfected several methods for the approximate solution of ordinary differential equations (1917).

In his mathematical education and his general view of mathematics, Krylov belonged to the Petersburg school of Chebyshev. Most representatives of this school, using concrete problems as their point of departure, developed primarily in a purely theoretical direction. Krylov, however, proceeded from theoretical foundations to the effective solution of practical engineering problems.

Krylov's practical interests were combined with a deep understanding of the ideas and methods of classical mathematics and mechanics of the seventeenth, eighteenth, and nineteenth centuries; and in the works of Newton, Euler, and Gauss he found forgotten methods that were applicable to the solution of contemporary problems.

BIBLIOGRAPHY

I. Original Works. Krylov's works are collected in *Sobranie trudov* ("Complete Works"), 11 vols. (Moscow–Leningrad, 1936–1951). His original development of Fourier's method appears in the article "Über die erzwungenen Schwingungen von gleichförmigen elastischen Stähen," in *Mathematische Annalen*, **61** (1905), 211–234; further work on Fourier and related series is found in *O nekotorykh differentsialnykh uravneniakh matematicheskoy fiziki, imeyushchikh prilozhenie v teknicheskikh voprosakh* ("On Several Differential Equations of Mathematical Physics Which Have Application in Engineering Problems"; St. Petersburg, 1912). For his work on the secular equation see "O chislennom reshenii uravnenia, kotorym v teknicheskikh voprosakh opredelaitsia chastoty malykh kolebanii materialnykh system" ("On the Numerical Solution of Equation by Which are Determined in Technical Problems the Frequencies of Small Vibrations of Material Systems"), in *Izvestiya Akademii nauk S.S.S.R.*, Otd. mat. nauk (1931), 491–539; see also *ibid.* (1933), 1–44.

Among his numerous works in the history of science, his Russian translation of Newton's *Principia* (*Matematicheskie nachala naturalnoy philosophii;* St. Petersburg, 1915) is especially noteworthy for its lucidity and for the depth of its scientific commentary.

II. Secondary Literature. A list of Krylov's works appears in N. A. Kryzhanovskaya, *Akademik A. N. Krylov, bibliografichesky ukazatel* ("Academician A. N. Krylov, Bibliographical Guide"; Leningrad, 1952). For a study of Krylov's life, see S. Y. Shtraykh, *Aleksei Nikolaevich Krylov, ego zhizn i deyatelnost* ("Aleksei Nikolaevich Krylov, His Life and Work"; Moscow–Leningrad, 1950).

A. T. Grigorian

KRYLOV, NIKOLAI MITROFANOVICH (*b.* St. Petersburg, Russia, 29 November 1879; *d.* Moscow, U.S.S.R., 11 May 1955), *mathematics.*

N. M. Krylov graduated from the St. Petersburg Institute of Mines in 1902. From 1912 to 1917 he was a professor there, and then from 1917 to 1922 he was a professor at Crimea University. In 1922 he was chosen a member of the Academy of Sciences of the Ukrainian S.S.R., and was appointed chairman of the mathematical physics department. In 1928 Krylov was elected an associate member of the Academy of Sciences of the U.S.S.R.; a year later he became a member. The rank of honored scientist of the Ukrainian S.S.R. was conferred on him in 1939.

Krylov's works relate mainly to problems of the theory of interpolation, of approximate integration of differential equations (applicable in mathematical physics), and of nonlinear mechanics. In his study on approximate integration, Krylov obtained extremely effective formulas for the evaluation of error in a field in which, prior to this work, one was limited either to proofs of existence or, at best, to proofs of the convergence of the method of approximation. Using the proof of Ritz's method of convergence, Krylov was the first to study—with the aid of the theory of infinite-order determinants—the general case of an arbitrary quadratic form standing under the variable integral sign. By using Ritz's method itself, he investigated the creation of more general methods which would be applicable to both the proof of the existence of a solution and the actual construction of the solution.

In 1932 Krylov began a study of actual problems of nonlinear oscillatory processes; in this work he succeeded in laying the foundation of nonlinear mechanics.

Krylov's work received wide application in many fields of science and technology. He published some 200 papers in mathematical analysis and mathematical physics.

BIBLIOGRAPHY

I. Original Works. A compilation is N. M. Krylov, *Izbrannye trudy* ("Selected Works"), 3 vols. (Kiev, 1961).

II. Secondary Literature. N. N. Bogolyubov, "Nikolai Mitrofanovich Krylov (k 70-letiyu so dnya rozhdenia)" ("Nikolai Mitrofanovich Krylov" [on his Seventieth Birthday]") in *Uspekhi matematicheskikh nauk*, **5**, no. 1 (1950); and O. V. Isakova, *Nikolai Mitrofanovich Krylov* (Moscow, 1945), includes material for bibliographical works of Soviet scientists.

A. T. Grigorian

KUENEN, JOHANNES PETRUS (*b.* Leiden, Netherlands, 11 October 1866; *d.* Leiden, 25 September 1922), physics.

Johan Kuenen was the son of Abraham Kuenen, professor of theology at the University of Leiden and Wiepkje Muurling, a daughter of W. Muurling, professor of theology at the University of Groningen. Kuenen began his studies of physics at the University of Leiden in 1884. In 1889 he was appointed assistant to H. Kamerlingh Onnes. He won his doctorate with a dissertation which, as a prize essay, had been awarded a gold medal in 1892. From 1893 until 1895 he served as conservator of the university physics laboratory; in the latter year, he went to England, where he worked with Sir William Ramsay in London and with Sir James Walker in Dundee, at the university of which town he was soon appointed professor of physics. He stayed at the University of Dundee until December 1906, when he was called back to Leiden to occupy a chair of physics. He held this post until his death.

Kuenen is known chiefly for his work on phase equilibria. The golden era of Dutch physics had been inaugurated in 1873 by J. D. van der Waals with his dissertation *Over de continuiteit van den gas- en vloeistoftoestand*, in which the famous equation of state had been introduced. The foundation for many pioneering investigations on critical phenomena and phase equilibria, it led Dutch physicists to such triumphs as the liquefaction and solidification of helium. Kuenen undertook his doctoral work to supply data for the theoretical research on binary mixtures that was then being conducted by van der Waals.

While pursuing this investigation, Kuenen discovered the phenomenon of retrograde condensation, which may be explained as follows. If the volume of a two- (or multi-) component gaseous system, kept at constant temperature and pressure below critical conditions, is gradually reduced, condensation will start when a certain volume is reached; the amount of condensate will gradually increase upon further reduction in volume, until finally the entire system is liquefied. If, however, the composition of such a system lies between the compositions defined by the so-called true and pseudo critical points, the condensate formed upon reduction of the volume will disappear on continued reduction of the volume. This disappearance of the condensate is called retrograde condensation. The phenomenon finds a practical application in the recovery of gasoline from gas wells.

Kuenen devoted his entire scientific life to the study of phase relations, designing many instruments for use in this study and contributing numerous data.

Although he excelled as an experimenter, he also found theoretical solutions to problems that arose from his own work or that of others. In addition, he contributed articles of a popular nature on the progress of physics and its implications to the monthly magazine *De Gids* and wrote a history of physics in the Netherlands.

BIBLIOGRAPHY

I. ORIGINAL WORKS. A list of Kuenen's scientific papers has been compiled by W. J. de Haas (see below).

His works include *Metingen betreffende het oppervlak van van der Waals voor mengsels van koolzuur en chloor methyl* (Leiden, 1892); *Theorie der Verdampfung und Verflüssigung von Gemischen und der fraktionierten Destillation*, vol. IV of G. Bredig, ed., *Handbuch der angewandten physikalischen Chemie* (Leipzig, 1906); *Die Zustandsgleichung der Gase und Flüssigkeiten und die Kontinuitätstheorie* (Brunswick, 1907); *Die Eigenschaften der Gase*, vol. III of W. Ostwald, ed., *Handbuch der allgemeinen Chemie* (Leipzig, 1919); and *Het aandeel van Nederland in de ontwikkeling der Natuurkunde gedurende de laatste 150 jaren* (*Gedenkboek van het Bataafsch Genootschap der Proefondervindelijke Wijsbegeerte te Rotterdam, 1769–1919*), privately printed (Rotterdam, 1919).

II. SECONDARY LITERATURE. On Kuenen and his work see W. J. de Haas, "Prof. Dr. J. P. Kuenen, ter nagedachtenis," in *Physica*, **2** (1922), 281–287; and "Lijst van verhandelingen van Professor Dr. J. P. Kuenen," *ibid.*, 342–344, a bibliography of Kuenen's scientific papers; J. Herderschee, "Abraham Kuenen," in *Nieuw Nederlandsch biografisch Woordenboek*, II (Leiden, 1912), 734–735; H. Kamerlingh Onnes, "Prof. J. P. Kuenen," in *Nature*, **110** (1922), 673–674; and H. A. Lorentz, "Kuenen als Natuurkundige," in *De Gids*, 4th ser., **86** (1922), 209–215; and "Johan Kuenen (1866–1922), grafrede," in P. Zeeman and A. D. Fokker, eds., *H. A. Lorentz, Collected Papers*, IX (The Hague, 1939), 404–406.

PETER W. VAN DER PAS

KÜHN, ALFRED (*b.* Baden-Baden, Germany, 22 April 1885; *d.* Tübingen, Germany, 22 November 1968), *zoology*.

Kühn began to study natural science in 1904 at the University of Freiburg im Breisgau, where he studied mainly with the zoologist A. Weismann and the physiologist Johannes von Kries. He received his doctorate in 1908 with a dissertation entitled *Die Entwicklung der Keimzellen in den parthenogenetischen Generationen der Cladoceren Daphnia pulex De Geer und Polyphemus pediculus De Geer*. In 1910 he qualified as a lecturer in zoology at the University of Freiburg, and in 1914 he was named extraordinary professor. After World War I Kühn worked for a short time under Karl Heider at the zoology laboratory of the University of Berlin. In 1920 he succeeded E. Ehlers in the chair of zoology at the University of Göttingen. He developed a zoology laboratory at Göttingen employing the most up-to-date concepts, and through his stimulating teaching and his widespread research activity he attracted a large number of students to it.

In 1937 he was appointed second director of the Kaiser Wilhelm Institute for Biology in Berlin-Dahlem, where he was able to devote himself completely to research. In 1951 the institute was moved to Tübingen—with the new name of the Max Planck Institute for Biology—and until 1958 Kühn was administrative director of the much expanded organization. Kühn continued to work intensively in his old laboratory as a scientific member of the institute until his death. From 1946 to 1951 he was also professor of zoology and director of the zoology laboratory at the University of Tübingen.

From the start of his career, Kühn's scientific work encompassed very varied fields. In Freiburg he simultaneously conducted investigations in embryology, cytology, and the physiology of sensation. These included studies on the ontogenesis and phylogenesis of the hydroids; the development of the cladocerans; processes of division among various protozoans, the physiology of the reptilian ear labyrinth, the spinal cord of the dove, and reflexes in crabs. His *Anleitung zu tierphysiologischen Grundversuchen* appeared in 1917 and *Die Orientierung der Tiere im Raum* in 1919. At Göttingen, Kühn also studied the physiology of sensation—the color vision of bees and cuttlefish.

Kühn soon concentrated his efforts almost entirely on investigating questions of genetics and early development, occasionally in collaboration with Karl Henke. He examined the flour moth *Ephestia kühniella Z*, which became the major object of study in genetics after *Drosophila;* later they were joined by the microlepidoptera *Ptychopoda seriata*. The formation of patterns and the effect of genes were the central problems of Kühn's research. Applying the genetic and developmental approaches to the design patterns on butterfly wings soon led to innovative concepts in genetics. Further investigations of eye color mutants resulted in the inclusion of the ommochrome pigments in the chain of biochemical processes set off by the genes. As a result of Kühn's collaborative work with Adolf Butenandt and his laboratory; the concept emerged that was to be the starting point for modern biochemical genetics—that genes achieve their effects by means of specific enzymes.

Kühn pursued detailed questions concerning only

a few species, yet hardly anyone else could rival his grasp of the entire field of genetics and developmental physiology. His knowledge of the field is illustrated by his outstanding *Vorlesungen über Entwicklungsphysiologie*, which appeared in a second edition in 1965. These lectures were preceded by two textbooks—*Grundriss der allgemeinen Zoologie* (17th ed., 1969), and *Grundriss der Vererbungslehre* (4th ed., 1965). The former work, the so-called "small Kühn," is in contrast to Kühn's contribution to the general section of the *Lehrbuch der Zoologie* (1932) by Claus, Grobben, and Kühn.

Kühn took a lively interest in the history of biology. He was the author of *Anton Dohrn und die Zoologie seiner Zeit* (1950), contributed to the collection Biologie der Romantik (1948), and wrote biographies of Gregor Mendel and Karl Ernst von Baer (1957).

BIBLIOGRAPHY

I. Original Works. A bibliography of Kühn's works is given by Viktor Schwartz in "A. Kühn 22 IV 1885–22 XI 1968," in *Zeitschrift für Naturforschung*, **248** (1969), 1–4. Kühn's autobiography to 1937 is in *Nova Acta Leopoldina*, **21** (1959), 274–280.

II. Secondary Literature. See Karl Henke, in *Naturwissenschaften*, **42** (1955), 193–199; and A. Egelhaaf, "Auf dem Weg zur molekularen Biologie. Alfred Kühn zum Gedenken," *ibid.*, **56** (1969), 229–232.

Hans Querner

KUHN, RICHARD (*b.* Vienna-Döbling, Austria, 3 December 1900; *d.* Heidelberg, Germany, 31 July 1967), *chemistry.*

Kuhn, the son of Richard Clemens and Angelika (Rodler) Kuhn, was taught by his mother until he entered the Döbling Gymnasium at the age of nine, where he remained for eight years until drafted into the army. Four days after his release on 18 November 1918, he entered the University of Vienna. After three semesters he proceeded to the University of Munich, where his experimental genius soon became evident. In 1922 he obtained his Ph.D. under Willstätter and not long afterward received his docentship; he became widely known as Willstätter's greatest discovery.

In 1926 Kuhn moved to Zurich to become professor of chemistry at the Eidgenössische Technische Hochschule. In 1928 he married Daisy Hartmann; they had two sons and four daughters. In 1929 he was appointed director of the new Chemistry Institute of the Kaiser Wilhelm Institute for Medical Research at Heidelberg and professor at the university, where he remained for the rest of his life, notwithstanding later offers from Berlin, Munich, Vienna, and the United States.

In 1937 Kuhn became director of the entire Kaiser Wilhelm Institute for Medical Research, and in 1946–1948 helped to transform the Kaiser Wilhelm Society for Scientific Research into the Max Planck Society for the Advancement of Science. He was a charter member of the society's senate and later served as vice-president under Otto Hahn and then Adolf Butenandt. His scientific papers numbered over 700, his students and collaborators over 150, and his distinctions over fifty, including the 1939 Nobel Prize in chemistry.

Kuhn demonstrated his major scientific traits—discipline and precision coupled with imagination and fantasy—in both his 1922 doctoral thesis, "Zur Spezifität von Enzymen imKohlenhydratstoffwechsel" ("On the Specificity of Enzymes in Carbohydrate Metabolism"), and 1925 *Habilitation* thesis, "Der Wirkungsmechanismus der Amylasen; ein Beitrag zum Konfigurations—Problem der Stärke" ("Mechanism of Action of Amylases"). They dealt with greatly improved enzyme adsorption and elution carrier materials and kinetic enzyme measurements applied to a variety of sugar derivatives (glycosides, oligosaccharides, and polysaccharides). Specificity problems led him inevitably to problems in optical stereochemistry that preoccupied him for the rest of his life. He began research on additions on ethylene bonds; thus, addition of hypochlorous acid to fumaric or maleic acid resulted in chloromalic acid and also in ring closure to form ethylene oxide dicarboxylic acid. This work led in Zurich to studies on inhibited rotation among diphenyls, especially ortho-substituted derivatives. Kuhn showed that benzidine was stretched out on a plane, with the two NH_2— groups about 10 Å apart, instead of angled back and only about 1.5 Å apart. Kuhn, like Pasteur, had early dominating experiences with different forms of isomerism, and, like van't Hoff, had a remarkable understanding of stereoisomerism.

With further studies on the activation energies of rotation among ortho-substituted diphenyls, Kuhn and his collaborators arrived at quantitative concepts of the spatial needs of particular groups, including the concept of "atropisomerism," long before the terms "conformation" and "constellation" gained acceptance and applicability not only to substituted compounds but also to totally unsubstituted ones as in trans-cycloocten.

Proceeding to the preparation of diphenylpolyenes containing conjugated double bonds $(-CH=CH-)_n$ added in unbroken order, Kuhn and his collaborators showed that the diphenyls were colorless when

$n = 1$ or 2, but colored for $n = 3$ to 15. They thereby definitely established the existence of colored hydrocarbons. Research on crocetin, bixin, and, most important, carotene (in crystallizable α, β, and γ forms) demonstrated that Kuhn-type polyenes are found in nature. Work on carotene proved that symmetrical provitamin A yields two molecules of water. Among the synthetic polyenes, it was found that the position of the longest wavelength maximum

$$\cdot\lambda_1 = K'n^{1/2}K''$$

helped to determine the structures of natural carotenoids.

Kuhn next attacked water-soluble vitamins. He showed vitamin B_2, isolated and crystallized from milk and named lactoflavin, to be part of Warburg's yellow enzyme; and he synthesized and structurally identified the vitamin. He prepared lactoflavin phosphate and, after combining it with the protein carrier of the yellow enzyme, he found it to be enzymatically identical with the reversible yellow oxidation enzyme (yellow flavin $\rightleftharpoons$ colorless leukoflavin), and that identifiable intermediate stages displayed free radical paramagnetism and dimerism. Kuhn then identified vitamin B_6 (adermin, pyridoxine), *p*-aminobenzoic acid, and pantothenic acid and synthesized numerous analogues and reversibly competitive inhibitors ("antivitamins"). It was for his work on carotenoids and vitamins that Kuhn was awarded the Nobel Prize.

From the 1950's on Kuhn worked on and identified various "resistance" factors effective against infection —nitrogenous oligosaccharides isolated from human milk; brain gangliosides; and potato alkaloid-glycoside (demissin), active against larvae of potato beetles. Of especial interest was the finding that lactaminyl (sialic acid) oligosaccharides could be split by influenza virus and also by the receptor-destroying enzyme (RDE) of cholera vibrio (α-ketosidase action). Lactaminyl oligosaccharide was recognized as a receptor for influenza virus; therefore, the virus-inhibiting action of human milk (compared with bovine, which does not contain lactaminyl oligosaccharide) was explained: cells that do not form lactaminyl-oligosaccharide structures on their surfaces show resistance to influenza virus. Some twenty-five papers on the isolation and synthesis of amino-sugar split products of N-oligosaccharides followed.

Like Pasteur of France and Virtanen of Finland, Kuhn of Germany and Austria developed over the years intense interest in applications of his academic researches to medicine and agriculture. At the end of World War II Kuhn demonstrated experimentally to a U.S. Army colonel's wife that she could turn green plants into red plants by adding triphenyltetrazolium chloride to the nutrient medium, and so the institute was spared molestation.

BIBLIOGRAPHY

The complete scientific works of Kuhn have not yet been published; the most complete review is in the 63-page supp. to the *Mitteilungen aus der Max-Planck-Gesellschaft zur Foerderung der Wissenschaften* (1968), which lists over 100 references to the most important works of Kuhn and his collaborators.

DEAN BURK

KUHN, WERNER (*b.* Maur, near Zurich, Switzerland, 6 February 1899; *d.* Basel, Switzerland, 27 August 1963), *physical chemistry.*

Kuhn began his studies in 1917 at the Eidgenössische Technische Hochschule in Zurich, where he earned his chemical engineering degree. He received his doctorate in 1924 for a work on the photochemical decomposition of ammonia. At a later date, Kuhn again investigated the interaction of electromagnetic radiation and matter. As a fellow of the Rockfeller Foundation he studied at the Institute for Theoretical Physics of the University of Copenhagen, where, like many others, he became a lifelong admirer of Niels Bohr. In his famous work on optical dispersion, "Über die Gesamtstärke der von einem Zustande ausgehenden Absorptionslinien" (Copenhagen, 1925), Kuhn derived the formula for the sum of the squares of the amplitudes of the electric moments belonging to all the transitions that start from the same energy level. Both Kuhn and W. Thomas, influenced by the work of H. A. Kramers and Bohr, worked in the same area, but published their results separately. The "*f*-summation theorem" of Kuhn and W. Thomas retained quantitative validity in the later matrix mechanics.

In 1927 Kuhn qualified as a lecturer at the Physico-Chemical Institute of the University of Zurich with a work on the anomalous dispersion of thallium and cadmium. From 1928 to 1930 he worked with K. Freudenberg in Heidelberg, where he furnished a model interpretation of natural optical activity. He then accepted the position of extraordinary professor at the Technical College in Karlsruhe.

Kuhn increasingly turned his attention to the study of macromolecules which, along with optical activity, became one of his chief fields of research. Taking rod-shaped molecules as the basis for the calculation of the viscosity of solutions, he arrived at results that contradicted those obtained by Hermann Staudinger.

From these calculations, Kuhn concluded that the molecules must have the form of a coiled chain. This model finally enabled him to understand the transformation of chemical energy into mechanical energy, as it occurs in the muscles, for example.

In 1936 Kuhn was appointed professor ordinarius at the University of Kiel. In 1939 he received an offer he had been hoping for—to return to his native Switzerland to assume a post at the Physico-Chemical Institute of the University of Basel. As early as 1932, together with Hans Martin, he had achieved by photochemical means a partial separation of the isotopes Cl^{35} and Cl^{37}. In Basel he soon developed an effective method (a hairpin countercurrent arrangement) of obtaining heavy water. Kuhn's theory of separation enabled him to understand important physiological processes—for example the mechanism of urine concentration in the kidney and the production of high gas pressure in the air bladder of the fish.

Kuhn was rector of the University of Basel in 1955–1956. From 1957 to 1961 he was president of the physical chemistry section of the International Union for Pure and Applied Chemistry.

BIBLIOGRAPHY

I. Original Works. An extensive bibliography of about 300 papers is given as an addendum to Hans J. Kuhn's memoir (see below), pp. 246–258.

II. Secondary Literature. On Kuhn's life, see W. Feitknecht, "Prof. Dr. Werner Kuhn 1899–1963," in *Verhandlungen der Schweizerischen naturforschenden Gesellschaft*, **143** (1963), 224–227; and Hans J. Kuhn, "Werner Kuhn 1899–1963 in Memoriam," in *Verhandlungen der Naturforschenden Gesellschaft in Basel*, **74** (1963), 239–246.

Armin Hermann

KÜHNE, WILHELM FRIEDRICH (*b.* Hamburg, Germany, 28 March 1837; *d.* Heidelberg, Germany, 10 June 1900), *physiology, physiological chemistry.*

Willy Kühne was the fifth of seven children. His father was a wealthy Hamburg merchant; his mother was interested in the arts and politics. He attended grammar school at Lüneburg, but preferred his own scientific experiments to the Latin grammar. In 1854, at the age of seventeen, he entered the University of Göttingen, where he studied chemistry under Wöhler, anatomy under Henle, and neurohistology under Rudolf Wagner. Two years later he obtained the Ph.D. with a thesis on induced diabetes in frogs. He then spent some time at the University of Jena, where he worked with Carl Gustav Lehmann on diabetes and sugar metabolism.

In 1858 Kühne worked in Berlin with E. du Bois-Reymond on problems of myodynamics and with Felix Hoppe-Seyler, who was in charge of the chemical laboratory at the institute of pathology directed by Rudolf Virchow. His other associates there included Julius Cohnheim, W. Preyer, F. C. Boll, Hugo Kronecker, and Theodor Leber. In the same year Bernard's report on sugar puncture prompted Kühne to go to Paris, where he received many suggestions regarding scientific methods. In 1860 he spent some time in Vienna working with Ernst Brücke and Carl Ludwig. Most probably it was Brücke who imparted to him his interest in the physiology of the protozoa and the nature of protoplasm. In 1861 he began his independent scientific career, succeeding Hoppe-Seyler as assistant in the chemical department of Virchow's institute. The latter gave him a completely free hand with his work. At the institute Kühne did not investigate strictly pathological questions but dealt with cytophysiological problems. In 1862, at the instigation of Albert von Bezold, Kühne was awarded an honorary M.D. by the University of Jena. During the Seven Weeks' War between Prussia and Austria in 1866 he was given charge of epidemic control. He left Berlin in 1868 to take over the chair of physiology at Amsterdam; Gustav Schwalbe and, later, Thomas Lauder-Brunton became his collaborators.

In 1871 Kühne succeeded Helmholtz in the chair of physiology at Heidelberg; his colleagues included E. Salkowski, J. N. Langley, and R. H. Chittenden. He married Helene Blum, the daughter of a Heidelberg mineralogist, and continued to teach and do research there until his retirement in 1899. In 1875, on his initiative, a new building for the institute of physiology was built. On the twenty-fifth anniversary of Kühne's appointment at Heidelberg (1896) a jubilee volume of the *Zeitschrift für Biologie* was published with contributions by E. Salkowski, J. von Uexküll, G. Schwalbe, Leon Asher, Lauder-Brunton, and Kronecker. A man of great versatility, Kühne was fond of good company and associated with artists as well as scientists. He died at the age of sixty-three of the sequelae of pneumonia.

Kühne's most outstanding gift was his ability to select significant problems, which he approached inventively, using a wide variety of technical devices. He substantially advanced research in the physiology of metabolism and digestion (sugar, protein, bile acids, trypsin), the physiology of muscle and nerves, the physiology of protozoa, and physiological optics. Many of his results and observations were immediately integrated into standard texts.

After presenting his doctoral thesis on induced diabetes in frogs, in 1856 Kühne elaborated a problem previously dealt with by Wöhler, who had stated that the benzoic acid ingested in food is found in the urine in the form of hippuric acid. Kühne proved that this compound is produced in the liver (although the same may also produce succinic acid, at least in carnivorous animals).

While collaborating with du Bois-Reymond in Berlin, Kühne began to study problems of myodynamics, simultaneously applying physiological, microscopical, and chemical methods to arrive at some essentially novel findings. He investigated histologically the nature of the motor nerve endings (1868) and found their terminal organs, which exist as motor end plates in warm-blooded animals. In preparations of the sartorius, a muscle rarely used until then for experimental purposes, he established the two-way conductibility of the nerve fiber (1859) and the direct excitability of the muscle fiber by both electric and chemical stimuli.

From the sartorius Kühne obtained coagulable myosin; assuming the heat rigor of the muscle to be a clotting process, accompanied by a clouding of the muscle substance, he was led to believe that the living muscle must be viscous in consistency. He later fortuitously proved this assumption when he observed a nematode creeping forward in a living muscle fiber (1863). Kühne also ascertained that the myohematin myoglobin is related to hemoglobin and, having suspected that contractile elements were involved in the creeping motions of the amoebas, proved the electric excitability of monocellular organisms (1864). Through his research, Kühne demonstrated the extraordinary usefulness of cytophysiological investigations for the solution of problems of general physiology.

Beginning with his stay at Amsterdam, Kühne again turned to problems of physiological chemistry, especially those of digestion. In particular he carried out investigations on the splitting of the large protein molecule during digestive fermentation. Since the stomach is not digested by its own pepsin, he concluded that such ferments have inactive protein precursors, which he called "zymogens"; he then traced the disintegration of proteins into albumoses and peptones. He further succeeded in demonstrating microscopic changes in living pancreatic cells during their activity. Having learned the technique of pancreatic fistula from Claude Bernard in Paris, Kühne pursued the study of the proteolytic effect of the pancreatic enzyme, which he called trypsin (1877).

Kühne's *Lehrbuch der physiologischen Chemie* appeared in 1868. It clearly and concisely presented the state of the science at that time. In 1877 Kühne took up a completely different topic. In 1876 Boll had established that the layer of rods of the retina contain a purple pigment that disappears on exposure to light. On this basis Kühne supposed that there was a primarily chemical process that preceded excitation of the optic nerves; he demonstrated that the retina works like a photographic plate, with light bleaching out the visual purple, which is regenerated in darkness. He succeeded in producing his famous "optograms"—the reproduction of the pattern of crossbars of a window on the chemical substance of the retina of a rabbit (1877–1878). Kühne was thus the first to perceive the migrating pigments in the living retina. His assumption that a photochemical process occurs prior to the excitation of the retina led him to investigate the electric processes in the eye, thus linking his research with the work of du Bois-Reymond and Holmgren. It may be noted that Kühne's research, as was typical of the work of German physiologists of the second half of the nineteenth century, always began with the formulation of a question rather than growing from a method.

Shortly before his death Kühne published a remarkable lecture on the responsibility of the physician to his patient and the relationship of medicine to science (1899). He postulated medical ethics as a subject in medical training and he believed that both a sympathetic heart and a clear, keen mind were necessary qualities for a physician. He regarded the concept of the immortality of the soul as important to the attitude of the physician and anticipated an era in which medical science would have to adapt more to the needs of society than it had done until then.

BIBLIOGRAPHY

I. ORIGINAL WORKS. Kühne's monographs include *Myologische Untersuchungen* (Leipzig, 1860); *Über die peripherischen Endorgane der motorischen Nerven* (Leipzig, 1862); *Untersuchungen über das Protoplasma und die Contractilität* (Leipzig, 1864); *Lehrbuch der physiologischen Chemie*, I, 3 pts. (Leipzig, 1866–1868); *Untersuchungen aus dem Physiologischen Institut der Universität Heidelberg*, 4 vols. (Heidelberg, 1878–1881); and *Über Ethik und Naturwissenschaft in der Medizin. Ein Auszug aus der Geschichte der Medizin* (Brunswick, 1899).

His articles include "Über künstlich erzeugten Diabetes bei Fröschen," in *Nachrichten von der Königlichen Gesellschaft der Wissenschaften und der Georg Augusta-Universität zu Göttingen* (1856), 217–219; "Über die Bildung der Hippursäure aus Benzolsäure bei fleischfressenden Thieren," in *Virchows Archiv für pathologische Anatomie und Physiologie*, **12** (1857), 386–396, written with W. Hallwachs; "Zur Metamorphose der Bernsteinsäure," *ibid.*, 396–401; "Beiträge zur Lehre vom Icterus. Eine physiologisch-chemische Untersuchung," *ibid.*, **14** (1858), 310–356; "Über die selbständige Reizbarkeit der Muskelfaser," in

Monatsberichte der K. preussischen Akademie der Wissenschaften zu Berlin (1859), 226–229; "Die Endigungsweise der Nerven in den Muskeln und das doppelsinnige Leitungsvermögen der motorischen Nervenfaser," *ibid.*, 395–402; "Sur l'irritation chimique des nerfs et des muscles," in *Comptes rendus hebdomadaires des séances de l'Académie des sciences*, **48** (1859), 406–409, 476–478; and "Recherches sur les propriétés physiologiques des muscles," in *Annales des sciences naturelles*, **14** (1860), 113–116.

Subsequent works are "Eine lebende Nematode in einer lebenden Muskelfaser beobachtet," in *Virchows Archiv*, **26** (1863), 222–224; "Über den Farbstoff der Muskeln," *ibid.*, **33** (1865), 79–94; "Über die Verdauung der Eiweissstoffe durch den Pankreassaft," *ibid.*, **39** (1867), 130–172; "Über das Trypsin (Enzym des Pankreas)," in *Verhandlungen des naturhistorisch-medizinischen Vereins zu Heidelberg*, **1** (1877), 194–198; "Zur Photochemie der Netzhaut," *ibid.*, 484–492; "Über den Sehpurpur," in *Untersuchungen aus dem Physiologischen Institut der Universität Heidelberg*, **1** (1878), 15–103; "Über die nächsten Spaltungsprodukte der Eiweisskörper," in *Zeitschrift für Biologie*, **19** (1883), 159–208, written with R. H. Chittenden; "Über elektrische Vorgänge im Sehorgan," in *Verhandlungen des naturhistorisch-medizinischen Vereins zu Heidelberg*, **3** (1886), 1–9, written with J. Steiner; "Über die Peptone," in *Zeitschrift für Biologie*, **22** (1886), 423–458; and "Zur Darstellung des Sehpurpurs," *ibid.*, **32** (1895), 21–28.

II. SECONDARY LITERATURE. There is no complete bibliography. Compilations are to be found in the *Index-Catalogue* of the Library of the Surgeon-General's Office, U.S. Army, 1st ser., VII (Washington, 1886), 569; 2nd ser., VIII (Washington, 1903), 872–873. See also the Royal Society *Catalogue of Scientific Papers*, III, 768–769; VIII, 133; X, 472–473; XII, 417; and XVI, 495–496.

Obituaries are by F. Hofmeister in *Berichte der Deutschen chemischen Gesellschaft*, **33** (1900), 3875–3880; Alois Kreidel, in *Wiener klinische Wochenschrift*, **13** (1900), 648–650; Franz Müller, in *Deutsche medizinische Wochenschrift*, **26** (1900), 440–441; Paul Schultz, in *Berliner klinische Wochenschrift*, **37** (1900), 606–608; J. von Uexküll, in *Münchener medizinische Wochenschrift*, **47** (1900), 937–939; Carl Voit, in *Zeitschrift für Biologie*, **40** (1900), i–viii.

See also Haberling, *et al.*, eds., *Biographisches Lexikon der hervorragenden Ärzte aller Zeiten und Völker*, 2nd ed., III (Berlin–Vienna, 1931), 627; Hugo Kronecker, "Ein eigenartiger deutscher Naturforscher. Zum Andenken an Willy Kühne," in *Deutsche Revue*, **32** (1907), 99–112; and Theodor Leber, "Willy Kühne," in *Heidelberger Professoren aus dem 19. Jahrhundert*, II (Heidelberg, 1903), 207–220.

K. E. ROTHSCHUH

KUMMER, ERNST EDUARD (*b.* Sorau, Germany [now Zary, Poland], 29 January 1810; *d.* Berlin, Germany, 14 May 1893), *mathematics.*

After the early death in 1813 of Kummer's father, the physician Carl Gotthelf Kummer, Ernst and his older brother Karl were brought up by their mother, the former Friederike Sophie Rothe. Following private instruction Kummer entered the Gymnasium in Sorau in 1819 and the University of Halle in 1828. He soon gave up his original study, Protestant theology, under the influence of the mathematics professor Heinrich Ferdinand Scherk and applied himself to mathematics, which he considered a kind of "preparatory science" for philosophy. (Kummer maintained a strong bent for philosophy throughout his life.) In 1831 he received a prize for his essay on the question posed by Scherk: "De cosinuum et sinuum potestatibus secundum cosinus et sinus arcuum multiplicium evolvendis."

In the same year Kummer passed the examination for Gymnasium teaching and on 10 September 1831 was granted a doctorate for his prize essay. After a year of probation at the Gymnasium in Sorau he taught from 1832 until 1842 at the Gymnasium in Liegnitz (now Legnica, Poland), mainly mathematics and physics. His students during this period included Leopold Kronecker and Ferdinand Joachimsthal, both of whom became interested in mathematics through Kummer's encouragement and stimulation. Kummer inspired his students to carry out independent scientific work, and his outstanding teaching talent soon became apparent. Later, together with his research work, it established the basis of his fame. His period of Gymnasium teaching coincided with his creative period in function theory, which began with the above-mentioned prize work. Its most important fruit was the paper on the hypergeometric series.[1] While doing his military service Kummer sent this paper to Jacobi. This led to his scientific connection with the latter and with Dirichlet, as well as to a corresponding membership, through Dirichlet's proposal, in the Berlin Academy of Sciences in 1839. After Kummer had thus earned a name for himself in the mathematical world, Jacobi sought to obtain a university professorship for him, in which endeavor he was supported by Alexander von Humboldt.

In 1840 Kummer married Ottilie Mendelssohn, a cousin of Dirichlet's wife. On the recommendation of Dirichlet and Jacobi, he was appointed full professor at the University of Breslau (now Wrocław, Poland) in 1842. In this position, which he held until 1855, he further developed his teaching abilities and was responsible for all mathematical lectures, beginning with the elementary introduction. During this period an honorary doctorate was bestowed on Ferdinand Gotthold Eisenstein; it had been proposed by Jacobi (probably on Humboldt's suggestion) and was carried out by Kummer despite considerable opposition. The second period of Kummer's research began about the time of his move to Breslau; it was dominated especially by number theory and lasted approximately twenty years. Not long after the death of his first wife in 1848, Kummer married Bertha Cauer.

When Dirichlet left Berlin in 1855 to succeed Gauss at Göttingen, he proposed Kummer as first choice for his Berlin professorship and Kummer was appointed. Kummer arranged for his former student Joachimsthal to become his successor at Breslau and hindered the chances of success for Weierstrass' application for the Breslau position, for he wanted the latter to be at the University of Berlin. This plan succeeded; Weierstrass was called to Berlin in 1856 as assistant professor. When Kronecker, with whom Kummer carried on an exchange of scientific views, also moved to Berlin in 1855, that city began to experience a new flowering of mathematics.

In 1861 Germany's first seminar in pure mathematics was established at Berlin on the recommendation of Kummer and Weierstrass; it soon attracted gifted young mathematicians from throughout the world, including many graduate students. It is permissible to suppose that in founding the seminar Kummer was guided by his experiences in Halle as a student in Scherk's Mathematischer Verein. Kummer's Berlin lectures, always carefully prepared, covered analytic geometry, mechanics, the theory of surfaces, and number theory. The clarity and vividness of his presentation brought him great numbers of students—as many as 250 were counted at his lectures. While Weierstrass and Kronecker offered the most recent results of their research in their lectures, Kummer in his restricted himself, after instituting the seminar, to laying firm foundations. In the seminar, on the other hand, he discussed his own research in order to encourage the participants to undertake independent investigations.

Kummer succeeded Dirichlet as mathematics teacher at the Kriegsschule. What would have been for most a heavy burden was a pleasure for Kummer, who had a marked inclination to every form of teaching activity. He did not withdraw from this additional post until 1874. From 1863 to 1878 he was perpetual secretary of the physics-mathematics section of the Berlin Academy, of which—on Dirichlet's recommendation—he had been a full member since 1855. He was also dean (1857–1858 and 1865–1866) and rector (1868–1869) of the University of Berlin. Kummer did not require leisure for creative achievements but was able to regenerate his powers through additional work.

In his third period, devoted to geometry, Kummer applied himself with unbroken productivity to ray systems and also considered ballistic problems. He retired at his own request in 1883 and was succeeded by Lazarus Fuchs, who had received his doctorate under him in 1858. Kummer spent the last years of his life in quiet retirement; his second wife and nine children survived him.

Kummer was first *Gutachter* for thirty-nine dissertations at Berlin. Of his doctoral students, seventeen later became university teachers, several of them famous mathematicians: Paul du Bois-Reymond, Paul Gordan, Paul Bachmann, H. A. Schwarz (his son-in-law), Georg Cantor, and Arthur Schoenflies. Kummer was also second *Gutachter* for thirty dissertations at Berlin. In addition, he was first referee when Alfred Clebsch, E. B. Christoffel, and L. Fuchs qualified for lectureships; and he acted as second referee at four other qualifying examinations. Kummer's popularity as a professor was based not only on the clarity of his lectures but on his charm and sense of humor as well. Moreover, he was concerned for the well-being of his students and willingly aided them when material difficulties arose; hence their devotion sometimes approached enthusiasm.

On Kummer's nomination Kronecker became a member of the Berlin Academy and Louis Poinsot, George Salmon, and Ludwig von Seidel became corresponding members. He himself became a correspondent of the Paris Academy of Sciences in 1860 and a foreign associate in 1868. This Academy had already awarded him its Grand Prix des Sciences Mathématiques in 1857 for his "Theorie der idealen Primfaktoren." Of his other memberships in scientific societies, that in the Royal Society as foreign member (1863) should be mentioned.

Kummer's official records reflect his characteristic strict objectivity, hardheaded straightforwardness, and conservative attitude. Thus it seems in keeping that during the revolutionary events of 1848, in which almost every important German mathematician except Gauss took an active role, Kummer was in the right wing of the movement, while Jacobi, for example, belonged to the progressive left. Kummer advocated a constitutional monarchy, not a republic. When, on the other hand, Jacobi, with his penchant for slight overstatement, declared that the glory of science consists in its having no use, Kummer agreed. He too considered the goal of mathematical research as the enrichment of knowledge without regard to applications; he believed that mathematics could attain the highest development only if it were pursued as an end in itself, independent of the external reality of nature. It is in this context that his rejection of multidimensional geometries should be mentioned.

Kummer's greatness and his limits lay in a certain self-restraint, manifested—among other ways—in his never publishing a textbook, but only articles and lectures. Weierstrass was led to state that, to some extent in his arithmetical period and more fully later, Kummer no longer concerned himself with

> . . . what was happening in mathematics. If you say to him, Euclidean geometry is based on an unproved axiom, he grants you this; but proceeding from this insight, the question now is phrased: How then does geometry look without this axiom? That goes against his nature; the efforts directed toward this question and the consequent general considerations, which free themselves from the empirically given or the presupposed, are to him idle speculations or simply a monstrosity.[2]

To be sure, the time at which this criticism was made must be considered: after Kummer, Kronecker, and Weierstrass had worked for twenty years in friendly, harmonious agreement and close scientific contact, an estrangement between Weierstrass and Kronecker took place in the mid-1870's which led to an almost complete break. Kummer's continuing friendship with Kronecker was not without its repercussions on Weierstrass' attitude toward Kummer. If, therefore, Weierstrass' evaluation of Kummer is to be taken with a grain of salt, it is nevertheless essentially correct.

Kummer's sudden decision to retire was another example of his inflexible principles. On 23 February 1882 he surprised the faculty by declaring that he had noticed a weakening of his memory and of the requisite ability to develop his thoughts freely in logical, coherent, and abstract arguments. On these grounds he requested retirement. No one else had detected such impairments, but Kummer could not be dissuaded and compelled the faculty to arrange for a successor.

Gauss and Dirichlet exerted the most lasting influence on Kummer. Each of Kummer's three creative periods began with a paper directly concerning Gauss, and his reverence for Dirichlet was movingly expressed in a commemorative speech on 5 July 1860 to the Berlin Academy.[3] Although he never attended a lecture by Dirichlet, he considered the latter to have been his real teacher. Kummer in turn had the strongest influence on Kronecker, who thanked him in a letter of 9 September 1881 for "my mathematical, indeed altogether the most essential portion of my intellectual life."

Today, Kummer's name is associated primarily with three achievements, one from each of his creative periods. From the function-theory period date his investigations, surpassing those of Gauss, of hypergeometric series, in which, in particular, he was the first to compute the substitutions of the monodromic groups of these series. The arithmetical period witnessed the introduction of "ideal numbers" in an attempt to demonstrate through multiplicative treatment the so-called great theorem of Fermat. After Dirichlet had pointed out to Kummer that the unambiguous prime factorization into number fields did not seem to have general validity, and after he had convinced himself of this fact, between 1845 and 1847, he formulated his theory of ideal prime factors.[4] It permitted unambiguous decomposition into general number fields and with its help Kummer was able to demonstrate Fermat's theorem in a number of cases.[5] It is again characteristic that Kummer elaborated his theory only to the extent required by those problems which interested him—the proof of Fermat's theorem and of the general law of reciprocity. Kummer's works were developed in the investigations of Richard Dedekind and Kronecker, thus contributing significantly to the arithmetization of mathematics.

The third result dates from Kummer's geometric period, in which he devoted himself principally to the theory of general ray systems, following Sir William Rowan Hamilton but treating them purely algebraically: the discovery of the fourth-order surface, named for Kummer, with sixteen isolated conical double points and sixteen singular tangent planes.[6] The number of other concepts connected with Kummer's name indicates that he was one of the creative pioneers of nineteenth-century mathematics.

NOTES

1. "Über die hypergeometrische Reihe . . .," in *Journal für die reine und angewandte Mathematik*, **15** (1836), 39–83, 127–172.
2. Gösta Mittag-Leffler, "Une page de la vie de Weierstrass," in *Comptes rendus du 2ᵉ Congrès international des mathématiciens* (Paris, 1902), pp. 131–153, see pp. 148–149, letter to Sonya Kovalevsky, 27 Aug. 1883.
3. "Gedächtnisrede auf G. P. L. Dirichlet," in *Abhandlungen der K. Preussischen Akademie der Wissenschaften* (1860), 1–36.
4. See "Über die Zerlegung der aus Wurzeln der Einheit gebildeten complexen zahlen in ihre Primfactoren," in *Journal für die reine und angewandte Mathematik*, **35** (1847), 327–367.
5. "Beweis des Fermatschen Satzes der Unmöglichkeit von $x^\lambda - y^\lambda = z^\lambda$ für eine unendliche Anzahl Primzahlen λ," in *Monatsberichte der K. Preussischen Akademie der Wissenschaften* (1847), 132–141, 305–319.
6. "Über die Flächen vierten Grades, auf welchen Schaaren von Kegelschnitten liegen," *ibid.* (1863), 324–338; "Über die Flächen vierten Grades mit sechzehn singulären Punkten," *ibid.* (1864), 246–260; "Über die Strahlensysteme, deren Brennflächen Flächen vierten Grades mit sechzehn singulären Punkten sind," *ibid.*, 495–499.

BIBLIOGRAPHY

I. ORIGINAL WORKS. Kummer published no books, nor has an edition of his works been published. His major writings appeared in *Journal für die reine und angewandte Mathematik*, **12–100** (1834–1886) and in *Monatsberichte* *der Königlichen Preussischen Akademie der Wissenschaften zu Berlin* (1846–1880).

Bibliographies of his writings may be found in Poggendorff, I, 1329–1330; III, 757; IV, 817; the Royal Society *Catalogue of Scientific Papers*, III, 770–772; VIII, 134–135; X, 475; XVI, 510; *Jahresbericht der Deutschen Mathematiker-vereinigung*, **3** (1894), 21–28; and Adolf Harnack, *Geschichte der Königlich Preussischen Akademie der Wissenschaften zu Berlin*, III (Berlin, 1900), 160–161.

Some of his correspondence appears in "Briefe Ernst Eduard Kummers an seine Mutter und an Leopold Kronecker," in Kurt Hensel, *Festschrift* ...(see below), 39–103.

MS material is in the archives of the Deutsche Akademie der Wissenschaften zu Berlin (D.D.R.) and of Humboldt University, Berlin (D.D.R.), and in the Deutsches Zentralarchiv, Merseburg (D.D.R.).

II. SECONDARY LITERATURE. For a bibliography of secondary literature, see Poggendorff, VIIa Suppl., 343–344. See also the following, listed chronologically: O. N.-H., "Eduard Kummer," in *Münchener allgemeine Zeitung*, no. 139 (20 May 1893); Emil Lampe, "Nachruf für Ernst Eduard Kummer," in *Jahresbericht der Deutschen Mathemátiker-vereinigung*, **3** (1894), 13–21; Leo Koenigsberger, *Carl Gustav Jacob Jacobi* (Leipzig, 1904); Wilhelm Ahrens, *Briefwechsel zwischen C. G. J. Jacobi und M. H. Jacobi* (Leipzig, 1907); Kurt Hensel, *Festschrift zur Feier des 100. Geburtstages Eduard Kummers* (Leipzig–Berlin, 1910), 1–37; Wilhelm Lorey, *Das Studium der Mathematik an den deutschen Universitäten seit Anfang des 19. Jahrhunderts* (Leipzig–Berlin, 1916); Leo Koenigsberger, *Mein Leben* (Heidelberg, 1919), 21, 24, 25, 27, 28, 31, 53, 114; Felix Klein, *Vorlesungen über die Entwicklung der Mathematik im 19. Jahrhundert*, I (Berlin, 1926), 167, 172, 199, 269, 282, 321–322; Kurt-R. Biermann, "Zur Geschichte der Ehrenpromotion Gotthold Eisensteins," in *Forschungen und Fortschritte*, **32** (1958), 332–335; Kurt-R. Biermann, "Über die Förderung deutscher Mathematiker durch Alexander von Humboldt," in *Alexander von Humboldt. Gedenkschrift zur 100. Wiederkehr seines Todestages* (Berlin, 1959), pp. 83–159; "J. P. G. Lejeune Dirichlet," in *Abhandlungen der Deutschen Akademie der Wissenschaften zu Berlin*, Kl. für Math., Phys. und Tech. (1959), no. 2; "Vorschläge zur Wahl von Mathematikern in die Berliner Akademie," *ibid.* (1960), no. 3; and "Die Mathematik und ihre Dozenten an der Berliner Universität 1810–1920" (Berlin, 1966), MS; and Hans Wussing, *Die Genesis des abstrakten Gruppenbegriffes* (Berlin, 1969).

Information on Kummer is also contained in Weierstrass' letters and in the secondary literature on him and on Leopold Kronecker.

KURT-R. BIERMANN

KUNCKEL, JOHANN (*b.* Hutten, Schleswig-Holstein, Germany, 1630 [possibly 1638]; *d.* Stockholm, Sweden [or nearby], 1702 [or 20 March 1703]), *chemistry*.

The details of Kunckel's life are obscure and are gleaned principally from his own writings. He was the son of an alchemist in the service of Duke Frederick of Holstein and learned chemistry from his father and practical chemistry from pharmacists and glassworkers. In his twenties (the year, as is usual with Kunckel, is uncertain) he was in the service of Dukes Franz Carl and Julius Heinrich of Sachsen-Lauenburg as pharmacist and chamberlain. About 1667 he became chemist and gentleman of the bedchamber to Johann Georg II, elector of Saxony, where there was an active alchemical circle. At Dresden he also learned the chemistry of glass manufacture.

After about ten years Kunckel lost his job (through the calumnies of his enemies, in his view); he says he then taught chemistry at Wittenberg. In 1679 he was invited by Frederick William, elector of Brandenburg, to become head of his chemical laboratory at Berlin and possibly director of the glassworks there. On the elector's death in 1688 he entered the service of King Charles XI of Sweden, as minister of mines; he was ennobled as Baron von Loewenstern in 1693, in which year he also became a member of the Academia Caesarea Leopoldina. He probably remained in Stockholm until his death.

Kunckel claimed always to be a follower of the experimental method, and his work that was best known outside Germany was his *Chymische Anmerckungen* of 1677, which received the Latin title *Philosophia chemica experimenti confirmata*, a name carried into its English title. His works are indeed filled with chemical fact, discovery, and observation, if not always with true experiment. His greatest theoretical interest was in promulgating the view that all fixed salts are the same, an opinion he carried over into his treatment of the manufacture of glass; he was of course correct in thinking that many plants produce potash but was unaware that seaside plants produce soda. Alkali salt he regarded as "the most universal Salt of all Metals and Minerals . . . the lock and key of all Metals" (*An Experimental Confirmation of Chymical Philosophy*, ch. 14). Thus mercury, he thought, was composed of "a Water and a Salt"; sulfur "first consists in a certain fatness of the Earth, which is a kind of combustible Oil, the like of which is found in all Vegetables; and then in a fix'd and volatile Salt, and a certain gross Earthiness." He did not think that sulfur was the principle of flame, for "where there is Heat, there is Acid, where there is Flame or Light, there is volatile salt."

Kunckel's views are patently a mixture derived from alchemy crossed with some rational natural philosophy and are not very far removed from those of his contemporary Becher. He belongs firmly to the late seventeenth-century German chemical tradition, and he was notable only for the keenness of his interest in practical and preparative chemistry. He evolved or adopted numerous recipes for the preparation of substances and displayed a good deal of common sense in discussing their probable nature. Like many men of his time he despised the alchemists for their mysticism while inclining to regard their aims as rational. Certainly he thought highly of *aurum potabile* as a medicine and believed that although nothing could be created *de novo*, yet base metals could be transformed or converted into gold. At the same time he pointed out the fallacy of the universal solvent: it would dissolve the vessel in which it is made. These severer views are found only in the posthumously published *Collegium physico-chymicum experimentale* and were perhaps the result of a lifetime's not altogether happy association with alchemists.

Kunckel's part in the discovery of phosphorus is not altogether clear, but it certainly enhanced his reputation. He described his own view of the affair in his *Collegium;* in 1678 he published an account of "his" phosphorus (*Oeffentliche Zuschrift von dem Phosphoro mirabili*) and its medical properties, but not of its method of preparation. It seems probable that Kunckel's statement that he saw Hennig Brand's phosphorus, got from him the hint that it was made from urine, and then proceeded on his own is true—it is very like the accounts of J. D. Krafft (who, at Kunckel's suggestion, bought the secret from Brand) and of Robert Boyle, who was the first to publish the method of preparation. It is interesting that Leibniz, in his *Historia inventionis phosphori* (1710), thought Kunckel's experiments more scientifically useful than Boyle's, presumably because they were more closely related to medicine. Kunckel's claim for priority was widely accepted on the Continent, and phosphorus was often associated with his name.

In 1679 Kunckel published a work which was much read for the next century and which comprised essentially a series of essays on aspects of glassmaking. This was a translation into German of Christopher Merret's Latin edition of Antonio Neri's *L'arte vetraria* of 1612. Kunckel preserved Merret's notes and added further notes as well as a section on the making of colored glass, together with several short treatises on related topics by other German writers. Its translation into French in the mid-eighteenth century added considerably to Kunckel's reputation.

Kunckel was clearly an able practical chemist, quick to seize upon new discoveries. His works had considerable appeal to his Germanic contemporaries. The reaction in England was respectful; when the elector of Brandenburg sent a copy of the controversial *Chymischer Probier-Stein* to the Royal Society, to which it was dedicated, in 1684, the Society asked Boyle to have a Latin abstract made; and it listened to a summary by Frederick Slare, which was perhaps what was printed in the *Philosophical Transactions.* But Boyle obviously thought poorly of it and delivered its author a stinging rebuke, declaring that the Society had not yet got to "framing systems" (T. Birch, *History of the Royal Society*, IV [London, 1757], 325–326, note). The only other work of Kunckel's to be translated into English is the short and rational *Chymische Anmerckungen.* But his main appeal was to those who admired German chemistry of the tradition to which he belonged. He cannot be said to rank high in the history of seventeenth-century chemistry.

BIBLIOGRAPHY

I. Original Works. Kunckel's works appeared originally in German; many were soon translated, usually into Latin, which complicates his bibliography. His earliest work was *Nützliche Observationes oder Anmerckungen von den Fixen und flüchtigen Salzen, Auro und Argento potabili spiritu mundi* (Hamburg, 1676), trans. as *Utiles observationes sive animadversiones de salibus fixis & volatilibus . . .* (London–Rotterdam, 1678). Next came *Chymische Anmerckungen: Darinn gehandelt Wird von denen Principiis chymicis, Salibus acidis und alkalibus, fixis et volatilibus . . .* (Wittenberg, 1677), trans. into Latin as *Cubiculari intimi et chymici philosophia chemica experimentis confirmata*, sometimes abbr. as *Philosophia chemica . . .* (London–Rotterdam, 1678; Amsterdam, 1694), and into English as "An Experimental Confirmation of Chymical Philosophy," in *Pyrotechnical Discourses* (London, 1705). His reputation was established by *Oeffentliche Zuschrift von dem Phosphoro mirabili und dessen leuchtenden Wunder-Pilulen* (Wittenberg, 1678; the BM copy is Leipzig, 1680 [?]), which was never translated. Next to be published was his trans. and ed. of Neri, with additions, *Ars vitraria experimentalis*, in German (Amsterdam–Danzig, 1679; Frankfurt–Leipzig, 1689; Nuremberg, 1743, 1756, 1785), trans. into French by Baron d'Holbach as *Art de la verrerie de Neri, Merret et Kunckel* (Paris, 1752). His *Epistola contra spiritum vini sine acido*, dated Berlin, 1681, may have been issued separately; it is usually found annexed to *Chymischer Probier-Stein de acido & urinoso, Sale calid. & frigid. Contra Herrn Doct. Voigts Spirit. Vini vindicatum* (Berlin, 1684, 1685 [?], 1686, 1696); there is a long review in *Philosophical Transactions of the Royal Society*, **15** (1685–1686), 896–914, with an English trans. of Kunckel's "Address to the Royal Society." The remainder of Kunckel's work was published posthumously in *Collegium physico-chymicum experimentale, oder Laboratorium chymicum* (Hamburg–Leipzig, 1716, 1722; Hamburg, 1738; Berlin, 1767). *V curiose chymische Tractätlein* (Frankfurt–Leipzig, 1721) contains *Chymische Anmerckungen, Nützliche Observationes, Epis-*

tola, *De Phosphoro mirabili*, and *Probier-Stein*, all in German.

II. Secondary Literature. There is a nearly complete bibliographical account with comment in J. R. Partington, *A History of Chemistry*, II (London, 1961), 361–377. Ferdinand Hoefer, *Histoire de la chimie* (Paris, 1866), II, 191–205, has a perhaps more balanced account, with useful quotations in French. Articles include Tenney L. Davis, "Kunckel and the Early History of Phosphorus," in *Journal of Chemical Education*, **4** (1927), 1105–1113; and "Kunckel's Discovery of Fulminate," in *Army Ordnance*, **7** (1926), 62. There is a biographical essay in Danish by Axel Helne, "Johann Kunckel von Lowenstern (1630–1702)," in *Tidsskrift vor Industri* (1912), with illustrations. More recent is H. Maurach, "Johann Kunckel: 1630–1703," in *Deutsches Museum Abhandlungen und Berichte*, **5**, no. 2 (1933), 31–64, with illustrations.

Marie Boas Hall

KUNDT, AUGUST ADOLPH (*b.* Schwerin, Germany, 18 November 1839; *d.* Israelsdorf, near Lübeck, Germany, 21 May 1894), *physics.*

Kundt was a leading experimental physicist in the nineteenth century. He studied at the University of Berlin, where he received his doctorate under G. Magnus in 1864 and qualified for lecturing in 1867. He obtained a professorship of physics in 1868 at the Polytechnikum in Zurich, at the University of Würzburg in 1869, and at Strasbourg in 1872. A follower of Helmholtz, he accepted the latter's chair at Berlin in 1888.

Kundt worked primarily in optics, acoustics, and gas theory; he became famous mainly for his method of determining the speed of sound in gases. By means of standing waves, which were produced in a tube closed at one end, the "Kundt tube," and made visible through finely divided lycopodium seeds—in what were known as "Kundt's dust figures"—he could determine the speed of sound in gases and compare it with that in solid bodies. In collaboration with E. G. Warburg, Kundt investigated friction and heat conduction in gases and established their pressure independence for certain regions. In 1876 Kundt and Warburg published their determination, which became famous, of the ratio of the specific heats of monatomic gases, such as mercury gas.

Kundt was, in addition, an expert on anomalous dispersion in fluids and thin metal layers. He demonstrated that this physical phenomenon appears in materials which absorb certain colors to a high degree. With W. C. Roentgen he demonstrated the Faraday effect—the magnetic rotation of the plane of polarization of light—for gases. Kundt also published investigations on the spectra of lightning and on a procedure for determining the thermo-, actino-, and piezoelectricity of crystals.

BIBLIOGRAPHY

I. Original Works. Bibliographies of Kundt's works are in Poggendorff, III, 757–758; and IV, 818; and A. Harnack, *Geschichte der Königlichen Preussischen Akademie der Wissenschaften zu Berlin*, III (Berlin, 1900), 162 (Kundt's academic writings only). He wrote more than 50 scientific papers, including the following, published in *Annalen der Physik und Chemie:* "Ueber eine besondere Art der Bewegung elastischer Körper auf tönenden Röhren und Stäben," **126** (1865), 513–527; "Ueber neue akustische Staubfiguren und Anwendung derselben zur Bestimmung der Schallgeschwindigkeit in festen Körpern und Gasen," **127** (1866), 497–523; "Untersuchung über die Schallgeschwindigkeit der Luft in Röhren," **135** (1868), 337–372, 527–561; "Ueber anomale Dispersion der Körper mit Oberflächenfarben," **142** (1871), 163–171; **143** (1871), 149–152, 259–269; **144** (1871), **145** (1872); "Reibung und Wärmeleitung verdünnter Gase," **155** (1875), 337–365, 525–550; **156** (1876), 177–211, written with E. G. Warburg; "Ueber die specifische Wärme des Quecksilbergases," **157** (1876), 353–369, written with E. G. Warburg; and "Ueber die electromotorische Drehung der Polarisationsebene des Lichtes in Gasen," n.s. **8** (1879), 278–298; n.s. **10** (1880), 257–265, written with W. C. Roentgen. He also wrote "Antrittsrede" (after his election as a member of the Prussian Academy of Sciences), in *Sitzungsberichte der Preussischen Akademie der Wissenschaften zu Berlin*, II (1889), 679–683; and "Gedächtnisrede auf Werner von Siemens," in *Abhandlungen der Preussischen Akademie der Wissenschaften* (1893), no. 2, 1–21.

II. Secondary Literature. See W. von Bezold, *August Kundt. Gedächtnisrede, gehalten in der Sitzung der Physikalischen Gesellschaft zu Berlin am 15. Juni 1894* (Leipzig, 1894); and E. du Bois-Reymond, "Antwort auf Kundts Antrittsrede," in *Sitzungsberichte der Preussischen Akademie der Wissenschaften zu Berlin* (1889), no. 2, 683–685.

Hans-Günther Körber

KUNKEL VON LÖWENSTERN. See **Kunckel, Johann.**

KURCHATOV, IGOR VASILIEVICH (*b.* Sim, Ufimskaya guberniya [now Ufimskaya oblast], Russia, 12 January 1903; *d.* Moscow, U.S.S.R., 7 February 1960), *physics.*

Kurchatov's father, Vasily Alekseevich Kurchatov, was a land surveyor; his mother, Maria Vasilievna, was a schoolteacher before her marriage. In 1911 the family moved to Simbirsk (now Ulyanovsk) and settled in Simferopol in 1912. Here Kurchatov graduated from the Gymnasium in 1920 and entered the mathematics section of the Faculty of Mathematics and Physics at the University of the Crimea. Financial hardships forced him to take various jobs while a student. In 1922 he became an assistant in the university physics laboratory. In 1923 Kurchatov graduated from the university, then became an observer at the magnetic-meteorological observatory in Pavlovsk (a suburb of Leningrad), where he completed his first scientific work on the radioactivity of snow.

In 1927 Kurchatov married, moved to Leningrad, and entered the Leningrad Physical-Technical Institute, which was headed by A. F. Joffe. During 1927–1929 he conducted a series of experiments on the physics of nonconductors (dielectrics), under Joffe's direct supervision. His first published works were on the conductance of solid bodies, the formation of an electric charge produced by the flow of an electric current through dielectric crystals, and the breakdown mechanism of solid dielectrics. With P. P. Kobeko he subsequently investigated the electrical characteristics of Rochelle (or Seignette's) salt in order to explain the nature of several anomalies described in the literature. In this investigation they discovered the far-reaching analogy between the dielectric characteristics of Rochelle salt and the magnetic characteristics of ferromagnetics. They called this new phenomenon "seignetto-electricity"; the established term for this in non-Soviet scientific literature is "ferroelectricity."

Beginning in 1932 Kurchatov made a gradual transition to studies of the atomic nucleus. He became the head of the large department of nuclear physics that was organized at the Leningrad Physical-Technical Institute. He devoted considerable attention to the creation of a high-voltage installation for the acceleration of ions and supervised the construction of what was then the largest cyclotron in Europe. He also studied nuclear reactions and in 1934 he established the branching of these reactions. In 1935 Kurchatov, with L. I. Rusinov, discovered nuclear isomers while irradiating the nucleus of bromine. In 1934 he was awarded a doctorate in physics and mathematical sciences. From 1935 to 1940 he conducted research in the physics of neutrons. In 1939 Kurchatov began to work on the problem of splitting heavy atoms and the possibility of obtaining a chain reaction. Under his supervision G. N. Flerov and K. A. Petrzhak proved experimentally the existence of spontaneous processes in the splitting of uranium nuclei.

During World War II, Kurchatov developed methods of defending ships against magnetic mines and tested them under battle conditions and was awarded the State Prize, first degree, in 1942 for his successful solution to this problem. At this time Kurchatov was appointed to head research on the development of atomic energy and of atomic weapons for defense. A large laboratory was set up in Moscow under his supervision, which later became the I. V. Kurchatov Institute of Atomic Energy.

In this undertaking Kurchatov's creative abilities were strikingly manifested, as were his brilliant talent for organization, his enormous capacity for work, and his exceptionally buoyant and good-natured personality. Toward the end of 1946 Kurchatov and his co-workers inaugurated the first atomic reactor in Europe, and in 1949 they developed and successfully tested the first Soviet atomic bombs. As a result of detailed investigation of methods for creating thermonuclear reactions, Kurchatov and his co-workers carried out the world's first hydrogen bomb test on 12 August 1953. Kurchatov paid particular attention to the peaceful uses of atomic energy. With his close collaboration a project for the world's first atomic electric station was devised, and the station was commissioned 27 June 1954. During the last years of his life Kurchatov worked a great deal on the results of controlled thermonuclear reactions. For service to his country he was thrice honored as Hero of Socialist Labor. His ashes are entombed in the wall of the Kremlin.

BIBLIOGRAPHY

I. Original Works. Kurchatov's writings include *Segnetoelektriki* ("Seignetto-electricity"; Leningrad–Moscow, 1933); *Rasshcheplenie atomnogo yadra* (*Problemy sovremennoy fiziki*) ("The Splitting of the Atomic Nucleus [Problems of Contemporary Physics]"), A. F. Joffe, ed. (Moscow–Leningrad, 1935); "Delenie tyazhelykh yader" ("The Division of Heavy Nuclei"), in *Uspekhi fizicheskikh nauk*, **25**, no. 2 (1941), 159–170; "O nekotorykh rabotakh Instituta atomnoy energii Akademii nauk SSSR po upravlyaemym termoyadernym reaktsiam" ("On Several Works of the Institute of Atomic Energy of the Academy of Sciences of the U.S.S.R. on Controlled Thermonuclear Reactions"), in *Atomnaya energia*, **5** (1958), 105–110; and "O nekotorykh rezultatakh issledovany po upravlyaemym termoyadernym reaktsiam, poluchennykh v SSR" ("On Several Results of Investigations of Controlled Thermonuclear Reactions Obtained in the U.S.S.R."), in *Uspekhi fizicheskikh nauk*, **73**, no. 4 (1961), 605–610.

II. Secondary Literature. See I. N. Golovin, *I. V. Kurchatov* (Moscow, 1967); "Igor Vasilievich Kurchatov," in *Uspekhi fizicheskikh nauk*, **73**, no. 4 (1961), 593–604, which includes a bibliography of Kurchatov's writings; and A. F. Joffe, "I. V. Kurchatov—issledovatel dielektrikov" ("I. V. Kurchatov—Investigator of Dielectrics"), *ibid.*, 611–614.

J. G. Dorfman

KURLBAUM, FERDINAND (*b.* Burg, near Magdeburg, Germany, 4 October 1857; *d.* Berlin, Germany, 29 July 1927), *physics, especially optics.*

As a result of scholastic difficulties, Kurlbaum did not enter the university until he was twenty-three. After only eight semesters, however, he began an important dissertation under the supervision of Heinrich Kayser in Helmholtz's laboratory. This work, completed in 1887, contained a new determination of the wavelengths of thirteen Fraunhofer lines of the solar spectrum. When Kayser was called to Hannover, Kurlbaum went with him as his assistant. In 1891 Kurlbaum moved to the Physikalisch-Technische Reichsanstalt in Berlin. At first he worked in the optical laboratory, which was directed by Otto Lummer. In 1893 Kurlbaum modified the bolometer in such a way that it was capable of making absolute measurements of radiation intensities.

In 1898 Lummer and Kurlbaum published, in the *Verhandlungen der Physikalischen Gesellschaft zu Berlin*, their famous work in which they described the radiation-containing hollow body in the form that is still customary: an electrically heated platinum cylinder, blackened on the inside with iron oxide and enclosed in an outer asbestos cylinder. In the same year Kurlbaum provided an absolute measure of blackbody radiation which was tantamount to determining the Stefan-Boltzmann constants to an accuracy of 5 percent.

In 1894 Kurlbaum was named an assistant at the Physikalisch-Technische Reichsanstalt; and in October 1899, at the proposal of the president, F. W. G. Kohlrausch, he received the title of "professor." At the invitation of Heinrich Rubens, Kurlbaum participated in the latter's measurements of the radiation intensity of the blackbody in the case of extremely long waves (residual radiation of fluorite and of rock salt). Lummer and Pringsheim had already questioned the validity of Wien's law of radiation; but the decisive break came, as Max Planck later expressly stated, only with Rubens' and Kurlbaum's long-wave measurements. Kurlbaum communicated the results at the meeting of the German Physical Society held in Berlin on 19 October 1900. It followed from their findings that at high temperatures the radiation intensity of the blackbody is proportional to the absolute temperature, as Rayleigh had already written in June 1900 in *Philosophical Magazine*. This result, which stood in obvious contradiction to Wien's law of radiation, was known to Planck on 7 October, through a verbal communication from Rubens; the discovery of his own radiation law followed in the middle of October. Directly after Kurlbaum gave his report, Planck presented his new formula: quantum theory had begun. Arnold Sommerfeld observed in 1911: "It will always remain a famous page in the history of the first decades of the Physikalisch-Technische Reichsanstalt that it erected one of the pillars of the quantum theory, the experimental bases of hollow-space radiation." One may add that Kurlbaum immortalized his name on this "famous page."

In 1901 Kurlbaum took charge of the high-voltage laboratory in the "second division" of the Physikalisch-Technische Reichsanstalt, where the principal activity was the testing of apparatus. His own endeavors, however, continued to be devoted primarily to the measurement of radiation. With Ludwig Holborn he constructed a filament pyrometer that was capable of measuring arbitrarily high temperatures in a brilliantly simple way and with great precision.

In the fall of 1904 Kurlbaum was appointed full professor at the Technical College of Berlin-Charlottenburg, as successor to C. A. Paalzow. He worked at first with Heinrich Rubens. After the latter's departure the main burden of teaching fell to Kurlbaum, and he carried out no further personal research at the Technical College.

BIBLIOGRAPHY

I. Original Works. Kurlbaum's writings include "Bestimmung der Wellenlänge Fraunhofer'scher Linien," in *Wiedemanns Annalen der Physik*, **33** (1888), 381–412; "Bolometrische Untersuchungen," *ibid.*, **46** (1892), 204–224, written with O. Lummer; "Über die Herstellung eines Flächenbolometers," in *Zeitschrift für Instrumentenkunde*, **12** (1892), 81–89, written with O. Lummer; "Notiz über eine Methode zur quantitativen Bestimmung strahlender Wärme," in *Wiedemanns Annalen der Physik*, **51** (1894), 591–592; "Der electrisch geglühte 'absolut schwarze' Körper und seine Temperaturmessung," in *Verhandlungen der Physikalischen Gesellschaft zu Berlin*, **17** (1898), 106–111, written with O. Lummer; "Über die Emission langwelliger Wärmestrahlen durch den schwarzen Körper bei verschiedenen Temperaturen," in *Sitzungsberichte der K. Preussischen Akademie der Wissenschaften zu Berlin* (1900), **2**, 929–941, written with H. Rubens; "Anwendung der Methode der Reststrahlen zur Prüfung des Strahlungsgesetzes," in *Wiedemanns Annalen der Physik*, **73** (1901), 649–666, written with H. Rubens; and "Über ein optisches Pyrometer," in *Drudes Annalen der Physik*, **10** (1903), 225–241.

II. Secondary Literature. See "Ferdinand Kurlbaum†," in *Zeitschrift für technische Physik*, **8** (1927), 525–527; F. Henning, "Ferdinand Kurlbaum," in *Physikalische Zeitschrift*, **29** (1928), 97–104; and Hans Kangro, *Vorgeschichte des Planckschen Strahlungsgesetzes. Messungen und Theorien der spektralen Energieverteilung bis zur Begründung der Quantenhypothese* (Wiesbaden, 1970).

Armin Hermann

KURNAKOV, NIKOLAI SEMYONOVICH (*b.* Nolinsk, Vyatka guberniya, Russia, 6 December 1860; *d.* Moscow, U.S.S.R., 19 March 1941), *chemistry.*

Kurnakov graduated from the St. Petersburg Institute of Mines in 1882. From 1893 he was a professor of inorganic chemistry there and also (1899–1908) professor of physical chemistry at the Institute of Electrical Engineering. From 1902 to 1930 he was professor of general chemistry at the St. Petersburg Polytechnic Institute. On 7 December 1913 Kurnakov was elected a member of the Academy of Sciences. In 1918, at Kurnakov's suggestion, the Academy created the Institute of Physical and Chemical Analysis; in 1934 it became the Institute of General and Inorganic Chemistry, and in 1944 it was renamed the N. S. Kurnakov Institute of General and Inorganic Chemistry.

Kurnakov's first scientific interest was in salts. In the 1890's he completed a series of experiments on the chemistry of coordination complexes that were generalized in his inaugural dissertation, "O slozhnykh metallicheskikh osnovaniakh" ("On Complex Metallic Bases," 1893). Kurnakov discovered the reaction of cis- and trans-isomers of divalent platinum with thiourea; this reaction is named for him. He showed that trans-isomers form a coordination complex with two molecules of thiourea in an internal sphere, while cis- compounds react with four particles of thiourea. This reaction permits question concerning the structure of derivatives of bivalent platinum.

Kurnakov's main scientific accomplishment was the creation of physicochemical analysis based on the study of equilibrium systems by measuring their characteristics in relation to changes in composition and by constructing appropriate phase diagrams. Beginning in 1898, Kurnakov carried on a systematic study of heterogeneous systems (initially of metallic alloys, later of organic and salt systems). Developing the work of Chernov, Le Châtelier, Osmond, and Roberts-Austen in thermal analysis, in 1900 Kurnakov developed the means of finding the composition of specific compounds in alloys by using the method of fusibility.

In 1903 Kurnakov invented a self-registering pyrometer—a device for recording heating and cooling curves—with the aid of which he significantly improved the methodology of thermal analysis. In 1906 he introduced the measurement of electroconductivity as a method of studying the change in the characteristics of a system in relation to changes in its composition. With S. F. Zhemchuzhny he demonstrated that in the formation of solid solutions of two metals there is a reduction in electroconductivity. They established the basic composition-electroconductivity diagrams for systems of two metals which form continuous solid solutions.

In 1906 Kurnakov established that technical alloys of high electric resistance, which are widely employed in the manufacture of rheostats and resistance boxes, consist of solid solutions. Expanding his research on alloys, Kurnakov applied the measurement of hardness and the determination of flow pressure as a new method of physicochemical analysis (1908–1912).

In 1912, while studying the viscosity of two-liquid systems in relation to their composition, Kurnakov found that the formation of specific compounds in the given systems corresponds to special points (which he called singular or Dalton points) on the phase diagrams. These results permitted Kurnakov to formulate the general conclusion that "a chemical individuum belonging to a specific chemical compound represents a phase, which possesses singular or Dalton points on the lines of its characteristics" (1914). The composition corresponding to these points remains constant even with a change in the facts of the system's equilibrium. Through investigation of tellurium-bismuth, iron-silicon, antimony, aluminum-iron, and lead-sodium systems, using all methods of physicochemical analysis, phases of variable composition were discovered that, according to Kurnakov, belonged to the berthollide type. These phases were not characterized by singular points on the phase diagrams. According to Kurnakov, the berthollides are crystal-phase nonstoichiometric compounds, the variable composition of which cannot be expressed by simple integer relationships.

Kurnakov ascribed great significance to the study of the genetic connection between daltonides and berthollides chemical compounds. "In equilibrium systems," he said, "discreteness and continuity are intercombined and coexist."

Kurnakov and his students proposed methods of physicochemical analysis for the study of a great variety of systems formed by salts, metals, and inorganic compounds. Without destroying the system under investigation, Kurnakov precisely specified the conditions of separation of the various phases for intermetallic compounds and for the hydrate forms of double salts; he also found the limits of their stable state—unobtainable with the common methods of chemical investigation then in use.

Investigation of the salt equilibrium in the system $2\,NaCl + MgSO_4 \rightleftharpoons Na_2SO_4 + MgCl_2$ (at 25° C. and 0° C.) enabled Kurnakov to clarify the mechanism of deposition of Glauber's salts in Kara-Bogaz-Gol Bay (1918); the results provided the scientific basis for the industrial exploitation of rich deposits of natural salts. In 1917 he published "Mestorozhdenia khloristogo kalia Solikamskoy solenosnoy tolshchi" ("Deposits of Potassium Chloride in the Solikamsk Saliferous Mass"), written with K. F. Beloglazov and M. K. Shmatko; this work played a conspicuous role in the industrial exploitation of the rich Solikamsk potassium deposits.

Studies of the salt lakes in the Crimea and in the Volga basin, of the Tikhvin bauxite deposits, and of other valuable sources of minerals are associated with Kurnakov.

Kurnakov was an active participant in the introduction of chemical processes into Soviet agriculture and helped to organize many scientific conventions and conferences. He also fostered a large school of inorganic chemists and metallurgists, including G. G. Urazov, S. F. Zhemchuzhny, N. I. Stepanov, and N. N. Efremov.

BIBLIOGRAPHY

I. Original Works. Kurnakov's writings include *Sobranie izbrannykh rabot* ("Collection of Selected Works"), 2 vols. (Leningrad, 1938–1939); *Vvedenie v fiziko-khimichesky analiz* ("Introduction to Physicochemical Analysis"), 4th ed. (Moscow, 1940); and *Izbrannye trudy* ("Selected Papers"), 3 vols. (Moscow, 1960–1963).

II. Secondary Literature. See L. Dlougatch, "N. S. Kournakow, sa vie, son oeuvre, son école," in *Revue de métallurgie*, **21** (1925), 650–662, 722–732; G. B. Kauffman and A. Beck, "Nikolai Semenovich Kurnakov," in *Journal of Chemical Education*, **36** (1962), 521; Y. I. Soloviev, *Ocherki istorii fiziko-khimicheskogo analiza* ("Essays on the History of Physicochemical Analysis"; Moscow, 1955); Y. I. Soloviev and O. E. Zvyagintsev, *Nikolay Semyonovich Kurnakov. Zhizn i deyatelnost* ("Nikolai Semyonovich Kurnakov. Life and Work"; Moscow, 1960); and G. G. Urazov, "Akademik N. S. Kurnakov—osnovatel fiziko-khimicheskogo analiza i glava nauchnoy shkoly" ("Academician N. S. Kurnakov—Founder of Physicochemical Analysis and Head of a Scientific School"), in *Izvestiya Sektora fiziko-khimicheskogo analiza, Institut obshchei i neorganicheskoi khimii*, **14** (1941), 9–35, which includes a bibliography of Kurnakov's papers.

Y. I. Soloviev

KÜRSCHÁK, JÓZSEF (*b.* Buda, Hungary, 14 March 1864; *d.* Budapest, Hungary, 26 March 1933), *mathematics.*

Kürschák's father, András Kürschák, an artisan, died when his son was six; the boy was very carefully brought up by his mother, the former Jozefa Teller. Kürschák's mathematical talent appeared in secondary school, after which he attended the Technical University in Budapest (1881–1886), which, although a technical school, also trained teachers of mathematics and physics. After graduating Kürschák taught for two years at Rozsnyó, Slovakia. In 1888 he moved to Budapest, where he worked toward the Ph.D., which he received in 1890. The following year he was appointed to teach at the Technical University, where he served successively as lecturer, assistant professor, and professor (1900) until his death. In 1897 he was elected a corresponding, and in 1914 an ordinary, member of the Hungarian Academy of Sciences.

Kürschák's mathematical interests were wide, and he had the ability to deal with various kinds of problems. His first paper (1887) concerned the extremal properties of polygons inscribed in and circumscribed about a circle and proved the existence of the extremum. Another paper (1902) showed, in connection with Hilbert's *Grundlagen der Geometrie*, the sufficiency of the ruler and of a fixed distance for all discrete constructions. Meanwhile, in extending a result of Julius Vályi's, Kürschák had turned to the investigation of the differential equations of the calculus of variations (1889, 1894, 1896), proved their invariance under contact (Legendre) transformations (1903), and gave the necessary and sufficient conditions—thereby generalizing a result of A. Hirsch's—for second-order differential expressions to provide the equation belonging to the variation of a multiple integral (1905). These investigations also furthered his interest in linear algebra, aroused by Eugen von Hunyady, an early exponent of algebraic geometry in Hungary, and led to a series of papers on determinants and matrices.

Kürschák's main achievement, however, is the founding of the theory of valuations (1912). Inspired by the algebraic studies of Julius König and by the fundamental work of E. Steinitz on abstract fields, as well as by K. Hensel's theory of *p*-adic numbers, Kürschák succeeded in generalizing the concept of absolute value by employing a "valuation," which made possible the introduction of such notions as convergence, fundamental sequence, distance function, and limits into the theory of abstract fields. He proved that any field with a valuation on it can be extended by the adjunction of new elements to a "perfect" (i.e., closed and dense in itself) field which is at the same time algebraically closed. Kürschák's valuation and his method were later developed, mainly by Alexander Ostrowski, into a consistent and highly important arithmetical theory of fields.

Above all, Kürschák was a versatile and thought-provoking teacher. One of the main organizers of mathematical competitions, he contributed greatly to the selection and education of many brilliant students and certainly had a role in the fact that—to use the words of S. Ulam—"Budapest, in the period of the two decades around the First World War, proved to be an exceptionally fertile breeding ground for scientific talent" (*Bulletin of the American Mathematical Society*, **64**, no. 3, pt. 2 [1958], 1). Among Kürschák's pupils were mathematicians and physicists of the first rank, the most brilliant being John von Neumann.

BIBLIOGRAPHY

I. ORIGINAL WORKS. Kürschák's longer works include "Propriétés générales des corps et des variétés algébriques," in *Encyclopédie des sciences mathématiques pures et appliquées*, I, pt. 2 (Paris–Leipzig, 1910), 233–385, French version of C. Landsberg's German article, written with J. Hadamard; *Analizis és analitikus geometria* (Budapest, 1920); and *Matematikai versenytételek* (Szeged, 1929), trans. into English by Elvira Rapaport as *Hungarian Problem Book*, 2 vols. (New York, 1963).

Articles are "Ueber dem Kreise ein- und umgeschriebene Vielecke," in *Mathematische Annalen*, **30** (1887), 578–581; "Über die partiellen Differentialgleichungen zweiter Ordnung bei der Variation der doppelter Integrale," in *Mathematische und naturwissenschaftliche Berichte aus Ungarn*, **7** (1889), 263–275; "Ueber die partielle Differentialgleichung des Problems $\delta \iint V(p, q)\, dx\, dy = 0$," in *Mathematische Annalen*, **44** (1894), 9–16; "Über eine Classe der partiellen Differentialgleichungen zweiter Ordnung," in *Mathematische und naturwissenschaftliche Berichte aus Ungarn*, **14** (1896), 285–318; "Das Streckenabtragen," in *Mathematische Annalen*, **55** (1902), 597–598; "Ueber die Transformation der partiellen Differentialgleichungen der Variationsrechnung," *ibid.*, **56** (1903), 155–164; "Über symmetrische Matrices," *ibid.*, **58** (1904), 380–384; "Über eine characteristische Eigenschaft der Differentialgleichungen der Variationsrechnung," *ibid.*, **60** (1905), 157–165; "Die Existenzbedingungen des verallgemeinerten kinetischen Potentials," *ibid.*, **62** (1906), 148–155; "Über Limesbildung und allgemeine Körpertheorie," in *Journal für die reine und angewandte Mathematik*, **142** (1913), 211–253, presented at the International Congress of Mathematicians (1912); "Ein Irreduzibilitätssatz in der Theorie der symmetrischen Matrizen," in *Mathematische Zeitschrift*, **9** (1921), 191–195; "On Matrices Connected With Sylvester's Dialytic Eliminant," in *Transactions of the Royal Society of South Africa*, **11** (1924), 257–260; and "Die Irreduzibilität einer Determinante der analytischen Geometrie," in *Acta mathematica Szeged*, **6** (1932–1933), 21–26.

II. SECONDARY LITERATURE. See G. Rados, "Kürschák József emlékezete," in *Magyar tudományos akadémia Elhunyt tagjai fölött tartott emlékbeszédek*, **22**, no. 7 (1934), 1–18; L. Stachó, "Kürschák József," in *Müszaki Nagyjaink*, III (Budapest, 1967), 241–282, with complete bibliography; T. Stachó, "Kürschák József 1864–1933," in *Matematikai és physikai lapok*, **43** (1936), 1–13; and S. Ulam, "John von Neumann, 1903–1957," in *Bulletin of the American Mathematical Society*, **64**, no. 3, pt. 2 (1958), 1–49.

L. VEKERDI

KUSHYĀR IBN LABBĀN IBN BĀSHAHRĪ, ABU-'L-ḤASAN, AL-JĪLĪ (*fl. ca.* 1000), *astronomy, trigonometry, arithmetic.*

Little is known about Kushyār's life. The word "al-Jīlī" added to his name refers to Jīlān, a region of northern Iran south of the Caspian Sea.

The earliest Arabic biographer to write about Kushyār is al-Bayhaqī (*d.* 1065), who states that Kushyār lived in Baghdad and died about A.H. 350 (A.D. 961). Later biographers copy al-Bayhaqī and add several attributes to Kushyār's name, including the title "al-kiya," which seems to mean "master." But ʿAlī ibn Aḥmad al-Nasawī, an arithmetician who flourished after 1029, is said to have been a student of Kushyār's. This makes 961 too early; and accordingly, Schoy, Suter, and Brockelmann state that Kushyār must have flourished between 971 and 1029. It may be pointed out, however, that Ibn al-Nadīm does not mention Kushyār. Ibn al-Nadīm completed the main bulk of his *Fihrist* about 987 but continued to make additions to it until about 995. It would be rather strange if *al-kiya* Kushyār, the prolific writer, had lived in the same city at the same time and remained unnoticed by Ibn al-Nadīm. In his study of Kushyār's *zījes*, Kennedy points out that most of them were probably written after 1000. Accordingly, until further evidence appears, it will be safe to state only that Kushyār ibn Labbān flourished around A.D. 1000.

The works attributed to Kushyār have survived, but of these only three have received scholarly attention, two *zījes* and an arithmetic. The two *zījes* are *al-Jāmiʿ*, "The Comprehensive," and *al-Bāligh*, "The Far-reaching." Each is in four sections: introductory notes, tables, explanations, and proofs. Of the *al-Bāligh* only the first two sections are extant in the Berlin manuscript. In his "Survey of Islamic Astronomical Tables," E. S. Kennedy refers to the doubts whether Kushyār actually wrote two distinct *zījes* and gives the impression that *al-Bāligh* is an abbreviated copy of *al-Jāmiʿ*.

Kushyār's arithmetic is *Uṣūl Ḥisāb al-Hind*, "Elements of Hindu Reckoning." There is a Hebrew commentary to this work written by ʿAnābī in the fifteenth century.

Kushyār also wrote *al-Lāmiʿ fī amthilat al-zīj al-Jāmiʿ*, "The Brilliant [Work] on the Examples Pertaining to *al-Jāmiʿ zīj*"; *Kitāb al-Asṭurlāb wa kayfiyyat ʿamalihī wa iʿtibārihī, . . .*, "A Book on the Astrolabe and How to Prepare It and Test It . . ."; *Tajrīd Uṣūl Tarkīb al-Juyūb*, "Extracts of the Principles of Building up Sine Tables"; *al-Madkhal* [or *al-Mujmal*] *fī ṣināʿat ahkām al-Nujūm*, "An Introduction [or Summary] of the Rules of Astrology [and

Astronomy]"; and *Risāla fī al-Abʿād wa al-Ajrām*, "A Treatise on Distances and Sizes," i.e., mensuration.

It is believed that Kushyār did not make any astronomical observations of his own; his *zījes* are classified with a few others called "al-Battānī's group." These take their elements from Muḥammad ibn Jābir ibn Sinān al-Battānī's *al-Zīj al-Ṣābi'*.

Kushyār is, however, credited with having developed the study of trigonometric functions started by Abu'l-Wafā' and al-Battānī. Abu'l-Wafā' gives sine tables, and al-Battānī gives sines and cotangents; but Kushyār's *zījes* contain sines, cotangents, tangents, and versed sines, together with tables of differences. In most of these tables the functions are calculated to three sexagesimal places and the angles increase in steps of one degree.

Kushyār's unique position in the development of Hindu arithmetic is not yet well understood. The Muslims inherited two arithmetical systems: the sexagesimal system, used mainly by astronomers, and finger reckoning, used by all. Finger reckoning contained no numeration. Numbers were stated in words, and calculations were done mentally. To remember intermediary results, calculators bent their fingers in distinct conventional ways; hence the name finger reckoning. Scribes were able to denote manually numbers from 1 to 9,999, one at a time.

The sexagesimal system used letters of the Arabic alphabet for numeration. Its fractions were always in the scale of sixty, but integers could be in the scale of sixty or of ten. We have arithmetic books that explain the concepts and practices of finger reckoning—the most important being the arithmetic written by Abu'l-Wafā' for state officials—but we do not have books that show how astronomers performed their calculations in the scale of sixty before the Indian methods began to exert their influence.

Almost every *zīj* starts by giving arithmetical rules stated rhetorically. Mainly these comprise rules of multiplication and division that may be expressed as

$$60^m \cdot 60^n = 60^{m+n}$$
$$60^m \div 60^n = 60^{m-n}$$

In books on finger reckoning and on Hindu arithmetic we find statements describing sexagesimal algorisms; these seem to bear Hindu influence. One is thus left with the impression that before they learned Hindu arithmetic, astronomers, like finger reckoners, made their calculations mentally, probably depending on finger reckoning as well as sexagesimal calculation. It should be stressed, however, that in Islam the system of finger reckoning used fractions to the scale of sixty and was very rich: it comprised algebra, mensuration, and the elements of trigonometry. It must therefore have been elaborated by the more gifted mathematicians. But whatever the case may be, there remains the question of whether there were any special manipulational methods devised for astronomical calculations.

Abu'l-Wafā' was more of an astronomer than an arithmetician; but since his arithmetic was expressly written for state officials, he may have deliberately avoided bothering his readers with material that he considered too advanced for them.

The importance of Kushyār's *Uṣūl Ḥisāb al-Hind* lies in his having written it to introduce the Hindu methods into astronomical calculations. Abū Ḥanīfa al-Dīnawarī, a lawyer, wrote on arithmetic to introduce these methods into business. ʿAlī ibn Aḥmad al-Nasawī, known to have been Kushyār's student, commented sarcastically on these two works because Abū Ḥanīfa's was lengthy and Kushyār's was compact; he said that the former proved to be for astronomers and the latter for businessmen. But al-Nasawī copied Kushyār freely in his work and showed no better understanding of the Hindu system.

Kushyār's *Uṣūl* is in two sections supplemented by a chapter on the cube root. The first section gives the bare rudiments of Hindu numeration and algorisms on the dust board, and in at least one place we find that Kushyār was not well informed on the new practice. Wishing to solve 5625 — 839, like any other computer he puts the array

$$\begin{array}{r}5625\\839\end{array}$$

on the dust board.

Other Arabic writers on Hindu artihmetic (excluding al-Nasawī) would start by subtracting 8 from 6; as this was not possible, they "borrowed" 1 from the 5, broke it down to 10, and thus subtracted 8 from 16. This process of borrowing and breaking down was not known to Kushyār (and his student). He subtracted 8 from 56 complete, obtaining 48. His next step was to subtract 3 from 82.

The second section of Kushyār's *Uṣūl* presents calculations in the scale of sixty, using Hindu numerals and the dust board. Here it seems, although it cannot always be proved decisively, that the author is dealing with concepts and manipulational schemes alien to both the Hindu system and what is found in books on finger reckoning. These must be schemes in the scale of sixty practiced by astronomers.

Briefly Kushyār states that the scale of sixty is indispensable because it is precise. In the absence of decimal fractions, which were added to the Hindu system by Muslim arithmeticians, Kushyār's statement is true. He then shows how to convert decimal integers to this scale; this is necessary for the methods he presents. His multiplication, division, and extraction of roots require the use of a multiplication table extending from 1×1 to 60×60, expressed in alphabetic numeration and in the sexagesimal scale. With this background he presents homogeneous methods of addition, subtraction, multiplication, division, and extraction of the square root (and, in the supplementary section, the cube root).

Kushyār certainly worked on the dust board and resorts to erasure and shifting of numbers from place to place; these were the distinguishing feature of the Hindu methods as they came to the Arabic world. But apart from these, he worked out $49°36' \div 12°25'$ as an endless operation of division in decimal fractions would be worked out. He obtained the answer 3°59′ and had 8′25″ left as a remainder. He added that the division could be continued if more precise results are desired.

The same applied to roots. Kushyār found the square root of 45°36′, obtaining the result 6°45′9″59‴, with a remainder. He added that the process could be continued to find the answer to higher degrees of precision.

From this treatment there is only one step to decimal fractions. Al-Uqlīdisī made that step in the tenth century, and al-Kāshī made it again in the fifteenth; but it was left to Stevin to establish it in his *La Disme* (1585).

BIBLIOGRAPHY

I. ORIGINAL WORKS. Kushyār's writings are *al-Bāligh* (Berlin 5751); *al-Jāmiʿ* (Leiden, Or. 1054), sec. 1 also available as Cairo MS 213; *Uṣūl Ḥisāb al-Hind* (Aya Sofya 4857), in M. Levey and M. Petruck, *Principles of Hindu Reckoning* (Madison, Wis., 1965), pp. 55–83, also edited with ample comparative notes in Arabic by A. S. Saidan in *Majallat Maʿhad al-Makhṭūṭāt* (Cairo, 1967); *al-Lāmiʿ fī amthilat al-zīj al-Jāmiʿ* (Fātiḥ 3418); *Kitāb al-Asṭurlāb wa kayfiyyat ʿamalihī wa iʿtibārihī* (Paris, BN 3487; Cairo, MS 138); *Tajrīd Uṣūl Tarkīb al-Juyūb* (Jarullah 1499/3); *al-Madkhal* or *al-Mujmal fī ṣināʿat ahkām al-Nujūm* (Berlin 5885; Escorial 972; British Museum 415); and *Risāla fī al-abʿād wa al-Ajrām* (Patna).

II. SECONDARY LITERATURE. See C. Brockelmann, *Geschichte der arabischen Literatur* (Leiden, 1943); E. S. Kennedy, "A Survey of Islamic Astronomical Tables," in *Transactions of the American Philosophical Society*, n.s. **46**, pt. 2 (1966); M. Krause, "Stambuler Handschriften islamischer Mathematiker," in *Quellen und Studien zur Geschichte der Mathematik, Astronomie und Physik*, Abt. B, Studien, **3** (1936). 472–473; C. A. Nallino, *Al-Battani sive Albattanii opus astronomicum*, 3 vols. (Milan, 1899–1907); C. Schoy, "Beiträge zur arabischen Trigonometrie," in *Isis*, **5** (1923), 364–399; and H. Suter, "Die Mathematiker und Astronomen der Araber und ihre Werke," in *Abhandlungen zur Geschichte der Mathematik*, **10** (1900); and "Nachträge und Berichtigungen zu 'Die Mathematiker und Astronomen . . .,' " *ibid.*, **14**, no. 2 (1902).

A. S. SAIDAN

KÜTZING, FRIEDRICH TRAUGOTT (*b.* Ritteburg, near Artern, Saxony, Germany, 8 December 1807; *d.* Nordhausen, Germany, 9 September 1893), *botany*.

Kützing was the eighth of the fourteen children of Johann Daniel Christoph Kützing, a miller, and the former Magdalene Zopf. He attended the village school of Ritteburg and from 1822 to 1832 served as apprentice and later as assistant to apothecaries in Artern, Aschersleben, Magdeburg, Schleusingen, and Tennstedt. During this period he studied chemistry and botany as well as pharmacology. He started a herbarium while at Aschersleben and began to study algae while at Schleusingen (1830–1831). At this time Kützing decided to pursue a career in pure science rather than in pharmacy and began to study subjects, such as Latin and Greek, that would qualify him for admission to a university. At Schleusingen he also started to prepare specimens for his *Algarum aquae dulcis Germanicarum*, which was issued in sixteen decades between 1833 and 1836. From May 1832 to October 1833 he attended the University of Halle, where the professor of pharmacology had given him an assistantship. Well-known biologists who were professors at Halle during that period were the zoologist Christian Ludwig Nitzsch, the botanist Kurt Sprengel, and the latter's successor, Dietrich Franz Leonhard von Schlechtendal. Kützing, who had been studying diatoms for about two years, was encouraged by Nitzsch, a diatomist, to continue investigating this then little-known group. A monograph resulted in 1833 in which Kützing pointed out the differences between diatoms and desmids, groups hitherto allied.

In the autumn of 1833 Kützing became an apothecary's helper in Eilenburg, where he remained through 1834. During this period he made the significant discovery that the walls of diatoms are silicified. He asked Alexander von Humboldt in Berlin to communicate his discovery to the Royal Prussian Academy and sent a manuscript to Poggendorff for publication in *Annalen der Physik und Chemie*. Although the paper never appeared (see *Die kieselschaligen Bacillarien oder Diatomeen*, pages 7–10), in 1835 Christian Gottfried Ehrenberg referred to this discovery.

The Royal Prussian Academy made a special award to Kützing, a grant of 200 talers to help him make a study trip to the Adriatic and Mediterranean seas.

The journey, financed in part by the sale of his *Algarum* decades, lasted from February to September 1835. In Vienna, where he met Endlicher, Diesing, and many other distinguished scientists, he demonstrated the siliceous nature of the diatom frustule.

In October 1835, Kützing was appointed teacher of chemistry and natural history (later he also taught geography) in the secondary school of Nordhausen, where he spent the remainder of his life, retiring in 1883. In October 1837 he married Maria Elisabeth Brose of Aschersleben; they had six children. In November 1837 he received the Ph.D. from the University of Marburg.

In the summer of 1839 Kützing made a study trip to the North Sea. In Hamburg he visited the senator Nicolaus Binder and the apothecary Otto Wilhelm Sonder, both of whom owned large, worldwide collections of seaweeds that were subsequently made available to Kützing for study.

In 1843 Kützing's first great algal work, *Phycologia generalis*, appeared. The 200 talers Kützing had received from the Royal Prussian Academy to help publish this work covered only a twelfth of the cost of the eighty plates, so Kützing decided to learn the art of engraving. The excellent results of his work and the knowledge that henceforth he would not be dependent upon favors from the Academy gave Kützing great satisfaction. As he remarked in his autobiography (p. 241), he was a "self-made man" and had practiced "help yourself" since the days of his youth. Especial attention was paid in this work to the physiology, anatomy, and development of representatives of all algal groups, as then recognized, except the diatoms, which formed the subject of a separate monograph that appeared in 1844. In the *Phycologia generalis* Kützing named the red and blue algal pigments phycoerythrin and phycocyanin, respectively, and also announced the discovery of starch granules (Floridian starch), the storage product in red algae. In recognition of the importance of his forthcoming *Phycologia*, in 1842 Kützing was made a royal professor.

Other major works by Kützing were *Die kieselschaligen Bacillarien oder Diatomeen* (1844), *Phycologia germanica* (1845), *Species algarum* (1849), and *Tabulae phycologiae* (1845–1871). The *Bacillarien*, which contained illustrations of 700 species, engraved by Kützing himself, received worldwide recognition. According to Kützing, this book made more friends for him than any of his others (it was reprinted in 1865). In the 1845 work Kützing proposed the class name Chlorophyceae, by which the green algae are now known. In the *Species algarum* diagnoses were given of more than 6,000 species of algae from all parts of the world. In the *Tabulae*, each volume of which contains 100 plates engraved by Kützing himself, 4,407 species and forms, exclusive of diatoms and desmids, were illustrated. Even today this work still continues to be the best reference on the habit of many species of algae.

In addition to his contributions to phycology, Kützing published articles and books on various other subjects, botanical and nonbotanical, including the discovery in 1837 that fermentation is a physiological process brought about by organisms; but his fame derives from his work on the algae. If it is remembered that Kützing never held a university professorship (his professional enemies prevented that) that would have allowed him some free time for research, and that he had to pursue his studies during his spare time from work in pharmacies or secondary school teaching his remarkable achievements become all the more a source of astonishment and admiration.

Kützing's diatom collection is now in the British Museum (Natural History) and the Natural History Museum of Antwerp; his algal collection is in the Rijksherbarium in Leiden.

BIBLIOGRAPHY

Kützing's bibliography is appended to his autobiography, *Friedrich Traugott Kützing 1807–1893 Aufzeichnungen und Erinnerungen*, R. H. Walther Müller und Rudolph Zaunick, eds. (Leipzig, 1960), with 2 portraits. His writings include *Algarum aquae dulcis Germanicarum*, 16 decades (Halle, 1833–1836); *Phycologia generalis* (Leipzig, 1843); *Die kieselschaligen Bacillarien oder Diatomeen* (Nordhausen, 1844); *Phycologia Germanica* (Nordhausen, 1845); *Tabulae phycologiae*, 19 vols. and index (Nordhausen, 1845–1871); and *Species algarum* (Leipzig, 1849).

Obituaries are W. Zopf, "Friedrich Traugott Kützing," in *Leopoldina*, **30** (1894), 145–151; and the unsigned "Friedrich Traugott Kützing," in *Hedwigia*, **32** (1893), 329–333.

GEORGE F. PAPENFUSS

KYLIN, JOHANN HARALD (*b*. Ornunga, Älvsborg, Sweden, 5 February 1879; *d*. Lund, Sweden, 16 December 1949), *botany*.

Kylin was the oldest son of five sons and five daughters of Nils Henrik Olsson, a farmer, and the former Johanna Augusta Johannesdotter. Since Olsson is a very common name in Sweden, the children adopted the name Kylin, derived from the name of the family farm (information supplied by Kylin's son, Dr. Anders O. Kylin of the University of Stockholm). In 1898 Kylin graduated from the Gymnasium in

Göteborg and entered the University of Uppsala, receiving the Ph.D. in 1907 under Frans Reinhold Kjellman. Following graduation Kylin remained as docent at Uppsala for thirteen years, during part of that time teaching in the Uppsala high school and the Uppsala teachers' college. In 1912–1913 he was an investigator in the laboratory of Wilhelm Pfeffer at Leipzig. In 1920 Kylin was appointed professor of botany (for anatomy and physiology) at the University of Lund, retiring in 1944. In 1924 he married Elsa Sofia Jacobowsky; they had a son and a daughter.

Kylin's doctoral dissertation was a study of the marine flora of the west coast of Sweden. In later years he and his family usually spent the summer vacations at Kristineberg on that coast, conducting research in the marine biological laboratory located there. Over the years he and his students published many papers dealing with the taxonomy, morphology, biochemistry, ecology, and physiology of the algae of this coast. Three of his last four major works constituted a taxonomic revision of the red (1944), brown (1947), and green (1949) algae of the Swedish west coast. Kylin also contributed significantly to knowledge of the marine algae of various other parts of the world: the west coast of Norway, which he visited in the summer of 1908 (1910); the red algae of the sub-Antarctic and Arctic (1919, with Carl Skottsberg); the red algae in the vicinity of Friday Harbor, Washington (1925); the Delesseriaceae (red algae) of New Zealand (1929); and the red algae of South Africa (1938) and of California (1941).

Kylin also studied the biochemistry and physiology of the algae. More than thirty papers dealing with pigments, storage products, pH relations, osmotic relations, and chemical composition of cell walls of various algae appeared between 1910 and 1946.

First and foremost a morphologist, Kylin, as he unraveled the step-by-step development of the vegetative and reproductive structures of the algae and details in their life histories, unearthed so much that was wrong with their taxonomy that he always was deeply involved in their systematics. In 1917 he revised the classification of the brown algae, basing his system largely on developmental and nuclear cycles. He recognized five orders. Utilizing the information that had accumulated since 1917, Kylin in 1933 erected a new system of classification of these algae, dividing them into three classes and twelve orders. This system has undergone much revision, but it served for a long time as a stimulating basis for research. In 1940 Kylin published an excellent taxonomic monograph on the brown algal order Chordariales.

Kylin's greatest contributions to phycology came from his outstanding morphological and systematic studies of the class Florideophycidae, which includes the bulk of the red algae. Between 1914 and 1923 he published several papers on the developmental morphology of six genera of red algae. In 1923 his very significant monograph on the morphology of twenty-five genera appeared. In this paper Kylin elaborated upon an earlier system of classification of the Florideophycidae (Friedrich Schmitz, 1883) based upon embryological details related to the mode of initiation and ontogeny of the generation developing from the fertilized egg, the carposporophyte.

Kylin saw that the ontogeny of practically every genus of red algae required a thorough investigation. To obtain properly prepared material he had to visit other parts of the world. In the summer of 1922, while his paper of 1923 was in press, he visited the United States, collecting at the Monterey Peninsula, La Jolla, Friday Harbor, and Woods Hole. In the summer of 1923 he visited the Isle of Man and Plymouth. In 1924 he returned to Friday Harbor to teach the laboratory's summer course on the algae. In the summers of 1927 and 1928 he collected at Roscoff, Guéthary, and Banyuls in France; in 1929 he collected at Naples.

Using material obtained at these various places and that in the Agardh Herbarium at Lund, the most important algal herbarium in the world, Kylin published a series of very important morphological and taxonomic monographs, culminating in one on the order Gigartinales in 1932. Through these studies he immensely advanced knowledge of the morphology and interrelationships of members of the large and diversified phylum Rhodophyta. Despite certain shortcomings, Kylin's system as outlined in his monograph of 1932 on the Gigartinales presents a much more natural arrangement of these algae than had previously been possible. His last great work, *Die Gattungen der Rhodophyceen*, which appeared posthumously in 1956 (Kylin's widow, herself a biologist, saw it through the press), is the standard reference on the red algae.

Kylin was elected to membership in the Royal Swedish Academy of Sciences and the Royal Danish Academy of Sciences, and was corresponding member of the Botanical Society of America and Societas pro Fauna et Flora Fennica. The Kungliga Fysiografiska Sällskapet of Lund awarded him its gold Linné Medal.

BIBLIOGRAPHY

Kylin's bibliography was compiled by John Tuneld, "Bibliografi över Harald Kylins Tryckta Skrifter," in *Botaniska Notiser* (1950), 106–116. His writings include "Studien über die Entwicklungsgeschichte der Florideen," in *Kungliga Svenska vetenskapsakademiens handlingar*, **63**, no. 11 (1923); "Entwicklungsgeschichtliche Florideenstudien," in *Acta Universitatis lundenis*, n.s., Afd. 2, **24**, no. 4 (1928); "Über die Entwicklungsgeschichte der Florideen," *ibid.*, **26**, no. 6 (1930); "Die Florideenordnung Gigartinales," *ibid.*, **28**, no. 8 (1932); "Über die Entwicklungsgeschichte der Phaeophyceen," *ibid.*, **29**, no. 9 (1933); "Anatomie der Rhodophyceen," in K. Linsbauer, *Handbuch der Pflanzenanatomie*, VI, pt. 2 (Berlin, 1937); and *Die Gattungen der Rhodophyceen* (Lund, 1956).

Obituaries are C. Skottsberg, in *Bihang till Göteborgs K. Vetenskaps- och vitterhets-samhälles Handlingar*, **69** (1950), 97–103; and Svante Suneson, in *Botaniska Notiser* (1950), 94–105.

GEORGE F. PAPENFUSS

LA BROSSE, GUY DE (*b.* Paris, France, *ca.* 1586; *d.* Paris, 1641), *botany, medicine, chemistry.*

The founder and first director (intendant) of the Royal Botanical Garden in Paris, Guy de La Brosse, was born during the reign of Henry III, probably in Paris, not in Rouen as is often claimed. He died—of Epicurean overindulgence, if his archenemy, Guy Patin, can be trusted—at his house in the Jardin des Plantes during the night of 30–31 August 1641.[1]

The La Brosse name was by no means uncommon, and it is not easy to sort out and identify his ancestors and relatives. But the La Brosse mentioned by the poet, historian, and chemist Jacques Gohory (*d.* 1576) as a learned "mathématicien du Roy," possessed of a fine botanical garden, may have been Guy's grandfather.[2] About the father we are on firmer ground: Isaïe de Vireneau, sieur de La Brosse, is described by his son as a respected physician and a fine medical botanist ("très bon simpliste").[3] Isaïe, who died about 1610, was long survived by Guy's stepmother, Judith de la Rivoire. These Christian names suggest that the family was originally Protestant,[4] although Guy was at least a nominal Catholic; he built a chapel in the Jardin des Plantes, where Mass was said on feast days and where he was eventually buried.

In his youth, Guy may have been a soldier;[5] in any case, his major book testifies to extensive travels in France. Yet by 1614 he had settled in Paris, had begun the study of chemistry, and was botanizing on Mont Valérien. Although we are ignorant about his medical training,[6] we know that by 1619 he was physician to Henry II de Bourbon, Prince de Condé, and that in 1626 he had become one of the physicians in ordinary to Louis XIII. Like many of the doctors of the royal household, a number of whom were products of the medical university of Montpellier, Guy was highly critical of the Paris medical faculty: its conservatism, its worship of Galen, its addiction to venesection, its relative neglect of botany and anatomy, and its distrust of the newly emerging, and highly controversial, field of medical chemistry.[7] By 1616 Guy had begun his efforts to secure the establishment in Paris of a royal botanical garden, not merely for the study of medicinal herbs, but where chemistry would be taught as a handmaiden to medicine.

Several small botanical gardens had been established in Paris by private persons—mostly physicians and apothecaries—during the sixteenth century. In Guy's youth the only Parisian garden of importance was the modest one of Jean Robin.[8] Guy hoped for something larger and more elaborate, and when he made his first overtures about 1616 to Louis XIII, through the good offices of Jean Hérouard, the chief body physician of the king,[9] his models were the botanical garden of Montpellier, established by Henry IV but recently fallen into ruin, and those of Padua and Leiden. What Guy envisaged was a teaching and research institution, designed to raise medical standards and advance the art. Besides a collection of living plants, Guy planned a herbarium of dried specimens and a *droguier*, or laboratory, where students could learn distillation and the preparation of herbal remedies. A royal edict of 6 January 1626 authorized the establishment, in one of the suburbs of Paris, of such a royal garden of medicinal plants, and designated Jean Hérouard as superintendent. Six months later, when the edict was registered by the Parlement, La Brosse received his appointment as intendant.

In the next two years, to assure support and financing for the project, Guy published a series of pleas to government officials, among them Richelieu, and brought out his *Advis défensif*, defending his plan and severely criticizing the Paris medical faculty. Most of these pamphlets were reprinted in 1628 in his major book, *De la nature, vertu et utilité des plantes.*

For several years there was no sign of progress; in the meanwhile Hérouard had died and Charles Bouvard succeeded him as superintendent of the proposed garden. At last, on 21 February 1633, there was purchased in the king's name, for the sum of 67,000 livres, a house and grounds in the Faubourg Saint-Victor. A year later Guy was able to show the king a plan of the new garden, where 1,500 species of plants were already growing. The act which detailed the organization and staffing of the Jardin des Plantes was a further royal edict of 15 May 1635. This specified that Guy, aided by a *sous-démonstrateur*, was to teach the "exterior" of plants, that is, their identification and taxonomic characteristics. Also authorized was the appointment of three demonstrators, to teach the "interior" of plants, in other words their pharmaceutical properties.

Meanwhile, to supervise the work, Guy moved into

LA BROSSE

the building that was later to serve as the zoological galleries. The ground was cleared and leveled, and garden plots and parterres were laid out. Many plants were provided by Vespasien Robin, the heir to his father's garden as well as to his title of *arboriste du Roi*, whom Guy appointed in 1635 as *sous-démonstrateur.*[10] Through active correspondence with botanists abroad, Guy obtained seeds and plants from foreign lands, notably from the East Indies and America. In 1636, when he published his *Description du Jardin royal des plantes médecinales* with two plans of the garden, he was able to list some 1,800 species and varieties under cultivation. Four years later, in 1640, came the formal opening of the institution, marked by the publication of a pamphlet of thirty-eight pages describing the foundation of the garden, comparing it with those of Padua, Pisa, Leiden, and Montpellier, printing some introductory remarks about the study of botany, and, finally, regulations for the students. The following year, the year of his death, Guy published a second catalogue of the plants growing in the garden, with a handsome perspective plan drawn and finely engraved by Abraham Bosse. Its appearance could hardly have changed much when the young John Evelyn recorded in his travel diary his visit to Guy's establishment in February 1644:

> The 8th I tooke Coach and went to see the famous Garden Royale, which is an Enclosure wall'd in, consisting of all sorts of varietys of grounds, for the planting & culture of Medical simples. It is certainely for all advantages very well chosen, having within it both hills, meadows, growne Wood, & Upland, both artificial and naturall; nor is the furniture inferiour, being very richly stord with exotic plants: has a fayre fountaine in the middle of the Parterre, a very noble house, Chapel, Laboratory, Orangerie & other accommodations for the Praesident, who is allwayes one of the Kings chiefe Physitians.[11]

We know something of Guy's friendships and his ties with the intellectual and free-thinking circles of his day. Descartes knew about him, and mentions in his letters Guy's refutation of the *Géostatique* of Jean Beaugrand.[12] Guy was at least an occasional visitor to the cell of the famous scientist and Minorite friar, Father Mersenne.[13] He was a familiar, too, of other learned circles like the "Cabinet" of the brothers Dupuy, and the "Tétrade" of Élie Diodati, Gabriel Naudé, Pierre Gassendi, and La Mothe le Vayer, who, while keeping up a discreet front as an early member of the French Academy, set forth his free-thinking views under the pseudonym of Orasius Tubero. Guy also formed part of the pleasure-loving group around François Lullier, financier and *maître des comptes*, and was perhaps closest of all to the libertine poet Théophile de Viau and his intimate and pupil Jacques Vallée, sieur des Barreaux, for Guy was Théophile's personal physician as well as friend. It was from Guy that the poet in his last illness received the narcotic pill that ended his life.[14]

Guy's earliest book was a short monograph on the causes of the plague, the *Traicté de la peste* (1623). In the following year (1624) appeared his *Traicté contre la mesdisance*, a work in which he defended various persons—among them, and perhaps notably, Théophile de Viau—who had been unjustly persecuted for their opinions.

In his work on the plague, Guy already showed his affinity with the Paracelsian doctors and his rejection of traditional medical theory. There are, he remarked, two different opinions about the efficient cause of the plague: (1) that it depends upon the active and passive qualities of the elements, a view held by all those who attempt to explain nature by the manifest qualities of things; and (2) that the cause is hidden, proceeding from agencies beyond the reach of our senses. Men in the first category follow Galen in making putrefaction the "principal and unique cause" of the plague. But Guy, opposing a philosophy "that knows the motions and changes of nature only through books," clearly preferred the second alternative, urging that the cause of the plague "is a venomous and contagious substance," which in turn is the cause of putrefaction. As Allen Debus has pointed out, Paracelsian doctors sought the causes of disease less in internal imbalances of fluids than in external factors.[15]

The major concern, however, of Guy de La Brosse was with medical botany, a field to which he was doubtless introduced by his father. Guy's *De la nature des plantes* is not only a defense of his project for creating a center for the study of medical botany, but is also a theoretical work about plants in general. In it he raises questions that would be meaningful today—about the generation, growth, and nutrition of plants—as well as asking whether plants have souls, a subject to which he devotes considerable space, discussing also the influence of the stars, yet criticizing the doctrine of signatures. His belief in the essential unity of plant and animal life led him to catalogue their similarities: growth is observed in both; motion is not peculiar to animals, for indeed some animals are motionless or sessile; both plants and animals suffer disease; both animals and plants hibernate and plants even sleep. Plants, he would have us believe, seem to have more vital force than animals, yet they are readily fatigued by the process of nutrition and by tempestuous weather. Extending this analogy further, he was convinced that plants, like animals, must differ in sex (the vernal rise of sap, he argued, "testifies to their amorous desires"); and he urged that an effort be made to examine plants closely to discover distinctive sexual features.

As to whether plants, like animals, display feeling and sensation, La Brosse disagreed with Aristotle and returned to, and cited, the earlier views of Empedocles and Anaxagoras. Although sense organs, he remarked, have not been observed in plants, several species—notably *Mimosa pudica*, the sensitive plant, which he claimed to have been the first to introduce into France—markedly display the quality of sensibility.[16] What the soul of the plant is, Guy did not pretend to know, although he could note its operations. The individualizing agent that he adduced as the cause of specific differences, what others would call the plant's "form," he called the "Artisan" or "esprit artiste."

Perhaps Guy's most interesting suggestion concerned plant nutrition. Plants, like animals, derive their nutriment not only from solid food (*viande*) drawn from the earth, and from aqueous liquid (*breuvage*) from water, but also from air. Air is as necessary to nourish and sustain plants as it is for the life of animals. Deprived of air, plants die; to be sure, they have no lungs, but in this they resemble insects, which nevertheless need air to live. It is not necessary to have lungs to draw in air; it suffices to be supplied with pores. If plants need earth, it is for the nitrous and saline juices it contains ("*la terre sans sels est inutile à la génération*"); manure is nothing but the salt of the urine of animals. Water by itself is not a nutrient (*pace* Van Helmont), but serves chiefly as the vehicle for the salts and the manna. Guy suggested an experiment to show that "pure" earth and distilled water cannot sustain the growth of plants: rich earth is leached with warm water and put into a large glass vessel; if seeds are then planted and watered with distilled water, they may sprout, but they will not grow. Similarly, it is for the *esprit*—the dew and the manna contained in it—that plants need air to live. That plants seek air, Guy attempted to prove by pointing to a shrub which, growing close to a wall or otherwise sheltered, sends its branches toward the open air. This was, of course, a misreading of the phenomenon of phototropism; such plants were seeking sunlight rather than air. Nevertheless, a century before Stephen Hales, although on wholly inadequate evidence, La Brosse argued for the nutritive role of air in plants.[17]

Guy's interest in chemistry was keen; the third of the five books into which *De la nature des plantes* is divided is devoted to the subject; Guy himself described it as "un traicté général de la Chimie." For Guy, chemistry was an important adjunct both to medicine and botany. It is through chemistry, rather than by ordinary dissection, that one learns the causes of the virtues of plants. Fire if guided by an experienced hand produces marvels, for it has the ability to disclose those things that are hidden from the senses.[18]

The fundamental assumption of the chemist is that every body can be reduced to those entities out of which it is formed; only when we have reduced substances to their principles and elements can we truly understand them. All natural compound bodies, Guy tells us, can be reduced into five simple bodies of different natures: into three principles—salt, sulfur, and mercury, the *tria prima* of the Paracelsians—and into two elements—water and earth.[19] Neither air nor fire (as in the old doctrine of the four elements) should be thought of as an element or principle.

Guy devotes an entire chapter to explaining why the chemist refuses to include air as one of the elements. This may seem odd, he admits, since air serves as an excellent and necessary food for man, for other animals, and indeed for plants, "which cannot live without respiring it." But the chemist would reply that when compound substances (*mixtes*) are dissected by fire, no air appears. Air, moreover, is not an element or simple body, but ought better to be called *chaos*, because of the great number of substances that it contains, and of which it is composed; atoms of earth, the vapor of water, and the three principles subtilized make up "that mixture of fine, subtile, and diaphanous substances that we respire." Air is the "magazine" of all the sensible substances which evaporate and are subtilized. Chemistry concerns itself only with sensible phenomena; it is from the senses that the chemist learns, Guy claims, that all compound bodies "contain and are made of Salt, Sulfur and Mercury," and that water and earth occur in the chemical dissection of all substances. Water and earth, however, are not to be considered principles, for without the capacity to produce the seeds which account for the specific forms and virtues of things, they are to be thought of as universal matrices, wombs, or "generous receptacles" found in all bodies "not as contained in them, but as containing them." Chemical change, in sum, is the result of the action of two agents: "the Form, or as we call it, the Artisan," and fire, "the universal instrument" or the "Great Artist," which in turn acts in some mysterious way upon the three seminal principles.

Even if Guy had not mentioned Paracelsus and his disciples in the *De la nature des plantes*, the influence of the Swiss doctor would be quite apparent. In his use of the word *chaos*—he spelled it "cahos"—to

LA BROSSE

describe aerial matter, and in his references to "dew" and "manna," Guy clearly echoed the Paracelsians. His "artisans," although they foreshadow the "plastic natures" imagined later in the century by Ralph Cudworth and John Ray, are the Paracelsian "Archei" under another name. Guy admired Paracelsus, not only as the enemy of medical bibliolatry and the first to dramatize an opposition to Aristotle and Galen, but because—as a revolutionary in the study of nature—he stressed experience and experiment, and because of his advocacy of chemistry as a key to nature and, through a knowledge of nature, to medical reform. The chemical doctrines of Paracelsus appealed strongly to Guy, but even in chemistry he cherished his independence and refused to follow blindly either the man "to whom first place is given in this excellent art" or Severinus, the best of the followers of Paracelsus. Rather than accept what they wrote, Guy insisted, "I have rather chosen to delve into the bowels of Nature," to test their assertions. "Using my hands," he continued, "I found that many of them wrote falsely; that even Paracelsus, at least if all the books bearing his name are his, was not always trustworthy . . . and that all the others did the same or even worse."[20]

Guy de La Brosse, it should be evident, shared a number of the attitudes and preconceptions we associate with the new learning of the seventeenth century: a distrust of authority, especially that of Aristotle; a preference, if one must choose, for the views of pre-Socratic philosophers; and a belief in the capacity of the natural sciences, guided by a critical use of human reason and a respect for experience, to move steadily forward. Pintard sees in one of Guy's basic doctrines—his trust in human reason—an anticipation of Descartes.[21] Like Descartes, Guy believed error to result not from some innate weakness of the human intellect but from the defective way the mind is used. If man can overcome in his thinking the influence of prejudice and the tyranny of "opinion," he may discover truth, "cette fille du temps." Like Descartes, and thirteen years before him, Guy announced his faith in that faculty of the human mind which, unhampered, allows men to distinguish truth from falsehood. This faculty—Descartes was to call it "le bon sens"—is found in all men and all climates; it works for the Parisian as well as for "the Indian, the Moor, the Chinese, the Jew, the Christian, the Mohammedan, even the Deist and the Atheist."

But the basis for the judgment of reason can only be experience, the real "maistresse des choses," and the only true foundation of the sciences. Here in his reiterated empiricism and eclecticism he diverges from Descartes and more closely resembles his friend Gassendi and Francis Bacon.

Guy's attitude toward medicine and science is neatly summed up in the handsome frontispiece of *De la nature des plantes*.[22] Four symmetrically arranged shields contain the portraits of Hippocrates, Dioscorides, Theophrastus, and Paracelsus, each accompanied by an appropriate motto. For Theophrastus, it is "Medicine is useless without plants." For Paracelsus, it is "Each thing has its heaven and its stars." Indeed the mottos of Hippocrates and Dioscorides pretty well sum up Guy's empiricist position: for Hippocrates, it is "From effects to causes"; and for Dioscorides, it is "From experience to knowledge." Good doctrine for a man who could write, "It is difficult to have conceptions of things which have not entered the understanding through the senses." In this and in other matters, Guy may have echoed the dictum of Aristotle, yet Galen and Aristotle are conspicuously absent from his frontispiece. And the reason is evident: centered at the top of the page is a radiant sun, and below it the legend, "Truth, not authority." At the bottom is Guy's own device, "De bien en mieux," which well epitomizes his faith in scientific progress.

Guy de La Brosse was in a number of respects a confused child of his time, echoing its aspirations and its intellectual discontents. His book—an odd mixture of the antiquated, the perverse, and the novel—cannot be said to have exerted a marked influence on scientific thought. The book was rarely cited.[23] Indeed Guy himself was largely forgotten, in a truly physical sense, for some two and a half centuries. In 1797, when the chapel he had built adjoining the main building of what had become the Muséum d'histoire naturelle was demolished to enlarge the zoological galleries (to accommodate, we must suppose, the collections of Cuvier and Lamarck), workers came upon the crypt containing the coffin of La Brosse, easily identified by a crude inscription written on the wall by his niece. The coffin was unceremoniously stored in a convenient basement; and if there were plans for a suitable reburial, they were long deferred. It was not until 1893, nearly a century later, that Guy was reinterred with seemly honors.[24]

NOTES

1. *Lettres de Gui Patin*, J. H. Reveillé-Parise, ed., 3 vols., (Paris, 1846), I, 81–82. See also Hamy, 1897, p. 1; and 1900, pp. 1–3. Patin's enmity was as an impassioned defender of the Paris Faculty of Medicine.
2. In his *Instruction sur l'herbe petum* (Paris, 1572), Gohory mentions a certain La Brosse "mathématicien du Roy"

and his "beau jardin garny d'une infinité de simples rares et de fleurs esquises" from which he obtained tobacco leaf for his experiments. See Hamy, 1899, pp. 4–5.

3. Guy de La Brosse, *De la nature des plantes*, p. 767.
4. See Hamy, 1900, p. 2. Cornelis de Waard (*Correspondance du P. Marin Mersenne*, V, 195) calls Guy a Calvinist.
5. Albrecht von Haller, without supporting evidence, describes Guy as "ex milite botanicus et medicus," in *Bibliotheca botanica*, I (1771), 440. *Cf.* Pintard, *Le libertinage érudit*, II, 605.
6. Hilarion de la Coste, in Tamisey de Larroque, ed., *Lettres écrites de Paris à Peiresc* (Paris, 1892), p. 59, mentions, as a visitor to Père Mersenne, a physician named La Brosse whom he describes as a "docteur de la Faculté de Montpellier." But there is no trace of Guy in the records of that medical university. Enemies called Guy an empiric, and doubted that he had ever received a medical degree.
7. See his "Advis défensif," in *De la nature des plantes*, pp. 754–799.
8. See M. Bouvet, "Les anciens jardins botaniques médicaux de Paris," in *Revue de l'histoire de la pharmacie* (Dec. 1947), 221–228. The garden of Jean Robin (1550–1629) first occupied, as Bouvet tells us (p. 226), "the western point of the Cité, where the Place Dauphine is located today." Late sixteenth-century plans of Paris show such a garden on that spot, but it must have moved to another location after the building (*ca.* 1607) of the Place Dauphine. Where it was located after that time is hard to determine.
9. For this Montpellier doctor whose name also appears as Héroard or Erouard, see Hamy, *Bulletin du Muséum* (1896), 171–176. For Guy's letter to Hérouard, see Denise, *Bibliographie*, no. 13.
10. The Robins introduced into Europe the first acacia tree (*Robinia*), which Vespasien planted in Guy's garden in 1636, and which still survives. For the younger Robin see Hamy, in *Nouvelles archives du Muséum d'histoire naturelle*, **8** (1896), 1–24.
11. John Evelyn, *Diary*, E. S. de Beer, ed., 6 vols. (Oxford, 1955), II, 1202. By "Praesident" Evelyn means the intendant, i.e., Guy de La Brosse.
12. For the Beaugrand episode see *Oeuvres de Descartes*, "Correspondance," I; and *Correspondance du P. Marin Mersenne*, V. See also Adrien Baillet, *Vie de M. Descartes* (Paris, 1691), bk. IV, ch. 12. For Beaugrand see Robert Lenoble, *Mersenne ou la naissance du mécanisme* (Paris, 1943), p. 472; and Lynn Thorndike, *A History of Magic and Experimental Science*, VII (New York, 1958), 437–438.
13. Hilarion de la Coste's list of visitors to Mersenne is printed *in extenso* in *Correspondance du P. Marin Mersenne*, I, xxx–xlii.
14. For Guy's *libertin* associations see Pintard, *Le libertinage érudit*, pp. 193–208.
15. Allen G. Debus, "The Medical World of the Paracelsians," to appear in essays in honor of Joseph Needham.
16. Mersenne wrote Descartes in 1638 about "l'herbe sensitive," he had seen "chez Mr. de la Brosse." *Correspondance du P. Marin Mersenne*, VIII, (1963), 56–57. For the discovery of *Mimosa pudica* and others of the genus see Charles Webster, "The Recognition of Plant Sensitivity by English Botanists in the Seventeenth Century," in *Isis*, **57** (1966), 5–23.
17. Guy's confidence in the role of air in plant nutrition surely has its origin in Paracelsian speculations. For this background, consult Allen G. Debus, "The Paracelsian Aerial Niter," in *Isis*, **55** (1964), 43–61.
18. The lower level of the chief building of the Jardin des Plantes was to be the laboratory "pour les distillations"; see *De la nature des plantes* ("Epistre au Roy"), p. 699. Distillation in early medical chemistry is described by Robert Multhauf, "Significance of Distillation in Renaissance Medical Chemistry," in *Bulletin of the History of Medicine*, **30** (1956), 329–346.
19. The five-element theory, which dominated the speculations of seventeenth-century chemists, was first set forth by Joseph Duchesne, or Quercetanus, with whose work Guy de La Brosse was familiar. See R. Hooykaas, "Die Elementenlehre der Iatrochemiker," in *Janus*, **41** (1937), 1–28, and Allen G. Debus, *The English Paracelsians* (New York, 1966), p. 90.
20. *De la nature des plantes*, "Argument du troisiesme livre" (inserted between pp. 288 and 289), fol. 2 v°.
21. Pintard, *Le libertinage érudit*, p. 196.
22. This frontispiece was designed by the artist and engraver Michel l'Asne (or Lasne). See Denise, *Bibliographie*, no. 39.
23. It was nevertheless referred to by William Harvey's disciple George Ent, in his *Apologia pro circulatione sanguinis* (London, 1641).
24. "Translation et inhumation des restes de Guy de La Brosse et de Victor Jacquement faites au Muséum d'histoire naturelle, le 29 November 1893," in *Nouvelles archives du Muséum d'histoire naturelle*, 3rd ser., **6** (1894), iii–xvi. On this occasion the principal discourse was delivered by the director of the Muséum, Henri Milne-Edwards.

BIBLIOGRAPHY

I. Original Works. The published works of Guy de La Brosse are the following:

Traicté de la peste (Paris, 1623); *Traicté contre la mesdisance* (Paris, 1624); and his most important *De la nature, vertu et utilité des plantes* (Paris, 1628). Several of Guy's previously published but undated pamphlets concerned with the proposed Jardin des Plantes Médecinales are reprinted in *De la nature des plantes*. These are *À Monseigneur le très révérend et le très-illustre cardinal, Monseigneur le cardinal de Richelieu;* the letters *Au Roy*, *À Monseigneur le garde des Sceaux*, *À Monseigneur le Superintendant des Finances de France;* the *Advis défensif du Jardin Royal des plantes médecinales à Paris;* and the *Mémoire des plantes usagères et de leurs parties que l'on doit trouver à toutes occurrences soit récentes ou sèches, selon la saison, au Jardin Royal des plantes médecinales.*

Also printed is the royal edict of January 1626 authorizing the establishment of the garden. But the earliest of Guy's pamphlets concerning the garden, his letter *À Monsieur Erouard, premier médecin du Roy* (n.p., n.d., but written *ca.* 1616), was not among those reprinted.

The following pamphlets are posterior to 1628 but published before the opening of the garden: *À Monsieur Bouvard, conseiller du Roy en ses conseils et son premier médecin* (n.p., n.d.); *Advis pour le Jardin royal des plantes médecinales que le Roy veut establir à Paris. Présenté à Nosseigneurs du Parlement par Guy de La Brosse, médecin ordinaire du Roy et intendant dudit jardin* (Paris, 1631); *Pour parfaictement accomplir le dessein de la construction du Jardin royal, pour la culture des plantes médecinales* (n.p., n.d.); *À Monseigneur le Chancelier* (n.p., n.d.). After the garden came into being, Guy published his *Description du Jardin royal des plantes médecinales estably par le Roy Louis le Juste à Paris; contenant le catalogue des plantes qui y sont de présent cultivées* (Paris, 1636) with an overall plan of the garden (by Scalberge) and a plan of the four great flower beds.

With a single exception, Guy's later publications all

dealt with the development of the Jardin des Plantes. The exception is his *Éclaircissement d'une partie des paralogismes ou fautes contre les loix du raisonnement et de la démonstration, que Monsieur de Beaugrand a commis en sa pretendue Demonstration de la première partie de la quatriesme proposition de son livre intitulé Geostatique. Adressé au mesme Monsieur de Beaugrand* (Paris, 1637).

Perhaps the two most important of his publications concerning the new garden are the following: *L'ouverture du Jardin royal de Paris pour la démonstration des plantes médecinales, par Guy de La Brosse, conseiller et médecin ordinaire du Roy, intendant du Jardin et démonstrateur de ses plantes, suivant les ordres de M. Bouvard, surintendant* (Paris, 1640), which summarizes the history of the garden, compares it with those of Padua, Pisa, Leiden, and Montpellier, refers to the acclimatizing of the *Mimosa pudica*, and prints the regulations for the students; and his *Catalogue des plantes cultivées à présent au Jardin royal des plantes médecinales estably par Louis le Juste, à Paris. Ensemble le plan de ce jardin en perspective orizontale. Par Guy de La Brosse, médecin ordinaire du roy et intendant dudit jardin* (Paris, 1641).

II. SECONDARY LITERATURE. There is no book-length biography of Guy de La Brosse, and he has been largely neglected by modern historians of botany and almost totally so by historians of chemistry. There are short sketches (not always reliable) in N. F. J. Eloy, *Dictionnaire historique de la médecine*, 4 vols. (Mons, 1778), I, 456–457; E. Gurlt, A. Wernich, and August Hirsch, *Biographisches Lexicon der hervorragenden Arzte*, 2nd ed. by W. Haberling *et al.*, 5 vols. (1929–1934), I, 715; Albrecht von Haller, *Bibliotheca botanica*, 2 vols. (Zurich, 1771–1772), I, 440–441; Curt Sprengel, *Historia rei herbariae*, 2 vols. (Amsterdam, 1807–1808), II, 111–112.

The common error that makes Rouen the birthplace of Guy is repeated by F. Hoefer in the *Nouvelle biographie générale;* by Théodore Lebreton, *Biographie normande*, 3 vols. (Rouen, 1857–1861), II, 316; and by Jules Roger, *Les médecins normands* (Paris, 1890), 36–39.

A series of articles by E. T. Hamy, professor of anthropology at the Muséum National d'Histoire Naturelle, has clarified a number of points about Guy's life. See especially his "La famille de Guy de la Brosse," in *Bulletin du Muséum d'histoire naturelle*, **6** (1900), 13–16, and his "Quelques notes sur la mort et la succession de Guy de la Brosse," *ibid.*, **3** (1897), 152–154.

The only study of Guy's botanical theories is by Agnes Arber, "The Botanical Philosophy of Guy de la Brosse," in *Isis*, **1** (1913), 359–369. See also her *Herbals, Their Origin and Evolution*, new ed., rev. (Cambridge, 1938), 144–145, 250, 255. Miss Arber remarks that Guy was deeply influenced by Aristotelian thought, although he "inveighed against the authority of the classics."

For Guy's associations with the *libertins* see René Pintard, *La Mothe de Vayer, Gassendi, Guy Patin* (Paris, n.d.), 23, 79, 128; and his *Le libertinage érudit*, 2 vols.-in-1, continuously paginated (Paris, 1943), 195–200, 437–441, 605–606.

For Guy's comments on the Paracelsians, and his interest in chemistry, see Henry Guerlac, "Guy de La Brosse and the French Paracelsians," to appear in Allen G. Debus, ed., *Science, Medicine and Society in the Renaissance: Essays to Honor Walter Pagel.*

Essential for any study of the garden founded by Guy de La Brosse is Louis Denise, *Bibliographie historique & iconographique du Jardin des plantes* (Paris, 1903), where the early pamphlets of Guy are listed and briefly described. For a short seventeenth-century description of the newly established garden, see Claude de Varennes, *Le voyage de France* (Paris, 1639, and later eds.). An early historical study of the garden, from its origins to the death of Buffon (1788), is that of the famous botanist Antoine-Laurent Jussieu, whose "Notices historiques sur le Muséum d'histoire naturelle," appeared in the *Annales du Muséum*, from 1802 to 1808; the first of these articles, covering the establishment of the Jardin and its development to 1643, was published in *Annales*, **1** (1802), 1–14.

Other accounts are by Gotthelf Fischer von Waldheim, *Das Nationalmuseum der Naturgeschichte zu Paris*, 2 vols. (Frankfurt am Main, 1802–1803), I, 21–42; and J. P. F. Deleuze, *Histoire et description du Muséum royal d'histoire naturelle*, 2 vols. (Paris, 1823).

For special aspects of the early history of the Jardin, see E. T. Hamy, "Recherches sur les origines de l'enseignement de l'anatomie humaine et de l'anthropologie au Jardin des Plantes," in *Nouvelles archives du Muséum d'histoire naturelle*, 3rd ser., **7** (1895), 1–29; "Vespasien Robin, arboriste du Roy, premier sous-démonstrateur de botanique du Jardin royal des plantes (1635–1662)," *ibid.*, **8** (1896), 1–24; and "Jean Héroard, premier superintendant du Jardin royal des plantes médecinales (1626–1628)," in *Bulletin du Muséum d'histoire naturelle*, **2** (1896), 171–176. Worth consulting is Jean-Paul Contant, *L'enseignement de la chimie au Jardin royal des plantes de Paris* (Cahors, 1952).

For early botanical gardens, see M. Bouvet, "Les anciens jardins botaniques médicaux de Paris," in *Revue d'histoire de la pharmacie* (Dec. 1947), 221–228. E. T. Hamy has corrected a persistent error that the garden of Jacques Gohory was located on the site of the labyrinth of the Jardin des Plantes, and that the garden of the man who may have been Guy's grandfather was close by. See Hamy, "Un précurseur de Guy de la Brosse. Jacques Gohory et le Lycium philosophal de Saint-Marceau-les-Paris (1571–1576)," in *Nouvelles archives du Muséum*, 4th ser., **1** (1899), 1–26.

The errors which originated with Gobet's *Anciens minéralogistes du royaume de France* (Paris, 1779) have been repeated by F. Hoefer, in *Histoire de la chimie*, 2nd ed., 2 vols. (Paris, 1869), II, 102–103, and in his article on Gohory in *Nouvelle biographie générale.* Hoefer, in turn, is relied upon by J. R. Partington, *A History of Chemistry*, II (London–New York, 1961), 162–163.

Miss Rio Howard, who is completing a Cornell University doctoral diss. on Guy de La Brosse, has helped the author of this article to avoid a number of errors.

HENRY GUERLAC

LACAILLE, NICOLAS-LOUIS DE (*b.* Rumigny, near Rheims, France, 15 March 1713; *d.* Paris, France, 21 March 1762), *astronomy, geodesy.*

The Abbé Lacaille was an immensely industrious observational astronomer whose career was climaxed by a scientific expedition to the Cape of Good Hope; his studies there made him "the father of southern astronomy," and his names for fourteen southern constellations remain as his most enduring monument.

His father, Louis de la Caille, was originally a gendarme and later served in various artillery companies; his mother was Barbe Rubuy. Both parents were descended from old and distinguished families; but since Lacaille believed that merit rested in the individual and not in his ancestors, he made no attempt to investigate his lineage.

The elder Lacaille recognized his son's scholastic ability and arranged for his education, first at Nantes and then, beginning in 1729, at the Collège de Lisieux in Paris. For two years the young Lacaille studied rhetoric, acquiring his lifelong habit of wide reading. The death of his father left him without resources; but his pleasant personality, hard work, and intelligence had impressed his teachers and it was arranged for the young man to receive support from the duke of Bourbon, an acquaintance of his father's. Sometime during this period he received the title of abbé, although he seems never to have practiced as a clergyman. After completing the course of philosophy, Lacaille transferred to the three-year theological course at the Collège de Navarre. There, by chance, he discovered Euclid and soon developed a keen but secret interest in mathematical astronomy, a subject in which he had no teacher and scarcely any books. He passed the examinations for the master's degree with honors; but at the traditional ceremony for conferring the hood Lacaille answered an already obsolete question of philosophy in a way that offended the vice-chancellor, who refused to award the hood. When the other examiners objected, the degree was grudgingly given. Although Lacaille had seemed destined for literature, the incident at his graduation fortified his resolve to study the mathematical sciences. Thus, rather than apply for the bachelor of theology degree, he spent the money on books.

In 1736 Lacaille contacted J.-P. Grandjean de Fouchy, soon to become permanent secretary of the Academy of Sciences, who was astonished at the young man's progress in astronomy in the absence of any formal teaching. Fouchy introduced Lacaille to Jacques Cassini, the leading astronomer at the observatory in Paris; and thereafter Lacaille received lodging there. He made his first astronomical observation in May 1737.

Throughout the eighteenth century, problems of geodetics were closely linked with astronomy, especially because of the growing requirements of navigation. Thus Lacaille was assigned the mapping of the seacoast from Nantes to Bayonne, and in May 1738 he left Paris with G.-D. Maraldi. Then, because of his demonstrated ability, he was assigned with Cassini de Thury to the verification of the great meridian of France, which extended by a series of triangles from Perpignan in the south to Dunkerque in the north. At that time the shape of the earth was the source of great controversy between the Cartesians and the Newtonians. Cassini actively defended the opinion that, according to the French geodetic measurements, the earth was a prolate spheroid, contrary to Newton's view of the earth with an equatorial bulge.

Lacaille took the leading role in the new measurements. He measured base lines at Bourges, at Rodez, and at Arles; and he established positions astronomically at Bourges, Rodez, and Perpignan. During the rigorous winter of 1740 he extended his triangles to the principal mountains of Auvergne in order to tie in with another newly measured base line at Riom. Soon he was able to improve Picard's measures of 1669, showing that Picard's base line near Juvisy was 1/1,000 too long. Lacaille's geodetic and astronomical measurements, continued north of Paris until the spring of 1741, enabled him to show that the degrees of terrestrial latitude increased in length toward the equator, a result in agreement with Newtonian theory but directly opposed to previous French results.

Because of his growing reputation, the twenty-six-year-old Lacaille was named, during his absence on the survey, to the chair in mathematics once held by Varignon at the Collège Mazarin. Two years later, in May 1741, in recognition of his work on the meridian and his resolution of the controversy over the shape of the earth, he was received into the Academy of Sciences as an adjoint astronomer. Once again in residence in Paris, he took his professorial duties seriously, publishing *Leçons élémentaires de mathématiques* in 1741. The prompt translations into Latin, Spanish, and English were an eloquent compliment to his book, which was destined to go through several French editions as well. In succession other elementary texts followed: *Leçons élémentaires de mécanique* (1743), *Leçons élémentaires d'astronomie géométrique et physique* (1746), and *Leçons élémentaires d'optique* (1756). These works were also translated into Latin and other foreign languages. In the same period Lacaille began the computation of the series *Éphémérides de mouvements célestes*, which eventually extended from 1745 to 1775; these were later continued by Lalande to 1800. Another impressive testimonial to his computational ability and intellectual discipline was his calculation of all the eclipses from the beginning of the Christian era through the year 1800 for the encyclopedic *L'art de vérifier les dates;* this he accomplished in five weeks, working fifteen hours per day. Because the work was done so quickly, the authors of the compendium assumed that Lacaille had calculated the eclipses long before and had merely recopied the tables.

In the 1740's Lacaille left his lodging at the Paris observatory, and in 1746 a new observatory became available for him at the Collège de Lisieux. Here he recorded a vast variety of celestial phenomena, including conjunctions, lunar occultations, and comets. The abbé Claude Carlier called him "an Argus who saw everything in the sky." Most important, at the Mazarin observatory he exploited transit instruments, which were scarely known and appreciated in France at that time.

Curiosity about the southern stars invisible from the latitude of Paris induced Lacaille to propose an expedition to the southern hemisphere. An endorsement was offered by the Academy of Sciences, which ensured government support; and on 21 October 1750 he departed from Paris on his southern journey. On 21 November he embarked on the *Glorieux*, a ship so badly constructed that it was necessary to stop at Rio de Janeiro (on 25 January 1751) to repair the leaks. The ship left Brazil a month later, arriving at the Cape of Good Hope on 30 March 1751; but the passengers were unable to disembark until 19 April. Lacaille was cordially received by the Dutch governor of the Cape and sent to lodge in one of the best houses of the town. His observatory, built in the yard, consisted of no more than a small room measuring about twelve feet square and erected on a heavy masonry foundation. In this room Lacaille had two piers for carrying instruments, a pendulum clock, and a bed. He had two sectors, each with a six-foot radius, one of them carrying two telescopes; a smaller quadrant; and a variety of telescopes, one fourteen feet long (which he used for observing Jupiter's satellites).

In seeking the support of the Academy, Lacaille had proposed to make observations for the determination of the parallaxes of the sun and moon, to determine the longitude of the Cape, and to chart all of the southern stars to the third or fourth magnitude. In spite of wretched seeing conditions caused by the southeast wind that blew steadily nearly half the year, and often made the stars look like comets, Lacaille far exceeded his planned program of observations.

Trigonometrical determinations of the distance to the moon or the scale of the solar system generally require as large a base line as possible. The Cape of Good Hope was ideally situated for parallax measurements because although it was far from Europe, it had the same longitude. While Lacaille made his observations at the Cape, simultaneous measurements were undertaken in Europe. It was on this occasion that the nineteen-year-old Lalande made his own astronomical reputation by observing the other end of the parallactic base line from Berlin. Lacaille observed for the lunar parallax from 10 May 1751 until October 1752. Observations for Venus were secured between 25 October 1751 and 15 November 1752, and for Mars from 31 August 1751 until 9 October 1751, while that planet was at a relatively favorable opposition. The value that he obtained for the solar parallax was 9.5 seconds of arc instead of 8.8 seconds, thus making the sun–earth distance roughly 10 percent too small.

When charting the southern skies, Lacaille's response to the bad seeing conditions was to use a small eight-power telescope, only twenty-eight inches long and one-half inch in diameter. In the field of this instrument he mounted a rhomboidal diaphragm. The telescope was rigidly attached to the mural quadrant so that it pointed to a chosen spot on the north–south meridian. As the stars in the 2.7-degree zone drifted through his field in their daily motion, Lacaille recorded the times when they entered and left the rhombus. The average of the two sidereal times for a star gave its right ascension, while the difference of the times was a function of its declination. With this instrument in the year beginning August 1751 he undertook 110 observing sessions of eight hours each, plus sixteen entire nights. In this fashion he mapped nearly 10,000 stars in the southern sky, an incredible achievement. Lacaille himself reduced the positions for only 1,942 of these stars for a preliminary catalog, and not until the 1840's was the entire catalog reduced in Edinburgh by Thomas Henderson and published under the direction of Francis Baily as *A Catalogue of 9766 Stars in the Southern Hemisphere* (1847). The magnitude of Lacaille's accomplishment can be compared with the only previous systematic attempt to map the southern skies, by Edmond Halley, who from the island of Saint Helena in 1677–1678 had cataloged 350 stars. Lacaille carried out his program in spite of continued fevers, rheumatism, and headaches exacerbated by his intemperate schedule.

In his work Lacaille completed the naming of the southern constellations, which had been begun by Dutch navigators around 1600. As an astronomer of the Enlightenment, Lacaille eschewed the mythology of classical antiquity and named his fourteen new constellations after modern tools of the arts and sciences: Sculptor, Fornax, Horologium, Reticulum Rhomboidalis, Caelum, Pictor, Pyxis, Antlia, Octans, Circinus, Norma, Telescopium, Microscopium, and Mons Mensa. Among these, the names of several of Lacaille's instruments take a prominent place.

A by-product of Lacaille's zone surveys was a catalog of forty-two nebulous objects. In describing this result to the Academy, Lacaille wrote:

> The so-called nebulous stars offer to the eyes of the observers a spectacle so varied that their exact and detailed description can occupy astronomers for a long time and give rise to a great number of curious reflections on the part of philosophers. As singular as those nebulae are which can be seen from Europe, those which lie in the vicinity of the south pole concede to them nothing, either in number or appearance [*Mémoires de l'Académie royale des sciences* (1755)].

The détour in the original journey to the Cape, plus the six weeks' delay while the observatory was being built, prevented Lacaille from completing his objectives within a year, as he had originally planned. Consequently, he extended his visit, which gave him more than enough time to meet the geodetic objective of his expedition. With assistance offered by the governor of the Cape he surveyed three-quarters of a degree along a north-south meridian. His eight-mile base line encompassed his observatory and a number of mountain peaks in the vicinity of Cape Town. Lacaille was troubled to find that his results supported the hypothesis that the earth was a prolate, not an oblate, spheroid. Although he partially rechecked the result, he could find no error and it remained a puzzle for some years. Apparently the result was due to the deviation of the plumb line at his southern station caused by the large mass of Table Mountain (the Mons Mensa of his constellation list).

While at the Cape, Lacaille collected many plants unknown in Europe for the royal botanical gardens in Paris. In addition he sent a great numbers of shells, rocks, and even the skin of a wild donkey to the cabinet of the royal gardens. His observations of the customs of "Hottentots and inhabitants of the Cape of Good Hope" were published posthumously in his *Journal historique du voyage fait au Cap de Bonne-Espérance* (Paris, 1776).

Before his return to France, Lacaille received instructions to establish the positions of two French islands in the Indian Ocean, Ile de France (Mauritius) and Ile de Bourbon (Réunion). He left the Cape for Mauritius on 8 March 1753 on the *Puisieulx;* en route he worked on the problem of determining longitude at sea from observations of the moon. He arrived on 18 April 1753 for a nine-month visit, during which he continued his astronomical observations as well as mapping the island. The following January he sailed to St.-Denis on Réunion. On 27 February 1754 he left on the *Achille* for France, stopping for five days in April on Ascension Island, whose position he determined. Lacaille arrived in Paris on 28 June 1754, after an absence of three years and eight months.

Upon his return to Paris, Lacaille found lavish praise awaiting him—he was even compared to a star returning to the horizon. With great modesty he refused all the fanfare. He wanted only to retire quietly to his observatory to reduce his observations; in fact, he dreamed of retiring to a southern province where he could once again observe the southern skies. He accepted an annual pension from the Academy but rejected all other means of advancing his fortune. Nevertheless, his fame spread and he was welcomed into membership in the academies of Berlin, St. Petersburg, Stockholm, Göttingen, and Bologna.

In 1757 Lacaille published *Astronomiae fundamenta*, a work now very scarce, apparently because it was privately distributed by the author in an edition of perhaps 120 copies. The book had two parts: the first contained tables for the reduction of true positions of stars to their apparent positions. In the second part of his work Lacaille gave the positions of 400 of the brightest stars. Appended to the work were observations of the sun made at the Cape and on Mauritius. The following year he published his detailed tables of solar positions; these included the effect of perturbations from the moon, Jupiter, and Venus. Another important contribution from his southern expedition was an extensive table of atmospheric refraction, showing the effects of both temperature and barometric pressure.

In this period Lacaille not only edited revisions of his own textbooks but also brought out a thoroughly revised edition of Bouguer's *Nouveau traité de navigation* and edited from manuscript Bouguer's *Traité d'optique sur la graduation de la lumière*. He initiated a project to be entitled *Les âges de l'astronomie*, in which he proposed to assemble and compare all the old astronomical observations. a work which later found partial fulfillment in Pingre's *Annales de l'astronomie*.

Lacaille's memoir on the Comet of 1759 (now known as Halley's Comet) not only described his particularly careful observations but also afforded the occasion to demonstrate his simplified method for finding the elements of a cometary orbit. Besides the observations that he regularly reported to the Academy, he made many others for his own star catalog. In 1760 he organized a plan to measure very accurately the positions of a number of zodiacal stars, and Lacaille's biographers are unanimous in attributing his early death to the rigors of his observational program. Not only did he spend many arduous hours observing the heavens; he even slept on the floor of the observatory. At the end of February 1762 the symptoms that he had previously suffered at the Cape returned: rheumatism, nosebleeds, and signs of indigestion. The doctors imposed the standard bloodletting procedures of the day, apparently not realizing the seriousness of his illness; and after an attack of particularly high fever, he died. He was only forty-nine.

Lacaille's deeply sincere modesty, his profound honesty, and his sustained devotion to his science impressed all who knew him. A younger colleague, Lalande, wrote that he had single-handedly made more observations and calculations than all the other astronomers of his time put together. Delambre added that although Lalande's statement appeared to be an exaggeration, it was literally true if only the twenty-seven years of Lacaille's astronomical career were considered.

BIBLIOGRAPHY

I. Original Works. The most extensive bibliography is found in Lacaille's posthumous *Coelum australe stelliferum* (Paris, 1763), pp. 20–24; a more readily accessible list is J. M. Querard's *La France littéraire* (Paris, 1830), pp. 353–354. *Catalogue général des livres imprimés de la Bibliothèque nationale, auteurs*, LXXXIV (Paris, 1925), cols. 942–948, tabulates many eds. of his books. A list of his memoirs may be found in *Table générale des matières contenues dans l'Histoire et les Mémoires de l'Académie royale des sciences* VI–VIII (1758–1774). In the octavo repr. of this work he is listed under "Caille." The principal books have been cited in the text; the memoir that contains the first plate of his new southern constellations is "Table des ascensions droites et des déclinations apparentes des étoiles australes renfermées dans le tropique du Capricorne; observés au Cap de Bonne-Espérance, dans l'intervalle du 6 août 1751, au 18 juillet 1752," *Mémoires ... présentés par divers sçavans* for 1752 (1756), 539–592.

Many of Lacaille's MSS are preserved at the Paris observatory; they are cataloged as C3.1–48 in G. Bigourdan, "Inventaire des manuscrits," in *Annales de l'observatoire de Paris. Memoires*, **21** (Paris, 1895), 1–60.

II. Secondary Literature. The most detailed biography, by the abbé Claude Carlier, is anonymously prefixed to the posthumous ed. of Lacaille's *Journal historique du voyage fait au Cap de Bonne-Espérance* (Paris, 1776). Other important sources are J.-P. Grandjean de Fouchy, "Éloge de Lacaille," in *Histoire de l'Académie royale des sciences pour l'année 1762* . . . (1767), 354–383 (octavo ed.); and J. B. Delambre, "Caille," in *Biographie universelle ancienne et moderne*, VI (Paris, after 1815), 350–354. A nineteen-page latin *vita* by G. Brotier introduces Lacaille's *Coelum australe stelliferum* (Paris, 1763). See also David S. Evans, "LaCaille: 10,000 Stars in Two Years," in *Discovery* (Oct. 1951), 315–319; and Angus Armitage, "The Astronomical Work of Nicolas-Louis de Lacaille," in *Annals of Science*, **12** (1956), 165–191.

Owen Gingerich

LACAZE-DUTHIERS, FÉLIX-JOSEPH HENRI DE (*b.* Château de Stiguederne, Lot-et-Garonne, Montpezat, France, 15[?] May 1821; *d.* Las-Fons, Dordogne, France, 21 July 1901), *zoology*.

Lacaze-Duthiers was the second son of Baron J. de Lacaze-Duthiers, a rather difficult and authoritarian man who was descended from an old Gascon family. After obtaining bachelor degrees in arts and in sciences, Lacaze-Duthiers went to Paris, despite paternal opposition, to undertake simultaneously medical and natural history studies. He became licentiate in 1845, doctor of medicine in 1851, and doctor of sciences at the Faculty of Sciences of Paris in 1853. He then departed on research excursions to the Balearic Islands and Brittany, where he commenced work on his scientific specialty, marine mollusks and zoophytes. In 1854, upon his return to Paris, his mentor Henri Milne-Edwards, for whom he had formerly acted as *préparateur*, obtained for him a professorship of zoology at the newly created Faculty of Sciences at Lille, where Louis Pasteur was dean of the faculty.

During this period of his career Lacaze-Duthiers made several more scientific excursions to the Atlantic and Mediterranean coasts. During his most important excursion, undertaken for the French government from 1860 to 1862, he studied coral fishing in Algeria. The results of this expedition, published as *Histoire naturelle du corail* (1864), won him the Prix Bordin from the Academy of Sciences in 1863. His work was especially valued for its accurate description of the generative organs and the phases of development of the coral and its polypary. In 1864 Lacaze-Duthiers left Lille for Paris, where he became professor of annelids, mollusks, and zoophytes at the Muséum d'Histoire Naturelle in 1865. He gave up this chair in 1869 to take one of the two chairs of zoology, anatomy, and comparative physiology at the Faculty of Sciences, the other chair being held by Milne-Edwards. In 1871 he was elected to the Academy of Sciences.

The earlier part of Lacaze-Duthiers's career was devoted to writing numerous long monographs on mollusks and zoophytes, in which he described with great accuracy and minuteness of detail the anatomy, histology, and embryogeny of his subjects, utilizing the results to classify what had appeared to be anomalous types of animals. The latter part of his career was characterized by activities directed toward assuring the progress of zoology and zoological education in France. Thus at the Faculty of Sciences he instituted laboratory work for students and set up a regularized three-year course in zoology for degree candidates. His journal, *Archives de zoologie expérimentale et générale*, founded in 1872, published the work of his students. The opening discourse, "Direction des études zoologiques," argues against the assertions of Claude Bernard and the French physiological school by insisting that zoology can and ought to be an experimental science, if one extends "experiment" to mean a prepared observation controlling an induction. Also to further zoological education, Lacaze-Duthiers founded and expended a great deal of effort developing two of the earliest marine zoological laboratories, one at Roscoff, on the coast of Brittany, and the other the Laboratory Arago, at Banyuls, on the Mediterranean.

In his work on marine invertebrates, Lacaze-Duthiers followed in the footsteps of his master, Henri Milne-Edwards, who also had carefully studied the anatomy and embryogeny of these animals. Repeatedly Lacaze-Duthiers stressed the importance for classification of studying marine animals in their natural habitat and of observing their embryogeny. He was a follower of the Cuvier school in that he held to a very empirical sort of science. Although he claimed to be "philosophical," he was extremely cautious about generalization and hypotheses. He believed, for example, that the problem of the origin of species was outside the domain of objective science. Lacaze-Duthiers did, however, often apply Geoffroy Saint-Hilaire's principle of connections in limited areas. His doctoral dissertation, for example, set up detailed homologies of the parts of the genital casing throughout the insects. A later memoir applying the principle of connections (1872) demonstrated that the auditory nerve of mollusks is always attached to the cerebroid ganglion and not sometimes to the pedal ganglion, as had been previously assumed.

Lacaze-Duthiers never married. Especially in his old age, he became rather rigid and suspicious. While very demanding of his students and reluctant to extend his confidences, he nevertheless felt a great need for the approval of others.

BIBLIOGRAPHY

I. Original Works. Lacaze-Duthiers's major publications are *Recherches sur l'armure génitale femelle des insectes* (Paris, 1853); *Voyage aux îles Baléares, ou Recherches sur l'anatomie et la physiologie de quelques mollusques de la Méditerranée* (Paris, 1857); *Histoire de l'organisation, du développement, des moeurs et des rapports zoologiques du dentale* (Paris, 1858); *Un été d'observations en Corse et à Minorque, ou Recherches d'anatomie et physiologie zoologiques sur les invertébrés des ports d'Ajaccio, Bonifacio et Mahon* (Paris, 1861); *Histoire naturelle du corail* (Paris, 1864); *Recherches de zoologie, d'anatomie et d'embryogénie sur les animaux des faunes maritimes de l'Algérie et de la Tunisée* (Paris, 1866); *Le monde de la mer et ses laboratoires* (Paris, 1888).

A chronological list of his books and memoirs can be found in *Archives de zoologie expérimentale et générale*, 3rd ser., **10** (1902), 64–78. It omits publications of memoirs in book form. The Royal Society's *Catalogue of Scientific Papers*, III, 787–789, VIII, 143–144, X, 485, and XVI, 534–535, contains his memoirs through 1900. Some of his MSS are at the Laboratoire Arago. An English trans. of part of Lacaze-Duthiers's introduction to the first vol. of his journal, "The Study of Zoology," is in William Coleman, *The Interpretation of Animal Form* (New York, 1967), 131–163.

II. Secondary Literature. There appears to be no full-length biography of Lacaze-Duthiers. Short accounts of his life and work can be found in Louis Boutan, "Henri de Lacaze-Duthiers," in *Revue scientifique*, 4th ser., **17** (1902), 33–40; Alfred Lacroix, "Les membres et correspondents ayant travaillé sur les côtes de l'Afrique du Nord et du Nord-Est," in *Mémoires de l'Académie des sciences de l'Institut de France*, **66** (1943), 17–27 (Lacroix gives the date of birth as 21 May instead of 15 May 1821); and G. Pruvot, "Henri de Lacaze-Duthiers," in *Archives de zoologie expérimentale et générale*, 3rd ser., **10** (1902), 1–46.

A recent article making use of MS material and helpful for probing the character of Lacaze-Duthiers is George Petit, "Henri de Lacaze-Duthiers (1821–1901) et ses 'carnets' intimes," in *Communications du Premier congrès international d'histoire de l'océanographie, Monaco, 1966*, II (1968), 453–465. For an account of Lacaze-Duthiers's professorship at Lille during Pasteur's tenure there as dean of the faculty, see Denise Wrotnowska, "Pasteur et Lacaze-Duthiers," in *Histoire des sciences médicales*, no. 1 (1967), 1–13.

Toby A. Appel

LACÉPÈDE, BERNARD-GERMAIN-ÉTIENNE DE LA VILLE-SUR-ILLON, COMTE DE (*b.* Agen, France, 26 December 1756; *d.* Épinay-sur-Seine, France, 6 October 1825), *zoology*.

Lacépède was the only son of Jean-Joseph-Médard, comte de La Ville, lieutenant general of the seneschalsy, and the former Marie de Lafond. The name Lacépède was taken from a maternal granduncle who made Lacépède his heir on condition that he take the name. He was educated by his father, soon a widower, with the help of the Abbé Carrière and M.

de Chabannes, bishop of Agen. At an early age Lacépède acquired a predilection for both science and music. He undertook a series of electrical experiments, began to compose an opera, and was soon corresponding with Buffon and Gluck. In 1777 he left Agen for Paris, where he was commissioned colonel of a German regiment (which he never saw), an appointment befitting his social position.

After an attempt at becoming a professional opera composer, Lacépède turned fully to science, attempting in his works to imitate the style of his master and model, Buffon. He published the *Essai sur l'électricité naturelle et artificielle* in 1781 and began a great treatise, *Physique générale et particulière*, of which only two volumes appeared (1782, 1784). According to Cuvier and Valenciennes, these works were rejected by the academicians for being too hypothetical. Buffon, however, was impressed with Lacépède and invited him to work on the continuation of Buffon's famous *Histoire naturelle*. Buffon had planned to write the natural history of all the vertebrates but at the time had completed only the viviparous quadrupeds and the birds. Reserving for himself the cetaceans, Buffon asked Lacépède to write the natural history of the reptiles and of the fishes. In order to facilitate this work, Buffon obtained for Lacépède the post of keeper and subdemonstrator at the Cabinet du Roi associated with the Jardin des Plantes. Lacépède's first volume, *Histoire naturelle des quadrupèdes ovipares*, appeared in 1788 and his second volume, *Histoire naturelle des serpents*, in 1789, after Buffon's death. Lacépède then completed Buffon's plan by publishing the *Histoire naturelle des poissons* in five volumes from 1798 to 1803 and the *Histoire naturelle des cétacés* in 1804.

During the Revolution, Lacépède occupied several political positions, becoming a deputy to both the Constituent Assemby and the Legislative Assembly. He helped formulate plans to reconstitute the Jardin des Plantes into the Muséum d'Histoire Naturelle but was forced to resign his position (March 1793) and leave Paris before the law instituting the Muséum was passed in June 1793. Under the constitution of the Muséum, Geoffroy Saint-Hilaire, who took Lacépède's place, became professor of vertebrate zoology in the newly established institution. Lacépède remained in exile at Leuville during the Terror, working on the *Histoire naturelle des poissons*. While at Leuville he married Anne Caroline Gauthier, the widow of a close friend, and adopted her son. When Lacépède returned to Paris after 9 Thermidor (1794), his former colleagues, who had been working all along to bring him back to the Muséum, induced the government to create a chair for him, a chair of zoology specializing in reptiles and fish. It was in this period that Lacépède attained a reputation among auditors of his lectures as "the successor of Buffon." When the Institut de France was established in 1795, Lacépède was chosen by the Directory as an original member of the section of anatomy and zoology.

After 1803 Lacépède no longer gave his course at the Muséum, allowing his aide, Constant Duméril, to perform most of the functions associated with his chair. With the completion in 1804 of Buffon's *Histoire naturelle*, Lacépède turned to public affairs and to moral and historical writings. Under Napoleon, whom he admired and fully supported, Lacépède took an active role in affairs of state, becoming a senator in 1799 and grand chancellor of the Legion of Honor in 1803. When Napoleon fell, Lacépède retired to Épinay; and although he served as a peer of France under the Restoration, he devoted most of the remaining years of his life to writing. During this period he published two novels, a natural history of man, and historical works on the development of civilization.

Buffon always remained Lacépède's chief mentor; but as his works in natural history progressed, Lacépède came to depend more and more on his close friend Daubenton, who after Buffon's death became the leading figure in French natural history. Under Daubenton's influence Lacépède paid increasing attention to exact anatomical description, mainly of external characters, and to the problems of taxonomy. In his history of reptiles Lacépède adopted the Linnaean binomial nomenclature and thereafter prefixed each of his works on natural history with a systematic table of orders, genera, and species and their characters. Lacépède believed that classification, although important, was nonetheless artificial; and he never claimed to be seeking a "natural system." Rather, he developed an artificial system based on convenient external characters and—merely for the sake of symmetry—often invented taxa for which no species actually existed. In all of his work on natural history Lacépède was able greatly to increase the number of known species by utilizing the resources of the Cabinet du Roi and information sent to the Muséum by voyagers.

However professional Lacépède became, he retained his desire to write in an elegant and elevated style in the manner of Buffon and to relate his knowledge of natural history to the development of man's material circumstances and moral nature. His broad views on natural history can be found in the discourses contained in each of his contributions to the *Histoire naturelle*, in the opening and closing lectures of his courses at the Muséum, and in his later historical works.

All of Lacépède's biographers have noted his reputation for good breeding, generosity, and fairness. Always generous in his judgment of others, he was popular among his colleagues.

BIBLIOGRAPHY

I. ORIGINAL WORKS. Apart from the scientific works mentioned in the text, Lacépède's *Discours d'ouverture et de clôture du cours d'histoire naturelle* (1798–1801) and his *Histoire naturelle de l'homme* (1827) ought to be noted. There are several nineteenth-century eds. of his major writings. The material has been somewhat rearranged, however, and important discourses have been left out of later eds. of the *Histoire naturelle des poissons*, notably "Troisième vue de la nature" and "Discours sur la durée des espèces." One should therefore consult the original eds. if possible. A fairly complete list of books published by Lacépède can be found in *Biographie universelle*, new ed., XXII, 345–346. This should be supplemented by the listings in the *Catalogue général de la Bibliothèque nationale*, LXXXIV, 1030–1043, esp. for later eds. and for Lacépède's political writings. A list of Lacépède's chief memoirs can be found in the Royal Society's *Catalogue of Scientific Papers*, III, 789–790.

II. SECONDARY LITERATURE. There is a full-length biography of Lacépède written by a later occupant of his chair at the museum: Louis Roule, *Lacépède et la sociologie humanitaire selon la nature* (Paris, 1932), vol. VI in Roule's *L'histoire de la nature vivante d'après l'oeuvre des grands naturalistes français*. An earlier version can be found under the title "La vie et l'oeuvre de Lacépède," in *Mémoires de la Société zoologique de France*, **27** (1917), 139–237. Earlier biographical sketches include Georges Cuvier's often inaccurate "Éloge historique de M. le Comte de Lacépède lu le 5 juin 1826," in *Mémoires de l'Académie des sciences de l'Institut de France*, **8** (1829), ccxii–ccxlviii; Achille Valenciennes, "Lacépède," in *Biographie universelle*, new ed., XXII, 336–346; and G.-T. Villenave, *Éloge historique de M. le Comte de Lacépède* (Paris, 1826), which is useful for exposing Lacépède's personality.

TOBY A. APPEL

LA CONDAMINE, CHARLES MARIE (*b.* Paris, France, 27 January 1701; *d.* Paris, 4 February 1774), *mathematics, natural history.*

For a detailed study of his life and work, see Supplement.

LACROIX, ALFRED (*b.* Mâcon, France, 4 February 1863; *d.* Paris, France, 16 March 1948), *mineralogy, petrology, geology.*

Lacroix's grandfather, Tony, and father, Francisque, were Paris-trained pharmacists; his grandfather's avocation was mineralogy. When he was twenty, Lacroix went to Paris to earn a pharmacist's diploma. Concurrently he attended courses in mineralogy at the Muséum d'Histoire Naturelle, where his advanced knowledge came to the attention of his professors, F. Fouqué and A. des Cloizeaux. Recognizing his interest and natural ability, they made funds available to enable him to visit classic localities and collections in Europe. After receiving the diploma in pharmacy in 1887, Lacroix chose to follow mineralogy and was appointed assistant to Fouqué, a post he held while earning his doctorate. He received it 31 May 1889 and then, having fulfilled Fouqué's prerequisite, on 6 June married Catherine, eldest of the Fouqué daughters. On 1 April 1893 Lacroix was named to succeed des Cloizeaux in the chair of mineralogy at the museum, a position he held until his official retirement on 1 October 1936. He continued a full schedule of work at his laboratory, walking there each day until four days before his death at the age of eighty-five.

Lacroix early realized the importance to research of an excellent mineral collection, and for fifty years one of his undeviating aims was to build up a systematic collection from around the world. As a corollary he envisaged an inventory of the minerals of France and its overseas possessions. He drew on acquaintances made during his early travels for exchange of specimens; he asked French colleagues in other sciences to collect while abroad; even administrators of French colonies found themselves collecting for Lacroix. This was particularly true of General Joseph Gallieni and his officers on Madagascar, which had come under French rule in 1895; they received detailed instructions for methodical collecting and responded so effectively that Lacroix visited the island in 1911 and undertook the detailed study that led to the publication of *Minéralogie de Madagascar*, a three-volume mine of information that has not been superseded. In like manner a systematic inventory of the minerals and rocks of the French volcanic islands of the South Pacific was completed and led to the recognition of two distinct petrologic domains characterized by mesocratic and melanocratic basalts.

Following in the footsteps of Fouqué, Lacroix made volcanology and volcanic rocks one of his specialties. These studies began in 1890; in May 1902 there was a major eruption on the French island of Martinique in the Caribbean. Appointed by the Académie des Sciences to head a mission to study the volcanism, Lacroix spent six months sending back a stream of letters which appeared in the Academy's *Comptes rendus.* These portended the vigorous exposition that he later published of the *nuée ardente*, or glowing cloud type of eruption, which he alone, of all the observers of Mt. Pelée, was perceptive enough to recognize as a newly observed phenomenon. First suggested by Fouqué in 1873, on the basis of tales of earlier eruptions in the Azores, an eruption of that type had never been witnessed by a scientist. Subsequent *nuée ardente* eruptions have been recorded on Martinique and in other active volcanic areas; and the term, in the French form, has been accepted and used throughout the world. Lacroix also derived new evidence on the origin and behavior of volcanic domes. The results of his observations and investigations are in the two-volume *La montagne Pelée*. On 11 January 1904, the Académie des Sciences elected Lacroix a member in recognition; his way of putting it was "Je suis entré à l'Institut sous l'irresistible poussée d'un volcan."

But this was not the only volcanic "poussée" to which Lacroix responded. Up to 1914 he was an eyewitness to outbursts of Vesuvius, which was then in almost constant eruption, of Etna, and of other volcanoes. His other travels were to study minerals in their natural settings in order to supplement his laboratory determinations. Lacroix had a broad interpretation of mineralogy, believing that the study of minerals should not be an end in itself but a means of learning their origins and relations to the rocks in which they occur, the structure of the rocks, and the entire terrain. Mineralogy as a point of union of chemistry, physics, mathematics, and natural history could make use of the techniques of those disciplines; but it should not be separated from geology or geophysics. Being on the scene when mineralogy was in transition from a purely descriptive to an interpretive science, Lacroix was a brilliant exponent of the new trend.

Lacroix was named *secrétaire perpetuel* of the Académie des Sciences on 8 June 1914. A lucid writer and an able administrator, he was admirably suited for the post and gave it meticulous attention during his thirty-four-year tenure. He reorganized the secretariat, enriched the archives, and lent needed support to an inventory of scientific materials in Paris libraries. Giving full play to his interest in the history of science and scientists, and with a wealth of facts at hand, he produced a series of biographies of French scientists of the seventeenth and eighteenth centuries. Written in a clear, fluent narrative style, they rank with the best in literature.

A kindly man of simple tastes who shunned any form of ostentation, Lacroix had the gift of creating an aura of goodwill among his students and colleagues. "La bienveillance, c'est quelque chose dans la vie des hommes," were his own words.

BIBLIOGRAPHY

I. Original Works. Lacroix wrote, either alone or with colleagues, about 650 papers, which were published primarily in French scientific journals between 1879 and 1946. His larger monographs include *Minéralogie de la France et de ses colonies*, 5 vols. (Paris, 1893–1913); *La montagne Pelée et ses éruptions* (Paris, 1904); *La montagne Pelée après ses éruptions* . . . (Paris, 1908); *Minéralogie de Madagascar*, 3 vols. (Paris, 1922–1923); *Figures de savants*, 4 vols. (Paris, 1932–1938); and *Le volcan actif de l'île de la Réunion et ses produits* (Paris, 1936), with supp. (1938).

II. Secondary Literature. At least twenty-two biographies of Lacroix have appeared in the scientific journals of several European countries and of the United States. A bibliography of 16 is in *Bulletin de la Société française de minéralogie et de cristallographie*, **73** (1950), 408.

Marjorie Hooker

LACROIX, SYLVESTRE FRANÇOIS (*b.* Paris, France, 28 April 1765; *d.* Paris, 24 May 1843), *mathematics.*

Lacroix, who came from a modest background, studied at the Collège des Quatre Nations in Paris, where he was taught mathematics by the Abbé Joseph François Marie. He became ardently interested in the exact sciences at a very young age; as early as 1779 he carried out long calculations on the motions of the planets, and 1780 he attended the free courses given by Gaspard Monge, who became his patron. The friendship between the two remained constant throughout the sometimes dramatic events of their lives. Monge, who was an examiner of students for the navy, in 1782 secured for Lacroix a position as professor of mathematics at the École des Gardes de la Marine at Rochefort. Following Monge's advice, Lacroix then began to concern himself with partial differential equations and with the calculus of variations. In 1785 Lacroix sent a memoir on partial differences to Monge which he reported on to the Académie des Sciences. In that same year Lacroix also sent new solar tables to the Academy.

Lacroix returned to Paris to substitute for Condorcet in his mathematics course at the Lycée, a newly founded free institution whose lectures attracted many members of the nobility and of the upper bourgeoisie. The course in pure mathematics had few auditors and was soon discontinued. Lacroix then taught astronomy and the theory of probability. He also shared a 1787 Academy prize (which they never received) with C. F. Bicquilley for a work on the theory of marine insurance. In the meantime he had married. Lacroix also succeeded d'Agelet first in the duties and eventually in the title to the chair of mathematics at the École Militaire in Paris, and began to gather material for his *Traité du calcul différentiel et du calcul intégral* (1797–1798).

Because the chair at the Lycée was abolished and the École Militaire closed in 1788, Lacroix once again left Paris; he took up a post as professor of mathematics, physics, and chemistry at the École Royale d'Artillerie in Besançon. In 1789 the Academy chose him to be Condorcet's correspondent, and in 1793 he succeeded Laplace as examiner of candidates and students for the artillery corps. In 1794 he became *chef de bureau* of the Commission Exécutive de l'Instruction Publique. He and Hachette later assisted Monge in the practical work connected with his course in descriptive geometry at the École Normale de l'An III. About this time Lacroix published his *Eléments de géométrie descriptive*, the materials for which had been assembled several years previously. The printing of his great *Traité* began during this period. Until 1791 Lacroix was a member of the admissions committee of the École Polytechnique.

Upon the creation of the Écoles Centrales, schools for intermediate education that were the forerunners of the modern lycée, Lacroix became professor of mathematics at the École Centrale des Quatres Nations. He then undertook to publish numerous textbooks, which further contributed to his fame. In 1799 he was elected a member of the Institute. He also succeeded to Lagrange's chair at the École Polytechnique, a position he held until 1809, when he became a permanent examiner.

The first volume of Lacroix's treatise on the calculus, in which he "united all the scattered methods, harmonized them, developed them, and joined his own ideas to them," appeared in 1797. It was followed by a second volume in 1798, and a third appeared in 1800 under the title *Traité des différences et des séries* (a second edition appeared in three volumes [1810, 1814, 1819]). This monumental work constituted a clear picture of mathematical analysis, documented and completely up to date. While Lacroix followed Euler on many points, he incorporated the various advances made since the middle of the eighteenth century. The treatise is a very successful synthesis of the works of Euler, Lagrange, Laplace, Monge, Legendre, Poisson, Gauss, and Cauchy, whose writings are followed up to the year 1819.

In his teaching, particularly at the École Polytechnique, Lacroix utilized his *Traité élémentaire du calcul différentiel et du calcul intégral*, which appeared in 1802. A work of enduring popularity, it was translated into English and German. From 1805 to 1815 Lacroix taught transcendental mathematics at the Lycée Bonaparte and, with the creation of the Facultés, he became dean of the Faculté des Sciences of Paris and professor of differential and integral calculus.

When Lacroix succeeded to the duties of Antoine Rémi Mauduit in 1812 at the Collège de France, he arranged for Paul Rémi Binet to succeed to his post at the Lycée Bonaparte. Upon Mauduit's death, Lacroix was appointed to the chair of mathematics at the Collège de France; he then definitively ended his connection with the Lycée Bonaparte, which in the meantime had become the Collège Bourbon.

Lacroix retired from his post as dean of the Faculté des Sciences in 1821 and a few years later from that of professor as well. In 1828 Louis Benjamin Francoeur succeeded to Lacroix's duties at the Collège de France, and beginning in 1836, Libri succeeded to these duties.

At the time of his visit to Paris in 1826, Abel found Lacroix "frightfully bald and remarkably old." Although he was only sixty-one, his astonishing activity since adolescence had affected his health.

Lacroix's mathematical work contained little that was absolutely new and original. His writings on analytic geometry, which refined ideas he derived from Lagrange and above all from Monge, served as models for later didactic works. It was he who actually proposed the term "analytic geometry": "There exists a manner of viewing geometry that could be called *géométrie analytique*, and which would consist in deducing the properties of extension from the least possible number of principles, and by truly analytic methods" (*Traité du calcul différentiel et du calcul intégral*, I [Paris, 1797] p. xxv).

A disciple of Condillac in philosophy, Lacroix brought to all his didactic works a liberal spirit, open to the most advanced ideas. He was particularly inspired by the pedagogical conceptions of Clairaut and, in addition, by those of the masters of Port-Royal and Pascal and Descartes. In this regard, the contrast is quite striking between Lacroix's *Éléments de géométrie* and the similar contemporary work by Legendre, which is a great deal more dogmatic.

Lacroix's sense of history is evident in all his writings. The preface to the first volume of the second edition of the great *Traité* (1810) is a model of the genre. He also wrote excellent studies, in particular those on Borda and Condorcet, for Michaud's *Biographie universelle*. In addition he participated in the editing of volume III of Montucla's *Histoire des mathématiques*, composed the section on mathematics for Delambre's report on the state of science in 1808, and prepared an essay on the history of mathematics, which unfortunately remained in manuscript form and now appears to have been lost.

His *Essais sur l'enseignement* (1805), a pedagogical

classic, display his acute psychological penetration, rich erudition, liberal cast of mind, and broad conception of education.

For more than half a century, through his writings and lectures, Lacroix thus contributed to an era of renewal and expansion in the exact sciences and to the training of numerous nineteenth-century mathematicians. The young English school of mathematics, formed by Babbage, Peacock, and Herschel, wished to breathe a new spirit into the nation's science, and one of its first acts was to translate the *Traité élémentaire du calcul différentiel et du calcul intégral.*

BIBLIOGRAPHY

I. Original Works. Lacroix's writings are *Essai de géométrie sur les plans et les surfaces* (Paris, 1795; 7th ed., 1840), also in Dutch trans. by I. R. Schmidt (1821); *Traité élémentaire d'arithmétique* (Paris, 1797; 20th ed., 1848), also in English trans. by John Farrar (Boston, 1818) and Italian trans. by Santi Fabri (Bologna, 1822); *Traité élémentaire de trigonométrie rectiligne et sphérique et d'application de l'algèbre à la géométrie* (Paris, 1798; 11th ed., 1863), also in English trans. by John Farrar (Cambridge, Mass., 1820) and German trans. by E. M. Hahn (Berlin, 1805); *Élémens de géométrie* (Paris, 1799; 19th ed., 1874); and *Complément des élémens d'algèbre* (Paris, 1800; 7th ed., 1863).

Subsequent works are *Traité du calcul différentiel et du calcul intégral*, 2 vols. (Paris, 1797–1798); 2nd ed., 3 vols. (Paris, 1810–1819); *Traité des différences et des séries* (Paris, 1800); *Traité élémentaire du calcul différentiel et du calcul intégral* (Paris, 1802; 9th ed., 1881), also in English trans. by C. Babbage, John F. W. Herschel, and G. Peacock (Cambridge, 1816) and German trans. (Berlin, 1830–1831); *Essais sur l'enseignement en général, et sur celui des mathématiques en particulier, ou manière d'étudier et d'enseigner les mathématiques* (Paris, 1805; 4th ed., 1838); *Introduction à la géographie mathématique et physique* (Paris, 1811); *Traité élémentaire du calcul des probabilités* (Paris, 1816); and *Introduction à la connaissance de la sphère* (Paris, 1828).

Works to which Lacroix was a contributor are *Lettres de M. Euler a une princesse d'Allemagne . . . avec des additions par M.M. le marquis de Condorcet et de la Croix*, new ed., 3 vols. (Paris, 1787–1789); *Élémens d'algèbre par Clairaut*, 5th ed. (Paris, 1797), with notes and additions drawn in part from the lectures given at the École Normale by Lagrange and Laplace and preceded by *Traité élémentaire d'arithmétique*, 2 vols. (Paris, 1797)—the notes, additions, and treatise are by Lacroix; J. F. Montucla, *Histoire des mathématiques*, 2nd ed., III (Paris, 1802), in which Lacroix revised ch. 33, pp. 342–352, on partial differential equations; *Rapport historique sur les progrès des sciences mathématiques depuis 1789 et sur leur état actuel, redigé par M. Delambre* (Paris, 1810)—in this report, presented in 1808, "everything concerning pure mathematics and transcendental analysis is drawn from a work of M. Lacroix, who submitted it to the assembled mathematics sections" (Delambre); and J. F. Montucla, *Histoire des recherches sur la quadrature du cercle*, 2nd ed. (Paris, 1831), which was prepared for publication by Lacroix.

II. Secondary Literature. On Lacroix and his work, see Carl Boyer, "Cartesian Geometry from Fermat to Lacroix," in *Scripta mathematica*, **13** (1947), 133–153; "Mathematicians of the French Revolution," *ibid.*, **25** (1960), 26–27; and *A History of Mathematics* (New York, 1968); Gino Loria, *Storia delle matematiche*, 2nd ed. (Milan, 1950), 771–772; Niels Nielsen, *Géomètres français sous la révolution* (Copenhagen, 1929), 134–136; Leo G. Simons, "The Influence of French Mathematicians at the End of the Eighteenth Century Upon the Teaching of Mathematics in American Colleges," in *Isis*, **15** (1931), 104–123; and René Taton, *L'oeuvre scientifique de Monge* (Paris, 1951), *passim;* "Sylvestre François Lacroix (1765–1843): Mathématicien, professeur et historien des sciences," in *Actes du septième congrès international d'histoire des sciences* (Paris, 1953), 588–593; "Laplace et Sylvestre Lacroix," in *Revue d'histoire des sciences*, **6** (1953), 350–360; "Une correspondance inédite: S. F. Lacroix–Quetelet," in *Actes du congrès 1953 de l'Association française pour l'avancement des sciences* (Paris, 1954), 595–606; "Une lettre inédite de Dirichlet," in *Revue d'histoire des sciences*, **7** (1954), 172–174; and "Condorcet et Sylvestre François Lacroix," *ibid.*, **12** (1959), 127–158, 243–262.

Jean Itard

LADENBURG, ALBERT (*b.* Mannheim, Germany, 2 July 1842; *d.* Breslau, Germany [now Wrocław, Poland], 15 August 1911), *chemistry.*

Ladenburg pioneered investigations of organic compounds of silicon and tin and advanced theories on the structure of aromatic compounds, but his chief contributions were the elucidation of the structure of alkaloids and their synthesis.

Ladenburg was one of eight children. His father, a prosperous attorney, objected to the classical education given at Gymnasium and sent him to a school where little Latin and no Greek were taught. At the Polytechnicium in Karlsruhe he emphasized study of mathematics and languages, attempting to fill in the gaps of his earlier education. In 1860 Ladenburg went to Heidelberg and, inspired by Bunsen's and Kirchhoff's lectures, decided on chemistry, taking the Ph.D., *summa cum laude*, in the spring of 1863. He stayed on for two years, working with Ludwig Carius, beginning a lifelong friendship with Erlenmeyer, and shifting his interest from inorganic to organic chemistry. Ladenburg worked under Kekulé at Ghent in 1865; but despite the scientific excitement he found life at Ghent dull, and after a visit with Frankland in London, he began working with Wurtz in Paris.

Late in 1866 Friedel invited Ladenburg to work with him at the École des Mines, where they began research on the compounds of silicon. Their concern centered on whether the new theories being developed about carbon compounds were applicable to the so-called inorganic elements and their compounds. Preliminary to synthesizing compounds containing two silicon atoms bonded to each other, they prepared dimethyldiethylmethane, the first known quarternary hydrocarbon, demonstrating that carbon can bond to four other carbon atoms as silicon was shown to do by Friedel and Crafts in their compound $Si(C_2H_5)_4$. They also prepared silicon analogues of carboxylic acids, ethers, ketones, and alcohols. Believing that compounds of the lower oxidation states of metals contained two atoms of the metal per molecule, Ladenburg began the study of tin compounds. He prepared various organotin compounds, including triethylphenyl tin, but soon realized that such compounds could not resolve the question.

Ladenburg qualified as a teacher in January 1868, set up a laboratory in Heidelberg, and taught a course on the history of chemistry, publishing the revised lectures in 1869. That same year he criticized Kekulé's formula for benzene and suggested alternatives, including his prism formula. Ladenburg argued that in Kekulé's formula the 1, 2- position is different from the the 1, 6- position and the 1, 3- position may be different from the 1, 5- position, whereas experimental evidence supported the identity of these positions. He moved to the University of Kiel in 1872 and, continuing his benzene researches, showed only three disubstitution products of benzene are possible, only one pentachlorobenzene exists, and mesitylene is symmetrical trimethylbenzene. In amassing evidence for the equivalence of the six benzene carbon atoms, Ladenburg recognized that he was weakening support for his prism formula and said in 1875 that there was no symbolic representation of benzene satisfying all requirements. In 1876 he summarized his views on benzene structure in *Die Theorie der aromatischen Verbindungen*, drawing attention to Körner's method for ascertaining the structure of benzene derivatives.

Ladenburg married Margarete Pringsheim, daughter of the professor of botany at Berlin, on 19 September 1876. Soon after, he began the study of alkaloids, concentrating on atropine and its derivatives. In 1884 Ladenburg began the work that resulted in the first synthesis of an alkaloid, coniine. He was the first to show that *d*, *l*- bases can be resolved by Pasteur's method for resolving *d*, *l*- acids. The acid tartrate was prepared and resolved, the coniine liberated, and the dextrorotatory form identified with naturally occurring coniine.

In 1889 Ladenburg became professor at Breslau, where he continued work on alkaloids and racemic compounds. He ascertained the formula of ozone as O_3 and accurately determined the atomic weight of iodine in seeking an answer to the question of its position relative to tellurium in the periodic table.

Ladenburg was a member of the Berlin and Paris academies of sciences. He was an honorary and foreign member of the Chemical Society of London and recipient of the Davy Medal of the Royal Society in 1907.

BIBLIOGRAPHY

I. Original Works. Ladenburg's first book was *Vorträge über die Entwicklungsgeschichte der Chemie in den letzten hundert Jahren* (Brunswick, 1869); there is an English trans. by Leonard Dobbin from the 2nd German ed., *Lectures on the History of the Development of Chemistry Since the Time of Lavoisier* (Edinburgh, 1900). Ladenburg's *Handwörterbuch der Chemie*, which he began compiling at Kiel in 1873, was published in 13 vols. (Breslau, 1882–1896). His controversial lecture, "Einfluss der Naturwissenschaften auf die Weltanschauung," delivered at Kassel, on 21 September 1903, caused bitter feelings between Ladenburg and his friends and colleagues; it was trans. into English by C. T. Sprague as *On the Influence of the Natural Sciences on Our Conceptions of the Universe* (London, 1908). His views on the nature of racemic compounds are summarized in *Über Racemie* (Stuttgart, 1903). Ladenburg wrote a short autobiography toward the end of his life which his son, Rudolph, had published as *Lebenserinnerungen* (Breslau, 1912). For a complete list of Ladenburg's publications see W. Herz, "Albert Ladenburg," in *Berichte der Deutschen chemischen Gesellschaft*, **45** (1912), 3636–3644.

II. Secondary Literature. Drawing from Ladenburg's *Lebenserinnerungen*, the following papers present comprehensive accounts of Ladenburg's life and work: F. S. Kipping, "Ladenburg Memorial Lecture," in *Journal of the Chemical Society*, **103** (1913), 1871–1895; and W. Herz, "Albert Ladenburg," in *Berichte der Deutschen chemischen Gesellschaft*, **45** (1912), 3597–3644. See also M. Delépine, ed., *La synthèse totale en chimie organique* (Paris, 1937), pp. 130–143, for memoirs of Ladenburg and others.

A. Albert Baker, Jr.

LADENBURG, RUDOLF WALTHER (*b.* Kiel, Germany, 6 June 1882; *d.* Princeton, New Jersey, 3 April 1952), *physics.*

Ladenburg was the second of three sons of the eminent chemist Albert Ladenburg and his wife, Margarete Pringsheim.

After his early education in the schools of Breslau, where his father was professor of chemistry in the university, Ladenburg went in 1900 to the University of Heidelberg. He returned to Breslau in 1901 and in 1902 went to Munich, where he took his Ph.D. in 1906 under Roentgen with a thesis on viscosity, a subject that held his interest until he began work in spectroscopy in 1908. From 1906 to 1924 he was on the staff of the University of Breslau, first as privatdocent and from 1909 as extraordinary professor. He married Else Uhthoff in 1911. He served in 1914 as a cavalry officer but later in the war did research in sound ranging. In 1924, at the invitation of Haber, he moved to the Kaiser Wilhelm Institute in Berlin as head of the physics division and remained there until going to Princeton in 1931.

Ladenburg is well known for his research in many fields of physics, but his most original work was done in the elucidation of anomalous dispersion in gases, in the period before 1931. To understand the originality of that work it is necessary to remember that the quantum theory had been proposed by Planck in 1901 but that its application to atomic phenomena did not become possible until after Bohr's paper on hydrogen in 1913. It was in the interim that Ladenburg started his important work.

A paper of considerable significance to an understanding of Ladenburg's mind must be discussed at this point. It is a sixty-page review of the photoelectric effect, published in 1909 in the *Jahrbuch der Radioaktivität und Elektronik*. Much of the experimental elucidation of that effect was the work of Erich Ladenburg, an older brother, whose tragic death by drowning in 1908 affected Rudolf very seriously. The paper is especially noteworthy because in it Ladenburg became one of the first physicists to accept Einstein's view of the constitution of radiation, at least as applied to the photoelectric effect. At that time there were two competing theories of that effect, the commonly accepted resonance theory due to Lenard and the theory which depended on the acceptance of Einstein's idea that radiation consisted of indivisible packets of energy $h\nu$. Ladenburg gives a masterly discussion of the two theories and shows conclusively that Lenard's theory is in difficulties in explaining several of the known facts, whereas Einstein's fits all of them in a quite natural way. At the same time, Ladenburg definitely recognized that Einstein's theory is in contradiction to the classical theory of radiation and gives no explanation of the mechanism of the release of the electrons. In other words, Ladenburg was face to face with the fundamental problem that led to the invention of quantum mechanics some sixteen years later.

The research undertaken by Ladenburg in the early part of his university career was published in five papers on viscosity, including his thesis. He then turned to spectroscopy, which was just becoming of major interest to physicists. His first effort in that new field was directed to a solution of the problem of whether electrically excited hydrogen could absorb the light of the Balmer lines, a question about which there was much conflicting evidence. Ladenburg solved it very quickly and definitely by an experiment in which he ensured that the source of the Balmer lines and the absorbing column of gas were both excited at the same instant, by simply having the two tubes in series in the same induction-coil circuit. He then proceeded to resolve the question by demonstrating that anomalous dispersion could be detected in the close neighborhood of the lines by placing the absorption tube in one arm of a Jamin interferometer. The only existing theory, based on classical electromagnetism, yielded a rough value of $\mathfrak{N}_e = 4 \times 10^{12}$ oscillators per cubic centimeter, equivalent to about one in 50,000 atoms. A further confirming experiment was carried out to measure the magnetic rotation of the plane of polarization of the light passing in the immediate neighborhood of the absorption lines. The small but measurable effect distinguished definitely between two current theories, confirming Voigt's theory based on the Hall effect. It also demonstrated that the oscillators were negatively charged particles. Ladenburg discussed these results in detail in a paper in *Annalen der Physik* (**38** [1912], 249–318).

Although Ladenburg's research had shown unequivocally that the Balmer lines were the seats of anomalous dispersion, the appearance of Bohr's theory of hydrogen in 1913 led surprisingly to attempts by theorists to base the theory of dispersion on the frequencies of the Bohr orbits rather than those of the lines. Ladenburg, of course, knew that the contrary was correct, and after World War I he set himself the task of finding the theoretical relations connecting the constant $\mathfrak{N}_e$ with the radically new way of describing emission and absorption. Using a correspondence-principle argument he found the following expression for $\mathfrak{N}_e$ (which is essentially correct except for the absence of a small correction term that was derived later by Kramers and that leads to negative dispersion):

$$\mathfrak{N}_e \equiv N_i \frac{g_k}{g_i} a_{ki} \frac{mc^3}{8\pi^2 l^2 v_{ik}^2},$$

in which g_k and g_i are the statistical weights of the upper and lower states and a_{ki} is the probability of a spontaneous transition from state k to state i. Ladenburg applied this formula to his observations of both hydrogen and sodium. In the latter element the density of oscillators appeared to be nearly equal to the number N of atoms per cubic centimeter. In hydrogen it was possible to find an approximate value of about four for the ratio of $\mathfrak{N}_e$ for H_α and H_β. This is a quantity that is not calculable from the Bohr theory. The equivalent ratio in sodium and other alkalies is very much greater.

With the advent of quantum mechanics a number of theoretical physicists developed a valid form of the equation for anomalous dispersion in gases involving in particular the factor mentioned above

$$\left(1 - \frac{N_k}{N_j}\frac{g_j}{g_k}\right)$$

which depends on the ratio of the number of atoms in the upper state to the number in the lower state. It appears that increasing excitation of a gas could increase that ratio and so produce a decrease in the refractive index, an effect usually referred to as negative dispersion.

During the years 1921–1928 Ladenburg performed a number of experiments with the aid of his students Kopfermann, Carst, and Levy that led to an important series of eight papers under the general title "Untersuchungen über die anomale Dispersion angeregte Gase." They are concerned mainly with the dispersion in electrically excited neon around the familiar strong yellow to red lines that result from transitions between levels p_1 to p_{10} and the lower levels s_2 to s_5. The new work was done with much improved techniques and gave results of outstanding importance.

The changing characteristics of the discharge through the absorption tube as the current was increased were studied in detail and revealed a close similarity between the curves for $\mathfrak{N}_e$ for all the lines with the same lower level up to a current of about 50 milliamperes. This showed that the population of the lower level was the dominant influence in that range of currents. However, with increasing current the curves of $\mathfrak{N}_e$ fell and diverged in a regular way, showing an unmistakable effect of the upper levels. This is exactly what should be expected from the factor

$$\left(1 - \frac{N_k}{N_j}\frac{g_j}{g_k}\right),$$

which gives the effect of negative dispersion. From the effects of that factor it is possible to derive a specific temperature for each level, all of which lie in the region of 20,000° C. for currents of about 700 milliamperes through the tube.

A further series of five papers, written with G. Wolfsohn, under the title "Untersuchungen über die Dispersion von Gasen und Dämpfen und ihre Darstellung durch die Dispersionstheorie" was published between 1930 and 1933. A new quartz Jamin interferometer made it possible to extend the measurements in mercury through the ultraviolet line 2536 Å using Rozhdestvensky's "hook" method, which had also been used in the work on neon. The value of $f = \mathfrak{N}_e/N$ was found to be constant at .0255 ± .0005 for all pressures from .01 to 200 millimeters of mercury in radical disagreement with previous measurements by R. W. Wood and others. A further paper carries measurements of the refractive index of mercury down to 1890 Å and makes possible an approximate calculation of the *f*-values at 1849 Å and 1413 Å which are the first and second members of the singlet principal series. New measurements were also made of the refractive index of oxygen down to 1920 Å, allowing the construction of a three-term formula depending on certain band frequencies as the centers of anomalous dispersion.

Ladenburg's experimental work on hydrogen and, especially, neon was very important for the theory of dispersion. In addition, it showed for the first time the possibility of obtaining definite results in the extremely difficult field of electrical discharges in gases. His successful formula for anomalous dispersion, which ignored the Bohr orbital frequencies in favor of the line frequencies, was of critical importance to the later development of the new quantum theory during the 1920's.

In 1931 Ladenburg was a visiting professor at Princeton and in 1932 accepted the appointment to the Cyrus Fogg Brackett professorship of physics, a position he held until his retirement in 1950. When he accepted the call to Princeton, he was probably insufficiently aware of the radical differences in the organization of departments of physics in Germany and the United States; furthermore, it may have come as a shock to find that the research professorship did not give him the control over graduate work that it would have given in Germany. The difficulty of reorientation was increased by the change of his chief interest to nuclear physics, a field that was then rapidly displacing spectroscopy.

For Ladenburg's use the Princeton physics department purchased, with funds from the Rockefeller Foundation, a transformer-rectifier set with a capacity of 400,000 volts and thirty milliamperes. It was considered that the high current would compensate for the low voltage, but unfortunately this hope was not realized because of the limitations in the available accelerating tubes and ion sources.

The apparatus was initially used for experiments on the light elements and eventually exclusively for the production of neutrons from the D-D reaction for a number of experiments including a determination of the relative cross sections for fission of uranium and thorium in the energy range 200 to 300 kilovolts.

World War II, which disrupted all physics research at Princeton, brought Ladenburg into contact with the Army Ballistic Laboratories, for which he developed a flash suppressor for rifles. This later led him to his postwar research on gas dynamics.

Refractivity changes accompanying density changes in a compressible flow field permit the use of optical methods in laboratory research. It was here that Ladenburg made the contribution for which he is best remembered by fluid dynamicists—the introduction in the United States of optical interferometry as a quantitative tool to map the density distribution in high-speed flows. He and his associates were the first to make systematic interferometric studies of over- and underexpanded supersonic jet flows and channel flows. The method was quickly adopted by many research laboratories for use with wind tunnels, shock tubes, ballistic ranges, and similar devices and in plasma dynamics. It is evident that the method will remain a major tool for the study of the physics of gases and gas flows.

In his social relationships Ladenburg was an extremely agreeable and hospitable person, but professionally he could be a very severe taskmaster, both to his students and to himself. An enthusiastic experimenter, he did not spare himself when great efforts were necessary. He was a very good teacher indeed and spent a great deal of time providing his graduate students with advice as well as instruction.

As early as 1933, the plight of German scientists under Hitler was causing great concern, and many had had to emigrate from Germany to other countries. The sympathy aroused in Ladenburg and E. P. Wigner by that situation led them to address an appeal to many physicists in American institutions for definite pledges of support for displaced colleagues, listing many by name. The work thus started was carried on by Ladenburg throughout the 1930's and 1940's by correspondence with scientists in many countries, including Germany. It was a work of the heart and showed that the human side of his nature was more important to him than the professional side. Einstein wrote after Ladenburg's death: "Ladenburg has been caught very suddenly by illness. He was a good human being who did not take things easily. During his last years he even fled from newspaper reading because he could not stand any more of the hypocrisy and mendacity" (*Albert Einstein–Max Born, Briefwechsel* 1916–1955 [Munich, 1969], p. 257).

BIBLIOGRAPHY

I. ORIGINAL WORKS. Ladenburg's name appears on 134 scientific papers; 34 other papers published under the names of his students are an integral part of his own principal researches. His doctoral thesis was "Die innere Reibung zäher Flüssigkeiten und ihre Abhängigkeit vom Druck," in *Annalen der Physik*, **22** (1907), 287–309. Also on viscosity are "Über den Einfluss von Wänden auf die Bewegung einer Kugel in einer reibenden Flüssigkeit," *ibid.*, **23** (1907), 447–458; "Einfluss der Reibung auf die Schwingungen einer Kugel," *ibid.*, **27** (1908), 157–185; and "Über die Reibung tropbares Flüssigkeiten," in *Jahresbericht der Schlesischen Gesellschaft für vaterländische Kultur* (1908), pp. 1–4.

Ladenburg's work on dispersion in gases is described in 58 papers, of which the most important are "Über die Dispersion des Leuchtenden Wasserstoffs," in *Physikalische Zeitschrift*, **9** (1908), 875–878; "Über die anomale Dispersion und die magnetische Drehung der Polarisationebene des leuchtenden Wasserstoffs, sowie über die Verbreitering von Spektrallinien," in *Annalen der Physik*, **38** (1912), 249–318; "Über selektive Absorption," *ibid.*, **42** (1913), 181–209; "Die Quantentheoretische Deutung der Zahl der Dispersionselektronen," in *Zeitschrift für Physik*, **4** (1921), 451–468; 8 papers (I to VIII) under the title "Untersuchungen über die anomale Dispersion angeregte Gase," written with H. Kopfermann, A. Carst, and S. Levy, of which no. VIII is *Zeitschrift für Physik*, **88** (1934), 461–468; and 5 papers under the title "Untersuchungen über die Dispersion von Gasen und Dämpfen und ihre Darstellung durch die Dispersionstheorie," written with G. Wolfsohn, of which no. V is *Zeitschrift für Physik*, **85** (1933), 366–372.

There are 12 papers on anomalous dispersion in sodium, of which the last is "Die Oszillatorenstärke der D-linien," in *Zeitschrift für Physik*, **72** (1931), 697–699.

During World War I Ladenburg did research on problems of sound; his last publication was "Experimentalle Beiträge zur Ausbreitung des Schalles in der freien Atmosphere," in *Annalen der Physik*, **66** (1921), 293–322, written with E. von Angerer.

In 1929 Ladenburg undertook work on the cleaning of effluent gases and published a number of papers, the last of which is "Elektrische Gasreinigung," in *Chemie-Ingenieur*, **1**, pt. 4 (1934), 31–81.

In nuclear physics, Ladenburg's principal papers are "On the Neutrons from the Deuteron-Deuteron Reaction," in *Physical Review*, **52** (1937), 911–918; "Study of Uranium Fission by Fast Neutrons of Nearly Homogeneous Energy," *ibid.*, **56** (1939), 168–170, written with M. H. Kanner, H. H. Barschall, and C. C. Van Voorhis; "Mass of the Meson by the Method of Momentum Loss," *ibid.*, **60** (1941), 754–761, written with J. A. Wheeler; and "Elastic and Inelastic Scattering of Fast Neutrons," *ibid.*, **61** (1942), 129–138, written with H. H. Barschall.

Ladenburg's interest in the atmosphere led to a number of papers of which the most important are "The Continuous Absorption of Oxygen Between 1750 Å and 1300 Å and Its Bearing Upon the Dispersion," *ibid.*, **43** (1933), 315–321, written with C. C. Van Voorhis; and "Light Absorption and Distribution of Atmospheric Ozone," in *Journal of the Optical Society of America*, **25** (1935), 259–269.

Ladenburg's early papers in gas dynamics are "Interferometric Study of Supersonic Phenomena," written with C. C. Van Voorhis and J. Winckler, in *Navy Department Bureau of Ordnance*, Washington, D.C. (1946), pt. 1, no. 69–46; (17 Apr. 1946), 1–84; pt. 2, no. 93–46; (2 Sept. 1946), 1–51; pt. 3, no. 7–47; (19 Feb. 1947), 1–22. Other papers of importance in this field are "Interferometric Studies of Faster than Sound Phenomena," in *Physical Review*, **73** (1948), 1359–1377 and **76** (1949), 662–677, written with C. C. Van Voorhis and J. Winckler; and "Interferometric Studies of Laminar and Turbulent Boundary Layers Along a Plane Surface at Supersonic Velocities," in *Symposium on Experimental Compressible Flow*, Naval Ordinance Laboratory Report, no. 1133 (1 May 1950), pp. 67–87, written with D. Bershader.

Ladenburg's most important monographs include "Die neueren Forschungen über die durch Licht- und Röntgenstrahlen hervorgerufene Emission negativer Elektronen," in *Jahrbuch der Radioaktivität und Elektronik*, **6** (1909), 425–484; "Bericht über die Bestimmung von Planck's elementaren Wirkungsquantum h," in *Jahrbuch der Radioaktivität und Elektronik*, **17** (1920), 93–145; and **17** (1920), 273–276; "Die Deutung der kontinuierlighen Absorptions- und Emissions-spektra von Atomen in Bohr's Theorie," *ibid.*, **17** (1920), 429–434; "Methoden zur h-Bestimmung und ihre Ergebnisse," in Geiger-Scheel, ed., *Handbuch der Physik* (Berlin), **23** (1926), 279–305; "Die Bestimmung der Lichtgeschwindigkeit in Ruhenden Körpern," in Wien-Harms, ed., *Handbuch der Experimentalphysik*, **18** (1928), 3–34; "Die Methoden zur h-Bestimmung und ihre Ergebnisse," in Geiger-Scheel, ed., *Handbuch der Physik*, 2nd ed., **23** (1933), 1; "Dispersion in Electrically Excited Gases," in *Review of Modern Physics*, **5** (1933), 243–256; "On Laminar and Turbulent Boundary Layer in Supersonic Flow," *ibid.*, **21** (1949), 510–515, written with D. Bershader; and "Interferometry," in *High Speed Aerodynamics and Jet Propulsion*, VII, sec. A3 (Princeton, N.J., 1954), 47–75, written with D. Bershader. Ladenburg was also the general editor of vol. VII.

II. SECONDARY LITERATURE. An obituary notice is Hans Kopfermann, "Rudolf Ladenburg," in *Naturwissenschaften*, **13** (1952), 289–290.

A. G. SHENSTONE

LAENNEC, THÉOPHILE-RENÉ-HYACINTHE (*b.* Quimper, France, 12 February 1781; *d.* Kerbouarnec, Brittany, France, 13 August 1826), *medicine.*

Laennec's mother died in her early thirties, probably from tuberculosis. His father, a lieutenant at the admiralty in Quimper, being unable to care for his children, Théophile was sent to his uncle, a physician at Nantes; there he was introduced to medical work. The French Revolution struck Nantes fiercely, and Laennec worked in the city's hospitals. In 1795 he was commissioned a third surgeon at the Hôpital de la Paix and shortly afterward at the Hospice de la Fraternité. It was at the latter that Laennec became acquainted with clinical work, surgical dressings, and treatment of patients. His health was not good, for he suffered from lassitude and occasional periods of pyrexia. He found consolation in music and spent his spare time playing the flute and writing poetry.

His father wished him to abandon the study of medicine, and during a period of indecision Laennec wasted time at Quimper, dancing, taking country rambles, and playing the flute, with occasional study of Greek. In June 1799 he returned to his medical studies and was appointed surgeon at the Hôtel-Dieu in Nantes. From there he entered the École Pratique in Paris and studied dissection in Dupuytren's laboratory. The following year, in June 1802, he published his first paper in *Journal de médecine*, "Observations sur une maladie de coeur"; it was followed in August, in the same journal, by "Histoire des inflammations du péritoine." His colleagues at this time were Gaspard Bayle, Xavier Bichat, Le Roux, and Corvisart. Bayle's early death from tuberculosis caused Laennec much sorrow; and this, coupled with his dislike for Dupuytren, nearly resulted in his leaving Paris. Bichat persuaded him not to go and together they published a number of papers on anatomy in *Journal de médecine* in 1802 and 1803.

In 1803 Laennec was awarded the prize for surgery and shared the prize for medicine awarded by the Grandes Écoles of Paris. His reputation increased, and he began to give private instruction in morbid anatomy to supplement his meager income. Although suffering from asthma, he worked hard and announced his classification of anatomical lesions into encephaloid and scirrhous types. He also found that the tubercle lesion could be present in all organs of the body and was identical with that which had previously been thought to be limited to the lungs; he did not, however, realize that the condition was infectious. His thesis, "Propositions sur la doctrine d'Hippocrate relativement à la médecine-pratique," was presented and accepted in July 1804; he thereby became an associate of the Société de l'École de Médecine.

Family troubles, the death of his uncle from tuberculosis, and financial difficulties, coupled with his break with Dupuytren, disturbed the continuity of Laennec's work and caused his health to fail. He recovered by going to Brittany, and on his return to Paris he became an editor-shareholder of the *Journal de médecine*. Private practice increased; but he was disappointed in not being appointed assistant physician at the Hôtel-Dieu in Paris, physician to the emperor's pages, or head of the department of anatomical studies. Taking the initiative himself, in 1808 he founded the Athénée Médical, which merged with the Société Académique de Médecine de Paris. Shortly afterward he was appointed personal physician to Cardinal Fesch, the uncle of Napoleon I, but the cardinal was exiled after the fall of Napoleon. After failing to be elected to the chair of Hippocratic medicine and rare cases, Laennec began preparing articles on pathological anatomy and ascarids for the *Dictionnaire des sciences médicales*. At this time (1812–1813) France was at war, and Laennec took charge of the wards in the Salpêtrière reserved for wounded Breton soldiers.

On the restoration of the monarchy, Laennec settled down to routine work but failed to obtain the chair of forensic medicine; reluctantly he accepted the post of physician to the Necker Hospital. It was here that he became interested in emphysema, tuberculosis, and physical signs of the chest. Although auscultation had been known since the days of Hippocrates, it was always done by the "direct" method, which often was very inconvenient. Laennec introduced what he called the "mediate" method, using a hollow tube for listening to the lungs and a solid wooden rod for heart sounds; by February 1818 he was able to present a paper on the subject to the Académie de Médecine. His poor health obliged him to live in Brittany as a gentleman farmer for some months; but by the end of 1818 he returned to Paris and soon was able to classify the physical signs of egophony, rales, rhonchi, and crepitations, which he described in detail in his book *De l'auscultation médiate*. The work was published in August 1819 and was acknowledged to be a great advance in the knowledge of chest diseases.

Again increasing asthma, headaches, and dyspepsia forced Laennec to return in October 1819 to his estates at Kerbouarnec, Brittany, where he assumed the role of a country squire; however the possibility of election to the chair of medicine in Paris led him back to his clinic at the Necker Hospital. Owing to personal animosities, it was not until July 1822 that he was appointed to the chair and a lectureship at the Collège de France. After this honors came rapidly to Laennec. In January 1823 he became a full member of the Académie de Médecine, and in August 1824 he was made a chevalier of the Legion of Honor. His private practice increased and included many distinguished persons. As a lecturer he became internationally famous; at times as many as fifty doctors awaited his arrival at the Charité Hospital, to which he had transferred his clinical work from the Necker.

Laënnec's health at this time was fairly good. Feeling the need of help in his domestic affairs, he engaged as housekeeper a Mme. Argon, whom he married in December 1824. The happy union helped him to publish a new edition of his book and enter the competition for the Montyon Prize in physiology, but the extra work caused a return of his chest symptoms and forced him to leave Paris in May 1826, never to return. The climate of Brittany brought a temporary improvement in his health, but he died on 13 August of that year.

BIBLIOGRAPHY

I. Original Works. Laennec's writings are "Note sur l'arachnoïde intérieure, ou sur la portion de cette membrane qui tapisse les ventricules du cerveau," in *Journal de médecine*, **5** (1803), 254–263; "Note sur une capsule synoviale située entre l'apophyse acromion et l'humerus," *ibid.*, 422–426; "Lettre sur des tuniques qui enveloppent certains viscères et fournissent des gaines membraneuses à leurs vaisseaux," *ibid.*, 539–575, **6** (1803), 73–90; *Propositions sur la doctrine d'Hippocrate* (Paris, 1804), his thesis presented on 11 June 1804 at the École de Médecine (repr., Paris, 1923); "Mémoire sur les vers vesiculaires et principalement sur ceux qui se trouvent dans le corps humain," in *Mémoires de la Société de la Faculté de médecine* (1804), 176; and "Mémoire sur le distomus intersectus, nouveau genre de vers intestins," *ibid.* (1809), 1–281.

De l'auscultation médiate, 2 vols. (Paris, 1819), appeared in many subsequent editions and translations; there is an English trans. of selected passages from the 1st ed., with biography by William Hale-White (New York, 1923).

II. Secondary Literature. Perhaps the most intimate biography is R. Kervran, *Laënnec: His Life and Times* (New York–Oxford, 1960), translated from the French by D. C. Abrahams-Curiel, which contains a comprehensive bibliography of the literature on Laennec.

Frederick Heaf

LA FAILLE, CHARLES DE (*b.* Antwerp, Belgium, 1 March 1597; *d.* Barcelona, Spain, 4 November 1652), *mathematics.*

The son of Jean Charles de La Faille, seigneur de Rymenam, and Marie van de Wouwere, Charles de La Faille received his early schooling at the Jesuit College of his native city. On 12 September 1613 he became a novitiate of the Jesuit order at Malines for two years. Afterward he was sent to Antwerp where he met Gregory of St. Vincent, who was renowned for his work on quadrature of the circle. La Faille was counted among Gregory's disciples, and in 1620 he was sent to France to follow a course of theology at Dôle, and to teach mathematics. After his return to Belgium in 1626, he taught mathematics at the Jesuit College of Louvain for the next two years. In 1629 he was appointed professor at the Imperial College in Madrid; he departed for Spain 23 March

1629. In 1644 Philip IV appointed him preceptor to his son Don Juan of Austria, whom he also accompanied on his expeditions to Naples, Sicily, and Catalonia. He died in Barcelona in 1652, a month after the capture of the town by Don Juan.

La Faille owed his fame as a scholar to his tract *Theoremata de centro gravitatis partium circuli et ellipsis*, published at Antwerp in 1632. In it the center of gravity of a sector of a circle is determined for the first time. In the first nine propositions each is established step by step. His procedure can be rendered as follows: If α is the angle of a given sector of a circle with radius R, and β is a sufficiently small angle of the same sector, the length

$$\frac{R\alpha^2}{\sin^2 \alpha}\left(1 - \frac{\sin \beta}{\beta}\right)$$

can be made arbitrarily small. For his proof, La Faille supposed that there can be constructed on one of the radii a triangle the area of which is equal to that of the sector. Of the next five propositions the first is especially interesting: If there are three lines AB, AC, and AD, and the straight line BCD cuts the lines given in such a way that $BD : BC =$ angle BAD : angle CAD, then $AD < AC < AB$, if $BAD < CAD$. The proof is based on the theorem which La Faille found in Clavius (*De sinibus*, prop. 10); it is also to be found in the first book of the *Almagest*.

In the next eight propositions the author proved that the centers of gravity of a sector of a circle, of a regular figure inscribed in it, of a segment of a circle, or of an ellipse lie on the diameter of the figure. These theorems are founded on a postulate from Luca Valerio's *De centro gravitatis solidorum* (1604). In his proofs, La Faille referred to Archimedes' *On the Equilibrium of Planes or Centers of Gravity of Planes* (book I). Propositions 23–31 lead to the proof that the distance between the center of gravity of a sector of a circle and the center of the circle is less than $\frac{2}{3}R$, but the difference between this distance and $\frac{2}{3}R$ can be made arbitrarily small by making the angle of the sector sufficiently small. Proposition 32, the main one of the work, can be rendered as follows: If A is the angle of a sector of a circle with radius R, the center of gravity lies on the bisector, and the distance d to the vertex of the angle of the sector is given by

$$d = \frac{2}{3}R\,\frac{\text{chord } A}{\text{arc } A}$$

Propositions 33–37 are consequences of 32, and 38–45 are an extension of the results on a sector and segment of an ellipse. La Faille ended his work with four corollaries which revealed his ultimate goal: an examination of the quadrature of the circle.

BIBLIOGRAPHY

According to C. Sommervogel, *Bibliothèque de la Compagnie de Jésus*, III (Brussels–Paris, 1897), cols. 529–530, there are some more works of La Faille in Spanish, but all of them are manuscripts and nothing is known about their contents. Moreover, there exists the correspondence of La Faille with the astronomer M. van Langren covering the period 20 Apr. 1634–25 Sept. 1645.

A very extensive biography was written by H. P. van der Speeten, "Le R.P. Jean Charles della Faille, de la Compagnie de Jésus, Précepteur de Don Juan d'Autriche," in *Collection de Précis Historiques*, **3** (1874), 77–83, 111–117, 132–142, 191–201, 213–219, and 241–246. Some information on his life and work can be found in A. G. Kästner, "Geschichte der Mathematik," **2** (Göttingen, 1797), 211–215; H. G. Zeuthen, "Geschichte der Mathematik im 16. und 17. Jahrhundert" (Leipzig, 1903), pp. 238–240.

See also H. Bosmans, "Deux lettres inédites de Grégoire de Saint-Vincent publiées avec des notes bibliographiques sur les oeuvres de Grégoire de Saint-Vincent et les manuscrits de della Faille," in *Annales de la Société Scientifique de Bruxelles*, **26** (1901–1902), 22–40; H. Bosmans, "Le traité 'De centro gravitatis' de Jean-Charles della Faille," *ibid.*, **38** (1913–1914), 255–317; H. Bosmans, "Le mathématicien anversois Jean-Charles della Faille de la Compagnie de Jésus," in *Mathésis*, **41** (1927), 5–11; and J. Pelseneer, "Jean Charles de la Faille (Anvers 1597–Barcelona 1652)," in *Isis*, **37** (1947), 73–74.

H. L. L. BUSARD

LAGNY, THOMAS FANTET DE (*b.* Lyons, France, 7 November 1660; *d.* Paris, France, 11 April 1734), *mathematics, computation.*

There are certain obscurities in our knowledge of Lagny's life, talented calculator though he was. Fantet was the name of his father, a royal official in Grenoble. It appears that Lagny studied with the Jesuits in Lyons and then at the Faculty of Law in Toulouse. In 1686 he appeared in Paris under the name of Lagny. He was a tutor in the Noailles family and the author of a study on coinage. His collaboration with L'Hospital and his first publications concerning the approximate calculation of irrationals (1690–1691) show that he was a good mathematician. He was living in Lyons when he was named an associate of the Académie Royale des Sciences on 11 December 1695. He stayed in Paris in 1696 and then, in 1697, through the Abbé Jean-Paul Bignon, obtained an appointment as professor of hydrography at Rochefort. This position assured him a salary but in a distant residence which allowed him only written contact with the Academy. His former pupil, the Maréchal Duc de Noailles, president of the Conseil des Finances of the regency, called upon him in 1716 to assume the deputy directorship of the Banque Générale founded by John Law. He resigned this job in 1718, at the time of the institution's transformation into the Banque Royale, and was not involved in the bankruptcy that shook the French state.

A *pensionnaire* of the Academy from 7 July 1719, Lagny finally earned his living from science, as he wished to do, but he was growing weaker and could barely revise his old manuscripts. His declining powers obliged him to retire in 1733, and the Academy completed the book that he planned to crown his work.

Lagny's work belonged to a type of computational mathematics at once outmoded and unappreciated. He lived during the creation of integral calculus without being affected by it. While the idea of the function was gaining dominance, he continued to approach mathematical problems—both ancient problems such as the solution of equations and new ones such as integration—with the aid of numerical tables. Employing with great skill the property possessed by algebraic forms of corresponding to tables in which the differences of a determined order are constant, he recognized the existence of transcendental numbers in the calculation of series.

Lagny made pertinent observations on convergence, in connection with the series that he utilized to calculate the first 120 decimal places in the value of π. He attempted to establish trigonometric tables through the use of transcription into binary arithmetic, which he termed "natural logarithm" and the properties of which he discovered independently of Leibniz.

In this regard his meeting with the inventor of the differential and integral calculus is interesting, but it was only the momentary crossing of very different paths. Lagny generally confined himself to numerical computation and practical solutions, notably the goniometry necessary for navigators. Nevertheless, his works retain a certain didactic value.

BIBLIOGRAPHY

I. Original Works. Lagny's writings include "Dissertation sur l'or de Toulouse," in *Annales de la ville de Toulouse* (Toulouse, 1687), I, 329–344; *Méthode nouvelle infiniment générale et infiniment abrégée pour l'extraction des racines quarrées, cubiques* ... (Paris, 1691); *Méthodes nouvelles et abrégées pour l'extraction et l'approximation des racines* (Paris, 1692); *Nouveaux élémens d'arithmétique et d'algèbre ou introduction aux mathématiques* (Paris, 1697); *Trigonométrie française ou reformée* (Rochefort, 1703), on binary arithmetic; *De la cubature de la sphère où l'on démontre une infinité de portions de sphères égales à des pyramides rectilignes* (La Rochelle, 1705); and *Analyse générale ou Méthodes nouvelles pour résoudre les problèmes de tous les genres et de tous les degrés à l'infini*, M. Richer, ed. (Paris, 1733).

Lagny addressed many memoirs to the Academy, and most of them were published. Perhaps the most important is "Quadrature du cercle," in *Histoire et Mémoires de l'Académie . . . pour 1719* (Amsterdam, 1723), pp. 176–189.

There is a portrait of Lagny in the Lyons municipal library, no. 13896.

II. Secondary Literature. See Jean-Baptiste Duhamel, *Regiae scientiarum academiae historia* (Paris, 1698), pp. 430–432; and B. de Fontenelle, "Éloge de M. de Lagny," in *Histoire et Mémoires de l'Académie . . . pour 1734* (Amsterdam, 1738), 146–155.

PIERRE COSTABEL

LAGRANGE, JOSEPH LOUIS (*b.* Turin, Italy, 25 January 1736; *d.* Paris, France, 10 April 1813), *mechanics, celestial mechanics, astronomy, mathematics.*

Lagrange's life divides very naturally into three periods. The first comprises the years spent in his native Turin (1736–1766). The second is that of his work at the Berlin Academy, between 1766 and 1787. The third finds him in Paris, from 1787 until his death in 1813.

The first two periods were the most fruitful in terms of scientific activity, which began as early as 1754 with the discovery of the calculus of variations and continued with the application of the latter to mechanics in 1756. He also worked in celestial mechanics in this first period, stimulated by the competitions held by the Paris Academy of Sciences in 1764 and 1766.

The Berlin period was productive in mechanics as well as in differential and integral calculus. Yet during that time Lagrange distinguished himself primarily in the numerical and algebraic solution of equations, and even more in the theory of numbers.

Lagrange's years in Paris were dedicated to didactic writings and to the composition of the great treatises summarizing his mathematical conceptions. These treatises, while closing the age of eighteenth-century mathematics, prepared and in certain respects opened that of the nineteenth century.

Lagrange's birth and baptismal records give his name as Lagrangia, Giuseppe Lodovico, and declare him to be the legitimate son of Giuseppe Francesco Lodovico Lagrangia and Teresa Grosso. But from his youth he signed himself Lodovico LaGrange or Luigi Lagrange, adopting the French spelling of the patronymic. His first published work, dated 23 July 1754, is entitled "Lettera di Luigi De la Grange Tournier." Until 1792 he and his correspondents frequently employed the particle, very common in France but quite rare in Italy, and named him in three words: de la Grange. The contract prepared for his second marriage (1792) is in the name of Monsieur Joseph-Louis La Grange, without the particle. In 1814 the *éloge* written for him by Delambre, the permanent secretary of the mathematics section of the Institut de France, was entitled "Notice sur la vie et les ouvrages de M. le Comte J. L. Lagrange"; and his death certificate designated him Monsieur Joseph Louis Lagrange, sénateur. As for the surname Tournier, he used it for only a few years, perhaps to distinguish himself from his father, who held office in Turin.

His family was, through the male members, of French origin, as stated in the marriage contract of 1792. His great-grandfather, a cavalry captain, had passed from the service of France to that of Charles Emmanuel II, duke of Savoy, and had married a Conti, from a Roman family whose members included Pope Innocent XIII.

His grandfather—who married Countess Bormiolo—was Treasurer of the Office of Public Works and Fortifications at Turin. His father and later one of his brothers held this office, which remained in the family until its suppression in 1800. Lagrange's mother, Teresa Gros, or Grosso, was the only daughter of a physician in Cambiano, a small town near Turin. Lagrange was the eldest of eleven children, most of whom did not reach adulthood.

Despite the official position held by the father—who had engaged in some unsuccessful financial speculations—the family lived very modestly. Lagrange himself declared that if he had had money, he probably would not have made mathematics his vocation. He remained with his family until his departure for Berlin in 1766.

Lagrange's father destined him for the law—a profession that one of his brothers later pursued—and Lagrange offered no objections. But having begun the study of physics under the direction of Beccaria and of geometry under Filippo Antonio Revelli, he quickly became aware of his talents and henceforth devoted himself to the exact sciences. Attracted first by geometry, at the age of seventeen he turned to analysis, then a rapidly developing field.

In 1754 Lagrange had a short essay printed in the form of a letter written in Italian and addressed to the geometer Giulio da Fagnano. In it he developed a formal calculus based on the analogy between Newton's binomial theorem and the successive differentiations of the product of two functions. He also communicated this discovery to Euler in a letter written in Latin slightly before the Italian publication. But in August 1754, while glancing through the scientific correspondence between Leibniz and Johann Bernoulli, Lagrange observed that his "discovery" was in fact their property and feared appearing to be a plagiarist and impostor.

This unfortunate start did not discourage Lagrange. He wrote to Fagnano on 30 October 1754 that he had been working on the tautochrone. This first essay is lost, but we know of two later memoirs on the same subject. The first was communicated to the Berlin Academy on 4 March 1767.[1] Criticized by the French Academician Alexis Fontaine des Bertins, Lagrange responded in "Nouvelles réflexions sur les tautochrones."[2]

At the end of December 1755, in a letter to Fagnano alluding to correspondence exchanged before the end of 1754, Lagrange speaks of Euler's *Methodus inveniendi lineas curvas maximi minimive proprietate gaudentes, sive solutio problematis isoperimetrici latissimo sensu accepti*, published at Lausanne and Geneva in 1744. The same letter shows that as early as the end of 1754 Lagrange had found interesting results in this area, which was to become the calculus of variations (a term coined by Euler in 1766).

On 12 August 1755 Lagrange sent Euler a summary, written in Latin, of the purely analytical method that he used for this type of problem. It consisted, he wrote in 1806, in varying the y's in the integral formula in x and y, which should be a maximum or a minimum by ordinary differentiations, but relative to another characteristic δ, different from the ordinary characteristic d. It was further dependent on determining the differential value of the formula with respect to this new characteristic by transposing the sign δ after the signs d and $\int$ when it is placed before. The differentials of δy under the $\int$ signs are then eliminated through integration by parts.

In a letter to d'Alembert of 2 November 1769 Lagrange confirmed that this method of maxima and minima was the first fruit of his studies—he was only nineteen when he devised it —and that he regarded it as his best work in mathematics.

Euler replied to Lagrange on 6 September 1755 that he was very interested in the technique. Lagrange's merit was likewise recognized in Turin; and he was named, by a royal decree of 28 September 1755, professor at the Royal Artillery School with an annual salary of 250 crowns—a sum never increased in all the years he remained in his native country.

In 1756, in a letter to Euler that has been lost, Lagrange applied the calculus of variations to mechanics. Euler had demonstrated, at the end of his *Methodus*, that the trajectory described by a material point subject to the influence of central forces is the same as that obtained by supposing that the integral of the velocity multiplied by the element of the curve is either a maximum or a minimum. Lagrange extended "this beautiful theorem" to an arbitrary system of bodies and derived from it a procedure for solving all the problems of dynamics.

Euler sent these works of Lagrange to his official superior Maupertuis, then president of the Berlin Academy. Finding in Lagrange an unexpected defender of his principle of least action, Maupertuis arranged for him to be offered, at the earliest opportunity, a chair of mathematics in Prussia, a more advantageous position than the one he held in Turin. This proposition, transmitted through Euler, was rejected by Lagrange out of shyness; and nothing ever came of it. At the same time he was offered a corresponding membership in the Berlin Academy, and on 2 September 1756 he was elected an associate foreign member.

In 1757 some young Turin scientists, among them Lagrange, Count Saluzzo (Giuseppe Angelo Saluzzo di Menusiglio), and the physician Giovanni Cigna, founded a scientific society that was the origin of the Royal Academy of Sciences of Turin. One of the main goals of this society was the publication of a miscellany in French and Latin, *Miscellanea Taurinensia ou Mélanges de Turin*, to which Lagrange contributed fundamentally. The first three volumes appeared at the beginning of the summers of 1759 and 1762 and in the summer of 1766, during which time Lagrange was in Turin. The fourth volume, for the years 1766–1769, published in 1773, included four of his memoirs, written in 1767, 1768, and 1770 and sent from Berlin.

The first three volumes contained almost all the works Lagrange published while in Turin, with the following exceptions: the courses he gave at the Artillery School on mechanics and differential and integral calculus, which remained in manuscript and now appear to have been lost; the two memoirs for the competitions set by the Paris Academy of Sciences in 1764 and 1766; and his contribution to Louis Dutens's edition of Leibniz' works.

In volume 1 of the *Mélanges de Turin* are Lagrange's "Recherches sur la Méthode de maximis et minimis,"[3] really an introduction to the memoir in volume 2 on the calculus of variations (dating, as noted above, from the end of 1754).

Another short memoir, "Sur l'intégration d'une équation différentielle à différences finies, qui contient la théorie des suites récurrentes,"[4] was cited by Lagrange in 1776 as an introduction to investigations on the calculus of probabilities that he was unable to develop for lack of time. There is also his unfinished and unpublished translation of Abraham de Moivre's *The Doctrine of Chances*, the third edition of which appeared in 1756. Lagrange mentioned this translation —which seems to have been lost—in a letter to Laplace of 30 December 1776.[5]

Recherches sur la nature et la propagation du son[6] constitutes a thorough and extensive study of a question much discussed at the time. In it Lagrange displays an astonishing erudition. He had read and pondered the writings of Newton, Daniel Bernoulli, Taylor, Euler, and d'Alembert; and his own contribution to the problem of vibrating strings makes him the equal of his predecessors.

Work of the same order is presented in "Nouvelles recherches sur la nature et la propagation du son"[7] and "Additions aux premières recherches,"[8] both published in volume 2 of the *Mélanges*. His most important contribution to this volume, though, is "Essai d'une nouvelle méthode pour déterminer les maxima et les minima des formules intégrales indéfinies,"[9] a rather brief memoir in which Lagrange published his analytic techniques of the calculus of variations. Here he developed the insights contained in his Latin letter to Euler of 1755 and added two appendixes, one dealing with minimal surfaces and the other with polygons of maximal area. Although published in 1762, the memoir and its first appendix were written before the end of 1759.

"Application de la méthode précédente à la solution de différens problèmes de dynamique"[10] made the principle of least action, joined with the theorem of *forces vives* (or *vis viva*), the very foundation of dynamics. Rather curiously, Lagrange no longer used the expression "least action," which he had employed until then, a minor failing due, perhaps, to the death of Maupertuis. This memoir heralded the *Mécanique analytique* of 1788 in its style and in the breadth of the author's views.

Volume 3 of the *Mélanges de Turin* contains "Solution de différens problèmes de calcul intégral."[11] An early section treats the integration of a general affine equation of arbitrary order. Lagrange here employed his favorite tool, integration by parts. He reduced the solution of the equation with second member to that of the equation without second member. This discovery dates—as we know from the correspondence with d'Alembert[12]—from about the end of 1764.

Lagrange's research also encompassed Riccati's equations and a functional equation, which he treated in a very offhand manner. Examining some problems on fluid motion, he outlined a study of the function later called Laplacian. He was following Euler, but with the originality that marked his entire career.

The consideration of the movement of a system of material points making only infinitely small oscillations around their equilibrium position led Lagrange to a system of linear differential equations. In integrating it he presented for the first time—explicitly—the notion of the characteristic value of a linear substitution.

Lagrange finally arrived at applications to the theory of Jupiter and Saturn. In September 1765 he wrote on this subject that, due to lack of time, he was contenting himself with applying the formulas he had just discovered to the variations in the eccentricity and position of the aphelia of the two planets and to those in the inclination and in the position of the nodes of their orbits. These were inequalities that "no one until now has undertaken to determine with all the exactitude" demanded.

Investigations of this kind were related to the prize questions proposed by the Paris Academy of Sciences. In 1762 it established a competition, for 1764, based on the question "Whether it can be explained by any physical reason why the moon always presents almost the same face to us; and how, by observations and by theory, it can be determined whether the axis of this planet is subject to some proper movement similar to that which the axis of the earth is known to perform, producing precession and nutation."

In 1763 Lagrange sent to Paris "Recherches sur la libration de la lune dans lesquelles on tâche de résoudre la question proposée par l'Académie royale des sciences pour le prix de l'année 1764."[13] In this work he provided a satisfactory explanation of the equality of the mean motion of translation and rotation but was less successful in accounting for the equality of the movement of the nodes of the lunar equator and that of the nodes of the moon's orbit on the ecliptic.

Lagrange also fruitfully applied the principle of virtual velocities, which is intimately and necessarily linked with his techniques in the calculus of variations. He also made it the basis of his *Mécanique analytique* of 1788. This principle has the advantage, over that of least action, of including the latter principle as well as the principle of *forces vives* and thus of giving mechanics a unified foundation. He had not yet achieved a unified point of view in the memoir published in 1762. Arriving at three differential equations, he demonstrated that they are identical to those relating to the precession of the equinoxes and the nutation of the earth's axis that d'Alembert presented in the *Mémoires* of the Paris Academy for 1754. Lagrange returned to this question and gave a more complete solution of it in "Théorie de la libration de la lune et des autres phénomènes qui dépendent de la figure non sphérique de cette planète," included in the *Mémoires* of the Berlin Academy for 1780 (published in 1782).[14] Laplace wrote to him on this subject on 10 February 1783: "The elegance and the generality of your analysis, the fortunate choice of your coordinates, and the manner in which you treat your differential equations, especially those of the movement of the equinoctial points and of the inclination of the lunar equator; all that, and the sublimity of your results, has filled me with admiration."

In 1763 d'Alembert, then on his way to Berlin, was not a member of the jury that judged Lagrange's entry. He had already been in correspondence with Lagrange but did not know him personally. Nevertheless, he had been able to judge of his ability through the *Mélanges de Turin.* In the meantime the Marquis Caraccioli, ambassador from the kingdom of Naples to the court of Turin, was transferred by his government to London. He took along the young Lagrange, who until then seems never to have left the immediate vicinity of Turin.

Lagrange departed his native city at the beginning of November 1763 and was warmly received in Paris, where he had been preceded by his memoir on lunar libration. He may perhaps have been treated too well in the Paris scientific community, where austerity was not a leading virtue. Being of a delicate constitution, Lagrange fell ill and had to interrupt his trip. His mediocre situation in Turin aroused the concern of d'Alembert, who had just returned from Prussia. D'Alembert asked Mme. Geoffrin to intercede with the ambassador of Sardinia at the court of Turin:

> Monsieur de la Grange, a young geometer from Turin, has been here for six weeks. He has become quite seriously ill and he needs, not financial aid, for Mr le marquis de Caraccioli directed upon leaving for England that he should not lack for anything, but rather some signs of interest on the part of his native country. . . . In him Turin possesses a treasure whose worth it perhaps does not know.

In the spring of 1765 Lagrange returned to Turin by way of Geneva and, without attempting to visit Basel to see Daniel Bernoulli, went on d'Alembert's advice to call on Voltaire, who extended him a cordial welcome. Lagrange reported: "He was in a humorous mood that day and his jokes, as usual, were at the expense of religion, which greatly amused the gathering. He is, in truth, a character worth seeing."

D'Alembert's intervention had had some success in Turin, where the king and the ministers held out great hopes to Lagrange—in which he placed little trust.

Meanwhile the Paris Academy of Sciences had proposed for the prize of 1766 the question "What are the inequalities that should be observed in the movement of the four satellites of Jupiter as a result of their mutual attractions . . . ." D'Alembert publicly objected to this subject, which he considered very poorly worded and incorrect, since the actions of the sun on these satellites were completely ignored. His stand on this matter led to a very sharp correspondence between him and Clairaut.

In August 1765 Lagrange sent to the Academy of Sciences "Recherches sur les inégalités des satellites de Jupiter . . .,"[15] which won the prize. He wrote to d'Alembert on 9 September 1765: "What I said there concerning the equation of the center and the latitude of the satellites appears to me entirely new and of very great importance in the theory of the planets, and I am now prepared to apply it to Saturn and Jupiter." He was alluding to the works published in volume 3 of the *Mélanges de Turin.*

The fine promises of the court of Turin had still not been fulfilled. In the autumn of 1765 d'Alembert, who was on excellent terms with Frederick II of Prussia, suggested to Lagrange that he accept a position in Berlin. He replied, "It seems to me that Berlin would not be at all suitable for me while Mr Euler is there." On 4 March 1766 d'Alembert notified him that Euler was going to leave Berlin and asked him to accept the latter's post. It seems quite likely that Lagrange would gladly have remained in Turin had the king been willing to improve his material and scientific situation. On 26 April, d'Alembert transmitted to Lagrange the very precise and advantageous propositions of the king of Prussia; and on 3 May, Euler, announcing his departure for St. Petersburg, offered him a place in Russia. Lagrange accepted the proposals of the Prussian king and, not without difficulties, obtained his leave at the beginning of July through the intercession of Frederick II with the king of Sardinia.

Lagrange left for Berlin on 21 August 1766, traveling first to Paris and London. After staying for two weeks with d'Alembert, on 20 September he arrived in the English capital, summoned there by Caraccioli. He then embarked for Hamburg and finally reached Berlin on 27 October. On 6 November he was named director of the mathematics section of the Berlin Academy. He quickly became friendly with Lambert and Johann III Bernoulli; but he immediately encountered the silent hostility of the undistinguished Johann Castillon, who stood sullenly aloof from the Academy when it passed him over for a colleague young enough "to be his son."

Lagrange's duties consisted of the monthly reading of a memoir, which was sometimes published in the Academy's *Mémoires* (sixty-three such memoirs were published there), and supervising the Academy's mathematical activities. He had no teaching duties of the sort he had had in Turin and would have again, although more episodically, in Paris. His financial compensation was excellent, and he never sought to improve on it during the twenty years he was there.

In September 1767, eleven months after his arrival, Lagrange married his cousin Vittoria Conti. "My wife," he wrote to d'Alembert in July 1769, "who is one of my cousins and who even lived for a long time with my family, is a very good housewife and has no pretensions at all." He also declared in this letter that he had no children and, moreover, did not want any. He wrote to his father in 1778 or 1779 that his wife's health had been poor for several years. She died in 1783 after a long illness.

The Paris Academy of Sciences had become accustomed to including Lagrange among the competitors for its biennial prizes, and d'Alembert constantly importuned him to participate. The question for 1768, like the one for 1764, concerned the theory of the moon. D'Alembert wrote to him: "This, it seems to me, is a subject truly worthy of your efforts." But Lagrange replied on 23 February 1767: "The king would like me to compete for your prize, because he thinks that Euler is working on it; that, it seems to me, is one more reason for me not to work on it."

The prize was postponed until 1770. Lagrange excused himself on 2 June 1769: "The illness that I have had these past days, and from which I am still very weakened, has completely upset my work schedule, so that I doubt whether I shall be able to compete for the prize concerning the moon as I had planned."

Lagrange did, however, participate in the competition of 1772 with his "Essai sur le problème des trois corps."[16] The subject was still the theory of the moon. In 1770 half the prize had been awarded to a work composed jointly by Euler and his son Johann Albrecht. The question was proposed again for 1772, and the prize was shared by Lagrange and Euler. On 4 April 1771 Lagrange wrote to d'Alembert: "I intend to send you something for the prize. I have considered the three-body problem in a new and general manner, not that I believe it is better than the one previously employed, but only to approach it *alio modo;* I have applied it to the moon, but I doubt very much that I shall have the time to complete the arithmetical calculations."

On 25 March 1772 d'Alembert announced to Lagrange: "You are sharing with Mr Euler the double prize of 5,000 livres, . . . by the unanimous decision of the five judges MM de Condorcet, Bossut, Cassini, Le Monnier, and myself. We believe we owe this recognition to the beautiful analysis of the three-body problem contained in your piece." In a note on this memoir J. A. Serret wrote:

> The first chapter deserves to be counted among Lagrange's most important works. The differential equations of the three-body problem . . . constitute a system of the twelfth order, and the complete solution required twelve integrations. The only knowns were those of the *force vive* and three from the principle of areas. Eight remained to be discovered. In reducing this number to seven Lagrange made a considerable contribution to the question, one not surpassed until 1873. . . .[17]

For the prize of 1774, the Academy asked whether it were possible to explain the secular equation of the moon by the attraction of all the celestial bodies, or by the effect of the nonsphericity of the earth and of the moon. Lagrange, who was equal to the scope of the subject, felt very stale and at the end of August 1773 withdrew from the contest. At d'Alembert's request Condorcet persuaded him to persevere. He was granted an extension and thanked the jury for this favor in February 1774. He took the prize with "Sur l'équation séculaire de la lune."[18]

The topic proposed for 1776 was the theory of the perturbations that comets might undergo through the action of the planets. Lagrange found the subject unpromising, withdrew, and wrote to d'Alembert on 29 May 1775: "I am now ready to give a complete theory of the variations of the elements of the planets resulting from their mutual action." These personal investigations resulted in three studies. One was presented in the *Mémoires* of the Paris Academy: "Recherches sur les équations séculaires des mouvements des noeuds et des inclinaisons des orbites des planètes."[19] Another appeared in the *Mémoires* of the Berlin Academy: "Sur le mouvement des noeuds des orbites planétaires."[20] The third was published in the Berlin *Ephemerides* for 1782: "Sur la diminution de l'obliquité de l'écliptique."[21]

It is understandable that Lagrange, having set out on his own path and being occupied with many other investigations, neglected to enter the competition on the comets. He excused himself by referring to his bad health and the inadequacy of the time allowed, and he pointed out to d'Alembert: "You now have young men in France who could do this work."

Only one entry, from St. Petersburg, was submitted, and the contest was adjourned until 1778. Lagrange "solemnly" promised to compete this time but sent nothing, and the prize was given to Nicolaus Fuss. The same subject was proposed for 1780, and in the summer of 1779 Lagrange submitted "Recherches sur la théorie des perturbations que les comètes peuvent éprouver par l'action des planètes,"[22] which won the double prize of 4,000 livres. This was the last time that he participated in the competitions of the Paris Academy.

Lagrange's activity in celestial mechanics was not confined solely to these competitions: in Turin it had often taken an independent direction. In 1782 he wrote to d'Alembert and Laplace that he was working "a little and slowly" on the theory of the secular variations of the aphelia and of the eccentricities of all the planets. This research led to the *Théorie des variations séculaires des éléments des planètes*[23] and the memoir "Sur les variations séculaires des mouvements moyens des planètes," the latter published in 1785.[24] A work on a related subject is "Théorie des variations périodiques des mouvements des planètes," the first part of which, containing the general formulas, appeared in 1785.[25] The second, concerning the six principal planets, was published in 1786.[26]

Lagrange's work in Berlin far surpassed this classical aspect of celestial mechanics. Soon after his arrival he presented "Mémoire sur le passage de Vénus du 3 Juin 1769,"[27] an occasional work that disconcerted the professional astronomers and contained the first somewhat extended example of an elementary astronomical problem solved by the method of three rectangular coordinates. He later returned sporadically to questions of pure astronomy, as in the two-part memoir "Sur le problème de la détermination des comètes d'après trois observations,"[28] published in 1780 and 1785, and in some articles for the Berlin *Ephemerides*. Furthermore, in 1767 he wrote "Recherches sur le mouvement d'un corps qui est attiré vers deux centres fixes,"[29] which generalized research analogous to that of Euler.

In October 1773 Lagrange composed *Nouvelle solution du problème du mouvement de rotation d'un corps de figure quelconque qui n'est animé par aucune force accélératrice*.[30] "It is," he wrote Condorcet, "a problem already solved by Euler and by d'Alembert . . .. My method is completely different from theirs. . . . It is, moreover, based on formulas that can be useful in other cases and that are quite remarkable in themselves." His method was constructed, in fact, on a purely algebraic lemma. The formulas he provided—with no proof—in this lemma pertain today to the multiplication of determinants.

A by-product of this study of dynamics was Lagrange's famous *Solutions analytiques de quelques problèmes sur les pyramides triangulaires*.[31] Starting from the same formulas as those of the lemma mentioned above, again asserted without proof, he expressed the surface, the volume, and the radii of circumscribed, inscribed, and escribed spheres and located the center of gravity of every triangular pyramid as a function of the lengths of the six edges. Published in May 1775, this memoir must have been written shortly after the preceding one, perhaps in the fall of 1773. It displays a real duality. Today it would be classed in the field of pure algebra, since it employs what are now called determinants, the square of a determinant, an inverse matrix, an orthogonal matrix, and so on.

From about the same period and in the same vein is "Sur l'attraction des sphéroïdes elliptiques,"[32] in which, after praising the solutions obtained by Maclaurin and d'Alembert with "the geometric method of the ancients that is commonly, although very improperly, called synthesis," Lagrange presented a purely analytic solution.

Lagrange had devoted several of his Turin memoirs to fluid mechanics. Among them are those on the propagation of sound. The study of the principle of least action, which appeared in volume 2 of the *Mélanges de Turin*, contained about thirty pages dealing with this topic; and "Solution de différens problèmes de calcul intégral" included another sixteen. He returned to this subject toward the end of his stay in Berlin with "Mémoire sur la théorie du mouvement des fluides,"[33] read on 22 November 1781 but not published until 1783. Laplace, before undertaking a criticism of this work, wrote to him on 11 February 1784: "Nothing could be added to the elegance and generality of your analysis."

Lagrange submitted to the *Mémoires de Turin* for 1784–1785 "Percussion des fluides."[34] In 1788 he published in the Berlin Academy's *Mémoires* "Sur la manière de rectifier deux endroits des principes de Newton relatifs à la propagation du son et au mouvement des ondes."[35] These works are contemporary with or later than the composition of his *Mécanique analytique*.

Lagrange began works of a very different sort as soon as he arrived in Berlin. They were inspired by Euler, whom he always read with the greatest attention.

First Lagrange presented in the *Mémoires* of the Berlin Academy for 1767 (published in 1769) "Sur la solution des problèmes indéterminés du second degré,"[36] in which he copiously cited his predecessor at the Academy and utilized the "Euler criterion." On 20 September 1768 he sent "Solution d'un problème d'arithmétique"[37] to the *Mélanges de Turin* for inclusion in volume 4. Through a series of unfortunate circumstances this second memoir was not published until October 1773. In it Lagrange alluded to the preceding memoir, and through a judicious and skillful use of the algorithm of continued fractions he demonstrated that Fermat's equation $x^2 - ay^2 = 1$ can be solved in all cases where x, y, and a are positive integers, a not being a perfect square and y being different from zero. This is the first known solution of this celebrated problem. The last part of this memoir was developed in "Nouvelle méthode pour résoudre les problèmes indéterminés en nombres entiers,"[38] presented in the Berlin *Mémoires* for 1768 but not completed until February 1769 and published in 1770.

On 26 August 1770 Lagrange reported to d'Alembert the publication of the German edition of Euler's *Algebra* (St. Petersburg, 1770): "It contains nothing of interest except for a treatise on the Diophantine equations, which is, in truth, excellent. . . . If you have the time you could wait for the French translation that they hope to bring out, and to which I shall be able to add some brief notes." The translation was done by Johann III Bernoulli and sent, with Lagrange's additions,[39] to Lyons for publication around May 1771. The entire work appeared in the summer of 1773.

In his additions Lagrange paid tribute to the works of Bachet de Méziriac on indeterminate first-degree equations and again considered the topics discussed in the memoirs cited above, at the same time simplifying the demonstrations. In particular he elaborated a great deal on continued fractions.

Meanwhile, in the Berlin *Mémoires* for 1770 (published 1772) he presented "Démonstration d'un théorème d'arithmétique."[40] On the basis of Euler's unsuccessful but nevertheless fruitful attempts, he set forth the first demonstration that every natural integer is the sum of at most four perfect squares.

On 13 June 1771 Lagrange read before the Berlin Academy "Démonstration d'un théorème nouveau concernant les nombres premiers."[41] The theorem in question was one developed by Wilson that had simply been stated in Edward Waring's *Meditationes algebraicae* (2nd ed., Cambridge, 1770). Lagrange was the first to prove it, along with the reciprocal proposition: "For n to be a prime number it is necessary and sufficient that $1 \cdot 2 \cdot 3 \cdots (n-1) + 1$ be divisible by n."

A fundamental memoir on the arithmetic theory of quadratic forms, modestly entitled "Recherches d'arithmétique,"[42] led the way for Gauss and Legendre. It appeared in two parts, the first in May 1775 in the Berlin *Mémoires* for 1773 and the second in June 1777, in the same periodical's volume for 1775.

Always timid before d'Alembert, whom he knew to be totally alien to this kind of investigation, Lagrange wrote to him regarding his memoirs recently published in Berlin: "The 'Recherches d'arithmétique' are the ones that caused me the most difficulty and are perhaps worth the least. I believe you never wished to find out very much about this material, and I don't think you are wrong. . . ." The encouragement that he vainly sought from his old friend was perhaps given him by Laplace, to whom he declared, when sending him the second part of his memoir on 1 September 1777: "I hastened to have it published only because you have encouraged me by your approval."

In any case, Lagrange was well aware of the value of his investigations—and posterity has agreed with his judgment. In the first part of the paper he stated: "No one I know of has yet treated this material in a direct and general manner, nor provided rules for finding a priori the principal properties of numbers that can be related to arbitrarily given formulas. As this subject is one of the most curious in arithmetic and particularly merits the attention of geometers because of the great difficulties it contains, I shall attempt to treat it more thoroughly than has previously been done." It may be said that Lagrange, who in many of his works is the last great mathematician of the eighteenth century, here opens up magnificently the route to the abstract mathematics of the nineteenth century.

On 20 March 1777 Lagrange read another paper before the Academy: "Sur quelques problèmes de l'analyse de Diophante."[43] It includes an exposition of "infinite descent" inspired by Fermat's comment on that topic, but this designation does not appear, since Fermat used it only in manuscripts that were unknown at the time. Lagrange writes: "The principle of Fermat's demonstration is one of the most fruitful in the entire theory of numbers and above all in that of the whole numbers. Mr Euler has further developed this principle." This memoir also contains solutions to several difficult problems in indeterminate analysis.

Lagrange's known arithmetical works end at this point, while he was still in Berlin. Yet "Essai d'analyse numérique sur la transformation des fractions,"[44] published at Paris in the *Journal de l'École polytechnique* (1797–1798), shows that Lagrange did not lose interest in problems of this type. But the main portion of his work in this area is concentrated in the first ten years of his stay in Berlin (1767–1777). The fatigue mentioned in the letter of 6 July 1775 (cited above) was probably real, for this pioneering work was obviously exhausting.

During these ten years Lagrange also tackled algebraic analysis—or, more precisely, the solution of both numerical and literal equations. On 29 October 1767 he read "Sur l'élimination des inconnues dans les équations"[45] (published in 1771), in which he employed Cramer's method of symmetric functions but sought to make it more rapid by use of the series development of $\log (1 + u)$. Nothing seems to remain of this "improvement" of Cramer's method.

Two important memoirs appeared in 1769 and 1770, respectively: "Sur la résolution des équations numérique" and "Addition au mémoire sur la résolution des équations numériques."[47] In them Lagrange utilized the algorithm of continued fractions, and in the "Addition" he showed that the quadratic irrationals are the only ones that can be expressed as periodic continued fractions. He returned to the question in the additions to Euler's *Algebra.* The two memoirs later formed the framework of the *Traité de la résolution des équations numériques de tous les degrés,*[48] the first edition of which dates from 1798.

On 18 January and 5 April 1770 Lagrange read before the Academy his "Nouvelle méthode pour résoudre les équations littérales par le moyen des séries."[49] The method was probably suggested to him by a verbal communication from Lambert. The latter had presented, in the *Acta helvetica* for 1758, related formulas for trinomial equations but had not demonstrated them. Lagrange's formula was destined to make a great impact. He stated it in a letter to d'Alembert of 26 August 1770, as follows: "Given the equation $\alpha - x + \varphi(x) = 0$, $\varphi(x)$ denoting an arbitrary function of x, of which p is one of the roots; I say that one will have $\psi(p)$ denoting an arbitrary function of p,

$$\psi(p) = \psi(x) + \varphi(x)\,\psi'(x) + \frac{d[\varphi(x)^2\,\psi'(x)]}{2dx} + \frac{d^2[\varphi(x)^3\,\psi'(x)]}{2\cdot 3dx^2} + \cdots +,$$

where

$$\psi'(x) = \frac{d\psi(x)}{dx},$$

provided that in this series one replaces x by α, after having carried out the differentiations indicated, taking dx as a constant."

Euler, his disciple Anders Lexell, d'Alembert, and Condorcet all became extremely interested in this discovery as soon as they learned of it. The "demonstrations" of it that Lagrange and his emulators produced were hardly founded on anything more than induction. Laplace later presented a better proof. Lagrange's formula occupied numerous other mathematicians, including Arbogast, Parseval, Servois, Hindenburg, and Bürmann. Cauchy closely examined the conditions of convergence, which had been completely ignored by the inventor; and virtually every analyst of the nineteenth century considered the problem.

On 1 November 1770 Lagrange communicated to the Academy the application of his series to "Kepler's problem."[50] But the culmination of his research in the theory of equations was a memoir read in 1771: "Réflexions sur la résolution algébrique des équations."[51] In November 1770 Vandermonde read before the Paris Academy an analogous but independent and perhaps more subtle study, published in 1774. These two memoirs constituted the source of all the subsequent works on the algebraic solution of equations. Lagrange publicly acknowledged the originality and depth of Vandermonde's research. As early as 24 February 1774 he wrote to Condorcet: "Monsieur de Vandermonde seems to me a very great analyst and I was very delighted with his work on equations."

Whereas Lagrange started from a discriminating critical-historical study of the writings of his predecessors—particularly Tschirnhausen, Euler, and Bezout—Vandermonde based his work directly on the principle that the analytic expression of the roots should be a function of these roots that can be determined from the coefficients alone. Yet each of these two memoirs reveals the appearance of the concept of the permutation group (without the term, which was coined by Galois), a concept which later played a fundamental role.

Two other memoirs on this subject should be mentioned. "Sur la forme des racines imaginaires des équations,"[52] ready for printing in October 1773, evoked the following response from d'Alembert: "Your demonstration on imaginary roots seems to me to leave nothing to be desired, and I am very much obliged to you for the justice you have rendered to mine, which, in fact, has the minor fault (perhaps more apparent than real) of not being direct, but which is quite simple and easy." D'Alembert was alluding to his *Cause des vents* (1747). According to the extremely precise testimony of Delambre in his biographical notice, the demonstration by François Daviet de Foncenex that appeared in the first volume of *Mélanges de Turin* was very probably at least inspired by Lagrange. It is known that in 1799 Gauss subjected these various attempts to fierce criticism.

The last memoir in this area that should be cited appeared in 1779: "Recherches sur la détermination du nombre des racines imaginaires dans les équations littérales."[53]

Lagrange's works in infinitesimal analysis are for the most part later than those concerned with number theory and algebra and were composed at intervals from about 1768 to 1787. More in agreement with prevailing tastes, they assured Lagrange a European reputation during his lifetime.

Returning, without citing it, to his letter to Fagnano of 1754, Lagrange presented in the Berlin *Mémoires* for 1772 (published in the spring of 1774) "Sur une nouvelle espèce de calcul relatif à la différentiation et à l'intégration des quantités variables."[54] This work, which is in fact an outline of his *Théorie des fonctions* (1797), greatly impressed Lacroix, Condorcet, and Laplace. Based on the analogy between powers of binomials and differentials, it is one of the sources of the symbolic calculuses of the nineteenth century. A typical example of Lagrange's thinking as an analyst is this sentence taken from the memoir: "Although the principle of this analogy [between powers and differentials] is not self-evident, nevertheless, since the conclusions drawn from it are not thereby less exact, I shall make use of it to discover various theorems. . . ."

On 20 September 1768 he sent to the *Mélanges de Turin*, along with "Mémoire d'analyse indéterminée," the essay "Sur l'intégration de quelques équations différentielles dont les indéterminées sont séparées, mais dont chaque membre en particulier n'est point intégrable."[55] In it Lagrange drew inspiration from some of Euler's works; and the latter wrote to him on 23 March 1775, when the essay finally came to his attention: "I was not sufficiently able to admire the skill and facility with which you treat so many thorny matters that have cost me much effort . . . in particular the integration of this differential equation:

$$\frac{m\,dx}{\sqrt{A + Bx + Cx^2 + Dx^3 + Ex^4}} = \frac{n\,dy}{\sqrt{A + By + Cy^2 + Dy^3 + Ey^4}}$$

in all cases where the two numbers m and n are rational."

With this essay, as with certain works of Jakob Bernoulli, Fagnano, Euler, Landen, and others, we are in the prehistory of the theory of elliptic functions, to which period belongs one other memoir by Lagrange. Included in volume 2 of the miscellany of the Academy of Turin for 1784–1785—this academy was founded in 1783 and Lagrange was its honorary president—it was entitled "Sur une nouvelle méthode de calcul intégral pour les différentielles affectées d'un radical carré sous lequel la variable ne passe pas le quatrième degré."[56] Lagrange here proposed to find convergent series for the integrals of this type of differential, which is frequent in mechanics. To this purpose he transformed these differentials in such a way that the fourth-degree polynomial placed under the radical separated into the factors $1 + px^2$ and $1 + qx^2$, the coefficients p and q being either very unequal or almost equal. Lagrange also utilized in this work the "arithmetico-geometric mean" and reduced the integration of the series

$$(A + A'U + A''U^2 + A'''U^3 + \cdots) V\,dx$$

to that of the differential $Vdx/(1 - aU)$. This memoir, which is difficult to date precisely, was written in the last years of his stay in Berlin, after the death of his wife.

We shall now consider some earlier works on differential and partial differential equations. About March 1773 Lagrange read before the Berlin Academy his study "Sur l'intégration des équations aux différences partielles du premier ordre."[57] The Berlin *Mémoires* for 1774 (published in 1776) contained the essay "Sur les intégrales particulières des équations différentielles."[58] In these two works he considered singular integrals of differential and partial differential equations. This problem had only been lightly touched on by Clairaut, Euler, d'Alembert, and Condorcet. Lagrange wrote: "Finally I have just read a memoir that Mr de Laplace presented recently. ... This reading awakened old ideas that I had on the same subject and resulted in the following investigations ... [which constitute] a new and complete theory." Laplace wrote on 3 February 1778 that he considered Lagrange's essay "a masterpiece of analysis, by the importance of the subject, by the beauty of the method, and by the elegant manner in which it is presented."

The Berlin *Mémoires* for 1776 (published 1778) included the brief study "Sur l'usage des fractions continues dans le calcul intégral."[59] The algorithm Lagrange proposed in it had, according to him, the advantage over series of giving, when it exists, the finite integral of a differential equation, while the other method can yield only approximations.

The memoir "Sur différentes questions d'analyse relatives à la théorie des intégrales particulières"[60] may have been written about 1780. In it Lagrange extended and deepened his studies of particular integrals. He demonstrated the equivalence of the integrations of the equation

$$\xi_1 \frac{\partial f}{\partial x_1} + \xi_2 \frac{\partial f}{\partial x_2} + \cdots + \xi_n \frac{\partial f}{\partial x_n} = 0$$

and the system

$$\frac{dx_1}{\xi_1} = \frac{dx_2}{\xi_2} = \cdots = \frac{dx_n}{\xi_n}.$$

Finally, just as he was leaving Prussia, Lagrange presented in the Berlin *Mémoires* for 1785 (published in 1787) "Méthode générale pour intégrer les équations partielles du premier ordre lorsque ces différences ne sont que linéaires."[61] This "general method" completed the preceding memoir.

Lagrange's contribution to the calculus of probabilities, while not inconsiderable, is limited to a few memoirs. We have cited one of them written before 1759 and mentioned his translation of de Moivre. Two others are "Mémoire sur l'utilité de la méthode de prendre le milieu entre les résultats de plusieurs observations . . .,"[62] composed before 1774, and "Recherches sur les suites récurrentes dont les termes varient de plusieurs manières différentes . . .,"[63] read before the Berlin Academy in May 1776. The latter memoir was inspired by two essays of Laplace, the reading of which recalled to Lagrange his first writing on the question, which predated 1759. He proposed to add to this early work and to Laplace's essays, and to treat the same subject in a manner at once simpler, more direct, and above all more general. Last we may mention, from the Paris period, "Essai d'arithmétique politique sur les premiers besoins de l'intérieur de la République,"[64] written in *an* IV (1795–1796).

The considerable place that mechanics, and more particularly celestial mechanics, occupied in Lagrange's works resulted in contributions that were scattered among numerous memoirs. Thinking it proper to present his ideas in a single comprehensive work, on 15 September 1782 Lagrange wrote to Laplace: "I have almost completed a *Traité de mécanique analytique*, based uniquely on [the principle of virtual velocities]; but, as I do not yet know when or where I shall be able to have it printed, I am not rushing to put the finishing touches on it."

The work was published at Paris. A. M. Legendre had assumed the heavy burden of correcting the proofs; and his former teacher, the Abbé Joseph-François Marie, was entrusted with the arrangements with the publishers, agreeing to buy up all the unsold copies. By the time the book appeared, at the beginning of 1788, Lagrange had settled in Paris.

About 1774 there was already talk of Lagrange's returning to Turin. In 1781, through the mediation of his old friend Caraccioli, then viceroy of Sicily, the court of Naples offered him the post of director of the philosophy section of the academy recently established in that city. Lagrange, however, rejected the proposal. He was happy with his situation in Berlin and wished only to work there in peace. But the death of his wife in August 1783 left him very distressed, and with the death of Frederick II in August 1786 he lost his strongest support in Berlin. Advised of the situation, the princes of Italy zealously competed in attracting him to their courts.

In the meantime Mirabeau, entrusted with a semiofficial diplomatic mission to the court of Prussia, asked the French government to bring Lagrange to Paris through an advantageous offer. Of all the candidates, Paris was victorious. France's written agreement with Lagrange was scrupulously respected by the public authorities through all the changes of regime. In addition, Prussia accorded him a generous pension that he was still drawing in 1792.

Lagrange left Berlin on 18 May 1787. On 29 July he became *pensionnaire vétéran* of the Paris Academy of Sciences, of which he had been a foreign associate member since 22 May 1772. Warmly welcomed in Paris, he experienced a certain lassitude and did not immediately resume his research. Yet he astonished those around him by his extensive knowledge of metaphysics, history, religion, linguistics, medicine, and botany. He had long before formulated a prudent rule of conduct: "I believe that, in general, one of the first principles of every wise man is to conform strictly to the laws of the country in which he is living, even when they are unreasonable." In this frame of mind he experienced the sudden changes of the Revolution, which he observed with interest and sometimes with sympathy but without the passion of his friends and colleagues Condorcet, Laplace, Monge, and Carnot.

In 1792 Lagrange married Renée-Françoise-Adélaïde Le Monnier, the daughter of his colleague at the Academy, the astronomer Pierre Charles Le Monnier. This was a troubled period, about a year after the flight of the king and his arrest at Varennes. Nevertheless, on 3 June the royal family signed the marriage contract "as a sign of its agreement to the union." Lagrange had no children from this second marriage, which, like the first, was a happy one.

Meanwhile, on 8 May 1790 the Constituent Assembly had decreed the standardization of weights and measures and given the Academy of Sciences the task of establishing a system founded on fixed bases and capable of universal adoption. Lagrange was naturally a member of the commission entrusted with this work.

When the academies were suppressed on 8 August 1793 this commission was retained. Three months later Lavoisier, Borda, Laplace, Coulomb, Brisson, and Delambre were purged from its membership; but Lagrange remained as its chairman. In September of the same year the authorities ordered the arrest of all foreigners born within the borders of the enemy powers and the confiscation of their property. Lavoisier intervened with Joseph Lakanal to obtain an exception for Lagrange, and it was granted.

The Bureau des Longitudes was established by the National Convention on 25 June 1795, and Lagrange was a member of it from the beginning. In this capacity he returned to concerns that had been familiar to him since his participation, with Johann Karl Schulze and J. E. Bode, among others, in the editing of the Berlin *Ephemerides*.

A decree of 30 October established an *école normale*, designed to train teachers and to standardize education. This creation of the Convention was short-lived. Generally known as the École Normale de l'An III, it lasted only three months and eleven days. Lagrange, with Laplace as his assistant, taught elementary mathematics there.

Founded on 11 March 1794 at the instigation of Monge, the École Centrale des Travaux Publics, which soon took the name École Polytechnique, still exists. Lagrange taught analysis there until 1799 and was succeeded by Sylvestre Lacroix.

The constitution of *an* III replaced the suppressed academies with the Institut National. On 27 December 1795 Lagrange was elected chairman of the provisional committee of the first section, reserved for the physical and mathematical sciences.

By the coup d'état of 18–19 Brumaire, *an* VIII (9–10 November 1799) Bonaparte replaced the Directory with the Consulate. A Sénat Conservateur, which continued to exist under the Empire, was established and included among its members Lagrange, Monge, Berthollet, Carnot, and other scientists. In addition Lagrange, like Monge, became a grand officer of the newly founded Legion of Honor. In 1808 he was made count of the Empire by a law covering all the senators, ministers, state councillors, archbishops, and the president of the legislature. He was named *grand croix* of the Ordre Impérial de la Réunion—created by Napoleon in 1811—at the same time as Monge, on 3 April 1813.

Lagrange was by now seriously ill. He died on the morning of 11 April 1813, and three days later his body was carried to the Panthéon. The funeral oration was given by Laplace in the name of the Senate and by Lacépède in the name of the Institute. Similar ceremonies were held in various universities of the kingdom of Italy; but nothing was done in Berlin, for Prussia had joined the coalition against France. Napoleon ordered the acquisition of Lagrange's papers, and they were turned over to the Institute.

With the appearance of the *Mécanique analytique* in 1788, Lagrange proposed to reduce the theory of mechanics and the art of solving problems in that field to general formulas, the mere development of which would yield all the equations necessary for the solution of every problem.

The *Traité* united and presented from a single point of view the various principles of mechanics, demonstrated their connection and mutual dependence, and made it possible to judge their validity and scope. It is divided into two parts, statics and dynamics, each of which treats solid bodies and fluids separately. There are no diagrams. The methods presented require only analytic operations, subordinated to a regular and uniform development. Each of the four sections begins with a historical account which is a model of the kind.

Lagrange decided, however, that the work should have a second edition incorporating certain advances. In the *Mémoires de l'Institut* he had earlier published some essays that represented a last, brilliant contribution to the development of celestial mechanics. Among them were "Mémoire sur la théorie générale de la variation des constantes arbitraires dans tous les problèmes de la mécanique,"[65] read on 13 March 1809, and "Second mémoire sur la théorie de la variation des constantes arbitraires dans les problèmes de mécanique dans lequel on simplifie l'application des formules générales à ces problèmes,"[66] read on 19 February 1810. Arthur Cayley later deemed this theory "perfectly complete in itself."

It was necessary to incorporate the theory and certain of its applications to celestial mechanics into the work of 1788. The first volume of the second edition appeared in 1811.[67] Lagrange died while working on the second volume, which was not published until 1816.[68] Even so, a large portion of it only repeated the first edition verbatim.

"Les leçons élémentaires sur les mathématiques données à l'École normale" (1795)[69] appeared first in the *Séances des Écoles normales recueillies par les sténographes et revues par les professeurs*, distributed to the students to accompany the class exercises and published in *an* IV (1795–1796). These lectures, which are very interesting from several points of view, included Lagrange's interpolation: If y takes the values P, Q, R, S, when $x = p, q, r, s$, then $y = AP + BQ + CR + DS$, with

$$A = \frac{(x-q)(x-r)(x-s)}{(p-q)(p-r)(p-s)}, \text{ and so on.}$$

(This interpolation had already been outlined by Waring in 1779.) The text of the "Leçons" as given in the *Oeuvres* is a much enlarged reissue that appeared in the *Journal de l'École polytechnique* in 1812.

Traité de la résolution des équations numériques de tous les degrés[70] was published in 1798. It is a reissue of memoirs originally published on the same subject in 1769 and 1770, preceded by a fine historical introduction and followed by numerous notes. Several of the latter consider points discussed in other memoirs whether in summary or in a developed form. In this work, which was republished in 1808, Lagrange paid tribute to the works of Vandermonde and Gauss.

Théorie des fonctions analytiques contenant les principes de calcul différentiel, dégagés de toute considération d'infiniment petits, d'évanouissants, de limites et de fluxions et réduits à l'analyse algébrique des quantités finies[71] indicates by its title the author's rather utopian program. First published in 1797 (a second edition appeared in 1813), it returned to themes already considered in 1772. In it Lagrange intended to show that power series expansions are sufficient to provide differential calculus with a solid foundation. Today mathematicians are partially returning to this conception in treating the formal calculus of series. As early as 1812, however, J. M. H. Wronski objected to Lagrange's claims. The subsequent opposition of Cauchy was more effective. Nevertheless, Lagrange's point of view could not be totally neglected. Completed by convergence considerations, it dominated the study of the functions of a complex variable throughout the nineteenth century.

Many passages of the *Théorie*, as of the "Leçons" (discussed below), were wholly incorporated into the later didactic works. This is true, for example, of the study of the tangents to curves and surfaces and of "Lagrange's remainder" in the expansion of functions by the Taylor series.

The "Leçons sur le calcul des fonctions,"[72] designed to be both a commentary on and a supplement to the preceding work, appeared in 1801 in the *Journal de l'École polytechnique* as the twelfth part of the "Leçons de l'École normale." A separate edition of 1806 contained two complementary lectures on the calculus of variations, and the *Théorie des fonctions* also devoted a chapter to this subject. In dealing with it and with all other subjects in these two works, Lagrange abandoned the differential notation and introduced a new vocabulary and a new symbolism: first derivative function f'; second derivative function, f''; and so on. To a certain extent this symbolism and vocabulary have prevailed.

Without having enumerated all of Lagrange's writings, this study has sought to make known their different aspects and to place them in approximately chronological order. This attempt will, it is hoped, be of assistance in comprehending the evolution of his thought.

Lagrange was always well informed about his contemporaries and predecessors and often enriched his thinking by a critical reading of their works. His close friendship with d'Alembert should not obscure the frequently striking divergence in their ideas. D'Alembert's mathematical production was characterized by a realism that links him with Newton and Cauchy. Lagrange, on the contrary, displayed in his youth, and sometimes in his later years, a poetic sense that recalls the creative audacity of Leibniz.

Although Lagrange was always very reserved toward Euler, whom he never met, it was the latter, among the older mathematicians, who most influenced him. That is why any study of his work must be preceded or accompanied by an examination of the work of Euler. Yet even in the face of this great model he preserved an originality that allowed him to criticize but above all to generalize, to systematize, and to deepen the ideas of his predecessors.

At his death Lagrange left examples to follow, new problems to solve, and techniques to develop in all branches of mathematics. His analytic mind was very different from the more intuitive one of his friend Monge. The two mathematicians in fact complemented each other very well, and together they were the masters of the following generations of French mathematicians, of whom many were trained at the École Polytechnique, where Lagrange and Monge were the two most famous teachers.

NOTES

All references are to volume and pages of Lagrange's *Oeuvres* cited in the bibliography.

1. II, 318–332.
2. III, 157–186.
3. I, 3–20.
4. I, 23–36.
5. XIV, 66.
6. I, 39–148.
7. I, 151–316.
8. I, 319–332.
9. I, 334–362.
10. I, 365–468.
11. I, 471–668.
12. XIII, 30.
13. VI, 5–61.
14. V, 5–123.
15. VI, 67–225.
16. VI, 229–324.
17. VI, 324.
18. VI, 335–399.
19. VI, 635–709.
20. IV, 111–148.
21. VII, 517–532.
22. VI, 403–503.
23. V, 125–207, pt. 1, published in 1783; V, 211–344, pt. 2, published in 1784.
24. V, 382–414.
25. V, 348–377.
26. V, 418–488.
27. II, 335–374.
28. IV, 439–532.
29. II, 67–121.
30. III, 581–616.
31. III, 661–692.
32. III, 619–649.
33. IV, 695–748.
34. II, 237–249.
35. V, 592–609.
36. II, 377–535.
37. I, 671–731.
38. II, 655–726.
39. VII, 5–180.
40. III, 189–201.
41. III, 425–438.
42. III, 695–795.
43. IV, 377–398.
44. VII, 291–313.
45. III, 141–154.
46. II, 539–578.
47. II, 581–652.
48. VIII, 13–367.
49. III, 5–73.
50. III, 113–138.
51. III, 205–421.
52. III, 479–516.
53. IV, 343–374.
54. III, 441–476.
55. II, 5–33.
56. II, 253–312.
57. III, 549–575.
58. IV, 5–108.
59. IV, 301–332.
60. IV, 585–635.
61. V, 544–562.
62. II, 173–234.
63. IV, 151–251.
64. VII, 573–579.
65. VI, 771–804.
66. VI, 809–816.
67. XI, 1–444.
68. XII, 1–340.
69. VII, 183–287.
70. VIII, 13–367.
71. IX, 15–413.
72. X, 1–451.

BIBLIOGRAPHY

I. ORIGINAL WORKS. *Oeuvres de Lagrange*, J. A. Serret, ed., 14 vols. (Paris, 1867–1892), consists of the following:

Vol. I (1867) contains the biographical notice written by Delambre and the articles from vols. **1–4** of *Mélanges de Turin*.

Vol. II (1868) presents articles originally published in vols. **4** and **5** of *Mélanges de Turin* and vols. **1** and **2** of *Mémoires de l'Académie des sciences de Turin*, and in the *Mémoires de l'Académie royale des sciences et belles lettres de Berlin* for 1765–1768. It should be noted that the Berlin *Mémoires* generally appeared two years after the date indicated.

Vol. III (1869) contains papers from the Berlin *Mémoires* for 1768 and 1769 and from the *Nouveaux mémoires de l'Académie de Berlin* for 1770–1773 (inclusive) and 1775.

Vol. IV (1869) reprints articles from the *Nouveaux mémoires de Berlin* for 1774–1779 (inclusive), 1781, and 1783.

Vol. V (1870) contains articles from the *Nouveaux mémoires de Berlin* for 1780–1783, 1785, 1786, 1792, 1793, and 1803.

Vol. VI (1873) consists of articles extracted from publications of the Paris Academy of Sciences and of the Class of Mathematical and Physical Sciences of the Institute.

Vol. VII (1877) contains various works that did not appear in the academic publications—in particular, the lectures given at the École Normale.

Vol. VIII (1879) is *Traité de la résolution des équations*

numériques de tous les degrés, avec des notes sur plusieurs points de la théorie des équations algébriques. This ed. is based on that of 1808.

Vol. IX (1881) is *Théorie des fonctions analytiques, contenant les principes du calcul différentiel dégagés de toute considération d'infiniment petits, d'évanouissants, de limites et de fluxions, et réduits à l'analyse algébrique des quantités finies*, based on the ed. of 1813.

Vol. X (1884) is *Leçons sur le calcul des fonctions*, based on the 1806 ed.

Vol. XI (1888) is *Mécanique analytique*, vol. I. This ed. is based on that of 1811, with notes by J. Bertrand and G. Darboux.

Vol. XII (1889) is vol. II of *Mécanique analytique*. Based on the ed. of 1816, it too has notes by Bertrand and Darboux. These two vols. have been reprinted (Paris, 1965).

Vol. XIII (1882) contains correspondence with d'Alembert, annotated by Ludovic Lalanne.

Vol. XIV (1892) contains correspondence with Condorcet, Laplace, Euler, and others, annotated by Lalanne.

Vol. XV, in preparation, will include some MSS that had been set aside by the commission of the Institute entrusted with publication of the collected works. This vol. will also present correspondence discovered since 1892 that has been published in various places or was until now unpublished—particularly the correspondence with Fagnano. It will also provide indexes and chronological tables to facilitate the study of Lagrange's works.

Opere matematiche del Marchese Giulio Carlo de Toschi di Fagnano, III (Milan–Rome–Naples, 1912), contains the correspondence between Lagrange and Fagnano: nineteen letters dated 1754–1756 and two from 1759.

There is also *G. G. Leibnitti opera omnia*, Dutens, ed., 6 vols. (Geneva, 1768).

II. Secondary Literature.

1. Sylvestre François Lacroix, "Liste des ouvrages de M. Lagrange," supp. to *Mécanique analytique* (Paris, 1816), pp. 372–378; and (Paris, 1855), pp. 383–389.

2. *Catalogue des livres de la bibliothèque du Comte Lagrange* (Paris, 1815).

3. Gino Loria, "Essai d'une bibliographie de Lagrange," in *Isis*, **40** (1949), 112–117, which is very complete.

4. Adolph von Harnack, *Geschichte der Königlich Preussischen Akademie der Wissenschaften*, 3 vols. in 4 pts. (Berlin, 1900), which includes (III, pt. 2, 163–165) the list of the memoirs Lagrange published in the *Mémoires* and *Nouveaux mémoires de Berlin* and (II, 314–321) the correspondence between the minister Hertzberg, Frederick William II of Prussia, and Lagrange on the subject of the latter's departure from Prussia and settling in Paris.

5. Honoré Gabriel Riquetti, comte de Mirabeau, *Histoire secrette de la cour de Berlin, ou correspondance d'un voyageur françois depuis le cinq juillet 1786 jusqu'au dixneuf janvier 1787*, 2 vols. (Paris, 1789).

6. Jean-Baptiste Biot, "Notice historique sur M. Lagrange," in *Journal de l'empire* (28 Apr. 1813), repr. in Biot's *Mélanges scientifiques et littéraires*, III (Paris, 1859), 117–124.

7. Carlo Denina, *La Prusse littéraire sous Frédéric II*, II (Berlin, 1790), 140–147.

8. Pietro Cossali, *Elogio di L. Lagrange* (Padua, 1813).

9. Jean Baptiste Joseph Delambre, "Notice sur la vie et les ouvrages de M. le Comte J. L. Lagrange," in *Mémoires de la classe des sciences mathématiques de l'Institut* for 1812 (Paris, 1816), repr. in *Oeuvres de Lagrange*, I, ix–li.

10. Frédéric Maurice, "Directions pour l'étude approfondie des mathématiques recueillies des entretiens de Lagrange," in *Le moniteur universel* (Paris) (26 Feb. 1814).

11. Frédéric Maurice, "Lagrange," in Michaud's *Biographie universelle*, XXIII (Paris, 1819), 157–175.

12. Dieudonné Thiebault, *Mes souvenirs de vingt ans de séjour à Berlin*, 5 vols. (Paris, 1804).

13. Julien Joseph Viery and [Dr.] Potel, *Précis historique sur la vie et la mort de Lagrange* (Paris, 1813).

14. Poggendorff, I, 1343–1346.

15. J. M. Quérard, "Lagrange, Joseph Louis de," in *La France littéraire*, IV (Paris, 1830), 429–432.

16. A. Korn, "Joseph Louis Lagrange," in *Mathematische Geschichte Sitzungsberichte*, **12** (1913), 90–94.

17. *Annali di matematica* (Milan), 3rd ser., **20** (Apr. 1913) and **21** (Oct. 1913), both published for the centenary of Lagrange's death.

18. Gino Loria, "G. L. Lagrange nella vita e nelle opere," *ibid.*, **20** (Apr. 1913), ix–lii, repr. in Loria's *Scritti, conferenze, discorsi* (Padua, 1937), pp. 293–333.

19. Gino Loria, *Storia delle matematiche*, 2nd ed. (Milan, 1950), pp. 747–760.

20. Soviet Academy of Sciences, *J. L. Lagrange. Sbornik statey k 200-letiyu so dnya rozhdenia* (Moscow, 1937), a collection of articles in Russian to celebrate the second centenary of Lagrange's birth. Contents given in *Isis*, **28** (1938), 199.

21. George Sarton, "Lagrange's Personality (1736–1813)," in *Proceedings of the American Philosophical Society*, **88** (1944), 457–496.

22. G. Sarton, R. Taton, and G. Beaujouan, "Documents nouveaux concernant Lagrange," in *Revue d'histoire des sciences*, **3** (1950), 110–132.

23. J. F. Montucla, *Histoire des mathématiques*, 2nd ed., IV (Paris, *an* X [1802]; repr. 1960). Despite some confusion the passages written by J. Lalande on Lagrange are interesting, especially those concerning celestial mechanics.

24. Charles Bossut, *Histoire générale des mathématiques*, II (Paris, 1810). Provides an accurate and quite complete description of Lagrange's *oeuvre*, particularly in celestial mechanics. The author was a colleague of Lagrange's at the Institute and had previously been one of his judges at the time of the competitions organized by the Academy of Sciences.

25. Moritz Cantor, ed., *Vorlesungen über Geschichte der Mathematik*, IV (Leipzig, 1908), *passim*. Very useful for situating Lagrange's work within that of his contemporaries.

26. Heinrich Wieleitner, *Geschichte der Mathematik*, II, *Von Cartesius bis zur Wende des 18 Jahrhunderts*, pt. 1, *Arithmetik, Algebra, Analysis*, prepared by Anton von Braunmühl (Leipzig, 1911), *passim*. Renders the same service as the preceding work.

27. Niels Nielsen, *Géomètres français sous la Révolution* (Copenhagen, 1929), pp. 136–152.

28. Maximilien Marie, *Histoire des sciences mathématiques et physiques*, IX (Paris, 1886), 76–234.

29. Nicolas Bourbaki, *Éléments d'histoire des mathématiques*, 2nd ed. (Paris, 1969). A study of the history of mathematics, including the work of Lagrange, from a very modern point of view.

30. Carl B. Boyer, *A History of Mathematics* (New York, 1968), pp. 510–543.

31. René Taton *et al.*, *Histoire générale des sciences*, II, *La science moderne*, and III, *La science contemporaine*, pt. 1, *Le XIXème siècle* (Paris, 1958–1961). Numerous citations from the work of Lagrange, which is considered in all its aspects.

32. Carl Ohrtmann, *Das Problem der Tautochronen. Ein historischer Versuch*; also trans. into French by Clément Dusausoy (Rome, 1875).

33. Robert Woodhouse, *A History of the Calculus of Variations in the Eighteenth Century* (Cambridge, 1810; repr. New York, 1965), pp. 80–109.

34. Isaac Todhunter, *A History of the Calculus of Variations in the Nineteenth Century* (Cambridge, 1861; repr. New York, n.d.), pp. 1–10.

35. C. Carathéodory, "The Beginning of Research in the Calculus of Variations," in *Osiris*, **3** (1938), 224–240.

36. Isaac Todhunter, *A History of the Mathematical Theory of Probability* (Cambridge, 1865; repr. New York, 1965), pp. 301–320.

37. René Dugas, *Histoire de la mécanique* (Neuchâtel–Paris, 1950), pp. 318–332. Includes an important study of the *Mécanique analytique* based on the 2nd ed. The development of Lagrange's thinking in mechanics is not considered; but his influence on his successors, such as Poisson, Hamilton, and Jacobi is well brought out.

38. Ernst Mach, *Die Mechanik in ihrer Entwicklung* (Leipzig, 1883), *passim*. Mach states that he found the original inspiration for his book in Lagrange's historical introductions to the various chapters of the *Mécanique analytique*. See especially ch. 4, "Die formelle Entwicklung der Mechanik."

39. Clifford Ambrose Truesdell, *Essays in the History of Mechanics* (Berlin–Heidelberg–New York, 1968), 93, 132–135, 173, 245–248. The author, who is more concerned with the origin and evolution of concepts than with personalities, nevertheless has a tendency to diminish the role of Lagrange in favor of Euler.

40. Julian Lowell Coolidge, *A History of Geometrical Methods* (Oxford, 1940; repr. New York, 1963). Provides several insights into Lagrange's work in geometry, in particular on minimal surfaces.

41. Gaston Darboux, *Leçons sur la théorie générale des surfaces*, I (Paris, 1887), 267–268, on minimal surfaces.

42. A. Aubry, "Sur les travaux arithmétiques de Lagrange, de Legendre et de Gauss," in *L'enseignement mathématique*, XI (Geneva, 1909), 430–450.

43. F. Cajori, *A History of the Arithmetical Methods of Approximation to the Roots of Numerical Equations of One Unknown Quantity*, Colorado College Publications, General Series, nos. 51 and 52 (Colorado Springs, Colo., 1910).

44. Leonard Eugene Dickson, *History of the Theory of Numbers*, 3 vols. (Washington, 1919–1923; repr. New York, 1952). Contains numerous citations from Lagrange throughout. Several of his memoirs are summarized.

45. Hans Wussing, *Die Genesis des Abstrakten Gruppenbegriffes* (Berlin, 1969). The author closely analyzes the 1771 memoir on the resolution of equations and describes Lagrange's role in the birth of the group concept.

JEAN ITARD

LAGUERRE, EDMOND NICOLAS (*b.* Bar-le-Duc, France, 9 April 1834; *d.* Bar-le-Duc, 14 August 1886), *mathematics.*

Laguerre was in his own lifetime considered to be a geometer of brilliance, but his major influence has been in analysis. Of his more than 140 published papers, over half are in geometry; in length his geometrical work represents more than two-thirds of his total output. He was also a member of the geometry section of the Academy of Sciences in Paris.

There was no facet of geometry which did not engage Laguerre's interest. Among his works are papers on foci of algebraic curves, on geometric interpretation of homogeneous forms and their invariants, on anallagmatic curves and surfaces (that is, curves and surfaces which are transformed into themselves by inversions), on fourth-order curves, and on differential geometry, particularly studies of curvature and geodesics. He was one of the first to investigate the complex projective plane.

Laguerre also published in other areas. Geometry led him naturally to linear algebra. In addition, he discovered a generalization of the Descartes rule of signs, worked in algebraic continued fractions, and toward the end of his life produced memoirs on differential equation and elliptical function theory.

The young Laguerre attended several public schools as he moved from place to place for his health. His education was completed at the École Polytechnique in Paris, where he excelled in modern languages and mathematics. His overall showing, however, was relatively poor: he ranked forty-sixth in his class. Nevertheless, he published his celebrated "On the Theory of Foci" when he was only nineteen.

In 1854 Laguerre left school and accepted a commission as an artillery officer. For ten years, while in the army, he published nothing. Evidently he kept on with his studies, however, for in 1864 he resigned his commission and returned to Paris to take up duties as a tutor at the École Polytechnique. He remained there for the rest of his life and in 1874 was appointed

 LAGUERRE

examinateur. In 1883 Laguerre accepted, concurrently, the chair of mathematical physics at the Collège de France. At the end of February 1886 his continually poor health broke down completely; he returned to Bar-le-Duc, where he died in August. Laguerre was pictured by his contemporaries as a quiet, gentle man who was passionately devoted to his research, his teaching, and the education of his two daughters.

Although his efforts in geometry were striking, all Laguerre's geometrical production—with but one exception—is now unknown except to a few specialists. Unfortunately for Laguerre's place in history, this part of his output has been largely absorbed by later theories or has passed into the general body of geometry without acknowledgment. For example, his work on differential invariants is included in the more comprehensive Lie group theory. Laguerre's one theorem of geometry which is still cited with frequency is the discovery—made in 1853 in "On the Theory of Foci"—that in the complex projective plane the angle between the lines a and b which intersect at the point O is given by the formula

$$\sphericalangle(ab) = \frac{R(a, b, OI, OJ)}{2i} \pmod{\pi}, \qquad (1)$$

where the numerator is the cross ratio of a, b and lines joining O to the circular points at infinity: $I = (i, 1, 0)$ and $J = (-i, 1, 0)$.

Actually, Laguerre proved more. He showed that if a system of angles $A, B, C, \cdots$ in a plane is related by a function $F(A, B, C, \cdots) = 0$, and if the system is transformed into another, $A', B', C', \cdots$, by a homographic (cross ratio-preserving) mapping, then $A', B', C', \cdots$, satisfies the relation

$$F\left(\frac{\log \alpha}{2i}, \frac{\log \beta}{2i}, \frac{\log \gamma}{2i}, \cdots\right) = 0,$$

where $\alpha, \beta, \gamma, \cdots$ are cross ratios, as in expression (1).

This theorem is commonly cited as being an inspiration for Arthur Cayley when he introduced a metric into the projective plane in 1859 and for Felix Klein when he improved and extended Cayley's work in 1871.[1] These assertions appear to be false. There is no mention of Laguerre in Cayley, and Cayley was meticulous to the point of fussiness in the assigning of proper credit. Klein is specific; he states that Laguerre's work was not known to him when he wrote his 1871 paper on non-Euclidean geometry.[2] Presumably the Laguerre piece was brought to Klein's attention after his own publication.

Nevertheless, Laguerre's current reputation rests on a very solid foundation: his discovery of the set of differential equations (Laguerre's equations)

$$xy'' + (1 + x)\,y' - ny = 0, \ (n = 0, 1, 2, \cdots) \qquad (2)$$

and their polynomial solutions (Laguerre's polynomials)

$$\sum_{k=0}^{n} \frac{n^2(n-1)^2 \cdots (n-k+1)^2}{k!}\, x^{n-k}. \qquad (3)$$

These ideas have been enlarged so that today generalized Laguerre equations are usually considered. They have the form

$$xy'' + (s + 1 - x)\,y' + ny = 0, \ (n = 0, 1, 2, \cdots) \qquad (4)$$

and have as their solutions the generalized Laguerre polynomials,

$$L_n^s(x) = \sum_{k=0}^{n} \frac{(-1)^k\, n!}{k!\,(n-k)!} \left(\prod_{j=0}^{k-1} (n + s - j)\right) x^{n-k}, \qquad (5)$$

which also are frequently written as

$$L_n^s(x) = (-1)^n\, x^{-s} e^{x} \frac{d^n}{dx^n} (x^{s+n} e^{-x}).$$

The alternating sign of (5) not present in (3) is due to the change of signs of the coefficients of the y and y' terms in (4). The Laguerre functions are defined from the polynomials by setting

$$\psi_n^s(x) = e^{-x/2} x^{s/2} L_n^s(x).$$

If $s = 0$, the notations $L_n(x)$ and $\psi_n(x)$ are often used. These functions and polynomials have wide uses in mathematical physics and applied mathematics—for example, in the solution of the Schrödinger equations for hydrogen-like atoms and in the study of electrical networks and dynamical systems.[3]

Laguerre studied the Laguerre equation in connection with his investigations of the integral

$$\int_x^{\infty} \frac{e^{-x}}{x}\, dx \qquad (6)$$

and published the results in 1879.

He started by setting

$$F(x) = \sum_{k=0}^{n-1} (-1)^k\, k!\, \frac{1}{x^{k+1}}, \qquad (7)$$

from which the relation

$$\int_x^{\infty} \frac{e^{-x}}{x}\, dx = e^{-x} F(x) + (-1)^n\, n! \int_x^{\infty} \frac{e^{-x}}{x^{n+1}}\, dx \qquad (8)$$

was obtained by integration by parts. Observe that as n increases beyond bound in (7), the infinite series obtained diverges for every x, since the nth term fails to go to zero. Nevertheless, for large-value x the first few terms can be utilized in (8) to give a good approximation to the integral (6).

Next, Laguerre set

$$F(x) = \frac{\varphi(x)}{f(x)} + \left\{\frac{1}{x^{2m+1}}\right\}, \quad (9)$$

where f is a polynomial, to be determined, of degree m, which is at most $n/2$; φ is another unknown polynomial; and $\{1/(x^{2m+1})\}$ is a power series in $1/x$ whose first term is $1/x^{2m+1}$. He then showed that f and $\varphi(x)$ satisfy

$$x[\varphi'(x)f(x) - f'(x)\varphi(x) - \varphi(x)f(x)] + f^2(x) = A,$$

where A is a constant. This was used to show that f is a solution of the second-order differential equation

$$xy'' + (x+1)y' - my = 0. \quad (10)$$

Another solution, linearly independent of f, is

$$u(x) = \varphi(x)e^{-x} - f(x)\int_x^\infty \frac{e^{-x}}{x}\,dx. \quad (11)$$

Substitution of f back into (10) shows, by comparison of coefficients, that it must satisfy (Laguerre's polynomial)

$$f(x) = x^m + m^2x^{m-1} + \frac{m^2(m-1)^2}{2!}x^{m-2} + \cdots + m!.$$

These results were combined by Laguerre to obtain the continued fraction representation for (6)

$$\int_x^\infty \frac{e^{-x}}{x}\,dx = \cfrac{e^{-x}}{x+1-\cfrac{1}{x+3-\cfrac{1}{\cfrac{x+5}{4}-\cfrac{1/4}{\cfrac{x+7}{9}-\cfrac{1/9}{\cfrac{x+9}{16}-\cfrac{1}{\cfrac{16}{x+\cdots}}}}}}} \quad (12)$$

Then Laguerre proved that the mth approximate of the fraction could be written as $e^{-x}[\varphi_m(x)/f_m(x)]$, where $f_m(x)$ is the Laguerre polynomial of degree m and φ_m is the associated numerator in expression (9). From this the convergence of the fraction in (12) was established.

Finally, Laguerre displayed several properties of the set of polynomials. He proved that the roots of $f_m(x)$ are all real and unequal, and that a quasi-orthogonality condition is satisfied, that is,

$$\int_{-\infty}^0 e^x f_n(x) f_m(x)\,dx = \delta_{mn}(n!)^2, \quad (13)$$

where $\delta_{mn} \neq 0$ if $m \# n$, 1 if $m = n$. Furthermore, from (13) he proved that if $\Phi(x)$ is "any" function, then Φ has an expansion as a series in Laguerre polynomials,

$$\Phi(x) = \sum_{n=0}^\infty A_n f_n(x). \quad (14)$$

The coefficients, A_n, are given by the formula

$$A_n = \frac{1}{(n!)^2}\int_{-\infty}^0 e^x\Phi(x)f_n(x)\,dx,\ n = 0, 1, 2, \cdots. \quad (15)$$

In particular,

$$x^m = (-1)^m m!\left[f_0(x) + \sum_{k=1}^m \frac{(-1)^k m(m-1)\cdots(m-k+1)}{(k!)^2} f_k(x),\right]$$

which led Laguerre to the following inversion: if (14) is symbolically written as $\Phi(x) = \theta(f)$, then $\theta(-x) = \Phi(-f)$.

This memoir of Laguerre's is significant not only because of the discovery of the Laguerre equations and polynomials and their properties, but also because it contains one of the earliest infinite continued fractions which was known to be convergent. That it was developed from a divergent series is especially remarkable.

What, then, can be said to evaluate Laguerre's work? That he was brilliant and innovative is beyond question. In his short working life, actually less than twenty-two years, he produced a quantity of first-class papers. Why, then, is his name so little known and his work so seldom cited? Because as brilliant as Laguerre was, he worked only on details—significant details, yet nevertheless details. Not once did he step back to draw together various pieces and put them into a single theory. The result is that his work has mostly come down as various interesting special cases of more general theories discovered by others.

NOTES

1. Arthur Cayley, "Sixth Memoir on Quantics" (1859), in *Collected Works*, II (Cambridge, 1898), 561–592; Felix Klein, "Uber die sogenannte nicht-Euklidische Geometrie," in *Mathematische Annalen*, **4** (1871), 573–625, also in his *Gesammelte mathematische Abhandlungen*, I (Berlin, 1921), 244–305.
2. Klein, *Gesammelte mathematische Abhandlungen*, I, 242.
3. V. S. Aizenshtadt, *et al.*, *Tables of Laguerre Polynomials and Functions*, translated by Prasenjit Basu (Oxford, 1966); J. W. Head and W. P. Wilson, *Laguerre Functions. Tables and Properties*, ITS Monograph 183 R (London, 1961).

BIBLIOGRAPHY

Laguerre's works were brought together in his *Oeuvres*, 2 vols. (Paris, 1898), with an obituary by Henri Poincaré. This ed. teems with errors and misprints.

See also Arthur Erdelyi, *et al.*, *Higher Transcendental Functions* (New York, 1953).

MICHAEL BERNKOPF

LA HIRE, GABRIEL-PHILIPPE (or **PHILIPPE II**) **DE** (*b.* Paris, France, 25 July 1677; *d.* Paris, 4 June 1719), *astronomy, geodesy, architecture.*

Son of the astronomer Philippe de La Hire and his first wife, Catherine Lesage, La Hire, whom his contemporaries most often called Philippe II, was educated at the Paris observatory, where he lived after 1682. Initiated from childhood into astronomy and the technique of meteorological and astronomical observations, he soon assisted his father in the regular work of observation, which led to his being named *élève-astronome* at the Academy of Sciences by 1694. (He became *associé* at the time of the reorganization of 1699 and succeeded his father as *pensionnaire* on 17 May 1718.) The first work of his own, establishing the *Ephémérides* for 1701, 1702, and 1703, involved him in a painful dispute with Jean Le Fèvre, *astronome pensionnaire* and editor of the *Connaissance des temps*, who accused La Hire and his father of plagiarism and incompetence. Severely censured by the Academy, Le Fèvre was expelled in January 1702 and also gave up the editorship of the *Connaissance des temps*. In 1702 La Hire published a new edition, with numerous additions, of Mathurin Jousse's *Le théâtre de l'art de charpenterie*.

Starting in 1703 La Hire presented short memoirs to the Academy of Sciences. Although they reveal no marked originality, their variety attests to the range of his interests: observational and physical astronomy (seven memoirs), meteorology and physics (seven), applied science (three), and medicine (two). His nomination on 25 January 1706 as member of the second class of the Royal Academy of Architecture led La Hire to consider several technical and architectural problems. His treatment of them is preserved in this academy's *Procès-verbaux*. In 1718 he succeeded his father as professor at this institution but filled this position for only a few months. In the same year La Hire participated in the geodesic operations carried out under the direction of Jacques Cassini to extend the meridian of Paris from Amiens to Dunkerque.

La Hire, his father's diligent collaborator and eventual successor, produced during his brief career a body of work almost as varied as the latter's, although of much more limited extent.

BIBLIOGRAPHY

I. ORIGINAL WORKS. In addition to the three fascicules of *Ephémérides* for 1701, 1702, and 1703, published under the auspices of the Académie des Sciences as *Regiae scientiarum academiae ephemerides ad annum 1701* . . . (Paris, 1700–1702), La Hire presented nineteen short memoirs to the Academy between 1703 and 1719; these were published in the annual volumes of the *Histoire de l'Académie royale des sciences*. A list of them is included in the *Tables générales des matières contenues dans l'Histoire et dans les Mémoires de l'Académie royale des sciences*, II and III (Paris, 1729), 318–319 and 169–170, respectively; incomplete lists are in J. M. Quérard, *La France littéraire*, IV (Paris, 1830), 447; and in Poggendorff, I, 1348–1349. La Hire also republished Mathurin Jousse's *L'art de charpenterie . . . corrigé et augmenté* . . . (Paris, 1702).

II. SECONDARY LITERATURE. Some biographical details are given by Weiss in Michaud's *Biographie universelle*, XXIII (Paris, 1819), 198–199; and by A. Jal, in *Dictionnaire critique de biographie et d'histoire*, 2nd ed. (Paris, 1872), pp. 730–731. Some information on La Hire's astronomical writings can be found in J. de Lalande, *Bibliographie astronomique* . . . (Paris, 1803), index; J. B. J. Delambre, *Histoire de l'astronomie moderne*, II (Paris, 1821), 683–685; and C. Wolf, *Histoire de l'observatoire de Paris* (Paris, 1902), index. The *Procès-verbaux de l'Académie royale d'architecture, 1697–1726*, H. Lemonnier, ed., III–IV (Paris, 1913–1915), and index to X (Paris, 1929), contain information on his architectural activity.

RENÉ TATON

LA HIRE, PHILIPPE DE (*b.* Paris, France, 18 March 1640; *d.* Paris, 21 April 1718), *astronomy, mathematics, geodesy, physics.*

La Hire was the eldest son of the painter Laurent de La Hire (or La Hyre) and Marguerite Cocquin. His father was a founder of and a professor at the Académie Royale de Peinture et de Sculpture and one of the first disciples of the geometer G. Desargues. Philippe de La Hire was educated among artists and technicians who were eager to learn more of the theoretical foundations of their trades. At a very early age he became interested in perspective, practical mechanics, drawing, and painting. Throughout his life La Hire preserved this unusual taste for the parallel study of art, science, and technology, which he undoubtedly derived from the profound influence of the conceptions of Desargues.

Following the death of his father, La Hire suffered, according to the testimony of Fontenelle, "very violent palpitations of the heart" and left for Italy in 1660, hoping that the trip would be as salutary for his health as for his art. During his four years' stay in Venice, he developed his artistic talent and also studied classical geometry, particularly the theory of the conics of

Apollonius. For several years after his return to France, he was active primarily as an artist, and he formed a friendship with Desargues's last disciple, Abraham Bosse. In order to solve, at the latter's request, a difficult problem of stonecutting, he developed, in 1672, a method of constructing conic sections, which revealed both his thorough knowledge of classical and modern geometry and his interest in practical questions.

His *Nouvelle méthode en géométrie pour les sections des superficies coniques, et cylindriques* (1673) is a comprehensive study of conic sections by means of the projective approach, based on a homology which permits the deduction of the conic section under examination from a particular circle. This treatise was completed shortly afterward by a supplement entitled *Les planiconiques*, which presented this method in a more direct fashion. The *Nouvelle méthode* clearly displayed Desargues's influence, even though La Hire, in a note written in 1679 and attached to a manuscipt copy of the *Brouillon projet* on Desargues's conics, affirmed that he did not become aware of the latter's work until after the publication of his own. Yet what we know about La Hire's training seems to contradict this assertion. Furthermore, the resemblance of their projective descriptions is too obvious for La Hire's not to appear to have been an adaptation of Desargues's. Nevertheless, La Hire's presentation, which was in classical language and in terms of both space and the plane, was much simpler and clearer. Thus La Hire deserves to be considered, after Pascal, a direct disciple of Desargues in projective geometry.

In 1685 La Hire published, in Latin, a much more extensive general treatise on conic sections, *Sectiones conicae in novem libros distributatae*. It was also inspired, but much less obviously, by the projective point of view, because of the preliminary study of the properties of harmonic division. It is primarily through this treatise that certain of Desargues's projective ideas became known. Meanwhile, in 1679, in his *Nouveaux élémens des sections coniques, les lieux géométriques*, La Hire provided an exposition of the properties of conic sections. He began with their focal definitions and applied Cartesian analytic geometry to the study of equations and the solution of indeterminate problems; he also displayed the Cartesian method of solving several types of equations by intersections of curves. Although not a work of great originality, it summarized the progress achieved in analytic geometry during half a century and contributed some interesting ideas, among them the possible extension of space to more than three dimensions. His virtuosity in this area appears further in the memoirs that he devoted to the cycloid, epicycloid, conchoid, and quadratures. This ingenuity in employing Cartesian methods was certainly what accounts for his hostility toward infinitesimal calculus in the discussions of its value raised in the Academy of Sciences starting in 1701. While he did not persist in ignoring the new methods, he nonetheless used them only with reservations. Having actively participated in the saving and partial publication of the mathematical manuscripts of Roberval and Frénicle de Bessy, La Hire was also interested in the theory of numbers, particularly magic squares.

Mathematics was only one aspect of La Hire's scientific activity, which soon included astronomy, physics, and applied mathematics. His nomination to the Academy of Sciences as *astronome pensionnaire* (26 January 1678) led him to undertake regular astronomical observations, a task which he pursued until two days before his death. In 1682 he moved into the Paris observatory where he was able to use rather highly developed equipment, in particular the large quadrant of a meridian circle that was installed in 1683. If the bulk of his observations have remained unpublished, at least he extracted from them numerous specific observations: conjunctions, eclipses, passages of comets, sunspots, etc. In 1687 and 1702, La Hire published astronomical tables containing his observations of the movements of the sun, the moon, and the planets; they were severely criticized by Delambre for their purely empirical inspiration. Furthermore, he studied instrumental technique and particular problems of observation and basic astronomy. As a result of his wide-ranging interests, he produced a body of work that was important and varied but that lacked great originality.

During these years La Hire also took part in many geodesic projects conducted by groups from the Paris observatory. From 1679 to 1682, sometimes in collaboration with Picard, he determined the coordinates of different points along the French coastlines in the hope of establishing a new map of France. In 1683 he began mapping the extension of the meridian of Paris toward the north. In 1684–1685 he directed the surveying operations designed to provide a water supply for the palace of Versailles. La Hire devoted several works to the methods and instruments of surveying, land measurement, and gnomonics. During his journeys, he made observations in the natural sciences, meteorology, and physics. In addition, he played an increasingly active role in the various regular observations pursued at the Paris observatory: terrestrial magnetism, pluviometry, and finally thermometry and barometry.

Appointed on 14 December 1682 to the chair of mathematics at the Collège Royal, which had been vacant since Roberval's death, La Hire gave courses in those branches of science and technology in which mathematics was becoming decisive—astronomy, mechanics, hydrostatics, dioptrics, and navigation. Although his lectures were not published, numerous memoirs presented to the Academy of Sciences preserve their outline. In the area of experimental science La Hire's efforts are attested by the description of various experiments—falling bodies, done with Mariotte in 1683, magnetism, electrostatics, heat reflected by the moon, the effects of cold, the physical properties of water, and the transmission of sound. He also studied the barometer, thermometer, clinometer, clocks, wind instruments, electrostatic machines, and magnets.

La Hire's work extended to descriptive zoology, the study of respiration, and physiological optics. The latter attracted him both by its role in astronomical observation and by its relationship to artistic technique, especially to the art of painting which La Hire continued to practice at the same time that he sought to grasp its basic principles.

La Hire was appointed, on 7 January 1687, professor at the Académie Royale d'Architecture, replacing F. Blondel. The weekly lectures that he gave until the end of 1717 dealt with the theory of architecture and such associated techniques as stonecutting. In the *Procès-verbaux de l'Académie royale d'architecture* there are many references to La Hire. In this regard he again appeared as a disciple of Desargues. Desargues's influence is confirmed by the manuscript of La Hire's course on "La pratique du trait dans la coupe des pierres pour en former des voûtes," which displays a generous use of the new graphic methods introduced by Desargues.

The important *Traité de mécanique* that La Hire published in 1695 represents a synthesis of his diverse theoretical and practical preoccupations. Although passed over by the majority of the historians of mechanics, this work marks a significant step toward the elaboration of a modern manual of practical mechanics, suitable for engineers of various disciplines. La Hire thus partially answered the wish expressed by Colbert in 1675 of seeing the Academy produce an exact description of all the machines useful in the arts and trades. On the theoretical plane, La Hire's treatise was already out of date at the time of its appearance because it ignored Newton's laws of dynamics and the indispensable infinitesimal methods. On the other hand, while La Hire did not tackle the problem of energy, he furnished useful descriptions and put forth the suggestion (already made in his *Traité des épicycloides* ... [1694]) following Desargues, of adopting an epicycloidal profile for gear wheels.

Associated with the leading scientists of the age, La Hire was, for nearly half a century, one of the principal animators of scientific life in France. Not satisfied with publishing a multitude of books and memoirs, he also edited various writings of Picard, Mariotte, Roberval, and Frénicle, as well as several ancient texts.

His family life was simple and circumspect. From his marriage with Catherine Lesage (*d. ca.* 1681), he had three daughters and two sons, one of whom, Gabriel-Philippe, continued his father's work in various fields. From a second marriage, with Catherine Nouet, he had two daughters and two sons; one of the latter, Jean-Nicolas, a physician and botanist, was elected an associate member of the Academy of Sciences.

It is difficult to make an overall judgment on a body of work as varied as La Hire's. A precise and regular observer, he contributed to the smooth running of the Paris observatory and to the success of different geodesic undertakings. Yet he was not responsible for any important innovation. His diverse observations in physics, meteorology, and the natural sciences simply attest to the high level of his intellectual curiosity. Although his rejection of infinitesimal calculus may have rendered a part of his mathematical work sterile, his early works in projective, analytic, and applied geometry place him among the best of the followers of Desargues and Descartes. Finally, his diverse knowledge and artistic, technical, and scientific experience were factors in the growth of technological thought, the advance of practical mechanics, and the perfecting of graphic techniques.

BIBLIOGRAPHY

I. Original Works. An exhaustive list of La Hire's numerous memoirs inserted in the annual volumes of the *Histoire de l'Académie royale des sciences* from the year 1699 to the year 1717 and, for the earlier period, in vols. **9** and **10** of the *Mémoires de l'Académie royale des sciences depuis 1666 jusqu'en 1699* is given in vols. I-III of M. Godin, *Table alphabétique des matières contenues dans l'histoire et les mémoires de l'Académie royale des sciences*; see vol. I, *1666–1698* (1734), 157–164; vol. II, *1699–1710* (1729), 306–317; vol. III, *1711–1720* (1731), 166–169 (cf. also J. M. Quérard, *La France littéraire*, IV [Paris, 1830], 445–447).

His principal works published separately are, in chronological order: *Observations sur les points d'attouchement de trois lignes droites qui touchent la section d'un cone* ... (Paris, 1672); *Nouvelle méthode en géométrie pour les sections des superficies coniques et cylindriques* (Paris, 1673); *Nouveaux élémens des sections coniques, les lieux géométriques, la construction ou effection des équations* (Paris, 1679; English trans., London, 1704); *La gnomonique* ...

(Paris, 1682; 2nd ed., 1698; English trans., 1685); *Sectiones conicae in novem libros distributae* . . . (Paris, 1685); *Tabularum astronomicarum* . . . (Paris, 1687); *L'école des arpenteurs* . . . (Paris, 1689; 4th ed., 1732); *Traité de mécanique* (Paris, 1695); *Tabulae astronomicae* . . . (Paris, 1702; 2nd ed., 1727; French ed., by Godin, 1735; German trans., 1735).

In addition, La Hire edited several works: J. Picard, *Traité du nivellement* (Paris, 1684); Mariotte, *Traité du mouvement des eaux* . . . (Paris, 1686); *Veterum mathematicorum Athenaei, Apollodori, Philonis, Bitonis, Heronis et aliorum opera*, with Sédillot and Pothenot (Paris, 1693). He also participated in the editing of the *Mémoires de mathématiques et de physique* . . ., published by the Académie des Sciences in 1692 and 1693.

Some of La Hire's manuscripts are preserved in the Archives of the Académie des Sciences de Paris and in the Library of the Institut de France (copy of the *Brouillon projet* of Desargues, "La pratique du trait dans la coupe des pierres").

II. Secondary Literature. The basic biographical notice is the one by B. Fontenelle in *Histoire de l'Académie royale des sciences pour l'année 1718*, *éloge* read on 12 Nov. 1718 (Paris, 1719), pp. 76–89. Other more recent ones are by E. Merlieux, in Michaud, *Biographie universelle*, XXIII (Paris, 1819), 196–198, new ed., XXII (Paris, 1861), 552–553; and F. Hoefer, *Nouvelle biographie générale*, XXVIII (Paris, 1861), cols. 901–904.

Complementary details are given by L. A. Sédillot, "Les professeurs de mathématiques et de physique générale au Collège de France," in *Bullettino di bibliografia e di storia delle scienze matematiche e fisiche*, **2** (1869), 498; A. Jal, *Dictionnaire critique de biographie et d'histoire*, 2nd ed. (Paris, 1872), pp. 730–731; J. Guiffrey, *Comptes des bâtiments du roi sous le règne de Louis XIV*, 5 vols. (Paris, 1881–1901)—see index; and H. Lemonnier, ed., *Procès-verbaux de l'Académie royale d'architecture, 1682–1726*, II–IV (Paris, 1911–1915)—see index in vol. X (1929).

La Hire's mathematical work is analyzed by J. F. Montucla, *Histoire des mathématiques*, 2nd ed., II (Paris, 1799), 169, 641–642; M. Chasles, *Aperçu historique* . . . (Brussels, 1837)—see index; R. Lehmann, "De La Hire und seine Sectiones conicae," in *Jahresberichte des königlichen Gymnasiums zu Leipzig* (1887–1888), pp. 1–28; N. Nielsen, *Géomètres français du XVIIIe siècle* (Copenhagen–Paris, 1935), pp. 248–261; J. L. Coolidge, *History of the Conic Sections and Quadric Surfaces* (Oxford, 1945), pp. 40–44; R. Taton, "La première oeuvre géométrique de Philippe de La Hire," in *Revue d'histoire des sciences*, **6** (1953), 93–111; and C. B. Boyer, *A History of Analytic Geometry* (New York, 1956)—see index.

The astronomical work is studied by J. B. Delambre, *Histoire de l'astronomie moderne*, II (Paris 1821), 661–685; C. Wolf, *Histoire de l'Observatoire de Paris* . . . (Paris, 1902)—see index; and F. Bouquet, *Histoire de l'astronomie* (Paris, 1925), pp. 381–383. On the technical work see M. Daumas, ed., *Histoire générale des techniques*, II (Paris, 1964), 285–286, 540–541.

René Taton

LALANDE, JOSEPH-JÉRÔME LEFRANÇAIS DE (*b.* Bourg-en-Bresse, France, 11 July 1732; *d.* Paris, France, 4 April 1807), *astronomy*.

Lalande's father was Pierre Le François, director of the post office at Bourg and also director of the tobacco warehouse. His mother was the former Marie-Anne-Gabrielle Monchinet. Lalande used the simple patronym Le François until 1752 when he began to write Le François de la Lande. With the abolition of noble titles during the Revolution he became simply Lalande. Apparently he had no brothers or sisters and was never married. His "nephew," Michel-Jean-Jérôme Lefrançais de Lalande, who became an astronomer under Lalande's tutelage, was actually a grandson of Lalande's uncle. Lalande also frequently referred to Michel's wife as his niece or daughter and occasionally employed her in the calculation of astronomical tables.

Lalande was extremely well known during his lifetime, partly because of the enormous bulk of his writings and partly because of his love for the limelight. Nothing pleased him more than to see his name in the public press, a weakness that he readily confessed: "I am an oilskin for insults and a sponge for praise." He was first and foremost a practical astronomer, a maker of tables and an excellent writer of astronomical textbooks. His enormous energy and active pen could never be confined to astronomy, however; and he also wrote on the practical arts, published travel literature, and was very active in the scientific academies.

Lalande was educated by the Jesuits at the Collège de Lyon and at first indicated an intention to join the order. His parents persuaded him to study law at Paris instead. During his student years he lived at the Hôtel de Cluny, where the astronomer Joseph-Nicolas Delisle had his observatory. Lalande followed Delisle's lectures at the Collège Royal and assisted him in his observations. He also attended the lectures of Pierre-Charles Le Monnier on mathematical physics, and it was Le Monnier who obtained for Lalande his first important assignment as an astronomer. In 1751 Abbé Nicolas de La Caille departed on an expedition to the Cape of Good Hope, one of the main purposes of which was to measure the lunar parallax. It was important that simultaneous measurements be made in Europe at some point on the same meridian. The most advantageous site was Berlin, which unfortunately lacked an adequate instrument. Le Monnier permitted Lalande to go in his place, entrusting to him his quadrant, which was generally considered to be the best in France. At Berlin, Lalande was admitted to the Prussian Academy, where he enjoyed the company of Maupertuis, Euler, and the marquis d'Argens. He published his observations in the *Acta eruditorum*, the *Histoire* of the Berlin Academy, and the *Mémoires* of the Paris Academy, which led almost immediately to his election to the latter on 4 February 1753 as *adjoint astronome*. He was promoted to *associé* in 1758 and became *pensionnaire* in 1772.

Lalande became involved in a series of controversies on astronomical questions. The first was with his teacher Le Monnier over the best way to correct for the flattening of the earth in calculating the lunar parallax. A commission appointed by the Academy to judge the dispute decided in Lalande's favor. His enthusiasm in pressing his claim caused ill will on the part of his former teacher and resulted in a rupture of their friendship.

A more important controversy arose over Alexis Clairaut's prediction of the return of Halley's comet. Halley had predicted that the comet of 1682 would return late in 1758 or early in 1759, but his prediction was based on the gravitational attraction of the sun alone, without considering the perturbations caused by the other planets. Clairaut determined to calculate the orbit more precisely and was aided in the extremely laborious calculations by Lalande and Mme. Lepaute, the wife of a famous French clockmaker. The comet appeared on schedule, as Clairaut predicted, and his feat was acclaimed in the popular press as a great vindication of Newton's law of gravitation.

The work of Clairaut and Lalande was made possible by the recently discovered mathematical methods of approximating solutions to the three-body problem. Clairaut, d'Alembert, and Euler had all been competing to solve this particular problem during the 1740's; and at one point it seemed that Newton's law would be shown to be in error, since the more precise calculations of the astronomers gave the wrong figure for the motion of the lunar apsides. A bitter controversy ensued between d'Alembert and Clairaut over the best method of approximation. D'Alembert was closely associated with Le Monnier and Lalande with Clairaut, and the recent rupture between Lalande and Le Monnier increased the hostility between the two camps. When the controversy was resumed over the return of Halley's comet, Lalande joined enthusiastically in the polemics. Many of the letters in the controversy were anonymous, however, and the extent of Lalande's involvement is difficult to determine. He published his account of the comet in his *Histoire de la comète de 1759*, which contained a new edition of Halley's planetary tables.

There followed new successes for Lalande. He was chosen to succeed G. D. Maraldi as editor of the astronomical almanac *Connaissance des temps*, which he greatly expanded during his years as editor from 1760 to 1776, adding accurate tables of lunar distances from the stars and the sun and other information of value for navigation. He also made it a chronicle of important astronomical events. During the Revolution, Lalande returned again to the *Connaissance des temps* and edited it from 1794 until his death in 1807. Also in 1760 he succeeded Delisle as professor of astronomy at the Collège Royale. Lalande was an excellent teacher and had many distinguished pupils during his forty-six years of service at the Collège Royale, including J. B. J. Delambre, G. Piazzi, P. Méchain, and his nephew, Michel Lalande.

Next to his indefatigable efforts to improve astronomical tables, Lalande's greatest contribution was as a writer of textbooks, the most important being his *Traité d'astronomie* of 1764, with subsequent editions in 1771 and 1792. It became a standard textbook and had the advantage over other texts of containing much practical information on instruments and methods of calculation. In 1793 he wrote *Abrégé de navigation historique, théorique, et pratique, avec des tables horaires*, for which the calculations were done by his niece, Mme. Lalande. Other major works are his enormous *Bibliographie astronomique* (1802), the last two volumes of Montucla's *Histoire des mathématiques* (1802), *Histoire céleste française contenant les observations de plusieurs astronomes français* (1801), *Traité des canaux de navigation* (1778), and numerous smaller works, including *Astronomie des dames* (1785, 1795, 1806) and annotated editions of works by earlier astronomers.

Lalande's leaning toward the spectacular attracted him to the most important astronomical event of the eighteenth century, the transits of Venus across the face of the sun, which occurred in 1761 and 1769. Astronomers believed that careful observations of the transits made from different places on the earth would provide a means for measuring very precisely the sun's parallax. Lalande's teacher Delisle had a major role in preparing for the transit of 1761 and benefited from Lalande's assistance. Before the second transit Lalande took a major organizational role and wrote to ministers and even to sovereigns of many countries in an attempt to coordinate a second international effort to send expeditions to the locations best suited for observing the transit. It was Lalande who constructed the mappemonde showing the portions of the world from which the transit could best be observed. He refused all offers to lead an expedition (he excused himself because of his extreme susceptibility to seasickness), but he regarded himself as the obvious person to compile the data and compute the solar distance. When Maximillian Hell refused to send his data from observations made at Wardhus in Lapland, Lalande intimated that Hell had failed to obtain satisfactory results and was concealing his ineptitude by refusing to send the data. Hell was later vindicated, and Lalande was forced to concede that his observation was one of the best. The most important calculations of the solar parallax from the transit of 1769 were those of Lalande (published in his *Mémoire sur le passage de Venus observé le 3 Juin 1769*) and Pingré.

Lalande caused another stir in 1773, when he discussed the possibility of a collision between the earth and a comet. His work on the perturbation of comets by the planets indicated that the orbit of a comet might be altered enough to make a collision with the earth possible. He realized that the likelihood of such a collision was extremely slight, but he failed to emphasize this point in summarizing his paper before the Academy. The result was a panic in Paris based on the rumor that Lalande had predicted the imminent destruction of the earth. Even prompt publication of the entire paper did not completely reassure the public.

Lalande also wrote lengthy accounts of his travels, the most important being his description of a trip to Italy in 1765 and 1766. The *Voyage d'un français en Italie* (1768) appeared in eight volumes and was the most complete guide available for the French traveler. Lalande went into great detail about prices, interesting places to visit, and other information of interest to the tourist. A similar description of a journey to England was never published but is of interest to the historian for the wealth of detail that it includes.

Another of Lalande's enthusiasms was for the practical arts, and he contributed a series of articles on technology to the collection of the Academy, eventually published as *Description des arts et métiers*. Lalande was not one of the original *encyclopédistes*; but he did contribute to the supplement and later rewrote the astronomical articles for the *Encyclopédie méthodique*, replacing d'Alembert's articles, which were drawn largely from Le Monnier's *Institutions astronomiques*, with material that he took from his own *Traité d'astronomie*.

Lalande had an important organizational role in many institutions of the *ancien régime*. He organized a literary society at Bourg in the winter of 1755–1756 which was active for over a year but was finally refused authorization after the attempt to assassinate the king in 1757. In 1783 Lalande renewed his efforts and obtained authorization to found a new Société d'Émulation et d'Agriculture de l'Ain. He was also a very active member of the Masonic order and founder of the famous Lodge of Nine Sisters at Paris. He had had an important part in the founding and early history of the Grand Orient de France in 1771 and wrote the short *Mémoire historique sur la Maçonnerie* (1777) as well as a new article "Franc-Maçon" for the supplement of the *Encyclopédie*. The Lodge of Nine Sisters was to be an "encyclopedic" lodge to bring together men of learning and talent. Originally conceived by Helvétius and Lalande, it was pursued by Lalande after the death of Helvétius in 1771. After some difficulty in getting permission from the Masonic hierarchy, the lodge was constituted in 1777. Membership in it was open only to those who were endowed with a specific talent in the arts or sciences and had already given public proof of that talent. The membership of this lodge (crowned by the initiation of Voltaire in 1778) reflected the essentially elitist ideas of Lalande. The most illustrious writers, scientists, artists, and political dignitaries became members.

In the Paris Academy, Lalande had little sympathy for those artisans, unskilled in mathematics, who complained about the autocratic manner of the Academy in dealing with their inventions. He supported the professional character of the Academy, a position that became increasingly unpopular after 1789. Lalande's political views were those of a cautious royalist, and he had to exercise great care during the Revolution. Nevertheless, he had the courage to hide the Abbé Garnier and Dupont de Nemours at the Paris observatory during the tempestuous days following 10 August 1792.

After Thermidor, Lalande worked to promote new scientific activity and to reestablish scientific organizations. On 21 November 1794 he gave a well-publicized speech at the Collège de France in which he attacked "Jacobin vandalism" of the sciences and described the reawakening of scientific activity in France. In February 1795 he founded a new scientific organization, the Réunion des Sciences, which, along with many other such societies, attempted to assume some of the functions of the old Academy of Sciences.

Lalande's desire for fame and his reputation as a freethinker led him into conflict with Napoleon. In 1803 he published a biographical notice on Sylvain Maréchal along with a supplement to Maréchal's *Dictionnaire des athées*. In a second supplement of 1805 he claimed that only philosophers could propagate science and thereby perhaps decrease the "number of monsters who govern and bloody the earth by war." Since Napoleon was busily at war and wished to retain cordial relations with the Church, he was greatly displeased and insisted that Lalande be censured before the entire Institut de France.

Throughout his life Lalande drew attention to himself by his numerous publications, by frequent letters to the Paris journals, by organizational activities, and by more bizarre episodes, such as a balloon ascent and a campaign to lessen the fear of spiders. (He ate several to prove his point.) He was an indefatigable worker, and the total volume of writing that flowed from his pen was prodigious. As a creative scientist he was not outstanding, but in the teaching and practical operations of astronomy he made major contributions. He remained an important figure in French astronomy until his death in 1807.

BIBLIOGRAPHY

Lalande's MSS are in the Bibliothèque Nationale, Paris (MS fr. 12271–12275); the archives of the Académie des Sciences, Paris, dossier Lalande; and in the archives of the Soviet Academy of Sciences. There is a diary by Lalande in the Bibliothèque Victor-Cousin, Paris, MS 99; and a MS of the Académie des Sciences "Collection de ses règlemens et déliberations par ordre de matière," annotated by Lalande, in Bibliotheca Medicea-Laurenziana, Florence, Ashburnham-Libri no. 1700.

Galina Pavlova has written a short biography in Russian, *Lalande, 1732–1807* (Leningrad, 1967), and a description of the Lalande letters in the Russian archives, "J. J. Lalande and the St. Petersburg Academy of Sciences," in *Proceedings of the Tenth International Congress of History of Science, Ithaca, N.Y.* (Paris, 1964), pp. 743–746. His *éloge* at the Académie des Sciences was given by J. B. J. Delambre and expanded for his *Histoire de l'astronomie au dix-huitième siècle* (Paris, 1827) and for the article on Lalande in Michaud's *Dictionnaire de biographie française*. The most complete description of Lalande's scientific work is that given by Delambre in his *Histoire*, but his evaluation is so hostile that it cannot be accepted uncritically. Valuable biographical information is contained in Louis Amiable, *Le franc-maçon Jérôme Lalande* (Paris, 1889); and in Constance Marie Salm-Reifferscheid-Dyck, "Éloge historique de M. de la Lande," in *Magasin encyclopédique* (April 1810).

Several articles on Lalande have appeared in the *Annales de la Société d'émulation et d'agriculture de l'Ain:* Joseph Bluche, "Jérôme Lalande," **37** (1904), 5–34; Denizet, "Lalande et l'art de l'ingénieur," **38** (1905), 232–261; and Charles E. H. Marchand, "Jérôme Lalande et l'astronomie au XVIIIe siècle," **40** (1907), 82–145, and **41** (1908), 313–417. Lalande's MS account of his English tour is described in Hélène Monod-Cassidy, "Un astronome philosophe, Jérôme de Lalande," in *Studies on Voltaire and the Eighteenth Century*, **56** (1967), 907–930. François Aulard described Lalande's conflict with Napoleon in "Napoléon et l'athée Lalande," in *Études et leçons sur la Révolution Française*, 4th ser. (Paris, 1904), pp. 303–316. Roger Hahn describes Lalande's activities at the Academy in *The Anatomy of a Scientific Institution; the Paris Academy of Sciences, 1666–1803* (Berkeley, Calif., 1971); and his involvement in the events of the transits of Venus are narrated by Harry Woolf in *The Transits of Venus. A Study of Eighteenth-Century Science* (Princeton, 1959).

THOMAS L. HANKINS

LALLA (*fl.* India, eighth century), *astronomy*.

The son of Trivikrama Bhaṭṭa and the grandson of Śâmba, Lalla was one of the leading Indian astronomers of the eighth century; the only other major figure known to us from that century is the author of the later *Pauliśasiddhânta*. Lalla adhered to the two traditions started by Āryabhaṭa I (*b.* 476); following the Āryapakṣa (see essay in Supp.), he wrote the *Śiṣyadhīvṛddhidatantra*, which is the most extensive extant exposition of the views of that school. It contains twenty-two chapters divided into two books—an arrangement which influenced Bhāskara II (*b.* 1115):

I. On the computation of the positions of the planets.

1. On the mean longitudes of the planets.
2. On the true longitudes of the planets.
3. On the three problems involving diurnal motion.
4. On lunar eclipses.
5. On solar eclipses.
6. On the syzygies.
7. On the heliacal settings and risings of the planets.
8. On the shadow of the moon.
9. On the lunar crescent.
10. On planetary conjunctions.
11. On conjunctions of the planets with the stars.
12. On the *pātas* of the sun and moon.
13. Conclusion.

II. On the sphere.

1. On graphical representations.
2. On the construction of the celestial sphere.
3. On the principles of mean motion.
4. On the terrestrial sphere.
5. On the motions and stations of the planets.
6. On geography.
7. On erroneous knowledge.
8. On instruments.
9. On certain (selected) problems.

A commentary on the *Śiṣyadhīvṛddhidatantra* was written by Bhāskara II.

In accordance with the teachings of the *ardharātrikapakṣa* (see essay in Supp.), which was also founded by Āryabhaṭa I, Lalla composed a commentary on the *Khaṇḍakhādyaka*, which had been written by Brahmagupta (*b.* 598) in 665; this commentary is no longer extant.

There does survive, however, in two or three manuscripts an astrological work by Lalla, the *Jyotiṣaratnakośa.* This was an extremely influential treatise on *muhūrtaśāstra,* or catarchic astrology, although it was later eclipsed by its shorter imitation, the *Jyotiṣaratnamālā* of Śrīpati (*fl.* 1040).

BIBLIOGRAPHY

The *Śiṣyadhīvṛddhidatantra* was edited by Sudhākara Dvivedin (Benares, 1886). There are brief notices concerning Lalla in Sudhākara Dvivedin, *Gaṇakataraṅgiṇī* (Benares, 1933), pp. 8–11, repr. from *The Pandit,* n.s. **14** (1892); and in Ś. B. Dīkṣita, *Bhāratīya Jyotiḥśāstra* (Poona, 1896; repr. 1931), 227–229. Fundamental for the problem of his date is the discussion by P. C. Sengupta in *The Khaṇḍakhādyaka* (Calcutta, 1934), pp. xxiii–xxvii.

DAVID PINGREE

LALOUVÈRE, ANTOINE DE (*b.* Rieux, Haute-Garonne, France, 24 August 1600; *d.* Toulouse, France, 2 September 1664), *mathematics.*

Lalouvère is often referred to by the Latin form of his name, Antonius Lalovera. Such a use avoids the problem of known variants; for example, Fermat wrote to Carcavi, on 16 February 1659, that the mathematician had a nephew who called himself Simon de La Loubère. Whatever the spelling, the family was presumably noble, since a château near Rieux bears their name.

Lalouvère himself became a Jesuit, entering the order on 9 July 1620, at Toulouse, where he was later to be professor of humanities, rhetoric, Hebrew, theology, and mathematics. The general of the order was at that time Guldin, a mathematician who may be considered, along with Cavalieri, Fermat, Vincentio, Kepler, Torricelli, Valerio—and indeed, Lalouvère—one of the precursors of modern integral calculus. That Lalouvère was on friendly terms with Fermat is evident in a series of letters; he further maintained a close relationship with Pardies in France and Wallis in England. His mathematics was essentially conservative; while modern analysis was alien to him, he was expert in the work of the Greeks, the Aristotelian-Scholastic tradition, and the commentators of antiquity. He depended strongly upon Archimedes.

Lalouvère's chief book is the *Quadratura circuli,* published in 1651, in which he drew upon the work of Charles de La Faille, Guldin, and Vincentio. His method of attack was an Archimedean summation of areas; he found the volumes and centers of gravity of bodies of rotation, cylindrical ungulae, and curvilinearly defined wedges by indirect proofs. He was then able to proceed by inverting Guldin's rule whereby the volume of a body of rotation is equal to the product of the generating figure and the path of its center of gravity. Thus, Lalouvère established the volume of the body of rotation and the center of gravity of its cross section; then by simple division he found the volume of the cross section.

By the time he published this work, Lalouvère was teaching Scholastic theology rather than mathematics, and believed that he had reached his goals as a mathematician. Indeed, he stated that he preferred to go on to easier tasks, more suited "to my advanced age." Nonetheless, he was drawn into the dispute with Pascal for which his name is best known.

In June 1658, Pascal made his conclusions on cycloids the subject of an open competition. The prize was to be sixty Spanish gold doubloons, and solutions to the problems he set were to be submitted by the following 1 October. Lalouvère's interest was attracted by the nature of the problems, rather than by the prize, and Fermat transmitted them to him on 11 July. Lalouvère returned his solutions to Pascal's first two problems only ten days later, having reached them by simple proportions rather than by calculation. The calculation of the volumes and centers of gravity of certain parts of cycloids and of the masses formed by their rotation around an axis was central to Pascal's problems, however; he did not accept Lalouvère's solutions, and Lalouvère himself later discovered and corrected an error in computation (although another remained undetected). The matter might have ended there had not Pascal, in his *Histoire de la roulette,* accused Lalouvère (without naming him) of plagiarizing his solutions from Roberval. Pascal's allegations were without foundation; Lalouvère asserted that he had reached all his conclusions independently, and became embittered, while Fermat, who might have helped to resolve the quarrel, chose instead to remain neutral. A second, incomplete solution to Pascal's problems was submitted by Wallis, and on 25 November 1658 the prize committee decided not to give the award to anyone.

Having returned to mathematics, Lalouvère went on to deal with bodies in free fall and the inaccuracies of Gassendi's observations in *Propositiones geometricae sex* (1658). He returned to problems concerning cycloids—including those posed by Pascal—in 1660, in *Veterum geometria promota in septem de cycloide libris.* In addition to these publications, Lalouvère maintained an active correspondence on mathematical subjects, several of his letters to Pascal being extant. Two of his letters to D. Petau may be found in the latter's *Petavii orationes*; the same work contains Petau's refutation of Lalouvère's views on the astronomical questions of the horizon and calculation of the calendar.

Lalouvère's work, rooted firmly in that of the ancients, was not innovative; nevertheless, he showed himself to be a man of substantial knowledge and clear judgment. He was a tenacious worker with a great command of detail. Montucla thought his style sufficient to keep "the most intrepid reader from straying."

BIBLIOGRAPHY

I. ORIGINAL WORKS. Lalouvère's writings are *Quadratura circuli et hyperbolae segmentorum ex dato eorum centro gravitatis* . . . (Toulouse, 1651); *Propositiones geometricae sex quibus ostenditur ex carraeciana hypothesi circa proportionem, qua gravia decidentia accelerantur* . . . (Toulouse, 1658); *Propositio 36ª excerpta ex quarto libro de cycloide nondum edito* (Toulouse, 1659); *Veterum geometria promota in septem de cycloide libris* (Toulouse, 1660), which has as an appendix Fermat's "De linearum curvarum cum lineis rectis comparatione dissertatio geometrica"; and *De cycloide Galilei et Torricelli propositiones viginti* (n.p., n.d.).

Works apparently lost are "Tractatus de principiis librae" and "De communi sectione plani et turbinatae superficiei ex puncto quiescente a linea recta per ellipsim . . .," both mentioned by Lalouvère in his works; and "Opusculum de materia probabile" and "Explicatio vocum geometricarum . . .," mentioned by Collins in a letter to Gregory of 24 Mar. 1671 (or 1672).

A reference to seven letters to D. Petau (1631–1644) at Tournon and Toulouse may be found in Poggendorff, II, col. 412.

Two letters from Petau to Lalouvère are in *Dionysii Petavii*, Aurelian Society of Jesus *Orations* (Paris, 1653).

II. SECONDARY LITERATURE. On Lalouvère or his work, see A. de Backer and C. Sommervogel, eds., *Bibliothèque de la Compagnie de Jesus*, V (Brussels–Paris, 1894), cols. 32–33; Henry Bosmans, in *Archives internationales d'histoire des sciences*, **3** (1950), 619–656; Pierre Costabel, in *Revue d'histoire des sciences et de leurs applications*, **15** (1962), 321–350, 367–369; James Gregory, *Tercentenary Memorial Volume*, Herbert Turnbull, ed. (London, 1839), p. 225; Gerhard Kropp, *De quadratura circuli et hyperbolae segmentorum des Antonii de Lalouvère*, thesis (Berlin, 1944), contains a long list of secondary literature; J. E. Montucla, *Histoire des mathématiques*, II (Paris, 1758), 56–57; and P. Tannery, "Pascal et Lalouvère," in *Mémoires de la Société des sciences physiques et naturelles de Bordeaux*, 3rd ser., **5** (1890), 55–84.

HERBERT OETTEL

LAMARCK, JEAN BAPTISTE PIERRE ANTOINE DE MONET DE (*b.* Bazentin-le-Petit, Picardy, France, 1 August 1744; *d.* Paris, France, 28 December 1829), *botany, invertebrate zoology and paleontology, evolution.*

Lamarck was the youngest of eleven children born to Marie-Françoise de Fontaines de Chuignolles and Philippe Jacques de Monet de La Marck. His parents were among the semi-impoverished lesser nobility of northern France; his father, following family tradition, served as a military officer. It was primarily economic and social considerations which led his parents to select the priesthood as his future career. Lamarck, at about age eleven, was sent to the Jesuit school at Amiens; he was not, however, interested in a religious career and much preferred the military life of his father and older brothers. When his father died in 1759, Lamarck left school in search of military glory. Within a few years he was fighting with a French army in the Seven Years' War. After the war was over, he spent five years (1763–1768) at various French forts on the Mediterranean and eastern borders of France. It was during this period that he began botanizing; his military transfers served to acquaint him with highly diverse types of French flora.

In 1768 Lamarck left military service because of illness and after several years found a job in a Paris bank. He subsequently studied medicine for four years and became increasingly interested in meteorology, chemistry, and shell collecting.

Lamarck's personal life was marked by tragedy and poverty. He had three or four wives and eight children. In 1777 he began a liaison with Marie Rosalie Delaporte, marrying her, fifteen years and six children later, as she was dying. In 1793 he married Charlotte Victoire Reverdy, by whom he had two children; she died in 1797. The following year he married again; Julie Mallet died childless in 1819. There is some indication that he married for a fourth time, but no documents to support this can be found. Lamarck's health began to fail in 1809, when he developed eye problems; in 1818 he became completely blind but was able to continue his work by dictating to one of his daughters. When he died in 1829, the family did not have enough money for his funeral and had to appeal to the Académie des Sciences for funds. His belongings, including his books and scientific collections, were sold at public auction; he left five children with no financial provisions. Of these, one son was deaf and another insane; his two daughters were single and without support. Only one child, Auguste, was financially successful as an engineer; he was the only one of the offspring to marry and have children.

Botany. Lamarck's recognition by the French

scientific community resulted from the publication of his *Flore françoise* in 1779 (not 1778 as the title page says). His innovation was the establishment of dichotomous keys to aid in the identification of French plants; by eliminating large groups of plants at each stage through the use of mutually exclusive characteristics, the given name of any plant could be rapidly determined. This "method of analysis," as Lamarck called it, was much easier to use in identifying plants than Linnaeus' artificial system of classification, which was based on sexual differences among plants or the natural methods of classification then developing in France with the work of Adanson, Bernard de Jussieu, and Antoine Laurent de Jussieu. Lamarck's new approach and his criticisms of Linnaeus impressed Buffon, who arranged to have the *Flore* published by the government. The first of the three volumes contained a theoretical "Discours préliminaire" which, among other things, explained the method of analysis and a lengthy exposition of the fundamentals of botany. The other two volumes listed all known French plants according to his method of classification and provided good descriptions of each species. The *Flore* was one of the first French works to include the Linnaean nomenclature as well as that of Tournefort. Written in French rather than Latin, the *Flore* was an immediate success and the first printing was sold out within the year; in 1780 the work was reprinted. Lamarck had various plans for a new edition but was unable to carry them out for lack of funds. Finally, in 1795, the *Flore* was reprinted; although it did not differ from the first edition, it was called a second edition. In 1802 Lamarck, who was too busy doing other things, turned the preparation of a new edition over to A. P. de Candolle, who published what is called the third edition in 1805. Candolle made major revisions, replacing Lamarck's system of classification with that of A. L. de Jussieu and revising the section on the fundamentals of botany to include new scientific discoveries. Ten years later this third edition was reprinted and another volume was added to include species previously unknown or overlooked.

Lamarck's other major work in botany was his contribution to the *Dictionnaire de botanique*, which formed part of the larger *Encyclopédie méthodique*. He wrote the first three and a half of eight volumes; they were published in 1783, 1786, 1789, and 1795. Lamarck composed a long "Discours préliminaire," articles on all aspects of botany including classification and the structures of plants, and articles describing specific plants and their classificatory groupings. The companion piece to the *Dictionnaire*, the *Illustration des genres*, appeared in three volumes in 1791, 1798, and 1800. It included about 900 plates, descriptions of genera arranged according to Linnaeus' system of classification, and a listing of all known species in these genera. Lamarck himself had identified several new genera and species; he published these discoveries as articles in various publications from 1784 to 1792.

In addition to devising a new and useful method for the identification of plants and doing systematic botany, Lamarck demonstrated a number of theoretical and philosophical interests in his botanical works. In the "Discours préliminaire" of the *Flore*, Lamarck made his first attempt to formulate a natural method of classification for the vegetable kingdom. His aim was to discover the position every vegetable species should occupy in a graduated unilinear chain of being on the basis of comparative structural relationships. Unable to achieve this, he had to settle for a natural order at the level of the genera and even this was very tentative. Although he shared a common assumption of the time, that a natural classification would begin with the most complex and descend to the simplest organism, he found that in practice it was easier to work in the opposite order. This order would later be an essential feature of his evolutionary theory. Lamarck intended to develop his natural method in a work which was to be entitled *Théâtre universel de botanique*. The proposed work was to include all members of the vegetable kingdom, not just those found in France; it was never written.

In the "Discours préliminaire" of the *Flore*, Lamarck showed the orientation of a naturalist philosopher concerned more with the broad problems than the little facts, as he called them. He conceived of nature as a whole composed of living and nonliving things, the former divided into plants and animals. It was the view of the whole, its processes, and interrelations which really interested him.

Lamarck, in the same work, demonstrated his awareness of the important influence of the environment, especially climate, on vegetable development. He noted that two seeds from the same plant growing in two very different environments would become two apparently different species. Lamarck was particularly conscious of the changes plants undergo in artificial cultivation and he referred to such changes as degradations, the term he first used in describing evolutionary processes in 1800. In 1779, however, Lamarck still believed in fixed species and thought of the environment as the factor responsible for the production of varieties; by 1800 he had extended these views on the production of varieties to the origin of all organisms below the level of classes.

In 1779 Lamarck also demonstrated his genetic approach to a subject; the present is understood by tracing the historical steps that produced it, beginning with the most primitive level and working up through time to the more complex. Lamarck shared with the Philosophes a belief in the idea of progress in human knowledge, which is clearly seen in his brief history of botany. Increasing progress over time is almost inevitable if circumstances are favorable. He was later to apply such a conception to natural as well as human history.

In the *Dictionnaire de botanique* Lamarck developed the theoretical and philosophical ideas he had advanced in the *Flore*. The "Discours préliminaire" was an expanded version of the history of botany from the *Flore* and it showed even more fully Lamarck's belief in the idea of progress. Lamarck himself had made some progress in his search for a natural method of vegetable classification. Following some suggestions from A. L. de Jussieu, he decided that a hierarchical arrangement could be established only for the larger groupings or classes of plants. A focus on classes rather than genera or species would later be an important part of his evolutionary theory. Lamarck's new views were set forth in the article "Classe," in which he listed the classes of plants, arranged from the most complex to the least complex; placement on the scale was determined by relative structural complexity. To complete the realm of living organisms, Lamarck presented a parallel series of descending classes for the animal kingdom. In pointing out the similarities between plants and animals, he laid the foundations for his biology. In another table the nonliving natural productions were also arranged in order of decreasing complexity. Lamarck held that all mineral substances were produced by organic beings as they and their waste products decayed over time and their debris underwent successive transformations until the simple element level was reached. The fact that Lamarck drew up these tables of comparison shows his concern with seeing nature as a whole.

During the 1790's, Lamarck's interests and studies turned away from botany to new fields. After 1800, when he began advocating his theory of evolution, he wrote only one work specifically dealing with botany. His two-volume *Introduction à la botanique* (1803) formed part of the fifteen-volume *Histoire naturelle des végétaux*; the rest of the work was written by Mirbel. This study of the vegetable kingdom was in turn part of the larger eighty-volume *Cours complet d'histoire naturelle pour faire suite à Buffon* edited by Castel. Lamarck's *Introduction* was his only botanical work to include his evolutionary theory. He stressed that for the vegetable kingdom a natural order of classification beginning with the simplest class and ending with the most complex class reflected the order which nature had followed in producing these groups in time. Although this was his last botanical work, Lamarck did not stop thinking about the vegetable kingdom. He discussed it in all of his evolutionary works and drew a number of examples from it.

Institutional Affiliations. Lamarck's election to the Académie des Sciences as an adjoint botanist was engineered by Buffon in 1779. He was promoted to an associate botanist in 1783 and became a pensioner in 1790. The Academy was suppressed in 1793, during the Terror; it was reorganized two years later as part of the Institut National des Sciences et des Arts. From 1795 until his death, Lamarck was a resident member of the botanical section. Until his health failed, he attended meetings regularly and prepared a number of reports on works submitted to the Academy.

In the 1790's Lamarck took an active role in the newly formed Société d'Histoire Naturelle, which included the prominent French naturalists of the time. He helped edit several of its publications and contributed a number of articles on botany and invertebrates to them.

Lamarck's most significant institutional affiliation was with the Jardin du Roi, which had become an important scientific center in the second half of the eighteenth century under the leadership of Buffon. From 1788 until 1793, Lamarck held various minor botanical positions there. During the French revolution, when all the institutions of the *ancien régime* were being subjected to critical examination, suggestions were made for the reorganization of the Jardin du Roi, among them a memoir by Lamarck. In 1793, when the academies were suppressed as privileged institutions of the old order, the Jardin du Roi was transformed into the Muséum National d'Histoire Naturelle. The botanical positions were filled by others and Lamarck was made a professor of zoology for the study of "insects and worms," a group of animals which he renamed "invertebrates." While this represented a rather drastic shift in fields for him, Lamarck was not unhappy about it, for he had been developing an interest in these animals. His new duties consisted of giving courses and classifying the large collection of invertebrates at the museum. He also took an active part in the administration of the new institution. Lamarck's own work benefited from contacts with his colleagues and their scientific investigations at the museum.

Chemistry. The first long works that Lamarck published after the reorganization of the museum dealt not with invertebrates but with chemistry, a subject in which he had been interested for many years. He had begun to study chemistry in the 1770's, when the four-element theory (earth, air, water, and fire) was generally accepted in France. He continued to believe in the four elements throughout his life despite the work of Lavoisier and the chemical revolution; for this reason his chemistry has often been dismissed as worthless speculation. Yet Lamarck took it very seriously, and it was an important part of his ideas about nature and evolution.

Lamarck's first work in the field, *Recherches sur les causes des principaux faits physiques*, was begun in 1776. It was submitted to the Académie des Sciences in 1780 and received an unfavorable report; it was finally published in 1794, after the Academy had been suppressed. Lamarck devoted two other full-length studies to chemistry: *Réfutation de la théorie pneumatique* (1796) and *Mémoires de physique et d'histoire naturelle* (1797). He also published two articles in 1799; they were reprinted at the end of his *Hydrogéologie* (1802), which contained a long chapter relating his chemical theories to his geological theories. Although Lamarck's chemical views were ignored, he continued to hold them; they appear with signs of increasing paranoia in his major evolutionary works. They play the most prominent role in *Recherches sur l'organisation des corps vivans* (1802), the first full-length exposition of his evolutionary theories.

In Lamarck's four-element theory, differences between compounds depended on both the number and proportion of the elements and the relative strengths of the bonds between the elements in the constituent molecules. Furthermore, each element had a natural state in which it demonstrated its real properties and several modified states in which it was present in compounds. The most important of the four elements in Lamarck's chemistry was fire, which existed in three main states: a natural one and two modified forms, which were fire in a state of expansion (or caloric fire) and fixed fire. Using these three main states and their many internal modifications, Lamarck attempted to account for a great number of chemical and physical phenomena such as sound, electricity, magnetism, color, vaporization, liquefaction, and calcination. Later, in his theory of evolution, he added life as another phenomenon to be explained by activity of fire. For Lamarck, fire not only explained many processes, it also was a constituent principle of compounds. He attempted to show how chemical substances in their various states depended on differing amounts of fixed fire. One temporary form of fixed fire was phlogiston.

Lamarck believed that only living beings could produce chemical compounds. Plants combined free elements directly to produce a number of substances of varying complexity. These in turn were elaborated by the different animals eating the plants, the more complex substances being produced by those animals with the most highly organized physiological structure. The process of compound formation involved modification of the elements away from their natural state and the more complex the substance, the greater the modification. Once the forces of life were removed, by death or the elimination of waste products, the compounds began to disintegrate. The natural tendency of all compounds, therefore, was to decompose until the elements returned to their natural state, in the process producing all known inorganic substances. For the mineral kingdom there was a chain of being with continous degradation from the most complex to the simplest; this chain was composed of individuals rather than species or types of minerals. Lamarck's first statement of his theory of evolution in 1800 showed a similar thought pattern: degradation and irrelevance of species.

In his chemistry, Lamarck showed a speculative orientation and an emphasis on nature as a whole with many interrelated parts and processes. His distinction between the living and the nonliving was crucial to his biology and his view of the mineralogical chain of being was basic to his geology. His chemistry was also later to be very important in his theory of evolution. It was used to provide a materialistic definition of life and to explain its maintenance, appearance (both through reproduction and through spontaneous generation), and the way in which living organisms gradually evolve, including the emergence of the higher mental faculties. Fire, as understood by Lamarck, was the key element in all of these explanations.

Meteorology. Lamarck's work in meteorology was similar in many respects to that in chemistry. Although one of his earliest scientific interests, he did not publish anything until the late 1790's; he experienced the same general lack of reception of his work in chemistry. Meteorology was the first scientific area in which Lamarck prepared a memoir and one which was well received by the Academy. The manuscript of this unpublished memoir (Muséum National d'Histoire Naturelle, Paris, MS 755–1) shows that as early as 1776, Lamarck was interested in the effects of climate on living organisms. It is highly probable that Lamarck's interest in chemistry resulted from his concern with certain aspects of meteorology. His general approach to science is also evident in this early manuscript: his emphasis is on the general principles, and he manifests disdain toward those devoted solely to the collection of little facts. The extent to which Lamarck saw his meteorology as part of his whole view of nature is indicated later in his

Hydrogéologie, in which he states that a terrestrial physics would include three subjects: meteorology, hydrogeology, and biology. He originally intended to write a work dealing with these areas but decided to postpone the sections on meteorology and biology until he had done further research.

Lamarck continued his study of meteorology in the 1780's, while he was involved in botanical studies; his awareness of the importance of climate for plants has already been mentioned. In 1797 he began publishing articles on meteorology and attempting to provide theoretical explanations for factors causing weather change. Three years later he started publishing the *Annuaire météorologique*. It has often been said that Lamarck wrote these volumes with the sole intention of earning money, but he showed too great an interest in them and in defending his theories for that to have been the case. It was surely no coincidence that he was assembling his meteorological studies at the same time he was elaborating his theory of evolution (the last *Annuaire* was published in 1810). Since climate was an important factor in his theory, it would be important to seek the laws regulating changes in climate and therefore perhaps be able to predict or understand changes in organisms more fully.

Lamarck's meteorology was devoted to a search for those laws of nature which regulate climatic change. The search was more important than the speculative theories he devised, for it indicates certain connections with the Enlightenment; he assumed that there must be simple discoverable laws governing weather changes. He also had grounds to think that such laws must exist because of Franklin's success in identifying lightning and terrestrial electricity; he greatly admired Franklin. In addition, a number of his contemporaries were studying meteorological phenomena, improving instruments, and theorizing. Lamarck was familiar with their work and used it as a point of departure for his own.

As in botany, Lamarck was attracted to the history and progressive development of meteorology. One might also say that he was concerned with a natural classification of meteorological phenomena. Finally, his theoretical considerations indicate an important thought pattern; he tried to explain all meteorological change as the result of one general cause (the moon) with irregularities produced by local circumstances.

Lamarck did have one public success with his meteorology. He recommended that the French government establish a central meteorological data bank. Following this suggestion, Chaptal established such a program in the ministry of the interior in 1800. One of Lamarck's concerns was that the daily observations from all parts of France be made in accordance with standardized procedures and instruments. The project and Lamarck's work in meteorology ended in 1810, when Napoleon ridiculed Lamarck's *Annuaire*.

Invertebrate Zoology and Paleontology. When the Muséum d'Histoire Naturelle was established in 1793 and Lamarck made professor of "insects and worms," he had the tasks of organizing the museum collection and giving courses, beginning in the spring of 1794. His only previous connection with the invertebrates was his interest in shell collecting. He was, however, a good friend of his colleague, Jean-Guillaume Bruguière, who was considered an expert on invertebrates, especially the mollusks. When Bruguière died in 1798, Lamarck finished his *Histoire des vers* for the *Encyclopédie méthodique*. Lamarck's published works in the field include a number of articles, in which he identified new genera and species and put forth some theoretical considerations, and books, the most important of which were *Système des animaux sans vertèbres* (1801) and his seven-volume major work, *Histoire naturelle des animaux sans vertèbres* (1815–1822).

Lamarck developed a system for the natural classification of invertebrates based on the anatomical findings of Cuvier. As in botany, the natural order consisted of classes arranged in a linear fashion from most complex to least complex. Such a series provided clear examples of degradation in anatomical structure and physiological function, as one system after another disappeared. Lamarck spoke of this degradation well before he advocated his theory of evolution, and that theory was first put forth as one of degradation. His study of invertebrates also helped him refine his definition of life, for the simplest organisms indicated the minimum conditions necessary for life. The origin (generation) of these simplest animals raised problems whose answer seemed to be spontaneous generation.

Burkhardt has pointed out (1972) that Lamarck came to be regarded as an expert in conchology and the successor to Bruguière. Lamarck made an important contribution to the classification of shells in his "Prodrome d'une nouvelle classification des coquilles" (1799), in which he established 126 genera. Any attempt to classify shells immediately raised the problem of what to do with fossil forms. According to Burkhardt, the pressing question of the late 1790's was whether there were any similarities between living and fossil forms. If the answer were no, the way was open to a belief in extinction, especially extinction brought about by some catastrophe. The issue being debated involved vertebrates as well. Cuvier and others were making impressive discoveries of large mammalian fossils which seemed to have no living analogues. There were some people at the museum, however, who were discovering analogues. Alexandre Brongniart and, slightly later, Faujas de Saint-Fond and Étienne Geoffroy Saint-Hilaire were investigating similarities between some fossil and living reptiles.

Several naturalists, including Faujas, expected Lamarck's investigations of shells from living and fossil forms to resolve the issue. His work did reveal a number of analogues. The question then became one of explaining the similarities and differences. Although the existence of analogues would rule out the possibility of a general catastrophe, more limited violent events could have produced some species extinction which would account for the failure to find analogues in many cases. There were, Burkhardt suggests, two other ways to explain the differences: migration or evolution. Lamarck, unable for philosophical reasons to entertain the possibility of extinction and unconvinced of migration as a plausible way to account for all the differences, chose an evolutionary explanation sometime in late 1799 or early 1800. Differences between living and fossil forms existed precisely because organisms had undergone change over time; Lamarck regarded this position as one of the strongest arguments against extinction. The study of fossil forms led Lamarck to conceive of nature as existing in time. Lamarck has often been called the founder of invertebrate paleontology. His most important work on the subject was *Mémoires sur les fossiles des environs de Paris* (1802–1806). The "Introduction" to this work discusses the significance of fossils for a theory of the earth.

Geology. Lamarck's geology was closely connected with his work in other fields. His *Hydrogéologie*, which grew out of a 1799 memoir presented to the Academy and his work in invertebrate paleontology, was published in 1802. He originally intended it to be a much broader work, as the manuscript shows (Muséum National d'Histoire Naturelle, Paris, MS 756–1, 2). It was to have been a terrestrial physics including meteorology, geology, and biology, a term he coined. He had not only a sense of the interrelation of fields but, within geology, a vision of the whole. He saw all of nature working according to similar principles: general natural tendencies producing gradual change over long periods of time, with local circumstances explaining the irregularities. His approach to geology was similar to that in other sciences: concern with the general principles and contempt for those who interested themselves too much with the specifics.

Although Lamarck's geological views were not original (he was strongly influenced by Buffon and Daubenton, among others), they were an important part of his conception of nature. His preoccupation with marine fossil shells had a decisive influence on his choice of geological theories. Since such shells had to have been laid down in water, he needed a theory to explain how this was possible. As in his meteorology, he used the moon as the main cause, in this case of a constant slow progression of the oceans around the globe. The main geological force was water acting according to uniformitarian principles over millions of years. The substances of the mineral kingdom were produced by the progressive disintegration of organic remains; water operated on these products to produce geological formations such as mountains. Lamarck's uniformitarianism and great geological time scale have led some to say that he was his own Lyell. Some historians have thought that Lamarck's perception of a slowly changing environment and the resulting necessity of organisms to change or become extinct (a possibility he could not accept) led him to his theory of evolution.

Theory of Evolution. Before beginning a discussion of Lamarck's theory of evolution, it is important to point out that he never used the term "evolution" but, rather, spoke of the path or order which nature had followed in producing all living organisms. "Evolution" is used here only as a shorthand form for Lamarck's longer phrases.

Lamarck's first public presentation of his theory of evolution was in his opening discourse for his course on invertebrates at the museum in 1800; it was published the following year at the beginning of his *Système des animaux sans vertèbres*. The evolutionary views sketched in the discourse leave much to be desired in terms of organization and explanation. They are, however, very much a part of a total view of nature, many aspects of which Lamarck had long accepted. Natural products consisted of living and nonliving things; in the two branches of living organisms, Lamarck pointed out the "degradations" in structural organization of the larger classificatory groupings or "masses" as one moved down the series from the most complex to the simplest. He indicated his lack of clarity about the new views he was proposing by using the term "complication" interchangeably with "degradation." Nature, after having formed the simplest animals and plants directly, produced all others from them with the aid of time and circumstances. In 1800 Lamarck did not explain how spontaneous generation occurred or how unlimited time and varied circumstances produced all other organisms. He did suggest that, for animals, changing circumstances and physical needs led to new responses which eventually produced new habits; these habits tended to strengthen certain parts or organs through use.

Gradually new organs or parts would be formed as acquired modifications were passed on through reproduction.

The great expansion of Lamarck's ideas occurred between 1800 and 1802, when the *Recherches sur l'organisation des corps vivans* was published. Much of the work was devoted to documenting the "degradations" in organization of the larger groupings of animals, from mammals to polyps. He was, however, more aware of the need to turn this series over so that it would correspond to the order of production in nature and time and thus be a really natural method of classification. Once he began to think in terms of increasing levels of complexity, he needed a mechanism to propel change; otherwise one would have only polyps in the world. He therefore began to talk of a natural tendency in the organic realms toward increasing complexity. Lamarck is not clear or consistent about the manner in which this natural tendency operates; often it seems to function like a moving escalator. At other times it can best be understood by a stairlike construction where descent is more directly tied to the historical past although Lamarck never suggests any dates marking the appearance of particular forms of life.

Lamarck's conception of a natural tendency toward increasing complexity provided a perfect complement to his views of the mineral kingdom with the opposite natural tendency. In both cases a long time span allowed nature to do her work and local circumstances explained irregularities. Among living beings, irregularities included all organisms below the level of the "masses," which usually meant classes but sometimes was extended to orders and families, never to genera and species.

In 1802 many of the additions were designed to explain in more detail why and how evolution happened; Lamarck's chemistry was essential to those explanations. He had to face the crucial problem of what happened at the lower ends of the vegetable and animal series. Rejecting the vitalist views of such contemporaries as Bichat, Lamarck defined life in physical terms. Life resulted from a particular kind of organization and a general tension maintained by the stimulation of the subtle fluids of his chemistry, especially modified forms of fire.

Spontaneous generation of animals, which Lamarck held was analogous to fertilization, occurred when heat (or caloric), sunlight, and electricity acted on small amounts of unorganized, moist, gelatinous matter to produce the simplest animals. He later specified that the simplest plants were spontaneously generated when the same physical substances organized moist, mucilaginous matter. The first traces of organic organization formed by the subtle fluids were simple structures capable of containing certain fluids, such as water, and more complex substances. From this point on, the natural tendency of organic movement toward increasing complexity could take over. In plants and simple animals, the physical cause of this tendency was the constant agitation of the contained fluids by the subtle fluids of the environment (especially the matter of heat), which were not containable and could penetrate the living organism. The result of this agitation was the gradual hollowing out of passages and tubes and the eventual formation of organs and then primitive systems.

Animals with circulatory systems were less directly dependent on the environment because, Lamarck believed, the matter of heat was constantly disengaging from the blood and thus providing an internal stimulus for greater development. Such was the materialistic explanation he gave for the origin of the larger groupings of plants and animals. All differences below that level were explained by variations in the movement of the containable and subtle fluids due to different circumstances, especially the temperature of the environment, and to changing life styles resulting in new habits. Through the use of parts, containable fluids were concentrated and the formation of new organs was accelerated. With new organs and systems, new faculties appeared. Acquired changes were preserved through inheritance.

In 1802 Lamarck dealt briefly with the upper limit of the animal series—man. He cautiously suggested that man was the result of the same processes that had produced all other living organisms. The major obstacle to the inclusion of man in the evolutionary process was his higher mental faculties. At the end of the *Recherches*, Lamarck offered a possible solution to this problem. Using his chemistry and comparative anatomy, he attempted to provide a materialistic explanation for the functioning of the nervous system. Following Haller's distinction, he maintained that while all animals exhibit irritability, only those with a nervous system experience sensibility or feeling. The degree to which an animal possessed the latter faculties depended on the level of complexity of the nervous system and the movement of the nervous fluid, which was a modified form of fire, similar to electricity. This subject of the evolution of the higher mental faculties underwent major development in the *Philosophie zoologique* (1809).

This work is the best-known and most extensive presentation of Lamarck's theory of evolution. An expanded version of the 1802 *Recherches*, it is divided into three sections. The first is a more elaborate analysis of the evidence for increasing levels of "complication" observed in the major classificatory groupings of animals and plants. It also presents in more detail Lamarck's two-factor theory of evolution: the natural tendency toward organic complexity as a way of explaining the hierarchical organization of the "masses" and the influence of the environment as the factor responsible for all variations from this norm. In the second part of the *Philosophie zoologique*, Lamarck developed his views on the physical nature of life, its spontaneous production resulting in simple cellular tissue, and its characteristics at the simplest level, the lower ends of the plant and animal series. While these two parts were very important in summarizing many of his evolutionary views, they do not differ significantly from the positions of 1802.

The third part contains the most important additions to the earlier theories. In this section Lamarck deals in great detail with the problem of a physical explanation for the emergence of the higher mental faculties. Some of the eighteenth-century materialists, such as Maupertuis, had attempted to avoid the question of emergence by making thought a property of matter. Some religious figures went to the other extreme and limited thought to man and his soul. Lamarck's breakthrough was tying a progressive development of higher mental faculties in a physical way to structural development of the nervous system. He had already advanced explanations for the evolution of new structures and systems, and the theories on the nervous system were an extension of these earlier views. Higher mental faculties could emerge precisely because they were a product of increased structural complexity, and in all this a physically defined nervous fluid was crucial. For Lamarck one of the most important events in the evolutionary process was the development of the nervous system, particularly the brain, because at that point animals began to form ideas and control their movements.

There has been great misunderstanding of Lamarck's concept of *sentiment intérieur*, or inner feeling, as a directing factor in the functioning and evolution of higher animals. Lamarck never believed that the giraffe has a longer neck because it consciously wanted one. Rather, he observed that higher animals were capable of voluntary motion which might become habit (as in a search for food or avoidance of danger) and of involuntary motion, or what we would call reflex action. Lamarck attempted to account for such behavior through the mechanism of the *sentiment intérieur*, an internal physical feeling resulting from agitation of the nervous fluid. The brain of an animal with an internal physical need, such as hunger, would direct the nervous fluid so as to cause muscular motion to satisfy that need. If this action were constantly repeated, new organs would eventually result. On the other hand, a sudden, strong stimulus, such as a loud noise, would produce a reflex action because of a particular perturbation of the nervous fluid.

The concept of the *sentiment intérieur* included not only the direct interaction with the physical world but also a more sophisticated level. It could be affected, particularly in human beings, by ideas or moral sensations. Such a view was in keeping with an extension of Condillac's sensationalist psychology and epistemology, especially as expressed by Cabanis and the Idéologues. Moral and aesthetic reactions were thus as physically caused as instinctive or reflex ones; the only difference was that between primary or secondary causation. It is not surprising that Lamarck has many references to Cabanis on the relationship between *physique* and *morale*. Lamarck felt he had provided a materialistic account for all the activities involving the nervous system, including instinct, will, memory, judgment, understanding, and imagination. He further developed these views in his last publication, *Système analytique des connaissances positives de l'homme* (1820).

Next to the *Philosophie zoologique*, Lamarck's best-known work dealing with evolution is the 1815 "Introduction" to his impressive seven-volume *Histoire naturelle des animaux sans vertèbres* (1815–1822). In this work he summarized his evolutionary views in four laws. The first law concerns his principle of the natural tendency toward increasing organic complexity as observed in the larger groupings of the plant and animal series. The other three laws explain how changes occurred and account for irregularities below the class level. The second law deals with the way new organs evolve by the indirect influence of the environment on an animal. The use-disuse principle, or third law, accounts for changes in the body as a result of new habits; this principle was not new with Lamarck but was generally accepted. The last law, dealing with the inheritance of acquired characteristics, was necessary after positing a slow, gradual evolution; without it Lamarck would have been unable to explain cumulative change and the emergence of new structures. Too much energy has been spent attacking this last law; because it represents an assumption not believed today, it has been said that this disproves Lamarck's whole theory of evolution. The historical context of Lamarck's thought has been forgotten. Most of his contemporaries believed in the inheritance of acquired characteristics, so much so that they rarely felt any need to offer proof of it. The above summary of the four 1815 laws shows that the basic features of Lamarck's evolutionary theory remained relatively unchanged from 1802.

Lamarck has been credited with introducing a branching family tree into evolutionary theory. It is true that in his major evolutionary works he often spoke of branchings below the level of the "masses," but he regarded these branchings as exceptions to the general rule of increasing structural complexity. In the *Philosophie zoologique*, Lamarck argued for a unilinear series of classes in each kingdom. In an addition at the end of the work, however, he presented a branching arrangement for animal classes in what is now a well-known diagram. It is significant that he was never able to integrate this new order with his evolutionary theory. The same discontinuity occurs in the 1815 "Introduction"; in the body of the work he assume a linear series of the "masses," and in a supplement he presents a diagram of branching classes (different from the 1809 version) which he labels as the presumed order of formation. These evolutionary trees were Lamarck's acknowledgment of advances in comparative anatomy and natural classification. That he could not include them with his two-factor theory shows his strong lifelong commitment to a philosophical idea: the chain of being and its modified version of a hierarchical unilinear series of classes in the two kingdoms.

The above example indicates a very important aspect of Lamarck's evolutionary theories. They were put forth by a philosopher–naturalist and not a positivist scientist. From his earliest scientific work, Lamarck was always more interested in the broad picture of nature and in general interrelations than in the details. While he did give scattered examples to support his theories, he was never systematic, always promising more evidence in a forthcoming work and never producing it. Lamarck felt that his theories were so obvious that they did not need extensive proof. In addition he was paranoid with respect to the French scientific community because of their attitudes toward his work in chemistry; he was convinced that he never could win over his enemies, so he did not try. With all his work in botany and invertebrate zoology, he would have had abundant examples if he had wanted them. He always separated his theories and his detailed classificatory work. Although he spent years carefully determining, describing, and classifying species, in his evolutionary views he maintained that species were almost irrelevant, exceptions to the general natural law of evolution.

Origins of Lamarck's Theory. Lamarck was fifty-five when, in 1800, he made his first public statement of evolution. Until the late 1790's he had believed in the fixity of species. Thus the question has always been why he changed his mind. Various answers have been and still are being put forth. They range from different influences from his own work to the particular influence of individuals. Among the former explanations, Lamarck's work in geology, invertebrate classification, paleontology and the problem of extinction, and chemistry have all been seen as the crucial factor. Individuals who have been held to have exerted the decisive influence on Lamarck's change of mind include Lacépède, Cabanis, Cuvier, and such earlier speculative thinkers as Buffon, Diderot, De Maillet, and Robinet. A study of the origins of Lamarck's theory of evolution must be broadened beyond the issue of the fixity of species. He also changed his views on other issues. Not only should we look at the changes, but we should also look at the continuities. Many aspects of Lamarck's evolutionary theory are found in his earlier works in different fields; some components of the theory, however, are new. Burkhardt, who recognizes the importance of continuities, has recently studied certain factors which he thinks were the immediate causes of Lamarck's evolutionary thought. In an article (1972) he presents a convincing argument to show that Lamarck changed his views on two important subjects (spontaneous generation and the mutability of species) between the spring of 1799 and the spring of 1800 and that in both cases the changes came about as a result of confronting the question of extinction. In the section on invertebrate paleontology, we summarized Burkhardt's position. Lamarck was faced with the extinction question in his study of fossil shells. Since an acceptance of extinction would have violated his view of nature, and since the migration theory was not really a satisfactory explanation, Lamarck was left with the choice that species change gradually with time.

Lamarck's acceptance of the possibility of spontaneous generation came via a more circuitous route but one which was also related to the study of invertebrates and to the extinction question. Burkhardt suggests that Lamarck's study of the invertebrates led him to a new definition of life. For Lamarck, the simplest organism demonstrated the minimum conditions necessary for life; lacking any specialized organs, they depended entirely on the movement of subtle fluids in the environment to maintain their organic movements. The next logical step was to move to a belief in spontaneous generation. If the subtle fluids could maintain this simple form of life, why could they not also create it when the circumstances were right? The extinction issue came in when Lamarck realized that these organisms were killed in bad weather. The only way he could account for their reappearance was by spontaneous generation. A belief in spontaneous generation was necessary for a theory of evolution, unless one wished to invoke a creator. Thus, according to Burkhardt, Lamarck's changing positions on these two issues were the crucial events which led to his theory of evolution.

While accepting the changes described above, we should also look at some of the continuities in Lamarck's thought. The continuities have been mentioned in each section of the article, but we will try to summarize them here. Lamarck had been interested in trying to develop a natural method of classification from the time of his earliest work in botany. Well before 1800, he had constructed a series of classes. In his theory of evolution, the natural method was the path nature had followed in producing the different groups of organisms. In Lamarck's work prior to 1800, we see his stress on nature as a whole whose processes and interrelations are more important than the details. The chemistry provides the key to these connections and perhaps a model for understanding the realm and order of living organisms. Finally, Lamarck's belief in the idea of progress may have prepared him for the application of such an idea to nature.

Lamarck's Reputation. When Lamarck died in 1829, he left few followers; generally he was ignored. The official eulogy prepared by Cuvier for the Academy condemned Lamarck's speculations and theories in all fields as being equally unacceptable; faint praise was offered for his contributions to biological classification. While he was ignored by his countrymen, he did receive some attention in England from the generation before Darwin. But it was really Darwin's theory of evolution which ensured Lamarck's fame. The question of the extent of Lamarck's influence on Darwin is still debated. It was mainly Darwin's enemies and detractors who revived Lamarck for a variety of reasons, ranging from scientific to religious to nationalistic (on the part of the French). Toward the end of the nineteenth century, a famous controversy developed between Darwinians and the so-called neo-Lamarckians; the latter used Lamarck's views selectively and often changed many of them to suit their purposes. Neo-Lamarckism had strong proponents in France, Germany, England, America, and more recently in the Soviet Union. With the wide acceptance of Darwinism as modified by modern genetic theory, much of Lamarckism has died out, although some still apply it to seemingly purposive biological behavior.

Aside from his legacies and the battles fought in his name, Lamarck deserves an important place in the history of science. He made significant contributions in botany, invertebrate zoology and paleontology, and developed one of the first thoroughgoing theories of evolution.

BIBLIOGRAPHY

I. Original Works. A more complete bibliography may be found in Landrieu (see below). Lamarck's most important published works are *Flore françoise*, 3 vols. (Paris, 1779; 2nd ed., 1795; 3rd ed., 1805, in collaboration with A. P. de Candolle, 4 vols.; repr. of 3rd ed. with 1 vol. supp., 1815); *Dictionnaire de botanique*, vols. I-III and first half of IV of 8 vols. (Paris, 1783–1795) in *Encyclopédie méthodique*, C. J. Panckoucke, ed., 193 vols. (Paris, 1782–1832); *Illustration des genres*, 3 vols. (Paris, 1791–1800), also in *Encyclopédie méthodique; Recherches sur les causes des principaux faits physiques*, 2 vols. (Paris, 1794); *Réfutation de la théorie pneumatique* (Paris, 1796); *Mémoires de physique et d'histoire naturelle* (Paris, 1797); "Prodrome d'une nouvelle classification des coquilles," in *Mémoires de la Société d'histoire naturelle*, I (Paris, 1799), 63–91; *Annuaires météorologiques*, 11 vols. (Paris, 1800–1810); *Système des animaux sans vertèbres précédé du 'Discours d'ouverture du cours de zoologie de l'an VIII'* (Paris, *an* IX [1801]); *Hydrogéologie* (Paris, *an* X [1802]), English trans. by A. V. Carozzi, *Hydrogeology* (Urbana, Ill., 1964); *Recherches sur l'organisation des corps vivans précédé du 'Discours d'ouverture du cours de zoologie, l'an X'* (Paris, *an* X [1802]); *Mémoires sur les fossiles des environs de Paris* (Paris, 1809), originally published as separate articles in *Annales du Muséum national d'histoire naturelle* (Paris, 1802–1806); see Landrieu for vol. and page references; *Introduction à la botanique*, 2 vols. (Paris, 1803), part of Lamarck and B. de Mirbel, *Histoire naturelle des Végétaux*, 15 vols. (Paris, 1803); *Philosophie zoologique*, 2 vols. (Paris, 1809; repr. Paris, 1830; New York, 1960), English trans., *Zoological Philosophy* (London, 1914; repr. New York–London, 1963); *Histoire naturelle des animaux sans vertèbres*, 7 vols. (Paris, 1815–1822); and *Système analytique des connaissances positives de l'homme* (Paris, 1820).

Posthumous works include "Discours d'ouverture des cours de zoologie donnés dans le Muséum d'histoire naturelle, an VIII (1800), an X (1802), an XI (1803), et 1806," A. Giard, ed., in *Bulletin scientifique de la France et de la Belgique*, **40** (1906), 443–595; *The Lamarck Manuscripts at Harvard*, William Wheeler and Thomas Barbour, eds. (Cambridge, Mass., 1933); "La biologie, texte inédit de Lamarck," Pierre Grassé, ed., in *Revue scientifique*, **82** (1944), 267–276; and *Inédits de Lamarck d'après les manuscrits conservés à la Bibliothèque centrale du Muséum national d'histoire naturelle de Paris*, Max Vachon, Georges Rousseau, and Yves Laissus, eds. (Paris, 1972).

Some unpub. MSS remain in the Muséum d'Histoire Naturelle's collection; consult Max Vachon, Georges Rousseau, and Yves Laissus, "Liste complète des manuscrits de Lamarck conservés à la Bibliothèque centrale du Muséum national d'histoire naturelle de Paris," in *Bulletin du Muséum national d'histoire naturelle*, 2nd ser., **40** (1969), 1093–1102.

II. Secondary Literature. See Jean-Paul Aron, "Les circonstances et le plan de la nature chez Lamarck," in *Revue générale des sciences pures et appliquées*, **64** (1957), 243–250; Franck Bourdier, "Esquisse d'une chronologie de la vie de Lamarck" (with the collaboration of Michel Orliac), unpub. memorandum, 3e section, École Pratique des Hautes Études, 22 June 1971; Richard W. Burkhardt, Jr., "Lamarck, Evolution and the Politics of Science," in *Journal of the History of Biology*, **3** (1970), 275–298; and "The Inspiration of Lamarck's Belief in Evolution," *ibid.*, **5** (1972); Leslie J. Burlingame, "The Importance of Lamarck's Chemistry for His Theories of Nature and Evolution or Transformism," in *Actes du XIIIe Congrès international d'histoire des sciences*, sect. IX A (Moscow); Dominique Clos, "Lamarck botaniste, sa contribution à la méthode dite naturelle," in *Mémoires de l'Académie des sciences, inscriptions et belles-lettres de Toulouse*, 9th ser., **8** (1896), 202–225; Georges Cuvier, "Éloge de M. de Lamarck," in *Mémoires de l'Académie Royale des sciences de l'Institut de France*, 2nd ser., **13** (1831), i–xxxi; Henri Daudin, *Cuvier et Lamarck. Les classes zoologiques et l'idée de série animale (1790–1830)*, 2 vols. (Paris, 1926), and *De Linné à Jussieu. Méthodes de la classification et idée de série en botanique et en zoologie* (1740–1790) (Paris, 1926), pp. 188–204; Charles C. Gillispie, "The Formation of Lamarck's Evolutionary Theory," in *Archives Internationales d'histoire des sciences*, **9** (1956), 323–338; John C. Greene, *The Death of Adam; Evolution and Its Impact on Western Thought* (Ames, Iowa, 1959), pp. 160–171; Emile Guyénot, *Les sciences de la vie aux XVIIe et XVIIIe siècles* (Paris, 1957), pp. 408–439; M. J. S. Hodge, "Lamarck's Science of Living Bodies," in *British Journal for the History of Science*, **5** (1971), 323–352; Marcel Landrieu, *Lamarck, le fondateur du transformisme, sa vie, son oeuvre* (Paris, 1909), which is **21** of *Mémoires de la Société Zoologique de France;* Ernst Mayr, "Lamarck Revisited," in *Journal of the History of Biology*, **5** (1972), 55–94; I. M. Poliakov, *Lamark i uchenia ob evolyutsii organicheskogo mira* ("Lamarck and the Theory of the Evolution of the Organic World") (Moscow, 1962); Georges Rousseau, "Lamarck et Darwin," in *Bulletin du Muséum national d'histoire naturelle*, **5** (1969), 1029–1041; Joseph Schiller, ed., *Colloque international Lamarck* (Paris, 1971); and "Physiologie et classification dans l'oeuvre de Lamarck," in *Histoire et biologie*, **2** (1969), 35–57; J. S. Wilkie, "Buffon, Lamarck and Darwin: The Originality of Darwin's Theory of Evolution," in P. R. Bell, ed., *Darwin's Biological Work; Some Aspects Reconsidered* (New York, 1959), pp. 262–307.

Highly recommended for background reading: Arthur O. Lovejoy, *The Great Chain of Being* (Cambridge, Mass., 1936); Jacques Roger, *Les sciences de la vie dans la pensée française du XVIIIe siècle* (Paris, 1963).

LESLIE J. BURLINGAME

LAMB, HORACE (*b.* Stockport, England, 29 November 1849; *d.* Manchester, England, 4 December 1934), *applied mathematics, geophysics.*

Lamb's father, John, was a foreman in a cotton mill who had a flair for inventing. Horace was quite young when his father died, and he was brought up by his mother's sister in a kindly but severely puritan manner. At the age of seventeen he qualified for admission to Queen's College, Cambridge, with a scholarship in classics but proceeded to a mathematical career. He gained major prizes in mathematics and astronomy and became second wrangler in 1872, when he was elected a fellow and lecturer of Trinity College. After three further years in Cambridge, he went to Australia as the first professor of mathematics at the University of Adelaide. He returned to England in 1885 as professor of pure mathematics (later pure and applied mathematics) at Owens College, Manchester, and held this post until his retirement in 1920. He married Elizabeth Foot; they had seven children.

Lamb was one of the world's greatest applied mathematicians. He was distinguished not only as a contributor to knowledge but also as a teacher who inspired a generation of applied mathematicians, both through personal teaching and through superbly written books. As a young man he was noted as a hard worker, shy and reticent; in later life he played a prominent part in academic councils. He also possessed considerable literary and general ability and enjoyed reading in French, German, and Italian. He liked walking and climbing and was one of the early climbers of the Matterhorn.

Like his teachers, Sir George Stokes and James Clerk Maxwell, Lamb saw from the outset of his career that success in applied mathematics demands both thorough knowledge of the context of application and mathematical skill. The fields in which he made his mark cover a wide range—electricity and magnetism, fluid mechanics, elasticity, acoustics, vibrations and wave motion, statics and dynamics, seismology, theory of tides, and terrestrial magnetism. Sections of his investigations in different fields are, however, closely linked by a common underlying mathematics. It was part of Lamb's genius that he could see how to apply the formal solution of a problem in one field to make profound contributions in another.

To the scientific world in general, Lamb is probably most widely known for his work in fluid mechanics, embodied in his book *Hydrodynamics*, which appeared first in 1879 as *A Treatise on the Motion of Fluids*, the title being changed to *Hydrodynamics* in the second, much enlarged, edition of 1895. Successive editions, to the sixth and last in 1932, showed a nice assimilation and condensation of new developments and increasingly included Lamb's own important contributions. The book is one of the most beautifully arranged and stimulating treatises ever written in a branch of applied mathematics—a model which modern scientific writers are often adjured to emulate.

In addition to solving numerous problems of direct hydrodynamical interest, as well as others of direct interest to electromagnetism and elasticity theory, Lamb applied many of the solutions with conspicuous success in geophysics. His much-quoted paper of 1904 gave an analytical account of the propagation, over the surface of an elastic solid, of waves generated by various assigned initial disturbances. The cases he studied bear intimately on earthquake wave transmission, and this paper is regarded today as one of the fundamental contributions to theoretical seismology. Modern attempts to interpret the finer details of earthquake records rest heavily on it. Another famous paper, published in 1882, analyzed the modes of oscillation of an elastic sphere. This paper is a classic in its completeness, and it recently rose to new prominence when free earth oscillations of the type Lamb had described were detected for the first time on records of the great Chilean earthquake of 1960. In 1903 he gave an analysis of two-dimensional wave motion which showed why the record of an earthquake usually has a prolonged tail.

Lamb's contributions to geophysics were by no means confined to seismology but extended to the theory of tides and terrestrial magnetism. In 1863 Lord Kelvin, using theory on fortnightly tides, came to the historic conclusion that the average rigidity of the earth exceeds the rigidity of ordinary steel. A significant point in Kelvin's theory, not well evidenced at the time, later came to be questioned. In 1895 Lamb gave an argument which placed the theory on a new basis and made Kelvin's conclusion inescapable. In 1915, in collaboration with Lorna Swain, he gave the first satisfactory account of the marked phase differences of tides observed in different parts of the oceans and seas, thereby settling a question which had been controversial since the time of Newton. In 1917 he worked out the deflection of the vertical caused by the tidal loading of the earth's surface.

In 1889 Arthur Schuster raised the question of the causes of diurnal variation of terrestrial magnetism. Lamb thereupon showed that the answer was immediately derivable from results he had published in 1883—that the variation is caused by influences outside the solid earth. He showed further that the magnitude of the variation is reduced by an increase in electrical conductivity below the earth's surface.

In addition to *Hydrodynamics* and numerous research papers, Lamb wrote texts, some of them still used today, on infinitesimal calculus, statics, dynamics, higher mechanics, and the dynamical theory of sound. His polished expositions led to his sometimes being called "the great artist" of applied mathematics.

Lamb continued to be active after his retirement; he was, for example, a key member from 1921 to 1927 of the Aeronautical Research Committee of Great Britain. He was elected a fellow of the Royal Society in 1884 and later received its Royal and Copley medals. He received many honors from overseas universities and academies and was knighted in 1931.

BIBLIOGRAPHY

Lamb wrote the following books: *A Treatise on the Motion of Fluids* (Cambridge, 1879)—the 2nd (1895) through 6th (1932) eds., greatly enl., are entitled *Hydrodynamics; Infinitesimal Calculus* (1897), *Statics* (1912), *Dynamics* (1914), *Higher Mechanics* (1920), all published by the Cambridge University Press, with several editions; and *The Dynamical Theory of Sound* (London, 1910). He contributed the article "Analytical Dynamics," in the *Encyclopaedia Britannica* supplement (London, 1902) and the article "Schwingungen elastischer Systeme, insbesondere Akustik," in *Encyclopädie der mathematischen Wissenschaften* (Leipzig, 1906).

Lamb's great contributions in the field of hydrodynamics are incorporated in his book bearing that title. His most important papers in other fields are: "On the Vibrations of an Elastic Sphere," in *Proceedings of the London Mathematical Society*, **13** (1882), 189–212; "On Electrical Motions in a Spherical Conductor," in *Philosophical Transactions of the Royal Society*, **174** (1883), 519–549; "On Wave-Propagation in Two Dimensions," in *Proceedings of the London Mathematical Society*, **35** (1903), 141–161; "On the Propagation of Tremors over the Surface of an Elastic Solid," in *Philosophical Transactions of the Royal Society*, **203** (1904), 1–42; "On a Tidal Problem" (with Lorna Swain), in *Philosophical Magazine*, **29** (1915), 737–744; "On the Deflection of the Vertical by Tidal Loading of the Earth's Surface," in *Proceedings of the Royal Society*, **93A** (1917), 293–312.

References to several further papers of Lamb and further details of his life and career may be found in *Obituary Notices of Fellows of the Royal Society*, **1** (1935), 375–392.

K. E. BULLEN

LAMBERT, JOHANN HEINRICH (*b.* Mulhouse, Alsace, 26 [?] August 1728; *d.* Berlin, Germany, 25 September 1777), *mathematics, physics, astronomy, philosophy.*

The Lambert family had come to Mulhouse from Lorraine as Calvinist refugees in 1635. Lambert's father and grandfather were tailors. His father, Lukas Lambert, married Elisabeth Schmerber in 1724. At his death in 1747, he left his widow with five boys and two girls.

Growing up in impoverished circumstances, Johann Heinrich had to leave school at the age of twelve in order to assist his father. But the elementary instruction he had received together with some training in French and Latin were sufficient to enable him to continue his studies without a teacher. He acquired all his scientific training and substantial scholarship by self-instruction—at night, when the tailoring in his father's shop was finished, or during any spare time left after his work as a clerk or private teacher.

Because of his excellent handwriting, Lambert was appointed clerk at the ironworks at Seppois at the age of fifteen. Two years later he became secretary to Johann Rudolf Iselin, editor of the *Basler Zeitung* and later professor of law at Basel University. There he had occasion to continue his private studies in the humanities, philosophy, and the sciences.

In addition to astronomy and mathematics, Lambert began to take a special interest in the theory of recognition. In a letter he reported:

> I bought some books in order to learn the first principles of philosophy. The first object of my endeavors was the means to become perfect and happy. I understood that the will could not be improved before the mind had been enlightened. I studied: [Christian] Wolf[f] "Of the powers of the human mind"; [Nicholas] Malebranche "Of the investigation of the truth"; [John] Locke "Thoughts of the human mind." [Lambert probably refers to the *Essay Concerning Human Understanding*.] The mathematical sciences, in particular algebra and mechanics, provided me with clear and profound examples to confirm the rules I had learned. Thereby I was enabled to penetrate into other sciences more easily and more profoundly, and to explain them to others, too. It is true that I was well aware of the lack of oral instruction, but I tried to replace this by even more assiduity, and I have now thanks to divine assistance reached the point where I can put forth to my lord and lady what I have learned.

In 1748 Lambert became tutor at Chur, in the home of the Reichsgraf Peter von Salis, who had been ambassador to the English court and was married to an Englishwoman. Lambert's pupils were von Salis' grandson, eleven-year-old Anton; Anton's cousin Baptista, also eleven; and a somewhat younger relative, Johann Ulrich von Salis-Seewis, seven years old. Lambert remained as tutor for ten years, a decisive period for his intellectual development. He was able to study intensively in the family library and to pursue his own critical reflections. He also met many of the friends and visitors of this noble Swiss family. Although Lambert became more refined in this cultivated atmosphere, he remained an original character who did not conform to many bourgeois conventions.

Lambert instructed his young charges in languages, mathematics, geography, history, and the catechism. The Salis family was very pious, and Lambert himself preserved his naturally devout attitude throughout his life. Later, when he lived in Berlin, this caused some embarrassment.

During the years at Chur, Lambert laid the foundation of his scientific work. His *Monatsbuch*, a journal begun in 1752 and continued until his death, lists his main occupations month by month. Besides theoretical investigations, Lambert carried out astronomical observations and constructed instruments for scientific experiments. Later, when he had access to improved instruments, he preferred to employ his simple homemade ones.

Lambert's spirit of inquiry did not remain unnoticed. He was made a member of the Literary Society of Chur and of the Swiss Scientific Society at Basel. On request of the latter he made regular meteorological observations, which he reported in 1755. His first of several publications in *Acta Helvetica* appeared in the same year in volume 2 and dealt with the measurement of caloric heat.

In 1756 Lambert embarked with Anton and Baptista on a *Bildungsreise*, or educational journey, through Europe. Their first stop was at Göttingen, where Lambert attended lectures in the faculty of law and studied works by the Bernoullis and Euler. He talked with Kaestner, with whom he continued to exchange books and letters until his death, and with the astronomer Tobias Mayer. He also participated in the meetings of the Learned Society at Göttingen and was elected corresponding member when he left the city after the French occupation in July 1757 during the Seven Years' War.

The greater part of the following two years was spent at Utrecht, with visits to all the important Dutch cities. Lambert visited the renowned physicist Pieter van Musschenbroek in The Hague, where his first book, on the path of light in air and various media, was published in 1758. Late in 1758 Lambert returned with his pupils to Chur via Paris (where he met d'Alembert), Marseilles, Nice, Turin, and Milan. A few months later he parted from the Salis family.

Seeking a permanent scientific position, Lambert at first hoped for a chair at the University of Göttingen. When this hope came to nothing, he went to Zurich, where he made astronomical observations with Gessner, was elected a member of the city's Physical Society, and published *Die freye Perspektive*. (It is also reported that the Zurich streetboys mocked him for his strange dress.) Lambert then spent some months with his family at Mulhouse.

During the following five years Lambert led a restless, peripatetic life. At Augsburg in 1759 he met the famous instrument maker Georg Friedrich Brander. (Their twelve-year correspondence is available in *Lamberts deutscher gelehrter Briefwechsel*, vol. III.) Lambert also found a publisher for his *Photometria* and his *Cosmologische Briefe*.

Meanwhile plans had been made for a Bavarian academy of sciences, based on the plan of the Prussian Academy at Berlin. Lambert was chosen a salaried member and was asked to organize this academy at Munich. But differences arose, and in 1762 he withdrew from the young academy. Returning to Switzerland, he participated as geometer in a resurvey of the frontier between Milan and Chur. He visited Leipzig in order to find a publisher for his *Neues Organon*, published in two parts in 1764.

In the meantime Lambert had been offered a position at the St. Petersburg Academy. Yet he hoped for a position at the Prussian Academy of Sciences in Berlin, of which he had been proposed as a member in 1761. He arrived in the Prussian capital in January 1764. Lambert was welcomed by the Swiss group of scientists, among them Euler and Johann Georg Sulzer, the director of the class of philosophy, but his strange appearance and behavior delayed his appointment until 10 January 1765. Frederick the Great is said to have exclaimed, after having seen Lambert for the first time, that the greatest blockhead had been suggested to him for the Academy, and at first he refused to instate him. Frederick changed his mind later and praised Lambert's "immeasurableness of insight." He raised his salary and made him a member of a new economic commission of the Academy, together with Euler, Sulzer, and Hans Bernard Merian. Lambert also was appointed to the committee for improving land surveying and building administration, and in 1770 he received the title *Oberbaurat*.

As a member of the physical class for twelve years, until his death at the age of forty-nine, Lambert produced more than 150 works for publication. He was the only member of the Academy to exercise regularly the right to read papers not only in his own class, but in any other class as well.

Of Lambert's philosophical writings only his principle works, *Neues Organon* and *Anlage zur Architectonic*, as well as three papers published in *Nova acta eruditorum*, appeared during his lifetime. Although the composition of the main books and papers was done during the period of his appointment to the Berlin Academy in the winter of 1764–1765, Lambert was occupied with philosophical questions at least as early as 1752, as his *Monatsbuch* testifies. During the last ten years of his life his interest centered on problems in mathematics and physics.

Lambert's philosophical position has been described in the most contradictory terms. R. Zimmermann in *Lambert, der Vorgänger Kants* (Vienna, 1879) tried to demonstrate the germs of Kant's philosophy everywhere in Lambert's writings. Two years later J. Lepsius' *J. H. Lambert* and *Das neue Organon und die Architektonik Lamberts* appeared in Munich. Although more reserved they still interpreted Lambert in terms of Kant. Otto Baentsch arrived at the opposite conclusion in his *Lamberts Philosophie und seine Stellung zu Kant* (Tübingen–Leipzig, 1902); he believed that Lambert might without harm be omitted from the history of critical philosophy. Kant himself recognized in Lambert a philosopher of the highest qualities; and he expected much from his critical attitude. He had drafted a dedication of the *Critique of Pure Reason* to Lambert, but Lambert's untimely death prevented its inclusion.

Lambert's place in the history of philosophy, however, should not be seen only in its relation to Kant. The genesis of his philosophical ideas dates from a time when Kant's major works had yet to be conceived. It was the philosophical doctrines of Leibniz, Christian Wolff, and Locke that exerted the more important influence—insofar as one can speak of influence in connection with a self-taught and wayward man such as Lambert. The Pietist philosophers Adolf Friedrich Hoffmann and Christian August Crusius, antagonists of the Wolffian philosophy, through their logical treatises also had some effect on his thinking.

The two main aspects of Lambert's philosophy, the analytic and the constructive, were both strongly shaped by mathematical notions; hence logic played an important part in his philosophical writing. Following Leibniz' ideas, Lambert early tried to create an *ars characteristica combinatoria*, or a logical or conceptual calculus. He investigated the conditions to which scientific knowledge must be subjected if it is to enjoy the same degree of exactness and evidence as mathematical knowledge. This interest was expressed in two smaller treatises, *Criterium veritatis* and *Über die Methode, die Metaphysik, Theologie und Moral richtiger zu beweisen*, published from manuscript in 1915 and 1918, respectively, by Karl Bopp (*Kantstudien,* supp. nos. 36 and 42). The second of these papers was composed with regard to the prize question posed for 1761 by the Berlin Academy:

> What is to be asked is whether metaphysical truths in general, and the first principle of natural theology and morality in particular are subject to the same evidence as mathematical truth; and in case they are not, what then is the nature of their certainty, how complete is it, and is it sufficient to carry conviction?

Lambert's paper, although fragmentary, firmly claimed that theorems and proofs in metaphysics can be given with the same evidence as mathematical ones.

In *Neues Organon, oder Gedanken über die Erforschung und Bezeichnung des Wahren und dessen Unterscheidung vom Irrthum und Schein* (Leipzig, 1764) these ideas are further developed. Lambert first dealt with the logical form of knowledge, the laws of thought, and method of scientific proof; he then exhibited the basic elements and studied the systematic character of a theory; he next developed (in the section headed "Semiotik") the idea of a characteristic language of symbols to avoid ambiguities of everyday language; and finally, in the most original part of his work called "Phänomenologie," he discussed appearance and gave rules for distinguishing false (or subjective) appearance from a true (or objective) one that is not susceptible to sensory illusions.

In his second large philosophical work, *Anlage zur Architectonic, oder Theorie des Einfachen und des Ersten in der philosophischen und mathematischen Erkenntnis* (Riga, 1771), Lambert proposed a far-reaching reform of metaphysics, stemming from discontent with the Wolffian system. Starting from a certain set of concepts which he analyzed, he turned to their a priori construction. Modeled on mathematical procedures, the body of general sciences so constructed was to be true both logically and metaphysically. Its propositions would be applied to experience. Each of the particular sciences would be founded on observations and experiments; the rules thereby abstracted would have to be joined with the propositions to give a foundation for truth. Leibniz' concept of a prestabilized harmony underlay Lambert's ideas, and Lambert followed Leibniz' belief in the best of all possible worlds. But Lambert's subtle discussions of basic notions, axioms, and elementary interrelations heralded the critical period in philosophy; and his logical analysis of a combinatorial calculus is particularly interesting in the development of mathematical logic.

Lambert's work in physics and astronomy must be seen in relation to his general philosophical outlook and his attempts to introduce mathematical exactness and certitude into the sciences. His interest in the paths of comets was stimulated by the appearance of a comet in 1744. While studying the properties of such paths, he discovered interesting geometrical theorems, one of which carries his name. It was later proven analytically by Olbers, Laplace, and Lagrange. In 1770 Lambert suggested an easy method of determining whether the distance between the earth and the sun is greater than the distance from the earth to a given comet.

Lambert's efforts to improve communication and collaboration in astronomy were noteworthy. He promoted the publication of astronomical journals and founded the *Berliner astronomisches Jahrbuch oder Ephemeriden*. Many of the articles that he contributed to it were not published until after his death. Lambert also suggested the publication of specialized trigonometrical and astronomical tables in order to reduce laborious routine work. Moreover, he proposed to divide the composition of such tables among several collaborating observatories. He also favored the founding of the Berlin observatory. These suggestions, in line with Leibniz' far-reaching plans for international cooperation of scientific societies, inaugurated a new period of scientific teamwork.

Of special interest among Lambert's astronomical writings—apart from applications of his physical doctrines (see below)—are his famous *Cosmologische Briefe über die Einrichtung des Weltbaues* (Augsburg, 1761). Not familiar with the similar ideas of Thomas Wright (published in 1750) and with Kant's *Allgemeine Naturgeschichte und Theorie des Himmels* (1755), Lambert had the idea (in 1749) that what appears as the Milky Way might be the visual effect of a lens-shaped universe. On this basis he elaborated a theory according to which the thousands of stars surrounding the sun constituted a system. Moreover, he considered the Milky Way as a large number of such systems, that is, a system of higher order. Certain difficulties and interpretations concerning the motions of Jupiter and Saturn had led him to conclude that a force must exist outside our planetary system, which must be but a small part of another, much larger system of higher order. These bold speculations, born of the Leibnizian belief in the most perfect of all possible worlds, far transcended astronomy. The whole universe, Lambert postulated, had to be inhabited by creatures like human beings. Hence collisions of heavenly bodies are not to be expected, and the widespread fear that comets (which Lambert also supposed to be inhabited) might destroy the earth was unfounded; the *Cosmologische Briefe* was a great sensation and was translated into French, Russian, and English. Only when William Herschel systematically examined the heavens telescopically and discovered numerous nebulae and "telescopic milky-ways" did it become obvious that Lambert's description was not mere science fiction but to a large extent a bold vision of the basic features of the universe.

Lambert's numerous contributions to physics center on photometry, hygrometry, and pyrometry. Many marked advances in these subjects are traceable to him. In Lambert's fundamental work in the sciences, he (1) searched for a basic system of clearly defined concepts, (2) looked for exact measurements (and often collected them himself), and (3) after establishing them tried to develop a mathematical theory that would comprise these foundations and would result in quantitative laws.

In his famous *Photometria sive de mensura et gradibus luminis, colorum et umbrae* (Augsburg, 1760), Lambert laid the foundation for this branch of physics independently of Bouguer, whose writings on the subject were unknown to him. Lambert carried out his experiments with few and primitive instruments, but his conclusions resulted in laws that bear his name. The exponential decrease of the light in a beam passing through an absorbing medium of uniform transparency is often named Lambert's law of absorption, although Bouguer discovered it earlier. Lambert's cosine law states that the brightness of a diffusely radiating plane surface is proportional to the cosine of the angle formed by the line of sight and the normal to the surface. Such a diffusely radiating surface does therefore appear equally bright when observed at different angles, since the apparent size of the surface also is proportional to the cosine of the said angle.

Lambert's *Hygrometrie* (Augsburg, 1774–1775) was first published in two parts in French as articles entitled *Essai d'hygrométrie*. A result of his meteorological studies, this work is mostly concerned with the reliable measurement of the humidity of the atmosphere. The instrument maker G. F. Brander constructed a hygrometer according to Lambert's description. Another product of Lambert's research in meteorology was his wind formula. Now discarded, it attempted to determine the average wind direction on the basis of observations made over a given period.

The *Pyrometrie oder vom Maasse des Feuers und der Wärme* (Berlin, 1779) was Lambert's last book, completed only a few months before his death; his first publication had also dealt with the question of measuring heat. It is characteristic of him that he dealt not only with radiation but also with reflection of heat, although the latter could not yet be demonstrated, and his results could only have been preliminary in nature. Lambert also took into consideration the sensory effect of heat on the human body and tried to give a mathematical formulation for it. Similarly, his work in acoustics, on speaking tubes, touched on the physical as well as on the psychophysical aspects of the problems.

In mathematics Lambert's largest publication was the *Beyträge zum Gebrauch der Mathematik und deren Anwendung* (3 pts. 4 vols., Berlin, 1765–1772). This is not at all a systematic work but rather a collection of papers and notes on a variety of many topics in pure and applied mathematics.

One of Lambert's most famous results is the proof of the irrationality of π and e. It was based on continued fractions, and two such fractions still bear his name. Of importance also is Lambert's series in which the coefficient 2 occurs only when the exponent is a prime number. Although it was expected that it might be useful in analytic number theory, it was not until 1928 that Norbert Wiener was able to give a proof of the prime number theorem employing this type of series. Lambert himself was interested in number theory and developed a method of determining the prime factors of a given (large) number and suggested a simplified arrangement for factor tables.

Many of Lambert's investigations were concerned with trigonometry and goniometry. He studied the hyperbolic functions and introduced them in order to reduce the amount of computation in trigonometric problems. He solved goniometric equations by infinite series, and he worked out a tetragonometry—a doctrine of plane quadrangles—corresponding to the common trigonometry. That he also discovered a number of theorems in the geometry of conic sections has already been mentioned in connection with his astronomical work.

Lambert's second book, *Die freye Perspektive, oder Anweisung, jeden perspektivischen Aufriss von freyen Stücken und ohne Grundriss zu verfertigen* (Zurich, 1759; 2nd ed. 1774), was also published in French (Zurich, 1759). Intended for the artist wishing to give a perspective drawing without first having to construct a ground plan, it is nevertheless a masterpiece in descriptive geometry, containing a wealth of geometrical discoveries. In this work Lambert proved himself a geometer of great intuitive powers. In the generality of his outlook, as in certain specific aspects, his work resembles that of Monge later on, usually considered the founder of descriptive geometry as a distinct branch of mathematics. Lambert's investigations of possible constructions by the use of a simple ruler, for example, constructing an ellipse from five given points, was similar in spirit to the program of Poncelet and J. Steiner two generations later.

Lambert contributed significantly to the theory of map construction. For the first time the mathematical conditions for map projections (to preserve angles and area) were stated, although analytically superior formulations were later given by Lagrange, Legendre, and Gauss. More important, Lambert made practical suggestions on how, for different purposes, either one of these contradicting conditions could best be satisfied. He also described constructions to determine the true distance between two places on a map drawn according to one of his projections. Lambert's map projections are still basic for the modern theory in this field.

One of Lambert's most important contributions to geometry was his posthumously published *Theorie der Parallel-Linien*. Here he returned to the famous question that had baffled mathematicians since Euclid: Is it not possible to give a proof for Euclid's axiom of parallels? As a starting point Lambert chose a quadrangle having three right angles, and assumed in turn the fourth angle to be a right angle, an obtuse angle, or an acute angle. He showed that the first assumption is equivalent to Euclid's postulate, and that the second leads to a contradiction (assuming each straight line to possess an infinite length). Having attempted to display a contradiction in case of the third assumption using the same line of argument unsuccessfully, he tried a different mode of attack but overlooked that his "proof" implicitly contained an assumption equivalent to his hypothesis. Obviously not satisfied with his investigations, Lambert did not publish them; yet he had already arrived at remarkable results. He had discovered that under the second and third assumption an absolute measure of length must exist and that the area of a triangle in these cases must be proportional to the divergence of its angle sum from two right angles. He noticed that the second assumption would correspond to the geometry of the sphere; and he speculated that the third assumption might be realizable on an imaginary sphere. An example for this last case was not given until the latter half of the nineteenth century by Beltrami, after this non-Euclidean geometry had been studied in particular by Lobachevsky. The quadrangle used as starting point by Lambert is called Saccheri's quadrangle. Whether Lambert was directly familiar with the investigations of this Italian mathematician is not known.

Lambert's work in non-Euclidean geometry had been overlooked until it was republished by F. Engel and Stäckel in their *Theorie der Parallellinien von Euklid bis auf Gauss* (Leipzig, 1895).

BIBLIOGRAPHY

I. Original Works. Since there exists a carefully prepared bibliography by M. Steck (see below), only some recent republications are mentioned here. Material for a scientific biography may be found in Max Steck, ed., *Johann Heinrich Lambert: Schriften zur Perspektive* (Berlin, 1943), which contains also a *Bibliographia Lambertiana. Johannis Henrici Lamberti opera mathematica*, Andreas Speiser, ed., 2 vols. (Zurich, 1946–1948), contains papers in analysis, algebra, and number theory. Further volumes did not appear. *Johann Heinrich Lambert: Gesammelte philosophische Werke*, H. W. Arndt, ed. (Hildesheim, 1967–), will comprise 10 vols.

II. Secondary Literature. Most important is Max Steck, *Bibliographia Lambertiana* (Berlin, 1943). An enl. 2nd ed. ("Neudruck") was issued shortly before the author's death (Hildesheim, 1970). It contains, apart from G. C. Lichtenberg's biography of Lambert, chronological bibliographies of all of Lambert's publications, including posthumous editions, translations, a survey of Lambert's scientific estate (formerly at Gotha, since 1938 in the university library at Basel), an incomplete chronological list of secondary literature, and reprints of two articles by Steck on Lambert's scientific estate and on his scientific correspondence. Steck also left in manuscript an index of Lambert's scientific manuscripts and correspondence.

The standard bibliography is still *Johann Heinrich Lambert nach seinem Leben und Wirken . . . in drei Abhandlungen dargestellt*, Daniel Huber, ed. (Basel, 1829). *Johann Heinrich Lambert—Leistung und Leben*, Friedrich Löwenhaupt, ed. (Mulhouse, 1943), is a collection of articles on various aspects of Lambert's life and work; its appendix contains a selection from the *Cosmologische Briefe*. About 90 references to Lambert are to be found in Clémence Seither, "Essai de bibliographie de la ville de Mulhouse des origines à 1798," in *Supplément au bulletin du Musée historique de Mulhouse*, **74** (1966), esp. 211–214, 250–253.

Some recent articles about Lambert which (with the exception of Berger) are not listed in the second edition of Steck's *Bibliographia* are (in chronological order): Peter Berger, "Johann Heinrich Lamberts Bedeutung in der Naturwissenschaft des 18. Jahrhunderts," in *Centaurus*, **6** (1959), 157–254; Wilhelm S. Peters, "Lamberts Konzeption einer Geometrie auf einer imaginären Kugel," in *Kantstudien*, **53** (1961–1962), 51–67 (short version of a dissertation [Bonn, 1961], 87 pp.); Roger Jaquel, "Vers les oeuvres complètes du savant et philosophe J.-H. Lambert (1728 à 1777): Velléités et réalisations depuis deux siècles," in *Revue d'histoire des sciences et de leurs applications*, **22** (1969), 285–302; Roger Jaquel, "Jean Henri Lambert (1728–1777) et l'astronomie cométaire au XVIIIe siècle," in *Comptes rendus du 92e Congrès national des Sociétés savantes (Strasbourg, 1967), Section des sciences*, **1** (Paris, 1969), 27–56; Roger Jaquel, "Le savant et philosophe mulhousien J. H. Lambert vu de l'étranger. Essai historiographique sur la façon dont les encyclopédies générales présentent le Mulhousien Lambert," in *Bulletin du Musée historique de Mulhouse*, **78** (1970), 95–130; Karin Figala and Joachim Fleckenstein, "Chemische Jugendschriften des Mathematikers J. H. Lambert (1728–1777)," in *Verhandlungen der Naturforschenden Gesellschaft Basel*, **81** (1971), 40–54; O. B. Sheynin, "J. H. Lambert's Work on Probability," in *Archive for History of Exact Sciences*, **7** (1971), 244–256.

Christoph J. Scriba

LAMÉ, GABRIEL (*b.* Tours, France, 22 July 1795; *d.* Paris, France, 1 May 1870), *mathematics.*

Like most French mathematicians of his time, Lamé attended the École Polytechnique. He entered in 1813 and was graduated in 1817. He then continued at the École des Mines, from which he was graduated in 1820.

His interest in geometry showed itself in his first article, "Mémoire sur les intersections des lignes et des surfaces" (1816–1817). His next work, *Examen des différentes méthodes employées pour résoudre les problèmes de géométrie* (1818), contained a new method for calculating the angles between faces and edges of crystals.

In 1820 Lamé accompanied Clapeyron to Russia. He was appointed director of the School of Highways and Transportation in St. Petersburg, where he taught analysis, physics, mechanics, and chemistry. He was also busy planning roads, highways, and bridges that were built in and around that city. He also collaborated with Pierre Dominique Bazaine on the text *Traité élémentaire du calcul intégral,* published in St. Petersburg in 1825. In 1832 he returned to Paris, where he spent the rest of his career and life.

For a few months after his return to Paris, Lamé joined with Clapeyron and the brothers Flachat to form an engineering firm. However, he left the firm in 1832 to accept the chair of physics at the École Polytechnique. He remained there until 1844.

Lamé always combined his teaching positions with work as a consulting engineer. In 1836, he was appointed chief engineer of mines. He also helped plan and build the first two railroads from Paris to Versailles and to St.-Germain.

In 1843 the Paris Academy of Sciences accepted him to replace Puissant in its geometry section. In 1844 he became graduate examiner for the University of Paris in mathematical physics and probability. He became professor of mathematical physics and probability at the university in 1851. In 1862, he went deaf, and resigned his positions. He was in retirement until his death in 1870. In spite of the unsettled and often troubled political climate, Lamé managed to lead a serene and quiet academic life. His sole and quite tenuous connection with politics was his *Esquisse d'un traité de la république* (1848).

Although Lamé did original work in such diverse areas as number theory, thermodynamics, and applied mechanics, his greatest contribution to mathematics was the introduction of curvilinear coordinates and their use in pure and applied mathematics. These coordinates were conceived as intersections of confocal quadric surfaces. By their means, he was able to transform Laplace's equation $\nabla^2 V = 0$ into ellipsoidal coordinates in a form where the variables were separable, and solve the resulting form of the generalized Laplace equation.

In 1836, Lamé had written a textbook in physics for the École Polytechnique, *Cours de physique de l'Ecole polytechnique.* In 1852 he published his text *Leçons sur la théorie mathématique de l'élasticité des corps solides*, in which he used curvilinear coordinates. This work resulted from his investigation into the conditions for equilibrium of a spherical elastic envelope subject to a given distribution of loads on the bounding spherical surfaces. He succeeded in the derivation and transformation of the general elastic surfaces.

As early as 1828, Lamé had shown an interest in thermodynamics in an article "Propagation de la chaleur dans les polyèdres," written in Russia. In 1837 his "Mémoire sur les surfaces isothermes dans les corps solides homogènes en équilibre de température" appeared in Liouville's *Journal.* In these articles Lamé used curvilinear coordinates and his elliptic functions, which were a generalization of the spherical harmonic functions of Laplace, in a consideration of temperatures in the interior of an ellipsoid. *Leçons sur les fonctions inverses des transcendentes et les surfaces isothermes* appeared in 1857, and *Leçons sur la théorie analytique de la chaleur* followed in 1861.

In *Leçons sur les coordonnés curvilignes et leurs diverses applications* Lamé extended his work in thermodynamics to the solution of various problems of a physical nature involving general ellipsoids, such as double refraction in the theory of propagation of light in crystals.

Lamé's investigations in curvilinear coordinates led him even into the field of number theory. He had begun with a study of the curves

$$\left(\frac{x_1}{a_1}\right)^n + \left(\frac{x_2}{a_2}\right)^n + \left(\frac{x_3}{a_3}\right)^n = 0,$$

which are symmetric with respect to a triangle (as well as the space analogs symmetrical with respect to a tetrahedron). When these equations are written in nonhomogeneous form, they appear as

$$\frac{x^n}{a^n} + \frac{y^n}{b^n} = 1;$$

and, when $a = b$, as $x^n + y^n = a^n$. This naturally led Lamé to study Fermat's last theorem.

In 1840 he was able to present a proof of the impossibility of a solution of the equation $x^7 + y^7 = z^7$ in integers (except for the trivial cases where $z = x$ or y, and the remaining variable is zero). In 1847 he developed a solution, in complex numbers, of the form $A^5 + B^5 + C^5 = 0$, and in 1851 a complete solution, in complex numbers, of the form $A^n + B^n + C^n = 0$.

Another result in number theory having nothing to do with Lamé's main interests and endeavors was the theorem: The number of divisions required to find the greatest common divisor of two numbers is never greater than five times the number of digits in the smaller of the numbers. This theorem is yet another example of the attraction that number theory has always seemed to have for mathematicians. This result and the "*Esquisse*" previously mentioned are the only examples of Lamé's work that were not devoted entirely to his main purposes.

Lamé was considered an excellent engineer. While in Russia, he wrote a number of articles that appeared in Gergonne's *Journal*, "Sur la stabilité des voûtes," on arches and mine tunnels (1822); "Sur les engrenages," on gears (1824); and studies on the properties of steel bridges. His work on the scientific design of built-up artillery was considered a standard reference and was much used by gun designers. His final text, *Cours de physique mathématique rationnelle* (1865), was a composite of practice and theory. All of his researches were undertaken with practical application in mind.

It is difficult to characterize Lamé and his work. Gauss considered Lamé the foremost French mathematician of his generation. In the opinion of Bertrand, who presented a eulogy of Lamé on the occasion of his demise, Lamé had a great capacity as an engineer. French mathematicians considered him too practical, and French scientists too theoretical.

Lamé himself once stated that he considered his development of curvilinear coordinates his greatest contribution to mathematical physics. True, his applications were all to physics—to the theory of elasticity, the thermodynamics of ellipsoids and other solids, ellipsoidal harmonics, among others. In his opinion, just as rectangular coordinates made algebra possible, and spherical coordinates made celestial mechanics possible, so general curvilinear coordinates would make possible the solution of more general questions of physics. Yet the work he began was generalized almost as soon as it appeared by such mathematicians as Klein, Bôcher, and Hermite. It now has a strictly mathematical format, being used in the study of ordinary and partial differential equations. We can conclude that Lamé's major work was in the field of differential geometry.

BIBLIOGRAPHY

A bibliography of Lamé's writings is in Poggendorff, I, pt. 1, 1359–1360; and in Royal Society *Catalogue of Scientific Papers*, III, 814–816; VIII, 152; and X, 501.

Works mentioned in the text are "Mémoire sur les intersections des lignes et des surfaces," in *Annales de mathématiques pures et appliquées*, **7** (1816–1817), 229–240; *Examen des différentes méthodes employées pour résoudre les problèmes de géométrie* (Paris, 1818); "Mémoire sur la stabilité des voûtes," in *Annales des mines*, **8** (1823), 789–836, written with Clapeyron; "Mémoire sur les engrenages," *ibid.*, **9** (1824), 601–624, written with Clapeyron; *Traité élémentaire du calcul intégral* (St. Petersburg, 1825); "Mémoire sur la propagation de la chaleur dans les polyèdres," in *Journal de l'École polytechnique*, cahier 22 (1833), 194–251; "Mémoire sur les surfaces isothermes dans les corps solides homogènes en équilibre de température," in *Journal de mathématiques pures et appliquées*, **2** (1837), 147–188; *Cours de physique de l'École polytechnique*, 2 vols. (Paris, 1836–1837); *Esquisse d'un traité de la république* (Paris, 1848); *Leçons sur la théorie mathématique de l'élasticité des corps* (Paris, 1852); *Leçons sur les fonctions inverses des transcendentes et les surfaces isothermes* (Paris, 1857); *Leçons sur les coordonnés curvilignes et leurs diverses applications* (Paris, 1859); *Leçons sur la théorie analytique de la chaleur* (Paris, 1861); and *Cours de physique mathématique rationnelle* (Paris, 1865).

SAMUEL L. GREITZER

LAMÉTHERIE, JEAN-CLAUDE DE (*b.* La Clayette [Mâconnais], France, 4 September 1743; *d.* Paris, France, 1 July 1817), *scientific editing and journalism, mineralogy, geology, biology, chemistry, natural philosophy.*

The son of François de Lamétherie, a doctor, and Claudine Constantin, Lamétherie turned to medicine, long a family tradition, after the death of his older brother in 1765. Before this he had prepared for an ecclesiastical career, first at a seminary in Thiers, then at the Sorbonne, and finally at the Seminary of Saint-Louis in Paris, where he took the four minor orders. After obtaining a medical degree he practiced medicine in La Clayette from about 1770 until 1780, when he abandoned the profession. Soon afterward he went to Paris to cultivate his interest in natural philosophy. Lamétherie's best-informed biographer regarded this action as the result of a natural preference for philosophical speculation and theoretical reflection over practical activity, an inclination he had demonstrated in his youth by reading such writers as Rollin and Pluche while other children engaged in more typical activities. At the age of sixteen, while he was in the seminary, Lamétherie had already begun, under the influence of Madame du Châtelet's *Institutions de physique*, to formulate the ideas of his *Principes de la philosophie naturelle*, completed in 1776 and published in 1778.

In 1785 he joined the editorial staff of the monthly journal *Observations sur la physique* (renamed the *Journal de physique* in 1794), and in May of 1785 became chief editor in place of Jean-André Mongez, who had left to participate in La Pérouse's ill-starred last voyage. Lamétherie continued in this position until the year of his death, although during his last five years, as his health failed, he was assisted by Blainville.

Lamétherie imposed a highly personal stamp upon the journal during his long tenure as editor. Not content merely to provide a forum for the publication of scientific papers, he sought to bring attention to many of his favorite ideas, and sometimes to rectify injustices supposedly done to unfortunate scientists—especially those committed by others of overmagnified reputation. Among those whose fame he attempted to diminish was Lavoisier, whom he regarded as a dictatorial force in French science. His animosity toward Lavoisier led Lamétherie to champion the little-known contributions of Pierre Bayen to the study of calcination and combustion, and he maintained a lasting hostility toward oxidation chemistry, never tiring of pointing out that the presumed "acid-forming" combustive principle was not present in certain acids. He frequently attempted to demonstrate that science was changing in ways contrary to the interests of his enemies, and he blocked publication of their work in his journal and often drew favorable attention to the work of foreign scientists. In this way the *Journal de physique* served in part as a vehicle for the introduction into France of foreign scientific ideas.

Not many honors and awards came Lamétherie's way in recognition of his own work. He refused to take steps to promote himself for candidacy in the Academy of Sciences, and did not possess adequate scientific merit to elicit a spontaneous invitation. He did hold membership in numerous provincial and foreign societies and academies. His sole teaching position came to him rather late in his career, at the Collège de France. In 1801, expecting to be named to succeed Daubenton as professor of natural sciences at the Collège, he was deeply hurt at being passed over in favor of Cuvier, but this wound was eased by his appointment that year in the same institution as *adjoint* professor of mineralogy and geology. He gave public courses in mineralogy, and was a pioneer in using field trips as part of the pedagogical method.

An annual feature of the *Journal de physique* from 1786 to 1817 was Lamétherie's lengthy "Discours préliminaire," a review of developments in science during the preceding year. Here, and also in the monthly "Nouvelles littéraires" and in editorial notes, Lamétherie did not refrain from recording his observations on topics of all sorts, including political developments, as if to illustrate his conviction that men of letters ought not to stand aside and passively watch worldly events taking their course. One can trace Lamétherie's attitudes toward the Revolution in his published remarks, beginning with an enthusiastically democratic optimism and then growing, when constitutional monarchy came to be threatened, into boldly outspoken opposition. His mood was sullen and foreboding throughout much of the 1790's. It no doubt required courage to place his name over some of his scathing denunciations of revolutionary excesses. Dolomieu, who described Lamétherie as his best friend, was right when he said that Lamétherie's "active imagination and hot courage do not allow him moderation in anything. He compromises neither with injustices nor with tyranny" (Alfred Lacroix, *Déodat Dolomieu*, I [Paris, 1921], 47). Excessive modesty was not one of Lamétherie's virtues, and he frequently published declarations agreeing with the latter part of this assessment, especially as with advancing age he came to cry out more shrilly against the critics of his management of his scientific journal.

Clearly he was a sensitive man whose spirit combined great generosity and an austere sense of obligation to humanity with extreme vanity and a measure of vindictiveness. He sacrificed his own resources to remedy the financial reversals of his younger brother Antoine (who had been a signer of the Tennis Court Oath as deputy of the Third Estate from Mâcon, and was a member of the *Assemblée constituante* and the *Corps législatif*), and depended to his last days on contributions made by Cuvier from his Collège de France income. He was a lifelong bachelor.

Lamétherie's writings range widely, often providing extensive summaries of the ideas of other authorities, but usually showing his distinctive viewpoint even on subjects treated more thoroughly by others. Even his defenders conceded that Lamétherie gave excessively free rein to his imagination and was apt to allow his enthusiasm for favorite notions to carry him too far. Frequently his commitments proved in the long run to land on the losing side of contemporary controversies; he maintained an Aristotelian four-element scheme in chemistry, denied the chemical decomposition of water into two gases, and persisted with a phlogistonist theory of combustion and calcination. Among his most cherished ideas were the possession of an innate force by every fundamental bit of matter, the universality of crystallization as the originating process in all ordered matter, and the reducibility of knowledge to the influence of sensation, which could be determined quantitatively. These ideas appear in his earliest work; after 1793 he developed the view that galvanic action is the basis of a vast range of phenomena.

Lamétherie's fundamental outlook on nature did not respect a firm boundary between the living and nonliving. As he put it, "An animal that exercises all its functions by the laws of physics alone is a machine that confounds all our ideas of mechanics; nonetheless it is nothing more than a simple machine . . ." (*Principes de la philosophie naturelle*, II [1787], 292). He held that attempts to distinguish clearly between different things fail ultimately because of the chain of being, according to which nature has created things by gradation. In Lamétherie's view, the basis of organic reproduction in the crystallization process served further to underline the basic similarity of all natural processes. Yet he did not deny that organic and inorganic things have differences. He admitted a vital force in life, attributing it in his later years to galvanic action, and he accepted spontaneous generation. Creation must have occurred in stages (for example, plants before animals), and might still be going on. But not all creations have been permanent; fossil evidence demonstrates extinction, perhaps the result of the incapacity of a being to sustain or reproduce itself. Lamétherie accepted as a fact transmutation of certain kinds, in both plants and animals. Breeding practices have changed the qualities of species, and similar changes have been effected naturally.

Lamétherie's influence may have been greater in mineralogy than in other sciences. His expanded French edition of Bergman's *Sciagraphia regni mineralis* was an important textbook for a generation of French scientists, and contributed to the acceptance of chemical composition as an important criterion in distinguishing minerals. Lamétherie's reliance on the significance of crystallization sustained his appreciation of Werner's geognosy, which he reported upon sympathetically, particularly on the occasion of Werner's visit to Paris in 1802. Taking a broad cosmogonical view of creation, Lamétherie regarded the major features of the earth as the result of the combined action of crystallization, moving water, and shifts in the planetary-motion characteristics of the earth. Major alterations in the crust, he believed, had not occurred since the main valleys and mountains were created by the primordial crystallization process. Mountain upheavals, violent floods, and other agents of change were generally rare and isolated events.

In true Enlightenment fashion, Lamétherie presumed that science illuminates all things, from the humblest to the highest production of nature—mankind. Confident that scientific knowledge could improve the nature and condition of man, Lamétherie held that his species, like any other, exhibits moral and intellectual qualities that result from its physical makeup. Man is a machine run by his nervous system. The complexity of human behavior can be understood in terms of the historical development of the race, whose ancestry ultimately merges with that of other animals. Self-interest is the central guide to human behavior, although men have become far subtler than other animals in their ability to project their self-interest into the future. If Lamétherie's assurance that he understood the main outlines of human biological and cultural development reflects on the historical credulity typical of his day, it can be said that he was an active participant in the movement to inject a historical element into natural science.

BIBLIOGRAPHY

I. Original Works. A bibliographical list, incomplete and riddled with errors, is included in Blainville's biographical sketch (see below), pp. 102–107. Lamétherie's major works include *Essai sur les principes de la philosophie naturelle* (Geneva, 1778), rev. and enl. as *Principes de la philosophie naturelle, dans lesquels on cherche à déterminer les degrés de certitude ou de probabilité des connoissances humaines*, 2 vols. (Geneva, 1787), and abridged as *De la nature des êtres existans, ou principes de la philosophie naturelle* (Paris, 1805); *Vues physiologiques sur l'organisation animale et végétale* (Amsterdam–Paris, 1780); *Essai analytique sur l'air pur et les différentes espèces d'air* (Paris, 1785; rev. 2nd ed., 2 vols., 1788); *Théorie de la terre*, 3 vols. (Paris, 1795), trans. into German by Christian Gotthold Eschenbach, enl. by Eschenbach and Johann Reinhold Forster, as *Theorie der Erde*, 3 vols. (Leipzig, 1797–1798), enl. 2nd ed., 5 vols. (Paris, 1797); *De l'homme considéré moralement; De ses moeurs, et de celles des animaux*, 2 vols. (Paris, 1802); *Considérations sur les êtres organisés*, 2 vols. (Paris, 1804); *Leçons de minéralogie, données au Collége de France*, 2 vols. (Paris, 1812); and *Leçons de géologie données au Collège de France*, 3 vols. (Paris, 1816). Lamétherie also produced an enlarged version of Jean-André Mongez's French translation of Torbern Bergman's mineralogy, *Manuel du minéralogiste; Ou sciagraphie du règne minéral, distribuée d'après l'analyse chimique*, 2 vols. (Paris, 1792). A large number of articles by Lamétherie are in *Observations sur la physique* (*Journal de physique* after 1794).

II. Secondary Literature. H. M. Ducrotay de Blainville, "Notice historique sur la vie et les écrits de J.-C. Delamétherie," in *Journal de physique*, **85** (1817), 78–107, makes use of autobiographical notes left by Lamétherie. Among other accounts, all dependent to some extent on Blainville, are Cuvier's biographical article on Lamétherie in Michaud, ed., *Biographie universelle*, XXVIII (1821), 461–463, and a biographical notice (probably by L. Louvet) in Hoefer, ed., *Nouvelle biographie générale*, XXIX (1859), cols. 209–212.

Kenneth L. Taylor

LA METTRIE, JULIEN OFFRAY DE (*b.* Saint-Malo, France, 19 December 1709; *d.* Berlin, Germany, 11 November 1751), *medicine, physiology, psychology, philosophy of science.*

The son of a prosperous textile merchant, La Mettrie studied medicine at the University of Paris from 1728 until 1733, when he transferred to Rheims to obtain the doctor's degree. He completed his training after another year at Leiden under the renowned Hermann Boerhaave, whose influence on him was decisive. From 1734 on, La Mettrie practiced medicine in the Saint-Malo district. Toward the end of 1742, however, he left abruptly for Paris and soon thereafter embarked on the adventurous and harried career that lasted until his death. Between 1743 and 1746 he served as an army doctor in the War of the Austrian Succession. Meanwhile, his first philosophical work, *Histoire naturelle de l'âme* (1745), which expounded a materialistic theory of the "soul," provoked a scandal and was officially condemned by the Paris Parlement. Despite this offense against orthodoxy, La Mettrie's professional ability was apparently esteemed enough for him to be promoted to the post of medical inspector of the armies in the field. But he imprudently turned, in *La politique du médecin de Machiavel* (1746), to ridiculing the incompetence, greed, and charlatanry of a gallery of prominent French physicians. This justified and successful attempt at medical satire was followed by *La faculté vengée* (1747) and his magnum opus in that vein, *L'ouvrage de Pénélope, ou Machiavel en médecine* (1748–1750).

La Mettrie's combined attacks against religion and the medical profession made him so many powerful enemies that, in order to escape arrest and imprisonment, he exiled himself to Holland in 1747. But unable to avoid trouble for long, he published there his most notorious book, *L'homme machine* (1748), the outspoken materialism and atheism of which raised a storm of protest even among the relatively tolerant Dutch. Its author, now regarded by the public as the most daring and dangerous of the Philosophes, was forced to flee again, this time to the court of Frederick II of Prussia, where he was appointed a member of the Royal Academy of Sciences, as well as reader and physician to the king. In this protected situation he continued to write tracts on scientific and philosophical subjects that shocked the conventional-minded. In particular his *Discours sur le bonheur* (1748), which denied that vice and virtue had any meaning within a deterministic view of human nature and, consequently, saw in remorse simply a morbid symptom to be got rid of, caused him (somewhat illogically) to be denounced as a debauched and cynical corrupter of morals. La Mettrie's querulous and mocking temper embroiled him in constant polemics, often of a mystifying sort, with his various adversaries, the most notable of these being the physiologist Albrecht von Haller. Even his death became an occasion for controversy, when his detractors, advertising that he had died by an act of gluttony, represented this as proof of the practical hazards of materialism and of the certainty of God's retribution.

La Mettrie's main service to medicine was his advocacy and propagation of Boerhaave's teaching. This he did by translating into French many of the master's works, in some cases appending to them commentaries of his own. The following translations from the Boerhaavian *corpus* deserve mention: *Système sur les maladies vénériennes* (Paris, 1735); *Aphorismes sur la connaissance et la cure des maladies* (Rennes, 1738); *Traité de la matière médicale* (Paris, 1739); *Abrégé de la théorie chimique de la terre* (Paris, 1741); and the monumental *Institutions de médecine* (Paris, 1743–1750), which included Haller's lengthy and valuable notes. La Mettrie's efforts to spread the lessons of Boerhaave had the positive result not only of prodding the rather sluggish medical science and practice in eighteenth-century France but also of bringing medical subject matter into the arena of philosophical discussion and intellectual history. In this respect there were two aspects of Boerhaavian doctrine that the zealous disciple was especially eager to have accredited by doctors and nondoctors alike. One was the emphasis on the empirical method and on clinical observation. The other was the aim of establishing medicine on as sound a theoretical basis as possible by linking it directly to anatomy, physiology, chemistry, and mechanics. La Mettrie thus became a leading expositor of the iatromechanistic philosophy of Boerhaave, to which he soon gave a radical application quite unintended by his teacher.

It is regrettable that the Boerhaavian methodology did not play a more noticeable role in the four treatises, long since forgotten, that La Mettrie wrote on venereal disease, vertigo, dysentery, and asthma. His personal contribution to medicine remained on the theoretical rather than the practical plane. Nevertheless, in his *Observations de médecine pratique* (1743) he gave some indication of the clinical ideal acquired from his days at Leiden. In particular he insisted on the importance of performing autopsies in order to verify diagnoses.

L'homme machine, which marked a culminating phase in the rise of modern materialism, was not merely the work of a doctor turned philosopher; it outlined a medical philosophy in the absolute sense of the term, springing as it did from the assumption

LA METTRIE

that reliable knowledge about man's nature was forthcoming only from the facts and theories that the medical sciences—anatomy, physiology, biology, pathology—could furnish. The human being was for La Mettrie a highly complex "living machine" of unique design that only those skilled in the investigation of the body's innermost secrets could hope eventually to explain (insofar, that is, as an explanation was possible, for the man-machine, rather than being a doctrinaire thesis, displayed heuristic and even skeptical features). Seen in historical perspective, such a position may be described as the final outcome both of the iatromechanistic tradition that had reached La Mettrie through Boerhaave and of the Cartesian automatist biology that had filtered down to him through numerous intermediaries who had already sought, in varying degrees, to extend its beast-machine concept to the study of human behavior.

The basic argument of *L'homme machine* was supported by different but complementary types of scientific evidence. La Mettrie cited many examples showing how particular psychological states derived from physical factors: illness, fatigue, hunger, diet, pregnancy, sexual stimulation, age, climate, and the use of drugs. Referring to data provided by comparative anatomy, he held that the great contrasts in the capabilities of the various animal species, including man himself, must be owing to the specific brain structure exhibited by each. He was astute enough to grasp, in relation to the man-machine idea, the theoretical value of the discoveries that Haller had just made concerning the irritable properties of muscle tissue. By generalizing the phenomenon of irritability, and combining it with related instances of reflex action, La Mettrie was able to picture the organism as a genuinely self-moving, inherently purposive mechanism. There were two distinguishable meanings present in this overall conception, even though its author would no doubt have regarded them as inseparable. On the primary level, the man-machine offered a strictly mechanistic interpretation of how living things are constituted and function; as such it served, in the eighteenth-century milieu, to express the counterpart of animistic or vitalistic theorizing in biology. On another and more original level, it claimed that all the mental faculties and processes in the human subject were products of the underlying bodily machine—more precisely, of its cerebral and neural components. In advancing this notion, La Mettrie was perhaps the earliest exponent of a school of psychology whose method of analysis would be consistently and rigorously physiological.

The technical documentation with which La Mettrie tried to prove his case was, to be sure, seriously limited by the knowledge then available concerning the life sciences. Even the term "machine," as he used it, suggested no definite mechanical model that might permit one to differentiate animate from inanimate systems. In describing man as a machine, what La Mettrie really meant was, first, that man was essentially a material being structured to behave automatically; and second, that this self-sufficing organic structure, together with the psychic activities it determined—consciousness, emotion, will, memory, intelligence, moral sense—ought to be explored and clarified with the aid of the same quantitative, mechanical principles that everyone had already recognized as operative in the realm of physics. He left it to his successors to fill in, as the progress of physiological psychology would allow, the concrete details of the mind-body correlation.

Several themes of interest to the history of science grew logically out of the man-machine thesis. One was the continuity it asserted between the mentality of man and that of those animals most resembling him. Supposing the observable differences in intelligent behavior among the various species to be a question merely of degree, La Mettrie ascribed these to the ascending order of complexity to be found in the central nervous apparatus of mammals from the lowliest up to man. It was his sharp awareness of the analogies between animal and human nature that led him, at one point, to entertain the experimental hope of instructing the anthropoid ape to speak. More generally, it prompted him to give a preponderant place to the instincts and other biologically conditioned needs in his evaluations of thought, feeling, and conduct. In accord with such an approach to psychology, La Mettrie envisioned a broad expansion of the ordinary limits set to the usefulness of medicine. He expected that medical science—in particular what is now called psychiatry—would someday be able, by modifying for the better the all-controlling state of the organism, to effect the ethical improvement of those who required it, thereby contributing to the well-being of society. A special instance of this concern was La Mettrie's proposal that many criminals be regarded as "sick" instead of "evil," and that they be turned over to competent doctors for diagnosis and treatment. But it must not be forgotten that the bond which he wished to forge between the practice of medicine and eudaemonistic or humanitarian ethics took for granted, on his part, a doctrine of physiological determinism that left no freedom to the individual, whose actions were held to be intrinsically amoral.

Among La Mettrie's other writings, the most important by far is the *Histoire naturelle de l'âme*, which anticipated closely, and corroborated with a richer accumulation of biological data and a greater reliance on sensationist psychology, the conclusions of *L'homme machine.* In that earlier treatise, however, he saw fit to set his demonstration of the materiality of the soul within the framework of a Scholastic type of metaphysics, somewhat blurring its import and leaving out of account the specifically mechanistic character of man that he was later to affirm so forcefully. The *Histoire naturelle de l'âme* was also, like *L'homme machine*, inspired in large part by an extra-scientific motive. This was La Mettrie's obvious desire, born of the freethinking and anticlerical tendencies of the period, to undermine religion by refuting, on the authority of biology and medicine, the dogma of the spiritual and immortal soul.

The *Système d'Epicure* (1750), an unsystematic group of reflections, gave to the naturalistic science of man sketched by La Mettrie an appropriate evolutionary dimension; more exactly, it represented the human race, no less than other animal races, as the final result of a long series of organic permutations in less perfect precursor species that had failed to survive. Although this work was among the earliest statements in the modern era of the idea of evolution, its exposition of that idea did not go much beyond the Lucretian background on which it freely drew. In *L'homme plante* (1748), a minor but curious work, La Mettrie sought to confirm his belief in a sort of universal organic analogy by pointing out, at times rather speciously, what he considered to be parallel organs and corresponding vital functions in plants and in the human body.

The influence of La Mettrie on the history of science, while difficult to fix with precision, may be said generally to have promoted the objectives of the mechanistic, as against the vitalistic, school of biology and, more significantly, to have militated in favor of a science of psychology based on the physiological method of investigating the mind and personality. Moreover, his deterministic interpretation of human behavior, and his likening of it to that of animals, foreshadowed two familiar tenets of present-day behaviorist psychology. Finally, one may rank among La Mettrie's more recent heirs those who have discovered in cybernetic technology not only the mechanical means of creating artificial thought but also a program for explaining how the brain itself thinks by assimilating its operations to the model of a computerized machine.

BIBLIOGRAPHY

I. Original Works. La Mettrie's philosophical, scientific, and literary writings have never been published together in a single ed. His philosophical texts alone were published numerous times in collected form, but not since the eighteenth century. A recent photo repr. (Hildesheim, 1968) reproduces the *Oeuvres philosophiques*, 2 vols. (Berlin, 1774). An anthology of selected materials can be found in Marcelle Tisserand, ed., *La Mettrie: Textes choisis* (Paris, 1954). There are critical presentations of two individual works: Francis Rougier, *L'homme plante*, repub. with intro. and notes (New York, 1936); and Aram Vartanian, *L'homme machine; a Study in the Origins of an Idea*, with an introductory monograph and notes (Princeton, 1960).

The only modern English trans. of La Mettrie is available in the now inadequate ed. by Gertrude C. Bussey: *Man a Machine; Including Frederick the Great's "Eulogy" . . . and Extracts From "The Natural History of the Soul"* (Chicago–London, 1927).

II. Secondary Literature. The following are useful studies of La Mettrie's scientific and philosophical thought: Raymond Boissier, *La Mettrie, médecin, pamphlétaire et philosophe, 1709–1751* (Paris, 1931); Emile Callot, "La Mettrie," in his *La philosophie de la vie au XVIIIe siècle* (Paris, 1965), ch. V, pp. 195–244; Keith Gunderson, "Descartes, La Mettrie, Language, and Machines," in his *Mentality and Machines* (New York, 1971), pp. 1–38; Günther Pflug, "J. O. de Lamettrie und die biologischen Theorien des 18. Jahrhunderts," in *Deutsche Vierteljahrsschrift für Literaturwissenschaft und Geistesgeschichte*, **27** (1953), 509–527; J. E. Poritzky, *Julien Offray de Lamettrie, sein Leben und seine Werke* (Berlin, 1900); and Guy-Francis Tuloup, *Un précurseur méconnu: Offray de La Mettrie, médecin-philosophe* (Dinard, 1938).

ARAM VARTANIAN

LAMONT, JOHANN VON (*b.* Braemar, Scotland, 13 December 1805, *d.* Bogenhausen, near Munich, Germany, 6 August 1879), *astronomy, physics.*

Lamont was the son of Robert Lamont, custodian of an earl's estate in Scotland. One of three sons of a second marriage, he displayed superior talents as early as primary school; but his education was placed in question when he was twelve by the death of his father. In 1817, however, he was accepted as a pupil at the St. Jacob Scottish Foundation in Regensburg. He never saw his family again.

In Regensburg, Lamont studied primarily German, Latin, and Greek, but was particularly fond of mathematics. He studied the works of Euler and other classics in the original language. In a mechanics workshop he acquired practical knowledge of great importance for his later work in constructing scientific measuring instruments. In 1827 he was sent to the astronomical observatory at Bogenhausen, near Munich, in order to develop his knowledge and abilities. His intelligence and dexterity won the full approval of the observatory's director, Soldner, and Lamont was consequently appointed an assistant at the observatory only one year later.

Following Soldner's death in 1833, Lamont provisionally took over the directorship of the observatory. In this capacity he displayed initiative and extraordinary scientific industry. In 1835, on the proposal of Schelling, who was then president of the Royal Bavarian Academy of Sciences, Lamont was named permanent director of the Bogenhausen observatory. He also became a full member of the Academy. Lamont's extremely varied scientific activity was directed toward astronomical, geodetic, meteorological, physical, and geophysical problems; and when it was necessary, he did not shrink from organizational endeavors.

To continue a promising astronomical investigation, Lamont began by equipping the observatory with better measuring apparatus and obtained funds for the publication of work that had already been completed. Hence he succeeded in publishing the observations that Soldner had made with the transit meridian during 1822–1827, after he himself had carried out the necessary reductions. The most valuable new instrument was a Fraunhofer refractor with a lens aperture of 10.5 Paris inches (approximately 11.25 inches) and a focal length of fifteen feet. With this telescope, which possessed the highest light-gathering power available at the time, Lamont observed the satellites of Saturn and Uranus and provided more exact data on their orbits. Moreover, he utilized the observations of the moons of Uranus to determine that planet's mass, which previously had been derived only from perturbations in the motion of Saturn. The new instrument also enabled Lamont to observe low-luminosity hazy objects, the study of which had been started by William Herschel around 1784. Lamont's exact measurement of the star cluster in Scutum constituted an important foundation for the later study of relative motions in star clusters.

Lamont energetically continued the work Soldner had begun with the meridian circle, and starting in 1838 he was assisted in this task by an observer. In 1840 he began to shift the emphasis of his activities to the observation of a broad zone of stars of the seventh to tenth magnitudes. Of the 80,000 observations in Lamont's zone catalog about 12,000 were of previously uncataloged stars. On two occasions he recorded the still undiscovered planet Neptune, without recognizing its planetary nature. Lamont's zone catalog ranks with those of Lalande, Bessel, Argelander, and Santini as among the most important undertakings of its kind in the nineteenth century.

Around 1850 Lamont introduced to Europe the method developed in American observatories of chronographically recording the transit times of stars across the meridian, and thereby contributed to the objectification of observational procedure. In 1867 an international project for measuring the earth's surface in Europe got under way, inspired by a suggestion of J. J. Baeyer. Lamont took charge of the geographical and astronomical work that was to be done in Bavaria. Beginning in 1878 Lamont turned his efforts chiefly to a thorough sorting out of his observations and to the publication of a general catalog of all the Munich observations, reduced to the year 1880. This work, however, was never completed.

Around 1840 Lamont had become interested in meteorological problems. Because of the great importance of atmospheric conditions for astronomical observation, astronomers had frequently considered problems related to this field; but the observational data, which were mostly sporadic, were still lacking in theoretical penetration. Lamont called for a network of meteorological stations in order to establish a systematic body of data. This project presupposed organized observational activity, however, something which Lamont hoped to bring about through the founding of a meteorological association (1842). The same need was evident for observations in the field of terrestrial magnetism. Lamont created an outlet for both disciplines in the *Annalen für Meteorologie und Erdmagnetismus* (1842–1844). Yet as a result of deficient financial support his progressive ideas regarding scientific organization did not have a chance to develop. Nevertheless, his ideas were influential; in particular, the recording and measuring apparatus that he devised proved their usefulness for decades. For more than forty years at Bogenhausen, Lamont carried out hourly meteorological recordings which were made from seven o'clock in the morning until six o'clock in the evening. The work was the foundation of meteorological science in Bavaria. In connection with his meteorological studies Lamont also investigated the phenomena of atmospheric electricity, in which widespread interest had been created by Franklin.

Lamont's most enduring achievements resulted from his research in terrestrial magnetism, which attracted the attention of John Herschel, Gauss, and Arago. Gauss, whose interest in such questions dated from the beginning of the nineteenth century, had created, with Wilhelm Weber, the Magnetische Verein, thus uniting many previously scattered efforts and furnishing the subject with a far-ranging and coherent methodology. Humboldt used his influence with foreign governments and learned societies to secure the establishment of a worldwide network of geomagnetic stations. Lamont received an official commission from the Bavarian government to take charge of these measurements, and the magnetic observatory erected expressly for this purpose received a special temporary subsidy from the private funds of Crown Prince Maximilian.

Lamont at first considered current methods of measurement and discovered a number of defects which he was able to avoid by employing his own instruments. For example, based on the most recent findings, he developed a magnetic theodolite for determining magnetic declination and horizontal intensity. He likewise devised a portable theodolite capable of meeting the demands made during scientific expeditions. The forty-five devices produced in the observatory's workshop found interested recipients throughout the world. The experience that Lamont gained in constructing these instruments was expressed in papers on the theory of magnetic measuring instruments, in which he demonstrated the influence of temperature on permanent magnets. He also created special temperature-compensated deflection magnets for his instruments.

The continuity of Lamont's activities is reflected in a series of observations of magnetic variations that were made with several assistants throughout the period 1841–1845 at one- or two-hour intervals, as well as at night. Later he also employed automatic recording apparatus of his own invention. From 1849 to 1855 Lamont established a magnetic survey of Bavaria by registering data at a total of 420 locations. In 1856 and 1857 he traveled with his measuring devices to France, Spain, and Portugal; and in 1858 he undertook an expedition to Belgium, Holland, and Denmark.

In his theory of terrestrial magnetism Lamont advocated the position that the earth possesses a solid magnetic core; but he always stressed that several possible conceptions were possible. A major result of his investigation of the earth's magnetism was the discovery that magnetic variations occur in periods of approximately ten years, which is the same time span that Schwabe found about 1843 for the appearance and frequency of sunspots. Lamont's discovery encouraged the study of the reciprocal effects of cosmic and terrestrial events. He set forth his experience and views in the field of magnetism in several comprehensive monographs.

Lamont did not undertake regular teaching duties at the University of Munich until 1852, when he assumed the chair of astronomy left vacant by the death of F. Gruithuisen, who had drawn much criticism for being a scientific visionary. In the framework of his university activity Lamont gave popular scientific lectures which attracted a large audience. He also published a popularized account of his work as *Astronomie und Erdmagnetismus.* He established, out of his own money, a foundation for gifted students of astronomy, physics, and mathematics and bequeathed to it the entire remainder of his considerable fortune.

Lamont was a member of many learned societies. His way of life was simple, and his efforts were dedicated exclusively to science. In an obituary notice in the *Astronomische Nachrichten* it is stated that his accomplishments "assure his name a lasting place in the history of the exact sciences" (**95** [1879], col. 253).

BIBLIOGRAPHY

I. Original Works. Lamont's writings are *Über die Nebelflecke* (Munich, 1837); *Handbuch des Erdmagnetismus* (Berlin, 1849); *Astronomie und Erdmagnetismus* (Stuttgart, 1851); *Der Erdstrom und der Zusammenhang desselben mit dem Erdmagnetismus* (Leipzig, 1862). The majority of his scientific findings are contained in the series of publications of the Bogenhausen observatory, especially *Observationes astronomicae in specula regia Monachiensi institutae* and *Annalen der Königlichen Sternwarte bei München.* Numerous publications can be found in the technical journals and in the publications of the Bavarian Academy of Sciences. See also Poggendorff, I (1863), col. 1361 and III (1898), pp. 768–769.

II. Secondary Literature. See S. Günther, in *Allgemeine deutsche Biographie*, XVII (Leipzig, 1883), 570–572; C. von Orff, in *Vierteljahrsschrift der Astronomischen Gesellschaft*, **15** (1880), 60–82; and the biography by Schafhäutl, in *Historisch-politische Blätter für das katholische Deutschland*, **85** (1880), 54–82. A short description of the Bogenhausen observatory and its instruments is given by G. A. Jahn in his *Geschichte der Astronomie vom Anfange des 19. Jahrhunderts bis zum Ende des Jahres 1842* (Leipzig, 1844), pp. 256–257.

Dieter B. Herrmann

LAMOUROUX, JEAN VINCENT FÉLIX (*b.* Agen, France, 3 May 1776; *d.* Caen, France, 26 May 1825), *natural history.*

The son of Claude Lamouroux and Catherine Langayrou, Lamouroux came from a well-to-do merchant family. He was first interested in botany as an amateur, under the guidance of F. B. de Saint-Amans, and traveled through southern France and Spain to broaden his knowledge of the subject. When the printed calico factory that his father managed suffered a severe reverse, Lamouroux had to plan on supporting himself. He went to Paris to complete his medical studies and received the M.D. in 1809. Named assistant professor of natural history at Caen in 1809, he became a full professor there at the Faculty of Science in 1810. Later he was elected a corresponding member of the Académie des Sciences. In 1818 he married Félicité de Lamariouze, by whom he had one son, Claude Louis Georges, (1819–1836), a midshipman who died at sea.

Lamouroux was attracted to the study of marine algae by his friend J. B. Bory de Saint-Vincent, and with the latter he was one of the first French botanists to take an interest in marine vegetation. In 1805 Lamouroux published his first memoir, which was illustrated with thirty-six engraved plates depicting several species of *Fucus* found along the coasts of Europe and in tropical regions. At that time *Fucus* was considered to include all marine algae that, when viewed by the naked eye or under a magnifying glass, did not exhibit a filamentous cellular structure (such as was seen among the articulated thallophytes). Moreover, both brown algae and red algae were indiscriminately attributed to this genus. Having recognized the heterogeneity of *Fucus*, Lamouroux described many new genera (*Dictyopteris*, *Amansia*, *Bryopsis*, *Caulerpa*, and others) in memoirs and in the first seven volumes of Bory de Saint-Vincent's *Dictionnaire classique d'histoire naturelle* (1822–1825).

In *Essai sur les genres de la famille des Thalassiophytes inarticulés* (1813) Lamouroux proposed a general classification of the marine algae, which he divided into Fucaceae, Florideae, Dictyoteae, Ulvaceae, Alcyonideae, and Spongodieae. Except for the last two, these groups have been maintained in present classifications, although with modifications regarding limits and hierarchy. For example, Lamouroux's Fucaceae include not only the current Fucales, Laminariales, and Desmarestiales but also certain Rhodophyceae (*Furcellaria*). The Florideae are more homogeneous, and in defining them Lamouroux employed a biochemical characteristic that has proved to be of fundamental value: the red color. Furthermore, he was the first to insist on the existence, among these algae, of two distinct types of reproductive organs: tubercles containing "seeds" (cystocarps) and capsules whose contents are almost invariably tripartite (tetrasporocysts). Until then it was assumed, following Dawson Turner and J. C. Mertens, that these two types of reproductive organs corresponded to different stages in the development of the same organ.

Lamouroux thus deserves credit for separating for the first time, even if imperfectly, the brown, red, and green algae, thus eliminating a good deal of confusion. Lamouroux considered the *Essai* of 1813 merely a preliminary exposition which he intended to extend to the nonarticulated thalassiophytes, but he died before he could do so. His ideas on the classification of the algae inspired those adopted by C. A. Agardh, and the two men may be considered the founders of modern phycology.

Lamouroux also wrote *Histoire des Polypiers coralligènes flexibles* (Caen, 1816), in which he described, besides such marine animals as hydrozoa and bryozoa, new genera of calcified algae previously joined with the polyparies (*Neomeris*, *Cymopolia*, *Halimeda*, *Liagora*, *Galaxaura*). By studying a great number of exotic algae brought back by scientific expeditions to the tropical seas and especially the Pacific Ocean, Lamouroux was able to describe many new species. In particular he furnished one of the first descriptions of the algae of Australia, including *Claudea elegans*, which he named for his father. Moreover, Lamouroux was the first to be concerned with the geographic distribution of marine algae, but the data upon which he attempted to establish its broad outlines were insufficient for the task.

BIBLIOGRAPHY

Dissertations sur plusieurs espèces de Fucus peu connues ou nouvelles avec leur description en latin et en français (Agen, 1805); "Mémoire sur trois nouveaux genres de la famille des Algues marines," in *Journal de botanique*, **2** (1809), 129–135; "Histoire des Polypiers coralligènes flexibles vulgairement appelés Zoophytes," in *Bulletin de la Société philomatique*, **3** (Caen, 1812–1816), 236–316; "Essai sur les genres de la famille des Thalassiophytes non articulées," in *Annales du Muséum d'histoire naturelle*, **20** (1813), 21–47, 115–139, 267–293; *Exposition methodique des genres de l'ordre des Polypiers* (Paris, 1821); "Mémoire sur la geographie des plantes marines," in *Annales des sciences naturelles*, **7** (1826), 60–82.

See also the many articles in Bory de Saint-Vincent, ed., *Dictionnaire classique de l'histoire naturelle* (Paris, 1822–1831).

J. FELDMANN

LAMY, BERNARD (*b.* Le Mans, France, June 1640; *d.* Rouen, France, 29 January 1715), *mathematics, mechanics.*

Lamy found his vocation at the Oratorian *collège* in Le Mans, where his parents, Alain Lamy and Marie Masnier, had sent him. As soon as his "Rhétorique" ended, he entered as a novice at the Maison d'Institution in Paris on 6 October 1658.

Lamy was both a product and a master of Oratorian pedagogy. In his principal work, *Entretiens sur les sciences*, the first edition of which appeared in 1683, he proposes an art of learning and teaching all the secular and religious disciplines. This book, admired later by Rousseau, is simultaneously an educational treatise, a discourse on method, and a guide to reading.

During his career Lamy taught almost all subjects. Following his novitiate (1658–1659) and two years of philosophical studies at the *collège* of Saumur, he became professor of classics at Vendôme (1661–1663) and at Juilly (1663–1668). In 1675, drawing on his knowledge of belles lettres, he composed *De l'art de parler*, which in 1688, became *La rhétorique ou l'art de parler*.

Ordained a priest in 1667, Lamy in 1669 finished his training at the École de Théologie de Notre-Dame des Ardilliers, at Saumur. There his teacher was Père André Martin, who found in Descartes support for his Augustinianism. Lamy's admiration for and attachment to Descartes were unwavering. When he became a professor of philosophy, it was Cartesianism that he taught, first at the *collège* of Saumur, and then, beginning in 1673, at the *collège* of Angers, which bore the title Faculté des Arts. This instruction was the cause of his misfortunes. Attacked and denounced for Augustinianism, Cartesianism, and antimonarchical opinions, Lamy was exiled by order of the king in Dauphiné at the beginning of 1676.

At first Lamy lived in a "solitude" at Saint-Martin de Miséré, but soon, thanks to the support of the bishop, Le Camus, he moved into the seminary in Grenoble, where he was again able to teach. During this period he published his principal scientific works: *Traitez de méchanique*, *Traité de la grandeur en général*, and *Les élémens de géométrie*.

These works were still those of a good teacher and not of a researcher; Lamy was more concerned with diffusion than with discovery. Connected with the small Oratorian group of mathematicians that his very good friend Malebranche inspired and animated, he asked of it more than he brought to it. He himself acknowledged his debt to his colleague Jean Prestet. Even when in 1687, in an appendix to the second edition of his *Traitez de méchanique*, Lamy stated, at the same time as Varignon, the rule of the parallelogram of forces, he did not see all of its implications and consequences. Despite Duhem's opinion, Varignon must be conceded the greater originality and awareness of novelty.

In 1686 Lamy obtained permission to live in Paris, but a work on the concordance of the evangelists provoked sharp polemics and his superior general judged it best to send him away again. Beginning in 1690 he lived in Rouen, where he remained until his death, occupied with historical and scriptural studies.

BIBLIOGRAPHY

I. Original Works. Lamy's writings include *Traitez de méchanique, de l'équilibre des solides et des liqueurs* (Paris, 1679); *Traitez de méchanique . . . Nouvelle Édition où l'on ajoute une nouvelle manière de démontrer les principaux théorèmes de cette science* (Paris, 1687); *Traité de la grandeur en général* (Paris, 1680); *Entretiens sur les sciences* (Grenoble, 1683), also in critical ed. by François Girbal and Pierre Clair (Paris, 1966); *Les élémens de géometrie* (Paris, 1685); and *Traité de perspective* (Paris, 1701).

II. Secondary Literature. See Pierre Costabel, "Varignon, Lamy et le parallélogramme des forces," in *Archives internationales d'histoire des sciences*, no. 74–75 (Jan.–June 1966), 103–124; Pierre Duhem, *Les origines de la statique* (Paris, 1906), II, 251–259; and François Girbal, *Bernard Lamy. Étude biographique et bibliographique* (Paris, 1964).

Joseph Beaude

LAMY, GUILLAUME (*b.* Coutances, France; *d.* Paris, France [?]), *philosophy, medicine.*

Lamy's dates of birth and death are completely uncertain; the only ones we can be sure of are those of his publications. It is known that he was born in the old Norman city of Coutances. None of his few biographers gives a birth date for him, and the destruction of the archives of Saint-Lô, in particular the baptismal registers, during the 1944 invasion makes research impossible. The notes of a scholar mention the marriage in Coutances of "Me. Guillaume Lamy, fils de feu Me. Bernard Lamy" on 16 May 1654. If he was then twenty years old, he would have been born in 1634. Accordingly, one is astonished to find that it was in 1672 (when, presumably, he was thirty-eight) that he received his doctorate from the Faculty of Medicine at Paris. Before this date—which is subject to doubt—he had already published (1668 and 1669) three works that are dated. In the first of these he is designated "Maistre aux Arts." That is, having completed the sequence of courses at the University of Paris and having defended a thesis in philosophy, he had the right to teach the humanities.

In 1669 Lamy published *De principiis rerum* (in octavo, 400 pages) that testifies to his vocation as a philosopher. In one of the opening sections of this book, Lamy sets forth the Peripatetic views on the definition of matter, the nature of substantial form, and the qualities of objects. His method is to develop each point of doctrine successively and to refute it immediately afterward. Book II of this work is a presentation and critique of Cartesian thought conducted in the same manner. Lamy attacks Descartes's methodic doubt and his proofs for the existence of God, and shows how Descartes's statements concerning the principles of natural things display a great affinity with the thought of Democritus and Epicurus. Book III is a systematic account of the thought of Epicurus as it is presented in Lucretius. In addition, Lamy shows on what points modern science has clarified and developed these ideas. His position is thus close to that of Gassendi, although he criticizes certain of the latter's opinions. Lamy's concern to harmonize philosophy and science is evident in two appendices, one devoted to the weight of the air and the vacuum and the other to fermentation.

Lamy's interest in the problems of life, however, led him to study medicine. Delaunay informs us that "in order to be inducted as a doctor of the very beneficient Faculty of Paris," one had to count on five to eight years of study. Since Lamy was admitted as a doctor in 1672, he must have commenced his medical studies no later than 1667. In that year he published *Lettre à M. Moreau contre l'utilité de la transfusion*, followed by a second letter in the same year. The year 1667 was decisive in the history of transfusion: the young Académie des Sciences carried out a series of experiments on dogs that led it to recommend against the practice. Lamy was one of the first who dared to contradict the advocates of transfusion. He asserted that this operation is more a means of tormenting the ill than of curing them.

After earning his degree Lamy concerned himself chiefly with medical questions. Between 1675 and 1682 he published three important works. The best-known of these works, *Discours anatomiques*, went through several editions (Paris, 1675 and 1685; Brussels, 1679). Lamy indicates that these discourses were written in conjunction with the presentation of a cadaver at the residence of a well-known surgeon. From precise details he ascends to the nobility of philosophical ideas. In a style now light and now grave he addresses to "Monsieur Notre Adversaire" profound words that appeal by their rationality and reveal a thinker who has deeply meditated on the phenomena of life.

The first discourse warns against the tendency to "exaggerate the nobility of man." Man, according to Lamy, receives from nature the same advantages and the same misfortunes as the beasts. His organization is not more perfect and he lacks apparatuses, such as wings, that a Galen would marvel at as a typically human attribute if man possessed them. Reasoning, Lamy affirms, must exclude finalism, even if one risks being accused of impiety—as Lamy in fact was. The study of the parts and membranes of the body constitutes the major portion of this discourse. The second discourse, devoted to the abdomen, is written from the same point of view. It is the arrangement of the atoms that defines the properties of matter: "Do not say that the eyes are made for seeing; we see because we have eyes." The third discourse is a study of the organs that convert the chyle into blood. The fourth treats the mammary glands and the milk, which Lamy thinks undoubtedly originates in the chyle. In this discourse he also considers the heart. It is enveloped, he asserts, in a useless membrane (pericardium). The heart causes the movement of the blood; the various liquors separate out in the organs that the blood passes through as it circulates. In the fifth discourse Lamy discusses the organs of generation. He is a convinced partisan of the Hippocratic theory of the double semen. Discourse VI is devoted to the brain. The soul, he affirms, does not know itself or the structure of its dwelling. Lamy declares that he is convinced of the immortality of the soul "by . . . Christian faith" and not at all "as a philosopher."

In 1677 Lamy published *Explication mécanique et physique des fonctions de l'âme sensitive*, in which he claimed that the mechanical explanation of the senses, of the passions, and of voluntary motion is necessarily a succession of risky hypotheses. The *Explication* was followed by *Discours sur la génération du laict* and *Dissertation contre la nouvelle opinion qui prétend que tous les animaux sont engendrés d'un oeuf*. Lamy stated that in this latter domain Harvey had not convinced him. Moreover, after criticizing the new views on logical grounds, he returned to the Hippocratic schema he had presented in the *Discours*; according to it the new being originates in the mixture of two seminal liquids.

Dissertation sur l'antimoine (1682) is the last work Lamy published. Renewing the "antimony war," ended fifteen years earlier by decrees of the Faculty of Medicine and of the Parlement condemning the thesis that rejected the use of antimony, Blondel, a former dean, and Douté, his brother-in-law, had challenged Lamy to write in favor of antimony. His work, approved by the highest medical authorities, sets forth the physical reasons for the harmlessness of antimony and the virtues of the preparations that can be made from it. Most notably it illustrates Lamy's wish that medicine benefit from the new discoveries made in anatomy and chemistry.

Lamy thus possessed the qualities of both the scientist who desires progress and the rationalist who is not afraid to dismiss novelties that do not satisfy the demands of reason.

Lamy expressed his opinions forcefully. He was in turn profound, witty, and ironic—Haller called him *impius homo*. His influence was considerable. Revéillé-

Parise, the only author to attempt a biography of him, wrote: "In the period in which he lived, his name resounded in all the Faculties of France and in the foreign universities." At the Faculty of Medicine of Paris his name is included in the list of the *honorandorum magistrorum nostrorum*, signed by Dean Le Moine in 1676. Popular with the public because of the originality of his remarks, Lamy had fervent disciples and impassioned enemies; he replied to the attacks of the latter on several occasions.

BIBLIOGRAPHY

I. ORIGINAL WORKS. In view of the almost complete absence of a real biography of Lamy, the titles of all works published by him have been given in the text, along with dates and an analysis.

II. SECONDARY LITERATURE. N. F. J. Eloy, *Dictionnaire historique de la médecine ancienne et moderne*, III (Mons, 1778), p. 8, contains a very short bibliography listing the principal publications and reflecting Portal's confusion of Lamy with Alain Amy, a physician born in Caen: Portal attributes Lamy's works to him. Joseph Henri Reveillé-Parise, "Étude biographique: Guillaume Lamy," in *Gazette médicale de Paris*, 3rd ser., **6** (1851), 497–502, gives a summary and of course incomplete biography; it includes an analysis of the anatomical discourses. Alphonse Pauly, *Bibliographie des sciences médicales* (Paris, 1874), in a section devoted to individual biographies, gives only Lamy's name and a reference to Reveillé-Parise's article. J. Levy-Valensi, *La médecine et les médecins français au XVIIe siècle* (Paris, 1933), does not devote an article to Lamy, but he is mentioned in the text in connection with the "antimony war." Jacques Roger, *Les sciences de la vie dans la pensée française du XVIIIe siècle*, thesis (Paris, 1963), mentions Lamy's name about thirty times and devotes several pages entirely to him.

L. PLANTEFOL

LANCHESTER, FREDERICK WILLIAM (*b.* Lewisham, England, 28 October 1868; *d.* Birmingham, England, 8 March 1946), *engineer.*

F. W. Lanchester, inventor, designer, and automotive engineer, was the son of Henry Jones Lanchester, an architect, and Octavia Ward. Educated at the Hartley Institute, Southampton, and at the Normal School of Science, Lanchester started work with Messrs. T. B. Barker, Birmingham, a manufacturing company. In 1899, five years after construction of the first Lanchester motor car began, the Lanchester Motor Company was formed with Lanchester as chief engineer and general manager. From 1904 to 1914 he served as the company's consulting engineer; from 1909 to 1929 he was consultant to the Daimler Company and the Birmingham Small Arms Company; and from 1928 to 1930 he was consulting engineer on Diesel engines for William Beardmore's, the manufacturing firm.

In 1894, Lanchester gave a talk before the Birmingham Natural History and Philosophical Society in which he is said to have stated his vortex theory of sustentation (lift). In 1897 a revised version of this paper (neither text exists) was rejected by both the Royal Society and the Physical Society. He was silent until 1907 when he published *Aerodynamics*, as volume I of *Aerial Flight*; volume II, *Aerodonetics*, followed in 1908.

This cavalier neglect by learned societies of an important concept is explainable only in the light of the low state of the theory and practice of hydrodynamics in the late nineteenth and early twentieth centuries. In addition, Lanchester's insistence on using his own terminology—"aerodonetics" for "aeronautics"—rather than that in common usage, and the novelty of his insights impeded the comprehension of his work. Lanchester's book had so considerable an influence on other investigators, however, that he merits a leading place, along with W. M. Kutta, Nikolai Zhukovsky, Ludwig Prandtl, Carl Runge, George H. Bryan, and Theodor von Karmán, in the history of the aeronautical sciences. Among his other works, *Aircraft in War* (1916), became a basic source for the development of the new science of operations analysis during World War II.

BIBLIOGRAPHY

For information on Lanchester's life and work, see "Frederick William Lanchester," in *The Daniel Guggenheim Medal for Achievement in Aeronautics* (New York, 1936); "Dr. F. W. Lanchester; Death of a British Pioneer in Aerodynamic Aerofoil Theory," in *Flight*, **49**, no. 1942 (14 Mar. 1946), 266; Raffaele Giacomelli, "In Memoria de Wilbur e Orville Wright, XII.—La Controversia sugli esperimenti di laboratorio dei fratelli Wright," in *Aerotecnica*, **29** (1949), 105–107; P. W. Kingsford, *F. W. Lanchester; A Life of an Engineer* (London, 1960); "F. W. Lanchester," sec. 2(4) in J. E. Allen, "Looking Ahead in Aeronautics," in *Aeronautical Journal*, **72**, no. 685 (1968), 6–7.

MARVIN W. MCFARLAND

LANCISI, GIOVANNI MARIA (*b.* Rome, Italy, 26 October 1654; *d.* Rome, 20 January 1720), *medicine.*

Lancisi was the son of a wealthy bourgeois family; his parents were Bartolomeo Lancisi and Anna Maria Borgiania. Following preparatory studies, Lancisi took courses in philosophy at the Collegio Romano, but soon realized that his real vocation lay in medicine and natural history. He therefore abandoned theology and entered the Sapienza to study medicine. He graduated in 1672 at the age of eighteen—young even for those times.

After obtaining his degree, Lancisi continued to study medicine independently and advanced rapidly in his career. In 1675 he was appointed doctor at the Hospital of Santo Spirito; in 1678 he was nominated to membership in the Collegio del Salvatore; and in 1684 he was appointed professor of anatomy at the Sapienza, where he taught for thirteen years. At the same time Lancisi became increasingly eminent in the papal court. In 1688 Pope Innocent IX made him pontifical doctor—a post he was to fill, if not always officially, under succeeding popes—and delegated him, as a representative of Cardinal Altieri, to head the pontifical committee for conferring degrees in the medical college.

In 1706 Pope Clement XI asked Lancisi to examine a mysterious increase in the number of sudden deaths, which had assumed the proportions of an epidemic. The following year Lancisi responded by publishing *De subitaneis mortibus*, in which he dealt in a masterly manner with the problems of cardiac pathology; he extended his study of the subject in a second book, *De motu cordis et aneurysmatibus*, published in 1728. Lancisi demonstrated in his first book that sudden deaths were often due to hypertrophy and dilatation of the heart, and to various kinds of valve defects. In the later book on aneurysms, he showed many heart lesions to be syphilitic in nature and gave a good clinical description of syphilis of the heart.

Lancisi also did important research on malaria, which was epidemic in Rome to such an extent that those who could fled the city during the hot months. Drawing upon the work of Fracastoro, Lancisi pointed out that the fevers afflicting Rome and the surrounding countryside were closely related to the presence of swamps, which encouraged the multiplying of mosquitoes. By a brilliant intuition Lancisi attributed the spread of the disease to these insects, and strongly advocated the draining of the swamps—unfortunately without success. He was more effective in bringing the then controversial treatment of malaria by cinchona bark into common practice. He made other significant epidemiological studies on influenza and cattle plague (rinderpest).

Lancisi was also successful in persuading Pope Clement XI to acquire Eustachi's anatomical tables, which had remained unpublished since the latter's death. Lancisi had them printed at his own expense, together with a comprehensive summary. During his life he himself collected a personal medical library of considerable size (well over 20,000 volumes) and interest, which he generously donated to the Hospital of Santo Spirito to be used for the education of the doctors and surgeons of that hospital. The Santo Spirito library was opened in 1716; now named for Lancisi, it constitutes a collection of basic importance for the history of medicine. By Lancisi's will all his own papers and manuscripts were also deposited in it.

BIBLIOGRAPHY

I. Original Works. Lancisi's books and monographs are *De subitaneis mortibus libri duo* (Rome, 1707); *Tabulae anatomicae clarissimi viri B. Eustachi . . . praefatione notisque illustravit Jo. Maria Lancisi* (Rome, 1714); *Dissertatio historica de bovilla peste ex Campaniae finibus* (Rome, 1715); *De noxiis paludum effluviis libri duo* (Rome, 1717); *Joannis Mariae Lancisi opera quae hactenus prodierunt omnia, dissertationibus nonnullis adhucdum locupletata* (Geneva, 1718); *De motu cordis et aneurysmatibus* (Rome, 1728); and *Consilia quadraginta novem posthuma* (Venice, 1744).

II. Secondary Literature. On Lancisi and his work see A. Bacchini, *La vita e le opere di G. M. Lancisi* (*1654–1720*) (Rome, 1920); G. Bilancioni, "La question della sede della cataratta e un carteggio inedito fra il Valsalva e il Lancisi," in *Rivista di storia critica delle scienze mediche e naturali*, **2** (1911), 1–10; "G. M. Lancisi e lo studio degli organi di senso," in *Giornale di medicina militare*, **68**, no. 9 (1920); G. Brambilla, *Un malariologo del Settecento: G. M. Lancisi* (Milan, 1912); A. Corradi, *Lettere di Lancisi a Morgagni e parecchie altre dello stesso Morgagni ora per la prima volta pubblicate* (Pavia, 1876); A. Giarola, M. Cantoni, and E. Magnone, "La dottrina esogena delle infezioni dall'antichità ai giorni nostri. IX: Un malariologo-igienista e un igienista-istorico del primo 700, G. M. Lancisi e L. A. Muratori," in *Rivista italiana di medicina e igiene della scuola*, **13** (1967), 296–311; F. Grondona, "La dissertazione di G. M. Lancisis sulla sede dell'anima razionale," in *Physis*, **7** (1965), 401–430; P. Piccinini, "Il concetto lancisiano degli studi medici," in *Atti del III congresso nazionale della Società italiana di storia delle scienze mediche e naturali* (Venice, 1925), pp. 29-30; and L. Stroppiana, "Giovanni Maria Lancisi," in *Scientia medica italica,* **8** (1959), 5-13.

Carlo Castellani

LANCRET, MICHEL ANGE (*b*. Paris, France, 15 December 1774; *d*. Paris, 17 December 1807), *differential geometry, topography, architecture.*

Son of the architect François Nicolas Lancret—who was the son of an engraver and nephew of the painter Nicolas Lancret—and Germaine Marguerite Vinache de Montblain, the daughter of a sculptor, Michel Ange Lancret was initiated into the plastic arts and architecture at a very early age. He entered the École des Ponts et Chaussées in 1793 and was sent as a student to the port of Dunkerque. Admitted on 21 November 1794 to the first graduating class of the École Polytechnique (at that time the École Centrale des Travaux Publics), he studied there for three years and, along with twenty-four of his fellow students—including J. B. Biot and E.-L. Malus—he served as monitor. After several months of specialization Lancret was named engineer of bridges and highways in April 1798, and in this capacity he was made a member of the Commission of Arts and Sciences attached to the Egyptian expedition. He reached Egypt on 1 July 1798 and was entrusted with important topographical operations, irrigation projects, and canal maintenance, as well as with archaeological studies, the description of the ancient monuments of the Upper Kingdom, and entomological studies.

On 4 July 1799 Lancret was named a member of the mathematics section of the Institut d'Égypte, where he presented several memoirs on his topographical work and communications from others, including one on the discovery of the Rosetta stone (19 July 1799) and Malus's first memoir on light (November 1800). Sent home at the end of 1801, he was soon appointed secretary of the commission responsible for the *Description de l'Egypte*, eventually succeeding Nicolas Conté as the official representative of the government in December 1805. The author of several memoirs on topography, architecture, and political economy, and of numerous drawings of monuments, he devoted himself passionately to this editorial assignment while continuing to do research in infinitesimal geometry.

In his first memoir on the theory of space curves, presented in April 1802, Lancret cites an unpublished theorem of Fourier's on the relationships between the curvature and torsion of a curve and the corresponding elements of the cuspidal edge of its polar curve. In addition he studied the properties of the rectifying surface of a curve and integrated the differential equations of its evolutes. In a second memoir (December 1806) he developed the theory of "développoïdes," cuspidal edges of developable surfaces which pass through a given curve and whose generating lines make a constant angle with this curve.

Although limited in extent, this work places Lancret among the most direct disciples of Monge in infinitesimal geometry.

BIBLIOGRAPHY

I. Original Works. Lancret's writings on Egypt appeared in the collection *Description de l'Égypte:* "Description de l'Ile de Philae," in no. 1, *Antiquité. Descriptions*, I (Paris, 1809), 1–60; "Mémoire sur le système d'imposition territoriale et sur l'administration des provinces de l'Égypte dans les dernières années du gouvernement des Mamlouks," in no. 71, *État moderne*, I (Paris, 1809), 233–260; "Notice sur la branche Canoptique," in no. 46, *Antiquité. Mémoires*, I (Paris, 1809), 251–254; "Mémoire sur le canal d'Alexandrie," in no. 90, *État moderne*, II (Paris, 1812), 185–194, written with G.-J. C. de Chabrol, previously pub. in *La décade égyptienne*, II (Cairo, 1799–1800), pp. 233–251; "Notice topographique sur la partie de l'Égypte comprise entre Rahmânich et Alexandrie et sur les environs du lac Maréotis," in no. 100, *État moderne*, II (Paris, 1812), 483–490, written with Chabrol; and "Description d'Héliopolis," in no. 28, *Antiquité. Descriptions*, II (Paris, 1818), 1–18, written with J. M. J. Dubois-Aymé. Architectural illustrations are in *Antiquité. Descriptions*, I, II, III, and V.

His mathematical writings include "Mémoire sur les courbes à double courbure," in *Mémoires présentés par divers savants* . . ., 2nd ser., **1** (1806), 416–454, an extract of which had appeared in *Correspondance sur l'École polytechnique*, **1**, no. 3 (Jan.–Feb. 1805), 51–52; and "Mémoire sur les développoïdes des courbes à double courbure et des surfaces développables," *ibid.*, **2** (1811), 1–79, extracts in *Nouveau Bulletin des Sciences par la Société philomatique de Paris*, 2nd series, **1** (1807), issues 56 and 57, and in *Correspondance sur l'École polytechnique*, **3**, no. 2 (May 1815), 146–149.

II. Secondary Literature. There are only a few brief and incomplete accounts of Lancret's life: G. Guémard, in *Bulletin de l'Institut d'Égypte*, **7** (1925), 89–90; J. P. N. Hachette, in *Correspondance sur l'École polytechnique*, **1**, no. 9 (Jan. 1808), 374; A. Jal, in *Dictionnaire critique de biographie et d'histoire* (Paris, 1872), pp. 734–735; A. de Lapparent, in *École polytechnique, livre du centenaire*, I (Paris, 1895), 91–92; A. Maury, in Michaud's *Biographie universelle*, new ed., XXIII (Paris, n.d.), 137–138; and F. P. H. Tarbé de Saint-Hardouin, in *Notices biographiques sur les ingénieurs des Ponts et Chaussées* . . . (Paris, 1884), pp. 123–124.

His mathematical work, on the other hand, has been analyzed quite thoroughly by M. Chasles, *Rapport sur les progrès de la géométrie* (Paris, 1870), pp. 10–13; J. L. Coolidge, *A History of Geometrical Methods* (Oxford, 1940), p. 323; N. Nielsen, *Géomètres français sous la Révolution* (Copenhagen, 1929), pp. 155–157; M. d'Ocagne, *Histoire abrégée des sciences mathématiques* (Paris, 1955), 199; and R. Taton, *L'oeuvre scientifique de Gaspard Monge* (Paris, 1951), see index.

René Taton

LANDAU, EDMUND (*b.* Berlin, Germany, 14 February 1877; *d.* Berlin, 19 February 1938), *mathematics.*

Landau was the son of the gynecologist Leopold Landau and the former Johanna Jacoby. He attended the "Französische Gymnasium" in Berlin and then studied mathematics, primarily also in Berlin. He worked mostly with Georg Frobenius and received his doctorate in 1899. Two years later he obtained the *venia legendi*, entitling him to lecture. He taught at the University of Berlin until 1909 and then became full professor at the University of Göttingen, suceeding Hermann Minkowski. David Hilbert and Felix Klein were his colleagues. Landau was active in Göttingen until forced to stop teaching by the National Socialist regime. After his return to Berlin he lectured only outside of Germany, for example, in Cambridge in 1935 and in Brussels in 1937, shortly before his sudden death.

Landau was a member of several German academies, of the academies of St. Petersburg (now Leningrad) and Rome, and an honorary member of the London Mathematical Society. In 1905 he married Marianne Ehrlich, daughter of Paul Ehrlich; they had two daughters and one son.

Landau's principal field of endeavor was analytic number theory and, in particular, the distribution of prime numbers. In 1796 Gauss had conjectured the prime number theorem: If $\pi(x)$ designates the number of prime numbers below x, then $\pi(x)$ is asymptotically equal to $x/\log x$, i.e., as $x \rightarrow \infty$, the quotient of $\pi(x)$ and $x/\log x$ approaches 1. This theorem was demonstrated in 1896 by Hadamard and de la Vallée-Poussin, working independently of each other. In 1903 Landau presented a new, fundamentally simpler proof, which, moreover, allowed the prime number theorem and a refinement made by de la Vallée-Poussin to be applied to the distribution of ideal primes in algebraic number fields. In his two-volume *Handbuch der Lehre von der Verteilung der Primzahlen* (1909), Landau gave the first systematic presentation of analytic number theory. For decades it was indispensable in research and teaching and remains an important historical document. His three-volume *Vorlesungen über Zahlentheorie* (1927) provided an extremely comprehensive presentation of the various branches of number theory from its elements to the contemporary state of research.

Besides two further books on number theory, Landau was author of *Darstellung und Begründung einiger neuerer Ergebnisse der Funktionentheorie*, which contains a collection of interesting and elegant theorems of the theory of analytic functions of a single variable. Landau himself discovered some of the theorems and demonstrated others in a new and simpler fashion. In *Grundlagen der Analysis* he established arithmetic with whole, rational, irrational, and complex numbers, starting from Peano's axioms for natural numbers. Also important is *Einführung in die Differentialrechnung und Integralrechnung*.

Written with the greatest care, Landau's books are characterized by argumentation which is complete, and as simple as possible. The necessary prerequisite knowledge is provided, and the reader is led securely, step by step, to the goal. The idea of the proof and the general relationships are, to be sure, not always clearly apparent, especially in his later works, which are written in an extremely terse manner—the so-called Landau style. Through his books and his more than 250 papers Landau exercised a great influence on the whole development of number theory in his time. He was an enthusiastic teacher and sought contact with fellow scientists. Harald Bohr and G. H. Hardy were often his guests in Göttingen.

BIBLIOGRAPHY

I. ORIGINAL WORKS. Landau was the author of more than 250 papers published in various journals. His books are *Handbuch der Lehre von der Verteilung der Primzahlen*, 2 vols. (Leipzig–Berlin, 1909); *Darstellung und Begründung einiger neuerer Ergebnisse der Funktionentheorie* (Berlin, 1916; 2nd ed., 1929); *Einführung in die elementare und analytische Theorie der algebraischen Zahlen und Ideale* (Leipzig–Berlin, 1918; 2nd ed., 1927); *Vorlesungen über Zahlentheorie*, 3 vols. (Leipzig, 1927); *Grundlagen der Analysis* (Leipzig, 1930); *Einführung in die Differentialrechnung und Integralrechnung* (Groningen, 1934); *Über einige Fortschritte der additiven Zahlentheorie* (Cambridge, 1937).

II. SECONDARY LITERATURE. A biography with portrait is in *Reichshandbuch der deutschen Gesellschaft*, II (Berlin, 1931), 1060; see also the obituaries in *Nachrichten von der Gesellschaft der Wissenschaften zu Göttingen* for 1937–1938, 10; by J. H. Hardy and Heilbronn in *Journal of the London Mathematical Society*, **13** (1938), 302–310; and by Konrad Knopp in *Jahresberichte der Deutschen Mathematiker-vereinigung*, **54** (1951), 55–62.

BRUNO SCHOENEBERG

LANDAU, LEV DAVIDOVICH (*b.* Baku, Russia, 22 January 1908; *d.* Moscow, U.S.S.R, 3 April 1968), *theoretical physics.*

Landau's father was a well-known petroleum engineer who had worked in the Baku oil fields. His mother received a medical education in St. Petersburg, where she did scientific work in physiology and later worked as a physician. When Landau finished school at thirteen, he was already attracted to the exact sciences. His parents thought him too young to enter the university, and he studied for a year at the Baku Economic Technical School. In 1922 he entered Baku University (now Kirov Azerbaydzhan State University), where he studied in the departments of physics-mathematics and chemistry. Although Landau did not continue his chemical education, he retained an interest in chemistry until his death.

In 1924 he transferred to the physics department of Leningrad University; three years later he published his first scientific work, on quantum mechanics. Also in 1927 he graduated from the university and became a graduate student at the Leningrad Institute of Physics and Technology. In his work devoted to *Bremsstrahlung* Landau first introduced the quantity later known as the density matrix (1927).

In 1929 Landau visited Germany, Switzerland, Holland, England, Belgium, and Denmark. There he became acquainted with Bohr, Pauli, Ehrenfest, and W. Heisenberg. Most important for Landau was his work in Copenhagen where theoretical physicists from Europe had gathered around Bohr. His participation in Bohr's seminar played an important role in Landau's development as a theoretical physicist. In 1930 Landau together with R. Peierls investigated a number of subtle problems in quantum mechanics. In the same year Landau did fundamental work in the field of the theory of metals, showing that the degenerate electron gas possesses diamagnetic susceptibility (Landau diamagnetism).

In 1931 he returned to Leningrad and worked in the Institute of Physics and Technology; in 1932 he transferred to Kharkov, where he became the scientific leader of the theoretical group of the newly created Ukrainian Institute of Physics and Technology. At the same time he occupied the chair of theoretical physics at the Kharkov Institute of Mechanical Engineering, and from 1935 he occupied the chair of general physics at Kharkov University.

In 1934 he was awarded the degree of Doctor of Physical and Mathematical Sciences without defending a dissertation, and in 1935 he received the title of professor. The foundation for his creation of an extensive Soviet school of theoretical physics was laid at Kharkov.

Landau's scientific work during this period dealt with various problems in the physics of solid bodies, the theory of atomic collisions, nuclear physics, astrophysics, general questions of thermodynamics, quantum electrodynamics, the kinetic theory of gases, and the theory of chemical reactions. Especially noteworthy is his well-known work on the kinetic equation for the case of Coulomb interactions, the theory of ferromagnetic domain structure and ferromagnetic resonance, the theory of the antiferromagnetic state, the statistical theory of nuclei, and the widely known theory of second-order phase transitions.

In 1937 Landau became director of the section of theoretical physics of the Institute of Physical Problems of the U.S.S.R. Academy of Sciences in Moscow, where he worked until the end of his life.

Landau's scientific work from 1937 to 1941 dealt especially with the cascade theory of electron showers and the intermediate state of superconductors. The physics of elementary particles and nuclear interactions began to occupy an ever greater place in his works. In 1941 he elaborated the basic features of the theory of the superfluidity of helium II. His work in the physics of combustion and the theory of explosions (1944–1945) is noteworthy, as is his research on the scattering of protons by protons and on the theory of ionization losses of fast particles in a medium. In 1946 Landau developed the theory of electron plasma oscillations.

From 1947 to 1953 Landau considered various questions in electrodynamics, the theory of viscosity of helium II, the new phenomenological theory of superconductivity and, the theory, of great importance in the physics of cosmic rays, of the multiple origin of particles in the collision of fast particles.

In 1954 Landau studied questions dealing with the principle of the quantum field theory. As a result of this work, in 1955 he and I. Y. Pomeranchuk obtained a significant argument suggesting that the perturbation series of quantum electrodynamics and the quantum field theory of strong interactions do not sum to a consistent solution.

From 1956 to 1958 Landau created a general theory of the so-called Fermi-liquid, to which liquid helium III and the electrons in metals are related. In 1957 he presented a new general law of modern physics, the law of CP conservation, to replace the law of the conservation of parity which appeared incorrect for weak interactions. In 1959 Landau advanced new principles of the structure of the theory of elementary particles. In a published article he noted a way to determine the basic properties of the so-called interaction amplitude of particles.

Landau's published textbooks for institutions of higher education and his monographs on theoretical physics are characterized by precision of exposition and richness of scientific material, combined with exceptional clarity and the presentation of profound physical ideas. His monographs on theoretical physics are widely known throughout the world. The first book of his course on theoretical physics, *Statisticheskaya fizika* ("Statistical Physics," 1938), was followed by *Mekhanika* ("Mechanics") and *Teoria polya* ("Field Theory").

In his last years Landau, together with E. M. Lifshits, continued to work on a course of theoretical physics. In 1948 a new book of this course appeared,

Kvantovaya mekhanika ("Quantum Mechanics"), as well as a revised edition of *Teoria polya.* In 1951 he published a completely new work on statistical physics and, in 1953, *Mekhanika sploshnykh sred* ("The Theory of Elasticity"). A course of lectures on general physics, given by Landau in the Moscow Institute of Physics and Technology was published in 1949, followed in 1955 by a course of lectures in the theory of the atomic nucleus written with Y. A. Smorodinsky. Another volume in this series, *Elektrodinamika sploshnykh sred* ("Electrodynamics of Continuous Media"), appeared in 1957. The authors' continuing revisions of these works were tantamount to the writing of a new book.

Landau created a very important scientific school. His students worked in the most varied fields of theoretical physics and became distinguished scientists. Among his students were E. M. Lifshits, I. Y. Pomeranchuk, I. M. Lifshits, A. S. Kompaneyts, A. I. Akhiezer, V. B. Berestetsky, I. M. Shmushkevich, V. L. Ginzburg, A. B. Migdal, Y. A. Smorodinsky, I. M. Khalatnikov, A. A. Abricossov, and K. A. Ter-Martirosian.

Landau's scientific achievements received wide recognition. He was elected to membership in the Academy of Sciences of the U.S.S.R. and was awarded the title of Hero of Socialist Labor. Landau received the State Prize of the U.S.S.R. three times, and in 1962 he was awarded the Lenin Prize.

International recognition was expressed by the award of the Nobel Prize in physics in 1962; he was also elected a member of many foreign academies and societies. In 1951 he was chosen a member of the Danish and, in 1956, the Netherlands academies of science. In 1959 he was elected a member of the British Physical Society and in 1960 of the Royal Society. In the same year he became a member of the U. S. National Academy of Sciences and the American Academy of Arts and Sciences and was awarded the F. London Prize (U.S.A.) for research in low-temperature physics and the Max Planck Medal (West Germany).

A tragic accident cut short Landau's scientific work. In January 1962 he sustained severe injuries in an automobile accident and for several months lingered between life and death. Through remarkable efforts the life of this great physicist was prolonged for six years.

BIBLIOGRAPHY

I. Original Works. Landau's writings include "Diamagnetismus der Metalle," in *Zeitschrift für Physik*, **64** (1930), 629; "Extension of the Uncertainty Principle to Relativistic Quantum Theory," *ibid.*, **69** (1931), 56, written with R. Peierls; "Eine mögliche Erklärung der Feldabhängigkeit der Suszeptibilität bei niedrigen Temperaturen," in *Soviet Physics*, **4**, no. 4 (1933), 675; "Struktur der unverschobenen Streulinie," *ibid.*, **5**, no. 1 (1934), 172, written with G. Platschek; "On the Theory of the Dispersion of Magnetic Permeability in Ferromagnetic Bodies," *ibid.*, **8**, no. 2 (1935), 153, written with E. Lifshits; "Zur Theorie der Schalldispersion," *ibid.*, **10**, no. 1 (1936), 34, written with E. Teller; "Die kinetische Gleichung für den Fall coulombscher Wechselwirkung," *ibid.*, **10**, no. 2 (1936), 154; "Zur Theorie der Supraleitfähigkeit," *ibid.*, **11**, no. 2 (1937), 129; and "K teorii fazovykh perekhodov" ("Toward a Theory of Phase Transitions"), in *Zhurnal eksperimentalnoi i teoreticheskoi fiziki*, no. 7 (1937), 19.

Subsequent works are "The Cascade Theory of Electronic Showers," in *Proceedings of the Royal Society*, **166A** (1938), 213, written with G. Rumer; *Statisticheskaya fizika* ("Statistical Physics"; Moscow–Leningrad, 1938), written with E. Lifshits; "Teoria sverkhtekuchesti gelia-2" ("Theory of the Superfluidity of Helium II"), in *Zhurnal eksperimentalnoi i teoreticheskoi fiziki*, no. 11 (1941), 592; *Teoria polya* ("Field Theory"; Moscow–Leningrad, 1941; rev. ed. 1951), written with E. Lifshits; "K teorii promezhutochnogo sostoyania sverkhprovodnikov" ("Toward a Theory of the Intermediate State of Superconductors"), in *Zhurnal eksperimentalnoi i teoreticheskoi fiziki*, 13 (1943), 377; "On the Theory of the Intermediate State of Superconductors," in *Fizicheskii zhurnal*, **7**, no. 3 (1943), 99; *Mekhanika sploshnykh sred* ("The Theory of Elasticity"; Moscow–Leningrad, 1944), written with E. Lifshits; and "On the Energy Loss of Fast Particles by Ionization," in *Fizicheskii zhurnal*, **8**, no. 4 (1944), 201.

Later writings are "On the Theory of Superfluidity of Helium II," in *Fizicheskii zhurnal*, **11**, no. 1 (1947), 91; *Kvantovaya mekhanika* ("Quantum Mechanics"; Moscow–Leningrad, 1948), written with E. Lifshits; "Asimptoticheskoe vyrazhenie dlya funktsii Grina elektrona v kvantovoy elektrodinamike" ("An Asymptotic Expression for Green's Function of the Electron in Quantum Electrodynamics"), written with A. Abricossov and I. Khalatnikov, in *Doklady Akademii nauk SSSR*, **95** (1954), 1177; "O tochechnom vzaimodeystvii v kvantovoy elektrodinamike" ("On Point Interaction in Quantum Electrodynamics"), *ibid.*, **102** (1955), 489, written with I. Pomeranchuk; *Lektsii po teorii atomnogo yadra* ("Lectures on the Theory of the Atomic Nucleus"; Moscow, 1955), written with Y. Smorodinsky; "On the Quantum Theory of Fields," in *Nuovo cimento*, supp. 3, no. 1 (1956), 80, written with A. Abricossov and I. Khalatnikov; "O zakonakh sokhranenia pri slabykh vzaimodeystviakh" ("On the Laws of Conservation in Weak Interactions"), in *Zhurnal eksperimentalnoi i teoreticheskoi fiziki*, **32**, no. 2 (1957), 405; and "Ob analiticheskikh svoystvakh vershinnykh chastey v kvantovoy teorii polya" ("On the Analytical Properties of the Vertex Function in Quantum Field Theory"), *ibid.*, **37**, no. 1 (1959), 62.

II. Secondary Literature. See V. B. Berestetsky, "Lev Davidovich Landau k 50-letiyu so dnya rozhdenia" ("Lev

Davidovich Landau on the Fiftieth Anniversary of His Birth"), in *Uspekhi fizicheskikh nauk*, **64**, no. 3 (1958), 615.

A. T. GRIGORIAN

LANDEN, JOHN (*b.* Peakirk, near Peterborough, England, 23 January 1719; *d.* Milton, near Peterborough, 15 January 1790), *mathematics.*

Landen was trained as a surveyor and from 1762 to 1788 was land agent to William Wentworth, second Earl Fitzwilliam. He lived a quiet rural life with mathematics as the occupation of his leisure, taking up those topics which caught his fancy. He contributed to the *Ladies' Diary* from 1744 and to the *Philosophical Transactions of the Royal Society*; he published his *Mathematical Lucubrations* in 1755 and the two-volume *Mathematical Memoirs* in 1780 and 1790; the latter volume was placed in his hands from the press the day before he died. He was elected a fellow of the Royal Society in 1766.

Landen wrote on dynamics, in which he had the temerity to differ with Euler and d'Alembert, and on the summation of series. He also tried to settle the arguments about the validity of limit processes used as a basis for the calculus by substituting a purely algebraic foundation.

Landen's name is perpetuated by his work on elliptic arcs (*Philosophical Transactions*, 1775). Giulio Carlo Fagnano dei Toschi had obtained elegant theorems about arcs of lemniscates and ellipses. Landen's development expressed the length of a hyperbolic arc in terms of lengths of arcs in two ellipses. The connection in size between these ellipses permits Landen's work to be seen as a relation between two elliptic integrals. In Legendre's notation, if

$$F(\phi, k) = \int_0^\phi \sqrt{(1 - k^2 \sin^2 \phi)}\, d\phi,$$

then in Landen's transformation

$$F(\phi, k) = \tfrac{1}{2}(1 + k_1)\, F(\phi_1, k_1),$$

where, writing $k' = \sqrt{(1 - k^2)}$ as usual, the new parameters ϕ_1, k_1 are expressed in terms of ϕ, k by the relations

$$\sqrt{(1 - k^2 \sin^2 \phi)} \cdot \sin \phi_1 = (1 + k') \sin \phi \cos \phi,$$
$$k_1 = (1 - k')/(1 + k').$$

By considering an iterated chain of such transformations, Legendre obtained a method for the rapid computation of elliptic integrals, of which Gauss's method of the arithmetico-geometric mean is another form. The Landen transformation can also be shown as a relation between elliptic functions; in the Jacobian notation,

$$\mathrm{sn}\{(1 + k')\, u, k_1\} = (1 + k')\, \mathrm{sn}(u, k)\, \mathrm{cd}(u, k).$$

An interest in integration, or "fluents," led Landen to discuss (*Philosophical Transactions*, 1760, and later) the dilogarithm

$$Li_2(z) = -\int_0^z \frac{\log(1 - z)\, dz}{z}$$

(the notation is modern). He obtained several formulas and numerical values that were found at almost the same time by Euler. In the first volume of the *Memoirs* he initiated discussion of the function (now sometimes called the trilogarithm)

$$Li_3(z) = \int_0^z \frac{Li_2(z)\, dz}{z},$$

deriving functional relations and certain numerical results, work followed up by Spence (1809) and Kummer (1840).

BIBLIOGRAPHY

I. ORIGINAL WORKS. Landen's books are *Mathematical Lucubrations* (London, 1755); and *Mathematical Memoirs*, 2 vols. (London, 1780–1790). Articles are "A New Method of Computing the Sums of Certain Series," in *Philosophical Transactions of the Royal Society*, **51**, pt. 2 (1760), 553–565, and "An Investigation of a General Theorem for Finding the Length of Any Conic Hyperbola . . .," *ibid.*, **65**, pt. 2 (1775), 283–289.

II. SECONDARY LITERATURE. A short biography is C. Hutton, "John Landen," in *A Mathematical and Philosophical Dictionary*, II (London, 1795), 7–9. For the life and work of Fagnano and of Landen, see G. N. Watson, "The Marquis and the Land-Agent," in *Mathematical Gazette*, **17** (Feb. 1933). Landen's transformation is discussed in any standard text on elliptic functions. For the dilogarithm and its generalizations, see L. Lewin, *Dilogarithms and Associated Functions* (London, 1958).

T. A. A. BROADBENT

LANDOLT, HANS HEINRICH (*b.* Zurich, Switzerland, 5 December 1831; *d.* Berlin, Germany, 15 March 1910), *chemistry.*

Landolt began his education in Zurich under Karl Löwig. He subsequently followed Löwig to Breslau, where he received his doctorate for work on arsenic ethyl. He then attended lectures by Rose and Mitscherlich in Berlin but found the laboratory facilities inadequate and soon moved to Heidelberg, where Bunsen's laboratory had become a center for chemical studies. Here he investigated the luminosity of gases produced in a Bunsen burner and, on the strength of his work, became a privatdocent at Breslau. A year later, in 1857, Landolt became associate professor at Bonn and in 1867 full professor. In 1869 he began to teach at Aachen, and in 1880 he moved to the Agricultural Institute in Berlin. He succeeded Rammelsberg in the Second Chemical Laboratory, Berlin, in 1891 remaining there until his retirement in 1905. He was elected to the Berlin Academy in 1882.

Primarily a physical chemist, Landolt centered his major work on molecular refractivity of organic compounds (specific refraction × molecular weight). In 1858 John Gladstone and Thomas Dale proposed an empirical formula which related the density and the refractive index of a substance. A second formula with a stronger theoretical basis was derived independently by H. A. Lorentz and Ludwig V. Lorenz in 1880. Berthelot, Gladstone, and Dale tried to correlate refractivity and chemical composition and suggested that molecular refractivity was an additive property. Landolt, studying fatty acids and esters, contributed to this view by arriving at values for the refraction of each element in a compound. In 1870 Gladstone showed that Landolt's values yielded erroneous results with such unsaturated compounds as benzene and the terpenes. Further work by Landolt's student Julius Wilhelm Brühl showed that molecular refractivity was not strictly an additive property but was affected by constitutive factors as well. Landolt later extended his research on molecular refractivity, using rays of various wavelengths.

Landolt also investigated the velocity of the reaction between iodic and sulfuric acid. Because of his appointments at technical schools, he was also interested in the design and industrial applications of polarimeters. His main publication, written in collaboration with Richard Börnstein, is *Physikalisch-chemische Tabellen* (1883). The book has been enlarged and reissued many times since Landolt's death.

BIBLIOGRAPHY

I. Original Works. Landolt wrote over forty articles, including "Ueber die Zeitdauer der Reaction zwischen Jodsäure und schwefliger Säure," in *Sitzungsberichte der Preussischen Akademie der Wissenschaften zu Berlin* (1885), pt. 1, 249–284; (1886), pt. 1, 193–219; pt. 2, 1007–1015; (1887), pt. 1, 21–37; and "Ueber den Einfluss der atomistischen Zusammensetzung C, H, und O-haltiger flüssiger Verbindingen auf die Fortpflanzung des Lichts," in *Annalen der Physik und Chemie*, **122** (1864), 545–563; **123** (1864), 595–628. With Richard Börnstein he wrote *Physikalisch-chemische Tabellen* (Berlin, 1883).

II. Secondary Literature. Works on Landolt include the following, listed chronologically: J. H. van't Hoff, "Gedächtniss Rede auf Hans Heinrich Landolt," in *Abhandlungen der K. Preussischen Akademie der Wissenschaften zu Berlin*, phys.-math. Kl. (1910), 67, English adaptation in *Journal of the Chemical Society of London*; **99** (1911), 1653; Richard Pribam, "Nekrolog auf H. Landolt," in *Berichte der Deutschen chemischen Gesellschaft*, **44** (1911), 3337; and A. Ihde, *Development of Modern Chemistry* (New York, 1964), pp. 265, 393.

Ruth Gienapp Rinard

LANDRIANI, MARSILIO (*b.* Milan, Italy, *ca.* 1751; *d.* Vienna, Austria, not later than 1816), *scientific instrumentation.*

We have no information about Landriani until 1775, when the *Ricerche fisiche intorno alla salubrità dell'aria* appeared. In 1776 he was appointed teacher of physics in the schools of higher education then being established in Milan. By appointment of the government, in 1787–1788 he made a long tour of the leading countries of Europe in order to study their scientific and technological development. In 1790 he was government adviser, and in this capacity he ordered the establishment of the Veterinary School of Milan. Toward the end of 1791 he was sent on a diplomatic mission to Dresden, where he continued to study physics, spreading knowledge of Galvani's recent electrophysiological discoveries. In 1794 he moved to Vienna, where he spent the remaining years of his life.

Landriani's name is repeatedly linked to Volta's inventions (from the electrophorus to the pile) and especially to the eudiometer. The term (derived from the Greek *eudia* ["fair weather"]) was first used by Landriani in the *Ricerche* to indicate the instrument he had devised to measure the purity of the air. The method had been introduced in 1772 by Joseph Priestley, who had proposed measuring the "different disposition of airs for breathing" by means of the $NO + O_2$ reaction: "nitrous air" (nitrogen bioxide) plus the gas of common air, which Priestley himself obtained in 1774 and called "dephlogisticated air" (later named oxygen by Lavoisier). By means of this reaction reddish vapors (higher oxides of nitrogen) are formed; being strongly water soluble, they are removed by water, in the presence of which the reaction is carefully performed. The reaction thus indicates the consumption of oxygen, or of part of common air. The greater the reduction in volume that the latter undergoes, the richer in oxygen it is and hence the healthier.

Volta very acutely defined the hygienic value of

the method and radically transformed the instrument, assigning it new tasks. In 1777 the eudiometer entered the history of science as a valued instrument for analyzing gases.

BIBLIOGRAPHY

I. ORIGINAL WORKS. Landriani's writings are *Ricerche fisiche intorno alla salubrità dell'aria* (Milan, 1775); *Physikalische Untersuchungen über die Gesundheit der Luft* (Basel, 1778; Bern, 1792); "Lettera al Signor D. Alessandro Volta," in *Scelta di opuscoli interessanti*, **19** (1776), 73–86, with a plate; *Opuscoli fisico-chimici* (Milan, 1781); *Dell' utilità dei conduttori elettrici* (Milan, 1784); *Abhandlung vom Nutzen der Blitzableiter* (Vienna, 1786); "Von einigen Entdeckungen in der thierischen Elektricität," in *Sammlung physikalischer Aufsätze, besonders die Böhmische Naturgeschichte betreffend, von einer Gesellschaft Böhmischer Naturforscher*, **3** (1793), 384–388, with 2 plates; and *Relazione sopra Basilea, Aarau e Bienne*, which follows the reprint of Pietro Moscati and M. Landriani, *Dei vantaggi della educazione filosofica nello studio della chimica*, Luigi Belloni, ed. (Milan, 1961).

II. SECONDARY LITERATURE. On Landriani and his work, see Luigi Belloni, "L'eudiometro del Landriani (contributo alla storia medica dell'eudiometria)," in *Actes du symposium international sur les sciences naturelles, la chimie et la pharmacie du 1630 au 1850, Florence–Vinci, 8–10 octobre 1960* (Florence, 1962), 130–151; "La salubrità dell'aria: l'eudiometro del Landriani," in Fondazione Treccani degli Alfieri, *Storia di Milano*, XVI (Milan, 1962), 946–947; and "La Scuola Veterinaria di Milano, Discorso celebrativo del 175° anniversario di fondazione della Scuola oggi Facoltà di Medicina Veterinaria letto il 14 ottobre 1966," in *Studium veterinarium mediolanense*, **1** (1969), 1–32.

LUIGI BELLONI

LANDSBERG, GEORG (*b.* Breslau, Germany [now Wrocław, Poland], 30 January 1865; *d.* Kiel, Germany, 14 September 1912), *mathematics.*

Landsberg spent his youth in Breslau. He studied at the universities of Breslau and Leipzig from 1883 to 1889, receiving his doctorate in mathematics from the former in 1890. He then went to the University of Heidelberg, where he became a privatdocent in mathematics in 1893 and extraordinary professor in 1897. He returned to Breslau in this capacity in 1904, but in 1906 he accepted an offer from the University of Kiel, where he was appointed professor ordinarius in 1911. He remained at Kiel until his death.

Landsberg investigated the theory of algebraic functions of two variables, which was then a hardly accessible subject that did not attain its major successes until much later. He also considered the theory of curves in higher dimensional manifolds and its connection with the calculus of variations and the mechanics of rigid bodies. In addition he studied theta functions and Gaussian sums. In this work he touched on the ideas of Weierstrass, Riemann, and Weber.

Landsberg's most important achievement lay in his contributions to the development of the theory of algebraic functions of one variable. In this field arithmetic, algebra, function theory, and geometry are most intimately related. In addition to Riemann's function-theoretical approach and the geometric approach favored by Italian mathematicians as an especially easy and sure access, there existed the arithmetical approach from Weierstrass. Landsberg's most important work in this area was his algebraic investigations of the Riemann-Roch theorem, which had been stated by Riemann in the context of his theory of algebraic functions and greatly extended by Roch. Landsberg provided a foundation for it within arithmetic theory, which then finally led to the modern abstract theory of algebraic functions.

BIBLIOGRAPHY

Landsberg's *Theorie der algebraischen Funktionen einer Variablen und ihre Anwendungen auf algebraische Kurven und abelsche Integrale* (Leipzig, 1902), written with Kurt Hensel, was a standard text for decades. A complete listing of Landsberg's articles can be found in Poggendorff, IV, 835; V, 706.

BRUNO SCHOENEBERG

LANDSBERG, GRIGORY SAMUILOVICH (*b.* Vologda, Russia, 22 January 1890; *d.* Moscow, U.S.S.R., 2 February 1957), *physics.*

Landsberg's father was a civil servant in a state forest preserve. The family first lived in Vologda, and then moved to Nizhniy Novgorod (now Gorky), where Landsberg graduated from the Gymnasium with a gold medal. In 1908 he entered the natural sciences section of the department of physics and mathematics of Moscow University, and after a year transferred to the mathematical section. He graduated in 1913 with a diploma of the first degree and remained at the university to prepare for a teaching career.

From 1913 to 1915 Landsberg was an assistant at the university; in 1915 he published with N. N. Andreev his first scientific work, on the manufacture of large electrical resistors. From 1918 to 1920 he was docent at Omsk Agricultural Institute.

In 1920 he returned to Moscow and became a scientific co-worker at the Institute of Physics and Biophysics. His interest in optics dates from this time. In 1925 L. I. Mandelshtam transferred to Moscow University, and from this time on Landsberg and Mandelshtam conducted joint research. Their first study was on Rayleigh scattering in crystals. A problem resulted from the presence in the crystals of internal defects, which caused an additional effect in the scattering of light. Using the fact that the intensity of the molecular scattering of light depends on temperature, Landsberg was able to separate the molecular scattering from the side effects. Landsberg and Mandelshtam subsequently began to study the spectral composition of light scattered by quartz crystal. It followed from theoretical considerations that a fine structure must be present in the scattered light, caused by the modulation of the Rayleigh line by heat waves distributed through the crystal.

In the fall of 1927 Landsberg and Mandelshtam discovered a new phenomenon: satellites were observed in the spectrum of scattered light from a crystal; but the changes in their wavelength from the primary light appeared considerably larger than those expected from the modulation by heat waves. It became obvious that these changes were caused by the modulation of light by the infrared vibrations of the molecules of the crystal. The new phenomenon was called "combination scattering." On 6 May 1928 the first communication on this discovery was submitted for publication; it contained not only experimental facts but also the theory of the new effect and a collection of experimental and computational data.

An analogous effect in liquids had been discovered simultaneously by C. V. Raman, who reported his discovery several weeks before Landsberg and Mandelshtam. Raman received the 1930 Nobel Prize in physics for his discovery, and the effect was named after him. After careful study of the new effect Landsberg and Mandelshtam continued their research on Rayleigh scattering in crystals, concentrating on the intensity and anisotropy of the light scattering. Through this research an incomplete theory was clarified, and under Landsberg's leadership a new theory was worked out. In 1931 Landsberg and Mandelshtam discovered a sharp intensification of the scattering near resonant spectral lines of atoms.

In 1932 Landsberg was elected corresponding member of the U.S.S.R. Academy of Sciences. His broad research in the area of emission spectral analysis and its applications began at this time. Landsberg and his co-workers developed a method of rapid identification of alloyed steels by spectral analysis. In 1934 Landsberg organized a large scientific research laboratory in the Lebedev Physical Institute of the U.S.S.R. Academy of Sciences; there Landsberg and his colleagues carried out investigations on combination scattering in organic substances, which permitted them to clarify a number of peculiarities in the hydrogen bond and the conditions of formation of associated complexes. Landsberg's development of methods and devices for spectral analysis played a considerable practical role during World War II, when Landsberg worked in Kazan. In 1940 Landsberg was awarded the State Prize for his work on spectral analysis, and in 1946 he was elected an active member of the U.S.S.R. Academy of Sciences. He subsequently carried out investigations of molecular scattering in viscous liquids and amorphous bodies.

Landsberg gave considerable attention to the teaching of physics. In 1929 with B. A. Vvedensky he wrote *Sovremennoe uchenie o magnetizme* ("Contemporary Theory of Magnetism"). In 1934 he published a basic course, *Optiki* ("Optics"), still widely used in Soviet higher educational institutions. On his initiative the three-volume *Elementarny uchebnik fiziki* ("Elementary Textbook of Physics") was created; it has been reprinted many times.

BIBLIOGRAPHY

Landsberg's selected works were published in Moscow in 1958. See *Uspekhi fizicheskikh nauk*, **63**, no. 2 (Oct. 1957), a commemorative issue that includes recollections of Landsberg by I. B. Tamm, pp. 287–288; a short sketch of his life and work by S. L. Mandelshtam, pp. 289–299; and a portrait by V. A. Fabrikant of Landsberg as author and editor of physics textbooks, pp. 455–460.

J. G. DORFMAN

LANDSTEINER, KARL (*b.* Vienna, Austria, or Baden [near Vienna], Austria, 14 June 1868; *d.* New York, N. Y., 26 June 1943), *medicine, serology, immunology.*

Landsteiner, who has been called the father of immunology, was the only son of Leopold Landsteiner, a well-known journalist and newspaper publisher, and Fanny Hess Landsteiner. He began his medical studies in 1885 and received his M.D. in 1891. In 1916 he married Helene Wlasto; their only child, Ernst Karl, was born the following year. Poor working conditions caused him to leave Vienna in 1919; but facilities in The Hague, where he was prosector at the RK Hospital for three years, were no better. He therefore accepted an offer from the Rockefeller Institute in New York, and went to the United States in 1922; he became an American citizen in 1929. Landsteiner was a modest, self-critical, rather timid man of science known for his wide reading. He was also an excellent pianist.

Although he had an M.D., Landsteiner's first scientific work was in chemistry, which he began to study in Ludwig's laboratory in Vienna while still a student. He continued these studies in Germany and Switzerland from 1892 to 1894. With Emil Fischer he synthesized glycolaldehyde at Würzburg in 1892. At Munich in 1892–1893, he learned the chemistry of benzene derivatives from Bamberger, and in Zurich in 1893–1894 he studied organic chemistry under Hantzsch.

Medicine, however, remained Landsteiner's chief interest. For a short time after receiving his M.D. he had worked with Kahler at the Second Medical University Clinic in Vienna; and from 1894 to 1895 he served with Eduard Albert at the First Surgical University Clinic. During 1896–1897 Landsteiner was assistant to Gruber in the newly established department of hygiene at the University of Vienna; and there his interest was awakened in serology and immunology.

Landsteiner's next teacher was Weichselbaum, whose assistant he was from 1897 to 1908. At that time Weichselbaum was director of the Pathological-Anatomical Institute of the University of Vienna. Under his supervision, Landsteiner conducted 3,639 postmortem examinations that gave him a comprehensive view of medicine and extensive experience as a pathological anatomist.

In 1900 Landsteiner published only one paper. But one of its footnotes contained information on one of his most important discoveries, namely, the interagglutination occurring between serum and blood cells of different humans as a physiological phenomenon, which he explained by individual differences. The following year, in the article "Über Agglutinationserscheinungen normalen menschlichen Blutes," Landsteiner described a simple technique of agglutination, whereby he divided human blood into three groups: A, B, and C (later O). Two of his inspired co-workers, the clinicians Decastello and Sturli, examined additional persons and found the fourth blood group, later named AB.

The blood grouping is done by mixing suspensions of red cells with the test sera anti-A and anti-B:

	O	A	B	AB
Serum Anti-A	−	+	−	+
Serum Anti-B	−	−	+	+

+ Agglutination
− No Agglutination

Blood group O is agglutinated by neither of the sera, AB by both, A by anti-A but not by anti-B, and B by anti-B but not by anti-A. The serum of group O has anti-A and anti-B antibodies, that of A has only anti-B, that of B has only anti-A, and that of AB has neither. According to Landsteiner's rule, serum contains only those antibodies (isoagglutinins) which are not active against their own blood group.

The discovery of blood groups made possible the safe transfusion of blood from one person to another, although several years passed before this knowledge was put to practical use. Richard Lewisohn's discovery in 1914 that adding citrates to blood prevented it from coagulating was the last prerequisite for the establishment of the modern blood bank, since blood could now be preserved for two- to three-week periods under refrigeration. Operations on the heart, lungs, and circulatory system, previously impracticable because of the magnitude of blood losses involved, were now feasible, as were complete blood exchanges in cases, for example, of intoxication or severe jaundice of the newborn.

Instead of pursuing further developments in blood groups, Landsteiner sought out other differences in human blood. He conceived the idea that the particularity of blood was reflected in antigen differences, and that these differences could be used to distinguish one person from another and to draw a serological "fingerprint." Today—if hereditary serum groups and enzyme groups are included—millions of combinations are possible; and Landsteiner's concept of the individuality of human blood, revealed serologically, has practically been realized. At first Landsteiner did not know that blood types were inheritable, for Mendel's laws of heredity had passed into oblivion. In 1900 the laws were rediscovered by Correns, De Vries, and Tschermak-Seysenegg. Ten years later Emil von Dungern and Hirszfeld postulated the first hypothesis for the inheritance of blood groups; this theory was corrected in 1924 by the mathematician B. A. Bernstein and was finally established. Serological genetics has existed since then and is applied in cases of disputed paternity. Today about 99 percent of paternity questions are settled by serological means.

During this period Landsteiner also worked on characterizing and evaluating the physiological meaning of cold agglutinations in human blood serum. In 1904 he and Donath described a test for the diagnosis of paroxysmal cold hemoglobinuria. In this disease, after the patient is exposed to cold, hemoglobin appears in his urine because some of the red blood cells have been lysed.

Ehrlich had also concerned himself with this problem. He originated a simple clinical diagnostic test, the so-called Ehrlich finger test. A finger to which a rubber tourniquet has been applied is put in ice water. After the dissolution of the congested material, hemoglobinuria occurs. Ehrlich erroneously attributed the phenomenon to a pathological change in the endothelium of the blood vessels. In opposition to Ehrlich, Landsteiner postulated that the disease-causing agent was found in the blood serum of the patient and that it was an antibody which, when exposed to cold, combines with the red cells and later, under warm conditions, causes their breakdown in the body. He demonstrated this process in a test tube and noted the lysis of the red cells (the Donath-Landsteiner test). Landsteiner also made important contributions to the etiology of meconium ileus in newborn children.

In 1905–1906 Landsteiner and Ernest Finger, then chief of the Dermatological Clinic in Vienna, were successful in infecting monkeys with syphilis. Experimentation with *Spirochaeta pallida*, the causal agent of the disease, was thereby made possible. The two investigators determined that infectious spirochetes were present in gummas. With the help of the venereologist Mucha they were able to demonstrate the syphilitic spirochetes in the dark field of the microscope and also describe their typical movements. In collaboration with the neurologist Poetzl and the serologist Mueller, they elucidated the previously unknown mechanism occurring in the Wassermann reaction. In 1907 Landsteiner also demonstrated that for this test, the extract (antigen) previously exclusively obtained from human organs could be replaced by a readily available extract of bovine hearts. This made possible the widespread use of the Wassermann test.

From 1908 to 1919, while he was prosector at the Royal-Imperial Wilhelminen Hospital in Vienna, Landsteiner concerned himself extensively with poliomyelitis. After conducting a postmortem examination of a child who had died of the disease, he injected a homogenate of its brain and spinal cord into the abdominal cavity of various experimental animals, including rhesus monkeys. On the sixth day following the injections, the monkeys showed signs of paralysis similar to those of poliomyelitis patients. The histological appearance of their central nervous systems also was similar to that of humans who had died of the disease. Since he could not prove the presence of bacteria in the spinal cord of the child who had died, Landsteiner postulated the existence of a virus: "The supposition is hence near, that a so-called invisible virus or a virus belonging to the class of protozoa, causes the disease." Between 1909 and 1912 he and Levaditi of the Pasteur Institute at Paris devised a serum diagnostic procedure for poliomyelitis and a method of preserving the viruses that cause it.

During the 1920's Landsteiner made further discoveries. In 1921, for instance, utilizing investigations dating as far back as 1904, he demonstrated the existence of hapten, a specific constituent of the antigens; this discovery was influential in the development of immunology. Landsteiner also differentiated various hemoglobins by means of chemical and serological techniques. In 1926 he and Philip Levine discovered the irregular agglutinins α_1 and α_2; the following year they found the blood factors M, N, and P. In 1934, with Strutton and Chase he described a blood factor found only in Negroes, which today is called the Hunter-Henshaw system.

Landsteiner and his co-workers Alexander Wiener and Philip Levine made an important discovery, reported in a paper (1940), describing a new factor in the human blood, the rhesus (Rh) factor. Levine was the first to see the connection between this factor and jaundice occurring in newborn children. A mother who does not have the Rh factor—that is, who is Rh-negative—can be stimulated by an Rh-positive fetus to form antibodies against the Rh factor. The red cells of the fetus are then destroyed by these antibodies, and the product of hemoglobin decomposition forms bilirubin which causes jaundice. Permanent brain damage can result, and the fetus or newborn child may die. By means of serological tests such cases can be recognized in time and saved by means of blood exchange transfusions.

The Rh factor is also of vital importance in blood transfusions, for Rh-positive blood must not be transfused into Rh-negative patients. If it is, Rh antibodies will be formed; and further transfusion of Rh-positive blood will lead to severe hemolytic reactions and the patient can die.

In the field of bacteriology, it should be noted that Landsteiner and Nigg were successful in 1930–1932 in culturing *Rickettsia prowazekii*, the causative agent of typhus, on living media.

Landsteiner's honors include honorary doctorates from the University of Chicago (1927), Cambridge (1934), the Free University of Brussels (1934), and Harvard (1936); the presidency of the American Association of Immunologists (1929); and the Nobel Prize in physiology or medicine (1930).

BIBLIOGRAPHY

I. Original Works. *The Specificity of Serological Reactions* (New York, 1962), a trans. of Landsteiner's major work, *Die Spezifitaet der serologischen Reaktionen*, contains, in addition to a new preface, a bibliography of Landsteiner's 346 scientific papers compiled by Merrill W. Chase.

II. Secondary Literature. On Landsteiner's life and

work, see H. Chiari, *Österreichische Naturforscher, Ärzte und Techniker* (Vienna, 1957); H. A. L. Degener, *Unsere Zeitgenossen* (Leipzig, 1914; Berlin, 1935); I. Fischer, *Biographisches Lexikon der hervorragenden Ärzte der letzten 50 Jahre*, vol. II (Berlin–Vienna, 1933); J. and R. Gicklhorn, *Die österreichische Nobelpreisträger* (Vienna, 1958); T. W. MacCallum and S. Taylor, *The Nobel Prize Winners and the Nobel Foundation 1901–1937* (Zurich, 1938); F. Oehlecker, *Die Bluttransfusion* (Berlin–Vienna, 1933); Peyton Rouse, "Karl Landsteiner," in *Obituary Notices of Fellows of the Royal Society of London*, **5** (1947), 295–324, with bibliography; L. Schönbauer, *Das Medizinische Wien* (Berlin–Vienna, 1944; 2nd ed. 1947); M. Schorr, *Zur Geschichte der Bluttransfusion im 19. Jahrhundert* (Basel–Stuttgart, 1956); G. R. Simms, *The Scientific Work of Karl Landsteiner* (Zurich, 1963); and P. Speiser, *Karl Landsteiner, Entdecker der Blutgruppen. Biographie eines Nobelpreisträgers aus der Wiener Medizinischen Schule* (Vienna, 1961), with a complete bibliography of Landsteiner's works.

A number of obituary notices appeared in a variety of medical journals at the time of his death; see especially *Journal of the American Medical Association*, **122** (1943), and *Wiener medizinische Wochenschrift*, **94** (1944).

PAUL SPEISER

DICTIONARY
OF
SCIENTIFIC BIOGRAPHY

PUBLISHED UNDER THE AUSPICES OF
THE AMERICAN COUNCIL OF LEARNED SOCIETIES

The American Council of Learned Societies, organized in 1919 for the purpose of advancing the study of the humanities and of the humanistic aspects of the social sciences, is a nonprofit federation comprising thirty-nine national scholarly groups. The Council represents the humanities in the United States in the International Union of Academies, provides fellowships and grants-in-aid, supports research-and-planning conferences and symposia, and sponsors special projects and scholarly publications.

MEMBER ORGANIZATIONS

AMERICAN PHILOSOPHICAL SOCIETY, 1743
AMERICAN ACADEMY OF ARTS AND SCIENCES, 1780
AMERICAN ANTIQUARIAN SOCIETY, 1812
AMERICAN ORIENTAL SOCIETY, 1842
AMERICAN NUMISMATIC SOCIETY, 1858
AMERICAN PHILOLOGICAL ASSOCIATION, 1869
ARCHAEOLOGICAL INSTITUTE OF AMERICA, 1879
SOCIETY OF BIBLICAL LITERATURE, 1880
MODERN LANGUAGE ASSOCIATION OF AMERICA, 1883
AMERICAN HISTORICAL ASSOCIATION, 1884
AMERICAN ECONOMIC ASSOCIATION, 1885
AMERICAN FOLKLORE SOCIETY, 1888
AMERICAN DIALECT SOCIETY, 1889
AMERICAN PSYCHOLOGICAL ASSOCIATION, 1892
ASSOCIATION OF AMERICAN LAW SCHOOLS, 1900
AMERICAN PHILOSOPHICAL ASSOCIATION, 1901
AMERICAN ANTHROPOLOGICAL ASSOCIATION, 1902
AMERICAN POLITICAL SCIENCE ASSOCIATION, 1903
BIBLIOGRAPHICAL SOCIETY OF AMERICA, 1904
ASSOCIATION OF AMERICAN GEOGRAPHERS, 1904
HISPANIC SOCIETY OF AMERICA, 1904
AMERICAN SOCIOLOGICAL ASSOCIATION, 1905
AMERICAN SOCIETY OF INTERNATIONAL LAW, 1906
ORGANIZATION OF AMERICAN HISTORIANS, 1907
COLLEGE ART ASSOCIATION OF AMERICA, 1912
HISTORY OF SCIENCE SOCIETY, 1924
LINGUISTIC SOCIETY OF AMERICA, 1924
MEDIAEVAL ACADEMY OF AMERICA, 1925
AMERICAN MUSICOLOGICAL SOCIETY, 1934
SOCIETY OF ARCHITECTURAL HISTORIANS, 1940
ECONOMIC HISTORY ASSOCIATION, 1940
ASSOCIATION FOR ASIAN STUDIES, 1941
AMERICAN SOCIETY FOR AESTHETICS, 1942
METAPHYSICAL SOCIETY OF AMERICA, 1950
AMERICAN STUDIES ASSOCIATION, 1950
RENAISSANCE SOCIETY OF AMERICA, 1954
SOCIETY FOR ETHNOMUSICOLOGY, 1955
AMERICAN SOCIETY FOR LEGAL HISTORY, 1956
SOCIETY FOR THE HISTORY OF TECHNOLOGY, 1958

DICTIONARY OF SCIENTIFIC BIOGRAPHY

CHARLES COULSTON GILLISPIE
Princeton University
EDITOR IN CHIEF

Volume 8

JONATHAN HOMER LANE – PIERRE JOSEPH MACQUER

CHARLES SCRIBNER'S SONS · NEW YORK

Editorial Staff

Panel of Consultants

Contributors to Volume 8

The following are the contributors to Volume 8. Each author's name is followed by the institutional affiliation at the time of publication and the names of articles written for this volume. The symbol † indicates that an author is deceased.

GIORGIO ABETTI
Istituto Nazionale di Ottica
LORENZONI

ERIC J. AITON
Didsbury College of Education
LEIBNIZ

LUÍS DE ALBUQUERQUE
University of Coimbra
LAVANHA; LISBOA

GARLAND E. ALLEN
Washington University
MCCLUNG

TOBY A. APPEL
Kirkland College
LEREBOULLET

CORTLAND P. AUSER
Bronx Community College, City University of New York
LARSEN

STANLEY L. BECKER
Bethany College
MCCOLLUM

WHITFIELD J. BELL, JR.
American Philosophical Society Library
LEA

LUIGI BELLONI
University of Milan
LARGHI

ENRIQUE BELTRÁN
Mexican Society of the History of Science and Technology
LICEAGA

RICHARD BERENDZEN
Boston University
MAANEN

ALEX BERMAN
University of Cincinnati
LESSON

OLEXA MYRON BILANIUK
Swarthmore College
LICHTENBERG

ASIT K. BISWAS
Department of Environment, Ottawa
LAUSSEDAT

MARGARET R. BISWAS
McGill University
LAUSSEDAT

L. J. BLACHER
Academy of Sciences of the U.S.S.R.
LAVRENTIEV

GEORGE BOAS
Johns Hopkins University
LOVEJOY

WALTER BÖHM
LOSCHMIDT

O. BORŮVKA
Czechoslovak Academy of Sciences
LERCH

GERT H. BRIEGER
Duke University
LOEFFLER

T. A. A. BROADBENT †
MACAULAY; MCCOLL; MACMAHON

THEODORE M. BROWN
City College, City University of New York
LOWER

STEPHEN G. BRUSH
University of Maryland
LENNARD-JONES

K. E. BULLEN
University of Sydney
LOVE

VERN L. BULLOUGH
California State University, Northridge
LIND; MACLEAN

IVOR BULMER-THOMAS
LEO; LEODAMAS OF THASOS

WERNER BURAU
University of Hamburg
LUEROTH

JOHN G. BURKE
University of California, Los Angeles
O. LEHMANN; LEONHARD

RICHARD W. BURKHARDT, JR.
University of Illinois
LATREILLE

H. L. L. BUSARD
State University of Leiden
VAN LANSBERGE; LE PAIGE

JEROME J. BYLEBYL
University of Chicago
LAURENS; LEONICENO

ANDRÉ CAILLEUX
Laval University
LAPPARENT; E. A. I. H. LARTET; L. LARTET

KENNETH L. CANEVA
University of Utah
A.-A. DE LA RIVE; C.-G. DE LA RIVE

JEFFREY CARR
University of Leeds
M. LISTER

ETTORE CARRUCCIO
Universities of Bologna and Turin
LORIA

JOHN CHALLINOR
C. LAPWORTH

SEYMOUR L. CHAPIN
California State University, Los Angeles
LE GENTIL DE LA GALAISIÈRE

MARSHALL CLAGETT
Institute for Advanced Study, Princeton
LEONARDO DA VINCI

T. H. CLARK
McGill University
W. E. LOGAN

M. J. CLARKSON
Liverpool School of Tropical Medicine
T. R. LEWIS

E. COUMET
E. LE ROY

MAURICE CRANSTON
London School of Economics
LOCKE

GLYN DANIEL
University of Cambridge
LEAKEY

MAURICE DAUMAS
Conservatoire National des Arts et Métiers
LANGLOIS

AUDREY B. DAVIS
Smithsonian Institution
LIEBERKÜHN

GAVIN DE BEER †
LANKESTER

SALLY H. DIEKE
Johns Hopkins University
J. W. LUBBOCK; C. S. LYMAN

HERBERT DINGLE
LOCKYER

CLAUDE E. DOLMAN
University of British Columbia
J. LISTER

J. G. DORFMAN
Academy of Sciences of the U.S.S.R.
P. P. LAZAREV; P. N. LEBEDEV

SIGALIA DOSTROVSKY
Barnard College
LISSAJOUS

STILLMAN DRAKE
University of Toronto
LE TENNEUR

J. M. EDMONDS
University of Oxford Museum
LHWYD

CONTRIBUTORS TO VOLUME 8

FRANK N. EGERTON III
University of Wisconsin-Parkside
LOVELL

JON EKLUND
Smithsonian Institution
W. LEWIS

V. A. ESAKOV
Academy of Sciences of the U.S.S.R.
M. P. LAZAREV

C. W. F. EVERITT
Stanford University
F. LONDON; H. LONDON

JOSEPH EWAN
Tulane University
LESQUEREUX

V. A. EYLES
MACCULLOCH

W. M. FAIRBANK
Stanford University
F. LONDON; H. LONDON

W. V. FARRAR
University of Manchester
A. LAPWORTH; LAWES AND GILBERT

GIOVANNI FAVILLI
University of Bologna
LUSTIG

IGNAZIO FAZZARI
University of Florence
LUNA

I. A. FEDOSEYEV
Academy of Sciences of the U.S.S.R.
LEPEKHIN; LOKHTIN

MARTIN FICHMAN
York University
J.-B. LE ROY

KARIN FIGALA
Technische Hochschule München
LONICERUS

N. FIGUROVSKY
Academy of Sciences of the U.S.S.R.
LOVITS

WALTHER FISCHER
LIESGANIG

DONALD FLEMING
Harvard University
J. LOEB

ROBERT FOX
University of Lancaster
N. LEBLANC

EUGENE FRANKEL
Trinity College
H. LLOYD

H.-CHRIST. FREIESLEBEN
LOHSE; LUDENDORFF; LUTHER

HANS FREUDENTHAL
State University of Utrecht
LIE; LOEWNER

BRUNO VON FREYBERG
University of Erlangen-Nuremberg
J. G. LEHMANN

DAVID J. FURLEY
Princeton University
LUCRETIUS

GERALD L. GEISON
Princeton University
J. N. LANGLEY; LOEWI; K. LUCAS

PATSY A. GERSTNER
Dittrick Museum of Historical Medicine
K. N. LANG

OWEN GINGERICH
Smithsonian Astrophysical Observatory
LEAVITT

P. GLEES
University of Göttingen
LEYDIG

THOMAS F. GLICK
Boston University
LEO THE AFRICAN

MARIO GLIOZZI
University of Turin
LEVI-CIVITA

GEORGE GOE
New School for Social Research
LUKASIEWICZ

J. B. GOUGH
Washington State University
C. LE ROY; LESAGE

I. GRATTAN-GUINNESS
Enfield College of Technology
M. P. H. LAURENT

SAMUEL L. GREITZER
Rutgers University
LEMOINE

NORMAN T. GRIDGEMAN
National Research Council of Canada
LOTKA; F.-E.-A. LUCAS

A. T. GRIGORIAN
Academy of Sciences of the U.S.S.R.
LEXELL; LEYBENZON; LYAPUNOV

HENRY GUERLAC
Cornell University
LAVOISIER

THOMAS L. HANKINS
University of Washington
P.-C. LE MONNIER

OWEN HANNAWAY
Johns Hopkins University
LE FEBVRE; L. LEMERY; N. LEMERY

THOMAS HAWKINS
Boston University
LEBESGUE

ROGER HEIM
Muséum National d'Histoire Naturelle
LE DANTEC

JOHANNES HENIGER
State University of Utrecht
LEEUWENHOEK

ARMIN HERMANN
University of Stuttgart
LAUE; LENARD; LUMMER

DIETER B. HERRMANN
Archenhold Observatory, Berlin
LINDENAU

ERWIN N. HIEBERT
Harvard University
MACH

JOEL H. HILDEBRAND
University of California, Berkeley
LATIMER

JOSEPH E. HOFMANN †
LEIBNIZ

HO PENG-YOKE
Griffith University
LI CHIH; LIU HUI

A. HOLLMAN
University College Hospital, London
T. LEWIS

FREDERIC L. HOLMES
University of Western Ontario
LIEBIG

I. B. HOPLEY
Clifton College
LIPPMANN

MICHAEL A. HOSKIN
University of Cambridge
MACLEAR

PIERRE HUARD
René Descartes University
LE DOUBLE; LIEUTAUD; LORRY

WŁODZIMIERZ HUBICKI
Marie Curie-Skłodowska University
LIBAVIUS

AARON J. IHDE
University of Wisconsin, Madison
LEVENE

MARIE JOSÉ IMBAULT-HUART
René Descartes University
LE DOUBLE; LIEUTAUD; LORRY

JEAN ITARD
Lycée Henri IV
P. A. LAURENT; LEGENDRE

R. V. JONES
University of Aberdeen
F. A. LINDEMANN

PAUL JOVET
Centre National de Floristique
L'ÉCLUSE; L'OBEL

SATISH C. KAPOOR
University of Saskatchewan
AUGUSTE LAURENT

B. M. KEDROV
Academy of Sciences of the U.S.S.R.
LENIN; LOMONOSOV

KENNETH D. KEELE
Research Fellow, Wellcome Institute
LEONARDO DA VINCI

A. G. KELLER
University of Leicester
LUSITANUS

CONTRIBUTORS TO VOLUME 8

MARTHA B. KENDALL
Vassar College
LESLEY

G. B. KERFERD
University of Manchester
LEUCIPPUS

H. C. KING
Royal Ontario Museum
LASSELL

MARC KLEIN
Louis Pasteur University
LAVERAN

FRIEDRICH KLEMM
Deutsches Museum
LINDE

H. KOBAYASHI
Hokkaido University
B. S. LYMAN

MANFRED KOCH
Bergbau Bucherei, Essen
LÖHNEYSS

R. E. KOHLER
University of Pennsylvania
G. N. LEWIS

ELAINE KOPPELMAN
Goucher College
M. LÉVY

SHELDON J. KOPPERL
Grand Valley State College
M. J. L. LE BLANC

FREDERICK KREILING
Brooklyn Polytechnic Institute
LEIBNIZ

A. D. KRIKORIAN
State University of New York at Stony Brook
LIVINGSTON

VLADISLAV KRUTA
Pediatric Research Institute, Brno
LEGALLOIS

GISELA KUTZBACH
University of Wisconsin
JOHN LECONTE; LOOMIS

HENRY M. LEICESTER
University of the Pacific
S. V. LEBEDEV; LE BEL; LE CHÂTELIER

CZESŁAW LEJEWSKI
University of Manchester
LEŚNIEWSKI

JACQUES R. LÉVY
Paris Observatory
LE VERRIER; LYOT

OLGA A. LEZHNEVA
Academy of Sciences of the U.S.S.R.
LENZ

STEN LINDROTH
University of Uppsala
LINNAEUS

ALBERT G. LONG
Hancock Museum
W. H. LANG

RUSSELL McCORMMACH
Johns Hopkins University
LORENTZ

ROBERT M. McKEON
Babson College
LE FÈVRE; J. LEMAIRE; P. LEMAIRE

H. LEWIS McKINNEY
University of Kansas
W. LAWRENCE; JOSEPH LECONTE

MICHAEL S. MAHONEY
Princeton University
LE POIVRE

J. C. MALLET
Centre National de Floristique
L'ÉCLUSE; L'OBEL

NIKOLAUS MANI
Bonn Institute for the History of Medicine
LANGERHANS

AUGUSTO MARINONI
LEONARDO DA VINCI

BRIAN G. MARSDEN
Smithsonian Astrophysical Observatory
LOWELL

KIRTLEY F. MATHER
University of New Mexico
LINDGREN

OTTO MAYR
Smithsonian Institution
LECORNU; H. LORENZ

A. A. MENIAILOV
Academy of Sciences of the U.S.S.R.
LEVINSON-LESSING

DANIEL MERRIMAN
Yale University
M'INTOSH

JÜRGEN MITTELSTRASS
University of Konstanz
LEIBNIZ

A. M. MONNIER
University of Paris
LAPICQUE; MACHEBOEUF

A. M. MONSEIGNY
Muséum National d'Histoire Naturelle
E. A. I. H. LARTET; L. LARTET

DON F. MOYER
Ripon College
S. P. LANGLEY; MACCULLAGH

AXEL V. NIELSEN †
LAU

J. D. NORTH
Museum of the History of Science, Oxford
MACMILLAN

KARL-GEORG NYHOLM
University of Uppsala
LOVÉN

HERBERT OETTEL
LINDELÖF

YNGVE ÖHMAN
Stockholm Observatory
LINDBLAD

OLIVIERO M. OLIVO
University of Bologna
LEVI

RICHARD G. OLSON
University of California, Santa Cruz
LESLIE

C. D. O'MALLEY †
LINACRE

A. PAPLAUSCAS
Academy of Sciences of the U.S.S.R.
LUZIN

JOHN PARASCANDOLA
University of Wisconsin
J. U. LLOYD

FRANKLIN PARKER
West Virginia University
L. LOEB

STUART PIERSON
Memorial University of Newfoundland
LEONHARDI; LYONET

MOGENS PIHL
University of Copenhagen
L. V. LORENZ

DAVID PINGREE
Brown University
LĀṬADEVA; LEO THE MATHEMATICIAN

A. F. PLAKHOTNIK
Academy of Sciences of the U.S.S.R.
LITKE

LUCIEN PLANTEFOL
University of Paris
L.-G. LE MONNIER

GIANLUIGI PORTA
LUCIANI

R. D. F. PRING-MILL
University of Oxford
LULL

P. RAMDOHR
University of Heidelberg
LOSSEN

NATHAN REINGOLD
Smithsonian Institution
LANE

LADISLAO RETI
LEONARDO DA VINCI

R. A. RICHARDSON
University of Western Ontario
MACALLUM

RUTH GIENAPP RINARD
Kirkland College
LIEBERMANN

S. DILLON RIPLEY
Smithsonian Institution
LEVAILLANT

CONTRIBUTORS TO VOLUME 8

PHILIP C. RITTERBUSH
Archives of Institutional Change, Washington, D.C.
LEIDY

J. M. ROBERTSON
University of Glasgow
K. LONSDALE

ABRAHAM ROBINSON
Yale University
L'HOSPITAL

PAUL G. ROOFE
University of Kansas
LASHLEY

GEORGE ROSEN
Yale University School of Medicine
LUDWIG

CHARLES E. ROSENBERG
University of Pennsylvania
LUSK

B. A. ROSENFELD
Academy of Sciences of the U.S.S.R.
LOBACHEVSKY

K. E. ROTHSCHUH
University of Münster/Westphalia
LOTZE

M. J. S. RUDWICK
University of Cambridge
W. LONSDALE

JULIO SAMSÓ
Universidad Autónoma de Barcelona
LEVI BEN GERSON

RALPH A. SAWYER
University of Michigan
T. LYMAN

WILLIAM L. SCHAAF
Florida Atlantic University
LEURECHON

H. SCHADEWALDT
University of Düsseldorf
LEUCKART

EBERHARD SCHMAUDERER
LUNGE

CECIL J. SCHNEER
University of New Hampshire
LEONARDO DA VINCI

BRUNO SCHOENEBERG
University of Hamburg
LIPSCHITZ

E. L. SCOTT
Stamford High School, Lincolnshire
R. LUBBOCK; D. MACBRIDE

J. F. SCOTT †
MACLAURIN

CAROL SHAMIEH
Boston University
MAANEN

DIANA M. SIMPKINS
Polytechnic of North London
E. W. MACBRIDE

N. SIVIN
Massachusetts Institute of Technology
LI SHIH-CHEN

W. A. SMEATON
University College, London
MACQUER

A. DE SMET
Bibliothèque Royale de Belgique
VAN LANGREN

H. A. M. SNELDERS
State University of Utrecht
H. F. LINK

E. SNORRASON
Rigshospitalet, Copenhagen
LANGE

Y. I. SOLOVIEV
Academy of Sciences of the U.S.S.R.
LUGININ

FRED SOMKIN
Cornell University
J. LUBBOCK

PIERRE SPEZIALI
University of Geneva
L'HUILLIER

FRANS A. STAFLEU
State University of Utrecht
L'HÉRITIER DE BRUTELLE

WILLIAM T. STEARN
British Museum (Natural History)
LINDLEY

C. G. G. J. VAN STEENIS
University of Leiden
LOTSY

LLOYD G. STEVENSON
Johns Hopkins University
MACLEOD

F. STOCKMANS
Free University of Brussels
LIGNIER; LOHEST

CHARLES SÜSSKIND
University of California, Berkeley
LANGMUIR; E. O. LAWRENCE; LODGE

F. SZABADVÁRY
Technical University, Budapest
R. LORENZ

RENÉ TATON
École Pratique des Hautes Études
LIOUVILLE

JEAN THÉODORIDÈS
Centre National de la Recherche Scientifique
LÉGER

D. N. TRIFONOV
Academy of Sciences of the U.S.S.R.
LEBEDINSKY

G. L'E. TURNER
University of Oxford
J. J. LISTER

GEORG USCHMANN
University of Jena
A. LANG

FRANCIS E. VAUGHAN
VEMCO Corporation
LAWSON

J. J. VERDONK
LA ROCHE

THÉODORE VETTER
Société Française d'Histoire de la Médecine
LE CAT

WILLIAM A. WALLACE, O.P.
Catholic University of America
LAX

RAY L. WATTERSON
University of Illinois
LILLIE

EUGENE WEGMANN
LUGEON

ADRIENNE R. WEILL-BRUNSCHVICG
LANGEVIN

DORA B. WEINER
Manhattanville College
LEURET

JOHN W. WELLS
Cornell University
LESUEUR

GEORGE W. WHITE
University of Illinois
MACLURE

LEONARD G. WILSON
University of Minnesota
LYELL

EDWIN WOLF II
Library Company of Philadelphia
J. LOGAN

A. E. WOODRUFF
Yeshiva University
LARMOR

DENISE WROTNOWSKA
Musée Pasteur
LEVADITI

H. WUSSING
Karl Marx University
C. L. F. LINDEMANN

J. WYART
University of Paris
S.-D. A. LÉVY

A. P. YOUSCHKEVITCH
Academy of Sciences of the U.S.S.R.
LEXELL

BRUNO ZANOBIO
University of Pavia
LUCIANI

DICTIONARY
OF
SCIENTIFIC BIOGRAPHY

DICTIONARY OF SCIENTIFIC BIOGRAPHY

LANE—MACQUER

LANE, JONATHAN HOMER (*b.* Genesee, New York, 9 August 1819; *d.* Washington, D.C., 3 May 1880), *physics.*

Lane left school at the age of eight but acquired enough learning by himself at home to teach in rural schools. In 1839 he entered Phillips Academy at Exeter, New Hampshire. While there, according to his own account, he became interested in what was to be his principal intellectual preoccupation, the experimental determination of absolute zero. He entered Yale, where he was apparently influenced by Denison Olmsted, and graduated in 1846.

After briefly teaching in Vermont, Lane came to Washington, D.C., in 1847 and was employed by the U.S. Coast Survey. He had already published the first of four articles on electricity. In 1848 Lane conferred with Joseph Henry about experiments to determine the speed of propagation of solar light and heat. In that same year, and with Henry's help, he was appointed an examiner in the U.S. Patent Office, where he was closely associated with Charles G. Page. Besides the articles on electricity, which were attempts to provide mathematical formulations for electrical phenomena, Lane published nothing during his years at the Patent Office (but see the comments in the bibliography below). In 1857, he was removed from the Patent Office by a spoils-minded Secretary of the Interior.

From 1857 to 1866, when Lane returned to Washington, his course is obscure, although the Abbe necrology (see below) asserts that he attempted to earn a living as a patent agent. He did attempt to develop his low-temperature apparatus, which was to utilize the expansion of gases for cooling. Failing to gain adequate backing, he went to Franklin, Venango County, Pennsylvania, in 1860, to live with his brother, a blacksmith. Lane expected his brother, who owed him money, to repay the debt by constructing the apparatus, an expectation that was not realized. Abbe also asserts that Lane made a handsome profit from the sale of oil lands in Pennsylvania, which enabled him to return to Washington. This seems unlikely, however, since Henry continued to send him odd computing jobs and even a small grant for the low-temperature experiments, apparently in the belief that Lane needed the money. In 1869 Lane joined the Office of Weights and Measures, the predecessor of the present National Bureau of Standards.

In 1869 Lane read a paper, "On the Theoretical Temperature of the Sun," before the National Academy of Sciences. It was printed in the *American Journal of Science* in the following year. His purpose was to test the adequacy of various current theories of heat by mathematical determinations of the temperature of the sun, on the assumption of a convection system, explicitly based upon James Espy's meteorological theories, for the movement of the photosphere. Lane concluded that none of the theories that assumed that heat was motion provided adequate explanations of his calculated values for the distribution of density, pressure, and temperature in the sun.

The paper gained a modest notoriety because of something it did not demonstrate but which Lane verbally proved to the satisfaction of both Kelvin and Newcomb. According to the latter's autobiography, Lane had given the proof to him prior to 1869 of the "law" by which a gaseous body contracts when it loses heat, but the heat generated by the contraction exceeds the heat lost in order to produce the contraction. Lane did not give the proof in his paper, although it was implicit in the presentation. Kelvin's interest stemmed in part from the possibility that "Lane's law" would contradict his calculation of the age of the earth by changing the quantity of energy available in the sun. Three years after Lane's paper August Ritter independently came to similar conclusions, including an explicit statement of the "paradoxical" law. Interestingly, Ritter also was applying meteorological theory to the study of the sun. Kelvin attempted to remove the supposed difficulty in 1887, for which he was subsequently criticized for "inexactitude" by Emden in 1907.

The principal significance of Lane's article, however, was not the unstated law nor even its testing of current theories of heat but the careful calculation of mass and heat relationships in the sun. The convection model, now discarded, proved useful for arriving at a good first approximation of the structure and energy distribution of the sun, while Lane's work on the structure of a star was a real contribution to the developing evidence of stellar evolution (leading to the Hertzsprung-Russell diagram of 1913).

It is not a wholly easy task to determine what other scientific work Lane actually performed. His bibliography contains only fifteen items, two of which are titles only. His contemporaries reported his unwillingness to rush into publication, this being the presumed reason for the absence of a published account of the work on absolute zero, supposedly completed by 1870. His personal papers in the U.S. National Archives contain several references to what may be published papers (perhaps in nonscientific journals), unpublished papers, or simply drafts or ideas for papers. In 1848, for example, Joseph Henry referred to a Lane paper on solar heat, which is now unknown.

This bibliographic uncertainty is particularly important because the surviving personal papers contain many fragments on proposed experiments and instruments, but since these are often incomplete or unclear, it is difficult to speak about Lane's scientific interests and accomplishments. Most of the fragments display a concern with precision instrumentation, while some clearly indicate Lane's mathematical aptitude. Few of the surviving manuscripts pertain to inventions in the popular sense, and Lane was never issued a patent.

In 1848, Henry called Lane a genuine mathematical physicist. He may indeed have been one. He may also have been a man of ability and overly strong fixations —a crank, as some of his other contemporaries saw him. It is, from the surviving fragments, difficult to decide if Lane was a man whose achievements did not live up to his potential—as his 1870 paper suggests—or a man whose notions surpassed his abilities.

If Lane did not, in fact, achieve his potential, was it his failure, or society's? Mathematical physicists were rare in antebellum America, a fact usually ascribed to a hostile national environment (although mathematical physicists were also relatively rare in European countries). Most of the positions available to Lane were in astronomy and meteorology, and he did work in both fields. Men of promise in his day did not automatically think of the university, nor did universities seek him out.

In Washington, Lane found a small, active, and congenial scientific community that recognized his talents. The Patent Office post, somewhat incongruously, to some extent offered an opportunity for men of scientific interests to earn an income—although they were not given either equipment or time to pursue research. But even time and equipment were provided to Lane, in part by Henry and by the cooperation of government agencies. Even without equipment, he could have pursued theoretical work. Moreover, Henry, Newcomb, Benjamin Peirce, and others provided him with a stimulating intellectual environment.

Newcomb and Cleveland Abbe, in their accounts of Lane, implicitly raise the matter of his personality as a cause for his limited productivity. He is described as gnomelike and living in a garret. His poverty is ascribed to his support of various relatives, but at his death his estate was valued at more than $10,000, of which $8,000 was in gold bonds. If his appearance and style of living were bedraggled, it was hardly due to lack of funds. In writing of him, Lane's contemporaries stressed his "hesitation in speaking," due, perhaps, to a speech defect or to shyness. At any rate, Lane had difficulty in socializing, even in the friendly environment of scientific Washington, where he remained somewhat of an outsider.

When Lane died, five sealed envelopes of priority claims, going back to 1850, remained in the Smithsonian Institution. S. P. Langley, on his arrival in Washington in 1887, prevailed upon Henry's successor, S. F. Baird, to authorize their opening. When this was done, in the presence of Langley and Newcomb, the claims were pronounced worthless. Lane was forgotten and his 1870 paper remained an inexplicable pioneering accident.

BIBLIOGRAPHY

I. Original Works. A small collection of Lane's personal papers is in the records of the National Bureau of Standards in the U.S. National Archives in Washington, D.C. These contain scientific notes, Lane's accounts of his life and work, and letters both personal and professional. Other records of the National Bureau of Standards contain documents on Lane's activities. Of these the most significant are the reports itemizing the records and papers found in Lane's office in 1879 at the start of his terminal illness. The sealed envelopes are in the "Clippings File" of the Smithsonian Institution Archives. The file account of Lane's estate is in the custody of the Register of Wills of the District of Columbia. Both the account file and the list of Lane records prepared by the Office of Weights and Measures are very useful but frustrating; omnibus listings rather than detailed itemizations occur at a few crucial points. The correspondence in the Lane papers simply adds to the mysteries about him. For example, two different letters indicate that Michael Faraday wrote to Lane, presumably on his electrical work, yet no Faraday letters are in the collection.

II. SECONDARY LITERATURE. Upon Lane's death, J. E. Hilgard, the superintendent of the U.S. Coast and Geodetic Survey issued a statement before the Philosophical Society of Washington on Lane's career, in *Bulletin of the Philosophical Society of Washington*, **3** (8 May 1880), 122–124. This statement and information from W. B. Taylor of the Smithsonian are apparently the principal sources for Cleveland Abbe's "Memoir of Jonathan Homer Lane, 1819–1880," in *Biographical Memoirs. National Academy of Sciences*, **3** (1895), 253–264. There is little evidence for contacts between Abbe and Lane. Abbe read his memoir in 1892, many years after Lane's death and the printed version is replete with errors and doubtful assertions. When Simon Newcomb published his autobiography, *The Reminiscences of an Astronomer* (Boston–New York, 1903); the chapter subheading on Lane was, "A Forgotten Scientist." So forgotten was Lane that the index to the book does not even list him, but the interested reader is directed to pages 245–249.

See also Kelvin, "The Sun's Heat," in the *Proceedings of the Royal Institution of Great Britain*, **12** (1889), 1–21; for the critique of Kelvin, see R. Emden, *Gaskugeln, Anwendungen der Mechanischen Wärmetheorie* . . . (Leipzig–Berlin, 1907), 462–469. August Ritter's work originally appeared in various numbers of the *Annalen der Physik und Chemie* and is summed up in his *Anwendungen der mechanischen Wärmetheorie auf kosmologische Probleme* (Leipzig, 1878).

N. REINGOLD

LANG, ARNOLD (*b.* Oftringen, Switzerland, 18 June 1855; *d.* Zurich, Switzerland, 30 November 1914), *zoology.*

Lang was the son of Adolf Lang, a cotton mill owner, who was also enthusiastically engaged in local politics. The family belonged to the Reformed Church.

Lang completed primary school in 1867 and district school in 1870. He then attended the cantonal school in Aarau until 1873. Starting in March 1873 he studied science, especially zoology, in Geneva and then, from 1874 to 1876, in Jena. After receiving the doctorate at Jena (1876), he qualified as privatdocent for zoology at Bern on 26 May 1876. In 1878 and 1879 he was the Swiss representative at the zoological station in Naples, where he remained as an assistant until 1885. In November of that year he went to Jena as privatdocent. In 1886, at the initiative of Ernst Haeckel, he was given the newly created post there of Ritter professor of phylogenetic zoology.

In 1889 he accepted an appointment as full professor of zoology and comparative anatomy at the University of Zurich. In addition, he became professor of zoology at the Eidgenössische Technische Hochschule, took over the directorship of the zoological collections, and founded a zoological institute. Along with his teaching duties, Lang assumed many further responsibilities, including membership on the Zurich school council, and played an active role in Swiss scientific societies. In the last years of his life he successfully campaigned for the rebuilding of the University of Zurich. Poor health forced him to retire on 15 April 1914.

In religious matters Lang characterized himself as an agnostic freethinker. In 1887 he married Jeanne Mathilde Bachelin. They had one son and two daughters. Lang worked intensively, without long periods of relaxation. As a result his arteriosclerotic heart complaint steadily worsened. Humorous and sociable, he was also musically and artistically gifted.

Lang became a corresponding member of the Société des Médecins et Naturalistes de Jassy in 1888 and of the Academy of Natural Sciences of Philadelphia in 1893. Furthermore, he was elected a member of the Royal Society of Sciences at Uppsala in 1901 and *socius extraneus* of the Swedish Academy of Sciences in 1910. Lang was honorary member of many learned societies and received honorary doctorates from both of Zurich's universities.

Lang's interest in zoology was awakened in Geneva by Karl Vogt, who in 1874 gave him a letter of introduction to Haeckel in Jena. At the latter's suggestion Lang translated Lamarck's *Philosophie zoologique* into German. In later writings he repeatedly discussed Lamarck's theory and questions pertaining to the history of the theory of evolution.

Lang's zoological works grew out of his research at Naples under the direction of Dohrn and were devoted to such topics as sessile crustaceans and the comparative anatomy and histology of the nervous system of the platyhelminths. He also wrote a monograph on the polyclads (marine turbellarians). In his popular *Lehrbuch der vergleichenden Anatomie der wirbellosen Tiere* (1888–1894), which was translated into English and French, Lang provided a critical account of the results of his own work and of other original papers on the subject. Through his studies on annelid phylogeny, and especially through his derivation of metamerism and his trophocoel theory of the formation of the entire alimentary canal, Lang participated vigorously in the debate over the problem of the origin of the bodily cavities in general. Moreover, his hybridization experiments with species of the genus *Helix* confirmed Mendel's results. Finally, he established an important basis for experimental genetics with his presentation of the "Anfangsgründe der Biometrik der Variation und Korrelation," which constituted a section of the compilation he published in 1914 under the title *Die experimentelle Vererbungslehre in der Zoologie seit 1900*.

BIBLIOGRAPHY

I. ORIGINAL WORKS. Lang's writings include *Die Polycladen (Seeplanarien) des Golfes von Neapel und der angrenzenden Meeresabschnitte* (Leipzig, 1884); *Mittel und Wege phylogenetischer Erkenntnis* (Jena, 1887); *Lehrbuch der vergleichenden Anatomie der wirbellosen Tiere*, 4 pts. (Jena, 1888–1894; 2nd ed. 1901); *Beiträge zu einer Trophocoeltheorie* (Jena, 1903); *Die experimentelle Vererbungslehre in der Zoologie seit 1900* (Jena, 1914); and "Aus meinem intimen Schuldbuch," in H. Schmidt, ed., *Was wir Ernst Haeckel verdanken*, II (Leipzig, 1914), 259–265.

II. SECONDARY LITERATURE. See E. Haeckel, K. Hescheler, and H. Eisig, *Aus dem Leben und Wirken von Arnold Lang* (Jena, 1916); and G. Uschmann, *Geschichte der Zoologie und der zoologischen Anstalten in Jena 1779–1919* (Jena, 1959), 112–113, 177–182.

GEORG USCHMANN

LANG, KARL NIKOLAUS (*b.* Lucerne, Switzerland, 18 February 1670; *d.* Lucerne, 2 May 1741), *paleontology*.

Lang was a collector of fossils who gave original descriptions of many of the fossils of Switzerland. He was categorically opposed to the idea of their organic origin and particularly argued against the conception of the diluvialists that fossils were animals destroyed in the Flood.

His principal concern was with marine fossils. Confronted with difficulties stemming both from the similarity of these fossils to living animals and their presence on land, especially in the mountains of Switzerland, Lang adopted a view similar to that of Lhwyd. According to Lang the fossils originated from tiny, seminal seeds of living marine animals that were scattered around the earth by the air. Once distributed in this manner the seeds were carried into and through the earth by water. The heat of the earth activated a plastic force inherent in each seed, and the *aura seminalis*, or seminal breeze, gave the seed shape. Because this force was particularly strong in the icy waters and snow of the mountain tops, the fossils were more common in these areas.

Lang's fossil descriptions were used and his theories discussed by Beringer, and Lang is said to have been a colleague of Scheuchzer. Yet the closeness of his relationship with Scheuchzer is open to question since the latter was a diluvialist. Lang's works were well known both in Great Britain and on the Continent but were often severely criticized by diluvialists, one of whom, John Woodward, successfully opposed Lang's membership in the Royal Society of London.

Lang, who studied medicine in Bologna and Rome, held many official medical positions in the forest cantons of Switzerland. He was married to Maria Anna Meyer of Altishofen on 5 November 1708. Lang suffered a stroke in 1733 from which he never fully recovered although he continued to work on his fossil collections with the aid of his son.

BIBLIOGRAPHY

I. ORIGINAL WORKS. For a complete list of Lang's works, see Bachmann (below). His main writings are *Historia lapidum figuratorum Helvetiae, ejusque viciniae, in qua non solum enarrantur omnia eorum genera, species et vires aeneisque tabulis repraesentantur, sed insuper adducuntur eorum loca nativa, in quibus reperiri solent, ut cuilibet facile sit eos colligere modo adducta loca adire libeat* (Venice, 1708); and *Tractatus de origine lapidum figuratorum in quo diffuse disseritur, utrum nimirum sunt corpora marina a diluvio ad montes translata, et tractu temporis petrificata vel an a seminio quodam e materia lapidescente intra terram generentur, quibus accedit accurata diluvii, ejusque in terra effectuum descriptio cum dissertatione de generatione viventium, testaceorum praecipue, plurimorumque corporum, a vi plastica aurae seminalis hinc inde delatae extra consuetum matricem productorum* (Lucerne, 1709).

II. SECONDARY LITERATURE. A biography with a descriptive list of works is Hans Bachmann, "Karl Nikolaus Lang Dr. Phil. et Med. 1670–1741," in *Geschichtsfreund*, **51** (1896), 167–280. Helpful references to Lang's works are in *The Lying Stones of Dr. Beringer, Being His Lithographiae Wirceburgensis*, translated and annotated by Melvin E. Jahn and Daniel J. Woolf (Berkeley–Los Angeles, 1963). A letter from Woodward to Scheuchzer is in Melvin E. Jahn, "Some Notes on Dr. Scheuchzer and on Homo diluvii testis," in Cecil J. Schneer, ed., *Toward a History of Geology* (Cambridge, Mass., 1969), pp. 193–194.

PATSY A. GERSTNER

LANG, WILLIAM HENRY (*b.* Withyham, Groombridge, Sussex, England, 12 May 1874; *d.* Storth, near Milnthorpe, Westmorland, England, 29 August 1960), *botany*.

Lang's father, Thomas Bisland Lang, was a doctor; he died at the age of thirty-four, when his son was only two. His widow, who originally came from Ireland, moved to Bridge of Weir, then a small village fourteen miles from Glasgow, where her husband's parents had lived. Here Lang attended the village school before proceeding to Denniston School in Glasgow and thence to the University of Glasgow in 1889, when he was fifteen. He graduated B.Sc. in 1894 with honors in botany and zoology, and in 1895 M.B., C.M. (medical qualification) with high commendation. Although he registered as a doctor, he never practiced but became junior assistant in botany under Frederick Bower. A year later Lang was awarded a Robert Donaldson scholarship, which enabled him to work under D. H. Scott, keeper of the Jodrell Laboratory at Kew. In the following year Lang returned to Glasgow to become senior assistant in botany, and in 1900 he graduated D.Sc. At that time D. T. Gwynne-Vaughan was the junior assistant. In 1909 Lang left Glasgow to become Barker professor of cryptogamic botany at the University of Manchester. In the following year he married his cousin Elsa Valentine, who died childless in 1957. After retirement in 1940 Lang continued for a time to reside in Manchester, working in the Manchester Museum, but eventually moved to Storth near Milnthorpe in Westmorland, where he died.

Lang's botanical interests were greatly influenced by F. O. Bower, D. H. Scott, and Robert Kidston. Bower occupied the chair of botany at Glasgow University for forty years (1885–1925). According to Lang, Bower was an inspiring teacher, and in a biographical memoir (1949) Lang wrote: "To enter Bower's class for the first time was an arresting experience, as I found in 1890." Both Bower and Scott were greatly influenced by W. C. Williamson, whom they visited in 1889 at Manchester to study his sections of Carboniferous fossil plants. Henceforth Scott was devoted to paleobotany, while Bower determined to study the existing Pteridophyta, especially with regard to the problems of alternation of generations and the origin of a land flora.

Lang's earliest research on apogamous reproduction in ferns was carried out at Kew but had begun in Glasgow as a result of Bower's kindred interest in apospory. Bower favored the view that the sporophyte was a new development from the zygote interpolated between two sexual generations and developed by progressive sterilization. He stated this theory in a paper entitled "On Antithetic as Distinct From Homologous Alternation of Generations in Plants" (1890). D. H. Scott ardently supported the homologous theory, and Lang likewise pointed out that the examples of apogamy he described suggested that the two generations were not as distinct as the antithetic theory supposed.

Lang's interest in alternation of generations led him to investigate apospory in *Anthoceros laevis* and the prothallia in Lycopodiales and Ophioglossales. To obtain specimens he visited Ceylon and Malaya in 1899 and brought back *Helminthostachys zeylanica*, *Ophioglossum pendulum*, and *Psilotum*. In these the prothallia are saprophytic by virtue of a symbiotic fungus, whereas in *Lycopodium cernuum* the prothallium has green photosynthetic lobes, a condition Lang regarded as being more primitive. In 1909 Lang published a paper on a theory of alternation of generations based on ontogeny, and as a result a discussion on "alternation" was organized at the Linnean Society.

Apart from two papers on the microsporangia and ovules of *Stangeria paradoxa* (1897, 1900) all of Lang's earlier research was on living cryptogams, including the cone structure of *Lycopodium cernuum* (1908) and the anatomy and morphology of *Botrychium lunaria* (1913) and of *Isoëtes lacustris* (1915). This earlier period of research was abruptly concluded by the unexpected death of D. T. Gwynne-Vaughan, at the age of forty-four, in 1915. The latter had collaborated with Kidston on a study of fossil Osmundaceae; and together they had commenced a series of papers on the Lower Carboniferous flora of Berwickshire, of which only part I (on *Stenomyelon*) was published in 1912. In that year William Mackie, erstwhile schoolmaster and then medical practitioner in Elgin, discovered the plant-bearing cherts of Rhynie in a dry-stone wall during one of his frequent geological excursions to central Aberdeenshire. Kidston took in hand further work to locate the chert bed *in situ* by having trenches dug under the supervision of David Tait of the Geological Survey. According to Crookall (1938), Lang visited Kidston at Stirling in 1915 to discuss the possibility of continuing the investigation of the Lower Carboniferous petrified plants; but it was decided to defer this in order to describe the silicified plants of the Rhynie chert. This they did in five classic papers (1917–1921), and these ancient vascular plants (probably Lower Devonian) have now become familiar to all students of botany under their generic names of *Rhynia*, *Hornea* (now *Horneophyton*), and *Asteroxylon*.

In *Rhynia* and *Hornea* the plants had no roots nor leaves and possessed terminal sporangia. In *Asteroxylon* simple leaves were present, but the vascular traces did not enter the leaves. A new order, Psilophytales, was created for them. Of this joint work D. H. Scott said: "Never was a great discovery more completely and wisely expounded"; and John Walton said it was "the most important contribution of the century to our knowledge of early plant life." Further papers by Kidston and Lang followed on *Hicklingia* and *Palaeopitys* (1923), and on *Nematophyton* and *Pachytheca* (1924).

After Kidston's death in 1924 Lang continued the investigation of pre-Carboniferous plants until about

LANG

1945. With Isabel Cookson he described vascular plants from Australia that were apparently of Silurian age (1935), and with W. N. Croft he described Lower Devonian plants from Wales (1942). His last publications were obituaries of J. E. Holloway (1947) and F. O. Bower (1949).

BIBLIOGRAPHY

I. Original Works. Lang's works include "Studies in the Development and Morphology of Cycadean Sporangia. I. The Microsporangia of *Stangeria paradoxa*," in *Annals of Botany*, **11** (1897), 421–438; "On Apogamy and the Development of Sporangia Upon Fern Prothalli," in *Philosophical Transactions of the Royal Society*, **190B** (1898), 187–238; "The Prothallus of *Lycopodium clavatum L.*" in *Annals of Botany*, **13** (1899), 279–317; "Studies in the Development and Morphology of Cycadean Sporangia. II. The Ovule of *Stangeria paradoxa*," *ibid.*, **14** (1900), 280–306; "On Apospory in *Anthoceros laevis*," *ibid.*, **15** (1901), 503–510; "On the Prothalli of *Ophioglossum pendulum* and *Helminthostachys zeylanica*," *ibid.*, **16** (1902), 23–56; "On a Prothallus Provisionally Referred to *Psilotum*," *ibid.*, **18** (1904), 571–577; "On the Morphology of *Cyathodium*," *ibid.*, **19** (1905), 411–426; "On the Sporogonium of *Notothylas*," *ibid.*, **21** (1907), 203–210; "A Theory of Alternation of Generations in Archegoniate Plants Based Upon the Ontogeny," in *New Phytologist*, **8** (1909), 3–12; "Discussion on 'Alternation of Generations' at the Linnean Society," *ibid.*, **8** (1909), 104–116; and "On the Interpretation of the Vascular Anatomy of the Ophioglossaceae," in *Memoirs and Proceedings of the Manchester Literary and Philosophical Society*, **56**, no. 12 (1912), 1–15.

Other papers include "Studies in the Morphology and Anatomy of the Ophioglossaceae. I. On the Branching of *Botrychium lunaria*, with Notes on the Anatomy of Young and Old Rhizomes," in *Annals of Botany*, **27** (1913), 203–242; "Studies in the Morphology and Anatomy of the Ophioglossaceae. II. On the Embryo of *Helminthostachys*," *ibid.*, **28** (1914), 19–37; "Studies in the Morphology and Anatomy of the Ophioglossaceae. III. On the Anatomy and Branching of the Rhizome of *Helminthostachys zeylanica*," *ibid.*, **29** (1915), 1–54; "Studies in the Morphology of Isoëtes. I. The General Morphology of the Stock of *Isoëtes lacustris*," in *Memoirs and Proceedings of the Manchester Literary and Philosophical Society*, **59**, pt. 1 (1915), 1–28; and "Studies in the Morphology of Isoëtes. II. The Analysis of the Stele of the Shoot of *Isoëtes lacustris* in the Light of Mature Structure and Apical Development," *ibid.*, **59**, pt. 2 (1915), 29–56.

"On Old Red Sandstone Plants Showing Structure From the Rhynie Chert Bed, Aberdeenshire," written with R. Kidston, was published in five parts in *Transactions of the Royal Society of Edinburgh;* they are: "Part I. *Rhynia Gwynne-Vaughani*," **51** (1917), 761–784; "Part II. Additional Notes on *Rhynia Gwynne-Vaughani*," **52** (1920), 605–627; "Part III. *Asteroxylon Mackiei*," **52** (1920), 643–680; "Part IV. Restorations of the Vascular Cryptogams, and Discussion of Their Bearing on the General Morphology of the Pteridophyta and the Origin of the Organization of Land Plants," **52** (1921), 831–854; and "Part V. The Thallophyta Occurring in the Peat-bed; the Succession of the Plants Throughout a Vertical Section of the Bed, and the Conditions of Accumulation and Preservation of the Deposit," **52** (1921), 855–902.

The following articles were published in the *Transactions of the Royal Society of Edinburgh:* "On *Palaeopitys Milleri* M'Nab," written with R. Kidston, **53**, pt. 2 (1923), 409–417; "Notes on Fossil Plants From the Old Red Sandstone of Scotland. I. *Hicklingia Edwardi*, K. and L.," **53** (1923), 405–407, written with R. Kidston; "Notes on Fossil Plants From the Old Red Sandstone of Scotland. II. *Nematophyton Forfarense*, Kidston sp., III. On Two Species of *Pachytheca* (*P. media* and *P. fasciculata*) Based on the Characters of the Algal Filaments," **53** (1924), 603–614, written with R. Kidston.

"Contributions to the Study of the Old Red Sandstone Flora of Scotland" was published in seven parts in *Transactions of the Royal Society of Edinburgh:* "I. On Plant Remains From the Fish-beds of Cromarty" and "II. On a Sporangium-bearing Branch-system From the Stromness Beds," **54** (1925), 253–279; "III. On *Hostimella (Ptilophyton) Thomsoni*, and Its Inclusion in a new Genus, *Milleria*," "IV. On a Specimen of *Protolepidodendron* From the Middle Old Red Sandstone of Caithness," and "V. On the Identification of the Large 'Stems' in the Carmyllie Beds of the Lower Old Red Sandstone as *Nematophyton*," **54** (1926), 253–279; "VI. On *Zosterophyllum Myretonianum* Penh., and Some Other Plant Remains From the Carmyllie Beds of the Lower Old Red Sandstone," and "VII. On a Specimen of *Pseudosporochnus* from the Stromness Beds," **55** (1927), 448–456; and "VIII. On *Arthrostigma*, *Psilophyton*, and Some Associated Plant-Remains From the Strathmore Beds of the Caledonian Lower Old Red Sandstone," **57** (1932), 491–521.

See also "Some Fossil Plants of Early Devonian Type From the Walhalla Series, Victoria, Australia," in *Philosophical Transactions*, **219B** (1930), 133–163, written with I. C. Cookson; "On the Spines, Sporangia, and Spores of *Psilophyton princeps*, Dawson, Shown in Specimens from Gaspé," *ibid.*, **219B** (1931), 421–442; "On a Flora, Including Vascular Land Plants, Associated With *Monograptus*, in Rocks of Silurian Age, From Victoria, Australia," *ibid.*, **224B** (1935), 421–449, written with I. C. Cookson; "On the Plant-remains From the Downtonian of England and Wales," *ibid.*, **227B** (1937), 245–291; and "The Lower Devonian Flora of the Senni Beds of Monmouthshire and Breconshire," *ibid.*, **231B** (1942), 131–168, written with W. N. Croft.

See also "Obituary. John Ernest Holloway 1881–1945," in *Obituary Notices of Fellows of the Royal Society*, **5** (1947), 425–444; "Obituary. Frederick Orpen Bower 1855–1948," *ibid.*, **6** (1949), 347–374.

II. Secondary Literature. For essential details of Lang's life and work, see E. J. Salisbury, "Obituary. William Henry Lang 1874–1960," in *Biographical Memoirs of the Fellows of the Royal Society*, **7** (1961), 147–160, with a complete bibliography; see also F. O. Bower, "On Antithetic as Distinct From Homologous Alternation of Generations in Plants," in *Annals of Botany*, **4** (1890), 347–370; R. Crookall, "The Kidston Collection of Fossil Plants," in *Memoirs of the Geological Survey of the United Kingdom*, **5** (1938); and R. Kidston and D. T. Gwynne-Vaughan, "On the Carboniferous Flora of Berwickshire. Part 1. *Stenomyelon tuedianum* Kidston," in *Transactions of the Royal Society of Edinburgh*, **48** (1912), 263–271.

ALBERT G. LONG

LANGE, CARL GEORG (*b*. Vordingborg, Denmark, 4 December 1834; *d*. Copenhagen, Denmark, 29 May 1900), *medicine, psychology.*

Lange's father, Frederik Lange, was professor of education at Copenhagen University; his mother, the former Louise Paludan-Müller, came from a learned family.

After graduating from the Copenhagen Metropolitan School in 1853, he studied medicine at Copenhagen University; after receiving the M.D. in 1859 he worked until 1867 as an intern in the medical departments of the Royal Frederiks Hospital and the Almindelig Hospital in Copenhagen. He published studies on ulcerous endocarditis and typhoid fever and an excellent description of the symptomatology and occurrence of rheumatic fever based on 1,900 cases. In 1863 Lange was sent to Greenland and reported on the widespread distribution of tuberculosis there. In 1867–1868 he studied histology in Zurich and experimental physiology in Florence with Moritz Schiff, who in 1856 had demonstrated the vasoconstrictor fibers in the cervical sympathetic segment. Schiff aroused Lange's interest in vasomotor reactions and in neurophysiology. At Florence, Lange published an experimental study concerning curare's influence on the nervous system.

After his return to Copenhagen, Lange was prosector at the Royal Frederiks Hospital and municipal health officer; he also had a private practice. In 1866 he had become coeditor of *Hospitalstidende*, in which many of his pioneer studies were published. In 1866 he was the first to describe acute bulbar paralysis; in 1870 he wrote on symptoms arising from cerebellar tumors; and in 1872 he demonstrated the secondary degeneration of the posterior columns of the spinal cord caused by spinal meningitis, thereby anticipating the later neuron doctrine. His discovery was unnoticed until 1894, when Jean Nageotte made the same findings; it was fully accepted in 1897 in C. W. Nothnagel's *Internal Pathology*. In 1873 Lange published his anatomical-clinical investigations on chronic myelitis, dividing the syndromes into those of the anterior horns with atrophy, in the lateral tracts with paraplegia and in the posterior tracts with root pains and ataxia. It was a very clear and really new point of view—but because it was written in Danish, it did not obtain the distribution and significance it deserved.

From 1869 to 1872 Lange lectured at Copenhagen University on pathology of the spinal cord. The lectures were published as *Forelaesninger over rygmarvens patologi*, which contains physiologically inspired descriptions of the various syndromes of paralysis, sensibility disturbances, and reflex phenomena. There are chapters on pain, hyperesthesia, and eccentric perceptions. His ideas of reflex pain, angina pectoris, and projected pain were later emphasized by Head and Wernøe.

In 1873 Lange failed to obtain the position of physician-in-chief in medicine at the Royal Frederiks Hospital, but 1875 he was appointed lecturer in pathological anatomy and in 1885 became professor of the subject at Copenhagen University. Despite very bad working conditions he continued scientific studies, most of them based on extensive clinical material from his private practice with nervous patients. In 1885 he published *Om Sindsbevaegelser*, a psychophysiological study on vasomotor disturbances and conditioned reflexes during periods of emotional stress. Excitement was the result of vasomotor manifestations and not of mental entities—a theory still known by psychologists as the James-Lange theory.

In 1886 Lange published *Periodiske depressionstilstande*, which separated periodic depressive conditions from the neurasthenic. Since he believed the depressions were caused by uric acid diathesis, his theory was attacked not only by psychiatrists but also by internists. His ideas found several defenders, however, especially in France. In 1899 Lange published *Bidrag til nydelsernes fysiologi*, a study of the pleasurable sensations in emotions. The book met with indignation—his explanations of vasomotor reactions during sympathetic reactions to the perception of beauty disturbed the ideas of aesthetes and philosophers.

As a member of several committees for public hygiene and hospital service and of the City Council, Lange procured reforms in vaccination, school hygiene, hospital buildings, and water supply. He was a member of the board of Medicinsk Selskab and a founder of Biologisk Selskab. As secretary-general of the International Congress of Physicians held at Copenhagen in 1884, he made possible the meeting of such people as Pasteur, Virchow, James Paget, and Donders with the rather provincial Danish medical profession.

BIBLIOGRAPHY

I. ORIGINAL WORKS. A full list of Lange's works is in Knud Faber, *Erindringer om C. Lange*, pp. 61–66. Translations include *Ueber Gemüthsbewegungen. Eine psychologisch-physiologische Studie*, Hans Kurella, trans. (Leipzig, 1887; 2nd ed., 1910); *Les émotions*, G. Dumas, trans. (Paris, 1895; 2nd ed., 1902); *Boden der harnsäure Diathese*, H. Kurella, ed. (Hamburg, 1896); *Sinnesgenüsse und Kunstgenüss*, L. Loewenfeld and H. Kurella, eds. (Wiesbaden, 1903); and *Psychology Classics*, William James, trans. (Boston, 1922).

II. SECONDARY LITERATURE. See Knud Faber, *Erindringer om C. Lange* (Copenhagen, 1927); Edvard Gotfredsen, *Medicinens Historie* (Copenhagen, 1964), pp. 427, 517, 522–523; and P. Bender Petersen, "La description de réflexes conditionnels par C. Lange," in E. Dein, *Sct Hans Hospital 1816–1966* (Copenhagen, 1966), pp. 188–192.

E. SNORRASON

LANGERHANS, PAUL (*b.* Berlin, Germany, 25 July 1847; *d.* Funchal, Madeira, 20 July 1888), *anatomy, pathology.*

Langerhans' father, for whom he was named, was a well-known physician in Berlin; and two younger stepbrothers were also physicians. One of them, Robert (1859–1904), was an assistant to Virchow and later became professor of pathology. Langerhans attended the Gymnasium zum Grauen Kloster in Berlin and graduated at the age of sixteen. From 1865 to 1866 he studied medicine at the University of Jena, where he was much impressed by Haeckel and Gegenbaur. He continued his medical studies in Berlin under K. Bardeleben, E. du Bois-Reymond, R. Virchow, J. Cohnheim, and F. T. von Frerichs. In 1869 he graduated M.D. At Berlin he was particularly influenced by Virchow and Cohnheim, and he became later a close friend of Virchow's. His first important research was done in Virchow's laboratory, where he discovered the cell islands of the pancreas named after him.

In 1870 Langerhans accompanied the geographer Heinrich Kiepert on an expedition to Egypt and Palestine. During the Franco-Prussian War he joined the German army as a physician and worked in a military hospital. After the conclusion of peace he went for a short while to Leipzig to see Ludwig's famous physiological institute and the obstetrical clinics of K. S. F. Crédé. In 1871 he was offered the position of prosector in pathology at the University of Freiburg im Breisgau, where he also became *Privatdozent* in the same year and, later, associate professor. In 1874 tuberculosis of the lung compelled Langerhans to interrupt his academic career. Attempts at cure in Switzerland, Italy, and Germany failed; and in 1875 he settled in Madeira. Its mild climate led to an improvement in his health. In Madeira he later practiced medicine in the capital, Funchal, where he died of a kidney infection.

Langerhans' main scientific achievements consist in his studies of human and animal microscopical anatomy. In this field he was among the first successful investigators to explore the new area of research with novel methods and staining techniques. Langerhans made his first scientific contributions as a medical student in Virchow's laboratory, under the guidance of Virchow and especially of Cohnheim. In 1868 he published a paper on the innervation of the skin; he had used gold chloride as a stain. He was able to show nerve endings in the Malpighian layer of the epidermis and described characteristic oblong bodies with branching processes in the Malpighian layer (later called Langerhans' cells). He believed that these bodies were probably nerve cells, but he did not exclude the possibility that they might be pigment cells (later called dendritic cells).

In his inaugural dissertation (1869) Langerhans immortalized his name by his discovery of characteristic cell islands in the pancreas named for him. The investigations for this work, "Beiträge zur mikroskopischen Anatomie der Bauchspeicheldrüse," were conducted in Virchow's pathology laboratory. The paper presented the first careful and detailed description of the microscopic structure of the pancreas.

Langerhans examined mainly the pancreas of rabbits. He studied fresh tissue and even sought to make microscopic observations of pancreatic tissue in the living animal. His main results were obtained through chemical fixation and maceration. Langerhans injected the pancreatic duct with Berlin blue in order to show the branching and the structure of the excretory system. Examining portions of the fresh gland, he distinguished three zones in the secreting cells: an apical granular zone, the zone of the cell nucleus, and a clear basal zone. In macerated pancreatic tissue he differentiated various types of cells, including small, irregularly polygonal cells without granules. These cells formed numerous spots scattered through the parenchyma of the gland measuring 0.1–0.24 mm. in diameter. Langerhans refrained from making any hypothesis as to the nature and significance of these cells. In 1893 the French histologist G. E. Laguesse named these cell spots "îlots de Langerhans"; the insulin-secreting function of these cells was established later.

Another important contribution, made with F. A. Hoffmann in Virchow's laboratory, dealt with the macrophage system. Langerhans and Hoffmann studied the intravital storage of cinnabar injected intravenously in rabbits and guinea pigs. After the injection they sought the presence of cinnabar in the blood and all the organs except the spleen. They were able to show that cinnabar was taken up by white blood corpuscles but never by the red. They also demonstrated the deposit of cinnabar in fixed cells of the bone marrow, in the capillary system, and in the connective tissue of the liver. This was one of the pioneer investigations that later led to Aschoff's concept of the reticuloendothelial system.

In 1873 Langerhans studied the structure of the skin and its innervation, staining the preparations with osmic acid and picrocarmine. He observed the nerve fibers branching in the interior of Meissner's corpuscles and described the oblong cells lying transversely in these corpuscles. In the report on this study Langerhans described his discovery of the granular cells in the exterior portion of the Malpighian layer (Langerhans' layer, *stratum granulosum*). In 1873 and 1874 Langerhans examined the microscopic structure of the cardiac muscle fibers in vertebrates and of the accessory genital glands of man. One of his last extensive histological papers (1876) was dedicated to the anatomy of the amphioxus.

During his trip to Palestine, Langerhans, stimulated by Virchow's interest in anthropology, made skull measurements and anthropological observations of the population of Palestine. In the last decade of his life, while practicing medicine on Madeira, Langerhans studied the etiology of tuberculosis and cautiously supported Virchow's dualistic view, which differentiated phthisis from tuberculosis. During that time he also published a handbook on Madeira dealing with the climatic and curative properties of that island.

BIBLIOGRAPHY

I. Original Works. A comprehensive list of Langerhans' anatomical works was compiled by K. Bardeleben, in *Anatomischer Anzeiger*, **3** (1888), 850–851. A partial list is by H. Morrison, in *Bulletin of the Institute of the History of Medicine*, **5** (1937), 266.

On microscopic anatomy see "Ueber die Nerven der menschlichen Haut," in *Virchows Archiv für pathologische Anatomie*, **44** (1868), 325–337; "Zur pathologischen Anatomie der Tastkörper," *ibid.*, **45** (1869), 413–417; "Ueber den Verbleib des in die Circulation eingeführten Zinnobers," *ibid.*, **48** (1869), 303–325; "Beiträge zur mikroskopischen Anatomie der Bauchspeicheldrüse," his inaugural dissertation (Berlin, 1869), repr., with English trans. and introductory essay by H. Morrison in *Bulletin of the Institute of the History of Medicine*, **5** (1937), 259–297; "Ein Beitrag zur Anatomie der sympathischen Ganglienzellen," his *Habilitationsschrift* (Freiburg im Breisgau, 1871); "Zur Histologie des Herzens," in *Virchows Archiv für pathologische Anatomie*, **58** (1873), 65–83; "Ueber mehrschichtige Epithelien," *ibid.*, 83–92; "Ueber die accessorischen Drüsen der Geschlechtsorgane," *ibid.*, **61** (1874), 208–228; "Ueber Tastkörperchen und Rete Malpighi," in *Archiv für mikroskopische Anatomie*, **9** (1873), 730–744.

Zoological works include "Zur Entwicklung der Gastropoda opisthobranchia," in *Zeitschrift für wissenschaftliche Zoologie*, **23** (1873), 171–179; and "Zur Anatomie des Amphioxus lanceolatus," in *Archiv für mikroskopische Anatomie*, **12** (1876), 290–348.

A work in pathology is "Zur Aetiologie der Phthise," in *Virchows Archiv für pathologische Anatomie*, **97** (1884), 289–306.

On anthropology see "Ueber die heutigen Bewohner des heiligen Landes," in *Archiv für Anthropologie*, **6** (1873), 39–58, 201–212; and *Handbuch für Madeira* (Berlin, 1885).

II. Secondary Literature. The most detailed biographical sketch of Langerhans and evaluation of his work is H. Morrison, in *Bulletin of the Institute of the History of Medicine*, **5** (1937), 259–267, with portrait. See also K. Bardeleben, "Paul Langerhans," in *Anatomischer Anzeiger*, **3** (1888), 850; W. R. Campbell, "Paul Langerhans 1847–1888," in *Canadian Medical Association Journal*, **79** (1958), 855–856; and G. Wolff, "Beiträge berühmter Studenten zur Erforschung des Zuckerstoffwechsels," in *Münchener medizinische Wochenschrift*, **102** (1960), 1203–1208.

Nikolaus Mani

LANGEVIN, PAUL (*b.* Paris, France, 23 January 1872; *d.* Paris, 19 December 1946), *physics.*

Langevin, the second son of Victor Langevin, an appraiser-verifier in the Montmartre section of Paris, very early displayed his liking for study. His mother, great-grandniece of the alienist Philippe Pinel, encouraged this inclination; and Langevin was always first in his class from the time he entered the École Lavoisier until he left the École Municipale de Physique et Chimie Industrielles de la Ville de Paris in 1891. (The latter school was established in 1881 by Paul Schützenberger to train engineers.) Langevin's enthusiasm was aroused by his contact with the school's director and by his laboratory work, which was supervised by Pierre Curie.

To further his knowledge Langevin attended the Sorbonne (1891–1893) while teaching a private course and learning Latin on his own. In 1893 he placed first in the competitive entrance examination for the École Normale Supérieure, but he did a year of military service before attending the school. At the École Normale Supérieure he heard the lectures of Marcel Brillouin and undertook research with Jean Perrin (then an *agrégé-préparateur*). Langevin placed first in the competition for the *agrégation* in physical sciences in 1897 and left for Cambridge to spend a year at the Cavendish Laboratory with J. J. Thomson. Under Thomson's direction, he worked on ionization by X rays, in the process discovering, independently of Sagnac, that X rays liberate secondary electrons from metals. Also while at the Cavendish he met J. Townsend, E. Rutherford, and C. T. R. Wilson; all of them soon became friends.

Upon returning to Paris, Langevin established a home (1898). He had four children: Jean (*b.* 1899), André (*b.* 1901)—both of whom became physicists—Madeleine (*b.* 1903), and Hélène (*b.* 1909). Still on scholarship, he was obliged to continue to give private lessons.

During this period the atmosphere in the Paris laboratories was one of intense excitement. At Jean Perrin's laboratory Langevin continued his investigations of the secondary effects of X rays and, on close terms with the Curies, he was present at the birth of the study of radioactivity. Langevin completed his doctoral dissertation in 1902 at the Sorbonne. It dealt with ionized gases and was based on investigations he had begun at Cambridge. After being named *préparateur* to Edmond Bouty at the Sorbonne, Langevin entered the Collège de France in 1902 to substitute for E. E. N. Mascart, whom he replaced in 1909. Meanwhile, he taught also at the École Municipale de Physique et Chimie, succeeding Pierre Curie (1904), and then at the École Nationale Supérieure de Jeunes Filles (Sèvres), replacing Marie Curie, who had been widowed (1906). Langevin loved teaching, and he excelled at it.

In his laboratory at the Collège de France, Langevin continued to study ions in gases, liquids, and dielectrics (1902–1913). In this work he was assisted by his students, including Edmond Bauer, Eugene Bloch, and Marcel Moulin. In his dissertation he had already given a method of calculating the mobility of both positive and negative ions during their passage through a condenser by considering their diffusion and recombination. Moreover, for the first time he communicated his results concerning secondary X rays (1898). Langevin was never in a hurry to publish; his written work is scanty in relation to the extent of his work. Whether it was a question of theory, of experimental results, or even of techniques or apparatus, he spent a long time seeking a simple and clear statement; often his publisher would snatch from him a manuscript filled with changes written in his clear, firm hand.

Langevin's position at the Collège de France was of particular importance to his development, for it freed him to lecture on subjects for which the standard French curriculum had little place. Although he continued after his arrival there to involve himself deeply with the experimental work of students, his own research and teaching turned increasingly to contemporary problems in theoretical physics. For most of the thirty years after he assumed the chair, he was the leading, and at times virtually the only, practitioner and expositor of modern mathematical physics in France. Einstein caught his role and status precisely when he wrote:

> Langevin's scientific thought displayed an extraordinary clarity and vivacity combined with a quick and sure intuition for the essential point. Because of those qualities, his courses exerted a decisive influence on more than a generation of French theoretical physicists. . . . It seems to me certain that he would have developed the special theory of relativity if that had not been done elsewhere, for he had clearly recognized its essential points.[1]

The last portion of that evaluation is in part a response to Langevin's first published theoretical papers, presented during 1904 and 1905. They dealt perceptively and authoritatively with a coherent set of current problems developed in the work of Lorentz, Larmor, and Abraham: the concept of electromagnetic mass, its rate of increase with velocity, and the related contraction hypotheses which suggested the impossibility in principle of determining the earth's motion through the ether. Both reports from his students and the speed with which he assimilated the special theory of relativity after 1905 suggest in addition that Langevin's own thoughts, at least on the relation between mass and energy, were developing along lines close to Einstein's before the latter's work appeared in 1905.[2]

That same year is the one in which Langevin published what was perhaps his most original and enduring contribution to physical theory, a quantitative account of paramagnetism and diamagnetism which demonstrated, he said, that it was "possible, using the electron hypothesis, to give precise meaning to the ideas [molecular models] of Ampère and Weber."[3]

To account for paramagnetism Langevin assumed that each molecule had a permanent magnetic moment m due to the circulation of one or more electrons. In the absence of an external field, thermal motion would orient the moments of individual molecules at random, so that there would be no net field. An external field, however, would tend to align molecular moments, the extent of the alignment depending both on the field strength and on the intensity of the thermal motion, the latter determined by temperature. Applying Boltzmann's techniques to the problem,

Langevin showed that for low fields the magnetic permeability of a gas should be given by $\mu = m^2N/3kT$, where N is the number of molecules per unit volume, k is Boltzmann's constant, and T is the absolute temperature.

The proportionality of susceptibility to the reciprocal of temperature was, Langevin emphasized, a result which Pierre Curie had found experimentally in 1895. Using the latter's measurement of the proportionality constant, he noted further that the observed susceptibility of oxygen could be accounted for by the orbital motion of even a single electron with velocity $2 \cdot 10^8$ cm/sec. At this time atoms were usually thought to consist of many hundreds of electrons. That so few were needed to explain magnetic properties suggested to Langevin that the electrons with this function might well be the superficial ones, i.e., the valence electrons, which were responsible also for chemical properties. Niels Bohr, whose route to the quantized version of Rutherford's atomic model was deeply influenced by Langevin, was to suggest precisely the same correlation.[4]

The even more puzzling phenomenon of diamagnetism Langevin explained in terms of molecules within which the orbital electronic motions canceled each other, so that no net molecular moment remained. An increasing magnetic field would, however, accelerate the electrons moving in one direction and retard those moving in the other, thus producing a small net moment in a direction opposed to the field. Again Langevin's treatment was quantitative. It predicted that, as Curie had found, diamagnetic susceptibility should be independent of temperature, and it permitted computation of plausible values for the radii of electronic orbits. As a tool for investigating both magnetism and molecular and atomic structure, Langevin's impressive theory was vigorously developed by a number of physicists, especially Pierre Weiss, and Langevin was himself invited to discuss its current state at the famous first Solvay Congress in 1911.

Three years before that, in 1908, Langevin, whose skill in kinetic theory had first been developed when he worked on ionic transport, turned briefly to the theory of Brownian motion developed by Einstein in 1905 and, via a more direct route, by Smoluchowski in 1906. The result was simplified, still-standard treatment which, unlike Smoluchowski's, produced precisely Einstein's formula for the mean-square displacement. Subsequently Langevin took up the subject of thermodynamics and reconsidered its basic notions, starting from theories of Boltzmann and of Planck ("the physics of the discontinuous") in 1913. At the same time he presented "the notions of time, space, and causality" with their relativistic significance. It required much difficult work to arrive at these elucidations; but he presented them simply, sometimes humorously (for example, Jules Verne's cannonball [1911] and Langevin's rocket or cannonball [1912]).

Although Langevin concerned himself with philosophical questions, he did not neglect the technical applications of his work. In 1914 he was called upon to work on ballistic problems and was later requested by Maurice de Broglie, his friend and former student, to find a way of detecting submerged enemy submarines. Lord Rayleigh and O. W. Richardson (1912) had thought of employing ultrasonic waves. In France a Russian engineer, Chilowski, proposed to the navy a device based on this principle; but its intensity was much too weak. In less than three years Langevin succeeded in providing adequate amplification by means of piezoelectricity. His team called the steel-quartz-steel triplet he developed a "Langevin sandwich." Functioning by resonance, it "finally played for ultrasonic waves the same roles as the antenna in radio engineering." Langevin continued to do important work in acoustics and ultrasonics after the war.

Langevin received many honors. In 1915 he was honored by the Royal Society of London, and in 1928 he became a member of that body (he had still not been elected to the Académie des Sciences). He was elected to many other foreign academies and to the Académie de Marine in Paris. His relativistic views still appeared revolutionary: he had invited Michelson and Einstein to speak at the Collège de France as early as 1922. Then came the theories of Louis de Broglie. Langevin, at first surprised, soon became their strongest advocate (1924). The Académie des Sciences elected Louis de Broglie in 1933 and Langevin in 1934. In writing his "Notice," Langevin relived forty years during which he had constantly contributed to deepening our understanding of the universe.

Internationally Langevin's influence became paramount in 1928, when he succeeded H. A. Lorentz as president of the Solvay International Physics Institute, of which he had been a member since 1921. On his initiative a message of sympathy was sent in 1933 to Einstein, who was already being persecuted by the Nazis.

As passionate in his concern for justice as in his quest for truth, in this period Langevin joined various movements supporting victims of fascism and, denouncing the horrors of war, actively participated in campaigns aimed at securing peace. To his mind the same enemy—reaction—opposed the new scientific theories, the modernization of teaching, individual liberty, and the spirit of brotherhood.

When war broke out, Langevin testified in favor of the forty-four Communist deputies excluded from their seats following the signing of the German-Soviet pact. In March–April 1940 he was invited by the navy to direct research on ultrasonic depth finders.

After the departure of the French government from Paris, Langevin again became director of the École Municipale de Physique et Chimie Industrielles (the duties of that office having already been delegated to him); but on 30 October 1940 he was arrested by the Wehrmacht. Dismissed by the Vichy government and imprisoned in Fresnes, he was finally placed under house arrest in Troyes. Messages of sympathy came to him from all over the world. Peter Kapitza invited Langevin to join him in the Soviet Union, but Langevin refused to leave France. Resigned, and surrounded by devoted friends, he resumed his calculations. He then learned of the execution of his son-in-law, the physicist Jacques Solomon (1942), and of the arrest and deportation of his daughter, Hélène Solomon-Langevin (1943). Fearing for his safety, his young friends Frédéric Joliot, H. Moureu, Denivelle, and P. Biquard persuaded Langevin to flee in May 1944. Warmly welcomed in Switzerland, he worked on educational reform for postwar France.

As early as 1904 Langevin had denounced the obstacles to progress found in scientific instruction in the form of "ossifying dogmas" that hinder the recognition of "fruitful principles"—such as rational mechanics vis-à-vis the atomic theory. Faithful to his own thought, he reprinted this article ("L'esprit de l'enseignement scientifique") in 1923. He took up the same theme again in the hope that liberated France would direct its youth along progressive paths in thought and action, and inculcate in them the notions indispensable to philosophers and technicians alike.

After returning to the École Municipale de Physique et Chimie Industrielles in October 1944, Langevin devoted his greatest efforts to educational reforms and to the support of his political friends. His daughter Hélène, returned from Auschwitz, sat in the Assemblée Consultative. He joined her as a member of the Communist Party—several members of which were also members of the government—in the hope of encouraging a brotherhood that capitalism had not succeeded in establishing.

Langevin died following a brief illness. The government, which had made him a grand officer of the Legion of Honor, accorded him a state funeral. His remains were transferred to the Pantheon in 1948, at the same time as those of Jean Perrin.

NOTES

1. A. Einstein, "Paul Langevin," in *La pensée*, **12** (May-June 1947), pp. 13–14.
2. E. Bauer, *L'électromagnétisme hier et aujourd'hui* (Paris, 1949), p. 156 n.
3. P. Langevin, "Magnétisme et théorie des électrons," in *Annales de chimie et de physique*, **5** (1905), 70–127; quotation from end of introduction.
4. Cf. John L. Heilbron and Thomas S. Kuhn, "The Genesis of the Bohr Atom," in *Historical Studies in the Physical Sciences*, **1** (1969), 211–290.

BIBLIOGRAPHY

I. Original Works. Langevin's writings on X rays and ionization of gas include "Recherches sur les gaz ionisés," his doctoral dissertation (1902), in *Annales de chimie et de physique*, **28** (1903), 289, 433; "Sur les rayons secondaires des rayons de Röntgen," his doctoral dissertation (1902), *ibid.*, p. 500; "Recherches récentes sur le mécanisme du courant électrique. Ions et électrons," in *Bulletin de la Société internationale des électriciens*, 2nd ser., **5** (1905), 615; "Recherches récentes sur le mécanisme de la décharge disruptive," *ibid.*, **6** (1906), 69; "Sur la recombinaison des ions dans les diélectriques," in *Comptes rendus . . . de l'Académie des sciences*, **146** (1908), 1011; "Mesure de la valence des ions dans les gaz," in *Radium*, **10** (1913), 113; and "Sur la recombinaison des ions," in *Journal de physique*, 8th ser., **6** (1945), 1.

In the following citations *Physique* refers to *La physique depuis vingt ans* (Paris, 1923). On ions in the atmosphere and particles in suspension, see "Interprétation de divers phénomènes par la présence de gros ions dans l'atmosphère," in *Bulletin des séances de la Société française de physique*, fasc. 4 (19 May 1905), 79; and "Électromètre enregistreur des ions de l'atmosphère," in *Radium*, **4** (1907), 218, written with M. Moulin.

Kinetic theory and thermodynamics are treated in "Sur une formule fondamentale de la théorie cinétique," in *Comptes rendus . . . de l'Académie des sciences*, **140** (1905), 35, also in *Annales de chimie et de physique*, 8th ser., **5** (1905), 245; "Sur la théorie du mouvement Brownian," in *Comptes rendus . . . de l'Académie des sciences*, **146** (1908), 530; and "La physique du discontinu," in *Les progrès de la physique moléculaire* (Paris, 1914), p. 1, also in *Physique*, p. 189.

On electromagnetic theory and electrons, see "La physique des électrons," in *Rapport du Congrès international des sciences et arts à Saint-Louis* (1904), also in *Physique*, p. 1; "La théorie électromagnétique et le bleu du ciel," in *Bulletin de la Société française de physique*, fasc. 4 (16 Dec. 1910), 80; "Les grains d'électricité et la dynamique électromagnétique," in *Les idées modernes sur la constitution de la matière* (Paris, 1913), p. 54, also in *Physique*, p. 70; and "L'électron positif," in *Bulletin de la Société des électriciens*, 5th ser., **4** (1934), 335.

Writings on magnetic theory and molecular orientation include "Magnétisme et théorie des électrons," in *Annales de chimie et de physique*, 8th ser., **5** (1905), 70; "Sur les biréfringences électrique et magnétique," in *Radium*, **7** (1910), 249; "La théorie cinétique du magnétisme et les magnétons," presented at the Solvay Conference in 1911, in *La théorie du rayonnement et les quanta* (Paris, 1913), also in *Physique*, p. 171; "Sur l'orientation moléculaire," a letter to M. W. Voigt, in *Göttingen Nachrichten*, no. 5 (1912), 589; and "Le magnétisme," in *Sixième Congrès de physique Solvay* (Paris, 1923), p. 352.

The principle of relativity and the inertia of energy are discussed in "Sur l'impossibilité de mettre en évidence le mouvement de translation de la terre," in *Comptes rendus . . . de l'Académie des sciences*, **140** (1905), 1171; "L'évolution de l'espace et du temps," in *Scientia*, **10** (1911), 31, also in *Physique*, p. 265; "Le temps, l'espace et la causalité dans la physique moderne," in *Bulletin de la Société française de philosophie*, **12**, no. 1 (Jan. 1912), 1, also in *Physique*, p. 301; "L'inertie de l'énergie et ses conséquences," in *Journal de physique*, 5th ser., **3** (1913), 553, also in *Physique*, p. 345; "Sur la théorie de la relativité et l'expérience de M. Sagnac," in *Comptes rendus . . . de l'Académie des sciences*, **173** (1921), 831; "La structure des atomes et l'origine de la chaleur solaire," in *Bulletin de l'Université de Tiflis*, **10** (1929); "La relativité," in *Exposés et discussions du Centre de synthèse* (Paris, 1932); "L'oeuvre d'Einstein et l'astronomie," in *L'astronomie*, **45** (1931), 277; "Déduction simplifiée du facteur de Thomas," in *Convegno di fisica nucleare* (Rome, 1931), p. 137; "Espace et temps dans un univers euclidien," in *Livre jubilaire de Marcel Brillouin* (Paris, 1935); and "Résonance et forces de gravitation," in *Annales de physique*, **17** (1942), 261.

On physical chemistry and radioactivity, see "Sur la comparaison des molécules gazeuses et dissoutes," in *Comptes rendus . . . de l'Académie des sciences*, **154** (1912), 594, also in *Procès-verbaux des commissions de la Société française de physique* (19 Apr. 1912), 54; "L'interprétation cinétique de la pression osmotique," in *Journal de chimie-physique*, **10** (1912), 524, 527; and "Sur un problème d'activation par diffusion," in *Journal de physique*, 7th ser., **5** (1934), 57.

Writings on magnitudes and units include "Notions géométriques fondamentales," in *Encyclopédie des sciences mathématiques*, IV, pt. 5, fasc. 1 (1912), 1; and "Sur les unités de champ et d'induction," in *Bulletin de la Société française de physique* (17 Feb. 1922), 33.

Classical and modern mechanics are discussed in "Sur la dynamique de la relativité," in *Procès-verbaux des commissions de la Société française de physique* (15 Dec. 1921), 97, also in *Exposés et discussions du Centre international de synthèse sur la relativité* (Paris, 1932); "Les nouvelles mécaniques et la chimie," in *L'activation et la structure des molécules* (Paris, 1929), p. 550; "La notion de corpuscules et d'atomes," in *Réunion internationale de chimie-physique* (Paris, 1933); and "Sur les chocs entre neutrons rapides et noyaux de masse quelconque," in *Annales de physique*, **17** (1942), 303, also in *Comptes rendus . . . de l'Académie des sciences*, **214** (1942), 517, 867, 889.

On acoustics and ultrasonics, see "Procédés et appareils pour la production de signaux sous-marins dirigés et pour la localisation à distance d'obstacles sous-marins . . .," in *Brevet français*, no. 502.913 (29 May 1916), written with M. C. Chilowski; no. 505.703 (17 Sept. 1918); no. 575.435 (27 Dec. 1923), written with M. C. Florisson; and no. 576.281 (14 Jan. 1924), 1st supp. no. (1 Mar. 1924) and 2nd supp. no. (16 Oct. 1924).

See also "Note sur l'énergie auditive," in *Publications du Centre d'études de Toulon* (25 Sept. 1918); "Émission d'un faisceau d'ondes ultra-sonores," in *Journal de physique*, 6th ser., **4** (1923), 537, written with M. C. Chilowsky and M. Tournier; "Utilisation des phénomènes piézo-électriques pour la mesure de l'intensité des sons en valeur absolue," *ibid.*, 6th ser., **4** (1923), 539, written with M. Ishimoto; "Sondage et détection sous-marine par les ultra-sons," in *Bulletin de l'Association technique maritime et aéronautique*, no. 28 (1924), 407; "La production et l'utilisation des ondes ultra-sonores," in *Revue générale de l'électricité*, **23** (1928), 626; "Sur le mirage ultra-sonore," in *Bulletin de l'Association technique maritime et aéronautique* (1929), 727; "Les ondes ultra-sonores," in *Revue d'acoustique*, **1** (1932), 93, 315; **2** (1933), 288; **3** (1934), 104, with notes by P. Biquard; and "Sur les lois du dégagement d'électricité par torsion dans les corps piézo-électriques," in *Comptes rendus . . . de l'Académie des sciences*, **200** (1935), 1257.

Various technical problems are treated in "Sur la production des étincelles musicales par courant continu," in *Annales des postes, télégraphes et téléphones*, 5th year, no. 4 (1916), p. 404; "Utilisation de la détente pour la production des courants d'air de grande vitesse," in *Procès-verbaux des commissions de la Société française de physique* (20 Feb. 1920), p. 21; "Note sur la loi de résistance de l'air," in *Mémorial de l'artillerie française* (1922), p. 253; "Note sur les effets balistiques de la détente des gaz de la poudre," *ibid.* (1923), p. 3; "Procédé et appareils permettant la mesure de la puissance transmise par un arbre," in *Brevet français* (22 Dec. 1927); "Banc piézo-électrique pour l'équilibrage des rotors," *ibid.* (19 Dec. 1927); "Procédé et dispositif pour la mesure des variations de pression dans les canalisations d'eau ou autre liquide," *ibid.* (6 Aug. 1927), written with R. Hocart; and "L'enregistrement des coups de bélier," in *Bulletin technique de la Chambre syndicale des entrepreneurs de couverture plomberie*, no. 23 (1927), p. 81.

On teaching and pedagogy, see "L'esprit de l'enseignement scientifique," in *L'enseignement des sciences mathématiques et des sciences physiques* (Paris, 1904), also in *Physique*; "Le théorème de Fermat de la loi du minimum de temps en optique géométrique," in *Journal de physique*, ser. 6, **1** (1920), 188; "La valeur éducative de l'histoire des sciences," in *Revue de synthèse*, **6** (1933), 5; "La réorganisation de l'enseignement public en Chine," in *Rapport de la mission d'experts de la Société des Nations*, written with C. H. Becker, M. Falski, and R. H. Tawney (Paris, 1932); "Le problème de la culture générale," in *Discours d'ouverture du Congrès international d'éducation nouvelle, Nice, July 1932*, also in *Full Report of the New Education Fellowship* (London, 1933), p. 73; "L'enseignement en Chine," in *Bulletin de la Société française de pedagogie*, no. 49 (Sept. 1933); and "La Réforme générale de l'enseignement

(Premier rapport sur les travaux de la Commission ministérielle)," in *Bulletin officiel de l'éducation nationale*, no. 23 (15 March 1945), p. 1461.

Other publications include "Notice sur les travaux de Monsieur P. Curie," in *Bulletin des anciens élèves de l'École municipale de physique et chimie industrielles* (Dec. 1904); "Henri Poincaré, le physicien" in *Henri Poincaré* (Paris, 1914), also in *Revue du Mois*, **8** (1913); "Paul Schutzenberger" in *Discours prononcé à l'occasion du centenaire de P. Schutzenberger* (1929); "L'orientation actuelle de la physique," in *L'orientation actuelle des sciences* (Paris, 1930), p. 29; "La physique au Collège de France," in *Volume du centenaire* (Paris, 1932), p. 61; "Ernest Solvay" in *Discours prononcé à l'inauguration du monument d'E. Solvay* (Brussels, 1932); "Paul Painlevé, le savant," in *Les Cahiers rationalistes*, no. 26 (Nov. 1933); "La valeur humaine de la science," preface to *l'Évolution humaine* (Paris, 1933); and "Discours prononcés à l'occasion du cinquantenaire de l'École municipale de physique et chimie industrielles" in *Cinquante années de science appliquée à l'industrie, 1882–1932*.

II. SECONDARY LITERATURE. See P. Biquard, *Paul Langevin, scientifique, éducateur, citoyen* (Paris, 1969), with preface by J. D. Bernal and a bibliography; Louis de Broglie, *Notice sur la vie et l'oeuvre de Paul Langevin* (Paris, 1947); S. Ghiseman, *Paul Langevin* (Bucharest, 1964); *La pensée* (Paris), no. 12 (May–June 1947), spec. no. "In memoriam"; O. A. Staroselskaya Nikitina, *Paul Langevin* (Moscow, 1962); and A. R. Weill, "Paul Langevin," in *Mémorial de l'artillerie française*, fasc. 4 (1946).

See also André Langevin, *Paul Langevin, mon père* (Paris, 1972).

ADRIENNE R. WEILL-BRUNSCHVICG

LANGLEY, JOHN NEWPORT (*b.* Newbury, England, 2 November 1852; *d.* Cambridge, England, 5 November 1925), *physiology, histology.*

Professor of physiology at Cambridge from 1903 until his death, Langley was renowned for his studies of glandular secretion and of the autonomic, or involuntary, nervous system. He was the second son of John Langley, a private schoolmaster at Newbury, who prepared him at home for Exeter Grammar School, where his uncle, Rev. H. Newport, was headmaster. His mother, Mary, was the eldest daughter of Richard Groom, assistant secretary to the Tax Office, Somerset House. In autumn 1871 Langley matriculated with a sizarship at St. John's College, Cambridge, where he later won a scholarship. During his first five terms he read mathematics and history as preparation for a place in the civil service, at home or in India. In May 1873 Langley enrolled in a newly created course in elementary biology offered by Michael Foster, who had come to Cambridge in 1870 as Trinity College praelector in physiology. Under Foster's inspiration Langley abandoned his former plans and began to read for the natural sciences tripos, in which he took first-class honors in 1874.

Langley graduated B.A. in 1875 and in 1876 succeeded Henry Newell Martin as Foster's chief demonstrator, his salary being paid by Trinity College, where he was elected to a fellowship in 1877. Also in 1877 Langley spent several months in the laboratory of Wilhelm Kühne in Heidelberg, where he studied salivary secretion in the cat. Langley received an M.A. from Cambridge in 1878 and an Sc.D. in 1896. From 1883 to 1903 he was university lecturer in physiology and lecturer in natural science at Trinity College. In 1900 he was named deputy professor to Foster and in 1903 succeeded him as professor of physiology. In 1914 a large new physiological laboratory was built at Cambridge under Langley's direction with funds provided by the Drapers' Company. Upon Langley's death in 1925 the chair passed to Joseph Barcroft.

Elected to fellowship in the Royal Society in 1883 and awarded its Royal Medal in 1892, Langley also served as a member of its Council (1897–1898) and as vice-president (1904–1905). He was president of the Neurological Society of Great Britain in 1893 and president of the physiological section of the British Association in 1899. He was awarded the Baly Medal of the Royal College of Physicians in 1903, the Andreas Retzius medal of the Swedish Society of Physicians in 1912, and honorary doctorates by the universities of Dublin, St. Andrews, Groningen, and Strasbourg. In 1902 he married Vera Kathleen Forsyth-Grant of Ecclesgreig, Kincardineshire, by whom he had one daughter.

Langley's career in research fell into two major phases. From 1875 to 1890 he devoted himself largely to glandular secretion, and from 1890 until his death he worked mainly upon the involuntary nervous system. At the very beginning of his career, however, Langley briefly joined the group of Cambridge researchers who were working in the area of Foster's special interest, the problem of the heartbeat. During the winter of 1874–1875 Foster gave Langley a small amount of the alcoholic extract of jaborandi and asked him to study its physiological action. By October 1875 Langley's focus had narrowed to the effects of the drug on the heart. His major conclusion —that jaborandi altered heart action not by acting on any nervous elements but, rather, by acting directly on the cardiac tissue itself—conformed nicely to Foster's own myogenic view of heart action.[1]

But Langley was quickly distracted from the problem of the heartbeat. One of the most striking effects of jaborandi was to evoke copious secretion from the submaxillary gland. Recognizing in this phenomenon an opportunity to clarify the nature of salivary secretion and of secretion in general, Langley undertook a systematic study of the secretory organs which occupied him for the next fifteen years. In a series of papers remarkable for technical ingenuity and theoretical caution, Langley challenged Rudolf Heidenhain's generally accepted views on the structural changes which take place in secreting glands and on the relationship between glandular secretion and nervous activity.

As early as 1879 Langley had begun to develop the view that gland cells become more granular during rest and less granular during secretion, a conclusion precisely the reverse of Heidenhain's.[2] In making this point Langley emphasized that the structural changes observed in secreting glands are progressive and gradual, with no dramatic turning points discernible in the underlying processes. By 1888 Langley was also prepared to dispute Heidenhain's theory that the salivary glands are supplied not only with secretory nerves but also with "trophic" (metabolic) nerves whose function is to increase the solubility of stored gland substance.[3] With increasing confidence Langley argued that the evidence adduced in support of "trophic" nerves was either unreliable or could more plausibly be ascribed to the effects of local vasomotor nervous action. To the end of his career Langley maintained that the salivary gland, at least, is supplied with but one kind of nerve, which can by itself produce all of the changes observed in glandular secretion.[4]

In the 1880's, although engaged chiefly in these studies on secretion, Langley also undertook studies of the normal anatomy of the dog's brain and of nervous degeneration in the decorticated dog. Although this work from its beginning was related to the celebrated controversy between David Ferrier and Friedrich Goltz over cerebral localization, Langley seems never to have taken a definitive position on the issue.[5] Also in the 1880's Langley devoted himself briefly to the fashionable and controversial subject of hypnotism. Reputedly an accomplished hypnotist himself, chiefly because of his remarkable steely blue eyes, Langley sought to subsume many of the hypnotic phenomena under the physiological concept of irradiated inhibition.[6]

Although Langley's work on glandular secretion had involved him from the outset with problems of sympathetic innervation, his subsequent concentration on the involuntary nervous system was not a direct outgrowth of his work on the secretory nerves. Rather, it had its origin in the discovery that nicotine could selectively interrupt nerve impulses at the sympathetic ganglia. In a joint paper announcing the discovery in 1889, Langley and William Lee Dickinson emphasized its immense potential as an analytic tool. In particular it seemed to offer an unimpeachable means of distinguishing those nerve fibers which ended in the nerve cells of a ganglion from those which merely ran through the ganglion without being connected with its nerve cells.[7] Although one alumnus of the Cambridge school has claimed in private correspondence that the discovery was really Dickinson's alone and not Langley's,[8] it was certainly Langley who most brilliantly and successfully exploited the new technique. Scarcely six months after the discovery, he had applied it with great success to extending and fortifying his earlier studies on the innervation of the salivary gland.[9] Thereafter the range of his studies expanded dramatically. By using nicotine in concert with more traditional techniques, such as nerve degeneration, Langley was able to map out the plan of much of the involuntary system in exquisite and unprecedented detail. In so doing he built upon the important earlier work of his Cambridge colleague W. H. Gaskell, and their combined contributions made Cambridge for some time the leading center for research on the involuntary nervous system.

Chiefly on histological grounds, Gaskell had established that the "visceral," or involuntary, nerves arise from the central nervous system in three distinct outflows: the cervicocranial, the thoracic, and the sacral. Gaskell's study of the distribution of the gray and white rami had also convinced him that the involuntary system is not essentially independent of the cerebrospinal system, as had been thought. While Langley's work strongly reinforced these fundamental generalizations, it also extended and modified Gaskell's results in several important ways. In the 1890's, at first with Charles Sherrington but mainly alone, Langley charted the distribution of the efferent sympathetic fibers to the skin and their relation to the afferent fibers of the corresponding spinal nerves. In so doing he not only clarified the basis for various pilomotor mechanisms (bristling of hair in cats and dogs, ruffling of feathers in birds) but also contributed importantly to clinical attempts to understand "referred" cutaneous pain in certain visceral diseases.

Moreover, these studies of the pilomotor nerves provided much of the evidence for three of Langley's most original contributions to the histology and physiology of the involuntary nervous system: (1) the conclusion that peripheral ganglia are not collected together according to function, as some had thought, but instead are associated with definite somatic areas, each ganglion being supplied with all of the sympathetic fibers of whatever function (whether inhibitory, motor, or other) which run to its associated area; (2) the discovery that every efferent nerve fiber passes through only one nerve cell on its way to the periphery, a discovery which overturned the hitherto rather casual assumption that such a fiber might run through several nerve cells (or even none), and which therefore had important physiological implications; and (3) his concept of "axon reflexes," developed in opposition to those French workers who followed Claude Bernard in their belief that peripheral ganglia are capable of true reflex action. The issues involved in this dispute are extremely complex, but Langley's basic conclusions were that the supposed ganglionic reflexes take place in the absence of afferent arcs and that they have their origin not in peripheral ganglia but in efferent preganglionic axon processes.

Another of Langley's influential contributions dating from the late 1890's was his successful introduction of a radically new nomenclature for what had been earlier (and loosely) designated the "organic," "vegetative," "sympathetic," "visceral," or "involuntary" nervous system. On the suggestion of Richard Jebb, professor of Greek at Cambridge, Langley proposed instead the term "autonomic,"[10] which still retains its preferred status. Langley thereafter confined the term "sympathetic" to the thoracic outflow of the autonomic system and introduced the term "parasympathetic" to designate its cranial and sacral outflows. By attaching this single name to both the cranial and sacral outflows, Langley wished to emphasize that these nerves belong to a somewhat separate system (which he sometimes called the "oro-anal" system), in that they resemble each other in their action and phylogenetic origin more than they resemble the thoracic (or "sympathetic") nerves. Langley gathered much of the evidence for the existence of this separate, parasympathetic system, although he perhaps tended to minimize the extent to which the idea was already present in Gaskell's earlier work. The importance of "merely" nomenclatural contributions tends to be greatly underrated, but Langley's terminological precision undeniably promoted clarity of thought and conciseness of expression in subsequent literature on the autonomic system. Still other contributions by Langley from the 1890's, perhaps less striking than those already described but characteristically thorough and valuable, were the studies he undertook with H. K. Anderson on the neuromuscular mechanisms of the iris and on the pattern of innervation for the pelvic organs.

Between about 1900 and 1905 Langley's most important new work centered on experimental cross unions of different kinds of nerve fibers. While confirming the functional distinction between efferent and afferent nerve fibers, these cross unions demonstrated the essential functional similarity of all efferent, preganglionic fibers and pointed to the conclusion that the final action of a peripheral nerve fiber depends not on the nature of the fiber itself, nor on the nature of the impulse which it transmits, but on the nature of the tissue in which it ends.

By 1905 Langley had entered perhaps the most fertile phase of his remarkably productive career. The subject to which he now turned was the mode by which nerve impulses or other stimuli are transmitted to various effector cells—whether muscular, glandular, or ganglionic. Here again nicotine and other drugs (especially curare and adrenaline) were utilized as fundamental analytic tools. The background for this new phase of Langley's work was formed in part by his earlier observation that nicotine could stimulate peripheral ganglia even after their preganglionic fibers had degenerated, but, more immediately, by the discovery of adrenalin in 1895 and the revelation soon afterward of its sympathomimetic effects on both normal and denervated muscle or gland cells. That nicotine and adrenalin produced the same effects on denervated as on normal cells raised doubts about the prevailing view that these and other drugs ordinarily act on nerve endings. These doubts were strikingly confirmed by Langley's discovery in 1905 that curare abolishes the slight tonic contractions produced in certain skeletal muscles of the fowl by injection of nicotine. Since this result occurs even after degeneration of the nerve fibers to these muscles, Langley argued that it can only be referred to the mutually antagonistic action of nicotine and curare on striated muscle. And this mutual antagonism could best be understood as the result of a competition between nicotine and curare for some constituent of the muscle substance with which each formed a specific chemical compound.

On grounds such as these Langley was able to suggest in 1906 that every cell connected with an efferent fiber contains a substance responsible for the chief function of that cell (whether contraction, or secretion, or—in the case of nerve cells—the discharging of nerve impulses) as well as other "receptive substances" capable of reacting specifically with chemical bodies (drugs) or sometimes with nervous stimuli. Although the tangled web of issues and influences has yet to be unraveled by historians, it seems clear that Langley's concept of "receptive substances" formed part of the background for later theories of the humoral transmission of nervous impulses[11] and that it provided a stimulus to many subsequent studies in neuromuscular physiology.

Langley himself elaborated his theory in several later studies.

With the advent of World War I, Langley and his depleted staff directed their energies and the facilities of their new laboratory toward studies of more immediate clinical relevance. Special attention was given to means of repairing denervated and atrophied muscle tissue, although Langley's typically cautious and thorough investigation offered little hope to those treating or suffering from such afflictions. His experiments on animals seemed to show that neither electrical stimulation nor massage nor any other physiotherapeutic measures could prevent atrophy in denervated muscle tissue. From the end of the war until his death, Langley worked chiefly on various aspects of vasomotor action, most notably the so-called "antidromic" vasodilatation produced in the hind limb by stimulating certain posterior nerve roots at their entry into the spinal cord. This phenomenon attracted considerable attention because it seemed to offer a unique exception to the Bell-Magendie law of the functional distinction between efferent and afferent fibers. In Langley's view, however, antidromic vasodilatation was produced by metabolites set free in muscle spindle cells by impulses passing down afferent fibers, and was therefore not an exception to the law.

In 1921 Langley gathered the results of much of his research in *The Autonomic Nervous System*, part I. (A projected second part never appeared.) Soon translated into French and German, this work was conceived in part as a rival to Gaskell's posthumous *The Involuntary Nervous System* (1916). Sensitive to suggestions that he had merely followed a path laid out for him by Gaskell, Langley in 1919 wrote a long letter to *Lancet* emphasizing the extent and manner in which his work had gone beyond Gaskell's and pointing to the latter's book as one possible factor impeding proper recognition of his own work. In this letter Langley also emphasized that all of his work on the autonomic system had evolved logically from the studies he had undertaken with Dickinson in 1889 on nicotine and related poisons.[12]

One final contribution of Langley's deserves mention. In 1894 he assumed editorial and financial control of the *Journal of Physiology*, which had acquired both a high reputation and a sizable debt since its founding by Foster in 1878. Besides arranging to pay off the debt and accepting the unsold stock, Langley substituted a policy of fiscal, verbal, and theoretical economy for Foster's rather genial and indulgent editorial policy. From 1894 until his death Langley invested enormous energy and talent in his editorial tasks, although he generously allowed Foster to retain the official title of editor until the latter's death in 1907.

Under Langley's direction the Cambridge School of Physiology preserved and enhanced the great reputation it had gained under Foster's initial inspiration. Before assuming the professorship Langley had taught only advanced students, but he thereafter followed Foster's example by taking responsibility for the introductory lectures. In the latter capacity he was not notably successful, being both too reserved and too exacting to carry most novices with him. In fact, even a few of his advanced students and colleagues apparently resented his reserve and his alleged egotism; and there is some evidence that personal relationships were consequently strained in the Cambridge physiological laboratory.[13] Like many members of the Cambridge school, Langley was an accomplished athlete; his performances in sprinting, rowing, tennis, and golf were overshadowed by his prodigious talent as a skater. For one of his less avid admirers, this talent was perfectly in keeping with the spirit of Langley's work and character—"brilliant technique on an icy background."[14]

NOTES

1. See J. N. Langley, "The Action of Jaborandi on the Heart," in *Journal of Anatomy and Physiology*, **10** (1876), 187–201; and G. Geison, *Michael Foster*, pp. 377–382.
2. J. N. Langley, "On the Structure of the Serous Glands in Rest and Activity," in *Proceedings of the Royal Society*, **29** (1879), 377–382.
3. See J. N. Langley, "On the Physiology of the Salivary Secretion. Part IV. The Effect of Atropin Upon the Supposed Varieties of Secretory Nerve Fibres," in *Journal of Physiology*, **9** (1888), 55–64, esp. 61–62; and J. N. Langley to E. A. Schäfer, 4 Feb. 1888, Sharpey-Schafer Papers, Wellcome Institute of the History of Medicine.
4. See, for example, J. N. Langley, "Note on Trophic Secretory Fibres to the Salivary Glands," in *Journal of Physiology*, **50** (1916), xxv–xxvi.
5. See E. Klein, J. N. Langley, and E. A. Schäfer, "On the Cortical Areas Removed From the Brain of a Dog, and From the Brain of a Monkey," *ibid.*, **4** (1883), 231–247; J. N. Langley, "The Structure of the Dog's Brain," *ibid.*, 248–285; and "Report on the Parts Destroyed on the Right Side of the Brain of the Dog Operated on by Prof. Goltz," *ibid.*, 286–309; J. N. Langley and C. S. Sherrington, "Secondary Degeneration of Nerve Tracts Following Removal of the Cortex of the Cerebrum in the Dog," *ibid.*, **5** (1884), 49–65; and J. N. Langley and A. S. Grünbaum, "On the Degeneration Resulting From Removal of the Cerebral Cortex and Corpora Striata in the Dog," *ibid.*, **11** (1890), 606–628. In private correspondence Langley did once refer to an apparently fleeting hypothesis of his on the issue of cerebral localization, but the content of this "hypothesis" remains obscure. See J. N. Langley to E. A. Schäfer, 3 Feb. 1884, Sharpey-Schafer Papers, Wellcome Institute of the History of Medicine.
6. See J. N. Langley, "The Physiological Aspect of Mesmerism," in *Proceedings of the Royal Institution of Great Britain*, **11** (1884), 25–43; and J. N. Langley and H. E. Wingfield, "A Preliminary Account of Some Observa-

tions on Hypnotism," in *Journal of Physiology*, **8** (1887), xvii–xxiv.

7. J. N. Langley and W. Lee Dickinson, "On the Local Paralysis of Peripheral Ganglia, and on the Connexion of Different Classes of Nerve Fibres With Them," in *Proceedings of the Royal Society*, **46** (1889), 423–431.
8. W. Langdon-Brown to Donal Sheehan, 11 Mar. 1940, copy deposited in the obituary files, Medical Historical Library, Yale University.
9. J. N. Langley, "On the Physiology of the Salivary Secretion. Part VI. Chiefly Upon the Connection of Peripheral Nerve Cells With the Nerve Fibres Which Run to the Sublingual and Submaxillary Glands," in *Journal of Physiology*, **11** (1890), 123–158.
10. See J. N. Langley, "The Sympathetic and Other Related Systems of Nerves," in *Text-Book of Physiology*, E. A. Schäfer, ed., 2 vols. (Edinburgh–London, 1898–1900), II, 616–696, esp. 659–660.
11. See, e.g., Walter B. Cannon, "The Story of the Development of Our Ideas of Chemical Mediation of Nerve Impulses," in *American Journal of the Medical Sciences*, n.s. **188** (1934), 145–159. According to Henry Dale, however, Langley reacted unfavorably toward, and diverted attention from, the hypothesis of the chemical transmission of nerve impulses as originally proposed by T. R. Elliott in 1904. See H. H. Dale, "T. R. Elliott," in *Biographical Memoirs of Fellows of the Royal Society*, **7** (1961), 53–74, esp. 63–64.
12. J. N. Langley, "The Arrangement of the Autonomic Nervous System," in *Lancet* (31 May 1919), p. 951.
13. Most notably in the letter of Walter Langdon-Brown to Donal Sheehan (see note 8). See also J. F. Fulton to Donal Sheehan, 27 Mar. 1940, copy deposited in obituary files, Medical Historical Library, Yale University. For a rather more favorable view of Langley's personality, see the obituary notices by W. M. Fletcher.
14. Langdon-Brown to Sheehan (note 8).

BIBLIOGRAPHY

I. Original Works. Of Langley's approximately 170 published research papers, the great majority appeared in the *Journal of Physiology*. A virtually complete bibliography, slightly flawed by occasional errors or omissions in pagination, is given by Walter Morley Fletcher in *Journal of Physiology*, **61** (1926), 16–27. Fletcher did overlook the important letter to *Lancet* cited in note 12 and the obituary notice on Lucas (cited below). The Royal Society *Catalogue of Scientific Papers*, X, 510–511; XVI, 595–596; lists Langley's papers up to 1900.

Langley's major contributions to the physiology of glandular secretion are contained in the six-part article "On the Physiology of the Salivary Secretion," in *Journal of Physiology*, **1** (1878), 96–103, 339–369; **6** (1885), 71–92; **9** (1888), 55–64; **10** (1889), 291–328; and **11** (1890), 123–158. The full measure of those contributions is made clear in Langley's admirable review article "The Salivary Glands," in *Text-Book of Physiology*, E. A. Schäfer, ed., 2 vols. (Edinburgh–London, 1898–1900), I, 475–530.

So extensive and rich was Langley's work on the autonomic nervous system that only a few of his major papers can be listed here. On his studies of the pilomotor nerves, see especially "On Pilomotor Nerves," in *Journal of Physiology*, **12** (1891), 278–291, written with C. S. Sherrington; and "The Arrangement of the Sympathetic Nervous System, Based Chiefly on Observations Upon Pilo-Motor Nerves," *ibid.*, **15** (1893), 176–244. Langley's work with H. K. Anderson on the innervation of the pelvic and adjoining viscera is described in a seven-part article, *ibid.*, **18** (1895), 67–105; **19** (1895), 71–84, 85–121, 122–130, 131–139; **19** (1896), 372–384; and **20** (1896), 372–406. On the concept of "axon reflexes," see "On Axon-reflexes in the Pre-ganglionic Fibres of the Sympathetic System," *ibid.*, **25** (1900), 364–398.

Langley's concept of "receptive substances" was first announced in "On the Reaction of Cells and of Nerve-Endings to Certain Poisons, Chiefly as Regards the Reaction of Striated Muscle to Nicotine and Curare," *ibid.*, **33** (1905), 374–413. More impressive for its scope and clarity is his Croonian lecture for 1906, "On Nerve-Endings and on Special Excitable Substances in Cells," in *Proceedings of the Royal Society*, **76B** (1906), 170–194. The concept was further developed in a four-part article, "On the Contraction of Muscle, Chiefly in Relation to the Presence of 'Receptive Substances,'" in *Journal of Physiology*, **36** (1907), 347–384; **37** (1908), 165–212, 285–300; and **39** (1909), 235–295.

For Langley's more general accounts of the autonomic nervous system, see his superb article "The Sympathetic and Other Related Systems of Nerves," in *Text-Book of Physiology*, II, 616–696; his presidential address to the physiological section of the British Association, in *Report of the Sixty-Ninth Meeting of the British Association for the Advancement of Science* (London, 1900), pp. 881–892; "The Autonomic Nervous System," in *Brain*, **26** (1903), 1–26; and, of course, his monograph *The Autonomic Nervous System*, pt. 1 (Cambridge, 1921). For a clear and succinct account of the rationale for Langley's nomenclatural innovations, see "The Nomenclature of the Sympathetic and of the Related System of Nerves," in *Zentralblatt für Physiologie*, **27** (1913), 149–152.

For examples of Langley's wartime research on denervated muscle, see "The Rate of Loss of Weight in Skeletal Muscle After Nerve Section With Some Observations on the Effect of Stimulation and Other Treatment," in *Journal of Physiology*, **49** (1915), 432–440, written with Toyojiro Kato; and "Observations on Denervated Muscle," *ibid.*, **50** (1916), 335–344. Langley's major postwar research projects are described in three two-part articles: "Vaso-Motor Centres," *ibid.*, **53** (1919), 120–134, 147–161; "The Secretion of Sweat," *ibid.*, **56** (1922), 110–119, 206–226; and "Antidromic Action," *ibid.*, **57** (1923), 428–446, **58** (1923), 49–69.

Langley also wrote useful obituary articles on four of his Cambridge colleagues—Michael Foster, in *Journal of Physiology*, **35** (1907), 233–246; W. H. Gaskell, in *Proceedings of the Royal Society*, **88B** (1915), xxvii–xxxvi; Arthur Sheridan Lea, *ibid.*, **89** (1916), xxv–xxvii; and Keith Lucas, in *Nature*, **98** (1916), 109. The obituary on Gaskell is noteworthy for its generous assessment of the work of his sometime rival. The range of Langley's interests is indicated in a historiographically undistinguished article, "Sketch of the Progress of Discovery in the Eighteenth Century as Regards the Autonomic Nervous System," in *Journal of Physiology*, **50** (1916), 225–258. Besides his monograph on the autonomic nervous system, Langley's published books were *Practical Histology* (London, 1901; 3rd ed., 1920); and, initially with Michael Foster, *A Course of Elementary Practical Physiology* (London, 1876; 7th ed., 1899); by the 7th ed., L. E. Shore was listed as coauthor.

There is apparently no central repository for Langley's letters or papers, and little seems to have survived. Some ten letters from him to E. A. Schäfer are to be found in the Sharpey-Schafer Papers at the Wellcome Institute of the History of Medicine, London.

II. SECONDARY LITERATURE. For valuable obituary notices, see W. M. Fletcher, in *Journal of Physiology*, **61** (1926), 1–27, which contains a full bibliography; and in *Proceedings of the Royal Society*, **101B** (1927), xxxiii–xli; and C. S. Sherrington, in *Dictionary of National Biography, 1922–1930*, pp. 478–481. For background on some of Langley's research and on his place in the Cambridge school, see Gerald L. Geison, "Michael Foster and the Rise of the Cambridge School of Physiology" (unpublished Ph.D. dissertation, Yale, 1970). His place in the recent history of pharmacological physiology and junctional transmission receives brief attention in Alfred Fessard, "Claude Bernard and the Physiology of Junctional Transmission," in *Claude Bernard and Experimental Medicine*, Francisco Grande and Maurice B. Visscher, eds. (Cambridge, Mass., 1967), pp. 105–123, esp. p. 116; and F. E. Shideman, "Drugs as Tools in the Elucidation of Physiological Mechanisms," *ibid.*, pp. 125–134, esp. pp. 125–127.

GERALD L. GEISON

LANGLEY, SAMUEL PIERPONT (*b.* Roxbury [now part of Boston], Massachusetts, 22 August 1834; *d.* Aiken, South Carolina, 22 February 1906), *astrophysics.*

Langley's parents, Mary Sumner Williams and Samuel Langley, a wholesale merchant of broad interests, traced their ancestry to old and prominent New England families. Langley attended private schools, Boston Latin School, and completed his formal education upon graduation from Boston High School in 1851. From his youth he read extensively in science, literature, and history. Langley had a Puritan morality but was an ardent seeker for religious truth and the nature of the soul, including study of metaphysics, psychology, and psychic phenomena. A bachelor, he was a large man with a florid countenance and robust health but a tendency to be melancholy. Because of a deep-seated shyness, he maintained a front of severe and haughty dignity. He was irascible and often gave offense, but to his intimates he was warmhearted, displaying great charm. Highly respected in scientific circles, he received several scientific honors and honorary degrees. He was a member of the National Academy of Sciences, on whose council he served; a foreign member of the Royal Society of London, the Royal Society of Edinburgh, and the Accademia dei Lincei of Rome; a correspondent of the Institut de France; and a president of the American Association for the Advancement of Science.

Langley worked as a civil engineer and architect from 1851 to 1864, ranging west to Chicago and St. Louis. After a tour of Europe with his brother John Williams Langley, a chemist, he was an assistant at the Harvard observatory for a year. In 1866 he was appointed assistant professor of mathematics at the Naval Academy and placed in charge of restoring its neglected observatory.

In 1867 Langley was appointed director of the Allegheny Observatory and professor of physics and astronomy at the Western University of Pennsylvania (later University of Pittsburgh). He pioneered in providing standard time signals to railroads; the income from this service helped support the observatory. Langley did his most original work in his twenty years at Pittsburgh.

In January 1887 Langley was appointed assistant secretary, and in November of that year secretary, of the Smithsonian Institution. He did not cease work at Allegheny or take up full-time residence in Washington until 1889, after the death of his Pittsburgh benefactor, William Thaw. Langley then fulfilled the long-held goal of a Smithsonian observatory by founding the Smithsonian Astrophysical Observatory, using private bequests and government support (1890). Besides giving new life to the physical sciences at the Smithsonian, Langley broadened the scope of the institution by beginning what became the National Zoological Park and the National Gallery of Art.

Langley began research on aerodynamics at Pittsburgh in 1887, investigating the pressure on a plane surface inclined to its direction of travel. The results were of minor importance. In 1891 Langley announced that "*Mechanical Flight is possible with engines we now possess*," on the basis of his prediction that a twenty-pound, one-horsepower steam engine could sustain the flight of a 200-pound airplane at a speed of forty-five miles per hour. In 1896 Langley successfully flew models over the Potomac.

Langley's models were developed in slow stages from bench tests and were designed to be inherently stable. When in 1898, at the request of the War Department, Langley began work on a man-carrying machine, he continued this evolutionary process. His one-quarter (of man-carrying size) model flew very successfully in 1903. The man-carrying airplane did not fly; on two attempts in 1906 the launching apparatus failed. The ridicule of the press, offended by Langley's aloofness, and the mismanaged combination of official secrecy but public testing ended Langley's expensive program.

It has been said that the work of the scientifically respected Langley brought experiments on flight out of the stage of ridicule and into a stage of science. His failure was the occasion for sharp ridicule, however; and the procedures of the Wright brothers were unlike those of Langley because they first mastered the art of flying inherently unstable gliders before adding power. The main line of development of flight followed the work of the Wright brothers and not that of Langley.

It is difficult to agree on what constitutes successful flight of powered, man-carrying, heavier-than-air craft; and there are many motivated by a desire to improve Langley's popular image or to discount the contributions of the Wright brothers in favor of supposed precursors. In 1914, with the support of Smithsonian officials, Glenn Curtiss, a patent opponent of the Wright brothers, barely flew Langley's airplane after significant but unreported modifications. The resulting controversy caused bitter feelings between the Wright brothers and the Smithsonian and did no service to Langley.

Langley's work in astrophysics and infrared spectroscopy was very much in the mainstream of those branches of physics. Although a keen visual observer, Langley made his contribution in the development and application of apparatus and techniques for the measurement of radiation.

Correctly recognizing the need for measurements of the energy of radiation as a function of wavelength, Langley developed a new instrument—the bolometer—between 1879 and 1881 to do this. The bolometer was an ingenious application of the principle—the temperature coefficient of the resistance of metals—being used by William Siemens for the measurement of temperature. It was a Wheatstone bridge with similar narrow platinum strips in opposite arms, one of which was exposed to radiation. With this Langley could measure a temperature difference of 10^{-5}° C. in a one-second exposure with an error of less than 1 percent. The bolometer allowed continuous and direct measurement of $E(\lambda)$ versus λ for the full known spectrum of radiation, through what had been delineated (because of the different techniques necessary for measurement before the bolometer) as actinic, visual, and thermal regions.

Langley developed the bolometer to study the solar constant by integrating the energy versus wavelength curves it gave; to study the selective absorption of the earth's atmosphere by repeating measurements for different thicknesses of the atmosphere; and to study the selective absorption of the sun's atmosphere by repeating measurements on various parts of the sun. On his expedition to Mt. Whitney to accomplish this in 1881, Langley discovered significant radiation beyond a wavelength of one micron, which had been thought to be the limit of solar radiation. The superior measurements by means of the bolometer, the newly discovered extent of the solar spectrum, and the new results for selective absorption of the earth's atmosphere were significant contributions to the study of the sun and its effects on the earth.

The dispersion and absorption of prisms depend on wavelength. Grating spectra, although without this disadvantage, are weak and overlap at large wavelengths. Langley's technique was to determine the dispersion and absorption of prisms, rock salt in particular, as a function of wavelength by explicitly using the overlap in grating spectra and the bolometer. The prism was then used in an automatically recording spectrobolometer, which allowed convenient measurement of infrared spectra well beyond previous capability.

The investigations of long wavelengths in the late 1890's which prompted Planck's interpolated radiation law were based, in part, on bolometer measurements. Langley extended measurement of spectra to a five-micron wavelength. The failure of Wein's law was seen beyond a ten-micron wavelength. Langley did some work measuring the spectrum of Leslie cubes—showing, for example, migration of the maximum of radiation intensity as a function of the temperature of the source—but did not study blackbody radiation as such.

In a paper titled "Laws of Nature," Langley expressed his empiricist creed: "[The] practical rule of life [is] that we must act with the majority where our faith does not compel us to do otherwise, [remembering that] we know nothing absolutely or in its essence [and science has authority in matters] only as far as they are settled by evidence and observation alone" ("The Laws of Nature," in *Report of the Board of Regents of the Smithsonian Institution* for 1901 [1902], p. 551).

BIBLIOGRAPHY

I. Original Works. There are bibliographies of Langley's published works appended to Walcott's memoir of Langley and to the report of the Langley Memorial Meeting (see below). Three major publications—"Researches on Solar Heat and Its Absorption by the Earth's Atmosphere, A Report of the Mount Whitney Expedition,"

Professional Papers of the Signal Service, no. 15 (1884); *Annals of the Astrophysical Observatory of the Smithsonian Institution*, **1** (1900); and "Langley Memoir on Flight," in *Smithsonian Contributions to Knowledge*, **27** (1911)—report most of his work.

II. SECONDARY LITERATURE. Langley prepared some biographical fragments which were used by his very close friend Cyrus Adler for his memoir, "Samuel Pierpont Langley," in *Bulletin of the Philosophical Society of Washington*, **15** (1907), 1–26. Adler contemplated a full biography of Langley, and there may be valuable materials among Adler's papers at the American Jewish Historical Society in New York.

Other notices of Langley are "Samuel Pierpont Langley Memorial Meeting," in *Smithsonian Miscellaneous Collections*, **49** (1907); C. G. Abbot, "Samuel Pierpont Langley," in *Astrophysical Journal*, **23** (1906), 271–283; George Brown Goode, "Samuel Pierpont Langley," in *The Smithsonian Institution: 1846 to 1896*, George Brown Goode, ed. (Washington, D.C., 1896); and C. D. Walcott, "Biographical Memoir of Samuel Pierpont Langley," in *Biographical Memoirs. National Academy of Sciences*, **7** (1917), 247–268.

Autobiographies of Langley's associates containing significant mention of Langley are Charles G. Abbot, *Adventures in the World of Science* (Washington, D.C., 1958); Cyrus Adler, *I Have Remembered the Days* (Philadelphia, 1941); and John A. Brashear, *The Autobiography of a Man Who Loved the Stars*, W. Lucien Scaife, ed. (Boston, 1925).

Among works on the Smithsonian the following, along with the Goode volume listed above, are valuable: Bessie Zaban Jones, *Lighthouse of the Skies, the Smithsonian Astrophysical Observatory: Background and History 1846–1955* (Washington, 1965); and Paul H. Oehser, *Sons of Science: The Story of the Smithsonian Institution and Its Leaders* (New York, 1949).

For Langley's place in the history of flight, see C. G. Abbot, "The 1914 Tests of the Langley 'Aerodrome,'" in *Smithsonian Miscellaneous Collections*, **103**, no. 8 (1942); and Charles H. Gibbs-Smith, *The Aeroplane: An Historical Survey of Its Origins and Development* (London, 1960).

Langley's bolometer work, concerning as it does instrumentation and technique, has not been itself the subject of historical scholarship, although there is perfunctory mention of Langley's contribution in works on astrophysics, infrared spectroscopy, and the prehistory of quantum mechanics.

DON F. MOYER

LANGLOIS, CLAUDE (*b.* France, *ca.* 1700; *d.* France, *ca.* 1756), *instrumentation.*

Documents concerning Langlois's life are rare and fragmentary, but from many accounts we know that between 1730 and 1756 he was the most highly regarded maker of scientific instruments in France and may be considered to have restored this art among the French. He and the other members of his trade provided scientists—at first astronomers—with the means of making highly precise observations during their many astronomical and geodesic missions. At the end of the seventeenth century the British instrument makers had become the undisputed leaders in the opinion of all European scientists, having surpassed Dutch workshops in their specialty—demonstration instruments for physics—and Italian workshops in the field of astronomical instruments (which the Italians had revitalized), and of instrumental optics (which they had created). French workshops were relatively numerous toward the beginning of the eighteenth century and furnished to a still small clientele all the types of instruments known. A few masters, like Nicolas Bion and Michael Butterfield, were better known than their colleagues, but none could be compared with the British competitors. Langlois, around the middle of the eighteenth century, was the first master to establish a firm tradition of precision workmanship in France. In 1730, the earliest date associated with his name, he received his first commission from the Paris observatory, for a wall quadrant of six-foot radius. He must therefore have been quite well known to French astronomers at this period to have merited the confidence represented by their commission. The signature on this instrument indicates that Langlois was established at the sign of the Niveau, probably on the Quai de l'Horloge, where most of the Parisian makers of scientific instruments were located. He was probably about thirty years old to have acquired not only the skill necessary for constructing large observatory wall instruments, but also the commercial foundation that could allow him to engage in such undertakings without serious financial risk. For no matter what price the purchaser paid, the instrument maker's investment in the form of tools, special equipment, and time was considerable in view of the very limited number of instruments of this size that a French workshop could hope to sell. Langlois was in fact the only French instrument maker to have produced such instruments for nearly thirty years. From his workshop came all the kinds of apparatus known at the time, for use in physics laboratories and schools, and by surveyors and navigators. Langlois's innovations in these areas included an improved pantograph that he submitted to the Académie des Sciences for approval.

The result of Langlois's activity seems to have been convincing enough for him to be chosen the official instrument maker of the French astronomers. About 1740 he was named *ingénieur en instruments de mathématiques* by the Academy and along with this title received lodgings at the Louvre, as did all officially designated "artists." This honor apparently reflected the maturing of Langlois's talent and his high reputation among astronomers, notably Cassini II, Cassini de Thury, Camus, Le Monnier, Maupertuis, and the Abbé de Lacaille. During the last twenty-five years of Langlois's life the Paris observatory ordered from him several instruments that were employed, along with English instruments, in the extensive triangulations of the first half of the century. In 1738 he constructed another wall quadrant of three-and-a half-foot radius and a six-foot sector; in 1742 a six-foot portable quadrant with certain features specified by Camus; a large sextant in 1750; and a three-foot portable quadrant in 1756, the year of his death.

In 1733 the Paris Academy decided to organize expeditions to Peru and Lapland to measure a meridian arc of one degree and the length of a seconds pendulum. Langlois was commissioned to construct five quadrants, of which three were taken to Peru and two to Lapland. In 1735 he furnished two standard *toises* established on the basis of the Chatelet *toise*, and each mission took along one standard. The Peru *toise*, also called the Academy *toise*, later replaced the Chatelet *toise* as the official standard. Langlois's successor, his nephew Canivet, was commissioned in 1766 to construct eighty copies, which were distributed to all the provincial *parlements*. Moreover, the Peru *toise* was used in 1793 to determine the length of the four platinum rulers constructed by Étienne Lenoir to measure the bases of Melun and Perpignan for the great meridian triangulation carried out by Delambre and Méchain. Several of Langlois's instruments, the six-foot sector of 1738 and a two-foot quadrant built in 1739, were employed in 1739 and 1740 by Cassini de Thury, Giovanni Domenico Maraldi, and the Abbé de Lacaille for the first series of triangulations undertaken to establish a new map of France. In 1744, Le Monnier entrusted Langlois with the restoration of the meridian gnomon of the Church of St. Sulpice in Paris, laid out twenty years earlier by the clockmaker Sully. The operculum, seventy-five feet above the south portal, was fitted with a lens having a focal length of eighty feet. Langlois placed the long copper plate representing the meridian on the ground; on the terminal obelisk he traced the curve and divisions of the solar pointer as a function of the calendar.

When Langlois died he was succeeded in his post at the Academy by Canivet, whom he had trained, and who in turn was succeeded by one of his own pupils, Lennel. Thus Langlois's workshop, established probably in the first quarter of the eighteenth century, pursued its activity until after 1785. Its founder applied to the construction of astronomical and geodesic instruments traditional methods, the insufficiencies of which were not overcome by English artisans until the 1740's and 1750's. Yet Langlois employed these methods with a skill superior to that of all his compatriots. Having inherited these qualities, his successors were fully prepared to adopt the new English methods. Thus a continuous tradition was established, linking Langlois with the brilliant generation of French instrument makers that flourished from 1780 to 1800 and participated in the establishment of the metric system.

BIBLIOGRAPHY

There are no written works by Langlois; his work survives only in his signed instruments. See G. Bigourdan, *Le système métrique* (Paris, 1901), p. 450; Maurice Daumas, *Les instruments scientifiques aux XVIIe et XVIIIe siècles* (Paris, 1953), p. 420; and C. Wolf, *Recherches historiques sur les étalons de poids et de mesures et les appareils qui ont servi à les construire* (Paris, 1882), p. 84; and *Histoire de l'observatoire* (Paris, 1902), p. 329.

MAURICE DAUMAS

LANGMUIR, IRVING (*b.* Brooklyn, New York, 31 January 1881; *d.* Falmouth, Massachusetts, 16 August 1957), *chemistry, physics.*

Irving Langmuir was the third of four children (all sons) of Charles Langmuir, a New York insurance executive of Scots ancestry, and Sadie Comings Langmuir, the daughter of a professor of anatomy and a descendant, on her mother's side, of the early English settlers who arrived in America on the *Mayflower*. Irving Langmuir alternately attended schools in New York and in Paris, where his father headed the European agencies of the New York Life Insurance Company. Irving was in his last year at the Pratt Institute's manual training high school in Brooklyn in 1898 when his father unexpectedly died at the age of fifty-four of pneumonia contracted during a winter transatlantic crossing. His mother lived to be eighty-seven. Fortunately, Charles Langmuir had provided well for his family, and Irving was able to enter Columbia University. Although deeply interested in science, he chose the more exacting curriculum of the School of Mines. "The course was strong in chemistry," he said later. "It had more physics than the chemical course, and more mathematics than the course in physics—and I wanted all three." He received the degree of metallurgical engineer (equivalent to bachelor of science) in 1903 and decided to enter on a course of postgraduate study in Germany. He hesitated between Leipzig and Göttingen and chose the latter, where he worked under

Walther Nernst on the dissociation of various gases by a glowing platinum wire, research that became the topic of his 1906 doctoral dissertation. The decision for Göttingen very likely had a profound influence on Langmuir's career, for not only was Nernst deeply engaged in the work on thermodynamics that was to earn him the Nobel Prize in 1920, but he was also favorably inclined toward applied research. Nernst had devised a new type of electric lamp that became a great commercial success, in a sense presaging Langmuir's own career in industrial research. At Leipzig, Langmuir would probably have come under the influence of Wilhelm Ostwald and might not have become interested in the problems of molecular and atomic structure that underlie much of his mature work, since Ostwald had little faith in that approach.

Armed with a doctorate from Göttingen, Langmuir returned to America and for a time taught chemistry at Stevens Institute of Technology in Hoboken, New Jersey, but the position gave little scope to his talents. After a summer job at the new research laboratory that the General Electric Company had established in 1901 in Schenectady, New York, he gratefully accepted an offer to stay there permanently. He remained there for forty-one years until his retirement in 1950 and continued as a consultant.

Langmuir's scientific work was extremely varied. During overlapping periods he made major contributions in at least seven fields: chemical reactions at high temperatures and low pressures (1906–1921); thermal effects in gases (1911–1936); atomic structure (1919–1921); thermionic emission and surfaces in vacuum (1913–1937); chemical forces in solids, liquids, and surface films (1916–1943); electrical discharges in gases (1923–1932); and atmospheric science (1938–1955). Much of his work led to technological developments of wide and lasting importance.

His work on chemical reactions at high temperatures and low pressures derived from his dissertation under Nernst, who wanted to understand the equilibrium conditions governing the formation of nitric oxide from air in the vicinity of a hot wire. The scope of the investigation presently came to include other gaseous equilibria. Langmuir was fascinated by the simplicity of the experimental setup, which resembled an incandescent light bulb. The electric lamp of the day was a relatively inefficient device containing a fragile carbon filament and was severely limited in lifetime and power. (Electric street lights used arc lamps.) Tantalum and tungsten filaments were being tried but did not last much longer even in the best available vacuum. Langmuir's first task at General Electric was to find the reason for the short life, which proved to be residual gases adhering to the glass envelope and other impurities. This research ultimately led to the recognition that filling the lamp with inert gases (a nitrogen-argon mixture worked best) greatly increased both lifetime and efficiency, a discovery that revolutionized the electric-lamp industry. The same studies led to the discovery and understanding of the formation of atomic hydrogen from molecular hydrogen and to a series of other results in the fields of heat transfer and low-pressure phenomena.

Continuing experiments on hot tungsten filaments in various gases, Langmuir turned to a study of heat losses and evaporation from hot filaments. He found that the laws governing the evaporation of tungsten into nitrogen resembled those of heat convection: thin wires lost as much tungsten per unit area as thick wires, but the latter were more efficient, a contradiction that led to the adoption of a lamp filament made from a thin wire tightly coiled into a helix. He also estimated the heat of dissociation of hydrogen and observed various properties of atomic hydrogen, including its adsorption on cold glass walls. The heating effects produced by the recombination of atomic hydrogen later led Langmuir to invent the atomic-hydrogen welding torch, in which an arc between tungsten electrodes in hydrogen produces hydrogen atoms that create heat when they recombine on the metal to be welded.

Of all his major contributions, Langmuir's sortie into atomic structure occupied him for the shortest period (1919–1921). Building upon the ideas of G. N. Lewis, Langmuir suggested a modification in the model of the atom proposed by his idol, Niels Bohr. Although the Lewis-Langmuir theory, called the "octet theory" of chemical valence, ultimately yielded to the quantum-mechanical concepts of chemical bonds, it proved to be astonishingly successful from a practical viewpoint, explaining a great variety of chemical phenomena by essentially classical methods. At this time, Langmuir apparently made a deliberate decision to concentrate his efforts on fields in which such classical methods were likely to yield new results and recognized that atomic structure was not one of them.

Concurrently with these efforts, he was engaged in investigations of thermionic emission and of surfaces in vacuum, two fields that proved of tremendous fundamental and practical importance. Elucidating the phenomenon of space charge (the cloud of charged particles that maintains itself in the interelectrode space) he independently derived for electrons the relationship established previously for ions by C. D. Child. Now known as the Child-Langmuir space-charge equation, the relationship shows that the current between the electrodes is proportional to the voltage raised to the 3/2 power, regardless of the shapes of the electrodes. This law, which underlies the design of a great variety of electron tubes and other devices, was elaborated by Langmuir and his co-workers for several geometrical configurations. At the same time, Langmuir observed that certain admixtures such as thoria, an oxide of thorium (ThO_2), greatly enhanced thermionic emission from tungsten, a result that he correctly ascribed to an effective lowering of the work function by the diffusion of a single layer of thorium atoms to the surface of the tungsten filament. His other studies in this field related to the changes in emission and in ionization produced by the presence of cesium, and were also destined to become of technological importance. A development of more immediate import was his invention of a condensation pump, which produces a good vacuum very quickly with the help of a refrigerant such as liquid nitrogen. This device rapidly came into widespread industrial and laboratory use and continued to dominate the field for decades.

The largest single topic to occupy Langmuir's attention was surface chemistry—the study of chemical forces at interfaces between different substances. He evolved a new concept of adsorption, according to which every molecule striking a surface remains in contact for a time before evaporating, so that a firmly held monolayer is formed. He developed a multitude of experimental techniques for studying surface films on liquids. Turning to solids, he developed the Langmuir adsorption isotherm, an expression for the fraction of surface covered by the adsorbed layer as a function of pressure and the temperature-dependent rates of surface condensation and evaporation. He also characterized the catalytic effect of an adsorbing surface by considering the chemical reaction as occurring in the film, a concept that served to explain many phenomena of surface kinetics that had not been previously understood. It was largely in recognition of this work that Langmuir was awarded the Nobel Prize in chemistry in 1932.

Possibly of even greater lasting significance was his work relating to electric discharges in gases, which derived from an interest in mercury-arc and other gaseous devices used in the control of heavy alternating currents. This effort led to a large body of experimental and analytical research concerned with electron-ion interactions, ionization, oscillations, and other aspects of a space-charge-free ionized gas, the richly complex and inherently unstable medium for which he coined the term "plasma." Subsequent work in such fields as electron physics, magnetohydrodynamics (MHD), and the control of thermonuclear fusion depends heavily on the basic results on plasma first reported in the papers of Langmuir and his associates. In these papers, the concept of electron "temperature" was introduced as well as a method of measuring both it and the ion density by a special electrode, now called the Langmuir probe.

The major field that occupied Langmuir into his retirement years was atmospheric science. Following a lifelong preoccupation with the weather (he was a great outdoorsman, a capable sailor, and an enthusiastic amateur flyer), he studied such phenomena as the regular formation of streaks of seaweed on the wind-blown surface of the sea (windrows), the formation of liquid particles of various sizes in air, and the nucleation of ice crystals in supercooled clouds by "seeding" with solid carbon dioxide particles. This last work, which was based on a discovery of Langmuir's associate V. J. Schaefer, led to the beginnings of artificial weather control, although early efforts at producing rain and diverting storms were marred by much scientific and public controversy regarding the efficacy of cloud seeding.

Langmuir was the first eminent scientist employed by an industrial laboratory to be honored by a Nobel Prize—an aspect of his career that had an impact on the worldwide development of scientific research and that lent a certain cachet to such endeavors. Forward-looking industries everywhere began to appreciate the importance of undirected basic research of a sort previously restricted to universities.

Langmuir's mature life was serene. At the age of thirty-one he married Marion Mersereau of Schenectady; they adopted two children, Kenneth and Barbara. He was close to his brothers and to his many nephews and nieces, several of whom also achieved distinction in scientific fields. Langmuir's elder brother Arthur was a successful industrial chemist, a fact that doubtless influenced young Irving's choice of a career. Moreover, Arthur's enthusiasm for science transmitted itself to Irving through their exceptionally close relationship. Langmuir's interests extended to such diverse fields as the Boy Scout movement, aviation, and music—he was a personal friend of Charles Lindbergh and of the conductor and sound-recording innovator Leopold Stokowski. He also concerned himself with problems of public policy such as the conservation of wilderness areas and control of atomic energy. In 1935 he stood for the city council of Schenectady, N. Y., but failed to be elected. He participated actively in America's scientific war efforts during both world wars: in antisubmarine defense and nitrate production during the first, and in aircraft de-icing and smoke-screen research (both outgrowths of his interest in meteorology) during the second. He worked best with small teams of collaborators. Outstanding among them was Dr. Katherine Blodgett, with whom he worked closely for over thirty years.

In addition to the Nobel Prize, Langmuir received many other honors, among them the Hughes Medal of the Royal Society (1918), the American Rumford Medal (1920), the American Chemical Society's Nichols Medal (1920) and Gibbs Medal (1930), the Franklin Medal (1934), and both Faraday Medals: that of the Chemical Society (1938) and of the Institution of Electrical Engineers (1943). He was granted fifteen honorary degrees. He was a member of the National Academy and a foreign member of the Royal Society and of other bodies. Mount Langmuir in Alaska is named after him, and in 1970 a residential college at the State University of New York at Stony Brook was named Irving Langmuir College in his honor.

BIBLIOGRAPHY

I. Original Works. Langmuir received 63 patents and published over 200 papers and reports between 1906 and 1956. Virtually all his papers appear in the 12-vol. memorial edition of *The Collected Works of Irving Langmuir*, C. G. Smits, ed. (London–New York, 1960–1962), which is devoted to the following topics: 1. Low-Pressure Phenomena, 2. Heat Transfer—Incandescent Tungsten, 3. Thermionic Phenomena, 4. Electrical Discharge, 5. Plasma and Oscillations, 6. Structure of Matter, 7. Protein Structures, 8. Properties of Matter, 9. Surface Phenomena, 10. Atmospheric Phenomena, 11. Cloud Nucleation, 12. The Man and the Scientist. Scattered throughout are memorial essays by scientists who knew him, including a book-length biography by Albert Rosenfeld, "The Quintessence of Irving Langmuir" in vol. XII, pt. 1, also published separately (New York–Oxford, 1966). In addition to a complete list of Langmuir's own papers, it contains an extensive bibliography of writings about him. Langmuir was also author of *Phenomena, Atoms and Molecules* (New York, 1950).

II. Secondary Literature. Tributes are by Sir Hugh Taylor in *Biographical Memoirs of Fellows of the Royal Society*, **4** (1958), 167–184; A. W. Hull in *Nature*, **181** (1958), 148; W. R. Whitney in *Yearbook. American Philosophical Society* (1957), 129–133; and Katherine B. Blodgett in *Vacuum*, **5** (1957), 1–3.

Charles Süsskind

LANGREN, MICHAEL FLORENT VAN (*b.* Mechlin or Antwerp, Belgium, *ca.* 1600; *d.* Brussels, Belgium, early May 1675), *engineering, cartography, selenography.*

Very little is known of Van Langren's life. He was the son of Arnold Florent Van Langren, archducal spherographer. His parents, who were Catholics, had left Holland before 1600 for the Spanish Netherlands and finally settled in Brussels in 1611. Like his father, Michael Florent became a royal cosmographer and mathematician. He married Jeanne de Quantere, by whom he had several children.

The primary goal of Van Langren's research was to discover a method of determining longitude at sea. As early as 1621 he attempted to do this by means of lunar observations, basing his procedure on the darkenings and illuminations of the lunar mountains; this approach obviously required good maps of the moon and a precise toponymy. In 1625, at Dunkirk, he presented his method to the archduchess Isabelle. The following year he requested and received assistance to travel to Spain in order to present his method to the king and to call attention to his books *Tábulas astronómicas y hydrográphicas*, which are now lost. The journey lasted from 1631 to 1634; the king promised to finance the publication of his observations. The *Advertencias de Miguel Florencio Van Langren . . . a todos los professores y amadores de la mathemática* . . . appeared in Madrid about 1634.

Van Langren continued his investigations in Brussels, publishing a *Calendarium perpetuum* . . . in 1636 and *La verdadera longitud por mar y tierra* . . . in 1644. Before 15 February 1645 he submitted to the Privy Council a manuscript map of the moon in support of his request for a privilege to publish it (entitled *Luna vel lumina Austriaca Philippica*; it is preserved in Brussels in the Archives Générales du Royaume, Cartes et Plans MS no. 7911). The map, engraved by Van Langren himself, appeared in May 1645. As is indicated by the title, *Plenilunii/Lumina Austriaca Philippica*, it presents the full moon and, as is stated in Van Langren's long legend, it was to be a part of a series of maps showing thirty phases of the moon. These maps, a veritable selenography, were prepared but were never published, owing to the death of Van Langren's patron, Erycius Puteanus, and to the wars. Van Langren also published a text on the comet of 1652, *Repraesentatio partis caeli quam cometa* . . . .

Van Langren also drew geographical maps, several of which have been lost. They included one of a canal from the Meuse to the Rhine (entitled *Fossa Eugeniana* in the first printings and *Fossa Sanctae Mariae* in the following ones [1628]), one of the archdiocese of Mechlin, one of Luxembourg, and one of the three parts of the former duchy of Brabant; the latter three appeared in the *Novus atlas* of W.J. Blaeu. In addition he provided maps and plans for projects for large public works.

In 1640, in Brussels, Van Langren published

Tormentum bellicum trisphaerium, concerning a three-shot cannon. His studies and projects for a port to be constructed at Mardyck, near Dunkirk, date from 1624; and in 1653 he published, again in Brussels, *Description particulière du . . . Banc de Mardijck . . .*, which ran to two editions. His work on the harbor-port of Ostend is set forth in *Profytelijcken middel om . . . de Zee-Haven van Oostende te verbeteren* (Brussels, 1650), in *Briefve description de la ville et havre d'Oostende . . .* (Brussels, 1659), and in *Copies de la VIe, XIe et XIIIe lettre que . . . don Juan d'Austriche a escrit . . . à Michel-Florencio Van Langren . . .* (1667). In 1661 he wrote a short treatise on the cleaning of the canals of Antwerp: *Bewys van de alder-bequaemste ende profytelyckste inventie . . . .* His works concerning Brussels dealt with fortifications, canals, and means of protecting the city from the flooding of the Senne: *Invention et proposition que Michel Florencio Van Langren . . . a faict . . . pour empescher . . . le debordement de la rivière de Senne . . .* (Brussels, 1644), *Eenighe middelen om . . . Brussel van de inondacien . . . te bevryden* (Brussels, 1648), and *Michael Florencio Van Langren . . . sprekende . . . stadt Brussel . . .* (Brussels, 1658).

Van Langren was in contact with most of the scientists of his time, including Erycius Puteanus, Gottfried Wendelin, the French astronomer Boulliau, Barthélémy Petit, the Jesuit Jean Charles de La Faille, and Huygens. He also knew the writings of Hevelius and Riccioli.

As an engineer, a cartographer, and an engraver, Van Langren occupied a leading place in the Spanish Netherlands. As a selenographer he was, like Hevelius, a pioneer. Circumstances did not permit him to develop his talents fully, and only a thorough study of his work will allow us to arrive at a just appreciation of his achievements.

BIBLIOGRAPHY

Eleven of Van Langren's published works are described in *Bibliotheca belgica*, 1st ser., XIII (Ghent–The Hague, 1880–1890), see index; and in *Bibliotheca belgica*, repub. under the direction of M. T. Lenger, III (Brussels, 1964), 674–677.

Secondary literature includes the following: D. Bierens de Haan, "Constantijn Huygens, als waterbouwkundige, Michael Florentz van Langren, in Bouwstoffen voor de geschiedenis der wis- en natuurkundige wetenschappen in de Nederlanden no. XXXIII," in *Verhandelingen der K. akademie van wetenschappen*, sec. 1, **2**, no. 1 (1893); H. Bosmans, "La carte lunaire de Van Langren conservée aux Archives générales du royaume, à Bruxelles," in *Revue des questions scientifiques*, 3rd ser., **4** (1903), 108–139, with facsimile; "Sur un pamphlet concernant les travaux à effectuer au port d'Ostende, publié en 1660 à Bruxelles," in *Revue des bibliothèques et archives de Belgique*, **1** (1903), 287–291; and "La carte lunaire de Van Langren conservée à l'Université de Leyde," in *Revue des questions scientifiques*, 3rd ser., **17** (1910), 248–264, with facsimile; L. Godeaux, *Esquisse d'une histoire des mathématiques en Belgique* (Brussels, 1943), pp. 19–20; J.-C. Houzeau, "Extrait des notes prises à la Bibliothèque royale, à Paris, en mars 1844," in *Bulletin de l'Académie royale . . . de Belgique. Classe des sciences*, **19**, pt. 3 (1852), 498–507; G. des Marez, "Notice sur les documents relatifs à Michel-Florent Van Langren . . . conservés aux Archives de la ville de Bruxelles," in *Revue des bibliothèques et archives de Belgique*, **1** (1903), 371–378, and **2** (1904), 23–31; W. Prinz, "L'original de la première carte lunaire de Van Langren," in *Ciel et terre*, **24** (1903–1904), 99–105, 149–155, with facsimile; Erycius Puteanus, *Honderd veertien Nederlandse brieven aan de astronoom Michael Florent Van Langren*, with intro. by J. J. Moreau (Antwerp–Amsterdam, 1957); A. Quetelet, *Histoire des sciences mathématiques et physiques chez les Belges*, new ed. (Brussels, 1871), pp. 247–253; A. Tiberghien, "Contribution à la bibliographie de M. F. Van Langren. Documents existant à la Bibliothèque royale de Belgique," in *Revue des bibliothèques et archives de Belgique*, **2** (1904), 1–14; and A. Wauters, "Langren (Michel-Florent Van)," in *Biographie nationale . . . de Belgique*, XI (Brussels, 1890–1891), cols. 276–292.

A. DE SMET

LANKESTER, EDWIN RAY (*b.* London, England, 15 May 1847; *d.* London, 15 August 1929), *zoology, natural history.*

The son of Edwin Lankester, M.D., Lankester was educated at St. Paul's School, London; Downing College, Cambridge; and Christ Church, Oxford. After graduating in zoology and geology, he studied at Vienna, Leipzig, and the Zoological Station at Naples. Elected a fellow of Exeter College, Oxford, in 1872, he was two years later appointed professor of zoology at University College, London, a post which he held until 1891, when he was appointed Linacre professor of comparative anatomy at Oxford. While in London he made the acquaintance of an art student at the Slade School, Edwin Stephen Goodrich, whom he inspired with zeal for comparative anatomy. He took him to Oxford; and, through Goodrich, his most distinguished pupil, he spread his teaching in zoology. In 1884 Lankester was foremost in promoting the foundation of the Marine Biological Association, whose laboratory at Plymouth has played a leading part in the training of British biologists. In 1898 Lankester was appointed director of the British Museum (Natural History), but not being suited to administrative work, he retired when he was knighted in 1907. He was elected fellow of the Royal Society in 1875 and corresponding member of the Paris Academy of Sciences in 1899.

Lankester's original researches extended over all major groups of living and fossil animals, from protozoa to mammals. His first paper, on *Pteraspis*, was published when he was sixteen. He demonstrated the fundamental similarities in structure, and therefore the affinities, between spiders, scorpions, and horseshoe crabs. He systematized the field of embryology and introduced the indispensable terms "stomodaeum," "blastopore," and "invagination." A firm supporter and friend of Charles Darwin and of the latter's theories on evolution and natural selection, he distinguished clearly between homology, homoplasy, and analogy of compared organs. Furthermore, he exposed the illogicality of Lamarck's speculations by showing that every character developed by every organism is a response by it to the conditions of its environment. No character is solely inherited or solely acquired; all are, to varying extents, both. Thus the phrase "inheritance of acquired characters" is meaningless.

The strength of Lankester's character and the forcefulness of his personality may be illustrated by his visit to the laboratory of the French neurologist Jean-Martin Charcot, who was experimenting on the power of electric currents (in wire conductors wound round the arms of the subjects) to reinforce suggestion and annul pain. On the order to make electric contact with a battery of storage cells, darning needles were pushed through the arms of young women, who felt nothing. Lankester arrived just before lunchtime and was allowed to wait for Charcot and his team in the laboratory. He employed his time in emptying the acid from all the storage cells, which he then filled with tap water, so that they gave no current. After lunch the experiments were renewed, contacts made, and needles pushed through the arms of the women, who still felt nothing. Lankester then confessed what he had done and, horrified, the team verified that no current flowed. After an awkward silence, Charcot embraced him.

BIBLIOGRAPHY

Lankester's scientific work was published in the journals of the Royal Society and the Palaeontographical Society, and in the *Quarterly Journal of Microscopical Science* (which he edited and brought to a standard of international repute). He contributed numerous articles to the *Encyclopaedia Britannica*, which were reprinted as *Zoological Articles*. He also edited *Treatise on Zoology* (1900–1909), of which 9 vols. appeared; it did much to help the teaching of zoology in English-speaking countries.

After his retirement, Lankester devoted himself to writing books on science for the lay reader: *Science From an Easy Chair* (1908); *Extinct Animals* (1909); *Diversions of a Naturalist* (1915); *Great and Small Things* (1923), which were a great success in introducing the public to scientific progress.

E. S. Goodrich, "Edwin Ray Lankester," in *Proceedings of the Royal Society* (1930), contains a portrait and a bibliography of his scientific papers.

GAVIN DE BEER

LANSBERGE, PHILIP VAN (*b.* Ghent, Belgium, 25 August 1561; *d.* Middelburg, Netherlands, 8 December 1632), *geometry, astronomy.*

On account of the religious troubles of those days his parents Daniel van Lansberge, lord of Meulebeke, and Pauline van den Honingh found themselves obliged to go to France in 1566 and afterward to England, where Philip studied mathematics and theology. He was already back in Belgium in 1579, where he received a call to be a Protestant minister in Antwerp in 1580. After the conquest of Antwerp by Spain on 16 August 1585, Philip left Belgium definitively to establish himself in the Netherlands. He went to Leiden, where he enrolled as a theological student. From 1586 to 1613 he was a Protestant minister at Goes in Zeeland, after which he went to Middelburg, where he died in 1632.

The Danish mathematician Thomas Finck had published an important work, the *Geometriae rotundi*, in Basel in 1583. It is this work that Van Lansberge seems to have followed very closely in his first mathematical study, *Triangulorum geometriae libri IV* of 1591. The first book is devoted to the definitions of the trigonometric functions. Van Lansberge, following Maurice Bressieu, used the term "radius" instead of "sinus totus." The second book contains the method of constructing the tables of sines, tangents, and secants, largely derived from those of Viète and Finck, and the tables themselves, which were used by Kepler in his calculations. The third book is devoted to the solution of plane triangles and is accompanied by numerical illustrations. Van Lansberge's statement and proof of the sine law differs very little from that given by Regiomontanus. The fourth book deals with spherical trigonometry; the first eleven items concern spherical geometry. In the solution of spherical triangles Van Lansberge employs a device similar to that of Bressieu in his *Metrices astronomicae* (Paris, 1581), the marking of the given parts of a triangle by two strokes. Van Lansberge's new proof for the cosine theorem for sides (Book IV, item 17) marks the first time that the theorem appeared in print for angles as well as sides.

But although Van Lansberge may lay claim to the discovery of the theorem for angles, sufficient evidence indicates that this theorem was known to Viète and to Tycho Brahe. On the whole Van Lansberge shows little originality in the content of his trigonometry, but his arrangement of definitions and propositions is less complicated and more systematic than that of Viète and Clavius.

In 1616 Van Lansberge published his *Cyclometriae novae libri II*, which was attacked the same year by Alexander Anderson in his *Vindiciae Archimedis, sive elenchus cyclometriae novae a Philippo Lansbergio nuper editae*. In this book Van Lansberge occupied himself with approximating the ratio between the circumference (Book I), the area (Book II), and the diameter of the circle. He carried the value of π to 28 decimal places by means of a method in which he seems to have joined the quadratrix of the ancients to trigonometric considerations. He thought that he had found a better approximation than that of Ludolf van Ceulen, who had used the Archimedean method of inscribed and circumscribed polygons and had carried the value of π to thirty-five decimal places in 1615. In his *Progymnasmatum astronomiae restitutae de motu solis* (Middelburg, 1619) Van Lansberge taught the probability of the earth's motion according to the Copernican doctrine; the same is true of *Bedenckingen op den dagelyckschen, ende jaerlyckschen loop van den aerdt-kloot* (Middelburg, 1629), translated into Latin by M. Hortensius as *Commentationes in motum terrae diurnum, et annuum* (Middelburg, 1630). Both works were attacked for their Copernican ideas by Morin in his *Famosi et antiqui problematis de telluris motu vel...* (Paris, 1631), and by Libert Froidmond in his *Anti-Aristarchus; sive orbis-terrae immobilis; liber I* (Antwerp, 1631). Although a follower of Copernicus, Van Lansberge did not accept the planetary theories of Kepler altogether. His *Tabulae motuum coelestium perpetuae* (Middelburg, 1632), founded on an epicyclic theory, were much used among astronomers, although they were very inferior to Kepler's Rudolphine tables.

BIBLIOGRAPHY

I. Original Works. Van Lansberge's works were published as *Philippi Lansbergii Astronomi Celebrium Opera Omnia* (Middelburg, 1663).

II. Secondary Literature. A very good biography is by C. de Waard in *Nieuw Nederlandsch biografisch woordenboek* (Leiden, 1912), cols. 775–782. See also the following (listed chronologically): J. E. Montucla, *Histoire des mathématiques*, II (Paris, 1799), 334; D. Bierens de Haan, "Notice sur quelques quadrateurs du cercle dans les Pays-Bas," in *Bullettino di bibliografia e di storia della scienze matematiche e fisiche*, **7** (1874), 120, 121; A. J. van der Aa, *Biographisch woordenboek der Nederlanden*, XI (Haarlem, 1876), 154–157; D. Bierens de Haan, "Bibliographie néerlandaise," in *Bullettino di bibliografia e di storia della scienze matematiche e fisiche*, **15** (1882), 229–231; A. Von Braunmühl, *Vorlesungen über Geschichte der Trigonometrie*, I (Leipzig, 1900), 175, 176, 192; C. de Waard, "Nog twee brieven van Philips Lansbergen," in *Archief vroegere en latere mededelingen ... in Middelburg* (1915), 93–99; H. Bosmans, "Philippe van Lansberge, de Gand, 1561–1632," in *Mathésis; recueil mathématique*, **42** (1928), 5–10; C. de Waard, ed., *Correspondance du M. Mersenne*, II (Paris, 1937), 36, 511–513; and M. C. Zeller, *The Development of Trigonometry From Regiomontanus to Pitiscus* (Ann Arbor., Mich., 1946), 87, 94–97.

H. L. L. Busard

LAPICQUE, LOUIS (*b.* Épinal, France, 1 August 1866; *d.* Paris, France, 6 December 1952), *physiology, anthropology.*

Lapicque's father, a veterinary surgeon, fostered his son's inclination toward natural sciences. At the Paris Medical School, Lapicque displayed an active interest in physics and chemistry, which in those days was rather unusual in medical circles. He was therefore entrusted by the celebrated clinician Germain Sée to organize a small chemical laboratory in the Hôtel-Dieu Hospital, where he investigated the circulation of iron in vertebrates. The amounts of metal to be measured being very small, he tried colorimetric tests. Such tests, despite their sensitivity, were then in disfavor because of their lack of accuracy. Lapicque succeeded in making the thiocyanate test for ferric iron a very precise tool that remained in general use for several decades. He showed that at birth iron is localized mainly in the liver, while there is none in the spleen. When pathological conditions bring severe destruction of red cells, iron accumulates in spleen and liver tissue in the form of a red pigment constituted essentially of colloidal ferric hydroxide. These results were published in his thesis for the doctorate in science. Lapicque's experimental work was then interrupted. A Mme Lebaudy, the wife of a sugar magnate, wanted her son to take a world cruise on her yacht *Semiramis* with a group of dedicated young scientists who might inspire him. The son failed to embark, but the cruise proceeded as planned. Lapicque proved to be an active anthropologist. In the lands of the Indian Ocean he endeavored to measure the anthropological characteristics of the various groups of Negritos. He devised a very useful index: the radius-pelvic ratio, which allowed him to conclude that in ancient times a single Negro race was distributed from Africa to Oceania.

Lapicque made an extensive study of the relation between brain and body weight, a relation discussed by Eugène Dubois, the discover of *Pithecanthropus erectus*. Lapicque, however, generalized this relation. The "exponent of relation" is the same not only between mammals but also between birds. The results appear most strikingly on logarithmic coordinates. The different animal species, even the fossil ones, are represented on parallel lines, the level of each defined by "the coefficient of cephalization," which represents the degree of nervous organization and, in a striking manner, the presumed intelligence of the species. Lapicque also found a precise relation between brain and eye weight.

Lapicque's essential work, however, was in the general physiology of the nervous system. From 1902 he investigated the time factor in nervous processes. From this study he soon derived a fundamental concept: the functional importance of the time factor of nervous electrical excitability. He made it possible to express excitability precisely in terms of "rheobase," an intensity factor, and of "chronaxie," a time factor. Chronaxie represents the functional rapidity of the tissue under investigation. Slow muscles and nerves are characterized by a long chronaxie and fast muscles and nerves by a short one. Furthermore, chronaxie measurements, for the first time, numerically expressed the effects of many agents: for instance, temperature, drugs, and anesthetics. The chronaxies of motor nerves in man could be obtained, thus permitting the evolution of degenerative or regenerative processes to be followed quantitatively. Chronaxie also reveals the fine details of motor organization.

As early as 1907 the concept of the functional role of the time factor of nervous processes, expressed by chronaxie, led Lapicque to theoretical speculations. He postulated that the activation of a chain of nerve cells depended upon the successive electrical stimulation of each cell by the impulse or action potential of the preceding one. By using an electrical stimulus having a shape similar to that of the nerve action potential, he showed that there is such a stimulus of optimal duration defined by the chronaxie of every tissue investigated. In other words, this is selective excitation realized by a tuning between duration of stimulus and chronaxie. Thus Lapicque proposed a theory of nervous processes that was analogous in its principle to the tuning or resonance between two oscillating radio circuits. The theory indicated that transmission of excitation between two nerve cells was optimal when the cells had the same chronaxie. When the second cell had a long chronaxie, its excitation required iterative activation of the first. In that case the numerical expressions derived by Lapicque are adequate regardless of whether transmission depends on electrical stimulation or, according to modern findings, depends upon liberation of successive "quanta" of a chemical mediator such as acetylcholine. Lapicque's theory was first met with widespread approval and later encountered strong criticism. Yet even in recent years it has provoked much research. It correlated many unexplained results at a time when nervous processes could be investigated only by the indirect methods of electrical stimulation. When the responses of nerve cells and fibers became directly accessible through modern electronics, several predictions based on Lapicque's theory were confirmed. For instance, in 1913 Lapicque and René Legendre had shown that chronaxie of motor fibers or axons is in inverse ratio to their diameter. Thus the velocity of their impulse was assumed to be proportional to their diameter, which was demonstrated by the cathode-ray oscillographic records obtained by Gasser and Erlanger in 1928.

Lapicque's activity was fruitful in many other areas. In nutritional physiology he added to the notion of isodynamic nutrients, developed by Rubner, a new concept: the margin of thermogenesis. His competence in that domain was amply used during World War I by the authorities responsible for the food supply of France. He was the first to advocate, in exceptional circumstances, the reduction of livestock to save grain that could be used for human consumption. This suggestion met with considerable opposition but was finally supported by the government. In connection with saving grain Lapicque found that certain seaweeds (*Laminaria*), collected in the proper season and correctly prepared, made an excellent oat substitute for the feeding of horses. He also investigated the physiology of marine algae. He discovered that ionic exchanges between the cell medium and the surrounding seawater proceed against an osmotic gradient, through the expenditure of metabolic energy. This process, involving an ionic "pump," he named "epictesis," akin to the modern concept of active transport.

As the field surgeon of an infantry regiment, Lapicque devised an adequate defense against the first attacks by poison gas: a filter of loose earth at the entrance of a trench shelter absorbed the gas. Unfortunately, this simple device was inefficient against the mustard gas later used by the German army. Lapicque always showed an unyielding and active opposition to any form of subjugation. In his late seventies he was jailed by the Gestapo for two months during the severe winter of 1941-1942.

Lapicque was an excellent sailor. Every summer he expertly navigated a twenty-ton yawl named *Axone* in the difficult waters off northern Brittany. The members of his laboratory were often on board, enjoying his teaching, which combined physiology and seamanship. He held prestigious teaching posts: associate professor at the Sorbonne in 1899, professor of general physiology at the Muséum d'Histoire Naturelle in 1911, and at the Sorbonne from 1919 to 1936. Both chairs had been Claude Bernard's. Lapicque's activity was not curtailed by his retirement; he studied many problems, such as the density of nerve cells in the various centers. Until his last days he gladly gave his extensive and diverse scientific experience to his students.

BIBLIOGRAPHY

I. Original Works. Lapicque's writings include *Observations et expériences sur la mutation du fer chez les vertébrés*, his *thèse de sciences* (Paris, 1897); "Unité fondamentale des races d'hommes à peau noire. Indice radio-pelvien," in *Comptes rendus de l'Académie des sciences*, **143** (1906), 81–84; "Échanges nutritifs des animaux en fonction du poids corporel," *ibid.*, **172** (1921), 1526–1529; "Mécanisme des échanges entre la cellule et le milieu ambiant," *ibid.*, **174** (1922), 1490–1492; *L'excitabilité en fonction du temps; la chronaxie, sa signification et sa mesure* (Paris, 1926); "Le poids du cerveau et l'intelligence," in G. Dumas, *Nouveau traité de psychologie*, I (Paris, 1930), 204; "Le problème du fonctionnement nerveux. Esquisse d'une solution," in G. Dumas, *Nouveau traité de psychologie*, I (Paris, 1930), 147–204; and *La machine nerveuse* (Paris, 1943), part of which was written during his imprisonment by the Germans.

II. Secondary Literature. See G. Bourguignon, *La chronaxie chez l'homme* (Paris, 1923); and A. Monnier, *L'excitation électrique des tissus. Essai d'interprétation physique* (Paris, 1934).

A. M. Monnier

LAPLACE, PIERRE SIMON, MARQUIS DE (*b.* Beaumont-en-Auge, Normandy, France, 28 March 1749; *d.* Paris, France, 5 March 1827), *mathematics, astronomy, physics.*

For a detailed study of his life and work, see Supplement.

LAPPARENT, ALBERT-AUGUSTE COCHON DE (*b.* Bourges, France, 30 December 1839; *d.* Paris, France, 4 May 1908), *geology.*

Lapparent was descended from a noble family of the Vendée. His paternal grandfather, Emmanuel, was a member of the first graduating class of the École Polytechnique (1794); and his father and uncle Henri were students there. Lapparent was admitted with highest standing to the École Polytechnique in 1858, after compiling a good record at the Lycée Bonaparte. He graduated first in his class and then entered the École des Mines, where he studied under Élie de Beaumont. He graduated in 1864, again first in his class.

Lapparent perfected his German and twice traveled in Germany in order to further his studies (1862, 1863). In 1863, in the town of Predazzo in the Tirol, he was the first to describe and name the rock called monzonite. During a stay in Rome in that same year he was enrolled in the Academy of St. Philip Neri, an institution concerned with religious, social, and economic questions. He became its secretary and excelled in summarizing the sometimes confusing discussions of the speakers. An excellent dancer, Lapparent frequented the Monday gatherings held by the Empress Eugénie in 1864. He became an assistant curator (*conservateur-adjoint*) at the École des Mines and, at the request of Achille Delesse, he agreed to edit the reviews of the memoirs on stratigraphy for the *Revue de géologie*. He carried out this task for fifteen years, aided by his knowledge of foreign languages, and acquired a profound knowledge of geology.

From 1865 to 1875 Lapparent supervised the surveying for six maps (scale of 1:80,000) of the Paris Basin. He established the stratigraphy and tectonics of the anticline of the Bray region, recording his findings on structural maps with contour lines. At Mortain he described faults "en échelons." He attributed the origin of the silt of the plateaus to a hypothetical alluvial process; more recent work, however, has tended to stress instead the role of wind (loess) and percolating water. On the other hand, in the controversy over the benches called "rideaux" in the fields of chalky regions, Lapparent presented arguments against their having had a natural origin and for an artificial one, such as ploughing. Since then aerial photographs and the investigations of M. Brochu and R. Agache have fully justified his views.

At the Paris Universal Exposition in 1867 Lapparent was the reporter for the public lectures on the standardization of weights and measures. Following this exposition the idea of a tunnel under the English Channel between England and France, unsuccessfully advanced as early as 1802, was again taken up and an Anglo-French committee was formed. Lapparent was one of the three French geologists assigned to study the project. A total of 7,671 dredgings were carried out at sea, 3,267 of which furnished samples. The continuity of the cretaceous beds on both sides of the Strait of Dover, the good quality of the sixty-meter-thick Cenomanian chalk, and the absence of faults were all demonstrated. The tunnel was possible. For his efforts he was made a member of the Legion of Honor.

Lapparent married in 1868. The marriage brought him happiness, nine children (six of whom survived), and financial independence.

In 1875 the Catholic University (later the Catholic Institute) was founded in Paris. A confirmed Catholic, Lapparent accepted the chair of geology and mineralogy and began his courses in January 1876. In 1879 he had to choose between a career in the Bureau of Mines, where he could look forward to a brilliant future, and remaining at the Catholic Institute, which had very few resources. He opted for the latter.

All the collections—minerals, rocks, fossils—had to be amassed. Lapparent took this task upon himself, without assistance; he was even obliged to write out the labels. He increased his knowledge of rocks considerably and also acquired a more developed taste for the nonabstract, which complemented very well the training—primarily in mathematics and physics—he had received at the École Polytechnique.

Lapparent's teaching was so successful that he received many offers from publishers, and he devoted the rest of his life to writing, especially textbooks. His *Traité de géologie* (Paris, 1882) was the first of its kind to be so well organized and so clear; it went through five editions, each carefully revised and brought up to date. Lapparent's *Cours de minéralogie* (Paris, 1883–1884; 4th ed. 1908) was based on the ideas, then underestimated, of Bravais and of Mallard. He also wrote *Abrégé de géologie* (Paris, 1886; 6th ed., 1907) and *La géologie en chemin de fer* (Paris, 1888), a geological description of the major railway routes of the Paris Basin. In 1889 he published *Précis de minéralogie*, a sixth edition of which came out in 1965. The *Lecons de géographie physique*, which appeared in 1896, is devoted to geomorphology; it is based on the ideas of W. M. Davis and of the American school, which it contributed to popularizing in France. *Le siècle de fer* is a useful collection of lectures and articles. *Science et apologétique* (1905), in which Lapparent attempts to reconcile science and Christianity, is also a plea in favor of science. In nature Lapparent saw, above all, unity, perfection, harmony, and finality.

Those who heard his course lectures and public addresses praised the liveliness of his demonstrations and his brilliant delivery. The successive editions of his treatises show, besides the qualities already mentioned, his care in constantly keeping his works up to date, his understanding, his lucidity, his mastery of the criticism of publications and ideas, his concern for precise observation, and his deep desire for order and harmony.

Lapparent received many honors. He was president of the Geological Society of France in 1880 and 1900, of the French Mineralogical Society in 1885, and of many other organizations. A member (1897) and later perpetual secretary (1907) of the Academy of Sciences, he was also a member or corresponding member of numerous other academies. He was awarded an honorary doctorate by Cambridge University.

BIBLIOGRAPHY

I. Original Works. In addition to the works mentioned above, all of which were published in Paris, Lapparent wrote *La formation des combustibles minéraux* (Paris, 1886); *Le niveau de la mer et ses variations* (Paris, 1886); *Les tremblements de terre* (Paris, 1887); and *Leçons de géographie physique* (3rd ed., Paris, 1907).

II. Secondary Literature. A. d'Ales, *Études religieuses* (Paris, 1908), p. 511; C. Barrois, "Albert de Lapparent et sa carrière scientifique," in *Revue des questions scientifiques* (July, 1909); A. Lacroix, "Notice historique sur Albert-Auguste de Lapparent," in *Académie des Sciences* (Paris, 1920), with photo; and E. de Margerie, "Albert de Lapparent," in *Annales de Géographie*, **17**, no. 94, 344–347.

André Cailleux

LAPWORTH, ARTHUR (*b.* Galashiels, Scotland, 10 October 1872; *d.* Manchester, England, 5 April 1941), *chemistry.*

Lapworth studied chemistry at Mason Science College (later the university), Birmingham, where his father was professor of geology. In 1893 he went to London as a research student at the City and Guilds College, working on naphthalene chemistry under H. E. Armstrong and collaborating with F. S. Kipping in studies on camphor. He, Kipping, and W. H. Perkin, Jr., became related by marrying three sisters. From 1895 to 1900 Lapworth was a demonstrator at the School of Pharmacy, then head of the chemistry department of Goldsmiths' College. In 1909 he left London for Manchester, where he spent the rest of his life, first as senior lecturer in inorganic and physical chemistry, then (1913) as professor of organic chemistry, finally (1922) occupying the senior chair in the department. He was made a fellow of the Royal Society in 1910. The last years of Lapworth's life, even before his retirement in 1935, were clouded by a long and painful illness. Until then he had been a man of many interests, including music, climbing, and what might broadly be called "natural history." He was a reticent and modest person of nonconformist convictions, remembered by his students as friendly although rather remote.

As may be seen from his appointments, Lapworth's knowledge was unusually wide. Although his name is not associated with any single great achievement, he did steady, meritorious work for nearly forty years. In his early work on camphor, he recognized an intramolecular change, related to the pinacol–pinacolone rearrangement, which made possible acceptance of Bredt's structure for camphor. A little later he began the study of reaction mechanisms, notably of cyanohydrin formation and the benzoin rearrangement, which entitles him to be regarded as one of the founders of modern "physical-organic chemistry." He was one of the first to emphasize that organic compounds could ionize either actually or incipiently and that different parts of an organic molecule behave as though they bear electrical charges, either permanently or at the moment of reaction.

With the development of theories of valence based on the electronic structure of the atom, Lapworth was able to refine some speculations about "alternate polarities" into a classification of reactive centers as "anionoid" and "cationoid," the charges being determined by the influence of a "key atom," usually oxygen. In the mid-1920's, when Lapworth elaborated on these concepts with his colleague Robert Robinson, a controversy ensued with C. K. Ingold and his school, who were developing a similar approach to the problems of organic reactivity using a different terminology ("nucleophilic" and "electrophilic" instead of anionoid and cationoid), which eventually gained general acceptance. The controversy, although occasionally sharp, was fruitful; and Lapworth's last paper (1931) bore Ingold's name as a coauthor.

BIBLIOGRAPHY

Lapworth worked by accumulation of detail rather than by publishing "important" papers, but special mention should be made of the following: "On the Constitution of Camphor," in *British Association Report for 1900*, 299–327, which finally pointed the way out of the labyrinth of camphor chemistry; and his two papers on "alternate polarities" in *Memoirs and Proceedings of the Manchester Literary and Philosophical Society*, **64** (1919–1920), no. 3, 16, and *Journal of the Chemical Society*, **121** (1922), 416–427.

The most authoritative notice of Lapworth is Sir Robert Robinson, in *Obituary Notices of Fellows of the Royal Society of London*, **5** (1945–1948), 555–572; repr. without portrait or bibliography, in *Journal of the Chemical Society* (1947), 989–996. See also G. N. Burkhardt, in *Memoirs and Proceedings of the Manchester Literary and Philosophical Society*, **84** (1939-1941), vi–x; and *Nature*, **147** (1941), 769.

W. V. FARRAR

LAPWORTH, CHARLES (*b.* Faringdon, Berkshire, England, 20 September 1842; *d.* Birmingham, England, 13 March 1920), *geology.*

Lapworth's early life was spent in Berkshire and Oxfordshire. In 1864 he became a schoolmaster in southern Scotland. Although his education and scholastic work were purely literary, he soon developed a genius for geological research, in which he engaged intensively during holidays in the surrounding countryside. In 1881 he was appointed to the newly established chair of geology at Mason College (now Birmingham University) and during the first two years there made long and arduous excursions into the northwest of Scotland. His health broke under the strain, and he never fully regained his strength. During his thirty years at Birmingham he taught, lectured, wrote, and advised with fruitful energy and made important contributions to geological knowledge of the English Midlands. He was awarded a Royal Medal by the Royal Society in 1891 and the Wollaston Medal by the Geological Society (its highest honor) in 1899.

Lapworth's investigations into the geology of the Southern Uplands of Scotland in the first phase of his career were epochal because they revolutionized the interpretation of the structure of that important region. A group of fossils, the graptolites, which he used for dating Lower Paleozoic rocks were the means by which he unraveled complicated regional structures. He combined large-scale geological mapping of critical areas with the exhaustive collection of graptolites from the black shale bands (to which they were confined) and the exact discrimination of species to establish his dating method. In the central area, around Moffat, which he studied first, Lapworth found that the numerous shale-band outcrops, which had been thought to represent separate bands at successive horizons in an enormously thick ascending series of strata, were in reality repetitions of a comparatively few bands that were a series of overfolds in which the limbs had become more or less parallel by compaction (isoclinal folding). He demonstrated that the strata, which could be shown to be middle or upper Ordovician and lower Silurian in age, here formed one condensed series only about 250 feet thick (1878). In the Girvan region to the west, the strata, though found to contain graptolite-bearing shale bands, were of quite different lithological types, containing shelly fossils and trilobites. By correlating these bands with the similar bands at Moffat, Lapworth showed that the Girvan rocks, spanning the same time range, were similarly complicated in structure and some twenty times as thick (1882).

From a study and comparison of his own collection of graptolites with collections made in other parts of the world, he tabulated a series of twenty graptolite zones (1879–1880) which seemed to be generally applicable and finally (1889) gave a comprehensive account of the stratigraphy of the Southern Uplands. These results were amplified and confirmed, respectively, in two major works on British paleontology and geology, the *Monograph* of British graptolites (1901–1918) and the Geological Survey *Memoir* on southern Scotland (1899). Lapworth's work inspired research into the working out of the detailed stratigraphy of other British regions by means of the graptolites, particularly by Nicholson and Marr in the English Lake District (1888), Gertrude Elles and Ethel Wood in the Welsh borderland (1900), and O. T. Jones in central Wales (1909).

Meanwhile, Lapworth had proposed (1879) a major classification of the Lower Paleozoic rocks into three systems: Cambrian, Ordovician (a new system), and Silurian. This proposal was so reasonable and convenient on general geological grounds, and provided so satisfactory a solution to the Sedgwick-Murchison controversy over "Cambrian" and "Silurian," that it was immediately and almost universally accepted.

Lapworth's investigations in the Northwest Highlands of Scotland helped to resolve the controversy that was beginning to rage as to the nature of the junction between the Cambrian-Ordovician and the Moinian, a question which affected the interpretation of the structure and history of the entire Scottish Highlands and possibly of other British regions. In the Northwest Highlands there are four main rock groups: gneiss (Lewisian, Pre-Cambrian, in modern terminology and age assignment), overlain unconformably by red sandstone (Torridonian, Pre-Cambrian), overlain unconformably by Cambrian-Ordovician quartzites and limestones, and schists (Moinian) physically overlying the Cambrian-Ordovician with about the same low dip to the east. It had been assumed by Murchison (1858–1860) that there was a conformable upward passage from the Cambrian-Ordovician into the Moinian, but faulting of various kinds had been suggested by Nicol (1861) and Callaway (1881). Lapworth, after painstaking researches (1883), found that the junction was a thrust and that the present character of the Moinian was undoubtedly produced by metamorphism at some post-Ordovician time. He left open the question of the age and character—and thus the correlation—of the original rocks that had become metamorphosed to form the Moinian. (It is now thought probable that these rocks were the Torridonian.) His interpretation was entirely confirmed and greatly extended by Peach and Horne of the Geological Survey, under the direction of Archibald Geikie. Lapworth's work in both the Uplands and the Highlands impressed him with the importance of tangential stress in the earth's crust throughout geological time.

In the last phase of his career Lapworth, with Birmingham as a center, and using the same insight that he had shown in his work in Scotland, investigated the surrounding rocks, directing his attention chiefly to the Lower Paleozoic inliers. Certain quartzites and sandstones as well as igneous rocks in these inliers were already strongly suspected of being, respectively, Lower Cambrian and Pre-Cambrian. Lapworth found similar rocks in other inliers and confirmed these ages by discovering unmistakable Lower Cambrian fossils, of which the *Olenellus* type of trilobite in Shropshire was particularly important. He mapped much of the Cambrian rocks in north Wales and the Pre-Cambrian, Cambrian, and Ordovician rocks in Shropshire; but his work on the old rocks of Wales and the Midlands was largely left to be carried on by others. Lapworth also did important work on other aspects of Midland geology: the Coal Measures (particularly concealed coalfields), the origin of Permian and Triassic breccias and conglomerates, glaciation, and river history.

BIBLIOGRAPHY

I. Original Works. Among Lapworth's more important works are "The Moffat Series," in *Quarterly Journal of the Geological Society of London*, **34** (1878), 240–346; "On the Tripartite Classification of the Lower Palaeozoic Rocks," in *Geological Magazine*, **16** (1879), 1–15; "On the Geological Distribution of the Rhabdophora," in *Annals and Magazine of Natural History*, 5th ser., **3** (1879), 245–257, 449–455, **4** (1879), 331–341, **5** (1880), 45–62, 273–285, 358–369, **6** (1880), 16–29, 185–207; "The Girvan Succession," in *Quarterly Journal of the Geological Society of London*, **38** (1882), 537–666; "The Secret of the Highlands," in *Geological Magazine*, **20** (1883), 120–128, 193–199, 337–344; "On the Close of the Highland Controversy," *ibid.*, **22** (1885), 97–106; "On the Discovery of the *Olenellus* Fauna in the Lower Cambrian Rocks of Britain," *ibid.*, **25** (1888), 484–487; "On the Ballantrae Rocks of the South of Scotland and Their Place in the Upland Sequence," *ibid.*, **26** (1889), 20–24, 59–69; "The Geology of South Shropshire," in *Proceedings of the Geologists' Association*, **13** (1894), 297–355, written with W. W. Watts; "A Sketch of the Geology of the Birmingham District," *ibid.*, **15** (1898), 313–416; and *A Monograph of British Graptolites* (London, 1901–1918), of which he was editor; "The Hidden

Coalfields of the Midlands," in *Transactions of the Institution of Mining and Metallurgy*, **33** (1907), 26–50.

II. SECONDARY LITERATURE. On Lapworth and his work see (listed chronologically) B. N. Peach and J. Horne, *The Silurian Rocks of Britain, Vol. I, Scotland* (*Memoirs of the Geological Survey of Great Britain*, 1899); Anon., "Eminent Living Geologists: Professor Charles Lapworth," in *Geological Magazine*, **38** (1901), 289–303; B. N. Peach, J. Horne *et al.*, *The Geological Structure of the North-West Highlands of Scotland* (*Memoirs of the Geological Survey of Great Britain*, 1907); W. W. W[atts], obituary notice in *Quarterly Journal of the Geological Society of London*, **77** (1921), lv–lxi; W. W. W[atts] and J. J. H. Teall, obituary notice in *Proceedings of the Royal Society*, **92B** (1921), xxxi–xl; W. W. Watts, "The Author of the Ordovician System: Charles Lapworth," in *Proceedings of the Birmingham Natural History and Philosophical Society*, **14** (1921), special suppl.; a revised version in *Proceedings of the Geologists' Association*, **50** (1939), 235–286; W. S. Boulton *et al.*, "The Work of Charles Lapworth," in *Advancement of Science*, **7** (1951), 433–442; and E. B. Bailey, *Geological Survey of Great Britain* (London, 1952).

JOHN CHALLINOR

LA RAMÉE, PIERRE DE. See **Ramus, Petrus.**

LARGHI, BERNARDINO (*b.* Vercelli, Italy, 27 February 1812; *d.* Vercelli, 2 January 1877), *surgery.*

Larghi was the son of Francesco Larghi and Maria de Giudice. He graduated in surgery from the University of Turin in 1833 and in medicine from the University of Genoa in 1836. Returning to Vercelli, he established a dispensary and a small infirmary in his house and began to practice surgery privately. He was so successful that in 1838 he was asked to operate at the St. Andrea Hospital in Vercelli, first in an honorary capacity and from 1844 as head surgeon.

Larghi is known chiefly for his performance of subperiosteal resection before the introduction of antisepsis. The reasons for Larghi's specific interest in this operation require further research; but it appears that earlier biological experiments on the power of the periosteum to regenerate bone were less influential than the brilliant practical results that he obtained. His surgical technique was marked by the extreme care that he took to respect soft paraosteal parts when separating the periosteum from the bone.

In 1845 Larghi performed a resection of the ribs—or actually, an extraction of their bony part—on a twelve-year-old boy suffering from abscess and central caries of the eighth and ninth ribs on the right side. Costal resection until then involved the total removal of the bone with its periosteal covering; it required that intercostal muscles, nerves, and vessels be cut and thereby introduced the danger of hemorrhage and pneumothorax from pleural perforation. Larghi considered such resection to be more a "demolition" than a surgical procedure. He maintained that the surgeon should not destroy the ribs but, rather, restore and rebuild them, since he believed that nature "strongly thickens the periosteum and for this purpose exudes a gelatinous humor which, working its way between the periosteum and the deteriorated bony part, separates them; this humor awaits only the expulsion of the deteriorated bony part, so that it can occupy the site more properly and be converted to new bone inside the case that generated it." The surgeon's task must therefore be to complete removal of the necrotic bone and to extract it from the thickened periosteal sheath.

Larghi's subsequent works, published in the *Giornale della R. Accademia medico-chirurgica di Torino*, reveal the progress of his techniques for subperiosteal resection and the improvement of his surgical instruments. In May 1852 he extracted the necrotic right hemimandible of a patient, removing the condyle but leaving the articular capsula—as well as the periosteum—in position. The neoarthrosis that formed between the neocondyle of the regenerated mandible and the glenoid cavity suggested to Larghi that other soft parts might be preserved, as shown in his most noted work, *Operazioni sottoperiostee e sottocassulari* . . . (Turin, 1855). After the introduction of antisepsis, subperiosteal resection developed substantially, especially through the work of L. Ollier.

BIBLIOGRAPHY

I. ORIGINAL WORKS. Larghi's main works are related to bone surgery and include "Estirpazione o rescissione delle ossa convertita nell'estrazione della loro parte ossea rigenerata dal periostio conservato," in *Giornale delle scienze mediche pubblicato dalla R. Accademia Medico-Chirurgica di Torino*, **28** (1847), 512–520; "Rescissione delle costole convertita nell'estrazione della loro parte ossea," *ibid.*, 521–529; "De l'extraction sous-périostée et de la réproduction des os; extraction sous-périostée des côtes en particulier," in *Gazette médicale de Paris*, **12** (1847), 434–437; "Taglio perpendicolare-longitudinale per le amputazioni delle membre umane," in *Giornale dell Accademia di medicina di Torino*, **5** (1849), 385–398; and "Estrazione e disarticolazione sottoperiostea della porzione destra della mascella inferiore affetta da necrosi," *ibid.*, **17** (1853), 49–62; and "Operazioni sottoperiostee e sottocassulari e guarigione delle malattie delle ossa ed articolazioni per il nitrato d'argento," *ibid.*, **24** (1855), 19–81, 131–169, 302–345, 410–461; **26** (1856), 350–378, 500–506; **27** (1856), 98–130; **31** (1858), 422–428, 467–482; **32** (1858), 225–248;

"Estrazione sotto-cassulo-periostea radio-carpea," *ibid.*, **31** (1858), 299–302; and "Supplemento alle operazioni sotto-periostee e sotto-cassulari," *ibid.*, **7** (1869), 242–356, 373–390, 463–475, 515–523; **8** (1869), 37–62.

II. SECONDARY LITERATURE. On Larghi's life, see Luigi Belloni, "Dalla osteogenesi periostale alla resezione sotto-periostale, Michele Troja (1775) e Bernardino Larghi (1847)," in *Simposi clinici*, **8** (1971), xxv–xxxii; Enrico Bottini, "Bernardino Larghi. Cenno necrologico," in *Giornale della R. Accademia di medicina di Torino*, 21 d.s. III (1877), 174–185; Carlo Dionisotti, *Notizie biografiche dei Vercellesi illustri* (Biella, 1862), 188–190; and Gabriele Stringa, "Bernardino Larghi," in *Archivio 'Putti' di chirurgia degli organi di movimento*, **5** (1954), 592–599.

LUIGI BELLONI

LA RIVE, ARTHUR-AUGUSTE DE (*b.* Geneva [then French], 9 October 1801; *d.* Marseilles, France, 27 November 1873), *physics.*

Arthur-Auguste de La Rive was the oldest son of Charles-Gaspard de La Rive and Marguerite-Adelaïde Boissier. In 1826 he married Jeanne-Mathilde Duppa, with whom he had two sons and three daughters. In 1855, five years after his first wife had died, he married the widow of his former colleague George Maurice.

Auguste received his elementary education at home and at the Collège Publique de Genève, passing thence to the Académie de Genève, where his father was an influential professor; from 1816 to 1823 he studied successively letters, philosophy, and law. Although his own interests were primarily scientific, he was still a student of law when, upon the retirement of P. Prevost in November 1822, the chair of physics at the Academy became vacant. Despite the fact that de La Rive had so far published only one minor paper, his father's influence secured him the appointment of professor of general physics on 27 October 1823 at the unusually young age of twenty-two. On 11 November 1823 he was elected to the influential post of secretary of the Sénat Académique, a post he held until 2 August 1836, although as of 4 March 1834 he had technically been secretary of the newly created Corps Académique; in both cases he was known informally simply as secretary of the Academy. When M. Pictet died, his chair was given to de La Rive, who on 1 June 1825 thereby became professor of experimental physics. This change gave him control, as director, of the Academy's physics laboratory, purchased the year before for 40,000 florins.

De La Rive was long a major force, both institutionally and personally, in the affairs of both the Academy and the government, which in oligarchic Geneva were closely connected. In fact it is as a leader of the Swiss scientific community rather than for his own work, which was devoted primarily to the chemical pile, that he is important. From 1832 to 1846 he sat on the Conseil Représentatif, where, as principal author and battle leader, he was instrumental in putting through the educational reforms of 27 January 1834 and 29 May 1835, which created a Conseil d'Instruction Publique to oversee all public education (in the process effectively eliminating the Church as an educational force) and which consolidated the control of the Corps Académique over the Academy. The so-called triumvirate of Auguste de La Rive, David Munier, and Abraham Pascalis effectively ran the Academy during the 1830's and 1840's. When Augustin-Pyramus de Candolle retired in 1835, Auguste became the Academy's chief spokesman. In addition, he was a leader of the antifederalist conservative faction in the Conseil Représentatif, where, especially after the Revolution of 22 November 1841, he found himself forced into an increasingly reactionary position. Similarly, since the late 1830's the Academy had been under severe attack as a reactionary oligarchy, and de La Rive was the opposition's prime target. He retired completely from academic and public life in December 1846, after the political changes in the wake of the Sonderbund Revolution brought into power a liberal government he had bitterly opposed.

He had been rector of the Academy twice, from 1837 to 1840 and from 1843 to 1844. A member since 1822, he was also twice president of the Société de Physique et d'Histoire Naturelle de Genève, in 1845 and 1865. The Paris Académie des Sciences named him one of its eight foreign associates on 11 July 1864; he had been a corresponding member since 6 December 1830.

For a dozen years after the start of his scientific career, de La Rive's favored journal of publication was the Sciences et Arts series of *Bibliothèque universelle des sciences, belles-lettres et arts*, except for his very longest papers, which appeared in *Mémoires de la Société de physique et d'histoire naturelle de Genève*. In 1836 he took over editorship of the former journal and abolished its separate series; the new *Bibliothèque universelle de Genève* was intended for a well-educated but general readership and avoided specialized or technical scientific papers. However, he soon desired a more strictly scientific journal in which to defend his chemical theory of the pile, and from 1841 to 1845 published *Archives de l'électricité. Supplément à la Bibliothèque universelle de Genève*. Finally, in 1846 the literary portion of the *Bibliothèque universelle* was given over to an editorial committee, and de La Rive limited himself to coediting the scientific portion, now published as *Archives des sciences physiques et naturelles. Supplément à la Bibliothèque universelle de Genève*, which enjoyed his collaboration until his death in 1873.

De La Rive made his scientific reputation and consolidated his power in the Academy roughly in the decade before 1835; for an equal length of time thereafter he was the most important figure in the Genevan scientific community, at least as regards local influence and contemporary fame. During this time he was the most powerful personage in the Academy, the editor of Geneva's leading scientific and cultural journal, and a leader of the conservative party in government. He was known as the friend of Ampère, Arago, and especially Faraday, with whom he maintained an extensive correspondence, and as the most dogged defender of the purely chemical theory of the pile.

It was as a critic of Volta's contact theory of the pile, which attributed the production of electricity in the pile to an electromotive force arising from the contact of heterogeneous substances, that de La Rive made his European reputation. He was a scientist of one tenaciously held but imprecisely conceived theory and in this long dispute made no significant original contributions. Neither his experiments, in which his lack of sense for the importance of quantitative measurements contributed to his poor control over his experimental variables, nor his arguments, which tended to be *ad hoc*, were cogent. He placed more weight on supposedly decisive experiments than on theoretical or philosophical considerations, and there is no evidence that, independently of Faraday, he thought to criticize the unreasonableness of the contact force from considerations of causality.

De La Rive's work must be understood in relation to two of Ampère's ideas that decisively shaped his thinking. The first was the absolute distinction between dynamic, or current, and static, or tension, electricity. Hence while a current flowed there could be no tension, and the contact theorists' electromotive force was fundamentally rejected. De La Rive went further than was required by this dichotomy and asserted the nonexistence of the electromotive force in the open pile, where no current flowed but where an electrostatic tension could be measured. (Becquerel and Stefano Marianini, two of his leading critics, defended a modified chemical theory that recognized the existence of an electromotive force but emphasized the necessity of chemical action for the continued production of electricity.) Second, after Ampère, de La Rive pictured the electric current as a series of decompositions and recompositions of a neutral fluid, the same aether whose greater or lesser vibrations were light and heat and whose positive and negative components were imagined to be variously associated with different chemical elements. These views, which held out the prospect of a unified explanation of electricity, heat, light, and chemical action, determined not only his theory of the pile but also his later work on the vibrations of bodies produced during the passage of electricity and on the light produced by electric arcs, which led to his electric theory of the aurora. In the development of these interconnections his work was largely derivative from that of Félix Savary and, especially, Becquerel, themselves also protégés of Ampère.

The contact theory regarded as its strongest proof experiments which showed that a sensitive electroscope could detect an electric tension between heterogeneous metals held in contact, even when all chemical activity was excluded. De La Rive quite gratuitously explained away these experiments by asserting that carelessness in the thoroughness with which air or water had been excluded had in every case permitted a chemical reaction to take place. The contact theory also made much of the nonproportionality between the apparent chemical activity in the pile and the electricity produced—a problem that bedeviled the chemical theory until Faraday's laws of electrochemical equivalence explained the precise relation between chemical activity and electricity. As a partial solution to this problem de La Rive argued the existence of "countercurrents" within the pile. That is, he allowed that greater chemical activity was accompanied by greater separation of the electric fluids but maintained that the increased conductivity of the more vigorously reacting liquid facilitated their immediate recombination within the pile, with the result that only a fraction of the originally produced electricity passed through the external circuit. In accordance with these views, his solution to the problem of which arrangement of plates and connecting wires produced the greatest current was to require that the resistance of the pile to the recomposition of the separated electricities be "just greater" than the resistance of the external circuit. De La Rive's theory could not explain the increase in tension with an increasing number of couples connected in series. Despite objections to his work—especially by C. H. Pfaff, an ardent contact theorist—he introduced only one significant modification into his theory in that he conceded, following Becquerel, that electricity could be produced not only by chemical means but also by mechanical and thermal actions, although he consistently played down the importance of all but the first.

BIBLIOGRAPHY

I. Original Works. The most complete bibliography is J. Soret, "Auguste de La Rive. Notice biographique," in *Archives des sciences physiques et naturelles. Supplément à la Bibliothèque universelle et revue suisse*, n.s. **60** (1877), 5–253; the bibliography (pp. 203–222) lists most of his non-scientific publications as well. For his scientific papers, see Royal Society, *Catalogue of Scientific Papers*, II, 212–217; VI, 636; VII, 507–509; and IX, 667. Less extensive but useful is Poggendorff, II, cols. 657–659; and III, 1126. De La Rive's major work was a series of three papers, "Recherches sur la cause de l'électricité voltaïque," in *Mémoires de la Société de physique et d'histoire naturelle de Genève*, **4**, pt. 3 (1828), 285–334; **6**, pt. 1 (1833), 149–208; and **7**, pt. 2 (1836), 457–517; also published separately (Geneva, 1836). The first paper was originally published in slightly different form in *Annales de chimie et de physique*, 2nd ser., **39** (1828), 297–324. Very useful is his "Esquisse historique des principales découvertes faites dans l'électricité depuis quelques années," in *Bibliothèque universelle des sciences, belles-lettres et arts.* Sciences et arts, **52** (1833), 225–264, 404–447; **53** (1833), 70–125, 170–227, 315–352; also published separately (Geneva, 1833). Apart from such separately published memoirs, de La Rive's only book was *Traité d'électricité théorique et appliquée*, 3 vols. (Paris, 1854–1858), translated into English by C. V. Walker as *A Treatise on Electricity, in Theory and Practice*, 3 vols. (London, 1853–1858). The Bibliothèque Publique et Universitaire de Genève preserves six and a half vols. of letters addressed to Auguste, plus a few other items.

II. Secondary Literature. The major biography, which also treats his scientific work, is by Soret, cited above, although it is unfortunately often vague with regard to names and dates. Also very useful is Jean-Baptiste-André Dumas, "Éloge historique d'Arthur-Auguste de La Rive," in *Mémoires de l'Académie des sciences de l'Institut de France*, **40** (1876), ix-lix; also published separately (Paris, 1874) and translated into English as "Eulogy on Arthur Auguste de La Rive," in *Smithsonian Annual Report for 1874*, 184–205. See also H. Deonna, "De la Rive," in *Dictionnaire historique et biographique de la Suisse*, II, 648. A great deal of information on his connections with the Academy can be found in Charles Borgeaud, *Histoire de l'Université de Genève*, II, *L'Académie et l'Université du XIXe siècle, 1814–1900* (Geneva, 1934), *passim.* For a detailed examination and estimation of his scientific work, see W. Ostwald, *Elektrochemie. Ihre Geschichte und Lehre* (Leipzig, 1896), *passim*, but especially ch. 12, "Der Kampf zwischen der Theorie der Berührungselektricität und der chemischen Theorie der galvanischen Erscheinungen," pp. 426–492.

Kenneth L. Caneva

LA RIVE, CHARLES-GASPARD DE (*b.* Geneva, 14 March 1770; *d.* Geneva, Swiss Confederation, 18 March 1834), *physics, chemistry, medicine.*

Nobile Charles-Gaspard de La Rive was the second son of Ami-Jean de La Rive and Jeanne-Elisabeth Sellon. A landowning noble family, the de La Rives had for centuries been prominent members of Geneva's Protestant patriciate. After receiving his primary education at the Collège de Genève, Gaspard entered the Académie de Genève in May 1789 to study law. Although he had studied some natural science, he was still a student of law in August 1794 when the first Genevan Tribunal Révolutionnaire sentenced him to a year's house arrest. However, the relaxation of the Terror after 9 Thermidor freed him to emigrate, whereupon he went to Edinburgh to study medicine. He received his doctorate in 1797 with the publication of his thesis, *Tentamen physiologicum de calore animali*, an unoriginal critique especially of Adair Crawford, in which de La Rive followed his teacher John Allen in attributing animal heat to the combustion in the blood of particles derived from food and oxygen absorbed through the lungs.

De La Rive returned to Geneva late in 1799 and shortly thereafter (probably in 1800) married the well-born Marguerite-Adelaïde Boissier, with whom he had two sons. He soon took charge of the Hospice des Aliénés, where he devoted some thirty years to improving the treatment of the insane. In November 1802 he was named honorary professor (without pay) of pharmaceutical chemistry at the Academy, filling one of the ten new honorary chairs created through the reform efforts of M. Pictet, who wanted a strong faculty of sciences within the Academy. In 1819, as Alexandre Marcet took over instruction in pharmaceutical chemistry, his title became that of honorary professor of general chemistry. He frequently gave series of lectures at the Academy and at the Musée Académique on medicine and, especially after 1818, on experimental chemistry. His well-equipped laboratory was outfitted at his own expense; its doors were always open to visiting scientific dignitaries such as Davy, Arago, and Ampère; and indeed his generosity and personal interest contributed significantly to his contemporary reputation.

On 13 December 1813 he participated in the creation of the reactionary Conseil Provisoire (renamed Conseil d'État in 1814) and remained an active member of the government as *conseiller d'état* until his resignation on 19 June 1818. He was elected *premier syndic* in 1817. On 4 November 1818 the Conseil d'État created an extraordinary seat for him on the Sénat Académique, where he used his considerable influence in support of reforms aimed at freeing the Academy from ecclesiastical control and in strengthening its faculty of science. He authored the *règlement* of 11 November 1818 providing for the conferring of academic degrees by the faculties of science and of letters. Later, during his term as rector of the Academy (November 1823–March 1826), he was the major force behind efforts to transform the Academy into a university and to secure for it the beginnings of state financial support. Most important were the two regulations of 28 November 1825, which consolidated the control of the individual faculties over their own affairs and provided for specialized instruction in the sciences and in letters. He was prominent in the founding and initial funding of the Jardin Botanique in 1817 and of the Société de Lecture in 1818, the latter organized at the urging of A.-P. de Candolle to supplement Geneva's meager libraries.

De La Rive's actual scientific contributions were negligible, both intrinsically and historically. Excepting his work on electricity, his sixteen publications between 1797 and 1833 were predominantly reviews, translations, extracts, and commentaries, with an occasional paper on a minor chemical or medical topic. Perhaps most valuable were his expositions, among the first on the Continent, of Davy's electrochemistry, Dalton's atomic theory, and Berzelius' theory of definite proportions. His work appeared in *Bibliothèque britannique*, Geneva's foremost scientific and literary journal, and in its successor, *Bibliothèque universelle*, both of which enjoyed his editorial collaboration.

De La Rive enjoyed modest contemporary renown as a critical defender of Ampère. He conceived a simple device—his "*flotteur électrique*"—which he believed exhibited the reasonableness of Ampère's theory of magnetism. It consisted of a strip each of copper and zinc, passed through a cork and fastened together at the top to form a ring; when floated on acidulated water near a magnet, the device demonstrated the action of the magnet on a simple current loop. De La Rive believed that certain experiments he performed with it led Ampère to modify his theory—by assuming that the molecular currents in a magnet were progressively more inclined the farther from the axis and the center of the magnet. There is no evidence, however, that the final version of Ampère's theory owed anything to de La Rive's criticisms.[1]

Ampère had shown that a rectangular current loop free to rotate about a vertical axis lying within the plane of the loop would assume a position perpendicular to the magnetic meridian, such that the direction of current in the lower portion of the rectangle was from east to west. Ampère initially attributed this effect to the interaction between east-west currents in the surface of the earth and those in the lower portion of his loop in accordance with his law for the attraction between like-directed currents. De La Rive varied this experiment by eliminating the lower horizontal segment. In terms of Ampère's earlier deficient explanation, he then expected that the action of the terrestrial currents—if, indeed, such exist—on the remaining upper horizontal portion would rotate the loop 180°; this effect, however, did not occur. The resolution of the difficulty lay in the analysis of the action, on the vertical sides of the loop, of the terrestrial currents—which Ampère now thought had to be located predominantly along the magnetic equator—and in a reconsideration of their action on the horizontal portions. It seems that Ampère had already recognized the deficiency of his original explanation, although de La Rive's criticism spurred him to develop this aspect of his theory in print before he otherwise would have done. De La Rive's experiment was carried somewhat further by his son Auguste, who owed his scientific debut to this controversy.[2]

NOTES

1. The confusion surrounding Gaspard's claims and Ampère's explanations is far out of proportion to the importance of the case. Be that as it may, the relevant sources are Gaspard de La Rive, "Notice sur quelques expériences électro-magnétiques," in *Bibliothèque universelle des sciences, belles-lettres et arts.* Sciences et arts, **16** (1821), 201–203; and "Mémoire sur quelques nouvelles expériences électro-magnétiques et en particulier sur celles de Mr. Faraday," *ibid.*, **18** (1821), 269–286. See also four letters of Ampère's—to S. S. van der Eyk, 12 Apr. 1822; to G. de La Rive, 12 June 1822; to F. Maurice, 6 July 1822; and to Faraday, 10 July 1822—published in *Correspondance du grand Ampère*, L. de Launay, ed., 3 vols. (Paris, 1936–1943), II, 579, 580–582; III, 927; and II, 588, respectively. The letter to de La Rive was originally published as "Extrait d'une lettre de Mr. Ampère au Prof. de La Rive sur des expériences électro-magnétiques," in *Bibliothèque universelle* . . ., **20** (1822), 185–192.
2. The relevant sources are Gaspard de La Rive, "Lettre du Professeur de La Rive à M. Arago (datée le 22 juin 1822), sur des courans galvaniques," in *Annales de chimie et de physique*, 2nd ser., **20** (1822), 269–275; a letter from Ampère to M. Pictet, dated 10 July 1822, published in the former's *Correspondance*, II, 583–585; Auguste de La Rive, "De l'action qu'exerce le globe terrestre sur une portion mobile du circuit voltaïque. (Mémoire lu à la Société de physique et d'histoire naturelle de Genève le 4 septembre 1822)," in *Bibliothèque universelle* . . ., **21** (1822), 29–48 (a footnote to the title adds that "Mr. Ampère, alors à Genève, assistoit à cette séance"). Pp. 29–42 are the memoir proper; pp. 42–47 are Auguste's version of Ampère's verbal explanation of the new experiments; pp. 47–48 describe two other experiments the two did together at Geneva. The version published in *Annales de chimie et de physique*, 2nd ser., **21** (1822), 24–48 ("Mémoire sur l'action qu'exerce . . ."), is the same except that the middle segment (pp. 39–46) has been modified slightly by Ampère; in addition, there follows a note by Ampère on two other new experiments (pp. 48–53).

   On the respective contributions of the two de La Rives, see also Gaspard's letter of 16 Sept. 1822 to Berzelius, in *Jac. Berzelius Bref*, H. G. Söderbaum, ed., 14 vols. in 6 (Uppsala, 1912–1932), VII (= *Strödda Bref (1809–1847)*, III, pt. 2), 60–61; and letters by Ampère to C. J. Bredin, [24] Sept. 1822, and to A. de La Rive, 11–31 Oct. 1822, in his *Correspondance*, II, 509 and 603–604, respectively.

BIBLIOGRAPHY

I. Original Works. Nearly complete bibliographies of Gaspard's scientific works are in Royal Society *Catalogue of Scientific Papers*, II, 217; Poggendorff, II, col. 657; and G. de Fère, "Charles-Gaspard de La Rive," in *Nouvelle biographie générale depuis les temps les plus reculés jusqu'à nos jours*, 46 vols. (Paris, 1853–1866), XXIX, cols. 603–606 (the bibliography, which follows a biography, is extensive but contains many errors). The Bibliothèque Publique et Universitaire de Genève preserves three vols. of MSS containing letters by and to Gaspard and a journal kept during the years 1794 to 1799.

II. Secondary Literature. The most important source is the anonymous "Notice biographique sur M. le prof. G. De La Rive," in *Bibliothèque universelle des sciences, belles-lettres et arts*. Sciences et arts, **55** (1834), 303–338, possibly written by A. Gautier, whose often identically worded obituary notice on Gaspard appeared in *Mémoires de la Société de physique et d'histoire naturelle de Genève*, **10** (1834), xii–xiv. In addition to the biography by de Fère cited above, mention should also be made of A. Maury, "Charles-Gaspard de Larive," in *Biographie universelle (Michaud) ancienne et moderne*, new ed. (Paris, n.d.), XXIII, 264–265. A wealth of information of Gaspard's activities at the Academy is in C. Borgeaud, *Histoire de l'Université de Genève*, II, "L'Académie de Calvin dans l'Université de Napoléon, 1798–1814," and III, "L'Académie et l'Université au XIXe siècle, 1814–1900" (Geneva, 1909–1934), *passim*.

Kenneth L. Caneva

LARMOR, JOSEPH (*b.* Magheragall, County Antrim, Ireland, 11 July 1857; *d.* Holywood, County Down, Ireland, 19 May 1942), *theoretical physics.*

Larmor was the eldest son in a large family. His father gave up farming to become a grocer in Belfast in 1863 or 1864. A shy, delicate, precocious boy, Larmor attended the Royal Belfast Academical Institution; received the B.A. and the M.A. from Queen's University, Belfast; and entered St. John's College, Cambridge, in 1877. In 1880 he was senior wrangler in the mathematical tripos (J. J. Thomson was second), was awarded a Smith's Prize, and was elected fellow of St. John's. For the next five years Larmor was professor of natural philosophy at Queen's College, Galway, then returned as a lecturer to St. John's in 1885. He succeeded Stokes as Lucasian professor in 1903 and retired from the position in 1932. His health deteriorating, he returned to Ireland to spend his final years. He never married.

Larmor became a fellow of the Royal Society in 1892 and served as a secretary from 1901 to 1912. From 1887 to 1912 he served on the council of the London Mathematical Society; he was president of this society in 1914–1915, having been at times vice-president and treasurer. The Royal Society awarded him its Royal Medal in 1915 and its Copley Medal in 1921, and he received the De Morgan Medal of the London Mathematical Society in 1914. Larmor was also awarded many honorary degrees and became a member of various foreign scientific societies. He was knighted in 1909. He represented Cambridge University in Parliament from 1911 to 1922. In his maiden speech in 1912 he defended the unionist position in the debate on Irish home rule. His major concern then and later was for education and the universities. Those who knew him report that Larmor was an unassuming, diffident man who did not readily form close friendships and whose numerous acts of generosity were performed without publicity. In the words of D'Arcy Thompson, "Larmor made few friends, perhaps; but while he lived, and they lived, he lost none."

Larmor's lectures and writings were often obscure, in that he would sketch the broad outlines of his thought without filling in the mathematical details, but this thought was stimulating and creative. He was concerned to stress the physical and geometrical characteristics of a problem rather than the analytical niceties. Of interest in this connection is his "Address on the Geometrical Method," delivered in 1896. In dynamics Larmor was a champion of the principle of least action. An early paper (1884) showed the analogies between diverse physical problems that it can bring to light. The use of the method of least action enables the compression of the basic assumptions involved in constructing a theory into a single function, from which results may be deduced with some guarantee of consistency and completeness. Larmor employed this method in his fundamental works, particularly in electron theory.

Larmor's scientific work centered on electromagnetic theory, optics, analytical mechanics, and geodynamics. As one of the great completers of the edifice of classical mathematical physics he bears comparison with H. A. Lorentz. Like Lorentz, his major work concerned electron theory, that is, the interaction of atomically charged matter and the electromagnetic field. Unlike Lorentz, Larmor did not participate to a large extent as a guide to the newer generation of physicists developing quantum theory and relativity. In general, he maintained a conservative, critical attitude toward the new ideas, particularly examining the possible limitations of the relativity theories.

Larmor's electron theory was a new fusion of electromagnetic and optical concepts. His first paper on electromagnetism, written in 1883, dealt with electromagnetic induction in conducting sheets and solids. In this work he encountered the problem of the effect of the motion of matter through the ether, the central problem leading to relativity and the key concern of his famous book, *Aether and Matter*. Larmor reported on the action of magnetism on light to the British Association for the Advancement of Science in 1893. In this report he discussed the dynamical theory of wave optics which the Irish physicist James MacCullagh had perfected in the 1830's. MacCullagh's treatment had avoided the flaws of other more or less contemporary theories, but MacCullagh had been unable to supplement his mathematical work with a specific mechanical model of the luminiferous ether. His expression for the action function of the ether corresponded to a medium possessing rotational elasticity, however, so that any element of it would resist rotation but otherwise would behave like a liquid. Kelvin's gyrostatic model of the ether, which had been the subject of an article by Larmor, removed the major objection to MacCullagh's theory on grounds of physical unrealizability. Furthermore, in 1880 G. F. FitzGerald had translated MacCullagh's analysis of optical reflection into the language of electromagnetic theory.

Inspired particularly by this last work, Larmor presented his electron theory in three important papers entitled "A Dynamical Theory of the Electric and Luminiferous Medium" in 1894, 1896, and 1898. He combined MacCullagh's type of ether with the electromagnetic field theory by identifying the magnetic force with the rate of displacement of the ethereal medium, and the electric displacement with the absolute rotation of the medium (the curl of the displacement of the ether). At first the permanent Amperian electric currents of material atoms were treated as vortex rings in the ether, thereby introducing Kelvin's vortex theory of the atom, while electric charge was not included integrally in the theory. Two months after the first article in the series was written, however, Larmor added a section incorporating "electrons" into the theory as mobile centers of rotational strain in the ether. In the MacCullagh type of ether such centers of strain would be permanent, possess inertia, and act upon one another as charged particles do.

The second article in the series (written in 1895) developed the theory of electrons foreshadowed in the addendum to the first. The only interaction between the ether and ordinary matter was assumed to be via the discrete electrons (of both signs of charge), and Larmor discussed the relation between a microscopic theory treating the dynamics of the electron and a macroscopic theory in which the current and other variables are treated as statistical averages. The influence of the motion of the matter through the ether on light propagation and the null result of the Michelson-Morley experiment were treated in a fashion similar to that of Lorentz in the same year. A standard of time varying from point to point was introduced, and it was shown that the FitzGerald-Lorentz contraction would arise out of the theory of the equilibrium of charges in a moving ether. Part 3 (written in 1897) dealt further with the effects involving material media, including motion through the ether, optical dispersion, and particularly electrical stresses. Much of this work was incorporated in *Aether and Matter* (published in 1900), which won the Adams Prize at Cambridge in 1898. This book concentrated mainly on the problem of motion of matter through the ether; here we find, perhaps for the first time, the full Lorentz transformations for space and time and for the electromagnetic field *in vacuo*.

Aside from his general version of the electron theory, constructed from a rotationally elastic ether, Larmor is noted for two specific contributions to electrodynamics. He introduced the Larmor precession, which orbiting charges undergo when subjected to a magnetic field, in 1897 in connection with a discussion of the Zeeman effect. In the same article he treated the radiation of an accelerating charge, obtaining the well-known nonrelativistic formula expressing the power radiated as proportional to the square of the product of charge and acceleration.

Larmor was interested in the dynamics of the earth's motion from 1896, when he published a work on the earth's free precession. In 1906 and 1915, with E. H. Hills, he analyzed possible causes of the irregular motion of the earth's axis; among his other articles one concerns irregularities in the earth's rotation and the definition of astronomical time (1915). Among the 104 articles included in Larmor's *Papers* is "Why Wireless Electric Rays Can Bend Round the Earth" (1924), which was of importance for radio communications. He edited several collections of scientific papers besides his own; and he contributed valuable biographical notices of scientists, particularly one of Kelvin (1908). Strongly interested in the history of his subject, he included in his longer papers and as appendixes to his *Papers* very interesting critical summaries of the work that preceded and led to his own research. His own work owed much to "that Scoto-Irish school of physics which dominated the world in the middle of the last century," particularly to W. R. Hamilton, J. MacCullagh, J. C. Maxwell, Kelvin, and G. FitzGerald; and there is little doubt that he considered himself the last follower of this tradition.

BIBLIOGRAPHY

I. Original Works. Most of Larmor's papers were published in his *Mathematical and Physical Papers*, 2 vols. (Cambridge, 1929). Not included are his later papers on relativity, which are listed in vol. II, and most of his biographical notices. His biography of Kelvin appears in *Proceedings of the Royal Society*, **81A** (1908), i–lxxvi. *Aether and Matter* (Cambridge, 1900) was published separately.

II. Secondary Literature. Obituary notices are by E. Cunningham, in *Dictionary of National Biography (1941–1950)*, pp. 480–483; A. S. Eddington, *Obituary Notices of Fellows of the Royal Society of London*, **4** (1942–1943), 197–207; W. B. Morton, in *Proceedings and Report of the Belfast Natural History and Philosophical Society*, 2nd ser., **2**, pt. 3 (1942–1943), 82–90; and D'Arcy W. Thompson, in *Yearbook of the Royal Society of Edinburgh*, **2** (1941–1942), 11–13.

A. E. Woodruff

LA ROCHE, ESTIENNE DE (*fl.* Lyons, France, *ca.* 1520), *arithmetic.*

La Roche, known as Villefrânche, was born in Lyons. A pupil of Nicolas Chuquet, he taught arithmetic for twenty-five years in the commercial center of his native town and was called "master of ciphers."

La Roche's *Larismetique*, published at Lyons in 1520, was long considered the work of an excellent writer who, early in the sixteenth century, introduced into France the Italian knowledge of arithmetic and useful notations for powers and roots. His fame decreased remarkably in 1880, when Aristide Marre published Chuquet's "Triparty," written in 1484 but preserved only in manuscript. The first part of *Larismetique* was then seen to be mostly a copy of the earlier work, with the omission of those striking features that established Chuquet as an algebraist of the first rank. Chuquet, for example, employed a more advanced notation for powers and he introduced zero as an exponent. It is not clear why La Roche failed to publish the "Triparty." He may have suppressed it in order to claim the credit for himself, or perhaps he thought it too far beyond the comprehension of prospective readers. Nevertheless, through *Larismetique* some of Chuquet's innovations influenced such French arithmeticians as Jean Buteo and Guillaume Gosselin.

The second, and greater, part of La Roche's work has, apart from some geometrical calculations at the end, a commercial character. The author states that as a basis he used "the flower of several masters, experts in the art" of arithmetic, such as Luca Pacioli, supplemented by his own knowledge of business practice. The result was a good but traditional arithmetic that presented an outstanding view of contemporary methods of computation and their applications in trade.

BIBLIOGRAPHY

Larismetique nouellement composee par maistre Estienne de la Roche dict Villefranche natif de Lyon sur le Rosne divisée en deux parties . . . (Lyons, 1520) was republished as *Larismetique et geometrie de maistre Estienne de la Roche dict Ville Franche, nouvellement imprimee et des faultes corrigee . . .* (Lyons, 1538).

See M. Cantor, *Vorlesungen über Geschichte der Mathematik*, II (Leipzig, 1913), 371–374; and N. Z. Davis, "Sixteenth-Century French Arithmetics on the Business Life," in *Journal of the History of Ideas*, **21** (1960), 18–48. On Chuquet and La Roche, see Aristide Marre, "Notice sur Nicolas Chuquet," in *Bullettino di bibliografia e di storia delle scienze matematiche e fisiche*, **13** (1880), esp. 569–580.

J. J. Verdonk

LARSEN, ESPER SIGNIUS, JR. (*b.* Astoria, Oregon, 14 March 1879; *d.* Washington, D.C., 8 March 1961), *geology.*

The son of a Danish immigrant who had settled in Oregon and had become the first Danish consul in Portland, Larsen attended the local schools. Upon graduation from high school, he did not enter college immediately but worked to ease the financial pressures on the family. Entering the University of California in 1902, Larsen came under the influence of A. C. Lawson and A. S. Eakle while an undergraduate. This led to his taking advanced courses in mathematics and chemistry and contributed to his ultimate decision to make geology and petrology his lifework. After receiving the B.S. degree in 1906, Larsen remained at the university to teach. He left in 1908 but returned to take his doctorate in 1918.

His early studies developed in Larsen the habit of extensive and detailed examination of specimens. He conducted advanced research first as an assistant petrologist in the geophysical laboratory of the Carnegie Institution in Washington, D.C. With H. E. Marwin he developed petrographic techniques, making investigations in optical crystallography and the immersion method of mineral analysis.

In 1909 Larsen was appointed assistant geologist for the U.S. Geological Survey, an association that was one of the most important in his entire professional life. He joined Whitman Cross, who had been studying the volcanic province of the San Juan Mountains of Colorado and New Mexico for fifteen

LARSEN

years. Their final report appeared forty-seven years later, in 1956. In Washington, Larsen was associated with F. E. Wright, who pioneered in the development of optical mineralogy. Their joint paper in 1909 was one of the earliest systematic efforts to establish criteria for geological thermometry.

In 1914 Larsen became a full geologist with the U.S. Geological Survey and served until 1923; for the last five years of this period, he headed the petrology section. Upon leaving the Survey, he became professor of petrography at Harvard University, where he remained until 1949. He then returned to the Survey until failing health forced him to reduce his activities in 1958.

Larsen's concern with optical mineralogy led him to assemble and tabulate the optical characteristics of more than 600 nonopaque minerals. Microscopy had been applied intensely to the study of rocks and minerals as early as 1850, with Sorby's development of the thin section technique. It was developed by Zirkel and Rosenbusch in the following years. Larsen extended and refined these early techniques, developing a hollow prism to measure directly the index of refraction of immersion liquids. These methods were refined for the measurement of mineral refraction indexes to three decimal places—until very recently this instrument served as the principal guide to mineral chemistry. Larsen's systematic methods of petrographic microscopy and his catalog-tables of the optical properties of minerals and the interrelationships of their properties with their chemistry were published in 1921. A revised handbook written with H. Berman, *The Microscopic Determination of the Nonopaque Minerals*, appeared in 1934 and remains the single indispensable handbook of optical crystallography.

Larsen's extensive field studies of the San Juans, the southern California batholith, the Idaho batholith, and later the Highwood Mountains of Montana made him a confirmed magmatist. His extensive microscopic studies of the specimens collected during the annual field seasons of nearly half a century reinforced the conclusions he drew in the field. Larsen strongly supported the laboratory investigations of minerals under pressures and high temperatures (phase equilibrium studies) associated with T. Vogt and N. L. Bowen.

Larsen's other major contributions were his development of the concepts of the petrographic province and the variation diagram. He studied the theory of thermal diffusion and applied it to the problem of the cooling of a batholith. He also developed a method of determining the age of igneous rocks using the lead in accessory minerals. His researches in California's San Diego County batholithic intrusive rocks had led him to examine the radioactivity of zircon. His retirement from Harvard in 1949 enabled him to devote full time in Washington to applying his methods for determining the age of rocks.

Larsen determined that lead would avoid zircon during the crystallization of magma. Therefore the trace amounts of lead found in zircon were of radiogenic origin, and their quantity was a function of the time since crystallization. He separated zircon from large quantities of igneous rock and proceeded to measure the lead in the zircon spectrographically and to determine the α activity. The formula

$$t = C\frac{\mathrm{Pb}}{\alpha}$$

is Larsen's: t = the age of the rock in millions of years; Pb, the lead concentration in parts per million; C = a constant between 2632 and 2013, depending upon the proportion of uranium to thorium; and α, the radioactivity in α counts per milligram of the mineral per hour. When a mineral contained unknown amounts of uranium and thorium, C was obtained by the measurement of the total α activity and a fluorescent analysis of the uranium.

In consequence of such contributions, Larsen was widely recognized as the foremost descriptive and theoretical petrologist in America and was awarded highly prized professional medals by both the Mineralogical and the Geological Societies of America. Perhaps his greatest contribution was in training many of the foremost petrologists of the age, not only in the classroom but also in the field and at the Geological Survey offices. His single-minded concentration on geology led to extreme absentmindedness; nonetheless his kindness and the grave consideration that he infallibly extended to those around him made him the object of an affectionate veneration on the part of his students. Among American geologists over the age of forty, the designation "The Professor" is immediately understood to refer to Larsen.

Larsen was survived by his wife of fifty-one years, the former Eva A. Smith, and for less than a year by the second of his two sons, the petrologist Esper S. Larsen III.

BIBLIOGRAPHY

I. Original Works. Larsen's writings include "Quartz as a Geologic Thermometer," in *American Journal of Science*, 4th ser., **27** (1909), 421–447, written with F. E. Wright; *The Microscopic Determination of the Nonopaque Minerals, Bulletin of the United States Geological Survey* no. 679 (1921), 2nd ed., written with Harry Berman, *ibid.*, no. 848 (1934); "The Igneous Rocks of the Highwood Mountains of Central Montana," in *Transactions of the American Geophysical Union 16th Annual Meeting*, pt. 1 (1935), 288–292, written with C. S. Hurlburt *et al.*; "Some New Variation Diagrams for Groups of Igneous Rocks," in *Journal of Geology*, **46** (1938), 505–520; "Petrographic Province of Central Montana," in *Bulletin of the Geological Society of America*, **51** (1940), 887–948; "Geochemistry," in *50th Anniversary Volume of the Geological Society of America* (Washington, D.C., 1941), pp. 393–413; "Time Required for the Crystallization of the Great Batholith of Southern and Lower California," in *American Journal of Science*, **243A** (1945), 399–416; "Batholith and Associated Rocks of Corona, Elsinore, and San Luis Rey Quadrangles, Southern California," in *Memoirs. Geological Society of America*, **29** (1948), 1–182; "Method for Determining the Age of Igneous Rocks Using the Accessory Minerals," in *Bulletin of the Geological Society of America*, **63** (1952), 1045–1052, written with N. B. Keevil and H. C. Harrison; "Geology and Petrology of the San Juan Region, Southwestern Colorado," in *Professional Papers. United States Geological Survey*, no. 258 (1956), 1–303, written with C. W. Cross; "Lead-Alpha Ages of the Mesozoic Batholiths of Western North America," in *Bulletin of the Geological Society of America*, no. 1070B (1958), 35–62, written with David Gottfried *et al.*; and "A Reconnaissance of the Idaho Batholith and Comparison With the Southern California Batholith," *ibid.*, no. 1070A (1958), 1–33, written with R. G. Schmidt.

II. SECONDARY LITERATURE. See C. S. Hurlburt, Jr., "Esper S. Larsen Jr.," in *American Mineralogist*, **47** (Mar. 1962), 450–459; and W. T. Pecora, "Esper Signius Larsen Jr. 1879–1961," in *Bulletin of the Geological Society of America*, **73** (Apr. 1962), 27–29, followed by a bibliography by Marjorie Hooker, 29–33. See especially the *Festschrift* volume, "Studies in Petrology and Mineralogy," *American Mineralogist*, **35** (1950), 619–958, and the remarkable dedication by J. C. Rabbitt.

CORTLAND P. AUSER

LARTET, ÉDOUARD AMANT ISIDORE HIPPOLYTE (*b.* St.-Guiraud, Gers, France, 15 April 1801; *d.* Seissan, Gers, France, 28 January 1871), *paleontology, prehistory.*

Descended from an old family of landed proprietors established for at least 500 years in the vicinity of Castelnau-Barbarens, Lartet spent his childhood on the family estate of Enpourqueron. While attending the *collège* in Auch, he received from Napoleon I, who was visiting the city, one of the medals awarded to the three most deserving pupils. Lartet went to Toulouse to study law and received his *licence* in 1820. His diploma was signed by Georges Cuvier, who was then a counsellor of state. In 1821 Lartet went to Paris as a probationary lawyer. Already attracted to the natural sciences, however, he attended courses at the Collège de France and visited the Muséum d'Histoire Naturelle. He also enjoyed browsing through the bookstalls along the banks of the Seine in the hope of discovering some rare book on natural history.

Having completed his probationary period, Lartet returned to Gers, where he had been called by his elderly father, who wished to divide his wealth among his children. From this time Lartet lived in the country, managing the property that he had inherited. He often gave free legal advice to the peasants who lived in the vicinity; aware of his interests, they repaid him by bringing him strange objects that they had found during the course of their work: medals, stone axes, shells, bones. One day a peasant brought a fossil tooth that Lartet recognized, after some research, as that of a Mastodon. His vocation was now settled. He hastened to acquire the books necessary to further his knowledge and to guide him in the paleontological research he wished to undertake. Lartet explored the Tertiary terrain of Gers; and toward the end of 1834 he came upon the rich fossil deposit of Sansan, where he was to discover more than ninety genera and species of fossil mammals and reptiles. His son reports: "For fifteen years he made excavations at his own expense and devoted late evenings to the study and classification of this material." Aware of the importance of these discoveries, at the beginning of 1834 Lartet sent Étienne Geoffroy Saint-Hilaire a letter on this subject, one later published in the *Bulletin de la Société géologique de France*.

It was the discovery in 1836 of the first anthropomorphic fossil ape, Pliopithecus, that revealed to Lartet the possibility of discovering human fossil remains. Encouraged by these first results, he continued his investigations with remarkable perseverance, gathering an abundant harvest of fossil remains, many of which were contributed to the Muséum d'Histoire Naturelle in Paris.

Lartet married in 1840 and had one son, Louis, who also became a scientist. In order to supervise his son's education more closely, Lartet decided in 1851 to move to Toulouse, where his son was attending the *lycée*. He remained there for only two years, having resolved to go to Paris, where he would be able to devote himself completely to the work he liked best. With A. Gaudry he investigated the identity of the fossil remains from Pikermi, Greece, and published notes on the paleontology of the Tertiary.

Lartet's memoir on the ancient migrations of the mammals of the Recent period (1858) presaged the turn his career was to take in 1860. He was particularly interested in the discoveries made by Boucher de

Perthes of chipped flint tools in the Somme Valley. Lartet himself had long been concerned with the antiquity of man. His belief in the continuity of life excluded the possibility of a sudden great upheaval that would have interrupted the regular succession of living creatures. His excavations in 1860 at the prehistoric sites of Massat (Ariège) and Aurignac (Haute-Garonne) yielded definite proof of the contemporaneity of man and extinct animal species and enabled him to establish, in 1861, a paleontological chronology based on the study of the large Quaternary mammals. Pursuing his research in this area, in 1863 he explored, along with the wealthy English collector Henry Christy, numerous grottoes in Périgord, particularly La Madeleine, Le Moustier, and Les Eyzies. The discovery of carved and sculpted objects furnished new proof of the existence of prehistoric art. Their finds later enriched the collections of the Musée des Antiquités Nationales in St.-Germain.

On the opening of this museum in 1867, Lartet was made an officer of the Legion of Honor. That same year he presided over the International Congress of Archaeology and Prehistoric Anthropology. In 1865 publication of the *Reliquiae aquitanicae* began in London. This magnificent work was the fruit of Lartet's collaboration with his friend Christy in the excavations in Périgord.

Lartet was named professor of paleontology at the Muséum d'Histoire Naturelle in March 1869; but already afflicted with the disease to which he succumbed two years later, he could not carry out his academic duties and returned to his estate in Gers.

BIBLIOGRAPHY

I. Original Works. A detailed bibliography of Lartet's writings is in *Vie et travaux de Édouard Lartet* (see below), pp. 77–80. The some 40 titles include *Notice sur la colline de Sansan* (Auch, 1851); "Sur la dentition des proboscidiens fossiles et sur la distribution géographique et stratigraphique de leurs débris en Europe," in *Bulletin de la Société géologique de France*, 2nd ser., **16** (1859), 469–515; "Nouvelles recherches sur la coexistence de l'homme et des grands mammifères fossiles réputés caractéristiques de la dernière période glacière," in *Annales des sciences naturelles*, 4th ser., **15** (1861), 177–253; "Sur des figures d'animaux gravées ou sculptées et autres produits d'art et d'industrie rapportables aux temps primordiaux de la période humaine," in *Revue d'archéologie* (1864), written with H. Christy; and *Reliquiae aquitanicae, Being Contributions to the Archeology and Palaeontology of Perigord . . .*, 2 vols. (London, 1865–1875), written with H. Christy.

II. Secondary Literature. *Vie et travaux de Édouard Lartet* (Paris, 1872) is a collection of notices published following his death; of particular interest are those by P. Fischer and E. T. Hamy. Also of value are G. Brégail, *Un éminent paléontologue gersois, Édouard Lartet* (Auch, 1948); Franck Bourdier, *L'art préhistorique et ses essais d'interprétation* (Paris, 1962); E. Cartailhac, "Édouard Lartet. Une féconde découverte à Aurignac," in *Revue du Comminges*, **33** (1918), 55–59; D. Dupuy, *Notice biographique sur Édouard Lartet* (Paris, 1873); and L. Méroc, "Édouard Lartet et son rôle dans l'élaboration de la préhistoire," in *Aurignac et l'aurignacien. Centenaire des fouilles d'Édouard Lartet* (Toulouse, 1963), pp. 7–18.

A. M. Monseigny
A. Cailleux

LARTET, LOUIS (*b.* Castelnau-Magnoac, Hautes-Pyrénées, France, 18 December 1840; *d.* Seissan, Gers, France, 16 August 1899), *geology, prehistory*.

Lartet was the son of the paleontologist and student of prehistory Édouard Lartet. After two years at the *lycée* in Toulouse, he continued his studies in Paris, where his family had moved. His father often entertained French and foreign scientists, both at his laboratory and at home; undoubtedly Lartet soon became familiar with the scientific problems discussed at these gatherings.

In 1862 Lartet was named *préparateur* at the Muséum d'Histoire Naturelle. In the same year he accompanied the geologist E. de Verneuil on one of his many trips to Spain. While there Lartet made interesting observations concerning geology and prehistory that were published in the *Bulletin de la Société géologique de France*. Immediately after receiving his *licence ès sciences* he was chosen by de Luynes to participate as a geologist on an expedition that took him from Lebanon to the Red Sea by way of the Jordan Valley and the Dead Sea (February–June 1864). The trip furnished Lartet with the subject of his doctoral dissertation, "Essai sur la géologie de la Palestine et des contrées avoisinantes," which he successfully defended in 1869.

In 1867 and 1868 Lartet was secretary of the International Congresses of Anthropology, held in Paris and London, respectively. Lartet's name, however, remains associated chiefly with the study of the Cro-Magnon deposit, the site of one of the most important discoveries in human paleontology. In April 1868 he was commissioned by the minister of education, Victor Duruy, to verify the authenticity of this discovery.

Apparently the Franco-Prussian War, in which Lartet had served as quartermaster-sergeant, and the death of his father in January 1871 dealt a serious blow to what had promised to be a fruitful career. At the end of the war, Lartet returned to Paris but remained there for only a short time. In 1873 he was named *suppléant* professor of geology at the Faculty of Sciences of Toulouse, and in July 1879 he became a full professor. Nevertheless, teaching apparently did not fulfill his aspirations. He wrote his friend E. T. Hamy in 1882: "I often wish, seeing that we have grown dull as college teachers, to throw off the official livery and return to Paris to resume, like my father, the disinterested study of science. . . ." He was not able to realize this desire. Apart from some observations concerning the geology and prehistory of the Pyrenees, Lartet spent the remainder of his career teaching. He had to retire prematurely because of poor health.

BIBLIOGRAPHY

I. Original Works. A bibliography of Lartet's writings by J. Canal is in *Bulletin de la Société d'histoire naturelle de Toulouse*, **45**, no. 2 (1912), 87–92. Among his works are "Sur le calcaire à Lychnus des environs de Segura (Aragon)," in *Bulletin de la Société géologique de France*, **20** (1863), 684–698, written with E. de Verneuil; "Mémoire sur une sépulture des anciens troglodytes du Périgord (Cro-Magnon)," in *Annales des sciences naturelles*, Zoologie, 5th ser., **10** (1868), 133–145; "Une sépulture des anciens troglodytes des Pyrénées," in *Comptes rendus hebdomadaires des séances de l'Académie des sciences*, **78** (1874), 1234–1236; and *Exploration géologique de la Mer Morte, de la Palestine et de l'Idumée, comprenant les observations recueillies par l'auteur durant l'expédition du duc de Luynes* . . . (Paris, 1877).

II. Secondary Literature. See E. Cartailhac, "Éloge de M. Louis Lartet," in *Mémoires de la Société archéologique du Midi*, **16** (1903–1908), 9–18; and A. Lavergne, "Louis Lartet," in *Revue de Gascogne*, **41** (1900), 177–182.

A. M. Monseigny
A. Cailleux

LASHLEY, KARL SPENCER (*b.* Davis, West Virginia, 7 June 1890; *d.* Poitiers, France, 7 August 1958), *psychology*, *neurophysiology*.

Karl Lashley's ancestry was English, primarily middle class. His father, Charles Gilpin Lashley, was a merchant in Davis, West Virginia, and its mayor and postmaster for many years. Lashley contended that he had received his native endowment from his mother, Maggie Blanche Spencer, a descendant of Jonathan Edwards. She exerted a profound influence upon her only child, encouraging his intellectual pursuits with her own library of 2,000 volumes. As a boy Karl collected various animal and plant specimens on long walks in the country. Throughout his life he formed assorted collections and kept unusual pets.

After receiving his B.A. from West Virginia University in 1910 he was awarded a teaching fellowship in biology at the University of Pittsburgh. He received his Ph.D. in 1914 from Johns Hopkins University under Jennings and Mast. In 1917 Lashley was instructor of psychology at the University of Minnesota; by 1923 he had been promoted to full professor. He remained at Minnesota until 1926, when he joined the staff at the University of Chicago as research psychologist; and he became professor of psychology in 1929. From 1935 to 1955 he was research professor of neuropsychology at Harvard University and, from 1942, director of the Yerkes Laboratories of Primate Biology.

In 1918 Lashley married Edith Ann Baker, an accomplished musician who died in 1948. A son born in 1919 died shortly thereafter. In 1957 he married Claire Imredy Schiller, widow of the Hungarian psychologist Paul Schiller.

Lashley was one of the world's foremost physiological psychologists. His scientific activities may be divided into four distinct periods. In the first (1912–1918) he laid a firm foundation in animal behavior working with lower forms, including their genetic make-up. In the second (1919–1929) his most noted contributions related to his penetrating analysis of the rat brain, in particular cerebral localization and interneuronal connections. His ideas of the learning process emerged from this analysis, culminating in *Brain Mechanisms and Intelligence* (New York, 1964). His theory of the equipotentiality of the cortex attracted world-wide attention and his article "In Search of the Engram" (*The Neuropsychology of Lashley*, pp. 478–505) made him famous. In the third period (1930–1942) he concerned himself primarily with vision, the chief sensory modality in learning. Few other scientific endeavors have equaled this thorough study of vision in relation to learning. In the final period (1941–1958) Lashley was a theorizer, especially on learning. He was also the world's sharpest critic of other theories, and especially of his own. He wrote: "My bricks won't hang together without speculative straw that I know is hooey."

BIBLIOGRAPHY

See Frank A. Beach's memoir of Karl Lashley in *Biographical Memoirs. National Academy of Sciences*, **35** (1961), 163–196. A full and complete bibliography of Lashley's works follows on pp. 196–204. *The Neuropsychology of Lashley*, F. A. Beach, D. O. Hebb, C. T. Morgan, and H. W. Nissen, eds. (New York, 1960), is a collection of Lashley's major publications.

Paul G. Roofe

LASSAR-COHN. See **Cohn, Lassar.**

LASSELL, WILLIAM (*b.* Bolton, Lancashire, England, 18 June 1799; *d.* Maidenhead, Berkshire, England, 5 October 1880), *astronomy.*

Lassell served a seven years' apprenticeship with a merchant at Liverpool and then became a brewer. About 1820 he began to construct reflecting telescopes and in 1840 installed a nine-inch Newtonian instrument in a private observatory near Liverpool.

Lassell was the first to design and use machines for surfacing mirrors of speculum metal in which the movement of the polisher closely imitated the circular motion used in polishing by hand. He was also the first to apply Fraunhofer's equatorial mounting to large reflecting telescopes. He mounted both the nine-inch telescope and a fine twenty-four-inch Newtonian, completed in 1846, in that way. With the latter he discovered Neptune's larger satellite, Triton, in 1846; detected Hyperion, the eighth satellite of Saturn (seen simultaneously by W. C. Bond at Harvard College Observatory) in 1848; and confirmed the existence of two satellites of Uranus, Ariel and Umbriel, in 1851.

In 1852 Lassell took the twenty-four-inch telescope to Valletta, Malta, where he expected that the climate would enable him to improve and extend his observational work. He then found that an even larger instrument was desirable, and in 1860 he completed a forty-eight-inch Newtonian telescope supported on a fork-type equatorial mounting. With this instrument, assisted by A. Marth, he searched unsuccessfully for new satellites, discovered 600 new nebulae, and monographed several others.

Upon his return to England in 1864, Lassell settled at Maidenhead, Berkshire. There he resumed his experiments with surfacing machines and constructed an improved form for polishing forty-eight-inch disks. Even so, he did not erect the forty-eight-inch telescope at Maidenhead but relied on the twenty-four-inch until failing eyesight forced him to abandon his studies.

Lassell became a fellow of the Royal Astronomical Society in 1839, received its gold medal in 1849, and was elected its president in 1870. He was elected a fellow of the Royal Society in 1849 and received the Royal Medal in 1858. He was an honorary member of the Royal Society of Edinburgh and the Royal Society of Sciences at Uppsala and held an honorary LL.D. conferred by the University of Cambridge.

BIBLIOGRAPHY

I. Original Works. For descriptions of machines for polishing specula, see *Monthly Notices of the Royal Astronomical Society*, **8** (1848), 197; **9** (1849), 29; **13** (1853), 43; and *Philosophical Transactions of the Royal Society*, **165** (1875), 303; Lassell's telescopes are described in *Memoirs of the Royal Astronomical Society:* the nine-inch reflector, **12** (1842), 265–272; the twenty-four-inch reflector, **18** (1850), 1–25; and the forty-eight-inch reflector, **35** (1866), 1–4. For an account of the Malta observations, see *Memoirs of the Royal Astronomical Society*, **35** (1866), 33–35.

II. Secondary Literature. Obituary notices include W. Huggins, in *Proceedings of the Royal Society*, **31** (1880–1881), vii–x; and M. L. Huggins, in *Observatory*, **3** (1880), 587–590. Lassell's activities in telescope-making are discussed in H. C. King, *The History of the Telescope* (London, 1955), pp. 218–224.

H. C. King

LĀṬADEVA (*fl.* India, *ca.* A.D. 505), *astronomy.*

Lāṭadeva, a pupil of Āryabhaṭa I (*b.* 476), was perhaps originally from Lāṭadeśa in southern Gujarat. He is known primarily through citations in the *Pañcasiddhāntikā* of Varāhamihira (sixth century) and in the commentary on the *Brāhmasphuṭasiddhānta* by Pṛthūdakasvāmin (*ca.* 864); the fragments are collected in O. Neugebauer and D. Pingree, *The Pañcasiddhāntikā of Varāhamihira.*

Lāṭadeva is said by Varāhamihira to have commented on the *Romakasiddhānta* and the *Pauliśasiddhānta*, which represent Greek and Greco-Babylonian astronomical techniques in Sanskrit (see essay in Supplement); he is evidently also responsible for a revision of the *Sūryasiddhānta* whereby it came to conform to the *ārddharātrikapakṣa* of his teacher, Āryabhaṭa I. His epoch for that work was 20/21 March 505.

BIBLIOGRAPHY

All the available material concerning Lāṭadeva and his works will be found in O. Neugebauer and D. Pingree, *The Pañcasiddhāntikā of Varāhamihira* (Copenhagen, 1970), pt. 1, pp. 14–15.

David Pingree

LATIMER, WENDELL MITCHELL (*b.* Garnett, Kansas, 22 April 1893; *d.* Oakland, California, 6 July 1955), *chemistry.*

Latimer entered the University of Kansas planning to become a lawyer; but finding that he enjoyed mathematics, he sought some subject to which he might apply it. His first contact with chemistry came during his third year at the university. The subject captured his interest, and he decided to become a chemist. He received the B.A. degree from the

University of Kansas in 1915 and served as instructor there from 1915 to 1917. The reputation of G. N. Lewis and his study of some of Lewis' papers led Latimer to Berkeley for graduate study, and he received the Ph.D. degree in 1919. His thesis research was concerned with low-temperature calorimetry and was conducted under the direction of G. E. Gibson. He was retained as a member of the staff and attained full professorship in 1931. Latimer was assistant dean of the College of Letters and Science from 1923 to 1924, dean of the College of Chemistry from 1941 to 1949, and chairman of the department of chemistry from 1945 to 1949.

His first paper, published in 1920 and entitled "Polarity and Ionization From the Standpoint of the Lewis Theory of Valence," written with W. H. Rodebush, was one of Latimer's most important. It contained the first clear recognition of the hydrogen bond as distinct from ordinary dipoles. The properties of numerous substances, including water, are largely determined by hydrogen bonding. The basic idea has found very wide and rapidly increasing application. Linus Pauling, who devotes some fifty pages in *The Nature of the Chemical Bond* to the discussion of examples of hydrogen bonding, states: "I believe that as the methods of structural chemistry are further applied to physiological problems it will be found that the significance of the hydrogen bond for physiology is greater than that of any other single structural feature."

Latimer's 108 scientific contributions are largely concerned with the application of thermodynamics to chemistry. However, he worked on such diverse subjects as dielectric constants, thermoelectric effect and electronic entropy, the ionization of salt vapors, radioactivity, and astrochemical processes involved in the formation of the earth.

Latimer was the first in the United States to liquefy hydrogen and to make measurements in that region of temperature. His leadership had a major influence on the subsequent work of W. F. Giauque, who extended the field to even lower temperatures. Latimer used low-temperature data in conjunction with the third law of thermodynamics to determine the entropies and free energies of aqueous ions. This work supplied much of the material included in his outstanding book, *The Oxidation States of the Elements and Their Potentials in Aqueous Solutions* (1938, 1952). He wrote *A Course in General Chemistry* with W. C. Bray and *Reference Book of Inorganic Chemistry* with J. H. Hildebrand.

Latimer was active on national defense research committees from 1941 to 1945 in the fields of oxygen production, chemical warfare, and plutonium research. He was director of a Manhattan Engineering District project in the University of California's department of chemistry, involving the chemistry of plutonium, from 1943 to 1947.

He was mainly responsible in the 1930's for initiating a seminar on nuclear chemistry that interested W. F. Libby, G. T. Seaborg, A. C. Wahl, and J. W. Kennedy and helped lay the foundation for the discovery of plutonium. The first separation and identification of plutonium depended on the relative oxidation potentials of the heaviest elements, and Latimer's availability for consultation contributed to the discovery of this extremely important element.

Latimer was a member of the National Academy of Sciences and chairman of the Chemistry Section in 1947–1950, of the American Chemical Society, the Electrochemical Society, the Faraday Society, the American Association for the Advancement of Science, and Sigma Xi. He received the Distinguished Service Award of the University of Kansas in 1948 and the Nichols Medal of the New York Section of the American Chemical Society in 1955. He was elected faculty research lecturer of the University of California in 1953.

Latimer received the Presidential Certificate of Merit for his contributions during World War II.

BIBLIOGRAPHY

A complete bibliography of Latimer's publications is in *Biographical Memoirs of the National Academy of Sciences*, **32** (1958), 230–237.

The following works may be cited from among his 108 publications: "Entropy Changes at Low Temperatures. I. Formic Acid and Urea. A Test of the Third Law of Thermodynamics," in *Journal of the American Chemical Society*, **42** (1920), 1533, written with G. E. Gibson and G. S. Parks; "Polarity and Ionization From the Standpoint of the Lewis Theory of Valence," *ibid.*, 1419, written with W. H. Rodebush; "The Mass Effect in the Entropy of Solids and Gases," *ibid.*, **43** (1921), 818; "A Revision of the Entropies of the Elements," *ibid.*, **44** (1922), 1008, written with G. N. Lewis and G. E. Gibson; "Thermoelectric Force, the Entropy of Electrons and the Specific Heat of Metals at High Temperatures," *ibid.*, 2136; and "The Electrode Potentials of Beryllium, Magnesium, Calcium, Strontium and Barium From Thermal Data," in *Journal of Physical Chemistry*, **31** (1927), 1267.

In the 1930's Latimer published *A Course in General Chemistry*, rev. ed. (New York, 1932; 3rd ed., 1940), written with W. C. Bray; "The Existence of Neutrons in the Atomic Nucleus," in *Journal of the American Chemical Society*, **54** (1932), 2125; "Bond Energies and Mass Defects in Atomic Nuclei," in *Journal of Chemical Physics*, **1** (1933), 133, written with W. F. Libby; "The Action of Neutrons on Heavy Water," in *Physical Review*, **47** (1935),

424, written with W. F. Libby and E. A. Long; "The Entropy of Aqueous Ions and the Nature of the Entropy of Hydration," in *Chemical Reviews*, **18** (1936), 349; "Silver Chromate: Its Heat Capacity, Entropy and Free Energy of Formation. The Entropy and Free Energy of Formation of Chromate Ion," in *Journal of the American Chemical Society*, **59** (1937), 2642, written with W. V. Smith and K. S. Pitzer; *The Oxidation States of the Elements and Their Potentials in Aqueous Solutions* (New York, 1938; 2nd ed., 1952); and "The Free Energy of Hydration of Gaseous Ions and the Absolute Potential of the Normal Calomel Electrode," in *Journal of Chemical Physics*, **7** (1939), 108, written with K. S. Pitzer and C. M. Slansky.

During the 1940's there appeared "The Entropies of Large Ions. The Heat Capacity, Entropy and Heat of Solution of Potassium Chlorplatinate, Tetramethylammonium Iodide and Uranyl Nitrate Hexahydrate," in *Journal of the American Chemical Society*, **62** (1940), 2845, written with L. V. Coulter and K. S. Pitzer; "Ionic Entropies and Free Energies and Entropies of Solvation in Water-Methanol Solutions," *ibid.*, 2019, written with C. M. Slansky; and "The Dielectric Constants of Hydrogen-Bonded Substances," in *Chemical Reviews*, **44** (1949), 59.

In the 1950's Latimer published "The Entropy of Aqueous Solutes," in *Journal of Chemical Physics*, **19** (1951), 1139, written with R. E. Powell; "Methods of Estimating the Entropies of Solid Compounds," in *Journal of the American Chemical Society*, **73** (1951), 1480; "Absolute Entropies in Liquid Ammonia," in *Journal of the American Chemical Society*, **75** (1953), 4147, written with W. L. Jolly; "Heats and Entropies of Successive Steps in the Formation of Al/F_6^{---}," *ibid.*, 1548, written with Jolly; "The Sign of Oxidation-Reduction Potentials," *ibid.*, **76** (1954), 1200; "Symposium on Hydration of Aqueous Ions, Introductory Remarks," in *Journal of Physical Chemistry*, **58** (1954), 513; and "The Complexing of Iron (III) by Fluoride Ions in Aqueous Solution: Free Energies, Heats and Entropies," in *Journal of the American Chemical Society*, **78** (1956), 1827, written with R. E. Connick *et al.*

JOEL H. HILDEBRAND

LATREILLE, PIERRE-ANDRÉ (*b.* Brive-la-Gaillarde, France, 29 November 1762; *d.* Paris, France, 6 February 1833), *entomology, zoology.*

Latreille was the natural son of J.-B.-J. Damazit de Sahuguet, baron d'Espagnac, a high-ranking officer in the French army. The identity of his mother is unknown. He was raised by foster parents, but his father provided for much of his education and in 1778 arranged for him to go to the Collège Cardinal Lemoine in Paris. Latreille received the degree of *maître ès arts* at the University of Paris in 1780. He then prepared for the priesthood and was ordained in 1786, but his intended ecclesiastical career was cut short by the Revolution.[1] In 1795 he attended the École Normale de Paris.

Frail in constitution throughout his life, Latreille had been encouraged at an early age to pursue natural history as a means of strengthening his health. He took up this activity eagerly, and by the 1790's he had achieved sufficient recognition as a naturalist to be elected an associate, or corresponding, member of the Société d'Histoire Naturelle de Paris (1791), the Société Philomathique de Paris (1795), and the First Class of the Institut de France (1798). Although he was considered one of the foremost entomologists of the day, Latreille spent most of his career in subordinate positions which earned him only a meager existence. In 1798 he went to the Muséum d'Histoire Naturelle in Paris as organizer of its entomological collection. There he officially assumed the role of *aide-naturaliste* in 1805, serving as demonstrator in Lamarck's course in invertebrate zoology. He eventually took over Lamarck's course in 1820, when Lamarck became completely blind. Ill health later forced Latreille to cede some of these responsibilities to Jean Victor Audouin. Following Lamarck's death in 1829, the chair of invertebrate zoology was divided into two, and Latreille became professor in the new chair of entomology. In addition to his work at the Muséum d'Histoire Naturelle, he taught briefly at the École Vétérinaire d'Alfort (1814–1815). He was named a member of the Institut de France in 1814 and received the decoration of the Légion d'Honneur in 1821. Although known primarily for his entomological studies, he wrote on a wide range of zoological subjects as well as ancient geography and chronology.

Latreille's major scientific contribution lay in applying the "natural method" to the classification of the insects, arachnids, and crustaceans. This task was undertaken at a time when entomological knowledge and collections were undergoing a spectacular growth (from an estimated 1,500 specimens in 1789 the collection at the Muséum d'Histoire Naturelle grew to roughly 40,000 specimens representing 22,000 species in 1823).[2] Linnaeus and Fabricius had constructed artificial systems for classifying the insects, the former relying primarily upon the number and configuration of the wings and the latter relying exclusively on the parts of the mouth. Latreille's goal in his first major work, his *Précis* of 1796, was to arrange the genera of insects in their "natural order" by taking numerous characters into consideration. His system was based essentially upon a combination of the characters that Linnaeus and Fabricius had employed.

Latreille's most significant work, the four-volume *Genera crustaceorum et insectorum* (1806–1809), was developed along the same lines. While Latreille praised Cuvier for having founded zoological classification on comparative anatomy, he based his own classificatory work primarily on the consideration of external characters. Nonetheless, as one of his successors remarked more than half a century after the appearance of the *Genera*, the work offered "such an exact exposition of the characters of the insects, arachnids, and crustaceans, and nearly always such a just appreciation of the natural affinities of these animals, that the most modern researches have only added rectifications of a second order."[3]

Unlike such notable predecessors as Réaumur and Fabricius, Latreille presented a balanced approach in his entomological studies, dealing with behavioral and taxonomic problems alike. He also became a self-conscious pioneer in the study of biogeography, observing that temperature alone is insufficient to explain animal distribution and calling attention to the way in which animal distribution is related to the distribution of food sources. Along other lines in the 1820's he advanced theoretical views displaying sympathy with the concept of the unity of plan in the animal kingdom.

Among such strong personalities as Cuvier, Lamarck, Geoffroy Saint-Hilaire, and Blainville, Latreille was a modest, unassuming figure. Although his concerns, as suggested here, were by no means limited to questions of insect classification, he was inclined to leave to others the formulation of broad systems and philosophies. With Étienne Geoffroy Saint-Hilaire he cosigned a report to the Institut that precipitated the famous Cuvier-Geoffroy debate in 1830. The report, on a memoir in which Laurencet and Meyranx had proposed an analogy between the cephalopods and the vertebrates, was, however, written by Geoffroy alone. Latreille, according to a letter he wrote to Cuvier, wished no credit for it, asserting that he himself had given up the idea he had once entertained of considering the invertebrates and vertebrates as constructed upon the same plan.[4]

On another speculative front Latreille's stand was less ambiguous. Although he was for many years Lamarck's assistant and friend, he did not share his mentor's evolutionary views. He found it impossible to conceive of the forms of insects, and in particular their instinctual behavior, as anything other than the evidence of divine wisdom and design. Furthermore, as he remarked on this view of creation shortly before he died: "If we are wrong, do not seek to destroy illusions which are useful rather than harmful to society and which make us happy or console us in the difficult pilgrimage of life."[5]

NOTES

1. Louis de Nussac, *Latreille à Brive*, pp. 25–43, presents in detail the often-told story of how Latreille, imprisoned during the Revolution as a priest who had not preached a sermon, gained his release through the attention focused on him by his chance discovery in his cell of a previously unknown species of beetle.
2. J. P. F. Deleuze, *Histoire et description du Muséum royal d'histoire naturelle* (Paris, 1823), p. 188.
3. Émile Blanchard, *Métamorphoses, moeurs et instincts des insectes* (Paris, 1877), p. 33.
4. Institut de France, MS 3060 (1).
5. Latreille, *Cours d'entomologie*, p. 21.

BIBLIOGRAPHY

I. Original Works. Latreille's publications total over seventy titles, excluding his contributions to various dictionaries of natural history. A reasonably accurate list of these appears in "Latreille," in *Nouvelle biographie générale* (Paris, 1862), XXIX, 851–854. The entomological titles alone have been compiled by A. Percheron, *Bibliographie entomologique* (Paris, 1837), I, 228–235. Among Latreille's more important works are *Précis des caractères génériques des insectes disposés dans un ordre naturel* (Brive, *an* v [1796]); *Histoire naturelle, générale et particulière des crustacés et des insectes. Ouvrage faisant suite aux oeuvres de . . . Buffon, rédigé par C. S. Sonnini . . .*, 14 vols. (Paris, 1802–1805); *Genera crustaceorum et insectorum secundum ordinem naturalem in familias disposita*, 4 vols. (Paris, 1806–1909); *Les crustacés, les arachnides et les insectes*, vol. III of Georges Cuvier, *Le règne animal distribué d'après son organisation, pour servir de base à l'histoire naturelle des animaux et d'introduction à l'anatomie comparée* (Paris, 1817)—also vols. IV and V of the 1829 ed.; and *Cours d'entomologie, ou de l'histoire naturelle des crustacés, des arachnides, des myriapodes et des insectes* (Paris, 1831).

The Société Entomologique de France has a collection of Latreille materials numbering 1,637 items, grouped in the following categories: (1) titles and distinctions; (2) MSS of various published scientific works; (3) unpublished works; and (4) scientific correspondence with other entomologists.

II. Secondary Literature. There is no full treatment of Latreille's life and scientific work. See Louis de Nussac, *Pierre-André Latreille à Brive de 1762 à 1798* (Paris, 1907); and "Le centenaire de Pierre-André Latreille," in *Archives du Muséum national d'histoire naturelle*, 6th ser., **11** (1934), 1–12; and A. J. L. Jourdan, "Latreille," in *Biographie universelle. Ancienne et moderne*, new ed. (Paris, n.d.), XXIII, 329–331. For a discussion of Latreille's classificatory contributions, see Henri Daudin, *Cuvier et Lamarck, les classes zoologiques et l'idée de série animale (1790–1830)* (Paris, 1926).

Richard W. Burkhardt, Jr.

LAU, HANS EMIL (*b.* Odense, Denmark, 16 April 1879; *d.* Usserød, Denmark, 16 October 1918), *astronomy*.

In 1906 Lau received the master's degree at the University of Copenhagen. During the following years he worked at the private Urania Observatory at Frederiksberg, a suburb of Copenhagen. About 1911–1912 he spent a year at the Treptow Sternwarte in Berlin, and after his return to Denmark he built a small private observatory.

In 1905 Lau was awarded a gold medal by Copenhagen University for a discussion of the observations of the bright new star of 1901 and of theories on new stars; he also published an interpretation of a nova spectrum. From his youth Lau worked on the problem of a trans-Neptunian planet (or two planets); and his conclusions must be characterized as sound.

During 1905–1911 Lau made a long series of visual measurements of double stars and did some pioneer work on photographic observations; he discussed the systematic errors of visual observations and made a close determination of the orbit of the system of 70 Ophiuchi; in connection with his reobservation of previously measured optical companions of bright stars he published several discussions of the proper motions and parallaxes of faint stars. He made observations of variable stars, of stars suspected of variability, and of colors of stars; he was also one of the first astronomers to become interested in photographic photometry.

Lau was always an eager observer of the planets, especially of Mars and Jupiter, making drawings of their surfaces and micrometrical determinations of the positions of details on them. His comprehensive descriptions of the yearly changes of the features of Mars and of the clouded world of Jupiter, based on observations collected from a great many observers, must be regarded as exhaustive contemporary discussions of these problems.

Lau never succeeded in obtaining a position at a scientific institution, but his contributions are those of an original and dedicated scientist.

BIBLIOGRAPHY

I. ORIGINAL WORKS. A listing of Lau's works in *Astronomische Nachrichten* is in Poggendorff, V, 711. His writings include "Sur la question des planètes transneptuniennes," in *Bulletin astronomique*, **20** (1903), 251–256; "Sur le spectre des étoiles nouvelles," *ibid.*, **23** (1906), 297–303; "Sur le système de 70 Ophiuchi," *ibid.*, **25** (1908), 139–141, and **26** (1909), 433–456; "Über die Rotation des Planeten Jupiter," in *Astronomische Nachrichten*, **195** (1913), 313–340; "La planète transneptunienne," in *Bulletin de la Société astronomique de France*, **28** (1914), 276–283; "Die periodischen Veränderungen auf dem Mars," in *Astronomische Nachrichten*, **204** (1917), 81–124; and "Untersuchungen über die Farben der Fixsterne," *ibid.*, **205** (1918), 49–70. Lau's measurements of visual double stars are published in a series of papers in *Astronomische Nachrichten*, **169–205** (1905–1918).

II. SECONDARY LITERATURE. There is an obituary by E. Strömgren in *Astronomische Nachrichten*, **208** (1919), 151–152.

AXEL V. NIELSEN

LAUE, MAX VON (*b.* Pfaffendorf, near Koblenz, Germany, 9 October 1879; *d.* Berlin, Germany, 24 April 1960), *theoretical physics*.

After passing the final secondary-school examination in March 1898, Laue began studying physics at the University of Strasbourg, although he was still in military service. In the fall of 1899 he transferred to Göttingen. Under the influence of Woldemar Voigt he chose to specialize in theoretical physics. At the same time he came to prefer optical problems, a preference that was strengthened by the lectures of Otto Lummer which he heard during his three semesters at the University of Berlin. In July 1903, Laue received his doctorate under Max Planck for a dissertation on the theory of interference in plane parallel plates. He then returned for two years to Göttingen, where he passed the state examination to qualify for teaching in the Gymnasiums.

The course that Laue's life was to take was decided in the autumn of 1905, when Planck offered him an assistantship. Laue became Planck's leading and favorite pupil, and the two formed a lifelong friendship. Laue introduced Planck's central concept, entropy, into optics and qualified as university lecturer in 1906 with a work on the entropy of interfering pencils of rays. In the winter semester of 1905–1906 Laue heard Planck's lecture at the Physical Colloquium on the special theory of relativity, which Einstein had just stated. After initial reservations Laue became one of the first adherents of the new theory and, as early as July 1907, presented a proof for it that he drew, characteristically, from optics.

In 1851, after many experiments, Fizeau had discovered a formula for the velocity of light in flowing water that could not be understood in terms of classical physics. Assuming light to be a wave phenomenon in the ether, one could either suppose that the ether does not contribute to the motion of the flowing water, in which case the velocity of light should be $u = c/n$; or one could postulate that the ether is carried along through the motion of the water, in which case the equation ought to be $u = c/n \pm v$. Yet, curiously, the experiments showed partial ether "drag" varying as a specific fraction of the velocity of water — $v(1 - 1/n^2)$—the Fresnel drag coefficient.

Einstein's special theory of relativity dispensed with the addition or subtraction of the velocities, hitherto assumed to be self-evident, and applied instead a special "addition theorem." In 1907 Laue demonstrated that this theorem readily yields Fizeau's formula with the previously enigmatic Fresnel drag coefficient: $u = c/n \pm v(1 - 1/n^2)$. Laue thereby furnished Einstein's theory with an important experimental proof, which, along with the Michelson-Morley experiment and arguments from group theory, contributed to early acceptance of the theory. Having thus proved himself an expert in relativity theory, in 1910 Laue wrote the first monograph on the subject. He expanded it in 1919 with a second volume on the general theory of relativity; the work went through several editions.

In 1909 Laue became a *Privatdozent* at the Institute for Theoretical Physics, directed by Arnold Sommerfeld, at the University of Munich. Here, in the spring of 1912, Laue had the crucial idea of sending X rays through crystals. At this time scientists were very far from having proved the supposition that the radiation that Roentgen had discovered in 1895 actually consisted of very short electromagnetic waves. Similarly, the physical composition of crystals was in dispute, although it was frequently stated that a regular structure of atoms was the characteristic property of crystals. Laue argued that if these suppositions are correct, then the behavior of X radiation upon penetrating a crystal should be approximately the same as that of light upon striking a diffraction grating; and interference phenomena had been studied by means of the latter arrangement since Fraunhofer. These ideas, which Laue expressed in a discussion with Peter Paul Ewald, were soon being talked about by the younger faculty members. Finally Walter Friedrich, an assistant of Sommerfeld's, and Paul Knipping, a doctoral candidate, began experiments in this field on 21 April 1912. The irradiation of a copper sulfate crystal yielded regularly ordered dark points on a photographic plate placed behind the crystal, the first of what are today called Laue diagrams. On 4 May 1912 Laue, Friedrich, and Knipping announced their success in a letter to the Bavarian Academy of Sciences.

Laue wrote in his autobiography:

> I was plunged into deep thought as I walked home along Leopoldstrasse just after Friedrich showed me this picture. Not far from my own apartment at Bismarckstrasse 22, just in front of the house at Siegfriedstrasse 10, the idea for a mathematical explanation of the phenomenon came to me. Not long before I had written an article for the *Enzyklopaedie der mathematischen Wissenschaften* in which I had to re-formulate Schwerd's theory of diffraction by an optical grating (1835), so that it would be valid, if iterated, also for a cross-grating. I needed only to write down the same condition a third time, corresponding to the triple periodicity of the space lattice, in order to explain the new discovery. . . . The decisive day, however, was the one a few weeks later when I could test the theory with the help of another, clearer photograph.

The awarding of the Nobel Prize in physics for 1914 to Laue indicated the significance of the discovery that Albert Einstein called one of the most beautiful in physics. Subsequently it was possible to investigate X radiation itself by means of wavelength determinations as well as to study the structure of the irradiated material. In the truest sense of the word scientists began to cast light on the structure of matter.

Laue was appointed associate professor at the University of Zurich in 1912 and full professor at Frankfurt in 1914. In the latter year Laue's father, an official in the military court system, was elevated to the hereditary nobility. Thus within a few years the unknown *Privatdozent* Max Laue became the world-famous Nobel Prize winner Professor Max von Laue.

During the war Laue worked with Willy Wien in Würzburg on developing electronic amplifying tubes for improving the army's communication techniques. In 1919 he arranged an exchange of teaching posts with Max Born: Born left Berlin to go to Frankfurt and Laue went to the University of Berlin, which he considered his true intellectual home. Here he was again able to be near Planck, his honored teacher and friend.

The new field of X-ray structural analysis that Laue established developed into an important branch of physics and chemistry. The leading researchers in the field were William Henry Bragg and William Lawrence Bragg. Laue himself, a true pupil of Planck's, was interested only in the "great, general principles" and did not concern himself with the study of the structure of individual substances; instead he continued to work on the fundamental theory. Following the preliminary investigations of Charles Galton Darwin and Peter Paul Ewald, Laue expanded his original geometric theory of X-ray interference into the so-called dynamical theory. Whereas the geometric theory dealt with only the interaction between the atoms of the crystal and the incident electromagnetic waves, the dynamical theory took into account the forces between the atoms as well. To be sure, the correction amounted to only a few seconds of arc, but deviations had appeared early in the course of the very precise X-ray spectroscopic measurements.

In the following decades the theory was developed in various directions. When Laue later undertook to provide a comprehensive view of only the principles in *Röntgenstrahl-Interferenzen* (1941), his account ran to 350 pages. After the discovery of electron interference, Laue included this phenomenon in his theory. He did not, however, otherwise participate in the creation or development of quantum theory; and, like Planck, Einstein, de Broglie, and Schrödinger, he was skeptical of the "Copenhagen interpretation."

In 1932 Laue received the German Physical Society's Max Planck Medal. In his acceptance speech Laue presented an important result in the field of superconductivity: the interpretation of a seemingly paradoxical measurement made by W. J. de Haas. Subsequently Laue engaged in a fruitful joint study of this topic with Walther Meissner. Meissner conducted the relevant experiments at the Physikalisch-Technische Reichsanstalt, and Laue acted as theoretical adviser to that institution. Whereas Werner Heisenberg, Fritz London, and Heinz London worked on a quantum theory of superconductivity, Laue characteristically remained within the framework of the classical theory. He applied the purely phenomenological Maxwellian theory to the superconductor and later worked on the thermodynamics of superconductivity.

Laue held positions of exceptional trust at an early age. In 1921, proposed by Max Planck, he became a member of the Prussian Academy of Sciences. Following the establishment of the Emergency-Association of German Science (later the German Research Association), the German physicists elected Laue the representative for theoretical physics. He was chairman of the physics committee and also a member of the electrophysics committee until 1934. Through his solid judgment he directed the available financial resources to the truly important projects and thereby played a not inconsiderable role in the continuance of the "golden age of German physics" even during the economic depression of the Weimar Republic.

Laue's scientific pride did not permit him passively to accept Einstein's dismissal following the Nazi seizure of power. Only two colleagues in the Prussian Academy joined in his protest. As the chairman of the German Physical Society, Laue took issue with the slandering of the theory of relativity as a "worldwide Jewish trick" and gave a highly regarded address at the opening of the physics congress in Würzburg on 18 September 1933. He likened Galileo, the champion of the Copernican world view, to Einstein, the founder of relativity theory, and openly expressed his hope and belief that, as the truth had once before won out against the Church's prohibition, this time it would win out against the National Socialist proscription: "No matter how great the repression, the representative of science can stand erect in the triumphant certainty that is expressed in the simple phrase: And yet it moves."

Although his defense of Einstein had been in vain, Laue did have one success at the end of 1933—in the Prussian Academy. Johannes Stark, Hitler's famous follower, who had become a rabid opponent of modern physical theories, was supposed to be admitted into the Academy at the request of the new regime; and a group of academicians was prepared—reluctantly—to consent to the election. Yet in the session of 11 December 1933 the objections to this choice were set forth so emphatically by Laue, Otto Hahn, and Wilhelm Schlenk that the sponsors withdrew the proposal and Stark was not admitted. On 23 March 1934, Einstein wrote to Laue: "Dear old comrade. How each piece of news from you and about you gladdens me. In fact I have always felt, and known, that you are not only a thinker, but a fine person too."

When Friedrich Schmidt-Ott, the elected president of the German Research Association, was dismissed by the Nazis and replaced by Johannes Stark, Laue was once again the spokesman for the physicists. He wrote to Schmidt-Ott on 27 June 1934: "I heard of your withdrawal from the presidency . . . with deep regret. The overwhelming majority of German physicists, especially the members of the physics committee, share this regret. . . . Under the present circumstances, moreover, I fear that the change in the presidency is the prelude to difficult times for German science, and physics will no doubt have to suffer the first and hardest blow."

In the Research Association, Laue's judgment was no longer asked for; he also lost his position as adviser to the Physikalisch-Technische Reichsanstalt. He continued, however, as professor at the University of Berlin and as deputy director of the Kaiser Wilhelm Institute for Physics. Following his early retirement from teaching in 1943, Laue moved to Württemberg-Hohenzollern; at this time the Kaiser Wilhelm Institute, busy with military research and now under the direction of Werner Heisenberg, was relocated in Hechingen. Although Laue did not participate in the uranium project for the production of atomic energy, he was interned after the war with the atomic physicists by the Allies.

From the beginning Laue stood at the forefront of the rebuilding of German science. In the fall of 1946, working in Göttingen, he created with former colleagues the German Physical Society in the British Zone and in 1950 took part in refounding the League of German Physical Societies, today known as the German Physical Society. Laue played an important part in the reestablishment of the Physikalisch-Technische Bundesanstalt in Brunswick (the successor to the Physikalisch-Technische Reichsanstalt in Berlin) and in the German Research Association, where he was reelected to the physics committee until 1955. At first Laue was active primarily in his former post of deputy director of the Kaiser Wilhelm Institute for Physics at Göttingen. In April 1951, at the age of seventy-one, he took over the directorship of the former Kaiser Wilhelm Institute for Chemistry and Electrochemistry in Berlin-Dahlem. Active up to the end, Laue died in his eighty-first year following an automobile accident. He was mourned by colleagues throughout the world.

BIBLIOGRAPHY

I. ORIGINAL WORKS. Laue's writings were collected in *Gesammelte Schriften und Vorträge*, 3 vols. (Brunswick, 1961), with his autobiography in vol. III. His works include *Die Relativitätstheorie*, 2 vols. (Brunswick, 1911–1919); *Korpuskular- und Wellentheorie* (Leipzig, 1933); *Röntgenstrahl-Interferenzen* (Leipzig, 1941); *Materiewellen und ihre Interferenzen* (Leipzig, 1944); *Geschichte der Physik* (Bonn, 1946); and *Theorie der Supraleitung* (Berlin, 1947).

II. SECONDARY LITERATURE. See the following, listed chronologically: R. Brill, O. Hahn, *et al.*, "Feierstunde zu Ehren von Max von Laue an seinem 80. Geburtstag," in *Mitteilungen aus der Max Planck-Gesellschaft*, **6** (1959), 323–366; Peter Paul Ewald, "Max von Laue," in *Biographical Memoirs of Fellows of the Royal Society*, **6** (1960), 135–156, with bibliography; James Franck, "Max von Laue (1879–1960)," in *Yearbook. American Philosophical Society* (1960), 155–159; Walther Meissner, "Max von Laue als Wissenschaftler und Mensch," in *Sitzungsberichte der Bayerischen Akademie der Wissenschaften zu München* (1960), 101–121; Max Päsler, "Leben und wissenschaftliches Werk Laues," in *Physikalische Blätter*, **16** (1960), 552–567; Peter Paul Ewald, *Fifty Years of X-Ray Diffraction* (Utrecht, 1962), *passim;* Friedrich Herneck, "Max von Laue," in *Bahnbrecher des Atomzeitalters* (Berlin, 1965), pp. 273–326; Paul Forman, "The Discovery of the Diffraction of X-Rays by Crystals," in *Archive for History of Exact Sciences*, **6** (1969), 38–71; and Armin Hermann, "Forschungsförderung der Deutschen Forschungsgemeinschaft und die Physik der letzten 50 Jahre," in *Mitteilungen der Deutschen Forschungsgemeinschaft*, **4** (1970), 21–34, also in *Physik in unserer Zeit*, **2** (1971), 17–23.

Copies of unpublished letters, almost all of them written to Laue, are available at the Deutsches Museum, Munich (Handschriftenabteilung) and at Stuttgart University, in the Department of the History of Science and Technology.

ARMIN HERMANN

LAURENS, ANDRÉ DU (LAURENTIUS) (*b.* Tarascon, near Arles, France, 9 December 1558; *d.* Paris, France, 16 August 1609), *anatomy, medicine.*

Laurens's father was a physician in Arles; his maternal uncle, Honoré Castellan, was an important royal physician. Although their father died when they were young, Laurens and his six brothers all enjoyed successful careers, two becoming archbishops and two becoming royal physicians.

After taking the M.D. at Avignon in 1578, Laurens went to Paris to study under Louis Duret. In 1583 he took another doctorate at Montpellier in order to qualify for the chair of medicine left vacant there in 1582 by the death of Laurent Joubert. After lecturing at Montpellier for about ten years, he left (apparently without relinquishing his chair) to serve as personal physician to the duchess of Uzès; she introduced him at the French court, where he was soon named one of the physicians of Henry IV. In 1596 he became a royal physician in ordinary and in 1600 was designated first physician to Henry's new queen, Marie de Médicis. In 1603 he became chancellor of the University of Montpellier, although he continued to reside at court and delegated his duties to a vice-chancellor. In 1606 he became first physician to the king; he died three years later. In 1601 he had married Anne Sanguin, by whom he had one son.

Laurens's first publication (1593) was a pamphlet in which he attacked the views of Simon Piètre, a prominent Parisian physician, on the foramen ovale of the fetal heart. According to Galen, this orifice permitted blood to pass from the vena cava, through the right and left auricles, and into the pulmonary vein, where it nourished the lungs. Piètre maintained (1593) that the actual function of the foramen was to bring blood from the vena cava into the left ventricle for distribution to the body through the arteries. In his pamphlet Laurens defended Galen's view. Piètre's response led to a further attack by Laurens, and a final defense by Piètre, whose view found little favor until its vindication by William Harvey in *De motu cordis* (1628).

The same Galenic orthodoxy is evident throughout Laurens's *Opera anatomica* (1593). Much of this work was incorporated into the more comprehensive, illustrated *Historia anatomica* (1600), which was one of the most widely used anatomical textbooks of the first half of the seventeenth century. The twelve books of the *Historia* include not only descriptions of the structures, actions, and uses of the parts, but also 178 "controversies" in which are discussed disputed questions generally of a theoretical nature, such as whether there is a natural spirit and whether the brain is the seat of a principal faculty. In his preface Laurens vowed to vindicate Galen from the innumerable calumnies of the moderns as far as truth would permit, and in fact he did support the Galenic position in most of the controversies, upholding, for example, the permeability of the cardiac septum against the pulmonary circuit and other alternatives and rejecting Falloppio's views on the action of the gall bladder.

Laurens's book contains little original material. His anatomical descriptions generally did not improve on those of his predecessors and at times fell short of them. His illustrations were, with few exceptions, borrowed from other works, especially those of Vesalius, as well as Coiter and Varolio; and even many of his controversies were based on earlier works written in the tradition of Peter of Abano's *Conciliator*. Nevertheless, as a textbook the *Historia anatomica* was quite successful. Its anatomical descriptions were concise and lucid, while the controversies provided a comprehensive survey of the various positions in disputed points of anatomy and physiology. Even the orthodoxy of Laurens's views often proved useful to those requiring a foil against which to develop alternative positions. The importance of the book is reflected in its numerous editions, both in Latin and in French, and in the frequency with which it was cited by contemporaries. Many of the controversies were translated into English in Helkiah Crooke's *Mikrokosmographia* (1615), and Theodore Colladon included a detailed critique of the *Historia* in his *Adversaria* (1615).

Laurens also published numerous other medical works, of which the most popular was *Discours de la conservation de la vue; des maladies mélancholiques; des catarrhes; et de la vieillesse* (1594). Intended for a lay audience, the work went through more than twenty editions, and was translated into English, German, Latin, and Italian.

BIBLIOGRAPHY

I. Original Works. The bibliography in Edouard Turner, "Bibliographie d'André Du Laurens . . . avec quelques remarques sur sa biographie," in *Gazette hebdomadaire de médecine et de chirurgie*, 2nd ser., **17** (1880), 329–341, 381–390, 413–435; and *Études historiques* (Paris, 1876–1885), pp. 209–243, is fairly comprehensive. He accepts *Historia anatomica humani corporis et singularum ejus partium, multis controversiis et observationibus novis illustrata* (Paris, 1600) as the 1st ed., instead of the undated Frankfurt edition, which he dates 1627 rather than 1599 and considers to be a reissue of the Frankfurt edition dated 1600. Turner's list includes the pamphlets exchanged by Piètre and Laurens in 1593. Laurens gives an account of the dispute, including a long extract from Piètre's first tract, in *Historia anatomica*, VIII, controversy 25. *A Discourse of the Preservation of Sight; of Melancholicke Diseases; of Rheumes, and of Old Age*, Richard Surphlett, trans. (London, 1599), was published in facs., with intro. by Sanford V. Larkey (London, 1938).

II. Secondary Literature. For biographical information, see Turner, cited above. For a critique of *Historia anatomica*, see M. Portal, *Histoire de l'anatomie et de la chirurgie*, II (Paris, 1770), 147–159.

Jerome J. Bylebyl

LAURENT, AUGUSTE (or **AUGUSTIN**) (*b.* St.-Maurice, near Langres, Haute-Marne, France, 14 November 1807; *d.* Paris, France, 15 April 1853), *chemistry*.

Jean Baptiste Laurent, who owned a small farm, married Marie-Jeanne Maistre, the daughter of a merchant from Burgundy. Augustin Laurent (who always signed his name as Auguste in later life) was the second of their four children. He received the traditional classical education at the local *collège* in Gray. He passed the entrance examination for the prestigious École des Mines in Paris, as an external student, on 9 December 1826, and received his engineering degree in June 1830. His first publication (1830), written jointly with Arrault, was a thesis submitted in partial fulfillment of the requirements for obtaining a degree; it was written on some of the techniques used in cobalt mines in Germany, which he had visited during his vacation the previous year.

At the beginning of the academic year 1830–1831 he was employed as a laboratory assistant at the École Centrale des Arts et Manufactures by J. B. Dumas. From 1832 to 1834 Laurent worked as chemist for the royal porcelain works at Sèvres, under the direction of Alexandre Brongniart, Dumas's father-in-law. There Laurent developed a method of analyzing silicates by the action of hydrofluoric acid that is still used by chemists.

In 1835 Laurent opened a private school in Paris where he taught chemistry to fee-paying adults. The school was closed after about a year, and its furnishings were sold to pay for the publication of Laurent's doctoral dissertation at the Sorbonne's Faculty of Sciences.

In December 1837 Laurent passed his oral examination and obtained the degree of *docteur-ès-sciences*. His main doctoral thesis developed the principal ideas of his theory of fundamental and derived radicals in organic chemistry.

In 1836–1837 Laurent worked as an analyst for a Paris perfumery owned by a certain Laugier, who regarded him as a partner and associated him with the firm's profits. Laurent received 10,000 francs when he left the perfumery, as his share of the profits.

Laurent was of a very sensitive and nervous disposition, easily discouraged and readily provoked into a quarrel over real or imagined insults. He was also generous and frank and had radical political commitments. A firm adherent of the left-wing republican tradition in France, he was impatient and suspicious of authority. Dumas, who epitomized conventional virtues of the authoritarian and traditionalist kind, was a natural target for Laurent's hostility. Although there is no positive evidence that Dumas tried to harm him, Laurent was persuaded of his senior's malevolence and bad faith. After receiving his doctorate, Laurent became persuaded that Dumas and other university authorities were opposed to the novelty of his ideas in organic chemistry and that he stood no chance of a university appointment in France. Consequently he contemplated abandoning a research career and in 1838 accepted a post as industrial chemist at a ceramics works in Luxembourg. The same year he married Anne-Françoise Schrobilgen, the daughter of an important Luxembourg dignitary.

Laurent's ideas were in fact so easily assimilated by French chemists that by 1839 both the younger chemists as well as their seniors active in research in organic chemistry, including Dumas, had accepted the essentials. Notwithstanding Liebig's exacerbated vanity and nationalism, by 1845 a large group of Germany chemists had also adopted Laurent's views. The same was true of England. Even Berzelius in Sweden, against whose views in organic chemistry Laurent directed most of his work, rendered homage to his genius for experimental work, while dismissing his theoretical assumptions. In his annual report on the development of chemistry during 1843, Berzelius devoted nearly half his account of organic chemistry to a highly complimentary account of Laurent's work; the previous year he had said that Laurent's work on indigo was the best piece of experimental research in vegetable chemistry since Liebig and Wöhler had discovered the benzoyl radical in 1832.

On 30 November 1838 Laurent was appointed to the newly created chair of chemistry at Bordeaux, in his native region. He returned from Luxembourg in early 1839 and occupied this post for the next six years. This was the most productive period of his life, during which he published nearly 100 papers.

During his annual summer vacation visit to his in-laws in 1843, Laurent went to see Liebig in Giessen, where he was well received and established important contacts with the younger German chemists, especially von Hofmann. In the fall of that year he met Gerhardt; by 1844 they were in close contact and their lifelong collaboration had started. Laurent became a *chevalier* of the Légion d'Honneur in 1844.

The first number of the journal jointly edited by Gerhardt and Laurent, *Comptes-rendus mensuels des travaux chimiques*, appeared in February 1845. On 11 August 1845, he was elected a corresponding member of the Académie des Sciences and later that month left Bordeaux for good; he was granted leave of absence with pay until his appointment in 1848 to the post he had long desired, assayer at the Paris Mint. In the meantime he worked at various laboratories at the Collège de France, École Normale, and École des Mines. In 1847 he published his book on crystallography.

Laurent presented his candidature for the chair of chemistry at the Collège de France left vacant by the resignation of Pelouze in November 1850. He was elected by thirteen votes to nine for his rival Balard. But the election had to be ratified by the Académie des Sciences, which preferred Balard (35 to 11). Presumably the hostility of the Academy was dictated by the political disturbances in France in 1851, which aggravated the antagonism of political moderates toward radicals such as Laurent. Having fallen seriously ill, in 1852 he went to recuperate in the south of France but died in Paris of consumption the following year. His family was awarded a state pension. He left the manuscript of *Méthode de chimie*, which was edited by J. Nicklès and published in 1854. Odling published an English version in 1855, and Kekulé offered to translate it into German.

Laurent's career coincided with the attempt to establish organic chemistry as a precise science, clearly distinguishable in its methods and fundamental principles from inorganic chemistry, which had been the main preoccupation of chemists during the previous half century. He was a central figure in the emergence of organic chemistry as a mature science.

A "philosophical" chemist, Laurent realized the need to pass beyond the narrow *ad hoc* theories of limited application adopted by his contemporaries, in order to establish a set of general theoretical principles valid for the whole domain of phenomena studied by organic chemistry. He insisted that an adequate scientific theory had to fulfill three requirements: explanation of all phenomena within the range of the theory; prediction of new and hitherto untested phenomena with its help; and enlargement of the scope of the scientific discipline by the imaginative application of theoretical principles to areas outside its original domain or subject matter.

His earlier contributions to organic chemistry were his thorough and meticulous experimental investigations of naphthalene and its derivatives. The subject had been chosen for him by Dumas, who was studying the reactions of the halogens upon various hydrocarbons (1832). Laurent adapted the existing method for the preparation of naphthalene (developed by Kidd) and succeeded in obtaining a very pure product, at low cost, by the fractional distillation of coal tar. By an extension of this early procedure, he became a major figure in the development of this branch of organic chemistry, the preparation and isolation of compounds by the distillation of coal tar. Using the same method, he and Dumas discovered anthracene (paranaphthalene) in 1832. Upon analysis both naphthalene and anthracene were found to be hydrocarbons containing the same relative proportions of carbon to hydrogen, namely 5:2 ($C = 6$).

The preparation of naphthalene was followed by a study of its reactions with chlorine, bromine, and nitric- and sulfuric-acid anhydrides. Laurent's practice of exhaustively investigating the compounds that a substance formed with each of a small number of reagents—rather than superficially spreading his researches over a very extensive range—was already formed and he never abandoned it. His most striking experimental discoveries were due to this method, such as the eventual preparation of nearly 100 new derivatives of naphthalene with the four reagents mentioned above.

While studying the reactions of naphthalene and its compounds with the halogens and nitric acid, Laurent was from the start characteristically concerned with the construction of an explanatory theory that would account for these phenomena. Like most creative scientists, he generalized his solution to a specific problem through the imaginative use of analogy, leading to the elaboration of the first comprehensive theory adequate for dealing with the whole domain of contemporaneous organic chemistry.

Several factors can be singled out in Laurent's earliest attempt to construct such a theory:

1. *Analogy*. Laurent based his theory upon at least four kinds of analogy drawn from the work of his contemporaries.

a) *Hydrocarbons:* Dumas had affirmed that a hydrocarbon such as ethylene (C_2H_4) acted as an organic radical and gave rise to a whole series of derivative compounds such as alcohol ($C_2H_4 + H_2O$), ether ($2C_2H_4 + H_2O$), and Dutch liquid ($C_2H_4 + Cl_2$). Laurent's study of naphthalene and anthracene, which appeared to contain the same relative proportions of carbon and hydrogen as ethylene and benzene, led him to generalize that all compounds could be understood as derivatives of hydrocarbons analogous to Dumas's series for ethylene.

b) *Substitutions and Additions:* Dumas had also shown that, in a large number of organic reactions, hydrogen was replaced by an equivalent amount of halogens, oxygen, etc. Laurent interpreted the action of the halogens upon naphthalene by applying Dumas's law of substitutions. But the most important demonstration of the truth of this law came from his study of naphthalic acid ($C_{40}H_8O_4 + O_4$; 4-volume formula, $C = 6$). This was formed by treating the following halogen derivative of naphthalene with nitric acid: $C_{40}H_{12}Cl_4 + H_4Cl_4$ (in modern notation $C_{10}H_8Cl_2$).

The reaction was explained by saying that H_4Cl_4 was removed from the radical directly derived from naphthalene (that is, $C_{40}H_{12}Cl_4$, where Cl_4 had replaced H_4 in the hydrocarbon: $C_{40}H_{16}$). But hydrochloric acid had further been added to the halogen derivative of naphthalene ($C_{40}H_{12}Cl_4 + H_4Cl_4$). This additional acid was shown by the reactions of naphthalic acid to remain outside the main radical corresponding to the hydrocarbon. Two analogies were thus involved: one with Dumas's theory of substitution and another based upon the explanation of the additional acid which remained outside of the substituted derivative of naphthalene. Laurent generalized from both and concluded that similar considerations applied to all organic reactions.

c) *Crystallography:* Laurent was greatly influenced by the work of Haüy in crystallography. Haüy had shown that the vast multiplicity of crystalline forms to be discovered in nature were in fact derived from five or six fundamental types of geometrical structure. By the application of a set of mathematical laws it was possible to reconstruct the basic form from which any given crystal was derived. Apart from the essential structure that could be discovered within a crystal, there were external accretions added to the crystal during its growth. A crystal was a unitary structure, and it was only by a feat of abstraction that different parts could be discovered within it. While developing his theory, Laurent closely followed an analogy with Haüy: the basic hydrocarbons from which all organic compounds were derived corresponded to the fundamental crystalline structure; the essential part of a crystal was like that portion of an organic molecule in which substitution occurred, while the additional parts, outside of the main radical, were assimilated to the accidental accretions to the crystal. Even more important than these detailed analogies was Laurent's general supposition, adopted from Haüy (via Baudrimont), that an organic molecule was a unitary structure that could not be interpreted in terms of the predominant dualistic theories of contemporary chemistry.

d) *Isomorphism:* Mitscherlich's law of isomorphism had shown that several substances which crystallized

LAURENT

identically possessed similar properties, even though their elementary components were different. From this Laurent concluded that in organic compounds an analogous situation was to be found: it was the position and arrangement of the atoms in a molecule that determined their properties—not their intrinsic natures. Likewise, the properties of organic compounds were dependent upon their position within a "series" into which all such substances were to be naturally classified.

2. *Models.* Laurent wanted to construct a model that would be both a visual aid in understanding his theory and an account of some of the more obvious features of the crystalline forms of the substances studied. There were two kinds of current chemical models upon which he wished to draw, although both were recognized as unsatisfactory in certain important respects. In the first model it was assumed that when two chemical compounds reacted with each other, their molecules retained their original forms and were simply juxtaposed in the resulting combination. The other model was based on the contrary assumption that the original molecules disintegrated completely during the reaction and gave rise to a completely new type of structure. Laurent recognized the force of reasoning behind this second position, especially since it could not be denied that during a reaction there was an internal movement of all the constituents. He suggested however that reactions would be more intelligible if a certain deformation rather than a destruction of the molecular structures of the reactants were assumed.

For organic substances he suggested a pyramidal model that functioned like Haüy's fundamental and derived crystalline structures. Since all organic compounds were ultimately derived from hydrocarbons, Laurent imagined that there was a right-angled pyramid at the center, having as many solid angles and edges as there were atoms of carbon and hydrogen composing the hydrocarbon. The carbon atoms were represented at the angles while the hydrogen atoms occupied the centers of the edges. Attached to the sides of this central structure, although not an integral part of it, there would be other pyramids representing additions to the hydrocarbon, such as water, in Dumas's account of the composition of alcohol or ether ($C_2H_4 + H_2O$ or $2C_2H_4 + H_2O$).

Substitution reactions were represented in this model by the replacement of a hydrogen atom from one of the edges by an equivalent atom, such as chlorine. This obviously meant that for Laurent an electronegative element like a halogen could play a role identical to that of an electropositive element like hydrogen. This idea led to serious personal difficulties with Berzelius. For in explaining a reaction such as the formation of naphthalic acid, where both substitution and addition occurred in Laurent's model, he assumed that two simultaneous modifications were involved: replacement of the atoms on the edges by their equivalents and attachment of new prisms to the bases. For instance, if four atoms of chlorine acted on the hydrocarbon $C_{12}H_{12}$, two chlorine atoms would replace two of the hydrogen atoms, while the latter combined with the other two chlorine atoms and formed hydrochloric acid, H_2Cl_2, which would attach itself to the bases.

When an equal number of equivalents replaced the atoms of a fundamental radical, the new substance formed had to have a similar formula and composition and the same fundamental properties as the original. In all these reactions it was emphasized that the central nucleus or pyramid retained its structure only as long as the carbon atoms were unaffected. If any of these were removed, the pyramid was destroyed and a wholly different type of product was formed having no relationship to the initial hydrocarbon. As long as the central pyramid was unaffected, the series of compounds belonged to the same family, of which the father was called the fundamental radical and the members to which it gave birth by substitution and addition were called derived radicals.

3. *Rationalism:* The third important influence on Laurent was his unquestioning faith in the rationality of nature. It found expression both in the assumption that nature always follows the simplest means to accomplish the most complicated ends and in the belief that natural phenomena embody mathematical principles. Thus, in his assertion that atoms always combine in simple numerical ratios to form organic compounds Laurent was also influenced by his belief in the uniformity of chemical principles that had to be equally valid both for organic and inorganic chemistry. Now if the laws of multiple proportions and of combining volumes in the latter were to be extended to organic chemistry then all combinations—and not only inorganic ones—occurred in simple numerical ratios. Concretely this meant that the fundamental hydrocarbons from which all organic substances were derived contained carbon and hydrogen in proportions of 1:1, 1:2, 1:3, 2:3, 3:5, · · · . Another consequence of this rationalistic thinking was Laurent's attempt to give a quasi-mathematical form to his theory, which was stated as a set of formal propositions, an idiosyncrasy that has often occurred in the history of chemistry.

Due to its analogy with Haüy's crystallography, the formal theory was called the theory of fundamental and derived radicals, the former being a hydrocarbon

LAURENT

and the latter its substitution and addition products. Among the theory's main tenets was the characterization of the hydrocarbons as neutral substances. The acidity was due to the existence of oxygen as a pyramid suspended outside the nucleus. Laurent maintained that alongside Dumas's law of substitution, the action of the halogens, oxygen, and nitric acid resulted in the formation of the corresponding halogen acid, or water, or nitrous acid, which were sometimes given out and sometimes combined with the new radical that was formed. He based this idea upon his view of acids as being hydrates. Laurent was also concerned to show why compounds containing large proportions of oxygen, such as sugar, gums, and carbon monoxide, were not acidic, surprising as this appeared to those who held that acidity was dependent upon the quantity of oxygen in a substance. This was even more difficult to reconcile with the fact that substances like stearic acid and margaric acids were distinctly acidic in their properties despite the tiny proportion of oxygen that they contained. The explanation Laurent offered was that in the former cases all the oxygen was contained within the fundamental nuclei, while in the latter cases the small quantity of this element was outside them. Thus, these acids were to be written as follows:

$$C_{70}H_{66}O_2 + O \quad \text{(margaric acid)} \qquad C_{140}H_{134}O_3 + O_2 \quad \text{(stearic acid)}$$

Both were derived from the hydrocarbon $C_{35}H_{35}$. Similar reasons led to the assertion that when the halogens or hydrogen were located outside the nucleus, the former generated acidic halogen compounds and the latter hydracids or hydrobases. These elements could be removed by the action of alkalis, heat, and other similar agents if they were outside the nucleus, but not when they were part of it. This furnished a simple experimental test for determining their positions.

Organic substances were classified into series defined by the relative numerical proportions of carbon and hydrogen. Given any such compound, it was possible to discover the series to which it belonged by imaginatively reconstructing the fundamental hydrocarbon from which it was derived. Examples of such series were:

a) C:H :: 1:1 which included cetene, tetrene, etherin, methylene, and their respective derivatives.
b) C:H :: 5:2 which included anthracene, naphthalene, and their derivatives.
c) C:H :: 2:1 which included cinnamene, benzogine, benzene, and their derivatives.
d) C:H :: 3:2 which included acetone, metacetone, and chloracetone.
e) C:H :: 5:4 which included pinic and silvic acids, camphene, citrene, and their derivatives.
f) C:H :: 10:7 which included camphor and its derivatives.

If a substance lost a carbon atom during its reactions, then it ceased to belong to the original series and gave rise to one or more compounds in other series.

Laurent's theory, formulated in 1835–1837, had thus succeeded not only in constructing an explanatory model and a set of general rules from which the formation of new organic compounds could be predicted by considering the numerous derivatives of a fundamental hydrocarbon nucleus but he had also furnished the first example of a comprehensive classification of organic compounds. The immediate experimental result that helped to test and confirm this theory was Laurent's discovery of two new hydrocarbons, pyrene and chrysene. Investigating the reactions of these and other similar hydrocarbons with nitric acid, Laurent also succeeded in preparing a new compound with anthracene, anthracenose (anthraquinone). In 1837 he investigated the preparation and properties of fatty acids and showed that the results coincided with those predicted on the basis of his theory.

In 1842 Laurent investigated the action of bromine on camphor and discovered that a compound was formed having a remarkable property: when this compound was heated or treated with an alkali it directly gave off bromine and not hydrobromic acid. Laurent had previously asserted that the theory of hydracids was correct in that when a halogen or oxygen acted upon a hydrocarbon, half of the element replaced an equivalent amount of hydrogen in the hydrocarbon nucleus. The other half united with the hydrogen given off during substitution, forming the corresponding acid or water; and this could combine with the derivative radical as an addition product located outside the central radical. It followed that, upon heating or reacting with potash, this acid or water was given off, a fact Laurent thought had been confirmed by all his previous work. But the behavior of the compound formed by camphor and bromine had disproved this, since bromine and not hydrobromic acid was given off. This observation was of special importance to him, for it involved modifying his theory in several important respects. First, he had to admit the correctness of the hydrogen theory of acids, due to Davy and Dulong, rather than the theory that the acids were hydrates, which he had hitherto maintained. Second, he had to abandon Dumas's theory of ethers and their halogen products. An ether did not contain water, nor did its halogen compounds

LAURENT

contain a halogen acid; in both cases the fundamental radical was directly combined to oxygen or a halogen.

This result was of such importance to Laurent, involving, as it did, a modification of some of his essential ideas, that he had to obtain further experimental confirmation. The new experiments upon naphthalene and its compounds fully confirmed the predictions based on the view that in superchlorates or superbromides the additional chlorine and bromine existed outside the central nucleus as halogens and not as hydracids. There was still the need to prove that oxygen existed as such and not as water in similar cases. The proof was furnished by experiments upon the benzoyl series.

Ammonium sulfide was made to react with bitter almond oil, and the new compound obtained corresponded to the sulfide of the oil (*hydrure de sulfobenzoïle*):

bitter almond oil	$C^{28}H^{12} + O^2$
new sulfide formed	$C^{28}H^{12} + S^2$.

Various products were obtained upon distillation of this sulfide of benzoyl, including a new hydrocarbon, first discovered by Laurent, which he named stilbene. It resembled naphthalene in its properties, and its composition was represented by bitter almond oil minus oxygen, $C^{28}H^{12}$, or in four volumes $C^{56}H^{24}$. On reacting with chromic acid, this substance gave the desired addition product in which the hydrocarbon nucleus combined directly with oxygen (located outside the center) rather than with water; bitter almond oil or benzoic acid were the oxidation products of stilbene. ($C^{56}H^{24}$ had thus formed $C^{28}H^{12} + O^2$, an oxide—not a hydrate.)

Alongside this discovery another important influence led Laurent to modify his original theory. This was his work on crystallography in which he attempted to prove that the compounds derived by substitution from a fundamental radical were all isomorphic. He had already suggested this for the naphthalene series in 1837 and had continued to collect experimental evidence for it. The influence of the arrangement and order of the atoms in an organic molecule thus appeared to be of far greater importance in determining its properties than Laurent had previously realized. He thus insisted that the substitution derivatives of a fundamental hydrocarbon radical were distinguishable not only by their composition but also by the order in which the elements were introduced into them. For instance, if the fundamental radical were $C^{32}H^{32}$, and four of its hydrogen atoms were replaced by two chlorine and two bromine atoms, the resulting products would then be two different derived radicals, $C^{32}H^{28}Br^2Cl^2$ and $C^{32}H^{28}Cl^2Br^2$.

Laurent also pointed out (a factor that was to have an important development under the designation "mixed types" in the theories of Williamson and others) that complicated fundamental radicals were in fact due to additions of simpler ones; conversely, simpler fundamental radicals could be obtained by successive subtractions from a more complicated one. In fact it was possible to discover the internal arrangement of simpler radicals that were combined to form a complex one. Here a change from Laurent's original model was involved; instead of a single pyramid at the center, the nucleus was a complicated structure of several pyramids. For example, the following series all contained the last fundamental radical ($C^{24}H^{12}$) or one of its derivatives as a constituent part:

coumalic series	C^{40}	benzoic series	C^{28}
anisic series	C^{32}	salicylic series	C^{28}
phthalic series	C^{32}	anthracenic series	C^{28}
cinnamic series	C^{32}	aniline series	C^{28}
hippuric series	C^{32}	chloranilic series	C^{24}
indigo series	C^{32}	benzic series	C^{29}
estragon oil	C^{40}	phenic series	C^{29}

It was possible to transform these series into simpler ones and thereby to discover the simpler radicals of which their fundamental radicals were formed. Estragon oil, for example (derived radical $C^{40}H^{24}O^2$), was composed of three simpler radicals: C^8H^8, $C^{24}H^{12}$, $C^8H^4O^2$.

The modified theory was accompanied by a new method of classification. Laurent criticized (1844) the prevalent schemes in organic chemistry, derived as they were mainly from the dualistic classification of compounds into acids, bases, and the salts they engendered. He said that the same organic substance, for example, the same vegetable oils, could in fact be simultaneously considered as an essence or a fatty body, and a base or a salt, depending upon the characteristics singled out. The best classification for him would have been one in which the only substances grouped together were those that could be mutually transformed into each other, for instance, acetic and chloracetic acids. Unfortunately, this principle of mutual generation could not be applied in the vast majority of cases, so that another, more indirect, set of criteria had to be sought.

Laurent suggested two such criteria that could serve as mutual checks upon the position of an organic compound within a classificatory scheme (constancy in the number of carbon atoms in all members of the same group) and the existence of a fundamental radical from which all compounds in the same group were to be derived. Likening his classificatory principles to those of botany, he predicted that while external characteristics were of no help in taxonomy, the need to classify plants according to their generating principles—from seed to tree, flower, fruit—would eventually lead to the discovery of an embryonic cell or a nucleus that was reduplicated within each member of a botanical family. Fundamental radicals were similar to such nuclei, since they were reduplicated as a stable structure in all the members of a series in organic chemistry. This analogy has led to Laurent's theory being referred to as the "nucleus" theory.

In this later classification Laurent pointed out that the fundamental nuclei in organic chemistry did not necessarily have to be hydrocarbons but could contain any number of primary constituents and carbon.

Organic chemistry for Laurent comprised five types of structures.

1. Fundamental radicals. These were groups of atoms which fulfilled the same functions as the nonmetallic elements. From these, by equivalent substitution, derived radicals were formed which also played the same role as nonmetallic elements. For example:

fundamental radical	$C_{32}H_{32} = R$
derived radical	$C_{32}H_{30}Cl_2 = R'$
derived radical	$C_{32}H_{28}Cl_4 = R''$
derived radical	$C_{32}H_{20}Cl_{12} = R'''$.

2. Derived and fundamental radicals combined with elements to form chlorides, oxides, sulfides, etc. represented by: aR, bR, cR, · · · ; aR′, bR′, cR′, · · · ; aR″, bR″, cR″, · · · .

3. An excess of oxygen transformed the radical into an acid, that is, under the influence of oxygen in excess, an equivalent of hydrogen underwent a modification of properties that made it easy for a metal to replace it. Depending upon the quantity of oxygen, various kinds of weak organic acids were formed when this element combined with the radical:

oxides OR, OR′, OR″, OR‴, · · · ;
monobasic acids O_2R, O_4R, O_4R', · · · ;
polybasic acids O_6R, O_6R', O_8R, · · · .

4. Organic metals. Hydrogen played the same role as a metal in organic chemistry, because the addition of hydrogen to a radical resulted in the formation of compounds that behaved identically to metals in inorganic chemistry.

5. Complex types. Radicals of two or more different types sometimes combined to form more complicated structures. For example, formiobenzoic acid, $C_{16}H_8 + O_6$, in spite of the six atoms of oxygen was monobasic, because it was formed by the combination of two different compounds: $C_{14}H_6 + O_6$ (bitter almond oil) + $C_2H_2 + O_4$ (monobasic formic acid).

Laurent wanted to introduce a consistent nomenclature into organic chemistry, rather similar to Lavoisier's reform of the nomenclature of inorganic chemistry. His attempt was eminently rational but came too soon to be effective. Although Laurent's ideas were to be the basis—acknowledged or unacknowledged—of later structural organic chemistry, his views needed to be supplemented by a clear recognition of the idea of valence and chemical bond before a complete reform of the basis of organic chemistry was to be possible.

Laurent tried various types of nomenclatures; this was necessitated by the growing number of organic compounds discovered by him and his contemporaries. Some idea of a later version of his attempted reform can be gathered from the following:

Hydrocarbons. These were to have names ending in "ene" for the fundamental radicals: benzene, stilbene, etc.

Derived radicals. If oxygen replaced the hydrogen in a fundamental radical, then the progressive substitution products had names ending in the same order as the vowels:

palène	C_4H_4
palase	C_4H_2O
palèse	C_4O_2.

If chlorine, bromine, · · · , were substituted then the prefixes chlo-, bro-, · · · , were to be added:

chlopalase	$C_4H_2Cl_2$
chlopalèse	C_4Cl_4.

Additional products. If the fundamental or derived radicals combined with an equivalent hydrogen, forming an organic metal, the ending "ene" was changed to "um":

palène	C_4H_4	*palum*	$H_4(C_4H_4)$
chlopalase	$C_2H_2Cl_2$	*chlopalasum*	$H_4(C_4H_2Cl_2)$

If oxygen combined with the organic metals to give acids, the ending was changed to "ique":

palène	C_4H_4	acid *palique*	$(C_4H_4)\ H_4 + O_4$
palase	C_4H_2O	acid *palasique*	$(C_4O)\ H_4 + O_4$

In *Méthode de chimie* Laurent gave a more detailed and systematic account of the ideas on theoretical chemistry and classification that he had been developing. Among other important discoveries the following two are especially noteworthy.

1. *Reform of atomic weights.* Laurent pointed out that the formulas of organic compounds needed to be reduced to half their accepted values, that is, two-volume formulas were to replace four-volume formulas. In arriving at this conclusion he had been

influenced by Gerhardt. Laurent also recognized that the molecules of the elements hydrogen, chlorine, etc. were biatomic.

2. *The benzene ring*. Chemical compounds, represented on a geometrical model, were taken to form complete polyhedra. In a substitution reaction with a simple element, such as the replacement of hydrogen by chlorine in naphthalene, one of the edges of the original polyhedron occupied by a hydrogen atom was supposed to be removed; and the resulting complex could be stable only if the original polyhedron was immediately reformed by the substituted chlorine edge. When the substitution occurred in a complex radical the situation was even more complicated. Here two complete polyhedra were originally present, as, for example, in the action of ammonia upon benzoyl chloride (C_7H_5ClO). During the reaction both polyhedra lost an edge, respectively due to the removal of hydrogen in ammonia and chlorine in benzoyl chloride, the amide (NH_2) and benzoyl radical being left in the end product. These were two incomplete polyhedra, each of which functioned as a missing edge for the other. The model taken by Laurent for these compounds was, significantly, a hexagon in each case and the substitution was represented thus: Bz = C_6H_5O; A = NH_2

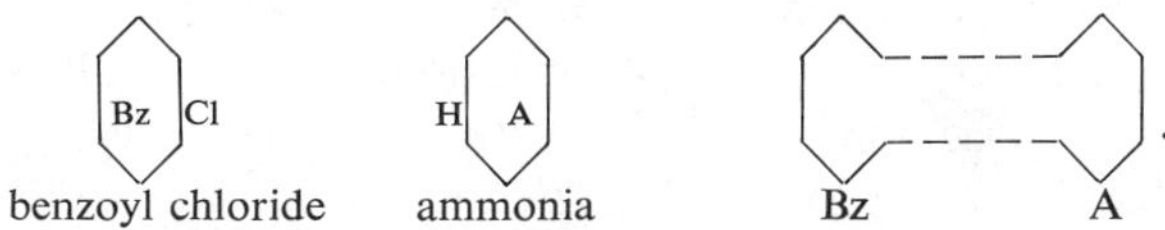

The two-faced Cl and H were removed during the reaction, and the polyhedron BzA was formed as a result.

Laurent constructed numerous other models of the same kind for different types of chemical reactions; but the fact that compounds of the benzene series were already envisaged as hexagonal, coupled with the view that the reinstatement of the edge was prompted by the requirement for chemical stability in the residual compounds, is especially significant for its obvious parallels with Kekulé's later views on the structure of benzene.

BIBLIOGRAPHY

An excellent bibliography, listing 216 of Laurent's publications, is J. Jacques, "Essai bibliographique sur l'oeuvre et la correspondance d'Auguste Laurent," in *Archives. Institut grand-ducal de Luxembourg*. Section des sciences naturelles, physiques et mathématiques, **22** (1955), 11–35. Laurent's doctoral diss. was published as *Recherches diverses de chimie organique. Sur la densité des argiles cuites à diverses températures* (Paris, 1837); a rare work, it may be consulted at the Faculté de Pharmacie and at the Bibliothèque de l'Institut, Paris. His two books are *Précis de cristallographie suivi d'une méthode simple d'analyse au chalumeau* (Paris, 1847); and *Méthode de chimie* (Paris, 1854). There is no worthwhile secondary literature.

SATISH C. KAPOOR

LAURENT, MATTHIEU PAUL HERMANN (*b.* Echternach, Luxembourg, 2 September 1841; *d.* Paris, France, 19 February 1908), *mathematics, higher education.*

Laurent's father, Auguste, was a noted chemist. After his father's death in 1853, Laurent was sent to the École Polytechnique in Paris and the École d'Application at Metz. He rose to the rank of officer but resigned in 1865 and took his *docteur-ès-sciences* at the University of Nancy with a thesis on the continuity of functions of a complex variable. In 1866 he became *répétiteur* at the École Polytechnique but returned to active military service during the Franco-Prussian War. He resigned to resume the post of *répétiteur* at the École Polytechnique, and in 1871 he was appointed actuary for the Compagnie d'Assurance de l'Union. In 1874 he married Berthe Moutard; in 1883 he became an *examinateur* at the École Polytechnique, and in 1889 he was appointed professor at the École Agronomique in Paris. In 1905 he was made an officer of the Legion of Honor.

Although Laurent's output comprised about thirty books and several dozen papers, his importance lies mainly in the teaching rather than in the development of mathematics. His papers dealt primarily with problems in analysis (especially relating to infinite series, elliptic functions, and Legendre polynomials), the theory of equations, differential equations and their solution, analytic geometry and the theory of curves, and the theory of substitutions and elimination. In the latter field he made one of his most useful contributions by extending some of the known techniques of eliminating variables to find the solutions to equations. Laurent was also noteworthy in developing statistical and interpolation formulas for the calculation of actuarial tables, annuities, and insurance rates; and he gave the name "chremastatistics" to subjects such as insurance and economics. He was a founding member of the Société des Actuaires Français and from 1903 was responsible for the mathematical section of the *Grande encyclopédie*, to which he contributed over 130 articles. He was a member of the editorial boards of the *Journal des mathématiques pures et appliquées* and of the *Nouvelles annales des mathématiques*, contributing articles especially to the latter.

Laurent's first textbook was the short *Traité des séries* (1862), one of the first works devoted entirely to the convergence as well as the summation of series; it was followed by *Traité des résidus* (1865). (It was Pierre Alphonse Laurent who was responsible, in 1843, for the Laurent series.) Of Laurent's later works, the comprehensive seven-volume *Traité d'analyse* (1885–1891) is notable. It is divided into two parts, of two and five volumes respectively, on the differential and integral calculus, and includes not only the standard treatments of the derivative and the integral and their applications to geometry but also substantial sections on the theory of functions, determinants, and elliptic functions. The last three volumes are devoted entirely to the solution and application of ordinary and partial differential equations. Laurent's other books deal with probability, arithmetic and algebra, rational mechanics, and statistics and its applications in economics. In 1895 he published *Traité d'arithmétique* under the names of his friends C. A. Laisant and Émile Lemoine, since arithmetic was part of the curriculum of the École Polytechnique, and he was not allowed as an examiner to publish the work under his own name.

BIBLIOGRAPHY

I. Original Works. In addition to works mentioned in the text, the following are worthy of note: *Traité du calcul des probabilités* (Paris, 1873); *Théorie élémentaire des fonctions elliptiques* (Paris, 1882); *Traité d'algèbre* (Paris, 1867; 5th ed., 1894); *Traité de mécanique rationnelle*, 2 vols. (Paris, 1889); *Théorie des jeux de hasard* (Paris, 1893); *Théorie des assurances sur la vie* (Paris, 1895); *Opérations financières* (Paris, 1898); *L'élimination* (Paris, 1900); *Petit traité d'économie politique* (Paris, 1902); *Traité de perspective* [intended for artists and draftsmen] (Paris, 1902); *Appendice sur les résidus* . . . (Paris, 1904); *Théorie des nombres ordinaires et algébriques* (Paris, 1904); *La géométrie analytique* (Paris, 1906); and *Statistique mathématique* (Paris, 1908), his last book. A substantial number of his papers on statistical and related matters were published in *Bulletin trimestriel de l'Institut des actuaires français*.

II. Secondary Literature. The most comprehensive source of bibliographical and biographical information is a 63-page pamphlet published in 1909 under the title *Hermann Laurent 1841–1908. Biographie—bibliographie*. This work is, however, extremely rare; and a fairly substantial amount of information may be gleaned from Poggendorff, IV, 844; V, 713. A notice on Laurent appeared in the *Grande encyclopédie*, XXI, 1038.

I. Grattan-Guinness

LAURENT, PIERRE ALPHONSE (*b.* Paris, France, 18 July 1813; *d.* Paris, 2 September 1854), *mathematics, optics.*

In 1832, after studying for two years at the École Polytechnique, Laurent graduated among the highest in his class, receiving the rank of a second lieutenant in the engineering corps. When he left the École d'Application at Metz he was sent to Algeria, where he took part in the Tlemcen and Tafna expeditions. He returned to France and participated in the study leading to the enlargement of the port of Le Havre. Laurent spent about six years in Le Havre directing the difficult hydraulic construction projects. His superiors considered him a promising officer; they admired his sure judgment and his extensive practical training.

In the midst of these technical operations Laurent composed his first scientific memoirs. Around 1843 he sent to the Académie des Sciences a "Mémoire sur le calcul des variations." The Academy had set the following problem as the subject of the Grand Prize in the mathematical sciences for 1842: Find the limiting equations that must be joined to the indefinite equations in order to determine completely the maxima and minima of multiple integrals. The prize was won by Pierre Frédéric Sarrus, then dean of the Faculty of Sciences of Strasbourg. A memoir by Delaunay was accorded an honorable mention. Laurent submitted his memoir to the Academy after the close of the competition but before the judges announced their decision. His entry presented great similarities to Sarrus's work. Although some of Laurent's methods could be considered more inductive than rigorous, Cauchy, in his report of 20 May, concluded that the piece should be approved by the Academy and inserted in the *Recueil des savants étrangers*. Laurent's work was never published, however, while Delaunay's memoir appeared in the *Journal de l'Ecole polytechnique* in 1843, and Sarrus's in the *Recueil des savants étrangers* in 1846.

A similar fate befell Laurent's "Extension du théorème de M. Cauchy relatif à la convergence du développement d'une fonction suivant les puissances ascendantes de la variable x." The content of this paper is known only through Cauchy's report to the Academy in 1843. Characteristically, Cauchy spoke more about himself than about the author. He stated his own theorem first:

> Let x designate a real or imaginary variable; a real or imaginary function of x will be developable in a convergent series ordered according to the ascending powers of this variable, while the modulus of the variable will preserve a value less than the smallest of the values

for which the function or its derivative ceases to be finite or continuous.

While carefully examining the first demonstration of this theorem, Laurent recognized that Cauchy's analysis could lead to a more general theorem, which Cauchy formulated in his report in the following way:

> Let x designate a real or imaginary variable; a real or imaginary function of x can be represented by the sum of two convergent series, one ordered according to the integral and ascending powers of x, and the other according to the integral and descending powers of x; and the modulus of x will take on a value in an interval within which the function or its derivative does not cease to be finite and continuous.

Cauchy thought Laurent's memoir merited approval by the Academy and inclusion in the *Recueil des savants étrangers*, but it too was not published. One can gain an idea of Laurent's methods by his study published posthumously in 1863.

Meanwhile Laurent abandoned research in pure mathematics and concentrated on the theory of light waves. The majority of his investigations in this area appeared in notes published in the *Comptes rendus hebdomadaires des séances de l'Académie des sciences.* Laurent summarized the principal ideas of his research in a letter to Arago. He declared that the theory of polarization was still at the point where Fresnel had left it. He criticized Cauchy's method of finding differential equations to explain this group of phenomena and asserted that the equations were purely empirical. He rejected the use of single material points in determining the equations of motion of light, and employed instead a system combining the spheroids and a system of material points. Cauchy responded vigorously to Laurent's claim that he had thereby proved that the molecules of bodies have finite dimensions.

When Jacobi, a correspondent of the Academy since 1830, was elected a foreign associate member in 1846, Cauchy nominated Laurent for Jacobi's former position, but he was not elected. A short time later Laurent was promoted to major and called to Paris to join the committee on fortifications. While carrying out his new duties, he continued his scientific research. He died in 1854 at the age of forty-two, leaving a wife and three children.

His widow arranged for two more of his memoirs to be presented to the Academy of Sciences. One, on optics, *Examen de la théorie de la lumière dans le système des ondes*, was never published, despite Cauchy's recommendation that it be printed in the *Recueil des savants étrangers*. The other did not appear until 1863, when it was published in the *Journal de l'Ecole polytechnique*. Designating the modulus of a complex number as X and its argument as p, Laurent proposed to study the continuous integrals of the equation

$$dF/dx = 1/(x\sqrt{-1}) \cdot dF/dp,$$

where F is a function of the form $\varphi + \psi\sqrt{-1}$ subject to the condition $F(x, \pi) = F(x, -\pi)$, and which, together with its first order partial derivatives, remains finite and continuous for all x and all p relative to the points of the plane included between two closed curves A and B, each of which encircles the origin of the system of coordinates. If C is a curve encircling the origin, contained between A and B and represented by the polar equation

$$X = P(p), \qquad \text{with} \quad P(\pi) = P(-\pi),$$

Laurent demonstrated that the integral

$$I = \int_{-\pi}^{+\pi} dp\, F(P, p)(\sqrt{-1} + 1/P \cdot dP/dp)$$

is independent of the curve C, that is, of the function P. He thus showed that the first variation of I is identically zero when C undergoes an infinitely small variation. He arrived at the same result, moreover, by calculating a double integral following a procedure devised by Cauchy. Laurent deduced his theorem from it by a method analogous to the one Cauchy had employed to establish his own theorem. He showed that if ρ_1 and ρ_2 are the radii of two circles centered at the origin and tangent respectively to the curves A and B, then at every polar coordinate point (X, p) situated in the annulus delimited by these circles:

$$F(X, p)$$
$$= \frac{1}{2\pi} \sum_0^\infty X^n e^{np\sqrt{-1}} \int_{-\pi}^{+\pi} dp'\, F(\rho_2, p') \frac{1}{\rho_2^n e^{np'\sqrt{-1}}}$$
$$+ \frac{1}{2\pi} \sum_0^\infty \frac{1}{X^n e^{np\sqrt{-1}}} \int_{-\pi}^{+\pi} dp'\, F(\rho_1, p')\, \rho_1^n e^{np'\sqrt{-1}}.$$

In the memoir he applied these results to the problem of the equilibrium of temperatures in a body and to the phenomenon of elasticity.

BIBLIOGRAPHY

I. ORIGINAL WORKS. Among Laurent's writings are "Mémoire sur la forme générale des équations aux différentielles linéaires et à coëfficients constants, propre à représenter les lois des mouvements infiniment petits d'un système de points matériels, soumis à des forces d'attraction ou de répulsion mutuelle," in *Comptes rendus hebdomadaires des séances de l'Académie des sciences*, **18** (1844), 294–297; "Equations des mouvements infiniment petits d'un système de sphéroïdes soumis à des forces d'attraction ou de répulsion mutuelle," *ibid.*, 771–774; "Sur la nature des forces répulsives entre les molécules," *ibid.*, 865–869; "Sur la rotation des plans de polarisation dans les mouvements infiniment petits d'un système de sphéroïdes," *ibid.*, 936–940; "Sur les fondements de la théorie mathématique de la polarisation mobile," *ibid.*, **19** (1844), 329–333; "Mémoire sur les mouvements infiniment petits d'une file rectiligne de sphéroïdes," *ibid.*, 482–483; and "Note sur les équations d'équilibre entre des forces quelconques, appliquées aux différents points d'un corps solide libre," in *Nouvelles annales de mathématiques*, IV (1845), 9–14.

See also "Note sur la théorie mathématique de la lumière," in *Comptes rendus hebdomadaires des séances de l'Académie des sciences*, **20** (1845), 560–563, 1076–1082, 1593–1603; "Observations sur les ondes liquides," *ibid.*, **20** (1845), 1713–1716; "Sur les mouvements atomiques," *ibid.*, **21** (1845), 438–443; "Sur les mouvements vibratoires de l'éther," *ibid.*, 529–553; "Recherches sur la théorie mathématique des mouvements ondulatoires," *ibid.*, 1160–1163; "Sur la propagation des ondes sonores," *ibid.*, **22** (1846), 80–84; "Sur les ondes sonores," *ibid.*, 251–253; and "Mémoire sur la théorie des imaginaires, sur l'équilibre des températures et sur l'équilibre d'élasticité," in *Journal de l'École polytechnique*, **23** (1863), 75–204.

II. Secondary Literature. For information on Laurent's work, see the following articles in Augustin Cauchy, *Oeuvres complètes*, 1st ser., VIII (Paris, 1893): "Rapport sur un mémoire de M. Laurent qui a pour titre: Extension du théorème de M. Cauchy relatif à la convergence du développement d'une fonction suivant les puissances ascendantes de la variable x (30 Octobre 1843)," pp. 115–117; "Rapport sur un mémoire de M. Laurent relatif au calcul des variations (20 Mai 1844)," pp. 208–210; and "Observations à l'occasion d'une note de M. Laurent (20 Mai 1844)," pp. 210–213.

See also "Mémoire sur la théorie de la polarisation chromatique (27 Mai 1844)," *ibid.*, XI (Paris, 1899), 213–225; and "Rapport sur un mémoire de M. Laurent, relatif aux équations d'équilibre et de mouvement d'un système de sphéroïdes sollicités par des forces d'attraction et de répulsion mutuelles (31 Juillet 1848)," *ibid.*, pp. 73–75; and "Rapport sur deux mémoires de M. Pierre Alphonse Laurent, chef de bataillon du Génie (19 Mars 1855)," *ibid.*, XII (Paris, 1900), pp. 256–258.

For further reference, see Joseph Bertrand, "Notice sur les travaux du Commandant Laurent, lue en Avril 1860 à la séance annuelle de la Société des Amis des Sciences," in *Éloges académiques* (Paris, 1890), pp. 389–393; and I. Todhunter, *A History of the Calculus of Variations* (1861), pp. 476–477.

Jean Itard

LAUSSEDAT, AIMÉ (*b.* Moulins, France, 19 April 1819; *d.* Paris, France, 18 March 1907), *photogrammetry*.

After graduating from the École Polytechnique in 1840, Laussedat joined the Engineer Corps of the French army. He spent some years working on the fortifications of Paris and along the Spanish border, then taught courses in astronomy and geodesy at the École Polytechnique. In 1856 he was made professor of geodesy, and in 1873 he became professor of geometry at the Conservatoire des Arts et Métiers. He was promoted to colonel in the Engineer Corps in 1874 and retired as an officer in 1879. Laussedat became the director of studies at the École Polytechnique in 1880, and in 1881 succeeded Hervé-Mangon as director of the Conservatoire des Arts et Métiers. He was a member of the council of the Paris observatory, and during the siege of Paris (1870–1871) presided over the commission charged with the establishment of a department of visual communication. In 1894 he was elected a member of the Académie des Sciences.

Laussedat's main contribution was in the field of photogrammetry. From about 1849 he carried out extensive investigations on the use of photography to prepare topographic maps. In 1858 he used a glass-plate camera, supported by a string of kites, for aerial photography. He also prepared some maps from photographs taken from balloons (during a photographic trip in a balloon called *L'Univers*, he broke his leg). Laussedat abandoned his experiments on aerial photography with kites and balloons about 1860, primarily because it was extremely difficult to take enough photographs from one air station to cover the entire area visible from that station.

Laussedat made considerable use of ground photography by means of a phototheodolite (a combination of theodolite and camera). In 1859 he announced to the Académie des Sciences that he had successfully prepared topographic maps from the photographs taken with this instrument. His new technique was critically examined by two members of the Academy, Pierre Daussy and P. A. Laugier, and was found to be completely satisfactory. Laussedat exhibited the first known phototheodolite at the Paris Exposition of 1867, along with a map of Paris he had prepared with it. The map compared favorably with those prepared with the aid of conventional surveying instruments.

In spite of the ridicule he received from many of his contemporaries, Laussedat conducted further research on the use of photographs to prepare topographic maps. He successfully developed a new method of mathematical analysis to convert overlapping perspective views into orthographic projections in one plane. Laussedat discussed the results of his many years of research in *Recherches sur les instruments, les méthodes et le dessin topographiques*. The book was highly acclaimed, and some of the principles he laid down are still in use. Because of his manifold contributions to the field of aerial photography, he is often called the "father of photogrammetry."

BIBLIOGRAPHY

Laussedat's writings include *Expériences faites avec l'appareil à mesurer les bases appartenant à la Commission de la carte d'Espagne* (Paris, 1860); *Leçons sur l'art de lever les plans* (Paris, 1861); *Recherches sur les instruments, les méthodes et le dessin topographiques*, 3 vols. (Paris, 1898–1903); and *La délimitation de la frontière franco-allemande* (Paris, 1901). See also "Sur le progrès de l'art de lever les plans à l'aide de la photographie en Europe et en Amérique," in *Comptes rendus . . . des séances de l'Académie des sciences*, Session of 6 Feb. 1896.

ASIT K. BISWAS
MARGARET R. BISWAS

LAVANHA, JOÃO BAPTISTA (*b.* Portugal, *ca.* 1550; *d.* Madrid, Spain, 31 March 1624), *cosmography, mathematics.*

Lavanha was appointed professor of mathematics at Madrid in 1583. (At that time Spain and Portugal were united under Philip II.) From 1587 he served as chief engineer and several years later became chief cosmographer; although provisionally named to this post in 1591, he did not actually assume its duties until 1596. Responsible for the technical aspects of navigation, he maintained in Lisbon a chair for the teaching of mathematics to sailors and pilots. He also inspected the maps and instruments used in navigation prepared in the cartography workshops and supervised the construction of astrolabes, quadrants, and compasses. In addition he was placed in charge of the examinations required of all who wished to become pilots, cartographers, or instrument makers for the navy.

The best-known of Lavanha's works is undoubtedly the *Regimento nautico* (1595; 2nd ed., 1606), which contains important texts for pilots, including rules for determining latitude and tables of the declination of the sun, corrected by Lavanha himself. Around 1600 Lavanha was the first to prepare tables of azimuths at rising and setting for the observation of magnetic declination by taking the rhumb line of the rising or setting of the sun with the magnetic compass.

As an engineer Lavanha carried out fieldwork for the tracing of topographical maps, the most important of which is the map of the kingdom of Aragon (1615–1618); to determine the coordinates of the fundamental points of the map Lavanha used a goniometer of his own design. This instrument was later perfected and proved very useful in surveying.

BIBLIOGRAPHY

I. ORIGINAL WORKS. The *Regimento nautico* underwent two eds. during the author's lifetime (see text). Lavanha's other works are "Tratado da arte de navegar," library of the National Palace, Madrid, MS 1910; still unpublished, this MS contains Lavanha's lectures given at Madrid in 1588; *Naufrágio de nau S. Alberto* (Lisbon, 1597); *Itinerário do reino de Aragon* (Zaragoza, 1895), which contains the notes taken by Lavanha during his travels in the course of preparing the topographical map of Aragon; "Descripción del universo," National Library, Madrid, MS 9251, a brief and carefully done treatise on cosmography dedicated to the crown prince and dating from 1613; *Quarta década da Ásia* (Madrid, 1615), a new ed. of the work of the same title by João de Barros, rev. and completed by Lavanha; and *Viagem da catholica real magestade del rey D. Filipe N. S. ao reyno de Portugal e relação do solene recebimento que nelle se lhe fez* (Madrid, 1622), an account of the festivities held in honor of the visit of Philip II to Portugal. Lavanha also wrote on history and the nobility.

II. SECONDARY LITERATURE. See Armando Cortesão, *Cartografia e cartógrafos portugueses dos séculos XVI e XVII*, II (Lisbon, 1935), 294–361; and Sousa Viterbo, *Trabalhos nauticos, dos Portugueses nos séculos XVI e XVII* (Lisbon, 1898), pp. 171–183.

LUÍS DE ALBUQUERQUE

LAVERAN, CHARLES LOUIS ALPHONSE (*b.* Paris, France, 18 June 1845; *d.* Paris, 18 May 1922), *medicine, biology, parasitology.*

Laveran studied medicine in Strasbourg, attending simultaneously the École Impériale du Service de Santé Militaire and the Faculté de Médecine, which in the 1860's were both well-known medical schools. His doctoral dissertation, defended at Strasbourg in 1867, dealt with experimental research on the regeneration of nerves; it is still worth consulting. Laveran served as military physician in the Franco-Prussian War. In 1874, after a competitive examination, he was named *professeur agrégé* at the École du Val de Grâce. From 1878 to 1883 he served in Algeria. During this period he carried out research on malaria, first at Bône and then at Constantine, where on 6 November 1880, at the military hospital, he discovered the living agent of malaria. Malarial fevers had afflicted men since antiquity, and their causes had been explained by the most varied and most contradictory hypotheses (see Hackett and Manson-Bahr). In his Nobel address he gave a precise and detailed account of his discovery.

Laveran, transferred to Paris, left the malaria-infested regions; and the discovery that its vector is the mosquito was made by Sir Ronald Ross, who always admitted having been put on this track by the work of Manson and of Laveran.

Laveran's discovery was greeted without enthusiasm. The military authorities did not recognize his merits, and he was not promoted. After having held the post of professor of military hygiene at the École du Val de Grâce from 1884 to 1894 and having filled several temporary administrative positions at Lille and Nantes, he left the army in 1896 to enter the Pasteur Institute, which "was proud to welcome independent spirits desiring to undertake disinterested research" (Roux). From 1897 Laveran carried out research on parasitic blood diseases in man and animals. In 1907 he was awarded the Nobel Prize for his work on pathogenic protozoans. He used the prize money to organize a laboratory at the Pasteur Institute for tropical medicine, which bears his name. In 1908 he founded the Société de Pathologie Exotique and was its president for twelve years. The Academy of Sciences, which had acknowledged his talents in 1889 with a major prize, elected him a member in 1901. He later became a member and president of the Academy of Medicine. He published the results of his numerous investigations not only in notes and memoirs but also in voluminous monographs remarkable for both the precision of their texts and the excellence of their illustrations.

Laveran was described by his friends Calmette and Émile Roux. He was cold and distant to those who did not know him and was a tenacious and solitary worker. Yet in his private life he was known for his cheerful disposition and was devoted to literature and the fine arts. His family consisted not only of his wife and sister, but also of his colleagues at the Pasteur Institute to whom he was attached by bonds of friendship.

BIBLIOGRAPHY

I. Original Works. Laveran's writings include "Recherches expérimentales sur la régénération des nerfs" (Strasbourg, 1867), his M.D. thesis; *Traité des maladies et épidémies des armées* (Paris, 1875); "Sur un nouveau parasite trouvé dans le sang de plusieurs malades atteints de fièvre palustre," in *Bulletin de l'Académie de Médecine*, 2nd ser., **9** (1880), 1268; "Deuxième note relative à un nouveau parasite trouvé dans le sang de malades atteints de la fièvre palustre," *ibid.*, p. 1346; "Description d'un nouveau parasite découvert dans le sang de malades atteints d'impaludisme," in *Comptes rendus hebdomadaires des séances de l'Académie des sciences*, **93** (1881), 627; *Nature parasitaire des accidents de l'impaludisme, description d'un nouveau parasite trouvé dans le sang des malades atteints de fièvre palustre* (Paris, 1881); *Nouveaux éléments de pathologie et de clinique médicales*, 2nd ed., rev. and enl., 2 vols. (Paris, 1883), written with S. Teissier—see "Paludisme," p. 92; *Du paludisme et de son hématozoaire* (Paris, 1891); "Protozoa as Causes of Diseases," in *Nobel Lectures, Physiology or Medicine 1901–1921* (Amsterdam, 1967), pp. 257–274; *Trypanosomes et trypanosomiases*, 2nd ed., rev. (Paris, 1912), written with F. Mesnil; and *Leishmanioses-Kala-Azar-Bouton d'Orient, Leishmaniose américaine* (Paris, 1917).

II. Secondary Literature. See A. Calmette, "Le professeur Laveran," in *Bulletin de la Société de pathologie exotique*, **15**, no. 6 (1922), 373–378; I. Fischer, *Biographisches Lexikon der hervorragenden Aerzte der letzten fünfzig Jahre*, II (Berlin, 1933), p. 873; L. W. Hackett, *Malaria in Europe, an Ecological Study* (Oxford, 1937), p. 109; P. Manson-Bahr, "The Story of Malaria: The Drama and Actors," in *International Review of Tropical Medicine*, **2** (1963), 329–390; R. Ross, *Nobel Lectures, Physiology or Medicine 1901–1921* (Amsterdam, 1967), pp. 19–119; J. L. Rouis, *Histoire de l'École impériale du Service de santé militaire instituée en 1856 à Strasbourg* (Nancy, 1898); E. Roux, "Jubilé de M. le pr. Laveran," in *Annales de l'Institut Pasteur*, **29** (1915), 405–414; and "A. Laveran," *ibid.*, **36**, no. 6 (1922), 459–461; and Edmond and Étienne Sergent and L. Parrot, *La découverte de Laveran* (Paris, 1929).

Marc Klein

LAVOISIER, ANTOINE-LAURENT (*b.* Paris, France, 26 August 1743; *d.* Paris, 8 May 1794), *chemistry, physiology, geology, economics, social reform.*

Remarkable for his versatility, as scientist and public servant, Lavoisier was first of all a chemist of genius, justly remembered for his discovery of the role of oxygen in chemical reactions and as the chief architect of a reform of chemistry, a reform so radical that he himself spoke of it early on as a "revolution" in that science. Yet Lavoisier also had a lifelong interest in geology and developed some original notions of stratigraphy; he was a pioneer in scientific agriculture, a financier of ability who holds a respected if minor place in the history of French economic thought, and a humanitarian and social reformer who used his position as a scientific statesman and landowner to alleviate the evils of society. His death on the guillotine in his fifty-first year, with creative powers still undiminished, has marked him,

 LAVOISIER

with the obvious exceptions of Louis XVI and Marie Antoinette, as the outstanding martyr to the excesses of the Reign of Terror during the French Revolution.

Lavoisier was a Parisian through and through and a child of the Enlightenment. Born in 1743, in the midreign of Louis XV and in the last year of the administration of Cardinal Fleury, his boyhood coincided with the flowering of the philosophic movement in France. The *Traité de dynamique*, which was to make d'Alembert's reputation, was published in the year of Lavoisier's birth. Voltaire's *Lettres philosophiques* had appeared ten years earlier (1733). Lavoisier was eight years old when the first volume of the *Encyclopédie* appeared and ten when Diderot, in his *Interprétation de la nature*, proclaimed that the future of science lay not in mathematical studies but in experimental physics—exemplified by the discoveries of Benjamin Franklin—and in chemistry.

Lavoisier's family, originally from Villers-Cotterets, a forest-encircled town some fifty miles north of Paris, was of humble, doubtless of peasant, origin.[1] His earliest recorded ancestor, a certain Antoine Lavoisier, was a courier of the royal postal service *(chevaucheur des écuries du roi)* who died in 1620; a son, another Antoine, rose to be master of the post at Villers-Cotterets, a position of some distinction. Three generations later Lavoisier's grandfather, still another Antoine, served as solicitor *(procureur)* at the bailiff's court of Villers-Cotterets and married Jeanne Waroquier, the daughter of a notary from nearby Pierrefonds. Their only son, Jean-Antoine, Lavoisier's father, was sent to Paris to study law. In 1741, still a mere fledgling in his profession, Jean-Antoine inherited the estate, as well as the *charge* of solicitor at the Parlement of Paris, of his uncle Jacques Waroquier. The following year he married Émilie Punctis, the well-dowered daughter of an attorney *(avocat)* at the Paris law courts. Their first child, Antoine-Laurent, the subject of this sketch, and a younger sister who was to die in her teens, were both born in the house in the cul-de-sac Pecquet (or Pecquay) which had been the residence of the old solicitor.[2] Here Lavoisier spent the first five years of his life, until the death of his mother in 1748. This bereavement led Jean-Antoine to move the two children to the house of his recently widowed mother-in-law, Mme Punctis, in the rue du Four near the church of St. Eustache.[3] In this house, tenderly cared for by an adoring maiden aunt, Mlle Constance Punctis, Lavoisier spent his childhood, his school days, and his young manhood until his marriage.

Lavoisier received his formal education in the Collège des Quatre Nations, a remarkable school—among its alumni were the physicist and mathematician d'Alembert, the astronomer J. S. Bailly, and the painter David—founded by the will of Cardinal Mazarin and so commonly called the Collège Mazarin. In the autumn of 1754, shortly after his eleventh birthday, Lavoisier was enrolled as a day student *(externe)* in the splendid building that today houses, under its gilded dome, the constituent academies of the Institut de France. At Mazarin, Lavoisier gained a sound classical and literary training, earned more than his share of literary prizes, and received the best scientific education available in any of the Paris schools. The course of study at Mazarin covered nine years; after the class of rhetoric, devoted to language and literature, and before the two years of philosophy, came a full year devoted to mathematics and the sciences under the tutelage of the distinguished astronomer Lacaille, famed for his expedition to the Cape of Good Hope and his charting of the stars of the southern hemisphere. We can gather something of Lacaille's teaching from his elementary books on mathematics, mechanics, optics, and astronomy. More than a trace of this training is to be found in Lavoisier's earliest scientific memoir, a highly precise description of an aurora borealis visible at Villers-Cotterets in October 1763. Observing it on a clear night, he showed his familiarity with Flamsteed's star table, carefully locating the streamers with respect to the visible stars and measuring their azimuth and altitude with a compass provided with an alidade.[4]

In 1761, rather than complete the class of philosophy leading to the baccalaureate of arts, Lavoisier transferred to the Faculty of Law; faithful, for the moment at least, to the family tradition, he received his baccalaureate in law in 1763 and his licentiate the following year.

Yet already his interest had turned to science, which he pursued in extracurricular fashion while carrying on his legal studies. Lavoisier may have received some further instruction from Lacaille in the intimacy of the latter's observatory at the Collège Mazarin, but only for a short while, for Lacaille died late in 1762. In the summer of 1763, and again the following year, Lavoisier accompanied the distinguished botanist Bernard de Jussieu on his *promenades philosophiques* in the Paris region.[5] Yet Jussieu's influence was probably slight—botany held little attraction for Lavoisier—whereas Lacaille's teaching helps account for Lavoisier's quantitative bent of mind, his lifelong interest in meteorological phenomena, and the barometric observations he made throughout his career.

Lavoisier's family and friends were soon aware that science had captivated the young man and was luring him away from the law. In an early letter (dated

LAVOISIER

March 1762) Lavoisier was addressed by a family friend as "mon cher et aimable mathématicien"; and the same correspondent expressed concern that his young friend's health might suffer from his intense application to the various branches of science.[6] It is better, he continued—and Lavoisier would surely have disagreed—to have a year more on earth than a hundred years in the memory of men. Clearly, Lavoisier already displayed the boundless energy and wide-ranging curiosity that was to characterize him all his life; indeed another family friend, the geologist Jean-Étienne Guettard, described him in these years as a young man whose "natural taste for the sciences leads him to want to know all of them before concentrating on one rather than another."[7] Yet it was Guettard who exerted the greatest influence upon Lavoisier and focused the young man's attention upon geology and mineralogy, and also, since it was an indispensable ancillary science, upon chemistry; for Guettard believed that to be "a mineralogist as enlightened as one can be," it was important to learn enough chemistry to be able to analyze, and so help to identify and classify, rocks and minerals.[8]

It was probably at Guettard's suggestion that Lavoisier attended the course in chemistry given by Guillaume François Rouelle.[9] A brilliant and flamboyant lecturer and a vivid popularizer, Rouelle filled his lecture hall at the Jardin du Roi with a mixed audience of students, young apothecaries, society folk, and such well-known men of letters as Diderot, Rousseau, and the economist Turgot. Besides Lavoisier, the leading French chemists of at least two generations were introduced to the subject by Rouelle. Included in Rouelle's course was a series of lectures on mineralogical problems; besides describing the physical and chemical properties of mineral substances, Rouelle touched upon his own rather special theory of geological stratification. Accordingly, persuaded that these lectures provided the best—if not the only—instruction in mineralogy and the chemistry of minerals, Lavoisier followed them faithfully, probably in 1762–1763. If he attended the showy lectures at the Jardin du Roi, he almost certainly followed the private—and perhaps more technical—course Rouelle taught in his apothecary shop on the rue Jacob, near St. Germain des Prés.

Under Guettard's guidance geology absorbed Lavoisier's attention in these years. As early as 1763 he began his collection of rock and mineral specimens; and with Guettard he explored the region around Villers-Cotterets, where he was accustomed to vacation with relatives.

For some time, indeed since 1746, Guettard had nurtured a plan for a geologic and mineralogic atlas of France. It did not receive official support until 1766, when it was commissioned and funded by the royal government; but even earlier Guettard enlisted young Lavoisier as a collaborator. In repeated excursions in northern France—to Mézières and Sedan, through Normandy to the coast at Dieppe—the two men collected specimens, noted outcroppings, drew sections, and described the principal strata, assembling material for the atlas. Their intensive exploration and mappings continued until 1770, by which time they had completed and printed sixteen regional quadrangles, using symbols to designate the rock formations and mineral deposits.

The most extensive of these joint expeditions, and the most venturesome, was a trip through the Vosges Mountains and parts of Alsace and Lorraine in 1767. Accompanied by Lavoisier's servant, Lavoisier and Guettard traveled on horseback for four months, often under trying, if not actually hazardous, conditions. The letters Lavoisier exchanged with his father and his adoring aunt during this long absence from Paris have largely survived; and Lavoisier's own warm, affectionate epistles are very nearly the only letters among the many extant in which we glimpse his personal feelings.

Lavoisier's special contribution to this venture, as to the other expeditions, was to add a quantitative character to their observations. He used the barometer systematically to measure the heights of mountains and the elevations and inclinations of strata, intending to use the results for drawing type sections; and he collected samples of mineral and spring waters to be analyzed in the field or shipped back to Paris. Nor was all this deemed of purely practical value: by 1766 he had outlined a research program of experiments and observations that should lead him, in the language of the day, to a "theory of the earth," that is, to an understanding of the changes that had altered the surface of the globe.

For the most part, Lavoisier's early theories of stratification were derived from Guettard and Rouelle or from his reading of Buffon's *Théorie de la terre* of 1749. Guettard, of a largely practical turn of mind, described the geology of France in terms of three *bandes* differing in their lithology. Rouelle distinguished the *terre ancienne*, the granitic and schistose formations in which fossils apparently were absent, from the generally horizontal, sedimentary strata rich in fossils, which he called the *terre nouvelle*. The strata of the *terre nouvelle*, he believed, had been deposited when the sea covered all or most of the continent of Europe. Accepting Rouelle's division (although often using Guettard's terms), Lavoisier at first assumed that there had been only a single

LAVOISIER

epoch in which the present continents were submerged. But he soon observed that the *terre nouvelle* was composed of two different kinds of strata: fine-grained, calcareous beds (pelagic beds), such as would result from a slow deposition in the open sea, and littoral deposits of rougher, abraded, and pebblelike material, formed at beaches and coastlines. These had been distinguished by his predecessors, but it was Lavoisier who noted that littoral and pelagic beds sometimes seemed to alternate with each other. As early as October 1766 he came to the radical conclusion—which greatly extended his conception of geologic time—that there may have been a succession of epochs marked by a cyclically advancing and retreating sea. In the regions he knew best there was not a single *terre nouvelle* but, rather, three different pelagic formations laid down at different times. Even the *terre ancienne* was not truly primitive rock; it was more likely, he remarked at one point, to be composed of littoral beds formed very long ago *(bancs littoraux beaucoup plus anciennement formés)*.

Lavoisier never lost interest in these geological problems, although after 1767 his opportunities for fieldwork markedly diminished; he continued, as far as innumerable distractions permitted, to be associated with the atlas project until 1777, the year when Antoine Monnet was officially put in charge of it. But it was not until 1788 that Lavoisier presented his theory of stratification to the Academy.

Lavoisier's earliest chemical investigation, his study of gypsum, was mineralogical in character; begun in the autumn of 1764, it was intended as the first paper in a series devoted to the analysis of mineral substances. This systematic inventory was to be carried out, not by the method of J. H. Pott—who exposed minerals to the action of fire—but by reactions in solution, by the "wet way." "I have tried to copy nature," Lavoisier wrote. "Water, this almost universal solvent . . . is the chief agent she employs; it is also the one I have adopted in my work."[10] Using a hydrometer, he determined with care the solubility of different samples of gypsum (samples of selenite, or *lapis specularis*, some supplied by Guettard and Rouelle). He made similar measurements with calcined gypsum (plaster of Paris). Analysis convinced him that this gypsum was a neutral salt, a compound of vitriolic (sulfuric) acid and a calcareous or chalky base. Not content with having shown by analysis the composition of the gypsum, Lavoisier completed his proof by a synthesis following, as he said, the way that nature had formed the gypsum. He further demonstrated that gypsum, when transformed by strong heating into plaster of Paris, gives off a vapor, which he showed to be pure water, making up about a quarter of the weight of gypsum. Conversely, when plaster of Paris is mixed with water and turns into a solid mass, it avidly combines with water. Using the expression first coined by Rouelle, he called this the "water of crystallization."

This first paper, which in so many respects embodies the quantitative methods Lavoisier was to employ in his later work, had in fact been largely anticipated by others, notably by Marggraf, who had already discovered the composition of gypsum and shown that it contained water *(phlegm)*. Yet Lavoisier's work was more thorough; and his paper, his first contribution to the Academy of Sciences (read to the Academy on 25 February 1765), appeared in 1768 in what was usually called the *Mémoires des savants étrangers*, an Academy organ which published some of the papers read to that body by nonmembers, often by those who, like Lavoisier, aspired to membership.[11]

At the age of twenty-one Lavoisier was already an aspirant; to be sure, the time had passed when men of that tender age were readily admitted to the Academy as *élèves* or *adjoints*. Yet Lavoisier had friends in the Academy, and his father and his aunt were not averse to pulling such strings as came to hand. Meanwhile, the young man laid siege to the Academy from another angle. He resolved in 1764 to compete for the prize offered by the lieutenant general of police, and to be judged by the Academy, for the best method of improving the street lighting of Paris. The effort he put into this inquiry was prodigious; he attempted, as A. N. Meldrum put it, to exhaust the subject—theoretically, practically, and even (as he was to do so often in later inquiries) historically. He sought the best available material for lamp wicks and the combustible; he determined what shape—whether parabolic, hyperbolic, or elliptical—made the best reflectors; he recommended how lanterns should be suspended to give the best illumination; and much else, although there was practically no chemistry involved. Lavoisier did not win the contest—indeed nobody did, for the prize was divided into smaller awards; but a gold medal, specially authorized by the king, was presented to him at the Easter public session of the Academy on 9 April 1766, a month after he had read to that body a second paper on gypsum.

Later that month Lavoisier's supporters in the Academy entered his name in the list of candidates for the place of *adjoint chimiste* that had fallen vacant. Needless to say, Lavoisier, at the age of twenty-two, was the youngest postulant. The winner, Louis Cadet de Gassicourt, like most of the others, was some ten years his senior. Nevertheless, Lavoisier's confidence was unabated; just prior to this election he drafted two similar letters, to be signed by one or more

LAVOISIER

of his friends in the Academy; one was addressed to Mignot de Montigny (the president of the Academy for that year), the other to the perpetual secretary, Grandjean de Fouchy. Both letters urged the Academy to create a new division or class of *physique expérimentale.*[12] If Lavoisier hoped to make room for himself in this rigidly structured and exclusive body, nothing came of this early example of his ambition and overwhelming self-confidence; the letters probably were never sent. But that Lavoisier was serious about the importance of experimental physics we need hardly doubt; he thought of himself, then and later, as a *physicien*, an experimental physicist, even more than as a chemist. Only when his investigations dealt with the reactive or combinatorial behavior of substances (as in the paper on gypsum) did Lavoisier speak of doing chemistry. When, on the other hand, he investigated the physical "instruments" that bring about or influence chemical change—water, elastic fluids like air, imponderable fluids like heat or electricity—this was physics; he was wont, characteristically, to refer to Boyle and Priestley as *physiciens*, not as chemists. There is further proof of his sincerity; in 1785, when the Academy of Sciences underwent its last reorganization before the Revolution—and this was largely Lavoisier's achievement—a *classe de physique générale* was finally established.

Lavoisier was not easily discouraged. In 1767, during his trip through the Vosges with Guettard, he received optimistic news from his father about his chances of election. Indeed, he seems to have taken his success for granted. On the title page of a copy of Agricola's *De re metallica*, which he purchased in Strasbourg in September, he wrote confidently: "Antonius Laurentius Lavoisier Regiae Scientiarum Academiae Socius anno 1767."[13]

In the spring of 1768 Lavoisier read to the Academy of Sciences a paper on hydrometry describing an accurate instrument of the constant-immersion type which he had designed. This was followed by the reading of a long paper on the analysis of samples of water he had collected in the course of his travels with Guettard. Then, as now, Europeans ascribed remarkable curative properties to the waters of various spas; and already there had grown up a large medicochemical literature on mineral waters. Before Lavoisier, these were the only waters commonly analyzed. The method employed was first to identify by qualitative analysis the principal salt in a given sample; the concentration of this salt was then determined by evaporating a certain volume of water to dryness and weighing the solid residue. Lavoisier distrusted this method, for the salt could be lost by decrepitation, spattering, volatilization, or in some cases by decomposition. His method was to determine the concentration of the characteristic salt by making specific-gravity measurements of the water sample with his improved hydrometer, a method which was, of course, limited to waters containing a single salt. Typically, Lavoisier did not concern himself with the analysis of mineral waters—which, he wrote, were important for only a small number of privileged persons—but with the potable waters used by society as a whole: waters from springs, wells, and rivers. He analyzed some samples in the field, but the best of his many measurements were made at the virtually constant temperature of the deep cellars of the Paris astronomical observatory.

Water, as we have seen, fascinated Lavoisier; he spoke of it as *l'agent favori de la nature* and as the chief agent in shaping the earth, producing all its crystal forms (including the diamond!). Analysis of waters ought to give some clue, he believed, to the kinds of hidden strata through which water flowed before emerging as a spring or river. The tables that accompanied this paper reveal his objective: the first table lists his analyses of water found in the inclined granitic and schistose strata of the Vosges Mountains, where all of his samples were more or less rich in Glauber's salt (hydrated sodium sulfate). The second table, analyzing water samples collected in the *bande calcaire*, disclosed a marked quantity of selenite or gypsum, with traces of sea salt.

Lavoisier's election to the Academy followed shortly upon the reading of this paper. The Academy's vote was divided between Lavoisier and Gabriel Jars, a mining engineer for the royal government, with Lavoisier receiving a slight majority. When, according to established practice, the two top names were submitted to the king's minister, the Comte de St. Florentin, Jars was designated *adjoint chimiste* but a special dispensation was accorded Lavoisier, who was admitted, to the great joy of his father and his aunt, as *adjoint chimiste surnuméraire*.

Shortly before his election, in March 1768, Lavoisier entered the Ferme Générale, a private consortium that collected for the government such indirect imposts as the tax on tobacco and on salt (the *gabelle*), as well as customs duties and taxes on produce entering Paris. Having inherited a considerable fortune from his mother, he now invested a substantial portion of his capital by assuming a third of the interest of the farmer-general, François Baudon, in a lease *(bail)* negotiated under the name of a certain Jean d'Alaterre. Some of Lavoisier's future colleagues in the Academy of Sciences feared these responsibilities would detract from his scientific work. Others, like the astronomer Lalande, supported Lavoisier by arguing that his increased wealth made it unnecessary for him to seek other ways of earning a living. Indeed, so the story goes, one mathematician remarked, "Fine. The dinners he will give us will be that much better."

For the next few years, much of Lavoisier's time was taken up with journeys, as a traveling inspector *(tourneur)*, on behalf of the Ferme Générale. One of these expeditions, from July to November 1769, took him across Champagne to inspect tobacco factories and check on the deployment of brigades assigned to catch smugglers. The early months of 1770 found him at Lille in Flanders; in August he was in Amiens, where he read a paper on the mineralogy of France (especially Picardy) at a public session of the Academy of Amiens. In the same year, again combining scientific work with his official travels, he presented a memoir on the water supply of Rouen to the academy of that city.

In 1771 Lavoisier married Marie Anne Pierrette Paulze, the only daughter of a farmer-general, Jacques Paulze. The discrepancy of age was notable: Lavoisier was twenty-eight, his bride not quite fourteen. Although the marriage was childless, it was happy and harmonious, a bourgeois marriage singularly devoid, it would seem, of anything other than fidelity and mutual esteem. Mme Lavoisier trained herself to be her husband's collaborator: she learned English (which he did not read), studied art with the painter David, and became a skilled draftsman and engraver. The thirteen copperplate illustrations in her husband's *Traité élémentaire de chimie* are her work; they are signed "Paulze Lavoisier sculpsit." In 1775, when Lavoisier was appointed a commissioner of the Royal Gunpowder Administration (Régie des Poudres et Salpêtres), to serve in effect as scientific director, the couple took up residence in the Paris Arsenal.[14] Here Lavoisier equipped a fine laboratory where most of his later scientific work was carried out. Mme Lavoisier often assisted him, recording the results of experiments; and as his hostess to visiting scientific celebrities, and at weekly gatherings of his scientific colleagues, she proved an indefatigable promoter of the "new chemistry" and her husband's renown.

As Lavoisier himself insisted, concern for the public welfare was at least as important in directing his attention to the problem of water supply for Paris as was scientific curiosity. One of his earliest publications (in the *Mercure de France* for October 1769) was a reply to a certain Father Félicien de St. Norbert who had opposed a much-discussed plan to bring to Paris, by an open canal, the waters of the Yvette, a tributary of the Orge River. The plan had been proposed by an engineer, Antoine de Parcieux, who in memoirs read to the Academy of Sciences defended the feasibility of the project. The potability of the water of the Yvette and its purity seemed attested by experiments performed under the separate auspices of the Academy of Sciences and the Faculty of Medicine. Since these analyses involved weighing the solid residue obtained by evaporation to dryness, the question arose at the Academy whether it was true that water, on distillation, could be in part transmuted into earth; if so—and several early chemists believed this to be the case—then the method of evaporating to dryness was undependable. The transmutation problem was discussed by the physicist J. B. Le Roy in a memoir presented at the Easter public meeting of the Academy in 1767. Le Roy did not believe that water could be changed into earth, but suggested instead that earth is somehow essential to the nature of water, or intimately associated with it, and that during distillation water and earth pass over together into the receiver. Nevertheless, the doctrine of transmutation had the authority of J. B. Van Helmont and of Robert Boyle, each of whom believed he had shown by experiment that the nutrition and growth of plants can only be attributed to water. Experiments like theirs were performed by various eighteenth-century scientists, chief among them the German physician Johann Theodor Eller, who found support for this doctrine by an experiment in which earth appeared to be formed when water was subjected to violent and prolonged shaking in a closed vessel and by experiments, performed with hyacinth bulbs, similar to the plant experiments of Van Helmont and Boyle.

This question, which had engaged so many distinguished scientists, Lavoisier was to solve, not long after his election to the Academy of Sciences, by means of an experiment that was to make his name widely known and satisfy that craving for public recognition which was already a marked aspect of his character. He began in the late summer of 1768 with attempts to obtain distilled water of the highest purity for use as a standard in his hydrometric measurements. Pure rain water, he found, when repeatedly distilled, always left behind a small solid residue, yet with no appreciable change in its specific gravity. It occurred to him that the solid matter might have been produced during the distillation, perhaps from the glass, some of which might have been dissolved by the boiling water. To settle the question, he placed the sample of water, which he had repeatedly distilled, in a carefully cleaned and weighed glass vessel called a pelican, a vessel so designed that water-vapor condensed in a spherical cap, then descended to the bottom through two handle-like tubes, only to be vaporized once more. Before placing the apparatus on a sand-bath, he weighed the pelican with its contained water (obtaining thereby, by difference, the weight of water). He began the refluxing on 24 October 1768, continuing it for 101 days. After a few weeks he noted the appearance of particles of solid matter. When he stopped the experiment on 1 February 1769, he weighed the apparatus, finding no appreciable change in weight. Then transferring the water and the solid matter to another glass container, Lavoisier dried the pelican, reweighed it and found it significantly lighter. When he weighed the earthy particles, and evaporated the water to dryness, weighing the solid residue, he found the total amount of the solid matter roughly equal to the loss in weight of the pelican. Clearly, the earthy material (silica) had been dissolved from the glass and was not the result of a transmutation of the water.

This well-known experiment demonstrates Lavoisier's experimental ingenuity and those gravimetric procedures that were to characterize his later work. The research was written up and initialed *(paraphé)* by the secretary of the Academy early in 1769; yet the paper was not read until November 1770. Lavoisier's discovery, presented in a public session, was promptly noted by the press; but impatient to see the results in print and unwilling to await the leisurely publication of the *Histoire et mémoires* of the Academy, Lavoisier profited from the establishment of a new scientific journal, the Abbé Rozier's *Observations sur la physique, sur l'histoire naturelle, et sur les arts.* A reference to Lavoisier's experiment appeared in a note in the first number of this journal (July 1771), and an extended summary was published in the second (August) number. Lavoisier's full account did not appear in the *Histoire et mémoires* until 1773.

This famous experiment may have had a practical aim, but its theoretical significance was probably foremost in Lavoisier's mind: water, he had shown, was not transmuted into earth and was probably an unchangeable element. We now know that as early as 1766 Lavoisier had begun to speculate about the nature of the traditional four elements, stimulated by two papers of J. T. Eller published in the *Memoirs* of the Berlin Academy. In these papers Eller rejected the doctrine of four elements and defended instead the notion that the true principles of nature were the "active" elements, fire and water. Water, as Eller believed he had shown by experiment, could produce earth. Lavoisier's experiment had pretty well disposed of this contention. But more interesting to Lavoisier was Eller's suggestion that air was the result of a combination of water with the matter of fire. This identification of water vapor with air can best be understood in the light of eighteenth-century theories about air and vapors. Air was the elastic body par excellence, playing its role (in combustion, for example) as a physical rather than a chemical agent. Vapors, on the other hand, were not thought to be inherently elastic but to be foreign particles dispersed and dissolved in air, just as particles of salt are dissolved in water. Nevertheless, scientists (Wallerius and Eller among them) had shown that water could evaporate in a vacuum, where there was no air into which the particles of water could be dissolved. Moreover, the increase in barometric pressure in the receiver of an air pump after the evaporation of water suggested to Eller that the water had been transformed into air by combining with the matter of fire.

In May 1766, Lavoisier jotted down two notes inspired by a rapid reading of Eller's papers; these, he noted, struck him as "very well done." In the first of his notes he recorded Eller's notion that air might not be an element but a combination of water with the matter of fire. In his second note he developed his own theory, ignoring Eller's identification of water vapor with air. Instead, he suggested that air, by being combined with the matter of fire, might be a fluid in a permanent state of expansion. The elements, Lavoisier wrote, enter into the composition of all bodies; but this combination does not take place in the same manner in all bodies. There is a great similarity between the aerial fluid and the igneous fluid. Both lose a part of their (characteristic) properties when they combine with bodies. It is known, for example, that air when combined ceases to be elastic and occupies a space infinitely less than when it was free.

This last comment is a clear allusion to the work of Stephen Hales, who, in a long chapter of his *Vegetable Staticks*, had demonstrated that air could exist "fixed" in a wide variety of animal, vegetable and mineral substances. Hales's book had been translated in 1735 by Buffon, and Lavoisier doubtless first learned of Hales's experiments on "fixed air" from the lectures of Rouelle. But he may have been impelled by a reference in one of Eller's papers to turn directly to the *Vegetable Staticks.* In the years that followed, Lavoisier was to puzzle over this problem of the elements, with his attention focused increasingly upon the fixation of air and the possibility that the aeriform state resulted from the combination of some base with the matter of fire.

A key idea drawn from his reading of Hales was that the effervescence produced in various chemical reactions was not merely the result of the thermal agitation accompanying the reaction, or "fermentation," but was instead the sudden release of "fixed air." In 1768 Lavoisier called attention to a phenomenon (observed long before by the chemist E. F. Geoffroy) that certain effervescent reactions produce a cooling effect.[15] This appeared to refute the belief that effervescences were simply the result of strong intestine motion of particles and that the heat, as some scientists believed, observed in other effervescent reactions was the result of friction. Moreover, it supported his notion that the air so released combined with the "matter of fire." Not long afterward, sometime between 1769 and 1770, Lavoisier became familiar with another phenomenon that supported his notion that air might be a combination of a base with the "matter of fire."

The observation of the Scots physician William Cullen (and of Lavoisier's compatriot Antoine Baumé) that when highly volatile liquids like ether or alcohol are vaporized, they produce a pronounced drop in temperature, was strong support for this theory, for the simplest explanation was that the change from liquid to vapor was accompanied by the absorption of heat or the "matter of fire." In 1771—as yet unaware of Joseph Black's unpublished discoveries concerning latent heat—Lavoisier observed that when ice melts, although heat is steadily applied to it, the temperature of an ice-water mixture remains unchanged. Clearly, "fire" becomes "fixed," unable to affect the thermometer.

All these scattered bits of evidence were given clear meaning when Lavoisier encountered—just when, we are not certain—a remarkable article entitled "Expansibilité" in the *Encyclopédie* of Diderot and d'Alembert. The author of this anonymous article was the economist, philosopher, and public servant Turgot. Turgot defined "expansibilité," a word he seems to have coined, to mean unlimited elasticity, like that of air. But to Turgot not merely air, but all vapors, are expansible, a property they acquire when heat, or the subtle matter of fire, enters bodies and weakens the attractive forces binding the particles together. Common air is simply a particular vapor, or what Lavoisier would have called an aeriform fluid; if it were possible to lower the temperature sufficiently, it should be possible to liquefy it. And Turgot made the remarkable suggestion, later taken up by Lavoisier, that all substances are in principle capable of existing in any of the three states of matter—as solid, liquid, or aeriform fluid (gas)—depending upon the amount of the matter of fire combined with them. Sometime before the summer of 1772 Lavoisier had drafted an incomplete but remarkable memoir which he called "Système sur les éléments" and which outlined a covering theory—if we may call it that—which was henceforth to guide his own investigations and his interpretation of the findings of others. The elements, notably water, air, and fire, can exist in either of two forms, fixed or free. In the crystals of certain salts, water is fixed as "water of crystallization." Air is not only fixed in many substances but even enters into the composition of mineral substances. The same is true of "phlogiston or the matter of fire." But how, he asks, can air, a fluid capable of such remarkable expansion, be fixed in solid substances and occupy such a small space? The answer is his *théorie singulière*, according to which water exists in the atmosphere as a vapor in which the particles are combined with the fine, elastic particles of the matter of fire; and the air we breathe is not a simple substance but a special kind of fluid combined with the matter of fire.

Up to this point, these were only private speculations which Lavoisier set down in several unpublished drafts. He had performed no experiments, and the theory derived entirely from his reading and the inferences he drew from the discoveries of others. But in the course of 1772 some new facts came to Lavoisier's attention and led to his famous first experiments on combustion, the dramatic first steps toward his "revolution in chemistry."

This revolution has often been summed up as Lavoisier's overthrow of the phlogiston theory (his new chemistry was later called the antiphlogistic chemistry), but this is only part of the story. His eventual recognition that the atmosphere is composed of different gases that take part in chemical reactions was followed by his demonstration that a particular kind of air, oxygen gas, is the agent active in combustion and calcination. Once the role of oxygen was understood—it had been prepared before Lavoisier, first by Scheele in Sweden and then independently by Priestley in England—the composition of many substances, notably the oxyacids, could be precisely determined by Lavoisier and his disciples. But the discovery of the role of oxygen was not sufficient of itself to justify abandoning the phlogiston theory of combustion. To explain this process, Lavoisier had to account for the production of heat and light when substances burn. It is here that Lavoisier's theory of the gaseous state, and what for a time was his theory of the elements, came to play a central part.

At this point, we may ask what Lavoisier could have known about the work of the British pneumatic chemists other than Stephen Hales. The answer would seem to be "very little." Of Joseph Black, often described as a major influence upon him, Lavoisier clearly knew nothing at first hand until 1773 or perhaps late 1772, for Black's famous *Essay on Magnesia Alba* was for long unavailable in French. Yet Black's major achievements were summarized in the French translation (1766) of David MacBride's *Experimental Essays on Medical and Philosophical Subjects*, a work strongly indebted to Hales and

LAVOISIER

Black, and chiefly concerned with the possible medical uses of Black's "fixed air" (carbon dioxide). Although Lavoisier had the book in his library, there is no evidence that he was impressed by it. On the other hand Jean Bucquet, a physician and chemist whom Lavoisier admired, and who later became his collaborator, had his interest in the chemistry of air aroused by MacBride's book. In his *Introduction à l'étude des corps naturels tirés du règne mineral* (2 vols., Paris, 1771), Bucquet remarks that air is found in almost all bodies in nature, in almost all minerals, although perhaps not in metals. He discusses the rival views of J. F. Meyer and MacBride on the causticity of alkalis and quicklime, and describes the precipitation of calcareous matter when "fixed air" is passed through limewater.

The famous paper of Henry Cavendish on "fixed air" long remained unknown, and it was some time before Joseph Priestley's work could serve as a stimulus to inquiries about different kinds of air. In March 1772, Priestley reported to the Royal Society the experiments that later appeared in the first volume of his *Observations on Different Kinds of Air*, a work only published late in 1772. Some faint echoes of his discoveries did reach the French chemists during the spring; what especially attracted their attention was Priestley's discovery that vitiated air is restored by the gaseous exchange of plants, and his report that an English physician had cured putrid fever by the rectal administration of "fixed air." It was not, however, these vague reports that made Frenchmen aware of Priestley's activity, but rather his first publication on gases, a modest pamphlet entitled *Directions for Impregnating Water with Fixed Air*, which appeared in June 1772 and was soon translated into French. Here again it was the presumed medical use of such artificial soda water that accounts for the impression it made. Yet we can only conclude that in 1772 the book by Stephen Hales was the predominant influence in interesting Lavoisier in the chemical role of air.

Early in 1772, at all events before June of that year, there appeared a book by the provincial lawyer and chemist Louis Bernard Guyton de Morveau, which conclusively proved that the well-known gain in weight of lead and tin when they are calcined is not a peculiarity of those metals, but that all calcinable metals become heavier when transformed into a calx; moreover, this increase has a definite upper limit characteristic of each metal. Guyton sought to explain these results by invoking a fanciful variation of the phlogistic hypothesis, but Lavoisier saw at once that the fixation of air might be the cause. By August he had devised an experiment to determine the role that air might play in chemical reactions involving metals. Was air absorbed by or released from a metal when exposed to the strong heat of a burning glass? Perhaps it would be possible to answer this question by using an apparatus devised by Stephen Hales (his pedestal apparatus) which enabled the amount of air released or absorbed to be measured. Soon after, however, Lavoisier learned that a Paris pharmacist, Pierre Mitouard, had reported that when phosphorus was burned to form the acid, air seemed to be absorbed.

In the early autumn of 1772 Lavoisier carefully verified this report. He burned weighed quantities of phosphorus and sulfur and discovered that the acids produced weighed more than the starting materials and absorbed, notably in the case of sulfur, "a prodigious amount of air." Soon after, using a modification of Hales's pedestal apparatus, he heated minium (red lead) in the presence of charcoal in a closed vessel and observed that a large amount of air was given off. This discovery struck him as "one of the most interesting of those that have been made since the time of Stahl."[16] Accordingly, he followed the common practice of the Academy of Sciences to assure priority and recorded these results in a famous sealed note *(pli cacheté)* deposited with the secretary of the Academy.

These experiments confirmed Lavoisier's suspicions that air, or some constituent of air, played an important role in the processes of combustion and calcination. In the autumn of 1772 he set himself to read everything that had been published on aeriform fluids. Translations were soon made available (probably through the indefatigable transmitter of scientific gossip Jean Hyacinthe Magellan) not only to Lavoisier but also to such of his fellow chemists in France as Bucquet, Hilaire-Marin Rouelle (the brother of Lavoisier's teacher), and the pharmacist Pierre Bayen, all of whom investigated, late in 1772 and in early 1773, the production and properties of Black's "fixed air."[17]

On 20 February 1773, Lavoisier opened a new research notebook (the first of his famous *registres de laboratoire*) with a memorandum announcing his intention to embark on a "long series of experiments" on the elastic fluid emitted from bodies during various chemical reactions and on the air absorbed during combustion. The subject was obviously so important—"destined," as he put it, "to bring about a revolution in physics and chemistry"—that he proposed in the succeeding months to repeat all the early experiments and to extend his own. The results, both of his reading and of his experimentation, were presented to the Academy of Sciences in the spring and summer of 1773—the *pli cacheté* was opened on 5 May—and formed the substance of his first book, the *Opuscules*

LAVOISIER

physiques et chimiques, which was published in January 1774.

Also in 1774 two French chemists, Pierre Bayen and Cadet de Gassicourt, investigated the peculiar behavior of red calx of mercury *(mercurius calcinatus per se).* From this substance, they claimed, the metal could be regenerated without the addition of a reducing agent rich in phlogiston, like charcoal, simply by heating it to a higher temperature. Was this true? And was this "red precipitate of mercury," as Lavoisier called it, a true calx?[18] Early in the autumn of 1774 the subject was discussed at the Academy of Sciences, and Lavoisier was made a member of a committee to examine this question. But before the inquiry could be made, there occurred an episode about which much ink has been spilt.

In October the English chemist Joseph Priestley, during a visit to Paris, dined with Lavoisier and a group of other French scientists. Priestley told the assembled guests that in the previous summer he had obtained a new kind of air by heating the "red precipitate of mercury." The air was poorly soluble in water, and in it a candle burned more brightly than in common air. Priestley had, in fact, prepared oxygen; but at the moment he thought he had found a species of nitrous air (nitric oxide), a gas he had discovered earlier.

Since Lavoisier, not long after, turned to investigate this new air, Priestley may be pardoned for believing that Lavoisier was simply following his clue. Many scholars have accepted Priestley's claim, but it is certainly not true.[19] Lavoisier and the other members of the Academy committee were already familiar with the "red precipitate of mercury" and reported in November that it was reduced without addition. If this substance was a true calx, Lavoisier did not need Priestley's disclosure to know that an air would be released when it was reduced. Indeed, Pierre Bayen, one of the first to call attention to the strange behavior of the red precipitate, had actually found that a gas was given off on reduction, although he erred in claiming that it was Joseph Black's "fixed air."

Lavoisier took up this question seriously in the early months of 1775. He collected the air produced by the reduction of mercury calx and tested it, not to see if it had the properties of Priestley's new air, which did not immediately occur to him, but to determine whether or not it was "fixed air," first by passing the gas through limewater (which it did not precipitate), then by seeing whether it would extinguish a flame. He recorded in his notebook that "far from being extinguished," the flame burned more brightly than in air.[20] He concluded that the air was "not only common air . . . but even more pure than the air in which we live."

At the public session of the Academy on 26 April 1775, Lavoisier described his discoveries in a paper entitled "Mémoire sur la nature du principe qui se combine avec les métaux pendant leur calcination, et qui en augmente le poids." Determined as usual to bring out his results as soon as possible and aware that the *Histoire et mémoires* of the Academy was several years behind schedule, Lavoisier published this classic paper (in its first version) in the May issue of Rozier's *Observations sur la physique.*

Meanwhile, Priestley was not idle. With a new sample of the red precipitate purchased in Paris, he again took up the investigation soon after his return to England. Various tests, carried out early in 1775, convinced him that he had not prepared nitrous air but had discovered a new gas "between four and five times as good as common air." A firm advocate of the phlogiston theory, he assumed—because this gas supported combustion so much better than common air—that it must be a better receptacle for phlogiston, had less phlogiston already in it, and should be called "dephlogisticated air." Priestley published these experiments and his conclusion in the second volume of his *Experiments and Observations on Different Kinds of Air,* which appeared before the end of 1775.

If Lavoisier owed little or nothing to Priestley's earlier disclosure, his indebtedness to Priestley at this point in time is undeniable. Advance sheets of Priestley's volume reached Paris by December,[21] and Lavoisier saw at once the significance of the new facts they contained. He set to work to confirm and extend Priestley's work; preparing once more the air from the mercury calx, he referred to it in his laboratory notebook as "the dephlogisticated air of M. Prisley [*sic*]."[22] He noted that calcining mercury would be a way of analyzing atmospheric air into its chief components. So in April he studied the residual air left after prolonged calcination of mercury, which of course was mainly nitrogen; and although he found that like Black's "fixed air" it did not support combustion, yet it did not precipitate limewater. He followed this analysis, according to his custom, by a synthesis: he mixed five parts of the residual air with one part of "dephlogisticated air" and found that a candle burned in it about as brightly as in common air. The highly respirable property of "dephlogisticated air" now commanded his attention. At the Château de Montigny, with his friend Trudaine de Montigny, Lavoisier carried out experiments on the respiration of birds, confirming that they lived much longer in the new gas than in common air and showing that in both cases they vitiated the air by producing Black's "fixed air."

LAVOISIER

When in 1778 Lavoisier reread to the Academy of Sciences his classic paper of three years earlier and published it in the Academy's *Mémoires*, he revised it without making it clear that he had changed his conclusions—and, indeed, recast the paper—because of what he had learned from Priestley's book and from his own subsequent experiments. Priestley, to be sure, is not mentioned. But Lavoisier no longer described the air produced by reducing mercury calx (without addition) as highly pure atmospheric air; now he called it "the purest part of the air" or "eminently respirable air." This air, he also showed, combines with carbon to produce Black's "fixed air," which he was soon to call, since it was known to be weakly acidic and to combine with alkalis and alkaline earths to form what we call the carbonates, "chalky aeriform acid" *(acide crayeux aériforme)*.

Lavoisier's activity during these years was prodigious; his scientific productivity was at its peak, and the experiments that he performed alone or with his collaborator, Jean Bucquet, contributed to round out his theory. Several main lines of inquiry were pursued simultaneously: the properties of aeriform fluids, the mystery of vaporization and heat, the role of "eminently respirable air" in combustion and respiration, and the formation of acids. This last was to add a new dimension to his oxygen theory.

Considerable progress had been made during the eighteenth century toward understanding the behavior of acids.[23] Rouelle, Lavoisier's teacher, had confirmed that salts were combinations of an acid with certain substances that served as a "base" (alkalis, earths, and metals); and he had shown that certain salts were genuinely neutral while others contained an excess of acid. New acids, too, were identified; beginning with his preparation of tartaric acid in 1770, the great Swedish chemist Carl Wilhelm Scheele added nearly a dozen organic acids to the roster of new chemical individuals, as well as preparing acids from such metals as arsenic and tungsten.

As to the cause of acidity, various theories were current—for example, Newton's doctrine that acids were substances "endued with a great attractive force, in which their activity consists," a theory favored by Macquer and Guyton de Morveau. Lavoisier, on the other hand, thought in more traditional terms and believed it possible to identify the "constituent principles" of acids and bases, just as his predecessors had analyzed salts into their two major constituents. One widely held notion was that there was one fundamental or "universal" acid, of which the individual acids were simply modifications. Various acids were nominated for this distinction, including vitriolic (sulfuric) acid and Black's "fixed air," which Bergman called the "aerial acid."

The sharp, fiery taste of acids led to a confusion of acidity with the causticity of such substances as corrosive sublimate or quicklime. Joseph Black had argued that the causticity of quicklime was an inhering property that was masked when the lime combined with "fixed air." A chemist and apothecary of Osnabrück, J. F. Meyer (1705–1765), attacked Black's theory and argued that mild alkalis become caustic when they take up an oily acid, *acidum pingue*, as he believed limestone does when it is burned to quicklime. The calxes of metals, Meyer thought, might contain *acidum pingue*. This imaginary protean substance, closely related to fire and light, was also the "universal primitive acid" of which all other acids were modifications.

As early as 1766, Lavoisier was familiar with Meyer's theory, for in his notes on Eller's memoirs he makes a passing reference to *acidum pingue*. When he wrote the *Opuscules* in 1773, Lavoisier had not definitely decided in favor of Black's theory rather than Meyer's; both theories of causticity were presented on their merits, although Lavoisier wrote of Meyer's book that it contained a "multitude of experiments, most of them well made and true," and noted that the theory of *acidum pingue* explains "in the most natural and simple fashion" the gain in weight of metals on calcination. Should Meyer's ideas be adopted, the result would be nothing less than a new theory directly contrary to that of Stahl.[24]

It is likely that Lavoisier at first shared the widespread notion that a universal acid, whether or not it was an *acidum pingue*, was to be found in the atmosphere and that this primitive acid gave rise to all the particular acids known to the chemist. His important step forward was to abandon the notion, essentially tautological, that an acid gave rise to acids and to suggest that an identifiable chemical substance, in fact an aeriform fluid or gas, played this universal role of acid-former. Yet before the autumn of 1772 Lavoisier did not suspect that air entered into the composition of acids. When air was released in the reaction between an acid and a metal, as in the production of "inflammable air" (hydrogen), Lavoisier believed that it came from the metal. In his August Memorandum of 1772 he wrote that "air enters into the composition of most minerals, even of metals, and in great abundance." Yet his experiments with phosphorus and sulfur convinced him that when they were burned to form the acids, air was absorbed. In the *Opuscules* of 1774 Lavoisier could remark that phosphoric acid was "in part composed of air, or at least of an elastic substance contained in air."

He did not return to the subject until two years later. On 20 April 1776, Lavoisier read to the Academy one of his most brilliant papers, a memoir entitled "Sur l'existence de l'air dans l'acide nitreux." Referring back to his experiments on phosphorus and sulfur, and his conclusion that air entered into the composition of the acids resulting from the combustion of these substances, he was led to consider the nature of acids in general and to conclude that all acids were in great part made up of air; that this substance was common to all of them; and, further, that they were differentiated by the addition of "principles" characteristic of each acid. When he "applied experiment to theory," Lavoisier was able to confirm his suspicion that it was the "purest portion of the air," Priestley's dephlogisticated air, that entered without exception into the composition of acids, and especially of nitric acid, which was particularly deserving of study because of the importance of saltpeter (potassium nitrate) in the making of gunpowder.

The experiments were as follows. When a known quantity of nitric acid was heated with a weighed amount of mercury, the resulting product was a white mercurial salt (mercuric nitrate); this decomposed to form the red oxide, yielding nitric oxide *(air nitreux)*. On further heating, the red oxide, as Lavoisier already knew, decomposed, producing metallic mercury and the "air better than common air," or "eminently respirable air." Lavoisier was careful to separate the different gaseous fractions by collecting the air under different bell jars over water. According to his practice, the analysis was followed by a synthesis: by bringing together the nitric oxide and the "pure air" in the presence of a small amount of water, he was able to regenerate nitric acid.

Almost exactly a year later, on 21 March 1777, Lavoisier read a memoir to the Academy of Sciences in which he showed that phosphorus burned in air combines with the "eminently respirable air" to form phosphoric acid. When this air is used up, the remaining air, which he called *mofette atmosphérique* (nitrogen), does not support combustion or sustain life. In the same paper he showed that sulfur takes up "eminently respirable air" in the formation of vitriolic acid.

In 1779 Lavoisier brought together the results of all this work in an important paper (published in 1781) entitled "Considérations générales sur la nature des acides." Here are set forth his conclusions that since "eminently respirable air" is a constituent of so many acids, it may play the role of "universal acid" or, rather, of the "acidifiable principle." When combined with carbonaceous substances or charcoal, this air forms chalky acid (carbon dioxide); with sulfur, vitriolic acid; with nitric oxide, nitric acid; with phosphorus, phosphoric acid. When combined with metals, on the other hand, it forms calxes (oxides). Nevertheless, it was the acid-forming characteristic that impressed him; he therefore proposed to call the acidifying principle or base of "eminently respirable air" the "oxygen principle" *(principe oxigine)*, that is, the "begetter of acids." What he soon called "vital air"[25] he described, faithful to his conception of aeriform fluids, as a combination of the "matter of fire" with a base, the "principe oxigine." By 1787, when the collaborative *Nomenclature chimique* was published, *oxigine* had become *oxygène*. In the meantime he had extended the application of his oxygen theory of acids and had discovered that there existed related acids that differed only in the proportion of oxygen they contained, as for example sulfurous and sulfuric acids; the higher the degree of oxygenation, he discovered, the stronger was the acid produced. In a paper published in 1785 Lavoisier made an equally important contribution by proving that the solution of metals in acids was a form of calcination: a calcination by the "wet way" *(par la voie humide)*. The metal combines with a quantity of the *principe oxigine* approximately equal to that which it is capable of removing from the air in the course of ordinary calcination.

Scholars have generally pointed to Lavoisier's theory of acids as perhaps his major error, or at least as a rash induction. From his analysis of a number of the oxyacids (carbonic, sulfuric, nitric) and certain organic acids (oxalic and acetic), all shown to contain oxygen, he argued that all acids must be so constituted, although he readily admitted that muriatic (hydrochloric) acid had not been shown to contain oxygen, which indeed it does not. Lavoisier explained this away by assuming that so far it had resisted further analysis.

Besides oxygen, other gases interested Lavoisier (the term "gas" was soon to replace the word "air" or the phrase "aeriform fluid"),[26] and none more than the "inflammable air" produced by the action of weak acids on iron or copper. In September 1777 with Bucquet he burned this "inflammable air" to see whether, as Bucquet expected, "fixed air" was produced, as was the case in the combustion of carbon. The limewater test was negative. What, then, was the product formed?

After the death of Bucquet, Lavoisier repeated this experiment in the winter of 1781–1782, using what he called a gasholder or pneumatic chest *(caisse pneumatique)* to store the oxygen and maintain a steady flow into a bottle filled with "inflammable air." Persuaded that some acid should be formed when the two gases were burned together, he was surprised when once

LAVOISIER

more he failed to detect any product or even a trace of acidity. Convinced that something must have been produced, since the basic article of scientific faith, on which his quantitative procedures were based, was that matter is neither created nor destroyed in chemical reactions, Lavoisier determined to carry out the experiment "with more precision and on a larger scale." To this end, assisted by his younger colleague, the mathematician Laplace, he designed a combustion apparatus in which streams of the two gases, each stored in a pneumatic chest, could be brought together in a double nozzle and burned together.

Before this new apparatus was completed, Lavoisier used the first pneumatic chest in a spectacular experiment in which a stream of oxygen, directed into a hollowed-out piece of charcoal, burned at such a high temperature that it melted platinum, a recently described metal that had resisted fusion even at the temperature of the great burning glass of Trudaine. Lavoisier reported his success at the public session of the Academy of Sciences on 10 April 1782. Early in June, and at considerable expense, he transported his equipment to the Academy and demonstrated, in the presence of a visiting Russian grand duke (traveling incognito as the Comte du Nord), the melting of a small mass of platinum by what Benjamin Franklin, another observer, called a fire "much more powerful than that of the strongest burning mirror," indeed the "strongest fire we yet know."[27]

Lavoisier's combustion apparatus was ready by June 1783 and was immediately put to historic use. In that month the Academy of Sciences was visited by the assistant of Henry Cavendish, Charles Blagden, a physician and scientist who the following year was to become the secretary of the Royal Society. Blagden, so the familiar account runs, informed Lavoisier and some other members of the Academy of Sciences that Cavendish had obtained pure water when he had detonated a mixture of "inflammable air" and "dephlogisticated air" in a closed vessel. As Blagden recalled this event, the French scientists replied that they had already heard about such experiments as repeated by Joseph Priestley but doubted that the weight of water, as Cavendish claimed, was equal to the weight of the gases used. They believed, rather, that the water was already contained in, or united to, the gases used. Nevertheless, Blagden's disclosure supported what they had already learned. Accordingly, Lavoisier and Laplace put into action the newly completed combustion apparatus. On 24 June 1783, in the presence of Blagden and several academicians, they burned substantial amounts of the dry gases and obtained enough of the liquid product so that it could be tested. It did not give an acid reaction with litmus or syrup of violets, precipitate limewater, or give a positive test with any of the known reagents; it appeared to be "as pure as distilled water." In reporting on these experiments, Lavoisier confessed that it was not possible to be certain of the exact quantity of the gases that were burned; but, he argued, since in physics, as in mathematics, the whole is equal to the sum of its parts, and since only water and nothing else was formed, it seemed safe to conclude that the weight of the water was the sum of the weights of the two gases from which it was produced.

With the return of the Academy in November from its vacation, Lavoisier read at the public meeting an account of his experiments, making the historic announcement that water was not a simple substance, an element, but a compound of the *principe oxigine* with what he proposed to call the "aqueous inflammable principle" *(principe inflammable aqueux)*, that is, hydrogen. A summary of these results appeared anonymously in the December issue of Rozier's journal.[28] A draft of this summary in Lavoisier's hand, recently identified in the archives of the Academy of Sciences, proves beyond a doubt that he was the author.[29] The fuller and more elaborate memoir, published after Lavoisier had begun extensive work with J. B. M. Meusnier, appeared in the Academy's *Mémoires* in 1784.[30]

The summer and autumn of 1783 riveted public attention, and that of the scientific community (including that shrewd American observer, Benjamin Franklin), upon the dramatic success of the first lighter-than-air flights. Soon after the pioneer demonstrations by the Montgolfier brothers came the historic exploit of the Marquis d'Arlandes and Pilatre de Rozier, who made the earliest manned free ascent. The hot air "aerostats," as the balloons were called, could be made to rise or descend only by varying the intensity of the fire of straw or other light combustible which caused hot air to rise into the open bottom of the balloon.

A quite different, and far more promising, solution was soon proposed and demonstrated by J. A. C. Charles, a free-lance teacher of physics. With his assistants, the brothers Robert, he launched an unmanned balloon that had been laboriously filled with "inflammable air." On 1 December, Charles and the younger Robert took off from the Tuileries in a basket *(nacelle)* lifted by a hydrogen-filled balloon twenty-six feet in diameter; they rose to the height of 300 fathoms: driven by a southeast wind, they covered a distance of nine leagues before an uneventful descent. Then Charles went aloft alone, rose to the height of nearly 1,700 fathoms, showing physicists, Lavoisier wrote, "how one can rise up to the clouds to study the causes of meteors."[31] What particularly impressed Lavoisier was the complete control Charles and Robert had over their machine, using bags of sand to slow the ascent of the balloon and descending by venting hydrogen.

Meanwhile, the Academy of Sciences appointed a standing committee to find measures for improving balloons, by determining the best shape, ways of maneuvering them, and above all by finding "a light gas, easy to obtain and always available, and which would be cheap" with which to fill the balloons. The committee included, besides the Academy's officers, physicists like J. B. Le Roy, Mathurin Brisson, and Coulomb, and chemists like Berthollet and Lavoisier, who—as he habitually did when serving on committees—became its secretary. Attached to this group was J. B. M. Meusnier, a young officer on leave from the Corps of Engineers. A graduate of the military engineering school at Mézières, he had been since 1776 a corresponding associate *(correspondant)* of the Academy of Sciences. The ease with which "inflammable air" had been obtained from the decomposition of water convinced Lavoisier that this was the route to follow in obtaining in quantity the cheap, light gas sought by the committee. In this inquiry, carried out in his laboratory at the Arsenal, Lavoisier had the invaluable assistance of Meusnier. In March 1784 they produced a small amount of the inflammable air by plunging a red-hot iron into water; later that month they successfully decomposed water by passing it drop by drop through an incandescent gun barrel. In this fashion Meusnier and Lavoisier prepared eighty-two pints of the light, inflammable air, and on 29 March they repeated this experiment in the presence of members of the standing committee. Meusnier presented the results of this joint effort to the Academy on 21 April; the full memoir was published soon after.

Steps were now taken to carry out with high precision a really large-scale decomposition of water into its constituent gases and its synthesis. Meusnier radically redesigned Lavoisier's gasholders *(caisses pneumatiques)* to allow the volume of gas and the rate of outflow to be measured with accuracy. These earliest "gasometers," as well as a combustion flask for the large-scale synthesis of water, were the work of the instrument maker Pierre Mégnié.

The experiments were carried out at the Arsenal on 27 and 28 February 1785, in the presence of members of a special evaluation committee of the Academy and other invited guests. For the decomposition experiment, water was percolated through a gun barrel filled with iron rings; the inflammable air was collected in bell jars over water in a pneumatic trough, from which it was transferred to a gasometer. In one experiment the volume of the gas (expressed in weight equivalents of water) was found to be equal to well over 335 *livres* of water. These experiments, in the opinion of Daumas and Duveen, exceeded all previous chemical investigations in "the perfection of the equipment used, the scale upon which the work was carried out, and the importance of the conclusions to be derived from the results."[32] On 19 March, Berthollet, who had followed these experiments closely, wrote to Blagden of the recent keen interest "in the beautiful discovery of Mr. Cavendish on the composition of water," remarking that "Mr. Lavoisier has tried to bring to this matter all the accuracy of which it is capable."[33] Water, all but a few intransigent opponents were obliged to admit, was not the irreducible "element" it had always been thought to be but a compound of the oxygen principle with the inflammable air that was soon to be baptized "hydrogen," the begetter of water.

The demonstration of the compound nature of water put the capstone on Lavoisier's oxygen theory of combustion. For a number of years Lavoisier had been reluctant to come out openly in opposition to the phlogiston theory, nor had he been in a hurry to publish his speculations about the nature of "aeriform fluids" or gases.[34] Not enough evidence was at hand to support his theory that gas was only a state which substances acquire when they are combined with a sufficient amount of the "matter of fire." His studies of heat were inspired by a desire to support what came to be called his "caloric" theory.

In the spring of 1777 Lavoisier had enlisted the cooperation of his young colleague at the Academy, Laplace, in experiments on the vaporization of water, ether, and alcohol in the evacuated receiver of an air pump.[35] These experiments convinced Lavoisier that under proper conditions of temperature and pressure, these fluids could be converted into vapors which, like air, existed "in a state of permanent elasticity." Emboldened by these results, Lavoisier presented in November 1777 the substance of one of his most famous papers, "Mémoire sur la combustion en général," the earliest announcement of his theory of combustion and his first, albeit cautious, assault on the phlogiston theory.

When Lavoisier resumed his collaboration with Laplace in 1781, the two soon undertook (1782–1783) a famous series of experiments, using a piece of apparatus—the ice calorimeter—and a technique, both suggested by Laplace.[36] With this contrivance the heat given off by hot bodies when they cooled, by exothermic chemical reactions, or by animals placed within the apparatus, was measured by the amount of water produced by the melting of ice.

LAVOISIER

With their calorimeter Lavoisier and Laplace determined the specific heats of various substances, the heats of formation of different compounds, and measured the heat produced by a guinea pig confined for several hours in their apparatus. Laplace read an account of these experiments to the Academy on 18 June 1783; their classic joint "Mémoire sur la chaleur" was published as a pamphlet later that summer and, essentially unaltered, in the *Histoire et mémoires* of the Academy in 1784. For Lavoisier, certain of these experiments on heat were convincing proof of his theory of vaporization, which held that gases owe their aeriform, elastic state to their combination with a large amount of the "matter of fire"; and so the heat and light given off during combustion must come from the fire matter released from the "vital air" (oxygen gas) when the oxygen principle combines with a combustible substance.

It has been argued that Lavoisier's caloric theory merely transferred the phlogiston from the combustible to the "vital air," and there is something to this view. Like phlogiston, the fire matter was a weightless fluid (or at least too tenuous to be weighed); nevertheless, unlike phlogiston, which defied measurement, both the intensity (the temperature) and the extensive measure of the fire (the heat produced in a given period of time) could be precisely measured. These calorimetric experiments, together with the work on water, seemed to Lavoisier to complete the evidence for his theory of combustion; and in 1786 he published his definitive attack on the old theory, "Réflexions sur le phlogistique." A brilliant dialectical performance, rightly called "one of the most notable documents in the history of chemistry,"[37] it is a closely reasoned refutation not only of Stahl but also of those latter-day phlogistonists, such as Macquer and Antoine Baumé, who had tried to modify Stahl's hypothesis in the light of the new experimental evidence.

The "Réflexions" was a personal manifesto, but it did not serve to convince all French scientists. Some, like Baumé, like Joseph Priestley, remained recalcitrant to the end. But the most tireless antagonist was the physician and naturalist Jean Claude de La Métherie. Lavoisier's exact contemporary (he was born in September 1743), he was a prolific author of generally wordy, often slovenly works on natural history, natural philosophy, geology, and chemistry. Never admitted to the Academy of Sciences, he yet occupied an influential position, for in 1785 he became editor of Rozier's old journal, the *Observations sur la physique*. La Métherie, a devoted and uncritical disciple of Priestley, was violently opposed to Lavoisier's theories; the *Observations*, therefore, which had once been so hospitable to Lavoisier, became—to a degree—the journal of the opposition.

The years 1785–1789 nevertheless brought a number of able chemists into Lavoisier's camp: Claude Berthollet, who had been converted when he closely followed the work on the decomposition and synthesis of water; Antoine de Fourcroy, the student of Lavoisier's former collaborator Bucquet; and Guyton de Morveau. These men, forming with Lavoisier and his wife what we might call the "antiphlogistic task force," set out to convert the scientific world to the "new chemistry." The instruments they employed were several and well-selected—intended to attract the young and the uncommitted.

First of all, to persuade a new generation of chemists to join their ranks and to complete what Lavoisier had envisaged since 1773—a revolution in chemistry—these men brought out a collaborative work, the *Méthode de nomenclature chimique* (1787). Originally suggested by Guyton de Morveau to eliminate the confused synonymy of chemistry, and prefaced by a memoir of Lavoisier, it emerged as a complete break with the past. In effect the scheme was based upon the new discoveries and theories, a fact that led the aging Joseph Black to complain that to accept the new nomenclature was to accept the new French theories. In a series of tables the *Nomenclature* listed the elements *(substances non décomposées)*, that is, those bodies that had not been, or perhaps could not be, decomposed. Fifty-five in number, these simple bodies included light and Lavoisier's "matter of fire," now called "caloric"; the elementary gases: oxygen, nitrogen *(azote)*, and "inflammable air," now called hydrogen; carbon, sulfur, and phosphorus; the sixteen known metals; a long list of organic "radicals" (i.e. acidifiable bases); and the as yet undecomposed alkaline earths and alkalis. Compounds were designated, as chemists have done ever since, so as to indicate their constituents: the metallic calxes were now called oxides; the salts were given names indicating the acid from which they are formed (sulfates, nitrates, carbonates, and so on). The *Nomenclature*—translated into English, German, Italian, and Spanish—was extremely influential and widely read.

A second expression of the collective effort was a French translation of Richard Kirwan's 1787 *Essay on Phlogiston*. This translation, made by Mme Lavoisier, was published in 1788, the year after the *Nomenclature chimique*, and was copiously annotated with critical notes by members of the task force: the chemists Guyton de Morveau, Berthollet, and Fourcroy; the physicists and mathematicians Monge and Laplace; and, of course, Lavoisier himself.

LAVOISIER

A third and most important instrument was the establishment of a new scientific journal, edited—and dominated—by the votaries of the "new chemistry." The first number of this journal, the *Annales de chimie*, appeared in 1789, the year of the Revolution. Its editors were, besides Lavoisier, his early disciples—Guyton, Berthollet, Fourcroy, and Monge—with the addition of three new recruits: the Strasbourg metallurgist the Baron de Dietrich, Jean-Henri Hassenfratz, and Pierre Auguste Adet.

Constituting still another, and much underestimated, vehicle for diffusing the new ideas were the later editions of Fourcroy's *Élémens d'histoire naturelle et de chimie.*[38] Of this immensely popular work a third edition, completely recast in terms of the "new chemistry," was published toward the end of 1788. A fourth edition appeared in 1791 and a fifth in 1793; it was translated into English, Italian, German, and Spanish. Its influence cannot be exaggerated, for it was one of the richest compendia of up-to-date chemical fact before the appearance of Thomas Thomson's *System of Chemistry* (1802). It was, moreover, the first work to present the whole of chemistry, as then known, in the light of Lavoisier's doctrines and according to the new nomenclature.

The last—and the best-known—of the propaganda instruments was Lavoisier's classic book, *Traité élémentaire de chimie* (1789). The culmination of Lavoisier's achievement, it grew out of the *Nomenclature*—indeed, it was a sort of justification of it. Neither a general reference work nor a technical monograph, this small work was a succinct exposition of Lavoisier's discoveries (and those of his disciples) and an introduction to the new way of approaching chemistry. Significantly, it begins, after the famous "Discours préliminaire," with an exposition of that theory of vaporization and the states of matter which had guided Lavoisier throughout much of his scientific career.

The "Discours préliminaire," a defense of the new chemistry with its new nomenclature, is also a short essay on scientific pedagogy, and on the proper method of scientific inquiry. The most widely read of Lavoisier's writings, it opens with quotations from the *Logique* (1780) of the Abbé de Condillac: language is the instrument of analysis; we cannot think without using words; the art of reasoning depends upon a well-made language *(l'art de raisonner se réduit à une langue bien faite)*. The *Traité*, Lavoisier remarks in this prefatory essay, grew out of his work on the nomenclature which, "without my being able to prevent it," developed into an element of chemistry. This must be taken with a grain of salt, for in notes set down in 1780–1781 Lavoisier had already outlined an elementary treatise, to be prefaced with two preliminary *discours*, one of which closely resembles the "Discours préliminaire" of 1789 and was intended to treat the application of logic to the physical sciences, especially chemistry. The notes for this preface, with page reference to the *Logique* in the margin, cite the same passages from Condillac that Lavoisier eventually used in print.

The epistemological and psychological assumptions Lavoisier outlines in the "Discours préliminaire" of 1789 are the commonplaces of eighteenth-century thought, found not only in Condillac, but in d'Alembert and in Rousseau's *Emile*: at birth, the mind is a *tabula rasa*; all our ideas come from the senses; analysis—that shibboleth of the Enlightenment—is the only way to truth; even a small child analyzes his experience; nature, through deprivation and pain, corrects his errors in judgment.

The beginning student of physical science must follow the path nature uses in forming the ideas of a child. His mind must be cleared of false suppositions, and his ideas should derive directly from experience or observation. But a scientific Emile is not promptly corrected by nature through the pleasure-pain principle; men are not punished for the hypotheses they invent, which their amour-propre leads them to elaborate and cling to. Because the imagination and reason must be held in check, we must create an artificial nature through the use of experiment.

Lavoisier gives numerous examples of the speculative notions that often hamper the progress of chemistry. He derides the order commonly followed in works on chemistry, which begin by treating the elements of bodies *(principes des corps)* and explaining tables of affinity. This latter branch of chemistry he acknowledges to be important (perhaps the best calculated to make chemistry eventually a true science), but it is not fully warranted by experiment, and is unsuited to an elementary treatise. As to the elements, all we can say about their nature and number amounts to mere metaphysical talk. And if by elements we mean those simple, indivisible particles *(molécules)* that bodies are made of, it is likely that we know nothing about them. Even the existence of his cherished caloric fluid, although it explains the phenomena of nature in a very satisfactory manner, is a mere hypothesis.

The chief operation of chemistry is to determine by analysis and synthesis the composition of the various substances found in nature; chemistry advances toward its goal by dividing, subdividing, and subdividing still again. If we mean by the elements of bodies the limit reached by this subdivision, then all the substances we are unable to decompose are to us as the elements out of which the other substances are made. In the *Traité* these substances are called *substances simples*; yet Lavoisier confesses: "We cannot be certain that what we regard today as simple is really so; all that we can say is that such a substance is the actual limit reached by chemical analysis, and that, in the present state of our knowledge, it cannot be further subdivided."

When we compare the table of *substances simples* in the *Traité* with that given in the *Nomenclature* two years earlier, we note some interesting changes. First, the list has been drastically reduced and now contains only thirty-three substances. Nineteen organic radicals and the three alkalis (potash, soda, and ammonia) have been eliminated, and with some justification. There was increasing evidence that the organic radicals could be broken down—in the *Traité* Lavoisier himself reports the composition of the "acetic radical"—and the case of the alkalis was similar. In 1785 Berthollet had shown that ammonia consists only of hydrogen and nitrogen, and presumably Lavoisier believed that potash and soda would also turn out to be compounds of familiar substances.

Both in the *Nomenclature* and the *Traité* the tables of undecomposed or simple substances are divided into subgroups, five in the first case, four (with the elimination of the alkalis) in the second. But in both tables the first subgroup is distinguished from the rest. In the *Nomenclature* this is made up of light, caloric, oxygen, and hydrogen. In the *Traité*, nitrogen *(azote)* has been added, and here for the first time these substances are described surprisingly as "simple substances which belong to the three kingdoms and which one can consider as the elements of bodies." It is curious that these five substances appear to Lavoisier to have more claim to the status of elements than other simple bodies, for example sulfur, phosphorus, carbon, or the metals.

The principal justification is revealed in a note Lavoisier set down in 1792 where he writes: "It is not enough for a substance to be simple, indivisible, or at least undecomposed for us to call it an element. It is also necessary for it to be abundantly distributed in nature and to enter as an essential and constituent principle in the composition of a great number of bodies." Wide distribution, and its presence in a great number of compounds, qualifies a substance to be called an element. Gold, on the other hand, is a simple substance, yet it is not an element. But there is something more to Lavoisier's conception of an element: he specifically notes that such a substance "enters as an essential and constituent principle" in compound bodies. In the older chemistry, from which Lavoisier is unable to free himself completely, the elements—the four elements of Aristotle, the *tria prima* of Paracelsus, and so on—were thought of as the bearers and the causes of the distinctive qualities of bodies into which they enter. As we saw in discussing his theory of acids, Lavoisier had strong ties to this "chemistry of principles." So in his first subgroup the substances deserving to be called elements act in this special way. Thus caloric is the "principle" that accounts for a body's physical state. Oxygen, of course, conveys acidity. Hydrogen is the producer of water, that essential substance found in all the realms of nature. But what of light and of nitrogen? Lavoisier never quite satisfied himself about the nature of light, but he knew from the work of Ingenhousz and Sennebier that light plays an essential part in the gaseous chemistry of vegetation, a matter to which he alluded in a remarkable speculative paper he published in 1788. Nitrogen, to be sure, is everywhere in the earth's atmosphere and is widely distributed in animal and vegetable substances. But was this sufficient reason for Lavoisier to elevate it to the dignity of an "element" of the first subgroup? Quite possibly he gave it this higher status because he tended to think, on the basis of Berthollet's analysis of ammonia, that nitrogen might be the "principle" of all alkalis, what Fourcroy would have liked to call "alkaligène."

Lavoisier's *Traité*, albeit an elementary textbook, contains important material much of which he had not previously published, notably his pioneer experiments on the combustion analysis of organic compounds and on the phenomenon of alcoholic fermentation, which impressed him as "one of the most striking and the most extraordinary" effects observed by the chemist. He was aware that the combustion of substances of vegetable origin, such as sugar and alcohol, yielded water and carbon dioxide; but he was convinced that these compounds did not exist preformed in the substances burned—which, however, do contain the elements hydrogen, oxygen, and carbon. Lavoisier's first combustion analyses involved burning various oils in a rather complicated apparatus shown in one of the plates of the *Traité*. A simpler contrivance was devised for the combustion of highly volatile substances like alcohol and ether. In his analysis of alcohol he determined the ratio of hydrogen to carbon by weight to be 3.6 to 1, not far from the correct value of 4 to 1.

Like his immediate predecessors—Macquer, for example—Lavoisier distinguished three kinds of fermentation: vinous or spirituous fermentation, acid or acetous fermentation (such as produces acetic acid), and putrid fermentation. He did not speculate about the underlying cause of these processes, although

LAVOISIER

he took for granted the existence of some sort of "ferment," but confined himself to the chemical changes. Vinous or alcoholic fermentation, marked by a violent intestine motion, he saw to be a chemical reaction involving rearrangement of the three essential organic "principles": oxygen, hydrogen, and carbon. He assumed an equality or "equation"[39] between the amount of these elements in the original sugar and the amount in the end products: alcohol, carbon dioxide, and acetic acid. His experiments seemed to bear out his assumption. His data, however, were wholly unreliable, although the end result was correct. As Arthur Harden put it: "The research must be regarded as one of those remarkable instances in which the genius of the investigator triumphs over experimental deficiencies." The numbers reveal grave errors, and "it was only by a fortunate compensation of these that a result as near the truth was attained."[40]

In describing his work on fermentation in chapter XIII of the *Traité*, Lavoisier stated most clearly his principle of the conservation of matter in chemical reactions:

> Nothing is created either in the operations of the laboratory, or in those of nature, and one can affirm as an axiom that, in every operation, there is an equal quantity of matter before and after the operation; that the quality and quantity of the principles are the same, and that there are only alterations and modifications. On this axiom is founded the whole art of making experiments in chemistry: we must suppose in all of them a true equality or equation between the principles of the body one examines, and those that we extract [*retire*] by analysis.[41]

Despite his prolonged interest in the chemistry of respiration, which had preoccupied him on and off since his earliest work on combustion, Lavoisier deliberately omitted all references to the subject in his *Traité élémentaire de chimie*. When in his earliest experiments on different kinds of air Lavoisier exposed small animals to these gases, it was less to explore the phenomenon of respiration than to characterize the gases. He was already aware that "fixed air" asphyxiated animals and that atmospheric air, "corrupted and infected" by the respiration of man or animals, takes on the character of "fixed air." In his *Opuscules physiques et chimiques*, where he summarized Priestley's experiments and recounted his own, Lavoisier advanced a physical, rather than a chemical, explanation of respiration. The role of atmospheric air, he then believed, is to inflate the lungs; "fixed air" is unable to do this because, being highly soluble in water, it is taken up by the moisture of the lungs "and suddenly loses its elasticity." In an atmosphere of "fixed air," therefore, the action of the lung is suspended and the animal suffocates.[42]

It was only after his first experiments with "dephlogisticated air" (oxygen) and his discovery that, far from killing animals, this gas "seemed, on the contrary, better fitted to sustain their respiration"[43] than ordinary air, that Lavoisier turned to the problem of respiration: it had now become a problem of physiological chemistry. His earliest experiments were carried out at the Château de Montigny with his friend Trudaine de Montigny during the Academy's vacation (October 1776).[44] He read the results at the Easter public session on 9 April 1777, under the title "Mémoire sur les changements que le sang éprouve dans les poumons et sur le mécanisme de la respiration."[45] By this time he could assert that respiration involved only the *air éminemment respirable* (oxygen), and that the remainder of the air is purely passive, entering and leaving the lung unchanged. In the course of respiration animals take in the *air éminemment respirable*, which is either converted into *acide crayeux aériforme* (carbon dioxide) or is exchanged in the lung, the oxygen being absorbed and an approximately equal volume of carbon dioxide being supplied by the lung. Lavoisier tended to favor the first alternative. Indeed, in that same year, in his "Mémoire sur la combustion en général," he suggests in a concluding paragraph that this reaction is similar to the combustion of carbon and must be accompanied by the release of a certain amount of the matter of fire. This, he points out, may account for the production of animal heat.[46] A persuasive argument seemed to him to be that, just as the calcination of mercury and lead results in red powders, so the absorption of oxygen gives a bright red color to arterial blood.

The calorimeter provided a new instrument for the quantitative study of animal heat and the verification of Lavoisier's combustion theory of respiration. In a famous experiment with Laplace they measured the heat produced by a guinea pig exhaling a given amount of carbon dioxide, comparing this figure with the heat of formation of the same amount of carbon dioxide produced by burning carbon in an atmosphere of oxygen. But the results showed a significant discrepancy: more heat was produced by the guinea pig than by the burning of carbon. It was soon discovered that the experimental animals gave off less carbon dioxide than the intake of oxygen had led them to expect. In a paper read before the Royal Society of Medicine in February 1785, Lavoisier argued that either some of the oxygen unites with the blood or it combines with hydrogen to form water.[47] Although not fully convinced, Lavoisier favored the second alternative;

and in his last experiments, carried out in cooperation with Armand Seguin, he assumed that both carbon dioxide and water were produced. The collaborative work was begun in 1789 and reported to the Academy late in 1790 and in 1791. These experiments were more genuinely physiological than Lavoisier's early ones. Lavoisier and Seguin found that the quantity of oxygen consumed increases with the temperature and is greater during digestion and exercise. The combustion, according to their findings, occurs in the lung, more specifically in its tubules, into which the blood secretes a "humor" composed of carbon and hydrogen. Understandably, the pressure of events during the Revolution left Lavoisier little time for scientific work; experiments begun collaboratively were in some cases left for Seguin to complete. Our most vivid record of these last experiments are the drawings made by Mme Lavoisier showing Seguin, his face covered by a mask, breathing air or oxygen to determine the amount consumed when at rest or at work.

The range of Lavoisier's activity is hard for lesser talents and less rigidly disciplined personalities to comprehend. To carry on his multifarious public responsibilities without neglecting his beloved science required a rigid and inflexible schedule. Mme Lavoisier tells us that he rose at six in the morning, worked at his science until eight and again in the evening from seven until ten. The rest of the day was devoted to the business of the Ferme Générale, the Gunpowder Administration, and meetings of the Academy of Sciences and of its numerous special committees. One day a week, his *jour de bonheur*, he devoted entirely to scientific experiments.[48]

Lavoisier rose steadily in the hierarchy of the Academy of Sciences. In August 1772 he was promoted to the rank of associate *(associé chimiste)*; he reached the top rank of *pensionnaire* in 1778. He served on a number of those committees set up at the request of the royal government to investigate matters of public concern. Among the most notable of these committees were those formed to investigate the condition of the prisons and hospitals of Paris and one, on which Lavoisier served with Benjamin Franklin, charged with investigating Mesmer's cures by what he called "animal magnetism."

Nor were his responsibilities confined to his official duties. Lavoisier's father, shortly before his death, purchased one of those honorific offices that carried with it the privilege of hereditary nobility. The title devolved on Lavoisier in 1775, and not long afterward he purchased the country estate of Fréchines, near Blois. Here he carried on agricultural experiments, chiefly concerned with the improvement of livestock, which he reported to the Royal Agricultural Society of Paris (to which he had been elected in 1783) and to the government's Committee on Agriculture, established soon after by Calonne, the minister of finance *(contrôleur général)*.

In sympathy with much of the criticism leveled against the *Ancien Régime*, Lavoisier shared many of the ideas of the *philosophes* and was closely associated with Pierre Samuel Dupont de Nemours and other physiocrats. Politically liberal, and almost certainly a Freemason, he took an active part in the events leading to the Revolution. In 1787 he was chosen a representative of the third estate at the provincial assembly of the Orléanais, the province in which he was a landowner. He served on the important standing committee *(bureau)* that dealt with social conditions and agriculture; he wrote reports on the *corvée*, on charitable foundations, and on trade; and he was responsible for many far-reaching proposals: schemes for old-age insurance, charity workshops to give employment to the poor, and tax reform.

When, after a lapse of nearly two centuries, the famous Estates-General was brought into being once again to deal with France's mounting problems, Lavoisier was chosen as alternate deputy for the nobility of Blois; as secretary, he drafted their bill *(cahier)* of grievances, a remarkably liberal document that embodied many of his earlier proposals. Although he never served in the National Assembly, multiple activities occupied him during the years of the Revolution. He was elected to the Commune of Paris and took his turn with the National Guard. He joined the most moderate of the revolutionary clubs, the short-lived Society of 1789. For this self-appointed planning group, whose members included Condorcet, the economist Pierre Samuel Dupont de Nemours, and the famous Abbé Sieyès, Lavoisier prepared his important paper on the *assignats*, pointing to the problems inherent in issuing paper money. Because of his mastery of financial matters, he was made a director of the Discount Bank (Caisse d'Escompte), established in 1776 by his friend Turgot; became an administrator of the national treasury; and in 1791 published a long report on the state of French finances. In the same year he brought out a classic statistical study of the agricultural resources of the country.[49] For the recently created Advisory Bureau for the Arts and Trades (Bureau de Consultation des Arts et Métiers) he helped draft, along with others, an elaborate scheme for the reform of the national educational system of France, based largely, to be sure, on ideas already put forward by Condorcet.

Unhappy at the leftward thrust of the Revolution, Lavoisier nevertheless remained loyal to his country.

LAVOISIER

In these critical years he was active in the Academy of Sciences, (he served as its treasurer from 1791 until the abolition of that body), and took an important part in the most enduring of its accomplishments: the establishment of the metric system of weights and measures.[50] Until his arrest he was engaged, in collaboration with the crystallographer René-Just Haüy, in establishing the metric unit of mass (the gram). Together they measured with high precision the weight of a given volume of distilled water at different temperatures.

Despite his eminence and his services to science and to France, Lavoisier was the target of increasingly violent attacks by radical journalists like Jean-Paul Marat. One after another the institutions with which Lavoisier had been associated changed form or were abolished. He was removed from his post in the Gunpowder Administration in 1791 but remained at the Arsenal until 1793, when he severed all connections with the *Régie*. That year brought the Reign of Terror, the abolition of the Discount Bank and the suppression of all the royal learned societies, including the Academy of Sciences.

In 1791 the National Assembly had abolished the unpopular Ferme Générale and set up a committee of experts from it (which did not include Lavoisier) charged with liquidating its affairs. This committee made slow progress and was accused of delaying tactics. On 24 December 1793 an order went out for the arrest of all the farmers-general. Lavoisier and his father-in-law were imprisoned with the others. All were tried by the Revolutionary Tribunal on the morning of 8 May 1794. They were convicted and executed on the guillotine that same afternoon. The story is told that Lavoisier appealed at his trial for time to complete some scientific work and that the presiding judge replied, "The Republic has no need of scientists." The story is apocryphal.[51] Authentic, however, is the remark attributed to Lagrange, the day after Lavoisier's execution: "It took them only an instant to cut off that head, and a hundred years may not produce another like it."

NOTES

1. For Lavoisier's genealogy, see Edouard Grimaux, *Lavoisier*, 2nd ed., p. 326.
2. Now the rue Pecquay. See Jacques Hillairet, *Dictionnaire historique des rues de Paris*, 3d ed. (Paris, 1966), II, 250. The Turgot plan of Paris (1739) shows the cul-de-sac "Pequet" opening from the rue des Blancs-Manteaux, not far from the Palais Soubise (the present Archives Nationales). By Piganiol de la Force it was called "Cul-de-sac Pequai ou de Novion dans la rue des Blancs-Manteaux," in *Description de Paris* (Paris, 1742), 427.
3. Grimaux erroneously called it the rue du Four Saint-Eustache; its name was simply rue du Four, or rue du Four-Saint Honoré, although to distinguish it from the rue du Four on the Left Bank, letters to residents were sometimes addressed "rue du Four St. Honoré près St. Eustache." It is now the rue de Vauvilliers, running from the rue Saint Honoré to the rue Coquillière.
4. Not published in Lavoisier's lifetime, the piece on the aurora is printed in *Oeuvres de Lavoisier*, IV, 1–7.
5. Lavoisier papers, archives of the Academy of Sciences, dossier 424. Cited by Rhoda Rappaport in "Lavoisier's Geologic Activities," in *Isis*, **58** (1968), 377.
6. *Oeuvres de Lavoisier—Correspondance*, fasc. I, 1–3.
7. Guettard, "Mémoire qui renferme des observations minéralogiques," cited by Rappaport, *loc. cit.*, p. 376.
8. Guettard, "Mémoire sur la manière d'étudier la minéralogie," cited by Rappaport, ibid.
9. For a general appraisal of Rouelle and his course in chemistry, see Rhoda Rappaport, "G.-F. Rouelle: An Eighteenth-Century Chemist and Teacher," in *Chymia*, **6** (1960), 68–101.
10. *Oeuvres de Lavoisier*, III, 112.
11. The full title of this publication is *Mémoires de mathématique et de physique, présentés à l'Académie royale des sciences, par divers savans et lûs dans ses assemblées.*
12. These unsigned letters, which have puzzled scholars, are printed in *Oeuvres de Lavoisier—Correspondance*, fasc. I, 7–10, repr. from *Oeuvres de Lavoisier*, IV, 561–563.
13. The Agricola title page, with Lavoisier's inscription, is reproduced as pl. I of Denis I. Duveen, *Bibliotheca alchemica et chemica* (London, 1949). The book itself is now in the Lavoisier Collection at Cornell University.
14. For the early history of the powder industry in France, see Régis Payan, *L'évolution d'un monopole: L'industrie des poudres avant la loi du 13 fructidor an V* (Paris, 1934). The *régie* was established by Turgot to replace the earlier *ferme*. See Douglas Dakin, *Turgot* (London, 1939), 164–166; and Edgar Faure, *La disgrâce de Turgot* (Paris, 1961), 108–110.
15. In a paper entitled "De la manière de composer des feux d'artifices colorés en bleu et en jaune," deposited as a sealed note on 14 May 1768. See *Oeuvres de Lavoisier—Correspondance*, fasc. I, 107–108. A significant reference by Lavoisier to the ideas of Hales about the fixation of air appeared in a paper of the same year. See J. B. Gough, "Lavoisier's Early Career in Science," in *British Journal for the History of Science*, **4** (1968), 52–57. More detailed early references to Hales's experiments are given in Lavoisier's "Sur la nature de l'eau," published in 1773, and in his *Opuscules physiques et chimiques* of 1774.
16. This famous phrase appears in Lavoisier's *pli cacheté* of 1 Nov. 1772. Repro. in Henry Guerlac, *Lavoisier—The Crucial Year* (Ithaca, N.Y., 1961), 227–228. In an important paper Robert E. Kohler, Jr., noting that Lavoisier's words suggest surprise, argues that the experiments on phosphorus and sulfur were simply to determine whether air enters into the composition of acids, as the Abbé Rozier had suggested, and were not guided by a primary interest in calcination. Only after his work on phosphorus and sulfur was Lavoisier led to wonder whether the addition of air might not explain the increase in weight of metallic calxes. This indeed is what Lavoisier tells us. See Kohler's "The Origin of Lavoisier's First Experiments on Combustion," in *Isis*, **63** (1972), 349–355.
17. The first of these translations, a *précis raisonné* of a memoir by a follower of Joseph Black, N. J. von Jacquin, was published in Rozier's *Observations sur la physique* in February 1773. Translations of Joseph Black's classic paper on magnesia alba, Joseph Priestley's "Observations on Different Kinds of Air," and Daniel Rutherford's paper on mephitic air (nitrogen) appeared in the spring and early summer in the same journal.
18. On this question, see a paper by C. E. Perrin, "Prelude to Lavoisier's Theory of Calcination—Some Observations on *mercurius calcinatus per se*," in *Ambix*, **16** (1969), 140–151.

19. In September 1774 the Swedish chemist Scheele communicated to Lavoisier a method of making oxygen by heating a preparation of silver carbonate. This letter was not published in *Oeuvres de Lavoisier—Correspondance*, fasc. II. See Uno Boklund, "A Lost Letter From Scheele to Lavoisier," in *Lychnos* (1957), 39–62. Boklund virtually accused the French scholars, including René Fric, of suppressing the letter so as to give all credit to Lavoisier, although he was aware that Grimaux had published this letter long before, as "Une lettre inédite de Scheele à Lavoisier," in *Revue générale des sciences pures et appliquées*, **1** (1890), 1–2. In any case Lavoisier seems to have been too preoccupied with other matters to follow up Scheele's suggestion.
20. M. Berthelot, *La révolution chimique*, 264–265.
21. See letters of Magellan (or Magalhaens) to Lavoisier in *Oeuvres de Lavoisier—Correspondance*, fasc. II, 504–508.
22. Berthelot, *op. cit.*, p. 271.
23. Much of this section is dependent upon Maurice Crosland's article "Lavoisier's Theory of Acidity," in press.
24. *Oeuvres de Lavoisier*, I, 482.
25. In a paper published in the *Histoire et mémoires* of the Academy in 1784, Lavoisier adopted, at the suggestion of Condorcet, the term *air vital* for what he had previously called (after Priestley) "dephlogisticated air" or "eminently respirable air." *Oeuvres de Lavoisier*, II, 263.
26. Originally a term used by Van Helmont to signify "a Spirit that will not coagulate," it was applied particularly to noxious vapors like firedamp. In the 1st ed. (1766) of his *Dictionnaire de chimie*, P. J. Macquer applied the term to "the volatile invisible parts which escape from certain bodies," for example, "the noxious vapors which rise from burning charcoal, and from matters undergoing spirituous fermentations." But in the 2nd ed. (1778) of the *Dictionnaire* he applied the word generally to aeriform fluids. See, for example, the articles "Gas ou air déphlogistiqué" and "Gas méphytique ou air fixe." In the same year Jean Bucquet included in his *Mémoire sur la manière dont les animaux sont affectés par différens Fluides Aériformes* an introductory section entitled "Histoire abrégé des différens Fluides aériformes ou Gas." The spelling used by both men is "gas," not "gaz." It is possible that this generalized use of the term "gas" may have been influenced by the second edition (1777) of James Keir's trans. of the first ed. of Macquer's *Dictionnaire* to which Keir appends *A Treatise on the Various Kinds of Permanently Elastic Fluids, or Gases*. A MS French translation of this treatise survives among Lavoisier's papers.
27. *Writings of Benjamin Franklin*, Albert Henry Smyth, ed., VIII (1906), 314.
28. *Observations sur la physique*, **23** (1783), 452–455. The paper is entitled "Extrait d'un mémoire lu par M. Lavoisier, à la séance publique de l'Académie royale des sciences, du 12 novembre, sur la nature de l'eau."
29. By C. E. Perrin. See his "Lavoisier, Monge and the Synthesis of Water," to appear in *British Journal for the History of Science*.
30. *Oeuvres de Lavoisier*, II, 334–359. Lavoisier notes that certain additions were made relative to the work done with Meusnier.
31. *Ibid.*, III, 733. See Benjamin Franklin's letters to Sir Joseph Banks, *Writings of Benjamin Franklin*, IX (1906), 79–85, 105–107, 113–117, and 119–121. On 16 September 1783 Franklin wrote to Richard Price: "All the Conversation here at present turns upon the Balloons fill'd with light inflammable Air, and the means of managing them, so to give men the Advantage of Flying." *Ibid.*, pp. 99–100.
32. Maurice Daumas and Denis I. Duveen, "Lavoisier's Relatively Unknown Large-Scale Decomposition and Synthesis of Water," in *Chymia*, **5** (1959), 126–127.
33. *Ibid.*, 127.
34. Two anonymous attacks on the phlogiston theory published in Rozier's *Observations sur la physique* in 1773 and 1774 have been widely attributed to Lavoisier—for example, by Berthelot, McKie, and Duveen. For arguments against Lavoisier's authorship and suggestions of likely alternatives, see Carleton Perrin, "Early Opposition to the Phlogiston Theory: Two Anonymous Attacks," in *British Journal for the History of Science*, **5** (1970), 128–144.
35. Henry Guerlac, "Laplace's Collaboration With Lavoisier," in *Actes du XIIe Congrès international d'histoire des sciences, Paris, 1968*, III B (Paris, 1971), 31–36.
36. Mme Lavoisier's drawing of the ice calorimeter is given in pl. VI of Lavoisier's *Traité élémentaire de chimie*. Two versions have survived and are in the collection of the Conservatoire des Arts et Métiers in Paris. It has not been possible to determine with certainty which of Lavoisier's artisans built the calorimeters, but it was probably a tinsmith named Naudin. See Maurice Daumas, *Lavoisier théoricien et expérimentateur* (1955), p. 142 and notes, and the illustrations in his "Les appareils d'expérimentation de Lavoisier," in *Chymia*, **3** (1950), 45–62, fig. 3.
37. Douglas McKie, *Antoine Lavoisier* (London–Philadelphia, 1935), p. 220.
38. For Fourcroy, and his relations with Lavoisier, see W. A. Smeaton, *Fourcroy, Chemist and Revolutionary, 1755–1809* (Cambridge, 1962).
39. J. R. Partington calls this the first use of the word "equation" in our modern sense of "chemical equation." Lavoisier's example is: must of grapes = carbonic acid + alcohol. See Partington's *History of Chemistry*, III, 480.
40. Arthur Harden, *Alcoholic Fermentation* (London, 1911), p. 3. Partington has remarked (*History of Chemistry*, III, p. 376) that Lavoisier "aimed at accurate results, but seldom achieved them." Guichard (*Essai historique sur les mesures en chimie* [Paris, 1937], 54–60), after examining Lavoisier's numerical results, judged his precision extremely variable; indeed in the later, more complicated experiments the accuracy was less than in the earliest ones. Yet Lavoisier repeatedly stressed the importance of quantitative accuracy. In the *Traité* he writes: "La détermination du poids des matières et des produits, avant et après les expériences, étant la base de tout ce qu'on peut faire d'utile et d'exact en chimie, on ne saurait y apporter trop d'exactitude" (*Oeuvres de Lavoisier*, I, 251); elsewhere he writes "Rien n'est supposé dans ces explications, tout est prouvé, le poids et la mesure à la main" (*ibid.*, V, 270).
41. *Oeuvres de Lavoisier*, I, 101. The axiom that matter is neither created nor destroyed originated with the atomists of antiquity. It was clearly stated by Mariotte in his *Essai de logique* (1678) and was a common assumption of eighteenth-century scientists. But Lavoisier deserves credit for applying it specifically to the operations of the chemist and for spelling out a law of the conservation of matter in chemical reactions.
42. *Ibid.*, 520–521, 625–627.
43. "Mémoire sur la nature du principe qui se combine avec les métaux pendant leur calcination, et qui en augmente le poids," in *Observations sur la physique*, **5** (1775), 433.
44. Berthelot, *op. cit.*, 290–291.
45. Cited from the *Procès-verbaux* of the Academy of Sciences by Maurice Daumas, in *Lavoisier, théoricien et expérimentateur*, p. 38. Daumas says this remained unpublished, yet it was almost certainly published, with the inevitable modifications, in 1780 as "Expériences sur la respiration des animaux." See *Oeuvres de Lavoisier*, II, 174–183.
46. *Ibid.*, 232.
47. "Altérations qu'éprouve l'air respiré," *ibid.*, 676–687.
48. Charles C. Gillispie, "Notice biographique de Lavoisier par Madame Lavoisier," in *Revue d'histoire des sciences et de leurs applications*, **9** (1956), 57.
49. *Résultats extraits d'un ouvrage intitulé: De la richesse territoriale du royaume de France* (Paris, 1791). Together with other of Lavoisier's economic writings this is printed in G. Schelle and E. Grimaux, *Lavoisier—statique agricole et*

projets de réformes (Paris, 1894). See also *Oeuvres de Lavoisier*, VI, 403–463.

50. For Lavoisier's role in the attempts to reform the Academy, and the political struggles within that body, see Roger Hahn, *The Anatomy of a Scientific Institution* (Berkeley–Los Angeles, 1971), especially chapters 8 and 9.
51. J. Guillaume, "Un mot légendaire: 'La République n'a pas besoin de savants,'" in *Révolution française*, **38** (1900), 385–399, and *Études révolutionnaires*, 1st ser. (Paris, 1908), pp. 136–155.

BIBLIOGRAPHY

I. ORIGINAL WORKS. For a bibliography of Lavoisier's papers—academic, scientific, and in popular periodicals—and his major and minor separate works, consult Denis I. Duveen and Herbert S. Klickstein, *A Bibliography of the Works of Antoine Laurent Lavoisier, 1743–1794* (London, 1954), and Duveen's *Supplement* (London, 1965).

Lavoisier's published articles, some hitherto unpublished papers, and two of his major books are printed in *Oeuvres de Lavoisier*, 6 vols. (Paris, 1862–1893). The first 4 vols. were edited by the distinguished chemist J. B. Dumas, the last 2 vols. by Edouard Grimaux, Lavoisier's principal biographer. Much, but by no means all, of Lavoisier's surviving correspondence through 1783 has appeared in *Oeuvres de Lavoisier—Correspondance*, 3 fascs. (Paris, 1955–1964), edited by the late René Fric under the auspices of the Académie des Sciences. A further fasc. completed before the death of M. Fric is due to appear. The editing of this work leaves much to be desired; many letters, some quite important, have been overlooked; for a scathing appraisal of fasc. I, see A. Birembaut, "La correspondance de Lavoisier," in *Annales historiques de la Révolution française*, **29** (Oct.–Dec. 1957), 340–351.

Lavoisier's major books are four in number. His *Opuscules physiques et chimiques* (Paris, 1774) was intended to be the first of a series of vols. containing the results of his investigations. The later vols. never materialized. A posthumous reissue of the 1st ed. was published in 1801; the *Opuscules* was trans. with notes and an appendix by Thomas Henry as *Essays Physical and Chemical* (London, 1776). A German trans. by C. E. Weigel appeared as vol. I of his 5-vol. collection entitled *Lavoisier physikalisch-chemische Schriften* (Greifswald, 1783–1794). The remaining 4 vols. contain German versions of a number of Lavoisier's scientific papers (the last 2 vols. were the work of H. F. Link).

Méthode de nomenclature chimique, proposée par MM. de Morveau, Lavoisier, Bertholet [sic] *& de Fourcroy. On y a joint un nouveau système de caractères chimiques, adaptés à cette nomenclature, par MM. Hassenfratz & Adet* (Paris, 1787). There is a 2nd printing of this 1st ed.; the 2nd ed. appeared in 1789, as vol. III of the 2nd ed. of Lavoisier's *Traité élémentaire de chimie*. It is entitled *Nomenclature chimique, ou synonymie ancienne et moderne, pour servir à l'intelligence des auteurs*. There is an English trans. of this collaborative work by James St. John (London, 1788); an Italian version by Pietro Calloud (Venice, 1790); and a German trans. by Karl von Meidinger (Vienna, 1793). Various abridgments and tabular versions of the new nomenclature appeared in English, German, and Italian. The chief agent for transmitting the new chemistry to Germany was Christoph Girtanner (1760–1800), who in 1791 brought out a German version of the *Nomenclature chimique* and the next year his *Anfangsgründe der antiphlogistischen Chemie* (Berlin, 1792). For the reception of Lavoisier's work in Italy, see Icilio Guareschi, "Lavoisier, sua vita e sue opere," in *Supplemento annuale alla Enciclopedia di chimica scientifica e industriale*, **19** (Turin, 1903), 307–469, esp. 452–455.

In 1778 Lavoisier proposed to devote a 2nd vol. of his *Opuscules* to setting forth a *système général* of chemistry expounded in a rational and deductive fashion, according to the *méthode des géomètres*. Such a work was never written; but this method was used in his most famous book, the *Traité élémentaire de chimie* (Paris, 1789), a book that in fact grew out of the work on the chemical nomenclature and first appeared in a single volume; a 2nd printing in 2 vols. appeared in the same year, as did a 2nd ed. An influential English trans. by Robert Kerr appeared as *Elements of Chemistry, in a New Systematic Order, Containing All the Modern Discoveries* (Edinburgh, 1790). There are several later eds. of Kerr's trans. The *Traité* appeared in German, translated by S. F. Hermbstadt (Berlin–Stettin, 1792); in Dutch, by N. C. de Fremery (Utrecht, 1800); in Italian, by Vincenzo Dandolo (Venice, 1791); and in Spanish, by Juan Manuel Munarriz (Madrid, 1798).

About 1792 Lavoisier planned a complete ed. of his memoirs in 8 vols.; also to be included were papers by some of his disciples. The posthumous *Mémoires de chimie*, printed in 2 vols. for Mme Lavoisier from incomplete proofs partially corrected by Lavoisier, is all that resulted from this project. The vols. are undated (there are no title pages, only half titles), and they were not commercially published; Mme Lavoisier presented copies to selected institutions and individuals. Although Duveen assigns the year 1805 to this work, it was circulating two years earlier—when, as J. R. Partington has shown, Berthollet quoted it by vol. and pg. The *Mémoires*, not repr. in the *Oeuvres*, includes reprints and revisions of previously published papers, as well as hitherto unpublished results, such as experiments carried out toward the close of his collaboration with Laplace in 1783–1784 on problems related to heat. There are several papers by Lavoisier's latter-day collaborator, Armand Seguin.

The chief repository of Lavoisier manuscripts is the Académie des Sciences in Paris, in whose archives one can consult Lavoisier's laboratory notebooks and some twenty *cartons* containing drafts of scientific papers, academy reports, and so on. There are important letters in the Collection de Chazelles in the city library of Clermont-Ferrand, and other manuscripts, notably letters, in the possession of Count Guy de Chabrol. Microfilms of the letters and other documents in the Fonds Chabrol are available in the archives of the Académie des Sciences, as is also a xerographic copy of the second laboratory notebook *(registre)* for the period 9 September 1773 to

5 March 1774. This had been given by François Arago to the library of Perpignan, and for a time was believed lost. See the *Comptes rendus des séances de l'Académie des sciences*, **135** (1902), 549–557 and 574–575. There are scattered documents in various provincial libraries in France, notably Orléans, and important material in the Archives Nationales, chiefly dealing with the period of the Revolution. The Bibliothèque Nationale has little to offer, except Fr. nouv. acq. 5153, which is an account of Lavoisier's repetition of his classic early experiments before an Academy committee. The Lavoisier Collection in the Olin Library of Cornell University was originally assembled by Denis I. Duveen. Besides a number of manuscripts, many books from Lavoisier's own library with his *ex libris*, this includes the most nearly complete assemblage of the books and pamphlets published by Lavoisier, and in nearly all the known editions and variants.

II. SECONDARY LITERATURE. The earliest biographical sketches of Lavoisier appeared in the early days of the Directory: Joseph Jérôme Lalande, "Notice sur la vie et les ouvrages de Lavoisier," in *Magasin encyclopédique*, **5** (1795), 174–188, English trans. in *Philosophical Magazine*, **9** (1801), 78–85; and Antoine François de Fourcroy, *Notice sur la vie et les travaux de Lavoisier* (Paris, 1796). Other early sketches are Thomas Thomson, "Biographical Account of M. Lavoisier," in *Annals of Philosophy*, **2** (1813), 81–92; and Georges Cuvier, "Lavoisier," in L. G. Michaud, ed., *Biographie universelle*, new ed. (Paris, 1854–1865), XXIII, 414–418, based in part on information, much of it inaccurate, supplied by Lavoisier's widow. See Charles C. Gillispie, "Notice biographique de Lavoisier par Madame Lavoisier," in *Revue d'histoire des sciences et de leurs applications*, **9** (1956), 52–61. The standard biography is Edouard Grimaux, *Lavoisier 1743–1795, d'après sa correspondance, ses manuscrits, ses papiers de famille et d'autres documents inédits* (Paris, 1888; 2nd ed., 1896; 3rd ed., 1899). No trans. was ever published, and most subsequent biographies have relied heavily upon Grimaux—for example, Mary Louise Foster, *Life of Lavoisier* (Northampton, Mass., 1926); J. A. Cochrane, *Lavoisier* (London, 1931); and Douglas McKie, *Antoine Lavoisier* (London–New York, 1952). For a general appraisal of this literature, see Henry Guerlac, "Lavoisier and His Biographers," in *Isis*, **45** (1954), 51–62. A readable study, which makes use of some recent research, is Léon Velluz, *Vie de Lavoisier* (Paris, 1966). Short but informative is Lucien Scheler, *Lavoisier et le principe chimique* (Paris, 1964).

There have been a number of general investigations of Lavoisier's work in chemistry. The pioneer study is surely the IVe Leçon of J.-B. Dumas, *Leçons sur la philosophie chimique professées au Collège de France* (Paris, n.d., but certainly 1836, the year the lectures were delivered). A second edition, an unaltered reprint, appeared in 1878. Still valuable, especially for its excerpts and paraphrases from Lavoisier's laboratory notebooks, is Marcelin Berthelot, *La révolution chimique—Lavoisier* (Paris, 1890, 2nd ed., 1902). Douglas McKie, *Antoine Lavoisier, the Father of Modern Chemistry* (Philadelphia, 1935), owes much to the pioneer work of Andrew Norman Meldrum (see below). In the same year appeared Hélène Metzger's classic *La philosophie de la matière chez Lavoisier* (Paris, 1935). Worth consulting is Sir Harold Hartley, "Antoine Laurent Lavoisier, 26 August 1743–8 May 1794," in *Proceedings of the Royal Society*, **189A** (1947), 427–456. Extremely valuable for the use made of the *procès-verbaux* of the Academy of Sciences and other unpublished documents relating to Lavoisier's work is Maurice Daumas, *Lavoisier, théoricien et expérimentateur* (Paris, 1955). An elaborate treatment of Lavoisier's achievements is given by J. R. Partington, *A History of Chemistry*, III, ch. IX. A general review of the state of Lavoisier studies is given by W. A. Smeaton, "New Light on Lavoisier: The Research of the Last Ten Years," in *History of Science*, **2** (1963), 51–69.

For eighteenth-century chemical theory before Lavoisier the classic studies are Hélène Metzger, *Les Doctrines chimiques en France du début du XVIIe à la fin du XVIIIe siècle* (Paris, 1923) and her *Newton, Stahl, Boerhaave et la doctrine chimique* (Paris, 1930). The technological background is described in Henry Guerlac, "Some French Antecedents of the Chemical Revolution," in *Chymia*, **5** (1959), 73–112. For Lavoisier's teacher of chemistry, see Rhoda Rappaport, "G. F. Rouelle: An Eighteenth-Century Chemist and Teacher," *ibid.*, **6** (1960), 68–101, and her "Rouelle and Stahl—The Phlogistic Revolution in France," *ibid.*, **7** (1961), 73–102. A different emphasis is found in Martin Fichman, "French Stahlism and Chemical Studies of Air, 1750–1779," in *Ambix*, **18** (1971), 94–122.

Studies of special aspects of Lavoisier's chemical research abound. For the early stages of his career, see A. N. Meldrum, "Lavoisier's Early Work on Science, 1763–1771," in *Isis*, **19** (1933), 330–363; **20** (1934), 396–425; and his "Lavoisier's Work on the Nature of Water and the Supposed Transmutation of Water Into Earth (1768–1773)," in *Archeion*, **14** (1932), 246–247. These studies have been corrected or amplified by Henry Guerlac, "A Note on Lavoisier's Scientific Education," in *Isis*, **47** (1956), 211–216; by Rhoda Rappaport, "Lavoisier's Geologic Activities, 1763–1792," *ibid.*, **58** (1968), 375–384; and by J. B. Gough, "Lavoisier's Early Career in Science, an Examination of Some New Evidence," in *British Journal for the History of Science*, **4** (1968), 52–57, who was the first to identify (correcting Daumas) Lavoisier's notes from Eller and to see their significance for Lavoisier's theory of heat and vaporization. Consult also W. A. Smeaton, "*L'avant-coureur*, the Journal in Which Some of Lavoisier's Earliest Research Was Reported," in *Annals of Science*, **13** (1957), 219–234. This article actually appeared in 1959. There is useful information in E. McDonald, "The Collaboration of Bucquet and Lavoisier," in *Ambix*, **13** (1966), 74–84.

For Lavoisier's first experiments on combustion, the pioneer work of A. N. Meldrum, *The Eighteenth Century Revolution in Science—The First Phase* (Calcutta–London–New York, n.d. [1930]), is classic. See too his "Lavoisier's Three Notes on Combustion: 1772," in *Archeion*, **14** (1932), 15–30; also Max Speter, "Die entdeckte Lavoisier-Note vom 20 Oktober 1772," in *Zeitschrift für angewandte Chemie*, **45** (1932); his "Kritisches über die Entstehung

LAVOISIER

von Lavoisiers System," *ibid.*, **39** (1926), 578–582; and his article "Lavoisier," in Gunther Bugge, ed., *Das Buch der grossen Chemiker* (Berlin, 1929), I, 304–333. For new light on the famous sealed note, see Henry Guerlac, "A Curious Lavoisier Episode," in *Chymia*, **7** (1961), 103–108. For the research leading up to Lavoisier's investigations in the autumn of 1772, see Guerlac, *Lavoisier—The Crucial Year* (Ithaca, N.Y., 1961), where the basic documents are reproduced in the app. For a reassessment see Robert E. Kohler, Jr., "The Origin of Lavoisier's First Experiments on Combustion," in *Isis*, **63** (1972), 349–355. The discovery of oxygen, and the influence of Priestley upon Lavoisier, are treated by Meldrum, *Eighteenth Century Revolution in Science*, chap. 5; by Sir Philip Hartog, "The Newer Views of Priestley and Lavoisier," in *Annals of Science*, **5** (1941), 1–56; and by Sidney J. French, "The Chemical Revolution—The Second Phase," in *Journal of Chemical Education*, **27** (1950), 83–88. C. E. Perrin, "Prelude to Lavoisier's Theory of Calcination—Some Observations on Mercurius Calcinatus per se," in *Ambix*, **16** (1969), 140–151, casts new light on Lavoisier's possible debt to Priestley and to Pierre Bayen.

For the stages in the preparation of Lavoisier's most famous book see Maurice Daumas, "L'élaboration du *Traité de Chimie* de Lavoisier," in *Archives internationales d'histoire des sciences*, **29** (1950), 570–590. For Lavoisier's attitude towards affinity theories see Daumas, "Les conceptions sur les affinités chimiques et la constitution de la matière," in *Thalès* (1949–1950), 69–80. Both these papers discuss unpublished notes for a work Lavoisier did not live to undertake seriously, to be entitled "Cours de chimie expérimentale" or "Cours de philosophie expérimentale." A provocative paper on the famous "Discours préliminaire" is Robert Delhez, "Révolution chimique et Révolution française: *Le Discours préliminaire* au *Traité élémentaire de chimie* de Lavoisier," in *Revue des questions scientifiques*, **143** (1972), 3–26.

Lavoisier's early theory of the elements and of the aeriform state of matter has been the focus of recent investigations. Guerlac, in "A Lost Memoir of Lavoisier," in *Isis*, **50** (1959), 125–129, attempts to reconstruct the character of a supposedly lost memoir by Lavoisier. This document was published with other *inédits* by René Fric, "Contribution à l'étude de l'évolution des idées de Lavoisier sur la nature de l'air et sur la calcination des métaux," in *Archives internationales d'histoire des sciences*, **12** (1959), 125–129, actually published 1960. Further contributions are Maurice Crosland, "The Development of the Concept of the Gaseous State as a Third State of Matter," in *Proceedings of the 10th International Congress of the History of Science* II (Paris, 1962), 851–854; J. B. Gough, "Nouvelle contribution à l'étude de l'évolution des idées de Lavoisier sur la nature de l'air et sur la calcination des métaux," in *Archives internationales d'histoire des sciences*, **22** (1969), 267–275, publishes a draft by Lavoisier entitled "De l'élasticité et de la formation des fluides élastiques," dated end of Feb. 1775. See also Robert Siegfried, "Lavoisier's View of the Gaseous State and Its Early Application to Pneumatic Chemistry," in *Isis*, **63** (1972), 59–78. The most extensive treatment of this aspect of Lavoisier's work is Gough's Ph.D. dissertation, "The Foundations of Modern Chemistry: The Origin and Development of the Concept of the Gaseous State and Its Role in the Chemical Revolution of the Eighteenth Century" (Cornell University, June 1971). Robert J. Morris has criticized, with some justice, Guerlac's narrow interpretation of Lavoisier's "Système sur les élémens" of the summer of 1772 but denies that it was in fact a system of the elements. See R. J. Morris, "Lavoisier on Air and Fire: The Memoir of July 1772," in *Isis*, **60** (1969), 374–377; and Guerlac's reply, *ibid.*, 381–382.

Closely related to the problem of the aeriform state is Lavoisier's caloric theory, a matter discussed cursorily in most accounts of his work. But recent work has reopened the question. Henry Guerlac, "Laplace's Collaboration With Lavoisier," in *Actes du XIIe Congrès international d'histoire des sciences; Paris, 1968*, III B (Paris, 1971), 31–36, is an abstract of a forthcoming study of the collaborative experiments of Lavoisier and Laplace on heat. Robert Fox, *The Caloric Theory of Gases, From Lavoisier to Regnault* (Oxford, 1971), has a long first chapter entitled "The Study of Gases and Heat to 1800." An important study is Robert J. Morris, "Lavoisier and the Caloric Theory," in *British Journal for the History of Science*, **6** (1972), 1–38.

For Lavoisier's instrument makers see Maurice Daumas, *Les instruments scientifiques aux XVIIe et XVIIIe siècles* (Paris, 1953). Descriptions of surviving pieces of Lavoisier's apparatus are given in P. Truchot, "Les instruments de Lavoisier. Relation d'une visite à La Canière (Puy-de-Dôme) où se trouve réunis les appareils ayant servi à Lavoisier," in *Annales de chimie et de physique*, 5th ser., **18** (1879), 289–319; also Maurice Daumas, "Les appareils d'expérimentation de Lavoisier," in *Chymia*, **3** (1950), 45–62; and chs. 5 and 6 of Daumas, *Lavoisier, théoricien et expérimentateur*; and Graham Luske, "Mementoes of Lavoisier, Notes on a Trip to Château de la Canière," in *Journal of the American Medical Association*, **85** (1925), 1246–1247. An attempt to evaluate the quantitative precision of Lavoisier's experiments is made by M. Guichard, *Essai historique sur les mesures en chimie* (Paris, 1937), 53–72.

On the synthesis and decomposition of water, the fullest account is in J. R. Partington, *History of Chemistry*, III, 325–338, 436–456. Maurice Daumas and Denis Duveen, "Lavoisier's Relatively Unknown Large-Scale Decomposition and Synthesis of Water, February 27 and 28, 1785," in *Chymia*, **5** (1959), 113–129, discuss in detail the experiments of Meusnier and Lavoisier that were never formally published, and were described only in the account printed in the obscure *Journal polytype des sciences et des arts* of 26 Feb. 1786; repr. in *Oeuvres de Lavoisier*, V, 320–339. C. E. Perrin, in an article in press entitled "Lavoisier, Monge and the Synthesis of Water," accounts for the apparent coincidence of the synthesis of water by Lavoisier and Laplace and by Gaspard Monge in the summer of 1783 by suggesting a knowledge (prior to the disclosure by Blagden of Cavendish's results) of Priestley's repetition of Cavendish's experiment.

LAVOISIER

For the pioneer balloon experiments, the most elaborate contemporary source is Barthélemy Faujas de Saint-Fond, *Description des expériences de la machine aérostatique de MM. Montgolfier . . . suivie de mémoires sur le gaz inflammable etc.*, 2 vols. (Paris, 1783–1784). Gaston Tissandier, *Histoire des ballons et des aéronautes célèbres* (Paris, 1887), is an admirable account, superbly illustrated. The report of the first commission on the Montgolfier balloons (composed of Le Roy, Tillet, Brisson, Cadet de Gassicourt, Lavoisier, Bossut, Condorcet, and Desmarest) appeared first as a pamphlet in 1784 and later in the *Mémoires* of the Academy of Sciences. It is repr. in *Oeuvres de Lavoisier*, III, 719–735, and in *Extraits des mémoires de Lavoisier concernant la météorologie et l'aéronautique* (Paris, 1926), published by the Office National Météorologique de France.

A brief account of Lavoisier's work on organic analysis is given in Ferenc Szabadváry, *History of Analytical Chemistry* (Oxford–London, 1966), pp. 284–287. His studies of alcoholic fermentation are described in P. Schutzenberger, *Les fermentations*, 5th ed. (Paris, 1889), pp. 14–15; Carl Oppenheimer, *Ferments and Their Actions*, trans. from the German by C. Ainsworth Mitchell (London, 1901), pp. 3–4; and Arthur Harden, *Alcoholic Fermentation* (London, 1911), pp. 2–3.

For Lavoisier's geology see Pierre Comte, "Aperçu sur l'oeuvre géologique de Lavoisier," in *Annales de la Société géologique du Nord*, **69** (1949), 369–375; A. V. Carozzi, "Lavoisier's Fundamental Contribution to Stratigraphy," in *Ohio Journal of Science*, **65** (1965), 71–85; Rhoda Rappaport, "The Early Disputes Between Lavoisier and Monnet, 1777–1781," in *British Journal for the History of Science*, **4** (1969), 233–244; "Lavoisier's Geologic Activities, 1763–1792," in *Isis*, **58** (1968), 375–384; "The Geological Atlas of Guettard, Lavoisier, and Monnet," in Cecil J. Schneer, ed., *Towards a History of Geology* (Cambridge, Mass., 1969), pp. 272–287; and "Lavoisier's Theory of the Earth," to appear in *British Journal for the History of Science*.

On the new nomenclature, see M. P. Crosland, *Historical Studies in the Language of Chemistry* (London, 1962), chs. 3–8. See also Denis I. Duveen and Herbert S. Klickstein, "The Introduction of Lavoisier's Chemical Nomenclature Into America," in *Isis*, **45** (1954), 278–292, 368–382; and H. M. Leicester, "The Spread of the Theory of Lavoisier in Russia," in *Chymia*, **5** (1959), 138–144. For Britain and Germany, see Crosland, *Historical Studies . . .*, pp. 193–206, 207–210.

Lavoisier's physiological writings, notably on respiration, are discussed in Charles Richet, "Lavoisier et la chaleur animale," in *Revue scientifique*, **34** (1884), 141–146; Alphonse Milne-Edwards, *Notice sur les travaux physiologiques de Lavoisier* (Paris, 1885); G. Masson, ed., *Lavoisier, la chaleur et la respiration, 1770–1789* (Paris, 1892); M. J. Rosenthal, "Lavoisier et son influence sur les progrès de la physiologie," in *Revue scientifique*, **47** (1891), 33–42; J. C. Hemmeter, "Lavoisier and the History of the Physiology of Respiration and Metabolism," in *Johns Hopkins Hospital Bulletin*, **29** (1918), 254–264; Graham Lusk, "A History of Metabolism," in L. F. Barker, ed., *Endocrinology and Metabolism* (New York, 1922), III, 3–78—a section of this history is devoted to Lavoisier, pp. 19–30; R. Foregger, "Respiration Experiments of Lavoisier," in *Archives internationales d'histoire des sciences*, **13** (1960), 103–106; and Everett Mendelsohn, *Heat and Life* (Cambridge, Mass., 1964), 134–139, 146–165, which discusses Lavoisier's work perceptively. Charles A. Culotta, "Respiration and the Lavoisier Tradition: Theory and Modification, 1777–1850," in *Transactions of the American Philosophical Society*, n.s. **62** (1972), 3–40, is the most recent study. See also John F. Fulton, Denis I. Duveen, and Herbert S. Klickstein, "Antoine Laurent Lavoisier's Réflexions sur les effets de l'éther nitreux dans l'économie animale," in *Journal of the History of Medicine and Allied Sciences*, **8** (1953), 318–323; W. A. Smeaton, "Lavoisier's Membership of the Société Royale de Médecine," in *Annals of Science*, **12** (1956), 228–244; and Denis I. Duveen and H. S. Klickstein, "Antoine Laurent Lavoisier's Contributions to Medicine and Public Health," in *Bulletin of the History of Medicine*, **29** (1955), 164–179.

For Lavoisier's role in the improvement of saltpeter production and gunpowder manufacture, see Robert P. Multhauf, "The French Crash Program for Saltpeter Production, 1776–94," in *Technology and Culture*, **12** (1971), 163–181. The gunpowder situation in America at the outbreak of the Revolution, and the aid received from France, is described by Orlando W. Stephenson, "The Supply of Gunpowder in 1776," in *American Historical Review*, **30** (1925), 271–281. More than half of vol. V of the *Oeuvres de Lavoisier* is devoted to his papers on saltpeter and the work for the Régie des Poudres. See also J. R. Partington, "Lavoisier's Memoir on the Composition of Nitric Acid," in *Annals of Science*, **9** (1953), 96–98. The relations between Franklin and Lavoisier are described by H. S. van Klooster, "Franklin and Lavoisier," in *Journal of Chemical Education*, **23** (1946), 107–109; and by Denis I. Duveen and H. S. Klickstein, "Benjamin Franklin (1706–1790) and Antoine Laurent Lavoisier (1743–1794)," in *Annals of Science*, **11** (1955), 103–128, 271–302, and **13** (1957), 30–46. See also Claude A. Lopez, "Saltpetre, Tin and Gunpowder: Addenda to the Correspondence of Lavoisier and Franklin," in *Annals of Science*, **16** (1960), 83–94. For a remarkable letter from Lavoisier to Franklin, see René Fric, "Une lettre inédite de Lavoisier à B. Franklin," in *Bulletin historique et scientifique de l'Auvergne*, **9** (1924), 145–152.

On Lavoisier as a financier and economic theorist, Eugène Daire and Gustave de Molinari, *Mélanges d'économie politique*, I (Paris, 1847), 577–580, contains a "Notice sur Lavoisier," followed by two of his essays in political economy. See also G. Schelle and E. Grimaux, *Lavoisier—statistique agricole et projets de réformes* (Paris, 1894); and R. Dujarric de la Rivière, *Lavoisier économiste* (Paris, 1949). For Lavoisier's relations with Pierre Samuel Dupont de Nemours and Pierre's son, see Bessie Gardner Du Pont, *Life of Éleuthère Irénée du Pont from Contemporary Correspondence* (11 vols., Newark, Delaware, 1923–

LAVOISIER

1926), *passim*, I, 141–145, and R. Dujarric de la Rivière, *E. I. Du Pont de Nemours, élève de Lavoisier* (Paris, 1954), esp. 157–158.

For Lavoisier's interest in agriculture, see Henri Pigeonneau and Alfred de Foville, *L'administration de l'agriculture au controle générale des finances (1785–1787). Procès-verbaux et rapports* (Paris, 1882); and Louis Passy, *Histoire de la Société nationale d'agriculture de France*, I, *1761–1793* (Paris, 1912), which is all that appeared. André J. Bourde, *The Influence of England on the French Agronomes, 1750–1789* (Cambridge, 1953); and *Agronomie et agronomes en France au XVIIIe siècle*, 3 vols. (Paris, 1967), provide the essential background with incidental references to Lavoisier. See also W. A. Smeaton, "Lavoisier's Membership of the Société Royale d'Agriculture and the Comité d'Agriculture," in *Annals of Science*, **12** (1956), 267–277.

Lavoisier's role as an officer of the Academy of Sciences, before and during the French Revolution, bulks large in Roger Hahn, *The Anatomy of a Scientific Institution, the Paris Academy of Sciences, 1666–1803* (Berkeley–Los Angeles, 1971). There are glimpses of Lavoisier and his wife in Arthur Young's *Travels in France*, the best (and fully annotated) ed. of which is the French trans. by Henri Sée, *Voyages en France en 1787, 1788 et 1789*, 3 vols. (Paris, 1931), espec. I, pp. 189–191; in Beatrix Cary Davenport, ed., *A Diary of the French Revolution by Gouverneur Morris, 1752–1816* (Boston, 1939); in V. A. Eyles, "The Evolution of a Chemist, Sir James Hall," in *Annals of Science*, **19** (1963), 153–182; and in J. A. Chaldecott, "Scientific Activities in Paris in 1791," *ibid.*, **24** (1968), 21–52.

Grimaux gives much useful information about Mme Lavoisier, her education, her marriage, and her contributions to her husband's career in his *Lavoisier*, 2nd ed., pp. 35–44; for the genealogy of her family and an account of her life after Lavoisier's execution, see the appendixes to the same work, pp. 330–336. Worth consulting is Denis I. Duveen, "Madame Lavoisier, 1758–1836," in *Chymia*, **4** (1953), 13–29. Mme Lavoisier's major contribution to the campaign for the new chemistry was her (anonymous) trans. of Richard Kirwan's *Essay on Phlogiston* (London, 1784); the French version, *Essai sur le phlogistique* (Paris, 1788), has notes by the translator and extensive critical commentary by Guyton de Morveau, Laplace, Monge, Berthollet, Fourcroy, and Lavoisier himself. Grimaux (*op. cit.*, p. 42) speaks of her "traductions inédites de Priestley, Cavendish, Henry, etc." Her last effort along these lines is her trans. of Kirwan's "Strength of Acids and the Proportion of Ingredients in Neutral Salts," in *Proceedings of the Royal Irish Academy*, **4** (1790), 3–89. It appeared in *Annales de chimie*, **14** (1792), 152, 211, 238–286.

Lavoisier's role in the French Revolution, his imprisonment, and his execution have produced a substantial, and sometimes controversial, literature. See, for example, Denis I. Duveen, "Antoine Laurent Lavoisier (1743–1794), a Note Regarding His Domicile During the French Revolution," in *Isis*, **42** (1951), 233–234; and "Antoine Laurent Lavoisier and the French Revolution," in *Journal of Chemical Education*, **31** (1954), 60–65; **34** (1957), 502–503; **35** (1958), 233–234, 470–471. See also Marguerite Vergnaud, "Un savant pendant la Révolution," in *Cahiers internationaux de sociologie*, **17** (1954), 123–139; Denis I. Duveen and Marguerite Vergnaud, "L'explication de la mort de Lavoisier," in *Archives internationales d'histoire des sciences*, **9** (1956), 43–50; Denis I. Duveen, "Lavoisier Writes to Fourcroy From Prison," in *Notes and Records. Royal Society of London*, **13** (1958), 59–60; and "Lavoisier Writes to His Wife From Prison," in *Manuscripts*, **10** (fall 1958), 38–39; and Denis I. Duveen and H. S. Klickstein, "Some New Facts Relating to the Arrest of Antoine Laurent Lavoisier," in *Isis*, **49** (1958), 347–348. On the problem of whether Fourcroy interceded on behalf of Lavoisier, see G. Kersaint, "Fourcroy a-t-il fait des démarches pour sauver Lavoisier?" in *Revue générale des sciences pures et appliquées*, **65** (1958), 151–152; and his "Lavoisier, Fourcroy et le scrutin épuratoire du Lycée de la rue de Valois," in *Bulletin. Société chimique de France* (1958), 259. For a comment on Kersaint's article, see Maurice Daumas, "Justification de l'attitude de Fourcroy pendant la Terreur," in *Revue d'histoire des sciences et de leurs applications*, **11** (1958), 273–274.

Other aspects of Lavoisier's career during the Revolution are mentioned by Henry Guerlac, "Some Aspects of Science During the French Revolution," in *Scientific Monthly*, **80** (1955), 93–101, repr. in Philipp G. Frank, ed., *The Validation of Scientific Theories* (Boston, 1957), 171–191; J. Guillaume, "Lavoisier anti-clérical et révolutionnaire," in *Révolution française*, **26** (1907), 403–423, repr. in *Études révolutionnaires*, 1st ser. (Paris, 1908), 354–379; Lucien Scheler, *Lavoisier et la Révolution française. II. Le journal de Fourgeroux de Bondaroy* (Paris, 1960), ed. with the collaboration of W. A. Smeaton. See also Smeaton's "The Early Years of the Lycée and the Lycée des Arts. A Chapter in the Lives of A. L. Lavoisier and A. F. de Fourcroy," in *Annals of Science*, **11** (1955), 257–267; **12** (1956), 267–277; and his "Lavoisier's Membership of the Assembly of Representatives of the Commune of Paris, 1789–1790," *ibid.*, **13** (1957), 235–248. For the text of letters seized at Lavoisier's house on 10 and 11 Sept. 1793, see Douglas McKie, "Antoine Laurent Lavoisier, F.R.S.," in *Notes and Records of the Royal Society*, **7** (1949), 1–41.

For the educational proposals attributed to Lavoisier, see Harold J. Abrahams, "Lavoisier's Proposals for French Education," in *Journal of Chemical Education*, **31** (1954), 413–416; and his "Summary of Lavoisier's Proposals for Training in Science and Medicine," in *Bulletin of the History of Medicine*, **32** (1958), 389–407. That Lavoisier was by no means the sole author of the *Réflexions sur l'instruction publique* has been shown by K. M. Baker and W. A. Smeaton, "The Origins and Authorship of the Educational Proposals Published in 1793 by the *Bureau de Consultation des Arts et Métiers* and Generally Ascribed to Lavoisier," in *Annals of Science*, **21** (1965), 33–46.

HENRY GUERLAC

LAVRENTIEV, BORIS INNOKENTIEVICH (*b.* Kazan, Russia, 13 August 1892; *d.* Moscow, U.S.S.R., 9 February 1944), *histology*.

Lavrentiev graduated from the Medical Faculty of Kazan University in 1914. From then until 1920 he was an army doctor, and in 1921 he became a prosector in the department of histology of Kazan University. In 1925 he was sent for a year to Utrecht to work in the neurohistological laboratory under J. Boecke. From 1927 to 1929 Lavrentiev was professor of histology at the Veterinary Faculty of the Zootechnical Institute in Moscow, then head of the department of histology at the First Moscow Medical Institute from 1929 to 1932. He headed the morphology department of the All-Union Institute of Experimental Medicine in Leningrad (it moved in 1934 to Moscow), at the same time holding the chair of histology at the Second Moscow Medical Institute. From 1941 to 1943, when the Institute of Experimental Medicine was evacuated to Tomsk, Lavrentiev was its acting director. In 1943 he continued his scientific work and teaching in Moscow. He was elected corresponding member of the Soviet Academy of Sciences in 1939.

The work of Lavrentiev and his collaborators was in the experimental study of the histophysiology of the autonomic nervous system. Their experiments with resection of ganglions and section of nerves, as well as their subsequent study by means of silver impregnation that adopted the Bielschowsky-Gross method (with Lavrentiev's modification), allowed investigation of the phenomena of degeneration and regeneration of the lost nerve connections and changes in synapses, autonomic ganglia, and the nerve endings in tissues and organs. On the basis of his research Lavrentiev supported the concept of the trophic effect of the nervous system and the theory of its neuronal structure. Lavrentiev demonstrated the existence and functions of interneuron bonds (synapses) not only on histological sections but also intravitally on the isolated heart of the frog, and he established that synapses ensure relative independence of neurons while connecting them in an integral system. In his work on the sensitive innervation of internal organs and the morphology of interoceptors, Lavrentiev suggested a classification of such receptors and determined the origin and localization of the sensitive neurons supplying internal organs.

BIBLIOGRAPHY

I. Original Works. Lavrentiev's writings include "Innervatsionnye mekhanizmy (sinapsy), ikh morfologia i patologia" ("Innervation Mechanisms [Synapses], Their Morphology and Pathology"), in *Transactions of the 1st Histological Conference* (Moscow, 1934); "Gistofiziologia innervatsionnykh mekhanizmov (sinapsov)" ("The Histology of Innervation Mechanisms [Synapses]"), in *Fiziologichesky zhurnal SSSR*, **21**, nos. 5–6 (1936), 858–859; "Nekotorye voprosy teorii stroenia nervnoy tkani" ("Some Problems of the Theory of the Structure of Nervous Tissue"), in *Arkhiv biologicheskikh nauk*, **48**, nos. 1–2 (1937), 194–210; "Chuvstvitelnaya innervatsia vnutrennikh organov" ("The Sensitive Innervation of Internal Organs"), in *Zhurnal obshchei biologii*, **4**, no. 4 (1943), 232–249; "Morfologia antagonisticheskoi innervatsii v avtonomnoi nervnoi sisteme" ("The Morphology of Antagonistic Innervation in the Autonomic Nervous System"), in *Morfologia avtonomnoy nervnoy sistemy* ("The Morphology of the Autonomic Nervous System"), 2nd ed. (Moscow, 1946); and "Chuvstvitelnaia innervatsia vnutrennikh organov" ("The Sensitive Innervation of Internal Organs"), in *Morfologia chuvstvitelnoy innervatsii vnutrennikh organov* ("Morphology of the Sensitive Innervation of Internal Organs") (Moscow, 1948), 5–21.

II. Secondary Literature. See N. G. Feldman, *B. I. Lavrentiev* (Moscow, 1970); A. N. Mislavsky, "Pamyati B. I. Lavrentieva" ("In Memory of B. I. Lavrentiev"), in *Zhurnal obshchei biologii*, **5**, no. 4 (1944), 199–204; E. K. Plechkova, "Boris Innokentievich Lavrentiev," in *Lyudi russkoy nauki. Biologia* . . . ("Men of Russian Science. Biology . . ."; Moscow, 1963), 448–456; and A. A. Zavarzin, "Pamyati B. I. Lavrentieva," in *Morfologia avtonomnoy nervnoy sistemy* ("Morphology of the Autonomic Nervous System"), 2nd ed. (Moscow, 1946), 7–12.

L. J. Blacher

LAWES, JOHN BENNET (*b.* Rothamsted, near Harpenden, Hertfordshire, England, 28 December 1814; *d.* Rothamsted, 31 August 1900); and **GILBERT, JOSEPH HENRY** (*b.* Hull, England, 1 August 1817; *d.* Harpenden, England, 23 December 1901), *agricultural chemistry*.

Lawes was educated at Eton and Oxford; he left without taking a degree, although he had cultivated an amateur's taste for chemistry and pharmacy. In 1834 he inherited the manor and estate of Rothamsted; and two years later, becoming interested in the problems of a neighboring landowner, he began his agricultural experiments. Having seen that ground bones, or "mineral phosphates," were highly effective as manures on some fields but useless on the majority, Lawes was able to devise a method—treatment with acids—that made them universally effective. (Acids convert insoluble tricalcium phosphate into more soluble monocalcium salt, a process that acidic soils perform naturally.) Despite the opposition of his mother, who was bitterly opposed to "trade," he acquired a factory at Deptford Creek and in the summer of 1843 began the manufacture of "superphosphate," using the profits to finance further experiments at Rothamsted. In the same year he engaged Gilbert to assist him in this work, especially in the performance of chemical analyses; and a rough laboratory was improvised in a converted barn. Gilbert, although partly blinded by a gun accident when a boy, had studied chemistry at Glasgow University and University College, London, and, for a short time, under Liebig at Giessen. The collaboration of Lawes and Gilbert lasted for more than fifty years, and together they built Rothamsted into a world-famous institution. In character they were complementary: Lawes was the organizer, the man of affairs, impatient of detail, dealing in large ideas and general principles; Gilbert was a laboratory chemist, methodical and fussily accurate, a diligent attender of scientific meetings.

In 1840, Liebig published *Die organische Chemie in ihre Anwendung auf Agricultur und Physiologie*—translated into English in the same year by Lyon Playfair—which expounded his belief that manures were effective only by virtue of their mineral content and that nitrogenous and humus-forming manures were unnecessary and wasteful. These views were set out with such clarity and force, and Liebig's authority was so great, that they were widely believed, although they ran counter to most farming experience and were based on the slenderest experimental evidence. Lawes and Gilbert, in a long series of carefully planned field experiments, were able to show by 1851 that, as far as nitrogen was concerned, Liebig was in error, although they did not solve the puzzle of the nitrogen balance of leguminous plants. During this study they evolved the "chessboard" system of random plots for field trials on the fields of Broadbalk and Barnfield.

During the 1850's the financial basis of Rothamsted was in danger because of the widespread pirating of the superphosphate patent. Lawes spent much of this decade in protracted lawsuits; but his eventual victory enabled him to extend his experiments to long-term studies of grassland economy, animal feeding, and water balance. His work on the agricultural utilization of sewage led to Lawes being appointed to membership of a Royal Commission in 1857.

As must be inevitable in such lengthy projects a certain stagnation eventually became obvious. This was due partly to Gilbert's innate conservatism, which, as he grew older, hardened into intolerance of younger chemists and new ideas. Rather than collaborate with younger colleagues, he built up over the years a corps of village boys, each highly skilled in one operation of analytical chemistry. His iron rule over this little empire fitted in well with Lawes's paternalistic supervision of his manor. But Lawes began to feel that he had to make innovations, and in 1876 he appointed Robert Warington as his personal assistant. Gilbert took this as a deliberate affront, and personal relations were never quite easy at Rothamsted again; for lack of cooperation, Warington achieved little.

Lawes sold the superphosphate business in 1872, although he then had other interests in the manufacture of tartaric and citric acids. In 1889 he put Rothamsted under the control of the Lawes Agricultural Trust, with an endowment of £100,000, so that its work would not cease with his death (unlike the parallel enterprise of Boussingault in France); it continues to the present day. He was elected fellow of the Royal Society in 1854 and was made a baronet in 1882; Gilbert was elected fellow of the Royal Society in 1860 and was knighted in 1893.

BIBLIOGRAPHY

I. Original Works. The work of Lawes and Gilbert was published entirely in the form of papers, under both joint and separate authorship, mostly in the *Journal of the Royal Agricultural Society*. After their deaths a summary of their work, incorporating much previously unpublished material, was written by A. D. Hall: *The Book of the Rothamsted Experiments* (London, 1905). Their collected papers (not quite complete) were published as *The Rothamsted Memoirs on Agricultural Chemistry and Physiology*, 7 vols. (London, 1893–1899).

II. Secondary Literature. There are several obituary notices, including (Lawes) *Proceedings of the Royal Society*, **75** (1905), 228; *Journal of the Chemical Society*, **79** (1901), 890; *Nature*, **62** (1900), 467; *Journal of the Royal Agricultural Society*, **61** (1900), 511; (Gilbert) *Proceedings of the Royal Society*, **75** (1905), 237; *Journal of the Chemical Society*, **81** (1902), 625; *Nature*, **65** (1902), 205; *Journal of the Royal Agricultural Society*, **62** (1901), 347. E. Grey, *Rothamsted Experiment Station: Reminiscences, Tales, and Anecdotes 1872–1922* (Harpenden, 1922), is exactly described by its title. The best account of their work is in E. J. Russell, *A History of Agricultural Science in Great Britain, 1620–1954* (London, 1966). Both of these books, and many of the obituary notices, include portraits.

W. V. Farrar

LAWRENCE, ERNEST ORLANDO (*b.* Canton, South Dakota, 8 August 1901; *d.* Palo Alto, California, 27 August 1958), *physics.*

Lawrence was the elder son of Carl Gustav Lawrence, a Wisconsin-born educator whose father,

Ole Hundale Lavrens (also a teacher), emigrated from Telemark in Norway to the Wisconsin Territory in 1846. His mother, Gunda Jacobson Lawrence, was also a teacher and of Norwegian ancestry. Lawrence's father was a graduate of the University of Wisconsin who successively served as city, county, and state superintendent of public education in South Dakota, and in 1919 became president of a teacher's college.

As a boy, Lawrence attended public schools in Canton, South Dakota, and in Pierre, the state capital. His best friend was Merle A. Tuve, another teacher's son who also became a well-known physicist (he measured, with Gregory Breit, the height of the ionosphere in 1925). Lawrence finished high school at sixteen and attended Saint Olaf College, a small Lutheran college in Northfield, Minnesota, on a scholarship for a year. He transferred to the University of South Dakota, where he first became interested in physics under the guidance of the professor of electrical engineering, Lewis E. Akeley, who recognized his student's unusual aptitude for science. After graduating with high honors in 1922, Lawrence enrolled at the University of Minnesota for postgraduate studies, mainly at the urging of Merle Tuve, who had preceded him to Minneapolis.

After earning a master's degree at Minnesota under the direction of W. F. G. Swann with an experimental confirmation of the theory of induction in an ellipsoid rotating in a magnetic field, Lawrence followed Swann to the University of Chicago, where he came into contact with A. A. Michelson, H. A. Wilson, Leigh Page, Arthur Compton, and Niels Bohr. In 1924 Lawrence followed Swann to Yale, where he completed his doctoral dissertation, a study of the photoelectric effect in potassium vapor. He remained at Yale as a research fellow and then assistant professor, quickly gaining a reputation as a brilliant experimenter, mainly in photoelectricity.

In 1928, at twenty-seven, Lawrence was offered an associate professorship at the University of California in Berkeley and startled fellow physicists by accepting—exchanging a famous old university for a little-known state university in the Far West. Its subsequent world renown as a center of research was due to a considerable extent to the primacy of its physics faculty, which came to rank with those of Cambridge and (in an earlier day) Göttingen among the world's finest. That phase began with the tenure of Lawrence and his contemporaries, among whom were Samuel Allison, R. B. Brode, and J. R. Oppenheimer. Within two years Lawrence, at twenty-nine, had become a full professor.

With his first graduate students, Niels E. Edlefsen and M. Stanley Livingston, Lawrence developed his invention of a circular accelerator, later called the cyclotron, in the shape of a flat circular can cut in two

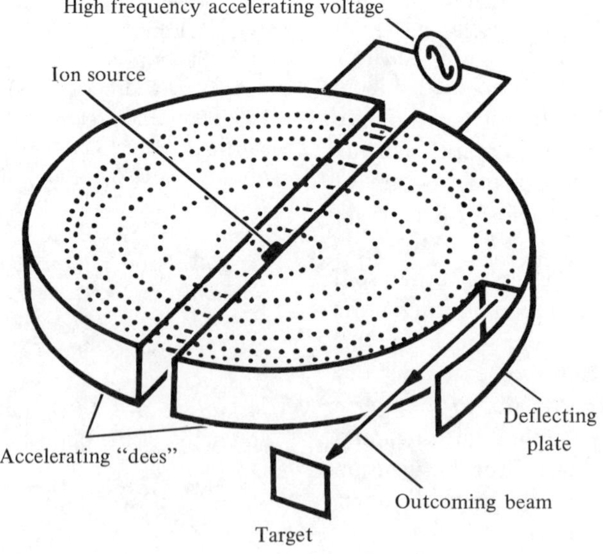

FIGURE 1. The circular accelerator.

along a diameter and placed inside a vacuum chamber. A high-frequency oscillator is connected between the two D-shaped halves, and charged particles are introduced near the center. The particles are constrained to travel in a circular path by a magnetic field along the axis of the can. With proper synchronization, the oscillating field serves to impart successive accelerations to each particle as it repeatedly crosses the boundary between the two halves, sending it on an ever-widening path with increasing velocity as it spirals outward. When it approaches the wall, it can be deflected through an opening toward a target, which it hits with a high velocity, producing nuclear disintegrations. Lawrence conceived the idea after seeing the illustrations for an article by the Norwegian-born engineer Rolf Wideröe elaborating a linear, rather than a circular, acceleration scheme proposed by the Swedish physicist Gustaf A. Ising.

The cyclotron was the first in a family of circular particle accelerators that made high acceleration energies available with relatively small instruments. Once the principle was proved, progress was swift. The unavailability of a sufficiently strong magnet was an obstacle that Lawrence characteristically overcame by persuading the Federal Telegraph Co. to donate an eighty-ton iron core originally intended for a radio arc generator but no longer needed. With this magnet and a cyclotron chamber 27.5 inches in diameter, Lawrence and his associates were able to produce energies of millions of electron volts. This achievement ushered in the era of high-energy physics and made possible the disintegration of atomic nuclei, artificial isotopes, and the discovery of new elements. Biomedical applications include radioactive tracers and uses in cancer therapy. One of the first results of major importance, the disintegration of lithium, was achieved at Berkeley as early as 1932, that *annus mirabilis* of modern physics which also saw the earlier disintegration of lithium by John Cockcroft and E. T. S. Walton, and the discovery of deuterium by H. C. Urey, of the neutron by James Chadwick, and of the positron by C. D. Anderson, each later honored by a Nobel Prize. In addition to lithium, many heavier nuclei were disintegrated in the Berkeley cyclotron, which was unique in that respect—no other machine could then do it. These disintegrations proved that nearly every nuclear reaction takes place if there is sufficient energy for it, a result of great importance in the development of nuclear physics. They also permitted accurate determination of the binding energy of various nuclei and, by comparison of the measured reaction energies (the masses were measured in a mass spectrograph), a complete verification of Einstein's law relating energy and mass.

World renown followed quickly. Lawrence was invited to his first Solvay Congress at thirty-two, was elected to the National Academy of Sciences at thirty-three, became director of his Radiation Laboratory at thirty-five, and in 1939 received the Nobel Prize, the first American associated with a state university to do so. (He had already received, in 1937, the Comstock Prize of the National Academy of Sciences and the Royal Society's Hughes Medal.) He was named director of what became known as the Radiation Laboratory ("Rad Lab") at Berkeley.

Lawrence was one of the American physicists who helped set up another "Radiation Laboratory" at the Massachusetts Institute of Technology in 1940; but that name disguised the laboratory's real function, the development of centimetric radar. He actively recruited young scientists for the laboratory but did not go himself, for with America's entry into World War II, an even larger project appeared on the horizon: the "Manhattan Project" to develop a nuclear or "A" (atomic) fission bomb.

Fearful that German scientists might be the first to develop such a weapon, other scientists, led by refugees from Nazism, proposed such a development to the U.S. government. The Berkeley Rad Lab, and Lawrence and Oppenheimer personally, were destined to play major roles in these endeavors. The Rad Lab helped devise a method of obtaining fissionable materials. A thirty-seven-inch cyclotron was converted into a mass spectograph for the purpose. The electromagnetic separation method devised at Berkeley was later used in a large laboratory (known as Y-12) at Oak Ridge, Tennessee, which provided the separated U^{235} for the Hiroshima atomic bomb. Oppenheimer became director of the laboratory at Los Alamos, New Mexico, where the first bombs were produced.

In 1945, with World War II ended by fission bombs dropped on Hiroshima and Nagasaki, the Rad Lab returned to predominantly scientific pursuits, including the construction of a 184-inch cyclotron (actually a new type of accelerator, the synchrocyclotron, based on a principle formulated by E. M. McMillan). The Los Alamos Scientific Laboratory, still managed by the University of California, continued in weapons research. Lawrence was actively involved in the subsequent controversy about the advisability of developing another, more destructive weapon, the thermonuclear-fusion or "H" (hydrogen) bomb. Its advocates (among whom the Hungarian-born physicist Edward Teller was the most prominent) prevailed. Oppenheimer was against it, and his long friendship with Lawrence ended over their deep disagreements on the desirable defense posture of the United States. The final break came when Oppenheimer, in a *cause*

LAWRENCE

célèbre, lost the security clearance he needed as a government consultant.

In addition to the laboratory at Los Alamos, a second laboratory for research on nuclear weapons was started at Livermore, California, under the sponsorship of Lawrence and Edward Teller. This laboratory was at first a branch of the Rad Lab. Research at the Berkeley site was limited to basic science. During these years, larger and more efficient accelerators were designed and constructed there, and for a long time the Rad Lab had the highest energies and a near-monopoly of the type of results depending on them. Machines that could accelerate particles to energies of billions of electron volts (BeV—hence the name of one of them, the "bevatron") were constructed. It was by means of the bevatron that the antiproton was discovered by Emilio Segrè and Owen Chamberlain, and the properties of mesons were explored in detail, which led to a general understanding of strongly interacting particles. Lawrence also created the style of "big science"; large-scale physics started in Berkeley and set the pattern around the world in such later organizations as the Brookhaven National Laboratory in New York, the Centre Européen Pour la Recherche Nucléaire (CERN) in Switzerland, and the National Accelerator Laboratory in Illinois. Lawrence remained personally involved throughout and suffered no lessening of his inventive genius (among other devices, he invented a novel type of color television tube at this time); but increasing demands on him as a government and industrial consultant often took him away from Berkeley. Among these endeavors was his participation, at the request of President Eisenhower, in the Conference of Experts to Study the Possibility of Detecting Violations of a Possible Agreement on Suspension of Nuclear Tests in Geneva in the summer of 1958. It proved to be his last contribution. Exhausted and plagued by recurrent ulcerative colitis, he was flown home for an operation, which he did not survive.

Lawrence married Mary (Molly) Kimberly Blumer, daughter of the dean of Yale's medical faculty, in 1932; he had first met her when she was a schoolgirl of sixteen. They had two sons and four daughters. He was very close to his younger brother John, a physician who was associated with him professionally as director of a biophysics laboratory founded at Berkeley for the purpose of exploiting biomedical applications of nuclear physics. Another close associate was E. M. McMillan, who succeeded to the directorship of the Rad Lab. Several other associates achieved great prominence, including Luis Alvarez, Owen Chamberlain, Glenn Seaborg, and Emilio Segrè, each of whom later received a Nobel Prize, as did McMillan.

Lawrence's name is commemorated in the two Rad Lab sites, now known as the Lawrence Berkeley Laboratory and the Lawrence Livermore Laboratory; and in the Lawrence Hall of Science, a Berkeley museum and research center devoted to the improvement of science education. Annual Lawrence Awards are given to young scientists honored by the U.S. Atomic Energy Commission (AEC), and the transuranium element 103, discovered at Berkeley, was named Lawrencium. Lawrence was a member of the National Academy of Sciences, a foreign member of the Swedish Academy, an honorary member of the Soviet Academy, and the recipient of countless honorary degrees and other awards from institutions all over the world.

BIBLIOGRAPHY

I. Original Works. A complete list of Lawrence's publications follows the article by Luis W. Alvarez in *Biographical Memoirs. National Academy of Sciences*, **41** (1970), 251–294. Lawrence's papers are in the Bancroft Library on the Berkeley campus of the University of California.

II. Secondary Literature. A biography commissioned by the University of California and based on hundreds of interviews and a large store of written materials is Herbert Childs, *An American Genius: The Life of Ernest Lawrence* (New York, 1968). Lawrence's role in the beginnings of the atomic age is described in several book-length studies, among which vol. I of the AEC-sponsored history by R. G. Hewlett and O. E. Anderson Jr., *The New World: 1939–1946* (New York, 1962), is the most detailed. There are also journalistic accounts, not uniformly flattering to Lawrence, such as those by Robert Jungk, *Heller als tausend Sonnen* (Stuttgart, 1956), trans. by James Cleugh as *Brighter Than a Thousand Suns* (New York, 1960); and Nuel Pharr Davis, *Lawrence and Oppenheimer* (New York, 1968).

For an authoritative account of the development of the cyclotron, see M. Stanley Livingston, "History of the Cyclotron (Part I)," in *Physics Today*, **12** (Oct. 1959), 18–23; and Edwin M. McMillan (Part II), *ibid.*, 24–34.

Charles Süsskind

LAWRENCE, WILLIAM (*b.* Cirencester, Gloucestershire, England, 16 July 1783; *d.* London, England, 5 July 1867), *anatomy, physiology, surgery.*

The son of William Lawrence, chief surgeon of Cirencester, and Judith Wood, Sir William Lawrence was a distinguished and gifted surgeon who was active at St. Bartholomew's Hospital in London for sixty-five years. His honors were numerous. In addition to serving as president of the Royal College of Surgeons for two terms (1846, 1855), he was also president of the Medical and Chirurgical Society of London (1831), surgeon extraordinary, and then sergeant-surgeon to Queen Victoria (1858). On 30 April 1867 a baronetcy was conferred upon him.

Lawrence was educated in a private school in his home town until 1799, when he was apprenticed to Dr. John Abernethy, assistant surgeon at St. Bartholomew's Hospital and one of the most popular medical teachers of his day. Only two years later, Lawrence was appointed Abernethy's demonstrator in anatomy, a post which he held for twelve years. Despite his youth, he soon acquired a reputation as an excellent teacher of anatomy and earned his diploma from the Royal College of Surgeons in 1805. He was elected assistant surgeon at St. Bartholomew's and fellow of the Royal Society in 1813. From 1814 to 1826 he was surgeon to the London Infirmary for Curing Diseases of the Eye (later the London Ophthalmic Infirmary and then the Royal Ophthalmic Hospital). He was appointed surgeon to the Royal Hospitals of Bridewell and Bethlehem in 1815 and was surgeon to St. Bartholomew's Hospital from 1824 to 1865.

After his nomination as professor of anatomy and surgery in 1815 by the Royal College of Surgeons, Lawrence delivered lectures to that body (1816–1818) on comparative anatomy, physiology, and the natural history of man. In 1829 he succeeded Abernethy at St. Bartholomew's as lecturer on surgery, a position he held until 1862. Although he had opposed the administration of the College of Surgeons, Lawrence was elected over two senior members to their council (1828) and later to the position of examiner (1840–1867), and was Hunterian orator in 1834 and 1846. His extensive private medical practice and lecturing at a private school of medicine on Aldersgate Street until 1829 were additional requirements on his time. He published many articles, particularly in *Lancet*, *Medical Gazette*, and *Transactions of the Medical and Chirurgical Society*. His *Lectures on Surgery Delivered in St. Bartholomew's Hospital* was published in 1863; unfortunately, by then his techniques had been superseded and the lectures obviously were out of date.

The earliest recognition of Lawrence's abilities occurred in 1806 when the College of Surgeons awarded him the Jacksonian Prize for his essay on ruptures. Published the following year as *A Treatise on Hernia*, the work became a standard reference and appeared in French (1818) and Italian (1820) translations. Dedicated to Abernethy, it is an excellent example of lucid exposition and thorough mastery of the subject.

Lawrence was one of the most distinguished English eye surgeons of his day. His private practice and his experiences as surgeon to the London Ophthalmic Infirmary laid the foundations for his *Treatise on the Venereal Diseases of the Eye* (1830). His lengthy *Treatise on the Diseases of the Eye* (1833) is an expanded version of his lectures on anatomy, physiology, and diseases of the eye delivered before the London Ophthalmic Infirmary and displays his outstanding abilities as an observer of diseases and as a practicing pathologist.

Although the Greeks had observed diseases of the eye and Celsus had summarized all examples known to his time, ophthalmology was largely in the hands of quacks and charlatans from the sixteenth to the eighteenth centuries. With the establishment of the Vienna school of ophthalmology in the latter half of the eighteenth century, the discipline entered its modern phase. In 1805 the London Ophthalmic Infirmary was established, and in 1810 students were admitted for observation and instruction. Lawrence was, therefore, one of its pioneering surgeons and produced one of the earliest reliable treatises in English on ophthalmology, "the equal of any work on the subject that appeared for many years."

Of a much more controversial nature were Lawrence's lectures presented before the Royal College of Surgeons from 1816 through 1818. In 1814, immediately preceding his presentation, Abernethy had delivered a series of anatomical lectures before that august body. The first two lectures, discussing his interpretation of John Hunter's views on life, were published in the same year. Reviewers were critical of Abernethy's somewhat obscure efforts, and Lawrence himself used his appointment to the college as a platform to criticize severely Hunter's vitalism as expounded by Abernethy. Dedicated to Blumenbach, whose *Short History of Comparative Anatomy* he had translated in 1807, Lawrence's *Lectures on Physiology* drew upon the works of various Continental authors, especially French, to expound his belief that the laws of physics and chemistry apply to organic nature. Nevertheless, physiological phenomena involve exceedingly complex systems: "We cannot expect to discover the true relations of things, until we rise high enough to survey the whole field of science, to observe the connexions of the various parts and their mutual influence." To compare the vital principle to magnetism, electricity, or galvanism, however, was unacceptable; extreme reductionism overlooked the complexities of the situation. Lawrence preferred a balanced view of life, encompassing as much scientific data as possible. Nevertheless, his belief that all mental activity was a function of the brain—not of an immortal, immaterial soul—cut to the heart of the theological establishment, which considered his "materialism" tantamount to atheism because he denied that the human soul existed beyond the body.

Lawrence's lectures, especially those published in 1819, generated severe criticism from Abernethy, theologians, physiologists, and his senior colleagues at the college. Although insisting he would not be silenced, he was compelled to withdraw his book within a month of its release by the threat of losing his position as surgeon at the Bridewell and Bethlehem hospitals. One contemporary observed, however, that Lawrence was not sincerely anxious for the matter to be dropped, for the book was still available from the publisher, "though at an advanced price." Indeed, the book went through ten subsequent editions (essentially printings), one of which was a pirated edition that James Smith brought out in 1822. Lawrence's attempt to prevent publication failed when Lord Eldon ruled in Chancery Court that parts of the work seemed to be "directed against the truth of the Scripture" and therefore could not be copyrighted. That various editions were subsequently printed suggests that Lawrence secretly urged their publication or was disinclined to go to court over the matter. It is clear, however, that his earlier suppression of his book was unrelated to the Chancery Court case.

While Lawrence has been described as an English medical evolutionist—another forerunner of Darwin—he maintained that man is "separated by a broad and clearly defined interval from all other animals," even the great apes. Although the evolutionist A. R. Wallace alluded to him with approbation in 1845, Lawrence had referred only to human variation which could explain the existence of different races of man: "The diversities of physical and moral endowment which characterize the various races of man, must be analogous in their nature, causes, and origin, to those which are observed in the rest of animal creation; and must therefore be explained on the same principles" (*Lectures on Physiology* [1822], p. 233). The different races of man had arisen through the ordinary process of variation observed in animals. Lawrence was no more an evolutionist, however, than many other English scientists like Blyth, Prichard, or Lyell, who also made highly suggestive remarks but failed to conclude that species evolve as well as vary. Lawrence's perceptive observations on inheritance do not make him a legitimate precursor of Darwin and Wallace.

Although an excellent lecturer and surgeon, and an outstanding member of the English medical community, Lawrence was not a creative thinker. In many ways his ideas were conservative. Had his *Lectures on Physiology* (1819) been published in France, instead of England, they would not have created the stir they did. We may conclude that his reputation did not rest on a single brilliant discovery, but rather on a lifetime of distinguished work in many areas of medicine and biology.

BIBLIOGRAPHY

I. Original Works. In addition to works cited in the text, Lawrence published *An Introduction to Comparative Anatomy and Physiology; Being the Two Introductory Lectures Delivered at the Royal College of Surgeons on the 21st and 25th of March, 1816* (London, 1816); *Lectures on Physiology, Zoology, and the Natural History of Man, Delivered at the Royal College of Surgeons* (London, 1819); and *The Hunterian Oration Delivered at the Royal College of Surgeons* (London, 1834); also his later lecture (London, 1846).

II. Secondary Literature. On Lawrence's life, see W. R. LeFanu, "Past Presidents: Sir William Lawrence, Bart.," in *Annals of the Royal College of Surgeons of England*, **25** (1959), 201–202; [W. S. Savory], "Sir William Lawrence, Bart.," in *St. Bartholomew's Hospital Reports*, **4** (1868), 1–18; and Norman Moore, *The History of St. Bartholomew's Hospital*, II (London, 1918), pp. 660–663.

For his work on ophthalmology see Burton Chance, "Sir William Lawrence in Relation to Medical Education with Special Reference to Ophthalmology in the Early Part of the Nineteenth Century," in *Annals of Medical History*, **8** (1926), 270–279.

On physiology see June Goodfield-Toulmin, "Blasphemy and Biology," in *Rockefeller University Review*, **4** (1966), 9–18; and "Some Aspects of English Physiology: 1780–1840," in *Journal of the History of Biology*, **2** (1969), 283–320; and Owsei Temkin, "Basic Science, Medicine, and the Romantic Era," in *Bulletin of the History of Medicine*, **37** (1963), 97–129.

For anthropology see C. D. Darlington, *Darwin's Place in History* (Oxford, 1959), pp. 19–24, 100–101; Peter G. Mudford, "William Lawrence and *The Natural History of Man*," in *Journal of the History of Ideas*, **29** (1968), 430–436; and Kentwood D. Wells, "Sir William Lawrence (1783–1867), A Study of Pre-Darwinian Ideas on Heredity and Variation," in *Journal of the History of Biology*, **4** (1971), 319–361.

H. Lewis McKinney

LAWSON, ANDREW COWPER (*b.* Anstruther, Scotland, 25 July 1861; *d.* San Leandro, California, 16 June 1952), *geology*.

The eldest of ten children of William Lawson and Jessie Kerr, Andrew was educated in public schools in Hamilton, Ontario. He received a bachelor's degree in 1883 from the University of Toronto, the master's degree in 1885, and the doctorate from the Johns Hopkins University in 1888. In 1890, after seven years with the Geological Survey of Canada, he joined the geology department of the University of California at Berkeley, where he was active for sixty years.

Lawson was raised a devout Calvinist. Through his studies he became an active opponent of dogma and gave up his religious beliefs, but he never effected a complete emotional adjustment to the change. In 1889 Lawson married Ludovika von Jansch, of Brünn, Moravia, who bore him four sons. She died in 1929. In 1931 he married Isabel R. Collins of Ottawa. He was survived by her and their one son and by two sons by his first wife. A man of powerful physique and dynamic drive, he continued working effectively well into his ninetieth year. Lawson served many learned and cultural societies with distinction and was, in 1926, president of the Geological Society of America. His achievements brought honors and recognition.

Lawson's studies of isostasy over the three decades 1920–1950 have drawn special attention to its importance in diastrophism. In a number of general theoretical papers he developed the logical consequences of isostatic adjustment as an important factor in orogenesis, indeed in some instances as the determinative agent in the elevation of lofty mountains and the depression of deep troughs and basins. Widely distributed field evidence in support of his findings is presented in these writings and others following, eighteen papers in all, showing the importance of isostasy in the development of such features as the Sierra Nevada, the Great Valley of California, the Mississippi delta, and the Cordilleran shield.

From time to time for some forty years Lawson turned his attention to the region northwest of Lake Superior, and in thirty-five papers he presented a great body of new information and revised the correlation of the pre-Cambrian rocks over a large part of North America. In his early studies about Lake of the Woods and Rainy Lake he found that the Laurentian, previously regarded as the oldest known formation and comprising metamorphosed sediments, was a granite intrusive into metamorphosed volcanic and sedimentary rocks which he called Keewatin. Underlying these he found another sedimentary series which he named Coutchiching and recognized as the oldest rocks in the region. His report of 1887 on these findings was published only after much controversy. Returning to this area in 1911, Lawson recognized two periods of batholithic invasion, the oldest being termed Laurentian and the younger Algoman, each followed by a great period of peneplanation. The resulting surfaces, termed the Laurentian peneplain and the Eparchean peneplain respectively, afforded references for the correlation of the invaded formations, and for the subsequently deposited sedimentary beds, over vast areas.

In Berkeley, Lawson initiated a systematic investigation of the geology of the western coastal area and established the *Bulletin of the Department of Geology*, for publishing papers relating to this work. Along with this undertaking, assisted by various students, Lawson carried out the preparation of the San Francisco Folio, which included five quadrangles in the San Francisco Bay region. This work was published by the U. S. Geological Survey in 1914. At his death, the *Bulletin* comprised twenty-eight volumes of from 400 to 600 pages each.

As chairman of the Department of Geology at the university for more than twenty years, Lawson was diligent in promoting its growth while continuing his own research work. For three years he was, at the same time, dean of the College of Mining, where he established instruction and research in petroleum engineering, the first program of the kind in any American university.

After the California earthquake of 18 April 1906, Lawson supervised the preparation of a report by the State Earthquake Investigation Commission. It included two volumes and an atlas comprising 500 pages and for many years it was the most informative treatise published on any earthquake; and it led to many new developments, one of the more important being the initiation of the elastic rebound theory of the dynamics of earthquakes by H. F. Reid. Lawson's continued activity in this field brought about the organization of the Seismological Society of America; the founding of the *Bulletin* of this society; the construction in Berkeley, in 1911, of the most complete seismograph station in America; the formulation of a plan, later carried out with the help of the Carnegie Institution of Washington, for an extensive network of seismograph stations in California; and the establishment at the University of California, in 1927, of the first professorship in seismology in the United States.

As a mining consultant Lawson not only visited most of the mining camps in North America, but worked in many from time to time over a period of years. At the same time, he was also a consultant in a number of construction engineering projects, including the Golden Gate bridge.

His sharp wit, dynamic personality, and sound thinking marked Lawson as a man feared in debate and revered in counsel. As an architect and skilled workman he put up several fireproof and earthquake-resistant buildings in the vicinity of his home. He collected paintings and wrote fifty-five poems expressing inner thoughts of considerable variety, from the light and cheerful to the profound and philosophical.

BIBLIOGRAPHY

A complete bibliography of all Lawson's 114 works is included in the biography by Francis E. Vaughan, *Andrew C. Lawson, Scientist, Teacher, Philosopher* (Glendale, Calif. 1970).

The following are representative: "The Archean geology of the region north-west of Lake Superior," in *Fourth International Geological Congress, London, 1888. Compte Rendu* (1891), 130–152; "The Cordilleran Mesozoic revolution," in *Journal of Geology*, **1** (1893), 579–586; "The Eparchean interval; a criticism of the use of the term Algonkian," in *University of California Publications, Department of Geology, Bulletin*, **3** (1902), 51–62; "The California earthquake of April 18, 1906," written with others, the report of the State Earthquake Investigation Commission, which is *Carnegie Institution of Washington Publication No. 87;* "The Archean geology of Rainy Lake restudied," which is *Geological Survey of Canada, Memoir 40;* "A standard scale for the pre-Cambrian rocks of North America," in *Twelfth International Geological Congress, Toronto, 1913. Compte Rendu* (1914), 349–370; "Ore deposition in and near intrusive rocks by meteoric waters," in *University of California Publications, Department of Geology, Bulletin*, **8** (1914), 219–242; "The spirit of science," in *University of California Chronicle*, **22** (1920), 143–161; "The geological implications of the doctrine of isostasy," in *National Research Council, Bulletin*, **8**, pt. 4, no. 46 (1924); "The classification and correlation of the pre-Cambrian rocks," in *University of California Publications, Department of Geological Sciences, Bulletin*, **19**, no. 11 (1930), 275–293.

FRANCIS E. VAUGHAN

LAX, GASPAR (*b.* Sariñena, Aragon, Spain, 1487; *d.* Zaragoza, Spain, 23 February 1560), *mathematics, logic, natural philosophy*.

After studying the arts and theology in Zaragoza, Lax taught in Paris in 1507 and 1508 at the Collège de Calvi, also known as the "Little Sorbonne." He transferred to the Collège de Montaigu, where he continued to teach and to study under the Scottish master John Maior; in 1517 he returned to the Collège de Calvi. Lax had an agile mind and an excellent memory, but he became so engrossed with the logical subtleties of the nominalist school that he was soon known as "the Prince of the Parisian *sophistae.*" A prolific writer, he turned out a series of ponderous tomes on *exponibilia, insolubilia*, and other topics of terminist logic. His student, the humanist Juan Luis Vives, later reported that he heard Lax and his colleague, John Dullaert of Ghent, moan over the years they had spent on such trivialities. Lax taught at Paris until 1523; possibly he returned to Spain in 1524, along with his countryman Juan de Celaya, when foreigners were asked to leave the university. In 1525 he taught mathematics and philosophy at the *studium generale* of Zaragoza. He remained there until his death, at which time he was vice-chancellor and rector.

Lax achieved greater fame as a mathematician than as a logician or as a philosopher. He published his *Arithmetica speculativa* and *Proportiones* at Paris in 1515 (Villoslada adds to these a *De propositionibus arithmeticis*, same place and date). The first is described by D. E. Smith as "a very prolix treatment of theoretical arithmetic, based on Boethius and his medieval successors" (p. 121). The *Proportiones* is a more compact and formalistic treatment of ratios, with citations of Euclid, Jordanus, and Campanus; unlike most sixteenth-century treatises on ratios, however, it does not deal with the velocities of motion in the Mertonian and Parisian traditions. Possibly Lax treated this interesting topic in his *Quaestiones phisicales,* printed at Zaragoza, 1527, according to early lists. (I have searched extensively for this, with no success.)

BIBLIOGRAPHY

None of Lax's works is translated from the Latin, and copies of the originals are rare; positive microfilms of the *Arithmetica speculativa* and various logical works are obtainable from the Vatican Library.

A brief general treatment of Lax is in R. G. Villoslada, S.J., *La universidad de Paris durante los estudios de Francisco de Vitoria (1507–1522)*, Analecta Gregoriana XIV (Rome, 1938), pp. 404–407 and *passim.* See also Hubert Élie, "Quelques maîtres de l'université de Paris vers l'an 1500," in *Archives d'histoire doctrinale et littéraire du moyen âge*, **18** (1950–1951), 193–243, esp. 214–216.

Lax's logical and philosophical thought is treated by Marcial Solana, *Historia de la filosofía española, Época del Renacimiento (siglo XVI)* (Madrid, 1941), III, 19–33. His arithmetic is described in D. E. Smith, *Rara arithmetica* (Boston–London, 1908), p. 121.

WILLIAM A. WALLACE, O.P.

LAZAREV

LAZAREV, MIKHAIL PETROVICH (*b.* Vladimir, Russia, 14 November 1788; *d.* Vienna, Austria, 23 April 1851), *navigation, oceanography.*

Lazarev was trained in both theoretical and practical navigation. From 1799 to 1803 he studied with P. Y. Gamaleya and G. A. Sarychev at the Naval Academy in St. Petersburg, then volunteered as a seaman in the British Navy. From 1808 he served on various military ships in the Baltic Sea. In 1813 he made the first of his three voyages around the world, commanding the *Suvorov*. This expedition lasted until 1813, and among its discoveries was a group of coral islands in the South Pacific that are now called the Suvorov atoll. In 1819 Lazarev was named commander of the sloop *Mirny*, the second ship of Bellinsgauzen's two-year expedition to the Antarctic. (For a complete account of this venture, which was organized to collect astronomical, hydrographical, and ethnographical data, and to map new lands and report on specifically polar phenomena, see the article on F. F. Bellinsgauzen, I, 594–595.) From 1822 to 1825 Lazarev was commander of the frigate *Kreyser*; he was promoted to admiral in 1843.

On all three of his voyages, Lazarev and his assistants were concerned with correcting navigational maps; they further determined precisely the locations of many islands discovered by themselves or by earlier explorers. Lazarev also made depth soundings of coastal waters (as, for example, those of Novo Arkhangel'sk and San Francisco Bay), observed local relief features, and determined the height of mountains. His measurements and observations were very accurate; on the voyage of the *Mirny*, he determined the height of Mt. Egmont in New Zealand to be 8,232 feet (it is actually 8,260 feet), while the meteorological data that he recorded on the *Kreyser* were of such high quality that they were published by the Naval Ministry in 1882.

Lazarev was an honorary member of the Russian Geographical Society and of Kazan University. One of the first Soviet Antarctic stations was named for him, as were a chain of Antarctic mountains, a great Antarctic trough, a group of mountains in the Tuamotu Archipelago, and several capes and harbors.

BIBLIOGRAPHY

I. ORIGINAL WORKS. Lazarev wrote *Meteorologicheskie nablyudenia, proizvodivshiesya vo vremya krugosvetnogo plavania fregata "Kreyser" pod komandoy kapitana II ranga Lazareva 1-go v 1822, 1823, 1824, 1825 godakh* ("Meteorological Observations, Made During the Voyage Around the World of the Frigate *Kreyser* Under the Command of Captain, Second Rank, Lazarev in 1822, 1823, 1824, 1825"; St. Petersburg, 1882); and *Dokumenty* . . ., 3 vols. (Moscow, 1952–1961).

II. SECONDARY LITERATURE. On Lazarev and his work see I, 595, and M. I. Belov, "Shestaya chast sveta otkryta russkimi moryakami (novye materialy . . .)" ("The Sixth Part of the World Discovered by Russian Sailors [New Materials . . .]"), in *Izvestiya Vsesoyuznogo geograficheskogo obshchestva*, **94**, no. 2 (1962), 105–114; V. A. Esakov *et al.*, *Russkie okeanicheskie i morskie issledovania v XIX-nach. XX v.* ("Russian Oceanic and Sea Investigations in the Nineteenth and Early Twentieth Centuries"; Moscow, 1964), pp. 53–69; M. I. Firsov, *Pervootkryvatel golubogo kontinenta* ("Discoverer of the Blue Continent"; Vladimir, 1963); K. I. Nikulchenkov, *Admiral Lazarev. 1788–1851* (Moscow, 1956); [P. M. Novoselsky], *Yuzhny polyus. Iz zapisok byvshego morskogo ofitsera* ("The South Pole. From the Notes of a Former Naval Officer"; St. Petersburg, 1853); B. Ostrovsky, *Lazarev* (Moscow, 1966), with bibliography pp. 174–175; A. V. Sokolov and E. G. Kushnarev, *Tri krugosvetnykh plavania M. P. Lazareva* ("M. P. Lazarev's Three Voyages Around the World"; Moscow, 1951); and N. N. Zubov, *Otechestvennye moreplavateli—issledovateli morey i okeanov* ("National Navigators and Investigators of the Seas and Oceans": Moscow, 1954), pp. 168–179; and "Mikhail Petrovich Lazarev," in *Otechestvennye fiziko-geografy i puteshestvenniki* ("National Physical Geographers and Travelers"; Moscow, 1959), pp. 193–197.

V. A. ESAKOV

LAZAREV, PETR PETROVICH (*b.* Moscow, Russia, 14 April 1878; *d.* Alma-Ata, U.S.S.R., 24 April 1942), *physics, biophysics, geophysics.*

Lazarev's father was a civil engineer. Upon graduation from the Gymnasium in Moscow, Lazarev entered the Medical Faculty of Moscow University, from which he graduated with distinction in 1901. In 1902 he passed the examinations for the academic degree of doctor of medicine and became an assistant in the ear clinic at Moscow University. Lazarev became deeply interested in the physical and mathematical sciences while he was still a student, and in 1903 he passed the examination for the entire course of study in the Faculty of Physics and Mathematics without ever having attended classes there.

While working in the ear clinic, Lazarev completed his first two research papers on the activity of the sensory organs. In the first paper he demonstrated experimentally that the phases of overtones do not influence the reception of sound. This helped to serve as an argument in favor of Helmholtz' resonance theory of hearing. In the second paper Lazarev demonstrated a distinct connection between auditory and visual sensations: the audibility of a sound is intensified by the simultaneous stimulation of the eyes by light.

In 1903–1905 Lazarev attended the weekly colloquiums of P. N. Lebedev and conducted several unofficial physical experiments in his laboratory. This work served to initiate a close friendship and scientific collaboration that lasted until Lebedev's death in 1912. In 1907 Lazarev passed his master's examinations and became an assistant professor at the Faculty of Physics and Mathematics of Moscow University. In 1911 he defended his master's dissertation on the temperature discontinuity at the gas-solid boundary. This topic, which arose from a research project conducted by Lebedev on the sensitivity of vacuum thermoelements, possessed great importance as a new experimental confirmation of the kinetic theory of gases. From that time on, Lazarev became Lebedev's chief assistant and the independent head of a group of researchers concerned with molecular and chemical physics.

In 1912 Lazarev defended his doctoral dissertation, "Vytsvetanie krasok i pigmentov v vidimom svete" ("The Fading of Colors and Pigments in Visible Light"). In this work Lazarev first established the connection between the quantity of matter decomposed under the influence of light and the light energy absorbed, as well as the dependence of the speed of a reaction upon the length of the light wave and the influence of oxygen on fading. This work, which is now acknowledged as one of the fundamental investigations of prequantum photochemistry, led to a series of investigations both abroad and in Russia.

In 1911 a significant number of professors and teachers left Moscow University as a protest against the reactionary actions of the minister of education, L. A. Kasso. Among them were Lazarev, Lebedev, and the majority of their colleagues. A short time later Lebedev's laboratory work was transferred to a small private institution, Shanyavsky City University. After Lebedev's death in April 1912, the supervision of his laboratory was entrusted to Lazarev. During the following years he began to concentrate his scientific interests on biophysics of the sense organs and the theory of nerve excitation. Guided by several experimental works of J. Loeb, Lazarev established his ion theory of nerve excitation. As a basis for his experimental law of excitation, Lazarev accepted Loeb's formula $C_1/(C_2 + b) = \text{constant}$ (where C_1 is the concentration of ions which give rise to the excitation of nerves and muscles, C_2 is the concentration of ions which depress this excitation, and b is a constant). In his theory Lazarev proceeded from the proposition that the coagulation of colloidal albuminous particles initiates the excitation process. The stability of the system is determined by the sign of the charge of the ions present in the solution. Lazarev produced many experimental research papers and, on the basis of extensive experimental material obtained by his collaborators, proposed the basic laws of ion theory for the excitation of living tissue. This theory has in significant measure retained its importance in modern biophysics.

Lazarev placed Loeb's formula at the base not only of nerve and muscle excitation theory but of peripheral vision theory as well. Lazarev was one of the first to raise the question of the photochemical interpretation of the phenomena of eye adaptation, and he produced a series of works in support of this conception. In proposing that ions are the products of the photochemical decomposition of visual purple, Lazarev constructed the mathematical theory of the laws of peripheral vision under various conditions. He also formulated on these principles a theory of hearing and taste. Lazarev occupied himself with these questions for the rest of his life.

Lazarev's activity was not limited to scientific research. From the beginning of World War I he participated in organizing medical aid for the front. He organized the production of medical thermometers, created the central diagnostic X-ray clinic for Moscow hospitals and a mobile X-ray section for the hospitals of the Moscow guberniya, and personally concerned himself with the training of physicians in roentgenology.

In 1917, shortly before the February Revolution, Lazarev was elected a full member of the Russian Academy of Sciences. He supported the October Revolution, and shortly afterward organized the first special scientific research institute for physics and biophysics in Russia. During the Civil War, Lazarev organized a laboratory in which he worked out the theoretical and practical problems of light and sound concealment and camouflage. In 1918 Lazarev, commissioned by the Soviet government, became the head of research into the Kursk Magnetic Anomaly. Before the Revolution, geomagnetic investigations in the region of the anomaly were conducted by E. J. Leyst. His investigations led him to suggest that enormous deposits of iron ore were to be found there, but geological prospecting did not support this view. The geomagnetic maps compiled by Leyst were stolen after his death.

Lazarev began all the work anew and established widespread geophysical and geological prospecting surveys which, in 1919, revealed the presence of enormous reserves of iron ore near the surface. He personally played a large part in implementing the geophysical calculations and initiated new investigations and geophysical methods of prospecting. He devoted much attention to scientific literature and was the founder of the journal *Uspekhi fizicheskikh nauk*.

BIBLIOGRAPHY

Many of Lazarev's more than 500 published works are in *Izbrannye sochinenia* ("Selected Works"), vols. II and III (Moscow–Leningrad, 1950)—vol. I has not yet been published.

On Lazarev or his work, see the following (listed chronologically): E. V. Shpolsky, "Petr Petrovich Lazarev (1878–1942)," in *Uspekhi fizicheskikh nauk*, **27** no. 1 (1945), 1–12; T. P. Kravets, "Tvorchesky put akademika P. P. Lazareva" ("The Creative Path of Academician P. P. Lazarev"), *ibid.*, 13–21; B. V. Deryagin, "O rabotakh akad. P. P. Lazareva v oblasti biologicheskoy fiziki" ("On the Works of Academician P. P. Lazarev in the Area of Biological Physics"), suppl. to *Issledovania po adaptatsii P. P. Lazareva* ("Studies in Adaptation by P. P. Lazarev"; Moscow–Leningrad, 1947); and S. V. Kravkov, "Petr Petrovich Lazarev (K desyatiletiyu so dnya smerti)" ("Petr Petrovich Lazarev [On the Tenth Anniversary of His Death]"), in *Uspekhi fizicheskikh nauk*, **46**, no. 4 (1952), 441–449.

J. G. DORFMAN

LEA, ISAAC (*b.* Wilmington, Delaware, 4 March 1792; *d.* Philadelphia, Pennsylvania, 8 December 1886), *malacology*.

The fifth son of James Lea, a Quaker merchant of Wilmington, Isaac attended the local academy with the idea of becoming a physician; but at the age of fifteen he went to work in the Philadelphia mercantile business of his brother John, who eventually made him a partner. In Philadelphia, Lea formed a lifelong friendship with Lardner Vanuxem, with whom he roamed the countryside collecting minerals, rocks, and fossils. They were both elected to the Academy of Natural Sciences of Philadelphia in 1815; and in 1818 Lea published in its *Journal* his first scientific paper, "An Account of the Minerals at Present Known to Exist in the Vicinity of Philadelphia." The study of geology led to the study of shells, and Lea spent several years studying the collections of freshwater mollusks that Major Stephen Long had sent to the Academy from the Ohio River and that his brother Thomas Lea had gathered for him near Cincinnati. In 1827 he published, in *Transactions of the American Philosophical Society*, "A Description of Six New Species of the Genus Unio." Thereafter he devoted most of his scientific attention to "this truly seducing branch of Nat: Hist:." His *Contributions to Geology* (Philadelphia, 1833), a descriptive catalogue principally of specimens of the Tertiary formations at Claiborne, Alabama, firmly established his reputation. Its text was distinguished by care and judgment; and its typography and illustrations, Roderick Murchison told the author, had "quite the stamp of being issued by one of the best workshops of Europe."

The subject was well chosen, for although mollusks abounded in American rivers, scarcely anyone had searched for them there. An acute and accurate observer, during a long lifetime Lea collected, identified, and described, chiefly in papers to the Academy of Natural Sciences and the American Philosophical Society, 1,842 species of some fifty genera of freshwater and terrestrial mollusks. "I sometimes fear I take up too much room in the Society's Transactions," he once half-apologized, "but it is a great matter to have new objects of Natural History well illustrated—it saves hundreds of mistakes which in time might otherwise occur." In addition to shells, Lea figured the embryonic forms of thirty-eight species of *Unio* and described the soft parts of another 234. His *Synopsis of the Family of Naïades* (Philadelphia–London, 1836; 4th ed. 1870) classified the synonymy of the Unionidae, until then in great confusion. He also investigated physiological questions, such as mollusks' sensitivity to light and differences due to sex. Lea ordered 250 reprints of each paper, bound them up from time to time with title page and introduction as *Observations on the Genus Unio* (13 vols., 1827–1874), and presented them to institutions and individuals. "Your intelligent perseverance with consummate skill & taste have reared a brilliant monument in this part of conchology," the elder Silliman told him in 1860 in acknowledging one of these volumes. Lea's bibliography, compiled in 1876, contained 279 titles.

Lea's personal collection was steadily augmented by gifts from other naturalists—among them Jeffries Wyman, Charles M. Wheatley, Bishop Stephen Elliott, Gerard Troost, John LeConte, and Dr. John Kirk, who accompanied Livingstone to Lake Nyassa. After seeing Lea's cabinet in 1846, Agassiz declared "that all which European naturalists have written on this subject must be revised." Captain Frederick Marryat pronounced it "certainly the most interesting [museum] I saw in the States" and a principal reason why Philadelphia, not Boston, must be considered "the most scientific city in the Union." In his later years Lea turned to the study of crystals and is said to have been the first in America to engage in microscopic mineralogy.

Unlike those of some contemporaries, Lea's disputes with fellow scientists were never acrimonious. He thought Rafinesque's descriptions "very imperfect"; and in a "rectification" of T. A. Conrad's errors and omissions, he asserted his own claims to priority in discoveries (1854). His report on the footprints of the reptile *Sauropus primaevus* in the Old Red Sandstone was disputed by Agassiz, who asserted that no air-breathing animals had existed earlier than the new Red Sandstone. Lea maintained his position in a carefully argued and illustrated monograph, which he ultimately reproduced in elephant folio, with a lithograph of the footprint in actual size. Of some importance then, the matter is of less concern now that fossils of air-breathers have been found in other formations.

Lea was a striking example of the self-taught amateur who by single-minded attention to a limited subject makes a comprehensive, basic, and lasting contribution to knowledge. He expressed the spirit that guided him in 1826: "To scrutinize the first cause is vain; we must make ourselves acquainted with the effects and compare them. In this we have ample room to engage all our faculties." Lea served as president of the Academy of Natural Sciences, vice-president of the American Philosophical Society (where he was also chairman of the finance and publications committees), and president of the American Association for the Advancement of Science. He received an honorary LL.D. from Harvard in 1852 and was a member or fellow of a score of foreign academies. On visits to England and Europe in 1832 and 1852–1853 he was warmly received by Alexander von Humboldt, William Buckland, Charles Lyell, Roderick Murchison, and Adam Sedgwick; and he inspected and named specimens of Unionidae in the British Museum, the Jardin des Plantes, and private collections.

Lea did his scientific work at the end of days filled with business and social obligations. Having joined Mathew Carey's publishing firm in 1821 after his marriage to Carey's daughter Frances Anne, he eventually became president, but retired in 1851. He was rich, with the obligations that wealth entails. Commenting on several unexpected family deaths that left him in 1845 responsible for the children of four relatives, he exclaimed to J. C. Jay, "God knows when I shall have leisure to pursue my wishes in our beautiful branch of Science." Of his three children, M. Carey Lea was a pioneer in photochemistry; and Henry Charles Lea, whose first published paper was on shells, became, after retiring from the publishing firm, the historian of the Inquisition. Lea bequeathed his collection, numbering nearly 10,000 specimens, to the National Museum, Washington.

BIBLIOGRAPHY

I. Original Works. Lea's contributions are analyzed in Newton P. Scudder, *Published Writings of Isaac Lea, LL.D., Bulletin. United States National Museum*, no. 23 (Washington, D.C., 1885). The Academy of Natural Sciences of Philadelphia and the American Philosophical Society Library possess correspondence used in this sketch. Lea's journals of foreign travel are available on film at the American Philosophical Society Library.

II. Secondary Literature. The principal biographical sketch is by Newton P. Scudder, introducing Lea's *Published Writings*, cited above. For other data see "A Sketch of the History of Conchology in the United States," in *American Journal of Science and Arts*, 2nd ser., **33** (1862), 161–180; W. H. Dall, "Isaac Lea, LL.D.," in *Science*, **8** (17 Dec. 1886), 556–558; Joseph Leidy, "Biographical Notice," in *Proceedings of the American Philosophical Society*, **24** (1887), 400–403; and W. J. Youmans, *Pioneers of Science in America* (New York, 1896), 260–269.

Whitfield J. Bell, Jr.

LEAKEY, LOUIS SEYMOUR BAZETT (*b.* Kabete, Kenya, 7 August 1903; *d.* London, England, 1 October 1972), *archaeology, human paleontology, anthropology.*

Leakey was the son of Canon Leakey of the Church Missionary Society in Kenya and was brought up with the native Kikuyu. After attending Weymouth College he went to St. John's College, Cambridge, from which he graduated in archaeology and anthropology and took his Ph.D. in African prehistory. He then became a research fellow of the college and, in 1966, an honorary fellow. His interests in prehistory and ethnology were stimulated by M. C. Burkitt and A. C. Haddon. He was a member of the British Museum East African Expedition to Tanganyika in 1924 and from 1926 led his own East African archaeological research expeditions. Leakey's important discoveries about the prehistory of East Africa and his discovery of early hominids were published in *The Stone Age Cultures of Kenya* (1931), *The Stone Age Races of Kenya* (1935), and *Stone-Age Africa* (1936).

His archaeological and paleontological work did not detract from his interest in Kenya and its politics, an interest which his close association with the Kikuyu made him particularly well equipped to pursue, as can be seen from his autobiographical *White African* (1937). The results of his research for the Rhodes Trustees into the customs of the Kikuyu tribe (1937–1939) are in press. At the outbreak of World War II he was in charge of special branch 6 of the Criminal Investigation Department in Nairobi; he continued as a handwriting expert to the department until 1951.

At the end of the war he returned to his archaeological and paleontological researches, as curator of the Coryndon Memorial Museum, Nairobi (1945–1961), and later as honorary director of the National Centre of Prehistory and Paleontology in Nairobi, and on behalf of various research foundations. He founded the Pan-African Congress on Prehistory, of which he was general secretary (1947–1951) and president (1955–1959). On periods of leave during the war Leakey and his second wife, the former Mary Douglas Nicol, discovered the Acheulean site of Olorgesailie in the Rift Valley. He continued his researches after the war. His work on the Miocene deposits of western Kenya produced among other discoveries the almost complete skull of *Proconsul africanus*, the earliest ape yet found.

Financed largely by the National Geographic Society of Washington, Leakey and his family, beginning in 1959, undertook large-scale work at Olduvai. There, in their first season, Mary Leakey found the skull of *Australopithecus (Zinjanthropus) boisei*; and in 1960 their son Jonathan discovered the first remains of *Homo habilis,* a hominid dated by the potassium-argon method at 1.7 million years. Also in 1960 Leakey discovered the skull of one of the makers of the Acheulean culture at Olduvai, which he named *Homo erectus*. These remarkable researches have been published and are still being published in a series of books entitled *Olduvai Gorge*. After his death his work was continued by his wife and his son Richard, who just before his father's death was able to show him the remains of a human being found on the shores of Lake Rudolf below a tufa dated at 2.6 million years.

Charles Darwin speculated that Africa might be the continent where man had emerged; and Leakey's fieldwork seems to have shown this guess to be a sound one. After his death the Kenya authorities established a museum and research institute which they propose to call The Louis Leakey Memorial Institute for African Prehistory.

A man of very wide interests and a great lover of both domestic and wild animals, Leakey was a trustee of the National Parks of Kenya and of the Kenya Wild Life Society and president of the East African Kennel Club. An enthusiastic and inspiring teacher, he was keen to demonstrate flint knapping, a technique he had learnt from Llewellyn Jewitt's account of the methods used by the nineteenth-century flint forger Edward Simpson, and from watching the knappers at Brandon in Suffolk. He traveled extensively, lecturing to large European and American audiences, and worked tirelessly to disseminate knowledge of his discoveries. His enthusiasm, it was claimed, often carried him to extremes. Although intolerant of opposing views that he considered ill-informed, he realized that, in his field, it was necessary to be a competent archaeologist, human paleontologist, zoologist, anatomist, and geologist—and hardly anyone could be expert in all of them. Many of his discoveries were controversial but his persistence and faith were amply justified. No one has hitherto contributed more to the direct discovery of early man and his ancient culture.

BIBLIOGRAPHY

A complete list of Leakey's published works will appear in his biography, now being written by Sonia Cole. On his early life, see *White African* (1937); a sequel is in press.

In addition to works cited in the text, Leakey's books include *Adam's Ancestors* (London, 1934); *Kenya: Contrasts and Problems* (London, 1936); *The Miocene Hominoidea of East Africa* (London, 1951), written with W. E. Le Gros Clark; *Olduvai Gorge* (Cambridge, 1951); *Mau Mau and the Kikuyu* (London, 1952); *Animals in Africa* (London, 1953), a book of photographs by Ylla to which Leakey contributed the text; and *Defeating Mau Mau* (London, 1954).

GLYN DANIEL

LEAVITT, HENRIETTA SWAN (*b.* Lancaster, Massachusetts, 4 July 1868; *d.* Cambridge, Massachusetts, 12 December 1921), *astronomy.*

Miss Leavitt was one of seven children of George Roswell Leavitt, a prominent Congregationalist minister, and Henrietta Swan Kendrick Leavitt. Both parents were of colonial stock, and their daughter held to the stern virtues of her Puritan ancestors.

Miss Leavitt found her first real opportunity to study astronomy at what is now Radcliffe College, from which she received an A.B. in 1892. She spent an additional year as an advanced student and, after a period of travel, became a volunteer research assistant at the Harvard College Observatory. In 1902, after receiving a permanent position there, she rapidly advanced from an assistant who measured the brightness of variable stars on photographic plates to chief of the photographic photometry department.

During the 1880's and 1890's Edward Pickering, the director of the Harvard College Observatory, had embarked on an extensive program for the determination of visual stellar magnitudes. By 1907 the importance of photographic magnitudes had become apparent, especially because by then it was realized that the photographic plate had a greater sensitivity to blue than the eye. Pickering consequently announced plans to establish a standard photographic sequence based on stars near the north celestial pole. According to S. I. Bailey (1922), Miss Leavitt carried out the plan with "unusual originality, skill and patience"; 299 plates from thirteen telescopes were compared to determine standards with a brightness range of six million (fourth to twenty-first magnitudes). This "north polar sequence" was eventually published as volume 27, number 3 of the *Annals of Harvard College Observatory* (1917). An extension of this research was given in number 4 of the same volume, in which Miss Leavitt supplied secondary magnitude standards for the forty-eight "Harvard standard regions" devised by Pickering. A similar work presented magnitude standards for the Astrographic Catalogue (*Annals of Harvard College Observatory*, **85,** no. 1, 1919; nos. 7 and 8, published posthumously, 1924–1926). The north polar sequence and its subsidiary magnitudes provided the standards for most statistical investigations of the Milky Way system until about 1940.

Miss Leavitt's most famous discovery, the period-luminosity relation of the Cepheid variable stars, originated in her study of the variables in the Magellanic Clouds, made on plates taken at the Harvard southern station in Arequipa, Peru. In the memoir "1777 Variables in the Magellanic Clouds" (*Annals of Harvard College Observatory*, **60,** no. 4 [1908]), she derived periods for sixteen of them and remarked, "It is worthy of notice . . . that the brighter variables have longer periods." In 1912 she extended the analysis to twenty-five stars and, in spite of considerable scatter in the data, found that the apparent magnitude decreased linearly with the logarithm of the period (*Circular. Astronomical Observatory of Harvard College*, no. 173). This relation was seized upon by Shapley at the Mount Wilson Observatory, who calibrated the absolute magnitudes of the Cepheid variables, thereby making it a basic tool for ascertaining the size of the Milky Way galaxy and its distance to other galaxies. To Miss Leavitt's contemporaries, however, her more obvious contribution was her discovery of 2,400 variable stars, about half of those known at the time.

Dorrit Hoffleit has written that although Miss Leavitt was not accorded the honors and publicity of her colleagues Williamina Fleming and Annie Cannon, she was fully as deserving. "Her most important work required greater understanding and even more meticulous care, and was more desperately needed by other astronomers of the time, even though it lacked the glamour and popular appeal of the newly opened field of stellar spectroscopy." Like Miss Cannon, she was extremely deaf. Deeply conscientious and religious, she was devoted to her work and her family, sharing her mother's home in Cambridge after her father's death in 1911. She died there of cancer in 1921.

BIBLIOGRAPHY

Miss Leavitt's principal contributions appeared in the *Annals of Harvard College Observatory* and *Circular. Astronomical Observatory of Harvard College*.

See the obituary by Solon I. Bailey in *Popular Astronomy*, **30** (1922), 197–199; and the biography by Dorrit Hoffleit in *Notable American Women* (Cambridge, Mass., 1971).

OWEN GINGERICH

LEBEDEV, PETR NIKOLAEVICH (*b.* Moscow, Russia, 8 March 1866; *d.* Moscow, 1 April 1912), *physics.*

Lebedev's father, Nikolay Vasilievich, a prosperous merchant, tried to guide his son into the same path by sending him to commercial school and then to a *Realschule*. After completing his secondary education, Lebedev entered the Moscow Highest Technical School. Convinced, however, that the career of an engineer was not for him, in 1887 he entered one of the best schools of physics in western Europe, at Strasbourg University, to work under A. Kundt. In 1888 Kundt moved to Berlin University; Lebedev could not follow him because he lacked the prerequisite diploma from a classical Gymnasium. At the suggestion of Friedrich Kohlrausch, Lebedev wrote "Ob izmerenii dielektricheskikh postoyannykh parov i o teorii dielektrikov Mossotti-Klauziusa" ("On the Measurement of the Dielectric Constants of Vapors and on the Theory of Dielectrics of Mossotti and Clausius"), in which he experimentally confirmed this theory. In the fall of 1891 he returned to Moscow; and at the invitation of A. G. Stoletov, then head of the department of physics at Moscow University, became a teacher there. In 1900 he defended his dissertation, "Eksperimentalnoe issledovanie ponderomotornogo deystvia voln na rezonatory" ("An Experimental Investigation of the Ponderomotive Action of Waves on Resonators"), for which he was awarded the doctorate in physical and mathematical sciences. Lebedev subsequently became professor of physics at Moscow University.

In his dissertation Lebedev posed the problem of studying the action of light waves on molecules in the simplest form. In view of the impossibility of investigating the effect on individual molecules, he tried to observe the influence of electromagnetic waves on a schematic molecule, using a resonator suspended from a twisted thread, which had its own period of vibration. Lebedev extended this research to hydrodynamic and acoustical waves. It appeared that the ponderomotive actions of waves completely different in nature were subject to the same laws. In investigating electromagnetic waves Lebedev encountered their similarity to light waves. He succeeded in constructing an extremely small vibrator, which allowed him to obtain waves from four to six millimeters long—100 times shorter than the waves studied by H. Hertz and ten times shorter than those studied shortly before by Righi. The small size of the apparatus allowed Lebedev to study, for the first time, the double refraction of electromagnetic waves in natural crystals of rhombic sulfur. His work "O dvoynom prelomlenii luchey elektricheskoy sily" ("On the Double Refraction of the Rays of Electric Force," 1895), established him as an excellent experimenter. Lebedev's success in obtaining short waves was not surpassed until thirty years later, by A. A. Glagoleva-Arkadieva and, independently, by M. A. Levitskaya.

As early as 1891 Lebedev became seriously interested in the pressure of light. He turned his attention to the fact that since the force of gravity is proportional to the volume of a body whereas light pressure must be proportional to its surface, it may be asserted that in a particle of cosmic dust the forces of light pressure pushing the particle away from the sun will be equal to the force of gravity attracting it toward the sun. Lebedev used this theory to explain why comets' tails always point away from the sun. His hypothesis was considered correct until the discovery of the solar wind, which creates substantially greater pressure than the sun's light.

Around 1898, Lebedev began experimental research on light pressure. Although its presence had been predicted by Maxwell's theory, it had not been detected experimentally before Lebedev. He first undertook research on the pressure of light on solid bodies. Because of the weakness of the effect itself and the considerable number of possible side effects, this experimental problem presented very great difficulties: if a body that is supposed to react to light pressure is placed in a gas, the warming of the body by the light will inevitably cause convection currents and thus set the body in motion. If the body is placed in a vacuum (in practice, in gas at very low pressure), the so-called radiometric effect will occur. As a result of the uneven warming of the front and back of the body, the molecules of gas hitting the body from the front will be repulsed more forcefully than those striking the back, thereby exerting greater pressure. By extremely ingenious methods Lebedev succeeded in completely eliminating these side effects and not only detected the pressure of light but also measured it and showed the correctness of Maxwell's quantitative theory. "Opytnoe issledovanie svetovogo davlenia" ("An Experimental Investigation of the Pressure of Light") was read by Lebedev at the International Congress of Physicists at Paris in 1899 and was published in 1901.

In connection with the study of processes in gas, Lebedev first considered the advantages of measuring radiant energy with vacuum thermocouples (1902).

He began the second, more difficult, phase of his research by studying the pressure of light on gases and completed this work in 1910. The attempts of other scientists to repeat Lebedev's research on light pressure were unsuccessful until the 1920's, when physicists developed new vacuum techniques.

Lebedev subsequently began to study the origin of terrestrial magnetism, seeking to relate it to the earth's rotation. He did not live to complete this and some other research.

Lebedev was one of many professors who in 1911 left Moscow University in protest against the illegal violation of the university's autonomy by the minister of education, L. A. Kasso. Deprived of his laboratory, he was offered chairs at a number of universities, including some abroad, and a campaign was begun to gather private funds to build him a laboratory. But the shock of these events affected his health and led to his sudden death.

Lebedev's activity was not limited to scientific research. He created an important school of Russian physics, which produced such scientists as P. P. Lazarev, T. P. Kravets, S. I. Vavilov, and V. K. Arkadiev, who have played a major role in the development of Soviet physics.

BIBLIOGRAPHY

Lebedev's writings have been brought together in three books: *Sobranie sochineny* ("Collected Works"; Moscow, 1913), with "Biografichesky ocherk" ("Biographical Sketch"), pp. vii–xxiii; *Izbrannye sochinenia* ("Selected Works"), A. K. Timiryazev, ed. (Moscow, 1949), with Timiryazev's foreword and biographical sketch, pp. 11–32; and *Sobranie sochineny*, in the series Klassikov Nauki (Moscow, 1963), which includes T. P. Kravets, "Petr Nikolaevich Lebedev," pp. 391–405; and N. A. Kaptsov, "Rol Petra Nikolaevicha Lebedeva v sozdanii nauchno-issledovatelskikh kadrov" ("The Role of Petr Nikolaevich Lebedev in the Development of Personnel for Scientific Research"), pp. 406–412.

J. G. DORFMAN

LEBEDEV, SERGEI VASILIEVICH (*b.* Lublin, Poland, 13 July 1874; *d.* Leningrad, U.S.S.R., 2 May 1934), *chemistry.*

Lebedev was the son of a priest. When he was eight, his father died and the family moved to Warsaw, where he received his Gymnasium training. In 1895 he entered the University of St. Petersburg, where he studied organic chemistry with Favorsky. Upon graduation in 1900 he was employed by the Institute of Communications to study the steel of railroad rails. However, wishing to return to academic work, in 1906 he spent some time at his own expense at the Institut Pasteur and the Sorbonne in Paris. After returning to St. Petersburg he began independent study of the chemistry of unsaturated hydrocarbons, a subject to which he devoted the rest of his life.

His extensive investigations of the conditions and products of polymerization of divinyl hydrocarbons led to his dissertation in 1913, for which he received the Tolstoy Prize. He became a docent at the university and, in 1915, professor of chemistry at the Women's Pedagogical Institute. In 1917 he was named professor of chemistry at the Military Medical Academy, where he remained until his death. He found the laboratory at this institution in very poor condition and succeeded in building it into one of the best organic laboratories in the Soviet Union. He continued his researches on many types of unsaturated compounds, investigating both their polymerization and hydrogenation. During World War I he began studies on the chemistry of petroleum. In 1925 he organized a petroleum laboratory at Leningrad University, where he found that the pyrolysis of petroleum hydrocarbons produced diethylene compounds such as he had previously studied.

In 1926, realizing that a severe shortage of rubber existed in the Soviet Union, he began the work for which he is best known. His earlier studies on polymerization had often yielded rubberlike polymers, and an ample source of divinyl was then available from petroleum. He gathered a group of seven chemists (five of them his students) to investigate the production of synthetic rubber. The work was at first carried on only in the chemists' spare time. They soon found a better method for producing divinyl from alcohol. In 1927 they obtained a form of synthetic rubber by polymerizing divinyl in the presence of sodium. In 1928 the petroleum laboratory at the university was converted to a laboratory for synthetic rubber. By 1930 the process was being carried out in an experimental plant. Full factory production began in 1932–1933. The polymer had a structure different from that of natural rubber, but Lebedev showed that it was a fully satisfactory substitute.

Lebedev received many honors in the Soviet Union. He was made a corresponding member of the Academy of Sciences in 1928 and a member in 1932. He received the Order of Lenin in 1931. A laboratory for the study of high molecular-weight compounds was founded under his direction at the Academy of Sciences, although his plans for its development were cut short by his death in 1934.

BIBLIOGRAPHY

I. ORIGINAL WORKS. There is a bibliography of Lebedev's scientific papers in *Zhurnal obshchei khimii*, **5** (1935), 15–17. The monograph reporting his early work and foreshadowing his work on rubber is *Issledovannie v oblasti polimerizatsii dvuetilenovykh uglevodorodov* ("Studies in the Field of the Polymerization of Diethylene Hydrocarbons") (St. Petersburg, 1913).

II. SECONDARY LITERATURE. A biography and critical evaluation of his scientific work is A. I. Yakubchik, "Akademik S. V. Lebedev," in *Zhurnal obshchei khimii*, **5** (1935), 1–14. Further biographical references are given in *Bolshaya sovetskaya entsiklopedia* ("Great Soviet Encyclopedia"), 2nd ed., XXIV (1953), 381–382.

HENRY M. LEICESTER

LEBEDINSKY, VYACHESLAV VASILIEVICH (*b.* St. Petersburg, Russia [now Leningrad, U.S.S.R.], 1 September 1888; *d.* Moscow, U.S.S.R., 12 December 1956), *chemistry.*

A student of Chugaev, Lebedinsky graduated in 1913 from St. Petersburg University and was retained by Chugaev in the department of inorganic chemistry to prepare for a teaching career. From 1920 to 1935 he was a professor at Petrograd (later Leningrad) University, and from 1935 to 1952 at the Moscow Institute of Fine Chemical Technology and at the Moscow Institute of Non-Ferrous Metals and Gold. Lebedinsky's scientific work was done in the Platinum Section of the Commission for the Study of Productive Forces (from 1916), at the Institute for the Study of Platinum and Precious Metals (1918–1934) in Leningrad, and at the Institute of General and Inorganic Chemistry of the Academy of Sciences in Moscow.

Lebedinsky's basic works were related to the study of the chemistry of complex compounds, chiefly rhodium and iridium. His studies of the complex compounds of rhodium won worldwide recognition. From the six possible forms of the ammonium derivative of trivalent rhodium he synthesized compounds corresponding to the four forms $Me_2[RhX_5NH_3]$, $Me[RhX_4(NH_3)_2]$, $[RhX_3(NH_3)_3]$, and $[RhX_2(NH_3)_4]X$, where X is an anion. His research on the ammonium compounds of rhodium and iridium led Lebedinsky to draw the important conclusion that the compounds containing an odd number of ammonia molecules in the inner sphere are the more stable and more easily synthesized; whereas, in the case of pyridine compounds, those containing an even number of molecules of ammonia are more easily synthesized. Study of the chemical compounds of rhodium enabled Lebedinsky to show that they possess a regularity of transinfluence.

Lebedinsky was one of the first to undertake work on the synthesis of complex compounds of rhenium and obtained its compound with ethylenediamine, with rhenium in the form of a cation.

Lebedinsky developed a new industrial method for obtaining pure rhodium and took part in the work of the refining commission. He carried out valuable investigations on processing wastes from refineries to extract precious metals, separating precious metals from weak solutions and salts, and developing the technology of extracting precious metals from copper-nickel sludge. In 1946 Lebedinsky was elected a corresponding member of the Soviet Academy of Sciences.

BIBLIOGRAPHY

I. Original Works. Lebedinsky's writings include "Novye kompleksnye soedinenia rodia" ("New Complex Compounds of Rhodium"), in *Zhurnal Russkago fiziko-khimicheskogo obshchestva*, **48** (1916), 1955, written with L. A. Chugaev; "Novy ryad ammiachnykh soedineny rodia" ("A New Series of Ammonia Compounds of Rhodium"), in *Izvestiya Instituta po izucheniyu platiny i drugikh blagorodnykh metallov*, no. 2 (1933); and "O nekotorykh novykh kompleksnykh soedineniakh renia" ("On Certain New Complex Compounds of Rhenium"), in *Zhurnal obshchei khimii*, **13** (1943), 253, written with B. N. Ivanov-Emin.

II. Secondary Literature. See O. E. Zvyagintsev, "Vyacheslav Vasilievich Lebedinsky," in *Zhurnal neorganicheskoi khimii*, **2** (1957), 1713.

D. N. Trifonov

LE BEL, JOSEPH ACHILLE (*b.* Pechelbronn, Bas-Rhin, France, 24 January 1847; *d.* Paris, France, 6 August 1930), *chemistry.*

Le Bel was one of those solitary workers who make a single outstanding contribution to science and then retire into semiobscurity for the rest of a long life. His contribution was the statement of the relation of stereochemical structure to optical activity in organic compounds.

He came from a wealthy family that controlled petroleum workings at Pechelbronn, and he himself was active in managing them during the first part of his life. He was educated at the École Polytechnique, and after serving as assistant to Balard, the discoverer of bromine, in 1873 he began to work in the laboratory of Wurtz. At this time van't Hoff was also working with Wurtz, and he and Le Bel announced the same theory almost simultaneously. Van't Hoff's paper appeared in Dutch in September 1874, and Le Bel's was published in French in November of the same year; however, there is no evidence that either was influenced by the other. Each arrived at his conclusion independently at this time probably because organic chemistry had reached a point at which such a theory had become essential to further progress. The structural theory, which had so well explained many types of isomerism, had failed completely to account for optical isomers.

Between 1848 and 1853 Pasteur had shown the significance of nonsuperposable mirror images in optically active crystals, but this idea had not been extended to molecular structures even after Kekulé had explained the structures of organic compounds in terms of the tetravalence of carbon. In 1874 Le Bel, starting from the views of Pasteur, and van't Hoff, starting from the more rigid ideas of Kekulé, arrived independently at the theory that when the four substituents around a carbon atom are different—that is, when the carbon compound is asymmetric—molecular mirror images must exist and they must show opposite optical activities. Both Le Bel and van't Hoff used this idea to explain many cases in which such isomerism did or did not occur. Van't Hoff subsequently shifted his interest to physical chemistry, in which he made his most important contributions; but Le Bel continued to work as an organic chemist. For a time he divided his work between his industrial activities at Pechelbronn and his chemical studies in Paris, but after 1889 he sold his petroleum interests and retired to carry on his favorite work privately in Paris. He continued to investigate optical activity. In 1891 he believed that he had found evidence for optical isomers in compounds of pentavalent nitrogen; such isomerism was later conclusively established by Pope in England.

Le Bel never held an academic appointment nor did he ever have any students. He served as president of the Société Chimique de France in 1892, and his work received general recognition in England. A conflict with members of the Académie des Sciences delayed his membership until 1929, near the end of his life. In his later years he devoted himself chiefly to paleontological and philosophical studies and to botanical work in his extensive garden.

BIBLIOGRAPHY

I. Original Works. Le Bel's works have been reprinted as *Vie et oeuvres de J.-A. Le Bel*, Marcel Delépine, ed. (Paris, 1949). His theory was explained in "Sur les relations qui existent entre les formules atomiques des corps organiques et le pouvoir rotatoire de leurs dissolutions," in *Bulletin de la Société chimique de Paris*, **22** (1874), 337–347.

II. Secondary Literature. Aside from the biography by Delépine listed above, there are short obituaries by W. J. Pope in *Journal of the Chemical Society* (1930), 2789–2791; and by E. Wedekind in *Zeitschrift für angewandte Chemie*, **43** (1930), 985–986.

HENRY M. LEICESTER

LEBESGUE, HENRI LÉON (*b.* Beauvais, France, 28 June 1875; *d.* Paris, France, 26 July 1941), *mathematics*.

Lebesgue studied at the École Normale Supérieure from 1894 to 1897. His first university positions were at Rennes (1902–1906) and Poitiers (1906–1910). At the Sorbonne, he became *maître de conférences* (in mathematical analysis, 1910–1919) and then *professeur d'application de la géométrie à l'analyse*. In 1921 he was named professor at the Collège de France and the following year was elected to the Académie des Sciences.

Lebesgue's outstanding contribution to mathematics was the theory of integration that bears his name and that became the foundation for subsequent work in integration theory and its applications. After completing his studies at the École Normale Supérieure, Lebesgue spent the next two years working in its library, where he became acquainted with the work of another recent graduate, René Baire. Baire's unprecedented and successful researches on the theory of discontinuous functions of a real variable indicated to Lebesgue what could be achieved in this area. From 1899 to 1902, while teaching at the Lycée Centrale in Nancy, Lebesgue developed the ideas that he presented in 1902 as his doctoral thesis at the Sorbonne. In this work Lebesgue began to develop his theory of integration which, as he showed, includes within its scope all the bounded discontinuous functions introduced by Baire.

The Lebesgue integral is a generalization of the integral introduced by Riemann in 1854. As Riemann's theory of integration was developed during the 1870's and 1880's, a measure-theoretic viewpoint was gradually introduced. This viewpoint was made especially prominent in Camille Jordan's treatment of the Riemann integral in his *Cours d'analyse* (1893) and strongly influenced Lebesgue's outlook on these matters. The significance of placing integration theory within a measure-theoretic context was that it made it possible to see that a generalization of the notions of measure and measurability carries with it corresponding generalizations of the notions of the integral and integrability. In 1898 Émile Borel was led through his work on complex function theory to introduce radically different notions of measure and measurability. Some mathematicians found Borel's ideas lacking in appeal and relevance, especially since they involved assigning measure zero to some dense sets. Lebesgue, however, accepted them. He completed Borel's definitions of measure and measurability so that they represented generalizations of Jordan's definitions and then used them to obtain his generalization of the Riemann integral.

After the work of Jordan and Borel, Lebesgue's generalizations were somewhat inevitable. Thus, W. H. Young and G. Vitali, independently of Lebesgue and of each other, introduced the same generalization of Jordan's theory of measure; in Young's case, it led to a generalization of the integral that was essentially the same as Lebesgue's. In Lebesgue's work, however, the generalized definition of the integral was simply the starting point of his contributions to integration theory. What made the new definition important was that Lebesgue was able to recognize in it an analytical tool capable of dealing with—and to a large extent overcoming—the numerous theoretical difficulties that had arisen in connection with Riemann's theory of integration. In fact, the problems posed by these difficulties motivated all of Lebesgue's major results.

The first such problem had been raised unwittingly by Fourier early in the nineteenth century: (1) If a bounded function can be represented by a trigonometric series, is that series the Fourier series of the function? Closely related to (1) is (2): When is the term-by-term integration of an infinite series permissible? Fourier had assumed that for bounded functions the answer to (2) is Always, and he had used this assumption to prove that the answer to (1) is Yes.

By the end of the nineteenth century it was recognized—and emphasized—that term-by-term integration is not always valid even for uniformly bounded series of Riemann-integrable functions, precisely because the function represented by the series need not be Riemann-integrable. These developments, however, paved the way for Lebesgue's elegant proof that term-by-term integration is permissible for any uniformly bounded series of Lebesgue-integrable functions. By applying this result to (1), Lebesgue was able to affirm Fourier's conclusion that the answer is Yes.

Another source of difficulties was the fundamental theorem of the calculus,

$$\int_a^b f'(x)\,dx = f(b) - f(a).$$

The work of Dini and Volterra in the period 1878–1881 made it clear that functions exist which have bounded derivatives that are not integrable in Riemann's sense, so that the fundamental theorem becomes meaningless for these functions. Later further classes of these functions were discovered; and additional problems arose in connection with Harnack's extension of the Riemann integral to unbounded functions because continuous functions with densely distributed intervals of invariability were discovered. These functions provided examples of Harnack-integrable derivatives for which the fundamental theorem is false. Lebesgue showed that for bounded derivatives these difficulties disappear entirely when integrals are taken in his sense. He also showed that the fundamental theorem is true for an unbounded, finite-valued derivative f' that is Lebesgue-integrable and that this is the case if, and only if, f is of bounded variation. Furthermore, Lebesgue's suggestive observations concerning the case in which f' is finite-valued but not Lebesgue-integrable were successfully developed by Arnaud Denjoy, starting in 1912, using the transfinite methods developed by Baire.

The discovery of continuous monotonic functions with densely distributed intervals of invariability also raised the question: When is a continuous function an integral? The question prompted Harnack to introduce the property that has since been termed absolute continuity. During the 1890's absolute continuity came to be regarded as the characteristic property of absolutely convergent integrals, although no one was actually able to show that every absolutely continuous function is an integral. Lebesgue, however, perceived that this is precisely the case when integrals are taken in his sense.

A deeper familiarity with infinite sets of points had led to the discovery of the problems connected with the fundamental theorem. The nascent theory of infinite sets also stimulated an interest in the meaningfulness of the customary formula

$$L = \int_a^b [1 + (f')^2]^{1/2}$$

for the length of the curve $y = f(x)$. Paul du Bois-Reymond, who initiated an interest in the problem in 1879, was convinced that the theory of integration is indispensable for the treatment of the concepts of rectifiability and curve length within the general context of the modern function concept. But by the end of the nineteenth century this view appeared untenable, particularly because of the criticism and counterexamples given by Ludwig Scheeffer. Lebesgue was quite interested in this matter and was able to use the methods and results of his theory of integration to reinstate the credibility of du Bois-Reymond's assertion that the concepts of curve length and integral are closely related.

Lebesgue's work on the fundamental theorem and on the theory of curve rectification played an important role in his discovery that a continuous function of bounded variation possesses a finite derivative except possibly on a set of Lebesgue measure zero. This theorem gains in significance when viewed against the background of the century-long discussion of the differentiability properties of continuous functions. During roughly the first half of the nineteenth century, it was generally thought that continuous functions are differentiable at "most" points, although continuous functions were frequently assumed to be "piecewise" monotonic. (Thus, differentiability and monotonicity were linked together, albeit tenuously.) By the end of the century this view was discredited, and no less a mathematician than Weierstrass felt that there must exist continuous monotonic functions that are nowhere differentiable. Thus, in a sense, Lebesgue's theorem substantiated the intuitions of an earlier generation of mathematicians.

Riemann's definition of the integral also raised problems in connection with the traditional theorem positing the identity of double and iterated integrals of a function of two variables. Examples were discovered for which the theorem fails to hold. As a result, the traditional formulation of the theorem had to be modified, and the modifications became drastic when Riemann's definition was extended to unbounded functions. Although Lebesgue himself did not resolve this infelicity, it was his treatment of the problem that formed the foundation for Fubini's proof (1907) that the Lebesgue integral does make it possible to restore to the theorem the simplicity of its traditional formulation.

During the academic years 1902–1903 and 1904–1905, Lebesgue was given the honor of presenting the Cours Peccot at the Collège de France. His lectures, published as the monographs *Leçons sur l'intégration . . .* (1904) and *Leçons sur les séries trigonométriques* (1906), served to make his ideas better known. By

1910 the number of mathematicians engaged in work involving the Lebesgue integral began to increase rapidly. Lebesgue's own work—particularly his highly successful applications of his integral in the theory of trigonometric series—was the chief reason for this increase, but the pioneering research of others, notably Fatou, F. Riesz, and Fischer, also contributed substantially to the trend. In particular, Riesz's work on L^p spaces secured a permanent place for the Lebesgue integral in the theory of integral equations and function spaces.

Although Lebesgue was primarily concerned with his own theory of integration, he played a role in bringing about the development of the abstract theories of measure and integration that predominate in contemporary mathematical research. In 1910 he published an important memoir, "Sur l'intégration des fonctions discontinues," in which he extended the theory of integration and differentiation to *n*-dimensional space. Here Lebesgue introduced, and made fundamental, the notion of a countably additive set function (defined on Lebesgue-measurable sets) and observed in passing that such functions are of bounded variation on sets on which they take a finite value. By thus linking the notions of bounded variation and additivity, Lebesgue's observation suggested to Radon a definition of the integral that would include both the definitions of Lebesgue and Stieltjes as special cases. Radon's paper (1913), which soon led to further abstractions, indicated the viability of Lebesgue's ideas in a much more general setting.

By the time of his election to the Académie des Sciences in 1922, Lebesgue had produced nearly ninety books and papers. Much of this output was concerned with his theory of integration, but he also did significant work on the structure of sets and functions (later carried further by Lusin and others), the calculus of variations, the theory of surface area, and dimension theory.

For his contributions to mathematics Lebesgue received the Prix Houllevigue (1912), the Prix Poncelet (1914), the Prix Saintour (1917), and the Prix Petit d'Ormoy (1919). During the last twenty years of his life, he remained very active, but his writings reflected a broadening of interests and were largely concerned with pedagogical and historical questions and with elementary geometry.

BIBLIOGRAPHY

Lebesgue's most important writings are "Intégrale, longueur, aire," in *Annali di matematica pura ed applicata*, 3rd ser., **7** (1902), 231–359, his thesis; *Leçons sur l'intégration et la recherche des fonctions primitives* (Paris, 1904; 2nd ed., 1928); *Leçons sur les séries trigonométriques* (Paris, 1906); and "Sur l'intégration des fonctions discontinues," in *Annales scientifiques de l'École normale supérieure*, 3rd ser., **27** (1910), 361–450. A complete list of his publications to 1922 and an analysis of their contents is in Lebesgue's *Notice sur les travaux scientifiques de M. Henri Lebesgue* (Toulouse, 1922). Lebesgue's complete works are being prepared for publication by l'Enseignement Mathématique (Geneva).

Biographical information and references on Lebesgue can be obtained from K. O. May, "Biographical Sketch of Henri Lebesgue," in H. Lebesgue, *Measure and the Integral*, K. O. May, ed., (San Francisco, 1966), 1–7. For a discussion of Lebesgue's work on integration theory and its historical background, see T. Hawkins, *Lebesgue's Theory of Integration: Its Origins and Development* (Madison, Wis., 1970).

THOMAS HAWKINS

LE BLANC, MAX JULIUS LOUIS (*b.* Barten, East Prussia [now Poland], 26 May 1865; *d.* Leipzig, Germany, 31 July 1943), *chemistry*.

Le Blanc received his chemical education at the universities of Tübingen, Munich, and Berlin, where in 1888 he became privatdocent with A. W. Hofmann in organic chemistry. In 1891 he went to Wilhelm Ostwald's institute at Leipzig and wrote his inaugural essay, "On the Electromotive Force of Polarization."

In 1896 he left his position as extraordinary professor at Leipzig to become the director of the electrochemical division of the Höchster Farbenfabriken. Five years later he accepted a call to the chair of physical chemistry at the technical institute at Karlsruhe, and in 1906 he became director of the physical chemical institute at Leipzig. He was personally responsible for much of the building and equipping of these two institutions. After his retirement in 1934 he remained a participant in scientific affairs mainly through his service to the Sächsische Akademie der Wissenschaften zu Leipzig.

Most of Le Blanc's chemical studies involved the phenomenon of electrochemical polarization: the nature of the decomposition voltage of aqueous solutions of electrolytes and the groundwork for the understanding of the processes occurring at the electrodes in electrolysis. Noting that the decomposition potentials of many aqueous solutions of salts were nearly the same (1.7 volts), he concluded that the electrode processes were identical and that hydrogen and oxygen gas were discharged at the electrodes.

In 1891 Le Blanc introduced the concept of overvoltage, the amount by which the decomposition potential is higher than the calculated value. This phenomenon has been most studied with hydrogen, where the effect is significant. The introduction of the hydrogen electrode is perhaps his most important contribution to science. In 1893 he observed that if a stream of hydrogen gas is allowed to flow over the surface of a platinized platinum electrode, it behaves like an electrode of hydrogen gas, with the platinum having no effect.

Interested in industrial chemistry, Le Blanc developed a process for the electrochemical regeneration of chromic acid, which found extensive use in the manufacture of dyes. During World War I he worked on a method of reclaiming rubber. He wrote several highly successful textbooks, including *Lehrbuch der Elektrochemie*.

BIBLIOGRAPHY

I. Original Works. Le Blanc's major textbook, *Lehrbuch der Elektrochemie* (Leipzig, 1895), went through twelve German eds., the latest appearing in 1925. This work was translated into French, and into English as *Elements of Electrochemistry*, Willis R. Whitney, trans. (London, 1896); and *A Textbook of Electrochemistry*, W. R. Whitney and John W. Brown, trans. (New York, 1907). *Die Darstellung des Chroms und seiner Verbindungen mit Hilfe des elektrischen Stroms* (Halle, 1902) appeared in English as *The Production of Chromium and Its Compounds by the Aid of the Electric Current*, Joseph W. Richards, trans. (Easton, Pa., 1904). Most of Le Blanc's important research on electrode processes is reported in "Die elektromotorischen Kräfte der Polarisation," in *Zeitschrift für physikalische Chemie*, **8** (1891), 299; **12** (1893), 333; the hydrogen electrode is discussed in the second paper.

II. Secondary Literature. A listing of Le Blanc's more than one hundred publications and a short biographical sketch is in M. Volmer, "Max Le Blanc als Forscher und Lehrer," in *Zeitschrift für Elektrochemie und angewandte physikalische Chemie*, **41** (1935), 309–314. Care must be taken in using this bibliography, as only the first of a series of several papers with the same title is listed. Karl Bonhoeffer has written two short biographical notes: "Max Le Blanc zum 26.5.1935," in *Forschungen und Fortschritte*, **11** (1935), 191–192; and "Max Le Blanc," *ibid.*, **19** (1943), 371–372. The two 1935 sketches honored Le Blanc on his seventieth birthday, and the third is a brief obituary notice.

Sheldon J. Kopperl

LEBLANC, NICOLAS (*b.* Ivoy-le-Pré, France, 6 December 1742; *d.* Paris, France, 16 January 1806), *chemistry*.

From 1751 Leblanc was brought up as an orphan in Bourges. Influenced by his guardian, who was a doctor, he went to the École de Chirurgie in Paris and in 1780 entered the service of the duc d'Orléans (the future Philippe Égalité) as a surgeon. The patronage of the Orléans family, which lasted until the duke was guillotined in November 1793, gave Leblanc the opportunity for research. He began with a study of crystallization, the results of which were later incorporated in his book *Cristallotechnie* (1802). But the most important work dating from this period was that which led him in 1789 to the discovery of the Leblanc process for the artificial preparation of soda. Although it was developed industrially, the discovery brought Leblanc neither happiness nor prosperity, and he eventually committed suicide.

During the late eighteenth century there were strong incentives to devise a new method for the preparation of soda. It was observed that it might soon become impossible to meet the growing industrial demand for soda from the traditional sources, such as potash, which was obtained from wood ashes, or from coastal shrubs like barilla, imported from Spain. In the 1770's and 1780's a number of alternative methods were developed, and between 1783 and 1788 the Académie des Sciences showed its interest by offering a prize for the best method of preparing soda from salt. Although no prize was awarded, by the time Leblanc made his discovery in 1789 several establishments were manufacturing artificial soda in France, if only as a by-product.

In Leblanc's method, sodium sulfate is prepared by the action of sulfuric acid on the salt from seawater and is then converted to soda by calcination with limestone and charcoal. It is not clear how Leblanc discovered the method, for there was no obvious precedent in the processes then in use; and the precise mechanism of the reaction was still obscure over a century later. In fact in 1810, in an article disputing Leblanc's claim to the discovery, J. J. Dizé even maintained that at first Leblanc had prepared sodium sulfide, mistaking it for soda, and that the correct process was developed only in subsequent researches undertaken by Leblanc and Dizé jointly, with guidance from Jean Darcet, the agent of the duc d'Orléans. Unfortunately there is no way of confirming or disproving Dizé's account, but it seems probable that the credit for the discovery should go to Leblanc. In any event, such was the decision of a commission specially appointed by the Académie des Sciences in 1856.

Once made, the discovery was quickly applied. In 1790 the duc d'Orléans, Leblanc, Dizé, and Henri Shée formed a company; and between 1791 and 1793, with capital provided by the duke, a factory was established at St.-Denis. But regular production never began. In 1793 work was abandoned because of the wartime shortage of sulfuric acid, and early in 1794 the factory was nationalized as property of the executed duc d'Orléans. Some months later details of Leblanc's process (with details of all the other known processes for the manufacture of soda) were published by the Committee of Public Safety in the interests of the war effort. Although Leblanc very soon began to seek compensation for his losses, it was several years before his many letters and petitions had any effect, and even when control of the factory was provisionally returned to him in 1800, pending a definitive ruling on his claims, he did not succeed in reviving it through lack of capital. His hope was always that the final settlement of his claims would give him the capital he needed. However, when a settlement was eventually made, in November 1805, the award announced fell far short of his expectations, and two months later, in despair and ill health, Leblanc shot himself.

BIBLIOGRAPHY

I. Original Works. Leblanc's only major publication is *De la cristallotechnie, ou sur les phénomènes de la cristallisation* (Paris, 1802). For references to the surviving MS material see the article by C. C. Gillispie cited below.

II. Secondary Literature. The standard biography, *Nicolas Leblanc, sa vie, ses travaux, et l'histoire de la soude artificielle* (Paris, 1884), is by Leblanc's grandson, Auguste Anastasi. Despite some useful documentation, it is uncritical and has set the tone for most subsequent biographical sketches, nearly all of which reflect Anastasi's undue reverence for Leblanc. The most useful of the derivative sketches are P. Baud, "Les origines de la grande industrie chimique en France," in *Revue historique*, **174** (1934), 1–18; and R. E. Oesper, "Nicolas Leblanc (1742–1806)," in *Journal of Chemical Education*, **19** (1942), 567–572, and **20** (1943), 11–20. An important reappraisal is C. C. Gillispie, "The Discovery of the Leblanc Process," in *Isis*, **48** (1957), 152–170, which uses MS material to show Leblanc as the victim less of misfortune than of his own difficult personality. The article also reexamines the circumstances of the discovery. A more favorable view of Leblanc appears in J. G. Smith, "Studies of Certain Chemical Industries in Revolutionary and Napoleonic France" (Leeds University, Ph.D. thesis, 1970). For the deliberations of the 1856 commission, see L. J. Thenard *et al.*, "Rapport relatif à la découverte de la soude artificielle," in *Comptes rendus hebdomadaires des séances de l'Académie des sciences*, **42** (1856), 553–578.

Robert Fox

LE BOUYER DE FONTENELLE, BERNARD. See **Fontenelle, Bernard Le Bouyer de.**

LE CAT, CLAUDE-NICOLAS (*b.* Blérancourt, Picardy, France, 6 September 1700; *d.* Rouen, France, 21 August 1768), *surgery, anatomy, physiology, biomechanics.*

Le Cat was the son of Claude Le Cat, a surgeon, and Anne-Marie Méresse, the daughter of a surgeon. Despite this family tradition, he himself was committed to an ecclesiastical career before taking up the study of surgery with his father. Following this early training he took courses under Winslow at the Écoles de Médecine in Paris and attended the lectures of Boudou at the Hôtel-Dieu, Le Dran at La Charité, and Duvernay at the Jardin Royal. In 1729 Le Cat was appointed physician and surgeon to the archbishop of Rouen, where he settled permanently, becoming chief surgeon at the Hôtel-Dieu there in 1731. On 29 January 1733 he received the M.D. at Rheims and then took the master's degree in surgery.

From the beginning of his surgical practice Le Cat devoted himself enthusiastically to operating for vesicular stones. He invented or perfected several instruments for lithotomy, one of which, a combined gorget and cystotome, he defended in a vigorous polemic with Jean Baseilhac, celebrated as a lithotomist under his name in religion, Frère Côme. Always open to technical innovation, he also operated on lachrimal fistula through the nasal passage and employed the method of reduction and Daviel's new method of extraction to treat cataracts. One of his procedures foreshadowed the modern method of diaphysectomy.

In 1736 Le Cat established a school for anatomy and surgery in Rouen. He was an early advocate of the union of medicine and surgery and adopted an anatomico-clinical method whereby his students participated in his researches, including animal vivisections. In 1737 he was appointed royal demonstrator in surgery, and gave courses in experimental physics as well as therapeutics, anatomy, and physiology.

Le Cat's medical investigations led him to a theory whereby malignant fevers represent the internalization of external diseases of a herpetic nature, while disease itself is directly attributable to a qualitative or quantitative loss of nervous fluid. Le Cat never fully formulated this hypothesis; like his theories on the cause of menstruation, prolonged pregnancy, and "spontaneous combustion" in human beings (and, indeed, like the works of many of his contemporaries), his theory of disease lacked coherence. More clearly realized was his biomechanic notion of the organism as a hydraulic machine, wherein he drew an analogy between the animal economy and a system of pipes and tubes. Such a view brought Le Cat into conflict

LE CAT

with Haller, especially in respect to the latter's ideas of irritability and sensibility, both of which Le Cat thought to be purely mechanical effects.

The *Traité des sens,* Le Cat's most important work, grew as much out of his researches in physics as in physiology. In it Le Cat presented a theory of the propagation of light contrary to that of Newtonian attraction. He further reported on the pigmented choroid coat of the eye and assigned it a common embryonic origin with the pigment of the skin. Other of his writings further demonstrate the breadth of his interests, which encompassed geophysics, astronomy, conchology, geology, embryology and teratology, electricity, literature, the fine arts, and archaeology. But in an age shaped by Descartes, Borelli, and Bellini, Le Cat must be remembered primarily, with Vaucanson and Quesnay, as one of the first adherents of a mechanistic approach to physiology.

Le Cat was a man of his times. Philosophically, he was a neo-Cartesian and believed in the union of the soul and the body. A devout Catholic throughout his life, he was a friend of the Abbé Nollet, Fontenelle, and Voltaire; he strongly opposed the ideas of Rousseau. He was a founding member of the Rouen Académie des Sciences, and a member of a number of other learned societies, both in France and elsewhere. Louis XV granted him titles of nobility and the rank of *écuyer* in recognition of his services.

In 1742, when he was forty-two years old, Le Cat married Marie-Marguerite Champossin, a girl of thirteen. Their only daughter, Charlotte-Bonne, married the surgeon Jean-Pierre David, who succeeded Le Cat in all his offices.

BIBLIOGRAPHY

I. Original Works. Among Le Cat's principal works that were published separately are his dissertation, *Sur le balancement d'un arc-boutant de l'église Saint-Nicaise* (Rheims, 1724); *Dissertation sur le dissolvant de la pierre, et en particulier sur celui de Melle Stephens* (Rouen, 1739); *Traité des sens en particulier* (Rouen, 1740; repr. Amsterdam, 1744); *Remarques sur les mémoires de l'académie de chirurgie. "Lettres d'un chirurgien de Paris à un chirurgien de province, son élève"* (Amsterdam, 1745); *Réfutation du discours du citoyen de Genève qui a remporté le prix de l'Académie de Dijon, en l'année 1750 sous le nom d'un académicien de la même ville* (London, 1751); *Éloge de M. de Fontenelle* (Rouen, 1759); *Nouveau système sur la cause de l'évacuation périodique du sexe* (Amsterdam, 1765); *Traité de l'existence, de la nature et des propriétés du fluide des nerfs et principalement de son action dans le mouvement musculaire suivi des dissertations sur la sensibilité des méninges, des tendons . . .* (Berlin, 1765); *Traité de la couleur de la peau humaine en général, de celle des nègres en particulier et de la métamorphose d'une de ces couleurs en l'autre, soit de naissance, soit accidentellement* (Amsterdam, 1765); *Parallèle de la taille latérale de M. Le Cat avec celle du lithotome caché suivi de deux dissertations: I. Sur l'adhérence des pierres dans la vessie; II. Sur quelques nouveaux moyens de briser la pierre* (Amsterdam, 1766); *Lettre de M. Le Cat à M. Maître ès arts et en chirurgie de Paris sur les avantages de la réunion du titre de Dr. en médecine avec celui de maître en chirurgie et sur quelques abus dans l'un et l'autre art, 11 juin 1762* (Amsterdam, 1766); *Traité des sensations et des passions en général et des sens en particulier* (La Chapelle, 1767); *La théorie de l'ouïe* (La Chapelle, 1768); *Cours abrégé d'ostéologie* (Rouen, 1768); and *Mémoire posthume sur les incendies spontanés de l'économie animale* (Paris, 1813).

The three-volume collected *Oeuvres physiologiques* (Paris, 1767–1768) comprise *Traité des sensations et des passions en général, Traité des sens,* and *La théorie de l'ouïe;* an early work, *Description d'un homme automate dans lequel on verra exécuter les principales fonctions de l'économie animale, la circulation, la respiration, les sécrétions, & au moyen desquels on peut déterminer les effets méchaniques de la saignée, & soumettre au joug de l'expérience plusieurs phénomènes interessants qui n'en paroissent pas susceptibles,* presented to the Académie de Rouen in 1744, appears to have been lost.

In addition to the above, Le Cat published a number of articles and memoirs in both French and foreign journals. His manuscripts may be found in the Bibliothèque Municipale, Rouen; Bibliothèque Municipale, Caen; Académie des Sciences, Belles-lettres, et Arts de Rouen, Rouen; Bibliothèque de la Faculté de Médecine, Paris; Collection of Dr. Jean, Rouen; Archives de l'Académie Royale de Chirurgie, Bibliothèque de l'Académie Nationale de Médecine, Paris; Bibliothèque Nationale, Paris; and Bibliothèque Municipale, Bordeaux.

II. Secondary Literature. On Le Cat and his work see the anonymous "Précis de la vie & des travaux de M. Le Cat, écuyer, docteur en médecine, & chirurgien en chef de l'hôtel-dieu de Rouen . . .," in *Journal des beaux-arts et des sciences,* **4** (1768), 312–340; and "Éloge de Monsieur Le Cat," in *Mercure de France* (2 Apr. 1769), 151–170; Louis Antoine, "Éloge de Monsieur le Cat," in E. Dubois, ed., *Éloges lus dans les séances publiques de l'Académie royale de chirurgie, de 1750 à 1792* (Paris, 1889), pp. 129–159; Ballière-Delaisment, *Éloge de Monsieur Le Cat* (Rouen, 1769); L. Boucher, "Notice sur les débuts de Claude-Nicolas Le Cat," in *Précis analytique des travaux de l'Académie des sciences, belles-lettres, et arts de Rouen* (1901); Valentin, *Éloge de Monsieur le Cat* (Rouen, 1769); A. Doyon and L. Liaigre, "Méthodologie comparée du biomécanisme et de la mécanique comparée," in *Dialectica,* **10** (1956), 292–335; André Girodie, "Le chirurgien Claude-Nicolas Le Cat de Blérancourt. Amateur d'art et chevalier de l'Arc," in *Bulletin de la Société d'histoire de Haute Picardie,* **6** (1928), 51–67; Merry Delabost, "Souvenirs épars. À propos de portraits de deux amis chirurgiens de l'Hôtel-Dieu, membres de l'Académie de Rouen," in *Précis analytique des travaux de l'Académie des sciences,*

 LE CHÂTELIER

belles-lettres, et arts de Rouen (1909–1910), 247–256; Gustave Pawlowski, "Claude-Nicolas Le Cat, célèbre chirurgien, ses lettres d'annoblissement et sa descendance," in *Revue d'histoire nobiliaire*, **1** (1882), 109–128; and Théodore Vetter, *Claude-Nicolas Le Cat (1700–1768). Chirurgien de province au siècle des lumières.*

THÉODORE VETTER

LE CHÂTELIER, HENRY LOUIS (*b.* Paris, France, 8 October 1850; *d.* Miribel-les-Échelles, Isère, France, 17 September 1936), *chemistry, metallurgy.*

On his father's side Le Châtelier came from a line of scientists and technologists, while his mother's ancestors were artists, sculptors, architects, and geographers. His maternal grandfather, Pierre Durand, operated lime kilns and was a friend of Louis Vicat, a specialist in synthetic cements. Two of Le Châtelier's uncles were engineers, one was an architect, and one a specialist in African affairs who helped to form French governmental policy in this field. His father Louis was inspector general of mines for France and the engineer responsible for building much of the French railway system. A strong republican, he resigned when Napoleon III became emperor and thereafter acted as advisor in the construction of railroads in Austria and Spain. He was associated with Deville in establishing an aluminum industry in France and with W. Siemens in building the first open-hearth steel furnace. The leading chemists of France frequently visited his home.

Thus young Le Châtelier grew up in an atmosphere in which science and technology met on equal terms. Even in his youth he was allowed to work for a time in Deville's laboratory. He later said that his contacts with these friends of his father were important influences in shaping his career and establishing his reputation as a chemist. The other significant influence on his life was his mother, who raised her children according to a strict schedule and instilled in them a sense of order and discipline reinforced by the somewhat rigid training obtained in French technical schools. It is not surprising that due to their father's activities all the children in the family were associated with scientific or technological employment throughout their lives or, in view of his mother's strictness, that Henry remained a scientific and political conservative. All his life he was more interested in confirming natural laws than in overthrowing them.

Le Châtelier attended a military academy in Paris for a short time before enrolling at the Collège Rollin, from which he received his Litt.B. in 1867 and his B.S. in 1868. In 1869 he entered the École Polytechnique. Even in his student days he showed his originality and his feeling for the importance of an experimental rather than a theoretical approach to science. With some of his fellow students he designed a course in physics based entirely on the positivism of Comte and eliminating all abstract entities such as force.

His studies were interrupted by the Franco-Prussian War, in which he served as a lieutenant in the army. He resumed his academic work in 1871, enrolling in the École des Mines, as he planned to make a career in government administration. After graduation in 1873 he spent several years traveling, chiefly in North Africa in connection with a government plan to create an inland sea in that region. In 1875 he took up the duties of a mining engineer at Besançon. He married Geneviève Nicolas in 1876, and their seven children carried on the family traditions. The three sons became engineers, the four daughters married into the same profession.

In 1877 Le Châtelier's career underwent a sharp change. Daubrée, the director of the École des Mines, did not know him personally, but he did know his father and had noted that the son had done well in chemistry while in the school. He therefore offered Le Châtelier the position of professor of general chemistry. He accepted the offer and remained at the institution until his retirement in 1919. At first he taught only chemistry but later added metallurgy to his subjects. He received the degree of Doctor of Physical and Chemical Science in 1887, at which time his title was changed to professor of industrial chemistry and metallurgy. In the same year he accepted the chair of chemistry at the Collège de France and retained the post until 1908. In 1907 he succeeded Moissan as professor of general chemistry at the Sorbonne and held the chair until becoming an honorary professor in 1925. In this post he was able to direct the researches of graduate students. During his active years between 1908 and 1922 he directed the work of more than a hundred students, twenty-four of whom received their doctorates under his supervision. His lectures were very popular, and several were afterward published as books.

Le Châtelier was always interested in the organization of science, especially in relation to its industrial applications. In particular, he attended many international congresses devoted to industrial problems. During his life he held increasingly important positions on a large number of commissions and boards to advise the government on scientific and technical questions. Among these were the Commission on Explosives, the National Science Bureau, the Commission on Weights and Measures, the Commission on Standardization of Metallic Products, the Commission on Inventions, and the Committee for the Control of French Monetary Circulation. In 1916 President Wilson appointed him advisor in the establishment of the National Research Council in the United States.

In 1907 he was appointed inspector general of mines and became a member of the Académie des Sciences. He received medals and honors from almost every French scientific and engineering society and from numerous foreign countries including Poland, Russia, England, and the United States. He also held honorary doctorates from the universities of Aix-la-Chapelle (Aachen), Manchester, Louvain, and Madrid and the Technical University of Denmark.

In his later years Le Châtelier devoted himself to writing on philosophical and social questions, an activity which he continued until his death at his country estate at the age of almost eighty-six.

The research activities of Le Châtelier were many and varied, and at first glance it would seem that there was little relation between some of them. Yet, as is often the case with scientists, there was continuity, and it is possible to see how one line of study led to another.

When he first assumed an academic position he was somewhat at a loss as to what research he should undertake. His grandfather had made a collection of different types of hydraulic cements, and he decided an investigation of these might be profitable. For simplicity he began first to study the setting of plaster of Paris, which had originally been investigated by Lavoisier and Payen. They believed that during the burning of gypsum, anhydrous calcium sulfate was formed, and that in setting, the gypsum was immediately regenerated. In his first paper on the subject (1882) Le Châtelier found that the gypsum could be burned at a lower temperature than Payen had called for and that the product formed was not the anhydrous salt but a hemihydrate. In the setting process it formed a supersaturated solution around the particles of gypsum, from which new crystals of gypsum precipitated to form an interlocking mass. This work showed him the importance of compound formation in complex mixtures, and much of his subsequent work in various fields involved the detection of such compounds.

He next turned his attention to the problem of the setting of cements composed of calcium silicates. He found it better first to study barium ortho and metasilicates which he could easily synthesize in their pure form. The anhydrous forms of these salts, formed by fusion, set with water to form a barium metasilicate hexahydrate. The same changes occurred with the calcium silicates, in all cases giving rise to the essential setting compound, which was $3CaO{\cdot}SiO_2$. Other components of the cements merely acted as fluxes. Since alumina was always present in cements, Le Châtelier next studied a series of calcium aluminates, preparing the pure salts synthetically. These compounds also showed the phenomenon of setting, but the products were less stable than were the silicates. A mixture of silicates and aluminates hydrolyzed to form the stable compound and calcium hydroxide. As was the case with plaster of Paris, a supersaturated solution of calcium salts precipitated the stable component in the form of interlocking crystals. Le Châtelier determined the effect of changing the ratio of lime or magnesia to silica or alumina and determined the conditions under which disintegration of the cement occurred. He also studied the setting of cements under varied conditions in air, water, or seawater. He carried out analogous studies on ceramics and glass. Published in 1887 as his doctoral thesis, the work on cements has been called a classic of inorganic chemistry.

In all these studies he was guided equally by the intrinsic scientific interest of the reactions and their practical value to users of cements. He never recognized a distinction between pure and applied science, and many of his most fruitful ideas came from industrial problems.

In the course of his studies on the silicates he had to use very high temperatures, the measurement of which was difficult with the instruments then available. Gas thermometers were inaccurate above 500°C. Platinum-iron and platinum-palladium thermocouples had been introduced, but Regnault, after careful study, had concluded that they gave widely varying results and should not be considered in any accurate work. Le Châtelier saw that the difficulty lay in the diffusion of one metal into the other at high temperatures and in lack of uniformity of the wires. After a series of studies he was able to show that a thermocouple consisting of platinum and a platinum-rhodium alloy gave accurate and reproducible results. He also introduced the custom of using the boiling points of naphthalene and sulfur and the melting points of antimony, silver, copper, gold, and palladium as standard fixed points in the calibration of his thermocouples. Since that time these thermocouples have been used successfully in all high-temperature work. He also made a number of studies of optical pyrometers which were useful in his day but have since been superseded by other methods.

A number of serious mine disasters led the French government in 1882 to ask the École des Mines to investigate their cause and prevention. In association

LE CHÂTELIER

with the professor of metallurgy, Mallard, Le Châtelier took up these problems. The two men first studied the temperature of ignition, speed of propagation of the flame, and conditions for explosion of mixtures of hydrogen, methane, and carbon monoxide with air. Later these same factors were investigated in other gases and gas mixtures. The study of the combustion of acetylene resulted in development of the oxy-acetylene torch and its use in welding. In the course of their study they determined accurately for the first time the specific heats of a number of gases at high temperatures. Their results subsequently proved very useful in the study of combustion in industrial furnaces. During his investigations of mining problems Le Châtelier made a number of improvements in miners' safety lamps and, together with Mallard, discovered several explosives that were safer than those in use at the time.

Le Châtelier then began to apply his earlier studies to events that occurred in the blast furnace. Industrialists had been puzzled by the reaction of iron oxides with carbon monoxide in the furnace. It was believed that the products should be iron and carbon dioxide. However, the gases emerging from the furnace still contained a considerable amount of carbon monoxide. Some industrialists believed that this was due to incomplete mixing of the reactants, but increasing the height of the furnace for better mixing did not help. Le Châtelier realized that the carbon monoxide formed carbon dioxide and carbon in a reversible reaction, for which the iron oxides acted as a catalyst.

His studies at high temperatures, both with the silicates and the combustion of gases, and the study of reactions in the blast furnace combined to interest Le Châtelier in the conditions needed for equilibrium in chemical reactions. His friend Deville had shown that in many reversible reactions dissociation occurred, and this emphasized the importance of studying equilibrium conditions. In his studies on the effect of temperature on such equilibria, van't Hoff had concluded that all equilibria between two different states of matter (systems) are displaced by a lowering of the temperature toward that of the system whose formation develops heat.

Le Châtelier took up the study of these phenomena and in 1884 was able to announce the famous generalization that bears his name, the Le Châtelier principle: "Every system in stable chemical equilibrium submitted to the influence of an exterior force which tends to cause variation either in its temperature or its condensation (pressure, concentration, number of molecules in the unit of volume) in its totality or only in some one of its parts can undergo only those interior modifications which, if they occur alone, would produce a change of temperature, or of condensation, of a sign contrary to that resulting from the exterior force." He stated it more simply in his summary article in the *Annales des mines* of 1888: "Every change in one of the factors of an equilibrium occasions a rearrangement of the system in such a direction that the factor in question experiences a change in the sense opposite to the original change." In this paper he illustrated his principle for a number of equilibria involving pressure, temperature, and electromotive force. As was usual with him, he applied the principle to practical studies, working out equilibrium conditions for gas furnaces and gaseous reactions.

Another result of his blast furnace studies was his investigation of the behavior of various gas mixtures. His student Boudouard worked out the conditions for the carbon monoxide reaction when catalyzed by iron oxide. Other gas mixtures were also tested. The most interesting results came from a study of the reaction of nitrogen and hydrogen in the presence of an iron catalyst in an attempt to synthesize ammonia. Le Châtelier worked out the theoretically correct conditions of temperature and pressure and the amount of catalyst needed as early as 1900. Unfortunately, when he tested his calculations in the laboratory, the gas mixture contained some air, and an explosion resulted. Unaware of the cause of this at the time, he abandoned the experiments, and it was left for Haber to accomplish the ammonia synthesis on a commercially practical scale after 1905.

Le Châtelier was greatly interested in thermodynamics. In the course of his work on equilibrium in reversible systems he discovered the work of J. W. Gibbs. He recognized that the highly abstract mathematical form of its presentation had prevented most chemists from understanding its chemical applications, and he did his best to spread knowledge of these in France. In spite of his efforts chemists did not accept it quickly, but he himself utilized it in the rest of his work. Some of his calculations anticipated the Nernst heat theorem. He applied his knowledge in this field to his studies of glasses and glazes, and investigated the properties of refractory materials and slags.

Le Châtelier's association with Mallard introduced him to metallurgy, and although he hesitated at first to work in this field, feeling that he knew too little of it, he gradually began to introduce lectures on the subject into his chemical course and was soon actively carrying on metallurgical research and lecturing on the subject, as he did for the rest of his life. Therefore, after 1887 he assumed the title of professor of metallurgy in addition to his other titles at the École des Mines.

Mallard asked him to study allotropic modifications in certain minerals. Having decided that a good approach was to investigate the expansion of the minerals with increasing temperature, he designed an efficient dilatometer for this purpose. With it he demonstrated that quartz, instead of expanding evenly with temperature, showed a break in its expansion curve at definite temperatures. These corresponded to formation of different allotropic modifications. He then applied the idea of using physical methods for detecting chemical changes to the study of metals and especially of alloys. Most chemists at this time believed that alloys were simply indefinite mixtures of the component metals. Le Châtelier disproved this by studies at higher temperature than had previously been used, employing his dilatometer and using measurements of electrical conductivity and freezing-point-composition diagrams. He was particularly successful in demonstrating the existence of intermetallic compounds by employing the latter method. In 1895 he pointed out that there were two parts to the study of alloys: determination of the chemical constitution, including formation of intermetallic compounds, solid solutions, and allotropic modifications; and the physical structure or arrangement of the several kinds of crystals in the alloys.

Much of his metallurgical work was concerned with the chemistry and metallurgy of iron and steel. It had been believed that the hardness of quenched steel was due to the retention of a hard variety of iron stable at high temperatures and retained by rapid cooling. Le Châtelier showed that the solid solution which was stable at high temperatures broke up upon cooling into two phases, iron and iron carbide. He pointed out that this process was analogous to crystallization of two phases from a liquid solution. Quenching prevented this separation. The hard form of iron was thus a solid solution of carbon in the low-temperature allotropic form of iron the formation of which was not prevented during the rapid cooling. The data from Le Châtelier's paper were later used by Roozeboom to construct the first equilibrium diagram for this series. This was the earliest application of the phase rule to a metallic system. Through Roozeboom's work the value of phase rule studies was finally recognized by industrial chemists.

Le Châtelier was always interested in the advancement of the science of metallurgy in France. Realizing the lack of a good textbook, he arranged for the committee on alloys that he had organized to publish a volume, *Contribution à l'étude des alliages,* containing papers by leading French metallurgists. Several of Le Châtelier's own contributions were included. For a long time it was the only really good textbook on metallurgy available in France. Le Châtelier was also aware that, while both England and Germany had special metallurgical journals, none existed in France. Therefore in 1904 he founded the *Revue de métallurgie*, which he edited until 1914. In the publication of this journal he was aided by his daughters, especially Geneviève, who died in 1923, the only one of his children who did not survive him.

One article printed in this journal concerned the high-speed steel tools developed in the United States by the industrial engineer F. W. Taylor. Thinking that they had been discovered accidentally, Le Châtelier entered into correspondence with Taylor and was relieved to discover that they were the result of a truly scientific study. He thus became aware of the work of Taylor, the father of scientific management, who stressed the application of scientific methods in the operation of factories and industrial concerns. Le Châtelier was greatly impressed by Taylor's ideas and devoted much time in his later years to advocating the introduction of the Taylor system into French industry. His book *Le Taylorisme*, published in 1928, was devoted to an effort to increase the efficiency of French workmen. As part of the desired program he recommended a more liberal status for factory workers, and he became interested in political economy.

During World War I, Le Châtelier served as advisor to the government on many military matters. He was influential in a metallurgical study of the heat treatment of shell cases. After the war he gradually withdrew from extensive laboratory work, although he continued to publish short notes on scientific subjects. His writings were always influential and clearly show his interest in the industrial applications of science. Thus *Leçons sur le carbone*, published from his lectures at the Sorbonne, opens with a discussion of the meaning of science and scientific laws. He then describes the various forms of carbon, leading up to a discussion of fuels and metallic carbides. The industrial uses of these materials are described. A survey of oxides of carbon and the carbonates leads to a consideration of the laws of equilibrium, including a description of his own work. The book concludes with a section on atomic and molecular weights, although Le Châtelier avoids descriptions in atomic terms. Historical examples are given throughout.

After World War I Le Châtelier became increasingly concerned with sociological and philosophical questions. In his lectures he had always stressed the importance of general principles rather than merely listing chemical compounds and their properties. He was sometimes accused of teaching physics rather than chemistry. Nevertheless he could not accept theories that appeared to him to lack experimental foundation. As he had avoided the concept of force in physics during his student days, so in later life he avoided the use of atomic theory as much as possible. Since he worked almost entirely with inorganic compounds, he was not concerned with questions of structure, and so was able to consider atoms merely as useful pedagogic tools without deciding whether or not they actually existed.

After his retirement he spent much time in consideration of problems of intellectual and moral education. He stressed the importance of literature and Latin as part of a general education and campaigned for restraint of political expenditures and better coordination of science and industry. Le Châtelier's last paper, on which he was working at the time of his death, was entitled "Morals and Human Affairs."

BIBLIOGRAPHY

I. ORIGINAL WORKS. There is a complete bibliography in *Revue de métallurgie*, **34** (1937), 145–160. A bibliography of Le Châtelier's scientific work is in *Bulletin de la Société chimique de France*, 5th ser., **4** (1937), 1596–1611. The complete form of the Le Châtelier principle is given in "Recherches expérimentales et théoriques sur les équilibres chimiques," in *Annales des mines et des carburants*, 8th ser., **13** (1888), 157–380. The studies on cements are described in the doctoral thesis *Recherches expérimentales sur la constitution des mortiers hydrauliques* (Paris, 1887).

II. SECONDARY LITERATURE. The most complete survey of Le Châtelier's life and work is the Jan. 1937 special issue, "À la mémoire de Henry Le Châtelier 1850–1936," of *Revue de métallurgie*, **34** (1937), 1–160. Full biographies with emphasis on the scientific work are C. H. Desch, "The Le Châtelier Memorial Lecture," in *Journal of the Chemical Society* (1938), 139–150; and P. Pascal, "Notice sur la vie et les travaux de Henry Le Châtelier," in *Bulletin de la Société chimique de France*, 5th ser., **4** (1937), 1557–1596. A shorter and more personal biography is R. E. Oesper, "The Scientific Career of Henry Louis Le Châtelier," in *Journal of Chemical Education*, **8** (1931), 442–461.

HENRY M. LEICESTER

LECLERC. See **Buffon, Georges-Louis Leclerc, Comte de.**

L'ÉCLUSE (CLUSIUS), CHARLES DE (*b.* Arras, France, 19 February 1526; *d.* Leiden, Netherlands, 1609), *botany.*

Known to botanists by the Latinized name of Clusius, L'Écluse was the son of Michel de l'Escluse, lord of Watènes and councillor at the provincial court of Artois, and of Guillièmine Quineault. The elder son in a rich and respectable family, he had a happy childhood and received a substantial education. Amiable and charming, he possessed a lively and penetrating intelligence and was a man of few words. He may have owed his serious nature and sad disposition to his delicate health. His interests were many: objets d'art, antiquities, the history of customs and of peoples. Although he lived in the company of the wealthy and powerful, L'Écluse remained unaffected and without prejudice. Firm in his philosophical and religious convictions, he suffered along with his family the effects of anti-Protestant persecutions. In 1571 he returned his share of the estate to his father, who had just been stripped of his fortune.

L'Écluse was trained as a lawyer and in 1548 received his *licence* in law from the University of Louvain. His interest in botany was not awakened until 1551, when he went to Provence and began gathering plants for the garden of his teacher, Guillaume Rondelet, a professor at the University of Montpellier. L'Écluse's excellent knowledge of Latin allowed him to write and to translate several works in natural science; he did not merely translate the works of his contemporaries but often expanded and even corrected them.

The first book on which L'Écluse collaborated, Rondelet's *De piscibus marinis libri XVIII*, which he wrote up entirely from the great ichthyologist's notes, appeared in 1554. Since his translations contain considerable original work, it is worth citing the most important of them:

Histoire des plantes (1557) is the French version of Dodoens' *Cruydeboeck;* this translation, mentioned by Gaspard Bauhin in his *Pinax*, can be considered a second edition of the original work.

Aromatum et simplicium aliquot medicamentorum apud iodos nascentium historia (1567) is taken from Garcia del Huerto's *Coloquios dos simples.*

For Jacques Amyot's *Vie des hommes illustres grecs et romains* (1568), L'Écluse translated the biographies of Hannibal and Scipio Africanus from Latin into French.

De simplicibus medicamentis ex occidentali India delatis quorum in medicina usus est (1574) is the translation of a work by the Seville physician Nicolas Monardes on the medicinal plants of the West Indies, to which L'Écluse added a supplement in 1582.

Aromatum et medicamentorum in orientali India nascentium liber (1582) is the translation of *Tractado de las drogas y medicinas de las Indias orientales* by Cristóbal Acosta.

A Latin translation of the accounts of Pierre Belon's voyage to the Orient and the Middle East appeared in 1589.

A complete edition of the three translations concerning the medicinal plants of the Indies appeared in 1593.

In addition, L'Écluse composed *Aliquot notae in garciae aromatum historiam* (1582) from the notes on the western coasts of America furnished by Francis Drake upon the return of his famous expedition to the Pacific.

Long periods spent in Provence, Spain, and Austria-Hungary enabled L'Écluse to make numerous botanical observations, on the basis of which he wrote accounts of the floras of these regions that are still valuable. L'Écluse's first entirely original work, *Rariorum aliquot stirpium per Hispanias observatarum historia* (1576), was prepared from notes collected during his travels in Spain in 1564–1565. It contains admirable engravings executed under the author's supervision, either from sketches or from specimens he had collected. These plates, paid for by the publisher Plantin, were also used to illustrate the works of Dodoens and Lobel. L'Écluse's flora of Hungary and Austria was published in 1583 and is still consulted for its descriptions of Alpine plants. *Rariorum plantarum historia* (1601) records approximately 100 new species; *Exoticorum libri decem* (1605) is an important work on exotic flora and includes everything that L'Écluse published on the subject. These two works contain all of L'Écluse's original contributions in botany and natural history and are still often consulted. They are sometimes found bound as a single work.

Honoring the promise that had been made to L'Écluse, Plantin's sons-in-law and successors published his works posthumously, in 1611, under the title *Curae posteriores*. There is also the 1644 folio volume of the works of Dodoens and L'Écluse, several copies of which contain colored plates.

The last sixteen years of L'Écluse's life brought him great satisfaction. Appointed in 1593 by the trustees of the University of Leiden to succeed Dodoens, he held the chair of botany until his death in 1609. It may justly be said that Leiden had become the botanical center for the whole of Europe. L'Écluse's communications to his many correspondents regarding exchanges of plants and documents allow us to reconstruct the state of botany in that period, especially in the Low Countries.

BIBLIOGRAPHY

I. ORIGINAL WORKS. L'Écluse's first original work was a "petit recueil auquel est contenue la description d'aucune gomme et liqueurs, provenant tant des arbres que des herbes . . .," appended to his trans. of Dodoens' *Cruydeboeck* as *Histoire des plantes* (Antwerp, 1557). It was followed by *Rariorum aliquot stirpium per Hispanias observatarum historia* (Antwerp, 1576), a flora of Spain that L'Écluse considered his first original work; *Aliquot notae in garciae aromatum historia* (Antwerp, 1582), collected notes on American flora; and *Rariorum aliquot stirpium, per Pannoniam, Austriam et vicinas quasdam provincias observatarum* (Antwerp, 1583). *Rariorum plantarum historia* (Antwerp, 1601) was an authoritative collection that included descriptions of European plants as well as of foreign ones—including the potato; a concluding treatise on mushrooms is also of interest.

L'Écluse's works were collected as *Exoticorum libri decem*, 3 vols. (Leiden, 1605), which also included 3 new chapters; a French trans. by Antoine Colin, as *Des drogues, espicèries et de medicamens simples*, was published at Lyons in 1602 (2nd ed., 1619). *Curae posteriores* (Leiden, 1611) was published posthumously by Plantin.

For bibliographical details of L'Écluse's translations, see *British Museum General Catalogue of Printed Books*, CXXXII, cols. 786–787; additional information may be found in *Nouvelle biographie générale*, XXX (Paris, 1862), cols. 220–221; and Michaud, *Biographie universelle*, new ed., XXIII, 534–535.

II. SECONDARY LITERATURE. On L'Écluse and his work, see L. Legré, *La botanique en Provence au XVIe siècle*, V (Marseilles, 1901); and C. J. É. Morren, *Charles de l'Escluse, sa vie et ses oeuvres* (Liége, 1875). See also C. F. A. Morren, in *Belgique horticole*, **3** (1853), v–xix; and *Bulletin. Société royale de botanique de Belgique*, **1** (1862), 14–15.

P. JOVET
J. C. MALLET

LECONTE, JOHN (*b.* near Savannah, Georgia, 4 December 1818; *d.* Berkeley, California, 29 April 1891), *physics, natural history.*

LeConte was the son of the Georgia plantation owner Louis LeConte and Ann Quarterman. The family provided a stimulating scientific environment. John Eatton LeConte, Louis' brother, and his son, John Lawrence, were important entomologists; Louis was a competent amateur botanist, and John LeConte and his brother Joseph, his closest scientific associate throughout his life, were among the most respected scientists of the South before the Civil War.

In 1835 LeConte entered Franklin College of the University of Georgia. After graduating in 1838 he studied medicine at the College of Physicians and Surgeons of New York and received his degree in 1841. Upon the death of his father he returned to Savannah with his wife, Josephine Graham of New York. After four years of a modest medical practice LeConte was appointed professor of chemistry and natural history at Franklin College. In 1855 he taught chemistry at the College of Physicians and Surgeons, and in 1856 he and his brother Joseph accepted professorships in physics and chemistry, respectively, at South Carolina College. During the Civil War both worked for the Confederate government in the production of niter and other chemicals. Choosing to leave the South after the war, they were appointed professors at the newly founded University of California in 1869 on the recommendation of Joseph Henry. As senior faculty member, LeConte became acting president in 1869 and again in 1875, and was president from 1876 to 1881. He held the chair of physics until his death.

Typical of early and mid-nineteenth-century scientists, he developed a wide-ranging interest in several branches of science. Most of his publications were in medicine and physics, but he also published on physiology, botany, astronomy, and geophysics. That he made no major contributions to scientific knowledge was perhaps partly due to this diversity. Many of his papers centered on criticizing the work of others. LeConte disliked and avoided experiments, collecting facts and observations for his studies mostly from the literature. In his career as a professor of physics he stressed general principles rather than their applications. This attitude won him respect among his colleagues, at a time when basic science was not generally appreciated. Preoccupation with administrative duties and personal problems severely hampered his scientific activity during his years at California.

Most important among his early papers was an article on the nervous system of the alligator. He promoted the use of statistics as a quantitative approach to medical research. His reputation as a physicist stemmed largely from his discovery of the sensitive flame in 1858, providing a method of visualizing the effects of sound waves. His 1864 paper on the velocity of sound clarified certain misconceptions concerning Laplace's theory. He also investigated the nature of sound shadows in water. Influenced by the work of John Tyndall and Jacques-Louis Soret he provided an explanation for the color of lakes in terms of selective reflection.

BIBLIOGRAPHY

I. ORIGINAL WORKS. LeConte's most important works include "Experiments Illustrating the Seat of Volition in the Alligator or Crocodilus Lucius of Cuvier," in *New York Journal of Medicine and the Collateral Sciences*, **5** (1845), 335–347; "Statistical Researches on Cancer," in *Southern Medical and Surgical Journal*, n.s. **2** (1846), 257–293; "On the Influence of Musical Sounds on the Flame of a Jet of Coal Gas," in *American Journal of Science*, 2nd ser., **23** (1858), 62–67; "On the Adequacy of Laplace's Explanation to Account for the Discrepancy Between the Computed and the Observed Velocity of Sound in Air and Gases," in *Philosophical Magazine*, 4th ser., **26** (1864), 1–32; "On Sound Shadows in Water," *ibid.*, 5th ser., **13** (1882), 98–113; and "Physical Studies of Lake Tahoe," in *Overland Monthly and Out West Magazine*, 2nd ser., **2** (1884), 41–52.

An almost complete MS for a college textbook on physics was destroyed during the Civil War along with most of his other early papers. The largest collection of his papers is the John LeConte Papers, Bancroft Library, University of California. This collection also contains an eight-page autobiography, "Sketch of John LeConte." Other MSS are at the South Carolina Library, University of South Carolina. The bulk of LeConte's letters are found in the papers of correspondents. See Lewis R. Gibbes Papers, Library of Congress, and Benjamin Peirce Papers, Harvard University Archives.

II. SECONDARY LITERATURE. A dissertation by J. S. Lupold, *From Physician to Physicist: The Scientific Career of John LeConte, 1818–1891* (University of South Carolina, 1970), 264, contains valuable bibliographical and biographical information. Joseph LeConte, "Biographical Memoir of John LeConte, 1818–1891," in *Biographical Memoirs. National Academy of Sciences*, **3** (1895), 371–393, includes a selected bibliography and an evaluation of his most important scientific papers. William LeConte Stevens, "Sketch of Professor John LeConte," in *Popular Science Monthly*, **36** (1889), 112–113, contains a complete bibliography of LeConte's works.

GISELA KUTZBACH

LECONTE, JOSEPH (*b*. Liberty County, Georgia, 26 February 1823; *d*. Yosemite Valley, California, 5 July 1901), *natural history, physiology, geology*.

LeConte was the fifth of seven children. His mother died when he was three, and the boy acquired his love of science from his father, Joseph LeConte, an erudite man who had once studied medicine. LeConte received his early education on his father's plantation and at a neighboring school in Liberty County before going (in January 1839) to Franklin College of the University of Georgia, from which he was graduated in 1841. LeConte studied medicine for a brief period in 1842 with a physician, Dr. Charles West, in Macon, Georgia. In 1843–1844, he took the intensive four-month winter medical course at the College of Physicians and Surgeons, New York, N.Y., but broke off his studies to undertake a trip to the headwaters of the Mississippi and to the Great Lakes. On this trip his interest in geology was intensified, although he subsequently finished his medical studies, taking his degree in 1845.

In 1850 LeConte abandoned the medical practice that he had established in Macon and went to Harvard to study zoology and geology with Louis Agassiz. With Agassiz he made scientific expeditions to Florida, to the fossiliferous areas of New York, and to the Massachusetts shores. In June 1851 he and three other students received the first degrees conferred by the newly established Lawrence Scientific School at Harvard. LeConte then began his academic career, teaching various sciences at Oglethorpe University, Georgia; Franklin College, University of Georgia; and South Carolina College, Columbia, South Carolina, a tenure that was interrupted by the Civil War, when LeConte served the Confederacy first as a government arbitrator and then as a chemist. In 1866 he returned briefly to teach at what had been renamed the University of South Carolina, but found Reconstruction policies to be intolerable. Consequently, he and his brother John joined the newly founded University of California in 1869. LeConte was appointed professor of geology and natural history and did his most important scientific work during his tenure there. LeConte became the first president of that institution.

LeConte published books and articles on a wide range of subjects, including various aspects of geology, binocular vision, physiology (especially glycogenic function of the liver), and evolution. Some of his most substantial contributions to geology included his views on the structure and origin of mountain ranges, the genesis of metals, crucial periods in the history of the earth, the demonstration of the great importance of the Ozarkian (Sierran) epoch, and the great lava flood of the northwest. His ideas became widely known through the publication of his *Elements of Geology* (1878), which offered a scholarly combination of principles (dynamical geology) and elements (historical and structural geology) with extensive treatment of American geology.

LeConte's work on vision culminated in his 1881 book *Sight*, which combined excellent experiments with perceptive observations (although devoid of mathematics). This, together with his physiological work on the liver, was further elaborated in his *Outlines of the Comparative Physiology and Morphology of Animals* (1900), which represents a summary of many of his biological ideas. LeConte himself described his work in evolution as "philosophical," since he contributed no new facts. Nevertheless, his books on the subject had a general, popular influence in reconciling science and religion, being designed to make evolution palatable to both clergymen and scientists who were frightened by the apparent materialism and possible impious implications of evolutionary doctrines.

LeConte enjoyed many academic honors, including an LL.D. from Princeton. In 1847 he married Bessie Nisbet; they had three daughters and one son. He died of a heart attack while on a geological expedition.

BIBLIOGRAPHY

I. Original Works. LeConte's scientific books are *Religion and Science. A Series of Sunday Lectures on the Relation of Natural and Revealed Religion, or the Truths Revealed in Nature and Scripture* (New York, 1873), which went into many editions; *Elements of Geology: A Text-Book for Colleges and for the General Reader* (New York, 1878; 5th ed., revised and partially rewritten by Herman LeRoy Fairchild, 1903); *Sight: An Exposition of the Principles of Monocular and Binocular Vision* (New York, 1881; 2nd ed., 1897); *A Compend of Geology* (New York, 1884; rev. ed., 1898); *Evolution and Its Relation to Religious Thought* (New York, 1888; 2nd ed. as *Evolution: Its Nature, Its Evidences, and Its Relation to Religious Thought*, 1891); and *Outlines of Comparative Physiology and Morphology of Animals* (New York, 1900).

See also his autobiographical writings, *The Autobiography of Joseph LeConte*, William Dallam Armes, ed. (New York, 1903); *'Ware Sherman. A Journal of Three Month's Personal Experiences in the Last Days of the Confederacy*, with "Introductory Reminiscence" by his daughter Caroline LeConte (Berkeley, 1937); *A Journal of Ramblings Through the High Sierra of California by the University Excursion Party* (San Francisco, 1875; repr. [1960]).

Manuscript materials are at the University of North Carolina Library, Southern Historical Collection (Elizabeth [Furman] Talley papers and Joseph LeConte papers); University of Michigan, Michigan Historical Collections (Alexander Winchell papers); Duke University Library (Benjamin Leonard Covington Wailes papers); University of South Carolina, South Caroliniana Library (Joseph LeConte papers); Henry E. Huntington Library, San Marino, California (Charles Russell Orcutt correspondence); and the University of California, Berkeley, Bancroft Library.

II. Secondary Literature. The most comprehensive account of LeConte and his work is Eugene W. Hilgard, "Biographical Memoir of Joseph LeConte, 1823–1901," in *Biographical Memoirs. National Academy of Sciences*, **6** (1909), 147–218, with bibliography, 212–218. A summary of some of his geological work is George P. Merrill, *The First One Hundred Years of American Geology* (New Haven, 1924), 344–345, 364–368.

H. Lewis McKinney

LECOQ DE BOISBAUDRAN. See **Boisbaudran, Paul Émile Lecoq de.**

LECORNU, LÉON FRANÇOIS ALFRED (*b.* Caen, France, 13 January 1854; *d.* St.-Aubin-sur-Mer, Calvados, France, 13 November 1940), *mechanics, mechanical engineering.*

Lecornu proved an exceptional student at the secondary school in Caen; in his final year he won the prize of honor in the annual *concours général* and easily passed the competitive entrance examinations for both the École Normale Supérieure and the École Polytechnique. He entered the latter in 1872 but two years later transferred to the École Supérieure des Mines. After graduating he joined the Corps des Mines and began to pursue two careers simultaneously —one professional, the other academic.

His first engineering assignments were in Rennes and in Caen. In 1893, when he returned to Paris, he was promoted to chief engineer in charge of the technical supervision of the railroads of western France. He rose in the Corps des Mines to the rank of inspector general first class and retired in 1924.

In 1880 Lecornu received a *doctorat ès sciences* from the Sorbonne with his dissertation, "Sur l'équilibre des surfaces flexibles et inextensibles." In 1881 he joined the faculty of Caen as associate professor. After his move to Paris he was appointed *répétiteur* at the École Polytechnique (1896) and professor of mechanics at the École des Mines (1900) and at the École Polytechnique (1904). After retiring from the latter in 1927, he continued to teach at the École Nationale Supérieure d'Aéronautique until 1934.

The tenor of Lecornu's original work in mechanics is the application of the classical methods to the analysis of a variety of engineering problems. In his dissertation he established the relationship between the stresses and the geometrical characteristics of deformed surfaces. Applying his general theory to an ellipsoidal surface, he was able to determine the conditions of rupture of air balloons. A similar approach was used to analyze the stability of rotating millstones. Other representative studies are his papers on pendulums of variable length, on inertial errors in the steam engine indicator, on speed regulation, and on the dynamics of gear teeth.

Much of Lecornu's writing was expository and didactic and was characterized by an informal, popular style. He wrote a number of textbooks on mechanics and mechanical engineering, and edited the multivolume *Encyclopédie de mécanique appliquée* (Paris, 1924). In 1910 Lecornu was elected a member of the mechanics section of the Académie des Sciences, which he served as president in 1930. He received the Prix Poncelet (1900) and the Prix Montyon (1909), and was a commander of the Legion of Honor.

Lecornu also had two minor avocations: geology (he published extensively while a mining engineer in Normandy) and the piano.

BIBLIOGRAPHY

I. Original Works. Lecornu's books are *Régularisation du mouvement dans les machines* (Paris, 1898); *Les régulateurs des machines à vapeur* (Paris, 1904); *Dynamique appliquée* (Paris, 1908); *Théorie des moteurs legers* (Paris, 1911); *Cours de mécanique, professé à l'École Polytechnique*, 3 vols. (Paris, 1914–1918); *La mécanique: les idées et les faits* (Paris, 1918); *Théorie mécanique de l'élasticité* (Paris, 1929); and *Propriétés générales des machines* (Paris, 1930). Incomplete bibliographies of his scientific papers are given in Poggendorff, IV, 855; V, 721; and VI, 1485; and in Royal Society *Catalogue of Scientific Papers*, X, 544–545; and XVI, 664–665.

II. Secondary Literature. The principal source on Lecornu is the obituary by H. Vincent, in *Comptes rendus hebdomadaires des séances de l'Académie des sciences*, **211** (1940), 493–498. Additional facts and a portrait are in N. Imbert, ed., *Dictionnaire national des contemporains*, I (Paris, 1936), 378. The catalog of the Bibliothèque Nationale lists also a *Notice sur les travaux scientifiques de M. Léon Lecornu* (Paris, 1896).

Otto Mayr

LE DANTEC, FÉLIX (*b.* Plougastel-Daoulas, France, 16 January 1869; *d.* Paris, France, 6 June 1917), *biology.*

Le Dantec, whose father was a naval physician and friend of Ernest Renan, was exceptionally precocious. He performed brilliantly on the entrance examination for the École Normale Supérieure; during his stay there he was influenced by C. Hermite and J. Tannery, who cultivated his penchant for mathematics. Suddenly attracted to the natural sciences, he made the acquaintance of A. Giard, who, through his unrelenting critical sense, freed Le Dantec from intellectual rigidity. Pasteur appointed him laboratory assistant at the Institut Pasteur in 1888, and sent him first to Laos and then to Brazil, where he founded a laboratory for the study of yellow fever. Le Dantec then gave free rein to the wide-ranging interests of an insatiably curious mind. In 1891 he defended a doctoral thesis on intracellular digestion in the protozoans, followed by numerous scientific and philosophical works that attest to the fruitfulness and brilliance of his intellect. Although an atheist, Le Dantec was always open to religious discussion. He was appointed lecturer at the University of Lyons in 1893, assistant lecturer in 1899, and full professor of general biology at the Sorbonne in 1908.

Le Dantec's gifts for generalization bore the mark of his mathematical training. His biological work began with a study of bacteria. He held that their physiological activity could be interpreted in the symbolic terms of a chemical equation and that elemental life was explicable by the existence of a specific substance that was only transmitted through heredity. He thus sought to reconstruct biology in accordance with the precise language of chemistry, eliminating any anthropomorphism. His logic led him to Lamarckian principles of adaptation and to the inheritance of acquired characteristics, which were the basis of his law of functional assimilation. For Le Dantec protoplasm grows by living; life and growth are a single phenomenon. Stasis produces erosion, destruction, and death. Starting from this conception he explained the law of natural selection and considered psychological and social questions. Conscience, he believed, is nonexistent: men are puppets, subject solely to the laws of mechanics.

Le Dantec's work survives only for its flashes of insight and its clarity. Although he constructed a precarious system based on facts obtained at second hand, his vigorous attacks on anthropomorphism, his passion for truth, his noble character, and his veneration of science explain his being described as a secular saint.

BIBLIOGRAPHY

I. Original Works. Among Le Dantec's more important works are *La matière vivante* (Paris, 1895); *Théorie nouvelle de la vie* (Paris, 1896); *La sexualité* (Évreux, 1899); *L'unité dans l'être vivant* (Paris, 1902); *Traité de biologie* (Paris, 1903); *Science et conscience* (Paris, 1908); and *La science de la vie* (Paris, 1912).

Among his philosophical works are *L'athéisme* (Paris, 1907); and his last work, *Le problème de la mort et la conscience universelle* (Paris, 1917).

II. Secondary Literature. On Le Dantec and his work, see G. Bonnet, *La morale de Félix Le Dantec* (Poitiers, 1930); C. Pérez, *Félix Le Dantec* (1918); and E. Rabaud, "Félix Le Dantec," in *Bulletin biologique* (1917).

Roger Heim

LE DOUBLE, ANATOLE FÉLIX (*b.* Rocroi, France, 14 August 1848; *d.* Tours, France, 22 October 1913), *anatomy.*

Le Double moved to Tours when still a child and never left that city; he died there, asphyxiated by carbon monoxide from a faulty stove. Le Double was a nonresident (1871) and a resident medical student (1873) in the Parisian hospitals and earned his medical degree at the University of Paris in 1873. He was a surgeon at the Hospice Général in Tours from 1879 to 1885, when he was appointed professor of anatomy. He then gave up the practice of medicine in order to devote his time to anatomical research. His career was crowned by his election as a corresponding, and then as a national associate member of the Académie de Médecine (1907).

Le Double was greatly interested in anomalies of the components of the human body including the muscles (1897), bones (1903), and hair. From his numerous statistics he attempted to derive general considerations valid for all observed cases. According to him, anomalies are not sports of nature occurring by mere chance. Defending the opposite thesis, he reaffirmed the conception of those anatomists who regarded anomalies as variations from a normal type and as subject to certain laws, including (1) the correlative and inverse balancing of the skull and face; (2) the correlation of cranial and brain volumes, especially the volume of the frontal lobes; (3) when several anomalies are observed in the same individual they are usually found in organs that have the same embryologic origin and that develop synchronously; (4) regardless of their type or location, variations in no way represent an anatomical indication of degeneracy, insanity, or criminality; (5) variations are indexes of human evolution: some are atavistic remnants of a previous evolutionary stage (reverse theromorphic variations), while others point to the form that such variations may take in adapting to a new environment (progressive or adaptive anomalous variations); (6) an anatomically aberrant organ is more subject to pathological processes than any other.

Between 1880 and 1883 Le Double published in Dechambre's *Dictionnaire encyclopédique des sciences médicales* twenty-four articles on the muscles of the trunk. In *Rabelais anatomiste et physiologiste* (1899), which was often plagiarized, Le Double asserted that Rabelais was the first before Fabricius Hildanus to have described tracheal deformations brought on by goiter and to have given the symptomatology of venereal sarcocele and of mercurial stomatitis. Le Double also indicated that Rabelais knew of the scabies-causing mite and exterminated it. However, not all his claims for Rabelais have been substantiated.

BIBLIOGRAPHY

I. Original Works. No complete bibliography of Le Double's works has been compiled. A list of 24 works may be found in the *Catalogue général* of the Bibliothèque

Nationale, Paris, XCII, cols. 407–410; the Royal Society *Catalogue of Scientific Papers*, X, 547; XVI, 668–669; lists 35 papers written up to 1900. In addition Le Double was author of *Fatale histoire* (Tours, 1892), under the pseudonym F. du Pleix; *Les velus* (Paris, 1912), written with F. Houssay; and *Bossuet, anatomiste et physiologiste* (Paris, 1913).

II. SECONDARY LITERATURE. On Le Double and his work, see (listed chronologically) *Gazette médicale du Centre*, **11** (1906), 129–133, with photograph; R. Mercier, "A. F. Le Double," in *Paris médical*, **12**, supp. (1912–1913), 935; *Semana médica* (Buenos Aires), **20**, pt. 2 (1913), 1611–1614; *Répertoire de médecine internationale*, **3** (1913), 24; L. Dubreuil-Chambardel, "Le professeur Le Double," in *Gazette médicale du Centre*, **18** (1913), 257–258; *Bulletin de l'Académie de médecine*, **70** (1913), 288–290; L. Dubreuil-Chambardel and Faix, "Le professeur Le Double," in *Aesculape*, **3** (1913), 271–272, with portrait; F. Baudoin, "Le professeur Le Double," in *Annales medico-chirurgicales du centre* (1913), 505–508, with photograph; C. de Varigny, "L'oeuvre du Dr. Le Double," *ibid.*, 172–176; 185–187; J. Renault, "Un anatomiste philosophe: Félix Anatole Le Double," in *Gazette médicale du Centre*, **18** (1914), 285–288; F. Houssay, "Anatole Félix Le Double et l'école tourangelle des variations anatomiques," *ibid.*, **19** (1914), 14–16; "Le Double," in *Biographisches Lexikon der hervoragenden Ärzte der letzen fünfzig Jahre*, II (Berlin–Vienna, 1933), 879–880; and C. Mentré, *Rabelais et l'anatomie* (Nancy, 1970), diss.

PIERRE HUARD
MARIE JOSÉ IMBAULT-HUART

LEEUWENHOEK, ANTONI VAN (*b.* Delft, Netherlands, 24 October 1632; *d.* Delft, 26 August 1723), *natural sciences, microscopy.*

Leeuwenhoek was the son of Philips Thoniszoon, a basket-maker, and Margriet Jacobsdochter van den Berch. He took his surname ("Lion's Corner") from the corner house near the Leeuwenpoort ("Lion's Gate") at Delft, which was owned by his father. The family belonged to the prosperous middle class of artisans, brewers, and lesser public officials, which was typical of the Golden Age of the Dutch Republic.

His father died at an early age, in 1638, and Leeuwenhoek was sent to the grammar school of Warmond, a village near Leiden. For some time afterwards he lived with a relative in Benthuizen, then, in 1648, moved to Amsterdam, where he was apprenticed to a cloth merchant. Returning to Delft, Leeuwenhoek began a career as a shopkeeper. On 29 July 1654, he married Barbara de Mey, daughter of Elias de Mey, a serge merchant from Norwich, England. They had five children, of whom only one, their daughter Maria, survived her father.

In 1660 Leeuwenhoek took up a new career as a civil servant with an appointment as usher to the aldermen of the municipality of Delft. In 1666 his wife died, and two years later Leeuwenhoek made one of the only two foreign trips that he took in his lifetime, visiting the chalk hills of Gravesend and Rochester in Kent, England. (His other occasion for travel abroad was a journey that he made to Antwerp in 1698 to see the Jesuit scholar Daniel Papenbroek.) Upon his return to Delft in 1669 he was appointed surveyor to the court of Holland.

Leeuwenhoek remarried on 25 January 1671. His second wife was Cornelia Swalmius, the daughter of Johannes Swalmius, a Calvinist minister at Valkenburg, near Leiden. She died in 1694; the one child of this marriage did not survive infancy. In the meantime Leeuwenhoek continued to advance in the service of the city of Delft, being made chief warden of the city in 1677 and, because of his mathematical skills, wine-gauger (or inspector of weights and measures) in 1679. The income and emoluments from these offices made him financially secure, especially in his old age, when the municipality, in gratitude for his scientific achievements, granted him a pension.

Leeuwenhoek's scientific life may be said to have begun in about 1671, when he was thirty-nine years old. At that time, developing the idea of the glasses used by drapers to inspect the quality of cloth, he constructed his first simple microscopes or magnifying glasses, consisting of a minute lens, ground by hand from a globule of glass, clamped between two small perforated metal plates. To this apparatus he fixed a specimen holder that revolved in three planes. From these beginnings Leeuwenhoek went on to grind about 550 lenses in his lifetime. Although always secretive about his technique, he achieved lenses of increasing quality, of which the best that survives, in the University Museum of Utrecht, has a linear magnifying power of 270 and a resolving power of 1.4 μ. (From his recorded observations it may be surmised that he must have actually made lenses of 500 power, with a resolution of 1.0 μ.) Whatever his methods, Leeuwenhoek's instruments were not surpassed until the nineteenth century.

To his work in lens-grinding Leeuwenhoek brought a good pair of eyes, mathematical exactitude, great patience, and even greater manual dexterity. These same qualities, together with a keen, practical intellect, served him in the exploration of the whole field of natural science that occupied him in fifty years of continuous work. Strictly speaking, his scientific training was incomplete—he never attended a university—and he was limited by his lack of skill in classical and foreign languages. He was, however, able to rely upon such friends as de Graaf and Constantijn Huygens (as well as upon professional translators) to aid him. He derived much of his scientific knowledge from Dutch authors—for example, Cornelis Bontekoe on medicine and Swammerdam on insects—and from Dutch translations of standard works, or, indeed, from the illustrations of books that he was not otherwise able to read (Hooke on microscopy, Grew on plant anatomy, and Redi on insects). He gained new information, too, from correspondence with the Royal Society (which had to be translated) and from conversations with visiting scholars.

Still, Leeuwenhoek remained in relative scientific isolation. One consequence of this was that he was not always fully acquainted with the researches and theories of his fellow scientists, and so could not incorporate their sometimes valuable suggestions into his own work. But he was thus able to work with full independence and to make a sharp distinction between the empiricism and speculation that marked the sometimes chaotic world of seventeenth-century science. Leeuwenhoek apparently regarded speculation as an academic occupation and therefore out of his domain, and this attitude permitted him a certain degree of detachment. He usually set out his observations fully, as facts, and only then, in a separate section, allowed himself to wonder what these facts might mean. (This caution is perhaps the reason that he did not publish a definitive work on microscopy.)

Leeuwenhoek's chief area of endeavor was the microscopical investigation of organic and inorganic structures. His most important contributions were made in the field of general biology; his studies in other branches of science were largely excursions in the interest of solving fundamental biological problems. Leeuwenhoek's method was based upon two presuppositions: that inorganic and organic nature are generally similar, and that all living creatures are similar in form and function. To demonstrate these similarities he attempted to study as many types as possible of any given group of organisms and then to generalize from this data; by drawing analogies between animals and plants, he sometimes succeeded in overcoming highly difficult problems in the interpretation of microscopic structures. Leeuwenhoek followed the Cartesian method in trying to apply his somewhat primitive knowledge of the laws of physics to living nature (and thereby exhibited some bias against chemical explanations).

Leeuwenhoek made his most important discovery early in his scientific career, in 1674, when he recognized the true nature of microorganisms. Starting from the assumption that life and motility are identical, he concluded that the moving objects that he saw through his microscope were little animals. He recorded these observations in his diary, and two years later, in a letter of 9 October 1676, communicated them to the Royal Society, where they caused a sensation. (Indeed, such was the disbelief of some of the fellows of that body that Leeuwenhoek felt obliged to procure written attestations to the reliability of his observations from ministers, jurists, and medical men.) Leeuwenhoek subsequently described, in about thirty letters to the Royal Society, many specific forms of microorganisms, including bacteria, protozoa, and rotifers, as well as his incidental discovery of ciliate reproduction. It was necessary to devise a scale by which to measure this formerly invisible new world, and Leeuwenhoek therefore developed a practical system of micrometry, utilizing as standards a grain of coarse sand (870 μ), a hair from his beard (100 μ), a human erythrocyte (7.2 μ), and bacteria in pepper-water (2–3 μ).

Microscopy, however, was only a tool that Leeuwenhoek put at the service of his two lifelong scientific concerns: his study of sexual reproduction, which was designed to refute the theory of spontaneous generation, and his study of the transport system of nutrients in plants and animals. He was drawn to the investigation of animal reproduction when, in 1677, a medical student, Johan Ham, of Arnhem, told him that he had seen animalcules in human seminal fluid. Ham presumed that these little animals had been generated by putrefaction; Leeuwenhoek, however, supposed them to be a normal component of semen throughout the animal kingdom. In the course of forty years, he described the spermatozoa of arthropods, mollusks, fishes, amphibians, birds, and mammals.

Such a persuasion led Leeuwenhoek to a new definition of the fertilization process. Having observed the spermatozoa, he was able to proceed beyond the prevalent notion that fertilization occurred from vapors arising out of the seminal fluid and to postulate that the spermatozoa actually penetrated the egg (although he was not able to observe this process). Assuming that the fast-moving spermatozoa were the origin of all new animal life (since he equated life with motility), he went on to state that the new life was nourished by the egg and the uterus. He thus denied any generative role to the motionless (and therefore lifeless) egg, and placed himself in direct opposition to those who, like Harvey in the *De generatione* of 1651, held the egg to be the source of all new life. Leeuwenhoek thus arrived at an animalculist theory of reproduction. A critical case in point was that of reproduction in mammals. The ovists thought that the follicle (discovered by de

LEEUWENHOEK

Graaf and named after him) was, in fact, the egg, while Leeuwenhoek pointed out that it was impossible for the entire follicle to pass through the narrow fallopian tube to the uterus. (It is interesting to note that the actual mammalian egg was not found until 1826, when K. E. von Baer discovered it, and that the ovist-animalculist controversy persisted until 1875, when Hertwig demonstrated that fertilization represents the fusion of the nuclei of the spermatozoon and the egg.)

A corollary of Leeuwenhoek's theory of reproduction was his denial, following in the footsteps of Swammerdam and Redi, of the Aristotelian theory of generation, whereby some animals were thought to originate from the putrefaction of organic matter. In particular the Aristotelians held that certain tiny animals—such as insects and intestinal worms—were very primitive in structure and were generated in this way. In a series of investigations Leeuwenhoek demonstrated the complex structure of mites, lice, and fleas, and described their copulation and life cycles. His studies took him as far as the problem of reproduction in the eel and in intestinal parasites.

Since he believed all living forms to be functionally similar to one another, Leeuwenhoek also made extensive investigations of reproduction in plants. In a letter of 12 October 1685 he drew upon his studies of the seeds of angiosperms to explain his theory of plant reproduction: since motionless plants could not copulate, each individual had to provide for its own propagation. The embryonic plant was therefore the source of new life; the endosperm was the primary nutritive uterus and the earth the secondary nutritive uterus for the seed of the plant. Since he considered the flower to be the beautiful but functionless ornament of the plant, Leeuwenhoek did not investigate the anthers and ovaries.

Leeuwenhoek endeavored to draw an analogy between the animal system and the plant system when he took up his second important area of investigation, the system of nutrient transport. He minutely analyzed the transport canals, particularly their walls, the transport media (as, for example, blood), and the nutritive matter to be moved. Having no concept of the cell, he imagined the system to be composed of pipes or vessels.

In his studies of animals, Leeuwenhoek distinguished among what he called air vessels, intestinal tubes, chyle vessels, blood vessels, and nerve tubes. He was, however, not always able to interpret their functions, and he was not always successful in drawing analogies within the animal systems. For example, although he was aware of the morphological similarity of the cartilaginous rings in the bronchia of mammals and the chitinous spiral in the trachea of insects, he chose to consider the trachea a blood vessel surrounding the intestinal tube.

Leeuwenhoek was particularly attentive to the blood vessels and the blood. Harvey had described the circulation of the blood in 1628, while Malpighi discovered the capillaries in 1661 and, in 1665, observed the corpuscles for the first time (although he wrongly identified them as fat globules). Leeuwenhoek, unaware of Malpighi's work, effectively rediscovered the blood corpuscles, in 1674, and the blood capillaries (in 1683). He went on to describe and measure the erythrocytes and their nuclei in fishes, amphibians, and mammals, and further investigated the walls of the blood vessels. (In the course of these studies Leeuwenhoek devised an ingenious apparatus —his famous "eel-spy"—to demonstrate the corpuscles and blood circulation in the tail of a live eel.) Nevertheless, Leeuwenhoek did not fully understand the function of the blood. Although he recognized it as the means whereby food particles were moved from the intestinal tube to the tissues, he rejected, on the basis of air-pump experiments, the notion that it might also serve to transport air.

Leeuwenhoek's work coincided again with that of Malpighi (as well as that of Grew) in the area of plant anatomy. Indeed, the three must be considered the founders of that study. Leeuwenhoek brought remarkable insight to the three-dimensional structure of the root, stem, and leaf, and illustrated his findings with radial, tangential, and cross sections of those organs. He considered plant tissue to be composed of a complex of tubes surrounded by other tissues made up of globules (cells), but his terminology is often unclear because he only rarely understood the function of the structures that he identified.

For example, in a long letter to the Royal Society of 12 January 1680, Leeuwenhoek described the elements that make up several kinds of wood. It is clear that he here considered all vertical and horizontal woody tracts to be sap-transporting vessels, in opposition to Grew, who distinguished between air vessels and sap vessels. Leeuwenhoek had observed the cross-walls of the rays and sieve tubes in plants on many occasions, and had drawn elaborate diagrams of them. He now interpreted them as being valves, analogous to those in animal veins, and was thus led to formulate a theory whereby the sap was moved from valve to valve by purely mechanical means. Twelve years later, however, in a letter of 12 August 1692, he changed his mind and, in a lengthy comparison of microscopic structures in plants and animals, arrived at something very close to Grew's theory. He now assumed the horizontal and vertical tracts in plants to be air tubes, and the secondary thickenings of these elements and the intercellular spaces between them to be sap vessels.

It was through letters—more than 300 of them, written to private scientists and amateurs in both Holland and other countries—that Leeuwenhoek made his work known. He wrote exclusively in Dutch, but had a few of his letters translated for the benefit of his correspondents. Primarily, though, he was indebted to the Royal Society of London for the publication of his views, and 190 of his letters are addressed to that body.

Leeuwenhoek's correspondence with the Royal Society began, through the intermediacy of de Graaf, one of the Society's Dutch correspondents, with his letter of 28 April 1673. This initial communication convinced the fellows of the great significance of his microscopic findings, and the Society prepared English translations or summaries of his letters for their *Philosophical Transactions* (1673–1724). It is apparent from the few letters to Leeuwenhoek that survive that he must have exchanged letters with the Royal Society on a regular basis, and that his London correspondents encouraged him to investigate new fields. Leeuwenhoek, in return, transmitted in each letter an exact account of the state of these studies. (It is sometimes perplexing to the modern reader, who may be unaware of the interaction of Leeuwenhoek and the Royal Society, to try to surmount the seeming lack of order in the sequence of subjects treated in the letters.)

Beginning in 1679, the French journal *Recueil d'expériences et observations sur le combat qui procède du mélange des corps* published Leeuwenhoek's letters, translating them from the *Philosophical Transactions*, and French summaries of them, drawn from the same source, appeared in the *Journal des sçavans*. Latin extracts were made and published in the *Acta eruditorum*, beginning in 1682.

Leeuwenhoek himself did not publish his work until 1684, when he brought out some of his letters in Dutch; from 1685 onward he also published Latin translations. He initially edited, reprinted, and reissued some of his letters separately or in groups of two or three, a practice that has resulted in some bibliographical confusion. From 1687 he adopted a more systematic course of publication, however, and in 1718 he brought out a collected edition of his letters in Dutch, followed in 1722 by a Latin edition. (In 1932, in commemoration of the tercentenary of Leeuwenhoek's birth, the Royal Academy of Sciences of Amsterdam and the *Nederlands Tijdschrift voor Geneeskunde* undertook a complete critical edition of all of his letters.)

Leeuwenhoek's scientific achievements were recognized during his lifetime by both his colleagues and the public. In 1680 he was elected a fellow of the Royal Society of London; in 1699 he was appointed a correspondent of the Paris Académie des Sciences; and in 1716 the Louvain College of Professors awarded him a silver medal. In addition to the pension that it gave him, the municipality of Delft made him special awards upon the publication of several of his books. The increasing number of learned and eminent visitors from several countries eventually caused Leeuwenhoek to demand introductory letters; his guests included kings and princes, among them Peter the Great, James II, Frederick the Great, Elector August II of Saxony, and the Grand Duke Cosimo III of Tuscany. In his old age, Leeuwenhoek became a legend; to his displeasure, his fellow townsmen reverently referred to him as a magician.

Four pictures of Leeuwenhoek made during his lifetime remain. He is a figure in Cornelis de Man's *Anatomy Lesson of Dr. Cornelis 'sGravesande*, painted in 1681 and now in the Old and New Hospital, Delft. In 1686 Johannes Verkolje made two portraits of him; one is a painting (now in the National Museum of the History of Science, Leiden), showing Leeuwenhoek holding his diploma of fellowship in the Royal Society, while the other is a mezzotint engraving in which the subject is seen handling one of his small microscopes. Another engraving was made by Jan Goeree in about 1707, and shows Leeuwenhoek in his old age.

BIBLIOGRAPHY

I. Original Works. A complete list of the many editions of Leeuwenhoek's works appears as an appendix to A. Schierbeek's biography, cited below.

His own collected Dutch edition, the *Brieven* or *Werken*, consists of four (sometimes five) volumes: I. *Brieven, Geschreven aan de Wyt-vermaarde Koninglijke Wetenschap-zoekende Societeit, tot Londen in Engeland*, 10 pts. (Leiden–Delft, 1684–1694), with a *Register* (Leiden, 1695)—the title page of each part gives a more substantial idea of its contents; II. *Vervolg Der Brieven, Geschreven aan de Wyt-vermaarde Koninglijke Societeit in Londen* (Leiden, 1688), *Tweede Vervolg Der Brieven* (Delft, 1689), *Derde Vervolg Der Brieven* (Delft, 1693), and *Vierde Vervolg Der Brieven* (Delft, 1694); III. *Vijfde Vervolg Der Brieven, Geschreven aan verscheide Hoge Standspersonen En Geleerde Luijden* (Delft, 1696), *Sesde Vervolg Der Brieven* (Delft, 1697), and *Sevende Vervolg Der Brieven* (Delft, 1702); IV. *Send-Brieven* (Delft, 1718).

His Latin edition also consists of four volumes: I. *Arcana Natura Detecta* (Leiden, 1722), *Continuatio Epistolarum* (Leiden, 1715), and *Continuatio Arcanorum Naturae*

Detectorum (Leiden, 1722); II. *Opera Omnia, Seu Arcana Naturae* (Leiden, 1722); III. *Epistolae ad Societatem Regiam Anglicam* (Leiden, 1719); IV. *Epistolae Physiologicae Super compluribus Naturae Arcanis* (Delft, 1719).

The only comprehensive English translation of the letters is Samuel Hoole, *The Select Works of Antony van Leeuwenhoek*, 2 vols. (London, 1798–1807), which, however, omits all references to animal reproduction.

The Collected Letters of Antoni van Leeuwenhoek, edited by the Leeuwenhoek Commission of the Royal Academy of Sciences of Amsterdam, presently comprises eight volumes (Amsterdam, 1939–1967) of 118 letters written between 1673 and 1692.

II. Secondary Literature. The most important recent works on Leeuwenhoek are C. Dobell, *Antony van Leeuwenhoek and His "Little Animals"* (London–Amsterdam, 1932; 2nd ed., New York, 1958); and A. Schierbeek, *Antoni van Leeuwenhoek. Zijn Leven en Zijn Werken*, 2 vols. (Lochem, 1950–1951), of which a concise English version is *Measuring the Invisible World. The Life and Works of Antoni van Leeuwenhoek FRS* (London–New York, 1959).

Johannes Heniger

LE FEBVRE, NICAISE (*b.* Sedan, France, *ca.* 1610; *d.* London, England, 1669), *pharmacy, chemistry.*

Nicaise Le Febvre (variously spelled Le Fèvre, Le Févre, Le Febure and sometimes incorrectly given the Christian name Nicolas) was born in the important Huguenot center of Sedan in the Ardennes. His father, Claude Le Febvre, was a Protestant apothecary from Normandy who had settled in Sedan during the Wars of Religion; his mother, Françoise de Beaufort, came from a local family of physicians and apothecaries. Following preliminary schooling in the pedagogium of the Calvinist academy of Sedan, Nicaise became an apprentice in his father's shop in 1625. Before he completed his training his father died, and the direction of his education was taken over by Abraham Duhan, a doctor of medicine and professor of philosophy at the academy. After qualifying as a master apothecary Le Febvre continued in his father's business until 1646–1647; he then moved to Paris, where he initially enjoyed the patronage of his coreligionist, the physician Samuel du Clos. Le Febvre soon began to offer private courses in pharmaceutical chemistry. In 1652 he obtained the *privilège* of a royal apothecary and distiller and was appointed demonstrator in chemistry at the Jardin du Roi after the departure of William Davison. Among those attending his courses were a number of English royalist emigrés; and following the restoration of Charles II in 1660, Le Febvre was invited to England as royal professor of chemistry and apothecary to the king's household. He was established in a laboratory in St. James's Palace, which he furnished with supplies from France. Le Febvre was admitted to the Royal Society in November 1661 on the nomination of Sir Robert Moray. His name also appears on the list of registered fellows drawn up at the reorganization of the Society in May 1663. He was an active member and in 1664 was appointed to the chemical committee of the society. Le Febvre died in London in the spring of 1669 and was buried in the parish of St. Martin-in-the-Fields.

Le Febvre's principal contribution to the literature of science is his textbook, the *Traicté de la chymie* (Paris, 1660). This work has a pivotal position in the series of French chemical teaching manuals of the seventeenth century. Its theoretical content, based largely on Paracelsus, Helmont, and Glauber, represents the culmination of the iatrochemical tradition in this series of French texts; its practical content and organization, however, were very closely followed by Nicolas Lemery in his *Cours de chymie* (1675), the most successful textbook of the late seventeenth century employing a corpuscular-mechanist interpretation. In his prefatory dedication to his fellow apothecaries, Le Febvre asserts the superiority of contemporary German pharmaceutical works to their French counterparts: in particular he cites the work of Johann Christian Schröder and Johann Zwelfer. Le Febvre's own text contains a significant infusion of mid-century German influences.

The theoretical introduction to the *Traicté* is an individualistic but interesting synthesis of chemical philosophy. For Le Febvre chemistry is a practical science of natural things with three main branches: philosophical chemistry, which seeks to understand the causes of natural and cosmological phenomena that are beyond the manipulative skill of the chemist to control or reproduce; iatrochemistry or medical chemistry, which seeks to apply the knowledge of natural phenomena acquired from philosophical chemistry to the practice of medicine; and pharmaceutical chemistry, which has as its sole object the preparation of remedies according to the directives of those versed in iatrochemistry. Thus chemistry is both science and art, "a knowledge of Nature itself reduced to operation."

Le Febvre's chemical theory is eclectic, drawing on Neoplatonic, Paracelsian, Helmontian, and Aristotelian ideas. The origin of all natural things is a homogeneous spirit, referred to as the universal spirit, which circulates throughout the cosmos. It is one in essence but threefold in nature. Being at once hot, moist, and dry, it is also known under the names of the three Paracelsian principles, sulfur, mercury, and salt. Individual chemical substances are generated and corporified out of this universal spirit by the action of specifying ferments present in the material matrices in which the spirit is entrapped. In this theoretical schema the four Aristotelian elements of fire, air, water, and earth are not considered as material components of chemical composition but as the four principal cosmogonic regions or matrices in which specification of the universal spirit takes place. Thus, fire is the matter of the heavens, and the stars are the specific ferments moving in this element whence the universal spirit always returns to take on its most perfect idea or form. Air is the first general matrix in which the universal spirit, emanating from the stars in the form of light, is corporified. This first corporeal form of the spirit is a hermaphroditic salt which is then absorbed successively by the other two matrices of water and earth; there it is specificated into the diverse forms of the mineral, vegetable, and animal kingdoms.

In addition to this theoretical discussion of principles and elements which belongs properly to philosophical, or "scientific," chemistry, Le Febvre recognizes five elements or principles which are the products of analysis by fire and belong to the art of chemistry. These elements are water or phlegm, spirit or mercury, sulfur or oil, salt, and earth. There is a limit to the resolution of mixt bodies, as further refinement would remove the specifying idea and corporifying matrix, thus resolving everything into the universal spirit. The art of chemistry resides precisely in refining and purifying the components of mixt bodies by repeated solutions and coagulations so that they may act more efficaciously—without simultaneously resolving everything into the universal spirit.

Following a brief description of the principal operations of the art, Le Febvre gives a description of apparatus illustrated with figures. The most notable illustration depicts an operator carrying out a calcination of antimony by means of a burning lens. In the text Le Febvre draws attention to the gain in weight following calcination which he attributes to the fixation of solar rays in the metal. This mode of calcining antimony had previously been described by Hamerus Poppius in *Basilica antimonii* (1618). Le Febvre's extensive list of chemical preparations, which forms the bulk of the text, are arranged under the headings of the three kingdoms of nature.

Le Febvre's other published work was a description of the preparation of Sir Walter Raleigh's great cordial, a polypharmaceutical preparation which, according to John Evelyn, he prepared before King Charles II in 1662.

BIBLIOGRAPHY

I. Original Works. The various eds. of the *Traicté de la Chymie* are listed in J. R. Partington, *A History of Chemistry*, III (London, 1962), 17–18. The 1st ed. (Paris, 1660) was followed by five eds. in French, two of which (1669, 1696) were published in Leiden. The last French ed. (1751) was enlarged by one Du Moustier, Apothicaire de la Marine (actually the Abbé Lenglet du Fresnoy). An English trans. by P. D. C. Esq. first appeared as *A Compendious Body of Chemistry* (London, 1662). Subsequent English eds. appeared in London in 1664 and 1670 as *A Compleat Body of Chemistry*. A German trans. entitled *Chymischer Handleiter* appeared at Nuremberg in 1672. Subsequent German eds. are reported for 1675, 1676, 1685, and 1688. J. Ferguson, *Bibliotheca chemica*, II (Glasgow, 1906), 18, reports a Latin ed. in 2 vols. (Besançon, 1737).

Le Febvre's discourse on Sir Walter Raleigh's cordial was *A Discourse upon Sir Walter Rawleigh's Great Cordial . . . Rendered into English by Peter Belon* (London, 1664). A French version appeared in London in 1665.

G. Goodwin, "Le Fevre," in *Dictionary of National Biography*, XXXII (London, 1892), 399; and the *British Museum General Catalogue* ("Lefèvre, Nicolas") wrongly attribute to him "Disputatio de myrrhata potione," in J. Pearson, *Critici sacri*, IX (1660). The author of this work was Nicolas Le Fèvre (*d.* 1612), a French theologian. The French translator of Sir Thomas Browne, *La réligion du médecin* (The Hague, 1688), is yet another Nicolas Le Fevre. On these false attributions see the Dorveaux articles below.

II. Secondary Literature. The "Avant-propos" of the *Traicté de la chymie* contains autobiographical reminiscences which are the principal source for Le Febvre's early life. Further background on his origins is in J. Villette, "Un procès entre un chirurgien et des médecins sedanais en 1646," in *Revue d'Ardenne et d'Argonne*, **10** (1903), 1–16. The most detailed and accurate biographical notice is P. Dorveaux, "L'apothicaire LeFebvre Nicaise, dit Nicolas," in *Proceedings of the Third International Congress of the History of Medicine, London, 1922* (Antwerp, 1923), 207–212. Substantially the same article appeared in *Bulletin de la Société d'histoire de la pharmacie*, no. 42 (May 1924), 345–356. See also J. P. Contant, *L'enseignement de la chimie au Jardin Royal des Plantes de Paris* (Paris, 1952), 88–90; J. R. Partington, *A History of Chemistry*, III (London, 1962), 17–24; and G. Goodwin (cited above). The best discussion of Le Febvre's textbook remains H. Metzger, *Les doctrines chimiques en France du début du XVIIe à la fin du XVIIe siècle* (Paris, 1923; repr. 1969), 62–82.

Owen Hannaway

LE FÈVRE (or **LE FÈBVRE), JEAN** (*b.* Lisieux, France, 9 April 1652[?]; *d.* Paris, France, 1706), *astronomy*.

Le Fèvre's origins remain uncertain, although tradition has it that he started life as a weaver.

Scholars have confused him with an instrument maker having the same surname. Not until the age of thirty did he appear in the French scientific community as the friend of a certain Father Pierre, an amateur astronomer slightly older than Le Fèvre who had studied humanities in Lisieux and was professor of rhetoric at the Collège de Lisieux in Paris. Father Pierre, an associate of Picard and of P. de La Hire, recommended his friend to Picard, who, after testing Le Fèvre's ability to calculate astronomical tables, assigned him the task of doing the calculations for the *Connaissances des temps*. Through Picard's influence and probably that of La Hire, Le Fèvre became a member of the Académie des Sciences a few months later. After Picard's death he continued the more pedestrian aspects of Picard's work by calculating astronomical tables, by publishing the *Connaissances des temps*, by making a few astronomical observations, and by assisting La Hire in surveying.

Soon after La Hire published his *Tabulae astronomicae*... (Paris, 1687), Le Fèvre accused him of stealing his astronomical tables and publishing them as his own. Le Fèvre's ill feeling toward his colleague and mentor smoldered until 1700, when it erupted again after La Hire's son, Gabriel-Philippe, had published *Ephemerides ad annum 1701* (Paris, 1700). In the original preface to the *Connaissances des temps* for 1701, Le Fèvre, who probably resented not having been named the Academy's official publisher of ephemerides, let his anger flame and treated the La Hires rudely. The French government, which took offense at this, had the original preface replaced by one praising the La Hires and transferred permission to publish the *Connaissances des temps* from Le Fèvre to the Académie des Sciences. In 1701 the government succeeded in having Le Fèvre excluded from the Academy for having failed to attend meetings regularly, as required by the rules. During the remaining few years of his life Le Fèvre seems to have continued publishing ephemerides under the pseudonym of J. de Beaulieu.

BIBLIOGRAPHY

I. Original Works. Le Fèvre published the *Connaissances des temps* from 1685 to 1701 and may also have published it during 1682–1684. J.-J. Le Français de Lalande, *Bibliographie astronomique* (Paris, 1803), pp. 341–343, gives the preface to the *Connaissances des temps* that Le Fèvre was obliged to delete. Le Fèvre also published *Ephémérides pour les années 1684 et 1685, calculées pour le méridien de Paris* ... (Paris, 1684). The following, published under the pseudonym J. de Beaulieu, may be by Le Fèvre: *Ephémérides des mouvements célestes pour l'an de grâce 1702* ... (Paris, 1702); *Ephémérides des mouvements célestes depuis l'an 1702 jusqu'à 1715* ... (Paris, 1703), republished with the title *Eph. ... 1703 ... 1714 ...;* and *État du ciel pendant l'année 1706* ... (Paris, 1706). Lalande, *op. cit.*, p. 349 identifies Beaulieu as being Charles Desforges; yet it seems reasonable to identify Beaulieu as Le Fèvre if a remark of the *Journal des sçavans* for 1703, p. 274, is interpreted in the context of the confrontation between Le Fèvre and the La Hires. Le Fèvre's MSS are at the Académie des Sciences.

II. Secondary Literature. Amédée Tissot, *Étude biographique sur Jean Le Fèvre* ... (Paris, 1872), gives the most complete biographical survey; M. J. Delacourtie, "L'astronome Jean Le Fèvre," in *Bulletin de la Société historique de Lisieux* (1951–1952), 43–48, accurately summarizes Tissot's study. Both studies fail to give scholarly references, although Tissot states that he used Le Fèvre's MSS, which he had found in the archives of the Académie des Sciences. They confuse a certain Le Fèvre, "Ingénieur pour les instruments de mathématique," who appeared in the Paris scientific community around 1702 and whose activity continued well after Le Fèvre's death, with Le Fèvre—Lalande and the *Journal des sçavans* (see its indexed table of contents) do not make this error. The *Histoire de l'Académie royale des sciences* for 1700, 2nd ed. (Paris, 1761), pp. 109–111, reports on the accuracy of Le Fèvre's tables—Tissot, *op. cit.*, p. 63, erroneously states that this occurred in 1703; the same work for 1686–1699 (Paris, 1733), p. 74, gives an account of a fireball observed by Le Fèvre. Lalande, *op. cit.*, pp. 312–314, details Le Fèvre's activities during the period 1682–1685. Michaud, *Biographie universelle* ..., XXIII (Paris, 1819), 546, gives a good account of Le Fèvre's life. Bonaventure d'Argonne's gossipy *Mélanges d'histoire et de littérature*, 4th ed. (Paris, 1725), I, 145, refers to Le Fèvre's humble origins. C. Wolf, *Histoire de l'observatoire de Paris* ... (Paris, 1902), pp. 221–222, mentions Le Fèvre in connection with the surveying of France. Lalande, *op. cit.*, pp. 341–344; J. Bertrand, *L'Académie des sciences et les académiciens* ... (Paris, 1869), pp. 55–58, 297–298; and *Histoire de l'Académie royale des sciences* for 1701, 2nd ed. (Paris, 1743), pp. 113–114, document Le Fèvre's quarrel with the La Hires. *Histoire de l'Académie royale des sciences* for 1700, 2nd ed. (Paris, 1761), pp. 129–130, recounts that the Academy commissioned G.-P. de La Hire to draw up new astronomical tables. *Journal des sçavans* for 1702, pp. 212–216, and for 1703, pp. 270–274 discusses ephemerides.

Robert M. McKeon

LEGALLOIS, JULIEN JEAN CÉSAR (*b.* Cherrueix, near Dol, France, 1 February 1770; *d.* Paris, France, 10 February 1814), *physiology*.

Legallois was the first of the great French physiologists who based their conclusions on animal experiments. (He was followed by Magendie, Flourens,

Claude Bernard, and Brown-Séquard, among others.) He is chiefly remembered as the first to localize the respiratory center in the medulla and as one of those who anticipated the internal secretions.

Legallois was the son of a Breton farmer who is said to have worked his land himself. The boy's mother died when he was very young, but his father secured him a good education. His father died when Legallois was thirteen and a student at the Collège de Dol; he received a modest inheritance that he used to further his education. He won first prize in rhetoric at Dol and then began to study medicine at Caen. His career, however, was interrupted once by illness and again, after he recovered, by his involvement as a student leader in an armed movement in support of the federalists. After the defeat of his party, Legallois returned home to escape punishment but was denounced by a relative and fled to Paris. There he remained unnoticed among the crowd of young men trying to learn medicine in the hospitals after the abolition of the medical faculties.

Denounced a second time, Legallois braved his fate, and presented himself to the Comité des Poudres et Salpêtres. He proved himself satisfactory to the committee and was sent to organize and direct the manufacture of gunpowder in his native district. The following year, after the foundation of the École de Santé, Legallois, who had distinguished himself by his accomplishments, was delegated by the district committee to go to Paris as a wage-earning student. Besides medicine he studied Greek, Italian, and English, graduating in 1801. His dissertation was addressed to the question of whether blood is identical in all vessels through which it flows. In it Legallois pointed out that although blood is identical in all the arteries, it changes as it passes through the various organs and thus one organ may by its products influence all the others. This dissertation revealed Legallois's interest in experimental physiology and was widely appreciated as a model of physiological discussion.

Even before his graduation Legallois had acquired great experience in practical medicine through working in hospitals for some ten years. While preparing his dissertation he realized the importance of experimental inquiry. For this reason he devoted much of his energy after he qualified to physiological research, although he had no official position in that field. For about ten years he served as physician to the poor of the twelfth civil district of Paris. He wrote two medico-historical surveys (one on the discovery of galvanism and one on the chronology of Hippocrates) and a review of data on the contagious nature of yellow fever. He then began a long series of physiological experiments to determine the conditions basic to maintaining the life of any part of the body or, indeed, of the organism as a whole—in short, what he called "the principle of life."

A clinical observation led him to speculate about how long a newborn, fully developed fetus might live without breathing. He was inspired in part by the experiments made by Lavoisier and Laplace in connection with the generation of animal heat and its relation to respiratory exchange, but soon he noticed that life closely depends on two functions, those of the nervous system and the circulation of the blood. Having noted that the interruption of circulation at some specific point led to the death of the part of the body affected, and that stoppage of the heart caused the death of the whole animal, Legallois formed his remarkable idea that it should be possible to maintain life by a sort of injection, "and if there be a continuous supply of arterial blood, natural or artificially prepared—provided such a procedure be possible—life of any portion could be indefinitely maintained. . ., even in the head after decapitation, with all the functions proper to the brain." He pointed out in this way life could be reestablished in a part reduced to a state of apparent death after the cessation of circulation. At a time when profuse bloodletting was the therapeutic procedure of choice, Legallois's idea of revival (or "resurrection," as he called it) through bringing arterial blood to the tissues was an outstanding one. Although he reached his conclusion from physiological experiments, the technique was developed and applied in practical medicine only much later.

In spite of his severe myopia and short, thick fingers, Legallois was very skillful in experiments on living animals. His general approach was equally important. He recognized both the possibilities and the limitations of the experimental method as well as the conditions that must be observed to reach valid results. He therefore stressed the importance of the choice of animals and the necessity of the controls being of the same species, sex, and age, writing that "The greatest harm to experimental physiology was the negligence of researchers in the choice of experimental animals [which] they took . . . as they fell into their hands without attention to . . . species or age, and compared the results of experiments made in this manner as though they had compared animals of the same species and of the same age." Legallois thought that life depended on the impression of blood on the brain and the spinal cord, or on a principle resulting from this impression; death was therefore due to the extinction of that principle and might be only partial if the extinction were incomplete.

It had already been well accepted that locomotion and motility in general depend on the brain, which determines and regulates all animal functions. Legallois, however, concluded from his experiments that the principle of sensation and motility could only reside in the spinal cord, and that the brain is not the unique origin of nervous power. One of his most important achievements was the demonstration of the metameric organization of the spinal cord, by which each segment serves as a center of a specific region, coordinating its sensory and motor activity. "If instead of destroying spinal cord we divide it into its transverse sections, parts of the body corresponding to each spinal-cord segment conserve sensation and motility, but without harmony and each one independent of the others as if the sections were made across the whole body of the animal. In short, there are as many centers of sensation as there are spinal segments." The brain thus regulates the movements of the body without supplying the immediate principle; this originates in the spinal cord. The brain acts on the spinal cord in the same way that the spinal cord acts on the muscles, the white matter of the brain being composed of filaments which end in the gray matter of the cord. This gray matter, the origin of the spinal nerves, is the seat of the principle of life.

The mechanics of respiration—those movements which make possible the entry of air into the lungs—depend immediately on the brain; since life is maintained by respiration, the life of the animal depends on the brain, through the medium of the spinal cord. Legallois then, in 1812, convincingly demonstrated that respiratory movements originate in a small area of the medulla oblongata near the occipital foramen and near the level of the origin of the vagus nerves. Indeed, if the cranium of a young rabbit is opened and the brain cut off in serial transverse slices, beginning at the stem, the whole of the cerebrum and of the cerebellum—as well as a part of medulla—can be removed; and respiration ceases suddenly only when the lesion reaches the level of the vagus nerves. This effect is not due to the elimination of the vagi, since in an adult animal respiration goes on (although in a somewhat modified form) for a long time after their transsection.

Legallois was the first to draw attention to variations according to age in the respiratory reactions to the excitation of the central nervous system of animals of the same species. "Repetition of the same experiments on animals of various ages could throw light on many questions in physiology," he wrote.

Legallois was cautious in drawing conclusions from his experiments. In dealing with complex phenomena he tried to isolate the elementary features and thus to determine the general laws of nature. His experiments were remarkable in the diversity and ingenuity of their design and arranged in a logical sequence. Legallois's unshakable belief in the supreme importance of observations and well-performed experiments, together with his reserve in interpreting his data, indicate that he was a scientist of great stature.

Legallois died prematurely, about a year after his appointment to the great hospital and asylum of Bicêtre in the outskirts of Paris. He used to walk there from his home every day and in winter, after one such walk, contracted peripneumonia. Well aware of the physiological role of the blood, he refused the usual treatment of the time, bleeding, objecting that his inflammation was of "adynamic" nature, a theory to which some of his biographers believed that he fell victim. Legallois's notion was, however, sound, although several decades had to pass before the disastrous practice of therapeutic bloodletting was abandoned.

BIBLIOGRAPHY

I. ORIGINAL WORKS. Legallois's works were posthumously collected and edited by his son Eugène as *Oeuvres de J. J. C. Legallois avec des notes de M. Pariset*, 2 vols. (Paris, 1824; 2nd ed., 1830). His most important and best-known work, *Expériences sur le principe de la vie, notamment sur celui des mouvements du coeur, et sur le siège de ce principe* (Paris, 1812), was translated into English by N. Nancrede and J. Nancrede as *Experiments on the Principle of Life and Particularly on the Principle of the Motion of the Heart, and on the Seat of This Principle* (Philadelphia, 1813).

His other works include *Le sang est-il identique dans tous les vaisseaux qu'il parcourt?* (Paris, 1802), his M.D. diss.; *Recherches chronologiques sur l'Hippocrate* (Paris, 1804); *Recherches sur la contagion de la fièvre jaune* (Paris, 1805); several articles in the *Dictionnaire des sciences médicales;* and the posthumous *Fragments d'un mémoire sur le temps durant lequel les jeunes animaux peuvent être, sans danger, privés de la respiration* (Paris, 1834).

II. SECONDARY LITERATURE. There is a biographical notice in Legallois's *Oeuvres*, I, 1–11; and an anonymous "Notice biographique sur M. Legallois," in *Bulletin et mémoires de la Société de la Faculté de médecine de Paris*, **4** (1814–1815), 105–109. See also the article on Legallois in *Dictionnaire des sciences médicales, biographie médicale*, V (Paris, 1822), 565–566; G. Canguilhem, *La formation du concept de réflexe aux XVIIe et XVIIIe siècles* (Paris, 1955); E. Clark and C. D. O'Malley, *The Human Brain and Spinal Cord* (Berkeley–Los Angeles, 1968); A. Dechambre, in *Dictionnaire encyclopédique des sciences médicales*, 2nd ser., II (Paris, 1876), 138–139; P. Huard, "César Legallois (1770–1814) découvre en 1811 le principe de la résuscitation," in *Histoire de la médecine*, **4** (1954), 23–25; M. Neuburger, *Die historische Entwicklung der experimentellen Gehirn- und Rückenmarksphysiologie von Flourens* (Stuttgart, 1897); and J. Sourry, *Système nerveux central. Structure et fonctions* (Paris, 1899).

The reports of the National Institute of France on *Expériences sur le principe de la vie* and on Legallois's other experiments are trans. in A. P. W. Philip, *An Experimental Inquiry Into the Laws of the Vital Functions* (London, 1817; Philadelphia, 1818).

VLADISLAV KRUTA

LEGENDRE, ADRIEN-MARIE (*b.* Paris, France, 18 September 1752; *d.* Paris, 9 January 1833), *mathematics.*

Legendre, who came from a well-to-do family, studied in Paris at the Collège Mazarin (also called Collège des Quatre-Nations). He received an education in science, especially mathematics, that was unusually advanced for Paris schools in the eighteenth century. His mathematics teacher was the Abbé Joseph-François Marie, a mathematician of some renown and well-regarded at court. In 1770, at the age of eighteen, Legendre defended his theses in mathematics and physics at the Collège Mazarin. In 1774 Marie utilized several of his essays in a treatise on mechanics.

Legendre's modest fortune was sufficient to allow him to devote himself entirely to research. Nevertheless he taught mathematics at the École Militaire in Paris from 1775 to 1780.

In 1782 Legendre won the prize of the Berlin Academy. The subject of its competition that year concerned exterior ballistics: "Determine the curve described by cannonballs and bombs, taking into consideration the resistance of the air; give rules for obtaining the ranges corresponding to different initial velocities and to different angles of projection." His essay, which was published in Berlin, attracted the attention of Lagrange, who asked Laplace for information about the young author. A few years later the Abbé Marie and Legendre arranged for Lagrange's *Mécanique analytique* (Paris, 1788) to be published and saw it through the press.

Meanwhile, Legendre sought to make himself better known in French scientific circles, particularly at the Académie des Sciences. He conducted research on the mutual attractions of planetary spheroids and on their equilibrium forms. In January 1783 he read a memoir on this problem before the Academy; it was published in the *Recueil des savants étrangers* (1785). He also submitted to Laplace essays on second-degree indeterminate equations, on the properties of continued fractions, on probabilities, and on the rotation of bodies subject to no accelerating force. As a result, on 30 March 1783 he was elected to the Academy as an *adjoint mécanicien*, replacing Laplace, who had been promoted to *associé.*

Legendre's scientific output continued to grow. In July 1784 he read before the Academy his "Recherches sur la figure des planètes." Upon the publication of this memoir, he recalled that Laplace had utilized his works in a study published in the *Mémoires de l'Académie des sciences* for 1782 (published in 1784) but written after his own, which allowed "M. de Laplace to go more deeply into this matter and to present a complete theory of it." The famous "Legendre polynomials" first appeared in these 1784 "Recherches."

The "Recherches d'analyse indéterminée" (1785) contains, among other things, the demonstration of a theorem that allows decision on the possibility or impossibility of solution of every second-degree indeterminate equation; an account of the law of reciprocity of quadratic residues and of its many applications; the sketch of a theory of the decomposition of numbers into three squares; and the statement of a theorem that later became famous: "Every arithmetical progression whose first term and ratio are relatively prime contains an infinite number of prime numbers."

In 1786 Legendre presented a study on the manner of distinguishing maxima from minima in the calculus of variations. The "Legendre conditions" set forth in this paper later gave rise to an extensive literature. Legendre next published, in the *Mémoires de l'Académie* for 1786, two important works on integrations by elliptic arcs and on the comparison of these arcs; here can be found the rudiments of his theory of elliptic functions.

In the works cited above, Legendre had marked off his favorite areas of research: celestial mechanics, number theory, and the theory of elliptic functions. Although he did take up other problems in the course of his life, he always returned to these subjects.

Legendre's career at the Academy proceeded without any setbacks. On the reorganization of the mechanics section he was promoted to *associé* (1785). In 1787, along with Cassini IV and Méchain, he was assigned by the Academy to the geodetic operations undertaken jointly by the Paris and Greenwich observatories. On this occasion he became a fellow of the Royal Society. His work on this project found expression in the "Mémoire sur les opérations trigonométriques dont les résultats dépendent de la figure de la terre." Here are found "Legendre's theorem" on spherical triangles:

> When the sides of a triangle are very small in relation to the radius of the sphere, it differs very little from a

rectilinear triangle. If one subtracts from each of its angles a third of the excess of the sum of the three angles over [the sum of] two right angles, the angles, diminished in this manner, may be considered those of a rectilinear triangle whose sides are equal in length to those of the given triangle.

This memoir also contains his method of indeterminate corrections:

> In these calculations there exist some elements that are susceptible to a slight uncertainty. In order to make the calculation only once, and in order to determine the influence of the errors at a glance, I have supposed the value of each principal element to be augmented by an indeterminate quantity that designates the required correction. These literal quantities, which are considered to be very small, do not prevent one from carrying out the calculation by logarithms in the usual manner.

In the *Mémoires de l'Académie* for 1787 Legendre published an important theoretical work stimulated by Monge's studies of minimal surfaces. Entitled "L'intégration de quelques équations aux différences partielles," it contains the Legendre transformation. Given the partial differential equation

$$R\frac{\partial^2 z}{\partial x^2} + S\frac{\partial^2 z}{\partial x\,\partial y} + T\frac{\partial^2 z}{\partial y^2} = 0.$$

Set $\partial z/\partial x = p$ and $\partial z/\partial y = q$. Suppose R, S, T are functions of p and q only. Legendre is concerned with the plane tangent to the surface being sought: $z = f(x, y)$. The equation of this plane being $pX + qY - Z - v = 0$, he shows that v satisfies the equation

$$R\frac{\partial^2 v}{\partial q^2} - S\frac{\partial^2 v}{\partial p\,\partial q} + T\frac{\partial^2 v}{\partial p^2} = 0.$$

Later, in volume II of his *Traité des fonctions elliptiques* (1826), Legendre employed an analogous procedure to demonstrate "the manner of expressing every integral as an arc length of a curve." If $\int p d\omega$ is the integral in question (p being a function of ω), he shows that the arc length of the envelope of the family of straight lines $x \cos \omega + y \sin \omega - p = 0$ is equal to $\int p d\omega + p'$, where p' is the derivative of p with respect to ω.

In 1789 and 1790 Legendre presented his "Mémoire sur les intégrales doubles," in which he completed his analysis of the attraction of spheroids; a study of the case of heterogeneous spheroids; and some investigations of the particular integrals of differential equations.

In April 1792 Legendre read before the Academy an important study on elliptic transcendentals, a more systematic account of material presented in his first works on the question, dating from 1786. The academies were suppressed in August 1793; consequently he published this study himself, toward the end of the same year, in a quarto volume of more than a hundred pages.

The times were difficult for everyone and for Legendre in particular. He may even have been obliged to hide for a time. In any case his "small fortune" disappeared, and he was obliged to find work, especially since he had married a girl of nineteen, Marguerite Couhin. He later wrote to Jacobi:

> I married following a bloody revolution that had destroyed my small fortune; we had great problems and some very difficult moments, but my wife staunchly helped me to put my affairs in order little by little and gave me the tranquillity necessary for my customary work and for writing new works that have steadily increased my reputation.

On 13 April 1791 Legendre had been named one of the Academy's three commissioners for the astronomical operations and triangulations necessary for determining the standard meter. His colleagues were Méchain and Cassini IV, who four years earlier had participated with him in the geodetic linking of the Paris and Greenwich meridians. On 17 March 1792, however, Legendre had requested to be relieved of this assignment. In ventôse *an* II (February–March 1794) the Commission of Public Instruction of the department of Paris appointed him professor of pure mathematics at the short-lived Institut de Marat, formerly the Collège d'Harcourt. During 1794 he was head of the first office of the National Executive Commission of Public Instruction (the second section, Sciences and Letters). He had eight employees under him and was expected to concern himself with weights and measures, inventions and discoveries, and the encouragement of science.

A tireless worker, during this same period Legendre published his *Éléments de géométrie.* This textbook was to dominate elementary instruction in the subject for almost a century. On 29 vendémiaire *an* II (20 October 1793) the Committee of Public Instruction, of which Legendre soon became senior clerk, commissioned him and Lagrange to write a book entitled *Éléments de calcul et de géométrie.* Actually his work must already have been nearly finished. He had probably been working on it for several years, encouraged perhaps by Condorcet, who stated in 1791, in his *Mémoires sur l'instruction publique:* "I have often spoken of elementary books written for children and for adults, of works designed to serve as guides for teachers. . . . Perhaps it is of some use to

LEGENDRE

mention here that I had conceived the project for these works and prepared the means necessary to execute them. . . ."

In the meantime the decimal system had been adopted for the measurement of angles; and the survey offices, under the direction of Prony, undertook to calculate the sines of angles in ten-thousandths of a right angle, correct to twenty-two decimal places; the logarithms of sines in hundred-thousandths of a right angle, correct to twelve decimal places; and the logarithms of numbers from 1 to 200,000, also to twelve decimal places. In 1802 Legendre wrote: "These three tables, constructed by means of new techniques based principally on the calculus of differences, are one of the most beautiful monuments ever erected to science." Prony had them drawn up rapidly through a division of labor that permitted him to employ people with low qualifications. The work was prepared by a section of analysts headed by Legendre, who devised new formulas for determining the successive differences of the sines. The other sections had only to perform additions. This collective work resulted in two completely independent copies of the tables, which were mutually verified by their identity. These manuscript tables were deposited at the Bureau des Longitudes. An explanatory article appeared in the *Mémoires de l'Institut* (1801).

Legendre figured neither among the professors of the *écoles normales* of *an* III nor among those of the École Polytechnique. He did, however, succeed Laplace, in 1799, as examiner in mathematics of the graduating students assigned to the artillery. He held this position until 1815, when he voluntarily resigned and was replaced by Prony. He was granted a pension of 3,000 francs, equal to half his salary. He lost it in 1824 following his refusal to vote for the official candidate in an election for a seat in the Institute.

Legendre was not one of the forty-eight scholars selected in August 1795 to form the nucleus of the Institut National, but on 13 December he was elected a resident member in the mathematics section. In 1808, upon the creation of the University, he was named a *conseiller titulaire*. A member of the Legion of Honor, he also obtained the title of Chevalier de l'Empire—a minor honor compared with the title of count bestowed on his colleagues Lagrange, Laplace, and Monge. When Lagrange died in 1813, Legendre replaced him at the Bureau des Longitudes, where he remained for the rest of his life.

We now return to Legendre's scientific publications. The first edition of his *Essai sur la théorie des nombres* appeared in 1798. In it he took up in a more systematic and more thorough fashion the topics covered in his "Recherches" of 1785.

His *Nouvelles méthodes pour la détermination des orbites des comètes* (1806) contains, in a supplement, the first statement and the first published application of the method of least squares. Gauss declared in his *Theoria motus corporum coelestium* (1809) that he himself had been using this method since 1795. This assertion, which was true, infuriated Legendre, who returned to the subject in 1811 and 1820.

Legendre had been dismayed once before by Gauss: in 1801 the latter attributed to himself the law of reciprocity of quadratic residues, which Legendre had stated in 1785. Later, in 1827, Legendre wrote to Jacobi: "How has M. Gauss dared to tell you that the majority of your theorems [on elliptic functions] were known to him and that he had discovered them as early as 1808? This excessive impudence is unbelievable in a man who has sufficient personal merit not to have need of appropriating the discoveries of others."

Both Legendre in his indignation and Gauss in his priority claims were acting in good faith. Gauss considered that a theorem was his if he gave the first rigorous demonstration of it. Legendre, twenty-five years his senior, had a much broader and often a hazier sense of rigor. For Legendre, a belated disciple of Euler, an argumentation that was merely plausible often took the place of a proof. Consequently all discussion of priority between the two resembled a dialogue of the deaf.

In "Analyse des triangles tracés sur la surface d'un sphéroïde" (read before the Institute in March 1806), in which he considered the triangles formed by the geodesics of an ellipsoid of revolution, Legendre generalized his theorem concerning spherical triangles. Gauss provided a much broader generalization of this theorem in 1827, in his "Disquisitiones generales circa superficies curvas."

Legendre brought out the second edition of *Essai sur la théorie des nombres* in 1808. He later wrote two supplements to it (1816, 1825).

The "Recherches sur diverses sortes d'intégrales définies" (1809) continued an early study of Eulerian integrals—the term is Legendre's—in particular of the "gamma function." The earlier work appeared in 1793, in *Mémoire sur les transcendantes elliptiques*. The investigations of 1809 were, in turn, revised, completed, and enlarged in later works devoted to elliptic functions (1811, 1816, 1817, 1826).

Volume I of the *Exercices de calcul intégral* (1811) contains a majority of the results that Legendre obtained in the study of elliptic integrals. Volume III was published in 1816. Volume II, which includes the important numerical tables of the elliptic integrals, appeared in 1817.

In 1823 Legendre published "Recherches sur quelques objets d'analyse indéterminée et particulièrement sur le théorème de Fermat" in *Mémoires de l'Académie.* It contains a beautiful demonstration of the impossibility of an integral solution of the equation $x^5 + y^5 = z^5$, followed by an examination of more complicated cases of the theorem. This memoir was reproduced as the second supplement to the *Essai sur la théorie des nombres* (1825).

In 1825 and 1826, in *Traité des fonctions elliptiques*, Legendre took up once more and developed the essential aspects of his 1811 work, including applications to geometry and mechanics. In 1827, however, Jacobi informed Legendre of his own discoveries in this area. The latter, inspired by the contributions of his correspondent and by those of Abel, published three supplements to his *Théorie des fonctions elliptiques* (the title of volume I of the *Traité*) in rapid succession (1828, 1829, 1832).

In May 1830 the third edition of Legendre's *Théorie des nombres* appeared. In this two-volume work he developed the material of the *Essai* of 1808, adding new thoughts inspired, to a large extent, by Gauss. Jacobi drew his attention to a weak point in his reasoning, and on 11 October 1830 Legendre presented a corrected memoir that appeared in 1832 in the Academy's *Mémoires*. Shortly before his death he referred to this memoir as a necessary complement and conclusion to his *Traité* and expressed the need for printing it at the end of that work. His wish was not fulfilled.

The "Réflexions sur différentes manières de démontrer la théorie des parallèles ou le théorème sur la somme des trois angles du triangle" appeared in the *Mémoires* in 1833. Legendre had already sent a separately printed copy of it to Jacobi at the end of June 1832.

Legendre died on 9 January 1833, following a painful illness. His health had been failing for several years. His wife, who died in December 1856, made a cult of his memory and until her death displayed a naïve, religious respect for everything that had belonged to him. She left to the village of Auteuil (now part of Paris) the last country house in which they had lived. They had no children.

We shall now return to the three main fields treated in Legendre's works: number theory, elliptic functions, and elementary geometry—more particularly, the theory of parallel lines.

Number Theory. In the introduction to the second edition of his *Théorie des nombres* (1808) Legendre exhibited an admirable concern for rigor. For example, he demonstrated the commutativity of the product of integers by a technique related to Fermat's method of infinite descent. A direct disciple of Euler and Lagrange, he, like them, made frequent use of the algorithm of continued fractions, as much to solve first-degree indeterminate equations as to show that Fermat's equation $x^2 - Ay^2 = 1$ always admits an integral solution.

Moreover, Legendre followed Lagrange step by step in the study of quadratic forms, a study that he completed in some respects. He showed, for example, that every odd number not of the form $8k + 7$ is the sum of three squares. (He had shown this imperfectly as early as 1785 and in a nearly satisfactory manner in 1798.) On the basis of this result, Cauchy, in 1812, demonstrated Fermat's theorem for the case of polygonal numbers.

Legendre's principal contribution was the law of reciprocity of quadratic residues. He stated it as early as 1785, when he produced a very long and imperfect demonstration of it. In 1801 Gauss subjected it to a thorough criticism and was able to declare that he was the first to have demonstrated the proposition rigorously. In 1808, while preserving his first exposition, improved in 1798, Legendre adopted the proof given by his young critic. In 1830 he added to it that of Jacobi, which he found superior.

We have already mentioned Legendre's contributions to the study of Fermat's great theorem concerning the impossibility of finding an integral solution of the equation $x^n + y^n + z^n = 0$. In this connection he had met Dirichlet, a young mathematician of great promise.

A very skillful calculator, Legendre furnished valuable tables listing the quadratic and linear divisors of quadratic forms and the least solutions of Fermat's equation $x^2 - Ay^2 = \pm 1$. For the latter table, published in 1798 and reproduced in a much abridged form in 1808, he later (1830) utilized the corrections made by the Danish mathematician C. F. Degen.

Legendre should be considered a precursor of analytic number theory. His law of the distribution of prime numbers, outlined in 1798 and made more precise in 1808, took the following form: If y is the number of prime numbers less than x, then $y = \frac{x}{\log x - 1.08366}$. He found it, he stated, by induction. In 1830 he pointed out again that it had been found through induction and that "it remains to demonstrate it a priori"; and he developed some observations on this subject in a manner similar to Euler's. About 1793 Gauss intuited the law of the asymptotic distribution of prime numbers, which could have occurred to any attentive reader of Euler. But it was Legendre who first drew attention to this remarkable law, which was not truly demonstrated until 1896 (by Charles de la Vallée Poussin and Jacques Hadamard).

On the other hand, Legendre thought he had demonstrated, as early as 1785, that in every arithmetic progression $ax + b$ where a and b are relative primes, there is an infinite number of primes. He even specified that in giving to b the $\varphi(a)$ values prime to a and less than this number ($\varphi(a)$ being Euler's indicatrix), the prime numbers are distributed almost equally among the $\varphi(a)$ distinct progressions. These propositions were first rigorously demonstrated by Dirichlet in 1837.

Giving a rather broad scope to number theory, Legendre sought in 1830 to present Abel's conceptions concerning the algebraic solution of equations. Legendre thought it had been convincingly proved that such a solution is in general impossible for degrees higher than the fourth. He was also interested in the numerical solution of equations. In particular he studied the separation of roots and their expansion as continued fractions. In 1808 he presented a demonstration of the fundamental theorem of algebra that was quite analogous to that given by J. R. Argand in 1806. These essentially correct analytical demonstrations required only a few restatements to be made rigorous.

It should be noted that in 1806 Legendre's attitude toward Argand was very understanding. He did not publicly adopt the latter's ideas on the geometric representation of complex numbers; but thanks to the letter that he wrote to François Argand concerning his brother's discovery, that discovery was not lost, and in 1813 it reached a large audience through Gergonne's *Annales*.

In note 4 of his *Éléments de géométrie*—a note included in the first editions of the work—Legendre, by employing the algorithm of continued fractions, established Lambert's theorem (1761): the ratio of the circumference of a circle to its diameter is an irrational number. He improved this result by showing that the square of this ratio is also irrational, and added: "It is probable that the number π is not even included in the algebraic irrationals, but it appears to be very difficult to demonstrate this proposition rigorously."

Much attracted throughout his life by number theory, Legendre was well aware of the difficulties it presents and in his last years experienced a sort of disenchantment with it. For example, in 1828 he wrote to Jacobi: "I would advise you not to give too much time to investigations of this nature: they are very difficult and are often fruitless."

Number theory, of course, does not constitute the most significant portion of his *oeuvre*. Instead, he should be considered the founder of the theory of elliptic functions. MacLaurin and d'Alembert had studied integrals expressible by arcs of an ellipse or a hyperbola. Fagnano had shown that to any given ellipse or hyperbola two arcs the difference of which equals an algebraic quantity can be assigned in an infinite number of ways (1716). He had also demonstrated that the arcs of Bernoulli's lemniscate $(x^2 + y^2)^2 = a^2(x^2 - y^2)$ can be multiplied and divided algebraically like the arcs of a circle. This was the first demonstration of the use of the simplest of the elliptic integrals, the one later designated by Legendre as $F(x)$ and considered by him to govern all the others.

In 1761 Euler had found the complete algebraic integral of a differential equation composed of two separate but similar terms, each of which is integrable only by arcs of conics: $\frac{dx}{R(x)} + \frac{dy}{R(y)} = 0$, R being the square root of a fourth-degree polynomial. This discovery, made almost by chance, allowed Euler to compare—in a more general manner than had ever been done previously—not only arcs of the same conic section or lemniscate but also, in general, all the transcendentals $\int \frac{Pdx}{R}$, where P is a rational function of x and R is the square root of a fourth-degree polynomial.

In 1768 Lagrange undertook to incorporate Euler's discovery into the ordinary procedures of analysis and, in 1784 and 1785, he presented a general method for finding the integrals $\int \frac{Pdx}{R}$ by approximation. Meanwhile, John Landen had demonstrated in 1775 that every arc of a hyperbola can be measured by two arcs of an ellipse. As a result of this memorable discovery the term "Landen's theorem" was applied not only to this result but also to the first known transformation of elliptic integrals.

Such was the state of research in the theory of elliptic transcendentals in 1786, when Legendre published his works on integration by elliptic arcs. The first portion of these, written before he had become aware of Landen's discoveries, contained new ideas concerning the use of elliptic arcs, notably a method of avoiding the use of hyperbolic arcs by replacing them with a suitably constructed table of elliptic arcs. He then gave a new demonstration of Landen's theorem and proved by the same method that every given ellipse is part of an infinite sequence of ellipses, related in such a way that by the rectification of two arbitrarily chosen ellipses the rectification of all the others is obtained. With this theorem it was possible to reduce the rectification of a given ellipse to that of two others differing arbitrarily little from a circle.

But this topic, and the theory of elliptic transcendentals in general, required a more systematic treatment. This is what Legendre, who was virtually alone in his interest in the problem, attempted to provide in his "Mémoire sur les transcendantes elliptiques" (1793). He proposed to compare all functions of this type, classify them into different species, reduce each one to the simplest possible form, evaluate them by the easiest and most rapid approximations, and create a sort of algorithm from the theory as a whole.

Later research having enabled him to perfect this theory in several respects, Legendre returned to it in the *Exercices* of 1811. On this occasion he gave his theory a trigonometric appearance. Setting $\Delta = \sqrt{1 - c^2 \sin^2 \varphi}$, with $0 \leqslant c \leqslant 1$, he calls c the modulus of the function. The integral being taken from 0 to φ, φ is called its amplitude; $\sqrt{(1 - c^2)} = b$ is the complement of the modulus. The simplest of the elliptic transcendentals is the integral of the first kind: $F(\varphi) = \int \frac{d\varphi}{\Delta}$. The integral of the second kind, which is representable by an elliptic arc of major axis 1 and eccentricity c, takes the form $E(\varphi) = \int \Delta d\varphi$. The integral of the third kind is

$$\Pi(\varphi) = \int \frac{d\varphi}{(1 + \eta \sin^2 \varphi) \Delta},$$

with parameter n. Every elliptic integral can be expressed as a combination of these three types of transcendentals.

Let φ and ψ be two variables linked by the differential equation

$$\frac{d\varphi}{\sqrt{1 - c^2 \sin^2 \varphi}} + \frac{d\psi}{\sqrt{1 - c^2 \sin^2 \psi}} = 0.$$

The integral of the equation is $F(\varphi) + F(\psi) = F(\mu)$, μ being an arbitrary constant. Euler's theorem gives

$$\cos \varphi \cos \psi - \sin \varphi \sin \psi \sqrt{1 - c^2 \sin^2 \mu} = \cos \mu.$$

Thus it allows μ to be found algebraically, such that $F(\varphi) + F(\psi) = F(\mu)$, and then $F(\varphi)$ to be multiplied by an arbitrary number, whole or rational. From these observations Legendre deduced many consequences for each of the three kinds of integrals.

Furthermore, by designating the integral of the first kind with modulus c and amplitude φ as $F(c, \varphi)$, Legendre was able, with the aid of Landen's theorem, to establish the transformation later called quadratic. Thus, if $\sin(2\varphi' - \varphi) = c \sin \varphi$ and if $c' = \frac{2\sqrt{c}}{1 + c}$, then $F(c', \varphi') = \frac{1 + c}{2} F(c, \varphi)$. Through repeated use of this transformation Legendre constructed and published (1817) his tables of elliptic functions. In 1825 he wrote:

> To render the theory wholly useful it remained to construct a series of tables by means of which one could find, in every given case, the numerical value of the functions. These tables have finally been constructed, after a multitude of investigations undertaken with a view toward discovering the methods and formulas most suitable to diminishing the length and difficulty of the calculations.

The work Legendre published in 1826, volume II of *Traité des fonctions elliptiques*, contains nine of these tables. The last one is "the general table of the functions F and E for each [sexagesimal] degree of the amplitude φ and of the angle of the modulus θ ($\sin \theta = c$) to ten decimal places for θ less than 45° and nine for θ between 45° and 90°." He wrote to Jacobi regarding these enormous computations that he carried out unassisted: "My goal has always been to introduce into calculation new elements that one can work with in arbitrary numbers, and I devoted myself to an exceedingly long and tedious task in order to construct the tables, a task I do not hesitate to believe is as considerable as that of Briggs's great tables."

Volume II of the *Traité* (1826) includes material on the construction of elliptic functions that is of the greatest interest from the point of view of numerical analysis. In particular Legendre presents a symbolic calculus, inspired by Lagrange, linking the expansion of a function by Taylor's formula with the calculation of its finite differences of various orders. Other investigations of this topic may be found in the analogous studies by Arbogast and Kramp, and by the students of Hindenburg. The same volume of the *Traité* lists—to twelve decimal places—the logarithms of the values of the gamma function $\Gamma(x)$ when x varies in thousandths of an integer and ranges from 1 to 2 inclusively.

In the meantime, around 1825, chance led Legendre to examine two functions F linked by relationships between their moduli, on the one hand, and their amplitudes, on the other—relationships that do not arise in the context of Landen's transformation. In generalizing these relationships Legendre discovered a new transformation closely related to the trisection of the function F. This trisection necessitated the solution of a ninth-degree algebraic equation. The new transformation reduced this solution to that of two third-degree equations.

In 1827, however, Jacobi (whose correspondence with Legendre, always of the greatest scientific and human interest, continued until 1832) communicated to Legendre his own discoveries and also informed him of Abel's. Legendre's attitude in the face of the discoveries of his young rivals was remarkable for its enthusiasm and forthrightness. In the foreword to volume III of his *Traité* he announced:

> A young geometer, M. Jacobi of Koenigsberg, who was not aware of the *Traité* [but who, it should be added, was familiar, like Abel, with the *Exercices* of 1811], has succeeded, through his own investigations, in discovering not only the second transformation, which is related to the number 3, but a third related to the number 5, and he has already become certain that there must exist a similar one for every given odd number.

Legendre also drew attention to Abel's memoirs and analyzed their content. Legendre's first two supplements are devoted primarily to Jacobi's works but also to those of Abel, which contained the first appearance of modern elliptic functions, the inverse functions of the Legendre integrals. Legendre discussed their extension to the complex domain and their double periodicity in his usual somewhat ponderous style. The third supplement deals mainly with Abel and his great theorem. Legendre concluded his work on 4 March 1832: "We have only touched the surface of this subject, but it may be supposed that it will be steadily enriched by the works of mathematicians and that eventually it will constitute one of the most beautiful parts of the analysis of transcendental functions."

In 1869 Charles Hermite made the following judgment concerning Legendre's writings: "Legendre, who for so many reasons is considered the founder of the theory of elliptic functions, greatly smoothed the way for his successors; it is the fact of the double periodicity of the inverse function, immediately discovered by Abel and Jacobi, that is missing and that gave such a restrained analytical character to his *Traité des fonctions elliptiques*."

Legendre's *Éléments de géométrie* long dominated elementary instruction in the subject through its numerous editions and translations. Quite dogmatic in its presentation, this work marked a partial return to Euclid in France. The notes that accompany and enrich the text still have a certain interest. The text itself, virtually unchanged since the first edition, does not take into account the various contributions made by Monge's disciples. The *Éléments* is above all a typical example of the difficulties encountered by the advocates of non-Euclidean geometries in their struggle to gain acceptance for their conceptions. The first published work of János Bolyai dates from 1832 and is thus contemporary with "Réflexions . . . sur la théorie des parallèles . . .," in which Legendre recalled the efforts (all unsuccessful although he was not convinced of this) that he made between 1794 and 1823 to demonstrate Euclid's postulate. Let us first mention two very positive achievements that date from 1800. First: "The sum of the three angles of a rectilinear triangle cannot be greater than two right angles," a proposition he arrived at, of course, by accepting all the axioms, theorems, and postulates preceding the parallel postulate in Euclid's *Elements*. Second: "If there exists a single triangle in which the sum of the angles is equal to two right angles, then in any triangle the sum of the angles will likewise be equal to two right angles."

Yet, aside from these beautiful theorems, demonstrated in an impeccable manner, hardly anything but paralogisms are to be found. Like all the disciples of Newton, Legendre believed in absolute space and in the "absolute magnitude" of the sides of a rectilinear triangle. Taking up again a favorite idea of Lagrange's, outlined by the latter in volume 2 of the *Mémoires de Turin* (1761), he utilized (1794) the "law of homogeneous magnitudes" to establish the theorem of the sum of the angles of a triangle. Suppose a triangle is given with a side a and the adjacent angles B and C. The triangle is then well defined. The third angle A is therefore a function of the given quantities: $A = \varphi(B, C, a)$. But A, B, and C are pure numbers and a is a length. Now, by solving the equation $A = \varphi(B, C, a)$ with a as the unknown, the equation $a = f(A, B, C)$ is obtained "from which it would result that the side a is equal to a pure number without dimension, which is absurd." The law of homogeneous magnitudes therefore requires that this length disappear from the formula at the start and thus that $A = \varphi(B, C)$. By considering a right triangle and its altitude it is easily found that $A + B + C = \pi$.

Until the end of his life Legendre remained convinced of the value of this reasoning, and his other attempts at demonstration held only a purely pedagogical interest for him. They all failed because he always relied, in the last analysis, on propositions that were "evident" from the Euclidean point of view. Among these are the following: two convex contours of opposed concavities intersect at a finite distance; from a point within an angle a straight line cutting the two sides can always be drawn; and through three nonaligned points a circumference of a circle can always be passed.

At the end of his *Réflexions*, Legendre even adopted the pseudodemonstration of Louis Bertrand involving infinite spaces of various orders; and he thought he had improved it by what was actually an even more obscure argument. His virtuosity in spherical geometry and spherical trigonometry did not free him from a blind belief in absolute Euclidean space. "It is nevertheless certain," he wrote in 1832, "that the theorem on the sum of the three angles of the triangle should be considered one of those fundamental truths that are impossible to contest and that are an enduring example of mathematical certitude."

Let us conclude this study by emphasizing the transitional character of Legendre's works, which, in time as well as in spirit, are neither completely of the eighteenth century nor of the nineteenth. His scientific activity, extending from about 1770 to the end of 1832, was divided equally between the two centuries. He was a first-rate disciple of Lagrange and above all of Euler. His writings, like theirs, treat both abstract mathematics and the application of mathematics to the system of the world. Yet his boundless confidence in the powers of abstract science bespeaks a certain naïveté. In 1808 he wrote: "It is remarkable that from integral calculus an essential proposition concerning prime numbers can be deduced." The law in question was that of the distribution of the primes, and his remark is very pertinent. He added: "All the truths of mathematics are linked to each other, and all means of discovering them are equally admissible." One can easily agree with him on this point. But he went further, and here he makes one smile: "Consequently we were led to consider functions in order to demonstrate various basic theorems of geometry and mechanics."

Number theory was a sound and difficult school of logic for Legendre, as for all mathematicians who have worked in that field. Yet on several occasions, as in his studies of Eulerian integrals, he employed disconcerting arguments. For example, having established for positive integral values a certain relationship in which the gamma function occurs, he declares that the relationship is true for every value of this variable because the two members of the relationship are continuous functions. Elsewhere he elaborates on his reasoning: the two members are reduced to the same expression, which he works out; it is an extremely divergent (in the modern sense of the term) series. Still another time he employs an "infinite constant."

Consequently Legendre's writings rapidly became obsolete. Nevertheless he remains a marvelous calculator, a skillful analyst, and, in sum, a good mathematician. In both the theory of elliptic functions and number theory he raised questions that were fruitful subjects of investigation for mathematicians of the nineteenth century.

BIBLIOGRAPHY

I. Original Works. Legendre's writings are *Theses mathematicae ex analysi, geometria, mecanica exerptae, ex collegio Mazarinaeo* (Paris, 1770); Abbé Marie, *Traité de mécanique* (Paris, 1774), which includes several passages by Legendre; *Recherches sur la trajectoire des projectiles dans les milieux résistants* (Berlin, 1782); "Recherches sur la figure des planètes," in *Mémoires de l'Académie des sciences* for 1784, pp. 370 ff.; "Recherches d'analyse indéterminée," *ibid.* for 1785, pp. 465-560; "Mémoire sur la manière de distinguer les maxima des minima dans le calcul des variations," *ibid.* for 1786, pp. 7–37, and 1787, 348–351; German trans. in Ostwalds Klassiker der exacten Wissenschaften, no. 47; "Mémoires sur les intégrations par arcs d'ellipse et sur la comparaison de ces arcs," in *Mémoires de l'Académie des sciences* for 1786, pp. 616, 644–673; "Mémoire sur l'intégration de quelques équations aux différences partielles," *ibid.* for 1787, pp. 309–351; "Mémoire sur les opérations trigonométriques dont les résultats dépendent de la figure de la terre," *ibid.* for 1787, pp. 352 ff.; "Mémoire sur les intégrales doubles," *ibid.* for 1788, pp. 454–486; "Recherches sur les sphéroïdes homogènes," *ibid.* for 1790; and "Mémoire sur les intégrales particulières des équations différentielles," *ibid.* for 1790.

Works published after the establishment of the republic are *Mémoire sur les transcendantes elliptiques* (Paris, 1793); *Éléments de géométrie, avec des notes* (Paris, 1794; 12th ed., 1823), new ed. with additions by A. Blanchet (Paris, 1845; 21st ed., 1876)—these eds. often depart from Legendre's text; English trans. by John Farrar (Cambridge, Mass., 1819) and other English eds. until 1890; German trans. by A. L. Crelle (Berlin, 1822); Romanian trans. by P. Poenaru (Bucharest, 1837); *Essai sur la théorie des nombres* (Paris, 1798; 2nd ed., 1808; with supps., 1816, 1825; 3rd ed., 2 vols., Paris, 1830) and German trans. by Maser as *Zahlentheorie* (Leipzig, 1893); *Méthodes analytiques pour la détermination d'un arc de méridien, par Delambre et Legendre* (Paris, 1799); "Nouvelle formule pour réduire en distances vraies les distances apparentes de la lune au soleil ou à une étoile," in *Mémoires de l'Institut national des sciences et arts*, **6** (1806), 30–54; and *Nouvelles méthodes pour la détermination des orbites des comètes* . . . (Paris, 1806).

Subsequent works are "Analyse des triangles tracés sur la surface d'un sphéroïde," in *Mémoires de l'Institut national des sciences et arts*, **7**, pt. 1 (1806), 130–161; "Recherches sur diverses sortes d'intégrales définies," *ibid.* (1809), pp. 416–509; "Méthode des moindres carrés pour trouver le milieu le plus probable entre les résultats de différentes observations," *ibid.*, pt. 2 (1810), 149–154, with supp. (Paris, 1820); "Mémoire sur l'attractions des ellipsoïdes homogènes," *ibid.*, pt. 2 (1810), pp. 155–183, read 5 Oct. 1812; *Exercices de calcul intégral*, 3 vols. (Paris, 1811–1817); "Sur une méthode d'interpolation employée par Briggs dans la construction de ses grandes tables trigonométriques," in *Connaissance des temps ou des mouvements célestes pour 1817*, **10** (Paris, 1815), 219–222; "Méthodes diverses pour faciliter l'interpolation des grandes tables trigonométriques," *ibid.*, pp. 302–331; and "Recherches sur quelques objets d'analyse indéterminée et particulièrement sur le théorème de Fermat," in *Mémoires de l'Académie des sciences*, n.s. **6** (1823), 1–60.

Traité des fonctions elliptiques et des intégrales eulériennes, avec des tables pour en faciliter le calcul numérique was published in 3 vols. (Paris, 1825–1828); vol. I contains the theory of elliptic functions and its application to various problems of geometry and mechanics; vol. II contains the methods of constructing the elliptical tables, a collection of these tables, the treatise on Eulerian integrals, and an appendix; and vol. III includes various supplements to the theory of Eulerian functions. The three supplements are dated 12 Aug. 1828, 15 Mar. 1829, and 4 Mar. 1832.

Legendre's last works are "Note sur les nouvelles propriétés des fonctions elliptiques découvertes par M. Jacobi," in *Astronomische Nachrichten*, **6** (1828), cols. 205–208; "Mémoire sur la détermination des fonctions Y et Z qui satisfont à l'équation: $4(x^n - 1) = (x - 1)(Y^2 \pm nZ^2)$," in *Mémoires de l'Académie des sciences*, **11** (1832), 81–99, read 11 Oct. 1830; and "Réflexions sur différentes manières de démontrer la théorie des parallèles ou le théorème sur la somme des trois angles du triangle," *ibid.*, **12** (1833), 367–410.

For Legendre's mathematical correspondence with Jacobi, see C. G. J. Jacobi, *Gesammelte Werke*, I (Berlin, 1881), 386–461. For his correspondence with Abel, see N. H. Abel, *Mémorial publié à l'occasion du centenaire de sa naissance*, pt. 2 (Oslo, 1902): pp. 77–79 (Legendre's letter to Abel, 25 Oct. 1828), pp. 82–90 (Abel's letter to Legendre, 25 Nov. 1828), and pp. 91–93 (Legendre's letter to Abel, 16 Jan. 1829).

II. SECONDARY LITERATURE. On Legendre and his work, see Élie de Beaumont, "Éloge historique d'Adrien-Marie Legendre, lu le 25 mars 1861," in *Mémoires de l'Académie des sciences*, **32** (1864), xxxvii–xciv; J. B. J. Delambre, *Rapport historique sur les progrès des sciences mathématiques depuis 1789, et sur leur état actuel* (Paris, 1810), pp. 7–10, 34, 46–96, 135–137, in which S. F. Lacroix was responsible for everything concerning pure mathematics; A. Birembaut, "Les deux déterminations de l'unité de masse du système métrique," in *Revue d'histoire des sciences et de leurs applications*, **12**, no. 1 (1958), 24–54; C. D. Hellman, "Legendre and the French Reform of Weights and Measures," in *Osiris*, **1** (1936), 314–340; Jacob, "A. M. Legendre," in Hoefer, ed., *Nouvelle biographie générale*, XXX (Paris, 1862), cols. 385–388; Gino Loria, *Storia delle matematiche*, 2nd ed. (Milan, 1950), pp. 768–770; Maximilien Marie, *Histoire des mathématiques*, X (Paris, 1887), 110–148; L. Maurice, "Mémoire sur les travaux et les écrits de M. Legendre," in *Bibliothèque universelle des sciences, belles-lettres et arts. Sciences et arts* (Geneva), **52** (1833), 45–82; and E. H. Neville, "A Biographical Note," in *Mathematical Gazette*, **17** (1933), 200–201.

See also N. Nielsen, *Géomètres français sous la Révolution* (Copenhagen, 1929), pp. 166–174; Maurice d'Ocagne, *Histoire abrégée des mathématiques* (Paris, 1955), pp. 182–187; Parisot, "Legendre," in Michaud, ed., *Biographie universelle ancienne et moderne*, XXIII, 610–615; Poggendorff, I, cols. 1406–1407; Denis Poisson, "Discours prononcé aux funérailles de M. Legendre . . .," in *Moniteur universel* (20 Jan. 1833), p. 162; J. M. Querard, "Legendre, Adrien-Marie," in *France littéraire*, **5** (1833), 94–95; A. Rabbe, *Biographie universelle et portative des contemporains*, III (Paris, 1834), 234–235; L. G. Simons, "The Influence of French Mathematicians at the End of the Eighteenth Century Upon the Teaching of Mathematics in American Colleges," in *Isis*, **15**, no. 45 (Feb. 1931), 104–123; D. E. Smith, "Legendre on Least Squares," in *A Source Book in Mathematics* (New York, 1929), pp. 576–579; I. Todhunter, *A History of the Progress of the Calculus of Variations* (Cambridge, 1861), ch. 9, pp. 229–253; A. Aubry, "Sur les travaux arithmétiques de Lagrange, de Legendre et de Gauss," in *Enseignement mathématique*, **11** (1909), 430–450; and Ivor Grattan-Guinness, *The Development of the Foundations of Mathematical Analysis from Euler to Riemann* (Cambridge, Mass., 1970), pp. 29, 36–41.

JEAN ITARD

LE GENTIL DE LA GALAISIÈRE, GUILLAUME-JOSEPH-HYACINTHE-JEAN-BAPTISTE (*b.* Coutances, France, 12 September 1725; *d.* Paris, France, 22 October 1792), *astronomy.*

Following early schooling in his native city, Le Gentil, the only son of a none-too-wealthy Norman gentleman, went to Paris to study theology. While pursuing that course he also attended Delisle's lectures at the Collège Royal on astronomy, to which he soon found himself more attracted. An introduction to Jacques Cassini at the observatory in 1748 definitely established this change in direction by bringing Le Gentil an offer of lodgings there and observational training under Cassini de Thury and G. D. Maraldi. He was soon engaged in a regular, well-rounded program of celestial observations.

Having obtained the support of important members of the Academy of Sciences and having demonstrated, as early as 1749, his acquired skill and promise through the discovery of a nebula, a drawing of which he presented to them, Le Gentil was elected to the Academy in 1753. During the succeeding six years his contributions to its *Mémoires* ranged from historical concerns with the saros, through observational and descriptive papers, to studies that combined his own observations with those of others to perfect existing theory and quantitative evaluation of orbit inclinations and the obliquity of the ecliptic.

The turning point in Le Gentil's career was his commission to observe the 1761 transit of Venus at Pondicherry, India. Because the English had captured that settlement just as he arrived (in 1760), Le Gentil was obliged to witness the transit from shipboard, without any possibility for scientifically significant observations. Since another transit of Venus was to take place in 1769, Le Gentil resolved to remain in the East in order to complete his mission. He used the intervening years to collect vast amounts of material on Indian astronomy and to make numerous excursions, from Madagascar to Manila, during the course of which he amassed observations on a broad spectrum of phenomena. Although his own calculations showed that the latter site would be excellent for observing the transit of 3 June 1769, the Academy ordered him back to Pondicherry for that purpose. The decision was unfortunate, since Manila was very clear that day, while at Pondicherry a cloud obscured the sun precisely during the crucial period. It was thus an extremely disappointed Le Gentil who returned to Paris two years later after an absence of eleven and a half years.

Problems of a different sort now confronted him. Le Gentil's relatives, believing him dead, had begun a division of his estate; and the Academy, some members of which apparently interpreted his absence as undertaken for personal enrichment, had relegated him to "veteran" status. Both problems were soon solved: the former by legal actions which, however, did not return stolen monies and left him responsible for court costs; and the latter by his reinstatement in 1772 as an associate and, a decade later, by his supernumerary promotion to the pensionary level. Le Gentil married and moved back to the observatory, and was soon dividing his time between attention to the only offspring of his marriage, a daughter, and to his writings, most of which were based on the materials that he had brought back from the East.

Le Gentil's major work was the two-volume *Voyage dans les mers d'Inde* . . . (1779–1781). The first volume was devoted to India and, after a discussion of the customs and religion of its inhabitants, dealt at considerable length with the history of Brahman astronomy. Le Gentil's largely conjectural contention of that science's great antiquity was disputed by many contemporaries and was rejected by virtually all later scholars. On the other hand, his personal astronomical observations—as well as those dealing with geography, meteorology, and physics—were securely based and of more lasting importance. His instruments having been verified at the outset, Le Gentil's latitude and longitude determinations inspired confidence; his newly calculated table of refractions for the torrid zone, even without barometer and thermometer readings, was a decided improvement over existing values and a remarkable anticipation of later ones. His solstitial observations confirmed his earlier conclusions about the diminution of the obliquity of the ecliptic.

The second volume of the *Voyage* was devoted to the Philippines, Madagascar, and the Mascarenes. Except for the difference in locale, the absence of a historical section, and greater emphasis on geography and navigation, it conveyed the same sort of useful information as did the first.

Some of these materials were abstracted and offered to the Academy as memoirs during the 1770's, while the *Voyage* was in preparation. Even after its publication, however, Le Gentil continued to exploit his Indian data, sometimes combining them with observations made after his return. Papers of the 1780's dealing with refraction and the obliquity of the ecliptic were representative of this type, and a group of historical offerings during the same period were largely restatements of his claims for Indian astronomy, occasionally buttressed by "evidence" from Gothic zodiacs in and near Paris. The most important of his totally new productions were a memoir on tides that Le Gentil had observed on the coasts of Normandy and a paper alleging certain advantages of binocular over monocular instruments. Neither work was responsible for any basic advance.

The reconstruction of the observatory displaced Le Gentil from his lodgings in 1787, and he never returned—nor did the Academy ever fill the vacancy created by his death in 1792.

BIBLIOGRAPHY

I. Original Works. Le Gentil's major work was *Voyage dans les mers de l'Inde (1760–1771), fait par ordre du roi, à l'occasion du passage de Vénus, sur le disque du soleil, le 6 Juin 1761, & le 3 du même mois 1769*, 2 vols. (Paris, 1779–1781). His only other non-Academic publication dealt with the transit of Venus and was written before his departure: "Mémoire de M. Le Gentil, au sujet de l'observation qu'il va faire, par ordre du roi, dans les Indes orientales, du prochain passage de Vénus pardevant le soleil," in *Journal des sçavans* (Mar. 1760), 137–139. Except for his 1749 "Mémoire sur une étoile nébuleuse nouvellement découverte à côté de celle qui est au-dessus de la ceinture d'Andromède," in vol. II (1755) of the so-called *Savants étrangers* series, all of his other publications were contributions to the *Mémoires*. These included a paper on the inequalities noticed in the movements of Jupiter and Saturn and tables of the oppositions of those planets with the sun that he had read to the Academy as a nonmember and that had been destined for the same *Savants étrangers* volume. Having attained membership before the latter was printed, however, he withdrew it from the printing office, made some changes in it, reread it to the Academy, and placed it in the *Mémoires* for 1754.

Inasmuch as Quérard provides a nearly complete chronological listing of forty papers in the *Mémoires* from 1752 through 1789 (with none between 1760 and 1770, because of Le Gentil's absence), there is no need to repeat long titles here. It has, however, been thought appropriate to indicate the years in which Le Gentil offered various types of memoirs. His historical studies began with two papers on the saros (1756), in which he rejected Halley's explanation of that period and substituted a fanciful scheme of his own; continued with three memoirs on various aspects of Indian astronomy (1772, pt. 2)—the antiquity of which he specifically insisted upon in a later work (1784); and ended by treating the origin of the zodiac and related matters (1782, two in 1785, two in 1788, 1789), again with frequent claims for Indian primacy.

Le Gentil's purely observational works of the 1750's dealt with determinations of the apparent diameter of the sun (1752, 1755) and of the earth's shadow during lunar eclipses (1755), and recorded such transient and permanent phenomena as nebulae (1759), a variable star (1759), rainbows (1757), lunar occultations of a star (1753) and Venus (1753), a lunar eclipse (1755), and inferior conjunctions of Venus (1753) and Mercury (1753). The latter, as with the observations of oppositions of Jupiter and Saturn (see above), were combined with remarks of theoretical interest.

More important efforts in the realm of theory were Le Gentil's researches into the obliquity of the ecliptic (1757) and his three-part study of the principal orbital elements of the superior planets (one part in 1757, two in 1758). In addition to the abstract items of the 1770's—the most important of which were two papers on horizontal refraction in the torrid zone (1774)—during that period he offered remarks on the temperature of the cellars of the Paris observatory (1774), an observation of a lunar eclipse (1773, with Bailly), and a memoir on the disappearance of Saturn's ring (1775). Although he thereafter reported both specifically astronomical and various general physical observations: for instance, on the intense cold of late 1783 and on the prevailing winds of Paris (both in 1784), the works noted in the text were far more important contributions: those on refraction (1789), the obliquity of the ecliptic (1783), tides (1782), and binocular instruments (1787).

II. SECONDARY LITERATURE. In his series of annual histories of astronomy from 1781 to 1802, J. J. de Lalande almost always offered brief notices of astronomers who died during those years; his succinct treatment of Le Gentil may be most conveniently consulted in his *Bibliographie astronomique* . . . (Paris, *an* XI [1803]), p. 722. During Condorcet's absence from the Academy in 1792, the writing of an "official" *éloge* fell to the vice-secretary, J. D. Cassini (IV); the product was published in his *Mémoires pour servir à l'histoire des sciences et à celle de l'Observatoire royal de Paris* (Paris, 1810), pp. 358–372. That rather laudatory and uncritical account should be supplemented by J. B. J. Delambre's trenchant analysis of Le Gentil's most important works in *Histoire de l'astronomie au dix-huitième siècle* (Paris, 1827), pp. 688–709; Delambre had earlier criticized Le Gentil's ideas about Indian astronomy in his *Histoire de l'astronomie ancienne* (Paris, 1817), I, 511–514.

As mentioned above, a nearly complete listing of Le Gentil's offerings in the *Mémoires*, although virtually nothing else, is available in J. M. Quérard, *La France littéraire* . . ., V (Paris, 1833), 95–96. The subsequent treatments in J. F. Michaud, ed., *Biographie universelle*, XXIII, 618–619; and in Niels Nielsen, *Géomètres français du dix-huitième siècle* (Paris, 1935), pp. 266–269, are flawed by minor errors. More specifically concerned with Le Gentil's voyage and its outcome are Alfred Lacroix, *Figures de savants*, III (Paris, 1938), 169–176; and, more important, Harry Woolf, *The Transits of Venus* (Princeton, 1959), esp. pp. 126–130, 151–156.

SEYMOUR L. CHAPIN

LÉGER, URBAIN-LOUIS-EUGÈNE (*b.* Loches, France, 7 September 1866; *d.* Grenoble, France, 7 July 1948), *zoology, protistology, hydrobiology.*

Léger, the son of Urbain-Louis Léger, a schoolteacher, and Marie-Eugénie Tretois, spent his childhood in Touraine. At the Faculty of Science of Poitiers he studied natural science, for which he received his *licence*, and, under the direction of Aimé Schneider, prepared his thesis for the *doctorat-ès-sciences* on the subject of the Gregarines (parasitic protozoans of invertebrates). He defended the thesis, for which he himself engraved the plates, in Paris on 27 February 1892 under the chairmanship of G. Bonnier. He was then appointed *préparateur* and then *chef de travaux* of zoology in Marion's laboratory at the Faculty of Sciences at Marseilles. He completed his medical studies and in 1895 defended his doctoral thesis in medicine on the histology of senile arteries before the Faculty of Medicine of Montpellier.

In 1898 Léger was appointed deputy lecturer and in 1904 professor of zoology at the Faculty of Sciences in Grenoble. He was appointed to the same post at the School of Medicine in 1910 and remained in Grenoble until his death.

While at Marseilles, Léger was married to Juliette Baraton, by whom he had three sons: Jacques, Jean, and Louis. He was the recipient of many honors: twice laureate of the Institute (Prix Serres, 1920; Prix Petit d'Ormoy, 1933); corresponding member of the Academy of Agriculture (1927); corresponding member of the Academy of Sciences (1928); officer of the Legion of Honor (1929); member of the Masaryk Academy of Prague; and member of the Romanian Academy of Sciences.

Léger's scientific output was considerable, comprising approximately 300 publications, which can be divided into two major groups: those on protistology and those on hydrobiology and pisciculture. In protistology Léger studied primarily parasitic species: Sporozoa (Gregarinida and Coccidia), Cnidosporidia, Flagellatae, and Ciliata. His researches on the

Gregarinida, done for the most part in collaboration with O. Duboscq, yielded data concerning their morphology, cytology, sexuality, life cycle, and their effect on the host. The papers on *Stylocephalus* and *Porospora* have become classics. Furthermore, Léger was interested in the Trichomycetes, curious parasites of arthropods which possess plantlike characteristics.

Léger began his piscicultural investigations in 1901. In 1909 he published a monograph on the fishes of Dauphiné, which was followed by numerous studies dealing with the physicochemical and biological factors of fresh waters and with the parasites and pathology of fishes. He also recommended attempts, which were successful, to introduce char in Alpine Lakes and rainbow trout in Madagascar.

At once a great researcher and a brilliant teacher, Léger was a cultivated artist—he was a watercolorist—and a direct and friendly man. His influence on his many pupils and his work in science give him a prominent place among French biologists of the first half of the twentieth century.

BIBLIOGRAPHY

I. ORIGINAL WORKS. Léger's principal works are *Recherches sur les Grégarines* (Poitiers, 1892); "Recherches sur les artères séniles," in *Annales de la Faculté des sciences de Marseille* (1895); "Nouvelles recherches sur les Polycystidées parasites des arthropodes terrestres, "*ibid.*, **6** (1896); "Essai sur la classification des Coccidies et description de quelques espèces nouvelles," in *Bulletin du Muséum d'histoire naturelle de Marseille*, **1** (1898), 71–123; "Les grégarines et l'épithélium intestinal chez les Trachéates," in *Archives de parasitologie*, **6** (1902), 377–473, with O. Duboscq; "Recherches sur les Myriapodes de Corse et leurs parasites," in *Archives de zoologie expérimentale et générale* (ser. 43), **1** (1903), 307–358, with Duboscq and H. Brolemann; "La reproduction sexuée chez les *Stylorhynchus*," in *Archiv für Protistenkunde*, **3** (1904), 303–357; "Nouvelles recherches sur les Grégarines et l'épithélium intestinal des Trachéates," *ibid.*, **4** (1904), 335–383, with Duboscq; "Étude sur la sexualité des Grégarines," *ibid.*, **17** (1909), 19–134, with Duboscq; "Poissons et Pisciculture dans le Dauphiné," in *Travaux du Laboratoire de pisciculture de l'Université de Grenoble*, **2** (1909); "Les Porospores et leur évolution," in *Travaux de la Station zoologique de Wimereux*, **9** (1925), 126–139, with Duboscq; and "Contribution à la connaissance des Eccrinides," in *Archives de zoologie expérimentale et générale*, **86** (1948), 29–144, with Duboscq and O. Tuzet.

II. SECONDARY LITERATURE. M. Caullery, "Notice nécrologique sur M. Louis-Urbain-Eugène Léger (1866–1948)," in *Comptes rendus hebdomadaires des séances de l'Académie des sciences*, **277** (1948), 101–102; A. Dorier, "Discours prononcé aux obsèques du Professeur Léger," in *Travaux du Laboratoire d'hydrobiologie et de pisciculture de l'Université de Grenoble*, **9–11** (1948), 38–40; and *Hommage à Louis Léger 1866–1948* (Grenoble, 1949).

J. THÉODORIDÈS

LEHMANN, JOHANN GOTTLOB (*b.* Langenhennersdorf, near Pirna, Germany, 4 August 1719; *d.* St. Petersburg, Russia [now Leningrad, U.S.S.R.], 22 January 1767), *medicine, chemistry, metallurgy, mineralogy, geology.*

Lehmann was the son of Martin Gottlob Lehmann, a prosperous gentleman farmer, and Johanna Theodora Schneider. His early education was largely at the hands of private tutors, although he did attend the Fürstenschule, in Schulpforta, for one semester in 1735 before ill health forced him to withdraw. In 1738 he matriculated as a medical student at the University of Leipzig; the following year he transferred to the University of Wittenberg, where he studied with the anatomist Abraham Vater. He received the M.D. in 1741, with a dissertation on the nervous papillae.

Lehmann then went to Dresden to practice medicine, but soon, having become acquainted with the natural scientists resident there, discovered his real field of interest to be mining and metallurgy. Saxony, a mining center, was an ideal place to take up that subject, and Lehmann became engrossed in all aspects of it. He was particularly concerned with the origins and distribution of ore deposits and with the chemical composition of various ores. He made field trips, among them one to Bohemia, to further his knowledge, and by 1750, the year in which he left Dresden for Berlin, he had become known for his writings on mines and mining.

Christlob Mylius, a scientist and a friend of Lehmann's, wrote that Lehmann initially came to Berlin in connection with the establishment of the state porcelain factory there. If such were the case, he did not participate in this enterprise for long, since shortly after his arrival he received an official commission to study mining procedures in the Prussian provinces and to make recommendations for their improvement. He spent several years on this project, mostly in the Harz. In August 1754 Lehmann was appointed *Bergrat;* he served as director of copper mining and of the Bureau of Mines in Hasserode, where he also established a smelter and factory for manufacturing blue pigment from cobalt from a neighboring mine. In 1755 and 1756 Lehmann traveled in Silesia in his official capacity.

In 1756, too, Lehmann settled in Berlin, having married Maria Rosina von Grünroth. He had been a member of the Royal Prussian Academy of Sciences since 1754, and his work in the capital was done largely under the auspices of that body. Lehmann thus entered into a period of extraordinary scientific activity, publishing his researches in chemistry, geology, and mineralogy. Of these, the most important and lasting were his geological studies. In geology Lehmann emphasized the importance of geological strata in determining the history of the earth; he was also one of the first to seek physico-chemical explanations for the origin of mineral deposits within the context of that history. (His new notion of the composition of the earth further enabled him to reach a theory of the propagation of earthquakes as being dependent on the "inner structure of the surface of the earth.")

In his *Versuch einer Geschichte von Flötz-Gebürgen* of 1756, Lehmann, in discussing sedimentary rocks, described and compared sequences of strata on the basis of his own novel observations. Recognizing that the older strata were formed by the action of water, he developed the laws governing the formation of mountains, making a distinction between what he called "Ganggebürge"—a mountain formed of veined rock—and "Flötzgebürgen," mountains formed of stratified rock. These are now called *Unterbau* and *Oberbau* (substructure and superstructure), respectively. Drawing upon his observations, Lehmann was to draw up the first geological profile. In it he demonstrated that rocks do not lie next to each other in a haphazard way, but rather are formed in historical sequence. He thus may be considered the founder of stratigraphy; his attempt to establish the laws underlying the formation of the earth provided the basis for modern geology.

Lehmann's work in chemistry—which, indeed, constituted the greatest part of his researches—is today primarily of historical value. In his studies of rocks and ores he sought to determine their composition and metallurgical properties, and he suggested a system of classification based upon chemical composition (in contrast to Agricola's classification by external characteristics and Linnaeus' classification by crystal form). He analyzed many minerals for the first time; these analyses led to the discovery of new metals, including cobalt and tungsten. Lehmann's *Cadmiologia* of 1760 dealt specifically with cobalt ores, treating their occurrence, mineralogy, and chemistry, and the technology necessary to their mining and commercial use. But because he was limited to the techniques of quantitative analysis, Lehmann was able to make no fundamental and lasting contribution to the development of chemistry.

Mining technology was the framework into which Lehmann fitted all his research. He consciously sought to introduce into mining new scientific findings from all the natural sciences to enrich the technology that had evolved over the centuries. An uncompromising empiricist, Lehmann brought this point of view to his work in the field, in the mine, and in the laboratory. Following in the steps of Karl Friedrich Zimmermann, Lehmann advocated the establishment of specialized research institutions; he envisioned a technical teaching and research institute that would be the equal in prestige of the established universities. He was rewarded in his efforts by the founding of the Freiberg Bergakademie in 1765.

In 1760 Lehmann was invited to St. Petersburg by the Imperial Academy of Sciences. He accepted the following year, and in July 1761 left Berlin for Russia, where he took up a post as professor of chemistry at the University of St. Petersburg and director of the Academy's natural history collection. He also continued his researches and, in the five and one-half years that remained to him, made thirty-seven reports to the Academy on the composition of minerals, the smelting of ores, the composition of soils and peats, the commercial manufacture of alum, fossil remains, and the geological structure of both specific regions and of the whole earth. He proposed the establishment of a governmental department to supervise the exploration of Russia's mineral resources through a cartographical survey of geological relationships. (In this Lehmann was ahead of his time; two years after his death Catherine the Great commissioned the first of several Siberian explorations, giving Pallas the particular charge of conducting geological and orographic observations.) Lehmann's death, at forty-seven, was caused by a bilious fever, and not by arsenic poisoning, as rumored.

BIBLIOGRAPHY

I. Original Works. Bruno von Freyberg gives a list of 112 titles in *Johann Gottlob Lehmann, ein Arzt, Chemiker, Metallurg, Bergmann, Mineraloge und grundlegender Geologe*, cited below. In particular, see *Dissertatio de consensu partium corporis humani occasione spasmi singularis in manu eiusque digitis ex hernia observati* (Wittenberg, 1741); "Sammlung einiger mineralischer Merkwürdigkeiten des Plauischen Grundes bey Dressden," in *Neue Versuche nützlicher Sammlungen* (1749, 580–597); *Abhandlung von Phosphoris* (Dresden–Leipzig, 1750); *Kurtze Nachricht vom Erbbereiten* (Dresden–Leipzig, 1750); *Kurtze Einleitung in einige Theile der Bergwerks-Wissenschaft* (Berlin, 1751); "Ohnmassgeblicher Vorschlag, auf was Art und Weise man zu einer genaueren Entdeckung der unter der Erde verborgenen Dinge, oder kurz zu sagen, zu einer unterirdischen Erdbeschreibung gelangen könne," in *Physika-*

lische Belustigungen, **2** (1752), 27–42; *Abhandlung von den Metallmüttern* (Berlin, 1753); *Versuch einer Geschichte von Flötz-Gebürgen* (Berlin, 1756); *Physicalische Gedanken von denen Ursachen derer Erdbeben* (Berlin, 1757); "Histoire du Chrysoprase de Kosemitz," in *Histoire de l'Académie de Berlin*, **11** (1757), 202–214; *Kurzer Entwurf einer Mineralogie* (Berlin, 1758); *Cadmiologia*, pt. 1 (Berlin, 1760); *Probierkunst* (Berlin, 1761); *Physikalisch-chymische Schriften* (Berlin, 1761); *Specimen orographiae generalis tractus montium primarios globum nostrum terraqueum pervagantes* (St. Petersburg, 1762); *Cadmiologia*, pt. 2 (Königsberg–Leipzig, 1766); *De nova minerae plumbi specie crystallina rubra* (St. Petersburg, 1766); "Historia et examen chymicum lapidis nephritici," in *Novi commentarii Academiae* [of St. Petersburg], **10** (1766), 381–412; "De vitro fossili naturali sive de Achate Islandico," *ibid.*, **12** (1768), 356–367; "De Cupro et Orichalco magnetico," *ibid.*, 368–390; and "Specimen Oryctographiae Stara-Russiensis et lacus Ilmen," *ibid.*, 391–402.

II. SECONDARY LITERATURE. Works about Lehmann include Bruno von Freyberg, *Die geologische Erforschung Thüringens in älterer Zeit* (Berlin, 1932); *Neues über Johann Gottlob Lehmann* (Erlangen, 1948); and *Johann Gottlob Lehmann, ein Arzt, Chemiker, Metallurg, Bergmann, Mineraloge und grundlegender Geologe*, vol. I of Erlanger Forschungen B, (Erlangen, 1955); W. von Gümbel, "Johann Gottlob Lehmann," in *Allgemeine deutsche Biographie*, XVIII (Leipzig, 1883), 140–141; and Hans Prescher, "Johann Gottlob Lehmann (1719–1767). Zum 200. Todestage des bedeutenden Bergmanns, Metallurgen und Begründers der modernen Erdgeschichtsforschung," in *Der Anschnitt*, **19** (1967), 9–18. See also Poggendorff, I, 1409–1416.

BRUNO VON FREYBERG

LEHMANN, OTTO (*b.* Konstanz, Germany, 13 January 1855; *d.* Karlsruhe, Germany, 17 June 1922), *crystallography, physics.*

Lehmann discovered liquid crystals; substances which behave mechanically as liquids but display many of the optical properties of crystalline solids.

Lehmann's father had been a professor of science and mathematics at the Gymnasium in Freiburg im Breisgau and was particularly interested in the mathematical manifestations of organic nature. He sought to develop mathematical formulas for such phenomena as the geometric forms of the leaves of plants, and his interests stimulated the scientific bent of his son. Lehmann received his doctorate from the University of Strasbourg in 1876 and taught in secondary schools at Freiburg im Breisgau and Mulhouse from 1876 to 1883. In the latter year he became *Dozent* and in 1885 associate professor of physics at the Technische Hochschule at Aachen; in 1888 he was named associate professor at the Technische Hochschule at Dresden; and in 1889 he succeeded Heinrich Hertz as professor of physics at the Technische Hochschule at Karlsruhe. He remained in this post until his death.

Lehmann's early scientific interest and experimentation were concerned with electric discharges in rarefied gases, but he soon turned his attention to the study of the fine structure of matter as revealed under the microscope. His first major work describing his studies was *Molekularphysik, mit besonderer Berüchtsichtigung mikroskopischer Untersuchung und Anleitung zu solchen* (1888–1889).

In 1888 the Austrian botanist Friedrich Reinitzer noticed that the solid compound cholesteryl benzoate seemed to have two distinct melting points, becoming a cloudy liquid at 145° C. and turning clear at 179° C. Reinitzer's observation came to Lehmann's attention, and he immediately began research on this and other organic substances displaying the same property. He determined in 1889 that the cloudy intermediate phase contained areas that possessed a molecular structure similar to that of solid crystals, and he called this phase "liquid crystal." In 1922 G. Friedel suggested the term "mesomorphic" to include Lehmann's liquid crystals as well as any state of matter intermediate between the amorphous and crystalline states; however, the term "liquid crystal" is still employed.

Lehmann published his results in two major works: *Flüssige Krystalle* (1904) and *Die neue Welt der flüssigen Krystalle und deren Bedeutung für Physik, Chemie, Technik und Biologie* (1911). His results astonished and perplexed the scientific world, since he demonstrated that the fluidity of many organic substances is not only equal to or greater than water but that they also display the double refracting properties of crystals, some being twice as birefringent as calcite.

Lehmann's work stimulated much research in this area as well as studies to find technical applications of the phenomenon, and these efforts are still continuing.

BIBLIOGRAPHY

I. ORIGINAL WORKS. Lehmann's chief publications are *Molekularphysik, mit besonderer Berüchtsichtigung mikroskopischer Untersuchung und Anleitung zu solchen*, 2 vols. (Leipzig, 1888–1889); *Krystallanalyse, oder die chemische Analyse durch Beobachtung der Krystallbindung mit Hülfe des Mikroskops* (Leipzig, 1891); *Elektrizität und Licht* (Brunswick, 1895); *Die elektrische Lichterschein* (Halle, 1898); *Flüssige Krystalle* (Leipzig, 1904); and *Die neue Welt der flüssigen Krystalle und deren Bedeutung für Physik, Chemie, Technik und Biologie* (Leipzig, 1911). In addition, he published approximately 120 articles in scientific journals.

II. SECONDARY WORKS. Obituary notices are in *Physikalische Zeitschrift*, **24** (1923), 289–291; and *Zeitschrift für technische Physik*, **4** (1923), 1.

JOHN G. BURKE

LEIBENSON, LEONID S. See **Leybenzon, Leonid S.**

LEIBNIZ, GOTTFRIED WILHELM (*b.* Leipzig, Germany, 1 July 1646; *d.* Hannover, Germany, 14 November 1716), *mathematics, philosophy, metaphysics.*

Leibniz was the son of Friedrich Leibniz, who was professor of moral philosophy and held various administrative posts at the University of Leipzig. His mother, Katherina Schmuck, was also from an academic family. Although the Leibniz family was of Slavonic origin, it had been established in the Leipzig area for more than two hundred years, and three generations had been in the service of the local princes.

Leibniz attended the Nicolai school, where his precocity led his teachers to attempt to confine him to materials thought suitable to his age. A sympathetic relative recognized his gifts and aptitude for self-instruction, and on the death of Friedrich Leibniz, in 1652, recommended that the boy be given unhampered access to the library that his father had assembled. By the time he was fourteen, Leibniz had thus become acquainted with a wide range of classical, scholastic, and even patristic writers, and had, in fact, begun that omnivorous reading that was to be his habit throughout his life. (Indeed, Leibniz' ability to read almost anything led Fontenelle to remark of him that he bestowed the honor of reading on a great mass of bad books.)

At the age of fifteen Leibniz entered the University of Leipzig, where he received most of his formal education, although that institution was at that time firmly entrenched in the Aristotelian tradition and did little to encourage science. In 1663 he was for a brief time a student at the University of Jena, where Erhard Weigel first taught him to understand Euclidean geometry. Leibniz continued his studies at Altdorf, from which he received the doctorate in 1666, with a dissertation entitled *Disputatio de casibus perplexis*. He was invited to remain at that university, but chose instead, during the second half of 1667, to undertake a visit to Holland.

Leibniz reached Mainz, where, through the offices of the statesman J. C. von Boyneburg, he met the elector Johann Philipp von Schönborn, who asked him into his service. Leibniz worked on general legal problems, developed his program for legal reform of the Holy Roman Empire, wrote (anonymously) a number of position papers for the elector, and began a vast correspondence that by 1671 had already brought him into contact with the secretaries of the Royal Society of London and the Paris Academy of Sciences, as well as with Athanasius Kircher in Italy and Otto von Guericke in Magdeburg. He also began work on his calculating machine, a device designed to multiply and divide by the mechanical repetition of adding or subtracting. In 1671 Pierre de Carcavi, royal librarian in Paris, asked Leibniz to send him this machine so that it could be shown to Colbert. The machine was, however, only in the design stage at that time (although a model of it was built in 1672 and demonstrated to the Academy three years later).

In the winter of 1671–1672, Leibniz and Boyneburg set forth a plan to forestall French attacks on the Rhineland. By its terms, Louis XIV was to conquer Egypt, create a colonial empire in North Africa, and build a canal across the isthmus of Suez—thereby gratifying his imperial ambitions at no cost to the Netherlands and the German states along the Rhine. Leibniz was asked to accompany a diplomatic mission to Paris to discuss this matter with the king. He never met Louis, but he did immerse himself in the intellectual and scientific life of Paris, forming a lifelong friendship with Christiaan Huygens. He also met Antoine Arnauld and Carcavi. The official mission came to nothing, however, and in December 1672 Leibniz' patron and collaborator Boyneburg died.

In January 1673 Leibniz went to London with a mission to encourage peace negotiations between England and the Netherlands; while there he became acquainted with Oldenburg, Pell, Hooke, and Boyle, and was elected to the Royal Society. The mission was completed, but the elector Johann Philipp had died, and his successor showed little interest in continuing Leibniz' salary, especially since Leibniz wanted to return to Paris. Leibniz arrived in the French capital in March 1673, hoping to make a sufficient reputation to obtain for himself a paid post in the Academy of Sciences. Disappointed in this ambition, he visited London briefly, where he saw Oldenburg and Collins, and in October 1676 left Paris for Hannover, where he was to enter the service of Johann Friedrich, duke of Brunswick-Lüneburg. En route, Leibniz stopped in Holland, where he had scientific discussions with Jan Hudde and Leeuwenhoek, and, at The Hague between 18 and 21 November, conducted a momentous series of conversations with Spinoza.

By the end of November Leibniz had arrived in Hannover, where he was initially a member of the duke's personal staff. He acted as adviser and librarian, as well as consulting on various engineering projects. (One of these, a scheme to increase the yield of the Harz silver mines by employing windmill-powered pumps, was put into operation in 1679, but failed a few years later, through no fault of the engineering principles involved.) Leibniz was soon formally appointed a councillor at court, and when Johann Friedrich died suddenly in 1679 to be succeeded by his brother Ernst August (in March 1680), he was confirmed in this office. Sophia, the wife of the new duke, became Leibniz' philosophical confidante; Ernst August commissioned him to write a genealogy of the house of Brunswick, *Annales imperii occidentes Brunsvicenses*, to support the imperial and dynastic claims of that family. Leibniz' researches on this subject involved him in a series of scholarly travels; his princely support opened the doors of archives and libraries, and he was enabled to meet and discuss science with eminent men throughout Europe.

Leibniz left Hannover in October 1687 and traveled across Germany; in Munich he found an indication that the Guelphs were related to the house of Este, an important point for his genealogy. In May 1688 he arrived in Vienna; in October of that year he had an audience with Emperor Leopold I, to whom he outlined a number of plans for economic and scientific reforms. He also sought an appointment at the Austrian court, which was granted only in 1713. He then proceeded to Venice and thence to Rome. He hoped to meet Queen Christina, but she had died; he did become a member of the Accademia Fisico-matematica that she had founded. In Rome, too, Leibniz met the Jesuit missionary Claudio Filippo Grimaldi, who was shortly to leave for China as mathematician to the court of Peking; Grimaldi awakened in Leibniz what was to become a lasting interest in Chinese culture. Returning north from Rome, Leibniz stopped in Florence for a lively exchange on mathematical problems with Galileo's pupil Viviani; in Bologna he met Malpighi.

On 30 December 1689 Leibniz reached Modena, his ultimate destination, and set to work in the ducal archives which had been opened to him. (Indeed, he threw himself into his genealogical research with such fervor that he afflicted himself with severe eyestrain.) He interrupted his work long enough to arrange a marriage between Rinaldo d'Este of Modena and Princess Charlotte Felicitas of Brunswick-Lüneburg (celebrated on 2 December 1695), but by February 1690 he was able to prove the original relatedness of the house of Este and the Guelph line, and his research was complete. He returned to Hannover, making various stops along the way; his efforts were influential in the elevation of Hannover to electoral status (1692) and earned Leibniz himself an appointment as privy councillor.

Elector Ernst August died in January 1698 and his successor, Georg Ludwig, although urging Leibniz to complete the history of his house, nevertheless declined to make any other use of his services. Leibniz found support for his project in other courts, however, particularly through the patronage of Sophia Charlotte, daughter of Ernst August and Sophia and electress of Brandenburg. At her invitation Leibniz went to Berlin in 1700, in which year, on his recommendation, the Berlin Academy was founded. Leibniz became its president for life. Sophia Charlotte died in 1705; Leibniz made his last visit to Berlin in 1711. He persisted in his efforts toward religious, political, and cultural reforms, now hoping to influence the Hapsburg court in Vienna and Peter the Great of Russia. In 1712 Peter appointed him privy councillor, and from 1712 to 1714 he served as imperial privy councillor in Vienna.

On 14 September 1714 Leibniz returned to Hannover; he arrived there three days after Georg Ludwig had left for England as King George I. Leibniz petitioned for a post in London as court historian, but the new king refused to consider it until he had finished his history of the house of Brunswick. Leibniz, plagued by gout, spent the last two years of his life trying to finish that monumental work. He died on 14 November 1716, quite neglected by the noblemen he had served. He never married.

FREDERICK KREILING

LEIBNIZ: Physics, Logic, Metaphysics

The special problems for any comprehensive treatment of the scientific investigations of Leibniz arise, on the one hand, from the fact that essential parts of his work have not been edited and, on the other hand, from the universality of his scientific interests. In view of this diversity of interests and the fragmentary, or rather encyclopedic, character of his work, the expositor is confronted with the task of achieving, at least in part, what Leibniz himself, following architectonic principles (within the framework of a *scientia generalis*), was unable to accomplish.

Leibniz is a striking example of a man whose universal interests (in his case ranging from physics through theory of law, linguistic philosophy, and historiography to particular questions of dogmatic theology) hindered rather than aided specialized scientific work. On the other hand, this broad interest, insofar as it remained oriented in architectonic principles, led to a concentration on methodological questions. In relation to the structure of a science, these are more important than concrete results. The position of Leibniz at the beginning of modern science is analogous to that of Aristotle at the beginning of ancient science. Leibniz' universality is comparable with that of Aristotle, different only in that it did not, as Aristotle's, remain grounded in essentially unchanged metaphysical distinctions but evolved by degrees from an encyclopedic multiplicity of interests. Consequently, in the strict sense in which there is an Aristotelian system, there is no Leibnizian system but rather a marked metaphysical and methodological concern that systematically expresses variations on the same theme in the various special fields (such as physics and logic) and underlies Leibniz' quest to establish a unified system of knowledge.

Leibniz' autobiographical statement in a letter to Rémond de Montmort in 1714 explains how, at the age of fifteen, though accepting the mechanical philosophy, his search for the ultimate grounds of mechanism led him to metaphysics and the doctrine of entelechies. This indicates the early orientation of Leibniz' thought towards the ideas of the *Monadologie*. Instead of setting out his philosophy systematically in a *magnum opus*, Leibniz presented piecemeal clarifications of his views in works that, in various ways, were inspired by the publications of others. After reading the papers of Huygens and Wren on collision and the *Elementorum philosophiae* of Hobbes, Leibniz composed his *Hypothesis physica nova*, consisting of two parts, *Theoria motus abstracti* and *Theoria motus concreti*, which in 1671 he presented respectively to the Paris Academy of Sciences and the London Royal Society. At this time, Leibniz owed more to Descartes, whose work he knew only at second hand, than he was later willing to admit. Closer study of the philosophy of Descartes led Leibniz to a more decisive rejection.

In 1686 he published in the *Acta eruditorum* a criticism of Descartes's measure of force, *Brevis demonstratio erroris memorabilis Cartesii et aliorum circa legem naturae*, which started a controversy with Catalan, Malebranche, and Papin lasting until 1691. Also in 1686 Leibniz sent to Arnauld, for his comments, an essay entitled *Discours de métaphysique*, in which he developed the ideas of the later *Théodicée*. The tracts entitled *De lineis opticis*, *Schediasma de resistentia medii*, and *Tentamen de motuum coelestium causis*, published in the *Acta eruditorum* in 1689, were hurriedly composed by Leibniz after he had read the review of Newton's *Principia* in the same journal, in an attempt to obtain some credit for results which he had derived independently of Newton. In 1692, at the instigation of Pelisson, Leibniz wrote an *Essay de dynamique*, which was read to the Paris Academy by Philippe de la Hire, and in 1695 there appeared in the *Acta eruditorum* an article entitled *Specimen dynamicum*, which contained the clearest exposition of Leibniz' dynamics.

Leibniz' *Nouveaux essais sur l'entendement humain*, completed in 1705 but not published during his lifetime, presented a detailed criticism of Locke's position. By adding *nisi ipse intellectus* to the famous maxim, *Nihil est in intellectu quod non prius fuerit in sensu* (wrongly attributed to Aristotle by Duns Scotus[1]), Leibniz neatly reversed the application of the principle by Locke. According to Leibniz, the mind originally contains the principles of the various ideas which the senses on occasion call forth.

In 1710 Leibniz published his *Essais de Théodicée sur la bonté de Dieu, la liberté de l'homme et l'origine du mal*, a work composed at the instigation of Sophia Charlotte, with whom Leibniz had conversed concerning the views of Bayle. In response to a request from Prince Eugene for an abstract of the *Théodicée*, Leibniz in 1714 wrote the *Principes de la nature et de la grâce fondées en raison* and the *Monadologie*. When in 1715 Leibniz wrote to Princess Caroline of Wales, criticizing the philosophical and theological implications of the work of Newton, she commissioned Samuel Clarke to reply. The ensuing correspondence, containing Leibniz' most penetrating criticism of Newtonian philosophy, was published in 1717.

Rational Physics (Protophysics). In his efforts to clarify fundamental physical principles, Leibniz followed a plan which he called a transition from geometry to physics through a "science of motion that unites matter with forms and theory with practice."[2] He sought the metaphysical foundations of mechanics in an axiomatic structure. The *Theoria motus abstracti*[3] offers a rational theory of motion whose axiomatic foundation *(fundamenta praedemonstrabilia)* was inspired by the indivisibles of Cavalieri and the notion of *conatus* proposed by Hobbes. Both the word *conatus* and the mechanical idea were taken from Hobbes,[4] while the mathematical reasoning was derived from Cavalieri. After his invention of the calculus, Leibniz was able to replace Cavalieri's indivisibles by differentials and this enabled him to apply his theory of *conatus* to the solution of dynamical problems.

The concept of *conatus* provided for Leibniz a path of escape from the paradox of Zeno. Motion is continuous and therefore infinitely divisible, but if motion is real, its beginning cannot be a mere nothing.[5] *Conatus* is a tendency to motion, a mind-like quality having the same relation to motion as a point to space (in Cavalieri's terms) or a differential to a finite quantity (in terms of the infinitesimal calculus). *Conatus* represents virtual motion; it is an intensive quality that can be measured by the distance traversed in an infinitesimal element of time. A body can possess several *conatuses* simultaneously and these can be combined into a single *conatus* if they are compatible. In the absence of motion, *conatus* lasts only an instant,[6] but however weak, its effect is transmitted to infinity in a plenum. Leibniz' doctrine of *conatus*, in which a body is conceived as a momentary mind, that is, a mind without memory, may be regarded as a first sketch of the philosophy of monads.

Mathematically, *conatus* represents for Leibniz accelerative force in the Newtonian sense, so that, by summing an infinity of *conatuses* (that is, by integration), the effect of a continuous force can be measured. Examples of *conatus* given by Leibniz are centrifugal force and what he called the solicitation of gravity. Further clarifications of the concept of *conatus* are given in the *Essay de dynamique* and *Specimen dynamicum*, where *conatus* is compared with static force or *vis mortua* in contrast to *vis viva*, which is produced by an infinity of impressions of *vis mortua*.

Physics (Mechanical Hypothesis). Leibniz soon recognized that the idea of *conatus* could not by itself explain the results of the experiments of Huygens and Wren on collision. Since in the absence of motion *conatus* lasts only an instant, a body once brought to rest in a collision, Leibniz explains, could not then rebound.[7] A new property of matter was needed and this was provided for Leibniz by the action of an ether. As conceived by Leibniz in his *Theoria motus concreti*,[8] the ether was a universal agent of motion, explaining mechanically all the phenomena of the visible world. This essentially Cartesian notion was adopted by Leibniz following a brief attachment to the doctrine of physical atomism defended in the works of Bacon and Gassendi. Leibniz did not, however, become a Cartesian, nor did he aim to construct an entirely new hypothesis but rather to improve and reconcile those of others.[9]

A good example of the way in which Leibniz pursued this goal is provided by the planetary theory expounded in his *Tentamen de motuum coelestium causis*. In this work, Leibniz combined the mechanics of *conatus* and inertial motion with the concept of a fluid vortex to give a physical explanation of planetary motion on the basis of Kepler's analysis of the elliptical orbit into a circulation and a radial motion. Leibniz' harmonic vortex accounted for the circulation while the variation in distance was explained by the combined action of the centrifugal force arising from the circulation and the solicitation of gravity. This solicitation he held to be the effect of a second independent vortex of the kind imagined by Huygens, to whom he described Newton's attraction as "an immaterial and inexplicable virtue," a criticism he made public in the *Théodicée* and repeated in the correspondence with Clarke.

Although Leibniz' planetary theory could be described as a modification of that of Descartes, he did not acknowledge any inspiration from this source. Attributing the idea of a fluid vortex to Kepler and also, but incorrectly, the idea of centrifugal force, Leibniz claimed that Descartes had made ample use of these ideas without acknowledgment. Leibniz had already, in a letter to Arnauld,[10] rejected the Cartesian doctrine that the essence of corporeal substance is extension. One reason that led him to this rejection was the theological problem of transubstantiation which he studied at the instigation of Baron Boyneburg,[11] but the most important dynamical reason was connected with the relativity of motion. As explained by Leibniz in the *Discours de métaphysique*,[12] since motion is relative, the real difference between a moving body and a body at rest cannot consist of change of position. Consequently, as the principle of inertial motion precludes an external impulse for a body moving with constant speed, the cause of motion must be an inherent force. Another argument against the Cartesian position involves the principle of the identity of indiscernibles. From this principle it follows that, besides extension, which is a property carried by a body from place to place, the body must have some intrinsic property which distinguishes it from others. According to Leibniz, bodies possess three properties which cannot be derived from extension: namely, impenetrability, inertia, and activity. Impenetrability and inertia are associated with *materia prima*, which is an abstraction, while *materia secunda* (the matter of dynamics) is matter endowed with force.

Since, for Leibniz, force alone confers reality to motion, the correct measure of force becomes the central problem of dynamics. Now Descartes had measured what Leibniz regarded as the active force of bodies, that is, the cause of their activity, by their quantity of motion. But, as Leibniz remarked in a letter of 1680,[13] Descartes's erroneous laws of collision implied that his basic principle of the conservation of motion was false. In 1686 Leibniz published his criticism of the Cartesian principle of the conservation of motion in his *Brevis demonstratio erroris mirabilis Cartesii*, thereby precipitating the *vis viva* controversy. According to Leibniz, Descartes had incorrectly generalized from statics to dynamics. In statics or virtual motion, Leibniz explains, the force is as the velocity but when the body has acquired a finite velocity and the force has become live, it is as the square of the velocity. As Leibniz remarked on several occasions, there is always a perfect equation between cause and effect, so that the force of a body in motion is measured by the product of the mass and the height to which the body could rise (the effect of the force). Using the laws of Galileo, this height was shown by Leibniz to be proportional to the square of the velocity, so that the force *(vis viva)* could be expressed as mv^2.

Since *vis viva* was regarded by Leibniz as the ultimate physical reality,[14] it had to be conserved throughout all transformations. Huygens had shown that, in elastic collision, *vis viva* is not diminished. The *vis viva* apparently lost in inelastic collision Leibniz held to be in fact simply distributed among the small parts of the bodies.[15]

Leibniz discovered the principle of the conservation of momentum, which he described as the "quantité d'avancement."[16] Had Descartes known that the quantity of motion is preserved in every direction (so that motion is completely determined, leaving no opportunity for the directing influence of mind), Leibniz remarks, he would probably have discovered the preestablished harmony. But in Leibniz' view, the principle of the conservation of momentum did not correspond to something absolute, since two bodies moving together with equal quantities of motion would have no total momentum. Leibniz' discovery of yet another absolute quantity in the concept of *action* enabled him to answer the Cartesian criticism that he had failed to take time into account in his consideration of *vis viva*. In his *Dynamica de potentia et legibus naturae corporeae*, Leibniz made an attempt to fit this new concept into his axiomatic scheme.

Although succeeding generations described the *vis viva* controversy as a battle of words, there can be no doubt that Leibniz himself saw it as a debate about the nature of reality. Referring to his search for a true dynamics, Leibniz remarked in 1689 that, to escape from the labyrinth, he could find "no other thread of Ariadne than the evaluation of forces, under the supposition of the metaphysical principle that the total effect is always equal to the complete cause."[17]

Scientia Generalis. According to the usual distinctions, the position that Leibniz took in physics, as well as in other fundamental questions, was rationalistic, and to that extent, despite all differences in detail, was related to Descartes's position. Evident confirmation of this may be seen in Leibniz' controversy with Locke; although he does not explicitly defend the Cartesian view, he uses arguments compatible with this position. It is often overlooked, however, that Leibniz was always concerned to discuss epistemological issues as questions of theoretical science. For example, while Locke speaks of the origins of knowledge, Leibniz speaks of the structure of a science which encompasses that particular field. Thus Leibniz sees the distinction between necessary and contingent truths, so important in the debate with Locke, as a problem of theoretical science which transcends any consideration of the historical alternatives, rationalism and empiricism. Neither an empiricist in the sense of Locke nor yet a rationalist in the sense of Descartes, Leibniz saw the refutation of the empiricist's thesis (experience as the nonconceptual basis of knowledge) not as the problem of a rational psychology as it was then understood (in Cartesian idiom, the assumption of inborn truths and ideas) but as a problem that can be resolved only within the framework of a general logic of scientific research. Nevertheless, he shares with Descartes one fundamental rationalistic idea, namely the notion (which may be discerned in the Cartesian *mathesis universalis)* of a *scientia generalis*. In connection with his theoretical linguistic efforts towards a *characteristica universalis*, this thought appears in Leibniz as a plan for a *mathématique universelle*.[18]

Inspired by the ideas of Lull, Kircher, Descartes, Hobbes, Wilkins, and Dalgarno, Leibniz pursued the invention of an alphabet of thought *(alphabetum cogitationum humanarum)*[19] that would not only be a form of shorthand but a formalism for the creation of knowledge itself. He sought a method that would permit "truths of reason in any field whatever to be attained, to some degree at least, through a calculus, as in arithmetic or algebra."[20] The program of such a *lingua philosophica* or *characteristica universalis* was to proceed through lists of definitions to an elementary terminology encompassing a complete encyclopedia of all that was known. Leibniz connected this plan with others that he had, such as the construction of a general language for intellectual discourse and a rational grammar, conceived as a continuation of the older *grammatica speculativa*.

While Leibniz' programmatic statements leave open the question of just how the basic language he was searching for and the encyclopedia were to be connected (the *characteristica universalis*, according to other explanations, was itself to facilitate a compendium of knowledge), a certain *ars combinatoria*, conceived as part of an *ars inveniendi*, was to serve in the creation of the lists of definitions. As early as 1666, in Leibniz' *Dissertatio de arte combinatoria*, the *ars inveniendi* was sketched out under the name *logica inventiva* (or *logica inventionis*) as a calculus of concepts in which, in marked contrast to the traditional

LEIBNIZ

theories of concepts and judgments, he discusses the possibility of transforming rules of inference into schematic deductive rules. Within this framework there is also a complementary *ars iudicandi*, a mechanical procedure for decision making. However, the thought of gaining scientific propositions by means of a calculus of concepts derived from the *ars combinatoria* and a mechanical procedure for decision making remained lodged in a few attempts at the formation of the "alphabet." Leibniz was unable to complete the most important task for this project, namely, the proof of its completeness and irreducibility, nor did he consider this problem in his plans for the *scientia generalis*, a basic part of which was the "alphabet," the *characteristica universalis* in the form of a *mathématique universelle*.

The *scientia generalis* exists essentially only in the "tables of contents," which are not internally consistent terminologically and thus admit of additions at will. Nevertheless, it is clear that Leibniz was thinking here of a structure for a general methodology, consisting, on the one hand, of partial methodologies concerning special sciences such as mathematics, and on the other hand, of procedures for the *ars inveniendi*, such as the *characteristica universalis;* taken together, these were probably intended to replace traditional epistemology as a unified conceptual armory. This was by no means impracticable, at least in part. For example, the analytical procedures in which arithmetical transformations occur independently of the processes to which they refer, employed by Leibniz in physics, may be construed as a partial realization of the concept of a *characteristica universalis*.

Formal Logic. Leibniz produced yet another proof of the feasibility of his plan for schematic operations with concepts. Besides the infinitesimal calculus, he created a logical calculus *(calculus ratiocinator, universalis, logicus,* or *rationalis)* that was to lend the same certainty to deductions concerning concepts as that possessed by algebraic deduction. Leibniz stands here at the very beginning of formal logic in the modern sense, especially in relation to the older syllogistics, which he succeeded in casting into the form of a calculus. A number of different steps may be distinguished in his program for a logical calculus. In 1679 various versions of an arithmetical calculus appeared that permitted a representation of a conjunction of predicates by the product of prime numbers assigned to the individual predicates. In order to solve the problem of negation—needed in the syllogistic modes—negative numbers were introduced for the nonpredicates of a concept. Every concept was assigned a pair of numbers having no common factor, in which the factors of the first represented the predicates and the factors of the second represented the nonpredicates of the concept. Because this arithmetical calculus became too complex, Leibniz replaced it in about 1686 by plans for an algebraic calculus treating the identity of concepts and the inclusion of one concept in another. The components of this calculus were the symbols for predicates, a, b, c, . . . *(termini)*, an operational sign $-$ *(non)*, four relational signs $\subset$, $\not\subset$, $=$, $\neq$ (represented in language by *est*, *non est*, *sunt idem* or *eadem sunt*, *diversa sunt)* and the logical particles in vernacular form. To the rules of the calculus *(principia calculi)*—as distinct from the axioms *(propositiones per se verae)* and hypotheses *(propositiones positae)* which constitute its foundation—belong the principles of implication and logical equivalence and also a substitution formula. Among the theses *(propositiones verae)* that can be proved with the aid of the axioms and hypotheses, such as $a \subset a$ (reflexivity of the relation $\subset$) and $a \subset b$ et $b \subset c$ implies $a \subset c$ (transitivity of $\subset$), is the proposition $a \subset b$ et $d \subset c$ implies $ad \subset bc$. This was called by Leibniz the "admirable theorem" *(praeclarum theorema)* and appears again, much later, with Russell and Whitehead.[21]

Leibniz extended this algebraic calculus in various ways, first with a predicate-constant *ens* (or *res*), which may be understood as a precursor of the existential quantifier, and secondly with the interpretation of the predicates as propositions instead of concepts. Inclusion between concepts becomes implication between propositions and the new predicate-constant *ens* appears as the truth value *(verum)*, intensionally designated as *possibile*. These discourses were concluded in about 1690 with two calculi[22] in which a transition is made from an (intensional) logic of concepts to a logic of classes. The first, originally entitled *Non inelegans specimen demonstrandi in abstractis* (a "plus-minus calculus"), is a pure calculus of classes (a dualization of the thesis of the original algebraic calculus) in which a new predicate-constant *nihil* (for *non-ens*) is introduced. The second calculus (a "plus calculus") is an abstract calculus for which an extensional as well as an intensional interpretation is expressly given. Logical addition in the "plus-minus calculus" is symbolized by $+$. In the "plus calculus," logical addition, as well as logical multiplication in the intensional sense, is symbolized by $\oplus$, while the relational sign $=$ *(sunt idem* or *eadem sunt)* is replaced by ∞ and the sign $\neq$ by *non* $A \infty B$. Furthermore, subtraction appears in the "plus-minus calculus," symbolized by $-$ or $\ominus$, and also the relation of incompatibility *(incommunicantia sunt)* together with its negation *(com-*

municantia sunt or *compatibilia sunt)*. For example, one of the propositions of the calculus states: $A - B = C$ holds exactly, if and only if $A = B + C$, and B and C are incompatible; in modern notation

$$a \llcorner b = c \wedge (b \subset a) \leftrightarrow a = (b \vee c) \wedge (b \mid c).$$

(The condition $b \subset a$ is implicit in Leibniz' use of the symbol $-$).

If we disregard a few syntactical details and observe that Leibniz' work gives an approximation to a complete interpretation of the elements of a logical calculus (including the rules for transformation), we see here for the first time a formal language and thus an actual successful example of a *characteristica universalis*. It is true that Leibniz did not make sufficient distinction between the formal structure of the calculus and the interpretations of its content; for example, the beginnings of the calculus are immediately considered as axiomatic and the rules of transformation are viewed as principles of deduction. Yet it is decisive for Leibniz' program and our appreciation of it that he succeeded at all, in his logical calculus, in the formal reconstruction of principles of deduction concerning concepts.

General Logic. Leibniz' general logical investigations play just as important a role in his systematic philosophy as does his logical calculus. Most important are his analytical theory of judgment, the theory of complete concepts on which this is based, and the distinction between necessary and contingent propositions. According to Leibniz' analytical theory of judgment, in every true proposition of the subject-predicate form, the concept of the predicate is contained originally in the concept of the subject *(praedicatum inest subiecto)*. The *inesse* relation between subject and predicate is indeed the converse of the universal-affirmative relation between concepts, long known in traditional syllogisms (*B inest omni A* is the converse of *omne A est B*).[23] Although it is thus taken for granted that subject-concepts are completely analyzable, it suffices, in a particular case, that a certain predicate-concept can be considered as contained in a certain subject-concept. Fundamentally, there is a theory of concepts according to which concepts are usually defined as combinations of partials, so that analysis of these composite concepts *(notiones compositae)* should lead to simple concepts *(notiones primitivae* or *irresolubiles)*. With these, the *characteristica universalis* could then begin again. When the predicate-concept simply repeats the subject-concept, Leibniz speaks of an identical proposition; when this is not the case, but analysis shows the predicate-concept to be contained implicitly in the subject-concept, he speaks of a virtually identical proposition.

The distinction between necessary propositions or truths of reason *(vérités de raisonnement)* and contingent propositions or truths of fact *(vérités de fait)* is central to Leibniz' theory of science. As contingent truths, the laws of nature are discoverable by observation and induction but they have their rational foundation, whose investigation constitutes for Leibniz the essential element in science, in principles of order and perfection. Leibniz replaces the classical syllogism, as a principle of deduction, by the principle of substitution of equivalents to reduce composite propositions to identical propositions. Contingent propositions are defined as those that are neither identical nor reducible through a finite number of substitutions to identical propositions.

All contingent propositions are held by Leibniz to be reducible to identical propositions through an infinite number of steps. Only God can perform these steps, but even for God, such propositions are not necessary (in the sense of being demanded by the principle of contradiction). Nevertheless, contingent propositions, in Leibniz' view, can be known *a priori* by God and, in principle, also by man. For Leibniz, the terms *a priori* and *necessary* are evidently not synonymous. It is the principle of sufficient reason that enables us (at least in principle) to know contingent truths *a priori*. Consequently, the deduction of such truths involves an appeal to final causes. On the physical plane, every event must have its cause in an anterior event. Thus we have a series of contingent events for which the reason must be sought in a necessary Being outside the series of contingents. The choice between the possibles does not depend on God's understanding, that is to say, on the necessity of the truths of mathematics and logic, but on his volition. God can create any possible world, but, being God, he wills the best of all possible worlds. Thus the contingent truths, including the laws of nature, do not proceed from logical necessity but from a moral necessity.

Methodological Principles. Logical calculi and the notions mentioned under "General Logic" belong to a general theory of foundations that also encompasses certain important Leibnizian methodological principles. The principle of sufficient reason *(principium rationis sufficientis*, also designated as *principium nobilissimum)* plays a special role. In its simplest form it is phrased "nothing is without a reason" *(nihil est sine ratione)*, which includes not only the concept of physical causality *(nihil fit sine causa)* but also in general the concept of a logical antecedent-consequent relationship. According to Leibniz, "a large part of metaphysics [by which he means rational theology], physics, and ethics" may be constructed on this proposition.[24] Viewed methodologically, this means that, in the principle of sufficient reason, there is a teleological as well as a causal principle; the particular import of the proposition is that both principles may be used in the same way for physical processes and human actions.

Defending the utility of final causes in physics (in opposition to the view of Descartes), Leibniz explained that these often provided an easier path than the more direct method of mechanical explanation in terms of efficient causes.[25] Leibniz had himself in 1682 used a variation of Fermat's principle in an application of his method of maxima and minima to the derivation of the law of refraction. Closely associated with the principle of sufficient reason is the principle of perfection *(principium perfectionis* or *melioris)*. In physics, this principle determines the actual motion from among the possible motions, and in metaphysics leads Leibniz to the idea of "the best of all possible worlds." The clearest expression of Leibniz' view is to be found in his *Tentamen anagogicum*,[26] written in about 1694, where he remarks that the least parts of the universe are ruled by the most perfect order. In this context, the idea of perfection consists in a maximum or minimum quantity, the choice between the two being determined by another architectonic principle, such as the principle of simplicity. Since the laws of nature themselves are held by Leibniz to depend on these principles, he supposed the existence of a perfect correlation between physical explanations in terms of final and efficient causes.

In relation to Leibniz' analytical theory of judgment and his distinction between necessary and contingent propositions, the principle of sufficient reason entails that, in the case of a well-founded connection between, for example, physical cause and physical effect, the proposition that formulates the effect may be described as a logical implication of the proposition that formulates the cause. Generalized in the sense of the analytical theory of truth and falsehood that Leibniz upholds, this means: "nothing is without a reason; that is, there is no proposition in which there is not some connection between the concept of the predicate and the concept of the subject, or which cannot be proved *a priori*."[27] This logical sense of the principle of sufficient reason contains also (in its formulation as a *principium reddendae rationis*) a methodological postulate; propositions are not only capable of being grounded in reasons (in the given analytical manner) but they must be so grounded (insofar as they are formulated with scientific intent).[28]

In addition to the principle of sufficient reason, the principle of contradiction *(principium contradictionis)* and the principle of the identity of indiscernibles *(principium identitatis indiscernibilium)* are especially in evidence in Leibniz' logic. In its Leibnizian formulation, the principle of contradiction, $\neg(A \wedge \neg A)$, includes the principle of the excluded middle, $A \vee \neg A$ *(tertium non datur)*: "nothing can be and not be at the same time; everything is or is not."[29] Since Leibniz' formulation rests on a theory according to which predicates, in principle, can be traced back to identical propositions, he also classes the principle of contradiction as a principle of identity. The principle of the identity of indiscernibles again defines the identity of two subjects, whether concrete or abstract, in terms of the property that the mutual replacement of their complete concepts in any arbitrary statement does not in the least change the truth value of that statement *(salva veritate)*. Two subjects s_1 and s_2 are different when there is a predicate P that is included in the complete concept S_1 of s_1 but not in the complete concept S_2 of s_2, or vice versa. If there is no such predicate, then because of the mutual replaceability of both complete concepts S_1 and S_2, there is no sense in talking of different subjects. This means, however, that the principle of the identity of indiscernibles, together with its traditional meaning ("there are no two indistinguishable subjects"[30]), is synonymous with the definition of logical equality ("whatever can be put in place of anything else, *salva veritate*, is identical to it"[31]).

Metaphysics (Logical Atomism). Since the investigations of Russell and Couturat, it has become clear that Leibniz' theory of monads is characterized by an attempt to discuss metaphysical questions within a framework of logical distinctions. On several occasions, however, Leibniz himself remarks that dynamics was to a great extent the foundation of his system. For example, in his *De primae philosophiae emendatione et de notione substantiae*, Leibniz comments that the notion of force, for the exposition of which he had designed a special science of dynamics, added much to the clear understanding of the concept of substance.[32] This suggests that it was the notion of mechanical energy that led to the concept of substance as activity. Again, it is in dynamics, Leibniz remarks, that we learn the difference between necessary truths and those which have their source in final causes, that is to say, contingent truths,[33] while optical theory, in the form of Fermat's principle, pointed to the location of the final causes in the principle of perfection.[34] Even the subject-predicate logic itself, which forms the rational foundation of Leibniz' metaphysics, seems to take on a biological image, such as the growth of a plant from a seed, when Leibniz writes to De Volder that the present state of a substance must

LEIBNIZ

involve its future states and vice versa. It thus appears that physical analogies very probably provided the initial inspiration for the formation of Leibniz' metaphysical concepts.

In the preface to the *Théodicée*, Leibniz declares that there are two famous labryrinths in which our reason goes astray; the one relates to the problem of liberty (which is the principal subject of the *Théodicée*), the other to the problem of continuity and the antinomies of the infinite. To arrive at a true metaphysics, Leibniz remarks in another place, it is necessary to have passed through the labyrinth of the continuum.[35] Extension, like other continuous quantities, is infinitely divisible, so that physical bodies, however small, have yet smaller parts. For Leibniz, there can be no real whole without real unities, that is, indivisibles,[36] or as he expresses it (repeating a phrase used by Nicholas of Cusa[37]), being and unity are convertible terms.[38] Now the real unities underlying physical bodies cannot be mathematical points, for these are mere nothings. As Leibniz explains, "only metaphysical or substantial points . . . are exact and real; without them there would be nothing real, since without true unities *[les véritables unités]* a composite whole would be impossible."[39] Within the narrower framework of physics, such unities can be understood as the concept of mass-points, but they are meant in the broader sense of the classical concept of substance, to the consideration of which Leibniz had in 1663 devoted his first philosophical essay, *Dissertatio metaphysica de principio individui*. Leibniz' metaphysical realities are unextended substances or monads (a term he used from 1696), whose essence is an intensive quality of the nature of force or mind.

Leibniz consciously adheres to Aristotelian definitions, when he emphasizes that we may speak of an individual substance whenever a predicate-concept is included in a subject-concept, and this subject-concept never appears itself as a predicate-concept. A concept fulfilling this condition may thus be designated as an individual concept *(notion individuelle)* and may be construed as a complete concept, that is, as the infinite conjunction of predicates appertaining to that individual. If a complete subject-concept were itself to appear as a predicate-concept, then according to the principle of the identity of indiscernibles, the predicated individual would be identical to the designated individual. This result of Leibniz is a logical reconstruction of the traditional ontological distinction between substance and quality.

The definition of individual concepts, according to which individual substances are denoted by complete concepts, leads also to the idea, central to the theory of monads, that each monad or individual substance represents or mirrors the whole universe. What Leibniz means is that, given a particular subject, all other subjects must appear, represented by their names or designations, in at least one of the infinite conjunction of predicates constituting the complete concept of that particular subject.

Leibniz describes the inner activity which constitutes the essence of the monad as perception.[40] This does not imply that all monads are conscious. The monad has perception in the sense that it represents the universe from its point of view, while its activity is manifested in spontaneous change from one perception to another. The attribution of perception and appetition to the monads does not mean that they can be sufficiently defined in terms of physiological and psychological processes, although Leibniz does compare them to biological organisms; he was, of course, familiar with the work of Leeuwenhoek, Swammerdam, and Malpighi on microorganisms, which seemed to confirm the theory of preformation demanded by the doctrine of monads. Once again, the logical basis for the theory of perception is that the concept of the monad's inner activity must occur in the concept of the individual monad; for everything that an individual substance encounters "is only the consequence of its notion or complete concept, since this notion already contains all predicates or events and expresses the whole universe."[41]

Within the framework of this conceptual connection between a theory of perception and a theory of individual concepts, sufficient room remains for physiological and psychological discussion and here Leibniz goes far beyond the level of debate in Locke and Descartes. In particular, Leibniz distinguishes between consciousness and self-consciousness, and again, between stimuli which rise above the threshold of consciousness and those that remain below it; he even observes that the summation of sub-threshold stimuli can finally lead to one that emerges over the threshold of consciousness, a clear hint of the existence of the unconscious.

Since, for Leibniz, the real unities constituting the universe are essentially perceptive, it follows that the real continuum must also be a continuum of perception. The infinite totality of monads represent or mirror the universe, of which each is a part, from all possible points of view, so that the universe is at once continuous and not only infinitely divisible but actually divided into an infinity of real metaphysical atoms.[42] As these atoms are purely intensive unities, they are mutually exclusive, so that no real interaction between them is possible. Consequently, Leibniz needed his principle of the preestablished harmony (which, he claimed, avoided the perpetual intervention of God involved in the doctrine of occasionalism) to explain the mutual compatibility of the internal activities of the monads. Leibniz thus evaded the antinomies of the continuum by conceiving reality not as an extensive plenum of matter bound by physical relations but as an intensive plenum of force or life bound by a preestablished harmony.

The monads differ in the clarity of their perceptions, for their activity is opposed and their perceptions consequently confused to varying degrees by the *materia prima* with which every created monad is endowed. As with the *materia prima* of dynamics, that of the monads is thus associated with passivity. Only one monad, God, is free of *materia prima* and he alone perceives the world with clarity, that is to say, as it really is.

Physical bodies consist of infinite aggregates of monads. Since such aggregates form only accidental unities, they are not real wholes and consequently, in Leibniz' view, not possessed of real magnitude. It is in this sense that Leibniz denies infinite number while admitting the existence of an actual infinite. When an aggregate has one dominant monad, the aggregate appears as the organic body of this monad. Aggregates without a dominant monad simply appear as *materia secunda*, the matter of dynamics. Bodies as such are therefore conceived by Leibniz simply as phenomena, but in contrast to dreams and similar illusions, well-founded phenomena on account of their consistency. For Leibniz then, the world of extended physical bodies is just a world of appearance, a symbolic representation of the real world of monads.

From this doctrine it follows that the forces involved in dynamics are only accidental, or derivative, as Leibniz terms them. The real active force, or *vis primitiva*, which remains constant in each corporeal substance, corresponds to mind or substantial form. Leibniz does not, however, reintroduce substantial forms as physical causes;[43] for in his view, physical explanations involve a *vis derivativa* or accidental force by which the *vis primitiva* or real principle of action is modified.[44] One monad, having more *vis primitiva*, represents the universe more distinctly than another, a difference that can be expressed by the terms active and passive, though there is, of course, no real interaction. In the world of phenomena, this relation is symbolized in the notion of physical causality. The laws of nature, including the principle of the conservation of *vis viva*, thus have relevance only at the phenomenal level, though they symbolize, on the metaphysical plane, an order manifested through the realization of predicates of individual substances in accordance with the preestablished harmony.

In the early stages of the formulation of his metaphysics, Leibniz located the unextended substances, that he later called monads, in points.[45] This presupposed a real space in which the monads were embedded, a view that consideration of the nature of substance and the difficulties of the continuum caused him soon to abandon. In his *Système nouveau*,[46] written in 1695, Leibniz described atoms of substance (that is, monads) as metaphysical points, but the mathematical points associated with the space of physics he described as the points of view of the monads. Physical space, Leibniz explains, consists of relations of order between coexistent things.[47] This may be contrasted with the notion of an abstract space, which consists of an assemblage of possible relations between possible existents.[48] For Leibniz then, space is an assemblage of relations between the monads. By locating the points of space in a quality of the monads, namely their points of view, Leibniz makes space entirely dependent on the monads. Space itself, however, is not a property of the monads, for points are not parts of space. In the sense of distance between points, space is a mere ideal thing, the consideration of which, Leibniz remarks, is nevertheless useful.[49] Distances between points are representations of the differences of the points of view of the monads. Owing to the preestablished harmony, these relations are compatible, so that space is a well-founded phenomenon, an extensive representation of an intensive continuum. Similarly time, as the order of noncontemporaneous events,[50] is also a well-founded phenomenon.

Leibniz' objections to the kind of absolute real space and time conceived by Newton are expressed most completely in his correspondence with Clarke. Real space and time, Leibniz argues, would violate the principle of sufficient reason and the principle of the identity of indiscernibles. For example, a rotation of the whole universe in an absolute space would leave the arrangement of bodies among themselves unchanged, but no sufficient reason could be found why God should have placed the whole universe in one of these positions rather than the other. Again, if time were absolute, no reason could be found why God should have created the universe at one time rather than another.[51]

From the ideal nature of space and time, it follows that motion, in the physical sense, is also ideal and therefore relative.[52] Against Newton, Leibniz maintains the relativity not only of rectilinear motion but also of rotation.[53] Yet Leibniz is willing to admit that there is a difference between what he calls an absolute real motion of a body and a mere change in its position relative to other bodies. For it is the body in which the cause of motion (that is, the active force) resides that is truly in motion.[54] Now it is evidently the *vis primitiva* that Leibniz has in mind, since consideration of the *vis derivativa (conatus* or *vis viva)* does not serve to identify an absolute motion, so that we may interpret Leibniz as saying that true absolute motion appertains to the metaphysical plane, where it can be perceived only by God. Indeed all bodies have an absolute motion in this sense, for since all monads have activity, all aggregates have *vis viva.* The world of phenomena is therefore conceived by Leibniz as a world of bodies in absolute motion (rest being a mere abstraction) but a world in which only relative changes of position can be observed.

The theory of monads may be seen as a sustained effort to present, in "cosmological" completeness, a systematic unified structure of knowledge on the basis of a logical reconstruction of the concept of substance. Insofar as the central assertion, namely, that because there are composite "substances," there must be simple substances, is not only a cosmological and metaphysical statement, but in addition, an assertion of the priority of synthetic over analytic procedures, his efforts retain their original methodological meaning. As revealed in his plan for a *characteristica universalis,* analysis, for Leibniz, implies synthesis, but a synthesis that must begin with irreducible elements. Where there are no such elements, neither analysis nor synthesis is possible within the framework of Leibniz' constructive methodology. This means that Leibniz, in the course of his protracted efforts to define an individual substance, moved from physical atomism to logical atomism in the modern sense, as represented by Russell. Leibniz believed that he had proved the thesis of an unambiguously defined world, in which the physico-theological theme running through the mechanistic philosophy of his age might once again be seen as a metaphysics in the classical sense.

Influence. The thought of Leibniz influenced the history of philosophy and science in two ways; first, through the mediocre systematization of Christian Wolff known as the "Leibnizo-Wolffian" philosophy, and secondly, through the significance of particular theories in the history of various sciences. While the tradition of the Leibnizo-Wolffian philosophy ended with Kant, the influence of particular theories of Leibniz lasted through the nineteenth and into the twentieth century.

The controversy with Newton and Clarke was not conducive to the reception of Leibniz' work in physics and hindered the objective evaluation of important contributions such as his law of radial acceleration. The *vis viva* controversy arose as a direct result of Leibniz' criticism of Descartes and concerned not only the measure of force but also the nature of force itself. While 'sGravesande and d'Alembert (in his *Traité de mécanique*) judged the dispute to be merely a semantic argument, Kant in 1747 *(Gedanken von der wahren Schätzung der lebendigen Kräfte*) made an ineffective attempt at reconciliation. Leibnizian dynamics was developed further by Bošković, who transformed the concept of dynamic force in the direction of a concept of relational force.

Within the framework of rational physics, Kant contributed some essential improvements, such as the completion of the distinction between necessary and contingent propositions by means of the concept of the synthetic *a priori* and clarification of the principle of causality. Leibniz' protophysical plan, however, remained intact. It was continued later by Whewell, Clifford, Mach, and Dingler, to mention only a few. Insofar as the fundamental concepts of space and time were concerned, the Newtonian ideas of absolute space and time at first prevailed over the Leibnizian ideas of relational space and time. Kant also tried here to mediate between the ideas of Leibniz and Newton, but his own suggestion (space as the origin of the distinction between a nonreflexible figure and its mirror image, such as a pair of gloves) strongly resembled the Newtonian concept of absolute space. Modern relativistic physics has turned the scales in favor of the ideas of Leibniz.

Among the methodological principles of Leibniz, only the principle of sufficient reason has played a prominent role in the history of philosophy. Wolff, disregarding Leibniz' methodological intentions, tried to prove it by ontological means *(Philosophia sive ontologia);* Kant reduced it essentially to the law of causality and in 1813 Schopenhauer drew on it for the elucidation of his four conditions of verification *(Über die vierfache Wurzel des Satzes vom zureichenden Grunde).* On the other hand, the methodological project of a *characteristica universalis* together with the ensuing development of logical calculi has played a most significant role in the history of modern logic. In 1896 Frege, recalling Leibniz, described his *Begriffsschrift* of 1879 as a *lingua characterica* (not just a *calculus ratiocinator*), thus distinguishing it from the parallel efforts of Boole and Peano. De Morgan and Boole tried to carry out what Scholz has described as the "Leibniz program" of the development of a logical algebra of classes. This connection between logic and mathematics, evident also in Peirce and Schröder, was once again weakened by Frege, Peano, and Russell, whose work (especially that of Frege) nevertheless bears the inescapable influence of Leibniz; for even where the differences are greatest,

LEIBNIZ

the development of modern logic can be traced back to Leibniz. In this connection, it is fortunate that (in the absence of any publications of Leibniz) there is a tradition of correspondence beginning with letters between Leibniz, Oldenburg, and Tschirnhaus. The emphasis here, as exemplified in the logic theories of Ploucquet, Lambert, and Castillon, is on the intensional interpretation of logical calculi.

While there is an affinity between the theory of monads and Russell's logical atomism, a direct influence of the more metaphysical parts of the theory of monads on the history of scientific thought is difficult to prove. Particular results, such as the biological concept of preformation (accepted by Haller, Bonnet, and Spallanzani) or the discovery of sensory thresholds, though related to the theory of monads in a systematic way, became detached from it and followed their own lines of development. Yet the term "monad" played an important role with Wolff, Baumgarten, Crusius, and, at the beginning, with Kant (as exemplified in his *Monadologica physica* of 1756), then later with Goethe and Solger as well. Vitalism in its various forms, including the "biological romanticism" of the nineteenth century (the Schelling school), embraced in general the biological interpretation of the theory of monads, but this did not amount to a revival of the metaphysical theory. It is more likely that vitalism simply represented a reaction against mechanism, a tradition to which Leibniz also belonged.

NOTES

1. Duns Scotus, *Questiones super universalibus Porphyrii* (Venice, 1512), Quest. 3.
2. L. Couturat, ed., *Opuscules et fragments inédits de Leibniz* (Paris, 1903; Hildesheim, 1966), p. 594.
3. G. W. Leibniz, *Sämtliche Schriften und Briefe*, VI, 2, pp. 258–276.
4. T. Hobbes, *Elementorum philosophiae* (London, 1655), sectio prima: de corpore, pars tertia, cap. 15, §2 and §3.
5. *Sämtliche Schriften und Briefe*, VI, 2, p. 264.
6. *Ibid.*, p. 266.
7. *Ibid.*, p. 231.
8. *Ibid.*, pp. 221–257.
9. *Ibid.*, p. 257.
10. *Sämtliche Schriften und Briefe*, II, 1, p. 172.
11. *Ibid.*, pp. 488–490.
12. G. W. Leibniz, *Die philosophischen Schriften*, C. I. Gerhardt, ed., IV, p. 444; cf. p. 369.
13. *Sämtliche Schriften und Briefe*, II, 1, p. 508.
14. P. Costabel, *Leibniz et la dynamique* (Paris, 1960), p. 106.
15. G. W. Leibniz, *Mathematische Schriften*, C. I. Gerhardt, ed., VI, pp. 230–231. The manuscript called *Essay de dynamique* by Gerhardt is not earlier than 1698.
16. P. Costabel, *op. cit.*, p. 105.
17. *Ibid.*, p. 12.
18. *Sämtliche Schriften und Briefe*, VI, 6, p. 487.
19. L. Couturat, *op. cit.*, p. 430.
20. *Die philosophischen Schriften*, VII, p. 32.
21. *Principia mathematica*, *3. 47.
22. *Die philosophischen Schriften*, VII, pp. 228–247. Cf. L. Couturat, *op. cit.*, pp. 246–270.
23. *Sämtliche Schriften und Briefe*, VI, 1, p. 183.
24. *Die philosophischen Schriften*, VII, p. 301.
25. *Ibid.*, IV, pp. 447–448.
26. *Ibid.*, VII, pp. 270–279.
27. *G. W. Leibniz, Textes inédits*, G. Grua, ed. (Paris, 1948), I, p. 287.
28. *Die philosophischen Schriften*, VII, p. 309. Cf. L. Couturat, *op. cit.*, p. 525.
29. L. Couturat, *op. cit.*, p. 515.
30. *Sämtliche Schriften und Briefe*, VI, 6, p. 230.
31. *Die philosophischen Schriften*, VII, p. 219.
32. *Ibid.*, IV, p. 469.
33. *Ibid.*, III, p. 645.
34. *Ibid.*, IV, p. 447.
35. *Mathematische Schriften*, VII, p. 326.
36. *Die philosophischen Schriften*, II, p. 97.
37. Nicholas of Cusa, *De docta ignorantia*, Bk. II, ch. 7.
38. *Die philosophischen Schriften*, II, p. 304.
39. *Ibid.*, IV, p. 483.
40. *Principes de la nature et de la grâce*, A. Robinet, ed., p. 27.
41. *Discours de métaphysique*, G. le Roy, ed., p. 50.
42. *Die philosophischen Schriften*, I, p. 416.
43. *Ibid.*, II, p. 58.
44. *Mathematische Schriften*, VI, p. 236.
45. *Die philosophischen Schriften*, II, p. 372.
46. *Ibid.*, IV, p. 482.
47. *Ibid.*, IV, p. 491. Cf. II, p. 450.
48. *Ibid.*, VII, p. 415.
49. *Ibid.*, VII, p. 401.
50. *Mathematische Schriften*, VII, p. 18.
51. *Die philosophischen Schriften*, VII, p. 364.
52. *Ibid.*, II, p. 270. Cf. *Mathematische Schriften*, VI, p. 247.
53. *Mathematische Schriften*, II, p. 184.
54. *Die philosophischen Schriften*, VII, p. 404. Cf. IV, p. 444 and L. Couturat, *op. cit.*, p. 594.

JÜRGEN MITTELSTRASS
ERIC J. AITON

LEIBNIZ: Mathematics

Leibniz had learned simple computation and a little geometry in his elementary studies and in secondary schools, but his interest in mathematics was aroused by the numerous remarks on the importance of the subject that he encountered in his reading of philosophical works. In Leipzig, John Kuhn's lectures on Euclid left him unsatisfied, whereas he received some stimulation from Erhard Weigel in Jena. During his student years, he had also cursorily read introductory works on Cossist algebra and the *Deliciae physicomathematicae* of Daniel Schwenter and Philipp Harsdörffer (1636–1653) with their varied and mainly practical content. At this stage Leibniz considered himself acquainted with all the essential areas of mathematics that he needed for his studies in logic, which attracted him much more strongly. The very modest specialized knowledge that he then possessed is reflected in the *Dissertatio de arte combinatoria*

(1666); several additions are presented in the *Hypothesis physica nova* (1671).

In accord with the encyclopedic approach popular at the time, Leibniz limited himself primarily to methods and results and considered demonstrations nonessential and unimportant. His effort to mechanize computation led him to work on a calculating machine which would perform all four fundamental operations of arithmetic.

Leibniz was occupied with diplomatic tasks in Paris from the spring of 1672, but he continued the studies (begun in 1666) on the arithmetic triangle which had appeared on the title page of Apianus' *Arithmetic* (1527) and which was well known in the sixteenth century; Leibniz was still unaware, however, of Pascal's treatise of 1665. He also studied the array of differences of the number sequences, and discovered both fundamental rules of the calculus of finite differences of sequences with a finite number of members. He revealed this in conversation with Huygens, who challenged his visitor to produce the summation of reciprocal triangular numbers and, therefore, of a sequence with infinitely many members. Leibniz succeeded in this task at the end of 1672 and summed further sequences of reciprocal polygonal numbers and, following the work of Grégoire de St.-Vincent (1647), the geometric sequence, through transition to the difference sequence.

As a member of a delegation from Mainz, Leibniz traveled to London in the spring of 1673 to take part in the unsuccessful peace negotiations between England, France, and the Netherlands. He was received by Oldenburg at the Royal Society, where he demonstrated an unfinished model of his calculating machine. Through Robert Boyle he met John Pell, who was familiar with the entire algebraic literature of the time. Pell discussed with Leibniz his successes in calculus of differences and immediately referred him to several relevant works of which Leibniz was not aware—including Mercator's *Logarithmotechnia* (1668), in which the logarithmic series is determined through prior division, and Barrow's *Lectiones opticae* (1669) and *Lectiones geometricae* (1670). (Barrow's works were published in 1672 under the single title *Lectiones opticae et geometricae*.)

Leibniz became a member of the Royal Society upon application, but he had seriously damaged his scientific reputation through thoughtless pronouncements on the array of differences, and still more through his rash promise of soon producing a working model of the calculating machine. Leibniz could not fully develop the calculator's principle of design until 1674, by which time he could take advantage of the invention of direct drive, the tachometer, and the stepped drum.

Through Oldenburg, Leibniz received hints, phrased in general terms, of Newton's and Gregory's results in infinitesimal mathematics; but he was still a novice and therefore could not comprehend the significance of what had been communicated to him. Huygens referred him to the relevant literature on infinitesimals in mathematics and Leibniz became passionately interested in the subject. Following the lead of Pascal's *Lettres de "A. Delonville"* [= *Pascal*] *contenant quelques unes de ses inventions de géométrie* (1659), by 1673 he had mastered the characteristic triangle and had found, by means of a transmutation—that is, of the integral transformation, discovered through affine geometry,

$$\frac{1}{2}\int_0^x \left[y(\bar{x}) - x\frac{dy}{dx}(\bar{x})\right] \cdot d\bar{x},$$

for the determination of a segment of a plane curve—a method developed on a purely geometrical basis by means of which he could uniformly derive all the previously stated theorems on quadratures.

Leibniz' most important new result, which he communicated in 1674 to Huygens, Oldenburg, and his friends in Paris, were the arithmetical quadrature of the circle, including the arc tangent series, which had been achieved in a manner corresponding to Mercator's series division, and the elementary quadrature of a cycloidal segment (presented in print in 1678 in a form that concealed the method). Referred by Jacques Ozanam to problems of indeterminate analysis that can be solved algebraically, Leibniz also achieved success in this area by simplifying methods, as in the essay on

$$x + y + z = p^2, x^2 + y^2 + z^2 = q^4$$

(to be solved in natural numbers). Furthermore, a casual note indicates that at this time he was already concerned with dyadic arithmetic.

The announcement of new publications on algebra provoked Leibniz to undertake a thorough review of the pertinent technical literature, in particular the Latin translation of Descartes's *Géométrie* (1637), published by Frans van Schooten (1659–1661) with commentaries and further studies written by Descartes's followers.

His efforts culminated in four results obtained in 1675: a more suitable manner of expressing the indices (*ik* in lieu of a_{ik}, for instance); the determination of symmetric functions and especially of sums of powers of the solutions of algebraic equations; the construction of equations of higher degree that can be represented by means of compound radicals; and ingenious attempts to solve higher equations algorithmically by means of radicals,

LEIBNIZ

attempts that were not recognized as fruitless because of the computational difficulties involved. On the other hand, Leibniz succeeded in demonstrating the universal validity of Cardano's formulas for solving cubic equations even when three real solutions are present and in establishing that in this case the imaginary cannot be dispensed with. The generality of these results had been frequently doubted because of the influence of Descartes. Through this work Leibniz concluded that the sum of conjugate complex expressions is always real (cf. the well-known example $\sqrt{1+\sqrt{-3}}+\sqrt{1-\sqrt{-3}}=\sqrt{6}$). The theorem named for de Moivre later proved this conclusion to be correct.

In the fall of 1675 Leibniz was visited by Tschirnhaus, who, while studying Descartes's methods (which he greatly overrated), had acquired considerable skill in algebraic computation. His virtuosity aroused admiration in London, yet it did not transcend the formal and led to a mistaken judgment of the new results achieved by Newton and Gregory. Tschirnhaus and Leibniz became friends and together went through the unpublished scientific papers of Descartes, Pascal, and other French mathematicians. The joint studies that emerged from this undertaking dealt with the array of differences and with the "harmonic" sequence $\cdots$, 1/5, 1/4, 1/3, 1/2, 1/1 and was treated by Leibniz as the counterpart of the arithmetic triangle. They then considered the succession of the prime numbers and presented a beautiful geometric interpretation of the sieve of Eratosthenes—which, however, cannot be recognized from the remark printed in 1678 that the prime numbers greater than three must be chosen from the numbers $6n \pm 1$.

When Roberval died in 1675 Leibniz hoped to succeed him in the professorship of mathematics established by Pierre de la Ramée at the Collège de France and also to become a member of the Académie des Sciences. Earlier in 1675 he had demonstrated at the Academy the improved model of his calculating machine and had referred to an unusual kind of chronometer. He was rejected in both cases because his negligence had cost him the favor of his patrons. Nevertheless, his thorough, critical study of earlier mathematical writings resulted in important advances, especially in the field of infinitesimals. He recognized the transcendence of the circular and logarithmic functions, the basic properties of the logarithmic and other transcendental curves, and the correspondence between the quadrature of the circle and the quadrature of the hyperbola. In addition, he considered questions of probability.

In the late autumn of 1675, seeking a better understanding of Cavalieri's quadrature methods (1635), Leibniz made his greatest discovery: the symbolic characterization of limiting processes by means of the calculus. To be sure, "not a single previously unsolved problem was solved" by this discovery (Newton's disparaging judgment in the priority dispute); yet it set out the procedure to be followed in a suggestive, efficient, abstract form and permitted the characterization and classification of the applicable computational steps. In connection with the arrangement in undetermined coefficients, Leibniz sought to clarify the conditions under which an algebraic function can be integrated algebraically. In addition he solved important differential equations: for example, the tractrix problem, proposed to him by Perrault, and Debeaune's problem (1638), which he knew from Descartes's *Lettres* (III, 1667); and which required the curve through the origin determined by

$$\frac{dy}{dx}=\frac{x-y}{a}.$$

He established that not every differential equation can be solved exclusively through the use of quadratures and was immediately cognizant of the far-reaching importance of symbolism and technical terminology.

Leibniz only hinted at his new discovery in vague remarks, as in letters to Oldenburg in which he requested details of the methods employed by Newton and Gregory. He received some results in reply, especially concerning power series expansions—which were obviously distorted through gross errors in copying—but nothing of fundamental significance (Newton's letters to Leibniz of June and October 1676, with further information on Gregory and Pell supplied by Oldenburg). Leibniz explained the new discovery to him personally, but Tschirnhaus was more precisely informed. He did not listen attentively, was troubled by the unfamiliar terminology and symbols, and thus never achieved a deeper understanding of Leibnizian analysis. Tschirnhaus also had the advantage of knowing the answer, written in great haste, to Newton's first letter, where Leibniz referred to the solution of Debeaune's problem (as an example of a differential equation that can be integrated in a closed form) and hinted at the principle of *vis viva*. Leibniz also included the essential elements of the arithmetical quadrature of the circle; yet it was derived not by means of the general transmutation but, rather, through a more special one of narrower virtue. Tschirnhaus did not know that the preliminary draft of this letter contained an example of the method of series expansion through gradual integration (later named for Cauchy [1844] and Picard [1890]) and, in any case, he would not have been able to understand and fully appreciate it. On the other hand, he did see and approve the definitive manuscript (1676) on the arithmetical quadrature of the circle. This work also contains the proof by convergence of an alternating sequence with members decreasing without limit; the rigorous treatment of transmutation and its application to the quadrature of higher parabolas and hyperbolas; the logarithmic series and its counterpart, the arc tangent series, and its numerical representation of $\frac{\pi}{4}$ (Leibniz series); and the representation of $\frac{ln\,2}{4}$ and $\frac{\pi}{8}$ through omission of members of the series

$$\sum_{n=2}^{\infty} \frac{1}{(n-1)(n+1)} = \frac{3}{4}.$$

The planned publication did not occur and subsequently the paper was superseded by Leibniz' own work as well as by that of others, particularly the two Bernoullis and L'Hospital.

Since there was no possibility of obtaining a sufficiently remunerative post in Paris, Leibniz entered the service of Hannover in the fall of 1676. He traveled first to London, where he sought out Oldenburg and the latter's mathematical authority, John Collins. He presented them with papers on algebra, which Collins transcribed and which were transmitted to Newton. A longer discussion with Collins was devoted primarily to algebraic questions although dyadics were also touched upon. Leibniz also made excerpts from Newton's letters and from the manuscript of his *De analysi per aequationes numero terminorum infinitas* (1669), which had been deposited with the Royal Society, and from the extracts procured by Collins of letters and papers of Gregory, only a small selection of which Leibniz had obtained earlier. He then went to Amsterdam, where he called on Hudde, who informed him of his own mathematical works.

In the intellectually limited atmosphere of Hannover there was no possibility of serious mathematical discussion. His talks with the Cartesian A. Eckhardt (1678–1679) on Pythagorean triangles with square measures and related questions were unsatisfying. The correspondence with Oldenburg provided an opportunity, in the early summer of 1677, to communicate the determination of tangents according to the method of the differential calculus, but this exchange ended in the autumn with Oldenburg's death. Huygens was ill and Tschirnhaus was traveling in Italy. Thus, in the midst of his multitudinous duties at court, Leibniz lacked the external stimulus needed to continue his previous studies on a large scale. Instead, he concentrated on symbological investigations, his first detailed draft in dyadics, and studies on pure geometric representation of positional relations without calculation, the counterpart to analytic geometry. Only after the founding of the *Acta eruditorum* (1682) did Leibniz present his mathematical papers to the public. In 1682 he published "De vera proportione circuli ad quadratum circumscriptum in numeris rationalibus" and "Unicum opticae, catoptricae et dioptricae principium," concise summaries of the chief results of the arithmetical quadrature of the circle and a hint regarding the derivation of the law of refraction by means of the extreme value method of the differential calculus. These revelations were followed in 1684 by the method of determining algebraic integrals of algebraic functions, a brief presentation of the differential calculus with a hint concerning the solution of Debeaune's problem by means of the logarithmic curve, and further remarks on the fundamental ideas of the integral calculus.

In 1686 Leibniz published the main concepts of the proof of the transcendental nature of $\int \sqrt{a^2 \pm x^2} \cdot dx$ and an example of integration, the first appearance in print of the integral sign (the initial letter of the word *summa*).

Yet Leibniz did not attract general attention until his public attack on Cartesian dynamics (1686–1688) by reference to the principle of the conservation of *vis viva*, with the dimensions mv^2. In the subsequent controversy with the Cartesians, Leibniz put forth for solution a dynamic problem that was also considered by Huygens (1687) and Jakob Bernoulli (1690): Under what conditions does a point moving without friction in a parallel gravitational field descend with uniform velocity? In this connection, Bernoulli raised the problem of the catenary, which was solved almost simultaneously by Leibniz, Huygens, and Johann Bernoulli (published 1691) and which introduced for debate a series of further subjects of increasing difficulty, stemming primarily from applied dynamics. Two are particularly noteworthy: The first was the determination, requested by Leibniz in 1689, of the isochrona paracentrica. Under what conditions does a point moving without friction under constant gravity revolve with uniform velocity about a fixed point? This problem was solved by Leibniz, Jakob Bernoulli, and Johann Bernoulli (1694). The second was the determination, requested by Johann Bernoulli, of the conditions under which a point moving without friction in a parallel gravitational field descends in the shortest possible time from one given point to another given point below it. This problem, called the brachistochrone, was solved in 1697 by Leibniz, Newton, Jakob Bernoulli, and Johann Bernoulli.

The participants in these investigations revealed only their results, not their derivations. The latter are found, in the case of Leibniz, in the posthumous papers or, in certain instances, in the letters exchanged with Jakob Bernoulli (from 1687), Rudolf Christian von Bodenhausen (from 1690), L'Hospital (from 1692), and Johann Bernoulli (from 1693). These letters, like the papers on pure mathematics that Leibniz published in the scientific journals, were usually hastily written in his few free hours. They were not always well edited and are far from being free of errors. Yet, despite their imperfections, they are extraordinarily rich in ideas. In part the letters were written to communicate original ideas, which were only occasionally pursued later; most of them, however, are drawn from earlier papers or brief notes. The following ideas in the correspondence should be specially mentioned:

(*a*) The determination of the center of curvature for a point of a curve as the intersection of two adjacent normals (1686–1692). Leibniz erroneously assumed that the circle of curvature has, in general, four neighboring points (instead of three) in common with the curve. It was only after some years that he understood, through the detailed explanation of Johann Bernoulli, the objections made by Jakob Bernoulli (from 1692). He immediately admitted his error publicly and candidly, as was his custom (1695–1696).

(*b*) The determination of that reflecting curve in the plane by which a given reflecting curve is completed in such a way that the rays of light coming from a given point are rejoined, after reflection in both curves, in another given point (1689).

(*c*) A detailed presentation of the results of the arithmetical quadrature of the circle and of the hyperbola, combined with communication of the power series for the arc tangent, the cosine, the sine, the natural logarithm, and the exponential function (1691).

(*d*) The theory of envelopes, illustrated with examples (1692, 1694).

(*e*) The treatment of differential equations through arrangement in undetermined coefficients (1693).

(*f*) The determination of the tractrix for a straight path and, following this, a mechanical construction of the integral curves of differential equations (1693), the earliest example of an integraph.

In 1694 Leibniz expressed his intention to present his own contributions and those of other contemporary mathematicians uniformly and comprehensively in a large work to be entitled *Scientia infiniti*. By 1696 Jakob Bernoulli had provided him with some of his own work and Leibniz had composed headings for earlier notes and selected some essential passages, but he did not get beyond this preliminary, unorganized collection of material. Much of his time was now taken up defending his ideas. Nieuwentijt, for example (1694–1696), questioned the admissibility and use of higher differentials. In his defense, Leibniz emphasized that his method should be considered only an abbreviated and easily grasped guide and that everything could be confirmed by strict deductions in the style of Archimedes (1695). On this occasion he revealed the differentiation of exponential functions such as x^x, which he had long known.

Even the originality of Leibniz' method was called into doubt. Hence, in 1691, Jakob Bernoulli, who was interested primarily in results and not in general concepts, saw in Leibniz' differential calculus only a mathematical reproduction of what Barrow had presented in a purely geometrical fashion in his *Lectiones geometricae*. He also failed to realize the general significance of the symbolism. In England it was observed with growing uneasiness that Leibniz was becoming increasingly the leader of a small but very active group of mathematicians. Moreover, the English deplored the lack of any public indication that Leibniz—as Newton supposed—had taken crucial suggestions from the two great letters of 1676. Representative material from these letters was in Wallis' *Algebra* (1685), and it was expanded in the Latin version (Wallis' *Opera*, II [1693]). Fatio de Duillier, who shortly before had seen copies of Newton's letters and other unpublished writings on methods of quadrature (1676), became convinced that in the treatment of questions in the mathematics of infinitesimals Newton had advanced far beyond Leibniz and that the latter was dependent on Newton. Since 1687 Fatio had been working with Huygens, who did not think much of Leibniz' symbolism, on the treatment of "inverse" tangent problems—differential equations—and in simple cases had achieved a methodical application of integrating factors.

In the preface to Volume I of Wallis' *Opera*, which did not appear until 1695, it is stated that the priority for the infinitesimal methods belongs to Newton. Furthermore, the words are so chosen that it could have been—and in fact was—inferred that Leibniz plagiarized Newton. Wallis, in his concern for the proper recognition of Newton's merits, was unceasing in his efforts to persuade Newton to publish his works on this subject. Beyond this, he obtained copies of several of the letters exchanged between Leibniz and Oldenburg in the years 1673 to 1677 and received permission from Newton and Leibniz to publish the writings in question. He included them in Volume III of his *Opera* (1699). The collection he assembled was based not on the largely inaccessible originals but, rather, on copies in which crucial passages were abbreviated. As a result, it was possible for the reader to gain the impression that Newton possessed priority in having obtained decisive results in the field of infinitesimals (method of tangents, power series in the handling of quadratures, and inverse tangent problems) and that Leibniz was guilty of plagiarism on the basis of what he had taken from Newton. Fatio pronounced this reproach in the sharpest terms in his *Lineae brevissimi descensus investigatio* (1699).

Leibniz replied in 1700 with a vigorous defense of his position, in which he stressed that he had obtained only results, not methods, from Newton and that he had already published the fundamental concepts of the differential calculus in 1684, three years before the appearance of the material that Newton referred to in a similar form in his *Principia* (1687). On this occasion he also described his own procedure with reference to de Moivre's theorem (1698) on series inversion through the use of undetermined coefficients. Leibniz made his procedure more general and easier to grasp by the introduction of numerical coefficients (in the sense of indices). The attack subsided because Fatio, who was excitable, oversensitive, and given to a coarse manner of expression, turned away from science and became a fervent adherent of an aggressive religious sect. He was eventually pilloried.

Wallis' insinuations were repeated by G. Cheyne in his *Methodus* (1703) and temperately yet firmly rebutted in Leibniz' review of 1703. Cheyne's discussion of special quadratures was probably what led Newton to publish the *Quadratura curvarum* (manuscript of 1676) and the *Enumeratio linearum tertii ordinis* (studies beginning in 1667–1668) as appendices to the *Optics* (1704). Newton viewed certain passages of Leibniz' review of the *Optics* (1705) as abusive attacks and gave additional material to John Keill, who, in a paper published in 1710, publicly accused Leibniz of plagiarism. Leibniz' protests (1711) led to the establishment of a commission of the Royal Society, which decided against him (1712) on the basis of the letters printed by Wallis and further earlier writings produced by Newton. The commission published the evidence, together with an analysis that had been published in 1711, in a *Commercium epistolicum* (1713 edition).

The verdict reached by these biased investigators, who heard no testimony from Leibniz and only superficially examined the available data, was accepted without question for some 140 years and was influential into the first half of the twentieth century. In the light of the much more extensive material now available, it is recognized as wrong. It can be understood only in the nationalistic context in which the controversy took place. The continuation of the quarrel was an embarrassment to both parties; and, since it yielded nothing new scientifically, it is unimportant for an understanding of Leibniz' mathematics. The intended rebuttal did not materialize, and Leibniz' interesting, but fragmentary, account of how he arrived at his discovery (*Historia et origo calculi differentialis* [1714]) was not published until the nineteenth century.

The hints concerning mathematical topics in Leibniz' correspondence are especially fascinating. When writing to those experienced in mathematics, whom he viewed as competitors, he expressed himself very cautiously, yet with such extraordinary cleverness that his words imply far more than is apparent from an examination of the notes and jottings preserved in his papers. For instance, his remarks on the solvability of higher equations are actually an anticipation of Galois's theory. Frequently, material of general validity is illustrated only by simple examples, as is the case with the schematic solution of systems of linear equations by means of number couples (double indices) in quadratic arrangement, which corresponds to the determinant form (1693).

In several places the metaphysical background is very much in evidence, as in the working out of binary numeration, which Leibniz connected with the creation of the world (indicated by 1) from nothingness (indicated by 0). The same is true of his interpretation of the imaginary number as an intermediate entity between Being and Not-Being (1702). His hope of being able to make a statement about the transcendence of π by employing the dyadic presentation (1701) was fruitless yet interesting, for transcendental numbers can be constructed out of infinite dual fractions possessing regular gaps (an example in the decimal system was given by Goldbach in 1729). The attempts to present $\sum_{n=1}^{r} \frac{1}{n}$ (reference in 1682, recognized as false in 1696) and $\sum_{n=1}^{\infty} \frac{1}{n^2}$ (1696) in closed form were unsuccessful, and the claimed rectification of an arc of the equilateral hyperbola through the quadrature of the hyperbola (1676) was based on an error in computation. Against these failures we may set the importance of the recognition of the correspondence between the multinomial theorem (1676) and the continuous differentiation and integration with fractional index that emerged from this observation. During this period (about 1696) Leibniz also achieved the general representation of partial differentiation; the reduction of the differential equation

$$a_{00} + a_{10}x + (a_{01} + a_{11}x)y' = 0$$

by means of the transformation

$$x = p_{11}u + p_{12}v,\ y = p_{21}u + p_{22}v;$$

the solution of

$$y' + p(x)y + q(x) = 0$$

and

$$(y')^2 + p(x)y' + q(x) = 0$$

through series with coefficients in number couples; and the reduction of the equation, which had originated with Jakob Bernoulli,

$$y' = p(x)y + q(x)y^n$$

to

$$y' = P(x)y + Q(x).$$

Leibniz' inventive powers and productivity in mathematics did not begin to slow until around 1700. The integration of rational functions (1702–1703) was, to be sure, an important accomplishment; but the subject was not completely explored, since Leibniz supposed (Johann Bernoulli to the contrary) that there existed other imaginary units besides $\sqrt{-1}$ (for example, $\sqrt[4]{-1}$) which could not be represented by ordinary complex numbers. The subsequent study, which was not published at the time, on the integration of special classes of irrational functions also remains of great interest. The discussion with Johann Bernoulli on the determination of arclike algebraic curves in the plane (1704–1706) resulted in both a consideration of relative motions in the plane and an interesting geometric construction of the arclike curve equivalent to a given curve. This construction is related to the optical essay of 1689 and to the theory of envelopes of 1692–1694, but it cannot readily be grasped in terms of a formula (1706). The remarks (1712) on the logarithms of negative numbers and on the representation of $\sum(-1)^n$ by 1/2 can no longer be considered satisfactory. The description of the calculating machine (1710) indicates its importance but does not give the important details; the first machines of practical application were constructed by P. M. Hahn in 1774 on the basis of Leibniz' ideas.

Leibniz accorded a great importance to mathematics because of its broad interest and numerous applications. The extent of his concern with mathematics is evident from the countless remarks and notes in his posthumous papers, only small portions of which have been accessible in print until now, as well as from the exceedingly challenging and suggestive influential comments expressed brilliantly in the letters and in the works published in his own lifetime.

Leibniz' power lay primarily in his great ability to distinguish the essential elements in the results of others, which were often rambling and presented in a manner that was difficult to understand. He put them in a new form, and by setting them in a larger context made them into a harmoniously balanced and comprehensive whole. This was possible only because in his reading Leibniz was prepared, despite his impatience, to immerse himself enthusiastically and selflessly in the thought of others. He was concerned with formulating authoritative ideas clearly and connecting them, as he did in so exemplary a fashion in the mathematics of infinitesimals. Interesting details were important—they occur often in his notes—but even more important were inner relationships and their comprehension, as this term is employed in the history of thought. He undertook the work required by such an approach for no other purpose than the exploration of the conditions under which new ideas emerge, stimulate each other, and are joined in a unified thought structure.

JOSEPH E. HOFMANN

BIBLIOGRAPHY

I. ORIGINAL WORKS. The following volumes of the *Sämtliche Schriften und Briefe*, edited by the Deutsche Akademie der Wissenschaften in Berlin, have been published: Series I *(Allgemeiner politischer und historischer Briefwechsel)*, vol. 1: 1668–1676 (1923; Hildesheim, 1970); vol. 2: 1676–1679 (1927; Hildesheim, 1970); vol. 3: 1680–1683 (1938; Hildesheim, 1970); vol. 4: 1684–1687 (1950); vol. 5: 1687–1690 (1954; Hildesheim, 1970); vol. 6: 1690–1691 (1957; Hildesheim, 1970); vol. 7: 1691–1692 (1964); vol. 8: 1692 (1970); Series II *(Philosophischer Briefwechsel)*, vol. 1: 1663–1685 (1926); Series IV *(Politische Schriften)*, vol. 1: 1667–1676 (1931); vol. 2: 1677–1687 (1963); Series VI *(Philosophische Schriften)*, vol. 1: 1663–1672 (1930); vol. 2: 1663–1672 (1966); vol. 6: *Nouveaux essais* (1962).

Until publication of the *Sämtliche Schriften und Briefe* is completed, it is necessary to use earlier editions and recent partial editions. The most important of these editions are the following (the larger editions are cited first): L. Dutens, ed., *G. W. Leibnitii opera omnia*, 6 vols. (Geneva, 1768); J. E. Erdmann, ed., *G. W. Leibniz. Opera philosophica quae extant latina gallica germanica omnia* (Berlin, 1840; Aalen, 1959); G. H. Pertz, ed., *G. W. Leibniz. Gesammelte Werke*, Part I: *Geschichte*, 4 vols. (Hannover,

1843–1847; Hildesheim, 1966); A. Foucher de Careil, ed., *Leibniz. Oeuvres*, 7 vols. (Paris, 1859–1875; Hildesheim, 1969); C. I. Gerhardt, ed., *G. W. Leibniz. Mathematische Schriften*, 7 vols. (Berlin–Halle, 1849–1863; Hildesheim, 1962). An index to the edition has been compiled by J. E. Hofmann (Hildesheim, 1971); C. I. Gerhardt, ed., *Die philosophischen Schriften von G. W. Leibniz*, 7 vols. (Berlin, 1875–1890; Hildesheim, 1960–1961); G. E. Guhrauer, ed., *G. W. Leibniz. Deutsche Schriften*, 2 vols. (Berlin, 1838–1840; Hildesheim, 1966); J. G. Eckhart, ed., *G. W. Leibniz. Collectanea etymologica* (Hannover, 1717; Hildesheim, 1970); A. Foucher de Careil, ed., *Lettres et opuscules inédits de Leibniz* (Paris, 1854; Hildesheim, 1971); A. Foucher de Careil, ed., *Nouvelles lettres et opuscules inédits de Leibniz* (Paris, 1857; Hildesheim, 1971); C. Haas, ed. and tr., *Theologisches System* (Tübingen, 1860; Hildesheim, 1966); C. L. Grotefend, ed., *Briefwechsel zwischen Leibniz, Arnauld und dem Landgrafen Ernst von Hessen-Rheinfels* (Hannover, 1846); C. I. Gerhardt, ed., *Briefwechsel zwischen Leibniz und Christian Wolff* (Halle, 1860; Hildesheim, 1963); E. Bodemann, ed., *Die Leibniz-Handschriften der Königlichen öffentlichen Bibliothek zu Hannover* (Hannover, 1889; Hildesheim, 1966); and *Der Briefwechsel des G. W. Leibniz in der Königlichen öffentlichen Bibliothek zu Hannover* (Hannover, 1895; Hildesheim, 1966); C. I. Gerhardt, ed., *Der Briefwechsel von G. W. Leibniz mit Mathematikern* (Berlin, 1899; Hildesheim, 1962); L. Couturat, ed., *Opuscules et fragments inédits de Leibniz* (Paris, 1903; Hildesheim, 1966); E. Gerland, ed., *Leibnizens nachgelassene Schriften physikalischen, mechanischen und technischen Inhalts* (Leipzig, 1906); H. Lestienne, ed., *G. W. Leibniz. Discours de métaphysique* (Paris, 1907, 2nd ed., 1929; Paris, 1952); I. Jagodinskij, ed., *Leibnitiana elementa philosophiae arcanae de summa rerum* (Kazan, 1913); P. Schrecker, ed., *G. W. Leibniz. Lettres et fragments inédits sur les problèmes philosophiques, théologiques, politiques de la réconciliation des doctrines protestantes (1669–1704)* (Paris, 1934); G. Grua, ed., *G. W. Leibniz. Textes inédits*, 2 vols. (Paris, 1948); W. von Engelhardt, ed. and tr., *G. W. Leibniz. Protogaea*, in *Leibniz. Werke* (W. E. Peuckert, ed.), vol. I (Stuttgart, 1949); A. Robinet, ed., *G. W. Leibniz. Principes de la nature et de la grâce fondés en raison. Principes de la philosophie ou monadologie. Publiés intégralement d'après les manuscrits de Hanovre, Vienne et Paris et présentés d'après des lettres inédits* (Paris, 1954); and *Correspondance Leibniz-Clarke. Présentés d'après les manuscrits originales des bibliothèques de Hanovre et de Londres* (Paris, 1957); G. le Roy, ed., *Leibniz. Discours de métaphysique et correspondance avec Arnauld* (Paris, 1957); O. Saame, ed. and tr., *G. W. Leibniz. Confessio philosophi. Ein Dialog* (Frankfurt, 1967); J. Brunschwig, ed., *Essais de Théodicée sur la bonté de Dieu, la liberté de l'homme et l'origine du mal* (Paris, 1969).

There are translations of single works, especially into English and German. English translations (including selections): *Philosophical Works*, G. M. Duncan, ed. and tr. (New Haven, 1890); *The Monadology and Other Philosophical Writings*, R. Latta, ed. and tr. (Oxford, 1898); *New Essays Concerning Human Understanding*, A. G. Langley, ed. and tr. (Chicago, 1916, 1949); *Discourse on Metaphysics*, P. G. Lucas and L. Grint, ed. and tr. (Manchester, 1953; 2nd ed. 1961); *Philosophical Writings*, M. Morris, ed. and tr. (London, 1934, 1968); *Theodicy*, E. M. Huggard and A. Farrer, ed. and tr. (London, 1951); *Selections*, P. P. Wiener, ed. (New York, 1951; 2nd ed. 1971); *Philosophical Papers and Letters*, L. E. Loemker, ed. and tr., 2 vols. (Chicago, 1956; 2nd ed. Dordrecht, 1969); *The Leibniz-Clarke Correspondence*, H. G. Alexander, ed. (Manchester, 1956); *Monadology and Other Philosophical Essays*, P. and A. Schrecker, ed. and tr. (Indianapolis, 1965); *Logical Papers*, G. H. R. Parkinson, ed. and tr. (Oxford, 1966); *The Leibniz-Arnauld Correspondence*, H. T. Mason, ed. and tr. (Manchester–New York, 1967); *General Investigations Concerning the Analysis of Concepts and Truths*, W. H. O'Brian, ed. and tr. (Athens, Ga., 1968).

German translations (including selections) are *Handschriften zur Grundlegung der Philosophie*, E. Cassirer, ed., A. Buchenau, tr., 2 vols. (Hamburg, 1904; 3rd ed. 1966); *Neue Abhandlungen über den menschlichen Verstand*, E. Cassirer, ed. and tr. (3rd ed., 1915; Hamburg, 1971); the same work, edited and translated by H. H. Holz and W. von Engelhardt, 2 vols. (Frankfurt, 1961); *Die Theodizee*, A. Buchenau, ed. and tr. (Hamburg, 2nd ed., 1968); *Schöpferische Vernunft*, W. von Engelhardt, ed. and tr. (Marburg, 1952); *Metaphysische Abhandlung*, H. Herring, ed. and tr. (Hamburg, 1958); *Fragmente zur Logik*, F. Schmidt, ed. and tr. (Berlin, 1960); *Vernunftprinzipien der Natur und der Gnade. Monadologie*, H. Herring, ed., A. Buchenau, tr. (Hamburg, 1960); *Kleine Schriften zur Metaphysik*, H. H. Holz, ed., 2 vols. (Frankfurt–Vienna, 1966–1967).

Bibliographical material on the works of Leibniz and the secondary literature can be found in E. Ravier, *Bibliographie des oeuvres de Leibniz* (Paris, 1937; Hildesheim, 1966), additional material in P. Schrecker, "Une bibliographie de Leibniz," in *Revue philosophique de la France et de l'Étranger*, **126** (1938), 324–346; Albert Rivaud, *Catalogue critique des manuscrits de Leibniz Fasc. II (Mars 1672–Novembre 1676)* (Poitiers, 1914–1924; repr. New York–Hildesheim, 1972); K. Müller, *Leibniz-Bibliographie. Verzeichnis der Literatur über Leibniz* (Frankfurt, 1967). Bibliographical supplements appear regularly in *Studia Leibnitiana. Vierteljahresschrift für Philosophie und Geschichte der Wissenschaften*, K. Müller and W. Totok, eds., **1** (1969); G. Utermöhlen, "Leibniz-Bibliographie 1967–1968," **1** (1969), 293–320; G. Utermöhlen and A. Schmitz, "Leibniz-Bibliographie. Neue Titel 1968–1970," **2** (1970), 302–320; A. Koch-Klose and A. Hölzer, "Leibniz-Bibliographie. Neue Titel 1969–1971," **3** (1971), 309–320.

II. SECONDARY LITERATURE. The literature on Leibniz' philosophy and science is so extensive that a complete presentation cannot be given. In the following selection, more recent works have been preferred, especially those which may provide additional views on the philosophy and science of Leibniz.

H. Aarsleff, "Leibniz on Locke and Language," in *American Philosophical Quarterly*, **1** (1964), 165–188; E. J. Aiton, "The Harmonic Vortex of Leibniz," in *The Vortex Theory of Planetary Motions* (London–New York, 1972); and "Leibniz on Motion in a Resisting Medium," in *Archive for History of Exact Sciences*, **9** (1972), 257–274; W. H. Barber, *Leibniz in France. From Arnauld to Voltaire* (Oxford, 1955); Y. Belaval, *Leibniz critique de Descartes* (Paris, 1960); and *Leibniz. Initiation à sa philosophie* (Paris, 1962; 3rd ed. 1969); A. Boehm, *Le "vinculum substantiale" chez Leibniz. Ses origines historiques* (Paris, 1938; 2nd ed. 1962); F. Brunner, *Études sur la signification historique de la philosophie de Leibniz* (Paris, 1950); G. Buchdahl, *Metaphysics and the Philosophy of Science. The Classical Origins—Descartes to Kant* (Oxford, 1969); P. Burgelin, *Commentaire du* Discours de métaphysique *de Leibniz* (Paris, 1959); H. W. Carr, *Leibniz* (London, 1929, 1960); E. Cassirer, *Leibniz' System in seinem wissenschaftlichen Grundlagen* (Marburg, 1902; Darmstadt, 1962); P. Costabel, *Leibniz et la dynamique. Les textes de 1692* (Paris, 1960); L. Couturat, *La logique de Leibniz* (Paris, 1901; Hildesheim, 1961); L. Davillé, *Leibniz historien. Essai sur l'activité et les méthodes historiques de Leibniz* (Paris, 1909); K. Dürr, *Neue Beleuchtung einer Theorie von Leibniz. Grundzüge des Logikkalküls* (Darmstadt, 1930); K. Dürr, "Die mathematische Logik von Leibniz," in *Studia philosophica*, **7** (1947), 87–102; K. Fischer, *G. W. Leibniz. Leben, Werke und Lehre* (Heidelberg, 5th ed. 1920, W. Kabitz, ed.); J. O. Fleckenstein, *G. W. Leibniz. Barock und Universalismus* (Munich, 1958); G. Friedmann, *Leibniz et Spinoza* (Paris, 2nd ed. 1946, 3rd ed. 1963); M. Gueroult, *Dynamique et métaphysique leibniziennes suivi d'une note sur le principe de la moindre action chez Maupertuis* (Paris, 1934, 1967); G. E. Guhrauer, *G. W. Freiherr von Leibniz. Eine Biographie*, 2 vols. (Wrocław, 1846; Hildesheim, 1966); G. Grua, *Jurisprudence universelle et théodicée selon Leibniz* (Paris, 1953); and *La justice humaine selon Leibniz* (Paris, 1956); H. Heimsoeth, *Die Methode der Erkenntnis bei Descartes und Leibniz*, 2 vols. (Giessen, 1912–1914); A. Heinekamp, *Das Problem des Guten bei Leibniz* (Bonn, 1969); K. Hildebrandt, *Leibniz und das Reich der Gnade* (The Hague, 1953); H. H. Golz, *Leibniz* (Stuttgart, 1958); K. Huber, *Leibniz* (Munich, 1951); J. Jalabert, *La théorie leibnizienne de la substance* (Paris, 1947); J. Jalabert, *Le Dieu de Leibniz* (Paris, 1960); J. Guitton, *Pascal et Leibniz* (Paris, 1951); M. Jammer, *Concepts of Force. A Study in the Foundations of Dynamics* (Cambridge, Mass., 1957); W. Janke, *Leibniz. Die Emendation der Metaphysik* (Frankfurt, 1963); H. W. B. Joseph, *Lectures on the Philosophy of Leibniz*, J. L. Austin, ed. (Oxford, 1949); W. Kabitz, *Die Philosophie des jungen Leibniz. Untersuchungen zur Entwicklungsgeschichte seines Systems* (Heidelberg, 1909); F. Kaulbach, *Die Metaphysik des Raumes bei Leibniz und Kant* (Cologne, 1960); R. Kauppi, *Über die leibnizsche Logik. Mit besonderer Berücksichtigung des Problems der Intension und der Extension* (*Acta Philosophica Fennica XII;* Helsinki, 1960); W. Kneale and M. Kneale, *The Development of Logic* (Oxford, 1962); L. Krüger, *Rationalismus und Entwurf einer universalen Logik bei Leibniz* (Frankfurt, 1969); D. Mahnke, "Leibnizens Synthese von Universalmathematik und Individualmetaphysik," in *Jahrbuch für Philosophie und phänomenologische Forschung*, **7** (1925), 305–612 (repr. Stuttgart, 1964); G. Martin, *Leibniz. Logik und Metaphysik* (Cologne, 1960; 2nd ed., Berlin, 1967; English tr. by P. G. Lucas and K. J. Northcott from the 1st ed. : *Leibniz. Logic and Metaphysics* (Manchester–New York, 1964); J. T. Mertz, *Leibniz* (Edinburgh–London, 1884; repr. New York, 1948); R. W. Meyer, *Leibniz and the 17th Century Revolution* (Glasgow, 1956); J. Mittelstrass, *Neuzeit und Aufklärung. Studien zur Entstehung der neuzeitlichen Wissenschaft und Philosophie* (Berlin, 1970); J. Moreau, *L'universe leibnizien* (Paris–Lyons, 1956); K. Müller and G. Krönert, *Leben und Werk von G. W. Leibniz. Eine Chronik* (Frankfurt, 1969); E. Naert, *Leibniz et la querelle du pur amour* (Paris, 1959); E. Naert, *Mémoire et conscience de soi selon Leibniz* (Paris, 1961); G. H. R. Parkinson, *Logic and Reality in Leibniz' Metaphysics* (Oxford, 1965); G. H. R. Parkinson, *Leibniz on Human Freedom* (*Studia Leibnitiana Sonderheft 2;* Wiesbaden, 1970); C. A. van Peursen, *Leibniz* (Baarn, 1966), tr. into English by H. Hoskins (London, 1969); H. Poser, *Zur Theorie der Modalbegriffe bei G. W. Leibniz* (*Studia Leibnitiana Suppl. VI;* Wiesbaden, 1969); N. Rescher, "Leibniz' Interpretation of His Logical Calculi," in *Journal of Symbolic Logic*, **19** (1954), 1–13; N. Rescher, *The Philosophy of Leibniz* (Englewood Cliffs, N.J., 1967); W. Risse, *Die Logik der Neuzeit II: 1640–1780* (Stuttgart, 1970); A. Robinet, *Malebranche et Leibniz. Relations personelles* (Paris, 1955); A. Robinet, *Leibniz et la racine de l'existence* (Paris, 1962); B. Russell, *A Critical Exposition of the Philosophy of Leibniz* (Cambridge, 1900; 2nd ed., London, 1937); H. Schiedermair, *Das Phänomen der Macht und die Idee des Rechts bei G. W. Leibniz* (*Studia Leibnitiana Suppl. VII;* Wiesbaden, 1970); H. Scholz, "Leibniz" (1942), reprinted in H. Scholz, *Mathesis universalis*, H. Hermes, F. Kambartel and J. Ritter, eds. (Basel, 1961); L. Stein, *Leibniz und Spinoza. Ein Beitrag zur Entwicklungsgeschichte der leibnizischen Philosophie* (Berlin, 1890); G. Stieler, *Leibniz und Malebranche und das Theodizeeproblem* (Darmstadt, 1930); W. Totok and C. Haase, eds., *Leibniz. Sein Leben, sein Wirken, seine Welt* (Hannover, 1966); A. T. Tymieniecka, *Leibniz' Cosmological Synthesis* (Assen, 1964); P. Wiedeburg, *Der junge Leibniz, das Reich und Europa*, pt. I, 2 vols. (Wiesbaden, 1962).

E. Hochstetter, ed., *Leibniz zu seinem 300 Geburtstag 1646–1946*, 8 pts. (Berlin, 1946–1952); G. Schischkoff, ed., *Beiträge zur Leibniz-Forschung* (Reutlingen, 1947); E. Hochstetter and G. Schischkoff, eds., *Zum Gedenken an den 250 Todestag von G. W. Leibniz* (*Zeitschrift für philosophische Forschung*, **20**, nos. 3–4 (Meisenheim, 1966), 377–658; and *Zum Gedenken an den 250 Todestag von G. W. Leibniz* (*Philosophia naturalis*, **10**, no. 2 (1968), 134–293); *Leibniz (1646–1716). Aspects de l'homme et de l'oeuvre (Journées Leibniz, organis. au Centre Int. de Synthèse, 28–30 mai 1966)* (Paris, 1968); *Studia Leibnitiana Supplementa*, vols. I-V *(Akten des Int. Leibniz-Kongresses Hannover, 14–19 November 1966)* (Wiesbaden, 1968–1970).

LEIDY, JOSEPH (*b.* Philadelphia, Pennsylvania, 9 September 1823; *d.* Philadelphia, 30 April 1891), *biology.*

A descriptive scientist of unusual range, Leidy made contributions of lasting value in human anatomy, parasitology, protozoology, and paleontology of mammals and reptiles.

Leidy was the second son and third child of Philip Leidy, a Philadelphia hatter and member of a family of mechanics and artisans who immigrated to Bucks County, Pennsylvania, from Wittenberg in the early eighteenth century, and Catherine Mellick, of Columbia County, Pennsylvania, following whose death in 1823 Philip Leidy married her cousin Christiana T. Mellick of Philadelphia. Joseph Leidy attended a private day school and was influenced to study natural history by a visiting lecturer on minerals. This early taste deepened under the guidance of a nearby horticulturist and during rambles in the countryside. A little book of drawings of shells accompanied by both scientific and common names, done by Leidy when he was ten, may be seen at the Academy of Natural Sciences of Philadelphia. Encouraged by his stepmother to become a student of anatomy, he matriculated at the University of Pennsylvania, where Paul B. Goddard introduced him to microscopy and supervised his dissertation on the comparative anatomy of the vertebrate eye, for which he was awarded the M.D. in 1844.

Leidy failed in an effort to build a private medical practice and seemed much better suited to close study. In the words of his biographer, W. S. W. Ruschenberger:

> It may truly be said that Dr. Leidy was born to be a naturalist. To his innate ability to perceive the minutest variations in the forms and color of things was united artistic aptitude of a high order. These natural faculties, in continuous exercise almost from his infantile days, and his love of accuracy, enabled him to detect minute differences and resemblances of all objects, and to correctly describe and portray them [p. 147].

His acuteness was exemplified by his noticing minute specks in a slice of pork he was eating, and thus discovering that *Trichinella spiralis*, known to be a parasite of man, infested the raw flesh of pigs and that it could be killed by boiling. In 1853 Leidy was appointed professor of anatomy at the University of Pennsylvania. He published a widely admired manual on human anatomy in 1861 and served during the Civil War as an army surgeon, performing many autopsies. In 1864 he married Anna Harden of Louisville, Kentucky; they adopted one daughter.

Leidy seems to have had an unusual degree of patience that allowed him to observe organisms for longer than was strictly necessary in order to publish a description. When writing up the discovery of an unusual planarian in a spring in Pennsylvania, he added notes on behavior and feeding habits and conducted experiments on its capacity for regeneration. In 1848 he commenced studies on the intestinal flora and fauna of healthy animals, an intricate and difficult subject which prompted reflections on spontaneous generation and the origin of life. Leidy found no difficulty in step-by-step reasoning from close study of intestinal microorganisms to considerations of life in the universe, and he offered a calmly worded suggestion that all life had evolved from comparably simple circumstances. His discovery in 1848 that the larval eye persisted into the adult stage of the Cirripedia attracted the attention of Charles Darwin and exemplified Leidy's capacity for detail, for these organs are very minute and rudimentary.

In 1848 Leidy was elected to the Academy of Natural Sciences of Philadelphia, then the foremost American institution for the deposit of paleontological specimens. In 1847 he described the fossil remains of the horse in America with the accuracy and minuteness that became his hallmarks. Charles Lyell urged Leidy to devote himself to paleontology. Spencer Baird turned over to him the vertebrate paleontology collections of the Smithsonian Institution and by 1855 Leidy had inspected most of the important finds from the American West. He proved to be a talented and perceptive comparative anatomist, writing meticulous descriptions of fossil mammals that have endeared him to subsequent workers. One typical appraisal is that he was "a paleontologist's paleontologist," whose monographs eschewed speculation and laid the foundation for understanding the diverse North American fossil vertebrates. "The Extinct Mammalian Fauna of Dakota and Nebraska" (1869), probably his greatest single work, was thought by Henry Fairfield Osborn to be "with the possible exception of Cope's *Tertiary Vertebrata*, the most important paleontological work which America has produced." The infant science of paleontology became embroiled in controversy among its promoters; and Leidy, who had no appetite for acrimony, gradually shifted his interests to protozoology.

Leidy devoted the years 1874 to 1878 to studies of freshwater Protozoa in a diverse array of habitats, ranging from the Uinta Mountains of Utah to the bogs of the Southeast. His monograph *Fresh-Water Rhizopods of North America* (1879), illustrated with superb lithographs of his field drawings, was the equal of any previous work in the field, constituting a remarkable portrayal of the diversity of unicellular life in microhabitats.

Leidy was a devoted worker who missed only five days in thirty-eight years as professor of anatomy, and his steadfastness contributed greatly to the strength of the institutions he served. He was president of the Academy of Natural Sciences for the last ten years of his life and in 1885 became president of the Wagner Free Institute of Science and director of its museum. His lectures there were popular discourses revealing an unaffected humanity—the quality that prompted him, while visiting the fish market in search of scientific specimens, to tell fish sellers the scientific names of uncommon varieties for them to copy onto their signs for the public.

His dislike for the controversies that beset vertebrate paleontology did not prevent Leidy from seeking the wider scientific implications of facts that he regarded as established. He drew connections from observations of parasites to concepts of contagion or primordial origins of life, for example, and exercised great powers of observation. His perceptions were marked by sober appreciation of the vastness and diversity of life, as in these comments on the Bridger Basin in Utah:

> The utter desolation of the scene, the dried-up watercourses, the absence of any moving object, and the profound silence which prevailed, produced a feeling that was positively oppressive. When I then thought of the buttes beneath my feet, with their entombed remains of multitudes of animals forever extinct, and reflected upon the time when the country teemed with life, I truly felt that I was standing on the wreck of a former world [*Contributions to the Extinct Vertebrate Fauna of the Western Territories*, pp. 18–19].

BIBLIOGRAPHY

I. Original Works. Leidy's writings include "A Flora and Fauna Within Living Animals," in *Smithsonian Contributions to Knowledge*, **5** (1853); "The Ancient Fauna of Nebraska, a Description of Extinct Mammalia and Chelonia From the Mauvaises Terres of Nebraska," *ibid.*, **6** (1854); "Cretaceous Reptiles of the United States," *ibid.*, **14** (1865); "The Extinct Mammalian Fauna of Dakota and Nebraska . . .," which is *Journal of the Academy of Natural Sciences of Philadelphia*, 2nd ser., **7** (1869); *Contributions to the Extinct Vertebrate Fauna of the Western Territories*, Report of the U.S. Geological Survey of the Territories (Washington, D.C., 1873); and *Fresh-Water Rhizopods of North America*, *ibid.* (Washington, D.C., 1879).

MS collections in the Academy of Natural Sciences, Philadelphia, include diaries and correspondence and a 373-page typescript of excerpts from these prepared by Joseph Leidy, Jr., covering the period 1823–1869. Medical correspondence is at the College of Physicians, Philadelphia.

II. Secondary Literature. See W. S. W. Ruschenberger, "A Sketch of the Life of Joseph Leidy, M.D., L.L.D.," in *Proceedings of the American Philosophical Society*, **30** (1892), 135–184, with bibliography; and Joseph Leidy, Jr., ed., "Researches in Helminthology and Parasitology by Joseph Leidy," in *Smithsonian Miscellaneous Collections*, **46**, no. 1477 (1904), with bibliography.

Philip C. Ritterbush

LEMAIRE, JACQUES (*fl.* Paris, France, 1720–1740), and **PIERRE** (*fl.* Paris, 1733–1760), *instrument making.*

A member of the Société des Arts and an associate of its president, the clockmaker Julien Le Roy, Jacques Lemaire built sundials, many of which Le Roy had designed. He invented the front-view reflecting telescope, which Herschel developed so advantageously in giant telescopes and which does away with the auxiliary mirror used to project the image onto the eyepiece of the Newtonian reflecting telescope. He developed a mechanical device that permits several screws to be advanced at once by turning a single screw.

Lemaire's son, Pierre, who carried on the tradition of excellence and inventiveness established in his father's workshop, and who had all sorts of scientific instruments made in the family shop, enjoyed the reputation of being the outstanding French compass maker. He was the first Frenchman to construct the Hadley octant and thereafter proudly used an image of it as his ensign, which had been a lodestone. He worked closely with the academicians Duhamel du Monceau and La Condamine, who inspired him.

A great variety of instruments possessed by the Académie des Sciences came from the Lemaire workshop, which, with that of Langlois, kept alive in France a tradition of quality precision-instrument making in spite of the impediments caused by the craft guild system of the *ancien régime*.

BIBLIOGRAPHY

I. Original Works. Jacques Lemaire's front-view reflecting telescope of 1728 is described in *Machines et inventions approuvées par l'Académie royale des sciences . . .*, VI (Paris, 1735), 61–63; his device of 1726 for advancing several screws is in *ibid.*, IV (Paris, 1735), 179–180, and in *Histoire de l'Académie des sciences* for 1726 (1728), p. 17.

Pierre Lemaire's work on compasses is described by Duhamel du Monceau, who did experiments with Pierre, in *Mémoires de l'Académie des sciences* for 1745 (1749), pp. 181–193; a description of his compass, inspired by Hadley's quadrant and by a proposal made in 1733 by La Condamine, is in *Histoire de l'Académie des sciences* for 1747 (1752), p. 126, and in *Machines et inventions approu-*

vées par l'Académie royale des sciences . . ., VII (Paris, 1777), 361–367.

Instruments constructed by the Lemaires may be found at the Conservatoire National des Arts et Métiers, Paris; at the Musées Royaux d'Art et d'Histoire, Brussels; in the Mensing Collection at the Adler Museum, Chicago; at the National Maritime Museum, Greenwich; at the Musée de la Marine, Paris; and at the British Museum, London. The inventory of instruments possessed by the Académie des Sciences before the French Revolution is at the Archives Nationales, Paris, F[14] 1274.

II. Secondary Literature. La Condamine's proposal and test of Pierre Lemaire's compass are found in *Mémoires de l'Académie des sciences* for 1733 (Paris, 1735), 446–456, and for 1734 (Paris, 1736), 597–599. Pierre Lemaire's address is mentioned in *ibid.* for 1734, 599; *ibid.* for 1745, 181; J. B. N. D. d'Apres de Mannevillette, *Le nouveau quartier anglois ou description et usage d'un nouvel instrument pour observer la latitude sur mer* (Paris, 1739), p. 46; H. Michel, *Introduction à l'étude d'une collection d'instruments anciens de mathématiques* (Antwerp, 1939), p. 96.

J.-A. Nollet, *Art des expériences* (Paris, 1770), I, 247–248, praises Pierre Lemaire's compasses. M. Daumas, *Les instruments scientifiques aux XVIIe et XVIIIe siècles* (Paris, 1953), offers a comprehensive survey that permits the reader to situate the Lemaires in the context of their times.

Robert M. McKeon

LEMERY, LOUIS (*b.* Paris, France, 25 January 1677; *d.* Paris, 9 June 1743), *chemistry*, *anatomy*, *medicine*.

Louis Lemery was the eldest son of the chemist Nicolas Lemery. Attracted at first to the legal profession, he finally gave in to his father's wishes and went on to a distinguished career in science and medicine which undoubtedly benefited in part from the reputation of his father. Educated at the Collège d'Harcourt, he proceeded to the Faculty of Medicine of Paris where he graduated M.D. in 1698. In 1700 he was admitted to the Academy of Sciences of Paris, as an *élève* of first the botanist Tournefort and then (from 1702) of his father. In 1712 he was elevated to the rank of *associé*, and in 1715, when his father took the *vétérance*, Louis succeeded him as *chimiste pensionnaire*. Between 1707 and 1710 he occasionally deputized for Guy-Crescent Fagon in the chemistry courses at the Jardin du Roi; he shared these duties with Claude Berger and E. F. Geoffroy. When Fagon resigned from the chemistry chair at the Jardin in 1712, however, Geoffroy was chosen over Lemery for the post. Lemery succeeded Geoffroy in this position in 1730.

Lemery was a physician at the Hôtel Dieu from 1710 to his death, a *médecin du roi* from 1722, and personal physician to Louis XV's cousin, the Princesse de Conti, in whose salon he composed many of his scientific works. He married Catherine Chapot in 1706, and one daughter survived from the marriage.

The bulk of Lemery's scientific writings were published in the *Mémoires de l'Académie royale des sciences*. They dealt mainly with problems of chemical analysis, especially of organic materials, and with fetal anatomy and the origin of monsters. In an early series of papers published between 1705 and 1708 he contested E. F. Geoffroy's view that iron could be synthesized from certain vegetable oils and mineral earths, and also that iron found in the ashes of plants was a product of the combustion.

Lemery's most important observations on organic analysis are contained in four papers published in 1719–1721. These represent a reevaluation of the Academy's long-standing project on the analysis of plants. In the first paper Lemery argued that fire is too destructive a tool for the analysis of organic materials, reducing them all to the same indiscriminate end products—a view his father had suggested in his textbook of 1675. In the subsequent papers, however, it became clear that he had no alternatives to offer to the traditional method of distillation analysis. Consequently he argued for a controlled use of this technique and stated that the important thing in such analysis is not to seek the ultimate principles of organic materials, but to seek only proximate principles. Lemery applied these considerations to the saline constituents of plants and animals, attempting to analyze them in terms of their acid and alkaline components. He concentrated on the ammoniacal and nitrate salts, drawing on the results of an earlier paper (1717) on the formation of niter, in which he had criticized the views of Mayow. In a 1709 paper on the nature of fire and light, he contested traditional Cartesian views and argued that fire and light were distinctive chemical fluids.

Lemery's anatomical papers dealt with the circulation of the blood in the fetal heart and with the origin of monsters. On this latter topic he disputed the emboîtement views of the Academy's anatomists, Duverney and Winslow, that monsters derived from monstrous embryos; Lemery argued instead that they were the product of purely accidental factors influencing the development of normal embryos in the womb.

In addition to his Academy memoirs, Lemery published two monographs. His *Traité des alimens* (1702) is a dictionary of edibles with a description of their nutritional value in terms of chemico-mechanical principles. His *Dissertation sur la nourriture des os* (1704) arose out of a protracted dispute with Nicolas Andry, whose *Traité de la génération des vers dans le corps de l'homme* Lemery had criticized severely in the

Journal de Trévoux. In this work Lemery maintained that part of the marrow had a nutritional function in young bones.

BIBLIOGRAPHY

I. Original Works. Lemery's works include *Traité des alimens où l'on trouve par ordre, et séparément la différence & le choix qu'on doit faire de chacun d'eux en particulier* (Paris, 1702; 2nd ed., Paris, 1705, repr. 1709; 3rd ed., 2 vols., Paris, 1755, trans. by D. Hay as *A Treatise of All Sorts of Foods* . . . (London, 1745); Lemery's three letters on N. Andry's book on the generation of worms appeared first in the *Journal de Trévoux*, beginning in November 1703; see also *Dissertation sur la nourriture des os: où l'on explique la nature & l'usage de la Moelle. Avec trois lettres sur le livre de la génération des vers dans le corps de l'homme* (Paris, 1704). Lemery's many presentations to the Academy of Sciences are to be found in the *Histoire* and *Mémoires* of the Academy between 1702 and 1740. A complete listing is given in J. M. Quérard, *La France littéraire*, V (Paris, 1833), 141–142; *Nouvelle biographie générale*, XIII (Paris, 1862), cols. 603–604, lists the more important papers, while Poggendorff, I, col. 1418, cites the principal chemical papers.

II. Secondary Literature. The principal source for Lemery's life is the *éloge* by Dortous de Marain in *Histoire de l'Académie royale des sciences pour l'année 1743–1746*, 195–208; see also J. P. Contant, *L'enseignement de la chimie au jardin royal des plantes de Paris* (Paris, 1952), 57–60; and J. A. Hazon, ed., *Notice des hommes les plus célèbres de la Faculté de Médecine en l'Université de Paris* (Paris, 1778), 195–198.

Lemery's contributions to organic analysis at the Academy of Sciences are discussed in F. L. Holmes, "Analysis by Fire and Solvent Extractions: the Metamorphosis of a Tradition," in *Isis*, **62** (1971), 129–148; some aspects of Lemery's chemistry are discussed in H. Metzger, *Les doctrines chimiques en France du début du XVII*[e] *à la fin du XVIII*[e] *siècle* (Paris, 1923, repr. 1969), pp. 341–420; and in J. R. Partington, *A History of Chemistry*, III (London, 1962), 41–42.

Owen Hannaway

LEMERY, NICOLAS (*b*. Rouen, France, 17 November 1645; *d*. Paris, France, 19 June 1715), *chemistry, pharmacy*.

Nicolas Lemery was the fifth of seven children born to Julien Lemery, a Protestant attorney in the Parlement of Normandy, and his second wife, Susan Duchemin. When Nicolas was eleven years old his father died, leaving a widow and four surviving children. Shortly before his fifteenth birthday he was indentured as an apprentice apothecary to his uncle Pierre Duchemin in Rouen. After serving six years with his uncle, in 1666 he went to Paris, where he became a boarding student of Christopher Glaser, then demonstrator in chemistry at the Jardin du Roi. Apparently Lemery did not find Glaser a compatible teacher, and this arrangement lasted only two months. He then embarked on a six-year period of travel and study. Few details survive of this period in his life. He is known to have visited Lyons and Geneva and to have spent a considerable part of the time between 1668 and 1671 in Montpellier, where he resided with a young Protestant master apothecary, Henri Verchant. In the summer of 1670 he was registered as a student of pharmacy in Montpellier and was permitted to attend the courses on "simples" and anatomy given at the Faculty of Medicine for such students. According to Fontenelle, he also taught chemistry courses to Verchant's students that attracted members of the medical faculty and other notables of the town.

In 1672 Lemery returned to Paris, where he associated with members of the household of Louis, prince of Condé (le Grand Condé). He attended the conferences of the Abbé Bourdelot, the prince's physician, and worked in the laboratory of the prince's apothecary Bernadin Martin. This connection no doubt introduced Lemery to the fashionable intellectual circles of Paris. In 1674 he secured his professional status by purchasing the office of apothecary to the king and *grand prévôt* of France, thus circumventing the legal obstacles in the path of a Protestant seeking admission to the guild of apothecaries of Paris. During the next seven years he established a highly successful pharmaceutical business, specializing in patent medicines. In addition, he gained a considerable reputation as a teacher of chemistry by his private courses. These courses not only catered to the professional needs of pharmacy apprentices but also attracted a large audience from fashionable Parisian society interested in semipopular scientific expositions. The textbook of his course, the *Cours de chymie* (1675), enjoyed unprecedented success for such a work, selling, as Fontenelle comments, like a work of romance or satire.

Beginning in 1681, however, increasing religious intolerance in France introduced a period of considerable anxiety in Lemery's life. He was required to dispose of his office as privileged apothecary to the king, and in 1683 proceedings were set in train to close his laboratory and shop. These events prompted him to go to England in 1683, perhaps in anticipation of securing a position there. Disappointed, however, he returned within the year to France, where he acquired an M.D. degree at the University of Caen in order to reestablish his professional status following the loss of his position as privileged apothecary. He continued his teaching in increasingly difficult circumstances until the revocation of the Edict of Nantes in 1685, when as a Protestant he lost all his professional and legal rights. Early in the following year he abjured his religion and was received, together with his family, into the Roman Catholic church. Shortly thereafter he was granted permission to reestablish his laboratory and shop on condition that he take no apprentices, following opposition from the Paris guild of apothecaries on the grounds that he had forfeited his right as an apothecary by qualifying as a physician. Lemery devoted the next twelve years largely to pharmacy, publishing at the end of this period his *Pharmacopée universelle* (1697) and the *Traité des drogues simples* (1698). In 1699 he was admitted to the reorganized Academy of Sciences as associate chemist, and in November of the same year he succeeded Claude Bourdelin as *chimiste pensionnaire*. His subsequent scientific work was associated almost exclusively with the Academy. He died in 1715.

Lemery's scientific career fell into three phases: the first (1674–1683) was dominated by his successful teaching career and the publication of his *Cours de chymie;* the second (1686–1698) saw the reestablishment of his pharmaceutical business following his conversion to Catholicism and culminated in the publication of his *Pharmacopée universelle* and *Traité des drogues simples;* and the third period (1699–1715) was dominated by his association with the Academy of Sciences, which resulted in the publication of several memoirs in the Academy's journal and a monograph on antimony entitled *Traité de l'antimoine* (1707).

Lemery's teaching and textbook on chemistry, the *Cours de chymie*, owed their success to his clear and entertaining presentation of chemistry in corpuscular-mechanist terms. His adoption of mechanical modes of explanation brought the French chemical teaching tradition out of its earlier Paracelsian-Helmontian inheritance into the mainstream of contemporary Cartesian natural philosophy. His originality, however, should not be exaggerated; his presentation of chemistry remained wedded to the pharmaceutical goals of the teaching tradition established at the Jardin du Roi, and in practical content and organization his text follows very closely the works of his predecessors Nicaise Le Febvre and Christopher Glaser. The sources of his mechanism are unclear, as there is no formal philosophical or methodological introduction to his chemistry. They were probably largely Cartesian in influence. He was a close friend of the Cartesian Pierre-Sylvain Régis, who lectured on Cartesian natural philosophy in Lemery's laboratory in 1680; but the atomism of Gassendi also probably influenced him, and he mentions François Bernier's redaction of Gassendi's philosophy in the 1690 edition of the *Cours de chymie*.

But Lemery cannot be said to have developed a thoroughgoing Cartesian or atomistic theory of matter. Rather he introduced his explanations of chemical reactions in terms of particle shape and movement on an *ad hoc* basis, appealing to a naive empiricism which stressed the visual imagination, bolstered in some instances by microscopic observation. Thus the best way to explain the nature of salts, according to Lemery, is to attribute shapes to their constituent particles which best answer to all the effects they produce. Acid salts must have sharp pointed particles because of their sharp taste and, even more convincingly, because they solidify in the form of sharp pointed crystals. Contrariwise, alkalis are composed of earthy solid particles whose interstitial pores are so shaped as to admit entry of the spiked particles of acid. For reaction to take place between a particular acid and alkali, there must be an appropriate relationship between the size of the acid spikes and alkaline pores. Effervescence is produced in some acid-alkaline reactions by the expulsion of fire particles entrapped in the pores of the alkalis. Lemery also deduced the shapes of particles from the alleged physiological action of chemical substances in conformity with then current iatrophysical doctrines.

As a common origin for all salts Lemery suggests the fossil or gem salt (common salt) which is formed from an acid liquor flowing in veins in the earth. The acid liquor insinuates itself into the pores of stones and after concoction for several years forms this primogenital salt. All salts are derived from this fossil salt, with the exception of saltpeter, which derives its acidity directly from acid particles in the atmosphere. He tentatively suggests, however, that the acid liquor responsible for the formation of fossil salt may derive its acidity, like saltpeter, from the acid particles in the atmosphere. Vegetable salts in their turn are derived from terrestrial salt by absorption into the plant. Lemery's discussion of vegetable salts leads him to an interesting critique of analysis by fire. He recognizes three species of vegetable salt: the acid or essential salt crystallized directly from the juice of plants; the volatile alkaline salt produced by distilling macerated and fermented seeds and fruits; and the alkaline fixed salt derived from the ashes of combusted plant materials. Of the three, only the first type is preexistent in plants; the other two are products of the action of fire. In discussing the production of the alkaline volatile salt of plants by heating, Lemery concludes that it must be admitted that fire destroys and confounds most things which it dissects, and there is no occasion to believe that it yields substances in their natural state. The probity of fire as a tool in vegetable analysis became a subject of much discussion and debate in the Academy of Sciences in Lemery's lifetime and subsequently.

Lemery, however, was not disposed to renounce analysis by fire entirely: he still finds a place in his chemistry for the five iatrochemical principles of salt, sulfur, mercury, water, and earth based on fire analysis. His retention of these principles reveals a curious tension in Lemery's chemistry between his innovative mechanist approach and its traditional iatrochemical framework. He states that he would like to believe that these principles are found in all mixt bodies but acknowledges that the existence of all five can be demonstrated only in animal and vegetable bodies: they cannot be separated so readily from minerals, and not even two of them can be extracted from gold and silver. Consequently, in spite of his initial discussion of them, the principles play a limited role in his subsequent exposition. Nevertheless, he is insistent that experimentally determined principles have a place in chemistry as an antidote to purely hypothetical mechanical theories of matter—one must proceed from the demonstrable products of chemical analysis to the shapes of particles and not vice versa. In spite of the drift of his arguments, Lemery, however, is unable or unwilling to formulate a new set of empirically determined principles based on a wider range of analysis than that of fire.

Lemery's chief contributions to pharmacy were his two complementary works, the *Pharmacopée universelle* and the *Traité des drogues simples.* These are alphabetically arranged lists of composites and simples respectively, giving the source, virtues, doses, and therapeutic action of the various medicaments. They represent a comprehensive dictionary of pharmaceuticals. Their chief rival was Pierre Pomet's *Histoire générale des drogues* (1694).

His last major work was *Traité de l'antimoine* (1707), which contained the results of his investigation into the properties and preparations of mineral antimony, his chosen topic of research on admission to the Academy of Sciences in 1699. It is a thorough and systematic collection of preparations of antimony arranged according to technique of preparation, but it does not depart in spirit or style from the section on antimony in the *Cours de chymie.* He also published in the *Mémoires* of the Academy papers on the physical and chemical explanations of subterranean fires, earthquakes, hurricanes, thunder, and lightning (1700), which he attributed to the spontaneous reaction of iron and sulfur, and on the analysis of camphor (1705), honey (1706), cows' urine (1707), and experiments on corrosive sublimate (1712). All are of minor importance, and his last researches in the Academy suffer comparison with the work of the other chemists, Guillaume Homberg, E. F. Geoffroy, and his son Louis Lemery.

The elder Lemery invites comparison with his older English contemporary Robert Boyle. Both were advocates of an eclectic mechanical philosophy and were firmly convinced of the contributions chemistry had to make to that philosophy. Lemery's vision of chemistry, however, was much more limited due to his professional commitment to pharmacy and to his inheritance of a well-established textbook tradition which was largely pharmaceutically oriented. He was perhaps inhibited by overriding professional concerns from attempting to integrate his chemistry more fully into the new scientific philosophy in the manner of Boyle. Also the opportunity to do so within the broader scientific environment of the Academy of Sciences came late in his life, and by that time his interests and outlook were set. Two of Lemery's sons followed their father's interest in chemistry, and both became members of the Academy. The elder, Louis Lemery (Lemery *le fils*, 1677–1743), succeeded his father as *chimiste pensionnaire* in 1715. A younger son, Jacques Lemery (Lemery *le jeune*, 1677/1678–1721), was an associate of the Academy from 1715, publishing several memoirs on phosphorus before his early death.

BIBLIOGRAPHY

I. Original Works. There is no wholly satisfactory guide to the numerous editions of Lemery's works; most useful are J. Roger, *Les médecins normands du XIIe au XIXe siècle* (Paris, 1890), 53–55; A. J. J. Vandevelde, "L'oeuvre bibliographique de Nicolas Lemery," in *Bulletin des sociétés chimiques Belges*, **30** (1921), 153–166; and J. R. Partington, *A History of Chemistry*, III (London, 1962), 29–31.

More than thirty eds. of the *Cours de chymie* (Paris, 1675) have been recorded. Eleven separate eds. appeared which claimed to have been revised and corrected by the author himself, the last of these being the 11th (Leiden, 1716), reissued in Paris in 1730. Other eds. in French were published at Amsterdam, Leiden, Lyons, Brussels, and Avignon between 1682 and 1751. The last French ed. was edited by Théodore Baron (Paris, 1756), who added many notes in an effort to update it in conformity with current phlogistic theory; this ed. was reissued in Paris in 1757. The *Cours* was also translated into Latin, German, Dutch, Italian, Spanish, and English. The English translations were by Walter Harris (London, 1677, 1686, from the 1st and 5th French eds. respectively) and by James Keill (London, 1698, 1720, from the 8th and 11th French eds. respectively).

The *Pharmacopée universelle* (Paris, 1697) appeared in at least five subsequent eds. from Paris, Amsterdam, and The Hague. The last ed. was in two vols., Paris 1763–1764. The *Traité universel des drogues simples* (Paris, 1698) appeared in at least five eds. to 1759. The 3rd (Amsterdam, 1716) and subsequent eds. have the title *Dictionnaire ou traité universel des drogues simples*. It was translated into Italian, German, Dutch, and English.

The *Traité de l'antimoine* (Paris, 1707) was translated into German and Italian.

A work entitled *Recueil de curiositez rares et nouvelles des plus admirables effets de la nature et de l'art* by a Sieur d'Emery, although sometimes attributed to Nicolas Lemery, is not generally considered to be by him.

Lemery's contributions to the Paris Academy of Sciences are to be found in its *Histoires* and *Mémoires* (1699–1712).

II. SECONDARY LITERATURE. Fontenelle's *éloge*, in *Histoire de l'Académie royale des sciences* for 1715 (Paris, 1717), pp. 73–82, remains an important source for Lemery's life. It is not without error, however, and should be supplemented by P. Dorveaux, "Apothicaires membres de l'Académie royale des sciences, VI. Nicolas Lemery," in *Revue d'histoire de la pharmacie*, **19** (1931), 208–219. J. Roger, *Les médecins normands du XIIe au XIX siècle* (Paris, 1890), 47–55, repeats several of Fontenelle's errors but includes two unpublished letters of Lemery from the 1660's. P.-A. Cap, *Études biographiques pour servir à l'histoire des sciences. Première série, chimistes-naturalistes* (Paris, 1857), 180–226, is a romanticized account of Lemery's life based largely on Fontenelle but gives useful genealogical information. See also E. and E. Haag, *La France protestante*, VI (Paris, 1856), 538–544. Lemery's chemistry is discussed in J. R. Partington, *A History of Chemistry*, III (London, 1962), 28–41; J. Read, *Humour and Humanism in Chemistry* (London, 1947), 116–123; and most fully by H. Metzger, *Les doctrines chimiques en France du début du XVIIe à la fin du XVIIIe siècle* (Paris, 1923; repr. 1969), 281–338, and *passim*.

OWEN HANNAWAY

LEMOINE, ÉMILE MICHEL HYACINTHE (*b.* Quimper, France, 22 November 1840; *d.* Paris, France, 21 December 1912), *mathematics.*

Lemoine can be characterized as an amateur mathematician and musician whose work was influential in both areas. Like a number of other famous French mathematicians, he was educated at the École Polytechnique in Paris, from which he was graduated in 1860. He taught there but was forced to resign after five or six years because of poor health. Subsequently he was a civil engineer, and he eventually became chief inspector for the department of gas supply in Paris. His avocations remained mathematics and music.

In 1860, while he was still at the École Polytechnique, Lemoine and other teachers formed a chamber music group, nicknamed "La Trompette." Camille Saint-Saëns wrote pieces for it.

Lemoine's major mathematical achievements were in geometry. He and John Casey are generally credited with having founded the newer geometry of the triangle.

In 1873, at the meeting of the Association Française pour l'Avancement des Sciences held in Lyons, Lemoine presented a paper entitled "Sur quelques propriétés d'un point remarquable du triangle." In this paper he called attention to the point of intersection of the symmedians of a triangle and described some of its more important properties. He also introduced the special circle named for him.

The point of concurrence of the symmedians of a triangle is called the Lemoine point (in France), the Grebe point (in Germany, after E. W. Grebe), or, most generally, the symmedian point. The last term was coined by the geometer Robert Tucker, of the University College School in London, in the interest of uniformity and amity. It is generally symbolized by *K*.

The symmedian point had appeared in the work of geometers before Lemoine, but his was the first systematic exposition of some of its interesting properties. Lemoine's concern with the problem of simplifying geometric constructions led him to develop a theory of constructions, which he called geometrography. He presented this system at the meeting of the Association Française pour l'Avancement des Sciences that was held at Oran, Algeria, in 1888.

Briefly, Lemoine reduced geometric constructions to five elementary operations: (1) placing a compass end on a given point; (2) placing a compass end on a given line; (3) drawing a circle with the compass so placed; (4) placing a straightedge on a given point; (5) drawing a line once the straightedge has been placed. The number of times any one of these five operations was performed he called the "simplicity" of the construction. The number of times operation (1), (2), or (4) was performed he called the "exactitude" of the construction. By a suitable examination of the operations involved in a construction, it is usually possible to reduce the simplicity. For example, it is possible to reduce the simplicity of the construction of a circle tangent to three given circles (Apollonius' problem) from over 400 steps to 199 steps.

This system had a mixed reception in the mathematical world. It appears to help reduce the number of steps required for constructions but generally requires more geometrical sophistication and ingenuity on the part of the constructor. It is now generally ignored.

Lemoine also wrote on local probability and on transformations involving geometric formulas. Concerning these transformations, he showed that it is always possible, by a suitable exchange of line segments, to derive from one formula a second formula of the same nature. Thus, from the formula for the radius of the incircle of a triangle, it is possible to derive formulas for the radii of the excircles of the same triangle.

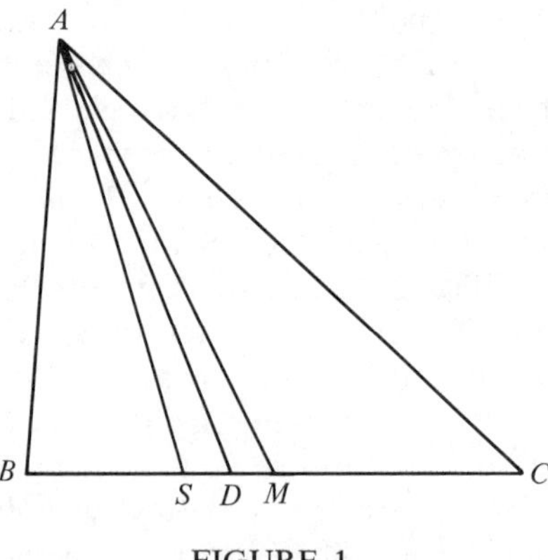

FIGURE 1

His mathematical work ceased after about 1895, but Lemoine's interest in the field continued. In 1894 he helped C. A. Laisant to found the periodical *Intermédiare des mathématiciens* and was its editor for many years.

Lemoine's reputation rests mainly on his work with the symmedian point. Briefly (see Figure 1), if, in triangle *ABC*, cevian *AM* is the median from vertex *A* and cevian *AD* is the bisector of the same angle, then cevian *AS*, which is symmetric with *AM* in respect to *AD*, is the symmedian to side *BC*. Among the results presented by Lemoine were that the three symmedians of a triangle are concurrent and that each symmedian divides the side to which it is drawn in the ratio of the squares of the other two sides.

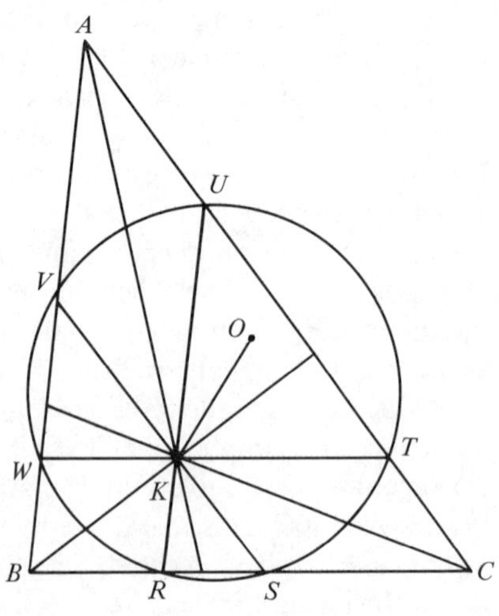

FIGURE 2

If (see Figure 2) lines are drawn through the given symmedian point *K* parallel to the sides of the given triangle, these lines meet the sides of the triangle in six points lying on a circle. This circle is called the first Lemoine circle. The center of the circle is the midpoint of the line segment, *OK*, joining the circumcenter and symmedian point of the triangle. The distances of *K* from the sides of the triangle are proportional to the lengths of the sides. The line segments cut from the sides by the first Lemoine circle have lengths proportional to the cubes of the sides. If antiparallels are drawn to the sides through *K*, they meet the sides of the triangle in six concyclic points. The circle thus determined, called the second Lemoine circle, has its center at *K*. Since their introduction by Lemoine, many properties of the symmedian point and the Lemoine circles have been discovered.

BIBLIOGRAPHY

I. ORIGINAL WORKS. Lemoine's papers are listed in Poggendorff, III, 793; IV, 864; V, 727–728; and in the Royal Society *Catalogue of Scientific Papers*, VIII, 200–201; X, 560–561; XVI, 699–700. See esp. "Note sur un point remarquable du plan d'un triangle," in *Nouvelles annales de mathématiques*, 2nd ser., **12** (1873), 364–366; "Quelques questions de probabilités résolues géométriquement," in *Bulletin de la Société mathématique de France*, **11** (1883), 13–25; and "Sur une généralisation des propriétés relative au cercle de Brocard et au point de Lemoine," in *Nouvelles annales de mathématiques*, 3rd ser., **4** (1885), 201–223, repr. as "Étude sur le triangle et certains points de géométrographie," in *Proceedings of the Edinburgh Mathematical Society*, **13** (1895), 2–25.

II. SECONDARY LITERATURE. On Lemoine and his work see *La grande encyclopédie*, XXI (Paris, n.d.), 1197; F. Cajori, *A History of Mathematics*, 2nd ed. (London, 1919), 299–300; J. L. Coolidge, *A History of Geometrical Methods* (Oxford, 1940), p. 58, and *passim;* and R. A. Johnson, *Modern Geometry* (Boston–New York, 1929), *passim*.

SAMUEL L. GREITZER

LE MONNIER, LOUIS-GUILLAUME (*b.* Paris, France, 27 June 1717; *d.* Montreuil, France, 7 September 1799), *botany, physics.*

Le Monnier was the son of Pierre Le Monnier, professor of philosophy at the Collège d'Harcourt and member of the Académie des Sciences; his older brother was the distinguished astronomer Pierre-Charles Le Monnier. Le Monnier's scientific career was quite unusual. He is considered a botanist because this was the category of his membership in the Académie Royale des Sciences and, later, of his associate membership in the Institute, and because he was a professor at the Jardin du Roi. Yet on the basis of his very slight written work, which dates mainly from his youth, the title of physicist would be much more appropriate.

In 1739, at the age of twenty-two, Le Monnier accompanied Cassini de Thury and Lacaille on a mission to extend the meridian of the observatory of Paris into the South of France. His publications from this time (in *Suite des Mémoires de l'Académie* [1740]) contain accounts of purely physical experiments (observations of mercury in barometers in the mountains of Auvergne, in the Canigou, and elsewhere), descriptions of mines and mineral springs, and the results of his botanizing.

It was with the publication of the manuscript containing his observations made in the South of France that Le Monnier started his botanical career. In the province of Berry he noted diseased wheat on which the growths of ergot were an inch and a half long. He collected plants and drew up floras, some of twenty pages: for Berry, the mountains of Auvergne, Roussillon, and the mountains of the diocese of Narbonne. He noted interesting ecological trends, finding similarities between the mountain plants and the plants of Lapland described by Linnaeus. Although he did not publish anything after 1752, he continued his work as a botanist; the evidence can be found in the archives of the Museum of Natural History and of the Academy of Sciences.

In 1741 Le Monnier presented a memoir to the Academy entitled "Sur le rapport de différents degrés de fluidité des liquides." Utilizing an original method, he presented measurements of the flow-times of a series of liquids through the same orifice. Le Monnier pointed out the importance of the "specific gravity" of these liquids. In 1744 and 1747 he carried out analyses—simultaneously physical, chemical, and medical—of the mineral waters of Mont d'Or and Barège.

In 1746, following the famous Leyden jar experiment, he published "Recherches sur la communication de l'électricité." Simplifying Musschenbroek's apparatus, he verified the facts previously reported and established important new results: conducting bodies become charged with electricity as a function of their surface area, not of their mass; water is one of the best conductors of electricity. In this field Le Monnier found himself the rival of the Abbé Nollet, who did not welcome any challenge to this authority. It was to Le Monnier, however, that Diderot assigned the two articles "Aimant" and "Aiguille aimantée" in the first volume of the *Encyclopédie.* Furthermore Le Monnier communicated to the Academy certain results "concerning the perfect resemblance recently observed between electric matter and that of thunder" ("Observations sur l'électricité de l'air" in *Mémoires de l'Académie royale des sciences* [1752]), and six months later published his method of studying storms and his own findings, which confirmed Franklin's.

In the *Mémoires de l'Académie royale des sciences* he signed his works "par M. Le Monnier, Médecin" to distinguish them from those of his father and of his brother, with whom he sat in the Academy for fourteen years. It was both as a physician and as a botanist that he published his observations, "Sur les pernicieux effets d'une espèce de champignons appelée par les botanistes, *Fungus mediae magnitudinis totus albus* Vaillant n° 17."

But all these efforts were, for him, peripheral: his own science was botany, and we should be no more astonished to find him in 1745 taking Linnaeus botanizing in the forest of Fontainebleau with Antoine and Bernard de Jussieu than to see him as a botanist at the Academy. If we are to believe G. Cuvier, "he could have been one of our most famous botanists," but, prevented by timidity, he wrote nothing; this timidity was due to the "unjust criticisms to which his first memoirs were subjected"—and Cuvier is undoubtedly referring to the criticisms of the Abbé Nollet.

Le Monnier first practiced medicine in the hospital of Saint-Germain-en-Laye. Subsequently he was chief physician of the Hanoverian army (1756) and then first physician in ordinary both to Louis XV and to Louis XVI. He became first physician to the latter in 1788. Through kindness he also dispensed medical care to the poor.

His practical work as a botanist is worth mentioning. While still young he met Louis de Noailles at the latter's garden. He enjoyed organizing de Noailles's park, which contained rare trees and a botanical garden. One day there he fainted from excitement in the presence of Louis XV, who gave him a post as botanist and placed him in charge of the botanical garden of the Trianon. Owing to his credit at court and at the Academy, Le Monnier obtained, through the help of diplomats and of botanists sent abroad on missions, a considerable number of plants which he raised in nurseries and which he gladly gave to plant-lovers. There are many examples of his work still in existence in Paris and he enriched French horticulture with many species of rhododendron and Kalmia, the long-flowered *Mirabilis Jalapa*, the pink-flowered locust, flowering dogwood (*Cornus florida*), and many other plants. He advocated the use of humus for plants liking an acid soil.

In Montreuil, near Versailles, Le Monnier lived on the property of the Comtesse de Marsan, governess of the royal children. He had saved her life in a serious illness and her interest in botany helped him in his career.

He also had an apartment in the Tuileries and on 10 August 1792, during the attack on the palace, his life was miraculously saved by an unknown individual. Destitute, he retired to Montreuil, where he lived a miserable existence on the income from an herbalist's shop which he opened. After the death of his wife in 1793 he was cared for with devotion by a niece, who eventually married him.

BIBLIOGRAPHY

I. Original Works. Le Monnier published the following papers: "Monnier (M. Le) Médecin. Ses observations d'histoire naturelle faites dans les provinces méridionales de la France pendant l'année 1739," in *Suite des Mémoires de l'Académie* (1740), 111–235; "Sur le rapport des différens degrés de fluidité des liquides," in *Histoire de l'Académie des Sciences* (1741), 11–17; "Examen des eaux minérales du Mont d'Or," in *Mémoires de l'Académie des Sciences* (1744), 157–169; "Recherches sur la communication de l'électricité," *ibid.* (1746), 447–464; "Examen de quelques fontaines minérales de la France," *ibid.* (1747), 259–271; "Sur les pernicieux effets d'une espèce de champignons, appelée par les botanistes, *Fungus mediae magnitudinis totus albus* Vaillant n° 17," *ibid.* (1749), 210–223; and "Observations sur l'électricité de l'air," *ibid.* (1752), 233–243. He also wrote the articles "Aiguille aimantée" and "Aimants" for *Encyclopédie ou Dictionnaire raisonné des Sciences, des Arts et des Métiers*, I (1751), 199–202, 214–223. A large number of MSS are preserved in the Bibliothèque Centrale of the Muséum d'Histoire Naturelle.

II. Secondary Literature. Two early biographies are G. Cuvier, "Notice sur la vie et les ouvrages du C^en^ Lemonnier," in *Mémoires de l'Institut National des Sciences et des Arts*, **3** (*an* IX), 101–117, and A. N. Duchesne, "Éloge historique du Cit. Lemonnier, Médecin," in *Magasin encyclopédique* (1799), 489–500.

L. Plantefol

LE MONNIER, PIERRE-CHARLES (*b.* Paris, France, 20 November 1715; *d.* Herils, Calvados, France, 3 April 1799), *astronomy*.

Le Monnier's father, for whom he was named, was professor of philosophy at the Collège d'Harcourt and a member of the Académie des Sciences. His brother was Louis-Guillaume Le Monnier, also a member of the Academy, professor of botany at the Jardin du Roi and *premier médecin ordinaire* to Louis XV. In 1763 he married a Mlle. Cussy from a prominent Norman family and had three daughters, one of whom married the mathematician Lagrange. A second daughter married his brother, Louis-Guillaume Le Monnier.

Le Monnier was presented to Louis XV by the duc de Noailles and remained a great favorite of the king, before whom he made important observations of the transits of Venus (1761) and Mercury (1753) across the face of the sun. The king in turn procured for him some of the best astronomical instruments in France. At that time the English instruments were superior to the French ones. He obtained a telescope by Short and two excellent quadrants by Bird and Sisson. He also constructed a great meridian at the church of Saint Sulpice in 1743.

Le Monnier began his astronomical career early with the assistance of his father. In 1732 J. P. Grandjean de Fouchy allowed him to observe at the Rue des Postes in Paris, and his first work was on the equation of the sun. In 1735 Le Monnier presented an elaborate lunar map to the Academy of Sciences and was admitted as *adjoint géomètre* on 23 April 1736, at the age of twenty. He rose to *pensionnaire* by 1746, became professor at the Collège Royal, and was admitted to the Royal Society, the Berlin Academy, and the Académie de la Marine.

Le Monnier's career was launched when he accompanied Clairaut and Maupertuis on their 1736 expedition to Lapland to measure a degree of an arc of meridian. Le Monnier had indicated his skill by observations made in 1731 on the opposition of Saturn, and his youth and vigor were an asset on such a hazardous and demanding voyage. In addition to surveying the degree of meridian, he made observations to calculate atmospheric refraction at various latitudes and in different seasons.

Of all his diverse interests, Le Monnier's work on lunar motion was the most extensive and the most important. In the first edition of the *Principia*, Newton had shown that the principal inequalities of the moon could be calculated from his law of universal gravitation; and in the second edition he applied these calculations to the observations of John Flamsteed. His methods, however, added little to the theory that Jeremiah Horrocks had suggested long before. Flamsteed calculated new tables based on Horrocks' theory incorporating Newton's corrections, but he did not publish them. They appeared for the first time in Le Monnier's *Institutions astronomiques*

(1746), his most famous work. It was basically a translation of John Keill's *Introductio ad veram astronomiam* (London, 1721) but with important additions and with new tables of the sun and moon. The book, the first important general manual of astronomy in France, later was largely replaced by the textbooks of Lalande and Lacaille.

Le Monnier supported the view of Edmond Halley that the irregularities of the moon's motion could be discovered by observing the moon regularly through an entire cycle of 223 lunations (the saros cycle of approximately eighteen years and eleven days), with the assumption that the irregularities would repeat themselves throughout each cycle. Le Monnier and James Bradley began such a series of observations, Le Monnier continuing his work for fifty years. During the 1740's Clairaut, d'Alembert, and Euler were working in competition to create a satisfactory mathematical lunar theory, which demanded an approximate solution to the three-body problem. In the ensuing controversy d'Alembert was seconded by Le Monnier, and he used Le Monnier's *Institutions astronomiques* as the basis for his tables. In 1746 Le Monnier presented a memoir to the Academy describing his observations of the inequalities in the motion of Saturn caused by the gravitational attraction of Jupiter, and he persuaded the Academy to choose as its prize question for 1748 the problem of providing a satisfactory theory to explain the observed inequalities. Euler entered the contest, used Le Monnier's data, and won the prize. His results were important for lunar theory, because an explanation of the perturbations of Saturn also required a treatment of the three-body problem.

The best lunar tables of the eighteenth century were those of Tobias Mayer, which were based on Euler's theory but with the magnitude of the predicted perturbations taken from observations. Le Monnier, however, adhered stubbornly to the task of correcting Flamsteed's tables even while confessing the superiority of Mayer's. Le Monnier's stubbornness is explained in part by his admiration for English methods in astronomy. In addition to his study of lunar motion he published *La théorie des comètes* (1743), which was largely a translation of Halley's *Cometography* but with several additions, constructed the first transit instrument in France at the Paris observatory, corresponded with James Bradley, and was the first to apply Bradley's discovery of the earth's nutation to correcting solar tables. He traveled to England in 1748 and observed an annular eclipse of the sun in Scotland. An indefatigable observer, Le Monnier undertook a star catalog and in 1755 produced a map of the zodiacal stars. His records show no fewer than twelve observations of Uranus, but he never identified it as a planet. He also wrote on navigation, magnetism, and electricity.

His most prominent pupil was Jérôme Lalande. Lalande attended Le Monnier's lectures in mathematical physics at the Collège Royale and with Le Monnier's strong recommendation and assistance went to Berlin in 1751 to make measurements of the moon simultaneous with those by Lacaille at the Cape of Good Hope, the purpose being to determine the lunar parallax. Le Monnier allowed Lalande to take his best quadrant on the journey. Soon afterward, however, the two men quarreled, partly because of Lalande's growing attachment to Lacaille, whom Le Monnier saw as a bitter foe. Le Monnier's irascibility led him into frequent controversy. He worked continually until 10 November 1791, when an attack of paralysis ended his career as a practicing astronomer.

BIBLIOGRAPHY

I. Original Works. The most important works by Le Monnier are *Histoire céleste ou recueil de toutes les observations astronomiques faites par ordre du roy, avec un discours préliminaire sur le progrès de l'astronomie* (Paris, 1741); *La théorie des comètes, où l'on traite du progrès de cette partie de l'astronomie avec des tables pour calculer les mouvements des comètes, du soleil et des principales étoiles fixes* (Paris, 1743); *Institutions astronomiques ou leçons élémentaires d'astronomie . . . précédées d'un essai sur l'histoire de l'astronomie* (Paris, 1746); *Observations de la lune, du soleil et des étoiles fixes pour servir à la physique céleste et aux usages de la navigation . . .*, 4 vols. (Paris, 1751–1773); *Nouveau zodiaque réduit à l'année 1755 avec les autres étoiles dont la latitude s'étend jusqu'à 10 degrés au nord et au sud de l'écliptique . . .* (Paris, 1755); *Astronomique nautique lunaire, où l'on traite de la latitude et de la longitude en mer de la période saros* (Paris, 1771); *Mémoires concernant diverses questions d'astronomie, de navigation, et de physique lus et communiqués à l'Académie royale des sciences*, 3 vols. (Paris, 1781–1788); and many contributions to the *Mémoires* of the Académie des Sciences.

II. Secondary Literature. A long and highly critical analysis of Le Monnier's works appears in J. B. J. Delambre, *Histoire de l'astronomie au dix-huitième siècle* (Paris, 1827), pp. 179–238. A more favorable account is in Jérôme Lalande, *Bibliographie astronomique avec l'histoire de l'astronomie depuis 1781 jusqu'à 1802* (Paris, 1803), pp. 819–826. Biographical information is in Michel Robida, *Ces bourgeois de Paris. Trois siècles de chronique familiale, de 1675 à nos jours* (Paris, 1955). An account of the expedition to Lapland is in Pierre Brunet, *Maupertuis, étude biographique*, I (Paris, 1929), pp. 33–64; Le Monnier's involvement in the controversy over lunar theory is described in an unpublished paper by Craig B. Waff, "The

Introduction of the Theory of Universal Gravitation Into Lunar Theory: Après le système du monde"; and in Thomas L. Hankins, *Jean d'Alembert: Science and the Enlightenment* (Oxford, 1970), pp. 28–42.

THOMAS L. HANKINS

LENARD, PHILIPP (*b.* Pressburg, Austria-Hungary [now Bratislava, Czechoslovakia], 7 June 1862; *d.* Messelhausen, Germany, 20 May 1947), *physics.*

Lenard was the son of a wealthy wine maker and wholesaler. His mother, the former Antonie Baumann, died young, and Lenard, an only child, was raised by an aunt who subsequently married his father.

Lenard was at first educated at home, but when he was nine he entered the cathedral school in Pressburg and later the Realschule. For him mathematics and physics were "oases in the desert" of other subjects, and he studied these two subjects by himself with the aid of college textbooks. In addition, he carried out chemistry and physics experiments on his own. He once devoted his summer vacation entirely to study of the new field of photography.

In his unpublished autobiography Lenard wrote, "When school days ended, a painful void came into my life." His father, who wanted him to carry on the wine business, permitted him, after long arguments, to continue his studies, but only at the Technische Hochschulen in Vienna and Budapest; moreover, he was to occupy himself primarily with the chemistry of wine. After a few dreary semesters Lenard joined his father's business.

In the summer of 1883, Lenard used the savings from a year of work for a trip to Germany. In Heidelberg he met Robert Bunsen, a figure who had long been a "secret object of worship." Bunsen's lectures were a revelation to Lenard, who was now firmly resolved to become a scientist; he matriculated at Heidelberg in the winter semester 1883–1884.

Lenard studied physics for four semesters at Heidelberg and two at Berlin. From Helmholtz he received the idea for his dissertation: "Über die Schwingungen fallender Tropfen." He began work on it at Berlin in the summer of 1885 and completed it at Heidelberg in 1886. He received his doctorate summa cum laude and thereupon became an assistant to Quincke at Heidelberg.

Since Lenard still returned to Pressburg on university vacations, he was now able to devote greater effort to the work he had begun while a student with his Gymnasium teacher Virgil Klatt. Their joint undertaking opened up for Lenard an important field of research, phosphorescence, in which he carried out investigations for more than forty years. As Klatt gradually withdrew from the work, Lenard played an increasingly larger role. When, toward the end of his life, three volumes of his planned four-volume *Wissenschaftliche Abhandlungen* were published, the entire second volume was devoted to this topic. As early as 1889 he had discovered that phosphorescence is caused by the presence of very small quantities of copper, bismuth, or manganese in what were previously thought to be pure alkaline earth sulfides.

After three years as an assistant at Heidelberg, Lenard spent half a year in England, where he worked in the electromagnetic and engineering laboratories of the "City and Guilds of the London Central Institution." Afterward he was an assistant for a semester at Breslau and then came to Bonn on 1 April 1891 to work in the same capacity under Hertz, who was already famous for his discovery of electromagnetic waves. Lenard was a great admirer of Hertz, yet his hypersensitivity, which later became pathological, caused him to feel neglected and pushed aside even by Hertz. When Hertz died unexpectedly on 1 January 1894, at the age of thirty-six, Lenard took charge of the publication of the three-volume *Gesammelte Werke*. He had to sacrifice much time for this task, especially for the *Prinzipien der Mechanik*, which, although complete, required editing. More than two decades later Lenard published a literary monument to Hertz in *Grosse Naturforscher*. In the meantime, however, Lenard had developed his racial ideology and he felt obliged to reveal that he had detected a split personality in Hertz, whose father's family was Jewish. Hertz's theoretical works, especially the *Prinzipien der Mechanik*, seemed to Lenard a characteristic product of his Jewish inheritance.

Lenard qualified as a lecturer in 1892 with a work on hydroelectricity, but he was principally engaged in continuing the cathode ray experiments that Hertz had begun. Cathode rays had been Lenard's favorite topic since 1880, when he had read Sir William Crookes's paper "Radiant Matter or the Fourth Physical State."

Lenard utilized Hertz's discovery that thin metal sheets transmit cathode rays, and at the end of 1892 he constructed a tube with a "Lenard window." With this device he was able to direct the rays out of the discharge space and into either open air or a second evacuated space, where they could be further examined independently of the discharge process. As Lenard said later, "For the first time completely pure experiments were now possible."

Like Jean Perrin, Willy Wien, and J. J. Thomson, Lenard established that cathode rays consist of negatively charged particles. In harsh priority disputes, especially with Thomson, Lenard claimed, on the basis of his 1898 publication "Über die electrostatischen Eigenschaften der Kathodenstrahlen," to have made the first "incontestable, convincing determination of what were soon called 'electrons.'"

In order to prove absorption Lenard utilized a phosphorescent screen that lit up as soon as it was struck by cathode rays:

> If we employ a phosphorescent screen, we find that it shines dazzlingly when placed right next to the window; . . . at a distance of about 8 cm the screen remains completely dark; obviously air at full atmospheric pressure is not at all permeable to cathode rays, not nearly as much as it is, for example, to light. Therefore, these rays must be extraordinarily short, so short that the molecular structure of matter, which vanishes in comparison with light waves—which are after all very small—becomes very appreciable in comparison with them. Naturally it then becomes possible, with the aid of these rays, to obtain information concerning the constitution of molecules and atoms [*Über Kathodenstrahlen*, p. 10].

Lenard was in fact able to infer from the absorption of the cathode rays by matter the correct conclusion that the effective center of the atom is concentrated in a tiny fraction of the atomic volume previously accepted in the kinetic theory of gases. Lenard's "dynamide" was an important predecessor of the atomic model of Rutherford, who in 1910–1911, on the basis of the deflections of α particles, drew the same conclusion as Lenard had earlier from the scattering of electrons.

Following Roentgen's discovery of X rays at the end of 1895, Lenard was deeply depressed that he, in the course of his constant experimenting with cathode ray tubes, had not achieved this success himself. Moreover, before the discovery was made, Lenard had provided Roentgen, at the latter's request, with a "reliable" tube and Roentgen did not reveal whether or not the discovery was made with this tube. Thus Lenard felt that his contribution should have been acknowledged in Roentgen's "Über eine neue Art von Strahlen." The affair left a permanent scar, and Lenard never used the term *Röntgenstrahlen*, although it became standard usage in Germany. He spoke only of "high frequency radiation."

At Hertz's untimely death, Lenard was also obliged to act as director of the laboratory. At the initiative of Friedrich Theodor Althoff, of the Prussian Ministry of Education, he obtained his first offer of a tenured position, an associate professorship at Breslau (October 1894). Since hardly any possibility of experimental work existed at Breslau, Lenard gave up this post after a year in favor of a nontenured lectureship at the Technische Hochschule in Aachen—as an assistant to Adolph Wüllner.

In October 1896 Lenard accepted an offer of an extraordinary professorship at Heidelberg. Two years later he went to Kiel—as a full professor and director of the physics laboratory. After a few years he was able to build a new laboratory. In 1902 Lenard succeeded in discovering important properties of the photoelectric effect. He found that as the intensity of the light increases the number of electrons set free rises, but their velocity remains unaffected: the velocity depends solely on the wavelength. The interpretation of this relationship was provided in 1905 by Albert Einstein's hypothesis of light quanta.

In 1905 Lenard received the Nobel Prize in physics for his cathode ray experiments; and in 1907 he succeeded Quincke as professor and director of the physics and radiology laboratory at the University of Heidelberg. Once more a new laboratory was constructed, on a splendid site on Philosophenweg, above the city; it was completed just before the outbreak of World War I. (The laboratory was renamed the Philipp Lenard Laboratory in 1935.)

In 1909, Adolf von Harnack, in the famous memorandum which led to the founding of the Kaiser Wilhelm Society, called Lenard the "most celebrated scientist" in the field of physical research. The initiated had a different opinion. In the summer of 1910, Einstein wrote to his friend Johann Jakob Laub, who was working as Lenard's assistant at Heidelberg: "Lenard must be very 'unevenly developed' in many things. His recent lecture on the abstruse ether [*Über Äther und Materie* (Heidelberg, 1910)] seems to me almost infantile. Moreover, the investigation that he has forced upon you . . . borders very closely on the ludicrous. I am sorry that you have to waste your time with such stupidities."

In August 1914 Lenard was swept along by the wave of patriotism and nationalism. Most scientists eventually found their way back to a more sober view, but Lenard persisted in his position of supernationalism. He composed the libelous *England und Deutschland zur Zeit des grossen Krieges*, in which he asserted that the work of German researchers was systematically hidden and plagiarized by their British colleagues. Congratulating Stark on the new advances on the effect that he had discovered, Lenard wrote (14 July 1915) with anger (which Stark shared) about the Dutch physicists whose sympathies were with the Allies: "Knowledge of the atoms is thus progressing well. That this now can happen among us cannot be put forth even by the 'neutralists' themselves as an English achievement."

 LENARD

After Germany's defeat Lenard felt himself called upon, as a teacher, to incite student youth against the ruling parties of the Weimar Republic, whom he despised for their democratic and antimilitaristic views. His final lecture for 1919 ended with the words: "We are a dishonorable nation, because we are a disarmed nation. He who does not offer resistance is worth nothing. To whom do we owe our dishonor? To the present rulers. Work, as I believe you will, so that next year we will have another government."

After the murder of Walther Rathenau, a state funeral and full cessation of work was decreed for 27 June 1922. Lenard protested by having the physics laboratory remain open. After the Heidelberg memorial service, where it became known that Lenard had stated he would not give his students time off because of a dead Jew, a crowd of hundreds of angry students and workers, led by the future Social Democratic Reichstag deputy, Carlo Mierendorff, stormed the laboratory. Lenard was taken into protective custody by police officials called to the scene and was released a few hours later. The incident attracted great attention throughout the country, and passionate stands were taken on both sides.

Lenard's anti-Semitism and nationalism increased. He attributed the turmoil in the newspapers about the general theory of relativity to an agreement between Einstein and the Jewish press. When the so-called Arbeitsgemeinschaft deutscher Naturforscher zur Erhaltung reiner Wissenschaft, founded by nationalistic and anti-Semitic demagogues, began a slander campaign against Einstein in Berlin in the summer of 1920, Lenard volunteered to head the movement.

The growing conflict broke into the open on 9 September 1920 at the eighty-sixth conference of the Deutsche Naturforscher und Ärzte in Bad Nauheim. The debate over the general theory of relativity turned into a dramatic duel between Einstein and Lenard. As Max Born recounted it, Lenard directed "sharp, malicious attacks against Einstein, with an unconcealed anti-Semitic bias." Fortunately, Max Planck, who was presiding over the debate, was able to prevent an uproar.

Along with personal antagonisms, another cause of Lenard's deep dissatisfaction—which often expressed itself in unusual aggression—was that mathematically difficult theories were attaining a decisive position in physics. Lenard, whose strength lay in experimentation, could not and did not want to follow this path. His anti-Semitism, joined with this disinclination for the new physical theories, led him to contrast the "dogmatic Jewish physics" with a pragmatic "German physics," in which experiments were paramount. In 1936–1937 Lenard published four volumes of experimental physics with the title *Deutsche Physik*, based on his lectures of the preceding decades. The preface begins with the author's war cry:

> "German physics?" one will ask.—I could also have said Aryan physics or physics of the Nordic man, physics of the reality explorers, of the truth seekers, the physics of those who have founded natural science.—"Science is and remains international!" someone will reply to me. He, however, is in error. In reality science, like everything man produces, is racially determined, determined by blood.

Only a very few of his colleagues (notably Stark) took Lenard's side, and his efforts remained, despite the support of the Third Reich, fruitless.

Lenard was among the ealiest adherents of National Socialism, and on 29 February 1924 he concluded his lectures for the winter semester with a reference to Hitler as the "true philosopher with a clear mind"; on 8 May his appeal to Hitler, cosigned by Stark, appeared in the *Grossdeutsche Zeitung;* and on 19 December of the same year he expressed the hope, also in a lecture, that Hitler would soon be released from prison. On 15 May 1926 Lenard traveled to a party meeting in Heilbronn, in order to meet Adolf Hitler in person. Further meetings followed, and throughout his life Hitler prized, even revered, this Nobel laureate who, unlike the other German scientists who treated Hitler as someone only half educated, had followed him unconditionally from the beginning.

Lenard became Hitler's authority in physics, teaching him, as Albert Speer reports in his memoirs, that in nuclear physics and relativity theory the Jews exercised a destructive influence. Hitler thus occasionally referred to nuclear physics as "Jewish physics," which resulted in delays in support for nuclear research. This ideological blindness and the expulsion of Jewish scientists from the Third Reich could only have had a damaging, perhaps fatal, effect on any German efforts to develop nuclear weapons—despite the early commanding lead of German physics.

After World War II, Lenard was not arrested (he was eighty-three), but he left Heidelberg, probably on orders, to live in the small village of Messelhausen, where he died.

BIBLIOGRAPHY

I. Original Works. For Lenard's writings, see *Über Kathodenstrahlen* (Leipzig, 1906; 2nd ed., Leipzig, 1920); *Über Äther und Materie* (Heidelberg, 1910; 2nd ed., Heidelberg, 1911); *England und Deutschland zur Zeit des grossen Krieges* (Heidelberg, 1914); *Über Relativitätsprinzip, Äther,*

Gravitation (Leipzig, 1918; 3rd ed., Leipzig, 1921); *Über Äther und Uräther* (Leipzig, 1921; 2nd ed., Leipzig, 1922); "Kathodenstrahlen," in *Handbuch der Experimentalphysik*, **14**, pt. 1 (Leipzig, 1927), written with August Becker; "Phosphoreszenz und Fluoreszenz," in *Handbuch der Experimentalphysik*, **23**, pt. 1 (Leipzig, 1928), written with Ferdinand Schmidt and Rudolf Tomaschek; and "Lichtelektrische Wirkung," in *Handbuch der Experimentalphysik*, **23**, pt. 2 (Leipzig, 1928), written with August Becker.

For further reference, see *Grosse Naturforscher. Eine Geschichte der Naturforschung in Lebensbeschreibungen* (Munich, 1929; 4th ed., Munich, 1941); *Deutsche Physik*, 4 vols. (Munich–Berlin, 1936–1937); *Wissenschaftliche Abhandlungen aus den Jahren 1886–1932*, 3 vols. (Leipzig, 1942–1944); and "Erinnerungen eines Naturforschers" (unpublished typewritten MS, completed in 1943. There is a copy in the Lehrstuhl für Geschichte der Naturwissenschaften und Technik der Universität Stuttgart); some of Lenard's instruments are in the possession of the Deutsches Museum in Munich. A list of them can be found in the Museum's library under no. 1962 B 350.

II. Secondary Literature. For information on Lenard's life and work, see August Becker, ed., *Naturforschung im Aufbruch. Reden und Vorträge zur Einweihungsfeier des Philipp Lenard-Institutes der Universität Heidelberg am 14. u. 15. Dezember 1935* (Munich, 1936), also including Johannes Stark, "Philipp Lenard als deutscher Naturforscher," pp. 10–15; *Philipp Lenard, der deutsche Naturforscher. Sein Kampf um nordische Forschung . . . Herausgegeben im Auftrag des Reichstudentenführers* (Munich-Berlin, 1937); "Philipp Lenard, der Vorkämpfer der Deutschen Physik," in *Karlsruher Akademische Reden*, **17** (Karlsruhe, 1937); and *Zum 80. Geburtstag von . . . Philipp Lenard und zum 70. Geburtstag von Reichspostminister Dr. Ing. e.h. Wilhelm Ohnesorge* (Vienna, 1942).

See also Ernst Brüche and H. Marx, "Der Fall Philipp Lenard—Mensch und 'Politiker,'" in *Physikalische Blätter*, **23** (1967), 262–267; Carl Ramsauer, *Physik-Technik-Pädagogik. Erfahrungen und Erinnerungen* (Karlsruhe, 1949), pp. 102, 106–119; "Zum zehnten Todestag. Philipp Lenard 1862–1947," in *Physikalische Blätter*, **13** (1957), 219–222; Charlotte Schmidt-Schönbeck, "300 Jahre Physik und Astronomie an der Universität Kiel," Ph.D. dissertation (University of Kiel, 1965); and Franz Wolf, "Zur 100. Wiederkehr des Geburtstages von Philipp Lenard," in *Naturwissenschaften*, **49** (1962), 245–247.

Armin Hermann

LENIN (ULYANOV), VLADIMIR ILYICH (*b.* Simbirsk, Russia [now Ulyanovsk, U.S.S.R.], 22 April 1870; *d.* Gorki, near Moscow, U.S.S.R., 21 January 1924), *politics, statesmanship, philosophy.*

Lenin's paternal grandfather was a peasant and a serf. His father, Ilya Nicolaevich, was an inspector of public schools in Simbirsk, and his mother, Maria Alexandrovna Blank, a well-educated woman. His character and outlook were strongly influenced by his family, by the political and social environment in which he was brought up, and by the progressive traditions of Russian literature. An elder brother, Alexander, was executed in 1887 for an attempt on the life of the Czar Alexander III. In 1887 Lenin graduated from the Gymnasium with a gold medal and entered the University of Kazan. He was soon expelled for participating in the student movement and arrested. He then began the study of Marxism, seeing in it the ideological weapon for liberation of the Russian proletariat. In 1891 he succeeded brilliantly in the examination for the faculty of law at the University of St. Petersburg.

This notice will be concerned only with the place of natural science and its philosophy in Lenin's work and thought. It is convenient to divide his career into three periods. The first (1893–1905) includes his work in St. Petersburg, his Siberian exile in the village of Shushenskoye, where he married Nadezhda Konstantinova Krupskaya, and his first residence abroad from 1900 to 1905. The second (1905–1917) includes his participation in the revolutionary events of 1905–1907, followed by his second and longer residence abroad from 1908 to 1917. The third (1917–1924) includes the February and October revolutions and the early years of the Soviet state.

Throughout the first period of his life, Lenin studied seriously about natural science with emphasis on its history and on the parallels to be made out between developments in natural and social science. He devoted particular attention to the Darwinian theory of evolution, to chemistry, and to psychology, with a view to substantiating and developing a scientific method applicable to the analysis and generalization of factual material. He adopted the name Lenin in 1901, as a literary pseudonym in the first instance.

It was after 1907, however, during his exile in Switzerland that Lenin gave himself most intensively to scientific reading. The positivist philosophies of Ernst Mach and Richard Avenarius were then arousing interest among the Russian intelligentsia. Lenin considered them to be merely varieties of subjective idealism, and directed a book, *Materialism and Empiriocriticism* (1909), against them and their Russian followers. The argument is a defense and a development of dialectical materialism, the philosophical theory of Marx and Engels.

In preparation for that work, Lenin studied hundreds of writings on natural science and philosophy, concentrating now on physics, and reading widely in the literature in English, French, German, and Italian, as well as Russian. He frequented the libraries of Geneva and Paris, and traveled to London to work, like Marx before him, in the reading room of the British Museum. In the course of his studies, he formed the view that a revolution in science had begun around the turn of the century, marked by Roentgen's discovery of X rays, the isolation of radium, the identification of the electron, and the transformation of chemical elements in early atomic physics. Traditional notions about the structure and physical properties of matter and the modes of motion had thus been vitiated. It was henceforth quite impossible to hold that atoms were indivisible, elements immutable, mass invariant, and the laws of classical mechanics universally valid. Experiment showed the contrary.

Nevertheless, changes in scientific comprehension of matter offered no justification for the views of idealists and the energeticists, who made the error of referring matter to electricity while separating motion from matter and declaring motion to be immaterial. Philosophical materialism, Lenin argued, does not presuppose admitting any specific physical theory of the structure of matter as a compulsory requirement. He showed that all physical notions are relative and changeable—and yet their relativity and changeability fails to lead to a conclusion that matter may disappear or be transferred into something spiritual or immaterial. That atoms turn out to be divisible is not to be taken as a derogation of the objective existence of matter. Indeed, Lenin predicted ever further penetration of science into matter, stating that the electron is as inexhaustible as the atom, and therefore could no more be regarded as the least structural unit of matter than could the atom itself. In Lenin's view, modern physics illustrated the dialectical nature of human cognition, and it substantiated materialism. Certain physicists who adopted idealistic notions in this crisis of the old views of matter failed to understand the point. Their errors, in Lenin's view, were a feature of the revolutionary disintegration of previous basic orthodoxies in physics. Lenin indicated that overcoming the difficulties and resolving the crisis required replacing obsolete metaphysical materialism with dialectical materialism.

During his years in Bern early in the war Lenin concentrated his studies on Hegel's philosophy as well as on the history of science and technology, which he thought the most pertinent aspect of history for verifying the operation of the dialectical method. His notes on these studies were the direct continuation of *Materialism and Empiriocriticism*. They were published after his death under the title *Philosophical Notebooks* (1929–1930), for once involved in the events of the Revolution in Russia, he had no time to finish the treatise on dialectics he had conceived and already begun.

As head of the Soviet state, Lenin took a direct part in the organization of science, higher education, and cultural life. Technological problems of the energy supply had interested him even during the years of exile, when among his scientific interests was a study of the possibility of converting mineral coal deposits into gas for urban and industrial use. Now in April 1918 he formed a plan according to which the Russian Academy of Sciences should take an active part in constructing a new socialist economy. His most immediate concerns were with seeking out new mineral resources within Soviet Russia and devising a scheme for electrifying the whole country.

The Civil War in Russia (1918–1920) impeded putting these ideas into effect for a time, but at the end of 1920 Lenin drew together leading scientists and engineers in preparing a detailed project for electrification under a famous slogan: "Communism is Soviet power plus the electrification of the whole country." His speeches and writings of that period emphasize the importance of developing science and technology for the benefit of all humanity, and not of a privileged class. It was his view that science is a force making for peace and welfare in a time of inevitable revolutionary change. He kept himself informed of developments in science, and never disputed technical findings or theory on dogmatic grounds. Lenin admired Einstein as one of the great reformers of natural science and considered the theory of relativity to be entirely consistent with the dialectical progress of knowledge. He ardently advocated union between progressive scientists and Marxist philosophers. His article "On the Significance of Militant Materialism" (1922) was dedicated to that subject.

BIBLIOGRAPHY

I. Original Works. Lenin's works have been collected in several editions, all published in Moscow: 1st ed., 20 vols. (1920–1926); 2nd and 3rd eds., 30 vols. (1925–1932); 4th ed., 45 vols. (1941–1957); and 5th ed., complete in 55 vols. (1958–1965). Translations include that into English of the 4th ed., as *Collected Works*, 45 vols. (Moscow–London, 1960–1970), with index to the complete works (1969–1970). Archive materials and manuscripts are listed in *Leninskye zborniki* ("Lenin Collections"), 37 collections, 39 vols. (Moscow, 1924–1970).

Lenin's works on science include *Materializm i empiriocrititsizm* ("Materialism and Empiriocriticism"), in several eds. (Moscow, 1909–1969), translated into several languages, including English (Moscow, 1970); *O nauke i vysshem obrazovanii* ("On Science and Higher Education"; 1st ed., Moscow, 1967; 2nd ed., 1971); and V. V. Adoratsky, ed., *Filosofskye tetradi*, which first appeared as vols. IX and XII (Moscow, 1929, 1930) of *Leninskye zborniki* (most recent separate edition, Moscow, 1969), and which was translated into several languages, including into French as *Cahiers philosophiques* (Paris, 1955).

II. SECONDARY LITERATURE. On Lenin and his scientific work, see the collections *Organizatsia nauki v pervye gody sovetskoy vlasti (1917–1925)*. M. S. Bastrakova *et al.*, eds., *Sbornik dokumentov* ("Organization of Science in the First Years of Soviet Power [1917–1925]. A Collection of Documents"; Leningrad, 1968); M. E. Omelyanovsky, ed., *Lenin i sovremenoye estestvoznanie* ("Lenin and the Contemporary Natural Sciences"; Moscow, 1969), English trans. in press; P. N. Pospelov, ed., *Lenin i Akademia nauk. Sbornik dokumentov* ("Lenin and the Academy of Sciences. A Collection of Documents"; Moscow, 1969); A. V. Koltsov, ed., *V. I. Lenin i problemi nauki* ("V. I. Lenin and Problems of Science"; Leningrad, 1969); and M. V. Keldysh *et al.*, eds., *Lenin i sovremennaya nauka* ("Lenin and Contemporary Science"; Moscow, 1970).

See also B. M. Kedrov, *Lenin i revolutsia v estestvoznanii XX veka* ("Lenin and the Revolution in the Natural Sciences in the Twentieth Century"; Moscow, 1969); and *Lenin i dialektika estestvoznania XX veka* ("Lenin and the Dialectics of the Natural Sciences in the Twentieth Century"; Moscow, 1971), trans. into French as Boniface Kedrov, *Dialectique, logique, gnoséologie: leur unité* (Moscow, 1970); and P. V. Kopnine, *Filosofskye idei V. I. Lenina i logika* ("Philosophical Ideas of V. I. Lenin and Logic"; Moscow, 1969); *Dialektika kak logika poznania. Opyt logiko-gnoseologitcheskogo issledovania* ("Dialectics as Logic and the Theory of Cognition. The Logical and Gnosiological Analysis"; Moscow, 1973).

Biographies include *Vladimir Ilyich Lenin. Biograficheskaya khronika 1870–1921* ("Vladimir Ilyich Lenin. A Biographical Chronicle 1870–1921"), written under the direction of G. N. Gorshkov, 3 vols. (Moscow, 1970–1972); *Vladimir Ilyich Lenin. Biografia*, written under the direction of P. N. Pospelov (1st ed., Moscow, 1960; 5th ed., 1972); K. A. Ostroukhova *et al.*, *V. I. Lenin. Kratkyi biograficheskiyi otcherk* ("V. I. Lenin. A Short Biography"; 1st ed., Moscow, 1960; 7th ed., written under the direction of G. D. Obytchkin, 1972); G. N. Golykov *et al.*, eds., *Vospominania o Vladimire Ilyiche Lenine* ("Recollections of V. I. Lenin"), 5 vols. (Moscow, 1968–1970); and, especially, N. K. Krupskaya, *O Lenine* ("On Lenin"; Moscow, 1971).

B. M. KEDROV

LENNARD-JONES, JOHN EDWARD (*b.* Leigh, Lancashire, England, 27 October 1894; *d.* Stoke-on-Trent, England, 1 November 1954), *theoretical physics, theoretical chemistry.*

As one of the first scientists to make a career of theoretical physics and theoretical chemistry, Lennard-Jones was an outstanding example of a type that has now become familiar in the physical sciences. His genius was not to make fundamental discoveries or to conduct crucial experiments but to realize quickly the potentialities of new theories proposed by other scientists and, by ingenious and painstaking calculations, to develop them to the stage where they could be applied quantitatively to a wide range of experimental data. His major contributions were in three closely related areas: the determination of interatomic and intermolecular forces; the quantum theory of molecular structure; and the statistical mechanics of liquids, gases, and surfaces.

Until the age of thirty-one his name was John Edward Jones. He studied mathematics at Manchester University, where he began his research career under the direction of Sir Horace Lamb, with work on the theory of sound. After serving in the Royal Flying Corps during World War I, he returned to Manchester as lecturer in mathematics. Jones became interested in the kinetic theory of gases through his contact with Sydney Chapman, who was then the professor of mathematics and natural philosophy at the university. In 1916 Chapman had published a new method for calculating the transport coefficients of a gas (viscosity, thermal conductivity, and diffusion) for various kinds of molecular force laws. In 1922 Jones extended this work to rarefied gases. He then went to Cambridge University, where he received his Ph.D. in 1924, and worked with R. H. Fowler, who was at that time developing quantum statistical mechanics. In 1925 he married Kathleen Mary Lennard and changed his own name to Lennard-Jones. In the same year he was elected reader in mathematical physics at the University of Bristol; he was later the first occupant of the chair of theoretical physics there.

The work for which Lennard-Jones is now best known—his determination of a semiempirical interatomic force law—was based on Chapman's gas theory and the Heisenberg-Schrödinger quantum mechanics, together with an analysis of experimental data on the properties of gases and solids. In 1924 he proposed to represent interatomic forces by the general formula

$$F(r) = -\frac{a}{r^n} + \frac{b}{r^m},$$

where r is the distance between the centers of two atoms. (Negative values of the function correspond to an attractive force, positive values to a repulsive force.) He attempted to determine the constants a, b, n, and m for various gases by comparing theoretical and experimental values of transport coefficients and virial coefficients (the latter being obtained from the pressure-volume-temperature relationships). In addition, X-ray diffraction data on the equilibrium interatomic distances in the solid state were used.

Lennard-Jones's first results led to values of about 5 for n and 14 or 15 for m, for the noble gases. Quantum-mechanical calculations by S. C. Wang, J. C. Slater, R. Eisenschitz, and F. London (1927–1930) indicated that n, the exponent for the long-range attractive force, should be 7 rather than 5. As a result Lennard-Jones adopted the value $n = 7$ and then, primarily for reasons of mathematical convenience, chose $m = 13$; the experimental data could be fitted equally well by several values of m. The potential energy function corresponding to this force law is

$$V(r) = -\frac{A}{r^6} + \frac{B}{r^{12}};$$

and this function, first proposed in 1931, is now generally known as the Lennard-Jones potential, or sometimes more specifically as the Lennard-Jones (6,12) potential, to distinguish it from the general form proposed earlier.

It was already evident by 1931 that quantum-mechanical calculations led to an exponential form for the short-range repulsive force; thus an alternative potential function, $V(r) = -Ar^{-6} + Be^{-kr}$, was advocated by J. C. Slater and later by R. A. Buckingham; it is generally known as the (exp, 6) potential. Extensive calculations of transport coefficients and virial coefficients of gases for these two potential functions led to the conclusion, by 1955, that both were about equally satisfactory. More recent research has shown that neither is really adequate for describing molecular and atomic interactions at very short or very long distances. Nevertheless, the Lennard-Jones potential continues to be used in many statistical mechanical calculations.

In 1932 Lennard-Jones was appointed to the new Plummer chair of theoretical chemistry at Cambridge University. As the first officially recognized "theoretical chemist" in England (or perhaps in the world), he quickly established himself as a leader in applying quantum mechanics to numerous aspects of the structure and properties of molecules. Many of his students are still actively pursuing research in this area, and Lennard-Jones must therefore be given much of the credit for introducing the discipline of quantum chemistry into Britain. His own research papers, in the 1930's and in the early 1950's, dealt with the molecular orbital method, in which calculations are based on trial wave functions chosen to reflect the structure and symmetry of the molecule as a whole, in contrast with the "valence bond" method, advocated by Linus Pauling and others, in which atomic wave functions are taken as the starting point. Lennard-Jones also pioneered the use of mechanical computers in quantum-mechanical calculations as early as 1939.

In 1937–1939 Lennard-Jones and A. F. Devonshire published a series of papers on critical phenomena and the equation of state of liquids and dense gases. Their theory was based on a "cell model" in which each molecule is assumed to move only within a small volume, where it is acted on by a force field determined by the average positions of its neighbors. While this model would seem to attribute a solidlike structure to the system, it was found to give approximately correct results for the behavior of gases and liquids near the critical point and thus stimulated considerable work by other theorists who attempted to improve the model.

Lennard-Jones used his considerable administrative talents not only in his academic positions but also as chief superintendent of armament research for the Ministry of Supply during World War II. In 1953 he accepted the position of principal at the newly founded University College of North Staffordshire, Keele, but he was prevented from carrying out his plans for that institution by his death in 1954.

BIBLIOGRAPHY

I. ORIGINAL WORKS. A partial list of Lennard-Jones's publications is given in the memoir by N. F. Mott (cited below). The following corrections should be made in the list:

1926. "The Forces Between Atoms and Ions." Pt. II of this paper has B. M. Dent as coauthor.

1927. "The Equation of State of a Gaseous Mixture." W. R. Cook is coauthor.

1934. "Energy Distribution in Molecules in Relation to Chemical Reactions." The discussion remark by Lennard-Jones begins on p. 242, not 239.

1936. "The Interaction of Atoms and Molecules With Solid Surfaces." Pts. III and IV have A. F. Devonshire as coauthor; pt. V is by Devonshire alone and therefore should not be listed here.

1937. (Same title). Pts. VI and VII have A. F. Devonshire as coauthor.

The following papers should be added to Mott's list: "Calculation of Surface-Tension From Intermolecular Forces," in *Transactions of the Faraday Society*, **36** (1940), 1156–1162, written with J. Corner; "Critical and Co-operative Phenomena. VI. The Neighbour Distribution Function in Monatomic Liquids and Dense Gases," in *Proceedings of the Royal Society*, **178A** (1941), 401–414, written with J. Corner; "The Education of the Man of Science (Discussion Remarks)," in *Advancement of Science*, **4** (1948), 318–319; "Implications of the Barlow Report (Discussion Remarks)," *ibid.*, **5**, no. 17 (1948), 7–8; "The

Molecular Orbital Theory of Chemical Valency. XIV. Paired Electrons in the Presence of Two Unlike Attracting Centers," in *Proceedings of the Royal Society*, **218A** (1953), 327–333, written with A. C. Hurley; and "New Ideas in Chemistry," in *Advancement of Science*, **11** (1954), 136–148, repr. in *Scientific Monthly*, **80** (Mar. 1955), 175–184.

II. SECONDARY LITERATURE. Short obituary notices on Lennard-Jones were published by C. A. Coulson in *Nature*, **174** (1954), 994–995; and by A. M. Tyndall in *Proceedings of the Physical Society of London*, **67A** (1954), 1128–1129, **67B** (1954), 916–917. An article by N. F. Mott, with bibliography and portrait, appeared in *Biographical Memoirs of Fellows of the Royal Society*, **1** (1955), 175–184. His work on interatomic forces and statistical mechanics is discussed by S. G. Brush, "Interatomic Forces and Gas Theory From Newton to Lennard-Jones," in *Archive for Rational Mechanics and Analysis*, **39** (1970), 1–29; and by J. O. Hirschfelder, C. F. Curtiss, and R. B. Bird, *Molecular Theory of Gases and Liquids* (New York, 1954).

STEPHEN G. BRUSH

LENZ, EMIL KHRISTIANOVICH (Heinrich Friedrich Emil) (*b.* Dorpat, Russia [now Tartu, Estonian S.S.R.], 24 February 1804; *d.* Rome, Italy, 10 February 1865), *physics, geophysics.*

Lenz's father, senior secretary of the Dorpat magistracy, died in 1817, leaving his widow and two sons in straitened circumstances.

After graduating from secondary school with highest honors in 1820, Lenz entered Dorpat University. At the university he studied chemistry under his uncle, J. E. F. Giese, and physics under G. F. Parrot, who had founded the physics department and its physical cabinet and was the first rector of the university. Soon Parrot recommended the nineteen-year-old student to Admiral I. F. Krusenstern as a geophysical observer during Kotzebue's second scientific voyage around the world (1823–1826) on the sloop *Predpriatie*.

In 1828 Lenz was elected junior scientific assistant of the St. Petersburg Academy of Sciences. In 1829–1830 he traveled in southern Russia, where he participated in an ascent of Mt. Elbrus in the Caucasus and determined its height, conducted magnetic observations at Nikolayev according to Humboldt's program of simultaneous observations, initiated accurate measurements of variations in the level of the Caspian Sea by setting up a surveying rod in Baku, and took samples of petroleum and natural gas. In 1830 he was elected associate academician and in 1834 a full academician.

In the spring of 1831 Lenz began his investigations of electromagnetism, which he continued until 1858. At the same time he lectured on physics at the Naval Military School (1835–1841), the Artillery Academy (1848–1861), the Central Pedagogical Institute (1851–1859), and the University of St. Petersburg (1836–1865), where from 1840 to 1863 he was dean of the physics and mathematics department and later the first rector to be elected by the professoriat in accordance with the new statutes of the university. Lenz died of an apoplectic stroke while on vacation. In 1830 he had married A. P. Helmersen, sister of the geologist G. P. Helmersen; they had seven children. Robert became a physicist and headed the physics department of the St. Petersburg Technological Institute. Lenz's students A. S. Savelyev, F. F. Petrushevsky, F. N. Shvedov, and M. P. Avenarius became university professors. His physics textbook for secondary schools ran to thirteen editions, and that on physical geography for higher military educational institutions ran to four.

Lenz's services as an expert were frequently enlisted to solve scientific and engineering problems. Many decorations were conferred on him, and he held the rank of privy councillor. He was an honorary member of many Russian universities, the Physical Society in Frankfurt am Main, and the Berlin Geographical Society, and was corresponding member of the Turin Academy of Sciences. His papers were published in the *Mémoires* and *Bulletin* of the St. Petersburg Academy of Sciences and, as a rule, in Poggendorff's *Annalen* at the same time. Some papers were also published in England, France, and Switzerland.

Lenz's name survives in the history of physics as the result of his discovery of two fundamental physical laws—soon seen as special cases of the law of conservation of energy—and a great many empirical quantitative relationships of electromagnetic, electrothermal, and electrochemical phenomena; his development of precise measuring methods, instruments, and standards; and his investigations of the theoretical principles of electrical engineering. All three of these aspects of his work can be found in almost any of his publications. His application of Ohm's law and Gauss's method of least squares and his graphical representation of various laws have distinguished his work from the scientific papers of most of his contemporaries.

In November 1833, Lenz read his paper ("Ueber die Bestimmung der Richtung der durch elektrodynamische Vertheilung erregten galvanischen Ströme") before the St. Petersburg Academy. It established Lenz's law, relating the phenomena of induction to those of the ponderomotive interaction of currents and magnets discovered by Oersted and Ampère. Lenz's law states that the induced current is in such a direction as to oppose, by its electromagnetic action, the motion of the magnet or coil that produces the induction. F. Neumann's derivation of the mathematical expression for the electromotive force of induction (1846) and Helmholtz's proof of the law of conservation of energy for electromagnetic phenomena (1847) were based on Lenz's law. This law also includes the principle of invertibility of motor and generator, which Lenz demonstrated on Pixii's magnetoelectric machine in 1838. The same law explains the phenomenon of armature reaction, discovered by Lenz in 1847 in his experiments with Störer's machine. This enabled Lenz to disprove W. Weber's incorrect hypothesis that the failure of current to increase with increased armature speeds was due to delay in the rate of magnetization of the iron, and to indicate the necessity of providing brushes at the neutrals of the machine. Continuing his experiments with Störer's machine, Lenz devised a method for plotting curves showing the phase relationship between current and magnetization (1853–1858).

Independently of Joule and with greater accuracy, in 1842–1843 Lenz established the law of the thermal action of a current, appreciating that the quantity of heat obtained was limited by the chemical processes of the battery. In 1838 he gave a visual and conclusive proof of J. C. A. Peltier's discovery by freezing water with an electric current on a layer of bismuth and antimony.

Lenz worked out the theoretical and practical aspects of the ballistic method of measuring electrical and magnetic quantities (1832) on the basis of his conception of the instantaneous, impactlike effect of induction current. Employing this method, he was able to make the first quantitative investigation of the induction phenomena themselves. (He established that the induced electromotive force in a coil is the sum of the electromotive force in each turn and does not depend upon the diameter of the turns, the thickness of the wire, or the metal of which it is made.) He also made a quantitative comparison of the resistivity of wire of different metals, established the laws of temperature dependence of resistivity in quadratic form for eight metals, derived the square-law variation of the attractive force of electromagnets with the magnetization current (under the conditions of the experiment, the magnetic field was proportional to the magnetic induction), and plotted curves showing the intensity distribution of magnetization along an iron core of a coil of finite length. The last of these measurements was made by Lenz as a member of a special committee which investigated the possibility of using Jacobi's electric motor for propelling ships. In their joint investigations (1837–1840) it was shown that the magnetization of electromagnets and, consequently, the available power of the motor depend upon the amount of zinc dissolved in the battery. Therefore all hopes of obtaining "free work" by means of electromagnets had to be abandoned.

In 1844 Lenz deduced the law of the branching of currents in a system of parallel-connected elements with arbitrary electromotive forces and resistances—four years before the publication of Kirchhoff's more general laws.

Lenz's most significant investigation in electrochemistry, conducted with A. S. Savelyev (1844–1846), established the additivity of electrode potentials in a galvanic cell; the existence of a series of electrode potentials at the metal-electrolyte boundary, similar to the Volta series; and the additivity laws of the electromotive forces of polarization on the cathode and anode, as well as of the electromotive forces of polarization and initial electrode potential of each electrode.

In the first half of the nineteenth century, data were being accumulated in geophysics; and the main problem was to ensure their reliability. This work was done by Lenz in his own observations and in the instructions he drew up for expeditions of the Russian Geographic Society, which he helped to organize in 1845.

During the voyage of the *Predpriatie*, Lenz's use of instruments, invented by Parrot for determining the specific gravity and temperature of seawater to a depth of about two kilometers, yielded results that were not excelled in accuracy by the *Challenger* expedition or on the voyages of S. O. Makarov on the *Vityaz* at the end of the nineteenth century.

Lenz also discovered and correctly explained the existence of salinity maximums in the Atlantic and Pacific oceans north and south of the equator, the greater salinity of the Atlantic in comparison with the Pacific, and the decreasing salinity of the Indian Ocean encountered in traveling east. He noted that at definite latitudes, water at the ocean's surface is warmer than the air above it; and he found two maximums and two minimums of barometric pressure in the tropics.

BIBLIOGRAPHY

I. Original Works. Some of Lenz's works were collected as *Izbrannie trudy* ("Selected Works"; Moscow, 1950), with articles by T. P. Kravetz, L. S. Berg, and K. K. Baumgart and a bibliography of 251 items. This collection does not include the following important works: "Ueber die Gesetze, nach welchen der Magnet auf eine Spirale einwirkt, wenn es ihr plötzlich genähert oder von ihr entfernt wird, und über die vortheilhafteste Konstruktion der Spiralen zu magneto-elektrischem Behufe," in *Mémoires de l'Académie impériale des sciences de St.-Pétersbourg*, 6th ser., **2** (1833), 427–457, also in R. Taylor, ed., *Scientific Memoirs*, I (London, 1837), 608–630; "Ueber einige Versuche im Gebiete der Galvanismus," in *Bulletin scientifique de l'Académie impériale des sciences de St.-Pétersbourg*, **3**, no. 21 (1838), cols. 321–326, also in *Annals of Electricity, Magnetism and Chemistry*, **3** (1838), 380–385; *Rukovodstvo k fizike, sostavlennoe dlya russkikh gimnazy* ("Manual of Physics, Prepared for Russian Secondary Schools"; St. Petersburg, 1839; 13th ed., Moscow, 1870); "Ueber galvanische Polarisation und elektromotorische Kraft in Hydroketten," in *Bulletin de la classe physico-mathématique de l'Académie impériale des sciences de St.-Pétersbourg*, **5** (1847), 1–28; and *Fizicheskaya geografia. S prilozheniem osobennogo atlasa* ("Physical Geography. With Special Atlas Appended"; St. Petersburg, 1851; 4th ed., 1865), also translated into Swedish (Kuopio, Finland, 1854).

II. SECONDARY LITERATURE. On Lenz or his work, see W. K. Lebedinsky, "E. K. Lenz kak odin iz osnovateley nauki ob elektromagnetisme" ("E. K. Lenz as One of the Founders of the Science of Electromagnetism"), in *Elektrichestvo*, nos. 11–12 (1895), 153–161; O. A. Lezhneva, "Die Entwicklung der Physik in Russland in der ersten Hälfte des 19. Jahrhunderts," in *Sowjetische Beiträge zur Geschichte der Naturwissenschaften* (Berlin, 1960); O. A. Lezhneva and B. N. Rzhonsnitsky, *Emil Khristianovich Lenz* (Moscow–Leningrad, 1952), which has a bibliography of 114 titles; and W. M. Stine, "The Contribution of H. F. E. Lenz to the Science of Electromagnetism," in *Journal of the Franklin Institute*, **155** (Apr.–May 1903), 301–314, 363–384.

OLGA A. LEZHNEVA

LEO (*fl.* Athens, first half of fourth century B.C.), *mathematics.*

Leo was a minor mathematician of the Platonic school. All that is known of him comes from the following passage in the summary of the history of geometry reproduced in Proclus' commentary on the first book of Euclid's *Elements*:[1]

> Younger than Leodamas [of Thasos] were Neoclides and his pupil Leo, who[2] added many things to those discovered by their predecessors, so that Leo was able to make a collection of the elements more carefully framed both in the number and in the utility of the things proved, and also found *diorismoi*, that is, tests of when the problem which it is sought to solve is possible and when not.

Proclus' source immediately adds that Eudoxus of Cnidus was "a little younger than Leo"; and since he has earlier made Leodamas contemporary with Archytas and Theaetetus, this puts the active life of Leo in the first half of the fourth century B.C. It is not stated in so many words that he lived in Athens; but since all the other persons mentioned belonged to the Platonic circle, this is a fair inference.

Leo had been preceded in the writing of *Elements* by Hippocrates of Chios. His book has not survived and presumably was eclipsed by Euclid's. He is the first Greek mathematician who is specifically said to have occupied himself with conditions of the possibility of solutions of problems; but the Greek word[3] does not necessarily mean that he discovered or invented the subject—and indeed it is clear that there must have been *diorismoi* before his time. From the earliest days it must have been realized that a triangle could be constructed out of three lines only if the sum of two was greater than the third, as is explicitly stated in Euclid I.22. This was certainly known to the Pythagoreans, and the more sophisticated *diorismos* in Euclid VI.28 is also probably Pythagorean: a parallelogram equal to a given rectilineal figure can be "applied" to a given straight line so as to be deficient by a given figure only if the given figure is not greater than the parallelogram described on half the straight line and is similar to the defect.[4] There is also a *diorismos* in the second geometrical problem in Plato's *Meno*; and if its latest editor, R. S. Bluck, is right in dating that work to about 386–385 B.C., it probably preceded Leo's studies.[5] Nevertheless, Leo must have distinguished himself in this field to have been so singled out for mention by Eudemus, if he is Proclus' ultimate source; it is probable that he was the first to recognize *diorismoi* as a special subject for research, and he may have invented the name.[6]

NOTES

1. Proclus, *In primum Euclidis*, Friedlein ed., pp. 66.18–67.1.
2. The Greek word is in the plural.
3. It is εὑρεῖν, which in its primary sense means no more than "to find." T. L. Heath, *A History of Greek Mathematics* (Oxford, 1921), I, 303, 319, takes it to mean that Leo "invented" *diorismoi*—which he rightly regards as an error—but this is to read too much into the word in this context.
4. The problem is equivalent to the solution of the quadratic equation
$$ax - (b/c)x^2 = A,$$
which has a real root only if
$$A \not> (c/b) \cdot (a^2/4).$$
See T. L. Heath, *The Thirteen Books of Euclid's Elements*, 2nd ed. (Cambridge, 1925; repr. Cambridge–New York, 1956), II, 257–265.
5. See R. S. Bluck, *Plato's Meno* (Cambridge, 1961), pp. 108–120, for the date and app., pp. 441–461, for a discussion of this much-debated problem with full documentation.
6. Primarily signifying "definition," διορισμός came to have two technical meanings in Greek mathematics: (1) the particular

enunciation of a Euclidean proposition, that is, a closer definition of the thing sought in relation to a particular figure; and (2) an examination of the conditions of possibility of a solution. Pappus uses the word in the latter sense only, but Proclus knows both meanings (*In primum Euclidis*, Friedlein ed., pp. 202.2–5, 203.4, 9–10). It is easy to see how one meaning merges into the other. See T. L. Heath, *The Thirteen Books of Euclid's Elements*, I, 130–131; and Charles Mugler, *Dictionnaire historique de la terminologie géométrique des grecs*, pp. 141–142.

IVOR BULMER-THOMAS

LEO THE AFRICAN, also known as **al-Ḥasan ibn Muḥammad al-Wazzān al-Zayyātī al-Gharnāṭī** (*b.* Granada, Spain, *ca.* 1485; *d.* Tunis, Tunisia, after 1554), *geography.*

Leo was born in Granada, some five years before the fall of the kingdom and his family's exile to Fez, where he was educated. His geographical knowledge was based on the medieval Islamic geographical corpus and on direct observation gleaned from four journeys (as reconstructed by Mauny): (1) from Fez to Constantinople in 1507–1508; (2) to Timbuktu in 1509–1510 or the following year; (3) to Timbuktu again, and thence to Egypt via Lake Chad in 1512–1514; and (4) a final sojourn (1515–1518) as ambassador from Morocco to the Ottoman court, which took him to Constantinople and then Egypt, Arabia, and Tripoli, where he was captured by Italian pirates, transported to Italy, and given as a slave to Pope Leo X. Leo was converted to Catholicism under the aegis of the pope, whose name he took upon baptism. In 1529 he returned to Tunis and was reconverted to Islam.

In Italy, Leo wrote his geographical treatise, *Della descrittione dell'Africa*, in Italian (although doubtless it was based on Arabic notes or drafts). The *Descrittione* is divided into nine books, with the first devoted to generalities about Africa and its people and the next five to a description of North Africa. Book VII added measurably to what little had been known of the upper Niger and other regions bordering the Sahara on the south. Book VIII describes Egypt. The ninth book (of most interest to historians of science because of fresh data presented by Leo and his discussion of the limitations of Pliny's knowledge of Africa) is an essay on the rivers, animals, minerals, and plants of Africa.

The *Descrittione* was a substantial addition to the impoverished knowledge of African geography inherited from the Middle Ages. A treatise of practical geography, giving distances between places in miles, it belongs to the traditional Islamic geographical genre of "routes and provinces" *(masālik wa-mamālik).* It was much consulted by makers of maps and portolanos (as in the maps of Africa of Ramusio [1554], Luchini [1559], and Ortelius [1570], and the Mediterranean portolano of W. Barentszoon [1596]) for 200 or more years after publication and was, in consequence, one of the few works composed in the postmedieval Islamic world which had influence in Europe.

BIBLIOGRAPHY

I. ORIGINAL WORKS. Leo's treatise, *Della descrittione dell'Africa et delle cose notabili che quivi sono*, was first published in G. B. Ramusio, ed., *Navigationi e viaggi*, I (Venice, 1550), 1–103. An inexact Latin trans. by Jean Fleurian, *De totius Africae descriptione* (Antwerp, 1556), was the source of John Pory's English trans., *A Geographical Historie of Africa* (London, 1600; facs. ed., Amsterdam–New York, 1969), which was reedited by Robert Brown for the Hakluyt Society: *The History and Description of Africa*, 3 vols. (London, 1896; repr. New York, 1963). The standard modern critical ed. is *Description de l'Afrique*, A. Epaulard, trans., 2 vols. (Paris, 1956).

II. SECONDARY LITERATURE. The fullest treatment of Leo's life and work is provided by Louis Massignon, *Le Maroc dans les premières années du XVIe siècle. Tableau géographique d'après Léon l'Africain* (Algiers, 1905), which is concerned solely with the Moroccan portion of his work. Two brief but important summaries are Massignon's entry in *Encyclopaedia of Islam*, III (Leiden, 1936), 22; and Angela Codazzi, "Leone Africano," in *Enciclopedia italiana di scienze, lettere ed arti*, XX (Rome, 1933), 899. For specific aspects of Leo's work, see Codazzi, "Dell'unico manoscritto conosciuto della 'Cosmografia dell'Africa' di Giovanni Leone l'Africano," in *International Geographical Congress (Lisbon, 1949). Comptes rendus*, IV (Lisbon, 1952), 225–226; Raymond Mauny, "Notes sur les 'Grands Voyages' de Léon l'Africain," in *Hespéris*, **41** (1954), 379–394; and Robert Brunschvig, "Léon l'Africain et l'embouchure du Chélif," in *Revue africaine* (Algiers), **79** (1936), 599–604.

THOMAS F. GLICK

LEO THE MATHEMATICIAN, also known as **Leo the Philosopher** (*b.* Constantinople[?], *ca.* 790; *d.* Constantinople[?], after 869), *mathematics, astronomy.*

Leo, who is often confused both by medieval and modern scholars with Emperor Leo VI the Wise (886–912) and with the patrician Leo Choerosphactes (*b. ca.* 845–850; *d.* after 919), belonged to a prominent Byzantine family, as is indicated by the fact that his cousin, John VII Morocharzianus the Grammarian, had been patriarch (837–843). The rather conflicting sources attest that Leo obtained a rudimentary

LEO THE MATHEMATICIAN

education in rhetoric, philosophy, and arithmetic from a scholar on the island of Andros but that his more advanced knowledge of these subjects, and of geometry, astronomy, and astrology was gained through his own researches among the manuscripts that he found in monastic libraries.

In the 820's Leo began to give private instruction at Constantinople; one of the students who had read Euclid's *Elements* with him was captured by the Arab army in 830 or 831, and his enthusiastic report of his master's accomplishments caused the caliph al-Ma'mûn (813–833) to invite Leo to Baghdad. The Byzantine emperor Theophilus (829–842), learning of this invitation, responded by charging Leo with the task of providing public education at the Church of the Forty Martyrs in Constantinople. While he was in Theophilus' service, Leo supervised the construction of a series of fire-signal stations between Loulon, located north of Tarsus and close to the Arab border, and the capital. A message could be transmitted over these stations in less than an hour; by establishing theoretically synchronized chronometers at either end, Leo was able to provide for the transmission of twelve different messages depending on the hour at which the first fire was lit.

From the spring of 840 until the spring of 843 Leo served as archbishop of Thessalonica, a post he received presumably because of his political influence rather than his holiness. He was promptly deposed when the iconodule Methodius I succeeded his iconoclast cousin John as patriarch in 843. Leo apparently returned to private teaching in Constantinople until about 855, when he was appointed head of the "Philosophical School" founded by Caesar Bardas (*d.* 866) in the Magnaura Palace, where arithmetic, geometry, and astronomy, as well as grammar and philosophy, were taught. The last notice of him is in a chronicle recording an astrological prediction that he made in Constantinople in 869.

From ninth-century manuscripts Leo is known to have been involved in the process of transcribing texts written in majuscule script into minuscule; this activity may well have included some editorial work, although its nature and extent are uncertain. In any case, he was connected with the transcription of at least some of the *Tetralogies* of Plato's works, of the larger part of the corpus of Archimedes' works, and of Ptolemy's *Almagest*; it is likely that he was also concerned with the collection of mathematical and astronomical writings known as the *Little Astronomy*. Arethas' copy of Euclid's *Elements* contains, at VI.5, a "school note" by Leo on the addition and subtraction of fractions, in which he uses the Greek alphabetical symbols for numbers as the denominators. Finally, from some of his poems preserved in the *Palatine Anthology*, it is known that he possessed copies of the *Mechanics* of Cyrinus and Marcellus, the *Conics* of Apollonius, the *Introduction* of the astrologer Paul of Alexandria, the romance of Achilles Tatius, as well as works of Theon on astronomy and Proclus on geometry.

His own surving scientific works are astrological. His "Scholia on the Hourly Motion," which claims to correct an error in an example given in Porphyrius' commentary on Ptolemy's *Apotelesmatics*, in fact refers to one in Pancharius' commentary on Ptolemy III 11, 8–11, as cited by Hephaestio of Thebes, *Apotelesmatics* II 11, 39–40 (Pingree, p. 124); his solution is absurdly taken from an entirely different example in the anonymous commentary on Ptolemy III 11, 10 (Wolf, p. 114). This proves that his technical mastery of astrology was very shaky indeed. His treatise "On the Solar Eclipse in the Royal Triplicity" is lifted from Lydus' *On Omens* 9 (Wachsmuth, pp. 19–21) and from Hephaestio I 21, 12–32 (Pingree, pp. 54–62). The short works on political astrology attributed to him in some manuscripts seem to be derived from the eleventh-century Byzantine translation of an Arabic astrological compendium of one Aḥmad, in which they are II 123–125. It is at present impossible to judge the authenticity of the other brief tracts on divination by thunder, earthquakes, the lords of the weekdays, and the gospels and psalms that are found in Byzantine manuscripts, but skepticism seems to be called for.

It remains, then, that Leo was important for his role in the transmission of Greek scientific literature and in the restoration of Byzantine learning after a long period of decline. He made few, if any, original contributions to science.

BIBLIOGRAPHY

I. ORIGINAL WORKS. 1. A homily delivered at Thessalonica on 25 March 842, is in V. Laurent, ed., *Mélanges Eugène Tisserant*, II (Vatican City, 1964), 281–302.

2. "Scholia on the Hourly Motion," F. Cumont, ed., in *Catalogus codicum astrologorum graecorum* (hereafter cited as *CCAG*), I (Brussels, 1898), p. 139.

3. "On the Solar Eclipse in the Royal Triplicity," F. C. Hertlein, ed., in *Hermes*, **8** (1874), 173–176; cf. F. Boll in *CCAG*, VII (Brussels, 1908), 150–151.

4. Twelve poems in the *Palatine Anthology* (IX 200–203, 214, 361, and 578–581; XV 12; and XVI [*Planudean Anthology*] 387 C). IX 1–358 was edited by P. Waltz (Paris, 1957), and XV by F. Buffière (Paris, 1970); the rest may be found in the ed. by F. Dübner, 2 vols. (Paris, 1864–1877).

5. A group of poems was attributed to Leo and edited by J. F. Boissonade, in *Anecdota graeca*, II (Paris, 1830), 469 ff.; the first, on old age, may indeed be his.

The following are of doubtful authenticity:

6–7. "How to Know the Lengths of the Reigns of Kings and Rulers, and What Happens in Their Reigns" and "On the Appearance of the Ruler," F. Cumont, ed., in *CCAG*, IV (Brussels, 1903), 92–93; it has been noted previously that these are taken from Aḥmad.

8. "Thunder Divination According to the Course of the Moon," an unpublished treatise found on fols. 1–2 of A 56 sup. in the Ambrosian Library, Milan. The index to *CCAG*, III (Brussels, 1901), incorrectly ascribes to Leo that text edited by A. Martini and D. Bassi on pp. 25–29.

9. "Divination From the Holy Gospel and Psalter," on fols. 28v–30 of Laurentianus graecus 86, 14, Florence.

10. "On a Sick Man," a treatise on medical astrology preserved on fols. 137v–138 of Vaticanus graecus 952.

11. A work on astrological predictions from the lords of the weekdays, accompanied by a "portrait" of Leo, on fols. 284v–285v of codex 3632 of the University Library, Bologna.

12. "Earthquake Omens," A. Delatte, ed., *CCAG*, X (Brussels, 1924), 132–135, is attributed to the Emperor Leo (Leo VI the Wise); but the attribution could possibly be a mistake for Leo the Mathematician.

13. "Gnomic Sayings," M. A. Šangin, ed., in *CCAG*, XII (Brussels, 1936), 105.

14. A poem edited by P. Matranga, *Anecdota graeca*, II (Rome, 1850), 559, is sometimes—and probably erroneously—ascribed to Leo; it follows some pieces by Leo's student Constantine, attacking him for studying pagan science, *ibid.*, 555–559.

II. SECONDARY LITERATURE. The best source is now P. Lemerle, *Le premier humanisme byzantin* (Paris, 1971), pp. 148–176; Lemerle gives references to almost all the earlier literature and is weak only in his discussion of Leo's astrological tracts.

DAVID PINGREE

LEO SUAVIUS. See **Gohory, Jacques.**

LEODAMAS OF THASOS (*b.* Thasos, Greece; *fl.* Athens, *ca.* 380 B.C.), *mathematics.*

Leodamas is treated as a contemporary of Plato in the summary of the history of geometry which Proclus reproduces in his commentary on the first book of Euclid's *Elements:*

> At this time also lived Leodamas of Thasos, Archytas of Tarentum and Theaetetus of Athens, by whom the theorems were increased in number and brought together in a more scientific grouping.[1]

Thasos, the birthplace of Leodamas, is an island in the northern Aegean off the coast of Thrace; but it would appear from the linking of his name with that of Archytas and Theaetetus that he spent his productive years in the Academy at Athens. This association with Archytas and Theaetetus suggests that he must have been a considerable mathematician; but the only other fact known about him is that Plato, according to Diogenes Laertius, "explained" (εἰσηγήσατο) to him the method of analysis[2] or, according to Proclus, "communicated" (παρεδέδωκεν) it to him.[3] Proclus describes analysis as carrying that which is sought up to an acknowledged first principle; he says it is the most elegant of the methods handed down for the discovery of lemmas in geometry, and he adds that by means of it Leodamas discovered many things in geometry.[4] These passages have led some to suppose that Plato invented the method of mathematical analysis, but this is possibly due to confusion with Plato's emphasis in the *Republic* on philosophical analysis. Mathematical analysis is clearly the same as reduction, of which Hippocrates had given a notable example in the reduction of the problem of doubling the cube to that of finding two mean proportionals; and the method must have been in use even earlier among the Pythagoreans.[5]

NOTES

1. Proclus, *In primum Euclidis*, Friedlein ed. (Leipzig, 1873, repr. Hildesheim, 1967), p. 66.14–18.
2. Diogenes Laertius, III.24. If this sentence is to be read in conjunction with the one preceding, Diogenes' authority is Favorinus.
3. Proclus, *op. cit.*, p. 211.21–22.
4. The fact that at this point Proclus waxes so eloquent about the merits of analysis is regarded by T. L. Heath, in *A History of Greek Mathematics*, I (Oxford, 1921), 120, as proof that he could not himself have been the author of the summary of geometry, and it is virtually certain that the ultimate source of this part of the summary is Eudemus.
5. In the same passage Diogenes Laertius, relying on Favorinus, attributes to Plato the first use of the word "element"—which is obviously untrue, since Democritus must have used it before him.

BIBLIOGRAPHY

See Kurt von Fritz, "Leodamas," in Pauly-Wissowa, Supp. VII (Stuttgart, 1940), cols. 371–373.

IVOR BULMER-THOMAS

LEONARDO DA VINCI (*b.* Vinci, near Empolia, Italy, 15 April 1452; *d.* Amboise, France, 2 May 1519), *anatomy, technology, mechanics, mathematics, geology.*

The reader may find helpful a preliminary word of explanation on the treatment being given the work of

LEONARDO DA VINCI

Leonardo da Vinci. The range of his knowledge was such as to recommend individual treatment of specific areas, but it is not that which is exceptional about the article that follows, for other articles in this *Dictionary* have been divided among several scholars specializing in appropriate disciplines. But the case of Leonardo is *sui generis* even in the context of the Renaissance, hospitable though its climate was to the growth of personal legends. It would be well to agree, before trying to penetrate Leonardo's sensibility, that it is anachronistic to ask whether he was a "scientist" and, although we may use the word "science" for convenience, it is largely irrelevant to wonder what he contributed toward its development. Strictly speaking, a thing cannot be a contribution unless it is known; and until the notebooks came to light, much was rumored but very little known of Leonardo's work except for his surviving paintings and (perhaps) certain features of his engineering practice, together with the well-founded tradition that he was learned in anatomy.

Rather than attributing this or that "discovery" to Leonardo, the interesting matter is to learn what Leonardo knew and how he knew it. It is to fulfill that purpose that the present article was composed. The task is important because it measures the scope of an extraordinary intellect and sensibility. Beyond that, it is rewarding because the study of Leonardo enables us to estimate what could be known at that particular juncture. Indeed, the opportunity is unique in the history of science, at least in its extent, for scientists and philosophers who advance their subjects by completing and communicating their work normally obscure in the process the elements with which they began. Not so Leonardo, whose art in drawing and artlessness in writing open windows to the knowledge latent in the civilization of the Renaissance.

The editors consider that they have been fortunate in persuading Dr. Kenneth D. Keele to write an introductory section on the lineaments of Leonardo's career together with a more detailed treatment of his studies in anatomy and physiology. Drs. Ladislao Reti, Marshall Clagett, Augusto Marinoni, and Cecil Schneer then develop in comparable detail the aspects of Leonardo's work that pertain to technology and engineering, to the science of mechanics, to mathematics, and to geology.

CHARLES C. GILLISPIE

LIFE, SCIENTIFIC METHODS, AND ANATOMICAL WORKS

Leonardo da Vinci was the illegimate son of Piero da Vinci, a respected Florentine notary, and a peasant girl named Caterina. The year that Leonardo was born, his father was rapidly married off to a girl of good family, Albiera di Giovanni Amadori. Genetically it is of some interest that Piero's youngest legitimate son, Bartolommeo, an enthusiastic admirer of his half-brother Leonardo, deliberately repeated his father's "experiment" by marrying a woman of Vinci and, as Vasari relates, "prayed God to grant him another Leonardo." In fact he produced Pierino da Vinci, a sculptor of sufficient genius to win himself the name of "Il Vinci" long before he died at the age of twenty-two.

Young Leonardo's education at Vinci was conventionally limited to reading and writing. His early manifested gifts for music and art induced his father to apprentice him, about 1467, to Andrea del Verrocchio, in whose workshop he studied painting, sculpture, and mechanics. During this period in Florence, Leonardo's activities appear to have been directed predominantly toward painting and sculpture; the earliest of his pictures still extant, the "Baptism of Christ," was painted in collaboration with Verrocchio in 1473. The "Adoration of the Magi" was still incomplete when he left for Milan in 1482, to enter the employ of Ludovico Sforza (Ludovico il Moro), duke of Milan. In his application to Ludovico, Leonardo revealed that a great deal of his attention had already been devoted to military engineering; only in concluding did he mention his achievements in architecture, painting, and sculpture, which could "well bear comparison with anyone else."

Leonardo lived at Milan in the duke's service until 1499. During these years his interest in the problems of mechanics and the physics of light grew steadily while his artistic output reached a peak in the fresco "The Last Supper" (1497) and in the clay model of his great equestrian statue of Francesco Sforza (1493). The notebooks of this period show the increase of his interest in mathematics, the physics of light, the physiology of vision, and numerous mechanical problems, including those of flight. Four projects for books—separate treatises on painting, architecture, mechanics, and the human figure—appear in his notes. These studies were to occupy him for the rest of his life. During his last years in Milan he collaborated with the mathematician Luca Pacioli on his *Divina proportione*; Leonardo drew the figures for the first book.

After the capture of the duke of Milan by the French, Leonardo left for Venice, eventually returning to Florence. He then entered the service of Cesare Borgia and was employed for about a year as chief inspector of fortifications and military engineer in the Romagna. Following this he was responsible for an un-

LEONARDO DA VINCI

successful attempt to divert the Arno near Pisa. While in Florence he began the portrait "Mona Lisa" and also the ill-fated fresco "Battle of Anghiari." During these years in Florence, from 1500 to 1506, Leonardo began his systematic researches into human anatomy. Mathematics and the mechanical sciences, too, increasingly occupied his time; and he began to couple his study of the problem of human flight with intensive research on bird flight and meteorology. His studies on the movement of water, a lifelong preoccupation, were later compiled into the *Treatise on the Movement and Measurement of Water*.

In June 1506 Leonardo returned to Milan, where Charles d'Amboise, the French governor, showed him the keenest appreciation he had yet experienced. During this period he produced most of his brilliant anatomical drawings, perhaps stimulated for a short while by the young Pavian anatomist Marcantonio della Torre. Leonardo had now come to feel that mathematics held the key to the "powers" of nature; and his work in hydrology, geology, meteorology, biology, and human physiology was increasingly devoted to a search for the geometrical "rules" of those powers through visual experience, experiment, and reason.

After the French were expelled from Milan in 1513, Leonardo left for Rome, hoping that the Medici Pope Leo X and his brother Giuliano would provide him with encouragement and a good working environment. Nothing came of this hope; and in 1516 he resumed the French liaison, this time with Francis I, with whom he traveled to Amboise in 1516. He died there after a stroke.

Problems of Evaluating Leonardo's Scientific Thought. There are many different opinions concerning Leonardo's stature as a scientist. The essential reason that this is so lies in the grossly abbreviated form in which his work has come down to posterity. Leonardo himself noted that "abbreviators of works do harm to knowledge . . . for certainty springs from a complete knowledge of all the parts which united compose the whole" (*Windsor Collection*, fol. 19084r, in I. A. Richter, *Selections . . .*, p. 3). Unhappily, his extant notes are both abbreviated and confused. It is therefore necessary to assess these defects before evaluating Leonardo's work in the history of science.

The Losses of Leonardo's Manuscripts. Although Leonardo clearly intended to write treatises on painting, architecture, mechanics, and anatomy, he brought none to publication. Two known works are published under his name, the *Treatise on Painting* and *Treatise on the Movement and Measurement of Water*; both were compiled after his death from his notes. These remain of great value, although their respective compilers, Francesco Melzi and Luigi Arconati, constructed them according to their own ideas, rather than Leonardo's. The great mass of surviving data consists of some 6,000 sheets of Leonardo's manuscript notes. It is difficult to estimate what proportion of the whole these represent. Reti has calculated that 75 percent of the material used by Melzi for his compilation of the *Treatise on Painting* has since been lost. If only a corresponding proportion of Leonardo's scientific notes are extant, this is indeed a severe truncation. The qualitative loss is probably even greater, since there is no trace of a number of "books" to which Leonardo frequently referred in his notes as "completed," and which he used as references, among them an "Elements of Mechanics," and a "Book on Water"—not to mention fifteen "small books" of anatomical drawings.

Sources of Confusion in the Notes. The notes that remain are in great confusion, in part because Leonardo himself made no effort to integrate them, and in part because after his death they underwent almost every conceivable kind of disarrangement and mutilation—as is illustrated by the great scrapbook of sheets composing the *Codex Atlanticus.* Thus the thousands of pages that do survive resemble the pieces of a grossly incomplete jigsaw puzzle.

FIGURE 1. The diversity of subjects on a single page of Leonardo's notes, of which this is an example, raises problems of interpretation (*Windsor Collection*, fol. 12283. Reproduced by gracious permission of Her Majesty Queen Elizabeth II).

Leonardo's mode of expressing himself often challenges interpretation. Apart from his "mirror" script, he habitually presented his thoughts visually, sometimes covering an astonishing range of phenomena with very few words, as in *Windsor Collection*, fol. 12283 (Figure 1). This page contains segments of circles, a geometrical study for the "rule of diminu-

LEONARDO DA VINCI

tion" of a straight line when curved, a study of curly hair (with a note on its preparation), and drawings of grasses curling around an arum lily, an old man with curly hair, trees, billowing clouds, rippling waves in a pool, a prancing horse, and a screw press. All can be seen as studies of curves, viewed from different aspects, each of which is developed in detail in various parts of his notes. The sheet provides a good example of Leonardo's visual approach to any problem—by integration—a mode which has until recently been frowned upon by orthodox science. Interpretation of the meaning of many of the drawings must necessarily be speculative; but it is always dangerous to call any drawing a "doodle," since some of Leonardo's most interesting scientific concepts appear as casual, small, inartistic figures.

A further source of confusion in Leonardo's notes derives from his habit of periodically returning to the same subject. This often resulted in incompatible statements in different notes, which Leonardo himself recognized, referring in 1508 to his notes as "a collection without order taken from many papers . . . for which O reader blame me not because the subjects are many, and memory cannot retain them . . . because of the long intervals between one time of writing and another" (*Codex Arundel*, fol. 1r).

Leonardo's thinking was intensely progressive during these "long intervals" of time. A note made in 1490 may therefore differ markedly from one on the same subject from 1500 or 1510. With this understanding, contradictory statements can often be transformed into sources of comprehension if they can be dated—a means of tracing the progress of Leonardo's thought that has only recently become available.

In 1936, with the facsimile production of the Forster codices, it was thought that publication of extant Leonardo notes was complete. Nevertheless, in 1967 two further codices, containing 340 folios, were found in Madrid. These still await publication; and one wonders how many more will be found.

Leonardo's Scientific Method. Leonardo lived at a time when theology was the queen of the sciences. Theological thinking emphasizes the divine incomprehensibility and mystery of natural phenomena; even the Neoplatonic thought of Leonardo's contemporary Marsilio Ficino focused on a direct, revealed link between the mind of man and God and was not concerned with the visual exploration of natural phenomena. Leonardo, on the other hand, declared that faith that is the foundation of all science: there is a logic in natural phenomena detectable by the senses and comprehensible to the mathematical logic of the human mind. Beyond this he postulated God as the creator of all.

Leonardo felt his way to his scientific outlook through his study of the theory of painting, which led him to analyze visual phenomena. As he declared,

> Painters study such things as pertain to the true understanding of all the forms of nature's works, and solicitously contrive to acquire an understanding of all these forms as far as possible. For this is the way to understand the Creator of so many admirable things, and this is the way to love such a great Inventor. In truth great love is born of great knowledge of the thing that is loved [*Treatise on Painting*, para. 80].

For Leonardo the world of nature was created by God, and the science or theory of painting "is a subtle invention which with philosophical and ingenious speculation takes as its theme all the various kinds of forms, airs and scenes, plants, grasses and flowers which are surrounded by light and shade. And this truly is a science and the true-born daughter of Nature" (MS *Ashburnham* 2038, fol. 20r, in E. MacCurdy, *Notebooks*, II, 229).

Science for Leonardo was basically visual: "The eye, the window of the soul, is the chief means whereby the understanding can most fully and abundantly appreciate the infinite works of Nature; and the ear is second" (MS *Ashburnham* 2038, fol. 20r, in E. MacCurdy, *Notebooks*, II, 227). He asserted his belief that the patterns of natural phenomena can conform with the patterns of form created in the human mind—"Painting compels the mind of the painter to transform itself into the very mind of Nature to become an interpreter between Nature and the art. It explains the causes of Nature's manifestations as compelled by its laws. . ." (*Treatise on Painting*, para. 55). These are "laws of necessity the artificer of nature—the bridle, the law and the theme" (*Codex Forster*, III, fol. 43v); while "Necessity constrains all effects to the direct result of their causes" (*Codex Atlanticus*, fol. 345v-b). For Leonardo these causes, effects, and laws could be expressed visually in the form of a geometry that includes the movement of things as well as their resting forms.

Leonardo's Mathematics of Science. Leonardo defined science as "that mental analysis which has its origin in ultimate principles beyond which nothing in nature can be found which is part of that science" (*Treatise on Painting*, para. 1). Geometry was such a science, in which "the point is that than which nothing can be smaller. Therefore the point is the first principle of geometry and nothing in nature or the human mind can be the origin of the point" (*ibid.*). He differentiated between the "natural" point, which has the characteristics of an atom, and the "conceptual" point of geometry: "The smallest natural point is larger than

LEONARDO DA VINCI

all mathematical points, because a natural point is a continuous quantity and as such is infinitely divisible, while the mathematical point is indivisible, having no quantity" (quoted from J. P. Richter, *Literary Works of Leonardo*, 44). In this way he linked geometry with physics and denied the term "science" to any investigation that is not capable of "mathematical" demonstration, such as the so-called sciences that "begin and end in the mind" and are "without the test of experience." He proclaimed the necessity for both geometry and experience, taking as his example astronomy, which was all "visual lines, which enclose all the different shapes of bodies created by nature, and without which the art of geometry is blind" (*Treatise on Painting*, para. 15). Thus, via geometry, human experience could interpret nature.

Leonardo divided his inquiry into three coincident parts: the geometry of vision, the geometry of nature, that is, physics or "natural philosophy," and pure geometry. His investigation of vision was the first of his physiological researches; it pertained primarily to perception and derived largely from the traditional concept of a cone of vision. (Pure geometry—that is, Euclidean geometry—is discussed in the mathematics section, below.)

The Geometry of Vision. Without the test of visual experience there could be no science for Leonardo, for "the observer's mind must enter into nature's mind" (*Treatise on Painting*, para. 40). The problem of vision involved his concept of the spread of light from its source. This he conceived of by analogy to the transverse wave-spread set up by dropping a stone into a still pond. "Just as a stone flung into water becomes the centre and cause of many circles . . . so

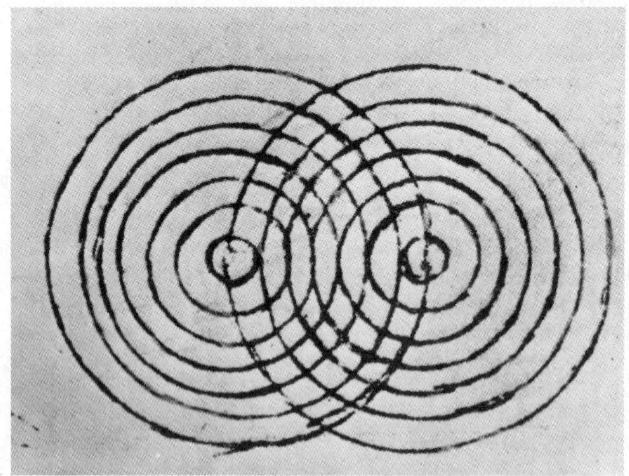

FIGURE 2. The intersection of waves produced by two stones dropped into a pond. Leonardo used this as an analogy for the spread of all powers produced by percussion, in which he included light and sound (MS *A*, fol. 61r).

any object placed in the luminous atmosphere diffuses itself in circles and fills the surrounding air with images of itself" (MS *A*, fol. 9r). These images spread out in circles from the surfaces of the object with steady diminution of power to the eye (Figure 2). As pond waves spread by the power of percussion, so the light wave varies in power with the force of percussion and inversely with its distance in the form of an infinite number of pyramids (like perspective), diminishing as they approach the eye (Figure 3). The

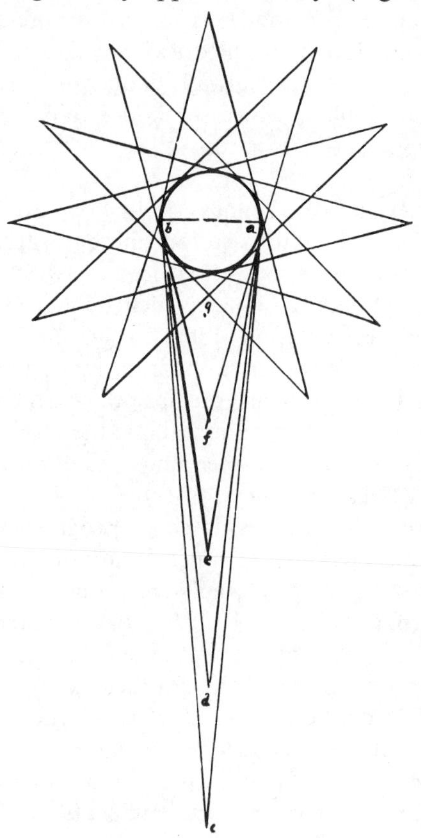

FIGURE 3. The pyramidal spread of light from a luminous object, *ab*, in all directions. "The circle of equidistant converging rays of the pyramid will give angles, and objects to the eye of equal size" (MS *Ashburnham* 2038, fol. 6b).

pyramidal form which Leonardo used to represent this decline of power has the characteristic that

> . . . if you cut the pyramid at any stage of its height by a line equidistant from its base you will find that the proportion of the height of this section from its base to the total height of the pyramid will be the same as the proportion between the breadth of this section and the breadth of the whole base [MS *M*, fol. 44r].

Leonardo thus represented the power of percussion by the base or diameter of a cone or pyramid, and its

LEONARDO DA VINCI

decline by its breadth at any height of the pyramid. When the spreading light image reaches the pupil, it finds not a point but a circle, which in its turn contracts or expands with the power of light reaching it and forms the base of another cone or pyramid of light rays directed by refraction through the lens system of the eye to the optic nerve (Figure 4). The image then travels via the central foramen up the nerve to the "imprensiva" of the brain, wherein the percussion is impressed. Here it joins with the impressions made by the "percussions" from the nerves that serve the senses of hearing, smell, touch, and pain—all of which are activated by similar percussive mechanisms following similar pyramidal laws.

Thus in the imprensiva a spatiotemporal verisimilitude of the external environment is produced, while a truly mathematical, geometrically conditioned model of reality is experienced by the "senso commune" in

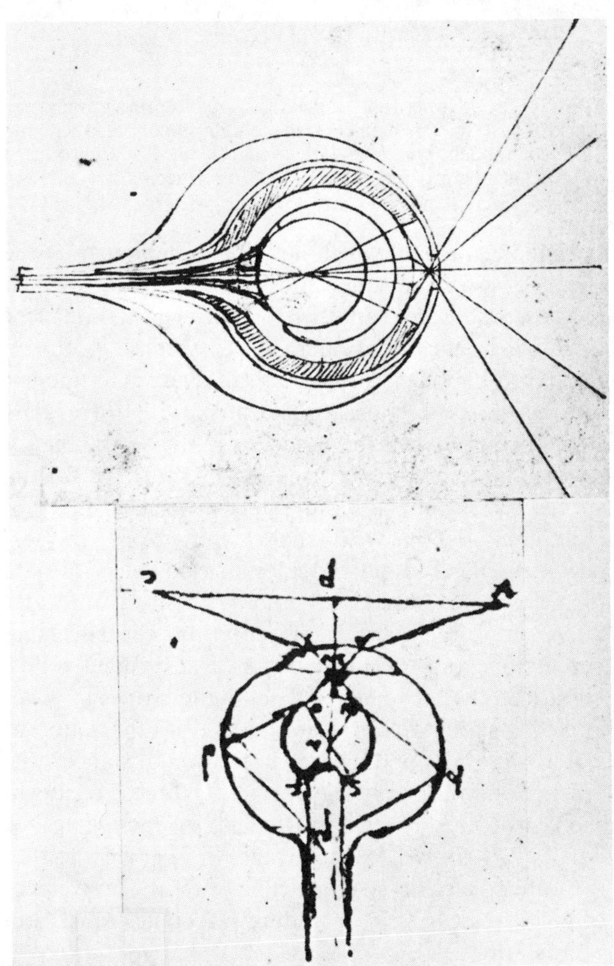

FIGURE 4. The dioptrics of the lens system of the eye. In the upper figure rays are shown crossing once at the pupil and again in the lens, to which the optic nerve is in apposition (*Codex Atlanticus*, fol. 337r-a). Below, rays are reflected from the retina, which acts as a mirror (MS *D*, fol. 10).

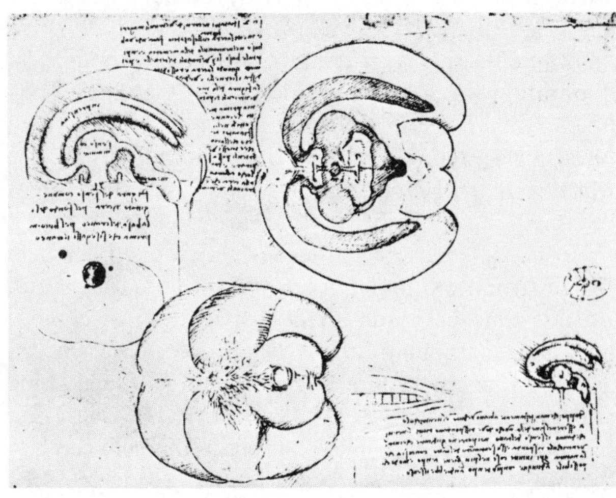

FIGURE 5. Wax casts of the cerebral ventricles. In the figure at left the lateral ventricle is labeled "imprensiva"; the third ventricle, "senso commune"; and the fourth ventricle, "memoria" (*Quaderni d'anatomia*, fol. 7r. Reproduced by gracious permission of Her Majesty Queen Elizabeth II).

the third ventricle where all these sensations come together (Figure 5). "Experience therefore does not ever err, it is only your judgment that errs in promising itself results which are not caused by your experiments" (*Codex Atlanticus*, fol. 154r). Leonardo later located the imprensiva in the lateral ventricles of the brain when he discovered them. "Memory" and "judgment" he located in the fourth and third ventricles, respectively. Thus his distinction between sensory "experience" and "judgment" have physiological as well as psychological connotations.

Leonardo recognized that errors of sensory observation can occur. He attributed these to "impediments"—the resulting experience "partaking of this impediment in a greater or less degree in proportion as this impediment is more or less powerful."

The Geometry of Nature. The pyramidal form that Leonardo imposed upon the propagation of visual images he further applied to the propagation of the four "powers" of nature: "movement, force, weight, and percussion." He saw it as perspectival, i.e., simple, pyramidal proportion. It was in this form, therefore, that he quantified these powers, whether dealing with light and vision, as above, or with other forms of the powers of nature, as, for example, weights, levers, the movements of water, and power transmission in machines, including the human heart or limbs. For example, he described the force of a spring as "pyramidal because it begins at a point or

LEONARDO DA VINCI

instant and with each degree of movement and time it acquires size and speed" (MS *G*, fol. 30r); he further described the movement of compound pulleys as "pyramidal since it proceeds to slowness with uniform diminution of uniformity down to the last rope" (*ibid.*, fol. 87v); see mechanics section, below. Such phrasing is reminiscent of the terminology of the Merton School and reminds one that the names "Suisset" and "Tisber" occur in Leonardo's notes.

Leonardo's close relationship with Luca Pacioli in Milan strongly stimulated his interest in mathematics. At this time he made an intensive study of algebra and the "manipulation of roots," describing algebra as "the demonstration of the equality of one thing with another" (MS *K*, II, fol. 27b). He does not, however, appear to have applied this concept of algebraic equations to his scientific work. On the other hand, he did apply square roots to the calculation of wingspan in relation to the weight of the human flying machine. His formula ran thus: "If a pelican of weight 25 pounds has a wingspan of 5 braccia, then a man of weight 400 pounds will require a wingspan of $\sqrt{400} = 20$ braccia" (*Codex Atlanticus*, fol. 320r-b). This calculation was made about the time of his final attempt at flight (1504). Leonardo never applied the inverse-square law to any "power" or "force."

Leonardo's Methods of Observation and Experiment. Leonardo's acquisition of "experience" was remarkable for three main methods: measurement, models, and markers. From about 1490, when he concluded that the "powers" of nature had geometrical relationships, Leonardo attempted to quantify his observations and experiments. Measurement entered into all his observations; measurements of weights, distances, and velocities. His notes from about this time contain descriptions of hodometers, anemometers, and hygrometers, as well as balances of various ingenious types.

In general one is struck, however, by the crudity of his measurements of both space and time. He rarely mentioned any spatial measurement other than the braccio (approximately twenty-four inches). For small intervals of time, such as the heart rate, he used musical tempi. Here he explained, "of which tempi an hour contains 1080" (*Quaderni d'anatomia*, II, 11r). For timing he used sandglasses and water clocks. Although he certainly did not invent the balance, he did use balances in a highly original way—for example, by placing a man in one pan and observing whether the downward movement of a levered wing to which he was attached would raise him or not, and by weighing objects at different temperatures.

Models were Leonardo's favorite form of experimental demonstration. He used them particularly for observations of the flow of water or blood currents. Perhaps his smallest model was a glass cast of an aorta with its valves (Figure 6); the largest was one of

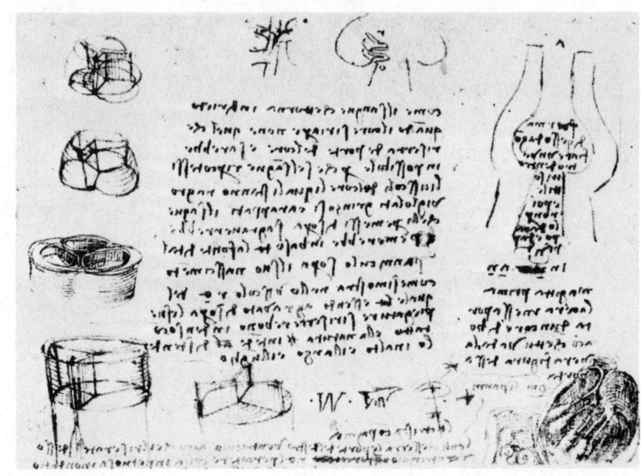

FIGURE 6. Upper right, a glass cast of an aorta with instructions for casting. The figures on the left margin are also schemes for aortic models (*Quaderni d'anatomia*, II, fol. 12r. Reproduced by gracious permission of Her Majesty Queen Elizabeth II).

the Mediterranean, which he built to demonstrate the effects of the rivers entering it.

Markers were another favorite method for visualizing the movements of water, both in Leonardo's models and in actual rivers. His models for this purpose consisted of tanks with three glass sides through which he could view the seeds, bits of paper, or colored inks used as markers; he had a special set of floats, each designed to be suspended at a different depth. With such observations and experiments he accumulated an enormous mass of data regarding the directions and velocities of water movements, their angles of reflection, the effects of percussion, and the movements of suspended solids. He traced the movements of sand and stones in water and their mode of deposition, as well as the action of water on surrounding surfaces—for instance, the erosion of river banks. He also applied these results to such problems as the movement of lock gates or aortic valves. He extended the marker method to the other "elements," particularly to the movement of solid bodies in air, including projectiles, dust, and "birds" (models).

Leonardo frequently advocated the repetition of experiments, "for the experiment might be false whether it deceived the investigator or not" (MS *M*, fol. 57r). If a confirmed experiment did not agree with a suggested mathematical "rule," he discarded

LEONARDO DA VINCI

the rule. By this means, and clearly to his surprise, he had to reject the Aristotelian "rule" that "if a power moves a mobile object with a certain speed, it will move half this mobile object twice as swiftly." By experiment Leonardo tested the rule, comparing the movements of "atoms" of dust in air with those of large objects fired from a mortar; and from his observations he drew a moral—mistrust those who have used nothing but their imagination and have not verified their statements by experiment. For Leonardo mathematical rules had to give way to experimental verification or refutation.

Leonardo applied his experimental or mathematical results only to visual space. Unlike such medieval antecedents as Oresme, he refused to quantify such abstract qualities as beauty or glory; he specifically stated that geometry and arithmetic "are not concerned with quality, the beauty of nature's creations, or the harmony of the world" (*Treatise on Painting*, para. 15). He honored the precept that these are fields for the creative arts rather than for the creative sciences.

Leonardo's "Rules." Not only did Leonardo advocate repetition of experiments, but he performed experiments in series, each repeated with one slight variation to resemble continuous change. For example, on investigating flight with model birds, he wrote: "Suppose a suspended body resembling a bird and that the tail is twisted to different angles. By means of this you will be able to derive a general rule as to the various movements of birds occasioned by bending their tails" (MS *L*, fol. 61v). Demonstrating the distribution of weight of a beam suspended by two cords, he wrote: "I make my figures so that you shall know all the cases that are placed under one simple rule" (*Codex Atlanticus*, fol. 274r-b). The statement is carried out in the following three pages of patient depiction of all the variable distributions of weight (*ibid.*, fols. 274r-b, r-a, v-b).

Thus by repeated observation, repeated experiment, and calculatedly varying his experiments, Leonardo built up quantitative data into limited generalized "rules." "These rules," he wrote, "are the causes of making you know the true from the false" (*ibid.*, fol. 119v-a). Since the rule is mathematical in form, he averred, "There is no certainty where one cannot apply any of the mathematical sciences" (MS *G*, fol. 96v); and clearly it follows that "No man who is not a mathematician should read the elements of my work" (*Windsor Collection*, fol. 19118v, in I. A. Richter, *Selections* . . ., p. 7).

The Four Powers. Leonardo's notes contain long debates on the nature of movement, weight, force, and percussion, all in the context of the Aristotelian elements of earth, air, fire, and water. From this debate emerged the statement: "Weight, force, together with percussion, are to be spoken of as the producers of movement as well as being produced by it" (*Codex Arundel*, fol. 184v). He finally saw weight as an accidental power produced by a displaced element "desiring" to return to its natural place in its own element. Thus "gravity" and "levity" were two aspects of the same drive. Force was caused by violent movement stored within bodies. Sometimes Leonardo called force "accidental weight," using the Aristotelian meaning of "accidental." Movement of an object from one point to another could be natural, accidental, or participating. Impetus was derived movement arising from primary movement when the movable thing was joined to the mover. He defined percussion as "an end of movement created in an indivisible period of time, because it is caused at the point which is the end of the line of movement" (*Codex Forster*, III, fol. 32r. in I. A. Richter, *Selections* . . ., p. 77).

This remarkable simplification of his experiences of "nature" was applied by Leonardo to all fields of investigation and practical creation, that is, to science and its derived art of invention. Clearly these four powers were both derived from, and most applicable to, mechanics. "Mechanics is the paradise of mathematical science because here one comes to the fruits of mathematics" (MS *E*, fol. 8v). This statement was generalized when Leonard asserted: "Proportion is not only found in numbers and measurements but also in sounds, weights, times, spaces, and whatsoever powers there be" (MS *K*, 497, Ravaisson-Mollien). From these two statements his essentially integrative approach to all problems can be glimpsed. In each field of investigation he built up, from quantitative observation and experiment, defined mathematical "rules" which he applied to the particular problem occupying him. For example, with regard to the flight of the bird, when discussing the problem of its being overturned, he set out the forces concerned and then stated: "This is proved by the first section of the Elements of Machines which shows how things in equilibrium which are percussed outside their center of gravity send down their opposite sides . . ." (*Codex on Flight* . . ., fol. 8r). He continued, "And in this way it will return to a position of equilibrium. This is proved by the fourth of the third, according to which that object is more overcome that is acted on by the greater forces; also by the fifth of the third, according to which a resistance is weaker the farther it is from its fixed point" (*ibid.*, fol. 9r). In this way Leonardo's "rules" were built into the structure of his thought about all the powers of nature, imparting a remarkable consistency in all fields, whether mechanics, light, sound, architecture, botany, or human physiology. Unfortunately, the books to which he refers in this way, showing how he reached these generalizations, have been lost.

Leonardo's Achievements in Science. Leonardo's achievements in art and science depended primarily on his remarkable acuity of vision, and his particular sensitivity to the geometrical consequences of that vision. These gifts he combined with the creative technology of the engineer and the artist. These creative aspects emerge in his achievements in scientific technology.

Mathematics. Leonardo's approach to mathematics was predominantly physical. Even when appearing abstract, as in his "Book on Transmutation" (of forms), his practical object was revealed by his reference to the metalworker or sculptor. Typical of his contribution to mathematics were proportional and parabolic compasses. His explorations of the properties of the pyramid were, as is to be expected, thorough. As a result of an experimental investigation of the center of gravity of a tetrahedron, he discovered that the center of gravity is at "a quarter of the axis of that pyramid" (MS *F*, fol. 51r). (For his development of pure mathematics, see the section on mathematics.)

Mechanics. (Here brief mention is made only of the main areas studied by Leonardo. For fuller development of these, see the section on mechanics.) Leonardo's point of departure for statics was Archimedes' work on the lever, of the principles of which he had some understanding. His appreciation that equilibrium of a balance depends arithmetically upon the weights and their distances from the fulcrum gave him many examples of pyramidal (arithmetical) proportion.

Leonardo never quite reached the concepts of mass or inertia, as opposed to weight. He still used the "halfway" concept of impetus when he said, "All moved bodies continue to move as long as the impression of the force of their motors remains in them" (*Codex on Flight* . . ., fol. 12r). This is as near as he came to Newton's first law. And although Leonardo disproved the Aristotelian relationship between force, weight, and velocity, he did not reach the concept of acceleration resulting from action of the force on a moving body. Newton's third law, however, Leonardo did state clearly in his concrete way and apply persistently. For example, "An object offers as much resistance to the air as the air does to the object" (*Codex Atlanticus*, fol. 381v-a). And after a repetition of this statement he added: "And it is the same with water, which a similar circumstance has shown me acts in the same way as air" (*Codex Atlanticus*, fol. 395r). One can see here the process by which he built up to one of his "rules"—Newton's third law.

Gravity for Leonardo was the force of weight which is exerted "along a central line which with straightness is imagined from the thing to the center of the world" (MS *I*, fol. 22v). In his investigations of it Leonardo dropped weights from high towers. For example, he dropped two balls of similar weight together and observed at the bottom whether they still touched (MS *M*, fol. 57r). From such experiments he noted: "In air of uniform density the heavy body which falls at each degree of time acquires a degree of movement more than the degree of preceding time . . . the aforesaid powers are all pyramidal, seeing that they commence in nothing and proceed to increase in degrees of arithmetical proportion" (*ibid.*, fol. 44r). He concluded that the velocity is proportional to the time of fall but, incorrectly, that the distance of fall is also proportional directly to the time, not to its square.

Leonardo appreciated and used the concepts of resolving and compounding of forces. He described the parallelogram of forces, although he did not draw it, in analyzing the flight of the bird (*Codex on Flight*, fol. 5r); and he resolved the movement of a weight down an inclined plane into two components (MS *G*, fol. 75r). Such movement led him to study friction, a force which, like so many others, he divided into simple and compound. Leonardo appreciated the importance of the pressure or weight factor and the nature of the surface and their independence of area. Thus, for "a polished smooth surface" he found a coefficient of friction of 0.25. He was aware that such friction produces heat.

Hydrodynamics. All the "rules" described above were carried into Leonardo's studies of the movement of water. Here two phenomena received particular attention: wave formation and eddies. He found the velocity of flow of water to be inversely proportional to the dimension of the passage, whether rivers or blood vessels are being discussed. Currents of water, as observed by his marker experiments, percussed so that the angles of incidence and reflection were equal, similar to those made by bouncing balls. The formation of the transverse wave from the percussion of a stone in water was similarly explained. The cohesion of a drop of water against the force of gravity led Leonardo to postulate a force "like that of a magnet," which was also responsible for capillary attraction. The extent of his studies of water is reflected in the *Codex Leicester*, where he draws "732 conclusions as to water" (*Codex Leicester*, fol. 26v).

Cosmology. Water, more than the other elements, led Leonardo to compare "the greater and the lesser world." The macrocosm–microcosm analogy was very real to him, since he believed that similar laws governed both the cosmos and the body of man. He considered at one time, for example, that the movement of the waters of the earth, particularly to the tops of mountains, was analogous to the movement of blood to the head, both being produced by heat. At first he postulated that water was drawn up by the heat of the sun; later he suggested that the heat arose from subterranean fires, from volcanos in the earth. He posited such subterranean fires as an analogy to the human heart, which he saw not only as the percussor of blood to the periphery but also as a source of heat to the body through the friction of the blood in its chambers (Figure 7).

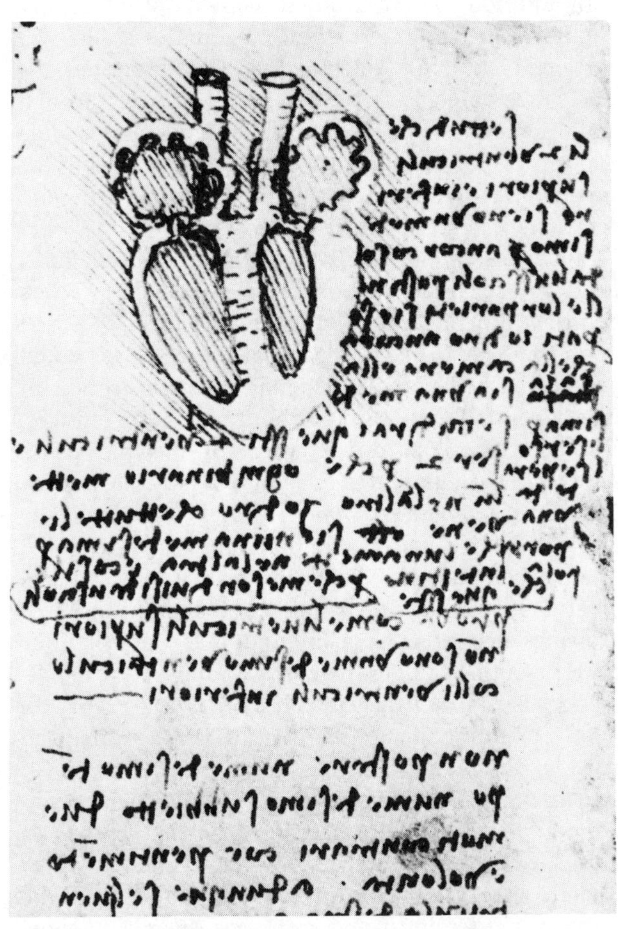

FIGURE 7. The heart as a percussion and heat-producing instrument. Blood from the atria, above, is propelled into the ventricles, below, from which part of the blood is propelled into the blood vessels and part back into the atria, producing heat by friction (*Quaderni d'anatomia,* II, fol. 3v. Reproduced by gracious permission of Her Majesty Queen Elizabeth II).

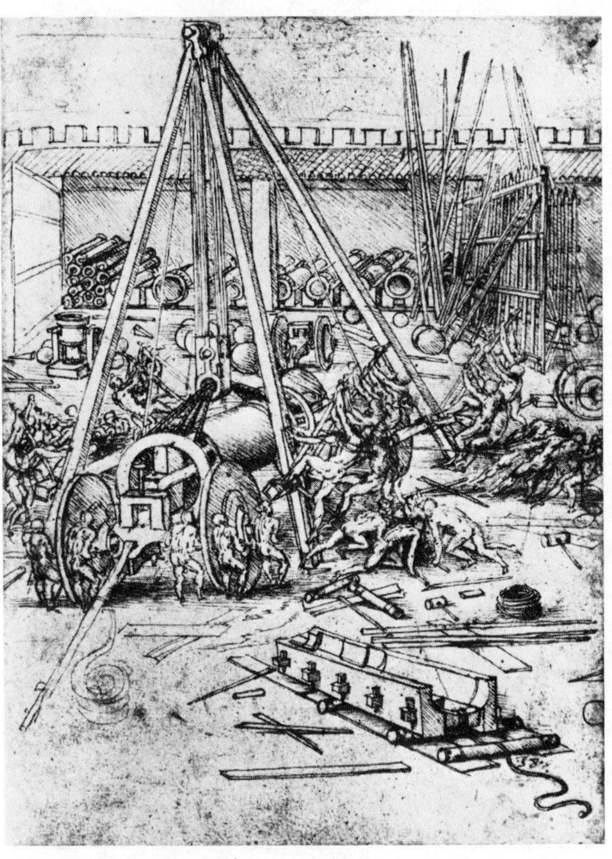

FIGURE 8. "The Arsenal," one of the many studies of the external powers of the human body, here being exerted to create a power transformation into the projectile force of a cannon (*Windsor Collection*, fol. 12647. Reproduced by gracious permission of Her Majesty Queen Elizabeth II).

Leonardo saw the land of the world as emerging from its surrounding element of water by a process of growth. Mountains, the skeleton of the world, were formed by destructive rain, frost, and snow washing the soft earth down into the rivers. Where large landslides occurred, inland seas were formed, from which the waters eventually broke through gorges to reach the sea, their "natural" level. No better summary of Leonardo's neptunist concept of these geological changes can be found than the background of his "Mona Lisa."

Biology. From about 1489 to 1500 Leonardo investigated the physiology of vision and the external powers of the human body. During this period he applied the principles of the "four powers"—movement, force, weight, and percussion—to every conceivable human activity. Many of the drawings of men in action found in the *Treatise on Painting* date from

 LEONARDO DA VINCI

this period, during which he also completed a book, now lost, on the human figure. He drew men sitting, standing, running, digging, pushing, pulling, and so on, all with such points of reference as the center of gravity or the leverage of the trunk and limbs. These simple movements he elaborated into studies of men at work and so to the transformation of human power into machines at work, the field of many of his technological inventions (Figure 8). (For Leonardo's development of this field, see the section on technology.)

Anatomy. Leonardo later investigated the internal powers of the human body. He described his approach to human anatomy and physiology in a passage headed "On Machines":

> Why Nature cannot give movement to animals without mechanical instruments is shown by me in this book, *On the Works of Motion Made by Nature in Animals*. For this I have set out the rules on the 4 powers of nature without which nothing can through her give local motion to these animals. Therefore we shall first describe this local motion, and how it produces and is produced by each of the other three powers. . . .

He then briefly defined these powers (*Quaderni d'anatomia*, I, fol. 1r) and added an admonition: "Arrange it so that the book of the elements of mechanics with examples shall precede the demonstration of the movement and force of man and other animals, and by means of these you will be able to prove all your propositions" (*Anatomical Folio A*, fol. 10r).

As usual, Leonardo gave intense thought to his methods of approaching the problem of man the machine. "I shall describe the function of the parts from every side, laying before your eyes the knowledge of the whole healthy figure of man in so far as it has local motion by means of its parts" (*Quaderni d'anatomia*, I, fol. 2r). He applied his gift of visual artistry and his concepts of the four powers to the human body with results that remain unique. The mechanics of the musculoskeletal system were displayed and explained, such complex movements as pronation and supination being correctly analyzed and demonstrated.

In his anatomical writings Leonardo, always fastidious, noted his intense dislike of "passing the night hours in the company of corpses, quartered and flayed, and horrible to behold."

Human Anatomy. On several occasions Leonardo laid out comprehensive plans for demonstrating the anatomy of the human body. In all of them he put great stress on the necessity for presenting the parts of the body from all sides. He stated, moreover, that each part must be dissected specifically to demonstrate vessels, nerves, or muscles:

> . . . you will need three [dissections] in order to have a complete knowledge of the veins and arteries, destroying all the rest with very great care; three others for a knowledge of the membranes, three for the nerves, muscles and ligaments, three for the bones and cartilages . . . . Three must also be devoted to the female body, and in this there is a great mystery by reason of the womb and its fetus.

Leonardo thus hoped to reveal "in fifteen entire figures . . . the whole figure and capacity of man in so far as it has local movement by means of its parts" (*Quaderni d'anatomia*, I, fol. 2r). His extant anatomical drawings show that he was in fact able to carry out a large part of this extensive program.

In addition to conventional methods of dissection, Leonardo brought to his study of anatomy his own particular skill in modelmaking. He discovered the true shape and size of the cerebral ventricles through making wax casts of them, and he came to appreciate the actions of the ventricles of the heart and the aortic valves by making casts of the cardiac atria and ventricles and glass models of the aorta.

Since his approach was primarily mechanical, Leonardo regarded the bones of the skeleton as levers and their attached muscles as the lines of force acting upon them. A firm knowledge of these actions could not be reached without a detailed demonstration of the shapes and dimensions of both bones and muscles. About 1510, Leonardo executed for this purpose a series of drawings, which constitute a large part of the collection *Anatomical Folio A*. Here he systematically illustrated all the main bones and muscles of the body, often accompanying the drawings with mechanical diagrams to show their mode of action. In these drawings Leonardo frequently adopted the technique of representing muscles by narrow bands, corresponding to what he called their "lines of force" (Figure 9), thus facilitating demonstration of their mechanical powers. Since an understanding of the mode of articulation of joints is clearly necessary to an appreciation of their movements in leverage, Leonardo illustrated these structures by an exploded view, whereby the joint surfaces are separated and the surrounding tendons of muscles are severed to show their exact lines of action in moving the joints. Such illustrations permitted Leonardo to demonstrate, for example, the subtleties of the movements of the upper cervical spine and the shoulder joint.

Possibly because the visual sense was so overwhelmingly important to him, Leonardo appears to have

LEONARDO DA VINCI

made his earliest anatomical studies on the optic nerves, cranial nerves, spinal cord, and peripheral nervous system. In the course of these studies he traced back the optic nerves, beautifully illustrating the optic chiasma and the optic tracts. In his attempts to elucidate the central distribution of sensory tracts in the brain, he injected the cerebral ventricles with wax, thereby ascertaining and demonstrating their approximate shape, size, and situation. On this discovery he based his own mechanical theory of sensation through percussion.

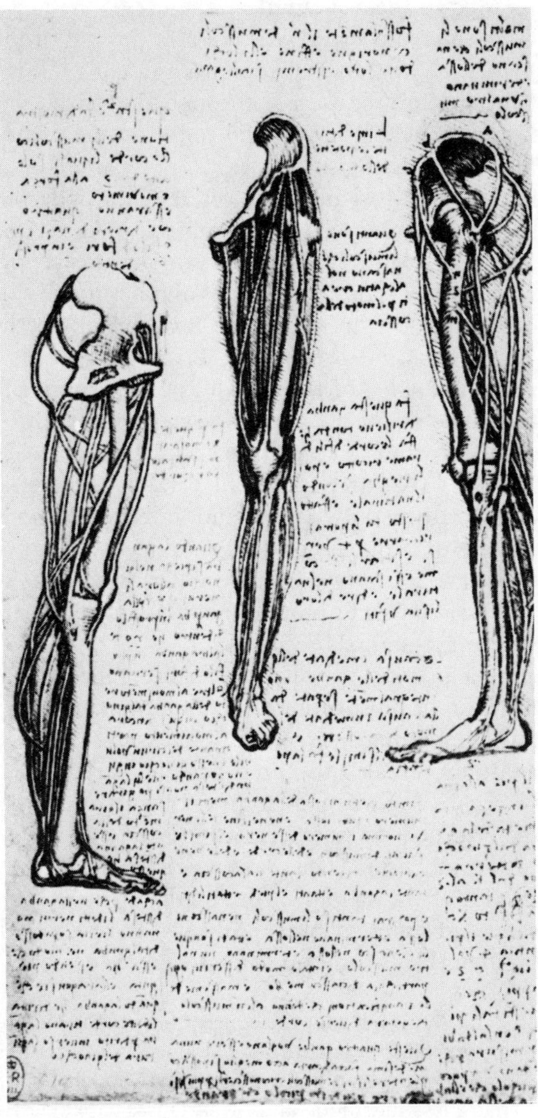

FIGURE 9. The muscles of the leg, one of the many studies of the internal powers of the human body; here the accurately dissected muscles are represented by their "central lines of force," showing their mechanical leverage on the joints (*Quaderni d'anatomia*, V, fol. 4r. Reproduced by gracious permission of Her Majesty Queen Elizabeth II).

Leonardo conceived the eye itself to be a camera obscura in which the inverted image, formed by light penetrating the pupil, is reinverted by the action of the lens. By constructing a glass model of the eye and lens, into which he inserted his own head with his own eye in the position corresponding to that of the optic nerve, he came to the conclusion that the optic nerve was the sensitive visual receptor organ. Here his own experimental brilliance deceived him.

On pithing a frog he noted, "The frog instantly dies when the spinal cord is pierced, and previous to this it lived without head, heart, bowels, or skin. Here therefore it would seem lies the foundation of movement and life" (*Quaderni d'anatomia*, V, fol. 21v). He here contradicted both Aristotelian and Galenic concepts, and opened the way for his own mechanical theory of sensation and movement.

Leonardo's interest in the alimentary tract arose with his dissection of the "old man" in the Hospital of Santa Maria Nuova in Florence. From this dissection he made drawings of the esophagus, stomach, liver, and gallbladder, and drew a first approximation of the distribution of the coils of small and large intestines in the abdomen. Perhaps most notable of all was a detailed sketch of the appendix (its first known representation) accompanied by the hypothesis that it served to take up superfluous wind from the bowel. The heat necessary for digestive coction of the food, Leonardo held, was derived from the heat of the heart—not from the liver, as the Galenists stated. The movement of food and digested products down the bowel was, in Leonardo's view, brought about by the descent of the diaphragm and the pressure produced by contraction of the transverse muscles of the abdomen. Since he did not vivisect, he did not observe peristalsis. Always aware of the necessity of feedback mechanisms to preserve the equilibrium of life, Leonardo was greatly concerned with the fate of superfluous blood, since blood was thought to be continuously manufactured by the liver. Superfluous blood, Leonardo thought, was dispersed through the bowel, contributing largely to the formation of the feces.

Leonardo understood in principle the mechanism of respiration, clearly described by Galen. Describing the movements of the ribs, he wrote, ". . . since there is no vacuum in nature the lung which touches the ribs from within must necessarily follow their expansion; and the lung therefore opening like a pair of bellows draws in the air" (*Anatomical Folio A*, fol. 15v). He considered, however, that the most powerful muscle of inspiration is the diaphragm, the "motor of food and air," as he called it, since by its descent it draws in air and presses food down the gut. After a long

LEONARDO DA VINCI

debate Leonardo decided that expiration is mostly passive, induced by the diaphragm rising with contraction of the abdominal muscles, and subsidence of the ribs.

Leonardo's studies of the heart and its action occupy more anatomical sheets than those of any other organ. By tying off the atria and ventricles and injecting them with air (later wax), he obtained an accurate idea of their shape and volume. He thus came to recognize that the atria are the contracting chambers of the heart which propel blood into the ventricles, a discovery to which he devoted pages of discussion and verification because it was entirely opposed to Galenic physiology. He applied the same technique to the root of the aorta to discover the aortic sinuses, subsequently named after the anatomist Antonio Maria Valsalva. Having established the shape and position of the valve cusps, Leonardo applied the same marker technique which he used so often in his studies of rivers and water currents. To a glass model of the aorta, containing the aortic valve ring and cusps of an ox, he attached the ventricle (or a bag representing it), which he squeezed so that water passed through the valve. The water contained the seeds of panic grass, which served to demonstrate the directional flow of the currents passing the valve cusps. From these experiments Leonardo showed that the aortic valve cusps close in vertical, not horizontal, apposition—that is, from the side, by pressure of eddying currents, not from above, by direct reflux. Although much modern evidence confirms this view, such action has not yet been directly visualized through angiocardiography.

Other features of Leonardo's study of the anatomy of the heart include his detailed demonstration of the coronary vessels and of the moderator band which some have named after him. By observation of the movements of the "spillo," the instrument that slaughterers used to pierce the heart of a pig, Leonardo deduced the relation of systole to the production of the pulse and apex beat. He considered that the force of cardiac percussion propelling the blood was exhausted by the time the blood reached the periphery of the body, however, and thus missed the Harveian concept of its circulation.

Although his drawings of the kidney, ureter, and bladder are relatively sophisticated, Leonardo's mechanistic physiology did not take him far beyond Galen's views on urology. He saw the kidney as acting as a kind of filter, as did Vesalius after him. His representation of the uterus as a single cavity marks a great advance, particularly in his drawing of a five-month *fetus in utero*, although even in this drawing Leonardo showed the cotyledons of the bovine uterus. His drawings of male and female genitalia that pertain to coitus show an austere emphasis on procreation that is perhaps reflected in his comment: "The act of procreation and the members employed therein are so repulsive that if it were not for the beauty of the faces and the adornments of the actors and the pent-up impulse, nature would lose the human species" (*Anatomical Folio A*, fol. 10r).

Comparative Anatomy. Although Leonardo concentrated his anatomical investigations on the human body, he by no means confined them to that field. From the artistic point of view, he was equally interested in the structural mechanics of the horse; and from the point of view of the ever-present problem of flying, the anatomy of the bird and bat took priority.

There are scattered through Leonardo's extant notebooks illustrations and notes on the anatomy of horses, birds, bats, oxen, pigs, dogs, monkeys, lions, and frogs. Most of these were studies undertaken to solve particular "power" problems. His anatomy of the horse, however, forms an important exception, for Lomazzo, Vasari, and Rubens all refer to the existence of Leonardo's "book on the anatomy of the horse," which has since been lost. Possible remnants of it remain in some of the drawings of the proportions of the horse in the Royal Library, Windsor, the *Codex Huygens*, and the musculoskeletal sketches in MS *K*, folios 102r and 109v. But perhaps the best-known comparative study of the hind limbs of the horse and the legs of man has come down to us in his drawing on *Quaderni d'anatomia*, V, fol. 22r, where he wrote: "Show a man on tiptoe so that you may compare man with other animals."

In a number of cases Leonardo repeated Galen's mistakes in substituting animal for human parts. These mistakes diminished as his increasing knowledge of anatomy revealed to him surprising variation of form in organs serving similar physiological functions. He then turned to such variations to gain knowledge of the function. Such was the basis of his numerous studies of the shapes of the pupils of the eye in men, cats, and different birds. And it was by dissection of the lion that Leonardo came to say:

> I have found that the constitution of the human body among all the constitutions of animals is of more obtuse and blunt sensibilities, and so is formed of an instrument less ingenious and of parts less capable of receiving the power of the senses. I have seen in the leonine species how the sense of smell forming part of the substance of the brain descends into a very large receptacle to meet the sense of smell. . . . The eyes of the leonine species have a great part of the head as their receptacle, so that the optic nerves are in immediate conjunction with the brain. With man the contrary is

LEONARDO DA VINCI

seen to be the case, for the cavities of the eye occupy a small part of the head, and the optic nerves are thin, long and weak [*Anatomical Folio B*, fol. 13v].

That Leonardo appreciated the importance of these studies in comparative anatomy is evinced in his note "Write of the varieties of the intestines of the human species, apes, and such like; then of the differences found in the leonine species, then the bovine, and lastly the birds; and make the description in the form of a discourse" (*Anatomical Folio B*, fol. 37r). His awareness of homologous structures was most pronounced in relation to the limbs, the form and power of which in both man and animals were of outstanding artistic and scientific importance to him. "Anatomize the bat, study it carefully and on this construct your machine," he enjoins on MS *F*, folio 41v. Comparing the arm of man and the wing of the bird, he pointed out: "The sinews and muscles of a bird are incomparably more powerful than those of man because the whole mass of so many muscles and of the fleshy parts of the breast go to aid and increase the movement of the wings, while the breastbone is all in one piece and consequently affords the bird very great power" (*Codex on Flight . . .*, fol. 16r). To elucidate this problem he embarked on a comparative mechanical anatomy of the proportions of the wings of the bat, the eagle, and the pelican, and the arm of man. He even reduced the forelimb to a three-jointed model of levers representing humerus, forearm, and hand, manipulating this artificial limb by pulley mechanisms. He was clearly using comparative anatomy as a means of solving the great problem of human flight.

Such comparisons led Leonardo to suggest repeatedly that he should represent the hands and feet of "the bear, monkey, and certain birds" in order to see how they differ from those of man. Examples of such completed work are scattered in the notes.

From such studies Leonardo came to realize the anatomical significance of "the movements of animals with four feet, amongst which is man, who likewise in his infancy goes on four feet, and who moves his limbs crosswise, as do other four-footed animals, for example the horse in trotting" (MS *E*, fol. 16r; *Codex Atlanticus*, fol. 297r).

From such steps Leonardo reached the generalization: "All terrestrial animals have a similarity of their parts, that is their muscles, nerves and bones, and these do not vary except in length and size, as will be demonstrated in the book of Anatomy. . . . Then there are the aquatic animals which are of many varieties, concerning which I shall not persuade the painter that there is any rule, since they are of almost infinite variety, as are insects" (MS *G*, fol. 5v). "Man differs from animals," concludes Leonardo, "only in what is accidental, and in this he is divine. . . ." (*Anatomical Folio B*, fol. 13v). After referring to skill in drawing, knowledge of the geometry of the four powers, and patience as being necessary for completing his anatomical researches, he finally wrote of his work: "Considering which things whether or no they have all been found in me, the hundred and twenty books which I have composed will give their verdict, yes or no. In these I have not been hindered either by avarice or negligence, but only by want of time. Farewell" (*Quaderni d'anatomia*, I, fol. 13r).

Botany. Leonardo's botanical studies developed, as it were, in miniature along lines similar to those of his anatomy, with the important difference that in this field he had no predecessors like Galen and Mondino to aid (or impede) his personal observations. These studies commenced in his early years with representation of the external forms of flowering plants, and it was not long before evidence appeared of his interest in plant physiology. About 1489, on a page containing two of his most beautiful designs for cathedrals, he presented a series of four botanical diagrams, one being described by the words: "If you strip off a ring of bark from the tree, it will wither from the ring upwards and remain alive from there downwards" (MS *B*, fol. 17v). Experiments and observations on the movement of sap and growth of plants and trees continued. They reached a climax about 1513, when he was in Rome. Many of these are gathered together in MS *G* and book VI of the *Treatise on Painting*. For many years Leonardo saw movement of sap in the plant as analogous to the movement of blood in the animal and water in the living earth. It is drawn upward by the heat of the sun and downward by its own "natural" weight:

> Heat that is poured into animated bodies moves the humours which nourish them. The movement made by this humour is the conservation of itself. The same cause moves the water through the spreading veins of the earth as that which moves the blood in the human species. . . . In the same way so does the water that rises from the low roots of the vine to its lofty summit and falling through the severed branches upon the primal roots mounts anew [*Codex Arundel*, fols. 234r, 235].

Incidentally, he postulated this same mechanism for the absorption of food from the human intestine by the portal veins.

Leonardo described an experiment with a gourd:

> The sun gives spirit and life to plants and the earth nourishes them with moisture. In this connection I once made an experiment of leaving only one very small

 LEONARDO DA VINCI

> root on a gourd and keeping it nourished with water and the gourd brought to perfection all the fruits that it could produce, which were about sixty. And I set myself diligently to consider the source of its life; and I perceived that it was the dew of the night which penetrated abundantly with its moisture through the joints of its great leaves to nourish this plant with its offspring, or rather seeds which have to produce its offspring. . . . The leaf serves as a nipple or breast to the branch or fruit which grows in the following year [MS *G*, fol. 32v].

Thus Leonardo reached a concept of the necessity of sunlight "giving spirit and life" through its leaves to a plant, while the moisture of its sap came from both root and leaf. He also perceived upward and downward movement of sap. He located the cambium when he stated: "The growth in thickness of trees is brought about by the sap which in the month of April is created between the 'camicia' [cambium] and the wood of the tree. At that time this cambium is converted into bark and the bark acquires new cracks" (*Treatise on Painting*, pt. VI, McMahon ed., p. 893).

Appreciating this seasonal growth of wood from the cambium, Leonardo came to recognize its part in the annual ring formation in trees. He wrote: "The age of trees which have not been injured by men can be counted in years by their branching from the trunk. The trees have as many differences in age as they have principal branches. . . . The circles of branches of trees which have been cut show the number of their years, and also show which years were wetter or drier according to their greater or lesser thickness . . ." (*ibid.*, p. 900). In these words Leonardo showed himself to be the originator of dendrochronology.

Leonardo's observation on the growth of the tree trunk and the patterns of its branches and leaves, combined with his views on plant nutrition, initiated his study of phyllotaxy:

> The lower trunks of trees do not keep their roundness of size when they approach the origin of their branches or roots. And this arises because the higher and lower branches are the organs [membra] by which the plants (or trees) are nourished; that is to say, in summer they are nourished from above by the dew and rain through the leaves and in winter from below through the contact which the earth has with their roots. . . . Larger branches do not grow toward the middle of the tree. This arises because every branch naturally seeks the air and avoids shadow. In those branches which turn to the sky, the course of the water and dew descends . . . and keeps the lower part more humid than the upper, and for this reason the branches have more abundant nourishment there, and therefore grow more [*ibid.*, p. 885].

Thus the trunk of the tree, being most nourished, is the thickest part of the tree. Leonardo saw here the mechanical implications of the force of winds on trees: "The part of the tree which is farthest from the force which percusses it is most damaged by this percussion because of its greater leverage. Thus nature has provided increased thickness in that part, and most in such trees as grow to great heights like pines" (*Codex Arundel*, fol. 277v). Leonardo asserted that the branch pattern of a plant or tree follows its leaf pattern: "The growth of the branches of trees on their principal trunk is like that of the growth of their leaves, which develop in four ways, one above the other. The first and most general is that the sixth leaf above always grows above the sixth below; and the second is for the third pair of leaves to be above the third pair below; the third way is that for the third leaf to be above the third below" (*Treatise on Painting*, McMahon ed., p. 889). This whole pattern of trunk, branch, and leaf Leonardo saw as designed to catch maximum sun, rain, and air—the leaves and branches were arranged "to leave intervals which the air and sun can penetrate, and drops that fall on the first leaf can also fall on the fourth and sixth leaves, and so on" (*ibid.*).

"All the flowers that see the sun mature their seed and not the others, that is, those that see only the reflection of the sun" (MS *G*, fol. 37v). And of seeds he wrote, "All seeds have the umbilical cord which breaks when the seed is ripe; and in like manner they have matrix [uterus] and secundina [membranes], as is shown in herbs and seeds that grow in pods" (*Quaderni d'anatomia*, III, fol. 9v). These words, written on a page of drawings of the infant in the womb, vividly reveal Leonardo's integrating mode of thought.

KENNETH D. KEELE

TECHNOLOGY

A statistical analysis of his extant writings suggests that technology was the most important of Leonardo's varied interests. Indeed, it is revealing to compare the volume of his technological writings with that of his purely artistic work. Of his paintings, fewer than ten are unanimously authenticated by art scholars. This evident disinclination to paint, which even his contemporaries remarked upon, contrasts strongly with the incredible toil and patience that Leonardo lavished upon scientific and technical studies, particularly in geometry, mechanics, and engineering.

Documentary evidence indicates that in his appointments Leonardo was always referred to not only as an artist but also as an engineer. At the court of Ludovico il Moro he was "Ingeniarius et pinctor," while Cesare Borgia called him his most beloved

"Architecto et Engegnero Generale" (1502). When Leonardo returned to Florence, he was immediately consulted as military engineer and sent officially "a livellare Arno in quello di Pisa e levallo del letto suo" (1503).[1] In 1504 he was in Piombino, in the service of Jacopo IV d'Appiano, working on the improvement of that little city-state's fortifications.[2] For Louis XII he was "notre chier et bien aimé Léonard de Vinci, notre paintre et ingénieur ordinaire" (1507).[3] When in Rome, from 1513 to 1516, his duties clearly included technical work, as documented by drafts of letters to his patron Giuliano de' Medici, found in the *Codex Atlanticus*. Even his official burial document refers to him as "Lionard de Vincy, noble millanois, premier peinctre et ingenieur et architecte du Roy, mescanichien d'Estat . . ." (1519).[4] The surviving notebooks and drawings demonstrate Leonardo's lifelong interest in the mechanical arts and engineering.

Leonardo's scientific and technological activities were well known to his early biographers, even if they did not approve of them. Paolo Giovio's short account on Leonardo's life (*ca.* 1527) contains a significant phrase: "But while he was thus spending his time in the close research of subordinate branches of his art he carried only very few works to completion."

Another biographer, the so-called "Anonimo Gaddiano" or "Magliabechiano," writing around 1540, said that Leonardo "was delightfully inventive, and was most skillful in lifting weights, in building waterworks and other imaginative constructions, nor did his mind ever come to rest, but dwelt always with ingenuity on the creation of new inventions."

Vasari's biography of Leonardo, in his *Lives of the Painters, Sculptors and Architects* (1550; 2nd ed., 1568), reflects the widespread sentiments of his contemporaries who were puzzled by the behavior of a man who, unconcerned with the great artistic gifts endowed upon him by Providence, dedicated himself to interesting but less noble occupations.

Vasari's testimony concerning Leonardo's widespread technological projects (confirmed by Lomazzo) is important in assessing the influence of Leonardo on the development of Western technology:

> He would have made great profit in learning had he not been so capricious and fickle, for he began to learn many things and then gave them up . . . he was the first, though so young, to propose to canalise the Arno from Pisa to Florence. He made designs for mills, fulling machines, and other engines to go by water. . . . Every day he made models and designs for the removal of mountains with ease and to pierce them to pass from one place to another, and by means of levers, cranes and winches to raise and draw heavy weights; he devised a method for cleansing ports, and to raise water from great depths, schemes which his brain never ceased to evolve. Many designs for these motions are scattered about, and I have seen numbers of them. . . . His interests were so numerous that his inquiries into natural phenomena led him to study the properties of herbs and to observe the movements of the heavens, the moon's orbit and the progress of the sun.

(It is characteristic of Leonardo's contemporary critics that Vasari, giving an account of Leonardo's last days, represents him as telling Francis I "the circumstances of his sickness, showing how greatly he had offended God and man in not having worked in his art as he ought.")

Lomazzo, in his *Trattato della pittura* (Milan, 1584), tells of having seen many of Leonardo's mechanical projects and praises especially thirty sheets representing a variety of mills, owned by Ambrogio Figino, and the automaton in the form of a lion made for Francis I. In his *Idea del tempio della pittura* (Milan, 1590), Lomazzo mentions "Leonardo's books, where all mathematical motions and effects are considered" and of his "projects for lifting heavy weights with ease, which are spread over all Europe. They are held in great esteem by the experts, because they think that nobody could do more, in this field, than what has been done by Leonardo." Lomazzo also notes "the art of turning oval shapes with a lathe invented by Leonardo," which was shown by a pupil of Melzi to Dionigi, brother of the Maggiore, who adopted it with great satisfaction.

Leonardo's actual technological investigations and work still await an exhaustive and objective study. Many early writers accepted all the ingenious mechanical contrivances found in the manuscripts as original inventions; their claims suffer from lack of historical perspective, particularly as concerns the work of the engineers who preceded Leonardo. The "inventions" of Leonardo have been celebrated uncritically, while the main obstacle to a properly critical study lies in the very nature of the available evidence, scattered and fragmented over many thousands of pages. Only in very recent times has the need for a chronological perspective been felt and the methods for its adoption elaborated.[5] It is precisely the earliest—and for this reason the least original—of Leonardo's projects for which model makers and general authors have shown a predilection. On the other hand, the preference for these juvenile projects is fully justified: they are among the most beautiful and lovingly elaborated designs of the artist-engineer. The drawings and writings of MS *B*, the *Codex Trivulzianus*, and the earliest folios of the *Codex Atlanticus* date from this period (*ca.* 1478–1490). Similar themes

LEONARDO DA VINCI

in the almost contemporary manuscripts of Francesco di Giorgio Martini offer ample opportunity to study Leonardo's early reliance on traditional technological schemes (Francesco di Giorgio himself borrowed heavily from Brunelleschi and especially from Mariano di Jacopo, called Taccola, the "Archimedes of Siena"); the same comparison serves to demonstrate Leonardo's originality and his search for rational ways of constructing better machines.

While he was still in Florence, Leonardo acquired a diversified range of skills in addition to the various crafts he learned in the workshop of Verrocchio, who was not only a painter but also a sculptor and a goldsmith. Leonardo must therefore have been familiar with bronze casting, and there is also early evidence of his interest in horology. His famous letter to Ludovico il Moro, offering his services, advertises Leonardo's familiarity with techniques of military importance, which, discounting a juvenile self-confidence, must have been based on some real experience.

Leonardo's true vocation for the technical arts developed in Milan, Italy's industrial center. The notes that he made during his first Milanese period (from 1481 or 1482 until 1499) indicate that he was in close contact with artisans and engineers engaged in extremely diversified technical activities—with, for example, military and civil architects, hydraulic engineers, millers, masons and other workers in stone, carpenters, textile workers, dyers, iron founders, bronze casters (of bells, statuary, and guns), locksmiths, clockmakers, and jewelers. At the same time that he was assimilating all available traditional experience, Leonardo was able to draw upon the fertile imagination and innate technological vision that, combined with his unparalleled artistic genius in the graphic rendering of the most complicated mechanical devices, allowed him to make improvements and innovations.

From about 1492 on (as shown in MS *A*; *Codex Forster*; *Codex Madrid*, I; MS *H*, and a great number of pages of the *Codex Atlanticus*), Leonardo became increasingly involved in the study of the theoretical background of engineering practice. At about that time he wrote a treatise on "elementi macchinali" that he returned to in later writings, citing it by book and paragraph. This treatise is lost, but many passages in the *Atlanticus* and *Arundel* codices may be drafts for it. *Codex Madrid*, I (1492–1497), takes up these matters in two main sections, one dealing with matters that today would be called statics and kinematics and another dedicated to applied mechanics, especially mechanisms.

Our knowledge of the technical arts of the fourteenth to sixteenth centuries is scarce and fragmentary. Engineers were reluctant to write about their experience; if they did, they chose to treat fantastic possibilities rather than the true practices of their time. The books of Biringuccio, Agricola, and, in part, Zonca are among the very few exceptions, although they deal with specialized technological fields. The notes and drawings of Leonardo should therefore be studied not only to discover his inventions and priorities, as has largely been done in the past, but also—and especially—for the insight they give into the state of the technical arts of his time. Leonardo took note of all the interesting mechanical contrivances he saw or heard about from fellow artists, scholars, artisans, and travelers. Speaking of his own solutions and improvements, he often referred to the customary practices. Thus his manuscripts are among the most important sources for medieval and Renaissance technology. In all other manuscripts and books of machines by authors of the periods both preceding and following Leonardo, projects for complete machines are presented, without any discussion of their construction and efficiency.[6] The only exception previous to the eighteenth-century authors Sturm, Leupold, and Belidor is represented by the work of Simon Stevin, around 1600.

As far as the evidence just mentioned shows, the mechanical engineering of times past was limited by factors of two sorts: various inadequacies in the actual construction of machines produced excessive friction and wear, and there was insufficient understanding of the possibilities inherent in any mechanical system. Leonardo's work deserves our attention as that of the first engineer to try systematically to overcome these shortcomings; most important, he was the first to recognize that each machine was a composition of certain universal mechanisms.

In this, as in several other respects, Leonardo anticipated Leupold, to whom, according to Reuleaux, the foundations of the science of mechanisms is generally attributed.[7] Indeed, of Reuleaux's own list of the constructive elements of machines (screws, keys or wedges, rivets, bearings, plummer blocks, pins, shafts, couplings, belts, cord and chain drives, friction wheels, gears, flywheels, levers, cranks, connecting rods, ratchet wheels, brakes, pipes, cylinders, pistons, valves, and springs), only the rivets are missing from Leonardo's inventories.

In Leonardo's time and even much later, engineers were convinced that the work done by a given prime mover, be it a waterwheel or the muscles of men or animals, could be indefinitely increased by means of suitable mechanical apparatuses. Such a belief led fatally to the idea of perpetual motion machines, on

LEONARDO DA VINCI

whose development an immense amount of effort was wasted, from the Middle Ages until the nineteenth century. Since the possibility of constructing a perpetual motion machine could not, until very recent times, be dismissed by scientific arguments, men of science of the first order accepted or rejected the underlying idea by intuition rather than by knowledge.

Leonardo followed the contemporary trend, and his earliest writings contain a fair number of perpetual motion schemes. But he gave up the idea around 1492, when he stated "It is impossible that dead [still] water may be the cause of its own or of some other body's motion" (MS *A*, fol. 43r), a statement that he later extended to all kinds of mechanical movements. By 1494 Leonardo could say that

> . . . in whatever system where the weight attached to the wheel should be the cause of the motion of the wheel, without any doubt the center of the gravity of the weight will stop beneath the center of its axle. No instrument devised by human ingenuity, which turns with its wheel, can remedy this effect. Oh! speculators about perpetual motion, how many vain chimeras have you created in the like quest. Go and take your place with the seekers after gold! [*Codex Forster*, II_2, fol. 92v].

Many similar statements can be found in the manuscripts, and it is worth noting that Leonardo's argument against perpetual motion machines is the same that was later put forth by Huygens and Parent.

Another belief common among Renaissance engineers was that flywheels (called "rote aumentative") and similar energy-storing and equalizing devices are endowed with the virtue of increasing the power of a mechanical system. Leonardo knew that such devices could be useful, but he also knew that their incorporation into a machine caused an increase in the demand of power instead of reducing it (*Codex Atlanticus*, fol. 207v-b; *Codex Madrid*, I, fol. 124r).

A practical consequence of this line of thought was Leonardo's recognition that machines do not perform work but only modify the manner of its application. The first clear formulation of this was given by Galileo,[8] but the same principle permeates all of Leonardo's pertinent investigations. He knew that mechanical advantage does not go beyond the given power available, from which the losses caused by friction must be deducted, and he formulated the basic concepts of what are known today as work and power. Leonardo's variables for these include force, time, distance, and weight (*Codex Forster*, II_2, fol. 78v; *Codex Madrid*, I, fol. 152r). One of the best examples is folio 35r of *Codex Madrid*, I, where Leonardo compares the performance of two lifting systems; the first is a simple windlass moved by a crank, capable of lifting 5,000 pounds; the second is also moved by a crank, but a worm gear confers upon it a higher mechanical advantage, raising its lifting capacity to 50,000 pounds. Leonardo affirms that operators of both machines, applying twenty-five pounds of force and cranking with the same speed, will have the load of 50,000 pounds raised to the same height at the end of one hour. The first instrument will raise its load in ten journeys, while the second will lift it all at once. The end result, however, will be the same.

In the same codex Leonardo established rules of general validity: "Among the infinite varieties of instruments which can be made for lifting weights, all will have the same power if the motions [distances] and the acting and patient weights are equal" (*Codex Madrid*, I, fol. 175r). Accordingly, "It is impossible to increase the power of instruments used for weight-lifting, if the quantity of force and motion is given" (*ibid.*, fol. 175v).

That Leonardo had an intuitive grasp of the principle of the conservation of energy is shown in many notes dispersed throughout the manuscripts. He tried to measure the different kinds of energy known to him (muscle power, springs, running and falling water, wind, and so forth) in terms of gravity—that is, using dynamometers counterbalanced by weights, anticipating Borelli and Smeaton. He even tried to investigate the energetic equivalent of gunpowder, weighing the propellant and the missile and measuring the range. The missile was then shot from a crossbow spanned with a given weight, which was then correlated with the quantity of gunpowder used in the first experiment (*ibid.*, fol. 60r).

Leonardo was aware that the main impediment to all mechanical motions was friction. He clearly recognized the importance of frictional resistance not only between solid bodies but also in liquids and gases. In order to evaluate the role of friction in mechanical motions, he devised ingenious experimental equipment, which included friction banks identical to those used by Coulomb 300 years later. From his experiments Leonardo derived several still-valid general principles—that frictional resistance differs according to the nature of the surfaces in contact, that it depends on the degree of smoothness of those surfaces, that it is independent of the area of the surfaces in contact, that it increases in direct proportion to the load, and that it can be reduced by interposing rolling elements or lubricating fluids between the surfaces in contact. He introduced the concept of the coefficient of friction and estimated that for "polished and smooth" surfaces the ratio F/P was 0.25, or one-fourth of the weight. This value is

LEONARDO DA VINCI

reasonably accurate for hardwood on hardwood, bronze on steel, and for other materials with which Leonardo was acquainted.[9]

Leonardo's main concern, however, was rolling friction. Realizing that lubrication alone could not prevent rapid wear of an axle and its bearing, Leonardo suggested the use of bearing blocks with split, adjustable bushings of antifriction metal ("three parts of copper and seven of tin melted together"). He was also the first to suggest true ball and roller bearings, developing ring-shaped races to eliminate the loss due to contact friction of the individual balls in a bearing. Leonardo's thrust bearings with conical pivots turning on cones, rollers, or balls (*ibid.*, fol. 101v) are particularly interesting. He also worked persistently to produce gearings designed to overcome frictional resistance. Even when they are not accompanied by geometrical elaborations, some of his gears are unmistakably cycloidal. Leonardo further introduced various new gear forms, among them trapezoidal, helical, and conical bevel gears; of particular note is his globoidal gear, of which several variants are found in the *Codex Atlanticus* and *Codex Madrid*, I, one of them being a worm gear shaped to match the curve of the toothed wheel it drives, thus overcoming the risk inherent in an endless screw that engages only a single gear tooth (*ibid.*, fols. 17v-18v). This device was rediscovered by Henry Hindley around 1740.

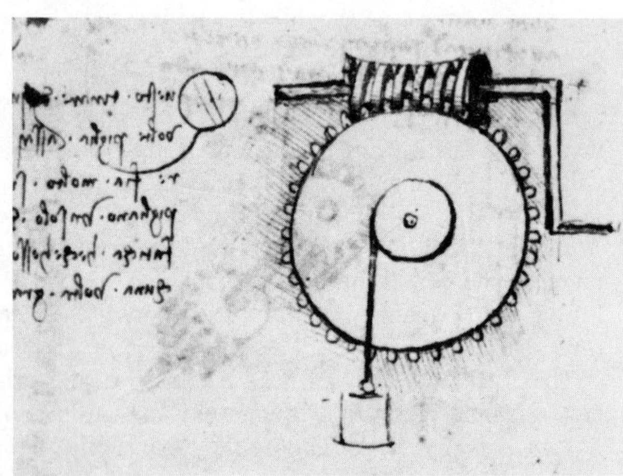

FIGURE 1. Worm gear, shaped to match the curve of the toothed wheel it drives. Designed by Leonardo in 1495 (*Codex Madrid*, I, 17v). It was reinvented by the English engineer Henry Hindley, *ca.* 1740.

Leonardo's development of complicated gear systems was not motivated by any vain hope of obtaining limitless mechanical advantages. He warned the makers of machines:

> The more wheels you will have in your instrument, the more toothing you will need, and the more teeth, the greater will be the friction of the wheels with the spindles of their pinions. And the greater the friction, the more power is lost by the motor and, consequently, force is lacking for the orderly motion of the entire system [*Codex Atlanticus*, fol. 207v-b].

Leonardo's contribution to practical kinematics is documented by the devices sketched and described in his notebooks. Since the conversion of rotary to alternating motion (or vice versa) was best performed with the help of the crank and rod combinations, Leonardo sketched hundreds of them to illustrate the kinematics of such composite machines as sawmills, pumps, spinning wheels, grinding machines, and blowers. In addition he drew scores of ingenious combinations of gears, linkages, cams, and ratchets, designed to transmit and modify mechanical movements. He used the pendulum as an energy accumulator in working machines as well as an escapement in clockwork (*Codex Madrid*, I, fol. 61v).

Although simple cord drives had been known since the Middle Ages, Leonardo's belt techniques, including tightening devices, must be considered as original. His manuscripts describe both hinged link chains and continuous chain drives with sprocket wheels (*ibid.*, fol. 10r; *Codex Atlanticus*, fol. 357r-a).

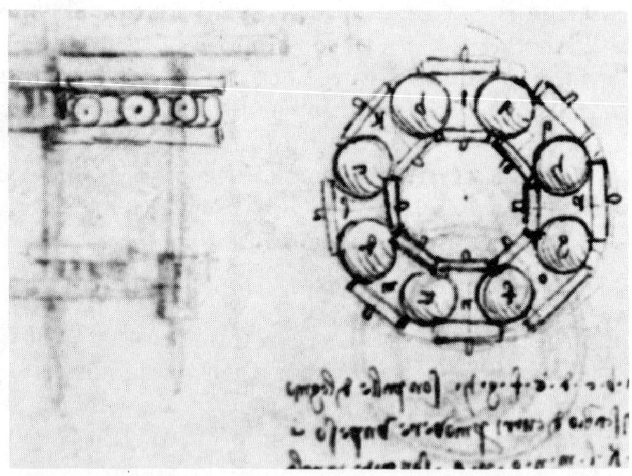

FIGURE 2. Antifriction balls designed by Leonardo for a ball bearing. Reinvented in 1920 (*Codex Madrid*, I, 101v).

Leonardo's notes about the most efficient use of prime movers deserve special attention. His particular interest in attaining the maximum efficiency of muscle power is understandable, since muscle power represented the only motor that might be used in a flying machine, a project that aroused his ambition as early as 1487 and one in which he remained interested until the end of his life. Since muscles were also the most common source of power, it was further important to establish the most effective ways to use them in performing work.

Leonardo estimated the force exerted by a man turning a crank as twenty-five pounds. (Philippe de La Hire found it to be twenty-seven pounds, while Guillaume Amontons, in similar experiments, confirmed Leonardo's figure; in 1782 Claude François Berthelot wrote that men cannot produce a continuous effort of more than twenty pounds, even if some authors admitted twenty-four.)[10] Such a return seemed highly unsatisfactory. Leonardo tried to find more suitable mechanical arrangements, the most remarkable of which employ the weight of men or animals instead of muscle power. For activating pile drivers (*Codex Leicester*, fol. 28v [*ca.* 1505]) or excavation machines (*Codex Atlanticus*, fols. 1v-b, 331v-a), Leonardo used the weight of men, who by running up ladders and returning on a descending platform, would raise the ram or monkey. Leonardo used the same system for lifting heavier loads with cranes, the counterweight being "one ox and one man"; lifting capacity was further increased by applying a differential windlass to the arm of the crane (*ibid.*, fol. 363v-b [*ca.* 1495]).

Until the advent of the steam engine the most popular portable prime mover was the treadmill, known since antiquity. Leonardo found the conventional type, in which men walk inside the drum, in the manner of a squirrel cage, to be inherently less efficient than one employing the weight of the men on the outside of the drum. While he did not invent the external treadmill, he was the first to use it rationally—the next scholar to analyze the efficiency of the treadmill mathematically was Simon Stevin (1586).

Leonardo also had very clear ideas about the advantages and the limitations of waterpower. He rejected popular hydraulic perpetual motion schemes, "Falling water will raise as much more weight than its own as is the weight equivalent to its percussion. . . But you have to deduce from the power of the instrument what is lost by friction in its bearings" (*ibid.*, fol. 151r-a). Since the weight of the percussion, according to Leonardo, is proportional to height, and therefore to gravitational acceleration ("among natural forces, percussion surpasses all others . . . because at every stage of the descent it acquires more momentum"), this represents the first, if imperfect, statement of the basic definition of the energy potential $Ep = mgh$.

Leonardo describes hydraulic wheels on many pages of his notebooks and drawings, either separately or as part of technological operations. He continually sought improvements for systems currently in use. He evaluated all varieties of prime movers, vertical as well as horizontal, and improved on the traditional Lombard mills by modifying the wheels and their races and introducing an adjustable wheel-raising device (MS *H*; *Codex Atlanticus*, fols. 304r-b, v-d [*ca.* 1494]).

In 1718 L. C. Sturm described a "new kind" of mill constructed in the Mark of Brandenburg, "where a lot of fuss was made about them, although they were not as new as most people in those parts let themselves be persuaded. . . ." They were, in fact, identical to those designed by Leonardo for the country estate of Ludovico il Moro near Vigevano around 1494.

It is noteworthy that in his mature technological projects Leonardo returned to horizontal waterwheels for moving heavy machinery (*Codex Atlanticus*, fol. 2r-a and b [*ca.* 1510]), confirming once more the high power output of such prime movers. His papers provide forerunners of the reaction turbine (*Codex Forster*, I_2) and the Pelton wheel (*Codex Madrid*, I, fol. 22v); drawings of completely encased waterwheels appear on several folios of the *Codex Atlanticus*.

There is little about wind power in Leonardo's writings, probably because meteorological conditions limited its practical use in Italy. Although it is erroneous to attribute the invention of the tower mill to Leonardo (as has been done), the pertinent sketches are significant because they show for the first time a brake wheel mounted on the wind shaft (MS *L*, fol. 34v). (The arrangement reappears, as do many other ideas of Leonardo's, in Ramelli's book of 1588 [plate cxxxiii].) Windmills with rotors turning on a vertical shaft provided with shield walls are elaborated on folios 43v, 44r, 74v, 75r, and 55v of *Codex Madrid*, II; and there can be no doubt that Leonardo became acquainted with them through friends who had seen them in the East.

In contrast with Leonardo's scant interest in wind power, he paid constant attention to heat and fire as possible sources of energy. His experiments with steam are found on folios 10r and 15r of the *Codex Leicester*. His approximate estimation of the volume of steam evolved through the evaporation of a given quantity of water suggests a ratio 1 : 1,500, the correct figure being about 1 : 1,700. Besson in 1569 still believed that the proportion was 1 : 10, a ratio raised to 1 : 255 in the famous experiments of Jean Rey; it was not until 1683 that a better estimate—1 : 2,000—was made by Samuel Morland. Leonardo's best-known contribution to the utilization of steam power was his "Architronito" (MS *B*, fol. 33r), a steam cannon. The idea is not as impractical as

LEONARDO DA VINCI

generally assumed, since steam cannons were used in the American Civil War and even in World War II (Holman projectors). It was Leonardo, and not Branca (1629), who described the first impulse turbine moved by a jet of steam (*Codex Leicester*, fol. 28v).

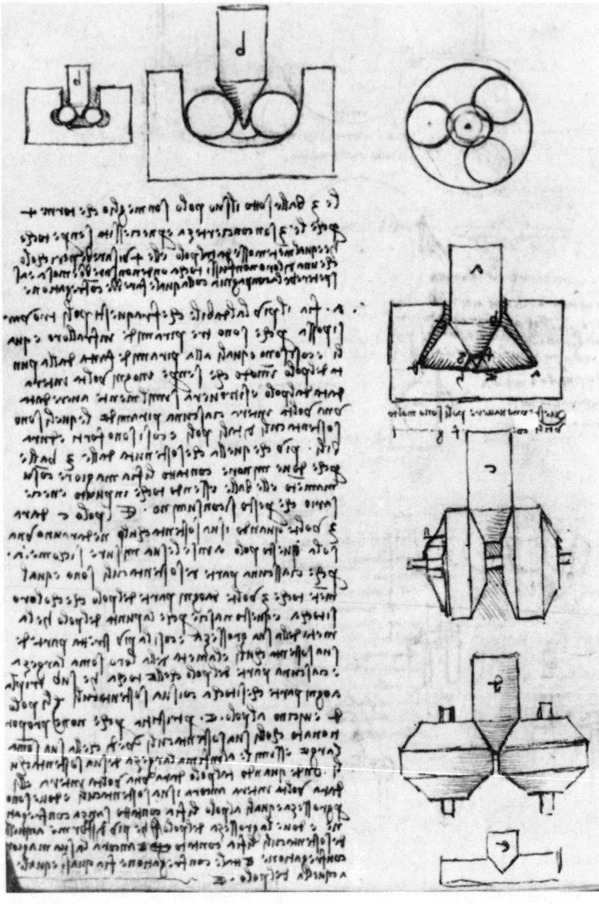

FIGURE 3. Ring-shaped race for a ball bearing with independent separation devices, *ca.* 1494 (*Codex Madrid*, I, 20v). Reinvented in the eighteenth century.

One of Leonardo's most original technological attempts toward a more efficient prime mover is the thermal engine drawn and described on folio 16v of MS *F* (1508–1509), which anticipates Huygens' and Papin's experiments. In 1690 Papin arrived at the idea of the atmospheric steam engine after Huygens' experiments with a gunpowder engine (1673) failed to give consistent results. Leonardo's atmospheric thermal motor, conceived "to lift a great weight by means of fire," like those of Huygens and Papin, consisted of cylinder, piston, and valve, and worked in exactly the same way.[11]

Leonardo's studies on the behavior and resistance of materials were the first of their kind. The problem was also attacked by Galileo, but an adequate treatment of the subject had to wait until the eighteenth century. One of Leonardo's most interesting observations was pointed out by Zammattio and refers to the bending of an elastic beam, or spring. Leonardo recognized clearly that the fibers of the beam are lengthened at the outside of the curvature and shortened at the inside. In the middle, there is an area which is not deformed, which Leonardo called the "linea centrica" (now called the neutral axis). Leonardo suggested that similar conditions obtained in the case of ropes bent around a pulley and in single as well as intertwisted wires (*Codex Madrid*, I, fols. 84v, 154v). More than two centuries had to pass before this model of the internal stresses was proposed again by Jakob Bernoulli.

Leonardo's notebooks also contain the first descriptions of machine tools of some significance, including plate rollers for nonferrous metals (MS *I*, fol. 48v), rollers for bars and strips (*Codex Atlanticus*, fol. 370v-b; MS *G*, fol. 70v), rollers for iron staves (*Codex Atlanticus*, fol. 2r-a and b), semiautomatic wood planers (*ibid.*, fol. 38v-b), and a planer for iron (*Codex Madrid*, I, fol. 84v). His thread-cutting machines (*Codex Atlanticus*, fol. 367v-a) reveal great ingenuity, and their principle has been adopted for modern use.

Not only did Leonardo describe the first lathe with continuous motion (*ibid.*, *fol.* 381r-b) but, according to Lomazzo, he must also be credited with the invention of the elliptic lathe, generally attributed to Besson (1569). Leonardo described external and internal grinders (*ibid.*, fols. 7r-b, 291r-a) as well as disk and belt grinding machines (*ibid.*, fols. 320r-b, 380v-b, 318v-a). He devoted a great deal of attention to the development of grinding and polishing wheels for plane and concave mirrors (*ibid.*, fols. 32r-a, 396v-f, 401r-a; *Codex Madrid*, I, fol. 163v, 164r; MS *G*, fol. 82v, for examples). Folio 159r-b of the *Codex Atlanticus* is concerned with shaping sheet metal by stamping (to make chandeliers of two or, better, four parts). The pressure necessary for this operation was obtained by means of a wedge press.

Leonardo's interest in water mills has already been mentioned. Of his work in applied hydraulics, the improvement of canal locks and sluices (miter gates and wickets) is an outstanding example. He discussed the theory and the practice of an original type of centrifugal pump (MS *F*, fols. 13r, 16r) and the best ways of constructing and moving Archimedean screws. Some of these methods were based on coiled pipes (MS *E*, fols. 13v, 14r) like those seen by Cardano at the waterworks of Augsburg in 1541.

Leonardo left plans for a great number of machines which were generally without parallel until the eighteenth and nineteenth centuries. Some of these are the improved pile drivers (*Codex Forster*, II, fol. 73v; MS *H*, fol. 80v; *Codex Leicester*, fol. 28v) later described by La Hire (1707) and Belidor (1737), a cylindrical bolter activated by the main drive of a grain mill (*Codex Madrid*, I, fols. 21v, 22r), and a mechanized wedge press (*ibid.*, 46v, 47r), which in the eighteenth century became known as the Dutch press. Particularly original are Leonardo's well-known canal-building machines, scattered through the *Codex Atlanticus*—in Parsons' opinion, "had Leonardo contributed nothing more to engineering than his plans and studies for the Arno canal, they alone would place him in the first rank of engineers for all time." The knowledge that most of those projects were executed in the Romagna during Leonardo's service with Cesare Borgia makes little difference.

Leonardo also designed textile machinery. Plans for spinning machines, embodying mechanical principles which did not reappear until the eighteenth century, are found on folio 393v-a of the *Codex Atlanticus*, as well as in the *Codex Madrid*, I (fols. 65v, 66r). The group represented in the *Codex Atlanticus* includes ropemaking machines of advanced design (fol. 2v-a and b), silk doubling and winding machines (fol. 36v-b), gig mills whose principle reappears in the nineteenth century (fols. 38r-a, 161v-b, 297r-a), shearing machines (fols. 397r-a, 397v-a), and even a power loom (fols. 317v-b, 356r-a, 356v-a).

Leonardo was also interested in the graphic arts and in 1494 presented details of the contemporary printing press, antedating by more than fifty years the first sensible reproduction of such an instrument. Even earlier (around 1480) he had tried to improve the efficiency of the printing press by making the motion of the carriage a function of the motion of the pressing screw. Leonardo's most interesting innovation in this field, however, was his invention of a technique of relief etching, permitting the printing of text and illustration in a single operation (*Codex Madrid*, II, fol. 119r [1504]). The technique was reinvented by William Blake in 1789 and perfected by Gillot around 1850.[12]

Leonardo's work on flight and flying machines is too well known to be discussed here in detail. The most positive part of it consists of his studies on the flight of birds, found in several notebooks, among them *Codex on Flight . . .*; MSS *K*, *E*, *G*, and *L*; *Codex Atlanticus*; and *Codex Madrid*, II, etc. Leonardo's flying machines embody many interesting mechanical features, although their basic conception as ornithopters makes them impractical. Only later did Leonardo decide to take up gliders, as did Lilienthal 400 years later. Although the idea of the parachute and of the so-called helicopter may antedate Leonardo, he was the first to experiment with true airscrews.

Leonardo's interest in chemical phenomena (for example, combustion) embraced them as such, or in relation to the practical arts. He made some inspired projects for distillation apparatus, based on the "Moor's head" condensation system that was universally adopted in the sixteenth century. Descriptions of water-cooled delivery pipes may also be found among his papers, as may a good description of an operation for separating gold from silver (*Codex Atlanticus*, fol. 244v [*ca.* 1505]). Some of the most original of the many practical chemical operations that are described in Leonardo's notebooks are concerned with making decorative objects of imitation agate, chalcedony, jasper, or amber. Leonardo began with proteins—a concentrated solution of gelatin or egg-white—and added pigments and vegetable colors. The material was then shaped by casting in ceramic molds or by extrusion; after drying, the objects were polished and then varnished for stability (MS *I*, fol. 27v; MS *F*, fols. 42r, 55v, 73v, 95v; MS *K*, fols. 114–118). By laminating unsized paper impregnated with the same materials and subsequently drying it, he obtained plates "so dense as to resemble bronze"; after varnishing, "they will be like glass and resist humidity" (MS *F*, fol. 96r). Leonardo's notes thus contain the basic operations of modern plastic technology.[13]

Of Leonardo's projects in military engineering and weaponry, those of his early period are more spectacular than practical. Some of them are, however, useful in obtaining firsthand information on contemporary techniques of cannon founding (*Codex Atlanticus*, fol. 19r-b) and also provide an interesting footnote on the survival, several centuries after the introduction of gunpowder, of such ancient devices as crossbows, ballistae, and mangonels. Leonardo did make some surprisingly modern suggestions; he was a stout advocate (although not the inventor) of breech-loading guns, he designed a water-cooled barrel for a rapid-fire gun, and, on several occasions, he proposed ogive-headed projectiles, with or without directional fins. His designs for wheel locks (*ibid.*, fols. 56v-b, 353r-c, 357r-a) antedate by about fifteen years the earliest known similar devices, which were constructed in Nuremberg. His suggestion of prefabricated cartridges consisting of ball, charge, and primer, which occurs on folio 9r-b of the *Codex Atlanticus*, is a very early (*ca.* 1480) proposal of a system introduced in Saxony around 1590.

Several of Leonardo's military projects are of importance to the history of mechanical engineering because of such details of construction as the racks used for spanning giant crossbows (*ibid.*, fol. 53r-a and b [*ca.* 1485]) or the perfectly developed universal joint on a light gun mount (*ibid.*, fol. 399v-a [*ca.* 1494]).

Leonardo worked as military architect at the court of Ludovico il Moro, although his precise tasks are not documented. His activities during his short stay in the service of Cesare Borgia are, however, better known. His projects for the modernization of the fortresses in the Romagna are very modern in concept, while the maps and city plans executed during this period (especially that of Imola in *Windsor Collection*, fol. 12284) have been called monuments in the history of cartography.

One of the manuscripts recently brought to light, *Codex Madrid*, II, reveals an activity of Leonardo's unknown until 1967—his work on the fortifications of Piombino for Jacopo IV d'Appiano (1504). Leonardo's technological work is characterized not only by an understanding of the natural laws (he is, incidentally, the first to have used this term) that govern the functioning of all mechanical devices but also by the requirement of technical and economical efficiency. He was not an armchair technologist, inventing ingenious but unusable machines; his main goal was practical efficiency and economy. He continually sought the best mechanical solution for a given task; a single page of his notes often contains a number of alternative means. Leonardo abhorred waste, be it of time, power, or money. This is why so many double-acting devices—ratchet mechanisms, blowers, and pumps—are found in his writings. This mentality led Leonardo toward more highly automated machines, including the file-cutting machine of *Codex Atlanticus,* folio 6r-b, the automatic hammer for gold-foil work of *Codex Atlanticus*, folios 8r-a, 21r-a, and 21v-a, and the mechanized printing press, rope-making machine, and power loom, already mentioned.[14]

According to Beck, the concept of the transmission of power to various operating machines was one of the most important in the development of industrial machinery. Beck thought that the earliest industrial application of this type is found in Agricola's famous *De re metallica* (1550) in a complex designed for the mercury amalgamation treatment of gold ores.[15] Leonardo, however, described several projects of the same kind, while on folios 46v and 47r of *Codex Madrid*, I, he considered the possibility of running a complete oil factory with a single power source. No fewer than five separate operations were to be performed in this combine—milling by rollers, scraping, pressing in a wedge press, releasing the pressed material, and mixing on the heating pan. This project is further remarkable because the power is transmitted by shafting; the complex thus represents a complete "Dutch oil mill," of which no other record exists prior to the eighteenth century.

Leonardo's interest in practical technology was at its strongest during his first Milanese period, from 1481 or 1482 to 1499. After that his concern shifted from practice to theory, although on occasion he resumed practical activities, as in 1502, when he built canals in the Romagna, and around 1509, when he executed works on the Adda River in Lombardy. He worked intensively on the construction of concave mirror systems in Rome, and he left highly original plans for the improvement of the minting techniques in that city (1513–1516). He was engaged in hydraulic works along the Loire during his last years.

NOTES

1. Luca Beltrami, *Documenti e memorie riguardanti la vita e le opere di Leonardo da Vinci* (Milan, 1919), nos. 66, 117, 126, 127.
2. *Codex Madrid*, II, *passim.*
3. Beltrami, op. cit., no. 189.
4. Beltrami, op. cit., no. 246.
5. G. Calvi, *I manoscritti di Leonardo da Vinci* (Bologna, 1925); A. M. Brizio, *Scritti scelti di Leonardo da Vinci* (Turin, 1952); Kenneth Clark, *A Catalogue of the Drawings of Leonardo da Vinci at Windsor Castle* (Cambridge, 1935; 2nd rev. ed., London, 1968–1969); C. Pedretti, *Studi Vinciani* (Geneva, 1957).
6. Kyeser (1405), Anonymous of the Hussite Wars (*ca.* 1430), Fontana (*ca.* 1420), Taccola (*ca.* 1450), Francesco di Giorgio (*ca.* 1480), Besson (1569), Ramelli (1588), Zonca (1607), Strada (1617–1618), Veranzio (*ca.* 1615), Branca (1629), Biringuccio (1540), Agricola (1556), Cardano (1550), Lorini (1591), Böckler (1661), Zeising (1607–1614).
7. F. Reuleaux, *The Kinematic of Machinery* (New York, 1876; repr. 1963); R. S. Hartenberg and J. Denavit, *Kinematic Synthesis of Linkages* (New York, 1964).
8. G. Galilei, *Le meccaniche* (*ca.* 1600), trans. with intro. and notes by I. E. Drabkin and Stillman Drake as *Galilei on Motion and on Mechanics* (Madison, Wis., 1960).
9. G. Canestrini, *Leonardo costruttore di macchine e veicoli* (Milan, 1939); L. Reti, "Leonardo on Bearings and Gears," in *Scientific American*, **224** (1971), 100.
10. E. S. Ferguson, "The Measurement of the 'Man-Day,'" in *Scientific American*, **225** (1971), 96. The writings of La Hire and Amontons are in *Mémoires de l'Académie . . .*, **1** (1699).
11. L. Reti, "Leonardo da Vinci nella storia della macchina a vapore," in *Rivista di ingegneria* (1956–1957).
12. L. Reti, "Leonardo da Vinci and the Graphic Arts," in *Burlington Magazine*, **113** (1971), 189.
13. L. Reti, "Le arti chimiche di Leonardo da Vinci," in *Chimica e l'industria*, **34** (1952), 655, 721.
14. B. Dibner, "Leonardo: Prophet of Automation," in C. D. O'Malley, ed., *Leonardo's Legacy* (Berkeley–Los Angeles, 1969).
15. T. Beck, *Beiträge zur Geschichte des Maschinenbaues* (Berlin, 1900), pp. 152–153.

LADISLAO RETI

LEONARDO DA VINCI

MECHANICS

Although Leonardo was interested in mechanics for most of his mature life, he would appear to have turned more and more of his attention to it from 1508 on. It is difficult to construct a unified and consistent picture of his mechanics in detail, but the major trends, concepts, and influences can be delineated with some firmness.[1] Statics may be considered first, since this area of theoretical mechanics greatly attracted him and his earliest influences in this field probably came from the medieval science of weights. To this he later added Archimedes' *On the Equilibrium of Planes*, with a consequent interest in the development of a procedure for determining the centers of gravity of sundry geometrical magnitudes.

In his usual fashion Leonardo absorbed the ideas of his predecessors, turned them in practical and experiential directions, and developed his own system of nomenclature—for example, he called the arms of balances "braccia" while in levers "lieva" is the arm of the lever to which the power is applied and "contralieva" the arm in which the resistance lies. Influenced by the Scholastics, he called the position of horizontal equilibrium the "sito dell'equalità" (or sometimes the position "equale allo orizzonte"). Pendent weights are simply "pesi," or "pesi attacati," or "pesi appiccati"; the cords supporting them—or, more generally, the lines of force in which the weights or forces act or in which they are applied—are called "appendicoli." Leonardo further distinguished between "braccia reali o linee corporee," the actual lever arms, and "braccia potenziali o spirituali o semireali," the potential or effective arms. The potential arm is the horizontal distance to the vertical (that is, the "linea central") through the center of motion in the case of bent levers, or, more generally, the perpendicular distance to the center of motion from the line of force about the center of motion. He often called the center of motion the "centro del circunvolubile," or simply the "polo."

The classical law of the lever appears again and again in Leonardo's notebooks. For example, *Codex Atlanticus*, folio 176v-d, states, "The ratio of the weights which hold the arms of the balance parallel to the horizon is the same as that of the arms, but is an inverse one." In *Codex Arundel*, folio 1v, the law appears formulaically as $W_2 = (W_1 \cdot s_1)/s_2$: "Multiply the longer arm of the balance by the weight it supports and divide the product by the shorter arm, and the result will be the weight which, when placed on the shorter arm, resists the descent of the longer arm, the arms of the balance being above all basically balanced."

A few considerations that prove the influence of the medieval science of weights upon Leonardo are in order. The medieval science of weights consisted essentially of the following corpus of works:[2] Pseudo-Euclid, *Liber de ponderoso et levi*, a geometrical treatment of basic Aristotelian ideas relating forces, volumes, weights, and velocities; Pseudo-Archimedes, *De ponderibus Archimenidis* (also entitled *De incidentibus in humidum*), essentially a work of hydrostatics; an anonymous tract from the Greek, *De canonio*, treating of the Roman balance or steelyard by reduction to a theoretical balance; Thābit ibn Qurra, *Liber karastonis*, another treatment of the Roman balance; Jordanus de Nemore, *Elementa de ponderibus* (also entitled *Elementa super demonstrationem ponderum*), a work existing in many manuscripts and complemented by several reworked versions of the late thirteenth and the fourteenth centuries, which was marked by the first use of the concept of positional gravity *(gravitas secundum situm)*, by a false demonstration of the law of the bent lever, and by an elegant proof of the law of the lever on the basis of the principle of virtual displacements; an anonymous *Liber de ponderibus* (called, by its modern editor, Version P), which contains a kind of short Peripatetic commentary to the enunciations of Jordanus; a *Liber de ratione ponderis*, also attributed to Jordanus, which is a greatly expanded and corrected version of the *Elementa* in four parts and is distinguished by a more correct use of the concept of positional gravity, by a superb proof of the law of the bent lever based on the principle of virtual displacements, by the same proof of the law of the straight lever given in the *Elementa*, and by a remarkable and sound proof of the law of the equilibrium of connected weights on adjacent inclined planes that is also based on the principle of virtual displacements—thereby constituting the first correct statement and proof of the inclined plane law—and including a number of practical problems in statics and dynamics, of which the most noteworthy are those connected with the bent lever in part III; and Blasius of Parma, *Tractatus de ponderibus*, a rather inept treatment of the problems of statics and hydrostatics based on the preceding works.

The whole corpus is marked methodologically by its geometric form and conceptually by its use of dynamics (particularly the principle of virtual velocities or the principle of virtual displacements in one form or another) in application to the basic statical problems and conclusions inherited from Greek antiquity. Leonardo was much influenced by the corpus' general dynamical approach as well as by particular conclusions of specific works. Reasonably conclusive

LEONARDO DA VINCI

evidence exists to show that Leonardo had read Pseudo-Archimedes, *De canonio*, and Thābit ibn Qurra, while completely conclusive evidence reveals his knowledge of the *Elementa*, *De ratione ponderis*, and Blasius. It is reasonable to suppose that he also saw the other works of the corpus, since they were so often included in the same manuscripts as the works he did read.

The two works upon which Leonardo drew most heavily were the *Elementa de ponderibus* and the *Liber de ratione ponderis*. The key passage that shows decisively that Leonardo read both of them occurs in *Codex Atlanticus*, folios 154v-a and r-a. In this passage, whose significance has not been properly recognized before, Leonardo presents a close Italian translation of all the postulates from the *Elementa* (E.01–E.07) and the enunciations of the first two propositions (without proofs), together with a definition that precedes the proof of proposition E.2.[3] He then shifts to the *Liber de ratione ponderis* and includes the enunciations of all the propositions of the first two parts (except R2.05) and the first two propositions of the third part.[4] It seems likely that Leonardo made his translation from a single manuscript that contained both works and, after starting his translation of the *Elementa*, suddenly realized the superiority of the *De ratione ponderis*. He may also have translated the rest of part III, for he seems to have been influenced by propositions R3.05 and R3.06, the last two propositions in part III, and perhaps also by R3.04 in passages not considered here (see MS *A*, fols. 1r, 33v). Another page in *Codex Atlanticus*, folio 165v-a and v-c, contains all of the enunciations of the propositions of part IV (except for R4.07, R4.11–12, and R4.16). The two passages from the *Codex Atlanticus* together establish that Leonardo had complete knowledge of the best and the most original work in the corpus of the medieval science of weights.

The passages cited do not, of course, show what Leonardo did with that knowledge, although others do. For example in MS *G*, folio 79r, Leonardo refutes the incorrect proof of the second part of proposition E.2 (or of its equivalent, R1.02). The proposition states: "When the beam of a balance of equal arm lengths is in the horizontal position, if equal weights are suspended [from its extremities], it will not leave the horizontal position, and if it should be moved from the horizontal position, it will revert to it."[5] The second part is true for a material beam supported from above, since the elevation of one of the arms removes the center of gravity from the vertical line through the fulcrum and, accordingly, the balance beam returns to the horizontal position as the center of gravity seeks the vertical. The medieval proof in E.2 and R1.02, however, treats the balance as if it were a theoretical balance (with weightless beam) and attempts to show that it would return to the horizontal position. Based on the false use of the concept of positional gravity, it asserts that the weight above the line of horizontal equilibrium has a greater gravity according to its position than the weight depressed below the line, for if both weights tended to move downward, the arc on which the upper weight would move intercepts more of the vertical than the arc on which the lower weight would tend to move. Leonardo's refutation is based on showing that because the weights are connected, the actual arcs to be compared are oppositely directed and so have equal obliquities—and thus the superior weight enjoys no positional advantage. In MS *E*, folio 59r, Leonardo seems also to give the correct explanation for the return to horizontal equilibrium of the material beam.

A further response to the *Liber de ratione ponderis* is in *Codex Atlanticus*, folio 354v-c, where Leonardo again translates proposition R1.09 and paraphrases proposition R1.10 of the medieval work: "The equality of the declination conserves the equality of the weights. If the ratios of the weights and the obliquities on which they are placed are the same but inverse, the weights will remain equal in gravity and motion" ("La equalità della declinazione osserva la equalità de' pesi. Se le proporzioni de' pesi e dell' obbliqua dove si posano, saranno equali, ma converse, essi pesi resteranno equali in gravità e in moto"). The equivalent propositions from the *Liber de ratione ponderis* were "R1.09. Equalitas declinationis identitatem conservat ponderis" and "R1.10. Si per diversarum obliquitatum vias duo pondera descendant, fueritque declinationum et ponderum una proportio eodum ordine sumpta, una erit utriusque virtus in descendendo."[6] While Leonardo preserved R1.09 exactly in holding that positional weight on the incline is everywhere the same as long as the incline's declination is the same, his rephrasing of R1.10 indicates that he adopted a different measure of "declination."

R. Marcolongo believed that Leonardo measured obliquity by the ratio of the common altitude of the inclines to the length of the incline (that is, by the sine of the angle of inclination), while Jordanus had measured obliquity by the length of the incline that intercepts the common altitude (that is, by the cosecant of the angle).[7] Thus, if p_1 and p_2 are connected weights placed on the inclined planes that are of lengths l_1 and l_2, respectively, with common altitude h, then, according to Marcolongo's view of Leonardo's method, $p_1/p_2 = (h/l_2)/(h/l_1)$, and thus $p_1/p_2 = l_1/l_2$,

LEONARDO DA VINCI

as Jordanus held. Hence, if Marcolongo is correct, Leonardo had absorbed the correct exposition of the inclined plane problem from the *Liber de ratione ponderis*. Perhaps he did, but it is not exhibited in this passage, for the figure accompanying the passage (Figure 1), with its numerical designations of 2 and 1

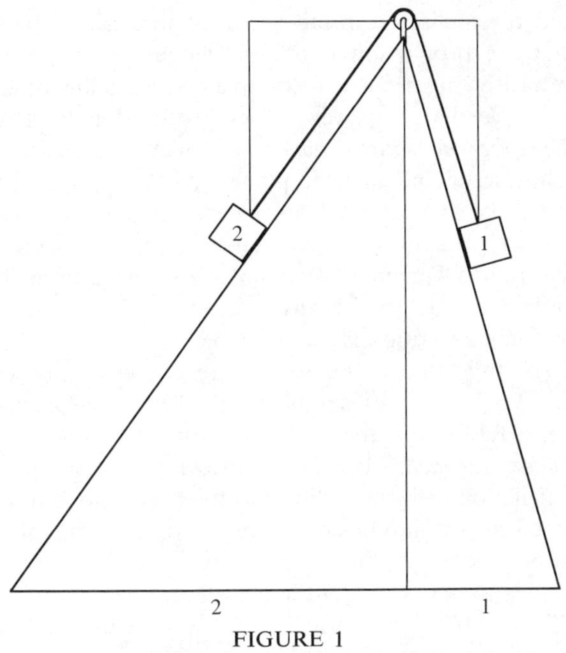

FIGURE 1

on the bases cut off by the vertical, shows that Leonardo believed the weights on the inclines to be inversely proportional to the tangents of the angles of inclination rather than to the sines—and such a solution is clearly incorrect. The same incorrect solution is apparent in MS *G*, folio 77v, where again equilibrium of the two weights is preserved when the weights are in the same ratio as the bases cut off by the common altitude (thus implying that the weights are inversely proportional to the tangents).

Other passages give evidence of Leonardo's vacillating methods and confusion. One (*Codex on Flight*, fol. 4r) consists of two paragraphs that apply to the same figure (Figure 2). The first is evidently an explanation of a proposition expressed elsewhere (MS *E*, fol. 75r) to the effect that although equal weights balance each other on the equal arms of a balance, they do not do so if they are put on inclines of different obliquity. It states:

> The weight *q*, because of the right angle *n* [perpendicularly] above point *e* in line *df*, weighs 2/3 of its natural weight, which was 3 pounds, and so has a residual force ["che resta in potenzia"] [along *nq*] of 2 pounds; and the weight *p*, whose natural weight was also 3 pounds, has a residual force of 1 pound [along *mp*] because of right angle *m* [perpendicularly] above point *g* in line *hd*. [Therefore, *p* and *q* are not in equilibrium on these inclines.]

The bracketed material has been added as clearly implicit, and so far this analysis seems to be entirely correct. It is evident from the figure that Leonardo has applied the concept of potential lever arm (implying static moment) to the determination of the component of weight along the incline, so that $F_1 \cdot dn = W_1 \cdot de$, where F_1 is the component of the natural weight W_1 along the incline, *dn* is the potential lever arm through which F_1 acts around fulcrum *d*, and *de* is the lever

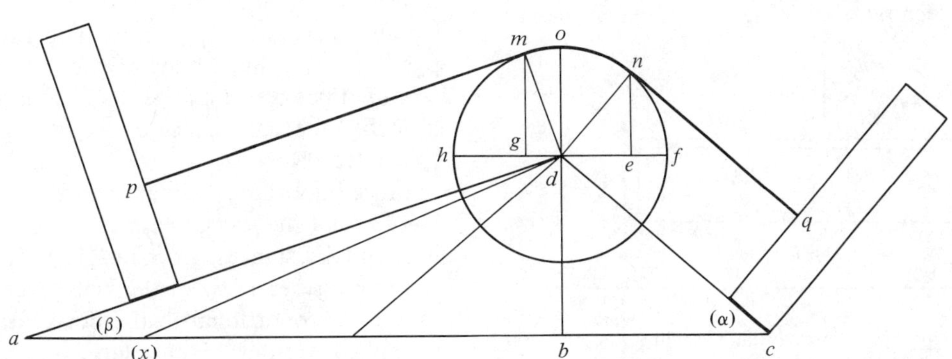

FIGURE 2

LEONARDO DA VINCI

arm through which W_1 would act when hanging from *n*. Hence, $F_1 = W_1 \cdot (de/dn) = W_1 \cdot \sin \angle dne$. But $\angle dne = \angle \alpha$, and so we have the correct formulation $F_1 = W_1 \cdot \sin \alpha$. In the same way for weight *p*, it can be shown that $F_2 = W_2 \cdot \sin \beta$. And since Leonardo apparently constructed $de = 2df/3$, $dg = hd/3$, and $W_1 = W_2 = 3$, obviously $F_1 = 2$ and $F_2 = 1$ and the weights are not in equilibrium.

In the second paragraph, which also pertains to the figure, he changes *p* and *q*, each of which was initially equal to three pounds, to two pounds, and one pound, respectively, with the object of determining whether the adjusted weights would be in equilibrium:

> So now we have one pound against two pounds. And because the obliquities *da* and *dc* on which these weights are placed are not in the same ratio as the weights, that is, 2 to 1 [the weights are not in equilibrium]; like the said weights [in the first paragraph?], they alter natural gravities [but they are not in equilibrium] because the obliquity *da* exceeds the obliquity *dc*, or contains the obliquity *dc*, $2\frac{1}{2}$ times, as is demonstrated by their bases *ab* and *bc*, whose ratio is a double sesquialterate ratio [that is, 5:2], while the ratio of the weights will be a double ratio [that is, 2:1].

It is abundantly clear, at least in the second paragraph, that Leonardo was assuming that for equilibrium the weights ought to be directly proportional to the bases (and thus inversely proportional to the tangents); and since the weights are not as the base lines, they are not in equilibrium. In fact, in adding line *d*[*x*] Leonardo was indicating the declination he thought would establish the equilibrium of *p* and *q*, weights of two pounds and one pound, respectively, for it is obvious that the horizontal distances *b*[*x*] and *bc* are also related as 2 : 1. This is further confirmation that Leonardo measured declination by the tangent.

Assuming that both paragraphs were written at the same time, it is apparent that Leonardo then thought that both methods of determining the effective weight on an incline—the method using the concept of the lever and the technique of using obliquities measured by tangents—were correct. A similar confusion is apparent in his treatment of the tensions in strings. But there is still another figure (Figure 3), which has accompanying it in MS *H*, folio 81(33)v, the following brief statement: "On the balance, weight *ab* will be as weight *cd*." If Leonardo was assuming that the weights are of the same material with equal thickness and, as it seems in the figure, that they are of the same width, with their lengths equal to the lengths of the vertical and the incline, respectively, thus producing weights that are proportional to these lengths, then he was indeed giving a correct example of the inclined-plane principle that may well reflect proposition R1.10 of the *Liber de ratione ponderis*. There is one further solution of the inclined-plane problem, on MS *A*, folio 21v, that is totally erroneous and, so far as is known, unique. It is not discussed here; the reader is referred to Duhem's treatment, with the caution that Duhem's conclusion that it was derived from Pappus' erroneous solution is questionable.[8]

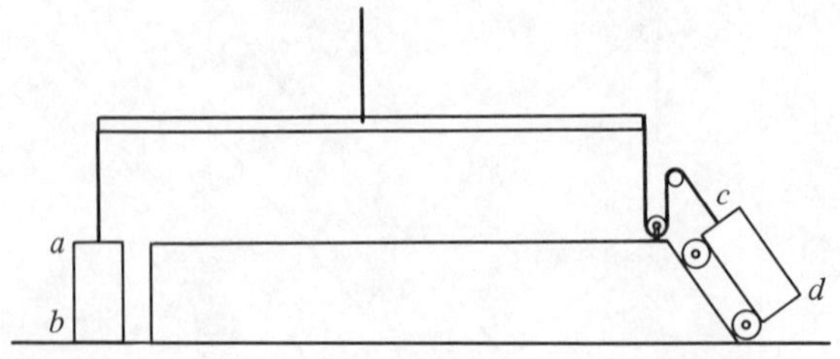

FIGURE 3. (Note: The horizontal base line and the continuation of the line from point *c* beyond the pulley wheel to the balance beam are in Leonardo's drawing but should be deleted.)

As important as Leonardo's responses to proposition R1.10 are his responses to the bent-lever proposition, R1.08, of the *Liber de ratione ponderis*. In *Codex Arundel*, folio 32v, he presents another Italian translation of it (in addition to the translation already noted in his omnibus collection of translations of the enunciations of the medieval work). In this new translation he writes of the bent-lever law as "tested": "Tested. If the arms of the balance are unequal and their juncture in the fulcrum is angular, and if their termini are equally distant from the central [that is, vertical] line through the fulcrum, with equal weights applied there [at the termini], they will weigh equally [that is, be in equilibrium]"—or "Sperimentata. Se le braccia della bilancia fieno inequali, e la lor congiunzione nel polo sia angulare, se i termini lor fieno equalmente distanti alla linea central del polo, li pesi appiccativi, essendo equali, equalmente peseranno," a translation of the Latin text, "R1.08. Si inequalia fuerint brachia libre, et in centro motus angulum fecerint, si termini eorum ad directionem hinc inde equaliter accesserint, equalia appensa, in hac dispositione equaliter ponderabunt."[9] Other, somewhat confused passages indicate that Leonardo had indeed absorbed the significance of the passage he had twice translated.[10]

More important, Leonardo extended the bent-lever law beyond the special case of equal weights at equal horizontal distances, given in R1.08, to cases in which the more general law of the bent lever—"weights are inversely proportional to the horizontal distances"—is applied. He did this in a large number of problems to which he applied his concept of "potential lever arm." The first passage notable for this concept is in

LEONARDO DA VINCI

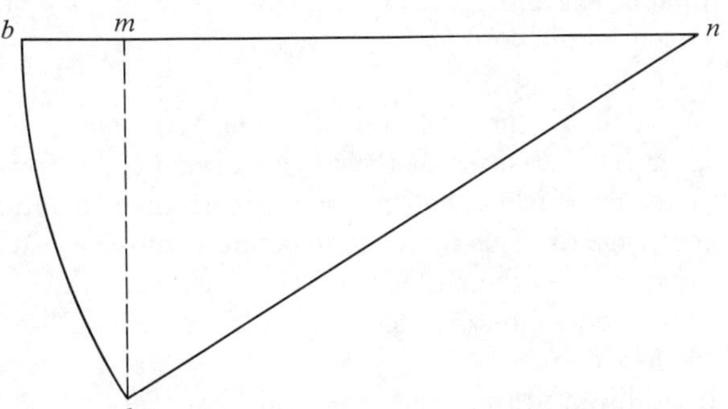

FIGURE 4. (Note: This is the essential part of a more detailed drawing.)

MS *E*, folio 72v, and probably derives from proposition R3.05 of the *Liber de ratione ponderis*.[11] Here Leonardo wrote (see Figure 4): "The ratio that space *mn* has to space *nb* is the same as the ratio that the weight which has descended to *d* has to the weight it had in position *b*. It follows that, *mn* being 9/11 of *nb*, the weight in *d* is 9/10 (! 9/11) of the weight it had in height *b*." In this passage *n* is the center of motion and *mn* is the potential lever arm of the weight in position *d*. His use of the potential lever arm is also illustrated in a passage that can be reconstructed from MS *E*, folio 65r, as follows (Figure 5). A bar *at* is pivoted at *a*; a weight *o* is suspended from *t* at *m* and a second weight acts on *t* in a direction *tn* perpendicular to that of the first weight. The weights at *m* and *n* necessary to keep the bar in equilibrium are to be determined. In this determination Leonardo took *ab* and *ac* as the potential lever arms (and so labeled them), so that the weights *m* and *n* are inversely proportional to the potential lever arms *ab* and *ac*. The same kind of problem is illustrated in a passage in *Codex Atlanticus*, folio 268v-b, which can be summarized by reference to Figure 6. Cord *ab* supports a weight *n*, which is pulled to the position indicated by a tangential force along *nf* that is given a value of 1 and there keeps *an* in equilibrium. The passage indicates that weight 1, acting through a distance of four units (the potential lever arm *an*), keeps in equilibrium a weight *n* of 4, acting through a distance of one unit (the potential counterlever *ac*). A similar use of "potential lever arm" is found in problems like that illustrated in Figure 7, taken from MS *M*, folio 40r. Here the potential lever arm is *an*, the perpendicular drawn from the line of force *fp* to the center of motion *a*. Hence the weights suspended from *p* and *m* are related inversely as the distances *an* and *am*. These and similar problems reveal Leonardo's acute awareness of the proper factors of horizontal distance and force determining the static moment about a point.

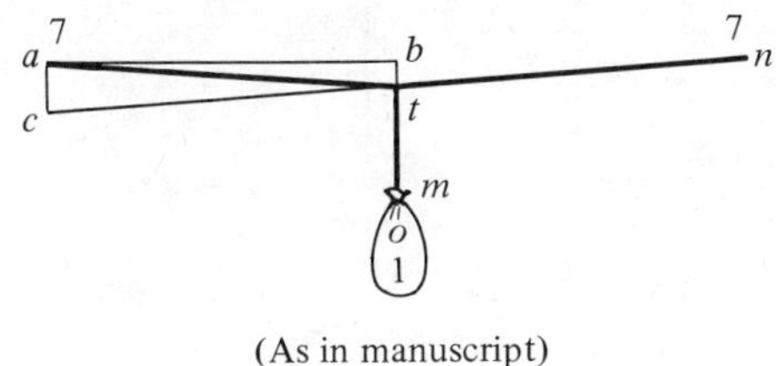

(As in manuscript)

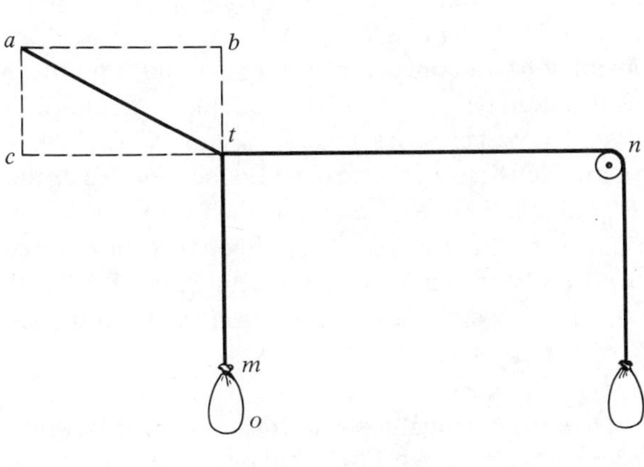

(As reconstructed)

FIGURE 5

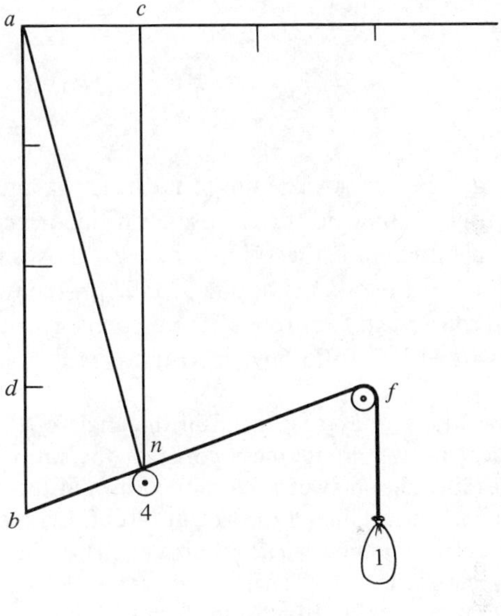

FIGURE 6

The concept of potential lever arm also played a crucial role in Leonardo's effort to analyze the tension

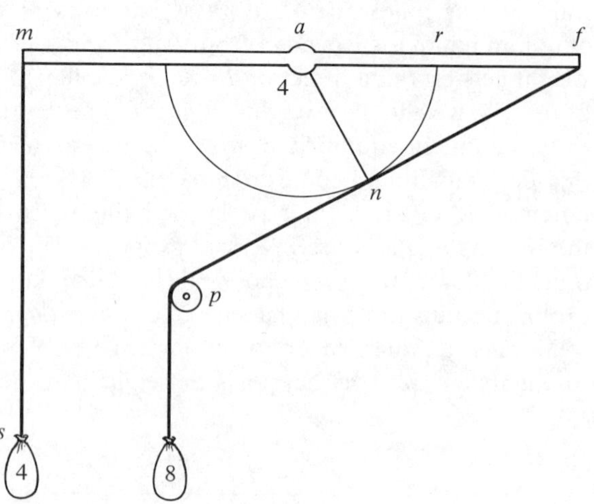

FIGURE 7

in strings. Before examining that role, it must be noted that Leonardo often used an incorrect rule based on tangents rather than sines, a rule similar to that which he mistakenly applied to the problem of the inclined plane. An example of the incorrect procedure appears in MS *E*, folio 66v (see Figure 8):

> The heavy body suspended in the angle of the cord divides the weight to these cords in the ratio of the angles included between the said cords and the central [that is, vertical] line of the weight. Proof. Let the angle of the said cord be *bac*, in which is suspended heavy body *g* by cord *ag*. Then let this angle be cut in the position of equality [that is, in the horizontal direction] by line *fb*. Then draw the perpendicular *da* to angle *a*

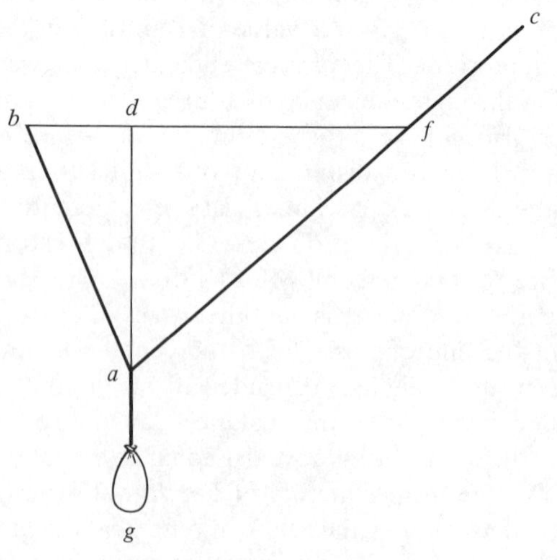

FIGURE 8

LEONARDO DA VINCI

> and it will be in the direct continuation of cord *ag*, and the ratio which space *df* has to space *db*, the weight [that is, tension] in cord *ba* will have to the weight [that is, tension] in cord *fa*.

Thus, with *da* the common altitude and *df* and *db* used as the measures of the angles at *a*, Leonardo is actually measuring the tensions by the inverse ratio of the tangents. This same incorrect procedure appears many times in the notebooks (for instance, MS *E*, fols. 67v, 68r, 68v, 69r, 69v, 71r; *Codex Arundel*, fol. 117v; MS *G*, fol. 39v).

In addition to this faulty method Leonardo in some instances employed a correct procedure based on the concept of the potential lever arm. In *Codex Arundel*, folio 1v, Leonardo wrote that "the weight 3 is not distributed to the real arms of the balance in the same [although inverse] ratio of these arms but in the [inverse] ratio of the potential arms" (see Figure 9).

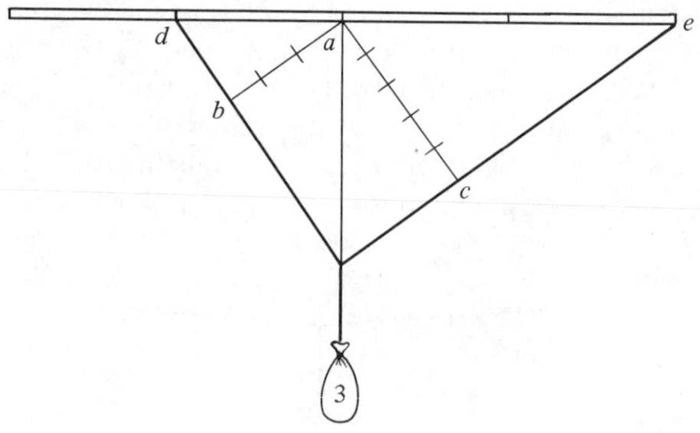

FIGURE 9

The cord is attached to a horizontal beam and a weight is hung from the cord. In the figure *ab* and *ac* are the potential arms. This is equivalent to a theorem that could be expressed in modern terms as "the moments of two concurrent forces around a point on the resultant are mutually equal." Leonardo's theorem, however, does not allow the calculation of the actual tensions in the strings. By locating the center of the moments first on one and then on the other of the concurrent forces, however, Leonardo discovered how to find the tensions in the segments of a string supporting a weight. In *Codex Arundel*, folio 6r (see Figure 10), he stated:

> Here the potential lever *db* is six times the potential counterlever *bc*. Whence it follows that one pound placed in the force line "appendiculo" *dn* is equal in power to six pounds placed in the semireal force line

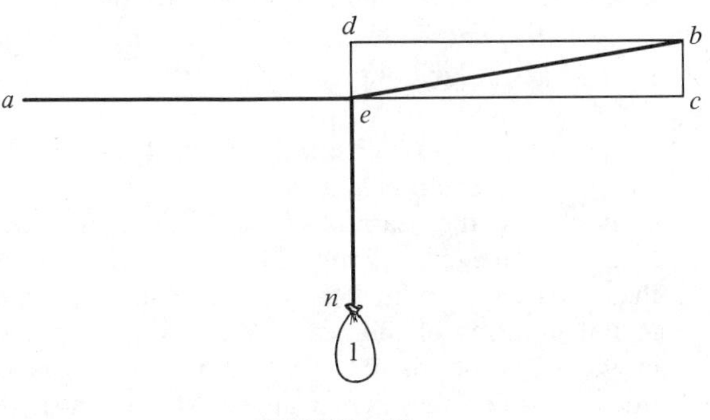

FIGURE 10

ca, and to another six pounds of power conjoined at *b*. Therefore, the cord *aeb* by means of one pound placed in line *dn* has the [total] effect of twelve pounds.

All of which is to say that Leonardo has determined that there are six pounds of tension in each segment of the cord. The general procedure is exhibited in MS *E*, folio 65r (see Figure 11). Under the figure is the caption "*a* is the pole [that is, fulcrum] of the angular balance [with arms] *ad* and *af*, and their force lines ['appendiculi'] are *dn* and *fc*." This applies to the figure on the left and indicates that *af* and *da* are the lever arms on which the tension in *cb* and the weight hanging at *b* act. Leonardo then went on to say: "The greater the angle of the cord which supports weight *n* in the middle of the cord, the smaller becomes the potential arm [that is, *ac* in the figure on the right] and the greater becomes the potential counter-arm [that is, *ba*] supporting the weight." Since Leonardo has drawn the figure so that *ab* is four times *ac* and has marked the weight of *n* as 1, the obvious implication is that the tension in cord *df* will be 4.

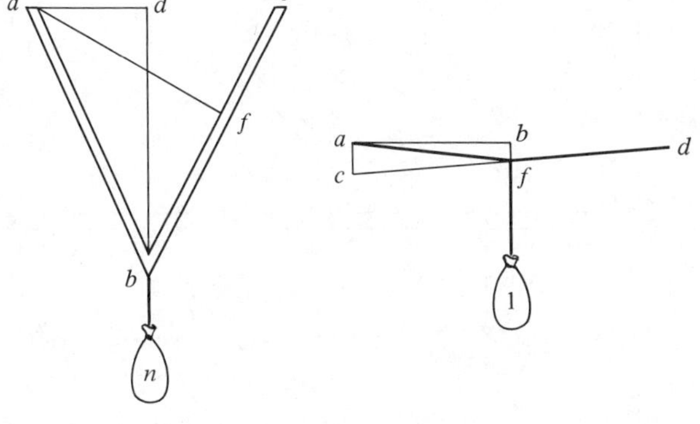

FIGURE 11

LEONARDO DA VINCI

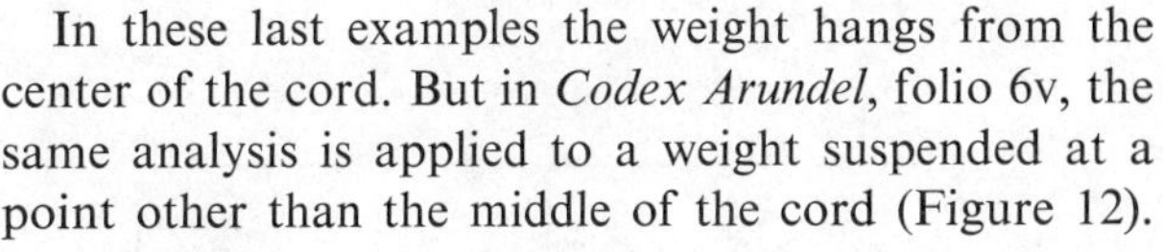
In these last examples the weight hangs from the center of the cord. But in *Codex Arundel*, folio 6v, the same analysis is applied to a weight suspended at a point other than the middle of the cord (Figure 12).

Here the weight *n* is supported by two different forces, *mf* and *mb*. Now it is necessary to find the potential levers and counterlevers of these two forces *bm* and *fm*. For the force [at] *b* [with *f* the fulcrum of the potential lever] the [potential] lever arm is *fe* and the [potential] counterlever is *fa*. Thus for the lever arm *fe* the force line ["appendiculo"] is *eb* along which the motor *b* is applied; and for the counterlever *fa* the force line is *an*, which supports weight *n*. Having arranged the balance of the power and resistance of motor and weight, it is necessary to see what ratio lever *fe* has to counterlever *fa;* which lever *fe* is 21/22 of counterlever *fa*. Therefore *b* suffers 22 [pounds of

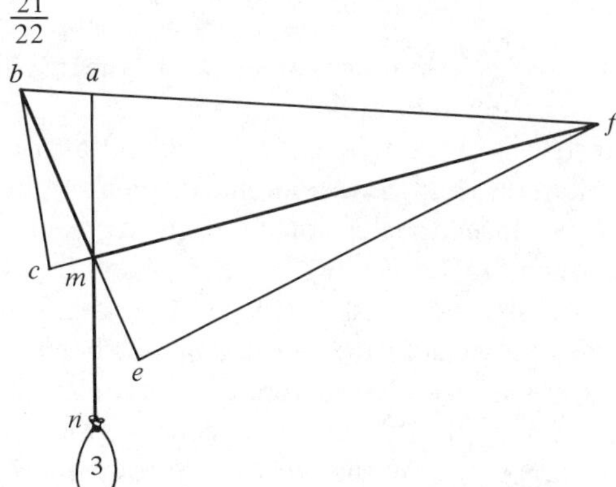

FIGURE 12. (Note: There is some discrepancy between Leonardo's text and the figure as concerns the relative length of the lines.)

tension] when *n* is 21. In the second disposition [with *b* instead of *f* as the fulcrum of a potential lever], *bc* is the [potential] lever arm and *ba* the [potential] counterlever. For *bc* the force line is *cf* along which the motor *f* is applied and weight *n* is applied along force line *an*. Now it is necessary to see what ratio lever *bc* has to counterlever *ba*, which counterlever is 1/3 of the lever. Therefore, one pound of force in *f* resists three pounds of weight in *ba;* and 21/22 of the three pounds in *n*, when placed at *b*, resist twenty-two placed in *ba*. . . . Thus is completed the rule for calculating the unequal arms of the angular cord.

In a similar problem in *Codex Arundel*, folio 4v, Leonardo determined the tensions of the string segments; in this case the strings are no longer attached or fixed to a beam but are suspended from two pulley wheels from which also hang two equal weights (Figure 13). Here *abc* is apparently an equilateral

LEONARDO DA VINCI

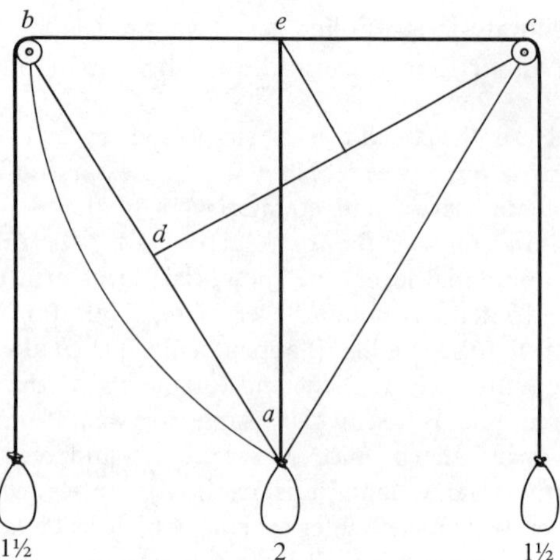

FIGURE 13

triangle, *cd* is the potential arm for the tension in string *ba* acting about fulcrum *c* and *ec* is the potential counterlever through which the weight 2 at *a* acts. If an equilateral triangle was intended, then, by Leonardo's procedure, the tension in each segment of the string ought to be $2/\sqrt{3} = 1.15$, rather than 1.5, as Leonardo miscalculated it. It is clear from all of these examples in which the tension is calculated that Leonardo was using a theorem that he understood as: "The ratio of the tension in a cord segment to the weight supported by the cord is equal to the inverse ratio of the potential lever arms through which the tension and weight act, where the fulcrum of the potential bent lever is in the point of support of the other segment of the cord." This is equivalent to a theorem in the composition of moments: "If one considers two concurrent forces and their resultant, the moment of the resultant about a point taken on one of the two concurrent forces is equal to the moment of the other concurrent force about the same point." The discovery and use of this basic concept in analyzing string tensions was Leonardo's most original development in statics beyond the medieval science of weights that he had inherited. Unfortunately, like most of Leonardo's investigations, it exerted no influence on those of his successors.

Another area of statics in which the medieval science of weights may have influenced Leonardo was that in which a determination is made of the partial forces in the supports of a beam where the beam itself supports a weight. Proposition R3.06 in the *Liber de ratione ponderis* states: "A weight not suspended in the middle [of a beam] makes the shorter part heavier according to the ratio of the longer part to the shorter part."[12] The proof indicates that the partial forces in the supports are inversely related to the distances from the principal weight to the supports. In *Codex Arundel*, folio 8v (see Figure 14), Leonardo arrived at a similar conclusion:

> The beam which is suspended from its extremities by two cords of equal height divides its weight equally in each cord. If the beam is suspended by its extremities at an equal height, and in its midpoint a weight is hung, then the gravity of such a weight is equally distributed to the supports of the beam. But the weight which is moved from the middle of the beam toward one of its extremities becomes lighter at the extremity away from which it was moved, or heavier at the other extremity, by a weight which has the same ratio to the total weight as the motion completed by the weight [that is, its distance moved from the center] has to the whole beam.

Leonardo also absorbed the concept, so prevalent in the medieval science of weights (particularly in the *De canonio*, in Thābit ibn Qurra's *Liber karastonis*, and in part II of the *Liber de ratione ponderis*), that a

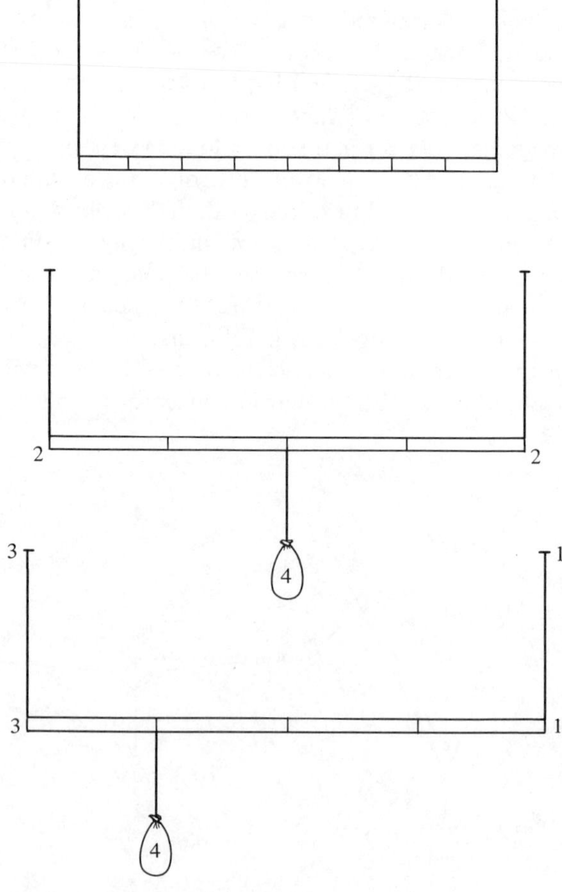

FIGURE 14

LEONARDO DA VINCI

segment of a solid beam may be replaced by a weight hung from the midpoint of a weightless arm of the same length and position. For example, see Leonardo's exposition in MS *A*, folio 5r (see Figure 15):

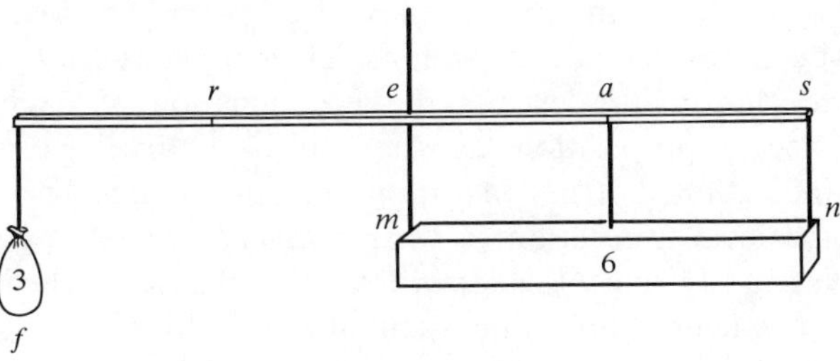

FIGURE 15

> If a balance has a weight which is similar [that is, equal] in length to one of its arms, the weight being *mn* of six pounds, how many pounds are to be placed in *f* to resist it [that is, which will be in equilibrium with it]? I say that three pounds will suffice, for if weight *mn* is as long as one of its arms, you could judge that it may be replaced in the middle of the balance arm at point *a;* therefore, if six pounds are in *a*, another six pounds placed at *r* would produce resistance to them [that is, be in equilibrium with them], and, if you proceed as before in point *r* [but now] in the extremity of the balance, three pounds will produce the [necessary] resistance to them.

This replacement doctrine was the key step in solving the problem of the Roman balance in all of the above-noted tracts.[13]

The medieval science of weights was not the only influence upon Leonardo's statics, however, since it may be documented that he also, perhaps at a later time, read book I of Archimedes' *On the Equilibrium of Planes*. As a result he seems to have begun a work on centers of gravity in about 1508, as may be seen in a series of passages in *Codex Arundel*. Since these passages have already been translated and analyzed in rather complete detail elsewhere,[14] only their content and objectives will be given here. Preliminary to Leonardo's propositions on centers of gravity are a number of passages distinguishing three centers of a figure that has weight (see *Codex Arundel*, fol. 72v): "The first is the center of its natural gravity, the second [the center] of its accidental gravity, and the third is [the center] of the magnitude of this body." On folio 123v of this manuscript the centers are defined:

> The center of the magnitude of bodies is placed in the middle with respect to the length, breadth and thickness of these bodies. The center of the accidental gravity of these bodies is placed in the middle with respect to the parts which resist one another by standing in equilibrium. The center of natural gravity is that which divides a body into two parts equal in weight and quantity.

It is clear from many other passages that the center of natural gravity is the symmetrical center with respect to weight. Hence, the center of natural gravity of a beam lies in its center, and that center would not be disturbed by hanging equal weights on its extremities. If unequal weights are applied, however, there is a shift of the center of gravity, which is now called the center of accidental gravity, the weights having assumed accidental gravities by their positions on the unequal arms. The doctrine of the three centers can be traced to Scholastic writings of Nicole Oresme, Albert of Saxony, Marsilius of Inghen (and, no doubt, others).[15] In all of these preliminary considerations Leonardo assumed that a body or a system of bodies is in equilibrium when supported from its center of gravity (be it natural or accidental). It should also be observed that Leonardo assumed the law of the lever as being proved before setting out to prove his Archimedean-like propositions.

The first Archimedean passage to note is in *Codex Arundel*, folio 16v, where Leonardo includes a series of statements on equilibrium that is drawn in significant part from the postulates and early propositions of book I of *On the Equilibrium of Planes*. His terminology suggests that when Leonardo wrote this passage, he was using the translation of Jacobus Cremonensis (*ca.* 1450).[16]

More important than this passage are the propositions and proofs on centers of gravity, framed under the influence of Archimedes' work, which Leonardo specifically cites in a number of instances. The order for these propositions that Leonardo's own numeration seems to suggest is the following:

1. *Codex Arundel*, folio 16v, "Every triangle has the center of its gravity in the intersection of the lines which start from the angles and terminate in the centers of the sides opposite them." This is proposition 14 of book I of *On the Equilibrium of Planes*, and Leonardo included in his "proof" an additional proof of sorts for Archimedes' proposition 13 (since that proposition is fundamental for the proof of proposition 14), although the proof for proposition 13 ignores Archimedes' superb geometrical demonstration. Depending, as it does, on balance considerations, Leonardo's proof is more like Archimedes' second proof of proposition 13. Leonardo's proof of proposition 14 is for an equilateral triangle, but at its end he notes that it applies as well to scalene triangles.

2. *Codex Arundel*, folio 16r: "The center of gravity of any two equal triangles lies in the middle of the

 LEONARDO DA VINCI

line beginning at the center of one triangle and terminating in the center of gravity of the other triangle." This is equivalent to proposition 4 of *On the Equilibrium of Planes*, but Leonardo's proof differs from Archimedes' in that it merely shows that the center of gravity is in the middle of the line because the weights would be in equilibrium about that point. Archimedes' work is here cited (under the inaccurate title of *De ponderibus*, since the Pseudo-Archimedean work of that title was concerned with hydrostatics rather than statics).

3. *Codex Arundel*, folio 16r: "If two unequal triangles are in equilibrium at unequal distances, the greater will be placed at the lesser distance and the lesser at the greater distance." This is similar to proposition 3 of Archimedes' work. Leonardo, however, simply employed the law of the lever in his proof, which Archimedes did not do, since he did not offer a proof of the law until propositions 6 and 7.

4. *Codex Arundel*, folio 17v: "The center of gravity of every square of parallel sides and equal angles is equally distant from its angles." This is a special case of proposition 10 of *On the Equilibrium of Planes*; but Leonardo's proof, based once more on a balancing procedure, is not directly related to either of the proofs provided by Archimedes.

5. *Codex Arundel*, folio 17v: "The center of gravity of every corbel-like figure [that is, isosceles trapezium] lies in the line which divides it into two equal parts when two of its sides are parallel." This is similar to Archimedes' proposition 15, which treated more generally of any trapezium. Leonardo made a numerical determination of where the center of gravity lies on the bisector. Although his proof is not close to Archimedes', like Archimedes he used the law of the lever in his proof.

6. *Codex Arundel*, folio 17r: "The center of gravity of every equilateral pentagon is in the center of the circle which circumscribes it." This has no equivalent in Archimedes' work. The proof proceeds by dividing the pentagon into triangles that are shown to balance about the center of the circle. Again Leonardo used the law of the lever in his proof, much as he had in his previous propositions. The proof is immediately followed by the determination of the center of gravity of a pentagon that is not equilateral, in which the same balancing techniques are again employed. Leonardo here cited Archimedes as the authority for the law of the lever, and the designation of Archimedes' proposition as the "fifth" is perhaps an indication that he was using William of Moerbeke's medieval translation of *On the Equilibrium of Planes* instead of that of Jacobus Cremonensis, in which the equivalent proposition is number 7. Although these exhaust those propositions of Leonardo's that are directly related to *On the Equilibrium of Planes*, it should be noted that Leonardo used the same balancing techniques in his effort to determine the center of gravity of a semicircle (see *Codex Arundel*, fols. 215r-v).

It should also be noted that Leonardo went beyond Archimedes' treatise in one major respect—the determination of centers of gravity of solids, a subject taken up in more detail later in the century by Francesco Maurolico and Federico Commandino. Two of the propositions investigated by Leonardo may be presented here to illustrate his procedures. Both propositions concern the center of gravity of a pyramid and appear to be discoveries of Leonardo's. The first is that the center of gravity of a pyramid (actually, a regular tetrahedron) is, at the intersection of the axes, a distance on each axis of 1/4 of its length, starting from the center of one of the faces. (By "axis" Leonardo understood a line drawn from a vertex to the center of the opposite face.) In one place (*Codex Arundel*, fol. 218v) he wrote of the intersection of the pyramidal axes as follows: "The inferior [interior?] axes of pyramids which arise from [a point lying at] 1/3 of the axis of their bases [that is, faces] will intersect in [a point lying at] 1/4 of their length [starting] at the base."

Despite the confusion of singulars and plurals as well as of the expression "inferior axes," the prop-

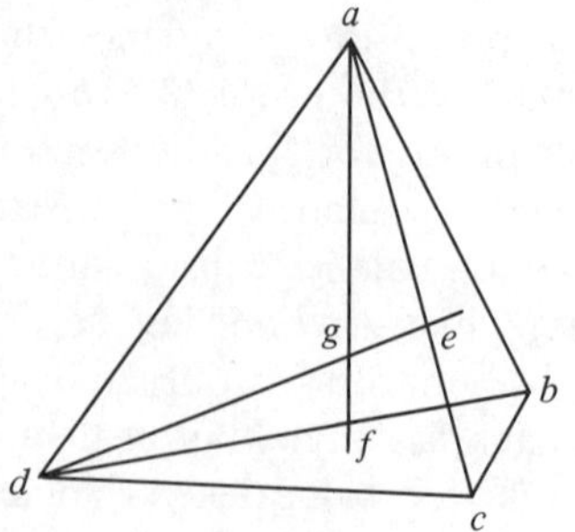

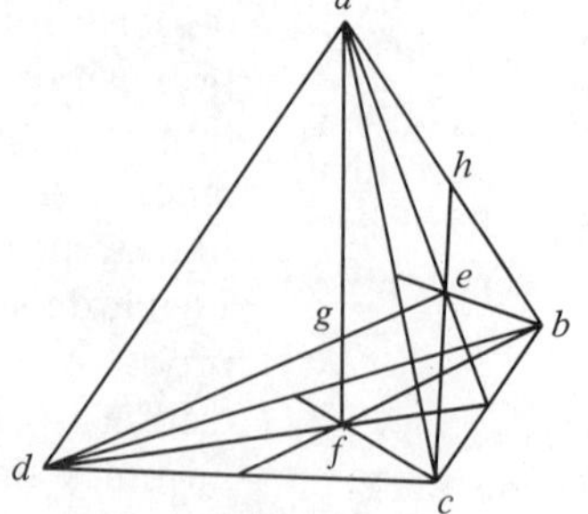

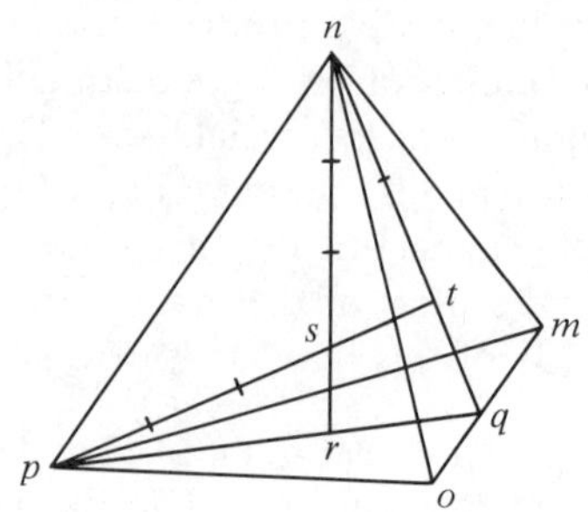

FIGURE 16

LEONARDO DA VINCI

osition is clear enough, particularly since Leonardo provided both the drawing shown in Figure 16 and its explanation and, in addition, the intersection of the pyramidal axes is definitely specified as the center of gravity in another passage (fol. 193v): "The center of gravity of the [pyramidal] body of four triangular bases [that is, faces] is located at the intersection of its axes and it will be in the 1/4 part of their length." The proof of this is actually given on the page of the original quotation about the intersection of the axes (fol. 218v), but only as a proof following a more general statement about pyramids and cones:

> The center of gravity of any pyramid—round, triangular, square, or [whose base is] of any number of sides—is in the fourth part of its axis near the base. Let the pyramid be *abcd* with base *bcd* and apex *a*. Find the center of the base *bcd*, which you let be *f*, then find the center of face *abc*, which will be *e*, as was proved by the first [proposition]. Now draw line *af*, in which the center of gravity of the pyramid lies because *f* is the center of base *bcd* and the apex *a* is perpendicularly above *f* and the angles *b*, *c* and *d* are equally distant from *f* and [thus] weigh equally so that the center of gravity lies in line *af*. Now draw a line from angle *d* to the center *e* of face *abc*, cutting *af* in point *g*. I say for the aforesaid reason that the center of gravity is in line *de*. So, since the center is in each [line] and there can be only one center, it necessarily lies in the intersection of these lines, namely in point *g*, because the angles *a*, *b*, *c* and *d* are equally distant from this *g*.

As noted above, the proof is given only for a regular tetrahedron, none being given for the cone or for other pyramids designated in the general enunciation. It represents a rather intuitive mechanical approach, for Leonardo abruptly stated that because the angles *b*, *c*, and *d* weigh equally about *f*, the center of gravity of the base triangle, and *a* is perpendicularly above *f*, the center of gravity of the whole pyramid must lie in line *af*. This is reminiscent of Hero's demonstration of the equilibrium of a triangle supported at its center with equal weights at the angles. This kind of reasoning, then, seems to be extended to the whole pyramid by Leonardo at the end of the proof in which he declared that each of the four angles is equidistant from *g* and, presumably, equal weights at the angles would therefore be in equilibrium if the pyramid were supported in *g*. It is worth noting that a generation later Maurolico gave a very neat demonstration of just such a determination of the center of gravity of a tetrahedron by the hanging of equal weights at the angles.[17] Finally, one additional theorem (without proof), concerning the center of gravity of a tetrahedron, appears to have been Leonardo's own discovery (*Codex Arundel*, fol. 123v):

> The pyramid with triangular base has the center of its natural gravity in the [line] segment which extends from the middle of the base [that is, the midpoint of one edge] to the middle of the side [that is, edge] opposite the base; and it [the center of gravity] is located on the segment equally distant [from the termini] of the [said] line joining the base with the aforesaid side.

Despite Leonardo's unusual and imprecise language (an attempt has been made to rectify it by bracketed additions), it is clear that he has here expressed a neat theorem to the effect that the center of gravity of the tetrahedron lies at the intersection of the segments joining the midpoint of each edge with the midpoint of the opposite edge and that each of these segments is bisected by the center of gravity. Again, it is possible that Leonardo arrived at this proposition by considering four equal weights hung at the angles. At any rate the balance procedure, whose refinements he learned from Archimedes, no doubt played some part in his discovery, however it was made.

Leonardo gave considerable attention to one other area of statics, pulley problems. Since this work perhaps belongs more to his study of machines, except for a brief discussion in the section on dynamics, the reader is referred to Marcolongo's brief but excellent account.[18]

Turning to Leonardo's knowledge of hydrostatics, it should first be noted that certain fragments from William of Moerbeke's translation of Archimedes' *On Floating Bodies* appear in the *Codex Atlanticus*, folios 153v-e, 153r-b, and 153r-c.[19] These fragments (which occupy a single sheet bound into the codex) are not in Leonardo's customary mirror script but appear in normal writing, from left to right. Although sometimes considered by earlier authors to have been written by Leonardo, they are now generally believed to be by some other hand.[20] Whether the sheet was once the property of Leonardo or whether it was added to Leonardo's material after his death cannot be determined with certainty—at any rate, the fragments can be identified as being from proposition 10 of book II of the Archimedean work. Whatever Leonardo's relationship to these fragments, his notebooks reveal that he had only a sketchy and indirect knowledge of Archimedean hydrostatics, which he seems to have drawn from the medieval tradition of *De ponderibus Archimenidis*. Numerous passages in the notebooks reveal a general knowledge of density and specific weight (for instance, MS *C*, fol. 26v, MS *F*, fol. 70r; MS *E*, fol. 74v). Similarly, Leonardo certainly knew that bodies weigh less in water than in air (see MS *F*, fol. 69r), and in one passage (*Codex Atlanticus*, fol. 284v) he proposed to measure the relative resistance of water as compared with air by

LEONARDO DA VINCI

plunging the weight on one arm of a balance held in aerial equilibrium into water and then determining how much extra weight must be added to the weight in the water to maintain the balance in equilibrium. See also MS *A*, folio 30v: "The weight in air exhibits the truth of its weight, the weight in water will appear to be less weight by the amount the water is heavier than the air."

So far as is known, however, the principle of Archimedes as embraced by proposition 7 of book I of the genuine *On Floating Bodies* was not precisely stated by Leonardo. Even if he did know the principle, as some have suggested, he probably would have learned it from proposition 1 of the medieval *De ponderibus Archimenidis*. As a matter of fact, Leonardo many times repeated the first postulate of the medieval work, that bodies or elements do not have weight amid their own kind—or, as Leonardo put it (*Codex Atlanticus*, fol. 365r-a; *Codex Arundel*, fol. 189r), "No part of an element weighs in its element" (cf. *Codex Arundel*, fol. 160r). Still, he could have gotten the postulate from Blasius of Parma's *De ponderibus*, a work that Leonardo knew and criticized (see MS *Ashburnham* 2038, fol. 2v). It is possible that since Leonardo knew the basic principle of floating bodies—that a floating body displaces its weight of liquid (see *Codex Forster* II_2, fol. 65v)—he may have gotten it directly from proposition 5 of book I of *On Floating Bodies*. But even this principle appeared in one manuscript of the medieval *De ponderibus Archimenidis* and was incorporated into John of Murs' version of that work, which appeared as part of his widely read *Quadripartitum numerorum* of 1343.[21] So, then, all of the meager reflections of Archimedean hydrostatics found in Leonardo's notebooks could easily have been drawn from medieval sources, and (with the possible exception of the disputed fragments noted above) nothing from the brilliant treatments in book II of the genuine *On Floating Bodies* is to be found in the great artist's notebooks. It is worth remarking, however, that although Leonardo showed little knowledge of Archimedean hydrostatics he had considerable success in the practical hydrostatic questions that arose from his study of pumps and other hydrostatic devices.[22]

The problems Leonardo considered concerning the hydrostatic equilibrium of liquids in communicating vessels were of two kinds: those in which the liquid is under the influence of gravity alone and those in which the liquid is under the external pressure of a piston in one of the communicating vessels. In connection with problems of the first kind, he expressly and correctly stated the law of communicating vessels in *Codex Atlanticus*, folio 219v-a: "The surfaces of all liquids at rest, which are joined together below, are always of equal height." Leonardo further noted in various ways that a quantity of water will never lift another quantity of water, even if the second quantity is in a narrower vessel, to a level that is higher than its own, whatever the ratio between the surfaces of the two communicating vessels (*Codex Atlanticus*, fol. 165v)—although he never gave, as far as can be seen, the correct explanation: the equality of the air pressure on both surfaces of the water in the communicating vessels. In some but not all passages (see *Codex Leicester*, fol. 25r; *Codex Atlanticus*, fol. 321v-a) Leonardo did free himself from the misapplication to hydrostatic equilibrium of the principle of the equilibrium of a balance of equal arms bearing equal weights (*Codex Atlanticus*, fol. 206r-a; *Codex Arundel*, fol. 264r). His own explanations are not happy ones, however. He also correctly analyzed the varying levels that would result if, to a liquid in a U-tube, were added a specifically lighter liquid which does not mix with the initial liquid (see Figure 17). In MS *E*, folio

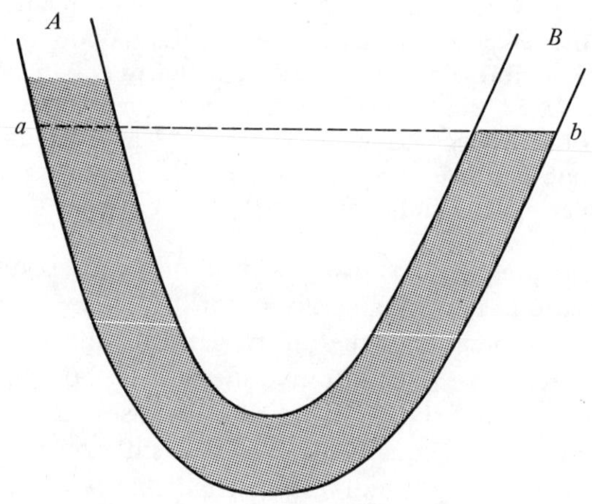

FIGURE 17

74v, he indicated that if the specific weight of the initial liquid were double that of the liquid added, the free surface of the heavier liquid in limb *B* would be at a level halfway between the level of the free surface of the lighter liquid and the surface of contact of the two liquids (the last two surfaces being in limb *A*).

In problems of the second kind, in which the force of a piston is applied to the surface of the liquid in one of the communicating vessels (Figure 18), despite some passages in which he gave an incorrect or only partially true account (see MS *A*, fol. 45r; *Codex Atlanticus*, fol. 384v-a), Leonardo did compose an entirely correct and generally expressed applicable

LEONARDO DA VINCI

statement or rule (*Codex Leicester*, fol. 11r). Assuming the tube on the right to be vertical and cylindrical, Leonardo observed that the ratio between the pressing weight of the piston and the weight of water in the tube on the right, above the upper level of the water in the vessel on the left, is equal to the ratio between the area under pressure from the piston and the area

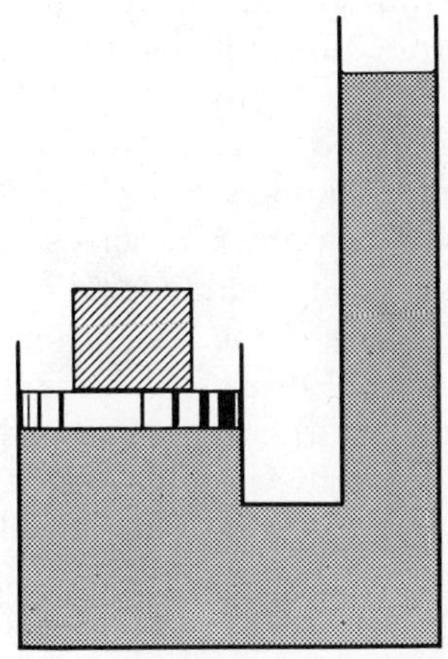

FIGURE 18

of the tube on the right (see *Codex Atlanticus*, fols. 20r, 206r-a, 306v-c; *Codex Leicester*, fol. 26r). And indeed, in a long explanation accompanying the rule in *Codex Leicester*, folio 11r, Leonardo approached, although still in a confused manner, the concept of pressure itself (a concept that appears in no other of his hydrostatic passages, so far as is known) and that of its uniform transmission through a liquid. Leonardo did not, however, generalize his observations to produce Pascal's law, that any additional pressure applied to a confined liquid at its boundary will be transmitted equally to every point in the liquid. Incidentally, the above-noted passage in *Codex Leicester*, folio 11r, also seems to imply a significant consequence for problems of the first kind: that the pressure in an enclosed liquid increases with its depth below its highest point.

Any discussion of Leonardo's knowledge of hydrostatics should be complemented with a few remarks regarding his observations on fluid motion. From the various quotations judiciously evaluated by Truesdell, one can single out some in which Leonardo appears as the first to express special cases of two basic laws of fluid mechanics. The first is the principle of continuity, which declares that the speed of steady flow varies inversely as the cross-sectional area of the channel. Leonardo expressed this in a number of passages, for example, in MS *A*, folio 57v, where he stated: "Every movement of water of equal breadth and surface will run that much faster in one place than in another, as [the water] may be less deep in the former place than in the latter" (compare MS *H*, fol. 54(6)v; *Codex Atlanticus*, fols. 80r-b, 81v-a; and *Codex Leicester*, fols. 24r and, particularly, 6v). He clearly recognized the principle as implying steady discharge in *Codex Atlanticus*, folio 287r-b: "If the water is not added to or taken away from the river, it will pass with equal quantities in every degree of its breadth [length?], with diverse speeds and slownesses, through the various straitnesses and breadths of its length."

The second principle of fluid motion enunciated by Leonardo was that of equal circulation. In its modern form, when applied to vortex motion, it holds that the product of speed and length is the same on each circle of flow. Leonardo expressed it, in *Codex Atlanticus*, folio 296v-b, as:

> The helical or rather rotary motion of every liquid is so much the swifter as it is nearer to the center of its revolution. This that we set forth is a case worthy of admiration; for the motion of the circular wheel is so much the slower as it is nearer to the center of the rotating thing. But in this case [of water] we have the same motion, through speed and length, in each whole revolution of the water, just the same in the circumference of the greatest circle as in the least. . . .

It could well be that Leonardo's reference to the motion of the circular wheel was suggested by either a statement in Pseudo-Aristotle's *Mechanica* (848A)—that on a rotating radius "the point which is farther from the fixed center is the quicker"—or by the first postulate of the thirteenth-century *Liber de motu* of Gerard of Brussels, which held that "those which are farther from the center or immobile axis are moved more [quickly]. Those which are less far are moved less [quickly]."[23] At any rate, it is worthy of note that Leonardo's statement of the principle of equal circulation is embroidered by an unsound theoretical explanation, but even so, one must agree with Truesdell's conclusion (p. 79): "If Leonardo discovered these two principles from observation, he stands among the founders of western mechanics."

The analysis can now be completed by turning to Leonardo's more general efforts in the dynamics and kinematics of moving bodies, including those that descend under the influence of gravity. In dynamics Leonardo often expressed views that were Aristotelian or Aristotelian as modified by Scholastic writers. His notes contain a virtual flood of definitions of gravity, weight, force, motion, impetus, and percussion:

1. MS *B*, folio 63r: "Gravity, force, material motion and percussion are four accidental powers

LEONARDO DA VINCI

with which all the evident works of mortal men have their causes and their deaths."

2. *Codex Arundel*, folio 37r: "Gravity is an invisible power which is created by accidental motion and infused into bodies which are removed from their natural place."

3. *Codex Atlanticus*, folio 246r-a: "The power of every gravity is extended toward the center of the world."

4. *Codex Arundel*, folio 37v: "Gravity, force and percussion are of such nature that each by itself alone can arise from each of the others and also each can give birth to them. And all together, and each by itself, can create motion and arise from it" and "Weight desires [to act in] a single line [that is, toward the center of the world] and force an infinitude [of lines]. Weight is of equal power throughout its life and force always weakens [as it acts]. Weight passes by nature into all its supports and exists throughout the length of these supports and completely through all their parts."

5. *Codex Atlanticus*, folio 253r-c: "Force is a spiritual essence which by accidental violence is conjoined in heavy bodies deprived of their natural desires; in such bodies, it [that is, force], although of short duration, often appears [to be] of marvelous power. Force is a power that is spiritual, incorporeal, impalpable, which force is effected for a short life in bodies which by accidental violence stand outside of their natural repose. 'Spiritual,' I say, because in it there is invisible life; 'incorporeal' and 'impalpable,' because the body in which it arises does not increase in form or in weight."

6. MS *A*, folio 34v: "Force, I say to be a spiritual virtue, an invisible power, which through accidental, external violence is caused by motion and is placed and infused into bodies which are withdrawn and turned from their natural use. . . ."

On the same page as the last Leonardo indicated that force has three "offices" ("ofizi") embracing an "infinitude of examples" of each. These are "drawing" ("tirare"), "pushing" ("spignere"), and "stopping" ("fermare"). Force arises in two ways: by the rapid expansion of a rare body in the presence of a dense one, as in the explosion of a gun, or by the return to their natural dispositions of bodies that have been distorted or bent, as manifested by the action of a bow.

Turning from the passages on "force" to those on "impetus," it is immediately apparent that Leonardo has absorbed the medieval theory that explains the motion of projectiles by the impression of an impetus into the projectile by the projector, a theory outlined in its most mature form by Jean Buridan and repeated by many other authors, including Albert of Saxony, whose works Leonardo had read.[24] Leonardo's dependence on the medieval impetus theory is readily shown by noting a few of his statements concerning it:

1. MS *E*, folio 22r: "Impetus is a virtue ["virtù"] created by motion and transmitted by the motor to the mobile that has as much motion as the impetus has life."

2. *Codex Atlanticus*, folio 161v-a: "Impetus is a power ["potenzia"] of the motor applied to its mobile, which [power] causes the mobile to move after it has separated from its motor."

3. MS *G*, folio 73r: "Impetus is the impression of motion transmitted by the motor to the mobile. Impetus is a power impressed by the motor in the mobile. . . . Every impression tends toward permanence or desires to be permanent."

In the last passage Leonardo's words are particularly reminiscent of Buridan's "inertia-like" impetus. On the other hand, Buridan's quantitative description of impetus as directly proportional to both the quantity of prime matter in and the velocity of the mobile is nowhere evident in Leonardo's notebooks. In some passages Leonardo noted the view held by some of his contemporaries (such as Agostino Nifo) that the air plays a supplementary role in keeping the projectile in motion (see *Codex Atlanticus*, fol. 168v-b): "Impetus is [the] impression of local motion transmitted by the motor to the mobile and maintained by the air or the water as they move in order to prevent a vacuum" (cf. *ibid.*, fol. 219v-a). In *Codex Atlanticus*, folio 108r-a, however, he stressed the role of the air in resisting the motion of the projectile and concluded that the air gives little or no help to the motion. Furthermore, in *Codex Leicester*, folio 29v, he gives a long and detailed refutation of the possible role of the air as motor, as Buridan had before him. And not only is it the air as resistance that weakens the impetus in a projectile; the impetus is also weakened and destroyed by the tendency to natural motion. For example, in MS *E*, folio 29r, Leonardo says: "But the natural motion conjoined with the motion of a motor [that is, arising from the impetus derived from the motor] consumes the impetus of the motor."

Leonardo also applied the impetus theory to many of the same inertial phenomena as did his medieval predecessors—for instance, to the stability of the spinning top (MS *E*, fol. 50v), to pendular and other kinds of oscillating motion (*Codex Arundel*, fol. 2r), to impact and rebound (*Codex Leicester*, fols. 8r, 29r), and to the common medieval speculation regarding a ball falling through a hole in the earth to the center and rising on the other side before falling back and oscillating about the center of the earth (*Codex Atlanticus*, fol. 153v-b). In a sense, all of these last are embraced by the general statement in MS

LEONARDO DA VINCI

E, folio 40v: "The impetus created in whatever line has the power of finishing in any other line."

In the overwhelming majority of passages, Leonardo applied the impetus theory to violent motion of projection. In some places, however, he also applied it to cases of natural motion, as did his medieval predecessors when they explained the acceleration of falling bodies through the continuous impression of impetus by the undiminished natural weight of the body. For example, in *Codex Atlanticus*, folio 176r-a, Leonardo wrote: "Impetus arises from weight just as it arises from force." And in the same manuscript (fol. 202v-b) he noted the continuous acquisition of impetus "up to the center [of the world]." Leonardo was convinced of the acceleration of falling bodies (although his kinematic description of that fall was confused); hence, he no doubt believed that the principal cause of such acceleration was the continual acquisition of impetus.

One last aspect of Leonardo's impetus doctrine remains to be discussed—his concept of compound impetus, defined in MS *E*, folio 35r, as that which occurs when the motion "partakes of the impetus of the motor and the impetus of the mobile." The example he gave is that of a spinning body moved by an external force along a straight line. When the impetus of the primary force dominates, the body moves along a simple straight line. As that impetus dies, the rotary motion of the spinning body acts with it to produce a composite-curved motion. Finally, all of the impetus of the original motion is dissipated and a simple circular motion remains that arises only from the spinning body. A series of passages (*Codex Arundel*, fols. 143r–144v) are further concerned with the relationship of transversal and natural motions, which, if the passages and diagrams have been understood correctly, concur to produce resultant composite motions. It appears later in the same manuscript (fol. 147v) that Leonardo thought that the first part of a projectile path was straight until the primary impetus diminished enough for the natural motion to have an effect:

> The mobile is [first] moved in that direction ["aspetto"] in which the motion of its motor is moved. The straightness of the transversal motion in the mobile lasts so long as the internal power given it by the motor lasts. Straightness is wanting to the transversal motion because [that is, when] the power which the mobile acquired from its motor diminishes.

A beautiful instance of compound motion upon which Leonardo reported more than once is that of an arrow or stone shot into the air from a rotating earth, which arrow or stone would fall to the ground with rectilinear motion with respect to the rotating earth because it receives a circular impetus from the earth. But with respect to a stationary frame, the descent is said to be spiral, that is, compounded of

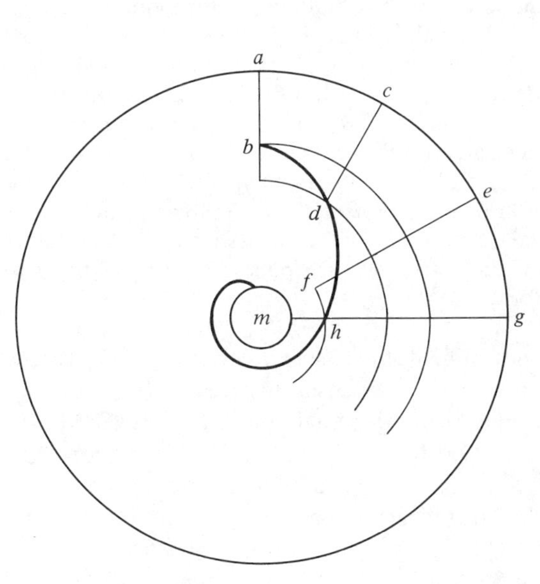

(As in manuscript)

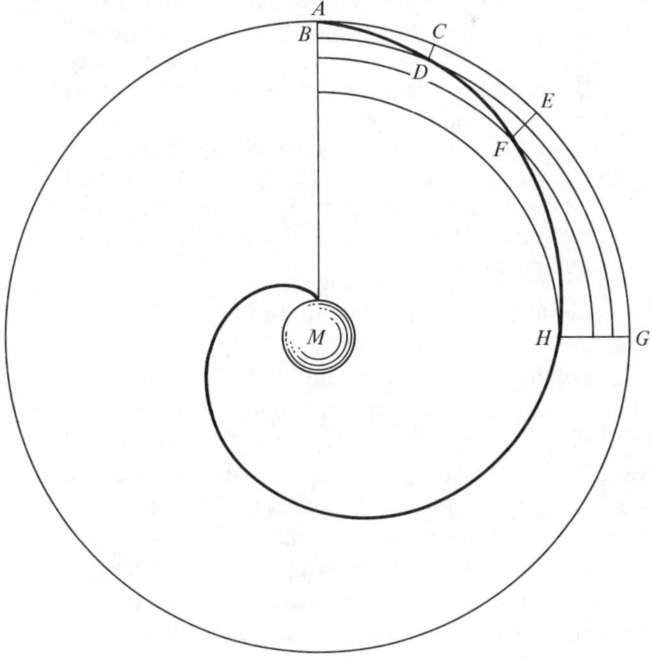

(Reconstructed)

FIGURE 19

LEONARDO DA VINCI

rectilinear and circular motions. The longer of the passages in which Leonardo described this kind of compound motion is worth quoting (MS *G*, fol. 55r; see Figure 19):

> On the heavy body descending in air, with the elements rotating in a complete rotation in twenty-four hours. The mobile descending from the uppermost part of the sphere of fire will produce a straight motion down to the earth even if the elements are in a continuous motion of rotation about the center of the world. Proof: let *b* be the heavy body which descends through the elements from [point] *a* to the center of the world, *m*. I say that such a heavy body, even if it makes a curved descent in the manner of a spiral line, will never deviate in its rectilinear descent along which it continually proceeds from the place whence it began to the center of the world, because when it departs from point *a* and descends to *b*, in the time in which it has descended to *b*, it has been carried on to [point] *d*, the position of *a* having rotated to *c*, and so the mobile finds itself in the straight line extending from *c* to the center of the world, *m*. If the mobile descends from *d* to *f*, the beginning of motion, *c*, is, in the same time, moved from *c* to *f* [!*e*]. And if *f* descends to *h*, *e* is rotated to *g;* and so in twenty-four hours the mobile descends to the earth [directly] under the place whence it first began. And such a motion is a compounded one.
>
> [In margin:] If the mobile descends from the uppermost part of the elements to the lowest point in twenty-four hours, its motion is compounded of straight and curved [motions]. I say "straight" because it will never deviate from the shortest line extending from the place whence it began to the center of the elements, and it will stop at the lowest extremity of such a rectitude, which stands, as if to the zenith, under the place from which the mobile began [to descend]. And such a motion is inherently curved along with the parts of the line, and consequently in the end is curved along with the whole line. Thus it happened that the stone thrown from the tower does not hit the side of the tower before hitting the ground.

This is not unlike a passage found in Nicole Oresme's *Livre du ciel et du monde.*[25]

In view of the Aristotelian and Scholastic doctrines already noted, it is not surprising to find that Leonardo again and again adopted some form of the Peripatetic law of motion relating velocity directly to force and inversely to resistance. For example, MS *F*, folios 51r-v, lists a series of Aristotelian rules. It begins "1^{0} if one power moves a body through a certain space in a certain time, the same power will move half the body twice the space in the same time. . . ." (compare MS *F*, fol. 26r). This law, when applied to machines like the pulley and the lever, became a kind of primitive conservation-of-work principle, as it had been in antiquity and the Middle Ages. Its sense was that "there is in effect a definite limit to the results of a given effort, and this effort is not alone a question of the magnitude of the force but also of the distance, in any given time, through which it acts. If the one be increased, it can only be at the expense of the other."[26] In regard to the lever, Leonardo wrote (MS *A*, fol. 45r): "The ratio which the length of the lever has to the counterlever you will find to be the same as that in the quality of their weights and similarly in the slowness of movement and in the quality of the paths traversed by their extremities, when they have arrived at the permanent height of their pole." He stated again (*E*, fol. 58v):

> By the amount that accidental weight is added to the motor placed at the extremity of the lever so does the mobile placed at the extremity of the counterlever exceed its natural weight. And the movement of the motor is greater than that of the mobile by as much as the accidental weight of the motor exceeds its natural weight.

Leonardo also applied the principle to more complex machines (MS *A*, fol. 33v): "The more a force is extended from wheel to wheel, from lever to lever, or from screw to screw, the greater is its power and its slowness." Concerning multiple pulleys, he added (MS *E*, fol. 20v):

> The powers that the cords interposed between the pulleys receive from their motor are in the same ratio as the speeds of their motions. Of the motions made by the cords on their pulleys, the motion of the last cord is in the same ratio to the first as that of the number of cords; that is, if there are five, the first is moved one braccio, while the last is moved 1/5 of a braccio; and if there are six, the last cord will be moved 1/6 of a braccio, and so on to infinity. The ratio which the motion of the motor of the pulleys has to the motion of the weight lifted by the pulleys is the same as that of the weight lifted by such pulleys to the weight of its motor. . . .

It is not difficult to see why Leonardo, so concerned with this view of compensating gain and loss, attacked the speculators on perpetual motion (*Codex Forster* II_2, fol. 92v): "O speculators on continuous motion, how many vain designs of a similar nature have you created. Go and accompany the seekers after gold."

One area of dynamics that Leonardo treated is particularly worthy of note, that which he often called "percussion." In this area he went beyond his predecessors and, one might say, virtually created it as a branch of mechanics. For him the subject included not only effects of the impacts of hammers on nails and other surfaces (as in MS *C*, fol. 6v; MS *A*, fol.

LEONARDO DA VINCI

53v) but also rectilinear impact of two balls, either both in motion or one in motion and the other at rest (see the various examples illustrated on MS *A*, fol. 8r), and rebound phenomena off a firm surface. In describing impacts in *Codex Arundel*, folio 83v, Leonardo wrote: "There are two kinds ["nature"] of percussion: the one when the object [struck] flees from the mobile that strikes it; the other when the mobile rebounds rectilinearly from the object struck." In one passage (MS *I*, fol. 41v), a problem of impacting balls is posed: "Ball *a* is moved with three degrees of velocity and ball *b* with four degrees of velocity. It is asked what is the difference ["varietà"] in such percussion [of *a*] with *b* when the latter ball would be at rest and when it [*a*] would meet the latter ball [moving] with the said four degrees of velocity."

In some passages Leonardo attempted to distinguish and measure the relative effects of the impetus of an object striking a surface and of the percussion executed by a resisting surface. For example, in *Codex Arundel*, folio 81v, he showed the rebound path as an arc (later called "l'arco del moto refresso") and indicated that the altitude of rebound is acquired only from the simple percussion, while the horizontal distance traversed in rebound is acquired only from the impetus that the mobile had on striking the surface, so that "by the amount that the rebound is higher than it is long, the power of the percussion exceeds the power of the impetus, and by the amount that the rebound's length exceeds its height, the percussion is exceeded by the impetus."

What is perhaps Leonardo's most interesting conclusion about rebound is that the angle of incidence is equal to that of rebound. For example, in *Codex Arundel*, folio 82v, he stated: "The angle made by the reflected motion of heavy bodies becomes equal to the angle made by the incident motion." Again, in MS *A*, folio 19r (Figure 20), he wrote:

> Every blow struck on an object rebounds rectilinearly at an angle equal ["simile"] to that of percussion. This proposition is clearly evident, inasmuch as, if you would strike a wall with a ball, it would rise rectilinearly at an angle equal to that of the percussion. That is, if the ball *b* is thrown at *c*, it will return rectilinearly through the line *cb* because it is constrained to produce equal angles on the wall *fg*. And if you throw it along line *bd*, it will return rectilinearly along line *de*, and so the line of percussion and the line of rebound will make one angle on wall *fg* situated in the middle between two equal angles, as *d* appears between *m* and *n*. [See also *Codex Atlanticus*, fol. 125r-a.]

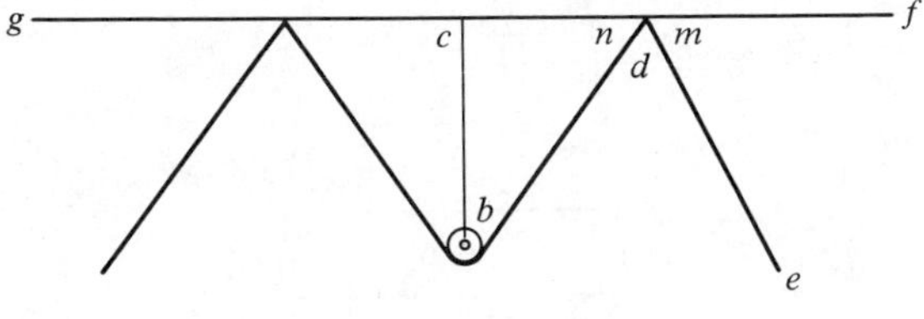

FIGURE 20

The transfer, and in a sense the conservation, of power and impetus in percussion is described in *Codex Leicester*, folio 8r:

> If the percussor will be equal and similar to the percussed, the percussor leaves its power completely in the percussed, which flees with fury from the site of the percussion, leaving its percussor there. But if the percussor—similar but not equal to the percussed—is greater, it will not lose its impetus completely after the percussion but there will remain the amount by which it exceeds the quantity of the percussed. And if the percussor will be less than the percussed, it will rebound rectilinearly through more distance than the percussed by the amount that the percussed exceeds the percussor.

Leonardo is here obviously groping for adequate laws of impact.

The last area to be considered is Leonardo's treatment of the kinematics of moving bodies, especially the kinematics of falling bodies. In *Codex Arundel*, folio 176v, he gave definitions of "slower" and "quicker" that rest ultimately on Aristotle:[27] "That motion is slower which, in the same time, acquires less space. And that is quicker which, in the same time, acquires more space." The description of falling bodies in respect to uniform acceleration is, of course, more complex. It should be said at the outset that Leonardo never succeeded in freeing his descriptions from essential confusions of the relationships of the variables involved. Most of his passages imply that the speed of fall is not only directly proportional to the time of fall, which is correct, but that it is also directly proportional to the distance of fall, which is not. In MS *M*, folio 45r, he declared that "the gravity [that is, heavy body] which descends freely, in every degree of time, acquires a degree of motion, and, in every degree of motion, a degree of speed." If, like Duhem, one interprets "degree of motion," that is, quantity of motion, to be equivalent not to distance but to the medieval impetus, then Leonardo's statement is entirely correct and implies only that speed of fall is proportional to time of fall. One might also interpret the passage in MS *M*, folio 44r, in the same way (see Figure 21):

> Prove the ratio of the time and the motion together with the speed produced in the descent of heavy bodies

LEONARDO DA VINCI

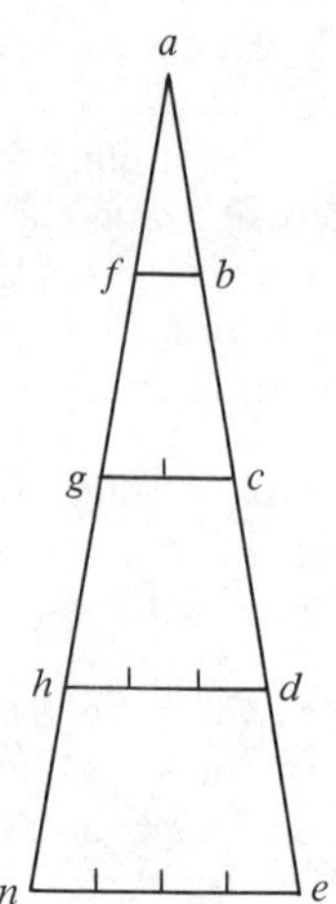

FIGURE 21

by means of the pyramidal figure, for the aforesaid powers ["potenzie"] are all pyramidal since they commence in nothing and go on increasing by degrees in arithmetic proportion. If you cut the pyramid in any degree of its height by a line parallel to the base, you will find that the space which extends from the section to the base has the same ratio as the breadth of the section has to the breadth of the whole base. You see that [just as] *ab* is 1/4 of *ae* so section *fb* is 1/4 of *ne*.

As in mathematical passages, Leonardo here used "pyramidal" where "triangular" is intended.[28] Thus he seems to require the representation of the whole motion by a triangle with point *a* the beginning of the motion and *ne* the final speed, with each of all the parallels representing the speed at some and every instant of time. In other passages Leonardo clearly coordinated instants in time with points and the whole time with a line (see *Codex Arundel*, fols. 176r-v). His triangular representation of quantity of motion is reminiscent of similar representations of uniformly difform motion (that is, uniform acceleration) in the medieval doctrine of configurations developed by Nicole Oresme and Giovanni Casali; Leonardo's use of an isosceles triangle seems to indicate that it was Casali's account rather than Oresme's that influenced him.[29] It should be emphasized, however, that in applying the triangle specifically to the motion of fall rather than to an abstract example of uniform acceleration, Leonardo was one step closer to the fruitful use to which Galileo and his successors put the triangle. Two similar passages illustrate this—the first (MS *M*, fol. 59v) again designates the triangle as a pyramidal figure, while in the second, in *Codex Madrid* I, folio 88v (Figure 22)[30] the triangle is divided into sixteen equally spaced sections. Leonardo explained the units on the left of the latter figure by saying that "these unities are designated to demonstrate that the excesses of degrees are equal." Lower on the same page he noted that "the thing which descends acquires a degree of speed in every degree of motion and loses a degree of time." By "every degree of motion" he may have meant equal vertical spaces between the parallels into which the motion is divided. Hence, this phrase would be equivalent to saying "in every degree of time." The comment about the loss of time merely emphasizes the whole time spent during the completion of the motion.

All of the foregoing comments suggest a possible, even plausible, interpretation of Leonardo's concept of "degree of motion" in these passages. Still, one should examine other passages in which Leonardo seems also to hold that velocity is directly proportional to distance of fall. Consider, for example, MS *M*, folio 44v: "The heavy body ["gravità"] which descends, in every degree of time, acquires a degree of motion more than in the degree of time preceding, and similarly a degree of speed ["velocità"] more than

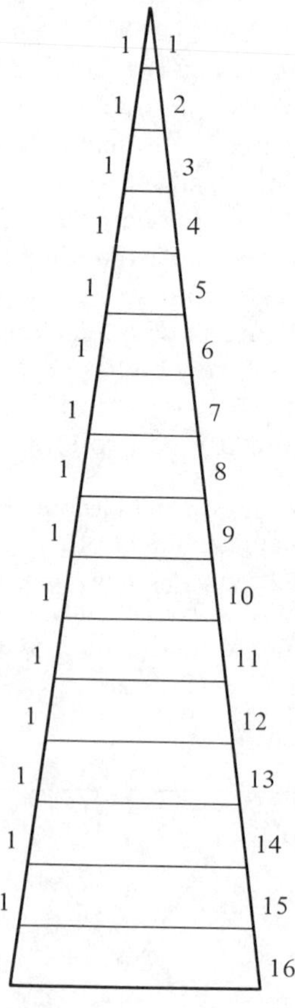

FIGURE 22

LEONARDO DA VINCI

1 2 3 4 5 6 7 8 9 10

FIGURE 23

in the preceding degree of motion. Therefore, in each doubled quantity of time, the length of descent ["lunghezza del discenso"] is doubled and [also] the speed of motion." The figure accompanying this passage (Figure 23) has the following legend: "It is here shown that whatever the ratio that one quantity of time has to another, so one quantity of motion will have to the other and [similarly] one quantity of speed to the other." There seems little doubt from this passage that Leonardo believed that in equal periods of time, equal increments of space are being acquired. One last passage deserves mention because, although also ambiguous, it reveals that Leonardo believed that the same kinds of relationships hold for motion on an incline as in vertical fall (MS *M*, fol. 42v): "Although the motion be oblique, it observes, in its every degree, the increase of the motion and the speed in arithmetic proportion." The figure (Figure 24) indicates that the motion on the incline is represented by the triangular section *ebc*, while the vertical fall is represented by *abc*. Hence, with this figure Leonardo clearly intended that the velocities at the end of both vertical and oblique descents are

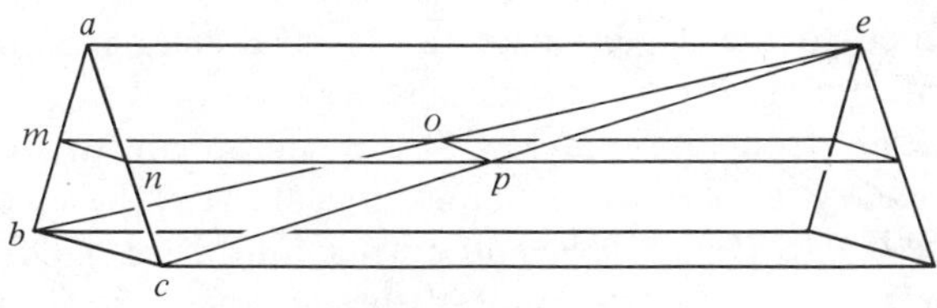

FIGURE 24

equal (that is, both are represented by *bc*) and also that the velocities midway in these descents are equal (that is, $mn = op$). The figure also shows that the times involved in acquiring the velocities differ, since the altitude of $\triangle ebc$ is obviously greater than the altitude of $\triangle$ *abc*.

So much, then, for the most important aspects of Leonardo's theoretical mechanics. His considerable dependence on earlier currents has been noted, as has his quite significant original extension and development of those currents. It cannot be denied, however, that his notebooks, virtually closed as they were to his successors, exerted little or no influence on the development of mechanics.

NOTES

1. The many passages from Leonardo's notebooks quoted here can be found in the standard eds. of the various MSS. Most of them have also been collected in Uccelli's ed. of *I libri di meccanica*. The English trans. (with only two exceptions) are my own.
2. The full corpus has been published in E. A. Moody and M. Clagett, *The Medieval Science of Weights* (Madison, Wis., 1952; repr., 1960). Variant versions of the texts have been studied and partially published in J. E. Brown, "The 'Scientia de ponderibus' in the Later Middle Ages" (dissertation, Univ. of Wis., 1967).
3. Moody and Clagett, *op. cit.*, pp. 128–131. Incidentally, the definition that precedes the proof is the sure sign that Leonardo translated the postulates and first two enunciations from the *Elementa* rather than from version P (where the enunciations are the same), since the definition was not included in Version P.
4. *Ibid.*, pp. 174–207.
5. *Ibid.*, pp. 130–131.
6. *Ibid.*, pp. 188–191. Leonardo's translation of these same propositions in *Codex Atlanticus*, fols. 154v-a–r-a, is very literal: "La equalità della declinazione conserva la equalità del peso. Se per due vie di diverse obliquità due pesi discendano, e sieno medesimo proporzione, se della d[eclinazione] de' pesi col medesimo ordine presa sarà ancora una medesima virtù dell'una e d[ell'altra in discendendo]." The bracketed material has been added from the Latin text.
7. Marcolongo, *Studi vinciani*, p. 173.
8. Duhem, *Les origines de la statique*, I, 189–190.
9. Moody and Clagett, *op. cit.*, pp. 184–187. Leonardo's more literal translation of R1.08 in *Codex Atlanticus*, fol. 154v-a, runs: "Se le braccia della libra sono inequali e nel centro del moto faranno un angolo, s'e' termini loro s'accosteranno parte equalmente alla direzione, e' pesi equali in questa disposizione equalmente peseranno."
10. Marcolongo, *op. cit.*, p. 149, discusses one such passage (*Codex Arundel*, fol. 67v); see also pp. 147–148, discussing the figures on MS *Ashburnham* 2038, fol. 3r; and *Codex Arundel*, fol. 32v (in the passage earlier than the one noted above in the text).
11. Moody and Clagett, *op. cit.*, pp. 208–211.
12. *Ibid.*, pp. 210–211.
13. *Ibid.*, pp. 64–65, 102–109, 192–193.
14. Clagett, "Leonardo da Vinci . . .," pp. 119–140.
15. *Ibid.*, pp. 121–126.
16. *Ibid.*, p. 126.
17. Archimedes, *Monumenta omnia mathematica, quae extant . . . ex traditione Francisci Maurolici* (Palermo, 1685), *De*

LEONARDO DA VINCI

momentis aequalibus, bk. IV, prop. 16, pp. 169–170. Maurolico completed the *De momentis aequalibus* in 1548.
18. Marcolongo, *op. cit.*, pp. 203–216.
19. See Clagett, "Leonardo da Vinci . . .," pp. 140–141. That account is here revised, taking into account the probability that the fragments were not copied by Leonardo.
20. *Ibid.*, p. 140, n. 65, notes the opinion of Favaro and Schmidt that the fragments are truly in Leonardo's hand. But Carlo Pedretti, whose knowledge of Leonardo's hand is sure and experienced, is convinced they are not. Arredi, *Le origini dell'idrostatica*, pp. 11–12, had already recognized that the notes were not in Leonardo's hand.
21. M. Clagett, *The Science of Mechanics in the Middle Ages* (Madison, Wis., 1959; repr., 1961), pp. 124–125.
22. Here Arredi's account is followed closely.
23. Clagett, *The Science of Mechanics*, p. 187.
24. *Ibid.*, chs. 8–9.
25. *Ibid.*, pp. 601–603.
26. Hart, *The Mechanical Investigations*, pp. 93–94.
27. Clagett, *The Science of Mechanics*, pp. 176–179.
28. Clagett, "Leonardo da Vinci . . .," p. 106, quoting MS *K*, fol. 79v.
29. M. Clagett, *Nicole Oresme and the Medieval Geometry of Qualities* (Madison, Wis., 1968), pp. 66–70.
30. I must thank L. Reti, editor of the forthcoming ed. of the Madrid codices, for providing me with this passage.

MARSHALL CLAGETT

MATHEMATICS

Leonardo's admiration for mathematics was unconditional, and found expression in his writings in such statements as "No certainty exists where none of the mathematical sciences can be applied" (MS *G*, fol. 96v). It is therefore useful to consider the development of his mathematical thought, drawing upon the whole of his manuscripts in order to reconstruct its principal stages; such a study will also serve to illustrate the sources for much of his work.

Leonardo particularly valued the rigorous logic implicit in mathematics, whereby the mathematician could hope to attain truth with the same certitude as might a physicist dealing with experimental data. (Students of the moral and metaphysical sciences, on the other hand, had no such expectation, since they were forced to proceed from unascertainable and infinitely arguable hypotheses.) His predilection did not, however, presuppose a broad mathematical culture or any real talent for calculation. Certainly, his education was that of an artisan rather than a mathematician, consisting of reading, writing, and the practical basics of calculus and geometry, together with the considerable body of practical rules that had been accumulated by generations of craftsmen and artists.

The effects of his early education may be seen in Leonardo's literary style (he preferred aphorisms and definitions to any prolonged organic development of ideas) and in his mathematics, which contains grave oversights not entirely due to haste. An example of the latter is his embarrassment when confronted with square and cube roots. In *Codex Arundel* (fol. 200r) he proposed, as an original discovery, a simple method for finding all roots "both irrational and rational," whereby the "root" is defined as a fraction of which the numerator alone is multiplied by itself one or two times to find the square or cubic figure. Thus the square root of 2 is obtained by multiplying 2/2 by 2/2 to find 4/2, or simply 2; while the cube root of 3 is reached by taking $3/9 \times 3/9 \times 3/9 = 27/9$, or 3. He applied this erroneous method several times in his later work; however absurd they might be, Leonardo seems to have been prouder of his discoveries in mathematics than in any other field.

When he was in his late thirties, Leonardo began to try to fill the gaps in his education and took up the serious study of Latin and geometry, among other subjects. At the same time he began to write one or more treatises. He wished to write for scientists, although scholars of the period were not prepared to admit the knowledge possessed by an artist as science. (An illustration of this occurs in Alberti's definition of the "principles" of geometry, in which he stated that he could not use truly scientific terms because he was addressing himself to painters.) It was only after Leonardo entered the Sforza service in Milan and became associated with philosophers and men of letters on a basis of mutual esteem that he took up his theory of the supremacy of painting. This theory was in part incorporated into the *Treatise on Painting* compiled by Leonardo's disciple Francesco Melzi, who, however, omitted Leonardo's definitions of the geometric "principles" of point, line, and area—definitions that Leonardo had drawn up with great care, since he wished them to be read by mathematicians rather than artists.

As a painter, Leonardo was, of course, concerned with proportion and also with "vividness of the actions" (this is how he translated Ficino's definition of beauty as "actus vivacitas"). In his pursuit of the latter, Leonardo inclined toward physics, seeking motions that could be studied experimentally. Perspective, too, had an important place in the treatise on painting, but Leonardo was more concerned with aerial perspective (chiaroscuro) than with linear. MS *C* (1490) is rich in geometrical drawings, although limited to the depiction of the projection of rays of light; it contains no organic system of theorems.

Aside from those in MS *C*, the pages dedicated to geometry in other early manuscripts are few; there are nine in MS *B* (*ca.* 1489) and nineteen in MS *A* (1492). The prevalent matter in all of these is the division of the circumference of the circle into equal parts, a prerequisite for the construction of polygons and for

LEONARDO DA VINCI

various other well-known procedures. The notes set down are not connected with each other and represent elementary precepts that were common knowledge to all "engineers." Because of these limitations, it is impossible to accept Caversazzi's thesis concerning MS *B*, folios 27v and 40r, wherein Leonardo explained how to divide a circumference into equal parts by merely "opening the compass"; Caversazzi sees this as an "exquisite geometric discovery," one that may be applied to solve Euclidean problems of the first and second degree. But since at this time Leonardo had not yet begun to study Euclid, it is probable that he was referring to precepts known to all draftsmen.

Seven pages are devoted to geometry in *Codex Forster* III (1493–1495) and, again, these contain only elementary formulas, expressed in language more imaginative than scientific (the volume of a sphere, for example, is described as "the air enclosed within a spherical body"). Only in *Codex Forster* II_1, written between 1495 and 1497, are there the first signs of a concentrated interest in geometrical problems, particularly those that were deeply to concern Leonardo in later years—lunes and the equivalence of rectilinear and curvilinear surfaces. An additional seventeen pages of this manuscript are devoted to the theory of proportion.

The two manuscripts that mark the close of Leonardo's first sojourn in Milan—MS *M* and MS *I*—are of far greater importance. The first thirty-six pages of MS *M* contain translations of Euclid ("Petitioni" and "Conceptioni") as well as derivations of propositions 1–42 (with a few omissions) of book I of the *Elements*, together with a group of propositions from the tenth book. Propositions 43–46 of book I appear in the first sixteen folios of MS *I* (which must for that reason be considered as being of a later date than MS *M*), as do propositions 1–4 and 6–10 of book II, selections from book III, and occasional references to book X. These evidences of Leonardo's systematic study of Euclid may be related to his friendship and collaboration with Luca Pacioli.

Leonardo must have read the *Elements* and acquired some deeper knowledge of geometry before undertaking the splendid drawings of solid bodies with which he illustrated the first book of Pacioli's *Divina proportione*. His interest in this work is probably reflected in the pages of *Codex Forster* II_1 on proportion; it would also explain his study of the tenth book of the *Elements*, the book least read because of its difficulty and its practical limitation to the construction of regular polygons. Pacioli was thus responsible for arousing Leonardo's enthusiasm for geometry and for introducing him to Euclid's work; indeed, he may have helped him read Euclid, since the text would have been extremely difficult for a man as relatively unlettered as Leonardo. (It is interesting to note that MS *I* also contains some first principles of Latin, copied from Perotti's grammar.) Pacioli's influence on Leonardo was probably also indirect, through his *Summa arithmetica* and perhaps through his translation of the *Elements*.

That Leonardo had at hand the Latin text of the *Elements* of 1482 or 1491 can be seen through the identity of some of his drawings with those texts as well as by certain verbal correspondences. At the same time, his method was not to transcribe sentences from the text he was studying but to attempt to present geometrical ideas graphically. Each page of his geometrical notes represents an aid to memorizing Euclid's text, rather than a compendium; hence it is not always easy to trace the specific passages studied. For example, Leonardo would often begin with a sketch, a number, or occasionally a word which he must have used initially to impress upon his mind some of the intricacies of Euclid's theses and later to recall the whole content. In MS *M*, folio 29v, three figures appear: a point and a line; the same point and line with an additional transversal line; and two parallel lines joined by the transversal, the original point having disappeared. The correspondence with the Euclidean thesis, together with the coincidence of similar notes on the preceding and following pages to propositions 27, 28, 29, 30, and 32, respectively, prove this to be Euclid's book I, proposition 31.

Notes on Euclid's first books also appear in certain folios of the *Codex Atlanticus*, where they are set out in better order and in a more complete form. On folio 169r-b, for instance, the "Petitioni" and "Conceptioni" are presented symbolically, while on folio 177v-a proposition 1.7 is transformed into a series of thirteen drawings.

Leonardo continued to work with Pacioli after they both left Milan in 1499. In the *Codex Atlanticus* he stated that he would learn "the multiplication of roots" from his friend "master Luca" (fol. 120r-d), and he had in fact transcribed all the rules for operations with fractions from the *Summa arithmetica* (fol. 69, a–b). Pietro da Novellara recorded in 1503 that Leonardo was neglecting painting in favor of geometry; this activity is reflected in MS *K*, *Codex Madrid*, II, and *Codex Forster* I_1, as well as in many folios of the *Codex Atlanticus*.

Two-thirds of MS K_1 (1504) are devoted to Euclid. From folio 15v to folio 48v Leonardo copied, in reverse order, almost all the marginal figures of books V and VI of the *Elements*. He transcribed none of the text, although the drawings are accompanied by

LEONARDO DA VINCI

unmistakable signs of his contemplation of the theory of proportion. MS K_2 contains notes similar to those in MS *M* and MS *I*; here they refer to the whole of the second book of the *Elements*, to a few propositions of books I, II, and III, and to nine of the first sixteen definitions of book V.

Leonardo's interest in the theory of proportion is further evident in *Codex Madrid*, II, folios 46v–50, in which he summarizes and describes the treatise "De proportionibus et proportionalitatibus" that is part of Pacioli's *Summa arithmetica.* A drawing on folio 78 of the same manuscript, graphically illustrating all the various kinds of proportions and proportionalities, is also taken from Pacioli's book, while Euclid reappears in the last five pages of the manuscript proper. In the latter portion Leonardo transcribed, in an elegant handwriting, the first pages of an Italian version of the *Elements*, the author of which is not known (the list of Leonardo's books cites only "Euclid, in Italian, that is, the first three books"). Folio 85r contains the ingenious illustration of Euclid's so-called algorithm (VII.2, X.2–3) that is repeated in more detail in *Codex Atlanticus*, folio 207r-b.

It is thus clear from manuscript sources that from the time that Leonardo began to collaborate with Pacioli on the *Divina proportione*, he concentrated on books I and II of Euclid—an indispensable base—and on books V and VI, the theory of proportion that had also been treated by Pacioli in the *Summa arithmetica.* The few references to book X deal with the ratios of incommensurate quantities; therefore there can be no doubt that the subject of proportions and proportionalities remained a constant center of Leonardo's interest.

There is no evidence of a similar study of the last books of Euclid, but it must not be forgotten that about two-thirds of Leonardo's writings have been lost. There is, moreover, no lack of practical applications of the propositions of the last books in Leonardo's writings. But *Codex Madrid,* II, is of particular importance because it demonstrates that Leonardo, after conducting a modest study of Euclid from approximately 1496 to 1504, began to conduct ambitious personal research. Having copied out the first pages of the *Elements* as, perhaps, a model, Leonardo wrote on folios 111r and 112r two titles indicative of the plan of his work, "The Science of Equiparation" and "On the Equality of Unequal Areas." For this new science he wrote on folio 107v a group of "Petitioni" and "Conceptioni" based upon Euclid's. Here Leonardo's chief concern lay in the squaring of curvilinear surfaces, which he generally divided into "falcates" (triangles with one, two, or three curved sides) and "portions" (circular segments formed by an arc and a chord, which could also be the side of a polygon inscribed in a circle). He carried out the transformation of these figures into their rectilinear equivalents by various means, some of which (such as superimposition and motion, or rotation) were mechanical.

The use of mechanical solutions, rarely accepted by Euclid, may perhaps be attributable to Leonardo's engineering background; he would almost seem to have recognized the unorthodoxy of his procedures in applying to himself Giordano's words: "This method is not simply geometrical but is subordinated to, and participates in, both philosophy and geometry, because the proof is obtained by means of motion, although in the end all mathematical sciences are philosophical speculations" (*Codex Madrid*, II, fol. 107r). Leonardo noted, however, that the squaring of curvilinear surfaces could be accomplished by more orthodox geometrical methods, provided the curved sides of the figure form parts of circles proportional among themselves (*Codex Atlanticus*, fol. 139v-a).

It is a short step from squaring falcates to squaring the circle, and Leonardo proposed several solutions to the latter problem. Again, some of them were mechanical and were suggested to him by studying Vitruvius. These consist in, for example, measuring the rectilinear track left by a wheel of which the width is one-quarter of the wheel's diameter (MS *G*, fol. 61r; also MS *E*, fol. 25v, where the width of the wheel that makes "a complete revolution" is mistakenly given as being equal to the radius). The circumference is obtained by winding a thread around the wheel, then withdrawing and measuring it (MS *K*, fol. 80r).

Leonardo was also well acquainted with the method for the "quadratura circuli per lunulas," which he knew from the *De expetendis et fugiendis rebus* of Giorgio Valla. A drawing corresponding to this method appears in MS *K*, folio 61r. (The identity of "Zenophont" or "Zenophonte," a mathematician criticized by Leonardo on the previous page of the same manuscript, is probably resolvable as Antiphon,

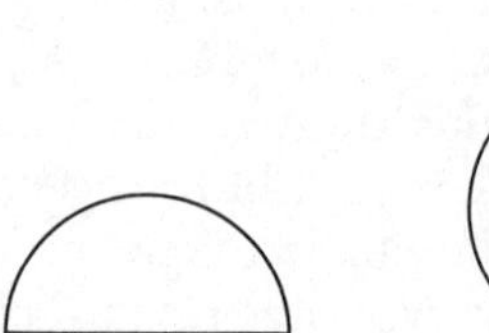

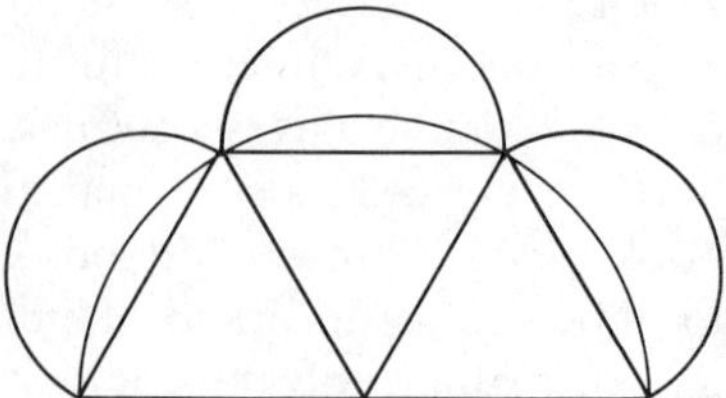

FIGURE 1. The quadratura circuli per lunulas.

LEONARDO DA VINCI

whose method of quadrature was also criticized by Valla.)

Leonardo of course knew Archimedes' solution to squaring the circle; Clagett has pointed out that he both praised and criticized it without ever having understood it completely. He accepted Archimedes' first proposition, which establishes the equivalence of a circle and a rectangular triangle of which the shorter sides are equal to the radius and the circumference of the circle, respectively; he remained unsatisfied with the third proposition, which fixes the approximate ratio between the circumference (taken as an element separating two contiguous classes of polygons) and the diameter as 22:7. Leonardo tried to take this approximation beyond the ninety-six-sided polygon; it is in this attempt at extension ad infinitum that the value of his effort lies.

Valla provided Leonardo with a starting point when, in describing the procedure of squaring "by lunes," he recommended "trapezium dissolvatur in triangula." Leonardo copied Valla's drawing, dividing the trapezium inscribed in a semicircle into three triangles; he extended the concept in MS *K*, folio 80r, in which he split the circle into sixteen triangles, rectified the circumference, then fitted the triangles together like gear teeth, eight into eight, to attain a rectangle equivalent to the circle. He took a further step in *Codex Madrid*, II, folio 105v, when departing from Euclid II.2 (circles are to each other as the squares of their diameters), he took two circles, having diameters of 1 and 14, respectively, and divided the larger into 196 sectors, each equivalent to the whole of the smaller circle.

In *Codex Atlanticus*, folio 118v-a, Leonardo dealt with circles of the ratio 1:1,000 and found that when the larger was split into 1,000 sectors, the "portion" (the difference between the sector and the triangle) was an "imperceptible quantity similar to the mathematical point." By logical extension, if each sector were "a millionth of its circle, it would be a straight portion" or "almost plane, and thus we would have carried out a squaring nearer the truth than Archimedes' " (*Codex Madrid*, II, fol. 105v). In *Codex Arundel*, folio 137v, Leonardo squared the two diameters, writing "one by one is one" and "one million by one million is one million"; his disappointment in seeing that the squares did not increase suggested the variant "four million by four million is sixteen million" (that he corrected his error in *Codex Atlanticus*, fol. 118r-a, writing "a million millions," reveals a poor grasp of calculation).

Leonardo was clearly proud of his discovery of the quadrature, and in *Codex Madrid*, II, folio 112r he recorded the exact moment that it came to him—on the night of St. Andrew's day (30 November), 1504, as hour, light, and page drew to a close. In actual fact, however, Leonardo's discovery amounted only to affirming the equivalence of a given circle to an infinitesimal part of another without calculating any measure; to have made the necessary calculations, he would have had to use Archimedes' formula 22:7, which he rejected. Leonardo characterized Archimedes' method of squaring as "ben detta e male data" (*Codex Atlanticus*, fol. 85r-a), that is, "well said and badly given." It would seem necessary to describe his own solution simply as "detta," not "data."

The presumed discovery on St. Andrew's night did, however, encourage Leonardo to pursue his geometrical studies in the hope of making other new breakthroughs. MS K_3, MS *F*, and a large number of pages in the *Codex Atlanticus* demonstrate his continuous and intense work in transforming sectors, "falcates," and "portions" into rectilinear figures. Taking as given that curved lines must have equal or proportional radii, Leonardo practiced constructing a series of squares doubled in succession, in which he inscribed circles proportioned in the same way; to obtain quadruple circles, he doubled the radii. To obtain submultiple circles, he divided a square constructed on the diameter of the circle into the required number of rectangles, transforming one into a square and inscribing the submultiple circle therein. In order to double the circle, however, instead of using the diagonal of the square constructed on the diameter, he turned to arithmetic calculus to increase proportionally the measure of the radius, adopting a ratio of 1:3/2, which is somewhat greater than the true one of $1:\sqrt{2}$. To obtain a series of circles doubling one another, he divided the greatest radius into equal parts, forgetting that arithmetic progression is not the same as geometric (*Codex Madrid*, II, fols. 117v, 132).

Alternating with the many pages in the *Codex Atlanticus* devoted to the solution of specific problems are others giving the fundamental rules of Leonardo's new science of comparison. Within the same sphere falls a group of pages in *Codex Madrid*, II, which are concerned with transforming rectilinear figures into other equivalent figures or solids into other solids. These sections of the two collections, presumably together with pages that are now lost, are preparatory to the short treatise on stereometry contained in *Codex Forster* I_1 (1505) and entitled "Book on Transformation, That Is, of One Body Into Another Without Decrease or Increase of Substance." In his study of this treatise, Marcolongo judges it to be suitable "for draftsmen more skilled in handling rule, square, and compass than in making numerical

LEONARDO DA VINCI

calculations"—Leonardo's viewpoint was often that of the engineer.

One of the problems to which Leonardo chose to apply his new science was that of how to transform a parallelepiped into a cube, or how to double the volume of a cube and then insert two proportional means between two segments. In the *Summa arithmetica*, Pacioli had solved the problem and had described how to find the cube root of 8 by geometrical methods, although he provided no geometrical demonstrations. Leonardo, in *Codex Atlanticus*, folio 58r, stated that if the edge of a cube is 4, the edge of the same cube doubled will be 5 plus a fraction that is "inexpressible and easier to make than to express." By 1504 Leonardo also possessed Valla's book, which gave the solution of the problem, with numerous demonstrations from the ancients. Leonardo copied various figures from Valla (in *Codex Arundel*, fols. 78r–79v, where the last one is, however, Pacioli's) and transcribed a vernacular translation of the part of Valla's book referring to the demonstrations of Philoponus and Parmenius. (That, having the Latin text at hand, Leonardo felt the need to translate or to have translated the two pages that interested him confirms his difficulty in reading Latin directly—Clagett and Marcolongo have both pointed out many mistakes in the translation of this passage which, if uncorrected, distort the text.)

Leonardo repeatedly expressed his dissatisfaction with the solution reported by Valla. He particularly objected to having to make the rule swing until the compass has fixed two points of intersection; this "negotiation," he stated (*Codex Atlanticus*, fol. 218v-b), seemed to him "dubious and mechanical." Since Valla attributed this procedure to Plato, Leonardo further stated, in MS *F* (1508), folio 59r, that "the proof given by Plato to the inhabitants of Delo is not geometrical." He continued to look for other, more truly "geometrical," means to double the cube or find the cube root.

In folios 50v–59v of MS *F* and in the similar folios 159r-a and b of the *Codex Atlanticus*, Leonardo tried, in opposition to the "ancient system," to resolve the problem by decomposing and rebuilding a cube; he further attempted to apply Pythagoras' theorem by substituting three cubes for three squares constructed on the sides of a right triangle. Since every square is equal to half the square constructed on its diagonal, he attempted to halve each cube with a diagonal cut, on the assumption that the face of a doubled cube could be obtained through manipulating the rectangular face formed by the cut. He arranged nine cubes in the form of a parallelepiped upon which he analyzed the diagonals that he supposed to represent the square and cube roots, respectively (*Codex Atlanticus*, fols. 159r-b, 303r-b; *Windsor Collection*, fol. 19128). He forgot, however, that the progression of squares and cubes does not correspond to the natural series of numbers. Although he studied the proportions between the areas that "cover" a cube and those of the same cube doubled, he realized that he could not apply "the science of cubes based on surfaces, but based on bodies [volumes]" (*Codex Arundel*, fol. 203r). He was thus aware that his formulation of the problem was inaccurate.

Leonardo finally arrived at a solution. He recorded in *Codex Atlanticus*, folios 218v-b, 231r-b, that in order to "avoid the difficulty of the mechanical system" taught by Plato and other ancients, he had

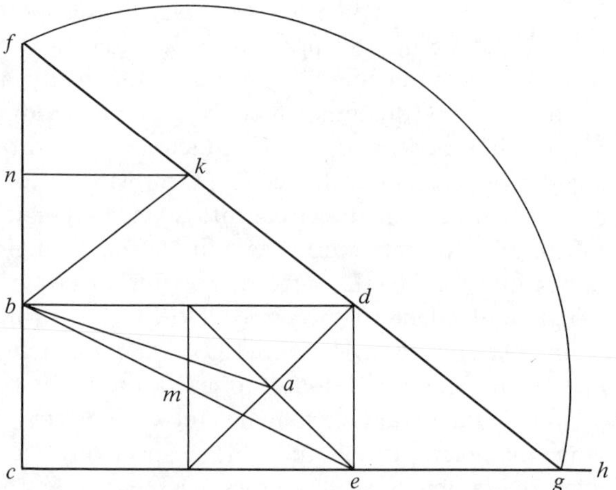

FIGURE 2. $bc : eg = eg : bf = bf : bd$
$eg = fk = kd = kb$
$ab = bf$ (*Codex Atlanticus*, 218v-b).

eliminated the compass and the imprecise movement of the rule in favor of placing two sides of the rectangle in the ratio $b = 2a$ (thereby unknowingly reestablishing the ratio used in the Greek text). Step by step, then, his procedure was to join two faces of a cube to make a rectangle, connect the upper-left angle to the center of the right square, then carry the line obtained to the upper extension of side *a*; from the extreme point thus reached, he drew a line that, touching the upper-right angle of the rectangle, cut the extension of its base. On this line Leonardo was able to determine the measure of the edge of the doubled cube, a measure that is to be found four times in the figure thus constructed. His results are reasonably accurate; with $a = 2$, Leonardo obtained a cube root for 16 as little as 0.02967 in excess of its true value.

Leonardo was unable to demonstrate the truth of his "new invention" save through criticism of the

LEONARDO DA VINCI

"mechanical test" of the compass, which after "laborious effort" helped the ancients to discover the proportional means—whereas now, Leonardo claimed, "without effort I use it to confirm that my experiment was confirmed by the ancients"—which, he added, "could not be done before our time." What Leonardo had discovered was that the third proportional number, used to determine the second, coincides with a line that can be more accurately constructed inside the rectangle; he thereby simplified the classical procedure without in any way discussing its scientific demonstrations, which he took for granted. He must, however, have checked his results experimentally and judged them with his unerring eye.

Other pages of the *Codex Atlanticus* contain summaries of problems to be solved, as, for example, folio 139r-a, which is entitled "Curvilinear Geometrical Elements." This, together with other titles—"On Transformation," folio 128r-a; "Book on Equation," folio 128r-a; and "Geometrical Play" ("De ludo geometrico"), folios 45v-a, 174v-b, 184v-c, and 259v-a —including the previously mentioned "Science of Equiparation"—might suggest that Leonardo had actually written systematic books and treatises that are now lost. But it is known that he cherished projects that he did not complete; and it is unlikely that he composed treatises using methods and forms different from those set out in, for example, *Codex Forster* I_2. The titles of treatises projected but unwritten do provide some record of the stages of his mathematical work, in addition to which Leonardo has recorded some specific dates for his successes.

Leonardo's discovery on St. Andrew's night has already been mentioned; in the same manuscript, *Codex Madrid*, II, on folio 118r, he further mentioned that a certain invention had been given to him "as a gift on Christmas morning 1504." In *Windsor Collection*, folio 19145, he claimed to have discovered, after prolonged research, a way of squaring the angle of two curved sides on Sunday, 30 April 1509. (Folio 128r-a of the *Codex Atlanticus*, the beginning of the "Book of Equation," is concerned with the squaring of a "portion" of a single curved side and promises a second book devoted to the new procedure.)

Although the last discovery is exactly dated, Leonardo's explanation of it is not very clear. (This is often the case in Leonardo's work, since the manuscripts are marked by incomplete and fragmentary discussions, geometrical figures unaccompanied by any explanation of the construction, and the bare statements of something proved elsewhere.) His accomplishment is, however, apparent—he had learned how to vary a "portion" in infinite figures, decomposing and rebuilding it in many different ways. The examples drawn on the *Windsor Collection* page cited reappear, with variations, in hundreds of illustrations scattered or collected in the *Codex Atlanticus*. The reader can only be dismayed by their complexity and monotony.

The mathematical pages of the *Codex Atlanticus* (to which all folio numbers hereafter cited refer) provide an interesting insight into the late developments of Leonardo's method. On folio 139r-a, among other places, he confirmed his proposal to vary "to infinity" one or more surfaces while maintaining the same quantity. It is clear that he received from the elaboration of his geometric equations the same pleasure that a mathematician derives from the development of algebraic ones. If, at the beginning of his research on the measure of curvilinear surfaces, his interest would seem to have been that of an engineer, his later work would seem to be marked by a disinterested passion for his subject. Indeed, the title of the last book that he planned— "Geometrical Play," or "De ludo geometrico"—which he began on folio 45v-a and worked on up to the last years of his life, indicates this.

For example, one of Leonardo's basic exercises consists of inscribing a square within a circle, then joining the resultant four "portions" in twos to get

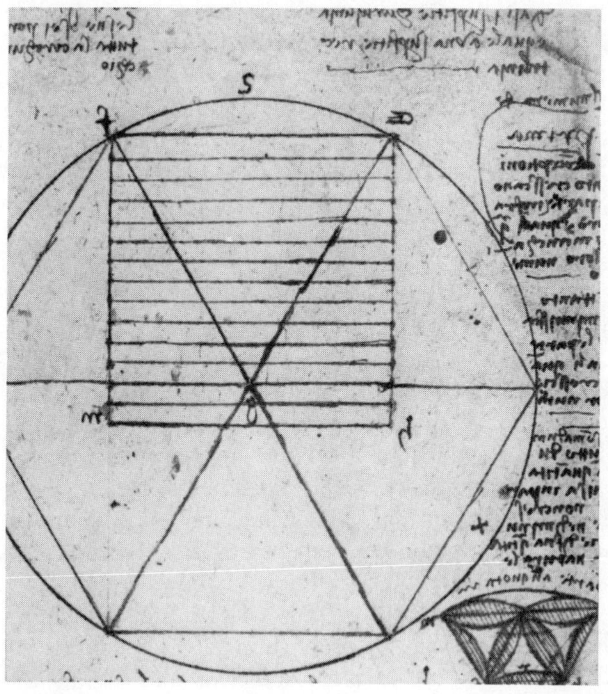

FIGURE 3. The square of proportionality (*Codex Atlanticus*, 111v-b).

LEONARDO DA VINCI

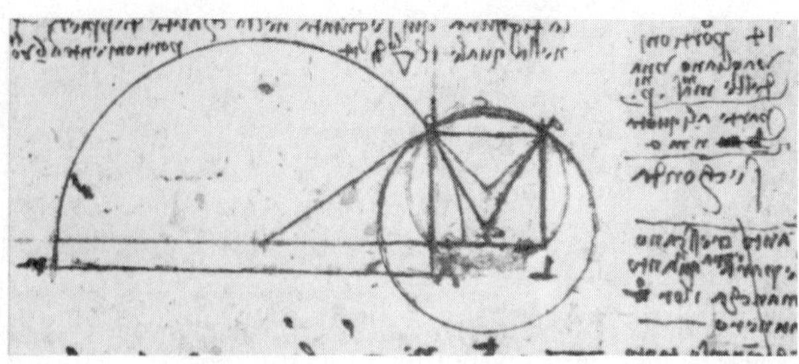

FIGURE 4. (*Codex Atlanticus*, 111v-b).

"bisangoli." The "bisangles" can then be broken up, subdivided, and distributed within the circumference to create full and empty spaces which are to each other as the inscribed square is to the four "portions" of the circumscribed circle. An important development of this is Leonardo's substitution of a hexagon for the square, the hexagon being described as the most perfect division of the circle (fols. 111v-a and b [Figs. 3, 4]). This leaves six "portions" for subsequent operations, and Leonardo recommended their subdivision in multiples of six. (On the splendid folio 110v-a [Fig. 5], he listed and calculated the first fifty multiples of six, but through a curious oversight the product of 34 × 6 is given as 104, rather than 204; and all the following calculations are thus incorrect.)

To obtain the exact dimensions of submultiple "portions," Leonardo devised the "square of proportionality" (fol. 107v-a), which he constructed on the side of the hexagon inscribed within the circle and then subdivided into the required number of equal rectangles. He next transformed one such rectangle into the equivalent square, the side of which is the radius of the submultiplied circle, the side of the corresponding inscribed hexagon, and the origin or chord of the minor "portions." These were then joined in pairs ("bisangoli"), combined into rosettes or stars, and distributed inside the circumference of the circle in a fretwork of full and empty spaces, of which the area of all full spaces is always equal to that of the six greater "portions" and the area of all the empty spaces to the area of the inscribed hexagon. The result is of undoubted aesthetic value and may be taken as the culmination of the geometrical adventure that began for Leonardo with Pacioli's *Divina proportione*.

Although Leonardo remained faithful throughout his life to the idea of proportion as the fundamental structure of reality, the insufficient and inconstant rigor of his scientific thesis precludes his being considered a mathematician in the true sense of the word. His work had no influence on the history of mathematics; it was organic to his reflections as an artist and "philosopher" (as he wished to call himself and as he was ultimately called by the king of France). In his geometrical research he drew upon Archimedes and the ancients' doctrine of lunes to develop the most neglected parts of Euclid's work—curvilinear angles, the squaring of curvilinear areas, and the infinite variation of forms of unaltered quantity. Thus, in Leonardo's philosophy, does nature build and infinitely vary her forms, from the simplest to the most complex; and even the most complex have a rational structure that defines their beauty. It was Leonardo's profound intuition that "Necessity . . . bridle and eternal rule of Nature" must have a mathematical foundation and that the infinite forms found in nature must therefore be the infinite variations of a fundamental "equation."

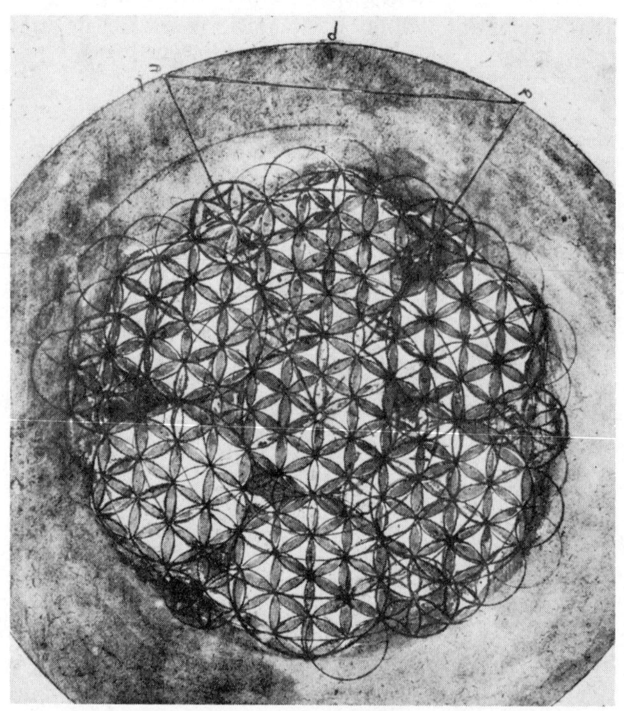

FIGURE 5. (*Codex Atlanticus*, 110v-a).

Leonardo thus expressed a Pythagorean or Platonic conception of reality, common to many artists of the Renaissance. It was at the same time a revolutionary view, inasmuch as it identified form as function and further integrated into the concept of function the medieval notion of substance. Leonardo did not formulate his conception of reality in the abstract terms of universal scientific law. Rather, as the purpose of his painting was always to render the most subtle

LEONARDO DA VINCI

designs of natural structures, his last geometric constructions were aimed at discovering the mathematical structure of nature.

AUGUSTO MARINONI

GEOLOGY

In the later years of his life Leonardo da Vinci described the configuration of the earth's crust as the result of actual processes, principally fluvial, operating over immense periods of time—a system of geology which Duhem described as "perhaps his most complete and lasting invention" (Duhem, II, p. 342).

For Leonardo, the study of the great world was related to the study of man. "Man is the model of the world" (*Codex Arundel*, fol. 156v; De Lorenzo, p. 8), he wrote, ". . . called by the ancients a microcosm . . ., composed like the earth itself, of earth, water, air and fire; as man contains within himself bones, the supports and armature of the flesh, the world has rocks, the supports of the earth: as man has in him the lake of blood which the lungs swell and decrease in breathing, the body of the earth has its oceanic sea which also swells and diminishes every six hours in nourishing the world" (MS *A*, fol. 55v).

Leonardo's geologic perceptions date back to his earliest apprenticeship in Florence. Horizontally stratified rocks in the foreground and pyramidal peaks above a background sea in the *Baptism of Christ* (Uffizi, Florence), which he worked on with his master Verrocchio in 1472, reflect the influence of Van Eyck on Florentine landscape conventions (Castelfranco, p. 472).

Leonardo's earliest dated work (1473), probably drawn in the field, is a sketch of the valley of the Arno (Uffizi). It displays similar characteristics—horizontal strata with waterfalls in the right foreground, and eroded hills above a broad alluvial valley behind.

In subsequent drawings and paintings, rocks may be thrown into cataclysmic contortions but at the same time are folded and fractured realistically. (Leonardo's infrequent mentions of earthquakes [*Codex Leicester*, fol. 10v], volcanoes [*Codex Arundel*, fol. 155r; and Richter, 1939], and internal heat betray a lack of firsthand familiarity with Italy from Naples south. His idea of catastrophe was of storm, avalanche, and flood, and his apocalyptic essays do not alter the essential actualism of his system. He pointedly neglects the plutonist orogenic ideas of Albertus Magnus for the gradualist-neptunism of Albert of Saxony [Duhem, II, p. 334].) The cliffs rising from a harbor in the background to the *Annunciation* (Uffizi) are not unlike views from above Lake Como or Lake Maggiore in the gathering mists. But the ultimate expression of Leonardo's visual apprehension of topography and the true measure of the extent of his journey from the first sketch of 1473 is in the Windsor drawings of the Alps above the plains of the Po (*Windsor Collection*, folios 12410, 12414). They were done in the final years of his life in Italy at about the same time that he formulated his geologic system (Calvi, p. xiii, places the composition of the geologic notes in Tuscany before his Alpine studies). In these notes Leonardo boldly rejected Judeo-Christian cosmology for the secular naturalism of the classical tradition as demonstrated by natural processes and by the actual configuration of the material world. Leonardo estimated that 200,000 years were required for the Po to lay down its plain ([Muentz, p. 34]; he had clearly abandoned the Judeo-Christian time scale for one that was two to three orders of magnitude greater [cf. Duhem, II, p. 335; and Richter, p. 915]).

In Leonardo's system, the highest peaks of the Alps and the Apennines were former islands in an ancient sea. Mountains are continually eroded by winds and rains. Every valley is carved by its river, which is proven by the concordance of the stratigraphic column across the valley walls (*Codex Leicester*, fol. 10r) and the proportionality of river size to valley breadth (*Codex Atlanticus*, fol. 321b). The foothills and plains made by alluvial deposition continually extend the area of land at the expense of the sea. Mountains made lighter by erosion rise slowly to maintain the earth's center of gravity at the center of the universe, bringing up petrified marine strata with their accompanying fauna and flora to be eroded into mountains in their turn.

Such great lakes as the Black Sea are impounded by the collapse of mountains and then, as streams breach the barriers, they drain down one into the other. In this way the Arno was seen to be cutting through its own flood plain, the Po to have filled in the great north Italian triangle from the Alps and Apennines to Venice, and the Nile from Memphis to Alexandria. The Mediterranean itself, the "greatest of rivers," is being filled by the expanding deltas of its tributaries. When the future extension of the Nile erodes through the barrier of the Pillars of Hercules, it will drain what remains of the Mediterranean. Its bottom, relieved of the weight of the superincumbent sea, will rise isostatically (not to be confused with the modern concept of isostasy) to become the summits of mountains.

Just as Leonardo turned to dissection to study anatomy, so he dissected the earth—first in Milan

LEONARDO DA VINCI

during the years that he spent with Ludovico Sforza (il Moro); later as architect and engineer-general to Cesare Borgia and in land reclamation in Tuscany and for Leo X—for a period of 33 years from 1482 until 1515. Excavations for canals, moats, and roadways in Lombardy, Tuscany, Emilia, and the Romagna were carried out under his direct supervision during most of his career. He constructed plans and relief maps requiring exact measurements as well as lithologic and structural insight. His designs for surveying and drafting instruments; his numerous sketches, notes, and calculations of costs, manpower, and time; and his meticulous plans of machines for excavation and hydraulic controls all attest to the extent of his occupation with practical geology. The maps themselves—"bird's-eye" topographic constructions, relief maps, and outline plans of drainage and culture—are the geological analogues of his anatomical drawings, transcending sixteenth-century technics and science.

Leonardo was familiar with classical geological traditions through the works of Albertus Magnus and Cecco d'Ascoli, Vincent of Beauvais, Ramon Lull, Isidore of Seville, Jan de Mandeville, and, above all, Albert of Saxony. Yet his demonstrations of the organic origins of fossils *in situ*, the impossibility of the biblical deluge, and the natural processes of petrifaction are based on meticulous observations—for example, of growth lines on shells (*Codex Leicester*, fol. 10a; Richter, p. 990).

"When I was casting the great horse at Milan. . . .," he wrote, "some countrymen brought to my workshop a great sack of cockles and corals that were found in the mountains of Parma and Piacenza" (*Codex Leicester*, fol. 9v). Was it his experience with the process of casting that led him ultimately to discuss the origins of the fossils and their casts; of worm tracks; *glossopetrae*; fragmented and complete shells; paired and single shells; leaves; tufa; and conglomerates? Did this casting experience lead to a discussion of the relationship of velocity of flow to the sedimentary gradation from the mountains to the sea, of coarse gravels and breccias to the finest white potter's clay?

Leonardo also observed and discussed turbidity currents (*Codex Leicester*, fol. 20r); initial horizontality (MS *F*, fol. 11v); the relationship of sedimentary textures to turbulence of flow, graded bedding, the formation of evaporites (*Codex Atlanticus*, fol. 160); and the association of folded strata with mountains (*Codex Arundel*, fol. 30b, Richter, 982). After numerous false starts, he arrived at an understanding of the hydrologic cycle (MS *E*, fol. 12a and Richter, 930); he dismissed the Pythagorean identification of the forms of the elements with the Platonic solids (MS *F* and Richter, 939); and he recorded the migration of sand dunes (MS *F*, fol. 61a and Richter, 1087).

After the period as engineer with Cesare Borgia, Leonardo had returned to Florence, then proceeded to Rome, and finally found sanctuary in Milan in 1508. Here he made the geologic notes of MS *F* under the heading *De mondo ed acque*. It is from this period also that the red chalk Windsor drawings of the Alps date, reflecting his interest in, and excursions into, the nearby Alps (Clark, p. 134).

Leonardo brought not only his experience and his scientific principles to his landscapes, but also his sense of fantasy (Castelfranco, p. 473). The foreground of the two versions of the *Virgin of the Rocks* (Louvre and National Gallery) appears to be alternately horizontal, vertical, and horizontal strata, with caves widening into tunnels so that the roofs form natural bridges, some of them falling in. Such hollowing out from underneath and falling in of the back is the same mechanism used by Nicholas Steno in his *Prodromus* of 1669 to account for the inclination of strata.

Duhem has argued cogently that Leonardo's ideas were transmitted through Cardano and Palissy to the modern world. There are many similarities which suggest a connection also with the highly influential *Telliamed* of Benoit de Maillet (1749). The notebooks were in part accessible well in advance of the modern development of a natural geology. G. B. Venturi's studies of the geologic material in the notebooks were published in 1797 when catastrophic views of earth history were dominant. This was the year of birth of Charles Lyell, whose *Principles of Geology* in 1830 first formally established actualism as geologic orthodoxy. (In later editions of the *Principles*, Lyell wrote that his attention was called to the Venturi studies by H. Hallam. G. Libri's notes on Leonardo's geology also date to the decade of Lyell's *Principles*. By contrast, the diluvial doctrine which Leonardo had demolished was seriously defended by William Buckland in his *Reliquiae diluvianae* as late as 1823, and catastrophism persisted well into the second half of the nineteenth century.)

Unfortunately, Leonardo rarely if ever sketched an indubitable fossil. He discussed but never illustrated the vivid Lake Garda *ammonitico rosso*—a Jurassic red marble with striking spiral ammonites used extensively in Milan. Leonardo's realistic strata, especially in the *Virgin and St. Anne* of the Louvre, closely resemble these limestones as they weather in the Milanese damp. How is it possible that his "ineffable left hand" traced no illustrations of his comments—those comments which in their freshness, their simplicity, and vivid detail are a warranty of firsthand

LEONARDO DA VINCI

observation and an actualistic geologic position not achieved again for centuries? "The understanding of times past and of the site of the world is the ornament and the food of the human mind," he wrote (*Codex Atlanticus*, fol. 365v).

CECIL J. SCHNEER

BIBLIOGRAPHY

Arbitrary Collections

1. *Codex Atlanticus*	1478–1518
2. *Windsor Collection*	1478–1518
3. *Codex Arundel*	1480–1518

Notebooks

4. *Codex Forster*, I_2 (fols. 41–55)	1480–1490
5. MS *B*	*ca.* 1489
6. *Codex Trivulzianus*	*ca.* 1489
7. MS *C*	1490
8. *Codex Madrid*, II (fols. 141–157)	1491–1493
9. MS *A*	1492
10. *Codex Madrid*, I	1492–1497
11. *Codex Forster*, II_1, II_2, III	1493–1495
12. MS *H*	1493–1494
13. MS *M*	*ca.* 1495
14. MS *I*	1495–1499
15. MS *L*	1497; 1502–1503
16. *Codex Madrid*, II (fols. 1–140)	1503–1505
17. MS K_1	1504
18. MS K_2	1504–1509
19. *Codex Forster*, I_1	1505
20. *Codex on Flight* . . .	1505
21. *Codex Leicester*	*ca.* 1506
22. MS *D*	*ca.* 1508
23. MS *F*	1508–1509
24. MS K_3	1509–1512
25. *Anatomical Folio A*	*ca.* 1510
26. MS *G*	1510–1516
27. MS *E*	1513–1514

Note: Not all scholars are in agreement as to the years given above. It is, however, a matter of plus or minus one or two years.

I. ORIGINAL WORKS. Published treatises (compilations) are *Treatise on Painting* (abr.), pub. by Rafaelle du Fresne (Paris, 1651), complete treatise, as found in *Codex Urbinas latinus* 1270, trans. into English, annotated, and published in facs. by Philip A. McMahon (Princeton, 1956); and *Il trattato del moto e misura dell'acqua* . . ., pub. from *Codex Barberinianus* by E. Carusi and A. Favoro (Bologna, 1923).

Published MSS are *Codex Atlanticus*, facs. ed. (Milan, 1872 [inc.], 1894–1904), consisting of 401 fols., each containing one or more MS sheets (*Codex* is in the Biblioteca Ambrosiana, Milan); MSS *A–M* and *Ashburnham* 2038 and 2037 (in the library of the Institut de France), 6 vols., Charles Ravaisson-Mollien, ed. (Paris, 1881–1891), consisting of 2,178 facs. reproducing the 14 MSS in the Institut de France and Bibliothèque Nationale, with transcription and French trans.; *Codex Trivulzianus*, transcription and annotation by Luca Beltrami (Milan, 1891), containing 55 fols.; *Codex on the Flight of Birds*, 14 fols. pub. by Theodore Sabachnikoff, transcribed by Giovanni Piumati, translated by C. Ravaisson-Mollien (Paris, 1893, 1946); *The Drawings of Leonardo da Vinci at Windsor Castle*, cataloged by Kenneth Clark, 2nd ed., rev. with the assistance of Carlo Pedretti (London, 1968), containing 234 fols.: repros. of all the drawings at Windsor, including the anatomical drawings (notes are not transcribed or translated where this has been done in other works—such as the selections of J. P. Richter, *Anatomical Folios A* and *B*, and the *Quaderni d'anatomia*; see below); *Dell'anatomia fogli A*, pub. by T. Sabachnikoff and G. Piumati (Turin, 1901); *Dell'anatomia fogli B*, pub. by T. Sabachnikoff and G. Piumati (Turin, 1901); *Quaderni d'anatomia*, 6 vols., Ove C. L. Vangensten, A. Fonahn, and H. Hopstock, eds. (Christiania, 1911–1916), all the anatomical drawings not included in *Folios A* and *B; Codex Leicester* (Milan, 1909), 36 fols., pub. by G. Calvi with the title *Libro originale della natura peso e moto delle acque* (*Codex* is in the Leicester Library, Holkham Hall, Norfolk); *Codex Arundel*, a bound vol. marked *Arundel 263*, pub. by the Reale Commissione Vinciana, with transcription (Rome, 1923–1930), containing 283 fols. (*Codex* is in British Museum); and *Codex Forster*, 5 vols., 304 fols., pub. with transcription by the Reale Commissione Vinciana (Rome, 1936) (*Codex* is in the Victoria and Albert Museum, London).

MSS are J. P. Richter, *The Literary Works of Leonardo da Vinci*, 2 vols. (London, 1970), containing a transcription and English translation of a wide range of Leonardo's notes—used as a reference work in Clark's *Catalogue of the Drawings at Windsor Castle;* and E. MacCurdy, *The Notebooks of Leonardo da Vinci*, 2 vols. (London, 1939; 2nd ed., 1956), the most extensive selection of Leonardo's notes.

II. SECONDARY LITERATURE. The following works can be used to obtain a general picture of Leonardo's scientific work: Mario Baratta, *Leonardo da Vinci ed i problemi della terra* (Turin, 1903); Elmer Belt, *Leonardo the Anatomist* (Lawrence, Kan., 1955); Girolamo Calvi, *I manoscritti di Leonardo da Vinci* (Rome, 1925); B. Dibner, *Leonardo da Vinci, Military Engineer* (New York, 1946), and *Leonardo da Vinci, Prophet of Automation* (New York, 1969); Pierre Duhem, *Études sur Léonard de Vinci* (Paris, 1906–1913; repr. 1955); Sigrid Esche-Braunfels, *Leonardo da Vinci, das anatomische Werk* (Basel, 1954); Giuseppe Favoro, *Leonardo da Vinci, i medici e la medicina* (Rome, 1923), and "Leonardo da Vinci e l'anatomia," in *Scientia*, no. 6 (1952), 170–175; Bertrand Gille, *The Renaissance Engineers* (London, 1966); I. B. Hart, *The World of Leonardo da Vinci* (London, 1961); L. H. Heydenreich, *Leonardo da Vinci* (Berlin, 1944), also trans. into English (London, 1954); K. D. Keele, *Leonardo da Vinci on the Movement of the Heart and Blood* (London, 1952); "The Genesis of Mona Lisa," in *Journal of the History of Medicine*, **14**

(1959), 135; and "Leonardo da Vinci's Physiology of the Senses," in C. D. O'Malley, ed., *Leonardo's Legacy* (Berkeley–Los Angeles, 1969); E. MacCurdy, *The Mind of Leonardo da Vinci* (New York, 1928); J. P. McMurrich, *Leonardo da Vinci, the Anatomist* (Baltimore, 1930); Roberto Marcolongo, *Studi vinciani: Memorie sulla geometria e la meccanica di Leonardo da Vinci* (Naples, 1937); and A. Marinoni, "The Manuscripts of Leonardo da Vinci and Their Editions," in *Leonardo saggi e ricerche* (Rome, 1954).

See also C. D. O'Malley, ed., *Leonardo's Legacy—an International Symposium* (Berkeley–Los Angeles, 1969); C. D. O'Malley and J. B. de C. M. Saunders, *Leonardo da Vinci on the Human Body* (New York, 1952); Erwin Panofsky, *The Codex Huygens and Leonardo da Vinci's Art Theory* (London, 1940); A. Pazzini, ed., *Leonardo da Vinci. Il trattato della anatomia* (Rome, 1962); Carlo Pedretti, *Documenti e memorie riguardanti Leonardo da Vinci a Bologna e in Emilia* (Bologna, 1953); *Studi vinciani* (Geneva, 1957); and *Leonardo da Vinci on Painting—a Lost Book (Libro A)* (London, 1965); *Raccolta vinciana*, "Commune di Milano, Castello Sforzesco," I–XX (Milan, 1905–1964); Ladislao Reti, "Le arti chimiche di Leonardo da Vinci," in *Chimica e l'industria*, **34** (1952), 655–721; "Leonardo da Vinci's Experiments on Combustion," in *Journal of Chemical Education*, **29** (1952), 590; "The Problem of Prime Movers," in *Leonardo da Vinci, Technologist* (New York, 1969); and "The Two Unpublished Manuscripts of Leonardo da Vinci in Madrid," *ibid.*; I. A. Richter, *Selections From the Notebooks of Leonardo da Vinci* (Oxford, 1962); Vasco Ronchi, "Leonardo e l'ottica," in *Leonardo saggi e ricerche* (Rome, 1954); George Sarton, *Léonard de Vinci, ingénieur et savant. Colloques internationaux* (Paris, 1953), pp. 11–22; E. Solmi, *Scritti vinciani*, papers collected by Arrigo Solmi (Florence, 1924); K. T. Steinitz, *Leonardo da Vinci's Trattato della pittura* (Copenhagen, 1958); Arturo Uccelli, *I libri di meccanica di Leonardo da Vinci nella ricostruzione ordinata da A. Uccelli* (Milan, 1940); Giorgio Vasari, *Lives of the Painters and Architects* (London, 1927); and V. P. Zubov, *Leonardo da Vinci*, trans. from the Russian by David H. Kraus (Cambridge, Mass., 1968).

The remainder of the bibliography is divided into sections corresponding to those in the text: Technology, Mechanics, Mathematics, and Geology.

Technology. On Leonardo's work in technology, the following should be consulted: T. Beck, *Beiträge zur Geschichte des Maschinenbaues*, 2nd ed. (Berlin, 1900), completed in *Zeitschrift des Vereines deutscher Ingenieure* (1906), 524–531, 562–569, 645–651, 777–784; I. Calvi, *L'architettura militare di Leonardo da Vinci* (Milan, 1943); G. Canestrini, *Leonardo costruttore di macchine e veicoli* (Milan, 1939); B. Dibner, *Leonardo da Vinci, Military Engineer* (New York, 1946); F. M. Feldhaus, *Leonardo der Techniker und Erfinder* (Jena, 1922); R. Giacomelli, *Gli scritti di Leonardo da Vinci sul volo* (Rome, 1936); C. H. Gibbs Smith, "The Flying Machine of Leonardo da Vinci," in *Shell Aviation News*, no. 194 (1954); *The Aeroplane* (London, 1960); and *Leonardo da Vinci's Aeronautics* (London, 1967); B. Gille, *Engineers of the Renaissance* (Cambridge, Mass., 1966); I. B. Hart, *The Mechanical Investigations of Leonardo da Vinci* (London, 1925; 2nd ed., with a foreword by E. A. Moody, Berkeley–Los Angeles, 1963), and *The World of Leonardo da Vinci* (London, 1961); *Léonard de Vinci et l'expérience scientifique au XVI siècle* (Paris, 1952), a collection of articles; *Leonardo da Vinci* (New York, 1967), a collection of articles originally pub. in 1939; R. Marcolongo, *Leonardo da Vinci artista-scienziato* (Milan, 1939); W. B. Parsons, *Engineers and Engineering in the Renaissance* (Baltimore, 1939; 2nd ed., Cambridge, Mass., 1968); L. Reti, "Leonardo da Vinci nella storia della macchina a vapore," in *Rivista di ingegneria* (1956–1957); L. Reti and B. Dibner, *Leonardo da Vinci, Technologist* (New York, 1969); G. Strobino, *Leonardo da Vinci e la meccanica tessile* (Milan, 1953); C. Truesdell, *Essays in the History of Mechanics* (New York, 1968); L. Tursini, *Le armi di Leonardo da Vinci* (Milan, 1952); A. Uccelli, *Storia della tecnica . . .* (Milan, 1945), and *I libri del volo di Leonardo da Vinci* (Milan, 1952); A. P. Usher, *A History of Mechanical Inventions*, rev. ed. (Cambridge, Mass., 1954); and V. P. Zubov, *Leonardo da Vinci* (Cambridge, Mass., 1968).

Mechanics. Particularly useful as a source collection of pertinent passages on mechanics in the notebooks is A. Uccelli, ed., *I libri di meccanica . . .* (Milan, 1942). The pioneer analytic works were those of P. Duhem, *Les origines de la statique*, 2 vols. (Paris, 1905–1906), and *Études sur Léonard de Vinci* (Paris, 1906–1913; repr. 1955). These works were the first to put Leonardo's mechanical works into historical perspective. Their main defect is that a full corpus of Leonardo's notebooks was not available to Duhem. They are less successful in interpreting Leonardo's dynamics. Far less perceptive than Duhem's work is F. Schuster's treatment of Leonardo's statics, *Zur Mechanik Leonardo da Vincis* (Erlangen, 1915), which also suffered because the sources available to him were deficient. A work that ordinarily follows Schuster (and Duhem) is I. B. Hart, *The Mechanical Investigations of Leonardo da Vinci* (London, 1925; 2nd ed., with foreword by E. A. Moody, Berkeley–Los Angeles, 1963). It tends to treat Leonardo in isolation, although the 2nd ed. makes some effort to rectify this deficiency. The best treatment of Leonardo's mechanics remains R. Marcolongo, *Studi vinciani: Memorie sulla geometria e la meccanica di Leonardo da Vinci* (Naples, 1937). It is wanting only in its treatment of Leonardo's hydrostatics and the motion of fluids. For a brief but important study of Leonardo's hydrostatics, see F. Arredi, *Le origini dell'idrostatica* (Rome, 1943), pp. 8–16. For an acute appraisal of Leonardo's mechanics in general and his fluid mechanics in particular, see C. Truesdell, *Essays in the History of Mechanics* (New York, 1968), pp. 1–83, esp. 62–79. Finally, for the influence of Archimedes on Leonardo, see M. Clagett, "Leonardo da Vinci and the Medieval Archimedes," in *Physis*, **11** (1969), 100–151, esp. 108–113, 119–140.

Mathematics. The most thorough study of Leonardo's mathematical works was carried out in the first part of the twentieth century by R. Marcolongo, whose most important writings are: "Le ricerche geometrico-meccaniche di Leonardo da Vinci," in *Atti della Società italiana delle scienze, detta dei XL*, 3rd ser., **23** (1929), 49–100; *Il trattato di Leonardo da Vinci sulle trasformazioni dei solidi* (Naples, 1934); and *Leonardo da Vinci artista-scienziato* (Milan, 1939). The cited article of C. Caversazzi, "Un'invenzione geometrica di Leonardo da Vinci," is in *Emporium* (May 1939), 317–323. Important articles are M. Clagett, "Leonardo da Vinci and the Medieval Archimedes," in *Physis*, **11** (1969), 100–151; and C. Pedretti, "The Geometrical Studies," in K. Clark, *The Drawings of Leonardo da Vinci at Windsor Castle*, 2nd ed., rev. (London, 1968), I, xlix–liii; and "Leonardo da Vinci: Manuscripts and Drawings of the French Period, 1517–1518," in *Gazette des beaux-arts* (Nov. 1970), 185–318.

The following studies by A. Marinoni provide a detailed analysis of some of Leonardo's works: "Le operazioni aritmetiche nei manoscritti vinciani," in *Raccolta vinciana*, XIX (Milan, 1962), 1–62; "La teoria dei numeri frazionari nei manoscritti vinciani. Leonardo e Luca Pacioli," *ibid.*, XX (Milan, 1964), 111–196; and "L'aritmetica di Leonardo," in *Periodico di matematiche* (Dec. 1968), 543–558. See also Marinoni's *L'essere del nulla* (Florence, 1970) on the definitions of the "principles" of geometry in Leonardo, and "Leonardo da Vinci," in *Grande antologia filosofica* (Milan, 1964), VI, 1149–1212, on the supposed definition of the principle of inertia.

Geology. On Leonardo's work in geology, see the following: Mario Baratta, *Leonardo da Vinci ed i problemi della terra* (Turin, 1903); *I disegni geografici di Leonardo da Vinci conservati nel Castello di Windsor* (Rome, 1941); Girolamo Calvi, *Introduction, Codex Leicester* (Rome, 1909); Giorgio Castelfranco, "Sul pensiero geologico e il paesaggio di Leonardo," in Achille Marazza, ed., *Saggi e Ricerche* (Rome, 1954), app. 2; Kenneth Clark, *Leonardo da Vinci* (Baltimore, 1963); Giuseppe De Lorenzo, *Leonardo da Vinci e la geologia* (Bologna, 1920); Pierre Duhem, *Études sur Léonard de Vinci*, 3 vols. (Paris, 1906–1913; repr. 1955); Eugène Muentz, *Leonardo da Vinci, Artist, Thinker, and Man of Science* (New York, 1898); and J. P. Richter, ed., *The Notebooks of Leonardo da Vinci* (New York, 1970).

LEONARDO OF PISA. See **Fibonacci, Leonardo.**

LEONHARD, KARL CÄSAR VON (*b.* Rumpenheim bei Hanau, Germany, 12 September 1779; *d.* Heidelberg, Germany, 23 January 1862), *mineralogy, geology.*

Early in his career, Leonhard adhered to the teachings of Werner but later deserted neptunism for volcanism. As the founding editor of the *Taschenbuch für die gesammte Mineralogie*, Leonhard earned a place among the foremost mineralogists of his time. His prolific writings contributed to the rise of popular interest in geology during the nineteenth century.

In 1797 Leonhard attended the University of Marburg; and in 1798 he went to the University of Göttingen, where his interest in mineralogy was awakened by Johann Friedrich Blumenbach. Because of his early marriage, however, he had to abandon his original intention to study mineralogy under Werner at Freiberg, and he took a position as an assessor in the bureau of land taxes in Hanau. Nevertheless, he corresponded with Werner, Voigt, and von Buch at Freiberg concerning geology and mineralogy, and he devoted his spare time to pursuing these studies. Beginning in 1803 he traveled frequently through Thuringen and Saxony to study the geology of these areas. From 1805 to 1810 he published his three-volume *Handbuch einer allgemeinen topographischen Mineralogie*, in which he took Werner's position on neptunism. Despite the Napoleonic campaigns of this period, Leonhard visited the Austrian Alps and the Salzkammergut, meeting Friedrich Mohs in Vienna and von Moll in Munich. In 1806 he collaborated with K. F. Marx and H. Kopp on *Systematisch tabellarische Übersicht und Charakteristik der Mineralien*, and in 1807 he originated the *Taschenbuch für die gesammte Mineralogie*. This journal soon attained prominence and received widespread support from German scientists, and it made Leonhard's name known throughout Europe. In 1830 the name of the journal was changed to the *Jahrbuch für Mineralogie, Geognosie, Geologie und Petrefaktenkunde*, and from 1833 to 1862 it appeared as the *Neues Jahrbuch für Mineralogie, Geognosie, Geologie und Petrekfaktenkunde*. It remains one of the foremost German scientific journals. From 1811 to 1821 Leonhard also edited the *Allgemeines Repertorium der Mineralogie*.

In 1809 Leonhard became a counselor and adviser in the bureau of mines of the Grand Duchy of Frankfurt, and in 1811 he was appointed the chief of land administration for this regime. In 1813 he was named inspector general and privy councillor of this portion of Napoleon's Confederation of the Rhine. Considered to be a friend of the French, Leonhard was stripped of his offices after the restoration and was forced to take another position as assessor. In 1815, however, he was called to Munich to teach in the academy, and in 1818 he was appointed professor of mineralogy at the University of Heidelberg, a position he held until his death. In an article concerning the instruction of science and medicine at Heidelberg, which appeared in Christian Karl André's *Hesperus* in 1831, Leonhard was described as having an astonishing fund of information concerning minerals and fossils and their classification. The article complained, however, that his approach was completely practical, that his knowledge was limited to the physical properties of minerals, that he completely neglected mathematics and chemistry, and that he used barbaric terminology in his lectures and gesticulated endlessly.

In 1817 Leonhard, together with J. K. Kopp and K. L. Gärtner, published *Propädeutik der Mineralogie*, considered at the time the most instructive source work in mineralogy. In 1818 he wrote *Zu Werners Andenken* as a tribute to his old friend; but that same year he also published *Zur Naturgeschichte der Vulkane*, in which he announced his change of allegiance from neptunism to volcanism because of the increasing evidence forthcoming from the study of basalt.

Leonhard's *Charakteristik der Felsarten* (1823) was the most complete work on petrology that appeared in the early nineteenth century. Although written primarily from a mineralogical point of view, the work also contained information concerning the occurrence of various kinds of rocks. Leonhard attempted to divide rocks into four classes: (1) rocks composed of unlike constituents; (2) rocks that are apparently uniform; (3) derivative or fragmented rocks; and (4) friable rocks. He based these distinctions on visible examination, so that his divisions were arbitrary and to a large extent unsatisfactory.

Leonhard's travels in Auvergne, Bohemia, and other volcanic areas, resulted in *Die Basaltgebilde* (1832). This work was a comprehensive study and conclusively proved the volcanic origin of basalt both with respect to its geological occurrence and to the appearance of the areas where the basalt contacted other rocks. The book aided substantially in the victory of volcanism.

Beginning in 1833 Leonhard turned his attention to the popularization of mineralogy and geology. His first effort, published that year, was *Geologie oder Naturgeschichte der Erde*, and in 1845–1847 he produced a three-volume work entitled *Taschenbuch für Freunde der Geologie* and, in 1846, *Naturgeschichte des Steinreichs*. His last scientific work, *Die Hüttenerzeugnisse als Stützpunkte geologischer Hypothesen*, appeared in 1858.

Leonhardite, $2(Ca_2Al_4Si_8O_{24} \cdot 7H_2O)$, an aluminosilicate of calcium, was named in honor of Leonhard by Blum in 1843.

BIBLIOGRAPHY

I. Original Works. Leonhard was a prolific writer. His chief works are *Handbuch einer allgemeinen topographischen Mineralogie*, 3 vols. (Frankfurt, 1805–1809); *Die Form-Verhältnisse und Grupperungen der Gebirge* (Frankfurt, 1812), written with P. E. Jasson; *Mineralogische Studien* (Nuremberg, 1812), written with C. J. Selb; *Propädeutik der Mineralogie* (Frankfurt, 1817), written with J. K. Kopp and K. L. Gärtner; *Zu Werners Andenken* (Frankfurt, 1818); *Zur Naturgeschichte der Vulkane* (Frankfurt, 1818); *Handbuch der Oryktognosie* (Heidelberg, 1821); *Charakteristik der Felsarten* (Heidelberg, 1823); *Die Naturgeschichte des Mineralreichs* (Heidelberg, 1825), 2nd ed. *Grundzüge der Geognosie und Geologie* (Heidelberg, 1831); *Agenda geognostica* (Heidelberg, 1829); *Die Basaltgebilde* (Stuttgart, 1832); *Geologie oder Naturgeschichte der Erde*, 5 vols. (Heidelberg, 1833); *Lehrbuch der Geognosie und Geologie* (Heidelberg, 1835); *Naturgeschichte des Steinreichs* (Heidelberg, 1846); *Taschenbuch für Freunde der Geologie*, 3 vols. (Heidelberg, 1845–1847); *Aus unserer Zeit in meinem Leben*, 2 vols. (Stuttgart, 1854–1856); and *Die Hüttenerzeugnisse als Stützpunkt geologischer Hypothesen* (Heidelberg, 1858). Leonhard also published about 30 articles in scientific journals.

II. Secondary Literature. See *Allgemeine deutsche Biographie*, XVIII (Leipzig, 1883), 308–311; *Aus der Geschichte der Universität Heidelberg und ihrer Fakültaten* (Heidelberg, 1961); and "Obituary, K. C. von Leonhard," in *American Journal of Science*, 2nd ser., **33** (1862), 453.

John G. Burke

LEONHARDI, JOHANN GOTTFRIED (*b.* Leipzig, Saxony, 18 June 1746; *d.* Dresden, Saxony, 11 January 1823), *chemistry, medicine, pharmacy.*

Leonhardi's father, a physician, inspired a love of learning in his young son. After studying at home under various tutors Leonhardi enrolled in the local lyceum, and there completed his training in the classical languages. In 1764 he entered the University of Leipzig, where in the course of his studies in philosophy and science he heard Johann Heinrich Winkler lecture on physics and Karl Wilhelm Pörner on chemistry. He began to read medicine under Christian Gottlieb Ludwig. He received his baccalaureate in 1767. Continuing his work in medicine, for which he had an "innate love," he delivered his "lectures for the License" in 1769 on the structure and function of the conglobate glands (that is, glands of simple structure, such as the lymphatics). He was awarded a master's degree in 1770 and the baccalaureate in medicine the following year with a dissertation *De resorptionis in corpore humano praeter naturam impeditae causis atque noxis* (Leipzig, 1771).

Remaining at Leipzig as *Privatdozent*, Leonhardi lectured on medicine and chemistry. He became extraordinary professor of medicine in 1781. In 1782 he moved to the University of Wittenberg (at that time still in Saxony) as professor of medicine; in 1791 he accepted the post of personal physician and councillor to the elector of Saxony, Friedrich Augustus III. Leonhardi was also given a seat in the Dresden public health council. Although he retained his professorship at Wittenberg, he arranged for a deputy to lecture in his place. In 1815 he became a knight of the Königliche Sächsische Civil-Verdienstorden. Despite almost constant ill health during the last decade or so of his life, Leonhardi managed to continue his favorite work of editing and translating.

Leonhardi did little original research. The bulk of his writings consists of summaries, compilations, and translations. He also wrote a number of short, academic, and unoriginal medical discourses. His chief accomplishment was *Chymisches Wörterbuch . . . übersetzt und mit Anmerkungen und Zusätzen vermehrt,* 6 pts. (Leipzig, 1781–1783), his authoritative translation of Pierre-Joseph Macquer's *Dictionnaire de chimie.* Leonhardi's well-received and widely used version amounted to a new, expanded edition, approximately two and a half times the length of the original, to which Leonhardi himself contributed about 150 new articles. The abundant annotation shows him to have been in complete command of the chemical literature of his day; the references are given exactly and the commentary in the notes is clear, factual, and comprehensive. Judging by these excellent notes, Leonhardi's primary interest appears to have been the qualitative analysis of inorganic substances, especially those of actual or potential industrial value. He followed the current controversies on the nature of combustion and the "airs" with great interest and published a summary of recent pneumatic discoveries as *Aerologie* (1781). The notational underbrush grows thickest in the less theoretical parts of the *Dictionnaire.* Like Macquer he was not a reductionist; he did not find it helpful to devise premature theories about the ultimate constituents of things. Leonhardi adhered to the phlogiston theory until the early 1790's. Leonhardi's was not a creative mind, but it was surely a disciplined and well-stocked one.

BIBLIOGRAPHY

I. Original Works. This list contains those of Leonhardi's works which are either (1) of special interest for chemistry or (2) not listed in the *British Museum Catalogue of Printed Books* or in the *Index Catalogue of the Surgeon General's Office, U.S. Army* (1st ser.). The latter catalogue has a nearly complete list of Leonhardi's medical works.

1. *Observationes quasdam chemicas proponit praelectionesque suas aestivas* . . . (Leipzig, 1775).

2. "Versuch einer Anleitung zu gemeinnütziger ökonomischen Prüfung der sämtlichen Gewässer in Sachsen," in *Schriften der Leipziger Oekonomischen Societät,* **6** (1784).

3. "Chymische Untersuchung des ächten Braunschweigischen Grüns," *ibid.*

4. *Vinorum alborum metallici contagii suspectorum docimasiae curae repetitae et novae* (Wittenberg, 1787).

5. *Programma de tubarum uterinarum morbis pauca quaedam* (Wittenberg, 1788).

6. *Physiologia muci primarum viarum* (Wittenberg, 1789).

7. *De succorum humanorum salibus dulcibus* (Leipzig, 1790).

8. Letters to the editor, Crell's *Chemische Annalen* (1789), pt. 2, 423–424; (1790), pt. 2, 126–128; (1794), pt. 1, 177–178.

Translations, editions, introductions:

9. D. P. Loyard, *Versuch über einen tollen Hundebiss,* with notes by Leonhardi (Leipzig, 1778).

10. *Herrn Peter Joseph Macquers Chymisches Wörterbuch, oder allgemeine Begriffe der Chemie nach alphabetischer Ordnung, aus dem Französischen nach der zweiten Ausgabe übersetzt und mit Anmerkungen und Züsatzen vermehrt,* 6 vols. (Leipzig, 1781–1783); 2nd ed., 7 vols. (Leipzig, 1788–1791); 3rd ed., prepared by J. B. Richter, 3 vols. (Leipzig, 1806–1809); separate publication of Leonhardi's notes and additions: *Neue Züsatze und Anmerkungen zu Macquers Chymischen Wörterbuch erster Ausgabe,* 2 vols. (Leipzig, 1792).

11. *Aerologie physico-chemicae recentioris primae linae* (Leipzig, 1781); expanded and translated as *Kurzer Umriss der neuern Entdekkungen über die Luftgattungen* (Graz, 1783).

12. *C. W. Scheeles . . . Chemische Abhandlung von der Luft und dem Feuer . . .,* 2nd ed. (Leipzig, 1782), contains notes by Leonhardi which were incorporated into the Baron de Dietrich's *Supplément au Traité Chimique de l'Air et du Feu de M. Scheele . . .* (Paris, 1785). C. G. Kayser, *Vollständiges Bücher-Lexicon* (Leipzig, 1834–1838), lists a third German ed. (1788).

13. *Schwedisches Apothekerbuch,* translated from the 2nd (Latin) ed. and annotated by Leonhardi (Leipzig, 1782).

14. Pierre Bayen, *Untersuchungen über das Zinn, und Beantwortung der Frage: Ob Man sich ohne Gefahr zu ökonomischen Gebrauche der zinnernen Gefässe bedienen könne ? . . . Aus dem Französischen übersetzt; herausgegeben und mit Anmerkungen begleitet D. Johann Gottfried Leonhardi . . .* (Leipzig, 1784).

15. A. F. L. Dörffurt, *Abhandlung über den Kampher, mit Vorrede von J. G. Leonhardi* (Wittenberg, 1793).

16. *Pharmacopoea Saxonica* (Dresden, 1820).

II. Secondary Literature. See J. Ferguson, *Bibliotheca chemica,* II (Glasgow, 1906), 28, and the literature cited there; L. F. F. Flemming, *De vita et meritis beati Joh. Gottfr. Leonhardi* (Dresden, 1823)—the author was unable to locate this in preparing the above article; Michaud, *Biographie universelle,* XXIV, 185–186; and J. R. Partington, *A History of Chemistry,* III (London, 1961), 81–82, 493–494, 616–619. Leonhardi's sketch of his early life is in A. G. Plaz, *Panegyrin medicam* (Leipzig, 1771), xii–xv. I am indebted to Dr. John B. Blake of the U.S. National

Library of Medicine for calling this reference to my attention. For reviews of the *Wörterbuch*, see *Annales de chimie*, **4** (1790), 289; **11** (1791), 204; **12** (1792), 110.

STUART PIERSON

LEONICENO, NICOLÒ (*b*. Vicenza, Italy, 1428; *d*. Ferrara, Italy, 9 June 1524), *medicine, philology*.

Leoniceno's father, Francesco, a physician, was from an ancient noble family of Vicenza; his mother, Madalena, was the daughter of Antonio Loschi, a humanist and secretary of Pope Alexander V. At Vicenza, Leoniceno received thorough training in Latin and Greek from Ognibene de' Bonisoli and then studied philosophy and medicine at Padua, taking his doctorate around 1453. Little is known of his activities during the following decade, except that he is said to have traveled to England; by 1462 he was back at Padua, possibly as a teacher. In 1464 he was called to teach at the University of Ferrara, where he remained for the rest of his ninety-six years, except for a year at Bologna (1508–1509). He did not marry.

Under the celebrated Guarino da Verona, Ferrara had become a major center for the study of classical literature, and Leoniceno took his place there as one of the leading Greek scholars of his age. (He is sometimes confused with the contemporary Hellenist Nicolò Leonico Tomeo.) Among Leoniceno's friends and correspondents were other major figures in the revival of ancient learning, including Pico della Mirandola, Giorgio Valla, Politian, Ermolao Barbaro, and Erasmus.

At Ferrara, Leoniceno taught mathematics, then Greek philosophy, and finally medicine, in which area he made his most important contribution. The teaching in European medical schools at this time was ultimately derived from the Greek physicians, especially Galen, but as interpreted and systematized by the Arabic authors. In the several stages of textual transmission, translation, and interpretation that separated European Arabist medicine from the original Greek sources, considerable distortion and corruption had been introduced; but it was obscured by the poor quality of such translations of genuine Galenic works as were generally available. Humanists had long ridiculed the Arabist medical tradition for its "barbarous" Latinity and the sterility of its scholastic disputations, but not until the late fifteenth century were serious attempts made to provide an alternative by reviving Greek medicine in its pristine form. Leoniceno was one of the chief pioneers in this effort to recover and edit the works of the Greek physicians and to prepare faithful Latin translations, and he was the first to develop this new approach into a serious rival to the established Arabist tradition. Under his leadership Ferrara became the main center for studying the revived Galenic medicine, and from there the movement spread to the other Italian medical schools and then to the rest of Europe. Among Leoniceno's students were Giovanni Manardi, Antonio Musa Brasavola, and Lodovico Bonacciolo; and others who were influenced by him included Giovanni Battista da Monte (Montanus), Thomas Linacre, and Leonhard Fuchs.

In accordance with the general humanist credo, Leoniceno placed great emphasis on the study of individual words and their meanings as the key to understanding the works of the Greek medical authorities. In his preface to his translations of Galen (1508) he cited examples of how an error in transcribing or translating a single word could distort the meaning of an entire passage; and in other works he sought to demonstrate that the words of the Greek physicians had been so often misconstrued by the Arabists as to make the resulting medical system a menace to human life. In his tracts on the errors of Pliny he cited numerous mistakes made by the Arabists in the identification of medicinal herbs described by the Greeks, and he also pointed out many corruptions, both terminological and factual, introduced into Galenic anatomy by Ibn Sīnā, Mondino de' Luzzi, and Benedetti. He based the latter criticisms in part on such Galenic treatises as the *Anatomical Procedures* and *Whether Blood is Naturally Contained in the Arteries* that were not generally available for another generation. In his treatise on syphilis he discussed similar errors in the naming and identification of diseases.

On the more constructive side, Leoniceno supplied the texts for the first genuine Galenic works to be published in Greek (1500) and published Latin translations of eleven Galenic treatises, beginning with the *Ars medica* (1508) and the *Commentary on the Aphorisms of Hippocrates*, together with the *Aphorisms* themselves (1509). He was very proud of these translations and in 1522 published a lengthy apologia against three men, each of whom had criticized his translation of only a single word or phrase. He also translated into Latin and Italian a number of works by nonmedical authors.

In 1497 Leoniceno published the first scholarly treatise on syphilis, a very influential work that went through numerous editions during the following century. He tried to show that although the disease had only recently appeared in Europe, it had occurred at various times in the remote past and was therefore not an essentially new disease.

Leoniceno's work *On the three ordered doctrines according to the opinion of Galen* (1508) was of some importance in the discussion of method during the sixteenth century. At the beginning of *Ars medica* Galen had stated that there are three ways of teaching a subject in an orderly manner: analysis, synthesis, and the explication of definitions. The medieval commentators had sought, with somewhat confusing results, to relate these three ways to the various methods of demonstration and dialectic referred to by Plato, Aristotle, and other authors. Leoniceno showed that as an exegesis of the *Ars medica* the entire discussion was pointless, because Galen was referring simply to didactic techniques and not to methods of philosophic inquiry or demonstration. His clarification of this point seems to have been widely accepted, although around 1517 a professor at Padua (possibly Ludovico Carenzio) defended the medieval authors, leading an anonymous Roman disciple of Leoniceno (or possibly Leoniceno himself) to publish a refutation.

In 1490 Leoniceno inaugurated a famous controversy on the errors of Pliny the elder. In that year he sent to Politian a critique of Ibn Sīnā, in which he noted in passing that Pliny seemed to have confused the two herbs ivy and cistus because of the similarity of their Greek names (*κισσός* and *κίσθος*); Politian commended Leoniceno's castigation of Ibn Sīnā but politely challenged his criticism of Pliny. Leoniceno responded with a tract, *On the errors of Pliny and others in medicine* (1492), in which he not only defended his original point but charged Pliny with many other errors stemming from verbal confusion. This work provoked an indirect response from Barbaro in support of Pliny and a direct attack on Leoniceno by the Ferrarese lawyer Pandolfo Collenuccio. Others joined in the fray on both sides, with Leoniceno himself contributing three additional tracts in 1493, 1503, and 1507, the latter in response to Alessandro Benedetti. (Leoniceno's tracts were apparently circulated in manuscript before being printed.)

This polemic was primarily an internal dispute among humanists, in contrast to Leoniceno's attacks on the Arabists. Pliny's *Natural History* was avowedly a compilation from earlier sources, chiefly Greek, and Leoniceno charged that he had often garbled the information that he transcribed. The controversy stemmed from the reluctance of some humanists to admit that an authentic Roman, who wrote in "eloquent" Latin and was fluent in Greek, could have made the same kinds of linguistic errors as the despised "barbarous" authors. To soften the blow Leoniceno sought to show that the Arabs had made even worse errors than Pliny in interpreting the Greeks and that the medieval Latin authorities were worse still. However, this argument did not placate his opponents, who thought that textual emendation and sympathetic interpretation would absolve Pliny of Leoniceno's charges. In this assumption they were partly justified, since some of the errors that Leoniceno singled out were in fact due to corruptions of the text he had used, and some of his interpretations of Pliny were unnecessarily harsh. Collenuccio also objected strongly to Leoniceno's assumption that the testimony of the Greek authorities, unconfirmed by observation, was an adequate standard of the truth of Pliny's statements, but the immediate outcome of the dispute, as of most of Leoniceno's work, was to enhance the authority of the Greeks at the expense of their Roman, Arabic, and medieval Latin followers. (Leoniceno eventually concluded that Celsus had similarly misinterpreted the Greek medical writers on whom he had relied.) Serious criticism of the Greeks themselves, based on independent observation and reasoning, was to begin a generation later, but it was to take many generations to overthrow completely the Hellenist tradition in medicine that Leoniceno had done so much to establish.

BIBLIOGRAPHY

I. Original Works. *De Plinii et aliorum in medicina erroribus* (Ferrara, 1492) is the first of Leoniceno's tracts on Pliny; *De Plinii, & plurimum aliorum medicorum in medicina erroribus* (Ferrara, 1509) contains all four, as do subsequent eds. His other works are *Libellus de epidemia, quam vulgo morbum gallicum vocant* (Venice–Milan–Leipzig, 1497); *De tiro seu vipera* and *De dipsade et pluribus aliis serpentibus* (Venice, *ca.* 1497), both repr. as *De serpentibus* (Bologna, 1518); *De virtute formativa* (Venice, 1506); *In libros Galeni e greca in latinam linguam a se translatos prefatio communis. Ejusdem in artem medicinalem Galeni . . . prefatio. Galeni ars medicinalis Nicolao Leoniceno interprete. . . . Ejusdem de tribus doctrinis ordinatis secundum Galeni sententiam opus* (Venice, 1508), repr. with *Galeni in aphorismos Hippocratis, cum ipsis aphorismis, eodem Nicolao Leoniceno interprete* (Ferrara, 1509); *Medici Romani Nicolai Leoniceni discipuli antisophisia* [*antisophista*](Bologna, 1519), of uncertain authorship; and *Contra obtrectatores apologia* (Venice, 1522). Most of Leoniceno's works went through one or more subsequent editions; all except his translations, the *Prefatio communis*, and the *Prefatio in artem medicinalem* are included in *Opuscula*, D. A. Leennium, ed. (Basel, 1532). A facs. of the first ed. of *De morbo gallico*, with intro., is included in K. Sudhoff, *The Earliest Printed Literature on Syphilis*, adapted by C. Singer, Monumenta Medica, III (Florence, 1925). Loris Premuda edited *De Plinii . . . erroribus*, with intro. and Italian trans. (Milan–Rome, 1958). R. J. Durling, "A Chronological Census of Renaissance Editions and Translations of Galen," in *Journal of the Warburg and Courtauld*

Institutes, **24** (1961), 230–305, lists Leoniceno's trans. of Galen, p. 297. P. O. Kristeller gives references to Italian archival material about Leoniceno in *Iter italicum*, 2 vols. (London–Leiden, 1963–1967).

II. SECONDARY LITERATURE. The only full-length study is D. Vitaliani, *Della vita e delle opere di Nicolò Leoniceno Vicentino* (Verona, 1892); see also Premuda's intro. to his ed. of *De Plinii . . . erroribus*. On the Galenic revival and Leoniceno's role in it, see Durling's intro. to his "Chronological Census," esp. p. 236. On Leoniceno's influence, esp. in Germany, see Luigi Samoggia, *Le ripercussioni in Germania dell' indirizzo filologico-medico leoniceniano della scuola Ferrarese per opera di Leonardo Fuchs*, Quaderni de Storia Della Scienza e Della Medicina, IV (Ferrara, 1964). On method, see N. W. Gilbert, *Renaissance Concepts of Method* (New York, 1960), pp. 102–104. On the errors of Pliny, see Lynn Thorndike, *A History of Magic and Experimental Science*, IV (New York, 1934), 593–610; A. Castiglioni, "The School of Ferrara and the Controversy on Pliny," in E. A. Underwood, ed., *Science, Medicine, and History* (London, 1953), I, 269–279; and F. Kudlien, "Zwei medizinisch-philologische Polemiken am Ende des 15. Jahrhunderts," in *Gesnerus*, **22** (1965), 85–92.

JEROME J. BYLEBYL

LE PAIGE, CONSTANTIN (*b.* Liège, Belgium, 9 March 1852; *d.* Liège, 26 January 1929), *mathematics.*

Le Paige, the son of Jeanne Jacques and Constantin Marie Le Paige, received his secondary education at Spa and Liège. After taking a course in mathematics with V. Falisse, he entered the University of Liège in 1869, where he attended the lectures of E. Catalan, to whom we owe a special form of determinants, the so-called circulants. After his graduation on 28 July 1875, Le Paige began teaching the theory of determinants and higher analysis at the university. He was appointed extraordinary professor of mathematics in 1882, and ordinary professor in 1885, which post he retained until his retirement in 1922; he was appointed professor emeritus in 1923. His wife was the former Marie Joséphine Ernst.

Le Paige began work at a time when the theory of algebraic forms, initiated by Boole in 1841 and developed by Cayley and Sylvester in England, Hermite in France, and Clebsch and Aronhold in Germany, had drawn the attention of the geometers. This theory studies the properties of such forms, which remain unchanged under linear substitutions. Le Paige's investigations touched mainly upon the geometry of algebraic curves and surfaces, and the theory of invariants and involutions. He coordinated and generalized the extensions which at that time had been tried. His best-known achievement was the construction of a cubic surface given by nineteen points. Starting from the construction of a cubic surface given by a straight line, three groups of three points on a line, and six other points, Le Paige comes to the construction of a cubic surface given by three lines and seven points. From this he proceeds to the construction of a cubic surface given by a line, three points on a line, and twelve other points, and by means of the construction of a cubic surface given by three points on a line and sixteen other points, he arrives at a surface given by nineteen points.

Steiner's theorem, that a conic section can be generated by the intersection of two projective pencils, had been extended by Chasles to plane algebraic curves. Le Paige also studied the generation of plane cubic and quartic curves, and the construction of a plane cubic curve given by nine points is presented in his memoir "Sur les courbes du troisième ordre" (*Mémoires de l'Académie royale de Belgique*, 43 [1881] and 45 [1882]), written with F. Folie.

Besides writing on algebraic geometry, Le Paige was also a historian of mathematics. He published the correspondence of Sluse, canon of Liège, with Pascal, Huygens, Oldenburg, and Wallis. His "Notes pour servir à l'histoire des mathématiques dans l'ancien pays de Liège" (*Bulletin de l'Institut archéologique liègeois*, 21 [1890]), devotes a large section to the Belgian astronomer Wendelin, and in "Sur l'origine de certains signes d'opération" (*Annales de la Société scientifiques de Bruxelles*, 16 [1891–1892]) Le Paige explains the origin of the symbols of operation. After his appointment as director of the Institut d'Astrophysique de Cointe-Sclessim in 1897 Le Paige wrote a number of astronomical treatises.

Le Paige was elected a member of the Royal Academy of Sciences of Belgium in 1885. His international reputation is attested by the following partial list of his affiliations: member of the Royal Society of Sciences of Liège (1878) and of the Royal Society of Bohemia; corresponding member of the Pontificia Accademia dei Nuovi Lincei (1881) and of the Royal Academy of Sciences of Lisbon (1883); and honorary member of the Mathematical Society of Amsterdam (1886).

BIBLIOGRAPHY

In addition to the works mentioned in the text, Le Paige published the following astronomical works: "Sur la réduction au lieu apparent. Termes dus à l'aberration," in *Mémoires de la Société royale des Sciences de Liège*, **3** (1901); and "Étude sur les visées au bain de mercure," *ibid.*

A fuller account of Le Paige's work and a complete list of his publications is given in L. Godeaux, "Notice sur Constantin le Paige," in *Annuaire de l'Académie royale de Belgique*, **105** (1939), 239–270.

H. L. L. BUSARD

LEPEKHIN, IVAN IVANOVICH (*b.* St. Petersburg, Russia [now Leningrad, U.S.S.R.], 21 September 1740; *d.* St. Petersburg, 18 April 1802), *geography, natural history.*

Lepekhin's father, Ivan Sidorovich Lepekhin, was a guardsman in the Semenovsky regiment. The soldiers of this regiment were entitled to send their children to the Gymnasium and University attached to the St. Petersburg Academy; and in the spring of 1751, the ten-year-old Lepekhin was accepted at the Gymnasium. He studied there, at state expense, until January 1760, when he was accepted as a student at the University. He remained at the University for two and a half years, studying the humanities and chemistry, but he wanted most of all to study the broader discipline of natural history which was not taught there. Lepekhin therefore asked for and received permission to study at a foreign university. His request was granted and in November 1762 he enrolled at the University of Strasbourg.

At Strasbourg, Lepekhin devoted himself to practical anatomy, physiological experiments on living animals, and medicine. In his free time, he collected and studied plants and insects in the vicinity of Strasbourg and compiled lists of local birds and fish. In the summer of 1766 he defended his dissertation on the formation of vinegar ("Specimen de acetificatione") and received the doctorate in medicine.

Lepekhin returned to St. Petersburg in the fall of 1767 and the following June was elected an adjunct of the Academy. In April of 1771 he was made academician in the department of natural history.

Lepekhin is known mainly as the leader of one of the five expeditions organized by the Academy in 1768 to study the natural resources of Russia and to collect ethnographic and economic data. In the first three years the expedition, which consisted of Lepekhin and six companions, traveled from Moscow to Simbirsk (Ulyanovsk), Samara (Kuibyshev), Saratov, Tsaritsyn (Volgograd), Gurev, Orenburg, Ekaterinburg (Sverdlovsk), and Tyumen. They were to proceed as far as Tobolsk and then to return to St. Petersburg. At Tyumen, where he spent the third winter, Lepekhin was granted permission to travel instead into northern European Russia—to the area around the White Sea. In 1772 the expedition left Tyumen and traveled to Archangel via Verkhoture and Veliki Ustyug. From Archangel they explored the White Sea coast, the Solovetskiye Islands, and the Kanin Peninsula. They then returned to St. Petersburg by way of Kholmogory, Kargopol, and Ladoga. In 1773 Lepekhin led a nine-month expedition to investigate Byelorussia.

From 1774 to 1802 Lepekhin was head of the Botanical Garden of the Academy of Sciences, and from 1777 to 1794 of the Gymnasium. From 1783 until his death he was the permanent secretary of the Russian Academy, created in the same year for work in the area of Russian literature (it was active until 1841, when it merged with the Petersburg Academy of Sciences as the Section of Russian Language and Literature). He participated in the compilation and publication of a six-volume *Dictionary of the Russian Academy* and directed its organization.

Along with his substantial administrative work in the two academies, Lepekhin systematically published articles devoted to the description of the new species of animals and plants he had discovered during his travels. He wrote forty-four articles (excluding the articles on physiology) in Latin, Russian, and French. But Lepekhin's chief scientific work was his four-volume *Diaries of a Journey Through Various Provinces of the Russian State* (1771–1805). The first three volumes were translated into German in 1774–1783, and published in 1784 in French in Lausanne. These publications enhanced Lepekhin's scientific reputation abroad. In 1776 he was elected a member of the Berlin Academy and several newly discovered plants and insects were named by German scientists in his honor.

The *Diaries* are a rich collection of natural science and ethnographic information relating to the Volga area, the Urals and adjacent northwestern Kazakhstan and western Siberia, and northern European Russia. Lepekhin gave special attention to the description of more than 300 species of animals and their habits and distribution. With great care he described more than 100 birds and 117 insect species. What is important about the *Diaries* is that they not only described different species but also gave whole zoological maps of broad areas, which was an innovation.

In the *Diaries* much attention was given also to plants. Many unknown plants of the Urals and the northern tundra were described, and much information was collected about plants useful to man and ways of cultivating them. Along with factual data the *Diaries* contain a number of interesting general observations. Concerning the question of the variability of the earth's surface, Lepekhin thoroughly rejected the widespread belief that changes in the earth's surface were the result of catastrophic floods. Like Lomonosov, with whose *O sloyakh zemnykh* ("On the Layers of the Earth") he became acquainted while still a student in Strasbourg, Lepekhin considered that the changes were the result of the slow influence of water and underground fire and were still taking place. To explain the presence of marine fossils in layers of rock, he said, "There was once ocean floor upon which now are built famous cities, and populous towns . . ." (*Diaries*, III, p. 36).

Speaking of the origin of the Ural Mountains, Lepekhin came to the correct idea of the hydrologic cycle in nature, aligning himself with those natural scientists "Who by ocean vapors, which are the source of rivers, determine the circulation of the atmosphere" (*Diaries*, II, p. 144). In a number of places in the *Diaries* Lepekhin closely approached an understanding of the possibility of change caused by external conditions. He described many deposits of useful fossils, and in particular he noted the phenomena of oil in several places between the Urals and the Volga. To Lepekhin's many contributions to science must be added his ten-volume Russian edition of Buffon's *Histoire naturelle* (St. Petersburg, 1789–1808). Lepekhin himself translated volumes V–X, and Volume I in collaboration with Rumovsky.

BIBLIOGRAPHY

I. ORIGINAL WORKS. See I. I. Lepekhin, *Dnevnye zapiski puteshestvia po raznym provintsiam Rossyskogo gosudarstva* ("Diaries of a Journey Through Various Provinces of the Russian State"; pts. 1–4, St. Petersburg, 1771–1805).

II. SECONDARY LITERATURE. For information on Lepekhin's life, see N. G. Fradkin, *Akademik I. I. Lepekhin i ego puteshestvia po Rossii v 1768–1773 gg* ("Academician I. I. Lepekhin and His Journeys Through Russia in 1768–1773"; Moscow, 1953); T. A. Lukina, *Ivan Ivanovich Lepekhin* (Moscow–Leningrad, 1965); N. Ya. Ozeretskovsky, "Zhizn I. I. Lepekhina" ("Life of I. I. Lepekhin"), in *Zhurnal departamenta narodnogo prosveshchenia*, no. 11 (1822), 281–288; V. A. Polenov, "Kratkoe zhizneopisanie I. I. Lepekhina" ("A Short Description of the Life of I. I. Lepekhin"), in *Trudy Rossyskoy Akademii*, pt. 2 (1840), 205–215; and M. I. Sukhomlinov, "I. I. Lepekhin," in *Istoria Rossyskoy Akademii*, no. 2 (1875), 157–299.

I. A. FEDOSEEV

LE POIVRE, JACQUES-FRANÇOIS (*fl.* France, early eighteenth century), *mathematics.*

A minor figure in the early history of projective geometry, Le Poivre is known only by his short treatise, *Traité des sections du cylindre et du cône considérées dans le solide et dans le plan, avec des démonstrations simples & nouvelles* (Paris, 1704). According to the review of this work in the *Journal des sçavans*, he lived in Mons and worked on the treatise for three years; nothing more is known about him.

Aimed both at presenting the conic sections in a form readily understandable to the novice and at offering new results to specialists, the *Traité* is divided into two parts. The first examines the ellipse by means of the parallel projection of a circle from one plane to another and, by inversion of the projection, within the same plane. Although primarily interested here in the tangent properties of the curve, Le Poivre also proves several theorems concerning its conjugate axes in a much simpler way than had been achieved earlier.

Part 2 is more interesting for its greater generality. Here Le Poivre generates the conic sections by means of the central projection of a circle. Taking a circle as base and a point S (at first assumed to be off the plane of the circle) as summit, he draws two parallel lines ab and DE in the plane of the circle (see Fig. 1).

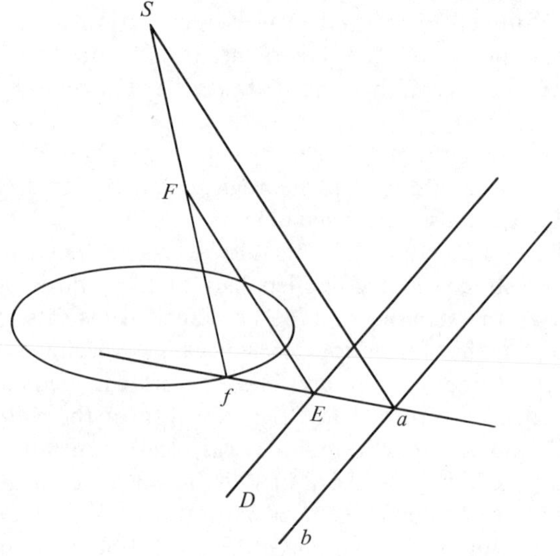

FIGURE 1

He then draws through any point f on the circle an arbitrary line intersecting DE at E and ab at a. Having drawn Sa, he draws EF parallel to it and intersecting Sf at F. Point F, he then shows, lies on a conic section, the precise nature of which depends on the location of line ab: if ab lies wholly outside the circle, the section is an ellipse; if ab is tangent to the circle, the section is a parabola; if ab cuts the circle, the section is a hyperbola. The theorems that he goes on to prove are nevertheless independent of the specific position of ab and hence apply to conic sections in general.

If one views ab and DE as the intersections of two parallel planes with the plane of the circle, and Sf as an element of the cone determined by point S and the circle, then Le Poivre's construction reduces to that of Apollonius. But the construction itself does not rely on that visualization of a solid, nor does it require that S in fact be off the plane of the circle.

Hence the theorems also hold for central projections in the same plane.

In reviewing Le Poivre's *Traité* for the *Acta eruditorum*, Christian Wolff spoke warmly of its elegance and originality. The anonymous reviewer of the *Journal des sçavans*, however, insisted on Le Poivre's omission of several important conic properties, in particular the focal properties, and on the extent to which he had failed to go beyond (or even to equal) the methods of Philippe de La Hire. In fact, Le Poivre's method of central projection is essentially that employed by La Hire in Part 2 of his *Nouvelle méthode en géométrie, pour les sections des superficies coniques et cylindriques* (Paris, 1673). Whether independent of La Hire or not, Le Poivre's work was apparently lost in his shadow, not to be brought to light again until Michel Chasles's survey of the history of geometry in 1837.

BIBLIOGRAPHY

Le Poivre's *Traité* is quite rare. The above description of it is taken from the reviews in the *Acta eruditorum* (Mar. 1707), 132–133 (the identification of Wolff as reviewer is from a contemporary marginal in the Princeton copy); and the *Journal des sçavans*, **32** (1704), 649–658. Chasles, who felt that Le Poivre belonged in a class with Desargues, Pascal, and La Hire, discusses his work in pars. 31–34 of ch. 3 of the *Aperçu historique sur l'origine et le développement des méthodes en géométrie, particulièrement de celles qui se rapportent à la géométrie moderne* (Brussels, 1837); German trans. by L. A. Sohncke (Halle, 1839; repr. Wiesbaden, 1968). The entry for Le Poivre in the *Nouvelle biographie générale*, XXX (Paris, 1862), 852, stems entirely from Chasles.

MICHAEL S. MAHONEY

LERCH, MATHIAS (*b.* Milínov, near Sušice, Bohemia, 20 February 1860; *d.* Sušice, Czechoslovakia, 3 August 1922), *mathematics.*

Lerch studied mathematics in Prague and, in 1884 and 1885, at the University of Berlin under Weierstrass, Kronecker, and Lazarus Fuchs. In 1886 he became a *Privatdozent* at the Czech technical institute in Prague and in 1896 full professor at the University of Fribourg. He returned to his native country in 1906, following his appointment as full professor at the Czech technical institute in Brno. In 1920 Lerch became the first professor of mathematics at the newly founded Masaryk University in Brno. He died two years later, at the age of sixty-two. In 1900 he received the grand prize of the Paris Academy for his *Essais sur le calcul du nombre des classes de formes quadratiques binaires aux coefficients entiers* [199].

Of Lerch's 238 scientific writings, some of which are quite comprehensive, 118 were written in Czech. About 150 deal with analysis and about forty with number theory; the rest are devoted to geometry, numerical methods, and other subjects. Lerch's achievements in analysis were in general function theory, general and special infinite series, special functions (particularly the gamma function), elliptic functions, and integral calculus. His works are noteworthy with regard to methodology. In particular he described and applied to concrete questions new methodological principles of considerable importance: the principle of the introduction of an auxiliary parameter for meromorphic functions [203] and the principle of most rapid convergence [210]. Lerch's best-known accomplishments include the Lerch theory on the generally unique solution φ of the equation $J(a) = \int_0^\infty \exp(-ax)\, \varphi(x)\, dx$—[73, 180]—which is fundamental in modern operator calculus; and the Lerch formula [101, 105, 116, 124], obtained originally from the theory of Malmsténian series, for the derivative of the Kummerian trigonometric development of log $\Gamma(v)$:

$$\sum_{k=1}^{\infty} \sin(2k+1)\, v\pi \cdot \log \frac{k}{k+1} = (\log 2\pi - \Gamma'[1]) \sin v\pi + \frac{\pi}{2} \cos v\pi + \frac{\Gamma'(v)}{\Gamma(v)} \sin v\pi \; (0 < v < 1).$$

BIBLIOGRAPHY

I. ORIGINAL WORKS. A complete, chronological bibliography of Lerch's scientific works was published by J. Škrášek in *Czechoslovak Mathematical Journal*, **3** (1953), 111–122. The numbers in text in square brackets refer to works so numbered in Škrášek's listing.

II. SECONDARY LITERATURE. An extensive discussion of Lerch's work in mathematical analysis was published by O. Borůvka *et al.* in *Práce Brněnské základny Československé akademie věd*, **29** (1957), 417–540. A detailed biography of Lerch can be found in an article by L. Frank in *Časopis pro pěstování matematiky*, **78** (1953), 119–137.

O. BORŮVKA

LEREBOULLET, DOMINIQUE-AUGUSTE (*b.* Épinal, France, 19 September 1804; *d.* Strasbourg, France, 5 October 1865), *zoology, embryology.*

Lereboullet was the son of an employee in the administration of indirect taxation at Épinal. After obtaining baccalaureates in letters and sciences at the Collège de Colmar, he spent a year at the University of Freiburg. There he studied German and was introduced to German science, a subject that he later analyzed, over a period of nearly thirty years, for the *Gazette médicale de Paris.* Returning to France, he studied medicine at Strasbourg, then in 1832 obtained permission to study the cholera epidemic at Paris. The data that he thus gathered was incorporated in the thesis with which he obtained the M.D. in the same year.

He immediately entered the Faculty of Sciences at Strasbourg as *préparateur* for G. L. Duvernoy, a friend and former collaborator of Cuvier. When in 1838 Duvernoy was appointed to Cuvier's chair at the Collège de France, Lereboullet (who had in 1837 received the licentiate and in 1838 the doctorate in natural science) succeeded him at Strasbourg in the chair of zoology and animal physiology.

Lereboullet had begun his zoological work by studying the comparative anatomy of vertebrates. His doctoral thesis dealt with the comparative anatomy of the respiratory system, while in another early work, which was awarded a prize by the Academy of Sciences, he examined the comparative anatomy of the genital organs in vertebrates. It was perhaps indicative of the German influence on his work that Lereboullet then took up the study, novel to France, of comparative histology. His memoir on the cellular structure of the liver, for which be was awarded the Portal Prize by the Academy of Medicine in 1851, described how cells are altered in the common pathological condition known as *foie gras.*

In 1849 Lereboullet began his celebrated research in comparative embryology. Having learned to fertilize fish eggs artificially, he was able to examine microscopically the earliest phases of development. In 1853 the Academy of Sciences proposed for competition the problem of comparing the development of a vertebrate with a mollusk or an articulate and Lereboullet chose as research subjects the pike, the perch, and the crayfish. Although he was paid for his work the Academy did not find it adequate to the problem, which was reset in 1856. This time Lereboullet studied the trout, the lizard, and the pond snail, and concluded his memoir with generalizations that satisfied the Academy, which awarded him the prize in 1857.

In his memoir Lereboullet had analyzed all phases of development, beginning with the unfertilized egg. He concluded that the unfertilized egg in all of his subjects appeared to be similar in the first three of Cuvier's four embranchments, but that soon after fertilization the paths of development separated, each embranchment manifesting its own proper characteristics. Finding himself in accord with von Baer and Henri-Milne Edwards, Lereboullet argued against the doctrine of unity of type and against E. R. A. Serres' recapitulation theory.

While studying the pike, Lereboullet noticed the frequent occurrence of monstrosities. Geoffroy Saint-Hilaire had previously claimed that monsters could be produced at will by the use of external agents, and the Academy set the problem of whether determined monstrosities could be experimentally produced by known physical agents for the subject of a competition in 1860. Lereboullet had begun to make teratological experiments in 1852. He subjected nearly 300,000 pike eggs to such disturbing agents as heat, cold, running water, and brushing, and carefully tabulated his results, noting that neither the fact nor the agent of perturbation had an effect on the number or type of abnormality that occurred. He concluded that abnormalities were most probably inherent in the structure of the germ, rather than caused by outside agents.

Lereboullet divided monsters into two groups (single and double) and seven classes, which he attributed to modifications in the blastodermic thickening and the embryonic bond. He recognized that the thickened edge of the blastoderm—the embryonic ring—gives rise to all embryonic structures and suggested that double monsters develop from two separate points in the blastodermic thickening. For these experiments the Academy gave Lereboullet the Alhumbert Prize, which he shared with Camille Dareste, although they had arrived at opposite theoretical conclusions.

In describing the earliest phases of animal development Lereboullet divided the elements constituting the fertilized egg into plastic elements (deriving from the germinative vesicle) that would eventually produce the cells of "animal life" and nutritive elements (deriving from the vitellus) that would produce the organs of "vegetative life." His work was precisely observed and factual. As a member of the *école de Cuvier*, he avoided theory and rejected the hypotheses of spontaneous generation, unity of type, recapitulation, multiple origin of the human species, and transformism.

Lereboullet was a modest and religious man. He was married; two of his sons later became doctors. He was active in several scientific societies, directed the Museum of Natural History of Strasbourg, and acted as dean of the Faculty of Sciences from 1861 until his death.

BIBLIOGRAPHY

I. Original Works. A complete list of Lereboullet's publications may be found in Hergott, *Notice sur le docteur*

Lereboullet (Strasbourg, 1866); an incomplete list is in Beaunis, "A. Lereboullet," in *Gazette médicale de Paris* (1865), 797–806. The Royal Society *Catalogue of Scientific Papers*, III, VIII, contains a list of his memoirs. His major publications were *Choléra-morbus observé à Paris et dans le département de la Meuse pendant l'année 1832* (Strasbourg, 1832), his M.D. diss.; *Anatomie comparée de l'appareil respiratoire dans les animaux vertébrés* (Strasbourg, 1838), his D.Sc. diss.; "Recherches sur l'anatomie des organes génitaux des animaux vertébrés," in *Nova acta academiae naturae curiosorum*, **23** (1851), 1–228, also published separately (Breslau, 1851); "Mémoire sur la structure intime du foie et sur la nature de l'altération connue sous le nom de *foie gras*," in *Mémoires de l'Académie de médecine*, **17** (1853), 387–501; "Recherches d'embryologie comparée sur le développement du brochet, de la perche et de l'écrevisse, Académie des Sciences," in *Mémoires présentés par divers savants étrangers*, **17** (1862), 447–805, also published separately (Paris, 1862); "Recherches d'embryologie comparée sur le développement de la truite, du lézard et du limnée," in *Annales des sciences naturelles*, 4th ser., **16** (1861), 113–196; **17** (1862), 89–157; **18** (1862), 87–211; **19** (1863), 5–130; **20** (1863), 5–58, also published separately (Paris, 1863); "Recherches sur les monstruosités du brochet observées dans l'oeuf et sur leur mode de production," *ibid.*, 4th ser., **20** (1863), 5th ser., **1** (1864). "Anatomie philosophique," in *Dictionnaire encyclopédique des sciences médicales*, contains a good summary of Lereboullet's scientific methodology and of his views on contemporary scientific theories.

II. Secondary Literature. There is no adequate or readily available biography of Lereboullet. The two most complete accounts of his life and work are listed above. For evaluations of Lereboullet's embryological work see J. Oppenheimer, "Historical Introduction to the Study of Teleostean Development," in *Osiris*, **2** (1936), 124–148, esp. 142–148; and "Some Historical Relationships Between Teratology and Experimental Embryology," in *Bulletin of the History of Medicine*, **42** (1968), 145–159, esp. 151–154.

Toby A. Appel

LE ROY, CHARLES (*b.* Paris, France, 12 January 1726; *d.* Paris, 10 December 1779), *physics, meteorology, medicine.*

Le Roy was a member of a distinguished family. His father, Julien, was a celebrated horologist. One of his older brothers, Jean-Baptiste, was an eminent physicist and a member of the Paris Academy of Sciences, and his younger brother, Julien-David, was a noted architect and historian of architecture. Charles received his medical education at the University of Montpellier. After traveling extensively in Italy, he returned to Montpellier as a member of the Faculty of Medicine.

Le Roy was a member of the Royal Academy of Sciences at Montpellier and a corresponding member of the Paris Academy of Sciences. In 1777 he returned to Paris, where he practiced medicine until his death in 1779.

Although a physician and teacher of medicine, Le Roy made his most important scientific contribution in physics. In 1751 he published a theory of evaporation that was to dominate discussions of the topic for more than forty years.

In the early eighteenth century, theories of evaporation were usually couched in the idiom of the prevailing mechanical philosophy. Many theories ascribed the "elevation" of vapors in the air to the specific legerity of water particles. Some scientists believed that the molecules of a liquid could be expanded by heat until their density became less than that of the atmosphere, at which point they would rise into the air rather like vacuum-filled balloons. Followers of Descartes thought that during evaporation "subtile matter" infiltrated particles of a liquid, thus reducing their specific gravity and causing them to rise into the ambient air. Other theories, especially those formulated by the followers of Newton, depended upon the concept of attractive forces acting at a distance. Desaguliers, for example, thought that evaporation was caused by atmospheric electricity attracting water particles into the air.

The problem with these early mechanical theories was that they were based upon untestable hypotheses concerning the behavior of submicroscopic particles. To invent an explanation of this kind, it was necessary only to conceive a plausible mechanism and present it as fact. To be sure, such mechanical conceptions could not be disproved—but they could not be proved either.

The value of Le Roy's theory was in its radically different approach to the problem of evaporation. Rather than attempting to deduce an explanation from an imagined account of the behavior of unseen particles, he presented instead a simple analogy between observable phenomena. Vapors, he said, are to the air as dissolved salts are to water. Le Roy made no attempt to explain the mechanism of solution itself, which he considered beyond the capabilities of the science of his day. It was sufficient merely to show that the phenomena of solution and evaporation were analogous and thus avoid useless debate over alternative mechanical hypotheses. All empirical generalizations concerning evaporation and vapors could be accounted for by referring to the analogous generalizations concerning the behavior of solution. For example, the fact that improved circulation and constant renewing of the air increases the rate of evaporation is analogous to the fact that stirring a liquid increases the rate at which salts dissolve in it. Heating will also increase the rates of both evaporation and dissolution, whereas cooling will retard both processes. Furthermore, just as a liquid may be saturated with a salt, so too may the atmosphere be saturated with water; and just as cooling a saturated solution will cause precipitation of the dissolved salt, so too will cooling a saturated atmosphere cause precipitation of water. In fact, the use of the word "precipitation" in meteorological contexts seems to be a linguistic remnant of Le Roy's once-popular theory.

The major weakness of Le Roy's solution theory of vapors was its inability to explain convincingly the widely known phenomenon of evaporation *in vacuo*. Clearly, liquids could not dissolve themselves in mere nothing. Le Roy's attempted explanation of this anomaly was farfetched. He knew from Stephen Hales's famous experiments that water contained a large quantity of dissolved air. Le Roy supposed that when water was placed in a vacuum apparatus, the air came out of it and dissolved the very water in which it had itself previously been dissolved.

There is nothing to indicate that Le Roy undertook any but the most casual experiments to substantiate his theory. He seems to have derived most of his empirical generalizations from everyday observation and from the writings of others. Indeed, the theory itself was probably anticipated by other scientists, some of whom Le Roy mentioned in his articles. Le Roy's version of the solution theory was, however, the most clearly and completely developed, and through its publication in the *Encyclopédie* it became widely known and accepted despite its one rather obvious weakness. Other versions of the theory were subsequently published in Britain by Benjamin Franklin and Henry Home, Lord Kames.

Although the solution theory of vapors has proved untenable, Le Roy nevertheless performed an invaluable service to science. By clearing away the arcane mechanisms that had been conceived by his predecessors and presenting instead a clear, testable hypothesis, he prepared the way for the important advancements in the theory of evaporation made by Lavoisier and Dalton. In particular, by focusing on the anomaly of evaporation *in vacuo*, he directed attention to a phenomenon that was to prove a key to the understanding of the gaseous state. In his insistence on the use of analogy instead of mechanism, he anticipated a mode of scientific explanation that was to prove remarkably fruitful in the hands of more thoughtful experimentalists, and notably in Lavoisier's. Finally, the theory itself was, for its time, a good one. It proved useful in most instances and provided an acceptable conceptual framework for dealing with many phenomena in the realms of meteorology, physics, and chemistry. The theory also provided a criterion for distinguishing true vapors from smokes and exhalations: true vapors, like true solutions, are perfectly transparent; whereas smokes and exhalations, like liquid suspensions, are cloudy.

In addition to his work on evaporation, Le Roy also communicated to the Academy a number of other observations. While traveling in Italy, he experimented on the asphyxiating gas produced in the Grotto del Cane, near Naples, and he attempted to provide an explanation for marine phosphorescence in the Mediterranean Sea. During his tenure at the Faculty of Medicine at Montpellier, he wrote extensive treatises on mineral waters, physiology, medicine, and chemistry.

BIBLIOGRAPHY

I. Original Works. Le Roy's principal works are "Mémoire sur l'élévation et la suspension de l'eau dans l'air," in *Histoire et mémoires de l'Académie Royale des Sciences* (1751), pp. 481–518; "Évaporation," in *Encyclopédie*, VI (Paris, 1756), pp. 123–130; and *Mélanges de physique et de médecine* (Paris, 1771).

II. Secondary Literature. For biographical references, see the *éloges* cited in *Nouvelle biographie générale*, XXX (Paris, 1862), 892; and L. Dulieu, "Un Parisien professeur à l'Université de Medecine de Montpellier: Charles Le Roy (1726–1779)," in *Revue d'histoire des sciences et leurs applications*, **6** (1953), 50–59. On eighteenth-century theories of evaporation, see S. A. Dyment, "Some Eighteenth Century Ideas Concerning Aqueous Vapor and Evaporation," in *Annals of Science*, **2** (1937), 465–473; and J. B. Gough, "Nouvelle contribution à l'évolution des idées de Lavoisier sur la nature de l'air et sur la calcination des métaux," in *Archives internationales d'histoire des sciences*, **22**, no. 89 (July-Dec. 1969), 267–275. On the solution theory, see Benjamin Franklin, "Physical and Meteorological Observations, Conjectures, and Suppositions," in *Letters and Papers on Philosophical Subjects* (London, 1769); and Lord Kames's article on evaporation in *Essays and Observations, Physical and Literary, Read Before a Society in Edinburgh*, III (Edinburgh, 1772), pp. 80 ff.

J. B. Gough

LE ROY, ÉDOUARD (*b.* Paris, France, 18 June 1870; *d.* Paris, 9 November 1954), *mathematics, philosophy.*

Le Roy's father worked for the Compagnie Transatlantique for several years and then established his own business outfitting ships in Le Havre. Le Roy studied at home under the guidance of a tutor and

LE ROY

was admitted in 1892 to the École Normale Supérieure. He became an *agrégé* in 1895 and earned a doctor of science degree in 1898; his thesis attracted the attention of Henri Poincaré. Until 1921 he taught mathematics classes that prepared students for the leading scientific schools. From 1924 to 1940 he was *chargé de conférences* at the Faculty of Sciences in Paris.

Although trained as a mathematician, Le Roy's enthusiastic discovery of the "new philosophy" of Bergson, to which he devoted a book in 1912, led him to teach philosophy. Bergson appointed Le Roy his *suppléant* in the chair of modern philosophy at the Collège de France (1914–1920). Named professor in 1921, Le Roy taught there until 1941; several of his published works are transcriptions of his courses. He was elected to the Académie des Sciences Morales et Politiques in 1919, and he entered the Académie Française as Bergson's successor in 1945.

A Catholic and a scientist, Le Roy had, in his youth, put forth theses in the philosophy of religion and the philosophy of science that provoked lively polemics. In "Qu'est-ce qu'un dogme?" (*La quinzaine*, 16 Apr. 1905) he emphasized the opposition between dogma and the body of positive knowledge. Dogma, he asserted, has a negative sense: it excludes and condemns certain errors rather than determining the truth in a positive fashion. Abóve all a dogma has a practical sense—that is its primary value.

Le Roy was thus involved in the quarrels precipitated by "modernist philosophy," which was condemned in the encyclical *Pascendi* by Pope Pius X in 1907, and his *Le problème de Dieu* (1929) was placed on the Index in June 1931. In the philosophy of science, Le Roy had proclaimed, in articles in *Revue de métaphysique et de morale* (1899, 1900), that facts are less established than constituted and that, far from being received passively by the mind, they are to some extent created by it. These paradoxical statements excited a certain interest through Poincaré's criticism of them in *La valeur de la science* (Paris, 1906, ch. 10).

While wishing to defend and justify the Bergsonian notion of creative evolution, Le Roy set forth his own philosophy of life, a "doctrine of authentically spiritualist inspiration" that sought a "restoration of finality" and respected the idea of creation. Le Roy's views were similar in many ways to those of his friend Teilhard de Chardin. According to Le Roy, there exists at the basis of life—as the major cause of its changes and progress—a psychic factor, a genuine power of invention. To recapture the activity of this factor one must consider the sole contemporary being —that is, Man—in which the power of creative evolution is still vital. Man must be observed in his capacity as inventor in order to return, by means of retrospective analogy, to the paleontological past. Biosphere and noosphere are the great moments of evolution. At the origin of the noosphere we must conceive a phenomenon *sui generis* of vital transformation affecting the entire biosphere: hominization. Humanity then appears as a new order of reality, sustaining a relationship with the lower forms of life analogous to that between these lower forms and inanimate matter. Man then no longer seems a paradoxical excrescence but becomes the key to transformist explanations. According to the "lesson that emerges from Christianity," man's intuition of a spiritual beyond and the ideal of an interior and mystic life show that we must form a concept of *Homo spiritualis* distinct from *Homo faber* and *Homo sapiens*.

In analyzing the history and philosophy of science, Le Roy likewise accorded a central position to the notion of invention, itself intimately linked to that of intuition, since intuitive thought is always, to some extent, inventive. True mathematical intuition is operative intuition; and the analyst should everywhere attempt to bring to consciousness the operative action perceived in its dynamic indivisibility. The transition from physics to microphysics clearly illuminates "the primary role of a factor of inventive energy in the innermost recesses of thought." Reason itself is in a state of becoming and must gradually be invented.

What will regulate this invention? An absolutely primary foundation must be sought which will be indisputable in every regard and, at the same time, dynamic and vital. Synthesizing twenty-five years of teaching, Le Roy in the *Essai d'une philosophie première*, his last course at the Collège de France, presented the major steps of such a search, the chief concern of which is to satisfy "the demands of idealism."

BIBLIOGRAPHY

I. Original Works. Le Roy's principal publications are "Sur l'intégration des équations de la chaleur," in *Annales scientifiques de l'École normale supérieure*, **14** (1897), 379–465, and *ibid.*, **15** (1898), 9–178, his doctoral diss.; *Dogme et critique* (Paris, 1906); *Une philosophie nouvelle: Henri Bergson* (Paris, 1912); *L'exigence idéaliste et le fait de l'évolution* (Paris, 1927); *Les origines humaines et l'évolution de l'intelligence* (Paris, 1928); *La pensée intuitive*, 2 vols. (Paris, 1929–1930); *Le problème de Dieu* (Paris, 1929); and *Introduction à l'étude du problème religieux* (Paris, 1944).

After Le Roy's death his son, Georges Le Roy, published the lectures his father had given at the Collège de France as *Essai d'une philosophie première, l'exigence idéaliste et l'exigence morale*, 2 vols. (Paris, 1956–1958); and *La pensée mathématique pure* (Paris, 1960). The "Notice bibliographique," in *Études philosophiques* (Apr. 1955), 207–210, records many of Le Roy's works.

II. SECONDARY LITERATURE. See the following, listed chronologically: S. Gagnebin, *La philosophie de l'intuition. Essai sur les idées de Mr Édouard Le Roy* (Paris, 1912); L. Weber, "Une philosophie de l'invention. M. Édouard Le Roy," in *Revue de métaphysique et de morale*, **39** (1932), 59–86, 253–292; F. Olgiati, *Édouard Le Roy e il problema di Dio* (Milan, 1929); J. Lacroix, "Édouard Le Roy, philosophe de l'invention," in *Études philosophiques* (Apr. 1955), 189–205; G. Maire, "La philosophie d'Edouard Le Roy," *ibid.*, **27** (1972), 201–220; M. Tavares de Miranda, *Théorie de la vérité chez Édouard Le Roy* (Paris–Recife, Brazil, 1957); and G. Bachelard, *L'engagement rationaliste* (Paris, 1972), 155–168.

Biographical information is in E. Rideau, "Édouard Le Roy," in *Études*, no. 245 (April 1945), 246–255; and in the addresses delivered at the Académie Française by A. Chaumeix at the time of Le Roy's reception on 18 Oct. 1945, in *Institut de France, Publications diverses*, no. 18 (1945); and by H. Daniel-Rops when he succeeded Le Roy at the Academy, *ibid.*, no. 8 (1956).

On Le Roy's religious philosophy, see the article by E. Rideau cited above and A.-D. Sertillanges, *Le christianisme et les philosophes*, II (Paris, 1941), 402–419.

In an app. to *Bergson et Teilhard de Chardin* (Paris, 1963), pp. 655–659, M. B. Madaule discusses Le Roy's friendship with Teilhard, the large area of agreement in their views, and their correspondence.

Concerning the polemic with Poincaré, the ways in which Le Roy weakened his first statements, and his final homage to Poincaré, see J. Abelé, "Le Roy et la philosophie des sciences," in *Études*, no. 284 (April 1955), 106–112.

M. Serres discusses Le Roy's *La pensée mathématique pure* in "La querelle des anciens et des modernes," in his *Hermès ou la communication* (Paris, 1968), pp. 46–77. While terming the work a "monument of traditional epistemology," he shows how Le Roy, in applying his method of analysis to a mathematics that was already outdated, failed completely to consider "modern mathematics."

E. COUMET

LE ROY, JEAN-BAPTISTE (*b.* Paris, France, 15 August 1720; *d.* Paris, 21 January 1800), *physics, scientific instrumentation.*

Son of the renowned clockmaker Julien Le Roy, Jean-Baptiste Le Roy was one of four brothers to achieve scientific prominence in Enlightenment France; the others were Charles Le Roy (medicine and chemistry), Julien-David Le Roy (architecture), and Pierre Le Roy (chronometry). Elected to the Académie Royale des Sciences in 1751 as *adjoint géomètre*, Le Roy played an active role in technical as well as administrative aspects of French science for the next half-century. He was elected *pensionnaire mécanicien* in 1770 and director of the Academy for 1773 and 1778, and became both a fellow of the Royal Society and a member of the American Philosophical Society in 1773.

Le Roy's major field of enquiry was electricity, a subject on which European opinion was much divided at mid-century. The most prominent controversy engaged the proponents of the Abbé Nollet's doctrine of two distinct streams of electric fluids (outflowing and inflowing) and the partisans of Benjamin Franklin's concept of a single electric fluid. This debate intensified in France in 1753 with an attack on Franklin's views by Nollet. Le Roy, later a friend and correspondent of Franklin, defended his single-fluid theory and offered considerable experimental evidence in support thereof. He played an important role in the dissemination of Franklin's ideas, stressing particularly their practical applications, and published many memoirs on electrical machines and theory in the annual *Histoires* and *Mémoires* of the Academy and in the *Journal de physique*.

A regular contributor to the *Encyclopédie*, Le Roy wrote articles dealing with scientific instruments. The most important of these included comprehensive treatments of "Horlogerie," "Télescope," and "Électromètre" (in which Le Roy claimed priority for the invention of the electrometer). He also promoted the use of lightning rods in France, urged that the Academy support technical education, and was active in hospital and prison reform. After the Revolutionary suppression of royal academies, Le Roy was appointed to the first class of the Institut National (*section de mécanique*) at its formation in 1795.

BIBLIOGRAPHY

I. ORIGINAL WORKS. Le Roy wrote no extended treatise on electricity. His more important publications include "Mémoire sur l'électricité, où l'on montre par une suite d'expériences qu'il y a deux espèces d'électricités, l'une produite par la condensation du fluide électrique, & l'autre par sa raréfaction," in *Mémoires de l'Académie royale des sciences* (1753), 447–474; "Mémoire sur un phénomène électrique intéressant et qui n'avoit pas encore été observé; ou sur la différence des distances auxquelles partent les étincelles entre deux corps métalliques de figures différentes, selon que l'un de ces deux corps est électrifié, & que l'autre lui est présenté," *ibid.* (1766), 541–546; "Mémoire sur les verges ou barres métalliques, destinées à garantir les édifices des effets de la foudre," *ibid.* (1770), 53–67; "Mémoire sur une machine à électriser d'une espèce nouvelle," *ibid.*, pt. 1 (1772), 499–512; and "Mémoire sur la forme des barres ou des conducteurs métalliques destinées à préserver les édifices des effets de la foudre, en transmettant son feu à la terre," *ibid.* (1773), 671–686.

For Le Roy's ideas on hospitals, see *Précis d'un ouvrage sur les hôpitaux, dans lequel on expose les principes résultant des observations de physique et de médecine qu'on doit avoir en vue dans la construction de ces édifices, avec un projet d'hôpital disposé d'après ces principes* (Paris, n.d.).

II. Secondary Literature. There is no biography of Le Roy. Details concerning his life and work can be found in Louis Lefèvre-Gineau, *Funérailles du citoyen le Roy* (Paris, 1801); and F. Hoefer, ed., *Nouvelle biographie générale*, XXX (1859), 891. Franklin's letters to Le Roy are in Albert Henry Smyth, ed., *The Writings of Benjamin Franklin*, 10 vols. (New York, 1905–1907), *passim*.

Martin Fichman

LESAGE, GEORGE-LOUIS (*b.* Geneva, Switzerland, 13 June 1724; *d.* Geneva, 9 November 1803), *physics.*

Lesage's father, also named George-Louis, was a distinguished teacher of mathematics and physics as well as a moral philosopher and theologian. Born in Conches, in Normandy, of a notorious Huguenot family, he was exiled to England at the age of eight to escape the religious persecutions in France. He married Anne-Marie Camp in 1722 and settled in Geneva shortly before the birth of his only son.

The elder Lesage instructed his son in the classics and the rudiments of mathematics and science. At the *collège* in Geneva Lesage studied physics with Jean-Louis Calandrani and mathematics with Gabriel Cramer.

Lesage wanted to become a philosopher; but his father, desiring that he should find more remunerative employment, determined instead that his son should become a physician. Lesage reluctantly attended medical school at Basel. After a brief sojourn in Paris, he returned to Geneva and attempted to take up the practice of medicine. The city authorities intervened—as the son of an immigrant, Lesage was not qualified. At first he tried to raise the money to purchase bourgeois status by entering the prize essay contest of the Paris Academy of Sciences. He failed and thus gave up medicine entirely and became a teacher of mathematics.

Lesage's reputation derived largely from his efforts to explain mechanistically the cause of gravitational phenomena. Although Newton's law had finally been accepted by Continental scientists, it still presented serious intellectual difficulties. Every particle of matter in the universe supposedly attracted every other inversely as the squares of the distances separating them and directly as the products of their masses. Many found it inconceivable, however, that lumps of inanimate matter could somehow divine the presence of their neighbors, measure the appropriate distances and masses, and attract each other across the intervening space. The absurdity of such a notion was manifest, especially to Continental scientists imbued with the precepts of the Cartesian mechanical philosophy. Yet Newton's law of gravity had been verified in innumerable instances and without exception. In the minds of most eighteenth-century scientists, the validity of the Newtonian law was unquestionable; but because of their commitment to the mechanical philosophy, they generally—albeit often tacitly—assumed that some underlying impulsive mechanism was responsible for the so-called Newtonian attraction. Newton himself had attempted to provide mechanical explanations of his law but without much success.

Inspired by Lucretius' *De rerum natura* and doubtless also by the pervasive climate of Cartesianism, Lesage set about the formidable task of explaining the Newtonian law of attraction in terms of the mechanical philosophy. His basic system rested upon his conception of "otherworldly particles" (*particules ultramondaines*), so-called because of their exemption from the law of gravity. These gravitational particles were presumed to be extremely small Lucretian atoms that moved in every direction and at very high velocity. Any isolated mass of ordinary matter (its atoms were presumed to be much larger than those composing the gravitational fluid) would not be moved by the impacts of the gravitational particles since they would impinge on it from all directions at once. (Lesage did allow for slight oscillation caused by temporary imbalance of forces, rather like the Brownian motion of particles in a gas or liquid.) Two masses of ordinary matter, however, would block some of the particles coming from either direction along the lines connecting the parts of one with those of the other. Each mass, in effect, would cast a kind of "gravitational shadow" on the other; and the resulting disequilibrium of force would impel the two bodies together, thus giving the illusion of an attraction between the masses. The greater the distance between the two bodies, the less intense would be the effect of the mutual gravitational shadow. Indeed, Lesage maintained that it would vary inversely as the square of the distance, in accordance with the Newtonian law. Similarly, the larger the mass of the body, the more gravitational particles would be intercepted and the greater would be the disequilibrium of force. To make the impulsive force vary with the mass rather than with the surface area of the body was always the most difficult problem faced by those who attempted to adduce mechanical explanations of the Newtonian law. By making his bodies extremely porous and quite large relative to their constituent atoms, Lesage was able at least to achieve an approximation of the proper relation between mass and gravitational force.

Using this ingenious mechanism, Lesage attempted to explain not only gravitation but also chemical affinity and corporeal cohesion. His system was not widely accepted among his contemporaries. It was too burdened with *ad hoc* assumptions, and because certain parameters (the size of the various atoms and the mean free path of the particles composing the gravitational fluid) could not be established, his theory was not subject to rigorous mathematical analysis. In short, his theory was not unlike the clever but untestable mechanical hypotheses that had burdened physics in the previous century. As a result Lesage's work was less criticized than neglected.

Despite this neglect and despite the fact that he published relatively little in his lifetime (although he did, in fact, write a great deal) Lesage acquired a fairly extensive reputation, largely through his correspondence with the great natural philosophers and mathematicians of his age. He was elected a correspondent of the Paris Academy of Sciences on 28 February 1761, and after the reorganization of the Academy became a first-class corresponding member on 12 September 1803.

BIBLIOGRAPHY

The only major work published during Lesage's lifetime was his *Essai de chimie mécanique* (Rouen, 1758), the main source for his explanation of Newtonian attraction.

On Lesage's life and work, see Pierre Prévost, *Notice de la vie et des écrits de George-Louis Le Sage de Genève* (Geneva, 1805), which contains an extensive and detailed biography, pp. 1–184; a list of his shorter published articles and letters, pp. 91–95; a list of his many unpublished large works and an account of their contents, pp. 95–101; "Lucrèce Neutonien," one of Lesage's many MS works on gravitation, with a conveniently succinct exposition of his system; and a portion of Lesage's large correspondence with contemporary scientists, philosophers, and mathematicians, pp. 189–502.

J. B. GOUGH

LESLEY, J. PETER (*b.* Philadelphia, Pennsylvania, 17 September 1819; *d.* Milton, Massachusetts, 1 June 1903), *geology.*

Fourth in a line of Peter Lesleys and first son of Peter III and Elizabeth Oswald Allen, Peter Lesley was a delicate, introverted, and bookish youngster. He entered the University of Pennsylvania at fifteen and graduated Phi Beta Kappa in 1838. Frail health frustrated his plans to continue directly in school studying for the Presbyterian ministry, so at his father's urging he sought outdoor employment. He joined the first state survey of Pennsylvania, directed by Henry Darwin Rogers, and passed the years 1838–1841 surveying the bituminous and anthracite coal regions.

Although he was untrained and inexperienced when he entered fieldwork, his insights into the complicated geology of the state won him Rogers' praise. When in 1841 the survey was abandoned for lack of funds, Lesley returned to his studies at Princeton Theological Seminary. He maintained contact with Rogers, however, and when the Seminary was in recess he helped him with the preparation of maps and sections for the final survey report.

Lesley was licensed to preach by the Philadelphia Presbytery in 1844 and took up pastoral work the following year after traveling in Europe. He began his brief career in the ministry as an avowed conservative on questions of religious doctrine. Subsequent exposure to the liberal opinions of Rogers, Agassiz, Emerson, Lyell, and other associates in Boston, where he had gone to assist Rogers, radicalized the young clergyman to such an extent that he gave up the ministry altogether and became a geologist in 1852.

His first book, *A Manual of Coal and Its Topography*, has been called "the most important matter of the decade, geographically considered" (Davis [1919], 183). It is an attempt to group the Appalachian coal beds systematically, to correlate them with coal measures in Europe and elsewhere, and to emphasize the importance of topographical geology in Pennsylvania. Lesley proffered a catastrophic explanation of Appalachian landforms, although in later works, when his break with the ministry was not so fresh, this approach was replaced by uniformitarianism.

His second major work, *The Iron Manufacturers' Guide* (1859), contains information and statistics on ironworks as well as a detailed geological discussion of iron deposits in Pennsylvania and elsewhere. The preface is notable for its intemperate personal attack on Rogers, who, Lesley felt, had failed to credit his young apprentices on the first state survey. Accusing Rogers of scientific theft and labeling him an "imposture," Lesley sought to rectify the indignity of being mentioned only briefly in Rogers' preface.

When in 1873 he himself became director of the second state survey, Lesley was careful to see that each fieldworker be given due credit for his work. He often altered field reports and maps or added footnotes to his apprentices' texts before sending them to press; he felt that doing so would clarify the material and protect the authors from unfavorable criticism. Lesley's practice caused understandable consternation among his staff. As director of the survey he examined every line, map, and illustration in the seventy-seven volumes of text and thirty-four atlases. Lesley attempted to distill all this for the final summary, but his health broke under the strain and he was forced to delegate responsibility to others.

Although christened Peter Lesley, he disliked his first name and in 1850 added the initial "J." from "Junior" and began signing his work "J. P. Lesley." Toward the end of his life he reverted to J. Peter Lesley.

BIBLIOGRAPHY

I. ORIGINAL WORKS. Lesley's most important contributions are *A Manual of Coal and Its Topography* (Philadelphia, 1856); *The Iron Manufacturers' Guide to Furnaces, Forges and Rollings Mills of the United States* (New York, 1859); and the *Second Geological Survey of Pennsylvania Report*: *A Historical Sketch of Geological Explorations in Pennsylvania and Other States* (Harrisburg, Pa., 1876); and *Final Report: A Summary Description of the Geology of Pennsylvania*, 3 vols. (1892–1895).

II. SECONDARY LITERATURE. A two-volume biography was edited by his eldest daughter, Mary Lesley Ames, as *Life and Letters of Peter and Susan Lesley* (New York, 1909). For biographical notices see W. M. Davis, "Biographical Memoir of J. P. Lesley," in *Biographical Memoirs. National Academy of Sciences*, **8** (1919), 152–240; P. Frazer, "J. Peter Lesley," in *American Geologist*, **32** (1903), 133–136; A. Geikie, "Notice of J. P. Lesley," in *Quarterly Journal of the Geological Society of London*, **60** (1904), xlix–lv; B. S. Lyman, "Biographical Sketch of J. Peter Lesley," in *Transactions of the American Institute of Mining Engineers*, **34** (1903), 726–739; G. P. Merrill, "Peter Lesley," in *Dictionary of American Biography;* and J. J. Stevenson, "Memoir of J. Peter Lesley," in *Bulletin of the Geological Society of America*, **15** (1904), 532–541.

MARTHA B. KENDALL

LESLIE, JOHN (*b.* Largo, Scotland, 16 April 1766; *d.* Coates, near Cupar, Fife, Scotland, 3 November 1832), *natural philosophy*.

Leslie was one of three sons born to a poor cabinetmaker, Robert Leslie, and his wife, Anne Carstairs. At the age of thirteen he entered the University of St. Andrews, where he studied mathematics under Nicholas Vilant, who also taught John Playfair and James Ivory. In 1785, Leslie went to Edinburgh to continue his studies with Joseph Black, John Robison, and the moral philosopher Dugald Stewart, from whom he derived a continuing interest in the philosophy and history of science. From 1790 to 1804, when his major work, *An Experimental Inquiry Into the Nature and Propagation of Heat*, appeared, Leslie supported himself by working as a science tutor to Thomas Wedgwood, son of the Etruria potter, Josiah, and by writing for the *Monthly Review*. Leslie became professor of mathematics at the University of Edinburgh in 1805, over the protests of the local clergymen that his acceptance of Hume's notions of causality made him an atheist; and in 1819 he was promoted to the chair of natural philosophy. Throughout his life Leslie refused membership in all British scientific societies because he felt slighted when the Royal Society of London refused his first communication in 1791, but he was made a corresponding member of the Paris Académie Royale des Sciences in 1820 and was knighted in 1832 for his scientific work.

Leslie's *Experimental Inquiry* (1804) established several fundamental laws of heat radiation: that the emissivity and absorptivity for any surface are equal, that the emissivity of a surface increases with the decrease of reflectivity, and that the intensity of heat radiated from a surface is proportional to the sine of the angle of the rays to the surface. The book also played a major role in the early nineteenth-century argument about whether heat was a form of matter or a mode of motion. Leslie's experiments showed that heat, unlike light, was not directly transmitted through transparent solids. Since Leslie embraced a corpuscular theory, he incorrectly interpreted the apparent blockage of heat radiation as evidence that heat was composed of particles much larger than those of light. He borrowed from James Hutton the basic notion that heat was a compound formed by the union of light particles with ordinary particles of matter. François Delaroche later showed that Leslie's failure to detect direct transmission of heat through solids was a result of using only low-temperature heat sources whose radiation was absorbed by the solid screens. In the meantime, Leslie's puzzling experimental results had stimulated further investigations of diathermancy and the nature of radiant heat.

Although Leslie is best known for his heat studies, his interests were wide-ranging. In 1791, for example, he wrote a paper which analyzed the leakage of static electricity from charged bodies through different conducting paths. First he theorized that the quantity of electricity conducted to ground per unit time I would be proportional to the instantaneous intensity of the source—that is, the quantity of electricity per unit volume at any moment V—where it joined the conductor. In addition, he argued that the amount communicated per unit time would be a function of the composition of the conductor ρ, that it would be proportional to the cross-sectional area of the conductor α, and that it would be inversely proportional to the length L of the conductor. In modern symbols, $I = V\rho(\alpha/L)$. Next he provided a limited confirmation of his theory by varying the length, composition, and cross sections of conductors discharging a bank of Leyden jars. Finally, he showed geometrically that one could replace a series of conducting segments of varying compositions, lengths, and cross sections with a single equivalent conductor, and he suggested that similar geometrical methods could be used to simplify "more intricate cases" such as those in which parallel conductors were used simultaneously to produce the discharge. Thus he prefigured many of Georg Ohm's considerations of voltaic electricity.

His systematic study, begun in 1793, of temperature-density relations in gases allowed him to propose a widely discussed formula for the decrease of temperature with increasing height in the atmosphere. In 1800, he announced the discovery of the wet- and dry-bulb hygrometer and provided an essentially correct theory of how the instrument operates. In 1802 he presented the first correct interpretation of capillary action, thus stimulating the work of Thomas Young and James Ivory. Finally, in 1810, Leslie showed that it was possible to attain very low temperatures by evaporating water in the presence of a desiccant in an evacuated receiver, thus providing the principle exploited by Ferdinand Carré in creating the first laboratory ice machines.

In his theoretical work Leslie emphasized the need for systematic explanations; he had little patience with the English love of what he called "dull empiricism." In conformity with this deep interest in theoretical schemes, Leslie was a principal proponent of the point atomist theory of Bošković.

BIBLIOGRAPHY

I. ORIGINAL WORKS. Leslie wrote ten book-length works, twenty-one journal articles, at least thirty-seven reviews, and sixteen encyclopedia articles. His most important books are *An Experimental Inquiry Into the Nature and Propagation of Heat* (London, 1804); *A Short Account of Experiments and Instruments Depending on the Relations of Air to Heat and Moisture* (Edinburgh, 1813); and *The Philosophy of Arithmetic* (Edinburgh, 1817). Leslie's important historical sketch of the exact sciences in the eighteenth century is most widely available in Dugald Stewart, Sir James Mackintosh, John Playfair, and John Leslie, *Dissertations on the History of Metaphysical and Ethical, and of Mathematical and Physical Science* (Edinburgh, 1835).

His more important articles include "Description of a Hygrometer and Photometer," in Nicholson's *Journal of Natural Philosophy*, **3** (1800), 461–467; "On Capillary Action," in *Philosophical Magazine*, **14** (1802), 193–205; "Méthode nouvelle de produire et d'entretenir la congelation," in *Annales de chimie et de physique*, **78** (1811), 177–182; "On Heat and Climate," in *Annals of Philosophy*, **14** (1819), 5–27, first read to the Royal Society of London in 1793; and "Observations on Electrical Theories," in *Edinburgh Philosophical Journal*, **11** (1824), 1–39, read to the Royal Society of Edinburgh in 1792.

II. SECONDARY LITERATURE. The only significant published biographical sketch is Macvey Napier, "Leslie, Sir John," in *Encyclopaedia Britannica*, 7th ed. (1842), VIII, 242–252. For more detailed information and bibliography see Richard Olson's thesis, "Sir John Leslie: 1766–1832; A Study of the Pursuit of the Exact Sciences in the Scottish Enlightenment" (Harvard, 1967).

RICHARD G. OLSON

LEŚNIEWSKI, STANISŁAW (*b.* Serpukhov, Russia, 18 March 1886; *d.* Warsaw, Poland, 13 May 1939), *philosophy of mathematics.*

After studying philosophy at various German universities, Leśniewski received the Ph.D. under Kazimierz Twardowski at Lvov in 1912. From 1919 until his death he held the chair of philosophy of mathematics at the University of Warsaw and, with Jan Łukasiewicz, inspired and directed research at the Warsaw school of logic. As a student Leśniewski studied the works of John Stuart Mill and Edmund Husserl, but through the influence of Łukasiewicz he soon turned to mathematical logic and began to study the *Principia* and the writings of Gottlob Frege and Ernst Schröder. A thorough and painstaking analysis of Russell's antinomy of the class of all those classes which are not members of themselves led Leśniewski to the construction of a system of logic and of the foundations of mathematics remarkable for its originality, elegance, and comprehensiveness. It consists of three theories, which he called protothetic, ontology, and mereology.

The standard system of protothetic, which is the most comprehensive logic of propositions, is based on a single axiom; and the functor of equivalence, "if and only if," occurs in it as the only undefined term. The directives of protothetic include (1) three rules of inference: substitution, detachment, and the distribution of the universal quantifier; (2) the rule of protothetical definition; and (3) the rule of protothetical extensionality. Protothetic presupposes no more fundamental theory, whereas all other deductive theories which are not parts of protothetic must be based on it or on a part of it.

Ontology is obtained by subjoining ontological axioms to protothetic, adapting the directives of protothetic to them, and allowing for a rule of ontological definition and a rule of ontological extensionality. The standard system of ontology is based on a single ontological axiom, in which the functor of singular inclusion, the copula "is," occurs as the only undefined term. Ontology comprises traditional logic and counterparts of the calculus of predicates, the calculus of classes, and the calculus of relations, including the theory of identity.

By subjoining mereological axioms to ontology and adapting the ontological directives to them, we obtain mereology, which is a theory of part-whole relations. The standard system of mereology, with the functor "proper or improper part of" as the only undefined term, can then be based on a single mereological axiom. No specifically mereological directives are involved. While ontology yields the foundations of arithmetic, mereology is the cornerstone of the foundations of geometry.

Leśniewski formulated the directives of his systems with unprecedented precision. In the art of formalizing deductive theories he has remained unsurpassed. Yet the theories that he developed never ceased for him to be interpreted theories, intended to embody a very general, and hence philosophically interesting, description of reality.

BIBLIOGRAPHY

I. Original Works. Leśniewski's writings include "O podstawach matematyki" ("On the Foundations of Mathematics"), in *Przegląd filozoficzny*, **30** (1927), 164–206; **31** (1928), 261–291; **32** (1929), 60–101; **33** (1930), 77–105, 142–170, with a discussion of the Russellian antinomy and an exposition of mereology; "Grundzüge eines neuen Systems der Grundlagen der Mathematik," in *Fundamenta mathematicae*, **14** (1929), 1–81, which gives an account of the origin and development of protothetic and contains the statement of protothetical directives; "Über die Grundlagen der Ontologie," in *Comptes rendus des séances de la Société des sciences et des lettres de Varsovie*, Cl. III, **23** (1930), 111–132, containing the statement of the directives of ontology; "Über Definitionen in der sogenannten Theorie der Deduktion," *ibid.*, **24** (1931), 289–309, with the statement of the rules of inference and the rule of definition for a system of the classical calculus of propositions; and *Einleitende Bemerkungen zur Fortsetzung meiner Mitteilung u.d. T. "Grundzüge eines neuen Systems der Grundlagen der Mathematik"* (Warsaw, 1938), which includes a discussion of certain problems concerning protothetic. The last two works are available in an English trans. in Storrs McCall, ed., *Polish Logic 1920–1939* (Oxford, 1967), pp. 116–169, 170–187. *Grundzüge eines neuen Systems der Grundlagen der Mathematik § 12* (Warsaw, 1938) offers the deduction of an axiom system of the classical calculus of propositions from a single axiom of protothetic.

II. Secondary Literature. See T. Kotarbiński, *La logique en Pologne* (Rome, 1959); C. Lejewski, "A Contribution to Leśniewski's Mereology," in *Polskie towarzystwo naukowe na obszyźnie. Rocznik*, **5** (1955), 43–50; "A New Axiom of Mereology," *ibid.*, **6** (1956), 65–70; "On Leśniewski's Ontology," in *Ratio*, **1** (1958), 150–176; "A Note on a Problem Concerning the Axiomatic Foundations of Mereology," in *Notre Dame Journal of Formal Logic*, **4** (1963), 135–139; "A Single Axiom for the Mereological Notion of Proper Part," *ibid.*, **8** (1967), 279–285; and "Consistency of Leśniewski's Mereology," in *Journal of Symbolic Logic*, **34** (1969), 321–328; E. C. Luschei, *The Logical Systems of Leśniewski* (Amsterdam, 1962), a comprehensive and reliable presentation of the foundations of the systems constructed by Leśniewski; J. Słupecki, "St. Leśniewski's Protothetics," in *Studia logica*, **1** (1953), 44–112; "S. Leśniewski's Calculus of Names," *ibid.*, **3** (1955), 7–76, which concerns ontology; "Towards a Generalised Mereology of Leśniewski," *ibid.*, **8** (1958), 131–163; B. Sobociński, "O kolejnych uproszczeniach aksjomatyki 'ontologji' Prof. St. Leśniewskiego" ("On Successive Simplifications of the Axiom System of Prof. S. Leśniewski's 'Ontology' "), in *Fragmenty filozoficzne* (Warsaw, 1934), pp. 144–160, available in English in Storrs McCall, ed., *Polish Logic 1920–1939* (Oxford, 1967), pp. 188–200; "L'analyse de l'antinomie Russellienne par Leśniewski," in *Methodos*, **1** (1949), 94–107, 220–228, 308–316; **2** (1950), 237–257; "Studies in Leśniewski's Mereology," in *Polskie towarzystwo naukowe na obczyźnie. Rocznik*, **5** (1955), 34–43; "On Well-Constructed Axiom Systems," *ibid.*, **6** (1956), 54–65; "La génesis de la escuela polaca de lógica," in *Oriente Europeo*, **7** (1957), 83–95; and "On the Single Axioms of Protothetic," in *Notre Dame Journal of Formal Logic*, **1** (1960), 52–73; **2** (1961), 111–126, 129–148, in progress; and A. Tarski, "O wyrazie pierwotnym logistyki," in *Przegląd filozoficzny*, **26** (1923), 68–89, available in English in A. Tarski, *Logic, Semantics, Metamathematics* (Oxford, 1956), pp. 1–23, important for the study of protothetic.

Czesław Lejewski

LESQUEREUX, LEO (*b.* Fleurier, Neuchâtel, Switzerland, 18 November 1806; *d.* Columbus, Ohio, 25 October 1889), *botany, paleontology.*

The earliest authority in the United States on fossil plants and its second-ranking bryologist, Lesquereux was an expatriate from the political revolution of

1847–1848, as were Louis Agassiz and Arnold Guyot. Lesquereux was of French Huguenot ancestry, the only son of V. Aimé and Marie Anne Lesquereux. His father was an unlettered manufacturer of watch springs. His well-read mother wished him to become a Lutheran minister. At the age of ten he suffered a near-fatal fall that brought on partial deafness. After two years at the *collège* of Neuchâtel, where he was a classmate of Guyot, he taught French at Eisenach with the intention of entering a university later. In 1830 he married Sophia von Wolffskel von Reichenberg, daughter of a general attached to the court of Saxe-Weimar; they had four sons and a daughter. Lesquereux discontinued his teaching because of his growing deafness and returned to Neuchâtel and the engraving of watchcases. It was probably while convalescing from a prolonged illness there that he became interested in mosses. He told Henry Bolander that he owed his enthusiasm for mosses to Wilhelm Schimper, whom he called *facile princeps bryologorum.* In an effort to enhance the nation's fuel supply the Swiss government offered a prize for an essay on the formation of peat bogs, and Lesquereux won. His essay, on the peat bogs of the Jura (1844), attracted the notice of Louis Agassiz, who was then at Neuchâtel.

Discouraged by the political ferment of the times and encouraged by Agassiz, Lesquereux and his family took steerage passage for Boston, arriving in September 1848. Agassiz at once employed him in classifying plants that he had collected on his 1848 expedition to Lake Superior. W. S. Sullivant, the country's leading bryologist and a well-to-do businessman of Columbus, Ohio, urged Lesquereux to work with him in the preparation of *Musci boreali-americani* (1856; 2nd ed., 1865) and *Icones muscorum* (1864). Lesquereux supplemented his income by a small jewelry business. In 1849 he toured the southern states for Sullivant and began writing a series of twenty-seven "Lettres écrites d'Amérique destinées aux émigrants," articles rich in commentary on American mores that were published in Neuchâtel's *Revue suisse.* Sullivant and Lesquereux began a *Manual of the Mosses of North America*, but it was interrupted by Sullivant's death in 1873 and by Lesquereux's failing eyesight. Lesquereux then enlisted Thomas Potts James to assist with the microscopic examinations; and the 447-page *Manual*, describing about 900 species, was published under their authorship (Boston, 1884) two years after James's death. Lesquereux's later years were spent naming fossils.

His first paleobotanical publication was a monograph of Carboniferous fossils of Pennsylvania (1854), followed by another on Illinois (1863), from which the locality of Mazon Creek became a classic. From 1867 to 1872 he organized the fossils that had accumulated at Harvard's Museum of Comparative Zoology; this was his only institutional affiliation. This engagement contributed to his monumental three-volume *Coal Flora of Pennsylvania* (1879–1884). Later R. D. Lacoe of Pittston, Pennsylvania, engaged Lesquereux as a semipensioner to organize his large private collection of fossils, which he bequeathed to the National Museum (now the National Museum of Natural History) in Washington, D.C.

BIBLIOGRAPHY

I. ORIGINAL WORKS. No thoroughly satisfactory bibliography of Lesquereux's publications has appeared; imperfect but useful are those by McCabe and Smith, cited below, and John M. Nickles, "Geologic Literature on North America, 1785–1918," in *Bulletin of the United States Geological Survey*, no. 746 (1923), 654–656. The essay that attracted Agassiz was published as part of "Quelques recherches sur les marais tourbeux en général," in *Neuchâtel Société des Sciences Naturelles, Mémoires*, **3** (1844), 1–138. *Lettres écrites d'Amérique* (Neuchâtel, 1849) was published in an enlarged ed. (Neuchâtel, 1853). The Lesquereux-James correspondence (1857–1881) of 270 letters, bound in 3 vols., is at Farlow Cryptogamic Laboratory, Harvard University; and 58 letters (1862–1872) to Henry Nicholas Bolander are at the Bancroft Library, Berkeley, California. An autobiographical letter was published as a pref. to Lesquereux's posthumous "Flora of the Dakota Group," in *Monographs of the U.S. Geological Survey*, **17** (1892), 1–400, reprinted in *Isis*, **34** (1942), 97–98. There is no catalog of his type specimens, figures, or cited specimens (2,460 in all), but according to W. C. Darrah, quoted in Sarton (see below), 1,415 specimens having the "status of types" are at the Botanical Museum, Harvard. His flowering plants, preserved in the British Museum (Natural History) and elsewhere, are numbered for example "Lx. 121" but lack date of collecting.

II. SECONDARY LITERATURE. Some authors give Charles as his first name, but he never used it in his letters or writings, as noted by G. P. Merrill, in *Dictionary of American Biography* XI (New York, 1933), 188–189. J. Peter Lesley, who met Lesquereux in 1851, wrote the most detailed account of his earlier life, in *Biographical Memoirs. National Academy of Sciences*, **3** (1895), 187–212, although the date of Lesquereux's death given there, 20 Oct., is an error. Heretofore unpublished facts regarding his parents, portraits, collections, and publications are presented with charm by George Sarton, in *Isis*, **34** (1942), 97–108; but Sarton's fig. 6 of a page of "MS preface to Manual of Mosses" is in the hand of Asa Gray and not of Lesquereux. Sereno Watson's handwriting appears on p. 18 ff. Both Gray and Watson edited the *Manual* MS. According to Charles R. Barnes, in *Botanical Gazette*, **15** (1890), 16–19, mementos of Lesquereux were presented to the Musée d'Histoire Naturelle at Neuchâtel. The sketch by Annie M. Smith, in *Bryologist*, **12** (1909), 75–78, includes a portrait; as does that of L. R. McCabe, in *Popular Science Monthly*, **30** (1887), 835–840, which is based on an interview with Lesquereux. Charles H. Sternberg, *Life of a Fossil Hunter* (New York, 1909), pp. 21–25, published Lesquereux's letter of 14 Apr. 1875 in facs. Additional biographical notes will be found in W. C. Darrah, "Leo Lesquereux," in *Harvard University Botanical Museum Leaflets*, **2** (1934), 113–119; and in Andrew Denny Rodgers, III, *Noble Fellow, William Starling Sullivant* (New York, 1940), pp. 191–204, and *passim*.

JOSEPH EWAN

LESSON, RENÉ-PRIMEVÈRE (*b.* Cabane-Carée, Rochefort, France, 20 March 1794; *d.* Rochefort, 28 April 1849), *natural history, scientific exploration.*

The son of a navy clerk of modest means, Lesson had had little formal education when in 1809, not yet sixteen years old, he entered the naval medical school of Rochefort. Lesson was largely self-taught in natural history, which became a lifelong passion. In 1811, he was conscripted into the navy as a third-class auxiliary surgeon, serving on several French ships and seeing action against the British. He qualified as *officier de santé* in 1816, competed successfully for third-class navy pharmacist that same year, and was promoted to second-class pharmacist in 1821. By this time, Lesson had also made a botanical survey of the Rochefort region, which was published much later (*Flore Rochefortine*, 1835). Lesson would probably have remained an obscure naturalist had he not embarked in 1822 on the corvette *Coquille* for a voyage of scientific exploration and discovery which dramatically altered his life and brought him into national prominence.

On 11 August 1822, the *Coquille* sailed from Toulon, commanded by Duperrey with J.-S.-C. Dumont d'Urville second in command and responsible for acquisitions in botany and entomology. The other two naturalists, Garnot and Lesson, also served as medical officers; Garnot's fieldwork covered mammals and birds, while Lesson was assigned fish, mollusks, crustaceans, zoophytes, and geology. Among the places visited by the *Coquille* were Tenerife, Brazil, the Falkland Islands, Chile, Peru, Tahiti, New Ireland, the Moluccas, and Australia, where Garnot was forced by illness to leave the expedition in January 1824, and Lesson assumed his scientific and medical duties. The *Coquille* proceeded to New Zealand, the Caroline Islands, New Guinea, Java, Mauritius, Réunion, and St. Helena, finally landing at Marseilles on 24 March 1825.

On 18 July 1825, Cuvier and Latreille reported to the Academy of Sciences on the expedition's zoological data and collections, which had been deposited at the Museum of Natural History in Paris. Lesson and Garnot were praised for bringing back hitherto unknown species of birds, reptiles, fish, mollusks, and crustaceans. Lesson was also cited for his remarkable colored illustrations of fish and mollusks and for his valuable aid to Dumont d'Urville for the insect collection. A later report on the voyage of the *Coquille*, made by Arago to the Academy on 22 August 1825, mentioned 330 geological specimens brought back by Lesson.

For Lesson, the four years of leave in Paris from 1825 to 1829 were his most productive scientifically. He wrote furiously, published the results of his voyage, studied, and made friends with outstanding naturalists and scientists of the capital. Upon his return to Rochefort he taught botany at the naval medical school and in 1831 was made professor of pharmacy. A succession of promotions culminated in 1835 with his appointment as the top-ranking navy pharmacist (*premier pharmacien en chef*) for Rochefort. In 1833 he was elected a corresponding member of the Academy of Sciences.

Lesson occupies a prominent place among French naturalist-voyagers of the period, namely F. Péron, Quoy, Gaimard, Garnot, Dumont d'Urville, and C. Gaudichaud-Beaupré. Lesson's numerous publications encompassed virtually all aspects of natural history, and included some archaeology, ethnography, and folklore, but his most important contributions were to zoology. Particularly significant was his work in ornithology, especially his writings on hummingbirds and birds of paradise.

BIBLIOGRAPHY

I. ORIGINAL WORKS. Among Lesson's more important publications are *Manuel de mammalogie, ou histoire naturelle des mammifères . . .* (Paris, 1827); *Manuel d'ornithologie, ou description des genres et des principales espèces d'oiseaux . . .*, 2 vols. (Paris, 1828); *Voyage médical autour du monde, exécuté sur la corvette du Roi la Coquille . . .* (Paris, 1829); *Zoologie du voyage autour du monde . . .* (Paris, 1829), written with P. Garnot, and with F.-E. Guérin-Méneville for the entomological part; *Histoire naturelle des oiseaux-mouches . . .* (Paris, 1829–1830); *Histoire naturelle des colibris . . .* (Paris, 1830–1831); *Les trochilidées, ou colibris et les oiseaux-mouches . . .* (Paris, 1830–1831); *Manuel d'histoire naturelle médicale et de pharmacographie . . .* (Paris, 1833); *Manuel d'ornithologie domestique, ou guide de l'amateur des oiseaux de volière . . .* (Paris,

1834); and *Histoire naturelle des oiseaux de paradis . . . et des épimaques* (Paris, 1835).

For additional listings of Lesson's publications, see *Catalogue général des livres imprimés de la Bibliothèque Nationale*, XCVI (Paris, 1929), 409–415; and British Museum, *General Catalogue of Printed Books*, CXXXV (London, 1962), 834–835. A comprehensive bibliography of Lesson's articles is given in Royal Society, *Catalogue of Scientific Papers (1800–1863)*, III, 971–975.

II. SECONDARY LITERATURE. A detailed account of Lesson's life and work is given by Louis Rallet, "Un naturaliste saintongeais: René-Primevère Lesson (1794–1849)," in *Annales de la Société des sciences naturelles de la Charente-Maritime*, n.s. **3** (May, 1953), 77–131.

Other sources include John Dunmore, *French Explorers of the Pacific*, vol. II, *The Nineteenth Century* (Oxford, 1969), 109–155, and *passim; Nouvelle biographie générale*, J. C. F. Hoefer, ed., XXIX, 972–974; and J. Léonard, *Les officiers de santé de la marine française de 1814 à 1835* (Paris, 1967), 129–133, 236–237, and *passim*. Brief biographies can be found in *Biographie universelle, ancienne et moderne*, L. G. Michaud and J. F. Michaud, eds., new ed., XXIV, 330–331; and *La Grande encyclopédie*, XXII, 103.

ALEX BERMAN

LESUEUR, CHARLES-ALEXANDRE (*b.* Le Havre, France, 1 January 1778; *d.* Le Havre, 12 December 1846), *natural history.*

Since his father was an officer in the Admiralty, Lesueur was readily accepted at the age of nine at the École Royale Militaire at Beaumont-en-Auge. From 1797 to 1799 he was *sous-officier* in the Garde Nationale at Le Havre. He had early shown remarkable talent in sketching, and in 1800 his great opportunity came when the expedition to Australia and Tasmania of the corvettes *Géographe* and *Naturaliste* (1800–1804), under the command of Nicolas Baudin, was fitted out and sailed from Le Havre. Lesueur shipped as apprentice helmsman but was soon placed in charge of drawing natural history subjects. His relations with the naturalist François Péron were close throughout the long voyage, and through their efforts more than 100,000 zoological specimens, including at least 2,500 new species, were brought back to Paris. The official report of the expedition, written by Péron and Lesueur, was completed later by Claude de Freycinet. Unfortunately Péron died in 1810 before the natural history results were completed, and few of the more than 1,500 of Lesueur's beautiful drawings were ever published. He later took them to America and eventually back to Le Havre, where they survived World War II.

In 1815 Lesueur met in Paris the geologist William Maclure, with whom he sailed for America on 15 August as a paid companion and naturalist. They spent a year in the West Indies on the way, where, among other novelties, Lesueur found many new species of fish, corals, and other marine organisms, some of which he described a few years later. In May 1816 they arrived in New York and immediately set out on a long trip through New England, New York, Pennsylvania, New Jersey, and Maryland. On the journey Maclure reexamined in more detail the geology he had studied some years earlier, and Lesueur assiduously collected fossils (which he probably understood better than anyone else in America at the time), fishes, mollusks, and insects.

For the next nine years Lesueur resided in Philadelphia, where he established himself as naturalist-engraver and teacher of drawing, earning a precarious living. He enjoyed his association with other Philadelphia naturalists, such as Thomas Say, Gerard Troost, and George Ord, to whom he brought his wide knowledge of tropical marine faunas. He soon became a member of the American Philosophical Society and an enthusiastic supporter of the newly founded Academy of Natural Sciences of Philadelphia.

In 1825 Lesueur left for Robert Owen's settlement at New Harmony, Indiana, in Maclure's "Boatload of Knowledge." Although New Harmony's utopian dream soon faded, Lesueur remained there many years, beginning a great work on the fishes of North America—one of several extensive projects that never materialized. An indifferent writer and a compulsive wanderer, Lesueur spent much time on long trips down the Ohio and took four long trips down the Mississippi to New Orleans. In 1837 he descended that river for the last time and returned to France on 27 July 1837, after an absence of twenty-two years. After eight years in Paris he was appointed curator of the newly established Muséum d'Histoire Naturelle du Havre, where he died on 12 December 1846.

In spite of his literary failings, Lesueur's explorations in the Mississippi valley contributed much to the zoology and paleontology of that poorly known area. He was the first to collect and study the fishes of the North American interior and published twenty-nine papers on them as well as others on reptiles, crustaceans, and other organisms. The record he left bears witness to "his quick and agile pencil and his clumsy and encumbered pen."

BIBLIOGRAPHY

I. ORIGINAL WORKS. A list of Lesueur's publications is appended to the memoir by Ord (below).

II. SECONDARY LITERATURE. See Gilbert Chinard, "The American Sketchbooks of C.-A. Lesueur," in *Proceedings of the American Philosophical Society*, **93** (1933), 114–118, with portrait, bust, and sketch; Jean Guiffry, ed., *Dessins de Ch.-A. Lesueur executés aux États-Unis de 1816 à 1837* (Paris, 1933), with 50 plates and reproductions of many of Lesueur's American sketches; E.-T. Hamy, "Les voyages du naturaliste Ch. Alex. Lesueur dans l'Amérique du Nord (1815–1837)," in *Journal de la Société des Américanistes de Paris*, **5** (1904), with 17 plates and 14 figures; and *The Travels of the Naturalist Charles A. Lesueur in North America, 1815–1837*, H. F. Raup, ed. (Kent, Ohio, 1968), with 26 illustrations and 4 maps; Waldo G. Leland, "The Lesueur Collection of American Sketches in the Museum of Natural History at Havre, Seine-Inférieure," in *Mississippi Valley Historical Review*, **10** (1923), 53–78, which lists American notes and sketches then in Le Havre; Mme Adrien Loir, *C.-A. Lesueur, artiste et savant français en Amérique de 1816–1839* (Le Havre, 1920), with 42 figures; André Maury, "C.-A. Lesueur, voyageur et peintre-naturaliste Havrais," in *French-American Review*, **1** (1948), 161–171, with portrait and 8 illustrations; George Ord, "A Memoir of Charles Alexandre Lesueur," in *American Journal of Science*, 2nd ser., **8** (1849), 189–216, a lengthy account of the *Géographe* and *Naturaliste* expedition by Lesueur's closest American friend, with a list of Lesueur's published writings.

JOHN W. WELLS

LE TENNEUR, JACQUES-ALEXANDRE (*b.* Paris, France; *d.* after 1652), *mathematics, physics.*

Described as a patrician of Paris on the title page of his principal book, little else is known of Le Tenneur, friend of Mersenne and correspondent of Gassendi. Probably a resident of Paris until the mid-1640's, he was at Clermont-Ferrand (near Puy-de-Dôme) late in 1646, and in 1651 he was counselor to a provincial senate. C. De Waard identifies him as counselor to the Cour des Aydes of Guyenne, but without indicating dates.

All modern authorities agree in attributing to Le Tenneur the *Traité des quantitez incommensurables*, whose author identified himself on its title page only by the letters I.N.T.Q.L.[1] Mersenne mentioned the *Traité* as in preparation on 15 January 1640 and sent a copy to Haak on 4 September, indicating that it was "by one of my friends." It was directed against Stevin's *L'arithmétique* (Leiden, 1585; ed. A. Girard, 1625), particularly opposing Stevin's treatment of unity as a number. Although Le Tenneur's arguments now seem elementary and conservative, they go to the heart of the foundations of algebra, standing as a final attempt to preserve the classical Greek separation of arithmetic from geometry that Descartes abandoned in 1637. The basic question is whether the unit may properly be considered as divisible. Against Stevin's affirmative answer, Le Tenneur took the view that this would in effect either merely substitute a different unit or deny the existence of any unit relevant to the problem at hand. The bearing of this analysis on the problems of indivisibles and infinitesimals that were then coming to the fore is evident; but a rapid and widespread acceptance of algebraic geometry doomed the classic distinctions, and the book was neglected. Aware of Viète's work and of the need to give symbolic treatment to incommensurables in the classic sense, Le Tenneur included a paraphrase of book X of Euclid's *Elements*, in which he gave the symbolic operations for each proposition. The book ends with an essay, addressed to the Académie Française, proposing that French should be used in science and outlining methods for the coining of terms, reminiscent of Stevin's earlier argument for the use of Dutch.

In January 1648 Mersenne wrote to Le Tenneur asking him to perform the barometric experiment at Puy-de-Dôme later done by F. Perier at Pascal's request. Le Tenneur, who had moved to Tours, replied that it could not be done in winter and expressed the view that, in any event, the level of mercury would not be changed by the ascent.

The importance of Le Tenneur to the history of science, however, depends on another book, *De motu naturaliter accelerato . . .* (1649), in which he showed himself to be the only mathematical physicist of the time who understood precisely Galileo's reasoning in rejecting the proportionality of speeds in free fall to the distances traversed. This subject was hotly debated in the late 1640's between Gassendi and two Jesuits, Pierre Cazré and Honoré Fabri. In 1647 Fermat, who had previously questioned the validity of Galileo's odd-number rule for distances traversed, wrote out for Gassendi a rigorous demonstration of the impossibility of space proportionality, believing that Galileo had deliberately withheld his own. Le Tenneur, who appears not to have seen Fermat's proof, perceived and illustrated the nature of Galileo's use of one-to-one correspondence between the speeds in a given fall from rest and those in its first half.

Le Tenneur's interest in the matter began with a request from Mersenne for a discussion of Cazré's *Physica demonstratio* (1645), which in turn had arisen from his letters to Gassendi opposing the latter's *De motu impresso a motore translato* of 1642. Le Tenneur's critique of Cazré took the form of a *Disputatio physico-mathematicus* sent in manuscript form to Mersenne, probably in mid-1646, and also sent to Cazré or forwarded to him by Mersenne. Gassendi published his own reply to Cazré as *De proportione qua gravia decidentia accelerantur* (1646) and sent a copy to Le Tenneur. Writing on 24 November 1646 to acknowledge this gift, Le Tenneur sent Gassendi a copy of his *Disputatio*. Gassendi replied on 14 December, and on 16 January 1647 Le Tenneur wrote him that Mersenne had mentioned some other Jesuit than Cazré. This was Honoré Fabri, whose *Tractatus physicus de motu locali* (1646) acknowledged the correctness of Galileo's rules for sensible distances only, explaining this in terms of the space–proportionality of insensible quanta of impetus. Baliani had (also in 1646) advanced a similar hypothesis. Two days later, Le Tenneur wrote that Cazré had meanwhile replied to his earlier *Disputatio* and sent the reply to Gassendi with his own rebuttal.

On 12 April 1647 Le Tenneur wrote out a refutation of Fabri's position in the form of a long letter to Mersenne. Fabri contended that the hypotenuse in Galileo's triangle was in ultimate reality a discrete step function (a denticulated line, as he called it) and that the space quanta traversed in true physical instants progressed as the positive integers rather than as the observed odd numbers. Le Tenneur opposed any analysis by physical instants (the medieval *minima naturalia*) on the grounds that if indivisible, such instants implied Galileo's law, whereas if divisible, they fell under Galileo's phrase "in any parts of time" and excluded Fabri's hypothetical increments of velocity added discretely (*simul*).

Mersenne appears to have communicated this letter to Fabri, who published it as an appendix to a new book in 1647, with critical comments. Finally, Le Tenneur wrote out a long reply to Fabri and composed a further treatise of his own in support of Galileo's laws against all his opponents. On 1 January 1649 he submitted it to Gassendi with the idea of publishing it together with his previous writings on the subject if Gassendi approved. Gassendi had meanwhile moved to Aix-en-Provence for his health and did not receive the material until Easter. His enthusiastic letter of endorsement was sent to Le Tenneur on 17 May 1649 and was included as the final item in *De motu accelerato*.

Of all the participants in this wordy dispute, only Le Tenneur correctly reconstructed Galileo's original argument. His book is of further interest for its inclusion of a strictly mathematical derivation of Galileo's odd-number rule by the young Christiaan Huygens, which had been sent by his father to Mersenne. Le Tenneur published it together with a postil predicting great things from Huygens. *De motu accelerato* is of further historical importance for its refutation of Fabri's dictum that no physical instant could be identical with a mathematical point in time, the last stand of orthodox Aristotelian physics and impetus theory against the continuity concept introduced by Galileo. Even Descartes had rejected the idea that in reaching any given speed from rest, a body must first have passed through every lesser speed. His letters to Mersenne on this point may have been the occasion for its having been called to the attention of Le Tenneur.

The only other books by Le Tenneur related to controversies with Jean-Jacques Chifflet over French history and royal genealogy. He sent one of these to Gassendi and in an accompanying letter seems to have considered himself more a historian than a scientist, despite his valuable if neglected contributions to fundamental issues of mathematics and physics in the seventeenth century.

NOTE

1. The attribution, although probably correct, is not yet certain. No early reference to this book includes more than the family name of the author, and the title page suggests the possibility that there may have been an I. N. (or J. N.) Le Tenneur; Gassendi mentioned a brother of Jacques-Alexandre who has not been identified. In the preface, the author expressed his intention of identifying himself after hearing comments on his book; if the same author wrote *De motu accelerato* (1649) mention of the earlier book would be expected in it. Moreover, the vigorous argument for the use of French hardly fits J.-A. Le Tenneur, who thereafter published only in Latin.

BIBLIOGRAPHY

I. Original Works. Le Tenneur's writings are *Traité des quantitez incommensurables* (Paris, 1640); *De motu naturaliter accelerato tractatus physico-mathematicus* (Paris, 1649), one section of which had appeared under the author's initials in H. Fabri (P. Mousnier, ed.), *Metaphysica demonstrativa* (Lyons, 1647); *Veritas vindicata adversus . . . Joan. Jac. Chifletii* (Paris, 1651); and *De sacra ampulla remensi tractatus apologeticus adversus Joan. Jac. Chifletum* (Paris, 1652). Some correspondence between Gassendi and Le Tenneur was published in Gassendi, *Opera omnia*, VI (Lyons, 1658). See Adam and Tannery, eds., *Oeuvres de Descartes*, V (Paris, 1903); L. Brunschvig and P. Boutroux, eds., *Oeuvres de B. Pascal*, II (Paris, 1908). Some of Le Tenneur's letters are preserved at Vienne, France, and at the Bibliothèque Nationale, Paris.

II. Secondary Literature. For references to and notes about Le Tenneur see *Correspondance du P. Marin Mersenne*, P. Tannery and C. De Waard, eds., IX–XI; most letters between the two men have not yet been published and were not accessible for this article. See also H. Brown, *Scientific Organization in 17th Century France* (New York, 1967), pp. 54–56; S. Drake in *Isis*, **49** (1958), 342–346; **49** (1958), 409–413; *British Journal of the History of Science*, **5** (1970), 34–36; and *Galileo Studies* (Ann Arbor, Mich., 1970), 235–236; W. E. K. Middleton, *History of the*

Barometer (Baltimore, 1964), 40, which gives references to papers on the Pascal controversy published by Felix Matthieu in the *Revue de Paris* (1906).

STILLMAN DRAKE

LE TONNELIER DE BRETEUIL. See **Châtelet, Gabrielle-Émilie Le Tonnelier de Breteuil, Marquise du.**

LEUCIPPUS (*fl.* Greece, fifth century B.C.), *philosophy*.

The first of the Greek atomists, Leucippus was probably the founder of the school of Abdera, whose most famous exponent was Democritus. Although at the end of the fourth century B.C. Epicurus denied that there had ever been any such person as Leucippus, the evidence of Aristotle is sufficient to establish that he existed and that he was earlier in date than Democritus. Aristotle treated his theories as providing a logical alternative to those of Parmenides, and a later tradition actually made him a pupil of Zeno, of the school of Parmenides. This, taken with the late chronology for Democritus, has led some scholars to date Leucippus' activity as late as 430 B.C. But neither the chronology of Democritus nor the relationship of Leucippus to Parmenides is in any way certain. Leucippus probably came from Miletus in Ionia and may have brought knowledge of the physical theories of the Ionians with him to Abdera in Thrace, either sometime after its refoundation as a colony about 500 B.C. (when it soon became an important city in the Persian system of government in Thrace) or sometime after its accession to the Athenian league in the period after 478.

By the fourth century B.C. Leucippus' basic doctrines and probably his writings as well seem to have become incorporated into a kind of corpus of atomist writings, the whole of which was attributed to his more famous pupil and successor, Democritus. Therefore, at many points it is no longer possible to distinguish what originated with Leucippus from what may have been added by Democritus. But Theophrastus attributed to Leucippus a *Megas diakosmos* ("Great World System") later listed among Democritus' writings, and one quotation is preserved from a work *On Mind.* (For the origins of atomism and the theories given in the Democritean corpus, see Democritus.)

As the last writer in a position to distinguish Leucippus' views from those of Democritus, Theophrastus makes it clear that the essentials of atomism were already held by Leucippus. Both matter and void have real existence. The constituents of matter are elements infinite in number and always in motion, with an infinite variety of shapes, completely solid in composition. We can be sure that a great deal more of Democritus' doctrine was also to be found in Leucippus, but we cannot say how much more.

On the other hand, we have preserved from Theophrastus a summary of Leucippus' account of the creation of physical worlds that is ascribed to him alone, so that it is likely to be correct in essentials. The earth and stars originated from a single whirl of colliding bodies, cut off from the infinite. The further details, which are strongly reminiscent of earlier Ionian cosmologies, include a doctrine of a surrounding membrane that has a typical early biological sound.

BIBLIOGRAPHY

The fragments and testimonia are collected in H. Diels and W. Kranz, *Die Fragmente der Vorsokratiker*, 6th ed., II (Berlin, 1952). For translation and discussion of the more important texts, see G. S. Kirk and J. E. Raven, *The Presocratic Philosophers* (Cambridge, 1957); and for more extended treatments, see C. Bailey, *The Greek Atomists and Epicurus* (Oxford, 1928); and W. K. C. Guthrie, *History of Greek Philosophy*, II (Cambridge, 1965), ch. 8.

G. B. KERFERD

LEUCKART, KARL GEORG FRIEDRICH RUDOLF (*b.* Helmstedt, Germany, 7 October 1822; *d.* Leipzig, Germany, 6 February 1898), *zoology, parasitology*.

Leuckart, whose father owned a printing plant, attended the Gymnasium in Helmstedt. His uncle, Friedrich Sigismund Leuckart, who was professor of zoology at Freiburg im Breisgau, made a strong impression on him and awakened his love for zoology. As a student he collected insects and soon decided to become a zoologist. At this period, however, he could do so only by obtaining a medical education, since there were no faculties of natural science. Consequently, in 1842 Leuckart began medical studies at the University of Göttingen. He soon came into contact with the distinguished zoologist Rudolf Wagner, with whom he formed a lifelong friendship. Wagner encouraged Leuckart to undertake independent research and made him an assistant in his institute in 1845, after Leuckart had passed the state medical examination. Leuckart's dissertation (*De monstris eorumque causis et ortu*), published in the same year, won a university prize; and Wagner immediately gave him an additional post as his lecture assistant. At the end of 1847 Leuckart qualified as a zoology lecturer at Göttingen with a lecture entitled "Naturgeschichte mit besonderer Berücksichtigung des Menschen und der Tiere." The following year Leuckart went on his first scientific expedition, to the German North Sea coast to study marine invertebrates. His investigations led to the establishment of the phylum Coelenterata.

In 1850, at the age of twenty-eight, Leuckart was appointed associate professor of zoology at the University of Giessen, to which Justus von Liebig and the anatomist and embryologist Theodor Bischoff were attracting many young researchers. Leuckart took his place among this illustrious company and later called his years in Giessen the happiest and most scientifically productive of his life. Here he met his wife, the daughter of a royal privy councillor, and conducted the investigations (1850–1869) on human and animal parasites that brought him world fame. In 1869 Leuckart accepted an offer from the University of Leipzig to succeed Eduard Pöppig, who had distinguished himself mainly as a geographer and world traveler. The facilities at his disposal were at first inadequate, but in 1880 he was given a new zoological institute constructed according to his plans and furnished with laboratories, a museum, and a library. Zoologists from all over the world came to the institute as guest researchers. Leuckart's outstanding skill as a teacher and his clear delivery brought him many students who later held chairs of zoology not only at German universities but also in England, France, Italy, Sweden, Russia, Switzerland, Japan, and the United States. The *Festschrift* published on his seventieth birthday includes contributions by more than 130 former students and co-workers.

Leuckart was named a privy councillor by the king of Saxony and was awarded an honorary Ph.D. by the University of Giessen in 1861. He was an honorary member of the Imperial Academy of Sciences of St. Petersburg and received the highest Prussian and Bavarian orders: Pour le Mérite für Künste und Wissenschaften and Maximilianorden. In 1873 he became dean of the philosophical faculty and in 1877–1878 was rector of the University of Leipzig. His powers gradually diminished, and the accidental death of his son Rudolf, a very promising chemist, and of a daughter, after a long illness, pained him deeply and affected his creative ability in his later life. He died of a stroke, following an attack of bronchitis.

Leuckart's first scientific work, which he began in 1847–1848, was the division of Cuvier's Radiata into Coelenterata and Echinodermata (terms of his own devising). The result marked the beginning of a new period of scientific zoology, that of animal systematics based on subtle morphological investigations. His division of the Metazoa into six principal phyla—Coelenterata, Echinodermata, Annelida, Arthropoda, Mollusca, and Vertebrata—is still considered classic, although it provoked considerable opposition when it was first proposed. For example, in 1848 the physiologist Carl Ludwig greeted with sarcasm the morphological point of view that Leuckart presented in his first book, *Über die Morphologie und die Verwandtschaftsverhältnisse der wirbellosen Tiere:* "It would have to be considered a good sign for German science if this book found no readers." Against this, the distinguished anatomist Heinrich Rathke wrote to Leuckart: "I hope not only that [your publication] finds many readers but also—should this happen—that it will be considered a sign of German science."

In *Die anatomisch-physiologische Übersicht des Tierreichs*, published jointly with Carl Georg Bergmann in 1852, Leuckart placed greater emphasis on the physiological approach that ultimately enabled him to discover the developmental process of many parasites. He was especially interested in the sexual organs of the lower animals and in parthenogenesis, as well as in polymorphism, a term he originated. This research was summarized in the article "Zeugung" in Rudolf Wagner's *Handwörterbuch der Physiologie* (IV, 1853). Leuckart's primary interest, however, was parasitology, a subject then being established. His studies of the *Pentastomum*, *Taenia*, liver fluke, and, most notably, *Trichina spiralis* were epoch-making. In opposition to Küchenmeister's view, Leuckart was able to show that *Taenia saginata* occurs only among cattle and *Taenia solium* only among pigs.

In 1860 Leuckart and Friedrich Albert von Zenker simultaneously discovered the *Trichina*, and an unpleasant priority dispute began. Although Leuckart's publication did appear shortly before Zenker's, there is no doubt today that Zenker made the decisive discovery first. Both works, following the initiative of Rudolf Virchow, led to the world's first meat inspection law. Leuckart carried out fundamental investigations of the Acanthocephalata and of *Onchocerca volvulus*, an organism responsible for *onchocerciasis*, which is still epidemic in central Africa. Leuckart also established the protozoan class Sporozoa and the order Coccidia. Many of his conclusions, derived from detailed investigations, were not confirmed until after his death; among these was the course of development of the roundworm *Ascaris lumbricoides* and the tapeworm *Diphyllobothrium latum*, which was first clarified in 1916 but had been correctly predicted by Leuckart.

The fruit of Leuckart's research in parasitology, the two-volume *Die menschlichen Parasiten und die von ihnen herrührenden Krankheiten*, was never completed; the aging Leuckart did not find the strength to finish it. Nevertheless, it has become a classic of parasitology, along with *Berichte über die Leistungen der niederen Thiere* (founded by Carl Theodor von Siebold), which Leuckart edited from 1848 to 1879 and which is a valuable source of information for mid-nineteenth-century zoological literature. Summing up his lifework, Leuckart wrote: "It is not possible for man, as a thinking being, to close his mind to the knowledge that he is ruled by the same power as is the animal world. Like the despised worm he lives in dependence upon external commands, and like the worm he perishes, even when he has shaken the world through the power of his ideas."

BIBLIOGRAPHY

I. Original Works. A list of the 115 zoological species, genera, and orders that Leuckart named, as well as a complete bibliography of his scientific publications, is in O. Taschenberg, "Rudolf Leuckart," in *Leopoldina* (1899), 63–66, 89–94, 102–112. His most important works are *Beiträge zur Kenntniss wirbelloser Thiere* (Brunswick, 1847), written with Heinrich Frey; *Über die Morphologie und die Verwandtschaftsverhältnisse der wirbellosen Thiere* (Brunswick, 1848); "Ist die Morphologie denn wirklich so ganz unberechtigt?" in *Zeitschrift für wissenschaftliche Zoologie*, **2** (1850), 271–275; *Über den Polymorphismus der Individuen oder die Erscheinungen der Arbeitstheilung in der Natur* (Giessen, 1851); *Anatomisch-physiologische Übersicht des Thierreichs* (Stuttgart, 1852), written with C. G. Bergmann; *Die Blasenbandwürmer und ihre Entwicklung* (Giessen, 1856); *Untersuchungen über Trichina spiralis* (Leipzig–Heidelberg, 1860; 2nd ed., 1866); *Die menschlichen Parasiten und die von ihnen herrührenden Krankheiten*, 2 vols. (Leipzig–Heidelberg, 1863–1876; 2nd ed., 1879–1894 [completed by Gustav Brandes]; 3rd ed., 1886–1901), also translated into English (Edinburgh–Philadelphia, 1886); "Organologie des Auges. Vergleichende Anatomie," in C. F. von Graefe and E. T. Saemisch, eds., *Handbuch der gesamten Augenheilkunde*, II (Leipzig, 1875), pp. 145–301; and "Zur Entwicklungsgeschichte des Leberegels," in *Archiv für Naturgeschichte*, **48** (1882), 80–119.

II. Secondary Literature. See R. Blanchard, "Notices biographiques: Rodolphe Leuckart," in *Archives de parasitologie*, **1** (1898), 185–190; O. Bütschli, "Die wichtigsten biographischen Daten aus dem Leben Rudolf Leuckarts," in *Zoologisches Zentralblatt*, **6** (1899), 264–266; J. V. Carus, "Zur Erinnerung an Rudolf Leuckart," in *Sitzungsberichte der K. Sächsischen Gesellschaft für Wissenschaften*, Math.-phys. Kl. (1898), 49–62; C. Grobben, "Rudolf Leuckart," in *Verhandlungen der Zoologische-botanischen Gesellschaft in Wien*, **48** (1898), 241–243; L. Grosse, "Leuckart in seiner Bedeutung für Natur- und Heilkunde," in *Jahresberichte der Gesellschaft für Natur- und Heilkunde in Dresden* (1898), 93–96; R. von Hanstein, "Rudolf Leuckart," in *Naturwissenschaftliche Rundschau*, **13** (1898), 242–246; A. Jacobi, "Rudolf Leuckart," in *Zentralblatt für Bakteriologie, Parasitenkunde, Infektionskrankheiten und Hygiene*, 1st Section, **23** (1898), 1073–1081; H. A. Kreis, "Rudolf Leuckart, der Begründer der modernen Parasitologie," in *CIBA-Zeitschrift* (Basel), **5** (1937), 1755–1757; G. Olpp, *Hervorragende Tropenärzte in Wort und Bild* (Munich, 1932), pp. 236–240, an especially detailed biography; and the unsigned "Rudolf Leuckart," in *Nature*, **57** (1898), 542.

H. Schadewaldt

LEURECHON, JEAN (*b.* Bar-le-Duc, France, *ca.* 1591; *d.* Pont-à-Mousson, France, 17 January 1670), *mathematics.*

Leurechon was a Jesuit who taught theology, philosophy, and mathematics in the cloister of his order at Bar-le-Duc, Lorraine. Very little is known of his personal life. In his earlier years he wrote several tracts on astronomy and an inconsequential work on geometry.

Leurechon is remembered chiefly for his collection of mathematical recreations, some of which were published under other names. Issued at a time when interest in recreational mathematics was rapidly rising, this work obviously appealed to popular fancy, for it passed through some thirty editions before 1700. Based largely on the work of Bachet de Méziriac, it included, besides many original problems, some taken from Cardano. It served in turn as a foundation for the works of Mydorge, Ozanam, Montucla, and Charles Hutton. For the most part Leurechon borrowed only Bachet's simpler and easier problems, completely bypassing the more significant sections. Leurechon's work was characterized by Montucla as "a pathetic jumble" and by D. E. Smith as "a poor collection of trivialities."

BIBLIOGRAPHY

I. Original Works. Leurechon's tracts on astronomy include *Pratiques de quelques horloges et du cylindre* (Pont-à-Mousson, 1616); *Ratio facillima describendi quamplurima et omnis generis horologia brevissimo tempore* (Pont-à-Mousson, 1618); and *Discours sur les observations de la comète de 1618* (Paris–Rheims, 1619). His work on geometry was *Selectae propositiones in tota sparsim mathematica pulcherrime propositae* (Pont-à-Mousson, 1622).

His collection of mathematical recreations, published under the pseudonym Hendrik van Etten, was *La récréation mathématique ou entretien facétieux sur plusieurs plaisants problèmes, en fait arithmétique* . . . (Pont-à-Mousson, 1624); 2nd ed., rev. and enl. (Paris, 1626). Many subsequent eds. and translations appeared: the first English trans., by William Oughtred, was *Mathematicall Recreations* (London, 1633).

II. SECONDARY LITERATURE. On Leurechon and his work, see Moritz Cantor, *Vorlesungen über Geschichte der Mathematik*, II (Leipzig, 1913), 673–674; 768–769; and Poggendorff, I, 1438. See also H. Zeitlinger, *Bibliotheca chemico-mathematica* (London, 1921), I, 61; II, 536.

WILLIAM L. SCHAAF

LEURET, FRANÇOIS (*b.* Nancy, France, 30 December 1797; *d.* Nancy, 6 January 1851), *psychiatry, public health, comparative anatomy.*

Leuret's education was achieved against the wishes of his father, a baker, who wanted his three sons to follow a trade. With his mother's persistent support, François was sent to a seminary. After an older brother, an army doctor, died in the Napoleonic Wars, François went to Paris to study medicine. In 1818, his father having cut off all funds, he enlisted in the army, a career that suited him ill. Stationed at St.-Denis, he walked to Paris daily to attend medical lectures, notably those of Jean-Étienne Esquirol at the Salpêtrière Hospital. His friend Ulysse Trélat secured a scholarship at Charenton, enabling Leuret to study further with Esquirol and with Spurzheim, and to work in the laboratories at Charenton and Alfort. He published several papers (1824, 1825) and earned his medical doctorate in 1826 with a thesis on changes in the blood. He spent the next twenty years in practice, research, writing, and as medical director of the private Maison de Santé du Gros Caillou in Paris. Late in 1839, Leuret was finally appointed psychiatrist in chief at Bicêtre, succeeding Guillaume Ferrus, whose assistant he had been since 1836. But Leuret was ill and died eleven years later of what seems to have been congestive heart failure. He attained only minor official distinctions and in an unsuccessful bid for the Paris Academy of Medicine in 1836 won only one vote out of 134. His decorations included the Legion of Honor and the Cholera Medal.

Leuret's early research in physiological chemistry and comparative anatomy resulted in a major book (1839), in which he elaborated the latest microscopic and differential staining techniques. The materials for a second volume were published by Gratiolet in 1857, on the initiative of Leuret's friend, the publisher J.-B. Baillière. After receiving his M.D. degree, Leuret practiced medicine in Nancy until 1829, when Esquirol appointed him editor of the newly founded *Annales d'hygiène publique et de médecine légale*. This excellent and influential journal attracted the contributions of concerned physicians, philanthropists, lawyers, and public officials. Leuret's editorial hand is evident in the sections "Correspondence," "Bibliography," and "Varia," and he contributed major papers on cholera (1831), on the Parisian poor (1836), and on his trip through northern Germany and to St. Petersburg where he visited poorhouses and insane asylums (1838).

Psychiatry was Leuret's major concern. Beginning with the comparative anatomy of the nervous system, he soon focused on psychosomatic and mental illness. His opposition to phrenology appears in two devastating book reviews in the *Annales* (1836) and a sarcastic account of Gall's visit to the Salpêtrière (1840). Like his heroes Philippe Pinel and Esquirol, Leuret was committed to "moral," that is, psychological, therapy. At Bicêtre, he introduced the teaching of academic subjects, dance, gymnastics, and music. He instituted a common dining hall and continued Pinel's practice of manual work in the fields. He argued somewhat paradoxically that, confronted with uncooperative patients, the psychiatrist may have to "attack and badger" in order to effect a cure ([1838], 553). He made frequent use of punitive cold showers and dousings, controversial methods that he justified at length in *Du traitement moral de la folie* (1840).

Leuret's major contribution as a teacher, therapist, and writer was his opposition to the prevalent tendency to find correlations between morbid brain anatomy and mental illness.

BIBLIOGRAPHY

I. ORIGINAL WORKS. A complete bibliography of Leuret's works is in Ulysse Trélat, "Notice sur François Leuret," in *Annales d'hygiène publique et de médecine légale*, **45** (1851), 241–263. Works cited in the text are *Mémoire sur l'altération du sang* (Paris, 1826); "Mémoire sur l'épidémie, désignée sous le nom de choléra-morbus," in *Annales d'hygiène publique et de médecine légale*, **6** (1831), 313–472; "Notice sur les indigents de la ville de Paris," *ibid.*, **15** (1836), 294–358; "Notice historique sur A. J. B. Parent-Duchatelet," *ibid.*, **16** (1836), v–xxi; "Notice sur quelques-uns des établissements de bienfaisance du nord de l'Allemagne et de St. Pétersbourg," *ibid.*, **20** (1838), 346–406; "Mémoire sur le traitement moral de la folie," in *Mémoires de l'Académie royale de médecine de Paris*, **7** (1838), 552–576; *Anatomie comparée du système nerveux considéré dans ses rapports avec l'intelligence*, 2 vols. (Paris, 1839–1857), vol. II edited by Pierre Gratiolet; and *Du traitement moral de la folie* (Paris, 1840).

II. SECONDARY LITERATURE. On Leuret and his work, see A. Brierre de Boismont, "Notice biographique sur François Leuret, médecin-en-chef de l'hospice de Bicêtre," in *Annales medico-psychologiques*, **3** (1851), 512–527; and U. Trélat, "Notice sur François Leuret," in F. Leuret and P. Gratiolet, *Anatomie comparée du système nerveux*, I (Paris, 1839), xiii–xxx.

See also the biographies in A. Dechambre, *Dictionnaire encyclopédique des sciences médicales*, 2nd ser., II (Paris, 1869), 403–404; A. Hirsch, *Biographisches Lexikon der hervorragenden Ärzte aller Zeiten und Völker*, 2nd ed., III (Berlin, 1931), 759–760; and R. Semelaigne, *Aliénistes et philanthropes* (Paris, 1912), 482–484.

DORA B. WEINER

LEVADITI, CONSTANTIN (*b.* Galaţi, Romania, 19 July 1874; *d.* Paris, France, 5 September 1953), *medicine, bacteriology.*

Orphaned at the age of eight, Levaditi was raised by an aunt in the port city of Galaţi. This stimulating environment heightened the sense of independence and initiative that he owed to his naturally lively intelligence. This aunt later entrusted him to his other aunt, a laundress at the Branconvan Hospital in Bucharest, where he completed his secondary schooling and medical studies. In 1895 he was accepted into the laboratory of Victor Babès. There he found his lifework: scientific research, particularly on the actinomycotic form of the tubercle bacillus. In 1898 Levaditi worked in Paris with Charles Bouchard and also with Albert Charrin at the Collège de France. Then, at Frankfurt am Main, he studied with Paul Ehrlich, whose views on chemotherapy profoundly influenced him. Admitted in 1900 to the Pasteur Institute, he served as assistant to Élie Metchnikoff in his work on phagocytosis and syphilis. Soon, however, he began to devote himself to his own research.

Levaditi received the M.D. at Paris in 1902. He became head of the laboratory of the Pasteur Institute in 1910 and in 1926 its *chef de service*. Throughout this period he trained numerous students and disciples. Entering World War I as a volunteer, he attained the rank of captain and was detached to serve with the ambulances of la Panne. In this capacity he worked on developing an antitetanus vaccine and studied streptococcus in wounds. After the war he resumed his work at the Pasteur Institute, adding new projects to those already in progress. After 1932 Levaditi divided his time between the Pasteur Institute and the Alfred Fournier Institute, which was the center for the French National League Against Venereal Diseases. After retiring from the Pasteur Institute in 1940, he devoted his efforts to the latter organization until his death at the age of seventy-nine.

In 1903 Levaditi married a Romanian woman who bore him two children. His son, Jean C. Levaditi, became head of the department of histopathology at the Pasteur Institute.

Levaditi became a naturalized French citizen in 1907. He was a member of many French and foreign academies, including the Académie de Médecine, to which he was named in 1928. He was commander of the Legion of Honor and received many prizes, including Cameron Prize (Edinburgh) in 1928, the Paul Ehrlich Prize in 1931, and eight prizes from the Académie des Sciences of Paris.

Levaditi published more than 1,200 notes, articles, and monographs. His penetrating intuition, enhanced by his power of reasoning and remarkable technical knowledge, enabled him to stay informed about new developments; by isolating these developments from their original context in order to set them against his own conceptions, he produced an immense body of work. Thus he noticed that the origin of a group of viral diseases whose agent he named ultravirus was still unknown. This group provided the connecting link for all his investigations, which were outwardly dissimilar (for instance, neurotropic skin diseases).

Levaditi dominated the study of syphilis in his time. After the discovery of the *Treponema pallidum* by Fritz Schaudinn and Paul-Erich Hoffmann, he studied it, using the silver-salt staining method, in the livers of newborn congenital syphilitics. He demonstrated that to supplement the Wassermann-Bordet reaction, a normal liver could be used for diagnosis, thus pioneering the study of antigens. Following Benjamin Sauton, he, Robert Sazerac, and Louis Fournier applied bismuth therapy to syphilis, and achieved results with metallotherapy. He was put in charge of Wassermann serum diagnosis at the Pasteur Institute. Levaditi also used stovarsol (Fourneau 190) in syphilis therapy, after its discovery by Fourneau, Jacques and Thérèse Tréfouël, and Navarro-Martin.

Inspired by Pasteur's work on rabies, Levaditi carried out important experiments on lethargic encephalitis. With Karl Landsteiner he established that the poliomyelitis virus is an ultravirus. During the epidemics in Sweden in 1913 and in Alsace in 1930 he was in charge of committees of assistance. In 1930 Louis Pasteur Vallery-Radot wrote that in France, Levaditi was "the instigator of all research on the etiology of poliomyelitis."

From 1936 to 1940 Levaditi studied the sulfamides and their derivatives. Beginning in 1946 he turned to antibiotics and their mechanisms, which he examined through experimentation on animals. He wrote several important monographs soon after the discovery of each new antibiotic. Levaditi continued his studies of viruses and of chemotherapeutic action on bacteria until his death. A major result of this dedicated effort was his discovery of what are today called the interference phenomena.

BIBLIOGRAPHY

I. Original Works. Bibliographies of monographs and articles by Levaditi are his *Travaux de médecine expérimentale, 1897–1931. Ectodermoses neurotropes, neuroprotozooses, syphilis, chimiothérapie et chimioprévention, phagocytose, immunité, érythème polymorphe, rhumatisme, ergostérol irradié* (Paris, 1931); and *Titres et travaux. Microbiologie, pathologie humaine et animale, chimiothérapie. 1897–1951* (Paris, 1952).

His writings include *La leucocytose et ses granulations* (Paris, 1902); *La nutrition dans ses rapports avec l'immunité* (Paris, 1904); *La réaction des anticorps syphilitiques dans la paralysie générale et le tabès* (Paris, 1906), written with A. Marie; *La syphilis* (Paris, 1909), written with F. Roché; *Traitement de la paralysie générale par injection du sérum salvarnisé dans la dure-mère cérébrale* (Paris, 1913), written with A. Marie et T. Martel; *Étude sur le tréponème de la paralysie générale* (Paris, 1919), written with A. Marie; *Les ectodermoses neurotropes, poliomyélite, encéphalite, herpès* (Paris, 1922), with preface by E. Roux; "Vaccine pure cérébrale, virulence pour l'homme," in *Comptes rendus hebdomadaires des séances de l'Académie des sciences*, **174** (1922), 248–252, written with S. Nicolau; *Étude de l'action thérapeutique sur la syphilis* (Paris, 1923), written with R. Sazérac; *Le bismuth dans le traitement de la syphilis* (Paris, 1924); *L'herpès et le zona* (Paris, 1926); *Travaux de microbiologie et de pathologie humaines et animales. 1897–1933* (Paris, 1933); *Prophylaxie de la syphilis* (Paris, 1936); *Traité des ultravirus des maladies humaines et animales*, 2 vols. (Paris, 1943–1948); *La pénicilline et ses applications thérapeutiques* (Paris, 1945); *Précis de virologie médicale* (Paris, 1945); *La streptomycine et ses applications thérapeutiques, principalement dans la tuberculose* (Paris, 1946); *Les antibiotiques autres que la pénicilline* (Paris, 1950); and *Le chloramphénicol et ses applications thérapeutiques* (Paris, 1951).

II. Secondary Literature. On Levaditi and his work, see R. Dujarric de La Rivière, *Souvenirs* (Périgueux [1962]), pp. 136–137; E. Iftimovici, *C. Levaditi* (Bucharest, 1960); P. Lépine, "Constantin Levaditi (1874–1953)," in *Presse médicale*, **71** (1953), 1455; and "C. Levaditi, 1874–1953," in *Annales de l'Institut Pasteur*, **85** (1953), 355; L. Pasteur Vallery Radot, "La poliomyélite épidémique," in *Revue des deux mondes*, **9** (1930), 899–914; E. Roux, "Rapport sur les travaux de M. le Docteur C. Levaditi. Sur les spirochètes en général et le *Treponema pallidum* en particulier," in *Comptes rendus hebdomadaires des séances de l'Académie des sciences*, **145** (1907), 1025.

Denise Wrotnowska

LEVAILLANT (LE VAILLANT), FRANÇOIS (*b.* Paramaribo, Netherlands Guiana [now Surinam], 1753; *d.* La Noue, near Sézanne, France, 22 November 1824), *natural history.*

The son of a wealthy trader from the vicinity of Metz who had become French consul in the Dutch colony, Levaillant presumably acquired a love of travel and exploration from his family. At the age of ten he went with them to Holland and later spent two years in Germany and seven in the French countryside, where he developed a love of hunting, began to study birds, and learned taxidermy. While visiting Paris in 1777 he became a devotee of the cabinets and collections of natural history and resolved to explore the remotest part of the globe. Having fixed on Africa as the least-known continent, he sailed from Holland for the Cape of Good Hope on 19 December 1779, arriving there on 29 March 1781. While he was hunting near the Bay of Saldanha, his boat was attacked by a British flotilla, and he lost his personal effects except for his gun, ten ducats, and the light clothes he was wearing. The colonists, however, outfitted him for his first tour of the interior, a six-month circular trip up from the east into the veld. His second trip, in 1783, lasted nearly a year and was attended with great physical difficulty as he traveled north along the left bank of the Orange River into land occupied by warring tribes. Protected by his faithful Hottentot retinue, Levaillant managed to hunt with the savage Hausa and to bring back his collections relatively unscathed. Returning to France in 1784 he was imprisoned for a time but survived the Revolution and retired to a small estate at La Noue, near Sézanne in Champagne. Here he wrote the books that have survived controversy and accusations of substitute authorship. Although his ornithological work consists essentially of magnificent illustrated books, devoid of scientific terminology, they were among the first to reveal to Frenchmen (and later Germans and Englishmen) the wonders of Africa and of the tropics. Levaillant was the first Frenchman to bring a giraffe to the Jardin des Plantes. Through his books and collections he popularized the wonders of the exotic fauna in which he had delighted. Decorated under the Empire with the Legion of Honor, he died at his country estate in 1824. In his best-known book, *Voyages de F. Le Vaillant dans l'intérieur de l'Afrique*, he mentions two of his African friends: Klaas, a Hottentot companion for whom Klaas' cuckoo is named; and "the fair" Narine, an African maiden (whose relationship to the explorer has always been thought romantic) for whom the lovely Narina trogon is named. Levaillant's ornithological production was great, though unscientific. His principal work was the magnificent *Histoire naturelle des oiseaux d'Afrique;* the general artistic supervisor is thought to have been Jean-Baptiste Audubert.

BIBLIOGRAPHY

I. ORIGINAL WORKS. Levaillant's writings are *Voyages de F. Le Vaillant dans l'intérieur de l'Afrique 1781–1785* (Paris, 1790); *Histoire naturelle des oiseaux d'Afrique*, 6 vols. (Paris, 1796–1812); *Histoire naturelle d'une partie d'oiseaux nouveaux et rares de l'Amérique et des Indes* (Paris, 1801); *Histoire naturelle des perroquets*, 2 vols. (Paris, 1801–1805), with 73 plates colored under the supervision of Barraband, supp. vols. added by Alexandre Bourjot-Saint-Hilaire (1837–1838) and Charles de Souancé (1857–1858); and *Histoire naturelle des oiseaux de paradis at des rolliers*, 2 vols. (Paris, 1801–1806).

II. SECONDARY LITERATURE. See Andrew Crichton, "Memoir of Le Vaillant," in William Swainson, *Birds of Western Africa*, pt. 2 (Edinburgh, 1845), pp. 17–31; and J. H. Ogilvie, *A Bibliography of Le Vaillant's Voyages and Oiseaux d'Afrique* (Johannesburg, 1962).

S. DILLON RIPLEY

LEVENE, PHOEBUS AARON THEODOR (*b.* Sagor, Russia, 25 February 1869; *d.* New York, N. Y., 6 September 1940), *biochemistry.*

An American biochemist who carried out extensive research on conjugated proteins, Levene was particularly noted for his pioneering work on nucleoproteins and their component nucleic acids. He was the second of eight children of a custom shirtmaker, Solom Michael Levene and the former Etta Brick. The family moved to St. Petersburg in 1873 so the children might attend private schools and the classical academy. Upon graduation from the latter in 1886, Phoebus was admitted to the Imperial Military Medical Academy in St. Petersburg, where chemistry was taught by Borodin and his son-in-law, Alexander Dianin. Because of growing anti-Semitism the Levene family emigrated to New York in 1891, but Phoebus returned to the medical academy and received his M.D. in the fall of that year.

He then undertook medical practice in New York's Lower East Side but was concurrently enrolled as a special student in the chemistry department of the Columbia University School of Mines. In 1896 he became associate in physiological chemistry at the recently opened laboratories of the Pathological Institute of the New York State Hospitals. He soon contracted tuberculosis and, while recovering in a sanitarium at Saranac Lake, decided to abandon the practice of medicine and devote his life to biochemical research. At various times in the next decade he served at the Pathological Institute and at the Saranac Lake Laboratory for the Study of Tuberculosis, where he worked on the chemistry of the tubercle bacillus. Interspersed were periods of study with Edmund Drechsel in Bern, Albrecht Kossel in Marburg, and Emil Fischer in Berlin. At Marburg he became interested in nucleic acids, and at Berlin he applied Fischer's new ester fractionation procedure to the determination of the amino acids in gelatin. In 1905 Simon Flexner selected him to head the biochemical studies at the newly created Rockefeller Institute for Medical Research, where he worked until his retirement in 1939.

Levene was an intense, hard-driving scientist. He was thin, of short stature, and had penetrating, dark-brown eyes, close-cropped mustache, and a stern expression. He liked to work in the laboratory and managed to do so while supervising a staff of post-doctoral students and assistants, many of them from foreign countries. He conferred daily with his associates, keeping in close touch with the progress of their research and making suggestions reflecting his broad familiarity with the literature. An accomplished linguist, he addressed his foreign students in their native tongues. He spoke excellent German and French, but his English always had a heavy Russian accent.

In 1919 he met Anna M. Erickson at Saranac Lake. She was born in Montana but was then a member of the Norwegian Lutheran Colony in Evanston, Illinois. They were married in 1920. There were no children. Their home reflected Levene's interest in music (he played the violin), contemporary art, and literature. Although a gracious host in a social setting, Levene never permitted social affairs to interfere with his scientific activities. His political views were liberal; he supported the Kerensky government, but was unsympathetic to the Bolsheviks.

Of particular importance in his career was his pioneering work on the nucleic acids. Although discovered in 1871, little was known about them in 1900 except that they were present in nucleoproteins, and contained phosphoric acid groups associated with nitrogenous and nonnitrogenous material. Levene showed the presence in cells of two principal types. One, obtained readily from yeast, he showed to be composed of four nucleosides in which he identified the hitherto unknown sugar D-ribose. The optical isomer, L-ribose, had recently been synthesized in Europe, and Levene showed his sugar to be identical except for direction of optical rotation. He also synthesized the hypothetical hexose sugars, D-allose and D-altrose, from D-ribose.

In subsequent work Levene identified four nucleotides from yeast nucleic acid and showed the presence of phosphoric acid, ribose, and a nitrogenous base (either a purine or a pyrimidine). He was able to show that nucleic acids are high polymers composed of the four nucleotides. Related work on thymus nucleic acid showed a similar composition, but the identity of the sugar remained elusive until Levene identified it in 1929, twenty years after the identification of ribose, as 2-deoxy-D-ribose. His laboratory established the sequence of units in the nucleotides, worked out the ring structures of the sugars, and established the position of attachment of the bases and phosphate units on the sugars. Levene died four years before RNA and DNA were recognized to play a principal role in transmission of hereditary information.

Besides his work on the ribose sugars, Levene undertook work on the glycoproteins in 1900, isolating nitrogenous sugars from the mucoids. The structures of these hexosamines could be established only after extensive synthetic studies on sugars and amino sugars. His identification of chondrosamine as 2-amino-2-deoxy-D-galactose and chitosamine as 2-amino-2-deoxy-D-glucose was later confirmed by the work of Haworth and his associates. In his studies of sugar Levene made important contributions to the understanding of the Walden inversion.

Other research dealt with the lipids. Levene showed that lecithin isolated from different parts of the body contains different fatty acids; that sphingomyelins prepared from various organs are identical; and that "kerasinic acid" from the cerebroside, kerasin, is identical with lignoceric acid.

Levene was a skilled laboratory worker who utilized simple apparatus with great effectiveness. Because of his medical training he carried out animal experiments with unusual skill. His more than 700 publications reveal a great diversity of research interests, yet they reflect an integration revealing a deep understanding of the problems involved in unraveling the chemistry of living processes. His early work resulted in the isolation of specific proteins. It was necessary to develop improved analytical procedures to unravel their composition. He complemented analysis with synthesis in attacking such problems. Early work on the racemization of synthetic diketopiperazines led him to conclude that protein structure was explainable by classical valence theory. While his isolation of propylglycine anhydride from a gelatin hydrolysate challenged the new polypeptide theory of protein structure, the problem was ultimately resolved. His work on protein structure quickly led to studies of the nonprotein constituents of conjugated proteins (nucleoproteins, lipoproteins, glycoproteins) and thereby to the studies of sugars, lipids, and nucleic acids.

Levene was a charter member of the American Society of Biological Chemists, the National Academy of Sciences, and of numerous other American and foreign societies. In 1931 he received the Willard Gibbs Medal of the American Chemical Society and in 1938 was awarded the William H. Nichols Medal of the New York section of the same society.

BIBLIOGRAPHY

I. Original Works. A bibliography of Levene's publications is appended to the biography by D. D. Van Slyke and W. A. Jacobs (see below). Most appeared in the *Journal of Biological Chemistry*. Levene published fragmentary results rapidly, rather than waiting until a major area of research was completed. His books are *Hexosamines, Their Derivatives, and Mucins and Mucoids* (New York, 1922); *Hexosamines and Mucoproteins* (New York, 1925); and *Nucleic Acids* (New York, 1931).

II. Secondary Literature. There are no extensive biographies of Levene. The best short biography of him as a person is Melville L. Wolfram, "Phoebus Aaron Theodor Levene," in Eduard Farber, ed., *Great Chemists* (New York, 1961), pp. 1313–1324. Wolfram served as a postdoctoral fellow in Levene's laboratory and obtained personal information about the family background from Alexander, the younger brother of Phoebus. He gives Aaron as the father's name and has much information on Levene's Russian name and its transliteration. There is also some personal material in D. D. Van Slyke and W. A. Jacobs, "Phoebus Aaron Theodor Levene," *Biographical Memoirs. National Academy of Sciences*, **23** (1945), 75–126. Other sketches are R. Stuart Tipson, "Phoebus Aaron Theodor Levene, 1869–1940," in *Advances in Carbohydrate Chemistry*, **12** (1957), 1–12; and Lawrence W. Bass, "American Contemporaries, P. A. Levene," in *Industrial and Engineering Chemistry. News edition*, **12** (1934), 105.

Aaron J. Ihde

LE VERRIER, URBAIN JEAN JOSEPH (*b.* Saint-Lô, France, 11 March 1811; *d.* Paris, France, 23 September 1877), *astronomy, celestial mechanics, meteorology*.

Le Verrier, whose family came from Normandy, attended secondary school in his native city and then in Caen. The family's finances were modest; and his father, an estate manager, sold his house in order to send Le Verrier to Paris to prepare for the École Polytechnique. Admitted in 1831, he graduated as *élève-ingénieur* of the state tobacco company and for two years studied at the specialized school of tobacco manufacture. He then resigned to pursue chemical research under Gay-Lussac.

In 1837 Le Verrier married the daughter of his former mathematics professor. He gave private lessons to support himself and applied for the position of *répétiteur de chimie* at the École Polytechnique. Having already published two well-received memoirs on combinations of phosphorus, he was accepted—as *répétiteur d'astronomie.*

Le Verrier was already interested in astronomy; his first publication (1832) had dealt with shooting stars and had led to his first contact with the astronomer royal, George Airy. Henceforth, Le Verrier devoted himself exclusively to astronomy. In 1837 he began to study the most general problem of celestial mechanics, the stability of the solar system. The perturbations of the major axes of the orbits had been treated, but Laplace had failed to obtain significant results for the eccentricities and inclinations. Extending Laplace's calculations by carrying the approximations much further and by making a more complete analytical study, Le Verrier derived in 1839 and 1840 precise limits for the eccentricities and inclinations of the seven planets, given the masses accepted at the time. For Jupiter, Saturn, and Uranus he demonstrated that stability is acquired without restriction.

Le Verrier next turned to the theoretical study of Mercury, the existing tables of which did not agree well with observation, and then to the identification of periodic comets. His progress in the analytic theory of perturbations was recognized by the Académie des Sciences; he became a member in January 1846, while engaged in the work that led him to the discovery of Neptune.

Until 1846 there was no theory of Uranus that permitted its movements to be represented satisfactorily. In 1821 Bouvard had constructed tables that, abandoning the older positions, adhered very closely to recent observations. Yet twenty years later a discrepancy of two minutes had already been observed, and several astronomers suggested that it might result from the attraction of an unknown planet. In 1845 Arago presented the problem to Le Verrier, who began by establishing a precise theory of Uranus. He then demonstrated that its observed perturbations could not be explained as the effect of the actions of Jupiter and Saturn, whatever modifications might eventually be made in the values assigned to the masses of those planets. He began to search for signs of an unknown disturbing planet. Finally, in a third memoir on the subject, appearing on 31 August 1846, Le Verrier fixed the exact position of the unknown planet and gave its apparent diameter.

Le Verrier communicated the result of his investigations to several astronomers who had powerful instruments at their disposal. Among them was J. G. Galle, at the Berlin observatory, who was notified by Le Verrier on 23 September. Two days later he wrote to Le Verrier: "The planet whose position you indicated *really exists.* The same day I received your letter I found a star of the eighth magnitude that was not recorded on the excellent Carta Hora XXI (drawn by Dr. Bremiker). . . . The observation of the following day confirmed that it was the planet sought." The map to which Galle referred had just reached him. The object observed was fifty-two minutes from the position predicted by Le Verrier. First called Le Verrier's planet, it was eventually named Neptune.

The English astronomer J. C. Adams had independently carried out a study essentially equivalent to Le Verrier's, although less elaborate, and had sent it to Airy. Having disregarded it until apprised of Le Verrier's second memoir on the subject, Airy then initiated a series of similar observations. Unfortunately, the Greenwich observatory did not possess a sufficiently accurate star map; thus identification of the planet required an effort that was not made until after the announcement of Galle's observation.

While Le Verrier, Airy, and Adams were establishing friendly relations at this period, a sharp polemic over the priority in this discovery was being conducted by French and English journalists. A short while later French and American scientists contested the role of Le Verrier's work and claimed that the discovery was due solely to chance. This assertion was based on the fact that at the end of several months, observations had shown that the elements of Neptune's orbit were quite different from those predicted by Le Verrier. Sir John Herschel explained the error of Le Verrier's opponents as follows: "The axis and excentricity are intellectual objects . . ., useful to represent to the mind's eye the general relations of the planet to space and time. The direct object of the enquiry was to say whereabouts in space at the present moment is the disturbing body. . . ." In fact, the perturbations of Uranus by Neptune are great only when these planets are at least approximately in heliocentric conjunction; and, for these times, their analysis furnishes the position then occupied by Neptune but not the totality of its orbit.

This successful application of mathematical analysis helped to make the public and governments aware of the importance of scientific research in general. Encouraged, in 1847 Le Verrier conceived the project "of embracing in a single work the whole of the planetary system," that is, of constructing theories and tables of the planets and determining their masses in a uniform manner while simultaneously taking into account all the mutual perturbations. This work, which he did not complete until a month before his death, occupies more than 4,000 pages of the *Annales de l'Observatoire de Paris*.

Le Verrier began by establishing the expansion of the perturbing force in its most general form—an expansion extended to the seventh power of the eccentricities and inclinations, and including 469 terms dependent on 154 special functions. This work, carried out in 1849, was the basis of later investigations. Le Verrier then turned to the first four planets, from Mercury to Mars, the theory and tables of which he completed in 1861. Between 1870 and 1877 he treated the other four planets.

Le Verrier's theories are literal and therefore permit interpretations of broad scope. But in elaborating them he did not adhere rigorously to his initial plan: the mutual perturbations of the big planets were too large for them all to undergo complete analytical expansions, and it was necessary to resort to interpolation procedures; for the other planets a uniform treatment was not possible, and certain constants had to be determined separately on the basis of isolated effects in which they played a preponderant role. On the whole, the ensemble represents a considerable advance in the determination of both the masses of the planets and their orbits.

Le Verrier hoped that his work would lead to further discoveries analogous to that of Neptune. His project of 1847 states that he wished to "put everything in harmony if possible and, if it is not, to declare with certainty that there exist still unknown causes of perturbations, the sources of which would then and only then be recognizable." He knew by 1843 that the observed motion of Mercury would not agree with theory. In 1859 he showed that Mercury moves as if an unknown agent produced an advance of its perihelion of about thirty-eight seconds per century. He then put forth the hypothesis of the perturbing action of an intramercurial planet—or, rather, of a group of such planets; the dimensions of a single perturbing body—on the order of those of Mercury—ruled out the possibility of this body having remained unobserved. The hypothesis was not confirmed by the observations conducted for this purpose. No satisfactory explanation was found until 1916, when Karl Schwarzschild, applying the theory of general relativity to celestial mechanics, demonstrated that the only notable correction in the orbits in relation to the Newtonian theory was an advance in the perihelia. This advance is very slight, except in the case of Mercury where it attains forty-three seconds per century. Hence the disagreement pointed out by Le Verrier ultimately became the most celebrated proof of the validity of Einstein's theory.

Le Verrier was one of the founders of modern meteorology. In 1854 the minister of war requested him to study the cyclone that struck the fleets besieging Sevastopol. His systematic inquiry in Europe and Asia enabled him to determine the path of the cyclone. Occupied during this period with the reorganization of the meteorological observation service of the Paris observatory, Le Verrier conceived "the project of a vast meteorological network designed to warn sailors of approaching storms." The greatest difficulty lay in securing the cooperation of the various telegraphic services. By 1857 the project was sufficiently advanced for distribution of a daily bulletin giving the atmospheric conditions at fourteen French and five foreign stations. In England, Robert Fitzroy took the lead in the practical application of the method, and succeeded in 1859 in establishing a storm warning and signaling center for British ports, something that Le Verrier obtained for France and continental Europe in 1863. The predictions were based on the examination of wind charts and isobars. The bulletin prepared by the Paris observatory, which became international in 1872, was sent to all European capitals; it was furnished twice a day when the situation was unsettled, and special telegrams were sent in case of severe storms. Le Verrier also organized a network of stations to report on thunderstorms. When he died, the various meteorological services were reorganized as the Bureau Central Météorologique de France.

Authoritarian and easily offended, but possessed of absolute integrity in scientific matters, Le Verrier conducted his projects like battles—in which, however, he displayed more emotion than strategy. As a result, in the course of his career he acquired the friendship and often the admiration of the greatest foreign scientists while quarreling with most of his French colleagues, who were in more direct contact with him. His dispute with Arago began, paradoxically, because the latter wished to name the new planet for Le Verrier.

Appointed director of the Paris observatory in 1854, following the death of Arago, Le Verrier successively lost the confidence of all its members. They reproached him especially for prohibiting them from taking the initiative and doing private work and for devoting to astronomy only a tenth of the available funds. His management was so contested that he was dismissed in 1870. Reinstated in 1873, after the death of his successor and enemy Charles-Eugène Delaunay, he was then prudent enough to devote all his activity to celestial mechanics. For the first time in his life he trained a student, J. B. Gaillot, who completed the theories of Jupiter and Saturn, which Le Verrier had not had the opportunity to treat thoroughly.

Le Verrier succumbed to the effects of a liver disease that caused him great discomfort during his last five years and affected his already difficult disposition. The fatal progress of this disease has been attributed to the intensity of his scientific work. His other responsibilities included a chair of celestial mechanics at the Sorbonne, created for him after the discovery of Neptune, and, beginning in 1849, a chair of astronomy. He occupied himself as well with the observatory of Marseilles, which was established at his initiative as a branch of the Paris observatory. In 1849 Le Verrier was elected deputy from La Manche, after a campaign in which he affirmed his Catholic faith and his conservative ideas. Three years later he was named to the Imperial Senate, and he remained a senator until the end of the Empire. He was a member of most of the scientific academies of Europe and had the distinction of twice receiving the gold medal of the Royal Astronomical Society of London (1868, 1876).

BIBLIOGRAPHY

I. Original Works. A complete list of Le Verrier's publications is in Institut de France, *Centenaire de la naissance de U. J. J. Le Verrier* (Paris, 1911), pp. 93–128. Most of his scientific work was presented in the form of notes (totaling 230) in the *Comptes rendus . . . de l'Académie des sciences* between 1835 and 1878. His studies in chemistry are "Sur les combinaisons du phosphore avec l'hydrogène," in *Annales de chimie*, **60** (1835), 174–194; and "Sur les combinaisons du phosphore avec l'oxygène," *ibid.*, **65** (1837), 18–35. The notes relating to the discovery of Neptune are in *Comptes rendus . . . de l'Académie des sciences*, **21** (1845), 1050–1055; **22** (1846), 907–918; **23** (1846), 428–438, 657–659; and **27** (1848), 208–210, 273–279, 304, 325–332. The organization of the meteorological investigations is set forth in *Historique des entreprises météorologiques de l'Observatoire impérial de Paris. 1854–1867* (Paris, 1868). All the memoirs on the theories and tables of the planets were published in *Annales de l'Observatoire impérial de Paris. Mémoires*, **1–9** (1855–1868), and then in the *Annales de l'Observatoire national de Paris. Mémoires*, **10** (1874) to **14**, no. 2 (1877).

II. Secondary Literature. See the following, listed chronologically: C. Pritchard, "Address . . . Gold Medal," in *Monthly Notices of the Royal Astronomical Society*, **28** (1868), 110–122; J. C. Adams, "Address . . . Gold Medal," *ibid.*, **36** (1876), 232–246; J. B. Dumas *et al.*, "U. J. Le Verrier. Discours prononcés aux funérailles," in *Comptes rendus . . . de l'Académie des sciences*, **85** (1877), 580–596; J. R. Hind, "Le Verrier," in *Monthly Notices of the Royal Astronomical Society*, **38** (1878), 155–168; J. Bertrand, "Éloge historique de Le Verrier," in *Annales de l'Observatoire de Paris. Mémoires*, **15** (1880), 3–22; F. Tisserand, "Les travaux de Le Verrier," *ibid.*, 23–43; E. Mouchez, F. Tisserand, and O. Struve, *Discours prononcés à l'occasion de la cérémonie d'inauguration de la statue de Le Verrier . . .* (Paris, 1889), pp. 15–41; Sir Robert Ball, *Great Astronomers* (London, 1895), pp. 335–353; Institut de France, *Centenaire de la naissance de U. J. J. Le Verrier* (Paris, 1911), and *Exposition Le Verrier et son temps. Catalogue* (Paris, 1946), both containing extensive documentation; A. Danjon, "Le Verrier créateur de la météorologie," in *La météorologie* (1946), 363–382.

See also F. Tisserand, "Notice sur les planètes intramercurielles," in *Annuaire publié par le Bureau des longitudes* for 1882, pp. 729–772; and, on the discovery of Neptune, R. Grant, *History of Physical Astronomy* (London, 1852), pp. 123–210.

Jacques R. Lévy

LEVI BEN GERSON (*b.* Bagnols, Gard, France, 1288; *d.* 20 April 1344), *mathematics, astronomy, physics, philosophy, commentary on the Bible and the Talmud.*

Levi was also called RaLBaG (a monogram of Rabbi Levi ben Gerson) by the Jewish writers and Gersoni, Gersonides, Leo de Bannolis or Balneolis, Leo Judaeus, and Leo Hebraeus (not to be confused with Leone Ebreo [*d.* 1521/1535], author of the *Dialoghi d'Amore*) by Latin writers. He lived in Orange and Avignon, which were not affected by the expulsion of the Jews from France in 1306 by the order of King Philip the Fair. He seems to have maintained good relations with the papal court—the Latin translations of *De sinibus, chordis et arcubus* and *Tractatus instrumenti astronomie* were dedicated to Clement VI in 1342—and with eminent French and Provençal personalities: his *Luḥot* and *De harmonicis numeris* were written at the request of a group of Jews and Christian noblemen and at the request of Philip of Vitry, bishop of Meaux, respectively. It has been insinuated, without foundation, that he embraced Christianity. It is possible that he practiced medicine, although the evidence is scant. He probably knew neither Arabic nor Latin and thus had to base his work on available Hebrew translations.

In 1321 or 1322 Levi finished the *Sefer ha mispar* ("Book of Number"), also called *Maʿaseh ḥosheb* ("Work of the Computer"; see Exodus 26:1). It deals with general principles of arithmetic and algebra and their applications to calculation: summations of series and combinatorial analysis (permutations and combinations). He used mathematical induction in his demonstrations earlier than Francesco Maurolico (1575) and Pascal (*ca.* 1654). He explained place

LEVI BEN GERSON

value notation, but instead of figures, he uses Hebrew letters according to their numerical value, as well as sexagesimal fractions.

In 1342 Levi wrote *De harmonicis numeris*, of which only the Latin translation survives. Its purpose was to demonstrate that, except for the pairs 1-2, 2-3, 3-4, and 8-9, it is impossible for two numbers that follow each other to be composed of the factors 2 and 3.

Levi's trigonometrical work is in *De sinibus, chordis et arcubus* (dated 1343) based on the Hebrew text of his *Sefer Tekunah*. He uses chords, sines, versed sines, and cosines but no tangents (known in Europe since 1126). Following Ptolemaic methods, he calculated sine tables with great precision. He also formulated the sine theorem for plane triangles; this theorem had been known in the Orient since the end of the tenth century, but it is not clear whether Levi rediscovered it independently or knew it through Jābir ibn Aflaḥ (twelfth century).

Levi wrote two geometrical works: a commentary on books I–V of Euclid's *Elements* that used the Hebrew translation by Moses ibn Tibbon (Montpellier, 1270)—an attempt to construct a geometry without axioms—and the treatise *Ḥibbur ḥokmat ha-tishboret* ("Science of Geometry"), of which only a fragment has been preserved.

The greatest work by Levi ben Gerson is philosophical in character. It is entitled *Milḥamot Adonai* ("The Wars of the Lord") and is divided into six books. The fifth book deals with astronomy and is composed of three treatises, the first of which is known as *Sefer Tekunah* ("Book of Astronomy"). Completed in 1328 and revised in 1340, it was translated into Latin in the fourteenth century. It is divided into 136 chapters and includes two works that are frequently found separately in the manuscripts: the *Luḥot* (chapter 99), astronomical tables calculated for the meridian of Ezob (Izop; Orange) in 1320, and his description of the construction and use of the instrument called the Jacob's staff (chapters 4-11). Levi treated this instrument not only in the chapters mentioned (which were translated into Latin by Peter of Alexandria in 1342) but also in two Hebrew poems. In these versions it is called *kelī* ("instrument") and *megalleh ʿamūqqōt* ("secretum revelator"; see Job 12:22); on the other hand, one of these two poems bears the subtitle "ʿal ha-maqel" ("Concerning the Staff") and in it he refers to Jacob's staff (Genesis 32:10). This is the origin of the expression *baculus Jacobi*, used in some Latin manuscripts, and of the misunderstanding that he wanted to attribute his invention to someone called Jacob. The instrument is also called *baculus geometricus, baculus astronomicus*, or *balestilha*. It consists of a graduated rod and a plate that moves along the rod perpendicular to it. In order to measure an angle, the observer must look at both ends of the plate. The angle α is determined solving triangle *ABC*. This instrument helped Levi in his attempts to determine the center of vision in the eye; unfortunately the experiment was not sufficiently precise and Levi concluded that the center of vision is in the crystalline lens, thus agreeing with Galen and Ibn al-Haytham.

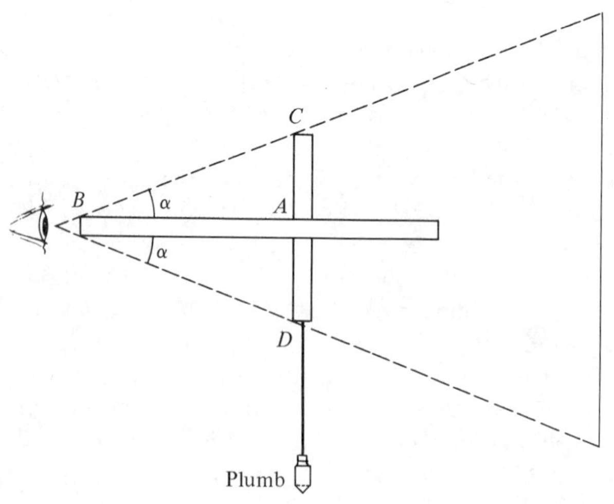

FIGURE 1. Jacob's staff.

Levi's astronomical system began with a critique of the *Almagest* and of al-Biṭrūjī's *Kitāb al-hay'a* (translated into Hebrew in the thirteenth century). His principal sources were al-Battānī (whom he could have known through Abraham bar Ḥiyya), the *Islāḥ al-Majisṭī* of Jābir ibn Aflaḥ (translated by Moses ibn Tibbon in 1274), Ibn Rushd, and Abraham ibn ʿEzra. He followed the doctrine of al-Farghānī (Hebrew translation by Jacob Anaṭoli, *ca.* 1231–1235) concerning the motion of the solar apogee. He rejected the planetary systems of Ptolemy and al-Biṭrūjī because he did not feel that they conformed with the data obtained by observation, as in the variations in the apparent sizes of the planets. Thus Levi observed that, according to Ptolemy, the apparent size of Mars must vary sixfold while, according to his own observations, it varies twice. The Ptolemaic epicycles also did not agree with observations (accepting the one of the moon would imply that we see both sides alternately). In the same way Levi rejected the theory of the trepidation of the equinoxes. This theoretical work was accompanied by many years of well-defined observation (1321–1339) in which Levi used his staff and the camera obscura. The use of the camera obscura in astronomy dates back to Ibn al-Haytham and was known in Europe in the second half of the thirteenth century; Levi is distinguished for his careful instructions on how the observations should be performed and for his explanation of the theoretical basis of the camera obscura.

The planetary system conceived by Levi contains a mixture of technical considerations with others of a metaphysical character. He postulates the existence of forty-eight spheres, some concentric with the earth and others not. The movement is transmitted from the innermost sphere to the outermost by means of an intermediate nonresistant fluid. The stars are affixed to the last sphere of their own system, and each sphere or system of spheres is moved by an immaterial intelligence. The number of spirits that move the heavens is, then, forty-eight or eight. Within this general framework, Levi investigated particularly the sun and the moon. His lunar model eliminated the use of epicycles; its results were practically equivalent to those of Ptolemy's model at syzygy and quadrature (where Levi believed them to be adequate enough) but an entirely new correction was introduced by him at the octants (where Ptolemaic theory could not be accepted because of systematic discrepancies with observation): there is no evidence to suggest a relationship between Levi and Tycho Brahe concerning the latter's discovery of variation (an inequality that reaches its maximum at octant). It is also clear that Levi's model represented the two first lunar inequalities better than Ptolemy's model did. Finally he eliminated practically all the variations in lunar distance from the earth, for which he rightly criticized Ptolemy's model.

The movements of the planets were less extensively dealt with (chapters 103–135). Levi carefully studied the controversial question concerning the places of Venus and Mercury with respect to the sun (chapters 129–135) without arriving at any definitive conclusion. Finally, one of his greatest contributions to medieval astronomy was his extraordinary enlargement of the Ptolemaic universe—for example, the maximum distance of Venus (Ptolemy, 1,079 earth radii; Levi, 8,971,112 earth radii); the distance of the fixed stars from the center of the earth (Ptolemy, 20,000 earth radii; Levi, $159 \times 10^{12} + 6{,}515 \times 10^{8} + 1{,}338 \times 10^{4} + 944$ earth radii); and the diameter of first-magnitude stars (Ptolemy, $4.5 + 1/20$ earth radii; Levi, more than 328×10^{8} earth radii).

A few astrological treatises of Levi have been preserved, including a prediction addressed to Pope Benedict XII in 1339 and his *Prognosticon de conjunctione Saturni et Jovis (et Martis) a.D. 1345*, left unfinished at his death (the Latin translation by Peter of Alexandria and Solomon ben Gerson, Levi's brother, is extant). This conjunction of the three superior planets was also the subject of predictions by John of Murs and Firminus de Bellavalle (Firmin de Beauval). At the outbreak of the black plague of 1348 the conjunction was believed, retrospectively, to be the celestial cause that had corrupted the air.

Levi's work was influential in Europe until the eighteenth century. In the fourteenth century his astronomical tables were used by Jacob ben David Yomtob in constucting his own tables (Perpignan, 1361). It influenced the astrological work of Symon de Covino (*d.* 1367), as well as the astronomical work of Immanuel ben Jacob of Tarascon (*fl. ca.* 1340–1377). In the fifteenth century the *Sefer Tekunah* was praised by Abraham Zacuto (*ca.* 1450–1510); and the treatise *De sinibus* was the model for the *De triangulis* of Regiomontanus (1464; published 1533). The latter, as well as his disciples Bernhard Walther (1430–1504) and Martin Behaim (*d.* 1507), used the Jacob's staff, and it continued to be widely used in navigation, with various improvements, until the middle of the eighteenth century. In the seventeenth century Kepler wrote to his friend Johannes Ramus, asking him to send a copy of the *Sefer Tekunah* to him. There is also a curious cosmographical treatise written in Hebrew by an unknown Roman Jew who cites Levi as being among the greatest authors of the past.

BIBLIOGRAPHY

I. ORIGINAL WORKS. Gerson Lange, *Die Praxis des Rechners* (Frankfurt, 1909), contains an edited German trans. of the *Ma'aseh Ḥosheb;* Joseph Carlebach, *Lewi als Mathematiker* (Berlin, 1910), has an ed. of the *De numeris harmonicis* on pp. 125–144. Maximilian Curtze, "Die Abhandlungen des Levi ben Gerson über Trigonometrie und den Jacobstab," in *Bibliotheca mathematica*, 2nd ser., **12** (1898), 97–112, is a Latin trans. of the *De sinibus* and the *Tractatus instrumenti astronomie*. The eds. of the *Milḥamot Adonai* do not include the *Sefer Tekunah*, which traditionally is considered as a separate work. On the latter see Ernest Renan, "Les écrivains juifs français du XIVe siècle," in *Histoire littéraire de la France*, XXXI (Paris, 1893), 586–644—see esp. pp. 624–641, the Hebrew and Latin texts of the intro. and table of contents of the *Sefer Tekunah;* see also Baldassarre Boncompagni, "Intorno ad un trattato d'aritmetica stampato nel 1478," in *Atti dell'Accademia Pontificia dei Nuovi Lincei* (1863), 741–753, a textual study of three Latin MSS from the above-mentioned work plus some short fragments. Goldstein ("Levi ben Gerson's Lunar Model" quoted below) gives an English translation of the significant passages concerning Levi's lunar theory. On the MSS of the *Sefer Tekunah*, as well as the remaining unpublished works by Levi, see Moritz Steinschneider, *Mathematik bei den*

Juden (Leipzig, 1893–1899; Frankfurt, 1901; Hildesheim, 1964).

II. SECONDARY LITERATURE. See F. Cantera Burgos, *El judío salamantino Abraham Zacut. Notas para la historia de la astronomía en la España medieval* (Madrid, n.d.), pp. 53, 55, 153, 189–190; Pamela H. Espenshade, "A Text on Trigonometry by Levi ben Gerson," in *Mathematics Teacher*, **60** (1967), 628–637; Bernard R. Goldstein, "The Town of Ezob/Aurayca," in *Revue des études juives*, **126** (1967), 269–271; "Preliminary Remarks on Levi ben Gerson's Contributions to Astronomy," in *Proceedings of the Israel Academy of Sciences and Humanities*, **3**, no. 9 (1969), 239–254; *Al-Biṭrūjī: On the Principles of Astronomy*, I (New Haven–London, 1971), 40–43; "Levi ben Gerson's Lunar Model," in *Centaurus*, **16** (1972), 257–283; and "Theory and Observation in Medieval Astronomy," in *Isis*, **63** (1972), 39–47; Isidore Loeb, "La ville d'Hysope," in *Revue des études juives*, **1** (1880), 72–82; D. C. Lindberg, "The Theory of Pinhole Images in the Fourteenth Century," in *Archives for History of Exact Sciences*, **6** (1970), 299–325, esp. pp. 303 ff.; José M. Millás Vallicrosa and David Romano, *Cosmografía de un judío romano del siglo XVII* (Madrid–Barcelona, 1954), pp. 20, 33, 65, 70, 76–77, 85, 87; B. A. Rosenfeld, "Dokazatelstva piatogo postulata Evklida srednevekovykh matematikov Khasana Ibn al-Khaisana i Lva Gersonida" ("The Proofs of Euclid's Fifth Postulate by the Medieval Mathematicians Ibn al-Haytham and Levi ben Gerson"), in *Istoriko-matematicheskie issledovaniya*, **11** (1958), 733–782; George Sarton, *Introduction to the History of Science*, III (Baltimore, 1947–1948), 594–607, and the bibliography given there—see also pp. 129, 886, 1116, 1516, 1518; and Lynn Thorndike, *A History of Magic and Experimental Science*, III (New York–London, 1934), 38, 303–305, 309–311.

JULIO SAMSÓ

LEVI, GIUSEPPE (*b.* Trieste, Austria-Hungary, 14 October 1872; *d.* Turin, Italy, 3 February 1965), *human anatomy, histology, embryology*.

Levi's father, Michele Levi, was a wealthy Jewish financier whose family had financially helped the Austrian administration; his mother, Emma Perugia, was from Pisa. In spite of being brought up conservatively Levi became an enthusiastic Irredentist. He completed his secondary education in Trieste; and when his father died, the family moved to Florence, where he studied medicine and surgery, graduating in 1895. He became an assistant at a psychiatric clinic in Florence (1896–1898); then in Berlin (1898–1899) at the Institute of Anatomy directed by O. Hertwig; and returned in 1900 to Florence, to the Institute of Anatomy, as an assistant to Chiarugi. Later that year Levi obtained the chair of anatomy at the University of Sassari, then at the University of Palermo in 1914, and finally that at the University of Turin, where he remained from 1919 until 1938. Dismissed from the university because of the fascist racial laws, he obtained a post at the University of Liège (1938–1941). He was reinstated into the Italian university system in 1945 and retired in 1947, subsequently becoming director of the Center for Studies on Growth and Senility of the National Research Council. Levi was married to Lidia Tinzi, a Catholic. The couple had five children.

Levi was a member, from 1926, of the Accademia dei Lincei; of the Accademia dei XL from 1933, of the Academy of Sciences of Bologna from 1945; and of many other Italian and foreign academies and scientific societies. He held honorary doctorates from the universities of Liège, Montevideo, and Santiago (Chile); he also received the royal prize of the Accademia dei Lincei in 1923 and a gold medal "dei Benemeriti" of the Scuola della Cultura e dell'Arte.

In 1896–1898, while an assistant in the psychiatric clinic, Levi was interested less in the patients than in morphology. He conducted studies on the structure of the nucleus of nerve cells, epoch-making studies that were fully understood and confirmed only fifty years later (H. Hydén, 1945). He also investigated the Nissl substance, which he did not consider exclusive to the nerve cell, a notion also confirmed much later.

While Levi was working with Chiarugi, he carried out important morphological and histogenetic studies on the comparative anatomy of the dorsal and ventral hippocampus nerve, completing previous observations by Elliot Smith, and on the morphology of the chondrocranium of man and other mammals. In a monograph on the spinal ganglia of over fifty vertebrate species, he illustrated the significance of the fenestrated apparatus of the ganglion cells and the size relationships between those cells and the body of the animal. He concluded that the size of neurons is proportional to the size of the body (Levi's law), in contrast with Driesch's law, which states that the size of the cells remains constant in both the variable and the stable elements. In 1921–1925 Levi further extended the study of the development mechanism which controls the dimensions and the number of supercellular units (the histomers of Heidenhain) during the growth of the body, in relation to the rate of growth and the length of life.

At Sassari, Levi studied the morphology of chondriosomes by normal histological methods; and later, using Harrison's method to study the behavior of mitochondria in cultured live cells, he concluded that there is no relation between mitochondria and secretion granules and that they do not transform themselves into paraplasm or myofibrils.

Levi was one of the pioneers in cultivating tissues *in vitro*, a line of research he pursued at the University of Palermo from 1914 to 1919. He confirmed the cytologic studies of Warren and Margaret Lewis and extended his observations to elements from various tissues, studying the changes that the tissues undergo while being placed in culture. He demonstrated the existence of neurofibrils in the live elements, a finding confirmed by the electron microscope (F. Schmitt, Fernandez Moran, Palay, and Bairati).

Levi continued to develop this line of research at the University of Turin and initiated a number of students into the field. In 1934 he published an extensive monograph, "Explantation," a documentation and discussion of the first twenty-five years of study on cultivation of tissues *in vitro*. Of particular interest are further observations by him and his students on the interdependence relationships of neurons grown *in vitro*, on the regeneration *per primam* of axons *in vitro*, and on the differentiation of neurons in the spinal ganglia of the chicken embryo. With this method he also confirmed the acute observations of Flemming and Boveri on the fibrillar structure of neurons. In 1927 he began an extensive plan of research on the morphological aspects of the aging of tissues and organs.

Levi's treatise on histology is marked by extreme individuality and progressiveness; it ran to four Italian (1927–1954) and two Spanish editions. Levi's knowledge of the field, his remarkable memory, his quick intuition, and his grasp of modern concepts and techniques made him for many years Italy's leading authority in biology. He was opposed to any racial or religious bigotry and did not approve of the Zionist movement. His sense of dignity and of independence, his social ideas, and his faith in the spiritual freedom of man made him a strenuous opponent of Fascism, under which he suffered persecution and imprisonment. He was strict, honest, loyal, intolerant of vulgarity and mediocrity; a severe and exacting, yet affectionate and generous teacher. Eight of his students became professors of human anatomy, histology, or embryology. Until the end of his life he retained his extraordinary memory and clarity of mind, and continued his keen interest in the progress of biological research. He died at the age of ninety-two, after a long and painful illness.

BIBLIOGRAPHY

I. Original Works. Levi's writings include "Su alcune particolarità di struttura del nucleo delle cellule nervose," in *Rivista di Patologia nervosa e mentale*, **1** (1896), 141–149; "Contributo alla fisiologia della cellula nervosa," *ibid.*, 169–180; "Ricerche sulla capacità proliferativa della cellula nervosa," *ibid.*, 385–386; "Ricerche citologiche comparate sulla cellula nervosa dei vertebrati," *ibid.*, **2** (1897), 1–43; "Sulla cariocinesi delle cellule nervose," *ibid.*, **3** (1898), 97–112; "Sulle modificazioni morfologiche delle cellule nervose di animali a sangue freddo durante l'ibernazione," *ibid.*, 443–459; "Beitrag zum Studium der Entwicklung des Primordialcranium des Menschen," in *Archiv für mikroskopische Anatomie und Entwicklungsmechanik*, **55** (1900), 341–414; "Ueber die Entwicklung und Histogenese der Amnioshornformation," *ibid.*, **64** (1904), 389–404; "Morfologia e minuta struttura dell'ippocampo dorsale," in *Archivio italiano di anatomia e di embriologia*, **3** (1904), 438–484; "Sull'origine filogenetica della formazione ammonica," *ibid.*, 234–247; the three-part "Studi sulla grandezza delle cellule. I. Ricerche comparative sulla grandezza delle cellule dei mammiferi," *ibid.*, **5** (1906), 291–358; "II. Le variazioni dell'indice plasmatico-nucleare durante l'intercinesi," *ibid.*, **10** (1911), 545–554; and "III. Le modificazioni della grandezza cellulare e nucleare e dell'indice plasmatico-nucleare durante i precoci periodi dell'ontogenesi dei mammiferi," in *Ricerche di biologia dedicate al Prof. A. Lustig* (Florence, 1914), pp. 1–26; "Cenni sulla costituzione e sullo sviluppo dell'uncus dell'ippocampo nell'uomo," in *Archivio italiano di anatomia e di embriologia*, **8** (1909), 535–562; "Contributo alla conoscenza del condrocranio dei mammiferi," in *Monitore zoologico italiano*, **20** (1909), 159–174; and "Il comportamento dei condriosomi durante i più precoci periodi dello sviluppo dei mammiferi," in *Archiv für Zellforschung*, **13** (1915), 471–524.

See also "Per la migliore conoscenza del fondamento anatomico e dei fattori morfogenetici della grandezza del corpo," in *Archivio italiano di anatomia e di embriologia*, supp. **18** (1921), 316–434; "Wachstum und Körpergrösse," in *Ergebnisse der Anatomie und Entwicklungsgeschichte*, **26** (1925), 87–342; *Trattato di istologia* (Turin, 1927; 4th ed., 1954); "Gewebezüchtung," in Tibor Péterfi, ed., *Methodik der wissenschaftlichen Biologie*, I (Berlin, 1928), pp. 494–558; *Fisiopatologia della vecchiaia*, I, *Istologia sieroterapico* (Milan, 1933); "Explantation, besonders die Struktur und die biologischen Eigenschaften der in vitro gezüchteten Zellen und Gewebe," in *Ergebnisse der Anatomie und Entwicklungsgeschichte*, **31** (1934), 125–707; and his revision and enlargement of Giulio Chiarugi's *Trattato di anatomia dell'uomo*, 9th ed. (Milan, 1959).

II. Secondary Literature. For further information on Levi's life and work consult the following biographical and bibliographical sources: Rodolfo Amprino, "Giuseppe Levi," in *Acta anatomica*, **66** (1967), 1–44, which includes a complete bibliography; and Oliviero M. Olivo, "Commemorazione del socio Giuseppe Levi," in *Atti dell'Accademia nazionale dei Lincei. Rendiconti*, 8th ser., **40** (1966), 954–972, with a bibliography. See also *Annuario Accademia Nazionale dei XL* (1961), pp. 49–66.

Oliviero M. Olivo

LEVI-CIVITA, TULLIO (*b.* Padua, Italy, 29 March 1873; *d.* Rome, Italy, 20 December 1941), *mathematics, mathematical physics.*

The son of Giacomo Levi-Civita, a lawyer who from 1908 was a senator, Levi-Civita was an outstanding student at the *liceo* in Padua. In 1890 he enrolled in the Faculty of Mathematics of the University of Padua. Giuseppe Veronese and Gregorio Ricci Curbastro were among his teachers. He received his diploma in 1894 and in 1895 became resident professor at the teachers' college annexed to the Faculty of Science at Pavia. From 1897 to 1918 Levi-Civita taught rational mechanics at the University of Padua. His years in Padua (where in 1914 he married a pupil, Libera Trevisani) were scientifically the most fruitful of his career. In 1918 he became professor of higher analysis at Rome and, in 1920, of rational mechanics. In 1938, struck by the fascist racial laws against Jews, he was forced to give up teaching.

The breadth of his scientific interests, his scruples regarding the fulfillment of his academic responsibilities, and his affection for young people made Levi-Civita the leader of a flourishing school of mathematicians.

Levi-Civita's approximately 200 memoirs in pure and applied mathematics deal with analytical mechanics, celestial mechanics, hydrodynamics, elasticity, electromagnetism, and atomic physics. His most important contribution to science was rooted in the memoir "Sulle trasformazioni delle equazioni dinamiche" (1896), which was characterized by the use of the methods of absolute differential calculus that Ricci Curbastro had applied only to differential geometry. In the "Méthodes de calcul différentiel absolus et leurs applications," written with Ricci Curbastro and published in 1900 in *Mathematische Annalen*, there is a complete exposition of the new calculus, which consists of a particular algorithm designed to express geometric and physical laws in Euclidean and non-Euclidean spaces, particularly in Riemannian curved spaces. The memoir concerns a very general but laborious type of calculus that made it possible to deal with many difficult problems, including, according to Einstein, the formulation of the general theory of relativity.

Although Levi-Civita had expressed certain reservations concerning relativity in the first years after its formulation (1905), he gradually came to accept the new views. His own original research culminated in 1917 in the introduction of the concept of parallelism in curved spaces that now bears his name; it furnishes a simple law for transporting a vector parallel to itself along a curve in a space of *n* dimensions. With this new concept, absolute differential calculus, having absorbed other techniques, became tensor calculus, now the essential instrument of the unitary relativistic theories of gravitation and electromagnetism. Two of the concept's many applications and generalizations were in the "geometry of paths," which extends the concept of Riemannian variety, and in the theory of spaces with affine and projective connections, which is used with the geometry for a complete representation of electromagnetic phenomena in the framework of general relativity.

In studying the stability of the phenomena of motion, Levi-Civita used a general method, which, by means of a periodic solution of a first-order differential system, restores stability or instability to the study of certain point transformations. He ascertained that periodic solutions, in a first approximation of apparent stability, prove instead to be unstable. Another of Levi-Civita's contributions to analytical mechanics was the general theory of stationary motions, in which moving bodies passing the same spot always do so at the same speed. The theory enabled him to find with a uniform method all the known cases of stationary motion and also to discover new ones.

From 1906 Levi-Civita's memoirs in hydrodynamics, his favorite field, deal with the resistance of a liquid to the translational motion of an immersed solid; he resolved the problem through the general integration of the equations for irrotational flows past a solid body, allowing for the formation of a cavity behind it. His general theory of canal waves originated in a memoir written in 1925.

In related memoirs (1903–1916) Levi-Civita contributed to celestial mechanics in the study of the three-body problem: the determination of the motion of three bodies, considered as reduced to their centers of mass and subject to mutual Newtonian attraction. In 1914–1916 he succeeded in eliminating the singularities presented at the points of possible collisions, past or future. His results furnished a rigorous solution to the classic problem—which, by indirect method and by transcending dynamic equations, Karl F. Sundmann had reached in 1912, as Levi-Civita himself admitted.

His research in relativity led Levi-Civita to mathematical problems suggested by atomic physics, which in the 1920's was developing outside the traditional framework: the general theory of adiabatic invariants, the motion of a body of variable mass, the extension of the Maxwellian distribution to a system of corpuscles, and Schrödinger's equations.

BIBLIOGRAPHY

I. Original Works. All of Levi-Civita's memoirs and notes published between 1893 and 1928 were collected in his *Opere matematiche. Memorie e note*, 4 vols. (Bologna, 1954–1960). Other works include *Questioni di meccanica classica e relativistica* (Bologna, 1924), also in German trans. (Berlin, 1924); *Lezioni di calcolo differenziale assoluto* (Rome, 1925); *Lezioni di meccanica razionale*, 2 vols. (Bologna, 1926–1927; 2nd ed., 1930), written with Ugo Amaldi; and *Fondamenti di meccanica relativistica* (Bologna, 1928).

II. SECONDARY LITERATURE. See the following, listed chronologically: Corrado Segre, "Relazione sul concorso al premio reale per la matematica, del 1907," in *Atti dell'Accademia nazionale dei Lincei. Rendiconti delle sedute solenni*, **2** (1908), 410–424; Albert Einstein, "Die Grundlage der allgemeinen Relativitätstheorie," in *Annalen der Physik*, 4th ser., **49** (1916), 769; "Tullio Levi-Civita," in *Annuario della Pontificia Accademia delle Scienze*, **1** (1936–1937), 496–511, with a complete list of his memberships in scientific institutions and of his academic honors; and Ugo Amaldi, "Commemorazione del socio Tullio Levi-Civita," in *Atti dell'Accademia nazionale dei Lincei, Rendiconti. Classe di scienze fisiche, matematiche e naturali*, 8th ser., **1** (1946), 1130–1155, with a complete bibliography.

MARIO GLIOZZI

LEVINSON-LESSING, FRANZ YULEVICH (*b.* St. Petersburg, Russia, 9 March 1861; *d.* Leningrad, U.S.S.R, 25 October 1939), *geology, petrography.*

An academician of the Soviet Academy of Sciences from 1925 (and a corresponding member since 1914), Levinson-Lessing was the son of a well-known doctor. He married Varvara Ippolitovna Tarnovskaya, who was his colleague in the Commission for Scientific Education, in 1919. Their son Vladimir, an art critic, became a well-known specialist on European art.

Levinson-Lessing spent his childhood in St. Petersburg and received his secondary education there. At St. Petersburg University he studied under the petrographer A. A. Inostrantsev. He graduated in 1883 and remained there to prepare for a teaching career. Undoubted influences on Levinson-Lessing in the first years of his scientific career were his teacher and colleague, the pioneer soil scientist V. V. Dokuchaev, and V. I. Vernadsky. Under this influence he concentrated on the chemistry of inorganic nature. Levinson-Lessing maintained lifelong ties with soil science, had a continuing interest in its problems, and headed various soil institutions.

Ten years after graduating from the university, during which period he had synthesized his work in regional petrography and marked out new paths in its theoretical structure, Levinson-Lessing became professor of mineralogy at the University of Yurev (Tartu).

Working from 1902 to 1930 at the Polytechnical Institute in Leningrad, Levinson-Lessing organized the first laboratory of experimental petrography in Russia, and then a geochemical section. From a division of mineralogy he created petrography, which has become an independent field of knowledge. His main areas of concentration were the analysis of the crust of the earth, the penetration of its depths, and the clarification of the structure of mountains and rocks and their accompanying ore deposits.

Levinson-Lessing developed the idea of the separation or differentiation of magma, placed it on a factual basis, and transformed it into a scientific theory. Having firmly established magma as a complex silicate solution differing from aqueous solutions by its considerable viscosity and susceptibility to intense supercooling, he advanced the doctrine of two ancestral magmas, granite and basalt, that had played a primary role in the creation of the rock of the earth's core. From them came all igneous rocks, the various compositions of which are caused by differing mixtures of these magmas, by the melting and assimilation of previously existing rocks, by fractional crystallization, and by gravity settling. All these processes together create the phenomenon of magmatic differentiation. Igneous rocks after the Precambrian era are primarily the result of the melting of particular parts of the earth's solid crust. Only older igneous rocks had the ancestral granite and gabbroic magmas as their sources.

Levinson-Lessing's regional petrographic works were the basis for the development of theoretical petrology. His expeditions in Karelia (now the Karelo-Finnish S.S.R.), the Urals, the Caucasus, Transcaucasia, the Crimea, the Khibiny Mountains, and eastern Siberia enabled him to approach the solution of the problems of petrographic formations and origin. In 1888 he advanced the idea of the Olonets diabasic formation, which was productively developed in the theory of formations. His trips to Italy and his ascent of Vesuvius with A. Rittmann provided an impetus for the creation in 1935 of the volcanological station at Kamchatka, where Soviet volcanology developed.

Classifying and systematizing igneous rocks occupied Levinson-Lessing throughout his career. In 1898 he proposed the first rational chemical classification of rocks. In his works he depended on physical and chemical research methods, using them to solve such problems as the differentiation of magma and the genesis and classification of rocks and ore deposits.

An active participant in the International Geological Congress, Levinson-Lessing corresponded with the most distinguished geologists and petrographers of western Europe and the United States. At the seventh session of the Congress, Levinson-Lessing was elected to the commission on the classification of igneous rock. For the eighth session, which took place in 1900, he prepared the *Petrographic Dictionary*. At the Paris meeting of the Commission on Petrographic Nomenclature he presented a report and his own additions. He also participated in the twelfth, fourteenth, and seventeenth sessions. At the last session the Permanent Commission on Petrography, Mineralogy, and Geochemistry was organized, with Levinson-Lessing as president.

Levinson-Lessing was a historian of petrography and the natural sciences whose works are well-known both in the Soviet Union and abroad; his monographs *Uspekhi petrografii v Rossii* ("The Progress of Petrography in Russia"; 1923) and *Vvedenie v istoriyu petrografii* ("Introduction to the History of Petrography"; 1936) and many biographical sketches of well-known scientists reveal a tireless researcher in the history of science. A historical approach to the solution of basic scientific problems characterized Levinson-Lessing and appeared in many of his works, particularly in the important "Problema genezisa magmaticheskikh porod i puti k ee razresheniyu" ("The Problem of the Genesis of Magmatic Rock and the Means of Solving It"; 1934).

Analyzing the phenomena of rock formation, the centuries-long changes in shorelines and volcanoes, Levinson-Lessing concluded that different parts of the earth's crust rise and fall simultaneously. Mountains were formed by these movements, and plastic or liquid magma was transferred at depth, serving as the source of magmatic and volcanic phenomena. Bringing together all that was known at the beginning of the twentieth century regarding the formation of mountains, he asserted that systems arose after prolonged sinking followed by slow uplift. Folded mountain chains were formed at the borders of the continents and seas. The sinking of some blocks in the sea was accompanied by rising of the adjoining lands. Sinking and rising were kept more or less equal by the warping of separate blocks of the earth's crust. Levinson-Lessing believed that volcanoes accompany uplifts of parts of the earth's crust and that a major factor in their eruption is the pressure on lava exerted by the sinking of other blocks. He considered the fluctuation of lava levels in the volcanoes of the Hawaiian Islands to be an especially clear example, believing that the lakes of lava form the upper parts of columns of liquid lava going down to subterranean lava reservoirs. These columns of lava would be in hydrostatic equilibrium.

For Levinson-Lessing science not only collects and describes facts but also compares them, penetrates into their essence, and structures them in broad generalizations. In his article "Rol fantazii v nauchnom tvorchestve" ("The Role of Fantasy in Scientific Creativity") he described scientific creativity as consisting of three main elements: the empirical, which provides a basis; scientific fantasy, or the creative idea; and the testing and investigation of the creative idea through logical analysis and experiment. All three elements are necessary, he believed, for scientific creativity and progress. The development and flowering of science requires a harmonious combination of observations, experiments, and ideas; induction and deduction together; a combination of concrete facts, perceived externally, and intuitive forms, which arise subjectively.

BIBLIOGRAPHY

I. Original Works. Many of Levinson-Lessing's writings are in *Izbrannye trudy* ("Selected Works"), 4 vols. (Moscow–Leningrad, 1944–1955). Among his works are "Olonetskaya diabazovaya formatsia" ("Olonets Diabasic Formation"), in *Trudy Imperatorskago S.-Peterburgskago obshchestva estestvoispytatelei*, Otdelenie geologii i mineralogii, **19** (1888); "Issledovania po teoreticheskoy petrografii v svyazi s ucheniem izverzhennykh porod tsentralnogo Kavkaza" ("Research in Theoretical Petrography in Connection with the Theory of Igneous Rock of the Central Caucasus"), *ibid.*, **26** (1898); "Sferolitovye porody Mugodzhar" ("Spherulite Rocks of Mugodzhar"), *ibid.*, **33**, no. 5 (1905); "Sushchestvuet li mezhdu intruzivnymi i effuzivnymi porodami razlichie v khimicheskom sostave" ("Is There a Difference in Chemical Composition Between Intrusive and Effusive Rocks?"), in *Izvestiya S.-Peterburgskago politekhnicheskago instituta Imperatora Petra Velikago* (1906), nos. 1–2; "Polveka mikroskopii v petrografii" ("A Half Century of Microscopy in Petrography"), *ibid.*, **10** (1908); "O samom yuzhnom mestopozhdenii platiny na Urale" ("On the Most Recent Platinum Deposits in the Urals"), *ibid.*, **13** (1910); and *Vulkany i lavy Tsentralnogo Kavkaza* ("Volcanoes and Lava of the Central Caucasus"; St. Petersburg, 1913).

Later works are *Uspekhi petrografii v Rossii* ("The Progress of Petrography in Russia"; Petrograd, 1923); "Rol fantazii v nauchnom tvorchestve" ("The Role of Fantasy in Scientific Creativity"), in *Peterburgskoe nauchnoe khimiko-tekhnicheskoe izdatelstvo* (1923), 36–50; "Professor Aleksandr Aleksandrovich Inostrantsev," in *Izvestiya Geologicheskago komiteta za 1919 g.*, **38**, no. 3 (1924); "Trappy Tuluno-Udinskogo i Bratskogo rayonov v vostochnoy Sibiri" ("Diabases of the Tulun-Udinsky and Bratsky Regions in Eastern Siberia"), in *Trudy Soveta po izucheniyu proizvoditelnykh sil*, Seriya sibirskaya, no. 1 (1932); "Problema genezisa magmaticheskikh porod i puti k ee razresheniyu" ("The Problem of the Genesis of Magmatic Rock and the Means of Solving It"), in *Izdatel-*

stvo Akademii nauk SSSR (1934); "O svoeobraznom tipe differentsiatsii v vartolite Yalguby" ("On a Peculiar Type of Differentiation in Variolite of Yalguba"), in *Trudy Petrina Akademii nauk SSSR*, no. 5 (1935); *Vvedenie v istoriyu petrografii* ("Introduction to the History of Petrography"; Leningrad, 1936); "Spornye voprosy sistematiki i nomenklatury izverzhennykh porod" ("Controversial Questions of Systematization and Nomenclature of Igneous Rocks"), in *Doklady Akademii nauk SSSR*, **21**, no. 3 (1938); and "Problemy magmy" ("Problems of Magma"), in *Uchenye zapiski Leningradskogo gosudartstvennogo universiteta*, **3**, no. 17 (1937), also in *Izvestiya Akademii nauk SSSR*, Seriya geologicheskaya, no. 1 (1939).

II. SECONDARY LITERATURE. See I. M. Asafova and O. V. Isakova, *Franz Yulevich Levinson-Lessing* (Moscow, 1941); D. S. Belyankin, "Akademik F. Y. Levinson-Lessing v trudakh ego po teoreticheskoy petrografii" ("Academician F. Yu. Levinson-Lessing in His Works on Theoretical Petrography"), in *Izvestiya Akademii nauk SSSR*, Seriya geologicheskaya (1945), no. 1, pp. 18–27; A. S. Ginzberg, "Znachenie petrograficheskikh rabot. F. Y. Levinson-Lessinga dlya russkoy i mirovoy nauki" ("The Significance of the Petrographical Work of F. Y. Levinson-Lessing in Russian and World Science"), *ibid.* (1952), no. 5, pp. 7–11; S. S. Kuznetsov, "Krupny russky ucheny F. Y. Levinson-Lessing" ("The Great Russian Scientist F. Y. Levinson-Lessing"), in *Vestnik Leningradskogo gosudarstvennogo universiteta* (1948), no. 5, pp. 128–144; P. I. Lebedev, *Akademik F. Y. Levinson-Lessing kak teoretik petrografii* ("Academician F. Y. Levinson-Lessing as a Theoretician of Petrography"; Moscow–Leningrad, 1947); B. L. Lichkov, "Idei F. Y. Levinson-Lessinga o vekovykh kolebaniakh zemnoy kory v svete sovremennykh vozzreny" ("The Ideas of F. Y. Levinson-Lessing on Secular Oscillations of the Earth's Crust in the Light of Contemporary Views"), in *Ocherki po istorii geologicheskikh znanii*, no. 5 (1965), pp. 248–259; A. A. Polkanov, "F. Y. Levinson-Lessing kak petrograf-myslitel" ("F. Y. Levinson-Lessing as a Petrographic Thinker"), in *Izvestiya Akademii nauk SSSR*, Seriya geologicheskaya, no. 4 (1950), pp. 25–27; and D. I. Shcherbakov, "Rol F. Y. Levinson-Lessinga v razvitii uchenia o rudnykh mestorozhdenia" ("The Role of F. Y. Levinson-Lessing in the Development of the Theory of Ore Deposits"), *ibid.*, no. 3 (1961), pp. 55–60.

A. A. MENIAILOV

LÉVY, MAURICE (*b.* Ribeauvillé, France, 28 February 1838; *d.* Paris, France, 30 September 1910), *mathematics, engineering.*

Maurice Lévy had a distinguished career as a practicing engineer, teacher, and researcher. He studied at the École Polytechnique from 1856 to 1858 and at the École des Ponts et Chaussées. After graduating he served for several years in the provinces and continued his studies. In 1867 he was awarded the *docteur ès sciences* and in 1872 was assigned to the navigation service of the Seine in the region of Paris. In 1880 he became chief government civil engineer and, in 1885, inspector general. At the same time Lévy pursued an academic career. From 1862 to 1883 he served as *répétiteur* in mechanics at the École Polytechnique. In 1875 he was appointed professor of applied mechanics at the École Centrale des Arts et Manufactures, and in 1885 he became professor of analytic and celestial mechanics at the Collège de France. Lévy was elected to the Académie des Sciences in 1883 and served as president of the Société Philomatique in 1880.

Lévy's research ranged over projective and differential geometry, mechanics, kinematics, hydraulics, and hydrodynamics. Of his two theses, the first, influenced by the work of Gabriel Lamé, deals with orthogonal curvilinear coordinates; the second is on the theory and application of liquid motion. He often returned to the question of elasticity, and his writings on the subject include a book published in 1873. His other works include theoretical treatises on energy and hydrodynamics, as well as more practical ones on boat propulsion.

Lévy considered his most significant work to be that on graphical statics (1874), which was an attempt to apply the methods of projective geometry, particularly the theory of geometric transformations, to problems of statics. It was directed toward engineers and was very popular, going through three editions.

BIBLIOGRAPHY

I. ORIGINAL WORKS. A complete bibliography is in Poggendorff, III, 804–805; IV, 876–877; and V, 737. Lévy's two theses are "Sur une transformation des coordonnées curvilignes orthogonales, et sur les coordonnées curvilignes comprenant une famille quelconque de surfaces du second ordre," in *Journal de l'École polytechnique*, **26** (1870), 157–200; and *Essai théorique et appliqué sur le mouvement des liquides*, published separately (Paris, 1867). His books are *Étude d'un système de barrage mobile* (Paris, 1873); *Application de la théorie mathématique de l'élasticité à l'étude de systèmes articulés* (Paris, 1873); *La statique graphique et ses applications aux constructions*, with atlas (Paris, 1874; 2nd ed., 4 vols., Paris, 1886–1888; 3rd ed., 1913–1918); *Sur le principe d'énergie* (Paris, 1890), written with G. Pavie; *Étude des moyens mécanique et électrique de traction des bateaux, Partie I: Halage funiculaire* (Paris, 1894); *Théorie des marées* (Paris, 1898); and *Éléments de cinématique et de mécanique* (Paris, 1902).

II. SECONDARY LITERATURE. There is very little available beyond the article in *La grande encyclopédie*, XXII,

148; and the obituary by Émile Picard in *Comptes rendus hebdomadaires des séances de l'Académie des sciences*, **151** (1910), 603–606.

ELAINE KOPPELMAN

LÉVY, SERVE-DIEU ABAILARD (called **ARMAND**) (*b.* Paris, France, 14 November 1795; *d.* le Pecq, near Saint-Germain, France, 29 July 1841), *mineralogy.*

Lévy's father, a Florentine Jew, was an itinerant merchant who, on one of his trips to France, married a Parisian Catholic, Céline Mailfert. Their son's birth records bear the unusual given names Serve-Dieu Abailard, which also appear on his death certificate, but Lévy called himself, more simply, Armand. It is under this name that he appears on the membership list of the Geological Society of London, beginning in 1824.

Lévy completed his secondary studies in Paris, at the Lycée Henri IV, and won the *concours général* in mathematics, a competition for the best students from all French *lycées* and *collèges.* He passed the entrance examination for the École Normale Supérieure in 1813 and graduated in 1816 with the title of *agrégé* in mathematical sciences. As it was difficult for a Jew to obtain a university position in France, he decided to leave the country and accept the post of professor at the Collège Royal of Île Bourbon (now Île de la Réunion) in the Indian Ocean. He embarked at Rochefort in 1818; the vessel was shipwrecked near Plymouth, an event that determined Lévy's scientific career. For the next two years he supported himself by giving mathematics lessons and in 1820 met a rich mineral merchant, Henry Heuland, who had just sold a large collection of minerals to Charles Hampden Turner. Lévy was hired to make an inventory of it and to produce a descriptive catalog. In 1822, in London, he married Harriet Drewet, who was nineteen; they had several children.

In 1827 the work for which Heuland had commissioned Lévy was finished; and it was decided that, for reasons of economy, the catalog would be printed in Belgium. Lévy therefore moved to Brussels to supervise the printing of the text and the engraving of the illustrations. He was also appointed reader at the University of Liège, teaching analytical mechanics, astronomy, crystallography, and mineralogy. On 3 April 1830 he was elected a member of the Académie des Sciences et Belles-lettres of Brussels. After the July Revolution of 1830 he returned to Paris as a lecturer in mathematics at the École Normale Supérieure and professor at the Collège Royal Charlemagne (1831).

A short while after his return to France, Lévy's wife died. In 1838 he married Amélie Rodriguez Henriquez, the sister of his close friend, the mathematician Olinde Rodriguez. They had two daughters, the elder of whom married Alexandre Bertrand, curator of the Musée de Saint-Germain and brother of the mathematician Joseph Bertrand.

Lévy was a man of great frankness and exquisite politeness. If he was sometimes touchy, even with his friends, it was because he had been the object of racial and religious prejudice. During his stay in England, Lévy became friendly with Wollaston and Herschel, and he was a member of the council of the Geological Society of London from 1826 to 1828. In France he had powerful defenders in Arago, Poisson, Charles Dupin, and the mineralogists Brochant, Brongniart, and Beudant. Des Cloizeaux considered himself Lévy's student. Lévy died of a ruptured aneurysm at the age of forty-five.

Lévy became a mineralogist apparently by necessity rather than by vocation. While a student at the École Normale Supérieure, he undoubtedly took the mineralogy courses given by Haüy. He used the latter's methods and perfected his system of notation in the illustrations of the Heuland collection catalog. The Haüy-Lévy notation or, as it is sometimes called, the Lévy crystallographic notation is still used.

From a morphological point of view, a crystal is characterized by its "primitive form," a parallelepiped the symmetry of which represents the symmetry of the crystal. Haüy had shown that all the plane surfaces of a crystal can be envisaged as "rational truncations" on the vertices and edges of the primitive solid. Lévy designated the vertices by vowels, the edges by consonants, and the planes by these vowels and consonants modified by a coefficient (a simple fraction) indicating their slope. For example, all the vertices of the cube are designated by the letter *a*, and the regular octahedron, which derives from the cube, by a^1. It is easy to pass from Lévy's notation to the crystallographic notation devised by W. H. Miller, which is now in general use.

In describing the Heuland collection and in measuring surface angles of its specimens with a goniometer, Lévy discovered important new mineral species: forsterite, babingtonite, brochantite, roselite, brookite, herschelite, phillipsite and beudantite. He also gave precise descriptions of the crystalline forms of eudialyte, wagnerite, and euclase. In addition, he named several minerals that were subsequently recognized as varieties of previously described species: turnerite, a variety of monazite; bucklandite, a variety of allanite; königite, a variety of brochantite; mohsite, a variety of ilmenite; and humboldtite, a variety of datolite.

While studying the zinc-bearing minerals of La Vieille-Montagne, in Belgium, Lévy discovered an important new zinc silicate which he named willemite. The results of his research were published posthumously in the *Annales des mines* (1843) through the efforts of Des Cloizeaux.

BIBLIOGRAPHY

See M. A. Lacroix, "A. Lévy (1795–1841)," in *Bulletin de la Société française de minéralogie et de cristallographie*, **42**, no. 3 (1919), 122.

J. Wyart

LEWIS, GILBERT NEWTON (*b.* West Newton, Massachusetts, 25 October 1875; *d.* Berkeley, California, 23 March 1946), *physical chemistry*.

Lewis received his primary education at home from his parents, Frank Wesley Lewis, a lawyer of independent character, and Mary Burr White Lewis. He read at age three and was intellectually precocious. In 1884 his family moved to Lincoln, Nebraska, and in 1889 he received his first formal education at the university preparatory school. In 1893, after two years at the University of Nebraska, Lewis transferred to Harvard, where he obtained his B.S. in 1896. After a year of teaching at Phillips Academy in Andover, Lewis returned to Harvard to study with the physical chemist T. W. Richards and obtained his Ph.D. in 1899 with a dissertation on electrochemical potentials. After a year of teaching at Harvard, Lewis made the pilgrimage to Germany, the center of physical chemistry, and studied with W. Nernst at Göttingen and with W. Ostwald at Leipzig. Upon his return to Harvard in 1901, he was appointed instructor in thermodynamics and electrochemistry. In 1904 Lewis was granted a leave of absence and became a chemist with the Bureau of Weights and Measures in Manila. After one year (a year he seldom spoke of) he joined the group of progressive young physical chemists around A. A. Noyes at the Massachusetts Institute of Technology.

Most of Lewis' lasting interests originated during his Harvard years. The most important was thermodynamics, a subject in which Richards was very active at that time. Although most of the important thermodynamic relations were known by 1895, they were seen as isolated equations, and had not yet been rationalized as a logical system, from which, given one relation, the rest could be derived. Moreover, these relations were inexact, applying only to ideal chemical systems. These were two outstanding problems of theoretical thermodynamics. In two long and ambitious theoretical papers in 1900 and 1901, Lewis tried to provide a solution. He proposed the new idea of "escaping tendency" or *fugacity* (a term he coined), a function with the dimensions of pressure which expressed the tendency of a substance to pass from one chemical phase to another. Lewis believed that fugacity was the fundamental principle from which a system of real thermodynamic relations could be derived. This hope was not realized, though fugacity did find a lasting place in the description of real gases.

Lewis' early papers also reveal an unusually advanced awareness of J. W. Gibbs's and P. Duhem's ideas of free energy and thermodynamic potential. These ideas were well known to physicists and mathematicians, but not to most practical chemists, who regarded them as abstruse and inapplicable to chemical systems. Most chemists relied on the familiar thermodynamics of heat (enthalpy) of Berthelot, Ostwald, and Van't Hoff, and the calorimetric school. Heat of reaction is not, of course, a measure of the tendency of chemical changes to occur, and Lewis realized that only free energy and entropy could provide an exact chemical thermodynamics. He derived free energy from fugacity; he tried, without success, to obtain an exact expression for the entropy function, which in 1901 had not been defined at low temperatures. Richards too tried and failed, and not until Nernst succeeded in 1907 was it possible to calculate entropies unambiguously. Although Lewis' fugacity-based system did not last, his early interest in free energy and entropy proved most fruitful, and much of his career was devoted to making these useful concepts accessible to practical chemists.

At Harvard, Lewis also wrote a theoretical paper on the thermodynamics of blackbody radiation in which he postulated that light has a pressure. He later revealed that he had been discouraged from pursuing this idea by his older, more conservative colleagues, who were unaware that W. Wien and others were successfully pursuing the same line of thought. Lewis' paper remained unpublished; but his interest in radiation and quantum theory, and (later) in relativity, sprang from this early, aborted effort. From the start of his career, Lewis regarded himself as both chemist and physicist.

A third major interest that originated during Lewis' Harvard years was valence theory. In 1902, while trying to explain the laws of valence to his students, Lewis conceived the idea that atoms were built up of a concentric series of cubes with electrons at each corner. This "cubic atom" explained the cycle of eight elements in the periodic table and was in accord with the widely accepted belief that chemical bonds were formed by transfer of electrons to give each atom a complete set of eight. This electrochemical theory of valence found its most elaborate expression in the work of Richard Abegg in 1904, but Lewis' version of this theory was the only one to be embodied in a concrete atomic model. Again Lewis' theory did not interest his Harvard mentors, who, like most American chemists of that time, had no taste for such speculation. Lewis did not publish his theory of the cubic atom, but in 1916 it became an important part of his theory of the shared electron pair bond.

Extremely bright and precocious as a young man, Lewis was also shy and lacking in self-confidence. His ideas were unorthodox and singular, perhaps owing in part to his unusual education. He was disappointed and resentful that his talents were not appreciated, especially by Richards. In 1928 he refused a call to Harvard and in 1929 refused an honorary degree.

Lewis remained at M.I.T. for seven years and there laid the foundations for his important work in thermodynamics. It was well-known that the equilibrium position of any chemical system could in theory be predicted from free-energy data—an invaluable aid to both pure and applied chemistry. But existing free-energy data were mostly unreliable, contradictory, and spotty. Existing methods for measuring free energies were imperfect. In practice, application of thermodynamics to chemistry was extremely difficult even for the specialist. Early in his career Lewis began to pursue this neglected opportunity. In Manila, he determined the oxygen-water electromotive potential, a key datum for many chemical reactions. At M.I.T. he systematically studied the free energy of formation of compounds of oxygen, nitrogen, the halogens, sulfur, and the alkali metals. Particularly important was his measurement of the free energy of formation of simple organic compounds, beginning in 1912 with ammonium cyanate and urea. Lewis realized that the most fruitful use of free-energy data would be in complex organic reactions, an insight that was borne out in the 1920's and 1930's.

In 1907 Lewis set forth a new system of thermodynamics based on the new concept of activity. A function with the dimensions of concentration, activity expresses the tendency of substances to cause change in chemical systems. Lewis derived activity by generalizing the idea of fugacity, but also emphasized that it could be derived directly from free energy, since change in free energy is proportional to change in activity. Lewis showed that in terms of activity, all the familiar thermodynamic equations for ideal systems became "perfectly exact and general" for real systems. He also defined the important new concept of partial molal properties and a more exact form of Nernst's equation for the potential of a single concentration cell. Both ideas proved very useful in treating real chemical systems.

Like fugacity, the conception of activity never played the central theoretical role that Lewis thought it would; but it proved indispensable for treating deviations from ideal behavior in real solutions. In general, Lewis' main contribution to thermodynamics was not in grand theory but, rather, in its practical applications to real systems. He made chemists aware of the importance of hydration of ions and clarified the theory of liquid boundary potentials and conductivity, all of which derived from the practical necessities of measuring free-energy data. His definition of ionic strength (1921) allowed the systematization of activity data. Lewis set new standards of experimental accuracy and reliability; to a fertile but unorganized field he began to bring new clarity and order.

Lewis' other theoretical interests also flourished at M.I.T. The publication of Einstein's theory of relativity (1905) and his mass-energy equation renewed Lewis' interest in his early speculations on radiation. He derived the mass-energy equation from his early idea of the pressure of light without using the principle of relativity (1908). This striking concurrence of his view with Einstein's convinced Lewis of the value of his youthful ideas and made him one of the very few early supporters of Einstein and relativity in America. A second paper with Richard Tolman (1912), deriving Einstein's equation from conservation laws and the principle of relativity, illustrates Lewis' delight in the bizarre paradoxes of relativity theory that most people found so profoundly disturbing: Lewis was an iconoclast and reveled in the overthrow of long-established ideas.

In 1912 Lewis accepted an offer to become dean and chairman of the College of Chemistry at the University of California at Berkeley. The chemistry department was badly run-down, and Lewis was given generous financial support and a free hand to recruit new faculty and to initiate reforms. With him he brought William Bray, Richard Tolman, and Joel Hildebrand, all of whom became distinguished teachers and authors; their textbooks trained the first generation of American chemists who rivaled the products of German universities. The most advanced ideas of German physical chemistry—above all, thermodynamics—still neglected at most American universities, were familiar at Berkeley.

Lewis himself taught no courses—he was always uneasy speaking before a large group—but his influence was felt everywhere. The curriculum was reformed, and introductory courses became models of clarity. Lewis kept the teaching load light and encouraged original research among both his colleagues and his students. No one was permitted to become a narrow specialist, and speculation and free discussion were encouraged. Lewis presided over weekly research conferences where the latest topics were discussed with the utmost freedom by professors and students alike. Lewis had missed this progressive, cooperative "spirit of research" in his student days at Harvard, and its results can be seen in the reputation that his department quickly attained and in the roster of fine chemists it turned out. Lewis' reform and modernization of chemical education set the standard for American chemistry and is one of his most important and enduring achievements.

At Berkeley, Lewis' work on thermodynamics grew more intense. In a long paper of 1913 he summarized and brought up to date the theory and methods for calculating free-energy data. The great utility of a complete table of free energies for predicting chemical behavior, he asserted, made the collection of such data "an imperative duty of chemistry." In the next seven years Lewis published a series of lengthy papers, many in collaboration with Merle Randall, systematically collecting and reworking all the known free-energy data for each element. Especially important is a compendium of entropy data (1917) and an empirical verification of Nernst's third law. All this material became the body of his book on thermodynamics (1923). Next to Nernst and Fritz Haber, Lewis was probably the most important figure in chemical thermodynamics at that time; and his clear, systematic organization of data was second to none.

Equally fruitful was Lewis' theoretical achievement in valence theory. In 1913 Bray and Lewis proposed a dualistic theory of valence which distinguished two distinctly different kinds of bond: the familiar polar bond formed by electron transfer, as in $Na^{+}Cl^{-}$, and a nonpolar bond that did not involve electron transfer. The polar theory, exemplified by J. J. Thomson's popular book *The Corpuscular Theory of Matter* (1907), was then at the peak of its popularity. Bray and Lewis were the first to challenge the view that all bonds, even those in the inert hydrocarbons, were polar; and their heresy was not well received. But other dissenters soon appeared. In 1914 Thomson himself postulated a nonpolar bond involving two electrons and two tubes of force. In 1915 Lewis saw in manuscript a paper by Alfred Parson, an English graduate student visiting Berkeley for a year, that postulated a two-electron nonpolar bond and also a cubic octet very similar to Lewis' cubic atom. This striking coincidence apparently revived Lewis' interest in his early speculations on atomic structure and valence. Lewis probably derived the shared electron pair bond from a combination of the novel and suggestive theories of Parson and Thomson with his own model of the cubic atom.

Early in 1916 Lewis published his germinal paper proposing that the chemical bond was a pair of electrons shared or held jointly by two atoms. The cubic atom was an integral part of Lewis' theory of molecular structure. In terms of cubic atoms, the single bond was represented by two cubes with a shared edge, or more simply by double dots (Figure 1), a convention that has been universally adopted.

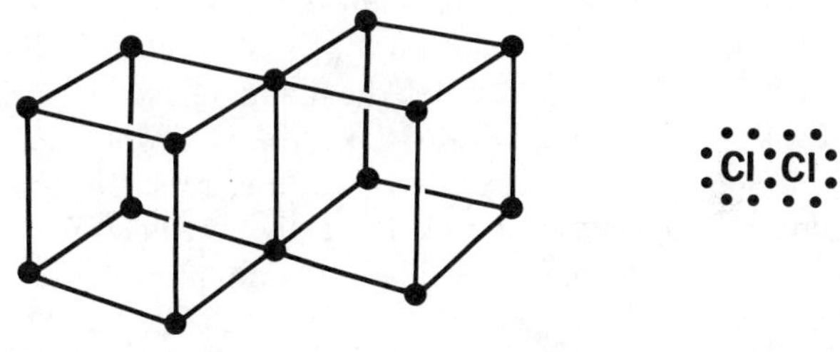

FIGURE 1. The Cl_2 molecules in terms of Lewis' cubic atoms and electron pairs.

According to the octet rule, for a molecule to be stable, each atom must be surrounded by four pairs of electrons that are either shared or free pairs. From this simple idea Lewis derived structures for the halogen molecules, the ammonium ion, and the oxyacids, all of which had proved insoluble for previous theories of valence. Lewis conceived polar bonds simply as unequally shared electron pairs. Since complete transfer of electrons was only the extreme case of polarity, Lewis abandoned his earlier dualistic view. The polar theory became a special case of Lewis' more general and unified theory.

Lewis' theory of the shared-pair bond received no notice in 1916, and he followed it up with only one further paper on color in molecules with "odd" or unpaired electrons (1916). His cubic or "static" atom appeared to be inconsistent with the physicists' view of the atom, based mainly on spectroscopy, which demanded moving electrons, as in Bohr's planetary model (1913). The physicists' "dynamic" atom, however, failed to explain the rigid stereochemistry of carbon compounds. This apparent paradox was much disputed in 1919–1923, and Lewis vigorously defended the static atom against Bohr's atom in a lecture to the Physical Society in 1916 and in his book *Valence* (1923).

By 1916 World War I had halted scientific work in Europe, and in January 1918 Lewis went to France as a major in the Chemical Warfare Service. There he organized the Gas Defense School to train gas officers, and proved so excellent an organizer that he was decorated for his service upon his return in September 1918.

In the spring of 1919 the cubic atom and the shared-pair bond were taken up by Irving Langmuir, who was already famous for his invention of the gas-filled electric lamp and his theory of surface absorption. His dramatic lecture on the new theory of the atom, delivered to the American Chemical Society in April 1919 and often repeated by request elsewhere, suddenly kindled the interest of American chemists. In a series of long papers and lectures in 1919–1921 Langmuir elaborated Lewis' theory so successfully that the Lewis-Langmuir theory, or the Langmuir theory as it was known to many, was talked of everywhere and soon was widely accepted. Lewis was resentful that Langmuir received so much of the credit for the ideas that he had originated. Langmuir always acknowledged his debt to Lewis but felt that he had added enough on his own to warrant the compound name. An exchange of polite but outspoken letters in 1919–1920 probably did not clear the air, but some years later Lewis and Langmuir were again on friendly terms.

Langmuir abruptly ceased publishing on valence in 1921, probably realizing the superiority of the increasingly sophisticated Bohr theory. Lewis, however, continued to support the static atom in a lecture to the Faraday Society in 1923 and in his *Valence and the Structure of Atoms and Molecules* (1923). The conflict between the static and dynamic atoms soon disappeared with the introduction of directed orbitals, and the cubic atom quickly became obsolete. But the shared-pair bond proved to be one of the most fruitful ideas in the history of chemistry. *Valence* became the textbook of the first generation of chemists for whom the chemical bond was more than a simple line. For the first time mechanisms of complex organic reactions could be explained in terms of shifting electron pairs; and in England a new school of physical-organic chemistry was formed by A. Lapworth, M. Lowry, R. Robinson, and others. In the late 1920's the shared-pair bond was the starting point for the new quantum chemistry of E. Schrödinger, H. London, L. Pauling, and others, which transformed Lewis' germinal idea into a quantum mechanical theory of molecular structure.

In 1923 Lewis also published, with Merle Randall, *Thermodynamics and the Free Energy of Chemical Substances.* This extremely influential textbook was for several generations the clearest and simplest presentation of chemical thermodynamics. Its summaries of reliable free-energy data made readily accessible to chemists, even to the novice, the powerful tools of thermodynamics. Once the luxury of specialists, after 1923 thermodynamics was increasingly regarded as an indispensable part of chemical education and research.

For Lewis 1923 marked an end to two of his most abiding interests. He had had enough of collecting free-energy data; and in organic and quantum chemistry, where the shared-pair theory proved most fruitful, Lewis was not at home. He thus found himself at loose ends; and for the next ten years he occupied himself almost exclusively with his third early interest, the theory of radiation and relativity. He tried to derive the laws of quantum radiation by thermodynamic reasoning from the law of microscopical reversibility. He proposed that light does not emit to all space but only to a receiver and that in the space-time manifold, emitter and receiver are in "virtual contact." Such entrancing paradoxes of space and time formed the bulk of his Silliman lectures of 1925 and the resulting book, *The Anatomy of Science* (1926). The decade 1923–1933 was certainly the least successful period of Lewis' career. He was a good enough mathematician to follow the contemporary developments in relativity and quantum theory, but he remained an amateur and an outsider. His novel ideas, stemming from youthful inspiration and his bent for thermodynamic reasoning, remained out of touch with the best professional thought. Lewis styled himself an *enfant terrible* and enjoyed shocking people with his unorthodox views. The profound revolution in physics in the 1920's encouraged his taste for paradoxes; it was clear that some cherished beliefs would be brought down. Anything seemed possible, and it was difficult to distinguish brilliant ideas from absurd ones. A letter from Einstein to Lewis suggests that even he took Lewis' ideas seriously; but in retrospect it is clear that Lewis was out of his depth.

Lewis probably sensed this, for in 1933 he abruptly abandoned theorizing to exploit an unexpected opportunity in a field quite new to him, the separation of isotopes. Deuterium had been discovered in 1932 by Harold Urey, who noted that it might be isolated on a large scale by fractional electrolysis of water. Lewis had been trying to separate oxygen isotopes when he realized that deuterium, being twice the size of ordinary hydrogen, would be easier and more interesting to obtain. In 1933 he succeeded in obtaining nearly pure heavy water and in the next two years rushed out, against intense competition, twenty-eight reports on deuterium chemistry, including several in collaboration with E. O. Lawrence on the nuclear reactions of deuterium in the cyclotron. Since deuterium was markedly different from hydrogen, Lewis foresaw a whole new chemistry of deutero compounds with distinct and unusual properties. But by 1934 these high hopes had apparently paled, for Lewis abruptly ceased work on heavy water. Covalent carbon-deuterium bonds are in fact not easy to make, and deutero compounds are not very different from ordinary compounds.

It is precisely these properties that make deuterium an ideal tracer for studying organic, and especially biochemical, reactions; but Lewis was not prepared to exploit this opportunity. He carried out several studies on the lethal effect of heavy water on germinating plant seeds and on living creatures, but he did not realize how deuterium could be used to study the microchemistry of living tissue. Lewis also tried to follow up the nuclear physics of deuterium in 1936–1937, but this attempt ended in failure when a report on the refraction of neutrons by wax had to be withdrawn as an experimental error. Again Lewis was not at home in either of the fields opened up by his own work.

In 1938 Lewis finally hit upon a fruitful combination of theory and experiment in photochemistry. He had long been interested in the theory of colored compounds. In 1914 he identified two different forms of the indicator methylene blue, and in 1916 his idea of the shared-pair bond led him to propose that color was due to the presence of "odd" electrons. (His work on color had won him the Nichols Medal in 1921.) The occasion for his return to this subject was an important lecture on the theory of acids and bases at the Franklin Institute in 1938. According to his generalized theory of acids and bases proposed in 1923, bases were molecules having free electron pairs to donate, whereas acids were electron-poor molecules that could accept an extra pair. The idea of a general base had been widely accepted; but because of the prevailing proton theory of acidity, Lewis' conception of acids had not. His 1938 lecture did much to make the Lewis acid an important part of chemical theory.

Many of the Lewis acids, such as the triphenylcarbonium salts, exist in a variety of different-colored forms. Lewis proposed that these forms were of two distinct "electromeric" types which differed only in the distribution of electrons in the molecule. Neutralization of one kind required energy of activation, while the other was neutralized spontaneously. Ironically, the concept of electromerism had been an important part of the abandoned electrochemical theory of valence, which Lewis' shared-pair bond and Pauling's theory of resonance hybrids had rendered obsolete. Despite its outmoded terminology, Lewis' theory of electromeric states proved extremely fruitful. Photochemistry was already becoming a popular field in the 1930's, but like thermodynamics in 1910 it lacked a solid theoretical foundation. In a long review in 1939 Lewis and Melvin Calvin summarized the known facts in terms of their elaborate theory of color. There followed a series of fine experimental papers on fluorescence and phosphorescence spectra and on photochemical reactions in rigid media.

Lewis separated two emission bands in phosphorescence spectra, which he associated with "electromeric" excited states. From one, emission was delayed, that is, it required energy of activation; from the other, emission was instantaneous. At first Lewis had consciously avoided quantum mechanical interpretations, preferring a more classical chemical approach. But by 1943 his kinetic and optical studies of phosphorescence had led him to believe that the two electromeric states were in fact the singlet and triplet states of quantum theory. This striking conclusion, which was confirmed in several important papers in 1944–1945 with Calvin and Michael Kasha, was the starting point for the rapid development of photochemistry.

These were Lewis' last papers. He died in 1946 while carrying out an experiment on fluorescence. Thus at the end of his career Lewis again found a field that combined theoretical interest and practical opportunities; to photochemistry he was able to bring the same rigor of experiment and vigor of imagination that characterized the best work of his early years.

BIBLIOGRAPHY

I. ORIGINAL WORKS. Lewis' principal writings are "Outlines of a New System of Thermodynamic Chemistry," in *Proceedings of the American Academy of Arts and Sciences*, **43** (1907), 259–293; "The Atom and the Molecule," in *Journal of the American Chemical Society*, **38** (1916), 762–785; *Thermodynamics and the Free Energy of Chemical Substances* (New York, 1923), written with M. Randall; *Valence and the Structure of Atoms and Molecules* (New York, 1923; repr. New York, 1965); "The Isotope of Hydrogen," in *Journal of the American Chemical Society*, **55** (1933), 1297–1300; "Acids and Bases," in *Journal of the Franklin Institute*, **226** (1938), 293–318; and "Phosphorescence and the Triplet State," in *Journal of the American Chemical Society*, **66** (1944), 2100–2109, written with M. Kasha.

Two boxes of Lewis' correspondence and some MS notes from his student years are in the Lewis Archive,

University of California, Berkeley. The Harvard Archives contain Lewis' Harvard records and a brief correspondence with T. W. Richards.

II. SECONDARY LITERATURE. J. H. Hildebrand, "Gilbert N. Lewis," *Biographical Memoirs. National Academy of Sciences*, **31** (1958), 209–235, includes a complete bibliography. See also W. F. Giauque, "Gilbert N. Lewis," in *Yearbook. American Philosophical Society* for 1946, pp. 317–322; A. Lachmann, *Borderland of the Unknown* (New York, 1956); and R. E. Kohler, "The Origin of G. N. Lewis's Theory of the Shared Pair Bond," in *Historical Studies in the Physical Sciences*, **3** (1971), 343–376. The relation between Lewis and Langmuir is discussed in R. E. Kohler, "Irving Langmuir and the Octet Theory of Valence," *ibid.*, **4** (1972).

R. E. KOHLER

LEWIS, THOMAS (*b.* Cardiff, Wales, 26 December 1881; *d.* Rickmansworth, England, 17 March 1945), *physiology, cardiology, clinical science.*

Thomas Lewis' parents were Welsh. His father, Henry Lewis, was a prominent coal mining engineer from a long line of Monmouthshire yeomen. The family was well-to-do, and Lewis was educated at home until the age of sixteen by his mother and a private tutor. His upbringing was Methodist and his father, to whose encouragement and stimulus he owed much, was a widely read and clearheaded man with a library of several thousand volumes. However, the young Lewis had no interest in books and spent every spare moment in sporting activities and in the local fields and woods.

He took up medicine because he hoped that by becoming a doctor he might emulate the skill of the family physician, who was an expert conjurer. After preclinical work at University College, Cardiff, where he was encouraged in critical thought and research by the physiologist Swale Vincent, in 1902 he entered University College Hospital, London, his medical home for the rest of his life. He graduated M.B., B.S. (university gold medal) in 1905, M.D. in 1907, and was elected fellow of the Royal College of Physicians in 1913. He became a fellow of the Royal Society of London in 1918 and was knighted in 1921. Of his numerous national and foreign distinctions none was more significant than the Copley Medal of the Royal Society (1941), which had previously been awarded to only one clinician, Lord Lister. In 1916 Lewis was appointed physician on the staff of the Medical Research Committee (later Council), which, as he later said, broke the ice in Great Britain as the first full-time research post in clinical medicine. Also in 1916 he married Lorna Treharne James of Merthyr Tydfil, Glamorgan, Wales; they had two daughters and one son.

A man of outstanding intellect with an enormous capacity for unrelenting hard work, Lewis remained in full mental vigor until the end of his life, although he sustained a myocardial infarct at the age of forty-five. At the time of a second attack eight years later he said to Sir Arthur Keith, "Another arrow from the same quiver, my friend, and one of them will get me in the end." He died from a fourth attack at the age of sixty-three.

Lewis' career in medical research began while he was still a student, with the publication of papers on the hemolymph glands that later became standard works. But his interest in cardiac physiology, stimulated by a year in E. H. Starling's laboratory, came from his work with Leonard Hill on the influence of respiration on the venous and arterial pulses. This association inevitably led him into contact with James Mackenzie, whose great work, *The Study of the Pulse*, had appeared in 1902. Their meeting in 1908 began a friendship that profoundly influenced Lewis' life. Mackenzie urged him to study the irregular action of the heart, and Lewis acquired the new Einthoven string galvanometer, set it up in the medical school basement, and plunged into his investigations on the spread of the excitatory processes in the dog heart, coupled with bedside and electrocardiographic studies on clinical arrhythmias. In 1909 he showed that Mackenzie's "irregular irregularity" of the heart was due to atrial fibrillation. Only two years later *The Mechanism of the Heart Beat* was published, and this scholarly scientific monograph soon became the bible of electrocardiography. In the foreword to the second edition (1920) Lewis set forth his scientific credo in a passage which illustrates the simplicity and clarity of his writing:

> Inexact method of observation, as I believe, is one flaw in clinical pathology to-day. Prematurity of conclusion is another, and in part follows from the first; but in chief part an unusual craving and veneration for hypothesis, which besets the minds of most medical men, is responsible. The purity of a science is to be judged by the paucity of its recorded hypothesis. Hypothesis is the heart which no man with right purpose wears willingly upon his sleeve.

Lewis' work from 1908 to 1925 gave him a complete mastery of all aspects of the electrophysiology of the heart, the human electrocardiogram, and clinical disorders of the heartbeat. It included his famous hypothesis of circus movement as the mechanism of atrial fibrillation. Much of this work, done with a succession of talented young men, of whom the first were Paul D. White of Boston and Jonathan Meakins of Montreal, was published in the journal *Heart*, which Lewis founded with Mackenzie's help in 1909.

World War I interrupted these studies in 1914, just after Lewis' return from America, where he had delivered the Herter lectures in Baltimore and the Harvey lecture in New York. Fortunately his talents were harnessed to a study of the condition known to the military as D.A.H. (deranged action of the heart), which was causing a serious loss of manpower at the front. His work at the Military Heart Hospitals led to its redefinition as the effort syndrome, with "a system of graded drills employed remedially and as a means of justly grading soldiers for supposed affections of the heart." This was the foundation of his later insistence on judging the state of the heart from the patient's exercise tolerance rather than on the then fashionable method of assessing cardiac murmurs. He was also involved in the problems of diagnosis and prognosis in organic heart disease and with R. T. Grant instituted an unsurpassed follow-up over ten years of 1,000 soldiers with valvular disease.

By 1925 Lewis had had enough of his elaborate electrocardiographic work and, believing that "the cream was off," he returned to observations on the cutaneous vessels which he had started during the war. He was also influenced by a growing belief that man, not animals, was the proper subject for hospital-based research. A series of observations on the vascular reaction of the skin to various injuries led to his description of the red line, flare, and wheal so produced as the triple response, with the postulate that the original stimulus acts by damaging the tissues, which produce a histaminelike compound that Lewis called the H-substance. This work was gathered together in 1927 in his monograph *The Blood Vessels of the Human Skin and Their Responses*. A notable feature of these experiments which became a hallmark of his later work was that they were carried out with the simplest apparatus, as was his brilliant elucidation of Trousseau's phenomenon in tetany using only two blood-pressure cuffs. In parallel with this laboratory work he was engaged in a study of Raynaud's disease and other maladies affecting blood flow to the limbs and digits.

Lewis' third, and last, period began with his work on pain. One suspects that this subjective phenomenon taxed even his unique research talents to the utmost. Certainly, the summary of his researches published as the monograph *Pain* (1942) lacks the sparkle of his earlier writings. Nevertheless, much was accomplished. The separateness of superficial and deep pain was emphasized. The double pain response to a single stimulus distinguished two systems of pain nerves in skin. Factor P was proposed as the cause of pain in ischemic limb muscle and possibly in heart muscle as well, in angina pectoris. The pain of visceral disease and referred pain in general were investigated experimentally by the injection of hypertonic saline into the interspinal ligaments. The curious hyperalgesia of damaged skin, "erythralgia," led to the uncharacteristically weak hypothesis of the nocifensor system of nerves.

Lewis' worldwide fame as a teacher came from a number of books written for a wider audience than was reached by his scientific papers and monographs. The first, in 1912, was *Clinical Disorders of the Heart Beat*, followed in 1913 by *Clinical Electrocardiography*. Both went through several editions and must have been a godsend to clinicians trying to understand the then novel methods of investigating the heart. However, Lewis grudged the time spent on them, writing only from a sense of duty to the profession. His best-known book, *Diseases of the Heart* (1933), was widely translated. *Clinical Science, Illustrated by Personal Experiences* (1934) recounted his methods of attack upon clinical problems such as intermittent claudication and bacterial endocarditis. *Vascular Disorders of the Limbs* (1936) is still a most useful acount. His last book, *Exercises in Human Physiology* (1945), demonstrates his interest in medical education and presents a detailed account of simple experiments that students may make on themselves.

At University College Hospital he worked constantly in the wards and in the out-patient department. His superb clinical teaching emphasized a meticulous examination of the patient followed by a penetrating analysis of the relationship of the symptoms and signs to the underlying hemodynamic and pathological disturbances. Few students other than his clinical clerks attended his rounds, however, the majority preferring didactic presentations. At his regular medical school lectures he displayed his interest in the history of medicine, especially the French physicians of the nineteenth century.

Lewis' most important contribution to medicine may have been his concept of clinical science, his insistence that progress in medicine would come chiefly from scientific studies on living men in health and disease, rather than from the basic science laboratories and from animal experimentation. With controlled investigation of patients now basic to much of medicine, it is difficult to relive the time when Lewis, almost alone, fought for the creation of established career posts in medical research, now numbered in the tens of thousands. This must have been the most difficult period of his life and he signaled it in 1930 by renaming his own department the Department of Clinical Research and by founding the Medical Research Society, and in July 1933 by changing the title of *Heart* to *Clinical Science.*

BIBLIOGRAPHY

A complete bibliography of Lewis' twelve books and 229 scientific papers is in the notice by A. N. Drury and R. T. Grant, in *Obituary Notices of Fellows of the Royal Society of London,* **5** (1945), 179–202. See also *American Heart Journal,* **29** (1945), 419–420; *British Medical Journal* (1945), **1**, 461–463, 498; *Lancet* (1945), **1**, 419–420; *University College Hospital Magazine,* **30** (1945), 36–39; and *British Heart Journal,* **8** (1946), 1–3.

A collection of papers by his former pupils and collaborators is in *University College Hospital Magazine,* **40** (1955), 63–74.

A. HOLLMAN

LEWIS, TIMOTHY RICHARDS (*b.* Llanboidy, Carmarthenshire, Wales, 31 October 1841; *d.* Southampton, England, 7 May 1886), *tropical medicine.*

Timothy Lewis, the second son of William Lewis, was born at Llanboidy but spent his early years at Crinow, a small village in Pembrokeshire, to which his family removed while he was still in his infancy. There he received his early education, first at the village school until he was ten years old, and later at a private grammar school recently opened in the locality by the Reverend James Morris, which he attended until he was fifteen. He then became apprenticed to a local pharmacist with whom he remained for four years, before moving to London. He became a compounder of medicines at the German Hospital in Dalston, London, where he became proficient in German. At the same time he attended lectures in human medicine at University College, London, where his skill in the laboratory as a chemist was quickly recognized. Subsequently, in the hospital's wards, his intelligence and attention to detail were so marked that he was awarded the Fellowes silver medal for clinical medicine in 1866. He completed his studies at the Medical School of the University of Aberdeen, from which he received the degree of M.B. with honors in 1867.

Immediately after qualifying he applied for a commission in the army medical department and was placed first in the entrance examination for the Army Medical School at Netley, near Southampton, his name again appearing at the head of the list at the completion of the four-month course. He was commissioned successively as assistant surgeon (31 March 1868), surgeon (1 March 1873), and surgeon major (31 March 1880), and for three years before his death in 1886 he held the post of assistant professor of pathology at the Army Medical School in Netley.

Little was known of the etiology of disease, and the imagination of many investigators ran riot in hypothesis and speculation. Lewis' work, collected and reprinted in 1888 in a commemorative volume, reveals the power of his critical examination and judgment, which enabled him to refute the wilder claims of some of his contemporaries, including such outstanding workers as Hallier, Koch, and Pasteur.

After completing the course at Netley, Lewis and D. D. Cunningham, the two best students of the session, were sent to study in Germany before proceeding to India to investigate the causation of cholera. They published a series of papers, together and singly, of which Lewis' first work, issued in 1870, on the microscopic objects found in the stools of patients with cholera, contains the first authentic account of amoebas from the human intestine. The delicate drawings (108 in all) demonstrate the care and exactitude of his observations. His interests were extremely wide, and over the next fourteen years his studies ranged over cholera, leprosy, fungal diseases, Oriental sore, bladder worms of pigs and cattle, and the diets of prisoners in Indian jails. His detailed descriptions of the blood parasites of man and animals constitute his most significant contribution to knowledge and exercised a fundamental influence over the attitude of his contemporaries to the etiology of disease. His two reports on the nematode hematozoa of man and animals (1874) contain the first account of microfilariae in human blood and revealed its connection with chyluria and elephantiasis. Lewis named the hematozoon *Filaria sanguinis hominis* (F.S.H.), later known as *Filaria bancrofti.* This work was originally published as an appendix to the annual report of the sanitary commissioner with the government of India and was not readily obtainable outside India. Manson, however, stumbled across it in the British Museum while on leave from China and, realizing its significance, carried out experiments on the life history of F.S.H. on his return to China in 1875. As a result he was able in 1877 to describe its development in mosquitoes.

In 1878 and 1884 Lewis published two reports on blood parasites of man and animals, including microfilariae, spirochetes in relapsing fever, and trypanosomes in the blood of rats, which was subsequently named *Trypanosoma lewisi* by Kent in 1880.

Two weeks before his death, Lewis was recommended by the council of the Royal Society for election as one of the fellows for 1886, but he did not live to be elected. He died of pneumonia contracted, his notebook suggests, by accidentally inoculating himself during his experiments.

Lewis' meticulous search for pathogenic organisms in blood, feces, and urine, and his knowledge of organisms which are normally present without producing disease, unquestionably bore fruit in the discoveries of subsequent workers. As Dobell rightly pointed out, "Lewis was, like Manson, a pioneer."

BIBLIOGRAPHY

Lewis' writings were brought together in W. Aitken, G. E. Dobson, and A. E. Brown, eds., *Physiological and Pathological Researches: Being a Reprint of the Principal Scientific Writings of the Late T. R. Lewis* (London, 1888). For his journal articles, see Royal Society *Catalogue of Scientific Papers*, X, 584; XII, 445; and XVI, 761.

Obituaries appeared in *British Medical Journal* (26 June 1886), 1242–1243; *Lancet* (22 May 1886), 993; and *Nature*, **34** (1886), 76–77, with bibliography. See also C. Dobell, "T. R. Lewis," in *Parasitology*, **14** (1922), 413–416.

M. J. CLARKSON

LEWIS, WILLIAM (*b.* Richmond, Surrey, England, 1708; *d.* Kingston, Surrey, England, 21 January 1781), *chemistry, pharmacy, chemical technology*.

Born into a brewing family in the environs of London, Lewis had an early opportunity to observe rudimentary chemical technology. By the time he had spent some five years at Oxford (1725–1730) and graduated M.B. from Cambridge in 1731, he had acquired a definite interest in the chemical arts. He was giving public lectures in chemistry "with a View to the Improvement of Pharmacy, Trades and the Art itself" at least as early as 1737. Within a few years he had established sufficient reputation to be elected a member of the Royal Society in October 1745, and he remained a part of the London scientific scene for the rest of his life.

Whatever Lewis learned at Newtonian Oxford and Cambridge was supplemented and shaped by later editorial labors on authors less devoted to the mechanical philosophy. His practical bent was nurtured by his 1746 edition of George Wilson's venerable *Compleat Course of Chemistry* and by his abridgment in the same year of the Edinburgh *Medical Essays*. The latter included many articles reflecting the long-term interests of the Edinburgh physicians in the practical chemistry associated with materia medica and mineral waters. Lewis' own interest in applied chemistry never waned, and, after he moved westward out of London proper to the village of Kingston in 1746, he continued to work on pharmacy and materia medica, gradually expanding his interests to include other areas of chemical technology.

That his plan for a broadly based attack on problems in applied chemistry was formed fairly early is suggested by his published pamphlet of 1748 proposing a periodical entitled *Commercium philosophico-technicum*. As planned, this new publication was to examine various manufacturing processes systematically, through laboratory studies where necessary, and to eliminate wasteful procedures and products. Lewis was far from alone in these interests. Many men of commerce and learning felt the need for the practical improvements in the arts that were promised but never delivered by the enthusiasts of the new mechanical philosophy. Indeed, this period saw the rise of several groups interested in such improvements, notably the Royal Society of Arts (founded 1754) and the Lunar Society of Birmingham. Lewis himself had a broad circle of acquaintances with similar outlooks, and he availed himself of their opinions of his projects on a number of recorded occasions.

While his broad plan for applied chemistry was thus being formulated, Lewis kept up his pharmaceutical activities. He had agreed to produce a new English pharmacopoeia and, perhaps as preparation for this task, translated the fourth Latin edition of the *Edinburgh Pharmacopoeia* and published it in 1748.

One of the most significant events in Lewis' life occurred in 1750, when he began his lifelong association with Alexander Chisholm. Searching for an intelligent and educated assistant for his numerous projects, Lewis found one, appropriately enough, in a London bookstore. Their careers, until Lewis' death, were inseparably entwined; to speak of one is to speak of the other. Chisholm, who in 1743 had graduated M.A. from Marischal College, Aberdeen, was an able linguist, competent in Latin and Greek and in the German and Swedish vernaculars. Indeed, it is believed that the translation of Caspar Neumann's chemical lectures was done by Chisholm.

Lewis' long-standing interest in materia medica and his studies of previous works in pharmacy, including his translation of the *Edinburgh Pharmacopoeia*, culminated in 1753 with the publication of *The New Dispensatory*, which was based on Quincy's *Complete English Dispensatory*. Careful and systematic, like all of Lewis' works, the *New Dispensatory* was highly praised by that archcritic William Cullen as the only English pharmacopoeia that "made any improvement on the Materia Medica . . . ." The work went through many editions before and after Lewis' death, and its spirit was retained in the form and arrangement of Andrew Duncan's *Edinburgh New Dispensatory* (1803), which survived well into the nineteenth century.

Perhaps no work better shows Lewis' abilities as an experimentalist than his series of researches on platinum during the 1750's. Although others had attacked this intractable metal before him, using a number of standard metallurgical techniques, Lewis succeeded where others had failed through his careful systematic approach and the clever use of chemical analogy. Well-versed in the trends of his time, Lewis quickly turned to acids to get the metal into solution. Noting certain similarities between platinum and gold, Lewis defined the new metal in terms of an old one whose chemistry was relatively well understood. In his knowledge of solution chemistry and his ability to employ it imaginatively, Lewis showed himself to be in the forefront of eighteenth-century experimental chemistry. For this effort and for his already substantial contributions to pharmacy and materia medica Lewis was awarded the Copley Medal of the Royal Society in 1754.

Attracted by the empiricism bias of the work of the German chemist and apothecary Caspar Neumann, Lewis published in 1759 a translation with both additions and abridgments of Neumann's chemical lectures. Neumann's approach presents a strong contrast to George Wilson's stark collection of recipes; for the former, although cautious as a theoretician, was a follower of Georg Ernst Stahl, and thus his work employed the complex of ideas associated with the "inflammable principle"—phlogiston. Much has been made of Lewis' description of Neumann as "biassed by no theory and attached to no opinions." That Lewis would hold such an opinion would literally mean that he had never read Neumann, for there is no possible doubt where the latter's theoretical sympathies lay. What Lewis undoubtedly objected to was the preconceived and—to him as well as to other practical chemists like Robert Dossie—ill-conceived notions of the corpuscularians. Like most of his contemporaries, Lewis felt free to accept or reject the various dicta associated with the phlogiston concept. His interests lay in the determination of chemical composition, a task that Stahl and such followers as Neumann were certainly better equipped to handle than the Newtonian chemists. Lewis himself used the other pillar of phlogiston chemistry—affinity—with ease, even though, like many others, he was not overly concerned with the corpuscular details of this theory.

Certainly Lewis' most ambitious work was the *Commercium philosophico-technicum*, which appeared in sections over the period 1763–1765. He divided the book into seven topics that by no means form a complete survey of chemical technology of this period. The topics included such diverse headings as the chemistry of gold, vitrification, expansion and contraction of bodies with change in temperature, blowing air into furnaces by the force of falling water, methods of producing the color black, and a summation of Lewis' studies on platinum. While these topics differed widely in their immediate applicability to technology, as a group they demonstrated forcefully the possibility of relating chemical knowledge to a wide variety of industrial problems.

Although Lewis was prolific, his general way of working was slow and methodical. He read extensively, took notes and made additions to them from his own experience, and even conducted new experiments to confirm or deny the claims of others. But it must not be assumed that his only virtue was his systematic methodology. His skepticism about the "mechanical philosophy" was in itself a philosophical position, one which was shared by a substantial number of his contemporaries. In his later works, and particularly in the highly empirical *Commercium philosophico-technicum*, he eschewed a simple empirical approach—such as descriptions of manufactures or histories of processes—for one that was based on the "invariable properties of matter." What Lewis avoided was a dependence on the sterile mechanical philosophy that promised so much and produced so little in chemistry.

BIBLIOGRAPHY

I. ORIGINAL WORKS. Lewis' works are *A Course of Practical Chemistry in Which are Contained All the Operations Described in Wilson's Complete Course of Chemistry* (London, 1746); *Medical Essays & Observations*, published by a society in Edinburgh, in six volumes, abridged and disposed under general heads (2 vols., London, 1746); *Pharmacopoeia of the Royal College of Physicians of Edinburgh* (London, 1748); *Proposals for Printing, by Subscription, Commercium philosophico-technicum or The Philosophical Commerce of the Arts* (London, 1748); *Oratio in theatro Sheldoniano, habilo idibus Aprilis* (Oxford, 1749); *An Answer to the Serious Inquiry Into Some Proceedings Relating to the University of Oxford* (London, 1751); *The New Dispensatory . . . Intended as a Correction and Improvement of Quincy* (London, 1753; 6th ed., 1799); "Experimental Examination of a White Metallic Substance . . .," in *Philosophical Transactions*, **48**, pt. 2 (1754), 638–689; "Experimental Examination of Platina," *ibid.*, **50**, pt. 1 (1757), 148–166; *The Chemical Works of Caspar Neumann* (London, 1759); *Experimental History of the Materia Medica* (London, 1761; 3rd ed., 1769); *Commercium philosophico-technicum or The Philosophical Commerce of Arts: Designed as an Attempt to Improve Arts, Trades and Manu-

factures (London, 1763–1765); *Experiments and Observations on American Potashes* (London, 1767); and the posthumous *A System of the Practice of Medicine, From the Latin of Dr. Hoffmann By the Late William Lewis* (London, 1783).

II. SECONDARY LITERATURE. See F. W. Gibbs, "William Lewis, M.B., F.R.S. (1708–1781)," in *Annals of Science*, **8** (1952), 122–151; Edward Kremers, "William Lewis," in *Journal of the American Pharmaceutical Association*, **20** (1931), 1204–1209; and Nathan Sivin, "William Lewis (1708–1781) As A Chemist," in *Chymia*, **8** (1962), 63–88.

JON EKLUND

LEXELL, ANDERS JOHAN (*b*. Äbo, Sweden [now Turku, Finland], 24 December 1740; *d*. St. Petersburg, Russia, 11 December 1784), *mathematics, astronomy.*

Lexell was the son of Jonas Lexell, a city councillor and jeweler, and his wife, Magdalena Catharina Björckegren. He graduated from the University of Äbo in 1760 as bachelor of philosophy and became assistant professor of Uppsala Nautical School in 1763 and professor of mathematics in 1766. Invited to work at the St. Petersburg Academy of Sciences on the recommendation of the Swedish astronomer P. W. Wargentin in 1768, he was appointed adjunct in 1769 and professor of astronomy in 1771. His research soon made him well-known. In 1775 the Swedish government offered him a professorship at the University of Äbo which would permit him to proceed with his work in St. Petersburg until 1780; but Lexell preferred to remain in Russia. He spent 1780–1782 traveling in Western Europe; and his letters to J. A. Euler, permanent secretary of the Academy, contain valuable information on scientific life in Germany, France, and England. (These letters are kept at the Archives of the U.S.S.R. Academy of Sciences, Leningrad.)

After Leonhard Euler's death in 1783, Lexell for a short time held Euler's professorship of mathematics, but Lexell himself died the following year. Lexell was a member of the Academy of Sciences in Stockholm, the Society of Sciences in Uppsala (1763), and a corresponding member of the Paris Académie des Sciences (1776).

At St. Petersburg Lexell immediately became one of the closest associates of Leonhard Euler. Under the famous mathematician's supervision he and W.-L. Krafft and J. A. Euler helped to prepare Euler's *Theoria motuum lunae, nova methodo pertractata* (1772) for publication. Euler's influence upon Lexell's scientific activity was considerable; but the latter's works were carried out independently.

On 3 June 1769 the St. Petersburg Academy conducted observations of the transit of Venus at many sites in Russia. Lexell took an active part in the organization and processing of the observations. Using L. Euler's method he calculated (1) the solar parallax to be 8″.68 (compared to today's value of 8″.80). No less interesting was Lexell's determination of the orbits of the 1769 comet (2) and especially of the comet discovered in 1770 by Messier (3). Lexell established the period of the latter on its elliptical orbit as five and a half years; this was the first known short revolution-period comet. Passing near Jupiter and its satellites the comet exerted no influence upon their motion; Lexell thus concluded that the masses of comets are rather small in spite of their enormous sizes. Lexell's comet, which had not been observed before 1770, has not been seen again; probably it lost its gaseous coat and become invisible. Still more important was Lexell's investigation of the orbit of the moving body discovered by W. Herschel on 13 March 1781 and initially regarded as a comet. Lexell's calculations showed that this heavenly body was a new planet (Uranus), nearly twice as far from the sun as Saturn (4). Moreover, Lexell pointed out that perturbations in the new planet's motion could not be explained by the action of the known members of the solar system and stated the hypothesis that they must be caused by another, more remote planet. This hypothesis proved correct when Neptune was discovered in 1846 on the basis of the calculations of J. Adams and U. Le Verrier.

Lexell's mathematical works are devoted to problems of analysis and geometry. Following Euler he elaborated the method of integrating factor as applied to higher order differential equations (1771). He gave a solution of linear systems of second order differential equations with constant coefficients (1778, 1783); suggested a classification of elliptic integrals (1778); calculated integrals of some irrational functions by reducing them to integrals of rational functions (1785).

Lexell for the first time constructed a general system of polygonometry (5), e.g., of trigonometrical solution of plane n-gons on the given $2n - 3$ sides and angles between them, provided that at least $n - 2$ elements are sides. These investigations formed a natural sequel to and generalization of the works on trigonometrical solution of quadrangles published shortly before by J.-H. Lambert (1770), J. T. Mayer (1773), and S. Björnssen (1780). Lexell based solutions of all the problems on two principal equations obtained when the sides of a polygon are projected on the two normal (to one another) axes situated in its plane, provided that one of them coincides with some side. From these equations he deduced others effective in the solution of triangles and quadrangles on certain given elements, suggested analogous formulae for pentagons, hexagons, and heptagons and stated considerations relevant to classification and solution of problems in the general case. He also considered, though in less detail, the problem of solution of *n*-gons on diagonals and on angles they form with sides. After Lexell polygonometry was also worked out by S. L'Huillier (1789).

Lexell considerably enriched spherical geometry and trigonometry. Especially brilliant is his theorem discovered no later than 1778 but published only in 1784: the geometric locus of the vertices of spherical triangles with the same base and equal area are arcs of two small circles whose extremities are points diametrically opposite to the extremities of the common base (6).

In his two articles which were published posthumously (7, 8), Lexell established other properties of small circles on the sphere and deduced a number of new propositions of spherical trigonometry in which he generalized Heron's and Ptolemy's theorems for the sphere and suggested elegant formulae for defining the radii of circumference of circles inscribed in spherical triangles or quadrangles and circumscribed about them. Works by St. Petersburg academicians N. Fuss and F. T. Schubert on spherical geometry and trigonometry were close to these investigations.

Five unpublished papers on geometry by Lexell are preserved in the Archives of the Academy of Sciences in Leningrad (18).

BIBLIOGRAPHY

I. Original Works.

1. *Disquisitio de investiganda vera quantitate parallaxeos Solis ex transitu Veneris ante discum solis anno 1769* (St. Petersburg, 1772).

2. *Recherches et calculs sur la vraie orbite elliptique de la comète de l'an 1769 et son temps périodique* (St. Petersburg, 1770), written with L. Euler.

3. *Réflexions sur le temps périodique des comètes en général et principalement sur celui de la comète observée en 1770* (St. Petersburg, 1772).

4. *Recherches sur la nouvelle planète découverte par Mr Herschel* (St. Petersburg, 1783).

5. "De resolutione polygonorum rectilineorum," in *Novi Commentarii Ac. Sc. Petropolitanae*, **19** (1774), 1775, 184–236; **20** (1775), 1776, 80–122.

6. "Solutio problematis geometrici ex doctrina sphaericorum," in *Acta Ac. Sc. Petropolitanae*, **5**, 1 (1781), 1784, 112–126.

7. "De proprietatibus circulorum in superficie sphaerica descriptorum," *ibid.*, **6**, 1 (1782), 1786, 58–103.

8. "Demonstratio nonnullorum theorematum ex doctrina sphaerica," *ibid.*, **6**, 2 (1782), 1786, 85–95.

II. Secondary Literature.

9. "Précis de la vie de M. Lexell," in *Nova acta Ac. Sc. Petropolitanae*, **2** (1784), 1788, 12–15.

10. J.-C. Poggendorf, *Biographisch-Literarisches Handwörterbuch zur Geschichte der exacten Wissenschaften*, I (Leipzig, 1863), 1444–1446, contains bibliography. (See also 13.)

11. *Nordisk Familjebok Konversationslexikon och realencyklopedi*, IX (Stockholm, 1885), 1189–1191.

12. M. Cantor, *Vorlesungen über Geschichte der Mathematik*, IV (Leipzig, 1908), see Index.

13. O. V. Dinze and K. I. Shafranovski, *Matematika v izdaniakh Akademii nauk* ("Mathematical Works Published by Academy of Sciences") (Moscow–Leningrad, 1936).

14. *Svenska män och kvinnor. Biografisk uppslagswork*, IV (Stockholm, 1948), 553.

15. *Istoria Akademii nauk SSSR* ("History of the Academy of Sciences of the U.S.S.R."), I (Moscow–Leningrad, 1958), see Index.

16. V. I. Lysenko, "Raboty po poligonometrii v Rossii XVIII v." ("Works on Polygonometry in Russia in the XVIII Century"), in *Istorico-matematicheskie issledovania*, **12** (1959), 161–178.

17. *Idem*, "O rabotakh peterburgskikh akademikov A. I. Lexella, N. I. Fussa u F. I. Shuberta po sfericheskoi geometrii i sfericheskoi trigonometrii" ("On the Works of Petersburg Academicians A. I. Lexell, N. I. Fuss, and F. T. Schubert on Spherical Geometry and Trigonometry"), in *Trudy Instituta istorii estestvoznania i tekhniki*, **34** (1960), 384–414.

18. *Idem*, "O neopublikovannykh rukopisiakh po geometrii akademikov A. I. Lexella i N. I. Fussa" ("On the Unpublished Works of Academicians A. I. Lexell and N. I. Fuss on Geometry"), in *Voprosy istorii estestvoznania i tekhniki*, **9** (1960), 116–120.

19. *Idem*, "Iz istorii pervoi peterburgskoi matematicheskoi shkoly" ("On the History of the First Mathematical School in St. Petersburg"), in *Trudy Instituta istorii estestvoznania i tekhniki*, **43** (1961), 182–205.

20. A. P. Youschkevitch, *Istoria matematiki v Rossii do 1917 goda* ("History of Mathematics in Russia Until 1917") (Moscow, 1968), see Index.

A. T. Grigorian
A. P. Youschkevitch

LEYBENZON, LEONID SAMUILOVICH (*b.* Kharkov, Russia, 26 June 1879; *d.* Moscow, U.S.S.R., 15 March 1951), *mechanical engineering*, *geophysics*.

The son of a physician, Leybenzon graduated in 1897 from the Tula classical Gymnasium. He then entered the physics and mathematics section of Moscow University. In 1901 he graduated from the university and in 1906 from the Moscow Higher Technical School. While still a student, he went, on the recommendation of his teacher Zhukovsky, as a mechanical engineer to the Aerodynamics Institute in Kuchino (1904). Here he helped to build the first wind tunnel in Russia and a machine for testing propellers. He also constructed two-component aerodynamic scales, built models, and worked out the first methods of aerodynamic and structural design of the airplane, following Zhukovsky's instructions.

In 1915 he defended his dissertation for a master's degree in applied mathematics at Moscow University, "K teorii bezbalochnykh pokryty" ("Toward a Theory of Ribless Covering"), and in 1916 was made professor of mechanics at Yurev University. In 1917 Leybenzon presented his dissertation for a doctorate in applied mathematics, "O prilozhenii metoda garmonicheskikh funktsy Tomsona k boprosu obvstoychivosti szhatyth sfericheskoy i tsilindricheskoy uprugikh obolochek" ("On the Application of the Method of Harmonic Functions of Thomson to the Question of the Stability of Compressed Spherical and Cylindrical Elastic Membranes"). From 1919 to 1929 Leybenzon was professor of applied mechanics at Tbilisi Polytechnical Institute and University.

In 1921 Leybenzon became professor at the Polytechnical Institute in Baku. At that time he began his remarkable activity in the science and technology of petroleum. In 1922 he returned to Moscow and was made head of the Department of Applied Mechanics at Moscow University. In 1933 Leybenzon was elected corresponding member of the Academy of Sciences of the U.S.S.R. and in 1943 active member.

The range of Leybenzon's scientific interests was extraordinarily wide. He did extensive research in aerodynamics, elasticity theory, hydraulics, and geophysics. In the theoretical section of the Central Aerohydrodynamic Institute from 1933 to 1936, Leybenzon was concerned with airplane design, with the theory of the border layer, and gas dynamics. His works on the theory of hydrodynamic lubrications and the theory of evaporation of liquid drops in a gas current are of great significance. The important transformations given by Leybenzon for the basic equations of gas dynamics of Chaplygin should also be noted.

Leybenzon contributed to elasticity theory and resistance of materials. He gave a method of determining the position of the center of a curve and presented an interesting theorem on the circulation of tangent stress on a curve. He also demonstrated a general method for softening or relaxing border conditions. He deserves credit for the development of methods of approximating the determination of the turning moment into the theory of torsion and the setting of upper and lower limits to its size.

Leybenzon's scientific research and his many years of teaching the theory of elasticity culminated in notable monographs and textbooks. Among them are *Kurs teorii uprugosti* ("Course in the Theory of Elasticity"; 1942), *Elementy matematicheskoy teorii plastichnosti* ("Elements of the Mathematical Theory of Plasticity"; 1943), and *Variatsionnye metody resheniya zadach teorii uprugosti* ("Variational Methods of Solving Problems in the Theory of Elasticity"; 1943).

Leybenzon's work on the production of petroleum was also of great significance. In 1932 he published a fundamental investigation, "Priblizhennaya dinamicheskaya teoria glubokogo nasosa" ("An Approximating Dynamic Theory of the Deep Pump"). His "Podzemnaya gidravlika vody, nefti i gaza" ("Underground Hydraulics of Water, Petroleum and Gas"; 1934) provided a scientific basis for the rational development of petroleum and gas deposits. His works laid the foundation for the development of the theory of filtration of aerated liquids. Leybenzon had a direct role in the planning and building of the first Soviet pipelines from Baku to Batumi and from Groznyy to Tuapse. A large part of his work was in geophysics, particularly the application of the elasticity theory to the study of the structure of the earth.

BIBLIOGRAPHY

I. Original Works. For Leybenzon's works, see *Sobranie trudov* ("Collected Works"), 4 vols. (Moscow, 1951–1955): I. *Teoria uprugosti* ("Theory of Elasticity"; 1951); II. *Podzemnaya gidrogazodinamika* ("Underground Gas Dynamics"; 1953); III. *Neftepromyslovaya mekhanika* ("Mechanics of the Petroleum Industry"; 1955); and IV. *Gidrodinamika. Geofizika* ("Hydrodynamics. Geophysics"; 1955).

II. Secondary Literature. For information on Leybenzon's life, see L. I. Sedov, "Osnovnye daty zhizni i deyatelnosti L. S. Leybenzona" ("Basic Facts in the Life and Work of L. S. Leybenzon"), in *Uspekhi matematicheskikh nauk*, **7** (1952), 127–134; and B. N. Yurev, "Leonid Samuilovich Leybenzon," in *Izvestiya Akademii nauk SSSR. Otdelenie tekhnicheskikh nauk* (1949), no. 8, 1138–1142.

A. T. Grigorian

LEYDIG, FRANZ VON (*b.* Rothenburg-ob-der-Tauber, Germany, 21 May 1821; *d.* Rothenburg-ob-der-Tauber, 13 April 1908), *comparative anatomy.*

One of three children, Leydig was the only son of Melchior Leydig, a Catholic and a minor public official, and his wife Margareta, a Protestant. Leydig shared his father's religion as well as his hobbies, the elder Leydig being a keen gardener and beekeeper. Leydig himself later recalled that these childhood interests established his lifelong concern with botany and zoology. At the age of twelve the boy acquired a simple microscope, at which he spent most of his free time. (A few years later, during his studies at Würzburg, a local doctor lent him a more sophisticated instrument.)

Beginning in 1840 Leydig studied medicine at the universities of Würzburg and Munich. He received his doctorate in medicine at Würzburg and became an assistant in the department of physiology, teaching also histology and developmental anatomy under Kölliker. He qualified as a university lecturer in 1849, and as extraordinary professor in 1855. In the winter of 1850–1851, Leydig made a trip to Sardinia, where he became aware of the rich marine life that was to become the subject of some of his most important researches. This journey, coupled with his early preoccupation with microscopy, determined the course of his life's work.

In 1857 Leydig was appointed full professor of zoology at the University of Tübingen and published his *Lehrbuch der Histologie des Menschen und der Tiere*, his outstanding contribution to morphology. In his introduction to the *Lehrbuch*, Leydig reviewed the crucial developments in the history of histology, including the discovery and definition of the cell by Purkyně, Valentin, and Schwann, the last of whom described it as a vesicle containing a nucleus (1839). Leydig paid further tribute to other contemporary anatomists, particularly Johannes Müller for his work on glands and for the emphasis that he properly placed upon the significance of the cellular doctrine for pathology. Leydig's book was published at about the same time as other general treatments of similar subjects—most notably Kölliker's *Handbuch der Gewebelehre des Menschen* (1852) and Gerlach's *Handbuch der allgemeinen und speciellen Gewebelehre des menschlichen Körpers . . .* (1848). The *Lehrbuch*, however, gives the best account of the rapid growth of comparative microscopical anatomy in the two decades following Schwann's discoveries.

In addition to its historical importance, Leydig's *Lehrbuch* is significant for his description in it of a large secretory cell, found in the epidermis of fishes and larval Amphibia. This mucous cell is peculiar in that it does not pour its secretions over the surface of the epithelium; Leydig believed that its function was to lubricate the skin, and the cell now bears his name.

Chief among Leydig's other discoveries is the interstitial cell, a body enclosed within a smooth endoplastic reticulum and containing lipid granules and crystals, that occurs in the seminiferous tubules and in the mediaseptum and connective-tissue septa of the testes. These cells are believed to produce the male hormone testosterone, which determines male secondary sexual characteristics. Leydig described the interstitial cells in his detailed account of the male sex organs, "Zur Anatomie der männlichen Geschlechtsorgane und Analdrüsen der Säugetiere," published in 1850:

> The comparative studies of the testis resulted in the discovery of cells surrounding the seminiferous tubules, vessels, and nerves. These special cells are present in small numbers where they follow the course of the blood vessels, but increase in mass considerably when surrounding seminiferous tubules. These cells are lipoid in character; they can be colorless or can be stained yellowish, and they have light vesicular nuclei [p. 47].

This description indicates clearly that Leydig recognized the specific morphology of these cells; their endocrine nature and ultrastructure have only recently been fully understood.

Leydig is also known for the discovery of the gland of Leydig (1892), a portion of the mesonephros in vertebrates, of which the secretions are thought to stimulate the movement of spermatozoa; and for describing large vesicular cells that occur in the connective tissue and in the walls of blood vessels in crustaceans (1883). Four different types of the latter have been determined.

Leydig became professor of comparative anatomy at the University of Bonn in 1875. He was made emeritus in 1887, and retired to the town of his birth. He had married Katharina Jaeger, the daughter of a professor of surgery at Erlangen, who survived him; they had no children. During his lifetime Leydig was granted many honors, including personal ennoblement, and an honorary doctorate of science from the University of Bologna. He was a member of a number of medical and scientific societies, among them the Royal Society of London, the Imperial Academy of Science of St. Petersburg, and the New York Academy of Sciences.

BIBLIOGRAPHY

I. Original Works. Leydig's most important writings include "Die Dotterfurchung nach ihrem Vorkommen in der Tierwelt und nach ihrer Bedeutung," in *Isis. Encyclopädische Zeitschrift, vorzüglich für Naturgeschichte, vergleichende Anatomie und Physiologie von* [*Lorenz*] *Oken*, pt. 3 (1848), cols. 161–193; "Zur Anatomie der männlichen Geschlechtsorgane und Analdrüsen der Säugetiere," in *Zeitschrift für wissenschaftliche Zoologie*, **2** (1850), 1–57; "Über Flimmerbewegung in den Uterindrüsen des

Schweines," in *Archiv für Anatomie, Physiologie und wissenschaftliche Medicin* (1852), 375–378; *Beiträge zur mikroskopischen Anatomie und Entwicklungsgeschichte der Rochen und Haie* (Leipzig, 1852); "Zum feinen Aufbau der Arthropoden," in *Archiv für Anatomie, Physiologie und wissenschaftliche Medicin* (1855), 376–480; "Über Tastkörperchen und Muskelstruktur," *ibid.* (1856), 150–159; *Lehrbuch der Histologie des Menschen und der Tiere* (Frankfurt, 1857); "Über das Nervensystem der Anneliden," in *Archiv für Anatomie und Physiologie* (1862), 90–124; "Neue Beiträge zur anatomischen Kenntnis der Hautdecke und Hautsinnesorgane der Fische," in *Festschrift zur Feier der 100jährigen Bestehens der Naturforschenden Gesellschaft zu Halle* (Halle, 1879), 129–186; *Zelle und Gewebe. Neue Beiträge zur Histologie des Tierkörpers* (Frankfurt, 1885); "Die riesigen Nervenröhren im Bauchmark der Ringelwürmer," in *Zoologischer Anzeiger*, **9** (1886), 591–597; "Das Parietalorgan der Wirbeltiere. Bemerkungen," *ibid.*, **10** (1887), 534–539; "Nervenkörperchen in der Haut der Fische," *ibid.*, **11** (1888), 40–44; "Das Parietalorgan der Reptilien und Amphibien kein Sinnesorgan," in *Biologisches Zentralblatt*, **8**, no. 23 (1889), 707–718; "Besteht eine Beziehung zwischen Hautsinnesorganen und Haaren?," *ibid.*, **13**, nos. 11–12 (1893), 359–375; "Zur Kenntnis der Zirbel- und Parietalorgane. Forgesetze Studien," in *Abhandlungen hrsg. von der Senckenbergischen naturforschenden Gesellschaft*, pt. 3 (1896), 217–278; "Der reizleitende Theil des Nervengewebes," in *Archiv für Anatomie und Physiologie*, Anatomische Abt. (1897), 431–464; "Zirbel und Jacobson'sche Organe einiger Reptilien," in *Archiv für mikroskopische Anatomie und Entwicklungsmechanik*, **50** (1897), 385–418; and "Bemerkung zu den 'Leuchtorganen' der Selachier," in *Anatomischer Anzeiger*, **22**, nos. 14–15 (1902), 297–301.

II. SECONDARY LITERATURE. Obituary notices are O. Boettger, in *Zoologischer Beobachter*, **50**, no. 1 (1909), 31; R. von Hanstein, in *Naturwissenschaftliche Rundschau*, **23**, no. 27 (1908), 347–351; M. Nussbaum, in *Anatomischer Anzeiger*, **32**, nos. 19–20 (1908), 503–506; and in *Kölnische Zeitung*, no. 520 (14 May 1908); O. Schultze, in *Münchener medizinische Wochenschrift*, **55**, no. 18 (1908), 972–973; O. Taschenberg, in *Leopoldina*, pt. 45 (1909), 82–88; and O. Zacharias, in *Archiv für Hydrobiologie*, **4**, pt. 1 (1908), 77–82.

P. GLEES

L'HÉRITIER DE BRUTELLE, CHARLES LOUIS

(*b.* Paris, France, 15 June 1746; *d.* Paris, 16 August 1800), *botany.*

L'Héritier was the son of a relatively well-to-do Roman Catholic family belonging to the court circles in Paris. Few details are known of his youth and upbringing. In 1772 he was appointed to the fairly high position of superintendent of the waters and forests of the Paris region, a sinecure which he took seriously. The agronomic duties connected with this position awakened an interest in botany that he developed by independent study and by seeking contact with the botanists of the Jardin du Roi. In 1775 he was appointed counselor at the Cour des Aides in Paris, a position that made him a member of high society. He was a friend of Malesherbes and the group around the *Encyclopédie.*

In 1785 L'Héritier started his botanical career with the publication of the sumptuous *Stirpes novae*, describing many plants that had recently been introduced into Paris gardens from throughout the world. He corresponded with the leading British naturalist Sir Joseph Banks, and worked at his private herbarium in 1786–1787, in order to learn from the collections made by Joseph Dombey. The right to publish on these South American plants was claimed by the Spanish government, and the French government wished to stop L'Héritier from studying the plants, which he had received from Dombey himself. The trip to England was made in order to be free to continue his botanical work but ultimately resulted in the publication of another folio volume, *Sertum anglicum*, dealing with plants newly introduced in London and Kew rather than with the Dombey material.

In December 1787 L'Héritier returned to Paris, where he resumed publication of the *Stirpes novae* and published monographs on *Cornus* and *Geranium* (in his *Geraniologia*). During the Revolution, L'Héritier lost his official position as well as most of his private fortune. Obliged to work in straitened circumstances, he could no longer publish his botanical works at his own expense. On 16 August 1800 he was murdered near his house; the crime was never solved.

L'Héritier married Thérèse Valère Doré in 1775; she bore him five children. He was a member of the Académie des Sciences and retained his seat after its reorganization. L'Héritier discovered the talents of the great botanical artist Pierre-Joseph Redouté, whom he employed to make drawings for the *Stirpes novae* and for subsequent works.

BIBLIOGRAPHY

I. ORIGINAL WORKS. L'Héritier's major works are *Stirpes novae aut minus cognitae, quae descriptionibus et iconibus illustravit*, 9 fascs. (Paris, 1784–1785 [published 1785–1805]); *Sertum anglicum, seu plantae rariores quae in hortis juxta Londinium . . .*, 4 fascs. (Paris, 1788 [published 1789–1792]; 2nd ed., 1788 [published *ca.* 1805]; facs. repr. of 1st ed., Pittsburgh, 1963, with intros. by W. Blunt, J. S. L. Gilmour *et al.*, and F. A. Stafleu); *Cornus. Specimen botanicum sistens descriptiones et icones specierum Corni minus cognitarum* (Paris, 1788 [published

1789]); "On the Genus of *Calligonum*, Comprehending *Pterococcus* and *Pallasia*," in *Transactions of the Linnean Society of London*, **1** (1791), 177; "On the Genus of *Symplocos*, Comprehending *Hopea*, *Alstonia*, and *Ciponia*," *ibid.*, 174; and *Geraniologia, seu Erodii, Pelargonii, Monsoniae et Grieli historia iconibus illustrata* (Paris, 1787–1788 [published Apr. 1792]), 44 plates—the text, unpublished, is at the Conservatoire Botanique, Geneva.

II. SECONDARY LITERATURE. See J. Britten and B. B. Woodward, "Bibliographical Notes, XXXV,—L'Héritier's Botanical Works," in *Journal of Botany, British and Foreign*, **43** (1905), 267–273, 325–329; and F. A. Stafleu, "L'Héritier de Brutelle: The Man and His Work," in G. H. M. Lawrence, ed., *Charles Louis L'Héritier de Brutelle, Sertum anglicum 1788, Facsimile With Critical Studies and a Translation* (Pittsburgh, 1963), with biography, commentaries, and unpublished sources; and *Taxonomic Literature* (Utrecht, 1967), pp. 266–268, which lists secondary literature.

FRANS A. STAFLEU

L'HOSPITAL (L'HÔPITAL), GUILLAUME-FRANÇOIS-ANTOINE DE (Marquis de Sainte-Mesme, Comte d'Entremont) (*b.* Paris, France, 1661; *d.* Paris, 2 February 1704), *mathematics.*

The son of Anne-Alexandre de L'Hospital and of Elizabeth Gobelin, L'Hospital served for a time as a cavalry officer but resigned from the army because of nearsightedness. From that time onwards he devoted his energies entirely to mathematics. He married Marie-Charlotte de Romilley de La Chesnelaye, who bore him one son and three daughters.

L'Hospital's mathematical talents were recognized when he was still a boy. It is reported that when he was only fifteen years of age he solved, much to the surprise of his elders, a problem on the cycloid which had been put forward by Pascal. Later he contributed solutions to several problems posed by Jean (Johann) Bernoulli, among them the problem of the brachistochrone, which was solved at the same time by three others—Newton, Leibniz, and Jacques (Jakob) Bernoulli. His memory has survived in the name of the rule for finding the limiting value of a fraction whose numerator and denominator tend to zero. However, in his own time, and for several generations after his death, his fame was based on his book *Analyse des infiniment petits pour l'intelligence des lignes courbes* (1st ed., 1696, 2nd ed. 1715). Following the classical custom, the book starts with a set of definitions and axioms. Thus, a *variable* quantity is defined as one that increases or decreases continuously while a *constant* quantity remains the same while others change. The *difference* (differential) is defined as the infinitely small portion by which a variable quantity increases or decreases continuously. Of the two axioms, the first postulates that quantities which differ only by infinitely small amounts may be substituted for one another, while the second states that a curve may be thought of as a polygonal line with an infinite number of infinitely small sides such that the angle between adjacent lines determines the curvature of the curve. Following the axioms, the basic rules of the differential calculus are given and exemplified. The second chapter applies these rules to the determination of the tangent to a curve in a given point. While many examples are given, the approach is perfectly general, that is, it applies to arbitrary curves or to the relation between two arbitrary curves. The third chapter deals with maximum-minimum problems and includes examples drawn from mechanics and from geography. Next comes a treatment of points of inflection and of cusps. This involves the introduction of higher-order differentials, each supposed infinitely small compared to its predecessor. Later chapters deal with evolutes and with caustics. L'Hospital's rule is given in chapter 9.

The *Analyse des infiniment petits* was the first textbook of the differential calculus. The existence of several commentaries on it—one by Varignon (1725)—attests to its popularity. The question of its intellectual ownership has been much debated. Jean Bernoulli, who is known to have instructed L'Hospital in the calculus about 1691, complained after L'Hospital's death that he (Bernoulli) had not been given enough credit for his contributions. L'Hospital himself, in the introduction to his books, freely acknowledges his indebtedness to Leibniz and to the Bernoulli brothers. On the other hand, he states that he regards the foundations provided by him as his own idea, although they also have been credited by some to Jean Bernoulli. However, these foundations can be found, less explicitly, also in Leibniz, although Leibniz made it clear that he did not accept L'Hospital's Platonistic views on the reality of infinitely small and infinitely large quantities.

At his death L'Hospital left the completed manuscript of a second book, *Traité analytique des sections coniques et de leur usage pour la résolution des équations dans les problèmes tant déterminés qu'indéterminés.* It was published in 1720. L'Hospital had also planned to write a continuation to his *Analyse des infiniment petits* which would have dealt with the integral calculus, but he dropped this project in deference to Leibniz, who had let him know that he had similar intentions.

L'Hospital was a major figure in the early development of the calculus on the continent of Europe. He advanced its cause not only by his scientific works but also by his many contacts, including correspondence with Leibniz, with Jean Bernoulli, and with Huygens. Fontenelle tells us that it was he who introduced Huygens to the new calculus.

According to the testimony of his contemporaries, L'Hospital possessed a very attractive personality, being, among other things, modest and generous, two qualities which were not widespread among the mathematicians of his time.

BIBLIOGRAPHY

L'Hospital's principal works are *Analyse des infiniment petits pour l'intelligence des lignes courbes* (Paris, 1696; 2nd ed., 1715); and the posthumous *Traité analytique des sections coniques et de leur usage pour la résolution des équations dans les problèmes tant déterminés qu'indéterminés* (Paris, 1720).

On his life and work, see the *éloge* by Fontenelle in the *Histoires* of the Paris Academy of Sciences for 1704, p. 125, and in Fontenelle's *Oeuvres diverses*, III (The Hague, 1729); J. E. Montucla, *Histoire des mathématiques*, II (Paris, 1758), 396; O. J. Rebel, *Der Briefwechsel zwischen Johann (I.) Bernoulli und dem Marquis de l'Hospital* (Heidelberg, 1932); and P. Schafheitlin, ed., *Die Differentialrechnung von Johann Bernoulli aus den Jahren 1691–1692*, Ostwalds Klassiker der Exacten Wissenschaften no. 211 (Leipzig, 1924).

ABRAHAM ROBINSON

L'HUILLIER (or **LHUILIER**), **SIMON-ANTOINE-JEAN** (*b.* Geneva, Switzerland, 24 April 1750; *d.* Geneva, 28 March 1840), *mathematics.*

L'Huillier, the fourth child of Laurent L'Huillier and his second wife, Suzanne-Constance Matte, came from a family of jewelers and goldsmiths originally from Mâcon. In 1691 they became citizens of Geneva, where they had found refuge at the time of the revocation of the Edict of Nantes. Attracted to mathematics at an early age, L'Huillier refused a relative's offer to bequeath him a part of his fortune if the young man consented to follow an ecclesiastical career. After brilliant secondary studies he attended the mathematics courses given at the Calvin Academy by Louis Bertrand, a former student of Leonhard Euler. He also followed the physics courses of Georges-Louis Le Sage, his famous relative, who gave him much advice and encouragement. Through Le Sage he obtained a position as tutor in the Rilliet-Plantamour family, with whom he stayed for two years. At Le Sage's prompting, in 1773 he sent to the *Journal encyclopédique* a "Lettre en réponse aux objections élevées contre la gravitation newtonienne."

Le Sage had had as a student and then as a collaborator Christoph Friedrich Pfleiderer, who later taught mathematics at Tübingen. In 1766, on the recommendation of Le Sage, Pfleiderer was named professor of mathematics and physics at the military academy in Warsaw recently founded by King Stanislaus II. He was subsequently appointed to the commission in charge of preparing textbooks for use in Polish schools. In 1775 he sent the commission's plans for a textbook contest to Le Sage, who tried to persuade L'Huillier to submit a proposal for a physics text, but the latter preferred to compete in mathematics. He rapidly drew up an outline, sent it to Warsaw, and won the prize. The king sent his congratulations to the young author, and Prince Adam Czartoryski offered him a post as tutor to his son, also named Adam, at their residence in Puławy.

L'Huillier accepted and spent the best years of his life in Poland, from 1777 to 1788. His pedagogical duties did not prevent him from writing his mathematics course, which he put in finished form with the aid of Pfleiderer, and which was translated into Polish by the Abbé Andrzej Gawroński, the king's reader. L'Huillier had an unusually gifted pupil and proved to be an excellent teacher. He had numerous social obligations arising from his situation (including hunting parties), but he still found time to compose several memoirs and to compete in 1786 in the Berlin Academy's contest on the theory of mathematical infinity. The jury, headed by Lagrange, awarded him the prize.

L'Huillier returned home in 1789 and found his native country in a state of considerable agitation. Fearing revolutionary disturbances, he decided to stay with his friend Pfleiderer in Tübingen, where he remained until 1794. Although offered a professorship of mathematics at the University of Leiden in 1795, L'Huillier entered the competition for the post left vacant in Geneva by his former teacher Louis Bertrand. In 1795 he was appointed to the Geneva Academy (of which he soon became rector) and held the chair of mathematics without interruption until his retirement in 1823. Also in 1795 he married Marie Cartier, by whom he had one daughter and one son.

Whereas the Poles found L'Huillier distinctly puritanical, his fellow citizens of Geneva reproached him for his lack of austerity and his whimsicality, although the latter quality never went beyond putting geometric theorems into verse and writing ballads on the number three and on the square root of minus one. Toward the end of his career Charles-François Sturm was among his students.

L'Huillier was also involved in the political life of Geneva. He was a member of the Legislative

Council, over which he presided in 1796, and a member of the Representative Council from its creation. His scientific achievements earned him membership in the Polish Educational Society, corresponding memberships in the academies of Berlin, Göttingen, and St. Petersburg, and in the Royal Society, and an honorary professorship at the University of Leiden.

L'Huillier's extensive and varied scientific work bore the stamp of an original intellect even in its most elementary components; and while it did not possess the subtlety of Sturm's writings, it surpassed those of Bertrand in its vigor. L'Huillier's excellent textbooks on algebra and geometry were used for many years in Polish schools. His treatise in Latin on problems of maxima and minima greatly impressed the geometer Jacob Steiner half a century later. L'Huillier also considered the problem, widely discussed at the time, of the minimum amount of wax contained in honeycomb cells. While in Poland he sent articles to the Berlin Academy, as well as the prize-winning memoir of 1786: *Exposition élémentaire des principes des calculs supérieurs.* Printed at the Academy's expense, the memoir was later discussed at length by Montucla in his revised *Histoire des mathématiques* and was examined in 1966 by E. S. Shatunova. In this work, which L'Huillier sent to Berlin with the motto "Infinity is the abyss in which our thoughts vanish," he presented a pertinent critique of Fontenelle's conceptions and even of Euler's, and provided new insights into the notion of limit, its interpretation, and its use. Baron J. F. T. Maurice recognized the exemplary rigor of L'Huillier's argumentation, although he regretted, not unjustifiably, that it "was accompanied by long-winded passages that could have been avoided."

In 1796 L'Huillier sent to the Berlin Academy the algebraic solution of the generalized Pappus problem. Euler, Fuss, and Lexell had found a geometric solution in 1780, and Lagrange had discovered an algebraic solution for the case of the triangle in 1776. L'Huillier based his contribution on the method used by Lagrange. More remarkable, however, were the four articles on probabilities, written with Pierre Prévost, that L'Huillier published in the *Mémoires de l'Académie de Berlin* of 1796 and 1797. Commencing with the problem of an urn containing black and white balls that are withdrawn and not replaced, the authors sought to determine the composition of the contents of the urn from the balls drawn. In this type of question concerning the probabilities of causes, they turned to the works of Jakob Bernoulli, De Moivre, Bayes and Laplace, their goal being clearly to find a demonstration of the principle that Laplace stated as follows and that L'Huillier termed the etiological principle: "If an event can be produced by a number n of different causes, the probabilities of the existence of these causes taken from the event are among themselves as the probabilities of the event taken from these causes." The four articles are of considerable interest, and Isaac Todhunter mentions them in his *History of the Mathematical Theory of Probability.*

The two-volume *Éléments raisonnés d'algèbre* that L'Huillier wrote for his Geneva students in 1804 was really a sequel to his texts for Polish schools. The first volume, composed of eight chapters, was concerned solely with first- and second-degree equations. One chapter was devoted to an account of Diophantine analysis. Volume II (chapters 9–22) treated progressions, logarithms, and combinations and went as far as fourth-degree equations. A chapter on continued fractions was based on the works of Lagrange and of Legendre; another concerned the method of indeterminate coefficients. Questions of calculus were discussed in an appendix. The main value of these two volumes lay in the author's clear exposition and judicious selection of exercises, for some of which he furnished solutions.

L'Huillier's last major work appeared in 1809 in Paris and Geneva. Dedicated to his former pupil Adam Czartoryski, who was then minister of public education in Russia, it dealt with geometric loci in the plane (straight line and circle) and in space (sphere). Between 1810 and 1813 L'Huillier was an editor of the *Annales de mathématiques pures et appliquées* and wrote seven articles on plane and spherical geometry and the construction of polyhedrons.

BIBLIOGRAPHY

I. Original Works. L'Huillier's writings include *Éléments d'arithmétique et de géométrie* . . . (Warsaw, 1778), partly translated by Gawroński as *Geometrya dla szkoł narodowych* (Warsaw, 1780) and *Algiebra dla szkoł narodowych* (Warsaw, 1782); "Mémoire sur le minimum de cire des alvéoles des abeilles, et en particulier un minimum minimorum relatif à cette matière," in *Mémoires de l'Académie Royale des sciences et belles-lettres de Berlin* (1781), 277–300; *De relatione mutua capacitatis et terminorum figurarum seu de maximis et minimis* (Warsaw, 1782); "Théorème sur les solides plano-superficiels," in *Mémoires de l'Académie Royale des sciences et belles-lettres de Berlin* (1786–1787), 423–432; *Exposition élémentaire des principes des calculs supérieures,* . . . (Berlin, 1787); "Sur la décomposition en facteurs de la somme et de la différence de deux puissances à exposants quelconques de la base des logarithmes hyperboliques . . .," in *Mémoires de l'Académie Royale des sciences et belles-lettres de Berlin* (1788–1789), 326–368; *Polygonométrie et abrégé d'isopérimétrie élémentaire* (Geneva, 1789); *Examen du mode d'élection proposé à la Convention nationale de France et adopté à Genève* (Geneva, 1794); and *Principiorum calculi differentialis et integralis expositio elementaris* (Tübingen, 1795).

See also "Solution algébrique du problème suivant: A un cercle donné, inscrire un polygone dont les côtés passent par des points donnés," in *Mémoires de l'Académie Royale des sciences et belles-lettres de Berlin* (1796), 94–116; "Sur les probabilités," *ibid.*, Cl. de math., 117–142, written with Pierre Prévost; "Mémoire sur l'art d'estimer les probabilités des causes par les effets," *ibid.*, Cl. de phil. spéc., 3–24, written with Pierre Prévost; "Remarques sur l'utilité et l'étendue du principe par lequel on estime la probabilité des causes," *ibid.*, 25–41, written with Pierre Prévost; "Mémoire sur l'application du calcul des probabilités à la valeur du témoignage," *ibid.* (1797), Cl. de phil. spéc., 120–152, written with Pierre Prévost; *Précis d'arithmétique par demandes et réponses à l'usage des écoles primaires* (Geneva, 1797); *Éléments raisonnés d'algèbre publiés à l'usage des étudiants en philosophie*, 2 vols. (Geneva, 1804); *Éléments d'analyse géométrique et d'analyse algébrique* (Geneva–Paris, 1809); and "Analogies entre les triangles, rectangles, rectilignes et sphériques," in *Annales de mathématiques pures et appliquées*, **1** (1810–1811), 197–201.

II. Secondary Literature. The first articles on L'Huillier, which appeared during his lifetime, were Jean Sénebier, *Histoire littéraire de Genève*, III (Geneva, 1786), 216–217; and J.-M. Quérard, *France littéraire*, V (Paris, 1833), 295. Shortly after his death there appeared Auguste de La Rive, *Discours sur l'instruction publique* (Geneva, 1840); and "Discours du prof. de Candolle à la séance publique de la Société des arts du 13 août 1840," in *Procès-verbaux des séances annuelles de la Société pour l'avancement des arts*, **4** (1840), 10–15. Brief articles are in Haag, *France protestante*, VII (Paris, 1857), 85; and A. de Montet, *Dictionnaire biographique des Genevois et des Vaudois*, II (Lausanne, 1878), 66–68.

The best account of L'Huillier's life and work is Rudolf Wolf, *Biographien zur Kulturgeschichte der Schweiz*, I (Zurich, 1858), 401–422. See also L. Isely, *Histoire des sciences mathématiques dans la Suisse française* (Neuchâtel, 1901), pp. 160–167.

More recent publications are Samuel Dickstein, "Przyczynek do biografji Szymona Lhuiliera (1750–1840)," in *Kongres matematyków krajów słowiańskich. sprawozdanie* (Warsaw, 1930), pp. 111–118; Émile L'Huillier, *Notice généalogique sur la famille L'Huillier de Genève* (Geneva, 1957); Emanuel Rostworowski, "La Suisse et la Pologne au XVIIIe siècle," in *Échanges entre la Pologne et la Suisse du XIVe au XIXe siècle* (Geneva, 1965), pp. 182–185; and E. S. Shatunova, "Teoria grani Simona Luilera" ("Simon L'Huillier's Theory of Limits"), in *Istoriko-matematicheskie issledovaniya*, **17** (1966), 325–331.

A. P. Youschkevitch discusses L'Huillier's 1786 prize-winning memoir in an essay, "The Mathematical Theory of the Infinite," in Charles C. Gillispie, *Lazare Carnot, Savant* (Princeton, 1971), 156–158.

Pierre Speziali

LHWYD, EDWARD (*b.* Cardiganshire, Wales, 1660; *d.* Oxford, England, 30 June 1709), *paleontology, botany, philology*.

Lhwyd (he used this spelling for his signature; but his name was also variously rendered Lhuyd, Llwyd, Lloyd, and Luidius) was the natural son of Edward Lloyd of Llanforda, near Oswestry, and of Bridget Pryse of Gogerddan, Cardiganshire. In 1682 he entered Jesus College, Oxford, where he studied for five years. In order to increase his limited means, he soon became assistant to Robert Plot, professor of chemistry and first keeper of the Ashmolean Museum. This museum, opened in 1683, was founded on the collections of John Tradescant father and son, augmented by the donor, Elias Ashmole. It was through his connection with the museum and with Plot, at that time secretary of the Royal Society, that Lhwyd was able to establish important scientific contacts.

During visits to north Wales, Lhwyd collected plants around the hill mass of Snowdon. He was the first to record, in Edmund Gibson's edition of *Camden's Britannia* (1695), that the mountains of Britain have a distinctive alpine flora and fauna. Lhwyd compiled a list of plants from Snowdon, which was published by John Ray in his *Synopsis methodica Stirpium Britannicarum* (1690), and Ray referred to these records as "the greatest adornment" of his book.

Lhwyd also assisted Martin Lister with lists of Oxfordshire species of mollusks and fossils, and some of his specimens were used in Lister's *Historiae sive synopsis methodice Conchyliorum* (1685–1692). By the time he succeeded Plot as keeper of the Ashmolean Museum in 1691, Lhwyd's interest in formed stones (fossils) had superseded his botanical interests. In 1686 he put before the Oxford Philosophical Society a new catalog of the shells in the Ashmolean Museum; and during the next few years he continued to add to it, with a view to publication. The work eventually appeared in 1699 in a limited edition of only 120 copies. The cost was subscribed by some of his patrons and friends, including Isaac Newton, Hans Sloane, and Martin Lister. Written in Latin and entitled *Lithophylacii Britannici ichnographia*, it consisted of a catalog of 1,766 localized items arranged systematically and was the first illustrated catalog of a public collection of fossils to be published in England.

As an appendix to his list Lhwyd printed six letters to friends, also in Latin, dealing with geological subjects. The most important of these was to John Ray on the problem of the origin of "marine bodies and mineral leaves." Although Lhwyd's theory was completely erroneous, his arguments illustrate the flexibility of thought which existed on this subject during the late seventeenth century. For Lhwyd, who had viewed the fossil content of the rocks in depth in the sea cliffs of south Wales and in limestone caves, there were obvious difficulties in accepting the generally held belief that all fossils had been laid down at the time of the universal deluge; and in his letter he set out several cogent arguments against the idea of a single inundation.

Lhwyd hoped that the considerable depth at which fossils are found could be explained by the hypothesis he put forward. He suggested a sequence in which mists and vapors over the sea were impregnated with the "seed" of marine animals. These were raised and carried for considerable distances before they descended over the land in rain and fog. The "invisible animalcula" then penetrated deep into the earth and there germinated; and in this way complete replicas of sea organisms, or sometimes only parts of individuals, were reproduced in stone. Lhwyd also suggested that fossil plants, known to him only as resembling leaves of ferns and mosses which have minute "seed," were formed in the same manner. He claimed that this theory explained a number of features about fossils in a satisfactory manner—the presence in England of nautiluses and exotic shells which were no longer found in neighboring seas; the absence of birds and viviparous animals not found by Lhwyd as fossils; the varying and often quite large size of the forms, not usual in present oceans; and the variation in preservation from perfect replica to vague representation, which was thought to indicate degeneration with time.

It does not seem that Lhwyd made any further effort to defend or propound his theory, and he assured Ray in his original letter that "They who have no other aim than the search of Truth, are no ways concerned for the honour of their opinions: and for my part I have always been led thereunto by your example, so much the less an admirer of hypotheses, as I have been a lover of Natural History."

Ray's later letters to Lhwyd reiterated his own opinion that fossils were remains of once-living organisms and that the strata in which they are found were once sediments of land floods or sea inundations. He did, however, generously allow a degree of possibility of his friend's theory; and certainly the republication of Lhwyd's letter, in English, in the third (posthumous) edition of Ray's *Miscellaneous Discourses* (1713) served to bring it to the notice of a much wider readership.

In 1695 Lhwyd contributed notes on the southern counties of Wales to *Camden's Britannia*, and out of this there arose an invitation to undertake a work on the natural history of Wales. Eventually it was proposed to include all the Celtic countries and to cover natural history, geology, history, archaeology, and philology. In May 1697, Lhwyd began a great tour which took him through Wales, Ireland, part of Scotland, Cornwall, and across into Brittany. He collected or transcribed many Welsh and Gaelic manuscripts; and when he returned to Oxford in 1701, he intended to publish his researches in two volumes, the first on linguistic studies, the second on archaeology and natural history. The first volume of *Archaeologia Britannica* appeared in 1707 and contained the first comparative study of the Celtic languages and an Irish Gaelic dictionary. Thus, Lhwyd can be considered the founder of comparative Celtic philology, but owing to his early death the remaining branches of his researches were not published.

BIBLIOGRAPHY

I. Original Works. In 1945 R. T. Gunther published *Life and Letters of Edward Lhwyd* as vol. XIV of his Early Science in Oxford series and included those letters preserved in the Martin Lister and John Aubrey collections at the Bodleian Library, Oxford, in addition to others previously published. Correspondence addressed to Lhwyd is also in the Bodleian.

Lhwyd's MS collections were originally very extensive; and at his death they were offered for purchase to the University of Oxford and to Jesus College, but the offers were not accepted. They were then sold to Sir Thomas Sebright of Beechwood, Hertfordshire. The Irish portion of the Celtic MSS was presented to Trinity College, Dublin, by Sir John Sebright in 1786. The remainder was sold at Sotheby's, London, in 1807—see *Gentleman's Magazine*, **77** (1807), 419—but it is believed that most of these were destroyed shortly afterward in a fire at a bookbinder's workshop.

Besides *Lithophylacii Britannici ichnographia* (London, 1699) and the first volume of *Archaeologia Britannica* (Oxford, 1707), Lhwyd made numerous contributions to the *Philosophical Transactions of the Royal Society*.

II. Secondary Literature. A memoir of Lhwyd was included in N. Owen's *British Remains* (London, 1777). R. T. Gunther's book, cited above, is a comprehensive work respecting Lhwyd's correspondence and fossil collections. J. L. Campbell and D. Thomson, *Edward Lhuyd in the Scottish Highlands* (Oxford, 1963), is devoted to his Gaelic studies.

M. E. Jahn has written on the editions of Lhwyd's catalogue, including the pirated Leipzig edition, in *Journal of the Society for the Bibliography of Natural History*, **6**, pt. 2 (1972), 86–97; other notes by Jahn on Lhwyd are in **4**, pt. 5 (1966), 244–248, **6**, pt. 1 (1971), 61–62.

J. M. Edmonds

LIBAVIUS (or **LIBAU), ANDREAS** (*b.* Halle, Saxony, *ca.* 1560; *d.* Coburg, 25 July 1616), *chemistry, medicine, logic.*

Libavius was the son of a poor linen weaver, Johann Libau, who, in search of work, went from Harz to Halle, where he settled. Libavius attended the Gymnasium in Halle during the rectorship of Johannes Rivius the Younger. In 1578 he entered the University of Wittenberg and in 1579 moved to the University of Jena, where in 1581 he received the Ph.D. and the title of poet laureate. In the same year he became a teacher at Ilmenau. In 1586 he was appointed "Stadt- und Raths-Schulen Rector" in Coburg.

At the beginning of 1588 Libavius enrolled in the University of Basel, where he received the M.D. after presenting a thesis entitled *Theses de summo et generali in medendo scopo*. The Medical Faculty at Basel was still strongly influenced by Thomas Erastus, who had died not long before (1583). He was an opponent of Paracelsus, whose position Libavius represented to the end of his life. Libavius' bond with the University of Basel was strong, however; among his friends were the professors Johannes Stupanus, Felix Platter, Gaspard Bauhin, and Jacob Zwinger.

In 1588 Libavius became professor of history and poetry at the University of Jena. Three years later he moved to Rothenburg, where he became municipal physician. From 1592 he was also inspector of schools. Endless quarrels with the rector of the town school, Elias Ehinger, led Libavius to return to Coburg on 21 February 1607; there he was named rector of the newly founded Gymnasium Casimirianum Academicum. From its very beginning the school was regarded as a university by its founder, Duke John Casimir of Coburg; but because it represented orthodox Lutheranism, it failed to receive the necessary imperial charter from Rudolf II. At the opening Libavius delivered the oration *Declamatio de discendi modis*.

There is little information about Libavius' private life, although it is known that he had two sons, Michael and Andreas; the latter was a physician who practiced for a time with Martin Ruhland the Younger in Prague and later was a physician in Moravia, and then a teacher in Leszno, Poland. Libavius also had two daughters. One of them, Susanna, was the wife of the physician Pancratius Gallus; the second, whose name is not known, was the wife of Philip Walther Seidenbecker, professor at the Gymnasium in Coburg.

Some light is thrown on Libavius' life by his correspondence (not yet published) with Leonhard Dolde, a medical doctor in Nuremberg (142 letters from 1600–1611 are extant), and the published letters to Sigismund Schnitzer of Bamberg. As can be seen from these letters, although Libavius traveled very little, he maintained contacts with many scientists and read a great deal. He paid little attention to medical practice.

Libavius' activities were many-sided, and his teaching is noteworthy. There are several printed works in various fields—disputations and exhibit-day lectures—which on the title page bear the note "Sub praeside A. Libavii." Between 1599 and 1601 Libavius published four volumes of *Singularium*, probably his lectures in natural science. Both the *Singularium* and his *Schulordnung* have considerable interest for the history of education.

An orthodox Lutheran, Libavius opposed the Catholics, especially the Jesuits (for instance, in *Gretserus triumphatus* [1604]). Toward the end of his life he also became an opponent of Calvinism. He wrote his theological treatises under the anagram Basilius de Varna. As a poet he wrote *Poemata epica* (1602). As a philosopher and logician he was known as the author of *Quaestionum physicarum* (1591), *Dialectica* (1593), *Exercitiorum logicorum liber* (1595), *Dialogus logicus* (1595), and *Tetraemerum* (1596). From the contents of his books it appears that he was a follower of Aristotle and opposed Petrus Ramus and —especially—his two British disciples, William Temple, of Cambridge, and James Martin, of Oxford. His correspondence with Dolde shows that Libavius was also interested in medicine, pharmacology, botany, mineralogy, zoology, and, above all, chemistry.

A man of exceptional industry and perhaps overweening self-confidence, Libavius underestimated others; consequently he fell easily into conflicts and lost friends. His voluminous works brought him recognition but also often attracted criticism. Joseph Duchesne (Quercetanus), in *Ad veritatem hermeticae medicinae* (1605), wrote of him: "Andreas Libavius Hallensis Sax. medicus, doctor celeberrimus rerumque naturalium perscrutator fidissimus et diligentissimus, verae Chymiae defensor acerrimus, cuius doctissima scripta . . .." In spite of this Libavius quarreled with Quercetanus a few years later. He wasted the greater part of his life on fruitless polemics, which earned him many enemies.

In 1591 Georgius an und von Wald (Amwald), a physician in Augsburg, wrote an account of the "universal medicine" he had discovered, *Kurzer Bericht . . . wie das Panacea am Waldiana . . . anzuwenden sei.* The book is an interesting example of the printed advertising of the time; half of it consists of testimonial letters from princes, counts, doctors, and other eminent men, while Rudolf II's imperial imprimatur emphasized its significance. Libavius, however, claimed that Amwald's drug contained not gold but mercury and presented harsh criticism of Amwald, his panacea, and his book in four publications: *Neoparacelsica* (1594), *Tractatus duo physici* (1595), *Gegenbericht von der Panacea Amwaldina* (1596), and *Panacea Amwaldina victa* (1596). The import of these works was that Amwald's medicine was quackery and his methods of healing worthless. At the same time Libavius took the opportunity to criticize the works and activities of the Paracelsian physicians Johann Graman of Erfurt, to whom he also devoted a separate critique, *Antigramania* (1595), and Joseph Michelius of Lucca. It is possible that these works all followed from Libavius' ire at not being invited to the court of Rudolf II, a well-known Maecenas of alchemy.

In 1600 Libavius issued *Variarum controversarium libri duo*, which was, as it were, a summing up of all the previous polemics. In 1607 he stated his opposition to Nicolas Guibert, a French physician who denied the possibility of the transmutation of metals, in a treatise in which he declared that the transmutation of base metals into gold was possible and that the secret of the philosophers' stone was known to many alchemists.

Libavius was also involved in the conflict between, on one side, the French Calvinist-Paracelsists Joseph Duchesne and Israel Harvet—whom he supported—and, on the other, the Galenist Catholic professor of medicine of the University of Paris, Jean Riolan. To an aggressive pamphlet by Riolan Libavius replied with the 926-page *Alchymia triumphans* (1607), demonstrating point by point the ignorance of his adversary.

Petrus Palmarius also became involved in the polemic and was answered by Libavius in *Syntagma arcanorum* (1613). This whole polemic, which took on an international character, was a battle of Paracelsists, anti-Paracelsists, Galenists, and Hermetics; in a word, everyone against everyone. Each accused the other of using ineffectual medicines, of not understanding the meaning of the words used by Paracelsus, Galen, or even the mythical Hermes. Entering the fray, the Medical Faculty of the University of Paris formally condemned Paracelsian chemical remedies.

Until 1607 Libavius was on good terms with Oswald Crollius, Théodore Turquet de Mayerne, Johannes Hartmann, Bernard Penotus, and Quercetanus, who were Calvinist-Hermetics. He visited Hartmann in Marburg and corresponded with Turquet de Mayerne and Quercetanus; the latter made him a present of his new publication *Pharmacopea* (1607). Some time in the period from 1608 to 1610 his friendship with these men turned to hatred, but the cause is difficult to state. One might suppose that it was caused either by Libavius' increasing religious orthodoxy or by professional disappointment. Libavius had maintained friendly contacts with Landgrave Maurice of Hesse, corresponded with him, and dedicated the *Commentationum metallicorum libri* to him. He therefore counted on obtaining a position at the court of Maurice, known for his liking for alchemy, at the University of Marburg, or at the Collegium Adelphicum Mauricianum in Kassel. But when a "chair of chymiatry" was founded in 1609, Hartmann was named professor.

Libavius gave vent to his bitterness in a letter to Leonard Dolde of 21 December 1609, in which he wrote that the Marburg school (the chair held by Hartmann) is a "pugnarum mare," a sea of impostures. He manifested his open enmity to the Hermetics in *Syntagmatis arcanorum chymicorum* (1613) and *Examen philosophiae novae* (1615). Of his former friend Hartmann he wrote: "Your philosophy is nothing but pure dung, sown with nonsense, impostures, the obscurest puzzles and allegories."

Libavius' last polemic is contained in two pamphlets against the Rosicrucians: *Analysis confessionis Fraternitatis de Rosae Cruce* (1615) and *Wohlmeinendes Bedencken von der Fama und Confessio der Brüderschaft des Rosen Creuzes* (1616). He hated things to be unclear and unintelligible and was greatly irritated by the anonymous fellowship of the Rosicrucians, which was much discussed throughout Europe at this time.

Libavius' main value to the history of science resides in his extraordinarily voluminous alchemical works, which represent a compendium of the chemical knowledge of his times. His first work was *Rerum chemicarum epistolica forma . . . liber* (two volumes in 1595 and the third in 1599), chemical lectures in the form of letters addressed to well-known physicians, including J. Stupanus, F. Platter, M. Ruland, and J. Camerarius. These lectures contain a definition of chemistry—"Chimia est a mineralium elaboratio"—a clarification of several obscure alchemical terms and phrases, a critique of Paracelsus, notes and considerations on the philosophers' stone, and chemical processes and preparations.

Libavius' chief work was *Alchemia*, which, together with the separately published *Commentationum metallicorum libri* (1597), appeared in a shortened form in German as *Alchymistische Practic* (1603) and in Latin as *Praxis alchymiae* (1604)—and, significantly enlarged, as *Alchymia* (1606). This last edition, with commentary, is considered the greatest and most beautiful (because of the numerous illustrations) of all books on chemistry in the seventeenth century. It is a folio edition with more than 200 designs and pictures of various sorts of chemical glassware, vessels, apparatuses, and furnaces, as well as architectural plans for the building of a chemical laboratory, "domus chymici."

In one sense, the *Alchemia* was completed by *Syntagma selectorum* (1611), *Syntagma arcanorum tomus secundus* (1613), and *Appendix philosophiae novae* (1615). These works, however, are devoted mainly to polemics. Although the *Alchymia* and its "continuation" comprise more than 2,000 folio pages, Libavius did not consider this to be everything that could be written on the subject and explained himself by saying: "Uni homini impossibile est chymiam condere et absolvere."

The *Alchymia* is unusually clear and highly systematic. The same cannot be said of the commentaries and supplements, especially if we consider Libavius' deliberations on the philosophers' stone, its contents, and the transmutation of metals. Libavius scrupulously cites the more than 200 authors whose works he used. He divided alchemy into two parts: "encheria" and "chymia." Encheria was the knowledge of chemical procedure and included furnaces, ovens, and vessels. Chymia meant the knowledge of how to prepare substances. Independent of these are two further divisions of alchemy: "ars probandi," the analysis of minerals, metals, and mineral waters, and "theoretical alchemy," knowledge concerning the philosophers' stone. Ars probandi (also ars probatoria), or assaying, was divided into "scevasia" and "ergastia." Scevasia was a kind of encheria: the technique of preparing crucibles, fluxes, and acids, the use of balances and weights, and the knowledge of alchemical symbols (Libavius gave examples of alchemical ciphers). Ergastia (or doecimasia) included assaying techniques. Libavius devoted a great deal of space to the analysis of mineral waters, "judicio aquarum mineralium."

In all practical recipes Libavius' style is extremely clear, in marked contrast to the bombastic verbosity of Paracelsus. It becomes obscure in the sections and fragments dealing with the philosophers' stone or transmutation of metals, in which he stated that these secrets were possessed by, to name a few, Pico della Mirandola, Giambattista della Porta, Edward Kelley, Alexander Seton, and Michael Seindivogius (Sendivogius).

Libavius tried to penetrate and understand works on the philosophers' stone, and commented broadly in his own way. He declared, for example, that the mysterious alchemical substance called "azoth" was really what he obtained and called "liquor [or spiritus] sublimati" (stannic chloride). The name "spiritus fumans Libavii" did not appear until the eighteenth century. Another secret alchemical substance, "lac virginis" (maiden's milk), was, according to Libavius, the product of the reaction of a solution of litharge in vinegar and salt brine or a solution of alum (lead chloride or sulfate).

This sort of interpretation caused Libavius to be criticized by many alchemists: J. Tancke, P. Palmarius, H. Scheunemann, and H. Khunrath declared that Libavius knew nothing of alchemy. It must be admitted that he indeed did not understand the chief ideas of the Hermetics, such as that of the existence in the air of an invisible life-giving substance. Nevertheless, Libavius can be ranked as a first-rate chemist on the basis of those parts of his book which can be considered truly chemical.

Alchemy, according to Libavius, was also—apart from the previous definitions of it—the art of extracting perfect magistracies and pure essences valuable in medicine. The term *magisterium* (magistracy) has various meanings. Sometimes it is a chemical species extracted from a given mixture, while at others it means the procedure, mystery, or secret of the nature or composition of substances.

Libavius was an exponent of the iatrochemical trend in medicine, and consequently the application of chemicals is stressed in his writings. The first part of *Syntagmatis*, for instance, is entitled "Alchymia pharmaceutica" and contains recipes for such substances as "tabulae perlatae," "elixir catharicum," and "balsamus sulphuris." He thought that drugs could be prepared in two ways: pharmaceutically, by gently infusing or cooking, and alchemically, by ennobling substances through the nature of fire.

Libavius had in his home (at both Rothenburg and Coburg) a chemical laboratory in which, either alone or with assistants, he carried on chemical experiments. It is difficult, however, to determine which of the compounds of which he wrote he obtained himself and which he merely tested from formulas he had been given.

He gave many recipes or remedies popular at that time, including "thurpethum minerale," the basic sulfate of mercury; "hepar sulphuris," potassium sulfide or polysulfide; "aurum potabile," potable gold (perhaps a colloidal gold solution); some antimony preparations which had purgative, emetic, or sudorific effects; and "sal prunellae mineralis," the contemporary fever remedy. Libavius' "butyra" (butters) were substances with the consistency of butter: butyrum antimonii ($SbCl_3$), butyrum arsenici (K_3AsO_4).

Libavius repeated Cesalpino's view that lead increases its weight while being calcined, this being caused by condensation of the smoke. He believed that iron turns into copper if immersed in vitriol water, although this was denied by Ercker, from whom Libavius had taken a great deal. Libavius also described the methods of distilling mineral acids and alcohol and gave a recipe for obtaining "quinta essentia Saturni" (acetone) by dry distillation of "saccharum Saturni" (lead acetate). Refluxing spirit of wine with oil of vitriol he obtained "oleum dulce" (ether).

Libavius can be regarded as one of the founders of chemical analysis, even though he took almost all his information from the books of Agricola, Ercker, and M. Fachs. He paid special attention to the analysis of mineral waters, investigating those in the environs of Rothenburg (*Singularium*, 1601) and Coburg (*Tractatus medicus physicus . . . und Historia Casimirianischen Sawer Brunnen*, 1610).

He gave quantitative methods of determining gold and silver in alloys and the analytical reactions of iron in water with an infusion of bile, a darkening of the blue of copper vitriol solution with the addition of "spiritus urinae" (ammonia). He was aware that water can yield a volatile acid (carbon dioxide). He also knew that a solution of lead, silver, or copper salts darkens when it comes in contact with sulfur vapors (H_2S) and that a solution of nitrate of mercury in nitric acid dyes the skin red (Millon's reaction).

Although Libavius had no pupils of his own, his books were used by many adepts of chemistry throughout most of the seventeenth century.

BIBLIOGRAPHY

I. Original Works. Libavius' most important books are cited in the text with shortened titles.

Libavius gives a list of his treatises in *Syntagmatis arcanorum chymicorum tomus secundus* (Frankfurt, 1613), pp. ix, x, where he writes that he would like to publish his correspondence with his friends in 3 vols. From this collection of Libavius' letters, those addressed to Sigismund Schnitzer were edited by Johann Hornung as *Cista medica* (Nuremberg, 1626).

A large collection of Libavius' letters to Wald, J. Camerarius the Younger, and especially to L. Dolde is in the Universitätsbibliothek, Erlangen, MS 1284. Letters addressed to Landgrave Maurice of Hesse are in Marbachsche und Landesbibliothek in Kassel, MS fol. 19, and in Universitätsbibliothek, Basel, e.g., "Jacob Zwingers Korrespondenz."

The fullest—but not complete—bibliography of Libavius' works is in E. Pietsch, A. Kotowski, and F. Rex, *Die Alchemie des Andreas Libavius* (Weinheim, 1964). It erroneously states, however, that *Syntagmatis selectorum . . . tomus primus* (no. 42) appeared in 1615, instead of 1611. An additional imperfection of this bibliography is that it does not give the sizes of Libavius' books. From 1606 almost all were printed in folio, while the earlier ones were in quarto or octavo. Folio books were rather rare in those times, so the fact that Libavius' works were printed in folio indicates how well his books sold.

The Pietsch-Kotowski-Rex *Die Alchemie des Andreas Libavius* is a very good trans. of the *Alchemia* (1597) into modern German. A kind of supplement is formed by the "Bildteil," 197 designs of chemical vessels, ovens, and instruments taken from the *Commentatorium alchymiae* (1606). The editorial commentary on the whole is good.

II. Secondary Literature. See the following, listed chronologically: G. Ludwig, *Ehre des Hoch-Fürstlichen Casimiriani Academici in Coburg*, I (Coburg, 1725), 72, 77, 84, 139–159; II (Coburg, 1729), 244–256; J. F. Gmelin, *Geschichte der Chemie*, I (Göttingen, 1797), 345–351; H. Kopp, *Beiträge zur Geschichte der Chemie*, III (Brunswick, 1875), 145–150; J. Ferguson, *Bibliotheca chemica*, II (Glasgow, 1906), 31–34; J. Ottmann, "Erinnerung an Libavius in Rothenburg ob der Tauber," in *Verhandlungen Gesellschaft der Naturforscher*, **65** (1894), 79–84; C. Beck, *Festschrift zur Feier des dreihundertjährigen Bestehens des Gymnasium Casimirianum in Coburg 1605–1905* (Coburg, n.d.), pp. 76–88; A. Schnizlein, "Andreas Libavius, der Stadt Rothenburg Physicus von 1591–1607 und gekrönter Poet," in *Die Linde, Beilage zum fränkischen Anzeiger 22. Juni 1912* (Rothenburg, 1912), pp. 21–24; and "Andreas Libavius und seine Tätigkeit am Gymnasium zu Rothenburg," in *Beilage zum Jahresbericht des Kgl. Progymnasiums Rothenburg ob der Tauber für das Schuljahr 1913/14* (Rothenburg, 1914); E. Darmstaedter, in G. Bugge, *Das Buch der grossen Chemiker*, I (Weinheim, 1955), 107–124; L. Thorndike, "Libavius and Chemical Controversy," in his *History of Magic and Experimental Science*, VI (New York, 1951), 238–253; R. P. Multhauf, "Libavius and Beguin," in E. Farber, ed., *Great Chemists* (New York–London, 1961), pp. 65–79; J. R. Partington, *History of Chemistry*, II (London, 1961), 244–270; and E. Pietsch and A. Kotowski, *Die Alchemie des Andreas Libavius, ein Lehrbuch der Chemie aus dem Jahre 1597* (Weinheim, 1964), plus supp. with illustrations and F. Rex, "Kommentarteil," pp. 77–136.

Włodzimierz Hubicki

LICEAGA, EDUARDO (*b.* Guanajuato, Mexico, 13 October 1839; *d.* Mexico City, Mexico, 14 January 1920), *medicine, public health.*

The son of a physician, Liceaga received his primary education at Guanajuato. After attending college in Mexico City and Guanajuato, he entered the School of Medicine in Mexico City, where he received the M.D. in 1866. After graduation he taught physics and natural history at the College of San Ildefonso. In 1868 he joined the staff of the School of Medicine as associate professor of surgery and became professor the following year. Appointed dean in 1902, he reformed the curriculum and resigned in 1911. Liceaga served on the staff of the San Andrés and maternity hospitals in addition to conducting a private practice. A member of the National Academy of Medicine, he was its president in 1878–1879 and 1906–1907.

Liceaga is better known as a hygienist and is considered the father of modern public health in Mexico. His interest was awakened by Chandler, chairman of the New York State Board of Health, while on a visit to the United States in 1883. As chairman of the Council of Public Health, which post he held until his resignation in 1914, Liceaga helped to organize public health work throughout Mexico. In 1891 he drafted the first sanitary code. Read at a meeting of the American Public Health Association, it was characterized by Baker as more advanced than any in the United States. Under Liceaga's influence the code was amended in 1894 and a new version enacted in 1902.

In 1887–1888 Liceaga visited Europe, inspecting public health works—mostly water supply, sewage, and fumigation facilities—in Paris, Vienna, Brussels, and Berlin. At Vienna he attended the International Congress of Hygiene and Demography; and in Paris he obtained rabies virus from the Pasteur Institute through Roux, which enabled him to administer the first human vaccination in Mexico in 1888.

Liceaga's successful campaigns against plague, yellow fever, and malaria and his efforts to arouse public interest in the fight against tuberculosis brought him international recognition. He organized the first and second Mexican Medical Congresses (1876, 1878) and was president of both. At the 1883 National Congress of Hygiene he presented a full program for public health legislation and administration. At the first Pan-American Sanitary Conference (Washington, 1902) he was influential in establishing the Pan-American Sanitary Bureau. Liceaga was president of the third convention, held in Mexico City in 1907. He also was president of the American Public Health Association in 1895.

In 1941 the Mexican government created the Eduardo Liceaga Medal (three classes) as the highest national distinction awarded in the field of public health.

BIBLIOGRAPHY

I. Original Works. Liceaga's autobiography is *Mis recuerdos de otros tiempos* (Mexico City, 1949). His articles include "Defensa de los puertos y ciudades fronterizas de México contra la epidemia de cólera que invadió Europa," in *Documentos e informes de la 20a. reunión de la Asociación americana de salud pública* (Mexico City, 1894), 257–266; "El combate contra la tuberculosis," in *Gaceta médica de México*, 3rd ser., **2** (1907), 117–147; and "El combate contra la fiebre amarilla y la malaria en la República mexicana," in *Memorias IV Congreso médico nacional mexicano* (Mexico City, 1910), pp. 579–587.

II. Secondary Literature. See A. Pruneda, "El Dr. Liceaga miembro de la Academia nacional de Medicina," in *Gaceta médica de México*, **70** (1940), 68, 74; B. Bandera, "El Dr. E. Liceaga, profesor y director de la Escuela nacional de Medicina," *ibid.*, 74–78; and M. E. Bustamante, "El doctor Liceaga higienista," *ibid.*, 79–91.

Enrique Beltrán

LI CHIH, also called **LI YEH** (*b.* Ta-hsing [now Peking], China, 1192; *d.* Hopeh province, China, 1279), *mathematics.*

Li Chih (literary name, Jen-ch'ing; appellation, Ching-chai) has been described by George Sarton as one of the greatest mathematicians of his time and of his race. His father, Li Yü, served as an attaché under a Jurchen officer called by the Chinese name Hu Sha-hu. Li Yü later sent his family back to his home in Luan-ch'eng, Hopeh province. Li Chih went alone to the Yüan-shih district in the same province for his education.

In 1230 Li Chih went to Loyang to take the civil service examination; after he passed, he was appointed registrar in the district of Kao-ling, Shensi province. Before he reported for duty, however, he was made governor of Chün-chou (now Yü-hsien), Honan province. In 1232 the Mongols captured the city of Chün-chou, and Li Chih was forced to seek refuge in Shansi province. The kingdom of the Jurchen fell into the hands of the Mongols in 1234. From that time on, Li Chih devoted himself to serious study, frequently living in poverty. It was during this period that he wrote his most important mathematical work, the *Ts'e-yüan hai-ching* ("Sea Mirror of the Circle Measurements").

About 1251 Li Chih, finding himself in an improved financial position, returned to the Yüan-shih district of Hopeh province and settled near Feng-lung, a mountain in that district. Although he continued to lead the life of a scholarly recluse, he counted Chang Te-hui and Yuan Yü among his friends; the three of them became popularly known as "the Three Friends of [Feng-]Lung Mountain." In 1257 Kublai Khan sent for Li Chih and asked him about the government of the state, the selection and deployment of scholars for civil service, and the reasons for earthquakes. Li Chih completed another mathematical text, the

I-ku yen-tuan ("New Steps in Computation") in 1259. Kublai Khan ascended the throne in 1260 and the following year offered Li Chih a government post, which was politely declined with the plea of ill health and old age. In 1264 the Mongolian emperor set up the Han-lin Academy for the purpose of writing the official histories of the kingdoms of Liao and Jurchen, and the following year Li Chih was obliged to join it. After a few months he submitted his resignation, again pleading infirmity and old age. He returned to his home near Feng-lung, and many pupils came to study under him.

Li Chih changed his name to Li Yeh at some point in his life because he wished to avoid having the same name as the third T'ang emperor, whose dynastic title was T'ang Kao-tsung (650–683). This circumstance has given rise to some confusion as to whether Li Yeh was a misprint for Li Chih.

Besides the *Ts'e-yüan hai-ching* and the *I-ku yen-tuan*, Li Chih wrote several other works, including the *Fan shuo*, the *Ching-chai ku-chin chu*, the *Wen chi*, and the *Pi-shu ts'ung-hsiao*. Before his death Li Chih told his son, Li K'e-hsiu, to burn all his books except the *Ts'e-yüan hai-ching*, because he felt that it alone would be of use to future generations. We do not know to what extent his wishes were carried out; but the *I-ku yen-tuan* survived the fire, and the *Ching-chai ku-chin chu* has also come down to us. His other works are now lost, although some passages from the *Fan shuo* are quoted in the *Ching-chai ku-chin chu*. Only the *Ts'e-yüan hai-ching* and the *I-ku yen-tuan* will be further described here, since the other extant work has neither mathematical nor scientific interest.

Originally called the *Ts'e-yüan hai-ching hsi-ts'ao* and completed in 1248, the *Ts'e-yüan hai-ching* was not published until some thirty years later, at about the same time as the *I-ku yen-tuan*. From a preface written by Wang Te-yüan, it appears that there was a second edition in 1287. In the late eighteenth century the *Ts'e-yüan hai-ching* was included in the imperial encyclopedia, the *Ssu-k'u ch'üan-shu*. It came from a copy preserved in the private library of Li Huang (*d.* 1811). A handwritten copy of the book was made by Juan Yuan (1764–1849) from the version in the *Ssu-k'u ch'üan-shu*. Later Ting Chieh presented a handwritten fourteenth-century copy of the *Ts'e-yüan hai-ching hsi-ts'ao* with the seal of Sung Lien (1310–1381) to Juan Yuan. This is probably the copy that is now preserved in the Peking Library. At the request of Juan Yuan, the Ch'ing mathematician Li Jui (1768–1817) collated the two versions in 1797. This has become the most widely circulated edition of the *Ts'e-yüan hai-ching* that exists today. In 1798 Li Jui's version was incorporated in the *Chih-pu-tsu-chai ts'ung-shu* collection, and in 1875 in the *Pai-fu-t'ang suan-hsüeh ts'ung-shu* collection. The modern reproduction in the *Ts'ung-shu chi-ch'eng* series is based on the version in the *Chih-pu-tsu-chai ts'ung-shu* collection.

The *Ts'e-yüan hai-ching* was studied by many eighteenth- and nineteenth-century Chinese mathematicians, such as K'ung Kuang-shen (1752–1786) and Li Shan-lan (1811–1882). A detailed analysis of the work was made by Li Yen (1892–1963), but it has not yet been translated.

The *I-ku yen-tuan* was completed in 1259 and was published in 1282. It has been regarded as a later version of a previous mathematical text, the *I-ku-chi*, which is no longer extant. The *I-ku yen-tuan* is incorporated in both the *Chih-pu-tsu-chai ts'ung-shu* and the *Pai-fu-t'ang suan-hsüeh ts'ung-shu* collections. The modern reproduction of the *I-ku yen-tuan* in the *Ts'ung-shu chi-ch'eng* series is based on the version in the *Chih-pu-tsu-chai ts'ung-shu* collection. It has been translated into French by L. van Hée.

Li Chih introduced an algebraic process called the *t'ien yüan shu* ("method of the celestial elements" or "coefficient array method") for setting up equations to any degree. The *t'ien yüan shu* occupied a very important position in the history of mathematics in both China and Japan. From the early fourteenth century until algebra was brought to China from the West by the Jesuits, no one in China seemed to understand this method. Algebra enabled Chinese mathematicians of the eighteenth century, especially Mei Ku-ch'eng, to recognize the algebra of the *t'ien yüan shu* and the *ssu yüan shu* of Chu Shih-chieh despite their unfamiliar notation. Knowing that algebra originally entered Europe from the East, some enthusiastic Chinese scholars of that time went so far as to claim that the *t'ien yüan shu* had gone from China to the West and there became known as algebra. The *t'ien yüan shu* also exerted a profound influence in Japan, where it became known as the *tengenjutsu*. The seventeenth-century Japanese mathematician Seki Takakazu (also known as Seki Kōwa), for example, developed from the algebra of Li Chih and Chu Shih-chieh a formula for infinite expansion, which is now arrived at by means of the infinitesimal calculus.

Li Chih did not claim to be the originator of the *t'ien yüan shu*. From his *Ching-chai ku-chin chu* it appears that he had copied the method from a certain mathematician in Taiyuan (in modern Shansi) named P'eng Che (literary name, Yen-ts'ai), of whom we know little. Chu Shih-chieh wrote in the early fourteenth century that one of Li Chih's friends, the

LI CHIH

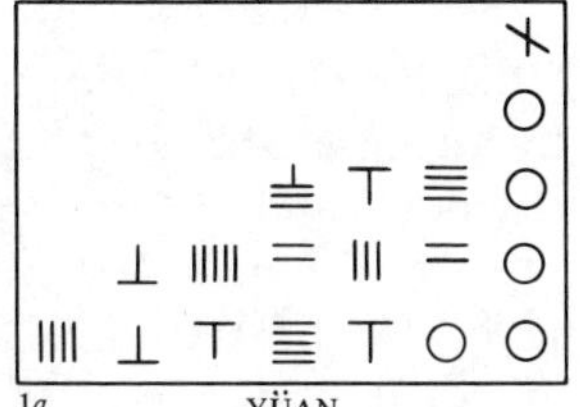

FIGURE 1*a*

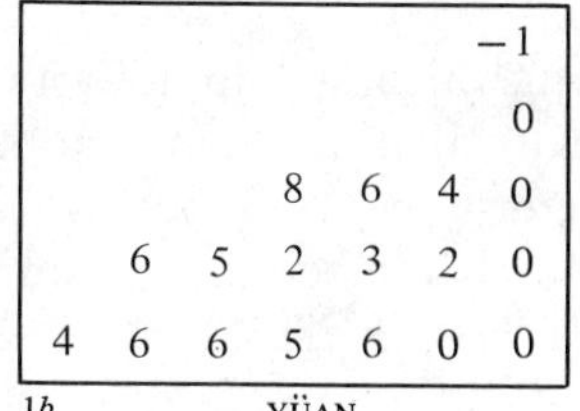

FIGURE 1*b*

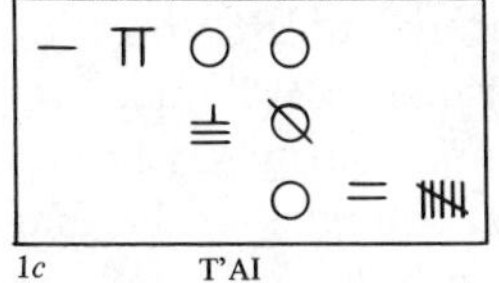

FIGURE 1*c*

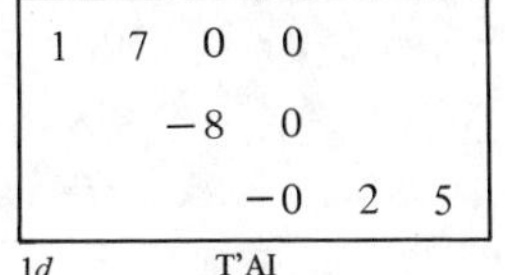

FIGURE 1*d*

famous poet Yuan Hao-wen, was also versed in the method of the celestial element. In the *t'ien yüan shu* method the absolute term is denoted by the character *t'ai* and the unknown by *yüan*, or element. An equation is arranged in a vertical column in which the term containing the unknown is set above the absolute term, the square of the unknown above the unknown, then the cube, and so on in increasing powers. Reciprocals or negative powers can also be placed in descending order after the absolute term. Thus the equation

$$-x^2 + 8640 + 652320x^{-1} + 4665600x^{-2} = 0$$

is represented on the countingboard as in Figure 1*a*, with its equivalent in Arabic numerals shown in Figure 1*b*. It is sufficient to indicate one position of the unknown and one of the absolute term by writing the words *yüan* and *t'ai*. It is curious that Li Chih reversed the process of expressing algebraic equations in his *I-ku yen-tuan* by writing the unknown below the absolute, the square below the unknown, and so on. For example, the equation $1700 - 80x - 0.25x^2 = 0$ is shown in Figure 1*c* and in Arabic numerals in Figure 1*d*. Li Chih's method was followed by later Chinese mathematicians.

Li Chih indicated negative quantities by drawing an oblique line over the final digit of the number concerned. He also used the zero symbol as his contemporary Ch'in Chiu-shao did, although there is no evidence that the two ever met or even heard of each other. It is likely that the zero symbol was used earlier in China, and it has even been suggested by Yen Tun-chieh that although the dot (*bindu*) was introduced from India in the eighth century and the circle for zero appeared in a magic square brought from the area of Islamic culture in the thirteenth century, the circle was nevertheless evolved independently in China from the square denoting zero sometime during the twelfth century.

Li Chih made use of numerical equations up to the sixth degree. He did not describe the procedure of solving such equations, which omission indicates that the method must have been well known in China during his time. This method must be similar to that rediscovered independently by Ruffini and Horner in the early nineteenth century, as described by Ch'in Chiu-shao. Li Chih stabilized the terminology used in connection with equations of higher degrees in the form

$$+ax^6 + bx^5 + cx^4 + dx^3 + ex^2 + fx + g = 0.$$

The absolute term g is called by the general term *shih*, or by the more specific terms *p'ing shih*, *fang shih*, *erh ch'eng fang shih*, *san ch'eng fang shih*, *shih ch'eng fang shih*, and *wu ch'eng fang shih* for linear equations, quadratic equations, cubic equations, quartic equations, and equations of the fifth and sixth degrees, respectively. The coefficient of the highest power of x (in this case, a) in the equation is called *yü*, *yü fa*, or *ch'ang fa*. The coefficient of the lowest power of x (in this case, f) is called *ts'ung* or *ts'ung fang*. All coefficients between the lowest and the highest powers are described by the word *lien*. For a cubic equation the coefficient of x^2 is known by the term *lien*. For a quartic equation the coefficient of x^2 is called *ti i lien*, that is, the first *lien*, and the coefficient of x^3 is *ti erh lien*, that is, the second *lien*. Hence, in the sixth-degree equation above we have e, *ti i lien* (first *lien*); d, *ti erh lien* (second *lien*); c, *ti san lien* (third *lien*); and b, *ti shih lien* (fourth *lien*).

All the above, except in the case of the absolute term, apply only to positive numbers. To denote negative numbers, Li Chih added either the word *i* or the word *hsü* before the terms applying to coefficients of the highest and lowest powers of x and before the word *lien*. He did not use different terms to distinguish between positive and negative absolute terms and, unlike Ch'in Chiu-shao, he did not make it a rule that the absolute term must be negative.

It is interesting to see how Li Chih handled the remainder in extracting a square root. An example is encountered in the equation

$$-22.5x^2 - 648x + 23002 = 0,$$

which occurred in the fortieth problem of his *I-ku yen-tuan*. He put $y = nx$, where $n = 22.5$, and transformed the equation into

$$-y^2 - 648y + 517545 = 0.$$

From the above, $y = 465$, and hence $x = 20^2/_3$. The same method was also used by Ch'in Chiu-shao.

The *Ts'e-yüan hai-ching* includes 170 problems dealing with various situations based on a circle inscribed in or circumscribing a right triangle. The same question is asked in all these problems and the same answer obtained. The book begins with a diagram showing a circle inscribed in a right triangle, ABC (Fig. 2). The square $CDEF$ circumscribing the circle lies along the base and height of $\triangle ABC$ and intersects the hypotenuse AB at G and H, from which perpendiculars GJ and HK are dropped on BC and AC, respectively. GJ and HK intersect at L. Through O, the center of the circle, $MNPOQ$ is drawn parallel to AC, meeting AB at M and BC at Q and cutting ED at N and HK at P. Also through O is drawn $RSTOU$, parallel to BC, meeting AB at R and AC at U and cutting EF at S and GJ at T. Finally, MV is drawn parallel to BC, meeting AC at V, and RW is drawn parallel to AC, meeting BC at W.

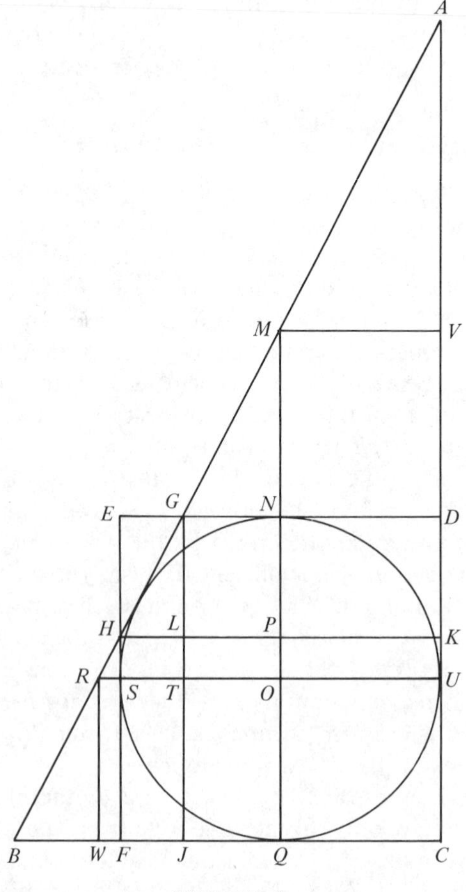

FIGURE 2

Special terms are then given for the three sides of each of the fifteen triangles in the diagram. These are followed by a list of relationships between the sides of some of these triangles and the circle. For example: the sides of $\triangle ABC$ and the diameter of the inscribed circle have the relationship $D = 2ab/(a + b + c)$; the sides of $\triangle ARU$ and the diameter of a circle with its center at one of the sides and touching the other two sides have the relationship $D = 2ar/(r + u)$; and the sides of $\triangle AGD$ and the diameter of an escribed circle touching the side ED and the sides AG and AD produced have the relationship $D = 2ag/(g + d - a)$. Similarly, for $\triangle MBQ$, $D = 2mb/(m + q)$; for $\triangle HBF$, $D = shb/(h + f - b)$; for $\triangle MRO$; $D = 2mr/o$; for $\triangle GHE$, $D = 2hg/(h + g - e)$; for $\triangle MGN$, $D = 2mg/(n - m)$; and for $\triangle HRS$, $D = 2hr/(s - r)$.

All the above are given in chapter 1 of the book. In the subsequent chapters Li Chih showed how these results can be applied to various cases. For example, the second problem in chapter 2 says:

> Two persons, *A* and *B*, start from the western gate [of a circular city wall]. *B* [first] walks a distance of 256 *pu* eastward. Then *A* walks a distance of 480 *pu* south before he can see *B*. Find [the diameter of the wall] as before.

The equation for $\triangle MBQ$ was then applied directly to give the diameter of the circular city wall.

Li Chih showed how to solve a similar problem by the use of a cubic equation. The fourth problem in chapter 3 says:

> *A* leaves the western gate [of a circular city wall] and walks south for 480 *pu*. *B* leaves the eastern gate and walks straight ahead a distance of 16 *pu*, when he just begins to see *A*. Find [the diameter of the city wall] as before.

Here Li Chih found x, the diameter of the city wall, by solving the cubic equation

$$x^3 + cx^2 - 4cb^2 = 0,$$

LI CHIH

where $c = 16$ *pu* and $b = 480$ *pu*. This obviously came from the quartic equation

$$x^4 + 2cx^3 + c^2x^2 - 4cb^2x - 4c^2b^2 = 0,$$

which can be derived directly from the equation for $\triangle MBQ$. In doing this Li Chih had discarded the factor $(x + c)$, knowing that the answer $x = -c$ was inadmissible. It is interesting to compare this with the tenth-degree equation used by Ch'in Chiu-shao for the same purpose.

All the 170 problems in the *Ts'e-yüan hai-ching* have been studied by Li Yen. To illustrate Li Chih's method of solving these problems, we shall follow step by step the working of problem 18 in chapter 11, which says:

> 135 *pu* directly out of the southern gate [of a circular city wall] is a tree. If one walks 15 *pu* out of the northern city gate and then turns east for a distance of 208 *pu*, the tree becomes visible. Find [the diameter of the city wall] as before. [The answer says 240 *pu* as before.]

Using modern conventions but following the traditional Chinese method of indicating the cardinal points in such a way that south is at the top, north at the bottom, east to the left, and west to the right, the problem is as illustrated in Figure 3.

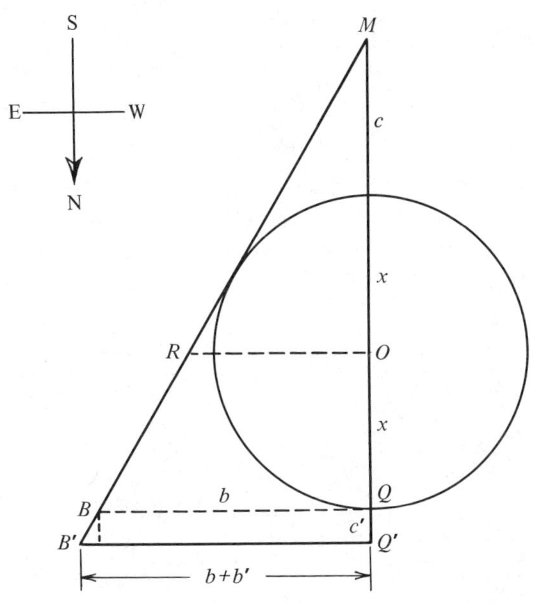

FIGURE 3

First, the method is given:

> Take the product of the distance to the east and that to the south [that is, $c(b + b')$], square it, and make it the *shih* [that is, $c^2(b + b')^2$ is taken as the absolute term]. Square the distance to the east, multiply it by the distance to the south, and double it to form the *ts'ung* [that is, the coefficient of x, the radius, is $2c(b + b')^2$]. Put aside the square of the distance to the east [that is, $(b + b')^2$]; add together the distance to the south and north, subtract the sum from the distance to the east, square the result, and subtract this from the distance to the east [that is, $(b + b')^2 - \{(b + b') - (c + c')\}^2$]; put aside this value. Again add together the distances to the south and north, multiply the sum first by the distance to the east and then by 2 [that is, $2(b + b')(c + c')$]; subtracting from this the amount just set aside gives the *tai i i lien* [that is, the coefficient of x^2 is $-2\{(b + b')(c + c')\} - [(b + b')^2 - \{(b + b') - (c + c')\}^2]$]. Multiply the distance to the east by 4 and put this aside [that is, $4(b + b')$]; add the distances to the south and north, subtract the sum from the eastward distance and multiply by 4 [that is, $4\{(b + b') + (c + c')\}$]; subtracting this result from the amount put aside gives the *tai erh i lien* [that is, the coefficient of x^3 is $-4(b + b') - 4\{(b + b') + (c + c')\}$]. Take 4 times the *hsü yü* [that is, $-4x^4$]. Solving the quartic equation gives the radius.

The procedure is as follows:

Set up one celestial element to represent the radius. (This is the *kao kou*.) Adding to this the distance to the south gives Figure 4*a* (below), which represents the

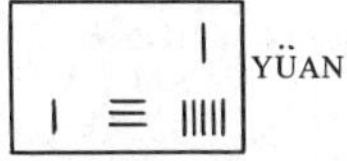

FIGURE 4*a*

kao hsien (that is, the vertical side *MO* in *MRO* in Figure 3 is $x + 135$). Put down the value for the *ta kou* (base of $MB'Q'$) of 208 and multiply it by the *kao hsien* $(x + 135)$; the result is $208x + 28080$ (Figure 4*b*).

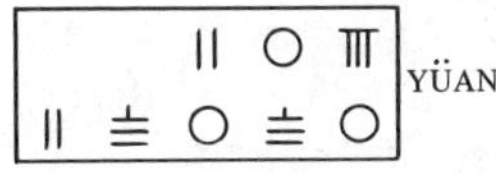

FIGURE 4*b*

Dividing this by the *kao kou* gives $208 + 28080x^{-1}$ (Figure 4*c*).

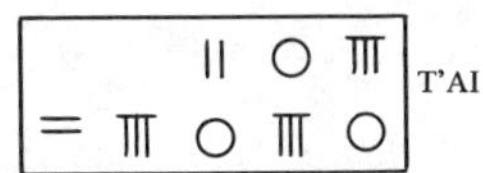

FIGURE 4*c*

LI CHIH

This is the *ta hsien* (the hypotenuse MB' of $\triangle MB'Q'$), and squaring this yields $43264 + 11681280x^{-1} + 788486400x^{-2}$ (Figure 4*d*).

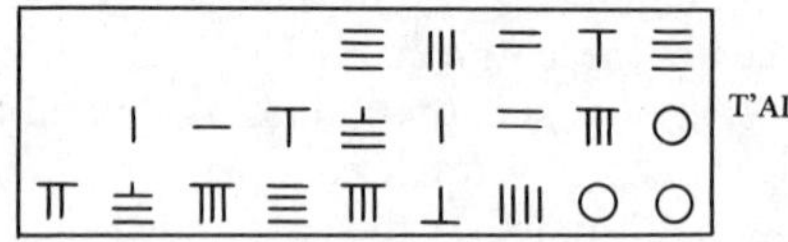

FIGURE 4*d*

The above is temporarily set aside. Take 2 as the celestial element and add this to the sum of the distances to the south and north, giving $2x + 150$ (Figure 4*e*),

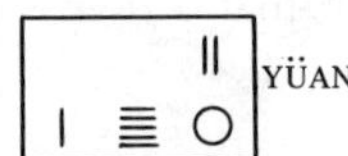

FIGURE 4*e*

or the *ta ku* (the vertical side MQ' of $\triangle MB'Q'$). Subtracting from this the value of the *ta kou*, 208, $2x - 58$ is obtained (Figure 4*f*).

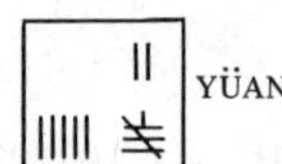

FIGURE 4*f*

This is known as the *chiao*. Squaring it gives the *chiao mi*, $4x^2 - 232x + 3364$, shown in Figure 4*g*.

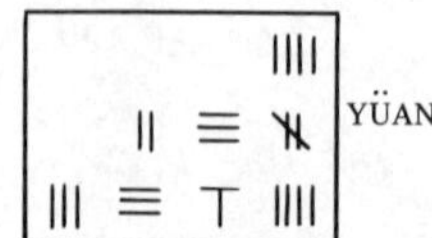

FIGURE 4*g*

Subtracting the *chiao mi* from the quantity set aside (Figure 4*d*) yields $-4x^2 + 232x + 39900 + 11681280x^{-1} + 788486400x^{-2}$ (Figure 4*h*).

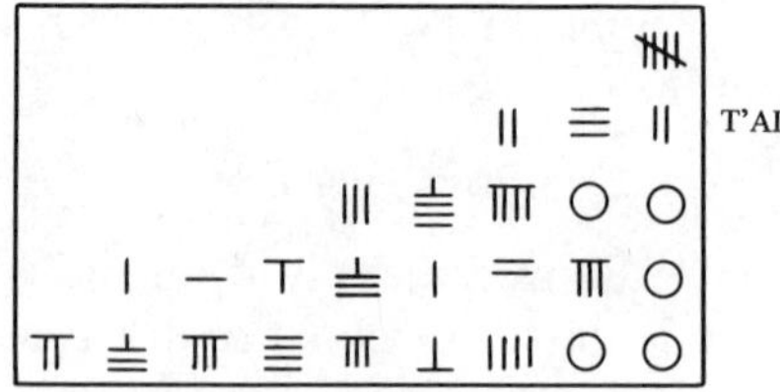

FIGURE 4*h*

This value, known as *erh-chih-chi*, is set aside at the lefthand side of the countingboard. Next multiply the *ta ku* (MQ') by the *ta kou* ($B'Q'$), which yields $416x + 31200$ (Figure 4*i*).

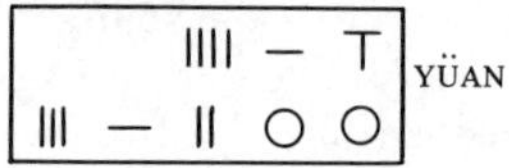

FIGURE 4*i*

This is the *chih chi*, which, when doubled, gives $832x + 62400$ (Figure 4*j*).

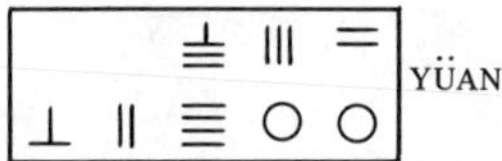

FIGURE 4*j*

This value is the same as that set aside at the left (that is, the value represented by Figure 4*h*). Equating the two values, one obtains $-4x^4 - 600x^3 - 22500x^2 + 11681280x + 788486400 = 0$ (Figure 4*k*).

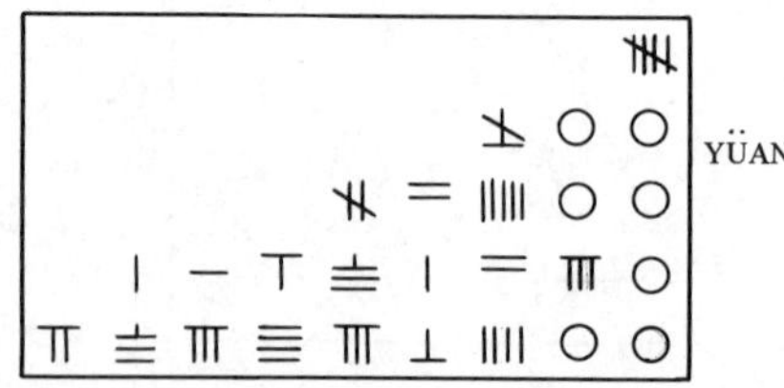

FIGURE 4*k*

Note that the position of the celestial element is no longer indicated here.

Solving this as a quartic equation, one obtains 120 *pu*, the radius of the circular city wall, which corresponds with the required answer.

Although written much later than the *Ts'e-yüan*

LI CHIH

hai-ching, the *I-ku yen-tuan* is considerably simpler in its contents. It is thought that Li Chih took this opportunity to explain the *t'ien yüan shu* method in a less complicated manner after finding his first book too difficult for people to understand. This second mathematical treatise has also been regarded as a later version of another work, *I-ku chi*, published between 1078 and 1224 and no longer extant. According to a preface in Chu Shih-chieh's *Ssu-yüan yu-chien*, the *I-ku chi* was written by a certain Chiang Chou of P'ing-yang. Out of a total of sixty-four problems in the *I-ku yen-tuan*, twenty-one are referred to as the "old method" (*chiu shu*), which presumably means the *I-ku chi*. Sixteen of these twenty-one problems deal with the quadratic equation

$$ax^2 + bx - c = 0,$$

where $a > 0$ or $a < 0$, $b > 0$ and $c > 0$. When $c > 0$ and $b > 0$, they are called by the terms *shih* and *ts'ung*, respectively. When $a > 0$, it is known as *lien* instead of the more general term *yü*, and when $a < 0$, it is also known as *lien*, but is followed by the words *chien ts'ung*.

Divided into three chapters, the *I-ku yen-tuan* deals with the combination of a circle and a square or, in a few cases, a circle and a rectangle. A full translation of the first problem in chapter 1 is given below:

> A square farm with a circular pool of water in the center has an area 13 *mou* and 7 1/2 tenths of a *mou* [that is, 13.75 *mou*]. The pool is 20 *pu* from the edge [1 *mou* = 240 square *pu*]. Find the side of the square and the diameter of the pool.
>
> Answer: Side of square = 60 *pu*, diameter of pool = 20 *pu*.
>
> Method: Put down one [counting rod] as the celestial element to represent the diameter of the pool. By adding twice the distance from the edge of the pool to the side of the farm, the side of the square farm is given by $x + 40$ [Figure 5*a*].
>
> 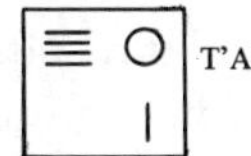
>
>
> FIGURE 5*a*
>
> The square of the side gives the area of the farm and the circular pool. That is, the total area is given by $x^2 + 80x + 1600$ [Figure 5*b*].
>
> Again, put down one [counting rod] as the celestial element to denote the diameter of the pool. Squaring the diameter, and multiplying the result by 3 [the ancient approximate value of π], then dividing the
>
> 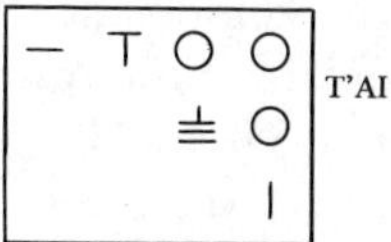
>
>
> FIGURE 5*b*
>
> result by 4, yields the area of the pool: $0.75x^2$ [Figure 5*c*].
>
> 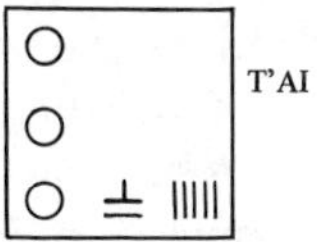
>
>
> FIGURE 5*c*
>
> Subtracting the area of the pool from the total area gives $0.25x^2 + 80x + 1600$ [Figure 5*d*].
>
> 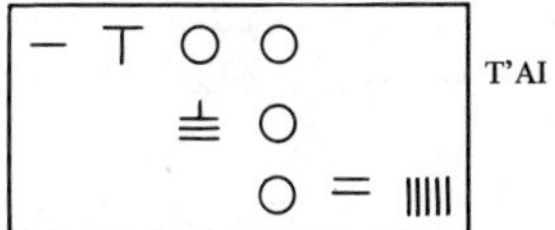
>
>
> FIGURE 5*d*
>
> The given area is 3,300 [square] *pu*. Equating this with the above yields $-0.25x^2 - 80x + 1700 = 0$ [Figure 5*e*].
>
> 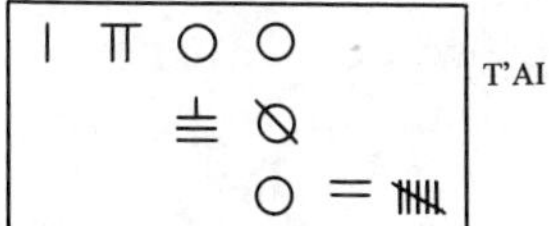
>
>
> FIGURE 5*e*
>
> Applying the method of solving quadratic equations shows the diameter of the pool to be 20 *pu*. If the distance from the side of the square to the edge of the pool is doubled and added to the diameter of the pool, the side of the farm is found to be 60 *pu*.

The first ten problems in the *I-ku yen-tuan* deal with a circle in the center of a square, each with different given parameters; the next ten problems are concerned with a square inside a circle. Problems 21 and 22, the last two problems in chapter 1, are concerned only with squares. In chapter 2, problems 23–29, we have the combination of a square with a circumscribed circle, while problem 30 gives the combination of two circles. Problem 31 concerns a rectangle with a circle in the center, and problems 32–37 give a circle with a rectangle at its center. In problem 38 two rectangles are given. Problems 39–42 treat various cases of a circle inside a rectangle. Problem 43, the first in chapter 3, deals with the three different values of π: the ancient value $\pi = 3$, the "close" value $\pi = 22/7$, and Liu Hui's value $\pi = 3.14$. Problem 44 is concerned with a trapezium, problem 45 with a square inside another square, problem 46 with a circle set outside but along the extended diagonal of a square, problem 47 with a square within a rectangle, and problem 48 with a rectangle within a square. In problems 49–52 a square is placed in the center of a larger square so that the diagonal of one is perpendicular to two sides of the other. In problems 53 and 54 the central square is replaced by a rectangle. Problems 55 and 56 are concerned with the annulus, and problems 57 and 58 with a rectangle inside a circle. In problem 59 a square encloses a circle, which in turn encloses another square at the center; and in problem 60 a circle encloses a square, which in turn encloses a circle at the center. Problem 62 concerns a square placed diagonally at a corner of another square. Problem 63 concerns a circle and two squares with another circle enclosed in one of them. The last problem, 64, has an annulus enclosed by a larger square.

Li Chih and Ch'in Chiu-shao were contemporaries, but they never mentioned each other in their writings. Li Chih lived in the north and Ch'in Chiu-shao in the south during the time when China was ruled in the south by the Sung dynasty and in the north first by the Jurchen Tartars and later by the Mongols. It is very likely that the two never even heard of each other. The terminology they used for equations of higher degree is similar but not identical. They also employed the so-called celestial element in different ways. Li Chih used it to denote the unknown quantity; but to Ch'in Chiu-shao the celestial element was a known number, and he never used the term in connection with his numerical equations. Ch'in Chiu-shao went into great detail in explaining the process of root extraction of numerical equations, but he did not describe how such equations were constructed by algebraic considerations from the given data in the problems. On the contrary, Li Chih concentrated on the method of setting out such equations algebraically without explaining the process of solving them. Thus Li Chih was indeed, as George Sarton says, essentially an algebraist.

BIBLIOGRAPHY

Works that may be consulted for further information on Li Chih and his writings are Ch'ien Pao-tsung, *Chung-kuo suan hsüeh-shih* ["History of Chinese Mathematics"] (Peking, 1932), pp. 116–124; Ch'ien Pao-tsung *et al.*, *Sung Yuan shu-hsüeh-shih lun-wen-chi* ["Collection of Essays on Chinese Mathematics in the Periods of Sung and Yuan"] (Peking, 1966), pp. 104–148; L. van Heé, "Li Yeh, mathématicien chinois du XIIIe siècle," in *T'oung Pao*, **14** (1913), pp. 537–568; Juan Yuan, *Ch'ou-jen chuan* ["Biographies of Mathematicians and Astronomers"] (Shanghai, 1935); Li Yen, *Chung-kuo shu-hsüeh ta-kang* ["Outline of Chinese Mathematics"], I (Shanghai, 1931), pp. 141–156; *Chung-kuo suan-hsüeh-shih* ["History of Chinese Mathematics"] (Shanghai, 1937), pp. 99–100; and "Chung-suan-shih lun-ts'ung," in *Gesammelte Abhandlungen über die Geschichte der chinesischen Mathematik*, **4**, no. 1 (1947), 15–251; Li Yen and Tu Shih-jan, *Chung-kuo ku-tai shu-hsüeh chien-shih* ["Concise History of Ancient Chinese Mathematics"], II (Peking, 1964), pp. 145, 147; Yoshio Mikami, *The Development of Mathematics in China and Japan* (Leipzig, 1913), pp. 79–84; Joseph Needham, *Science and Civilisation in China*, III (Cambridge, 1959), esp. pp. 40–41; George Sarton, *Introduction to the History of Science*, 3 vols. (Baltimore, 1927–1947), esp. vol. 2, pp. 627–628; Sung Lien *et al.*, *Yuan shih* ["Official History of the Yuan Dynasty"] (*ca.* 1370), ch. 160; and Alexander Wylie, *Notes on Chinese Literature* (Shanghai, 1902), p. 116.

HO PENG-YOKE

LICHTENBERG, GEORG CHRISTOPH (*b.* Oberramstadt, near Darmstadt, Germany, 1 July 1742; *d.* Göttingen, Germany, 24 February 1799), *physics.*

Lichtenberg was the seventeenth child—the fifth to survive—of a Protestant pastor. From his father he received his early schooling, including mathematics and natural science, for which subjects he developed an early predilection. A permanent spinal deformity in his childhood perhaps enhanced his propensity for scholarly work. Upon graduation from the secondary school at Darmstadt, Lichtenberg was accorded the patronage of his sovereign, Ludwig VIII, the duke of Hesse-Darmstadt, and he continued his studies at the University of Göttingen.

At the university Lichtenberg studied a wide range of subjects, particularly literature under Christian Gottlob Heyne, history under Johann Christoph Gatterer, and natural sciences under the witty Abraham Gotthelf Kästner. He studied avidly and with such thoroughness that he frequently found himself digressing into cognate fields. As he himself put it: ". . . I have covered the path which leads toward science like a dog accompanying his master on a walk . . .. I have covered it over and over again in all directions . . ..” As a result he became the leading German expert in a number of scientific fields, including geodesy, geophysics, meteorology, astronomy, chemistry, statistics, and geometry, in addition to his foremost field and prime interest—experimental physics. To all these areas he contributed respectably for his time, gaining the admiration and friendship of such contemporaries as Volta, F. W. Herschel, Kant, Goethe, Humboldt, and George III of England, with whom he became well acquainted during one of his visits to that country. Lichtenberg was appointed *professor extraordinarius* at the University of Göttingen in 1769 and was made *professor ordinarius* in 1775. He was given the title of royal British privy councillor in 1788 and in 1793 was elected to membership in the Royal Society (London) and in 1795 to membership in the Petersburg Academy of Sciences.

In geodesy Lichtenberg carried out a precise determination of the geodetic coordinates of Hannover, Stade, and Osnabrück. These measurements were performed at the request of George III (who, besides being the king of England, was also elector of Hannover) for purposes of military cartography, and also to help verify the concept advanced by Christiaan Huygens and Isaac Newton that the earth is an oblate spheroid.

Lichtenberg was particularly interested in volcanology. Among his writings on the subject is a calculation of the volume of lava ejected from Vesuvius during its eruption of 1784.

Also concerned with meteorology, Lichtenberg in 1780 was the first to erect in Germany a correct version of Benjamin Franklin's lightning rod. (A year earlier a lightning rod had been installed at Hamburg by the physician J. A. H. Reimarus, but without the essential connection between rod and ground.) In 1796 Lichtenberg wrote a brilliant monograph in defense of Jean André Deluc's theory of rain formation.

Along with Joseph Priestley and Carl Wilhelm Scheele, Lichtenberg was one of the last notable holdouts against the “French and new chemistry” of Lavoisier. Convinced in the end that Lavoisier was right, Lichtenberg capitulated by admitting that the new chemistry was a “magnificent structure.” A fusible metal of 50 percent bismuth, 30 percent lead, and 20 percent tin, having a melting point of 91.6° C., is known as Lichtenberg's alloy.

In mathematics Lichtenberg attempted to clear up the controversy between Daniel Bernoulli and d'Alembert regarding the probabilities in the “St. Petersburg problem.”[1] He sided with the former but admitted that an element of paradox remained in the solution. It was not until 1928 that an acceptable resolution of this paradox was suggested by Thornton C. Fry. This resolution was formalized in 1945 by William Feller, who cleared up the question of what constitutes a “fair” game and showed the Petersburg game to be “fair” in the classical sense.

Lichtenberg edited and published works of the great German astronomer Johann Tobias Mayer, the founder of the astronomical observatory of Göttingen. He also prepared for engraving and published Mayer's detailed map of the moon, which was highly appreciated by contemporary astronomers. In 1795 he published a biography of Copernicus. Himself an active observer, Lichtenberg sighted and described a comet, studied the fall of meteorites, and observed the transit of Venus on 19 June 1769. In 1807 the astronomer Johann Hieronymus Schröter gave to a feature on the moon the “unforgettable name of the great naturalist Lichtenberg.” Later the selenographer J. H. Mädler reassigned the name of Lichtenberg to a much more prominent feature—a first-order ring plain north of the Ocean of Storms (−67° 5′3″ long., +31° 25′20″ lat.).[2]

Yet it was in physics that Lichtenberg produced his greatest scientific achievements. First and foremost he was a teacher. The first German university chair of experimental physics was established for him. As professor at the University of Göttingen, he was enormously popular with students, and his lecture-demonstrations attracted an extremely large number of auditors. Along with the scrupulous presentation of facts, Lichtenberg offered his students a view of physics which would not be out of tune with the attitudes of the twentieth century. He combined bold imagining with radical scientific skepticism and left a legacy of maxims, many of which are as valid today as they were in his time. He wrote: “Almost everything in physics must be investigated anew, even the best-known things, because it is precisely here that one least suspects something new or incorrect.” Accordingly, he was among the first to question the validity of the postulates of Euclidean geometry, in particular the postulate that only one straight line can pass through two points. He questioned the usefulness of the concept of ether because of the absence of any measurable effects that could be attributed to it. At the same time he recognized the importance of “experimenting with ideas,” provided they are based on fact and the resulting theory is verifiable experimentally. He was contemptuous of dogmatists and scornful of those who confuse “facts” and “dreams.”

Lichtenberg saw as the purpose of his teaching the “coherent exposition of physical relationships as preparation for a *future* science of nature.” He used hypotheses very much as we use “models” in physics of the twentieth century. He wrote, "I see such hypotheses in physics as nothing else but convenient *pictures* that facilitate the conception of the whole." The interconnectedness of things, the wholeness of nature, was a postulate that guided all his work. In his quest for the "conception of the whole" Lichtenberg expressed some ideas which must have sounded rather farfetched to his contemporaries. For example, commenting on the controversy concerning the Huygens-Euler undulatory theory and the Newton-Kant corpuscular theory of light, Lichtenberg wrote, "Wouldn't one perhaps accomplish most by unifying the two theories . . . considering the limitation of our knowledge; both deserve respect, and both may indeed be right."[3] In addition to his penetrating philosophical insight, Lichtenberg displayed an experimental skill and thoroughness of the highest order. Guided by the precept that "repeating an experiment with larger apparatus is tantamount to looking at the phenomenon through a microscope," Lichtenberg constructed a gigantic electrophorus and, while experimenting with it, discovered in 1777 the basic process of xerographic copying. In the words of Chester F. Carlson, the inventor of modern xerography, "Georg Christoph Lichtenberg, professor of physics at Göttingen University and an avid electrical experimenter, discovered the first electrostatic recording process, by which he produced the so-called 'Lichtenberg figures' which still bear his name."

Lichtenberg must be considered the most significant early observer of the subconscious. Concepts of repression, compensation, subconscious motivation, and sublimation are all in his writings. Lichtenberg is quoted over a dozen times by Sigmund Freud. In particular, Lichtenberg examined the interrelation between moods and facial expressions and gestures. The term "pathognomy"—which he coined to describe outward signs of emotion—is still used in psychology of expression. At the same time Lichtenberg combated the mystical aspects of Johann Kaspar Lavater's "physiognomic theory"—fashionable in those times—which maintained that anatomical structure was the outward expression of the soul.[4]

Lichtenberg's insight into human nature, coupled with an elegant and lucid style, made him a leading literary figure and earned him a secure place in German literature. His aphorisms and his commentaries on the engravings of Hogarth are unsurpassed. This facet of Lichtenberg is even more impressive than his scientific work. It is for his role as a "heretic" and an "antifaust," who in his superb satiric and aphoristic writings stood out strongly against "metaphysical and romantic excesses," that Lichtenberg is known best.

NOTES

1. A player pays the "bank" an *entrance fee* of X dollars. A coin is then tossed until heads shows up, say at the n-th toss. The player is now paid by the bank the amount of 2^n dollars. *Example:* Suppose the entrance fee is 7 dollars. If heads shows up at the first toss ($n = 1$) the player gets 2 dollars, losing 5 dollars. If heads shows up at $n = 2$, the player gets $2^2 = 4$ dollars, losing 3 dollars. If heads shows up at $n = 3$, the player gets $2^3 = 8$ dollars, winning 1 dollar. If heads does not show up until the 8th toss, the player gets $2^8 = 256$ dollars, winning 249 dollars. The problem is to determine what the "fair" *entrance fee* should be. Classical theory of probability gives

$$X = \sum_{n=1}^{\infty} 2^n \left(\frac{1}{2}\right)^n = \infty$$

which is not only paradoxical but absurd as well.
2. The selenographic longitudes are measured in the plane of the moon's equator *positive* toward the *west* in the sky and negative toward the east in the sky. The reference axis (0° longitude) is the radius of the moon passing through the mean center of moon's visible disk. The latitudes are measured from the moon's equator *positive* toward the *north* in the sky and negative toward the south.
3. It may be of interest to compare this statement to one made by Niels Bohr in 1928 in a similar context: "However contrasting such phenomena may at first appear, it must be realized that they are complementary, in the sense that taken together they exhaust all information about the atomic object which can be expressed in common language without ambiguity."
4. The differences between Lavater's "physiognomic theory" and Lichtenberg's "pathognomy" were best summarized by Franz H. Mautner (in his *Lichtenberg—Geschichte seines Geistes*, Berlin, 1968, p. 188): "Lavater was concerned primarily about revelation, Lichtenberg about learning; Lavater about principle, Lichtenberg about practical applicability; Lavater about religion, Lichtenberg about science; Lavater mainly about symbolism, Lichtenberg about scientific phenomenologic research."

BIBLIOGRAPHY

I. ORIGINAL WORKS. Collections of Lichtenberg's writings are *Georg Christoph Lichtenbergs vermischte Schriften*, Ludwig Christian Lichtenberg and Friedrich Kries, eds., 9 vols. (Göttingen, 1800–1806); and *Georg Christoph Lichtenberg, Schriften und Briefe*, Wolfgang Promies, ed., 4 vols. (Munich, 1967–1972). A third collection, *Lichtenberg: Schriften*, Franz H. Mautner, ed., is to be published.

Other writings and letters by Lichtenberg are listed in the following, along with rich biographical and secondary bibliographical material (presented chronologically): F. Lauchert, *G. Chr. Lichtenbergs schriftstellerische Tätigkeit in chronologischer Übersicht dargestellt* (Göttingen, 1893); Albert Schneider, *G. C. Lichtenberg, précurseur du romantisme* (Nancy, 1954); J. P. Stern, *Lichtenberg: A Doctrine of Scattered Occasions* (Bloomington, Ind., 1959); W. Promies, *Georg Christoph Lichtenberg in Selbstzeugnissen und Bilddokumenten*, Rowohlt Monographien no. 90 (Reinbek bei Hamburg, 1964); Franz H. Mautner, *Lichtenberg: Geschichte seines Geistes* (Berlin, 1968); and Anacleto

Verrecchia, *Georg Christoph Lichtenberg: L'eretico dello spirito tedesco* (Florence, 1969).

Lichtenberg's anonymous contributions to *Göttingische Anzeigen von gelehrten Sachen* are identified by Karl S. Guthke in *Libri*, **12** (1963), 331–340.

II. SECONDARY LITERATURE. "Lichtenberg figures" are discussed in the following and in references therein (listed chronologically): Karl Przibram, "Die elektrischen Figuren," in *Handbuch der Physik*, **14** (1927), 391–404; C. E. Magnusson, "Lichtenberg Figures," in *Journal of the American Institute of Electrical Engineers*, **47** (1928), 828–835; F. H. Merrill and A. von Hippel, "The Atomphysical Interpretation of Lichtenberg Figures and Their Application to the Study of Gas Discharge Phenomena," in *Journal of Applied Physics*, **10** (1939), 873–887; A. Morris Thomas, "Heat Developed and Powder Lichtenberg Figures and the Ionization of Dielectric Surfaces Produced by Electrical Impulses," in *British Journal of Applied Physics*, **2** (1951), 98–109; J. M. Meek and J. D. Craggs, *Electrical Breakdown of Gases* (Oxford, 1953), 215–222; and Chester F. Carlson, "History of Electrostatic Recording," in John H. Dessauer and Harold E. Clark, eds., *Xerography and Related Processes* (London–New York, 1965), 15–49.

Lichtenberg's other scientific activity is discussed in the following and in references therein (listed chronologically): Herbert Pupke, "Georg Christoph Lichtenberg als Naturforscher," in *Naturwissenschaften*, **30** (1942), 745–750; P. Hahn, "Lichtenberg und die Experimentalphysik," in *Zeitschrift für physikalischen und chemischen Unterricht*, **56** (1943), 8–15; F. H. Mautner and F. Miller, "Remarks on G. C. Lichtenberg, Humanist-Scientist," in *Isis*, **43** (1952), 223–232; D. B. Herrmann, "Georg Christoph Lichtenberg und die Mondkarte von Tobias Mayer," in *Mitteilungen der Archenhold-Sternwarte Berlin-Treptow*, no. 72 (1965), 2–6; D. B. Herrmann, "Georg Christoph Lichtenberg als Herausgeber von Erxlebens Werk 'Anfangsgründe der Naturlehre,'" in *NTM, Schriftenreihe für Geschichte der Naturwissenschaften, Technik und Medizin*, **6**, no. 1 (1970), 68–81, and **6**, no. 2 (1970), 1–12; and Eric G. Forbes, "Georg Christoph Lichtenberg and the *Opera Inedita* of Tobias Mayer," in *Annals of Science*, **28** (1972), 31–42.

For a discussion of the "Petersburg game," see Thornton C. Fry, *Probability and Its Engineering Uses* (New York, 1928), 197, and W. Feller, "Note on the Law of Large Numbers and 'Fair' Games," in *Annals of Mathematical Statistics*, **16** (1945), 301–304; and his *An Introduction to Probability Theory and Its Applications*, 2nd ed. (New York, 1957), 233–237.

Breakdowns of references on Lichtenberg according to his areas of activity (science, philosophy, literature, and so on) are in the bibliographical sections of the books by J. P. Stern and A. Verrecchia (see above).

A definitive bibliography is *Lichtenberg-Bibliographie*, prepared by Rudolf Jung, in the series Repertoria Heidelbergensia, II (Heidelberg, 1972).

OLEXA MYRON BILANIUK

LIE, MARIUS SOPHUS (*b.* Nordfjordeide, Norway, 17 December 1842; *d.* Christiania [now Oslo], Norway, 18 February 1899), *mathematics.*

Sophus Lie, as he is known, was the sixth and youngest child of a Lutheran pastor, Johann Herman Lie. He first attended school in Moss (Kristianiafjord), then, from 1857 to 1859, Nissen's Private Latin School in Christiania. He studied at Christiania University from 1859 to 1865, mainly mathematics and sciences. Although mathematics was taught by such people as Bjerknes and Sylow, Lie was not much impressed. After his examination in 1865, he gave private lessons, became slightly interested in astronomy, and tried to learn mechanics; but he could not decide what to do. The situation changed when, in 1868, he hit upon Poncelet's and Plücker's writings. Later, he called himself a student of Plücker's, although he had never met him. Plücker's momentous idea to create new geometries by choosing figures other than points—in fact straight lines—as elements of space pervaded all of Lie's work.

Lie's first published paper brought him a scholarship for study abroad. He spent the winter of 1869–1870 in Berlin, where he met Felix Klein, whose interest in geometry also had been influenced by Plücker's work. This acquaintance developed into a friendship that, although seriously troubled in later years, proved crucial for the scientific progress of both men. Lie and Klein had quite different characters as humans and mathematicians: the algebraist Klein was fascinated by the peculiarities of charming problems; the analyst Lie, parting from special cases, sought to understand a problem in its appropriate generalization.

Lie and Klein spent the summer of 1870 in Paris, where they became acquainted with Darboux and Camille Jordan. Here Lie, influenced by the ideas of the French "anallagmatic" school, discovered his famous contact transformation, which maps straight lines into spheres and principal tangent curves into curvature lines. He also became familiar with Monge's theory of differential equations. At the outbreak of the Franco-Prussian war in July, Klein left Paris; Lie, as a Norwegian, stayed. In August he decided to hike to Italy but was arrested near Fontainebleau as a spy. After a month in prison, he was freed through Darboux's intervention. Just before the Germans blockaded Paris, he escaped to Italy. From there he returned to Germany, where he again met Klein.

In 1871 Lie was awarded a scholarship to Christiania University. He also taught at Nissen's Private Latin School. In July 1872 he received his Ph.D. During this period he developed the integration theory of partial differential equations now found in many textbooks, although rarely under his name.

Lie's results were found at the same time by Adolph Mayer, with whom he conducted a lively correspondence. Lie's letters are a valuable source of knowledge about his development.

In 1872 a chair in mathematics was created for him at Christiania University. In 1873 Lie turned from the invariants of contact transformations to the principles of the theory of transformation groups. Together with Sylow he assumed the editorship of Niels Abel's works. In 1874 Lie married Anna Birch, who bore him two sons and a daughter.

His main interest turned to transformation groups, his most celebrated creation, although in 1876 he returned to differential geometry. In the same year he joined G. O. Sars and Worm Müller in founding the *Archiv för mathematik og naturvidenskab.* In 1882 the work of Halphen and Laguerre on differential invariants led Lie to resume his investigations on transformation groups.

Lie was quite isolated in Christiania. He had no students interested in his research. Abroad, except for Klein, Mayer, and somewhat later Picard, nobody paid attention to his work. In 1884 Klein and Mayer induced F. Engel, who had just received his Ph.D., to visit Lie in order to learn about transformation groups and to help him write a comprehensive book on the subject. Engel stayed nine months with Lie. Thanks to his activity the work was accomplished, its three parts being published between 1888 and 1893, whereas Lie's other great projects were never completed. F. Hausdorff, whom Lie had chosen to assist him in preparing a work on contact transformations and partial differential equations, got interested in quite different subjects.

This happened after 1886 when Lie had succeeded Klein at Leipzig, where, indeed, he found students, among whom was G. Scheffers. With him Lie published textbooks on transformation groups and on differential equations, and a fragmentary geometry of contact transformations. In the last years of his life Lie turned to foundations of geometry, which at that time meant the Helmholtz space problem.

In 1889 Lie, who was described as an open-hearted man of gigantic stature and excellent physical health, was struck by what was then called neurasthenia. Treatment in a mental hospital led to his recovery, and in 1890 he could resume his work. His character, however, had changed greatly. He became increasingly sensitive, irascible, suspicious, and misanthropic, despite the many tokens of recognition that were heaped upon him.

Meanwhile, his Norwegian friends sought to lure him back to Norway. Another special chair in mathematics was created for him at Christiania University, and in September 1898 he moved there. He died of pernicious anemia the following February. His papers have been edited, with excellent annotations, by F. Engel and P. Heegaard.

Lie's first papers dealt with very special subjects in geometry, more precisely, in differential geometry. In comparison with his later performances, they seem like classroom exercises; but they are actually the seeds from which his great theories grew. Change of the space element and related mappings, the lines of a complex considered as solutions of a differential equation, special contact transformations, and trajectories of special groups prepared his theory of partial differential equations, contact transformations, and transformation groups. He often returned to this less sophisticated differential geometry. His best-known discoveries of this kind during his later years concern minimal surfaces.

The crucial idea that emerged from his preliminary investigations was a new choice of space element, the contact element: an incidence pair of point and line or, in n dimensions, of point and hyperplane. The manifold of these elements was now studied, not algebraically, as Klein would have done—and actually did—but analytically or, rather, from the standpoint of differential geometry. The procedure of describing a line complex by a partial differential equation was inverted: solving the first-order partial differential equation

$$F\left(x, x_1, \cdots, x_{n-1}, \frac{\partial x}{\partial x_1}, \cdots, \frac{\partial x}{\partial x_{n-1}}\right) = 0$$

means fibering the manifold $F(x, x_1, \cdots, x_{n-1}, p_1, \cdots, p_{n-1}) = 0$ of $(2n - 1)$-space by n-submanifolds on which the Pfaffian equation $dx = p_1 dx + \cdots + p_{n-1}\, dx_{n-1}$ prevails. This Pfaffian equation was interpreted geometrically: it means the incidence of the contact elements $\ulcorner x, x_1, \cdots, x_{n-1}, p_1, \cdots, p_{n-1}\urcorner$ and $\ulcorner x + dx, x_1 + dx_1, \cdots, x_{n-1} + dx_{n-1}, p_1 + dp_1, \cdots, p_{n-1} + dp_{n-1}\urcorner$. This incidence notion was so strongly suggested by the geometry of complexes (or, as one would say today, by symplectic geometry) that Lie never bothered to state it explicitly. Indeed, if it is viewed in the related $2n$-vector space instead of $(2n + 1)$-projective space, incidence means what is called conjugateness with respect to a skew form. It was one of Lie's idiosyncrasies that he never made this skew form explicit, even after Frobenius had introduced it in 1877; obviously Lie did not like it because he had missed it. It is another drawback that Lie adhered mainly to projective formulations in $(2n - 1)$-space, which led to clumsy formulas as soon as things had to be presented analytically; homogeneous formulations in $2n$-space are more elegant and make the ideas much clearer, so they will be used in the sequel such that the partial differential equation is written as $F(x_1, \cdots, x_n, p_1, \cdots p_n) = 0$, with $p_1dx_1 + \cdots + p_ndx_n$ as the total differential of the nonexplicit unknown variable. Then the skew form (the Frobenius covariant) has the shape $\sum(\delta p_i dx_i - dp_i \delta x_i)$.

A manifold $z = f(x_1, \cdots, x_n)$ in $(n+1)$-space, if viewed in the $2n$-space of contact elements, makes $\sum p_i dx_i$ a complete differential, or, in geometrical terms, neighboring contact elements in this manifold are incident. But there are more such n-dimensional *Elementvereine:* a k-dimensional manifold in $(n+1)$-space with all its n-dimensional tangent spaces shares this property. It was an important step to deal with all these *Elementvereine* on the same footing, for it led to an illuminating extension of the differential equation problem and to contact transformations. Finding a complete solution of the differential equation now amounted to fibering the manifold $F = 0$ by n-dimensional *Elementvereine*. In geometrical terms the Lagrange-Monge-Pfaff-Cauchy theory (which is often falsely ascribed to Hamilton and Jacobi) was refashioned: to every point of $F = 0$ the skew form assigns one tangential direction that is conjugate to the whole $(2n - 1)$-dimensional tangential plane. Integrating this field of directions, or otherwise solving the system of ordinary differential equations

$$\frac{dx_i}{dt} = \frac{\partial F}{\partial p_i}, \frac{dp_i}{dt} = -\frac{\partial F}{\partial x_i},$$

one obtains a fibering of $F = 0$ into curves, the "characteristic strips," closely connected to the Monge curves (touching the Monge cones). Thus it became geometrically clear why every complete solution also had to be fibered by characteristic strips.

Here the notion of contact transformation came in. First suggested by special instances, it was conceived of as a mapping that conserves the incidence of neighboring contact elements. Analytically, this meant invariance of $\sum p_i dx_i$ up to a total differential. The characteristic strips appeared as the trajectories of such a contact transformation:

$$(F,\cdot) = \sum\left(\frac{\partial F}{\partial p_i}\frac{\partial}{\partial x_i} - \frac{\partial F}{\partial x_i}\frac{\partial}{\partial p_i}\right).$$

Thus characteristic strips must be incident everywhere as soon as they are so in one point. This led to a geometric reinterpretation of Cauchy's construction of one solution of the partial differential equation. From one $(n - 1)$-dimensional *Elementverein* on $F = 0$, which is easily found, one had to issue all characteristic strips. But even a complete solution was obtained in this way: by cross-secting the system of characteristics, the figure was lowered by two dimensions in order to apply induction. Solving the partial differential equation was now brought back to integrating systems of ordinary equations of, subsequently, $2, 4, \cdots, 2n$ variables. In comparison with older methods, this was an enormous reduction of the integration job, which at the same time was performed analytically by Adolph Mayer.

With the Poisson brackets $(F,\cdot)$ viewed as contact transformations, Jacobi's integration theory of systems

$$F_j(x_1, \cdots, x_n, p_1, \cdots p_n) = 0$$

was reinterpreted and simplified. Indeed, $(F,\cdot\ F_j)$ is nothing but the commutator of the related contact transformations. The notion of transformation group, although not yet explicitly formulated, was already active in Lie's unconscious. The integrability condition $(F_i, F_j) = \sum \rho_{ij}^k F_k$ (where the ρ_{ij}^k are functions) was indeed closely connected to group theory ideas, and it is not surprising that Lie called such a system a group. The theory of these "function groups," which was thoroughly developed for use in partial differential equations and contact transformations, was the last stepping-stone to the theory of transformation groups, which was later applied in differential equations.

Lie's integration theory was the result of marvelous geometric intuitions. The preceding short account is the most direct way to present it. The usual way is a rigmarole of formulas, even in the comparatively excellent book of Engel and Faber. Whereas transformation groups have become famous as Lie groups, his integration theory is not as well known as it deserves to be. To a certain extent this is Lie's own fault. The nineteenth-century mathematical public often could not understand lucid abstract ideas if they were not expressed in the analytic language of that time, even if this language would not help to make things clearer. So Lie, a poor analyst in comparison with his ablest contemporaries, had to adapt and express in a host of formulas, ideas which would have been said better without them. It was Lie's misfortune that by yielding to this urge, he rendered his theories obscure to the geometricians and failed to convince the analysts.

About 1870 group theory became fashionable. In 1870 C. Jordan published his *Traité des substitutions*, and two years later Klein presented his *Erlanger Programm.* Obviously Klein and Lie must have discussed group theory early. Nevertheless, to name a certain set of (smooth) mappings of (part of) n space, depending on r parameters, a group was still a new way of speaking. Klein, with his background in the theory of invariants, of course thought of very special groups, as his *Erlanger Programm* and later works prove.

Lie, however, soon turned to transformation groups in general—finite continuous groups, as he christened them ("finite" because of the finite number of parameters, and "continuous" because at that time this included differentiability of any order wanted). Today they are called Lie groups. In the mid-1870's this theory was completed, although its publication would take many years.

Taking derivatives (velocity fields) at identity in all directions creates the infinitesimal transformations of the group, which together form the infinitesimal group. The first fundamental theorem, providing a necessary and sufficient condition, tells how the derivatives at any parameter point $a_1, \cdots, a_r$ are linearly combined from those at identity. The second fundamental theorem says that the infinitesimal transformations will and should form what is today called a Lie algebra,

$$[X_i, X_j] = X_iX_j - X_jX_i = \sum_k c_{ij}^k X_k,$$

with some structure constants c_{ij}^k. Antisymmetry and Jacobi associativity yield the relations

$$c_{ij}^k + c_{ji}^k = 0,$$

$$\sum_k (c_{ij}^k c_{kl}^m + c_{jl}^k c_{ki}^m + c_{li}^k c_{kj}^m) = 0$$

between the structure constants. It cost Lie some trouble to prove that these relations were also sufficient.

From these fundamental theorems the theory was developed extensively. The underlying abstract group, called the parameter group, showed up. Differential invariants were investigated, and automorphism groups of differential equations were used as tools of solution. Groups in a plane and in 3-space were classified. "Infinite continuous" groups were also considered, with no remarkable success, then and afterward. Lie dreamed of a Galois theory of differential equations but did not really succeed, since he could not explain what kind of *ausführbare* operations should correspond to the rational ones of Galois theory and what solving meant in the case of a differential equation with no nontrivial automorphisms. Nevertheless, it was an inexhaustible and promising subject.

Gradually, quite a few mathematicians became interested in the subject. First, of course, was Lie's student Engel. F. Schur then gave another proof of the third fundamental theorem (1889–1890), which led to interesting new views; L. Maurer refashioned the proofs of all fundamental theorems (1888–1891); and Picard and Vessiot developed Galois theories of differential equations (1883, 1891). The most astonishing fact about Lie groups, that their abstract structure was determined by the purely algebraic phenomenon of their structure constants, led to the most important investigations. First were those of Wilhelm Killing, who tried to classify the simple Lie groups. This was a tedious job, and he erred more than once. This made Lie furious, and according to oral tradition he is said to have warned one of his students who was leaving: "Farewell, and if ever you meet that s.o.b., kill him." Although belittled by Lie and some of his followers, Killing's work was excellent. It was revised by Cartan, who after staying with Lie wrote his famous thesis (1894). For many years Cartan—gifted with Lie's geometric intuition and, although trained in the French tradition, as incapable as Lie of explaining things clearly—was the greatest, if not the only, really important mathematician who continued Lie's tradition in all his fields. But Cartan was isolated. Weyl's papers of 1922–1923 marked the revival of Lie groups. In the 1930's Lie's local approach gave way to a global one. The elimination of differentiability conditions in Lie groups took place between the 1920's and 1950's. Chevalley's development of algebraic groups was a momentous generalization of Lie groups in the 1950's. Lie algebras, replacing ordinary associativity by Jacobi associativity, became popular among algebraists from the 1940's. Lie groups now play an increasingly important part in quantum physics. The joining of topology to algebra on the most primitive level, as Lie did, has shown its creative power in this century.

In 1868 Hermann von Helmholtz formulated his space problem, an attempt to replace Euclid's foundations of geometry with group-theoretic ones, although in fact groups were never explicitly mentioned in that paper. In 1890 Lie showed that Helmholtz's formulations were unsatisfactory and that his solution was defective. His work on this subject, now called the Helmholtz-Lie space problem, is one of the most beautiful applications of Lie groups. In the 1950's and 1960's it was reconsidered in a topological setting.

BIBLIOGRAPHY

I. Original Works. Lie's collected papers were published as *Gesammelte Abhandlungen*, F. Engel and P. Heegaard, eds., 6 vols. in 11 pts. (Leipzig–Oslo, 1922–1937). His writings include *Theorie der Transformationsgruppen*, 3 vols. (Leipzig, 1888–1893), on which Engel collaborated; *Vorlesungen über Differentialgleichungen mit bekannten infinitesimalen Transformationen* (Leipzig, 1891), written with G. Scheffers; *Vorlesungen über continuierliche Grup-*

pen mit geometrischen und anderen Anwendungen (Leipzig, 1893), written with G. Scheffers; and *Geometrie der Berührungstransformationen* (Leipzig, 1896), written with G. Scheffers.

II. SECONDARY LITERATURE. Works on Lie or his work are F. Engel, "Sophus Lie," in *Jahresbericht der Deutschen Matematiker-Vereinigung*, **8** (1900), 30–46; and M. Noether, "Sophus Lie," in *Mathematische Annalen*, **53** (1900), 1–41.

HANS FREUDENTHAL

LIEBER, THOMAS. See **Erastus, Thomas.**

LIEBERKÜHN, JOHANNES NATHANAEL (*b.* Berlin, Germany, 5 September 1711; *d.* Berlin, 7 December 1756), *anatomy.*

One of the most skillful German anatomists of the early eighteenth century, Lieberkühn was the son of Johannes Christianus Lieberkühn, a goldsmith. His father insisted that Johannes and his brother plan for careers in theology; thus prior to attending the academy at Jena, the boy was sent to the Halle Magdeburg Gymnasium.

At Jena, Lieberkühn studied mathematics, mechanics, and natural philosophy, coming under the influence of the physician-iatromathematician G. E. Hamberger (1697–1755). His interest in medicine was sharpened by this experience and he went on to study chemistry, anatomy, and physiology with Hermann Friedrich Teichmeyer (1685–1744) and Johann Adolph Wedel (1675–1747). Lieberkühn left Jena in 1733 and joined his brother at Rostock as a candidate to become a preacher. Here he delivered only a few sermons, choosing instead to continue studies of what most interested him. Johann Gustav Reinbech (1683–1741), the noted Protestant theologian, recognized Lieberkühn's aptitude for scientific studies and introduced him to the Prussian king, Frederick William I. After interviewing Lieberkühn, the king released him from the career set by his father, who had died in the meantime, so that he could devote full time to science and medicine.

Even before Lieberkühn returned to Jena in 1735, the Berlin Academy of Sciences enrolled him as a fellow as a result of his earlier work at Jena. Subsequent to his second period at Jena, Lieberkühn traveled and studied in other centers, including the Imperial Natural Sciences Academy in Erfurt, where its president, A. E. Buchner, made him a fellow.

He pursued further medical study, especially of anatomy and chemistry, at Leiden under Boerhaave, B. S. Albinus, J. D. Gaub, and Swieten. Lieberkühn's Leiden tutelage culminated in the award of a medical degree in 1739. His dissertation, *De valvula coli et usu processus vermicularis*, was commended by Boerhaave and Swieten. Another dissertation written at this time, "De plumbi indole," was not published and apparently is no longer extant.

Lieberkühn's fascination with anatomical structures and their mechanisms expressed itself in his *De fabrica et actione villorum intestinorum tenuium hominis* (1745). Here, for the first time, were described, in greatest detail, the structure and function of the numerous glands attached to the villi, appropriately called Lieberkühnian glands, as well as the structure and function of the villi found in the intestines.

All of these were made comprehensible by the meticulous and skillful injections of a mixture of wax, turpentine, and colophony or dark resin.

To explain the flow of fluids into these intestinal components Lieberkühn constructed a model. By means of an open curved brass tube, cone-shaped at both ends, with two outlet tubes placed toward the narrow center, each of which drained into a separate vessel, he demonstrated the flow of chyle from the arterioles to the villi and the ascent of the lymph from the villi into the small veins. He used tinted water to show the flow of lymph from the villus into the veins and plain water to emulate the flow of chyle to the villus. As a result of his excellent demonstrations of the intestinal contents of an experimental animal before the members of the Royal Society of London, Lieberkühn was made a fellow in 1740.

In the tradition of Hamberger, his first medical teacher at Jena, Lieberkühn explored the circulatory vessels, devising special microscopes to view in greater detail the intricacies of fluid motion within the living animal. One of these was the anatomical microscope used for viewing the circulation in frogs. The specimen was attached to the body of the microscope, which consisted of two thin silver plates between which was placed a small lens and around which were arranged hooks to hold and manipulate the animal. The part of the animal to be observed was fixed over the lens.

En route to Leiden, Lieberkühn had visited Amsterdam, where he saw a solar microscope similar to the one Fahrenheit made in 1736. Another microscope, for the invention of which Lieberkühn has been given credit, to be used in illuminating opaque objects, was based on the principle of Fahrenheit's solar microscope. It consisted of a small, concave, highly polished silver speculum, later termed a Lieberkühn, that provided intense reflection of the sun's rays directly upon the object. Although Descartes had shown a solar microscope in his *Dioptrique* as early as 1637, it was probably due to Lieberkühn's revival of interest in it that the speculum arrangement on this microscope received his name. Lieberkühn's solar microscope is illustrated in William Carpenter's *The Microscope and Its Revelations* (7th ed. [Philadelphia, 1891], p. 138). The noted English microscope maker John Cuff (*ca.* 1708–1772) later adapted Lieberkühn's model by adding a mirror to it which provided better control by reflecting the sun's rays to the speculum and then to the object.

After settling in Berlin in 1740, Lieberkühn practiced medicine until his death at the age of 45. His most significant contribution during these last sixteen years of his life grew out of his desire to expose and preserve the vascular tissues of various animals. He assembled a collection of over 400 items which consisted of three main types of anatomical preparations: those preserved in a transparent liquid; dry specimens injected and hardened; and injected preparations of minute pieces of tissue (especially lung) to be viewed under the microscope. After his death this collection was advertised by a Paris dealer named Mettra and eventually was broken up and sold to several museums, as were other instruments he had constructed, including pneumatic pumps, pyrometers, and air guns.

Lieberkühn was survived by his wife, the former Catherine Dorothy Neveling, and a son and a daughter.

Lieberkühn combined a gift for observation with technical facility, enabling him to perfect the instruments and injections required to explore the most minute vascular structure of both living and preserved organisms.

BIBLIOGRAPHY

I. Original Works. A collected ed. of Lieberkühn's works appeared under the title *Dissertationes quatuor, omnia nunc primum in unum collecta . . .*, John Sheldon, ed. (London, 1782). This ed. contains *Memoria* extracted from the Berlin Academy *Memoirs* (1758), "Sur les moyens propres à découvrir la construction des viscères," which appeared originally in the *Memoirs* of 1748, and "Description d'un microscope anatomique," which appeared originally in the *Memoirs* of 1745. Other writings are *De fabrica et actione villorum intestinorum tenuium hominis* (Leiden, 1739); and *De valvula coli et usu processus vermicularis*, his dissertation for the medical degree (Leiden, 1745).

II. Secondary Literature. The best sources for an account of Lieberkühn's life are the *Memoria* published in the Sheldon ed. and the *éloge* by Formey published in the Berlin Academy *Memoirs* in 1756. The most explicit description of his anatomical preparations appears in *L'année littéraire* (1764), 136–140.

Audrey B. Davis

LIEBERMANN, CARL THEODORE (*b.* Berlin, Germany, 23 February 1842; *d.* Berlin, 28 December 1914), *chemistry.*

Liebermann was the son of Benjamin Liebermann, a cotton manufacturer. As a child he attended a private boys' school and then the Gray Cloister Gymnasium, from which he graduated in 1859. His father wanted him to work in his factory and eventually assume its direction, but Liebermann preferred to continue his studies at the university. Since he chose to study chemistry—which might have been useful in the cotton industry—his father encouraged him. In 1861 he went to Heidelberg to study spectral analysis with Bunsen. The following year he returned to Berlin. Although laboratory space was scarce, he found a place with F. Sonnenschein. In 1863 he moved to Baeyer's laboratory and two years later completed a paper on propargyl ether.

Liebermann then joined the firm of Koechlin, Baumgarten and Co. as a chemist. Here he learned about the dye industry while also discovering that his real interest lay in pure rather than applied science. After spending several months in the family factory to satisfy his father, he returned to Baeyer's laboratory in 1867. After Graebe's departure from the laboratory, Liebermann became Baeyer's assistant. He soon formed a close personal relationship with Baeyer and qualified himself to teach at the technical school in Berlin. In 1870 he also became a privatdocent at the University of Berlin. When Baeyer left Berlin in 1873, Liebermann became professor ordinarius at the Gewerbeakademie. He remained at the academy, which later became the Charlottenburg Technical High School, for the rest of his career. During this time he directed the research of dozens of students.

In 1868 Liebermann became a member of the Deutsche Chemische Gesellschaft. After 1870 he continually held a position of leadership in the society and was president twice. He helped to found the Hofmann Haus and to promote an association of laboratory students. He was a member of the scientific academies of Uppsala, Christiania, and Göttingen, the Manchester Literary and Philosophical Society, and the Chemical Society of London. In 1906 he received the Perkin Medal. He married Tony Reichenheim in 1869 and they had one daughter, Else.

Liebermann produced more than 350 scientific papers and directed the work of many students. Most of his work was in the area of aromatic organic chemistry. He became interested in this area as a student when he synthesized alizarin with Graebe. They used Baeyer's zinc dust method to show that alizarin is derived from anthracene; they also showed that purpurin, a second colored substance of madder, was trihydroxyanthraquinone.

With this introduction to anthracene compounds, Liebermann prepared halogen derivatives and investigated the hydrogen addition products. He studied the monocarbonic acids of anthracene and prepared phenol derivatives. As early as 1868 Graebe and Liebermann had suggested that anthracene was a ring structure. Heinrich Limpricht's synthesis made both

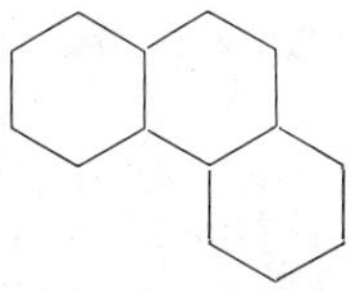

and

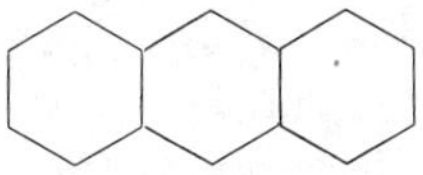

possible formulas; Liebermann felt through his work on the placement of oxygen in alizarin that the latter formula was the most probable. Liebermann's formula proved to be correct when W. A. van Dorp in 1872 produced anthracene from benzyl toluol in Liebermann's laboratory. While studying the hydroxyl isomers of anthraquinones, Liebermann found a way of exchanging the hydroxyl with an amide and eliminating the latter through a diazo reaction.

Liebermann then turned to the methyl homologs of anthraquinones; he investigated chrysophanic acid found in rhubarb and showed it to be methyldihydroxyanthraquinone. He proposed that the hydroxyanthraquinones were part of a reduction series of anthracene. Between 1876 and 1881 he was able to produce most of the compounds for which he had proposed structural formulas. In his work on reduction series of anthracene, Liebermann also became interested in oxyanthraquinones. One of these compounds, Goa powder or chrysarobin, was of pharmacological interest. Later researchers extended his work on the relationship between hydroxyl placement and therapeutic effectiveness.

Liebermann also undertook an extensive study of naphthalene compounds. He wanted to understand the basis for the location of substitution derivatives of naphthalene and to produce a whole series of β-naphthalene derivatives. He was successful, for he produced β-naphthylamine from α-naphthylamine with Scheidung. He also found retene, chrysene, and picene, and soon began to study other highly colored condensed aromatics.

Liebermann also studied the dye cochineal, or carminic acid. Paul Schützenberger and Warren de la Rue had both investigated this substance and each had assigned to it a different formula. Liebermann and van Dorp finally showed that it contains ruficoccin and ruficarmin, derivatives of anthracene and α-naphthaquinone. Other natural and artificial dyes on which Liebermann worked include β-rosaniline, hematoxylin, and xanthorhamnin. He was interested in spectrographic work which might relate color production in plants to molecular constitution.

Liebermann and his students investigated the alkaloids of coca. They devised technical methods for the production of cocaine and related compounds and studied the isomeric cinnamic acids.

Since his work on alizarin in 1868, Liebermann had tried to relate the structure of a compound with its color and its usefulness as a dye. Colors were generally produced by reduction. He felt that the color must therefore be due to unsaturated valences or an inner storage capacity in a molecule which exceeded the minimum force necessary to hold the molecule together. Compounds must have a specific composition to be mordants. In alizarin he found that the placement of the two hydroxyl groups influenced the color. Further study on the significance of the ortho position was done by Alfred Werner with organometallic compounds.

BIBLIOGRAPHY

I. Original Works. Works by Liebermann include "Ueber Alizarin und Anthracene," in *Berichte der Deutschen chemischen Gesellschaft*, **1** (1868), 49–51, written with K. Graebe; "Ueber den Zusammenhang zwischen Molecularconstitution und Farbe bei organischen Verbindungen," *ibid.*, **1** (1868), 106–108, written with Graebe; "Ueber kunstliche Bildung von Alizarin," *ibid.*, **2** (1869), 332–334, written with Graebe; "Ueber die isomeren α- und β-Derivate des Naphtalins," *ibid.*, **6** (1873), 945–951, written with A. Dittler; "Zur Kenntniss des Cochenillefarbestoffs," *ibid.*, **4** (1871), 655–658, written with W. A. van Dorp; and "Reduktionsversuche am Anthrachinon," *ibid.*, **13** (1880), 1596–1603.

II. Secondary Literature. For information on Liebermann's life, see James Partington, *A History of Chemistry*, IV (New York, 1961), 790; and O. Wallach and P. Jacobsen, "C. Liebermann," in *Berichte der Deutschen chemischen Gesellschaft*, **51**, pt. 2 (1918), 1135–1204.

Ruth Gienapp Rinard

LIEBIG, JUSTUS VON (*b.* Darmstadt, Germany, 12 May 1803; *d.* Munich, Germany, 18 April 1873), *chemistry.*

Justus Liebig was the second of the nine children of Johann Georg and Maria Karoline Moserin Liebig. He taught chemistry at the University of Giessen from 1825 until 1851, and for the remainder of his life at the University of Munich. In 1826 he married Henriette Moldenhauer, who survived him. There were five children: Georg, Hermann, Agnes, Johanna, and Marie. In 1845 he was made a baron. Liebig achieved prominence in general analytical, organic, agricultural, and physiological chemistry. He wrote many books and articles popularizing chemistry, and sought to turn the results of his research to social use by marketing such products as a meat extract and artificial milk. He played a leading role in the formation of scientific laboratories and agricultural experiment stations. His scientific reputation, however, rests mainly on his role in the development of organic chemistry between 1829 and 1839.

As a child Liebig learned to perform chemical operations in the small laboratory his father maintained to supply a family drug and painting materials business. He reproduced many of the experiments described in the chemical books available in the town library. After watching an itinerant trader make the explosive silver fulminate, Liebig produced the same compound. He was much less interested in ordinary school subjects. Recognizing his preferences, his father apprenticed him to an apothecary. Within a year, however, he returned home to continue his experiments. In 1820 he went to the University of Bonn to study chemistry under Wilhelm Gottlob Kastner. He followed Kastner to Erlangen, where he received the doctorate in 1822. Liebig found Kastner unskilled in analyses and unable to provide comprehensive chemical instruction, and he continued his own investigation of fulminate of mercury. Liebig soon realized he would have to go abroad to complete his education, and he obtained from Grand Duke Louis I of Hesse a grant to study in Paris, where he remained from 1822 until early 1824.

Liebig attended the lectures of Gay-Lussac, Thenard, and Dulong, where he encountered a rigorous, quantitative, experimental chemistry unlike anything he had found in Germany and learned for the first time some of the general principles connecting his knowledge of particular compounds and processes. Continuing his own investigations of the fulminates, he submitted a memoir on the subject to the Académie des Sciences in December 1823. He argued that silver and mercury fulminate were salts of a peculiar acid which could not be separated from the metals except by combining with other metals, with which it formed a number of new salts. Among those impressed by his paper was Humboldt, who arranged for Liebig to work with Gay-Lussac. For Liebig this opportunity was the most important of his life, for with Gay-Lussac he not only mastered methods of analysis, but learned to pursue investigations systematically.

Their collaboration soon resulted in another memoir on the fulminates, a paper making apparent that Liebig had learned from Gay-Lussac to pay stricter attention to precise determinations of the quantitative elementary composition of substances. Among their new experiments was a combustion analysis of the percentages of carbon, hydrogen, and oxygen in the fulminates. Although Liebig had produced silver fulminate entirely from inorganic reagents, he and Gay-Lussac applied to it a modification of the method for organic substances that Gay-Lussac had helped to develop. Thus, although working in inorganic chemistry, Liebig was introduced to the general problem of organic analysis to which he afterward made major contributions.

Laboratory at Giessen. Humboldt once again gave timely aid by recommending to Louis I of Hesse that he provide Liebig with an academic position, and in 1824 Liebig was appointed extraordinary professor at the University of Giessen. He returned to Germany determined to make opportunities similar to that afforded him by his work with Gay-Lussac available to a larger number of students. Together with a mineralogist and a mathematician, he proposed that a pharmaceutical training institute be established. Their proposal was turned down, but they were permitted to set up the institution by themselves in a recently evacuated military barracks. By 1827 chemistry dominated the instruction, and the twenty places in the laboratory were soon filled.

Liebig's first laboratory consisted of one unventilated room, containing a large coal oven in the center and benches around the walls. Liebig had to buy most of the supplies and pay his assistant from his own modest salary. Nevertheless this laboratory saw the beginning of a whole new mode of training scientists. It was the first institution deliberately designed to enable a number of students to progress systematically from elementary operations to independent research under the guidance of an established scientist. A carefully planned program of exercises led the students from one stage to the next. Liebig's success derived in part from his own analytical skill and his magnetic qualities as a teacher, but in part also from the state of knowledge in chemistry. The science had attained to general principles which were technically demanding and which invited application to a nearly limitless number of particular cases. For the first time it offered great scope for clearly delimited research projects under the leadership of one man in command of the field. Even as he trained others, he was able to extend his own investigations far beyond what he could perform with his own hands. If other chemists wished to compete with him, they had little choice but to imitate his approach, and the example soon spread to other experimental sciences.

Students began by learning a system of qualitative analysis, that is, an orderly way to separate and identify the acids and bases present in a solid or solution of unknown composition. To develop this system Liebig drew upon a great knowledge of the chemical properties of the acids, alkalies, alkaline earths, metals, and salts in order to standardize a sequence of operations by which the presence of any of these substances in combination with any others in a given material could most effectively be determined. The procedure began with application of general reagents which separated the components of a solution into groups that could then be further separated and distinguished by more specific reactions. Liebig believed that mastery of these reactions was the first requirement for a chemist. In developing his analytical system he drew on the traditions of distinguished analysts both before and after Lavoisier. He later turned over the elementary teaching of this system to assistants, while he gave his attention to advanced students. Two of his assistants, Fresenius and Will, published textbooks outlining the methods which in modified form have been used by students ever since.

During his first five years in Giessen, Liebig carried out many investigations concerning methods for preparing known inorganic substances. When Wöhler published an analysis of silver cyanate giving the same composition that he and Gay-Lussac had found for silver fulminate, Liebig confirmed the unusual result. He drew from it the important new idea that two chemical compounds with entirely different properties could have the same elementary compositions, differing only in the manner in which the elements were joined. In 1829 he completed an extensive study of the decomposition of a variety of chemical combinations by means of chlorine. Among the compounds with which he reacted the gas was silver cyanide. He was attempting to convert the cyanic acid which Wöhler had previously investigated into a similar acid which Georges Serullas had produced from cyanogen chloride, but which seemed to contain more oxygen. Learning that Wöhler was pursuing a similar investigation, Liebig wrote to him, and they collaborated to settle the question of the relation between the two acids. After encountering a series of difficulties, Liebig found that Serullas's analysis of his cyanic acid was incorrect, that its percentage composition was the same as that of Wöhler's acid. Wöhler inferred that the two compounds were isomers, according to Berzelius' recent definition of that term, and renamed them, respectively, cyanuric acid and cyanic acid. Liebig and Wöhler continued the cooperation begun in this study, became close friends, and made several important investigations in common.

Liebig also utilized chlorine in an effort to determine the composition of uric acid. The reaction produced cyanic acid, ammonia, and oxalic acid. Probably hoping to emulate Wöhler's synthesis of urea, Liebig tried unsuccessfully to compose uric acid from oxalic acid, potassium cyanate, and ammonia. Despite the indecisive result, his interest in uric acid carried him from inorganic into organic chemistry. He thought he might find a clue to the composition of the acid in an earlier discovery by Fourcroy and Vauquelin that horse urine yields benzoic acid. Liebig soon ascertained that the acid in the urine was distinct from benzoic acid but yielded the latter on distillation. He named the new compound hippuric acid.

Liebig's study of hippuric acid confronted him with some general difficulties in the elementary analysis of organic compounds. The customary combustion analysis, he found, gave unreliable results for nitrogen, because of the very small proportion of that element in the acid. He was able to reduce the error by using a method that he and Gay-Lussac had devised earlier for burning the compound in a vacuum. To determine the hydrogen content more accurately, he measured it in a separate analysis. This method enabled him to burn a larger amount of the compound than he could use in measuring the carbon, for the necessity to handle the carbonic acid gas that the latter element produced limited the quantity of the compound that could be analyzed. Liebig applied the same procedures to other organic acids, and entered the discussions arising then over the general question of the best method for elementary analysis. The procedures worked out by Gay-Lussac, Thenard, and Berzelius between 1810 and 1820 had become the customary method, but chemists were still seeking to improve on it. In 1830 Liebig wrote a critique of several recent efforts, arguing that the techniques he himself had worked out for hippuric acid best solved the crucial problems.

When Liebig turned to the analysis of alkaloids, he met further obstacles. He was attracted to these distinctive plant substances because an investigation of ten of them by Pelletier and Dumas had given the paradoxical result that the nitrogen in the various compounds, the supposed source of their alkaline properties, was not proportional to the quantities of acid with which each could combine. Liebig set out to improve the measurement of nitrogen, but he soon found that carbon also posed special difficulties. Because of the unusually large molecular weights of the alkaloids, errors of the order of 1 percent in the determination of carbon could lead to incorrect empirical formulas. The solution, he thought, would be to make it possible to analyze larger amounts of the organic compound, just as he had done earlier for hydrogen. In order to achieve this, Liebig made a crucial technical innovation. In his apparatus the combustion gases passed through a tube containing calcium chloride, which absorbed the water vapor. Then they entered a tube bent into a triangle, with five bulbs blown into it. A caustic potash solution in this tube completely absorbed the carbonic acid. He could then simply weigh the increase in the weight of the tube and dispense with the collection of gases pneumatically for non-nitrogenous compounds. He could now analyze in one operation ten times the amount of organic compound as by the older methods. In addition he made a number of small refinements, resulting in a much simplified, reliable procedure that soon became standard. He could, however, find no such satisfactory solution for the problem of nitrogen. He was able to improve the results Dumas and Pelletier had published for the nitrogen in the alkaloids, but he did this mostly by sheer persistence, repeating his analyses many times. The corrections he made partially resolved the theoretical problem with which he had begun, for he showed that within one class of the plant bases the nitrogen was strictly proportional to the quantities of sulfuric acid which would combine with them.

Although Liebig's new method for non-nitrogenous compounds did not immediately produce results strikingly better than had previously been attained, the simplifications he introduced eliminated much of the tedium and extraordinary skill formerly demanded. Not only could he analyze many more compounds in far shorter time than his predecessors, but he could entrust analyses to his students. Because he had many students, his laboratory began turning out hundreds of analyses annually. Liebig could now determine the elementary compositions not only of the most important organic compounds, but of all of the products of the reactions encountered in most of his investigations. His combustion apparatus became a symbol of the new era of organic chemistry that he helped to establish.

Liebig's organic research soon involved him in a question which was growing increasingly important as a consequence of the ability to depict the numbers of atoms of each element in more and more organic compounds. How are these atoms ordered in the molecules? The discoveries of different compounds with the same elementary proportions added urgency to the problem. In 1830 Berzelius showed that tartaric acid and racemic acid have identical compositions but distinguishable properties. On the basis of this and several previously reported examples, he predicted that such situations were commonplace, and named the compounds so related isomeric bodies. Liebig and Wöhler from their experiences with fulminic acid and cyanic acid were well prepared to appreciate the importance of isomerism. Impressed with Berzelius' investigation, they treated the results of their work on cyanic acid and cyanuric acid similarly and inferred that these compounds could be explained only "by a different arrangement of their molecules."

Dumas had already proposed one way to approach the problem of the ordering of atoms in compounds. An investigation of ethers (esters) in 1827 and 1828 had led him and Polydore Boullay to believe that olefiant gas, which they called *hydrogène bicarboné*, C_2H_2, acted as a base and combined with water, an acid, or both to form the various ethers and alcohol. [*Note:* Each chemist's original names and formulas are used here because the substitution of modern equivalents would be misleading. Dumas and most French chemists at this time adopted the atomic weight for carbon proposed by Gay-Lussac. Liebig followed Berzelius in taking a weight for carbon double Gay-Lussac's value. Consequently Dumas's formulas depict twice as many carbon atoms as those of Berzelius and Liebig. Berzelius' barred symbols indicating double atoms, dots for oxygen atoms, and variant symbols for elements such as chlorine and nitrogen have not been used.]

Dumas and Boullay represented a series of ethers as a sequence of compounds containing ammonia in place of olefiant gas. Their discussion implied that the elementary particles in organic molecules form persistent subgroups exchanged as units in reactions, just as ammonia is; and they assumed in their formulas that the groupings were in accordance with the electrochemical theory of Berzelius. Their theory attracted a great deal of attention, but Liebig and Wöhler dismissed it as an ingenious idea with no empirical foundation. They considered their own finding, i.e., that the same elements are combined in different ways to form cyanic acid and cyanuric acid, to be a sufficient refutation of Dumas's view. But Dumas for his part could see the same phenomena as additional support for his theory.

Despite his skepticism about Dumas's attempts to define groups of atoms which function as units in organic compounds, Liebig soon began to apply similar reasoning to other substances. In 1831 he analyzed camphor and camphoric acid. He depicted the composition of camphor as $6(2C + 3H) + O$, and that of camphoric acid as $5(2C + 3H) + 5O$, and he suggested that the unit $2C + 3H$ "acts as an elementary body." He admitted that his opinion could only be regarded as a hypothesis. A year later he and Wöhler were able to provide a far more impressive illustration of their case for a unit appearing in a series of compounds.

In 1830 Pierre Robiquet and Antoine Boutron-Charlard had converted the oil of bitter almonds to benzoic acid by oxidation. They had also formed from the oil a neutral compound which decomposed to benzoic acid and another crystalline substance which formed the same acid. They named the crystalline material amygdalin, and inferred that the oil was a "benzoic radical." During the summer of 1832, Liebig and Wöhler repeated and extended the experiments of the French pharmacists. They were able to show that the conversion of oil of bitter almonds to benzoic acid can be represented as the replacement of two atoms of hydrogen by one of oxygen. They also produced a series of related compounds containing bromine, iodine, sulfur, and cyanide. From their elementary analyses they deduced that a "benzoyl radical," $C^{14}H^{10}O^{2}$, persisted unchanged through all of these reactions. Although unable to isolate the radical, they presented persuasive evidence that it actually existed. When Berzelius saw their result, he praised it as the most important step yet made in plant chemistry.

Liebig had several times undertaken studies of the action of chlorine on alcohol and on ether, but so many compounds had formed that he had given up the investigations. Late in 1831 he had more success. He devised a new method to decompose the alcohol, passing the chlorine gas through it and periodically driving off the hydrochloric acid that formed by heating the solution. This process left behind a new compound which he named chloral and which he wrote $C^{9}Cl^{12}O^{4}$. He discovered that chloral decomposed with alkalies to yield another compound containing only carbon and chlorine $C^{2}Cl^{5}$. He emphasized that these compositions satisfied not only the elementary analyses but the relationship of chloral to its decomposition products.

The hydrocarbon radical which Dumas and Boullay had defined in 1828 continued to be influential. The boldness of their generalization combined with the uncertainty of their evidence invited others to challenge their views, in part because their theory involved an explanation of one of the most important reactions in organic chemistry—the conversion of alcohol to ether by the action of sulfuric acid. The reaction was unusually complex because, depending on the conditions, a number of other products also formed. By 1827 chemists had become particularly interested in a compound known as "sulfovinic acid," which appeared to accompany the formation of ether. The acid contained sulfur and a substance composed of carbon and hydrogen. Dumas and Boullay believed that the sulfur was in the form of sulfurous acid, and that the other substance was "oil of wine," another product of the etherification reaction that they had discovered and given the formula $C^{4}H^{3}$. Henry Hennell determined, however, that the hydrocarbon contained equal portions of carbon and hydrogen, and that therefore it was probably Dumas's and Boullay's *hydrogène bicarboné*. Unlike Dumas and Boullay, Hennell considered sulfovinic acid an essential intermediary in etherification, supporting his position with several elegant and persuasive experiments. Georges Serullas also analyzed the compounds associated with etherification, and concluded that the hydrocarbon in sulfovinic acid was *hydrogène bicarboné*, but combined with water to form ether. The conflict of opinions drew Liebig and Wöhler to study the acid in 1831.

The Ether Question. Liebig and Wöhler were able to show easily that Dumas's and Boullay's formula, $C^{4}H^{3}$, was based on incorrect elementary proportions. It was more difficult to decide whether the olefiant gas in sulfovinic acid was combined with water, in the form of ether, or directly joined with anhydrous sulfuric acid. Since they could not isolate the acid itself, they tried to deduce its composition indirectly by analyzing barium sulfovinate. Their results fitted either the interpretation that sulfovinic acid contains anhydrous sulfuric acid combined with alcohol or, as they thought more probable, hydrated sulfuric acid combined with ether. Either view, they thought, lent itself to a satisfactory explanation of the intermediary action of sulfovinic acid in the formation of ether.

Even though Serullas, Hennell, and Liebig and Wöhler had overturned certain of Dumas's and Boullay's conclusions, they had not undermined the general theory that alcohol, ethers, and related compounds were combinations of the same hydrocarbon base. The critics envisioned these compounds in the same way, and even strengthened the theory by making sulfovinic acid and oil of wine fit within the scheme more convincingly than its authors had done. Berzelius was not persuaded that the compounds were necessarily constituted in this way, but he acknowledged that such a representation placed so many phenomena in a simple and convenient order that the idea was worth attention. He believed, however, that the radical was not olefiant gas itself, but a multiple of it, which he named etherin.

New research soon forced further reassessments of the olefiant gas radical. In 1833 Gustav Magnus published the results of an investigation of sulfovinic acid, in which he claimed that the formula contained one molecule of water more than Liebig and Wöhler had assigned it. From arguments based on analogy and on the conditions under which sulfovinic acid was formed, Magnus inferred that two of the three atoms of water were attached to the hydrocarbon in the compound, forming alcohol with it. By reacting anhydrous sulfuric acid with absolute alcohol, he obtained a new acid containing only one molecule of water; he named it ethionic acid. In this acid, he asserted, anhydrous sulfuric acid is combined with ether in place of the alcohol in sulfovinic acid. Magnus depicted both alcohol and ether as compounds of the etherin radical, but he disputed Dumas and Boullay's view that the radical acts as a base analogous to ammonia. Meanwhile Pelouze was studying an acid similar to sulfovinic acid, but formed by the action on alcohol of phosphoric acid instead of sulfuric acid. Phosphovinic acid was more stable, so that Pelouze was able to dry and analyze it directly. The compound yielded carbon and hydrogen in the proportions necessary to form alcohol, from which finding Pelouze concluded that it was constituted from that substance combined with phosphoric acid. He too challenged the theory that the etherin radical plays the part of a base in such compounds. Liebig examined a sample of Pelouze's barium phosphovinate and saw that the substance absorbed water so rapidly that Pelouze's combustion analyses had shown an excess of hydrogen and carbon over the true composition of the dry salt. Liebig therefore analyzed the crystalline salt and subtracted from the measured carbon and hydrogen the elements of the water of crystallization. The results, he concluded, fitted the view that the salt was composed of phosphoric acid and ether.

The new studies of phosphovinic, sulfovinic, and ethionic acid induced Berzelius to change his views concerning the first-order compound radicals in organic chemistry. None of these acids could contain etherin and water, he thought, because the water ought then to be removable, like the water of crystallization in the corresponding salts of ordinary bases. Furthermore, if ether consisted of etherin plus one atom of water, and alcohol of etherin and two atoms of water, then sulfovinic acid ought to be convertible to ethionic acid by driving off some of the water. Neither did that occur, however, and therefore Berzelius asserted that alcohol and ether could not be hydrates of the same radical. Discarding the etherin radical, Berzelius proposed that alcohol and ether were oxides of different radicals composed of carbon and hydrogen. Ether he considered to be $C^4H^{10} + O$ and alcohol to be $C^2H^6 + O$. The compounds were, he thought, equivalent to the oxides of inorganic chemistry.

When Berzelius expressed these views to Liebig in a letter (May 1833), Liebig wrote back enthusiastically that they were the only satisfactory explanation of all the phenomena. He had, he said, been thinking along similar lines himself. Several months later he wrote an article for a chemical dictionary in which he took up the principal evidence for Dumas and Boullay's theory, and systematically refuted it. The first argument was that olefiant gas was supposed to combine directly with sulfuric acid to produce sulfovinic acid. Using olefiant gas specially purified of the ether and alcohol vapor which he believed usually to be mixed with it, Liebig found that the reaction no longer took place; he therefore asserted that it was not the olefiant gas but the ether ordinarily accompanying it that actually combined with sulfuric acid. The second case was a compound Zeise had recently produced by reacting alcohol with platinum chloride. Zeise had concluded that the resulting salt contained carbon and hydrogen in the proportions for olefiant gas, and no oxygen. According to Liebig, however, Zeise's own data fit better with a formula including oxygen, joined with carbon and hydrogen to form ether rather than olefiant gas. The most difficult evidence to counter involved a compound which Dumas and Boullay had produced by reacting dry ammonia with oxalic ether. From the fact that the resulting salt, which they named ammonium oxalovinate, yielded carbonic acid and nitrogen in the ratio of 8:1; and from some indirect considerations, they inferred that it was composed of oxalic acid, ammonia, and the radical *hydrogène bicarboné*. On repeating the reaction, Liebig found that the product he obtained was oxamide, with a ratio of carbonic acid to nitrogen of only 2:1. This outcome seemed to him to eliminate the last foundation of the etherin theory, for that radical could not be assigned to the composition of oxamide. Liebig believed that in eliminating Dumas and Boullay's theory he had firmly established that of Berzelius. He had no doubt that the radical of ether, C^4H^{10}, would soon be isolated, even though he failed in his first attempt.

Although Liebig supported Berzelius' conception of the ether radical, he could not accept the latter's opinion that alcohol was the oxide of a different radical. Some of the reactions of sulfovinic acid and phosphovinic acid, as well as the conversions of alcohol to compound ethers, involved principally the removal of water—so he argued—and were best explained by assuming that alcohol is the hydrate of ether. By thus rejecting one aspect of Berzelius' theory, Liebig attained much greater generality for Berzelius' ether radical. Giving it the name ethyl, and the symbol E, Liebig presented rational formulas based on it for over twenty compounds. Among them were the following, compared here with their interpretation according to the etherin theory:

	Ethyl Radical	Etherin Radical
Radical	$E = C^4H^{10}$	C^4H^8
Ether	$E+O$	$C^4H^8+H^2O$
Alcohol	$EO+H^2O$	$C^4H^8+2H^2O$
Hydrochloric ether (ester)	$E+Cl^2$	C^4H^8+2HCl
Sulfovinic acid	$(EO+H^2O)+2SO^3$	$C^4H^8+2SO^3+2H^2O$

Controversy With Dumas. Dumas strongly defended his own theory during the following years. He maintained his position that under the circumstances in which he had carried out the reaction of ammonia with oxalic acid, he did obtain ammonium oxalovinate, and he supported his view of its constitution with a fuller analysis of its elementary composition. Searching for a situation in which he could test the consequences derived from his theory, he noted that, according to his view, alcohol contains hydrogen in two states. Four volumes of hydrogen are combined with oxygen in water, and eight volumes with carbon in the hydrocarbon radical. It ought, therefore, to be possible to distinguish these two states in the reactions of the compound ethers. He found a verification of this prediction in the formation of chloral, the compound recently isolated by Liebig. Reexamining the reactions and products Liebig had investigated, Dumas detected a small amount of hydrogen, which Liebig had missed, both in chloral and in the decomposition product Liebig had described. Dumas renamed the latter chloroform. He gave the two compounds, respectively, the formulas $C^8H^2Cl^6O^2$ and C^2HCl^3. Stressing that in the reaction producing chloral from alcohol, ten volumes of hydrogen were replaced by only six of chlorine, Dumas explained the anomaly in terms of his view of the composition of alcohol, and of two substitution rules that he postulated. He then showed that similar rules could account for the conversion of alcohol to acetic acid. Later in the same year, Dumas and Eugene Peligot discovered a new compound "isomorphic" with alcohol, and a series of compounds analogous to those formed by alcohol. They interpreted these in terms of a new hypothetical radical, methylene, C^4H^4, corresponding to the *hydrogène bicarboné* radical. These investigations won Liebig's grudging admiration, in spite of his continuing dissatisfaction with Dumas's theoretical deductions.

In 1835 Liebig began deliberately to seek reactions which might contradict Dumas's ether theory. One such opportunity arose when he repeated some of the experiments of Magnus which had corrected his own work with Wöhler on sulfovinic acid. From the composition and reactions of Magnus' ethionic acid, Liebig inferred that during the formation of that compound one atom of oxygen from the sulfuric acid in the process must combine with two atoms of hydrogen from the ether to form water. Liebig considered that reaction to be proof that ether is not a hydrate of olefiant gas, for the water would not form at the expense of the elements of the ether and sulfuric acid if it were already present in the ether.

Shortly afterward Liebig discovered an important new compound which he thought revealed another fatal flaw in Dumas's theory. Since 1831, when Döbereiner had sent him an "ether-like" fluid obtained from the oxidation of alcohol, Liebig had tried several times to identify the resultant compounds. Döbereiner maintained that the fluid contained an "oxygen ether." In 1833 Liebig found two distinct compounds present. One of them, Döbereiner's oxygen ether, he examined more thoroughly and renamed acetal. Early in 1835 he was able to isolate the other substance. It turned out to be an unknown and extraordinarily volatile compound, which he called aldehyde. By analyzing aldehyde and its ammonium salt, he determined its composition to be $C_4H_8O_2$, and concluded that it is formed from alcohol by the loss of four hydrogen atoms. In keeping with his view of alcohol as the hydrate of ether, $C_4H_{10}O + H_2O$, he postulated that aldehyde is $C_4H_6O + H_2O$, and described it as the hydrate of C_4H_6O, an unknown oxide of a hypothetical hydrocarbon, C_4H_6. Liebig pointed out that if Dumas's view of the composition of alcohol as $C_4H_8 + H_4O_2$ was accepted, then aldehyde might be considered an oxide containing the same hydrocarbon radical, or $C_4H_8 + 2O$. This interpretation required fewer new hypothetical compounds. Liebig argued, however, that it would be impossible in these terms to account for the conversion of alcohol to aldehyde. It would have to be assumed either that the hydrogen in the water of the alcohol is oxidized, or that the alcohol gives up all its water at the same time that two atoms of oxygen are added to its radical. He regarded both possibilities as absurd.

At the same time Liebig drew a refutation of another feature of Dumas's theory from work done under his direction by Regnault. Dumas had listed the "oil of the Dutch chemists" first among the binary combinations of the *hydrogène bicarboné* radical. It was a substance obtained by reacting olefiant gas with chlorine, which he represented as C^8H^8, Cl^4. Regnault produced from the oil a new ether-like substance of composition $C_4H_6Cl_2$ (in Liebig's notation). Regnault concluded that in the reaction half of the chlorine of the Dutch oil was separated, together with enough of the hydrogen to form hydrochloric acid. It followed, he asserted, that the chlorine in the Dutch oil was combined in two different ways, so that Dumas's view of the compound as a simple combination of chlorine and olefiant gas could not be correct. Liebig triumphantly concluded that the "first limb" of Dumas's ether theory had been proven completely false.

In the fall of 1836 Liebig appended a lengthy footnote attacking Dumas's ether theory to a joint memoir reporting investigations he had made with Pelouze. First he countered the idea that olefiant gas is the basis of the composition of ether by rejecting supporting arguments that Dumas had derived from the reactions of analogous compounds. Then he enumerated ways in which the behavior of ether differed from that of hydrates and resembled that of oxides. Finally, he tried to prove that Dumas's rules of substitution, applied to his own formulas for ether and alcohol, failed to give results in accord with the reactions these compounds undergo, and that Dumas had used the rules inconsistently. He showed in detail how the conversion of alcohol to acetic acid, either directly or through the intermediary of aldehyde, and the formation of chloral, led to contradictions if Dumas's formulas and rules were applied strictly. Liebig believed that his note thoroughly destroyed the credibility of Dumas's ether theory and his rules of substitution.

As he confronted Dumas's theory, Liebig had presumed that he and Berzelius stood united in support of a shared ether theory, for he still regarded his view of alcohol as the hydrate of ether to be only an improvement on Berzelius' original theory. It was therefore very disappointing to find on reading Berzelius' *Jahresbericht* for 1836 that he was depicted as disputing Berzelius' own theory of the composition of ether and alcohol. Berzelius discounted Liebig's arguments and maintained his previous opinion that alcohol was too unlike ether to be merely a hydrate of it. In February 1837 Liebig wrote Berzelius to try to overcome his objections; he was anxious, he said, to reach an agreement, lest their division encourage the opponents of "the new ether theory." Later in the year Liebig found a new occasion to strike a blow at this opposition and simultaneously to try to dispel Berzelius' reservations. Recently Zeise had defended his earlier conclusion that the double salt formed by alcohol and platinum chloride was a combination of the etherin radical. Liebig asserted that even though Zeise's new analyses appeared more accurate than the older ones, the fit between observation and theory was fortuitous; his formula was based on elementary analyses alone, unsupported by relationships between the compound and its decomposition products. Zeise was, according to Liebig, basing his interpretation mostly on a prior and misguided allegiance to the etherin theory. Reactions which could remove the oxygen and hydrogen of the water of hydration of alcohol demonstrated that that hydrogen was in a different state from the rest of the hydrogen in the compound. There were no grounds at all, however, for distinguishing any of the hydrogen atoms in ether from the others, so that ether must be a simple oxide of the hydrocarbon, rather than a hydrate. Finally Liebig tried to counter various objections that defenders of the etherin theory had raised against treating ether as an oxide.

Liebig's adamant defense of his ether theory and his relentless opposition to that of Dumas seems hard to reconcile with the attitude that both he and Dumas professed, that all of their theories were necessarily provisional. In principle both recognized that no theory yet available was general enough to encompass all the relevant data. They believed that they must formulate broad theories even though these might prove wrong, in order to organize a burgeoning mass of analytical results. Like Dumas and Berzelius, Liebig repeatedly said that debates over divergent theories were beneficial because the defenders of each were led to discover and analyze new compounds while searching for support for their views. In this judgment they were correct, for almost every investigation conducted to advance a particular theoretical goal revealed compounds and reactions which helped expand the foundations of chemistry. The stubbornness with which Liebig sought to advance his theory and to discredit that of his rival probably owed less to the issues than to the personalities.

As Dumas's influence grew after 1832, Liebig began to perceive him as the leader of the French school of organic chemistry, a school which he thought stood in opposition to himself and Berzelius. He believed that Dumas was using his dominant position to impose his views on the other chemists in Paris. As an influential independent voice, Liebig felt he had a duty to oppose these tendencies and to rescue French chemistry from the "false route" on which Dumas was directing it. Liebig was also resentful because he thought that the younger French chemists still acted as though no important investigations were going on outside Paris. In seeking to overthrow Dumas's ether theory Liebig was in part trying to make the French acknowledge him and his followers as a leading force in organic chemistry. Besides these calculated reasons for treating his scientific differences with Dumas as political struggles, there were strong emotional elements moving Liebig. Since 1830 he and Dumas had opposed each other repeatedly on both theoretical and experimental questions, and during the course of their published debates each had allowed himself to interject into his arguments personal criticisms which wounded the pride of the other. Liebig was particularly prone to give free rein to the expression of heated reproaches which caused the disputes to take on a harshness out of proportion to the depth of their intellectual differences. Their scientific differences were not fundamental, and in 1837 it appeared possible that the two men might be able to reconcile them.

Since 1831 Pelouze had been Liebig's closest friend and colleague in Paris. They corresponded frequently, exchanging scientific news and research ideas. In 1836 Pelouze spent the summer in Giessen working with Liebig. When Liebig attached his attack on Dumas's ether theory to the joint memoir reporting these investigations, he was intentionally linking Pelouze with himself as the representative of his school in Paris, hoping that Pelouze could act as a counterbalance to Dumas's position there. No sooner had he done this, however, than it appeared that the alliance might have been constructed at Pelouze's expense; Pelouze had become a candidate for the Academy of Sciences, where Dumas had much influence. Liebig reflected that if he could reach a reconciliation with Dumas, Pelouze would have a better chance. Accordingly he wrote Dumas in May 1837, explaining away as best he could his motives for combating Dumas's theory and offering to put their quarrels behind them so that they could cooperate in the future. Dumas seemed pleased at the prospect of ending the hostilities, which he acknowledged had been a burden, and accepted Liebig's suggestion that they publish together. Liebig visited Dumas in October on his return from a trip to England; and the meeting went so well that Liebig left Paris feeling that all of the points in dispute had been settled and that he had converted Dumas to his ether theory. Dumas composed a note "Sur l'état actuel de la chimie organique," which he presented in both their names to the Academy. The note suggests that Dumas had not necessarily accepted Liebig's views, for their common beliefs were stated in propositions so general as to skirt the points of contention. He proclaimed their faith that the laws of combination and reactions in organic chemistry were the same as those of inorganic chemistry, but that the role played in the latter by elements was represented in the former by radicals. Further, the characterization of these radicals had been the constant task of both men for ten years, and they regretted that they had debated their differences of opinion so heatedly. They announced their new alliance and appealed to all chemists to join their effort to classify all organic compounds according to the radicals they contained. This "manifesto," as it was soon called, was most unfortunate. Instead of resolving Dumas's and Liebig's disagreements, it suppressed them. Moreover it publicly committed them to a doctrine which both were already coming to doubt. Liebig had publicly and privately questioned whether fixed radicals really exist. Instead of uniting other chemists, the document divided them. Older chemists got the impression that Dumas and Liebig claimed to have originated organic chemistry and were dismissing the contributions of their predecessors. Younger chemists saw the program as an attempt by the two to set themselves up as privileged directors of the work of everyone else. Sensing the hostile reactions, Liebig sought to dissociate himself from the note and prevented its publication in German.

In 1837 Liebig regarded the defense of his ether theory as one of his principal tasks, and by the end of that year he was confident that he had eliminated all opposition to it. Yet two years later he acknowledged that the opposing views were basically the same. He now represented the compounds involved in the earlier debates as combinations of an acetyl radical, $Ac = C_4H_6$, and claimed that this system satisfied the aims of both of the earlier theories. This sudden switch in his attitude resulted from a complex blend of scientific and personal developments.

The origins of the acetyl radical were in Liebig's 1835 paper on aldehyde, which he had described as a combination of the hypothetical hydrocarbon C_4H_6. Shortly afterward Regnault produced two new compounds by reacting alcohol with bromine and iodine; he assigned to them the formulas C_4H_6Br and C_4H_6I. He thought these compounds and the Dutch oil, C_4H_6Cl, could best be regarded as combinations of the hypothetical radical C_4H_6, which he named aldehyden because he considered it the same as the radical which Liebig had discovered in aldehyde. Berzelius renamed Regnault's aldehyden "acetyl." From measurements of gas densities Berzelius confirmed the application of Regnault's formulas to acetic acid and aldehyde, and by 1838 he wrote in his *Jahresbericht* that there was a firm basis for the acetyl theory.

In his new system of 1839 Liebig extended the acetyl combinations to include the compounds formerly interpreted by the etherin or ethyl theories. Thus, if acetyl, C_4H_6, $= Ac$;

then $AcH_2 =$ olefiant gas;
$AcH_4 =$ ethyl;
$AcH_4O =$ ether;
and $AcH_4O + H_2O =$ alcohol.

Liebig added a long list of related compounds. Neither etherin nor ethyl appeared any longer as fundamental radicals, but as combinations of acetyl with hydrogen. The formulas were compatible both with his own view that ether is an organic oxide and with the view of Dumas and Boullay that the compounds of ether resemble those of ammonia. The man who had recently contrasted the emptiness of the etherin theory with the soundness of his own, now stated that from his new standpoint "both . . . theories have the same foundation." He could now compromise on the issue partly because he could appreciate more fully than before that the composition of ether had been less important for itself than as a testing ground for general principles of organic composition. He had already been to some degree aware of that. He had written in 1837 that the question of the constitution of ether embraced the general question of whether organic materials which contain oxygen were oxides of a complex radical or combinations of a radical with a compound body. He believed, therefore, that in settling the problem of ether he would orient all future investigations in organic chemistry. By 1839 he had recognized that so much could not depend on one case. The existence of organic oxides was so well established in other situations that he no longer needed to defend that concept by insisting that ether was the oxide of the ethyl radical. Liebig was also more flexible because he had become far less confident that there was any satisfactory single way to envision organic composition. He was readier to accommodate his views with those of Dumas because he was coming to see that Dumas's theoretical flexibility was more viable than the fixed principles that Berzelius maintained. Within this period the man whom Liebig had regarded since 1830 as his mentor came to appear to him as an increasingly rigid voice of the past. These shifts in Liebig's attitude grew largely out of the problem of organic acids.

The Problem of Acidity. By 1830 the unifying conception derived from Lavoisier's oxygen theory of acids was no longer intact. The discovery that muriatic and other acids contained no oxygen had led to the division of acids into two categories, oxacids and hydracids. When a hydracid reacted with an inorganic base it was thought not to combine directly but to decompose the base. The hydrogen formed water with its oxygen, while the metal of the base replaced the hydrogen of the acid. Oxacids were thought to be combined with a quantity of water which, unlike the water of crystallization, could not be removed by drying. These hydrated acids combined directly with bases, which displaced the combined water. According to Berzelius, a given acid, when reacted with different bases, always combined with quantities of the base containing the same amount of oxygen. If the resulting salt were neutral, the acidic oxide had the same number of atoms of oxygen as did the basic oxide. A few oxacids could be obtained both in the hydrous and anhydrous states, but in most cases the anhydrous acid was hypothetical. Its composition was inferred by subtracting the quantity of water separated during the formation of a neutral salt from the formula determined by analyzing the free hydrated acid. In this system the organic acids formed the class of oxacids of compound radicals. For the case of acetic acid these considerations led to the formulas $C_4H_6O_3$ for the theoretical anhydrous acid; $C_4H_6O_3 + H_2O$ for the aqueous acid; and $C_4H_6O_3 + BaO$ for the barium salt.

Most of the organic acids fitted very well into this scheme, but two anomalies developed. Oxalic acid contained so little hydrogen that the subtraction of a molecule of water from its formula left the anhydrous acid with no hydrogen at all. This peculiarity caused Dulong to propose in 1816 that oxalic acid was a hydracid, $H_2 + C_4O_4$. The second difficulty concerned citric acid. From the analysis of its lead salt Berzelius had given the anhydrous acid the formula $C_4H_4O_4$. During the summer of 1832 Jules Gay-Lussac, who was studying chemistry in Liebig's laboratory, performed an analysis of copper citrate, the results of which seemed to fit the formula for the hydrate of the acid rather than the anhydrous acid. Consequently Liebig asked Berzelius to reexamine his earlier analyses. A few months later Berzelius reported back that his new results were quite puzzling. The proportions of acid and base in the lead salt varied according to the acidity or alkalinity of the solution in which it formed, a phenomenon which he ascribed vaguely to the propensity of the acid to form acidic and basic salts. The sodium salt formed only a single combination of fixed proportions, but the amount of removable water of crystallization was equivalent to 2 1/3 atoms. The dry salt could be represented as $NaC_4H_4O_4 + H_2O$, but when heated further it lost one-third of an atom more than the formula indicated it contained. Berzelius could not give a satisfactory explanation for these results. It occurred to Liebig that the contradictions might be resolved if citric acid were considered to be $C_6H_6O_6$ or $C_3H_3O_3$. He suggested for the sodium salt the formula

$$2(C_6H_6O_6) + 3NaO + 11H_2O \text{ (or } 12H_2O),$$

which he thought accounted fairly closely for the amount of water lost in drying the salt. Berzelius objected, however, that the error would still be too large, and that it was contrary to experience for a neutral salt to have more atoms of base than of acid. Both men agreed that the problem could not be solved until organic chemistry had progressed further.

During the same years the number of organic acids under investigation began to grow rapidly. Some were isolated from natural materials; others, known as "pyrogenic" acids, were produced by heating the natural acids. In 1832 Liebig interpreted the conversion of meconic acid to "metameconic acid" as the removal of one atom of carbonic acid from each atom of meconic acid, leaving one-half atom of metameconic acid. Soon afterward Pelouze learned to produce simple controlled decompositions systematically, by distilling various acids at moderate, constant temperatures. In this way he found a sublimate of lactic acid which contained two equivalents of water less than the ordinary lactic acid. He converted tannic acid to gallic acid, interpreting the reaction as the loss of one equivalent each of carbonic acid and water. He treated similarly the conversion of malic acid to fumaric acid, and tartaric acid to pyrotartaric acid. Liebig followed these investigations with close interest and judged Pelouze's new approach to be one of the most important discoveries in organic chemistry.

Following the prevalent opinion of the time, Pelouze assumed that the pyrogenic acids were oxacids. Liebig, however, already considered the division between oxacids and hydracids to be unnatural. He and Wöhler had thought for a while of interpreting benzoic acid as a hydracid in 1832. A year later he speculated briefly that tartaric acid and malic acid might be hydracids. Early in 1836 he still seemed uncertain of his position on acids, but by the end of that year he was trying to persuade Berzelius that several organic acids were hydracids. He also asserted that there are organic acids that neutralize one, two, and three atoms of base. This break with Berzelius' general conception of salts was inspired by Thomas Graham's recent discovery that there are three types of phosphoric acid containing respectively three atoms, two atoms, and one atom of water replaceable by bases. Liebig believed that a number of organic acids are analogous to phosphoric acid. According to Berzelius, potassium cyanurate is $C_3N_3H_3O_3 + KO$. Liebig contended, however, that the so-called neutral potassium salt is $C_6N_6H_2O_4 + 2KO$, whereas the so-called acid salt is $C_6N_6H_4O_5 + KO$. He discovered a new silver salt which he depicted as $C_6N_6O_3 + 3AgO$. Thus in the second case one atom of base replaced one atom of water; in the first, two atoms replaced two of water; and in the third, three atoms replaced three of water. With meconic acid he obtained a silver salt containing two atoms of silver oxide and another containing three of silver oxide with no water. The same principles could explain the anomalies in the composition of citric acid.

A new analysis of silver citrate gave Liebig slightly more silver and less water than Berzelius had found, enabling Liebig to substitute for Berzelius' formula, $C_4H_4O_4 + AgO$, one that required three atoms of silver. Liebig wrote it $C_{12}H_{10}O_{11} + 3AgO$. Tripling the formula eliminated the fractional atoms Berzelius had encountered and favored the interpretation that citric acid, too, can neutralize three atoms of base. Liebig also argued that tartar emetic has four atoms of replaceable water. From Liebig's first letter, Berzelius thought that the replacement of water in these reactions represented the ordinary removal of water from acids and salts by heating. In a second letter Liebig emphasized that he had dried silver citrate and the salts of cyanuric acid at room temperature. His results, therefore, must directly involve the constitution of the acids themselves.

Liebig especially wanted to clarify that point because it was one of the reasons that he did not think his results could be interpreted by an adaptation of the oxacid theory to the idea of multiple bases. He was now convinced that alkali and metallic bases did not simply replace water already present in the acid, but that the oxygen of the bases combined with hydrogen to form the water. His other reason was that silver oxide, a weak base, could displace all the water, whereas potash, a strong base, could displace only part of it. This was anomalous under the old theory, but predictable in terms of hydracids. Silver is easily reducible and would therefore more readily give up its oxygen to the hydrogen. Consequently cyanuric acid was not $3Cy_2O + 3Ag$, but $C_6N_6O_6 + 6H$. He thought that potassium sulfate might similarly be regarded as $K + SO_4$ rather than $KO + SO_3$. Sulfates would then be exactly analogous to chlorides, and all compounds would fit into a simple, harmonious system.

Liebig returned to these ideas nine months later, as he set out on the reformulation of organic chemistry that he and Dumas had decided upon during their meeting. On 16 November 1837 he sent Dumas a letter proposing that they begin their collaboration by extending to the rest of the organic acids the consequences of his hypothesis that all acids are hydracids. Dumas hesitated and cautioned Liebig to be prudent about proposing theories. Although annoyed by this reaction, Liebig wrote again, explaining his ideas more fully, and succeeded in overcoming Dumas's reservations. In December Dumas presented a short note, "Sur la constitution de quelques acides," to the Academy summarizing the views on organic acids upon which they agreed. The contents came mostly from Liebig's two letters. The note summarized the arguments for considering citric acid, meconic acid, and cyanuric acid as tribasic, and stressed that by tripling the formula for citric acid the authors had resolved the dilemma that Berzelius' research had posed. They asserted that the phenomena could be envisioned in a simpler, more direct manner, if these acids were regarded as hydracids. Using tartaric acid as an example, they showed that in the old system complicated double formulas were necessary to represent salts such as cream of tartar. But if tartaric acid were a hydracid, $C_8H_4O_{12}$, H_8 then all of its compounds would appear as ordinary substitutions of different amounts of metal for equivalent amounts of the hydrogen. Thus:

$$C_8H_4O_{12}\left\{\begin{matrix}K_2\\H_4\end{matrix}\right\}\text{ neutral potassium tartrate}$$

$$C_8H_4O_{12}\left\{\begin{matrix}K\\H_6\end{matrix}\right\}\text{ cream of tartar.}$$

The anhydrous acid assumed in the old theory would not exist. Relying entirely on the advantage of simplicity in representation as justification, and presenting only a few of the examples Liebig had worked out, the note did not reveal the full scope of his argument.

Even as he exchanged letters with Dumas concerning how they would proceed in their new venture, Liebig was becoming increasingly anxious about Berzelius' reaction to his acid theories. He wrote Berzelius repeatedly, expanding and developing his earlier arguments, and pleading for considerate, forbearing criticism. Berzelius tactfully referred to the analyses underlying Liebig's ideas as "highly interesting discoveries," but could not accept the explanation that the acids were hydracids. By February 1838 Liebig's hope that he could win Berzelius over was waning, and the rapprochement with Dumas had proved to be only temporary. When Pelouze read Dumas and Liebig's note on acids in the *Comptes rendus de l'Académie des sciences*, he saw that Dumas had referred to the fact that many citrates in addition to those investigated by Berzelius lose one-third of an atom of water. Pelouze was convinced that Dumas had appropriated without acknowledgment his own research on citrates, work which he had mentioned to Dumas in the fall of 1837. Pelouze therefore wrote Liebig expressing his displeasure. At first Liebig did not consider the problem serious, but during the exchange of letters and the events that followed he once again became suspicious of Dumas's motives. Moreover, Liebig had moved ahead with his research on the organic acids while, as far as he could see, Dumas had done no significant work on the problem. He persuaded himself, therefore, that Dumas was only seeking to share in the credit. By mid-February, Liebig was nearly ready to publish his own results; he wrote Dumas that he had decided to give up their collaboration and that he would continue independently.

Liebig's massive article on organic acids, "Über die Constitution der organischen Säuren," appeared in April 1838. The first part contained his analyses of the acids and their salts and the remainder treated three related theoretical questions. Liebig gave detailed arguments for considering that some organic acids neutralize more than one base and that all acids are hydrogen acids. In addition he tried to account for the molecular rearrangements involved in the conversion of some of the organic acids to their related pyrogenic acids. In part he was extending the approach of Pelouze, but he was attempting also to explain in terms of changes in composition the decrease in the number of atoms of base which the resulting acids could neutralize.

The interpretive portion began with a discussion of phosphoric acid based on Graham's work. After reviewing the three series of phosphoric acid salts, Liebig proposed various ways in which the isomeric changes in the acid might be envisioned to account for the differences in the number of bases with which the three forms could combine. He also pointed out that the property of the acid of combining with more than one base revealed the uncertainty underlying the customary conception of neutrality. He suggested that the property that most clearly distinguished acids like phosphoric from the majority of acids was the ability to form true double salts such as sodium–potassium phosphate; for this capacity was the same as that of combining with more than one atom of a single base.

Liebig next applied the principles he had drawn from the example of phosphoric acid to the organic acids which embodied similar combining properties. From this viewpoint he depicted the constitution of cyanuric, meconic, citric, tartaric, mucic, malic, gallic, and aspartic acids, their salts, and their pyrogenic products. The case of meconic acid will illustrate his approach:

$$C_{14}H_2O_{11} + 3H_2O\quad\text{dried acid}$$

$$C_{14}H_2O_{11} + 3AgO\quad\text{silver salt}$$

$$C_{14}H_2O_{11} + \left.\begin{matrix}H_2O\\2AgO\end{matrix}\right\}\text{ silver salt}$$

$$C_{14}H_2O_{11} + \left.\begin{matrix}2H_2O\\KO\end{matrix}\right\}\text{ so-called acid potassium salt}$$

$$C_{14}H_2O_{11} + \left.\begin{matrix}H_2O\\2KO\end{matrix}\right\}\text{ so-called neutral potassium salt}$$

Liebig showed that the pyrogenic acids related to meconic, citric, tartaric, and gallic acids usually neutralized only one or two atoms of base. For example, heat and concentrated acid transformed meconic acid into comenic acid, which could combine with two bases. The constitution of comenic acid he considered to be

$C_{12}H_4O_8 + 2H_2O$ crystalline acid

$C_{12}H_4O_8 + \left.\begin{matrix} H_2O \\ KO \end{matrix}\right\}$ so-called potassium salt, etc.

In the formation of comenic acid two atoms of carbonic acid were removed. That did not account, however, for the loss of one-third of the saturation capacity, a change caused, according to Liebig, by the incorporation of one atom of the previously separable water into the acid radical itself, where it could no longer be exchanged for another base. The further conversion of comenic acid to pyromeconic acid involved the separation of another atom of carbonic acid and the transfer of another atom of water to the radical. The resulting constitution, $C_{10}H_6O_5 + H_2O$, accorded with the capacity of the acid to neutralize only one atom of base. Liebig described in a similar manner the conversions of other acids to their pyrogenic products, although some of the changes were more complex, and in some cases unsolved difficulties prevented him from accounting fully for the reactions.

In the last section of his article, Liebig noted that he had built the preceding interpretations on current conceptions of acids and bases. For certain of the phenomena, however, he could find no explanation in that system. Besides the previously mentioned paradox that silver oxide displaced more water than did potash, there was another question. Why, if the saturation capacity of an acid depends on water it contains, do those acids which lose part of their saturation capacity in the modifications that displace some of the water not regain both when they are placed again in water? Such situations suggested that it might be profitable to consider generalizing the view that some acids are hydracids. He ascribed the strangeness of resulting formulas such as $SO_4 + K$ mostly to habit, and argued that the hydrogen theory was as valid a way as the conventional theory to envision acids containing oxygen. The close similarity of various reactions of bases with the halogen acids and with oxygen acids such as sulfuric made it unlikely that water was formed in the first type and merely separated in the second. The hydrogen theory led to some deeper insights into the nature of compounds, which the oxygen theory could not provide. A comparison of the acids formed by the various oxidation states of chlorine showed that their saturation capacities depended only on the amount of hydrogen present, not on the composition of the radicals. Liebig thought he could apply the same principle to numerous organic acids; various complicated combinations of elements could be added to their radicals without changing their saturation capacity. After completing a few more general arguments, Liebig sketched only briefly with a single example what the constitution of organic acids would be like in the framework of the hydrogen theory.

$C_{12}H_4O_{10} + H_4 =$ meconic acid

$C_{10}H_6O_6 + H_2 =$ pyromeconic (comenic) acid

Liebig carefully presented his theory as a hypothesis which might not necessarily express the true constitution of the compounds. He maintained only that it had guided him in his investigations, that it brought certain relations into a simple and general form, closing gaps that the prevailing theory could not close, and that he was deeply persuaded that the theory would lead chemists to important discoveries.

The paper on organic acids was one of Liebig's finest achievements, reflecting the best of the attributes that had marked his previous work. He based his position on precise analyses of numerous compounds. Some concerned substances he had discovered, but many were refinements of analyses done by others. He had not originated the theories he defended, but had greatly extended approaches drawn from Davy, Graham, Pelouze, and others. Through his extensive knowledge of compounds and reactions, he was able to amass impressive evidence for his inferences. He displayed a realistic sense of the value and limits of theoretical conceptions. He utilized flexibly such currently accepted foundations of reasoning as the radical theory. He was able to weld these elements into a comprehensive, unifying whole. He took a major step in one of the most important revisions in general chemical theory since the acceptance of Lavoisier's system of chemistry: a revision completed a few years later in the more universal statement of the hydrogen theory of acids by Liebig's former student, Gerhardt.

For Liebig his investigation of the organic acids only continued to be a source of personal turmoil. In part this was inevitable, for the new system he was building entailed the disruption of that which had guided Berzelius for many years and which had been the foundation for their scientific partnership. Recognizing the seriousness of the breach between them, both men made strenuous efforts to repair it. Liebig tried again and again to overcome Berzelius' disapproval of his new ideas. Diplomatically he portrayed them as generalizations of interpretations Berzelius himself had given in specific cases, and he appealed to Berzelius to join him in exploring the consequences of his theory. Berzelius studied Liebig's views with care, but he could not subscribe to them. Within the framework in which he viewed organic composition, Liebig's formulas were incomprehensible. Both men acknowledged that no single theory was exclusively true, which was a step toward reconciling their intellectual differences, but at the same time nonrational factors were making harmony difficult. Liebig had recently published the *Anleitung zur Analyse organischer Körper*, in which he had included several criticisms of details of Berzelius' methods. Berzelius was hurt by what seemed to him an abrupt dismissal of his earlier contributions, and began to feel that Liebig was trying to enhance his own reputation at the expense of his colleagues. Liebig expressed regrets at his thoughtlessness, and the two men appeared reconciled; but they were estranged again by Pelouze's quarrel with Dumas.

In January 1838 Dumas had assured Pelouze that his claims concerning the citrates would be given full credit in an article that he and Liebig would publish as soon as Berzelius had confirmed Liebig's analysis of silver citrate. By March, Pelouze felt that he could wait no longer, and he wrote Berzelius to find out if Berzelius actually intended to do the analyses in question. If not, Pelouze said, he planned to present his own case at the Academy of Sciences. In his letter Pelouze described Dumas's conduct in very harsh terms. Berzelius, who had long distrusted Dumas, wrote a letter to the Academy in which he not only supported Pelouze's assertion that the explanation given in the joint note of Dumas and Liebig of the loss of water in citrates was the same as that which Pelouze had written him, but attacked Dumas's theory of substitutions and the joint theory of organic acids as well. After Pelouze read this letter at the Academy, he and Dumas engaged in an acrimonious debate over what had actually happened. Liebig was soon drawn into the fracas. Pelouze and Dumas both sought to enlist Liebig's support, each writing to persuade him that honor and self-interest required Liebig to join him in defense against the accusations of the other. For a time Liebig sided with Dumas and unsuccessfully attempted to persuade Pelouze to drop the affair. Berzelius' open attack on his theory of organic acids greatly angered Liebig; and, in a letter to Dumas, he was led in turn to criticize Berzelius' acid theory. He authorized Dumas to read the scientific portions of the letter at the Academy.

As the dispute continued, however, Liebig's old suspicions about Dumas reasserted themselves. On the basis of a partial knowledge of what was happening in Paris, reflected especially in an inaccurate report in the daily press, Liebig decided that Dumas had violated his confidence and was manipulating him for personal advantage. By June 1838 he had made a "total break" with Dumas and thought of him as an "implacable" enemy. With Berzelius, however, he tried hard to repair the damage the events of the preceding weeks had done to their relationship. Patiently the two men renewed their efforts to reconcile their views about organic acids. Yet they were inexorably drifting apart, and Liebig was beginning to see Berzelius as a spent force in chemistry. Despite the personal enmity between Liebig and Dumas, they were coming to share a general conviction that organic chemistry was so much more complex than inorganic that new rules were required for it, and that whatever promised to simplify and order some of the phenomena deserved consideration. Like it or not, during the next two years Liebig was drawn toward an intellectual alliance with his most formidable rival directed against his most revered friend.

During the years that Liebig was preoccupied with the ether theory and with organic acids, he also carried out two important investigations with Wöhler. In October 1836 Wöhler wrote that he had discovered a way to transform amygdalin to oil of bitter almonds and hydrocyanic acid, by distilling it with manganese and sulfuric acid, and he invited Liebig to join in pursuing the topic. Two days later he made a more remarkable discovery. It had occurred to him that perhaps the transformation of amygdalin could be effected by the albumin in the almonds, in a manner similar to the action of yeast on sugar. He was able to confirm his expectation, for either crushed almonds or an aqueous emulsion derived from them produced the reaction. Wöhler suspected that the decomposition was an example of what Berzelius had recently defined as catalysis. Liebig and Wöhler then divided up the detailed examination of the properties and composition of amygdalin. They precipitated from the emulsion of almonds a substance which when redissolved retained its action. They named the active substance "emulsin." Its effectiveness in very small quantities confirmed that it acted like yeast. Liebig wrote up the memoir "Über die Bildung der Bittermandelöls," which they published in 1837, but the critical ideas and experiments came from Wöhler. Their characterization of emulsin, following that of diastase by Payen and Persoz, and of pepsin by Theodor Schwann, helped to generalize the conception of specific fermenting actions, which became a central theme in organic and physiological chemistry during the next decade.

Wöhler also provided the initial impulse for their other major joint investigation. In June 1837 he wrote Liebig that he had found a way to identify the constituents of uric acid by decomposing it in water with lead peroxide as the oxidizing agent. The products formed were urea, carbonic acid, and a colorless crystalline substance. Invited by Wöhler to take up the investigation with him, Liebig immediately determined the elementary composition of the new substance and identified it with allantoin, a compound long known to be present in the allantoic fluid of calves. The composition of uric acid was $C_5N_4H_4O_3$ and that of allantoin $C_4N_4H_6O_3$. Both men then started to examine other means of oxidizing uric acid. When Liebig attended the meeting of the British Association for the Advancement of Science later in the summer, he presented a preliminary report on the new work. He depicted the reaction which formed allantoin as:

$$\underbrace{C_{10}N_4H_4O_6}_{\text{1 atom uric acid}} + \underbrace{Pb_2O_4}_{\text{2 atoms peroxide of lead}}$$

$$= \underbrace{CO + O_2 + 2PbO}_{\text{oxalate of lead}} + \underbrace{4Cy + 3H_2O}_{\text{allantoin}} + \underbrace{Ur.}_{\text{1 atom urea}}$$

On the basis of the reaction and the conventional assumption that a compound containing four elements must be formed by a union of binary compounds, Liebig interpreted uric acid as a compound of urea with a peculiar unknown acid, the latter containing the radical of oxalic acid combined with cyanogen. He represented this formulation as $4(CO + Cy)$ + urea. Wöhler came to Giessen at Christmas so that he and Liebig could complete the project together; but they found such an astonishing number of new compounds that they were unable to finish. Liebig continued the experiments through the spring, and the results were finally ready for publication in June 1838.

The paper, "Über die Natur der Harnsäure," described the conversion of uric acid into allantoin and then the products of its decomposition by nitric acid. In dilute nitric acid and ammonia, uric acid produced ammonium oxalurate, $C_6N_4H_8O_8$. In more concentrated cold nitric acid it yielded crystals of a water-soluble substance which they named alloxan and gave the formula $C_8N_4H_8O_{10}$. If an excess of ammonia were present the reaction produced instead the ammonium salt of purpuric acid, a remarkable compound identified earlier by William Prout. Liebig and Wöhler renamed it murexid and considered its composition to be $C_{12}N_{10}H_{12}O_8$. Under other circumstances uric acid produced a substance they named alloxantin, the composition of which, $C_8N_4H_{10}O_{10}$, was closely related to that of alloxan. Alloxan could readily be converted to alloxantin by reducing agents, and back again with oxidizing agents. The diverse products of the direct decomposition of uric acid gave rise in turn to numerous other products. Among those that Liebig and Wöhler discovered and named were thionuric acid, uramil, uramilic acid, dialuric acid, alloxic acid, mesoxalic acid, and mycomelinic acid.

The complexity of the reactions made it difficult for Liebig and Wöhler to draw unequivocal conclusions concerning the rational constitution of the compounds involved. They elaborated essentially the same interpretation of the relation between uric acid and allantoin that Liebig had summarized the previous summer. They considered, as an admittedly provisional and "prejudiced" view, that urea preexisted in uric acid. Assuming this to be so, they subtracted the formula for urea from that for uric acid and derived a hypothetical body, $C_8N_4O_4$, which they named uril. They then proposed interpretations of reactions such as the conversion of uric acid to alloxan based on the supposed separation of uric acid into urea and uril. In general they recognized that the diversity of decomposition reactions made it impossible to deduce from the products that the same ones preexisted in the original compounds. Often two or more explanations of a reaction seemed equally conformable to the result. The most general conclusion they reached was that uric acid demonstrated that organic compounds undergo "innumerable metamorphoses," a characteristic setting them quite apart from inorganic compounds.

Further research on the decomposition products of uric acid occupied Liebig's laboratory during 1839, although he began to spend more of his time writing textbooks. He was now supreme in his field, and at the age of thirty-six at the peak of his experimental productivity. His grasp of theoretical issues was firmer yet more flexible than ever before, and he headed the largest scientific laboratory and school in the world. By August of that year additions to his laboratory and improved financial support enabled him to expand his training and research programs. Yet within a year Liebig had nearly given up those experimental areas in which he excelled and was devoting his energy to applications of chemistry in agriculture and physiology. To some extent his switch was a response to new opportunities; but it reflected also a weariness with the work he had been doing, and especially with the disputes which pervaded organic chemistry. In 1839 and 1840 the debates over Dumas's substitution theory reached a climax, as Dumas responded to new objections by Berzelius. Liebig supported the basic conception of substitution, and dissociated himself from Berzelius' alternative explanations of the reactions Dumas had used to support his theories. Early in 1840, however, Pelouze joined the opposition to Dumas and urged Liebig also to join in refuting the substitution theory. As he saw himself in danger of being pulled into further arguments dominated by personalities more than by issues, Liebig began to feel an aversion toward the activities of chemistry. He therefore turned to new subjects, in part to escape the impasse he had reached in his old field.

If Liebig's significance were to be judged in terms of his lasting tangible contributions to organic chemistry, the emphasis would have to be on the methods he devised or refined, and the many compounds and reactions he discovered or described. Few of his theories were highly original, and none of them definitive. Yet a summation of Liebig's achievements in terms of discoveries and general theories is bound to underrate his stature, for that is measurable more by the way in which he participated in a crucial stage in the development of chemistry. The manner in which he did organic chemistry was as important as the tabulation of his results. Through the reliability of his analyses, the thoroughness of his examination of all the products and reactions related to any particular problem, the quality of his reasoning, and the soundness of his judgment about most theoretical issues, Liebig helped his contemporaries and successors to see more clearly how the science ought to be pursued. Similarly, his new way of training chemists was as important as the many excellent chemists he himself taught. He helped to set standards in another way in his role as the critical editor of the *Annalen der Chemie und Pharmacie*, a journal he turned into the preeminent publication in chemistry. Liebig did as much as any one person to bring about the era of large-scale research, in which the ability to organize men became as critical as the ability to conceive and carry out experiments. Between 1830 and 1840 he was at the very center of the rapidly growing field of chemistry and he gave it an impetus that was felt long after he had ceased actively to participate in it.

Agricultural Chemistry. Liebig moved into his new areas of interest at his usual pace. Within four months after he began systematically studying the relation of organic chemistry to agriculture and physiology, he had produced the first version of one of the most important books in the history of scientific agriculture. In a letter to Berzelius in April 1840, he wrote that he had derived his general arguments from several surprising discoveries. First, from analyses of straw, hay, and fruits, he reached the "very remarkable result" that a given area of land, whether cultivated field or forest, produces each year the same total quantity of carbon in the composition of whatever plants grow on it. That inference became the starting point for his argument that the carbon must derive from the atmosphere rather than from humus in the soil. The problem of the source of nitrogen had "long occupied" him, he said, but after he found ammonia in the sap of every plant he investigated he became persuaded that the nitrogen must come from ammonia dissolved in rainwater, and he found that all of it contained determinate amounts of ammonia. Next Liebig examined the alkalies and alkaline earths in plants. Because they were always present, he presumed that they were essential, and set out to explain why they nevertheless varied according to the soil in which the plants grew. He surmised that the total equivalent quantity of base must be constant for a given species of plant, but that one type of base might substitute for another. He verified this prediction by reference to analyses that other chemists had made of the ashes of evergreens. Pine trees grown in one locality contained magnesia, while those grown in another place did not, but the total oxygen content (and thus by current acid-base theory the saturation capacity) of the potash, lime, and magnesia was in both cases the same. Fir trees grown in different regions also yielded different proportions of the specific alkalies, but the same total, a total which differed from that of the spruce ashes. Liebig thought these uniformities could not be accidental, and he explained them by postulating that the bases served to neutralize the organic and inorganic acids which were characteristic, essential constituents of each type of plant. Liebig drew a last key observation from the sad experience of a farmer near Göttingen who had grown absinthe in his fields to use as a source of potash; he found afterward that he could not raise ordinary crops in the field because the potash of the soil was exhausted. From all these considerations Liebig concluded that the true purpose of a fertilizer is to supply ammonia and such salts as potassium silicate, calcium phosphate, and magnesium phosphate to plants. He wrote Berzelius that a large number of practical applications for farmers could be developed from his results, and he would describe them in a forthcoming book.

Liebig's letter implied that he had derived his views on agricultural chemistry chiefly from investigations he had carried out himself. No doubt he felt he had, and the observations and calculations he made may well have been decisive in shaping his opinions. Yet in the short time he had devoted to the subject he could not have mastered the whole field of investigation, and the full treatment he published shortly afterward shows that he drew most of his arguments from the work of others and from strongly deductive reasoning. He had ably and impressively reviewed an area of activity that was new to him but that had been pursued with varying intensity for several decades. He joined the scattered conclusions of his predecessors into the most comprehensive picture of the problems of plant nutrition that had ever been presented, using his own findings at certain crucial points to decide between alternative theories.

Liebig's *Die organische Chemie in ihre Anwendung auf Agricultur und Physiologie* (Brunswick, 1840; English trans., *Organic Chemistry in Its Applications to Agriculture and Physiology* [London, 1840]) begins with a discussion of the role of carbon in plant nutrition. He refuted the widely held theory that humus, the product of the decay of plant matter, formed the main nutrient substance for plant growth and supported the view that the source of the carbon assimilated into plant substances is the atmosphere. The stability of the carbonic acid content of the air, despite the continual exhalation of that compound by animals, required that something else must be continually removing it from the atmosphere. Even though the percentage of carbonic acid in the air was very small, the atmosphere contained an ample amount to supply all of the plant material on the surface of the earth. The most important plant function, Liebig asserted, was to separate the carbon and oxygen of carbonic acid, releasing the oxygen and assimilating the carbon into compounds such as sugar, starch, and gum. In a later section he admitted uncertainty as to whether carbon was separated and joined with the elements of water to produce these compounds, or whether the plant in fact separated the oxygen from the water and combined its hydrogen with carbonic acid. He favored the latter alternative, because it was evident that in the formation of waxes, oils, and resins, the hydrogen in excess of the proportions of water must come from the decomposition of water.

Liebig denied the view that the consumption of oxygen and the exhalation of carbonic acid by plants at night constituted a "true respiration." That transformation was, he thought, a purely chemical decomposition. He believed that the starch, sugar, and gum formed by the primary nutritive process were transported throughout the plant and converted by numerous further metamorphoses into the special constituents of such parts as the flowers and fruit. Liebig's general picture of plant physiology was not original; he had obtained most of it from the earlier investigations of Priestley, Sénebier, Ingen-Housz, and Saussure. Through his knowledge of recent developments in organic chemistry, however, Liebig was able to give many more examples of chemical processes of the kind that might occur in plants. From the elementary compositions of various plant compounds he could calculate the amounts of carbonic acid and water consumed, and of oxygen released, in the formation of a given quantity of each compound. Although he had not investigated directly the chemical processes in plants, he could depict impressively the types of metamorphoses which he supposed took place by likening them to the metamorphoses he had produced in the laboratory. The creation in the laboratory of such important organic compounds as formic acid, oxalic acid, and urea enabled him to predict that vegetative processes would be explained through chemistry. One of his main purposes was to persuade botanists and physiologists that they must pay more attention to chemistry if they were to make further progress.

This glimpse of a chemical understanding of the internal processes of vegetables was, however, not what gave Liebig's book its most forceful impact. Whatever explanations he gave of the necessity for certain nutrient materials in terms of their role in plant growth, his views on the external sources of the substances he deemed essential were what aroused immediate interest, for these conclusions directly impinged on agricultural practices. The source of hydrogen was no problem, since all agreed that it was water. Nitrogen was a more pressing question. To support his claim that plants are supplied with nitrogen by means of ammonia washed out of the atmosphere, Liebig had to dispose of the fact that the ammonia in the air is so small in amount as to be undetectable, whereas nitrogen gas constitutes about 80 percent of the atmosphere. His confirmation that ammonia exists in rainwater provided him with an important supporting argument, but his conviction was based on his general chemical experience. Atmospheric nitrogen is one of the least reactive of substances, whereas ammonia enters readily into many organic compounds, a large number of which Liebig had investigated. Therefore he considered it most probable that ammonia is also the medium through which nitrogen is introduced into the constitution of plants. His reasoning was typical of the approach which Liebig and other organic chemists brought to the discussion of physiological questions; that is, they expected to predict the behavior of the substances within living organisms from their knowledge of chemical properties of elements and compounds in laboratory reactions. The separation of carbon from oxygen by plants was the one great and mysterious exception to this rule: "a force and capacity for assimilation which we cannot match with an ordinary chemical action, even the most powerful."

Turning to the inorganic constituents of plants, Liebig elaborated his view that plants need alkalies and alkaline earths to neutralize their essential acids. Analyses of ash contents, he contended, form the basis for determining the requirements of different plants for bases. Because these contents differ from one plant to another, some plants can flourish in soils that will scarcely support others. After finishing his discussion of plant nutrients, Liebig drew the appropriate lessons for agriculture. Cultivation of the same crop year after year gradually diminishes the fertility of the soil as the inorganic constituents essential to the growth of that crop are depleted. Rotation of crops can extend the resources of a given field because one plant can utilize those particular bases that the other did not require, but eventually the soil will become totally infertile unless these elements are returned to it. Allowing land to lie fallow slowly restores its fertility because the weathering of minerals gradually restores the acids and bases. A rational agriculture, however, would sustain fertility more effectively by supplying artificially the necessary elements. Liebig defined fertilizer as the means for adding to the soil the nutritional requirements of crops not supplied naturally from the atmosphere. The composition of a good fertilizer therefore varied according to the ash content of each specific crop. Fertilizers should ordinarily be composed of bases such as lime, potash, and magnesia, but since all plants contain phosphoric acid that compound should be included also. Liebig considered the best source of phosphoric acid to be pulverized animal bones. The bones should be dissolved in sulfuric acid and the phosphoric acid would thus be freed to combine with the bases in the soil. The best available sources of the other inorganic salts were plant ashes and human and animal excrement. But Liebig looked forward to the day when chemical industries would be able to produce salts to make up specific fertilizers for each crop.

Liebig was then rather ambivalent about the need for nitrogen in fertilizers. The ammonia derived from rainwater was sufficient, he argued, for natural plant growth, but plants could convert added ammonia into greater yields of nitrogenous organic compounds, and therefore increase the supply of the important nitrogenous constituents of human and animal diets. Gypsum, calcium chloride, and sulfuric acid were effective in increasing growth because they transferred ammonia from ammonium carbonate to less volatile salts so that the ammonia remained in the soil until absorbed by plants.

Liebig's book excited an unexpectedly intense interest among practical agriculturalists, especially in England, where his former student, Lyon Playfair, acted as translator and propagandist. Playfair arranged for large-scale tests of Liebig's ideas on several farms. In America some of his ideas also aroused a lively response. Liebig thus became more deeply involved in the application of his general ideas to actual farming practices. The changing titles that he gave his book symbolized his shifting conception of the scope of the topic with which he was dealing. The work appeared first in French (April 1840) as volume one of a *Traité de chimie organique*, with the explanation that Liebig was publishing it separately because the material lay outside of chemistry proper. The German edition, four months later, reflected that situation directly in its title. In later editions the adjective "organic" disappeared.

Although he had taken up the subject as an application of organic chemistry, which he regarded as the chemistry of materials of biological origin, the problem had proved to have different boundaries; the analysis of inorganic compounds had in fact become dominant. The main additions Liebig made in the second edition were efforts to specify more concretely the types of fertilizers that would be suitable for particular crops in given localities. Thus he discussed the mineral constituents characterizing the major crops in England as a guide to the selection of inorganic fertilizers. Anticipating the substitution of artificial for natural fertilizers, he suggested a method for converting the ammoniacal liquid left over from industrial production of coal gas into a purified ammonium sulfate. Stressing that knowledge of the composition of soils was the basis for the whole system of rational agriculture, he appended a long list of soil analyses carried out earlier by Kurt Sprengel. In the third edition (1843), Liebig revised his previous opinion that agriculturists should supplement the natural supply of nitrogen by adding or fixing ammonium salts. The addition of nitrogenous fertilizers alone, he said, could not augment the fertility of a field, for its productivity increased or decreased in direct proportion to the mineral nutrients provided in its fertilizer. In keeping with his greater emphasis on the mineral requirements of plants, he added analyses performed in his laboratory of the ash contents of various crops.

In 1845 Liebig took another step toward the practical implementation of his theories; he devised instructions for making artificial fertilizers formulated entirely of mineral salts that were combined in proportions corresponding to the ash contents of various crops. Muspratt and Co. of Liverpool put them into production. The venture proved a severe setback for Liebig's views, for nowhere did the fertilizers cause increased yields sufficient to cover costs. Surprised and dismayed, Liebig purchased a plot of sterile grazing land outside Giessen and began to investigate why his fertilizer was ineffective. Between 1845 and 1849 he managed to convert the plot into a fertile field, but he remained puzzled by the slowness of the action of the fertilizer. Not until over a decade later did he identify the cause of the difficulty. In order to prevent the potash from being washed out of the soil, he had fused it with calcium carbonate into a highly insoluble combination, and its insolubility had prevented the plant roots from absorbing the alkali. By this time he realized that he need not have worried about soluble salts being removed by rainwater, because the topsoil itself absorbed and held them. Meanwhile the error had threatened to discredit the general principles he had espoused.

At Rothamsted, in England, Lawes and Gilbert tested one of Liebig's fertilizers on wheat and found no noticeable increase in production, whereas ammonium salts added annually brought about significant improvements year after year in the harvests. Lawes and Gilbert regarded these results as a refutation of Liebig's entire "mineral theory," and other agriculturists came to regard it as a hastily conceived mistake. The success with guano, a substance rich in ammonia, seemed further to contradict Liebig's views. Liebig defended himself ably, however, in a series of polemic tracts published during the 1850's. Lawes and Gilbert's experiments were not decisive, he argued, because they had conducted their investigation under conditions which precluded a fair test of his fertilizer. From their description of the fields they had used, Liebig inferred that the soil already contained such ample quantities of the essential minerals that the provision of more thorough fertilizers could not significantly affect its productivity. Furthermore, he asserted that the English experimenters had misrepresented him, for he had not claimed that agricultural yields are dependent solely on the addition of mineral constituents to the soil, nor that it is never useful to add ammonia. He had only stated that in most cases it is superfluous to supplement the natural supply of ammonia and that fertilizers cannot be evaluated by their nitrogen contents. The current fashion for fertilizing with nitrogenous salts, he believed, would simply cause the mineral stores of the soil to be more rapidly exploited by quicker plant growth; if not supplemented by proportionate quantities of mineral fertilizers that practice would only deplete the soil sooner. Farmers, he contended, must think beyond the objective of obtaining the largest possible crop in a given year. In order to protect their capital, they must preserve the productivity of their land for future years by restoring to it all that their harvests removed. Nitrogen is replenished from the atmosphere but minerals come from the soil alone, and therefore it was the minerals with which they must be chiefly concerned.

While defending his theories, Liebig also clarified them. The discrepant effects of fertilizers in different situations led him to perceive that the addition of any one constituent to a field has value only in relation to the availability of the other required constituents of a given crop. If one is lacking, even though all others are in excess, plants will still not thrive. The addition of a single constituent will increase the crop only if a particular soil can deliver the other necessary constituents in greater quantities as well. Depending on the circumstances, therefore, any of the essential minerals might become the controlling factor. This generalization became known as Liebig's "Law of the Minimum."

In 1862 Liebig completed an expanded seventh edition of his *Chemistry in Its Applications to Agriculture and Physiology;* the sixth edition had appeared in 1847. The new book was the most comprehensive statement of his views, backed with the most extensive analytical and field data, and it was also his last major scientific endeavor. His theories remained controversial and vulnerable. Some of the issues dividing him from his critics could not be settled at the time, for the question of the source of nitrogen depended on phenomena they did not yet understand. Liebig's role in agricultural chemistry was far different from his earlier role in organic chemistry. In both areas he displayed an acute perception of central issues and was a powerful advocate for his own opinions; but in organic chemistry he had also been the most distinguished experimentalist of his time. He could support his theoretical positions with analyses he knew to be more reliable than those of most of the chemists who disagreed. In agricultural chemistry he did carry out important analyses of plant composition and tested fertilizers on a small scale, but he was aware that for the large-scale investigations which would decide the key questions he was dependent on other people. He could act only as catalyst and critic, providing the chemical principles and the theory which he claimed would lead agriculturists to achieve significant investigations. When these agriculturists reached conclusions which appeared to contradict his views, he could not improve on their experiments, but only complain that they had failed to follow his reasoning. He could not direct the emergence of a new science from the inside, as he had done in organic chemistry, but only draw on the influence he had attained through his earlier leadership to induce others to follow the program he had laid out.

In agricultural chemistry, Liebig also indulged in bolder speculative conclusions than he had done in his previous investigations. In organic chemistry his deep knowledge of the properties and reactions of compounds had helped to control his theoretical bent. He acquired his knowledge of agriculture more hastily and largely at secondhand, and his reasoning from chemical principles was less checked by anticipations of complicating physiological or physiographical factors. From the rebuffs of his initial views, he learned to be a good deal more cautious. Yet he justly claimed that even though he had made mistakes, even though there were in the first edition of his book the most extraordinary inconsistencies and the greatest disorder, its appearance had completely changed the nature of the problem of scientific agriculture. Before 1840 it was generally believed that both plant and animal life were dependent on the circulation of an organic, previously living material. Now, whatever opinion individuals held on specific points, they agreed that the nutrient substances of plants were inorganic. That change had transformed the objectives of agriculture, for under the older conception the potential production of foodstuffs would seem to have a fixed limit, whereas in the new view an unbounded increase in organic life appeared possible.

Liebig's students and followers provided much of the means for the rigorous scientific study of agriculture that he envisioned, as they began to set up experiment stations in Europe and the United States. In that way they gradually removed the questions involved from the control of the scientifically unsophisticated farmers who had caused Liebig so much trouble. His abrupt change of subjects in 1840 thus marked a crucial step in the emergence of modern scientific agriculture.

Physiology. After finishing his first treatise on agricultural chemistry in 1840, Liebig entered with equal intensity into the study of animal chemistry. He brought to this topic the same confidence that his knowledge of the chemical properties of organic compounds would enable him to infer the transformations occurring within living organisms; the application of quantitative chemical methods would solve the problems that physiologists had failed to solve. He had already become interested in nutrition in 1838 when G. J. Mulder found that the elementary compositions of plant albumin, animal fibrin, casein, and albumin were identical. During the next three years Liebig's students verified and extended Mulder's results. From that basis Liebig concluded that animals receive the chief constituents of their blood already formed in their nutrients. He developed the idea that the nitrogenous plant substances were assimilated in the blood and organized tissues of animals, while starch, sugar, and other nonnitrogenous compounds were consumed in respiration. During 1841 and 1842 Liebig expressed his physiological ideas in three articles, then elaborated them more extensively in his book *Die Thierchemie oder die organische Chemie in ihrer Anwendung auf Physiologie und Pathologie* (Brunswick, 1842; English trans., *Animal Chemistry or Organic Chemistry in Its Applications to Physiology and Pathology* [London, 1842]). He maintained that animal heat is produced solely by the oxidation to carbon dioxide and water of the carbon and hydrogen of the nutrient compounds. The idea dated from Lavoisier's experiments on combustion and respiration, but during the intervening years some physiologists had proposed other views, and the most thorough experimental investigations undertaken had cast some doubt on it. Liebig had no new evidence to prove Lavoisier correct, but from his understanding of chemical reactions it seemed to him that there was no other possibility. Liebig broadened the combustion theory by perceiving that the respiratory oxidations were net reactions, encompassing the cumulative results of all of the transformations that nutrients may undergo while in the body. He also examined the mutual proportionalities between the internal nutritive metamorphoses within an animal, its respiratory exchanges, intake of food, excretions, and production of heat or work.

Beyond establishing such general principles, Liebig attempted to deduce the actual chemical transformations which the three classes of nutrients undergo within the body; and in order to depict these more concretely, he constructed hypothetical chemical equations similar in form to those he and his colleagues had been using to interpret the reactions of organic compounds in their laboratories.

Liebig's *Animal Chemistry* aroused sharply divergent reactions. Some accepted it uncritically as a revelation of the true inner workings of animals and humans. Others were so antagonized by its speculative excesses that they refused to credit the important insights it contained. Even those who reacted against it, however, began to view the chemical phenomena of life differently than they had before, for Liebig had provided one of the first comprehensive pictures of the overall meaning of the ceaseless chemical exchanges which form an integral part of the vital processes. A few men adopted it as a guide for further investigation, and Bischoff and Voit based the field of energy metabolism on Liebig's ideas. As with his agricultural chemistry, Liebig's physiological writings provided an impetus which outlasted the refutation of some of his specific theories.

By transferring his investigations into agriculture and animal chemistry, Liebig had hoped to avoid further disputes with his old rival Dumas, but to his dismay Dumas's interests were moving in the same direction. In August 1841 Dumas published a lecture which expressed nutritional ideas similar in general outline to those Liebig had already presented in his agricultural chemistry, and also to those on animal chemistry which he had begun to teach in Giessen. Liebig now accused Dumas of plagiarism, a charge that could only exacerbate the earlier bitterness between them. Over the following years Liebig engaged Dumas and his associates Boussingault and Payen in a debate over the most prominent difference between their respective physiological views. Liebig had asserted that animals can convert dietary starch into fats, whereas the French chemists insisted that the source of all animal constituents, including fats, is in plant nutrients. The French chemists carried out extensive feeding experiments on bees, geese, hogs, and cattle, which were expected to demonstrate the sufficiency of dietary fat, but which instead proved Liebig to be correct. The outcome enhanced the prestige of Liebig's physiological views.

Industrial Chemistry. Liebig did not directly involve himself in industrial chemistry as he did in agricultural chemistry, but his indirect influence was nearly as great. Many of his students founded or worked in the new large-scale chemical factories, and a number of the analytical methods he developed for research purposes found important industrial applications. Most decisive, however, was his effect on the dye industry. In 1843 Ernest Sell, a former student who had recently set up a plant to distill coal tar, sent Liebig a sample of a light oil which he had produced. Liebig asked one of his assistants, August Wilhelm Hofmann, to analyze the oil. Hofmann separated two organic bases, one of which, afterward named aniline, reacted with concentrated nitric acid to give a deep blue fluid that became yellow and then deep scarlet when heated. Liebig, who had already been interested in colored organic compounds as potential sources of dyes, predicted that aniline would have important effects on the dye industry. At first, Hofmann was more interested in using the new substance to help solve a theoretical question previously raised by Liebig: organic bases contain a hypothetical compound, amid (NH_2), combined with an organic radical, so that the radical substituted for one of the hydrogens of ammonia. Hofmann, who had found by the usual elementary analyses that aniline has the formula $C_{12}H_7N$, interpreted its composition in the light of Liebig's view, as

$$\left.\begin{array}{r} H \\ H \\ C_{12}H_5 \end{array}\right\} N.$$

Hofmann then began to try to substitute other organic radicals for the remaining hydrogen atoms. He followed this general approach over the next two decades, and it eventually yielded many of the compounds which formed the basis for the synthetic dye industry that burgeoned after 1860. Thus Liebig provided the institutional, analytical, and conceptual framework within which Hofmann began the study that helped give birth to the first industry based entirely on the application of the results of systematic scientific research.

In 1839 Liebig advanced the theory that fermentation is caused by the decomposition of a nitrogenous material, the reaction of which evokes a similar decomposition in another compound such as sugar. Although formulated mainly because of his dissatisfaction with Berzelius' definition of catalysis, which in Liebig's view provided no explanation for the phenomena it designated, the theory placed Liebig in the middle of the debate over whether fermentation is a vital or a purely chemical process. Ultimately his position drew him into a famous controversy with Pasteur, who defended with great resourcefulness the belief that living microorganisms are essential for fermentation.

In 1852 Liebig left Giessen for Munich. His old laboratory had continued to attract students from all over Europe, and increasingly from the United States, but he himself had grown weary of the unremitting burdens of his teaching. In Munich he had an institute built to his specifications, with better research facilities than in Giessen and a large lecture hall. He no longer carried on a formal program of instruction, but allowed only a few selected assistants and students to work in his laboratory. He himself continued to do some experiments, mostly following up certain aspects of his earlier research; but he devoted most of his time to writing, especially in support of his agricultural ideas. In the livelier milieu of Munich he also became more involved in society than he had been. He instituted a popular series of evening talks on scientific topics of broad interest. Although he remained quite active until his last illnesses, it was for Liebig a more quiet time than those twenty-eight hectic years when he was one of the most formidable figures of a formidable scientific age.

BIBLIOGRAPHY

The articles by Liebig and his contemporaries on which this text is based are mostly found in Poggendorff's *Annalen der Physik und Chemie*, Liebig's *Annalen der Pharmacie*, and the *Annales de Chimie et de Physique*. Indispensable sources are *Berzelius und Liebig Ihre Briefe von 1831–1845*, Justus Carrière, ed. (Munich, 1893; repr. Wiesbaden, 1967); and *Aus Justus Liebig's und Friedrich Wöhler's Briefwechsel in der Jahren 1829–1873*, A. W. Hofmann, ed. (Brunswick, 1888). The correspondence of Berzelius and Pelouze in *Jac. Berzelius Brev*, H. G. Söderbaum, ed. (Uppsala, 1941), supp. 2, contains useful references to Liebig.

Convenient descriptions of the analytical methods used in Liebig's laboratory are Justus Liebig, *Instructions for the Chemical Analysis of Organic Bodies*, William Gregory, tr. (Glasgow, 1839), and Henry Will, *Outlines of the Course of Qualitative Analysis Followed in the Giessen Laboratory* (London, 1841).

Carlo Paoloni, *Justus von Liebig. Eine Bibliographie sämtlicher Veröffentlichungen* (Heidelberg: Carl Winter, 1968), contains a full bibliography of Liebig's articles and books and published collections of letters, and a list of secondary literature concerning Liebig.

The most important collections of MSS of Liebig are in the Bibliothek der Liebig-Museum-Gesellschaft and the Universitätsarchiv of the Justus Liebig Universität in Giessen, which contain especially documents concerning Liebig's laboratory; and the Bayerische Staatsbibliothek in Munich, which has an extensive collection of correspondence.

The discussions in the present article based on unpublished letters between Liebig, Pelouze, and Dumas are derived from the following sources: the letters from Liebig to Pelouze and to Dumas are preserved in the Dumas dossier at the Archives of the Académie des Sciences in Paris; letters to Liebig from Pelouze and from Dumas are in the Bayerische Staatsbibliothek. I am grateful to these two institutions for making available to me microfilms of these documents.

Liebig's laboratory in Giessen is preserved as a museum in the state in which it existed after the additions of 1839.

The most extensive biography of Liebig is Jakob Volhard, *Justus Von Liebig* (Leipzig, 1909). A. W. Hofmann, *The Life-Work of Liebig*, Faraday lecture for 1875 (London, 1876), gives a laudatory but valuable discussion of Liebig's personality and contributions to organic chemistry. Many of the chemical investigations discussed above are summarized in J. R. Partington, *A History of Chemistry*, IV (London, 1964). For Liebig's agricultural chemistry, see F. R. Moulton, ed., *Liebig and After Liebig: a Century of Progress in Agricultural Chemistry*, Publications of the American Association for the Advancement of Science, no. 16 (Washington, 1942); and Margaret W. Rossiter, *Justus Liebig and the Americans: a Study in the Transit of Science, 1840–1880* (unpub. diss., Yale Univ., 1971).

For Liebig's animal chemistry, see F. L. Holmes, "Introduction," to Justus Liebig, *Animal Chemistry* (New York, 1964); and *Claude Bernard and Animal Chemistry* (Cambridge, Mass., 1974); and Timothy O. Lipman, "Vitalism and Reductionism in Liebig's Physiological Thought," in *Isis*, **58** (1967), 167–185.

Liebig's role in the controversies over fermentation is treated in Joseph S. Fruton, *Molecules and Life* (New York, 1972). Liebig's laboratory at Giessen is discussed most recently in J. B. Morrell, "The Chemist Breeders: the Research Schools of Liebig and Thomas Thomson," in *Ambix*, **19** (1972), 1–47. For Liebig's influence on the dye industry, see John J. Beer, *The Emergence of the German Dye Industry* (Urbana, Ill., 1959).

F. L. HOLMES

LIESGANIG, JOSEPH XAVER (*b.* Graz, Austria, 13 February 1719; *d.* Lemberg, Galicia [now Lvov, U.S.S.R.], 4 March 1799), *astronomy, geodesy.*

Liesganig was the son of Wolfgang Lisganig, *Hofmeister* of Graz, and his wife Rosalie. He entered the Society of Jesus in 1734 and received his education at the Jesuit College in Vienna. From 1742 to 1751 he served the order in various academic and ecclesiastical posts (having been ordained in 1749) in Austria, Hungary, and Slovakia; in 1752 he returned to the Vienna Jesuit College as *professor matheseos.* At the same time he was attached to the Jesuit astronomical observatory (of which he was appointed prefect in 1756, a position he held until the order was suppressed in 1773) and professor of mathematics at the University of Vienna. He was made dean of the university's philosophical faculty in 1771.

Liesganig's scientific career may be said to have begun with his prefecture of the observatory. In 1757, at the suggestion of Bošković, Liesganig constructed with Joseph Ramspoeck a ten-foot zenith sector that indicated tangents instead of angles. He also designed a quadrant (built by Ramspoeck) with a radius of two and one-half feet; it was provided with micrometer screws for accuracy in reading. Another of Liesganig's instruments consisted of a Graham clock fitted with a gridiron compensation pendulum.

In 1758 Liesganig had used his zenith sector to determine the latitude of Vienna to be 48° 12′ 34.5″.

In 1760 Maria Theresia, at Bošković's urging, commissioned Liesganig to survey the environs of the city. Liesganig for this purpose obtained from Canivet a copy of the Paris toise, on which he calibrated the Vienna fathom. During the following years, Liesganig established a series of bases near Wiener Neustadt and in the Marchfeld region; he incorporated these into a triangulation system with the azimuth of Mt. Leopold, near Vienna, and calculated the geographic latitudes of Warasdin, Brünn (Brno), and Graz. (A later survey, made in 1803, confirmed the exactness of Liesganig's measurements of the Graz–Vienna–Brünn triangle, but found other errors in his survey which were attributed to defects in his azimuth zenith.) He also carried out experiments with a seconds pendulum (1765) to determine the length of the Vienna fathom, and made computations with a sphere in which the reduction to sea level represented 2.3 fathoms per degree of meridian.

In 1769 Liesganig undertook a survey in Hungary between Szeged and Peterwardein (now Petrovaradin, Yugoslavia). For this purpose he constructed bases near Kistelek and Csuróg and linked them, by a chain of twenty-six triangles, to the meridian of Budapest. In connection with this project Liesganig wrote a manual on map-making methods, preserved in manuscript in the Vienna war archives.

From 1772 until 1774 Liesganig was engaged in a trigonometric survey of East Galicia and Lodomeria (now Vladimir-Volynsky, Ukrainian S.S.R.), which had just become a part of that province. He made use of five brass quadrants that had been built to his specifications by Ramspoeck and the Viennese clockmaker Schreibelmayer; each was equipped with a movable telescope having a cross-hair micrometer at the focal point and a fixed telescope having a micrometer at the focal point of a microscope. (These instruments could also be used as spirit levels.) During the same period Liesganig observed solar azimuths in Lemberg, near Cracow, and in Rzeszów, reducing all the measurements to the meridian and clock of the Lemberg observatory. At the same time he held the position of Imperial and Royal Building Inspector in that city and from 1775 was professor of mechanics at the Collegium Nobilium there, as well as provincial director of military engineering and navigation.

Liesganig made six circular maps from his data of the Lemberg area, although these lacked place names and topographical information. In 1784 he reviewed the cadastral survey of the area of Gutenbrunn, in Lower Austria; the following year he took over the direction of the cadastral survey of Galicia and prepared for it a manual that was translated into Polish, Czech, Slovene, and Italian. In this work he recommended that measurements be made in local linear units to permit their comparison to the Vienna fathom.

BIBLIOGRAPHY

I. Original Works. Liesganig's writings include *Tabulae memoriales praecipue arithmeticae tum numericae tum literalis cum tabulis tribus figurarum* (Vienna, 1746); *Prolusio ad auditores matheseos* (Vienna, 1753); *Tabulae memoriales praecipue arithmeticae tum numericae tum literalis, geometriae, etiam curvarum et trigonometriae, atque utriusque architecturae elementa complexae* (Vienna, 1754); "A Short Account of the Measurement of Three Degrees of Latitude Under the Meridian of Vienna," in *Philosophical Transactions of the Royal Society*, **58** (1767), 15–16; and *Dimensio graduum meridiani Viennensis et Hungarici, Augustorum jussu et auspiciis suscepta* (Vienna, 1770).

II. Secondary Literature. For early biographical literature, see Constant von Wurzbach, *Biographisches Lexikon des Kaiserthums Österreich*, XV (Vienna, 1866), 179–180. Important details are in Franz Freiherr von Zach, "Über die Sternwarte von Lemberg," in *Monatliche Correspondenz zur Beförderung der Erd- und Himmelskunde*, **4** (1801), 547–550.

See also "Zusätze des Herausgebers: Über die Lemberger Sternwarte, über die trigonometrische Aufnahme von Galizien und Lodomerien und die darauf gegründete Karte dieser Länder, und über die geographische Bestimmung von Lemberg," *ibid.*, 550–558; "Nachschrift des Herausgebers zu D. Seetzen's Reise-Nachrichten," *ibid.*, **7** (1803), 24–36; "Geographische Bestimmung einiger Orte in Ungarn, aus Liesganig's Ungarischer Gradmessung, nebst einem Verzeichniss aller in Ungarn astronomisch und trigonometrisch bestimmten Orte," *ibid.*, 37–48 (including commentary on 50–52 attributed in most biographies to Liesganig); "Beweis, dass die Oesterreichische Gradmessung des Jesuiten Liesganig sehr fehlerhaft, und zur Bestimmung der Gestalt der Erde ganz untauglich war," *ibid.*, **8** (1803), 507–527, and **9** (1804), 32–38, 120–130.

More recent treatments of Liesganig are Günther, in *Allgemeine deutsche Biographie*, XVIII (Leipzig, 1883), 637; A. Westphal, "Basisapparate und Basismessungen," in *Zeitschrift für Instrumentenkunde*, **5** (1885), 257–274, 333–345, 373–385, 420–432 (esp. 342–345, 374); Ernst Nischer, *Österreichische Kartographen. Ihr Leben und Wirken* (Vienna, 1925), esp. pp. 76–80; Richard Krauland, "Legales und internationales Meter in Österreich," in *Österreichische Zeitschrift für Vermessungswesen*, **37** (1949), 30–42; K. Lego, "Abbé Joseph Liesganig zur 150. Wiederkehr seines Todestages," *ibid.*, 59–62; Paula Embacher, "Die Liesganig'sche Gradmessung," *ibid.*, **39** (1951), 17–22, 51–55; and Herwig Ebner, "Joseph Liesganig. Ein Beitrag zu seiner Biographie," in *Blätter für Heimatkunde*, **36** (1962), 129–131.

Walther Fischer

LIEUTAUD, JOSEPH (*b.* Aix-en-Provence, France, 21 June 1703; *d.* Versailles, France, 6 December 1780), *anatomy, medicine.*

Lieutaud was raised by his uncle, the physician and botanist Pierre-Joseph Garidel. He received a degree in medicine at Aix and a later, honorary one from Paris in 1752. Lieutaud moved to Paris and, with the help of Sénac, chief physician to Louis XV, began a career as a royal physician. He held successive appointments as physician to the royal children (in 1755), chief physician to Monsieur and to the Comte d'Artois, and finally chief physician to Louis XV. An essentially modest man, Lieutaud then asked J. M. F. de Lassone to assume his duties at court so that he might devote his time to his own research. By so doing he further stimulated the professional activity of the other court physicians, as well as that of the doctors of the Versailles infirmary, of which he was more or less the director.

Lieutaud's lifework was strongly oriented toward practical medicine, a notion that was then being defined. Practical medicine as such rejected all theoretical systems and speculative etiologies; in the schools it sought to replace Latin with the vernacular to benefit physicians who should "read little, see a great deal, and do a great deal," as Fourcroy later put it. Such medicine was to be concerned with facts, not opinions, and to be learned at the bedside and in the autopsy room. While it might be illuminated by the "preliminary sciences"—that is, by "most of mathematics, experimental physics, chemistry, anatomy, natural history, etc."—erudition alone was insufficient to the discipline. Lieutaud, as one of the leaders of the new school, "reproached those whom he saw occupied solely with accumulating books" (although he himself possessed a splendid library, in part inherited from his uncle).

For these reasons Lieutaud devoted a great deal of attention to postmortem observations. From 1750 on, he examined more than 1,200 cadavers at the Versailles infirmary, as well as attending the autopsies of the son of the dauphin (1761), the dauphin (1765), and the dauphine (1765). He used postmortem materials to study both normal anatomy and pathology; he further undertook to determine the correlation between the symptoms recorded during the life of a patient and the physical lesions found after his death. Lieutaud thus established a method based upon complete and precise observation of the patient which was carried on by his students, including Pierre Edouard Brunyer, his successor at the infirmary. He may thus be considered one of the French pioneers in the pathological anatomy and anatomico-clinical concept of disease that gradually replaced the various medical systems of the eighteenth century.

Lieutaud's clinical practice included the inoculation of his patients against smallpox, using Sutton's method. Lassone reported to the Académie des Sciences the successful results of a series of inoculations that Lieutaud performed on 18 June 1774 at the Château of Marly; this communication contributed greatly toward modifying the generally conservative French attitude toward inoculation. Lieutaud himself reported to the Academy on a number of subjects, including the pathology of the gall bladder; empyema of the frontal sinus; hydrocephalus; osteoma of the cerebellum; pathological anatomy of the spleen, heart, and stomach; hydatids of the thyroid; laryngeal polyps; and the correlation between stomach contraction and splenic hypertrophy (or, conversely, between stomach distension and splenic hypotrophy). This last finding was accepted by Haller and Soemmerring, but criticized severely by Bichat.

Like his clinical work, Lieutaud's writings were directed toward practical medicine. His *Essais anatomiques* of 1742 was widely read for thirty years, and was given a scholarly second edition, as *Anatomie historique et pratique*, by A. Portal in 1776–1777. Although Lieutaud's intent was to place basic practical knowledge of both normal and pathological anatomy (which he considered to be two branches of the same field) at the disposal of the practitioner, his book and the techniques presented in it surpassed this aim. The first part of the book is devoted to descriptive anatomy, while the second gives precise instructions for dissecting both mature and fetal human bodies, ending with a "recapitulation" that is actually an essay on topographical anatomy in which Lieutaud demonstrates the disposition of the organs in one system after another. (It is typical of Lieutaud's thoroughness that in the second section he devoted a full seven pages to the technique for dissecting the middle and inner ear.)

In *Précis de la médecine pratique* of 1759, Lieutaud offered a classification of diseases as being those of the general population or those affecting men, women, or children, respectively. He set out the symptomatology of each disease and discussed the therapy for it; although he treated surgical therapy only in broad outline, his prescriptions of medical treatments are fully detailed. The book includes a number of case histories drawn from Lieutaud's own observations and from the literature.

Précis de la matière médicale (1766) is a catalog, with an index in both French and Latin, in which Lieutaud classified drugs by their pharmaceutical complexity and gave directions for their compounding and their therapeutic indications. In it he recommended the use of simple, "domestic" (or "Galenic") medicines, of which the effects are well known, and denounced the polypharmacy of the Arabs and the pyrotechnical remedies that had been fashionable in the preceding century. Lieutaud's standards for inclusion in his materia medica were, indeed, so strict that he here presented "only a twentieth part" of the contents of other contemporary manuals.

Lieutaud's work earned him admission to the Académie des Sciences as a corresponding member in 1731; in 1751 he became adjunct anatomist and, in 1759, associate anatomist. He was also a member of the Royal Society. He remained a grateful friend of the Paris Faculty of Medicine, and did not hesitate to support it against its opponents—as when on 22 December 1777 he arranged to present at court Sigault, who had, against the unanimous advice of the Académie Royale de Chirurgie, just performed a symphysiotomy on a woman named Souchot. Although he actively opposed the founding of the Société Royale de Médecine, which was nonetheless accomplished by Lassone and Vicq d'Azyr in 1778, he later accepted its presidency.

Lieutaud died five days after contracting pneumonia; faithful to his principles, he refused all remedies. He was by royal favor buried in the Ancienne Église near Notre-Dame de Versailles. He had previously sold his library to the Comte de Provence, who had allowed him to continue to use it; at the death of the Comte de Provence Lieutaud's books were taken to Versailles, where they now constitute the bulk of the medical collection of the municipal library. His name is perpetuated in the medical literature in Lieutaud's sinus, Lieutaud's uvula, and, especially, Lieutaud's body—the triangular area, limited by the interureteric fold and by the uvula of the bladder, that he isolated and described.

BIBLIOGRAPHY

I. ORIGINAL WORKS. Lieutaud's communications to the Académie des Sciences, all published in *Histoire de l'Académie royale des sciences*, include "Observations sur la vésicule du fiel, sur une quantité très considérable de pus dans les sinus frontaux, sur deux livres au moins de sérosité très claire, trouvées dans les ventricules du cerveau," in *Histoire . . . année 1735* (1738), 16–22; "Observation sur un corps osseux . . . trouvé dans le cervelet d'un jeune homme de 18 ans . . .," in *Histoire . . . année 1737* (1740), 51; "Observation sur la grosseur naturelle de la rate," in *Histoire . . . année 1738* (1740), 39; "Sur une maladie rare de l'estomac, sur le vomissement et sur l'usage de la rate, sur un écu de six livres avalé, sur une maladie singulière . . .," in *Histoire . . . année 1752* (1756), 45–49, 71–75; "Relation d'une maladie rare de l'estomac avec quelques observations concernant le méchanisme du vomissement et l'usage de la rate," *ibid.*, 223–232; "Observations anatomiques sur le coeur," *ibid.*, 244–265, 308–322; "Observations anatomiques sur la structure de la vessie," in *Histoire . . . année 1753* (1757), 1–26; "Observations anatomiques sur le coeur. Troisième mémoire contenant la description particulière des oreillettes du trou oval et du canal artériel," in *Histoire . . . année 1754* (1759), 369–381; and "Observations sur les suites d'une suppression et sur les hydatides formées dans la glande thyroïde, sur un polype en forme de grappe, situé immédiatement au dessous du larynx," *ibid.*, 70–74.

Lieutaud's books, most of which were published in several editions, are *Essais anatomiques contenant l'histoire exacte de toutes les parties qui composent le corps de l'homme avec la manière de disséquer* (Paris, 1742); 2nd ed. by A. Portal, as *Anatomie historique et pratique. Nouvelle edit. augmentée de diverses remarques historiques et critiques et de nouvelles planches*, 2 vols. (Paris, 1776–1777); *Elementa physiologiae, juxta solertiora, novissimaque physicorum experimenta et accuratiores anatomicorum observationes concinnata* (Amsterdam, 1749); *Précis de la médecine pratique, contenant l'histoire des maladies et de la manière de les traiter avec des observations et remarques critiques sur les points les plus intéressants* (Paris, 1759); *Synopsis universae praxeos medicae in binas partes divisa* (Amsterdam, 1765); English trans. by E. A. Atlee as *Synopsis of the Universal Practice of Medicine. Exhibiting a Concise View of all Diseases, Both Internal and External; Illustrated With Complete Commentaries* (Philadelphia, 1816); *Précis de la matière médicale, contenant les connaissances les plus utiles sur l'histoire, la nature, les vertus et les doses des médicaments, tant simples qu'officinaux, usités dans la pratique actuelle de la médecine avec un grand nombre de formules éprouvées* (Paris, 1766); and A. Portal, ed., *Historia anatomica medica, sistens numerosissima cadaverum humanorum extispicia, quibus in apricum venit generina morborum sedes, horumque reserantur causae, vel patent effectus. Recensuit suas observationes numero plures adjecit* (Paris, 1767).

II. SECONDARY LITERATURE. On Lieutaud and his work see the article under his name in *Biographie universelle ancienne et moderne*, XXIV (Paris, 1819), 470–471; that by A.-L. Bayle and A.-J. Thillaye, in their *Biographie médicale*, II (Paris, 1855), 321–322; and A. Chéreau, in *Dictionnaire encyclopédique des sciences médicales Dechambre*, 2nd ser., II (Paris, 1876), 555–556.

See also M. Bariéty, "Lieutaud et la méthode anatomoclinique," in *Mémoires de la Société française d'histoire de la médecine*, **3** (1947), 13–16; P. Brassart, "Contribution à l'étude du monde médical versaillais sous le règne de Louis XVI et pendant la Révolution," medical thesis, University of Rennes (1965); P. Delaunay, "Le monde Médical parisien au XVIIIème siècle," medical thesis, University of Paris (1906), 443–453; and C. Fiessinger, "La thérapeutique de Joseph Lieutaud," in his *Thérapeutique des vieux maîtres* (Paris, 1897), 250–256.

Éloges are that published anonymously in *Histoire de la Société royale de médecine . . . année 1779* (1782), 94–117; and A. de Condorcet, in *Mémoires de l'Académie royale des sciences . . . année 1780* (1784), 46–59.

PIERRE HUARD
MARIE JOSÉ IMBAULT-HUART

LIGNIER, ÉLIE ANTOINE OCTAVE (*b.* Pougy, Aube, France, 25 February 1855; *d.* Paris, France, 19 March 1916), *botany.*

His father, who was probably a farmer, is described on Octave's birth certificate as a landowner. Nothing is known of the botanist's primary and secondary schooling. He obtained his bachelor of arts degree in Paris in 1873, and a baccalaureate in science at Lille in 1880.

As teaching assistant in the Faculté des Sciences at Lille, he worked under the direction of the paleobotanist Charles Eugène Bertrand and helped to organize the Botanical Teaching and Research Laboratory. He subsequently helped to reorganize the Botanical Gardens, which had been moved from Lille to La Louvière. At the same time, he was working toward his licentiate, which he was awarded in 1882, having studied under the geologist Gosselet and the zoologist Giard, as well as Bertrand. In 1887 he received his doctorate in natural sciences from the University of Paris. In the same year he was appointed assistant lecturer in the Faculté des Sciences at Caen, where in 1889 he was made full professor.

When he first arrived in Caen, the town had only a few collections of dried plants, a Municipal School of Botany, some greenhouses, and a park. Lignier had to compile a catalog, which he published with B. Le Bey, and later with M. Lortet. Through sheer persistence he succeeded in having a botanical institute built. His teaching covered all aspects of botany, including paleobotany. He published some 110 writings of his own and coauthored others. His work encompassed both existing and fossil plants.

His anatomical research was directed particularly at the floral organs, seeds, and fruit of the Myrtaceae, Oenothera, Cruciferae, and Fumariaceae, and at the vascular bundles of the stems and leaves of many families, but only part of his research was ever published. He regarded the Gnetales as angiosperms.

He published some remarkable studies of the Bennettitales of Normandy. He was wrong, though, in believing the *Bennettites morierei* to be parthenogenetic, which he inferred from the closed micropyles that he had found in them, and he assumed that the loss of sexual reproduction was responsible for the disappearance of progeny.

He was also interested in paleozoic materials, specifically the *Radiculites reticulatus*, which he first placed in the Taxodiaceae and then assigned to the *Poroxylon*, close to the *Cordaites*. Finally, he described for the first time with rare precision fossil woods that came from Normandy.

But what attracted the attention of the world's leading paleobotanists were Lignier's theories on evolution and proposed classifications. Among his theories were the cauloid concept, involving an undifferentiated fundamental element with dichotomous ramification, which underlies the theory of the present telome, and the meriphyte concept, which attributes such great importance to the fibrovascular system of the leaf in phanerogams.

He contributed greatly to our understanding of the group of Articuleae, a name which he invented to cover various genera. He became interested also in the *Stauropteris* because of their relevance to theory.

Concerning the hypotheses advanced by E. A. N. Arber and J. Parkin on the Euanthostrobiles and the Proanthostrobiles to explain the origin of the angiosperms, Lignier was generally inclined to go along with them, although he did suggest some modifications. He disagreed in particular with the conclusions about the Bennettitales. Notwithstanding G. R. Wieland's opinion, he regarded their fruiting as an inflorescence and not as a flower or part of a flower.

BIBLIOGRAPHY

I. ORIGINAL WORKS. For a listing of Lignier's writings, see *Titres et travaux scientifiques de M. Octave Lignier* (Laval, 1914).

II. SECONDARY LITERATURE. For further information on Lignier's work, see L. Emberger, *Les plantes fossiles dans leurs rapports avec les végétaux vivants* (Paris, 1968). pp. 157–158, 382, 651; C. Houard, and M. Lortet, "Rapport annuel pour 1916 sur l'Institut botanique et les collections botaniques de Caen," in *Bulletin de la Société linnéenne de Normandie*, 6th ser., **9** (1919), 237; E. C. Jeffrey, "Octave Lignier," in the *Botanical Gazette*, **62** (1916), 507–508; D. H. Scott, *Studies in Fossil Botany*, I (London, 1920), 334–337, 413 ff.; II (1923), *passim;* and A. C. Seward, *Fossil Plants* (Cambridge, 2, 1910; 3, 1917; 4, 1919).

F. STOCKMANS

LILLIE, FRANK RATTRAY (*b.* Toronto, Canada, 27 June 1870; *d.* Chicago, Illinois, 5 November 1947), *embryology, zoology.*

Lillie's father, George Waddell Lillie, accountant and partner in a wholesale drug company, was a man of exceedingly upright character, but without special intellectual interests. His mother, the former Emily Ann Rattray, was the daughter of a Scottish tobacco merchant who later became a Congregational minister and amateur astronomer.

An indifferent student at a boys' grammar school, in high school Lillie was most interested in extracurricular scientific activities. High school was a time of emotional stress and religious doubts, but he was converted to the Church of England and decided to make religion his lifework. During his last year in high school he met Alexander J. Hunter, who taught him to collect, identify, and arrange insects and fossils. Both entered the University of Toronto expecting to study for the ministry. They majored in natural sciences, believing that science posed a threat to religion. During their last two years they worked with minimum guidance and literally lived in the laboratory except when discussing science and evolutionary theory, or conducting their collecting expeditions. Lillie was increasingly disillusioned by many churchmen. A. B. Macallum showed him how he could devote his life to science, a choice Lillie made during his senior year, and also transmitted to him his physiological point of view. R. Ramsey Wright imbued Lillie with his lifelong interest in embryology and recommended him, upon his graduation with honors, to work as a fellow with C. O. Whitman at the Marine Biological Laboratory at Woods Hole, Massachusetts, and afterward (1891–1892) at Clark University.

In 1891, when Lillie arrived at Woods Hole, E. B. Wilson, E. G. Conklin, A. B. Mead, and L. Treadwell were conducting cell-lineage studies in an attempt to determine the exact contribution of each blastomere to the formation of larval organs. Local ponds contained abundant supplies of freshwater clams bearing eggs and embryos in their gills; Whitman suggested that Lillie study the cell lineage of these lamellibranchs for his doctoral dissertation. In 1892 Whitman, accompanied by Lillie and his other students, moved to the newly established University of Chicago as chairman of the zoology department.

During 1892–1893 Lillie was a fellow in morphology at Chicago and the following year a reader in embryology. He received his doctorate in zoology *summa cum laude* in 1894 with an outstanding thesis which received wide praise and established him as a young investigator of great promise. From 1894 to 1899 he was instructor in zoology at the University of Michigan and from 1899 to 1900 professor of biology at Vassar College; he then returned to the University of Chicago as assistant professor of embryology. He became associate professor in 1902, professor in 1906, chairman of the department of zoology in 1910, Andrew MacLeish distinguished service professor of embryology and dean of the newly established division of biological sciences in 1931, and emeritus professor in 1935.

In 1895 Lillie married Frances Crane, who in the summer of 1894 had been a student in his embryology course at Woods Hole. They had five daughters and one son and adopted three sons. Their Chicago and Woods Hole homes offered warmth and hospitality to a wide circle of friends, although life with Mrs. Lillie was anything but peaceful because of her extensive social work and marked changes in religious preference. The Lillies gave the Whitman Laboratory of Experimental Zoology to the University of Chicago at a time when that institution would permit solicitation of funds only for its new medical school.

Until 1916 Lillie taught an embryology course for medical students and until 1924 another for advanced undergraduate and beginning graduate students. His undergraduate teaching resulted in publication of a useful laboratory manual (1906), twice revised under slightly different titles (1919, 1923). More important, it resulted in publication of his classic book on chick embryology (1908), revised by Lillie (1919) and later by H. L. Hamilton (1952), one of Lillie's "scientific grandchildren." This textbook has been used by students throughout the world and remains one of the best accounts of bird development.

Lillie offered graduate seminars on the physiology of development, problems of fertilization, and the biology of sex. His approach to research training was rigorous. He suggested a problem, usually related to his own current investigations, briefly discussed possible approaches, let the student know that conscientious application and results were expected, then left him largely to his own resources until he was prepared to discuss his results and present a preliminary interpretation of their meaning.

Upon Lillie's return to Chicago (1900) the operation of the department of zoology devolved increasingly on him as Whitman became progressively more engrossed in his own research. In 1910 Lillie was named second chairman of the department. Although adhering to Whitman's emphasis upon original research and graduate student training as the primary mission of the department, Lillie's administration differed from Whitman's in two ways: the department functioned more democratically, and the scope and importance of the undergraduate program in zoology were greatly strengthened, with all members of the department teaching courses specifically for undergraduates.

Lillie resigned his chairmanship in 1931 to become dean of the newly established division of biological sciences, which included not only the usual biological science departments but also the departments of psychology, home economics, physical culture, and preclinical and clinical medicine. He was faced with amalgamating established preclinical departments with newly established clinical ones at a time when there was much disagreement about the advisability of attempting to operate both Rush Medical College and the university's new medical school and about the basic organization and management of the latter. By the time of his retirement four years later he had succeeded.

Lillie is most widely known for his vital roles in development of the Marine Biological Laboratory, one of the strongest influences on the rapid development of the broad field of biology in America. He first arrived at the laboratory at the beginning of its fourth summer session. There was only one small building and no equipment besides a steam launch and a few skiffs. Accommodations existed for only a few investigators. There was no endowment; each summer ended with a deficit, covered initially by members of the board of trustees. Lillie returned to Woods Hole every summer thereafter for fifty-five years and during his lifetime became the leader in every aspect of the growth and development of the laboratory.

Lillie was assistant in the embryology course at Woods Hole when Whitman started it in 1893; he became course director the following year. At the same time operation of the entire laboratory began to devolve increasingly on Lillie, while the board of trustees resisted expansion and became increasingly reluctant to meet annual deficits. In 1900 Lillie was named assistant director. He immediately interested Charles R. Crane, his wife's favorite brother, in the operation and support of the laboratory and firmly backed Whitman in convincing the corporation and board of trustees to reverse their decision to transfer the laboratory to the Carnegie Institution of Washington as its permanent marine laboratory. Lillie was named director of the laboratory in 1910, and held that position until 1926. During this period, coinciding closely with the period in which Crane served as president of the board of trustees (1904–1924) and contributed much to the support of the laboratory, Lillie and Crane convinced financiers and foundations that the laboratory deserved their strong support. Consequently, Woods Hole became a great institution for marine biology. Lillie served as president of the corporation and board of trustees from 1925 to 1942 and then as president emeritus.

In 1915, a year before the National Research Council was established, Lillie was elected to the National Academy of Sciences. In 1919, as representative of the American Society of Zoologists to the Division of Biology and Agriculture of the National Research Council, he was made a member of the executive committee and immediately received their unqualified support in his efforts to obtain increased financial support for the marine biological laboratory. He was vice-chairman and chairman of the division in 1921 and 1922, respectively. He was further active in the establishment, in 1922, of the Union of Biological Sciences, which assumed full responsibility for publication of *Biological Abstracts* and became chairman of the board administering the newly established National Research Council fellowships in the biological sciences (1923), later serving on the natural sciences fellowship board when the separate programs in physical and biological sciences were combined. He served effectively on the Division of Medical Sciences of the National Research Council Committee for Research in Problems of Sex, established in 1921 to administer grants and offer advice to investigators in this area, and wrote the first chapter of the 1932 and 1939 editions of *Sex and Internal Secretions*.

The year he became emeritus professor, Lillie was unanimously elected president of the National Academy of Sciences (1935–1939) and chairman of the National Research Council (1935–1936), the first man to hold these two key positions simultaneously. He was given this extraordinary joint responsibility in an attempt to eliminate serious controversies concerning the activities rightfully belonging to each of these two organizations. He did so with such dispatch that the task was accomplished in one year. He simultaneously strengthened relations of the Academy and Council with agencies of the federal government.

For some years Lillie had discussed with Wickliffe Rose, president of the General Education Board—and the latter with John C. Merriam, president of the Carnegie Institution of Washington, and Vernon Kellogg, chairman of the National Research Council—the need for an oceanographic institution on the East Coast. In large part the result of his labors, the Woods Hole Oceanographic Institution was formally incorporated on 6 January 1930. Financing was so well organized in advance that the building was occupied 15 June 1931 and its research ship, *Atlantis*, was available 31 August 1931. Lillie was president of the Oceanographic Institution from 1930 to 1939 and in 1940 received the Agassiz Medal of the National Academy in recognition of his efforts.

Lillie's research contributions were as remarkable as his administrative accomplishments. His chronological bibliography (Willier, 1957) contains 109 titles,

LILLIE

sixty-six of which are original works dealing with varied but always fundamental problems of development; some were published ten years after his retirement. He only rarely published research papers with his students, although after his retirement he did publish jointly with his research associates, Mary Juhn and Hsi Wang.

Lillie's research was masterfully planned and executed. His work invariably opened up new areas of research. He was adept at creative interpretation and constantly strove to tie details together by broad interpretive concepts because he firmly believed that sound speculation was the only true stimulus for future research, whether or not it survived additional probing. That he never gave up active participation in research is exemplified by his service as managing editor or member of the editorial boards of several journals in biology and experimental zoology.

Beginning with his doctoral dissertation (1895), a superb descriptive study on cell lineage of the freshwater mussel *Unio*, Lillie's early work (1892–1909) dealt primarily with problems of mechanisms of cleavage and of early development of eggs of invertebrates, notably those of *Unio* and of the marine annelid *Chaetopterus*. His emphasis was on how the special features of cleavage in each species were adapted to the needs of the future larva, rather than on the common features of cleavage even in different phyla, which were emphasized by many others. His studies on eggs with rigid mosaic cleavage (with developmental fates of cleavage cells already determined in early cleavage stages) led him to explore in detail the cytology of such eggs in an attempt to find some material basis for cleavage patterns, polarity, bilateral symmetry, and so on. His light microscope studies on the organization of egg protoplasm in normal and centrifuged eggs and on the extent to which differentiation without cleavage is possible in activated *Chaetopterus* eggs led to the conclusion that the control of early development must reside in the architecture of the ground substance of the egg, that is, on its physical and chemical properties. Thus, along with E. B. Wilson, he focused attention on the necessity for studying the ultramicroscopic organization of the egg; but even with modern ultrastructural techniques there is still a long way to go in understanding the basic nature of egg organization, established in the mother's ovary and important in control of early development.

During this period Lillie deviated briefly from his major line of investigation. One paper concerned regeneration in the protozoan *Stentor*; two others, regeneration and regulation in planarians; one, the effect of temperature on animal development. Of greater import was the appearance of two papers in 1903 and two in 1904 involving operations on living chick embryos *in ovo*. The first two described and analyzed experimentally the formation of the amnion (which fills with amniotic fluid, thus enclosing the embryo in its own miniature aquarium). By the use of heated needles or electrocautery he produced anamniote embryos which could develop normally for five to six days, and by localized destruction of specific sectors of the amniotic folds he could demonstrate convincingly certain causal relationships between them and the role of their progressive fusion in elevating amniotic folds at adjacent levels.

The second two papers primarily demonstrated the inability of parts of the embryo posterior to the vitelline arteries to regenerate following cauterization of the areas or rudiments that should have formed them; they further showed the inability of the right wing bud to regenerate following its extirpation. Although his own experimental analysis of chick embryo development was limited to these four publications, Lillie was convinced that chick embryos were suitable material for almost any type of experimental analysis of embryological problems.

From 1910 to 1921 Lillie's research centered primarily on the morphology and physiology of fertilization in the annelid *Nereis* and the sea urchins *Arbacia* and *Strongylocentrotus*, resulting in the publication of eighteen papers, one book (1919), and a chapter in a book (1924). Lillie hypothesized the existence of specific combining groups in sperm and in eggs and developed a far-reaching, comprehensive "fertilizin theory" of the role of union (linkage) of such combining groups in fertilization phenomena, including activation of the egg by sperm or parthenogenetic agents, species-specificity of fertilization, prevention of polyspermy, inhibition of fertilization by what he believed to be a blood inhibitor from adult sea urchins, and so on; these results are summarized schematically in ". . . The Mechanism of Fertilization in *Arbacia*" (1914, fig. 1, p. 579). Lillie was the first to visualize the linkage of his hypothetical specific combining groups in lock-and-key fashion, analogous with hypothetical antigen-antibody or immunological reactions. He thus applied entirely new ideas to the physiology of fertilization. Although many aspects of his basic theory have had to be modified in the light of increased knowledge and understanding of the molecules and combining groups involved, many investigators still believe that interactions between specific combining groups in antifertilizin molecules on the sperm surface and specific combining groups in the fertilizin molecules of the egg cortex play essential roles in the fertilization process (see Metz, 1957, 1967, for the continuing impact of Lillie's ideas in this research area).

Lillie's first publication concerning sexual differentiation appeared in 1907. His interest in this research area soared in 1914, when the manager of his large Buffalo Creek farm, where he kept a herd of purebred cattle in which the birth of freemartins occurred, sent him a pair of calf fetuses still within their fetal membranes. H. H. Newman, because of his interest in twins, was first offered this material for study but he was too heavily committed to other tasks to examine it. Therefore Lillie tackled it and immediately became intensely interested in analyzing the factors responsible for development of the freemartin, which, largely on the basis of his work, was subsequently defined as a sterile genetic female born twin to a normal genetic male. The cooperative foreman of the Swift and Company abattoir at the nearby Union Stockyards agreed to notify Lillie whenever he found a pregnant uterus (preferably with both ovaries attached) containing twins as young as possible.

By 24 February 1916 Lillie had carefully analyzed forty-one cases of bovine twins and announced his theory of the freemartin. His full analysis, based on fifty-five pairs of fetal twins, was published the following year (1917). This work is usually considered his most significant and enduring contribution to research (see Burns, 1961, for an authoritative review and impact of this theory), although Lillie's explanation of causal relationships in freemartin development has been criticized by Short (1970).

Briefly, Lillie's findings were as follows:

1. Cattle twins are two-egg or nonidentical twins (each ovary contains an ovulation site or corpus luteum).

2. The outermost extraembryonic membrane of bovine fetuses, the chorion, is greatly elongated and, in the case of twin fetuses, the two chorions (and their contained allantoic or umbilical blood vessels, especially the arteries) usually fuse at an early stage, so that blood is freely exchanged between the twin fetuses.

3. Differentiation of the gonads of genetic males into testes occurs early in fetal life, as does the differentiation of the interstitial cells of the testes, known to secrete male hormone in adult males.

4. Under such circumstances the gonads of the genetic female twin differentiate as ovotestes or as rudimentary testes, which are usually sterile.

5. The mesonephric or Wolffian ducts, and some of the mesonephric tubules, of the genetic female twin persist and develop into typical male vasa deferentia and vasa efferentia (instead of degenerating, as they would in genetic females).

6. The Müllerian ducts of the genetic female twin usually degenerate (as they normally do in genetic males) instead of forming oviducts, uterus, and possibly the vagina (as they normally do in genetic females).

7. The degree of masculinization of the gonads and reproductive ducts in the genetic female twin is greater the earlier and the more complete the fusion of the extraembryonic blood vessels of the twin fetuses.

8. The development of the genetic male reproductive system is always normal under such circumstances.

9. In those rare instances when the extraembryonic blood vessels of the heterosexual twins fail to fuse during fetal life, the reproductive systems of both members of the heterosexual twin pair develop normally.

Lillie concluded that male hormone produced by the fetal testes of the genetic male twin enters the genetic female twin via the fused extraembryonic circulations and in the genetic female inhibits development of gonads into ovaries, stimulates development of mesonephric ducts and tubules into male-type sperm passageways, and inhibits development of Müllerian ducts into components of the female reproductive system. An almost identical theory was proposed independently by two Austrian investigators in an obscure publication (see in *Science*, 1919).

This theory to account for development of the freemartin condition led directly to the concept that fetal sex hormones play a comparable role in normal sex differentiation in mammals and other vertebrates, once gene control of sex differentiation causes the gonads to develop into testes or ovaries. Thus in the normal sex differentiation of mammals the duct system develops in the male direction in genetic males because of the presence of fetal testes producing male hormone during fetal life, whereas the duct system develops in the female direction in genetic females because of the absence of fetal testes producing male hormone during fetal life. Thus fetal castration of genetic male and genetic female mammals, such as rabbits, should result in development of typical female-type reproductive systems regardless of genetic sex. This result has been amply demonstrated (see Jost, 1954, 1961, 1971).

Lillie's freemartin research introduced biologists to the problem of the nature, origin, and action of sex hormones at a time when almost nothing was known about them. It stimulated attempts to duplicate the circumstances producing freemartins in cattle by parabiosis of amphibian larvae of opposite genetic sex, or by grafting gonads of donor chick embryos of one genetic sex onto the choriallantoic membrane or into the coelom of host chick embryos of the opposite sex. The freemartin research led Lillie directly into collaborative work with members of the department of biochemistry that resulted in the isolation, chemical analysis, and synthesis of sex hormones.

The first known androgens were isolated in the laboratory of F. C. Koch at the University of Chicago with Lillie as colleague. These pure chemical substances were then available for injection into experimental animals to determine to what extent development of the reproductive systems of the injected animals could be modified. This work opened up for investigation the field of the biology of sex, the physiology of sex hormones and, more recently, the role of sex hormones in animal behavior. It also opened up for investigation the role in sex hormone production of gonadotrophic hormones from the anterior pituitary gland and, later, of specific gonadotrophin-releasing factors from special neurosecretory cells of the hypothalamus, the nature of feedback mechanisms, and so on.

After his retirement Lillie turned to a series of investigations on the physiology of development of regenerating feathers, which, in certain breeds of fowl, such as the Brown Leghorn, are especially sensitive to estrogen, thyroxin, and similar hormones. In a series of papers published between 1932 and 1947, mostly in collaboration with Mary Juhn and Hsi Wang, Lillie added much to our understanding of the highly complex development of a regenerating feather, whose definitive structure, according to Cohen, is "... the sum of the debris of some 1.5×10^7 to 5×10^9 cells, exactly organized and engineered to the utmost precision" (1965, p. 9). The regenerating feather forms from the feather papilla or feather germ, a morphogenetic system in miniature.

Lillie and his co-workers studied the role of induction and of the processes of twinning in feather development, attempting especially to reveal some of the factors responsible for the consistent morphogenetic and color reactions of parts of the individual feather papilla, and of feather papillae in different feather tracts, to different concentrations of hormones. Although Cohen is sharply critical of some of Lillie's general interpretations about mechanisms of feather development and reaction to hormones, his criticisms are relatively minor when viewed against Lillie's overall contributions to our understanding of development of the regenerating feather, examination of whose definitive structure is, according to Cohen, "a humbling process."

Lillie received the Sc.D. from the University of Toronto (1920), Yale University (1932), and Harvard University (1938), and the LL.D. from Johns Hopkins University (1942). He was a member of several learned societies, both American and foreign.

BIBLIOGRAPHY

I. Original Works. Lillie's writings include "The Embryology of the *Unionidae*. A Study in Cell-Lineage," in *Journal of Morphology*, **10** (1895), 1–100; "Experimental Studies on the Development of the Organs in the Embryo of the Fowl *(Gallus domesticus)*," in *Biological Bulletin, Marine Biological Laboratory, Woods Hole, Mass.*, **5** (1903), 92–124; "Experimental Studies on the Development of Organs in the Embryo of the Fowl *(Gallus domesticus)*. II. The Development of Defective Embryos, and the Power of Regeneration," *ibid.*, **7** (1904), 33–54; *Laboratory Outlines for the Study of the Embryology of the Chick and the Pig* (Chicago, 1906), 2nd ed., rev., *A Laboratory Outline and Manual for the Study of Embryology* (Chicago, 1919), written with C. R. Moore, 3rd ed., rev., *A Laboratory Outline of Embryology With Special Reference to the Chick and the Pig* (Chicago, 1923), written with C. R. Moore; "The Biological Significance of Sexual Differentiation—A Zoological Point of View," in *Science*, **25** (1907), 372–376; *Development of the Chick. An Introduction to Embryology* (New York, 1908, 2nd ed., New York, 1919); 3rd ed., rev., *Lillie's Development of the Chick* (New York, 1952), by H. L. Hamilton; "Studies of Fertilization. VI. The Mechanism of Fertilization in *Arbacia*," in *Journal of Experimental Zoology*, **16** (1914), 523–590; "The Theory of the Freemartin," in *Science*, **43** (1916), 611–613; "The Free-Martin; a Study of the Action of Sex Hormones in the Foetal Life of Cattle," in *Journal of Experimental Zoology*, **23** (1917), 371–452; *Problems of Fertilization* (Chicago, 1919); "Tandler and Keller on the Freemartin," in *Science*, **50** (1919), 183–184; "Fertilization," in E. V. Cowdry, ed., *General Cytology* (Chicago, 1924), pp. 449–536, written with E. E. Just; "General Biological Introduction," in E. Allen, ed., *Sex and Internal Secretions* (Baltimore, 1932), pp. 1–11, also in E. Allen, C. H. Danforth, and E. A. Doisy, eds., *Sex and Internal Secretions*, 2nd ed. (Baltimore, 1939), pp. 3–14; "The Builders of the Marine Biological Laboratory," in *Collecting Net*, **13** (1938), 107–108; and *The Woods Hole Marine Biological Laboratory* (Chicago, 1944).

An undated 33-p. MS entitled "My Early Life" provides much information on his ancestry, youth, and early education. It is filed with the F. R. Lillie reprint collection at the Marine Biological Laboratory.

II. Secondary Literature. See "Appreciations of Professor Frank R. Lillie on the Occasion of his Sixtieth Birthday," in *Collecting Net*, **5** (anniversary supp., 28 June 1930), 1–12; Mary Prentice Lillie Barrows, *Moon out of the Well. Reminiscences* (1970), copy filed with F. R. Lillie reprints at the Marine Biological Laboratory; and *Frances Crane Lillie (1869–1958). A Memoir* (1970), copy provided through the courtesy of B. H. Willier; R. K. Burns, "Role of Hormones in the Differentiation of Sex," in W. C. Young, ed., *Sex and Internal Secretions*, 3rd ed. (Baltimore, 1961), pp. 76–158; J. Cohen, "Feathers and Patterns," in M. Abercrombie and J. Brachet, eds., *Advances in Morphogenesis* (New York, 1966), V, 1–38; E. G. Conklin, "Science and Scientists at the M. B. L.

Fifty Years Ago," in *Collecting Net*, **13** (1938), 101–106; and "The Contributions of Dr. Frank R. Lillie to Oceanography," in *Collecting Net*, **15** (1940), 30–31 (on the occasion of the presentation of the Agassiz Medal for Oceanography by the National Academy of Sciences); G. Fankhauser, "Memories of Great Embryologists," in *American Scientist*, **60** (1972), 46–55; and R. J. Harrison, "Frank Rattray Lillie (1870–1947)," in *Yearbook. American Philosophical Society* (1947), pp. 264–270.

See also A. Jost, "Hormonal Factors in the Development of the Fetus," in *Cold Spring Harbor Symposia on Quantitative Biology*, XIX, *The Mammalian Fetus: Physiological Aspects of Development* (Cold Spring Harbor, N.Y., 1954), 167–181; "The Role of Fetal Hormones in Prenatal Development," in *Harvey Lectures*, **55** (1961), 201–226; and "Hormones in Development: Past and Prospects," in M. Hamburgh and E. J. W. Barrington, eds., *Hormones in Development* (New York, 1971), pp. 1–18; C. B. Metz, "Specific Egg and Sperm Substances and Activation of the Egg," in A. Tyler, R. C. von Borstel, and C. B. Metz, eds., *The Beginnings of Embryonic Development* (Washington, D.C., 1957), pp. 22–69; and "Gamete Surface Components and Their Role in Fertilization," in C. B. Metz and A. Monroy, eds., *Fertilization: Comparative Morphology, Biochemistry and Immunology*, I (New York, 1967), 163–236; C. R. Moore, "Frank Rattray Lillie 1870–1947," in *Science*, **107** (1948), 33–35; and "Frank Rattray Lillie. 1870–1947," in *Anatomical Record*, **101** (1948), 1–4; H. H. Newman, "History of the Department of Zoology in the University of Chicago," in *Bios*, **19** (1948), 215–239; A. N. Richards, "Dr. Frank R. Lillie," a typewritten memorial read at a dinner of the National Academy of Sciences, November, 1947, two weeks after Lillie's death (filed with the F. R. Lillie reprint collection at the Marine Biological Laboratory); R. V. Short, "The Bovine Freemartin: A New Look at an Old Problem," in *Philosophical Transactions of the Royal Society*, **B259** (1970), 141–147; B. H. Willier, "Frank Rattray Lillie," in *Anatomical Record*, **100** (1948), 407–410; and "Frank Rattray Lillie 1870–1947," in *Biographical Memoirs. National Academy of Sciences*, **30** (1957), 179–236; B. H. Willier, R. J. Harrison, H. B. Bigelow, and E. G. Conklin, "Addresses at the Lillie Memorial Meeting Woods Hole, August 11, 1948," in *Biological Bulletin. Marine Biological Laboratory, Woods Hole, Mass.*, **95** (1948), 151–162; B. H. Willier, "The Work and Accomplishments of Frank R. Lillie at Chicago," pp. 151–153; R. G. Harrison, "Dr. Lillie's Relations With the National Academy of Sciences and the National Research Council," pp. 154–157; H. B. Bigelow, "Dr. Lillie and the Founding of the Woods Hole Oceanographic Institution," pp. 157–158; and E. G. Conklin, "Frank R. Lillie and the Marine Biological Laboratory," pp. 158–162; and W. C. Young, ed., *Sex and Internal Secretions*, 3rd ed., 2 vols. (Baltimore, 1961).

RAY L. WATTERSON

LINACRE, THOMAS (*b.* Canterbury, England, 1460[?]; *d.* London, England, 1524), *medicine*.

Linacre received his early education at the school of Christ Church Monastery, Canterbury, under the direction of William de Selling, later prior. It appears at the age of twenty Linacre went to Oxford, where he learned some Greek and in 1484 was elected fellow of All Souls College. About 1487 he accompanied Selling to Italy. He studied Greek with Demetrius Chalcondylas in Florence, met Hermolaus Barbarus in Rome, became acquainted with the printer Aldus Manutius (and was involved in the editing of the Aldine edition of the Greek *Opera omnia* of Aristotle [1495–1498]) in Venice, and graduated M.D. from Padua in 1496. He then engaged in further study with the humanist-physician Nicolò Leoniceno in Vicenza and returned to England by way of Geneva, Paris, and Calais.

Back at Oxford, Linacre was incorporated M.D. in consequence of his Paduan degree. He is said to have given some public lectures on medicine and seems also to have taught Greek privately, one of his students being Sir Thomas More. During this period he also made a Latin translation of Proclus' *De sphaera*, which was published by Aldus in a collection of ancient Greek astronomical works (1499). About 1501 Linacre became tutor to Prince Arthur (who died in the following year) and in 1509 was chosen as one of the physicians to Henry VIII at a salary of £50 a year. From this time onward he lived chiefly in London, where his patients and friends included Cardinal Wolsey, Archbishop Warham, Bishop Fox, and such eminent scholars as Colet, More, Erasmus, and William Lily.

At about the same time Linacre, in search of greater leisure, took holy orders. His highly placed friends found him a succession of ecclesiastical livings, which he either sold or deputized; he was thus enabled to devote most of his efforts to scholarship (although he remained physician to the king, in which post his duties were nominal). Linacre was especially concerned in translating Galen's writings into Latin, beginning with the treatise of hygiene, *De sanitate tuenda*. Since there was then no printer in England sufficiently able or willing to assume the financial risk of producing this work for the English market, the tract, dedicated to Henry VIII, was published in Paris in 1517. Linacre's second translation, Galen's *Methodus medendi*, was also published in Paris (1519). It, too, was dedicated to the English king.

Linacre's next translation of Galen, *De temperamentis*, was, however, published in England, from the press of John Siberch in Cambridge (1521). In 1522, Linacre saw two Galenic translations by his former teacher, Nicolò Leoniceno, through the press of Richard Pynson in London; Pynson also published Linacre's own subsequent translations, *De usu pulsuum* (1522), *De facultatibus naturalibus* (1523), and *De symptomatis differentiis* (1524).

Linacre was a medical humanist as well as one of the finest Greek scholars of his day; his major effort, therefore, was directed toward bringing to English physicians a series of classical medical texts that he considered essential and that were, in fact, superior to other medical writings published in England at that time. In addition to the works that he brought to publication, he is known to have translated yet others, but with the one exception, a brief extract from Paul of Aegina, these were either lost or destroyed after his death. His very considerable Continental reputation, especially in Greek medical scholarship, was clearly recognized by Erasmus: "Medicine has begun to make herself heard in Italy through the voice of Nicolaus Leonicenus . . . and among the French by Guillaume Cop of Basel; while among the English, owing to the studies of Thomas Linacre, Galen has begun to be so eloquent and informative that even in his own tongue he may seem to be less so." Such commendation remained unheeded in England, however, and Linacre's translations were not republished there, although there were approximately forty Continental reprintings of them between 1524 and 1550.

Linacre himself may have come to realize that more than the Greek medical classics were necessary to improve the state of English medicine. Just prior to his death he arranged that funds from his considerable estate should be used to establish the Linacre lectures at Oxford and Cambridge. Although he did not specifically declare that the lectures must be devoted to medical subjects, there nonetheless seems little doubt that such was his intention. He may have hoped that the oral presentation of classical medicine would prove attractive to physicians and medical students who might resist the printed word. He did not live to see the misuse of this endowment. The lectures failed through the conservatism and lethargy of the universities, and whatever potential they might have had for revitalizing the medical curriculum or promoting a more advanced medicine in England was unrealized.

Linacre was nevertheless responsible for an important contribution to English medicine in his work toward founding the College of Physicians of London. Through his influence the College was granted, from its foundation in 1518, the power to license physicians in London and for seven miles surrounding. The right to practice medicine had previously been conferred only upon medical graduates of Oxford or Cambridge or to such men as were licensed by bishops (or in London by the bishop or by the dean of St. Paul's). The more stringent professional license enhanced the prestige of the relatively small group who met the College's standards. This result, desirable from the physicians' standpoint, was recognized and envied by those practicing outside the area of the College's supervision, where licensing remained beyond the control of the medical profession.

Linacre's sense of the dignity of medicine was paramount. As the College served to promote that dignity, so it also maintained it. Although the medical profession in Continental cities was able to rally around the faculty of medicine of a university, this was not possible in England, since the universities were in relatively small and inaccessible towns. For this reason the College of Physicians of London became the focus of physicians in London and, to some degree, elsewhere. The College, which Linacre directed until his death, established the character of English medicine for centuries to follow. It thus represents his most enduring work.

Apparently it was of the complications of a calculary disorder that Linacre died.

BIBLIOGRAPHY

I. Original Works. The original editions of Linacre's translations of Galen are very rare, but one of them, *De temperamentis*, was published in facsimile with a biographical introduction by J. F. Payne (Cambridge, 1881).

II. Secondary Literature. On Linacre's life and work see Josephine W. Bennett, "John Morer's Will: Thomas Linacre and Prior Sellying's Greek Teaching," in *Studies in the Renaissance*, **15** (New York, 1968), 70–91; George Clark, *A History of the Royal College of Physicians of London*, I (Oxford, 1964), 37; John Noble Johnson, *The Life of Thomas Linacre, Doctor of Medicine*, Robert Graves, ed. (London, 1835); C. D. O'Malley, *English Medical Humanists. Thomas Linacre and John Caius* (Lawrence, Kans., 1965); and William Osler, *Thomas Linacre* (Cambridge, 1908).

C. D. O'Malley

LIND, JAMES (*b.* Edinburgh, Scotland, 4 October 1716; *d.* Gosport, Hampshire, England, 13 July 1794), *medicine*.

The discoverer of citrus fruit as a cure for scurvy, Lind was the son of Margaret Smelum (alternately spelled Smellum or Smellome) and James Lind, a substantial Edinburgh merchant. (Many early biographical sketches confuse him with his cousin, also named James Lind, a physician to the household of George III.) After attending grammar school in Edinburgh, in December 1731 he was apprenticed to George Langlands, an Edinburgh physician; and in 1739 he entered the British navy as a surgeon's mate. He was promoted to surgeon in 1747. Most of his naval service was aboard ships patrolling the English Channel during the War of Austrian Succession (1740–1748), and his longest cruise was on the fourth-rated ship H.M.S. *Salisbury* (August–October 1746) in the English Channel. It was on a somewhat shorter cruise on this ship in 1747 that he carried out his classic experiments on scurvy.

Lind left the navy in 1748 and obtained his M.D. from the University of Edinburgh in the same year. His doctoral thesis on venereal lesions, *De morbis venereis localibus*, gives evidence of having been written in haste and is a trivial work but was sufficient to win him a license to practice in Edinburgh. In 1750 Lind became a fellow of the Royal College of Physicians of Edinburgh and served as treasurer from December 1756 to 1758, when he was appointed first chief physician to the new Royal Naval Hospital at Haslar. Lind was also a member of the Philosophical and Medical Society of Edinburgh and an original fellow of the Royal Society of Edinburgh. In 1753 he published *A Treatise of the Scurvy*, which created little stir in Edinburgh but apparently attracted the attention of Lord Anson, then first Lord of the Admiralty, to whom it was dedicated. It is believed that Anson was later influential in securing Lind's appointment to the Royal Naval Hospital.

This post gave Lind additional opportunity for study. In the first two years of his administration the hospital received 5,734 admissions, of which 1,146 were cases of scurvy. The last edition of his *Treatise* includes additions derived, said the author, from four large notebooks of case histories and observations made at Haslar. Lind served as chief physician until 1783, when he was succeeded by his son John, who had been his assistant. On his retirement his portrait, now lost, was painted by Sir George Chalmers; a print of an engraving made from the picture is extant. Lind was buried in St. Mary's Church, Gosport. The church register mentions his wife, Isobel (Isabel) Dickie, whom Lind must have married while practicing in Edinburgh.

Lind's *An Essay on the Most Effectual Means of Preserving the Health of Seamen in the Royal Navy* (1757) throws much light on the living conditions and the diet of seamen. In *Two Papers on Fevers and Infections* (1763), which contains most of what is known of the medical history of the Seven Years War, Lind emphasized that the "number of seamen in time of war who died by shipwreck, capture, famine, fire or sword" was only a small proportion compared to those who died by "ship diseases and the usual maladies of intemperate climates." (During the Seven Years War 133,708 men were lost to the navy by disease or desertion; 1,512 were killed in action.) Lind's last major work, *An Essay on Diseases Incidental to Europeans in Hot Climates* (1768), was written primarily for the use of travelers and emigrants to tropical colonies and remained the standard work on tropical medicine in English for half a century.

Although Lind is now recognized as the greatest of naval doctors and as the "father of nautical medicine," in his own day he was not elected a fellow of the Royal Society. He has also come to be regarded as one of the first modern clinical investigators and clearly was one of the first to have complete faith in the validity of his own observations and the logic of his conclusions. Lind's scurvy cure was not officially introduced into the diet of the Royal Navy until after his death, and this delayed recognition has aroused considerable historical interest. Herbert Spencer, in his *Study of Sociology*, cited the navy's neglect of Lind's findings as the most flagrant example of administrative apathy on record. More recent investigators such as Sir Humphrey Rolleston and A. P. Meiklejohn regarded it as yet another example of the prejudice which greets many innovators.

Lind's basic experiments were carried out in May 1747 while he was serving as surgeon aboard the *Salisbury*. Lind selected twelve scurvy patients with the most similar cases. He divided them into six groups of two, and although all received the same basic diet, he added different supplements to each one. Two he gave a quart of cider a day; two others, twenty-five drops of elixir vitrio three times a day; two were given spoonfuls of vinegar; two were treated with seawater; two, with an electuary of garlic, mustard seed, horseradish, and other ingredients; and two were given two oranges and a lemon every day. Lind observed the most sudden and visible improvement in those fed citrus fruit: one was pronounced fit for duty after six days, and the other was reported to have made the best recovery of all. Lind held that cider had the next best effect (although only small improvements were noticeable) and that the other treatments were useless. The question is why Lind's remedy was not quickly adopted. Meiklejohn held that it was because Lind was confronted with tenaciously held current medical practices and concepts about scurvy; Christopher Lloyd has given a somewhat more complex explanation.

Lloyd pointed out that while the voyages of Captain James Cook between 1768 and 1780 are assumed to have demonstrated the effectiveness of citrus fruits in preventing scurvy, this is not exactly the case. Cook himself emphasized that the absence of scurvy on his voyages was due to a diet of sauerkraut, soup, and malt. On his second voyage Cook also insisted on obtaining fresh water and vegetables at every opportunity, and while they undoubtedly served to prevent scurvy, no genuine experiments were made to determine their therapeutic effectiveness. Cook's surgeons in their report attributed the absence of scurvy to the inclusion of malt, and this idea dominated naval thinking for a time.

The outbreak of the American revolution forced the British navy to reexamine this supposition, since in 1780, despite the inclusion of malt, one-seventh of the fleet, some 2,400 cases, were landed for treatment of scurvy. The high cost of Mediterranean lemons was also a major factor in the delay of their use by the navy. By the last decade of the eighteenth century the navy had finally adopted citrus fruits as a remedy. By 1796 it was generally accepted that scurvy could be prevented by fresh vegetables and cured effectively by lemons or their juice, preserved effectively by boiling; and since that time scurvy has played little part in medical history. (One of the seamen's demands during the mutiny of 1797 was a ration of lemon juice.) Limes replaced lemons in the British navy in the mid-nineteenth century because the West Indian lime was cheaper and easier to obtain than the Mediterranean lemon. However, there were still serious outbreaks of scurvy on several Arctic expeditions, and it was not until 1917 that preserved lime juice was discovered to have only one-third the antiscorbutic value of lemon. Unfortunately the same dosages had usually been followed.

Second only to his treatise on scurvy in importance to naval medicine was his general effort to improve the health of seamen. In his book on the subject, Lind included recommendations for the practice of shipboard hygiene, comments upon diseases in northern latitudes (especially scurvy), advice on the prevention of tropical fevers, and a means for procuring a wholesome water supply at sea. Others had suggested ways for distilling seawater, but Lind's method did not require the addition of chemicals as did those of his predecessors or even of his contemporaries. Lind's importance in medical history lies in his willingness to experiment and to break with past medical tradition when it conflicted with his own observations. Although his ideas were very much of his own time, he modified them where necessary. Lind's younger contemporaries and ardent supporters, Thomas Trotter and Sir Gilbert Blane, helped carry on his tradition; and the triumvirate is now regarded as responsible for the founding of the modern naval health service.

BIBLIOGRAPHY

I. ORIGINAL WORKS. *A Treatise of the Scurvy* (Edinburgh, 1753) went through three eds.; the second ed. (London, 1757) is little more than a reprint, but the third edition, retitled *A Treatise on the Scurvy* (London, 1772)—the title most often cited—contains some additions based on Lind's observations at Haslar and an updated bibliography. The French trans., *Traité du scorbut*, 2 vols. (Paris, 1756; repr. 1783), was published with a treatise on the same subject by Boerhaave. An Italian ed. was published at Venice in 1766, and a German ed. at Leipzig in 1775. The first ed. was reprinted as *Lind's Treatise on Scurvy* (Edinburgh, 1953) with additional notes and comments by C. P. Stewart, D. Guthrie, and others as part of observances commemorating the bicentenary of its original publication.

An Essay on the Most Effectual Means of Preserving the Health of Seamen in the Royal Navy (London, 1757) was reprinted in 1762; the third ed. (1779) includes *Two Papers on Fevers and Infections*, originally read before the Philosophical and Medical Society of Edinburgh and first published separately in 1763 in London. The *Essay* was translated into Dutch (Middleburgh, 1760) and was probably also published in French and German. *Two Papers on Fevers and Infections* was translated into French (Lausanne, 1798). Most of the *Essay* as well as the papers on fevers and infections were reprinted by the Navy Records Society as *The Health of Seamen*, C. Lloyd, ed. (London, 1965). *An Essay on Diseases Incidental to Europeans in Hot Climates, With the Method of Preventing Their Fatal Consequences* (London, 1768; 6th ed., 1808; Philadelphia, 1811), Lind's most popular work during his lifetime, was translated into German (Leipzig, 1773), Dutch (Amsterdam, 1781), and French (Paris, 1785).

Two letters addressed to Sir Alexander Dick in Lind's handwriting are extant, one resigning his office of treasurer of the Royal College, and a second describing his new place of work, Haslar Hospital; for text of both and photograph of one, see *Lind's Treatise on Scurvy* (1953), pp. 390–391.

II. SECONDARY LITERATURE. L. H. Roddis, *James Lind* (New York, 1950), is a rather naïve biography, valuable primarily for the author's attempt to locate all eds. of Lind's works. Much biographical information was gathered in Sir H. D. Rolleston, "James Lind, Pioneer of Naval Hygiene," in *Journal of the Royal Naval Medical Service*, **1** (1915), 181–190. For explanations of the delay in adopting Lind's cure see A. P. Meiklejohn, "The Curious Obscurity of Dr. James Lind," in *Journal of the History of Medicine and Allied Sciences*, **9** (1954), 304–310; and esp. C. Lloyd, "The Introduction of Lemon Juice as a Cure for Scurvy," in *Bulletin of the History of Medicine*, **35** (1961), 123–132. See also E. A. Hudson and A. Herbert, "James Lind: His Contribution to Shipboard Sanitation," in *Journal of the History of Medicine*, **11** (1956), 1–12; C. Lloyd and J. L. S. Coulter, *Medicine and the Navy*, III (Edinburgh, 1957); and the various articles on Lind, naval medicine, and scurvy in *Lind's Treatise on Scurvy*, cited above.

VERN L. BULLOUGH

LINDBLAD, BERTIL (*b.* Örebro, Sweden, 26 November 1895; *d.* Stockholm, Sweden, 25 June 1965), *astronomy*.

Lindblad's father, Lieutenant Colonel Birger Lindblad, was from Askersund, where four generations of his family had been merchants and active members of the magistrates court. Bertil Lindblad's paternal grandmother, Jenny Hybbinette, belonged to a distinguished Walloon family that migrated to Sweden around 1630 to work in the iron industry. His mother, Sara Gabriella Waldenström, belonged to an intellectual Swedish family whose members for several generations had distinguished themselves in theology and medicine.

Lindblad graduated in 1914 from the Karolinska Läroverket in Örebro. That same year he began studies in mathematics, physics, and astronomy at Uppsala University. Östen Bergstrand, director of the Uppsala observatory, soon realized that Lindblad was an unusually gifted student. Lindblad became interested in the study of the colors of stars, and he was encouraged in his early research not only by Bergstrand himself but also by Hugo von Zeipel, associate professor of astronomy.

In 1920 Lindblad defended his doctoral thesis, an important contribution to the theory of radiative transfer in the solar atmosphere with special application to the phenomenon of solar limb darkening. In the same paper he presented a first approach to the problem of determining absolute magnitudes of stars by combining two color indices from different parts of the spectrum. During a research period (1920–1922) at the Lick and Mount Wilson observatories, Lindblad was able to establish new, important spectroscopic criteria of stellar luminosity in addition to those which had been found somewhat earlier by Kohlschütter and Adams. He thus found that absorption bands of the cyanogen molecule CN are considerably stronger in the spectra of giants than in the spectra of dwarfs for stars with the same temperature as the sun or with a somewhat lower temperature. Compared with Kohlschütter's and Adams' criteria, the new criteria had the great advantage of being visible also in spectra for very low dispersion. Thus, faint stars could also be studied with this new spectroscopic method, and it soon became a most efficient tool in stellar research.

Upon his return to Uppsala in 1922 Lindblad continued his work with one of the observatory astrographs which had been especially equipped with an objective prism giving short dispersion spectra. He also included other principal spectrum lines and bands in his analysis and was able to use this astrograph for a first survey of faint stars in selected regions of the sky. In order to improve the calibration of his luminosity criteria he selected regions where the proper motions were known down to the limiting magnitude. In this way he improved considerably the basis of stellar statistics, and derived more accurately the density distribution of stars and the state of motion. This work gave a great stimulus to Swedish astronomy, and many young astronomers joined in the development of this field of research, which remains a major program at the observatories of Uppsala, Stockholm, and Lund.

Lindblad married Dagmar Bolin in 1924. The couple had four children.

In 1927 Lindblad was appointed director of the Stockholm observatory and was thus in charge of building the new observatory in Saltsjöbaden. The instruments for the new institution were selected to allow further applications of the spectroscopic criteria of stellar luminosity.

During his last years in Uppsala, Lindblad became increasingly interested in stellar dynamics and he introduced new concepts which could explain the asymmetric drift of high velocity stars. He advanced the fundamental idea that the galactic system is rotating around a distant center. In his detailed discussion he introduced a model of the galactic system consisting of a number of subsystems of different speeds of rotation and with different degrees of flatness and velocity dispersion.

Lindblad's ideas were soon confirmed by J. H. Oort with his discovery of the differential rotation. Following that discovery, Lindblad became increasingly engaged in stellar dynamics and soon became a leading authority. He devoted much time to the difficult problem of explaining the spiral structure of rotating stellar systems assuming gravitational forces only. His first results indicated that spiral arms probably open up in their motion, but important numerical calculations made on his suggestion by his son Per Olof Lindblad (appointed director of the Stockholm observatory in 1967) have shown that trailing arms probably are the most frequent ones. This led Lindblad to a further development of his theory. He now introduced the concepts "dispersion orbits" and "density waves," which were followed up with considerable success by C. C. Lin. Lindblad also found ways of explaining how spiral features may be preserved.

Despite his many official duties, Lindblad continued his scientific research; and even in his later years he found time to do observational work on eclipses and photometric studies of nebulae.

Lindblad was very active internationally; he was president of the International Astronomical Union from 1948 to 1952, when he became President of the International Council of Scientific Unions. A few months before his death, he was elected president of the council of the European Southern Observatory.

While director of the Stockholm observatory, Lindblad was also professor of astronomy at Stockholm University. He was elected to the Royal Swedish Academy of Sciences in 1928 and served as president in 1938 and 1960 and as vice-president for many years. He served as chairman of the Swedish Natural Science Research Council from 1951. Lindblad was honored with the Gold Medal of the Royal Astronomical Society in 1948, and he was a foreign member of academies in Denmark, Finland, Belgium, Portugal, Italy, France, Germany, Poland, and the United States. At his death he was chairman of the Nobel Foundation.

BIBLIOGRAPHY

A comprehensive list of publications (about 100) has been given by J. H. Oort in his obituary notice printed in *Quarterly Journal of the Royal Astronomical Society*, **7** (1966).

YNGVE ÖHMAN

LINDE, CARL VON (*b.* Berndorf, Germany, 11 June 1842; *d.* Munich, Germany, 16 November 1934), *engineering, cryogenics.*

Linde was the third of nine children of Friedrich Linde and his wife Franziska Linde, the daughter of a businessman from Neuwied. In Kempten, where his father, a minister, had been transferred, the young Linde pursued classical studies at the Gymnasium. To his literary and cultural interests he soon added a great enthusiasm for technical matters which led him to study machine construction. Thus from 1861 to 1864 he attended the famous Eidgenössische Polytechnikum in Zurich, where he studied science and engineering with Clausius, Zeuner, and F. Reuleaux, as well as aesthetics with F. T. Vischer and art history with W. Lübke. From 1864 to 1866 Linde received practical training in the workshop and drafting studios of, among others, the locomotive and machine factory of A. Borsig in Tegel, near Berlin. In 1866 he became the head of the technical department of the newly founded Krauss and Company, locomotive manufacturers in Munich. There he was able to implement a series of his own ideas, including those on braking arrangements.

On the founding of the Munich Polytechnische Schule (later the Technische Hochschule) in 1868, Linde became extraordinary professor, and in 1872, full professor of theoretical engineering. Here, in 1875, he established an engineering laboratory, the first of its kind in Germany. He lectured on a number of subjects—motors in general; the theory of steam engines, crank motion, and water wheels; turbines; hot-air engines; locomotives; and steamships—and carried out practical laboratory projects. Among his colleagues at the Polytechnische Schule he was especially close to Felix Klein, who taught mathematics there from 1875 to 1880.

In 1870 Linde started to investigate refrigeration. His research on heat theory led, from 1873 to 1877, to the development of the first successful compressed-ammonia refrigerator. Refrigerators existed before Linde's, but his was especially reliable, economical, efficient. He stressed that refrigerators should be useful not only for the making of ice, but also for the direct cooling of liquids. For these reasons breweries were particularly interested in his device.

Linde left his teaching position in 1879 and founded the Gesellschaft für Linde's Eismaschinen in Wiesbaden to develop his process industrially. Because of the many applications of artificial cooling the undertaking was an international success. In order to explore the subject of low temperatures at a research station (founded in 1888) and to teach again on a reduced schedule, Linde retired from the management of the business in 1891; at that time twelve hundred refrigerators of his construction were already in operation.

In 1895 Linde succeeded in liquefying air with the help of the Joule-Thomson effect and by the application of the counterflow principle, and the foundations were thereby laid for the maintenance of low temperatures. In London Hampson arrived at a similar process shortly after Linde (1896). Linde also successfully developed devices for obtaining pure oxygen by means of rectification (1902), for the production of pure nitrogen through the use of the nitrogen cycle process (1903), and for producing hydrogen from water gas by means of partial condensation of carbon monoxide (1909). The production of pure oxygen was of great importance for oxyacetylene torches used in metalworking, as was that of pure nitrogen for the large-scale production of calcium nitrate, ammonia, and saltpeter. Linde founded a group of enterprises in Europe and overseas to utilize his processes.

In the second phase of his teaching, beginning in Munich in 1891, Linde dealt primarily with the theory of refrigeration machines. It was at his instigation that the first laboratory for applied physics in Germany was founded at the Munich Technische Hochschule in 1902. Linde was also active on numerous scientific and technical committees. He was a member of the Bavarian Academy of Sciences and a corresponding member of the Vienna Academy. In addition, he belonged to the board of trustees of the Physikalisch-Technische Reichsanstalt and to boards of directors of the Verein Deutsche Ingenieure and of the Deutsches Museum.

In 1866 Linde married Helene Grimm; they had six children. In 1897 he was made a nobleman. The exceedingly fortunate combination of scientific, technical, and entrepreneurial abilities met and were developed in this simple man of strong moral character and an uncommon capacity for work.

BIBLIOGRAPHY

I. ORIGINAL WORKS. Linde's manuscript material is in the Archiv des Polytechnischen Vereins in Bayern (now in the MS collection of the library of the Deutsches Museum in Munich) and at his company, Linde A. G. in Wiesbaden.

Linde's important published works are: *Über einige Methoden zum Bremsen der Lokomotiven und Eisenbahnzüge* (Munich, 1868); "Wärmeentziehung bei niedrigen Temperaturen durch mechanische Mittel," in *Bayerisches Industrie- und Gewerbeblatt*, **2** (1870), 205; "Eine neue Eis- und Kühlmaschine," *ibid.*, **3** (1871), 264; "Theorie der Kälteerzeugungsmaschinen," in *Verhandlungen des Vereins zur Beförderung des Gewerbefleisses*, **54** (1875), 357 and **55** (1876), 185; "The Refrigerating Machine of Today," in *Transactions of the American Society of Mechanical Engineers*, **14** (1893), 1414; and "Refrigerating Apparatus," in *Journal of the Society of Arts*, **42** (1894), 322.

Also see his "Erzielung niedrigster Temperaturen," in *Annalen der Physik und Chemie*, n.s. **57** (1896), 328; "Process and Apparatus for Attaining Lowest Temperatures," in *Engineer* (London), **82** (1896), 485; "Kälteerzeugungsmaschine," in *Luegers Lexicon der gesamten Technik*, V (Stuttgart, 1897), 353; "Über die Veränderlichkeit der specifischen Wärme der Gase," in *Sitzungsberichte der Bayerischen Akademie der Wissenschaften zu München*, **27** (1897), 485; "Die Entwicklung der Kältetechnik," in *Festschrift 71. Versammlung der Gesellschaft Deutscher Naturforscher und Ärzte* (Munich, 1899), 189. Among Linde's later works are: "Zur Geschichte der Maschinen für die Herstellung flüssiger Luft," in *Berichte der Deutschen Chemischen Gesellschaft*, **32**, no. 1 (1899), 925; "Über Vorgänge bei Verbrennung in flüssiger Luft," in *Sitzungsberichte der Bayerischen Akademie der Wissenschaften zu München*, **29** (1899), 65; "Sauerstoffgewinnung mittels fraktionierter Verdampfung flüssiger Luft," in *Zeitschrift des Vereins Deutscher Ingenieure*, **46** (1902), 1173; "Die Schätze der Atmosphäre," in *Deutsches Museum. Vorträge und Berichte*, no. 1 (1908); and "Physik und Technik auf dem Wege zum absoluten Nullpunkt," in *Festrede in der Bayerischen Akademie der Wissenschaften* (Munich, 1912); and *Aus mein Leben und von meiner Arbeit* (Munich, 1916).

II. SECONDARY LITERATURE. Works about Linde are (in chronological order) *50 Jahre Kältetechnik, 1879–1929. Geschichte der Gesellschaft für Linde's Eismaschinen A.G., Wiesbaden* (Wiesbaden, 1929); "Carl von Linde zum 90 Geburtstag," in *Abhandlungen und Berichte des Deutschen Museums*, **4** (1932), 55; R. Plank, "Carl von Linde und sein Werk," in *Zeitschrift für die gesamte Kälte-Industrie*, **42** (1935), 162; H. Mache, "Carl von Linde," in *Almanach der Akademie der Wissenschaften in Wien*, **85** (1936), 272. W. Meissner, "Carl von Lindes wissenschaftliche Leistungen," in *Zeitschrift für die gesamte Kälte-Industrie*, **49** (1942), 101; R. Plank, "Geschichte der Kälteerzeugung," in *Handbuch der Kältetechnik*, **1** (1954), 1; *75 Jahre Linde* (Munich, 1954); and *Technische Hochschule München, 1868–1968* (Munich, 1968), 102.

FRIEDRICH KLEMM

LINDELÖF, ERNST LEONHARD (*b.* Helsingfors, Sweden [now Helsinki, Finland], 7 March 1870; *d.* Helsinki, 4 June 1946), *mathematics.*

Lindelöf was the son of the mathematician Leonard Lorenz Lindelöf, who was a professor in Helsingfors from 1857 to 1874. He studied in Helsingfors from 1887 to 1900, and spent the year 1891 in Stockholm, 1893 and 1894 in Paris, and 1901 in Göttingen. He passed his university examination in 1895 in Helsingfors and then gave courses in mathematics as a docent. In 1902 he became assistant professor and in 1903 full professor of mathematics. From 1907 he belonged to the editorial board of the *Acta Mathematica*, and he was a member of many learned academies and societies. He received honorary doctoral degrees from the Universities of Uppsala, Oslo, and Stockholm, and he was also named honorary professor by the University of Helsinki. He retired in 1938.

At the beginning of his career Lindelöf published a remarkable work on the theory of differential equations, in which he investigated the existence of solutions ("Sur l'intégration de l'équation differentielle de Kummer," in *Acta Societatis Scientiae Fennicae*, 19 [1890], 1). He soon turned his attention to function theory, and in this area solved some fundamental problems in the theory of analytic functions. His primary field of interest, however, was in entire functions. He considered the mutual dependency between the growth of the function and the coefficients of the Taylor expansion. He also treated the behavior of analytic functions in the neighborhood of a singular point. The investigations concern questions which arise from Picard's problem. Together with Phragmén he developed a general principle that he applied to function theory. His works are characterized by clarity and purity of method as well as elegance of form.

Lindelöf's investigations of analytic continuation with the help of summation formulas had far-reaching results, which are set down in his excellent book *Le calcul des résidus.* In it he examines the role which residue theory (Cauchy) plays in function theory as a means of access to modern analysis. In this endeavor he applies the results of Mittag-Leffler. Moreover, he considers series analogous to the Fourier summation formulas and applications to the gamma function and the Riemann function. In addition, new results concerning the Stirling series and analytic continuation are presented. The book concludes with an asymptotic investigation of series defined by Taylor's formula. The method of successive correction and the examination of this procedure following the studies of Picard and Schwarz is characteristic of Lindelöf's work. His last works dealt with conformal mapping.

Lindelöf early abandoned creative scientific research and devoted himself enthusiastically to his duties as a professor; he was always available to researchers and colleagues. He laid the foundations for the study of the history of mathematics in Finland and trained many students. In an interesting festschrift published in honor of Lindelöf's sixtieth birthday fourteen authors presented papers on function theory, diophantine equations, correlation theory, and number theory.

Lindelöf devoted his last years mainly to publishing textbooks which were noted for their lucidity and comprehensible style. Of his publications, eight deal with the theory of differential equations, twenty-three with function theory, four with the theory of error in harmonic analysis, and five with other fields.

BIBLIOGRAPHY

A listing of Lindelöf's mathematical works may be found in P. J. Myrberg, "Ernst Lindelöf in memoriam," in *Acta Mathematica*, **79** (Uppsala, 1947), i–iv. Lindelöf's major textbooks are *Le calcul des résidus et ses applications à la théorie des fonctions* (Paris, 1905; New York, 1947); *Inldening til högre Analysen* (Stockholm, 1912), translated into Finnish as *Johdatus korkeampaan analysiin* (Helsinki, 1917, new ed., 1956), and German as *Einführung in die höhere Analysis* (Berlin, 1934); *Differentiaali- ja integraalilaska ja sen sovellutukset* ("Differential and Integral Calculus and their Application"), 4 vols. (Helsinki, 1920–1946); and *Johdatus funktioteoriaan* ("Introduction to Function Theory"; Helsinki, 1936).

The festschrift, *Commentationes in honorem Ernesti Leonardi Lindelöf. Die VII Mensis Martii A MCMXXX Sexagenarii*, edited by his students, is in *Suomalaisen tiedeakatemiam toimituksia*, ser. A, **32** (Helsinki, 1929).

HERBERT OETTEL

LINDEMANN, CARL LOUIS FERDINAND (*b.* Hannover, Germany, 12 April 1852; *d.* Munich, Germany, 6 March 1939), *mathematics.*

Lindemann studied in Göttingen, Erlangen, and Munich from 1870 to 1873. He was particularly influenced by Clebsch, who pioneered in invariant theory. Lindemann received the doctorate from Erlangen in 1873, under the direction of F. Klein. His dissertation dealt with infinitely small movements of rigid bodies under general projective mensuration. He then undertook a year-long academic journey to Oxford, Cambridge, London, and Paris, where he met Chasles, J. Bertrand, C. Jordan, and Hermite. In 1877 Lindemann qualified as lecturer at Würzburg. In the same year he became assistant professor at the University of Freiburg. He became full professor in 1879. In 1883 he accepted an appointment at Königsberg and finally, ten years later, one at Munich, where he was for many years also active on the administrative board of the university. He was elected an associate member of the Bavarian Academy of Sciences in 1894, and a full member in 1895.

Lindemann was one of the founders of the modern German educational system. He emphasized the development of the seminar and in his lectures communicated the latest research results. He also supervised more than sixty doctoral students, including David Hilbert. In the years before World War I, he represented the Bavarian Academy in the meetings of the International Association of Academies of Sciences and Learned Societies.

Lindemann wrote papers on numerous branches of mathematics and on theoretical mechanics and spectrum theory. In addition, he edited and revised Clebsch's geometry lectures following the latter's untimely death (as *Vorlesungen über Geometrie* [Leipzig, 1876–1877]).

Lindemann's most outstanding original research is his 1882 work on the transcendence of π. This work definitively settled the ancient problem of the quadrature of the circle; it also redefined and reanimated fundamental questions in the mathematics of its own time.

In the nineteenth century mathematicians had realized that not every real number was necessarily the root of an algebraic equation and that therefore non-algebraic, so-called transcendental, numbers must exist. Liouville stated certain transcendental numbers, and in 1873 Hermite succeeded in demonstrating the transcendence of the base e of the natural logarithms. It was at this point that Lindemann turned his attention to the subject.

The demonstration of the transcendence of π is based on the proof of the theorem that, except for trivial cases, every expression of the form $\sum_{i=1}^{n} A_i e^{a_i}$, where A_i and a_i are algebraic numbers, must always be different from zero. Since the imaginary unity i, as the root of the equation $x^2 + 1 = 0$, is algebraic, and since $e^{\pi i} + e^0 = 0$, then πi and therefore also π cannot be algebraic. So much the less then is π a root, representable by a radical, of an algebraic equation; hence the quadrature of the circle is impossible, inasmuch as π cannot be constructed with ruler and compass.

Lindemann also composed works in the history of mathematics, including a "Geschichte der Polyder und der Zahlzeichen" (in *Sitzungsberichte der mathematisch-physikalischen Klasse der Bayrischen Akademie der Wissenschaften* [1896]). He and his wife collaborated in translating and revising some of the works of Poincaré. Their edition of his *La science et l'hypothèse* (as *Wissenschaft und Hypothese* [Leipzig, 1904]) contributed greatly to the dissemination of Poincaré's ideas in Germany.

BIBLIOGRAPHY

I. Original Works. In addition to the works cited in the text, see especially "Über die Ludolphsche Zahl," in *Sitzungsberichte der Preussischen Akademie der Wissenschaften zu Berlin*, math.-phys. Klasse, **22** (1882); and "Über die Zahl π," in *Mathematische Annalen*, **25** (1882).

II. Secondary Literature. See "Druckschriften-Verzeichnis von F. Lindemann," in *Almanach der Königlich Bayrischen Akademie der Wissenschaften zum 150. Stiftungsfest* (Munich, 1909), 303–306; "F. von Lindemanns 70. Geburtstag," in *Jahresberichte der Mathematikervereinigung*, **31** (1922), 24–30; and C. Carathéodory, "Nekrolog auf Ferdinand von Lindemann," in *Sitzungsberichte der mathematisch-naturwissenschaftlichen Abteilung der Bayrischen Akademie der Wissenschaften zu München*, no. 1 (1940), 61–63.

H. Wussing

LINDEMANN, FREDERICK ALEXANDER, later **Lord Cherwell** (*b.* Baden-Baden, Germany, 5 April 1886; *d.* Oxford, England, 3 July 1957), *physics.*

Lord Cherwell's father was A. F. Lindemann, a wealthy Alsatian engineer who became a British subject after the Franco-Prussian War; his mother, Olga Noble, was American and of Scottish descent. He was educated in Scotland and Germany, and took his Ph.D. with Nernst in 1910. Together they formulated the Nernst-Lindemann theory of specific heats, and at the same time Lindemann derived his formula relating the melting point of a crystal to the amplitude of vibration of its atoms.

He returned to Britain in August 1914 and joined the Royal Aircraft Establishment to work on problems of flight. Aircraft at that time were prone to spin uncontrollably, often crashing fatally. Lindemann worked out the relevant theory, showed how to recover an aircraft from a spin, and made the first tests of the theory himself. Despite defective eyesight, he had already learned to fly.

In 1919, backed by his friend Henry Tizard, whom he had met in Nernst's laboratory, Lindemann was appointed Dr. Lee's professor of experimental philosophy at Oxford and head of the Clarendon Laboratory. The laboratory was then moribund, for almost no research had been done during its fifty years of existence. Lindemann started the long fight to build it up, often against the prejudice of the classical tradition at Oxford; and in 1933 he strengthened it by giving hospitality to Jewish emigrés from Germany, led by Franz Simon.

During this period he had become the friend of Winston Churchill, and from 1932 the two fought to awaken Britain to the German threat and to the need—and hope—for effective air defense. Here, unhappily, he quarreled with Tizard, who had returned to this field. Ostensibly the quarrel was over priorities, but there was a deeper conflict of personalities, and from 1933 to 1939 Lindemann was in eclipse.

The return of Churchill to the Admiralty brought Lindemann forward as his scientific adviser, a relationship maintained after Churchill became prime minister in 1940. Lindemann advised him over the entire fields of science and economics. Although the two men were apparently very different (Lindemann was aloof, a vegetarian, and a total abstainer), each respected the complementary qualities in the other and they had essentials in common—courage, humor, and a love of language. And while Churchill had excelled at polo, Lindemann was outstanding at tennis; he had won the European championship in Germany in 1914 and later competed at Wimbledon.

Lindemann's wartime advice was sometimes controversial, but the fact that it was sought and accepted signified a changed attitude in the British government. For the first time since the brilliant exception of Lyon Playfair in the nineteenth century, a scientist had a direct voice in national affairs. This change was effected mutually by Churchill and Cherwell, who had been ennobled in 1941 and appointed paymaster general in 1942.

Cherwell returned to Oxford in 1945 but was again with Churchill as paymaster general from 1951 to 1953 and accompanied him to summit meetings. He was primarily responsible for establishing the Atomic

Energy Authority, and when this was assured he returned to Oxford. He retired from his chair in 1956 but continued to influence Churchill's thought, especially regarding technology. The two wanted to establish in England an institution similar to MIT; but tradition proved too strong and a compromise was reached in founding a new college at Cambridge, Churchill College, intended especially to promote technology.

Lindemann's wide interests in physics were reflected in the Lindemann electrometer (in which he perceived the advantages of reducing instrument size), Lindemann glass for transmitting X rays, the Dobson-Lindemann theory of the upper atmosphere, indeterminacy, chemical kinetics, and, with Aston in 1919, the separation of isotopes. While serving as paymaster general during World War II, he proved the prime-number theorem by a new argument. The task of building up the Clarendon Laboratory in a difficult environment took effort that would otherwise have earned him a greater place in personal research; but his ideas illuminated many fields of physics at the pioneering stage. Beyond this, his achievements lie in the success of the laboratory and in the help that he gave Churchill.

BIBLIOGRAPHY

See *The Physical Significance of the Quantum Theory* (London, 1931).

For information on Lindemann's work, see Lord Birkenhead, *The Prof in Two Worlds* (London, 1961); R. F. Harrod, *The Prof* (London, 1959); and G. P. Thomson, "Frederick Alexander Lindemann, Viscount Cherwell 1886–1957," in *Biographical Memoirs of Fellows of the Royal Society*, **4** (1958), 45.

Both Birkenhead and Thomson give complete bibliographies.

R. V. JONES

LINDENAU, BERNHARD AUGUST VON (*b.* Altenburg, Germany, 11 June 1779; *d.* Altenburg, 21 May 1854), *astronomy.*

Lindenau was the son of Johann August Lindenau, a regional administrator *(Landschaftsdirektor).* He began the study of law and mathematics at Leipzig in 1793; and beginning in 1801 he worked at the astronomical observatory directed by F. X. von Zach located at Seeberg, near Gotha. For this purpose he procured a leave of absence from government service. Following Zach's departure (1804) Lindenau became temporary, and in 1808 official, director of the Gotha observatory. There he also was editor of the important technical journal *Monatliche Correspondenz zur Beförderung der Erd- und Himmelskunde.* In 1816, after a temporary absence required by the "War of Liberation" against Napoleon, Lindenau founded, with J. G. F. von Bohnenberger, the *Zeitschrift für Astronomie und verwandte Wissenschaften.* In 1818 he was forced to return to the civil service. His successor at Gotha was J. F. Encke.

Lindenau rapidly advanced from vice-president of the Altenburg board of finance to chief minister of Saxe-Gotha-Altenburg (1820). Later he became minister of the interior of Saxony. In 1848 he was elected to the Frankfurt Parliament. His political activity was comparable with the reforming efforts of Baron Karl vom und zum Stein in Prussia.

An art collector, during his lifetime he established a museum in Altenburg, today known as the Staatliche Lindenau-Museum, which contains one of the most important collections of early Italian panel painting.

Aside from his editorial activity (and considerable assistance in the publication of Bessel's *Fundamenta astronomiae*), Lindenau's importance for astronomy lies mainly in the tables of planets he produced for the calculation of the ephemerides of Mercury, Venus, and Mars. He also published improved values for the constants of aberration and nutation, as well as papers on the history of astronomy. Regrettably, however, astronomy represented only an episode in his life.

BIBLIOGRAPHY

I. ORIGINAL WORKS. Many of Lindenau's publications are listed in J. G. Galle, ed., *Register zu von Zach's Monatlicher Correspondenz zur Beförderung der Erd- und Himmelskunde* (Gotha, 1850), pp. 104–105. Among his writings are *Tabulae Veneris novae et correctae ex theoria gravitatis cl. de Laplace et ex observationibus recentissimis in specula astronomica Seebergensi habitis erutae* (Gotha, 1810); *Tabulae Martis novae et correctae ex theoria gravitatis cl. De la Place et ex observationibus recentissimis erutae* (Eisenberg, 1811); *Investigatio nova orbitae a Mercurio circa solem descriptae cum tabulis planetae* (Gotha, 1813); and "Beitrag zur Geschichte der Neptuns-Entdeckung," in *Ergänzungsheft zu "Astronomische Nachrichten"* (Altona, 1849), pp. 1–32.

II. SECONDARY LITERATURE. Biographies are P. von Ebart, *Bernhard August v. Lindenau* (Gotha, 1896); F. Volger, *Bernhard von Lindenau* (Altenburg, n.d.); and Pasch, "Lindenau," in *Allgemeine Deutsche Biographie*, XVIII (Leipzig, 1883), 681–686. On his political activity, see G. Schmidt, *Die Staatsreform in Sachsen in der ersten Hälfte des 19. Jahrhunderts* (Weimar, 1966), especially pp. 110 ff. Lindenau as an art collector is discussed in H. C. von der Gabelentz and H. Scherf, *Das Staatliche Lindenau-Museum, seine Geschichte, seine Sammlungen* (Altenburg, 1967).

On Lindenau's scientific work, see W. Gresky, "Aus Bernhard v. Lindenaus Briefen an C. F. Gauss," in *Mitteilungen der Gauss-Gesellschaft*, no. 5 (1968), 12–46; D. B. Herrmann, "Lindenaus Abschied von der Astronomie," in *Die Sterne*, **42** (1966), 66–72; and "Bernhard August von Lindenau und die Herausgabe der *Fundamenta astronomiae* von F. W. Bessel," in *Vorträge und Schriften der Archenhold-Sternwarte Berlin-Treptow*, no. 20 (1968); and F. Lessig, "Der Astronom Bernhard von Lindenau," in *Abhandlungen des Naturkundlichen Museums "Mauritianum" Altenburg*, **2** (1960), 29–34.

DIETER B. HERRMANN

LINDGREN, WALDEMAR (*b.* Kalmar, Sweden, 14 February 1860; *d.* Brookline, Massachusetts, 3 November 1939), *geology*.

The leading contributor in his generation to the science of ore deposition, Lindgren stressed the dominance of igneous processes in ore formation, developed a still-useful classification of the metasomatic processes in fissure veins, and was one of the first to identify contact-metamorphic ore bodies in North America. He pioneered in the application of the petrographic microscope to the study of ores and their constituent minerals. The unified study of the geology and ore deposits of extensive regions, which he inaugurated, led to important ideas concerning metallogenetic epochs and laid the foundations for modern concepts of metallogenetic provinces.

The son of Johan Magnus Lindgren, a judge of a provincial court, and Emma Bergman Lindgren, whose ancestors had been important in the history of Sweden for more than three centuries, Waldemar Lindgren displayed a keen interest in mines and minerals when he was young. Encouraged by his parents, he entered the Royal Mining Academy at Freiberg, Saxony, at the age of eighteen. Graduating in 1882 with the degree of mining engineer, he remained there for postgraduate studies in metallurgy and chemistry until, in June 1883, he sailed for America with the hope of participating in the mining and geological activity then flourishing in the western United States. Letters from professors at Freiberg introduced him to Raphael Pumpelly and George F. Becker, and in November 1884 he began an association with the U.S. Geological Survey that continued for thirty-one years.

At first Lindgren was assigned to fieldwork that took him from one western mining district to another; then, in 1905, he became head of the section in the Division of Mineral Resources devoted to the nonferrous metals, and in 1908 he was made chief of the Division of Metalliferous Geology. Appointed chief geologist of the Survey in 1911, he resigned from that post in 1912 to become William Barton Rogers professor of geology and head of the department of geology at the Massachusetts Institute of Technology. There he served with great distinction, well past the normal retirement age of seventy, until he became professor emeritus in 1933.

A founder of *Economic Geology* in 1905, Lindgren continued as an associate editor of that journal for the rest of his life. While he was chairman of the Division of Geology and Geography of the National Research Council in 1927 and 1928, he established the *Annotated Bibliography of Economic Geology* and subsequently supervised the abstracting of an enormous volume of current literature, foreign as well as American. He also found time to serve as consulting geologist for mining companies in Canada, Mexico, Chile, Bolivia, and Australia. In 1886 he married Ottolina Allstrim of Goteborg, Sweden, who was his inseparable companion until her death in 1929. They had no children.

Lindgren was a member, fellow, or corresponding member of many scientific societies in the United States, Canada, Sweden, Belgium, England, and the Soviet Union. He was president of the Mining and Metallurgical Society of America in 1920, of the Society of Economic Geologists in 1922, and of the Geological Society of America in 1924. The latter society awarded him its Penrose Medal in 1933, and the Geological Society of London bestowed its Wollaston Medal upon him in 1937. The American Institute of Mining and Metallurgical Engineers made Lindgren an honorary life member in 1931, and in 1933 he was honorary chairman of the Sixteenth International Geological Congress.

BIBLIOGRAPHY

I. ORIGINAL WORKS. A "classified list" of Lindgren's scientific writings prior to 1933 (228 titles) accompanies the biographical sketch by L. C. Graton (see below). Many of these writings are in the geologic atlases, annual reports, professional papers, and bulletins of the U.S. Geological Survey; others are contributions to technical journals. To Graton's bibliography should be added 12 titles published in 1933–1938. Among the latter, two are of prime importance: "Differentiation and Ore Deposition, Cordilleran Region of the United States," in *Ore Deposits of the Western States* (New York, 1933), pp. 152–180; and "Succession of Minerals and Temperature of Formation in Ore Deposits of Magmatic Affiliations," in *Technical Publications, American Institute of Mining and Metallurgical Engineers*, no. 713, also in *Transactions of the American Institute of Mining and Metallurgical Engineers*, **126** (1937), 356–376. Notable also is Lindgren's *Mineral Deposits*

(New York, 1913), with rev. and enl. eds. in 1919, 1928, and 1933, which was the standard textbook in economic geology for at least a third of a century.

A considerable file of Lindgren's personal records, correspondence, and memorabilia is in the archives of the Massachusetts Institute of Technology, where the library of the department of earth and planetary sciences is named for him.

II. SECONDARY LITERATURE. The most inclusive and perceptive article about Lindgren is L. C. Graton's "Life and Scientific Works of Waldemar Lindgren," in *Ore Deposits of the Western States, Lindgren Volume* (New York, 1933), pp. xiii–xxxii. Among the several obituaries published soon after his death are those by Hans Schneiderholm in *Zentralblatt für Mineralogie*, Abt. A, no. 3 (1940), 65–69; and M. J. Buerger, in *American Mineralogist*, **25** (1940), 184–188.

KIRTLEY F. MATHER

LINDLEY, JOHN (*b.* Catton, near Norwich, England, 5 February 1799; *d.* Turnham Green, Middlesex, England, 1 November 1865), *botany, horticulture.*

A man endowed with an extraordinary capacity for work and a restless, aggressive, untiring intellect, who attained distinction in all his varied activities, Lindley was among the most industrious, many-sided and productive of the nineteenth-century botanists. As administrator, professor, horticulturist, taxonomist, editor, journalist, author, and botanical artist he used to the full his time, his abundant energy, and his remarkable talents, with lasting beneficial results in many fields of botany and horticulture. His major botanical contribution was to the study of orchids.

His father, George Lindley, a skilled but financially unsuccessful nurseryman, could not afford to buy his son an officer's commission in the army or a university education but gave him a good schooling in Norwich to the age of sixteen. Young Lindley then went to Belgium as a British seedsman's representative. He early displayed his remarkable powers of sustained work by translating into English at one sitting L. C. M. Richard's *Démonstrations botaniques, ou Analyse du Fruit* (1808), published in 1819 as *Observations on the Structure of Seeds and Fruits.* In 1818 or 1819 he entered the employment of Sir Joseph Banks as an assistant in the latter's rich library and herbarium, working there for eighteen months with Robert Brown. Banks died in 1820. The Horticultural Society of London had commissioned Lindley in that year to draw some single roses, and in 1822 he entered its service as assistant secretary of its newly established Chiswick garden, thus beginning an association of forty-three years. His early publications, for which Banks's library and herbarium provided facilities then unrivaled, included *Rasarum monographia* (1820), *Digitalium monographia* (1821), *Collectanea botanica* (1821–1825) and a survey of the Rosaceae subfamily Pomoideae (Pomaceae), published in *Transactions of the Linnean Society of London* (**13** [1821], 88–106), in which he established the genera *Chaenomeles, Osteomeles, Eriobotrya, Photinia, Chamaemeles,* and *Raphiolepis,* all still accepted. Together with contributions to the *Botanical Register* (beginning with volume 5, plate 385, August 1819), they quickly won him an international reputation.

These youthful publications displayed remarkable taxonomic judgment, detailed observation, and precision of language in both English and Latin. In 1828, despite his lack of a university education, Lindley was elected Fellow of the Royal Society of London and appointed professor of botany in the newly founded University of London, giving his inaugural lecture in April 1829. He did not, however, relinquish his employment by the Horticultural Society, of which he became general assistant secretary in 1827 and secretary in 1858; indeed, he carried a heavy load of responsibility and made important innovations during the society's troubled years. Late in 1832 the University of Munich, at the instigation of Martius, enterprisingly conferred an honorary Ph.D. upon Lindley. In 1838 he prepared the report on the management of the royal gardens at Kew which led ultimately to the foundation of the Royal Botanic Gardens, Kew, as a national botanical institution.

Time-consuming though his official duties and public activities certainly were, Lindley nevertheless managed to prepare the specific characters for the 16,712 species of flowering plants and cryptogams included in John Loudon's *Encyclopaedia of Plants* (1829) and to produce a series of well-documented, clearly written, authoritative educational publications, including *An Introduction to Botany* (1832; 2nd ed., 1835; 3rd ed., 1839; 4th ed., 1848), of permanent value for its botanical vocabulary (reprinted in W. T. Stearn, *Botanical Latin* [1966], pp. 314–353, and elsewhere), *Flora medica* (1838), *School Botany* (1839; 12th ed., 1862), and *The Theory of Horticulture* (1840; 2nd ed., entitled *The Theory and Practice of Horticulture,* 1855), which Lindley himself considered his best book. He also managed to engage in research, notably in paleobotany, to which Lindley and Hutton's *Fossil Flora of Great Britain* (3 vols., 1831–1837) bears witness, and in orchidology. During Lindley's lifetime European penetration into the moist tropics abounding with Orchidaceae, the employment of professional plant collectors by European nurseries, swifter transport by sea, improved methods of greenhouse construction and management, and the social prestige associated with orchid-growing by the aristocracy and gentry of Britain, who spent vast sums on this, led to the introduction and successful cultivation of orchids in unprecedented quantity and diversity. They became Lindley's major botanical specialty, and he became the leading authority on their classification. Ultimately he established more than 120 genera of Orchidaceae, among them *Cattleya*, *Cirrhopetalum*, *Coelogyne*, *Laelia*, *Lycaste*, and *Sophronitis*; described many hundreds of new species; and produced three major works: *Genera and Species of Orchidaceous Plants* (1830–1840), *Sertum orchidaceum* (1838), and *Folia orchidacea* (1852–1855), as well as many articles in periodicals.

As a young man Lindley campaigned vigorously against the artificial "sexual system" of classification of plants introduced by Linnaeus and in favor of a more natural system, as propounded by A. L. de Jussieu and A. P. de Candolle and improved in detail by Robert Brown. On his appointment as a professor, he immediately prepared for the use of students *A Synopsis of the British Flora, Arranged According to the Natural Orders* (1829; 2nd ed., 1835; 3rd ed., 1841), the second account of British plants thus classified. In 1830 he published *Introduction to the Natural System of Botany*, which was the first work in English to give descriptions of the families (then called "natural orders") on a worldwide basis; it embodied detailed, firsthand observations of their representatives in the garden and herbarium. Uninfluenced by theories of evolution, and hence without thought of phylogeny, Lindley regarded the characters of plants as "the living Hieroglyphics of the Almighty which the skill of man is permitted to interpret. The key to their meaning lies enveloped in the folds of the Natural System." This he continuously sought to unfold, with but partial success. He took the view that "the investigation of structure and vegetable physiology are the foundation of all sound principles of classification," that within the vegetable kingdom "no sections are capable of being positively defined, except as depend upon physiological peculiarities," and that "physiological characters are of greater importance in regulating the natural classification than structural."

This emphasis led Lindley astray and resulted in major classifications which he himself never found wholly satisfactory, since he changed them from work to work, and which other botanists accepted only in part. Because, however, he also believed "that the affinities of plants may be determined by a consideration of all the points of resemblance between their various parts, properties and qualities; and that thence an arrangement may be deduced in which these species will be placed next each other which have the highest degree of relationship," he gave attention to a much wider range of characters than did many of his contemporaries. Such information, derived from Lindley's profound and extensive observation of plants and a thorough study of available literature, made his *Introduction to the Natural System* (374 pages, 1830) and its enlarged successors, *A Natural System of Botany* (526 pages, 1836) and *The Vegetable Kingdom* (908 pages, with more than 500 illustrations, 1846; 3rd ed., 1853), reference works long unrivaled for matters of detail. In the 1836 work Lindley introduced a nomenclatural reform by proposing that divisions of the same hierarchical standing should have names formed in the same distinctive way, with terminations indicative of these divisions. Thus he consistently used the termination "-aceae" for names of natural orders (now called "families"), replacing, for example, "Umbelliferae" by "Apiaceae" (from *apium*, celery) and "Leguminosae" by "Fabaceae" (from *faba*, broad bean), and the termination "-ales" for alliances (now called "orders"); this became the internationally followed procedure.

In his youth Lindley gallantly but unwisely assumed responsibility for his father's heavy debts, and their redemption burdened him for many years. Hence, driven partly by financial necessity, he took on ever more tasks and duties without relinquishing those he already had. In 1826, for example, he became *de facto* editor of *Botanical Register*; in 1836, superintendent of the Chelsea Physic Garden; in 1841, horticultural editor of *Gardeners' Chronicle*. Between 1833 and 1840 he completed Sibthorp and Smith's magnificent *Flora Graeca* (see *Taxon*, **16** [1967], 168–178); he also wrote innumerable botanical articles in the *Penny Cyclopaedia*. For the London exhibition of 1862 he took charge of the Colonial Department, but the load of manifold onerous activities had now become too great even for a man of his sturdy constitution and stubborn, determined mind. In 1862 his health declined, and Lindley had reluctantly to give up posts he had so honorably and industriously held over many years. In 1865 he died, within a few months of his lifelong friends William Jackson Hooker and Joseph Paxton. His orchid herbarium was acquired by the Royal Botanic Gardens, Kew; his general herbarium by the department of botany, University of Cambridge. His private library, very rich in botanical tracts and pamphlets, became the foundation of the Lindley Library of the Royal Horticultural Society of London, of which he had been so long an efficient servant.

BIBLIOGRAPHY

I. ORIGINAL WORKS. Lindley's books are listed in the text. There is a more detailed list in the Royal Horticultural Society, *The Lindley Library Catalogue* (London, 1927), pp. 256–257. His contributions to periodicals other than the numerous articles in the *Botanical Register* and the *Gardeners' Chronicle* are listed in the Royal Society of London, *Catalogue of Scientific Papers 1800–1863*, IV, 31–32. An unpublished "Bibliography of the Published Works of John Lindley," compiled by J. M. Allford in 1953, lists 236 publications (including eds.) by Lindley. His first publication, "a most ingenious and elaborate description by Mr Lindley, junior" of *Maranta zebrina*, appeared in *Botanical Register*, **5**, pl. 385 (Aug. 1819) and was followed by the text to pls. 397, 404, 419, 420, 425, 430, 431 (1819–1820). He was officially editor of the *Botanical Register* from vol. **16** (1829) to vol. **33** (1847) but contributed most of the articles from vol. **11** (1825) on. Many letters to and from Lindley are at the Royal Botanic Gardens, Kew.

II. SECONDARY LITERATURE. The major sources of biographical information are the obituary in *Gardeners' Chronicle* (1865), 1058–1059, 1082–1083; W. Gardener, "John Lindley," *ibid.*, **158** (1965), 386, 406, 409, 430, 434, 451, 457, 481, 502, 507, 526; and F. Keeble, "John Lindley," in F. W. Oliver, ed., *Makers of British Botany* (London, 1912), 164–177. J. R. Green, *A History of Botany in the United Kingdom* (London, 1914), pp. 336–353, gives a fair assessment of Lindley's scientific work. H. R. Fletcher, *The Story of the Royal Horticultural Society 1804–1968* (London, 1969), contains many references to Lindley in connection with the Society's affairs. Lindley's part in the 1838 committee of inquiry into the management of the Royal Gardens at Kew is summarized by W. T. Stearn, "The Self-Taught Botanists Who Saved the Kew Botanic Garden," in *Taxon*, **14** (Dec. 1965), 293–298.

WILLIAM T. STEARN

LINK, HEINRICH FRIEDRICH (*b.* Hildesheim, Germany, 2 February 1767; *d.* Berlin, Germany, 1 January 1851), *zoology, botany, chemistry, geology, physics.*

Link studied medicine and natural sciences at the University of Göttingen, from which he received the M.D. in 1789 with a dissertation entitled *Florae Göttingensis specimen, sistens vegetabilia saxo calcareo propria* (published in Usteri's *Delectus opusculorum botanicorum*, I [1790], 299–336). In 1792 he was appointed full professor of zoology, botany, and chemistry at the University of Rostock; in 1811, of chemistry and botany at the University of Breslau; and in 1815, of botany at the University of Berlin. In Berlin he was also director of the university's botanic garden.

Link published a great many articles on topics in botany, zoology, geology, and chemistry. His primary interest was in chemistry and physics, and he conducted experiments on adhesion, solution, crystallization, and double salts and acid salts. In his dissertation Link declared himself an adherent of Lavoisier's oxidation theory; he published a critical discussion on phlogiston, "Bemerkungen über das Phlogiston" (1790), as well as papers on the general concept of chemical affinity (1791) and Berthollet's theory of chemical equilibrium (1807). In 1814 Link published an article on the action of sulfuric acid on vegetable material, and in 1815 one on chemical reactions of solids produced by trituration. In 1806 he wrote a textbook on antiphlogistic chemistry: *Die Grundwahrheiten der neuern Chemie.*

Link also published on botany and geology. In 1797–1799 he traveled with Count J. C. von Hoffmannsegg through Portugal, where he studied the flora as well as the geology, agriculture, and industry. In his *Versuch einer Anleitung zur geologischen Kenntniss der Mineralien* (1790), Link declared himself an adherent of neither the neptunist nor the plutonic geological school; rather, he defended the general view that inorganic nature can produce the same materials by different means. In botany, Link published on morphology, classification, anatomy, and physiology of plants; and in zoology, a work on mollusks (1806) which was used by Lamarck. His last zoological publication (1830) dealt with zoophytes.

Link was always deeply interested in the philosophical foundations of the natural sciences. Throughout his life he followed critically the various philosophical currents in Germany. His philosophical thought formed no complete system, and he sought a position intermediate between the prevailing more or less speculative philosophies. In particular, Link was strongly influenced by Kant's dynamical theories of matter; the influence of Schelling's more speculative dynamical concepts was much less, and that of Hegel was negligible. Like the majority of the scientists of his day, Link was greatly interested in the idea of unity in all natural sciences, but he rejected emphatically the speculative *Naturphilosophie* of Schelling and his adherents.

In his first philosophical work, *Ueber Naturphilosophie* (1806), Link started from the concept that it is absolutely necessary for both physicists and chemists to study philosophy. He declared himself against a speculative natural-philosophical mode of thought but praised the dynamical theory of matter given by Kant, although he mentioned many objections against it. It remained the primary task of the natural scientist constantly to examine all natural phenomena. This critical and—to the then-ruling philosophical systems—negative attitude was undoubtedly the reason why Link's views met with great approval among the opponents of *Naturphilosophie* but, on the other hand, could not convince its adherents. The same is true of Link's other philosophical publications, of which *Propyläen der Naturkunde* (1836–1839) is the most important.

From his first publications, Link was averse to an atomic theory of matter. In 1798 he already accepted the infinite divisibility of matter. Like many of his contemporaries, however, in practice Link worked with little particles from which matter could be built up, since he found it possible to explain more phenomena in this way than with the dynamical construction. Link was also very interested in the problem of the nature *(Wesen)* of the liquid and solid states. He took the liquid state as the original form of matter and tried to derive the solid state from it. Many mineralogists, most notably J. J. Bernhardi and J. J. Prechtl, accepted Link's theory as a starting point for a theory of crystallization. Link applied his ideas to the explanation of such phenomena as solution, formation of crystals, chemical reactions, flexibility, and elasticity.

BIBLIOGRAPHY

Link's most important philosophical works are *Ueber Naturphilosophie* (Leipzig–Rostock, 1806); *Ideen zu einer philosophischen Naturkunde* (Breslau, 1814); and *Propyläen der Naturkunde*, 2 vols. (Berlin, 1836–1839).

Among his other publications are *Versuch einer Anleitung zur geologischen Kenntniss der Mineralien* (Göttingen, 1790); *Beyträge zur Physik und Chemie*, 3 vols. (Rostock, 1795–1797); *Beyträge zur Naturgeschichte*, 2 vols. (Rostock–Leipzig, 1797–1801); *Grundriss der Physik* (Hamburg, 1798); *Die Grundwahrheiten der neuern Chemie, nach Fourcroy's Philosophie chimique herausgegeben, mit vielen Zusätzen* (Leipzig–Rostock, 1806); *Grundlehren der Anatomie und Physiologie der Pflanzen* (Göttingen, 1807); *Flore portugaise ou Description de toutes les plantes, qui croissent naturellement en Portugal*, 2 vols. (Berlin, 1809–1820), written with J. C. von Hoffmannsegg; *Elementa philosophiae botanicae* (Berlin, 1824); *Handbuch der physikalischen Erdbeschreibung*, 2 vols. (Berlin, 1826–1830); and *Ueber die Bildung der festen Körper* (Berlin, 1841).

On Link, see C. F. P. von Martius, *Denkrede auf Heinrich Friedrich Link* (Munich, 1851).

H. A. M. SNELDERS

LINNAEUS (or **VON LINNÉ**), **CARL** (*b.* Södra Råshult, Småland, Sweden, 23 May 1707; *d.* Uppsala, Sweden, 10 January 1778), *botany, zoology, geology, medicine.*

During his lifetime, Linnaeus exerted an influence in his fields—botany and natural history—that has had few parallels in the history of science. Driven by indomitable ambition and aided by an incredible capacity for work, he accomplished the tremendous task that he had set for himself in his youth: the establishment of new systems for the three kingdoms of nature to facilitate the description of all known animals, plants, and minerals.

His father, Nils Linnaeus, a country parson in the south Swedish province of Småland, settled in 1709 in Stenbrohult, where Linnaeus grew up. The father was a great lover of flowers and laid out a beautiful garden in the parsonage grounds. He also introduced his son to the mysteries of botany. In 1716 Linnaeus entered the Latin School in the nearby cathedral city of Växjö, and from that time natural history remained his favorite study. An average boy, he was set apart only by his delight in botanizing and learning about plants.

Of great importance for Linnaeus' future development was an outstanding teacher during his last years at Växjö, Johan Rothman, who encouraged his aptitude for botany and taught him Tournefort's system. In 1727 he entered the University of Lund to study medicine. Medical instruction was poor there, but during his stay Linnaeus undertook extensive botanical excursions in the surrounding area and met Kilian Stobaeus, one of the many patrons who helped him in his youth. A medical doctor and a learned polymath, Stobaeus placed his rich library and his collection of natural specimens at Linnaeus' disposition.

In the fall of 1728 Linnaeus went on to the University of Uppsala, lured by its greater reputation. On the whole he must have been disappointed, for the medical faculty was not much better than at Lund. Although he came to rely mostly on his own, independently acquired abilities as a botanist and natural historian, the atmosphere of Uppsala was in many ways useful. The botanical garden, although somewhat neglected, contained rare foreign plants, and he found influential promoters: the elderly professor of medicine Olof Rudbeck the younger, formerly an outstanding botanist and zoologist, and the learned Olof Celsius, with whom Linnaeus studied the flora of the surrounding region.

Around 1730 Linnaeus matured as a researcher and began delineating the fundamental features of his botanical reform. In a series of manuscripts, unpublished at the time, he noted the results of his observations and thoughts. He had already found the tool with which to construct a new system of plant taxonomy—the new theory of plant sexuality. At this time Linnaeus' sexual system first appeared, still incomplete, in a manuscript version of his *Hortus uplandicus* (1730).

Linnaeus barely supported himself during this period by substituting for Rudbeck in conducting demonstrations at the botanical garden and by giving private lessons. In the spring of 1732, sponsored by the Uppsala Society of Science, he made a trip to Lapland that lasted until the fall. His account of the journey (first published in English translation in 1811) gives vivid evidence of his joy in the unfamiliar world of the Scandinavian mountains, in which he studied the unknown alpine flora and Lapp customs. Two years later (1734) he undertook another, more modest study trip in the Dalarna region of central Sweden. He constantly revised his manuscripts and widened his teaching activities with courses in mineralogy and assaying. In the spring of 1735 Linnaeus left Sweden—with a Lapp costume, a troll-drum, and unpublished manuscripts in his luggage—to obtain the M.D. in Holland. At the University of Harderwijk he defended a thesis on the causes of ague, received his degree, and went on to Leiden.

Linnaeus' sojourn in Holland lasted until 1738. It was a period of uninterrupted success in approaching his goal of being acknowledged as Europe's outstanding botanist. Learned and wealthy patrons continued to help him: Boerhaave, J. F. Gronovius in Leiden, and the botanist Jan Burman in Amsterdam fostered Linnaeus' plans and helped to see his works through to publication. During a short visit to England in 1736 he met Sir Hans Sloane and J. J. Dillenius. Most of the time he stayed with George Clifford, a wealthy merchant who established an incomparable botanical garden at the Hartekamp, near Haarlem; as its superintendent Linnaeus could concentrate on his botanical work.

Linnaeus' literary production during his three years in Holland is astonishing. Most of the works had been nearly finished in Sweden; but others, such as *Hortus Cliffortianus* (1737), a beautiful folio on the plants in Clifford's garden and greenhouse, were newly composed. Linnaeus' key work, and the first to be published in Holland, was *Systema naturae* (1735). On seven folio leaves he presented in a schematic arrangement his new system for the animal, plant, and mineral kingdoms; the plants were arranged according to the sexual system that this work first made known to the scientific world. In the small *Fundamenta botanica* (1736) Linnaeus put forward with dictatorial authority his theory for systematic botany, and in *Bibliotheca botanica* he listed the botanical literature in systematic fashion. Other works of this period are *Flora lapponica* (1737), in which he described his botanical collections from the Lapland journey; *Critica botanica* (1737), with rules for botanical nomenclature; *Classes plantarum* (1738), in which he reviewed the various plant systems going back to Cesalpino; and *Genera plantarum* (1737), which, next to *Systema naturae*, is his outstanding early work and contains short descriptions of all 935 plant genera known at the time.

Linnaeus then set out for Sweden via Paris, where he visited the brothers Jussieu, and by June 1738 he was back in his native country. His foreign triumphs had not gone unnoticed; but no academic position awaited him, and he was obliged to become a practicing physician in Stockholm. He married Sara Elisabeth Moraea, the daughter of a wealthy city physician. In 1739 he was appointed physician to the admiralty and was one of the moving powers in the founding of the Swedish Academy of Science in Stockholm, of which he became the first president and eventually the most outstanding member.

A new phase of Linnaeus' life began in the spring of 1741. After a rather scandalous fight for tenure he was appointed professor of practical medicine at the University of Uppsala; in 1742, through a sort of job exchange, he took over the more suitable chair of botany, dietetics, and materia medica. For the rest of his life Linnaeus remained in Uppsala as professor of medicine, while his fame as premier botanist spread throughout the world.

One of his first tasks was to renovate the university's neglected botanical garden, founded in 1655 by Rudbeck the elder. Remodeled in French style with formal parterres and an ornamental orangery, it was filled over the years with innumerable plants from Europe and distant parts of the world. Linnaeus spent the rest of his life in the professorial residence in a corner of the garden.

As a teacher and supervisor Linnaeus was incomparable. Students flocked to his lectures, which were characterized by humor and the presentation of unusual ideas, and which included the whole of botany and natural history. He liked dealing with what he called "the natural diet," that is, man's way of living under the pressure of civilization. In private seminars he gave deeper insights into his botanical thought. His botanical excursions held near Uppsala were immensely popular; these *herbationes* were organized according to a fixed ritual: on the return home the participants marched in closed formation to the music of French horns and drums.

Linnaeus took an unselfish and devoted interest in his more mature students, supporting and encouraging them. He persuaded an imposing number of them to obtain the doctorate, but he usually wrote their theses himself (collected under the title *Amoenitates academicae*, I–VII [1749–1769]; VIII–X [1785–1790]).

From 1759 Linnaeus was assisted by his son Carl von Linné the younger, who had been appointed botanical demonstrator and who succeeded his father. Linnaeus occasionally worked for the royal court: he described the valuable natural history collections at Ulriksdal and at the castle of Drottningholm, and charmed the discerning queen, Lovisa Ulrika, by his unaffected manner. On the whole he had an unerring instinct for profitable social connections; he counted among his friends and admirers the country's outstanding cultural leaders, including the president of the chancellery, Carl Gustaf Tessin.

Linnaeus was in contact with the world's botanists and collectors of natural history specimens. He corresponded with them and received seeds for his garden and plants for his herbarium, which became very extensive. He became a member of innumerable foreign scientific societies (he was one of the eight foreign members of the Paris Academy in 1762) and received distinctions and honors in Sweden; in 1747 he was appointed court physician and in 1762 he was elevated to the nobility as von Linné. In later years he traveled every summer to Hammarby, his small estate outside Uppsala; there, free from academic chores, he devoted himself to botany or taught selected students.

Linnaeus' research work during his mature years began with trips to various Swedish provinces. By order of the parliament, which wanted an inventory of all the natural resources of the country, during three summers in the 1740's Linnaeus traveled through selected areas to describe them and to search for dyestuffs, minerals, clay, and other economically useful substances. His reports of these expeditions were published as *Ölandska och gothländska resa* (1745), *Västgöta resa* (1747), and *Skånska resa* (1751), all written in Swedish. Nothing escaped his attention on his travels on horseback—plants and insects, runic stones and other ancient remnants, farmers working in the fields and meadows, the changes in the weather. His prose style was simple and strong, sometimes rising to lyrical outbursts or spiced with effective similes.

But Linnaeus' main mission was to complete his reform of botany. In the work produced during his stay in Holland he had established the principles, maintained more or less unchanged for the rest of his life, but they still had to be developed and put into practice. In 1751 he published *Philosophia botanica*, his most influential work but actually only an expanded version of *Fundamenta botanica*. In it Linnaeus dealt with the theory of botany, the laws and rules that the botanist must follow in order to describe and name the plants correctly and to combine them into higher systematic categories. At the same time he struggled with the enormous undertaking of cataloging all of the world's plant and animal species and giving each its correct place in the system.

Linnaeus started modestly with his native land: *Flora suecica*, containing 1,140 species, appeared in 1745 and *Fauna suecica* followed a year later. The remaining inventory of the world's flora was arduous; surrounded by herbarium sheets and folios, Linnaeus occasionally felt unable to continue. Yet before long he completed the monumental *Species plantarum* (1753), the work he valued the most. In its two volumes he described in brief Latin diagnoses about 8,000 plant species from throughout the world. He also continued to update his *Systema naturae*, which, as the bible of natural history, was to present the current state of knowledge and which he periodically expanded and revised. The definitive tenth edition (1758–1759), which included only animal and plant species, was a two-volume work of 1,384 pages.

Linnaeus had to pay the price his ambitions demanded. After suffering a fit of apoplexy in the spring of 1774 that forced him to retire from teaching, his mind became increasingly clouded; another stroke followed in the fall of 1776, and he died early in 1778.

A born taxonomist and systematist, Linnaeus had a deep-rooted, almost compulsive need for clarity and order; everything that passed through his hands he cataloged and sorted into groups and subgroups. This characteristic typifies all of his writings, including his private correspondence. His view of the universe as a gigantic nature collection that God had given him to describe and to fit into a framework was applied first to his reform of botany. There was at that time an urgent need for a simple and easy-to-grasp system for the plant kingdom. A stream of previously unknown plants was reaching Europe, and botanists felt chaos threatening. Since the time of Cesalpino many botanists had tried to create a useful system; and toward the eighteenth century some of these systems had begun to win considerable support, especially those of John Ray and Tournefort, based upon the appearance of the corolla. But none was sufficiently practical, and the various systems were reciprocally competitive, thus increasing the confusion. Only Linnaeus' system, based on sexuality, had the requisite of being generally adoptable.

Around 1720 the sexuality of plants was still being disputed by many botanists. Linnaeus had already learned about it from his teacher Rothman at the Växjö Gymnasium, following the publication of Sébastien Vaillant's *Sermo de structura florum* (1717). As a student at Uppsala he had pursued the subject through his own investigations and soon was con-

LINNAEUS

vinced of its truth. In the small *Praeludia sponsaliorum plantarum* (1730), written in Swedish, he announced that the stamens and pistils were the sexual organs of plants. At the same time he began to investigate whether the stamens and pistils could be used to construct a new botanical system. After a short period of doubt he was sure that they could, and in *Systema naturae* he presented the sexual system in its definitive form. The various plants are grouped into twenty-four classes according to the number of stamens and their relative order; each class is then divided into certain orders, mostly according to the number of pistils. The practical applicability of the sexual system made botany an easy and pleasant science; after one look at the organs of fertilization, any plant could quickly be placed in the proper class and order. To be sure, the system was attacked: in Germany by J. G. Siegesbeck and Heister, in Switzerland by Haller, and in France, where Tournefort's system had long been generally accepted. But acceptance of the sexual system could not be halted; it gained footing almost everywhere, especially in England, beginning in the 1760's.

Linnaeus was aware that the sexual system was an artificial structure. It was founded with rigid consistency upon a single principle of division and therefore represented the natural affinities only partially. Linnaeus tried throughout his life to replace the sexual system with a *methodus naturalis*, a botanical system that would express the "natural" relations. In *Classes plantarum* (1738) he described sixty-five plant orders that he considered natural, and he returned to the problem in private lectures toward the end of his life—but he never succeeded in solving it. A natural plant system presented theoretical difficulties that he was unable to overcome; in practice he seems to have depended on an intuitive knowledge of the general similarity (facies) of plants. Therefore Linnaeus and his pupils continued to use the sexual system, and it remained for the French (Adanson, A. L. de Jussieu) to lay the foundation of a natural botanical system in the modern sense of the word.

Linnaeus' other achievements as the reorganizer of botany were as fundamental as the sexual system. He made a deep study of the principles of taxonomy and with masterly clarity worked out fixed rules for the differentiation of the lowest systematic categories—genera and species—both of which were, according to him, "natural." Although Tournefort had already delineated the genus with great precision, it remained for Linnaeus, inspired by Ray, to be the first to work with species as a clearly defined concept. The species were the basic units of botany, and for a long time he considered them to be fixed and unchangeable: "We can count as many species now as were created at the beginning." He united closely related species in genera according to the appearance of their sexual organs.

Linnaeus dictated with inexorable logic in *Philosophia botanica* how botanists should proceed in practice. Every species had to be differentiated from all other species within the genus through a Latin *differentia specifica*, that is, a short diagnosis (twelve words at most) including the characteristic scientific name or *nomen specificum*. Furthermore, there had to be a longer, complete description of the whole plant. Linnaeus himself laid down the morphological terms that every taxonomist should use for each part of the plant, in order to avoid confusion.

Latin species names notwithstanding, the precision ordered by Linnaeus was nevertheless found to be unmanageable, and he therefore took a decisive step—the introduction of binomial nomenclature. Heralded in his bibliographical writings, it is effectively used for the first time in his dissertation on the plants eaten by animals, "Pan suecicus" (1749), and was subsequently applied in *Species plantarum* (1753). The binomial nomenclature implied that every plant species would be provided with only two names: one for the genus and one for the species. Linnaeus was delighted by the idea; it "was like putting the clapper in the bell." Yet at the beginning he did not realize its full import. He considered the new species names only as handy labels, calling them "trivial names"; they could never replace the scientific "specific" names. Only gradually did he understand the practical advantages offered by the new nomenclature, and he then decreed that the "trivial name" should be unchangeable. Since then his own *Species plantarum* has been the solid basis for botanical nomenclature.

Linnaeus' botanical ideas changed over the years on an important point: he was forced to relinquish confidence in the constancy of the number of species. In *Peloria* (1744) he described a monstrous form of *Linaria vulgaris* that he wrongly interpreted as a hybrid between *Linaria* and a completely different plant. He thereafter held that new plant species could develop through hybridization (see *Plantae hybridae* [1751]) and reached the daring conclusion that within every genus only one species had originally been created, and that new species had developed in time through hybridization of the mother species with species of other genera. This concept was based upon a peculiar theory, traceable to Cesalpino, concerning the "marrow" and the "bark": Every organism was supposed to consist of a marrow substance (medulla), which was inherited from the mother, and a bark substance (cortex), inherited from the father.

All species within the same genera had a common marrow covered by bark substances deriving from different fathers. Around 1760 Linnaeus took the full step and asserted that for every natural plant order only one species had been created at the beginning; the various genera and species came into existence through cross-fertilization at different levels. The theory, although interesting as an attempt to approach a genetic kinship concept, has nothing to do with a theory of evolution in the modern, Darwinian sense of the word.

As a botanist Linnaeus was monumentally one-sided. To order the plants in a system *(divisio)* and to name them *(denominatio)* was for him the real task of botany. He scorned anatomy and physiology and had little understanding of the then developing experimental biological investigations. To be sure, many acute observations in various areas of botany can be found in his works, especially in his academic dissertations. They deal in large part with ecology. Linnaeus investigated the dormancy of plants and the pollination process, undertaking some simple experiments in connection with the latter. With a sharp eye he distinguished different climatic and phytogeographical regions and dealt successfully with problems of dispersion (wind dissemination of seeds and fruits, and so on).

Worked out in *Oeconomia naturae* (1749) and *Politia naturae* (1760), Linnaeus' views on the harmony or order in nature are of great interest. Here his perspective broadens; he sees organic nature as an eternal cyclic course of life and death in which every plant and animal species fulfills its destined task in the service of the whole. The domination of the insects prevents the plants from filling the world; the birds contain the insects within fixed limits. Intense competition prevails in nature, but precisely because of it a delicate balance is maintained that permits the work of creation to function. Both Erasmus and Charles Darwin later profitably adopted some of Linnaeus' views on the hidden struggle in nature.

Linnaeus' real mastery, in botanical systematics, was more problematic than generally realized. He was an empiricist and a brilliant observer whom nothing escaped. As an exact depicter of natural phenomena he scarcely had an equal. His herbarium comprised dried plants from all parts of the world, and he himself scrutinized more plant species than any of his predecessors. But he was ultimately forced by his special talents to abandon empiricism and instead stamped his work as taxonomist and systematist with a dogmatic and philosophical spirit. Order was paramount, and therefore he resorted to an artificial structure in his effort to place nature's swarming multitudes in a simple, clear scheme. He violated nature, as in the sexual system he devised. Linnaeus' scientific method implied that with the aid of certain rules, once and for all determined by himself, the three kingdoms of nature would be organized. He considered the *Philosophia botanica* to be a statute book spelling out the terminology and concepts to be used by botanists. Logic and definitions of concepts were the true tools of botany.

The fixed basis of Linnaeus' system was derived from Aristotelian logic, and as a taxonomist he strictly applied its procedure *per genus et differentiam.* Significantly, he gratefully acknowledged his dependence on the neo-Aristotelian Andrea Cesalpino. As pointed out by Julius Sachs (1875), Linnaeus' reformation of botany undoubtedly carried a Scholastic stamp, the problems that he wanted to solve with his systematic work being of a logical nature. The Scholastic feature is also noticeable in his conviction that the sexual organs are the real "essences" of plants and therefore must be made the basis for the classification system. He preferred to give his writing a Scholastic form and argued *pro* and *contra* like a medieval doctor. All of his work shows a tension between empirical and logical demands, and in situations of conflict the latter almost always wins.

Linnaeus' classification of the animal and mineral kingdoms does not display the same rigorous consistency as his plant system. Presented and revised in the various editions of *Systema naturae*, his animal system was of considerable influence. It is somewhat forced and artificial; but Linnaeus, who here had Ray as principal forerunner, nevertheless sought a natural arrangement and often succeeded brilliantly. He did not use, as in botany, a single, universally valid basis for division but tried to divide the six animal classes he had distinguished by means of a specific organ: mammals according to teeth, birds according to bills, fishes according to fins, and insects according to wings. Added to these are the reptiles (Amphibia) and the "worms" (Vermes), the large group of animals without backbone. In the first edition of *Systema* Linnaeus described 549 animal forms; the last edition included 5,897 species.

In details he was rather unfortunate as a zoological taxonomist, particularly with the "worms" (Vermes), which he arranged elaborately according to misleading exterior features. He united pigs, armadillos, and moles in the same mammalian order and considered the rhinoceros to be a rodent. But as a whole Linnaeus' animal system signified a considerable advance. He unhesitatingly united man with the apes and was the first to recognize whales as mammals (1758). Linnaeus was at his best in entomology, which from his youth had been his favorite field apart from botany; the insect orders he recognized are essentially still valid. In the important tenth edition of *Systema* (1758–1759), he used the "trivial name" consistently in the animal kingdom. As a result it became the standard text for all zoological nomenclature.

Linnaeus' system of the mineral kingdom had little influence but was not without its interesting features. Although he attached great importance to crystal structure for classification and thus was one of the pioneers in crystallography (*De crystallorum generatione* [1747]), he had no feeling for the chemical composition of minerals and consequently stood outside of the fruitful contemporary trends in mineral systematics. More important were his contributions to paleontology and historical geology: he described many fossils, including trilobites (entomolithus), which he correctly placed among the arthropods. In his odd speech on "The Growth of the Earth" ("Oratio de telluris habitabilis incremento" [1743]) and in later works Linnaeus gave his conception of the development of the planet. Inspired by the Christian doctrine of paradise and the old concept of the golden age, he imagined paradise as an island on the equator, from which animals and plants, with the decrease of the water, spread out over the earth. In the 1740's he contributed observations in the field to the lively debates in Sweden about "water reduction," i.e., the uplift of the Swedish plateau in postglacial time. He believed that the sedimentary strata had been deposited over enormous time periods but, mindful of ecclesiastical orthodoxy, he was wary of expressing his heretical views on the age of the world and the length of the geological epochs.

As a university teacher Linnaeus worked within the medical faculty and published many medical works of varying value. His passion for order and classification extended to his medical work, and through his *Genera morborum* (1763) he figures with François Boissier de Sauvages as the founder of systematic nosology. To a great extent inspired by Sauvages, with whom he maintained a lively correspondence, Linnaeus methodically grouped diseases into classes, orders, and genera according to their symptoms. His theory, explained in *Exanthemata viva* (1757), which holds that certain diseases are caused by invisible living organisms *(acari)*, is interesting but not original; he cites A. Q. Bachmann (Rivinus) as precursor. Linnaeus applied his botanical knowledge in the three-volume *Materia medica* (1749–1763) and sought medical profundity in the peculiar *Clavis medicinae* (1766). Filled with number mysticism and based upon speculations about "marrow" and "bark," *Clavis medicinae* is almost incomprehensible in its classification of diseases according to complaints deriving from maternal marrow or paternal bark substance.

Linnaeus had a tendency toward unrestrained, impulsive speculation; for him, vague fancies often became eternal truths that needed no close examination. Similarly, he sometimes put blind faith in fables and folk tales; during his entire life he remained convinced that swallows slept on the bottom of lakes during the winter. He was also rather old-fashioned and simplistic in his general theory of nature, which had a strong Stoic-Aristotelian basis. The peculiar theological remarks that Linnaeus later in life collected under the title *Nemesis divina* (first published in 1848) also belong in this category. They deal, in melancholic fatalism, with divine retribution as the law ruling human existence.

Linnaeus attained worldwide influence not only through his writings but also through his students. A great many foreign students came to him in Uppsala to learn the foundations of systematic natural science. They included the Scandinavians Peter Ascanius and J. C. Fabricius, the entomologist; the Germans Johann Beckmann, Paul D. Giseke, J. C. D. Schreber; the Russian Barons Demidoff; and the American Adam Kuhn, who became professor in Philadelphia. All of them helped to spread the Linnaean doctrine in their own countries.

The world was Linnaeus' sphere of activity also in another sense. He sent many of his students abroad, rejoicing in the plants and other natural specimens that he received from them but grieving bitterly when they died of disease or hardship. Peter Kalm traveled in North America and Fredrik Hasselquist in Egypt, Syria, and Palestine; Peter Osbeck and others sailed on East Indian merchant ships to China. Peter Forsskål went to Arabia as a member of a Danish-German expedition, while Pehr Loefling, Linnaeus' favorite pupil, died in Venezuela. Daniel Solander sailed with Joseph Banks on Captain Cook's first global circumnavigation, and later Anders Sparrman sailed under Cook's command across the Pacific. The most industrious of all Linnaean travelers, Karl Peter Thunberg, managed after many trips to gain entry into Japan. Linnaeus not only received botanical specimens from his pupils but also, on occasion, published posthumously their travel descriptions, such as Hasselquist's *Iter Palaestinum* (1757) and Loefling's *Iter Hispanicum* (1758).

Linnaeus was a complicated man. His enormous scientific production was supported by a self-esteem almost without parallel. He considered his published works to be unblemished masterpieces; no one had ever been a greater botanist or zoologist. Unable to accept criticism, he sulked like a child when he encountered it. He persecuted those, like Buffon, who did not accept him. Linnaeus looked upon himself as a prophet called by God to promulgate the only true dogma; botanists who did not follow the rules and regulations in *Philosophia botanica* clearly were "heretics." Of unstable temperament and essentially naïve, he vacillated between overweening pride and brooding despair. His peculiarities often made him difficult; many were repelled by his egocentric behavior, and over the years he became increasingly isolated.

Yet Linnaeus also could radiate an overpowering charm, and almost all of his students loved him. In a sunny mood, as one sees him in his letters and other relaxed moments, he was irresistible. In playful moods he interpreted nature in the light of classical mythology or personified it. His view of nature was deeply religious; central to all his work was God's omnipotence. He never deviated from the devout physicotheology that was widespread in the eighteenth century, but he stamped it with his peculiar sense of mystery and wonder. "I saw," he wrote in the introduction to the later editions of *Systema naturae*, "the infinite, all-knowing and all-powerful God from behind as He went away and I grew dizzy. I followed His footsteps over nature's fields and saw everywhere an eternal wisdom and power, an inscrutable perfection."

BIBLIOGRAPHY

I. Original Works. Linnaeus' library, herbarium, archives, and MSS were left to his only son, Carl von Linné the younger, who succeeded his father as professor of botany. After the son's death in 1783, the collections were sold for 900 guineas to the young English physician James Edward Smith, who in 1788 founded the Linnean Society of London. Since then the Linnean Society has been the keeper of all Linnaeus' collections. Swedish libraries have only a few of the more important MSS, such as *Praeludia sponsaliorum plantarum* and *Nemesis divina*, both in the library of Uppsala University.

Two complete modern bibliographies are J. M. Hulth, *Bibliographia Linnaeana*, I, pt. 1 (Uppsala, 1907); and B. H. Soulsby, *A Catalogue of the Works of Linnaeus* (London, 1933), which includes works about Linnaeus. Two modern facsimile eds. are *Genera plantarum*, which is Historiae Naturalis Classica, III (Weinheim, 1960); and *Species plantarum*, 2 vols. (London, 1957–1959), both with important intros. by W. T. Stearn.

Praelectiones in ordines naturales plantarum was published by P. D. Giseke (Hamburg, 1792). Collections of letters from Linnaeus or to him include D. H. Stöver, ed., *Collectio epistolarum quas . . . scripsit Carolus a Linné* (Hamburg, 1792); J. E. Smith, ed., *A Selection of the Correspondence of Linnaeus and Other Naturalists*, 2 vols. (London, 1821); C. N. J. Schreiber, ed., *Epistolae ad Nicolaum Josephum Jacquin* (Vienna, 1841); and *Lettres inédites de Linné à Boissier de la Croix de Sauvages* (Alais [Alès], 1860). Publication of the whole of Linnaeus' correspondence was never completed: T. M. Fries, J. M. Hulth, and A. H. Uggla, eds., *Bref och skrifvelser af och till Carl von Linné*, 10 vols. (I, 8 vols.; II, 2 vols.) (Stockholm–Uppsala, 1907–1943). E. Ährling, *Carl von Linnés brefvexling* (Stockholm, 1885), contains a complete list of the letters to Linnaeus in the Linnean Society.

A critical ed. of Linnaeus' five preserved autobiographies is in E. Malmeström and A. H. Uggla, *Vita Caroli Linnaei* (Stockholm, 1957).

Works from Linnaeus' youth, including those on the Lapland and Dalarna trips, are in E. Ährling, *Carl von Linnés ungdomsskrifter*, 2 vols. (Stockholm, 1888–1889). The Lapland voyage, *Iter lapponicum*, was translated by J. E. Smith as *Lachesis lapponica or a Tour in Lappland*, 2 vols. (London, 1811); a critical ed. is in T. M. Fries, ed., *Skrifter af Carl von Linné*, V (Uppsala, 1913). There is also A. H. Uggla, ed., *Dalaresa* (with *Iter ad exteros* and *Iter ad fodinas*) (Stockholm, 1953), with commentary.

See also A. H. Uggla, ed., *Diaeta naturalis 1733* (Uppsala, 1958); A. O. Lindfors, ed., "Linnés dietetik," in *Uppsala universitets årsskrift* (1907); Einar Lönnberg, ed., *Föreläsningar öfver djurriket* (Uppsala, 1913); E. Malmeström and T. Fredbärj, eds., *Nemesis divina* (Stockholm, 1968), the first complete ed.

II. Secondary Literature. The literature on Linnaeus is very comprehensive; a survey is given in Sten Lindroth, "Two Centuries of Linnaean Studies," in *Bibliography and Natural History* (Lawrence, Kans., 1966), pp. 27–45.

The first biographies appeared in the eighteenth century: Richard Pulteney, *A General View of the Writings of Linnaeus* (London, 1781); and D. H. Stöver, *Leben des Ritters Carl von Linné*, 2 vols. (Hamburg, 1792), also translated into English (London, 1794). Another older biography is J. F. X. Gistel, *Carolus Linnaeus. Ein Lebensbild* (Frankfurt, 1873). T. M. Fries, *Linné*, 2 vols. (Stockholm, 1903), translated into English in an abridged version by Benjamin Daydon Jackson as *Linnaeus: The Story of His Life* (London, 1923), is the indispensable modern biography for all Linnaean research. Among later, shorter popular biographies are Knut Hagberg, *Carl Linnaeus* (Stockholm, 1939; rev. ed., 1957), also translated into English (London, 1952); Norah Gourlie, *The Prince of Botanists: Carl Linnaeus* (London, 1953); Elis Malmeström, *Carl von Linné* (Stockholm, 1964); Heinz Goerke, *Carl von Linné*, which is Grosse Naturforscher no. 31 (Stuttgart, 1966); and Wilfrid Blunt, *The Compleat Naturalist. A Life of Linnaeus* (London, 1971). For Linnaean iconography, see Tycho Tullberg, *Linnéporträtt* (Uppsala, 1907).

A. J. Boerman has dealt with *Carolus Linnaeus als middelaar tussen Nederland en Zweden* (Utrecht, 1953). Articles on Linnaeus' life and work are being published in *Svenska Linnésällskapets årsskrift* (Uppsala, 1918–). *Carl von Linné's Bedeutung als Naturforscher und Arzt* (Jena, 1909), published by the Swedish Academy of Science in a collection of papers, reviews Linnaeus' scientific achievements. Nils von Hofsten deals with Linnaeus' conception of nature in *Kungliga Vetenskaps-societetens årsbok 1957* (Uppsala, 1958), pp. 65–105. An attempt to evaluate Linnaeus in terms of the history of ideas and psychology is Sten Lindroth, "Linné-legend och verklighet," in *Lychnos* (1965–1966), pp. 56–122.

Valuable insights on Linnaeus as a botanist are in Julius Sachs, *Geschichte der Botanik* (Munich, 1875). The best modern summary of Linnaeus' botanical ideas is Gunnar Eriksson, *Botanikens historia i Sverige intill år 1800*, which is Lychnos-Bibliotek, XVII, 3 (Uppsala, 1969). His taxonomic and systematic principles are investigated in J. Ramsbottom, "Linnaeus and the Species Concept," in *Proceedings of the Linnean Society of London*, **150** (1938), 192–219; H. K. Svenson, "On the Descriptive Method of Linnaeus," in *Rhodora*, **47** (1945), 273–302, 363–388; C. E. B. Bremekamp, "Linné's Views on the Hierarchy of the Taxonomic Groups," in *Acta botanica neerlandica*, **2** (1953), 242–253; A. J. Cain, "Logic and Memory in Linnaeus's System of Taxonomy," in *Proceedings of the Linnean Society of London*, **169** (1958), 144–163; W. T. Stearn, *Three Prefaces on Linnaeus and Robert Brown* (Weinheim, 1962); and James L. Larson, *Reason and Experience. The Representation of Natural Order in the Work of Carl von Linné* (Berkeley, 1971).

On the binomial nomenclature, see W. T. Stearn, "The Background of Linnaeus's Contributions to the Nomenclature and Methods of Systematic Biology," in *Systematic Zoology*, **8**, no. 1 (1959), 4–22; and John L. Heller, "The Early History of Binomial Nomenclature," in *Huntia*, **1** (1964), 33–70. Linnaeus' relations to other botanists are discussed in Heinz Goerke, "Die Beziehungen Hallers zu Linné," in *Sudhoffs Archiv für Geschichte der Medizin und der Naturwissenschaften*, **38** (1954), 367–377; and Hans Krook, "Lorenz Heister och Linné," in *Svenska Linnésällskapets årsskrift*, **31** (1948), 57–72.

On Linnaeus' zoological system, see Nils von Hofsten, "Linnés djursystem," in *Svenska Linnésällskapets årsskrift*, **42** (1959), 9–43; and "A System of 'Double Entries' in the Zoological Classification of Linnaeus," in *Zoologiska bidrag från Uppsala*, **35** (1963), 603–631. In addition there is Thomas Bendyshe, "On the Anthropology of Linnaeus," in *Memoirs Read Before the Anthropological Society of London*, **1** (1863–1864), 421–458. On Darwin and Linnaeus, see Robert C. Stauffer, "Ecology in the Long Manuscript Version of Darwin's *Origin of Species* and Linnaeus," in *Proceedings of the American Philosophical Society*, **104** (1960), 235–241. Linnaeus' geological thought is examined in Tore Frängsmyr, *Geologi och skapelsetro. Föreställningar om jordens historia från Hiärne till Bergman*, which is Lychnos-Bibliotek, XXVI (Uppsala, 1969), ch. 4; and his nosological system in Fredrik Berg, "Linné et Sauvages: Les rapports entre leurs systèmes nosologiques," in *Lychnos* (1956), pp. 31–54; and *Linnés Systema morborum* (Uppsala, 1957).

Elis Malmeström considers Linnaeus as theological thinker in *Carl von Linnés religiösa åskådning* (Stockholm, 1926); and "Die religiöse Entwicklung und die Weltanschauung Carl von Linnés," in *Zeitschrift für systematische Theologie*, **19** (1942), 31–58. See also Elof Ehnmark, "Linnaeus and the Problem of Immortality," in *Kungliga Humanistiska vetenskapssamfundet i Lund, Årsberättelse* (1951–1952), pp. 63–93. Linnaeus' view of man is analyzed in K. R. V. Wikman, *Lachesis and Nemesis*, which is Scripta Instituti Donneriani Aboensis, no. 6 (Stockholm, 1970).

For the spread of Linnaean botany, see Frans A. Stafleu, *Linnaeus and the Linnaeans. The Spreading of Their Ideas in Systematic Botany (1735–1789)* (Utrecht, 1971).

STEN LINDROTH

LIOUVILLE, JOSEPH (*b.* St.-Omer, Pas-de-Calais, France, 24 March 1809; *d.* Paris, France, 8 September 1882), *mathematics.*

Liouville is most famous for having founded and directed for almost forty years one of the major mathematical journals of the nineteenth century, the *Journal de Liouville*. He also made important contributions in pure and applied mathematics and exerted a fruitful influence on French mathematics through his teaching.

The few articles devoted to Liouville contain little biographical data. Thus the principal stages of his life and career must be reconstructed on the basis of original documentation. There is no exhaustive list of Liouville's works, which are dispersed in some 400 publications—the most nearly complete is that in the Royal Society's *Catalogue of Scientific Papers*. His work as a whole has been treated in only two original studies of limited scope, those of G. Chrystal and G. Loria. On the other hand, certain of Liouville's works have been analyzed in greater detail, such as those on geometry. In view of the limited space available, this study cannot hope to provide a thorough account of Liouville's work but will attempt instead to present its major themes.

Life. Liouville was the second son of Claude-Joseph Liouville (1772–1852), an army captain, and Thérèse Balland, both originally from Lorraine. Liouville studied in Commercy and then in Toul. In 1831 he married a maternal cousin, Marie-Louise Balland (1812–1880); they had three daughters and one son. Liouville lived a calm and studious life, enlivened by an annual vacation at the family house in Toul. His scientific career was disturbed only by a brief venture into politics, during the Revolution of 1848. Already known for his democratic convictions, he was elected on 23 April 1848 to the Constituent Assembly as one of the representatives from the department of the Meurthe. He voted with the moderate democratic party. His defeat in the elections for the Legislative

Assembly in May 1849 marked the end of his political ambitions.

Admitted to the École Polytechnique in November 1825, Liouville transferred in November 1827 to the École des Ponts et Chaussées, where, while preparing for a career in engineering, he began original research in mathematics and mathematical physics. Between June 1828 and November 1830 he presented before the Académie des Sciences seven memoirs, two of which dealt with the theory of electricity, three with the analytic theory of heat, and two with mathematical analysis. Although Academy reporters expressed certain reservations, these first works were on the whole very favorably received; and their partial publication in the *Annales de chimie et de physique*, in Gergonne's *Annales*,[1] and in Férussac's *Bulletin* gained their author a certain reputation. In order to secure as much freedom as possible to pursue his research, Liouville soon thought of changing professions. In 1830, upon graduating from the École des Ponts et Chaussées, he refused the position of engineer that he was offered, hoping that his reputation would permit him to obtain a teaching post fairly soon.

In November 1831 Liouville was selected by the Council on Instruction of the École Polytechnique to replace P. Binet as *répétiteur* in L. Mathieu's course in analysis and mechanics. This was the beginning of a brilliant career of some fifty years, in the course of which Liouville taught pure and applied mathematics in the leading Paris institutions of higher education.

In 1838 Liouville succeeded Mathieu as holder of one of the two chairs of analysis and mechanics at the École Polytechnique, a position that he resigned in 1851, immediately after his election to the Collège de France. From 1833 to 1838 Liouville also taught mathematics and mechanics, but at a more elementary level, at the recently founded École Centrale des Arts et Manufactures. In March 1837 he was chosen to teach mathematical physics at the Collège de France as *suppléant* for J. B. Biot. He resigned in March 1843 to protest the election of Count Libri-Carrucci to the chair of mathematics at that institution.

Liouville did not return to the Collège de France until the beginning of 1851, when he succeeded Libri-Carrucci, who had left France.[2] This chair had no fixed program, and so for the first time Liouville could present his own research and discuss current topics. He took advantage of this to present unpublished works, some of which were developed by his students before he himself published them.[3] Appreciating the interest and flexibility of such teaching, he remained in the post until 1879, when he arranged for O. Bonnet to take over his duties.

Liouville also wished to teach on the university level. Toward this end he had earned his doctorate with a dissertation on certain developments in Fourier series and their applications in mathematical physics (1836). He was therefore eligible for election, in 1857, to the chair of rational mechanics at the Paris Faculty of Sciences, a position vacant since the death of Charles Sturm. In 1874, he stopped teaching there and arranged for his replacement by Darboux.

Parallel with this very full academic career, Liouville was elected a member of the astronomy section of the Académie des Sciences in June 1839, succeeding M. Lefrançois de Lalande; and in 1840 he succeeded Poisson as a member of the Bureau des Longitudes. From this time on, he participated regularly in the work of these two groups. For forty years he also passionately devoted himself to another particularly burdensome task, heading an important mathematical journal. The almost simultaneous demise, in 1831, of the only French mathematical review, Gergonne's *Annales de mathématiques pures et appliquées*, together with one of the principal science reviews, Férussac's *Bulletin des sciences mathématiques, astronomiques, physiques et chimiques*, deprived French-language mathematicians of two of their favorite forums. Liouville understood that the vigor of French mathematical writings demanded the creation of new organs of communication. Despite his youth and inexperience in the problems of editing and publishing, he launched the *Journal de mathématiques pures et appliquées* in January 1836. He published the first thirty-nine volumes (the twenty of the first series [1836–1855] and the nineteen of the second series [1856–1874]), each volume in twelve fascicles of thirty-two to forty pages. Finally, at the end of 1874, he entrusted the editorship to H. Résal.

Open to all branches of pure and applied mathematics, the publication was extremely successful and was soon called the *Journal de Liouville*. Its first volume contained articles by Ampère, Chasles, Coriolis, Jacobi, Lamé, V.-A. Lebesgue, Libri-Carrucci, Plücker, Sturm, and Liouville himself. Although not all the tables of contents are as brilliant as that of the first, the thirty-nine volumes published by Liouville record an important part of the mathematical activity of the forty years of the mid-nineteenth century. In fact, Liouville secured regular contributions from a majority of the great mathematicians of the era and maintained particularly warm and fruitful relations with Jacobi, Dirichlet, Lamé, Coriolis, and Sturm. At the same time he sought to guide and smooth the way for the first works of young authors, notably Le Verrier, Bonnet, J. A. Serret, J. Bertrand, Hermite, and Bour. But his outstanding qualities as an editor are most clearly displayed in his exemplary effort to assimilate as perfectly as possible the work of Galois. He imposed this task on himself in order to establish in an irreproachable fashion the texts he published in volume 11 of the *Journal de mathématiques*.

During the first thirty years of his career Liouville, while maintaining a very special interest in mathematical analysis, also did research in mathematical physics, algebra, number theory, and geometry. Starting in 1857, however, he considerably altered the orientation of his studies, concentrating more and more on particular problems of number theory. Departing in this way from the most fruitful paths of mathematical research, Liouville saw his influence decline rapidly. Yet it was the abandonment of the editorship of his *Journal* in 1874 that signaled the real end of his activity. His publications, which had been appearing with decreasing frequency since 1867, stopped altogether at this time. Simultaneously he gave up his courses at the Sorbonne, where his *suppléants* were Darboux and then Tisserand. He still attended the sessions of the Académie des Sciences and of the Bureau des Longitudes and still lectured at the Collège de France; but he no longer really participated in French mathematical life.

Mathematical Works. Liouville published thirty-nine volumes of the *Journal de mathématiques*, republished a work by Monge, and published a treatise by Navier; but he never composed a general work of his own. He did write nearly 400 memoirs, articles, and notes on a great many aspects of pure and applied mathematics. Despite its great diversity, this literary output is marked by a limited number of major themes, the majority of which were evident in his first publications. Liouville also published numerous articles correcting, completing, or extending the results of others, especially of articles that had appeared in the *Journal de mathématiques*. Other articles by him were summaries of or extracts from courses he gave at the Collège de France.

The divisions used below tend to obscure one of the main characteristics of Liouville's works, their interdisciplinary nature. Yet they are indispensable in the presentation of such a diverse body of work. References generally will be limited to the year of publication; given that information, further details can rapidly be found by consulting the *Catalogue of Scientific Papers*.

Mathematical Analysis. It was in mathematical analysis that Liouville published the greatest number and the most varied of his works. Composed mainly from 1832 to 1857, they number about one hundred. It is thus possible to mention only the most important and most original of these contributions, including some that were virtually ignored by contemporaries. To grasp their significance fully, it must be noted that these apparently disordered investigations were actually guided—for lack of an overall plan—by a few governing ideas. It is also important to view each of them in the context of the studies carried out at the same time by Gauss, Jacobi, Cauchy, and Sturm. It should be remembered that in his courses at the Collège de France, Liouville treated important questions that do not appear among his writings and that he inspired many disciples.

Certain of Liouville's earliest investigations in mathematical analysis should be viewed as a continuation of the then most recent works of Abel and Jacobi. The most important are concerned with attempts to classify all algebraic functions and the simplest types of transcendental functions, with the theory of elliptic functions, and with certain types of integrals that can be expressed by an algebraic function.

Following the demonstration by Abel and Galois of the impossibility of algebraically solving general equations higher than the fourth degree, Liouville devoted several memoirs to determining the nature of the roots of algebraic equations of higher degree and of transcendental equations. One of the results he obtained was that according to which the number e cannot be the root of any second- or fourth-degree equation (1840). In 1844 he discovered a specific characteristic of the expansion as a continued fraction of every algebraic number, and showed that there are continued fractions that do not possess this characteristic. This discovery ("Liouville's transcendental numbers") is set forth in a memoir entitled "Sur des classes très-étendues de nombres dont la valeur n'est ni algébrique, ni même réductible à des irrationnelles algébriques" (*Journal de mathématiques*, **16** [1851], 133–142). It was not until 1873 that G. Cantor demonstrated the existence of much more general transcendental numbers.

In 1844 Liouville took up the study of all the functions possessing—like the elliptic functions—the property of admitting two periods. He set forth his theory to Joachimsthal and Borchardt (1847) before presenting it at the Collège de France (1850–1851). Thus, although he published little on the subject, the theory itself rapidly became widely known through the publications of Borchardt and the *Théorie des fonctions doublement périodiques* (1859) of Briot and Bouquet. Liouville made important contributions to the theory of Eulerian functions. He also sought to develop the theory of differential equations and of partial differential equations. In addition to numerous memoirs treating particular types of differential equations, such as Riccati's equation, and various technical questions, Liouville also worked on general problems, such as the demonstration (1840) of the impossibility, in general, of reducing the solution of differential equations either to a finite sequence of algebraic operations and indefinite integrations or—independently of Cauchy—to a particular aspect of the method of successive approximations (1837–1838). His contributions in the theory of partial differential equations were also of considerable value, even though a large portion of his "discoveries" had been anticipated by Jacobi. Much of his research in this area was closely linked with rational mechanics, celestial mechanics, and mathematical physics.

Pursuing the pioneer investigations of Leibniz, Johann Bernoulli, and Euler, Liouville devoted a part of his early work (1832–1836) to an attempt to enlarge as far as possible the notions of the differential and the integral, and in particular to establish the theory of derivatives of arbitrary index. Assuming a function $f(x)$ to be representable as a series of exponentials

$$f(x) = \sum_{n=0}^{n=+\infty} (c_n e^{a_n x}),$$

he defined its derivative of order s (s being an arbitrary number, rational or irrational, or even complex) by the series

$$D_s f(x) = \sum_{n=0}^{n=+\infty} c_n a_n^s \, e^{a_n x}.$$

This definition does in fact extend the ordinary differential calculus of integral indexes, but its generality is limited: not every function admits an expansion of the type proposed; and the convergence and uniqueness of the expansion are not guaranteed for all values of s. Despite its weaknesses and deficiencies, Liouville's endeavor, one of the many efforts that led to the establishment of functional calculus, shows his great virtuosity in handling the analysis of his time.

Liouville and Sturm published an important series of memoirs in the first two volumes of the *Journal de mathématiques* (1836–1837). Bourbaki (*Éléments d'histoire des mathématiques*, 2nd ed., pp. 260–262) shows that these investigations, which extended the numerous earlier works devoted to the equation of vibrating strings, permitted the elaboration of a general theory of oscillations for the case of one variable. He also shows that they are linked to the beginning of the theory of linear integral equations that contributed to the advent of modern ideas concerning functional analysis.

This constitutes only a broad outline of Liouville's chief contributions to analysis. The principal works mentioned below in connection with mathematical physics should for the most part be considered direct applications—or even important elements—of Liouville's analytical investigations.

Mathematical Physics. Under the influence of Ampère and of Navier, whose courses he had followed at the Collège de France and at the École des Ponts et Chaussées from 1828 to 1830, Liouville directed his first investigations to two areas of mathematical physics of current interest: the theory of electrodynamics and the theory of heat. After these first studies he never returned to the former topic. On the other hand he later (1836–1838, 1846–1848) devoted several memoirs to the theory of heat, but only for the purpose of employing new analytic methods, such as elliptic functions. Similarly, it was to demonstrate the power of analysis that he considered certain aspects of the theory of sound (1836, 1838) and the distribution of electricity on the surface of conductors (1846, 1857).

Celestial Mechanics. Liouville came in contact with celestial mechanics in 1834, through a hydrodynamical problem: finding the surface of equilibrium of a homogeneous fluid mass in rotation about an axis. He showed that with Laplace's formulas one can demonstrate Jacobi's theorem concerning the existence among these equilibrium figures of an ellipsoidal surface with unequal axes. Between 1839 and 1855 Liouville returned to this topic on several occasions, confirming the efficacy of the Laplacian methods while verifying the great power of the new procedures. From 1836 to 1842 he wrote a series of memoirs on classical celestial mechanics (perturbations, secular variations, the use of elliptic functions). Subsequently, although he was active in the Bureau des Longitudes, he devoted only a few notes to the problem of attraction (1845) and to particular cases of the three-body problem (1842, 1846).

Rational Mechanics. Rational mechanics held Liouville's interest for only two short periods. From 1846 to 1849, influenced by the work of Hamilton and Jacobi, he sought to erect a theory of point dynamics. He did not return to related subjects until 1855 and 1856, when he published, respectively, two studies of the differential equations of dynamics and an important note on the use of the principle of least action.

Algebra. Although Liouville wrote only a few memoirs dealing with algebraic questions, his contributions in this field merit discussion in some detail.

In 1836, with his friend Sturm, Liouville demonstrated Cauchy's theorem concerning the number of complex roots of an algebraic equation that are situated in the interior of a given contour, and in 1840 he gave an elegant demonstration of the fundamental theorem of algebra. From 1841 to 1847 he took up, on new bases, the problem of the elimination of an unknown from two equations in two unknowns, and applied the principles brought to light in this case to the demonstration of various properties of infinitesimal geometry. In 1846 he presented a new method for decomposing rational functions, and in 1863 he generalized Rolle's theorem to the imaginary roots of equations.

Liouville's most significant contribution by far, however, was concerned with the theory of algebraic equations and group theory. He published very little on these questions, but from 1843 to 1846 he conducted a thorough study of Galois's manuscripts in order to prepare for publication, in the October and November 1846 issues of his *Journal de mathématiques*, the bulk of the work of the young mathematician who tragically had died in 1832. Liouville's brief "Avertissement" (pp. 381–384) paid fitting homage to the value of Galois's work, but the second part of the *Oeuvres*, announced for the following volume of the *Journal*, was not published until 1908 by J. Tannery. According to J. Bertrand, Liouville invited a few friends, including J. A. Serret, to attend a series of lectures on Galois's work.[4] Although he did not publish the text of these lectures, Liouville nevertheless contributed in large measure to making known a body of work whose extraordinary innovations were to become more and more influential. He thus participated, indirectly, in the elaboration of modern algebra and of group theory.

Geometry. Liouville's work in geometry, the subject of an excellent analysis by Chasles (*Rapport sur les progrès de la géométrie* [Paris, 1870], pp. 127–130), consists of about twenty works written over the period from 1832 to 1854. The majority of them are applications or extensions of other investigations and are more analytic than geometric in their inspiration.

In 1832 Liouville showed that his "calculus of differentials of arbitrary index" could facilitate the study of certain questions of mechanics and geometry. In 1841 and 1844 he demonstrated and extended, by employing elimination theory, numerous metric properties of curves and surfaces that were established geometrically by Chasles. In 1844 Liouville published a new method for determining the geodesic lines of an arbitrary ellipsoid, a problem that Jacobi had just reduced to one involving elliptic transcendentals. In 1846 he proposed a direct demonstration of the so-called Joachimsthal equation and generalized the study of polygons of maximal or minimal perimeter inscribed in or circumscribed about a plane or spherical conic section to geodesic polygons traced on an ellipsoid. Finally, in 1847, prompted by a note of W. Thomson concerning the distribution of electricity on two conducting bodies, Liouville undertook an analytic study of inversive geometry in space, which he called "transformation by reciprocal vector rays." He showed that the inversion is the only conformal nonlinear spatial transformation and pointed out its applications to many questions of geometry and mathematical physics.

When Liouville published the fifth edition of Monge's *Application de l'analyse à la géométrie* (1850), he added to it Gauss's famous memoir *Disquisitiones generales circa superficies curvas*. He also appended to it seven important related notes of his own; these dealt with the theory of space curves, the introduction of the notions of relative curvature and geodesic curvature of curves situated on a surface, the integration of the equation of geodesic lines, the notion of total curvature, the study of the deformations of a surface of constant curvature, a particular type of representation of one surface on another, and the theory of vibrating strings.

Liouville returned to certain aspects of the general theory of surfaces in later works and in his lectures at the Collège de France—for instance, the determination of the surface the development of which is composed of two confocal quadrics (1851). Although they treat rather disparate topics, these works on infinitesimal geometry exerted a salutary influence on research in this field, confirming the fruitfulness of the analytic methods of Gauss's school.

Number Theory. Liouville entered the new field of number theory in 1840 with a demonstration that the impossibility of the equation $x^n + y^n = z^n$ entails that of the equation $x^{2n} + y^{2n} = z^{2n}$. He returned to this field only occasionally in the following years, publishing a comparison of two particular quadratic forms (1845) and a new demonstration of the law of quadratic reciprocity (1847). In 1856, however, he began an impressive and astonishing series of works in this area, abandoning—except for some brief notes on analysis—the important investigations he was carrying out in other branches of mathematics. At first he took an interest in quite varied questions, including the sum of the divisors of an integer; the impossibility, for integers, of the equation $(p-1)! + 1 = p^m$, p being a prime number greater than 5; and the number $\varphi(n)$ of integers less than n that are prime to n. From 1858 to 1865, in eighteen successive notes published in the *Journal de mathématiques* under the general title "Sur quelques formules générales qui peuvent être utiles dans la théorie des nombres," Liouville stated without demonstration a series of theorems—a list is given in volume II of L. E. Dickson, *History of the Theory of Numbers*—that constitute the foundations of "analytic" number theory.

But from 1859 to 1866 Liouville devoted the bulk of his publications—nearly 200 short notes in the *Journal de mathématiques*—to a monotonous series of particular problems in number theory. These problems are reducible to two principal types: the exposition of certain properties of prime numbers of a particular form and their products (properties equivalent to the existence of integral solutions of equations having the form $ax + by = c$, the numbers a, b, c being of a given form); and the determination of different representations of an arbitrary integer by a quadratic form of the type $ax^2 + by^2 + cz^2 + dt^2$, where a, b, c, and d are given constants.

L. E. Dickson mentions the examples Liouville studied in this series and the demonstrations that were published later: Liouville's notes contain only statements and numerical proofs. One may well be astonished at his singular behavior in abandoning his other research in order to amass, without any real demonstration, detailed results concerning two specialized topics in number theory. He never explained this in his writings. Hence one may wonder, as did G. Loria, whether he may have justified his apparently mysterious choice of examples in the course of his lectures at the Collège de France. Bôcher, more severe, simply considers these last publications to be mediocre.

Four hundred memoirs, published mainly in the *Journal de mathématiques* and the *Comptes rendus hebdomadaires des séances de l'Académie des sciences*, and 340 manuscript notebooks[5] in the Bibliothèque de l'Institut de France, Paris—this is the mass of documents that must be carefully analyzed in order to provide an exhaustive description of Liouville's work. To complete this account, it would also be necessary to trace more exactly his activity as head of the *Journal de mathématiques*, to assemble the greatest amount of data possible on the courses he taught, above all at the Collège de France, and to ascertain his role in the training of his students and in their publications.

NOTES

1. This, Liouville's second publication—he was then twenty-one—appeared in *Annales de mathématiques pures et appliquées* (**21**, nos. 5 and 6 [Nov. and Dec. 1830]). The second of these issues also contains two short notes by Evariste Galois, whose work Liouville published in *Journal de mathématiques pures et appliquées*, **11** (1846). But this circumstance obviously is not sufficient to prove that the two knew each other. A note by Gergonne, following Liouville's memoir, severely criticizes its style and presentation. The editor of the *Annales* denied all hope of a future for the young man who, five years later, succeeded him by creating the *Journal de mathématiques*. See also Liouville's "Note sur l'électro-dynamique," in *Bulletin des sciences mathématiques . . .*, **15** (Jan. 1831), 29–31, which is omitted from all bibliographies.
2. In 1843 the election for the chair of mathematics left vacant by the death of S. F. Lacroix saw three opposing candidates, all members of the Académie des Sciences: Cauchy, Liouville, and Libri-Carrucci. At the time Liouville was engaged in a violent controversy with Libri-Carrucci, whose abilities he publicly disputed. He therefore resigned as soon as he learned that the Council of the Collège de France was recommending that the Assembly vote for Libri-Carrucci. In 1850, when this chair became vacant, following Libri-Carrucci's dismissal, Cauchy and Liouville again were candidates. The particular circumstances of the votes that resulted in Liouville's finally being chosen (18 Jan. 1851) gave rise to vigorous protests by Cauchy.
3. Thus the works on doubly periodic functions that Liouville had presented privately (1847) or in public lectures (1851) were used by Borchardt and Joachimsthal and also by Briot and Bouquet. Similarly, Lebesgue published in his *Exercices d'analyse numérique* (1859) Liouville's demonstration, unpublished until then, of Waring's theorem for the case of biquadratic numbers.
4. J. Bertrand (*Éloges académiques*, II [Paris, 1902], 342–343) wrote that it was out of deference to Liouville, who had taught him Galois's discoveries, that J. A. Serret mentioned these discoveries in neither the first (1849) nor the second edition (1859) of his *Cours d'algèbre supérieure*, but only in the third (1866).
5. A rapid examination of some of these notebooks shows that they consist mainly of outlines and final drafts of the published works, but the whole collection would have to be gone over very thoroughly before a definitive evaluation could be made.

BIBLIOGRAPHY

I. Original Works. Liouville published only two of his works in book form: his thesis, *Sur le développement des fonctions ou parties de fonctions en séries de sinus et de cosinus, dont on fait usage dans un grand nombre de questions de mécanique et de physique . . .* (Paris, 1836); and a summary of a course he gave at the École Centrale des Arts et Manufactures in 1837–1838, *Résumé des leçons de trigonométrie. Notes pour le cours de statique* (Paris, n.d.). He edited and annotated Gaspard Monge, *Application de l'analyse à la géométrie*, 5th ed. (Paris, 1850); and H. Navier, *Résumé des leçons d'analyse données à l'École polytechnique . . .*, 2 vols. (Paris, 1856).

Liouville published the first 39 vols. of the *Journal de mathématiques pures et appliquées*: 1st ser., **1–20** (1836–1855); 2nd ser., **1–19** (1856–1874). He also wrote approximately 400 papers, of which the majority are cited in the Royal Society's *Catalogue of Scientific Papers*, IV (1870), 39–49 (nos. 1–309 for the period 1829–1863, plus three articles written in collaboration with Sturm); VIII (1879), 239–241 (nos. 310–377 for the period 1864–1873); X (1894), 606 (nos. 378–380 for the period 1874–1880). See also Poggendorff, I, cols. 1471–1475 and III, p. 818.

The list of Liouville's contributions to the *Journal de mathématiques* is given in the tables of contents inserted at the end of vol. **20**, 1st ser. (1855), and at the end of vol. **19**, 2nd ser. (1874). (The articles in the *Journal de mathématiques* signed "Besge" are attributed to Liouville by H. Brocard, in *Intermédiaire des mathématiciens*, **9** [1902], 216.) The memoirs and notes that Liouville presented to the Académie des Sciences are cited from 1829 to July 1835 in the corresponding vols. of the *Procès-verbaux des séances de l'Académie des sciences* (see index) and after July 1835 in the vols. of *Tables des Comptes rendus hebdomadaires de l'Académie des sciences*, published every 15 years.

J. Tannery edited the *Correspondance entre Lejeune-Dirichlet et Liouville* (Paris, 1910), a collection of articles published in the *Bulletin des sciences mathématiques*, 2nd ser., **32** (1908), pt. 1, 47–62, 88–95; and **33** (1908–1909), pt. 1, 47–64.

Liouville's MSS are in the Bibliothèque de l'Institut de France, Paris. They consist of 340 notebooks and cartons of various materials, MS 3615–3640. See also the papers of Evariste Galois, MS 2108, folios 252–285.

II. SECONDARY LITERATURE. Liouville's biography is sketched in the following articles (listed chronologically): Jacob, in F. Hoefer, ed., *Nouvelle biographie générale*, XXXI (Paris, 1872), 316–318; G. Vapereau, *Dictionnaire universel des contemporains*, 5th ed. (Paris, 1880), 1171–1172; H. Faye and E. Laboulaye, in *Comptes rendus hebdomadaires des séances de l'Académie des sciences*, **95** (1882), 467–471, a speech given at his funeral; David, in *Mémoires de l'Académie des sciences, inscriptions et belles-lettres de Toulouse*, 8th ser., **5** (1883), 257–258; A. Robert and G. Cougny, *Dictionnaire des parlementaires français*, II (Paris, 1890), 165; H. Laurent, in *Livre du centenaire de l'École polytechnique*, I (Paris, 1895), 130–133; and L. Sagnet, in *La grande encyclopédie*, XXII (Paris, 1896), 305; and *Intermédiaire des mathématiciens*, **9** (1902), 215–217; **13** (1906), 13–15; **14** (1907), 59–61.

The principal studies dealing with Liouville's work as a whole are G. Chrystal, "Joseph Liouville," in *Proceedings of the Royal Society of Edinburgh*, **14** (1888), 2nd pagination, 83–91; and G. Loria, "Le mathématicien J. Liouville et ses oeuvres," in *Archeion*, **18** (1936), 117–139, translated into English as "J. Liouville and His Work," in *Scripta mathematica*, **4** (1936), 147–154 (with portrait), 257–263, 301–306.

Interesting details and comments concerning various aspects of Liouville's work are in J. Bertrand, *Rapport sur les progrès les plus récents de l'analyse mathématique* (Paris, 1867), pp. 2–5, 32; M. Chasles, *Rapport sur les progrès de la géométrie* (Paris, 1870), 127–140; N. Saltykow, "Sur le rapport des travaux de S. Lie à ceux de Liouville," in *Comptes rendus hebdomadaires des séances de l'Académie des sciences*, **137** (1903), 403–405; M. Bôcher, "Mathématiques et mathématiciens français," in *Revue internationale de l'enseignement*, **67** (1914), 30–31; F. Cajori, *History of Mathematics*, rev. ed. (New York, 1919), see index; L. E. Dickson, *History of the Theory of Numbers*, 3 vols. (Washington, D.C., 1919–1923), see index; and N. Bourbaki, *Éléments d'histoire des mathématiques*, 2nd ed. (Paris, 1964), 260–262.

RENÉ TATON

LIPPMANN, GABRIEL JONAS (*b.* Hollerich, Luxembourg, 16 August 1845; *d.* at sea 12 July 1921), *physics, instrumental astronomy.*

Lippmann's parents were French and settled in Paris while he was still a boy. There he attended first the Lycée Napoléon and later the École Normale Supérieure. In the early stages of his career he collaborated with Bertin in the publication of *Annales de chimie et de physique* by abstracting German papers. This work led him to develop an interest in contemporary research in electricity. While on a scientific mission to Germany he was shown by Wilhelm Kühne, professor of physiology at Heidelberg, an experiment in which a drop of mercury, covered by dilute sulfuric acid, contracts on being touched with an iron wire and regains its original shape when the wire is removed. Lippmann realized that there was a connection between electrical polarization and surface tension and obtained permission to conduct a systematic investigation of this phenomenon in Kirchhoff's laboratory.

These experiments resulted in the development of the capillary electrometer, a device sensitive to changes in potential of the order of 1/1,000 of a volt. The instrument consists essentially of a capillary tube, which is inclined at a few degrees from the horizontal, containing an interface between mercury and dilute acid. If the potential difference between the two liquids is varied, the effective surface tension at the meniscus is altered and, as a result, the meniscus moves along the capillary tube. Although the device could be calibrated, it was more usually employed in potentiometer and bridge measurements as a null instrument. It possessed the advantage, significant at the time, of being independent of magnetic and electric fields. Lippmann's researches on electrocapillarity received recognition in the form of a doctorate awarded by the Sorbonne in 1875. In a paper published in 1876, Lippmann investigated the application of thermodynamic principles to electrical systems.

In 1883 Lippmann was appointed professor of mathematical physics at the Faculty of Sciences in Paris. In 1886 he succeeded Jules Jamin as professor of experimental physics and later became director of the research laboratory, which was subsequently transferred to the Sorbonne. He retained this position until his death.

Lippmann made numerous contributions to instrumental design, particularly in connection with astronomy and seismology. His most notable contribution was the invention of the coelostat. In this instrument the light from a portion of the sky is reflected by a mirror which rotates around an axis parallel to its plane once every forty-eight sidereal hours. This axis is arranged parallel to the axis of the earth. The light from the mirror enters a telescope fixed to the earth. The arrangement ensures that a region of the sky, and not merely one particular star, may be photographed without apparent movement. In this respect it represented an improvement on the siderostat. Lippmann later put forward a suggestion for the accurate adjustment of a telescope directed at its zenith by reflecting light from a pool of mercury.

Lippmann also devised a number of improvements on observational technique by the introduction of photographic or electrical methods of measurement. These include a method for the measurement of the difference in longitude between two observatories by means of radio waves and photography and an improvement on the method of coincidences for measuring the difference between the periods of two pendulums. The pendulum measurements involve the use of a high-speed flash photograph to determine the change in phase over a short interval of time. Lippmann later investigated the problem of maintaining a pendulum in continuous oscillation. He showed that the period will be unaffected if the maintaining impulse is applied at the instant the pendulum swings through the point of zero displacement. The necessary impulse could, he suggested, be obtained by alternately charging and discharging a capacitor through coils mounted on either side of the pendulum. This arrangement ensures that the impulse is of short duration and is independent of wear on the pendulum contacts.

Lippmann's contributions to seismology include a suggestion for the use of telegraph signals to give early warning of earth tremors and for the measurement of their velocity of propagation. He also proposed a new form of seismograph intended to give directly the acceleration of the earth's movement.

In 1908 Lippmann was awarded the Nobel Prize in physics "for his method, based on the interference phenomenon, for reproducing colours photographically." In the same year he was elected a fellow of the Royal Society of London. In Lippmann's color process the sensitive emulsion, which is relatively thick, is backed by a reflecting surface of mercury. As a result the incident light is reflected back toward the source, and the incident and reflected beams combine to produce stationary waves. After development the film is found to contain reflecting planes of silver separated by distances of half a wavelength. When the film is viewed by reflected light, the color corresponding to the original beam is strongly reinforced by reflection from successive planes.

BIBLIOGRAPHY

I. Original Works. Lippmann's writings are "Relation entre les phénomènes électriques et capillaires," in *Comptes rendus . . . de l'Académie des sciences*, **76** (1873), 1407; "Sur une propriété d'une surface d'eau électrisée," *ibid.*, **81** (1875), 280—see also *ibid.*, **85** (1877), 142; "Extension du principe de Carnot à la théorie des phénomènes électriques," *ibid.*, **82** (1876), 1425; "Photographies colorées du spectre, sur albumine et sur gélatine bichromatées," *ibid.*, **115** (1892), 575; "Sur la théorie de la photographie des couleurs simples et composées par la méthode interférentielle," *ibid.*, **118** (1894), 92; "Sur un coelostat," *ibid.*, **120** (1895), 1015; "Sur l'entretien du mouvement du pendule sans perturbations," *ibid.*, **122** (1896), 104; "Méthodes pour comparer, à l'aide de l'étincelle électrique, les durées d'oscillation de deux pendules réglés sensiblement à la même période," *ibid.*, **124** (1897), 125; "Sur l'emploi d'un fil télégraphique pour l'inscription des tremblements de terre et la mesure de leur vitesse de propagation," *ibid.*, **136** (1903), 203; "Appareil pour enregistrer l'accélération absolue des mouvements séismiques," *ibid.*, **148** (1909), 138; "Sur une méthode photographique direct pour la détermination des différences des longitudes," *ibid.*, **158** (1914), 909; and "Méthode pour le réglage d'une lunette en collimation," *ibid.*, 88.

II. Secondary Literature. Obituary notices are in *Annales de physique*, **16** (1921), 156; and *Proceedings of the Royal Society*, ser. A, **101** (1922). See also H. H. Turner, "Some Notes on the Use and Adjustment of the Coelostat," in *Monthly Notices of the Royal Astronomical Society*, **56** (1896), 408.

I. B. Hopley

LIPSCHITZ, RUDOLF OTTO SIGISMUND (*b.* near Königsberg, Germany [now Kaliningrad, R.S.F.S.R.], 14 May 1832; *d.* Bonn, Germany, 7 October 1903), *mathematics.*

Lipschitz was born on his father's estate, Bönkein. At the age of fifteen he began the study of mathematics at the University of Königsberg, where Franz Neumann was teaching. He then went to Dirichlet in Berlin, and he always considered himself Dirichlet's student. After interrupting his studies for a year because of illness, he received his doctorate from the University of Berlin on 9 August 1853. After a period of training and teaching at the Gymnasiums in Königsberg and Elbing, Lipschitz became a *Privatdozent* in mathematics at the University of Berlin in 1857. In the same year he married Ida Pascha, the daughter of a neighboring landowner. In 1862 he became an associate professor at Breslau, and in 1864 a full professor at Bonn, where he was examiner for the dissertation of nineteen-year-old Felix Klein in 1868. He rejected an offer to succeed Clebsch at

Göttingen in 1873. Lipschitz was a corresponding member of the academies of Paris, Berlin, Göttingen, and Rome. He loved music, especially classical music.

Lipschitz distinguished himself through the unusual breadth of his research. He carried out many important and fruitful investigations in number theory, in the theory of Bessel functions and of Fourier series, in ordinary and partial differential equations, and in analytical mechanics and potential theory. Of special note are his extensive investigations concerning n-dimensional differential forms and the related questions of the calculus of variations, geometry, and mechanics. This work—in which he drew upon the developments that Riemann had presented in his famous lecture on the basic hypotheses underlying geometry—contributed to the creation of a new branch of mathematics.

Lipschitz was also very interested in the fundamental questions of mathematical research and of mathematical instruction in the universities. He gathered together his studies on these topics in the two-volume *Grundlagen der Analysis.* Until then a work of this kind had never appeared in German, although such books existed in French. The work begins with the theory of the rational integers and goes on to differential equations and function theory. The foundation of mathematics is also considered in terms of its applications.

In basic analysis Lipschitz furnished a condition, named for him, which is today as important for proofs of existence and uniqueness as for approximation theory and constructive function theory: If f is a function defined in the interval $\langle a, b\rangle$, then f may be said to satisfy a Lipschitz condition with the exponent α and the coefficient M if for any two values x, y in $\langle a, b\rangle$, the condition

$$|f(y) - f(x)| \leqslant M\,|y - x|^{\alpha}, \qquad \alpha > 0,$$

is satisfied.

In his algebraic number theory investigations of sums of arbitrarily many squares, Lipschitz obtained certain symbolic expressions from real transformations and derived computational rules for them. In this manner he obtained a hypercomplex system that is today termed a Lipschitz algebra. In the case of sums of two squares, his symbolic expressions go over into the numbers of the Gaussian number field; and in that of three squares, into the Hamiltonian quaternions. Related studies were carried out by H. Grassmann.

Lipschitz's most important achievements are contained in his investigations of forms on n differentials which he published, starting in 1869, in numerous articles, especially in the *Journal für die reine und angewandte Mathematik.* In this area he was one of the direct followers of Riemann, who in his lecture of 1854, before the Göttingen philosophical faculty, had formulated the principal problems of differential geometry in higher-dimension manifolds and had begun the study of the possible metric structures of an n-dimensional manifold. This lecture, which was not exclusively for mathematicians and which presented only the basic ideas, was published in 1868, following Riemann's death. A year later Lipschitz published his first work on this subject. With E. B. Christoffel, he was one of the first to employ cogredient differentiation; and in the process he created an easily used computational method. He showed that the vanishing of a certain expression is a necessary and sufficient condition for a Riemannian manifold to be Euclidean. The expression in question is a fourth-degree curvature quantity. Riemann knew this and had argued it in a work submitted in 1861 to a competition held by the French Academy of Sciences. Riemann did not win the prize, however; and the essay was not published until 1876, when it appeared in his collected works.

Lipschitz was especially successful in his investigations of the properties of Riemannian submanifolds V_m of dimension m in a Riemannian manifold V_n of dimension n. He showed flexure invariants for V_m in V_n, proved theorems concerning curvature, and investigated minimal submanifolds V_m in V_n. He was also responsible for the chief theorem concerning the mean curvature vector that yields the condition for a minimal submanifold: A submanifold V_m of V_n is a minimal submanifold if and only if the mean curvature vector vanishes at every point. Thus a manifold V_m is said to be a minimal manifold in V_n if, the boundary being fixed, the variation of the content of each subset bounded by a closed V_{m-1} vanishes. The definition of the mean curvature vector of a submanifold V_m is based on the fundamental concept of the curvature vector of a nonisotropic curve in V_m. This vector can be described with the help of Christoffel's three index symbols. Its length is the first curvature of the curve and vanishes for geodesics in V_m.

Lipschitz's investigations were continued by G. Ricci. The latter's absolute differential calculus was employed, beginning in 1913, by Einstein, who in turn stimulated interest and further research in the differential geometry of higher-dimension manifolds.

BIBLIOGRAPHY

Lipschitz's books are *Grundlagen der Analysis*, 2 vols. (Bonn, 1877–1880); and *Untersuchungen über die Summen von Quadraten* (Bonn, 1886). He also published in many German and foreign journals, especially the *Journal für die reine und angewandte Mathematik* (from 1869). There is a bibliography in Poggendorff, IV, 897.

A biographical article is H. Kortum, "Rudolf Lipschitz," in *Jahresberichte der Deutschen Matematiker-Vereinigung*, **15** (1906), 56–59.

BRUNO SCHOENEBERG

LISBOA, JOÃO DE (*b.* Portugal; *d.* Indian Ocean, before 1526), *navigation.*

It has been claimed that Lisboa, a Portuguese pilot, accompanied Vasco da Gama on the voyage on which the route to the Indies was discovered (1497–1499). While this claim is probably false, there are documents proving that he navigated the coast of Brazil (at an unknown date), that he participated in an expedition to the fortress of Azamor in northern Africa (1513), and that he sailed to the Indies at least three times: in 1506, as pilot of the fleet led by Tristão da Cunha; in 1518, in the squadron of Diogo Lopes de Sequeira; and in the squadron commanded by Filipe de Castro in 1525. The last fleet had a difficult crossing from Madagascar to the Curia-Muria Isles; and it is likely that Lisboa, then advanced in age, did not survive this arduous passage.

In the course of his first documented visit to the Indies, Lisboa and the pilot Pêro Anes carried out observations to determine the magnetic declination with the aid of the constellations Ursa Minor and the Southern Cross. The method is set forth in the *Tratado da agulha de marear*, a small book written by Lisboa in 1514. It contains the first description of the nautical "dip" compass and the first presentation of a method for measuring the magnetic declination. Present knowledge of the text derives from copies that are incomplete and faulty but numerous enough to allow one, through comparison, to determine its complete and correct form. The treatise consists of an introduction and ten chapters. The first three are concerned with the construction of the compass, the following three with the observation of the magnetic declination with the aid of stars, and the seventh with the method of magnetizing the needle. In the last three chapters Lisboa presents a theory without any experimental support; he assumes that the declination of the compass undergoes variations proportional to the longitude and is convinced that this "law" can enable sailors to determine this coordinate.

BIBLIOGRAPHY

Portions of João de Lisboa's treatise can be found in the following MSS:

"Livro de marinharia," attributed to Lisboa himself (but completed about 1550, well after his death), in the National Library, Lisbon. The text was published by Brito Rebelo (Lisbon, 1903).

"Livro de marinharia" by Bernardo Fernandes, in the Vatican Library, Codex Borg. lat. 153. An ed. was published by Fontoura da Costa (Lisbon, 1940).

"Livro de marinharia" by André Pires in the Bibliothèque Nationale, Paris, MS portugais, 40. Published by L. de Albuquerque (Coimbra, 1963).

"Livro de marinharia" by Gaspar Moreira, in the Bibliothèque Nationale, Paris, still unpublished.

"Rotero de navigación," written in Portuguese despite the title, in the library of the Academia Real de la Historia, Madrid, Cortes 30-2165. This compilation is anonymous and is unpublished.

LUÍS DE ALBUQUERQUE

LI SHIH-CHEN[1] (*tzu* Tung-pi,[2] *hao* P'in-hu[3]) (*b.* Wa-hsiao-pa,[4] near Chichow [now Hupeh province], China, 1518; *d.* autumn 1593), *pharmacology.*

Li Shih-chen was the son of Li Yen-wen,[5] the educated and comparatively successful offspring of several generations of medical practitioners who, as was common in China, also practiced pharmacy. The father was at one time a medical officer of lower rank in the Imperial Medical Academy and, at another time, a client in the house of the philosopher and administrator Ku Wen.[6] He wrote five treatises, no longer extant, on diagnosis, smallpox, sphygmology, ginseng, and the *Artemisia* of Hupeh.

Li Shih-chen was bright and inquisitive about nature in his youth. His father encouraged him to prepare for the civil service examinations. At fourteen he passed the preliminary examinations but, despite three attempts, was unable to advance further. Thus ineligible for a bureaucratic career, he obtained his father's permission to devote himself to medicine. His education was, typically, not institutional but combined clinical experience in his father's practice and study of the literature of theoretical medicine, therapeutics, and materia medica. Li's skill as a physician won him considerable renown in his lifetime and even official posts in medical administration—which provided him access to rare drugs and books—for two short periods. He wrote at least twelve books, which, in addition to medical works, include a treatise on the rules of prosody and a collection of poetry and short prose writings. Reports also mention writings of unknown scope, size, and title on astrology, geomancy, and divination.

Li's greatest work, known to every educated Chinese even today as the culmination of the pharma-

LI SHIH-CHEN

cognostic tradition, is the *Pen-ts'ao kang mu* ("Systematic Pharmacopoeia"). At about the age of thirty Li assumed the enormous task of producing a comprehensive and up-to-date encyclopedia of pharmaceutical natural history, taking into account the need for correcting the many mistakes of identification, classification, and evaluation in the previously standard Sung dynasty series of pharmacognostic treatises, which begins with T'ang Shen-wei's *Ching shih cheng lei pei chi pen-ts'ao* ("Pharmacopoeia for Every Emergency, With Classifications Verified From the Classics and Histories," drafted in 1082–1083) and reaches its apogee in Chang Ts'un-hui's *Ch'ung hsiu Cheng-ho ching shih cheng lei pei yung pen-ts'ao* ("Revised Pharmacopoeia of the Cheng-ho Reign Period for Every Use . . .," 1249), which incorporates part of the brilliant collection of critical notes of K'ou Tsung-shih, *Pen-ts'ao yen i* ("Dilatations Upon the Pharmacopoeias," 1116). By Li's time many new substances, some of them imported, had been introduced into therapeutic use. Lacking the imperial patronage usual for works of this magnitude, Li over a span of thirty years incorporated in three successive drafts the contributions of over 800 books; and, from 1556, traveled widely in major drug-producing provinces (Honan, Kiangsu, Anhwei), collecting specimens and studying the natural occurrences of minerals, plants, and animals. After completion of the final draft in 1587 he visited Nanking to arrange for publication. When, in 1590, an agreement with a printer was concluded, Li's eminent literary friend Wang Shih-chen[7] wrote a preface in which he praised Li for his lucidity as "a scholar unique south of the Great Bear." The first printed edition (the so-called Chin-ling xylograph) did not appear until 1596, after Li's death.

That the *Pen-ts'ao kang mu* is far more than a pharmacopoeia is immediately evident from its introduction. There Li affirms his intention of discussing everything that has been recorded about the simples under discussion, whether strictly concerned with medical practice or not; thus the book became a comprehensive treatise on mineralogy, metallurgy, botany, and zoology. His critical approach to the factual record led him to cite previous accounts with clear ascription and in chronological order. In both of these respects he follows well-established precedent. The catholicity of Li's book goes back to the beginning of the pharmacognostic tradition in China, and his citations of early literature follow the lead of T'ang Shen-wei. Neither author is extremely accurate in quoting, and critical historians of science do not rely upon them for citations from early works.

The articles on drugs in the body of the book are arranged in a taxonomic order based on that of T'ang Shen-wei but somewhat more systematic. This classification is far from rigorous; the rubrics indifferently reflect habitat, physical characteristics, and use. Nevertheless it is more detailed, and based less on purely verbal criteria, than any which preceded it. As in modern practice, the earliest name given in the literature to any drug is made the basis for classification; and varieties are subgrouped into families, which correspond in a very rough way to the Linnaean genera. The *t'ung*[8] trees, for instance, are characterized by specific differences, and names which correspond to varieties are meticulously distinguished from mere synonyms. Location of a particular item depends both upon its morphology (a *yin* feature, because static) and its qualitatively defined energy (a *yang* feature, because dynamic). In principle the latter is defined largely in terms of physiological action, according to a conceptual scheme closer to that of Renaissance medicine than to that of modern pharmacodynamics.

Li's most obvious modification of T'ang Shen-wei's taxonomy, as this outline of the *Pen-ts'ao kang mu*'s rubrics shows, is rearrangement according to a hierarchy of being from the inorganic world to man (compare Aristotle's *scala naturae*):

1. The substances corresponding to the Five Phases (*wu hsing*,[9] sometimes misleadingly translated "five elements"). The Five Phases, which in every field of traditional science were used analytically to represent divisions of generalized dynamic processes (or, to a lesser extent, of static configurations), were named for wood, fire, earth, metal, and water. Although in this section Li gives special consideration to the namesakes of the Phases, every medical substance was understood to correspond to one of the five as part of the definition of its specific energy. Actually only four of the five substances are represented here; wood, which never needed so abstract a rationale for inclusion in the materia medica, is treated separately in section 8.

   Natural varieties of water, both celestial and terrestrial

   Natural varieties of fire

   Natural varieties of earth (including clay, ink, and ash)

   Natural varieties of metal (including ores and corrosion products)
2. Jades (including coral and quartzes)
3. Inorganic substances

   Minerals

   Salt and salt derivatives

LI SHIH-CHEN

4. Herbs
   Mountain herbs
   Aromatic herbs
   Moist herbs
   Toxic herbs
   Creeping herbs
   Aquatic herbs
   Herbs of stony habitat
   Mosses
5. Grains
   Hemp, wheat, rice
   Millet
   Legumes
   Fermented grains
6. Vegetables
   Aromatic vegetables
   Soft and slippery vegetables
   Ground vegetables (gourds, eggplant, and such)
   Aquatic vegetables
   Fungi
7. Fruit
   Classic edible fruits
   Mountain fruits
   Fruits of foreign origin
   Spices
   Ground fruits (melons, grapes, sugarcane, and such)
   Aquatic fruits
8. Arboreal drugs
   Aromatic woods
   Woods from trees
   Woods from shrubs
   Parasitic plants (including amber)
   Canes (including bamboo)
   Miscellaneous woods
9. Furnishings and implements
   Cloth and clothing
   Articles of use (from paper to privy buckets)
10. Insects and insect products
   Oviparous insects
   Insects engendered by transformation
   Insects born of moisture (including frogs)
11. Scaly creatures and their products
   Dragons
   Snakes
   Fish
   Fish without scales (including eels)
12. Creatures with shells
   Turtles and tortoises
   Oysters and clams
13. Birds
   Aquatic birds
   Plains birds
   Forest birds
   Mountain birds
14. Animals
   Domestic animals
   Wild animals
   Rodents
   Parasites and prodigious beasts (including apes)
15. Man (including hair, meat, bone, and assorted secretions and excrements)

In addition to this basic taxonomic arrangement, Li incorporated in his preliminary chapters a second classification according to diseases cured. This had been done often by earlier writers, but Li's carefully appended notes on the physiological functions of each drug with respect to each disease (phrased in the abstract terminology of rational medicine) were his own innovation. There is elsewhere in his treatise a more condensed classification according to the theoretical variables of physiological action, without reference to specific disease, probably generalized from a much more partial schema of Ch'en Ts'ang-ch'i[10] (early eighth century). Finally, Li noted in each major entry, for drugs which had appeared in the first classic of materia medica, the *Shen-nung pen-ts'ao* ("Canonic Pharmacopoeia," attributed to the legendary emperor Shen-nung, probably first century), where the substance belonged in the tripartite order of that book—whether it was assigned to the lower class, which merely cured disease; the middle class, which maintained health; or the higher class, which conferred immortality. These rubrics were still functional as secondary divisions in the *Pen-ts'ao p'in hui ching yao* ("Classified Essential Pharmacopoeia"), compiled for use in the imperial palace less than a century earlier, but apparently had only antiquarian interest for Li Shih-chen.

The internal arrangement of the articles is also a step forward in terms of system. The general name of the substance is followed by an etymologic explanation of variant names (*shih ming*[11]); corrections of errors in earlier pharmacopoeias (*cheng wu*[12]); a pastiche of quotations, with Li's own comments added, about the habitat, varieties, and qualities of the substance, with tests for the genuine article (*chi chieh*[13]); instructions for processing and storing the drug (*hsiu chih*[14]); a specification of the physiological activity of the drug and of diseases cured (*chu chih*[15]); a statement of essential energetic qualities in terms of sapidity (correlated with the Five Phases), warming or cooling activity, and toxicity (*ch'i wei*[16]); quotations which account for these qualities theoretically (*fa ming*[17]); and a number of prescriptions in which the drug is employed (*fu fang*[18]).

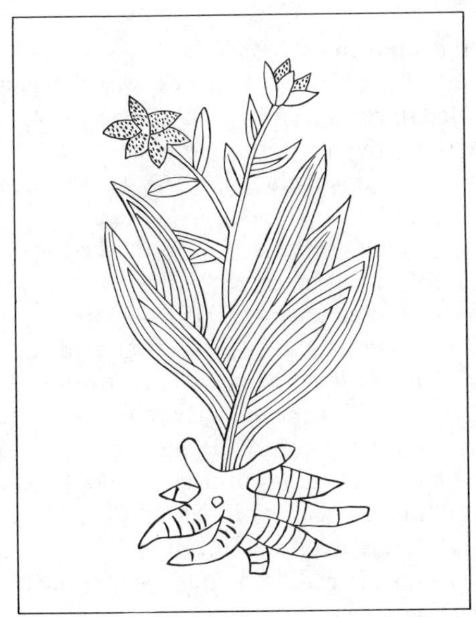

FIGURE 1

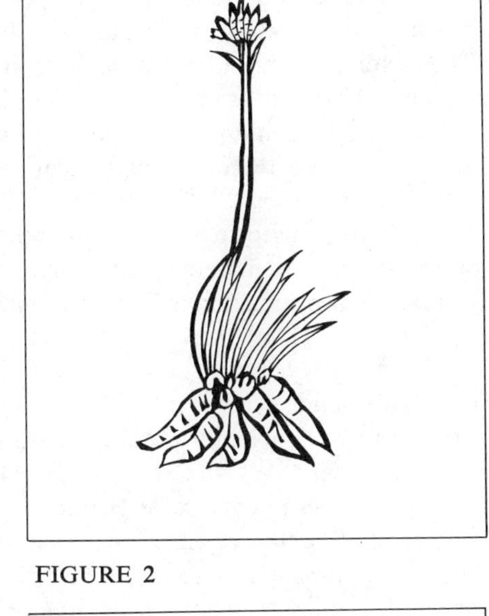

FIGURE 2

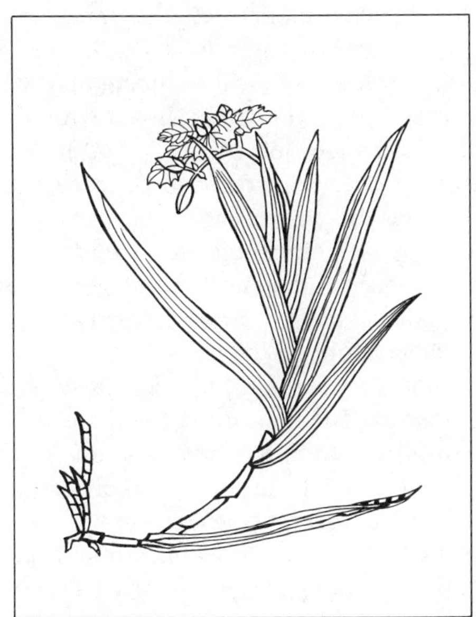

FIGURE 3

FIGURE 4

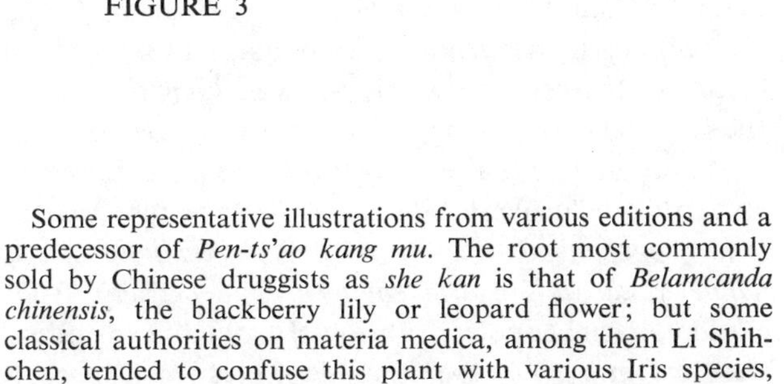

Some representative illustrations from various editions and a predecessor of *Pen-ts'ao kang mu*. The root most commonly sold by Chinese druggists as *she kan* is that of *Belamcanda chinensis*, the blackberry lily or leopard flower; but some classical authorities on materia medica, among them Li Shih-chen, tended to confuse this plant with various Iris species, especially *Iris tectorum maxim*. The latter, ordinarily called *yuan wei*, is still found on the market in Szechuan province as *she kan*.

FIGURE 1. Reproduced from the 1929 Wuchang reproduction of the 1302 edition of the *Ta-kuan pen-ts'ao* (see Bibliography II, item 5).

FIGURE 2. From the first printed edition of *Pen-ts'ao kang mu* (1596).

FIGURE 3. From the 1640 edition of *Pen-ts'ao kang mu* as reproduced in 1712 by the Edo Karahonya Seibei, Tokyo. The branching of the blossoms unambiguously suggests the blackberry lily, although in other respects the limning is defective.

FIGURE 4. From the 1885 Anui edition and a good likeness of an iris. Because this illustration was copied from *Chih wu ming shih t'u k'ao*, a compilation oriented toward botanical philology rather than medical therapy, it is uninformative about the appearance of the root, the part used in medicine.

LI SHIH-CHEN

One finds in the *Pen-ts'ao p'in hui ching yao* an even more systematic scheme of presentation, but this book had no important influence in Chinese medicine. Li never saw this manuscript; and it was not printed until 1936. Its framework, while clearly superior for concise presentation of factual information, did not provide for the depth of critical discussion unique to Li's treatise.

The systematic order used in this imperial pharmacopoeia of 1505 is set out here as a summary of the most fully elaborated parameters of Chinese pharmacognosy.

1. Name of substance
2. Specification of toxicity (an indication of potency)
3. Curative properties (a concise description quoted from or patterned after that of earlier compendiums)
4. Variant names (*ming*[19])
5. Description of plant (for botanicals) (*miao*[20])
6. Habitat (*ti*[21])
7. Times of growth and gathering (*shih*[22])
8. Processing and storage procedures (*shou*[23])
9. Part (or physical state in which) used (*yung*[24])
10. Physical characteristics of part, form, or type used (generally comparative) (*chih*[25])
11. Color (five possibilities, corresponding to the five phases) (*se*[26])
12. Sapidity (five possibilities) (*wei*[27])
13. Nature (nine possibilities) (*hsing*[28])
14. Strength of configurational energy (*ch'i*), *yin-yang* activity (*ch'i*[29])
15. Odor (five possibilities) (*hsiu*[30])
16. Physiological action (*chu*[31])
17. Circulation subsystems affected (*hsing*[32])
18. Ancillary action of other drugs (*chu*[33])
19. Antagonistic substances (*fan*[34])
20. Preparation for ingestion (that is, prescriptions) (*chih*[35])
21. Specific diseases cured (*chih*[36])
22. Use in combination with other drugs to cure specific diseases (*ho chih*[37])
23. Conditions in which administration is prohibited, limits on dosage (*chin*[38])
24. Substitutes for the drug in question, or its use as a substitute for other drugs (*tai*[39])
25. Dietary prohibitions when the drug is taken (*chi*[40])
26. Poisons neutralized (*chieh*[41])
27. Adulterants (*yen*[42])

The fifty-two chapters of *Pen-ts'ao kang mu* contain 1,892 entries, of which 275 are mineral and 444 animal. (Another 921 substances were added two centuries later in the supplement, *Pen-ts'ao kang mu shih-i.*) Li himself introduced 374 substances into the literature and credited thirty-nine more, not previously recorded, to doctors of the preceding four centuries. Of the 11,096 prescriptions included, 8,161 were recorded or created by Li.

The more than 1,100 illustrations in the first edition (see Figure 2) are not, in general, up to the level of those in the Sung series of pharmacopoeias which began five centuries earlier. Lu Gwei-djen has compared Li's illustrations, on the whole, to those in late fifteenth-century European herbals (such as Arnald of Villanova), and those of the Sung compilations to those of the sixteenth-century Occident (such as Fuchs); but in many cases Li's leave little to be desired in terms of accuracy of identification. There were two major revisions of the illustrations, one for the printing of 1640 (Figure 3) and the other for the Anhwei edition of 1885 (Figure 4). The latter, from which the pictures in most modern reprints are descended, borrows considerably from Wu Ch'i-chün's[43] fastidiously illustrated botanical compendium *Chih wu ming shih t'u k'ao*[44] ("Illustrated Investigations of the Names and Identities of Plants," 1848). Thus, in addition to an overall increase in delicacy there are many improvements in precision.

Although it is impossible to specify the first European influence of the *Pen-ts'ao kang mu*, we know that a copy was taken by Portuguese or Dutch traders to Japan, where about 1607 it came into the possession of the shogun Ieyasu and his scholarly secretary Hayashi Dōshun. The first of many Japanese editions appeared in 1637. In 1783 there was an immensely influential Japanese translation, amply annotated, by the great naturalist Ono Ranzan.

Europeans first learned of China's botanical riches in the Polish Jesuit Michael Boym's *Flora sinensis* (Vienna, 1656), which describes twenty-two plants in detail. About forty years later the physician G. E. Rumpf, deputy governor of Amboina, sent to Holland a copy of the first edition of *Pen-ts'ao kang mu*, which subsequently was long preserved in Germany. In 1735 J. B. du Halde cited *Pen-ts'ao kang mu* extensively as part of the general account of China in his *Description géographique, historique ... de l'empire de la Chine et de la Tartarie chinoise*. The extracts from earlier pharmacopoeias, abridged from the introduction to *Pen-ts'ao kang mu*, made it clear to European readers that Li's compilation belonged to a long and cumulative tradition. Du Halde also provided French renderings of the table of contents, Li's systematic account of the principles and types of prescriptions, and about fifteen articles on simples of special interest, including ginseng, tea, musk, and rhubarb.

LI SHIH-CHEN

Because du Halde was soon translated into English (1736, 1741), we can be reasonably confident that Li's method of classification, reflected in his table of contents, was known to the great seventeenth-century taxonomists and to Linnaeus. Linnaeus also knew indirectly of the work on Chinese plants grown in the Philippines carried out by the Jesuit Georg Josef Kamel, after whom he named the genus *Camellia.*

Considering the greater lapse of time, it is not surprising that Li Shih-chen's influence on Charles Darwin is much more easily documented. In *Variation of Animals and Plants Under Domestication* (London, 1868; I, 247), Darwin notes that seven breeds of fowls "including what we should now call jumpers or creepers, and likewise fowls with black feathers, bones and flesh" are described in a "Chinese Encyclopaedia published in 1596." In his account of goldfish he cites "an old Chinese work" to the effect that "fish with vermilion scales were first raised in confinement during the Sung dynasty (which commenced in A.D. 960) and now they are cultivated in families everywhere for the sake of ornament"—a literal rendition from *Pen-ts'ao kang mu*, chüan 44. The sinologist W. F. Mayers was led by Darwin's book to make a study published in China as "Gold Fish Cultivation" (*Notes and Queries on China and Japan*, 2 [1868], 123–124). Mayers in turn was quoted in Darwin's account of the secondary sexual characteristics of fishes in *The Descent of Man* (London, 1881; p. 343). Thus was Li Shih-chen's work woven into the fabric of modern science.

BIBLIOGRAPHY

I. LI SHIH-CHEN. The major documents for the life of Li Shih-chen are the many pertinent remarks in *Pen-ts'ao kang mu* and the document (*su*[45]) composed by his son Li Chien-yuan[46] to accompany the presentation of the book to the government. The latter is printed as part of the front matter of most editions of *Pen-ts'ao kang mu.* The "official" life in *Ming shih*[47] ("Standard History of the Ming Period"), Po-na ed., ch. 299, pp. 19*b*–20*a*, is derived from the latter. Other early biographical sketches, such as that in Ku Ching-hsing's[48] collected works, *Pai mao t'ang chi*[49], ch. 38; and that found in the collected prose of the great historiographer Chang Hsueh-ch'eng,[50] *Chang shih i shu,*[51] Chia yeh t'ang ed. (1922), ch. 25, pp. 44*a*–47*a*, are considerably less reliable.

The most notable modern studies of Li Shih-chen and his work are Chang Hui-chien,[52] *Li Shih-chen* (Shanghai, 1954), a severely positivistic popular account, also published in English (Peking, 1956); Lu Gwei-djen, "China's Greatest Naturalist, a Brief Biography of Li Shih-chen," in *Physis*, **8**, no. 4 (1966), 383–392, the most authoritative to date; Alfred Mosig and Gottfried Schramm, *Der Arzneipflanzen- und Drogenschatz Chinas und die Bedeutung des Pen-ts'ao Kang-mu als Standardwerk der chinesischen Materia Medica*, supp. to *Pharmazie*, no. 4 (1955), a rather slovenly jaunt through traditional pharmacology; Ts'ai Ching-feng,[53] "Shih lun Li Shih-chen chi ch'i tsai k'o-hsueh shang ti ch'eng-chiu"[54] ("An Essay on Li Shih-chen and His Scientific Achievements"), in *K'o-hsueh-shih chi-k'an*, **7** (1964), 63–80; Watanabe Kōzō,[55] "Ri Ji-chin no Honzo komoku to sono hampon"[56] ("Li Shih-chen's *Pen-ts'ao kang mu* and Its Editions"), in *Tōyōshi kenkyū*, **12**, no. 4 (1953), 333–357, which details Japanese as well as Chinese printed eds.; and the important series of evaluative essays by Japanese specialists which constitutes pp. 147–325 of Yabuuchi Kiyoshi[57] and Yoshida Mitsukuni,[58] eds., *Min-Shin jidai no kagaku gijutsu shi*[59] ("History of Science and Technology in the Ming and Ch'ing Periods"; Kyoto, 1970).

To this day there has not been a complete translation into a Western language of even one article of *Pen-ts'ao kang mu.* A great deal of material on botanicals has been summarized in E. V. Bretschneider, *Botanicon sinicum. Notes on Chinese Botany From Native and Western Sources*, **16** (1881), **25** (1893), and **29** (1895) of the *Journal of the North China Branch of the Royal Asiatic Society;* reprinted at London (1882–1895) and Tokyo (1937). An index to other treatises on Chinese simples is Bernard E. Read, *Chinese Medicinal Plants From the Pen Ts'ao Kang Mu . . . 1596*, 3rd ed. (Peking, 1936). On animal drugs see B. E. Read, *Chinese Materia Medica* (Peking, 1931–1941); for a detailed list of fascicles see J. Needham *et al.*, *Science and Civilisation in China* (Cambridge, 1959), III, 784–785. B. E. Read and C. Pak, *A Compendium of Minerals and Stones Used in Chinese Medicine From the Pen Ts'ao Kang Mu . . . [of] Li Shih-Chen . . . 1597 A.D.* (2nd ed., Peking, 1936), is not a translation but a raw juxtaposition of data from heterogeneous sources. Its historical usefulness is therefore somewhat limited, although its general accuracy is high. For complete and literal translations of all prescriptions given in *Pen-ts'ao kang mu* for a number of medicinal substances see William C. Cooper and N. Sivin, "Man as a Medicine. Pharmacological and Ritual Aspects of Traditional Therapy Using Drugs Derived From the Human Body," in S. Nakayama and N. Sivin, eds., *Chinese Science;* M.I.T. East Asian Science Series, I (Cambridge, Mass., 1972).

II. LANDMARKS OF THE PHARMACOGNOSTIC TRADITION. This list of basic works is provided for the reader not only because of their pertinence to the development of materia medica but also because they are essential to the historical study of botany, zoology, mineralogy, and natural history in China. The selection encompasses only a few of the many works belonging to what might be called the main line of development of the pharmacognostic tradition, as well as examples of intrinsically important but less influential or more specialized books on materia medica. The "main line" is defined most objectively as the series of pharmacopoeias spanning the millennium and a half from *Shen nung pen-ts'ao* to *Pen-ts'ao kang mu*, each of which largely incorporates, amplifies, and improves

 LI SHIH-CHEN

upon its predecessors. This succession has been sketched by Okanishi Tameto[66] in his preface to *Hsin hsiu pen-ts'ao;* and a much fuller flow chart is provided by Kimura Kōichi[81] in his explanatory essay appended to the reprint cited under *Ching shih cheng lei pei chi pen-ts'ao.* Other bibliographical sources are noted in part III.

1. *Shen nung pen-ts'ao ching*[60] ("Canonic Pharmacopoeia of the Emperor Shen-nung"; the last word of the title is often omitted in citations).

Anonymous and almost certainly compiled in the first or second century, this work covers 365 medicinal substances in tripartite arrangement (as noted above). It is no longer extant but is quoted fully in later pharmacopoeias because of its canonical status. Of the seven published reconstructions, that by Sung Hsing-yen and Sun P'ing-i[61] (*ca.* 1800, many reliable editions) is superior, although far from thorough in its use of sources.

2. *Pen-ts'ao ching chi chu*[62] ("Canonic Pharmacopoeia [of Shen-nung] With Collected Commentaries").

An annotated version of *Shen nung pen-ts'ao ching* compiled by the physician and Taoist magus T'ao Hung-ching, probably shortly after 500. The number of medically active substances was doubled; and detailed descriptions of drugs and notes on their gathering, preparation, and use were added. A copy of a MS of the preface dated 718 was published under the title *Pen-ts'ao ching chi chu ts'an chüan*[63] in the Chi shih an collection *(Chi shih an ts'ung-shu*[64]*)* of 1917 and was reprinted at Peking in 1955. The 1849–1852 Mori manuscript reconstruction of the complete text has been supplemented and reproduced by Okanishi Tameto[65] (Osaka, 1972).

3. *Hsin hsiu pen-ts'ao*[66] ("New Pharmacopoeia").

Compiled under imperial sponsorship by Su Ching and others, and completed in 659. Compared with the previous item, on which it was based, the number of substances included in this book was not much increased. Coverage was revised to take into account the wider resources of a reunited China, and illustrations were provided. Classification was by mineral, vegetable, or animal origin, in only seven categories (and one more containing drugs no longer in use), with the threefold system of the earlier pharmacopoeias retained for subdivisions.

Slightly over half of the text is preserved in four fragments, which have been used in the preparation of a definitive reconstruction of the whole text with variorum notes by Okanishi Tameto, *Ch'ung chi hsin hsiu pen-ts'ao*[67] (Ch'ing-t'an, Taiwan, 1964).

4. *Shih liao pen-ts'ao*[68] ("Dietary Pharmacopoeia").

Compiled by Meng Shen,[69] probably shortly after 700, and expanded by Chang Ting (between 720 and 740[?]) to include 227 substances of value in dietary hygiene for both health and illness. According to Nakao Manzō,[70] Chang changed the title from *Pu yang fang*[71] ("Restorative and Tonic Prescriptions"); see "Tonkō sekishitsu hakken Shokuryō honzō zankan kō"[72] ("Study of a Fragment of the *Shih liao pen-ts'ao* Discovered in a Tun-huang Grotto"), in *Shanhai shizen kagaku kenkyūsho ihō* (Shanghai), **1**, no. 3 (Feb. 1930), 9–18. *Shih liao pen-ts'ao* is perhaps the most important and most often cited book in the tradition of dietary regulation and therapy, which goes back to the beginnings of pharmacology in China.

The manuscript fragment just mentioned was found in northwest China in 1907; written in 933, it comprises about a tenth of the original book. Nakao published it as "Shokuryō honzō no kōsatsu"[73] ("A Study of the *Shih liao pen-ts'ao*"), *ibid.*, pp. 79–216, repr. as *Tun-huang shih shih ku pen-ts'ao ts'an chüan*[74] ("Fragment of an Ancient Pharmacopoeia From a Tun-huang Grotto"; Shanghai, 1937; repr. Taipei, 1970).

5. *Ching shih cheng lei pei chi pen-ts'ao*[75] ("Pharmacopoeia for Every Emergency, With Classifications Verified From the Classics and Histories").

Compiled by T'ang Shen-wei,[76] this work was drafted in 1082–1083 and, after conflation with the *Ch'ung kuang pu-chu pen-ts'ao*[77] ("Reamplified Pharmacopoeia With Added Commentary"; 1092) of Ch'en Ch'eng[78] was printed in 1108 under the title *Ching shih cheng lei Ta-kuan pen-ts'ao*[79] (usually abbreviated as *Ta-kuan pen-ts'ao;* Ta-kuan is the name of the reign period in which it was published). This book contains more than twice as many substances (*ca.* 1,750) as *Hsin hsiu pen-ts'ao.* The principle of classification was unchanged, but the number of rubrics was slightly increased. The series which this book begins comprises more than forty printed revisions and recensions in China, Japan, and Korea. See Watanabe Kōzō, "Tō Shimbi no Keishi shōrui bikkyū honzō to sono hampon"[80] ("On the Series Descended From T'ang Shen-wei's *Ching shih cheng lei pei chi pen-ts'ao*, and Its Editions"), in *Tōhō gakuhō* (Kyoto), **21** (1952), 160–206.

The earliest extant printed edition of the *Ta-kuan pen-ts'ao* is that of 1195, which had been further augmented by appending the *Pen-ts'ao yen i.* The recent reduced reprint with an index and a valuable historical introduction, *Ching shih cheng lei Ta-kuan pen-ts'ao*, Kimura Kōichi and Yoshizaki Masao,[81] eds. (Tokyo, 1970), is based on a 1904 repro. of the ed. of 1215.

6. *Pen-ts'ao yen i*[82] ("Dilatations Upon the Pharmacopoeias").

A collection of critical notes on 472 medical substances by K'ou Tsung-shih, printed in 1119. The original title was apparently *Pen-ts'ao kuang i*,[83] but the necessity to avoid a taboo on use of the personal name of Emperor Ning-tsung (1195–1201) led to the minor change in wording. This work, which made influential contributions to pharmacological theory, takes *Hsin hsiu pen-ts'ao* as its point of departure. K'ou, a minor medical official, did not have access to *Ching shih cheng lei pei chi pen-ts'ao* or its early successors. Submission of his work to the central government led to his promotion but not to imperial sponsorship of the book's publication.

The earliest surviving version was printed in 1195. A critical text based on surviving editions and citations was published at Peking in 1937 (repr. 1957).

7. *Ch'ung hsiu Cheng-ho ching shih cheng lei pei yung pen-ts'ao*[84] ("Revised Pharmacopoeia of the Cheng-ho Reign Period for Every Use, With Classifications Verified From the Classics and Histories").

This is a revision by Chang Ts'un-hui,[85] printed in 1249,

LI SHIH-CHEN

of the *Cheng-ho hsin hsiu ching shih cheng lei pei yung pen-ts'ao* ("Pharmacopoeia of the Cheng-ho Reign Period . . ."), which had been published under imperial auspices by Ts'ao Hsiao-chung[86] and others in 1116. Aside from minor discrepancies in content from *Ching shih cheng lei pei chi pen-ts'ao* (the net increase is only two substances), the major difference is that in this book the contents of *Pen-ts'ao yen i* are distributed among the articles to which they are related.

The exceptional usefulness of this member of the *Cheng lei pen-ts'ao* series results largely from the existence of the printed edition of 1249, which has been photographically reproduced in a widely distributed edition (Peking, 1957). The earlier edition in the Ssu pu ts'ung k'an collection (Shanghai, 1929), represented as based on the 1249 ed., is actually a reprint of the version of 1468.

8. *Chiu huang pen-ts'ao* ("Famine Relief Pharmacopoeia").

By Chu Hsiao,[87] fifth son of the Ming Emperor T'ai-tsu, preface dated 1406. Includes 440 substances which could be used to sustain life in time of famine, with illustrations and systematic notes on habitat, preparation, and ingestion. For a detailed description and historical appreciation see Lu Gwei-djen and Joseph Needham, "The Esculentist Movement in Mediaeval Chinese Botany; Studies on Wild (Emergency) Food Plants," in *Archives internationales d'histoire des sciences*, **21** (1968), 226–248; and Amano Motonosuke,[88] "Mindai ni okeru kyūkō sakumotsu chojutsu kō,"[89] ("A Study of the Writing of Works on Famine Relief in the Ming Period"), in *Tōyō gakuhō*, **47**, no. 1 (1964), 32–59.

The oldest extant printed edition is the xylograph 2nd ed. of 1525 (facs. repr., Shanghai, 1959). The most accessible edition is *Nung cheng ch'üan shu*[90] ("Complete Writings on Agricultural Administration"; completed 1628; repr. Peking, 1956), chüan 46–59. Later separate versions of *Chiu huang pen-ts'ao* (1837, 1856; Japan, 1718) are extracts from this agricultural handbook.

9. *Pen-ts'ao p'in hui ching yao*[91] ("Classified Essential Pharmacopoeia").

Compiled by a board including Liu Wen-t'ai[92] and others in 1505 but not printed until 1936, probably because of the death of its patron, Emperor Hsiao-tsung, soon after its completion. Since the MS was kept in the imperial palace, its historic influence was negligible. This is a work of great sophistication, on the scale of *Pen-ts'ao kang mu*, but based more narrowly on written sources.

The whereabouts of the MS are now unknown. While still in the palace library, it was published in a typeset edition under the title *Tien pan*[93] *pen-ts'ao p'in hui ching yao* (Shanghai, 1937). This included a supplement of 1701 (which incorporated material from *Pen-ts'ao kang mu*) but did not reproduce the fine color illustrations of the MS. On the illustrations see Giuliani Bertuccioli, "A Note on Two Ming Manuscripts of the Pên-Ts'ao P'in-Hui Ching-Yao," in *Journal of Oriental Studies*, **3** (1956), 63–68.

10. *Pen-ts'ao kang mu*[94] ("Systematic Pharmacopoeia").

Of the many available editions, the most convenient for scholarly use is the 2-vol. reprint (Peking, 1957) of the Anhwei recension of 1885. It adds not only variorum notes but also indexes of drug names and synonyms. The *Pen-ts'ao kang mu* and Chao Hsueh-min's[95] useful supplement, *Pen-ts'ao kang mu shih i*[96] ("Gleanings for the *Pen-ts'ao kang mu*"; compiled and revised between 1760 and 1803 or later, first printed 1871), are available in a punctuated typeset edition with index according to the four-corner system (Shanghai, 1954–1955).

III. Important Reference Tools for the History of Pharmacognosy. The standard descriptive guide to the extant pharmacopoeias of China and to Japanese pharmacopoeias in the Chinese tradition is Lung Po-chien,[97] *Hsien ts'un pen-ts'ao shu lu*[98] ("Bibliography of Extant Pharmacopoeias"; Peking, 1957). An invaluable pastiche of bibliographical data, prefaces and colophons, critical discussions, and analytic information is provided for the entire Chinese medical literature, extant or lost, up to the late thirteenth century in Okanishi Tameto, *Sung i-ch'ien i chi k'ao*[99] ("Studies of Medical Books Through the Sung Period"; Peking, 1958), which devotes nearly 200 pages to the classics of pharmacognosy. Less detailed information of a similar kind is provided for later works as well in Tamba Mototane[100] (*nom de plume* of Taki Gen'in[101]), *Chung-kuo i chi k'ao*[102] ("Studies of Chinese Medical Books"; Peking, 1956). Liou Ho and Claudius Roux, *Aperçu bibliographique sur les anciens traités chinois de botanique, d'agriculture, de sériciculture et de fungiculture* (Lyons, 1927), reflects very limited study.

There is no reliable general historical study of Chinese materia medica in any Western language. Pierre Huard and Ming Wong, "Évolution de la matière médicale chinoise," in *Janus*, **47** (1958), reviews the main pharmacopoeias, but because of carelessness and excessive reliance on uncritical modern Chinese secondary sources this article must be used with caution. The historical data in vol. I of Bretschneider's book (cited above), because less ambitious, is perhaps more useful. Needham *et al.*, *Science and Civilisation in China*, VI, can be expected to fill this gap.

In Chinese there is as yet nothing on medicine in general to replace Ch'en Pang-hsien,[103] *Chung-kuo i-hsueh shih*[104] ("History of Chinese Medicine"; Shanghai, 1920), almost completely bibliographical in approach; and there is no monograph on the history of materia medica. In Japanese the survey of Liao Wen-jen,[105] *Shina chūsei igaku shi*[106] ("History of Medieval Chinese Medicine"; Kyoto, 1932), is similar in style and value to that of Ch'en; and scholarship on Chinese pharmacognosy is scattered through a great many articles on special subjects. Among the most informative of these is Okanishi, "Chūgoku honzō no dentō to Kin-Gen no honzō,"[107] ("The Chinese Pharmacognostic Tradition and the Pharmacopoeias of the Chin and Yuan Periods"), in Yabuuchi, ed., *Sō-Gen jidai no kagaku gijutsushi*[108] ("History of Science and Technology in the Sung and Yuan Periods"; Kyoto, 1967), pp. 171–210.

The major guides to the secondary literature of Chinese medicine, and thus of materia medica, are Chi Hung and

Wu Kuan-kuo,[109] *Chung-wen i-hsueh wen-hsien fen-lei so-yin*[110] ("Classified Index of Medical Contributions in Chinese"; Peking, 1958), esp. pp. 8–14 on history; A. R. Ghani, *Chinese Medicine and Indigenous Medicinal Plants*, Pansdoc Bibliography no. 396 (Karachi, 1965), a random collection of citations; and Medical History Research Subcommittee, Shanghai Municipal Research Committee on Chinese Materia Medica, ed., *Chung-wen i-shih lun-wen so-yin*[111] ("Index to Articles in Chinese on the History of Medicine"), 3 vols. (Shanghai, 1957–1958).

This article is, with some additions and omissions and a new bibliography, adapted with permission from the study of Lu Gwei-djen noted above. The aid of Miyasita Saburō, librarian for Takeda Chemical Industries, Osaka, Japan, especially in procuring the illustrations, is gratefully acknowledged.

N. SIVIN

NOTES

1. 李時珍
2. 東壁
3. 瀕湖
4. 瓦硝羂
5. 李言聞
6. 顧問
7. 王世貞
8. 桐
9. 五行
10. 陳藏器
11. 釋名
12. 正誤
13. 集解
14. 脩治
15. 主治
16. 氣味
17. 發明
18. 附方
19. 名
20. 苗
21. 地
22. 時
23. 收
24. 用
25. 質
26. 色
27. 味
28. 性
29. 氣
30. 臭
31. 主
32. 行
33. 助
34. 反
35. 製
36. 治
37. 合治
38. 禁
39. 代
40. 忌
41. 解
42. 贋
43. 吳其濬
44. 植物名實圖考
45. 疏
46. 李建元
47. 明史
48. 顧景星
49. 白茅堂集
50. 章學誠
51. 章氏遺書
52. 張慧劍
53. 蔡景峯
54. 試論李時珍及其在科學上的成就
55. 渡邊幸三
56. 李時珍の本草綱目とその版本
57. 藪内清
58. 吉田光邦
59. 明清時代の科学技術史
60. 神農本草經
61. 孫星衍，孫馮翼
62. 本草經集注
63. 殘卷
64. 吉石盒叢書
65. 岡西爲人
66. 新修本草
67. 重輯新脩本草
68. 食療本草
69. 孟詵
70. 中尾万三
71. 補養方
72. 敦煌石室發見食療木草殘卷考
73. 食療本草の考祭
74. 敦煌石室古本草殘卷
75. 經史證類備急本草
76. 唐愼微
77. 重廣補注本草
78. 陳承
79. 經史證類大觀本草
80. 唐愼微の經史證類備急本草の系統とその版本
81. 木村康一，吉崎正雄
82. 本草衍義
83. 廣義
84. 重修政和新修經史證類備用本草
85. 張存惠
86. 曹孝忠
87. 朱橚
88. 天野元之助
89. 明代における救荒作物著述考
90. 農政全書
91. 本草品彙精要
92. 劉文泰
93. 殿板
94. 本草綱目
95. 趙學敏
96. 拾遺
97. 龍伯堅
98. 現存本草書錄
99. 宋以前醫籍考
100. 丹波元胤
101. 多紀元胤
102. 中國醫籍考
103. 陳邦賢
104. 中國醫學史
105. 廖溫仁
106. 支那中世醫學史
107. 中国本草の伝統と金元の本草
108. 宋元時代の科学技術史
109. 吉鴻，吳观國
110. 中文醫學文献分類索引
111. 中文醫史論文索引

LISSAJOUS, JULES ANTOINE (*b.* Versailles France, 4 March 1822; *d.* Plombières, France 24 June 1880), *physics.*

Lissajous developed an optical method for studying vibration and was generally interested in the physics of wave motion. "Lissajous figures" are the curves in the xy plane generated by the functions $y = a \sin(w_1 t + q_1)$ and $x = b \sin(w_2 t + q_2)$, where w_1 and w_2 are small integers. The curves are today easily produced on an oscilloscope screen; but Lissajous obtained them in the context of acoustics, from the superposition of the vibrations of tuning forks. He entered the École Normale Supérieure in 1841 and received the *agrégé* in 1847. He then became professor of physics at the Lycée Saint-Louis. In 1850 he presented his thesis, *Sur la position des noeuds dans les lames qui vibrent transversalement*, to the Faculty of Sciences. In 1874 he became rector of the academy of Chambéry and, in the following year, of the academy of Besançon. Lissajous was a candidate for the physics section of the Paris Academy in 1873 but was only elected corresponding member in 1879. In 1873 he received the Lacaze Prize, primarily for his work on the optical observation of vibration.

Like some other physicists of the time, Lissajous was interested in demonstrations of vibration that did not depend on the sense of hearing. Most of his experiments involved visual manifestations of vibrations: in the thesis on vibrating bars he used Chladni's sand-pattern method to determine nodal positions; he studied the waves produced by tuning forks in contact with water; and he did experiments on the popular phenomenon of "singing flames." Lissajous's most important research, first described in 1855, was the invention of a way to study acoustic vibrations by reflecting a light beam from the vibrating object onto a screen. He introduced his discussion of this topic with the assertion that, although sound vibrations are too rapid for direct observation, he could provide a visual demonstration of the wave form and obtain precise tuning without using the ear. (Lissajous was concerned with the problem of tuning—in 1855 he strongly recommended defining a standard frequency for the tuning of musical instruments.) He wrote that he thought of projecting the motion of vibration by reflecting light because he wanted to avoid the mechanical linkage present in some graphic devices (such as Duhamel's). He claimed that the use of rapidly rotating mirrors in some contemporary experiments (such as that of his friend Léon Foucault, on the velocity of light) had influenced his thinking.

Lissajous produced two kinds of luminous curves. In the first kind, light is reflected from a tuning fork (to which a small mirror is attached), and then from a large mirror that is rotated rapidly. When viewed on a screen, the beam shows the trigonometric form of the displacements, because the vibrations have been "spread out." The second kind of curve, named the "Lissajous figure," is more useful. The light beam is successively reflected from mirrors on two forks that are vibrating about mutually perpendicular axes. Persistence of vision causes various curves, whose shapes depend on the relative frequency, phase, and amplitude of the forks' vibrations, to be seen on the screen. For example, forks vibrating with the same amplitude and frequency produce ellipses the parameters of which depend on the phase difference between them.

Lissajous was interested in using his superposition curves to measure vibration parameters and to analyze more complicated acoustical problems. If one of the forks is a standard, the form of the curve enables an estimate of the parameters of the other. As Lissajous said, they enable one to study beats (the ellipses rotate as the phase difference changes). "Lissajous figures" have been, and still are, important in this respect. For further use in research Lissajous invented the "phonoptomètre," a vibrating microscope in which a tuning fork is attached to the objective lens. The vibrations of the object being observed combine with those of the lens and can therefore be analyzed in terms of the Lissajous figures produced. Helmholtz used this instrument in his investigations of string vibration (Helmholtz, *On the Sensations of Tone*, Dover reprint, p. 80).

Lissajous's optical method of observing vibration and the vibrating microscope were shown at the Paris exhibition of 1867 (they are described in *Reports of the United States Commissioners to the Paris Universal Exposition 1867* [Washington, D.C., 1870], III, 507–509). The French physicists awarded the Lacaze Prize to Lissajous for his "beautiful experiments," and both Rayleigh and Tyndall discussed his work in their treatises on sound. Scientists were enthusiastic about the work because there were still not many ways of demonstrating and measuring the parameters of vibration. Lissajous's optical experiments are simple, but the reasoning behind them depends on the principle of superposition; it is probably for this reason that they were done only in the middle of the nineteenth century.

BIBLIOGRAPHY

Lissajous's most important papers are "Note sur un moyen nouveau de mettre en évidence le mouvement vibratoire des corps," in *Comptes rendus hebdomadaires des séances de l'Académie des sciences*, **41** (1855), 93–95; "Note sur une méthode nouvelle applicable à l'étude des mouvements vibratoires," *ibid.*, 814–817; "Mémoire sur l'étude optique des mouvements vibratoires," in *Annales de chimie*, 3rd ser., **51** (1857), 147–231 (this is the most substantial article on the optical method); "Sur l'interférence des ondes liquides," in *Comptes rendus hebdomadaires des séances de l'Académie des sciences*, **58** (1868), 1187; "Sur le phonoptomètre, instrument propre à l'étude optique des mouvements périodiques ou continus," *ibid.*, **76** (1873), 878–880; and "Notice historique sur la vie et les travaux de Léon Foucault," in J. B. L. Foucault, *Recueil des travaux scientifiques*, II (Paris, 1878). For other papers by Lissajous, see the Royal Society's *Catalogue of Scientific Papers*, IV, 52; VIII, 244.

On Lissajous's work, see "Prix Lacaze, physique," in *Comptes rendus hebdomadaires des séances de l'Académie des sciences*, **79** (1874), 1607–1610.

SIGALIA DOSTROVSKY

LISTER, JOSEPH (*b.* Upton, Essex, England, 5 April 1827; *d.* Walmer, Deal, Kent, England, 10 February 1912), *surgery.*

The antiseptic doctrines and practices developed by Joseph Lister in the mid-Victorian era transformed the ancient craft of surgery into an enlightened art governed by scientific disciplines. His methodical, conscientious determination to reduce the appalling mortality rates resulting from traumatic and post-operative sepsis in the surgical wards of hospitals, allied with extraordinary probity and charm of character, captured the devotion of adherents in many countries and eventually silenced opponents. During his lifetime surgery expanded tremendously in resources and scope; the terrified despair of prospective victims gave place to faith and confidence that injury could be remedied and suppuration averted; and Lister himself became widely acknowledged as one of mankind's greatest benefactors.

Lister was the fourth child and second son of Quaker parents, Joseph Jackson Lister and the former Isabella Harris, who had four sons and three daughters. Several generations of the Lister family had lived in Yorkshire when Joseph Lister, the illustrious surgeon's great-grandfather (whose parents had joined the Society of Friends), went to London about 1720 and set up as a tobacconist. His youngest son, John, acquired a vintnery from Stephen Jackson, his father-in-law. John's only son, Joseph Jackson, left school at fourteen to become apprenticed to the prospering wine business. In his early thirties he married a schoolteacher about ten years his junior, daughter of the headmistress of Ackworth Quaker School. In 1826 he purchased Upton House, a mansion set on seventy acres about five miles east of London. Here Joseph Lister was born. The estate remained the family home until his father's death in 1869.

Lister's pensive, handsome mother was an unfailing source of affection, guidance, and instruction to the young boy. His remarkable father had artistic talent and, despite meager schooling, became a good Latin scholar and achieved distinction in mathematics, particularly in application to optics. Joseph Jackson Lister's work led to production of the achromatic microscope objective and to his election to fellowship in the Royal Society of London in 1832. The close-knit household was happy and lively, enjoying occasional fun. Amusements were restrained, however, and relaxations purposeful; and the high-mindedness of parents and neighboring Friends was reflected by scientists who occasionally visited Upton House for group discussions. Lister was introduced by his father to microscopy and to pursuits in natural history that gave him lifelong pleasure. As a child he macerated bones, dissected animals and articulated their skeletons, and announced his intention to become a surgeon. A close sympathy developed between father and son, the latter habitually writing home about his affairs and appreciating the shrewd paternal comments and counsel.

Lister attended two private schools, the first at Hitchin, where he was ahead of his year, especially in classics. At about thirteen he went to the Quaker school of Grove House, Tottenham, which emphasized mathematics, natural science, modern languages, and the writing of formal essays on various topics. (Lister's subjects included chemistry, human osteology, and laughing gas.) He left school at seventeen with a well-rounded education that included considerable familiarity with the classics, some facility in French and German, and a punctiliousness in speech and writing that became his hallmark.

In 1844 Lister entered the nonsectarian University College, London. It presented no obstacles to registration on religious grounds; possessed a modern hospital and a distinguished medical faculty; and was dedicated to sober study. In three years he received his B.A. degree; but soon after he began to study medicine, he contracted smallpox. Premature return to work brought on nervous depression, necessitating a long holiday in 1848. He resumed his studies in London that autumn. In those days students interspersed their classes in botany, physics, and chemistry (subsequently termed "premedical" courses) with physiology and anatomy ("preclinical" courses), and with ward visits, as the opportunity offered. Lister's painstaking system of learning and his retentive memory allowed him to profit from this system. He always showed talent for integrating and applying knowledge acquired from manifold sources.

In his first years as a medical student, Lister worked very seriously under austere living conditions. A more cheerful and wholesome life supervened when he became a hospital resident and met at close quarters intelligent young men of diverse backgrounds, some destined for prominent careers. Participating in the debating society and the hospital medical society, he attacked the homeopaths and read papers (never published) on hospital gangrene and on the use of the microscope in medicine. After being house physician to the cardiologist W. H. Walshe, he was house surgeon for nine months in 1851 to J. (later Sir John) Erichsen. Awarded many examination honors, in 1852 he received the M.B. degree of the University of London and the fellowship of the Royal College of Surgeons.

In his final years at University College, Lister was influenced particularly by two professors: the ophthalmic surgeon Wharton Jones—well-known for researches on inflammation—and the eminent physiologist William Sharpey. Under their guidance and example, and with his father's practical encouragement, he launched some histological investigations. These studies, conducted before modern methods of section cutting and tissue staining were available, evoked the technical enterprise and tenacity that marked Lister's subsequent microbiological researches and antiseptic practices. The first published report,

"Observations on the Contractile Tissue of the Iris" (1853), confirmed R. A. von Kölliker's claim that the iris comprises involuntary muscle and demonstrated that pupillary size is controlled by two distinct muscles. Another report soon appeared, "Observations on the Muscular Tissue of the Skin" (1853), dealing mainly with involuntary muscles of the scalp. Both papers, illustrated by delicate camera lucida drawings, attracted favorable attention at home and abroad. Lister's first experimental inquiry, begun shortly afterward (but unreported for four years), concerned the flow and absorption of chyle in the mesenteric lacteals of mice given indigo in their feed.

Although these microscopic excursions fascinated Lister, his interest in surgery wavered only briefly. He had been present at University College Hospital in December 1846, when Robert Liston performed the first major operation under ether anesthesia in Britain. Several years elapsed before the availability of anesthetics persuaded surgeons to broaden the scope and reduce the haste of their procedures—thus increasing the mortality from postoperative sepsis. Prolonged suffering and death often followed treatment of a minor disability. A conscientious surgeon-to-be might well be appalled at the unpredictable prospect for his patients, and saddened by the grim mortality associated with his calling. But Lister was undaunted; after nine years of studies at University College he was now twenty-six, and it was time to establish himself. Sharpey advised that after spending a month with James Syme, professor of clinical surgery at the University of Edinburgh—perhaps the most original and thoughtful surgeon of his day—Lister should broaden his experience by visiting famous Continental medical centers.

In September 1853 Lister was warmly received at Edinburgh by the brilliant, opinionated Syme, who entrusted him with so much responsibility at the infirmary that before long Lister decided to stay the winter. Syme appointed him supernumerary house surgeon and made him welcome in his household. Early in 1854 Lister became his resident house surgeon, a post that brought special privileges: twelve dressers to supervise, a wide choice of hospital patients to operate upon, and opportunities of assisting in his chief's private practice. The close relationship between the two men was enhanced when Lister became engaged in July 1855 to Syme's eldest daughter, Agnes. He resigned his membership in the Society of Friends (but continued to use the Quaker form of address to his parents and siblings) and later joined the Scottish Episcopal Church. The marriage took place in April 1856 and, although childless, was very happy. For nearly forty years his wife proved a devoted companion, an understanding helpmate, and a patient and competent amanuensis.

Late in 1854, an assistant surgeoncy at the Royal Infirmary fell vacant, and Lister was urged to apply for it. The duties of the appointment included giving a course on the principles and practice of surgery, which he painstakingly prepared and began to give in the autumn of 1855, although he was not elected to the post until a year later, shortly after returning with his wife from a prolonged wedding tour. (At such renowned medical centers as Pavia, Padua, Vienna, Prague, Würzburg, Leipzig, and Berlin he had met many well-known figures, particularly in the field of ophthalmic surgery, to which he was then attracted.) The new appointee zestfully undertook a second lecture course, on surgical pathology and operative surgery. He also conducted "public" operations at the Royal Infirmary—sometimes applauded by students—and extended his microscopic researches.

After presenting a paper entitled "On the Minute Structure of Involuntary Muscular Fibre" to the Royal Society of Edinburgh, Lister intensified his inquiries into inflammation, launched the previous year following "a most glorious night" at the microscope. He read three reports to the Royal Society of London in June 1857; they were published in its *Philosophical Transactions* (1858), the most important being "On the Early Stages of Inflammation." This records the earliest vascular and tissue changes induced in the frog's web by such irritants as hot water and mustard. He concluded and thereafter taught that "the primary lesion in inflammatory congestion" is a "suspension of function or temporary abolition of vital energy," characterized by "adhesiveness" of the blood corpuscles. Led thence to investigate blood coagulation, he summarized current knowledge of this phenomenon in his Croonian lecture of 1863.

Lister's reputation as an original and thorough investigator brought him election to fellowship of the Royal Society in 1860, when he was only thirty-three years old. His scrupulous concern to verify conclusions and give full weight to conflicting evidence is revealed in letters he wrote in 1857–1858 to William Sharpey. Lister's genuine modesty and obvious integrity, and a capacity to inspire students with enthusiasm and respect for their calling, made him a natural candidate for the regius professorship of surgery at the University of Glasgow, which became vacant late in 1859. Syme persuaded him to apply and canvass for the post, and his appointment was confirmed early in 1860. Nearly seven formative years at Edinburgh thus closed with regretful congratulations from students and a testimonial dinner from colleagues.

Lister was broadly trained, well reputed, courageous and conciliatory, and in the prime of physical and mental vigor. Glasgow's medical school was prosperous, with a talented and congenial faculty. Lister was initially handicapped, however, by lack of a hospital appointment, for the professorial chair did not automatically involve the surgeoncy at the Royal Infirmary. Nevertheless, he assumed heavy teaching and administrative duties, including the preparation of a daily lecture in systematic surgery for a class of 182 enthusiastic students. Characteristically, he renovated the students' desks and redecorated the lecture theater at his own expense. He set up a meticulously detailed and troublesome system of marking each question in his written and oral examinations; became actively interested in arrangements for removal of the university to a new locality; and served as secretary of the medical faculty. Lister also wrote lengthy and valuable chapters on amputations and anesthetics for T. Holmes's comprehensive *System of Surgery* (1860–1864), to which he was sole contributor from outside London. These undertakings left little time for the demands of private practice.

Lister worked even harder when elected in 1861 to take charge of the surgical wards at the Royal Infirmary. He explored the greater opportunities, more conservative procedures, and lessened shock offered by the advent of anesthesia. (He favored chloroform, discovered in 1847 by Sir James Y. Simpson, an eminent gynecologist at Edinburgh—who later attacked the antiseptic system.) During this period Lister invented several ingenious instruments, including a needle for silver-wire sutures, a hook for extracting small objects from the ear, a slender-bladed sinus forceps, and a screw tourniquet for compressing the abdominal aorta. Later he devised many kinds of dressings, ligatures, and drains for wounds. He was a cautious operator, deliberate and thorough rather than spectacular, but alert and resolute in an emergency. Broad-shouldered and of powerful physique, with very strong hands, Lister was yet extremely gentle; and the most delicate manipulations, such as removal of a cataract or a urethral stone, were performed unfalteringly. His innovations—such as radical mastectomy for breast cancer, the wiring of fractured patellae and pegging of un-united fractures, and the revival of supra-pubic cystotomy—were a consequence of greater surgical boldness following the introduction of antiseptic techniques. Improved methods of amputating through the thigh or at the hip dated from the pre-antiseptic era, however, as did Lister's practice of rendering the site of operation bloodless by elevating the involved limb before applying the tourniquet.

The manifold commitments and heavy routine were incompatible with Lister's research aspirations. Moreover, tenure of his hospital appointment was limited to ten years. Hence, when the chair of systematic surgery at Edinburgh became vacant in 1864, he was persuaded to apply for it. His disappointment when James Spence was chosen became mingled with grief at the almost simultaneous death of his mother; but he involved himself in helping to select a site for Glasgow's new College Hospital (the future Western Infirmary) and in preparing a pioneer paper, "Excision of the Wrist for Caries" (1865). This describes how, out of fifteen cases of tuberculosis of the wrist joint, at least ten hands were spared from amputation and rendered functional.

Lister was dismayed that surgical progress should be blocked by the threat of erysipelas, septicemia, pyemia, or hospital gangrene. Admittance to the New Surgical Hospital portion of the Royal Infirmary meant courting unpredictable catastrophe. The mortality from amputations, as in many other well-known hospitals, was around 40 percent. Abdominal surgery was seldom contemplated and was limited to ovariotomy, and the thoracic and cranial cavities were practically sacrosanct. Lister's wards never assumed the frightful conditions that necessitated closure of other parts of the building—probably because he was unusually fastidious, demanded strict attention to rules of cleanliness, ensured good air circulation, and refused to tolerate overcrowding. Yet the sufferings of his patients were so heartrending that it sometimes seemed "a questionable privilege to be connected with the institution."

In the early 1860's Lister began declaring wound suppuration a form of decomposition. The prevailing medical doctrine about the cause of putrefaction derived from Liebig's dictum (1839) that organic substances in the moist state and in the presence of oxygen undergo a peculiar form of combustion. Liebig rode roughshod over the observations in 1837 of the physicist Charles Cagniard de la Tour and the physiologist Theodor Schwann that fermentative phenomena resulted from the multiplication of yeast cells. A subsequent generation of misguided surgeons had treated wounds systematically on the supposition that they should be shielded from the effects of atmospheric oxygen. One system encouraged scab formation through application of powders, caustics, or fragrant balsams. Every surgeon had his favorite dressing, dry or wet. The alternative system entailed mechanical occlusion of the wound with agents such as collodion, adhesive plaster, and goldbeater's skin.

Lister realized that oxygen could not be excluded from wounds, and he soon doubted its responsibility for provoking suppuration. His 1865 paper referred to the oral administration of potassium sulfite to a patient who developed fatal pyemia, "with the view of counteracting the poisonous effect of any septic matter already introduced into the circulation." Further, when two persistent sores became gangrenous in a patient whose wrist he had successfully excised, the process was "checked by the application of carbolic acid. . . ."

Thomas Anderson, professor of chemistry at Glasgow University, drew Lister's attention in 1865 to the work and writings of Louis Pasteur. In several notable reports, especially "Mémoire sur les corpuscules organisés qui existent dans l'atmosphère. Examen de la doctrine des générations spontanées" (1861) and "Recherches sur la putréfaction" (1864), Pasteur had claimed that putrefaction was a fermentative process caused by living microorganisms carried on dust particles and transported by the air; that the air could be freed of these agents by filtration, heat, and other means; that certain body fluids, such as blood and urine, would keep indefinitely without decomposing if collected and stored under sterile conditions; and that spontaneous generation was a myth. These claims revealed startlingly to Lister the causes of wound sepsis and provided the key to banishment of hospital diseases. He always gladly acknowledged his indebtedness to Pasteur, as in the 1867 address "On the Antiseptic Principle in the Practice of Surgery" to the British Medical Association:

> When it had been shown by the researches of Pasteur that the septic property of the atmosphere depended not on the oxygen or any gaseous constituent, but on minute organisms suspended in it, which owed their energy to their vitality, it occurred to me that decomposition in the injured part might be avoided without excluding the air, by applying as a dressing some material capable of destroying the life of the floating particles.

In 1864 Lister learned that carbolic acid treatment of the sewage at Carlisle had rendered neighboring irrigated lands odorless and had destroyed entozoa that infected cattle grazing there. Procuring a sample of crude "German creosote" from his colleague Anderson, he used it (unsuccessfully) in March 1865 in the treatment of a compound fracture of the leg. The impure material, immiscible with water, was superseded by the less irritating crystalline carbolic acid—newly manufactured at Manchester by F. C. Calvert. This was soluble up to 5 percent in water but dissolved readily in such organic fluids as olive oil or linseed oil. Between August 1865 and April 1867, by means of these agents, Lister obtained results exceeding all expectations: of eleven cases of compound fractures of limbs, nine recovered. The main features of his antiseptic treatment were thorough cleansing of the wound with carbolic acid and its protection from airborne germs by a dressing soaked in the acid. The site was covered with a molded sheet of tin to diminish evaporation, around which absorbent material was packed to collect discharge. The dressings were changed daily and the tenacious crust of carbolized blood at the site of injury retouched with the acid.

Lister adapted his principle and technique to drainage of indolent abscesses (including a large psoas abscess), associated with tuberculous bones or joints. Until opened, such lesions generally contained no "septic organisms," so it was unnecessary to introduce antiseptic into them. To protect against environmental living particles during and after surgical intervention, however, the overlying skin was dressed with rag soaked in a 1 : 4 solution of crystalline carbolic acid in boiled linseed oil. Under this "antiseptic curtain" the abscess was incised and its contents evacuated. To prevent persistent discharge from such abscesses becoming contaminated, Lister prepared an antiseptic putty by mixing ordinary whiting with carbolic acid solution in linseed oil. This was spread thickly upon tinfoil and secured over the site with adhesive plaster.

Lister described these remarkable advances in a classic series of reports in *The Lancet*, "On a New Method of Treating Compound Fracture, Abscess, etc., With Observations on the Conditions of Suppuration" (1867), which included two novel observations on the healing capacities of tissues rendered uninfected by antiseptic treatment. First, a carbolized blood clot, left undisturbed, became organized into living tissue by ingrowth of cells and blood vessels from surrounding parts. Second, in an aseptic wound, portions of dead bone were absorbed by adjacent granulation tissue. Encouraged by these results, Lister began to apply the antiseptic principle to surgical wounds toward the end of 1866. In August 1867, at the Dublin meeting of the British Medical Association, he announced that during the last nine months his wards—previously "amongst the unhealthiest in the whole surgical division of the Glasgow Royal Infirmary"—had been entirely free from hospital sepsis.

Lister's accomplishments were viewed by his university colleagues with sympathetic interest and by his assistants with unfeigned admiration; but many fellow surgeons, unaccustomed to weighing scientific evidence and prone to polemics (in which Lister never engaged), were indifferent, skeptical, or even hostile. For example, the chief of an adjacent surgical ward at the Royal Infirmary evinced no interest whatever in Lister's patients or methods. Others found his system too exactingly detailed or, overlooking crucial points of technique, performed ambitious operations with disastrous results. His detractors deplored the frequent changes in technique and questioned the significance of such unsettled rituals. They did not realize that every modification followed exhaustive tests, often conducted far into the night in Lister's home laboratory. His "protective" oiled-silk dressing, which allowed secretions to escape while sparing skin and tissues from carbolic acid irritation, took great time and trouble to develop. For three decades Lister sought improvements in the catgut ligature—a quest that began in 1868 at Upton, where he ligated a calf's carotid artery during the last Christmas holidays before his father died. The animal was killed one month later and the catgut absorption process examined microscopically.

Some misunderstandings arose because Lister chose compound fractures for preliminary trials of his antiseptic method, in view of their frequency in heavily industrialized Glasgow and of the extremely high mortality rate resulting from them. To his distress, carbolic acid often was looked upon as a specific nostrum against sepsis, rather than as a potent protective barrier behind which injured parts could exercise their natural recovery powers. He was blamed both when the agent failed through tardy or inadequate usage and when its careless or excessive application caused tissue devitalization or carbolic acid poisoning.

Although he disdained recrimination, Lister felt obliged to refute the allegation of a formidable opponent, Sir James Simpson, that he had plagiarized the work of French and German surgeons, especially of Jules Lemaire, a pharmaceutical chemist of Paris, whose book *De l'acide phénique* (1863) described the medical and surgical uses of carbolic acid. Simpson's bitter attack began with a letter signed "Chirurgicus," published in the *Edinburgh Daily Review* of 23 September 1867. This was reproduced unworthily in *The Lancet*, which a few weeks later published an article by Simpson, "Carbolic Acid and Its Compounds in Surgery," insinuating that Lister was culpably ignorant of medical literature and had merely transplanted an established Continental system of treatment. The latter had never heard of Lemaire but with some difficulty located a copy of his book. In a letter to *The Lancet*, Lister disavowed ever having claimed the first usage of carbolic acid in surgery and dismissed Lemaire's work by stating: "The principles and practice which he mentions are such as sufficiently to explain the insignificance of the results."

Simpson was Syme's rival for leadership of the Edinburgh medical profession; but he resented the antiseptic doctrine chiefly because it could render superfluous his proposal to replace existing large hospitals with small disposable units—in order to combat "hospitalism" or hospital sepsis—as well as his "acupressure" method of arresting hemorrhage, which would remove a common cause of postoperative infection by dispensing with ligatures. According to Lister, efficient antisepsis permitted ligatures to be cut short and left within wounds, instead of dangling purulently outside.

Favorable reports in medical journals now began to counterbalance the opposition. Visitors, including emissaries from European professors, although sometimes shocked by Lister's surgical temerity, were impressed by the results. Students, house surgeons, and assistants, such as Hector Cameron, were his most enthusiastic disciples and increasingly promoted the doctrine. In April 1868, in a well-received review of the antiseptic system for the Medico-Chirurgical Society of Glasgow, illustrated by clinical cases, Lister advocated the transient germicidal potency of the 5 percent aqueous solution of carbolic acid for initially cleansing a wound; the bland, more retentive oily preparation was reserved for external dressing. On this occasion he presented his first modified Pasteurian demonstration of the airborne nature of putrefactive germs. Several months previously he had boiled for five minutes three bent-necked flasks containing fresh urine and a similar flask with attenuated and shortened vertical neck. In the former flasks the urine remained unaltered, but in the other decomposition had occurred and microorganisms abounded. This classic demonstration was used repeatedly to illustrate that putrefaction was not due to atmospheric gases alone but to mechanically arrestive, viable particles floating in the air. (Lister wrote subsequently to his father, describing attempts to exclude airborne germs by sheathing amputation stumps in thin rubber.) Meanwhile, his Royal Infirmary colleagues remained aloof, supported by a management that attributed the lower surgical mortality to improvements in hygiene, diet, and nursing.

In 1866 Lister unsuccessfully applied for the chair of surgery at his alma mater, University College, London. Three years later he submitted his candidacy as successor to Syme (who had suffered a severe stroke), and in August 1869 was elected professor of clinical surgery at Edinburgh. Within a month his father developed a fatal illness. Syme died the following summer, a few weeks after Simpson.

Although Lister left Glasgow amid expressions of regret, shortly after his arrival in Edinburgh a discordant note was struck by publication of his article "On the Effects of the Antiseptic System of Treatment Upon the Salubrity of a Surgical Hospital" (1869). He contended that the prevailing deplorable conditions at the Glasgow Royal Infirmary, aggravated on the ground floor by adjacent pit burials of victims of the 1849 cholera epidemic, should be contrasted with the healthy conditions brought about in his men's accident ward on that floor by antiseptic treatment. On this and other occasions when Lister's ingenuous candor and reforming zeal bruised the sensitivities of others, he was apologetic and sought to make amends without sacrifice of principle or dilution of precept.

The next eight years in Edinburgh were the happiest of his life. He was summoned and warmly welcomed by faculty and students as the exponent of a liberating new doctrine; both his wisest counselor and fiercest opponent had died; and he had assured respect and independence. Comparatively wealthy since his father's death, he bought a costly house in the finest square and eventually built up an extensive practice. Less onerous university duties allowed Lister to devote himself to perfecting and propagating his antiseptic system. This demanded unstinting personal attention to patients of every class, besides greatly expanded laboratory researches. Most of his important work on pure and applied bacteriology was done during these Edinburgh years.

In his introductory lecture Lister reviewed the researches of Pasteur and his precursors on atmospheric germs, emphasizing their role in putrefaction without mentioning antiseptic surgery. He thus challenged the claims of the professor of medicine at Edinburgh, John Hughes Bennett, a microscopist who (like Sir James Simpson) ridiculed microbes as "mythical fungi." Lister's belief in atmospheric bacteria as the principal source of wound contamination, reinforced by John Tyndall's essay "On Dust and Disease," led him in 1871 to recommend carbolic spray as a means of pervading the vicinity of surgical wounds with antiseptic. During the next decade various types of atomizers were added to the operating room paraphernalia. Hand sprays evolved into foot sprays, both types requiring relays of perspiring assistants. The "donkey engine," a long-handled model mounted on a tripod, was easier to manipulate but very cumbersome. Eventually, a portable "steam spray" was manufactured in large numbers. These devices kept the patient, surgeon, and assistants in intimate contact with droplets of 1:40 aqueous solution of carbolic acid, which were inevitably inhaled, sometimes with toxic effects, although Lister denied that any of his patients suffered carbolic acid poisoning. His own skin was delicate, yet it tolerated the acid well. The operator's hand went white and numb, however, and the ultrasensitive found the regimen impossible. Although Lister never tested the spray's action experimentally, from the first he regarded it as a necessary evil. Eventually growing skeptical, he abandoned it in 1887.

Lister's time and ingenuity were now devoted mainly to the improvement of wound dressings and to various bacteriological researches. Because carbolic acid was evidently an irritant, liable to interfere with tissue healing, he developed an impermeable "protective" of oiled silk covered with copal varnish for direct application to the wound. This was covered by an antiseptic plaster of four parts shellac and one part carbolic acid. He summarized the underlying principle thus in 1870: "An antiseptic to exclude putrefaction, with a protective to exclude the antiseptic, will by their joint action keep the wound from abnormal stimulus." Soon afterwards he replaced the nonabsorbent lac plaster with carbolized, absorbent muslin gauze, which remained in vogue for many years. Besides the spray and dressings, in 1871 Lister introduced into British practice the rubber drainage tube, invented by P. M. E. Chaussaignac in France twelve years before. Shortly after becoming surgeon in ordinary in Scotland to Queen Victoria, he treated her at Balmoral for an axillary abscess, obtaining good results after inserting a small drainage tube into the incision.

The elaborate safeguards for wounds, devised on the supposition that "after being exposed even for a second to the influence of septic air, putrefaction would be pretty certain to occur," contrasted with Lister's simple personal precautions in the operating room. Removing his coat and rolling up his sleeves, he pinned a large towel (clean but unsterilized) over his waistcoat and trousers. He wore neither gown, nor mask, nor gloves. Two strengths of carbolic acid solution, 1:20 and 1:40, were kept in trays and basins, the stronger being used for preliminary hand washing, for cleansing the patient's skin at the operative site, and for immersing instruments. The weaker lotion, held in saturated sponges, served for frequent hand-dipping during the operation.

Many of Lister's microbiological studies, recorded on more than 400 foolscap sheets closely written by his wife or himself, were tests of his antiseptic system and methods, not intended for publication. Besides his 1868 address at Glasgow and the inaugural lecture "On the Causation of Putrefaction and Fermentation" (1869), his *Collected Papers* include six reports under the heading "Bacteriology," two dating from this

Edinburgh period. His lengthy address to the Royal Society of London in 1873, on the germ theory of putrefaction and other fermentative changes, was largely provoked by John Burdon-Sanderson's report "On the Origin and Distribution of Microzymes (Bacteria) in Water . . ." (1871), which claimed that bacteria are conveyed by water but not by air, and are killed by simple drying at 100° F. These contentions, if true, would have nullified Lister's efforts to provide an "antiseptic atmosphere" in surgical practice. In unboiled human urine taken with antiseptic precautions—a better medium than Pasteur's 10 percent sucrose and yeast ash solution employed by Burdon-Sanderson—bacteria grew when drops of tap water were added and also after this medium had been exposed for some hours to room air. The resulting aerial flora included a yeast and a filamentous fungus. Lister mistakenly concluded that the latter could develop into either the yeast or the bacteria.

He renewed this pleomorphist interpretation when subsequently reporting the behavior of two bacterial species and a filamentous fungus that developed in an exposed milk sample. One of the bacterial species soured and curdled milk. Early in 1874 Lister sent his article to Pasteur, who had noted this phenomenon in 1857. The resulting correspondence initiated a lifelong mutual admiration. Pasteur's delicately suggested explanation of these findings was duly acknowledged: "Next to the promulgation of new truth, the best thing, I conceive, that a man can do, is the recantation of a published error."

Some false starts notwithstanding, Lister made several fruitful contributions to bacteriology. He noted morphological and fermentative variations undergone by microorganisms in different nutrient media, and he observed that bacterial metabolites of a proteinaceous fluid could be odorless, or devoid of putrefactive smell. He devised novel apparatus—lidded glassware, a sterilizable syringe-pipette, and the "hot box," a pioneer autoclave providing diffuse dry heat of 300° F. By confirming, extending, and unequivocally sponsoring Pasteur's fundamental contentions, he induced many British colleagues to reappraise the germ theory or to launch fresh inquiries.

Meanwhile, antiseptic surgery gained more adherents abroad than at home. In England its chief proponents were younger surgeons at hospitals in large provincial cities, such as Liverpool, Birmingham, and Manchester. In Scotland, Listerian techniques yielded highly successful results at Glasgow Royal Infirmary in the mid-1870's, when Hector Cameron and William Macewen became surgeons there. Alexander Ogston of Aberdeen, who later discovered the pyogenic staphylococcus, was a staunch supporter. At Edinburgh the chief sympathizer among the senior staff was Thomas Keith, who with T. Spencer Wells of London had achieved a remarkably low mortality rate in ovariotomy, mainly through scrupulous attention to surgical cleanliness. After he adopted antiseptic techniques, Keith's results improved. In London, with few exceptions, apathy had given place to opposition. This trend, encouraged by the influential *The Lancet*, stemmed from various factors, including plain prejudice, misconceptions about details of the antiseptic ritual, Florence Nightingale's campaign to improve hospital hygiene, and H. Charlton Bastian's revival of the spontaneous-generation obsession.

In Germany, Karl Thiersch of Leipzig successfully adopted Lister's system as early as 1867. A simplified form of antiseptic treatment for battle wounds in the Franco-Prussian War gave disappointing results; but Richard von Volkmann became a doughty devotee after 1872, when his hospital at Halle, overcrowded with wounded soldiers and so dreadfully infected that its closure was imminent, obtained astonishing benefits from Listerian techniques. Other prominent supporters included J. von Nussbaum of Munich, A. Bardeleben and A. W. Schultze of Berlin, and Friedrich von Esmarch of Kiel. The particularly strong support for Lister's methods in Germany was attributed by some British surgeons to that country's less advanced state of sanitary science.

Among the earliest foreign visitors to become enthusiastic disciples were M. H. Saxtorph of Copenhagen and J. Lucas-Championnière of Paris (author of the first manual of antiseptic surgery). Other European adherents included Theodor Kocher in Bern and J. W. R. Tilanus in Amsterdam. Enrico Bottini's use of carbolic acid as a surgical antiseptic at Novara, Italy, reported in "Dell'acido fenico nella chirurgia pratica e nella tassidermia" (1866), was unknown to Lister and was generally ignored.

The reality of Lister's fame abroad became obvious in 1875, during a much feted European tour with his wife, his brother Arthur, and other relatives. *The Lancet* termed his progress through the university towns of Germany "a triumphal march." In America antiseptic surgery made slow progress until Lister personally expounded his doctrine, as president of the Surgical Section, to the 1876 International Medical Congress at Philadelphia. Again accompanied by his wife and brother, he paid pre-Congress visits to Montreal and Toronto, subsequently crossed the United States to San Francisco, and met enthusiastic receptions in Boston and New York.

Meanwhile, Lister campaigned against the antivivisectionists, whose allegations caused the appointment of a royal commission in 1875. To the queen's plea that he should condemn vivisection, he responded by declaring "legislation on this subject is wholly uncalled for." He deplored the inconsistency of those who approve fox-hunting but brand as cruel the animal experimenter who takes "every care to render the pain as slight as is compatible with the high object in view"; and he reaffirmed his stand in evidence before the commission. Appointed to the General Medical Council in 1876, Lister was designated chairman of its committee to report on the Cruelty to Animals Bill before Parliament. The final act contained relaxations for which Lister was largely responsible.

Between 1870 and 1876 Lister published ten papers on antiseptic surgery, including a major address to the British Medical Association at Plymouth in 1871, and presented masterful demonstrations before that organization at Edinburgh in 1875. Notwithstanding these evangelistic efforts and successful European and North American missions, the persistent hostility of many London surgeons represented a challenge to Lister (himself a Londoner) to carry his gospel to the metropolis. The opportunity came early in 1877, when he was approached as possible successor to Sir William Fergusson, titular professor of surgery at King's College, who had died. His students, hearing rumors, presented a complimentary memorial with 700 signatures, begging him to remain. An impromptu response, in which Lister incautiously called the London system of teaching clinical surgery "a mere sham" that largely neglected "magnificent opportunities of demonstrative teaching" appeared the next day in the Edinburgh and London newspapers. When castigated by *The Lancet*, Lister temperately explained that his words were not intended as a personal affront. Negotiations were resumed, and in June he was elected to a newly created chair of clinical surgery.

Lister's brilliant interlude in Edinburgh ended in his fiftieth year, without fuss or fanfare. In London he settled in Park Crescent, near Regent's Park and the Botanical Gardens, where he liked to roam and meditate. A nursing home for private patients was also close. Among the four-man team accompanying him were Watson Cheyne and John Stewart of Halifax, Nova Scotia, who as house surgeons disconsolately faced two dozen empty beds at King's College Hospital instead of six wards and up to seventy patients at Edinburgh. The nursing sisters of St. John were uncooperative because the Listerian regimen conflicted with their time-honored routine. Eager lecture audiences of more than 400 auditors had dwindled to fewer than twenty listless students who feared examination penalties for airing antiseptic doctrines.

Without complaining, Lister overcame student apathy, nursing obstructiveness, and much professional opposition. James Spence, his former Edinburgh rival, and Robert Lawson Tait, the Birmingham gynecologist, remained antagonistic; but the latter's truculence was largely neutralized by support from another expert ovariotomist, Spencer Wells. During a debate on antiseptic surgery at St. Thomas' Hospital in December 1879, Lister fulfilled William Savory's demands, echoed in *The Lancet*, for statistical data justifying his claims. Thereafter the few loyal but inarticulate followers of "Listerism" in London, backed by the *British Medical Journal*, were reinforced by the outspoken advocacy of such leading surgeons as William MacCormac, Jonathan Hutchinson, John Wood, and (a late convert) Sir James Paget.

Lister's complete sincerity, pertinacious but conciliatory approach, and dramatic surgical achievements—such as the successful wiring of a broken kneecap—largely account for this transformation. Advances in bacteriology also strengthened the antiseptic doctrine. His introductory address at King's College, "The Nature of Fermentation," inferentially clinched the microbic etiology of putrefaction by linking the lactic acid fermentation of milk to growth of a specific bacillus, which he termed *Bacterium lactis*. With small inocula of this microorganism, prepared in pure culture by a novel dilution method, he curdled boiled milk at will. An expanded lecture-demonstration on the lactic fermentation before the Pathological Society of London late in 1877 was his last major contribution to bacteriological research. For many years thereafter Lister maintained his laboratory for testing antiseptics, read the foreign literature punctiliously, and occasionally reviewed developments in microbiology; but he found restrictions on animal experiments a handicap and left the field to specialists.

His prestige at home was further enhanced by his celebrity abroad. Lister visited Paris in 1878 as president of the jury on medical matters at the Universal Exhibition, lectured (in French) at the Academy of Medicine, and met Pasteur. At the 1879 International Medical Congress in Amsterdam, he was acclaimed with unprecedented enthusiasm. He received honorary doctorates from Oxford and Cambridge in 1880. The queen appointed him surgeon-in-ordinary in 1878 and conferred a baronetcy on him in 1883. Among the earliest foreign honors awarded him were the Boudet Prize (1881), for his application of Pasteur's researches to the healing art, and the Prussian Ordre pour le Mérite (1885).

Robert Koch's discoveries on the etiology of anthrax and wound infections (1876–1878) evoked Lister's admiration. Correspondence ensued, and the *Wundinfectionskrankheiten* monograph was translated into English by Cheyne in 1880. Lister summarized this work, and Pasteur's early experiments on immunity against fowl cholera, in masterly addresses on the relation of microorganisms to disease and inflammation, given at Cambridge and at the 1881 International Medical Congress in London. The Surgical Section of the Congress sponsored a symposium on the treatment of wounds, in which Lister stressed the defensive powers of the blood and tissues and spoke diffidently about the carbolic spray, lately declared superfluous in an article by P. Bruns of Tübingen, arrestingly titled "Fort mit dem Spray!" Both Pasteur and Koch attended the congress and at Lister's instigation met at King's College, where the latter demonstrated his gelatin medium for cultivating and isolating pure cultures.

Koch's initial report on disinfection (1881) revealed that the antiseptic supremacy of carbolic acid was overestimated, at least as regards anthrax spores, on which mercuric chloride had far greater "disinfectant" (bactericidal) and "antiseptic" (bacteriostatic) actions. In 1884 Lister adopted external dressings impregnated with mercuric chloride; but these proved too irritating, and after five years of laboratory tests and manufacturing problems he recommended that the outer gauze dressing should contain the double cyanide of mercury and zinc. The frequent modifications in Listerian technique were now approved by *The Lancet*, as evidence that the paramount issue was the principle of antisepsis, rather than the peculiar virtues of any given antiseptic. Lister kept faith in his favorites, however; an address in 1893, "The Antiseptic Management of Wounds," extolled the manifold virtues of 1:20 carbolic acid.

Developments in some German hospitals and clinics took another direction. Further studies from Koch's laboratory, evaluating the bactericidal power of dry heat and the greater efficiency of steam sterilization for inert objects, coincided with increased awareness (anticipated earlier by Lister) that chemical antiseptics are liable to damage natural healing mechanisms, notably the phagocytic phenomenon first described in 1883 by Elie Metchnikoff. Berlin's leading surgeon, Ernst von Bergmann, and his assistant Curt Schimmelbusch became apostles of "aseptic" surgery. They proclaimed great respect for Lister, although the ritualistic details and expensive paraphernalia published by Schimmelbusch in a text on aseptic surgery (1892) tended to disparage Listerism as outmoded.

To satisfy himself that the full-fledged aseptic tenets were feasible, Lister removed a tumor successfully "without contact of any antiseptic material with the wound." He deplored the pretense that aseptic surgery completely obviates antiseptics and contended that its techniques, to achieve results comparable with his own, entailed greater care and trouble, as well as compelling hospitalization. He considered the terms "antisepsis" and "asepsis" to be interchangeable and kept aloof from extremist viewpoints, but he unwaveringly upheld the basic doctrine common to both systems. On details of antiseptic technique he was more pliable. The spray was renounced when he realized that the great majority of atmospheric microbes were neither pathogenic nor eliminated by carbolic acid droplets. In his address to the 1890 International Medical Congress in Berlin, "The Present Position of Antiseptic Surgery," he commented: "As regards the spray, I feel ashamed that I should ever have recommended it. . . ." The tributes paid him at this congress nevertheless proved "almost overpowering." Next year, at the International Congress of Hygiene in London, Lister presided over the exceptionally well-attended Bacteriology Section; and a galaxy of delegates dined at Park Crescent. In 1892 he reached retirement age and relinquished the King's College chair, but agreed to retain charge of his wards for another year. That December he represented the Royal Societies of London and Edinburgh at the Sorbonne ceremonies commemorating Pasteur's jubilee. The artist J. A. Rixens captured the memorable scene as Lister stepped forward to embrace Pasteur after completing his eulogy.

Lister and his wife took frequent holidays in the British Isles and on the Continent. They enjoyed collecting flowers and studying wildlife, and kept diaries about these interludes. In the spring of 1893, at Rapallo, Lady Lister contracted pneumonia and died four days later, leaving her husband bereft of all intimate daily companionship and overwhelmed by sudden grief. He fulfilled the commitment to King's College Hospital; but his private practice almost vanished, his laboratory experiments languished, and he generally avoided social gatherings. Lucy Syme, a sister-in-law who had often stayed with them, lightened the solitude by keeping house for him. Although Lister never ceased to mourn his wife, personal sadness was not allowed to interfere with public duty. In the following decade he filled many highly responsible offices with appropriate dignity and learning.

Elected foreign secretary of the Royal Society late in 1893, Lister became president two years later. His wide reading, logical exposition, and acute grasp of public health issues were apparent in five annual presidential addresses, which dealt mainly with developments in medical and veterinary sciences, particularly in microbiology. In 1896 Lister's presidential address to the British Association, "The Interdependence of Science and the Healing Art," included his first public explanation of Pasteur's scientific influence upon his surgical practices, and also his last major allusion to antisepsis.

Lister's concern for public health found expression in his persistent efforts to secure establishment of a British Institute of Preventive Medicine. Despite irrational opposition and studied indifference, it was incorporated in 1891 with Lister as chairman; and by 1895 it was producing diphtheria antitoxin under Armand Ruffer's direction. In 1897, the centenary of the discovery of vaccination, Jenner's name was attached to the Institute. Since this caused conflict with a vaccine lymph manufacturer, in 1903 it was finally named after Lister, who was its governing board's first chairman and president for several years.

He was raised to the peerage in 1897, assuming the title Baron Lister of Lyme Regis, a small Dorsetshire town where he and his brothers had purchased a seaside house many years before. That autumn Lister attended the annual meetings of the British Association in Toronto and the British Medical Association in Montreal, afterward traveling across Canada by special railroad car with his brother and nieces. His final journey abroad was a voyage to South Africa in the winter of 1901–1902; but he continued to visit Lyme Regis, Buxton, and other health resorts in Britain, hoping to relieve his increasing rheumatic afflictions. Now a greatly venerated figure, he emerged from retirement only on such special occasions as that of his last great public address, the third Huxley lecture (1900)—a sweeping retrospect of his early physiological and pathological researches—or for his resourceful chairmanship of the session of the Second Tuberculosis Congress in London (1901), when Koch asserted that bovine tuberculosis was a negligible hazard to human health. Disappointed at Koch's premature disclosure of tuberculin ten years before, Lister unhesitatingly revealed flaws in his present argument.

In 1903, at Buxton, Lister apparently suffered a slight stroke, which hampered walking and mental effort; but he recovered sufficiently to deal with matters of personal import. For example, in 1906 he refuted the allegation that his antiseptic system was derived from the work of I. P. Semmelweis; and in 1907 he actively cooperated with friends in editing his *Collected Papers* to memorialize his eightieth birthday. This anniversary was celebrated in many countries and Lister received innumerable congratulatory messages. His last public appearance was at the Guildhall a few weeks later, to receive the freedom of the City of London.

Lister and his sister-in-law moved in 1908 to the small town of Walmer, on the Kentish coast. The bracing air, however, failed to renew his strength; his sight and hearing became impaired; and he fell into a gradual, prolonged decline. The end came almost imperceptibly. A widespread desire that Lister should be buried in Westminster Abbey was overridden by his own wish to be interred beside his wife in West Hampstead Cemetery. An impressive funeral service was nonetheless held in the Abbey, where a medallion by Sir Thomas Brock, one of the finest of Lister's many portraits and busts, commemorates his fame.

During later years Lister's multiplying tributes included the freedom of Edinburgh, Glasgow, and London; honorary doctorates from many British and foreign universities; corresponding or honorary membership in some sixty scientific and medical societies in various countries; and the Copley and several other medals. He became sergeant-surgeon to Queen Victoria in 1900 and, at his accession, to King Edward VII, who appointed Lister to the newly instituted Order of Merit and also to the Privy Council.

In analyzing Lister's personality, his chief biographer, Sir Rickman Godlee, was restrained by his uncle's expressed desire for a simple record of his contributions to science and surgery. His faults were trivial and easily explained. He was unpunctual; he often greatly exceeded the allotted time in addresses and discussions; and he embarrassed patients and professional colleagues by refusing to specify his fees. He epitomized himself in a letter to his father soon after first arriving in Edinburgh: "I am by disposition very averse to quarrelling and contending with others," he wrote, "but at the same time I do love honesty and independence." The rare nobility of his character inspired one of his ward patients, the poet W. E. Henley, to express his admiration in the sonnet "The Chief," which compared him to Hercules, "Battling with custom, prejudice, disease. . . ." A fellow surgeon, Sir Frederick Treves, provided a fitting epitaph for his accomplishments:

> Lister created anew the ancient art of healing; he made a reality of the hope which had for all time sustained the surgeon's endeavours; he removed the impenetrable cloud which had stood for centuries between great principles and successful practice, and he rendered possible a treatment which had hitherto been but the vision of the dreamer.

BIBLIOGRAPHY

I. Original Works. The only ed. of Lister's works is *The Collected Papers of Joseph, Baron Lister*, 2 vols. (Oxford, 1909). This was prepared by a committee consisting of Hector C. Cameron, W. Watson Cheyne, R. J. Godlee, C. J. Martin, and D. Williams, in response to the widely expressed desire, on Lister's eightieth birthday, for some appropriate memorial of "a life so rich in benefits to mankind." The committee was advised by Lister on the selection of papers and addresses—about half of all his publications—which he himself thought possessed permanent interest and importance. Vol. I contains 16 contributions to physiology (pt. 1), and 9 to pathology and bacteriology (pt. 2). In vol. II are 26 publications on the antiseptic system (pt. 3), and 6 on surgery (pt. 4), while pt. 5 comprises 4 addresses and communications on miscellaneous topics, including an obituary tribute to his father.

Lister's publications were incompletely compiled by C. W. W. Judd in his prize essay on the life and work of Lister (see below); by J. Chiene in the Lister no. of the *British Medical Journal* (1902), **2**, 1853–1854; and by C. N. B. Camac in *Epoch-Making Contributions to Medicine, Surgery and the Allied Sciences* (London, 1909), appended to a repr. of "On the Antiseptic Principle in the Practice of Surgery," pp. 9–22. A more extensive bibliography, supplemented by biographical articles, precedes (pp. 9–27) the repr. of three of Lister's papers on antisepsis in *Medical Classics*, **2** (1937–1938), 28–101. The *Souvenir Handbook of the Lister Centenary Exhibition at the Wellcome Historical Medical Museum* (London, 1927), pp. 155–166, offers the most complete record of his published work, numbering more than 120 items. None of these sources is free from errors.

Reports on researches in physiology and pathology include "Observations on the Contractile Tissue of the Iris," in *Quarterly Journal of Microscopical Science*, **1** (1853), 8–17; "Observations on the Muscular Tissue of the Skin," *ibid.*, 262–268; "On the Flow of Lacteal Fluid in the Mesentery of the Mouse," in *Report of the 27th Meeting of the British Association for the Advancement of Science* (Dublin, 1857), 114; "On the Minute Structure of Involuntary Muscle Fibres," in *Transactions of the Royal Society of Edinburgh*, **21** (1857), 549–557; "On the Early Stages of Inflammation," in *Proceedings of the Royal Society*, **8** (1857), 581–587; "An Enquiry Regarding the Parts of the Nervous System Which Regulate the Contractions of the Arteries," in *Philosophical Transactions of the Royal Society*, **148** (1858), 607–625; "On the Cutaneous Pigmentary System in the Frog," *ibid.*, 627–643; "Spontaneous Gangrene From Arteritis and the Causes of Coagulation of the Blood in Diseases of the Blood-Vessels," in *Edinburgh Medical Journal*, **3** (1858), 893–907; "Preliminary Account of an Enquiry Into the Functions of the Visceral Nerves, With Special Reference to the So-Called 'Inhibitory System,'" in *Proceedings of the Royal Society*, **9** (1859), 367–380; "Some Observations on the Structure of Nerve-Fibres," in *Quarterly Journal of Microscopical Science*, **8** (1860), 29–32, written with W. Turner but with "Supplementary Observations" by Lister, 32–34; "Notice of Further Researches on the Coagulation of the Blood," in *Edinburgh Medical Journal*, **5** (1860), 536–540; "Coagulation of the Blood," in *Proceedings of the Royal Society*, **12** (1863), 580–611, the Croonian lecture; and "Address on the Value of Pathological Research," in *British Medical Journal* (1897), **1**, 317–319.

Lister's earlier papers on the antiseptic treatment in surgery include "On a New Method of Treating Compound Fracture, Abscess, etc., With Observations on the Conditions of Suppuration," in *Lancet* (1867), **1**, 326–329, 357–359, 387–389, 507–509, and **2**, 95–96; "On the Antiseptic Principle in the Practice of Surgery," *ibid.*, 353–356; "On the Antiseptic Treatment in Surgery," in *British Medical Journal* (1868), **2**, 53–56, 101–102, 461–463, 515–517, and (1869), **1**, 301–304—these three reports are repr. in German in Karl Sudhoff's Klassiker der Medizin, no. 17 (Leipzig, 1912); "Observations on Ligature of Arteries on the Antiseptic System," in *Lancet* (1869), **1**, 451–455; "On the Effects of the Antiseptic System Upon the General Salubrity of a Surgical Hospital," *ibid.* (1870), **1**, 4–6, 40–42; "The Glasgow Infirmary and the Antiseptic Treatment," *ibid.*, 210–211, letter to the editor; "Remarks on a Case of Compound Dislocation of the Ankle, With Other Injuries, Illustrating the Antiseptic System of Treatment," *ibid.*, 404–406, 440–443, 512–513; "Further Evidence Regarding the Effects of the Antiseptic Treatment Upon the Salubrity of a Surgical Hospital," *ibid.*, **2**, 287–289; "A Method of Antiseptic Treatment Applicable to Wounded Soldiers in the Present War," in *British Medical Journal* (1870), **2**, 243–244; "On Recent Improvements in the Details of Antiseptic Surgery," in *Lancet* (1875), **1**, 365–367, 401–402, 434–436, 468–470, 603–605, 717–719, 787–789; and "Clinical Lecture Illustrating Antiseptic Surgery," *ibid.* (1879), **2**, 901–905.

Lister's chief publications on pure and applied bacteriology are "Introductory Lecture" (delivered at the University of Edinburgh, 8 Nov. 1869), in *British Medical Journal* (1869), **2**, 601–604; "The Address in Surgery" (at the 39th annual meeting of the British Medical Association, Plymouth), *ibid.* (1871), **2**, 225–233; "A Further Contribution to the Natural History of Bacteria and the Germ Theory of Fermentative Changes," in *Quarterly Journal of Microscopical Science*, n.s. **13** (1873), 380–408; "A Contribution to the Germ Theory of Putrefaction and Other Fermentative Changes, and to the Natural History of Torulae and Bacteria," in *Transactions of the Royal Society of Edinburgh*, **27** (1875), 313–344; "On the Nature of Fermentation," in *Quarterly Journal of Microscopical Science*, n.s. **18** (1878), 177–194; "On the Lactic Fermentation, and Its Bearings on Pathology," in *Transactions of the Pathological Society of London*, **29** (1878), 425–467; "On the Relation of Microorganisms to Disease," in *Quarterly Journal of Microscopical Science*, n.s. **21** (1881), 330–342; "On the Relations of Minute Organisms to Unhealthy Processes Arising in Wounds, and to Inflammation in General," in *Transactions of the 7th International*

Medical Congress, I (London, 1881), 1311–1319; and "The Causes of Failure in Obtaining Primary Union in Operation Wounds, and on the Methods of Treatment Best Calculated to Secure It" (discussion), *ibid.*, II, 369–383. (Many of the reports in this and the preceding paragraph appeared in French in G. Borginon, trans., *Oeuvres réunies de chirurgie antiseptique et théorie des germes* [Paris, 1882].

Reports of progress in antiseptic techniques include "Corrosive Sublimate as a Surgical Dressing," in *Lancet* (1884), **2**, 723–728; "A New Antiseptic Dressing," *ibid.* (1889), **2**, 943–947; "On Two Cases of Long-Standing Dislocation of Both Shoulders Treated by Operation; With Further Observations on the Cyanide of Zinc and Mercury," *ibid.* (1890), **1**, 1–4; "An Address on the Present Position of Antiseptic Surgery," in *British Medical Journal* (1890), **2**, 377–379; "An Address on the Antiseptic Management of Wounds," *ibid.* (1893), **1**, 161–162, 277–278, 337–339; "On Early Researches Leading up to the Antiseptic System of Surgery," in *Lancet* (1900), **2**, 985–993, the Huxley lecture; "Note on the Preparation of Catgut for Surgical Purposes," *ibid.* (1908), **1**, 148–149; and "Remarks on Some Points in the History of Antiseptic Surgery," *ibid.*, 1815–1816.

Important and characteristic writings on surgery are "Report of Some Cases of Articular Disease Occurring in Mr. Syme's Practice, Illustrating the Advantages of the Actual Cautery," in *Monthly Journal of Medical Science*, **19** (1854), 134–137; "On Excision of the Wrist for Caries," in *Lancet* (1865), **1**, 308–312, 335–338, 362–364; "Clinical Lecture on a Case of Excision of the Knee-Joint, and on Horsehair as a Drain for Wounds; With Remarks on the Teaching of Clinical Surgery" (delivered at King's College Hospital, 10 Dec. 1877), in *Lancet* (1878), **1**, 5–9; "A Case of Multiple Papillomatous Growths in the Larynx, Extirpated by Complete Laryngotomy; Removal of the Whole Length of Both True and False Vocal Cords; Preservation of the Voice; Co-existence of Thoracic Aneurism," in *Transactions of the Clinical Society of London*, **11** (1878), 104–113, written with J. B. Yeo; "An Address on the Treatment of Fracture of the Patella," in *British Medical Journal* (1883), **2**, 855–860; "Amputation," in Holmes's *System of Surgery*, 3rd ed., III (London, 1883); "Anaesthetics" (pt. 1 written in 1861, pt. 2 in 1870, pt. 3 in 1882), *ibid.*; "Remarks on the Treatment of Fractures of the Patella of Long Standing," in *British Medical Journal* (1908), **1**, 849–850.

Miscellaneous writings of lasting interest include "Obituary Notice of the Late Joseph Jackson Lister, F.R.S., With Special Reference to His Labours in the Improvement of the Achromatic Microscope," in *Monthly Microscopical Journal*, **3** (1870), 134–143; "On the Coagulation of the Blood in Its Practical Aspects," in *British Medical Journal* (1891), **1**, 1057–1060; "On the Relations of Clinical Medicine to Modern Scientific Development," *ibid.* (1896), **2**, 733–741; "Presidential Address Before the British Association for Advancement of Science," in *Science*, n.s. **4** (1896), 409–429; "Presidential Address at the Anniversary Meeting of the Royal Society," in *Year-Book of the Royal Society* (London, 1900), 144–156; "On Recent Researches With Regard to the Parasitology of Malaria," in *British Medical Journal* (1900), **2**, 1625–1627; and the intro. to Stephen Paget's *Experiments on Animals* (London, 1900), pp. xi–xii.

Many of Lister's papers were published also in pamphlet form, or in more than one journal. Some appeared in foreign medical periodicals, trans. into German, French, or Italian. Handwritten letters from him are heirlooms in many families and are treasured by numerous institutions. Interesting small collections of his letters are in the Royal Faculty of Physicians and Surgeons, Glasgow, and the Osler Library, McGill University, Montreal. The former includes 14 unpub. letters (1897–1906) to Peter Paterson (later professor of surgery, Glasgow Royal Infirmary) about induced immunity to tuberculosis in experimental animals. Of the Osler Library letters, written to A. E. Malloch of Hamilton, Ontario, a former house surgeon at Glasgow Royal Infirmary, the earliest, dated 10 Sept. 1868, gives detailed instructions on treatment of a ward case and is reproduced in Godlee's biography of Lister. The others, written during his last decade of life, thank Malloch for Canadian apples sent each Christmas to his former chief. Lister's letters (1857–1864) to W. Sharpey, professor of physiology at University College, mostly concerning researches on inflammation, were published by C. R. Rudolf, "Eight Letters of Joseph (Lord) Lister to William Sharpey," in *British Journal of Surgery*, **20** (1932), 145–164, 459–466.

Memorabilia of Lister are in many institutions, especially Edinburgh and Glasgow universities, and King's College Hospital, London. The richest collections are at the Wellcome Institute of the History of Medicine, comprising some of his holiday diaries, manuscript notes on clinical surgery, and a volume of about 100 autograph letters; and at the Royal College of Surgeons of England, where the Lister Cabinet contains his "Common Place Book" (3 vols.), case notes, and other MSS, along with surgical instruments.

II. Secondary Literature. Tributes to Lister's work during his lifetime include the editorials "Professor Lister in Germany," in *Lancet* (1875), **1**, 868; "The Surgical Use of Carbolic Acid," *ibid.*, **2**, 234–235; and "Lord Lister and the Antiseptic Method," *ibid.* (1909), **1**, 1617–1620; J. Finlayson, "Lord Lister and the Development of Antiseptic Surgery," in *Janus*, **5** (1900), 1–5, 57–63; L. L. Hill, "Some Personal Reminiscences of Lord Lister," in *Medical Record*, **80** (1911), 327–329; C. C. W. Judd, "The Life and Work of Lister," in *Bulletin of the Johns Hopkins Hospital*, **21** (1910), 293–304, the Lister Prize essay; W. W. Keen, "Lister on the Use of Animals in Research," in *Journal of the American Medical Association*, **68** (1917), 53; H. Tillmanns, "Sir Joseph Lister," in *Nature*, **54** (1896), 1–3, followed by an editorial on Lister, pp. 3–5; and Frederick Treves, "The Progress of Surgery," in *Practitioner*, **58** (1897), 619–630.

The Lister jubilee no. of the *British Medical Journal* (1902), **2**, 1817–1861, commemorates the fiftieth anniversary of his entering the medical profession. Among the more noteworthy contributions are T. Annandale, "Early

Days in Edinburgh," pp. 1842–1843; O. Bloch, "On the Antiseptic Treatment of Wounds," pp. 1825–1828; Hector C. Cameron, "Lord Lister and the Evolution of Modern Surgery. Glasgow, 1861–1869," pp. 1844–1848; W. W. Cheyne, "Listerism and the Development of Operative Surgery," pp. 1851–1852; J. Chiene, "Edinburgh Royal Infirmary, 1869–1877," pp. 1848–1851; the editorial "The Listerian System," pp. 1854–1861; J. Lucas-Championnière, "An Essay on Scientific Surgery. The Antiseptic Method of Lister in the Present and in the Future," pp. 1819–1821; and A. Ogston, "The Influence of Lister Upon Military Surgery," pp. 1837–1838.

Obituaries include W. Watson Cheyne, "Lord Lister, 1827–1912," in *Proceedings of the Royal Society*, **86B** (1912–1913), i–xxi; J. Stewart, "Lord Lister," in *Canadian Journal of Medicine and Surgery*, **31** (1912), 323–330; Frederick Treves, "Lister," in *London Hospital Gazette*, **15** (1911–1912), 171–172; and also (unsigned) "Death of Lord Lister," in *Boston Medical and Surgical Journal*, **166** (1912), 301–302; "Lord Lister," in *British Medical Journal* (1912), **1**, 397–402; "Lister," in *Glasgow Medical Journal*, **77** (1912), 190–196; and "Lord Lister, O.M.," in *Lancet* (1912), **1**, 465–472. Accounts of the obsequies are "Funeral of Lord Lister," in *British Medical Journal* (1912), **1**, 440–446, followed by the editorial "The Maker of Modern Surgery," pp. 447–448; and William Osler, "The Funeral of Lord Lister," in *Canadian Medical Association Journal*, **2** (1912), 343–344.

Among foreign-language obituaries are H. Coenen, "Joseph Lister † 1827–1912," in *Zeitschrift für Ärtzliche Fortbildung*, **9** (1912), 161–164; P. Daser, "Lord Lister," in *Münchener medizinische Wochenschrift*, **59** (1912), 480–481; A. Fraenkel, "Gedenkrede auf Lord Josef Lister," in *Wiener klinische Wochenschrift*, **25** (1912), 381–386; O. Lanz, "Joseph Lister †," in *Nederlandsch Tijdschrift voor Geneeskunde* (1912), **1**, 425–428; J. Lucas-Championnière, "Mort de Lord Lister," in *Journal de médecine et chirurgie pratique*, **83** (1912), 129–135; and F. Trendelenburg, "Zur Erinnerung an Joseph Lister," in *Deutsche medizinische Wochenschrift*, **38** (1912), 713–716.

Reminiscences and posthumous tributes include those of Hector C. Cameron, *Lord Lister, 1827–1912. An Oration Delivered in the University of Chicago on Commemoration Day, 23rd June, 1914* (Glasgow, 1914); W. Watson Cheyne, "Lister, the Investigator and Surgeon," in *British Medical Journal* (1925), **1**, 923–926; A. E. Malloch, "Personal Reminiscences of Lister," in *Canadian Medical Association Journal*, **2** (1912), 502–506; J. Stewart, "First Listerian Oration," *ibid.*, **14** (1924), 1011–1040; and St. Clair Thomson, "A House-Surgeon's Memories of Joseph Lister (Born April 5, 1827. Died February 10, 1912)," in *Annals of Medical History*, **2** (1919), 93–108.

The centenary of Lister's birth occasioned many biographical articles in medical and scientific journals, including A. P. C. Ashhurst, "Centenary of Lister: Tale of Sepsis and Antisepsis," in *Annals of Medical History*, **9** (1927), 205–221; John Bland-Sutton, "The Conquest of Sepsis," in *Lancet* (1927), **1**, 781; W. Bulloch, "Lord Lister as a Pathologist and Bacteriologist," *ibid.*, 744–746; D. Cheever, "Lister: 1827–1927," in *Boston Medical and Surgical Journal*, **196** (1927), 984–993; E. L. Gilcreest, "Lord Lister and the Renaissance of Surgery," in *Surgical Clinics of North America*, **7** (1927), 1117–1123; W. W. Keen, "Some Personal Recollections of Lord Lister," in *Surgery, Gynecology and Obstetrics*, **45** (1927), 861–864; C. J. Martin, "Lister's Contribution to Preventive Medicine," in *Nature*, **119** (1927), 529–531; C. F. Painter, "Sir Joseph Lister. An Historical Sketch," in *Boston Medical and Surgical Journal*, **196** (1927), 1093–1096; Charles S. Sherrington, "Lister's Contributions to Physiology," in *Lancet* (1927), **1**, 743–744; and "Listerian Oration, 1927," in *Canadian Medical Association Journal*, **17** (1927), 1255–1263; J. Tait, "Lister as Physiologist," in *Science*, **66** (1927), 267–272; C. J. S. Thompson, "Surgical Instruments Designed by Lord Lister," in *Boston Medical and Surgical Journal*, **196** (1927), 946–951; and A. Young, "Lord Lister's Life and Work," in *Janus*, **31** (1927), 318–335. Among articles in foreign journals commemorating Lister's centenary are J. P. zum Busch, "Zur Erinnerung an Joseph Lister," in *Deutsche medizinische Wochenschrift*, **53** (1927), 583–585; A. Eiselberg, "Zu Josef Lister's 100. Geburtstage," in *Wiener klinische Wochenschrift*, **40** (1927), 461–465; B. F., "Le centenaire de Lister," in *Presse médicale*, **35** (1927), 523–524; and M. von Gruber, "Lord Lister und Deutschland," in *Münchener medizinische Wochenschrift*, **74** (1927), 592–593.

Other biographical accounts and articles on miscellaneous aspects of Lister's life and work, mostly published after 1927, include: E. Archibald, "The Mind and Character of Lister," in *Canadian Medical Association Journal*, **35** (1936), 475–490, fifth Listerian oration; C. Ballance, "Lister and His Time," in *Lancet* (1933), **1**, 815–816; H. C. Cameron, "Lord Lister and the Catgut Ligature," in *British Medical Journal* (1912), **1**, 579 (correspondence); Johnson and Johnson, *Lister and the Ligature. A Landmark in the History of Modern Surgery* (New Brunswick, N. J., 1925); C. Martin, "Lister's Early Bacteriological Researches and Origin of His Antiseptic System," in *Medical Journal of Australia* (1931), **2**, 437–444, the Listerian oration; Lord Moynihan, "Lister—the Idealist," in *Canadian Medical Association Journal*, **23** (1930), 479–488, the third Listerian oration; R. Muir, "The Fourth Listerian Oration," *ibid.*, **29** (1933), 349–360; J. Riera, "The Dissemination of Lister's Teaching in Spain," in *Medical History*, **13** (1969), 123–153; I. M. Thompson, "Lister's Early Scientific Background," in *Manitoba Medical Review*, **20** (1940), 95–100; I. Veith, "Lord Lister and the Antivivisectionists," in *Modern Medicine of Canada*, **16** (1961), 43–57; and A. O. Whipple, "A Consideration of Recent Advances in Medical Science in the Light of Lord Lister's Studies," in *Canadian Medical Association Journal*, **41** (1939), 323–331, the sixth Listerian oration.

Writings that provide background and perspective to Listerian doctrines and methods are J. P. Arcieri, "Enrico Bottini and Joseph Lister in the Method of Antisepsis. Pioneers of Antiseptic Era," in *Alcmaeon*, **1** (1939), 2–13, repr. by Istituto di Storia della Medicina dell'Università di Roma (Rome, 1967); E. Bottini, "Dell'acido fenico nella chirurgia pratica e nella tassidermica," in *Annali universali di medicina*, **198** (1866), 585–636; A. K. Bowman, *The Life and Teaching of Sir William Macewen* (London, 1942); J. Burdon-Sanderson, "The Origin and Distribution of Microzymes (Bacteria) in Water, and the Circumstances Which Determine Their Existence in the Tissues and Liquids of the Living Body," in *Quarterly Journal of Microscopical Science*, n.s. **11** (1871), 323–352; W. W. Cheyne, *Antiseptic Surgery* (London, 1882); D. J. Ferguson, "Paréian and Listerian Slants on Infections in Wounds," in *Perspectives in Biology and Medicine*, **14** (1970), 63–68; S. Gamgee, "The Present State of Surgery in Paris," in *Lancet* (1867), **2**, 392–393, 483–484; T. W. Jones, *Report on the State of the Blood and the Blood Vessels in Inflammation* (London, 1891); T. Keith, "Fifty-one Cases of Ovariotomy," in *Lancet* (1867), **2**, 290–291; and "Second Series of Fifty Cases of Ovariotomy . . .," *ibid.* (1870), **2**, 249–251; J. Lemaire, *De l'acide phénique* (Paris, 1863); J. Lucas-Championnière, *Chirurgie antiseptique* (Paris, 1876), of which the 2nd ed., trans. by F. H. Gerrish, appeared as *Antiseptic Surgery. The Principles, Modes of Application and Results of the Lister Dressing* (Portland, Me., 1881); W. MacCormac, *Antiseptic Surgery* (London, 1880); V. Manninger, *Der Entwicklungsgang der Antiseptik und Aseptik* (Breslau, 1904); L. Munster, "Ein vergessener Vorkämpfer der Parasitenlehre: Agostino Bassi aus Lodi . . .," in *Janus*, **37** (1933), 221–246; J. N. Ritter von Nussbaum, *Leitfaden zur antiseptischen Wundbehandlung insbesondere zur Lister'schen Methode* (Stuttgart, 1881); L. Pasteur, "Recherches sur la putréfaction," in *Comptes rendus hebdomadaires des séances de l'Académie des sciences*, **56** (1863), 1189–1194; and related papers in *Oeuvres de Pasteur*, II (Paris, 1922); A. Sabatier, *Des méthodes antiseptiques chez les anciens et chez les modernes* (Paris, 1883); C. Schimmelbusch, *Anleitung zur aseptischen Wundbehandlung* (Berlin, 1892), trans. by F. J. Thornbury as *A Guide to the Aseptic Treatment of Wounds* (New York, 1895); T. Spencer Wells, "Some Causes of Excessive Mortality After Surgical Operations," in *British Medical Journal* (1864), **2**, 384–388; J. Tyndall, "On Dust and Disease," in *Fragments of Science* (London, 1871); and *Essays on the Floating-Matter of the Air, in Relation to Putrefaction and Infection* (London, 1881); and B. A. G. Veraart, "Semmelweis and Lister," in *Nederlandsch Tijdschrift voor Geneeskunde*, **74** (1930), 1762–1775.

Skeptical or critical articles include those of P. Bruns, "Fort mit dem Spray!," in *Berliner klinische Wochenschrift*, **17** (1180), 609–611; the editorials "Antiseptic Surgery," in *Lancet* (1875), **2**, 565–566, 597–598; and "The Debate on Antiseptic Surgery," *ibid.*, 744–747; J. Y. Simpson, "Carbolic Acid and Its Compounds in Surgery," *ibid.*, 546–549; the editorial "Professor Lister," *ibid.* (1877), **1**, 361; and letters to the editor by T. Bryant, "Professor Lister and Clinical Surgery in the London Hospitals," *ibid.* (1877), **1**, 367–368; W. A. Leslie, *ibid.*, 368; and T. Smith, *ibid.*, with Lister's reply appearing as "Clinical Surgery in London and Edinburgh," *ibid.*, 475–477.

Biographies of Lister are J. R. Bradford, in *Dictionary of National Biography* (1912–1921 supp.), pp. 339–343; Hector C. Cameron, *Joseph Lister the Friend of Man* (London, 1949); W. Watson Cheyne, *Lister and His Achievement* (London, 1925); J. D. Cromie, *History of Scottish Medicine* (London, 1932), II, 597–600, 635–639, 669–671; Cuthbert Dukes, *Lord Lister* (London, 1924); Rickman J. Godlee, *Lord Lister* (London, 1917); D. Guthrie, *Lord Lister, His Life and Doctrine* (Edinburgh, 1949); J. R. Leeson, *Lister as I Knew Him* (New York, 1927); A. E. Maylard, J. A. Morris, and L. W. G. Malcolm, *Lister and the Lister Ward* (Glasgow, 1927); J. G. Mumford, *Surgical Memoirs* (New York, 1908), pp. 100–113; J. Pagel and W. Haberling, "Lister, Lord Joseph," in *Biographisches Lexikon der hervorragender Aerzte*, III (Berlin–Vienna, 1929–1935), 803–805; H. Sigerist, "Joseph Lister (1827–1912)," in *Grosse Ärzte* (Munich, 1932), pp. 277–280; St. Clair Thomson, "Joseph Lister (1827–1912), the Founder of Modern Surgery," in D'Arcy Power, ed., *British Masters of Medicine* (London, 1936), ch. XVII, pp. 137–147; R. Truax, *Joseph Lister, Father of Modern Surgery* (London, 1947); A. L. Turner, ed., *Joseph, Baron Lister, Centenary Volume 1827–1927* (Edinburgh–London, 1927); Kenneth Walker, *Joseph Lister* (London, 1956); M. E. M. Walker, "Lord Lister, O.M., F.R.S., 1st Baron of Lyme Regis, 1827–1912," in *Pioneers of Public Health* (Edinburgh, 1930), pp. 154–165; and G. T. Wrench, *Lord Lister, His Life and Work* (London, 1913).

CLAUDE E. DOLMAN

LISTER, JOSEPH JACKSON (*b.* London, England, 11 January 1786; *d.* West Ham, Essex, England, 24 October 1869), *optics.*

Lord Lister (1827–1912) has taken most of the glory of the family name, but the contribution to the perfection of the objective lens systems of the microscope by his father, Joseph Jackson Lister, marks a spectacular turning point in the development of that important and ubiquitous instrument. At the time of the elder Lister's death, at the age of eighty-three, he was described as "the pillar and source of all the microscopy of the age."

Lister was the third child and only son of John Lister and the former Mary Jackson, Quakers who were wine merchants in the City of London. At the age of fourteen he joined his father in the firm, later becoming a partner. In the summer of 1818 he married Isabella Harris; they lived above the business at 5 Tokenhouse Yard until 1822, moving then to Stoke Newington and shortly after to Upton House, West Ham. It was here that Lister occupied himself with his microscopy and optical experiments. When the Listers vacated Tokenhouse Yard, their place as resident partner was taken by Richard Low Beck, the son of Lister's sister Elizabeth. Beck had just married

Rachel Lucas; and two of their many children, Richard and Joseph, formed the optical instrument-making firm still known as R. & J. Beck Ltd.

Although Lister's approach to his investigations was that of an amateur, the standard of his work was to the highest degree professional. His interest in optics originated in his boyhood; of his schoolmates only he owned a telescope. Nevertheless, he did not become an innovator until his thirty-eighth year. Among the papers he left to his son is one that gives a brief survey of his involvement with the microscope from 1824 to 1843:

> I had been from early life fond of the compound microscope, but had not thought of improving its object-glass till about the year 1824, when I saw at W. Tulley's an achromatic combination made by him at Dr. Goring's suggestion, of two convex lenses of plate glass, with a concave of flint glass between them, on the plan of the telescopic objective. They were thick and clumsy. I showed him this by a tracing with a camera lucida, which I had attached to my microscope, and the suggestions resulted in "Tulley's 9/10", which became *the* microscopic object-glass of the time. But the subject continued to engage my thoughts, and resulted in the paper "On the Improvement of Compound Microscopes" read before the Royal Society, Jan. 21st, 1830, announcing the discovery of the existence of two aplanatic foci in a double achromatic object-glass [Joseph Lister, "Obituary. . .," p. 141].

In this paper Lister reported that an achromatic combination of a negative flint-glass lens with a positive crown-glass lens has two aplanatic focal points. For all points between these foci the spherical aberration is overcorrected; for points outside, it is undercorrected. If, then, a doublet objective is formed that is composed of two sets of achromatic lens combinations, spherical aberration is avoided if the object is at the shorter aplanatic focus of the first lens pair, which then passes the rays on to the longer aplanatic focus of the second element. This design removed, for the first time, the fuzziness of the image caused by both chromatic and spherical aberrations and, in addition, nullified coma. The new principle elevated the making of microscope objectives from the traditional trial-and-error procedure of Tulley, Chevalier, and Amici to a scientific one, and it continues as the basis for the design of low-power objectives. As a result of this paper Lister was elected a fellow of the Royal Society on 2 February 1832.

Having established a principle, Lister naturally wished to continue experimenting; but he found that Tulley was too busy to make lenses for him. In November 1830, therefore, Lister began the grinding and polishing of lenses in his own home. The result was, he said in a letter to Sir John Herschel, beyond his expectations: ". . . without having ever before cut brass or ground more than a single surface of a piece of glass, I managed to make the tools and to manufacture a combination of three double object-glasses, without spoiling a lens or altering a curve, which fulfilled all the conditions I proposed for a pencil of 36 degrees" (*ibid.*, p. 140). Some of Lister's optical lathe chucks, polishing sticks, and experimental lenses are extant.

The optical instrument-makers in London did not immediately adopt Lister's ideas in designing their objectives; but in 1837 he gave details for the construction of an objective of 1/8-inch focal length to Andrew Ross, one of the foremost optical instrument-makers of the time. In 1840 Lister instructed James Smith in the techniques for constructing 1/4-inch objectives, which were for a long time known as "Smith's quarters" among microscopists. Such new objectives, commercially available in quantity for the first time, turned the microscope into a serious scientific instrument; and it continued to develop very rapidly until the 1880's, when the limit of resolution of the light microscope was reached. Richard Beck, the grandnephew of Lister, was apprenticed to James Smith and joined him in partnership in 1847. When Smith retired, Joseph Beck joined his brother; and the firm traded as R. & J. Beck, with their factory in Holloway appropriately known as the Lister Works.

When first working with Tulley's improved objective, Lister realized that the mounting of the compound body tube of his microscope was not sufficiently stable for the increased magnification that could now be used. He therefore designed a completely new form of stand, which was fabricated by James Smith in 1826. Throughout his life Lister maintained his interest in microscopical observations. He published a paper in 1827 with Thomas Hodgkin on red blood cells, giving their true form and the most accurate measurement of their diameter that had yet been achieved. His patient work on zoophytes and ascidians was read to the Royal Society in 1834. His research into the resolving power of the human eye and of the microscope, which anticipated some of the findings of Abbe and Helmholtz, was not published during his life, even though it was prepared for press. The manuscript was found among his son's papers and was published by the Royal Microscopical Society in 1913.

Lister died at Upton House in 1869, having received few honors. That he was responsible for changing the microscope from a toy to a vital scientific instrument can be fully appreciated only in historical perspective.

BIBLIOGRAPHY

I. Original Works. Lister's writings include "Notice of Some Microscopic Observations of the Blood and Animal Tissues," in *Philosophical Magazine*, n.s. **2** (1827), 130–138, written with T. Hodgkin; "On the Improvement of Achromatic Compound Microscopes," in *Philosophical Transactions of the Royal Society*, **120** (1830), 187–200; "Some Observations on the Structure and Functions of Tubular and Cellular Polypi, and of Ascidiae," *ibid.*, **124** (1834), 365–388; and "On the Limit to Defining-power, in Vision With the Unassisted Eye, the Telescope, and the Microscope" (dated 1842–1843), in *Journal of the Royal Microscopical Society*, **33** (1913), 34–35. Uncataloged MS material is in the possession of the Royal Microscopical Society, as are some tools and experimental lenses. Four of Lister's microscopes, including the first made to his design by James Smith, are in the possession of the Wellcome Institute of the History of Medicine, London.

II. Secondary Literature. See William Beck, *Family Fragments Respecting the Ancestry, Acquaintance and Marriage of Richard Low Beck and Rachel Lucas* (Gloucester, 1897); A. E. Conrady, "The Unpublished Papers of J. J. Lister," in *Journal of the Royal Microscopical Society*, **33** (1913), 27–33; "Obituary Notice of the Late Joseph Jackson Lister, F.R.S., Z.S., With Special Reference to His Labours in the Improvement of the Achromatic Microscope . . .," in *Monthly Microscopical Journal*, **3** (1870), 134–143; *Lister Centenary Exhibition at the Wellcome Historical Medical Museum. Handbook. 1927*, 133 f., 138; G. L'E. Turner, "The Microscope as a Technical Frontier in Science," in *Historical Aspects of Microscopy*, S. Bradbury and G. L'E. Turner, eds. (Cambridge, 1967), pp. 175–199; and "The Rise and Fall of the Jewel Microscope 1824–1837," in *Microscopy*, **31** (1968), 85–94.

G. L'E. Turner

LISTER, MARTIN (christened Radclive, Buckinghamshire, England, 11 April 1639; *d.* Epsom, England, 2 February 1712), *zoology, geology*.

Born into a landed family with estates in the North and Midlands, Lister was educated at St. John's College, Cambridge (B.A. 1658, M.A. 1662), becoming a fellow, at the Restoration, by royal mandate; his uncle, Sir Matthew Lister, had been physician to the new king's mother. He studied medicine at Montpellier from 1663 to 1666 but did not graduate. He resigned his fellowship in 1669 and began to practice medicine in York, where, in comparative isolation, he carried out and published pioneer studies in several fields of invertebrate zoology.

Lister's work on mollusks was at first confined to natural history and taxonomy. The latter, although conventionally artificial, attempted to be comprehensive; his spider classification was, for its date, masterly, and agrees remarkably well with a modern system. Based *ex moribus et vita* and on a wide range of characteristics, it includes, for example, exemplary descriptions of the eye arrangement in each group. He was aware of intraspecific variation and came close to a biological definition of the species. Lister's systematic attention to field observation is noteworthy; some of his notes on courtship behavior and on the early stages of some species have never been repeated. This work was not appreciated fully at the time, even by other zoologists; and Thomas Shadwell clearly had Lister in mind when he created Sir Nicholas Gimcrack.

This early enthusiasm is also seen in Lister's work on the life histories of several parasites—gall wasps, ichneumons, gut worms, and horsehair worms—which had been used by others as evidence of spontaneous generation. Lister was very close to giving a complete account of the life histories of these animals, but his enthusiasm declined about 1676. It revived in the early 1680's, but in a different direction. In 1684, after receiving an Oxford M.D.—largely because of his donations to the Ashmolean Museum—he moved to London. A fellow of the Royal Society since 1671, he now regularly attended its meetings and was vice-president from January 1685. Two years later, however, he was involved in some personal controversy at the Society and ceased attending. He became a fellow of the College of Physicians in 1687.

Lister's work was now concentrated on mollusk anatomy and taxonomy. His best-known work, the *Historia . . . conchyliorum* of 1685–1692, consists entirely of engravings by his wife and daughter, with no real text or even titles. Because of the popularity of conchology, the work became well-known; but the three illustrated anatomical supplements were of greater scientific value. These were the first attempt to cover the morphology of a whole invertebrate group in detail. Each contained detailed descriptions of a small number of types, with briefer notes on the structure of a number of other species. Although not of Swammerdam's standard, Lister's dissections were reasonably competent. Not surprisingly, he had difficulty with the complex mollusk reproductive system, and he suffered from the contemporary tendency to overanalogize; thus, he assumed that the "gill" of a snail must receive blood directly from the heart, as in the fish, and thus believed the blood to circulate in what is in fact the wrong direction. On the other hand, he used sound comparative methods to show the true nature of the mollusk "liver."

His concern with mollusk classification brought Lister into the controversy on the nature of fossils. Many ideas on the origin of these "shell stones" had been suggested, ranging from the supernatural theories of Paracelsus to those of writers such as Palissy and Leonardo da Vinci, who accepted their animal origin. The problem could not become of fundamental importance, however, until the second half of the seventeenth century, when the rejection of the idea of spontaneous generation caused a clear distinction to be made between the living and the nonliving. The controversy was centered in England, where the interest in natural theology made important any evidence for the Noachian flood and the interest in natural history encouraged the collection of fossils and the gathering of reliable information on them. In the period 1660–1690 an animal origin was generally accepted in England for formed stones; but those natural philosophers accepting this idea, such as Robert Hooke and John Ray, were not themselves collectors and systematizers of fossils. The men with greatest firsthand knowledge of the subject—Martin Lister, Edward Lhwyd, Robert Plot, John Beaumont, and William Coles—found the difficulties of explaining the distribution of fossils too great for them to accept their dispersal by a universal flood; and, having a nonevolutionary outlook, they were convinced by the differences in detail between extant and fossil shells that there could be no direct link between them. It is likely that Lister's criticism of his ideas encouraged Hooke's suggestions on the mutability of these specific characteristics. Lister was in fact the center of this group of collectors, and his arguments were worked out in most detail. He noticed that the distribution of fossil shells is correlated with the distribution of rocks, and he believed that this was an argument for their geological origin. In tracing the distribution of one particular fossil through a certain rock formation across half of England, he came close to a stratigraphical use for these formed stones (being interested in the classification and distribution of rock types, Lister in 1684 made the first suggestions for the compilation of geological maps). He explained the growth of fossils in rock as a complex crystallization from lapidifying juices found naturally in the earth. Living mollusks were also able to secrete such juices, from which, by a nonvital process, their shells crystallized; in fact he tried to grow such shells from the body juices of mollusks.

Lister's energies were, from the middle 1690's, concerned mainly with the College of Physicians, of which he was censor in 1694. In 1698 he accompanied Lord Portland as paid physician on his embassy to Paris; his account of the city, satirized at the time for its attention to detail, is now a valuable source book. In 1702 he was appointed one of Queen Anne's physicians, apparently largely through the influence of his niece, Sarah Churchill. This influence, and his philosophical activities, appear to have helped to make Lister unpopular among his fellow physicians. He was, however, a difficult man in any case; and the only close friends he ever had appear to have been John Ray in the 1670's and Edward Lhwyd in the 1690's.

After 1700 Lister almost ceased scientific activity, although he did publish some medical works. His attempt at a comprehensive physiology, *Dissertatio de humoribus* (1709), is extremely speculative, containing little observation or experiment. It was old-fashioned in its reliance on humors, and Lister was unsympathetic to the mathematical physiologists of his day—Keill, Friend, and Pitcairne. The book completes the course of Lister's work, from the diligent and original fieldworker of 1670, through the laboratory anatomist and systematist of 1690, to the armchair philosopher of 1709. The superficiality of much of Lister's thought, largely concealed by his early enthusiasm, was now obvious.

BIBLIOGRAPHY

I. Original Works. *Historiae animalium Angliae tres tractatus* (London, 1678), despite the title, has four sections, covering spiders, land snails, freshwater and saltwater mollusks, and fossil shells; the latter has a separate preface and all have individual title pages. An appendix to this work was issued in 1685, bound in with the Latin ed. by Goedart. A German trans. of the spider section by J. A. E. Goeze was published as *Naturgeschichte der Spinnen* (Quendlingburg–Blankenburg, 1778).

De fontibus medicatis Angliae, exercitatio nova et prior (York, 1682) is an account of medical mineral waters and includes an outline of Lister's physiological system. Rev. and enl. eds. were published as *De thermis et fontibus medicatis Angliae* (London, 1684); and *Exercitationes et descriptiones thermarum et fontium medicatorum Angliae* (London, 1685; 1689).

Johannes Geodartius of Insects. Done Into English and Methodized. With the Addition of Notes (York, 1682) has Lister's notes as a substantial part of the whole and new plates by F. Place. A Latin version was published by the Royal Society as *J. Goedartius de insectis in methodum redactum* (London, 1685).

Letters and Divers Other Mixt Discourses in Natural Philosophy (York, 1683) is a collection of papers, almost all of which had been published in the *Philosophical Transactions of the Royal Society*.

Historia sive synopsis methodica conchyliorum (London, 1685–1692; 2nd ed., London, 1692–1697) is bibliographically extremely complex. It was published in parts, and few copies appear to be identical. A number of bound sets of samples of the earlier sheets, issued with the title *De cochleis*, about 1685, survive; it is debatable whether they should be looked upon as a separate work. See G. L. Wilkins, *Journal of the Society for the Bibliography of Natural History*, **3**, no. 4 (1957), 196–205. A 3rd ed., edited by G. Huddesford, was published at Oxford in 1770; and a 4th ed., L. W. Dillwyn, ed., at Oxford in 1823, with a correlation of Lister's arrangement with the Linnaean system. This last ed. bears the words *editio tertia*.

Exercitatio anatomica in qua de cochleis maxime terrestribus et limacibus agitur (London, 1694).

Exercitatio anatomica altera de buccinis fluviatilibus et marinis (London, 1695) was issued bound with *Exercitatio medicinalis de variolis*. Some copies of pt. I were issued separately as *Dissertatio anatomica altera* . . . (London, 1695), and pt. II was issued as *Disquisitio medicinalis de variolis* (London, 1696).

Conchyliorum bivalvium utriusque aquae exercitatio anatomica tertia huic accedit dissertatio medicinalis de calculo humano (London, 1696), with the two *Exercitatio anatomica*, was intended as an anatomical supplement to the *Historia conchyliorum*.

Sex exercitationes medicinales de quibusdam morbis chronicis . . . (London, 1694; a rev. and enl. ed. published as *Octo exercitationes medicinales*, 1697).

A Journey to Paris in the Year 1698 (London, 1699) had 2 further eds. in the same year. Repr. in Pinkerton's *General Collection of the Best and Most Interesting Voyages and Travels* . . . (London, 1809); and there is a rev. ed. by George Henning, *An Account of Paris at the Close of the Seventeenth Century* . . . (London, 1823). Henning's ed. was trans. into French as *Voyage de Lister à Paris* . . . (Paris, 1873), and there is a facs. repr. of the 3rd ed. with notes by R. P. Stearns (Urbana, Ill., 1967).

Other works are *S. sanctorii de statica medicina . . . cum commentario* (London, 1701; new ed., 1728); *Commentariolus in Hippocratem* (London, 1702), repub. as part of *Hippocratis aphorismi cum commentariolo* (London, 1703); *De opsoniis et condimentis sive arte coquininaria* (London, 1705; 2nd ed., 1709), Lister's ed. of Apicius Caelius' work; and *Dissertatio de humoribus in qua veterum ac recentiorum medicorum ac philosophorum opiniones et sententiae examinantur* (London, 1709; new ed., Amsterdam, 1711)—"As full and compleat a system of the animal oeconomie . . . as I could contrive."

Lister's *De scarabaeis Britannicus* was printed as part of John Ray's publication of Francis Willughby's *Historia insectorum* (London, 1710).

Lister was a frequent contributor to the *Philosophical Transactions of the Royal Society*, submitting papers on insects, spiders, parasites, mollusks, birds, plants, physiology (particularly on the lymphatics), medicine, geology, meteorology and archaeology. There are in all 51 papers by Lister, from vol. **4** (1669) to vol. **22** (1701). In addition, 31 letters sent to Lister were passed on by him for publication in the *Transactions* in the same period.

The Lister MSS at the Bodleian Library, Oxford, form a set of 40 vols. of mixed letters and papers. The letters, although from a large number of correspondents, are incomplete—containing, for example, not a single letter from John Ray. In general, they are of slight scientific interest. The other papers include drafts of parts of his published works; sketches for unfinished geological works; papers on fossils, geology, and barnacles; diaries and account books; and so on. Letters to or from Lister can be found in MSS Ashmole 1816, 1829, and 1830 (particularly concerning Lhwyd) and in MSS Smith 51 and 52 (Thomas Smith, the Cotton librarian, was arranging Lister's papers but died with the work incomplete). There are a few letters in the British Museum, MSS Sloane and Stowe. The correspondence between Lister and Ray has been published in E. R. Lankester, ed., *The Correspondence of John Ray* (London, 1848); and R. T. Gunther, ed., *The Further Correspondence of John Ray* (London, 1928). Letters to and from Robert Plot and Edward Lhwyd have been published by Gunther in *Early Science in Oxford*, vols. XII and XIV (Oxford, 1939 and 1945).

II. SECONDARY LITERATURE. There is no full-length study of Lister's life or works. Of several articles in English local journals, the only one which can be recommended is Davies, in *Yorkshire Archaeological Journal*, **2** (1873), 297–320. The eds. of *A Journey to Paris* by Henning and Stearns (see above) both contain a biographical introduction. The author of this article has a thesis on Lister in progress at the University of Leeds.

JEFFREY CARR

LITKE, FYODOR PETROVICH (*b.* St. Petersburg, Russia, 17 September 1797; *d.* St. Petersburg, 8 October 1882), *earth sciences, geography*.

Litke's parents died when he was young. He was poor during his childhood and thus could not obtain advanced education. In 1812, however, he was accepted as a sailor in the Baltic fleet; and his resourcefulness and boldness led to his rapid promotion. Litke's participation in the round-the-world voyage (1817–1819) of the sloop *Kamchatka*, commanded by the well-known geographer and traveler V. M. Golovnin, decisively determined his future. Soon after his return from this voyage Litke, on the recommendation of Golovnin, was appointed commander of an expedition to survey the shores of Novaya Zemlya. In 1821–1824 the expedition made four attempts to circumnavigate Novaya Zemlya from the north. Although exceptionally heavy ice prevented it from succeeding, the expedition contributed much to science. Important mistakes on maps in the position of the western coast of Novaya Zemlya, Matochkin Shar, Proliv, Strait, Kanin Nos, the eastern coast of the mouth of the White Sea, and the Murmansk coast of the Barents Sea were corrected; a number of bays of the Barents Sea were investigated with the aim of determining the possibility of using them for anchorage; and extensive areas of the Barents Sea were studied, including its northern part, between Spitsbergen and Novaya Zemlya.

In 1826–1829 Litke commanded the sloop *Senyavin*, on a round-the-world voyage designed to survey the little-known islands of the central Pacific and the coast of the Bering Sea.

In the Bering Sea, Litke made astronomical determinations of the most important points of the coast of Kamchatka north from the Avachinskaya gulf; he measured the height of many hills; he described in detail the hitherto unknown Karaginskiy Pribilof Ostrova, and Matveyev Islands and also the Chukchi coast from East Cape almost to the mouth of the Anadyr River. In the central Pacific he made a detailed investigation of the Carolines, discovering twelve previously unknown islands, describing the entire archipelago in detail, and placing it on the map. The expedition gathered materials on the botany, zoology, ethnography, geophysics, and oceanography of the regions explored.

After the voyage of the *Senyavin*, Litke became well known and was considered to be one of the most famous travelers of the first half of the nineteenth century. Soon afterward Litke became involved in organizing the Russian Geographical Society and was elected its president at the first meeting, in 1845.

Led by Litke, the Russian Geographical Society conducted a number of important expeditions to the outlying areas of the country and beyond its borders. In 1873, citing his advanced years, he asked to be relieved of the presidency; the society agreed, and in recognition of his exceptional services it established a gold medal in his name to honor especially distinguished geographical discoveries and research.

From 1864 to 1881 Litke was president of the St. Petersburg Academy of Sciences, retiring only a few months before his death. As president he did much to make the Academy flourish, helping various scientific institutions and societies to develop rapidly. In particular, through his efforts the Pulkovo observatory, the main physical observatory, and the Pavlov magnetic-meteorological observatory substantially broadened their activities.

Litke was an honorary member of the Naval Academy, of Kharkov and Dorpat universities, and of the geographical societies of London and Antwerp, as well as a corresponding member of the Paris Academy of Sciences.

BIBLIOGRAPHY

I. Original Works. Litke's most important published writings are *Chetyrekhkratnoe puteshestvie v Severny Ledovity okean na voennom brige "Novaya Zemlya," v 1821–1824 godakh* ("Four Journeys to the Northern Arctic Ocean on the Military Brig *Novaya Zemlya*, in 1821–1824"; St. Petersburg, 1828; 2nd ed., Moscow, 1948); and *Puteshestvie vokrug sveta na voennom shlyupe "Senyavin" v 1826–1829 godakh* ("A Trip Around the World in the Military Sloop *Senyavin* in 1826–1829"), 3 vols. (St. Petersburg, 1834–1836; 2nd ed., Moscow, 1948).

II. Secondary Literature. The most important publications on Litke are the following, listed chronologically: O. V. Struve, *Ob uchenykh zaslugakh grafa F. P. Litke* ("On the Scientific Contributions of Count F. P. Litke"; St. Petersburg, 1883); V. P. Bezobrazov, "Graf Fyodor Petrovich Litke," in *Zapisok Akademii nauk*, **57** (1888), app. 2; F. Wrangel, "Graf Fyodor Petrovich Litke," in *Izvestiya Russkogo Geograficheskogo obshchestva*, **33** (1897), 326–347; and A. D. Dobrovolsky, *Plavania F. P. Litke* ("Voyages of F. P. Litke"; Moscow, 1948).

For further reference, see B. P. Orlov, *Fyodor Petrovich Litke . . . (k 150-letiyu so dnya rozhdenia)* ("Fyodor Petrovich Litke . . . [for the 150th Anniversary of His Birth]"; Moscow, 1948); M. Marich, *Zhizn i plavania flota kapitan-leytenanta Fyodora Petrovicha Litke* ("Life and Voyages of the Fleet of Captain-Lieutenant Fyodor Petrovich Litke"; Moscow–Leningrad, 1949); A. E. Antonov, *F. P. Litke* (Moscow, 1955); and N. N. Zubov, *Fyodor Petrovich Litke*, in the series Otechestvennye Fiziko-Geografy i Puteshestvenniki ("Native Physical Geographers and Travelers"; Moscow, 1959).

A. F. Plakhotnik

LIU HUI (*fl.* China, *ca.* A.D. 250), *mathematics.*

Nothing is known about the life of Liu Hui, except that he flourished in the kingdom of Wei toward the end of the Three Kingdoms period (A.D. 221–265). His mathematical writings, on the other hand, are well known; his commentary on the *Chiu-chang suan-shu* ("Nine Chapters on the Mathematical Art") has exerted a profound influence on Chinese mathematics for well over 1,000 years. He wrote another important, but much shorter, work: the *Hai-tao suan-ching* ("Sea Island Mathematical Manual").

Some scholars believe that the *Chiu-chang suan-shu*, also called the *Chiu-chang suan-ching* ("Mathematical Manual in Nine Chapters"), was already in existence in China during the third century B.C. Ch'ien Pao-tsung, in his *Chung-kuo suan-hsüeh-shih*, and Chang Yin-lin (*Yenching Hsüeh Pao*, **2** [1927], 301) have noted that the titles of certain officials mentioned in the problems date from Ch'in and earlier (third and early second centuries B.C.). There are also references which must indicate a taxation system of 203 B.C. According to Liu Hui's preface, the book was burned during the time of Emperor Ch'in Shih-huang (221–209 B.C.); but remnants of it were later recovered and put in order. In the following two centuries, commentaries on this book were written by Chang Ts'ang (*fl.* 165–142 B.C.) and Keng Shou-ch'ang (*fl.* 75–49 B.C.). In a study by Ch'ien Pao-tsung (1963) it is suggested, from internal textual evidence, that the *Chiu-chang suan-shu* was written between 50 B.C. and A.D. 100 and that it is doubtful whether Chang Ts'ang and Keng Shou-ch'ang had anything to do with the book. Yet Li Yen and Tu Shih-jan, both colleagues of Ch'ien Pao-tsung, still believed Liu Hui's preface when they wrote about the *Chiu-chang suan-shu* in the same year.

During the seventh century both the *Chiu-chang suan-shu* and the *Hai-tao suan-ching* (A.D. 263) were included in *Suan-ching shih-shu* ("Ten Mathematical Manuals," A.D. 656), to which the T'ang mathematician and astronomer Li Shun-feng (602–670) added his annotations and commentaries. These works then became standard texts for students of mathematics; official regulations prescribed that three years be devoted to the works of Liu Hui. Liu Hui's works also found their way to Japan with these ten mathematical manuals. When schools were established in Japan in 702 and mathematics was taught, both the *Chiu-chang suan-shu* and the *Hai-tao suan-ching* were among the prescribed texts.

According to Ch'eng Ta-wei's mathematical treatise, the *Suan-fa t'ung-tsung* ("Systematic Treatise on Arithmetic"; 1592), both the *Chiu-chang suan-shu* and the *Hai-tao suan-ching* were first printed officially in 1084. There was another printed version of them by Pao Huan-chih in 1213. In the early fifteenth century they were included, although considerably rearranged, in the vast Ming encyclopedia, the *Yung-lo ta-tien* (1403–1407). In the second part of the eighteenth century Tai Chen (1724–1777) reconstructed these two texts after having extracted them piecemeal from the *Yung-lo ta-tien*. They were subsequently included by K'ung Chi-han (1739–1787) in his *Wei-po-hsieh ts'ung-shu* (1773). Three years later Ch'ü Tseng-fa printed them separately with a preface by Tai Chen.

Other reproductions based on Tai Chen's reconstruction in the *Wei-po-hsieh ts'ung-shu* are found in the *Suan-ching shih-shu* ("Ten Mathematical Manuals") of Mei Ch'i-chao (1862) and in the *Wan-yu-wen-k'u* (1929–1933) and *Ssu-pu ts'ung-k'an* series (1920–1922; both of the Commercial Press, Shanghai). Two nineteenth-century scholars, Chung Hsiang and Li Huang, discovered that certain passages in the text had been rendered incomprehensible by Tai Chen's attempt to improve on the original text of the *Chiu-chang suan-shu*. A fragment of the early thirteenth-century edition of the *Chiu-chang suan-shu*, consisting of only five chapters, was found during the seventeenth century in Nanking, in the private library of Huang Yü-chi (1629–1691). This copy was seen by the famous Ch'ing scholar Mei Wen-ting (1633–1721) in 1678, and it later came into the possession of K'ung Chi-han (1739–1784) and then Chang Tun-jen (1754–1834); finally it was acquired by the Shanghai Library, where it is now kept. In 1684, Mao I (1640–after 1710) made a handwritten copy of the original text found in the library of Huang Yü-chi. This copy was later acquired by the emperor during the Ch'ien-lung reign (1736–1795). In 1932 it was reproduced in the *T'ien-lu-lin-lang ts'ung-shu* series.

In 1261 Yang Hui wrote the *Hsiang-chieh chiu-chang suan-fa* ("Detailed Analysis of the Mathematical Rules in the Nine Chapters") to elucidate the problems in the *Chiu-chang suan-shu*. Ch'ien Pao-tsung in 1963 collated the text of the *Chiu-chang suan-shu* from Tai Chen's version, the fragments of the late Sung edition as reproduced in the *T'ien-lu-lin-lang ts'ung-shu* series, and Yang Hui's *Hsiang-chieh chiu-chang suan-fa*.

As for the *Hai-tao suan-ching*, only the reconstructed version by Tai Chen remains. It was reproduced in the *Wu-ying-tien* palace edition (before 1794), the "Ten Mathematical Manuals" in K'ung Chi-han's *Wei-po-hsieh ts'ung-shu*, and the appendix to Chü Tseng-fa's *Chiu-chang suan-shu*.

The *Chiu-chang suan-shu* was intended as a practical handbook, a kind of aide-mémoire for architects, engineers, officials, and tradesmen. This is the reason for the presence of so many problems on building canals and dikes, city walls, taxation, barter, public services, etc. It consists of nine chapters, with a total of 246 problems. The chapters may be outlined as follows:

(1) *Fang-t'ien* ("Land Surveying") contains the rules for finding the areas of triangles, trapezoids, rectangles, circles, sectors of circles, and annuli. It gives rules for addition, subtraction, multiplication, and division of fractions. There is an interesting but inaccurate formula for the area of the segment of a circle, where the chord c and the sagitta s are known, in the form $s(c + s)/2$. This expression later appeared during the ninth century in Mahāvīra's *Gaṇitasāra-sangraha*.

Of special interest is the value of the ratio of the circumference of a circle to its diameter that Liu Hui used. The ancient value of π used in China was 3, but since the first century Chinese mathematicians had been searching for a more accurate value. Liu Hsin (*d.* A.D. 23) used 3.1547, while Chang Heng (78–139) gave $\sqrt{10}$ and 92/29. Wang Fan (219–257) found 142/45, and then Liu Hui gave 3.14. The most important names in this connection are, however,

LIU HUI

those of Tsu Ch'ung-chih (430–501), a brilliant mathematician, astronomer, and engineer of the Liu Sung and Ch'i dynasties, and his son, Tsu Cheng-chih. Tsu Ch'ung-chih gave two values for π, first an "inaccurate" one (*yo lü*), equal to 22/7, given earlier by Archimedes, and then a "more accurate" one (*mi lu*), 355/113 (3.1415929). He even looked for further approximations and found that π lies between 3.1415926 and 3.1415927. His method was probably described in the *Chui Shu*, which he and his son wrote but is now lost. Tsu Ch'ung-chih's value of 355/113 for π disappeared for many centuries in China until it was again taken up by Chao Yu-ch'in (*fl. ca.* 1300). Liu Hui obtained the accurate value 3.14 by taking the ratio of the perimeter of a regular polygon of ninety-six sides to the diameter of a circle enclosing this polygon. Let us begin with a regular hexagon of side L_6. The ratio of the perimeter of the hexagon to the diameter of the circle enclosing it is 3. If we change the hexagon to a regular polygon of twelve sides, as shown in Figure 1—noting that $L_6 = r$, the radius of the circumscribed circle—then the side of the twelve-sided polygon is given by

$$L_{12} = \sqrt{2r\left[r - \sqrt{r^2 - \left(\frac{L_6}{2}\right)^2}\right]}.$$

Hence, if L_n is known, then L_{2n} can be found from the expression

$$L_{2n} = \sqrt{2r\left[r - \sqrt{r^2 - \left(\frac{L_n}{2}\right)^2}\right]}.$$

Taking $r = 1$, the following values can be found: $L_6 = 1$; $L_{12} = 0.517638$; $L_{24} = 0.261052$; $L_{48} = 0.130806$; $L_{96} = 0.065438$.

The perimeter of a regular polygon of $n = 96$ and $r = 1$ is $96 \times 0.065438 = 6.282048$. Hence $\pi = 6.282048/2 = 3.141024$, or approximately 3.14. Liu Hui also used a polygon of 3,072 sides and obtained his best value, 3.14159.

(2) *Su-mi* ("Millet and Rice") deals with percentages and proportions. Indeterminate equations are avoided in the last nine problems in this chapter by the use of proportions.

(3) *Ts'ui-fen* ("Distribution by Progression") concerns distribution of properties among partners according to given rates. It also includes problems in taxation of goods of different qualities, and others in arithmetical and geometrical progressions, all solved by use of proportions.

(4) *Shao-kuang* ("Diminishing Breadth") involves finding the sides of a rectangle when the area and one of the sides are given, the circumference of a circle

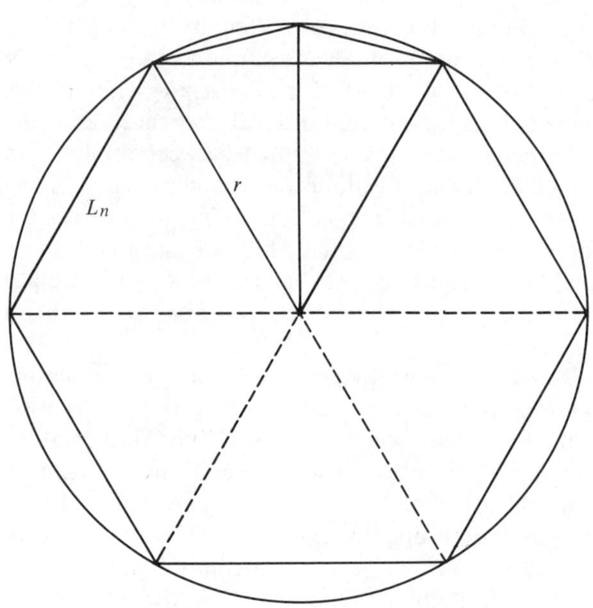

FIGURE 1

when its area is known, the side of a cube given its volume, and the diameter of a sphere of known volume. The use of the least common multiple in fractions is shown. It is interesting that unit fractions are used, for example, in problem 11 in this chapter. The given width of a rectangular form is expressed as

$$1 + 1/2 + 1/3 + 1/4 + 1/5 + 1/6 + 1/7 + 1/8 + 1/9 + 1/10 + 1/11 + 1/12.$$

The problems in this chapter also lead to the extraction of square roots and cube roots; problem 13, for example, involves finding the square root of 25,281. According to the method given in the *Chiu-chang suan-shu*, this number, known as the *shih* (dividend), is first placed in the second row from the top of the counting board. Next, one counting rod, called the preliminary *chieh-suan*, is put on the bottom row of the counting board in the farthest right-hand digit column. This rod is moved to the left, two places at a time, as far as it can go without overshooting the farthest left digit of the number in the *shih* row. With its new place value this rod is called the *chieh-suan*. It is shown in Figure 2*a*.

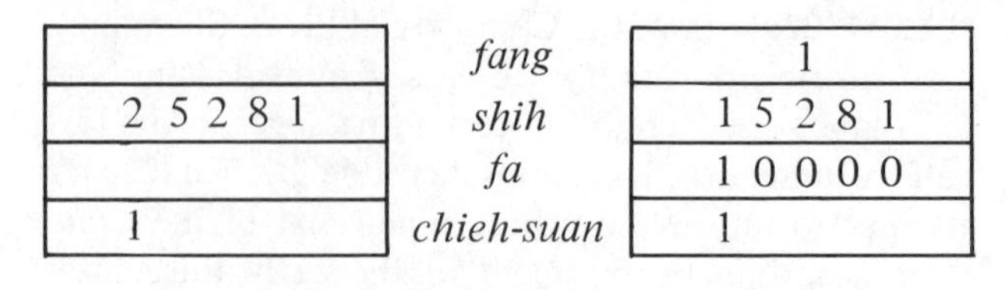

Figure 2a		Figure 2b
	fang	1
2 5 2 8 1	*shih*	1 5 2 8 1
	fa	1 0 0 0 0
1	*chieh-suan*	1

FIGURE 2*a* FIGURE 2*b*

LIU HUI

The first figure of the root is found to lie between 100 and 200. Then 1 is taken as the first figure of the root and is placed on the top row in the hundreds column. The top row is called *fang*. The *chieh-suan* is multiplied by the first figure of the root. The product, called *fa*, is placed in the third row. The *shih* (25,281) less the *fa* (10,000) leaves the "first remainder" (15,281), which is written on the second row, as shown in Figure 2*b*. After the division has been made, the *fa* is doubled to form the *ting-fa*. This is moved one digit to the right, while the *chieh-suan* is shifted two digits to the right, as shown in Figure 2*c*.

Figure 2c		Figure 2d
1	*fang*	
1 5 2 8 1	1st/2nd re.	2 7 8 1
2 0 0 0	*ting-fa*	2 5 0 0
1	*chieh-suan*	1

FIGURE 2*c* FIGURE 2*d*

The second figure, selected by trial and error, is found to lie between 5 and 6. The tens' digit is therefore taken to be 5 and will be placed in its appropriate position on the top row in Figure 2*e*. The *chieh-suan* (which is now 100) is multiplied by this second figure and the product is added to the *ting-fa*, which becomes 2,500. The *ting-fa* multiplied by 5 is subtracted from the first remainder, which gives a remainder of 2,781 ($15{,}281 - 2{,}500 \times 5 = 2{,}781$), as shown in Figure 2*d*. The *ting-fa* is next shifted one digit to the right and the *chieh-suan* two places (see Figure 2*e*). The third figure, again selected by trial and error, is found to be 9. This unit digit is placed in its appropriate position on the top row. The *chieh-suan*, which is now 1, is multiplied by this third figure and the product is added to the *ting-fa*, which becomes 259. The second remainder is divided by the *ting-fa*, which leaves a remainder of zero ($2{,}781 \div 259 = 9 + 0$). Hence the answer is 159 (see Figure 2*f*).

Figure 2e		Figure 2f
1 5	*fang*	1 5 9
2 7 8 1	2nd/last re.	0
2 5 0	*ting-fa*	2 5 9
1	*chieh-suan*	1

FIGURE 2*e* FIGURE 2*f*

(5) *Shang-kung* ("Consultations on Engineering Works") gives the volumes of such solid figures as the prism, the pyramid, the tetrahedron, the wedge, the cylinder, the cone, and the frustum of a cone:

(*a*) Volume of square prism = square of side of base times height.

(*b*) Volume of cylinder = 1/12 square of circumference of circle times height (where π is taken to be approximately 3).

(*c*) Volume of truncated square pyramid = 1/3 the height times the sum of the squares of the sides of the upper and lower squares and the product of the sides of the upper and lower squares.

(*d*) Volume of square pyramid = 1/3 the height times the square of the side of the base.

(*e*) Volume of frustum of a circular cone = 1/36 the height times the sum of the squares of the circumferences of the upper and lower circular faces and the product of these two circumferences (where π is taken to be approximately 3).

(*f*) Volume of circular cone = 1/36 the height times the square of the circumference of the base (where π is taken to be approximately 3).

(*g*) Volume of a right triangular prism = 1/2 the product of the width, the length, and the height.

(*h*) Volume of a rectangular pyramid = 1/3 the product of the width and length of the base and the height.

(*i*) Volume of tetrahedron with two opposite edges perpendicular to each other = 1/6 the product of the two perpendicular opposite edges and the perpendicular common to these two edges.

(6) *Chün-shu* ("Impartial Taxation") concerns problems of pursuit and alligation, especially in connection with the time required for taxpayers to get their grain contributions from their native towns to the capital. It also deals with problems of ratios in connection with the allocation of tax burdens according to the population. Problem 12 in this chapter says:

> A good runner can go 100 paces while a bad runner goes 60 paces. The bad runner has gone a distance of 100 paces before the good runner starts pursuing him. In how many paces will the good runner catch up? [Answer: 250 paces.]

(7) *Ying pu-tsu* or *ying-nü* ("Excess and Deficiency"). *Ying*, referring to the full moon, and *pu-tsu* or *nü* to the new moon, mean "too much" and "too little," respectively. This section deals with a Chinese algebraic invention used mainly for solving problems of the type $ax + b = 0$ in a rather roundabout manner. The method came to be known in Europe as the rule of false position. In this method two guesses, x_1 and x_2, are made, giving rise to values c_1 and c_2, respectively, either greater or less than 0. From these we have the following equations:

(1) $$ax_1 + b = c_1$$

(2) $$ax_2 + b = c_2.$$

LIU HUI

Multiplying (1) by x_2 and (2) by x_1, we have

$$(1') \qquad ax_1x_2 + bx_2 = c_1x_2$$
$$(2') \qquad ax_1x_2 + bx_1 = c_2x_1$$
$$(3) \qquad \therefore b(x_2 - x_1) = c_1x_2 - c_2x_1.$$

From (1) and (2),

$$(4) \qquad a(x_2 - x_1) = c_2 - c_1$$
$$\therefore \frac{b}{a} = \frac{c_1x_2 - c_2x_1}{c_2 - c_1}.$$

Hence

$$(5) \qquad x = -\frac{b}{a} = \frac{c_1x_2 - c_2x_1}{c_1 - c_2}.$$

Problem 1 in this chapter says:

> In a situation where certain things are purchased jointly, if each person pays 8 [units of money], the surplus is 3 [units], and if each person pays 7, the deficiency is 4. Find the number of persons and the price of the things brought. [Answer: 7 persons and 53 units of money.]

According to the method of excess and deficiency, the rates (that is, the "guesses" 8 and 7) are first set on the counting board with the excess (3) and deficiency (−4) placed below them. The rates are then cross multiplied by the excess and deficiency, and the products are added to form the dividend. Then the excess and deficiency are added together to form the divisor. The quotient gives the correct amount of money payable by each person. To get the number of persons, add the excess and deficiency and divide the sum by the difference between the two rates. In other words, x and a are obtained using equations (5) and (4) above.

Sometimes a straightforward problem may be transformed into one involving the use of the rule of false position. Problem 18 in the same chapter says:

> There are 9 [equal] pieces of gold and 11 [equal] pieces of silver. The two lots weigh the same. One piece is taken from each lot and put in the other. The lot containing mainly gold is now found to weigh less than the lot containing mainly silver by 13 ounces. Find the weight of each piece of gold and silver.

Here two guesses are made for the weight of gold. The method says that if each piece of gold weighs 3 pounds, then each piece of silver would weigh 2 5/11 pounds, giving a deficiency of 49/11 ounces; and if each piece of gold weighs 2 pounds, then each piece of silver would weigh 1 7/11 pounds, giving an excess of 15/11 ounces. Following this, the rule of false position is applied.

(8) *Fang-ch'eng* ("Calculation by Tabulation") is concerned with simultaneous linear equations, using both positive and negative numbers. Problem 18 in this chapter involves five unknowns but gives only four equations, thus heralding the indeterminate equation. The process of solving simultaneous linear equations given here is the same as the modern procedure for solving the simultaneous system

$$a_1x + b_1y + c_1z = d_1$$
$$a_2x + b_2y + c_2z = d_2$$
$$a_3x + b_3y + c_3z = d_3,$$

except that the coefficients and constants are arranged in vertical columns instead of being written horizontally:

$$\begin{matrix} a_1 & a_2 & a_3 \\ b_1 & b_2 & b_3 \\ c_1 & c_2 & c_3 \\ d_1 & d_2 & d_3. \end{matrix}$$

In this chapter Liu Hui also explains the algebraic addition and subtraction of positive and negative numbers. (Liu Hui denoted positive numbers and negative numbers by red and black calculating rods, respectively.)

(9) *Kou-ku* ("Right Angles") deals with the application of the Pythagorean theorem. Some of its problems are as follows:

> A cylindrical piece of wood with a cross-section diameter of 2 feet, 5 inches, is to be cut into a piece of plank 7 inches thick. What is the width? [problem 4]
> There is a tree 20 feet high and 3 feet in circumference. A creeper winds round the tree seven times and just reaches the top. Find the length of the vine. [problem 5]
> There is a pond 7 feet square with a reed growing at the center and measuring 1 foot above the water. The reed just reaches the bank at the water level when drawn toward it. Find the depth of the water and the length of the reed. [problem 6]
> There is a bamboo 10 feet high. When bent, the upper end touches the ground 3 feet away from the stem. Find the height of the break. [problem 13]

It is interesting that a problem similar to 13 appeared in Brahmagupta's work in the seventh century.

Problem 20 has aroused even greater interest:

> There is a square town of unknown dimension. A gate is at the middle of each side. Twenty paces out of the north gate is a tree. If one walks 14 paces from the south gate, turns west, and takes 1,775 paces, the tree will just come into view. Find the length of the side of the town.

LIU HUI

The book indicates that the answer can be obtained by evolving the root of the quadratic equation

$$x^2 + (14 + 20)x = 2(1775 \times 20).$$

The method of solving this equation is not described. Mikami suggests that it is highly probable that the root extraction was carried out with an additional term in the first-degree coefficient in the unknown and that this additional term was called *tsung*, but in his literal translation of some parts of the text concerning root extractions he does not notice that the successive steps correspond closely to those in Horner's method. Ch'ien Pao-tsung and Li Yen have both tried to compare the method described in the *Chiu-chang suan-shu* with that of Horner, but they have not clarified the textual obscurities. Wang Ling and Needham say that it is possible to show that if the text of the *Chiu-chang suan-shu* is very carefully followed, the essentials of the methods used by the Chinese for solving numerical equations of the second and higher degrees, similar to that developed by Horner in 1819, are present in a work that may be dated in the first century B.C.

The *Hai-tao suan-ching*, originally known by the name *Ch'ung ch'a* ("Method of Double Differences"), was appended to the *Chiu-chang suan-shu* as its tenth chapter. It was separated from the main text during the seventh century, when the "Ten Mathematical Manuals" were chosen, and was given the title *Hai-tao suan-ching*. According to Mikami, the term *ch'ung ch'a* was intended to mean double, or repeated, application of proportions of the sides of right triangles. The name *Hai-tao* probably came from the first problem of the book, which deals with an island in the sea. Consisting of only nine problems, the book is equivalent to less than one chapter of the *Chiu-chang suan-shu*.

In its preface Liu Hui describes the classical Chinese method of determining the distance from the sun to the flat earth by means of double triangulation. According to this method, two vertical poles eight feet high were erected at the same level along the same meridian, one at the ancient Chou capital of Yan-ch'eng and the other 10,000 *li* (1 *li* = 1,800 feet) to the north. The lengths of the shadows cast by the sun at midday of the summer solstice were measured, and from these the distance of sun could be derived. Liu Hui then shows how the same method can be applied to more everyday examples. Problem 1 says:

> A sea island is viewed from a distance. Two poles, each 30 feet high, are erected on the same level 1,000 *pu* [1 *pu* = 6 ft.] apart so that the pole at the rear is in a straight line with the island and the other pole. If one moves 123 *pu* back from the nearer pole, the top of the island is just visible through the end of the pole if he views it from ground level. Should he move back 127 *pu* from the other pole, the top of the island is just visible through the end of the pole if viewed from ground level. Find the elevation of the island and its distance from the [nearer] pole. [Answer: The elevation of the island is 4 *li*, 55 *pu*. The distance to the [nearer] pole is 102 *li*, 150 *pu* (300 *pu* = 1 *li*).]

The rule for solving this problem is given as follows:

> Multiply the height of the pole by the distance between the poles and divide the product by the difference between the distances that one has to walk back from the poles in order to view the highest point on the island. Adding the height of the pole to the quotient gives the elevation of the island. To find the distance from the nearer pole to the island, multiply the distance walked back from that pole by the distance between the poles. Dividing the product by the difference between the distances that one has to walk back from the poles gives that distance.

Problem 7 is of special interest:

> A person is looking into an abyss with a piece of white rock at the bottom. From the shore a crossbar is turned to lie on the side that is normally upright [so that its base is vertical]. If the base is 3 feet and one looks at the surface of the water [directly above the rock] from the tip of the base, the line of sight meets the height of the crossbar at a distance of 4 feet, 5 inches; and when one looks at the rock, the line of sight meets the height of the crossbar at a distance of 2 feet, 4 inches. A similar crossbar is set up 4 feet above the first. If one looks from the tip of the base, the line of sight to the water surface [directly above the rock] would meet the height of the crossbar at a distance of 4 feet; and if one looks at the rock, it will be 2 feet, 2 inches. Find the depth of the water.

In Figure 3, if P is the water surface above the white rock, R, and BC and FG are the two crossbars, then $BC = FG = 3$ feet; $GC = 4$ feet; $AC = 4$ feet, 5 inches; $DC = 2$ feet, 4 inches; $EG = 4$ feet; and $HG = 2$ feet, 2 inches. The depth of the water, PR, is sought. To obtain the answer, Liu Hui gives the following rule:

$$PR = GC\,\frac{EG(DC - HG) - HG(AC - EG)}{(DC - HG)(AC - EG)}.$$

Liu Hui has not taken into account here the refractive index of water. The rule given is an extension of that used in solving problem 4, which uses the same method for determining the depth of a valley:

> A person is looking at a deep valley. From the edge of the valley a crossbar is turned to lie on the side that is normally upright [so that its base is vertical]. The base

LIU HUI

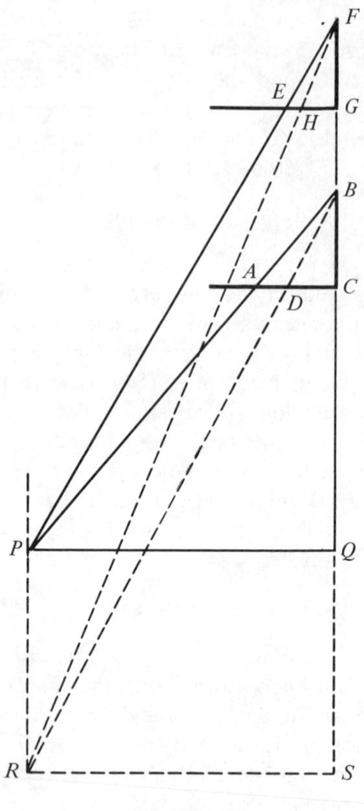

FIGURE 3

is 6 feet long. If one looks at the bottom of the valley from the edge of the base, the line of sight meets the vertical side at a distance of 9 feet, 1 inch. Another crossbar is set 30 feet directly above the first. If the bottom of the valley is observed from the edge of the base, the line of sight will meet the vertical side at a distance of 8 feet, 5 inches. Find the depth of the valley.

If we refer again to Figure 3, ignoring the broken lines, we have $CB = GF = 6$ feet; $CG = 30$ feet; $AC = 9$ feet, 1 inch; $EG = 8$ feet, 5 inches; and CQ is the depth. From similar triangles ABC and PBQ,

$$QB \cdot AC = PQ \cdot CB;$$

and from similar triangles EFG and PFQ,

$$QF \cdot EG = PQ \cdot GF.$$

Since $CB = GF$, and $QF = QB + BF$,

$$QB \cdot AC = (QB + BF)\, EG,$$

$$QB(AC - EG) = BF \cdot EG = GC \cdot EG,$$

that is,

$$(CQ + CB)(AC - EG) = GC \cdot EG.$$

Hence,

$$CQ = \frac{GC \cdot EG}{AC - EG} - CB.$$

In problem 7 one can also obtain the distance from the bank to the bottom of the abyss (CS in Figure 3) from the expression

$$CS = \frac{GC \cdot HG}{(DC - HG)} - CB.$$

PR is derived from the difference between CS and CQ.

As for the other problems, problem 2 concerns finding the height of a tree on a hill; problem 3 deals with the size of a distant walled city; problem 5 shows how to measure the height of a tower on a plain as seen from a hill; problem 6 gives a method for finding the width of a gulf seen from a distance on land; problem 8 is a case of finding the width of a river seen from a hill; and problem 9 seeks the size of a city seen from a mountain.

BIBLIOGRAPHY

A modern ed. of the *Chiu-chang suan-shu* is vol. 1121 in the *Ts'ung-Shu Chi-Chêng* series (Shanghai, 1936).

Works dealing with Liu Hui and his writings are Ch'ien Pao-tsung, *Suan-ching shih-shu* ("Ten Mathematical Manuals"), 2 vols. (Peking, 1963), 83–272; and *Chung-kuo suan-hsüeh-shih* ("History of Chinese Mathematics") (Peking, 1964), 61–75; L. van Hée, "Le Hai Tao Suan Ching de Lieou," in *T'oung Pao*, **20** (1921), 51–60; Hsü Shun-fang, *Chung-suan-chia te tai-shu-hsüeh yen-chiu* ("A Study of Algebra by Chinese Mathematicians") (Peking, 1955), 1–8; Li Yen, *Chung-kuo shu-Hsüeh ta-kang* ("Outline of Chinese Mathematics"), I (Shanghai, 1931); and *Chung-kuo suan-hsüeh-shih* ("History of Chinese Mathematics") (Shanghai, 1937; rev. ed., 1955), 16, 19, 21; Li Yen and Tu Shih-jan, *Chung-kuo ku-tai shu-hsüeh chien-shih* ("Brief History of Ancient Chinese Mathematics"), I (Peking, 1963), 45–77; Yoshio Mikami, *The Development of Mathematics in China and Japan* (New York, 1913); Joseph Needham, *Science and Civilisation in China*, III (Cambridge, 1959), 24–27; George Sarton, *Introduction to the History of Science*, 3 vols. (Baltimore, 1927–1947), esp. I, 338; Wang Ling, "The Chiu Chang Suan Shu and the History of Chinese Mathematics During the Han Dynasty," a doctoral diss. (Cambridge Univ., 1956); Wang Ling and Joseph Needham, "Horner's Method in Chinese Mathematics; Its Origins in the Root-Extraction Procedure of the Han Dynasty," in *T'oung Pao*, **43** (1955), 345–401; and Alexander Wylie, *Chinese Researches* (Shanghai, 1897; repr. Peking, 1936, and Taipei, 1966), 170–174.

Some important special studies on the *Chiu-chang suan-shu* are E. I. Berezkina, "Drevnekitaysky traktat matematika v devyati knigach" ("The Ancient Chinese Mathematical Treatise in Nine Books"), in *Istoriko-matematicheskie issledovaniya*, **10** (1957), 423–584, a Russian trans. of the

Chiu-chang suan-shu; Kurt Vogel, *Neun Bücher arithmetischer Technik* (Brunswick, 1968), a German trans. and study of the work; and A. P. Youschkevitsch, *Geschichte der Mathematik im Mittelalter* (Leipzig, 1964), 1–88 ("Die Mathematik in China"), translated from the Russian.

Access to old biographical notes and bibliographical citations concerning mathematical works are Hu Yü-chin, *Ssu-K'u-T'i-Yao Pu-Chêng* ("Supplements to the *Ssu-K'u-T'i-yao*"), 2 vols. (Taipei, 1964–1967); and Ting Fu-pao and Chou Yün-ch'ing, *Ssu-Pu-Tsung-Lu Suan-Fa-Pien* ("Bibliography of Mathematical Books to Supplement the *Ssu-K'u-Ch'uan-Shu* Encyclopedia"; Shanghai, 1956).

More information on the *Suan-Ching Shi-Shu* can be found in Needham, *Science and Civilisation in China*, III, 18; and in A. Hummel, *Eminent Chinese of the Ch'ing Period* (Washington, 1943), p. 697.

The two extant volumes of the *Yung-Lo Ta-Tien* encyclopedia have been reproduced photographically (Peking, 1960); they show that the arrangement was according to mathematical procedures and not by authors.

HO PENG-YOKE

LIVINGSTON, BURTON EDWARD (*b.* Grand Rapids, Michigan, 9 February 1875; *d.* Baltimore, Maryland, 8 February 1948), *plant physiology, physiological ecology.*

The son of Benjamin and Keziah Livingston, Burton was raised in a home and environment that gave him early contact with and love of flora and natural history. Since his father was a contractor in the street paving and sewer construction business, he became familiar with tools and machinery. Moreover, a large home library and access to simple microscopes gave him considerable familiarity with science before he entered school. In high school Livingston became proficient in languages and began a herbarium. Upon graduation he worked in a Short Hills, New Jersey, nursery and then entered the University of Michigan in 1894, receiving ten hours of advanced credit in botany. He was awarded the B.S. degree in 1898 and received the Ph.D. from the University of Chicago in 1901. From 1899 to 1904 he was assistant in plant physiology at Chicago, then became a soil expert for the U.S. Bureau of Soils, Department of Agriculture, from 1905 to 1906. He was a staff member of the department of botanical research at the desert laboratory of the Carnegie Institution of Washington at Tucson, Arizona, from 1906 to 1909; professor of plant physiology at Johns Hopkins University from 1909 to 1940; and director of the laboratory of plant physiology at Johns Hopkins from 1913 to 1940. His marriage to Grace Johnson in 1905 ended in divorce in 1918; he married Marguerite Anna Brennan MacPhilips in 1921. He had no children.

Livingston's early interests and training prepared him for a research career in plant physiology at a time when that science was not given academic standing at most American universities. It was mainly through his efforts at Johns Hopkins that plant physiology was gradually given full standing elsewhere. His interests were broad but centered mainly on the water relations of plants. He studied soil moisture and the evaporative power of the environment, with special emphasis on the effects of light, temperature, wind, and other factors on foliar transpiring power. His careful compilations of data on the daily course of transpiration, the seasonal soil moisture conditions, the water-supplying power of soils in relation to wilting, and the oxygen-supplying ability of soils in relation to seed germination contributed immensely to the foundations of modern physiological plant ecology.

Livingston invented the porous cup atmometer, an instrument for measuring the evaporating capacity of air; the auto-irrigator, for automatic control of soil moisture of potted plants; water-absorbing points for measuring the water-supplying power of soil; and rotating tables for assuring equal exposure of plant cultures to environmental conditions. These inventions or modifications of them are still in use.

Livingston's laboratory became a mecca for plant physiologists, and more than 300 papers were published with his students and colleagues. His book *The Role of Diffusion and Osmotic Pressure in Plants* (1903) is still an authoritative summary of the early work. In 1921 he published with Forrest Shreve *The Vegetation of the United States as Determined by Climatic Conditions.* His translation, with extensive editorial comment, of V. I. Palladin's *Plant Physiology* in 1918 gave the work the impetus it needed to become the standard American textbook in plant physiology for many years. It ran to three editions, the last appearing in 1926.

BIBLIOGRAPHY

See D. T. MacDougal, "Burton Edward Livingston (1875–1948)," in *Yearbook. American Philosophical Society*, **12** (1948), 278–280; and an article on Livingston in *National Cyclopedia of American Biography*, **36** (1950), 334; and C. A. Shull, "Burton Edward Livingston (1875–1948)," in *Science*, **107** (1948), 558–560; and "Burton Edward Livingston (1875–1948)," in *Plant Physiology*, **23** (1948), iii–vii.

A. D. KRIKORIAN

LI YEH. See **Li Chih.**

LLOYD, HUMPHREY (*b.* Dublin, Ireland, 16 April 1800; *d.* Dublin, 17 January 1881), *physics.*

An active member of the Church of Ireland and provost of Trinity College, Dublin, Lloyd was known among physicists as an expert in optics and a leader in the British program to map the earth's magnetic field. His career closely paralleled that of his father, the Reverend Bartholomew Lloyd, who was professor of mathematics, natural philosophy, Greek, and divinity at Trinity College, as well as provost of the college, and president of the Royal Irish Academy. Humphrey displayed his abilities while young, placing first on the entrance examination for Trinity and, upon his graduation in 1819, receiving a gold medal for excellence in science. He became a member of the faculty and rose through the ranks: junior fellow (1824), professor of natural and experimental science (1831), senior fellow (1843), vice provost (1862), and provost (1867). During this period he received his Master of Arts and Doctor of Divinity. He was president of the Royal Irish Academy from 1846 to 1851, and a member of the Royal Societies of London and Edinburgh and of the British Association for the Advancement of Science. At the age of forty he married Dorothea Bulwer, the daughter of a clergyman.

Lloyd's first important original research was in optics. In 1832 W. R. Hamilton deduced from Fresnel's equations for double refraction in a biaxial crystal the consequence that under certain conditions a ray of light within the crystal would split into an infinite number of rays forming a conic surface. Hamilton asked Lloyd to investigate this "conical refraction" experimentally, and Lloyd succeeded in giving a demonstration of the phenomenon and finding the direction of polarization of the rays making up the luminous cone. This achievement gave new support to the wave theory of light and established Lloyd's reputation as an expert in optics.

In 1834, while examining Fresnel's classic experiment in which an interference pattern is produced by light from a single source reflected by two mirrors, Lloyd discovered that a similar effect could be produced by using a single reflecting surface. Arranging a light source, mirror, and screen as shown in Figure 1, he caused reflected light to interfere with direct light from the source in a limited area on the screen. The location of the bands of this pattern could be deduced by imagining they were caused by a second source of light, located at the mirror image of the first and producing an interference pattern. This treatment enabled Lloyd to establish two important properties of reflection: first, that the intensity of light reflected at a 90° incidence is equal to that of direct light (as Fresnel had predicted), and second, that a half-wavelength phase acceleration takes place upon reflection by a higher-density medium.

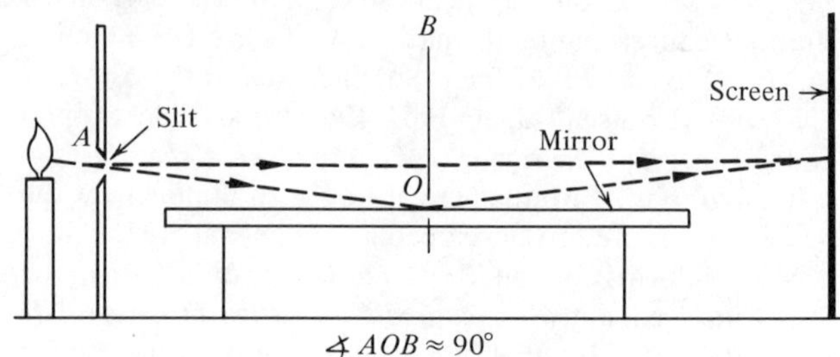

FIGURE 1

Lloyd's other contributions to optics include a report to the British Association on the state of the subject in 1833, two textbooks expounding the wave theory, and a study of reflection and refraction by thin plates. The bulk of his remaining scientific work was devoted to terrestrial magnetism.

In 1831–1832 the British Association proposed that a series of observations of the intensity of the earth's magnetic field be made in various parts of the kingdom. It appointed a committee to coordinate this activity and Lloyd was made a member. Shortly thereafter he undertook a series of magnetic observations in Ireland with Sir Edward Sabine. During the course of this project Lloyd made measurements in twenty-four areas of his native land and invented a new technique for the simultaneous determination of magnetic inclination and intensity.

By 1838 the discovery of temporal variations in the earth's magnetic field created the need for a set of permanent observatories capable of making simultaneous measurements in different parts of the world. The British Association, with government financial support, undertook the construction of these stations and assigned Lloyd the important job of drawing up instructions for the observers and teaching the officers in charge the use of the instruments. Lloyd's own observatory at Trinity College, which had been constructed under his supervision in 1837, served as a model for the other stations and received the results of their observations.

Lloyd's contributions to the understanding of terrestrial magnetism did not end with his role in setting up the British magnetic survey. He remained active in the field throughout his life and succeeded in demonstrating the existence of currents of electricity in the earth's crust and in calculating their effect on the daily variation in the magnetic field.

BIBLIOGRAPHY

I. ORIGINAL WORKS. Lloyd's articles include "On the Phenomena Presented by Light in Its Passage Along the Axes of Biaxial Crystals," in *Philosophical Magazine*, **2** (1833), 112–120, 207–210; "Report on the Progress and Present State of Physical Optics," in *British Association Reports* (1834), 295–413; "An Attempt to Facilitate Observations of Terrestrial Magnetism," in *Transactions of the Royal Irish Academy*, **17** (1837), 159–170; "Further Developments of a Method . . .," *ibid.*, 449–460; "On a New Case of Interference of the Rays of Light," *ibid.*, 171–178; "On the Light Reflected and Transmitted by Thin Plates," *ibid.*, **24**, pt. 1 (1860), 3–15; and "On Earth-Currents, and Their Connexion With the Phenomena of Terrestrial Magnetism," in *Philosophical Magazine*, **22** (1861), 437–442. His most important scientific papers are reprinted in his *Miscellaneous Papers Connected With Physical Science* (London, 1877), and his other published works are listed in *Dictionary of National Biography*, XI, p. 1304. His textbooks are *The Elements of Optics* (Dublin, 1849) and *Elementary Treatise on the Wave Theory of Light* (Dublin, 1857).

II. SECONDARY LITERATURE. The best biographical material on Lloyd is in the *Proceedings of the Royal Society*, **31** (1881), 23. A bibliography of his scientific papers is in Royal Society, *Catalogue of Scientific Papers*, IV, 62–63.

EUGENE FRANKEL

LLOYD, JOHN URI (*b.* West Bloomfield, New York, 19 April 1849; *d.* Van Nuys, California, 9 April 1936), *pharmacy, chemistry.*

Lloyd was the son of Nelson Lloyd, a civil engineer and schoolteacher, and the former Sophia Webster. When the boy was four, the family moved to Kentucky, where his parents taught school in their home. Most of his early education was obtained at home, and in 1863 he was apprenticed to a pharmacist in Cincinnati. After serving for several years as an apprentice and drug clerk, Lloyd was approached by the eclectic physician John King with an offer to join the pharmaceutical firm of H. M. Merrell of Cincinnati as a chemist. In 1871 he joined this firm, which specialized in eclectic remedies; and eventually he and two brothers gained control of the company, which became known as Lloyd Brothers. Although he did not possess a college education, Lloyd served as a professor of chemistry at the Cincinnati College of Pharmacy from 1883 to 1887 and at the Eclectic Medical Institute from 1878 to 1907.

Lloyd was a man of small physical stature and was possessed of a lively and versatile mind. He was a prolific writer whose publications included scientific articles, historical works, and novels. He and his brothers also created the Lloyd Library in Cincinnati, which contains significant collections in pharmacy, botany, medicine, natural history, and allied fields. Among the honors he received in his lifetime were the Remington Medal of the American Pharmaceutical Association (1920), the Procter International Award of the Philadelphia College of Pharmacy and Science (1934), and several honorary degrees. He was married twice and had three children by his second wife.

His commitment to eclecticism, which stressed the use of botanical drugs, led Lloyd to concentrate his scientific studies on plants. He investigated the medicinal and chemical properties of numerous indigenous plants and marketed tinctures, fluid extracts, and other products derived from these plants. The so-called "specific medicines" of Lloyd Brothers, although intended for eclectic practitioners, also found favor with many other physicians and pharmacists. In the course of this work Lloyd developed and patented several useful techniques, such as his "cold still," which extracted the soluble constituents of plants with a minimum of heat, and Lloyd's reagent, hydrous aluminum silicate, which adsorbed alkaloids from solutions. He also wrote several important pharmaceutical textbooks and reference works, such as *Drugs and Medicines of North America.*

Lloyd was a pioneer in the application of physical chemistry to pharmaceutical techniques. In a series of papers published in 1879–1885, he discussed such physical phenomena as adsorption and capillarity and their relationship to the preparation of fluid extracts. Wolfgang Ostwald, one of the founders of the field of colloid chemistry, later found that these works contained much of interest to those studying colloidal phenomena. He felt that they were significant and original enough to merit republication in his journal, *Kolloidchemische Beihefte*, in 1916.

BIBLIOGRAPHY

I. ORIGINAL WORKS. There does not appear to be a published bibliography of Lloyd's writings, which number in the thousands. His most important scientific books are *The Chemistry of Medicines* (Cincinnati, 1881); *Elixirs . . .* (Cincinnati, 1883); *Drugs and Medicines of North America*, 2 vols. (Cincinnati, 1884–1887), written with his brother C. G. Lloyd; and *Origin and History of All the Pharmacopeial Vegetable Drugs, Chemicals and Preparations. I. Vegetable Drugs* (Cincinnati, 1921). His novels largely concern Kentucky life and folklore, the best-known being *Stringtown on the Pike* (Cincinnati, 1901). *Etidorpha* (Cincinnati, 1897), a fantasy involving a trip to the center of the earth, is a curious mixture of scientific and metaphysical speculations.

The six articles selected by Ostwald for republication (see text) were originally published in *Proceedings of the*

American Pharmaceutical Association, **27–33** (1879–1885), under the general title "Precipitates in Fluid Extracts" (except for the first paper, which was entitled "On the Conditions Necessary to Successfully Conduct Percolation"). Lloyd also published related papers on adsorption, on solvents, and on physics in pharmacy in *Journal of the American Pharmaceutical Association*, especially in the period 1916–1936. Several of these papers bear the name of Wolfgang Ostwald, among others, as a coauthor. For Lloyd's first detailed report of the reagent bearing his name, see "Discovery of the Alkaloidal Affinities of Hydrous Aluminum Silicate," in *Journal of the American Pharmaceutical Association*, **5** (1916), 381–390, 490–495. For his views on colloids in pharmacy and on the methods of extracting active principles of plants, see "Colloids in Pharmacy," in Jerome Alexander, ed., *Colloid Chemistry*, II (New York, 1928), 931–934.

Lloyd was a frequent contributor to *Eclectic Medical Journal;* and numerous essays by him on pharmacy, eclecticism, and other subjects can be found in this publication.

The Lloyd Library in Cincinnati possesses correspondence and other MS materials of Lloyd's. A significant amount of his correspondence and other documents related to him are in the Kremers Reference Files of the University of Wisconsin School of Pharmacy in Madison.

II. Secondary Literature. The life and work of John Uri Lloyd have not yet been satisfactorily evaluated in detail. Corinne Miller Simons, *John Uri Lloyd, His Life and Works, 1849–1936, With a History of the Lloyd Library* (Cincinnati, 1972), provides a significant amount of useful biographical information, but the author herself describes the book in her preface as a "chronicle" intended "merely to preserve and enumerate the authentic facts of his life while they are still fresh in the memories of those who knew him"; she acknowledges that she has left for later generations the task of writing biographies of Lloyd "that may be philosophical, analytical or interpretative of his influence and place in contemporary history." For biographical sketches, see the article by Corinne Miller Simons in *Dictionary of American Biography*, XXII, supp. 2 (1958), 389–390; Roy Bird Cook, "John Uri Lloyd: Pharmacist, Philosopher, Author, Man," in *Journal of the American Pharmaceutical Association, Practical Pharmacy Edition*, **10** (1949), 538–544; and George Beal, "Lloyds of Cincinnati," in *American Journal of Pharmaceutical Education*, **23** (1959), 202–206. *Eclectic Medical Journal*, **96**, no. 5 (May 1936), is a memorial issue which reprints several biographical articles on Lloyd. *National Eclectic Medical Quarterly*, **41**, no. 2 (Dec. 1949), also contains several short articles about Lloyd.

On the history of the Lloyd Library, as well as for biographical information on its founders, see Caswell Mayo, *The Lloyd Library and Its Makers* (Cincinnati, 1928); and Corinne Miller Simons, "John Uri Lloyd and the Lloyd Library," in *National Eclectic Medical Quarterly* (Mar. 1951), 1–5.

On eclecticism, see Alex Berman, "The Impact of the Nineteenth Century Botanico-Medical Movement on American Pharmacy and Medicine" (Ph.D. dissertation, University of Wisconsin, 1954), pp. 251–316, esp. pp. 307–311. See also Berman's "Wooster Beach and the Early Eclectics," in *University of Michigan Medical Bulletin*, **24** (1958), 277–286; and "A Striving for Scientific Respectability: Some American Botanics and the Nineteenth-Century Plant Materia Medica," in *Bulletin of the History of Medicine*, **30** (1956), 7–31. Also of interest in this connection is Harvey Felter, *History of the Eclectic Medical Institute, Cincinnati, Ohio, 1845–1902* (Cincinnati, 1902)—see pp. 130–132 for a biographical sketch of Lloyd.

John Parascandola

LLWYD, EDWARD. See **Lhwyd, Edward.**

LOBACHEVSKY, NIKOLAI IVANOVICH (*b.* Nizhni Novgorod [now Gorki], Russia, 2 December 1792; *d.* Kazan, Russia, 24 February 1856), *mathematics.*

Lobachevsky was the son of Ivan Maksimovich Lobachevsky, a clerk in a land-surveying office, and Praskovia Aleksandrovna Lobachevskaya. In about 1800 the mother moved with her three sons to Kazan, where Lobachevsky and his brothers were soon enrolled in the Gymnasium on public scholarships. In 1807 Lobachevsky entered Kazan University, where he studied under the supervision of Martin Bartels, a friend of Gauss, and, in 1812, received the master's degree in physics and mathematics. In 1814 he became an adjunct in physical and mathematical sciences and began to lecture on various aspects of mathematics and mechanics. He was appointed extraordinary professor in 1814 and professor ordinarius in 1822, the same year in which he began an administrative career as a member of the committee formed to supervise the construction of the new university buildings. He was chairman of that committee in 1825, twice dean of the department of physics and mathematics (in 1820–1821 and 1823–1825), librarian of the university (1825–1835), rector (1827–1846), and assistant trustee for the whole of the Kazan educational district (1846–1855).

In recognition of his work Lobachevsky was in 1837 raised to the hereditary nobility; he designed his own familial device (which is reproduced on his tombstone), depicting Solomon's seal, a bee, an arrow, and a horseshoe, to symbolize wisdom, diligence, alacrity, and happiness, respectively. He had in 1832 made a wealthy marriage, to Lady Varvara Aleksivna Moisieva, but his family of seven children and the cost of technological improvements for his estate left him with little money upon his retirement from the university, although he received a modest pension. A worsening sclerotic condition progressively affected his eyesight, and he was blind in his last years.

Although Lobachevsky wrote his first major work, *Geometriya*, in 1823, it was not published in its original form until 1909. The basic geometrical studies that it embodies, however, led Lobachevsky to his chief discovery—non-Euclidean geometry (now called Lobachevskian geometry)—which he first set out in "Exposition succincte des principes de la géométrie avec une démonstration rigoureuse du théorème des parallèles," and on which he reported to the Kazan department of physics and mathematics at a meeting held on 23 February 1826. His first published work on the subject, "O nachalakh geometrii" ("On the Principles of Geometry"), appeared in the *Kazanski vestnik*, a journal published by the university, in 1829–1830; it comprised the earlier "Exposition."

Some of Lobachevsky's early papers, too, were on such nongeometrical subjects as algebra and the theoretical aspects of infinite series. Thus, in 1834 he published his paper "Algebra ili ischislenie konechnykh" ("Algebra, or Calculus of Finites"), of which most had been composed as early as 1825. The first issue of the *Uchenye zapiski* ("Scientific Memoirs") of Kazan University, founded by Lobachevsky, likewise carried his article "Ob ischezanii trigonometricheskikh strok" ("On the Convergence of Trigonometrical Series"). The chief thrust of his scientific endeavor was, however, geometrical, and his later work was devoted exclusively to his new non-Euclidean geometry.

In 1835 Lobachevsky published a long article, "Voobrazhaemaya geometriya" ("Imaginary Geometry"), in the *Uchenye zapiski*. He also translated it into French for Crelle's *Journal für die reine und angewandte Mathematik*. The following year he published, also in the *Uchenye zapiski*, a continuation of this work, "Primenenie voobrazhaemoi geometrii k nekotorym integralam" ("Application of Imaginary Geometry to Certain Integrals"). The same period, from 1835 to 1838, also saw him concerned with writing *Novye nachala geometrii s polnoi teoriei parallelnykh* ("New Principles of Geometry With a Complete Theory of Parallels"), which incorporated a version of his first work, the still unpublished *Geometriya*. The last two chapters of the book were abbreviated and translated for publication in Crelle's *Journal* in 1842. *Geometrische Untersuchungen zur Theorie der Parallellinien*, which he published in Berlin in 1840, is the best exposition of his new geometry; following its publication, in 1842, Lobachevsky was, on the recommendation of Gauss, elected to the Göttingen Gesellschaft der Wissenschaften. His last work, *Pangéométrie*, was published in Kazan in 1855–1856.

Lobachevskian Geometry. Lobachevsky's non-Euclidean geometry was the product of some two millennia of criticism of the *Elements*. Geometers had historically been concerned primarily with Euclid's fifth postulate: If a straight line meets two other straight lines so as to make the two interior angles on one side of the former together less than two right angles, then the latter straight lines will meet if produced on that side on which the angles are less than two right angles (it must be noted that Euclid understood lines as finite segments). This postulate is equivalent to the statement that given a line and a point not on it, one can draw through the point one and only one coplanar line not intersecting the given line. Throughout the centuries, mathematicians tried to prove the fifth postulate as a theorem either by assuming implicitly an equivalent statement (as did Posidonius, Ptolemy, Proclus, Thābit ibn Qurra, Ibn al-Haytham, Saccheri, and Legendre) or by directly substituting a more obvious postulate for it (as did al-Khayyāmī, al-Ṭūsī, and Wallis).

In his early lectures on geometry, Lobachevsky himself attempted to prove the fifth postulate; his own geometry is derived from his later insight that a geometry in which all of Euclid's axioms except the fifth postulate hold true is not in itself contradictory. He called such a system "imaginary geometry," proceeding from an analogy with imaginary numbers. If imaginary numbers are the most general numbers for which the laws of arithmetic of real numbers prove justifiable, then imaginary geometry is the most general geometrical system. It was Lobachevsky's merit to refute the uniqueness of Euclid's geometry, and to consider it as a special case of a more general system.

In Lobachevskian geometry, given a line a and a point A not on it (see Fig. 1), one can draw through A more than one coplanar line not intersecting a. It

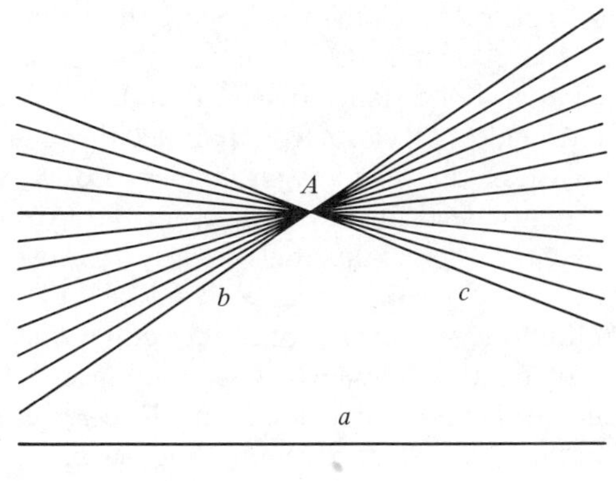

FIGURE 1

LOBACHEVSKY

follows that one can draw infinitely many such lines which, taken together, constitute an angle of which the vertex is A. The two lines, b and c, bordering that angle are called parallels to a, and the lines contained between them are called ultraparallels, or diverging lines; all other lines through A intersect a. If one measures the distance between two parallel lines on a secant equally inclined to each, then, as Lobachevsky proved, that distance decreases indefinitely, tending to zero, as one moves farther out from A. Consequently, when representing Lobachevskian parallels in the Euclidean plane, one often draws them conventionally as asymptotic curves (see Fig. 2). The

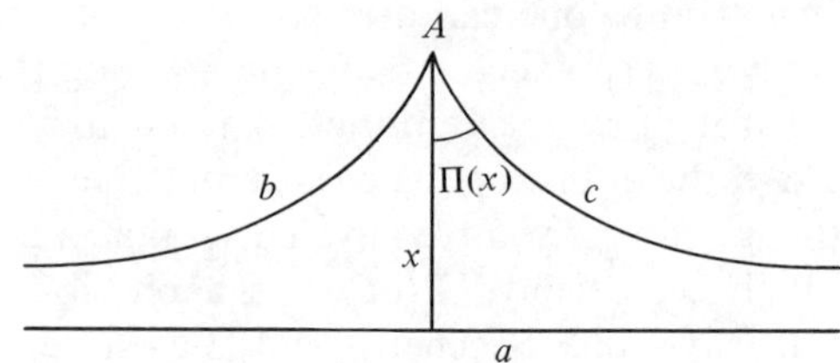

FIGURE 2

angle $\Pi(x)$ between the perpendicular x from point A to line a and a parallel drawn through A is a function of x. Lobachevsky showed that this function, named after him, can be expressed in elementary terms as

$$\cot \frac{\Pi(x)}{2} = e^{x};$$

clearly, $\Pi(0) = \pi/2$, and, for $x > 0$, $\Pi(x) < \pi/2$. Lobachevsky later proved that the distance between two diverging lines, where distance is again measured on a secant equally inclined to each, tends to infinity as one moves farther out from A; the distance has a minimum value when the secant is perpendicular to each line, and this perpendicular secant is unique.

A comparison of Euclidean and Lobachevskian geometry yields several immediate and interesting contrasts. In the latter, the sum of the angles of the right triangle of which the vertices are the point A, the foot of the perpendicular x on a, and a point on a situated at such a distance from x that the hypotenuse of the triangle lies close to a parallel through A is evidently less than two right angles. Lobachevsky proved that indeed for all triangles in the Lobachevskian plane the sum of the angles is less than two right angles.

In addition to pencils of intersecting lines and pencils of parallel lines, which are common to both the Euclidean and Lobachevskian planes, the latter also contains pencils of diverging lines, which consist of all perpendiculars to the same line. In both planes, the circle is the orthogonal trajectory of a pencil of intersecting lines, but the orthogonal trajectories of the other pencils differ in the Lobachevskian plane. That of a pencil of parallel lines (see Fig. 3) is called

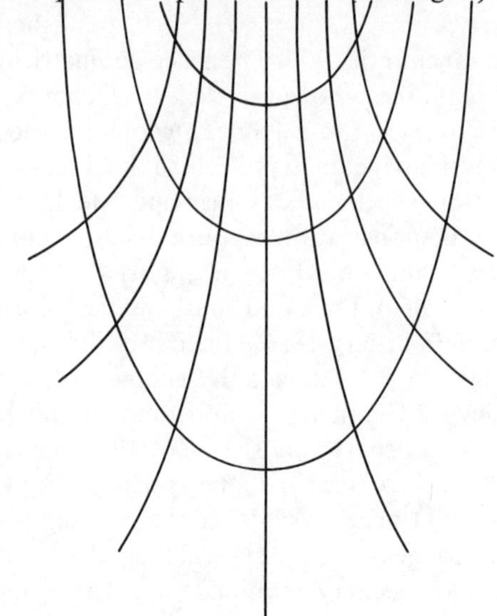

FIGURE 3

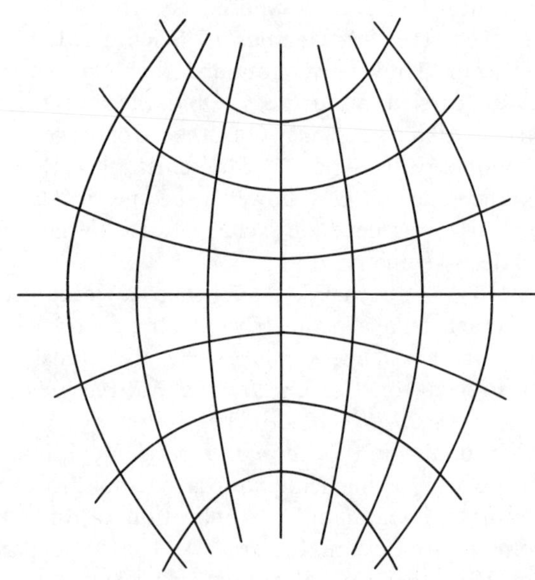

FIGURE 4

a horocycle, or limit circle; as the name implies, it is the curve toward which the circular trajectory of a pencil of intersecting lines tends as the intersection point tends to infinity. The orthogonal trajectory of a pencil of diverging lines (see Fig. 4) is called an equidistant, or hypercycle; it is the locus of points of the Lobachevskian plane that are equally distant from the common perpendicular of the diverging lines in the pencil, and this perpendicular is called the base of the equidistant.

LOBACHEVSKY

On the basis of these orthogonal trajectories, Lobachevsky also constructed a space geometry. By rotating the circle, the horocycle, and the equidistant about one of the orthogonal lines, he obtained, respectively, a sphere, a horosphere (or limit sphere), and an equidistant surface, the last being the locus of the points of space equally distant from a plane. Although geometry on the sphere does not differ from that of Euclidean space, the geometry on each of the two sheets of the equidistant surface is that of the Lobachevskian plane, the geometry on the horosphere that of the Euclidean plane.

Working from the geometry (and, hence, trigonometry) of the Euclidean plane on horospheres, Lobachevsky derived trigonometric formulas for triangles in the Lobachevskian plane. In modern terms these formulas are:

$$\cosh\frac{a}{q} = \cosh\frac{b}{q}\cosh\frac{c}{q} - \sinh\frac{b}{q}\sinh\frac{c}{q}\cos A,$$

$$\frac{\sinh\frac{a}{q}}{\sin A} = \frac{\sinh\frac{b}{q}}{\sin B} = \frac{\sinh\frac{c}{q}}{\sin C},$$

$$\cos A = -\cos B\cos C + \sin B\sin C\cosh\frac{a}{q}, \qquad (1)$$

where q is a certain constant of the Lobachevskian plane.

Comparing these formulas with those of spherical trigonometry on a sphere of radius r, that is,

$$\cos\frac{a}{r} = \cos\frac{b}{r}\cos\frac{c}{r} + \sin\frac{b}{r}\sin\frac{c}{r}\cos A,$$

$$\frac{\sin\frac{a}{r}}{\sin A} = \frac{\sin\frac{b}{r}}{\sin B} = \frac{\sin\frac{c}{r}}{\sin C},$$

$$\cos A = -\cos B\cos C + \sin B\sin C\cos\frac{a}{r}, \qquad (2)$$

Lobachevsky discovered that the formulas of trigonometry in the space he defined can be derived from formulas of spherical trigonometry if the sides a, b, c of triangles are regarded as purely imaginary numbers or, put another way, if the radius r of the sphere is considered as purely imaginary. Indeed, (2) is transformed into (1) if it is supposed that $r = qi$ and the correlations $\cos ix = \cosh x$ and $\sin ix = \sinh x$ are employed. In this Lobachevsky saw evidence of the noncontradictory nature of the geometry he had discovered. This idea can be made quite rigorous by introducing imaginary points of Euclidean space, producing the so-called complex Euclidean space; defining in that space the sphere of purely imaginary radius; and considering a set of points on that sphere that have real rectangular coordinates x and y and purely imaginary coordinate z. (Such a set of points in the complex Euclidean space was first considered in 1908 by H. Minkowski, in connection with Einstein's special principle of relativity; it is now called pseudo-Euclidean space.) The sphere of the imaginary radius in this space thus appears to be a hyperboloid of two sheets, each of which is a model of the Lobachevskian plane, while construction of this model proves it to be noncontradictory in nature.

Lobachevskian geometry is further analogous to spherical geometry. The area of triangle S in Lobachevsky's plane is expressed through its angles (in radian measure) by the formula

$$S = q^2(\pi - A - B - C), \qquad (3)$$

whereas the area of spherical triangle S is expressed through its angles (by the same measure) as

$$S = r^2(A + B + C - \pi). \qquad (4)$$

Formula (3) is transformed into formula (4) if $r = qi$.

Lines of the Lobachevskian plane are represented on the sphere of imaginary radius of the pseudo-Euclidean space by sections with planes through the center of the sphere. Intersecting lines can then be represented by sections with planes intersecting on a line of purely imaginary length; parallel lines by sections with planes intersecting on lines of zero length (isotropic lines); and diverging lines by sections with planes intersecting on lines of real length. Circles, horocycles, and equidistants of the Lobachevskian plane are represented by sections of the sphere of imaginary radius with planes that do not pass through its center.

If the sides a, b, c of the triangle are very small or the numbers q and r very large, it is necessary to consider only the first members in the series

$$\cosh x = 1 + \frac{x^2}{2} + \frac{x^4}{4!} + \cdots,$$

$$\sinh x = x = \frac{x^3}{3!} + \frac{x^5}{5!} + \cdots,$$

$$\cos x = 1 - \frac{x^2}{2} + \frac{x^4}{4!} - + \cdots,$$

$$\sin x = x - \frac{x^3}{3!} + \frac{x^5}{5!} - + \cdots. \qquad (5)$$

If it is assumed for formulas (1) and (2), above, that $\sinh x = x$ and $\cosh x = \cos x = 1$, $\sin x = x$ or $\cosh x = 1 + \frac{x^2}{2}$ and $\cos x = 1 - \frac{x^2}{2}$, these formulas become

$$a^2 = b^2 + c^2 - 2bc\cos A,$$

$$\frac{a}{\sin A} = \frac{b}{\sin B} = \frac{c}{\sin C},$$

$$A + B + C = \pi, \qquad (6)$$

in the Euclidean plane. Euclidean geometry may then be considered as a limiting case of both spherical and Lobachevskian geometry as r and q tend to infinity, or as a particular case for $r = q = \infty$. In analogy to the number $\frac{1}{r^2}$, which represents the curvature of the sphere, the number $-\frac{1}{q^2}$ is called the curvature of the Lobachevskian plane; the constant q is called the radius of curvature of this plane.

Lobachevsky recognized the universal character of his new geometry in naming it "pangeometry." He nevertheless thought it necessary to establish experimentally which geometry—his or Euclid's—actually occurs in the real world. To this end he made a series of calculations of the sums of the angles of triangles of which the vertices are two diametrically opposed points on the orbit of the earth and one of the fixed stars Sirius, Rigel, or 28 Eridani. Having established that the deviation of these sums from π is no greater than might be due to errors in observation, he concluded that the geometry of the real world might be considered as Euclidean, whence he also found "a rigorous proof of the theorem of parallels" as set out in his work of 1826. In explaining his calculations (in "O nachalakh geometrii" ["On the Principles of Geometry"]), Lobachevsky noted that it is possible to find experimentally the deviation from π of the sum of the angles of cosmic triangles of great size; in a later work (*Novye nachala geometrii s polnoi teoriei parallelnykh* ["New Principles of Geometry With a Complete Theory of Parallels"]) he moved to the opposite scale and suggested that his geometry might find application in the "intimate sphere of molecular attractions."

In his earlier papers Lobachevsky had defined imaginary geometry on an a priori basis, beginning with the supposition that Euclid's fifth postulate does not hold true and explaining the principle tenets of his new geometry without defining it (although he did describe the results of his experiment to prove his theorem of parallels). In "Voobrazhaemaya geometriya" ("Imaginary Geometry"), however, he built up the new geometry analytically, proceeding from its inherent trigonometrical formulas and considering the derivation of these formulas from spherical trigonometry to guarantee its internal consistency. In the sequel to that paper, "Primenenie voobrazhaemoi geometrii k nekotorym integralam" ("Application of Imaginary Geometry to Certain Integrals"), he applied geometrical considerations in Lobachevskian space to the calculation of known integrals (in order to make sure that their application led to valid results), then to new, previously uncalculated integrals.

In *Novye nachala geometrii s polnoi teoriei parallelnykh* ("New Principles of Geometry With a Complete Theory of Parallels"), Lobachevsky, after criticizing various demonstrations of the fifth postulate, went on to develop the idea of a geometry independent of the fifth postulate, an idea presented in his earliest *Geometriya* (of which a considerable portion was encompassed in the later work). The last two chapters of the book—on the solution of triangles, on given measurements, and on probable errors in calculation—were connected with his attempts to establish experimentally what sort of geometry obtains in the real world. Lobachevsky's last two books, *Geometrische Untersuchungen* and *Pangéométrie*, represent summaries of his previous geometrical work. The former dealt with the elements of the new geometry, while the latter applied differential and integral calculus to it.

At the same time as Lobachevsky, other geometers were making similar discoveries. Gauss had arrived at an idea of non-Euclidean geometry in the last years of the eighteenth century and had for several decades continued to study the problems that such an idea presented. He never published his results, however, and these became known only after his death and the publication of his correspondence. Jànos Bolyai, the son of Gauss's university comrade Farkas Bolyai, hit upon Lobachevskian geometry at a slightly later date than Lobachevsky; he explained his discovery in an appendix to his father's work that was published in 1832. (Since Gauss did not publish his work on the subject, and since Bolyai published only at a later date, Lobachevsky clearly holds priority.)

It may be observed that Lobachevsky's works in other areas of mathematics were either directly relevant to his geometry (as his calculations on definite integrals and probable errors of observation) or results of his studies of foundations of mathematics (as his works on the theory of finites and the theory of trigonometric series). His work on these problems again for the most part paralleled that of other European mathematicians. It is, for example, worth noting that in his algebra Lobachevsky suggested a method of separating roots of equations by their repeated squaring, a method coincident with that suggested by Dandelin in 1826 and by Gräffe in 1837. His paper on the convergence of trigonometric series, too, suggested a general definition of function like that proposed by Dirichlet in 1837. (Lobachevsky also gave a rigorous definition of continuity and differentiability, and pointed out the difference between these notions.)

Recognition of Lobachevskian Geometry. Lobachevsky's work was little heralded during his lifetime. M. V. Ostrogradsky, the most famous mathematician of the St. Petersburg Academy, for one, did not understand Lobachevsky's achievement, and published an uncomplimentary review of "O nachalakh geometrii" ("On the Principles of Geometry"); the magazine *Syn otechestva* soon followed his lead, and in 1834 issued a pamphlet ridiculing Lobachevsky's paper. Although Gauss, who had received a copy of the *Geometrische Untersuchungen* from Lobachevsky, spoke to him flatteringly of the book, studied Russian especially to read his works in their original language, and supported his election to the Göttingen Gesellschaft der Wissenschaften, he never publicly commented on Lobachevsky's discovery. His views on the new geometry became clear only after the publication, in 1860–1865, of his correspondence with H. C. Schumacher. Following this, in 1865, the English algebraist Cayley (who had himself paved the way for the theory of projective metrics in 1859 with his "Sixth Memoir Upon Quantics") brought out his "A Note Upon Lobachevsky's Imaginary Geometry," from which it is evident that he also failed to understand Lobachevsky's work.

The cause of Lobachevskian geometry was, however, furthered by Hoüel, one of its earliest proponents, who in 1866 brought out a French translation of *Geometrische Untersuchungen*, with appended extracts from the Gauss-Schumacher correspondence. The following year he also published Bolyai's appendix on non-Euclidean geometry, which was translated into Italian by Battaglini in 1867. Hoüel's own *Notices sur la vie et les travaux de N. I. Lobachevsky* appeared in 1870. In the meantime, Lobachevsky's *Geometrische Untersuchungen* had been translated into Russian by A. V. Letnikov; it was published in 1868 in the newly founded Moscow magazine *Matematichesky sbornik*, together with Letnikov's article "O teorii parallelnykh linii N. I. Lobachevskogo" ("On the Theory of Parallel Lines by N. I. Lobachevsky").

These translations and reviews were soon augmented by extensions of Lobachevskian geometry itself. In 1858 Beltrami published in the *Giornale di matematiche* his "Saggio di interpretazione della geometria non-euclidea," in which he established that the intrinsic geometry of the pseudosphere and other surfaces of constant negative curvature coincides with the geometry of part of the Lobachevskian plane and that an interpretation of the whole Lobachevskian plane can be constructed in the interior of a circle in the Euclidean plane. This interpretation can be derived by projecting a hemisphere of imaginary radius in pseudo-Euclidean space from its center onto a Euclidean plane tangent to this hemisphere. The article was translated into French by Hoüel in 1869 for publication in the *Annales scientifiques de l'École Normale Supérieure*.

In 1870 Weierstrass led a seminar on Lobachevsky's geometry at the University of Berlin; the young Felix Klein was one of the participants. Weierstrass' own contribution to the subject, the so-called Weierstrass coordinates, are essentially rectangular coordinates of a point on a hemisphere of imaginary radius in pseudo-Euclidean space. Klein compared Lobachevskian geometry with Cayley's projective metrics to establish that Lobachevsky's geometry is in fact one of Cayley's geometries, which Cayley himself (although he was acquainted with Lobachevskian geometry) had failed to notice. The Lobachevskian plane can be regarded as the interior domain of a conic section on a projective plane; when this conic section is a circle, representation of the Lobachevskian plane on the projective plane coincides with Beltrami's interpretation. It thus follows that motions of the Lobachevskian plane are represented in Beltrami's interpretation by projective transformations of the plane mapping a circle into itself. Lines of the Lobachevskian plane are represented in Beltrami's interpretation by chords of the circle, while parallel lines are represented by chords intersecting on the circumference, and diverging lines by non-intersecting chords. There are analogous interpretations for three-dimensional and multidimensional Lobachevskian spaces, in which Beltrami's circle or Klein's conic section is replaced by a sphere or oval quadric, e.g., ellipsoid.

Poincaré made two important contributions to Lobachevskian geometry at somewhat later dates. In 1882 he suggested an interpretation of the Lobachevskian plane in terms of a Euclidean semiplane; in this intrepretation, motions of the Lobachevskian plane are represented by inversive transformations, mapping the border of the semiplane into itself, while Lobachevskian lines are mapped by perpendiculars to this border or by semicircles with centers on it. An important property of this interpretation is that angles in the Lobachevskian plane are mapped conformally. Taking as given that inversive transformations on the plane are represented as fractional linear functions of a complex variable, Poincaré used such an interpretation for the theory of automorphic functions.

Poincaré's interpretation is often described in other terms, as when a semiplane is mapped by inversive transformation into the interior of a circle. An interpretation in terms of the circle can also be presented by projecting the hemisphere of imaginary radius in pseudo-Euclidean space from one of its points onto the Euclidean plane perpendicular to the radius through this point. This projection is analogous to stereographic projection, through which the preservation of angles in Poincaré's interpretation is explained. In 1887 Poincaré suggested a second interpretation of the Lobachevskian plane, this one in terms of a two-sheeted hyperboloid coinciding with that on the hemisphere of imaginary radius as already defined.

A further important development of Lobachevsky's geometry came from Riemann, whose address of 1854, *Uber die Hypothesen, welche der Geometrie zu Grundliegen*, was published in 1866. Developing Gauss's idea of intrinsic geometry of a surface, Riemann presented a notion of multidimensional curved space (now called Riemannian space). Riemannian space of constant positive curvature (often called Riemannian elliptic space) is represented by the geometry on a sphere in fourth-dimensional or multidimensional Euclidean space or in the space with a projective metric in which an imaginary quadric plays the part of a conic section (or Klein's quadric). It subsequently, then, became clear that Lobachevskian space (often called hyperbolic space) is the Riemannian space of constant negative curvature. It was with this in mind that Klein, in his memoir of 1871, "Über die sogenannte nicht-euklidische Geometrie," chose to consider all three geometries—Euclidean, elliptic, and hyperbolic—from a single standpoint, whereby groups of motions of all three are subgroups of the group of projective transformations of projective space.

Here the extension of the concept of space, which was introduced by Lobachevsky and Riemann, merges with the concept of group, which was introduced by Galois, a synthesis that Klein developed further in 1872 in *Vergleichende Betrachtungen über neuere geometrische Forschungen* (also known as the *Erlanger Programm*) in which he arrived at a general view of various geometries as theories of invariants of various continuous groups of transformations. A new stage in the development of mathematics thus began, one in which the mathematics of antiquity and the Middle Ages (that is, the mathematics of constant magnitudes) and the mathematics of early modern times (that is, the mathematics of variable magnitudes) were replaced by modern mathematics—the mathematics of many geometries, of many algebras, and of many mathematical systems having no classical analogues. A number of these systems found applications in modern physics—pseudo-Euclidean geometry, one of the projective metrics, figures in the special theory of relativity; Riemannian geometry appears in the general theory of relativity; and group theory is significant to quantum physics.

The presence of specifically Lobachevskian geometry is felt in modern physics in the isomorphism of the group of motions of Lobachevskian space and the Lorentz group. This isomorphism opens the possibility of applying Lobachevskian geometry to the solution of a number of problems of relativist quantum physics. Within the framework of the general theory of relativity, the problem of the geometry of the real world, to which Lobachevsky had devoted so much attention, was solved; the geometry of the real world is that of variable curvature, which is on the average much closer to Lobachevsky's than to Euclid's.

BIBLIOGRAPHY

I. Original Works. Lobachevsky's writings have been brought together in *Polnoe sobranie sochinenii* ("Complete Works"), 5 vols. (Moscow–Leningrad, 1946–1951): vols. I–III, geometrical works; vol. IV, algebraical works; vol. V, works on analysis, the theory of probability, mechanics, and astronomy. His geometrical works were also collected as *Polnoe sobranie sochinenii po geometrii* ("Complete Geometrical Works"), 2 vols. (Kazan, 1883–1886): vol. I, works in Russian; vol. II, works in French and German.

Translations include J. Hoüel, *Études géométriques sur la théorie des parallèles* (Paris, 1866; 2nd ed., 1900); G. B. Halsted, *Geometrical Researches on the Theory of Parallels*, Neomonic ser. no. 4 (Austin, Tex., 1892; repr. Chicago–London, 1942); and *New Principles of Geometry With Complete Theory of Parallels*, Neomonic ser. no. 5 (Austin, Tex., 1897); F. Engel, *Zwei geometrische Abhandlungen* (Leipzig, 1898); F. Mallieux, *Nouveaux principes de la géométrie avec une théorie complète des parallèles* (Brussels, 1901); and H. Liebmann, *Imaginäre Geometrie und Anwendungen der imaginären Geometrie auf einige Integrale* (Leipzig, 1904).

II. Secondary Literature. Among works on Lobachevsky are the following: A. A. Andronov, "Gde i kogda rodilsya N. I. Lobachevsky?" ("Where and When Was Lobachevsky born?"), in *Istoriko-matematicheskie issledovaniya*, **9** (1956); R. Bonola, *La geometria non-euclidea: Esposizione storico-critico del suo sviluppo* (Bologna, 1906); translated into German as *Die nicht-euklidische Geometrie. Historisch-kritische Darstellung ihrer Entwicklung*, 3rd ed. (Leipzig, 1921); translated into Russian as *Neevklidova geometriya. Kritiko-istoricheskoe issledovanie e razvitiya* (St. Petersburg, 1910); translated into English as *Non-Euclidean Geometry. A Critical and Historical Study of Its Developments* (New York, 1955); J. L. Coolidge, *The Elements of Non-Euclidean Geometry* (Oxford, 1909); H. S. M. Coxeter, *Non-Euclidean Geometry*, 3rd ed. (Toronto, 1957); V. M. Gerasimova, *Ukazatel literatury po geometrii Lobachevskogo i razvitiyu e idei* ("A Guide to Studies on Lobachevskian Geometry and the Development of Its Ideas"; Moscow, 1952); B. V. Gnedenko, "O rabotakh N. I. Lobachevskogo po teorii veroyatnostei" ("On Lobachevsky's Works on the Calculus of Probabilities"), in *Istoriko-matematicheskie issledovaniya*, **2** (1949); N. I. Idelson, "Lobachevskii-astronom" ("Lobachevsky as an Astronomer"), *ibid.*

Also of value are V. F. Kagan, *Lobachevsky*, 2nd ed. (Moscow–Leningrad, 1948); and *Osnovaniya geometrii* ("Foundations of Geometry"), 2 vols. (I, Moscow–Leningrad, 1949; II, Moscow, 1956); E. K. Khilkevich, "Iz istorii rasprostraneniya i razvitiya idei N. I. Lobachevskogo v 60-70-kh godakh XIX stoletiya" ("The History of the Spread and Development of N. I. Lobachevsky's Ideas in the 60th to 70th Years of the Nineteenth Century"), in *Istoriko-matematicheskie issledovaniya*, **2** (1949); F. Klein, *Vorlesungen über nichteuklidische Geometrie*, 3rd ed. (Berlin, 1928), also translated into Russian as *Neevklidova geometriya* (Moscow–Leningrad, 1936); B. L. Laptev, "Teoriya parallelnykh pryamykh v rannikh rabotakh Lobachevskogo" ("Theory of Parallel Lines in Lobachevsky's Early Works"), in *Istoriko-matematicheskie issledovaniya*, **4** (1951); H. Liebmann, *Nicht-euclidische Geometrie*, 3rd ed. (Berlin, 1923); G. L. Luntz, "O rabotakh N. I. Lobachevskogo po matematicheskomu analizu" ("On N. I. Lobachevsky's Work in Mathematical Analysis"), in *Istoriko-matematicheskie issledovaniya*, **2** (1949); M. B. Modzalevsky, *Materialy dlya biografii N. I. Lobachevskogo* ("Materials for N. I. Lobachevsky's Biography"; Moscow–Leningrad, 1948); V. V. Morozov, "Ob algebraicheskikh rukopisyakh N. I. Lobachevskogo" ("On Algebraic Manuscripts of N. I. Lobachevsky"), in *Istoriko-matematicheskie issledovaniya*, **4** (1951); and A. P. Norden, ed., *Ob osnovaniyakh geometrii. Sbornik klassicheskikh rabot po geometrii Lobachevskogo i razvitiyu e idei* ("On Foundations of Geometry. Collection of Classic Works on Lobachevskian Geometry and the Development of Its Ideas"; Moscow, 1956).

See also B. A. Rosenfeld, *Neevklidovy geometrii* (Moscow, 1955); "Interpretacii geometrii Lobachevskogo" ("Interpretations of Lobachevskian Geometry"), in *Istoriko-matematicheskie issledovaniya*, **9** (1956); and *Neevklidovy prostranstva* ("Non-Euclidean Spaces"; Moscow, 1968); D. M. Y. Sommerville, *Bibliography of Non-Euclidean Geometry* (London, 1911); and *The Elements of Non-Euclidean Geometry* (London, 1914); S. A. Yanovskaya, "O mirovozzrenii N. I. Lobachevskogo" ("On N. I. Lobachevsky's Outlook"), in *Istoriko-matematicheskie issledovaniya*, **4** (1951); and A. P. Youschkevitch and I. G. Bashmakova, " 'Algebra ili vychislenie konechnykh' N. I. Lobachevskogo" (" 'Algebra or Calculus of Finites' by N. I. Lobachevsky"), in *Istoriko-matematicheskie issledovaniya*, **2** (1949).

B. A. Rosenfeld

L'OBEL (or **LOBEL), MATHIAS DE** (*b.* Lille, France, 1538; *d.* Highgate, near London, England, 3 March 1616), *botany.*

L'Obel's name is ingeniously alluded to in the coat of arms on his books: two poplar trees (in French "aubel" means "abele," or "white poplar") constitute part of the design. Few details of his life are known. He seems to have had a powerful personality—rough, violent, passionate, tireless—worthy of a leader among the sixteenth-century restorers of learning. At the age of sixteen he was already attracted to botany and materia medica. According to Édouard Morren he undoubtedly practiced medicine, although there is no known evidence that he ever received a medical degree. In 1565 (he was then twenty-seven) he became the favorite student of Guillaume Rondelet, professor at the University of Montpellier. There he met Jacques Uitenhove of Ghent, and a young Provençal, Pierre Pena, who became his friend, his companion in study and in botanizing, and his collaborator.

It is very probable that at their first meeting Rondelet discerned L'Obel's intellectual superiority, for at his death, on 20 July 1566, he bequeathed the latter all his botanical manuscripts. L'Obel then spent three more years in Montpellier, writing a book with Pierre Pena, which appeared in 1570 (or 1571) under the title *Stirpium adversaria nova.* Although one may criticize L'Obel's inelegant, even barbarous, Latin, this work is nevertheless one of the milestones of modern botany. A collection of notes and data on 1,200–1,300 plants that he himself had gathered and observed in the vicinity of Montpellier, in the Cévennes, in the Low Countries, and in England, it also furnished precise information on other subjects, including the making of beer and the cultivation of exotic plants and of chicory.

L'Obel held that the thorough and exact observation of men and things constituted the basis of the two sciences that interested him: medicine and botany. This conviction is clearly evident in his book. Before contemporary botanists he divided the plants into groups according to the form of their leaves (whole, divided, more or less compound). It seems that he had an inkling of natural families and that he attempted to use, if not the terms, at least the concepts of genus and family.

The second edition of *Adversaria* (1576) was published with an appendix including many engravings larger than those in the 1571 edition. *Stirpium observationes*, a sort of complement to the *Adversaria*, was joined to it under the title *Plantarum seu stirpium historia* (1576). In 1581, Christophe Plantin published *Kruydtboeck*, a Flemish translation of the *Stirpium*, and, in a separate volume, the illustrations that L'Obel prefaced with a brief table—"Elenchus plantarum fere congenerum"—which presented the plants according to affinities. Linnaeus subsequently used the table in his *Species plantarum.*

Finding it impossible to work in a country torn by civil war, L'Obel left the Low Countries, where he had been living since 1571, following the assassination of William of Orange in July 1584. He went to England, his wife's native country, where he remained for the rest of his life.

Despite his conceit and unpolished language, which earned him many enemies (he was involved in an unfortunate dispute with Mattioli), L'Obel was one of the great pre-Linnaean botanists. His name was perpetuated by Plumier, who in 1702 dedicated to him the genus *Lobelia*—which Linnaeus later called *Scaevola*, at the same time giving the name "Lobelia" to a genus of aquatic plant.

BIBLIOGRAPHY

I. Original Works. L'Obel's first work was *Stirpium adversaria nova* (London, 1570), written with Pierra Pena; the colophon is dated 1571, and the work may have been printed at Antwerp by Plantin. *Plantarum seu stirpium historia*, 2 vols. (Antwerp, 1576), comprising the 2nd ed. of *Adversaria* together with *Stirpium observationes*, contained 1,486 engravings, most of which are not original. *Kruydtboeck* (Antwerp, 1581), a Flemish trans. of *Stirpium historia*, was followed the same year by *Icones stirpium*, a 2-vol. work printed at Antwerp that included the "Elenchus" and all engravings published to date; a 2nd ed. appeared in 1591. *Balsami, opobalsami, carpobalsami et xylobalsami* (London, 1598) was followed by a new ed. of the *Adversaria* entitled *Dilucidae simplicium medicamentorum explicationes et stirpium adversaria* (London, 1605) that included several new treatises. *Stirpium illustrationes* was published posthumously (London, 1655).

II. Secondary Literature. On L'Obel and his work, see B. C. Dumortier, "Discours sur les services rendus par les belges à la botanique," in *Bulletin. Société r. de botanique de Belgique*, **1** (1862), 16; L. Legré, *La botanique en Provence au XVIe siècle*, I, *Pierre Pena et Mathias de Lobel* (Marseilles, 1899); C. F. A. Morren, "Prologue à la mémoire de L'Obel," in *Belgique horticole*, **2** (1852), v–xviii; C. J. É. Morren, *Mathias de L'Obel. Sa vie et ses oeuvres* (Liége, 1875); and J. É. Planchon, *Rondelet et ses disciples* (Montpellier, 1866).

J. C. Mallet
P. Jovet

LOCKE, JOHN (*b*. Wrington, Somersetshire, England, 29 August 1632; *d*. Oates, Essex, England, 28 October 1704), *philosophy*.

John Locke was the most important British philosopher of the Age of Reason. If the modern Western world has been shaped by scientists, merchants, statesmen, and industrialists, Locke was the first philosopher to expound their view of life, articulate their aspirations, and justify their deeds. No philosopher has exercised a greater influence. Yet it could be said that Locke was not a philosopher at all; he was certainly not a metaphysician in the same sense as his contemporaries Leibniz or Spinoza. Locke offered no all-embracing system to explain the nature of the universe. On the contrary, he tried to show that human understanding is so limited that such comprehensive knowledge is beyond man's reach. Although he did not have the answers to the philosophical problems he formulated, Locke was able to establish the importance of science as an object of philosophical analysis.

Two powerful streams in seventeenth-century thought, the semiskeptical rational theorizing of Descartes and the *ad hoc* scientific experimentation of Bacon and the Royal Society, merged in Locke. Because the streams were so different, their union was imperfect, but his mind was the meeting point that marked a new beginning, not only in theoretical philosophy, but in man's approach to the problems of practical life. Locke, one might almost say, had the first modern mind. Descartes was still in many ways a medieval thinker, with his philosophy bound to theology; and even Gassendi, who anticipated much of Locke, did not free himself completely from its hold. By divorcing philosophy from theology, Locke placed the study of philosophy within the boundaries of man's experience: "Our portion," he wrote, "lies only here in this little spot of earth, where we and all our concernments are shut up."

Yet Locke was not an atheist, and in his capacity as theologian he expounded his own thoughts about God. He quarreled with bishops and with the orthodox of most denominations. He maintained that a Christian need believe no more than the single proposition "that Christ is the Messiah"; but to that minimal creed he clung with the firmest assurance. He had a quiet and steady faith in the immortality of the soul and in the prospect of happiness in the life to come. He had no belief in miracles and no patience with people who had mystical experiences or visions of God. He detested religious enthusiasm, but in an unemotional way he was, like Newton, a deeply religious man.

Locke was born at Wrington, Somersetshire, in western England. His grandfather, Nicholas Locke, was a successful clothier in that county. His father, also named John Locke, was a less prosperous lawyer and clerk to the local magistrates. Baptized by Samuel Crook, a leading Calvinist intellectual, Locke was brought up in an atmosphere of austerity and discipline. He was ten when the Civil War broke out, and his father was mounted as a captain of Parliamentary horse by Alexander Popham, a rich local magistrate-turned-colonel. Apart from demolishing some images in Wells Cathedral, the two officers saw little action, but a grateful Popham became the patron of his captain's eldest son. A few years later, when Westminster School was taken over by the Parliament, Popham found a place for his protégé in what was then the best boarding school in the country.

LOCKE

At Westminster, Locke came under the influence of the Royalist headmaster Richard Busby, whom the Parliamentary governors had imprudently allowed to remain in charge of the school. Locke left Westminster in 1652 to become an undergraduate at Christ Church, Oxford, where he once again came under Puritan influence. Although he did not wholly enjoy the university, he received the M.A. degree and was appointed to a teaching position, lecturing on such subjects as natural law, although he was more interested in medicine.

In the summer of 1666 Locke became friends with Anthony Ashley Cooper, then Lord Ashley and later the first earl of Shaftesbury. Although not yet the leader of a party, Shaftesbury was already the outstanding politician of liberalism and the most forceful champion of religious toleration. In 1667, at the age of thirty-four, Locke went to live at Shaftesbury's house in London. His Oxford career had not been particularly distinguished: he had been a temporary lecturer and a censor at Christ Church; he had become friends with Robert Boyle, one of the founders of the Royal Society, and helped him by collecting scientific data; and he had studied medicine. But he had done no important laboratory work and had failed to get a medical degree. Even so, it was as a domestic physician that Locke entered Shaftesbury's household, and he soon proved himself an able doctor by saving his patron's life (as Shaftesbury believed) when it was threatened by a suppurating cyst of the liver. Convinced that Locke was far too great a genius to be spending his time on medicine alone, Shaftesbury encouraged other pursuits, and under his patronage Locke discovered his own capabilities.

Locke first became a philosopher. At Oxford he had been, like Hobbes before him, bored and dissatisfied with the medieval Aristotelian philosophy that was taught there. Through Descartes, Locke became acquainted with the "new philosophy," and discussions with Shaftesbury and other London friends led him to write, in his fourth year under Shaftesbury's roof, the earliest drafts of his masterpiece, *An Essay Concerning Human Understanding*. In London, Locke also met the physician Thomas Sydenham, who introduced him to the new clinical method that he had learned at Montpellier. Shaftesbury himself introduced Locke to the study of economics and gave him his earliest experience in political administration.

Locke stayed with Shaftesbury intermittently until the latter's death in 1683. In those fifteen years Shaftesbury's Protestant zeal carried him to the point of organizing a rebellion over the Catholic James's legitimate right of succession. But the plot was nipped, and Shaftesbury withdrew to Holland, where he died a month later. Locke followed the example of Shaftesbury and fled to Holland, remaining in exile until William of Orange invaded England in 1688 and reclaimed the country for Protestantism and liberty. Locke returned to England in 1689 and devoted his remaining fifteen years to scholarship and public service.

While in Holland, Locke had completed *An Essay Concerning Human Understanding* and his first *Letter for Toleration* and may have done some work on his *Civil Government*. The *Letter for Toleration* was published in Locke's original Latin at Gouda in 1689 and an English translation made by the Socinian William Popple—"without my privity" Locke later said—was published in London a few months later. The *Essay* and the *Civil Government* were also brought out by various London booksellers in the winter of 1689–1690. Only the *Essay* bore Locke's name, but its success was so great that he became famous throughout Europe.

In the "Epistle to the Reader," which begins the *Essay*, Locke says that in an age of such "master builders" as Boyle, Sydenham, Huygens, and "the incomparable Mr. Newton" it is, for him, "ambition enough to be employed as an under-labourer in clearing the ground a little and removing some of the rubbish that lies in the way of knowledge." Despite this modest explanation of purpose Locke provided, among other things, the first modern philosophy of science, in which the Cartesian "idea" was a recurrent theme. Locke says that we have ideas in our minds not only when we think but when we see, hear, smell, taste, and feel as well. The core of his epistemology is the notion that the objects of perception are not things but ideas derived in part from things in the external world and dependent, to some extent, on our own minds for their existence. Locke defines an "idea" as the "object of the understanding," whether it is a notion, an entity, or an illusion; perception is for him a "species of understanding."

Locke continues his *Essay* with an attack on the currently established opinion that certain ideas are innate. He claims that they have been considered innate only because people cannot remember first having learned them. Locke believed that we are born in total ignorance and that even our theoretical ideas of identity, quantity, and substance are derived from experience. He says that a child gets ideas of black and white, of sweet and bitter before he gets an idea of abstract principles such as identity or impossibility. "The senses at first let in particular ideas, and furnish the yet empty cabinet . . . ." The mind later abstracts these theoretical ideas and so "comes to be furnished with ideas and language, the materials about which to exercise the discursive faculty." In the child's development "the use of reason becomes daily more visible as these materials that give it employment increase."

Locke had thus to defend the assumption that everything which he calls an idea is derived from sensation, although he admits that an idea may also be produced by reflection—"remembering, considering, reasoning." He classifies ideas as simple (those which the mind receives passively) and complex (those produced by the exercise of the mind's own powers). In the chapters on simple ideas he sets out the main lines of his theory of perception. Most people, if asked what it is that they see, smell, hear, taste, or touch, would answer "things," although they might add "but sometimes illusions, chimeras, mirages, which are not real things." They would probably maintain that there are two elements in an act of perception: the observer and the object. Locke differs from this view in two respects. First, he claims that what we perceive is always an idea, as distinct from a thing; second, that there are not two but three elements in perception: the observer, the idea, and the object that the idea represents.

The reasoning that led Locke to this conclusion is not difficult to appreciate. We look at a penny. We are asked to describe it. It is round, brown, and of modest dimensions. But do we really see just this? We think again and realize that more often than not what we see is elliptical, not circular; in some lights it is golden, in others black; close to the eye it is large, seen from afar it is tiny. The actual penny, we are certain, cannot be both circular and elliptical, both golden all over and black all over. So we may be led to agree that there must be something which is the one and something which is the other, something which changes and something which does not change, the elliptical "penny" we see and the real circular penny, or, in Locke's words, the "idea" in the mind of the observer and the material "body" itself.

Today Locke's theory of perception is defended on the basis of the physicist's description of the structure of the universe and the physiologist's description of the mechanism of perception. To a child a penny is something that looks brown, feels warm, and tastes sharp; to the scientist it is a congeries of electrons and protons. The scientist speaks of certain light waves striking the retina of the observer's eye while waves of other kinds strike different nerve terminals, producing those modifications of the nervous system that are called "seeing," "feeling," or "tasting" a penny. Neither the electrons nor the protons, neither the external waves nor the internal modifications are brown or warm or astringent. The scientist then, in a certain sense, differentiates what Locke called the secondary qualities—color, taste, sound—which Locke said depended on the observer's mind for their existence. At the same time science seems to accept the objective existence of the qualities that Locke called primary and that he thought belonged to material bodies themselves—impenetrability, extension, figure, mobility, and number. Although the language of primary and secondary qualities was used by earlier theorists, it is Locke's analysis of the distinction which makes it of significance both to science and to common sense. Most people would probably agree that they could imagine an object divested one by one of its qualities of taste, smell, and so forth; but they could not imagine a body divested of impenetrability, shape, size, or position in space. A body without primary qualities would not exist at all.

While there is much to be said for Locke's epistemology, it is not without its faults. If we are aware in our perceptual experience only of ideas (which represent objects) and never of objects themselves, there can be no means of knowing what, if anything, is represented by those ideas. The human predicament, according to Locke's account, is that of a man permanently imprisoned in a sort of diving bell, receiving some signals from without and some from within his apparatus but having no means of knowing which, if any, signals come from outside; hence he has no means of testing their authenticity. Man, therefore, cannot have any definite knowledge whatever of the external world.

In later chapters of the *Essay* Locke places an even heavier emphasis on human ignorance. Our knowledge, he says, "is not only limited to the paucity and imperfections of the ideas we have, and which we employ about it," but is still more circumscribed. Our knowledge of identity and diversity in ideas extends only as far as our ideas themselves; our knowledge of their coexistence extends only a little way, because knowledge of any necessary connection between primary and secondary qualities is unattainable. However, with the area of certainty thus diminished, Locke does not deny the possibility of an assurance which falls short of perfect knowledge. We can have probable knowledge, even though we cannot have certain knowledge. Moreover, unlike most of his successors in empiricist philosophy, Locke admits the existence of substance, which he says is somehow present in all objects even though we do not see or feel it. What we see and feel are the primary and secondary qualities, which are propped up by substance. Beyond that the subject must necessarily remain a mystery:

> It seems probable to me that the simple ideas we receive from sensation and reflection are the boundaries of our thoughts; beyond which the mind, whatever efforts it would make, is not able to advance one jot, nor can it make any discoveries, when it would pry into the nature and hidden causes of those ideas.

The tone of the *Essay* is at once moral and pragmatic. Its style is homely rather than elegant, its construction informal and even amateurish. The "pursuit of Truth," Locke says, "is a duty we owe to God . . . and a duty we owe also to ourselves"; here utility is at one with piety. Truth, as Locke defines it, is the "proper riches and furniture of the mind," and he does not claim to have added to that stock, but rather to have shown the conditions under which the mind could acquire truth:

> We have no reason to complain that we do not know the nature of the sun or the stars, that the consideration of light itself leaves us in the dark and a thousand other speculations in nature, since, if we knew them, they would be of no solid advantage, nor help to make our lives the happier, they being but the useless employment of idle or over-curious brains. . . .

Locke's theory of knowledge has obvious implications for a theory of morals. The traditional view in Locke's time was that some sort of moral knowledge was innate in the human person. Locke thought otherwise. What God or Nature had given man was a faculty of reason and a sentiment of self-love. Reason in combination with self-love produced morality and could discern the general principles of ethics, or natural law; and self-love should lead men to obey those principles.

For Locke, Christian ethics was natural ethics. The teaching of the New Testament was a means to an end, happiness in this life and in the life to come. Loving one's neighbor and otherwise obeying the precepts of the Saviour was a way to that end. The reason for doing what Christ said was not simply that He had said it, but that by doing it one promoted one's happiness. There was no need to ask why anyone should desire happiness, because all men were impelled by their natural self-love to desire it.

Wrongdoing was thus for Locke a sign of ignorance. People did not always realize that long-term happiness could usually be bought only at the cost of short-term pleasures. Folly drove them to destroy their own well-being. If people were enlightened, if they used their own powers of reason, they would be good; if they were prudent, reflective, calculating, instead of moved by the transitory winds of impulse and emotion, they would have what they most desired. There is perhaps in this system of morals something rather naïve and commonplace; but Locke was in many ways a very ordinary thinker, inspired by a prophetic common sense.

BIBLIOGRAPHY

I. Original Works. See H. O. Christopherson, *A Bibliographical Introduction to the Study of John Locke* (Oslo, 1930). Collected editions and selections of Locke's work are *The Works*, 3 vols. (London, 1714), repr. throughout the eighteenth century; *The Works*, Edmund Law, ed., 4 vols. (London, 1777), the best collected ed.; *The Philosophical Works*, J. A. St. John, ed. (1843); and *Locke on Politics, Religion and Education*, M. Cranston, ed. (1965).

Separate works include *Epistola de tolerantia ad clarissimum virum* . . . (Gouda, 1689), English trans. (London, 1689); *Two Treatises of Government* (London, 1690); P. Laslett, ed. (Cambridge, 1960), a critical ed. with original text, intro., and commentary; *An Essay Concerning Human Understanding* (London, 1690), enl. (1694, 1700); A. C. Fraser, ed., 2 vols. (Oxford, 1894), prepared from a collation of 4 eds. published in Locke's lifetime; John W. Yolton, ed. (London–New York, 1961), the best recent text; see also *An Early Draft of Locke's Essay*, R. I. Aaron and J. Gibbs, eds. (Oxford, 1936), prepared from one of three surviving MSS; *A Second Letter Concerning Toleration* (London, 1690); *Some Considerations of the Consequences of the Lowering of Interest, and Raising the Value of Money* (London, 1692); *A Third Letter For Toleration* (London, 1692)—the letters on toleration have been reprinted together several times, see the eds. by A. Millar (London, 1765) and A. Murray (London, 1870); *Some Thoughts Concerning Education* (London, 1693), enl. (London, 1695); R. H. Quick, ed. (London, 1880); the most useful ed. is in James L. Axtell, ed., *Educational Writings* (Cambridge, 1968); *Locke's Travels in France 1675–1679*, J. Lough, ed. (Cambridge, 1953), Locke's French travel diaries published for the first time; *Essays on the Laws of Nature*, W. von Leyden, ed. (Oxford, 1954), an early Latin text found among the Lovelace papers by the editor, who includes a trans. and an intro. of exceptional interest; and *Two Tracts on Government*, Philip Abrams, ed. (Cambridge, 1965), drawn from early MSS.

For Locke's correspondence, see T. Forster, ed., *Original Letters of John Locke, Algernon Sydney and Lord Shaftesbury* (London, 1830); H. Ollion, ed., *Lettres inédites de John Locke* (The Hague, 1912); and B. Rand, ed., *The Correspondence of John Locke and Edward Clarke* (London, 1927). A definitive ed. of Locke's letters is being prepared by E. S. De Beer.

II. Secondary Literature. Some bibliographical and critical studies (in chronological order) are Lord King, *The Life of John Locke With Extracts From His Correspondence, Journals and Commonplace Books*, 2 vols. (London, 1829); H. R. F. Bourne, *The Life of John Locke*, 2 vols. (London, 1876), a substantial Victorian work that has not lost its utility despite later discoveries; J. Gibson, *Locke's Theory of Knowledge* (Cambridge, 1917), a useful guide to the *Essay;* S. Lamprecht, *The Moral and Political Philosophy of John Locke* (London, 1918); A. I. Aaron, *John Locke* (Oxford, 1937), the best-known modern commentary on Locke's philosophy; Willmoore Kendall, *John Locke and the Doctrine of Majority Rule* (New Haven, 1941), which depicts Locke as a forerunner of modern progressive democratic ideas; J. W. Gough, *John Locke's Political Philosophy: Eight Studies* (Oxford, 1952), a well-arranged commentary; M. Cranston, *John Locke: A Biography* (London, 1957), largely based on the earl of Lovelace's inherited collection of Locke's MSS acquired by the Bodleian Library, Oxford, in 1948; R. H. Cox, *Locke on War and Peace* (Oxford, 1961), a distinctive reinterpretation that stresses Locke's affinity with Hobbes; C. B. MacPherson, *The Political Theory of Progressive Individualism* (Oxford, 1962), a critique of Marxian inspiration; Martin Seliger, *The Liberal Politics of John Locke* (London, 1968), a defense of the liberal interpretation of Locke's politics; and John W. Yolton, ed., *John Locke: Problems and Perspectives* (Cambridge, 1969), a series of critical essays by English and American scholars.

MAURICE CRANSTON

LOCKYER, JOSEPH NORMAN (*b.* Rugby, England, 17 May 1836; *d.* Salcombe Regis, England, 16 August 1920), *astrophysics.*

Lockyer came from a middle-class family, derived, it was believed, from early Celtic immigrants from France into England. His father, Joseph Hooley Lockyer, was a surgeon-apothecary with broad scientific interests, and his mother, Anne Norman, was a daughter of Edward Norman, the squire of Cosford, Warwickshire. His formal education, at schools in the English Midlands, tended to concentrate on the classics, but the scientific atmosphere at home, supplemented by travel in Switzerland and France, served to broaden his interests and to produce the type of mind that was later to show a marked versatility. His earliest employment was as a civil servant in the English War Office, which he entered in 1857. He remained there until his striking success as an amateur astronomer led, via a period of temporary service as secretary of the duke of Devonshire's commission on scientific instruction, to a permanent post under the Science and Art Department, which culminated in his appointment as director of the Solar Physics Observatory established at South Kensington. There he remained until, in 1911, the observatory was transferred, much to his disappointment, to Cambridge, whereupon he retired.

In 1858 he married Winifred James, who died in 1879, leaving seven children, and in 1903 he married Thomasine Mary, the younger daughter of S. Woolcott Browne and widow of Bernhard E. Brodhurst, F.R.C.S., who survived him. After leaving the Solar Physics Observatory he established an observatory near his home in Salcombe Regis, Devonshire, known at first as the Hill Observatory and after his death as the Norman Lockyer Observatory—an institution now attached to the University of Exeter. Lockyer's interests, though very wide, were dominated by his belief in science and its potentialities for the welfare of the human race. Nevertheless, he published, with the collaboration of his daughter Winifred, *Tennyson as a Student and a Poet of Nature*—inspired by his friendship with the poet. He took more than a nominal interest in his membership of the Anglican Church, while on the lighter side he was stimulated to write *The Rules of Golf.* He received numerous honors from scientific bodies throughout the course of a long and exceptionally active life.

The direction of Lockyer's early scientific investigations was determined by the fact that the spectroscope was just beginning to reveal its great possibilities as an almost miraculous means of probing the secrets of nature. It was in 1859 that Kirchhoff and Bunsen laid the foundations of astrophysics by showing how the composition of the heavenly bodies could be determined—a problem not long before selected by Auguste Comte as a type of the permanently insoluble. Lockyer, thrilled by the new prospect thus opened up, obtained a spectroscope, which he attached to his $6\frac{1}{4}$-inch refracting telescope, and began a series of observations which, in 1868, brought him his first major success—the observation, at times other than during a total solar eclipse, of the solar prominences. The dramatic circumstances attending this discovery doubtless added to the attention which it received, but the ingenuity of the method and the importance of its implications would alone have entitled it to full credit.

The observation depended on the possibility of using a dispersion large enough to weaken the spectrum of the diffused sunlight in the atmosphere sufficiently to make visible the bright lines of the prominence spectrum. Lockyer conceived the idea as early as 1866, but, through a series of delays in obtaining a satisfactory instrument, it was not until 20 October 1868 that he was able to observe the prominence spectrum. It happened that on 18 August 1868 a total eclipse of the sun was visible in India, at which the brilliance of the prominence spectrum lines suggested to Jules Janssen, who observed them, the same idea that had occurred long before to Lockyer; and on the following day he successfully applied it. Both Lockyer and Janssen transmitted the news of the discovery to a meeting of the French Academy of Sciences, and by a remarkable coincidence the messages were received within a few minutes of one another. The French government commemorated the event by striking a medal bearing the portraits of the two astronomers.

Following up the observations, Lockyer soon observed a yellow line in the prominence spectrum, and in that of the outer atmosphere of the sun (which he similarly discovered and named the "chromosphere"), which had not been produced in the laboratory. This suggested to him the existence in the sun of an unknown element, which he named helium. In this he ran counter to the general opinion, which was that the line was due to a familiar element under exceptional conditions of excitation; and it was not until 1895 that William Ramsay, by producing the line from terrestrial sources, verified Lockyer's early conclusion. The constitution of the sun remained a leading interest with Lockyer, and he conducted many solar eclipse expeditions with a view to establishing ideas which he had formed from his studies of laboratory spectra.

One of the most conspicuous of such ideas was the "dissociation hypothesis," which much later acquired a special significance through developments in the theory of spectra. Lockyer observed in his experiments that the spectrum of an element varied with the intensity of the stimulus used to produce it. This appeared incompatible with the general view that the spectrum of any element was an invariable characteristic of the atoms (or molecules) of the element itself. It was recognized that the spectra of atoms and of molecules, even of the same element, differed; and Lockyer, following this clue, postulated that a stimulus greater than that necessary to break up the molecule into atoms would break up the atoms themselves and produce subatoms having their own characteristic spectra. The observation that, as he believed, the spectra of different elements contained lines in common led to the hypothesis that the atoms of what were known to the chemist as elements were themselves groupings of smaller constituents, the common lines being due to such constituents obtained by the dissociation of atoms of different elements.

This idea, like many of Lockyer's speculations, met with almost universal opposition, but although, in its original form, it is now believed to be groundless, it is recognized that it contains more than a germ of truth. It is now held that an increase in the energy of stimulus does indeed produce a dissociation of the atoms, and that atoms of different elements are indeed composed of different associations of the same more elementary components; but the apparent coincidences of lines from different elements are found, with the greater accuracy of measurement now possible, to be illusory. Although each element can yield a succession of spectra (silicon, for instance, to take one of Lockyer's own examples, under continuously increasing stimulus yields spectra known as Si I, Si II, Si III, Si IV), these are all different from any other spectrum that can be produced from anything at all. The data accumulated by Lockyer and, under his direction, by his students and assistants proved of the greatest value when a more tenable explanation became possible.

Another idea of Lockyer's connected with his observations of the sun, to which he devoted much attention and which likewise has not—at least in its original form—been strongly substantiated, was the supposed connection between the sunspot cycle and terrestrial meteorology. The eleven-year cycle of sunspots has been correlated, with greater or less plausibility, with various terrestrial phenomena, but what was peculiar to Lockyer's view of the matter was that he attempted to connect it with his idea of dissociation. He observed that certain lines in the spectra of sunspots varied in width during the course of a cycle, and this to him denoted a change of physical conditions in the sun that manifested itself in parallel effects in the spectra of the spots and in the weather on the earth. These phenomena were thus related not as cause to effect but as effects of the same cause which he postulated as pulsations in the sun. This work came to an abrupt end with the transfer of the Solar Physics Observatory to Cambridge and has not had any significant development.

Perhaps the most far-reaching of Lockyer's ideas was his meteoritic hypothesis, which is one of the most comprehensive schemes of inorganic evolution ever devised and which, though almost none of it now survives, led him to conceptions far in advance of those prevalent in his day. It may be summed up in his own words: "All self-luminous bodies in the celestial spaces are composed either of swarms of meteorites or of masses of meteoritic vapour produced by heat." The idea arose gradually in his mind from a variety of circumstances, but the crucial evidence, as he believed, lay in the examination of the spectra of meteorites under varying conditions. He believed that, by enclosing meteorites in vacuum tubes and subjecting them to gradually increasing degrees of temperature and electrical excitation, he could retrace the course of celestial evolution, and his endeavors for many years were concentrated on the establishment of a parallelism between the various types of meteoritic spectra and the spectra of the several kinds of heavenly body.

Lockyer had the good fortune to witness the very striking meteoric display of 1866, which appears to have left an indelible impression on his mind and predisposed him to give special attention to meteorites when later he began to speculate on the possibility of determining the course of celestial evolution in the light of spectroscopic evidence. Observations of Coggia's comet, which appeared in 1874, produced a further impulse in this direction—a close association between comets and meteors was already known—and he naturally associated the successive changes in the spectrum of the comet with the changes of laboratory spectra that had created in his mind the idea of dissociation. The ultimate result was an all-comprehensive scheme of the following character.

In the beginning space was occupied by a more or less uniform distribution of meteorites which, through their rapid motions and chance collisions, tended to accumulate in groups. These developed in various ways according to the conditions, giving rise to the various types of heavenly body observed. Very large loose groups would constitute nebulae, within which further condensations would occur, destined to become stars. A star, beginning as a meteoritic swarm, would rise in temperature as it condensed, until a stage was reached at which the mass, then completely vaporized, would lose heat by radiation at a rate equal to its generation by condensation; thereafter it would cool down toward its final state as a cold solid body. An exceptionally rapid conglomeration would constitute a nova, or new star, of which Lockyer examined some specimens that appeared during his working life; it would rise rapidly to its maximum temperature, and then decline in the normal stellar manner. Original groupings too small to become nebulae would form comets, while a number of isolated meteors would remain to enter the earth's atmosphere and appear as meteor swarms and occasionally as a meteorite large enough to reach the earth's surface. The final state of the universe would be that of a number of cold, dark bodies moving about aimlessly forever.

The correlation of the various types of celestial spectra with the succession of meteoritic spectra observed in the laboratory was carried out by Lockyer with great ingenuity but, it must be admitted, with insufficient critical power and too great a tendency to special pleading. A band spectrum arising from molecules presents the appearance, under small dispersion, of bright strips, each sharp at one edge and gradually fading out at the other, until, after a short dark gap, the sharp edge of the next appears. Now it is possible to regard this, in spectra so imperfect as those then obtainable from faint celestial sources, in the opposite way—as a dark band with a sharp edge, gradually growing into brightness in the reverse direction and ending abruptly at the beginning of the next dark band. In that case it would be an absorption instead of an emission spectrum, the substances present revealing themselves by the absorption of light from a strip of continuous radiation produced by a remoter source, and the identification of the substances would accordingly be quite different from that which the assumption of an emission spectrum would yield. Lockyer felt himself free to adopt either explanation, and to allow liberally for errors of measurement, according to the needs of his hypothesis, and it is not surprising that the conclusions he reached failed to convince the more critical of his contemporaries.

One of the most striking features of the hypothesis, however, that aroused some of the strongest opposition, was its implication that a star began as a cold body, rose to a maximum temperature, and then cooled down again, so that the coolest stars observable belonged to two groups, young and old. It was generally held that stars were born hot and cooled continuously throughout their lives, and the succession of stellar spectra was held to support this. According to Lockyer's idea, however, spectra of very young and very old stars, though very similar because their temperatures, the main determining characteristic, were the same, should differ in detail because of the much greater density of the old stars compared with the young. His chief activity in astronomical research during his later years was devoted to the classification of the cooler stars according to criteria which he devised.

It is a striking fact that, before his death, the theory of stellar evolution advanced by Henry Norris Russell and widely accepted, though having no connection with the meteoritic hypothesis, required just such a life history, so far as temperature and density were concerned, as that envisaged by Lockyer; and a really satisfactory criterion was discovered—which Lockyer just missed—for distinguishing young and old (giant and dwarf, as they were called) cool stars from one another. This partial vindication of his ideas gave him great satisfaction in his closing years. The fact that the whole problem now appears much more complex than either Lockyer or Russell could have conceived does not detract from the value of their work as a stimulus to further progress.

Another activity of Lockyer's restless mind was the astronomical interpretation of eastern temples and prehistoric stone circles, of which many examples are to be found in the British Isles. His idea was that these structures were connected with sun and star worship, their orientation being toward the points of rising or setting of conspicuous heavenly bodies. If that were so, a method existed, he conceived, of dating these erections, because our knowledge of the precession of the equinoxes enabled us to determine the precise azimuths of such risings and settings at different epochs, and if a particular temple axis pointed reasonably close to such an azimuth for a particular bright star, it could be assumed that the temple was erected when that star would be on the horizon at the point where the temple axis met it. More cautious archaeologists, while evincing considerable skepticism concerning the fundamental idea, were deterred also by the fact that the erosion of centuries, or even millennia, would have changed the visible horizon too drastically for the method to have much value; but Lockyer exhibited the same fertility of imagination in overcoming such difficulties as he had shown in connection with celestial spectra, and he built up an imposing system of dates associated with a few stars which he regarded as having had special religious significance.

Perhaps the most lasting of Lockyer's achievements was his creation of the scientific journal *Nature*, which, beginning in 1869, he edited for the first fifty years of its existence. *Nature* is now, by common consent, the world's leading general scientific periodical, but for many years it had to struggle to keep alive, and no small credit is due to Lockyer, and the publishers of that time, Messrs. Macmillan, whose faith in its ultimate success led them to persevere in the face of what at the earlier stages must have appeared almost insuperable obstacles. When Lockyer retired from the editorship in 1919 he was succeeded by his former assistant, Sir Richard Gregory, who held the office for the next twenty years.

This was perhaps the most outstanding example of Lockyer's lifelong concern for the recognition of science as a most potent agent in the general progress of civilization. Of his many efforts toward this end it is impossible to omit mention of one which arose from his election as president of the British Association for the Advancement of Science in 1903. His address on "The Influence of Brain Power on History" was intended to stimulate the Association to an extension of its activities from the pursuit of pure science into the realization of another of the objects of its founders —"to obtain a more general attention to the objects of science and a removal of any disadvantages of a public kind which impede its progress." Being unable to carry the council with him to the extent that he desired, he thereupon founded a new body, the British Science Guild, with this as its main object. This organization continued to function until after his death, when the gradual recognition by the British Association that this duty indeed properly lay upon it, led to the Guild relinquishing its activities to the Association.

Lockyer is an outstanding example of the adventurous rather than the critical scientist. It is easy to find faults in his advocacy of his ideas; it is not so easy to estimate the influence of those ideas on those who were stimulated to oppose them. Just as his speculations aroused either enthusiastic support or, more often, violent dissent but never indifference, so did his person: he had devoted friends and implacable enemies. His effect on the course of science is impossible to assess, but it is undoubtedly greater than the fate of his particular speculations would suggest.

BIBLIOGRAPHY

Lockyer wrote many papers and books on astrophysics. Most of the papers were published in the *Proceedings of the Royal Society* or *Philosophical Transactions of the Royal Society*. His major books are *Contributions to Solar Physics* (London, 1874), containing an account of his early observations of the sun by his new method; *Chemistry of the Sun* (London, 1887), in which the dissociation hypothesis and the evidence for it are presented; *The Meteoritic Hypothesis* (London, 1890); *The Dawn of Astronomy* (London, 1894), explaining the method of dating ancient structures and the results of its application; and *Inorganic Evolution* (London, 1900), which describes the most developed form of the hypothesis and its significance for celestial spectroscopy.

The main source of information concerning Lockyer is the biography by his widow and daughter, T. Mary Lockyer and Winifred L. Lockyer: *The Life and Work of Sir Norman Lockyer* (London, 1928). See also A. J. Meadows, *Science and Controversy: a Biography of Sir Norman Lockyer* (London, 1972).

HERBERT DINGLE

LODGE, OLIVER JOSEPH (*b.* Penkhull, Staffordshire, England, 12 June 1851; *d.* Lake, near Salisbury, England, 22 August 1940), *physics.*

Lodge was a member of an uncommonly vigorous and prolific clan. He was the eldest of nine children —eight sons and a daughter— of the merchant Oliver Lodge and grandson of the clergyman and schoolmaster Oliver Lodge, who had twenty-five children; his mother was Grace Heath, likewise descended from educators and clergymen. Lodge and his wife Mary Marshall had twelve children, ten of whom survived them; the youngest son, Raymond, was killed in World War I.

Lodge was educated privately until he was fourteen and then entered the business of his father, who was a supplier of materials used in the pottery industry. His interest in science was periodically sparked by visits to London, where he heard John Tyndall and others lecture at the Royal Institution. He resumed his education at the age of twenty-two at the Royal College of Science and at University College in London, where for a time he served as a demonstrator in physics for George Carey Foster. He received the D.Sc. in 1877 and began publishing papers on electricity, mechanics, and allied topics. In 1881 he became the first professor of physics at the new University College in Liverpool, where he remained for nineteen years. It was during those years that he made his principal contributions, mainly in two areas: theory of the ether and electromagnetic propagation.

The nineteenth-century concept of the ether left numerous questions unresolved, one of which was whether the ether in the vicinity of moving matter moved along with it. The Michelson-Morley experiment of 1887 was thought to have answered the question in the affirmative. But in 1893—by an ingenious experiment involving the interference between two opposing light rays traveling around the space between a pair of rapidly rotating parallel steel disks—Lodge showed that the ether was *not* carried along. The apparent contradiction helped to discredit the theory of the ether and to set the stage for the theory of relativity.

Lodge is also remembered for his experiments on electromagnetic radiation, in which he came close to anticipating Heinrich Hertz's discovery of propagating waves, and for his participation in the beginnings of radiotelegraphy. Lodge's 1887–1888 discovery that oscillations associated with the discharge of a Leyden jar result in waves and standing waves along conducting wires, with measurable wavelengths and other characteristics predicted by Maxwell's theory, was overshadowed by the more spectacular results that Hertz obtained in free space in the same year. Lodge's interest in these phenomena extended to improved methods of detecting electromagnetic waves, in one of which he utilized the observation that electromagnetic irradiation of loosely connected metals makes them stick together. This principle, simultaneously observed and elaborated by others, formed the basis for an early detector of radio waves, a container of loose metal particles subjected to mechanical vibration (so as continually to restore the original conductivity) that he named the coherer.

On 1 June 1894, in a lecture to commemorate the untimely death of Hertz five months earlier, Lodge spoke at the Royal Institution on "The Work of Hertz," stressing the experimental aspects of the work, including the importance of "syntony" (resonant tuning) in obtaining good results. The lecture was published and subsequently incorporated in a book; it had a widespread influence on the development of radiotelegraphy, inspiring experimenters in Germany, Italy, Russia, and other countries. With an associate, Alexander Muirhead, Lodge formed a syndicate to exploit one of his ideas, the resonant antenna circuit, and obtained some important patents.

In 1900 the University of Birmingham became the first British civic institution to receive a charter as a full-fledged university, and Lodge was appointed its first principal. He held the post until 1919, devoting himself increasingly to administrative work and the leadership of professional societies. He was president of the Physical Society in 1900 and of the British Association for the Advancement of Science in 1913. He received many prizes and honors, including the Rumford Medal of the Royal Society (1898) and the Faraday Medal of the Institution of Electrical Engineers (1932). He was knighted in 1902.

Beginning in 1883, Lodge also became interested in psychic research—telepathy, telekinesis, and communication with the dead—an interest that was intensified after his son's death and his own retirement in 1919. On two occasions he served as president of the Society for Psychical Research. He lived to the age of eighty-nine.

BIBLIOGRAPHY

I. Original Works. Lodge's Royal Institution lecture was published in *Electrician*, **33** (1894), 153 ff., and reprinted by that journal with addenda as *The Work of Hertz and His Successors* (London, 1894, 1898); the third and fourth eds., retitled *Signalling Through Space Without Wires* (London, 1900, 1908), also contained reprints of several other papers and various correspondence. Among his other books were texts and more popular works, all published in London: *The Ether of Space* (1909), *Talks About Wireless* (1925), *Advancing Science* (1931), and *My Philosophy* (1933), as well as a work on psychical research, *Raymond: Or Life and Death* (1916). He also wrote an autobiography, *Past Years* (London, 1931).

II. Secondary Literature. J. A. Hill compiled and annotated *Letters From Sir Oliver Lodge* (London, 1932); and Theodore Besterman assembled *A Bibliography of Sir Oliver Lodge* (London, 1935). There is an entry by Allan Ferguson in the 1931–1940 supp. to *Dictionary of National Biography*, pp. 541–543; and there are obituaries in *Obituary Notices of Fellows of the Royal Society*, **3** (1941), 551–574; *Proceedings of the Physical Society*, **53** (1941), 54–65; and *The Times* (23 Aug. 1940), p. 7. A well-documented chronological account of Lodge's contributions to tuned radiotelegraphy was prepared in support of his petition for extension of his patent by S. P. Thompson, *Notes on Sir Oliver Lodge's Patent for Wireless Telegraphy* (London, 1911).

Charles Süsskind

LOEB, JACQUES (*b.* Mayen, Rhine Province, Prussia, 7 April 1859; *d.* Hamilton, Bermuda, 11 February 1924), *physiology, biology.*

The apostle of mechanistic conceptions in biology, Loeb was the elder son of a prosperous Jewish importer. His brother Leo later became professor of pathology at Washington University, St. Louis. His father, an ardent Francophile, encouraged his son to read the classics of eighteenth-century thought. The mature Loeb always sought to harness his empirical researches to the wide-ranging political, social, and philosophical concerns of the *philosophes.* There is no direct evidence that he was influenced by the cognate tradition of the German-speaking medical materialists, led by Jacob Moleschott, Ludwig Büchner, and Carl Vogt, who published their major pronunciamentos in the 1850's. But it would be extraordinary if Loeb had not been acquainted with a celebrated essay by a kindred spirit of the medical materialists—"Die mechanistische Auffassung des Lebens" by Rudolf Virchow.

When he prepared to enter the University of Berlin in 1880, Loeb was not interested in science as a career. Having become intoxicated by reading Schopenhauer and Hartmann, the philosophers of the will, he resolved to become a philosopher in order to discover if freedom of the will existed—preferably to establish that it did not. To his bitter disillusionment, he felt the professors of philosophy at Berlin to be mere "wordmongers," uninterested in resolving the issues they posed and incapable of it.

Rebounding from this disappointment, he turned to science. His quest remained the same—to find a "starting point for an experimental analysis of the will." He thus enrolled at Strasbourg in 1880 to study the localization of function in the brain under Goltz. Loeb presumably conceived of brain surgery upon experimental animals as forcing their wills. He took his M.D. in 1884 with a thesis written under Goltz on blindness induced by injury to the cerebral cortex. Loeb regarded his five years at Strasbourg as totally wasted for his chosen purpose.

Loeb finally found the concatenation of men and impulses that he was looking for at Würzburg in 1886, where he had become assistant to the professor of physiology Adolf Fick, one of the chief pioneers in the application of physics to biology and medicine. The decisive contact of Loeb's career was his intimate friendship with Julius von Sachs, from whom he learned of plant tropisms, obligatory movements elicited by physical stimuli such as light and gravity. The concept was ultimately traceable to T. A. Knight and Candolle but had been greatly amended and enhanced in interest by Sachs's own research. Loeb's task in science suddenly became clear: to establish a concept of tropisms in animals by which they too could be shown to be irresistibly driven by external stimuli and impotent to interpose their wills. He could simultaneously allay his metaphysical anxieties and prosecute a clear-cut program of empirical researches.

Loeb's final period of scientific initiation came in the winter of 1889–1890 at the marine biological station at Naples, where he found himself in an intellectual environment dominated by Wilhelm Roux's *Entwicklungsmechanik.* In this context, he first encountered a group of brilliant young American embryologists and cytologists, including Thomas Hunt Morgan. In 1890 Loeb married an American, Anne Leonard, who had taken a Ph.D. in philology at Zurich; they settled in the United States in 1891. He later wrote that he "could not live in a regime of oppression such as Bismarck had created" and was also disturbed by reading Heinrich von Treitschke, "the court historian of the King of Prussia," on the superiority of the German race. In the United States, Loeb taught successively at Bryn Mawr College, the University of Chicago, and the University of California at Berkeley, and from 1910 until his death at the Rockefeller Institute in New York. He spent his summers at the marine biological laboratories at Pacific Grove, California, or Woods Hole, Massachusetts. Two of his three children, Leonard and Robert, became scientists.

Loeb's publications on animal tropisms began in 1888. Certain caterpillars on emerging in the spring had long been observed to climb to the tips of branches and to feed upon the buds; and this behavior had been attributed to an infallible instinct for self-preservation. Loeb showed, however, that if the only source of light was in the opposite direction from food, the caterpillars would move toward the light and starve to death. The alleged instinct, Loeb maintained, was merely a product of heliotropism: the caterpillars were photochemical machines enslaved to the light. He demonstrated by ingenious experiments that they strove to achieve a frontal fixation upon a light source, with the plane of symmetry bisecting the source of stimulation. Loeb thought that he had found the solution to the problem of will by applying the appropriate tropic stimulus, "forcing, by external agencies, any number of individuals of a given kind of animals to move in a definite direction by means of their locomotor apparatus." He showed that certain organisms, thought to be exempt from tropistic behavior, could be oriented in a single direction by such simple expedients as adding carbonic acid to the medium.

Although some organisms indubitably manifest tropistic behavior, Loeb's conception of animal tropisms as generally applicable soon came under severe attack. The classic refutation was given by the American biologist Herbert Spencer Jennings in his *Behavior of the Lower Organisms* (1906). He showed that some apparently tropistic responses were merely "avoiding reactions," unspecifically elicited by many stimuli and totally unaffected by the direction from which the stimulus was applied. More generally, Jennings invoked the concept, originated by C. Lloyd Morgan and Edward Lee Thorndike, of trial and error reactions to unfamiliar stimuli, leading by chance to rewarded behavior which tended to recur in the future.

Undaunted by such criticism, Loeb saw himself as having suppressed one form of biological mysticism by substituting tropisms for instincts. In the field of development mechanics to which he had been introduced at Naples, he thought that the equivalent achievement would be "the substitution of well-known physico-chemical agencies for the mysterious action of the spermatozoon." To test the hypothesis that salts acted upon the living organism by the combination of their ions with protoplasm, Loeb immersed fertilized sea urchin eggs in salt water the osmotic pressure of which had been raised by the addition of sodium chloride. When replaced in ordinary seawater, they underwent multicellular segmentation. T. H. Morgan then subjected unfertilized eggs to the same process and found that they too could be induced to start segmentation, although without producing any larvae. It was Loeb who first succeeded, in 1899, in raising larvae by this technique—the first notable triumph in achieving artificial parthenogenesis. Although Loeb greatly relished the philosophical implications of artificial parthenogenesis, he also saw it as an addition to the repertory of experimental embryologists.

The third of Loeb's major lines of research, from 1918 until his death in 1924, was summarized in *Proteins and the Theory of Colloidal Behavior* (1922). He demonstrated that proteins are amphoteric electrolytes, capable of reacting chemically either as an acid or as a base, and that many observed properties of proteins and protein solutions are explicable in terms of the creation of a Donnan equilibrium between two solutions separated by a semipermeable membrane. This view seemed to many to be an epoch-making advance in the unification of physiology and physical chemistry. As the Donnan theory postulates the existence of salt solutions, Loeb's concern with these issues was directly linked to his earlier research on artificial parthenogenesis.

Few scientists of Loeb's generation were as well known to the American public. As a materialist in philosophy, a mechanist in science, and a socialist in politics, he offended against the prevalent American orthodoxies. But he was correspondingly idolized by the dissenters and debunkers—including Veblen, Mencken, and Sinclair Lewis—who increasingly set the intellectual tone. Lewis, in his satirical delineation of the Rockefeller Institute in *Arrowsmith* (1925), drew significantly upon Loeb for the only senior scientist consistently exempted from the author's scorn and derision.

Loeb's principal statement of his basic philosophy for a lay audience was his famous address "The Mechanistic Conception of Life," delivered before the First International Congress of Monists in Hamburg in September 1911 and published as the title piece of his most widely read book (1912). In this work Loeb argued that the mechanistic conception had made colossal strides in the first decade of the twentieth century, largely through his own researches: the activation of the egg "completely reduced to a physicochemical explanation"; the instincts of men equated with tropisms; and the field of heredity, traditionally "the stamping ground of the rhetorician and metaphysician," now transformed by Mendelism into "the most exact and rationalistic part of biology." The one great remaining task was to explain the origin of life, but he did not expect this problem to be insurmountable.

The long-term impact of Loeb's work and orientation was substantial. Jennings, his chief critic on tropisms, always affirmed his indebtedness to Loeb as the great pioneer in the objective analysis of behavior, the man who first made the study of behavior a rigorously experimental science. While a graduate student at the University of Chicago, J. B. Watson, the founder of behaviorism in psychology, chose Loeb as his thesis director but was dissuaded by other professors who told him that Loeb was not "safe" for a student to work with. It is a fair inference, however, that Loeb had already instilled in him a latent hostility toward subjective modes of analysis in psychology. Loeb may justly be described as Watson's greatest precursor in exemplifying and proselytizing for this attitude.

BIBLIOGRAPHY

An exhaustive bibliography of Loeb's publications compiled by Nina Kobelt is appended to the standard biographical account, W. J. V. Osterhout, "Jacques Loeb," in *Journal of General Physiology*, **8** (1928), ix–xcii. The other extended account of Loeb is Donald Fleming, "Introduction," in Jacques Loeb, *The Mechanistic Conception of Life*, repr. ed. (Cambridge, Mass., 1964), vii–xli. On Loeb's relationship to Jennings and Watson, see

LOEB

Donald D. Jensen, "Foreword," in H. S. Jennings, *Behavior of the Lower Organisms*, repr. ed. (Bloomington, Ind., 1962), ix–xvii.

DONALD FLEMING

LOEB, LEO (*b*. Mayen, Germany, 21 September 1869; *d*. St. Louis, Missouri, 28 December 1959), *medicine, pathology, cancer research*.

Loeb's mother, Barbara Isay Loeb, died when he was three; his father, Benedict Loeb, died of tuberculosis when the boy was six. He lived with his maternal grandfather in Trier and at age ten moved to Berlin to live with a maternal aunt and uncle, whose daughter Helene married Albert Schweitzer. Tuberculosis and other ailments interrupted Loeb's schooling and recurred occasionally during his life. After 1889 he attended for short periods the universities of Heidelberg, Berlin, Freiburg (where he studied with August Weismann) and Basel (where he studied with Bunge and Miescher). His disapproval of rising German nationalism and militarism led him to take up medicine at the University of Zurich in 1890–1892. He did his clinical work at the University of Edinburgh and the medical school of London Hospital, occasionally attended lectures at other London medical schools, and returned to Zurich to complete his medical studies in 1895–1897. In his work toward the M.D., which required a thesis, he did skin transplantation experiments on guinea pigs under the direction of the pathologist Hugo Ribbert.

Loeb received the M.D. in 1897, then went to Chicago, where his older brother Jacques was a physiologist at the University. (He had previously visited his brother at Woods Hole, Massachusetts, in 1892 and 1894.) During five years in Chicago, Loeb practiced briefly near the University of Chicago, where he was physician to John Dewey's experimental school and was adjunct professor of pathology at Rush Medical College (later affiliated with the University of Illinois). In a rented room behind a drugstore he did experimental research on the healing of skin wounds of guinea pigs; he extended this research during a brief stay at Johns Hopkins School of Medicine, where he met William Osler, W. S. Thayer, and L. F. Barker (internal medicine); W. S. Halsted (surgery); W. H. Welch and Simon Flexner (pathology); and Mall and Harrison (anatomy). He also did research during summers at the Marine Biological Laboratory at Woods Hole.

Loeb next held a research fellowship under Adami at McGill University in 1902–1903. He was assistant professor of experimental pathology at the University of Pennsylvania in 1904–1910; directed laboratory research at Barnard Skin and Cancer Hospital, St. Louis, 1910–1915; and was professor of comparative pathology at Washington University, St. Louis, in 1915–1924, succeeding Eugene L. Opie as Mallinckrodt professor of pathology and department chairman in 1924–1937. On 3 January 1922, at the age of fifty-three, Loeb married Georgianna Sands, a physician in Port Chester, New York. He became emeritus professor of pathology in 1937 but continued to work as research professor (endowed by the Oscar Johnson Institute) until his final retirement in 1941 at the age of seventy-two.

Loeb continued to do research during summers at the Marine Biological Laboratory until 1950, when he nearly died of tuberculosis. Remaining thereafter in St. Louis, he worked on two books, one on the causes and nature of cancer and the other on the psychical factors of human life; both were unfinished at his death.

Among Loeb's honors were appointments to two endowed professorships; the John Phillips Memorial Prize (1935), awarded annually to an outstanding physician by the American College of Physicians; an annual lectureship in his name, endowed by his students at Washington University; an honorary D.Sc. from Washington University in 1948; and election as a member and officeholder in national and international medical and scientific organizations.

Of his contributions to science, his biographer, Ernest W. Goodpasture, wrote, "Although Loeb did not perfect *in vitro* culture of cells, he conceptually paved the way." Placing Loeb among the pioneers in studying the compatibility reactions of hosts toward transplanted tissues of the same and different species, Goodpasture wrote, "Loeb's histological studies of the fate of transplanted tissue, both normal and tumorous, were probably the first, certainly the most detailed, investigations of this kind."

His chief research writings were on tissue ai tumor growth, tissue culture, pathology of circulation, venom of Heloderma, analysis of experimental amoebocyte tissue, internal secretions, and the biological basis of individuality. Philip A. Shaffer cited Loeb as having helped inaugurate the experimental approach to the study of cancer in the United States.

BIBLIOGRAPHY

Loeb's major books, mainly collaborations, are *The Venom of Heloderma*, Carnegie Publication no. 177 (Washington, D.C., 1913); *Edema* (Baltimore, 1924); and *The Biological Basis of Individuality* (Springfield, Ill., 1945). His "Autobiographical Notes" appeared in *Perspectives in Biology and Medicine*, **2**, no. 1 (Autumn 1958), 1–23.

Philip A. Shaffer's "Biographical Notes on Dr. Leo Loeb" precedes a comprehensive (over 400 entries) "Bibliography of Writings of Dr. Leo Loeb From 1896 to 1949," in *Archives of Pathology* (Chicago), **50**, no. 6 (Dec. 1950), 661–675.

See also *New York Times* (Dec. 30, 1959), 21; Ernest W. Goodpasture, "Leo Loeb, September 21, 1869–December 28, 1959," in *Biographical Memoirs, National Academy of Sciences*, **35** (1961), 205–219; and W. Stanley Hartroft, "Leo Loeb, 1869–1959," in *Archives of Pathology* (Chicago), **70**, no. 2 (Aug. 1960), 269–274.

FRANKLIN PARKER

LOEFFLER (LÖFFLER), FRIEDRICH AUGUST JOHANNES (*b.* Frankfurt an der Oder, Germany, 24 June 1852; *d.* Berlin, Germany, 9 April 1915), *microbiology, medicine.*

Loeffler's father, Gottfried Friedrich Franz Loeffler, was a distinguished physician who rose to the rank of *Generalarzt* in the Prussian army. He wrote books on military medicine and from 1867 until his death in 1874 was assistant director of the Friedrich Wilhelm Institut for military doctors in Berlin, where he did much to raise the status of army medical officers. With this advantage the younger Loeffler began his own studies at the French Gymnasium of Berlin and became fluent in French, an important skill for those who were then pursuing microbiology. He attended the medical school of the University of Würzburg before transferring to his father's school just before the Franco-Prussian War. After serving as a hospital assistant in that conflict, he was awarded the medical degree in 1874. Loeffler then became assistant physician at the Charité Hospital in Berlin, where for a year and a half he came in contact with some of the best clinicians of the time. From 1876 to 1879 he served as military surgeon and public health officer in Hannover and Potsdam, and in October 1879 he was assigned to the newly established Kaiserliches Gesundheitsamt in Berlin.

The assignment proved to be the turning point of his career. Nine months later Robert Koch arrived to set up a bacteriological laboratory, which, modest at first, soon became a leading center of research and discovery. Already famous for his work on anthrax, the development of a solid medium for cultivating bacteria, and a monograph on wound infection, Koch chose Loeffler and Georg Gaffky as his assistants. The next four years were to be highly productive for all the members of Koch's group. As Loeffler later recalled:

> The memory of those days, when we still worked in this room, Koch in the center and we about him, when almost daily new wonders in bacteriology arose before our astounded vision, and we, following the brilliant example of our chief, worked from morning to evening and scarcely had regard to our bodily needs—the memory of that time will remain unforgettable to us. Then it was that we learnt what it means to observe and work accurately and with energy to pursue the problem laid before us [in F. Nuttall, "Biographical Notes," 235].

During Loeffler's four years at the imperial health office laboratories he pursued a number of bacteriological problems, including a series of experiments to determine effective means of disinfection. A significant practical result of these studies was the identification of the infectious agent of glanders, a disease seen mainly in horses. With the identification of the bacterium came the practical consequence of prevention. Although several others had described the occurrence of microorganisms in cases of glanders, no positive identification of an etiological agent had been made.

As with most of Loeffler's subsequent bacteriological work, the glanders studies were models of the new experimental methods and criteria as applied to bacteriology by Koch. Loeffler took materials from glanders nodules in the lung and spleen of a horse and cultured them in test tubes containing blood serum. Here the bacteria were successfully grown in succeeding generations and then inoculated into a healthy horse, which began to show typical symptoms of glanders within forty-eight hours. At autopsy Loeffler found fresh nodules which again were productive of bacteria. He now clearly demonstrated them by staining with methylene blue. These bacteria, injected into rabbits, mice, and guinea pigs, produced the signs and symptoms suggestive of glanders. Bacilli cultivated from the small experimental animals, when injected into horses, produced the typical disease of glanders. Thus, in this sequence of experiments Loeffler satisfied the so-called Koch's postulates, in 1882, the year Koch enunciated them: to isolate the organism, to grow it in pure culture, to infect experimental animals with the cultured organism, and then to again isolate the bacteria from the experimental animals.

Loeffler's best-known work is the elucidation of the characteristics of the diphtheria bacillus and its growth in pure culture (1884), which again clearly revealed the imprint of Koch, with whom he was then so closely associated. For the first time bacteriologists could work with single microbial species even though the original specimen taken from the throat of a patient, for instance, might be teeming with myriads of organisms of different species.

Diphtheria, a disease known since antiquity, to the history of which Loeffler also devoted much time, was particularly feared because it produced a false membrane in the throat that could suffocate its victims, especially children. In 1871 Max Oertel, of Munich, showed that the false membrane could be produced in rabbits by swabbing their throats with secretions from human patients. In 1875 Edwin Klebs postulated a fungus as the cause, but at the German Medical Congress of 1883 Klebs presented new information pointing to a specific bacterium that could be seen, after staining, in the throat membranes of diphtheria patients. The task remained to differentiate the several bacteria that were implicated in the disease and to grow in pure culture the one responsible for causing it.

One of the difficulties Loeffler faced in isolating the agent of diphtheria was that the throats of diphtheria patients carried many microorganisms, one of which, the Streptococcus, had already led to much confusion. In a series of twenty-seven cases of fatal throat inflammation, twenty-two had been diagnosed as diphtheria, five as scarlatinal diphtheria. In the latter, Loeffler found that the Streptococcus was the dominant organism. It is now known that scarlet fever is accompanied or preceded by a streptococcal throat infection. In the case of diphtheria, Loeffler reasoned that these chains of cocci played a secondary role.

In the case of typical diphtheria Loeffler observed that the bacteria described by Klebs were easily demonstrated in about half the cases he studied. Loeffler found these bacilli, which stained markedly with methylene blue, in the deeper layers of the false membrane but never in the deeper tissues or other internal organs, although these organs may have been greatly damaged. Loeffler still had to culture both the Klebs bacillus, never grown before, as well as the Streptococcus to prove or disprove either one as the cause of diphtheria. The Streptococci were easily grown on the solid medium of peptone and gelatin devised by Koch. Inoculation into animals produced generalized infections but never a disease resembling human diphtheria.

The bacillus implicated by Klebs—and now strongly suspected by Loeffler as well—as the diphtheria-causing organism was difficult to culture on the usual gelatin plates because it would not grow at the low temperatures required to keep the gelatin solid. The Streptococci, on the other hand, grew well at temperatures below 24° C., needed to keep the medium from liquefying. At this point Loeffler's innovative and experimental skills come out most clearly. He developed a new solid medium using heated blood serum rather than gelatin as the means of solidifying. This medium could now be incubated at 37° C., or body temperature. The Klebs bacilli grew well under these conditions. When they were injected into animals, Loeffler found that the guinea pig developed tissue lesions very similar to those of human diphtheria. Bacilli could be easily recovered from the infection produced at the site of inoculation, but they were never recovered from the damaged internal organs. Loeffler thus postulated that this, too, was similar to human diphtheria, in which the bacteria were confined to the throat membrane. He reasoned that perhaps the bacteria released a poisonous substance that reached other parts of the body through the bloodstream. This supposition was soon proved correct by the work of Émile Roux and Yersin, who did much to reveal the nature of the diphtheria toxin. In practical terms the toxin theory soon bore fruit in the work of Behring and others who developed an effective antitoxin to counter the effects of the soluble poison produced by the bacillus.

One further test carried out by Loeffler in this series of experiments to identify and isolate the agent of diphtheria was an attempt to culture the organisms from healthy children. Much to his surprise he was able to isolate the bacillus from one of the twenty subjects under study. He thereby called attention to the fact that not all people infected by the diphtheria bacillus or the tubercle bacillus had the disease diphtheria or tuberculosis. This concept of a healthy carrier had immense public health significance, especially in the period when medicine was making a headlong rush to ascribe all diseases to bacterial agents and when physicians too often simply equated the presence of a bacillus with a particular disease. The host factors thus had to come under study as well.

With the publication in 1884 of these wide-ranging studies of diphtheria, which in their process revealed not only the identifiable agent but the equally important phenomena of toxin production and the healthy-carrier state, Loeffler may be considered the principal discoverer of the diphtheria bacillus. He thus capped the findings of numerous clinical investigators of diphtheria and those of Klebs, who first observed the bacteria.

In 1884 Loeffler was transferred from the laboratory at the Gesundheitsamt to become director of the hygienic laboratory at the First Garrison Hospital in Berlin, where he continued his bacteriological studies and lectured on sanitation. At this time he discovered the cause of swine erysipelas and of swine plague. In 1886 he was appointed to the Faculty of Hygiene at the University of Berlin, and the following year, with his bacteriological colleagues, R. Leuckart and O. Uhlworm, he founded the *Centralblatt für Bakteriologie und Parasitenkunde* (later the *Zentralblatt für Bakteriologie, Parasitenkunde, Infektionskrank-*

heiten und Hygiene), one of the most influential journals in the field.

In 1888 Loeffler, renowned as a bacteriologist, researcher, and teacher, received calls to chairs in hygiene at the universities of Giessen and Greifswald. He accepted the latter offer and spent twenty-five years as a distinguished professor, serving as rector of the university from 1903 to 1907. In 1905 he was raised to the rank of *Generalarzt* in the army.

Early in his years at Greifswald, Loeffler began work on the problem of mouse typhoid and its bacteriological cause, *Salmonella typhi-murium*. This led to the first attempt to use bacteria for controlling the spread of an unwanted animal population. Pasteur had earlier suggested this be done to reduce the rabbit population in Australia. Having learned of Loeffler's paper describing a new bacteriological method for controlling field mice, in 1892 the Greek government asked him to assist it with control of the rodents that threatened the harvests on the plain of Thessaly. In his laboratory studies Loeffler had shown that *Salmonella typhi-murium* was infectious for mice but not for other farm animals, and he planned to spread the disease among mice by contaminating their food sources. With Rudolf Abel, his assistant at the university, Loeffler journeyed to Greece in April 1892. After determining that the Thessalian vole was as susceptible to the bacteria as his laboratory animals, he began to culture huge quantities of the Salmonella necessary to spread in the fields. The dispersed bacteria killed many field mice, but there was some debate as to the real, practical effectiveness of Loeffler's efforts at biological control. The subsequent realization that the bacteria may not be innocuous to man precluded further study.

In 1897 Loeffler became chairman of a German commission to investigate foot-and-mouth disease of cattle. The following year, working with his assistant, Paul Frosch, Loeffler made yet another fundamental discovery. The first real proof of the existence of a pathogenic organism so small that it was invisible under the microscope came from the work of Dmitry Ivanowsky, who in 1892 demonstrated its activity in tobacco mosaic. The first knowledge that these agents, known as filterable viruses because they passed through porcelain filters that stopped the larger bacteria, were also capable of causing disease in animals resulted from the work of Loeffler and Frosch (1898). They discovered that aphthous fever (foot-and-mouth disease) in cattle was caused by such a virus.

Lymph from infected animals, diluted and filtered so as to be entirely free of bacteria, produced the typical disease after injection into test animals. The question now was whether the lymph contained an effective toxin or a previously undiscovered agent. Loeffler and Frosch believed the latter possibility, that the activity of the lymph in causing further disease was due to an organism that could multiply yet was invisible with existing microscopes. Shortly after their work became known, William H. Welch in Baltimore called it to the attention of Walter Reed, who soon elucidated the viral nature of yellow fever and its transmission by mosquitoes.

Through his discoveries of several microorganisms as well as through his teaching and methods of work, Loeffler contributed greatly to the so-called golden age of bacteriology. Besides his improved culture media, bacteriological technique is also indebted to him for the popularization of the alkaline methylene blue stain as well as a specific stain for cilia and flagella. He introduced the standard medium for cultivation of the typhoid bacillus and improved the methods of isolating and differentiating the numerous fecal bacteria. He is also known for work on disinfection, hygiene of milk and water, and sewage disposal.

In 1913 Loeffler was recalled to Berlin to succeed Gaffky as director of the Institut für Infektionskrankheiten. He was active at the beginning of World War I in planning hygiene programs for the army. His health began to fail in December 1914, and despite surgical intervention he died on 9 April 1915. He was buried in Greifswald.

BIBLIOGRAPHY

I. Original Works. Loeffler's memoir on the diphtheria bacillus is "Untersuchungen über die Bedeutung der Mikroorganismen für die Entstehung der Diphtherie," in *Mittheilungen aus dem kaiserlichen Gesundheitsamt*, **2** (1884), 421–499; see also "Berichte der Kommission zur Erforschung des Maul und Kanlenseuche," in *Centralblatt für Bakteriologie und Parasitenkunde*, **23**, Abt. 1 (1898), 371–391, written with Paul Frosch. His many other medical and bacteriological papers appeared mainly in the *Mittheilungen aus dem K. Gesundheitsamte*, the *Deutsche medizinische Wochenschrift*, and the *Centralblatt für Bakteriologie*. One of the first books dealing with the history of bacteriology was his *Vorlesungen über die geschichtliche Entwicklung der Lehre von den Bacterien* (Leipzig, 1887).

Loeffler's papers available in English trans. are "On the Bacillus of Glanders"; "Investigations as to the Significance of Microorganisms in the Production of Diphtheria in Man"; and "Disinfection by Steam," written with R. Koch and G. Gaffky, in W. Watson Cheyne, ed., *Recent Essays by Various Authors on Bacteria in Relation to Disease* (London, 1886); and "The History of Diphtheria," in G. H. F. Nuttall and G. S. Graham-Smith, eds., *The Bacteriology of Diphtheria* (Cambridge, 1908), pp. 1–52.

II. SECONDARY LITERATURE. The following works (listed chronologically) contain useful biographical data: F. Goldschmidt, "Die Begründer der modernen Diphtheriebehandlung," in *Münchener medizinische Wochenschrift*, **42** (1895), 335–336; and "Professeur Friedrich Loeffler," in *Revue d'hygiène*, **22** (1900), 652–655; G. Gaffky, "Friedrich Loeffler," in *Zentralblatt für Bakteriologie, Parasitenkunde, Infektionskrankheiten und Hygiene*, **76** (1915), 241–245; George H. F. Nuttall, "Biographical Notes Bearing on Koch, Ehrlich, Behring, and Loeffler . . .," in *Parasitology*, **16** (1924), 214–238; Dexter H. Howard, "Friedrich Loeffler and the Thessalian Field Mouse Plague of 1892," in *Journal of the History of Medicine and Allied Sciences*, **18** (1963), 272–281; and H. A. Lechevalier and M. Soltorovsky, *Three Centuries of Microbiology* (New York, 1965), pp. 122–136, 282–286.

GERT H. BRIEGER

LOEWI, OTTO (*b.* Frankfurt am Main, Germany, 3 June 1873; *d.* New York, N.Y., 25 December 1961), *pharmacology, physiology*.

In 1936 Loewi shared with Henry Dale the Nobel Prize in physiology or medicine for work relating to the chemical transmission of nervous impulses. He was the first child and only son of Jakob Loewi, a wealthy wine merchant, and his second wife, Anna Willstädter. At the Frankfurter Städtisches Gymnasium, where Loewi studied from 1882 to 1891, his record was much better in the humanities than in physics and mathematics, and he hoped to pursue a career in the history of art. For pragmatic reasons, however, his family wanted him to study medicine, and in 1891 he matriculated as a medical student at the University of Strasbourg. Loewi found the preclinical studies there uninteresting and preferred during his first two years to attend the lectures of the philosophical faculty. After barely passing his first medical examination (*Physikum*) in 1893, Loewi spent the next two semesters at Munich and then returned to Strasbourg, where Bernhard Naunyn's clinical lectures now attracted his attention. Nevertheless, with little or no preparation in the subject, Loewi chose to write his dissertation on a topic in pharmacology. Under the direction of the renowned teacher and pharmacologist Oswald Schmiedeberg, he worked on the effects of various drugs on the isolated heart of the frog. Although quite typical of the work done in Schmiedeberg's laboratory, this choice of topic adds force to Loewi's own retrospective testimony that his interest in basic science had been aroused in part by reading Walter Holbrook Gaskell's Croonian lecture of 1883 on the isolated heart of the frog.[1]

After graduating M.D. from Strasbourg in 1896, Loewi took a course in analytical inorganic chemistry from Martin Freund in Frankfurt and then studied physiological chemistry for several months in Franz Hofmeister's laboratory in Strasbourg. In 1897–1898 Loewi served as Carl von Noorden's assistant in internal medicine at the city hospital in Frankfurt. Assigned to wards for advanced tuberculosis and epidemic pneumonia, Loewi was discouraged by the lack of therapy and resultant high mortality for these cases and later claimed that it was because of this experience that he chose basic research over a clinical career.[2] From 1898 to 1904 Loewi was assistant to Hans Horst Meyer at the pharmacological institute of the University of Marburg. His *Habilitationschrift*, which brought him the title of *Privatdozent* in 1900, dealt with nuclein metabolism in man. During 1902–1903 Loewi spent several months in England, which he believed had by then replaced Germany as the world's leading center for physiology.[3] At E. H. Starling's laboratory in London, where he spent most of this time, and during a brief visit at Cambridge, he met and exchanged ideas with the leading English physiologists of the day, including Walter Gaskell, John Newport Langley, Thomas Renton Elliott, and Dale. Despite the brevity of this trip, it clearly exerted an important influence on Loewi's later research interests and achievements.

In 1904 Loewi was appointed assistant professor at Marburg and briefly succeeded Meyer as director of the pharmacological institute there. In 1905 he followed Meyer to the University of Vienna to serve again as his assistant. He was appointed assistant professor at Vienna in 1907 and in 1909 accepted a post as professor and head of pharmacology at the University of Graz. Despite offers from more famous universities, including Vienna, Loewi remained at Graz until his expulsion by the Nazis in 1938. Like other male Jewish citizens of Graz, Loewi and two of his four children were imprisoned in March of that year. He was released two months later and in September was allowed to leave for London, but only after the Gestapo had forced him to transfer his Nobel Prize money from a bank in Stockholm to a bank under Nazi control.[4] Meanwhile the Nazis detained his wife, Guida Goldschmidt (whom he had married in 1908), while seeking to dispossess her of some real estate in Italy. Stripped of all property and means, Loewi managed during the next year or so to secure invitations to work at the Franqui Foundation in Brussels and then at the Nuffield Institute, Oxford. From 1940 until his death he was research professor of pharmacology at the College of Medicine, New York University. In 1941 his wife was at last able to rejoin him, and in 1946 Loewi became a naturalized citizen of the United States. While in his adopted country, he spent his summers at the Marine Biological Laboratory in Woods Hole, Massachusetts, where his remains are buried.

Although the range of Loewi's research interests and studies was vast, his most influential work fell into two broad categories: (1) protein and carbohydrate metabolism; and (2) the autonomic nervous system, especially the cardiac nerves. Other major interests were the pharmacology and physiology of the kidney and the physiogical role of cations. Loewi produced work in all of these areas quite early in his career, and the later fame of his work on the chemical transmission of nerve impulses ought not to obscure his continued interest in other areas.

Loewi's concern with the problems of metabolism emerged while he was still a medical student at Strasbourg. Naunyn was himself a pioneer in the area of metabolism, as were two members of his staff, Adolph Magnus-Levy and Oscar Minkowski, who had just begun his famous experimental studies on the relationship between the pancreas and diabetes mellitus. Loewi later included one of Minkowski's contributions in a list of the five papers that had most inspired his own career.[5] While working in Schmiedeberg's laboratory, Loewi was introduced to Johann Friedrich Miescher's classic papers on the metabolism of the Rhine salmon during its freshwater phase. It was this work to which Loewi later variously referred as his "scientific bible," as jointly responsible (with Gaskell's paper of 1883) for arousing his interest in basic science, and as the chief influence on his choice of physiology in particular as a calling.[6] Another probable influence on Loewi's interest in metabolism was van Noorden, whose own clinical studies focused on pathologic metabolism.

In his earliest publications on metabolism Loewi argued that of the components of urinary secretion, only the uric acid depended on diet. These studies were followed by papers on phlorizin-induced glycosuria and on the question whether fat could be converted into sugar in dogs; Loewi concluded that it could not. His greatest contribution to metabolic studies, dealing with protein synthesis in the animal body, appeared as early as 1902. Since no one had previously been able to maintain nitrogen balance in animals by feeding them with the degradation products of protein in place of the original protein itself, it was supposed that animals were incapable of protein synthesis. By his own account Loewi became interested in the question after reading a paper by the biochemist Friedrich Kutscher, a colleague at Marburg, who in 1899 showed that extended trypsin digestion of a pancreas could yield end products none of which displayed any longer the chemical reactions characteristic of protein.[7] This result somehow led Loewi to believe that animals might be able to synthesize protein if their diet consisted of the degradation products of a whole organ rather than of an isolated protein, and he immediately undertook a series of nutritional experiments to test his idea. Although the dogs he used found this diet unpalatable at first, Loewi persisted and eventually established that animals were indeed capable of synthesizing proteins from their degradation products, even from the most elementary end products, the amino acids. This work established Loewi's early reputation and greatly influenced later work in nutrition, including perhaps that of Frederick Gowland Hopkins, who met Loewi during his visit to England and who was to win the Nobel Prize in 1929 for his discovery of "accessory food factors," or vitamins.[8]

Between 1902 and 1905, Loewi—in collaboration with Walter Fletcher, Nathaniel Alcock, and Velyian Henderson—produced a series of five papers on the function of the kidney and the action of diuretics. In the classic monograph, *The Secretion of the Urine* (London, 1917), Arthur Cushny made important use of these studies while criticizing Loewi's early attempt (1902) to strike a compromise between the filtration and vital theories of kidney function. Nonetheless, Loewi's ultimate position, as represented in the last paper of this series, seems to conform quite closely to Cushny's own modified filtration theory.

From 1907 to 1918, Loewi published six additional studies on metabolism, chiefly with regard to the function of the pancreas and glucose metabolism in diabetes[9]; and after Frederick Banting and Charles Best discovered insulin in 1921, he undertook with H. F. Haüsler a long series of studies on the mode of action of insulin and its antagonists. By 1929 Loewi was forced to disavow some of the results of his earlier work on insulin and thereby to withdraw support from his earlier hypothesis that insulin promoted the binding of glucose to erythrocytes. In the 1930's Loewi and his co-workers claimed to have discovered a proprioceptive metabolic reflex ("a mechanism through which the central nervous system . . . is informed of the state of the metabolism in individual organs")[10] and pursued some of the issues raised by the work of Bernardo Houssay on the role of the pituitary gland in carbohydrate metabolism.

By his own account, Loewi "imported" from England (and particularly from Gaskell, Langley, and Elliott) his interest in the "vegetative," or autonomic, nervous system.[11] His earliest publications in the field, all of which postdate his visit to England, concern vasomotor action, salivary secretion, and the action

LOEWI

of adrenalin, all topics under active investigation at Cambridge during his visit there. Loewi's first really original contribution in this area was his demonstration with Alfred Fröhlich in 1910 that cocaine increases the sensitivity of autonomically innervated organs to adrenalin, a response so specific that it came to be used as a test for the latter drug.[12] In 1912 Loewi published a series of three communications on the action of the vagus nerve on the heart, dealing particularly with the effects on vagal action of various drugs and of variations in the calcium ion concentration of the nutrient saline solution. Between 1913 and 1921 Loewi pursued these interests in a series of three publications on the role in heart action of physiological cations (especially calcium) and in a series of six publications on the physiological relationship between calcium ions and the series of digitalis drugs, especially with respect to their action on the frog's heart. In the latter studies Loewi emphasized that the digitalis series affects the heart as it does chiefly because digitalis sensitizes the heart to calcium, a conclusion with clinical consequences that Loewi sought to specify.

Thus, for several years before 1921 Loewi had been turning increasingly to the physiology and pharmacology of the frog's heart and its nerves. But none of this research led directly to the work for which he won the Nobel Prize. In fact, there is an element of mystery and drama in the way Loewi came to demonstrate experimentally the chemical transmission of nervous impulses. By the time he did so, in 1921, the hypothesis of chemical transmission was nearly twenty years old. Credit for the hypothesis is usually given to Elliott, who published the suggestion in 1904.[13] In 1929, after Loewi's work had become well known, another Cambridge physiologist, Walter Fletcher, recalled that Loewi had independently proposed the chemical transmission hypothesis in 1903, in a private conversation with Fletcher, who was then working in Loewi's laboratory at Marburg.[14] Before this reminder from Fletcher, Loewi had completely forgotten the conversation; but he thereafter attached considerable significance to it, undoubtedly to emphasize the independence of his own work. However, as Dale suggested more than once, it is hard to believe that Loewi's mind had not been at least somewhat prepared for the idea by his meeting with Elliott and in general by his visit to Cambridge, where the meaning of the neuromimetic effects of drugs was then a topic of intense interest and discussion.[15]

In any case, by 1920 no decisive experimental evidence for chemical transmission had yet been found, and the hypothesis had fallen into rather general discredit. Loewi revived it with an elegant experiment which, by his own account, occurred to him in the midst of sleep on two successive nights:

> The night before Easter Sunday of [1921] I awoke, turned on the light, and jotted down a few notes on a tiny slip of thin paper. Then I fell asleep again. It occurred to me at six o'clock in the morning that during the night I had written down something most important, but I was unable to decipher the scrawl. The next night, at three o'clock, the idea returned. It was the design of an experiment to determine whether or not the hypothesis of chemical transmission that I had uttered seventeen years ago was correct. I got up immediately, went to the laboratory, and performed a simple experiment on a frog heart according to the nocturnal design. I have to describe briefly this experiment since its results became the foundation of the theory of the chemical transmission of the nervous impulse.
>
> The hearts of two frogs were isolated, the first with its nerves, the second without. Both hearts were attached to Straub canulas filled with a little Ringer solution. The vagus nerve of the first heart was stimulated for a few minutes. Then the Ringer solution that had been in the first heart during the stimulation of the vagus was transferred to the second heart. [This second heart] slowed and its beats diminished just as if its vagus had been stimulated. Similarly, when the accelerator nerve was stimulated and the Ringer from this period transferred, the second heart speeded up and its beats increased. These results unequivocally proved that the nerves do not influence the heart directly but liberate from their terminals specific chemical substances which, in their turn, cause the well-known modifications of the function of the heart characteristic of the stimulation of its nerves.[16]

So dramatic an account of a scientific achievement naturally arouses skepticism, but Dale has testified that Loewi told him essentially the same story in 1921 and repeated it virtually unchanged several times during the next four decades.[17] Some of the mystery was dispelled when Loewi eventually realized that the method he used to demonstrate chemical transmission for the cardiac nerves was essentially the same method he had used in two slightly earlier studies "also in search of a substance given off from the heart."[18] For Loewi the remarkable nocturnal experimental design then became nothing more than a sudden unconscious association between the old hypothesis and the recent method.[19]

Loewi's classic experiment was described in a four-page article in *Pflügers Archiv* in 1921, the first in a series of fourteen papers on humoral transmission which Loewi and his several collaborators published in the same journal during the next fifteen years. The initial paper is considerably more modest in its claims than the retrospective accounts of Loewi and others might suggest: although it clearly emphasized chemical mediation of nerve impulses as one plausible interpretation of the experimental results, it also raised the alternative possibility that vagomimetic and accelerator-mimetic substances might be products of altered cardiac activity and thus not directly products of nerve stimulation. In the second paper of the series (1922) Loewi rejected the latter interpretation in favor of direct chemical mediation. The next immediate task, pursued in several papers in the series, was to determine the chemical nature of the substances released by stimulation of the vagus and accelerator nerves. The fact that the action of the vagus-transmitting substance (*Vagusstoff*) was abolished by atropine greatly simplified the search for its chemical equivalent; and as early as the second paper Loewi argued that the *Vagusstoff* must be a choline ester. But not until the tenth and eleventh papers (both published in 1926) did Loewi and his collaborator E. Navratil positively identify the *Vagusstoff* with acetylcholine. For some, notably Dale, this identification seemed to be approached with excessive caution in view of what was already known about acetylcholine, much of it as a result of Dale's own work a decade earlier.

But Loewi's caution may well reflect the skepticism that his work aroused in others, particularly Leon Asher, against whose criticisms some of the papers in the series were specifically directed. Nor was all of the criticism misguided: however simple on the surface, Loewi's famous experiment is not always easy to reproduce, and it is said that he often failed in his own attempts to do so.[20] As Loewi pointed out in the third paper of the series (1924), success depended on achieving a combination of several intricate experimental conditions, including a test heart of appropriate responsiveness (which in turn could depend on species differences and the season of the year). Perhaps because of such difficulties, Loewi was invited to demonstrate his experiment before the Twelfth International Congress of Physiology at Stockholm in 1926. Although he apparently accepted the invitation with some reluctance, Loewi executed the demonstration with great success.[21] By 1934 W. B. Cannon could describe the support for Loewi's work as "conclusive,"[22] a judgment confirmed by G. Liljestrand in his presentation speech at the Nobel Prize ceremonies in 1936.[23]

If Loewi proceeded cautiously toward the identification of *Vagusstoff* with acetylcholine, he moved even more cautiously toward an identification of the substance released by stimulation of the accelerator nerve. Even though Elliott had proposed the hypothesis of chemical transmission in the first place because of the sympathomimetic effects of adrenalin, it was not until 1936 that Loewi positively identified the *Acceleransstoff* or *Sympathicusstoff* with adrenaline (epinephrine). Like many others, Loewi apparently did assume immediately that his results for the cardiac nerves would apply as well to all other peripheral autonomic nerve fibers, and one of the earliest and most important pieces of evidence for this extension was produced in Loewi's laboratory by E. Engelhart.[24] But the suggestion that chemical transmitters were also released by ordinary voluntary motor fibers or across other nonautonomic synaptic junctions aroused in Loewi what Dale characterized as "almost obstinate scepticism."[25] Chiefly because nervous impulses produce their effects so much more rapidly in voluntary and central nervous processes than in autonomic, Loewi doubted that chemical transmission could be involved in the former. In the end, as Loewi himself emphasized, it was Dale and his associates who were mainly responsible for establishing the existence of chemical transmission outside the postganglionic autonomic nervous system.[26]

In a sense, however, Dale's ability to make this bold extension depended in part on Loewi's caution. For in the course of their careful approach toward a definitive identification of *Vagusstoff*, Loewi and Navratil not only found that the effects of *Vagusstoff* faded rapidly because it was speedily metabolized by an enzyme but also that this enzyme could be inhibited (and *Vagusstoff* thereby protected) by the alkaloid physostigmine (eserine). These results, reported in the tenth and eleventh papers of the famous series, were important for several reasons. (1) Since acetylcholine was also speedily metabolized by the same enzyme (a specific cholinesterase) and also selectively protected by eserine, the identity of *Vagusstoff* and acetylcholine was virtually assured. (2) By this work Loewi and Navratil introduced the concept that the pharmacologic action of an alkaloid could be defined in terms of its inhibition of an enzyme, and thus helped to explain why alkaloids could be effective even in minute doses. (3) The ability of eserine to prevent the destruction of acetylcholine made it possible to develop new methods for detecting the transmitter in tissues where its low concentration or rapid destruction had kept it hitherto undetected. It was chiefly by the use of eserinized solutions that Dale and others were able to establish the existence of chemical transmission in warm-blooded vertebrates and in parts of the nervous system where Loewi himself had doubted its presence. (4) The knowledge that a specific cholinesterase was responsible for the destruction of the acetylcholine transmitter made it easier to understand and to determine why the speed and duration of nervous effects might vary in different species and in different parts of the nervous system. Among other things, these differences might be traced to differences in the amount of the cholinesterase normally present in different tissues. Along lines such as these, an explanation can be proposed for the difficulties encountered in the early attempts to reproduce Loewi's original experiment. According to William Van der Kloot, the heart of the Hungarian frogs that Loewi used contains only a small amount of the cholinesterase, so that the acetylcholine released by vagus stimulation persists unusually well in it.[27] In the era before eserinized solutions, those who used other species would naturally have experienced greater difficulty in detecting the vagus transmitter.

With the publication in 1936 of the fourteenth and final paper in the series on humoral transmission, the truly creative phase of Loewi's research career came to an end. Especially after 1938, when the Nazis forced him out of Graz, Loewi's main role was that of critic, reviewer, and guide for those who more actively pursued the new lines of research opened up by the discovery of the chemical transmission of nervous impulses. This discovery in fact inaugurated a conceptual revolution in neurophysiology. Besides offering an entirely new mode of thinking about such phenomena as inhibition and summation, the concept of chemical transmission had clinical implications as well, particularly with regard to certain symptoms of neurological hyperactivity formerly regarded as purely reflex in nature. Loewi himself often reflected on these questions as well as on the implications of chemical transmission for general physiology and biology. For him, the existence of chemical transmission seemed to give further support to the organismic conception of the living body as an adaptive, regulated, delicately coordinated, and peculiarly biological mechanism. In his inclination toward overtly teleological thinking, in his abiding love of music, art, and culture in general, and in his profound moral and humanitarian sensitivity, Loewi maintained to the end a kinship with that young aspiring historian of art who had been deflected by circumstance into a richly creative career in science.

NOTES

1. Loewi, "Prefatory Chapter," p. 2.
2. Loewi, "An Autobiographic Sketch," p. 7.
3. *Ibid.*, p. 10.
4. *Ibid.*, p. 21.
5. Loewi, "Prefatory Chapter," p. 2.
6. *Ibid.*, p. 2; and Loewi, "An Autobiographic Sketch," p. 6.
7. Loewi, "An Autobiographic Sketch," p. 9.
8. See Dale, "Otto Loewi," p. 70.
9. Of the work done during these two decades, Loewi himself chose to emphasize the following: "One of these studies proved that the preference of pancreatectomized dogs for fructose rather than glucose, as had been demonstrated by Minkowski, is not specific for this deficiency but is shared by dogs deprived of their glycogen by other means, e.g., by phosphorus poisoning. It was further shown that the heart, in contrast to the liver, cannot utilize fructose. It was finally discovered that epinephrine injections into rabbits completely depleted by starvation of their liver glycogen brought the glycogen back to almost normal values in spite of continued starvation." See Loewi, "An Autobiographic Sketch," pp. 12–13.
10. *Ibid.*, p. 15.
11. *Ibid.*, p. 13.
12. *Ibid.*, p. 13. Cf. Loewi, "The Ferrier Lecture," p. 300.
13. See T. R. Elliott, "On the Action of Adrenalin," in *Journal of Physiology*, **31** (1904), xx–xxi; and *ibid.*, **32** (1905), 401–467.
14. See Loewi, *From the Workshop of Discoveries*, p. 33; "An Autobiographic Sketch," p. 17; and the sources cited in note 15 below.
15. See H. H. Dale, "T. R. Elliott," in *Biographical Memoirs of Fellows of the Royal Society*, **7** (1961), 64; and "Otto Loewi," pp. 70–73.
16. Loewi, "An Autobiographic Sketch," p. 17. In an obvious oversight, Loewi here places the nocturnal event in 1920 rather than 1921. For a virtually identical version of the story, but with the correct date, see Loewi, *From the Workshop of Discoveries*, pp. 32–33.
17. Dale, "Otto Loewi," pp. 76–77.
18. Loewi, "An Autobiographic Sketch," p. 18. Loewi does not identify the two studies he has in mind, but the likeliest candidates are "Über Spontanerholung des Froschherzens bei unzureichender Kationenspeisung. II. Mitt. Ein Beitrag zur Wirkung der Alkalien aufs Herz," in *Pflügers Archiv für die gesamte Physiologie des Menschen und der Tiere*, **170** (1918), 677–695; and "III. Mitt. Quantitative mikroanalytische Untersuchungen über die Ursache der Calciumabgabe von seiten des Herzens," *ibid.*, **173** (1919), 152–157, written with H. Lieb.
19. Loewi, "An Autobiographic Sketch," p. 18.
20. Letter from William G. Van der Kloot to G. L. Geison, 2 June 1972.
21. Loewi, "An Autobiographic Sketch," p. 19; Dale, "Otto Loewi," pp. 77–78. Incidentally, neither account corroborates a remarkable claim which Loewi made many years later to William Van der Kloot, who was chairman of the pharmacology department at New York University during Loewi's last years there. According to Van der Kloot (see note 20) Loewi told him that during this demonstration in Stockholm he had been "obliged to stand at one end of the room and simply give instructions, so that the possibility of his secreting some chemical under his fingernails and dropping it on the preparation could be eliminated." This claim also appears to conflict with a photograph taken of Loewi at the time of the demonstration (see Holmstedt and Liljestrand, *Readings in Pharmacology*, p. 194).
22. Walter B. Cannon, "The Story of the Development of Our Ideas of Chemical Mediation of Nerve Impulses," in *American Journal of the Medical Sciences*, n.s. **188** (1934), 149.
23. See *Nobel Lectures Including Presentation Speeches and Laureates' Biographies: Physiology or Medicine 1922–1941* (Amsterdam, 1965), pp. 397–401, esp. p. 399.
24. See e.g., Otto Loewi, "Chemical Transmission of Nervous Impulses," in George A. Baitsell, ed., *Science in Progress*, 4th ser. (New Haven, 1945), p. 102.
25. Dale, "Otto Loewi," p. 79.
26. See, e.g., Otto Loewi, "Salute to Henry Hallet Dale," in *British Medical Journal* (1955), **1**, 1356–1357.
27. Van der Kloot to Geison (see note 20).

LOEWI

BIBLIOGRAPHY

I. Original Works. Lembeck and Giere (see below) give an apparently complete bibliography of more than 170 items by Loewi, arranged both topically and chronologically. In their topical bibliography, a summary is also given of the contents of most of his research papers. Appended to Henry Dale's biographical memoir of Loewi (see below) is a bibliography of 150 items; Dale omits many of the essentially nonscientific papers included by Lembeck and Giere.

The most important of Loewi's papers on metabolism is "Über Eiweisssynthese im Thierkörper," in *Archiv für experimentelle Pathologie und Pharmakologie*, **48** (1902), 303–330. For the early series of five papers on the kidney, see "Untersuchungen zur Physiologie und Pharmakologie der Nierenfunction," *ibid.*, **48** (1902), 410–438; **50** (1903), 326–331; **53** (1905), 15–32, 33–48, 49–55. Loewi abandoned his earlier hypothesis of the mode of action of insulin in "Insulin und Glykämin," in *Klinische Wochenschrift*, **8** (1929), 391–393. On the concept of the proprioceptive metabolic reflex, see "Über den Glykogenstoffwechsel des Muskels und seine nervöse Beeinflussung," in *Pflügers Archiv für die gesamte Physiologie des Menschen und der Tiere*, **233** (1933), 35–56. For Loewi and Frölich's work on cocaine and adrenaline, see "Untersuchungen zur Physiologie und Pharmakologie des autonomen Nervensystems. II. Mitt. Über eine Steigerung der Adrenalinempfindlichkeit durch Cocain," in *Archiv für experimentelle Pathologie und Pharmakologie*, **62** (1910), 159–169.

For Loewi's work on the heart between 1912 and 1921, see "Untersuchungen zur Physiologie und Pharmakologie des Herzvagus," *ibid.*, **70** (1912), 323–343, 343–350, 351–368; "Über Beziehungen zwischen Herzmittel- und physiologischer Kationenwirkung," *ibid.*, **71** (1913), 251–260; **82** (1918), 131–158; **83** (1918), 366–380; *Pflügers Archiv*, **187** (1921), 105–122, 123–131; **188** (1921), 87–97; and "Über die Spontanerholung des Froschherzens bei unzureichender Kationenspeisung," *ibid.*, **157** (1914), 531; **170** (1918), 677–695; and **173** (1919), 152–157. For the celebrated series of fourteen papers on humoral transmission, see "Über homorale Übertragbarkeit der Herznervenwirkung," *ibid.*, **189** (1921), 239–242; **193** (1922), 201–213; **203** (1924), 408–412; **204** (1924), 361–367, 629–640; **206** (1924), 123–134, 135–140; **208** (1925), 694–704; **210** (1925), 550–556; **214** (1926), 678–688, 689–696; **217** (1927), 610–617; **225** (1930), 721–727; **237** (1936), 504–514. See also Otto Loewi, "Kritische Bemerkungen zu L. Ashers Mitteilungen," *ibid.*, **212** (1926), 695–706.

Loewi's more general views on chemical transmission and his role as historian, reviewer, critic, and guide are best revealed in a number of his lectures given between about 1930 and 1945. Of these, the most important and valuable are "The Humoral Transmission of Nervous Impulse," in *Harvey Lectures*, **28** (1934), 218–233; "The Ferrier Lecture on Problems Connected with the Principle of Humoral Transmission of Nervous Impulses," in *Proceedings of the Royal Society*, **118B** (1935), 299–316; "Die chemische Übertragung der Nervwirkung," in *Schweizerische medizinische Wochenschrift*, **67** (1937), 850–855; "Die chemische Übertragung der Herznervenwirkung," in *Les Prix Nobel en 1936* (Stockholm, 1936), translated into English in *Nobel Lectures Including Presentation Speeches and Laureates' Biographies: Physiology or Medicine 1922–1941* (Amsterdam, 1965), pp. 416–429; and "The Edward Gamaliel Janeway Lectures: Aspects of the Transmission of Nervous Impulse," in *Journal of the Mount Sinai Hospital*, **12** (1945), 803–816, 851–865. See also the lecture cited in note 24 above.

For useful autobiographical material, see Otto Loewi, "An Autobiographic Sketch," in *Perspectives in Biology and Medicine*, **4** (1960), 1–25, reprinted in Lembeck and Giere, pp. 168–190; "Prefatory Chapter: Reflections on the Study of Physiology," in *Annual Review of Physiology*, **16** (1954), 1–10, reprinted in *The Excitement and Fascination of Science: A Collection of Autobiographical Essays* (Palo Alto, Calif., 1965), pp. 269–278; and *From the Workshop of Discoveries* (Lawrence, Kans., 1953). Lembeck and Giere have also published two posthumous MSS by Loewi, "Meaning of Life" and "The Organism as a Unit" (pp. 191–203 and 204–217, respectively); as well as a dozen brief ceremonial addresses given by Loewi between 1942 and 1952 (pp. 218–241). Several of these sources emphasize Loewi's organismic philosophy and approach; a few relate to the Jewish question and to more general political and humanitarian concerns.

According to Lembeck and Giere, Loewi's MSS and papers are deposited in thirteen file drawers at the Royal Society of London. In the obituary files of the Medical-Historical Library, Yale University, is a letter from Loewi to Harvey Cushing (27 Nov. 1937) in which Loewi seeks aid for a victim of Nazi persecution. Coming just a few months before his own imprisonment by the Nazis, it is both a reminder of the horror of Nazism and a testimony to the solidarity of its victims.

II. Secondary Literature. The basic source for Loewi's life and work is Fred Lembeck and Wolfgang Giere, *Otto Loewi: Ein Lebensbild in Dokumenten* (Berlin, 1968). Although not a scientific biography of Loewi, this remarkable book reprints or provides references to almost all the material that a biographer could conceivably use. Besides the reprinted material and bibliographies mentioned above, there is a brief biographical sketch (pp. 2–16), a list of his honors and the professional associations to which he belonged (pp. 19–21), a bibliography of more than fifty sketches and obituaries (pp. 22–24), and a lengthy section containing extensive and impressive documentation of his life (pp. 27–91).

Of the many biographical sketches of Loewi, the most valuable is that by his Nobel colaureate Henry Dale, "Otto Loewi," in *Biographical Memoirs of Fellows of the Royal Society*, **8** (1962), 67–89, with portrait and bibliography. For short biographical sketches, brief attempts to place Loewi's work in historical perspective, and excerpts in English from his writings, see Lloyd G. Stevenson, *Nobel Prize Winners in Medicine and Physiology, 1901–1950* (New York, 1953), pp. 186–195; B. Holmstedt and G. Liljestrand, *Readings in Pharmacology* (New York,

1963), pp. 190–196; and Edwin Clarke and C. D. O'Malley, *The Human Brain and Spinal Cord: A Historical Study Illustrated by Writings from Antiquity to the Twentieth Century* (Berkeley, Calif., 1968), pp. 250–255.

GERALD L. GEISON

LOEWINSON-LESSING. See **Levinson-Lessing, Frants Yulevich.**

LOEWNER, CHARLES (KARL) (*b.* Lany, Bohemia [now Czechoslovakia], 29 May 1893; *d.* Stanford, California, 8 January 1968), *mathematics.*

Loewner was the son of Sigmund and Jana Loewner. He studied mathematics with G. Pick at the German University of Prague and received the Ph.D. in 1917. From 1917 to 1922 he was an assistant at the German Technical University of Prague; from 1922 to 1928, assistant and *Privatdozent* at the University of Berlin; from 1928 to 1930, extraordinary professor at the University of Cologne; and from 1930 to 1939, full professor at Charles University in Prague, which he left when the Nazis occupied Czechoslovakia. From 1939 to 1944 he was lecturer and assistant professor at the University of Louisville, Kentucky; from 1945 to 1946, associate professor, and from 1946 to 1951, full professor at Syracuse University; and from 1951 until his retirement in 1963, full professor at Stanford University.

Loewner was married to Elizabeth Alexander, who died in 1955; they had one daughter. Of short stature, soft-spoken, modest, shy (but exceedingly kind to his acquaintances), he had a large number of research students, even after his retirement. His knowledge of mathematics was broad and profound, and included significant parts of mathematical physics. His originality was remarkable; he chose as his problems far from fashionable topics.

One idea pervades Loewner's work from his Ph.D. thesis: applying Lie theory concepts and methods to semigroups, and applying semigroups to unexpected mathematical situations. This led him in 1923 to a sensational result (4): the first significant contribution to the Bieberbach hypothesis. (A *schlicht* function $f(z) = \sum a_n z^n$ in the unit circle with $a_0 = 0$, $a_1 = 1$, has $|a_n| \leqslant n$—the case $n = 2$ was Bieberbach's and Loewner proved it for $n = 3$; in its totality the problem is still open.) In 1934 Loewner defined nth-order real monotonic functions by the property of staying monotonic if extended to nth-degree symmetric matrices (7) and characterized ∞th-order monotonic functions as functions which, analytically extended to the upper half-plane, map it into itself. The semigroups of first- and second-order monotonic mappings are infinitesimally generated; this property breaks down for orders greater than 2 (28, 32). The infinitesimally generated closed subsemigroup of monotonic mappings of infinite order is characterized by *schlicht* extensions to the upper half-plane (21). Loewner studied minimal semigroup extensions of Lie groups; for the group of the real projective line there are two: that of monotonic mappings of infinite order, and its inverse (21). In higher dimensions the question becomes significant under a suitable definition of monotony (30). Loewner also studied semigroups in a more geometrical context: deformation theorems for projective and Moebius translations (19), and infinitesimally generated semigroups invariant under the non-Euclidean or Moebius group (19), particularly if finite dimensionality and minimality are requested (22).

Among Loewner's other papers, many of which deal with physics, one should be mentioned explicitly: his non-Archimedean measure in Hilbert space (8), which despite its startling originality (or rather because of it) has drawn little attention outside the circle of those who know Loewner's work.

BIBLIOGRAPHY

Loewner's works are

(1) "Untersuchungen über die Verzerrung bei konformen Abbildungen des Einheitskreises, die durch Funktionen mit nichtverschwindender Ableitung geliefert werden," in *Berichte über die Verhandlungen der Sächsischen Akademie der Wissenschaften zu Leipzig*, Math.-phys. Klasse, **69** (1917), 89–106;

(2) "Über Extremumsätze bei der konformen Abbildung des Äusseren des Einheitskreises," in *Mathematische Zeitschrift*, **3** (1919), 65–77;

(3) "Eine Anwendung des Koebeschen Verzerrungssatzes auf ein Problem der Hydrodynamik," *ibid.*, 78–86, written with P. Frank;

(4) "Untersuchungen über schlichte konforme Abbildungen des Einheitskreises," in *Mathematische Annalen*, **89** (1923), 103–121;

(5) "Bemerkung zu einem Blaschkeschen Konvergenzsatze," in *Jahresbericht der Deutschen Mathematikervereinigung*, **32** (1923), 198–200, written with T. Rado;

(6) Chapters 3 and 16 in P. Frank and R. von Mises, eds., *Die Differential- und Integralgleichungen der Mechanik und Physik* (Brunswick, 1925), ch. 3, 119–192, and ch. 16, 685–737;

(7) "Über monotone Matrixfunktionen," in *Mathematische Zeitschrift*, **38** (1934), 177–216;

(8) "Grundzüge einer Inhaltslehre im Hilbertschen Raume," in *Annals of Mathematics*, **40** (1939), 816–833;

(9) "A Topological Characterization of a Class of Integral Operators," *ibid.*, **49** (1948), 316–332;

(10) "Some Classes of Functions Defined by Difference or Differential Inequalities," in *Bulletin of the American Mathematical Society*, **56** (1950), 308–319;

(11) *A Transformation Theory of the Partial Differential Equations of Gas Dynamics*, NACA Technical Report no. 2065 (New York, 1950);

(12) "Generation of Solutions of Systems of Partial Differential Equations by Composition of Infinitesimal Baecklund Transformations," in *Journal d'analyse mathématique*, **2** (1952–1953), 219–242;

(13) "On Generation of Solutions of the Biharmonic Equation in the Plane by Conformal Mappings," in *Pacific Journal of Mathematics*, **3** (1953), 417–436;

(14) "Conservation Laws in Compressible Fluid Flow and Associated Mappings," in *Journal of Rational Mechanics and Analysis*, **2** (1953), 537–561;

(15) "Some Bounds for the Critical Free Stream Mach Number of a Compressible Flow Around an Obstacle," in *Studies in Mathematics and Mechanics Presented to Richard von Mises* (New York, 1954), 177–183;

(16) "On Some Critical Points of Higher Order," Technical Note no. 2, Air Force Contract AF 18(600)680 (1954);

(17) "On Totally Positive Matrices," in *Mathematische Zeitschrift*, **63** (1955), 338–340;

(18) "Continuous Groups," mimeographed notes, University of California at Berkeley (1955);

(19) "On Some Transformation Semigroups," in *Journal of Rational Mechanics and Analysis*, **5** (1956), 791–804;

(20) "Advanced Matrix Theory," mimeographed notes, Stanford University (1957);

(21) "Semigroups of Conformal Mappings," in *Seminar on Analytic Functions, Institute for Advanced Study*, I (Princeton, N.J., 1957), 278–288;

(22) "On Some Transformation Semigroups Invariant Under Euclidean and non-Euclidean Isometries," in *Journal of Mathematics and Mechanics*, **8** (1959), 393–409;

(23) "A Theorem on the Partial Order Derived From a Certain Transformation Semigroup," in *Mathematische Zeitschrift*, **72** (1959), 53–60;

(24) "On the Conformal Capacity in Space," in *Journal of Mathematics and Mechanics*, **8** (1959), 411–414;

(25) "On Some Compositions of Hadamard Type in Classes of Analytic Functions," in *Bulletin of the American Mathematical Society*, **65** (1959), 284–286, written with E. Netanyahu;

(26) "A Group Theoretical Characterization of Some Manifold Structures," Technical Report no. 2 (1962);

(27) "On Some Classes of Functions Associated With Exponential Polynomials," in *Studies in Mathematical Analysis and Related Topics* (Stanford, Calif., 1962), 175–182, written with S. Karlin;

(28) "On Generation of Monotonic Transformations of Higher Order by Infinitesimal Transformations," in *Journal d'analyse mathématique*, **11** (1963), 189–206;

(29) "On Some Classes of Functions Associated With Systems of Curves or Partial Differential Equations of First Order," in *Outlines of the Joint Soviet-American Symposium on Partial Differential Equations* (Novosibirsk, 1963);

(30) "On Semigroups in Analysis and Geometry," in *Bulletin of the American Mathematical Society*, **70** (1964), 1–15;

(31) "Approximation on an Arc by Polynomials With Restricted Zeros," in *Proceedings of the Koninklijke Nederlandse Akademie van Wetenschappen*, Section A, **67** (1964), 121–128, written with J. Korevaar;

(32) "On Schlicht-monotonic Functions of Higher Order," in *Journal of Mathematical Analysis and Applications*, **14** (1966), 320–326;

(33) "Some Concepts of Parallelism With Respect to a Given Transformation Group," in *Duke Mathematical Journal*, **33** (1966), 151–164;

(34) "Determination of the Critical Exponent of the Green's Function," in *Contemporary Problems in the Theory of Analytic Functions* (Moscow, 1965), pp. 184–187;

(35) "On the Difference Between the Geometric and the Arithmetic Mean of *n* Quantities," in *Advances in Mathematics*, **5** (1971), 472–473, written with H. B. Mann; and

(36) *Theory of Continuous Groups*, with notes by H. Flanders and M. H. Protter (Cambridge, Mass., 1971).

HANS FREUDENTHAL

LOEWY, ALFRED (*b.* Rawitsch, Germany [now Rawicz, Poznan, Poland], 20 June 1873; *d.* Freiburg im Breisgau, Germany, 25 January 1935), *mathematics.*

Loewy studied from 1891 to 1895 at the universities of Breslau, Munich, Berlin, and Göttingen. He earned his Ph.D. in 1894 at Munich; was granted his *Habilitation* as *Privatdozent* at Freiburg in 1897; and became extraordinary professor in 1902. Not until 1919 was he appointed a full professor at Freiburg. In 1935 he was forced into retirement because he was a Jew. From about 1920 he was troubled with poor eyesight and he died totally blind.

Loewy published some seventy papers in mathematical periodicals and a few books. He edited German translations of works by Abel, Fourier, and Sturm; and the greater portion of the first part of Pascal's *Repertorium* was his work. His publications were concerned mainly with linear groups, with the algebraic theory of linear and algebraic differential equations, and with actuarial mathematics.

BIBLIOGRAPHY

I. ORIGINAL WORKS. Loewy's writings include *Versicherungsmathematik*, Göschen collection, 180 (Leipzig, 1903); *Lehrbuch der Algebra* (Leipzig, 1915); and *Mathematik des Geld- und Zahlungsverkehrs* (Leipzig, 1920).

Translations edited by Loewy or to which he contributed are N. H. Abel, *Abhandlung über eine besondere Klasse algebraisch auflösbarer Gleichungen*, Ostwald's Klassiker no. 111 (Leipzig, 1900); J. B. Fourier, *Die Auflösung der bestimmten Gleichungen*, Ostwald's Klassiker no. 127 (Leipzig, 1902); C. Sturm, *Abhandlung über die Auflösung der numerischen Gleichungen*, Ostwald's Klassiker no. 143 (Leipzig, 1904); and E. Pascal, ed., *Repertorium der höheren Mathematik*, P. Epstein and M. E. Timerding, eds., 2nd ed., I (Leipzig, 1910).

II. Secondary Literature. See S. Breuer, "Alfred Loewy," in *Versicherungsarchiv*, **6** (1935), 1–5; and A. Fraenkel, "Alfred Loewy," in *Scripta mathematica*, **5** (1938), 17–22, with portrait.

Hans Freudenthal

LOGAN, JAMES (*b.* Lurgan, Ireland, 20 October 1674; *d.* Germantown, Pennsylvania, 31 October 1751), *dissemination of knowledge.*

The most intellectually capable scientist of colonial America, Logan published works in botany and optics. His major contribution, however, lay in his acknowledged competence in science and in the help and advice that he gave to Thomas Godfrey, John Bartram, Benjamin Franklin, and Cadwallader Colden.

Logan was the son of Patrick Logan, a Scotch Quaker schoolmaster who moved in 1690 from Lurgan to Bristol, where he taught school assisted by his son. James Logan learned Latin and Greek as well as a number of other languages, including Arabic and Italian. He taught himself mathematics by mastering William Leybourn's *Cursus mathematicus*, only one work in his boasted first library of 700 volumes, which he later sold in order to invest in the linen trade. Having failed to so establish himself, in 1699 Logan accepted William Penn's invitation to accompany him to Pennsylvania as his secretary.

On his return to England in 1701, Penn left the twenty-seven-year-old Logan in charge of his affairs. For over four decades—as administrator, land agent, and merchant—he represented, not without dispute, the interests of Penn and his heirs. He was first secretary to and then for most of his life a member of the provincial council, mayor of Philadelphia in 1722–1723, chief justice of the Supreme Court of Pennsylvania in 1731–1739, and acting governor of the colony in 1736–1737. He negotiated treaties with the Indians and foresaw a clash with the French on the frontiers. He designed the Conestoga wagon to transport trade goods to Indian country and furs for export back to Philadelphia. He became rich through personal investments in land and an active mercantile trade.

Logan's copy of the *Principia*, which he bought in 1708 while he was in London, was the first known in America. On the same trip Logan saw Newton perform an experiment on the velocity of falling bodies under the dome of St. Paul's. He also made the acquaintance of a number of learned English men, including Flamsteed, and haunted the bookshops of London. Self-taught from books—which he bought in increasing quantities as he grew older and richer—Logan became highly knowledgeable in mathematics, natural history, and astronomy, making observations with his own telescope. Unhappily alone on the frontier of civilization without intellectual peers to converse with, he entered into extensive correspondence with Robert Hunter and William Burnet, successive governors of New York; the Quaker scholar Josiah Martin; Johann Albrecht Fabricius; William Jones, the British mathematician; John Fothergill, the Quaker physician; and Peter Collinson.

On his second trip to England in 1723–1724, Logan procured a set of sheets of Edmond Halley's still unpublished moon tables and added to them an explanation of Halley's method and an account of his own observation of the solar eclipse of 11 May 1724, which he watched at Windsor with members of the Penn family. Many of his astronomical books, some of which he procured from the widow of Johann Jacob Zimmermann, the German scientist and pietist, are full of notes, corrections, and comments, including Flamsteed's suppressed calculations of stars which Logan inserted in his own copy of Hevelius' *Prodromus Astronomiae*.

Thomas Godfrey, a glazier, was introduced to higher mathematics by Logan and became so proficient that he invented an improved mariner's quadrant. In 1732 Logan vigorously defended Godfrey's claim to the invention before the Royal Society against the simultaneous announcement of a similar instrument by John Hadley. His letters on behalf of Godfrey were supplemented by his own communications to the Royal Society, including in 1735 a preliminary essay on the generation of Indian corn, and in 1736 observations on a Hebrew shekel (which he sent as a gift), further notes on his corn experiments, a theory to explain the crooked aspect of lightning, and a hypothesis to account for the huge appearance of the moon on the horizon. He was, however, never made a fellow of the Royal Society.

In 1727 Logan read of the theories of seminal preformation and aerial pollination and began his own experiments with Indian corn. After his initial letter to the Royal Society, he worked to improve his techniques and to make his conclusions more explicit. Although the British scientists took little notice of his work, it was received enthusiastically on the Continent. Linnaeus hailed Logan as one of the heroes of botany, and Johann Friedrich Gronovius saw that the essay was published in full at Leiden in 1739 as *Experimenta et meletemata de plantarum*. Others had noted the sexuality of plants and pollination; Logan, by covering the flowers of corn with linen bags and by cutting off the tassels, demonstrated how the pollen generates kernels, or seeds, by ascending the tassels. Fothergill translated the work into English as *Experiments and Considerations on the Generation of Plants*, and it was published in London in 1747. Logan's work, a pioneer step toward plant hybridization, was referred to throughout the century. Within a year of the publication of Linnaeus' *Systema naturae* in 1735 Logan had acquired a copy. He introduced the "natural" naturalist, John Bartram, to the advances in taxonomy and taught him enough Latin to be able to read Linnaeus. It was through his encouragement and introductions that Bartram became known to botanists abroad.

Logan's scientific virtuosity found other expression. His simplification of Christiaan Huygens' method of finding the refraction of a lens appeared at the end of his *Experimenta* as "Canonum pro inveniendis refractionum." A second work in optics was even bolder. Starting with Huygens and going beyond Newton, he tried to show that the laws of spherical aberration could be more briefly expressed mathematically otherwise than geometrically. His treatise *Demonstrationes de radiorum lucis in superficies sphaericas* was reprinted at Leiden in 1741 with the help of Gronovius, whose colleague Pieter van Musschenbroek read the proofs. Such was Logan's reputation at home that Franklin brought him the accounts of his experiments in electricity to read before he sent them off to England.

Early in 1728 Logan slipped on ice and broke the head of his thighbone; it never healed, and he was lame for the rest his life. In 1730, after moving to his country house, Stenton, in Germantown, he tried with only moderate success to free himself of administrative responsibilities so that he could devote himself to his books. His translation of the pseudo-Cato's *Moral Distichs* was printed by Franklin in 1735, and his more famous version of Cicero's *Cato major*, written in 1733, was issued in 1744, also from Franklin's press. These, as well as his scientific articles, were minor works. During the second half of the 1730's Logan worked in bursts of energy upon a major philosophical treatise, "The Duties of Man Deduced from Nature." The recently rediscovered manuscript contains three completed chapters, one almost complete, and two only begun. In the section "Of the Exterior Senses" Logan brought to bear his knowledge of mathematics, harmonics, and medicine. The work, although imperfect, is the only surviving nontheological tractate on moral philosophy written in colonial America.

In 1742 Logan decided to bequeath his extensive library for public use and to spend his remaining years increasing it. By a codicil to his will, canceled because of a disagreement with his son-in-law, Isaac Norris, who had been named one of the trustees, he had left his books to be installed in a building at Sixth and Walnut Streets in Philadelphia. After his death his children established the Bibliotheca Loganiana, then consisting of about 2,600 volumes, as a trust in accordance with their father's unfulfilled intentions. In 1792 the trust was transferred to the Library Company of Philadelphia, where the books have remained ever since. The finest library in British America, it is now the only collection of its kind that has been preserved virtually intact. As he wrote for Franklin in 1749, Logan's library contained "A good Collection of Mathematical Pieces, as Newton in all three Editions, Wallis, Huygens, Tacquet, Deschales, &c. in near 100 Vols. in all Sizes." He had in addition all the works of the ancient mathematicians, a large collection of astronomical works, and as good a representation of books on natural history as could be found in the colonies. His impact as a scientist was limited, for, with few exceptions, he read, worked, and experimented for his own intellectual enjoyment.

BIBLIOGRAPHY

I. Original Works. Logan's writings have been mentioned in the text; he also published a number of political pamphlets and broadsides. His philosophical treatise, "The Duties of Man Deduced From Nature," in the Historical Society of Pennsylvania, is still unpublished. The principal materials on Logan's life are in the extensive Logan Papers in the Historical Society of Pennsylvania, which include MSS of many of his writings, a virtually complete run of letter books and other documents by and about him, but no corpus of letters received.

II. Secondary Literature. The only adequate biography of Logan is Frederick B. Tolles, *James Logan and the Culture of Provincial America* (Boston, 1957) but it is too brief. Logan's experiments on Indian corn are treated by Conway Zirkle, *The Beginnings of Plant Hybridization and Sex in Plants* (Philadelphia, 1935). His mathematical knowledge is noted, somewhat inaccurately, by Frederick E. Brasch, "James Logan, a Colonial Mathematical Scholar, and the First Copy of Newton's *Principia* to Arrive in the Colonies," in *Proceedings of the American Philosophical Society*, **86** (1942), 3–12. The best general appraisal of Logan as a scientist is Frederick B. Tolles, "Philadelphia's First Scientist, James Logan," in *Isis*, **47**

(1956), 2–30. The *Catalogus bibliothecae Loganianae* (Philadelphia, 1760) lists his library as it existed at the time of his death. A detailed catalogue of that library, noting his extensive annotations and his correspondence about the books, is being compiled by the author.

EDWIN WOLF II

LOGAN, WILLIAM EDMOND (*b.* Montreal, Canada, 20 April 1798; *d.* Llechryd, Wales, 22 July 1875), *geology.*

Logan's grandfather, James Logan, emigrated from Stirling, Scotland, to Montreal in 1784 and soon developed a prosperous bakery business, which passed upon his retirement to his eldest son, William. The latter married Janet Edmond of Stirling; and their second son, William Edmond, was born in Montreal. Logan's education began at Skakel's Private School in Montreal and continued at the Edinburgh High School (1814–1816) and Edinburgh University (1816–1817), where he studied chemistry, mathematics, and logic. He spent the years 1818–1831 in his uncle Hart's bank in London, becoming its manager upon his uncle's retirement in 1827. Later (1831–1838) he joined the management of a copper-smelting and coal-mining venture near Swansea, Wales, in which his uncle was interested, remaining there until his uncle's death. He soon found that chemistry and geology were essential to the success of the business and embarked upon a geological study of the local Glamorganshire coalfield—ultimately, in 1838, producing a memoir, with maps and sections. Its excellence was recognized by the director of the Geological Survey of Great Britain, Sir Henry De la Beche, who with Logan's permission incorporated it *in toto* in the Survey's report on that region. From that time on, Logan devoted himself exclusively to geology, particularly to the coal formations.

Logan's work on underclays with fossil *Stigmaria* in South Wales coalfields was to weigh heavily on the side of the *in situ* theory of the origin of coal, and, with his papers on the packing of ice in the St. Lawrence River, soon established him as a geologist of note. In 1842 the appointment of a provincial geologist was approved by the Canadian government under Sir Charles Bagot, who set about finding a suitable candidate. Logan obtained "a mass of testimonials," including letters from four of the most influential British geologists of the time: De la Beche, Roderick Murchison, Adam Sedgwick, and William Buckland. As a consequence he was offered, and accepted, the directorship of the newly created Geological Survey of Canada, a post which he held until 1869. For twenty-seven years he and his assistants traveled in all reachable parts of Canada from the Great Lakes to the Maritime Provinces; they also issued reports of progress, of which "Report on the Geology of Canada" (1863), his *magnum opus*, provided a compilation of twenty years of research. After more than a century, it is still a reservoir of important information.

Logan was fortunate in the choice of his assistants for both fieldwork and office work. Alexander Murray was his first and most important field geologist until he resigned to become director of the Geological Survey of Newfoundland in 1864. T. Sterry Hunt, his chemist, was responsible for hundreds of analyses of minerals, rocks, and ores. Elkanah Billings, his paleontologist, examined all fossils collected by field geologists and provided Logan with information invaluable for the correct identification of the age and the stratigraphic position of rock formations. Others included the geologists James Richardson and Robert Bell and the draftsman Robert Barlow. Later, Edward Hartley, Thomas Macfarlane, Charles Robb, and H. G. Vennor joined the Survey.

A twelve-hour day in the field was the rule for Logan. He was usually alone, carrying all necessary equipment together with the day's collection of specimens; he recorded his progress by means of pacing and compass in regions of which, for the most part, there were no reliable maps. If at the end of a day in the bush his plotting of his traverse showed an error of more than two chains, he was disappointed. In the evenings he wrote up his notes and completed his maps. Logan's notebooks, preserved in Ottawa, are marvels of simplicity, perspicacity, and brevity, here and there embellished by illuminating pen sketches of the country covered. He was equally tireless during the winters, composing his reports of progress, revising those of his assistants, and above all seeking adequate governmental financial support. Thousands of pounds of his own resources were poured into the early ill-supported organization.

Following the publication of the 1863 report one can detect a slight but increasing diminution of Logan's powers, which in 1869 he recognized had reached a point where a younger man was needed to carry the burden of a vigorous and growing organization. As a consequence, in that year he resigned as director and divided his time between an estate he had bought in Wales and exploration, at his own expense, in Canada, designed to settle certain vexatious problems which had been left unsolved at the time of his resignation. While preparing for a summer's fieldwork in the eastern townships of Quebec, he became ill;

and following a short illness he died in 1875. He was buried in the churchyard at Llechryd, Wales.

Logan's bibliography is not extensive and consists mostly of progress reports to the government concerning the work of the Survey. Many of these reports, sixteen in all, he wrote in his own hand—some in quadruplicate. The most important, and nearly the last, was his 1863 report, which provided the first complete coverage, according to information then available, of the geology of Canada from the Great Lakes to the Atlantic seaboard. In this remarkable compilation Logan was ably assisted by Sterry Hunt, whose work as chemist provided the foundation on which much of the information concerning the rocks, minerals, and ores of Canada was based. Early articles on underclay, the Glamorganshire coalfield, and ice packing have been mentioned. Others, mostly short notes, recorded his observations on the copper-bearing rocks of Lake Superior, animal tracks in the Potsdam sandstone, the supposed fossil *Eozoön*, subdivision of the Precambrian rocks of Canada, and remarks on the Taconic question, in which he avoided controversy by using the term "Quebec group" for equivalent rocks in Canada. Although he was the first to publish the discovery of *Eozoön*, and exhibited specimens of it during his visits to England, Logan later became noncommittal as the battle was waged between those who saw it as a fossil and those who advocated its metamorphic origin.

Among Logan's achievements was his recognition of an anomalous structural condition in which rocks of the Quebec Group lay structurally above younger (Middle and Late Ordovician) beds of the St. Lawrence Lowland. This he explained by proposing "an overturn anticlinal fold, with a crack and dislocation running along its summit, by which the [Quebec] group is made to overlap the Hudson River [Upper Ordovician] formation." He traced this thrust fault from Alabama to the Canadian border, and thence to the tip of the Gaspé peninsula. Although made up of a multitude of imbricating faults the structure is still referred to as Logan's Line. The earliest use of that term is not known.

Logan was never directly connected with university affairs, although as a result of his regard for Sir John William Dawson he donated $19,000 to found the Logan chair in geology and lesser amounts for Logan medals. Both McGill University in Montreal (1856) and the University of Bishop's College in Lennoxville, Quebec (1855), conferred honorary degrees on him. The excellence of his display of Canadian rocks and minerals at the London exhibition of 1851 led to his election as fellow of the Royal Society; he was sponsored by the most prominent contemporary British geologist, Sir Roderick Murchison. A similar exhibit at Paris in 1855 earned him the Grand Gold Medal of Honor from the Imperial Commission and an investiture as chevalier of the Legion of Honor in the same year. The following year he was knighted by Queen Victoria and also received the Wollaston Palladium Medal from the Royal Society. He was a fellow of the Geological Society of London (1837) and of the Royal Society of Edinburgh (1861), and a member of the Academy of Natural Sciences of Philadelphia (1846), the American Academy of Arts and Sciences, Boston (1859), and the American Philosophical Society (1860).

BIBLIOGRAPHY

I. Original Works. Logan's complete bibliography is in John M. Nickles, "Geologic Literature of North America 1785–1918," in *Bulletin of the United States Geological Survey*, no. 746, pt. 1 (1923), 671–672. His most important work was "Report on the Geology of Canada," in Geological Survey of Canada, *Report of Progress to 1863* (Ottawa, 1863), a summation of all his previous annual reports. His bibliography also includes a dozen short reports and articles on various topics.

II. Secondary Literature. See Robert Bell, *Sir William E. Logan and the Geological Survey of Canada* (Ottawa, 1877); B. J. Harrington, "Sir William Edmond Logan," in *American Journal of Science and Arts*, 3rd ser., **11** (1876), 81–93; *Canadian Naturalist and Geologist*, **8** (1876), 31–46, with portrait; and Geological Survey of Canada, *Report of Progress* for 1875–1876 (Ottawa, 1877), 8–21; and *Life of Sir William E. Logan, Kt.* (London, 1883), with portrait; J. M. Harrison and E. Hall, "William Edmond Logan," in *Proceedings of the Geological Association of Canada*, **15** (1963), 33–42; G. P. Merrill, *The First One Hundred Years of American Geology* (New Haven, 1924), 237, 411–416, 636; and W. Notman and Fennings Taylor, "Sir William Edmond Logan, LL.D., F.R.S., F.G.S.," in *Portraits of British Americans With Biographical Sketches*, II (Montreal, 1867), 133–145.

T. H. Clark

LOHEST, MARIE JOSEPH MAXIMIN (called **MAX**) (*b.* Liège, Belgium, 8 September 1857; *d.* Liège, 6 December 1926), *geology, mineralogical sciences, paleontology.*

Lohest was the son of Joseph Lohest, doctor of laws and merchant. After completing his secondary studies at the Collège des Jésuites, he enrolled first in the Faculty of Philosophy and Letters, and then in the School of Mines, of the University of Liège. In 1883 he received an honorary engineering diploma from the Technical Faculty. In the following year Gustave Dewalque engaged him as an assistant for a geology course. Having obtained the rank of *agrégé spécial*, he was put in charge of a course on deposits of fuels and phosphates; on 15 June 1893 he was given an optional course in geology instead. In 1897 Lohest succeeded Dewalque and thereby became professor of the general geology course. In his teaching he eliminated inessential detail to concentrate on geological principles and their possible applications. He took part, along with several colleagues from Liège, in the creation of the rank of engineer-geologist in 1900. During World War I he was involved in the organization of the School of Anthropology.

Lohest was elected a corresponding member of the Académie Royale des Sciences de Belgique in 1904 and a full member in 1910. In 1908 he was awarded the decennial prize in mineralogical science. He achieved membership in numerous Belgian and foreign scientific societies; of these his nomination as corresponding member of the Geological Society of London brought him the most pleasure. Lohest was on the board of directors of the Belgian Geological Commission. In 1919 he became a member of the Geological Council and was named to the commission set up to study the waters at Spa.

Like most scientists of the period, Lohest had many interests. Paleontology, especially the study of Paleozoic fishes, was the subject of his first scientific studies. Mere description of fossils was insufficient for him; from his observations he drew conclusions concerning the mode of life of the organisms and the milieu in which they evolved. In 1888 J. S. Newberry, who was visiting Liège, recognized bony plates from the head of a fish of the Dinichtys genus that until then had been reported only in America. Lohest found, in addition, resemblances—of other bones, fins, and dorsal plates—to fishes of the Upper Devonian of Canada.

Lohest also made mineralogical and petrographic studies. In particular he was interested in tourmaline, the conglomerates (pudding stones) of the Gedinnian, and certain anthracites. He investigated the specific conditions under which the deposits occur. Thus, having discovered in 1884, with G. Rocour, phosphate rocks in Hesbaye, he studied them extensively in 1885 and envisaged their possible extent in 1890. He also devoted a note to the age and origin of the plastic clay in the vicinity of Andenne and another to the Tertiary deposits of upper Belgium.

Stratigraphy and tectonics especially attracted Lohest, particularly those of the Paleozoic formations of eastern Belgium and he carried out a detailed survey of the highway and railway cuts in Belgium. His researches on the Dinantian led him to consider the presence of dolomites to be a local phenomenon unsuitable for stratigraphic reference. He found that the gray breccia encountered at various levels of the Visean were of tectonic origin due to fracturing of the limestone during the formation of the synclines. As for the very different red breccia of Waulsort and Landelies, he erroneously associated them with emersion phenomena. This opinion met with lively opposition in 1912 from one of his former students, Victor Brien.

Lohest was also interested in the details of the sedimentation of the Famennian, such as gullying and intraformational conglomerates. The Cambrian of the Stavelot Massif produced a revision in his views; in collaboration with Henri Forir, following the profound modifications that J. Gosselet had made within the stratigraphy of this system, he demonstrated the soundness of earlier surveys done by André Hubert Dumont.

Opposing Gosselet's thesis concerning the paleogeography of the Paleozoic of the Belgian Ardennes, Lohest defended the purely tectonic origin of anticline zones, stating that the partition into basins was simply a result of the Hercynian deformation.

After having been concerned with the relative age of certain faults of the coal-bearing formation, Lohest turned to those affecting both Mesozoic and Cenozoic formations. He sought to find out whether they indicate the disposition of the Paleozoic substratum; in this regard he attempted to apply the concepts of the American geologist Joseph Le Conte and the French geologist Marcel Bertrand. He also played a prominent role in guiding research on the westward extension into Belgium of the Dutch Limburg coalfields.

In the memoir "Les grandes lignes de la géologie des terrains primaires de la Belgique," Lohest showed himself an ardent partisan of actualism—neglecting, however, certain facts accepted by his contemporaries. Although tectonics was his chief interest—he even attempted to demonstrate its principal features experimentally—he devoted several publications to metamorphism in the Ardennes. He originated the term "boudinage," applied to the pinch and swell of quartz seams within sandstone beds. Various explanations have been offered for this phenomenon since it was first observed. In considering the seismic phenomena of the region around Liège (which are, of course, infrequent and of little consequence), Lohest accepted Élie de Beaumont's view that all geological phenomena are due to the contractions of the earth's crust caused by the cooling of the globe, and he defended this position all his life.

"Introduction à. . . la géologie. La vie de l'écorce terrestre," which appeared in 1924, is a condensation of Lohest's ideas in 224 pages. He wrote of recurrences, of the cyclic evolution of inanimate matter, and even of the succession of species toward an ideal of progress and perfection. Lohest became interested in the Quaternary in 1886, when, with Julien Fraipont and Marcel de Puydt, he discovered a characteristic Neanderthal fossil man—one dated, for the first time, by accompanying evidence of a Mousterian industry—in the terrace of Spy Cavern.

Lohest also devoted some of his time to physical geography; and although he never went to the Congo, he understood the importance of this territory for Belgian engineers. In 1897 he published "Notions sommaires de géologie à l'usage de l'explorateur du Congo." In Belgium he made several surveys on a scale of 1:40,000 and was requested to advise on important questions of applied hydrology, notably those concerning preliminary plans for water catchment and for the extension of water distribution.

BIBLIOGRAPHY

I. Original Works. There is complete bibliography in Fourmarier (see below). The most important works are "Recherches sur les poissons des terrains paléozoiques de Belgique. Poissons des psammites du Condroz, famennien supérieur," in *Annales de la Société géologique de Belgique*, **15** (1888), 112–203; "De l'origine des failles des terrains secondaires et tertiaires et de leur importance dans la détermination de l'allure souterraine des terrains primaires," *ibid.*, **20** (1893), 275–287; "Les grandes lignes de la géologie des terrains primaires de la Belgique," *ibid.*, **31** (1904), 219–232; "Introduction à l'étude de la géologie. La vie de l'écorce terrestre," in *Mémoires de la Société royale des sciences* (Liège), 3rd ser., **21** (1924); "Stratigraphie du massif cambrien de Stavelot," in *Mémoires de la Société géologique de Belgique*, **25** (1900), 71–119, written with H. Forir; "La race humaine de Néanderthal ou de Canstadt, en Belgique. Recherches ethnographiques sur des ossements humains découverts dans les dépôts quaternaires d'une grotte à Spy et détermination de leur âge géologique," in *Bulletin de l'Académie royale des sciences*, 3rd ser., **12** (1886), 741–784, also in *Archives de biologie* (Ghent), **7** (1887), 587–757, written with J. Fraipont; "Exploration de la grotte de Spy. Notice préliminaire," in *Annales de la Société géologique de Belgique*, **13** (1886), 34–39, written with M. de Puydt.

II. Secondary Literature. See P. Fourmarier, "Notice sur Max Lohest, membre de l'Académie," in *Annuaire. Académie royale de Belgique*, **119** (1953), 279–386, with portrait and complete bibliography.

F. Stockmans

LÖHNEYSS (or **LÖHNEIS** or **LÖHNEYSSEN**), **GEORG ENGELHARDT VON** (*b.* Witzlasreuth, Fichtelgebirge, Germany, 7 March 1552; *d.* Remlingen, near Wolfenbüttel, Germany, 1625[?]), *mining, metallurgy.*

Löhneyss, who came from a noble family, was brought up in Würzburg and Coburg. Before he was twenty, he entered the service of the prince of Ansbach. In 1575 he went to the court of Elector Augustus I of Saxony as equerry. Beginning in 1583 he held the same position under Augustus' son-in-law, Heinrich Julius of Brunswick and Wolfenbüttel, who in 1589 made him director of Brunswick's mines and foundries in the Upper Harz. In 1596 Löhneyss was appointed inspector general of the mines of Zellerfeld, Clausthal, and Andreasberg. He carried out his official duties first in Wolfenbüttel and on the estate at Remlingen; and from 1613 he worked in Zellerfeld. He was inspector general until 1622. During this time he kept many mines and foundries from being shut down.

Löhneyss was highly regarded as a writer by his contemporaries. From 1596, after a quarrel with his publisher, he brought out his works himself in printing shops he established in Remlingen and Zellerfeld. To illustrate his books he hired a woodblock engraver and a copperplate engraver. Along with *Della cavalleria*, both the text and the illustrations of which command respect, and *Aulico-politica oder Hof-, Staats- and Regierungskunst*, one should mention above all his *Bericht vom Bergkwerck*, which appeared in 1617. The intrinsic value of this book, which treats technical as well as economic and administrative aspects of mining and metallurgy, lies solely in the economic sections. Otherwise, *Bericht vom Bergkwerck* contributes nothing new; and even though it has frequently been called an outstanding description of mining and metallurgy, it in no way merits this praise. The technical material is shameless plagiarism: without citing his sources Löhneyss summarizes Agricola's *De re metallica libri XII* (1556) and copies word for word Lazarus Ercker's *Beschreibung allerfürnemisten mineralischen Ertzt und Berckwercksarten . . .* (1574), then combines them with the mining regulations of Brunswick to form a book.

BIBLIOGRAPHY

I. Original Works. Löhneyss's major works are *Della cavalleria, grundtlicher Bericht von allem was zu der löblichen Reiterei gehörig und einem Cavallier davon zu wissen geburt* (Remlingen, 1609–1610; 2nd ed., 1624; 3rd ed., Nuremberg, 1729); and *Bericht von Bergkwerck, wie man dieselben Bawen und in guten Wolstandt bringen soll, sampt allen darzu gehörigen Arbeiten, Ordnung und rechtlichen Process* (Zellerfeld, 1617); undated new eds. appeared between 1650 and 1670; in 1690 there appeared an ed. entitled *Gründlicher und ausführlicher Bericht von Bergwercken, wie man dieselbigen nützlich und fruchtbarlich bauen, in glückliches Auffnehmen bringen, und in guten Wolstand beständig erhalten.*

II. SECONDARY LITERATURE. See "Georg Engelhard[t] von Löhneyss," in *Allgemeine deutsche Biographie*, XIX (Leipzig, 1884), 133–134; W. Serlo, *Männer des Bergbaus* (Berlin, 1937), 99–100; H. Dickmann, "Das grösste Plagiat im berg- und hüttenmännischen Schrifttum," in *Das Werk* (Düsseldorf), **16** (1936), 572; W. Hommel, "Berghauptmann Löhneysen, ein Plagiator des 17. Jahrhunderts," in *Chemikerzeitung*, **36** (1912), 137–138; and "Über den Berghauptmann Löhneysen," *ibid.*, 562; M. Koch, *Geschichte und Entwicklung des bergmännischen Schrifttums* (Goslar, 1963), 60–62; and "Berghauptmann Georg Engelhardt von Löhneyss, Bergbauschriftsteller und Plagiator," in *Glückauf!*, **100** (1964), 49–50.

M. KOCH

LOHSE, WILHELM OSWALD (*b.* Leipzig, Germany, 13 February 1845; *d.* Potsdam, Germany, 14 May 1915), *astronomy*.

Lohse, the son of a tailor, attended schools in his native town and then, in 1859–1862, took the preliminary courses of the engineering college in Dresden. In the latter year he entered the University of Leipzig to study chemistry, and in 1865 he received the Ph.D. After some years of practical technical activity at Salzmunde, near Halle, and at Leipzig, he was interested in astronomy by Vogel, who at this time became director of the private observatory of the Kammerherr von Bülow at Bothkamp, near Kiel. Vogel engaged Lohse as a collaborator. Both men made spectroscopic observations; Lohse concentrated on solar observation and celestial photography, for which he prepared the plates himself. In 1874 Vogel and Lohse moved to the new astrophysical observatory at Potsdam, where Lohse worked for more than forty years—from 1882 as observer and from 1901 as head observer. Soon after his arrival Lohse installed the chemical and photographic workrooms and laboratories. At the Berlin observatory he observed *Nova Cygni* 1876 spectroscopically. In 1877, continuing his solar observations at Potsdam, Lohse had the necessary instruments constructed. At the same time he observed Jupiter and Mars. A little later he made photographs of stars and star clusters, and in 1899 he began photographing double stars. By comparing photographs made within twenty-five years Lohse tried to derive proper motions of faint stars; and through investigations of metallic spectra he contributed to the increased accuracy of spectroscopic measurements, using both prisms and a Rowland concave grating.

After the death of Vogel in 1907, Lohse headed the astrophysical observatory for two years, until advanced age and melancholia led him to resign. He was still healthy, however, and worked in his private workshop and in his garden until the end of 1914, when his health began to fail rapidly. At his death in May 1915 Lohse left his wife, two sons, and two daughters.

BIBLIOGRAPHY

Lohse's books include *Meteorologische und astronomische Beobachtungen in Bothkamp 1871–1873*, 3 vols. (Leipzig, 1872–1874); and *Tafel für numerisches Rechnen mit Maschinen* (Leipzig, 1909). The following articles appeared in *Potsdamer astrophysikalische Publikationen:* "Physische Beschaffenheit des Jupiter, Beobachtungen des Mars" (1882), 1–76; "Abbildungen von Sonnenflecken nebst Bemerkungen über astronomische Zeichnungen" (1883), 1–109; "Heliograph" (1889), 1–15; "Beobachtungen des Mars," **8** (1889), 1–40; "Beobachtungen des südlichen Polarflecks des Mars und der Elemente des Mars-Äquators," **11** (1898), 1–25; "Funkenspektren einiger Metalle," **12** (1902), 1–208; "Doppelsterne," **20** (1909), 1–168; and "Physische Beschaffenheit des Jupiter," **21** (1911), 1–11. He also published many short notes in *Astronomische Nachrichten*, vols. **82, 83, 85, 96, 97, 111, 115, 142, 146,** and **165.**

An obituary by P. Kempf is in *Vierteljahrsschrift der Astronomischen Gesellschaft* (1915), 160–169.

H.-CHRIST. FREIESLEBEN

LOKHTIN, VLADIMIR MIKHAYLOVICH (*b.* St. Petersburg, Russia, 1849; *d.* Petrograd, Russia, 1919), *hydrotechnology, hydrology*.

In 1875 Lokhtin graduated from the St. Petersburg Institute of Communications Engineers, after which he participated in surveys of the tributaries of the Kama River. In 1882 he was appointed head of a party to survey on the Dniester River, and after two years he was put in charge of improving navigation conditions on that river. From 1892 to 1899 Lokhtin, as head of the Kazan Communication District, led major research and corrective work on the Volga River near Nizhni Novgorod (now Gorki) and on a number of its shoals. In the early twentieth century he studied economic problems of water transport, and 1904 he participated in the study of the ice conditions of the Neva River. From 1907 he was inspector of macadam roads of Petersburg Province, and from 1915 he was a member of the Committee of State Construction and edited the journal *Vodnye puti i shosseynye dorogi* ("Waterways and Macadam Roads").

Lokhtin was one of the founders of the hydrology of rivers. The development of his theory on the formation of a riverbed may be followed from his work *Reka Chusovaya* ("The Chusovaya River"; 1878) to the monograph *O mekhanizme rechnogo rusla* ("On the Mechanism of a Riverbed"; 1895). In the first of these works Lokhtin, examining the peculiarities of the Chusovaya as a mountain stream, also touched on several general properties of rivers, the character of the changes in their velocities, their slopes, and the movement of high water. He turned his attention to the change in the transverse profile of the surface of the river during rises and falls in its level; on the basis of his observations, he asserted that a rise in level of the river produces a distended or convex profile while a fall in level leads to formation of a depression or concave profile.

Reports given by Lokhtin at the Russian Technical Society, *Sovremennoe polozhenie voprosa o sposobakh uluchshenia rek* ("The Present State of Methods of Improving Rivers"; 1883) and *Sovremennoe sostoyanie voprosa ob izuchenii svoystv rek* ("The Present Condition of the Study of the Properties of Rivers"; 1884), represented a substantial contribution to the development of the theory of riverbed processes. In the former Lokhtin spoke of the incorrectness of the then prevalent idea of the parallel stream movement of water in rivers and advanced the theory of the inner displacement of water within the current, pointing out that such a displacement explains the spiral movement of water, which determines the form of the riverbed. He did not, however, develop the question of the spiral form of movement of a liquid. In this report Lokhtin spoke of the three elements—climatic conditions, the character of the soil, and the topography of the basin—that determine the peculiarities of each river.

Lokhtin developed his original ideas of the processes which take place in riverbeds, particularly the formation of shoals, spits, and islands, in a work published in 1886, "Reka Dnestr, ee sudokhodstvo, svoystva i uluchshenie" ("The River Dniester, Its Navigation, Properties, and Improvement"). He connected the redistribution of slope between the reach and the shoal portions at the times of low and high water and the formation of shoals with three factors: the longitudinal profile of the riverbed (the influence of rocky "strong points"), the arrangement of the river in the plane (the influence of curvature), and the width of the river (influence of local widening). Referring to the shoal-reach form of the river, Lokhtin wrote that the low-water reaches and shoals are the result of the action of energetic high water—a result so considerable that weak and brief low waters are not strong enough to erase it or modify it. Lokhtin explained the formation of spits, and then of islands, as the result of the filling up of excess widths by sediment at the time of high water.

The fullest expression of Lokhtin's views on riverbed processes occurs in his monograph *O mekhanizme rechnogo rusla* ("On the Mechanism of a Riverbed"; 1895). It should be noted that in their theories other authors of works on the formation of riverbeds did not take into account the essential consideration that the discharge and level of water in rivers is not constant and can change within considerable limits. They made their generalizations fit chiefly the low-water situation. Moreover, a large range of variation in level is characteristic of the Russian rivers of the plains, which are fed primarily by melted snow. (During spring floods the majority of large Russian rivers rise more than three meters: on the Volga the rise reaches ten–thirteen; on the Don, seven–ten; on the Dnieper, six–eight.)

According to Lokhtin, the character of each river is determined by the following elements, each of which is independent of the others: (1) the amount of water, determined by sedimentation in the basin, and the soil conditions of the basin; (2) the slope or steepness, determined by the topography of a local cross section of the river; and (3) the degree of erosion or stability of the river channel, which depends on the properties cut into it by the flow of the bottom layers. Since all these elements can be present in different degrees and in different combinations, studying the character of a given river requires a knowledge of the physical and geographical conditions of the basin. The equipment which Lokhtin had at his disposal did not permit him to discover the inner structure of the current, determined by the turbulence and action of the centrifugal and Coriolis forces; he regarded the current as affected only by the action of gravity and as counterbalancing the resistance arising from the friction of water against the bottom.

Having examined and compiled longitudinal profiles of the free-water surfaces of the Volga, Dniester, and Garonne rivers at high-water and low-water levels, Lokhtin noted that the slope of the water surface at the transition from high water to low water changes at the same places in such a way that in the reaches it becomes less, and at the shoals more, than it was at high water. From this it follows that at low water, as a result of the increase in the longitudinal slope at the shoals—and thus also of the velocity of the current—

the deposits which accumulate at times of high water or flood are eroded and carried to lower-lying reaches, where at low water the slope and velocity of the current decrease. Lokhtin wrote:

> Having in its fall a single force for carrying away the obstructions which constantly enter the river and sensing an inadequacy in this force, in comparison with the resistance of the deposits, the river, so to speak, economizes the force, concentrating it at the place where, considering this, it is at a given time most necessary. At high water the slope is concentrated at the reaches, so that, cleaning them of deposits, it can go over, when the water level falls, to the shoals and begin carrying away the deposits which were temporarily left there at high water because of a lack of strength. Thus both parts of the riverbed, reaches and shoals, are inevitable and necessary in the attraction of deposits and are, in addition, definitely limited and constantly keep their places corresponding to the organically essential local conditions of the riverbed [*O mekhanizme rechnogo rusla*, p. 33].

Lokhtin classified rivers as stable or unstable. He considered a stepped longitudinal profile of the water surface to be characteristic of stable rivers. Between stable and unstable rivers are those of intermediate character. Unstable rivers, according to Lokhtin, are characterized by a uniform slope of the longitudinal profile of the water surface and by continuous movement of bottom sediments along the whole river. For a measure of the stability of the river, Lokhtin proposed the expression

$$k = \frac{d}{\Delta H},$$

representing the ratio of the mean diameter of the section of the bed to the fall of the water surface. Lokhtin concentrated his analysis of riverbed phenomena on the stable river bed. He believed that rivers could be improved through engineering methods based on the degree of stability of the river. (Although several inadequacies are now apparent in Lokhtin's discussions of separate river phenomena, the basic elements of his theory of formation of riverbeds are still significant.)

Lokhtin's research on the formation of ground ice is of considerable interest. He ascertained the presence of ice crystals at all depths of the current, which showed that they are formed in the open portions of the river as a result of the contact of flowing water with frosty air.

BIBLIOGRAPHY

I. Original Works. Lokhtin's writings include *Reka Chusovaya* ("The Chusovaya River"; St. Petersburg, 1878); "Reka Dnestr, ee sudokhodstvo, svoystva i uluchshenie" ("The Dniester River, Its Navigation, Properties, and Improvement"), in *Inzhener* (1886), no. 9–10, 410–441, no. 11–12, 485–546; *Sovremennoe polozhenie voprosa o sposobakh uluchshenia rek* ("The Present State of Methods of Improving Rivers"; St. Petersburg, 1886); *O mekhanizme rechnogo rusla* ("On the Mechanism of a Riverbed"; Kazan, 1895), also translated into French (Paris, 1909); and *Ledyanoy nanos i zimnie zatory na r. Neve* ("Ice Deposits and Winter Ice Blocks on the Neva River"; St. Petersburg, 1906).

II. Secondary Literature. See I. A. Fedoseyev, *Razvitie gidrologii sushi v Rossii* ("Development of the Hydrology of Dry Land in Russia"; Moscow, 1960), pp. 105–106, 168–172; and A. K. Proskuryakov, *V. M. Lokhtin i N. S. Lelyavsky. Osnovateli uchenia o formirovanii rusla* ("V. M. Lokhtin and N. S. Lelyavsky. Founders of the Theory of the Formation of the Riverbed"; Leningrad, 1951).

I. A. Fedoseyev

LOMONOSOV, MIKHAIL VASILIEVICH (*b.* Mishaninskaya, Arkhangelsk province, Russia, 19 November 1711; *d.* St. Petersburg, Russia, 15 April 1765), *chemistry, physics, metallurgy, optics.*

Lomonosov's father, Vasily Dorofeevich, owned several fishing and cargo ships; his mother, Elena Ivanovna Sivkova, was the daughter of a deacon. A gifted child, Lomonosov learned to read and write at an early age, and by the time he was fourteen was studying M. Smotritsky's Slavonic grammar and Magnitsky's arithmetic (which also dealt with other sciences and with technology).

In December 1730 Lomonosov received permission from local authorities to go to Moscow, where he entered the Slavonic, Greek, and Latin Academy the following month. He displayed brilliant linguistic abilities and soon became an accomplished Latinist. In 1734 he went to Kiev to work in the archives of the Religious Academy; he then returned to Moscow. At the beginning of 1736 he was sent, as one of the Moscow Academy's best pupils, to study at the University of St. Petersburg; and in the fall of that year he went to the University of Marburg, where he studied for three years with Christian Wolff. Lomonosov and Wolff respected each other's abilities but held few scientific views in common, although in 1745 Lomonosov did do a translation of Wolff's work on physics into Russian. That book attempted to combine the views of Newton with the ideas of Leibniz and Descartes and to reconcile the continuous ether with atomic theory. From Wolff, however, Lomonosov acquired a schematic style of scientific description that served him throughout his life.

At Marburg, Lomonosov also studied the humanities, on his own initiative. His studies abroad were in general oriented toward mathematics and chemistry (which he studied with Duising), mining, natural history, physics, mechanics, hydraulics, and hydrotechnics. In the summer of 1739 he traveled to Freiburg to study with Johann Henckel, a specialist in mining. Henckel was an Aristotelian and an opponent of the mechanistic interpretation of chemical phenomena that Lomonosov, for his part, always supported. The teacher and his new student differed sharply, and Lomonosov soon departed, having nevertheless acquired in less than a year much knowledge of mineralogy and metallurgy. By the spring of 1740 Lomonosov was traveling extensively in Germany and Holland. Later that year, in Marburg, he married Elizabeth Zilch; their daughter Elena was born in 1749. In 1741 Lomonosov returned with his wife to St. Petersburg, where he spent the rest of his life.

Lomonosov's two chief interests, poetry and science, had already come to the fore in the course of his foreign studies. In 1739 he composed an ode on the Russian capture of the Turkish fortress of Khotin, to which he appended theoretical considerations on the reform of Russian versification. In two student dissertations in physics (1738–1739), "On the Transformation of a Solid Body Into a Liquid" and "On Distinguishing Mixed Bodies That Consist of Chains of Corpuscles," Lomonosov established the basis of his future atomic-kinetic conceptions. "Corpuscles are different," he wrote, "if they are distinguished by mass or by figure, or by both at the same time." In 1756 he recalled, "From the time when I read Boyle I had a fervent desire to study the smallest particles. I reflected on them for eighteen years." While in Marburg he had also developed the idea of applying algebra to theoretical chemistry and physics, a notion that he began to implement on his return to St. Petersburg. A fully mature scientist, he had turned completely from theology and ancient languages to the natural sciences and technology, and to the Russian language and its poetry. His subsequent life and career may be divided into three distinct stages.

Theoretical Physics (1741–1748). At the beginning of 1742 Lomonosov was named adjunct of the St. Petersburg Academy of Sciences in the class of physics, in which post he remained for three and a half years. He was commissioned to compile a catalog of minerals and fossils, which was followed by "First Principles of Mining Science." His first independent scientific work, "Elements of Mathematical Chemistry," is marked by bold hypotheses and speculations. Defending the unity of theory and experiment, he wrote that "A true chemist must be both a theorist and a practical worker . . . as well as a philosopher." His notion of the difference between a compound, which is composed of corpuscles, and an element, has been said to resemble the laws of definite and multiple proportion later established in chemistry. The hypothetical nature of Lomonosov's approach to the physical sciences was further conditioned by the circumstances of his time: in the mid-eighteenth century experimental data on the quantitative chemical composition of various substances were not sufficient for testing his suppositions. Thus he was obliged to proceed on the basis of hypothesis.

In the period 1741–1743 he outlined the topics of his future research, compiling 276 notes on physics and corpuscular philosophy. They include his observations that

> When it is warm sound is more intense than when it is cold because the corpuscles move faster and strike each other more forcefully. . . . We must not think of many reasons when one is sufficient; thus corpuscular motion suffices to explain heat . . . there is no need to look for other reasons . . . . Nature strongly adheres to her laws and is everywhere the same . . . . The continuous formation and destruction of bodies speaks sufficiently for corpuscular motion.

Other notes have an ethical character and reflect his scientific outlook: "I will not attack for their errors people who have served the republic of science; rather, I will try to use their good thoughts for useful work."

In 1743–1744 he developed an atomic theory in a series of papers: "On the Intangible Physical Particles That Constitute Natural Substances," "On the Adhesion of Corpuscles," "On the Adhesion and Position of Physical Monads," and "On the Intangible Physical Particles That Constitute Natural Substances, in Which Substances the Sufficient Basis for Specific Qualities is Contained." His theory of matter, which was atomistic in principle, was further developed in his "On the Weight of Bodies" (1748), in which he imagined a materialistic monadology to oppose Leibniz' idealistic picture. Lomonosov's monads were corporeal rather than spiritual, having form, weight, and volume; and he employed them to explain the nature of heat and the elasticity of gases.

In 1745 Lomonosov read his paper "Reflections on the Reason for Heat and Cold" to the St. Petersburg Academy. He considered it to be one of his most important works for its argument against phlogiston in particular and against the theory of weightless fluids in general. He explained heat in terms of the velocity of motion (rotation) of material particles.

Cold was a diminution of motion, and with the full cessation of particle motion the greatest degree of cold was achieved. In "Attempt at a Theory of Elasticity of the Air" (1748) Lomonosov considered the nature of heat more fully. Inasmuch as the particles themselves occupy a certain volume of space, he predicted a deviation from Boyle's law in air subjected to very great pressure. In a letter to Euler of 5 July 1748 Lomonosov set down the general law to which these researches had led him: "All changes that we encounter in nature proceed so that . . . however much matter is added to any body, as much is taken away from another . . . since this is the general law of nature, it is also found in the rules of motion."

Continuing Boyle's line of thought, Lomonosov based his theory of heat on the mechanical action of bodies in contact, rather than upon the dynamics of Newtonian forces. (For this reason the external form of the material monad is important.) But Lomonosov employed corpuscular mechanics in chemical explanations more extensively than Boyle had done. Treating chemical compounds as particles in adhesion, he held that "adhesion is eliminated and renewed by means of motion . . . since no change in a body can take place without motion." He attempted to apply these theories to chemical phenomena—although he was limited to speculation—in papers on the action of chemical solvents in general (read in 1745) and on "metallic brilliance."

Lomonosov's work during this first period was not, however, confined to the physical sciences. In 1744 he described a comet that had appeared that year and, until 1748, he kept a record of the phenomena of thunderstorms. He wrote on electrical experiments in 1745—a subject to which he was to return—and in 1746 sought a method for measuring temperature at the bottom of a frozen sea. Returning to his mineralogical studies, he published a memoir on the wave motion of air observed in mines and conducted chemical analyses of salts, ores, and other rocks sent to the St. Petersburg Academy. He also compiled a syllabus of lectures on physics to be delivered in Russian and, in the summer of 1746, gave the first public lecture on that subject ever to be presented in that language.

Meanwhile, Lomonosov continued to combine science with poetry. In 1743 he wrote two major philosophical poems, "Morning Reflection on the Greatness of God" and "Evening Reflection on the Greatness of God on the Occasion of the Great Northern Lights." In the latter he asked, "What is the ray that surges through the clear night? What is the fine flame that strikes the firmament? . . . How is it possible that the frozen vapor of winter engendered fire?" He competed with the poets Tretyakovsky and Sumarokov in translating a Biblical psalm into Russian and composed a brief guide to rhetoric. In an ode on "The Delight of Earthly Kings and Kingdoms" (1747) he glorified science and the peaceful flowering of Russia. Addressing a patriotic challenge to youth, he summoned them "to show that the Russian land can give birth to its own Platos and quick-witted Newtons."

Despite these accomplishments, Lomonosov became embroiled in a number of heated disputes within the Academy. In 1743 he was arrested and imprisoned for eight months as a result of these encounters with Academy bureaucrats, led by Schumacher, whose interests were far removed from science. Nevertheless, in 1745 he was named professor of chemistry at the Academy. In 1747 Schumacher, hoping for an unfavorable review, sent a copy of Lomonosov's works to Euler; but his malice was rewarded by Euler's complete approval of Lomonosov's "Reflections on the Reason for Heat and Cold." The explanations of physical and chemical problems were so sound, Euler stated, that he was fully convinced of the accuracy of Lomonosov's proofs. Euler greeted with equal enthusiasm other works of Lomonosov that were sent to him from the Academy. It was largely through Euler's good opinion that the pioneering nature of Lomonosov's work was first recognized by that body. All the works that Lomonosov presented for publication were included in the first volume of the Academy's *Novye kommentarii* ("New Commentaries"), issued in 1748.

Experimental Chemistry (1748–1757). On assuming his duties as professor of chemistry, Lomonosov began to plan the construction of the first scientific chemical laboratory in Russia, which was opened in October 1748. Its equipment included balances, so that quantitative methods could be introduced into chemistry and the general law of conservation proven experimentally. Although Lomonosov did not completely suspend work in theoretical physics, he began to turn his interest to the experimental chemistry that he was just learning to do. His first chemical work, on the origin and nature of saltpeter (1749), presents the results of laboratory experiments together with theoretical speculation on the nature of mixed bodies (chemical compounds) and of chemical affinity. The latter were based on Lomonosov's kinetic interpretation of heat. In a paper on the usefulness of chemistry read to the Academy in 1751, he spoke of the problems of chemistry and of training chemists, noting that the discipline "requires a highly skilled practical worker and a profound mathematician in the same person." Thus, Lomonosov worked toward elevating chemistry to the level of a genuine theoretical, rather than a purely empirical, science. Pointing to the practical importance of chemistry, he challenged the dogma that useful minerals—especially precious metals in rocks—do not exist in northern countries.

In 1752 Lomonosov implemented his ideas on the training of chemists by drawing up a program of instruction in physical chemistry designed for young students. In an introductory note he wrote, "The study of chemistry has a dual purpose: advancing the natural sciences and improving the general welfare." He later set forth in detail the theoretical and empirical aspects of this science, considering that physical chemistry explains "on the basis of the ideas and experiments of physics what takes place in mixed bodies under chemical operations."

Lomonosov's surviving laboratory notes and journals testify to the number and variety of experiments that he himself performed. In his journal for 1751, for example, he reported on the results of seventy-four reagents and on their mutual interactions with various solvents, on his experiments on the production of glass, on his work with various powders, and on his investigations of a large number of chemical reactions. From 1752 to 1756 he took notes on physical-chemical experiments with salts and liquids and on the freezing (crystallization) of liquids. In 1756, following up Boyle's experiments on the heating of metals in closed containers, he found that when air is not admitted into the vessel, the total weight of the vessel and its contents remains constant—another confirmation of the general law of conservation as it applies to the total weight of chemically reacting substances. "My chemistry," he wrote in the same year, "is physical."

Lomonosov returned to the study of electrical phenomena in 1753, when he resumed experiments on atmospheric electricity. With G. V. Richmann he attempted to discover methods of conducting lightning and wrote "A Word on Atmospheric Phenomena Proceeding From Electrical Force." Richmann was killed while conducting experiments during a thunderstorm, but Lomonosov continued his researches and drew up a syllabus for further study. In 1756 he compiled 127 notes on the theory of light and electricity, presented a mathematical theory of electricity, and read a paper on the origin of light and on a new theory of colors that constitute it to a public meeting of the Academy. His reflections on the relation between mass and weight (1757) led him to the idea that another concept of measurement, perhaps that of weight, should be introduced as an expression of mass.

Lomonosov was also busy with practical projects. Having undertaken research on the production of glass, he turned to the revival of mosaic as an art form. In 1752 he presented a work on this subject to Czarina Elizabeth and introduced into the Duma a proposal to establish mosaic factories in Russia. In the same year he wrote a poem on the usefulness of glass, in which he contrasted glass and objects made of it with man's lust for gold. In 1753 Lomonosov received permission to build factories "for making varicolored glass and beads" and was given an estate near Moscow for this purpose. He built a mosaic workshop, with an attached chemical and optical laboratory, in St. Petersburg in 1756 and between 1761 and 1764 designed a large mosaic mural, *The Battle of Poltava*. Executed after his death, it is now in the Academy of Sciences in Leningrad. In 1754 he demonstrated at the St. Petersburg Academy a model of an "aerodrome machine" that he had invented and sent to I. I. Shuvalov his project for creating a university in Moscow, which was opened at the beginning of 1755.

During this period Lomonosov was especially active in history, philosophy, and literature. He presented a severe criticism of the historian G. Miller's "Norman theory" of the origin of Russia. In 1750 his dramatic tragedy, *Tamira and Selim*, was presented in St. Petersburg, and the following year an edition of his collected poems and prose works was published by the Academy of Sciences. His Russian grammar (1755–1757) was an important reform of the Russian language.

Lomonosov's Wide Practical Interests (1757–1765). At the beginning of 1757 Lomonosov was named member of the academic chancellery and, a year later, head of the geographical department of the St. Petersburg Academy. Occupied in scientific administration, he was less able to devote attention to physics and chemistry. Although his energies were drawn increasingly to practical matters, his surviving chemical and optical notes indicate that he nevertheless continued to do a considerable amount of pure research. In a paper on the hardness and fluidity of bodies (1760) he once again stated the general law of the conservation of matter and motion.

During the late 1750's Lomonosov became interested in exploration and, extending his earlier work on mining and metallurgy, in the exploitation of Russia's natural resources. Between 1757 and 1763 he wrote three works on mining and metallurgy; "Metals and minerals are not found lying on the doorstep," he stated in one of them. "Eyes and hands must search for them." Interested in navigation, especially of the northern seas, in 1759 he invented a number of instruments for astronomy and navigation, including a self-recording compass, and reflected on the precise determination of a ship's route. In 1761 he communicated to the Swedish Academy of Sciences a paper on the origin of icebergs in the northern seas and, two years later, described various voyages in the northern seas and discussed a possible approach to East India through the Siberian Sea. These works contain the first classification of ice and introduced the ideas of fossil ice and the presence of a huge ice drift.

As head of the geography department of the Academy, Lomonosov attempted to serve the general national interest. In addition to his works on exploiting natural resources, in 1761 he wrote Shuvalov a letter "on the propagation and preservation of the Russian people," in which he considered a broad range of social, economic, and political problems.

Lomonosov's works in literature and linguistics during the last period of his life include a foreword on the usefulness of church books, in the first volume of his collected works, published in 1758 by Moscow University. In it he opposed the current tendency to restore Church Slavonic and established the basis for a Russian scientific language in which the literary idiom approached the vernacular. In 1759–1760 he compiled a brief Russian chronicle with genealogy, wrote a memoir on the need for reforming the Academy, and began a heroic poem on Peter the Great. In 1761 all academic institutions in his region were entrusted to his sole management. His ideas for a pictorial map of Russian history were set forth in 1764.

Lomonosov was elected an honorary member of the Swedish Academy of Sciences (1760), the St. Petersburg Academy of Arts (1763), and a member of the Bologna Academy of Sciences (1764). His strong-willed and independent character created continuing difficulties with the czarist government; and in 1763 the czarina, having granted him the rank of state councillor, ordered his retirement. Several days later she herself rescinded the order.

Shortly before his death Lomonosov intended to produce two major generalizing philosophical works that would embody the whole of his atomic-kinetic principles and would illuminate his concept of the unity of nature. The works, to have been entitled "System of All Physics" and "Micrology," survive in outline notes and in theses that include

> Forces are miracles of harmony, a harmonious structure of causes. . . . The harmony and concord of nature. . . . The voice of nature everywhere in tune. . . . Unanimity and assent. . . . The concord of all causes is the most constant law of nature. . . . Everything is linked by a common force and the concord of nature.

The theses also reflect Lomonosov's ethical character, and many are autobiographical. Death prevented him from carrying out this projected work.

Lomonosov's last years were marred by illness. He appeared at the St. Petersburg Academy of Sciences for the last time early in 1765 and died a few months later. He was buried at the Alexander Nevsky Monastery.

Unique in the history of Russian culture and science, Lomonosov had an encyclopedic education. The first great Russian scientist, he united in himself knowledge not only of every basic area of the science of his time but of history, languages, poetry, literary prose, and art. His outlook on natural science and philosophy was frequently presented in verse. He was a distinguished teacher and social reformer, an enlightened humanist who worked to develop his country's productive forces. Pushkin called him Russia's first university. His scientific creativity consisted especially in his theoretical union of two basic concepts—the atomic (recognition of the discrete structure of matter) and the kinetic (recognition that particles of matter are endowed with motion). It was by basing this theory on the most general concept of the law of conservation of matter and motion that Lomonosov demonstrated experimentally the conservation of matter. A number of predictions in physics and chemistry derived from the combination of these three concepts were not verified until many years after his death.

Lomonosov was described as having "a cheerful temperament, [he] spoke tersely and wittily and loved to use pointed remarks in conversation; he was faithful to his country and friends, protected and encouraged novices in the literary arts; his manner was, for the most part, gentle but was passionate and hot-tempered withal."

Lomonosov was long considered to have been primarily a poet and man of letters, and little was known of his scientific works, which remained in manuscript. At the beginning of the twentieth century, B. N. Menshutkin discovered a great quantity of his unpublished material in the archives of the Academy of Sciences. Their publication revealed Lomonosov's importance as a physicist and chemist to the scientists of our time.

BIBLIOGRAPHY

I. Original Works. Lomonosov's complete works were published by the Soviet Academy of Sciences, under the general editorship of S. I. Vavilov, as *Polnoe sobranie sochineny*, 10 vols. (Moscow–Leningrad, 1950–1959). An earlier, 8-vol. ed. of his collected works, *Sobranie sochineny* (Moscow–Leningrad, 1934–1948), includes his poetry; works on linguistics and literature; articles on the natural sciences, physics, optics, chemistry, astronomy, and metallurgy; and his correspondence.

Important translations of his works (listed in chronological order) are *Physikalisch-chemische Abhandlungen*, Ostwalds Klassiker der Exacten Wissenschaften no. 178 (Leipzig, 1910), with works dating from 1741 to 1752; *Ausgewählte Schriften*, 2 vols. (Berlin, 1961), with works on natural science, history, and linguistics and collected letters; and *Michail Vasilievitch Lomonosov on the Corpuscular Theory*, trans. and with intro. by Henry M. Leicester (Cambridge, Mass., 1970).

II. Secondary Literature. It is impossible here to offer more than a brief indication of the vast literature on Lomonosov and various aspects of his career. The following works, therefore, listed chronologically within each subheading, represent only a small portion of existing nineteenth- and twentieth-century sources in Russian and other languages.

Biographies. See V. I. Lamansky, *Lomonosov* (St. Petersburg, 1864); V. I. Pokrovsky, *Mikhail Vasilievich Lomonosov* (Moscow, 1905); M. de Lur-Saluces, *Lomonossoff. Le prodigieux moujik* (Paris, 1933); G. Shtorm, *Lomonosov* (Moscow, 1933); J. D. Bernal, "M. V. Lomonosov," in *Nature*, no. 3688 (1940), 16–17; B. N. Menshutkin, *Zhizneopisanie Mikhaila Vasilievicha Lomonosova* (Moscow–Leningrad, 1947), the app. by L. B. Modzalevsky is a biographical guide to the basic literature on Lomonosov's life and work; and *Russia's Lomonosov, Chemist, Courtier, Physicist, Poet* (Princeton, 1952); A. A. Morosov, *M. W. Lomonosov* (Berlin, 1954); and *Mikhail Vasilievich Lomonosov* (Moscow, 1955), with bibliography, pp. 909–924; S. I. Vavilov, *Mikhail Vasilievich Lomonosov* (Moscow, 1961), a collection of articles and lectures—see esp. "Zakon Lomonosova" ("Lomonosov's Law"); G. S. Vasetsky, *Mirovozzrenie M. V. Lomonosova* (Moscow, 1961), on his world view; G. T. Korovin, *Biblioteka Lomonosova* (Moscow–Leningrad, 1961), with a catalog of Lomonosov's personal library; B. G. Kuznetsov, *Tvorchesky put Lomonosova* ("Lomonosov's Creative Path"; Moscow, 1961); G. E. Pavlova, ed., *Lomonosov v vospominaniakh i kharakteristikakh sovremennikov* ("Lomonosov Recalled and Described by His Contemporaries"; Moscow–Leningrad, 1962); L. Langevin, *Lomonosov, sa vie, son oeuvre* (Paris, 1967); and G. Vasetsky, *Lomonosov's Philosophy* (Moscow, 1968).

Lomonosov and the Academy of Sciences. Recent works include G. A. Knyazev, ed., *Rukopisi Lomonosova v Akademii nauk SSSR* ("Lomonosov's Manuscripts at the Soviet Academy of Sciences"; Moscow–Leningrad, 1937); V. F. Gnucheva, *Geografichesky departament Akademii nauk* ("The Geography Section of the Academy of Sciences"; Moscow–Leningrad, 1946), pp. 66–86, 178–199; M. I. Radovsky, *Lomonosov i Peterburgskaya Akademia nauk* ("Lomonosov and the St. Petersburg Academy of Sciences"; Moscow–Leningrad, 1961); and E. S. Kulebyako, *Lomonosov i uchebnaya deyatelnost Peterburgskoy Akademii nauk* ("Lomonosov and the Educational Activity of the St. Petersburg Academy of Sciences"; Moscow–Leningrad, 1962).

Lomonosov as Physicist, Chemist, and Astronomer. Useful sources on particular aspects of Lomonosov's career include B. N. Menshutkin, *Trudy Lomonosova po fizike i khimii* ("Lomonosov's Works on Physics and Chemistry"; Moscow–Leningrad, 1936); B. E. Raykov, *Ocherki po istorii geliotsentricheskogo mirovozzrenia v Rossii* ("Sketches in the History of the Heliocentric View in Russia"; Moscow–Leningrad, 1947); P. G. Kulikovsky, *Lomonosov—astronom i astrofizik* ("Lomonosov—Astronomer and Astrophysicist"; Moscow, 1961); N. M. Raskin, *Khimicheskaya laboratoria Lomonosova* ("Lomonosov's Chemistry Laboratory"; Moscow–Leningrad, 1962); and B. M. Kedrov, *Tri aspekta atomistiki*, II, *Uchenie Daltona* ("Three Aspects of Atomic Theory, II, Dalton's Theory"; Moscow, 1969), see pp. 178–200 for Lomonosov's discoveries of the law of conservation of matter, pp. 219–263 on his atomic-kinetic conceptions, and app., pp. 269–274, 290–292.

Other Aspects. On Lomonosov's other scientific and literary activity, see K. S. Aksakov, *Lomonosov v istorii russkoy literatury i russkogo yazyka* ("Lomonosov in the History of Russian Literature and the Russian Language"; Moscow, 1846); V. I. Vernadsky, *O znachenii trudov Lomonosova v mineralogii i geologii* ("The Significance of Lomonosov's Work in Mineralogy and Geology"; Moscow, 1900); A. P. Pavlov, *Znachenie Lomonosova v istorii pochvovedenia* ("Lomonosov's Importance in the History of Soil Science"; Moscow, 1911); P. N. Berkov, *Lomonosov i literaturnaya polemika ego vremeni, 1750–1765* ("Lomonosov and the Literary Polemic of His Time"; Moscow–Leningrad, 1936); V. P. Lystsov, *Lomonosov o sotsialno-ekonomicheskom razvitii Rossii* ("Lomonosov on the Social and Economic Development of Russia"; Voronezh, 1939); M. A. Bezborodov, *Lomonosov i ego rabota po khimii i tekhnologii stekla* ("Lomonosov and His Work in the Chemistry and Technology of Glass"; Moscow–Leningrad, 1948), published on the 200th anniversary of the founding of the first scientific chemical laboratory in Russia; V. A. Perevalov, *Lomonosov i arktika* ("Lomonosov and the Arctic"; Moscow–Leningrad, 1949); S. I. Volfkovich and V. V. Kozlov, *Tekhnicheskaya khimia v tvorchestve Lomonosova* ("Technical Chemistry in Lomonosov's Creative Work"; Moscow, 1961), with bibliography, pp. 75–81; and P. N. Berkov *et al.*, eds., *Literaturnoe tvorchestvo Lomonosova. Issledovania i materialy* ("Lomonosov's Creative Literary Works. Studies and Material"; Moscow–Leningrad, 1962).

Collections. Collected material on Lomonosov's life and activity include *Lomonosovsky sbornik* ("Lomonosov Collection"; St. Petersburg, 1911); N. A. Golubtsova, ed., *Lomonovsky sbornik* (Arkhangelsk, 1911); and A. I. Andreyev, L. B. Modzelevsky, S. I. Vavilov, V. L. Chenakal, *et al.*, eds., *Lomonosov. Sbornik statey i materialov* ("Lomonosov. A Collection of Articles and Materials"), 6 vols. (Moscow–Leningrad, 1940–1965).

B. M. Kedrov

LONDON, FRITZ (*b.* Breslau, Germany [now Wrocław, Poland], 7 March 1900; *d.* Durham, North Carolina, 30 March 1954), *physics, theoretical chemistry.*

Fritz London and his younger brother, Heinz, were the only children of Franz London, *Privatdozent* in mathematics at Breslau and later professor at Bonn, and Luise Hamburger, daughter of a linen manufacturer. The home atmosphere was that of a cultivated and prosperous liberal German-Jewish family, disturbed only by the premature death of Franz London in 1917. London attended high school at Bonn, where he received a classical education, and the universities of Bonn, Frankfurt, and Munich. His first academic interest was philosophy: in 1921 he received a doctorate *summa cum laude* at Munich for a dissertation on the theory of knowledge based on the symbolic methods of Peano, Russell, and Whitehead and their followers. He had done the work without supervision and entered it for the degree at the suggestion of G. Pfänder, to whom he had shown it for criticism. The philosophic bent remains noticeable throughout London's work, which is characterized by a constant search for general principles and thorough exploration of the logical foundations of his chosen subjects. He was never a mere calculator. In 1939 he published jointly with Ernst Bauer a short monograph in French on the theory of measurement in quantum mechanics.

During the three years following publication of his dissertation London wrote two further philosophical papers and spent some time as a high school teacher in various parts of Germany, but in 1925 he returned to Munich to work in theoretical physics under Sommerfeld. He then held in succession appointments at Stuttgart with P. P. Ewald and at Zurich and Berlin with Schrödinger. In 1933 the Nazi persecution forced both London brothers to leave Germany. London spent two years at Oxford and two years in Paris at the Institut Henri Poincaré. In 1939 he was appointed professor of theoretical chemistry (later chemical physics) at Duke University, where he remained until his death. In 1929 he married the artist Edith Caspary. They had two children.

Between 1925 and 1934 London's interests centered on spectroscopy and the new quantum mechanics, chiefly as applied to the chemical bond. His first scientific paper, written jointly with H. Hönl, was on the intensity rules for band spectra. In 1927 he and W. Heitler produced their classic quantum-mechanical treatment of the hydrogen molecule, "the greatest single contribution to the chemist's conception of valence made since G. N. Lewis' suggestion in 1916 that the chemical bond between two atoms consists of a pair of electrons held jointly by the two atoms."[1] Heitler and London set themselves the problem of determining the energy of a quantum mechanical system having two electrons in motion about two charged nuclei (protons). At large distances, with one electron orbiting each nucleus, the system is just a pair of hydrogen atoms; but when the nuclei are brought closer the forces between all four particles must be taken into account. It had long been recognized that two atoms A, B, may combine as a stable molecule with separation d if the interatomic forces vary so as to make the total energy a minimum at that distance. A complete quantum mechanical calculation of the forces requires precise knowledge of the probability function for the distribution of electrons, which is exceedingly complicated, but Heitler and London were able to approximate the solution by means of an analytical technique developed in the theory of sound by Lord Rayleigh. In a vibrating body, such as a bell, the energy associated with the lowest normal mode of vibration is a minimum. Any motion of slightly different form will have higher energy, but since it is near a minimal point, the increase in energy due to first-order deviations in motion is of the second order of small quantities. A rather good estimate of the energy of vibration may therefore be made from quite a crude approximation to the actual motion; and of any two approximate solutions the one yielding lower energy is nearer the truth. Rayleigh's method supplies a guide for guesswork in the study of complex vibrating systems; equivalent ideas apply to the solution of Schrödinger's wave equation in quantum mechanics. The first trial wave function used by Heitler and London for the hydrogen molecule was simply the known expression for one electron orbiting each nucleus, that being the exact solution at large distances where the interactions between the atoms are negligible. In this case the function probability of the two electrons 1, 2 being found in the same volume element of space is (by ordinary statistical theory) the product of their separate probabilities, and hence their joint wave function $\psi_{AB}(1,2)$ from which the probabilities are determined is

$$\psi_{AB}(1,2) = \psi_A(1)\,\psi_B(2), \qquad (1)$$

where $\psi_A(1)$ is the wave function for electron 1 orbiting atom A, and $\psi_B(2)$ for electron 2 orbiting atom B. Insertion of (1) into Schrödinger's equation does yield an attractive force between the atoms at large distances and an energy minimum when they are separated by about 0.9 Å, but the calculated binding energy is far from the experimental value.

The next step hinged on a concept put forward in

another connection by Heisenberg about a year earlier. Since in quantum mechanics the two electrons are indistinguishable, they may be supposed to change places without altering the system in any way, and a better approximation to the wave function will be

$$\psi_{AB}(1, 2) = \psi_A(1)\,\psi_B(2) \pm \psi_A(2)\,\psi_B(1). \qquad (2)$$

From (2) with the positive sign Heitler and London obtained an expression for the total energy W_M of the molecule having the form

$$W_M = 2W_H + \frac{e^2}{r_{AB}} + \frac{2J + J' + 2K\triangle + K'}{1 + \triangle^2}, \qquad (3)$$

where W_H is the ground state energy of an isolated hydrogen atom, e^2/r_{AB} the electrostatic repulsion between the two nuclei, and J, J', K, K', and $\triangle$ are various integrals, of which J can be interpreted as the net attraction between each electron cloud and the nucleus of the other atom, J' the repulsion between the electron clouds, and K and K' are the so-called exchange integrals whose meaning will be discussed below. The energy is substantially reduced, so by Rayleigh's principle the function (2) is indeed a closer approximation than (1) to the real distribution. Figure 1 reproduces the calculated curve, which has a

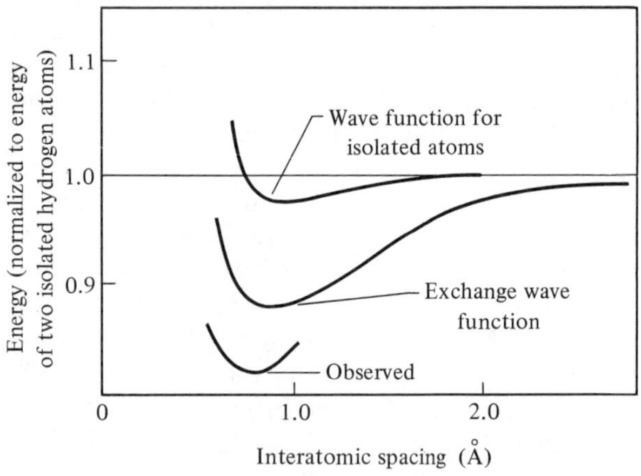

FIGURE 1. Binding energy for the hydrogen molecule.

pronounced minimum at an interatomic spacing of 0.79 Å and gives a binding energy of 3.14 electron volts, which may be compared with the experimental values 0.74 Å and 4.72 ev. Finally, extending London's earlier interest in the band spectra of molecules Heitler and London determined a vibrational frequency for the molecule from the curvature of the potential near the minimum and obtained a value of 4800 cm^{-1} as compared with the observed spectral frequency of 4318 cm^{-1}.

Improvements in detail to the Heitler-London solution were soon made by others, chief being corrections for screening of the nuclei by the electron cloud (S. C. Wang), polarization of the charge cloud (N. Rosen), and the addition to equation (2) of terms describing ionic structures (S. Weinbaum). In 1933 H. M. James and A. S. Coolidge gave a laborious investigation introducing the interelectronic distance r_{12} explicitly into the analysis and comparing results for variation functions with five, eleven, and thirteen terms. The calculations converged on values for the binding energy, vibrational frequency, and interatomic spacing extraordinarily close to experiment, and equally good results have since been found for other properties of molecular hydrogen such as its magnetic and electric susceptibilities. All in all, the description of the hydrogen molecule by quantum mechanics achieves a level of success which makes it one of the most compelling pieces of evidence for the theory itself. The remaining major issue is the interpretation of the exchange integrals. Heisenberg had originally illustrated the idea of a quantum mechanical system oscillating between two states by analogy with classical resonance phenomena, for example, the periodic interchange of energy between two tuning forks on a common base. Quantum mechanical resonance, however, differs from its classical analogues by lowering the total energy of the system. Since a reduced energy implies an attraction, contributions to the interaction arising from the use of wave functions like (2) are often spoken of as exchange forces, pictured as the result of a frequent switching of positions of the two electrons in the molecule—a quite unnecessary piece of mystification. All the forces binding the molecule together are electrical in origin: the exchange integral is merely one of a number of contributions to the electromagnetic energy, and there is moreover no intrinsic need to divide the energy in this particular way. Shortly after the Heitler-London paper, another, very different treatment of molecular structures, the molecular-orbital method, was developed by E. U. Condon, F. Hund, R. S. Mulliken and others, which builds up the solution by modifying the wave function for the helium ion rather than those for two hydrogen atoms. It gives results of comparable accuracy without exchange terms, using instead a quite different pair of resonant structures to find the wave function for the ion. A third mode of division appears in James and Coolidge's general variational solution. With these qualifications it must be added that the idea of treating complex molecules as resonating between a number of simple structures, which grew out of the work of Heitler and London and their contemporaries, has been amazingly fruitful. London later used to recall Schrödinger's comment that while he had a high opinion of his equation, expecting it to describe the entire field of chemistry as well as physics was more even than he would have dared.[2]

London continued for several years to work on molecular theory. In 1928 he formulated a description of chemical reactions as activation processes, and in 1930 with R. Eisenschitz he turned his attention to the quantum theory of intermolecular forces. The modern burgeoning of this branch of physics dates effectively from 1875, when van der Waals applied corrections to the ideal gas equation to describe the phenomena of critical points and the condensation of liquids and showed that the additional terms could be attributed to the combination of a long-range attraction and a shorter-range repulsion between molecules. These must, of course, be distinguished from the more powerful chemical forces. Analysis of their effects gradually advanced during the next forty years, and by 1926, largely through the work of J. E. Lennard-Jones on crystal structures and the transport properties of gases, some knowledge had been gained of their magnitude and distance dependence. Meanwhile, P. J. W. Debye and also W. Keesom in 1921 had given elementary calculations of forces due to the electric dipole moments of molecules and had found an attraction varying inversely as the seventh power of the distance. This attraction, however, was far too weak to account for the observed phenomena. Following two false starts by Born and Heisenberg, the corresponding quantum mechanical calculation given in 1927 with numerical errors by Wang was then corrected and extended by Eisenschitz and London in their paper. They found that fluctuation processes cause a dipole-dipole interaction much greater than Debye's, because in effect the orbital motions of the electrons in one molecule create large time-varying fields at the other, and the force is proportional not to the mean but to the mean-square value of the field. Such forces are customarily called dispersion forces because they are determined mainly by the outer electrons, which are also responsible for the dispersion of light. The interaction energy is

$$W = \frac{6e^2 a_o^5}{r^6}, \tag{4}$$

where e is the electron charge and a_o the Bohr radius of the atom. This was the form finally adopted by Lennard-Jones for the attractive part of his potential: it leads to forces varying inversely as the seventh power of the distance. Eisenschitz and London were able to correlate the forces with molecular polarizabilities, and they were later applied to many other matters. More elaborate analyses yielded higher-order corrections. For polar molecules there are additional terms of comparable size, which also obey the inverse-seventh-power law but vary with temperature. A further correction was given in 1946 by H. B. G. Casimir and D. Polder, who discovered that at short distances the retardation of the field introduces phase shifts in the motions of the electrons which make the forces vary more nearly as the inverse eighth power. Effects of this kind are observed in surface phenomena, such as the equilibrium thickness of very thin helium films.

Toward the end of 1932 London completed the manuscript of a book on molecular theory. His agreement for its publication was broken by Springer, the German publisher, after his departure from the country. In England he attempted to arrange a translation, but although several persons offered to help, he was unable to strike a working relationship with any of them. At the time of his death he was planning to rework and translate the material himself. The manuscript is preserved at Duke University Library. London's departure from Germany coincided with a general shift in his scientific interest. Only on one problem in molecular physics, the diamagnetism of the aromatic compounds, did he spend much time thereafter. While studying the benzene ring in 1937 he began to form the ideas about long-range order that became central in his work on superconductivity.[2]

In 1932 London's brother, Heinz, started a Ph.D. thesis on superconductivity in the low-temperature group at Breslau directed by F. E. Simon. Since Kammerlingh Onnes' discovery in 1911 that the electrical resistance of mercury vanishes at very low temperatures, the behavior of superconductors with direct currents had been extensively studied, but nothing was known of the effects of high-frequency alternating currents. Following a suggestion by W. Schottky of Siemens, H. London looked for high-frequency losses in superconductors by attempting to detect the Joule heating caused by currents from a 40 MHz radio source. His search did not lead to anything for several years, but early on he formed some strikingly original ideas about the superconducting state. He decided that a.c. effects, if they exist, probably occur through the existence of two groups of electrons, one subject to losses, the other not. Direct currents then flow only in the superconducting electrons; alternating currents couple inductively to both groups in parallel and so cause dissipation. The same two-fluid model was independently advanced by C. J. Gorter and H. B. G. Casimir in 1934 to account for certain thermodynamic properties of superconductors, and since Heinz London gave the idea only in his thesis, it is commonly associated with their names. He then wrote an acceleration equation for the superconducting electrons,

$$\Lambda \dot{\mathbf{J}} = \mathbf{E}, \tag{5}$$

where **E** is the internal field, **J** the current, and Λ a constant equal to m/ne^2, where m and e are the mass and charge of the electrons and n their number density. The constant Λ, known as the London order parameter, is often attributed to Fritz London but is in fact exclusively due to his brother. Combining (5) with Maxwell's electromagnetic equations, H. London then concluded that the currents in any superconductor are confined to a shallow surface layer characterized by a penetration depth $\lambda = \sqrt{\Lambda/4\pi}$. This idea is evidently analogous in some degree to the well-known a.c. penetration depth derived by Maxwell for alternating currents in normal metals. The quantity λ was the first of a number of characteristic distances important in the theory of superconductivity. Again it was due exclusively to Heinz London. Similar ideas about the acceleration equation and penetration depth were developed independently by R. Becker, G. Heller, and F. Sauter in an interesting paper which also evaluated magnetic effects in a rotating superconductor.

In 1933 shortly before Heinz London joined his brother at Oxford, W. Meissner and R. Ochsenfeld made a startling discovery. It was well known that currents in superconductors flow in such a way as to shield points inside the material from changes in the external magnetic field. This indeed is an obvious property of any resistanceless medium, fully discussed by Maxwell in 1873 long before the discovery of superconductivity.[3] But a superconductor does more. Whereas a zero resistance medium only counteracts changes in the field, it actually tends to expel the field present in its interior before cooling. The distinction between the two cases is illustrated in Figure 2. The Londons quickly saw its implications and in 1935 published a joint paper on the electrodynamics of superconductors, in which they replaced (5) by a new phenomenological equation connecting the current with the magnetic rather than the electric field,

$$\operatorname{curl} \Lambda \mathbf{J} = -\frac{1}{c}\mathbf{H}. \tag{6}$$

Formally, (6) is nothing other than the integral of the equation obtained by inserting (5) into Maxwell's equation, curl $\mathbf{E} = -(1/c)\dot{\mathbf{H}}$, and taking the constant of integration as zero. However, in contrast to many physical processes where constants of integration are trivial, this choice has profound significance: it represents a preferred magnetic state of the superconductor independent of past history. In the paper the logical relation between the propositions is set forth with Fritz London's philosophical clarity. As the authors observe, equations (5) and (6) stand roughly at the same level of generality; but (6) embraces more in respect to the Meissner effect, yet less in another respect, since it implies not (5) but the weaker condition curl $(\Lambda\mathbf{J} - \mathbf{E}) = 0$ or, after integration,

$$\Lambda \mathbf{J} = \mathbf{E} + \operatorname{grad} \mu, \tag{7}$$

where μ is an arbitrary scalar quantity. The virtue of the new description is that it covers neither more nor less than the known facts. In the remaining sections the Londons went on to determine boundary conditions, examine the covariant properties of the equations, and give detailed solutions for a sphere and a cylindrical wire which again showed that the currents are restricted to a penetration layer d of order $c\sqrt{\Lambda/4\pi}$.

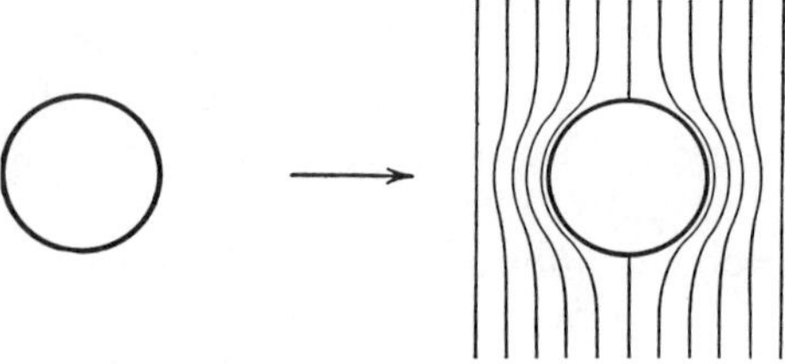

Magnetic field applied to zero resistance medium after cooling in zero field.

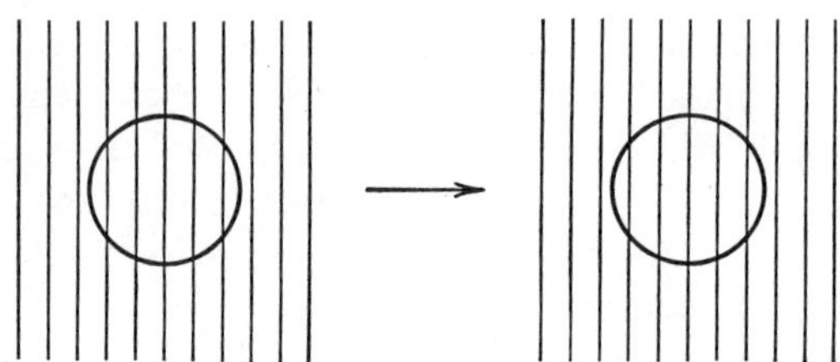

Zero resistance medium cooled in finite field.

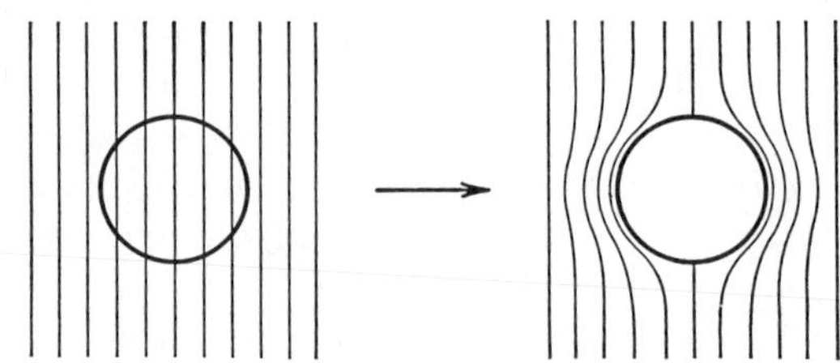

Superconductor cooled in finite field.

FIGURE 2. The Ochsenfeld-Meissner effect.

London continued to study superconductivity for many years, along with the parallel phenomena of superfluidity in liquid helium which were discovered soon afterward and on which he wrote an important paper in 1938. He expounded his ideas on both subjects in various articles and in the two-volume *Superfluids* (1950–1954). He gradually came to see a deeper meaning to equation (6). In ordinary electrodynamics the canonical momentum of the electron is given by

$$\mathbf{p} = m\mathbf{v} + e\mathbf{A}/c, \tag{8}$$

where $m\mathbf{v}$ is the ordinary momentum and $e\mathbf{A}/c$ is a kind of effective momentum associated with the magnetic field, **A** being the vector potential of the field. The electric current is determined by **v**, but the quantity determining the de Broglie wavelength of the electron is **p**. Writing $\mathbf{p}_s$ for the average momentum of the superconducting electrons and using Maxwell's relation $\mathbf{B} = \text{curl } \mathbf{A}$ and the definition of Λ, London discovered that (6) is equivalent to

$$\text{curl } \mathbf{p}_s = 0. \tag{9}$$

In the special case of an ideal solid superconductor the solution of (9) is $\mathbf{p}_s = 0$, while for a wire of constant cross section fed at its ends by a current from an external source, $\mathbf{p}_s$ is uniform over the length and breadth of the wire. This last result is of extraordinary interest. It implies that although the superconducting current is concentrated at the surface of the wire, the wave functions of the electrons extend uniformly throughout the material. From this London came to a radically new concept of the nature of the superconducting state. Ordinary substances when cooled to low temperatures lose the kinetic energy of heat and become progressively more ordered in position: they solidify. For particles subject to classical mechanics, that indeed is the only possibility. For quantum mechanical systems, however, the kinetic energy does not quite vanish at the absolute zero; there is a residual zero-point energy because the positions and momenta of the particles are subject to Heisenberg's uncertainty relation $\triangle x \triangle p \sim h$. Hence, London conjectured, there might in some circumstances be an advantage in energy for a collection of particles to condense with respect to momentum rather than position. This would account for the uniformity of $\mathbf{p}_s$ throughout the wire and would imply that the current constitutes a macroscopic quantum state, describable by a single wave function, like a gigantic molecule. A striking corollary is that the magnetic flux through a superconducting ring should be quantized, since there must be an integral number of waves around the ring.[4] This conjectural condition on the entire current resembles the quantum condition on a single electron in the de Broglie picture of the atom. London's unit of magnetic flux was hc/e or about 4×10^{-7} gauss cm^2. Some years later the existence of flux quantization was demonstrated independently in Germany and the United States,[5] but the unit was found to be half London's value, a result explained in the microscopic theory of superconductivity by the idea that electrons interact in pairs.

London's concept of long-range order in momentum space transformed the issue of what the theory of superconductivity was about. It no longer had to explain vanishing resistance but, rather, why superconducting electrons acquire order of this particular kind. This opened the way of escape from an embarrassing theorem due to F. Bloch,[6] who had rigorously established that the lowest energy state of a system of electrons isolated from external magnetic fields is one of no current. Now if superconductivity were merely an absence of resistance, fluctuation effects would inevitably disturb the current and cause it to decay. In London's viewpoint no such problem arises, because the current is an excited metastable state, with energy barriers that may (as was later shown) be related to the quantized flux condition.[7] The goal for a microscopic theory of superconductivity is then to find the mechanism for cooperative behavior of the electrons. This was achieved by H. Fröhlich; J. Bardeen, L. N. Cooper, and J. R. Schrieffer; and N. Bogoliubov[8] in the years immediately after London's death. The crucial idea, due to Cooper, is that pairs of electrons moving in opposite directions with opposite spins become briefly paired through interaction with the ionic lattice of the metal. In the general case the London equations apply only in one particular limit, and the currents are better described by more elaborate phenomenological equations given in 1952 by V. L. Ginzburg and L. D. Landau. Many of the clues to the solution came through interpreting experiments on the microwave properties of superconductors, which followed the work of Heinz London.

London's volume of superconductivity contained much else of value on the thermodynamics of superconductors, the intermediate state, and other special topics. One interesting result was that a superconductor spinning with angular velocity $\boldsymbol{\omega}$ should generate a magnetic field $\mathbf{H}_L$ equal to $(2mc/e)\boldsymbol{\omega}$ gauss, where m and e are the mass and charge of the electron. The corresponding calculations for a zero-resistance medium spun from rest had been given by Becker, Heller, and Sauter, but London predicted the field would also spontaneously appear as the spinning body is cooled through its transition temperature, in analogy with Meissner and Ochsenfeld's discovery. The effect was demonstrated experimentally in 1960.[9] An interesting point, which has been little explored, is that $\mathbf{H}_L$ appears to depend on the rotation of the superconductor relative to the universe as a whole. There then appears to be some constraint on the underlying quantum condition closely related to that determining rotation in gravitational theory. Besides their theoretical implications, the ideas discussed by London have led to many interesting advances in technology, for example in the application of quantized flux devices to measurement of very low magnetic fields. The authors and their colleagues are planning to use the magnetic moment of a rotating superconductor for readout of a high-precision gyroscope to perform a new test of general relativity in a satellite suggested by L. I. Schiff.

In 1908 Kammerlingh Onnes liquefied helium at a temperature of 4.2° K. During the 1920's evidence began to accumulate that at 2.19° K. the liquid undergoes a peculiar transition, marked by discontinuities in the specific heat and density curves, into another phase which became known as liquid helium II; but not until 1938 were its extraordinary superfluid properties discovered simultaneously by P. Kapitsa and by J. F. Allen and A. D. Misener. The liquid passed through the finest of capillary channels with no measurable resistance. Indications of abnormally high thermal conductivity had previously been found by W. H. and A. P. Keesom. Allen with several colleagues then established that the heat current in the helium II is not linear with temperature and that a pressure gradient proportional to heat input exists in the fluid. The latter effects are often called the fountain pressure, since Allen and J. Jones demonstrated it in an especially striking way by creating a helium fountain. Clearly the mass flow and heat transfer equations of helium II must differ from those of any other fluid; and in addition, the nature of the phase transition—indeed, even the existence of two distinct liquid phases—was puzzling. Thinking over the problem, London was reminded of a strange feature of quantum statistics derived in 1924 by Einstein. In 1923 S. N. Bose had shown that Planck's well-known radiation formula might be derived from the hypothesis of light quanta (photons), by assuming that the photons are subject to a different kind of statistics from classical Maxwell-Boltzmann particles, with the statistical specification based not on the number of particles but on the number of particle states. Bose communicated his result privately to Einstein, who arranged for its publication and then wrote two papers extending the statistics to other quantum gases. One of Einstein's conclusions was that at low temperatures such a gas would have more particles than the number of available states and would then undergo a peculiar kind of condensation with the surplus molecules concentrated in the lowest possible quantum state. Applying the new statistics to ordinary molecules, Einstein then argued that the condensation would have perceptible effects in gases at low temperatures; the viscosity of helium gas, for example, might be expected to fall off rapidly below 40° K. All this seemed rather farfetched: it was not obvious that the new statistics applied to gases. Indeed, a year later E. Fermi and P. A. M. Dirac showed that electrons obey another, quite different kind of quantum statistics, and in 1927 G. Uhlenbeck claimed that Einstein had mistakenly approximated a sum by an integral and that in the correct calculation no condensation would occur. There things rested until London revived the argument and made the even more radical suggestion that Bose-Einstein condensation might occur in liquids as well as gases and thus might account for the peculiar behavior of liquid helium II.

London's suggestion was controversial and also difficult to develop owing to the unsatisfactory state of the theory of liquids. The next advance was due to L. Tisza. Partly through London's idea and partly through considering the results of experiments with oscillating disks, which unlike those on capillary flow gave a finite viscosity, he suggested that helium II may be conceived of as a mixture of two liquids, superfluid and normal, analogously with the two-fluid model of superconductivity. In 1939 Heinz London made a further important contribution by proving from thermodynamic arguments that the superfluid does not carry entropy. In this form the two-fluid model provided a framework for many experimental discoveries over a number of years. Meanwhile, a different approach to the fundamental theory was advanced by Landau, and the idea of Bose-Einstein condensation was heavily criticized. However, London stuck to his guns. An important test, as he pointed out, was to search for superfluidity in the rare isotope He^3 which, unlike ordinary He^4, contains an odd number of particles and may be expected to obey Fermi-Dirac rather than Bose-Einstein statistics. In fact no superfluidity has been observed in He^3 down to 10^{-2}° K. In volume II of *Superfluids* London gave an approximate treatment of a Bose-Einstein liquid, following E. Guggenheim's smoothed-potential model of the liquid state, and estimated the condensation temperature at roughly 3.13° K. as compared with the experimental value of 2.19° K. He also continued to emphasize the close analogy between superfluidity and superconductivity. Here he was in more difficulty than he cared to admit. Electrons obey Fermi-Dirac statistics, and there seemed to be no grounds for linking superconductivity with Bose-Einstein condensation. With the emergence of the microscopic theory, however, the tables were turned on London's critics: paired electrons do behave as Bose particles, and his interpretation of the condensation extends to superconductivity also.

There is a perspective in London's work that has made many of his ideas become progressively more influential with the passage of time. One of his favorite opinions was that some concept of long-range order would eventually prove important to the understanding of biological systems. Although the suggestion remains in the nascent phase it has already had interesting consequences. W. A. Little has proposed that a particular class of long-chain organic molecules might be expected to have superconductive properties, with transition temperatures around room temperature.[10] Although organic superconductors have not yet been made, the idea has raised many interesting questions of the kind that would have appealed to London.

NOTES

1. L. Pauling and E. B. Wilson, *Introduction to Quantum Mechanics* (New York, 1935), p. 340.
2. Personal recollections of conversations between London and W. M. Fairbank.
3. J. C. Maxwell, *Treatise on Electricity and Magnetism*, II (London, 1873), secs. 654, 655.
4. F. London, in *Physical Review*, **74** (1948), 570, is the earliest statement of flux quantization.
5. B. S. Deaver, Jr. and W. M. Fairbank, in *Physical Review Letters*, **7** (1961), 43; R. Doll and M. Näbauer, *ibid.*, **7** (1961), 51.
6. L. Brillouin, in *Proceedings of the Royal Society*, **75** (1949), 502, crediting theorem to Bloch.
7. N. Byers and C. N. Yang, in *Physical Review Letters*, **7** (1961), 46.
8. H. Fröhlich, in *Physical Review*, **79** (1950), 845; J. Bardeen, L. N. Cooper, and J. R. Schrieffer, *ibid.*, **106** (1957), 162; **108** (1957), 1175; and N. Bogoliubov, in *Zhurnal eksperimentalnoi i teoreticheskoi fisiki*, **34** (1958), 58; translated in *Soviet Physics JETP*, **7** (1958), 41.
9. A. F. Hildebrandt, in *Physical Review Letters*, **12** (1964), 190; A. King, Jr., J. B. Hendricks, and H. E. Rorschach, Jr., in *Proceedings of the Ninth International Conference on Temperature Physics* (New York, 1965), p. 466; M. Bol and W. M. Fairbank, *ibid.*, p. 471.
10. W. A. Little, in *Physical Review*, **134** (1964), 1416.

BIBLIOGRAPHY

I. Original Works. London's published books are *La théorie de l'observation en mécanique* (Paris, 1939), written with E. Bauer (a copy of this rare work is in Brown University Library); *Superfluids:* Vol. I, *Macroscopic Theory of Superconductivity;* Vol. II, *Macroscopic Theory of Superfluid Helium* (New York, 1950–1954); 2nd ed., rev., reprinted in the Dover series with additional material by other authors, 2 vols. (New York, 1961–1964), with bibliography of London's scientific papers, by Edith London, I, xv–xviii. A large collection of notebooks and MSS is preserved at the Duke University Library, Durham, N. C., together with two bound vols. containing London's complete published papers. Some MSS of scientific interest remain in Mrs. London's possession.

II. Secondary Literature. Personal biographical material by E. London and a sketch of London's scientific work by L. W. Nordheim are in *Superfluids*, 2nd ed., I, v–xviii.

On the chemical bond the monographs by L. Pauling, *Nature of the Chemical Bond*, 3rd ed. (Ithaca, N. Y., 1960); and C. A. Coulson, *Valence* (Oxford, 1960), are useful. On dispersion forces, see S. G. Brush, "Interatomic Forces and Gas Theory From Newton to Lennard-Jones," in *Archives of Rational Mechanics and Analysis*, **39**, no. 1 (1970), 1–29.

On superconductivity, see D. Shoenberg, *Superconductivity* (Cambridge, 1950); J. Bardeen and J. R. Schrieffer, in C. J. Gorter, ed., *Progress in Low Temperature Physics*, III (Amsterdam, 1961), 170–287; and P. de Gennes, *Superconductivity of Metals and Alloys* (New York, 1966). On superfluidity, see K. R. Atkins, *Liquid Helium* (Cambridge, 1958); and J. Wilks, *Properties of Liquid and Solid Helium* (Oxford, 1967).

C. W. F. Everitt
W. M. Fairbank

LONDON, HEINZ (*b.* Bonn, Germany, 7 November 1907; *d.* Oxford, England, 3 August 1970), *physics.*

Like his brother, Fritz London, who exercised close parental influence on him after their father's early death, Heinz received a classical education but was also interested in mathematics and science, especially chemistry. He attended the University of Bonn from 1926 to 1927 and after six months with the W. C. Heraeus chemical company completed undergraduate studies in Berlin and Munich. His graduate work on superconductivity was done at Breslau under F. E. Simon, and his Ph.D., issued early in 1934, was one of the last awarded to a Jew in Nazi Germany. Later in the same year he joined Simon and other members of the Breslau group at Oxford, where they were establishing the first center of low-temperature research in England since the time of James Dewar. Fritz London was already in Oxford, and Heinz lived there with the family for two years. In 1936 he moved to Bristol, where he remained until 1940, when he was for a brief time interned as an alien. After his release he worked with Simon and others on the British atomic bomb project and then spent two years at Birmingham before transferring to Harwell in 1946, where he continued to work until his death. In 1939 he married Gertrude Rosenthal, but the marriage broke up soon afterward. In 1946 he married Lucie Meissner; they had four children.

The origin of London's work on alternating-current losses in superconductors has been described in the preceding entry, along with his early contributions to the theory, culminating in the joint paper with Fritz London on the electrodynamics of the superconducting state. An interesting feature of the theory outlined there is the indeterminate parameter μ in equation (7). Evidently one solution would be $\mu = 0$, in which case the acceleration equation (5) remains valid, although its justification is different from that originally given. A superconductor would then, like any other metal, exhibit surface charges in an electric field. An equally plausible alternative, discussed by the Londons, was to make $\mu = \Lambda c^2 \rho$, where ρ is charge density. This had the advantage of allowing a symmetrical four-dimensional representation of the equations; it implied, however, that the electric field penetrates a superconductor to the same depth as the magnetic field and therefore that the capacity of a condenser with superconducting plates should change appreciably on cooling through its transition from the normal to the superconducting state. In 1935 London tried this experiment at Oxford, showing conclusively that there is no change in capacity and hence that the solution $\mu = 0$ must be accepted. About the same time he published an investigation of the equilibrium between superconducting and normal phases of a metal, based on a thermodynamic analysis of the London equations. He found that the observed transition to the normal state in high magnetic fields depends on the existence of a threshold value for current density in the superconductor. He also discovered that unless a superconductor has a positive surface energy sufficient to counteract the field energy in the penetration layer, it will split into a finely divided mixture of normal and superconducting regions in high fields. Since the superconductivity of most pure metals disappears entirely above the critical field, London concluded that their surface energy is indeed positive. He conjectured however that some hysteresis effects just then observed in superconducting alloys might arise from a negative surface energy. These ideas were later entirely confirmed, and during the 1950's they were elaborated into the distinction between type I and type II superconductors. The special properties of type II superconductors proved of great technical importance in obtaining high field superconducting magnets.

While at Oxford, London continued to search for alternating-current losses in superconductors, but had no success, even though he raised the operating frequency to 150 MHz. After his move to Bristol he took the subject up once more and finally in 1939 succeeded in demonstrating the occurrence of high-frequency losses at 1500 MHz. Figure 1 reproduces the experimental data from London's paper, which shows significant alternating-current dissipation at points well below the ordinary transition temperature. An unexpected by-product was the discovery of the "anomalous skin effect" in normal metals. At low temperatures the resistivity of tin in the normal state, as deduced from Joule heating at 1500 MHz, was several times larger than the measured direct-current resistivity, although at room temperatures the two figures were identical. London tentatively attributed the effect to the electrons' having a mean free path considerably greater than the skin depth, so that relatively few of them contribute to conduction.

London's work on high-frequency losses was terminated by World War II, but the subject was revived in 1946 by several other investigators, using resonant cavity techniques from radar research. A. B. Pippard in particular reached some highly significant results by measuring changes in resonant frequency of cavities above and below the transition temperature. He concluded that the wave functions were indeed coherent over distances greater than the penetration depth but that the coherence length was finite rather than infinite as in the London theory. Pursuing London's idea about the electronic mean free path, he introduced impurities into the material, which acted as scattering centers to reduce both the free path and the coherence length. When this was done the penetration depth increased, the product of the Pippard coherence length l and the penetration depth λ being approximately constant. Other measurements led to the hypothesis of an energy gap in the distribution of conducting particles. These developments provided the basis of ideas for the Bardeen-Cooper-Schrieffer microscopic theory of superconductivity. The Pippard length was then identified with the pairing distance of the electrons.

At Bristol, London also investigated superconductivity of thin metallic films in collaboration with E. T. S. Appleyard, A. D. Misener, and J. R. Bristow. From measurements of critical fields they were able to determine the penetration depth experimentally. Measurements on the magnetic susceptibilities of superconducting particles by D. Shoenberg about the same time gave comparable results. This work on thin films was another example, like the experiments on alternating-current losses, of work begun by London which became a major field of research studied by

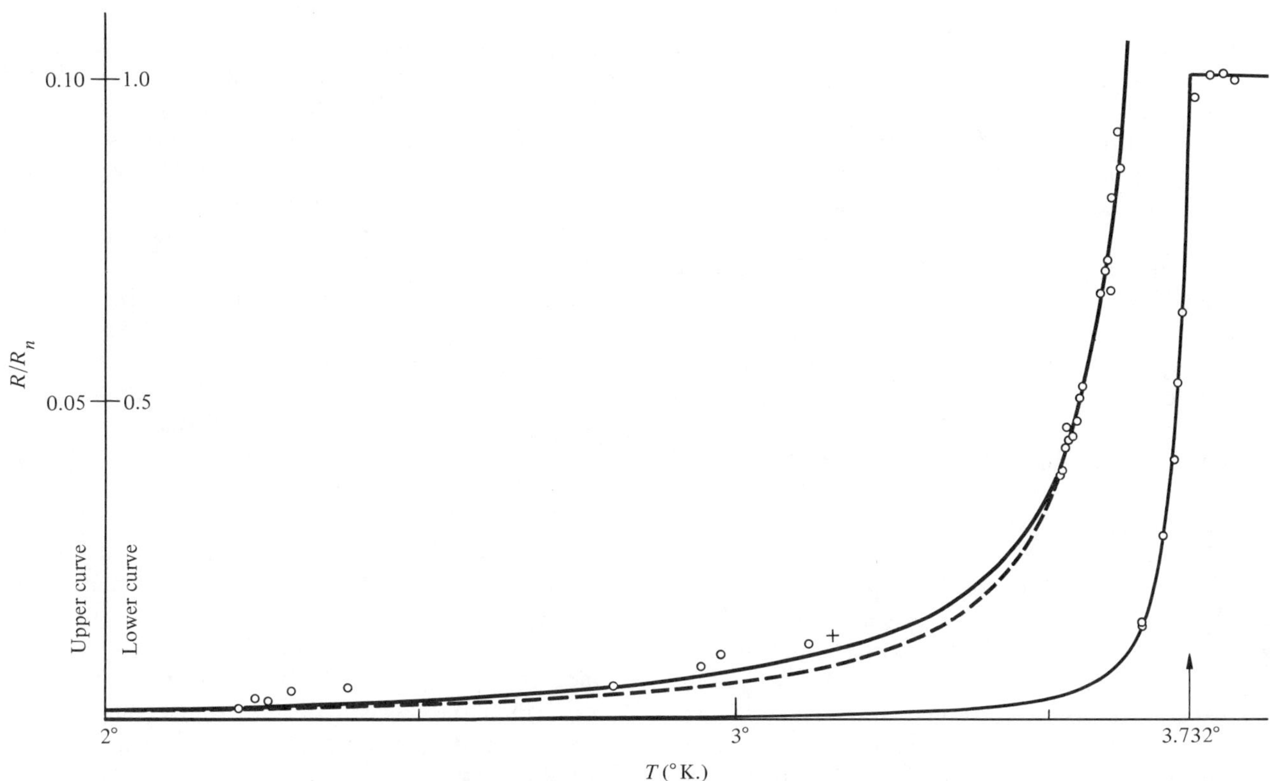

FIGURE 1. Alternating field losses in superconducting tin at 1,500 MHz.

hundreds of other investigators. A third subject of study from the same period was his work on helium, which included an experiment that helped establish the two-fluid model, and the thermodynamic analysis mentioned in the previous entry, which proved that the superfluid component carries no entropy. A by-product of the analysis was London's prediction of the mechanocaloric or inverse fountain effect, a temperature difference generated by the moving superfluid.

Following the work of Fritz London and Tisza much effort was applied to the theory of liquid helium II. In 1941 Lev Landau formulated the problem in quite a different way, describing the normal fluid in terms of quantized excitations, phonons and rotons, and giving new hydrodynamical equations to describe superfluidity. One of his conclusions was that the superfluid is subject to the condition curl $\mathbf{v}_s = 0$, making it incapable of rotation. Actually the same condition for the superfluid was also strongly advocated by Fritz London in the London-Tisza model, along with the analogous condition on curl $\mathbf{p}_s = 0$ in a superconductor, as evidence that each phenomenon constitutes a macroscopic quantum state. In 1946 Heinz London proposed an experiment to distinguish the behavior of a fluid subject to constraints of this kind from mere absence of viscosity. The idea was to set a cylindrical container of helium I in slow uniform rotation about its axis and then cool the system slowly to a temperature well below the lambda point. If the superfluidity is simply absence of viscosity, the angular momentum of the fluid should be unaffected by the transition; but if the equilibrium state is one of constrained vorticity, the superfluid created by cooling should stop rotating and the container has to rotate more rapidly to conserve the total annular momentum. London then made a further conjecture. In 1941 A. Bijl, J. de Boer, and A. Michels had pointed out that the critical velocity v_c (the maximum velocity for superfluid flow) in a film of helium II seems to obey the condition $v_c \sim h/4\pi md$, where m is the atomic mass of helium and d the film thickness. London suggested that the angular velocity Ω of the helium II might be subject to a similar quantum condition, so that in a container of radius R, rotation stops only if

$$\Omega < h/2\pi mR^2. \quad (1)$$

For a vessel one cm. in diameter the rotational period obtained from (1) is about one day, so the experiment was a difficult one. Experiments at higher speeds tried several times from 1950 onward disclosed no difference between the rotating superfluid and classical rigid-body rotation. Various suggestions about the state of motion were advanced by Fritz London and others; then Lars Onsager and later, independently, Richard Feynman assumed that the circulation around an arbitrary path in the fluid is a multiple of h/m consistent with London's condition (1), but that the rotating fluid as it gains more angular momentum will break up into an array of vortices, each with quantized circulation h/m. Thus by a different path the suggestion of a macroscopic quantum condition entered the theory of helium II as well as the theory of superconductivity; it is remarkable that one suggestion originated with Heinz London and the other with Fritz London. The existence of individual quantized vortices in helium II was first demonstrated experimentally by W. F. Vinen in 1961 through the coupling of vorticity to a vibrating wire;[1] since then they have been observed in several other ways. The rotating-bucket experiment suggested by London was finally performed successfully in 1965.[2] The measurements confirmed that at low angular velocities the superfluid stops rotating and the bucket rotates faster on cooling through the lambda point. Thus the conjecture that curl $\mathbf{v}_s = 0$ is an equilibrium state for a superfluid was confirmed. A notable feature was that while quantized vortices of the predicted size were indeed observed, the angular velocity below which rotation ceased was about fifteen times higher than condition (1). The last result is explained by the large amount of energy required to form a single vortex core. It is the equivalent for the superfluid of the Ochsenfeld-Meissner effect in a superconductor.

London's work on the atomic bomb project had been on methods of separating uranium 235 by ionic migration and liquid thermal diffusion. At Harwell after the war he continued to work on isotope separation, concentrating on the production of carbon 13 as a stable tracer element for medical research. He developed a method of low-temperature distillation and designed a fractionating column using carbon monoxide for the enrichment of C^{13} and O^{18}. This machine has operated successfully for many years and currently supplies all the C^{13} used in Great Britain and the United States.

During the last fifteen years of his life London worked in collaboration with various colleagues on three main topics, neutron production and neutron-scattering experiments in liquid helium, techniques for producing high field superconducting magnets, and, most important, the He^3-He^4 dilution refrigerator, one of the most ingenious and useful contributions to cryogenic technology in many years. He had suggested the essential idea of the dilution refrigerator in 1951 at the Oxford Conference on Low-Temperature Physics. He pointed out that since He^3 behaves as a Fermi gas below 1° K., and He^4 as a Bose gas, then if He^3 is allowed to mix with He^4 it will in effect expand. Now in this region He^4, being practically pure superfluid, has negligible entropy; whereas He^3 has entropy proportional to $n^{-2/3}T$, where n is the atomic concentration. Dilution to one part in a thousand would therefore cool a bath of helium to 10^{-2}° K., while still retaining appreciable specific heat. In 1951 the idea seemed quite impractical in view of the scarcity of He^3, but in 1955 London revived it in collaboration with E. Mendoza and G. Clarke. They first measured the osmotic pressure of He^3 dissolved in He^4 and attempted adiabatic dilution starting at 0.8° K. but obtained no cooling. After G. K. Walters and W. M. Fairbank had discovered that He^3 and He^4 separate at 0.87° K. into two phases, one concentrated and one dilute, London and Mendoza at first concluded that dilution cooling would be ineffective below about 0.8° K. They later realized that it could be achieved in another way, by exploiting the latent heat of transition from the concentrated to the dilute phase.[3] The essential principle of the dilution refrigerator as conceived by London and Mendoza is illustrated in Figure 2. Refrigeration occurs in the pot A, where the lighter He^3-rich phase floats on top of the dilute

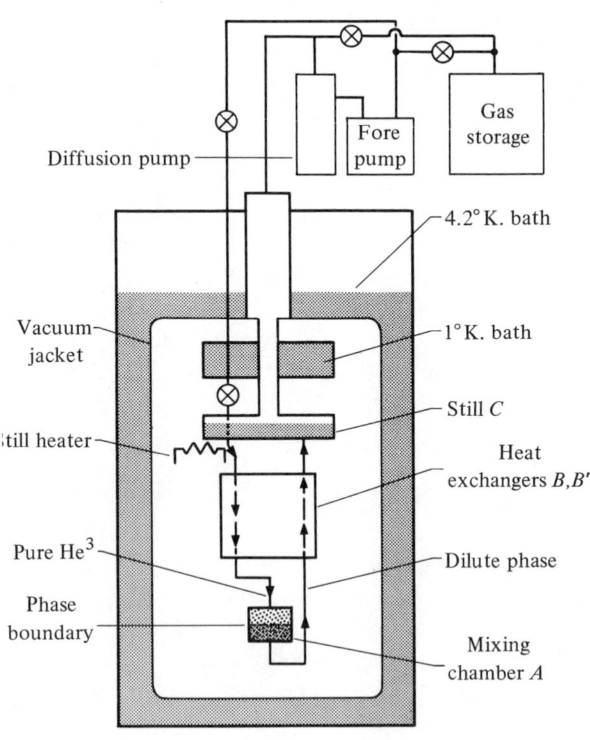

FIGURE 2. The He^3–He^4 dilution refrigerator.

phase and expansion proceeds downward. The diluted mixture is then taken out to a still, *C*, where the He^3 evaporates and is recycled through heat exchangers *B*, *B′*. The effectiveness of the cooling cycle depends on the fortunate coincidence of two special properties of He^3. First, the phase boundary for the mixtures has a shape such that significant proportions of He^3 enter the dilute solution down to the absolute zero. Second, He^3 has a higher vapor pressure than He^4 and is therefore evaporated preferentially in the still. It is customary to use the refrigerator with conventional He^3 cooling as a first stage. After various experiments by Mendoza at Manchester, a working model was built in 1965 by H. E. Hall, P. J. Ford, and K. Thompson.[4] Improved versions with operating temperatures down to 0.005° K. have since been made and marketed in several countries, and the device is now widely used.

The two London brothers form a study of unusual psychological interest. Fritz's influence on Heinz in the critical years after their father's death was strong, yet despite their close professional relationship they are in many ways contrasting figures. Fritz was well-organized, Heinz shockingly untidy. It is customary to think of Fritz as a theoretician and Heinz as an experimentalist, but actually neither fits exactly into ordinary categories. Of Heinz, D. Shoenberg has said: "Though he spent much time on experiments his most valuable contributions have been ideas and inventions. Perhaps he might be described as a cross between a theoretical physicist and an inventor."[5] In their theoretical contributions Fritz's genius consisted in identifying the deep conceptual issues, Heinz's in the vital and far from simple task of bringing experiment into fruitful contact with theory. As an experimentalist he belonged to that not undistinguished class of workers, among whom was J. J. Thomson, whose junior colleagues wisely protect them by various subterfuges from too close contact with apparatus. His clumsiness was a byword. Both brothers brought a remorseless thoroughness to everything they did. Heinz's attitude to physics had also an element of spiritual passion. "For the second law of thermodynamics," he proclaimed vehemently to one colleague, "I would die at the stake."[6]

NOTES

1. W. F. Vinen, "The Detection of Single Quanta of Circulation in Liquid Helium II," in *Proceedings of the Royal Society*, **260** (1961), 218.
2. G. B. Hess and W. M. Fairbank, "Measurement of Angular Momentum in Superfluid Helium," in *Physical Review Letters*, **19** (1967), 216.
3. Many interesting details are given in an unpublished paper by E. Mendoza, "Early History of the He^3-He^4 Dilutions Refrigerator: Some Personal Impressions."
4. H. E. Hall, P. J. Ford, and K. Thomson, "A Helium-3 Dilutions Refrigerator," in *Cryogenics*, **6** (1966), 80.
5. D. Shoenberg, obituary of H. London, in *Biographical Memoirs of Fellows of the Royal Society*, **17** (1971), 441–461.
6. H. Montgomery, quoted in 5.

BIBLIOGRAPHY

A complete bibliography of London's published papers with an admirable biographical sketch is D. Shoenberg, in *Biographical Memoirs of Fellows of the Royal Society*, **17** (1971), 441–461. His unpublished MSS are in Mrs. London's possession. See also K. Mendelsohn, in *Review of Modern Physics*, **36** (1964), 71.

For further sources on particular areas of low-temperature physics, see the bibliography for Fritz London.

C. W. F. Everitt
W. M. Fairbank

LONGOMONTANUS. See **Severin, Christian.**

LONICERUS (LONITZER), ADAM (*b.* Marburg, Germany, 10 October 1528; *d.* Frankfurt, Germany, 29 May 1586), *botany.*

Lonicerus was the son of Johann Lonitzer, a philologist and professor at Marburg. He received his baccalaureate in 1540 and his master's degree in 1545. In the latter year he began teaching at the Gymnasium in Frankfurt, but he returned to Marburg because of disorders caused by war. He studied medicine there and later in Mainz, where he was a private tutor in the home of a Dr. Osterod. In 1553 Lonicerus became professor of mathematics at Marburg, and in 1554 he received his medical degree. Also in 1554 he married the daughter of the Frankfurt printer Egenolph Magdalena; and following the death of Graff, the municipal physician of Frankfurt, in that year, he was appointed to the post. Lonicerus worked as a proofreader in the printing shop of his father-in-law, who specialized in the revision of old herbals (for example, those of Eucharius Röslin and Dioscorides).

Lonicerus wrote extensively in many fields, including botany, arithmetic, history of medicine, and medicine, particularly public health books such as regulations for controlling the plague (1572) and regulations for midwives (1573). His herbals were so influential that in 1783 at Augsburg—almost 250 years after the first edition—*Adam Lonicers Kräuter-Buch* was still published. In addition, Linnaeus immortalized his name in the genus *Lonicera.*

Lonicerus based the first, Latin edition of his herbal on Röslin's revision of the *Ortus sanitatis* (1551), which contained many illustrations, most of them borrowed from Bock. The popularity of Lonicerus' herbal is shown by the many, steadily enlarged editions he brought out. Although the provision of plant names in German, Latin, Greek, French, Italian, and Spanish lends the herbal a scientific air, the inclusion of fabulous stories betrays its late medieval character. (For example, the formation of bezoars is attributed to the hardening of the tears of stags!) The herbal also lists animal and metallic medicaments and contains one of the earliest descriptions of local flora. In addition, the book distinguishes the deciduous trees from the conifers; the group composed of the yew, the cypress, the juniper, and the savin is contrasted with that containing the spruce and the fir. Lonicerus' son Johann Adam (*b.* 1557) edited his father's writings.

BIBLIOGRAPHY

I. ORIGINAL WORKS. Lonicerus' writings include *Naturalis historiae* . . ., 2 pts.: I, *Naturalis historiae opus novum, in quo tractatur de natura arborum, fructicum, herbarum* . . . (Frankfurt, 1551), II, *Naturalis historiae de plantarum potissimum quae circa Francofurtum nascuntur, descriptione et virtute* . . . (Frankfurt, 1555); *Aphorismoi Hippocratis* (Frankfurt, 1554); *Kreuterbuch, neu zugericht, künstliche Conterfeytunge der Bäume . . . Item von fürnembsten Gethieren der Erden, Vögeln und Fischen; auch von Metallen* . . ., a German version of *Naturalis historiae* (Frankfurt, 1557; later eds. at Frankfurt, 1560–1650; Ulm, 1679–1770; Augsburg, 1783)—this is simply a revision of a work by Cube (municipal physician in Frankfurt 1484–*ca.* 1503), which was first issued by Röslin: only the medicinal plants were described in pt. I, local flora in pt. II; *Botanicon. Plantarum historiae cum earundem ad vivum arteficiose expressis iconibus tomi duo* . . . (Frankfurt, 1565); *Brevis et utilis arithmetices introductio* . . . (Frankfurt, 1570); *Beschreibung der Arzney und fürnehmsten Compositionen* (Frankfurt, 1572); *Hebammenbüchlin, Empfengnuss und Geburt dess Menschen* . . . (Frankfurt, 1582), a revision of Röslin's work; *Omnium corporis humani affectuum explicatio methodica* . . . (Frankfurt, 1594); and *De purgationibus libri III, ex Hippocrate, Galeno Aetio et Mesue depromti . . . foras dati per Teuc. Annaeum Privatum, C. Adami Loniceri . . . filium* . . . (Frankfurt, 1596).

A partial list of his works is in *Catalogue général de la Bibliothèque nationale*, XCIV (Paris, 1930), cols. 925 f.

II. SECONDARY LITERATURE. See A. Arber, *Herbals, Their Origin and Evolution*, 2nd ed. (Cambridge, 1953), 72, 134, 221; A. von Haller, *Bibliotheca botanica*, 2 vols. (Zurich, 1771–1772; repr. Hildesheim, 1969): I, 309 f., II, 451; W. Kallmorgen, *700 Jahre Heilkunde in Frankfurt am Main* (Frankfurt, 1936), 342 (portrait p. 146); Linnaeus, *Bibliotheca botanica* (Amsterdam, 1736), 14, 25; M. Möbius, *Geschichte der Botanik*, 2nd ed. (Stuttgart, 1968), 30, 138; G. A. Pritzel, *Thesaurus literaturae botanicae* (Leipzig, 1872), 195 f.; F. W. E. Roth, "Botaniker Eucharius Rösslin, Theod. Dorsten u. Adam Lonicer (1526–1586)," in *Zentralblatt für Bibliothekswesen*, **19** (1902), 271–286, 338–345; H. Schelenz, *Geschichte der Pharmazie* (Berlin, 1904), 432, 446; and G. Stricker, in *Allgemeine deutsche Biographie*, XIX (1884), 157.

KARIN FIGALA

LONITZER, Adam. See **Lonicerus, Adam.**

LONSDALE, DAME KATHLEEN YARDLEY (*b.* Newbridge, Ireland, 28 January 1903; *d.* London, England, 1 April 1971), *crystallography, chemistry, physics.*

Kathleen Lonsdale was the tenth and youngest child of Harry Frederick Yardley and Jessie Cameron. Her father was postmaster at Newbridge, but the family later moved to England and Kathleen won scholarships which took her to the County High School for girls at Ilford and later to Bedford College for Women, which she entered at the age of sixteen. There she studied mathematics and physics and received a B.Sc. with highest honors at the age of nineteen. She took first place in the University of London list and one of her examiners, W.H. Bragg, was so impressed that he offered her a post in his research team to work on the crystal structure of organic compounds by X-ray analysis. Thus her scientific career began in 1922, first at University College and then at the Royal Institution in London. It continued throughout her life, with only brief interruptions following her marriage to Thomas Lonsdale in 1927, when she moved to Leeds for three years before returning once more to the Royal Institution.

In 1945 she became the first woman to be admitted as a Fellow of the Royal Society. In 1946 she was appointed reader in crystallography at University College, London, and professor of chemistry in 1949. She was appointed Dame Commander of the British Empire in 1956, received the Davy Medal of the Royal Society in 1957 and was its vice-president from 1960–1961. In 1966 she was president of the International Union of Crystallography and president of the British Association from 1967 to 1968. During her long and intensely active scientific career Kathleen Lonsdale raised two daughters and one son, had many outside interests, and traveled widely in most countries of the world. She was a member of the Society of Friends, and worked hard for peace and for prison reforms. But her many and varied interests were always closely integrated with her scientific work.

X-ray crystallography is primarily concerned with deducing the positions of atoms in space from a study of the diffraction patterns obtained when a beam of X rays is passed through a crystal. Kathleen Lonsdale began her work in 1922, ten years after the original discovery of the diffraction effect by Laue. In that time considerable advances in the theory and practice of the method had been made, and the structure of many simple and some quite complex inorganic substances had been deduced. But little was known about the dimensions and exact metrical structure of organic molecules.

As a physicist and mathematician by training, Kathleen Lonsdale's first major contribution to this subject was a profound and systematic study of the theory of space groups, methods for their determination, and the possibilities of molecular symmetry that are involved. This work was published with W. T. Astbury in 1924. To some extent it duplicated the work of Wyckoff and Niggli (of which Astbury and Kathleen Lonsdale were ignorant), but it went further and was much more useful for those whose work was concerned with organic molecular crystals. A later continuation and extension of this work consisted of the structure factor tables published in 1936 in photo-litho-printed form from the handwritten manuscript, and the many volumes of the International Tables for X-Ray Crystallography published from 1952 onwards. These volumes, of which Kathleen Lonsdale was principal editor, are in constant use today and form an essential tool for crystal structure determination.

Kathleen Lonsdale's main task at the Royal Institution was to try to solve the problem of organic crystal structures. Although she confessed that she "knew no organic chemistry and very little of any other kind," she was destined to make one of the most outstanding advances in that subject for many decades. Her work on hexamethylbenzene, published in 1929, and hexachlorobenzene demonstrated for the first time that the benzene ring was hexagonal and planar, and she gave its precise dimensions. Of this result C. K. Ingold wrote, "one paper like this brings more certainty into organic chemistry than generations of activity by us professionals."

Her later work, however, was more concerned with the physics of crystals than with chemistry. She made an intensive study of the magnetic anisotropy of crystals, and by measuring the diamagnetic susceptibilities in and perpendicular to the planes of a number of aromatic molecules, she was able to show that, while the σ electronic orbits were of atomic dimensions, the π orbits were of molecular dimensions. The far-reaching importance of this result was that it established the reality of the concept of molecular orbitals.

Later she was responsible for the development of divergent-beam X-ray photography of crystals, a technique which gives information about the texture and perfection of the crystal, and can also be used to make precise measurements of lattice constants or wavelength. By this means she was able to measure the C—C distance in individual diamonds to seven significant figures. She had a special interest in the structure of diamonds and made many other important contributions to their study.

Another important field in which she made outstanding contributions was the study of diffuse X-ray reflection by single crystals. She showed that this diffuse scattering is directly related to the elastic constants of the crystal. It is also dependent on the crystal structure and can be used to determine molecular orientation. In general, the thermal motion of atoms and molecules in crystals was one of her special interests.

In her later work she returned again to more chemical interests, as in her study of the solid-state transformation of anthracene to anthraquinone and anthrone, and to medical problems concerned with the mode of action of drugs and the composition of endemic bladder stones.

Kathleen Lonsdale had a profound influence on the development of X-ray crystallography and related fields in chemistry and physics. Very few have made so many important advances in so many different directions.

BIBLIOGRAPHY

Among Lonsdale's more important works are "Tabulated Data for the Examination of the 230 Space-groups by Homogeneous X-Rays," in *Philosophical Transactions of the Royal Society*, **224A** (1924), 221–257, written with W. T. Astbury; "The Structure of the Benzene Ring in $C_6(CH_3)_6$," in *Proceedings of the Royal Society*, **123A** (1929), 494–515; "An X-Ray Analysis of the Structure of Hexachlorobenzene, Using the Fourier Method," *ibid.*, **133A** (1931), 536–552; *Simplified Structure Factor and Electron Density Formulae for the 230 Space Groups of Mathematical Crystallography* (London, 1936), and "Magnetic Anisotropy and Electronic Structure of Aromatic Molecules," in *Proceedings of the Royal Society*, **159A** (1937), 149–161.

With H. Smith, she wrote "An Experimental Study of Diffuse X-Ray Reflection by Single Crystals," *ibid.*, **179A** (1941), 8–50. Also see "Extra Reflections From the Two Types of Diamond," in *Proceedings of the Royal Society*,

179A (1942), 315–320; "X-Ray Study of Crystal Dynamics: An Historical and Critical Survey of Experiment and Theory," in *Proceedings of the Physical Society*, **54** (1942), 314–353; "Divergent Beam X-ray Photography of Crystals," in *Philosophical Transactions of the Royal Society*, **240A** (1947), 219–250; *Crystals and X-rays* (London, 1948); "Geiger Counter Measurements of Bragg and Diffuse Scattering of X-rays by Single Crystals," in *Acta crystallographica*, **1** (1948), 12–20; and "Vibration Amplitudes of Atoms in Cubic Crystals," *ibid.*, **1** (1948), 142–149.

J. M. ROBERTSON

LONSDALE, WILLIAM (*b.* Bath, England, 9 September 1794; *d.* Bristol, England, 11 November 1871), *geology*.

Lonsdale was the younger son of William Lonsdale of Bath. He obtained an army commission in 1812, served in the Peninsular War, and was present at the battle of Waterloo. When peace was restored in 1815, he retired from the army on half-pay and settled near Bath. He never married. Lonsdale's interests inclined toward natural history, and in 1826 he was appointed a curator at the Bath Museum. He was elected to the Geological Society of London in 1829, and soon afterward was appointed curator and librarian to the society. Lonsdale worked hard at cataloging the society's rapidly growing collections; in 1838 his post was redefined, and as assistant secretary and librarian he undertook an increasing administrative and editorial burden. In 1842, poor health forced him to resign, and he retired to southwest England for the rest of his life. In 1846 he was given the society's highest award, the Wollaston Medal, for his work on fossil corals.

Shortly before his election to the Geological Society, Lonsdale contributed an excellent, although conventional, account of the stratigraphy and paleontology of the area around Bath. Once he was in the society's employment, he had little time for writing. His work, however, gave him an unrivaled knowledge of British fossils, which led to his most important contribution to geology.

In the late 1830's the geology of Devonshire became fiercely controversial because of the assertion by Roderick Murchison that the government's official geological surveyor, Henry De la Beche, had misinterpreted the entire sequence of strata and had overlooked the presence of a large area of strata of Coal Measure (that is, late Carboniferous) age. Most geologists agreed, however, that the older strata of Devonshire were pre-Silurian because of their lithological character. In 1837, however, it was asserted instead that the fossils in these rocks indicated a Mountain Limestone (that is, an early Carboniferous) date. Lonsdale immediately suggested, on the basis of his own studies of the fossil corals, that the strata were neither Carboniferous nor Silurian, but intermediate in age, since some fossil species were known from the Silurian, some from the Carboniferous, and some from neither. This implied that the Devon rocks were the lateral equivalents of the Old Red Sandstone, despite their totally different lithology. This seemed so improbable that Lonsdale's suggestion was virtually forgotten until 1839, when it was adopted by Murchison and Adam Sedgwick and became the basis of their Devonian system. The validity of Lonsdale's hypothesis was then quickly confirmed by the recognition of the same fossil fauna in the expected position in clearer sequences of strata elsewhere in Europe and in North America.

The Devonian controversy involved much more than a technical or local problem. The argument turned on the reliability of fossils for the estimation of the relative geological age of strata in the absence of clear evidence from superposition and in the face of clear contrary evidence from lithology. Even more fundamentally, this methodological argument depended on conceptions of biological history. Lonsdale's hypothesis was based on his perception of the intermediate character of the Devonian corals; it was valid only if the fauna had been transformed in the course of time by the gradual production of new species and extinction of old species and in a roughly synchronous manner over a very wide area, with anomalies in only exceptional circumstances (as, for example, the unusual Old Red Sandstone environment). The vindication of Lonsdale's hypothesis was therefore taken as confirming strongly the validity of correlating strata by means of fossils, even on an intercontinental scale. It was thus highly influential indirectly in the development of stratigraphical geology.

BIBLIOGRAPHY

Lonsdale's principal scientific publications were "On the Oolitic District of Bath," in *Transactions of the Geological Society of London*, 2nd ser., **3**, pt. 2 (1832), 241–276; and "Notes on the Age of the Limestones of South Devonshire," *ibid.*, **5**, pt. 3 (1840), 721–738. The history of the Devonian problem is summarized in M. J. S. Rudwick, "The Devonian System: A Study in Scientific Controversy," in *Actes du XIIe Congrès international d'Histoire des Sciences*, **7** (Paris, 1971), 39–43. Lonsdale's work for the Geological Society is referred to in Horace B. Woodward, *The History of the Geological Society of London* (London, 1907).

M. J. S. RUDWICK

LOOMIS, ELIAS (*b.* Willington, Connecticut, 7 August 1811; *d.* New Haven, Connecticut, 15 August 1889), *meteorology, mathematics, astronomy.*

Loomis graduated from Yale College in 1830, and was a tutor there from 1833 to 1836. He became professor of mathematics and natural philosophy at Western Reserve College, in Hudson, Ohio, in 1836, although he spent the first year of his appointment in further study in Paris. From 1844 to 1860 he was professor at the University of the City of New York. In 1860 he accepted a call to Yale, where he stayed until the end of his life. After the early death of his wife, Loomis led a rather isolated life, centered around his work, his only diversion being the compilation of an extensive genealogy of the Loomis family. He was member of the National Academy of Sciences and of scientific societies in the United States and Europe.

Loomis' interest was divided among several branches of science. Most of his scientific achievements were of a practical, rather than a theoretical, nature. His work, reflecting his conviction that the laws governing natural phenomena can be uncovered only by studying observed data, was carried out with utmost precision, and his research, although not always original, was highly valued because of its great reliability. In his own time, however, Loomis was better known for the publication of a large number of textbooks on mathematics, astronomy, and meteorology than for his scientific investigations.

Loomis made his most important contributions in the field of meteorology. In 1846 he published the first "synoptic" weather map, a new method of data representation that in the following decades exerted a profound influence on the formulation of theories of storms. This method became of fundamental importance in the development of weather prediction. The weather maps presented in Loomis' paper brought clarification to the heated controversy between J. P. Espy and W. C. Redfield concerning the surface wind pattern in storms. Later, Loomis essentially followed Espy in regarding thermal convection, reinforced by the latent heat released during condensation of water vapor, as the chief factor in storm formation. As soon as weather maps began to be published on a daily basis in the United States, in 1871, Loomis embarked on a long series of meticulous statistical investigations of cyclones and anticyclones. His results effectively supported the convection theory of cyclones.

Throughout his life Loomis was strongly interested in geomagnetism. In 1833–1834 he conducted a series of hourly observations of the earth's magnetic field and mapped his results for the United States. In 1860 Loomis prepared the first map of the frequency distribution of auroras and pointed out that the oval belt of most frequent auroras was not centered on the geographic pole but approximately paralleled the lines of equal magnetic dip.

Loomis devoted much of his time to astronomical investigations. These studies dealt mainly with the observation of meteors and the determination of longitude and latitude of various localities. Together with D. Olmsted he rediscovered Halley's comet on its return in 1835 and computed its orbit.

BIBLIOGRAPHY

I. Original Works. Loomis published a large number of his papers in the *American Journal of Science and Arts*, including his investigations on the aurora borealis (1859–1861) and a series of twenty-three papers entitled "Contributions to Meteorology" (1874–1889). A complete bibliography of Loomis' works is to be found in H. A. Newton, "Biographical Memoir of Elias Loomis," in *Biographical Memoirs. National Academy of Sciences*, **3** (1895), 213–252. Collections of Loomis' papers and correspondence are to be found in the Manuscripts and Archives Department of the Beinecke Rare Book and Manuscript Library at Yale University.

II. Secondary Literature. The most complete biographical notice is H. A. Newton, cited above. See also F. Waldo, "Some Remarks on Theoretical Meteorology in the United States, 1855 to 1890," in *Report of the International Meteorological Congress, Chicago*, August 21–24, 1893, edited by O. L. Fassig as *Bulletin. Weather Bureau, United States Department of Agriculture* **2**, pt. 2 (1895), 318–325.

Gisela Kutzbach

LORENTZ, HENDRIK ANTOON (*b.* Arnhem, Netherlands, 18 July 1853; *d.* Haarlem, Netherlands, 4 February 1928), *theoretical physics.*

Lorentz' father, Gerrit Frederik Lorentz, owned a nursery near Arnhem. His mother, the former Geertruida van Ginkel, died when he was young; and his father married Luberta Hupkes when the boy was nine. Lorentz attended primary and secondary schools in Arnhem and was always the first in his class. From an early age he was drawn to physical science, although he also read widely in Victorian novels and Reformation history. He was unusually quick with foreign languages, inferring grammar and idiom from context. Although he grew up in Protestant circles, he was a freethinker in religious matters; he regularly attended the local French church to improve his French.

By the time Lorentz matriculated at the University of Leiden in 1870, his primary interests were mathematics and physics. He became close friends with the astronomy professor, Frederick Kaiser, and followed his lectures on theoretical astronomy with keen interest. He also attended the lectures of Pieter Leonhard Rijke, the only professor of physics at the time. After a year and a half Lorentz passed his candidate's examination in mathematics and physics and returned to Arnhem to prepare for his doctoral examination and to write his dissertation. He lived at home for seven years, studying on his own and teaching in the evening high school. It was at this time that he bought an edition of Fresnel's collected works, the first reference volume he owned beyond the usual textbooks. He admired Fresnel above any of the early physical authors (as he came to admire Hertz above any of the modern ones), being impressed by his logical clarity and by the sure physical intuition that enabled him to overcome a limited mathematical skill. Following his French master, Lorentz cultivated clarity and physical insight as his own special talents.

Theoretical physics tended to be an isolated activity for the early specialists in the discipline; Lorentz, like his contemporary J. W. Gibbs and like Einstein somewhat later, carried out highly original researches on the central problems of physics while removed from direct contact with other active researchers. Lorentz needed only to make the short trip to Leiden's physics laboratory to borrow Maxwell's works and to keep in touch with the current specialist physics literature. His linguistic fluency was a marked asset in his Arnhem years, enabling him to command a wide range of physical writings in French, German, and English. Trained in Leiden and working in Arnhem, his receptivity to outside influences was not inhibited by any dominant local school of physics. He passed his doctoral examination *summa cum laude* in 1873 and received his doctorate in 1875, at age twenty-one; his dissertation was on physical optics, the subject Fresnel had made his own but which Lorentz treated from the viewpoint of Maxwell, as interpreted by Helmholtz.

Lorentz remained at Arnhem after receiving his doctorate, uncertain whether to follow a career in physics or mathematics. His uncertainty was in part a reflection of the professional condition of physics. Although he did some optical and electromagnetic experiments at Arnhem, he was strongly drawn to the theoretical aspects of physics; at this time, however, theoretical physics was only starting to be organized as an independent discipline within physics; and the prospects for a career in the discipline were problematic. The importance of Lorentz' thesis was early recognized in his own country, as was his academic promise generally; and in 1877 he was offered a chair of mathematics at the University of Utrecht. That same year the University of Leiden offered him its new chair of theoretical physics, which had originally been intended for J. D. van der Waals. Lorentz accepted the appointment, resolving his career uncertainty; he was not yet twenty-five. The Leiden theoretical physics chair was the first of its kind in the Netherlands, and one of the first in Europe. Lorentz holds a major place in the history of physics for his part in shaping the new role of the discipline as well as for his scientific contributions.

In 1881 Lorentz married Aletta Kaiser, a niece of his former astronomy teacher. Their first daughter, Geertruida Luberta, was born in 1885; their second, Johanna Wilhelmina, in 1889. Their first son died in infancy; their second, Rudolf, was born in 1895. Lorentz' elder daughter remembers him as constantly at his writing desk, yet never seeming to be engaged in hard work. He had none of the mannerisms of the eccentric genius nor of the bookish ascetic. He was a disciplined scholar of regular habits and disposition. He was social, with a marked sense of humor and a gift for conversation; he always enjoyed a cigar and glass of wine with friends. To all who knew him, he seemed a man of remarkable inner harmony.

At Leiden, Lorentz developed his most original contribution to theoretical physics, his celebrated electron theory. He announced and articulated the theory in a series of publications beginning in 1892. In his fifteen years at Leiden prior to 1892, Lorentz had not been uncommonly productive by contemporary standards. He had published an average of one paper a year; in addition he had written two widely used textbooks in Dutch, one on the calculus for physical scientists in 1882 and one on elementary physics in 1888. Lorentz' most productive period began in 1892, when he was almost forty. Between 1892 and 1904—the time during which he largely completed his electron theory—he produced an average of three to four research publications a year. For Lorentz, as for others, the electron theory opened up a vast number of new experimental and theoretical directions. Much of the physics community now looked to Leiden for guidance.

Lorentz' influence on theoretical physics was exerted not only through his writings but also through the personal impression he made on young physicists who came to Leiden from all over the world to hear his lectures. His influence was unwitting in the sense that he did not interfere with others and did not inspire a school of the usual kind. He followed the work of students and younger physicists, but he did not try to influence the direction of their interests; his relations with them were in keeping with his basically private nature, at once kindly and aloof. Einstein and other theorists of the younger generation venerated him, making frequent journeys to Leiden to visit him and hear his thoughts on their latest ideas. They prized him for his intellectual daring and mastery, his thorough familiarity with all areas of physics, and his nonmanipulative, natural leadership. Einstein said Lorentz had been the greatest influence in his life.

In 1905 Lorentz considered moving to the University of Munich as professor of theoretical physics. It was only after the University of Leiden relieved him from the chore of delivering introductory lectures by appointing a third professor of physics that he agreed to stay. Lorentz resigned his Leiden chair in 1912, after thirty-five years of lecturing on all aspects of theoretical physics. He moved to Haarlem, where he served as curator of the physical cabinet in Teyler's Stichting, a museum housing science, art, and coin collections, and as secretary of the Hollandsche Maatschappij der Wetenschappen, an organization to promote private patronage for science. The museum contained a laboratory, which was remodeled after that of the Royal Institution of London for Lorentz's use. For the first time Lorentz had his own laboratory, something the University of Leiden had promised him but failed to deliver, to his great disappointment.

For years Lorentz gave popular physics lectures, a task he liked and took seriously; but chiefly he used the greater freedom of his new post to pursue his theoretical researches. He retained a connection with the University of Leiden as an honorary professor, in which capacity he delivered his famous Monday morning lectures on current problems in physics. In the latter part of his career he performed various services for the government. He was keenly interested in education and served on the government board of education from its founding in 1919 until 1926, helping to reform university examinations and to redistribute professorial chairs among the universities; from 1921 he was president of the department of higher education. He also worked on practical applications of physics for the government; in 1920 he took charge of the calculations for the height of the dike closing the Zuider Zee, devoting years to the completion of the problem.

Lorentz lived all his life in a few close-lying Dutch towns—Arnhem, Leiden, Haarlem—yet he was the most cosmopolitan of physicists. For the first twenty years of his career, his cosmopolitanism was of a restricted, literary kind. After that he began to move outside his Leiden study and lecture room to make personal contact with physicists abroad. He did this at the time his electron theory was securing him a commanding position in physics, which was also the time of a new phase of international awareness of the science. International physics meetings are essentially a twentieth-century phenomenon, although before then physicists of one country sometimes attended physics meetings in another. Lorentz was first drawn out in 1898, when he accepted Boltzmann's invitation to address the physics section of the Düsseldorf meeting of the German Society of Natural Scientists and Physicians. In 1900 he addressed the International Congress of Physics in Paris, a truly worldwide assembly of physicists.

Lorentz' most important international activity in physics was his regular presidency of the Solvay Congresses for physics from their inception in 1911 through 1927, the last meeting before his death. For a quarter-century Lorentz was a kind of institution at these and other international gatherings, where he presided as the willing and acknowledged leader of the physics discipline. Everyone remarked on his unsurpassed knowledge, his great tact, his ability to summarize lucidly the most tangled argument, and above all his matchless linguistic skill. In addition to presiding at congresses, he did a good deal of invited lecturing, further evidence of the increasing internationalism of physics. He spoke on the electron theory and other topics of current physical interest in Berlin in 1904, in Paris in 1905, in New York in 1906, and later in Göttingen, Pasadena, and elsewhere.

After World War I, Lorentz' cosmopolitanism took on a political cast. As president of the physics section of the Royal Netherlands Academy of Sciences and Letters from 1909 to 1921, he used his influence to persuade his countrymen to join the postwar international scientific organizations created by the Allies. He sought to repeal the clauses excluding the Central Powers from the organizations and to restore true internationalism to science. In 1923 he became one of the seven members of the International Commission on Intellectual Cooperation of the League of Nations, succeeding Henri Bergson as its president. Lorentz' efforts to restore internationalism in science made little headway against the powerful nationalisms supporting the scientific boycott and counterboycott of the Central Powers. The continuing divisiveness of science was a source of great unhappiness to him.

In 1902 Lorentz shared the Nobel Prize in physics with his countryman Pieter Zeeman. He also received most of the other honors that ordinarily come to one of his scientific stature: he was awarded the Royal Society's Rumford and Copley medals; he received honorary doctorates from the universities of Paris and

Cambridge; and he was elected a foreign member of the German Physics Society and the Royal Society.

When Lorentz died in 1928, at age seventy-four, he was honored as the greatest cultural figure the Netherlands had produced in recent times. On the day of his funeral the Dutch telegraph and telephone services were suspended for three minutes in tribute. Representatives of the Dutch royalty and government attended his funeral in Haarlem, as did representatives of scientific academies from around the world. Einstein, a leader of the second generation of professional theoretical physicists and representative of the Prussian Academy of Sciences, spoke at the graveside, referring to Lorentz as the "greatest and noblest man of our times."

One of Lorentz' marked characteristics was an uncommon openness to new trains of thought. Again and again Einstein, Erwin Schrödinger, and other theorists sought his opinion, which was at once sympathetic, informed, and critical. Lorentz' openness was rooted partly in temperament and partly in his view of the proper work of the theoretical physicist. In his inaugural lecture at the University of Leiden in 1878, he explained that the object of all physical research was to find simple, basic principles from which all phenomena can be deduced. He warned against placing undue importance on the mental images associated with basic principles or hoping that the principles themselves can be further explained. He believed that we cannot penetrate deeply into the nature of things and that it is therefore unthoughtful to advocate any given approach as the only valid one. Various basic theoretical approaches—such as atoms and distance forces, contiguous action in a material plenum, and vortex-ring atoms—should be explored at the same time by different investigators, for only then could physicists compare approaches and decide which one leads to the simple basic principles. Much of Lorentz' lifework was spent in critically examining others' theories and helping them seek out simple principles, an activity that conformed to his understanding of the critical function of the theoretical physicist. It should be remarked that in his activities as critic and as constructor of physical theories, Lorentz was strongly sympathetic to certain approaches. However, he had an uncommon ability to distinguish between sympathy and critical judgment in his appraisal of theories. This ability—basically a trait of temperament—underlay in large part his seemingly contradictory roles as strongly committed theorist and openminded critic.

Lorentz' most important work was in optical and electromagnetic theory; to follow its bearing on the development of physics, we shall have to look at the state of the fields as he found them in the 1870's. Optical thinking was then dominated by the elastic-solid theory of the luminiferous ether, which had been developed by Fresnel, Cauchy, Neumann, Stokes, and others. The theory had major difficulties. One was the presence of longitudinal as well as transverse waves in elastic solids, whereas in optical phenomena only transverse waves were known. Another was the failure of elastic solids to yield Fresnel's fraction for the light reflected at the interface of two optical media. These and other difficulties resulted in an accumulation of hypotheses. The longitudinal wave could be disregarded if its velocity of propagation were assumed to be infinite. Certain properties of the passage of light through optical media could be explained by assuming different densities for the ether inside and outside of matter; certain others, by assuming different elasticities for the ether inside and outside matter. The state of optical theory in the 1870's was, in a word, unsatisfactory.

The state of electrodynamics in the 1870's was at least as unsatisfactory as that of optics. There was little agreement on the proper theoretical principles for developing electrodynamic theory. The two most successful electrodynamic theories in Germany were those of Weber and Neumann. Both theories had originated in the mid-1840's, and Weber's especially had been extensively developed since then. Encompassing all electric and magnetic phenomena, Weber's theory was based on the hypothesis that an electric current consisted of two fluids of electric particles of opposite signs moving in opposite directions. It was also based on the hypothesis of a central, instantaneous, action-at-a-distance force between pairs of electric particles. The force was analogous to the gravitational attraction between material particles, except that it depended on the relative motion as well as on the separation of the particles and was attractive or repulsive according to the signs of the particles. Weber's motion-dependent force had an unclear relation to the energy principle, and was challenged for this reason. Neumann's theory of induced electric currents was based on the hypothesis of a position- and motion-dependent force between two elements of electric current rather than between pairs of electric particles. Both theories followed Ampère's in referring magnetism to elementary circular electric currents.

Several other electrodynamic theories were subsequently put forward in Germany. Riemann in 1858 and Neumann in 1868 advanced theories modeled after Weber's; they differed in that they postulated a finite velocity of propagation of electric action, introducing retarded potentials for that purpose. Encouraged by the closeness of the values for the speed of light and the ratio of the electromagnetic and electrostatic measures of electricity—a ratio that had units of velocity and that entered all Weber-type force laws—Riemann identified light with the propagation of electric action. Neumann, however, was more impressed by the differences between light and the finitely propagated potential of his theory, rejecting the suggestion that they were the same thing. Clausius introduced another variant of Weber's theory in the mid-1870's, assuming only a single mobile electric fluid and a force between two electric particles that was noncentral and that depended on the absolute, rather than relative, motion of the particles.

A prominent non-German electrodynamic theory was that proposed by the Danish physicist Ludwig Lorenz in 1867. From Kirchhoff's equations—which stemmed from Weberian electrodynamics—for the motion of electricity, Lorenz, with the use of retarded potentials, showed that periodic electric currents behave like the vibrations of light, concluding that light itself consists of electric currents. Another prominent non-German Continental theory was proposed by the Swedish physicist Erik Edlund in 1871. Edlund derived Weber's fundamental law by assuming a single, particulate electric fluid—which he thought was probably the luminiferous ether—and by assuming that electric actions are finitely propagated. Whereas Lorenz broke with action at a distance, Edlund did not; whereas Edlund assumed an electric ether, Lorenz did not. The disagreements multiplied.

The several Continental electrodynamic theories resulted in a maze of incompletely tested hypotheses, and no immediate experimental decision seemed likely. The conflicting hypotheses centered on several issues: the number—whether one or two—of mobile electric fluids; the existence of an electric ether; the identity of light with electric motions; the nature of the forces between electric fluid particles, to which belonged the questions of central or noncentral forces, and of the relative or absolute motion that entered the force laws; and the relation of motion-dependent forces to the principles of dynamics, especially to the energy principle.

Contributing further to the profusion of electrodynamic principles was the very different theoretical tradition that had developed in Britain during the same period. Drawing on Faraday's and William Thomson's work, Maxwell proposed a mathematical theory of electricity and magnetism in papers in the 1850's and 1860's and in his comprehensive *Treatise on Electricity and Magnetism* in 1873. Rejecting action at a distance and the associated particulate electric fluids of Continental electrodynamics, Maxwell based his theory on the concept of an electromagnetic medium. The essential point of the theory was that the electromagnetic medium supported transverse vibrations that propagated at the speed of light. The theory promised to be fruitful, especially in the prediction it shared with certain Continental theories that light is an electric phenomenon. Maxwell took seriously the optic implications of his theory, investigating a number of optical problems: notably, the pressure exerted by light, the propagation of light in crystals, the relation between electric conductivity and optical opacity of metals, and the magnetic rotation of the plane of polarization of light. He failed, however, to derive the laws of the reflection and refraction of light, the problem that had proved so embarrassing to the elastic-solid optical theories.

For all its promise, Maxwell's theory—in the form, or various forms, in which he left it—was a source of great perplexity for many of its close readers. Ehrenfest recalled that Maxwell's *Treatise* seemed a "kind of intellectual primeval forest, almost impenetrable in its uncleared fecundity." The theory was unclear in precisely those basic features that differentiated it from most Continental theories. Maxwell did not elucidate the nature of electric charge, a concept that had received clear, if not unanimous, interpretations in terms of particulate electric fluids in the action-at-a-distance theories. Indeed, Maxwell refused to commit himself to any position on the nature of electricity. Sometimes he spoke of it as the process of dielectric polarization; in reverse fashion he spoke at other times of polarization as a motion of electricity. He sometimes likened the motion of electricity to that of an incompressible fluid, without intending that the reader should think that electricity is an incompressible fluid. In the *Treatise* he spoke of a "molecule of electricity"; he added, however, that this way of speaking was "out of harmony with the rest of this treatise."

Conceding only that electricity was a physical quantity, Maxwell cautioned against assuming that "it is, or is not, a substance, or that it is, or is not, a form of energy, or that it belongs to any known category of physical quantities." It is small wonder that his Continental interpreters found the task of clarifying the concept of charge to be both essential and difficult. They found Maxwell's concept of the electromagnetic field equally difficult to grasp. Maxwell did not distinguish between the roles of the medium and ordinary matter in electromagnetic processes. The clear distinction in the Continental theories between distance forces—the counterpart of Maxwell's field—and the electricity they arise from and act upon is entirely absent from Maxwell's theory.

The eventual clarification of the conceptual basis of Maxwell's theory was aided immeasurably by a paper that Helmholtz published in 1870. The purpose of his paper was to bring order to electrodynamics by examining the major rival theories for their agreement with dynamical principles and by exposing their differing, testable consequences. To do this he generalized Neumann's electrodynamic potential between two current elements to encompass Maxwell's and Weber's theories; each of the three theories was characterized by a numerical parameter entering the formulas of the general theory. Because of the recent interest in the velocity of propagation of electric actions and because of its great significance for physics, Helmholtz applied his general induction law to the case in which an electrically and magnetically polarizable medium is present. In the spirit of action-at-a-distance theories, he assumed that the electric interaction of bodies depends in part on direct distance forces and in part on the polarization of the medium, which in turn depends on the distance forces and is the source of new distance forces. He showed that the electric displacement in a dielectric obeys the same wave equations as the displacement of ponderable particles in an elastic solid.

Helmholtz thought that the resulting "remarkable analogy" between the motions of electricity and the motions in the luminiferous ether were consequences of the older action-at-a-distance electrodynamics no less than they were of Maxwell's special hypotheses. He concluded that finitely propagated transverse waves of polarization in a dielectric medium were possible without basic changes in the foundations of action-at-a-distance electrodynamics. His general theory also yielded longitudinal waves, which in the special case of Maxwell's theory were propagated with infinite velocity. From his critique of the rival electrodynamic theories, Helmholtz argued that Weber's theory was untenable and that Maxwell's theory, as he had interpreted it, should be taken seriously. Many Continental physicists approached Maxwell's theory through Helmholtz; his action-at-a-distance formulation of the theory rendered it intelligible to physicists nurtured in the Continental tradition, and his favorable judgment assured it serious attention.

Helmholtz recognized that resolution of the difficulties of optical theory might be directly related to resolution of the difficulties of electrodynamics, and this clear recognition was the starting point for Lorentz' optical researches. Lorentz chose his dissertation topic as a result of a footnote in Helmholtz' 1870 paper, which pointed to the superiority of Maxwell's electromagnetic theory of light over the older elastic-solid theories in yielding the crucial interface conditions required by Fresnel's laws of reflection and refraction.

In 1875 Lorentz submitted to the University of Leiden his doctoral dissertation on electromagnetic optics, "Sur la théorie de la réflexion et de la réfraction de la lumière" (translated from the original Dutch). Although he had studied Maxwell's papers, his starting point was Helmholtz' action-at-a-distance general theory. He was not opposed in principle to Maxwell's contiguous-action point of view, but he felt that Maxwell's theory was incompletely developed and that in its present form it relied on more special and unconfirmed hypotheses than Helmholtz' did. Lorentz' dissertation reveals his characteristic blend of incisive clarity and unifying vision. He opened with a comparative critique of the older wave theories of light and of the new electromagnetic theory of light. He then reproduced Helmholtz' derivation of the wave equation for the propagation of variations in the state of polarization, applying it successively to the reflection and refraction of light by isotropic media, crystal optics, total reflection, and the interaction of light and metals.

The main result of Lorentz' dissertation was his electromagnetic derivation of Fresnel's amplitudes for light reflected at a dielectric interface. He did this without having to make ad hoc assumptions comparable with those of the elastic-solid theory concerning the density and elasticity of the ether. He also showed that in a more natural way than the elastic-solid theories the electromagnetic theory dispensed with embarrassing, unobserved longitudinal vibrations. He concluded that because of its greater simplicity and comprehensiveness, Maxwell's electromagnetic theory of light was to be preferred over the elastic-solid ones. Lorentz closed his dissertation with a prophetic vision of the vast unifying potentiality for physics implicit in the combination of Maxwell's electromagnetic theory of light with the molecular theory of matter, a prospect that proved to be an adumbration of his own future research.

Apart from being the first systematic treatment of electromagnetic optics, one that succeeded where the elastic-solid theories failed most conspicuously, Lorentz' dissertation was significant as a first step toward distinguishing the electromagnetic field from matter and thus clarifying the physical basis of Maxwell's theory. He pointed to Boltzmann's recent experimental investigations of the predicted relation of the specific inductive capacity of gases to their index of refraction as strongly confirming Maxwell's electromagnetic theory of light. He interpreted Boltzmann's results to mean that the ether is the seat of polarizations and that gaseous molecules exert only a small, secondary influence on the specific inductive capacity of the ether pervading the intermolecular spaces; in this sense he viewed the ether as the only proper dielectric.

In a sequel paper in 1878, on the optical theory of matter, Lorentz began the research program he had sketched at the close of his dissertation; and in so doing he further strengthened his distinction between the roles of matter and the ether. He developed there a theory of the dispersion of light by assuming that material molecules contain charged harmonic oscillators and that the ether everywhere—except possibly in the immediate vicinity of the molecules—has the same properties that it does in a vacuum. Lorentz accounted for the dispersion of light in a body in the following way: incident light waves in the ether cause the electric particles in the body to vibrate; the vibrating particles send out secondary waves in the ether which interfere with the incident ones, thus making the velocity of propagation of light through the body depend on frequency. In his 1878 paper Lorentz predicted a relation between the density of a body and its index of refraction, a relation that became known as the Lorentz-Lorenz formula for its independent development by the Danish physicist in 1869.

In the 1880's Lorentz' continuing interest in electromagnetism was reflected in a paper on the force between two current elements regarded from the point of view of Continental electrodynamics, and in another on the Hall effect and the electromagnetic rotation of the plane of polarization of light. His continuing interest in the ether was reflected in a study of the aberration of light in 1886, in which he concluded that Fresnel's view of the luminiferous ether was superior to Stokes's. Unlike Stokes, Fresnel in his theory of aberration assumed that the ether near the earth did not participate in its motion. Lorentz thought that the hypothesis of the complete transparency of matter to the ether was implicit in Fresnel's whole theory.

Lorentz' chief interest in the 1880's was not, however, electromagnetism and optics but the molecular-kinetic theory of heat. The science of thermodynamics had been largely completed by his day, and he did not participate in its construction. But the molecular-kinetic interpretation of thermodynamics was far from complete, and he worked on a number of its problems; for one, he corrected Boltzmann's proof of the *H*-theorem. Lorentz was strongly receptive to the kinetic theory, as he was to the molecular viewpoint in physics in general, the theme he had enlarged on in his inaugural lecture at the University of Leiden.

From the 1890's, the electron theory dominated Lorentz' researches. In constructing a theory of electrons and fields, he brought together a number of insights he had accumulated over several years of varied researches. The particulate concept of electricity that lay at the basis of his electron theory was conditioned by his long familiarity with Continental electrodynamics and its assumption of particulate electric fluids. It was also conditioned by the same molecular orientation in physics that had prompted his intensive work on the molecular-kinetic theory and, before that, on the interaction of light and molecular matter. Equally fundamental to his electron theory was the concept of the stationary ether, which he had reached through his studies of optical aberration. Finally, his pioneering studies of Maxwellian optics in the 1870's prepared the way for the full, explicit separation of the roles of ether and matter in his electron theory.

The immediate occasion for Lorentz to return to the foundations of Maxwell's theory was Hertz's researches in electromagnetism. In the late 1880's, in response to Helmholtz' program for bringing about an empirical decision between the rival electrodynamic theories, Hertz carried out his influential experiments on Maxwellian electric waves. Following his experimental demonstration of electric waves in air, Hertz made theoretical studies in 1890 of the Maxwellian electrodynamics of stationary and moving bodies. Having abandoned the Helmholtzian action-at-a-distance interpretation of Maxwell's theory with which he had begun his experimental researches, Hertz in his theoretical studies accepted Maxwell's contiguous-action interpretation of electromagnetic processes in the ether, together with his denial of particulate electric fluids. Hertz based his study of the electrodynamics of moving bodies on the concept of an ether that was completely dragged by ponderable bodies. Fully realizing that a dragged ether was in conflict with optical facts, he justified its application to narrowly electrodynamic phenomena on the grounds that it was compatible with such phenomena and that it facilitated a systematic presentation of the theory, the sole purpose of his theoretical study. By assuming a dragged ether, Hertz needed only one set of electric and magnetic vectors instead of the two required by an ether that was separable from matter. He also based his study on his version of Maxwell's equations, which he postulated rather than deriving from mechanical models and principles.

Hertz's experimental and theoretical researches generated widespread interest in Maxwell's theory among Continental physicists. Of the major theoretical statements of Maxwellian electrodynamics following Hertz's researches, several advanced a molecular view of electricity together with a stationary ether. Such theories—soon to be called electron theories—were proposed independently in the early 1890's by Lorentz, by Wiechert, and by Larmor. Although the three theories bore large areas of resemblance, they also differed in major respects. Whereas Larmor, for example, viewed electrons as structures in the ether and the ether as the sole physical reality, Lorentz treated electrons and the ether as distinct entities. Of the three theories, Lorentz' gained the greatest authority on the Continent, in part because of its clear, if ultimately unsatisfactory, dualism of electron and field.

In an address to the Dutch Congress of Physics and Medicine in 1891, Lorentz, following Hertz, declared his conversion to contiguous action in electromagnetic processes. Lorentz objected, however, to two features of Hertz's electrodynamics, which prompted his first memoir in 1892 on the electron theory: "La théorie électromagnétique de Maxwell et son application aux corps mouvants." He criticized Hertz's totally dragged ether on well-known optical grounds; and although he admired the clear, succinct mathematical form that Hertz and Heaviside had given to Maxwell's theory, he criticized Hertz's bare postulation of the field equations. Referring to "one of the most beautiful" chapters in Maxwell's *Treatise*, in which Maxwell applied Lagrangian mechanics to elucidate electromagnetism without assuming a detailed mechanism, Lorentz in 1892 continued the mechanical exploration begun by Maxwell. He remarked that "one has always tried to return to mechanical explanations," and properly so, in his opinion. After laying down his own hypotheses concerning the stationary ether and electrons, he applied d'Alembert's principle to derive the field equations and the equations of motion of an electron in the field. Characteristically, Lorentz spelled out at the start the physical suppositions of his theory; it was a constructive effort of exemplary clarity in a science chronically subject to obscurity.

Lorentz incorporated into his electron theory Fresnel's view that the ether penetrates matter (Lorentz held more tenaciously to a stationary ether than even Fresnel, who admitted some dragging), but he rejected Fresnel's further view that the density of the ether varies from substance to substance as well as Neumann's view that the ether differs in elasticity in different substances. For Lorentz the ether had identical properties everywhere, and he was little inclined to speculate about its ultimate nature. He completely separated the ether and ordinary matter; however, since ether and matter were observed to interact, he needed to reestablish their connection. He assumed that their sole connection was provided by positive and negative electrons, which he regarded as small, ponderable, rigid bodies. He assumed that electrons are contained in all molecules of ordinary matter (Lorentz wrote of "charged particles" in 1892, of "ions" in 1895, and of "electrons" only after 1899; in this article the term "electrons" is used throughout, without suggesting that Lorentz had an unchanging view of the properties of electrons during these years).

In contrast with Maxwell and Hertz, Lorentz provided a clear, simple interpretation of electric charge and current and of their relation to the electromagnetic field. A body carries a charge if it has an excess of one or the other kind of electrons. An electric current in a conductor is a flow of electrons; a dielectric displacement in a nonconductor is a displacement of electrons from their equilibrium positions. The electrons create the electromagnetic field, the seat of which is the ether; the field in turn acts ponderomotively on ordinary matter through the electrons embedded in material molecules.

Because Lorentz completely separated ether and matter, he needed only one pair of directed magnitudes—one electric and one magnetic—to define the field at a point, and this was so whether or not matter was present at the point; in this way he answered Hertz's formal objection to a stationary ether that it required two sets of directed magnitudes to define the field at a point, one for matter and one for ether. Lorentz also answered Hertz's objection that particulate electric fluids belonged to action-at-a-distance, not contiguous-action, electrodynamics. By means of his concepts of a stationary ether and of electrons transparent to it, Lorentz constructed a consistent electrodynamics that at the same time rejected action at a distance and retained particulate electric fluids.

For this reason Lorentz characterized his theory as a fusion of Continental and Maxwellian electrodynamics. He retained the clear understanding of electricity of the theories of Weber and Clausius, and at the same time he accepted the crux of Maxwell's theory: the propagation of electric action at the speed of light. In the electrodynamics of Weber and Clausius, two electric particles act on each other with a force that depends on their separation and their motion at the instant of the action. In Lorentz' theory the force on an electron also depends on the position and motion of other electrons, but at earlier instants, owing to the finite propagation of electric disturbances through the mediating ether.

From his understanding of ether and electricity in 1892, Lorentz explained the interaction of light and matter in a way close to that of his optical theory of the 1870's. Light is turned and retarded in its passage through a body owing to the electrons in the molecules of the body; the incident light sets up vibrations in the electrons, which in turn produce light waves that interfere with the original ones and with one another. With this conception Lorentz in 1892 derived the Fresnel drag coefficient, a measure of the motion that a moving transparent body communicates to light passing through it. The coefficient had been confirmed by Fizeau's 1851 experiment on the motion of light in a moving column of liquid and Michelson and Morley's repetition of it in 1886. Among the physical interpretations that had been given to the coefficient, one was that the moving body partially drags the ether with it. Lorentz demonstrated that the drag coefficient resulted from the interference of light and thus did not imply a true, partial dragging of the ether. This demonstration was the single most impressive achievement of his first memoir on the electron theory, inspiring Max Born to refer to it, rightly, as one of the "most beautiful examples of the might of mathematical analysis in the physical world."

Although in 1892 Lorentz believed that it was important to relate electromagnetic processes to mechanical laws, he had not constructed a wholly mechanistic electrodynamics. In certain respects his ether was inherently a nonmechanical substance. Since his ether occupies the same space as electrons and material molecules, and its properties are unaffected by their coextension, it has no mechanical connection with ordinary matter.

Lorentz' next major exposition of his electron theory was *Versuch einer Theorie der electrischen und optischen Erscheinungen in bewegten Körpern* (1895). He no longer derived the basic equations of his theory from mechanical principles, but simply postulated them. He wrote the equations for the first time in compact vector notation; in electromagnetic units the four equations that describe the electromagnetic field in a vacuum are

$$\begin{aligned} \operatorname{div} \mathbf{d} &= \rho, \\ \operatorname{div} \mathbf{H} &= 0, \\ \operatorname{rot} \mathbf{H} &= 4\pi(\rho\mathbf{v} + \dot{\mathbf{d}}), \\ -4\pi c^2 \operatorname{rot} \mathbf{d} &= \dot{\mathbf{H}}, \end{aligned}$$

where $\mathbf{d}$ is the dielectric displacement, $\mathbf{H}$ the magnetic force, $\mathbf{v}$ the velocity of the electric charge, ρ the electric charge density, and c the velocity of light. A fifth and final equation describes the electric force of the ether on ponderable matter containing electrons bearing unit charge:

$$\mathbf{E} = 4\pi c^2\mathbf{d} + \mathbf{v} \times \mathbf{H}.$$

The first four equations embody the content of Maxwell's theory; the fifth equation is Lorentz' own contribution to electrodynamics—known today as the Lorentz force—connecting the continuous field with discrete electricity.

In 1892 Lorentz had briefly discussed the problem of the effects of the earth's motion through the stationary ether; in the *Versuch* he systematically went over the whole problem. Since the ether is not dragged, a moving body such as the earth has an absolute velocity relative to it. The question arises whether or not the earth's absolute velocity is detectable through optical or electromagnetic effects of the accompanying ether "wind." The magnitude of the effects of the wind is measured theoretically by the ratio of the speed of the earth's motion v to the speed of light c. The ratio is small for the earth, but not so small as to be beyond the reach of observation.

The effects of the wind, however, were not observed; and for his theory to be credible Lorentz had to explain why. He showed that, according to the theory, an unexpected compensation of actions eliminates all effects of the ether wind to first-order approximation (neglecting terms involving the very much smaller second and higher powers of v/c). He analyzed the absence of first-order effects of the ether wind in phenomena such as reflection, refraction, and interference with the aid of a formal "theorem of corresponding states." The theorem asserted that to first-order accuracy no experiments using terrestrial light sources could reveal the earth's motion through the ether. By introducing transformations for the field magnitudes, spatial coordinates, and a "local time," Lorentz showed that in first-order approximation the equations describing an electric system in a moving frame were identical with those describing the corresponding electric system in a frame at rest in the ether (for which Lorentz assumed Maxwell's equations to hold exactly).

Lorentz' first-order approximation accounted for nearly all experience with the optics and electrodynamics of moving bodies. Second-order experiments had been performed, however, for which theoretically no compensating actions should occur. The most important second-order experiments were Michelson's interferometer experiment in 1881 and his and Morley's more accurate repetition of it in 1887. Their experiments failed to produce evidence of the second-order ether wind effects expected from Lorentz' theory. The only solution Lorentz could think of was the famous contraction hypothesis that he and Fitzgerald proposed independently at about the same time. Lorentz published an approximate form of his hypothesis in 1892. In 1895 he published an exact form, which stated that the arms of the interferometer contract by a factor of $\sqrt{1 - v^2/c^2}$ in the direction of the earth's motion through the ether. Lorentz regarded the hypothesis as dynamic rather than kinematic, requiring that the molecular forces determining the shape of the interferometer arms are propagated through the ether analogously to electric forces.

In 1899 Lorentz published another important statement of his theory, "Théorie simplifiée des phénomènes électriques et optiques dans des corps en mouvement." It was a response to Alfred Liénard's contention that according to Lorentz' theory, Michelson's experiment should yield a positive effect if the light passes through a liquid or solid instead of air. Lorentz believed that the positive effect was improbable, and he simplified and deepened his theory to support his belief. He now treated his dynamical contraction hypothesis mathematically, as though it were a general coordinate transformation on a par with the local time transformation. Except for an undetermined coefficient, the resulting transformations for the space and time coordinates were equivalent to those he published in his better-known 1904 article: the "Lorentz transformations." In 1899 Lorentz discussed for the first time the extension of his corresponding-states theorem to second-order effects. In connection with this extension he introduced, without thoroughly exploring it, an idea that would soon move to the center of electron theory concerns: the second-order transformations imply that the mass of an electron varies with velocity, and that the mass is different for motions parallel and perpendicular to the direction of translation through the ether. Although he discussed only electrons, his analysis implies that all mass, charged or otherwise, should vary with velocity, an implication that he made explicit in his more comprehensive 1904 paper.

Shortly before 1900 it was empirically established that cathode rays are negatively charged particles and that they are the same regardless of how and from what source they are produced. Since the particulate properties of cathode rays agreed with Lorentz' hypothesis about the nature of electricity, he confidently identified the cathode ray particles with the electrons of his theory. The empirical confirmation of the discrete unit of electricity lent great authority to Lorentz' and others' electron theories. In 1896 Zeeman observed a broadening of the spectral lines when a sodium flame is placed between the poles of a magnet. Lorentz immediately explained the effect on the basis of his electron theory, a result which pointed to his theory as a promising tool for unraveling the complexities of atomic structure.

Lorentz' application of the theory to the Zeeman effect showed that the vibrating electrons in sodium atoms are negative. Moreover, it yielded a value for the ratio of charge to mass of negative electrons which agreed with subsequent estimates by J. J. Thomson, Walther Kaufmann, and others, found by deflecting cathode rays in electric and magnetic fields. The success of Lorentz' analysis of the Zeeman effect further enhanced the standing of his theory. Indeed, at the turn of the century Lorentz' theory could count many solid accomplishments, such as explanations of Fizeau's experiment, of normal and anomalous dispersion, and of Faraday's rotation of light. For both its accomplishments and its promise, Lorentz' electron theory became widely adopted, especially on the Continent around 1900.

In 1904 Lorentz published "Electromagnetic Phenomena in a System Moving With Any Velocity Smaller Than That of Light." He was motivated to make another, nearly final, reformulation of his electron theory by several theoretical and experimental developments. First, Poincaré in 1900 had urged Lorentz not to make ad hoc explanations for each order of null effects, but to adopt instead a single explanation that excluded effects of the earth's motion of all orders. Second, in 1902 Max Abraham had published a rival electron theory based on the hypothesis of a rigid electron. Third, Kaufmann had begun publishing experimental data on the variable mass of moving electrons that seemed to favor Abraham's theory. Finally, new second-order experiments had been performed in 1902–1904 by Lord Rayleigh, D. B. Brace, Frederick Trouton, and H. R. Noble; and Lorentz wanted to discuss their null results in light of his theory. In his 1904 paper Lorentz refined his corresponding-states theorem to hold for all orders of smallness for the case of electromagnetic systems without charges, which meant that no experiment, however accurate, on such systems could reveal the translation of the apparatus through the ether. He also showed that his theory agreed with Kaufmann's data as well as Abraham's theory did. With this paper Lorentz all but solved the problem of the earth's motion through the stationary ether as it was formulated at the time. Poincaré in 1905 showed how to extend Lorentz' corresponding-states theorem to systems that included charges and to make the principle of relativity, as Poincaré understood it, more than approximation within the context of Lorentz' theory.

Lorentz' solution—developed over the years since 1892—entailed a number of radical departures from traditional dynamics; these he spelled out explicitly in 1904. First, the masses of all particles, charged or not, vary with their motion through the ether according to a single law. Second, the mass of an electron is due solely to its self-induction and has no invariant mechanical mass. Third, the dimensions of the electron itself, as well as those of macroscopic bodies, contract in the direction of motion, the physical deformation arising from the motion itself. Fourth, the molecular forces binding an electron and a ponderable particle or binding two ponderable particles are affected by motion in the same way as the electric force. Finally, the speed of light is the theoretical upper limit of the speed of any body relative to the ether; the formulas for the energy and inertia of bodies become infinite at that speed. Thus, to attain a fully satisfactory corresponding-states theorem, Lorentz had to go far beyond the domain of his original electron theory and make assertions about all bodies and all forces, whether electric or not.

The new dynamics of Lorentz' 1904 electron theory was compatible with the leading physical thought of the time. That it was shows how deeply the electron theory—Lorentz' form of it and others'—had changed the foundations of prerelativistic classical physics. To understand the change, we need to look more closely at the place of the electron theory in physics in the years just prior to 1904. In part because of the large and growing number of phenomena encompassed by the electron theory, physicists around the turn of the country extrapolated from it a new physical world view.

Lorentz, Larmor, Wien, Abraham, and others anticipated that mechanics would be replaced by electrodynamics as the fundamental, unifying branch of physics. They believed that mechanical concepts and laws were probably special cases of those of the electron theory. They foresaw a physical universe consisting solely of ether and charged particles, or possibly of ether alone. They were encouraged in their world view anticipations by the mounting experimental evidence for the electromagnetic nature of the electron mass. Lorentz wrote on electromagnetic mass in 1900, urging that experiments be made to determine the exact dependency of the electron mass on velocity. The problem of mass was central not only for its bearing on electron dynamics, but also for its bearing on the question of the electromagnetic foundation of physics and the related question of the connection of the ether and electricity with ordinary matter. Although Lorentz spoke more cautiously than other advocates of the electromagnetic program for physics, he gave it powerful impetus through his own applications of the electron theory. He showed in 1900 that if matter were constituted solely of charged particles, it was possible to give an electron-theoretical interpretation of gravitation as a finitely propagated, etherborne force. In that same year, 1900, he spoke of his belief that the electron theory could be extended to spectroscopy, atomic structure, and chemical forces.

Those physicists seeking an electromagnetic foundation for all of physics recognized the immensity of the reductionist tasks awaiting them. The tasks included the demonstration that all matter is electrical and the further demonstration that molecular and gravitational forces are explicable on wholly electromagnetic grounds. Even within electrodynamic theory itself, however, there were elements of incompleteness, which undermined the classical electromagnetic program for physics. The problem of securing a purely electromagnetic foundation for electrodynamics seemed increasingly unrealizable. The contractile electron at the core of Lorentz' theory presented a major difficulty in this regard.

Following Abraham, Lorentz by 1904 recognized that such an electron contains an energy of deformation that is necessarily nonelectromagnetic. Abraham's theory avoided this nonelectromagnetic dependence, but at the price of postulating a rigid scaffolding for the electron; other electron theorists posed alternative conceptions of the electron's shape, structure, and charge distribution. Another closely related difficulty for the electromagnetic program arose from the need for the electron to be finitely extended in order to avoid infinite self-energy, which in turn required nonelectromagnetic forces or constraints to prevent the finite electron from disintegrating through the electric repulsion of its parts. Yet another formidable difficulty emerged as a result of one of Lorentz' applications of the electron theory. In connection with his electron theory of metals, Lorentz in 1903 derived a formula for the energy distribution in blackbody radiation. He was able to calculate from his theory only the long wavelength limit of the energy spectrum. He recognized that Planck's 1900 quantum theory of blackbody radiation comprehended the entire spectrum; he also recognized that Planck's energy quanta were foreign to the foundations of the electron theory.

In 1908 Lorentz spoke out in favor of Planck's quantum theory as the only theory capable of explaining the complete spectrum of blackbody radiation. He was one of the first to do so and to emphasize the deep antithesis between the quantum hypotheses and those of the electron theory. The discussions of the first Solvay Congress in 1911—which Lorentz chaired—pointed to the conclusion that only through radical reform could the electron theory and the molecular-kinetic theory be made compatible with Planck's theory. Einstein had already long sought a revised electron theory to incorporate light quanta. Bohr's 1913 quantum theory of atoms and molecules was based on an explicit limitation of the validity of the classical electron and mechanical theories in the domain of atomic processes. Lorentz' career was divided roughly in half by the transition from classical to quantum physics. Although he was intensely interested in the new physics, and although he worked with the new quantum axioms and developed their consequences, his heart was never fully in it. He deeply regretted the passing of classical physics.

Einstein's 1905 special relativity paper provided Lorentz' theory with a physical reinterpretation. Reversing Lorentz' logic, Einstein made relativity a principle rather than a problem (Lorentz did not speak of a "principle of relativity" in connection with his own theory until after Einstein's 1905 paper). Likewise, for Einstein the constancy of the velocity of light for all observers was a principle, whereas for Lorentz the constancy was only apparent, a consequence of his theory. From the principles of relativity and the constancy of the velocity of light, Einstein deduced the Lorentz transformations and other results that had first been made known through Lorentz' and others' electron theories. Einstein's and Lorentz' interpretations of the formalisms were, however, very different. For Lorentz, time dilation in moving frames was a mathematical artifice; for Einstein, measures of time intervals were equally legitimate in all uniformly moving frames. For Lorentz the contraction of length was a real effect explicable by molecular forces; for Einstein it was a phenomenon of measurement only.

Einstein argued in 1905 that the ether of the electron theory and the related notions of absolute space and time were superfluous or unsuited for the development of a consistent electrodynamics. Lorentz admired, but never embraced, Einstein's 1905 reinterpretation of the equations of his electron theory. The observable consequences of his and Einstein's interpretations were the same, and he regarded the choice between them as a matter of taste. To the end of his life he believed that the ether was a reality and that absolute space and time were meaningful concepts.

Einstein's special relativity principle began to be widely accepted around 1910. It weakened the goal of a completely electromagnetic physics by questioning the need for an ether. It weakened it even more by questioning inferences from the variation of mass with velocity to the electromagnetic nature of mass. The relativity principle led to the same testable dynamical conclusions as Lorentz' theory, but without relying on the assumptions about electron structure that had been central to electron physics.

The origins of another leading twentieth-century theoretical development, the theory of general relativity, were closely tied to the classical electron theory. Beginning in 1907 and especially from 1911 on, Einstein sought a gravitational and, subsequently, a unified gravitational and electromagnetic field theory in the context of his general relativity theory. Lorentz was strongly sympathetic to the general relativity theory, making fundamental contributions to it in 1914–1917. Lorentz, and Einstein too, regarded the physical space of general relativity as essentially fulfilling the role of the ether of the older electron theory. A principal objective of Einstein's field theoretical proposals was to relate electrons and material particles to the total field in a more satisfactory way than the earlier electron theories had. Einstein wanted above all to remove the radical dualism of continuous field and discrete particle that had characterized Lorentz' electron theory.

The classical electron theory was related through highly complex conceptual developments to later fundamental theories in physics. The Schrödinger and Dirac equations of the 1920's formed the mathematical bases of electron physics refashioned on nonclassical suppositions. These equations, together with the tradition of earlier electron theory investigations of the self-induction and electromagnetic mass of the electron, formed part of the basis for the development of modern elementary particle theory. Further, problems explored in the context of the electron theory have had a continuous history into the nonclassical period; for instance, the problem of the electron structure of atoms and the electron properties of metals.

Although the classical electron theory did not fulfill the anticipations it stirred at the turn of the century of an electromagnetic world view, it worked a profound change in the thinking of physicists and contributed to new world view perspectives. For Lorentz' immediate followers—those who had struggled to comprehend the obscurities of Maxwell's theory in the aftermath of Hertz's confirmation of electric waves—his electron theory was immensely clarifying. The younger generation of European theoretical physicists who learned much of their electrodynamics from Lorentz—Einstein, Ehrenfest, A. D. Fokker—agreed that Lorentz' great idea was the complete separation of field and matter. Einstein called Lorentz' establishment of the electromagnetic field as an independent reality distinct from ponderable matter an "act of intellectual liberation." The act at the same time determined the outstanding problem for future research, one that is still at the heart of theoretical physics. It is the problem of reestablishing the connection of field and particle in a way that does not contradict experience.

Einstein grasped the need to reject the wave theory of light and to modify the concepts of space and time in part from his analysis of the essential duality of Lorentz' theory: its joint foundation in the continuous electromagnetic field and molecular dynamics. Lorentz and his co-workers in electron theory carried classical physics to its furthest stage of clarity, precision, and unity. In doing so, Lorentz identified the strong limitations of the electron theory, defining the problem areas from which ongoing transformations of the foundations of physics arose. He brought classical physics to a state of development that made the need for reform in its fundamental principles evident and urgent to Einstein and other followers; this is the essential historical significance of Lorentz' work.

BIBLIOGRAPHY

I. Original Works. Lorentz' dissertation, both in Dutch and in French trans., his 1895 *Versuch*, and most of his important papers are conveniently repub. in *H. A. Lorentz, Collected Papers*, P. Zeeman and A. D. Fokker, eds., 9 vols. (The Hague, 1935–1939). Vols. I–VIII contain technical writings, all in English, French, or German. Vol. IX is a collection of Lorentz' nonmathematical writings, mostly in Dutch; his 1878 inaugural lecture is included in English trans. as well as in the original Dutch. There are both systematic and chronological bibliographies at the end of vol. IX. The bibliographies list all of the eds. and translations of Lorentz' textbooks, all the places of publication of each paper, and all the books, papers, and essays that are not repub. in the *Papers*.

Lorentz' two textbooks are *Leerboeck der differentiaal- en integraalrekening en van de eerste beginselen der analytische meetkunde* (Leiden, 1882) and *Beginselen der Natuurkunde* (Leiden, 1888); both are available in German trans. His lectures were published in Dutch in 1919–1925, in German in 1927–1931, and in English, as *Lectures on Theoretical Physics*, A. D. Fokker *et al.*, eds., 8 vols. (London, 1931). Separate vols. treat radiation, quanta, ether theories and models, thermodynamics, kinetic theory, special relativity, entropy and probability, and Maxwell's theory.

The best introduction to Lorentz' electron theory in English is *Theory of Electrons* (Leipzig, 1909), an expanded version of his 1906 Columbia University lectures. A more mathematical treatment of the electron theory in German is his 1903 encyclopedia article, "Weiterbildung der Maxwellschen Theorie. Elektronentheorie," in *Encyklopädie der mathematischen Wissenschaften*, V, pt. 2 (Leipzig, 1904), 145–280, one of several articles he wrote for the encyclopedia on the electromagnetic theory of light.

Lorentz began an early ed. of his writings but finished only the first vol., *Abhandlungen über theoretische Physik* (Leipzig, 1907). This vol. contains material Zeeman and Fokker did not republish. Other books by Lorentz are *Sichtbare und unsichtbare Bewegungen* (Brunswick, 1902); *Les théories statistiques en thermodynamique* (Leipzig, 1912); *The Einstein Theory of Relativity: A Concise Statement* (New York, 1920); and *Problems of Modern Physics. Lectures at the Institute of Technology at Pasadena* (Boston, 1927).

Writings on the electron theory by Lorentz in separate English trans. include the discussion of Michelson's interference experiment in his 1895 *Versuch*, in the Dover ed. of *The Principle of Relativity*, edited and translated by W. Perrett and G. B. Jeffery (New York, n.d.), pp. 3–7; and the intro. to his 1895 *Versuch* and his entire 1899 paper, "Simplified Theory of Electrical and Optical Phenomena in Moving Systems," in Kenneth F. Schaffner, *Nineteenth-Century Aether Theories* (Oxford, 1972), pp. 247–254 and 255–273, respectively.

Some Lorentz correspondence has been pub. in *Einstein on Peace*, O. Nathan and H. Norden, eds. (New York, 1960); A. Hermann, "H. A. Lorentz—Praeceptor physicae, sein Briefwechsel mit dem deutschen Nobelpreisträger Johannes Stark," in *Janus*, **53** (1966), 99–114; and *Letters on Wave Mechanics: Schrödinger, Planck, Einstein, Lorentz*, K. Prizbram, ed., translated by M. J. Klein (New York, 1967).

There is a vast amount of unpublished Lorentz material, chiefly letters. The catalog of the Archive for History of Quantum Physics, T. S. Kuhn, *et al.*, *Sources for History of Quantum Physics* (Philadelphia, 1967), lists correspondence between Lorentz and Bohr, Kramers, Schrödinger, and Sommerfeld. Copies of the documents cataloged in *Sources* are deposited in the Niels Bohr Institute, Copenhagen; the library of the University of California, Berkeley; and the American Philosophical Society library, Philadelphia. Subsequent to the publication of *Sources*, the Archive acquired twenty microfilm reels of Lorentz correspondence covering the period 1888–1928. The originals of this correspondence are in the Rijksarchief, The Hague. Other locations of Lorentz papers not included in the Archive are listed in *Sources*.

II. Secondary Literature. There is no large-scale study of Lorentz' life and work. His private life is most fully detailed in the memoir by his daughter, G. L. de Haas-Lorentz, in the vol. she edited, *H. A. Lorentz. Impressions of His Life and Work* (Amsterdam, 1957). Other contributors to this volume are Einstein, W. J. de Haas, A. D. Fokker, B. van der Pol, J. T. Thijsse, P. Ehrenfest, and H. B. G. Casimir. Especially valuable are Fokker's essay on Lorentz' scientific work, and Einstein's tribute, "H. A. Lorentz, His Creative Genius and His Personality." Other useful accounts by Lorentz' contemporaries are P. Ehrenfest, "Professor H. A. Lorentz as Researcher 1853–July 18, 1923," in *Paul Ehrenfest: Collected Scientific Papers*, M. J. Klein, ed. (Amsterdam, 1959), pp. 471–477; J. Larmor, "Hendrik Antoon Lorentz," in *Nature*, **111** (1923), 1–6; W. H. Bragg, *et al.*, "Prof. H. A. Lorentz, For. Mem. R. S.," *ibid.*, **121** (1928), 287–291; O. W. Richardson, "Hendrik Antoon Lorentz—1853–1928," in *Proceedings of the Royal Society*, **121A** (1928), xx–xxviii; M. Born, "Antoon Lorentz," in *Nachrichten von der Königlichen Gesellschaft der Wissenschaften zu Göttingen. Geschäftliche Mitteilungen* (1927–1928), 69–73; and M. Planck, "Hendrik Antoon Lorentz," in *Naturwissenschaften*, **16** (1928), 549–555.

The outstanding historical study of Lorentz is T. Hirosige, "Origins of Lorentz' Theory of Electrons and the Concept of the Electromagnetic Field," in *Historical Studies in the Physical Sciences*, **1** (1969), 151–209. Other pertinent writings by Hirosige are "A Consideration Concerning the Origins of the Theory of Relativity," in *Japanese Studies in the History of Science*, no. 4 (1965), 117–123; "Electrodynamics Before the Theory of Relativity, 1890–1905," *ibid.*, no. 5 (1966), 1–49; and "Theory of Relativity and the Ether," *ibid.*, no. 7 (1968), 37–53.

For other useful historical discussions, see Stanley Goldberg, "The Lorentz Theory of Electrons and Einstein's Theory of Relativity," in *American Journal of Physics*, **37** (1969), 982–994; G. Holton, "On the Origins of the Special Theory of Relativity," *ibid.*, **28** (1960), 627–636; R. McCormmach, "H. A. Lorentz and the Electromagnetic View of Nature," in *Isis*, **61** (1970), 459–497; and "Einstein, Lorentz, and the Electron Theory," in *Historical Studies in the Physical Sciences*, **2** (1970), 41–87; and K. F. Schaffner, "The Lorentz Electron Theory and Relativity," in *American Journal of Physics*, **37** (1969), 498–513; and "Interaction of Theory and Experiment in the Development of Lorentz' Contraction Hypothesis," in *Actes du XIIe Congrès internationale de l'histoire des sciences, 1968*, V, 87–90. The standard older discussion is the closing chapter, "Classical Theory in the Age of Lorentz," in E. Whittaker, *History of the Theories of Aether and Electricity. 1: The Classical Theories* (London, 1910).

RUSSELL MCCORMMACH

LORENZ, HANS (*b.* Wilsdruff, Germany, 24 March 1865; *d.* Sistrans, near Innsbruck, Austria, 4 July 1940), *mechanics, mechanical engineering.*

The son of a teacher, Lorenz grew up in Leipzig, then studied (1885–1889) mechanical engineering at the Dresden Polytechnic Institute, where Gustav Zeuner was his principal teacher. His first professional experience (1890–1893) was with the Augsburg firm of L. A. Riedinger, developing a pneumatic power distribution system, and with the Escher-Wyss Company of Zurich, improving refrigeration compressors (1893–1894).

In 1894 Lorenz established himself in Munich as an independent consulting engineer. He founded, and for the first five years edited, the *Zeitschrift für die gesamte Kälteindustrie*, which quickly became the leading international journal of refrigeration technology. In 1894 he received a Ph.D. from the University of Munich—a rare feat then for an engineer—with a dissertation on the thermodynamic limits of energy conversion.

Lorenz' academic career began with brief appointments as extraordinary professor of applied science at Halle (1896) and Göttingen (1900). In Göttingen he was also director of the Institute for Technical Physics, which had recently been established upon recommendation of the mathematician Felix Klein, and which was to become world famous under Lorenz' successor, Ludwig Prandtl. In 1904 Lorenz was appointed to the chair of mechanics at the newly founded Technische Hochschule of Danzig, a position that he held for the rest of his active career. He also served as director of its materials testing laboratory (from 1909) and as rector (1915–1917). After his retirement in 1934, Lorenz lived in Munich and Sistrans.

Whether any of Lorenz' research efforts will be remembered in history is difficult to judge. Without doubt, however, he was an important figure in his own time. He was active in three areas of engineering: practice, teaching, and research. He was sought as a consultant in the burgeoning refrigeration industry, and his professional leadership was recognized in his election to presidencies and honorary memberships of several scientific societies. As an engineering teacher he was a leading proponent of a distinctive style of scientific engineering that flourished in early twentieth-century Germany. It was his belief that a creative engineer combined mathematical and scientific competence of a high order with the ability to translate actual problems into simple, tangible models (and was occasionally willing to sacrifice mathematical rigor for the sake of practical results). For undergraduates, Lorenz was a difficult teacher, but his advanced courses, as many of his former graduate students (prominent among whom are W. Hort, R. Plank, and A. Pröll) have testified, were unforgettable.

Lorenz' original research is characterized by unbounded versatility. His topics derived from his own engineering practice (pneumatic transmission lines, refrigeration), from his work as head of the materials testing laboratory (buckling, plastic deformation), and from ongoing scientific debates (vibrations, gyroscopes, turbomachinery, turbulent flow), to the war effort (ballistics). In his later years he published a considerable amount of work on astrophysics and astronomy.

BIBLIOGRAPHY

I. ORIGINAL WORKS. An almost complete list of Lorenz' scientific papers (approximately 130 items) is given in Poggendorff. His books are *Neuere Kühlmaschinen* (Munich, 1896); *Dynamik der Kurbelgetriebe* (Leipzig, 1901), with O. Schlick; *Lehrbuch der technischen Physik*, 4 vols. (Munich, 1902–1913); *Neue Theorie und Berechnung der Kreiselräder* (Munich, 1906); *Einführung in die Elemente der höheren Mathematik und Mechanik* (Berlin, 1910); *Ballistik* (Munich, 1917); *Technische Anwendungen der Kreiselbewegungen* (Berlin, 1919); *Einführung in die Technik* (Leipzig, 1919); and *Das Verhalten fester Körper im Fliessbereich* (Leipzig, 1922). Lorenz left unpublished MSS of an autobiography, "Praxis, Lehre, und Forschung: Akademische Erinnerungen und Erfahrungen" (*ca.* 1935) and of a book on astronomy (*ca.* 1938).

II. SECONDARY LITERATURE. Biographical writings on Lorenz consist of birthday tributes and obituaries, the most extensive being by Rudolf Plank, "Hans Lorenz zum 70. Geburtstag," in *Zeitschrift für die gesamte Kälteindustrie*, **42** (1935), 42–46. See also G. Cattaneo, "Was ich Hans Lorenz verdanke," *ibid.*, **47** (1940), 114; Wilhelm Hort, "Hans Lorenz zum 70. Geburtstag," in *Zeitschrift für technische Physik*, **16** (1935), 93–95; R. Plank, "Hans Lorenz zum 75. Geburtstag," *ibid.*, **47** (1940), 33; and "Hans Lorenz," in *Zeitschrift des Vereins deutscher Ingenieure*, **84** (1940), 638; and A. Pröll, "Zu Geheimrat Lorenz' 70. Geburtstag," in *Zeitschrift für angewandte Mathematik und Mechanik*, **15** (1935), 183–184.

OTTO MAYR

LORENZ, LUDWIG VALENTIN (*b.* Elsinore, Denmark, 18 January 1829; *d.* Copenhagen, Denmark, 9 June 1891), *physics.*

Lorenz, a civil engineer by training, graduated from the Technical University of Denmark in Copenhagen; he taught at the Danish Military Academy and at a teacher's training college in Copenhagen. During the last years of his life he was financially supported by the Carlsberg Foundation to such an extent that he could devote all his time to research.

Lorenz was an eminent physicist, although the content and range of his achievements were not fully recognized by his contemporaries. This lack of recognition was due mainly to his great difficulties in presenting his ideas and mathematical calculations in intelligible form, but it was also due to his publishing some of his important papers only in Danish.

Lorenz' researches in physics, mainly optics and heat and electricity conductivities of metals, were carried out with equal emphasis on theoretical and experimental methods. The basis of his research in optics was the propagation of light waves conceptualized in terms of the theory of elasticity. Finally, however, he was convinced of the incompatibility of the boundary conditions of the theory of elasticity with Fresnel's formulas for reflection and refraction; he then concentrated his efforts on finding a phenomenological description of the propagation of light waves instead of speculating on the nature of light. Independently of the Irish physicist MacCullagh, he showed in 1863 that the partial differential equation for the light vector $\vec{u}$ should be (in modern notation)

$$-\operatorname{curlcurl}\vec{u} = \frac{1}{a^2}\frac{\partial^2\vec{u}}{\partial t^2},$$

where the phase velocity a is a constant that is characteristic for the homogeneous optical medium under consideration. He also gave the correct boundary conditions for the light vector when it is passing from one medium to another.

Lorenz then showed that if a is not considered as a constant but as a periodic function of space, his wave equation led to the theory of double refraction; where periods of a are small in comparison with the wavelength. This theory, by use of the same mathematical technique, has again been used recently.

A further remarkable result of Lorenz' optical researches on the basis of his fundamental wave equation was the well-known formula (Lorentz-Lorenz formula) for the refraction constant R, according to which

$$R = \frac{n_\infty^2 - 1}{n_\infty^2 + 2}\, v = \text{constant},$$

where n_∞ is the index of refraction for an infinite wavelength and v is the specific volume. He also showed that for a mixture

$$R = \sum_i \frac{n_i^2 - 1}{n_i^2 + 2}\, v_i,$$

where the index $_i$ refers to the components of the mixture. His first paper on the refraction constant, in which he also gave an experimental verification of his formula in the case of water, dates from 1869. In 1870 H. A. Lorentz arrived at the same result, independently of Lorenz.

In 1890, Lorenz published in *Det kongelige danske Videnskabernes Selskab Skrifter* a paper on the diffraction of plane waves by a transparent sphere, again using his fundamental wave equation with the relevant boundary conditions; he thereby anticipated later calculations by Mie and Debye. Unfortunately, this very important paper was published only in Danish, and thus it had no influence on later developments.

This paper contains the first determination of the number of molecules in a given volume of air, based upon the scattering of sunlight in the atmosphere. It is the first fairly accurate estimation of the order of magnitude of Avogadro's number, usually ascribed to Lord Rayleigh, who published it in 1899, nine years after the publication of Lorenz' paper and independently of Lorenz.

Most impressive of all Lorenz' achievements in optics is his electromagnetic theory of light, developed in a relatively unknown paper of 1867, two years after Maxwell's famous paper on the same subject. At that time Lorenz did not know Maxwell's theory, and his own approach was quite different. Lorenz' electromagnetic theory of light can be described briefly as an interpretation of the light vector as the current density vector in a medium obeying Ohm's law. This paper contains the fundamental equations for the vector potential and the scalar potential or —for the first time—the corresponding retarded potentials expressed in terms of the current density vector and the electrical charge density. The concept of retarded potentials had already been introduced in an earlier paper by Lorenz in connection with research on the theory of elasticity. He found that the differential equation for the current density vector was the same as his fundamental wave equation for the light vector, completed with a term which explains the absorption of light in conducting media, and that his theory led to the correct value for the velocity of light.

Lorenz' most important contribution in heat and electricity conductivities of metals was the Lorenz law, according to which the ratio between the conductivities of heat and electricity is proportional to the absolute temperature, where the constant of proportionality is the same for all metals. It had already been shown by R. Franz, who worked with Wiedemann, that this ratio is the same for all metals at the same temperature; and Lorenz then proved, through very accurate measurements of the two conductivities at 0° C. and 100° C. that this ratio is proportional to the absolute temperature.

It is characteristic of Lorenz' mastery of both theoretical and experimental methods that when, in the determination of the conductivity of heat, he encountered the problem of the cooling of a body by the air through heat convection, he solved the problem by calculations based on Navier's and Stokes's equations of hydrodynamics. It is a very difficult problem in mathematical physics, but he succeeded in solving it by using quite modern methods of similarity. Having finished his calculations, he found them confirmed by experiments.

In the determination of the specific conductivities for electricity, Lorenz used an absolute method of his own invention, which later was applied by himself and others, such as Rayleigh, to fix the international unit of resistance, the ohm.

In an unpublished paper of great interest, Lorenz developed a complete theory of currents in telephone cables, showing that the attenuation of the current along the cable can be reduced by increasing the inductance per unit of length. He therefore proposed the use of cables wound with a covering of soft iron, an idea also proposed by Heaviside, independently of Lorenz, that later was taken up by the Danish electrotechnician C. E. Krarup. This improvement is now known as continuous loading.

Lorenz was also deeply interested in problems of pure mathematics, such as the distribution of prime numbers; but here he encountered severe criticism from Danish mathematicians, who did not appreciate the undoubted lightness—if not recklessness—with which he applied the methods of mathematics. It is, however, probable that there also are problems of interest in some of his mathematical works.

Lorenz's contemporaries in the scientific circles of Copenhagen no doubt realized that he was an eminent scientist, and by the standard of his time his work was well supported. But although he was much respected, few understood what he actually did. This lack of deeper understanding, combined with his own tendency to isolation and his rather polemical attitude toward some of his colleagues, made him a somewhat lonely philosopher.

BIBLIOGRAPHY

Lorenz' writings were brought together in *Oeuvres scientifiques de L. Lorenz*, 2 vols., revised and annotated by H. Valentiner (Copenhagen, 1898–1904).

Secondary literature includes Kirstine Meyer, in *Dansk Biografisk Leksikon*, XIII (Copenhagen, 1937), 464–469; and Mogens Pihl, *Der Physiker L. V. Lorenz* (Copenhagen, 1939); and "The Scientific Achievements of L. V. Lorenz," in E. C. Jordan, ed., *Electromagnetic Theory and Antennas*, I (New York, 1963), xxi–xxx.

MOGENS PIHL

LORENZ, RICHARD (*b.* Vienna, Austria, 13 April 1863; *d.* Frankfurt am Main, Germany, 23 June 1929), *physical chemistry.*

Lorenz' father was the historian Ottokar Lorenz; his mother was the daughter of the philosopher and educational reformer Franz Lott. Lorenz studied medicine at the universities of Vienna and Jena, but his main interest was chemistry. He received his doctorate at Jena for work on the valence of boron. After completing his studies he was for a short time an assistant in the physiology institute at the University of Rostock; he left in order to work with Otto Wallach in the chemistry laboratory of the University of Göttingen. Despite this beginning he did not devote his career to organic chemistry, Wallach's principal field of interest. Instead, he took up the study of physical chemistry. Nernst had just been appointed professor at Göttingen, in the first chair ever dedicated to that subject, and in 1892 Lorenz became an assistant and *Privatdozent* under him.

Lorenz then began to explore the field in which he worked for the rest of his life—the electrochemistry of substances in the fused state. In 1896 he was invited to the Technische Hochschule in Zurich to join the newly founded electrochemical laboratory. In Zurich he married Lili Heusler, who bore him two children. Following the death of his wife, Lorenz accepted an offer in 1910 from the Frankfurt Academy to direct its physics and chemistry institute. The academy soon became a university, and Lorenz pursued his work there until his retirement, which occurred shortly before his death. For thirty years he was editor of the *Zeitschrift für anorganische und allgemeine Chemie*.

Lorenz' most important scientific achievement was his demonstration that Faraday's law is completely valid for fused salts. He discovered the cause of earlier results which contradicted it in the formation of the so-called metallic fog, which occurs in the reaction of the precipitated metals with the melt. He then found methods which inhibited this fog formation. With M. Katajama he developed the thermodynamic relations to calculate the electromotive force of galvanic cells consisting of fused salts alone or of the latter joined with metals. By determining the particle sizes in disperse systems, which had been suggested by Einstein, he was able to compute the ion sizes in the fused electrolytes from the mobility and the rate of diffusion.

Lorenz' work contributed to the clarification of many other properties of fused salts, for example internal friction, capillary electricity, and conductivity. In his last years he established that the law of mass action is not valid in highly concentrated fusions. He also developed van der Waals's equation of state taking this fact into account and, with the aid of the concept of thermodynamic potential, developed a law of mass action for condensed systems.

BIBLIOGRAPHY

I. Original Works. Lorenz wrote about 250 articles; a fairly extensive listing may be found in Poggendorff VI and in the obituary in the *Zeitschrift für Elektrochemie* but there is no complete list. His most important books are *Die Elektrolyse geschmolzener Salze*, 3 vols. (Halle, 1905); *Elektrochemie geschmolzener Salze* (Leipzig, 1909), written with F. Kaufler; *Raumerfüllung und Ionenbeweglichkeit* (Leipzig, 1922); and *Das Gesetz der chemischen Massenwirkung, seine thermodynamische Begründung und Erweiterung* (Leipzig, 1927).

II. Secondary Literature. See W. Fraenkel, "Richard Lorenz," in *Zeitschrift für angewandte Chemie und Zentralblatt für technische Chemie*, **42** (1929), 801–802; G. Hevesy, "Richard Lorenz. Erinnerungen aus den Zürcher Jahren," in *Helvetica chimica acta*, **13** (1930), 13–17; and A. Magnus, "Richard Lorenz," in *Berichte der Deutschen chemischen Gesellschaft*, Abt. A, **62** (1929), 88–90; and "Richard Lorenz," in *Zeitschrift für Elektrochemie und angewandte physikalische Chemie*, **35** (1929), 815–822.

F. Szabadváry

LORENZONI, GIUSEPPE (*b.* Rolle di Cisón, Treviso, Italy, 10 July 1843; *d.* Padua, Italy, 7 July 1914), *astronomy, astrophysics, geodesy*.

The son of an elementary school teacher, Lorenzoni received his degree in engineering at the University of Padua in 1863. In the same year, he entered the Padua Observatory as assistant to the director, Giovanni Santini. In 1872 he was appointed professor of astronomy at the University of Padua, where he also held the professorship of geodesy from 1869 to 1885. In 1878 he became the director of the Padua Observatory.

Lorenzoni's first observations and research dealt with astrophysics. He began these studies with spectral analysis applied to the physical study of celestial bodies. As early as 1871, he collaborated with A. Secchi and P. Tacchini in establishing the Società degli Spettroscopisti Italiani and in the publication of its *Memorie*. During the total solar eclipse of 1870, which he observed in Terranova (now Gela), Sicily, he noted many brilliant lines of emission of the prominences, which led him to search for lines of emission besides those commonly visible in the chromosphere in full sunlight. Through this he discovered a new line that proved to be the 4471 A of the diffused series of the neutral helium of which he had measured the wavelength. Later he found that the same line had been independently observed by C. H. Young in the United States. He then published in the *Memorie della Società degli spettroscopisti italiani* important considerations on the spectroscopic visibility of monochromatic images. Among his writings on a variety of subjects published in the *Memorie* is an article dealing with theoretical optics, in which it is possible to foresee the invention of the spectroheliograph (1892) and the interferential filter (1929).

After abandoning spectroscopy, Lorenzoni turned to classical astronomy and the related science of geodesy, which occupied the greater part of his life. A keen observer who used instruments of very modest dimension, Lorenzoni observed several comets and small planets; and using Starke's meridian circle, he continued and completed the stellar catalogs of his predecessors.

His numerous publications on classical astronomy discuss the formulas of precession and of nutation, in which the Euler equations for the rotation of a body around a given point are given by trigonometric and geometric methods.

Lorenzoni also dealt with fundamental formulas of spherical trigonometry used for calculating the parallax in the coordinates of a planet and for determining the angular coordinates by means of astronomical instruments.

In 1878 Lorenzoni was appointed a member of the Commissione Italiana per la Misura del Grado, which was then carrying on intensive geodesic activity. He made important contributions to the commission's work on latitude and azimuth observations in various Italian localities. He also collaborated with Schiaparelli in Milan, with Oppolzer in Vienna, and with other foreign astronomers on telegraphic longitude determinations. His works on classical astronomy are found in the publications of the Commissione Geodetica Italiana.

Realizing the importance of measuring the force of gravity in Italy, especially in collaboration with scientific organizations established in other countries, the commission in 1880 aroused Lorenzoni's interest in this problem. He soon presented a concrete plan of studies and the history of research to determine the length of the simple seconds pendulum. After having participated in a conference at Munich, where the subject was discussed, Lorenzoni was able to begin work in Italy. He had at his disposal a Repsold pendulum that he placed at the base of the Padua Observatory. His patient and accurate measurements and discussions of the problem were presented in a scholarly memoir published by the Accademia dei Lincei in 1888. It is divided into three parts; the first is theoretical; the second deals with the place of observation, instruments, and errors; the third presents the results of observation and reduction.

The value that Lorenzoni computed at Padua (sixty feet above sea level) of the length of the seconds pendulum agreed completely with the value that he had obtained by reducing, for Padua, the values obtained at five other stations. From this he concluded that there were no great anomalies in the gravity at Padua. The site of these experiments was officially named the *Sala di gravità di Giuseppe Lorenzoni* and became the goal of geodetic scholars who use Padua for gravimetric reference and as a meeting place.

Lorenzoni's fundamental research in gravity and the effect of support flection on the time of pendulum oscillation led to the design and construction of a bipendular support in a vacuum.

BIBLIOGRAPHY

A fairly complete listing of Lorenzoni's papers is in Royal Society *Catalogue of Scientific Papers*, VIII, 260; X, 632; and XVI, 870. The articles cited in the text are "Di un mezzo atto a rendere visibile tutta in una volta una immagine monocromatica completa della cromosfera e delle protuberanze solare al lembo con la costruzione di lenti ipercromatiche," in *Memorie della Società degli spettroscopisti italiani*, **3** (1874), 61–64, on theoretical optics; and "Relazione sulle esperienze istituite ... per determinare la lunghezza del pendolo semplice ..., in *Atti dell'Accademia nazionale dei Lincei. Memorie*, **5** (1888), 41–281.

Obituaries are by A. Abetti, in *Vierteljahrsschrift der Astronomischen Gesellschaft*, **49** (1914), 232–238; A. Antoniazzi, in *Astronomische Nachrichten*, **198** (1914), col. 486; and E. Millosevich, in *Atti dell'Accademia nazionale dei Lincei*, **23**, pt. 2 (1914), 442–446.

GIORGIO ABETTI

LORIA, GINO (*b.* Mantua, Italy, 19 May 1862; *d.* Genoa, Italy, 30 January 1954), *mathematics, history of mathematics.*

After graduating in 1883 from Turin University, where higher geometry was taught by Enrico D'Ovidio, Loria attended a postgraduate course at Pavia University taught by Beltrami, Bertini, and F. Casorati. Loria became D'Ovidio's assistant in November 1884; on 1 November 1886 he was appointed extraordinary professor of higher geometry at Genoa University, becoming full professor in 1891. He held the chair of higher geometry at Genoa for forty-nine years, until he retired in 1935; he also taught the history of mathematics, descriptive geometry, and mathematics education. He wrote a number of treatises on descriptive geometry that were appreciated for their simplicity, elegant constructions, and generality. In them he introduced elements of photogrammetry. He was a member of the Accademia Nazionale dei Lincei and of various other academies.

Loria's geometrical research, at first carried out in cooperation with Corrado Segre, was especially concerned with new applications of algebraic concepts to the geometry of straight lines and spheres, with hyperspatial projective geometry, with algebraic correspondence between fundamental forms, and with Cremona transformations in space. But the field in which Loria's scientific activity was most extensive was the history of mathematics from antiquity down to his own time.

Special mention should be made of *Le scienze esatte nell'antica Grecia* (1914), a reworking of the articles which appeared in the *Memorie della Regia Accademia di scienze, lettere e arti in Modena* in 1893–1902. It constitutes a rich source of information on Greek mathematics. *Storia della geometria descrittiva* (1921) shows Loria's liking for the subject. *Guida allo studio della storia delle matematiche* (last ed., 1946) is methodological in character and places at scholars' disposal the rich source materials and collections on the subject with which he had worked for years. Full bibliographical information is in *Il passato e il presente delle principali teorie geometriche*, updated in the editions of 1897, 1907, and 1931.

Loria devoted two works to special algebraic and transcendental curves which have attracted the attention of mathematicians. The first of these works was published in German in 1902, and only much later (1930–1931) in Italian, with the title *Curve piane speciali algebriche e trascendenti*; the other is *Le curve sghembe speciali algebriche e trascendenti* (1925). As F. Enriques has pointed out, the curves studied by Loria are of special interest because the study of particular cases often leads to general results.

Many of Loria's writings are biographical, notably those on Archimedes, Newton, Cremona, Beltrami, and Tannery. With G. Vassura he edited the largely unpublished writings of Torricelli (E. Torricelli, *Opere*, 4 vols. [Faenza, 1919–1944]). In 1897 Loria recognized, in one of Torricelli's manuscripts, the first known rectification of a curve: the logarithmic spiral.

Some of Loria's works have been collected in the volume *Scritti, conferenze e discorsi sulla storia delle matematiche* (1937); and an overall view of the evolution of mathematical thought, from ancient times to the end of the nineteenth century, is provided in his *Storia delle matematiche* (2nd ed., 1950).

Loria's library, containing many works on the history of mathematics, was left to the University of Genoa.

BIBLIOGRAPHY

I. Original Works. A list of Loria's works is in the commemorative address delivered by A. Terracini at the Accademia dei Lincei (see below) and includes 386 writings.

Some of his more important articles are: "Sur les différentes espèces de complexes du second degré des droites qui coupent harmoniquement deux surfaces du second ordre," in *Mathematische Annalen*, **23** (1883), 213–235, written with C. Segre; "Sulle corrispondenze proiettive fra due piani e fra due spazi," in *Giornale di matematiche*, **22** (1883), 1–16; "Ricerche intorno alla geometria della sfera e loro applicazione allo studio ed alla classificazione delle superficie di quarto ordine aventi per linea doppia il cerchio immaginario all'infinito," in *Memorie della Reale Accademia delle scienze di Torino*, 2nd ser., **36** (1884), 199–297; "Sulla classificazione delle trasformazioni razionali dello spazio, in particolare sulle trasformazioni di genere uno," in *Rendiconti dell'Istituto lombardo di scienze e lettere*, 2nd ser., **23** (1890), 824–834; "Della varia fortuna d'Euclide in relazione con i problemi dell'insegnamento della geometria elementare," in *Periodico di matematica per l'insegnamento secondario*, **8** (1893), 81–113; "Evangelista Torricelli e la prima rettificazione di una curva," in *Atti dell'Accademia nazionale dei Lincei, Rendiconti*, 5th ser., **6**, no. 2 (1897), 318–323; "Eugenio Beltrami e le sue opere matematiche," in *Bibliotheca mathematica*, 3rd ser., **2** (1901), 392–440; "Commemorazione di L. Cremona," in *Atti della Società ligustica di scienze naturali e geografiche*, **15** (1904), 19; "Necrologio. Paolo Tannery," in *Bollettino di bibliografia e storia delle scienze matematiche pubblicato per cura di Gino Loria*, **8** (1905), 27–30; and "La storia della matematica vista da un veterano," in *Bollettino dell'Unione matematica italiana*, 3rd ser., **5** (1950), 165–170.

His books include *Spezielle algebraische und transscendente ebene Curven* (Leipzig, 1902), trans. by author as *Curve piane speciali algebriche e trascendenti*, 2 vols. (Milan, 1930–1931); *Le scienze esatte nell'antica Grecia* (Milan, 1914); *Newton* (Rome, 1920); *Storia della geometria descrittiva dalle origini sino ai nostri giorni* (Milan, 1921); *Le curve sghembe speciali algebriche e trascendenti*, 2 vols. (Bologna, 1925); *Archimede* (Milan, 1928); *Storia delle matematiche*, 2 vols. (Torino, 1929–1931; 2nd ed., Milan, 1950); *Il passato e il presente delle principali teorie geometriche* (Padua, 1931), an updating of arts. In *Memorie della Reale Accademia delle scienze di Torino*, sers. 2, **38** (1887); *Scritti, conferenze e discorsi sulla storia delle matematiche* (Padua, 1937); and *Guida allo studio della storia delle matematiche* (Milan, 1946).

II. Secondary Literature. See A. Terracini, "Gino Loria. Cenni commemorativi," in *Atti dell'Accademia delle scienze di Torino*, **88** (1953–1954), 6–11; and "Commemorazione del socio Gino Loria," in *Atti della Accademia Nazionale dei Lincei, Rendiconti*, 8th ser., Classe di scienze fisiche, matematiche e naturali, **17** (1954), 402–421, with bibliography.

Ettore Carruccio

LORRY, ANNE CHARLES (*b.* Crosnes, France, 10 October 1726; *d.* Bourbonne-les-Bains, France, 18 September 1783), *medicine.*

The son of a well-known professor of law in Paris, Lorry received an excellent education supervised by

Charles Rollin. Soon after receiving his medical degree (1748), Lorry was presented by the chief royal physician L. G. Le Monnier, to Marshal Noailles. He was thus introduced to Paris high society and became the physician of the Richelieus, of the Fronsacs, of Mlle de Lespinasse, and of Voltaire, all of whom received him warmly but paid him less enthusiastically. He attended the autopsy of the young duke of Burgundy, the dauphin's son, in 1761, and was called in as a consultant on 29 April 1774, during Louis XV's last illness, at the request of the Duc d'Aiguillon, the friend of the Comtesse du Barry. The king (who died on 10 May) spoke to Lorry several times and his baptismal name was, one evening, the password given to the captain of the guards.

Lorry's worldly successes were numerous. He was consequently taken to task by La Mettrie in his *Politique du médecin de Machiavel* (1746) and was a character in a play by Poinsinet. Indeed, according to Paul Delaunay, the elegant Lorry was the darling of the salons and of the ladies. He was extravagant with money and unconcerned about his health; thus he became gouty. A serious stroke, around 1782, caused him to lose the use of his legs; and many of his patients turned to other physicians. He fell into financial difficulties, but obtained a royal pension that enabled him to go to Bourbonne-les-Bains, where he died.

Lorry's worldly concerns did not prevent him from studying with Astruc and Ferrein, nor from doing important scientific work. In 1760 he presented two memoirs to the Académie Royale des Sciences on the then fashionable subject of medullo-cerebral physiology. In the first he studied the normal movements of the encephalon, while in the second he established that compression of the cerebellum produces sleep and that a puncture in the upper part of the spinal cord between the second and third cervical vertebrae is immediately followed by death. It was thus that he placed in the spinal column the seat of the soul—the *sensorium commune*—the location of which had already been discussed by Vieussens, Boerhaave, Astruc, François de La Peyronie, Haller, and Antoine Louis, and later by Soemmerring. This "noeud vital" was later studied by Legallois (1808–1812) and by Flourens (1827); they showed that it was actually a regulatory center of the respiratory system.

In other publications Lorry discussed hygiene, anatomy, physiology, the history of medicine, medical pathology, epidemiology, clinical medicine, and smallpox inoculation, a method he defended in 1763 against Astruc. He is now primarily known, however, for his work in dermatology and mental disease.

Lorry's treatise on dermatology, which he dedicated to his friend E. L. Geoffroy, was the first French monograph devoted to diseases of the skin and, according to L. T. Morton, the first modern work that attempted to classify them according to their physiological, pathological, and etiological similarities. He divided skin diseases into two groups. The first consisted of those he considered to be the external expression of an internal disease; these were the general or *dépuratoires* diseases, which can either invade the entire cutaneous surface or be localized. The second group consists of diseases that originate in the skin itself and do not affect other parts of the body. They can—like those of the first group—cover either small areas or the general surface of the skin. Lorry's book is a source of precise observations, but it is spoiled by an obsolete Galeno-Arabic nomenclature and archaic medical conceptions.

English and French physicians in the eighteenth century were much concerned with depression; the word "suicide" was first used in France in 1737 by Y. Pelicier. In England one spoke of "melancholy" (R. Burton, 1621), of "spleen" (R. Blackmore, 1725), or "English malady" (G. Cheyne, 1733); and in France, of "nostalgie" (J. Hoffer, 1688; La Caze, 1763; Lecat, 1767), "affection vaporeuse" (J. Raulin, 1758; P. Pomme, 1763), or "mélancolie" (C. Andry, 1785). Lorry recognized two types of melancholy, one a consequence of an alteration of the solid parts ("mélancolie nerveuse"), the other originating in the humors ("mélancolie humorale"). The latter could result in hysteria in women and, in men, hypochondria. Lorry stressed spasmodic episodes: torticollis, pyloric spasm with vomiting, pharyngeal spasm with difficulties in swallowing. Although he assigned a physical etiology to melancholy, Lorry sometimes contradicted himself by citing the roles of fear and anguish.

Lorry left the Paris Faculty of Medicine, where he taught surgery (1753), and was one of the founders of the Société Royale de Médecine, of which he became the director and vice-president. It was for his efforts in this capacity that Vicq d'Azyr dedicated to him one of his best memorial addresses.

BIBLIOGRAPHY

I. Original Works. Lorry's writings include *An causa caloris in pulmone aeris actione temperetur?* (Paris, 1746); *Essai sur les aliments, pour servir de commentaire aux livres diététiques d'Hippocrate*, 2 vols. (Paris, 1754–1757), reiss. as *Essai sur l'usage des aliments*, 2 vols. (Paris, 1781), translated into Italian as *Saggio sopra gli alimenti per servire di commentario ai libri dietetici d'Ippocrate*, 2 vols. (Venice, 1787); *De melancholia et morbis melancholicis*, 2 vols. (Paris, 1765); *Tractatus de morbis cutaneis* (Paris, 1777), trans. into German by C. F. Held as *Abhandlung von den Krankheiten der Haut*, 2 vols. (Leipzig, 1779); and *De praecipuis morborum mutationibus et conversionibus tentamen medicum*, J. N. Hallé, ed. (Paris, 1784).

Editions and translations include *Hippocratis aphorismi graece et latine* (Paris, 1759); *Mémoires pour servir à l'histoire de la Faculté de médecine de Montpellier par feu M. Jean Astruc*, revised and published by Lorry (Paris, 1767); J. Barker's *Essai sur la conformité de la médecine ancienne et moderne dans le traitement des maladies aigües*, translated by Schomberg, new ed., revised by Lorry (Paris, 1768); *Sanctorius . . . de medicina statica aphorismi*, comments and notes added by Lorry (Paris, 1770); *Richardi Mead Opera*, translated by Lorry (Paris, 1751); and *Hippocratis aphorismi, Hippocratis et Celsi locis parallelis illustrati*, edited by Lorry (Paris, 1784).

Many writings by Lorry are in the *Mémoires de l'Académie royale des sciences* (1760) and the *Mémoires de la Société royale de médecine* (1776–1779).

II. Secondary Literature. See A. Kissmeyer, "Autour d'Anne-Charles Lorry," in *Bulletin de la Société française de dermatologie et de syphiligraphie*, **38** (1931), 1524–1527; E. Bongrand, "Lorry," in *Dictionnaire encyclopédique des sciences médicales de Dechambre*, 2nd ser., III (Paris, 1870), 112–113; P. Delaunay, *Le monde médical parisien au XVIII° siècle* (Paris, 1906), *passim;* R. Desgenettes, "Lorry," in *Biographie médicale*, VI (Paris, 1824), 102–109; "Éloge de Lorry," in *Journal de médecine militaire*, **3** (1784), 379–387; Arne Kissmeyer, *Anne-Charles Lorry et son oeuvre dermatologique* (Paris, 1928); "Lorry," in A. L. J. Bayle and A. G. Thillaye, *Biographie médicale*, II (Paris, 1855), 503–504, with portrait; "Lorry," in *Biographisches Lexicon*, III (Berlin–Vienna, 1962), 843; "Lorry," in Dezeimeris, *Dictionnaire historique de la médecine ancienne et moderne*, III (Paris, 1836), 479–480; "Lorry," in N. F. J. Eloy, *Dictionnaire historique de la médecine* (Mons, 1778), pp. 101–102; Richter, "Geschichte der Dermatologie," in *Handbuch der Haut- und Geschlechtskrankheiten*, **14**, no. 2 (1928), 178a, 180; C. Saucerotte, "Lorry," in *Nouvelle biographie générale*, XXXI (Paris, 1852), 688–690; and F. Vicq d'Azyr, "Éloge de M. Lorry," in *Histoire de la Société royale de médecine années 1782–1783* (Paris, 1787), V, pt. 1, 25–59.

Pierre Huard
Marie José Imbault-Huart

LOSCHMIDT, JOHANN JOSEPH (*b.* Putschirn, near Carlsbad, Bohemia [now Karlovy Vary, Czechoslovakia], 15 March 1821; *d.* Vienna, Austria, 8 July 1895), *physics, chemistry.*

Loschmidt was the oldest of four children of a poor peasant family. The young Loschmidt showed more aptitude for schoolwork than work in the fields and was enrolled in the parochial school in Schlackenwerth (near Carlsbad) in 1833 with the help of the village priest; in 1837, again with the assistance of the Catholic clergy, he entered Humanistic Gymnasium in Prague. From 1839 he studied classical philology and philosophy at the German University in Prague.

Loschmidt became a lecturer under Franz Exner, who aided the impoverished young man and advised him to concentrate on mathematics. He commissioned Loschmidt to enlarge on Herbart's attempts at a mathematical treatment of psychology and to put it on a more solid basis. However, despite intense efforts, Loschmidt achieved only negative results, primarily because of difficulties encountered in the measurement of the intensities of psychological experiences.

In 1841 Loschmidt joined Exner in Vienna to study the natural sciences, though his favorite field remained the border area between physics and philosophy. Influences of Herbartian philosophy were evident in his thought throughout his life. Loschmidt was unsuccessful in obtaining a teaching post when he graduated in 1843 and he therefore turned to industry. He worked in Anton Schrötter's laboratory until the end of 1846. With his friend Benedikt Margulies, Loschmidt discovered a process for converting sodium nitrate into potassium nitrate, used for manufacturing gunpowder. They could not know, of course, that the same process had been discovered half a century earlier by Thaddäus Haenke, who had actually turned it to practical use in Peru.[1] From then until 1854 Loschmidt tried again and again to establish businesses but the drastic social, political, and financial upheavals of Europe in the mid-nineteenth century made it almost impossible for any businessman to succeed, and in 1854 he went into bankruptcy.

Discouraged by his many failures, Loschmidt decided to return to a career in science. Early in 1856 he passed, with excellent marks, his examinations to qualify as a teacher. In September of that year he obtained a post at a Vienna Realschule, where he taught chemistry, physics, and algebra and used his free time for scientific research. At that time Vienna had become the center of crystallography, and Loschmidt concentrated his studies on the chemistry of crystals.[2] During this period he met J. Stefan, the young director of the Institute of Physics of the University of Vienna, and they became close friends. Stefan soon recognized Loschmidt's talents and offered him the facilities of the institute's laboratory and library, and Loschmidt, who had had enough of practical work, turned to theoretical research.

Various attempts were being made in those days to represent symbolically and graphically molecular formulas in analogy to the efforts at a unified and internally consistent system of atomic weights for the chemical elements. In 1861, at the age of forty, Loschmidt published in Vienna, at his own expense, his first scientific work, *Chemische Studien*, I (there never was a part II). Loschmidt was the first to represent graphically the double and triple bonds of polyvalent atoms by means of connecting lines. He also was probably the first who viewed cane sugar as an "ether-like compound," and he expressed the opinion that ozone consisted of three oxygen atoms. Above all, Loschmidt was the first to envision a ring-shaped chain formula of carbon atoms for benzene, but he expressly rejected the assumption of double bonds, preferring to hold such a decision in abeyance.

In his representation of benzene derivatives, for which he gave 121 graphical formulas, Loschmidt represented the benzene as a hexavalent nucleus by means of a circle, erroneously assuming that the carbon atoms and their valences were unevenly distributed about the circle. This model of the benzene nucleus impeded his understanding of the isomeric relationships in polysubstitution derivatives, whereas the strength of Kekulé's benzene model, by contrast, showed up at just this point. Nevertheless, Loschmidt correctly recognized toluene as methyl benzene, and thus accurately explained the isomerism of cresol with benzyl alcohol. He was the first to state that in alcohols with several OH groups each C atom can bind no more than one OH group.

Loschmidt also assumed that an element could have multiple valences (he used the terms "pollency" and "capacity"). Sulfur, for example, could have a valence of 2, 4, or 6. Thus, Loschmidt arrived at the proper structural formula for sulfuric acid, giving sulfur a valence of 6. According to him, nitrogen had a valence of 3 or 5, carbon always had a valence of 4, oxygen always had a valence of 2, and hydrogen of 1. Exceptions were NO, NO_2, and CO.

Loschmidt's book apparently had little influence on other chemists, except perhaps on Kekulé, who published his textbook *Lehrbuch der organischen Chemie* (Stuttgart) in different volumes between 1859 and 1887, and Crum-Brown, also an organic chemist who was involved with the representation of chemical formulas. Priority disputes and mutual influences among the three men are still open to investigation, but resolving the problem would require the analysis and evaluation of much published and unpublished material. Loschmidt's book remained largely ignored for over fifty years until it was republished in Ostwald's *Klassiker* in 1913,[3] with the subtitle *Constitutions-Formeln der organischen Chemie in graphischer Darstellung*.

Herbart's metaphysical speculations on the antinomies of the real continuum had led him to construct reality out of simple entities, which he considered spherical, and which attract and repel each other by partial penetration. Loschmidt, in *Zur Constitution des Aethers* (Vienna, 1862), extended this Herbartian metaphysics of elementary quanta into the realm of physics in line with Lamé's elasticity and ether theory. He attempted to formulate a mathematical theory for luminiferous ether very similar to that of Lord Kelvin thirty years later. According to Loschmidt, every molecule is surrounded with an envelope of ether, the density of which falls off inversely with distance from the molecule's center. Matter and ether attract each other; the ether particles repel each other. Ether is the carrier of energy exchange in the universe; and the attracting and repelling forces of ether spheres summarized under the term "affinity" are, in a manner of speaking, the souls of the atoms. All natural phenomena were to be derived from the reciprocal action of the atoms by means of their energy spheres. He contrasted the prevalent atomic theory—which ascribed to each atom a sharply limited nucleus of impenetrable matter—with the concept of an energy sphere surrounding the atom, itself perhaps a conglomerate of the smallest ether particles.

Loschmidt's other scientific achievement was the first accurate estimations of the size of air molecules.[4] This linked directly with the speculations of Clausius and Maxwell. Clausius assumed at the outset that all molecules of a gas are at rest and that a mass point moves through them.[5] The probability that this mass point penetrates a layer of thickness x without colliding with the molecules $= a^x$, where a is the probability for thickness 1. Thereby a is a function of the radius ρ of the molecular energy spheres and of the mean distance λ of nearby molecules. With $a = e^{-\alpha}$ (e being the base of the natural logarithms), α has to be determined. For small layer thicknesses δ there results $W_\delta = e^{-\alpha\delta} = 1 - \alpha\delta$. Furthermore, the probability that the point penetrates the layer is equal to the relationship between the superficial area of the layer not covered by the molecular cross sections to the total surface. For a layer of thickness λ the fraction $\pi\rho^2/\lambda^2$ of the total surface is covered with molecular cross sections. For a layer of thickness δ this must be multiplied by δ/λ, that is, $(\pi\rho^2/\lambda^3)\delta$. Thus, the probability that the point penetrates a layer of thickness δ is $W_\delta = 1 - (\pi\rho^2/\lambda^3)\delta$. Equating the two expressions for W_δ, $\alpha = \pi\rho^2/\lambda^3$. For any layer thickness x with this value for α the following formula for the penetration possibility is obtained:

$$W = e^{-\frac{\pi\rho^2}{\lambda^3}x}.$$

Accordingly, out of Z mass points the layer thickness $x + dx$ is penetrated by

$$Z \cdot e^{-\frac{\pi\rho^2}{\lambda^3}(x+dx)} = Z \cdot e^{-\frac{\pi\rho^2}{\lambda^3}x} \cdot \left(1 - \frac{\pi\rho^2}{\lambda^3}dx\right).$$

Accordingly, $Z \cdot e^{-(\pi\rho^2/\lambda^3)x} \cdot (\pi\rho^2/\lambda^3)\,dx$ points are retained in layer dx. Multiplication by x gives the total length of path of all these points together. Then, by integration from $x = 0$ to $x = \infty$ and division by Z, the mean free length of path to $l' = \lambda^3/\pi\rho^2$ is obtained for one point. In the case of molecules with radius ρ, instead of mass points, the value of this radius must be doubled—that is, replaced with diameter s. However, if there are no molecules at rest but all molecules move at the same mean speed, this expression must also be multiplied by 0.75, so that

$$l = \frac{3}{4}\frac{\lambda^3}{\pi s^2}.$$

The proof is as follows. If, according to Maxwell,[6] v represents the speed of a molecule and $1/\lambda^3 = N$, the number of molecules per volume unit, then the number of collisions per minute for a moving molecule, with the remaining molecules at rest, is:

$$v/l' = v \cdot \pi s^2 N,$$

with $s = 2\rho$. But if the other molecules are also moving, then the v on the right side of the equation must be replaced by the mean speed r of the molecule under consideration relative to the others. In that case, one obtains for the number of collisions per minute of the molecule under consideration: $v/l = r \cdot \pi s^2 N$. Thus

$$v/l : v/l' = v : r, \qquad l' : l = v : r.$$

Now, according to Clausius, the relative speed between one molecule moving at speed v and another at speed u is: $\sqrt{u^2 + v^2 - 2uv\cos\theta}$, if θ is the angle between the directions of movement of the two molecules.[7] Since all angles occur with equal frequency, the number of those molecules whose lines of motion make angles between θ and $\theta + d\theta$ with the line in which the first molecule moves will have the same ratio to the whole number of molecules as the corresponding spherical zone to the total spherical surface—that is, as $2\pi \cdot \sin\theta \cdot d\theta : 4\pi$. Accordingly, the number of these molecules per volume unit is $N \cdot 1/2 \sin\theta \cdot d\theta$. Thus, the mean speed r of the particular molecule relative to the other molecules in motion is

$$r = \frac{1}{2}\int_0^\pi \sqrt{u^2 + v^2 - 2uv\cos\theta} \cdot \sin\theta d\theta.$$

On the average, $u = v$; and in this case the pertinent integral is $r = 4v/3$, *q.e.d.*

Thus $1/l = \frac{4}{3}\pi s^2 N$. As Loschmidt showed, estimates of the relative size of the gas molecules can be obtained from this equation. He initially transformed it to some extent:

$$\frac{1}{N} = \frac{16}{3} \cdot \frac{\pi l s^2}{4}.$$

$1/N$ is that part of the volume unit assigned to one molecule, if the molecules are considered to be evenly distributed. Loschmidt termed it the "molecular gas volume."

Since, according to Avogadro's principle, N is equal for all gases, the same applies to $1/N$. On the right side of the equation, $\pi l s^2/4$ indicates the cylindrical path traveled by the molecule covering the distance l, that is, the "molecular distance volume." Accordingly, the equation states that the latter quantity also is the same for all gases. $\pi s^3/6$ is the volume of a molecule. Its Nfold is that part of the unit of volume which is taken up by the molecules assumed at rest; its ratio to the unit is designated as the condensation coefficient ϵ of the gas. Consequently, the above equation results in $s = 8\epsilon l$. The mean free length of path l for air had already been calculated by Maxwell and O. E. Meyer. The condensation coefficient ϵ can be approximated from increases and decreases in volume due to evaporation and condensation, if the plausible assumption is made that in a liquefied state the action spheres of the molecules almost touch. The regularities indicated by Hermann Kopp also permit an estimation of the condensation coefficient of air from those of condensable gases. With these estimated values for l and ϵ, Loschmidt obtained from the last equation the molecule diameter $s = 8 \times 0.000866 \times 0.000140 = 0.000000970$ mm—that is, the correct size of somewhat less than 10^{-7} cm. He also stressed that this was to be considered only as a rough approximation, although this value certainly was not ten times too large or too small.

The value of 0.000140 mm for l is that given by Meyer. Loschmidt felt that this value was preferable to the older value of 0.000062 given by Maxwell. But had Loschmidt preferred Maxwell's value, he would have obtained for the molecular diameter the more accurate value of $s = 0.0000004295$ mm, that is, approximately 0.5×10^{-7} cm. Thus he could have calculated directly from his first formula a likewise accurate value of $2 \cdot 10^{19}$ molecules per cubic centimeter of gas. But Loschmidt calculated the number of molecules in living organisms, as 13.10^{19}. For the number of gas molecules, in his lecture "Die Weltanschauung der modernen Naturwissenschaft" (1867)[8a], he indicated—generally without basis, although referring to his work of 1865—the figure of 866 billion molecules per cubic millimeter, which is about thirty times too small. He recommends 10^{-18} of a milligram as a suitable unit for atomic weights, which is about 100 or 1,000 times too large. (According to modern measurements, a hydrogen atom weighs 1.67×10^{-24}g.) In conclusion, Loschmidt mentioned the hypothesis that atoms consist of even smaller particles and raised the question of whether his hypothetical ether envelopes could be used to explain the life processes.

Three years after Loschmidt, G. Johnstone Stoney attempted to compute the number of molecules per unit volume on the basis of Clausius' rough estimates. Thereby he obtained a value 100 times too large.[8] In 1870, William Thomson (Lord Kelvin), unaware of Loschmidt's work, published in *Nature* (issue of 31 March, pp. 551 ff.) a fundamental work in which he calculated the sizes of atoms by various methods. Initially he stated that as early as thirty years previously Cauchy had frightened natural scientists with the daring statement that the well-known prism colors indicated that the sphere of molecular action was comparable with the wavelength of light. Eight years previously he himself had published notes on experiments with electrically charged, mutually attractive thin copper and zinc membranes, which enabled him—through a method also outlined there in principle—to estimate a magnitude of $3.3 \cdot 10^{-9}$ cm for atoms.[9] He also obtained from experiments with liquid lamellae a molecular diameter of $\geqslant 5 \cdot 10^{-9}$ cm. From the kinetic theory of gases he calculated, in principle by the same method as Loschmidt, a molecular size not below $2 \cdot 10^{-9}$ cm and, for the number of molecules per cubic centimeter, $\leqslant 6 \cdot 10^{21}$. Thus the values obtained by Thomson were ten times too small for the molecular diameters and 100 times too large for the number of molecules.

Later, Thomson himself acknowledged Loschmidt's priority. In 1890, Lord Rayleigh directed attention to a forgotten work by Thomas Young, printed half a century before Loschmidt's, in the *Supplement* to the fourth edition (1810) of the *Encyclopaedia Britannica*.[10] In it Young had estimated the range of molecular attraction from the surface tension of water, thus obtaining a value between 10^{-10} and $5 \cdot 10^{-10}$ inch for the size of water molecules. Furthermore, from the ratio between water density and steam density he obtained the value 10^{-8} cm for the distance between steam particles. These figures are about 100 times too small.

To date, many methods have been invented for determining these figures. With each method, roughly the same values are obtained: about $0.5 \cdot 10^{-7}$ cm for the molecular diameter. For the distance between molecules the figure was about $3 \cdot 10^{-7}$ cm and, for the number of molecules per cubic centimeter of gas at 0° C. and 760 mm mercury, $2.7 \cdot 10^{19}$. Thus, one mole of any gas—22.414 liters—contains $6.03 \cdot 10^{23}$ molecules. This figure is called Loschmidt's figure or, sometimes, Avogadro's constant, although Avogadro never performed any pertinent numerical calculations.

In 1868 Loschmidt was appointed to the new position of associate professor of physical chemistry at the University of Vienna, and he received an honorary Ph.D. several months later. He then received modest grants for experimentation and investigated the diffusion of gas in the absence of a porous membrane.[11] This involved the experimental confirmation of certain conformities (diffusion speeds) already theoretically derived by Maxwell from the kinetic gas theory, and a more accurate determination of the mean length of path from diffusion instead of from interior friction, as done previously.

Loschmidt also worked on the electromagnetic wave theory and even attempted to demonstrate experimentally the Kerr and Hall effects, although without success, because of the inadequate equipment. Also, as Hertz did later, he worked on the production of electrical resonances and came close to inventing the dynamo. Jokingly, he proposed the founding of a Viennese journal for unsuccessful experiments.

Loschmidt also continued his theoretical work. In the course of a controversy with Boltzmann, he sought a way to escape the heat death resulting from the kinetic theory ("reverse argument"), although without success.[12] These speculations, however, resulted in Loschmidt's first application of the second law of thermodynamics to the theory of solutions and chemical compounds.[13] Thus he became a forerunner of Horstmann and Gibbs. This also led to a modification of Maxwell's homogeneous-distribution axiom in the case of perceptible gravity effect.

Loschmidt let gases escape into a vacuum in order to observe the effects on their temperature (in accordance with the kinetic theory, the temperature must not change for ideal gases). He also speculated on the manner of propagation of sound in air: "Deduktion der Schallgeschwindigkeit aus der kinetischen Gastheorie."[14] He gave a simpler derivation of the equation of a point system.[15] He attempted to derive the Weber-Ampère law from that of Coulomb, and, in accordance with Kirchhoff, to derive Ohm's law from hydrodynamic flow laws, analogous to Poiseuille's law.[16] Finally, he attempted to calculate, on the basis of Lamé's elasticity theory and his own atomistic concept, the existence of spectral lines from the vibrations of ether spheres surrounding the atoms.[17]

In 1867 Loschmidt was named corresponding member of the Imperial Academy of Sciences in

Vienna, and in 1870 a full member. In 1869 he and Josef Stefan founded the Chemical-Physical Society. Loschmidt became the first director of the Physical-Chemical Laboratory (today the Second Institute of Physics). He was dean of the Philosophical Faculty in 1877–1878 and acting dean in 1878–1879. In 1887 he married his housekeeper; they had a son who died of scarlet fever at the age of ten. Loschmidt retired in 1890 and was decorated with the Order of the Iron Crown, third class. He turned over his institute to his pupil Franz Serafin Exner, the son of his professor. Other important pupils were Gustav Jäger and Ludwig Boltzmann.

NOTES

1. Compare Gicklhorn, in *Sudhoffs Archiv für Geschichte der Medizin und der Naturwissenschaften*, **32** (1939–1940), 337–370; and Donald, in *Annals of Science*, **1** (1936), 29–47.
2. See *Sitzungsberichte der Akademie der Wissenschaften in Wien*, Math.-natwiss. Kl., **51**, Abt. II (1865), and **52**, Abt. II (1866).
3. The problem is treated in detail by the organic chemist and textbook ed. and reviser, R. Anschütz, in *August Kekulé*, 2 vols. (Berlin, 1929), especially pp. 271–306. Also see his "Anmerkungen" in *Ostwalds Klassikern der exakten Wissenschaften* no. 190 (Leipzig, 1913); and *Chemische Studien* in the same ed. for Loschmidt's formulas.
4. "Zur Grosse der Luftmolekule," in *Sitzungsberichte der Akademie der Wissenschaften zu Wien*, Math.-natwiss. Kl., **52**, Abt. II (1866), 395–413.
5. *Philosophical Magazine*, 4th ser., **17** (1859), 81–91.
6. *Ibid.*, **19** (1860), 19–32.
7. *Ibid.*, pp. 434–436.
8. *Ibid.*, **36** (1868), 132–141.

8a. In *Schriften des Vereins zur Verbreitung naturwissenschaftlicher Kenntnisse in Wien*, **8** (1867–1868), 41–106, esp. 53.

9. William Thomson (Lord Kelvin), *Reprint of Papers on Electrostatics and Magnetism* (London, 1972), 400.
10. *Philosophical Magazine*, 5th ser., **30** (1890), 474.
11. "Diffusion von Gasen ohne poröse Scheidewände," in *Sitzungsberichte der Akademie der Wissenschaften in Wien*, Math.-natwiss. Kl., **61**, Abt. II (1870), 367–380; **62**, Abt. II (1870), 468–478.
12. *Sitzungsberichte der Akademie der Wissenschaften in Wien*, Math.-natwiss. Kl., **73**, Abt. II (1876), 128–142, 366–372; **75**, Abt. II (1877), 287–288; **76**, Abt. II (1877), 209–225.
13. *Ibid.*, **59**, Abt. II (1869), 395–418.
14. *Ibid.*, **54**, Abt. II (1866), 646–666.
15. *Ibid.*, **55**, Abt. II (1867), 523–538.
16. *Ibid.*, **58**, Abt. II (1868), 7–14.
17. *Ibid.*, **93**, Abt. II (1886), 434–446.

BIBLIOGRAPHY

Loschmidt's works are cited in the text of the article. A secondary source is Hubert de Martin, "Johann Joseph Loschmidt" (Vienna, 1948), a dissertation in typescript, poorly written but thorough and with a complete bibliography.

WALTER BÖHM

LOSSEN, KARL AUGUST (*b.* Kreuznach, Germany, 5 January 1841; *d.* Berlin, Germany, 24 February 1893), *geology, petrology.*

Lossen was the son of a doctor and his wife Charlotte Mayer. He was educated at the high school and Gymnasium in Kreuznach. He married his cousin Therese Lossen and had two daughters and one son. Lossen began work as a mining engineer in the siderite deposits of the Siegerland in Westphalia and in the coal mines of Saarbrücken. He broadened his studies in geology, mineralogy, and microscopic petrography, which was then just developing, under the guidance of Beyrich, Girard, Rammelsberg, G. Rose, and J. Roth. In 1866, after graduating from the University of Halle, he was introduced by Dechen at the Geological Survey of Prussia, and started working under the direction of Hauchecorne and Beyrich.

Lossen's main work was the mapping and description of the very complicated geology of the Harz Mountains, which were then still rather unknown, but which have since, because of Lossen's work, become one of the classic regions for geological study. He worked there part of every year from 1866 to 1892 and completed a geological map of the area in the scale of 1 : 100,000, and, incidentally, maps of smaller areas in the scale of 1 : 25,000 as by-products. He produced significant papers on the Devonian period, especially the Lower Devonian, and the early Carboniferous period in that region, which had previously been mistaken for Silurian.

Lossen also did essential work in microscopical petrology, and with this, along with geological observations in the field, became one of the first to report the influence of tectonic movements on metamorphism (dynamometamorphism).

By 1866 Lossen could show that gneissic and phyllitic rocks of the Hunsrück Mountains were Lower Devonian sediments. He discovered similar things in the Ostharz, where he also observed transitions between contact metamorphic and dynamometamorphic rocks. For metamorphic tuffs he coined the name "porphyroid."

In 1870 Lossen became a lecturer at the University of Berlin. He was appointed an associate professor in 1882 and full professor of petrology in 1886. While in Berlin, he redefined the old name "Hercynian," which he then used only for the lower part of the Lower Devonian.

Many of Lossen's works concern the magmatic rocks of the Saar-Nahe basin. From these were drawn numerous definitions, in part now completely misused, especially of rocks of the melaphyre-porphyrite-tholeite family.

BIBLIOGRAPHY

A list of Lossen's publications may be found in the obituary by E. Kayser in *Neues Jahrbuch fur Mineralogie, Geologie und Paläontologie*, **2** (1893), App. 1–18.

P. RAMDOHR

LOTKA, ALFRED JAMES (*b.* Lemberg, Austria [now Lvov, Ukrainian S.S.R.], 2 March 1880; *d.* Red Bank, New Jersey, 5 December 1949), *demography, statistics.*

Born of American parents, Lotka received his primary education in France and Germany, took science degrees at the University of Birmingham, England, and subsequently attended graduate courses in physics at Leipzig and Cornell. At various times he worked for the General Chemical Company, the U.S. Patent Office, the National Bureau of Standards, and from 1911 to 1914 was editor of the *Scientific American Supplement*. In 1924, after having spent two years at Johns Hopkins University composing his magnum opus, *Elements of Physical Biology*, he joined the statistical bureau of the Metropolitan Life Insurance Company in New York. There he spent the remainder of his working life (eventually becoming assistant statistician). In 1935 he married Romola Beattie; they had no children. During 1938–1939 he was president of the Population Association of America, and in 1942 he was president of the American Statistical Association.

Lotka's cardinal interests lay in the dynamics of biological populations, and this stemmed from his being struck, as a young physicist, by the similarities between chemical autocatalysis and the proliferation of organisms under specified conditions. He soon developed an "analytical" theory of population that was a function of birthrate, deathrate, and age distribution. His model was more realistic than earlier models; nevertheless he was admirably cautious about its predictive value. He also paid attention to interspecies competition. Building on some mathematical work of Volterra's (in 1926), he formulated a growth law for two competing populations, expressing the process in terms of two interlocking differential equations.

Lotka took a broad culture-conscious view of the quantification of biological change and devoted much thought to the ecologic influence of industrial man. The *Elements of Physical Biology* is an ambitious work, treating the whole biological world from a mathematico-physical standpoint. Later, during his service in the insurance business, he collaborated in writing about life expectancy and allied topics.

BIBLIOGRAPHY

Lotka's most important work, *Elements of Physical Biology* (Baltimore, 1925), was reissued with the title *Elements of Mathematical Biology* (New York, 1956). This ed. carries a full 94-item bibliography of Lotka's scientific and technical papers. With Louis I. Dublin he published three books: *The Money Value of a Man* (New York, 1930); *Length of Life* (New York, 1936); and *Twenty-Five Years of Health Progress* (New York, 1937). The most thorough exposition of his views on population change and evolution is to be found in *Théorie analytique des associations biologiques* (Paris, 1934; 1939), an English trans. of which he was working on at the time of his death. Almost all modern books on population theory mention Lotka's contributions.

NORMAN T. GRIDGEMAN

LOTSY, JAN PAULUS (*b.* Dordrecht, Netherlands, 11 April 1867; *d.* Voorburg, Netherlands, 17 November 1931), *botany.*

Born into a patrician family, Lotsy devoted himself to biological research against the wishes of his father. He studied botany with Beyerinck at the Wageningen Agricultural College and continued his studies at the University of Göttingen, from 1886 to 1890. His doctoral thesis was on German lichens. From 1890 to 1895 he was in the United States as a lecturer at Johns Hopkins University. He showed his versatility in working on the food supply of the adult oyster, the nitrogen assimilation of mustard, a study on the root formation of the swamp cypress, toxic substances of the pear-blight bacteria, the staining of diatoms, the fixing of cells of red algae, the significance of herbaria for botany, and a taxonomical revision of some Euphorbiaceae from Guatemala.

From 1895 to 1900 Lotsy was in Java, where he worked on the localization, physiology, and secretion of quinine in Cinchona species at the Cinchona Experiment Station (for which he received a gold medal). He was later invited by Treub to work in the Botanic Gardens at Buitenzorg (Bogor) and the mountain garden connected with it at Tjibodas. While there, he wrote a basic work on the life history of the genus *Gnetum*, the detailed anatomy and taxonomy of some Balanophoraceae, and sketches of Javanese forest plants. After an attack of malaria he returned to the Netherlands and from 1901 on lived at Leiden.

On his initiative the Association Internationale de Botanistes was founded, for which he edited the journal *Progressus rei botanicae*, which was comparable in scope to the present *Botanical Review*. This association also bought the *Botanisches Centralblatt* (1902), of which Lotsy was also the editorial secretary. In 1904 he was appointed lecturer of plant systematics at Leiden University and in 1906, in addition, director of the Rijksherbarium. He held these posts until 1909, when he entered a new phase. This was a time of great ferment and reform in biological research owing to the rediscovery of Mendel's laws, the publication of De Vries' mutation theory, the recognition of the role of chromosomes, and the close study of haploid and diploid generations in cormophytes. He viewed systematics in a wide sense, and thought that it should be combined with genetics to produce a phylogeny based on evolution. As early as 1903 Lotsy had given a lecture on the implications of the mutation theory, genetics, and hybridization for plant breeding. For teaching he composed in an incredibly short time two major handbooks—lectures on descent theories (1906–1908) and lectures on botanical phylogeny (1907–1911)—which were frequently used for university lectures.

After concluding that knowledge of dicotyledons was insufficient to establish their ancestry, Lotsy switched to a third phase of scientific research, the experimental approach to the mechanism of evolution. At Leiden he wanted a new herbarium with a garden attached for experimental taxonomy. Unfortunately the House of Commons rejected his plan, partly because of the opposition of De Vries, who saw it as a competitive threat. Lotsy and his close friend J. W. C. Goethart then offered to construct the garden from private means, but to no avail. Disgusted by this obstruction he resigned in 1909 from his official functions and with his own funds erected experimental gardens at Haarlem and later at Velp.

About 1912 Lotsy became convinced that species originated and evolved because of hybridization, and the development and advocacy of this theory occupied him for the rest of his life. He defined more precisely the concepts *jordanon*, *linneon*, and *syngameon* to tie systematics to genetics, and thus became a pioneer of modern research. Possibly he and his close collaborator Goethart obtained experimentally allopolyploid new species of *Antirrhinum* and *Scrophularia*. Though the creation of an official center of experimental taxonomy in the Netherlands was frustrated, Lotsy created a forum for genetics with the journal *Genetica* (1919), later adding *Resumptio genetica* and *Bibliographia genetica*. In the years 1920 to 1930 he traveled to North America, New Zealand, South Africa, and Egypt to disseminate his ideas and to test his theories, thus stimulating much research in experimental taxonomy.

Lotsy's writings were prolix and vague. His alternative hybridization theory and equally untenable ideas found in his writings—*Homo sapiens* as a genus; linneons are homozygous—arose largely as a critical response to De Vries' mutation theory. Although Lotsy was not held in high esteem in the Netherlands, abroad he had many friends—William Bateson, Erwin Baur, Heribert Nilsson, Karl von Goebel, Richard von Wettstein, Leonard Cockayne—who admired him for his many-sided initiatives, energy, and stimulating spirit.

BIBLIOGRAPHY

There are 135 books and papers of Lotsy listed by W. A. Goddijn in his "In Memoriam," in *Genetica*, **13** (1931), i–xx. Lotsy's major works are: *Vorlesungen über Deszendenztheorien*, 2 vols. (Jena, 1906; 1908); *Vorträge über botanische Stammesgeschichte*, 3 vols. (Jena, 1907–1911); *Van den Atlantischen Oceaan naar de Stille Zuidzee in 1922* ('s Gravenhage, 1922); and "Voyages of Exploration to Judge of the Bearing of Hybridisation Upon Evolution. I. South Africa," in *Genetica*, **10** (1928), 1–315, written with W. A. Goddijn. Also important are his shorter works: *De kruisingstheorie, een nieuwe theorie over het ontstaan der soorten* (Leiden, 1914); *Evolution by Means of Hybridisation* ('s Gravenhage, 1916); *Over Oenothera lamarckiana als type van een nieuwe groep van organismen, die der kernchimeren* ('s Gravenhage, 1917); *De wereldbeschouwing van een natuuronderzoeker* ('s Gravenhage, 1917); *Het evolutievraagstuk* ('s Gravenhage, 1921); *Evolution Considered in the Light of Hybridisation* (n.p., 1925); "Versuche über Artbastarde und Betrachtungen über die Möglichkeit einer Evolution trotz Artbeständigkeit," in *Zeitschrift für induktive Abstammungs-Vererbungslehre*, **8** (1912), 325–333; "La théorie de croisement," in *Archives néerlandaises des sciences exactes et naturelles*, ser. IIIB, **2** (1914), 1–61; "Qu'est ce qu'une espèce," in *Archives néerlandaises des sciences exactes et naturelles*, sér. IIIB, **3** (1916), 57–110; "Evolution im Lichte der Bastardierung betrachtet," in *Genetica*, **7** (1925), 365–470; "Kreuzung und Deszendenz," in *Ber. Bot. Ges. Zürich 1924/26* (1926), 16–47; and the Cawthorn Lecture: "A Popular Account of Evolution" (Nelson, New Zealand, 1927), 1–22.

C. G. G. J. VAN STEENIS

LOTZE, HERMANN RUDOLPH (*b.* Bautzen, Germany, 21 May 1817; *d.* Berlin, Germany, 1 July 1881), *theoretical biology, metaphysics.*

Lotze was the son of the military doctor Carl Friedrich Lotze and Christiane Caroline Noak of Dresden. He had an older sister and brother, Natalie and Carl Robert. He spent his Gymnasium years in Zittau from 1824 to 1834 and at seventeen graduated first in his class. In the summer semester of 1834 he began to study philosophy and natural science at the University of Leipzig. C. H. Weisse, an idealistic and Hegelian philosopher, made a strong impression at first on Lotze's thinking. A counterbalance was provided by his scientific studies, for example, of physiology under E. H. Weber, of anatomy under A. W. Volkmann, and of physics under Fechner. He completed his philosophical studies with a work in French on Descartes and Leibniz, for which he received his Ph.D. on 1 March 1838. The ideas of Leibniz, moreover, were influential in his own philosophy.

Lotze earned his M.D. on 7 July 1838 with the dissertation *De futuris biologiae principiis philosophicis.* This youthful work contained the essence of his later philosophical system. In it he required that philosophy take careful account of scientific knowledge and reciprocally that physics and the life sciences subject their basic principles to the purifying scrutiny of contemporary philosophical analysis. In particular, he ruled out explanations that resorted to the concept of a life force, although the validity of this principle was then largely unquestioned, and he resolved the mind-body problem by a combination of Leibnizian monadology with the older occasionalism.

After Lotze had practiced medicine in Zittau for about a year (1838–1839), he qualified at Leipzig to lecture in medicine in the fall of 1839 and in philosophy in May 1840. With this combination he hoped to be fully equipped to consider the basic questions of biology, medicine, and anthropology. In that same year (1840) he published a volume of idealistic poems, which were contemplative yet lyrical and dealt with basic religious and aesthetic questions. In the period 1840 to 1844 he lectured on a broad field of subjects—general pathology, anthropology, philosophy of medicine, psychology, logic, and theological medicine—which he attempted to unify in his later works. In 1843 he became an assistant professor of philosophy in Leipzig, and in 1844, at the age of twenty-seven, he obtained, on the recommendation of the anatomist Rudolph Wagner, a full professorship in philosophy in Göttingen, there succeeding the famous Herbart.

In September 1844 Lotze married Ferdinande Hoffmann, a pastor's daughter; they had four sons. In the following decades Lotze devoted all his energies to his extensive scientific work. He maintained friendly relations with Wagner, and especially with members of the Freitagsverein, including the ophthalmologist Carl Ruete, the surgeon Baum, Carl Stumpf, and Karl Ewald Hasse. He remained in Göttingen for thirty-seven years, refusing a call to return to the University of Leipzig in 1859. Finally in April 1881, at the age of sixty-four, he accepted a new offer, this time for the chair of philosophy at Berlin. A few weeks later, on 1 July 1881, he died in Berlin of pneumonia.

Lotze sought to classify nature, soul, mind, history, and culture in a great unifying concept, and to do so, moreover, with careful consideration of the great advances in the inorganic sciences, biology, and medicine, as well as in psychology. His principal writings appeared in rapid succession beginning in 1841 and reached their high point in his comprehensive three-volume work *Mikrokosmus* (1856–1864). This work is an architectonic one, and forms a kind of keystone joining Herder's *Ideen zur Philosophie der Menschheit* (1784–1791) and Humboldt's *Kosmos.* Here too Lotze displays his intention of explaining the nature of microcosm and macrocosm philosophically through the concepts of soul, mind, and culture, thus creating a philosophic-medical anthropology.

In his *Metaphysik* of 1841 Lotze attempted for the first time, by completely accepting a mechanistic causality in the organic as well as the inorganic world, to inquire into the real meaning and teleology of the cosmos, and therefore into the governing foundation of all phenomena. He concluded that mechanisms (natural laws) are the means by which the immanent goals of the universe are realized.

The relations between things can be causes, grounds, or purposes. Not everything is determined solely by causes. In the *Allgemeine Pathologie* (1842) he teaches that living things are subject to the same natural laws as inanimate nature; only the arrangement of the causally determined parts is different. There is no life force and no natural healing power, yet all life processes function harmoniously as a means toward the goal of survival. Between body and mind there exists a reciprocal cooperation.

In the history of physiology and the interpretation of living matter, Lotze's essay "Leben, Lebenskraft" (1843) played an important role, for no one up to that time had so penetratingly demonstrated the senselessness of the then generally accepted concept of a life force. To him a life force as a single cause of a multitude of consequences was unimaginable. How could it activate or alter the extraordinary number of the most varied mechanisms? The organism is simply a combination of mechanical processes that are in harmony with natural goals. The vital properties, or "life force," arise only from the concatenation of the mechanical properties of the entire organism. Nonetheless, each individual organism emerges not through chance, but through creation. The union of mind and body indicates a higher purposefulness, and the communication between body and soul occurs by means of a psychophysical mechanism with a preestablished occasionalistic harmony. Thus emerges a philosophical physiology which seriously investigates the scientific relation of causality but at the same time only briefly touches upon but does not develop the closely related question of goals.

In the article "Seele und Seelenleben" (1846) Lotze emphasized the limits of mechanism in an organism and discussed the three essential aspects of the concept of the mind: consciousness, unity of experience, and freedom. These go beyond the organic and constitute the separate domain of ethics and values.

In 1851 he discussed again, in a more profound manner, the subject of the interpretation of existence in *Allgemeine Physiologie des körperlichen Lebens.* He stated that we must recognize a kind of ensoulment in all things and repeated the justification of teleological views alongside of mechanical conceptions. He again dismisses vitalism with the observation that irritability of organisms is no "life" force but a phenomenon explaining nothing, which must itself be explained. In the system of ordered existence the means and forces are organized in a particular way characteristic of each organism. The *Medizinische Psychologie* (1852) continues the investigation of 1846 on the limits of mechanism. Lotze had by then assumed a completely spiritualistic view, believing that the central focus of the world was spiritual, with matter secondary and dependent. The mind is independent of all material events; it influences the body, but is not influenced by it. He also considers the fate of the soul and believes that immortality is granted only to the most distinguished souls.

Lotze's three-volume masterpiece, the *Mikrokosmus* (1856–1864), presents everything effectively and skillfully in a unified survey. His structure arches from inanimate nature, through animate, to the mind and to man, and encompasses man's history and culture. Lotze's fundamental idea remains the same, namely, that throughout the inanimate, animate, and spiritual worlds, hidden purposes are active, of which science, with its analysis of causality, investigates and explains only the instrumentation of the causal relations. But the key to the understanding of the world is to be sought in ideas and values, not in these mechanical processes. Even the formation and development of a seed are supramechanical and determined by the immanence of an infinite being.

Lotze, however, had found it necessary to publish *Streitschriften* (1857) in order to offset the impressions that he was a materialist, a misconception which arose from an imperfect understanding of his anti-vitalistic stance. His writings received much attention from the natural scientists of the nineteenth century, especially because of his attack on vitalism and his disavowal of the philosophical uses of deductive reasoning and the dialectic method. This break with tradition stands at the beginning of the debate that was carried on particularly by the students of Johannes Müller, namely Schwann, Emil du Bois-Reymond, Helmholtz, Brücke, and also Ludwig, and which was concluded with a definitive victory about 1850.

Lotze was a small, gaunt man, and exceptionally taciturn. He was very formal, stiff, and inclined to melancholy. His style of lecturing was plain, even dull, and lacking high points, but very precise, logical, and convincing. His industry is evident from his abundant writings. The anti-materialists of the nineteenth century saw in him the defender of a view of the world nobler and more beautiful than that offered by Carl Vogt, Moleschott, Friedrich Büchner, and the monists. Lotze saw his goal as uniting in a consistent fashion the results of scientific research with an ethical and religious world view, and in demonstrating a harmony between natural laws and the world of values. Hence his philosophy bordered on theosophy. As A. Krohn said in his article on Lotze, "He sought in that which should be, the basis of that which is."

BIBLIOGRAPHY

I. Original Works. Lotze's original works are: *Metaphysik* (Leipzig, 1841; Paris, 1883); *Allgemeine Pathologie und Therapie als mechanische Naturwissenschaften* (Leipzig, 1842; 2nd ed., 1848); *Allgemeine Physiologie des körperlichen Lebens* (Leipzig, 1851); *Medizinische Psychologie oder Physiologie der Seele* (Leipzig, 1852); *Mikrokosmus. Ideen zur Naturgeschichte und Geschichte der Menschheit. Versuch einer Anthropologie*, 3 vols. (Leipzig, 1856–1864); *System der Philosophie*, 2 vols. (Leipzig, 1874–1879); and *Geschichte der deutschen Philosophie seit Kant. Dictate aus den Vorlesungen* (Leipzig, 1882).

Three important articles Lotze wrote are: "Leben und Lebenskraft," in R. Wagner, ed., *Handwörterbuch der Physiologie*, **1** (Brunswick, 1842), ix–lviii; "Instinkt," *ibid.*, **2** (1844), 191–209; and "Seele und Seelenleben," *ibid.*, **3** (1846), 142–264. Lotze's shorter works were edited by D. Peipers, *Kleine Schriften*, 3 vols. (Leipzig, 1885–1891).

II. Secondary Literature. Works about Lotze and his works are: R. Falkenberg, *H. Lotze. Teil I. Leben und Entstehung der Schriften nach den Briefen* (Stuttgart, 1906); A. Krohn, "Zur Erinnerung an Hermann Lotze," in *Zeitschrift für Philosophie und Philosophische Kritik*, n.s. **80** (1882), 56–93; L. Seibert, *Lotze als Anthropologe* (Wiesbaden, 1900); M. Wentscher, *Hermann Lotze. I. Band. Lotzes Leben und Werke* (Heidelberg, 1913); E. Wentscher, *Das Kausalproblem in Lotzes Philosophie* (Halle, 1903); and S. Witkowski, "Über den Zusammenhang von Lotzes medizinisch-physiologischer Anschauung mit seiner Auffassung von Entstehen und Fortleben der Seele," Ph.D. diss., Giessen University, 1924.

Short bibliographical sketches of Lotze may be found in: W. G. M. Haberling, F. Hubotter, H. Vierordt, eds., *Biographisches Lexikon der hervorragenden Ärzte aller Zeiten und Völker*, **3** (Berlin–Vienna, 1931), 846–847; *Meyers Grosses Konversations-Lexikon*, 6th ed., **12** (Leipzig–Vienna, 1909), 737–738; and *Allgemeine Deutsche Biographie*, **52** (Leipzig, 1906), 93–97.

K. E. ROTHSCHUH

LOVE, AUGUSTUS EDWARD HOUGH (*b.* Weston-super-Mare, England, 17 April 1863; *d.* Oxford, England, 5 June 1940), *applied mathematics, geophysics.*

Love was one of four children and the second son of John Henry Love, a surgeon of Somersetshire. He was educated at Wolverhampton Grammar School, and his subsequent career owed much to his mathematical master, the Reverend Henry Williams.

He entered St. John's College, Cambridge, in 1882. He was a fellow of St. John's College from 1886 to 1889 and held the Sedleian chair of natural philosophy at Oxford from 1899 on. He was elected a fellow of the Royal Society of London in 1894. Love was secretary of the London Mathematical Society for fifteen years and president in 1912–1913. He was noted as a quiet, unassuming, brilliant scholar, with a logical and superbly tidy mind. He liked traveling, was interested in music, and played croquet. He never married; a sister, Blanche, kept house for him.

Love's principal research interests were the theory of deformable media, both fluid and solid, and theoretical geophysics. He also contributed to the theory of electric waves and ballistics, and published books on theoretical mechanics and the calculus.

Love's first great work, *A Treatise on the Mathematical Theory of Elasticity*, appeared in two volumes in 1892–1893. A second edition, largely rewritten, appeared in 1906 and was followed by further editions in 1920 and 1927. This treatise, translated into several foreign languages, served as the world's standard source on the subject for nearly half a century. It is a masterpiece of exposition and stands as a classic in the literature of mathematical physics. It continues to be much referred to by workers in the field.

While Love's contribution to the pure theory of elasticity rests principally on his expository powers, his excursions into theoretical geophysics led to far-reaching discoveries about the structure of the earth. His second work, *Some Problems of Geodynamics*, won the Adams Prize at Cambridge in 1911. The work includes contributions on isostasy, tides of the solid earth, variation of latitude, effects of compressibility in the earth, gravitational instability, and the vibrations of a compressible planet. Many of his contributions are basic in current geophysical research, especially Love waves and Love's numbers, the latter being key numbers in tidal theory.

Developing the theory of Love waves was probably his greatest contribution. Formal theory on the transmission of primary (*P*) and secondary (*S*) waves in the interior of an elastic body had been worked out by Poisson and Stokes (1830–1850). In 1885 Rayleigh had shown that waves (Rayleigh waves) could be transmitted over the surface of an elastic solid. Rayleigh's theory concerned a semi-infinite, uniform, perfectly elastic, isotropic solid with an infinite plane boundary over which the waves travel.

According to Rayleigh the only permissible surface waves under these conditions are polarized so that the *SH* component of the particle motions is absent; this is the component which lies in the plane of the surface and is at right angles to the direction of wave advance. A second property of the waves is that for any general initial disturbance they advance unchanged in form: there is no dispersion—no spreading out into sine wave constituents over time.

When surface seismic waves were first detected in studies of earthquake records (some time after 1900), they were found to be discordant with the above two properties of Rayleigh waves. Love set out to investigate a suggestion that the earth's crust is responsible for the discordances. He examined a mathematical model consisting of Rayleigh's uniform medium overlain by a uniform layer of distinct elastic properties and density, and found that this model both permits the transmission of *SH* waves and requires the waves to be dispersed. These waves are now called Love waves. Love was thus the first to satisfy the general observational requirements of seismology with respect to surface waves.

Love's analysis also supplied a relation between periods and group velocities of surface waves, which became a powerful tool in estimating crustal thicknesses in various geographical regions of the earth; it led *inter alia* to the first evidence of the large differences in crustal structures below continents and oceans.

BIBLIOGRAPHY

Love published forty-five research papers over the period 1887–1929. A full list is given in the *Obituary Notices of Fellows of the Royal Society*, **3** (1939–1941), 480–482.

Love's books are *A Treatise on the Mathematical Theory of Elasticity*, 2 vols. (Cambridge, 1892–1893; 2nd ed., 1906; 3rd ed., 1920; 4th ed., 1927); *Theoretical Mechanics* (Cambridge, 1897; 2nd ed., 1906; 3rd ed., 1921); *Elements of the Differential and Integral Calculus* (Cambridge, 1909); *Some Problems of Geodynamics* (Cambridge, 1911).

K. E. BULLEN

LOVEJOY, ARTHUR ONCKEN (*b.* Berlin, Germany, 10 October 1873; *d.* Baltimore, Maryland, 30 December 1962), *epistemology, history of ideas.*

Lovejoy was the son of an American father, the Reverend W. W. Lovejoy, and a German mother, Sara Oncken. He was educated at the University of California (Berkeley) and did graduate work at Harvard, where he came mainly under the influence of William James. After teaching for brief periods at Stanford, Washington University, and the University of Missouri, he went to Johns Hopkins in 1910 and remained there until his retirement in 1938.

Lovejoy's epistemology was based on the premise that experience is irreducibly temporal. He held that the time series was irreversible and dates absolute. From this premise he argued that if two apparently similar (or for that matter different) objects have different dates, they are existentially dual, whatever their causal relations may be. Therefore, the sensory impressions which presumably arise in the human brain must be existentially different from their "objects." Stars, for instance, are seen at dates later than the date at which the light rays, which cause our vision, left them. The star that we see is thus not the star of the date at which we think we see it. This illustration would be true for any visual object and our impression of it. Lovejoy's epistemological dualism was expounded in detail in *Revolt Against Dualism* (1930), along with critical analyses of various forms of monism. Because of his temporalism, he attempted to refute some of the inferences drawn from the special theory of relativity; later, in conversation with the author, he repudiated the papers in which the refutations appeared.

Lovejoy's epistemological dualism was accompanied by a firm belief in the causal efficacy of ideas. He argued against all forms of anti-intellectualism and favored freedom of speech and conscience. He was one of the organizers of the American Association of University Professors and chairman of its committee on academic freedom for several years. A pronounced intellectualist, he became interested in the history of ideas. In collaboration with Philip Wiener, Lovejoy founded the *Journal of the History of Ideas* in 1940. His historiographic program consisted in analyzing ideas into their component elements and then looking for a given elemental idea in various fields, regardless of the context in which it first appeared. He argued that a given idea—evolution, for instance—might begin as a theory of biology, but could turn up in theories of art, religion, or social organization.

Lovejoy also believed that ideas often begin as simple descriptive labels but take on eulogistic connotations as time goes on; the historian must become aware of these connotations and of their influence on thought. His most famous example was probably his analysis of the meanings of "nature" and its derivatives. He pointed out that authors are frequently unaware of the ambiguity of their ideas—ambiguities that have accumulated over the centuries—and thus fall into unconscious inconsistencies of thinking. His most influential contribution to the history of ideas is undoubtedly *The Great Chain of Being* (1936), preceded by certain chapters of *Primitivism and Related Ideas in Antiquity* (1935). He realized that to complete a detailed history of any idea would require the collaboration of many scholars.

BIBLIOGRAPHY

I. ORIGINAL WORKS. Lovejoy's works include "The Thirteen Pragmatisms," in *Journal of Philosophy, Psychology and Scientific Methods*, **5** (1908), 1–12, 29–39; "Reflections of a Temporalist on the New Realism," *ibid.*, **8** (1911), 589–599; "Some Antecedents of the Philosophy of Bergson," in *Mind*, **22** (1913), 465–483; "On Some Novelties of the New Realism," in *Journal of Philosophy, Psychology and Scientific Methods*, **10** (1913), 29–43; *Bergson and Romantic Evolutionism* (Berkeley, 1914); "On Some Conditions of Progress in Philosophical Inquiry," in *Philosophical Review*, **26** (1917), 123–163; "The Paradox of the Thinking Behaviorist," *ibid.*, **31** (1922), 135–147; *The Revolt Against Dualism: An Inquiry Concerning the Existence of Ideas* (La Salle, Ill., 1930); *Primitivism and Related Ideas in Antiquity* (Baltimore, 1935), written with G. Boas; *The Great Chain of Being: A Study of the History of an Idea* (Cambridge, Mass., 1936); and *Essays in the History of Ideas* (Baltimore, 1948), which includes a complete bibliography up to 1947.

See also "Buffon and the Problem of Species," in Bentley Glass, Owsei Temkin, and William L. Straus, Jr., eds., *Forerunners of Darwin* (Baltimore, 1959), 84–113; "Kant and Evolution," *ibid.*, 173–206; "Herder: Progressionism Without Transformation," *ibid.*, 207–221; "The Argument for Organic Evolution Before the *Origin of Species*, 1830–1858," *ibid.*, 356-414; "Schopenhauer as an Evolutionist," *ibid.*, 415–437; "Recent Criticism of the Darwinian Theory of Recapitulation: Its Grounds and its Initiator," *ibid.*, 438–458; *Reflections on Human Nature* (Baltimore, 1961); and *The Reason, the Understanding, and Time* (Baltimore, 1961).

II. Secondary Literature. See George Boas, "A. O. Lovejoy as Historian of Philosophy," in *Journal of the History of Ideas*, **9**, no. 4 (1948), 404–411; Maurice Mandelbaum, "Arthur O. Lovejoy and the Theory of Historiography," *ibid.*, 412–423; W. P. Montague, "My Friend Lovejoy," *ibid.*, 424–427; Marjorie H. Nicholson, "A. O. Lovejoy as Teacher," *ibid.*, 428–438; Theodore Spencer, "Lovejoy's *Essays in the History of Ideas*," *ibid.*, 439–446; H. A. Taylor, "Further Reflections on the History of Ideas: An Examination of A. O. Lovejoy's Program," in *Journal of Philosophy*, **40** (1943), 281–299; and Philip P. Wiener, "Lovejoy's Role in American Philosophy," in *Studies in Intellectual History* (Baltimore, 1953), 161–173.

George Boas

LOVELL, JOHN HARVEY (*b.* Waldoboro, Maine, 21 October 1860; *d.* Waldoboro, 2 August 1939), *botany*, *entomology*, *apiology*.

Lovell was the son of Harvey H. Lovell, a sea captain, and of Sophronia Caroline Bulfinch Lovell. An interest in natural history led him to study science at Amherst, where he was elected to Phi Beta Kappa. After graduating in 1882 he taught school for several years, but he resigned when his father became seriously ill. His father died in 1898, leaving him a fine house on the Medomac River and enough wealth to provide him with an income for life.

In 1899 he returned to Amherst and earned a master of arts degree, and in the same year he married Lottie Magune. She took an interest in his work and often assisted in collecting, labeling, and cataloging his insects. They had two sons, Harvey Bulfinch Lovell and Ralph Marston Lovell. Harvey developed interests similar to his father's and received a Ph.D. in zoology from Harvard in 1933. They collaborated on six papers on flower pollination from 1932 to 1939.

The particular focus of John Lovell's researches was established when he read Hermann Müller's *Fertilization of Flowers*. The correlation which Müller had described between the colors of flowers and the kinds of insects that were attracted to them was repudiated by Felix Plateau in 1895. One of the main goals of Lovell's early studies on the colors of flowers and color preferences of insects was to substantiate Müller's findings.

Lovell found that the identification of wild bees was often difficult, and he decided to make a special study of them. He collected over 8,000 Apoidea specimens and described, sometimes with the assistance of Theodore D. A. Cockerell, thirty-two new species.

Lovell's interest in bees and pollination soon led him to the study of apiculture and the photography of flowers. He combined these interests in *The Flower and the Bee* (1918) and *Honey Plants of North America* (1926), each of which contains more than 100 of his plant photographs.

Many of his photographs were also reproduced with his articles in the *American Bee Journal* and in about a thousand articles on plants which he wrote, beginning in 1926, for the *Boston Globe* and other newspapers.

BIBLIOGRAPHY

I. Original Works. Lovell's two books are *The Flower and the Bee* (1918) and *Honey Plants of North America* (1926). Beginning in 1913 he wrote seventy-eight articles for the four eds. of Amos Ives Root and E. R. Root, *The ABC and XYZ of Bee Culture* (1923–1940). He wrote forty-one articles and notes for the *American Bee Journal*, which appeared from March 1913 to October 1937 (see indexes). Seventeen botanical articles by Lovell are listed in the Torrey Botanical Club *Index to American Botanical Literature*, 1886–1966, 4 vols., III (Boston, 1969), 111–112.

Eleven articles which he wrote describing Apoidea are listed in the bibliography of Covell (cited below). Of his insect collections, 156 type specimens are now part of the collections of the U. S. National Museum and the remainder are located in the Lovell Insect Museum, University of Louisville.

II. Secondary Literature. There is a discussion and summary of his Apoidea descriptions in Charles V. Covell, Jr., "A Catalog of the J. H. Lovell Types of Apoidea (Hymenoptera), with Lectotype Designations," in *Proceedings of the Entomological Society of Washington*, **74** (1972), 10–18. For an outline of the history of apiculture in America, see John E. Eckert, Frank R. Shaw, and Everett F. Phillips, *Beekeeping* (New York, 1960), 7–10, 453–458. For a more recent understanding of pollination ecology, see K. Faegri and L. van der Pijl, *The Principles of Pollination Ecology* (Oxford–New York, 1966). For the history of pollination ecology down to 1873, see Hermann Müller, *The Fertilization of Flowers*, trans. into Eng. by D'Arcy Wentworth Thompson (London, 1883), 1–29. A useful biographical sketch is Frank C. Pellett, "John H. Lovell: Notes on the Life and Writings of the Maine Naturalist," in *American Bee Journal*, **79** (1939), 568–570, which includes two photographs of Lovell.

Frank N. Egerton III

LOVÉN, SVEN (*b.* Stockholm, Sweden, 6 January 1809; *d.* near Stockholm, 3 September 1895), *marine biology*.

Lovén was the son of Christian Lovén, the mayor of Stockholm. He received his early education in a private school, then in 1824 entered the University of Lund. His teachers included the botanist Agardh and the zoologist Sven Nilsson, with the latter of whom Lovén made a trip to Norway in 1826. The subject of Lovén's first paper, the geographical distribution of birds, shows Nilsson's influence. Lovén took the M.A. in 1829; in 1830 he went to Berlin for a year's further study. In Germany he learned microscope technique and studied with Ehrenberg and Rudolphi, who, together with the Swedish anatomist Anders Retzius, persuaded him to give up ornithology in favor of marine biology.

In 1835 Lovén published a treatise on the plankton crustacean *Evadne nordmanni* and a work entitled *Contribution to the Knowledge of the Genera Campanularia and Syncoryne*, in which he traced the life-cycles of these genera from the egg to the formation of new colonies. He supplemented his knowledge of marine fauna with a number of field excursions, of which the most important was a seventeen-month trip to Spitsbergen and northern Norway that he undertook in 1836–1837. On this voyage Lovén studied marine life, especially plankton, and assembled a rich collection of specimens. He also recorded significant observations on Paleozoic fossils and Quaternary geological formations.

In 1841 Lovén was appointed curator of the invertebrate section of the Museum of Natural History in Stockholm. At the same time, he was engaged in research on the anatomy and evolution of mollusks; his *Index molluscorum* was published in 1846. Lovén then took up the study of the molluscean radula, noting that this structural detail is useful in classification by genera and species. In 1848, while studying the embryological development of mollusks, Lovén (simultaneously with Friedrich Müller) discovered the polar bodies that appear during the maturation of the egg. He incorporated this discovery, together with several other important results, in his *Contribution to the Knowledge of the Development of Mollusca Acephala Lamellibranchiata*, a study of the metamorphic and embryonic stages in a number of different bivalves which was published the same year.

Lovén's interest in fossil forms led him to take up the question of the evolution of the Scandinavian peninsula, a problem that had also been discussed by Celsius and Swedenborg. Lovén studied shell-banks—deposits of fossil shells at altitudes considerably above sea level—on the west coast of Sweden and found such arctic forms as *Pecten islandicus*. These finds corroborated his theory that an arctic sea had once covered much of the present Scandinavian land; he also found that the proportion of southern fossil forms was significantly higher near the present shore, demonstrating a gradually warming climate. He further investigated the peculiar distribution of certain glacial marine forms which were common in the Baltic Sea and in several large inland lakes, but completely lacking on the west coast of Scandinavia. His *On Several Crustaceans Found in Lakes Vättern and Vänern*, published in 1860, offers a geological explanation of this zoological phenomenon.

Although his work on mollusks was most widely recognized, Lovén also published excellent anatomical studies on other species. His *Études sur les Echinoidées*, published in 1874, is of lasting value to students of the Echinodermata.

Lovén remained active at the Museum of Natural History until his retirement in 1892. He made a further permanent contribution to science in his work toward the foundation of the Kristineberg zoological station on the west coast of Sweden. With Berzelius, he began the series *Summary of the Transactions of the Royal Academy of Science* in 1844. In 1903 the Academy issued a memorial medal in his honor.

BIBLIOGRAPHY

A bibliography of eighty-nine works is supplied by Hjalmar Théel, "Sven Ludwig Lovén," in *Kungliga Svenska vetenskapakademien lefnadsteckning*, **4,** no. 3 (1903), 75–82.

KARL-GEORG NYHOLM

LOVITS (LOWITZ), JOHANN TOBIAS (in Russian, **Tovy Yegorovich)** (*b.* Göttingen, Germany, 25 April 1757; *d.* St. Petersburg, Russia, 17 December 1804), *chemistry*.

Lovits' mother died when he was very young; his father was Georg Moritz Lovits, a cartographer and instrument maker, who was also, from 1762, a professor at Göttingen University. In 1768 father and son went to St. Petersburg and thence on an ill-fated expedition to the Caspian steppes. The elder Lovits was executed there by the Cossack revolutionary Pugachev, and in 1774, following that event, his orphaned son was placed in the Academy Gymnasium in St. Petersburg. Lovits left the Gymnasium after two years and entered the main pharmacy of St. Petersburg as a student. By 1779 he had become a journeyman pharmacist, and the following year he was sent to Göttingen to continue his education. His health had been precarious throughout his school years, and his university career was interrupted by a serious illness. He attempted to regain his strength by a long foot trip throughout Europe, and succeeded so well that by 1784 he was able to return to St. Petersburg, and to his studies at the main pharmacy there. The study of chemistry by that time occupied all his free time. In 1787 Lovits was made court apothecary; in 1790 he was elected an adjunct, and in 1793 a full member of the St. Petersburg Academy of Sciences.

Lovits became known as a talented and inventive experimenter. In 1785, while trying to obtain a crystalline form of tartaric acid, he noticed that powdered charcoal which accidentally contaminated the acid solution effectively removed visible impurities. He thus began his research on adsorption, at first attributing the adsorptive qualities of charcoal to its "dephlogistic action." He set up an extensive program to test the effects of charcoal on a number of different substances, employing carbons of wood, bone, and even pure tartaric acid. Having discovered the efficacy of charcoal for removing color from organic products, he recommended that it be used as a purifying agent for vodka, sugar syrup, and drinking water. He made further investigations of gas adsorption by charcoal, and noted its deodorizing action. In the 1790's, Lovits adapted Lavoisier's theory and explained the phenomenon of adsorption chemically (unlike Klaproth and Green, who remained committed to a mechanical interpretation).

Lovits did further research in the crystallization of substances from solutions, introducing the concepts of supercooling and supersaturation. He obtained the crystal hydrates $NaCl \cdot 2H_2O$ and $KOH \cdot 2H_2O$, and distinguished between forced crystallization and spontaneous crystallization, noting that the same substance could thus yield crystals of different form and composition. He explained the role of seeding in crystal growth, and used this method to obtain Rochelle salt crystals "of unusual size" from hot, lightly saturated solutions. He further employed the seeding technique to separate the crystal forms of salts, which he used in a number of chemical analyses. Lovits was the first to describe convection currents in the crystallization process. He also developed several formulas for cooling mixtures; fabricated 288 wax models of crystals; and used a microscope to observe the crystalline patterns left on glass after evaporation of the mother solution (thereby laying the foundations of microchemical analysis).

In analytical chemistry Lovits investigated strontium, chrome, titanium, manganese, columbium (niobium), and their salts. Independently of A. Crawford, he isolated strontium from barite, examined the properties of its salts in detail, and developed a method of distinguishing it from barium and calcium according to its solubility in alcohol. Independently of Vauquelin, Lovits, in 1798, derived chrome from Siberian lead ore and established its crystalline form.

Lovits designed new methods to use in his analytical work. He dissolved silicates in heated caustic alkalies, instead of melting them, and in 1794 described a technique for titrating acetic acid with potassium tartrate. (The appearance of sediment consisting of undissolved potassium bitartrate marked the end of the process.) In his research on the intermediate and acid salts of carbonic and sulfuric acids, he obtained crystalline bicarbonates and bisulfates to demonstrate that the excess of acids in acid salts represented a chemical, rather than a mechanical, mixture.

Lovits was also the first to isolate a number of organic substances in the pure state. Through the use of a technique combining the principles of crystallization, distillation, and adsorption, he was able to prepare frozen acetic acid, anhydride alcohol, pure sulfuric ether, and many organic acids. He was the first to isolate glucose from honey, and he obtained two new acids—dichloracetic and trichloracetic—by subjecting acetic acid to the action of chlorine.

Lovits was active scientifically for only twenty years. He was often seriously ill, occasionally as a result of his experiments, and survived a number of accidents. One of the most dedicated of chemists, he found satisfaction and joy only in his research.

BIBLIOGRAPHY

I. Original Works. A collection is *T. E. Lovits. Izbrannye trudy po khimii i khimicheskoy tekhnologii* ("T. E. Lovits. Selected Works in Chemistry and Chemical Technology"; Moscow, 1955), ed. and with notes by N. Figurovsky. A partial list of individual works is given by Poggendorff.

II. Secondary Literature. A biography is Figurovsky, *op. cit.*, 405–514; see also his *Leben und Werk des Chemikers Tobias Lowitz* (Berlin, 1959); A. N. Scherer, *Worte der Erinnerung an das Leben und die Verdienste von Tobias Lowitz* (St. Petersburg, 1820); and P. Walden, "Tobias Lowitz, ein vergessener Physikochemiker," in Diergart, ed., *Beiträge zur Geschichte der Chemie* (Leipzig, 1909), 533–544.

N. Figurovsky

LOWELL, PERCIVAL (*b.* Boston, Massachusetts, 13 March 1855; *d.* Flagstaff, Arizona, 12 November 1916), *astronomy.*

More than any other astronomer of his generation, Percival Lowell had a profound influence on the general public. His thesis that the planet Mars was the abode of intelligent life continued to excite the public mind decades after his death; although the idea never gained the acceptance of his colleagues, it was not until after the Mariner flights to Mars during the late

1960's that it could firmly and finally be banished from all consideration.

Lowell's name is also forever linked with Pluto; although he did not live to see that distant planet (and while subsequent evidence has revealed its discovery to be accidental), there is no doubt that his inspiration advanced the date of its detection by many years. His most important contribution to astronomy, however, was his realization that superior observational work can be conducted only where atmospheric conditions are superior. He was a pioneer in constructing an observatory far from the lights and smoke of civilization.

Lowell was descended from some of the most prominent New England families. His father, Augustus Lowell, was closely identified with the cultural life of Boston. His mother, Katharine Bigelow Lawrence, was the daughter of Abbott Lawrence, sometime United States minister to Great Britain. A great-great-grandfather, John Lowell, was a member of the Continental Congress; his son, Lowell's great-grandfather, also John Lowell, represented Boston in the Massachusetts legislature. Both were actively interested in horticulture, and the younger John Lowell had on his farm some of the first greenhouses in the United States to be built on truly scientific principles. John Amory Lowell, Lowell's grandfather, collected a valuable herbarium and botanical library, the latter now forming an important part of the Gray Herbarium. Lowell's younger brother, Abbott Lawrence Lowell, became president of Harvard University; and his youngest sister, Amy Lowell, was well known as a poet and critic. James Russell Lowell, a first cousin of his father, was the foremost American man of letters of his time.

Lowell's early schooling was at Miss Fette's dame school. He subsequently attended Noble's School, where he developed interests in many fields, including astronomy. He graduated from Harvard University in 1876 with distinction in mathematics. After a year in Europe he went into his grandfather's business (cotton mills and trust and utility companies) and in 1883 traveled to Japan to study its people, customs, and language. While in Japan he was invited to serve as foreign secretary and general counsellor to the first diplomatic mission from Korea to the United States. He subsequently visited Korea and wrote of his experiences there in *Chosön—the Land of the Morning Calm—A Sketch of Korea* (1885). For several years he journeyed throughout the Far East and learned much of the oriental character, which he portrayed in *The Sound of the Far East* (1888). *Noto* (1891) is a straightforward but exciting account of a trip to a remote part of Japan. An earlier interest in Shintoism was renewed by his visit to the sacred mountain of Ontake, and he described his studies of the Shinto trances in *Occult Japan, or the Way of the Gods* (1895).

But Lowell's thoughts were turning more and more to astronomy, and on his final voyage to Japan he took a small telescope with him. He left the Orient for the last time in 1893. A favorable opposition of Mars was due toward the end of 1894, and Lowell resolved to study that planet then under the best possible conditions. At the 1877 opposition Mars had been extensively studied by Giovanni Schiaparelli, whose widely publicized observations of the "canals" had opened a new era in the investigation of that planet. Lowell took up a suggestion by W. H. Pickering that the steadiest air in North America was to be found in the Arizona Territory, and as a result of tests in the spring of 1894 he selected an observing site on the eastern edge of the mesa to the west of Flagstaff, at an altitude of some 7,000 feet. He acquired an eighteen-inch and a twelve-inch telescope and began observations. In *Mars* (1895) Lowell concluded

> . . . that the broad physical conditions of the planet are not antagonistic to some form of life; secondly, that there is an apparent dearth of water upon the planet's surface, and therefore, if beings of sufficient intelligence inhabited it, they would have to resort to irrigation to support life; thirdly, that there turns out to be a network of markings covering the disk precisely counterparting what a system of irrigation would look like; and, lastly, that there is a set of spots placed where we should expect to find the lands thus artificially fertilized, and behaving as such constructed oases should. All this, of course, may be a set of coincidences, signifying nothing; but the probability points the other way.

By 1896 Lowell had replaced the two borrowed telescopes with a twenty-four-inch telescope and had resumed observations of Mars by night and of Mercury and Venus by day. He had also tested sites in the Sahara and in Mexico and South America. During the winter of 1896–1897 he temporarily transferred his telescope to Tacubaya, Mexico, where it was believed—erroneously, as it turned out—that observing conditions would be better than in Flagstaff.

Overwork and insufficient sleep took their toll, and Lowell's health broke down. He was kept from astronomy for four years, except for his participation in an eclipse expedition to Tripoli in 1900. Observations at the Lowell observatory continued, however, and the first two volumes of the observatory's *Annals* appeared. Lowell was back in Flagstaff observing Mars in 1901, 1903, and 1905, and he expanded his earlier ideas about that planet in *Mars and Its Canals*

(1906) and *Mars as the Abode of Life* (1908). Lowell was not the first to regard the Martian bright areas as deserts and the dark areas as vegetation, but he studied in unprecedented detail the progressive "wave of darkening" of the dark areas, from pole to equator, as the seasons advanced from late winter, through spring, and into summer. As water was released from the melting polar caps, plant life would be revived; the accompanying increased prominence of the canals would indicate that water was flowing through them.

The premise that the dark areas are vegetation was almost universally accepted until the late 1950's. Then astronomers began to suspect that the light and dark areas are equally barren, and that the changes in the latter are simply due to light-colored dust being blown across them by winds. Lowell had himself suggested that the polar caps were due to hoarfrost, although astronomers now believe them to arise more from solid carbon dioxide than from solidified water. Observations from the Mariner flights have confirmed the barren nature of the Martian surface. The canals exist, although not as the fantastic system of hundreds of straight lines depicted by Lowell; most, if not all, of them are in fact merely chance alignments of dark patches.

While it was Mars that received the greatest part of Lowell's attention, he did not ignore the other planets. He "confirmed" Schiaparelli's result that Mercury rotates on its axis in the same time it takes to orbit the sun, although radar studies have revealed the true rotation period to be only two-thirds as long. With V. M. Slipher he made, in 1911, the first reliable determination of the rotation period of Uranus—ten and three-quarter hours. He also made a critical investigation of the structure of Saturn's rings, particularly when they were presented edgewise to the earth in 1907. And he made extensive studies of the "cloud formations" on Jupiter.

A nonresident professor at the Massachusetts Institute of Technology, Lowell gave a series of lectures there in 1902, later published as *The Solar System* (1903). Another series of lectures there led to the publication of *The Evolution of Worlds* (1909). In the latter work, as in *Mars as the Abode of Life*, Lowell adopted the Chamberlin-Moulton hypothesis that the solar system arose as the result of the encounter of the sun and another star. He then discussed various stages in the evolution of a planet; as cooling set in, a crust formed, conditions became suitable for the development of life, and finally the oceans and atmosphere disappeared and life ceased. He supposed that Mars had evolved much more completely than the earth, and that the moon had reached its final stage. *The Evolution of Worlds* concludes with a vivid description of how life on the earth could end, as its inhabitants became aware of a dark star steadily approaching, finally to collide with the sun.

Lowell was also interested in purely theoretical studies of the solar system, and in particular that of the significance of the resonances among the planets. He maintained that, after each of the planets was formed, there would be a tendency for the next one to collect at a point where its period of revolution bore some simple relationship to that of its predecessor: thus, after the formation of Jupiter, Saturn was formed with a period just two-and-a-half times longer; then Uranus with a period three times that of Saturn; and Neptune with a period twice as long again. There was a similar relationship among the periods of the inner planets.

The planets are not now observed to be exactly at these resonances, and Lowell discussed a mechanism whereby each planet was subsequently perturbed somewhat in toward the sun. Much of this theory is described in Lowell's final book, *The Genesis of the Planets* (1916). The process cannot now be given serious consideration, but it led Lowell to anticipate the existence of another planet orbiting the sun with a revolution period twice that of Neptune. He had previously come to a similar conclusion by studying the distribution of the aphelion distances of the periodic comets; there are a large number of cometary aphelia near the orbit of Jupiter, a few near the orbits of Saturn and Uranus, several near that of Neptune, but apparently none in between.

The concept of "cometary families" has since been largely discredited (except for the Jupiter family), but more distant clusters of cometary aphelia could suggest the presence of at least one planet beyond Neptune. For the last eight years of his life Lowell paid particular attention to the problem of finding a trans-Neptunian planet. Recognizing that the orbit of Neptune was too imperfectly known for his purpose, he analyzed the residuals remaining between the observed and calculated positions of Uranus, as J. C. Adams and Le Verrier had done before him; this time, the perturbations by Neptune were taken into account. Lowell had C. O. Lampland photograph the region of the sky in which the new planet X was expected to lie. Subsequent refinement of the calculations led to considerable changes in the prediction, but X still eluded detection.

The search was continued at the Lowell observatory long after its founder's death, and X was finally identified by Clyde Tombaugh in 1930. Named Pluto, with its symbol ♇ depicting also the initials of Percival Lowell, the new planet was announced to the world on the seventy-fifth anniversary of Lowell's birth. Pluto was found to travel in an orbit remarkably close to that predicted, although with a period only three-halves (not twice) that of Neptune. Pluto was considerably fainter than had been anticipated. Since it has now been established that Pluto has a mass scarcely greater than that of Mars (and possibly no greater than that of Mercury), it is quite clear that the discovery was due to an incredible coincidence.

Lowell married Constance Savage Keith in 1908. He was a brilliant speaker and was always in demand for lecture tours. He enjoyed all nature and derived from his ancestors a great love of botany. He found near Flagstaff a number of plants not previously known to prevail so far north, and he also discovered a new species of ash tree that now bears his name. Lowell was an honorary member of the Royal Astronomical Society of Canada, and he received medals from the national astronomical societies of France and Mexico.

BIBLIOGRAPHY

I. ORIGINAL WORKS. In addition to the books mentioned in the text many of Lowell's writings are contained in the *Annals of Lowell Observatory*, **1–3** (1898–1905) and the *Bulletin of Lowell Observatory*, **1–2** (1903–1916). Other works include "On the Capture of Comets by Jupiter," in *Science*, **15** (1902), 289; "Expedition for the Ascertaining of the Best Location of Observatories," in *Monthly Notices of the Royal Astronomical Society*, **63** (1902), 42–43; "A Standard Scale for Telescopic Observations," *ibid.*, **63** (1902), 40–42; "On the Variable Velocity of Zeta Herculis in the Line of Sight," in *Astronomical Journal*, **22** (1902), 190; "On the Kind of Eye Needed for the Detection of Planetary Detail," in *Popular Astronomy*, **13** (1905), 92–94; and "Chart of Faint Stars Visible at the Lowell Observatory," *ibid.*, 391–392.

See also his "Comparative Charts of the Region Following Delta Ophiuchi," in *Monthly Notices of the Royal Astronomical Society*, **66** (1905), 57; "Planetary Photography," in *Nature*, **77** (1908), 402–404; "The Tores of Saturn," in *Popular Astronomy*, **16** (1908), 133–146; "The Revelation of Evolution: A Thought and Its Thinker," in *Atlantic Monthly*, **104** (1909), 174–183; "Planets and Their Satellite Systems," in *Astronomische Nachrichten*, **182** (1909), 97–100; "The Plateau of the San Francisco Peaks in Its Effect on Tree Life," in *Bulletin of the American Geographical Society of New York* (1909); "On the Limits of the Oblateness of a Rotating Planet and the Physical Deductions From Them," in *Philosophical Magazine*, 6th ser., **19** (1910), 710–712; "Saturn's Rings," in *Astronomische Nachrichten*, **184** (1910), 177–182; "The Hood of a Comet's Head," in *Astronomical Journal*, **26** (1910), 131–134; "On the Action of Planets Upon Neighboring Particles," *ibid.* (1911), 171–174; and "Libration and the Asteroids," *ibid.*, **27** (1911), 41–46.

See also "The Sun as a Star," in *Popular Astronomy*, **19** (1911), 283–287; "The Spectroscopic Discovery of the Rotation Period of Uranus," in *Observatory*, **35** (1912), 228–230; "Sur la Désintégration des Comètes," in *Bulletin astronomique*, **29** (1912), 94–100; "Precession and the Pyramids," in *Popular Science Monthly*, **80** (1913), 449–460; "Precession of the Martian Equinoxes," in *Astronomical Journal*, **28** (1914), 169–171; "Mimas and Enceladus," in *Popular Astronomy*, **22** (1914), 633; "Memoir on a Trans-Neptunian Planet," in *Memoirs of the Lowell Observatory*, **1**, no. 1 (1915); "Memoir on Saturn's Rings," *ibid.*, no. 2 (1915); "Measures of the Fifth Satellite of Jupiter Made at the Lowell Observatory in September 1915," in *Astronomical Journal*, **29** (1916), 133–137; and "Our Solar System," in *Popular Astronomy*, **24** (1916), 419–427.

II. SECONDARY LITERATURE. The most complete biographical information is to be found in A. L. Lowell, *Biography of Percival Lowell* (New York, 1935), and in an account issued by the Lowell Observatory and published in *Popular Astronomy*, **35** (1917), 219–223. L. Leonard, *Percival Lowell: An Afterglow* (Boston, 1921) is a delightful character sketch and contains many extracts from his letters.

For recent information on Mars, see S. Glasstone, *The Book of Mars* (Washington, D.C., 1968) and R. B. Leighton, N. H. Horowitz, B. C. Murray, R. P. Sharp, A. G. Herriman, A. T. Young, B. A. Smith, M. E. Davies, and C. B. Leovy, "Mariner 6 Television Pictures: First Report," in *Science*, **165** (1969), 684–690.

For material on Pluto and its discovery, see E. W. Brown, "On a Criterion for the Prediction of an Unknown Planet," in *Monthly Notices of the Royal Astronomical Society*, **92** (1931), 80–101; P. K. Seidelmann, W. J. Klepczynski, R. L. Duncombe, and E. S. Jackson, "Determination of the Mass of Pluto," in *Astronomical Journal*, **76** (1971), 488–492; and C. W. Tombaugh, "Reminiscences of the Discovery of Pluto," in *Sky and Telescope*, **19** (1960), 264–270.

BRIAN G. MARSDEN

LOWER, RICHARD (*b.* Tremeer, near Bodmin, Cornwall, England, 1631; *d.* London, England, 17 January 1691), *medicine, physiology.*

Lower came from an old, affluent family with an interesting history. His grandmother, Mary Nicholls, was related to Anthony Nicholls, a member of the Long Parliament. His mother, Margery Billing, was of Hengar, the largest house in the district; and when Lower married Elizabeth, daughter of John Billing of Hengar, in 1666, the house came into the Lower family. Lower's father, Humphry, inherited Tremeer, the Lower family estate, and bequeathed it to Edward, Richard's older brother. Richard's younger brother, Thomas, a physician, was later imprisoned with the

Quaker leader George Fox for his religious beliefs. Lower was also related to the dramatist Sir William Lower.

Lower was admitted from Westminster School to Christ Church, Oxford, in 1649. He took his B.A. in February 1653 and his M.A. in June 1655. In April 1663 he wrote to Boyle that he had been put out of his place "above a year and a half since for not being in orders," and in June 1665 he took both the B.M. and doctor of physic degrees by accumulation. By this time Lower had spent several years at Oxford in close association with its famous circle of science devotees. He worked particularly closely with Thomas Willis, appointed Sedleian professor of natural philosophy in June 1660, whom he served for many years as research assistant.

In 1666 Lower moved from Oxford to London, where Willis had recently moved, principally to establish a medical practice. He settled at first in Hatton Garden, but during the next decade moved several times, in each case to a more fashionable location. He was admitted candidate of the Royal College of Physicians on 22 December 1671 and fellow on 29 July 1675. After Willis' death in 1675, Lower's successful medical career flourished even more. According to Anthony Wood he was "esteemed the most noted physician in Westminster and London, and no man's name was more cried up at court than his." Lower, having strong Protestant and anti-Popish sentiments, was closely identified with the Whig party, and his later career followed its fortunes. In the 1680's, with the discreditation under Charles II of the Whigs and the ascendance and accession of James II, Lower fell into some disrepute; he lost his court appointment and steadily lost much of his practice. Nevertheless, he carried on for several years, probably spending much of his time in Cornwall, until his death in 1691. In his will he left money to St. Bartholomew's Hospital in London and to the French and Irish Protestant refugees.

Lower was invited to join the Royal Society early in his residence in London, and for a few years he was closely associated with the society and its various scientific activities. After receiving mention several times in the minutes for 1666, Lower himself was "introduced" to the society by Robert Boyle on 2 May 1667. He was formally admitted on 17 October 1667, and in November, after having been considered as early as June, he was invited to take up the post of curator. Lower refused this offer, but on 21 November he was appointed to a committee to audit the society's accounts.

The sorry state into which the society was falling despite Lower's efforts is reflected in an entry in the minutes for 29 June 1668, in which Lower was "desired . . . to make a list of particulars necessary for the making of anatomical experiments." Lower himself dropped from regular participation in the society's affairs by March 1669, probably to attend more fully to his medical practice; in 1678 he formally resigned his fellowship.

During his few, intensely active years with the Royal Society, Lower did much of the work that established his reputation as perhaps the best seventeenth-century English physiologist after Harvey. He was concerned principally with two areas of investigation: transfusion and cardiopulmonary function. His interest in both problems can be traced to his days at Oxford, but the fame of his investigations and many of his most fruitful results owed a great deal to his association with the Royal Society.

Apparently transfusion was attempted at Oxford in the late 1650's. There, according to later accounts, Christopher Wren tried to convey certain medicinal liquors directly into the bloodstream using quills and special bladders. Familiar with these earlier attempts, Lower in 1661 expressed interest in using similar procedures to transmit broth and other nutritive fluids directly into the bloodstream. In a letter to Boyle dated 18 January 1661, Lower expressed his "fancy to try, how long a dog may live without meat, by syringing into a vein a due quantity of good broth" and described his intended procedure as follows: "I shall try it in a dog, and I shall get a tin pipe made, about two inches long, and about the usual bigness of a jugular vein, and hollow, which I may put into the vein. . . ."[1]

By 8 June 1664, Lower was able to write to Boyle in London about a more daring experiment: he intended to "get two dogs of equal bigness [and] let both bleed into the others vein. . . ."[2] As Lower was to explain retrospectively in his *Tractatus de corde* (1669), he was led from the broth experiment to the transfusion attempt by observing how harmoniously the blood of different animals mixed with various injected substances. It was natural to "try if the blood of different animals would not be much more suitable and would mix together without danger or conflict."[3] It is quite possible that Lower was influenced as well by reports of discussions at the Royal Society late in 1663. At one of these, Timothy Clarke had described his method of infusing certain medicinal preparations directly into the veins of dogs, and an unnamed fellow of the society proposed "to let the blood of a lusty dog into the veins of an old one, by the contrivance of two silver pipes fastened to the veins of such two dogs."[4]

With his ideas crystallized, it took Lower only a few months to perfect the requisite experimental technique.

He performed the first successful transfusion at Oxford late in February 1665, transfusing blood "from an artery of one animal into a vein of a second." The Royal Society soon heard of these results, and in early 1666, after several months' interruption due to plague and the London fire, society members were busy making their own investigations into transfusion. In June 1666 John Wallis, who had been present at Lower's successful experiment at Oxford the previous February, reviewed Lower's success; and the society, through Boyle, requested a full account from Lower. This was officially received in September, replicated at the society in November, and printed in *Philosophical Transactions* (December 1666). By mid-1667 Lower had joined the society.

Meanwhile, in Paris, Denis—without proper citation—had appropriated Lower's techniques and applied them to human transfusion. The Royal Society was outraged and Lower, sensitive to his colleagues' concerns, tried and succeeded at human transfusion. On 12 December 1667 the procedure was firmly established in England with its second successful trial, this a public one before a large crowd.

Lower's second major area of physiological investigation was cardiopulmonary function. Again, his interest can be traced to his Oxford days. Already in 1658 Lower and Willis were looking into the fundamental problem from which, when solved, all of Lower's principal results were to derive: the reason for the perceived difference in color between venous and arterial blood.

Willis formulated his own answer to this problem in his *Diatribae duae* (Oxford, 1659), a two-part essay on fermentation and fevers. Willis assumed that the blood, composed of five chemical principles, is normally in a state of gentle fermentation. But when the blood reaches the chambers of the heart the already fermenting fluid effervesces further. As the blood passes through the heart, "its mixture is very much loosned, so that the Spirits, together with the Sulphureous Particles, being somewhat loosned, and as it were inkindled into a flame, leap forth, and are much expanded, and from thence they impart by their deflagration, a heat to the whole."[5]

Lower himself at first accepted this theory of a sudden, energetic enkindling of the blood in the closed chambers of the heart as an explanation for the lighter, more vivid appearance of arterial blood. He made it the basis of an essay written in defense of Willis, the *Diatribae Thomae Willisii . . . de febribus vindicatio adversus Edmundum De Meara* (1665). A few years after writing the *Vindicatio*, however, Lower's own ideas were to change substantially. The changes derived from certain subtle unorthodoxies he permitted himself in 1665 and, even more, from his positive, fruitful association with the Royal Society, which allowed his doubts to develop into open disagreement with Willis' ideas.

Lower's early departures were evident in a few scattered passages of the *Vindicatio*. Thus, while elaborately defending Willis' theory of an accension of the blood in the heart, Lower nevertheless introduces an idea nowhere evident in the *Diatribae duae*: that the lungs (which were given no clear role by Willis) serve not only to discharge the soot resulting from the "fire" in the heart, but likewise serve to impregnate the blood passing through them with the "nitrous pabulum" of the air.[6] Willis, like Lower after him, had alluded to a "nitrosulphureous ferment implanted in the heart," but only Lower explicitly referred to a pabulum in the lungs that "impregnated" the blood.

Furthermore, in other passages in the *Vindicatio*, Lower refers in detail to the analogies between accension in the heart and air-controlled combustion.[7] He even mentions an experiment in which the heart of a vivisected animal is revived by blowing in air, with the mouth, through a tube into the chyle vessels or vena cava.[8] Finally, in a slightly different context, Lower describes active expulsion of blood from the heart during cardiac systole;[9] Willis, by contrast, had spoken merely of blood being "wheeled about after a constant manner, as it were in a water Engine."[10]

These minor unorthodoxies apparently played on Lower's mind for several years. Then, at the Royal Society, Lower began to collaborate with Hooke, who for some time had been interested in respiratory physiology. Already in 1664 Hooke had tried to determine whether the efficacy of the lungs depended on their motive effect on the blood or on the role they played in admitting air to the body. In 1667 he perfected an experimental technique that allowed him to keep the lungs motionless and inflated, even with the chest cavity fully exposed. Hooke pricked holes in the lungs and blew a continuous stream of air through the motionless lungs of a dog by a bellows arrangement.

In October 1667 Hooke and Lower began to collaborate on cardiopulmonary experiments. Upon Hooke's suggestion, they attempted together to determine the exact effect on the blood of its passage through the lungs; and to accomplish this by altering the normal route, bypassing the lungs, and transmitting blood directly from the pulmonary artery to the aorta through an air-free channel.

Preliminary results from these experimental techniques had an enormous impact on Lower's thinking. He was now able for the first time to observe closely the appearance of the blood, both as it left the right ventricle of the heart and before it entered the lungs, and as it returned from the inflated lungs via the pulmonary vein. Lower could now see clearly that only exposure to the air in the lungs altered the blood's appearance; whereas blood still appeared venous as it left the right ventricle, it was already arterial as it moved from the lungs to the left ventricle. Thus the bright, scarlet color of arterial blood depended on contact with the air and not on effervescence in the heart (which should have occurred as surely in the right as the left ventricle). Lower could also observe closely the continual, rapid, and vigorous pumping action of the heart, which now seemed clearly to be unrelated to a postulated effervescence in the cardiac chambers.

An intense period of reconceptualization and experimentation followed this work at the Royal Society, and led Lower to the publication of his *Tractatus de corde* (London, 1669), which comprised five chapters. The fourth chapter quickly reviews the history of transfusion and presents Lower's principal results; the fifth surveys Lower's thoughts on chyle and its transformation into blood.

The first three chapters of the *De corde* are exclusively concerned with the heart, the blood, respiration, and circulation. Lower begins with detailed descriptions of the muscular anatomy and nerve supply of the heart, comparisons of the fibrous structure of the human heart with that of other animals, and an account of the heart's contractive and expulsive movements. He now denies that there is any nitrosulfureous ferment in the heart and explains that the blood is "too inert to effervesce so violently and suddenly in the Heart or its Vessels."[11] Lower stresses the movement of the blood by the active cardiac systole (an early emphasis of his) and pointedly asks, "If the blood moves through its own power, why does the Heart need to be so fibrous and so well supplied with Nerves?"[12]

Lower also reports experiments he performed to determine the velocity of circulation. In one of these, by taking the average capacity of the left ventricle, counting the number of beats per hour, and estimating the total quantity of blood in the body, he finds that all the blood passes through the heart thirteen times per hour. In another experiment Lower drains almost all the blood from a dog, through the cervical arteries, within three minutes. Both experiments clearly suggest that there can be no profound difference between venous and arterial blood, since overall circulatory velocity is too great to allow drastic transformation.

Finally, Lower comes to a series of experiments on the influence of the air on blood color. He reports that he once held different views on this matter, confessing that he had "relied more . . . on the authority and preconceived opinion of the learned Dr. Willis than my own experience."[13] Acknowledging his indebtedness to Hooke, Lower presents the new experimental basis for his greatly modified thinking. He reports his observation that blood withdrawn from the pulmonary artery "is similar in all respects to venous blood,"[14] and claims that this would not be so if a cardiac effervescence were responsible—as he once believed—for the different appearance of venous and arterial blood. Exposure of the blood to the air is the crucial occurrence, which, he explains, can be further established by blocking of the trachea and noting the still-venous appearance of blood that passes through the air-blocked lungs. This proves definitively that the bright red color of arterial blood is "due to the penetration of particles of air into the blood"; it is the "nitrous spirit of the air" that normally enters the blood in the lungs.

The *Tractatus de corde* was quickly recognized as a major new work on physiology. Warmly reviewed in the *Philosophical Transactions* for 25 March 1669, it won plaudits both in England and abroad and in a short time dramatically changed the thinking of many about the role of the heart and the lungs. Even Willis was deeply impressed. In 1670 Willis published an essay, *De sanguinius accensione*, in which he abandoned his old views for Lower's new ones and applauded "the most learned Doctor Lower" for making him see the light.

Mayow, whose *Tractatus quinque* of 1674 was to represent the next major breakthrough in respiratory physiology, was also greatly influenced. The favorable reception proved enduring, and for many years Lower's *Tractatus de corde* continued to be cited as an important and authoritative work, clearly in the high Harveian tradition of "anatomical experiment." In the next century, too, Lower continued to receive credit. His accomplishments were enthusiastically acknowledged by Senac in his *Traité de la structure de coeur* (Paris, 1749) and by the influential Boerhaave. The *De corde* itself was published in a Leiden edition as late as 1749.

After the *De corde*, Lower published only one minor work on anatomical physiology, the *Dissertatio de origine catarrhi* (London, 1672). Here Lower codified the ideas and experiments he and Willis had performed to show that no passage existed for discharging serous fluids from the cerebral ventricles through the palate and nose; Willis had already reported similar results in chapter twelve of his *Cerebri anatome* (1664), as had Lower himself in the

Tractatus de corde. Other medical works bearing Lower's name appeared in later years, but none of these can be definitively traced to Lower's pen. It seems clear that after the *De corde* Lower's active interest in physiology gave way to a vigorous cultivation of his medical practice, and it was as a physician that he concluded his career.

NOTES

1. *The Works of the Honourable Robert Boyle*, Thomas Birch, ed., V (London, 1744), 518.
2. *Ibid.*, 525–526.
3. Lower, *Tractatus de corde* (London, 1669), 173.
4. Thomas Birch, *The History of the Royal Society*, I (London 1756–1757), 303.
5. Thomas Willis, *The Remaining Medical Works of . . . Dr. Thomas Willis* (London, 1681), 59.
6. *Diatribae Thomae Willisii . . . de febribus vindicatio adversus Edmundum De Meara* (London, 1665), p. 116.
7. *Ibid.*, 117–119.
8. *Ibid.*, 125–126.
9. *Ibid.*, 59.
10. Willis, *Remaining Medical Works*, 64.
11. Lower, *Tractatus de corde*, 62.
12. *Ibid.*, 64.
13. *Ibid.*, 163.
14. *Ibid.*, 164.

BIBLIOGRAPHY

I. Original Works. Lower's principal works have been referred to above. Two of the medical works that are variously attributed to him are *Bromographia* (Amsterdam, 1669) and *Receipts* (London, 1700). *Diatribae Thomae Willisii . . . de febribus vindicatio adversus Edmundum De Meara* exists only in contemporary Latin eds. *Tractatus de corde* (London, 1669), however, has been reprinted in facs. and translated by K. J. Franklin as vol. IX in R. T. Gunther, *Early Science in Oxford* (Oxford, 1932). *De origine catarrhi* has also been reproduced in facs. and translated, this by Richard Hunter and Ida Macalpine (London, 1963).

Lower also published several minor papers in *Philosophical Transactions*. Among them are "On Making a Dog Draw His Breath Like a Broken-winded Horse," **2** (1667), 544–546; and "An Observation Concerning a Blemish in an Horses Eye . . .," **2** (1667), 613–614.

Many references to Lower's work and opinions can be found in Thomas Birch, *The History of the Royal Society*, 4 vols. (London, 1756–1757), and in A. Rupert and M. B. Hall, eds., *The Correspondence of Henry Olderling* (Madison, Wisc., 1966–). Correspondence between Lower and Boyle is found in Thomas Birch, ed., *The Works of the Honourable Robert Boyle*, V (London, 1744), 517–529.

II. Secondary Literature. There is no comprehensive study of Richard Lower, but a number of works are of considerable interest and usefulness. Among these are John F. Fulton, *A Bibliography of Two Oxford Physiologists: Richard Lower (1631–1691), John Mayow (1643–1679)* (Oxford, 1935), which is much more than a bibliography; Francis Gotch, *Two Oxford Physiologists: Richard Lower and John Mayow* (Oxford, 1908); Ebbe C. Hoff and Phebe M. Hoff, "The Life and Times of Richard Lower, Physiologist and Physician," in *Bulletin of the History of Medicine*, **4** (1936), 517–535; and K. J. Franklin, "Biographical Notice," which accompanies his trans. of Lower's *De corde* (see above).

Of great help for particular aspects of Lower's work are Franklin, "Some Textual Changes in Successive Editions of Richard Lower's *Tractatus de corde*," in *Annals of Science*, **4** (1939), 283–294; "The Work of Richard Lower," in *Proceedings of the Royal Society of Medicine*, **25** (1931), 7–12; Leonard G. Wilson, "The Transformation of Ancient Concepts of Respiration in the Seventeenth Century," in *Isis*, **51** (1960), 161–172; and T. S. Patterson, "John Mayow in Contemporary Setting," *ibid.*, **15** (1931), 47–96, 504–546.

Also worth consulting are Michael Foster, *Lectures on the History of Physiology* (Cambridge, 1901); Marjorie Hope Nicolson, *Pepys' Diary and the New Science* (Charlottesville, Va., 1963); Kenneth Dewhurst, "An Oxford Medical Quartet," in *British Medical Journal*, **2** (1963), 857–860; and Hebbel E. Hoff and Roger Guillemin, "The First Experiments on Transfusion in France," in *Journal of the History of Medicine and Allied Sciences*, **18** (1963), 103–124.

Theodore M. Brown

LOWITZ, J. T. See **Lovits, Johann Tobias.**

LUBBOCK, SIR JOHN (LORD AVEBURY) *(b.* London, England, 30 April 1834; *d.* Kingsgate Castle, Kent, England, 28 May 1913), *entomology, anthropology, botany.*

Lubbock was the eldest son of a baronet, Sir John William Lubbock. His father was a banker and mathematician who did work on probability and the theory of tides, and was treasurer of the Royal Society. After three years at Eton, young Lubbock was removed from school before he was fifteen and taken into the family bank, where he shortly assumed the responsibilities of an adult partner. Thereafter his education was largely self-directed according to a rigorous schedule, with emphasis upon natural history. He trained himself to shift from one subject to another at short intervals with entire concentration, a habit he retained through life. Of crucial importance for the development of his interests was the presence at Down

House, close by the Lubbock estate in Kent, of Charles Darwin. From the time of his settlement there in 1842, Darwin gave the boy encouragement and direction, beginning a friendship that continued for forty years. In his treatise on barnacles, Darwin utilized Lubbock's talent for drawing, and Lubbock's earliest scientific papers were on zoological specimens from the *Beagle*. His careful work earned the notice and respect of Lyell, T. H. Huxley, Joseph Hooker, and Tyndall, all of whom became his friends. In 1855 Lyell proposed him for the Geological Society. In the same year Lubbock discovered the first fossil remains of a musk-ox to be unearthed in Britain, evidence of a glacial age. Lubbock's account of the methods of reproduction in *Daphnia*, which Darwin submitted to the Royal Society for him, led to his election as a fellow in 1858; three years later he was made a member of the council. A convinced natural selectionist from the beginning, after the release of the Darwin-Wallace papers Lubbock's work in microanatomy, such as his notice of the irregularity in the central ganglion of *Coccus hesperidum*, pointed to the high degree of variation in nature. Despite his youth Lubbock was one of the handful of men whose opinions mattered to Darwin, and he took a prominent part in the controversy that followed the appearance of the *Origin of Species*.

Lubbock first gained an international reputation by his provision of an evolutionary framework for the accumulated archaeological remains bearing on human beginnings. The tools collected from French river gravels by Boucher de Perthes had long indicated an origin for culture antedating the geologically recent past. The final acceptance of this view by leading British men of science in the late 1850's generated an enthusiasm for the reconstruction of man's prehistory. Lubbock had already taken part in the furtherance of Lyell's search for fossil gaps in the geological record. Between 1860 and 1864, he traveled to the Somme Valley and the Dordogne Caves, the Swiss lake village sites, and the tumuli, kitchen middens, and museums of Denmark. He went over the ground with the investigators, studied the finds, and read the reports and literature, even those in Danish.

The series of articles Lubbock wrote aroused wide interest and formed the basis for a pioneer work, *Pre-Historic Times* (1865), that consolidated the data on the life of prehistoric man in Europe and North America. He coined the terms *Neolithic* and *Paleolithic* to distinguish the later and earlier Stone Age periods. Here and in a sequel, *The Origin of Civilisation* (1871), Lubbock identified prehistoric cultures as evolutionary precursors of modern civilization and contended for the independent origination of cultural inventions as against diffusion or borrowing. He saw in a common creative mind for mankind a promise of the general evolutionary movement toward civilization and happiness. Lubbock rejected Bishop Whately's theory of degeneration and categorized savage tribes as comparable to the opossums of the natural world. By studying contemporary primitives he sought clues to the function of ancient implements. The popularity of these books led to their reissue in new editions for over a generation, even after their simplistic evolutionism had become outmoded. But Lubbock never modified these first conclusions.

Anthropology did not interfere with Lubbock's research on insects, and in 1873 the Ray Society published his standard *Monograph on the Collembola and Thysanura*. Beautifully illustrated with his own plates, this book of over two hundred pages separated the springtails from the bristletails on the basis of the ventral tube, and named the new order Collembola. About this time he began the seminal studies in insect behavior that were reported in *Ants, Bees, and Wasps* (1882) and *On the Senses, Instincts, and Intelligence of Animals* (1888). It was not only the normal habits of his subjects that he set out to investigate, but their powers of sensing, learning, and what seemed to be calculated response. For this purpose he devised the "Lubbock nest," as it became known, in which ant colonies are confined in moistened earth between two panes of glass. Stacked in series and attached to a post, these could be lowered to a platform surrounded by a moat and the ants let out for excursions. Previously, ant nests had never been kept under observation for more than several months, and the life-span of an ant was thought to be a year. Lubbock was able to keep some workers alive for as long as seven years, and two queens for twice that time. No one as yet knew how an ant nest started, but Lubbock watched queens of *Myrmica ruginodis* rear larvae and establish a colony. He observed, and described for the first time, that aphid eggs laid in the fall are taken into ant nests over the winter, and in the spring the newly hatched young are transported out to feed on plant shoots. In the nests Lubbock discovered a new mite, *Uropoda formicariae*, and two parasitic dipterons, *Platyphora Lubbocki* and *Phora formicarum*.

While trying to make ants respond to sounds, Lubbock located in their legs a chordontal organ known until then only in Orthoptera, and suggested correctly that it was a sort of hearing instrument. An imaginative and ingenious experimenter, Lubbock introduced specificity into the study of insect behavior by marking individuals with paint for identification, a practice that later became common. He also used obstacles and mazes to test the intelligence of ants, thus anticipating animal psychologists like Kohler.

Lubbock's experiments on insect vision and color sense were of special significance. On a table with movable concentric rings, designed for him by Francis Galton, Lubbock found that some ants were partly influenced in their sense of direction by the angle of the light, a discovery important for homing. By using colored glass and solutions with spectral light, he established that ants and *Daphnia* could not only distinguish colors, but were especially sensitive to ultraviolet light. After tabulating the color preferences of bees, and training them to return to colors associated with honey after the honey was removed, Lubbock concluded that bees could see colors. He did not test them with spectral light, however, to eliminate the possibility that they were attracted merely by different degrees of brightness. With modifications, Karl von Frisch utilized Lubbock's procedures, but it was not until the 1920's that A. Kühn's use of the spectrum delineated the range of color vision in bees and revealed their sensitivity to ultraviolet light.

In his lifetime Lubbock was one of the best-known men in England. He began a long career as a Liberal in Parliament by sponsoring the famous Bank Holidays Act (1871) which established what came to be called St. Lubbock's Days. He went on to sponsor over two dozen other bills, including acts regulating the health professions, the Wild Birds Protection Act (1880), Open Spaces Act (1880), Ancient Monuments Act (1882), and acts requiring the limitation of shop hours and the provision of seats for employees. In 1900 he was made a peer.

He published some twenty-five books, over a hundred scientific papers, and gave lectures on subjects ranging from free trade to the forces that formed the Alps, the hearing of Crustacea, and the pleasures of life. He was never merely a popularizer; his widely read books on the scenery of Switzerland and Britain were also fascinating treatises on geology by an expert. In volumes on British flowers he dealt with the questions of the relation of a flower's parts to each other and to insects.

Essentially, Lubbock was a great public educator, perhaps the foremost of his time. As an exponent of Darwinism, he was as active as Huxley, without his truculence. A vice-chancellor of London University, head of its Extension Society, and president of the Working Men's College, he helped widen educational opportunities and the spread of scientific literacy. An 1887 address to the college on "The Hundred Best Books" had a far-reaching effect on publishing in both England and the United States. Possessed of charm and a conciliatory manner, he helped smooth the proceedings of the many scientific societies he headed at various times. He married twice, the second time to the daughter of Pitt-Rivers the ethnologist. He presided over a large household, and often drafted family members and servants alike to assist and appreciate his always ongoing investigations of nature. Lubbock's mind was not subtle or deep, but he had an organized intelligence and great energy, and he was a magnificent amateur in the old sense. In the world of politics and commerce he was an arch-representative of science; to the intelligent public, during a transition period in intellectual history, he stood for the harmony of experimental truth with idealism and social progress.

BIBLIOGRAPHY

I. Original Works. An extensive, though not exhaustive, classified bibliography of Lubbock's works appears in Ursula Grant Duff, ed., *The Life-work of Lord Avebury (Sir John Lubbock, 1834–1913)* (London, 1924), but it does not list the numerous revised eds. of his principal books.

II. Secondary Literature. The authorized biography is Horace G. Hutchinson's *Life of Sir John Lubbock, Lord Avebury*, 2 vols. (London, 1914). Uncritical and lacking scientific expertise, it is useful for facts and the many letters and memoirs reprinted. Appreciative yet authoritative assessments of Lubbock's contributions in a number of fields, including anthropology, geology, entomology, zoology, and botany, were made by a group of specialists who collaborated on the Grant Duff volume cited above.

Insight into the difficulties Lubbock's anthropology met with in his later years is provided by Andrew Lang, "Lord Avebury on Marriage, Totemism, and Religion," in *Folklore*, **22** (1911), 402–425. Especially valuable for its updating of Lubbock's results in the light of later experimentation and the excerpts it provides from the relevant literature, particularly from W. M. Wheeler on ants and Karl von Frisch on bees, is the reissue of the seventeenth ed. of *Ants, Bees, and Wasps*, edited and annotated by J. G. Myers (London–New York, 1929).

A vigorous defense of Lubbock's scientific achievements and an explanation for their persistent denigration was presented by a zoologist, R. J. Pumphrey, F.R.S., in "The Forgotten Man—Sir John Lubbock, F.R.S.," in *Notes and Records of the Royal Society of London*, **13**, no. 1 (June 1958), 49–58. One aspect of Lubbock's relation to Darwin is treated in Fred Somkin's "The Contributions of Sir John Lubbock, F.R.S. to the *Origin of Species:* Some Annotations to Darwin," *ibid.*, **17**, no. 2 (Dec. 1962), 183–191.

Fred Somkin

LUBBOCK, JOHN WILLIAM (*b.* London, England, 26 March 1803; *d.* High Elms, near Farnborough, England, 20 June 1865), *astronomy.*

By profession a banker, Lubbock spent his leisure hours in scientific pursuits. He was among the first to check theory against extensive observations so as to get better predictions of the tides; and for most of his life he worked on the problem of simplifying and extending Laplace's methods for calculating perturbations of lunar and planetary orbits.

Lubbock was named for his father, from whom he inherited (1840) both his mercantile bank (Lubbock & Co.) and his baronetcy. His mother's maiden name was Mary Entwisle. He attended Eton before enrolling in 1821 at Trinity College, Cambridge, where he received a B.A. in 1825 and an M.A. in 1833. In the latter year he married Harriet Hotham; the eldest of their eleven children was the fourth Sir John Lubbock and first Baron Avebury, who wrote many books on natural history.

Lubbock was introduced to Laplace's powerful mathematical techniques during a visit to Paris in 1822, three years before the final volume of the *Mécanique céleste* had appeared. In 1828 Lubbock wrote an article on annuities, applying Laplace's probability theory. By 1831 he was making progress both with predicting the exact position of the moon in the sky and also—thanks to business connections with the chairman of the London Dock Company—with the assembling of many years of water level readings, from which to determine the average time high tide lagged behind the moon (an interval commonly known as the establishment of the port).

For his work on the tides Lubbock received from the Royal Society one of the two medals awarded in 1833; the other went to the geologist Charles Lyell.

Lubbock joined the Royal Astronomical Society in 1828. He became a fellow of the Royal Society in 1829, and served twice as its treasurer and vice-president (1830–1835, 1838–1847). Rather late in life (1848) he was elected fellow of the Geological Society and contributed to its *Quarterly Journal* in 1849 a suggestion as to how a change in the axis of the earth might have accounted for the redistribution of water and land areas.

BIBLIOGRAPHY

I. Original Works. Lubbock's first published work, written in 1828, was "On the Calculation of Annuities, and on Some Questions in the Theory of Chances," in *Transactions of the Cambridge Philosophical Society*, **3** (1830), 141–155. His papers on the tides began with "On the Tides on the Coast of Great Britain," in *Philosophical Magazine*, **9** (1831), 333–335; "On the Tides in the Port of London," in *Philosophical Transactions of the Royal Society*, **121** (1831), 379–416; and "Report on the Tides," in *Report of the British Association for the Advancement of Science* (1831–1832), 189–195. His approach to the problem of the tides is summarized in his Bakerian lecture, "On the Tides at the Port of London," in *Philosophical Transactions of the Royal Society*, **126** (1836), 217–266; with an extension, "On the Tides," *ibid.*, **127** (1837), 97–140.

"On Change of Climate Resulting From a Change in the Earth's Axis of Rotation" appeared in *Quarterly Journal of the Geological Society of London*, **5** (1849), 4–7. Lubbock's last published paper, "On the Lunar Theory," in *Memoirs of the Royal Astronomical Society*, **30** (1862), 1–53, provides a summary of his work on the motions of the moon, with comparison to the results of other workers.

The Royal Society *Catalogue of Scientific Papers*, IV, 105–106, lists seventy-four articles by Lubbock, one of which was published jointly with William Whewell. He also wrote, with John Drinkwater Bethune, *A Treatise on Probability* (London, 1835) and at least nine other books or pamphlets; see his entry in *Dictionary of National Biography*, XXXIV, 227–228.

II. Secondary Literature. Three obituaries appeared: two unsigned, in *Monthly Notices of the Royal Astronomical Society*, **26** (1866), 118–120; and in *Proceedings of the Royal Society*, **15** (1866–1867), xxxii–xxxvii; and one by William John Hamilton, in *Quarterly Journal of the Geological Society of London*, **22** (1866), xxxi–xxxii.

Sally H. Dieke

LUBBOCK, RICHARD (*b.* Norwich, England, 1759[?]; *d.* Norwich, 2 September 1808), *chemistry.*

Lubbock's place in the history of chemistry depends solely on his M.D. dissertation (1784), which embodies one of the earliest examples of the rejection of the phlogiston theory in Britain. He was a pupil of Joseph Black, who, according to Lubbock, had long been teaching the phlogiston theory without confidence. Lubbock implies that Black had recently abandoned it, although Black's acceptance of Lavoisier's theory is not generally thought to have occurred as early as 1784. In a letter to Lavoisier in October 1790 Black mentions his original "aversion to the new system" and says that although the approval by older chemists of the new ideas might be prevented by power of habit, "the younger ones will not be influenced by the same power" (see D. McKie, "Antoine Laurent Lavoisier, 1743–1794," in *Notes and Records. Royal Society of London*, **7** [1950], 1–41). Lubbock said that his own theory was partly met by that of Lavoisier but that the latter left some important points unexplained, such as the nature of heat and light.

Little is known of Lubbock's life. He was educated at Norwich Free School and the University of Edinburgh, practiced medicine in Norwich, and was physician at the Norfolk and Norwich Hospital from 1790 to 1808. His younger son, the Reverend Richard Lubbock, wrote a classic work on the natural history of Norfolk.

In his *Dissertatio . . . de principio sorbili*, Lubbock said that it had long been accepted that the air contained a principle which maintained life and which governed calcination, combustion, and other natural processes. It had been given a variety of names—Priestley's "dephlogisticated air" and Lavoisier's "eminently respirable air"—Lubbock favored the term "pure air," which had also been used. According to him it was a compound of an absorbable part (*principium sorbile*) and of *principium aëri proprium*; the latter was the matter of light and heat and was without mass. Metals and combustible substances such as charcoal, sulfur, and phosphorus were true elements: on calcination or combustion they combined with principium sorbile and thereby gained weight, the principium aëri proprium being liberated as heat and light. The variation in the evolution of the latter depended on the extent to which the principium sorbile was removed; as this increased, so the amount of heat manifested increased; if the removal of the principle was extensive, light appeared. Lubbock contended that Lavoisier's theory did not explain why air in which combustion had taken place was vitiated. His idea was that "phlogisticated air"—Lavoisier's "azotic gas" (nitrogen)—was a compound of principium sorbile with a greater proportion of principium aëri proprium than exists in "pure air."

Lubbock's exposition, which was preceded by a critical examination of Stahl's ideas and some prevailing variants of them, was illustrated by numerous experiments. His work was known in Europe: *principe sorbile* is given as a synonym for *oxigène* in Lavoisier's *Méthode de nomenclature chimique* (1787), and extracts from his dissertation were quoted by Christoph Girtanner in an early textbook based on the antiphlogistic system (1792).

BIBLIOGRAPHY

Lubbock's unique work is *Dissertatio physico-chemica, inauguralis, de principio sorbili, sive communi mutationum chemicarum causa, quaestionem, an phlogiston sit substantia, an qualitas, agitans; et alteram ignis theoriam complectens* . . . (Edinburgh, 1784). His ideas are examined by D. McKie and J. R. Partington, in *Annals of Science*, **3** (1938), 356–361, as part of their "Historical Studies on the Phlogiston Theory," *ibid.*, **2** (1937), 361–404; **3** (1938), 1–58, 337–371 (with portrait of Lubbock); **4** (1940), 113–149. See also J. R. Partington, *A History of Chemistry*, III (London, 1962), 489, 627.

E. L. SCOTT

LUCAS, FRANÇOIS-ÉDOUARD-ANATOLE (*b.* Amiens, France, 1842; *d.* Paris, France, 3 October 1891), *number theory, recreational mathematics.*

Educated at the École Normale in Amiens, he was first employed as an assistant at the Paris Observatory. After serving as an artillery officer in the Franco-Prussian War, he became professor of mathematics at the Lycée Saint-Louis and the Lycée Charlemagne, both in Paris. He was an entertaining teacher. He died as a result of a trivial accident at a banquet; a piece of a dropped plate flew up and gashed his cheek, and within a few days he succumbed to erysipelas.

In number theory his research interest centered on primes and factorization. He devised what is essentially the modern method of testing the primality of Mersenne's numbers, his theorem being as follows: The number $M_p = 2^p - 1$, in which p is a prime, is itself prime if and only if $S_{p-1} \equiv 0 \bmod M_p$, where S belongs to the sequence $S_1 = 4$; $S_n = S_{n-1}^2 - 2$. Using this he was able in 1876 to identify $2^{127} - 1$ as a prime.

The first new Mersenne prime discovered in over a century, it is the largest ever to be checked without electronic help. He loved calculating, wrote on the history of mechanical aids to the process, and worked on plans (never realized) for a large-capacity binary-scale computer. He did some highly original work on the arithmetization of the elliptic functions and on Fibonacci sequences, and he claimed to have made substantial progress in the construction of a proof of Fermat's last theorem.

Lucas' many contributions to number theory were balanced by extensive writings on recreational mathematics, and his four-volume book on the subject remains a classic. Perhaps the best-known of the problems he devised is that of the tower of Hanoi, in which n distinctive rings piled on one of three pegs on a board have to be transferred, in peg-to-peg single steps, to one of the other pegs, the final ordering of the rings to be unchanged.

BIBLIOGRAPHY

I. ORIGINAL WORKS. A comprehensive bibliography of 184 items is appended to Duncan Harkin, "On the Mathematical Works of François-Édouard-Anatole Lucas," in *Enseignement mathématique*, 2nd ser., **3** (1957), 276–288.

Only the 1st vol. of Lucas' projected multivolume *Théorie des nombres* (Paris, 1891) was published. See also *Récréations mathématiques*, 4 vols. (Paris, 1891–1894; repr. Paris, 1960).

II. SECONDARY LITERATURE. Lucas' most important contributions to number theory are synopsized in L. E. Dickson, *History of the Theory of Numbers*, 3 vols. (Washington, 1919–1923)—see esp. vol. I, ch. 17. Harkin's article (see above) is also informative in this respect.

NORMAN T. GRIDGEMAN

LUCAS, KEITH (*b.* Greenwich, England, 8 March 1879; *d.* over Salisbury Plain, near Aldershot, England, 5 October 1916), *physiology*.

Lucas was the second son of Francis Robert Lucas, an inventor and engineer who supervised the laying of the early intercontinental submarine telegraph cables and who ultimately became managing director of the Telegraph Construction and Maintenance Company. Under his father's influence, Lucas early displayed great mechanical ingenuity, remarkable manual dexterity, and a deep interest in science and engineering. His mother was the former Katherine Mary Riddle, granddaughter of Edward Riddle and daughter of John Riddle, successive directors of a school for sons of naval officers in Greenwich and both renowned as teachers of navigation and nautical astronomy.

From the Reverend T. Oldham's preparatory school at Blackheath, Lucas went to Rugby School on a classical scholarship in 1893 and then on a minor classical scholarship to Trinity College, Cambridge, where he matriculated in 1898. Lucas had already decided to study science; and he began at once to read for the natural sciences tripos, in part I of which he took a first class in 1901. In keeping with his interests, and despite two full years of support on a classical scholarship, Lucas had as his director of studies at Trinity the noted physiologist Walter Morley Fletcher. In 1901, under the strain of his studies and the death of an old school friend in the Boer War, Lucas suffered a breakdown in his health and left Cambridge for two years. Part of this period he spent in New Zealand, where he carried out a bathymetric survey of a number of lakes.

By the time he returned to Cambridge in the autumn of 1903, Lucas had decided to devote himself to a career in physiological research. He had in fact already devised, at home and on his own, an impressive photographic recording method for tracing the events of muscle contraction. So clearly had he formulated his research interests that he was allowed to forgo the usual routine of organized course work and examinations for part II of the natural sciences tripos and was immediately given a place in the crowded physiological laboratory at Cambridge, initially in a sort of anteroom passageway and later in a small cellar.

In 1904 Lucas won the Gedge Prize and the Walsingham Medal and was elected fellow of Trinity College. He was appointed additional university demonstrator in physiology in 1907 and lecturer in natural science at Trinity College in 1908. In addition to the B.A. degree in 1901, Lucas received the M.A. from Cambridge in 1905 and the D.Sc. in 1911. He was elected fellow of the Royal Society in 1913, having been Croonian lecturer the year before. From 1906 to 1914 Lucas was a director of the Cambridge Scientific Instrument Company. A member of the Trinity College Council, he also helped to plan the new physiological laboratory built at Cambridge in 1914.

Upon the outbreak of World War I, Lucas enlisted and in September 1914 was assigned to the experimental research department of the Royal Aircraft Factory at Farnborough. There he applied his graphical recording methods to an analysis of roll, pitch, and yaw in airplanes. He also helped to design an accurate bombsight and a new magnetic compass which greatly improved aerial navigation and which was granted a War Office secret patent in July 1915. His contributions in this area were recognized in the published report of the Advisory Committee for Aeronautics. In September 1916, after repeated requests, he was allowed to attend the Central Flying School, where he quickly qualified as a pilot. He was killed, while flying solo, in a midair collision. He was survived by three sons and his wife, the former Alys Hubbard, eldest daughter of the Reverend C. E. Hubbard, whom he had married in 1909. After his death, his wife and sons changed their names by deed poll to "Keith-Lucas."

Lucas' career in physiological research was devoted wholly to investigating the properties of nerve and muscle, especially the characteristics in each of waves of excitation (what Lucas called "propagated disturbances"). The dominant and most valuable features of his work in general were clarity, precision, and enormous methodological originality. To the design of physiological instruments he brought the same basic principle that he had already applied at Rugby to the design of a bicycle and an especially admirable microscope:[1] the design itself should ensure the accurate operation of the device even in the face of shoddy worksmanship and wear and tear. On this basis he designed an instrument for the analysis of photographic curves from the capillary electrometer, a method of drawing fine glass tubes of uniform dimensions for the capillary electrometer, and a photographic time marker along the lines of Einthoven's string galvanometer. To reduce the distortion produced by the inertia of recording levers, he used light levers or photographic methods where feasible. So integral to Lucas' achievements were these and other instruments of his own design that he has been described as "essentially an engineer."[2]

In addition to these important methodological contributions, Lucas produced a series of valuable experimental results and established at least one fundamental principle: the "all or none" law for ordinary skeletal muscle. Since the work of Henry P. Bowditch in 1871, it had been known that cardiac muscle follows an "all or none" rule: a given stimulus either evokes the maximum possible contraction, or it evokes no contraction at all; if a cardiac contraction does occur, its strength is independent of the exciting stimulus. Some indirect evidence existed that this principle applied as well to skeletal muscle, as Francis Gotch had explicitly suggested in 1902; but direct support for the suggestion was lacking. This support was provided by Lucas in two papers published in 1905 and 1909.

In the first paper, Lucas showed that when the frog's *cutaneus dorsi* muscle is stimulated directly by electrical currents, the resulting contraction increases in discrete, discontinuous steps as the stimulus is increased. Whether the preparation is the *cutaneus dorsi* muscle as a whole (consisting of 150 to 200 individual muscle fibers) or only a small section of the muscle, these discrete steps are always fewer than the number of individual muscle fibers in the preparation. This result suggested that the individual muscle fibers fall into distinct groups according to their excitability and that each discrete step of increased contraction marks the excitation of an additional fiber or small group of fibers of similar excitability.

In the paper of 1909, Lucas confirmed these earlier results under more normal conditions, for he now evoked contraction not by direct stimulation of the *cutaneus dorsi* muscle but through stimulation of its motor nerve fibers. Persuasive evidence was thus produced that for each skeletal muscle fiber, as for cardiac muscle, contraction (if evoked at all) is maximal regardless of the strength of the exciting stimulus. A stronger stimulus increases contraction in an ordinary many-fibered muscle only because it activates a larger number of the individual constituent fibers; and the "submaximal" contraction of such a muscle is merely the maximal contraction of less than all of its fibers. The absence of submaximal contractions in the heart is not to be ascribed to any fundamental differences in the functional capacities of skeletal and cardiac muscle cells but, rather, to the fact that cardiac muscle is functionally continuous, its fibers being in connection with one another, while skeletal muscle fibers are separated by their sarcolemma. By 1914 the "all or none" law had been extended to motor nerve fibers by Lucas' most celebrated student, Edgar Douglas Adrian, later a Nobel laureate in physiology or medicine.

In extending to another tissue a property previously established only for cardiac muscle, Lucas followed a course which is at least implicit in most of his work. Upon determining the temperature coefficient for nerve conduction, for example, Lucas pointed out that his mean value of 1.79 for the ten degrees between 8° and 18° C. was similar to values already obtained for muscle conduction, suggesting to him that the conduction process is fundamentally the same in nerve and muscle. In both kinds of tissue, he believed, conduction takes place in two stages: a preliminary local excitatory process at the seat of stimulation, and the subsequent production of a partially independent "propagated disturbance." His attempt to show that the propagated disturbance in muscle is an electric and not a contractile disturbance[3] probably reflects his search for underlying similarities among all excitable tissues.

Similarly motivated was Lucas' comparison of conduction across the myoneural junction, across the A.V. bundle in the heart, and across the synapse in the central nervous system—in all these cases, he argued, impulses pass through a region of diminished conductivity, a "region of decrement."[4] In general, Lucas considered it folly to ascribe special properties to certain kinds of tissue unless and until every attempt had been made to explain the phenomena in terms of known properties common to all excitable tissues. This attitude is evident in his posthumous monograph, *The Conduction of the Nervous Impulse* (1917), in which the phenomena of the peripheral nerves are offered as a guide to processes in the central nervous system.

With respect to the refractory period, inhibition, and the summation of stimuli, Lucas conceived of them in similar terms and attributed them to similar causes, whether they took place in muscle tissue, peripheral nerves, or the central nervous system. In all excitable tissues, he believed, inhibition results when successive stimuli arrive too frequently for the tissue to recover from the impaired conductivity of the refractory period; summation, on the other hand, results when successive stimuli are so timed as to coincide with a temporary postrecovery phase of supernormal excitability and conductivity. This theory of summation, developed by Lucas in association with Adrian, differed essentially from the prevailing theory of Max Verworn and F. W. Fröhlich.

In addition to the "all or none" law and the Lucas-Adrian theory of summation, Lucas contributed to the physicochemical theory of excitation, although his role in this area was more that of an effective and suggestive critic than of a creative pioneer. His departure point here was Walther Hermann Nernst's influential theory of the local excitatory process, according to which the threshold of excitation is reached when a certain difference of ionic concentration is produced at a semipermeable membrane contained in the excitable tissue. Drawing in part on the work of A. V. Hill, another young Cambridge physiologist and future Nobel laureate, Lucas argued that Nernst's theory, properly conceived and appropriately modified, could account for virtually all the phenomena of local excitation. In his Croonian lecture of 1912, Lucas emphasized in typical fashion that what difficulties the theory did present "seem to dovetail into one another, being probably expressions of a common property of the tissues."[5]

Lucas' last research, published posthumously, dealt with the neuromuscular physiology of the crayfish. Besides discovering that the crayfish claw is innervated by two sets of nerve fibers—one responsible for the slow, prolonged contraction of the claw and the other for a brief twitch—Lucas also found that the phase of supernormal excitability is much more pronounced in the crayfish than in the frog, a circumstance that enabled him to produce further support for the Lucas-Adrian theory of summation.[6] But, as Adrian has suggested,[7] the choice of the crayfish as an experimental animal may reflect a shift of emphasis in Lucas' thought far more significant than any of these immediate results. It may be that Lucas had decided to move explicitly and actively in the direction of comparative and evolutionary physiology, a deep and long-standing interest occasionally evident even during his earlier concentration on the frog's neuromuscular tissue.

Almost from the beginning of his career Lucas had noticed that different excitable tissues do possess a few fundamentally different properties despite all attempts to emphasize their similarities. In particular, he found that the value of optimal exciting stimuli varies for different tissues in the same animal and for the same excitable tissues in different animals. So persistent, distinct, and quantifiable are these differences that Lucas was led to postulate the existence of three distinct "excitable substances." Each of these substances was associated with a particular kind of tissue—one with muscle, one with nerve, and one with a hypothetical myoneural "junctional tissue"—and each responded differently to the same exciting currents, especially those of short duration. Moreover, for any one of these excitable substances, the optimal stimulus differed in different animals. In attaching the term "substances" to these characteristic differences in the excitability of different tissues, Lucas was probably influenced by the theory of "receptive substances" then being developed by John Langley, professor of physiology at Cambridge. Certainly Langley had helped to convince Lucas that the region between nerve and muscle contains something whose physiological properties differ markedly from those of ordinary nerve and muscle.[8] That Lucas supposed this something to be a special kind of tissue, rather than a chemical substance, reminds us forcefully that his work preceded the general adoption of the humoral (or chemical) theory of nervous transmission.

In any case, the concept of "excitable substances" was for Lucas only one of the implications to be drawn from the existence of fundamental differences in the excitability of different tissues. From another point of view, even these very basic differences could be referred to a common principle, for they represented "a point of obvious interest in the evolutionary history of the excitable tissues."[9] According to Adrian, Lucas developed his interest in the evolution of function while studying zoology at Cambridge and, under the impulse of this interest, "gave a course of lectures in the Zoology department on the comparative physiology of muscle."[10] An even more direct expression of this interest is provided in a two-part paper of 1909 on the evolution of function. In this paper Lucas both deplores and seeks to explain the lack of interaction between evolutionary concepts and physiological research. His explanation is historical, tracing the separation between function and evolution to the famous debate in the 1820's between Georges Cuvier and Geoffroy Saint-Hilaire. Under Geoffroy's influence, Lucas argues, physiology was excluded from the problem of animal classification and thereby lost the impetus to become a properly comparative and evolutionary science. In advocating a reunion of physiology and evolutionary concepts, Lucas emphasized that "the primary problem of comparative physiology—a problem whose investigation is wholly necessary for the understanding of the evolutionary process . . . [is] the question to what extent and along what lines the functional capabilities of animal cells have been changed in the course of evolution."[11] Lucas' premature death prevents us from knowing to what extent, and with what success, he might have pursued this line of work. As it is, his work stands as a monument of clarity and precision in that era of neurophysiology which preceded the adoption of the chemical theory of nervous transmission and the development of important new methods of amplifying bioelectrical activity.[12]

NOTES

1. About 1925, Lucas' microscope, an unorthodox arrangement with differential screws, was placed among the exhibits of the South Kensington Science Museum. See Col. F. C. Temple, "At Rugby," in *Keith Lucas*, p. 48, *n*. 1.
2. See H. H. Turner, "Ancestry," *ibid*., p. 12.
3. "On the Relation Between the Electric Disturbance in Muscle and the Propagation of the Excited State," in *Journal of Physiology*, **39** (1909), 207–227, esp. 208–210.
4. *The Conduction of the Nervous Impulse*, pp. 68–73.
5. "Croonian Lecture," p. 516.
6. "On Summation of Propagated Disturbances in the Claw of Astacus, and on the Double Neuro-Muscular System of the Adductor," in *Journal of Physiology*, **51** (1917), 1–35.
7. See E. D. Adrian, "Cambridge 1904–1914," in *Keith Lucas*, pp. 105–106.
8. *The Conduction of the Nervous Impulse*, pp. 67–68.
9. "Croonian Lecture," p. 517.
10. Adrian, *Keith Lucas*, p. 105.
11. "The Evolution of Animal Function," pt. II, p. 325.
12. On the importance of the new amplifying methods, see J[oseph] B[arcroft], "Keith Lucas," in *Nature*, **134** (1934), 475.

BIBLIOGRAPHY

I. Original Works. A complete bibliography of Lucas' publications is given in *Keith Lucas* (see below), pp. 129–131. His posthumous monograph, *The Conduction of the Nervous Impulse* (London, 1917), was based on the Page May memorial lectures, delivered by Lucas at University College, London, in 1914. Revised for publication by E. D. Adrian, the monograph emphasizes the work of Lucas and his students, especially Adrian. In addition to this monograph, Lucas published 32 papers, nearly all of them in *Journal of Physiology*. For insight into his general approach and guiding principles, the most valuable are "The Evolution of Animal Function," in *Science Progress*, pt. I, no. 11 (1909), 472–483; pt. II, no. 14 (1909), 321–331; and "Croonian Lecture: The Process of Excitation in Nerve and Muscle," in *Proceedings of the Royal Society*, **85B** (1912), 495–524.

Lucas established the "all or none" law for skeletal muscle in "On the Gradation of Activity in a Skeletal Muscle Fibre," in *Journal of Physiology*, **33** (1905), 125–137; and "The 'All-or-None' Contraction of the Amphibian Skeletal Muscle Fibre," *ibid*., **38** (1909), 113–133. Other important papers by Lucas include those cited in notes 3 and 6; "The Excitable Substances of Amphibian Muscle," in *Journal of Physiology*, **36** (1907), 113–135; "The Temperature Coefficient of the Rate of Conduction in Nerve," *ibid*., **37** (1908), 112–121; "On the Transference of the Propagated Disturbance From Nerve to Muscle With Special Reference to the Apparent Inhibition Described by Wedensky," *ibid*., **43** (1911), 46–90; and "On the Summation of Propagated Disturbances in Nerve and Muscle," *ibid*., **44** (1912), 68–124, written with E. D. Adrian.

II. Secondary Literature. The basic source for Lucas' life is *Keith Lucas* (Cambridge, 1934), a series of sketches by men who knew Lucas well at various stages of his life. Written in 1916, when Walter Fletcher planned to write a memoir of Lucas, these sketches were finally published in their original form after Fletcher's death. The sketches, and their authors, are as follows: H. H. Turner, "Ancestry" and "Earliest Years"; Col. F. C. Temple, "At Rugby"; Sir Walter Fletcher, "Undergraduate Days" and "Return to Cambridge"; G. L. Hodgkin, "New Zealand"; E. D. Adrian, "Cambridge 1904–1914"; and Col. Mervyn O'Gorman, Bertram Hopkinson, and Maj. R. H. Mayo, "Wartime." For an evaluation of Lucas as scientist, the chapters by Fletcher and Adrian are the most valuable.

For other sketches of Lucas' life and work, see Horace Darwin and W. M. Bayliss, "Keith Lucas," in *Proceedings of the Royal Society*, **90B** (1917–1919), xxxi–xlii; C[harles] S. S[herrington], "Keith Lucas," in *Dictionary of National Biography*, supp. 1912–1921, p. 347; and John Langley, "Keith Lucas," in *Nature*, **98** (1916), 109. For a very brief attempt to place Lucas' work in historical perspective, with excerpts from his writings, see Edwin Clarke and C. D. O'Malley, *The Human Brain and Spinal Cord: A Historical Study Illustrated by Writings From Antiquity to the Twentieth Century* (Berkeley, 1968), pp. 218–221.

Gerald L. Geison

LUCIANI, LUIGI (*b*. Ascoli Piceno, Italy, 23 November 1840; *d*. Rome, Italy, 23 June 1919), *physiology*.

Luciani was the son of Serafino Luciani and Aurora Vecchi. In 1860 he completed secondary studies in Ascoli Piceno and then—after a period spent in independently pursuing politics, literature, and philosophy—in 1862 began to study medicine at the University of Bologna. He also studied in Naples for a time, returning to Bologna to graduate in 1868. Following his graduation he remained at the university to work in the physiology laboratory directed by Luigi Vella.

Luciani spent March 1872 to November 1873 in Leipzig, at the physiological institute directed by Carl Ludwig. From that time he regarded Ludwig as his real master and this period of residence in Germany as the main epoch of his scientific life. It did, in fact, leave a deep imprint on his mind. Returning to Bologna Luciani became a lecturer in general pathology; after teaching this subject at Bologna (1873–1874) and Parma (1875–1880), he became professor of physiology in Siena (1880–1882), Florence (1882–1893), and finally Rome (1893–1917), where he died, as professor emeritus, of a urinary disease. Luciani was a Senator of the Kingdom and a member of Italian and foreign academies. He was an excellent teacher (many of his pupils rose to university chairs), a man of vast general and philosophical culture, and an able experimenter.

Luciani not only investigated all phases of physiology—general, human, comparative, and linguistic—but also conducted research in many other fields. (He made observations on silkworms, studied experimental phonetics, and took up the theory of self-intoxication as an effect of experimental removal of the thyroid-parathyroid complex in animals, for example.) His treatise *Fisiologia dell'uomo* went through a number of editions in several languages.

Luciani had made a study of the activity of the cardiac diastole in 1871, before going to Ludwig's laboratory. Once there he took up the work of H. F. Stannius and carried out experimental research on the genesis of the automatic activity of the heart (that is, of the periodic cardiac rhythm, now known as Luciani's phenomenon). To this end he used a graphic method, obtaining tracings of three distinct, characteristic phenomena (access, periodic rhythm, and crisis) which can be interpreted as three different phases of cardiac activity prior to its exhaustion. Luciani subsequently took up the related question of the automatic activity of the respiratory centers. He then carried out studies on cerebral localizations; made contributions to the doctrine of the cortical pathogenesis of epilepsy (1878); performed experimental extirpations of the various regions connected with the sensory functions; and did research on the mixed sensory and motor nature of the cortical excitation.

Luciani also carried out significant research on the physiology of fasting, in which he determined the various changes that, in man, the great organic functions undergo. On the basis of his observations he was able to distinguish three stages in fasting—an initial, or hunger, period; a period of physiological inanition; and a final period of morbid inanition or crisis.

Luciani's most important scientific work, however, was his research on the physiology of the cerebellum (1891). Through skillful vivisection, he succeeded in experimentally removing the cerebellum in the dog and the monkey, thereby making a fundamental contribution to the development of knowledge of the nervous system—and, in fact, illuminating the peculiar function of the cerebellum within the framework of the whole nervous system.

Luciani held that the multiform symptomatology consequent upon decerebellation is a series of discrete phenomena, and distinguished among three clinical stages following decerebellation. The first is characterized by the presence of dynamic signs, the second by motor deficiencies, and the third by compensatory phenomena. The limits between the various stages are not, however, sharply defined. He considered cerebellar ataxy to be based on the triad—atonia, asthenia, astasia—now named for him. Clinical observation largely confirmed this triad as valid in human physiopathology. Luciani also asserted that atonia and asthenia are constant components of cerebellar deficiency in the experimental animal.

BIBLIOGRAPHY

I. Original Works. Luciani published about seventy works, some of them written with collaborators, which appeared in both Italian and foreign journals, or as separate publications, between 1864 and 1917. Baglioni, cited below, provides a complete list. See also Luciani's posthumously published "Cenni autobiografici," in *Archivio di fisiologia*, **19** (1921), 319–349.

II. Secondary Literature. On Luciani's life and work, see esp. Silvestro Baglioni, "Luigi Luciani," in Aldo Mieli, ed., *Gli scienziati italiani dall'inizio del medio evo ai giorni nostri*, I (Rome, 1921), 336–343, with a bibliography of Luciani's publications, works published on the occasion of jubilees, and literature on Luciani. See also Paolo Crepax, "The First Italian Contributions to the Study of Cerebellar Functions and the Work of Luigi Luciani. II. Luigi Luciani," in Luigi Belloni, ed., *Essay on the History of Italian Neurology* (Milan, 1963), 225–236, with bibliography.

Bruno Zanobio
Gianluigi Porta

LUCRETIUS (*b.* Italy, *ca.* 95 B.C.; *d. ca.* 55 B.C.), *natural philosophy.*

Lucretius followed the Epicurean maxim "Live unnoticed" so well that almost nothing is known for certain about him. His "poemata" were known to Cicero and his brother in 54 B.C.[1] St. Jerome reports that he was born in 94 (or 93 or 96, according to other manuscripts), that he was driven mad by an aphrodisiac, that in the intervals between fits of madness he wrote several books, which Cicero afterwards edited or corrected (*emendavit*), and that he took his own life at the age of forty-four.[2] In the life of Vergil attributed to Donatus, it is said that Lucretius died on the same day that Vergil took the *toga virilis*, but the date of this event is stated ambiguously and may be either 55 or 53.

Lucretius was probably a Roman, and probably of aristocratic family, although other views have been advanced. His only work, the poem *De rerum natura*, shows familiarity with Roman life and with Roman literature. It is written for "Memmius," probably C. Memmius, a Roman aristocrat who was praetor in 58 B.C. and later governor of Bithynia. The poem is written in Latin hexameters and divided by the poet into six "books." The title itself is a translation of Περὶ φύσεως, the name of many Greek works of natural philosophy, including a hexameter poem by Empedocles and a long prose treatise by Epicurus, of which only mutilated fragments survive. There is some indication that the poem was not finished—there are repetitious passages (although some may be deliberate), some books are more polished than others, and the end of book VI does not seem to be the intended end of the whole poem.

Books I and II contain an exposition of the elements of the atomic theory. Books III and IV are about the soul. Book V, which is the longest, describes the origin of the cosmos and the natural growth of living creatures and of civilization. Book VI is on sundry natural phenomena of the sky and earth.

The core of Epicurean natural philosophy is given in books I, II, and V, and there are signs that these were the first books in order of composition. There is some correspondence between the order and proportion of these three books and Epicurus' outline of the subject in the extant *Letter to Herodotus*. In the absence of Epicurus' major work, *On Nature*, it is impossible to say how closely Lucretius follows him, but there is perhaps little or nothing in the natural philosophy of Lucretius that was not in that of Epicurus: The poem shows astonishingly little consciousness of such post-Epicurean philosophies as Stoicism. The Italian imagery and the animadversions on Roman life and morality are no doubt Lucretius' own.

The existence of this long poem on the Epicurean world picture is surprising in itself. The extant work of Epicurus shows a prosaic mind; and the rather impoverished utilitarianism that he taught gives little obvious encouragement to a poet, although there is evidence that Epicureans were interested in poetry and music in the fragments of work by Lucretius contemporary Philodemus. But for all one knows, it was lonely poetic genius that produced this poem—one of the greatest in the Latin language.

If one compares Lucretius' poem with the extant writings of Epicurus and with the general picture of Epicurus' philosophy given by ancient critics, certain changes of emphasis are noticeable. Lucretius devoted two whole books, set in the middle of the exposition of the atomic cosmology, to the structure and functions of the soul—a much higher proportion than in Epicurus' *Letter to Herodotus*. Theory of knowledge or "canonic" received little attention from Lucretius. Moreover, the moral lessons, although they are certainly not omitted, occur mainly in the introductions to each book, or in the form of satire on Roman life; there is little preaching in the manner of Epicurus' *Letter to Menoeceus*. For all these differences, the *De rerum natura* remains purely Epicurean in that the acknowledged motive is not disinterested scientific curiosity, but peace of mind.

Lucretius' atomic theory can be understood best by comparing it with the world picture of Plato and Aristotle. In contrast with the finite, single world of Platonic and Aristotelian theory, later adopted by Christian philosophers, the atomists argued that the universe is infinitely large and contains an unlimited number of *cosmoi*, past, present, and future.[3] The single Aristotelian cosmos was without beginning or end in time; there was thus no problem for Aristotelians about the origin of its parts and of their order. The atomic theory, on the other hand, was committed to explaining the origin of everything in the world. Atoms and void were ungenerated and indestructible, but the earth, sea, air, and stars were compounds that came into being at a particular time; their origin had to be explained.

Plato and Aristotle preferred teleological explanations in their natural philosophies. Plato's *Timaeus* offers some grounds for this preference in the notion that the cosmos was created by an intelligent craftsman; the Christian world picture was similar in this respect. Aristotle's eternal cosmos had no creator; the evidence of order and design showed that the cosmos is an ordered whole, but did not point to a designer. Epicurean atomism rejected this teleology, and sought to explain everything by mechanical causes.[4] It was therefore necessary to find plausible explanations for apparently purposive or highly ordered phenomena, and some of Lucretius' most earnest arguments were devoted to this end. The hypothesis of the infinity of the universe was a vital premise in these arguments: in infinite time an infinite supply of atoms moving in infinite space will produce everything that atoms can possibly produce by their combinations.[5]

The theory held that all atoms are too small to be perceived individually. They are all made of the same material, and have no properties other than shape, size, and weight; their weight is the cause of their natural motion downward through the void. This motion could be interrupted by two causes: an unexplained swerve (*clinamen*)—postulated by Epicurus to explain the formation of compounds and to free the movements of the soul from "fate"—and collisions with other atoms.[6] The perceptible qualities of compounds were explained by relating them to atomic shapes and to the proportion of void in the compound. Atoms in compounds move continually, with very frequent collisions in dense compounds, but more freely in compounds containing a greater proportion of void.[7]

Lucretius presented a simple causal theory of perception: all compounds throw off "films" of atoms (*simulacra*) from their surfaces, and these somehow mark their pattern on the soul atoms by direct contact.[8] The soul itself is a mixture of atoms of four kinds—something like heat, something like *pneuma*, something like air, and a fourth unnamed kind.[9] All the soul atoms are highly mobile, but they are not themselves alive; like other atoms they possess no properties other than shape, size, and weight.[10] At death they simply disperse into the air (the theory so confidently rebutted in Plato's *Phaedo*, 70a ff.).[11]

Book V of Lucretius' poem puts the elements of the atomic theory into action, so to speak, to show how the cosmos and all things in it originated from atoms moving in the void, without any divine plan or direction. First the world masses were formed, then the earth began to grow vegetation spontaneously, and finally animal species also emerged from the earth. Once having come from earth, a species survived if it were so adapted as to nourish and reproduce itself; otherwise it simply died out as soon as the earth, growing old, ceased to be spontaneously productive. (The theory thus includes the origin of species and the survival of the fittest, but not the evolution of the species.) It was this account of the natural origin of everything, denying Providence and divine creation, that, along with the denial of the survival of the soul, most antagonized Christian philosophers.[12]

The fatal weakness of the ancient atomic theory, considered as a framework on which to build explanations of natural phenomena, was that it could not offer any laws describing the elementary interactions of atoms. The whole theory depended on the effects of shape, size, and weight of atoms when they collided with each other; but the mechanics of the theory relied on simple intuitions and analogies, such as the vague principle of "like to like" taken over from Democritus. In the ancient world the teleological explanation adopted by Platonists, Aristotelians, and Stoics was clearly more satisfying, especially in biology.

During the early part of the Christian era, Lucretius was first attacked for denying Providence, then, later, forgotten except by grammarians and lexicographers. His poem may have been known in Padua in the thirteenth century,[13] but it was Gian Francesco Poggio's rediscovery of a manuscript early in the fifteenth century that began the great revival of interest in his work. Both the anti-Aristotelian cosmology and the beauty of Lucretius' poetry attracted writers of the Italian Renaissance. Ficino studied him, but rejected his doctrines. Bruno found support in Lucretius' doctrine of the infinite universe and plurality of worlds, and wrote Latin poems in imitation of him. With the full revival of atomism in Western Europe, especially as manifested by Gassendi, it became harder to distinguish the influence of Lucretius from that of Democritus, who was known mainly through Aristotle's polemics, and from that of Epicurus.

NOTES

1. Cicero, *Ad Q. Fratrem*, 2, 9, 3.
2. St. Jerome, on Eusebius' *Chronicle*, 94.
3. *De rerum natura*, I, 951–1082; II, 1023–1089.
4. *Ibid.*, IV, 822–876; V, 110–234.
5. *Ibid.*, I, 1021–1051; II, 1058–1063; V, 419–431.
6. *Ibid.*, II, 62–293.
7. *Ibid.*, II, 332–521; 730–864.
8. *Ibid.*, IV, 26–857.
9. *Ibid.*, III, 177–369.
10. *Ibid.*, II, 865–1022.
11. *Ibid.*, III, 417–829.
12. E.g., Lactantius, *De ira dei*, 10.
13. Billanovich, "Veterum vestigia vatum."

BIBLIOGRAPHY

I. Original Works. The *editio princeps* of *De rerum natura* was that of T. Ferrandus (Brescia, 1473). The best modern English ed. is that of C. Bailey (Oxford, 1947, repr. with corrections, 1949), although the commentary is weak on philosophical matters. Other particularly interesting eds. are those of H. A. J. Munro (Cambridge, 1864) and C. Giussani (Milan, 1896–1898). There is a useful commentary by A. Ernout and L. Robin (Paris, 1925–1926). A reliable but dull English trans. in prose is that of R. E. Latham (Harmondsworth, 1951); a more exciting one, in verse, is by Rolfe Humphries, deplorably entitled *Lucretius: The Way Things Are* (Bloomington, Ind., 1968).

II. Secondary Literature. E. Bignone, *Storia della letteratura Latina*, II (Florence, 1945), 456–462; and C. Bailey and D. E. W. Wormell in M. Platnauer, ed., *Fifty Years (and Twelve) of Classical Scholarship* (Oxford, 1968) contain good secondary bibliographies.

On Lucretius as a natural philosopher see V. E. Alfieri, *Atomos Idea* (Florence, 1953); Anne Amory, "*Obscura de re lucida carmina:* Science and Poetry in *De Rerum Natura*," in *Yale Classical Studies*, **21** (1969), 145–168; C. Bailey, *The Greek Atomists and Epicurus* (Oxford, 1928; repr. New York, 1964); and E. Bignone, cited above. Also see G. Billanovich, "Veterum vestigia vatum," in *Italia Medioevale e Umanistica*, **1** (1958), 155–243; P. Boyancè, *Lucrèce et l'Épicurisme* (Paris, 1963); P. De Lacy, "Limit and Variation in the Epicurean Philosophy," in *Phoenix*, **23** (1969), 104–113; and D. R. Dudley, ed., *Lucretius*

(London, 1965; seven essays by different authors). B. Farrington's *Science and Politics in the Ancient World* (London, 1935) was reviewed by A. Momigliano in *Journal of Roman Studies* (1941), 149 ff.

Other studies include D. J. Furley, "Lucretius and the Stoics," in *Bulletin of the Institute of Classical Studies*, **13** (1966), 13–33; and "Variations on Themes From Empedocles in Lucretius' Proem," *ibid.*, **17** (1970), 55–64; G. D. Hadzits, *Lucretius and His Influence* (New York, 1935); Gerhard Müller, *Die Darstellung der Kinetik bei Lukrez* (Berlin, 1959); G. Santayana, *Three Philosophical Poets: Lucretius, Dante and Goethe* (Cambridge, 1910); F. Solmsen, "Epicurus and Cosmological Heresies," in *American Journal of Philology*, **62** (1951), 1 ff.; "Epicurus on the Growth and Decline of the Cosmos," *ibid.*, **64** (1953), 34 ff.; and W. Spoerri, *Späthellenistische Berichte über Welt, Kultur und Götter* (Basel, 1959).

DAVID J. FURLEY

LUDENDORFF, F. W. HANS (*b.* Thunow, near Koszalin, Germany, 26 May 1873; *d.* Potsdam, Germany, 26 June 1941), *astronomy.*

Ludendorff was the youngest son of Wilhelm Ludendorff, a farmer, and his wife, the former Clara von Tempelhof, a descendant of the old Prussian military family. After an elementary education at home from a tutor, Ludendorff attended the Falk Realgymnasium in Berlin, then he qualified for admission to the University of Berlin in 1892. He studied astronomy, mathematics, and physics and received the Ph.D. in 1897. His main teachers were Förster and Bauschinger. After a year at the Hamburg observatory under Rümcker, Ludendorff joined the astrophysical observatory at Potsdam in 1898 and remained there until his retirement in 1939. In 1905 he became observer, and in 1915, chief observer; in 1921–1939 he was director of the famous institute. Ludendorff was elected a member of the Preussische Akademie der Wissenschaften at Berlin in 1921, a position that carried with it the title of professor and the right to lecture at the University of Berlin.

During his student years, at Hamburg, and early in his stay at Potsdam, Ludendorff was active in both theoretical and measurement astronomy. In 1899 the Potsdam part of the *Photographische Himmelskarte* was completed. Ludendorff assisted in this work and also in the evaluation of the plates until 1900. However, from 1901 to 1921 he worked only in spectrography, to some extent with Eberhard. After becoming director of the observatory, Ludendorff could no longer spend the nights observing and gave up that work. At the beginning of his spectroscopic studies Ludendorff investigated the radial velocities of many stars. He later determined the orbits of spectroscopic binaries, calculated their masses, and regularly published catalogs of their orbital elements. Thus he found that the masses of the B stars are three times greater than the masses of the stars of classes A–K. He was also occupied with the statistics of spectroscopic binaries and classified helium stars. Attempting to treat the δ Cephei and ζ Geminorum variables as binaries, Ludendorff found in 1913 that they could not be double stars. Subsequently he investigated the relationship of these stars and of the planetary nebulae to the helium stars, problems which led him more and more to variable star investigations. Among Ludendorff's spectroscopic work his measurements of radial velocities of the Ursa Major group, especially of star ϵ, should be mentioned.

The last two decades of Ludendorff's life were devoted to the variable stars and to writing. In particular he investigated ϵ Aurigae, *R* Coronae Borealis, and such long-period variables as o Ceti (Mira). Here he did not carry out observations but did critical sifting. He also proposed a new classification of variable stars. With G. Eberhard and A. Kohlschütter he edited the *Handbuch für Astrophysik*, writing parts of volumes VI and VII. He also helped to revise some popular astronomy books.

From the time of a visit to Mexico to observe a solar eclipse in 1923—he observed the solar corona, photometering its spectrum—and to Bolivia, where he erected a branch observatory for the spectrographic *Durchmusterung* of the southern sky in 1926, Ludendorff was interested in the astronomical knowledge of the older American civilizations, especially of the Mayans. He passionately investigated various Mayan inscriptions, relating them to conspicuous constellations long before the Christian era.

From 1933 to 1939 Ludendorff was president of the Astronomische Gesellschaft and presided over three of its congresses. He was also a member of many scientific societies, including the Royal Astronomical Society.

Ludendorff was the youngest brother of General Erich Ludendorff, well-known for his service during World War I and later for his extremist political ideas, of which Hans never approved. Hans Ludendorff served for only a short time as military meteorologist, and though a lance-corporal, he sometimes dined with Hindenburg and his brother Erich. In 1907 Ludendorff married Käthe Schallehn; they had two sons and one daughter.

BIBLIOGRAPHY

Ludendorff's writings include *Die Jupiter-Störungen der kleinen Planeten vom Hecuba-Typus* (Berlin, 1896), his dissertation; "Untersuchungen über die Kopien des Gitters Gautier No. 42 und über Schichtverzerrungen auf photographische Platten," in *Publikationen des Astrophysikalischen Observatoriums zu Potsdam*, **49**, pt. 15 (1903); "Der grosse Sternhaufen im Hercules Messier 13," *ibid.*, **50**, pt. 15 (1905); "Der Veränderliche Stern R Coronae Borealis," *ibid.*, **57**, pt. 19 (1908); "Erweitertes System des Grossen Bären," in *Astronomische Nachrichten*, **180** (1908), 183; "Verzeichn. d. Bahnelem. spektroskop. Doppelsterne," in *Vierteljahrschrift der Astronomische Gesellschaft*, **45–50**, **62** (1910–1915, 1927); "Über die Bezieh. d. verschied. Klass. Veränderl. Sterne," in *Seeliger Festschrift* (Berlin, 1924), pp. 89–93; "Radialgesch. v. ε Aurigae," in *Sitzungsberichte der Preussischen Akademie der Wissenschaften zu Berlin* (1924), 49–69; "Spektroskop. Untersuchungen über d. Sonnenkorona," *ibid.* (1925), 83–113; "Über d. Abhängigkeit d. Form d. Sonnenkorona v. d. Sonnenfleckenhäufigkeit," *ibid.* (1928), 185–214; 13 papers with the general title, "Untersuchungen zur Astronomie der Maya," *ibid.* (1930–1937); and "Über d. Lichtkurven d. Mira-Sterne," *ibid.* (1932), 291–325.

In addition, Ludendorff was editor of *Handbuch der Astrophysik* from 1920 to 1930.

A biography is P. Guthnik, "Hans Ludendorff," in *Vierteljahrschrift der Astronomischen Gesellschaft*, **77** (1942).

H. CHRIST. FREIESLEBEN

LUDWIG, CARL FRIEDRICH WILHELM (*b.* Witzenhausen, Germany, 29 December 1816; *d.* Leipzig, Germany, 27 April 1895), *physiology.*

Ludwig's father, Friedrich Ludwig, was a cavalry officer during the Napoleonic wars, who in 1816 retired from his military post and entered the civil service of the Electorate of Hessen as a *Landrezeptor* at Witzenhausen; he was promoted to *Rentmeister* in 1821, and in 1825 moved to Hanau. Ludwig completed his schooling at the Hanau Gymnasium in 1834, then enrolled as a medical student at the University of Marburg, where he was a stormy petrel among his fellow-students. (A deep scar on his upper lip bore witness in later years to his youthful turbulence and dueling.) He was compelled to leave the university because of conflicts with its authorities, possibly a result of his political activities. After studying at Erlangen and at the surgical school in Bamberg, he was allowed to return to Marburg. He received his medical degree in 1840 and the following year became prosector of anatomy under his friend Ludwig Fick. In 1842 he obtained the *venia legendi*, the right to teach in the medical faculty, with a dissertation on the mechanism of renal function.

In 1846 Ludwig was appointed associate professor (extraordinarius) at Marburg, where he continued his teaching and research until 1849, when he received an appointment as professor of anatomy and physiology at Zurich. Although most of his time was taken up with dissections and lectures, he continued the investigations of the physiology of secretion that he had initiated earlier, and showed that the secretion of the salivary glands is dependent on nerve stimulation and not on the blood supply (1850). During this time he was also working on his *Lehrbuch der Physiologie*, of which the first volume appeared in 1852 and the second in 1856.

In 1855 Ludwig went to Vienna as professor of anatomy and physiology at the Josephinum, the Austrian military-medical academy, which had been organized the preceding year. At this institution he carried on the work that he had begun in Marburg and Zurich, particularly his investigations of the blood gases. In 1859, in a paper published by his student I. M. Sechenov, he described his invention of the mercurial blood pump. This instrument enabled him to separate the gases from a given quantity of blood taken directly *in vivo*. Over the next three decades, Ludwig and his students, who were growing in number, developed these early researches in a variety of directions, eventually embracing the entire subject of respiratory exchange.

When the new chair of physiology was instituted at Leipzig in 1865, Ludwig was offered the post and accepted it. His first problem was the planning of a physiological institute, and he created a model for others to follow. It is probable that Liebig's chemical laboratory at Giessen and Bunsen's physical laboratory at Marburg served as examples. Ludwig's institute, however, was organized in keeping with his broad concept of physiology. He sought to explain vital phenomena in terms of mechanics, that is, through laws of physics and chemistry. It is clear that a visit to Berlin in 1847, in the course of which he met Helmholtz, Brücke, and du Bois-Reymond—men who were to be his lifelong friends and companions in the creation of modern physiology—was influential in the development of his thought.

Ludwig hoped to elucidate physiological problems by combining the study of the anatomy of an organ with a knowledge of the physicochemical changes that occur in its functioning. To this end he created physical, chemical, and anatomical (including histology) divisions in his laboratory, which was built in the form of a capital "E." One wing was devoted to histology, another to physiological chemistry, and the third, main section was equipped for the physical study of physiological problems.

When Ludwig began his career, there was an almost complete lack not only of physiological laboratories but also of experimental instruments. The circumstances under which he worked, particularly in his earlier years, fixed the direction of his ideas and methods. He was both an anatomist and a student of physics, conversant with the newest methods and developments of his science. Bunsen was his intimate friend and influenced him greatly in regard to chemistry and physics. Ludwig's combination of ingenuity, resourcefulness, and knowledge of physical science enabled him to become one of the greatest experimenters in the history of physiology.

In designing the experiments, Ludwig invented numerous methods and instruments. In 1846, he described the kymograph, which, in its subsequent modification by Marey and Chauveau, became a standard tool for the graphic recording of experimental results. His work on circulation led him to devise the mercurial blood pump (1859), the stream gauge (1867), and the method of maintaining circulation in isolated organs (1865).

Ludwig's teaching was as important as his research achievements. The entire work of the institute was characterized by its complete unity of purpose, a quality well illustrated by his lectures. They were usually given at four in the afternoon, and were addressed to the mature student; all those working under Ludwig's direction at the institute attended them, and the undergraduates often had a difficult time (most of them heard the lectures several semesters before being examined). Ludwig attacked his lectures in the same thorough, enthusiastic manner that characterized all his work. The focal point was the illustrative experiments, often the result of recent work done in the laboratory, which Ludwig prepared each day, assisted only by his laboratory aide, Salvenmoser.

The spirit of the laboratory was compounded of equal shares of hard work and enthusiasm and the students sensed it immediately. Ludwig's capacity to judge the abilities of his students led to a useful division of labor. He found something for each worker to do, and he collaborated in the experimentation as well as in the publication of the final results. His role as a teacher was enhanced still further by his *Lehrbuch*, the first modern text on physiology. The book was meant for students and made no attempt to define life or to elucidate "vital forces" or "biological principles." In accordance with Ludwig's basic approach, all explanations of living processes were given in terms of physics and chemistry and the research results that he presented bore testimony to his use of the methods of these sciences.

Ludwig's desire to investigate and to explain vital processes in physicochemical terms was apparent as early as 1842 in his Marburg dissertation (published the following year as *Beiträge zur Lehre vom Mechanismus der Harnsekretion*). In it Ludwig developed a physical theory of renal secretion, suggested to him by the structure of the kidney glomeruli, whereby the first stage in renal excretion represented a diffusion of liquid through a membrane because of a difference in pressure between the two sides. He sought to support this theory in further experiments published in 1849 and 1856; these researches are fundamental to present knowledge of diffusion through membranes. In 1851 Ludwig discovered through another series of experiments that salivary secretion depends on initiation by the glandular nerves, the pressure of the secretion being independent of the blood pressure.

Ludwig was interested, too, in all the aspects of circulation. He studied the fluidity of the circulating blood and its lateral pressure, as well as the dependence of these functions on cardiac activity, on the muscles, and on various other factors. He investigated the physiology of respiration, including gas exchange in the lungs, the respiratory movements, and tissue respiration. He also worked on the lymphatic system, making quantitative as well as histological determinations. His treatise on the structure of the kidneys is a classic, and a large part of what is now known about the mechanism of cardiac activity is based upon his work.

Ludwig belonged to that remarkable group of German physiologists and teachers who in the latter half of the nineteenth century created modern physiology. In 1874, on the twenty-fifth anniversary of his professorship, students who had worked with him at Marburg, Zurich, Vienna, and Leipzig presented him with a distinguished collection of original papers. On his seventieth birthday, in 1886, a second *Festgabe* was presented to him by former students. At the time of his death, almost every physiologist who was active had at some juncture studied with him.

In 1849 Ludwig married Christiane Endemann, the daughter of a Marburg law professor. They had two children, one of whom, a daughter, survived him, as did his wife. He died at the age of seventy-nine, having been ill with bronchitis for seven weeks.

BIBLIOGRAPHY

I. Original Works. For an understanding of Ludwig's work, see *Lehrbuch der Physiologie des Menschen*, I (Heidelberg, 1852), II (Heidelberg, 1856). His other publications, including those of his students, are listed in Heinz Schröer, *Carl Ludwig, Begründer der messenden Experimentalphysiologie 1816–1895* (Stuttgart, 1967), 294–312, which also contains a number of Ludwig's letters. For the correspondence between Ludwig and du Bois-Reymond, see Estelle du Bois-Reymond, ed., *Zwei Grosse Naturforscher des 19. Jahrhunderts. Ein Briefwechsel zwischen Emil Du Bois-Reymond und Carl Ludwig* (Leipzig, 1927), with foreword and annotations by Paul Diepgen; other letters are included in E. Ebstein, "Briefe von Du Bois-Reymond, Claude Bernard and Johannes Müller an Karl Ludwig," in *Medizinische Klinik*, **40** (1925), 1517.

II. SECONDARY LITERATURE. For information on Ludwig's life and work, see T. N. Bonner, *American Doctors and German Universities* (Lincoln, Nebraska, 1963), 120–129; J. Burdon-Sanderson, "Carl Friedrich Wilhelm Ludwig," in *Proceedings. Royal Society of London*, **59** pt. 2 (1895–1896), 1–8; Simon Flexner and James Thomas Flexner, *William Henry Welch and the Heroic Age of American Medicine* (New York, 1941), 84 ff.; Morton H. Frank and Joyce J. Weiss, "The 'Introduction' to Carl Ludwig's Textbook of Human Physiology," in *Medical History*, **10** (1966), 76–86; Wilhelm His, "Carl Ludwig and Karl Thiersch," in *Popular Science Monthly*, **52** (1898), 338–353; Hugo Kronecker, "Carl Friedrich Wilhelm Ludwig," in *Berliner klinische Wochenschrift*, **25** (1895), 466; Erna Lesky, "Zu Carl Ludwigs Wiener Zeit, 1855–1865," in *Sudhoffs Archive*, **46** (1962), 178; and Warren P. Lombard, "The Life and Work of Carl Ludwig," in *Science*, **44** (1916), 363–375.

See also N. S. R. Maluf, "Carson, Ludwig, and Donders and the Negative Interpleural Pressure," in *Bulletin of the History of Medicine*, **16** (1944), 417–418; "How a Physiologist Anticipates a Physical Chemist," *ibid.*, **14** (1943), 352–365; George Rosen, "Carl Ludwig and His American Students," *ibid.*, **4** (1936), 609–650; K. E. Rothschuh, "Carl-Ludwig-Portraits in der bildenden Kunst," in *Archiv für Kreislaufforschung*, **33** (1960), 33; William Stirling, "Carl Ludwig, Professor of Physiology at the University of Leipzig," in *Science Progress*, **4** (1895), 155–176; Robert Tigerstedt, "Karl Ludwig, Denkrede," in *Biographische Blätter*, **1**, pt. 3 (1895), 271–279; Gerald B. Webb and Desmond Powell, *Henry Sewall, Physiologist and Physician* (Baltimore, Md., 1946), 46–50.

GEORGE ROSEN

LUEROTH (or LÜROTH), JAKOB (*b.* Mannheim, Germany, 18 February 1844; *d.* Munich, Germany, 14 September 1910), *mathematics.*

Lueroth's first scientific interests lay in astronomy. He began making observations while he was still in secondary school, but since he was hampered by bad eyesight he took up the study of mathematics instead. He attended the universities of Heidelberg, Berlin, and Giessen from 1863 until 1866; he had already, in 1865, written his doctoral dissertation on the Pascal configuration. In 1867 he became *Privatdozent* at the University of Heidelberg, and two years later, when he was still only twenty-five years old, he was appointed professor ordinarius at the Technische Hochschule in Karlsruhe. From 1880 until 1883 he taught at the Technische Hochschule of Munich, and from the latter year until his death, at the University of Freiburg.

Lueroth's first mathematical publications were concerned with questions in analytical geometry, linear geometry, and theory of invariants, a development of the work of his teachers Hesse and Clebsch. His name is associated with three specific contributions to science. The first of these, a covariant of a given ternary form of fourth degree, is called the "Lueroth quartic," and Lueroth discovered it when he examined, following Clebsch, the condition under which a ternary quartic form may be represented as a sum of five fourth-powers of linear forms. In 1876 he demonstrated the "Lueroth theorem," whereby each uni-rational curve is rational—Castelnuovo in 1895 proved the analogous but more difficult theorem for surfaces. Finally, the "Clebsch-Lueroth method" may be employed in the construction of a Riemann surface for a given algebraic curve in the complex plane.

In addition, Lueroth worked in other areas of mathematics far removed from algebraic geometry. He obtained partial proof of the topological invariance of dimension (proved in 1911 by L. Brouwer) and, following the work of Staudt, did research in complex geometry. He was also involved in the logical researches of his friend Schröder and published two books in applied mathematics and mechanics. These were *Grundriss der Mechanik*, in which he used the vector calculus for the first time, and *Vorlesungen über numerisches Rechnen.* Lueroth collaborated in editing the collected works of Hesse and Grassmann.

BIBLIOGRAPHY

I. ORIGINAL WORKS. A complete bibliography of Lueroth's works may be found in the Brill and Noether obituary (see below). The papers containing his main discoveries are "Einige Eigenschaften einer gewissen Gattung von Kurven 4. Ordnung," in *Mathematische Annalen*, **2** (1869), 37–53; "Das Imaginäre in der Geometrie und das Rechnen mit Würfen," *ibid.*, **8** (1875), 145–214; "Beweis eines Satzes über rationale Kurven," *ibid.*, **9** (1876), 163–165; and "Über die kanonischen Querschnitte einer Riemannschen Fläche," in *Sitzungsberichte der physikalisch-medizinischen Sozietät in Erlangen*, **15** (1883), 24–30. He also published *Grundriss der Mechanik* (Munich, 1881), and *Vorlesungen über numerisches Rechnen* (Leipzig, 1900).

II. SECONDARY LITERATURE. An obituary is A. Brill and M. Noether, "Jacob Lueroth," in *Jahresberichte der Deutschen Mathematikervereinigung*, **20** (1911), 279–299.

WERNER BURAU

LUGEON, MAURICE (*b.* Poissy, France, 10 July 1870; *d.* Lausanne, Switzerland, 23 October 1953), *geology.*

Lugeon was the youngest of the five children of David Lugeon, a sculptor and burgher of Chevilly, Vaud, Switzerland, and Adèle Cauchois, who came from Normandy. The elder Lugeon had been a member of the staff brought together by Viollet-le-Duc to restore the cathedrals of France; the family returned to Switzerland in 1876, when he was commissioned to take part in the restoration of the cathedral in Lausanne. The family had limited means, and Lugeon began studying for a career in banking or industry, entering a bank as an apprentice after six years of school. (It is interesting to note that later in life he served on several boards of directors, continuing his early financial inclination.)

Lugeon's interest in geology was awakened on a school trip to Mt. Salève, near Geneva, when he was ten years old. By the time he was twelve, he was a regular visitor to the geology collections of the Lausanne university museum, where he met Eugene Renevier and his assistant, Théophile Rittener, both of whom took an interest in him. When Lugeon was fifteen, Rittener took him on a summer mapping trip in the mountains of Savoy, and Lugeon succeeded him as Renevier's assistant.

Having attained a monthly salary of seventy-five francs in his new post, Lugeon decided to advance his education by taking final school-leaving examinations. This was only the first step in his aspiration to earn an academic degree, and his circumstances remained difficult. He held a number of different jobs during this period, among them collecting and classifying lichens for the botanical department of the museum and working as a reporter. He also found time to publish several papers on paleontology.

In 1891 Lugeon accompanied Renevier to the mountains of Savoy to do mapping for the Société Géologique de France. On this expedition he received intensive instruction in stratigraphy—Renevier's chief interest—which became the basis for much of his later work. Renevier sent Lugeon to Munich, where there was a center for paleontological research, directed by Zittel, in the winter of 1893–1894; Lugeon thus received further training in this subject. The following spring, Lugeon returned to Savoy and in December 1895 he presented his thesis, "La région de la brèche du Chablais (Haute-Savoie)," to the University of Lausanne. In 1897—the same year in which he married Ida Welti, a grandniece of Oswald Heer—Lugeon became a lecturer at the university; he was made titular professor in 1898, and in 1906 he succeeded Renevier as head of the department of geology.

Lugeon's first inaugural lecture, on drainage and physiographic problems, was an important contribution to alpine geology, but it was overshadowed by the more fundamental (and more decisive to the development of geology) paper that he delivered to the Société Géologique de France in 1901. The latter, entitled "Les grandes nappes de recouvrement des Alpes du Chablais et de la Suisse," was the product of seven years' fieldwork and, more important, exhibited Lugeon's ability to organize a multitude of observations within a new framework—that of the new science of tectonics, for which the memoir became a source book. In it, Lugeon described the macrostructure of the whole Alpine chain, and showed for the first time the interrelationships of a vast series of recumbent folds and successive thrust sheets—the *nappes de charriage* (or *nappes de recouvrement*) or *Decken*—within the historical dynamics of the Alps. His radical view of the enormous overturned folds of the Valais or Pennine Alps was later dramatically confirmed in the building of the Simplon Tunnel, and the tectonic theories set out in his paper have since been applied to all parts of the world.

Lugeon's work opened a new era in the systematic alpine geology that had been developing since the time of Saussure and Buch. The stratigraphic succession of the Alps had gradually been defined, and they had been geologically mapped for the first time in 1853; a new set of sheets, to a scale of 1 : 100,000, was published between 1865 and 1887. With this increasingly detailed data, however, a greater number of irregularities or abnormalities in the stratigraphic succession became apparent, and new explanations were necessary. J. J. Scheuchzer had depicted folds and Escher von der Linth had suspected both recumbent folds and nappes or overthrusts (although he had been afraid to publish such heretical views); Escher's successor Albert Heim had explored and described the famous Glarus double fault, and thereby laid the foundation of the study of rock deformations. In 1883 Marcel Bertrand reinterpreted Heim's findings and provided one of the most important arguments for the unilateral movement that Suess had already proposed in 1875. In 1893 Hans Schardt had recognized the exotic origin of the Prealps, and had explained it by gravity gliding.

But these were exceptional analyses, and the structure of the Alps as a whole was still understood only as a chaos of many different fold directions and interfering faults. The hypotheses then available were inadequate to a satisfactory explanation of the broad range of observations, and the value of Lugeon's pioneering paper lay in the synthesis that it offered. Following the presentation of his memoir in 1901,

Lugeon conducted the annual field meeting of the Société Géologique de France through the Savoy Prealps and convinced many of the participants of the allochthony of those mountains.

Two years later, in 1903, Lugeon, who had never visited the Carpathians, studied published observations of them and pointed out that they, like the Alps, were formed by recumbent folds and thrust masses. His interpretations, drawn as they were from documentary evidence only, were opposed by some geologists, but the results obtained by the Vienna Geological Congress, during an excursion to the Tatras, vindicated his ideas. In 1907 V. Uhlig, the leading authority on Carpathian geology, accepted Lugeon's views, while in 1909 Suess drew upon them in his *Das Antlitz der Erde*. In addition to introducing a whole new interpretative principle, Lugeon was responsible for both the concepts and terminology of such tectonic staples as autochthony, allochthony, window, and involution.

Lugeon had thus attained an international reputation by the time he was thirty. When he became head of the department of geology at the University of Lausanne in 1906, scientists and students came to him from all over the world. Many of the latter became distinguished geologists, most notably, perhaps, Émile Argand, who began to explore the Pennine Alps with his teacher, and who later developed Lugeon's methods further. Lugeon not only helped, but often collaborated directly with his fellow scientists, as in the analysis of the periodic variations of the Swiss glaciers that he carried out with F. A. Forel and the studies of the northern Alps that he made with Emile Haug. In a less direct way, his ideas stimulated Termier to demonstrate windows in the eastern Alps.

Lugeon made numerous scientific trips, on which he made such significant observations as the nappes of Sicily. He generally preferred, however, to remain in France and Switzerland and make more and more detailed studies of areas that he knew well. From 1900 he was engaged in highly refined mapping of the high calcareous Alps between the Sanetsch Pass and the valley of the Kander. This map was published in 1910, with four books of explicative descriptions, sectional maps, and panoramas that brilliantly corroborated his theories. He continued mapping this area for the next several decades; this cartographic work, marked as it is by the careful analysis of local structures and the integration of this analysis into a comprehensive theory, remains a major contribution to the tectonic literature of the first half of the twentieth century.

Lugeon also served as an expert consultant in applied geology, especially in the determination of dam sites. These activities took him and his staff all over the world, and a report of their work was published in 1932. Lugeon's services were in such demand that he soon had to limit himself to taking up only the most difficult problems, although he continued to work as a consultant for the rest of his life. He had once defined the relationship between theoretical and applied geology for his students: "Faites de la bonne géologie, on peut toujours l'appliquer."

Despite his other projects, structural geology in all its facets—from local details to the grand problems of orogeny—remained his chief interest. In 1941 he published, with his successor Gagnebin, a memoir in which he took up the idea of gravity gliding to explain the emplacement of the Prealps and offered demonstrations of many new consequences of this hypothesis. Between 1941 and 1948 he completed a geological relief map, of which several copies are extant, of the high calcareous Alps that embodied his own cartographic results.

Lugeon was witty and ironical, and a lover of practical jokes (about which there are numerous stories, some of them apocryphal, many authentic). In addition to his scientific works, he wrote on food and wines and he was especially proud of his *La fondue Vaudoise*. Although his last years were not untroubled, he maintained a vivid scientific curiosity and interest in the work of his friends and colleagues. He died in his home, "Les Préalpes," and was buried in the cemetery of Chevilly, where his family had lived for generations.

BIBLIOGRAPHY

I. Original Works. A complete bibliography of Lugeon's works may be found in the biographies by H. Badoux, in *Actes Société helvétique des sciences naturelles*, **133** (1953), 327–341, with a list of his honors and a portrait; P. Fallot, in *Bulletin de la Société géologique de France*, 6th ser., **4** (1954), 303–340, with portrait; trans. into English in *Proceedings. Geological Society of America* (1955), 122–127.

Individual writings include "La région de la brèche du Chablais (Haute-Savoie)," in *Bulletin des Services de la carte géologique de la France et des topographies souterraines*, **7** (1896); "Leçon d'ouverture du cours de géographie physique professé à l'Université de Lausanne. (La loi des vallées transversales des Alpes occidentales; l'histoire de l'Isère; le Rhône était-il tributaire du Rhin?)," in *Bulletin de la Société vaudoise des sciences naturelles*, **33** (1897), 49–78; "Les dislocations des Bauges (Savoie)," in *Bulletin des Services de la carte géologique de la France et des topographies souterraines*, **11**, no. 77 (1900), 359–470; "Réunion extraordinaire de la Société géologique de

France à Lausanne et dans le Chablais. Programme et comptes rendus des excursions et des séances, avec notes diverses," in *Bulletin de la Société géologique de France*, 4th ser., **1** (1901), 678–722; and "Les grandes nappes de recouvrement des Alpes du Chablais et de la Suisse," *ibid.*, **1** (1901), 723–825, with a letter from A. Heim.

See also "Sur la coupe géologique du Simplon," in *Comptes rendus hebdomadaires des séances de l'Académie des sciences*, **134** (1902), and in *Bulletin de la Société vaudoise des sciences naturelles*, **38** (1902), 34–41; "Les nappes de recouvrement de la Tatra et l'origine des klippes des Carpathes," *ibid.*, **39** (1903), and in *Bulletin des laboratoires de géologie, géographie physique, minéralogie, paléontologie, géophysique et du Musée géologique de l'Université de Lausanne*, no. 4; "Sur l'existence, dans le Salzkammergut, de quatre nappes de charriage superposées," in *Comptes rendus hebdomadaires des séances de l'Académie des sciences* (1904), written with E. Haug; "Sur la grande nappe de recouvrement de la Sicile," *ibid.* (1906), written with E. Argand; "La fenêtre de Saint-Nicolas," in *Bulletin de la Société vaudoise des sciences naturelles*, **43** (1907), 57; and "Sur les relations tectoniques des Préalpes internes avec les nappes helvétiques de Morcles et des Diablerets," in *Comptes rendus hebdomadaires des séances de l'Académie des sciences* (1909).

See further "Carte géologique des Hautes-Alpes calcaires entre la Lizerne et la Kander. Échelle 1 : 50,000," in *Beiträge zur geologischen Karte der Schweiz;* "Sur deux phases de plissements paléozoiques dans les Alpes," in *Comptes rendus hebdomadaires des séances de l'Académie des sciences* (1911); "Sur l'existence de grandes nappes de recouvrement dans le bassin du Sebou," *ibid.* (1918), written with L. Gentil and L. Joleaud; and "Observations et vues nouvelles sur la géologie des Préalpes romandes," in *Bulletin des laboratoires de géologie, géographie physique, minéralogie, paléontologie, géophysique et du Musée géologique de l'Université de Lausanne*, no. 72 (1941), and *Mémoires de la Société vaudoise des sciences naturelles*, **47** (1941), 1–90, written with Elie Gagnebin.

II. SECONDARY LITERATURE. For further information on Lugeon's life, see the obituary notices by C. Jacob, in *Comptes rendus hebdomadaires des séances de l'Académie des sciences*, **237** (1953), 1045–1050; A. Lombard, in *Revue de l'Université de Bruxelles*, **2** (1954); and E. Wegmann, in *Geologische Rundschau*, **42**, no. 2 (1954), 311–314. See also L. Seylaz, *Les Alpes* (1953), pp. 177–178.

EUGENE WEGMANN

LUGININ, VLADIMIR FEDOROVICH (*b.* Moscow, Russia, 2 June 1834; *d.* Paris, France, 26 October 1911), *chemistry.*

Luginin graduated from the Mikhailov Artillery School in St. Petersburg in 1853 and the Artillery Academy in 1858.

In 1854–1856 Luginin took part in the Crimean War as an artillery officer. In the 1860's he was active in politics. He was a member of Velikoruss, a secret revolutionary-democratic organization, and was in close contact with Alexander Herzen, N. P. Ogarev, and Mikhail Bakunin. At one time Herzen suggested that Luginin should join the staff of the publication *Kolokol* ("The Bell"). In 1867 Luginin abandoned all political activity and devoted himself entirely to science. From 1874 to 1881 he worked in his own laboratory in St. Petersburg, and from 1882 to 1889 in Paris. Between 1889 and 1905 Luginin's scientific and educational activities were centered around Moscow University, where with his own funds he established the first thermochemical laboratory in Russia, which today bears his name.

Luginin's interest in thermochemistry was influenced by Berthelot, with whom he published a number of articles. Their thermochemical research on the heats of neutralization of citric and phosphoric acids (1875) proved these acids to be tribasic. In 1879 he investigated the effect of exchanging the hydrogen atoms with electronegative groups as well as with NH_2 groups at the heat of neutralization of substances belonging to various classes of organic chemistry. From 1880 Luginin engaged in important research to determine the heats of combustion of organic compounds.

In 1894, Luginin published a monograph entitled *Opisanie razlichnykh metodov opredelenia teplot gorenia organicheskikh soedineny* ("Description of Various Methods for Determining the Heats of Combustion of Organic Compounds"). This led the development of methods and techniques in thermochemical research. During 1894–1900, Luginin investigated the specific heats and latent heats of evaporation of various compounds in liquid form. He supplied a significant amount of exact data for the calculation of the heats of formation of organic substances. His work demonstrated the limits of applicability of Trouton's rule.

BIBLIOGRAPHY

Luginin's writings include "Étude thermochimique sur l'effet produit par les substitutions de Cl et de NO_2 et de NH_2 dans les corps de différentes groupes de la chimie organique," in *Annales de chimie et de physique*, 5th ser., **17** (1879), 222–268; "Sur la mesure de chaleur de combustion des matières organiques," *ibid.*, **27** (1882), 347–374; *Opisanie razlichnykh metodov opredelenia teplot gorenia organicheskikh soedineny* (Moscow, 1894); and *Méthodes de calorimétrie, usitées au laboratoire thermique de l'Université de Moscou* (Geneva, 1908), written with A. Shukarev.

A biography is Y. I. Soloviev and P. I. Starosselsky, *Vladimir Fedorovich Luginin* (Moscow, 1963).

Y. I. SOLOVIEV

ŁUKASIEWICZ, JAN (*b.* Lvov, Austrian Galicia [now Ukrainian S.S.R.], 21 December 1878; *d.* Dublin, Ireland, 13 February 1956), *mathematical logic.*

Łukasiewicz' father, Paul, was a captain in the Austrian army; his mother, the former Leopoldine Holtzer, was the daughter of an Austrian civil servant. The family was Roman Catholic, and the language spoken at home was Polish. Young Łukasiewicz studied mathematics and philosophy at the University of Lvov, earning his doctorate *sub auspiciis imperatoris*, a rare honor (1902). At the same institution he received his *Habilitation* (1906) and lectured in logic and philosophy, as *Privatdozent* until 1911, then as extraordinary professor. In 1915 Łukasiewicz accepted an invitation to lecture at the University of Warsaw, then in German-occupied territory.

Between the world wars, as a citizen of independent Poland, Łukasiewicz was minister of education (1919), professor at the University of Warsaw (1920–1939), twice rector of that institution, an active member of scientific societies, and the recipient of several honors. He and Stanislaw Leśniewski founded the Warsaw school of logic, which A. Tarski helped make world famous. Viewing mathematical logic as an instrument of inquiry into the foundations of mathematics and the methodology of empirical science, Łukasiewicz succeeded in making it a required subject for mathematics and science students in Polish universities. His lucid lectures attracted students of the humanities as well.

The sufferings endured by Łukasiewicz and his wife (the former Regina Barwinska) during World War II are poignantly recalled in an autobiographical note. (See Sobociński's "In Memoriam," cited below.) In 1946 Łukasiewicz, then an exile in Belgium, accepted a professorship at the Royal Irish Academy, Dublin, where he remained until his death.

After some early essays on the principles of noncontradiction and excluded middle (1910), Łukasiewicz arrived by 1917 at the conception of a three-valued propositional calculus. His subsequent researches on many-valued logics is regarded by some as his greatest contribution. He viewed these "non-Aristotelian" logics as representing possible new ways of thinking, and he experimented with interpreting them in modal terms and in probability terms. The nonstandard systems he developed have value independently of the philosophy that inspired them or of the usefulness of those interpretations. Łukasiewicz created the elegant "Łukasiewicz system" for two-valued propositional logic and the parenthesis-free "Polish notation."

The metalogic (a term he coined on the model of Hilbert's terminology) of propositional calculi, notably the theory of their syntactic and semantic completeness, owes much to Łukasiewicz and his school. He regarded these studies as a prelude to analogous investigations for the rest of logic, which were then carried out by Tarski.

Using modern formal techniques, Łukasiewicz reconstructed and reevaluated ancient and medieval logic. Through his work in this area, we have changed our view of the history of logic.

During his last years in Ireland, Łukasiewicz published important studies on modal and intuitionistic logic, and he again made logical history with a detailed and novel study of Aristotle's syllogistic. Essentially he interpreted syllogisms in Aristotle to be theorems of logic, not rules of derivations.

BIBLIOGRAPHY

I. Original Works. Most of Łukasiewicz' contributions were first presented in short notes, often in Polish, or in his university lectures. A list of all, or almost all, of his publications is appended to Andrzej Mostowski's "L'oeuvre scientifique de Jan Łukasiewicz dans le domaine de la logique mathématique," in *Fundamenta mathematicae*, **44** (1957), 1–11. His following writings present important results systematically: *Elementy logiki mathematycznej* ("Elements of Mathematical Logic"; Warsaw, 1929; 2nd ed., 1958), translated by Olgierd Wojtasiewicz as *Elements of Mathematical Logic* (New York, 1963); "Philosophische Bemerkungen zu mehrwertigen Systemen des Aussagenkalküls," in *Comptes rendus des séances de la Société des sciences et des lettres de Varsovie*, Cl. III, **23** (1930), 51–77, written with Alfred Tarski; "Untersuchungen über den Aussagenkalkül," *ibid.*, 30–50; "Zur Geschichte der Aussagenlogik," in *Erkenntnis*, **5** (1935–1936), 111–131; "A System of Modal Logic," in *Journal of Computing Systems*, **1** (1953), 111–149; and *Aristotle's Syllogistic From the Stand-Point of Modern Formal Logic*, 2nd ed. (Oxford, 1957).

II. Secondary Literature. The following two articles jointly constitute a valuable survey of Łukasiewicz' lifework as a logician, philosopher, and historian of logic: L. Borkowski and J. Słupecki, "The Logical Works of Jan Łukasiewicz," in *Studia logica*, **8** (1958), 7–56; and Tadeusz Kotarbiński, "Jan Łukasiewicz's Works on the History of Logic," *ibid.*, 57–62. Shorter general treatments of Łukasiewicz's work are the Mostowski article cited above; Bołeslaw Sobociński, "In Memoriam Jan Łukasiewicz," in *Philosophical Studies* (Maynooth, Ireland), **6** (1956), 3–49, which contains an autobiographical note, "Curriculum vitae of Jan Łukasiewicz," and a bibliography; and Heinrich Scholz, "In Memoriam Jan Łukasiewicz," in *Archiv für mathematische Logik und Grundlagenforschung*, **3** (1957), 3–18, which contains an excellent summary of the technical aspects of Łukasiewicz' contributions.

Łukasiewicz' exegesis of Aristotle's syllogistic is disputed in Arthur N. Prior, "Łukasiewicz's Symbolic Logic," in *Australasian Journal of Philosophy*, **30** (1952), 33–46, and is discussed in Gunther Patzig, *Die Aristotelische Syllogistik: Logisch-philologische Untersuchungen über das Buch A der "Ersten Analytiken"* (Göttingen, 1959); English trans. by J. Barnes, *Aristotle's Theory of the Syllogism: A Logico-philological Study of Book A of the Prior Analytics* (Dordrecht, 1968), *passim*, esp. 196–202.

For a general evaluation of Łukasiewicz' philosophical and logical ideas, see Henryk Skolimowsky, *Polish Analytical Philosophy: A Survey and a Comparison with British Analytical Philosophy* (New York, 1967), 56–72.

GEORGE GOE

LULL, RAMON (*b.* Ciutat de Mallorques [now Palma de Mallorca], *ca.* 1232; *d.* Ciutat de Mallorques [?], January/March [?] 1316), *polymathy.*

A Catalan encyclopedist, Lull invented an "art of finding truth" which inspired Leibniz's dream of a universal algebra four centuries later. His contributions to science are understandable only when examined in their historical and theological context. The son of a Catalan nobleman of the same name who participated in the reconquest of Mallorca from the Moors, Lull was brought up with James the Conqueror's younger son (later crowned James II of Mallorca), whose seneschal he became. About six years after his marriage to Blanca Picany (1257) he was converted from a courtly to a religious way of life, following a series of visions of Christ crucified. He never took holy orders (although he may have become a Franciscan tertiary in 1295), but his subsequent career was dominated by three religious resolutions: to become a missionary and attain martyrdom, to establish colleges where missionaries would study oriental languages, and to provide them with "the best book[s] in the world against the errors of the infidel."[1]

Lull's preparations lasted a decade; his remaining forty years (from 1275, when he was summoned by Prince James to Montpellier, where he lectured on the early versions of his Art) were spent in writing, preaching, lecturing, and traveling (including missionary journeys to Tunis in 1292; Bougie, Algeria, in 1307; and Tunis late in 1315), and in attempts to secure support from numerous kings and four successive popes for his proposed colleges. During Lull's lifetime only James II of Mallorca established such a foundation (1276, the year of his accession); when he lost Mallorca to his elder brother, Peter III of Aragon, the college at Miramar apparently was abandoned (*ca.* 1292). In Lull's old age his proposals were finally approved by the Council of Vienne (1311–1312); and colleges for the study of Arabic, Hebrew, and Chaldean were founded in Rome, Bologna, Paris, Salamanca, and Oxford after Lull's death. Pious tradition has it that he died after being stoned by Muslims in Bougie (January 1316[?]), although his actual death is variously said to have occurred in Bougie, at sea, or in Mallorca; modern scholars doubt the historicity of his martyrdom. As for his third resolution, it led to the various versions of Lull's Art—and all his scientific contributions were by-products of this enterprise.

James the Conqueror's chief adviser, the Dominican Saint Ramon de Penyafort, dissuaded Lull from studying in Paris, where his age and lack of Latin would have told against him; he therefore studied informally in Mallorca (1265[?]–1273[?]). His thought was thus not structured at the formative stage by the Scholastic training which molded most other late medieval Christian thinkers; this fostered the development of his highly idiosyncratic system by leaving his mind open to numerous non-Scholastic sources. These included cabalism (then flourishing in learned Jewish circles in both Catalonia and Italy), earlier Christian writers discarded by Scholasticism (for instance, John Scotus Eriugena, whose ninth-century *De divisione naturae* influenced Lullian cosmological works, notably the *Liber chaos*, either directly or indirectly—and hence also his Art), and probably also Arabic humoral medicine and astrology. The Augustinian Neoplatonism of the Victorines also proved important, partly because of its continuing prominence but mainly because its marked coincidences with both Islamic and cabalistic Neoplatonism favored the creation of a syncretistic system which was firmly grounded in doctrines equally acceptable to Christians, Jews, and Muslims.

This fusion occurred after the eight years Lull spent in Mallorca studying Latin, learning Arabic from a slave, reading all texts available to him in either tongue, and writing copiously. One of his earliest works was a compendium of the logic of al-Ghazālī in Arabic (1270[?]); it has since been lost, although two later compendia with similar titles survive—one in Latin, the other in Catalan mnemonic verse. In all, Lull wrote at least 292 works in Catalan, Arabic, or Latin over a period of forty-five years (1270–1315); most of them have been preserved, although no Arabic manuscripts have yet been traced and many Catalan and Latin works remain unpublished. His initial awkwardness in Latin, coupled with his desire that knowledge be made available to non-Latin-speaking sectors of society, made Lull the first person to mold Catalan into a literary medium. He used it not only in important mystical works, poetry, and allegorical novels (none of which concerns us here) but also to deal with every learned subject which engaged his attention: theology and philosophy; arithmetic, geometry, and astronomy (often mainly astrology), which, together with music, formed the quadrivium (the higher division of the seven liberal arts); grammar, rhetoric, and logic (the trivium); law; and medicine. Thus, Lull created a fully developed learned vocabulary in Catalan almost a century before any other Romance vernacular became a viable scholarly medium. Almost all Lull's works in such nonliterary fields were connected in some way with his Art, because the "art of finding truth" which he developed to convert "the infidel" proved applicable to every branch of knowledge. Lull himself pioneered its application to all subjects studied in medieval universities—except for music—and also constructed one of the last great medieval encyclopedias, the *Arbor scientiae* (1295–1296), in accordance with its basic principles.

Yet the Art can be understood correctly only when viewed in the light of Lull's primary aim: to place Christian apologetics on a rational basis for use in disputations with Muslims, for whom arguments *de auctoritate* grounded on the Old Testament—widely used by Dominicans in disputations with the Jews—carried no weight. The same purpose lay behind the *Summa contra gentiles* of Aquinas, written at the request of his fellow Dominican Penyafort, whose concern for the conversion of all non-Christians (but particularly those in James the Conqueror's dominions) thus inspired the two chief thirteenth-century attempts in this direction; the *Summa contra gentiles* was finished during the interval between Lull's discovery of his own calling and his interview with Penyafort. But whereas Aquinas distinguished categorically between what reason could prove and that which, while not contrary to reason, needed faith in revelation, Lull advanced what he called necessary reasons for accepting dogmas like the Trinity and the Incarnation. This gave his Art a rationalistic air that led to much subsequent criticism. Lull himself described his Art as lying between faith and logic, and his "necessary reasons" were not so much logical proofs as reasons of greater or lesser congruence which could not be denied without rejecting generally accepted principles. In this respect they were not appreciably more "rationalistic" than Aquinas's "proofs" that the truths of faith were not incompatible with reason. But the differences between the two apologetic systems are far more striking than their resemblances.

Lull regarded his Art as divinely inspired and hence infallible (although open to improvement in successive versions). Its first form, the *Ars compendiosa inveniendi veritatem* or *Ars maior*[2] (1273–1274[?]), was composed after a mystical "illumination" on Mount Randa, Mallorca, in which Lull saw that everything could be systematically related back to God by examining how Creation was structured by the active manifestation of the divine attributes—which he called Dignities and used as the absolute principles of his Art. Examining their manifestations involved using a set of relative principles; and both sets could be visualized in combinatory diagrams, known as Figures *A* and *T*. The original Figure *A* had sixteen Dignities, lettered *BCDEFGHIKLMNOPQR*; the original Figure *T* had five triads, only three of which (*EFG* + *HIK* + *LMN*) were strictly principles of relation, the others being sets of subjects (God + Creature + Operation, *BCD*) and possible judgments (Affirmation + Doubt + Negation, *OPQ*). All early versions had a proliferation of supplementary visual aids, which always included diagrams showing the four elements, and—with the obvious exception of Figure *T*—most features of the system were grouped into sets of sixteen items, lettered like the Dignities.

This quaternary base seems to provide the key to the origins of the Art's combinatory aspect, apparently modeled on the methods used to calculate combinations of the sixteen elemental "grades" (four each for fire, air, water, and earth) in both astrology and humoral medicine. A major simplification in the *Ars inventiva* (*ca.* 1289) eliminated the elemental features, reduced the diagrams to four (unchanged thereafter), reduced Figure *T* to the nine actual relative prin-

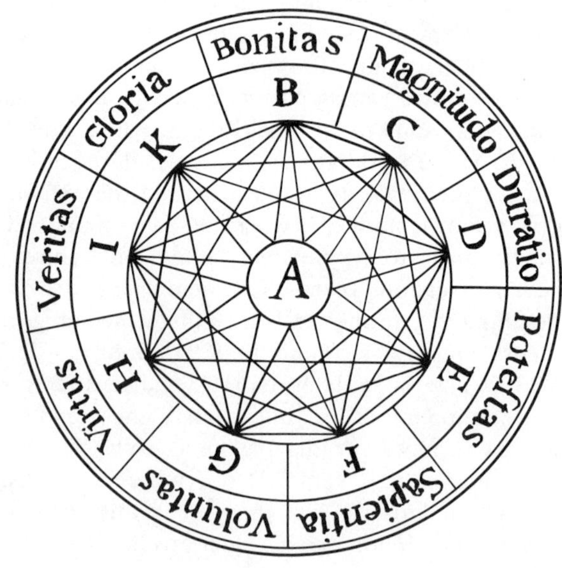

FIGURE 1. The Lullian "Dignities."

		A (1. Essentia, 2. Unitas, 3. Perfectio)	B	C	D	E	F	G	H	I	K
Alphabetum seu principia huius artis sunt aut	Praedicata	Absoluta	Bonitas	Magnitudo	Æternitas seu Duratio	Potestas	Sapientia	Voluntas	Virtus	Veritas	Gloria
		T. Relata seu respectus	Differentia	Concordantia	Contrarietas	Principium	Medium	Finis	Maioritas	Æqualitas	Minoritas
	Q. Quaestiones		Vtrum?	Quid?	De quo?	Quare?	Quantum?	Quale?	Quando?	Vbi?	Quomodo? Cum quo?
	S. Subiecta		Deus	Angelus	Coelum	Homo	Imaginatio	Sensitiua	Vegetatiua	Elementatiua	Instrumentatiua
	V. Virtutes		Iustitia	Prudentia	Fortitudo	Temperantia	Fides	Spes	Charitas	Patientia	Pietas
	V. Vitia		Auaritia	Gula	Luxuria	Superbia	Acidia	Inuidia	Ira	Mendacium	Inconstantia

FIGURE 2. The "alphabet" of Lull's *Ars brevis* (1303).

ciples (Difference + Concordance + Contradiction, Beginning + Middle + End, and Majority + Equality + Minority) and the sixteen original Dignities to the nine shown in Figure 1. In still later versions the symbolic letters *BCDEFGHIK* acquired up to six meanings that were ultimately set out in the gridlike "alphabet" of the *Ars generalis ultima* and its abridgment, the *Ars brevis* (both 1308), from which Figure 2 is reproduced. The traditional seven virtues and seven vices have been extended to sets of nine, to meet the requirements of the ternary system; the last two of ten *quaestiones* (a series connected with the ten Aristotelian categories) had to share the same compartment, since the set of fundamental questions could not be shortened and still be exhaustive.

The most distinctive characteristic of Lull's Art is clearly its combinatory nature, which led to both the use of complex semimechanical techniques that sometimes required figures with separately revolving concentric wheels—"volvelles," in bibliographical parlance (see Figure 3)—and to the symbolic notation of its alphabet. These features justify its classification among the forerunners of both modern symbolic logic and computer science, with its systematically exhaustive consideration of all possible combinations of the material under examination, reduced to a symbolic coding. Yet these techniques taken over from nontheological sources, however striking, remain ancillary, and should not obscure the theocentric basis of the Art. It relates everything to the exemplification of God's Dignities, thus starting out from both the monotheism common to Judaism, Christianity, and Islam and their common acceptance of a Neoplatonic exemplarist world picture, to argue its way up and down the traditional ladder of being on the basis of the analogies between its rungs—as becomes very obvious in Lull's *De ascensu et descensu intellectus* (1305). The lowest rung was that of the elements, and Lull probably thought that the "model" provided by the physical doctrines of his time constituted a valid "scientific" basis for arguments projected to higher levels. Since this physical basis would be accepted in the scientific field by savants of all three "revealed religions," he doubtless also hoped that the specifically

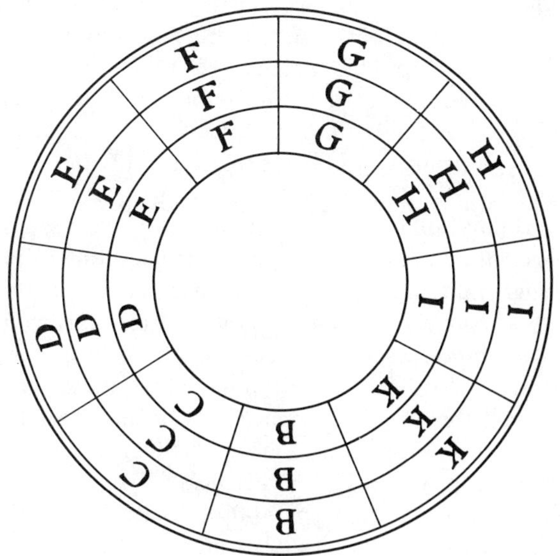

FIGURE 3. The fourth figure of the later Art: the inner wheels rotate independently, allowing all possible ternary combinations of the letters *BCDEFGHIK* to be read off.

Christian conclusions which he drew in the apologetic field would be equally acceptable. It even seems likely that what hit him with the force of a divine "illumination" on Mount Randa was his sudden recognition of such a possibility.

There is no evidence that Lull's Art ever converted anybody, but his application of the combinatory method to other disciplines (begun in the four *Libri principiorum, ca.* 1274–1275) was followed by numerous later Lullists; the Art's function as a means of unifying all knowledge into a single system remained viable throughout the Renaissance and well into the seventeenth century. As a system of logical inquiry (see Lull's *Logica nova* [1303] for the strictly logical implications, disentangled from other aspects), its method of proceeding from basic sets of preestablished concepts by the systematic exploration of their combinations—in connection with any question on any conceivable subject—can be succinctly stated in terms taken from the *Dissertatio de arte combinatoria* (1666) of Leibniz, which was inspired by the Lullian Art: "A proposition is made up of subject and predicate; hence all propositions are combinations. Hence the logic of inventing [discovering] propositions involves solving this problem: 1. given a subject, [finding] the predicates; 2. given a predicate, finding the subjects [to which it may] apply, whether by way of affirmation or negation."[3]

Recent research has concentrated on the clarification of Lull's ideas, the identification of their sources, and the nature of their influence on later thinkers—especially Nicholas of Cusa and Giordano Bruno. Major advances in all these fields have taken place since the 1950's, but much more research is still required. The specific origins of Lull's doctrines regarding the elements, whose importance has been fully recognized only since 1954 (see Yates), are particularly significant. A proper exploration of the antecedents of his *Opera medica* is a prerequisite for establishing Lull's final place in the history of Western science. In this connection it must be mentioned that although Lull himself was opposed to alchemy (but not to astrology, a "science" he sought to improve in the *Tractatus novus de astronomia* [1297]), his methods had obvious applications in the alchemical field—and they were so applied in a host of pseudo-Lullian alchemical works, most of them composed more than fifty years after his death. These works explain the traditional (but false) "scientific" view which made him "Lull the Alchemist."

NOTES

1. *Vida coëtanea.* The Latin (dictated by Lull [?], probably 1311) says "book," which doubtlessly agreed with Lull's original resolve; the plural, in the fourteenth-century Catalan text (modernized in *Obres essencials*, see I, 36), would better fit the series of "improved" versions of the Art itself, which first took shape almost ten years after Lull's conversion.
2. References to an *Ars magna* in later centuries are usually either to the definitive *Ars magna generalis ultima* (1308) or to Lull's system in general. The alternative title of the first version recalls Roger Bacon's *Opus maius* (1267); the connections between Lull and Bacon have yet to be investigated, but many resemblances may well be due to common Arabic sources.
3. "Propositio componitur ex subiecto et praedicato, omnes igitur propositiones sunt combinationes. Logicae igitur inventivae propositionum est hoc problema solvere: 1. Dato subiecto praedicata. 2. Dato praedicato subiecta invenire, utraque cum affirmative, tum negative" (G. W. Leibniz, *op. cit.* [in text], no. 55, in *Sämtliche Schriften und Briefen*, 2nd ser., I [Darmstadt, 1926], 192).

BIBLIOGRAPHY

I. Original Works. The larger of the standard eds. in Catalan is *Obres de Ramon Lull*, 21 vols., with 9 more planned (Palma de Mallorca, 1901–), cited below as *ORL. Obres essencials*, 2 vols. (Barcelona, 1957–1960), cited below as *Ob es*, contains the chief literary works; the modernized text has many errata, but there are excellent general introductions and a select bibliography.

Collections in Latin are Beati Raymundi Lulli, *Opera omnia*, I. Salzinger, ed., vols. I-VI, IX (VII and VIII unpublished) (Mainz, 1721–1742), contains 48 works, repr. Minerva GMBH, cited below as *MOG;* reprint of *Quattuor libri principiorum* from *MOG* I (Paris–The Hague, 1969), cited below as *QLP; Opera parva*, 3 vols. (Palma, 1744–1746), 15 works, 12 of which are not in *MOG;* and *Opera medica* (Palma, 1742)—it and *Opera parva* are to be issued in Gerstenberg Reprints. A modern critical ed. is *Raymundi Lulli Opera latina*, 5 vols., of 30 planned (Palma, 1961–), thus far previously unpublished works, cited below as *ORL*.

Major scientific works and versions of the Art are *Ars compendiosa inveniendi veritatem/Ars maior* (1273–1274[?]), *MOG* I; *Ars Universalis* (1274–1275[?]), *MOG* I; *Liber principiorum theologiae, Liber principiorum philosophiae, Liber principiorum juris, Liber principiorum medicinae* (1274–1275[?]), *MOG* I, *QLP; Liber chaos* (1275–1276[?]), *MOG III*; *Ars compendiosa medicinae* (1275–1276[?]), *Opera medica; Logica Algazalis* (*ca.* 1281), which is related to the Catalan verse *Logica del Gatzel* (after 1282[?]), *ORL* XIX; *Ars inventiva [veritatis]* (1289[?]), *MOG* V; *De levitate et ponderositate elementorum* (1293), *Opera medica; Arbor scientiae* (29 Sept. 1295–1 Apr. 1296), in various rare eds., in Catalan as *Arbre de ciencia* (simultaneous[?]), *ORL* XI–XIII, *Ob es* I; *Tractatus novus de astronomia* (1297; unpublished), Latin text discussed in Yates (1954; see below), Catalan MS is British Museum, Add. 16434, to appear in *ORL;* and *Liber de geometria [nova et compendiosa]* (1299), ed. by J. M. Millás-Vallicrosa (Barcelona, 1953), considered unreliable—see R. D. F. Pring-Mill, "La geometría luliana," in *Estudios Lulianos*, **2** (1958), 341–342.

Other works by Lull are *Liber de natura* (1301) (Palma, 1744); *Logica nova* (Palma, 1744); *Liber de regionibus sanitatis et infirmitatis* (1303), *Opera medica; De ascensu et descensu intellectus* (1305) (Palma, 1744); *Ars brevis* (1308), *Opera parva* I; *Ars generalis ultima/Ars magna [generalis ultima]* (1305–1308), in L. Zetzner, *Raymundi Lulli opera ea quae ad . . . artem . . . pertinent* (Strasbourg, 1598, 1609, 1617, 1651); and *Vita coëtanea* (dictated [?] 1311), B. de Gaiffier, ed., in *Analecta Bollandiana*, **48** (1930), 130–178, in Catalan as *Vida coëtania, Ob es.*

II. SECONDARY LITERATURE. See E. Colomer, S.J., *Nikolaus von Kues und Raimund Lull* (Berlin, 1961); C. E. Dufourcq, *L'Espagne catalane et le Maghreb aux XIIIe et XIVe siècles* (Paris, 1966); J. N. Hillgarth, *Ramon Lull and Lullism in Fourteenth-Century France* (Oxford, 1971); Erhard-Wolfram Platzeck, O.F.M., *Raimund Lull, sein Leben, seine Werke, die Grundlagen seines Denkens*, 2 vols. (Düsseldorf, 1961–1964), vol. II contains the fullest bibliography of Lull's writings and Lullian studies; and Frances A. Yates, "The Art of Ramon Lull: An Approach to It Through Lull's Theory of the Elements," in *Journal of the Warburg and Courtauld Institutes*, **17** (1954), 115–173; and "Ramon Lull and John Scotus Erigena," *ibid.*, **23** (1960), 1–44; but see R. D. F. Pring-Mill, in *Estudios Lulianos*, **7** (1963), 167–180.

See also *Estudios Lulianos*, which began publication in 1957, and E. W. Platzeck's comprehensive survey of research from 1955 to 1969 in *Antonianum*, **45** (1970), 213–272.

R. D. F. PRING-MILL

LUMMER, OTTO RICHARD (*b.* Gera, Germany, 17 July 1860; *d.* Breslau, Germany [now Wrocław, Poland], 5 July 1925), *optics.*

After completing his dissertation in 1884, Lummer became an assistant to Helmholtz; and in 1887 he followed the latter to the newly founded Physikalisch-Technische Reichsanstalt (PTR) in Berlin. Helmholtz, his revered teacher and his constant model, was for Lummer, as Gehrcke recalled, the "absolute standard as a research worker and as a man." In 1889 Lummer became a member of the PTR, and in 1894 he was given the title of professor. Lummer did not qualify for lecturing until 1901, at the University of Berlin.

As early as 1849 Haidinger had announced the existence of interference fringes that occur on mica plates and that, unlike Newton's rings, "do not move when the plates producing them are displaced." These fringes of equal inclination, caused by interference between rays emerging after multiple internal reflections from a parallel-sided plate, were rediscovered (for the third time, after Haidinger and Mascart) by Lummer in 1884 in Helmholtz' laboratory at the University of Berlin; they became known as Lummer fringes. Helmholtz, who had not perceived the phenomenon because of his nearsightedness, was not willing to accept the existence of the interference effects. He was soon convinced, however, and called Lummer's dissertation "an unusually good work."

Since Lummer fringes are the result of differences in path length of many wavelengths, Lummer arrived at the idea, in 1901, of developing the plane parallel plates into a spectroscope of the highest resolution. This device had the advantage of possessing greater resolving power than the interferometer produced in 1897 by Fabry and Perot. The considerable drawback of low luminous intensity, caused by the glancing incidence of the light, was eliminated in 1902 by Gehrcke, who cemented a prism to the plate with Canada balsam. The new apparatus, for which Lummer proposed the name Lummer–Gehrcke interference spectroscope, proved to be an excellent tool for spectroscopy and superior to the simple line grating.

At the PTR, Lummer had the task of working out the bases for a suitable international primary standard of luminosity. In 1889 he constructed, with Brodhun, an exact photometer, the Lummer-Brodhun cube. The new instrument fulfilled "through optical arrangements all the conditions of an ideal grease spot," something that the previously employed real grease-spot photometer could not do. In photometry it became necessary to hold constant the light sources that were being compared. The existing bolometer was not exact enough for Lummer's needs. In 1892, with F. Kurlbaum, he constructed a surface bolometer, which superseded all the previous types. Lummer thus gradually approached the field in which he was to have his greatest successes: thermal radiation.

In 1898, after preliminary work concerning, among other things, the production of the blackbody, Lummer tackled the problem of determining the Kirchhoff function depending only on wavelength and temperature—that is, he was seeking the emissive power of the blackbody. With Pringsheim he confirmed Wien's displacement law and also, with greater precision, Wien's radiation law, which Wien had stated in 1896 and Paschen had experimentally ascertained. In 1900, however, Lummer and Pringsheim discovered the "nonvalidity" of this law, which they called the Wien-Planck spectral equation: It "is thereby demonstrated that the Wien-Planck spectral equation does not yield the blackbody radiation that we measured in the region of 12μ to 18μ."

The competing team of Rubens and Kurlbaum verified this important result and derived from their measurements a conclusion that went even further:

"The intensity of a monochromatic beam at high temperatures is proportional to the temperature." This was the decisive stimulus leading to Planck's radiation formula, which he communicated on 19 October 1900. Lummer and Pringsheim subsequently confirmed this law with great precision, thereby establishing an unassailable foundation for the emerging quantum theory.

In 1904 Lummer was appointed a full professor at the University of Breslau. At first it was not easy for this specialist in optics to cover the whole field of physics. But his lectures soon became, as Clemens Schaefer wrote, "revelations of a mind in which the divine spark glowed."

BIBLIOGRAPHY

I. ORIGINAL WORKS. Lummer's writings include "Über eine neue Interferenzerscheinung an planparallelen Glasplatten und eine Methode, die Planparallelität solcher Gläser zu prüfen (Inauguraldissertation)," in *Annalen der Physik und Chemie*, n.s. **23** (1884), 49–84; "Über eine neue Interferenzerscheinung," *ibid.*, 513–548; "Photometrische Untersuchungen," in *Zeitschrift für Instrumentenkunde*, **9** (1889), 41–50, 461–465; **10** (1890), 119–133; **12** (1892), 41–50, written with Eugen Brodhun; "Über die Herstellung eines Flächenbolometers," *ibid.*, **12** (1892), 81–89; "Über die Strahlung des absolut schwarzen Körpers und seine Verwirklichung," in *Naturwissenschaftliche Rundschau*, **11** (1896), 65–68, 81–83, 93–95; "Die Vertheilung der Energie im Spectrum des schwarzen Körpers," in *Verhandlungen der Deutschen Physikalischen Gesellschaft*, **1** (1899), 23–41, written with Ernst Pringsheim; "Le rayonnement des corps noirs," in *Rapports présentés au Congrès international de physique réuni à Paris en 1900*, II (Paris, 1900), 41–99; *Die Lehre von der strahlenden Energie (Optik)*, II, pt. 3 of Müller–Pouillet's *Lehrbuch der Physik und Meteorologie*, 10th ed. (Brunswick, 1909); and *Grundlagen, Ziele und Grenzen der Leuchttechnik* (Munich, 1918).

II. SECONDARY LITERATURE. See Ernst Gehrcke, "Erinnerungen an Lummer," in *Physikalische Blätter*, **11** (1955), 315–317; Hans Kangro, *Vorgeschichte des Planckschen Strahlungsgesetzes* (Wiesbaden, 1970), esp. pp. 192–200; Fritz Reiche, "Otto Lummer," in *Physikalische Zeitschrift*, **27** (1926), 459–467; and Clemens Schaefer," Otto Lummer zum 100. Geburtstag," in *Physikalische Blätter*, **16** (1960), 373–381.

ARMIN HERMANN

LUNA, EMERICO (*b.* Palermo, Italy, 6 January 1882; *d.* Palermo, 4 December 1963), *anatomy.*

Born into a socially prominent family, Luna studied at the Institute of Histology of the University of Palermo. He received his degree with honors and became an assistant at the Institute of Normal Human Anatomy. He was made a full professor in 1922. Luna founded the Italian Anatomical Society and served as secretary until his death.

Luna was among the first to suture the heart successfully and to introduce the teaching of radiological anatomy. He was also active in politics and served as a deputy from Palermo and as a city councillor in Palermo. A major interest was expanded medical care for the poor, and he introduced legislation toward this end.

Luna's earliest studies dealt with the morphology of the cerebellum and with the projection on its surface of the cerebellar nuclei, in order to avoid them in the successive experimental extirpation of sections of the lobulus paramedianus, of the formatio vermicularis and of the crus secundum. By applying Marchi's method he demonstrated the course of the degenerated bundles and was able to localize the motor center in the cortex of the internal segment of the anterior lunated lobulus; he further demonstrated that for movements of the neck in the lobulus simplex the fibers follow the cerebrospinal fasciculus, the presence of which he confirmed in the dog.

In later investigations Luna demonstrated the nerve paths which proceed from the intermediate nerve and the vago-glosso-pharyngeal complex to the cerebellum and the spino-olivari fibers that terminate in the reticular substance of the medulla oblongata. He also carried out investigations on the development and morphology of the intercalated nucleus of Staderini, and he described the existence of a nucleus situated ventrally to that of the hypoglossus in *Sus schropha* (1). These studies preceded those of other investigators on the tegmental rhombencephalic centers.

In histology Luna studied the fine structure and nature of the reticular tissue and published a masterly and still authoritative monograph (2). He also studied the relation between the diameter of the capillaries and the size of the subject, the lymphatics of the lung, the morphology of the suprarenals, and muscular and vascular anomalies. He wrote many distinguished books, both literary and scientific, among which are the magnificent volume on the nervous system (3) in Bertelli's *Trattato di anatomia umana* and on regional clinical anatomy (4).

BIBLIOGRAPHY

Luna's writings include (1) "Ricerche istologiche sopra un nucleo riscontrato nel rombencefalo di *Sus schropha*," in *Folia neuro-biologica*, **5**, no. 1 (1911); (2) "Studio sul tessuto reticolare," in *Ricerche di morfologia*, **1** (1921);

(3) vol. V of the *Trattato di anatomia umana*, Bertelli, ed., 2nd ed. (Milan, 1938); (4) *Trattato di anatomia clinica regionale* (Florence, 1948).

IGNAZIO FAZZARI

LUNGE, GEORG (*b.* Breslau, Prussia [now Wrocław, Poland], 15 September 1839; *d.* Zurich, Switzerland, 3 January 1923), *chemistry, chemical technology.*

Lunge was the son of a merchant. At the age of seventeen he graduated at the head of his class from a Gymnasium in Breslau. Subsequently, at the University of Breslau, he studied the natural sciences, especially chemistry, and in 1859 obtained his doctorate *magna cum laude* with a dissertation entitled *De fermentatione alcoholica*. He first studied under Löwig and Cohn, then with Bunsen and Kirchhoff at Heidelberg. In 1860 Lunge joined the chemical industry, initially as a chemist at a Silesian fertilizer factory. In 1862 he opened his own plant for the production of potassium ferrocyanide, ammonia, lead salts, and tartaric acid; but the venture was unsuccessful.

Lunge then went to England, where he first worked in a factory producing coal-tar products. In 1865 he joined the newly founded, small alkali works of Baily, Bowron & Co., in South Shields, first as chemist and subsequently as manager. Lunge married a Miss Bowron, daughter of one of his employers. He participated in the founding and activities of the Newcastle Chemical Society; established contacts with representatives of science and industry; and established a fine reputation as a specialist in chemical technology. On the recommendation of Heinrich Caro, Lunge was invited in 1875 to become professor of technical chemistry at the Eidgenössische Polytechnikum Schule in Zurich. Subsequently he was granted Swiss citizenship. He resigned his teaching post at Zurich in 1907 but continued to write and do laboratory work until 1916. He retired at the age of seventy-seven.

At a time when most chemists worked in organic chemistry, Lunge was among the few who chose inorganic chemistry as a field of concentration. He was eminently pragmatic; Haber said of him that he had awakened spirits and conquered problems in applied science which previously had been dealt with only by technicians. Lunge worked toward the control of manufacturing processes by means of refined technical-chemical investigations and analyses and by the quantitative analysis of synthetic compounds. His vast technical knowledge based on practical experience made him a master in applying scientific results to economic-technical problems.

Lunge published an enormous number of papers. A bibliography he compiled himself—although not complete—lists eighty-six books and pamphlets as well as 590 articles. Practitioners acclaim Lunge's works as particularly dependable, detailed, precise, and practical, the general approach being coupled with the explanation of the scientific principles involved. Lunge's *Handbuch der Sodaindustrie*, his *Steinkohlenteer und Ammoniak*, and his *Chemisch-technische Untersuchungsmethoden* (written with Ernst Berl) were for decades considered standard works for the chemical raw-materials industry.

Experimental work by Lunge and by many of his pupils treated all problems of chemical manufacturing, especially that of soda, sulfuric acid, chlorine, and chloride of lime, nitric acid, phosphates, coal-tar products, explosives, cellulose, and synthetic fibers. It resulted in several inventions, such as Lunge's plate towers (which used countercurrents) for the acid industry and Heumann's synthesis of indigo.

Lunge also personally directed a large chemical laboratory—in addition to his lecturing—in which students learned to test raw materials and to supervise manufacturing processes. The newly constructed chemical laboratory complex (built in 1883 in association with Victor Meyer) became the model for chemical-technical research and teaching facilities.

BIBLIOGRAPHY

I. ORIGINAL WORKS. Lunge's writings include *Handbuch der Sodaindustrie und ihrer Nebenzweige für Theorie und Praxis*, 2 vols. (Brunswick, 1879; 3rd ed., 3 vols., 1903–1909); *Theoretical and Practical Treatise on the Manufacture of Sulfuric Acid and Alkali*, 3 vols. (London, 1879–1880; 3rd ed., 2 vols. in 4 pts., 1903–1911; supp. to vol. I, London, 1917); *Die Industrie der Steinkohlenteer—Destillation und Ammoniakwasser-Verarbeitung* (Brunswick, 1882), also translated into English (London, 1882; 5th ed., 1912); *Die chemischen Laboratorien des Eidgenössischen Polytechnikums* (Zurich, 1889); *Chemisch-technische Untersuchungsmethoden*, 3 vols. (Berlin, 1899–1900; 6th ed., 4 vols., 1910–1911; 8th ed., 5 vols., 1931–1934), also translated into English by C. A. Keane and others, 3 vols. in 5 pts. (London, 1908–1914), written with Berl; *Zur Geschichte der Entstehung und Entwicklung der chemischen Industrien in der Schweiz* (Zurich, 1901); and *Das Zusammenwirken von Chemie und Ingenieurwesen in der Technik* (Vienna, 1907).

II. SECONDARY LITERATURE. See E. Bosshard, "Georg Lunge 1839–1923," in *Verhandlungen der Schweizerischen naturforschenden Gesellschaft*, **104**, pt. 2 (1923), "Anhang, Nekrologe und Biographien," 25–43, with a bibliography; and "Georg Lunge," in Günther Bugge, ed., *Das Buch der grossen Chemiker*, II (Weinheim, 1965); and Poggendorff, III, 843–844; IV, 926–927, VI, pt. 2, 1591.

EBERHARD SCHMAUDERER

LUSITANUS, AMATUS (RODRIGUES, JOÃO) (*b.* Castelo Branco, Portugal, 1511; *d.* Salonika, Greece (then Turkish), 21[?] January 1568), *medicine, botany.*

No more is known of the parents of João Rodrigues than that they were Marranos, probably of Spanish origin, for they sent him to study medicine at Salamanca University, where he graduated in 1529 or 1530. He seems to have studied surgery longer than was common for medical students of the time, possibly after completing his main course, for he stayed in Spain until 1532. He practiced only briefly in his native country, for in 1533 he migrated to Antwerp, then under Spanish rule, which had a large Portuguese colony. He later mysteriously explained his exile as due to an expectation that something would happen which subsequently did. (It was possibly the intensified harassment of the Marranos.) While in Flanders he made the acquaintance of leading humanists and published in 1536 his first book, *Index Dioscoridis*, a commentary on the first book of the *De materia medica*, intended to be eruditely philological as well as useful to apothecaries. In 1540 he was appointed professor of anatomy at the University of Ferrara, where he collaborated with Canano in his dissections. In the course of research they discovered (probably in 1547) the valve in the azygos vein. Amatus points out that "if you blow air into the lower part of the azygos vein, the vena cava will not be inflated because of a valve at the orifice where the two join." His is the first report of this observation, but argument continues as to who was the active partner. Neither appreciated the real significance of their discovery.

In the summer of 1547 Amatus moved to Ancona, where he established an extensive practice, traveling to Florence, Rome, and Venice to treat highly placed patients. While in Ancona he began to write up his case histories; a hundred of these, starting from his brief practice in Portugal, were published in 1551 as *Curationum medicinalium centuria.* Six similar collections appeared over the next decade, the last in 1561. In further commentaries on Dioscorides he concerned himself more with descriptions of plants and included his own opinions on their identification and supposed medicinal properties, claiming his own successful experience; he also added new Asian plants, such as the coconut palm. Unfortunately for him, he criticized Mattioli and made a bitter and dangerous enemy. When, in 1555, the new pope, Paul IV (the previous pope, Julius III, had been his patient), began persecuting the Marranos, Amatus was forced to flee the papal dominion. His papers, including the fifth *centuria*, were retrieved for him by friends. He escaped first to Pesaro and then in 1556 to Ragusa, perhaps as municipal physician there—he had long had hopes of securing this post. But in 1559 he left Christian territory altogether and went to Salonika, where he returned openly to Judaism. He died there during an outbreak of the plague in 1568, having, it is said, contracted the disease from his patients.

Although his work on Dioscorides includes some interesting observations and information on contemporary economic botany, his lifework is to be found mainly in his *Curationum medicinalium.* For each case history he gives the patient's name or social position, age, physiognomy, and build; outlines the symptoms and course of the disease; and records his diagnosis, prescription, and the result. There is little order, and he ranges over most branches of medicine. Local climate is held responsible for at least the specific characters of epidemics. Amatus is at his best when discussing the social origin of stress diseases and mental disorders, for which he recommended empirically well-founded diets. Although usually concerned, as a good humanist, to show that his treatment was founded on Hippocratic or Galenic principles, he was ready to innovate, especially in surgery. In operating for empyema, for example, he proposed the third intercostal space, rather than higher up, for the puncture site, in order not to damage the diaphragm. He was also a pioneer in the use of such devices as the obturator in cases of cleft palate and the breast reliever.

BIBLIOGRAPHY

I. Original Works. Amatus' botanical works are *Index Dioscoridis, en, candide lector, historiales Diosocoridis campi exegemataque simplicium* . . . (Antwerp, 1536), the only work to appear under his own name rather than the humanistic nom de plume Amatus Lusitanus; and *In Dioscoridis de materia medica libros quinque enarrationes* (Venice, 1553; illustrated ed., Lyons, 1558). *Curationum medicinalium centuria* (Florence, 1551), was followed by *Curationum medicinalium centuria secunda* (Paris, 1552). Publication history of the remaining five *centuriae* is quite complicated as they usually appeared with reissues of earlier ones. The first complete ed. of all seven was published at Venice from 1565 to 1570. There were also several posthumous eds. (Lyons, 1580; Bordeaux, 1620).

II. Secondary Literature. See (listed chronologically) M. Salomon, "Amatus Lusitanus in seine Zeit," in *Zeitschrift für klinische Medizin* 41–2 (1901); M. de Lemos, *Amato Lusitano, a sua vida e a sua obra* (Porto, 1907); H. Friedenwald, "Amatus Lusitanus," in *Bulletin of the Institute of History of Medicine, Johns Hopkins University*, **4** (1937); J. Lopes-Dias, "Dr. João Rodrigues de Castelo Branco. Amato Lusitano. Ensaio Bio-bibliográfico," in

Congresso do mundo português. Publicacões, **13** (1940). A commemorative volume is *IV Centenario de João Rodrigues de Castelo Branco–Amato Lusitano (Estudios Castelo Branco 1968)*.

On the *Curationum*, see J. O. Leibovitz, "A Probable Case of Peptic Ulcer as Described by Amatus Lusitanus," in *Journal of the History of Medicine and Allied Sciences*, **27** (1953), 212–216; "Amatus Lusitanus and the Obturator in Cleft Palates," in *Bulletin of the History of Medicine*, **13** (1958), 492–503; and F. Segret, "Amatus Lusitanus, Témoin de son temps," in *Sefarrad*, **23** (1968), 285–309.

A. G. KELLER

LUSK, GRAHAM (*b*. Bridgeport, Connecticut, 15 February 1866; *d*. New York, N. Y., 18 July 1932), *physiology, nutrition, clinical investigation*.

Lusk was the son of William Thompson Lusk, a prominent New York teacher and practitioner of obstetrics, and Mary Hartwell Chittenden. Partially deaf, Lusk did not pursue a family interest in clinical medicine, but chose instead a laboratory science career. He received a bachelor's degree from the Columbia University School of Mines (1887); then—as his father had a generation before—he traveled to Germany for advanced study. Lusk came under the influence of Voit and in 1891 received a Ph.D. from Munich, with a dissertation on the influence of carbohydrates in protein metabolism. Upon his return to the United States in 1891, Lusk taught physiology at the Yale Medical School, then moved in 1898 to New York City and the Bellevue Hospital–New York University College of Medicine. In 1909, he was appointed professor of physiology at Cornell Medical College, where he remained until his retirement and death in 1932.

Inspired by Voit's work, particularly in nutrition and metabolism, Lusk devoted his career to the illumination of metabolic processes through the use of the calorimeter in the analysis of intake and output, in both normal and pathological states. While at Yale, Lusk began an important series of studies on diabetes. He administered the chemical phlorhizin to dogs to produce glycosuria and demonstrated a constant ratio between the amount of dextrose and nitrogen excreted in the urine. Later experiments indicated a differential yield of carbohydrate from the metabolism of particular amino acids. His demonstration that carbohydrates were not normally formed in the metabolism of fats held important clinical implications. He also did research on the respiratory quotient and the significance of surface area in the determination of basal metabolism rates (as opposed to those physiologists who emphasized the importance of body weight).

In 1912 Lusk was appointed scientific director of the Russell Sage Institute of Pathology, newly affiliated with the Cornell division of Bellevue. Working in conjunction with Eugene F. DuBois (clinical director of the program) Lusk had, for the first time, adequate financial and clinical resources to utilize and maintain a calorimeter for work in human physiology and pathology. Much of Lusk's most important work resulted from two decades of collaboration with DuBois and other clinicians, resulting in more than 1,400 calorimeter experiments completed before Lusk's retirement. Lusk probably exerted his broadest scientific influence through his synthetic work, *Elements of the Science of Nutrition*, which appeared in four increasingly detailed editions between 1906 and 1928 and which played an important role in popularizing the findings of calorimeter research.

With Atwater and Benedict, Lusk was a leader in metabolism studies. Like Benedict, however, Lusk's scientific view was firmly oriented toward research methods he had learned in the 1890's; measuring energy relationships with the calorimeter was a basic and unquestioned approach to the understanding of nutrition. As late as 1932, he still dismissed vitamin studies briefly in a chapter on dietetics in his work *Nutrition*. Lusk never appreciated the significance of the vitamins and especially their implications for further work in enzymatic and intracellular processes.

Few of his American contemporaries were more active than Lusk in making the medical profession a truly scientific one. When Lusk arrived in New Haven in 1891, suffused with the German research ideal and devoted to the reshaping of clinical medicine in this mold, he found Yale—like almost every American medical school—dominated by clinicians who could not appreciate his laboratory methods. Since his annual salary was $300, only a private income enabled Lusk to pursue the research demanded by his European training. Lusk soon became part of a group of young American scientists and clinical investigators—frequently German-trained—who actively sought to raise standards of medical research and education in the United States. Genuine scientific training, they argued, would have to become part of the medical school curriculum and part of the intellectual equipment and job qualifications of a new type of clinician. Lusk's family background and his interest in pathological phenomena made him an ideal spokesman for such ideas. His collaboration over two decades with DuBois, moreover, and the practical implications of their work on diabetes had a particularly telling effect in the practice-oriented world of American medicine. Family wealth and his marriage to a daughter of Louis C. Tiffany provided the means for Lusk to become part of New York's social and medical establishment.

A member of numerous scientific societies, including the National Academy of Sciences and the Royal Society, Lusk was founder and organizer of the Society for Experimental Biology and Medicine (1903), American Society of Biological Chemists (1906), and New York's Harvey Society (1905). During World War I, Lusk served as adviser in matters of nutrition for the U. S. government.

BIBLIOGRAPHY

I. Original Works. There is no full-length biography of Lusk and no extensive collection of his papers. Useful in understanding the background of Lusk's calorimeter work are his *Elements of the Science of Nutrition* (Philadelphia, 1906; 2nd ed., 1909; 3rd ed., 1917; 4th ed., 1928), and his *Nutrition* (New York, 1933). This latter work, edited and published after Lusk's death, concludes with a discussion of basal metabolism; vitamins are mentioned briefly in a chapter on dietetics.

II. Secondary Literature. For useful biographical sketches of Lusk, see *Addresses Given at a Memorial Meeting for Graham Lusk at the New York Academy of Medicine on December 10, 1932* (Baltimore, 1933); the editorial "Graham Lusk," in *Journal of Biological Chemistry*, **98** (1932), vii; and Russell H. Chittenden, "Graham Lusk," in *Dictionary of American Biography*, supp. I, 517–518. The most detailed sketch is by Eugene F. DuBois, "Biographical Memoir of Graham Lusk, 1886–1932," in *Biographical Memoirs. National Academy of Sciences*, **21** (1941), 95–142; "Graham Lusk," in *Science*, **76** (1932), 113–115; and John R. Murlin, "Graham Lusk. A Brief Review of His Work," in *Journal of Nutrition*, **5** (1932), 527–538. Compare also C. Culotta, "Francis G. Benedict," in *Dictionary of Scientific Biography*, I, 325–326; and C. E. Rosenberg, "Wilbur Olin Atwater," *ibid.*, I, 609–611. For a useful comparison with Lusk's work, see E. V. McCollum, *A History of Nutrition. The Sequence of Ideas in Nutrition Investigations* (Boston, 1957), esp. ch. 10, "Respiration and Calorimetry."

Charles E. Rosenberg

LUSTIG, ALESSANDRO (*b.* Trieste, Austria [now Italy], 5 May 1857; *d.* Marina di Pietrasanta, Lucca, Italy, 23 September 1937), *pathology, bacteriology.*

Lustig received his medical degree at Vienna in 1882, then was an assistant at the Institute of Physiology in Vienna and, in 1884, at the Institute of Physiology at Innsbruck. He next went to Turin, where he obtained an Italian medical degree; soon afterward he started his scientific work in the research laboratory of the Mauriziano Hospital in Turin. In 1886, during a cholera epidemic, Lustig tended the victims in lazarettos; his investigations on the disease carried out at that time are among the earliest contributions to its study after the discovery of *Vibrio cholerae* by Koch (1884). He was appointed professor of general pathology at the University of Cagliari in Sardinia in 1889, and in 1890 at Florence, where he worked until his retirement in 1932.

Lustig began his scientific activity in histology and physiology. He dealt with the nerve ends of smooth muscles, epithelial degeneration of the olfactory membrane after destruction of the olfactory lobe, myocardial changes after section of extracardial nerves, the effects of celiectomy, the etiology of endemic goiter, and experimental acetonuria.

From 1891 Lustig's activity was mainly in bacteriology, immunology, and the prevention of infections. After his investigations on cholera, he devoted further study to the bacteriology of this infection and published an excellent monograph on the detection of water bacteria. In 1897, with his pupil G. Galeotti, Lustig extracted from *Bacterium pestis* a substance having the chemical characteristics of nucleoproteids, and demonstrated its immunizing properties: these investigations stand, in the history of bacteriology and immunology, as one of the first attempts at chemical identification of a bacterial antigen and at using chemically defined antigens to induce active immunity.

These substances were used by Lustig and his co-workers as chemical vaccines to immunize animals and obtain immune sera that were employed for serum therapy in cases of plague. Later, Lustig and his pupils Galeotti and G. Polverini were invited to India to apply the results of their research. In 1897 and 1898 they were in Bombay, where they made epidemiological observations on plague and used a serum prepared in the Laboratory of General Pathology at Florence to treat a considerable number of plague victims. Having verified the efficacy of the serum, they started prophylactic vaccination with nucleoproteids and created in Bombay a laboratory for the preparation of nucleoproteids and antiplague sera. This activity is documented in numerous papers by Lustig and his co-workers, which were published in Italian, English, and German journals from 1897 to 1903. In the Italian medical journal *Morgagni*, Lustig published "I metodi razionali di immunizzazione per mezzo di sostanze chimicamente estratte dai microrganismi patogeni" (1903), in which the premises for future developments of both bacteriological chemistry and immunology are clearly seen.

In 1901 Lustig published *Trattato di patologia generale*, which went through eight editions. Starting with the third edition he had the collaboration of his pupils G. Galeotti, P. Rondoni, and G. Guerrini. While general pathology had until then been considered as the propaedeutic part of morbid anatomy, in that book Lustig and his collaborators clearly indicated the new dimension of this discipline, including in it immunology and physiopathology. This extension proved to be essential for a complete understanding of fundamental pathogenetic mechanisms.

In 1913, with several collaborators, Lustig published *Le malattie infettive dell'uomo e degli animali*. In 1931, in collaboration with G. C. Rovida and G. Ferraloro, he wrote *Fisiopatologia e clinica dei gas da combattimento*, which he dedicated "to the memory of the eight thousand Italian soldiers who died in World War I following the first gas attack on the Austrian front."

Later, Lustig's interest turned more and more to social medicine and hygiene: this is attested by his many papers on combating malaria, tuberculosis, and tumors.

BIBLIOGRAPHY

Lustig's writings include "Studi sul colera asiatico," in *Resoconto sanitario dello Spedale civile di Trieste* (1886); "Ueber die Aetiologie des endemischen Kropfes," in *Verhandlungen des Internationalen Kongresses für Medizin* (1890); *Diagnostik der Bakterien des Wassers* (Jena, 1893); "Risultati delle ricerche fatte in India negli animali e nell'uomo intorno alla vaccinazione preventiva contro la peste bubbonica," in *Pubblicazioni del R. Istituto di studi superiori . . . in Firenze* (1897); "Versuche mit Pestschutzimpfungen bei Thieren," in *Deutsche medizinische Wochenschrift*, **23** (1897), 227, written with G. Galeotti; *Trattato di patologia generale* (Milan, 1901); "I metodi razionali di immunizzazione per mezzo di sostanze chimicamente estratte dai microrganismi patogeni," in *Morgagni*, **45** (1903), 721; "Ueber Bakteriennukleoproteide," in W. Kolle and A. von Wassermann, eds., *Handbuch der pathogenen Mikroorganismen*, II (Jena, 1913), 1362; sections of *Le malattie infettive dell'uomo e degli animali* (Milan, 1913); and *Fisiopatologia e clinica dei gas da combattimento* (Milan, 1931), written with G. C. Rovida and G. Ferraloro.

GIOVANNI FAVILLI

LUTHER, C. ROBERT (*b.* Schweidnitz, Germany [now Swidnica, Poland], 10 April 1822; *d.* Düsseldorf, Germany, 15 February 1900), *astronomy*.

Luther was the son of F. H. A. August Luther and the former Wilhelmine von Ende. In 1831–1841 he attended the secondary school of his native town. From 1841 to 1843 he studied natural sciences, especially astronomy, at the University of Breslau and continued these studies at Berlin in the fall of 1843. As a student of Encke's, Luther soon began doing computations for the *Berliner astronomisches Jahrbuch*, which published his work from 1849 to 1899. From 1847 he was an observer at the Berlin observatory; in 1848 he reduced Encke's meridian circle observations; and beginning in 1850 he succeeded Galle at the nine-inch refractor. In 1851 he followed Brünnow as director of the small observatory at Bilk, a suburb of Düsseldorf, remaining in Düsseldorf until his death.

Luther's lifework was the planetoids; he discovered twenty-four with priority and many others which had been discovered independently about the same time. For several of these planetoids he calculated the ephemerides continuously; these ephemerides are among the best and most accurately calculated ones for planetoids. The work deserves to be appreciated because at first Luther had only a telescope with a six-foot focal length at Bilk (in 1877 he acquired a larger one) and made all his observations with a nonilluminated ring micrometer. In the last years of his life Luther restricted his work to the reduction of observations that, because of his defective vision, were made by his son Wilhelm.

Luther received many honors, including an honorary doctorate from the University of Bonn (1855). The *Astronomische Nachrichten*, from volume 21 (1844) to volume 152 (1900), contains many, mostly short, notes by Luther.

Luther married Caroline Märcker in 1859; they had a son, Wilhelm.

BIBLIOGRAPHY

Luther's writings, besides the notes in *Astronomische Nachrichten*, include *Hora O Berliner akademische Sternkarten* (Berlin, 1857); and "Neuberechnung der Barker' schen Hilfstafel," in Encke's new ed. of Olbers' *Methode eine Kometenbahn zu berechnen* (Weimar, 1847).

An obituary is V. Knorre, in, *Vierteljahrsschrift der Astronomischen Gesellschaft*, **17** (1900), 191–200.

H.-CHRIST. FREIESLEBEN

LUZIN, NIKOLAI NIKOLAIEVICH (*b.* Tomsk, Russia, 9 December 1883; *d.* Moscow, U.S.S.R., 28 February 1950), *mathematics*.

Luzin was the son of a trade official, Nikolai Mitrofanovich Luzin, and Olga Nikolaievna. He completed his secondary schooling at Tomsk in 1901 and entered the mathematics division of the Physics-Mathematics Faculty of Moscow University. Here he became an active member of the circle of science students headed by N. E. Zhukovsky and studied mathematics under the guidance of D. F. Egorov. He spent the winter and spring of 1906 in Paris, where he attended lectures at the Sorbonne and the Collège de France. At the end of 1906 Luzin completed the course at Moscow University and remained there to prepare for a professorship. In 1910, after passing the examinations for the master's degree, he was appointed assistant professor at Moscow University but did not take up his duties there; he was sent to Göttingen and Paris for further study in those fields that had interested him for several years—the theory of functions of a real variable, integration theory, and the theory of trigonometric series. Luzin's first memoirs, published in 1911–1913 in *Matematicheskii sbornik* and the *Comptes rendus* of the Paris Academy of Sciences, immediately attracted the attention of the scientific world. In the spring of 1914, upon his return to Moscow, Luzin started lecturing, one of his courses being on function theory; he also organized a special research seminar on the subject. In 1915 he submitted a monograph entitled *Integral i trigonometrichesky ryad* ("Integrals and the Trigonometric Series"), published in *Matematicheskii sbornik*, and defended it at the Physics-Mathematics Faculty as his master's thesis. On the basis of its outstanding merits, the scientific council awarded Luzin the doctorate in pure mathematics, and in 1917 he became a professor.

During the period 1914–1924 Luzin, a brilliant lecturer and scientific organizer, was the center of a Moscow school of function theory which greatly influenced the subsequent development of mathematics, both in the Soviet Union and abroad. Such outstanding mathematicians as P. S. Alexandrov, A. Y. Khinchin, D. E. Menshov, M. Y. Suslin, A. N. Kolmogorov, N. K. Bari, and P. S. Novikov were his pupils.

Luzin was a member of the Moscow Mathematical Society, of the Moscow Society of Explorers of Nature, and of the Cracow Academy of Sciences; an honorary member of the Calcutta Mathematical Society and of the Belgian Mathematical Society; and was elected vice-president of the International Mathematical Congress held at Bologna in 1928. In 1927 Luzin was elected a corresponding member of the Soviet Academy of Sciences, and in 1929 he became an academician.

After 1930 Luzin devoted less time to teaching and worked mainly at the Soviet Academy of Sciences: the Mathematical Institute (1929–1936; 1941–1950) and the Institute of Automatics and Telemechanics (1936–1950).

Luzin's mathematical creativity relates mostly to the theory of functions of a real variable in its two branches, metric and descriptive. Insofar as the first is concerned, Luzin's thesis (Moscow, 1915) is paramount, since it contains important results on the structure of measurable sets and functions, on primitive functions, on convergence of trigonometric series, and on representability of functions by trigonometric series. Cited in the thesis are examples of power series with coefficients tending to zero and nevertheless diverging everywhere on the boundary of the circle of convergence (1912), as well as a trigonometric series with coefficients tending to zero that nevertheless diverges almost everywhere (1912). These examples stimulated many later investigations. Luzin later showed that every measurable function can, in a certain sense, be represented as a trigonometric series that may be summable almost everywhere to that function by the methods of Poisson and Riemann.

Another important result obtained by Luzin on absolute convergence resides in the fact that when a trigonometric series converges absolutely at two points, the distance between which is incommensurable with respect to π, this series converges absolutely over an everywhere dense set and therefore will either converge everywhere or diverge everywhere.

The investigation of the structure of measurable sets and functions led Luzin to prove the so-called *C* property: that is, that every measurable function which is almost everywhere finite over a given segment can be made continuous over the segment by varying its value on a set of arbitrarily small measure (1912). In Luzin's hands this theorem was a most important means of investigation; by resorting to it he managed to solve completely a number of cardinal problems: finding a primitive function, the representability of a function as a trigonometric series, and finding a harmonic function that is holomorphic inside a circle and takes given values on the circumference.

In the theory of integration, Luzin solved the problem of how to distinguish the Lebesgue integral or the Denjoy integral from the other primitives of the given function. For this purpose he introduced a new concept of complete variation of a function for a perfect set. The metric theory of functions was applied by Luzin to the study of boundary properties of analytical functions.

After 1915 Luzin turned to the descriptive theory of functions. Having investigated the so-called *B* sets studied by Borel, Baire, and Lebesgue, Luzin raised questions concerning the power of B sets and the construction of sets that are not B sets without resorting to Zermelo's axiom, to which many mathematicians objected. Both of the problems were solved by Luzin's pupils; M. Y. Suslin constructed a more extended class of sets than that of B sets, called analytical or A sets. Luzin introduced new definitions of both A sets and B sets, as well as the "sieving process," which became very important in the theory of A sets. Through the use of this process, Luzin also showed that a segment may be represented as a sum of pairwise disjoint B sets. It is to date the strongest result in the theory of sets that is not based upon Zermelo's axiom. Luzin also created a theory of "projective" sets (1925), which are obtained from B sets by successively performing the operation of projection and taking a complement. Through the introduction of this concept, some difficulties of a mathematical and logical nature came to light and were studied by his successors.

Luzin also provided decisive results in the solution of the problem of bending on the main base (1938) that dated back to K. M. Peterson, and the evaluation of the convergence of S. A. Chaplygin's method of approximate solution of differential equations (1932). He also wrote brilliant articles on Euler and Newton (1933, 1943). Luzin's many manuals and textbooks on mathematical analysis and the theory of functions of a real variable went through many editions and are still used by students in the Soviet Union.

BIBLIOGRAPHY

I. ORIGINAL WORKS. Luzin's writings were brought together in *Sobranie sochinenii* ("Collected Works"), 3 vols. (Moscow, 1953–1959). Individual works include *Leçons sur les ensembles analytiques et leurs applications* (Paris, 1930); and *Integral i trigonometrichesky ryad* ("Integrals and Trigonometric Series"), new ed. (Moscow–Leningrad, 1951), with biography of Luzin by V. V. Golubev and N. K. Bari, pp. 11–31; an analysis of his work by Bari and L. A. Lusternik, pp. 32–45; detailed commentary by Bari and D. E. Menshov, pp. 389–537; and a bibliography, pp. 538–547.

II. SECONDARY LITERATURE. See N. K. Bari and L. A. Lusternik, "Raboti N. N. Luzina po metricheskoy teorii funktsy" ("The Works of N. N. Luzin in the Metric Theory of Functions"), in *Uspekhi matematicheskikh nauk*, **6**, no. 6 (1951), 28–46; V. S. Fedorov, "Trudi N. N. Luzina po teorii funktsy kompleksnogo peremennogo" ("The Works of N. N. Luzin in the Theory of Functions of a Complex Variable"), *ibid.*, **7**, no. 2 (1952), 7–16; V. K. Goltsman and P. I. Kuznetsov, "Raboti N. N. Luzina po differentsialnim uravneniam i po vychislitelnim metodam" ("The Works of N. N. Luzin in Differential Equations and in Computational Methods"), *ibid.*, 17–30; L. V. Keldysh and P. S. Novikov, "Raboti N. N. Luzina v oblasti deskriptivnoy teorii mnozhestv" ("The Works of N. N. Luzin in the Area of the Descriptive Theory of Sets"), *ibid.*, **8**, no. 2 (1953), 93–104; and A. P. Youschkevitch, *Istoria matematiki v Rossii* ("The History of Mathematics in Russia"; Moscow, 1968), 565–577.

A. B. PAPLAUSCAS

LYAPUNOV, ALEKSANDR MIKHAILOVICH (*b.* Yaroslavl, Russia, 6 June 1857; *d.* Odessa, U.S.S.R., 3 November 1918), *mathematics, mechanics.*

Lyapunov was a son of the astronomer Mikhail Vasilievich Lyapunov, who worked at Kazan University from 1840 until 1855 and was director of the Demidovski Lyceum in Yaroslavl from 1856 until 1863; his mother was the former Sofia Aleksandrovna Shilipova. Lyapunov's brother Sergei was a composer; another brother, Boris, was a specialist in Slavic philology and a member of the Soviet Academy of Sciences.

Lyapunov received his elementary education at home and later with an uncle, R. M. Sechenov, brother of the physiologist I. M. Sechenov. With R. M. Sechenov's daughter Natalia Rafailovna (whom he married in 1886), he prepared for the Gymnasium. In 1870 Lyapunov's mother moved to Nizhny Novgorod (now Gorky) with her children. After graduating from the Gymnasium in Nizhny Novgorod in 1876, Lyapunov enrolled in the Physics and Mathematics Faculty of St. Petersburg University, where P. L. Chebyshev greatly influenced him.

Upon graduating from the university in 1880, Lyapunov remained, upon the recommendation of D. K. Bobylev, in the department of mechanics of St. Petersburg University in order to prepare for a professorial career. In 1881 he published his first two scientific papers, which dealt with hydrostatics: "O ravnovesii tyazhelykh tel v tyazhelykh zhidkostyakh, soderzhashchikhsya v sosude opredelennoy formy" ("On the Equilibrium of Heavy Bodies in Heavy Liquids Contained in a Vessel of a Certain Shape") and "O potentsiale gidrostaticheskikh davleny" ("On the Potential of Hydrostatic Pressures").

In 1882 Chebyshev posed the following question: "It is known that at a certain angular velocity ellipsoidal forms cease to be the forms of equilibrium of a rotating liquid. In this case, do they not shift into some new forms of equilibrium which differ little from ellipsoids for small increases in the angular velocity?" Lyapunov did not solve the question at the time, but the problem led him to another that became the subject of his master's dissertation, *Ob ustoychivosti ellipsoidalnykh form ravnovesia vrashchayushcheysya zhidkosti* ("On the Stability of Ellipsoidal Forms of Equilibrium of a Rotating Liquid"; 1884), which he defended at St. Petersburg University in 1885.

In the autumn of that year Lyapunov began to teach mechanics at Kharkov University as a *Privatdozent*. For some time he was completely occupied by the preparation of lectures and by teaching, but in 1888 his papers on the stability of the motion of mechanical systems having a finite number of degrees of freedom began to be published. In 1892 he published the classic *Obshchaya zadacha ob ustoychivosti dvizhenia* ("The General Problem of the Stability of Motion"), and in the same year he defended it as his doctoral dissertation at Moscow University; N. E. Zhukovsky was one of the examiners. Lyapunov studied this field until 1902.

In 1893 Lyapunov became a professor at Kharkov. In addition to mechanics he taught mathematics courses. He was also active in the Kharkov Mathematical Society. From 1891 until 1898 he was its vice-president, and from 1899 until 1902 its president and the editor of its *Soobshchenia* ("Reports").

At Kharkov, Lyapunov conducted investigations in mathematical physics (1886–1902) and the theory of probability (1900–1901) and obtained outstanding results in both. At the beginning of 1901 he was elected an associate member of the St. Petersburg Academy of Sciences, and at the end of that year he became an academician in applied mathematics, a chair which had remained vacant for seven years, since the death of Chebyshev.

In St. Petersburg, Lyapunov devoted himself completely to scientific work. He returned to the problem that Chebyshev had placed before him and, in an extensive series of papers which continued until his death, developed the theory of figures of equilibrium of rotating heavy liquids and of the stability of these figures.

In 1908 Lyapunov attended the Fourth International Congress of Mathematicians in Rome. He was involved in the publication of the complete collected works of Euler *(L. Euleri Opera omnia)* and was an editor of volumes XVIII and XIX of the first (mathematical) series, which appeared in 1920 and 1932. Lyapunov's scientific work received wide recognition. He was elected an honorary member of the universities of St. Petersburg, Kharkov, and Kazan, a foreign member of the Accademia dei Lincei (1909) and of the Paris Academy of Sciences (1916), and a member of many other scientific societies.

In the summer of 1917 Lyapunov went to Odessa with his wife, who suffered from a serious form of tuberculosis. He began to lecture at the university: but in the spring of the following year his wife's condition rapidly deteriorated and she died on 31 October 1918. On that day Lyapunov shot himself and died three days later, without regaining consciousness. In accordance with a wish stated in a note that he had left, he was buried with his wife.

Lyapunov and A. A. Markov, who had been schoolmates at St. Petersburg University and, later, colleagues at the Academy of Sciences, were Chebyshev's most prominent students and representatives of the St. Petersburg mathematics school. Both were outstanding mathematicians and both exerted a powerful influence on the subsequent development of science. Lyapunov concentrated on three fields: the stability of equilibrium and motion of a mechanical system having a finite number of degrees of freedom; the stability of figures of equilibrium of a uniformly rotating liquid; the stability of figures of equilibrium of a rotating liquid.

Lyapunov's papers on the stability of systems having a finite number of degrees of freedom, among which his doctoral dissertation occupies a central position, belong equally to mathematics and to mechanics. They contain thorough analyses of a great many problems in the theory of ordinary differential equations.

The mathematical formulation of a problem closely related historically to investigations in celestial mechanics of the eighteenth and nineteenth centuries (the problem of the stability of the solar system) is the following: Given a system of n ordinary, first-order differential equations for n functions $x_k(t)$ of the independent variable t (time). It is assumed that the equations are solved with respect to the derivative dx_k/dt on the left side and that the right sides are power series with respect to $x_1, x_2, \cdots, x_n$ without free members, such that the equations have the obvious null solution $x_1 = x_2 = \cdots = x_n = 0$. The coefficients of the series can depend on t. The solution of the system is completely defined by assignment of the values of the unknown functions x_k for some value $t = t_0$. The stability of the null solution, according to Lyapunov, is stability in the infinite time interval $t \geqslant t_0$, with respect to the initial data. In other words, stability consists in the fact that, for $t \geqslant t_0$, the solution of the system $x_k(t)$ will be sufficiently small in absolute value for sufficiently small absolute values of the initial data $x_k(t_0)$. The mechanical systems described above, often called dynamic systems, play a fundamental role in dynamics.

When the corresponding system of differential equations is integrated and its solution is found in simple form, the investigation of stability presents no difficulty. As a rule, however, this integration is impracticable. Therefore, mathematicians generally use the approximation method, which consists in replacing the right sides of the equations by the system of linear members of their expansions into power series. In this manner the task devolves into a study of the stability of a linear system of differential equations; this substantially simplifies the problem, especially when the coefficients of the linear system are constants. It remained unclear, however, whether the replacement of the given system by a linear one was valid and, if valid, under what conditions. Use of a second- or somewhat higher-order approximation (that is, the retention of second- or somewhat higher-order members on the right side) enables one to improve the accuracy of knowledge about functions $x_k(t)$ in a finite time interval but gives no new basis for any conclusions about stability in the infinite interval $t \geqslant t_0$. As Lyapunov noted, the only attempt at a rigorous solution of the question had been made by Poincaré a short time earlier in the special cases of second-order and, in part, third-order systems.

With the aid of new methods that he had created, Lyapunov himself solved, for extremely general assumptions, the question of when the first approximation solves the problem of stability. He thoroughly examined the cases, especially important in practice, when the coefficients of the series on the right side of the equations are constants (the "established motion") or are periodic functions of time t, having one and the same period. For example, if, given constant coefficients, the real parts of all the roots of the system's characteristic equation (an nth degree, "secular" algebraic equation) are negative, the solution of the initial system is stable; if, however, there is among the roots one having a positive real part, then the solution is unstable. The first approximation, however, does not permit one to solve the problem of stability if the characteristic equation, while not having roots with positive real numbers, has roots the real parts of which are equal to zero. Here, the cases when the characteristic equation has one root equal to zero, or when it has two purely imaginary conjugate roots, are of special interest; these cases were exhaustively investigated by Lyapunov. In the case of periodic coefficients Lyapunov examined the possibilities arising in two especially interesting instances: when one of the roots of the characteristic equation is equal to unity and when two imaginary conjugate roots have a modulus equal to unity.

Among many other results is the proof of the theorem of the instability of motion if the force function of the forces acting on the system is not a maximum. Several of Lyapunov's articles dealing with a detailed analysis of the solution of homogeneous linear second-order equations having periodic coefficients (1896–1902) have the same orientation.

The ideas of this cycle of papers, especially Lyapunov's doctoral dissertation, are related to Poincaré's investigation. Specific results obtained by both these scholars coincide, but they do not deal with the basic content or the basic methods of their works. In particular, Poincaré made wide use of geometrical and topological concepts, while Lyapunov used purely analytical methods. Both Poincaré's papers and Lyapunov's works are fundamental to the qualitative theory of ordinary differential equations. At first Lyapunov's theory of the stability of mechanical systems did not receive the wide response given to Poincaré's more general ideas. But, from the early 1930's the number of papers directly related to Lyapunov's investigations increased very rapidly, especially with the growth of the significance of problems concerning the stability of motion in modern physics and engineering, primarily because of the study of fluctuations in various mechanical and physical systems. Problems of stability arise in the determination of the work regimen of various machines, in the construction of airplanes, electrical engineering, and ballistics.

Lyapunov studied figures of equilibrium of a uniformly rotating liquid over a period of thirty-six years. In the lecture "O forme nebesnykh tel" ("On the Shape of Celestial Bodies"; 1918) he said:

> According to a well-known hypothesis, each such body was initially in a liquid state; it took its present form before solidification, having previously received an unchanging form as the result of internal friction. Assuming this, the shape of a celestial body must be one of those which can be assumed by a rotating liquid mass, the particles of which mutually attract one another according to Newton's law, or, at least, must differ little from such a figure of equilibrium of a rotating liquid [*Izbrannye trudy*, p. 303].

Lyapunov mentioned the mathematical difficulty of studying equilibrium figures, a study which entails the solution of nonlinear integral equations.

It was Newton, his eighteenth-century successors Maclaurin and d'Alembert, and others who established that ellipsoids of rotation could be figures of equilibrium of homogeneous rotating liquids. Later, Jacobi demonstrated that certain triaxial ellipsoids could also be such figures. Other scholars also studied this problem. When, in 1882, Chebyshev placed before Lyapunov the question concerning the possibility of the existence of other equilibrium figures that are close to ellipsoidal, Lyapunov could solve the problem only in the first approximation. Believing it impossible to judge the existence of new figures according to the first approximation, he put off a definitive solution to the question. In his master's thesis of 1884 he guardedly mentioned that certain algebraic surfaces which were close to earlier-known ellipsoids of equilibrium, satisfy the conditions of equilibrium in the first approximation. On the other hand, in this thesis he examined the problem of the stability of the Maclaurin and Jacobi ellipsoids, injecting clarity and rigor into the statement of the problem, defining for the first time the concept of stability for a continuous medium.

In a series of papers written between 1903 and 1918, Lyapunov moved deeply into the investigation of Chebyshev's problem and related questions. In *Recherches dans la théorie de la figure des corps célestes* and *Sur l'équation de Clairaut* . . . he proved the existence of nearly spherical figures of equilibrium for a sufficiently slowly rotating nonhomogeneous liquid, and investigated the solutions to the integral-differential equations arising from this, which contain the unknown function both under the integral sign and under the sign of the derivatives; the first of these equations is "Clairaut's equation." In *Sur un problème de Tchébycheff* and *Sur les figures d'équilibre peu différentes des ellipsoids* . . . it was established that among ellipsoids of equilibrium there are "ellipsoids of bifurcation" and that, in addition to ellipsoidal figures close to these, there also exist close, nonellipsoidal figures of equilibrium. A number of methods are entailed in the consistent determination of the equations of these figures. Finally, in the posthumously published *Sur certaines séries de figures d'équilibre d'un liquide hétérogène en rotation,* it is proved that each Maclaurin and Jacobi ellipsoid differing from the ellipsoids of bifurcation engenders new figures of equilibrium that are close to the Maclaurin and Jacobi ellipsoids in form; their density is not considered constant but, rather, as weakly varying.

In *Problème de minimum dans une question de stabilité des figures d'équilibre d'une masse fluide en rotation*, Lyapunov further developed and made more precise the theory of stability stated in his master's thesis and investigated the stability of the new, nearly ellipsoidal, figures of equilibrium that he had previously found.

In solving all these problems and the corresponding nonlinear integral and integral-differential equations, Lyapunov had to overcome great mathematical difficulties. To this end he devised delicate methods of approximation, the convergence of which he proved with the rigor of contemporary mathematics; generalized the concept of the integral (in the direction of the Stieltjes-Riemann integral); and proved a number of new theorems on spherical functions.

Lyapunov's works again approach those of Poincaré. At a certain angular velocity, figures of equilibrium that Poincaré called "pear-shaped" branch off from Jacobi ellipsoids. The astronomer G. H. Darwin encountered the problem of the stability of pear-shaped forms in his hypothesis concerning the origin of double stars arising from the division of a rotating liquid mass into two bodies. Poincaré, who examined the question within the limits of the second approximation, stated the hypothesis that these forms were stable; and Darwin, using Poincaré's general theory, seemingly confirmed this opinion, which was indispensable to his cosmogonical hypothesis, by calculations. Lyapunov's calculations, which were based on precise formulas and evaluations, led him to the opposite conclusion—that the pear-shaped figures were unstable. In 1911 Poincaré stated that he was not sure of the correctness of his prior opinion, but that to solve the question would require one to begin the very complex computations again. In 1912 Lyapunov published the necessary calculations in the third part of *Sur certaines séries de figures d'équilibre* . . ., but no one sought to verify them. In 1917 Sir James Jeans confirmed Lyapunov's results, having discovered the defect in Darwin's computations.

During the period 1886–1902, Lyapunov devoted several works to mathematical physics; *Sur certaines questions qui se rattachent au problème de Dirichlet* (1898) is fundamental among these. Here, for the first time, a number of the basic properties of the potentials of simple and double layers were studied with utter rigor and the necessary and sufficient conditions, under which the function that solves Dirichlet's problem within a given range has normal derivatives over the limiting range of the surface, were indicated. These investigations created the foundation of a number of classic methods for solving boundary-value problems. In addition, Lyapunov's works on mathematical physics brought that area of analysis to the attention of a number of Kharkov mathematicians, especially V. A. Steklov, Lyapunov's student.

Finally, in two works that arose from a course in the theory of probability taught by Lyapunov, he substantially generalized Laplace's limit theorem in application to sums of random independent values. Chebyshev gave the first such generalization of this theorem in 1887, indicating the possibility of its proof, in the form given by him, by the method of moments; and Markov carried out the full proof on this basis in 1898. In "Sur une proposition de la théorie des probabilités" (1900) and "Nouvelle forme du théorème sur la limite de probabilité" (1901), Lyapunov proved the central limit theorem by the method of characteristic functions, which has subsequently assumed a fundamental place in the theory of probability; moreover, he did so under much broader conditions than Markov had used. Some time later, however, Markov proved the central limit theorem under Lyapunov's conditions by using the method of moments. These works by Lyapunov also served as the starting point for many later investigations.

BIBLIOGRAPHY

I. ORIGINAL WORKS. A complete list of Lyapunov's works is in *Aleksandr Mikhailovich Lyapunov. Bibliografia*, compiled by A. M. Lukomskaya, V. I. Smirnov, ed. (Moscow–Leningrad, 1953). Collections of his writings are *A. M. Lyapunov, Izbrannye trudy* ("Selected Works"; Moscow–Leningrad, 1948), with a bibliography compiled by A. M. Lukomskaya; and his *Sobranie sochineny* ("Collected Works"), 5 vols. (Moscow, 1954–1965), which includes Russian translations of all works written in French.

Individual works are *Ob ustoychivosti ellipsoidalnykh form ravnovesia vrashchayushcheysya zhidkosti* ("On the Stability of Ellipsoidal Forms of Equilibrium of a Rotating Liquid"; St. Petersburg, 1884), his master's thesis, translated into French as "Sur la stabilité des figures ellipsoidales d'équilibre d'un liquide animé d'un mouvement de rotation," in *Annales de la Faculté des sciences de l'Université de Toulouse*, 2nd ser., **6** (1904), 5–116; *Obshchaya zadacha ob ustoychivosti dvizhenia* ("The General Problem of the Stability of Motion"; Kharkov, 1902; 2nd ed., Moscow–Leningrad, 1935; 3rd ed., 1950), his doctoral dissertation, translated into French as "Problème générale de la stabilité du mouvement," in *Annales de la Faculté des sciences de l'Université de Toulouse*, 2nd ser., **9** (1907), 203–474, reprinted as no. 17 in the Annals of Mathematics series (Princeton, 1947); *Recherches dans la théorie de la figure des corps célestes* (St. Petersburg, 1903); *Sur l'équation de Clairaut et les équations plus générales de la théorie de la figure des planètes* (St. Petersburg, 1904); *Sur un problème de Tchebycheff* (St. Petersburg, 1905); *Sur les figures d'équilibre peu différentes des ellipsoids d'une masse liquide homogène douée d'un mouvement de rotation*, 4 vols. (St. Petersburg, 1906–1914); and *Problème de minimum dans une question de stabilité des figures d'équilibre d'une masse fluide en rotation* (St. Petersburg, 1908).

Published posthumously were *Sur certaines séries de figures d'équilibre d'un liquide hétérogène en rotation*, 2 vols. (Leningrad, 1925–1927); and *Raboty po teorii potentsiala* ("Papers on the Theory of Potentials"; Moscow–Leningrad, 1949), with a biographical essay by V. A. Steklov.

II. SECONDARY LITERATURE. See Y. L. Geronimus, *Ocherki o rabotakh korifeev russkoy mekhaniki* ("Essays on the Works of the Leading Figures of Russian Mechanics"; Moscow, 1952); A. T. Grigorian, *Ocherki istorii mekhaniki v Rossii* ("Essays on the History of Mechanics in Russia"; Moscow, 1961), 139–148, 219–225; V. I. Smirnov, "Ocherk nauchnykh trudov A. M. Lyapunova" ("Essay on the Scientific Works of A. M. Lyapunov"), in V. I. Smirnov, ed., *A. M. Lyapunov, Izbrannye trudy* (see above), 495–538; "Ocherk zhizni A. M. Lyapunova" and "Obzor nauchnogo tvorchestva A. M. Lyapunova" ("Essay on the Life of A. M. Lyapunov" and "A Survey of the Creative Scientific Work of A. M. Lyapunov"), in *Aleksandr Mikhailovich Lyapunov. Bibliografia* (see above), 469–478, 479–532; and "Iz perepiski P. Appelya, Z. Adamara, G. Burkkhardta, V. Volterra, P. Dyugema, K. Zhordana, A. Puankare i N. Rado s akademikom A. M. Lyapunovym" ("From the Correspondence of P. Appell, J. Hadamard, H. Burckhardt, V. Volterra, P. Duhem, C. Jordan, H. Poincaré, and N. Rado With Academician A. M. Lyapunov"), in *Trudy Instituta istorii estestvoznania i tekhniki. Akademia nauk SSSR*, **19** (1957), 690–719; V. A. Steklov, *Aleksandr Mikhailovich Lyapunov. Nekrolog* (Petrograd, 1919), repr. in *Obshchaya zadacha ustoychivosti dvizhenia*, 2nd ed. (see above), 367–388 and in Lyapunov's *Raboty po teorii potentsiala* (see above), 7–32; and A. P. Youschkevitch, *Istoria matematiki v Rossii do 1917 goda* ("The History of Mathematics in Russia Before 1917"; Moscow, 1968), 448–458.

A. T. GRIGORIAN

LYELL, CHARLES (*b.* Kinnordy, Kirriemuir, Angus, Scotland, 14 November 1797; *d.* London, England, 22 February 1875), *geology, evolutionary biology.*

Charles Lyell was the eldest child of Charles Lyell, Esq., of Kinnordy and the former Frances Smith, daughter of Thomas Smith of Maker Hall, Swaledale, Yorkshire. Both of Lyell's parents had been born in London, where both his grandfathers were connected with the British navy. His paternal grandfather, also Charles Lyell, served as a purser on various ships and in 1770 entered into a partnership to supply His Majesty's ships on the coast of North America. In 1778 he became secretary to Admiral John Byron during the latter's disastrous expedition to America and the West Indies. In 1782 he purchased the estate of Kinnordy at Kirriemuir, in what was then the county of Forfarshire, Scotland, and retired to lead the life of a Scottish laird.

Lyell's father matriculated in 1786 at Saint Andrews University and in 1787 at Peterhouse College, Cambridge, from which he was graduated B.A. in 1791. He thereafter studied law at Lincoln's Inn until he inherited the Kinnordy estate in Scotland on his father's death in 1796. On 4 October 1796 he married Frances Smith at Sidmouth, Devonshire.

When Charles Lyell was less than a year old, his father left Scotland for the south of England, where, in the autumn of 1798, he took a long lease on Bartley Lodge at Lyndhurst, on the border of the New Forest in Hampshire. There Lyell spent his boyhood. Strongly interested in botany, his father collected rare plants in the New Forest and corresponded with James Sowerby, Dawson Turner, William Hooker, and other botanists. He also made regular trips to Scotland to supervise his estate.

In 1805 Lyell, aged seven, was sent to school at Ringwood, Hampshire, and in 1808 at Salisbury. In December 1808 he became severely ill with pleurisy, and during his convalescence at Bartley he began to collect insects and study their habits. In 1810 Lyell and his younger brother Thomas were sent to school at Midhurst, in Sussex. He left Midhurst in June 1815 and in February 1816 matriculated at Exeter College, Oxford, as a gentleman commoner. At Oxford he received the customary classical education based largely on the reading of Aristotle and other ancient authors, but in 1817 he took the course of lectures in mineralogy given by William Buckland and in 1818 attended Buckland's lectures in geology.

Since boyhood Lyell had been an enthusiastic amateur entomologist, and now Buckland aroused his interest in geology. In July 1817 Lyell visited his father's friend Dawson Turner at Yarmouth, Norfolk, where he studied the effects of the interaction of the Yare River with the sea in forming the delta on which Yarmouth stood. In September 1817, while in Scotland with his father, he went with two college friends on horseback through the Highlands, along Loch Awe to Oban and thence by water to the island of Mull and to Staffa, famous for its columnar basalt formation. On Staffa, Lyell confirmed for Buckland that there were the broken ends of basalt columns in the roof of Fingal's Cave, a fact which showed that the cave had not been formed by the erosion of an intrusive dike of softer lava, as had been suggested by Leopold von Buch. During the summer of 1818 Lyell accompanied his parents and sisters on a carriage tour through France, Switzerland, and the north of Italy, crossing the Jura and the Alps. In the Alps he observed the effects of glaciers and the destruction produced by mountain torrents.

On 19 March 1819 Lyell was elected to the Geological Society of London and in the same year to the Linnean Society. He was entered at Lincoln's Inn, where in the autumn of 1819 he began to study law, reading in the office of John Patteson, the pleader. During the spring of 1819, however, when Lyell was studying for his degree examinations at Oxford, his eyes had begun to give him a great deal of pain and he found that he could not sustain the prolonged and intense reading needed for his legal studies. He got on as best he could with the help of a reading clerk, but in August 1820 set out with his father on a long tour of the Continent. They traveled through Belgium to Cologne and thence up the Rhine valley and south into Italy, where they visited Ravenna and Rome. Along the Adriatic coast of Italy Lyell observed how the rivers descending from the Apennines had created a coastal plain, and cities which were ports during Roman times were now five miles from the sea. After his return to London in November 1820, Lyell continued his legal studies, but in a more relaxed fashion.

During the summer of 1821 Lyell visited his old school at Midhurst, Sussex, and, becoming curious about the geology of Sussex, called on Gideon Mantell, a surgeon in Lewes, who had formed a collection of fossils from the chalk of the South Downs and the underlying strata of the Weald. They became close friends and pursued a common interest in the geology of southeast England. In April 1822 Lyell visited Winchelsea and Romsey, on the southeast coast, to study the series of strata below the chalk, since they were visible in the cliffs along the English Channel. In late June 1822 Lyell also visited the Isle of Wight, where he discovered that the sequence of strata below the chalk was the same on the Isle of Wight as on the coast of Sussex and that the formation on the Isle of Wight called the greensand by Thomas Webster was identical with the Reigate firestone and the marl rock of western Sussex. The formation in England called greensand by G. B. Greenough and W. D. Conybeare was identical, however, with a lower series of beds which Webster had called the ferruginous sand on the Isle of Wight. Lyell outlined the distinction between the formations now known as the upper and lower greensand in a letter to Mantell of 4 July 1822. He did not publish his findings, and the same observations were first published in a general study of the greensand by William Fitton in 1824, after he had seen Lyell's letter to Mantell on the subject.

On 25 June 1823 Lyell arrived in Paris, where he spent the next two months. He studied French and attended free lectures on mining, geology, chemistry, and zoology. He met many French scientists, including Alexandre Brongniart, Georges Cuvier, and Constant Prévost, and the great German scientist and traveler Alexander von Humboldt, who was then living in Paris. With Prévost, Lyell studied the geology of the Paris basin and was influenced by Prévost to believe that the alternation of freshwater and marine formations could be explained naturally, without resort to catastrophes of unknown magnitude or cause. Prévost had found places where marine and freshwater fossils were mixed together, and he suggested that the basin had been at times a freshwater lake and at other times a great bay of the sea into which various rivers had emptied. The change from one state to another would have required only a relatively minor geological change in the barrier separating the Paris basin from the open sea, or in the level of the basin in relation to sea level. Prévost's explanation of geological changes in the Paris basin was radically different from that of Cuvier, who had accounted for the alternation of marine and freshwater formations by suggesting violent incursions of the sea over the land. Prévost, a student of Lamarck, also held to the principle of "induction to deduce what has been from what is."[1]

In Paris, Lyell took a broad and discriminating interest in French culture and institutions, but the scientific influence of Paris on him was most profound. He became aware of the depth and seriousness of French studies on natural history and geology, and in particular he became interested in freshwater formations.

In June 1824 Lyell entertained Constant Prévost when he visited England, and they went together on a geological tour of the southwest of England. During the late summer of 1824 Lyell worked at the geological mapping of his native county of Forfarshire in Scotland and made a detailed study of the marl deposits formed in its small freshwater lakes. This study provided the material for Lyell's first scientific paper, "On a Recent Formation of Freshwater Limestone in Forfarshire . . . ," which he began to read to the Geological Society on 17 December 1824. He found that modern freshwater marls were similar in their physical characteristics and contained the same types of fossils as the ancient freshwater marls studied by Cuvier and Brongniart in the Paris basin. In particular Lyell found that in certain circumstances a hard, dense limestone deposited in the Scottish lakes was exactly like certain ancient freshwater limestones of the Paris basin. Cuvier and Brongniart had believed that under modern conditions only soft, friable deposits formed in lakes; and Lyell showed that the distinction they had drawn between ancient and modern freshwater sediments was not valid.

In 1822 Lyell was called to the bar; and from 1825 to 1827 he was a practicing barrister, traveling on the Western Circuit, which included assizes at Winchester, Salisbury, Dorchester, Taunton, Exeter, and Bodmin. Although Lyell never derived any significant income from his legal practice, he valued his professional status as a lawyer because it freed him from many social demands upon his time. On circuit he had opportunities to study the geology of the southwest of England, and it was while on circuit during the summer of 1827 that he read Lamarck's *Philosophie zoologique*. From 1823 to 1826 Lyell served as one of the secretaries of the Geological Society, and in 1826 he was elected a fellow of the Royal Society. In 1825 he published a paper on a dike of serpentine cutting through sandstone strata in Forfarshire.[2] He also began to write articles for the *Quarterly Review*. His first article, published in December 1825, was a review of Thomas Campbell's *Letter to Mr. Brougham on the Subject of a London University;* and the writing of it aroused Lyell's interest in the state of university education in England. In June 1826 he published his second article, "On Scientific Institutions," in the *Quarterly Review;* and in September 1826 his third article appeared, a review of the first volume of the second series of the *Transactions of the Geological Society of London*. The last article was in effect a review of the state of geology in 1826. Lyell discussed current knowledge of the regular succession of stratified formations and the many recent discoveries of fossils, especially of the skeletons of fossil vertebrates. In considering the forces that had brought about geological change, he thought it possible that, given sufficient time, geological causes now active, such as earthquakes and volcanic action, could have raised sedimentary strata from the bottom of the sea to the places they now occupy in hills and mountains. He thought it premature to consider that any unknown force would have been needed and that to postulate more powerful disturbing forces in earlier geological periods would tend to discourage geologists from studying those changes now going on. Lyell showed that older strata had been deposited during long periods of tranquil and orderly conditions, just as younger strata had, and that the elevation and bending of strata proceeded in much the same way in successive geological periods.

In June 1827 the *Quarterly Review* published Lyell's fourth article, "State of the Universities," an extensive and profound comparison of universities in England with those in Scotland and on the Continent. Lyell described the concentration of teaching in the colleges at Oxford and Cambridge; their attention to classical studies to the exclusion of modern languages, modern history, economics, political science, and natural sciences; and the fact that although the dons taught nothing but classics, they made almost no contribution to classical scholarship. He followed this original and comprehensive analysis of English university education with various suggestions for reform, some of which were implemented much later in the nineteenth century.

During the summer of 1827 Lyell wrote a review of George Scrope's *Memoir on the Geology of Central France* (London, 1827), which was published in the *Quarterly Review* for October 1827. Scrope's book aroused Lyell's interest in the Auvergne district of France. In 1822 the discovery by Gideon Mantell of a river delta formation in the Tilgate forest beds beneath the chalk in Sussex had suggested to Lyell the magnitude of the earth movements which could sink the land surface, represented by the Tilgate forest beds, beneath the deep ocean in which the chalk was deposited, and then elevate it again to form the high ridges of the Weald. He also realized that the outlines of the ancient land surfaces, even those contemporary with the younger Tertiary strata of the Hampshire and London basins, had been completely obliterated. In contrast, the Auvergne, with its succession of ancient volcanic formations, was a geographically definable area since an early period in geological time. In Auvergne there were lava flows from at least three successive periods of volcanic activity. The most recent extinct volcanoes still retained their cones and craters intact, with unbroken sheets of lava extending into the river valleys. Beneath these layers of lava were others which bore no relation to the existing cones. Still older were isolated masses of basalt capping outlying hills. In addition to volcanic rocks of successive ages there were in Auvergne thick formations of freshwater strata containing fossils of extinct animals. Thus, through geological periods long enough for great changes in organic life, the Auvergne had remained a region of volcanic mountains, valleys, and freshwater lakes. In its geological formations was an extraordinarily long and continuous record of geological history.

In May 1828 Lyell set off for Auvergne to see its geology for himself. In Paris he joined Roderick Murchison and Mrs. Murchison, and they traveled south in Murchison's light open carriage. During the next two months they traversed the volcanic mountains and freshwater formations of central France. Lyell was struck by the resemblance of the ancient freshwater sediments of Auvergne to the modern deposits he had studied in Scotland. Both contained such fossils as the seedpods of *Chara* (gyrogonites), *Cypris*, caddis fly larva cases (*Phryganea*), and freshwater shells. But he was also astonished by a Tertiary freshwater limestone in the Cantal, near Aurillac, which at first sight looked exactly like the English chalk, a much older marine formation. Lyell and Murchison observed the action of rivers in Auvergne in reexcavating river valleys after they had been filled by successive lava flows.

From the Auvergne, Lyell and Murchison descended into the Rhone valley at Montélimar and went south. On 9 August they were at Nice, where they paused for some weeks to rest and to allow Murchison to recover from an attack of fever. The effect of the tour thus far had been to suggest to Lyell many analogies between conditions during the geological past and those of the present. His enthusiasm for geology was raised to the highest pitch. Murchison now suggested to Lyell that he go on to the south of Italy and Sicily, where he would see younger strata and many instances of modern changes produced by the volcanoes of Vesuvius and Etna. Before he left London, Lyell had begun the manuscript of a geological book in which he pointed out analogies between ancient and modern conditions. The tour through France had provided him with numerous instances to amplify this work, and he decided to go on to gather even more information.

In Nice, Lyell learned from Giovanni Risso that the fossil seashells in the local marine Tertiary strata included about 18 percent of species still living in the Mediterranean. From Nice, Lyell and Murchison traveled east along the coast to Genoa. They then returned to Savona and crossed the Apennines to Turin, where Franco Bonelli, director of the museum of zoology, told them that the fossil shells of the nearby hill of Superga differed from those of the subapennine beds but corresponded closely to the shells of the Bordeaux Tertiary formation in France. They differed also in the proportion of fossil shells belonging to still-living species. Lyell now formed the idea of classifying Tertiary formations according to the proportion of living species among their fossils. Furthermore, because he considered that volcanoes and earthquakes were associated with the uplift of strata, he thought that as he approached the active volcanoes of Vesuvius and Etna, he would find younger and more recently uplifted strata containing larger proportions of still-living species.

Murchison and Lyell traveled together as far as Padua, where they parted, Lyell going south through Italy while Murchison turned north on his way back to England. Lyell examined collections of fossils at Parma, Bologna, Florence, and Siena and made his own collection of subapennine shells. From Naples he went to the island of Ischia, where he collected some thirty species of fossil shells, all belonging to species living in the Mediterranean, from marl strata at an elevation of 700 feet on the side of the extinct volcano of Epomeo. Up to this point Lyell's discoveries entirely fulfilled his expectations. He thought the strata on Ischia had been elevated very recently and represented more or less contemporary deposits of the Mediterranean sea bottom. He assumed that soft marls containing only living species of shells must be very recent. He had found that the strata in the vicinity of volcanoes, and in an area where earthquakes were frequent, contained fossil shells of the same species as were still living in the Mediterranean. In Naples, Lyell studied Vesuvius and the temple of Serapis at Pozzuoli, which showed evidence of both elevation and subsidence of the land during historic times.

On 16 November 1828 Lyell crossed to Messina, Sicily, and rode on muleback south along the coast to Taormina. He visited the Val del Bove and from Nicolosi climbed Mount Etna. He was deeply impressed by the enormous size and great age of Etna, which he was convinced had been built up by a long series of volcanic eruptions like those still going on. His first doubts about the necessary recentness of strata containing living species of fossils may have arisen from a bed of clay above Catania that contained many modern Mediterranean shells. When he looked back at Etna from the ridge above Primosole, he saw that the enormous mass of the volcano appeared to rest, as though on a platform, on the plateau formed by the clay hills above Catania, so that the clay strata appeared older than the mountain.

In Lentini, Lyell saw for the first time the hard white limestone of the Val di Noto; it puzzled him because, while he thought it ought to be recent, it contained only the casts of shells, like an ancient Secondary limestone. As he rode down the valley of the Anapo River toward Syracuse, Lyell saw the white limestone exposed in the cliffs to a thickness of more than 800 feet. At the foot of the white limestone escarpment at Floridia, near Syracuse, were strata of calcareous sandstone. On the south side of the harbor of Syracuse, Lyell found, apparently beneath the white limestone—and therefore still older—strata of soft marl containing many perfect fossil shells, some still retaining their colors and all belonging to species still living in the surrounding Mediterranean. He was, however, uncertain of the relation of the marl to the white limestone until he reached Girgenti (now Agrigento), where he found strata of blue clay containing shells overlain by strata of calcareous sandstone identical with that at Floridia.

Astonished at what he was finding, and doubting whether he had observed correctly, Lyell explored the interior of Sicily until at Castrogiovanni (now Enna) he found all the strata of the Sicilian formation, from the white limestone of Lentini down through sandstones and marls to the blue clay of Syracuse, containing living species of shells with their original colors, all exposed in order in one immense escarpment. For Lyell the discovery that the white limestone, containing only casts and impressions of shells, was a very recent formation, had the force of revelation. It meant that the characteristics of a rock stratum were determined not by its geological age but by the conditions under which it was laid down, and these conditions had recurred again and again in successive geological periods. In Sicily, Lyell saw too that volcanic activity had gone on gradually and uniformly, interspersed with periods of rest, through an immense extent of time. Finally, he saw that the assemblage of shell species now living in the Mediterranean constituted a very old fauna, which had changed very little during the immense time required to deposit the stratified rocks of Sicily and to elevate them into hills of rounded and worn outline.

Because Lyell saw in Sicily a continuity between the fauna living in the Mediterranean and its fossil ancestors preserved in the rocks, he realized that the conditions under which both lived must be analogous. Therefore conditions during past geological ages must have been essentially similar to modern conditions on the earth's surface, and the forces which brought about geologic changes must have been the result of processes similar to those going on at the present time. Fired with such new ideas, Lyell set out for England in January 1829, determined to rewrite his book to see how far he could "explain the former changes of the earth's surface by causes now in operation." He was now certain that the Tertiary formations represented a much greater period of time than he had dreamed and that the strata of the Paris basin, of the hill of Superga at Turin, of the subapennine hills, and of Sicily were not parallel and contemporary formations but represented different and successive periods of geological time.

In Turin, Lyell discussed his Sicilian results with Bonelli, and in Geneva he obtained from Augustin de Candolle information on the geographical distribution of plants. In Paris, which he reached on 14 February 1829, Lyell called on Jules Desnoyers, who in 1825 had shown that the Tertiary formation of the Loire valley was younger than that of the Paris basin. When Lyell told Desnoyers of his Sicilian discoveries and his opinion that the various Tertiary formations were of different ages and could be arranged according to the proportion of living species among their fossils, Desnoyers said that he himself had come to similar conclusions and that Gérard Paul Deshayes, who possessed a large collection of fossil shells, had also decided that Tertiary formations might be arranged in a chronological series on the basis of differences in their characteristic groupings of fossil shell species. Lyell immediately consulted Deshayes, who told him that on the basis of their fossil shells, Tertiary formations might be distinguished into three successive geological periods. Lyell then asked Deshayes to prepare a catalog of fossil shells from all the Tertiary formations of Europe.

Lyell returned to London on 24 February 1829, and from then until the beginning of June 1830 he wrote and saw through the press the first volume of his *Principles of Geology*. It was published by John Murray in July 1830 (not January, as stated in later editions of the *Principles*) and created an immediate sensation. In it Lyell first discussed the historical development of geology and then treated the principles of geological reasoning. He argued that the order of nature in the past was uniform with that in the present and, therefore, that the geologist should always try to explain geological phenomena by analogy with modern conditions. The greatest difficulty he had to overcome was the clear evidence for a warmer climate during past geological epochs even in high northern latitudes, such as those of Great Britain. To account for the changes in climate which had occurred between past geological epochs and the present, Lyell studied the factors which determine climate in different parts of the world at the present time. He showed how not only local climate, but even worldwide climatic conditions, depend on the pattern of distribution of land and sea and would, therefore, be altered by changes in their distribution. An increase in the proportion of land near the equator, and of ocean area toward the poles, would tend to create a warmer world climate and vice versa.

Lyell surveyed the full range of processes which at present are altering the earth's surface—the eroding effects of running water in streams and rivers and of waves along the seacoast, the accumulation of sediments in deltas and on the sea bottom, and the cumulative effects of earthquakes and volcanoes in elevating the land. He showed that even the largest volcanoes, such as Vesuvius and Etna, were the product of a long series of eruptions distributed through immense periods of time, and the eruptions were never greater nor more frequent than in historic times. Lyell emphasized repeatedly that the magnitude of the geological changes which had occurred during the past was not a reason to postulate extraordinary convulsions or catastrophes. The greatest changes could be accomplished by ordinary geological processes acting gradually, if they were given sufficient time.

Lyell sought to separate geology from cosmology and refused to consider the question of the origin of the earth, or its state when first formed. Furthermore, he considered it unscientific to attempt to explain otherwise puzzling geological phenomena by reference to the origin of the world. In the succession of geological formations and geological changes Lyell saw an unending vista of uniform conditions on the earth's surface, as stable as those that exist today, and of geological processes ceaselessly at work. The principle of the uniformity of the "course of Nature" throughout the history of the earth gave significance to every feature of the present natural world for the understanding of events in the past. The geologist therefore must study all changes now occurring on the earth, because these changes, daily at work, may enable him to understand otherwise puzzling geological phenomena of the past.

During the summer of 1830 Lyell made a geological expedition to the Pyrenees, followed by several weeks of study of Tertiary shells with Deshayes at Paris. When he returned to London on 9 November, half the edition of the first volume of the *Principles* was already sold. During 1831 he wrote the second volume, which was published in January 1832. In this volume Lyell considered the changes which had occurred in the living world through geological time. His studies on Tertiary shells had already shown him that as one proceeded from older to younger strata, the proportion of fossils belonging to extinct species declined gradually while that of living species increased. Thus, throughout the Tertiary period species of shells had become extinct, one by one, to be replaced, one by one, by new species. The continuity thus revealed in the living world, accompanied by gradual change, was in sharp distinction to the apparently abrupt changes in the forms of animal life observed in the secondary strata of France and England. Lyell pointed out that all the species of mammals living when the present assemblage of marine mollusks had become established in the sea had since become extinct. Thus the land mammals, which were subject to many more vicissitudes than marine forms, became extinct more rapidly.

To account for extinction Lyell showed that the life of each species was dependent upon the continuance of a certain set of physical conditions in its environment. Geological processes, however, were steadily tending to alter these conditions, both locally and over wide regions. Thus the possible habitats for a species were steadily shifting and sometimes were obliterated. Lyell also showed that the life of a species was at the same time dependent upon a multitude of relationships with other species living in the same area. There was severe competition between species for living space, so that sometimes a species might become extinct simply because it could not contend successfully with others. And the extinction of one species would drag in its train the extinction of others which were dependent upon it. Similarly, the increase and spread of a successful species would force many others into extinction. Yet the geographical distribution of species showed that every species had tended to spread outward from its geographical center of origin. Lyell thus showed that the living world of plants and animals was in a state of dynamic balance and that the fluctuations of the balance of nature would themselves steadily tend to produce the extinction of species.

Lyell's argument required the assumption that species were real, definite, and stable entities. If species were infinitely plastic, or had no real existence, as Lamarck had supposed, they need never become extinct but could simply adapt themselves endlessly to varying conditions. Lyell therefore began his discussion of species with a systematic examination and criticism of Lamarck's theory.

Lyell also used evidence drawn from the geographical distribution of plants and animals to show that each species appeared to have arisen in a single center, from which it had spread. The area of its distribution was limited by geographical barriers of different kinds. Places with similar altitude and climate on two isolated continents have entirely different sets of species, although the two sets of species may have adapted in similar ways to similar conditions. Lyell showed that each species had arisen at a particular place and at a particular time in geologic history, and had persisted for a certain period, at the end of which it became extinct and was replaced by a new species.

Lyell left the species problem at this point. He had shown that the emergence of new species was a steady process in geologic history that was needed to compensate for the equally steady process of extinction, but he did not attempt to say how new species emerged. His readers were thus left in suspense. Privately, Lyell had pondered the question of the origin of species and was certain that they emerged by some natural process, but he had no explanation to offer. Although he was fascinated by Lamarck's theory, he felt certain, on scientific grounds, that it was false.

In 1831 Lyell was appointed professor of geology at King's College, London, and in May and June 1832 delivered his first course of lectures on geology, a course which was widely attended by members of the public. On 12 July 1832, at Bonn, Germany, Lyell married Mary Elizabeth Horner, aged twenty-three, daughter of Leonard Horner, a fellow of the Geological Society and from 1828 to 1830 warden of London University. On their return to London after a wedding trip up the Rhine valley and through Switzerland and France, Lyell took a house at 16 Hart Street. Mary Lyell, who read German and French fluently, assisted Lyell in his reading and geological work at home and accompanied him on almost all his geological travels abroad. She became an accomplished conchologist and, because of the weakness of his eyes, often wrote Lyell's letters for him.

In the third volume of the *Principles*, published in 1833, Lyell attempted to answer some of the criticisms of his doctrines provoked by the first two volumes. The opening chapter is a splendid defense of the point of view to which William Whewell had already attached the name "uniformitarian." Lyell argued that the natural order of the past was uniform with that of the present; the same physical laws held true and the same kinds of processes occurred. The geologist should, therefore, attempt to explain geological phenomena by analogy with modern processes. The apparent difficulty in doing so was chiefly a result of ignorance of the nature and extent of geological changes now going on. The remainder of the volume was an application of the doctrine of uniformity and the use of modern analogies to the problems of Tertiary geology. Lyell introduced his classification of Tertiary formations into four successive epochs and the terms Eocene (Gr. *ἠως*, dawn; *καίνος*, recent), Miocene (Gr. *μειων*, less; *καίνος*, recent), and the older and newer Pliocene (Gr. *πλειον*, more, larger; *καίνος*, recent) to designate them. In the coinage of these terms he was assisted by William Whewell, who suggested the common root *καινος*. Lyell's classification showed that the Tertiary formations made up a series of strata enormously greater than had previously been suspected and represented a period of time correspondingly great. Lyell included, as an appendix to the third volume, Deshayes's tables of more than 3,000 species of Tertiary fossil shells.

Besides establishing a new era in geology, the *Principles* became a lifework for Lyell. He continued to revise it and brought it out in twelve successive editions. The twelfth edition was corrected by him but appeared after his death. The second edition consisted of a corrected version of the first and second volumes, published in May 1832 and January 1833, respectively, before the appearance of the third volume in May 1833. In May 1834 he published a new edition of the whole work, called the third edition, in four small volumes, dividing the material in the three volumes of the first edition into four books but omitting Deshayes's tables of Tertiary shells. While continuing to revise and add to his text, Lyell retained the format of four small volumes for the fourth edition, which appeared in June 1835, and the fifth edition, which appeared in March 1837.

In July 1838 Lyell published his *Elements of Geology* in one volume. Although the *Elements* was intended to be a brief descriptive work and not theoretical, in writing it Lyell took for granted the theoretical viewpoint he had already established in the *Principles*. Thus Lyell's *Elements* was the first modern textbook of geology written on the assumption that geological phenomena could be explained completely in terms of natural and knowable causes.

When Lyell came to prepare the sixth edition of the *Principles*, he removed from it the discussion of Tertiary formations in book IV, in order to transfer all purely descriptive geology to a second edition of the *Elements*. He also transferred four chapters dealing with various theoretical questions from book IV to book I in order to retain them in the *Principles*. Lyell was anxious that the *Principles* and the *Elements* should not overlap in their subject matter. In addition he developed, in chapter 10 of book I, a theory that erratic boulders had been transported by icebergs; and in chapter 11 he presented evidence that volcanic action had occurred in all geological periods with about the same intensity as at present. Lyell thus established in the sixth edition of the *Principles*, published in two volumes in June 1840, the plan of organization to which he would adhere through the seventh edition (February 1847), the eighth edition (May 1850), and the ninth edition (June 1853). The seventh through ninth editions were each a single large octavo volume. After 1853 Lyell did not revise the *Principles* until 1865, when he undertook a major reorganization of the book.

The second edition of the *Elements* appeared in July 1841. With the inclusion of the chapters on Tertiary formations and other additions, it had grown from twenty-five to thirty-six chapters in two volumes. In the third edition, published in January 1851, Lyell changed the title to *A Manual of Elementary Geology* but did not alter the basic plan of the book. It remained the *Manual* in the fourth edition of January 1852 and the fifth edition of February 1855. For the sixth edition, however, published in January 1865, Lyell returned to the original *Elements of Geology*.

Lyell took great trouble to provide both the *Principles* and the *Elements* with maps, diagrams, and illustrations. He had numerous drawings of fossils made and transferred to woodcuts or engravings that were often of unusual beauty. As a rule each new edition of his work contained additional illustrations. Lyell took equal pains with his prose. The *Elements* is a model of simple, lucid exposition, while in the *Principles* he cultivated a disarming and entrancing style of great persuasive power. Lyell used visual images, analogies, and metaphors to help the reader envision the past history of the earth.

From 1834 to 1836 Lyell was president of the Geological Society of London. During June and July 1834 he traveled alone to Sweden to determine whether the rising of the land there, a phenomenon reported by a number of writers, were real or not. Lyell found that the evidence for a gradual and imperceptible, but steady, rise of the land in northern Sweden was clear and inescapable. On 27 November and 18 December 1834 he presented the results of his Swedish tour to the Royal Society of London as the Bakerian lecture.

In October 1836 Charles Darwin returned from the *Beagle* voyage, and a warm friendship soon sprang up between him and Lyell. Darwin had taken Lyell's *Principles* with him on the *Beagle*, and in South America he had discovered that the whole southern part of the continent had undergone a recent elevation in gradual and successive stages. On 7 March 1838 Darwin presented to the Geological Society a paper in which he connected the phenomena of earthquakes and volcanic action in South America to show that the elevation of the Andes and of the whole continent had been brought about gradually by the cumulative effect of repeated earthquakes and volcanic eruptions. Such a large-scale example of the action of modern causes provided Lyell with very welcome support for the theory of geological uniformity, for which he was still contending against catastrophist opponents. Lyell was also deeply impressed by Darwin's theory of the origin of coral reefs and atolls by gradual subsidence. Darwin and Lyell met most frequently from 1836 to 1841, when Darwin was living in London. After Darwin moved to Downe, Kent, he and Lyell corresponded regularly and met occasionally in London. From time to time the Lyells went to Downe to spend several days with the Darwins, and during these visits Lyell usually took work with him.

In 1837 Lyell became involved in a controversy with Edward Charlesworth over the classification of the crag, a Tertiary formation of Norfolk. In 1835 Charlesworth had correctly distinguished the red crag from the older coralline crag; but Lyell had thought that because there were 150 species of shells common to the two formations, they must belong to the same geological period. He suggested that the two deposits had been laid down contemporaneously in different parts of the same sea. Charlesworth then attacked the basis for Lyell's classification of Tertiary formations. He pointed out the uncertainty in the discrimination of species and cited the opinion of some naturalists that almost all species of crag fossils were extinct. Because of the uncertainty whether fossil shells were identical with living forms or belonged to extinct species, Charlesworth argued that Lyell's system of classification based on the proportion of living to extinct species in a formation was unreliable.

To find answers to the questions raised by Charlesworth, Lyell made three geological tours of the crag, in April 1838 to Suffolk and in July 1838 and June 1839 to Suffolk and Norfolk. He found that Charlesworth was right in considering the coralline crag to be an older formation than the red crag, but after an exhaustive study of crag fossil shells he also found that the percentage of living species in the various crag formations confirmed Charlesworth's geological findings instead of contradicting them. The coralline crag contained only 19 percent living species, while the red crag contained 30 percent and the Norwich crag, which Charlesworth considered a still younger formation, contained between 50 and 60 percent living species of shells. Ironically, therefore, Charlesworth's conclusions were upheld by the very system he had attacked. The crag question also caused Lyell to study intensively the range of variation, through time, of species of Tertiary fossil shells and to compare the opinions of the two leading fossil conchologists of Europe, Heinrich Beck and Deshayes, on the identity of crag shell species.

On 20 July 1841 the Lyells sailed for America, where they remained until August 1842. After landing at Boston, Lyell visited the geologists of the New York State Geological Survey at Albany and the Lyells traveled through western New York, accompanied by James Hall. Deeply impressed by Niagara Falls, Lyell collected evidence to show that the falls were receding steadily toward Lake Erie and that the Niagara gorge had been produced by the gradual recession of the falls from the face of the escarpment at Queenston, Ontario. During October and November 1841 Lyell delivered the Lowell lectures in Boston. In December the Lyells traveled to the South, where Lyell studied the Tertiary formations of the coastal plain in Virginia, the Carolinas, and Georgia. On their return northward Lyell gave courses of geological lectures in Philadelphia and New York, attended the meeting of the American Association of Geologists and Naturalists at Boston in April 1842, and then made a tour of the Ohio valley, Lake Erie, Lake Ontario, and the St. Lawrence River valley. He studied the succession of terraces around Lake Erie and Lake Ontario; and in Montreal and Quebec and at Lake Champlain he collected, from deposits of clay several hundred feet above sea level, fossil seashells belonging to living Arctic species. During July and August 1842 Lyell spent a month studying the coal formation of Nova Scotia. He showed that the gypsum-bearing strata were part of the lower Carboniferous formation rather than the upper.

At Pictou, Nova Scotia, Lyell met John William Dawson, then aged twenty-one and newly returned from a year of study at the University of Edinburgh, who accompanied him on several geological expeditions. Lyell encouraged Dawson to prepare papers to send to the Geological Society of London and thereby launched Dawson on his career as a geologist. Lyell was a sympathetic and perceptive observer of the United States and Canada and made a number of warm friends among Americans, who were impressed by Lyell's geological knowledge and universal curiosity but sometimes bewildered by his absentmindedness and odd mannerisms. Invariably they were charmed by the beauty and winning manners of Mary Lyell, who was equally at ease in mansion or log cabin, in steamboat, railway car, or stagecoach, and always cheerful and uncomplaining, whatever mishaps or hardships she might suffer. On her part Mary Lyell developed a strong affection for America. During the Civil War, when the United States was very unpopular in England, she defended the cause of the Union and befriended the United States ambassador, Charles Francis Adams, when he was subjected to social ostracism in London.

In 1845 Lyell published his *Travels in North America*, a work unusual among descriptions of America by British travelers in its fairness and insight, and incorporating much discussion of the geology of North America. He included a geological map of eastern North America from the Gulf of St. Lawrence to Florida and Texas, based on the published works of twenty-six authors as well as on his own travels. In discussing American colleges Lyell was also led to discuss English university education because, he said, in trying to explain it to Americans he himself had become more aware of its unusual nature. Lyell followed up his criticisms of the English universities by helping to obtain the establishment of a royal commission on the universities.

In September 1845 Lyell returned to the United States, again to give the Lowell lectures in Boston. He and his wife first traveled to Maine and through the White Mountains of New Hampshire. After Lyell had delivered his course of lectures, the Lyells set off in December on a long tour through the South, stopping for two weeks in January 1846 with Hamilton Couper at Hopeton plantation, on the Altamaha River in Georgia. From Savannah, Georgia, they went to Claiborne, on the Alabama River, where Lyell collected fossils, then by steamboat to Mobile. From Mobile they went up the Tombigbee River to Tuscaloosa, to visit the Alabama coal field. Returning to Mobile, they continued by steamboat to New Orleans, where Lyell made an expedition to the mouth of the Mississippi River to study the growth of its delta. From New Orleans the Lyells went up the Mississippi and Ohio rivers by steamboat, and overland across Pennsylvania to Philadelphia. On 1 June 1846 they sailed for England. In 1849 Lyell published an account of this tour in his *Second Visit to the United States.*

In October 1846 the Lyells moved from 16 Hart Street to a larger house at 11 Harley Street (now 73 Harley Street). In 1848 Lyell was knighted, and in 1849 he was reelected president of the Geological Society of London. In this capacity he gave evidence before the Oxford University Commission, and by this and other means he was active in the reform of the universities and the introduction of science into university education. His books on America attracted the attention of Prince Albert, and he served as one of the royal commissioners for the Great Exhibition of 1851.

In August 1852 the Lyells sailed for Halifax, Nova Scotia, where Lyell went with J. W. Dawson to study the great fossil cliff at South Joggins. On this occasion they found inside the trunk of one of the upright fossil *Sigillaria* trees of Joggins the skeleton of a reptile, later described by Richard Owen and named *Dendrerpeton acadianum.* Since the first air-breathing animal was discovered in the coal only in 1844, *Dendrerpeton* was one of but three or four fossil reptiles then known from the Carboniferous period. From Nova Scotia the Lyells went to Fredericton, New Brunswick, where they were guests of Sir Edmund Head, the governor, and thence to Boston, where Lyell delivered the Lowell Lectures during the autumn of 1852. In December they returned to London. In June 1853 Lyell made a fourth visit to the United States, to represent, with Lord Ellesmere, the Royal Commission for the Exhibition of 1851 at the New York Industrial Exhibition of 1853; he returned to London at the beginning of August.

In December 1853 the Lyells sailed to Madeira, accompanied by Mr. and Mrs. Charles Bunbury, to spend several months studying the volcanic geology of Madeira and the Canary Islands. On Madeira, Lyell met a young German naturalist, George Hartung, who had come there to recover from consumption. Hartung joined Lyell in his geological expeditions and in the collection of specimens. Lyell was deeply impressed by the distinctive flora of these oceanic islands. On Madeira, at an elevation of about 1,000 feet above sea level, he discovered a bed of lignite containing fossilized leaves of plants similar to those of the modern forests of Madeira. The fossil plants showed that Madeira had existed as a land surface, and probably as an isolated oceanic island, since the Miocene. In the light of their long geological history as isolated islands, the zoology of the islands was also remarkable. They possessed no native mammals but contained an unusual number of insect species, especially of beetles; and many of the species were peculiar to individual islands. The species of land shells were even more remarkable. There were seventy-seven species on Madeira and fifty-five species on Pôrto Santo, but only nine species were common to the two islands, although they are only thirty miles apart.

The Lyells returned to England in March 1854. Until February 1855 Lyell was occupied in preparing the fifth edition of the *Manual.* He then visited Berlin and during the summer made a long visit to Scotland. In November 1855 Hartung came to London, and Lyell and he began to go over their collections from Madeira and the Canary Islands. The remarkable number of distinct species of insects and land shells —a pattern of geographical distribution of species similar to that which Charles Darwin had observed in the Galápagos Islands twenty years earlier—impressed Lyell as bearing directly on the question of the origin of species. On 26 November 1855 Lyell read an article by A. R. Wallace, "On the Law Which Has Regulated the Introduction of New Species," in the *Annals and Magazine of Natural History* for September 1855. Wallace presented evidence from paleontology and from the geographical distribution of plants and animals to show that "Every species has come into existence co-incident in both space and time with a pre-existing closely allied species." Wallace's paper seemed to express very aptly the meaning of the distribution of both fossil and living species in Madeira and the Canary Islands; and Lyell made extensive notes on it in a notebook devoted solely to the species question, the first of what would ultimately be seven such notebooks. Through the winter of 1856 Lyell continued to study Madeiran and Canary species, and he began to investigate the question of how species might migrate across areas of ocean to colonize islands.

On 13 April 1856 the Lyells went to Downe, Kent, to spend four days with the Charles Darwins. On the morning of 16 April, Darwin explained to Lyell his theory of the origin of species by natural selection. Lyell had been aware for many years that Darwin believed in the transmutation of species; but he now learned, apparently for the first time, the details of Darwin's theory to account for the process of transmutation. Lyell immediately urged Darwin to publish his theory, a step which Darwin was reluctant to take because he thought that it could not be understood, or appreciated fully, except within the framework of a broad array of connected biological and geological evidence. During the summer of 1856, however, Darwin began to write on the relationship of the geographical distribution of species to geological history. In October 1856 Darwin decided that he could not publish a brief sketch of his theory and began to write the large work on which he was still engaged in June 1858, when he received from Alfred Russel Wallace the manuscript of Wallace's paper containing the theory of natural selection.

The species of Madeira and the Canary Islands, together with Darwin's theory of natural selection, caused Lyell to accept at least tentatively the possibility of the transmutation of species. In 1857 a series of discoveries of fossil mammals in the Purbeck strata showed that a rich mammalian fauna had lived during the Secondary period, in what had been considered exclusively an age of reptiles. This caused Lyell again to doubt whether there had been a genuine progressive development of animal life through geological time. In 1858 the publication of the Darwin-Wallace papers on natural selection renewed Lyell's interest in the species question. From 1858 to 1861, when he began to write the *Antiquity of Man*, Lyell made many entries in his species journals and corresponded frequently with Darwin.

In October 1857 Lyell revisited southern Italy and Sicily, where he had not been since 1828, to make a fresh study of the structure of Vesuvius and Etna. In 1853–1854, in his study of Madeira and the Canaries, made with Hartung, Lyell had become convinced that the evidence on which Leopold von Buch had founded his theory of craters of elevation (1825) was completely fallacious. According to Buch's theory, the volcanic rocks forming such islands as Tenerife and Palma had been formed originally as horizontal sheets of lava which had later been upheaved in a great convulsion to create the cones and craters of their modern volcanoes. Buch believed that liquid lavas could not solidify to form sheets of solid rock on the steep slopes where they now occurred. Therefore, they must have been solidified in a horizontal position and later upheaved. In 1834 Élie de Beaumont, in a paper on Mount Etna, argued that the solid beds of lava in the Val del Bove, inclined at angles of 28° and more, resembled portions of modern lavas that flowed over ground almost level or inclined at no more than 3°. Geologists who accepted Élie de Beaumont's conclusions were forced to believe that all modern volcanoes had acquired their conical form by later upheaval of their beds of lava.

In 1853–1854, on Madeira and Palma, Lyell had seen modern lavas forming sheets of solid rock and inclined at angles of 15° to 20° but showing no sign of any disturbance in position. In 1855 Hartung had also observed on Lanzarote in the Canaries a solid basaltic lava on a slope of 30°. In 1857 Lyell found on Etna lavas which had solidified on steep slopes of from 15° to 40° in inclination. The lavas formed continuous sheets of rock alternating with layers of loose scoriae above and beneath them. Lyell reported these results to the Royal Society on 10 June 1858, and in September and October 1858 he returned to Etna to make a more thorough study of its structure. He found that Etna had neither a linear axis nor a single center of upheaval. Instead, he found two earlier centers of eruption in the Val del Bove. From each center the beds of lava sloped away in all directions, but beds of lava arising from Etna's modern center of eruption had flowed over and buried those from the earlier centers. Lyell considered this to be decisive evidence against the crater-of-elevation theory because "although one cone of eruption may envelope and bury another cone of eruption, it is impossible for a cone of upheaval to mantle round and overwhelm another cone of upheaval so as to reduce the whole mass to one conical mountain."[3] He concluded that the conical form of Etna, as of all volcanoes, was entirely the result of the long-continued process of volcanic eruption.

In June 1858 Charles Darwin consulted Lyell about Alfred Russel Wallace's paper on natural selection, which he had just received, and Lyell and Joseph Hooker together presented Darwin's and Wallace's papers on natural selection to the Linnean Society on 1 July 1858. In May 1859 Joseph Prestwich announced to the Royal Society that he had found, at St. Acheul near Abbeville, France, a flint implement, the product of human workmanship, embedded in place in ancient river gravel. Prestwich's discovery, which established the geological age of the flint implements described earlier from Abbeville by Boucher de Perthes, was clear evidence that man had lived in Europe as far back as the later Pliocene and, therefore, far longer than had previously been thought. In July 1859 Lyell went to Abbeville and Amiens to see for himself the places where flint implements had been found and continued to pursue the question of the antiquity of man, which he saw was related to that of the origin of species.

During 1860 and 1861 Lyell studied evidence of the early existence of man in Europe; and in 1862 he began to write *The Geological Evidences of the Antiquity of Man*, which was published in February 1863. In the *Antiquity of Man* Lyell discussed the state of knowledge concerning the prehistory of man, including the discovery of stone tools, the definition of the Bronze and Stone ages, the shell mounds of Denmark, and the Swiss lake dwellings. He gave a particularly thorough account of the discovery in 1857 of the skeleton of a man in a cave at Neanderthal, Germany, because the skull of the Neanderthal man, although unquestionably human, showed heavy eyebrow ridges like those of apes. In considering the origin of man, Lyell argued that if all races of man were descended from a common ancestry, as he believed they were, then it would have required a very long period of time for the gradual formation of such distinct races as the Caucasian, Mongolian, and Negro. If the different races of man were descended from a common ancestry, however, Lyell thought it equally probable that closely related species of animals and plants were likewise descended from a common parentage. He then reviewed his own 1832 analysis of Lamarck's theory and said that while he had disagreed with Lamarck's view of the transmutation of species, he had agreed with Lamarck "in believing that the system of changes now in progress in the organic world would afford, when fully understood, a complete key to the interpretation of all the vicissitudes of the living creation in past ages."[4] Lyell described Darwin's and Wallace's theory of the indefinite modification of species by natural selection and gave a brief abstract of some of Darwin's arguments in the *Origin of Species.* He accepted Darwin's theory as providing a clear explanation of many biological and geological phenomena previously puzzling, but did not make any declaration of his own adherence to the theory.

In the *Antiquity of Man*, however, Lyell, in the manner of a barrister, presented to the reader a broad array of evidence which indicated that man had evolved gradually from lower animals over an immense period of time, that species had been modified one into another, and that the modification had probably been produced by natural selection. He compared the gradual divergence of the various European languages from a common root language to the gradual divergence of species. By presenting evidence without drawing conclusions from it, Lyell kept the reader's attention on the evidence and required him to draw his own conclusions. Lyell thus made the question of the origin of man and other species one of evidence rather than of doctrine. He kept his readers in suspense to the end of the book and induced them to consider the meaning of the evidence for themselves. Lyell thus gave Darwin powerful support in the controversy following the publication of the *Origin of Species*. Darwin, however, expressed disappointment that Lyell did not make a clear statement of his own opinion on the origin of species.[5] In reply Lyell said that he had expressed his opinion fully "so far as my reason goes, and farther than my imagination and sentiment can follow."[6] Lyell seems to have felt a particular revulsion toward the idea that man was descended from animals and to have thought that the origin of the human intellect required a special explanation. At Darwin's urging he amended his statements concerning species in a later edition of the *Antiquity of Man* but remained hesitant.

On 30 November 1864, at the anniversary meeting of the Royal Society at which the Copley Medal was awarded to Charles Darwin in absentia, Lyell announced publicly that he had changed his mind concerning species and declared his faith in Darwin's theory of the indefinite modification of species by natural selection. Between 1865 and 1868 Lyell rewrote and revised the *Principles* for the tenth edition, so as to make it conform in every part with Darwin's theory and to express his wholehearted conviction of the truth of the theory of evolution by natural selection. In the second volume of the tenth edition, published in 1868, Lyell devoted the first ten chapters of book III to various aspects of the species question. In chapter 37 he discussed the theory of natural selection and in four succeeding chapters showed how the complex facts of the geographical distribution of species were explicable in terms of natural selection but not by any other hypothesis. The tenth edition of the *Principles* was Lyell's last great feat of scientific writing. The eleventh edition of 1872 and the twelfth edition, published in 1875 after his death, included only minor revisions.

In 1864 Lyell was elected president of the British Association for the Advancement of Science and delivered the presidential address at the annual meeting at Bath. Also in 1864 he was created a baronet by Queen Victoria. Since their marriage in 1832 Lady Lyell had helped her husband with his correspondence, writing many of his letters from dictation; but in 1865 Lyell engaged as his secretary Arabella Buckley, and his letters after that were usually written in her hand. After 1869 Lyell's health began to fail. His eyesight became gradually worse, until he was blind, and his strength diminished. Yet he continued to make geological expeditions to various places in the British Isles; and in April 1872 he and Lady Lyell, accompanied by T. McKenna Hughes, went to the south of France to visit the caves of Aurignac.

On 25 April 1873 Lady Lyell died unexpectedly; and the loss, after forty years of so close and constant a companion, was a severe blow to Lyell. After some months of failing health he died on 22 February 1875 and was buried in Westminster Abbey.

In addition to his knighthood and baronetcy Lyell received many honors. The Royal Society awarded him the Copley Medal and the Royal Medal; the Geological Society, the Wollaston Medal. He was elected a member of the American Philosophical Society and a corresponding member of the Institut de France and of the Royal Academy of Sciences at Berlin, in addition to membership in many other scientific and learned societies.

Lyell established geology as a science. He applied to the subject the strictest discipline of rigorous reasoning and expunged it of all that was merely fanciful and speculative. His criteria for admissible causes in geology stand today more strongly than ever; and the kind of detailed analogy which he drew between ancient and modern conditions in oceans, lakes, streams, and estuaries foreshadowed the modern development of paleoecology. In certain theoretical interpretations Lyell was clearly wrong. Until 1857 he defended the view that erratic boulders had been transported by icebergs, whereas for the most part they had been transported by continental glaciation. Lyell also exaggerated the role of the sea in shaping the form of the land and did not fully appreciate the immense scale of subaerial denudation in determining landforms. For many years he mistakenly upheld the fixity of species. Yet Lyell's theoretical positions, even when wrong, were always carefully reasoned; and he showed an extraordinary capacity even into old age to understand the meaning of new evidence and to change his mind.

After Lyell's death his reputation, together with that of Darwin, suffered a serious decline—principally because, in the controversy over the age of the earth initiated in 1865 by Lord Kelvin, Kelvin and other physicists had asserted that the earth could not be older than 10 to 25 million years. Within these time limits geologists could not account for the long series of geological changes which had occurred, in terms of the uniform action of natural causes. Consequently there was a tendency for geologists to revert to catastrophic interpretations of past geological changes. Lord Kelvin's estimate of the age of the earth was based on calculations of the rate of heat loss by the earth, assuming that it had begun as an incandescent molten mass, approximately at the temperature of the sun, and had gradually cooled according to the laws of radiation. Kelvin had admitted that his calculations would be invalidated if any steady source of heat were to be discovered within the earth.

As early as 1830 Lyell had concluded that the occurrence of volcanic activity throughout all periods of geological history required a steady source of heat within the earth, and he had rejected the concept of a gradually cooling earth. In 1904 Ernest Rutherford pointed out that a steady source of heat within the earth had been found in the discovery of radioactivity. Estimates of the age of the earth were rapidly revised upward and the modern estimates, which are in excess of 4,000 million years, are great enough to allow the uniform action of gradual causes to produce all the geological changes that have occurred, just as they are also adequate to allow the gradual action of natural selection to produce all the results of evolution in biology. Lyell's faith in geological explanation in terms of the uniform action of observable natural processes has been vindicated by the progress of his science.

NOTES

1. C. Prévost, "De l'importance de l'étude des corps organisés vivants pour la géologie positive," in *Mémoires de la Société d'histoire naturelle de Paris*, **1** (1823), 259–268, see 262.
2. "On a Dike of Serpentine, Cutting Through Sandstone, in the County of Forfar," in *Edinburgh Journal of Science*, **3** (1825), 112, 126.
3. "On the Structure of Lavas Which Have Consolidated on Steep Slopes; With Remarks on the Mode of Origin of Mount Etna, and on the Theory of 'Craters of Elevation,' " in *Philosophical Transactions of the Royal Society*, **148** (1858), 703–786, see 761.
4. *The Antiquity of Man*, 2nd ed. (London, 1863), p. 393.
5. Darwin to Lyell, 6 Mar. 1863, in *The Life and Letters of Charles Darwin* (London, 1888), III, 11–13.
6. Lyell to Darwin, 11 Mar. 1863, in *Life, Letters and Journals of Sir Charles Lyell Bart.* (London, 1881), II, 363–364.

BIBLIOGRAPHY

I. Original Works. There is a fairly complete list of Lyell's scientific papers in the Royal Society's *Catalogue of Scientific Papers* and of all his writings in appendix E to *Life, Letters and Journals of Sir Charles Lyell Bart.*, Katherine M. Lyell, ed., 2 vols. (London, 1881). For Lyell's publications up to 1841, the bibliography in Leonard G. Wilson, *Charles Lyell, the Years to 1841: The Revolution in Geology* (New Haven–London, 1972), contains some items omitted from the other two lists and complete references to his early articles in the *Quarterly Review*.

Selected papers are "On a Recent Formation of Freshwater Limestone in Forfarshire, and on Some Recent Deposits of Freshwater Marl; With a Comparison of Recent With Ancient Freshwater Formations; and an Appendix on the Gyrogonite or Seed Vessel of the Chara" (1824), in *Transactions of the Geological Society of London*, 2nd ser., **2** (1829), 73–96; "The Bakerian Lecture. On the Proofs of a Gradual Rising of the Land in Certain Parts of Sweden," in *Philosophical Transactions of the Royal Society*, **125** (1835), 1–38; "On the Relative Ages of the Tertiary Deposits Commonly Called 'Crag' in the Counties of Norfolk and Suffolk," in *Magazine of Natural History*, n.s. **3** (1839), 313–330; "On the Upright Fossil Trees Found at Different Levels in the Coal Strata of Cumberland, Nova Scotia" (1843), in *Proceedings of the Geological Society of London*, **4** (1842–1845), 176–178; "On Craters of Denudation, with Observations on the Structure and Growth of Volcanic Cones," in *Quarterly Journal of the Geological Society of London*, **6** (1850), 207–234; "The Tertiary Strata of Belgium and French Flanders," *ibid.*, **8** (1852), 277–371; and "On the Structure of Lavas Which Have Consolidated on Steep Slopes; With Remarks on the Mode of Origin of Mount Etna, and on the Theory of 'Craters of Elevation,' " in *Philosophical Transactions of the Royal Society*, **148** (1858), 703–786.

Lyell's books were *Principles of Geology. Being an Attempt to Explain the Former Changes of the Earth's Surface by Reference to Causes Now in Operation*, 3 vols. (London, 1830–1833; 2nd ed., 2 vols., 1832–1833; 3rd ed., 4 vols., 1834; 4th ed., 1835; 5th ed., 1837; 6th ed., 3 vols., 1840; 7th ed., 1 vol., 1847; 8th ed., 1850; 9th ed., 1853; 10th ed., 2 vols., 1867–1868; 11th ed., 1872; 12th ed., 1875); *Elements of Geology* (London, 1838; 2nd ed., 2 vols., 1841; 3rd ed., pub. as *A Manual of Elementary Geology*, 1 vol., 1851; 4th ed., 1852; 5th ed., 1855; 6th ed., 1865); *Travels in North America, in the Years 1841–2; With Geological Observations on the United States, Canada and Nova Scotia*, 2 vols. (London–New York, 1845); *A Second Visit to the United States of North America*, 2 vols. (London, 1849); *The Geological Evidences of the Antiquity of Man With Remarks on Theories of the Origin of Species by Variation* (London, 1863; 2nd ed., 1863; 3rd ed., 1863; 4th ed., 1873); and *The Student's Elements of Geology* (London, 1871).

Many of Lyell's letters and travel diaries were published in the *Life, Letters and Journals of Sir Charles Lyell Bart.*, Katherine M. Lyell, ed., 2 vols. (London, 1881). The original copies of many of Lyell's letters to members of his family, his scientific journals, and his notebooks are the property of Lord Lyell of Kinnordy, Kirriemuir, Angus, Scotland. Lyell's letters to other scientists are scattered among libraries and MS collections, both public and private, throughout the world.

Between 1855 and 1861 Lyell filled seven notebooks with notes and references on the species question; these have been published as *Sir Charles Lyell's Scientific Journals on the Species Question*, Leonard G. Wilson, ed. (New Haven–London, 1972).

II. SECONDARY LITERATURE. In addition to K. M. Lyell's *Life, Letters and Journals*, three brief popular biographies of Lyell have been published: Sir Edward Bailey, *Charles Lyell* (New York, 1963); Thomas George Bonney, *Charles Lyell and Modern Geology* (London, 1895); and F. J. North, *Sir Charles Lyell* (London). Although all of these biographies are heavily dependent on the *Life, Letters and Journals* and on Lyell's published writings, each was written by a professional geologist who contributed useful insights into Lyell's work.

Leonard G. Wilson, *Charles Lyell, the Years to 1841: The Revolution in Geology* (New Haven–London, 1972), is the first of three volumes of a detailed, large-scale biography of Lyell based on extensive study of both published and MS sources.

There are references to Lyell among the lives and letters of many eminent Victorians, both British and American. Of particular value are Katherine M. Lyell, ed., *Memoir of Leonard Horner, F.R.S., F.G.S., Consisting of Letters to His Family and From Some of His Friends*, 2 vols. (London, 1890); and Frances Joanna Bunbury, ed., *Life, Letters and Journals of Sir Charles J. F. Bunbury Bart.*, 2 vols. (London, 1894).

The intellectual and scientific background to Lyell's thought is discussed in a lively study by Charles Coulston Gillispie, *Genesis and Geology* (Cambridge, Mass., 1951). Various aspects of Lyell's concept of uniformity are discussed in the following: Walter F. Cannon, "The Uniformitarian–Catastrophist Debate," in *Isis*, **51** (1960), 38–55; and "The Impact of Uniformitarianism. Two Letters From John Herschel to Charles Lyell 1836–1837," in *Proceedings of the American Philosophical Society*, **105** (1961), 301–314; R. Hooykaas, *Natural Law and Divine Miracle, A Historical-Critical Study of the Principle of Uniformity in Geology, Biology and Theology* (Leiden, 1959), an attack on the principle of uniformity, inspired by religious doctrines; M. J. S. Rudwick, "A Critique of Uniformitarian Geology: A Letter From W. D. Conybeare to Charles Lyell, 1841," in *Proceedings of the American Philosophical Society*, **111** (1967), 272–287; and "Lyell on Etna, and the Antiquity of the Earth," in Cecil J. Schneer, ed., *Toward a History of Geology* (Cambridge, Mass., 1969), pp. 288–304; and Leonard G. Wilson, "The Development of the Concept of Uniformitarianism in the Mind of Charles Lyell," in *Proceedings of the Tenth International Congress of the History of Science* (Paris, 1964), pp. 993–996; and "The Origins of Charles Lyell's Uniformitarianism," in *Uniformity and Simplicity* (New York, 1967), pp. 35–62.

LEONARD G. WILSON

LYMAN, BENJAMIN SMITH (*b.* Northampton, Massachusetts, 11 December 1835; *d.* Philadelphia, Pennsylvania, 30 August 1920), *geology.*

In 1852–1855 Lyman studied law at Harvard University. He spent the field seasons of 1856 and 1857 assisting his uncle, J. P. Lesley, with the geological survey of Pennsylvania. He went to Philadelphia in 1856, and the following year he traveled to the middle Atlantic and southern states to collect statistics on iron manufacturing for the Iron and Steel Association. In 1858 Lyman joined the Geological Survey of Iowa as assistant to James Hall. Under Hall's influence he decided to become a geologist. In September of that year he entered the School of Mines at Paris, and in 1861 he transferred to the Freiberg Mining Academy. Lyman returned to the United States in 1862, subsequently working, again with his uncle, on geological surveys of Pennsylvania, Virginia, Nova Scotia, California, and Alabama. About 1866 Lyman devised the method of representing subsurface geological structure by the use of structural contour lines. In 1870 he was employed by the government of India to carry out a geological survey of oil fields in the Punjab.

The Meiji government of Japan invited experts from the West to assist in transforming Japanese society. The American contingent, led by Horace Capron, included Lyman. In 1872 he contracted with the Hokkaido *kaitakushi* (development board) as geologist and mining engineer. He undertook a geological survey of Yesso (the ancient name for Hokkaido) in 1873–1875. Beginning in 1876 he prospected for oil in Japan, and he returned to America in 1881. He worked as a geologist for the Geological Survey of Pennsylvania from 1887 to 1895.

In 1873–1875 Lyman and his co-workers, to whom he taught mathematics, physics, and geology, undertook the geological survey of the Kayanuma and Poronai coalfields. They published a geological map on a scale of 1:5,000. In the same period Lyman visited other areas on Hokkaido; results of these visits are *Geological Sketch Map of the Island of Yesso, Japan* (1876) and *A General Report of the Geology of Yesso* (1877).

Lyman had hoped to return to America after finishing the geological survey of Hokkaido; but instead he undertook a geological survey for oil for the Ministry of Industry. He visited the Niigata oil field and western Japan in 1876–1881. The article "Geological and Topographical Maps of the Oil Lands in Japan," written during this period, includes a geological map of the Niigata oil field on the scale 1:60,000.

Lyman made lasting contributions to Japan by establishing the first geological map of an extensive area, making structure contour maps of the coalfields and oil fields, and teaching practical American geology to thirteen young Japanese geologists. The future geologists, managers, and administrators of the Japanese mining industry were among Lyman's protégés, for example, E. Yamagiwa and J. Shimada, who found the great coalfield at Ikushumbetsu (1880), and I. Ban, who found the Oyubari coalfield (1888).

Lyman returned to the United States and from 1887 to 1895 Lyman was vice-director of the Pennsylvania Geological Survey. He had a deep knowledge of literature, philosophy, law, art, and archaeology. A lifelong bachelor, Lyman was also a vegetarian.

BIBLIOGRAPHY

I. ORIGINAL WORKS. Lyman's books and papers have been collected at the Forbes Library, Northampton, Mass., but there is no complete bibliography of the more than 150 publications attributed to him. His most important writings are the *Geological Survey of Hokkaido*, 5 vols. (Tokyo, 1875–1877); *Geological Sketch Map of the Island of Yesso, Japan (1:2,000,000)* (Tokyo, 1876); and *A General Report on the Geology of Yesso* (Tokyo, 1877). Other of his works on Japanese geology are included in Horace Capron, *Reports and Official Letters to the Kaitakushi* (Tokyo, 1875) and *Geological Survey of Japan, Reports of Progress for 1878, 1879* (Tokyo, 1879). Works on American geology, mostly that of Pennsylvania, are listed in J. M. Nickles, "Geologic Literature on North America, 1785–1918," in *Bulletin of the United States Geological Survey*, no. 746 (1922). See also E. S. Dunkman, "Notes on the Benjamin Smith Lyman Collection," MSS, Forbes Library, Northhampton, Mass.

II. SECONDARY LITERATURE. See M. L. Ames, *Life and Letters of Peter and Susan Lesley* (New York, 1909), which contains correspondence between Lyman and the Lesleys; Gonpei Kuwada, *Biography of Benjamin Smith Lyman* (Tokyo, 1937); and F. B. Sanborn, *Recollections of Seventy Years* (Boston, 1909), which includes recollections of Lyman, with a photograph.

H. KOBAYASHI

LYMAN, CHESTER SMITH (*b*. Manchester, Connecticut, 13 January 1814; *d*. New Haven, Connecticut, 29 January 1890), *astronomy*, *geology*.

Lyman is credited with being the first person to obtain reliable evidence for an atmosphere surrounding the planet Venus: when the dark side of Venus was toward the earth in 1866, and again in 1874, he saw it surrounded by a complete bright circle, which he correctly attributed to refraction of sunlight in the Cytherean atmosphere.

The son of a miller, Chester Lyman, and his wife Mary Smith, Lyman gave evidence of an interest in astronomy while still quite young. But the career he chose for himself was the ministry, and so, after an elementary education in public schools, he taught school for several years to finance a college education. At age nineteen he entered Yale College; here he helped found the *Yale Literary Magazine* and had access to the astronomical observatory as a perquisite of his job as assistant to the professor of natural philosophy. He graduated B.A. in 1837.

After two years as superintendent of the Ellington Academy, Lyman entered the Union Theological Seminary in 1839, but transferred after a year to the theological school at Yale, from which he graduated B.D.

His first pastorate, that of the First Congregational Church in New Britain, Connecticut, was also his last: in 1845 he resigned, for reasons of health, and set off for the South Seas. He went around Cape Horn to Hawaii (then called the Sandwich Islands), where he worked sporadically as a missionary, schoolteacher, and surveyor. He also visited the volcano of Kilauea and recorded geological details of the surrounding region.

In 1847 Lyman moved on to California, arriving shortly before it was ceded to the United States by Mexico, and just in time to have his surveying activities interrupted by the Gold Rush. In 1850 he returned to Connecticut, married Delia Williams Wood, and settled down in New Haven for the rest of his life.

Lyman was appointed professor of industrial mechanics and physics in the Sheffield Scientific School of Yale College in 1859. It was during this tenure that he made his observations of the planet Venus. His professorship was altered in 1871 to include only astronomy and physics, and again in 1884 to cover only astronomy. He served as president of the Connecticut Academy of Sciences from 1859 to 1877, and was granted an M.A. degree by Beloit College in 1864.

BIBLIOGRAPHY

I. Original Works. Lyman published almost exclusively in Silliman's *American Journal of Science and the Arts:* letters from him describing California and gold mining—said to have been the first reliable accounts received on the East Coast of this bonanza—appeared in 2nd ser., **6** (1848), 270–271 (includes mention of the discovery of gold at Sutter's Fort on the Sacramento River); 2nd ser., **7** (1849), 290–292 (his first visit to the gold fields); *ibid.*, 305–309 (how he himself dug for gold, and a description of conditions in the mining camps); 2nd ser., **8** (1849), 415–419 (describes the geology of placers); and 2nd ser., **9** (1850), 126–127 (the discovery of gold-bearing veins).

Lyman's description of the volcano Kilauea also appeared in Silliman's *Journal*, 2nd ser., **12** (1851), 75–82; as did his observations of Venus, 2nd ser., **43** (1867), 129–130, and 3rd ser., **9** (1875), 47–48.

The journal that Lyman kept during his travels, *Around the Horn to the Sandwich Islands and California, 1845–50* (New Haven, 1924), was prepared for publication by Frederick John Teggert; it includes a portrait and an introduction by Lyman's daughter, Delia Lyman Porter.

Thirteen articles by Lyman are listed in the *Royal Society. Catalogue of Scientific Papers*, **4** (London, 1870), 141; **8** (London, 1879), 285; and **10** (London, 1894), 665.

II. Secondary Literature. Facts about Lyman's early life can be found in the annual *Catalogue of the Officers and Students in Yale College*, for the years 1833–1834 through 1842–1843, and in *Historical Register of Yale University* (New Haven, 1939), p. 367.

For contemporary evaluations, see "Sketch of Chester S. Lyman," in *Popular Science Monthly*, **32** (1887–1888), 116–121, with portrait facing p. 1, and an obituary notice in Silliman's *American Journal of Science and the Arts*, 3rd ser., **39** (1890), 245–246. Both are unsigned.

Sally H. Dieke

LYMAN, THEODORE (*b.* Boston, Massachusetts, 23 November 1874; *d.* Brookline, Massachusetts, 11 October 1954), *experimental physics.*

Lyman came from an old and wealthy Massachusetts family. His great-grandfather was a very successful Boston merchant. His father, also Theodore, was a marine biologist who served one term in Congress. His mother, the former Elizabeth Russell, was the granddaughter of a U.S. minister to Sweden. Theodore Lyman never married and spent his life in his grandfather's mansion on a large estate in Brookline.

He received the B.A. degree in 1897 and the Ph.D. in 1900 from Harvard. After a year at the Cavendish Laboratory in Cambridge and a summer at Göttingen he returned to Harvard as an instructor in physics. His entire career was spent at Harvard, where he was director of the Jefferson Physical Laboratory from 1910 to 1947 and held the Hollis professorship (the oldest endowed scientific chair in the United States) from 1921 until he retired.

Lyman's doctoral dissertation was devoted to the problem of applying the concave grating to the measurement of spectral lines in the extreme ultraviolet, a region where the rays cannot pass through air. The technical problems were great and consumed six years. In the work Lyman found false lines in the spectrum which he was able to explain as due to periodic errors in the grating ruling. The clarification of these "Lyman ghosts" in 1900 constituted his doctoral thesis. His first published measurement of wavelengths in the "Lyman region" (1906) gave the first accurate measurements below 2,000 Å. and extended the known extreme ultraviolet region significantly. Viktor Schumann in Germany had used a fluorite prism to disperse the light that did not permit wavelength determination.

Lyman's later scientific work was devoted to measuring various spectra and the optical properties of various materials in the region and to extending the ultraviolet spectrum to the final limit of 500 Å., which he attained in 1917. In 1914 he announced the discovery of the fundamental series of hydrogen, which was an essential part of the foundation on which Bohr developed the quantum theory of the atom.

Never in robust health, Lyman retired from his professorship in 1925 but continued as director of the Jefferson Physical Laboratory until 1947. His last paper was published in 1935, although he continued to direct doctoral dissertations until 1942.

He was a member of the National Academy of Sciences, the American Philosophical Society, the American Academy of Arts and Sciences (president, 1924–1927), the American Physical Society (president, 1921–1922); an honorary member of the Optical Society of America and the Royal Institution of Great Britain; and a fellow of the Royal Geographical Society.

He received the Rumford Medal of the American Academy of Arts and Sciences, the Cresson Medal of the American Philosophical Society, and the Ives Medal of the Optical Society of America.

BIBLIOGRAPHY

I. Original Works. Lyman published about forty scientific papers. His most important work was *The Spectroscopy of the Extreme Ultra-Violet* (New York, 1914; rev. ed., 1928). Important papers include: "An Explanation of False Spectra From Diffraction Gratings," in *Physical Review*, **16** (1903), 257–266; and "The Spectrum of Helium in the Extreme Ultra-Violet," in *Astrophysical Journal*, **60** (1924), 1–14.

II. Secondary Literature. See the discussion of Lyman and his work by P. W. Bridgman, in *Biographical Memoirs. National Academy of Sciences*, **30** (1957), 237–250, which includes a complete bibliography.

Ralph A. Sawyer

LYONET, PIERRE (*b.* Maastricht, Netherlands, 21 July 1706; *d.* The Hague, Netherlands, 10 January 1789), *entomology*.

Lyonet's family came originally from northeastern France. They were Calvinists, and their search for religious peace took them to Metz, the Palatinate, and finally Switzerland. Lyonet's father, Benjamin Lyonet, migrated at the beginning of the eighteenth century to Holland, where in 1704 he married Marie le Boucher, the daughter of Huguenot refugees. He was a Presbyterian pastor and was given a parish in Heusden. Lyonet, the elder of two sons (four other children died in infancy), was a frail child. He later wrote: "I was always so quiet that I was thought to be deprived of reason; it was thus with no mild joy that my parents heard me, at two years of age, cry for the first time."

Little is known about Lyonet's early education, but it is likely that he studied Latin, Greek, and Hebrew under the tutelage of his father, who intended his son for the clergy. Lyonet entered Leiden University (1724), accordingly, as a student of theology. In view of the congeniality that existed in those years between science and theology (at least in Protestant countries), it is not suprising that Lyonet studied mathematics, Newtonian physics, military architecture, and anatomy during his first two years at Leiden. In 1728, a year after completing his course, he was admitted to the pastorate. He soon found, however, that he did not have a real vocation, and he persuaded his reluctant father to allow him to return to Leiden to study law (1730). He graduated after a year's work, with a thesis entitled *De justo quaestionis usu* (on the just use of torture), and in the same year set himself up as an *avocat* in The Hague.

Although Lyonet was apparently successful at the bar, he discovered his life's work elsewhere, through two discrete but probably related activities. He began to study insects and became a translator (1738) and then a cipher clerk for the government. As a clerk, working on his own initiative, he broke the code used by the Prussian ambassador to the United Provinces (cracking it took him eighteen months), then other codes used by diplomats in the country, and even those of diplomats in London whose letters passed through The Hague. After convincing his initially skeptical government of the usefulness of decoding, he was allowed to continue the secret work, and did so until his death. He fought for nearly ten years (from 1753) for recognition of his services in the only form acceptable to him: the title of Secretary of Ciphers, which was granted him in 1762.

Lyonet, like Bonnet, first became interested in the study of insects through reading Pluche's *Natural History*, but he was inspired to study them seriously by the first volume (on "chenilles et papillons") of Réaumur's *Mémoires*, which appeared in 1734. Réaumur's influence on Lyonet was decisive, as it had been on Bonnet and Trembley. The magisterial volumes of *Mémoires* served as models to emulate and bases to build upon. Lyonet, who had respect for few naturalists, praised only Swammerdam and Réaumur. From Réaumur he learned method and the importance of being exceedingly careful.

Lyonet began systematic observation on insects in 1736. In 1738 he undertook to correct and expand a translation of F. C. Lesser's *Insectotheologia*. Lyonet's annotations to the translation indicate that he was familiar with the subject and that his ideas on the general biological problems of his time were already formed; his ideas on classification and generation found in the notes to Lesser also appear in the *Traité anatomique* of 1760 and in the posthumously published *Recherches*.

The *Traité anatomique de la chenille qui ronge le bois de saule* was begun in 1745. Lyonet had originally planned a treatise on all the insects in the vicinity of The Hague (and indeed the *Recherches* consist of that part of this project which he was able to complete). When he saw the fame that Trembley and Bonnet had achieved, however (Trembley had created a stir with the demonstration of regeneration of hydras in his *Mémoires pour servir à l'histoire d'un genre de polypes d'eau douce*, largely illustrated by Lyonet, while Bonnet had observed the parthenogenesis of aphids), he decided to establish his own reputation in microanatomy. He examined the common goat moth caterpillar (*Cossus ligniperda*) and the anatomy of its chrysalis and imago. He intended to delineate the metamorphosis of the insect in subsequent volumes, but was prevented from doing so by an affliction of the eyes, which after 1767 made it impossible for him to do close work. *Traité anatomique* is devoted wholly to the anatomy of the caterpillar (except for a short initial chapter on its lifecycle), and the plates, drawn and engraved by Lyonet, portray the muscles, nerves, bronchia, heart, viscera, silk vessels, and the internal parts of the head with astonishing precision.

For Lyonet the principal enemies of sound thinking in natural history were *l'esprit de système* and mechanistic ideas, both exemplified by Buffon. Lyonet rejected both animalculist and ovist preformation theories of generation and believed that spontaneous ("equivocal") generation was an illusion fostered by sloppy experimental technique. All animals come from eggs, he said, echoing Harvey (by way of Swammerdam), except those few which reproduce by budding (Trembley's polyps). Lyonet did not formulate a theory of generation, but the logic of his position would have led him to a theory of epigenesis, if he had pursued it. His declarations on generation could have been approved by Aristotle: in reproduction, both the female and male "principles" play a role.

Lyonet considered himself an empiricist, like Swammerdam and Réaumur, who examined insects, to be sure, but who saw only what any unbiased and respectful observer would see. In a way, it was so: the *Traité anatomique* is a triumph of the eye. But his empiricism was less than pure, since two assumptions controlled every line he wrote or drew. He believed, first, that the world and all its creatures are a vast cipher and, second, that the duty of man is to decode it. He found the natural world to be as intricately, precisely, and richly designed as a work by a Dutch artist; and he believed that the more this great design was elucidated, the greater would be man's reverence for the Designer. The tasks of breaking the code by tracing the design to its last perfect detail was therefore perhaps the one supremely worthwhile thing to do.

Beginning about 1767 Lyonet, whose eyesight was still sufficient for everyday purposes, began to collect paintings, including those of Vermeer (widely thought to have been neglected in the eighteenth century). He spent his later years in preparing his notes on insects in the vicinity of The Hague for publication. Although he had a manuscript ready for the printer by 1787, various delays kept it from the presses until 1832, when W. de Haan published it as the *Recherches*.

BIBLIOGRAPHY

I. Original Works. A complete list of Lyonet's writings and drawings, published and unpublished, may be found in W. H. Van Seters, *Lyonet* (see below), 185–199. A sketchbook of Lyonet's which turned up after the biography was published is described by Van Seters, "Lyonet's Kunstboek," in *Medical and Biological Illustration*, **13** (1963), 255–264. Lyonet's works referred to in the text are *Théologie des insectes ou démonstration des perfections de Dieu dans ce que concerne les insectes, traduit de l'allemand de Mr Lesser avec des remarques de Mr P. Lyonet*, 2 vols. (The Hague, 1742); *Traité anatomique de la chenille qui ronge le bois de saule* (The Hague, 1760); and *Recherches sur l'anatomie et les métamorphoses de différentes espèces d'insectes, ouvrage posthume de Pierre Lyonet, publié par M. W. de Haan . . .* (Paris, 1832).

II. Secondary Literature. A complete biographical source is W. H. Van Seters, *Pierre Lyonet, 1706–1789, sa vie, ses collections de coquillages et de tableaux, ses recherches entomologiques* (The Hague, 1962); but see also Émile Hublard, "Le naturaliste hollandais Pierre Lyonet. Sa vie et ses oeuvres 1706–1789," in *Mémoires et publications de la Société des sciences, des arts, et des lettres du Hainaut*, **61** (1910), 1–159, also published separately (Mons, 1910), especially for a number of Lyonet's letters. For the milieu, see J. R. Baker, *Abraham Trembley of Geneva* (London, 1952); R. S. Clay and T. H. Court, *The History of the Microscope* (London, 1932); F. J. Cole, *Early Theories of Sexual Generation* (Oxford, 1930); L. C. Miall, *The Early Naturalists, Their Lives and Work (1530–1789)* (London, 1912); Jean Torlais, *Réaumur* (Paris, 1936; new ed., 1961); Maurice Trembley, ed., *Correspondance inédite entre Réaumur et Abraham Trembley* (Geneva, 1943); Aram Vartanian, "Trembley's Polyp, La Mettrie, and Eighteenth-Century French Materialism," in *Journal of the History of Ideas*, **11** (1950), 259–286, repr. in P. P. Wiener and Aaron Noland, eds., *Roots of Scientific Thought* (New York, 1957), 497–516.

Stuart Pierson

LYOT, BERNARD (*b.* Paris, France, 27 February 1897; *d.* Cairo, Egypt, 2 April 1952), *solar and planetary astronomy, optics.*

Lyot, whose father, a surgeon, died when the boy was seven, studied to be an engineer. After graduating from the École Supérieure d'Électricité in 1917, he obtained a post at the École Polytechnique as assistant to the physicist A. Pérot, with whom he studied maritime and aerial radio guidance devices for the army. Pérot put him in contact with the observatory at Meudon, where he himself was working. Lyot joined this observatory in 1920 and, having earned his *licence ès sciences*, successively became assistant (1925), astronomer (1930), and chief astronomer (1943).

The first problem that Lyot studied—which also was the subject of his doctoral dissertation in 1929—concerned the polarization of light reflected from the surfaces of the planets. In order to detect this polarization he had to construct a polariscope ten times more sensitive than the existing instruments. From 1923 to 1930 he determined, for each planetary surface, the polarization curve as a function of the angle of vision and, by comparison with laboratory measurements of various substances, he obtained information on the structure of the surfaces observed. He showed, notably, that the lunar soil behaves like volcanic dust and that Mars experiences sandstorms.

The polarization of the solar corona could not be studied during eclipses because the latter's duration was too short. The observation of the corona when there was no eclipse—conducted with the aid of a screen masking the image of the disk—was thought to be impossible because of the light diffused by the atmosphere. Lyot showed that the "parasite" light was caused by the apparatus and could be eliminated. Thus in 1930 he invented the coronograph, which has become the classic instrument for observing the solar corona.

In order to be able to use all the information contained in the image given by the coronograph, it was necessary to work in strictly monochromatic light and to avoid introducing absorption. These two apparently contradictory conditions were satisfied in the monochromatic filter that Lyot devised in 1933. He employed a property of birefringent crystalline laminae, which in polarized light produce interferences, the periods of which vary with the thickness of the lamina: by stacking up a suitable series of laminae, one obtains well-separated passbands, the width of which does not exceed one angstrom. This filter is used in the network of automated heliographs that today assure the permanent observation of solar eruptions and rapidly evolving chromospheric phenomena. Conjoined with the coronograph, the filter has permitted the taking of films showing the movements of solar prominences; they are among the most popular astronomical documents. Lyot himself took the first of these films in 1939.

Lyot spent his life studying phenomena which seemed to be impossible to detect. His skill as a physicist was such that no major advance has been made in the types of instruments that he invented. His work was conceived in accordance with an initial overall plan; for instance, as early as 1923 he stated the principle of the photoelectric polarimeter, which the electron multiplier enabled him to realize in 1950. This instrument, which can detect polarized light when it constitutes as little as 1/10,000 of the total flux, has made possible the observation of coronal rays without the coronograph and without the need for high-altitude stations.

Lyot was elected to the Académie des Sciences in 1939 and in the same year received the gold medal of the Royal Astronomical Society. He was awarded the gold medal of the Astronomical Society of the Pacific in 1947. A pleasant and unpretentious person, Lyot divided his time between his laboratory at Meudon, mountain observations, and his family. He enjoyed skiing, swimming, and sailing and appeared astonishingly youthful. He died of a heart attack upon his return from an exhausting mission to Khartoum to observe a solar eclipse.

BIBLIOGRAPHY

I. Original Works. Lyot's investigations were primarily in four fields.

On polarization he wrote "Recherches sur la polarisation de la lumière des planètes et de quelques substances terrestres," in *Annales de l'Observatoire, Section de Meudon*, **8** (1929), 1–161; and six notes in *Comptes rendus . . . de l'Académie des sciences*, **189** (1929), 425–426; **191** (1930), 703–705, 834–836; **198** (1934), 249–251, 774–777.

On the coronograph and its applications, see three notes, *ibid.*, **191** (1930), 834–836; **193** (1931), 1169–1173; **194** (1932), 443–446; two articles in *Astronomie:* **45** (1931), 248–253; **46** (1932), 272–287; "Études de la couronne solaire en dehors des éclipses," in *Zeitschrift für Astrophysik*, **5** (1932), 73–95; and "A Study of the Solar Corona and Prominences Without Eclipses," in *Monthly Notices of the Royal Astronomical Society*, **99** (1939), 578–596.

The monochromatic filter is the subject of two notes in *Comptes rendus*, **197** (1933), 1593–1595, **212** (1941), 1013–1017; and "Le filtre monochromatique polarisant et ses applications en physique solaire," in *Annales d'astrophysique*, **7** (1944), 31–79.

On the photoelectric polarimeter, see a note in the *Comptes rendus*, **226** (1948), 25–28.

Three other papers should be mentioned: "Une nouvelle méthode d'observation de la couronne solaire," *ibid.*, **231** (1950), 461–464; "Étude spectroscopique de la rotation de la couronne solaire," *ibid.*, **233** (1951), 1529–1532, written with A. Dollfus; and "Étude des défauts d'homogénéité de grands disques de verre," in *Revue d'optique théorique et instrumentale*, **29** (1950), 499–512, written with M. Françon.

II. SECONDARY LITERATURE. See the following, listed chronologically: F. Robbins, "The Gold Medallist of the Royal Astronomical Society," in *Journal of the British Astronomical Association*, **49** (1939), 259–263; A. Chevalier, "Notice nécrologique," in *Comptes rendus*, **234** (1952), 1501–1505; A. Danjon, "Bernard Lyot," in *Annales d'astrophysique*, **15** (1952), 75–78; and L. d'Azambuja, "L'oeuvre de B. Lyot," in *Astronomie*, **66** (1952), 265–277.

JACQUES R. LÉVY

MAANEN, ADRIAAN VAN (*b.* Sneek, Netherlands, 31 March 1884; *d.* Pasadena, California, 26 January 1946), *astronomy.*

Descended from a long line of aristocrats, van Maanen was the son of Johan Willem Gerbrand and Catharina Adriana Visser van Maanen. He received his B.A. (1906), M.A. (1909), and Sc.D. (1911) from the University of Utrecht. From 1908 to 1911 he worked at the University of Groningen, where he met J. C. Kapteyn. In 1911 van Maanen joined the Yerkes Observatory as a volunteer assistant; and in 1912, on Kapteyn's recommendation, he was appointed to the staff of the Mt. Wilson Observatory. His job there —to measure the proper motions and parallaxes of stars—employed the skills he had acquired while working on his thesis, "The Proper Motions of the 1418 Stars in and Near the Clusters h and χ Persei."

From its completion in 1914, van Maanen used the sixty-inch telescope at the eighty-foot Cassegrain focus for parallax determinations, the first such use of a reflector for such delicate measurements. In the 1920's he also began using the 100-inch reflector at the forty-two-foot Newtonian focus. To measure either parallax or proper motion from a photograph, van Maanen employed a stereocomparator. After superimposing sets of comparison stars, he measured the distance separating the two images in question with a movable micrometer thread.

Van Maanen's study of the parallaxes of more than 500 stars can be divided into two parts. He used one group, comprising stars of apparent magnitude $+5$ to $+7$ with moderate proper motions, as a distance standard for measuring spectroscopic parallaxes of more distant objects. Another group, made up of faint stars with great proper motions, led to better understanding of the luminosity function. During these observations he discovered the second known white dwarf (now named "van Maanen's star"), which yielded new information about such objects. A skillful determiner of parallaxes, his average probable error in those measurements was given as $\pm 0\rlap{.}''006$.

Van Maanen studied the proper motions of planetary nebulae, globular and open clusters, faint stars in or near the Orion nebula, near bright stars with large proper motions, faint stars in forty-two of Kapteyn's Selected Areas, and spiral nebulae. These studies yielded important fundamental information about several then puzzling phenomena, among them measures of the distances and absolute magnitudes of planetary nebulae, identification of stars as members of the Orion system and of the Pleiades and h Persei clusters, and rudimentary distances to 125 Cepheid variables.

In 1916 van Maanen published the results that he derived from the displacements of eighty-seven nebular points on two pairs of plates of M101. He detected a rotation rate of $0\rlap{.}''02$ per year at a distance of 5′ from the nucleus, a finding he checked by having Seth Nicholson, a meticulous observer, measure half the points. All nine of the spirals that van Maanen eventually investigated seemed to show motions outward along their arms. When Knut Lundmark in 1927 measured van Maanen's plates of M33, he found a rotational component only one-tenth as large as van Maanen's; moreover, van Maanen himself later obtained rotations only half as large as his earlier ones. But although other information strongly indicated that spirals are remote (and hence could not be rotating as fast as he had found), he continued to trust his calculations. In the famous Shapley-Curtis debate in 1920, Shapley, van Maanen's lifelong friend, cited van Maanen's findings as proof that spirals are relatively nearby. Fifteen years later, after plates of several spirals had been taken at longer intervals, Edwin Hubble and van Maanen published papers in *Astrophysical Journal* stating that van Maanen's results on rotations had been incorrect, apparently because of systematic errors.

Van Maanen also attempted to measure the general solar magnetic field, an undertaking begun in 1908 by G. E. Hale. The technique involved measuring the weakly polarized components of Zeeman-split lines. Several staff members at Mt. Wilson worked on the project, but the reduction of the observations was specifically van Maanen's responsibility. Initially he found an overall field strength of roughly fifty gauss, which was later revised to about twenty. Recently J. O. Stenflo analyzed van Maanen's plates by computer and found that van Maanen's visual measurements apparently had again involved systematic errors. It should be noted that the field strengths being measured are so slight (about 1 gauss) that no one was able to determine them reliably until new techniques were introduced in about 1952.

Van Maanen's entire career dealt with visually measuring almost imperceptible changes on photographic plates. In parallax observations he ranked high, but his results on the proper motions of spirals and the magnetic field of the sun were largely incorrect. Numerous explanations have been offered as to why he made such errors. Whatever the reason, the fundamental fact remains that the changes he was attempting to measure were at the very limits of precision of his equipment and techniques. Ironically, even though some of his findings caused confusion for over a decade, they also provided a healthy catalyst for research and discussion and demonstrated the ever-present dangers of systematic error, for even a careful observer.

BIBLIOGRAPHY

I. Original Works. Van Maanen published more than 150 papers, about fifty of them in the *Contributions from the Mount Wilson Solar Observatory*. Aside from research papers he also wrote popular pieces and biographical sketches and collaborated in the publication of *Hemel en Dampkring*. Many of his most important papers appeared in *Astrophysical Journal*, among them "List of Stars With Proper Motion Exceeding 0″.50 Annually," **41** (1915), 187–205; "Preliminary Evidence of Internal Motion in the Spiral Nebula Messier 101," **44** (1916), 210–228; and "The General Magnetic Field of the Sun. Apparent Variations of Field-Strength With Level in the Solar Atmosphere," **47** (1918), 206–254, written with G. E. Hale, F. H. Seares, and F. Ellerman.

II. Secondary Literature. Obituaries of van Maanen include Dorrit Hoffleit, in *Sky and Telescope*, **5** (1946), 10; Alfred H. Joy, in *Popular Astronomy*, **54** (1946), 107–110; and Frederick H. Seares, in *Publications of the Astronomical Society of the Pacific*, **58** (1946), 88–103, with portrait.

Van Maanen's errors in the determination of the rotations of spirals have been discussed by numerous investigators, including Knut Lundmark, in *Uppsala Astron. Obs. Medd.*, no. 30 (1927); Edwin Hubble, "Angular Rotations of Spiral Nebulae," in *Astrophysical Journal*, **81** (1935), 334–335; Walter Baade, in C. Payne-Gaposchkin, ed., *Evolution of Stars and Galaxies* (Cambridge, Mass., 1963), 28–29; Harlow Shapley, *Through Rugged Ways to the Stars* (New York, 1969), 50, 55–57, 80; and J. D. Fernie, "The Historical Quest for the Nature of the Spiral Nebulae," in *Publications of the Astronomical Society of the Pacific*, **82** (1970), 1212–1213, 1218–1219.

Commentaries on the difficulties of measuring the general solar magnetic field and van Maanen's attempts to do so are given by K. O. Kiepenheuer, in G. P. Kuiper, ed., *The Sun* (Chicago, 1953), 361 ff.; C. de Jager, in *Handbuch der Physik*, II (Berlin, 1959), 340 ff.; Einar Tandberg-Hanssen, *Solar Activity* (Waltham, Mass., 1967), 75 ff.; and Jan Olof Stenflo, "Hale's Attempts to Determine the Sun's General Magnetic Field," in *Solar Physics*, **14** (1970), 263–273.

Correspondence with and about van Maanen is contained in "The George Ellery Hale Papers, 1882–1937" (Pasadena, Calif., 1968), microfilm ed., California Institute of Technology.

Richard Berendzen
Carol Shamieh

MACALLUM, ARCHIBALD BYRON (*b*. Belmont, Ontario, Canada, 1858; *d*. London, Ontario, 5 April 1934), *biochemistry, physiology*.

Macallum was one of several children of Alexander and Annie (née McAlpine) McCallum. His father emigrated from Kilmartin, Argyllshire, Scotland, in the early nineteenth century, settling in southwestern Ontario. The family was of Gaelic-speaking Presbyterian background, and many members achieved distinction in the professions in Canada and the United States.

Macallum graduated in 1880 from the University of Toronto, winning a silver medal in the natural sciences. For some time afterward he taught high school in Cornwall, Ontario, where he married Winifred Isobel Bruce. In 1883 he was appointed lecturer in biology at Toronto. He obtained his Ph.D. from Johns Hopkins in 1888 and his M.B. from Toronto in 1889. He received many other degrees and academic honors and was a member of numerous learned societies and professional groups, serving at various times as an officer in several of these. He was elected a fellow of the Royal Canadian Society in 1901 and fellow of the Royal Society in 1906.

From 1890 to 1916 Macallum was professor of physiology and later of biochemistry at Toronto. After a three-year period devoted to the founding of the National Research Council of Canada, he joined the faculty of McGill University as professor of biochemistry. Under the auspices of the Rockefeller Foundation he was visiting professor in 1921 at the Peking Union College. Upon his retirement in 1928 he returned to London, Ontario, continuing his research on a part-time basis at the Medical School of the University of Western Ontario, of which his son Archibald Bruce became dean (1928–1934).

Macallum is perhaps best known today for his theory that there is a significant relationship between the inorganic composition of vertebrate blood plasma and that of the ancient oceans. He argued that specific concentrations of inorganic ions encountered in modern vertebrate blood plasma constitute an heirloom from primeval sea life, preserved by the emergence of multicellular organisms having closed circulatory systems, and by the evolution of the kidney. In a 1918 paper he ascribed to the kidney the important function of regulating the inorganic composition of the body fluids. His most mature statement on the supposed paleochemistry of the body fluids appeared in 1926. Despite obvious difficulties of conclusive demonstration, this theory is still referred to in modern texts of general physiology.

Some of his other important scientific papers dealt with the microchemical determination and localization of the inorganic constituents of plant and animal tissues. More recent investigations have shown a number of errors in some of his conclusions due to the limitation of techniques available in his time.

Macallum may be considered one of the pioneers of medical and biological science in Canada, both for his own scientific researches and for his many efforts on behalf of higher education, especially the development of the study of the biosciences in the Canadian premedical curriculum.

BIBLIOGRAPHY

I. Original Works. Macallum's writings include "Contributions to the Morphology and Physiology of the Cell," in *Transactions of the Royal Canadian Institute*, **1** (1891), 247–278; "On the Demonstration of the Presence of Iron in Chromatin by Micro-Chemical Methods," in *Proceedings of the Royal Society*, **50** (1892), 277–286; "On the Absorption of Iron in the Animal Body," in *Journal of Physiology*, **16** (1894), 268–297; "On the Detection and Localisation of Phosphorus in Animal and Vegetable Tissues," in *Proceedings of the Royal Society*, **63** (1898), 467–479; "On the Cytology of Non-nucleated Organisms," in *Transactions of the Royal Canadian Institute*, **6** (1899), 439–506; "On the Inorganic Composition of the Medusae, *Aurelia flavidula* and *Cyanea arctica*," in *Journal of Physiology*, **29** (1903), 213–241; "The Palaeochemistry of the Ocean in Relation to Animal and Vegetable Protoplasm," in *Transactions of the Royal Canadian Institute*, **7** (1904), 535–562; "On the Distribution of Potassium in Animal and Vegetable Cells," in *Journal of Physiology*, **32** (1905), 95–128; "On the Nature of the Silver Reaction in Animal and Vegetable Tissues," in *Proceedings of the Royal Society*, **76B** (1905), 217–229; and "On the Distribution of Chlorides in the Nerve Cells and Fibres," *ibid.*, **77B** (1906), 165–193, written with M. L. Menten.

Later works are "The Scientific Spirit in Medicine," in *Montreal Medical Journal*, **37** (1908), 1–21; "Die Methoden und Ergebnisse der Mikrochemie in der biologischen Forschung," in *Ergebnisse der Physiologie*, **7** (1908), 552–652; "On the Origin of Life on the Globe," in *Transactions of the Royal Canadian Institute*, **8** (1910), 423–441; "The Inorganic Composition of the Blood in Vertebrates and Invertebrates, and its Origin," in *Proceedings of the Royal Society*, **82B** (1910), 602–624; "Oberflächenspannung und Lebenserscheinungen," in *Ergebnisse der Physiologie*, **11** (1911), 598–659; an English trans., "Surface Tension and Vital Phenomena," appears in *University of Toronto Studies. Physiological Series*, no. 8 (1912), 1–82; "*Acineta tuberosa:* A Study on the Action of Surface Tension in Determining the Distribution of Salts in Living Matter," in *Proceedings of the Royal Society*, **86B** (1913), 527–550; "The Ancient Factors in the Relations Between the Blood Plasma and the Kidneys," in *American Journal of the Medical Sciences*, **156** (1918), 1–11; "Paleochemistry of the Body Fluids and Tissues," in *Physiological Reviews*, **6** (1926), 316–357; and "The Significance of Ketogenesis," in *Canadian Medical Association Journal*, **22** (1930), 3–11.

II. Secondary Literature. For obituaries of Macallum see *Obituary Notices of Fellows of the Royal Society of London*, **1** (Dec. 1934), 287–291; and *Proceedings and Transactions of the Royal Society of London*, 3rd ser., **28** (1934), xix–xxi.

R. A. Richardson

MACAULAY, FRANCIS SOWERBY (*b.* Witney, England, 11 February 1862; *d.* Cambridge, England, 9 February 1937), *mathematics.*

The son of a Methodist minister, Macaulay was educated at Kingswood School, Bath, a school for the sons of the Methodist clergy, and at St. John's College, Cambridge. After graduating with distinction, he taught mathematics for two years at Kingswood and, from 1885 to 1911, at St. Paul's School, London, where he worked with senior pupils who were preparing to enter a university. He was remarkably successful: two of his many pupils who became eminent mathematicians were G. N. Watson and J. E. Littlewood. In *A Mathematician's Miscellany*, Littlewood gives a vivid picture of Macaulay's methods: there was little formal instruction; students were directed to read widely but thoroughly, encouraged to be self-reliant, and inspired to look forward to pursuing research in mathematics.

In recognition of his own researches, which he had steadily carried on despite his heavy teaching responsibilities, in 1928 Macaulay was elected a fellow of the Royal Society, a distinction very seldom attained by a schoolmaster. Apart from some elementary articles in the *Mathematical Gazette* and a school text on geometrical conics, he wrote some fourteen papers on algebraic geometry and a Cambridge tract on modular systems, otherwise polynomial ideals. The earlier papers concerned algebraic plane curves, their multiple points and intersections, and the Noether and Riemann-Roch theorems. This work led to later papers on the theory of algebraic polynomials and of modular systems. Much of this was pioneering work with an important influence on subsequent research in algebraic geometry, and it was directed toward the construction of a firm and precise basis of algebra on which geometrical theorems could be safely erected.

BIBLIOGRAPHY

I. ORIGINAL WORKS. Macaulay's books are *Geometrical Conics* (Cambridge, 1895); and *The Algebraic Theory of Modular Systems*, Cambridge Mathematical Tracts, no. 19 (Cambridge, 1916). A list of Macaulay's papers follows the obituary notice by H. F. Baker cited below.

II. SECONDARY LITERATURE. See H. F. Baker's notice of Macaulay in *Journal of the London Mathematical Society*, **13** (1938), 157–160; and J. E. Littlewood, *A Mathematician's Miscellany* (London, 1953), 66–83—the paragraphs relevant to Macaulay are quoted in Baker's notice.

T. A. A. BROADBENT

MACBRIDE, DAVID (*b.* Ballymoney, Antrim, Ireland, 26 April 1726; *d.* Dublin, Ireland, 28 December 1778), *medicine, chemistry.*

Macbride's father and grandfather were both Presbyterian ministers; his brother John rose to the rank of admiral in the British navy. Macbride chose a career in medicine and became apprenticed to a local surgeon after leaving the village school. He served as a surgeon in the navy during the War of the Austrian Succession (1740–1748) and then spent some time in Edinburgh and London, learning more about his profession from well-known teachers, before establishing a practice in Ballymoney in 1749. In 1751 he moved to Dublin.

His career seems to have met with only a limited success until after the publication of *Experimental Essays* (1764), which secured him a doctorate from Glasgow (prior to this he had no degree) and a European reputation. The work dealt with various aspects of a theory that Macbride had developed from Stephen Hales's concept of air. Hales's observations of the expulsion of "air" from all kinds of materials, during heating and fermentation, led him to regard air as an essential constituent of all bodies. Its apparently dual nature—elastic in the free state yet capable of fixation in solids—prevented the component particles from coalescing into a "sluggish lump," while at the same time it contributed to their union (Hales, *Vegetable Staticks* [London, 1727], pp. v–vi, 313–314).

Stressing the latter property (air as a "cementing principle"), Macbride attempted to apply the idea to medicine. He believed that it had been supported by Pringle's experiments and confirmed by his own, which appeared to show that "air" freed from fermenting mixtures could counteract putrefaction. He believed that any mixture of animal and vegetable substances with water would ferment, hence the value of a mixed diet. The need for vegetable matter in the diet led to his advocacy of wort (an infusion of malt) in the treatment and prevention of scurvy. This method was favorably reported on by Captain Cook and others.

In one of his essays Macbride established the precipitation of chalk from limewater as a test for "fixed air"—(a phenomenon that confirmed his theory)—but he was unable to decide whether "fixed air" was a substance quite distinct from atmospheric air or simply a part of the latter modified by its fixation in solids.

His experiments with limewater led him to observe that it is more efficacious than plain water in extracting tannin from oak bark; and he petitioned the Irish parliament for recompense if he divulged his "secret" to Irish tanners, claiming that considerable time and money would be saved by his method. There is no record that his petition was successful; but the innovation was reported on favorably by the Dublin Society, and the discovery was eventually published.

Macbride was apparently a skilled surgeon and obstetrician, and his practice became lucrative. His medical treatise was accepted as authoritative and was translated into several European languages.

BIBLIOGRAPHY

I. ORIGINAL WORKS. Macbride's writings include *Experimental Essays* (London, 1764, 1767); and *A Methodical Introduction to the Theory and Practice of Physic* (London, 1772), 2nd ed., enl., *A Methodical Introduction to the Theory and Practice of the Art of Medicine*, 2 vols. (Dublin, 1777). *Some Account of a New Method of Tanning* (Dublin, 1769) gives the advantages but not the secret of the method; the latter is given in "An Improved Method of Tanning Leather," in *Philosophical Transactions of the Royal Society*, **68** (1778), 111–130.

II. SECONDARY LITERATURE. The main biographical source is A. Smith, "David Macbride, M.D.," in *Dublin Quarterly Journal of Medical Science*, **3** (1847), 281–290.

Macbride's theory of air and some of its consequences are dealt with in E. L. Scott, "The 'Macbridean Doctrine' of Air . . .," in *Ambix*, **17** (1970), 43–57.

E. L. Scott

MacBRIDE, ERNEST WILLIAM (*b.* Belfast, Ireland, 12 December 1866; *d.* Alton, Hampshire, England, 17 November 1940), *embryology*.

The eldest son of Samuel MacBride and Mary Jane Browne, MacBride was educated at Belfast and Cambridge and in Germany. He graduated from Cambridge in 1891 and worked for a year under Anton Dohrn at the Marine Biological Station in Naples before returning to Cambridge. In 1897 he became the first professor of zoology at McGill University, Montreal, where he built up a strong school before returning in 1909 to the Imperial College of Science and Technology, remaining there until his retirement in 1934. In 1902 he married Constance Harvey; they had two sons. He was elected fellow of the Royal Society in 1905, vice-president of the Zoological Society of London in 1913, and was active on several committees concerned with marine biology and fisheries.

MacBride's *Textbook of Embryology*, Volume I, *Invertebrata* (1914) was the standard work for many years and reflects his wide knowledge and interest in comparative embryology. His particular interest, the development of echinoids, began in Naples and continued throughout his professional life. He studied the problem of metamorphosis from the bilateral larva to the radial adult and used it to elucidate phylogenetic affinities. He found that metamorphosis occurs in Asterina during a previously unnoticed fixed stage, which discovery enabled him to postulate a fixed stage in the phylogeny of this group. There had already been some suggestion by Bateson of affinities between the echinoderms and chordates, for which hypothesis MacBride provided support from embryological evidence, mainly in the origin of the nervous system and coelom, in echinoderms and in amphioxus. His *Textbook* includes the Protochordata and analyzes relationships.

From his earliest years MacBride was increasingly a supporter of some form of Lamarckian inheritance of acquired characters. As he expressed it, ". . . habit is response to environment and inherited structure is nothing but the crystallisation of the habits of past generations."

BIBLIOGRAPHY

I. Original Works. The Royal Society obituary lists a selected bibliography of thirty items, among them the series of papers on the embryology of echinoderms and his books. Worth noting in supplement to that list are MacBride's contribution of "Zoology" to *Evolution in the Light of Modern Knowledge: A Collective Work* (London, 1924), 211–261; and his intro. to his own trans. of E. Rignano, *Biological Memory* (London, 1926), 1–16. He also wrote the section on larvae of echinoderms for *Natural History Reports. British Antarctic Terra Nova Expedition, 1910. Zoology*, **4**, pt. 3 (1920), 83–94; and contributed numerous letters to *Nature* and *The Times*.

II. Secondary Literature. The only comprehensive analysis of MacBride's life and work is by W. T. Calman, in *Obituary Notices of Fellows of the Royal Society of London*, **3** (1940), 747–759, with portrait. An anonymous obituary appeared in *Nature*, **146** (1940), 831–832; and a printed pamphlet was prepared by MacBride, *Application for the Chair of Zoology in University College London, with Testimonials* (1906).

Diana M. Simpkins

McCLUNG, CLARENCE ERWIN (*b.* Clayton, California, 5 April 1870; *d.* Philadelphia, Pennsylvania, 17 January 1946), *cytology*, *zoology*.

Through his studies on the accessory chromosome in insects, Clarence E. McClung contributed significantly to the establishment of the chromosome theory of inheritance. In 1901 and 1902 he pointed out that the accessory chromosome (or X chromosome as it is sometimes called) was possibly the nuclear element responsible for determining sex. Thus he was among the first to offer evidence that a given chromosome carried a definable set of hereditary traits.

Of Scotch-Irish descent (i.e., Scots who moved to Ireland), McClung's ancestors had come to the United States in 1740 and settled in Lancaster County, Pennsylvania. His father, Charles Livingston McClung, was a civil and mining engineer, and his mother, Annie H. Mackey, the daughter of a physician. Because his father's business required the family to move about (mostly through the West and Midwest), young McClung's schooling was sporadic. By the time he was ready to enter high school, however, the family had settled in Columbus, Kansas (in the mid-1880's) and McClung's intellectual abilities began to show. He became especially interested in science; from his father he learned surveying, and from working in an uncle's drug store, he learned pharmacy. Following this latter interest, he entered the University of Kansas School of Pharmacy in Lawrence (1890), and completed the pharmacy course in two years (receiving his Ph.G. in 1892). For a year he taught chemistry and pharmacy at the university, and in the fall of 1893 enrolled in the College of Liberal Arts.

As an undergraduate his interests shifted from chemistry to zoology, particularly through the influence of S. W. Williston, who encouraged McClung's natural mechanical bent and allowed him to work in the histology laboratory learning special techniques. Williston also entrusted McClung with teaching part of his course in histology during one semester. McClung received his B.A. in 1896, and immediately entered graduate school at Kansas, receiving his M.A. in 1898, and Ph.D. in 1902. During his graduate years he spent one semester at Columbia University with the cytologist Edmund Beecher Wilson, and one summer (1898) working with William Morton Wheeler at the University of Chicago. Both of these men were interested in the nature and behavior of chromosomes, and it was through their influence that McClung's attention was directed to these nuclear elements.

While still a graduate student at Kansas he was appointed assistant professor of zoology, and later of histology and animal morphology (1898–1900); in 1901 he was made associate professor of zoology and head of the zoology department. He served as curator of the university's paleontological collections (1902–1912), and was acting dean of the Medical School (1902–1906). In 1906 he became professor of zoology, a post he held until 1912 when he accepted a call to become head of the zoological laboratories at the University of Pennsylvania, where he remained until his retirement in 1940. During the academic year 1940–1941 he served as acting chairman of the department of zoology at the University of Illinois, and in 1943–1944 was acting chairman of the biology department at Swarthmore College (Pennsylvania). Always interested in teaching, McClung introduced a variety of pedagogical innovations during his career. At Kansas, where he taught introductory biology, he deemphasized memorization and encouraged students to think through problems on their own. And at Pennsylvania, he served on many university committees concerned with curriculum reform and matters of basic educational policy.

McClung's biological work covers several distinct areas of interest: paleontology (on which he wrote some early papers in 1895, 1905, and 1908), technical microscopy and microscopy techniques, and studies on chromosomes. The latter area includes his most important contributions.

By the 1890's it had become clear to most biologists that the chromosomes were somehow involved in the processes of heredity. Both Wilhelm Roux and August Weismann had suggested theoretical roles for the chromosomes as hereditary determiners, while a host of microscopists had carefully detailed the movements of chromosomes in the mitotic and meiotic divisions. The constancy of chromosome number for any species was recognized; and the number of chromosomes for the cells of higher animals and plants was found to be even, suggesting that equal numbers come from the egg and from the sperm. It was also known that the chromosomes divide longitudinally at each somatic division, and that in the formation of gametes (sperm or egg cells) the chromosome number is reduced by one-half (each sperm or egg receiving only half the total number characteristic of the species). Yet there were also a number of facts now part of our common understanding of heredity which were unknown at that time. It was not known until after 1901 or 1902 that the chromosomes exist in definite pairs (homologs). Furthermore, it was thought that when the chromosomes reappear at the end of interphase (just before the cell begins its division cycle), they are linked together end-to-end as one continuous thread (called a spireme). As the division cycle begins, the thread was thought to break up into the number of chromosomes characteristic of the species. The most basic question still unanswered was, in terms of the hereditary information they carry, does each chromosome differ qualitatively from the others, or are they all basically the same, each bearing a full complement of hereditary information? If it were possible to identify a given hereditary trait (or set of traits) with any specific chromosome, this basic question would be answered. The chromosomes would then appear to have individuality, i.e., each would appear to differ from the others, and would control a different trait or traits.

In 1891 Hermann Henking had noted the peculiar meiotic behavior of an unusual chromosome in spermatocytes of the small fire wasp (Hemipteran) *Pyrrhocoris*. This chromosome, which Henking called a "chromatin body" and labeled "X," did not seem to pair up with a partner chromosome at prophase I of meiosis, thus seeming to lead a peculiarly separate existence. Furthermore, Henking noted that the male of the species had an uneven number of chromosomes, while the female had an even number. In spermatogenesis one-half of the sperm received this "X" element (thus ending up with twelve chromosomes), while the other half did not receive it (thus ending up with only eleven chromosomes). Similar "accessory chromosomes" were reported for other species of insects, including the long-horned grasshopper, *Xiphidium fasciatum*, as observed by McClung in 1899. It was not until McClung's suggestive papers of 1901 and 1902, however, that the significance of these unpaired "X" elements was fully realized. Most important, McClung's work had a crucial, though

McCLUNG

indirect significance for the question of the individuality of the chromosomes.

When McClung was in Chicago in 1898, Wheeler suggested that he study spermatogenesis in the grasshopper *Xiphidium*; Wheeler himself had been working on oogenesis in this species, and wanted someone to do the complementary studies on males. In 1899 McClung published his first set of observations on the "X" element which he observed clearly in grasshopper spermatocytes; because it was unpaired he coined the term "accessory chromosome" to replace Henking's rather vague designation of "X." In two longer papers (1901 and 1902) McClung went beyond the cytological observations of his predecessors. He saw in the behavior of the accessory chromosome a clear mechanism for understanding how sex could be determined. McClung started with two notions (1) that the chromosomes were the bearers of hereditary information, and (2) that sex was to some extent at least determined by heredity. Thus it was logical to conclude that sex must in some way be determined by the chromosomes. The accessory chromosome, McClung pointed out, fits all the requirements for being a sex determiner: it is present in all cells of the organism until they are fully formed, thus serving to control the development of sexual characteristics in the adult; and it is present in one-half the gametes of the sex that bears it (e.g., males). Thus, at fertilization, one-half the zygotes will contain the accessory, and one-half will lack it. Now, McClung reasoned, sex is the only hereditary characteristic which divides a species into two equal groups. It is quite possible, then, that the accessory chromosome is the hereditary element which determines this difference.

Although his theory proposed that sex was determined at the moment of fertilization, McClung did not hold that this was necessarily true for all species, or that the environment did not sometimes influence the development of the hereditary predisposition. In humans, he felt sex was determined at conception, as evidenced by the fact that identical twins are always of the same sex. But in other species this might not be the case. He took a broad (and from today's perspective too indulgent a view of the environmental theory) approach to this question when he wrote in 1902:

> Sex, then, *is* determined sometimes by the fact of fertilization and can not be subsequently altered. But between this extreme and the other of marked instability [*i.e.*, totally dependent on the environment] there may be found all degrees of response to environment. It must accordingly be granted that there is no hard-and-fast rule about the determination of sex, but that specific conditions have to be taken into account in each case.

We know today that environmental factors do have some influence on sex ratios in some species; but they do not have the fundamental role which McClung and some of his contemporaries often assigned to them.

It should be noted that McClung made two additional errors in his theory of sex determination. Because he gave such prominence to the influence of environmental conditions, he believed in a theory popular at the time called "selective fertilization"—the idea that the egg could "choose" whether it was fertilized by a sperm bearing an accessory, or by one lacking the accessory. We know today that the egg is incapable of carrying out any such selectivity. Furthermore, through a miscount of the number of chromosomes in the female of the insects he was studying, McClung thought that a sperm bearing an accessory chromosome (the X) would be male-determining. In fact it is, as we know now, female-determining (a sperm bearing either no accessory or a Y-chromosome is male-determining).

McClung posed his ideas about the role of the accessory only as a working hypothesis, tentative and perhaps suggestive, but not as a proof of the chromosomal nature of sex determination. The evidence was largely circumstantial, resting upon the observed parallelism between the presence or absence of the accessory and the sexual differentiation into males or females. His hypothesis proved to be very stimulating, however. Within a few years (1905) E. B. Wilson and Nettie M. Stevens, independently, showed that for most groups of animals males carried one accessory chromosome (the genotype being designated as X, XO, or XY, "Y" indicating a second element, the Y-chromosome, in many species found associated with the X), and females two accessory chromosomes (the genotype being designated XX). In some species they found that the relationship was just reversed: males had two accessories (XX) and females one (XO or XY). Their work provided a sound and systematic basis for the chromosomal theory of sex determination.

McClung's work, and subsequently that of Wilson and Stevens, substantiated the concept of the individuality of the chromosomes. By associating the inheritance of one set of traits (sexual) with one particular chromosomal element, the McClung theory suggested that each chromosome was different from the others, governing one specific set of characteristics. At just about the same time, the concept of chromosomal individuality received considerable support from other quarters: particularly from the work of Boveri with polyspermy in the sea urchin (1902), and that of T. H. Montgomery (1901) and W. S. Sutton (1902, 1903).

Through the concept of chromosomal individuality, McClung's work led directly to the correlation of the chromosome theory with Mendel's newly publicized laws of heredity (rediscovered in 1900). The parallelism between the separation of maternal and paternal members of each chromosome pair in meiosis and Mendel's postulate of segregation of heritable factors was picked up by both Montgomery (1901) and Sutton (1902, 1903). Sutton had been a graduate student of McClung's at Kansas in 1901, and gained both interest and insight into the chromosome question from his teacher. At the time he wrote his own important papers on Mendelism, however, Sutton had moved on to Columbia University as a student of E. B. Wilson. McClung's studies on the accessory chromosome came at a fortuitous time in the history of genetics. They pulled together disparate information by suggesting a purpose for the otherwise inexplicable movements of the chromosomes in general and of the accessory in particular during spermatogenesis. They gave a new and fruitful means for understanding the inheritance of sex by providing a mechanism for explaining the 1 : 1 male–female ratio observed for most species. And lastly, they set the stage for relating Mendel's abstract "factors" to real material bodies (chromosomes) in the cell nucleus.

Like many biologists at the time, McClung saw heredity and development as inextricably linked. Heredity operates, he maintained, by controlling the cell's metabolic functioning. And since that functioning depends upon environmental circumstances (for availability of raw materials, for example), it was clear that an individual's ontogeny was molded by hereditary potentialities interacting with specific environmental conditions. This was a consciously epigenetic view, conditioned in part by reaction to the prevalence of particulate theories of heredity in the past (such as those of Haeckel or Weismann) which placed all emphasis on heredity. To McClung, the most important arena in which further understanding of the relationships between heredity and development could be worked out was cytology. Biochemical knowledge was too scanty at the time to make his approach a feasible one for understanding how cells grew and differentiated during embryonic development. But cytology was accessible, and should be used to explore these fundamental questions:

> The apprehension of large principles of organization should therefore be our aim, and I have no doubt that once an understanding of the cytological changes in the body of an animal during its ontogeny is reached, we shall have solved, as far as it is possible for us to do, some of the larger problems of heredity and development that have become our scientific inheritance [1908].

In addition to his studies on chromosomes, McClung also made contributions to techniques of biological staining and to the design of microscopes. Particularly noted was the "McClung model" microscope, which had an improved mechanical design. He was also the author of a series of papers on photomicrography and a *Handbook of Microscopical Technique* (1929), and he pioneered in the use of simple, new designs in laboratory equipment for introductory biology courses.

Besides his teaching and research, McClung served in a number of administrative posts outside the university. In 1913 he was appointed a trustee of the Marine Biological Laboratories, Woods Hole, Massachusetts (he had taught part of the embryology course there in the summer of 1903), and in 1914 a member of the investigative staff. In 1917 he became chairman of the Zoology Committee of the Division of Biology and Medicine of the National Research Council (NRC), and in 1919 the first chairman of its newly created Division of Biology and Agriculture (NRC). In the latter post he initiated plans for a comprehensive biological abstracting service, which eventually led to the publication of *Biological Abstracts* (published by the Union of American Biological Societies, of which McClung became president of the board of trustees, 1925–1933). As chairman of the Division of Biology and Agriculture, he also initiated plans for standardizing biological stains, eventually giving rise to the Biological Stain Commission. He served as the managing editor of the *Journal of Morphology* from 1920 until 1946, and the associate editor of *Cytologia* from 1930 onward.

McClung held memberships in all of the important zoological societies. These include American Morphological Society (1901), American Association for the Advancement of Science (1902; fellow, 1908); American Society of Zoologists (secretary, Central Branch, 1905; president, Central Branch, 1910; president, national organization, 1914); American Philosophical Society (1913); American Society of Naturalists (1913; president, 1927); Philadelphia Academy of Sciences (1914); Wistar Institute of Anatomy and Biology, advisory board (1914); American Association of Anatomists (1914); and the Washington Academy of Sciences (1920).

In addition, he received many honors, including membership in several honorary societies. The latter include Sigma Xi (national president, 1917–1921), Beta Beta Beta (president, 1936), and the National Academy of Sciences (1920). He was also the United States representative at the International Biological Congress in Montevideo, Uruguay, in 1930, a goodwill scientific ambassador (sponsored by the Rockefeller

Foundation) to Japan (1933–1934), a recipient of the Distinguished Service Citation from the University of Kansas (1941), and a D.Sc. from Franklin and Marshall College (1942).

McClung was known as a congenial and friendly person by both his colleagues and students. He had a broad range of interests involving athletics, English literature, music, photography, and dramatics. Noted as a sensitive and meticulous worker, McClung was described by one student as "an artist in everything he does." He was married to Anna Adelia Drake of Lawrence, Kansas, on 31 August 1899; they had two children, Ruth Cromwell and Della Elizabeth.

BIBLIOGRAPHY

I. Original Works. The most complete bibliography of McClung's writings is at the conclusion to the biographical sketch by Wenrich mentioned below; however, this covers only up until 1940. His most important papers include: "The Spermatocyte Divisions of the Acrididae," in *Kansas University Quarterly*, **9** (1900), 73–100; "Notes on the Accessory Chromosome," in *Anatomischer Anzeiger*, **20** (1901), 220–226; "The Accessory Chromosomes: Sex Determinant?" in *Biological Bulletin*, **3** (1902), 43–84; "Cytology and Taxonomy . . .," in *Kansas University Science Bulletin*, **4** (1908), 199–215; "A Comparative Study of Chromosomes in Orthopteran Spermatogenesis," in *Journal of Morphology*, **25** (1914), 651–749; and "The Cell Theory—What of the Future?" in *American Naturalist*, **74** (1939), 47–53.

II. Secondary Literature. The only substantial biographical treatment of McClung was prepared by D. H. Wenrich on the occasion of McClung's seventieth birthday: "Clarence Erwin McClung," in *Journal of Morphology*, **66** (1940), 635–688. Wenrich also prepared several shorter treatments after McClung's death: *American Naturalist*, **80** (1946), 294–296; and *Yearbook. American Philosophical Society* (1946), 322–325. These are all, however, truncated versions of the more complete biographical sketch. Further information can be found in a series of short sketches devoted to McClung in *Bios*, **11** (1940), 141–155, by various colleagues.

Garland E. Allen

McCOLL, HUGH (*b.* 1837; *d.* Boulogne-sur-Mer, France, 27 December 1909), *mathematical logic.*

McColl's contributions to mathematical logic and its symbolic expression helped to clarify the subject in the particular period which may be said to begin with Boole's *An Investigation of the Laws of Thought*. . . (1854) and reach a climax in the *Principia Mathematica* (1950) of Whitehead and Russell.

The logical calculus of propositions has a certain analogy to that of classes, with implication in the former corresponding to inclusion in the latter. Thus, for propositions p, q, r, we have that if p implies q and q implies r, then p implies r, while for classes A, B, C, the dual statement is that if A is contained in B and B is contained in C, then A is contained in C. But the duality is not complete. If p implies q or r, then p implies q or p implies r, but if A is contained in B or C, with "or" in its usual inclusive sense, then we cannot say that either A is contained in B or A is contained in C. The ambiguity is to be seen in Boole's *Laws of Thought*, where he is aware of the duality but is not always quite clear about which interpretation of his symbolic calculus he is using. Since the duality is not perfect, the question arises as to which calculus is the more basic. In his papers, chiefly in the period 1880–1900, discussing many points of the symbolic logic then in process of formation, McColl takes the view, which has much to commend it, that implication and propositions have a more fundamental character than inclusion and classes. His arguments are forceful; but his logical position would have been clearer had he distinguished between a propositional function containing an indeterminate such as "x is a prime number," and a proposition, which is the form assumed by a propositional function when the indeterminate receives a specific value. The proposition is then a statement which is either true or false, whereas no truth-value can be assigned to a propositional function. The distinction was hinted at by Peano, but seems to have been first clearly drawn by Russell.

BIBLIOGRAPHY

McColl's main writings on symbolic logic are "On the Calculus of Equivalent Statements," in *Proceedings of the London Mathematical Society*, 1st ser. **9**, **10**, **11**, **13**; "Symbolic Reasoning," in *Mind* (1880, 1897, 1900); and "La logique symbolique et ses applications," in *Bibliothèque du Congrès International de Philosophie*, III (Paris, 1901).

T. A. A. Broadbent

McCOLLUM, ELMER VERNER (*b.* near Fort Scott, Kansas, 3 March 1879; *d.* Baltimore, Maryland, 15 November 1967), *organic chemistry, nutrition.*

Elmer V. McCollum originated the first white rat colony in the United States devoted solely to the purpose of experimentation in nutrition. The outcome of this endeavor was his personal discovery and codiscovery of a number of the vitamins. Although he worked with other aspects of nutrition, the major portion of his professional career was devoted to vitamins and other trace nutrients.

McCollum received his bachelor's degree from the University of Kansas in 1903 and completed his doctorate at Yale University in 1906. Although his primary field was organic chemistry, circumstances led him to work for a period of time under Thomas B. Osborne at the Connecticut Agricultural Experiment Station where he acquired a strong interest in what was then called agricultural chemistry (biochemistry). In 1907 he was employed by the Wisconsin Agricultural Experiment Station to conduct chemical analyses on the food and excreta of dairy cattle, part of an experiment inaugurated a year earlier to determine the effect of various cereal grains upon the health and reproductive capacity of cattle. McCollum, despairing of the long and tedious procedures entailed in the use of such large animals, instituted a study which resulted in his setting up the albino rat colony. Under the most adverse conditions, including the hostility of the dean of the College of Agriculture, he conducted experimental work of such a nature that he was able to report in 1913 that rats fed on a diet deficient in certain fats resumed normal growth when fed "the ether extract of egg or of butter." Furthermore, he was able to transfer this "growth-promoting factor" to otherwise nutritionally inert fat or oil which then exhibited growth-promoting activity in rats.

Within two years McCollum also demonstrated that certain water-soluble substances were necessary for normal health in rats, and he consequently named these substances, present in foods in relatively small quantities, "fat-soluble A" and "water-soluble B," thus initiating the alphabetical nomenclature for vitamins. He at first thought that there existed one fat-soluble A and one water-soluble B, but further work in his laboratory and at other institutions soon indicated that there were numerous chemical entities involved.

In 1917 McCollum left Wisconsin to become the first biochemist of the School of Hygiene and Public Health of Johns Hopkins University, where he continued his studies of the vitamins and where, in collaboration with members of the medical school, he aided in the elucidation of what is now known as vitamin D (the antirachitic factor). Another outcome of McCollum's work both at Wisconsin and at Johns Hopkins was the development of the use of the living animal as an analytical tool. As far as nutrition was concerned, the only means by which the presence of curative or preventive substances could be detected was the animal feeding experiment, that is, biological analysis.

Because of his outstanding contributions in the field of nutrition, McCollum received many awards and he was invited to serve as a member of numerous national and international organizations devoted to public health. He was involved with the World Health Organization and the Nutrition Board of the National Research Council, as well as being an American fellow of the Royal Society of London.

McCollum's interests extended beyond the laboratory to the public domain via lectures, magazine articles, and books. His career indicates he was a man of tenacious character who, even after his retirement in the early 1940's, retained an active interest in nutrition and related health fields, evidenced by the fact that he published his book *A History of Nutrition* in 1957, and maintained an influence upon the science he had pioneered many years before.

BIBLIOGRAPHY

A detailed biography and bibliography of McCollum is in *Biographical Memoirs of Fellows of the Royal Society*, **15** (1969), 159–171. His works include *The Newer Knowledge of Nutrition* (1918; 5th ed. 1939); with E. Simmonds, *Food, Nutrition and Health* (1926); *A History of Nutrition* (1957); and *Text Book of Organic Chemistry for Medical Students*. His autobiography is *From Kansas Farm Boy to Scientist* (Lawrence, Kans., 1964). McCollum also contributed chapters or sections to a number of textbooks such as *Endocrinology and Metabolism* (1922); and *Human Biology and Racial Welfare* (1930). He published over 200 papers in the major scientific periodicals, as well as numerous popular magazine articles.

McCollum's early years of nutritional investigations (1907–1915) are most thoroughly dealt with in Stanley L. Becker, "The Emergence of a Trace Nutrient Concept Through Animal Feeding Experiments" (Ph.D. diss., 1968; Univ. of Wisconsin Microfilms), esp. chaps. 3–5.

STANLEY L. BECKER

MacCULLAGH, JAMES (*b.* near Strabane, County Tyrone, Ireland, 1809; *d.* Dublin, Ireland, 24 October 1847), *physics.*

There is little reliable biographical information concerning MacCullagh. He was the son of a poor farmer and apparently inherited some money from a wealthy grandfather. At any rate, he entered Trinity College, Dublin, as a pensioner in 1824 and went on to take every undergraduate honor in classics and science. In 1833 he became professor of mathematics at Trinity and in 1842 he succeeded Humphrey Lloyd as professor of natural philosophy. MacCullagh developed the school of mathematical physics that Lloyd had founded to such an extent that in 1845 the *Cambridge Mathematical Journal* became the *Cambridge and Dublin Mathematical Journal* as part of the campaign conducted by Kelvin (its editor), Stokes, and others to upgrade mathematical physics in Britain. In 1833 MacCullagh became a member of the Royal Irish Academy. He held successively higher offices, and in 1838 received the Academy's first gold medal. In 1842 he received the Copley Medal of the Royal Society of London; he became a fellow in the following year.

MacCullagh gave seven courses of lectures on mathematical physics. In one he independently duplicated Poinsot's work on rigid bodies; in another he exploited potential theory. His papers on geometry are generally characterized by depth, taste, and elegance, while a paper on curves of the second order was also highly original.

Pursuing the goal of giving rigor to Fresnel's wave optics, MacCullagh contributed valuable geometrical constructions. He worked steadily toward clarifying the problem posed by Fresnel's theory, and by 1837 had settled on four assumptions: (1) the vibrations of the wave are parallel to the plane of polarization; (2) the density of the ether is constant, but the elastic properties vary; (3) the *vis viva* of the wave is preserved at interfaces; and (4) the vibrations are continuous across interfaces. With these as given, and assuming Fresnel's form for the wave surface, in 1837 MacCullagh set out a phenomenological theory in a paper entitled "On the Laws of Crystalline Reflexion and Refraction." (The simultaneous development of a similar theory by F. E. Neumann led to considerable nationalistic controversy.)

William Rowan Hamilton, in presenting the gold medal of the Royal Irish Academy to MacCullagh for this paper, called MacCullagh's method "mathematical induction" as opposed to "dynamical deduction," indicating that MacCullagh had sought and found a mathematical generalization of the phenomena rather than having sought to derive the phenomena from established dynamical principles. John Herschel seconded Hamilton's praise of the paper and predicted that physical optics was "on the eve of some considerable improvement . . . [by] searching among the phenomena for laws simple in their geometrical enunciation, . . . *without (for a while) much troubling ourselves how far those laws may be in apparent accordance with . . . general principles in dynamics.*"

Three theories of physical optics were published in 1839: Cauchy's second theory, Green's elastic solid theory, and MacCullagh's dynamical theory (the first two being read on the same day). Seeking independently to deduce the phenomena from dynamical principles, MacCullagh and Green each proceeded by finding a potential for the presumed luminiferous medium and then deriving a wave equation by means of the Lagrangian variational principle. MacCullagh, in "An Essay Towards a Dynamical Theory of Crystalline Reflexion and Refraction," called this method "dynamical reasoning." Such "dynamical" theories would be "mechanical" if the potential were in turn derived from a complete mechanical structure rather than being postulated from general considerations. MacCullagh, however, admitted failure in his search for such a mechanical basis.

Stokes in 1862 led the way in preferring Green's linear displacement potential to MacCullagh's rotational potential, since the latter involved unbalanced couples. Although both Green's and MacCullagh's theories were more natural than Cauchy's—in that they avoided Cauchy's disposable constants—no theory prevailed until about 1900, when the electromagnetic theory achieved dominance.

In 1880 FitzGerald showed (in "On the Electromagnetic Theory of Reflection and Refraction of Light") that MacCullagh's formulation

$$\delta \int \ddot{\vec{R}}\, d^3r = -\delta \int (\vec{C} \cdot \vec{\nabla} X \vec{R})^2\, d^3r$$

could be translated into an electromagnetic one,

$$\delta \int \frac{1}{2} \vec{E} \cdot \vec{B}\, d^3r = \delta \int \frac{1}{8\pi} \vec{B} \cdot \vec{H}\, d^3r.$$

(Both MacCullagh's and FitzGerald's formulations are given in vector notation for clarity.) The wave equation studied by MacCullagh, FitzGerald, and later Larmor was for the displacement variable R (which was translated by FitzGerald as being equivalent to $\int \vec{H}\, dt$).

In a set of papers on "A Dynamical Theory of the Electric and Luminiferous Medium" (1893–1897), Larmor attempted with little success to combine MacCullagh's theory, Maxwell's electrodynamics, and Kelvin's rotationally elastic ethers. (The problem is more tractable with a purely electromagnetic approach, which also avoids the search for a mechanical basis for MacCullagh's displacement variable.)

MacCullagh's 1839 paper was a prototype of the techniques of mathematical physics exploited so successfully by the British in the latter half of the nineteenth century. It is an early example of the growing importance of rotational terms in the dynamics of continuous mediums, of which the curl relations in Maxwell's electrodynamics and Kelvin's theories of vortex motion provide prime later illustrations.

MacCullagh was an Irish nationalist and a strict adherent of the doctrines of the Roman Catholic church. He was modest and sternly moral; despite a keen relish for society, he never married. His disappointment at losing a parliamentary election, in which he had stood as a nationalist candidate, coupled with overwork, resulted in severe dyspepsia and an aggravation of earlier mental illness. He committed suicide at the age of thirty-eight.

BIBLIOGRAPHY

I. Original Works. MacCullagh's writings are gathered in John H. Jellett and Samuel Haughton, eds., *The Collected Works of James MacCullagh* (Dublin, 1880).

II. Secondary Literature. Chief biographical sources are *Encyclopaedia Brittanica*, 9th and 11th eds.; *Proceedings of the Royal Irish Academy*, **4** (1847–1850), 103–116; and *Abstracts of the Papers Communicated to the Royal Society*, **5** (1843–1853), 713–718.

On MacCullagh's work and milieu, see especially W. R. Hamilton, "Address of the President," in *Proceedings of the Royal Irish Academy*, **1** (1836–1839), 212–221, on the occasion of the award of the Academy's gold medal to MacCullagh. See also George F. FitzGerald, "On the Electromagnetic Theory of Reflection and Refraction of Light," in *Philosophical Transactions of the Royal Society*, **171** (1880), 691–711; Richard T. Glazebrook, "Report on Optical Theories" in *Report of the British Association for the Advancement of Science* (1885), 157–261, a thorough summary, with references, of the theories of optics available in the late nineteenth century; Joseph Larmor, *Mathematical and Physical Papers*, 2 vols. (Cambridge, 1921), of which the abstract and introduction to the first paper contain valuable historical material and commentary on method consistent with British usage of the period; his *Aether and Matter* (Cambridge, 1900) is a revision of these papers with a greater emphasis on electron theory and relativity; and George G. Stokes, "Report on Double Refraction," in *Report of the British Association for the Advancement of Science* (1862), 253–267 (in his obituary notice of Stokes, in *Scientific Papers*, V [New York, 1964], 180, Rayleigh held that Stokes had treated MacCullagh's theory unjustly). The foregoing represent the main sources for the determination of MacCullagh's place in the historical development of physical optics.

Don F. Moyer

MACCULLOCH, JOHN (*b.* Guernsey, Channel Islands, 6 October 1773; *d.* Poltair, Cornwall, England, 20 August 1835), *geology, chemistry.*

Macculloch was descended from a Scottish family, the Maccullochs of Ardwell, Galloway. His father, James Macculloch, after being engaged in business in Brittany, retired to Cornwall, where Macculloch attended school. He entered Edinburgh University to study medicine and graduated M.D. in 1793. His interest in geology arose during his stay in Edinburgh, possibly because he attended the lectures of John Walker, professor of natural history, who included geology and mineralogy in his course.

Macculloch's first appointment was as assistant surgeon to the Royal Regiment of Artillery, a branch of the army controlled by the Master General and Board of Ordnance. In 1803 he was appointed chemist to the Board of Ordnance. From 1807 he also practiced at Blackheath, near London. He became a member of the newly formed Geological Society of London in 1808, and its president in 1816.

In 1811 he gave up his medical practice and thereafter was employed by the Board of Ordnance in tasks requiring a knowledge of geology. He spent the summers of 1811 to 1813 in Scotland, investigating what rocks could be used safely in mills for grinding gunpowder. In 1814 the Board of Ordnance appointed him geologist to the Trigonometrical Survey then being carried out in Scotland. This involved two tasks. One was to choose sites geologically suitable for setting up the zenith sector used to determine the meridians; the other was to select a mountain geologically suitable for determining the earth's density (a previous determination made by the astronomer Maskelyne had in 1774 proved unsatisfactory for geological reasons).

While engaged in this work Macculloch used his spare time to record additional geological information with the intention of constructing ultimately a geological map of the whole of Scotland; and he tried to persuade his employers that official support for the preparation of such a map would be in the national interest. His employment in Scotland was terminated in 1820. During this period Macculloch also lectured in chemistry during the winter at the Royal Military Academy at Woolwich and later held a similar post at the East India Company's College at Addiscombe, where he also taught geology. He was elected a fellow of the Royal Society in 1820.

In the next few years Macculloch spent some time in completing for the Board of Ordnance reports on the work he had carried out in Scotland. In 1826 his post as chemist was abolished. He was then informed that for the purpose of completing a geological map of Scotland he could transfer his services to the Treasury, which would pay him a salary and expenses.

From 1826 until about 1832 Macculloch visited Scotland every summer to complete his geological survey. The question of whether his geological map should be published caused some delay, but eventually the Treasury sanctioned publication. Macculloch died as a result of a carriage accident in 1835; the map was published, posthumously, in 1836. He married a Miss White in 1835.

Macculloch was the author of a large number of books and papers of geological, mineralogical, and chemical interest. His most important work was that carried out in Scotland. Here his knowledge of mineralogy and chemistry proved invaluable in studying the igneous and metamorphic rocks, which occur over about three-quarters of the country. He was an acute observer of geological phenomena, and the value of his published work was often enhanced by his careful sketches of the field relations between the different types of rock.

In some respects Macculloch was conservative in outlook. He was unwilling to accept new geological ideas unless supported by evidence acceptable to him. He examined the lithological and mineralogical characteristics of the sediments carefully, and accepted that their contained fossils yielded information about the physical conditions existing at the time they were formed. He believed, however, that contemporaneous knowledge of fossil forms and their distribution was inadequate. He felt that their use as stratigraphical indexes and for correlation was unreliable and likely to cause confusion, even though by about 1820 they were quite widely used for this purpose. Contemptuous of stratigraphical paleontologists, Macculloch once described them as "namby pamby cockleologists and formation men." Nevertheless, in *A Description of the Western Isles of Scotland* (1819), he recorded the occurrence of Jurassic fossils in the islands of Skye and Raasay, and remarked on their similarity to those found in Jurassic rocks in Somerset and Gloucestershire in England.

Macculloch also found for the first time organic remains (*Serpulites maccullochii*) in the Cambrian quartzite in the northwest highlands of Scotland. Though he did not recognize the significance of this discovery, it later proved of great importance in working out the complicated overthrust rock succession in this area.

His *Description of the Western Isles of Scotland* was Macculloch's most important book. The numerous islands described, large and small, many of which had not previously been examined by geologists, contain rocks ranging in age from Precambrian to Tertiary, including many igneous rocks. His descriptions of the igneous rocks and the sketches and maps in the accompanying atlas promoted a true understanding of the nature and origin of igneous rocks at a time when the mistaken views of Werner on their origin had not yet been eradicated.

Macculloch's *Geological Classification of Rocks . . . Comprising the Elements of Geology* (London, 1821) describes his system of classifying rocks. He divided rocks into two main classes: Primary and Secondary. He subdivided Primary into unstratified (granite) and stratified (mainly gneisses and schists, with some sediments). The secondary included all the younger sediments and a few "unstratified" rocks, which his descriptions clearly indicate were igneous intrusions, although he did not specify them as such. Macculloch also described in detail the occurrences of basalt lava ("trap"), the remains of the extensive Tertiary basalt lava plateau off the west of Scotland. Although he realized they were analogous in appearance and composition to recent lava flows, he seemed unwilling to commit himself to the view that the "trap" rocks were true lava flows. He regarded the problem as an obscure one.

Macculloch's rock classification was based on that of Werner—as a matter of convenience because it was still widely used—but this classification did not imply acceptance of Werner's ideas about the origin of rocks. He omitted Werner's "transition" group, regarding it as an unjustifiable complication.

Macculloch's last geological book was his *System of Geology, With a Theory of the Earth* (London, 1831). His "theory of the earth" was largely Huttonian in concept, modified by the suggestion—perhaps derived from Cuvier—that a succession of revolutions might have brought about the extinction of some forms of life. An unusual feature of the work is the inclusion, in an appendix, of advice on the qualifications required of a geologist; on describing geological observations; on the instruments required; and on constructing geological maps. Much of this information is still useful.

Macculloch's geological map of Scotland, on a scale of four miles to the inch, was the first large-scale geological map of the country. It differentiated eighteen rock types and was largely, if not entirely, based on his own observations. It was a remarkable achievement for one man, especially considering the difficulties of access to the remoter parts of the country and the inaccuracies of the only topographical map available. Though gradually superseded by other maps during the nineteenth century, some of the areas he described had still not received detailed examination even as late as the mid-twentieth century.

Macculloch's observations greatly advanced general knowledge of the varied rock formations in Scotland, especially that of the igneous rocks. Though his work is now largely forgotten, it was appreciated in his time by Lyell, who made several references to it in both his *Principles of Geology* and his *Elements of Geology*. In an obituary notice, Lyell recorded "that as an original observer Macculloch yields to no other geologist of our time, and he is perhaps unrivalled in the wide range of subjects on which he displayed great talent and profound knowledge," and he added that he had "received more instruction from his labours in geology than from those of any living writer."

BIBLIOGRAPHY

I. ORIGINAL WORKS. Macculloch's geological works include *A Description of the Western Isles of Scotland . . . Comprising an Account of Their Geological Structure*, 3 vols., with an atlas of plates and maps in one vol. (London, 1819); *A Geological Classification of Rocks . . . Comprising the Elements of Geology* (London, 1821); *A System of Geology, With a Theory of the Earth*, 2 vols. (London, 1831); *A Geological Map of Scotland by Dr. Macculloch, F.R.S. &c.*, published by order of the Lords of the Treasury by S. Arrowsmith (London, 1836). There are four different issues of the map, with different titles, dates, and publishers. The geology is the same in all issues. None is dated 1836, but independent evidence establishes this as the date on which the map was first published. *Memoirs to His Majesty's Treasury Respecting a Geological Survey of Scotland* (London, 1836), published posthumously, accompanied the geological map.

A complete list of Macculloch's scientific papers is in the Royal Society *Catalogue of Scientific Papers* (1800–1863), IV (1870), 153–155. Of the total of seventy-nine, forty-six are of geological or mineralogical interest. The earlier (1811–1824) are in the *Transactions of the Geological Society of London;* later papers (1819–1830) are mainly in the *Quarterly Journal of Literature, Science and the Arts*. Macculloch published several books on subjects other than geology which are listed in the usual reference works.

II. SECONDARY LITERATURE. There is an obituary notice by Lyell in *Proceedings of the Geological Society*, **2** (1838), 358–359; see also T. G. Bonney, *Dictionary of National Biography*, XII, 461–463; V. A. Eyles, "John Macculloch, F.R.S., and His Geological Map: an Account of the First Geological Survey of Scotland," in *Annals of Science*, **2** (1937), 114–129; and "Macculloch's Geological Map of Scotland: an Additional Note," *ibid.*, **4** (1939), 107.

V. A. EYLES

MACELLAMA. See **Mashallah.**

ERNST MACH (*b.* Chirlitz-Turas near Brno, Moravia [now Chrlice-Tuřany, Czechoslovakia], 18 February 1838; *d.* Vaterstetten, near Haar, Germany, 19 February 1916), *physics, physiology, psychology.*

Ernst Mach spent his entire life, until the final three years, in the Austro-Hungarian empire. It was a time of increasing national self-determination, affirmations of linguistic identity among the non-Germanic peoples, and physical and intellectual emancipation of various ethnic groups. His parents provided an environment which nurtured unrestrained, critical, and stubborn scientific inquisitiveness and skepticism.

Mach was the first of three children of Johann Mach and Josephine Lanhaus. His father had received an excellent classical education that included two years' study in the philosophy faculty of the University of Prague. In 1840 he settled with his family on a farm in Untersiebenbrunn, near Vienna. Johann Mach was a practical and inventive idealist, a stark individualist who, with the collaboration of members of his more or less secluded household, distributed his efforts between improving methods of silkworm cultivation, tending his orchard, observing behavior in animals and children, reading Greek and Latin classics, and private tutoring. Mach's mother was raised within the tradition of a family engaged in law and medicine. She was a woman of tender character and artistic disposition, absorbed in instilling in her children a love of music and poetry.

Except for a year at the Benedictine Gymnasium in Seitenstetten, Mach was instructed at home by his father until he was fourteen. In 1848, at age ten, a year that remained fixed in Mach's memory as one dominated by parental sympathies for the Hungarian revolution against the Hapsburg monarchy, he entered the first class of the Gymnasium. He enjoyed the geography lessons, but he was so turned away by the grammar of ancient languages and pious aphorisms that his religious mentor recommended a handicraft or career in business for the untalented student. Accordingly, Mach's father again assumed the supervision of the boy's education. The mornings were taken up with studies in the prescribed Gymnasium subjects—Greek and Latin grammar and literature, history, algebra, and geometry—liberally punctuated with observational and experimental interludes in house, garden, and woods. The afternoons were devoted in part to manual labor in the father's agricultural ventures and in part to an apprenticeship with a cabinetmaker.

Mach entered the sixth class of the public Piarist Gymnasium in Kremsier (Kroměříž) in 1853 at the age of fifteen. The religious exercises displeased him but were more than compensated for by his enthusiasm for the way in which the natural sciences were presented. In later years Mach at various times paid special tribute to his teacher of natural history and geography, F. X. Wessely. Apparently Wessely presented Lamarck's theory of evolution (before Darwin's *Origin of Species* was published) so convincingly that Mach never managed to rid himself of an underlying evolutionary epistemology grounded in an appeal to the inheritance of acquired characteristics.

After five years of mathematics, physics, and philosophy at the University of Vienna, in 1860 Mach received the doctorate with a dissertation on electrical discharge and induction. While at the university, working as a *Privatdozent* in the laboratory of his teacher, Andreas von Ettingshausen (Doppler's successor in the chair of physics), Mach carried out a series of experimental investigations designed to provide theoretical support for Doppler's controversial law relating changes of musical pitch and optical frequency to the relative motion of signal and receiver. Furthermore, he presented a number of papers to the Academy of Sciences in Vienna in which he employed the idea of intermolecular vibrations to account for gaseous spectra. "Molecular functions" was the term Mach used to explain the phenomena of resonance in mechanically vibrating systems, the behavior of fluids in relation to density and viscosity, capillary phenomena, and the principal radii of curvature of liquid surfaces.

Although Mach derived most of his income in Vienna from popular scientific lectures on optics, musical acoustics, and psychophysics, he also presented formal university lectures on the principles of mechanics and designed a special course in physics for medical students. The latter formed the basis for his *Compendium der Physik für Mediciner* (1863). This work notably demonstrated that early in his career Mach had adopted a thoroughly mechanistic interpretation of natural phenomena and accepted the atomic-molecular theory and the kinetic theory of gases without reservation—at least as a working model and hypothesis.[1] In this instance Mach simply was following the philosophically atomistic account of physics that was then in fashion among physicists. Even so, in the preface and at the end of the *Compendium*, Mach explicitly discusses the inadequacies of the atomic theory. He remarks there that whatever metaphysical conception of matter may be put forward in future, the results obtained according to the atomic theory will always be capable of being translated into another conception—just as formulas in polar coordinates may be expressed in rectangular coordinates.

Before leaving Vienna Mach's scientific interests had begun to shift from physics to the physiology and psychology of sensation and to the new discipline of psychophysics. In the reflections on his life that appear in his *Leitgedanken* (1910) Mach says that he was pressed into the field of the psychology of sensation because he lacked the means for physical investigations. "Here, where I could observe my sensations, and against their environmental circumstances, I attained, as I believe, a natural *Weltanschauung* freed from speculative, metaphysical ingredients."[2] When Mach began to explore psychophysical problems associated with vision, audition, and variations in blood pressure, he gradually but surely reached the conclusion that his mechanistic and atomistic approach had netted him very little. This reorientation of Mach's scientific attention is manifestly visible in the "Vorträge über Psychophysik" that appeared toward the end of 1863. For him this shift represented an escape from metaphysical questions, a retreat from mechanism and physical reductionism, and the intentional avoidance of hypotheses. He wrote: "For the value of a hypothesis consists mainly herein, that by a kind of *regula falsi* it always leads closer and closer to the truth."[3] By 1863 the atomic hypothesis was already for Mach a kind of *regula falsi.*

In 1864 Mach accepted a full professorship in mathematics at the University of Graz. Having received no promise of an institute or resources for scientific equipment, he used his own funds to secure the necessary research facilities. By 1866 he was given the title of professor of physics, although he continued to pursue physiological and psychophysical problems on aural accommodation, the sense of time, and spatial vision. Most important at this juncture was Mach's discovery of what later came to be known as Mach's bands, a phenomenon that relates the physiological effect of spatially distributed light stimuli to visual perception. A Mach band is observed when a spatial distribution of light results in a sharp change in illumination at some point. A negative change corresponds to a band brighter than its surroundings in the region of sharp change. A positive change corresponds to a band darker than its surroundings in the region of sharp change. This phenomenon, a physiological effect that has no physical basis, was the subject of five papers published between 1865 and 1868. A final paper appeared in 1906. The effect was essentially rediscovered in the 1950's and since has been the subject of considerable investigation.[4]

In 1867 Mach married Ludovica Marussig in Graz. He had already accepted the professorship of experimental physics at Charles University in Prague. He produced most of his important work during the twenty-eight years he spent in that chair and saw the publication of over a hundred scientific papers. Although he was deeply involved throughout in the theoretical reformation of his views on mechanics and thermodynamics, his journal publications chiefly reflect his experimental endeavors.

Prominent among the psychophysical investigations during the Prague years are Mach's studies on the changes of kinesthetic sensation and equilibrium associated with physical movement, acceleration, and change of orientation in the human body. To these may be added a steady stream of research papers in which Mach resolutely continued to probe problems already initiated in Vienna and Graz: experiments with spatially distributed retinal stimuli (Mach bands), monocular stereoscopy, the anatomy and function of the organs of auditory perception, and aural accommodation.

Within the discipline of physics proper Mach's output was no less impressive. The more conventional physical studies carried out in Prague conspicuously include a great variety of optical experiments connected with refraction, interference, polarization, and spectra. He investigated the wave motion associated with mechanical, electrical, and optical phenomena and notably clarified the longitudinal-wave propulsion characteristics of stretched and unstretched glass rods and quartz rods. He also studied the mechanical effects resulting from spark discharge within solids and on surfaces.

Between 1873 and 1893 Mach and various collaborators, including his son Ludwig, devised and perfected optical and photographic techniques to study sound waves and the wave propulsion and gas dynamics of projectiles, meteorites, explosions, and gas jets. Stimulated by the remarks of the Belgian artillerist Henri Melsens, in 1881 Mach undertook to study the flight of projectiles by means of photographic techniques that he had already devised for other experiments in his Prague laboratory.[5] His celebrated 1887 paper on supersonics was published jointly with P. Salcher of the Marine Academy of Fiume (now Rijeka, Yugoslavia) in the *Sitzungsberichte* of the Academy of Sciences in Vienna.[6] The experiments described in this classic paper were carried out in Fiume with the support of the Royal Austrian Navy. In this paper, the angle α, which the shock wave surrounding the envelope of an advancing gas cone makes with the direction of its motion, was shown to be related to the velocity of sound ν and the velocity of the projectile ω as $\sin\alpha = \nu/\omega$ when $\omega > \nu$. After 1907, following the work of Ludwig Prandtl at the Kaiser Wilhelm Institut für Strömungsforschung in Göttingen, the angle α was called the Mach angle.[7]

Recognizing that the value of ω/ν (the ratio of the speed of an object to the speed of sound in the undisturbed medium in which the object is traveling) was becoming increasingly significant in aerodynamics for high-speed projectile studies, J. Ackeret in his inaugural lecture in 1929 as *Privatdozent* at the Eidgenössische Technische Hochschule, Zürich, suggested the term "Mach number" for this ratio.[8] The Mach number was introduced into the literature in English by the late 1930's and since the end of World War II has taken on considerable importance in theoretical and fluid dynamics. The work that has come to be most closely associated with Mach represents, in all probability, contributions that Mach would have conceived to be almost inconsequential when compared with his criticisms of classical mechanics and his experimental innovations in psychophysiology.

Mach also published while in Prague numerous popular scientific lectures and essays of historical and educational import and major treatises and monographs on conservation of energy (1872), spectral and stroboscopic investigations of musical tones (1873), the theory of the sensation of motion (1875), a critical history of mechanics (1883), and a volume on the analysis of sensations (1886). From 1882 to 1884 he was rector of the university during the difficult days when it separated into a German and a Czech faculty. In 1887 Mach and Johann Odstrčil published, with the help of a number of collaborators, the first of a series of physics textbooks, which ran to some twenty editions and were used with varying success in Germany and Austria for about four decades. Similarly, demonstration apparatus designed in Mach's laboratory was in use in Prague, Vienna, and Leipzig.

In 1895 Mach moved to the University of Vienna to assume a teaching position in philosophy, with the title professor of the history and theory of the inductive sciences. His *Popular Scientific Lectures*, a later edition of which was dedicated to William James, was first published in English in 1895. His *Principien der Wärmelehre*, dedicated to J. B. Stallo, appeared in 1896. In 1897 Mach suffered a stroke which left the right side of his body paralyzed. After a period of recuperation, he resumed lecturing and writing. He officially retired from his professorship in 1901, the year of his appointment to the upper chamber of the Austrian parliament. His *Erkenntnis und Irrtum* appeared in 1905 and his *Space and Geometry* in 1906.

In 1913 Mach moved with his wife to the country home of his son Ludwig in Vaterstetten, Germany. His *Kultur und Mechanik* was published in 1915, the year before he died. Mach's *Die Principien der physikalischen Optik* was published posthumously in 1921.

A physicist by training, Mach wished to be recognized as such. All the same, he was deeply engaged for most of his life in investigating problems in physiology, psychology, and the history and philosophy of science. An overall examination of his lifework reveals that he was an inventive experimentalist, an acute and imaginative critic of scientific theory, and was unusually sensitive to the importance of formulating problems in areas where physics, physiology, and psychology intersect. Mach's broad fascination with the phenomenal world of his immediate environment was leavened with an unusual and solicitous curiosity about nature in its pristine form. He recognized a strong desire for self-enlightenment and realized that he wanted to be a physicist, but a physicist unconstrained by the conventional barriers of the specialists with whom he came in contact.

In Einstein's obituary for Mach in 1916 we read: "The unmediated pleasure of seeing and understanding, Spinoza's *amor dei intellectualis,* was so strongly predominant in him that to a ripe old age he peered into the world with the inquisitive eyes of a carefree child taking delight in the understanding of relationships."[9] Einstein believed that even in those cases in which Mach's scientific inquiries were not founded on new principles, his work at all times displayed "extraordinary experimental talent." Even where Mach's philosophy intruded upon his science his colleagues nevertheless praised his intuition and skill in scientific research. In 1927 Wilhelm Ostwald wrote: "So clear and calculated a thinker as Ernst Mach was regarded as a visionary *[Phantast],* and it was not conceivable that a man who understood how to produce such good experimental work would want to practice nonsense *[Allotria]* which was so suspicious philosophically."[10] Arnold Sommerfeld spoke of Mach as a brilliant experimentalist but a peculiar theoretician who, in seeking to embrace the "physiological" and "psychical" in his physics, had to relegate the "physical" to a less pretentious level than physicists were accustomed to expect from a colleague.[11]

Fundamental among Mach's theoretical contributions and reflections were his explanations of visual, aural, and kinesthetic sensation; his views on mechanics, thermodynamics, optics, and molecular spectroscopy; and the ideas associated with his wave propulsion studies. Generally speaking, at least during his lifetime, Mach's theoretical opinions were judged to be amalgamated with too many hypercritical, obstreperous, and extrascientific remarks. In other instances, either the subject matter of his inquiries or his approach to the problems, or both, were too far afield to interest professional physicists. Mach's uncompromising rejection of the atomic theory, for example, far surpassed that of his contempories who, even when noncommittal about atoms and molecules as existential entities, never questioned its extraordinary usefulness as a powerful hypothesis. Mach became less denunciatory about the theory as he became older, but it is doubtful that he ever abandoned his dogmatic antiatomistic position, even after the discoveries at the turn of the century had furnished rather convincing evidence that he was wrong.

Mach looked upon the atomic theory and "the artificial hypothetical atoms and molecules of physics and chemistry" as "traditional intellectual implements" of the discipline. He wrote: "The value of these implements for their special, limited purposes is not one whit destroyed. As before, they remain economical ways of symbolizing experience. But we have as little right to expect from them, as from the symbols of algebra, more than we have put into them, and certainly not more enlightenment than from experience itself."[12]

The most severe critic of Mach's position on thermodynamics was Max Planck,[13] for whom the principle of conservation of energy was a true law of nature, a reality independent of man's existence. For Mach the principle took the position of a maxim or convention for organizing a large class of natural phenomena and was rooted in an anthropomorphic sanction related to a biologically determined economy of effort conducive to survival. "Energy" was for Mach no more than a plausible and powerful concept like force, space, or temperature. Thus it is wrong to include Mach among energeticists such as Wilhelm Ostwald and Georg Helm—as is often done.[14]

As for the second law of thermodynamics, Boltzmann felt that neither Planck nor Mach had penetrated its statistical essence. After 1900 Planck, but not Mach, accepted Boltzmann's interpretation. In fact, Mach continued to challenge, on philosophical grounds, the atomic-molecular-mechanical explanation of the laws of thermodynamics, preferring characteristically to give them a more simple and purely postulational status.[15] In his *Wärmelehre* (1896) Mach stated,

> The mechanical conception of the Second Law through the distinction between *ordered* and *unordered* motion, through the establishment of a parallel between the increase of entropy and the increase of unordered motion at the expense of ordered, seems quite *artificial.* If one realizes that a real analogy of the *entropy increase* in a purely mechanical system consisting of absolutely elastic atoms does not exist, one can hardly help thinking that a violation of the Second Law—and without the help of any demon—would have to be possible if such a mechanical system were the *real* foundation of thermal processes. Here I agree with

F. Wald completely, when he says: "In my opinion the roots of this (entropy) law lie much deeper, and if success were achieved in bringing about agreement between the molecular hypothesis and the entropy law, this would be fortunate for the hypothesis, but not for the entropy law."[16]

While thermodynamics at present enjoys autonomy as a discipline independent of mechanics, at the beginning of the century the statistical interpretation thoroughly permeated thermodynamics because of the great achievements and general reception of the atomic-molecular theory. It is thus even more difficult today to see how Mach could have been so blind to the steady advance of what became one of the most powerful theories for all of the physical sciences.

Mach's critical reflections on mechanics gave rise to a spirited discussion of the scientific, historical, and philosophical foundations of classical physics. His perspicuity in these matters is admirably demonstrated by the richness of the responses elicited from Hertz, Pearson, Boltzmann, Föppl, Love, Stallo, Clifford, Picard, Poincaré, Duhem, Seeliger, Vailati, and Jourdain. Even so, as Einstein wrote in 1916, referring to Mach's *Die Mechanik* of 1883, "There you will find set forth brilliantly ideas which by no means as yet have become the common property of physicists."[17]

According to Mach, Newton possessed the two characteristics necessary for greatness in a scientist: an imaginative grasp of the essential elements of experience of the world and the intellectual power of generalization. In *Die Mechanik*, Mach presented Newton's mechanical views in considerable detail. He offered generous praise for the clarity of presentation of the *Principia* along with some forceful arguments for the rational reformulation of some of its concepts. A case in point was Mach's conception of inertial mass, which he treated not as an intrinsic property of an object but as an entity specified by the dynamical coupling between the object and the rest of the universe. He proposed that Newton's conception of mass, as quantity of matter, be replaced by an "arbitrarily established definition," namely, that "all those bodies are bodies of equal mass, which, mutually acting on each other, produce in each other equal and opposite accelerations."[18] Such a definition, Mach contended, would render superfluous Newton's special enunciation of the principle of reaction, a conclusion favored for reasons of economy of thought.

Newton's views on absolute time, space, and motion were challenged in *Die Mechanik* on the grounds that they could in no way be related to experimental observations. Mach suggested the elimination of all propositions from which observables cannot be deduced and further proposed that the motions of bodies be considered relative to all observable matter in the universe at large: "When we reflect that we cannot abolish isolated bodies . . ., that is, cannot determine by experiment whether the part they play is fundamental or collateral, that hitherto they have been the sole and only competent means of the orientation of motions and of the descriptions of mechanical facts, then it will be found expedient provisionally to regard all motions as determined by these bodies."[19] On Mach's terms a body in an empty universe has no inertia. The inertia of a system is reduced to a functional relationship between the system and the rest of the universe, including the most distant parts of the interacting material system.

To call attention to this principle, Einstein, in a four-page paper on general relativity (1918), introduced the expression "Mach principle" *(Machsches Prinzip)* to emphasize a generalization of Mach's claim that the inertia of an isolated body can have no meaning;[20] that inertia must be reduced to the reciprocal action of bodies; that the inertial frame is determined by the mass distribution in the universe; and that the inertial force on a body is the interaction of distant matter on the body. What is entailed is the choice, even if it be provisional, of a material system that mathematically approximates absolute space.

Mach's critique of Newtonian mechanics, interpreted by Einstein within the context of Riemannian field theory, served as one of the strongest incentives for the development of Einstein's gravitational theory—although Einstein eventually discovered that the Mach principle did not hold for his new theory. What it did show was that the metric of space-time could be determined by the distribution of matter and energy; that the curvature of space, and from this the motion of bodies, could be determined from matter in space. Thus the motion of bodies was seen to be due to the influence of the surrounding masses (including the stars) and not, as Newton had supposed, to any effort on the part of bodies to maintain their direction of motion in absolute space.

Einstein hoped that he would be able to give mathematical expression to the Mach principle; admittedly he was not completely successful. He found support for the Mach principle by showing that his field equations turned out to have no solution, no metric, in matter-free space. He managed to assimilate the reconcilable aspect of Mach's principle into the general theory of relativity, based on the equivalence principle and the conception of covariance under general transformations of space-time coordinates, but he was not content to accept the qualification of restriction to finite space-time boundary conditions that is implied in the Mach principle. In fact, he discovered that he could write field equations that gave the correct solutions only by adding the so-called cosmological term.[21]

The scientific and philosophical literature on the reinterpretations and formulations of the Mach principle is huge.[22] Demonstrations of its compatibility with gravitational theory and other cosmological models continue to be proposed, reformulated, and rejected. Attempts to show where the principle does or does not make sense continue; as do objections to the principle itself or to parts of a particular formulation of it. Suffice it to say that there is little agreement on the mathematical formulation of the Mach principle. The expression is highly anachronistic as a collective term and has little in common with Mach's original conception of the problem. Still, the logical connection between Mach's views on mechanics and Einstein's theory of relativity, even now, is not a meaningless issue.

In his *Autobiographical Notes* (1946) Einstein saw "Mach's greatness in his incorruptible skepticism and independence." Although he there referred to Mach's epistemological position as one "which today appears to me to be essentially untenable," he also recognized that Mach had influenced his thought during his early years.[23] Einstein on several occasions mentioned that he had drawn inspiration from Mach's *Mechanik*, and it is therefore something of an enigma that Mach so categorically rejected the theory of relativity. In the preface to *Die Principien der physikalischen Optik*, written in 1913 but not published until 1921, Mach complained that he was "gradually becoming regarded as the forerunner of relativity" and that philosophers and physicists were carrying on a crusade against him: "I have repeatedly observed that I was merely an unprejudiced rambler, endowed with original ideas, in various fields of knowledge. I must, however, as assuredly disclaim to be a forerunner of the relativists as I withhold from the atomistic belief of the present day. The reason why, and the extent to which, I reject the present-day relativity theory, which I find to be growing more and more dogmatical, together with the particular reasons which have led me to such a view—the considerations based on the physiology of the senses, the theoretical ideas, and above all the conceptions resulting from my experiments—must remain to be treated in the sequel."[24] The sequel—apparently an attack on Einstein's theory of relativity—never materialized.

In considering Mach's contribution to the history of science, it is helpful to understand that all of his historicocritical writings lend strength to the thesis that the history of a scientific discipline, concept, or theory was for him a means to interpret and illuminate epistemological problems in the philosophy of science that puzzled him as a physicist.[25] Mach never undertook to write history with the intent merely of reconstituting the development of the subject as an end in itself. He wrote no history of areas in which he recognized no problem. By the same token he made no attempt to treat any aspect of the history of science in a systematic fashion. He was not predisposed to examine the nature of scientific revolutions or to relate science to its organizational, institutional, or sociocultural framework. He put forward no systematic philosophy of science.

As a young man Mach was given to considerable introspection and puzzlement about the nature of physics and the obligations of the physicist toward the new branches of science opening up alongside physics. By means of historical studies and the critical examination of the foundations of physics, he hoped to achieve insights into the direction that his own work might take. Mach's writings exerted considerable influence on late nineteenth- and early twentieth-century philosophic thought. The characteristic epistemological questions that Mach formulated were the outgrowth of his preoccupation with physics as seen from within the context of an obstinate inquisitiveness concerning the historical tradition inherited by physics.

Mach was first attracted to the history of science for what it might teach him about physics. In *Die Mechanik* he spelled this out, saying that

> . . . not only a knowledge of the ideas that have been accepted and cultivated by subsequent teachers is necessary for the historical understanding of a science, but also . . . the rejected and transient thoughts of the inquirers, nay even apparently erroneous notions, may be very important and very instructive. The historical investigation of the development of a science is most needful, lest the principles treasured up in it become a system of half-understood prescripts, or worse, a system of *prejudices*. Historical investigation not only promotes the understanding of that which now is, but also brings new possibilities before us, by showing that which exists to be in great measure *conventional* and *accidental*. From the higher point of view at which different paths of thought converge we may look about us with freer vision and discover routes before unknown.[26]

Mach's historical perspective was biased toward the investigation of specific problems and concepts. To this end, history was not conceived chronologically, biographically, or even topically. The documents that formed the basis of his historical analyses were almost exclusively limited to scientific writings. There is ample evidence that Mach had a good command of

Latin, Greek, French, Italian, and English and that he examined the primary sources. In addition he was familiar with a very broad secondary literature and, as his personal correspondence shows, exchanged ideas with an international group of scholars.

In 1863 Mach was already persuaded that his scientific curiosity might be nurtured by a serious study of the history of science, if only as an aid in the teaching of science. For he believed that students should not be expected to adopt propositions as self-evident which had developed over several thousand years. In his first major historical work, *Die Geschichte und die Wurzel des Satzes von der Erhaltung der Arbeit* (1872), he conjectured that there is "only one way to [scientific] enlightenment: historical studies!" The investigation of nature, he believed, should be founded on a special classical education that "consists in the knowledge of the historical development . . . of science." It was not the logical analysis of science but the history of science that would encourage the scientist to tackle problems without engendering an aversion to them. The scientist might follow two paths in order to become reconciled with reality: "Either one grows accustomed to the puzzles and they trouble one no more, or one learns to understand them with the help of history and to consider them calmly from that point of view."[27]

The youthful considerations included in the small (fifty-eight-page) *Geschichte und Wurzel* touched upon themes and points of view that continued to be explored by Mach for the rest of his life: the meaning and function of scientific theories, the epistemological importance of the physiology and psychology of sensation for the natural sciences in general, the principle of economy of thought, the inadequacy of Newtonian mechanics, the sterility of the atomic theory, and the sharp criticism of classical causality, physical reductionism, mechanism, materialism, and all forms of metaphysical speculation. In his historico-critical treatises on mechanics, optics, and heat theory, and in his *Analyse der Empfindungen* (1886) and *Erkenntnis und Irrtum* (1905), Mach returned to these same issues. It is instructive to try to identify some of the recurrent themes and cardinal epistemological questions raised in these works and then to discover how they are formulated within the context of Mach's historical outlook.

The pivotal questions that Mach's historicocritical analyses sought to elucidate may be formulated as follows: How did we inherit our current scientific concepts and theories? Why are they given to us in the way that we have become accustomed to accept them, rather than in some other form that may be logically more plausible or aesthetically more commendable? What factors can we identify as contributory to the adoption of preferred modes of reasoning and analogical adaptation from other domains? At any given time in history, what was considered to constitute evidence, verification, or a conclusive proof for a scientific theory?

With these questions, formulated and elucidated with examples drawn from the history of science, the essential framework of Mach's conception of the scientific enterprise takes on a fairly predictable form: Physics comprises only a small part of the body of scientific knowledge: its intellectual concepts and implements are too specialized. The natural sciences are inexpressibly richer than physics, and their history is inexhaustibly diversified. In any particular science that is being actively cultivated the ideas are generally undergoing metamorphosis. Since scientific concepts, laws, and theories are to some extent perennially obsolete, it is meaningless to harp on their anchored truth status. The conceptual creations of science, always tentative and at best incomplete, take on a configuration at any time that reveals the attendant historical circumstances and the convergence of interest and attention of those scientific investigators at work on the problems—now physicists, now physiologists, now psychologists.

Implied here is the notion that the nature and form given to a scientific construct depend in large part on the whims of history and on the environmentally conditioned process of cognitive organization employed by scientists. The concept of matter, for example, is seen to be bound, in its content, to a particular historically determined state of scientific development rather than to any fixed reality. Necessarily subjoined to this cognitive activity—biologically joined, in fact—is a mandate for abstraction and the exercise of economy of thought. Given that there exists no conceivable rock bottom of information potentially relevant for understanding the ways of nature, and recognizing the many alternatives for relating and organizing the facts, the scientific investigator has no choice but to fall back on drastic abstraction and generalization.

Mach, of course, understood that scientists commonly strive to achieve logical consistency, simplicity and elegance of formulation, inclusiveness, and range of applicability. Actually, scientific concepts, laws, and hypotheses, when studied within the context of their genesis, do not display so prominently the elegant logical traits and clear features that characterize the orthodox image of science. Instead, their most visible feature is their scientific extrinsicness. With the passage of time, that which is historically acquired comes to be philosophically affirmed. Mach underlined the need for scientists to acknowledge and neutralize the subtle way in which scientific constructs take on a status of philosophical necessity rather than historical contingency.

In *Die Mechanik*, for example, Mach sought to demonstrate how the historical study of the development of the principles of statics and dynamics—from historical contingency to philosophical necessity—could furnish the tactics for developing the analytical and methodological desiderata for interpreting and purifying the conceptual components of science. By means of the critical, historical, and psychological exposure of the roots of science, Mach intended, above all, to expose metaphysical obscurities, inherited anthropomorphisms, and ambiguities and to demonstrate the artificiality of the mechanical interpretation of the sciences. To this end he explored the function of instinctive knowledge, the role of memory, the correlation between the origin of an idea and its status in science at a given time, the psychology of discovery in contradistinction to the logic of discovery, and the authority that induction and deduction derive from experience.

The disclosure of the mental steps of proof adopted by different investigators is informative. For what reason, by what authority, did Stevin in his consideration of the principle of the inclined plane (based on the condition of equilibrium of an endless uniform chain on a triangular prism) argue from the general to the specific case? Why did Archimedes, on the other hand, in his treatment of the principle of the lever argue from the specific to the general case? What constituted proof is by no means unambiguously clear in these two cases.

Again, when Daniel Bernoulli purported to have demonstrated that the proposition for composition of forces is a geometrical truth independent of experience, Mach accepted this "proof" as no more than a dubious reduction of what is easier to observe from what is more obscure and difficult to observe, a demonstration in which it is possible to conceal the nature of the so-called proof. The significant feature in this case, as Mach saw it, was to search for the historical reasons why Bernoulli's demonstration might have been accepted as proof. Was it not the case, here as in other examples, that the investigator had built his demonstration on an appeal to instinctive knowledge that is supposed to be self-evident and that claims to be basically different from experiential knowledge? For Mach so-called instinctive knowledge, no matter how heuristically valuable, was assumed to contain important elements of experience hidden in its premises. He tried, wherever possible, to buttress this experience-oriented point of view with examples drawn from the history of science.

Mach's overall objective, narrow as it may seem, was to rid science of concepts that have no parallel in experience. This emphasis frequently has been characterized in the literature as the "Mach criterion," according to which only those propositions from which statements about observable phenomena can be deduced should be employed in theory. Scientific "proofs" that are not tied to experience, Mach felt, serve either as a cover-up for counterfeit rigor or as an appeal to the so-called higher authority of purely instinctive cognitions. In the case of a family of mechanical principles (exclusion of the *perpetuum mobile*, center of gravity, virtual work, *vis viva*, d'Alembert's principle, Gauss's principle of least constraint, Maupertuis's principle of least action, Hamilton's principle) Mach demonstrated to his own satisfaction that each of these can be derived from any other—given sufficient mathematical ingenuity and perseverance.

Clearly, Mach reasoned, these principles are all related in such a way that, if any one is taken to be true, then all the rest can be deduced by mathematical and logical reasoning. But what affords the rationale for accepting any specific principle as true in the first place? It can only be that the principle provides the right answers to problems in statics and dynamics. If any one of the above principles can be translated by deduction into any other, it nevertheless remains to be shown how the validity of at least one of them can be established. Mach's answer, given historically, was that the roots of the most primitive of such principles derive from experience. An example might be the experimental impossibility of designing a machine that furnishes unlimited quantities of mechanical work without the input of effort. Mach was dogmatic about this point—that the behavior of nature cannot be divined exclusively on the basis of so-called self-evident suppositions, unless the suppositions are themselves drawn from experience.

Mach analyzed at some length the history of the treatment of problems in statics (using both static and dynamic arguments) and in dynamics (using both dynamic and static arguments). Mach found that, historically considered, statics was prior to dynamics; whereas logically or conceptually understood, statics could be reduced to a limiting case of dynamics. Mach hoped in this way to discover by analysis what clear meaning, if any, was to be attached to the statement that one scientific principle is more fundamental than another. The wider generality of one principle compared to another, Mach felt, was not necessarily a sufficient warrant to accept it as more basic. Nor should the historical priority of discovery or enunciation of one principle over that of an alternative formulation dictate logical status. In short, "basic" and "fundamental" had for Mach no fixed significance when severed from the context of the scientific problem. The practical solution to understanding mechanical principles, Mach felt, was to learn by experience in scientific problem solving.[28]

Mach's emphasis in *Die Mechanik* is neatly summarized as follows:

> The most important result of our reflections is . . . *that precisely the apparently simplest mechanical theorems are of a very complicated nature; that they are founded on incomplete experiences, even on experiences that never can be fully completed; that in view of the tolerable stability of our environment they are, in fact, practically safeguarded to serve as the foundation of mathematical deduction; but that they by no means themselves can be regarded as mathematically established truths, but only as theorems that not only admit of constant control by experience but actually require it.*[29]

Mach's position on mechanics was an ambiguous one. On the one hand he played upon the crucial importance of a critical analysis of physical principles. On the other hand, he concluded that from a logical, economical, and practical point of view there is no persuasive rationale for defending either the historical legacy or the relevance of mechanics for the development of other scientific disciplines. While the study of the history of science abundantly demonstrates the prestigious place of mechanics, Mach conjectured that its applicability for other scientific domains was severely limited. While mechanics, Mach thought, offered little means of understanding heat theory, electricity, and light, its application within psychology was unqualifiably pernicious. The resolute thrust that characterized Mach's efforts to exploit experimental physiology and psychology—in part, by means of physical techniques—was indicative of the genuineness with which he was searching for novel problems and perspectives beyond the conventional physics of his day, without ever being able to relinquish his hold on professional physics or to put aside experimental physics as such.

Mach's philosophy of science was governed by the impulse to explore in depth the epistemological roots of science; but he was certain that this exploration could not be undertaken by examining the scientist's work without analyzing his behavior. Hence we recognize Mach's perseverance in demonstrating the relevance of the analysis of sensations and of the importance of psychology and physiology as a corrective to the prevailing mechanistic physicalism. Mach's philosophical point of view and, in particular, his emphasis on scientific laws as summary statements of observation led to his being identified with the positivists. In 1909 Lenin published, in Moscow, his *Materializm i empirio-kritisizm. Kriticheskie zametki ob odnoy reaktsionnoy filosofii* ("Materialism and Empirio-Criticism; Critical Comments on a Reactionary Philosophy"). In this work—much commented upon since then and essential in the development of Soviet Marxism—Lenin scathingly denounced the young Machists. Along with Mach, Avenarius, and Berkeley, they were accused of supporting a pernicious positivism and subjective idealism antithetical to Marxist dialectial materialism.

In addition to being associated with radical and empirical skepticism, positivism, subjective idealism, and the renunciation of truth in any transcendental sense, Mach is known to have supported a theory of knowledge based on the study of biological behavior. The ideas presented in Mach's *Beiträge zur Analyse der Empfindungen* (1886; rev. ed., 1900) sprang from the profound conviction, as explained in the preface, "that the foundations of science as a whole, and of physics in general, await their next greatest elucidations from the side of biology, and especially from the analysis of sensations."[30]

In the various revisions of this work, the 1901 edition of which was dedicated to Pearson, Mach explicitly clarified his views on how psychological, physiological, and physical experience, taken as sensations, provide man's sole source of knowledge concerning the world. The so-called elements of sensation that Mach had in mind—colors, sounds, temperatures, pressures, spaces, times, and so forth—are functionally connected in many ways and are associated with mind, feelings, and volition. Some of these associations (adaptations) stand forth more prominently than others, are engraved on our memory, and are expressed in language. Phenomena are represented as a complex of interrelated sensations embracing the object and subject of the sensations, the sensed, and the senser, such that any dualism between subject and object disappears.

The elements of sensation are not entirely independent, but for the sake of analysis can be separated into three groups: The external *(äussere)* elements, $A, B, C, \cdots$, characterize physical bodies and define the discipline of physics; the internal *(innere)* elements $K, L, M, \cdots$, characterize human bodies and define the discipline of physiology; the interior *(innerste)* elements, $\alpha, \beta, \gamma, \cdots$, characterize our ego (memory, volitions) and define the discipline of psychology. Since the properties of one and the same physical object, therefore, appear modified and conditioned by our own bodies, our sense organs under different circumstances produce different sensations and perceptions. The gulf that separates physical from psychological investigation resides not in a difference of subject matter but in different modes of investigation in the two domains. For example, to refer to that which is physical in an object being examined is but to designate one method of cognitive organization of which there are many. It is not the facts but the points of view that distinguish the disciplines.

From a psychophysiological point of view, the process of cognition includes the adaptation of thoughts to facts *(Beobachtung)* and thoughts to other thoughts *(Theorie)*. Mach asserted that the world of the percipient is composed entirely of sensations and not any illusive, protean, pseudophilosophical, Kantian *Ding-an-sich* that is supposed to remain after all of the qualities of a body are taken from it. Things, bodies, matter—the so-called objects of our experience—are thought symbols for combinations of elements and complexes of sensations and therefore are nothing apart from the totality of their attributes. A more or less complete description of phenomena should be accompanied by a more or less complete statement of the relevant functional relationships. To suit his purposes an investigator can analyze that which is given in experience, but he cannot by any manipulation of the empirically given data produce new information about how nature will behave in a new situation. He can, of course, predict how nature will behave in the empirically unknown, but it is always experience that serves as final judge about how nature in fact does behave in the unknown.

Mach's so-called principle of psychophysical parallelism as a guide in the investigation of sensations follows from the assumption that all sensations can be investigated from the point of view of physics, physiology, or psychology. There are no sensations uniquely physical or physiological or psychological. There are only sensations to be investigated from different points of view and according to the choice of connections between elements isolated for examination.

To strengthen his principle of psychophysical parallelism and to encourage the broadest possible interchange within the physical and biological field, Mach condoned the use of provisional teleological motives because of their heuristic value in formulating significant questions. More important for Mach's doctrine of how sensations become the basis for behavior or action are his views on biological evolution. He was convinced that all of science had its origins in the needs of life. The animal body is a relatively constant sum of touch and sight sensations, and all physical knowledge can only mentally represent and anticipate compounds of the elements of sensation.

Mach's argument can be briefly reconstructed as follows: The external phenomenal world is hopelessly heterogeneous and complex. Biological organisms cannot adapt to this environment, cannot survive, unless the given heterogeneous sensations as stimuli are converted into some focused form of homogeneous behavior that favors self-preservation. Over long periods of time biological organisms have developed a mechanism of accepting (scanning, rejecting) heterogeneous stimuli and correlating them and integrating them into homogeneous, purposeful behavior. How? By a sensor mechanism—tactile, aural, thermal—that excites but one nerve path; this in turn connects with the brain and gives rise to voluntary and, by habituation, involuntary behavior. The net process from unorganized stimuli to anatomically and symmetrically organized response (an example is the need for rapid locomotion) is a movement toward the biological organism's evolutionary acquisition of control over nature's stimuli.

It might be suggested that Mach's principle of economy of thought was also rooted in evolutionary arguments that championed those survival acts that relate to minimizing effort. Mach's views on *Denkökonomie* were first spelled out in his 1868 lecture on "The Form of Liquids" and in *Die Geschichte und die Wurzel* (1872). In the former he compared the principle of least surface for liquids to the miserly but intelligent mercantilist principle of a tailor working with the greatest saving of material, adding: "But why, tell me, should science be ashamed of such a principle? Is science . . . as a maximum or minimum problem . . . itself anything more than—a business? Is not its task to acquire with the least possible work, in the least possible time, with the least possible thought, the greatest possible part of eternal truth?"[31] Elsewhere, in his 1882 lecture "On the Economical Nature of Physical Inquiry" and in a special chapter in *Die Mechanik*, Mach reiterated and reemphasized the principle of *Denkökonomie* and its connection with mathematical reasoning, causality, teleology, evolution, and psychic phenomena.[32] The transparent omnipresence of this principle in Mach's writings led Peirce to say, "Dr. Ernst Mach who has one of the best faults a philosopher can have, that of riding his horse to death, does just this with his principle of Economy in science."[33]

As a physicist given to championing psychophysiology as the science of the future, Mach was not at all persuasive in his obscure position. Was he suggesting that concepts, methods, and distinctive points of view within the domain of psychophysiology might prove effective for investigating problems in the physical sciences that had hitherto been the exclusive concern of professionals in the discipline? Or should the fields of psychology and physiology and their attendant problems be opened up for investigation by the methods developed within the physical sciences? The former position, a rather radical one, calls for an extraordinarily sweeping psychophysiological reconstruction of physical principles—whatever that might entail. The second position, psychology as auxiliary to physics, either approximates physical reductionism, toward which Mach was ill-disposed, or else borders on psychophysics in the style of Fechner, whose approach considerably influenced Mach. In any case, Mach apparently chose to exploit either possibility, as the occasion demanded: psychophysiology to provide new directives for physics, and physics enriched through the focus of attention on psychophysiological problems. He seems to have leaned philosophically toward the former but to have organized his own research according to the latter.

While Mach made no claim to be a philosopher, much has been written about his scientific philosophy; he was more of a scientist's philosopher than a philosopher's philosopher. In 1886 he wrote in his *Analysis of Sensations*, "I make no pretension to the title of philosopher. I only seek to adopt in physics a point of view that need not be changed immediately on glancing over into the domain of another science; for, ultimately, all must form one whole."[34]

In the preface to *Erkenntnis und Irrtum* (1905) Mach remarked that the scientist, without claiming to be a philosopher, has a strong urge to satisfy an almost insatiable curiosity about the origins, structure, process, and conceptual roots of science. Averse to being labeled a philosopher, Mach nevertheless was not timid about raising and discussing methodological and epistemological questions. His progressive drift from physics to the philosophy of science via the history of science is nowhere more clearly evident than in *Erkenntnis und Irrtum*. The most comprehensive analytical and systematic treatment of Mach's philosophical views, the work is seldom read and almost never cited by historians of science. When his critics referred to his writings as *die Machsche Philosophie*, he replied:

> Above all there is no Machist philosophy. At most [there is] a scientific methodology and a psychology of knowledge [*Erkenntnispsychologie*]; and like all scientific theories both are provisional and imperfect efforts. I am not responsible for the philosophy which can be constructed from these with the help of extraneous ingredients. . . . The land of the transcendental is closed to me. And if I make the open confession that its inhabitants are not able at all to excite my curiosity, then you may estimate the wide abyss that exists between me and many philosophers. For this reason I already have declared explicitly that I am by no means a philosopher, but only a scientist. If nevertheless occasionally, and in a somewhat noisy way, I have been listed among the former then I am not responsible for this. Of course, I also do not want to be a scientist who blindly entrusts himself to the guidance of a single philosopher in the way that Molière's physician expected and demanded of his patients.[35]

In 1872 Mach wrote, "The object of science is the connection of phenomena; but the theories are like dry leaves which fall away when they have long ceased to be the lungs of the tree of science."[36] Science was for Mach primarily "the compendious representation of the actual." It is on this crucial point that his thought exhibits a gross miscalculation about the strength and reach, especially within recent science, of the highly artificial conceptual tools that scientists employ in their exploration of nature and even more of the exploration of that aspect of nature that is a product of the scientists' own creation.

Superimposed on Mach's ambitious program to relate the sciences to their historical and philosophical implications was his unflinching drive to unmask the theological, animistic, and metaphysical elements of science as he saw them. For this reason (among others that touch upon the long-range implications of his antirealistic, anticausal, antimechanistic, antimaterialistic, and antiatomistic world view) it was virtually impossible for contemporary scientists, philosophers, and historians to avoid taking sides *vis-à-vis* Mach. Positions of neutrality or indifference were rare during his lifetime—and still are. The deeper he delved into his historical studies the more he became engulfed by philosophical questions. Much as he eschewed the title of philosopher, his involvement with such questions ultimately forced him to carve out a scientific metaphysics of his own—although he would have denied that.

If at times Mach's historical position and emphasis seem superficial, erroneous, and misdirected, the caliber and import of his work can still be evaluated against the state of the art of historical and philosophical discussions in the sciences in his day.

As Einstein wrote, in the year of Mach's death, "I even believe that those who consider themselves to be adversaries of Mach scarcely know how much of Mach's outlook they have, so to speak, absorbed with their mothers' milk."[37]

NOTES

1. Erwin Hiebert, "The Genesis of Mach's Early Views on Atomism," in R. S. Cohen and R. J. Seeger, eds., *Ernst Mach: Physicist and Philosopher* (1970), 79–106.
2. "Die Leitgedanken . . .," in *Scientia*, **8** (1910), 234.
3. "Aus Dr. Mach's Vorträgen über Psychophysik," in *Österreichische Zeitschrift für praktische Heilkunde*, **9** (1863), 366.
4. Ratliff, *Mach Bands: Quantitative Studies on Neural Networks in the Retina* (1965).
5. Cf. Mach's 1897 lecture: "On Some Phenomena Attending the Flight of Projectiles," in *Popular Scientific Lectures* (1943).
6. E. Mach and P. Salcher, "Photographische Fixirung der durch Projectile in der Luft eingeleiteten Vorgänge," in *Sitzungsberichte der Akademie der Wissenschaften in Wien*, **95** (1887), 764–780. This paper and four others are conveniently reproduced in Joachim Thiele, ed., *Ernst Mach. Arbeiten über Erscheinungen an fliegenden Projectilen* (Hamburg, 1966).
7. Ludwig Prandtl, "Neue Untersuchungen über die strömende Bewegung der Gase und Dämpfe," in *Physikalische Zeitschrift*, **8** (1907), 23–30.
8. J. Ackeret, "Der Luftwiderstand bei sehr grossen Geschwindigkeiten," in *Schweizerische Bauzeitung*, **94** (1929), 179–183.
9. Albert Einstein, "Ernst Mach," in *Physikalische Zeitschrift*, **17** (1916), 101–104.
10. Wilhelm Ostwald, *Lebenslinien, Eine Selbstbiographie*, II (Berlin, 1927), 171.
11. Arnold Sommerfeld, "Nekrolog auf Ernst Mach," in *Jahrbuch der bayerischen Akademie der Wissenschaften* (1917), 58–67.
12. *Beiträge* (1886), pp. 142–143; English ed. (1959), p. 311.
13. For a discussion and evaluation of the differences in philosophical outlook between Mach and Planck, see Erwin Hiebert (1968).
14. Cf. Erwin Hiebert, "The Energetics Controversy and the New Thermodynamics," in Duane H. D. Roller, ed., *Perspectives in the History of Science and Technology* (Norman, Okla., 1971), 67–86.
15. Cf. Hiebert, *The Conception of Thermodynamics in the Scientific Thought of Mach and Planck* (Freiburg im Breisgau, 1968), pp. 104–107.
16. *Wärmelehre* (1896), p. 364.
17. Einstein, *loc. cit.* (1916), p. 102.
18. *Die Mechanik* (1889), p. 203; English ed. (1960), p. 266.
19. *Ibid.*, pp. 215–216 and 283 resp.
20. Albert Einstein, "Prinzipielles zur allgemeinen Relativitätstheorie," in *Annalen der Physik*, 4th ser., **55** (1918), 241–244.
21. See *Dictionary of Scientific Biography*, IV, 312–333.
22. Cf. Hong-Yee Chiu and W. F. Hoffmann, *Gravitation and Relativity* (New York, 1964).
23. P. A. Schilpp, ed., *Albert Einstein: Philosopher-Scientist* (Evanston, Ill., 1949), p. 21.
24. *Die Principien der physikalischen Optik* (1921), pp. viii–ix; English ed. (1926), p. vii–viii. See also Joseph Petzoldt, "Das Verhältnis der Machschen Gedankenwelt zur Relativitätstheorie," in the 8th ed. (1921) of *Die Mechanik;* and Ludwig Mach's pref. to the 9th ed. (1933).
25. Cf. Erwin Hiebert, "Mach's Philosophical Use of the History of Science," in Roger H. Stuewer, ed., *Historical and Philosophical Perspectives of Science* (Minneapolis, 1970), pp. 184–203.
26. *Die Mechanik* (1889), pp. 237–238; English ed. (1960), pp. 316–317.
27. *Geschichte und Wurzel* (1872), pp. 1–4; English ed. (1911), pp. 15–18.
28. For Mach's conception of the role of physical experiments and thought experiments in scientific problem solving, see Erwin Hiebert, "Mach's Conception of Thought Experiments in the Natural Sciences," in *S. Sambursky Symposium* (in press).
29. *Die Mechanik* (1889), pp. 221–222; English ed. (1960), pp. 289–290.
30. *Beiträge* (1886), p. v; English ed. (1959), pp. xxxv–xxxvi.
31. *Popular Scientific Lectures* (1943), p. 16.
32. *Ibid.* (1943), pp. 186–213; and "Die Oekonomie der Wissenschaft," in *Die Mechanik* (1889), pp. 452–466; English ed. (1960), pp. 577–595.
33. *Collected Papers of Charles Sanders Peirce*, C. Hartschorne and P. Weiss, eds., I (Cambridge, 1931), par. 122.
34. *Beiträge* (1886), p. 21; English ed. (1959), p. 30.
35. *Erkenntnis und Irrtum* (1905), pp. vii–viii.
36. *Geschichte und Wurzel* (1872), p. 46; English ed. (1911), p. 74.
37. Albert Einstein, "Ernst Mach," in *Physikalische Zeitschrift*, **17** (1916), 102.

BIBLIOGRAPHY

I. ORIGINAL WORKS. There is no collected edition of Mach's works. The Royal Society *Catalogue of Scientific Papers* lists 91 papers to the year 1900; Poggendorff is useful but incomplete. The most complete Mach bibliography, for both original works and secondary literature is Joachim Thiele, "Ernst Mach—Bibliographie," in *Centaurus*, **8** (1963), 189–237. The section on original works in this bibliography (without the list of secondary works) was revised and updated in Thiele (see below).

The library of the Ernst-Mach-Institut der Fraunhofer-Gesellschaft zur Förderung der angewandten Forschung, Freiburg im Breisgau, owns the largest collection of correspondence and other Mach documents: about 2,500 letters (mostly addressed to Mach) dated 1860–1916, some MSS, and other personalia and information by and about family members. The institute also houses a large but far from complete collection of Mach's publications and related secondary literature. Other major repositories for Mach correspondence and documents are the Morris Library of Southern Illinois University, Carbondale, Ill.; the Burndy Library, Norwalk, Conn.; the Akademie-Archiv of the Deutsche Akademie der Wissenschaften zu Berlin; and the University Library of the Technische Universität, Berlin-Charlottenburg.

The following chronological list of Mach's published books does not duplicate separately published pamphlets and lectures contained in those volumes.

1. *Compendium der Physik für Mediciner* (Vienna, 1863).

2. *Einleitung in die Helmholtz'sche Musiktheorie. Populär für Musiker dargestellt* (Graz, 1866).

3. *Die Geschichte und die Wurzel des Satzes von der Erhaltung der Arbeit* (Prague, 1872), includes Mach's "Ueber die Definition der Masse," in *Repertorium für physikalische Technik*, **4** (1868), 355–359. 2nd ed. (Leipzig, 1909) unchanged except for added "Bemerkungen zum zweiten Abdruck," pp. 59–60. English trans. and annotations by Philip E. B. Jourdain (Chicago–London), 1911.

4. *Optisch-akustische Versuche. Die spectrale und stroboskopische Untersuchung tönender Körper* (Prague, 1873).

5. *Grundlinien der Lehre von den Bewegungsempfindungen* (Leipzig, 1875); facs. ed. (Amsterdam, 1967).

6. *Die Mechanik in ihrer Entwickelung historisch-kri-*

tisch dargestellt (Leipzig, 1883). The 8th ed. (1921) includes Joseph Petzoldt, "Das Verhältnis der Machschen Gedankenwelt zur Relativitätstheorie." The 9th ed. (1933) contains a pref. by Ludwig Mach; facs. (Darmstadt, 1963). English trans. of 9th German ed. by Thomas J. McCormack (Chicago–La Salle, Ill., 1942). 6th American ed., with new intro. by Karl Menger (La Salle, Ill., 1960).

7. *Beiträge zur Analyse der Empfindungen* (Jena, 1886); 2nd ed., rev. and enl., *Die Analyse der Empfindungen und das Verhältnis des Physischen zum Psychischen* (Jena, 1900); 9th ed., 1922. English trans. of the 1st German ed. by C. M. Williams (Chicago, 1897); rev. and supp. from the 5th German ed. by Sidney Waterlow (Chicago, 1914). Paperback ed., with new intro. by Thomas S. Szasz (New York, 1959).

8. *Popular Scientific Lectures*, Thomas J. McCormack, trans. (La Salle, Ill., 1895); 5th ed., 1943. *Populärwissenschaftliche Vorlesungen* (Leipzig, 1896), 5th ed., 1923.

9. *Die Principien der Wärmelehre. Historisch-kritisch entwickelt* (Leipzig, 1896); 4th ed., with added index of persons, 1923. No known translations to date.

10. *Erkenntnis und Irrtum. Skizzen zur Psychologie der Forschung* (Leipzig, 1905); 2nd ed., 1906. The 3rd (1917) through 5th (1926) eds. contain Friedrich Jodl, "Ernst Mach und seine neueste Arbeit: 'Erkenntnis und Irrtum,' " from *Neue Freie Presse*, 24 Sept. 1905. English trans. in process for *Vienna Circle Collection* to be published by D. Reidel, with intro. and commentary by E. Hiebert.

11. *Space and Geometry in the Light of Physiological, Psychological and Physical Inquiry*, repr. from *Monist* (La Salle, Ill., 1906; 1960).

12. *Kultur und Mechanik* (Stuttgart, 1915).

13. *Die Leitgedanken meiner naturwissenschaftlichen Erkenntnislehre und ihre Aufnahme durch die Zeitgenossen. Sinnliche Elemente und naturwissenschaftliche Begriffe* (Leipzig, 1919), a pamphlet repr. of two essays originally published in *Scientia*, **8** (1910), 227–240; and *Pflügers Archiv für die gesamte Physiologie des Menschen und der Tiere*, **136** (1910), 263–274.

14. *Die Principien der physikalischen Optik. Historisch und erkenntnis-psychologisch entwickelt* (Leipzig, 1921). English trans. by John S. Anderson and A. F. A. Young (London, 1926; London–New York, 1956).

15. *Grundriss der Naturlehre für die unteren Classen der Mittelschulen* (Prague, 1887), written with Johann Odstrčil. Reissued (1887–1913) by Mach, K. Habart, and G. Wenzel in various eds. specifically designed for secondary school instruction in Austria and Germany.

16. *Grundriss der Physik für die höheren Schulen des Deutschen Reiches* (Leipzig, 1890), edited with the help of F. Harbordt and M. Fischer.

17. *Leitfaden der Physik für Studierende* (Prague–Vienna–Leipzig, 1891), written with G. Jaumann. 2nd ed., 1891.

II. SECONDARY LITERATURE.

1. [Maria Mach], *Erinnerungen einer Erzieherin. Nach Aufzeichnungen von ***, mit einem Vorwort herausgegeben von Professor Ernst Mach*, 2nd ed. (Vienna, 1913). Maria Mach was Ernst Mach's sister.

2. Friedrich Herneck, "Ueber eine unveröffentliche Selbstbiographie Ernst Machs," in *Wissenschaftliche Zeitschrift der Humboldt-Universität Berlin*, Math.-nat. Reihe, **6** (1956–1957), 209–220, from a thirteen-page MS of 1913 in the archives of the Deutsche Akademie der Wissenschaften zu Berlin.

3. Joachim Thiele, *Die Bedeutung Ernst Machs für die Wende von der klassischen zur modernen Physik. Ein Beitrag zur vergleichenden Geschichte wissenschaftstheoretischer Systeme* (Hamburg, 1959), an unpublished doctoral diss.

4. Theodor Ackermann Antiquariat, *Bibliothek Ernst Mach*, 2 pts. (Munich, 1959–1960), a list of some 3,700 works from Mach's working library.

5. K. D. Heller, *Ernst Mach. Wegbereiter der modernen Physik. Mit ausgewählten Kapiteln aus seinem Werk* (Vienna–New York, 1964).

6. Floyd Ratliff, *Mach Bands: Quantitative Studies on Neural Networks in the Retina* (San Francisco, 1965), contains in English trans. Mach's six papers on this subject (1865–1906).

7. Frank Kerkhof, ed., *Symposium zum Anlass des 50. Todestages von Ernst Mach* (Freiburg im Breisgau, 1966).

8. Erwin Hiebert, *The Conception of Thermodynamics in the Scientific Thought of Mach and Planck* (Freiburg im Breisgau, 1968).

9. J. Hintikka, ed., "A Symposium on Ernst Mach," in *Synthese*, **18** (1968), 132–301.

10. Joachim Thiele, ed., *Ernst Mach Abhandlungen: Die Geschichte und die Wurzel des Satzes von der Erhaltung der Arbeit (1872). Zur Geschichte des Arbeitbegriffes (1873). Kultur und Mechanik (1915)* (Amsterdam, 1969), a facs. ed. with foreword and bibliography of Mach's works.

11. Robert S. Cohen and R. J. Seeger, eds., *Ernst Mach: Physicist and Philosopher*, Boston Studies in the Philosophy of Science, no. 6 (Dordrecht, 1970).

12. J. Bradley, *Mach's Philosophy of Science* (London, 1971).

13. John T. Blackmore, *Ernst Mach. His Work, Life, and Influence* (Berkeley, Calif., 1972).

ERWIN N. HIEBERT

MACHEBOEUF, MICHEL (*b.* Châtel-Guyon, France, 19 October 1900; *d.* Paris, France, 20 August 1953), *biochemistry.*

Born to a sturdy Auvergnese family and son of Elie Macheboeuf, a distinguished physician, Macheboeuf studied medicine at Clermont-Ferrand and at Paris. He soon entered the Pasteur Institute, working in Gabriel Bertrand's laboratory (he succeeded Bertrand as head of the department of biochemistry in 1942). His teaching abilities were recognized by professorships at the University of Bordeaux and at the Sorbonne. His early scientific work, as Rockefeller fellow, was done under S. P. L. Sørensen in Copenhagen. They jointly investigated the structure of egg albumin and, in 1928, demonstrated that it is a mixture of phosphoproteins to which phosphorus is strongly bound. Shortly afterward Macheboeuf made his fundamental discovery of the linkage between the lipides and the proteins of blood plasma: the protidolipidic "cenapse." Blood plasma contains large amounts of lipides, which, when isolated, are insoluble in water. In the blood plasma they form soluble cenapses or lipide-protein complexes. Lipide and protein molecules are then strongly bound by their homologous carbon chains—that is, by their water-repellent groups. These groups are somewhat mutually neutralized, and thus the lipide-protein cenapse becomes water-soluble. This concept is now universally accepted. It had immediate important clinical application: the lipide-albumin index, the amount of lipides that remain in aqueous solution with 100 grams of albumin. In normal blood this index is about 12, but it reaches 60 or more in cases of lipoidic nephrosis. Macheboeuf also showed that alteration of the normal amount of cenapses modifies the exchange of water and electrolytes between blood and tissues and thus is one of the major causes of edema.

Macheboeuf made a major contribution to the immunochemistry of the tubercle bacillus. He found in this bacillus different groups of important substances: a precipitatory factor which he identified as a specific phosphatidic acid, strong antigens, and toxic substances which cancel out the effect of the strong antigens because they inhibit the activity of the antibody-producing cells. Thus the essential difficulties of antituberculosis vaccine manufacture were precisely defined.

Macheboeuf was among the first to investigate the effect of ultrahigh pressures. He found that bacteria and viruses are permanently inactivated above 6,000 atmospheres, and enzymes resist up to 12,000, while bacterial spores are killed only above 22,000 atmospheres.

Macheboeuf headed an active laboratory that produced fruitful results in many areas: plasma lipoproteins, mitochondrial enzymes, and the effect of antibiotics upon bacterial metabolic enzymes. Thus Macheboeuf was a source of inspiration to biochemists, French and foreign alike, until his untimely death (of lung ailment). He married Simone Bezou; they had three daughters.

BIBLIOGRAPHY

I. Original Works. Macheboeuf's most important works include "Études sur les effets biologiques des ultra-pressions. Résistance des bactéries, des diastases et des toxines aux pressions très élevées," in *Comptes rendus hebdomadaires des séances de l'Académie des sciences*, **195** (1932), 1431, written with J. Basset; "Études chimiques sur le bacille tuberculeux. I. Essais préliminaires d'extraction et de fractionnement des substances lipoïdiques de corps bacillaires tués par la chaleur," in *Bulletin de la Société de chimie biologique*, **16** (1934), 355, and in *Annales de l'Institut Pasteur*, **52** (1934), 277, written with G. Levy, N. Fethke, J. Dieryck, and A. Bonnefoi; "Études sur les effets biologiques des ultra-pressions. Modification de la spécificité antigénique des sérums sous l'influence des pressions très élevées," in *Comptes rendus hebdomadaires des séances de l'Académie des sciences*, **200** (1935), 496, written with J. Basset and J. Perez; "Sur l'état des lipides et du cholestérol dans le sérum sanguin. Destruction de certaines cénapses lipidoprotéidiques et libération de leurs substances lipoïdiques par un savon," *ibid.*, **206** (1938), 860, written with F. Tayeau; *L'état des lipides dans la matière vivante des cénapses et leur importance biologique*, 7 vols., Hermann and co-editors (1937).

Subsequent works include "Nouvelles recherches sur la nature et la stabilité des liaisons unissant les lipides aux protéides dans le sérum sanguin," in *Bulletin de la Société de chimie biologique*, **23** (1941), 31, written with F. Tayeau; "Recherches sur les phosphoaminolipides du sérum sanguin. Nature des phospholipides liés aux albumines du sérum de Cheval à l'état de cénapses acido-précipitables," *ibid.*, **25** (1943), 358, written with J.-L. Delsal; "Action des savons à cation actif sur les protéines. I. Précipitation de la sérum-albumine par un cation gras. II. Précipitation des protéines du sérum par un cation gras," in *Annales de l'Institut Pasteur*, **74** (1948), 196, written with J. Polonovski; "Recherches biochimiques sur le mode d'action de la streptomycine dans le métabolisme d'une bactérie : *Clostridium sporogenes*," *ibid.*, **75** (1948), 242, written with F. Gros and S. Jaulen; "Action des antibiotiques sur le métabolisme protidique d'une bactérie : *Clostridium sporogenes*. I. Recherche sur le catabolisme des protéines et des peptides," *ibid.*, **75** (1948), 320, written with F. Gros and P. Lacaille; "II. Recherches sur le métabolisme des amino-acides," *ibid.*, 446, written with U. Rambech and F. Gros; "Études des cénapses lipoprotéiques par relargage et électrophorèse," in *Bulletin de la Société de chimie biologique*, **33** (1951), 998, written with P. Rebeyrotte; and "Some Aspects of the Influence of Hydrostatic Pressure on Reactions Catalysed by Enzymes," in *Journal of Colloid Science*, supp. 1 (1954), written with G.-P. Talwar and J. Basset.

II. Secondary Literature. See the notice by J. Polonovski, in *Bulletin de la Société de chimie biologique*, **35** (1953), 1279–1286.

A. M. Monnier

M'INTOSH, WILLIAM CARMICHAEL (*b.* St. Andrews, Scotland, 10 October 1838; *d.* St. Andrews, 1 April 1931), *marine zoology.*

William M'Intosh was the only son of Baillie John M'Intosh, a builder and magistrate in St. Andrews.

He credited his mother, Eliza Mitchell, with an early stimulus toward marine descriptive zoology. His youngest sister, Roberta, who married Albert C. L. Günther (keeper of zoology at the British Museum of Natural History), illustrated many of his works.

He was educated at St. Andrews (1853–1857) and Edinburgh, where he received his M.D. in 1860. He specialized in mental diseases and was superintendent of the Perthshire Asylum at Murthly from 1863 to 1882. During this time, the results of his marine zoological studies appeared regularly in the scientific literature.

Volume I of his *Monograph of the British Marine Annelids* (*The Nemerteans*, published in two parts [1873–1874]) brought him wide recognition; hence the Annelid collections from the *Porcupine* and *Challenger* expeditions were assigned to him for description. The *Monograph* was not continued until 1900, when volume II, *Polychaetes*, appeared; this work was concluded with volume IV in 1923.

M'Intosh assumed the chair of natural history at St. Andrews in 1882; he retired from it in his eightieth year. In 1883 he was appointed to conduct a royal commission scientific inquiry into the effects of beam-trawling on the Scottish sea fisheries. A knowledge of the early life histories of a variety of marine species was needed, and requisite facilities became available when he established the Gatty Laboratory (1884), the first marine laboratory in Great Britain, at St. Andrews. M'Intosh and his co-workers first described in detail the pelagic eggs and larvae of the major British marine food-fishes, a critical aspect of fishery biology.

M'Intosh received the Royal Medal in 1899, the Linnean Medal in 1924, and was president and active in the affairs of the Ray Society in London from 1913 through March 1931. He never married.

BIBLIOGRAPHY

Among M'Intosh's writings are: *Monograph of the British Marine Annelids*, 4 vols. (London, 1873–1923); "On the Annelida of the Porcupine Expeditions of 1869 and 1870," in *Transactions of the Zoological Society of London*, **9** (1876), 395–416; *Report on the Annelida polychaeta collected by H. M. S. Challenger during the years 1873–1876*, Challenger Expedition, Zool., XII (London, 1885); *Report on Cephalodiscus dodecalophus, M'Intosh, a New type of the Polyzoa, procured on the voyage of H. M. S. Challenger during the years 1873–1876*, Challenger Expedition, Zool., XX (London, 1887); with E. E. Prince, "On the Development and Life-Histories of the Teleostean Food- and Other Fishes," in *Transactions of the Royal Society of Edinburgh*, **35** (1890), 665–946; *The Gatty Marine Laboratory* . . . (Dundee, 1896); with A. T. Masterman, *The Life-Histories of the British Marine Food-Fishes* (London, 1897); and *The Resources of the Sea* . . . (London, 1899).

Obituary notices appeared in the 2 Apr. 1931 issues of *The New York Times* and the *Times* (London); W. T. C., "Prof. W. C. M'Intosh, F. R. S.," in *Nature*, **127** (2 May 1931), 673–674; W. T. C., "William Carmichael M'Intosh—1838–1931," in *Proceedings of the Royal Society*, **110B** (1932), xxiv–xxviii, with portrait. Also see *List of Works, Memoirs, and Papers* (Dundee, n.d.).

DANIEL MERRIMAN

MACLAURIN, COLIN (*b.* Kilmodan, Scotland, February 1698; *d.* Edinburgh, Scotland, 14 January 1746), *mathematics.*

Maclaurin was the youngest of the three sons of John Maclaurin, minister of the parish of Kilmodan and a man of profound learning. John, the eldest son, followed in his father's footsteps and became a noted divine. The father died when Maclaurin was only six weeks old and after the death of his mother nine years later, Maclaurin was cared for by an uncle, Daniel Maclaurin, a minister of Kilfinnan.

In 1709 Maclaurin entered the University of Glasgow where he read divinity for a year. At Glasgow he became acquainted with Robert Simson, professor of mathematics. Simson, who tried to revive the geometry of the ancients, particularly the *Elements* of Euclid, stimulated Maclaurin's interest in this aspect of mathematics.

In 1715 Maclaurin defended the thesis "On the Power of Gravity," for which he was awarded a master of arts degree. It led to his appointment, in 1717, as professor of mathematics at Marischal College, Aberdeen, although he was still only in his teens. This appointment marked the beginning of a brilliant mathematical career which was to continue without interruption until the end of his life.

In 1719 Maclaurin visited London, where he was well received in the scientific circles of the capital and where he met Newton. On a second visit he met and formed a lasting friendship with Martin Folkes, who became president of the Royal Society in 1741, Maclaurin was meanwhile actively working on his *Geometrica organica*, which was published in 1720 with Newton's imprimatur.

Geometrica organica, sive descriptio linearum curvarum universalis dealt with the general properties of conics and of the higher plane curves. It contained proofs of many important theorems which were to be found, without proof, in Newton's work, as well as a considerable number of others which Maclaurin had discovered while at the university. Following traditional geometrical methods, Maclaurin showed that the higher plane curves, the cubic and the quartic, could be described by the rotation of two angles about their vertices. Newton had shown that the conic sections might all be described by the rotation of two angles of fixed size about their vertices *S* and *C* as centers of rotation. If the point of intersection of two of the arms lie on a fixed straight line, the intersection of the other two arms will describe a conic section which will pass through *S* and *C*.

In 1722 Maclaurin left Scotland to serve as companion and tutor to the son of Lord Polwarth, plenipotentiary of Great Britain at Cambrai. They visited Paris, then went on to Lorraine, where Maclaurin, during a period of intense mathematical activity, wrote *On the Percussion of Bodies*. It won him the prize offered by the French Academy of Sciences in 1724.

In the same year, the sudden death of Maclaurin's pupil caused him to return to Aberdeen. As a result of three years' absence, however, his chair had been declared vacant. Maclaurin then moved to Edinburgh, where he acted as deputy for the elderly James Gregory. There is no doubt that Maclaurin owed his appointment to a strong recommendation from Newton, who wrote to him:

> I am very glad to hear that you have a prospect of being joined to Mr. James Gregory in the Professorship of the Mathematics at Edinburgh, not only because you are my friend, but principally because of your abilities, you being acquainted as well with the new improvements of mathematics as with the former state of these sciences. I heartily wish you good success and shall be very glad of hearing of your being elected.

In a further letter to the lord provost of Edinburgh, Newton wrote:

> I am glad to understand that Mr. Maclaurin is in good repute amongst you, for I think he deserves it very well: And to satisfy you that I do not flatter him, and also to encourage him to accept the place of assisting Mr. Gregory, in order to succeed him, I am ready (if you will please give me leave) to contribute twenty pounds per annum towards a provision for him till Mr. Gregory's place becomes void, if I live so long.

Maclaurin was appointed to the Edinburgh chair when it fell vacant. He assumed its duties in 1725, lecturing on a wide range of topics that included twelve books of Euclid, spherical trigonometry, the conics, the elements of fortification, astronomy, and perspective, as well as a careful exposition of Newton's *Principia*.

Maclaurin was elected a fellow of the Royal Society in 1719. He was also influential in persuading the members of the Edinburgh Society for Improving Medical Knowledge to widen its scope. The society's name was thus changed to the Philosophical Society and Maclaurin became one of its secretaries. (In 1783 the organization was granted incorporation by George III as the Royal Society of Edinburgh.) In 1733 Maclaurin married Anne, daughter of Walter Stewart, solicitor general for Scotland. Of their seven children, two sons and three daughters survived him.

Maclaurin's *Treatise of Fluxions* (1742) has been described as the earliest logical and systematic publication of the Newtonian methods. It stood as a model of rigor until the appearance of Cauchy's *Cours d'analyse* in 1821. Maclaurin, a zealous disciple of Newton, hoped to silence criticism of the latter's doctrine of "prime and ultimate ratios," which proved something of a stumbling block to even Newton's staunchest supporters. In the *Treatise* Maclaurin tried to provide a geometrical framework for the doctrine of fluxions; in this way he hoped to refute his critics, the most vociferous of whom was George Berkeley, Bishop of Cloyne. In 1734 Berkeley had published *The Analyst. A Letter Addressed to an Infidel Mathematician*, in which he derided Newton's conception of "prime and ultimate ratios." (The "infidel mathematician" himself was Halley, who had piloted the first edition of the *Principia* through the press.) Berkeley maintained that it was not possible to imagine a finite ratio between two evanescent quantities. "What are these fluxions?" he asked. "The velocities of evanescent increments? They are neither finite quantities, nor quantities infinitely small, nor yet nothing. May we not call them the ghosts of departed quantities?"

In the preface to the *Treatise*, Maclaurin gave his reasons for replying to Berkeley:

> A Letter published in the year 1734 under the Title of *The Analyst*, first gave occasion to the ensuing Treatise, and several Reasons concurred to induce me to write on the subject at so great length. The Author of that piece had represented the Method of Fluxions as founded on false Reasoning, and full of Mysteries. His objections seemed to have been occasioned by the concise manner in which the Elements of this method have been usually described, and their having been so much misunderstood by a Person of his abilities appeared to me to be a sufficient Proof that a fuller account of the grounds of this was required.

He took up the question of fluxions almost immediately and defended Newton's methods:

> In explaining the notion of Fluxions I have followed Sir Isaac Newton in the First Book imagining there can be no difficulty in conceiving Velocity wherever there is

> motion, nor do I think I have departed from his sense in the Second Book, and in both I have endeavoured to avoid several expressions which though convenient, might be liable to exceptions and perhaps occasion disputes. . . .
>
> There were some who disliked the making much use of infinites and infinitesimals in geometry. Of this number was Sir Isaac Newton (whose caution was almost as distinguishing a part of his character as his invention) especially after he saw that this liberty was growing to so great a height. In demonstrating the grounds of the method of fluxions he avoided them, establishing it in a way more agreeable to the strictness of geometry.

Maclaurin followed Newton in abandoning the view that variable quantities were made up of infinitesimal elements and in approaching the problem from kinematical considerations. Moreover, he consistently followed the Newtonian notation, although the Leibnizian notation was by this time well established on the Continent. Thus he wrote (in the *Treatise*, p. 738), "The fluxion of xy is $x\dot{y} + \dot{x}y$."

The *Treatise* is otherwise noteworthy for the solution of a great number of problems in geometry, statics, and the theory of attractions. It contains an elaborate discussion on infinite series, including Maclaurin's test for convergence, as well as a remarkable investigation of curves of quickest descent and various isoperimetrical problems. It describes his series for the expansion of a function of x, namely,

$$f(x) = f(0) + xf'(0) + \frac{x^2}{2!}f''(0) + \frac{x^3}{3!}f'''(0) + \cdots .$$

Maclaurin also elaborated many of the principles enunciated by Newton in the *Principia* in this work, including problems in applied geometry and physics, founded on the geometry of Euclid.

Maclaurin's discussion of the attraction of an ellipsoid on an internal point is particularly significant. His interest in this subject began in 1740 when he submitted an essay "On the Tides" (*De causa physica fluxus et refluxus maris*) for a prize offered by the Académie des Sciences. Maclaurin shared the award with Daniel Bernoulli and Euler; all three men based their work upon proposition 24 of the *Principia*, on the flux and reflux of the sea. Maclaurin's original essay was hastily assembled, but he developed his ideas much further in the *Treatise* (II, article 686). He showed that a homogeneous fluid mass revolving uniformly about an axis under the action of gravity must assume the form of an ellipsoid of revolution. Clairaut was so impressed with Maclaurin's exposition that in his *Théorie de la figure de la terre* (1743), he abandoned analytical techniques and attacked the problem of the shape of the earth by purely geometrical methods.

In the *Treatise* Maclaurin presented for the first time the correct theory for distinguishing between maximum and minimum values of a function; he further indicated the importance of this distinction in the theory of multiple points of curves. In Chapter IX of Volume I, article 238, "Of the Greatest and Least Ordinates, of the points of contrary flexion and reflexion of various kinds, and of other affections of curves that are defined by a common or by a fluxional equation," he wrote that "There are hardly any speculations in Geometry more useful or more entertaining than those which relate to the *Maximum* and *Minimum*."

Maclaurin's persistent defense of the Newtonian methods was not without harmful consequences for the progress of mathematics in Great Britain. National pride induced Englishmen to follow the geometrical methods which Newton had employed in the *Principia*, and to neglect the analytical methods which were being pursued with such conspicuous success on the Continent. As a result, English mathematicians came to think that the calculus was not really necessary. This unfortunate neglect persisted for a century or more. It was said that during the eighteenth century Maclaurin and Matthew Stewart, who succeeded him in the mathematical chair at Edinburgh, were the only prominent mathematicians in Great Britain. Writing toward the end of the century, J. Lalande, in his *Life of Condorcet*, maintained that in 1764 there was not a single first-rate analyst in the whole of England.

Maclaurin's advice was sought nevertheless on many topics, not all of them mathematical. He was a skilled experimentalist, and he devised a variety of mechanical appliances. He made valuable astronomical observations and did actuarial computations for the use of insurance societies. He also took an active part in improving maps of the Orkney and Shetland Islands, with a view to discovering a northeast polar passage from Greenland to the southern seas, and prepared an extensive memorial upon this subject for the government. (Since the government was at that time primarily interested in finding a northwest passage, the matter was dropped.)

When a Highland army marched upon Edinburgh in the uprising of 1745, Maclaurin wholeheartedly organized the defenses of the city. With tireless energy, he planned and supervised the hastily erected fortifications, and, indeed, drove himself to a state of exhaustion from which he never recovered. The city fell to the Jacobites and Maclaurin was forced to flee to England. He reached York and sought refuge with Thomas Herring, the archbishop. He returned to

Edinburgh once it became clear that the Jacobites were not going to occupy the city, but the rigors he had endured had very severely undermined his health. He died soon after, at the age of forty-eight. Only a few hours before his death he dictated the concluding passage of his work on Newton's philosophy, in which he affirmed his unwavering belief in a future life.

At the meeting of the university following Maclaurin's death, his friend, Alexander Munro, professor of anatomy at the University of Glasgow, paid tribute to him: "He was more nobly distinguished from the bulk of mankind by the qualities of the heart: his sincere love of God and men, his universal benevolence and unaffected piety together with a warmth and constancy in his friendship that was in a manner peculiar to himself."

BIBLIOGRAPHY

I. Original Works. Maclaurin's works are *Geometrica organica, sive descriptio linearum curvarum universalis* (London, 1720); *The Treatise of Fluxions*, 2 vols. (Edinburgh, 1742); and *A Treatise of Algebra* (1748), a somewhat elementary posthumous work on the application of algebra to geometry, to which is joined a Latin tract, "De linearum geometricarum proprietatibus generalibus" (1756), printed from a manuscript written and corrected in Maclaurin's own hand. *An Account of Sir Isaac Newton's Philosophical Discoveries* (London, 1748) has a prefatory memoir on Maclaurin, "An Account of the Life and Writings of the Author," by Patrick Murdoch.

Maclaurin's papers published in the *Philosophical Transactions of the Royal Society* are: "Of the Construction and Measure of Curves," no. 356 (1718); "A New Universal Method of Describing All Curves of Every Order by the Assistance of Angles and Right Lines," no. 359 (1719); "A Letter . . . to Martin Folkes Esq. Concerning Equations with Impossible Roots (May 1726)," no. 394; "A Second Letter . . . to Martin Folkes Concerning the Roots of Equations, With the Demonstration of Other Rules in Algebra," no. 408 (1729); "On the Description of Curve Lines With an Account of Further Improvements, and a Paper Dated at Nancy, 27 Nov. 1722," no. 439; and "An Abstract of What Has Been Printed Since the Year 1721, as a Supplement to a Treatise Concerning a Description of Curve Lines, Published in 1719, and to Which the Author Proposes to Add to That Supplement," **39**.

Further papers published in the *Philosophical Transactions* are "An Account of the Treatise of Fluxions," no. 467 and continued in no. 469; "A Rule for Finding the Meridional Parts to a Spheroid With the Same Exactness as a Sphere," no. 461 (1711); "A Letter From Mr. Colin Maclaurin . . . to Mr. John Machin Concerning the Description of Curve Lines. Communicated to the Royal Society December 21. 1732"; "An Observation of the Eclipse of the Sun, on February 18, 1737 Made at Edinburgh, in a Letter to Martin Folkes," **40**, 177; "Of the Basis of Cells Wherein Bees Deposit Their Honey," no. 471 (1743).

Maclaurin also left a large number of manuscripts and unfinished essays on a variety of subjects, mathematical and nonmathematical.

II. Secondary Literature. Further information on Maclaurin's life and work may be found in W. W. R. Ball, *A Short Account of the History of Mathematics* (1912), 359–363; Florian Cajori, *History of Mathematics* (1919); Moritz Cantor, *Vorlesungen über Geschichte der Mathematik* (1884–1908); J. P. Montucla, *Histoire des Mathématiques*, 4 vols. (1799–1802); H. W. Turnbull, "Colin Maclaurin," in *American Mathematical Monthly*, **54**, no. 6 (1947).

J. F. Scott

MACLEAN, JOHN (*b.* Glasgow, Scotland, 1 March 1771; *d.* Princeton, New Jersey, 17 February 1814), *chemistry.*

Maclean's father, John Maclean, a surgeon, and his mother, Agnes Lang Maclean, both died when he was young. He was raised by George Macintosh, the father of Charles Macintosh, the inventor of the waterproofed cloth named after him.

Maclean attended Glasgow Grammar School and entered the University of Glasgow when he was about thirteen. At the university, under the influence of Charles Macintosh, who was four years older, he joined the Chemical Society and presented several papers. In 1787 he left Glasgow to travel and study at Edinburgh, London, and Paris. At Paris he made contact with Lavoisier and Berthollet, who won him over to the antiphlogistic theory of the "new chemistry." He returned to Glasgow in 1790, and on 1 August 1791 he received his diploma authorizing him to practice surgery and pharmacy. The same day he became a member of the faculty of physicians and surgeons.

In April 1795 Maclean left Scotland for America, at least in part because of his sympathy for the United States. Benjamin Rush advised him to settle in Princeton, seat of the College of New Jersey. After delivering a course of lectures in Princeton during the summer of 1795, Maclean was chosen professor of chemistry and natural history at the college. He is recognized as the first professor of chemistry, outside of a medical school, in any American college. He was also appointed professor of mathematics and natural philosophy; such comprehensiveness was necessary since there was only one other professor, who also served as president.

Maclean's *Two Lectures on Combustion*, published at Philadelphia in 1797, was important in the overthrow of the phlogiston theory. Publication of the work led to a discussion by Maclean, Priestley, James Woodhouse, and Samuel Mitchill in the New York *Medical Repository*.

In his teaching at Princeton, Maclean divided chemistry into "dead" and "living" matter, foreshadowing the later division into organic and inorganic. His dead matter was further divided into "simple" and "compound" substances. Of the thirty-seven simple substances he discussed, some twenty are now recognized as elements, including oxygen, hydrogen, aluminum, tungsten, platinum. The rest would now be classified as compounds—lime, soda, potash, and a number of acids.

In 1802 Maclean met with Benjamin Silliman, who, although he was not yet qualified for that role, had been appointed professor of chemistry at Yale. Maclean prepared a reading list for him.

Maclean resigned from the College of New Jersey in 1812 and shortly after joined the faculty of the College of William and Mary in Williamsburg, Virginia. Because of ill health he stayed there only a little over a year. He returned to Princeton, where he died soon after.

Maclean married Phoebe Bainbridge, the sister of Commodore William Bainbridge, on 7 November 1798. They had two daughters and four sons. The eldest son, John, became the tenth president of Princeton.

BIBLIOGRAPHY

Maclean's published work was *Two Lectures on Combustion* (Philadelphia, 1797). The chief source of biographical information is a memoir by his son, John Maclean, *A Memoir of John Maclean, M.D., the First Professor of Chemistry*, printed privately (Princeton, 1876). See also William Foster, "John Maclean—Chemist," in *Science*, **60** (3 Oct. 1924), 306–308; John F. Fulton and Elizabeth H. Thomson, *Benjamin Silliman* (New York, 1947), and Theodore Hornberg, *Scientific Thought in the American Colleges 1638–1800* (Austin, 1945). Princeton University Library has several of Maclean's letters.

VERN L. BULLOUGH

MACLEAR, THOMAS (*b*. Newtown Stewart, Tyrone, Ireland, 17 March 1794; *d*. Mowbray, near Cape Town, South Africa, 14 July 1879), *astronomy*.

Defying his father's wish that he become a clergyman, Maclear studied medicine at Guy's and St. Thomas's hospitals in London. In 1815 he joined the staff of Bedford Infirmary, where his interest in astronomy was aroused by his friendship with Admiral William Henry Smyth. From 1823 he practiced medicine in Biggleswade, Bedfordshire, where he set up a small observatory in his garden; he was lent the Wollaston telescope by the Royal Astronomical Society to observe occultations of Aldebaran which he had himself calculated. In 1833 he was appointed Her Majesty's astronomer at the Cape of Good Hope, where he arrived on 5 January 1834, shortly before John Herschel.

Until his retirement, Maclear devoted himself to the work of the observatory and to matters of public welfare in the colony. From 1837 to 1847 he carried out the remeasurement and extension of Lacaille's arc as a basis for the survey of the colony, and this earned him the Lalande Prize in 1867 and a Royal Medal in 1869. In 1839–1840 and 1842–1848 he observed α Centauri to confirm the parallax derived by Thomas Henderson. He made numerous observations of comets and double stars, and furnished materials for successive Cape Catalogues. His observations of Mars during the opposition of 1862 were used by others in determinations of the sun's distance.

Maclear also promoted a wide range of applied sciences in the colony. He assembled meteorological, magnetic, and tidal data, and in 1860 he established time signals for Port Elizabeth and Simonstown. He sponsored the construction of lighthouses, promoted sanitary improvements, and took part in a commission on weights and measures. He was keenly interested in African exploration and was a friend of Livingstone and Stanley.

Maclear was knighted in 1860. He retired in 1870 and went blind in 1876, but he continued to interest himself in public affairs at the Cape until his death.

BIBLIOGRAPHY

I. ORIGINAL WORKS. Maclear's writings include *The Verification and Extension of Lacaille's Arc of the Meridian at the Cape of Good Hope*, 2 vols. (London, 1866), and numerous papers, mostly in the *Memoirs* and *Monthly Notices* of the Royal Astronomical Society. Lists of these papers are given in the Royal Society *Catalogue of Scientific Papers*, IV (1870), 166–168, and VIII (1879), 299–300; and in Poggendorff, III, 852–853.

II. SECONDARY LITERATURE. The most important memoir on Maclear's astronomical work at the Cape is ch. 7, "Les travaux de Maclear au cap de Bonne-Espérance," of "Tableau de l'astronomie dans l'hémisphère austral et dans l'Inde," by É. Mailly, in *Mémoires couronnés et mémoires publiés par l'Académie royale des sciences, des lettres et des beaux-arts de Belgique*, **23** (1873), 77–109. Extensive biographical sketches are by David Gill in

Monthly Notices of the Royal Astronomical Society, **40** (1880), 200–204, and by Agnes M. Clerke in *Dictionary of National Biography*, XII, 648–649.

MICHAEL A. HOSKIN

MACLEOD, JOHN JAMES RICKARD (*b.* Cluny, near Dunkeld, Perthshire, Scotland, 6 September 1876; *d.* Aberdeen, Scotland, 16 March 1935), *physiology*.

Macleod was a distinguished teacher, author, investigator, and administrator. He was the son of Rev. Robert Macleod, who shortly after the boy's birth was called to Aberdeen; Macleod received his education at the Aberdeen Grammar School and Aberdeen University. He studied medicine at Marischal College, and graduated with distinction as M.B., Ch.B in 1898, winning the Matthews Duncan and Fife Jamieson medals; he was also awarded the Anderson traveling scholarship and spent a year in the Physiology Institute at Leipzig, where he studied biochemistry under Siegfried and Burian. In 1900 Macleod joined the London Hospital Medical College as demonstrator in physiology under Sir Leonard Hill. He became lecturer in biochemistry at the school in 1902 and was also selected as Mackinnon research scholar of the Royal Society. He married Mary Watson McWalter in 1903; there were no children.

During his short stay in London Macleod published an account of experiments on intracranial circulation and on caisson disease, carried out in conjunction with Hill; he retained his interest in the problems of respiration throughout his life, publishing many papers on the control of respiration between 1902 and 1922. In the early years of the twentieth century Macleod continued his postgraduate studies at Cambridge, where he took his diploma in public health. In 1903 he published a text entitled *Practical Physiology*, and in the same year he was appointed professor of physiology at Western Reserve University, Cleveland (now Case Western Reserve University). Here he remained as a teacher and researcher for fifteen years. In 1918 he became professor of physiology at the University of Toronto and not long afterward published, with collaborators, a textbook of nearly 1,000 pages, *Physiology and Biochemistry in Modern Medicine*. This text, which reached its seventh edition the year before Macleod's death, was widely read and consulted. Its title reveals it as one of the last of such broad scope and, at the same time, as the precursor of other, more recent wide-ranging treatises appealing both to the special student and to the clinician.

In his first years at Western Reserve Macleod published a series of papers on the carbamates and one on purine metabolism. In 1907 there appeared the first of a long series of "Studies in Experimental Glycosuria" in the *American Journal of Physiology*, studies on the breakdown of liver glycogen, whether produced by piqûre, stimulation of the splanchnics, reflexly by asphyxia, or by injection of adrenaline. He conceived of the problem as one fundamentally involving the access of the diastatic enzyme to the stored glycogen. In 1913, nearly ten years before the discovery of insulin, Macleod wrote a book on diabetes and its pathological physiology, an expansion of lectures which he had delivered during the summer of 1912 at the University of London. Although he published on surgical shock (1918–1920) and other subjects, his first years in Toronto were devoted chiefly to studying the peculiarities of respiration in decerebrate animals and of the effects of anoxemia and of excess oxygen. Later, in 1921, he made a thorough examination of the control of the blood-sugar level in the normal and in the depancreatized animal and of the roles played by the liver, the muscles, and the pancreas in the metabolism of sugar.

His return to work on carbohydrate metabolism had been stimulated by the initial successes of Banting and Best. As J. B. Collip subsequently observed, Macleod had already attained, by the time of his arrival in Toronto (1918), "an outstanding position in the field of carbohydrate metabolism, and it was both appropriate and fortunate that the discovery of insulin should have been made in his laboratory." It was early in 1921 that Macleod agreed to receive Frederick G. Banting, a young surgeon, into his department to carry out investigations aimed at determining the true function of the pancreatic islets; C. H. Best, a member of the professor's senior class in physiology, was assigned to be Banting's assistant. Banting and Best began their research on 16 May. The general pattern of their work, following Banting's conception of how the islets might be freed from the acinar tissue of the gland and then extracted, was worked out with Macleod; but its first results were obtained in midsummer, when Macleod had gone to Scotland. On his return, he discontinued his work on anoxemia and turned all the resources of his laboratory to the new work. J. B. Collip joined the team; and usable preparations of insulin were ready, and were used with success, early in 1922.

Macleod was president of the American Physiological Society at the time of the discovery of insulin and had received many honors; now they multiplied. In 1923 Banting and Macleod shared the Nobel Prize; Banting divided his share with Best, and

Macleod divided with Collip. Macleod became fellow of the Royal Society in 1923 and was awarded honorary degrees by Toronto, Western Reserve, Aberdeen, and other universities. In 1928 he returned to Scotland as regius professor of physiology at Aberdeen. Productive work there, and at the Rowett Institute, continued to add to knowledge of carbohydrate metabolism. The metabolism of the decerebrate eviscerated animal, with special reference to the respiratory quotient, occupied much of his time until crippling arthritis put an end to his laboratory work; even then he continued to direct the activities of his department. Macleod's last important publication marked his return to an earlier problem, the nervous control of the glycogenic function of the liver.

BIBLIOGRAPHY

I. ORIGINAL WORKS. Macleod's writings include *Practical Physiology* (1903); "Studies in Experimental Glycosuria," in *American Journal of Physiology* (1907); "Teaching and Research Positions in Medical Science," in *Western Reserve University Bulletin*, **11**, no. 6 (1908), 129–150; and *Physiology and Biochemistry in Modern Medicine* (St. Louis, 1918), written with Roy G. Pearce and others.

II. SECONDARY LITERATURE. See C. H. Best, "The Late John James Rickard Macleod, M.B., Ch.B., LL.D., F.R.C.P.," in *Canadian Medical Association Journal*, n.s. **32** (1935), 556; J. B. C[ollip], "John James Rickard Macleod (1876–1935)," in *Biochemical Journal*, **29**, no. 1 (1935), 1253–1256; "John J. R. Macleod, F.R.S., D.Sc., M.D. Aberd., F.R.C.P. Lond.," in *Lancet*, no 228 (1935), 716–717; and L. G. Stevenson, *Sir Frederick Banting*, rev. ed. (Toronto, 1947).

LLOYD G. STEVENSON

MACLURE, WILLIAM (*b.* Ayr, Scotland, 27 October 1763; *d.* St. Angelo, Mexico, 23 March 1840), *geology.*

Maclure, the son of David and Ann Maclure, was educated by private tutors. He came to the United States in 1782 to establish "mercantile arrangements" and returned to London as a partner in the house of Miller, Hart and Company, where he soon amassed a fortune. He traveled back and forth between Europe and America, and spent several years traveling through Europe, including Scandinavia and Russia, observing geological features and seeking instruction from geologists in several countries. In Paris he met Count Volney, who discussed American geology with him.

Returning to the United States in 1796, Maclure examined the geology of the country as he traversed it from New England to the far southeast, "crossing the dividing lines of the principle formations in 15 or 20 different places." He published his report and colored map, "Observations on the Geology of the United States, Explanatory of a Geological Map" (in *Transactions of the American Philosophical Society*, **6** [1809], 411–428).

In 1815 Maclure visited France and met Lesueur the painter-naturalist, whom he persuaded to return to America with him as his personal cartographer-naturalist. They spent the winter of 1815–1816 examining the geology and natural history of the West Indies, then made repeated traverses of the Allegheny Mountains, collecting specimens and revising the geological map. From December 1817 to April 1818, Maclure organized an expedition to Georgia and Spanish Florida with Lesueur, Thomas Say, Titian Peale, and George Ord.

Maclure developed an estate in Spain and lived there from 1820 to 1824. In 1824, following a revolution and the confiscation of his property, he returned to America. Here he joined Robert Owen in his New Harmony venture, in which he invested $82,000. He visited Mexico in 1827 and later moved there.

Maclure's best known work is the *Observations on the Geology of the United States.* Although preceded by Johann D. Schöpf's work in German (1787) and Volney's in French (1803), Maclure's articles and book are the first connected account originally written in English on the geology of the United States. The expanded and revised work and map were published as a separate volume in 1817 and in the *Transactions of the American Philosophical Society* in 1818. In the text and map, Maclure divided the country into areas of "primitive rocks," "transition rocks," "floetz and secondary rocks," and "alluvial rocks." The primitive rocks are crystalline rocks in the area that extends from New England southwest to Alabama. The transition rocks occur in the folds of the Appalachians; the secondary rocks are the flat-lying strata west of the Appalachians. The alluvial rocks are found in the valley of the Mississippi and in the coastal plain from Cape Cod southeast to the mouth of the Mississippi River.

In 1817 Maclure published a report on his detailed examination of the West Indies (*Journal of the Academy of Natural Sciences of Philadelphia*, **1**, 134–149). In this report he discussed the uplifted coral limestone in some of the islands and the various kinds of volcanic materials, including volcanic breccia, which he interpreted as volcanic mud flows. His "Essay on the Formation of Rocks . . ." (1818) is important for his explanation of his terminology and for his theories of the origin of rocks. In a series of eleven papers (1819–1831) published in the *American*

Journal of Science, Maclure recorded observations on the geology of the United States and other parts of the world and expanded upon his ideas of rock origin and geological processes.

The short paper "Genealogy of the Earth–Geological Observations" (1838) is the most philosophical of his geological writings. In it he espoused Lamarckian evolution and clarified his belief in organic and inorganic development through a gradual series of minute changes, rather than by large jumps or catastrophes.

Maclure is usually regarded as a follower of Werner because he used that terminology, but he specifically disclaimed his belief in that school's theory of crystalline rock origin. He was convinced of the igneous origin of basalt and stated that sedimentary rocks were deposited in separate basins and were therefore not continuous in an "onionskin" fashion around the world. He was uncertain about the origin of primitive rocks, and thought that they might be diverse in origin. From his studies of various areas of the world, he had concluded that sedimentary rocks and lava flows alternated with each other. Thus he speculated that the primitive rocks may have originated as sediments and lava, later becoming "smelted, roasted and liquified," almost the concept of granitization in present-day terminology.

Maclure was convinced that basalt and related rocks developed from lava flows, but he did not discuss granites as intrusive rocks, nor did he mention the explanations of Hutton and Playfair. He urged more attention to the "primitive formations" because of their great age and the changes that they had undergone. He disclaimed support, however, of the aqueous theory of the origin of primitive rocks. "The geology of the United States . . . is a strong argument against the Wernerian system—all these theories have had their day, and are fast going out of fashion" (1822).

Maclure recognized the role of streams in the formation of valleys. Following S. L. Mitchill and many others, he thought that the region west of the Appalachian Mountains had once been the site of a great lake, which had drained out through the Hudson and broken out across the mountains in several other valleys. Maclure was one of the first men in America to notice and speculate on the origin of the granite erratics in the northern part of the country. He stated that the granite erratics between Lake Erie and the Ohio River had been transported southward by "large pieces of floating ice" in the lake that he believed had occupied the trans-Appalachian area (1823).

Maclure's work on metallic deposits and the connection between metals and volcanic regions are of interest. In writing about Mexico (1831), he proposed that "precious metals have been converted into vapor, that would penetrate through chinks that would not permit lava to pass." Thus he realized that lava was intrusive until it reached the surface. This explanation for veins is directly contrary to that of Werner, who supposed that veins had been formed by infilling of cracks from the surface.

Maclure's influence was not only great as an observer and writer on geology and as a maker of important maps, but also through his support of scientists and scientific organizations. One of the founders of the Academy of Natural Sciences of Philadelphia, he was president from 1817 to his death.

BIBLIOGRAPHY

I. Original Works. Maclure's map and *Observations* of 1817 have been produced in facsimile (New York, 1962). The 1809 map has been reproduced by Merrill (New Haven, 1924). His papers on his travels are in the early volumes of the *American Journal of Science* and the *Journal of the Academy of Natural Sciences of Philadelphia.* Among them are "Essay on the Formation of Rocks . . .," in *Journal of the Academy of Natural Sciences of Philadelphia*, **1** (1818), and published separately (Philadelphia, 1818; New Harmony, 1832).

The three-volume work *Opinions on Various Subjects Dedicated to the Industrious Producers* (New Harmony, 1831–1838) sets forth in detail his ideas on economic, political, and related areas; it contains little science except for one important paper, "Genealogy of the Earth—Geological Observations," *ibid.*, III, 175–178.

II. Secondary Literature. There is as yet no full biography of Maclure. S. G. Morton, *A Memoir of William Maclure, Esq.* (Philadelphia, 1841), with a portrait, reprinted in *American Journal of Science*, **47** (1844), 1–17, is the source of the most immediate information. A. E. Bestor, "Education and Reform at New Harmony: Correspondence of William Maclure and Marie Duclos Fretageot, 1820–1833," in *Indiana Historical Society*, **15**, no. 3 (1948), discusses Maclure's association with New Harmony and contains some of the letters from Europe and Mexico preserved at the Working Men's Institute at New Harmony; Charles Keyes, "William Maclure: Father of American Geology," in *Pan-American Geologist*, **44** (1925), 81–94, is a fulsome account with some information; G. P. Merrill, *First One Hundred Years of American Geology* (New Haven, 1924; New York, 1962), 31–37, 46–47, has summarized Maclure's geological contributions; and Jessie Poesch, *Titian Ramsey Peale, 1799–1885 . . .* (Philadelphia, 1961), gives information on Maclure's expedition to Georgia and Florida based on original correspondence.

Lesueur's connection with Maclure is given in some detail by R. W. G. Vail, *The American Sketchbooks of Charles Alexander Lesueur 1816–1837* (Worcester, 1938). A portrait of Maclure and many references to his association with Thomas Say are in H. B. Weiss and G. M. Zeigler, *Thomas Say, Early American Naturalist* (Springfield–Baltimore, 1931). William E. Wilson, *The Angel and the Serpent, the Story of New Harmony* (Bloomington, Indiana, 1964), gives extensive consideration to Maclure at New Harmony.

All of Maclure's geological papers are being prepared for publication in facsimile, edited and annotated by George W. White.

GEORGE W. WHITE

MACMAHON, PERCY ALEXANDER (*b.* Malta, 26 September 1854; *d.* Bognor Regis, England, 25 December 1929), *mathematics.*

MacMahon was the son of Brigadier General P. W. MacMahon. He entered the army in 1871, rising to the rank of major in 1889. He served as instructor in mathematics at the Royal Military Academy from 1882 to 1888, as assistant inspector at Woolwich arsenal to 1891, and as professor of physics at the Artillery college until his retirement in 1898. Afterward, he was deputy warden of standards at the Board of Trade from 1906 to 1920. A fellow of the Royal Society in 1890, he was president of the London Mathematical Society in 1894–1896, and president of the Royal Astronomical Society in 1917.

MacMahon was a master of classical algebra, who had remarkable insight into algebraic form and structure, together with a power of rapid and precise calculation. His early work dealt with invariants, following the studies of Cayley and Sylvester. He noticed that the partial differential equation for semi-invariants is fundamentally the same as that for general symmetric functions. MacMahon made use of the concept of generating functions, and of U. Hammond's symbolic calculus of differential operators in connection with symmetric functions. His power of calculation helped him in the work of tabulation and enumeration.

The study of symmetric functions led to MacMahon's interest in partitions and to the enumeration of Euler's Latin squares. His presidential address to the London Mathematical Society gave a survey of combinatorial analysis, and his two-volume treatise of 1915–1916 is a classic in this field. It identified and clarified the master theorems, and indicated a wealth of applications. An introductory version was published in 1920.

MacMahon's interest in repeating patterns and space-filling solids began in his childhood with observations of piles of shot found in his military environment. He revived this interest in latter years, writing a book on mathematical pastimes.

BIBLIOGRAPHY

I. ORIGINAL WORKS. The presidential address on combinatorial analysis is printed in the *Proceedings of the London Mathematical Society*, 1st ser., **28** (1897). Some ninety research papers are listed in the obituary notice by H. F. Baker, cited below. Works by MacMahon were *Combinatorial Analysis*, I–II (Cambridge, 1915–1916); *An Introduction to Combinatorial Analysis* (Cambridge, 1920); and *New Mathematical Pastimes* (Cambridge, 1921).

II. SECONDARY LITERATURE. The obituary notice by H. F. Baker in the *Journal of the London Mathematical Society*, **5** (1930), 305–318, gives a brief sketch of MacMahon's life and a substantial account and analysis of his mathematical work.

T. A. A. BROADBENT

MACMILLAN, WILLIAM DUNCAN (*b.* La Crosse, Wisconsin, 24 July 1871; *d.* St. Paul, Minnesota, 14 November 1948), *astronomy, mathematics.*

William Macmillan was the son of Duncan D. Macmillan and Mary Jean MacCrea. He attended Lake Forest College, the University of Virginia, and Fort Worth University, from which he received the B.A. in 1898. He then went to the University of Chicago, where he spent most of his working life. He took the M.A. there in 1906 and the Ph.D. in 1908. A pupil of F. R. Moulton, he became a research assistant, first in geology (1907–1908) and then in mathematics and astronomy (1908–1909). He held a succession of posts in astronomy at the university, becoming professor emeritus in 1936.

Macmillan's interests centered around cosmogony and related topics in applied mathematics. Probably his most widely known works were his textbooks of theoretical mechanics. He made a number of original contributions to potential theory, the theory of differential equations with periodic coefficients, and the theory of automorphic functions. He took an active part in the then-controversial discussions of the theory of relativity, contributing to *A Debate on the Theory of Relativity* (New York, 1927) with R. D. Carmichael, H. T. Davis, and others.

One of Macmillan's most influential pieces of work was an attempt to remove the supposed paradox of P. L. de Cheseaux (1744) and H. W. M. Olbers (1823), whereby, with the hypothesis of an infinite and uniform distribution of stars throughout space, the night sky would shine with a brightness corresponding to their average surface brightness. In 1918 and 1925 Macmillan proposed a form of continual material creation (*Astrophysical Journal*, **48** (1918), 35, and *Science*, **62** (1925), 63–72, 96–99, 121–127). His main concern was with the formation of the planets and stars. Among his so-called postulates he included two according to which the universe maintains a steady state, and another according to which the energy of a large region of the universe, supposedly unbounded, is conserved. He acknowledged that matter is converted to energy in stellar interiors, and explained away the De Cheseaux-Olbers paradox as a disappearance or dissipation of the radiation traversing empty space. (This radiation was to reappear in the form of hydrogen atoms.) Subsequently, R. A. Millikan, one of Macmillan's colleagues at Chicago, used his theory to account for the origin of cosmic rays, but by 1935 A. H. Compton proved that it could not account for the high energies of much cosmic radiation. The theory was then abandoned. It should be noted that, unlike more recent steady-state theories, Macmillan's identified a source from which the mass or energy of the created particle was drawn. He did not suggest creation *ex nihilo*.

BIBLIOGRAPHY

Macmillan collaborated with F. R. Moulton, C. S. Slichter, *et al.*, in *Contributions to Cosmogony and the Fundamental Problems of Geology. The Tidal and Other Problems* (Washington, D.C., 1909), and with Moulton, F. R. Longley, *et al.*, in *Periodic Orbits*, Carnegie Institution of Washington publ. no. 161 (Washington, D.C., 1920). His best-known work was *Theoretical Mechanics*, 3 vols. (New York, 1927–1936). On the background to Macmillan's writings on cosmogony, see J. D. North, *The Measure of the Universe* (Oxford, 1965), esp. 18, 198–199, and 260–261.

J. D. NORTH

MACQUER, PIERRE JOSEPH (*b.* Paris, France, 9 October 1718; *d.* Paris, 15 February 1784), *chemistry.*

Best known as the author of a widely read textbook and of the first chemical dictionary, Macquer made no lasting contributions to chemical theory and discovered few new substances. Nevertheless he was an influential member of the Paris scientific community and did much important but unpublished research behind the scenes as a government scientific adviser. The descendant of a Stuart supporter[1] who in 1688 had accompanied James II into exile, he was the elder son f Joseph Macquer and Marie-Anne Caillet. An early erest in science was encouraged by one of his ers, Charles Le Beau, a well-known historian; decided to devote himself to research. However, ts insisted on his first qualifying in a profession, and he chose medicine. After graduating from the Paris Faculty of Medicine in 1742 he practiced for only a few years, being at one time doctor for the poor of the parishes of St.-Nicolas and St.-Sauveur,[2] near his home in the rue St.-Sauveur. He studied chemistry under G. F. Rouelle and soon began to do research.

Rouelle had started lecturing at the Jardin du Roi in 1742, and his influence on French chemistry was yet to be felt. Since there were few experienced and able chemists competing for places in the Academy of Sciences, Macquer was elected in 1745, before reading his first published paper, on the solubility in alcohol of different oils. He was hoping to relate the changes that occurred in oils on rectification to the solubilities of various products, but his experiments and theoretical discussion contributed little to this difficult subject. He was on firmer ground in a later investigation of the solubilities of carefully dehydrated inorganic salts in alcohol (1766, 1773); his quantitative results were of value to chemists who used alcohol extraction in the analysis of residues from evaporated mineral waters.

The most important of Macquer's early researches was his study in 1746 and 1748 of white arsenic (arsenious oxide), which he found to react with niter (potassium nitrate) to form a previously unknown crystalline salt (potassium arsenate) that was quite different from the compound (potassium arsenite) obtained by dissolving white arsenic in potash. The new salt formed precipitates with metallic salts, and Macquer recognized that in them the metal was combined with the "arsenical part" of his compound. He did not, however, go so far as to assert that arsenic formed an acid although, by heating white arsenic with oil of vitriol (sulfuric acid), he had, in fact, prepared its anhydride as an "arsenical glass" that gave an acid solution in water. The true nature of arsenic acid and its salts was recognized in 1775 by C. W. Scheele.

Pastel and indigo were the only fast blue dyes used before 1749, when Macquer discovered a method of dyeing with Prussian blue, an insoluble substance used as an artists' pigment. It was prepared by adding potash that had been calcined with animal matter to a solution of alum and green vitriol (ferrous sulfate), and Macquer therefore boiled skeins of flax, cotton, silk, and wool with alum and green vitriol, and dipped them into a solution of the specially treated potash. The specimens were dyed a dull blue, which brightened on rinsing in dilute sulfuric acid and proved to be fast for wool and silk. Macquer then attempted to analyze Prussian blue (1752). Although he did not solve the problem of its constitution, he found that alum was not an essential constituent. Again, his research was later extended by Scheele. These researches on arsenic and Prussian blue were praised by Macquer's contemporaries, but it was through his books that he became widely known and influenced the development of chemistry.

The first textbook, *Élémens de chymie théorique*, appeared in 1749, when there was a real need for an up-to-date book by a French author; Nicolas Lemery's *Cours de chymie* had not been printed in Paris since 1730 (an edition revised by Baron was not published until 1756), and the anonymous *Nouveau cours de chymie* (generally attributed to J. B. Senac) had not been printed since 1737. The *Nouveau cours* included an account of the phlogiston theory, which was being taught by Rouelle in the 1740's, but it is unlikely to have been available to the many people whose interest in chemistry he was stimulating. Macquer wrote his *Élémens* for people with no previous chemical knowledge. He started with an account of the four elements, air, water, earth, and fire (of which phlogiston was a modification), briefly discussed affinity, and then considered compound substances in order of their increasing complexity, mineral, vegetable, and animal. He was careful not to mention any substance before acquainting the reader with it: for example, acids, alkalis, and salts were described before metals, so in the account of acids their power of dissolving metals was not mentioned. This "synthetic" treatment, as Macquer called it, seems to have been popular with beginners but was of little value in laboratory work, where an experimenter would normally be concerned with the analysis of naturally occurring substances that were compounds or mixtures. Macquer gave a thorough account of compound substances in *Élémens de chymie pratique* (1751), which was no mere laboratory manual. The two books were complementary and were reprinted together in 1756, with slight modifications. There were no further editions in France,[3] but the combined work enjoyed a great vogue in Britain, where Andrew Reid's 1758 translation was reprinted as late as 1777—by which time it was badly out of date.

Textbooks are generally written by teachers, but Macquer's teaching experience began only in 1752, when he was elected professor of pharmacy in the Paris Faculty of Medicine for a year.[4] His partnership with Antoine Baumé was more lasting. In 1757 they gave a joint course in chemistry that was attended by fifty or sixty people[5] and proved to be the first of sixteen annual winter courses. The outline in their *Plan d'un cours de chymie* . . . (1757) shows that they followed roughly the same order as in Macquer's *Élémens*, and they demonstrated 2,000 experiments. Their joint research included a study of one of the first specimens of platinum to reach France (1758); they discovered no new chemical properties but showed that small portions could be melted with the aid of a powerful burning lens.

By 1773, when the joint courses ceased, Macquer had started lecturing at the Jardin du Roi. L. C. Bourdelin had been professor of chemistry since 1743, but his health was poor and up to 1769 the lectures were often given by P. J. Malouin. Buffon, the intendant, then granted Macquer the reversion of the chair. He gave his first course in 1770 and, after Bourdelin's death in 1777, became the titular professor and continued to lecture every summer until 1783. The professor's lectures were always followed by experiments, performed and explained by the demonstrator. H. M. Rouelle (brother of Macquer's chemistry teacher) had been demonstrator since 1768, and Macquer seems to have left him to plan the course, for the vegetable and animal kingdoms were treated before the mineral kingdom, as in Rouelle's own private course. When A. L. Brongniart (1742–1804) succeeded Rouelle, Macquer revised his course and from 1779 taught the mineral kingdom first, as in his *Élémens* and in a private course given by Brongniart. Macquer occasionally introduced new material into his lectures but, surprisingly, said little about the chemistry of gases, explaining in an introductory lecture that he preferred not to deal with modern and controversial topics in a short course intended for beginners. His lectures were praised by Condorcet, who attended them, but his audiences were deprived of an up-to-date account of important new developments.

Macquer's principal book, the *Dictionnaire de chymie* (1766), was one of a series that included J. C. Valmont de Bomare's *Dictionnaire raisonné universel d'histoire naturelle* (1764) and an anonymous *Dictionnaire portatif des arts et métiers* (1766) compiled by Macquer's brother Philippe.[6] Like the latter, Macquer's work also appeared anonymously, because Macquer had been commissioned to write it hastily and was worried about his reputation.[7] Some articles to which he gave cross-references do not, in fact, appear, but he need not have worried, for the *Dictionnaire* was an immediate success. It was reprinted in 1766 and again in 1769, and Macquer soon started work on an enlarged and revised edition which was published—no longer anonymously—in 1778. There were several Swiss editions, and German, English, Danish, and Italian translations.

The *Dictionnaire* contained about 500 articles in alphabetical order, ranging in length from two-line definitions to long essays like that on "salt," which filled more than seventy pages. Most articles were revised in the second edition, and Macquer added some new ones, notably "Gas," a 168-page account of a topic that hardly existed when he was writing the first edition.

In the preface of 1766 Macquer confessed that when asked to write a chemical dictionary—for which there was no precedent—he had misgivings because he thought that all parts of chemistry were interdependent and also that the subject did not seem suited to alphabetical treatment. However, he had come to see certain advantages—the dictionary had taken the form of a collection of essays and therefore could be studied by each reader in the order he preferred; and, while listing topics in alphabetical order, Macquer had thought of some that might not otherwise have occurred to him. Cross-references were freely given in both editions, and the utility of the second was greatly increased by the inclusion of a subject index of more than 600 pages. In his 1778 preface Macquer stated clearly that the book was intended not for beginners but for those who had attended a course of lectures and had read a textbook, and he suggested the order in which such a student should read about 280 of the main articles. Essays on general theoretical topics were to be followed by those on the mineral kingdom, and then by articles on the vegetable and animal kingdoms—the ordering of the *Élémens* and of his later courses.

Like most of his contemporaries, Macquer accepted the four-element theory of matter and expounded on it in his books and lectures, but he constantly tried to reconcile it with new discoveries. He followed Boerhaave in believing that earths such as lime and silica were modifications of the pure elementary earth which, he thought, must be a hard, transparent, crystalline substance. In 1766 he identified it with diamond; but in 1772, in collaboration with Lavoisier and others, he showed that diamond was combustible, and so by 1778 he had concluded that pure rock crystal was the element. A metal was composed of a kind of earth (the calx) combined with phlogiston (which in 1749 he regarded as elementary fire in combination with some other principle) and possibly also with the mercurial earth of Becher and Stahl. But by 1766 he had serious doubts about the mercurial earth, for most calxes could be converted to metal by the addition of phlogiston alone, in the form of charcoal. He therefore concluded that a metal was composed only of calx [illegible]d phlogiston. It was difficult to expel the last traces [illegible]hlogiston by heating, and Macquer thought that [illegible] could be achieved all metals might be found to [illegible] the same earth, which would open up the [illegible] of transmutation of metals. Macquer also made the perceptive suggestion that mineral earths like lime and magnesia were of essentially the same nature as metallic calxes[8] and might one day be combined with phlogiston to form new metals. In 1773, with other chemists of the Academy, he used a large burning glass to heat magnesia and alumina with ivory black, a form of charcoal; this experiment may have been an attempt to combine them with phlogiston.

When the second edition of the *Dictionnaire* was written, Lavoisier was challenging the phlogiston theory, but in 1778 the antiphlogistic theory was still far from complete, and Macquer was able to suggest a compromise. In 1773 he had praised Lavoisier's early experiments on gases,[9] and he now agreed that vital air (oxygen) was absorbed during combustion and calcination. He believed this effect to be accompanied by the emission of phlogiston from the combustible or metal. He now identified phlogiston with light and still regarded it as a modification of elementary fire; he did not think of heat as a material substance but preferred the Baconian theory that it was a phenomenon caused by the motion of particles of matter. Lavoisier's discovery of the composition of water in 1783 greatly strengthened the antiphlogistic theory, and his calorimetric researches with Laplace tended to support the material theory of heat. In December 1783, when discussing C. L. Morozzo's experiments on calcination, Macquer said that on this subject Lavoisier had "an absolutely new idea, to which he has already given much probability by many fine experiments."[10] Soon afterward Macquer was reported to be preparing a new edition of the *Dictionnaire*, with some revised articles,[11] and it seems likely that he was very near to accepting the antiphlogistic theory just before his death.

The compromise theory proposed by Macquer in 1778 resembled one advocated by L. B. Guyton de Morveau, with whom he corresponded regularly. He welcomed Guyton's reform of chemical nomenclature in 1782 for he had himself criticized the old unsystematic nomenclature as early as 1749, although he offered only tentative proposals for reform and generally retained the old names in his books. He was, however, responsible for reintroducing Van Helmont's word "gas," for he was reluctant to describe as an "air" any substance that differed from the air of the atmosphere.

Macquer's books helped to spread knowledge of chemical affinity. While admitting that progress in chemistry could have led to its extension, he reprinted E. F. Geoffroy's affinity table of 1718 unaltered in the *Élémens*, arguing that it included all the principal affinities;[12] he did, however, discuss criticisms to which it had been subjected and added some new ones.

In 1749 he believed that there was an affinity between two substances when they resembled each other, but in 1766 he defined affinity simply as the tendency of constituent or integrant parts of substances to combine and as the force that makes them adhere when united. He did not speculate about the causes of affinity but referred his readers to the works of Newton, Keill, and others who had attempted to illuminate this obscure subject. Both the *Élémens* and the *Dictionnaire* contained sets of propositions showing the ways in which affinity could manifest itself: simple affinity, when two parts of the same substance adhered or two substances combined to form a new one; and various kinds of complex affinity, when, for example, a compound of two components was decomposed by a third substance that had a greater affinity for one of them, or when double decomposition took place between two compounds, each of two components. These distinctions had been recognized by Geoffroy and others, but Macquer expressed them systematically for the first time.

A conscientious and respected member of the Academy of Sciences, Macquer served as its director in 1774. Although only briefly a practicing physician, he had become interested in the applications of chemistry to medicine, and in 1776 was a founding member of the Société Royale de Médecine. He prepared many reports on memoirs and books submitted to these learned societies, and from 1750 examined books on chemistry, medicine, and natural history in his capacity as a royal censor. He became even more deeply involved in literary work in 1768, when he succeeded J. H. Macquart as a scientific and medical member of the editorial board of the *Journal des sçavans*, a state-owned monthly journal that published book reviews and occasional original memoirs.[13] Macquer's importance in the scientific life of Paris was therefore greater than is indicated by his few published papers, which were mainly concerned with applications of chemistry; his theoretical discussions and some original experimental work were included in his books.

With J. Hellot and M. Tillet he carefully studied the ancient method of assaying silver and gold by cupellation (1763); they standardized the procedure to obtain consistent results and disproved the old belief that silver was formed during the process by transmutation of lead. He investigated rubber—recently introduced from South America—and found that it dissolved in rectified ether and could be formed into flexible tubes (1768); but he was unsuccessful in his efforts to make flint glass (1773).

Dyeing and porcelain manufacture were the techniques that benefited most from Macquer's researches. After his success with Prussian blue, he was appointed to assist Hellot as government inspector of the dyeing industries. A leading silk dyer who allowed him to visit his workshop explained the processes, and this led to the publication of Macquer's *Art de la teinture en soie* (1763) in the Academy's series, Descriptions des Arts et Métiers. In 1768 he found that cochineal, a scarlet dye previously used only for wool, could be applied to silk with the aid of a mordant of tin dissolved in aqua regia. He also became interested in the theory of dyeing. Hellot advocated a mechanical theory, believing that particles of dye were retained in the pores of the fabric. While not entirely denying this, Macquer concluded in the *Dictionnaire* (1778) that there was adhesion, or perhaps even chemical combination, among the fabric, the mordant, and the dye. This chemical theory was later elaborated by Berthollet.

From 1751 to 1766 Hellot was scientific adviser to the Royal Porcelain Works at Sèvres, and Macquer assisted him for much of that time. Most of his technical research was not published, so there is only fragmentary information about his work on ceramics. With Baumé he examined more than 800 specimens of clay from many parts of France and studied the effect of heating them with various earthy substances in a powerful charcoal furnace of his own design (1758). The much-admired oriental porcelain had been imitated at Meissen in 1715, but the secret eluded the French until after 1765, when the necessary kaolin was found near Limoges. From August to November 1768 Macquer traveled in the southwest[14] and verified the presence of large deposits, and within a year the craftsmen at Sèvres developed the necessary techniques and produced porcelain as good as any in the world.

After Hellot's death in 1766 Macquer succeeded him as inspector of dyeing and, with E. Mignot de Montigny, as a scientific adviser at Sèvres.[15] His influence on French technology may have been considerable through his post as consultant to the government Bureau de Commerce, which controlled French trade and industry.[16] This well-paid post was no sinecure, and in 1781 Macquer told Bergman that much of his time was spent investigating problems, some of them secret, concerning the saltpeter and other industries as well as pottery and dyeing.[17]

Commissioned by the government to write a comprehensive treatise on dyeing,[18] by the end of 1783 he found it necessary to ask for an assistant to relieve him of some of his work for the *Journal des sçavans*.[19] His health was already failing, and he died a few weeks later. He left a widow, whom he had married in 1748, and two daughters.[20]

NOTES

1. In their *éloges* of Macquer (see Bibliography), Condorcet described him as descended from a noble Scottish family, and Vicq d'Azyr said that his great-grandfather accompanied James I (presumably a misprint for James II) to France. Evidence produced by D. McKie, "The Descent of Pierre Joseph Macquer," in *Nature*, **163** (1949), 627, suggests that he was descended from an Irish family, the Maguires, barons of Enniskillen.
2. Macquer disclosed his motives for studying medicine in the draft of a letter to D. F. R. Mesnard de Chousy, 24 Mar. 1774, Bibliothèque Nationale, MS français 9134, fol. 100. He referred to his work as *médecin des pauvres malades* in an unpublished memoir on the accidental poisoning of wine and milk with lead and copper, Bib. Nat., MS français 9132, fol. 112.
3. In a letter to T. O. Bergman, 18 July 1772, Macquer said that he was working on new eds. of both the *Élémens* and the *Dictionnaire;* see F. Carlid and J. Nordström, eds., *Torbern Bergman's Foreign Correspondence*, I (Stockholm, 1965), 243. An incomplete MS of a revised ed. of *Élémens*, which internal evidence shows to have been written after 1768, is Bib. Nat., MSS français 9131, fols. 1–18 (chs. 1–2, pp. 1–36), and 9133, fols. 216–254 (chs. 3–5, pp. 37–113). This rev. ed. was never completed. A 1775 ed. mentioned by J. R. Partington, *A History of Chemistry*, III (London, 1962), 80, seems not to exist.
4. He is named as professor of pharmacy in *Almanach royal* (Paris, 1753), 347. At this time a new professor was elected each year.
5. The attendance is recorded in a note by Macquer, Bib. Nat., MS français 9134, fols. 131–132.
6. For an account of Philippe Macquer (1720–1770), advocate, historian, and man of letters, see the anonymous "Éloge de M. Macquer, avocat en parlement," in *Le nécrologe des hommes célèbres de la France* (Paris, 1771), 187–206. His *Dictionnaire portatif des arts et métiers* is discussed by J. Proust, "Deux encyclopédistes hors de l'Encyclopédie: Philippe Macquer et l'abbé Jaubert," in *Revue d'histoire des sciences*, **11** (1958), 330–336. At one time he edited the scientific part of the weekly *Avant-coureur;* see a letter from P. J. Macquer to Bergman, 22 Feb. 1768, Carlid and Nordström, *op. cit.*, 230.
7. He explained why the *Dictionnaire* was published anonymously in a letter to Bergman, 22 Feb. 1768, in Carlid and Nordström, *op. cit.*, 229.
8. Bergman later suggested that baryta might be the calx of an unknown metal; see W. A. Smeaton, in *Dictionary of Scientific Biography*, II (1970), 8. Lavoisier expressed the same idea in antiphlogistic terms, when he conjectured that all earths might be metals oxidized to a certain degree (Lavoisier, *Traité élémentaire de chimie* [Paris, 1789], 195).
9. See W. C. Ahlers, "P. J. Macquer et le rapport sur les *Opuscules physiques et chimiques de Lavoisier*," in *Actes du XIIe Congrès international d'histoire des sciences*, VI (Paris, 1971), 5–9.
10. Review by Macquer of *Lettre de M. le Comte Morozzo à M. Macquer, sur la décomposition du gas méphitique & du gas nitreux* (Turin, 1783), in *Journal des sçavans* (1783), 864–867 (quoted on p. 867).
11. M. Landriani to Bergman, 13 Feb. 1784, in Carlid and Nordström, *op. cit.*, 201.
12. The affinity table in Macquer's *Élémens* is an exact copy of Geoffroy's; careful comparison shows that it was printed from the same copper plate as that in Geoffroy's posthumous *Traité de matière médicale*, I (Paris, 1743). Perhaps it was because this plate was available that Macquer did not draw up a new table. He included no affinity table in the *Dictionnaire*.
13. Macquer's appointment as editor is mentioned in the anonymous "Éloge de Jacques-Henri Macquart," in *Journal des sçavans* (1768), 635–638. For the history of the *Journal des sçavans*, see E. Hatin, *Histoire politique et littéraire de la presse en France*, II (Paris, 1859), 151–217; a brief account is in Hatin's *Bibliographie historique et critique de la presse française* (Paris, 1866), 28–32.
14. Letters from Macquer to his brother and other correspondents, written from the southwest between 23 Aug. and 4 Nov. 1768, are in Bib. Nat., MSS français 9134, fols. 86–96; 9135, fols. 65–96.
15. Macquer and De Montigny (*d.* 1782) are described as *commissaires de l'Académie Royale des Sciences, pour les recherches* in the official list of personnel at the Manufacture de Porcelaines de France à Sève [sic] that was published in the *Almanach royal* from 1776 onward. There is no entry under this heading in earlier years.
16. See E. Lelong's, "Introduction," in Pierre Bonnassieux, ed., *Conseil de commerce et Bureau de Commerce. Inventaire analytique des procès-verbaux* (Paris, 1900), xxvii–xxviii.
17. Macquer to Bergman, 7 July 1781, in Carlid and Nordström, *op. cit.*, 253–254.
18. Macquer to Bergman, 7 Mar. 1782, *ibid.*, 255. The proposed work, described in an eight-page prospectus (see below), was never completed.
19. Correspondence concerning Macquer's request is in Bib. Nat., MSS français 9134, fols. 116, 120, and 12306, fols. 66, 78.
20. Macquer's brother Philippe (see note 6) had died in 1770. Another relation, who died in 1782, was a cousin called Macquer (other names unknown). He was coauthor, with H. G. Duchesne, of *Manuel du naturaliste* (Paris, 1770), frequently but incorrectly attributed to P. J. Macquer. Information about the cousin is given by W. C. Ahlers in his doctoral thesis (see Bibliography).

BIBLIOGRAPHY

I. Original Works. Macquer's *Élémens de chymie théorique* (Paris, 1749) was reprinted with only minor alterations as a new ed. in 1753. *Élémens de chymie pratique*, 2 vols. (Paris, 1751), appeared as a 2nd ed. in 1756, and *Chymie théorique* was again reprinted then. The 1756 eds. were together translated into English by Andrew Reid as *Elements of the Theory and Practice of Chemistry*, 2 vols. (London, 1758). D. McKie, in *Endeavour*, **16** (1957), 135, records 4 subsequent English eds. published in London (1764; 1775) and Edinburgh (1768; 1777). German (1752, 1758), Italian (1781), and Russian (1791) translations are recorded by H. C. Bolton, *A Select Bibliography of Chemistry* (Washington, 1893), 646. Earlier Russian (1774–1775) and later German (1768) eds. and a Dutch trans. (1773, 1775), are mentioned by J. Ferguson, *Bibliotheca chemica*, II (Glasgow, 1906), 60, but he often gives incorrect dates.

Macquer and Antoine Baumé gave a detailed account of their joint lecture course in *Plan d'un cours de chymie, expérimentale et raisonnée, avec un discours historique sur la chymie* (Paris, 1757). Macquer's *Art de la teinture en soie* (Paris, 1763), was published in the series of Descriptions des Arts et Métiers of the Académie Royale des Sciences. An English trans. appeared of *The Art of Dying [sic] Wool, Silk and Cotton. Translated from the French of M. Hellot, M. Macquer and M. Le Pileur d'Apligny* (London, 1789), 263–381, and on pp. 233–339 of the reprint (London, 1901). The book was also translated into Spanish (Madrid, 1771), German (Leipzig, 1779), and Arabic

(Boulāq, 1823). Macquer's method of dyeing silk with cochineal, published in the *Mémoires* of the Academy for 1768 (see below), was also described in a pamphlet, *Méthode pour teindre la soie en plusieurs nuances de rouge vif de cochenille et autres couleurs, par M. Macquer* (Paris, 1769). His projected general treatise on dyeing was described in *Prospectus et plan d'une description générale de l'art de la teinture . . . par M. Macquer* (Paris, 1782).

Macquer was coauthor, with J. M. F. de Lassone, E. Gourley de la Motte, A. L. de Jussieu, J. B. Carburi, and L. C. Cadet, of a short medical work, *Traitement contre le ténia ou ver solitaire, pratiqué à Morat en Suisse, examiné et éprouvé à Paris* (Paris, 1775; repr. 1776); summarized in *Journal des sçavans* (1776), 606–611; English trans. by S. Foart Simmons, *An Account of the Tenia or Long Tape Worm* . . . (London, 1777).

The breadth of Macquer's interests is shown by the full title of his *Dictionnaire de chymie, contenant la théorie & la pratique de cette science, son application à la physique, à l'histoire naturelle, à la médecine & à l'économie animale*, first published anonymously in 2 small octavo vols. (Paris, 1766); a reimpression, with slightly different pagination, appeared later in 1766 and another in 1769. There was also a Swiss printing (3 vols., Yverdon, 1767). The words *à l'économie animale* in the title were replaced by *aux arts dépendans de la chymie* in the 2nd ed., which was published, with Macquer named as author, in both octavo in 4 vols. (Paris, 1778) and quarto in 2 vols. (Paris, 1778). A 3-vol. octavo reprint, incorrectly dated 1777 was not published before 1778. There were also 3 Swiss printings (Yverdon, 1779; "en Suisse," 1779–1780; Neuchâtel, 1789). James Keir's English trans., *A Dictionary of Chemistry*, first appeared in 2 vols. (London, 1771); a 2nd English ed., required before the 2nd French ed. was ready, contains supplementary material by Keir (3 vols., London, 1777). The 1st ed. was also translated into German and Danish; the 2nd ed. into German and Italian. Full details of the translations are given by J. R. Partington, *A History of Chemistry*, III (London, 1962), 81–82.

The following memoirs by Macquer appeared in the *Histoire [H.] et Mémoires [M.] de l'Académie Royale des Sciences* (the date of publication in parenthesis follows the nominal date): "Sur la cause de la différente dissolubilité des huiles dans l'esprit de vin," *M.*, 1745 (1749), 9–25; "Recherches sur l'arsenic. Premier mémoire," *M.*, 1746 (1751), 233–236; "Observations sur la chaux et sur le plâtre," *M.*, 1747 (1752), 678–696; "Second mémoire sur l'arsenic," *M.*, 1748 (1752), 35–50; "Mémoire sur une nouvelle espèce de teinture bleue, dans laquelle il n'entre ni pastel ni indigo," *M.*, 1749 (1753), 255–265; "Examen chymique du bleu de Prusse," *M.*, 1752 (1756), 60–77; "Mémoire sur une nouvelle méthode de M. le Comte de la Garaye pour dissoudre les métaux," *M.*, 1755 (1761), 25–35; "Recherches sur la nature de la teinture mercurielle de M. le Comte de la Garaye. Premier mémoire," *M.*, 1755 (1761), 531–546 (there was no further memoir on this topic); "Mémoire sur un nouveau métal connu sous le nom d'or blanc ou de platine," *M.*, 1758 (1763), 119–133, written with A. Baumé; "Mémoire sur les argiles et sur la fusibilité de cette espèce de terre avec les terres calcaires," *M.*, 1758 (1763), 155–176; "Examen chimique de l'eau de la rivière d'Yvette," *M.*, 1762 (1764), 376–380, written with J. Hellot; "Mémoire sur les essais des matières d'or et d'argent," *M.*, 1763 (1766), 1–14, written with J. Hellot and M. Tillet; "Mémoire sur l'action d'un feu violent de charbon, appliqué à plusieurs terres, pierres et chaux métalliques," *M.*, 1767 (1770), 298–314; "Sur une source minéral trouvé à Vaugirard," *H.*, 1768 (1770), 69–75, written with S. F. Morand and L. C. Cadet; "Mémoire sur un moyen de teindre la soie en un rouge vif de cochenille et de lui faire prendre plusieurs autres couleurs plus belles et plus solides qu'on a faites jusqu'à présent," *M.*, 1768 (1770), 82–90; "Mémoire sur un moyen de dissoudre la résine caoutchouc connue présentement sous le nom de résine élastique de Cayenne, et de la faire reparoitre avec toutes ses qualités," *M.*, 1768 (1770), 209–217; "Recherches sur la composition du flint-glass, avec des vues pour le perfectionner," *M.*, 1773 (1777), 502–511; "Premier essai du grand verre ardent de M. Trudaine établi au jardin de l'Infante au commencement du mois d'octobre de l'année 1774," *M.*, 1774 (1778), 62–72, written with J. C. P. Trudaine de Montigny, L. C. Cadet, A. L. Lavoisier and M. J. Brisson; and "Analyse de l'eau du lac Asphaltite," *M.*, 1778 (1781), 69–72, written with A. L. Lavoisier and B. G. Sage.

Macquer's research on the solubility of salts in alcohol was published as "Mémoire sur la différente dissolubilité des sels neutres dans l'esprit de vin, contenant des observations particulières sur plusieurs de ces sels," in *Mélanges de philosophie et de mathématique de la Société Royale de Turin*, **3** (1766), 1–30; and "Second mémoire . . . de ces sels," *ibid.*, **5** (1770–1773), 173–190; the first memoir was reprinted in *Introduction aux observations sur la physique*, **1** (1777), 461–472, 559–568; and the second in *Observations sur la physique*, **9** (1777), 182–193. Other memoirs by Macquer include "Lettre . . . au sujet des expériences qui se font au foyer de la grande lentille à liqueur de M. de Trudaine," in *Journal des sçavans* (1776), 561–563; "Observations sur la dissolution des sédimens et incrustations pierreuses que forme l'urine dans les vaisseaux où elle séjourne," *ibid.*, 613–616; "Mémoire sur les savons acides, et sur les avantages qu'on en pourroit retirer dans la pratique de la médecine," in *Mémoires de la Société royale de médecine* (1776), 379–386; "Réflexions sur la magnésie du sel d'Epsom," in *Histoire de la Société royale de médecine* (1779), 235–243. Minor reports written by Macquer jointly with other academicians are not included in this bibliography; details of those of which Lavoisier was a coauthor are in D. I. Duveen and H. S. Klickstein, *A Bibliography . . . of Lavoisier* (London, 1954).

Eleven bound vols. of Macquer's papers are at the Bibliothèque Nationale, Paris: MSS français 9127–9135 contain MSS of published and unpublished books, memoirs, and lecture notes as well as drafts of letters written by him; MSS français 12305–12306 contain letters received by him from many correspondents. An inventory and index of these vols. (except MS 9135, which contains documents relating to porcelain), compiled in 1943 by

Paul M. Bondois, is in Bib. Nat., MS français, nouvelle acquisition 13260; it is useful but not completely accurate. A small collection of Macquer's correspondence is at the library of the Muséum National d'Histoire Naturelle, Paris, MS 283; some of his notebooks and other documents concerning porcelain are in the archives of the Manufacture Nationale de Sèvres, Y 57–60. In a doctoral thesis (see below) W. C. Ahlers lists the principal contents of all these MSS and draws attention to Macquer material in other French archives. The eighteen letters from Macquer to Bergman at the archives of the University of Uppsala are printed with an additional letter in G. Carlid and J. Nordström, eds., *Torbern Bergman's Foreign Correspondence*, I (Stockholm, 1965), 229–255, 437–438.

II. SECONDARY LITERATURE. The best contemporary account of Macquer's life and work is F. Vicq d'Azyr, "Éloge de M. Macquer," in *Histoire de la Société royale de médecine* for 1782–1783 (1787), pp. 69–94, repr., but without the valuable footnotes, in his *Éloges historiques*, I (Paris, 1805), 277–303. Less informative is Condorcet, "Éloge de M. Macquer," in *Histoire de l'Académie Royale des Sciences* for 1784 (1787), pp. 20–30, repr. in his *Oeuvres*, III (Paris, 1847), 125–138. A recent study, containing much information from MS sources, is the unpublished doctoral thesis by Willem C. Ahlers, "Un chimiste du XVIIIe siècle. Pierre Joseph Macquer (1718–1784). Aspects de sa vie et de son oeuvre," Université de Paris, Faculté des Lettres et Sciences Humaines, thèse de troisième cycle, 1969. L. J. M. Coleby, *The Chemical Studies of P. J. Macquer* (London, 1938), is a convenient source of information about Macquer's published work. Useful bibliographical data are given by D. McKie, "Macquer, the First Lexicographer of Chemistry," in *Endeavour*, **16** (1957), 133–136. Macquer's joint lecture course with Baumé is described by R. Davy, *L'apothicaire Antoine Baumé* (Cahors, 1955), 38–40; and the text of their agreement is printed in J. P. Contant, *L'enseignement de la chimie au Jardin Royal des Plantes de Paris* (Cahors, 1952), 65–69. Contant says little about Macquer's lectures; a short account is given by W. A. Smeaton, "P. J. Macquer's Course of Chemistry at the Jardin du Roi," in *Actes du Xe Congrès international d'histoire des sciences, Ithaca, 1962* (Paris, 1964), 847–849.

Macquer's views on the nature of the elements are discussed by W. A. Smeaton, "Macquer on the Composition of Metals and the Artificial Production of Gold and Silver," in *Chymia*, **11** (1966), 81–88. A good account of his ideas on affinity is given by A. M. Duncan, "Some Theoretical Aspects of Eighteenth-century Tables of Affinity," in *Annals of Science*, **18** (1962), 177–194, 217–232. His contributions to the reform of nomenclature are discussed by M. P. Crosland, *Historical Studies in the Language of Chemistry* (London, 1962), 120–122, 134–138.

The importance of Macquer and Baumé's research on platinum is assessed by D. McDonald, *A History of Platinum* (London, 1960), 32–34; and the analysis of an iron ore by Macquer and E. Mignot de Montigny is described by R. G. Neville, " 'Observations sur la mine de fer de Bagory': An Unpublished Manuscript by P. J. Macquer," in *Chymia*, **8** (1962), 89–96. Macquer's technical research is set in its historical context by H. Guerlac, "Some French Antecedents of the Chemical Revolution," in *Chymia*, **5** (1959), 73–112; further information is given by W. C. Ahlers, "P. J. Macquer, pionnier de la recherche fondamentale dans le domaine de l'industrie chimique," in *Actes du XIIIe Congrès international d'histoire des sciences, Moscow/Leningrad, 1971* (in press).

References to additional secondary sources are given in the notes above.

W. A. SMEATON